	1B	2B	3A	4A	5A	6A	7A	8A
								2 He 4.0026
			5 B 10.81	6 C 12.011	7 N 14.0067	8 O 15.9994	9 F 18.998	10 Ne 20.179
			13 Al 26.9815	14 Si 28.085	15 P 30.974	16 S 32.06	17 Cl 35.453	18 Ar 39.948
Ni 58.69	29 Cu 63.546	30 Zn 65.39	31 Ga 69.72	32 Ge 72.59	33 As 74.922	34 Se 78.96	35 Br 79.904	36 Kr 83.80
Pd 106.42	47 Ag 107.868	48 Cd 112.41	49 In 114.82	50 Sn 118.71	51 Sb 121.75	52 Te 127.60	53 I 126.905	54 Xe 131.29
Pt 195.08	79 Au 196.967	80 Hg 200.59	81 Tl 204.383	82 Pb 207.2	83 Bi 208.98	84 Po (209)	85 At (210)	86 Rn (222)

Amine [illegible] Ch. 23.5

Gd 157.25	65 Tb 158.925	66 Dy 162.50	67 Ho 164.930	68 Er 167.26	69 Tm 168.934	70 Yb 173.04	71 Lu 174.967
Cm (247)	97 Bk (247)	98 Cf (251)	99 Es (252)	100 Fm (257)	101 Md (258)	102 No (259)	103 Lr (260)

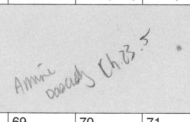

Organic Chemistry

SECOND EDITION

Organic Chemistry

SECOND EDITION

G. Marc Loudon

Purdue University

The Benjamin/Cummings Publishing Company, Inc.

Menlo Park, California • Reading, Massachusetts
Don Mills, Ontario • Wokingham, U.K. • Amsterdam • Sydney
Singapore • Tokyo • Madrid • Bogota • Santiago • San Juan

Sponsoring Editor: *Diane Bowen*
Production Supervisor: *Karen Gulliver*
Production and Art Coordination: *Pat Waldo and Deborah Gale/Partners in Publishing*
Copy Editor: *Carol S. Kinney*
Developmental Art Editor: *Audre Newman*
Text and Cover Designer: *Michael Rogondino*
Illustrators: *Linda Harris-Sweezy, J & R Technical Services, George Klatt, Linda McVay, Carol Verbeeck*
Proofreader: *Steve Sorensen*
Compositor: *York Graphic Services*

Library of Congress Cataloging-in-Publication Data

Loudon, G. Marc.
Organic chemistry.

Includes index.
1. Chemistry, Organic. I. Title.
QD251.2.L68 1988 547 87-29996
ISBN 0-8053-6643-1

Permission for publication herein of Sadtler Standard Spectra® has been granted, and all rights are reserved, by Sadtler Research Laboratories, Division of Bio-Rad Laboratories, Inc.

Additional credits are listed starting on page C-1.

7 8 9 10 -VH- 95 94 93 92 91

The Benjamin/Cummings Publishing Company, Inc.
2727 Sand Hill Road
Menlo Park, California 94025

To Judy and
to Kyle and Chris,
who can't remember when
Dad wasn't writing.

Preface

This book was written because I have for several years felt the need for a text that corresponds more closely than other texts to the course I teach. Although it is organized along tried-and-true functional-group lines, the book contains some unique features that have served me well in both my teaching and my learning of organic chemistry.

I have had four major concerns in both the initial writing and the revision of this text: *readability, type of presentation, organization,* and *scientific accuracy.*

READABILITY

In addition to the writing style, some of the devices used to enhance readability are:

1. **Use of common analogies** to aid in the understanding of difficult topics. Two of the many examples can be found on pages 437 and 606.

2. **Use of numerous figures and diagrams** to illustrate important concepts.

3. **Frequent cross-referencing** to assist students in finding the initial discussion of seminal topics. This technique is illustrated on page 963, just below Eq. 22.93.

4. **Framing of problems in humorous or "real-life" contexts.** Problem 31 on page 32 and Problem 32 on page 577 demonstrate this approach.

5. **Guidance on how to solve problems.** This takes several forms: worked-out examples, as on pages 407–408, 500, and 513; hints as in Problem 35, page 362; and detailed lists of leading questions that show students the path to an answer without revealing the answer outright, as in the box at the top of page 177. This guidance is extended in the Study Guide–Solutions Manual, in which the problem-solving approaches parallel those in the text.

6. **Frequent summaries and recapitulations of key ideas.** These also take several forms. There are the end-of-chapter "Key Ideas" summaries, which bring together important concepts from within the chapter. In a number of sections, important points are brought together with numbered lists, as on page 354. And there are appendices at the end of the text that summarize rules of nomenclature, key spectroscopic information, acidity and basicity data, and synthetic methods. Because I try to discourage students (not always successfully) from memorizing tables of reactions, I have moved reaction summaries out of the text and into the Study Guide–Solutions Manual supplement. I encourage students to make their own summaries and check them against those in the supplement.

7. **Use of a second color within chemical equations** to show contrasts and changes. The scheme on page 943 is illustrative of the literally hundreds of examples of this technique within the text. Labeling with color the groups that change in a reaction draws a student's attention to the point of the equation. Yet this is not done in every equation, because students must learn to look for themselves. To avoid confusion, we intentionally avoided the use of a four-color presentation within equations, compound names, and spectra.

8. **Presentation of historical sketches in anecdotal style,** as opposed to dry biographical data. Examples of such sketches are found at the bottom of pages 782 and 936. The purpose of these is to stress that chemistry is a human endeavor, and that the road to discovery is often paved with serendipity and humor. Screened boxes have been used to set these "asides" apart from the text.

During the writing of this book, I have tried to anticipate the questions of a student studying organic chemistry for the first time. The book has benefitted from the questions that my own students have posed to me. I have tried to challenge the student to think about the subject rather than simply memorizing it. I have endeavored to write under the conviction that we must continue to confront our students not only with the factual material of organic chemistry, but also with the considerable intellectual beauty and challenge of the subject.

TYPE OF PRESENTATION

A number of elements of the presentation used in this text are worth special comment.

1. I have chosen a mechanistic approach within the overall functional-group framework of the text. This emphasis is consonant with the recognition by many teachers of organic chemistry that the only way for students to truly learn the subject is for them to see the unifying elements that connect what at first appear to be unrelated phenomena. Thus, students learn a given reaction more easily when its mechanistic connection to an earlier reaction is apparent. The benefit of this approach is that students can understand why organic reactions occur as they do. They can make reasonable predictions about the outcome of unknown reactions. In contrast, students who view each reaction as an isolated entity court a nearly impossible task in trying to learn organic chemistry by virtual rote memorization. To assist with this mechanistic approach, I have not only presented, but also *thoroughly explained,* the "curved-arrow" formalism, I have required that students master it, and I have *used it consistently* throughout the text. I have also stressed both Lewis and Brønsted acid-base properties of organic compounds, not only because these topics are important in their

own right (how many organic reactions incorporate acid-base steps in their mechanisms?), but also because these concepts can be logically extended in many cases to the prediction of chemical reactivity.

2. One of the key elements in the presentation of organic chemistry by any textbook is the *mechanistic centerpiece*—the reaction used to introduce the notions of mechanism, such as multi-step reactions, reaction free-energy diagrams, reactive intermediates, rate-determining step, and the like. The traditional vehicle for this purpose has been free-radical halogenation of alkanes. I have felt that since most of the common organic reactions encountered by students are polar reactions, the mechanistic centerpiece should also be a polar reaction. I have chosen to use for this purpose polar additions to alkenes, because an unsymmetrical alkene can in principle undergo two competing addition reactions. We can evaluate the relative merits of the two reaction pathways by a direct comparison of carbocation stabilities and, invoking Hammond's postulate, transition-state stabilities; the issue is not complicated by relative stabilities of starting materials, because the starting materials are the same for both pathways.

Free-radical reactions are not ignored, but only postponed until students have a chance to master the essentials of polar reactions. At the proper time, free-radical reactions (and the corresponding "fishhook" formalism) are introduced and thoroughly discussed.

3. More than 1300 problems of both the in-text and end-of-chapter variety are provided. These range in difficulty from simple drill-type problems to problems that will challenge the best students.

4. There is a thorough discussion of stereochemistry. The subject is introduced early and strongly reinforced throughout the text, both in discussions of reactions and in problems. The application of stereochemistry to chemical reactions (Chapter 7), and group equivalence and nonequivalence (Chapter 10) are two stereochemical topics whose treatments are particularly unique to this text.

5. I have introduced an important topic too-often ignored in undergraduate texts: solvents in organic chemistry, and the relationship of gas-phase chemistry to solvent effects. The level of this discussion is introductory, appropriate for the beginning undergraduate.

6. I have presented a thorough approach to defining and understanding oxidation and reduction in organic chemistry.

7. There are biological applications of appropriate chemistry, not set apart in "special topics" chapters, but instead included in sections adjacent to the related laboratory chemistry. In these examples, I have not forgotten that this is a chemistry text, not a biochemistry text; and the underlying theme of these sections is not the details of the biology involved, but rather the close analogy of biological chemistry to laboratory reactions.

8. Nomenclature is treated thoroughly in this book because I believe that after students finish a first course in organic chemistry, they should be able to construct a systematic name for any simple organic compound.

9. Finally, I have tried to indicate the important role of organic chemistry in today's economy, and have discussed some of the key social issues surrounding organic chemistry (pollution and chemical carcinogenesis, to name two). Since we teach so many non-majors, I believe that they must leave our classes convinced that organic chemistry is a potent economic force, and that embodied within the discipline is the capacity to solve a number of social problems, not just to create them.

ORGANIZATION

In this edition, I have grouped reactions of alkenes, alkyl halides, alcohols, ethers, epoxides, thiols, and sulfides together in early chapters. I have two reasons for this strategy: first, the chemistry of these groups is strongly interrelated; and second, a substantial amount of nonhydrocarbon chemistry can be covered in the early part of the course. Following an interlude dealing with spectroscopy (which could be placed anywhere with little adjustment in a course using this text), I return to concepts of resonance and aromaticity by considering dienes, aromatic compounds, and allylic/benzylic reactivity. Then comes carbonyl chemistry, where I have consolidated a discussion of enols, enolate ions, and condensation reactions in a special chapter. This is followed by amine and heterocycle chemistry. The text closes with pericyclic reactions, amino acid and peptide chemistry, and finally, sugar and nucleic acid chemistry.

SCIENTIFIC ACCURACY

Each topic in this book was researched back into the original or review literature. In the preface to the first edition I stated that it would be presumptuous to state that this book is free of factual errors—and I am glad I made that statement! I have endeavored to correct the errors that I, my students, and my colleagues have found, and I am indebted to the many people who have sent concrete suggestions for sharpening the accuracy of the text.

CHANGES TO THE SECOND EDITION

How is the second edition different from the first? To begin with, the book is considerably shorter. Using the format of the first edition as a frame of reference, the equivalent of about three hundred pages of material has been cut. About a hundred pages of these savings have been expended in design considerations—in opening up the layout to give it a less "dense" look. The result is a text that is nearly two hundred pages shorter and more visually digestible than the first edition—and this including a new chapter on pericyclic reactions. What was deleted from the first edition? First, I cut a number of the more difficult problems. (This edition still has more problems than any other text.) Second, I deleted several spectra that I viewed as redundant. Third, I consolidated several topics. For example, the section on mass spectrometry has been included in the chapter with infrared spectroscopy, and the level of presentation has been reduced somewhat. Finally, a few topics have been deleted. I became personally convinced from my own teaching that these changes would be beneficial, and I believe that they have enhanced the usability of the text without altering the positive qualities of the first edition.

The text has seen some reorganization, much of which was discussed above. In addition, much of the art has been totally redesigned, with the introduction of considerably more airbrushed drawings, and the use of ball-and-stick models instead of line formulas in situations calling for stereochemical detail.

The treatment of organic synthesis has been reorganized. Instead of concentrating a single discussion within a detailed chapter near the end of the text, I have considered the strategy of organic synthesis much earlier, in Chapters 10 and 11, points at which students have accumulated just enough reactions in their repertoire that they can begin to construct relatively simple multistep syntheses. The concepts of these sections are continually reinforced within several later chapters in the discussions of reactions that are particularly useful in synthesis. Some of the deletions in this edition are of synthetically redundant reactions. For example, I have deleted the reaction of carboxylic acids with organolithium reagents—a perfectly useful reaction, but one which, at the beginner's level of expertise, offers no advantages over other methods of ketone synthesis.

SUPPLEMENTS

Another important change is that we have provided a completely redesigned and rewritten *Study Guide–Solutions Manual.* Assembled into an attractive format with modern desktop-publishing technology, this supplement contains glossaries, conceptual outlines, and summary tables, as well as a solution to every problem in the text. As with the previous edition, *transparencies* of key diagrams from the text are available for classroom teaching purposes.

ACKNOWLEDGMENT

There are many whom I wish to acknowledge for their assistance in the preparation of this text. I am indebted to my Department Chairman, Professor John M. Cassady, for providing encouragement as well as an environment in which this text could be completed expeditiously. I am grateful to my departmental colleagues Mark Cushman, John Schwab, Del Knevel, and Joe Stowell, as well as to chemistry faculty Harry Morrison, Bob Benkeser, and Jim Brewster for providing corrections, consultations, and suggestions for improvement. I am indebted to Professor John Pinzelik, Chemistry Librarian at Purdue, and to Professor Theodora Andrews, and their staffs for frequent assistance throughout the preparation of both editions. Several reviewers were helpful in this revision, particularly Professor Charles Wilcox of Cornell University; Professor Maitland Jones of Princeton University; Professor John Wiseman of the University of Michigan; Professor Jay Bardole of Vincennes University; Robert Belloli of California State University–Fullerton; Jed Fisher of the University of Minnesota; John Hogg and Tammy Tiner of Texas A & M University; Michael Rathke of Michigan State University; Grant Taylor of the University of Louisville; and most especially, Professor Ron Magid of the University of Tennessee, whose constant attention to accuracy and detail from start to finish, not to mention good humor, were very important to me throughout the project. In addition, I am indebted to the many students who used the first edition and offered constructive advice—including those who called from other universities.

The relationship between author and publisher has been exceptionally gratifying. I would particularly acknowledge David Chelton for his extremely useful guidance as Developmental Editor; Audre Newman, who helped develop the art program; Steve Mautner, my editor who guided the revision through the manuscript phase; Diane Bowen, my present editor at Benjamin-Cummings, who has been a valued fountain of advice and encouragement; and Pat Waldo, Deborah Gale, and Mimi Hills of "Partners in Publishing," who guided the second edition through the production phase with expertise and good humor.

I am indebted to the authors and publishers acknowledged separately in the credits section for permission to reproduce copyrighted materials.

Finally, and most importantly, I acknowledge the love and support of my family, to whom this book is dedicated.

My wish for this text is that students and professors will enjoy using this text as much as I have enjoyed writing it!

West Lafayette, Indiana G. M. L.
October 1987

About the Author

Marc Loudon received his B. S. (Magna Cum Laude) in chemistry from Louisiana State University in 1964 and his Ph.D. in organic chemistry from the University of California, Berkeley, where he worked with Professor Donald S. Noyce. After two years of postdoctoral work with Professor Daniel E. Koshland in the Biochemistry Department at Berkeley, Dr. Loudon joined the faculty of the Chemistry Department at Cornell University, where he taught organic chemistry to both preprofessional students and science majors. He received the Clark Award for Distinguished Teaching in 1976. Since 1977, Dr. Loudon has been at Purdue University, where he is presently professor of Medicinal Chemistry. At Purdue, Dr. Loudon has twice won the Professor Henry Heine Award for his teaching of organic chemistry.

Dr. Loudon's research interests are in the application of organic chemistry to biological problems, particularly in the peptide/protein field; development of antimetastatic agents for cancer chemotherapy; and the mechanisms of organic reactions, particularly those in aqueous solution.

Dr. Loudon is an accomplished organist and has performed professionally in the San Francisco Area, at Cornell University, and in Indiana. He also enjoys playing competitive tennis and has participated in several tournaments with no success whatsoever.

Reviewers

Elizabeth Armstrong
Skyline College

Jay Bardole
Vincennes University

Robert Belloli
California State University–Fullerton

Jed Fisher
University of Minnesota

Norbert Hepfinger
Rensselaer Polytechnic Institute

John Hogg
Texas A & M University

Maitland Jones
Princeton University

Joseph Kurz
Washington University

Dennis Liotta
Emory University

Ronald Magid
University of Tennessee–Knoxville

Kenneth Marsi
California State University–Long Beach

Michael Rathke
Michigan State University

K. Grant Taylor
University of Louisville

B. S. Thyagarajan
University of Texas–San Antonio

Tammy Tiner
Texas A & M University

Charles Wilcox
Cornell University

John Wiseman
University of Michigan

Contents in Brief

Contents in Detail

CHAPTER 8 Introduction to Alkyl Halides, Alcohols, Ethers, Thiols and Sulfides 272

CHAPTER 9 Substitution and Elimination Reactions of Alkyl Halides 319

Organic Chemistry

SECOND EDITION

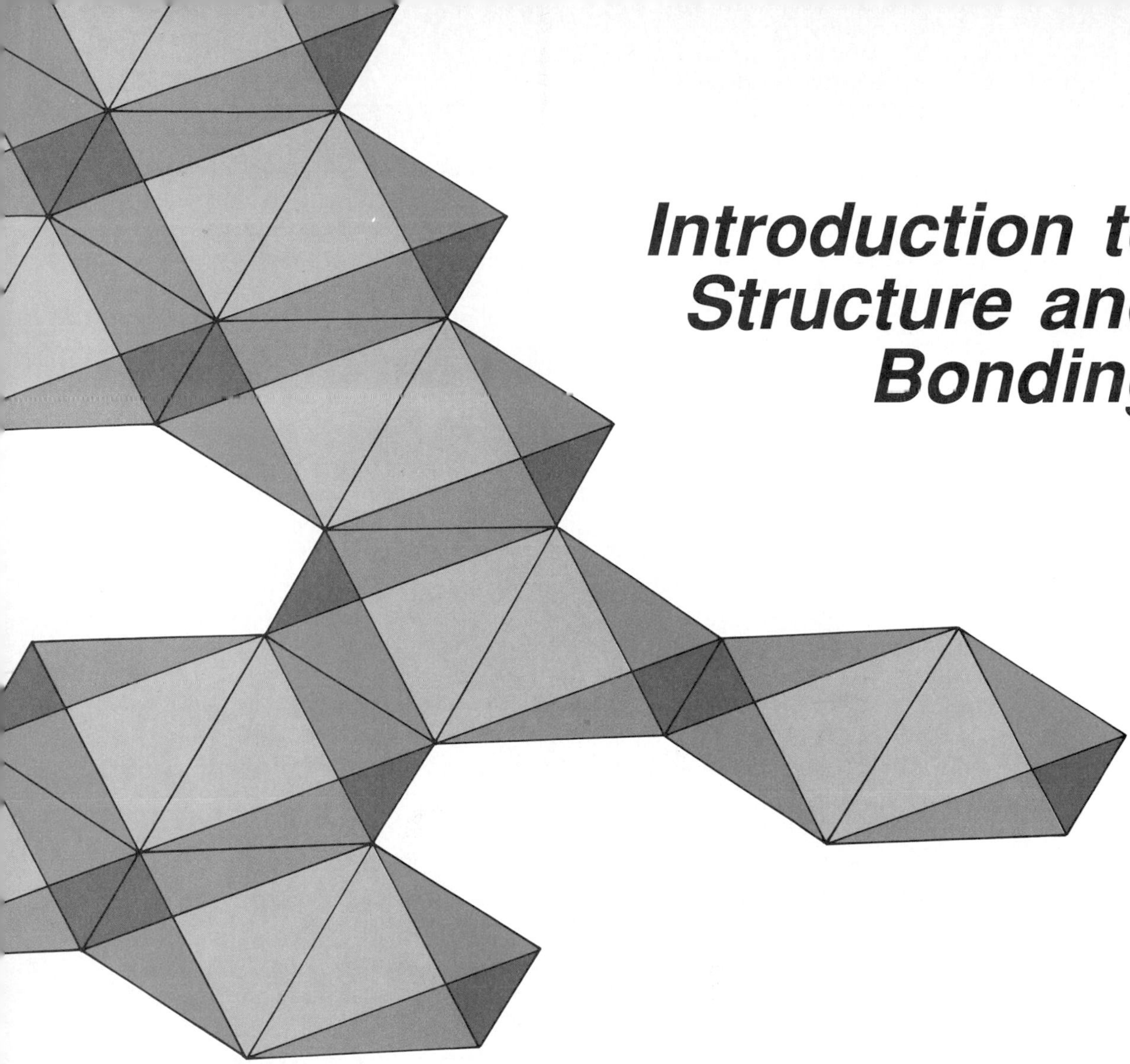

1 Introduction to Structure and Bonding

1.1 INTRODUCTION

A. Why Study Organic Chemistry?

The vitality of organic chemistry as a branch of science is unquestioned. Organic chemistry is characterized by continuing development of new knowledge, a fact evidenced by the large number of journals devoted exclusively or in large part to the subject. Organic chemistry lies at the heart of a substantial fraction of modern chemical industry and therefore contributes to the economy of many nations.

Many students who take organic chemistry nowadays do not intend to become professional chemists. Organic chemistry courses are populated largely with students majoring in the biological sciences or in the allied health disciplines such as pharmacy or medicine. In fact, the subject is immensely important to the biological and health sciences, and its importance there is sure to increase. One need only open a modern textbook or journal of biochemistry or biology to appreciate the sophisticated organic chemistry that is central to these fields.

Even for those who do not plan a career in any of the sciences, a study of organic chemistry is important. We live in a technological age that is made possible in large part by applications of organic chemistry to industries as diverse as plastics, textiles, communications, transportation, food, and clothing. In addition, problems of energy, pollution, and depletion of resources are all around us. If organic chemistry has played a part in creating these problems, it will almost surely have a role in their solution.

As a science, organic chemistry lies at the interface of the physical and biological sciences. Research in organic chemistry is a mixture of sophisticated logic and empirical observation. At its best, it can assume artistic proportions. We can use the study of organic chemistry to develop and apply basic skills in problem solving while at the same time learning a subject of immense practical value. Thus, to develop as a chemist, to remain in the mainstream of a health profession, or to be a well-informed citizen in a technological age, the study of organic chemistry is appropriate.

In this text we have several objectives. Of course, we shall learn the "nuts and bolts"—the nomenclature, classification, structure, and properties of organic compounds. We shall study the principal reactions and the syntheses of organic molecules. But more than this, we shall develop underlying principles that will allow us to understand, and sometimes predict, reactions rather than simply memorizing them. We shall see some of the organic chemistry that is industrially important. Finally, we shall examine some of the beautiful applications of organic chemistry in biology—for example, how organic chemistry is executed in nature and how the biological world has provided the impetus for much of the research in organic chemistry.

B. Emergence of Organic Chemistry

As early as the sixteenth century scholars seem to have had some realization that the phenomenon of life has chemical attributes. Theophrastus Bombastus von Hohenheim, a Swiss physician and alchemist (ca. 1493–1541), who was also known as Paracelsus, sought to deal with medicine in terms of its "elements" mercury, sulfur, and salt. An ailing person was thought to be deficient in one of these elements and therefore in need of supplementation with the missing substance. Ascribed to Paracelsus were some dramatic "cures." Paracelsus died in exile after quarreling with a local cleric over a fee for professional services.

By the eighteenth century chemists were beginning to recognize the chemical aspects of life processes in a modern sense. Antoine Laurent Lavoisier (1743–1794) recognized the similarity of respiration to combustion in the uptake of oxygen and expiration of carbon dioxide. Lavoisier, charged with watering soldiers' tobacco, was executed under the orders of a revolutionary tribunal with the comment, "The Republic has no need of scientists; justice must take its course."

Certain compounds were clearly associated with living systems; it was observed that these compounds generally contain carbon. They were thought to have arisen from, or to be a consequence of, a "vital force" responsible for the life process. The term *organic* was applied to substances isolated from living things by Jöns Jacob Berzelius (1779–1848), the great Swedish analyst. Somehow, the fact that these chemical substances were organic in nature was thought to put them beyond the scope of the experimentalist. The logic of the time seems to have been that life is not understandable; organic compounds spring from life; therefore, organic compounds are not understandable.

The barrier between organic (living) and inorganic (nonliving) chemistry was penetrated in 1828 by an accidental discovery of Friedrich Wöhler (1800–1882), a German analyst originally trained in medicine. (Many noteworthy discoveries—perhaps more than we might care to admit—have been due to similar "accidents." This phenomenon of accidental discovery has been so widely noted that it has been given a name: *serendipity*.) When Wöhler heated ammonium cyanate, an inorganic compound, he isolated urea, a known urinary excretion product of mammals.

$$\underset{\text{ammonium cyanate}}{\overset{+}{N}H_4NCO^-} \xrightarrow{\text{heat}} \underset{\text{urea}}{H_2N{-}CO{-}NH_2} \tag{1.1}$$

Wöhler recognized that he had synthesized this biological material "without benefit of a kidney, a bladder, or a dog." Not long thereafter followed the synthesis of acetic acid by Kolbe in 1845 and the preparation of acetylene and methane by Berthelot in the period 1856–1863. Although "vitalism" was not so much a textbook theory as an intuitive idea that there might be something special and beyond human grasp about the chemistry of living things, Wöhler did not identify his urea synthesis with the demise of the vitalistic idea; rather, his work signaled the start of a period in which the synthesis of so-called organic compounds was no longer regarded as something outside the province of laboratory investigation. **Organic chemistry** is now recognized as the branch of science that deals generally with compounds of carbon, regardless of origin. Today organic chemists investigate not only molecules of biological importance, but also intriguing molecules of bizarre structure and purely theoretical interest. Through these investigations, chemists learn the limitations of their science and, sometimes through failures in their original goals, learn new and interesting chemistry. Wöhler seems to have anticipated these developments when, a little more than a century ago, he wrote to his mentor Berzelius, "Organic chemistry appears to be like a primeval tropical forest, full of the most remarkable things."

1.2 CLASSICAL THEORIES OF CHEMICAL BONDING

To understand organic chemistry, it is necessary to have some understanding of the **chemical bond**—the forces that hold atoms together within molecules. In this chapter we shall review some of the older or classical ideas of chemical bonding—ideas that, despite their vintage, remain useful today. In Chapter 2 we shall consider more modern ways of describing the chemical bond.

Through the work of Davy and others, it was known in the early nineteenth century that many compounds can undergo **electrolysis:** separation into their component elements through the action of electric current. Thus it seemed reasonable that chemical bonds are basically electrical in nature. As the twentieth century opened, this idea, along with the discovery of the electron, gave rise to complementary models of the chemical bond: the **ionic bond** and the **covalent bond.**

A. The Ionic Bond

To appreciate the theories of the chemical bond we should first review the organization of the periodic table (inside front cover). The shaded elements are of greatest importance in organic chemistry; knowing their atomic numbers and relative positions will be valuable later on. For the moment, however, the following details of the periodic table should be considered, for these were important in the development of concepts of bonding.

The hydrogen atom, atomic number 1, has one nuclear proton and one electron. Each successive element of atomic number Z is characterized by Z protons and Z electrons. (We can forget about neutrons for the moment.) The organization of the periodic table lent itself to the idea that electrons reside in layers, or *shells,* about the nucleus. The outermost shell of electrons in an atom is called its **valence shell,** and the electrons in this shell are called **valence electrons.** *The number of valence electrons for any neutral atom in an A group of the periodic table* (except helium) *equals its group number.* Thus lithium, sodium, and potassium (Group 1A) have one valence electron; carbon (Group 4A) has four valence electrons, the halogens (Group 7A) have seven valence electrons, and the noble gases (except helium) have eight valence electrons; helium has two valence electrons. The German Walther Kossel noted in 1916 that *stable ions are formed when atoms gain or lose valence electrons in order to have the same number of electrons as the noble gas of closest atomic number.* Thus, potassium, with one valence electron (and 19 total electrons), tends to lose an electron to become K^+, the potassium ion, which has the same number of electrons (18) as the nearest noble gas, argon. Chlorine, with seven valence electrons (and 17 total electrons) tends to accept an electron to become the 18-electron chloride ion, Cl^-, which also has the same number of electrons as argon.

Since the noble gases have an octet of electrons (that is, eight electrons) in their valence shells, the tendency of atoms to gain or lose valence electrons to form ions with the noble-gas configuration has been known as the **octet rule.** According to the octet rule, *an atomic species tends to be especially stable when its valence shell contains eight electrons.* (The corresponding rule for elements near helium, of course, is a "duet rule.")

It is not equally easy for all atoms to achieve the valence-electron configuration of a noble gas. It is found that *the ease with which neutral atoms lose electrons to form positive ions increases to the left and toward the bottom of the periodic table.* Thus, the alkali metals, on the extreme left of the periodic table, have the greatest tendency to exist as positive ions; and within the alkali metals, cesium has the greatest tendency to form a positive ion. *The ease with which neutral atoms gain electrons to form negative ions increases to the right and toward the top of the periodic table.* The halogens, which are on the extreme right of the periodic table, have the greatest tendency to form negative ions; and within the halogens, fluorine has the greatest tendency to form a negative ion. (The noble gases are written on the right of the periodic table for convenience, but as we know, these atoms do not easily form ions, and are not considered in the above trends.)

A common type of chemical compound is one in which the component atoms exist as ions. Such a compound is called an **ionic compound.** Potassium chloride, KCl, is a common ionic compound. Because potassium and chlorine come from the extreme left and right, respectively, of the periodic table, it is not surprising that potassium readily exists as the positive ion K^+ and chlorine as the negative ion Cl^-. The electronic configurations of these two ions obey the octet rule. The attraction between these two ions in the compound KCl is an example of an **ionic bond.** Let us examine some of the characteristics of this bond.

In crystalline KCl (Fig. 1.1), the potassium and chloride ions are arranged in a regular array about one another, and the ions are essentially immobile. The attraction between each pair of ions, to a useful approximation, is given by the **electrostatic law** of physics—the law that describes the interaction of two charged particles.

$$E = \frac{q_1 q_2}{\epsilon r_{12}} \tag{1.2}$$

According to this law, the energy E of attraction of two charges depends directly on the magnitudes of the charges, q_1 and q_2, and inversely on the distance between them r_{12}. By convention, the energy is negative (attractive) when the charges have opposite signs. The **dielectric constant,** ϵ, is a property of the medium (solvent) in which the charges are imbedded and is a measure of the ability of the medium to shield charges from each other. The higher ϵ is, the lower is the energy of attraction or repulsion between charges at a given distance. For crystalline KCl, there is only empty space between the ions, and ϵ is unity.

In KCl, a definite charge is associated with each potassium and chloride ion, +1 and −1, respectively. The ions are held together by the electrostatic attraction described by Eq. 1.2. In effect, this equation is a description of the ionic bond. A potassium ion has the same attraction for each of the nearest-neighbor chloride ions that surround it because all are located at the same distance; for the same reason, a chloride has the same attraction for each of its nearest-neighbor potassium ions.

When an ionic compound such as KCl dissolves in water, it dissociates into free ions (each surrounded by water). Each potassium ion moves around in solution more or less independent of each chloride ion. The conduction of electricity by KCl solutions shows that the ions are present. Thus, the ionic bond is broken when KCl dissolves in water. Therefore, we see that the ionic bond is an electrostatic attraction between ions; it is the same in all directions; and it is readily broken when an ionic compound dissolves in water. Atoms at opposite sides of the periodic table are most likely to form ionic bonds.

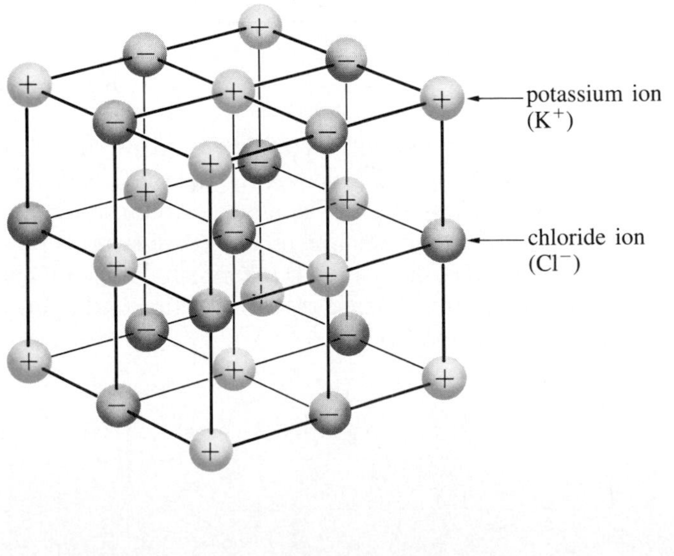

Figure 1.1 Crystal structure of **KCl.** The potassium and chlorine are present in this substance as *ions,* the attractive forces between the potassium and chlorine are purely electrostatic.

Problems

1 When two different species have the same number of electrons they are said to be *isoelectronic*. Name the species that satisfies each of the following criteria:
(a) the negative ion isoelectronic with neon
(b) the dipositive ion isoelectronic with argon
(c) the negative ion isoelectronic with helium
(d) the neon species that is isoelectronic with neutral fluorine

2 How many valence electrons are found in each of the following species?
(a) Na (b) O^{2-} (c) Ca (d) Si (e) Ca^+

B. The Covalent Bond

Consider the compound methanol, CH_4O. Like KCl, this compound is highly soluble in water. In fact, it forms solutions when mixed in any proportions with water. However, neither pure methanol nor its aqueous solution conducts electricity. This behavior, in marked contrast to that of KCl, indicates that methanol is not an ionic compound. How are the forces that hold the atoms together in methanol different from those in KCl? In 1916 G. N. Lewis, an American physical chemist, proposed an electrical model for bonding in nonionic compounds. According to this model, a chemical bond consists of an electron pair that is *shared* between bonded atoms. This type of bond is called a **covalent bond.**

Lewis Structures One of the simplest examples of a covalent bond is the bond between the two hydrogen atoms in the hydrogen molecule.

$$H : H \qquad H{-}H$$

covalent bond

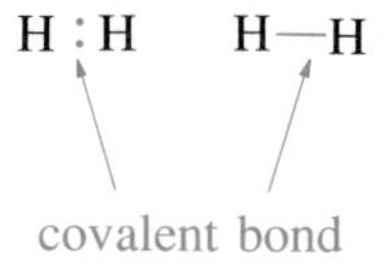

The symbols : and — are both used to denote a covalent bond. Molecular structures that use this notation for the electron-pair bond are called **Lewis structures.** The electron-pair bond holds the two hydrogen atoms together in a stable molecule. Formally, the bond arises from the pairing of the valence electrons of two hydrogen atoms.

$$H \cdot + H \cdot \longrightarrow H : H \qquad\qquad (1.3)$$

Both electrons in the covalent bond are shared equally between the hydrogen atoms. Even though electrons are mutually repulsive, bonding occurs because the electron of one hydrogen atom is attracted to the nucleus (proton) of the other.

Another example of a covalent bond is provided by the simplest stable organic molecule, methane, CH_4. Formally, methane results from the pairing of each of the four carbon valence electrons with a hydrogen valence electron to make four C—H electron-pair bonds.

$$\cdot \overset{\textstyle .}{\underset{\textstyle .}{C}} \cdot + 4H \cdot \longrightarrow H : \overset{\textstyle ..}{\underset{\textstyle ..}{C}} : H \quad \text{or} \quad H{-}\overset{\textstyle H}{\underset{\textstyle H}{C}}{-}H \quad \text{or} \quad CH_4 \qquad (1.4)$$

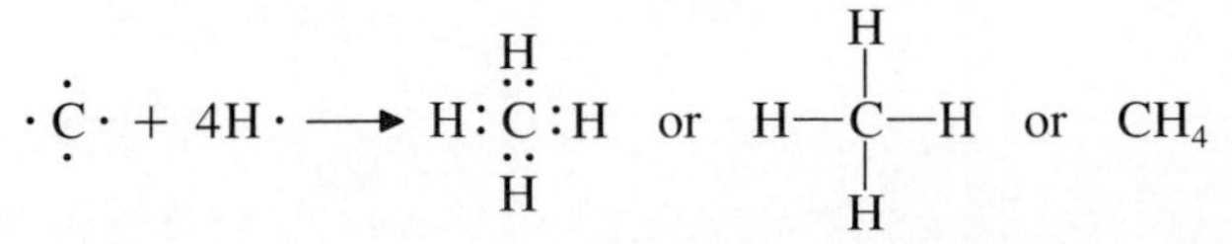

If we count *all* of the electrons involved in the bonds to a particular atom in CH_4 or H_2, we find that each atom is sharing a noble-gas number of electrons. Thus, in both methane and the hydrogen molecule, each hydrogen shares a total of two electrons (remember that the number of electrons for the noble gas helium is two). In methane, the carbon shares a total of eight electrons—two for each electron-pair bond.

Water, H_2O, is another useful example of a covalent compound. Oxygen has six valence electrons. Two of these combine with hydrogens to make an O—H electron-pair bond; four of the oxygen valence electrons are left over. These are represented in the Lewis structure of water as two **unshared pairs** of electrons.

Although we often write water as H—O—H, or even H_2O, it is a good habit to indicate with paired dots all unshared valence electrons until we remember intuitively that they are there.

These examples—H_2, CH_4, and H_2O—illustrate the point that if we count all shared and unshared electron pairs around a given atom, *the octet rule is often obeyed in covalent bonding.*

Atoms in covalent compounds may be bound by more than one covalent bond. The following compounds are well-known examples of this phenomenon:

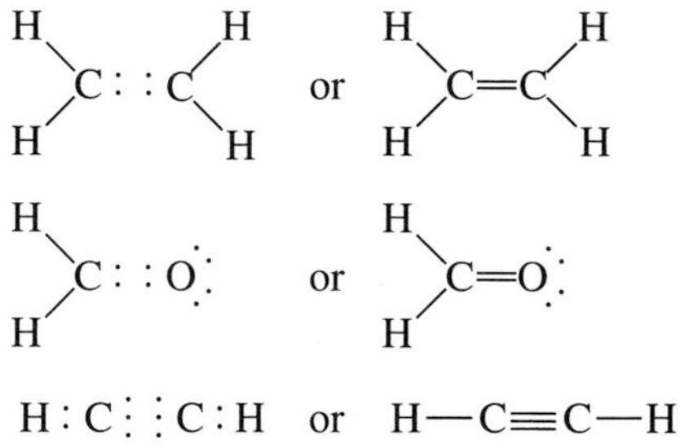

ethylene

formaldehyde

acetylene

Ethylene and formaldehyde each contain a **double bond**—a bond consisting of two electron pairs. Acetylene contains a **triple bond**—one with three electron pairs.

The covalent bond is especially important in organic chemistry because *all organic molecules contain covalent bonds.*

Formal Charge The Lewis structures considered in the previous discussion are those of neutral molecules. However, many familiar ionic species such as SO_4^{2-}, NH_4^+, and BF_4^- also contain covalent bonds. Consider the tetrafluoroborate anion, which contains covalent B—F bonds:

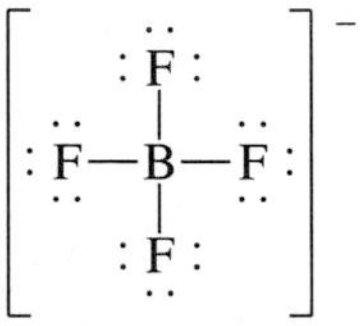

tetrafluoroborate ion

Since the ion bears a negative charge, one or more of the atoms within the ion must be charged—but which one(s)? The rigorous answer is that the charge is shared by all of the atoms. However, there is a useful formalism for electronic bookkeeping that assigns the charge to specific atoms. In this method, we assign a **formal charge** to each atom of the ion; the sum of the formal charges of each atom must, of course, equal the net charge on the ion.

To assign formal charge to an atom, first assign a *valence electron count* to that atom by adding *all unshared* valence electrons on the atom and *half* the electrons in the covalent bonds to the atom. Subtract this electron count from the group number of the atom. The group number is used because it is the number of valence electrons in the neutral atom. The difference between this number and the actual number of valence electrons is the formal charge.

This procedure can be illustrated by applying it to a fluorine atom of BF_4^-.

Group number of fluorine: 7
Valence electron count: 7
(Unshared pairs contribute 6 electrons; chemical bond contributes 1 electron)
Formal charge on fluorine:
Group number − valence electron count = 7 − 7 = 0

Since all fluorine atoms of BF_4^- are the same, they all have zero formal charge. Thus, it follows that the boron must bear the formal negative charge. Let us compute it to be sure. For the boron:

Group number of boron: 3
Valence electron count: 4
(Four chemical bonds contribute 1 electron each)
Formal charge on boron: $3 - 4 = -1$

Because the formal charge of boron is -1, the structure of BF_4^- is written with the minus charge assigned to boron:

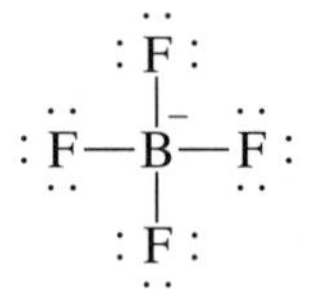

Rules for Writing Lewis Structures The previous two sections can be summarized in the following guidelines for writing Lewis structures:

1. Hydrogen can share no more than two electrons.
2. The sum of all bonding electrons and unshared pairs for atoms in the first full row of the periodic table—the row beginning with lithium—will under no circumstances be greater than eight (octet rule). These atoms may, however, have fewer than eight electrons.
3. In some cases, atoms in the A groups of the periodic table below the first row may have more than eight electrons. However, these cases occur so infrequently that we should also follow Rule 2 for all atoms in the A groups of the periodic table until the exceptions are discussed later in the text. (Unfortunately, these rules are not so easily applied to the transition elements, the elements in the B groups of the periodic table.)

4. Nonvalence electrons are not shown in Lewis structures.
5. The formal charge on each atom is computed by the formalism described in Sec. 1.2B and, if not equal to zero, is indicated with a plus or minus sign.

Notice that we have used two types of electron counting. When we wish to see whether an atom has a complete octet, we count all unshared valence electrons and *all bonding electrons* (Rule 2 above). When we count to determine formal charge, we count all unshared valence electrons and *half of the bonding electrons.*

Problems

3 Draw the Lewis structure for each of the following species. Show all unshared pairs and the formal charges, if any. Assume that the bonding follows the octet rule in all cases.
(a) ammonia, NH_3 (b) ammonium ion, NH_4^+ (c) HCl (d) $CHCl_3$ (e) H_3O^+ (f) BF_3 (g) $[AlCl_4]^-$ (h) methanol, CH_4O

4 Write two reasonable Lewis structures corresponding to the formula C_2H_6O. Assume the molecule adheres to the octet rule; no atoms should have formal charges.

5 Draw a Lewis structure for each of the following compounds. Each contains at least one multiple bond to carbon and no formal charges.
(a) Allene, C_3H_4, contains only double bond(s) to carbon.
(b) Acetonitrile, C_2H_3N, contains a carbon–nitrogen triple bond.

6 Compute the formal charge on each atom of the following complex of BH_3 and NH_3; all valence electrons are shown. What is the charge on the entire complex?

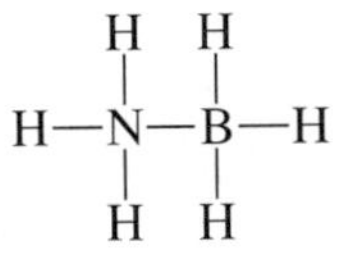

C. The Polar Covalent Bond

In some covalent bonds the electrons are not shared equally between two bonded atoms. Consider, for example, the covalent compound hydrogen chloride, HCl. (Although HCl dissolves in water to form H_3O^+ and Cl^- ions, in the pure gaseous state HCl is a covalent compound.) The electrons in the H—Cl covalent bond are unevenly distributed between the two atoms; they are polarized, or "pulled," toward the chlorine and away from the hydrogen. When electrons in a covalent bond are unequally shared, as in HCl, the bond is called a **polar bond.** (We might think of a polar bond as a covalent bond that is trying to become ionic!) The tendency of an atom to attract the electrons of a covalent bond is indicated by its **electronegativity.** A number of electronegativities are given in Table 1.1. We see from this table that chlorine, for example, has a higher electronegativity than hydrogen. *A polar bond is a covalent bond between two atoms of differing electronegativity.*

The uneven electron distribution in a compound like HCl is measured by a quantity called the **dipole moment.** A dipole moment occurs when a positive and

TABLE 1.1 Average Pauling Electronegativities of Some Main Group Elements

			H **2.20**			
Li **0.98**	Be **1.57**	B **2.04**	C **2.55**	N **3.04**	O **3.44**	F **3.98**
Na **0.93**	Mg **1.31**	Al **1.61**	Si **1.90**	P **2.19**	S **2.58**	Cl **3.16**
K **0.82**	Ca **1.00**	Ga **2.01**	Ge **2.01**	As **2.18**	Se **2.55**	Br **2.96**
Rb **0.82**	Sr **0.95**	In **1.78**	Sn **1.96**	Sb **2.05**		I **2.66**
Cs **0.79**	Ba **0.89**	Tl **2.04**	Pb **2.33**	Bi **2.02**		

negative charge are separated by a distance. A quantitative statement of this definition is Eq. 1.5, in which $\boldsymbol{\mu}$ is the dipole moment and $\mathbf{r}$ is the distance between two charges of absolute magnitude q.

$$\boldsymbol{\mu} = q \cdot \mathbf{r} \qquad \mathbf{r}\ \overset{+q}{\underset{-q}{\downarrow}} \quad \text{direction of } \mathbf{r} \text{ and } \boldsymbol{\mu} \tag{1.5}$$

The dipole moment is a vector quantity; thus, it has not only magnitude but also direction. The direction of the dipole moment vector $\boldsymbol{\mu}$ is the same as that of the $\mathbf{r}$ vector, which by convention runs from the positive charge to the negative charge. The debye is the standard dipole moment unit; a dipole moment of one debye (1 D) results when opposite charges of 1×10^{-10} electrostatic unit (esu) are separated by one angstrom (Å). One debye therefore equals, by Eq. 1.5, 10^{-10} esu-Å.

Let us apply Eq. 1.5 to the calculation of the charge separation in HCl. It is found experimentally that HCl has a dipole moment of 1.08 D ($\boldsymbol{\mu}$ in Eq. 1.5). The bond length of HCl ($\mathbf{r}$ in Eq. 1.5) is 1.274 Å. Solving for q, we find that q is 0.85×10^{-10} esu. Thus, the polarity of the covalent bond in HCl is equivalent to the separation of two opposite charges of this magnitude by the length of the HCl bond. The meaning of the dipole moment becomes clearer when we express the charge units in terms of the charge on the electron, 4.8×10^{-10} esu. The amount of extra charge on the Cl of HCl is

$$\frac{0.85 \times 10^{-10} \text{ esu}}{4.8 \times 10^{-10} \text{ esu/electron}} = 0.18 \text{ electron} \tag{1.6}$$

Thus, H—Cl behaves as if it has an excess of 0.18 electron on the chlorine and a deficiency of 0.18 electron (that is, 0.18 proton charge) on the hydrogen. We shall sometimes indicate such an electron distribution in a qualitative way by structures such as the following:

$$\overset{\delta+}{\text{H}}\text{——}\overset{\delta-}{\text{Cl}}$$

The delta (δ) is read *partially,* so that the hydrogen of HCl is *partially positive,* and the Cl is *partially negative.*

A molecule such as HCl that possesses a permanent dipole moment is called a **polar molecule.** The magnitude of the dipole moment is a measure of the polarity of a molecule.

The fact that the dipole moment is a vector quantity has important consequences when a molecule contains more than one polar bond. Consider, for example, the molecule carbon dioxide, CO_2.

bond dipoles

$$\overset{\longleftarrow\;+\quad+\;\longrightarrow}{O=C=O}$$

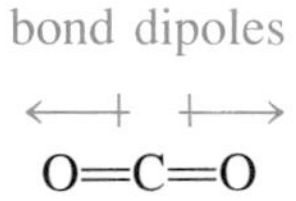

Because the oxygens, which are more electronegative than the carbon (Table 1.1), attract electrons, there is a dipole moment contribution aligned from carbon to oxygen along each C=O bond. Such dipole moment contributions from individual bonds are called **bond dipoles.** Despite the presence of C=O bond dipoles, CO_2 has a zero dipole moment, because CO_2 is a linear molecule, and the bond dipoles are oriented in opposite directions (more precisely, aligned at 180° to each other). The dipole moment is the vector sum of all bond dipoles within a molecule. Just as two identical forces oriented in opposite directions sum to zero by vector addition, the two oppositely oriented C=O bond dipoles cancel to give a net dipole moment of zero for CO_2. Thus, CO_2 has polar bonds, but it is not a polar molecule!

The polarity of a molecule can significantly affect its chemical and physical properties. For example, its polarity may give some indication how a molecule reacts chemically. Returning to the HCl molecule, we know that HCl in water dissociates to its ions in a manner suggested by its bond polarity.

$$\overset{\delta+\;\;\delta-}{H-Cl} \xrightarrow{H_2O} H^+ + Cl^- \tag{1.7}$$

We shall find many similar examples in organic chemistry in which bond polarity provides a clue to chemical reactivity.

Problems

7 Analyze the polarity of each bond in the following organic compound. Which bond, other than the C—C bond, is the least polar one in the molecule? Which carbon has the most partial positive character?

$$\begin{array}{c} H \quad\; O \\ | \quad\;\; \| \\ H-C-C-Cl \\ | \\ Cl \end{array}$$

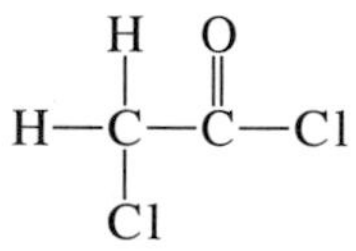

8 What is the direction of the N—H bond dipoles in the ammonium ion, NH_4^+? Does the assignment of the formal charge in this ion agree with your analysis? (Read on for a resolution to this problem.)

We should notice that our procedure for specifying formal charge sometimes appears to place the formal charge on the wrong end of a polar bond (see Problem 8). In the ammonium ion, NH_4^+, nitrogen has a formal charge of +1 and the hydrogen has a formal charge of zero; yet nitrogen is the more electronegative of the two atoms. In reality the hydrogens bear significant positive charge and the charge on the nitrogen is less than +1. Similarly, in the BH_4^- anion, the formal negative charge is placed on boron, but in reality the hydrogens bear a significant fraction of the negative charge. We must remember that formal charge is merely a device for electronic bookkeeping; it ensures that all charges have been properly accounted for. That it does not always correspond to the real electronic situation does not lessen its utility.

1.3 LEWIS ACIDS AND BASES: THE ARROW FORMALISM FOR LEWIS ACID–BASE REACTIONS

Although we can safely assume that none of the atoms in the compounds we shall encounter in our early studies possesses more than an octet of electrons, some compounds may contain atoms with fewer than an octet of electrons. Three common examples of these are the following:

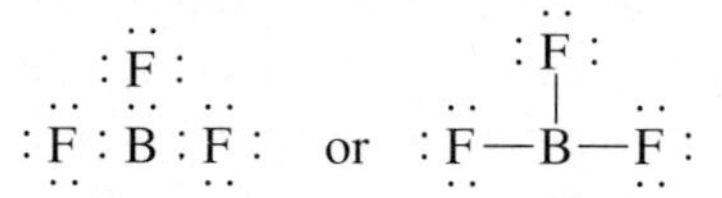

boron trifluoride, BF_3

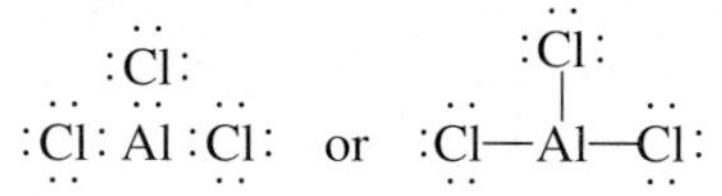

aluminum chloride, $AlCl_3$

H^+ the proton, or hydrogen ion

The boron of BF_3 and the aluminum of $AlCl_3$ both have a sextet of electrons in their valence shells; the proton has no electrons. Each of these species is two electrons short of the noble-gas valence-shell configuration. Species such as this show a common tendency to undergo chemical reactions that complete their valence-shell octets. In other words, *there is a driving force in the chemical reactions of covalent compounds toward an electronic octet about each atom* (or a "duet" for hydrogen). This is comparable to the similar drive in the elements to form ions with complete valence shells. Two examples of chemical reactions that illustrate this idea are shown in the following equations:

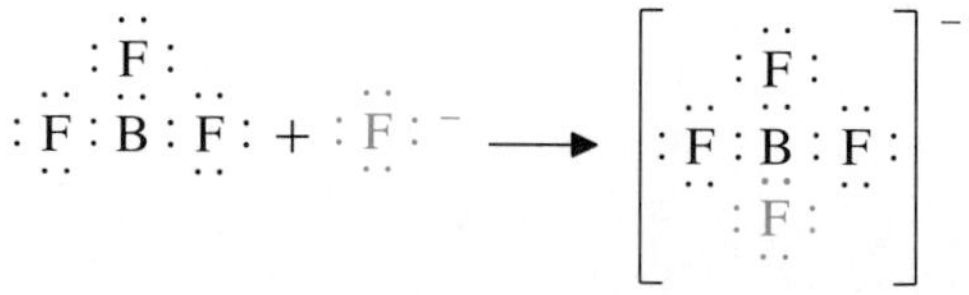

tetrafluoroborate ion

(1.8)

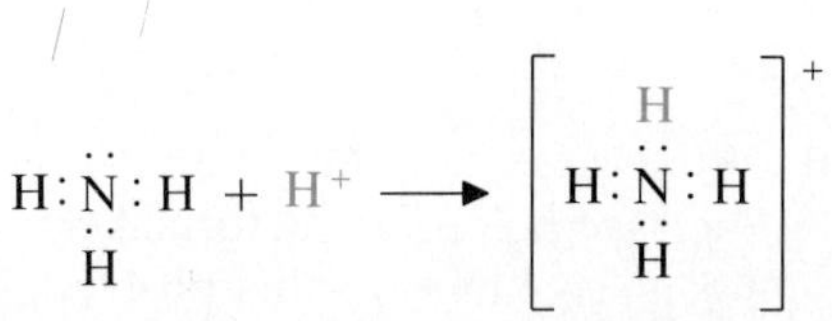

ammonium ion

(1.9)

In Eq. 1.8 boron, with an incomplete valence shell in BF_3, has an octet in the BF_4^- ion. Similarly, in Eq. 1.9 the proton fills out its incomplete valence shell when it reacts with NH_3 to achieve the inert-gas configuration in the NH_4^+ ion. In both reactions one species donates an electron pair to another species with an incomplete octet.

Species that are short of an electronic octet by one or more electron pairs are called **Lewis acids.** Species that react with Lewis acids by donating an electron pair are called **Lewis bases.** In Eqs. 1.8 and 1.9 above, BF_3 and H^+ are Lewis acids; F^- and NH_3 are Lewis bases. Organic chemists commonly use other names for Lewis acids and bases. Lewis acids are frequently called **electrophiles** (*phile* = loving; *electrophile* = electron-loving); Lewis bases are called **nucleophiles** (nucleus-seeking, or "lover of electron-deficient centers"). Thus, Lewis acid–Lewis base reactions are reactions between electrophiles and nucleophiles.

Methane, CH_4, cannot act as a Lewis acid, or electrophile, because its carbon already has a complete electronic octet and every hydrogen has two electrons. It cannot act as a Lewis base, or nucleophile, because it cannot donate an electron pair.

There is a peculiarity in the octet counting formalism evident in Eq. 1.8. The fluoride ion has an octet. When it donates an electron pair to BF_3, the fluoride still has the octet in the product BF_4^-. Students frequently ask, "How can fluorine have an octet both before and after it donates electrons?" The answer is that we count unshared pairs of electrons in the fluoride ion, and in BF_4^-, we count as fluorine electrons unshared pairs as well as *both* electrons in the newly formed chemical bond. The electrons in this bond are also counted as part of the boron octet. There is an apt analogy to this situation in the marriage of a poor gentleman (we might call him Lewis A.) to a wealthy lady. Before marriage, the gentleman is poor and the lady is wealthy; after marriage, the lady is still wealthy, but the gentleman is wealthy by marriage! The justification for this practice of counting electrons twice is that it provides an extremely useful framework for predicting chemical reactivity. The ultimate justification for any theory is that it explains old phenomena and predicts new phenomena; no other justification is needed. Note again that the procedure used in counting electrons for the octet differs from the one used in calculating formal charge (Sec. 1.2B).

Lewis acid–base reactions are extremely common in all of chemistry, including organic chemistry. For this reason organic chemists have developed a symbolic device for keeping track of electrons in these reactions, called the **curved-arrow formalism.** In this formalism, the formation of a chemical bond is described by a formal "flow" of an electron pair *from an electron donor to an electron acceptor.* This electron flow is indicated by a curved arrow drawn *from the electron source to the position of the newly formed bond at the electron acceptor.* This formalism is illustrated as follows for the reaction of Eq. 1.8:

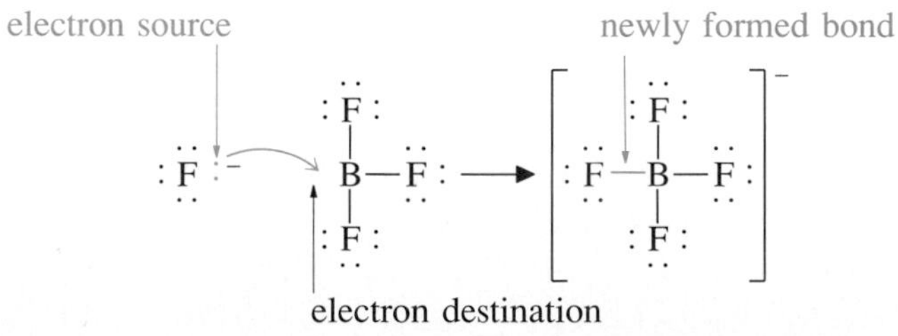

(1.10)

The colored arrow indicates that the electron pair on the fluoride ion formally becomes the shared electron pair in the newly formed bond of the tetrafluoroborate anion. The arrow formalism for the reverse reaction is as follows:

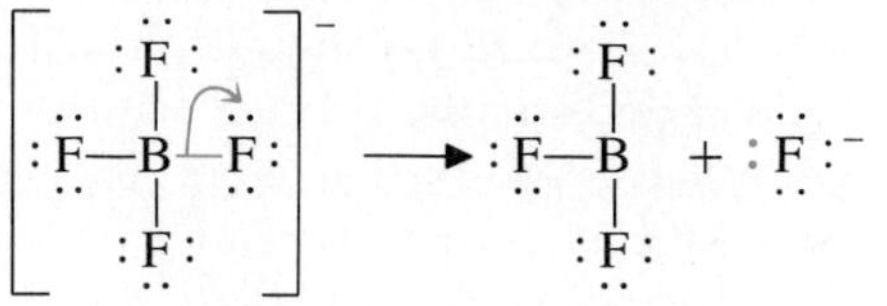
(1.11)

In this case, the arrow is drawn *from the position of the breaking bond to the atom that will bear the newly formed unshared electron pair.* This indicates that the electrons in a B—F electron-pair bond have become an unshared pair on fluorine.

Problem

9 (a) Suggest a structure for the product of each of the following Lewis acid–base reactions. Be sure to assign formal charges

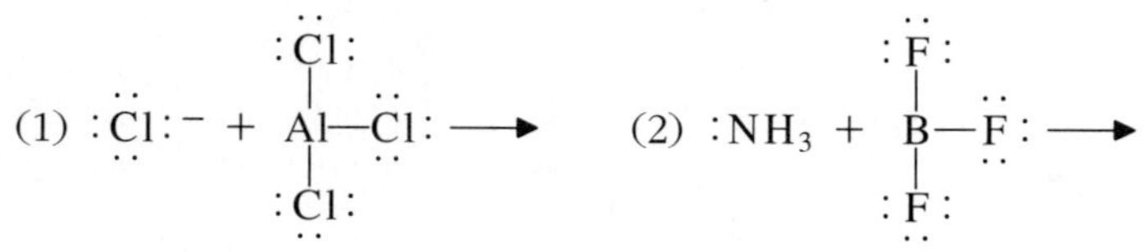

(b) Draw the arrow formalism for the forward and reverse of each of these reactions.

1.4 BRØNSTED–LOWRY ACIDS AND BASES: THE ARROW FORMALISM FOR DISPLACEMENT REACTIONS

The Lewis acid–base concept is but one of several definitions of acids and bases. Perhaps the most familiar idea of acids and bases is the **Brønsted–Lowry acid–base concept.** This definition of acids and bases was published in 1923, the same year that Lewis formulated his ideas of acidity and basicity. A **Brønsted acid** is any species that can donate a proton; a **Brønsted base** is any species that can accept a proton. Many Brønsted bases are also Lewis bases.

The reaction of ammonium ion with hydroxide ion is a simple Brønsted acid–base reaction.

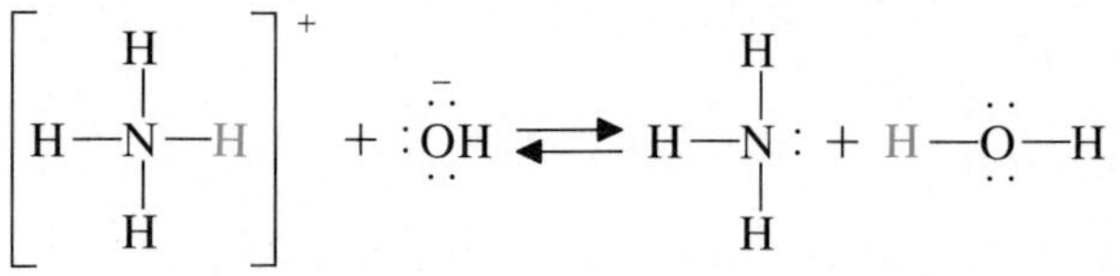
(1.12)

On the left side of this equation, NH_4^+ is acting as a Brønsted acid and OH^- is acting as a Brønsted base. Looking at the equation from right to left, H_2O is acting as a Brønsted acid, and NH_3 is acting as a Brønsted base. In this equation, NH_4^+ and NH_3 are a **conjugate acid–base pair;** an acid loses a proton to form its conjugate base; a base gains a proton to form its conjugate acid. Thus, NH_4^+ acts as an acid on one side of the equation to form, by loss of a proton, its conjugate base NH_3 on the other side of the equation. Likewise, H_2O and OH^- are a conjugate acid–base pair. Note that the conjugate acid–base relationship is *across* the equilibrium arrows; that is, NH_4^+ and OH^- are *not* a conjugate acid–base pair, because one is not converted into the other by loss of a proton.

A given compound may act as a base in one reaction and as an acid in another. A common example of such behavior is water. In Eq. 1.12, for example, water is the conjugate acid of the acid–base pair H_2O/OH^-; in the following reaction, water is the conjugate base of the acid–base pair H_3O^+/H_2O:

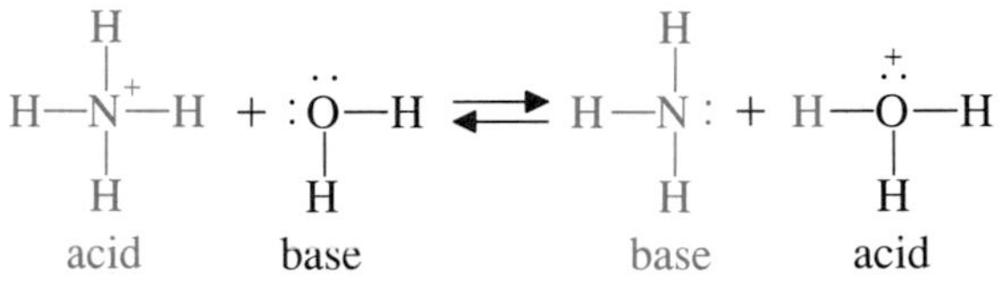

$$\text{(1.13)}$$

acid base base acid

The Brønsted acid–base reaction is one of the most general reactions of chemistry—including organic chemistry. Many organic reactions are made possible by the transfer of a proton to or from an organic molecule. For this reason it is important to master the ideas of Brønsted–Lowry acidity and basicity. But there is another reason for the importance of this concept. *A large number of organic reactions can be readily understood as simple analogies of acid–base reactions.* A specific example of this idea is the reaction of methyl chloride, CH_3—Cl, with the hydroxide ion.

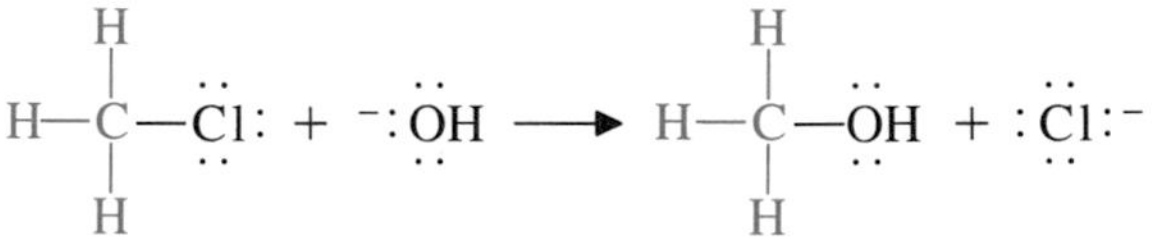

$$\text{H}-\overset{\overset{\displaystyle\text{H}}{|}}{\underset{\underset{\displaystyle\text{H}}{|}}{\text{C}}}-\overset{..}{\underset{..}{\text{Cl}}}:\ +\ ^-:\overset{..}{\underset{..}{\text{O}}}\text{H} \longrightarrow \text{H}-\overset{\overset{\displaystyle\text{H}}{|}}{\underset{\underset{\displaystyle\text{H}}{|}}{\text{C}}}-\overset{..}{\text{O}}\text{H}\ +\ :\overset{..}{\underset{..}{\text{Cl}}}:^- \qquad \text{(1.14a)}$$

A simple acid–base analogy is the reaction of OH^- with HCl.

$$\text{H}-\overset{..}{\underset{..}{\text{Cl}}}:\ +\ ^-:\overset{..}{\underset{..}{\text{O}}}\text{H} \longrightarrow \text{H}-\overset{..}{\underset{..}{\text{O}}}\text{H}\ +\ :\overset{..}{\underset{..}{\text{Cl}}}:^- \qquad \text{(1.14b)}$$

Notice that the formal replacement of the hydrogen of HCl with a methyl (CH_3) group has no apparent effect on the outcome of the reaction. In fact, the equilibria in both reactions lie well to the right. This example illustrates that *one can learn much about what to expect in organic reactions merely by examining familiar acid–base analogies.*

There is also a curved-arrow formalism for Brønsted acid–base reactions. This can be illustrated by the reaction of Eq. 1.12. The hydroxide ion, OH^-, reacts with a proton of NH_4^+. Because this proton in NH_4^+ already has its full complement of two electrons, it cannot accept electrons from OH^- without losing some in return. Consequently, the Brønsted acid–base reactions of OH^- with NH_4^+ is shown as a *displacement* of electrons from the proton using the arrow formalism.

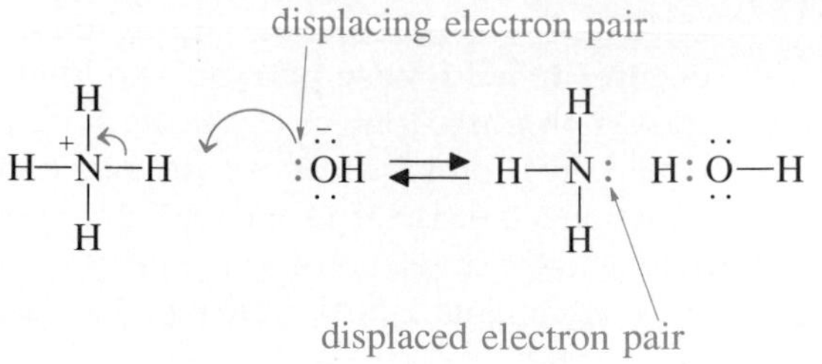

$$(1.15)$$

Notice that the curved arrow shows the movement of electrons, not atoms. The arrow, as in the Lewis arrow formalism, is drawn *from the source of the electrons to their destination,* and the displaced electrons are indicated to move away from the site of displacement onto the nitrogen. Note also the accompanying changes in formal charge.

Another instructive example of the displacement arrow formalism is provided by the reaction of BH_4^- with H_2O.

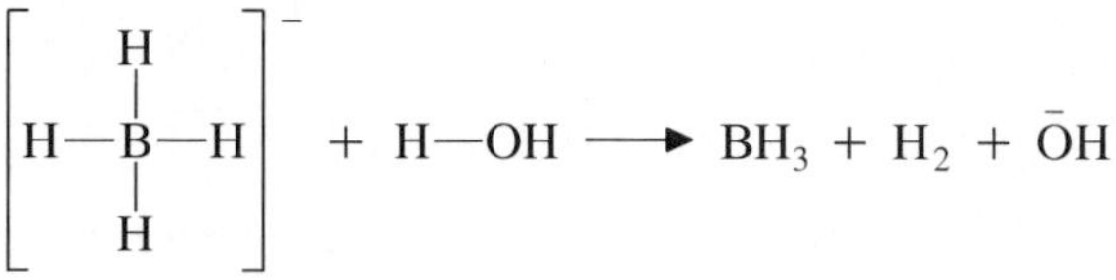

$$\left[\begin{array}{c} H \\ | \\ H-B-H \\ | \\ H \end{array} \right]^- + \ H-OH \longrightarrow BH_3 + H_2 + \bar{O}H \qquad (1.16)$$

In this case, the displacing electrons are derived not from an electron pair, but from a chemical bond—the B—H bond.

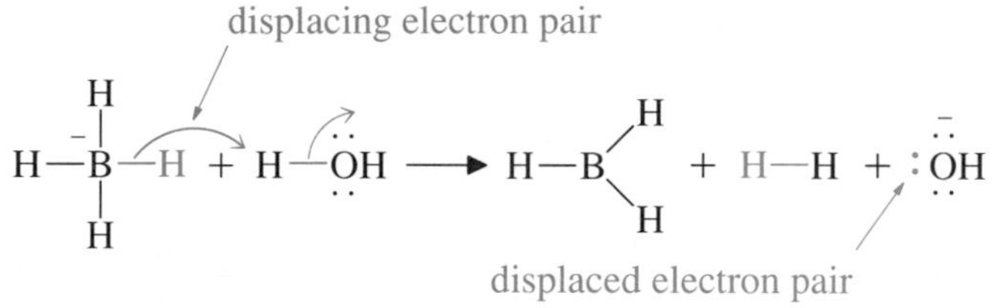

$$(1.17)$$

Does the arrow formalism tell us anything fundamental about how electrons move in chemical reactions? The answer is no. However, it does represent an extremely simple device for electronic bookkeeping, and it enables us to see quickly how the different atoms in a chemical reaction come and go between reactants and products. This formalism is widely used by practicing organic chemists, and will be used throughout this book.

Now that we have considered Lewis and Brønsted acids separately, it is instructive to contrast their behavior. The methyl cation is a Lewis acid; the ammonium ion is a Brønsted acid.

methyl cation ammonium ion

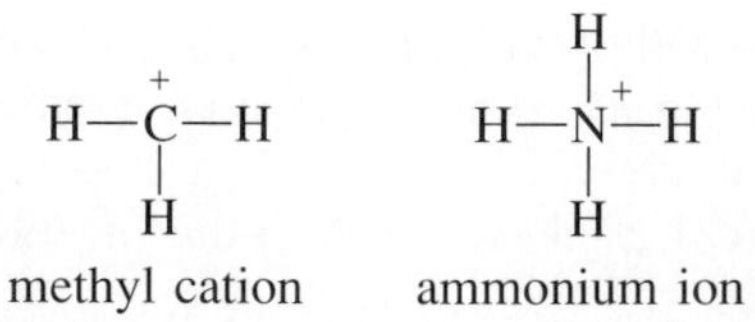

Both species are cations, but they behave chemically in fundamentally different ways. The carbon of the methyl cation lacks an electronic octet, so it can react with Lewis bases (nucleophiles).

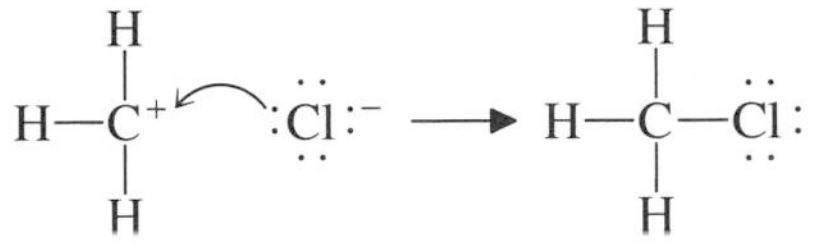

methyl chloride
(a covalent compound)

(1.18)

The nitrogen of the ammonium ion is also positively charged; however, this atom has a complete octet, so it *cannot* be attacked by nucleophiles. Such a reaction would violate the octet rule for nitrogen (Rule 2, Sec. 1.2B).

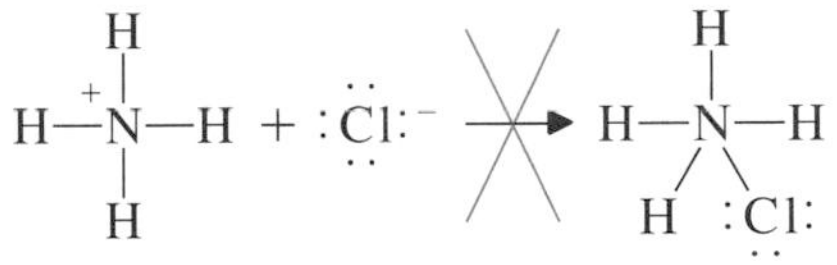

(1.19)

The ammonium ion does, however, react as a Brønsted acid by losing a proton to Brønsted bases as shown, for example, in Eq. 1.12.

Problems

10 In the following reactions, label the conjugate acid–base pairs. Then draw the arrow formalism for these reactions in the left-to-right direction.

$$\ddot{N}H_3 + :\ddot{O}H^{-} \rightleftharpoons :\ddot{N}H_2^{-} + H_2\ddot{O}:$$

$$\ddot{N}H_3 + \ddot{N}H_3 \rightleftharpoons :\ddot{N}H_2^{-} + \overset{+}{N}H_4$$

11 Write a Brønsted acid–base reaction in which $H_2\ddot{O}/:\ddot{O}H^{-}$ and $CH_3\ddot{O}H$ (methanol)/$CH_3\ddot{O}:^{-}$ act as conjugate acid–base pairs.

12 Give the arrow formalism for the reaction shown in Eq. 1.14a.

1.5 A REVIEW OF THE ARROW FORMALISM

The arrow formalism will help us follow the course of complex reactions only if we learn to apply it properly. A student who understands the application of the arrow formalism to simple Lewis or Brønsted acid–base reactions has all the tools necessary to understand the application of the formalism to more complex reactions.

Given a compound and the associated arrow formalism for the change in bonding associated with a chemical reaction, we want to be able to write the structure of the product. Consider, for example, the reaction of the following two compounds.

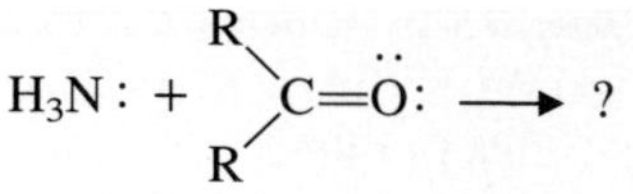

$$H_3N: + \quad \begin{matrix} R \\ \diagdown \\ C=\ddot{O}: \\ \diagup \\ R \end{matrix} \longrightarrow \quad ? \tag{1.20}$$

At this stage of our study of organic chemistry, we cannot know in advance how these two molecules will react; but *given the arrow formalism,* we should be able to complete the reaction.

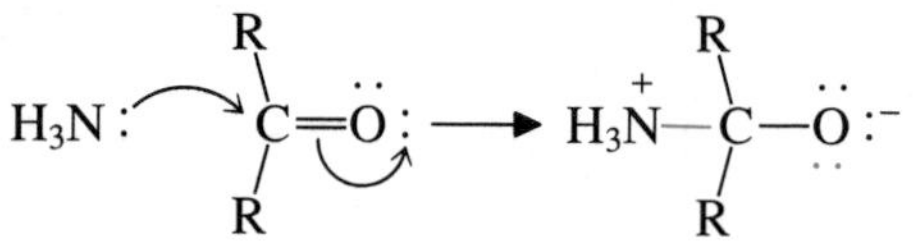

$$H_3N: \quad \overset{R}{\underset{R}{\diagup}} C=\ddot{O}: \longrightarrow H_3\overset{+}{N}-\overset{R}{\underset{R}{C}}-\ddot{O}:^- \qquad \begin{matrix} - & \text{new bond} \\ \\ : & \text{new unshared} \\ & \text{electron pair} \end{matrix} \tag{1.21}$$

The bonds or unshared electron pairs of the reactant that are no longer in the same relative position in the product are those at the tails of the arrows. The heads of the arrows point to the places in which new bonds or unshared electron pairs exist in the product. Using this example, let us examine in detail the steps that should be followed in using the arrow formalism.

1. Redraw all atoms just as they were in the reactant.

$$H_3N \quad \begin{matrix} R \\ \\ R \end{matrix} \quad C \quad O$$

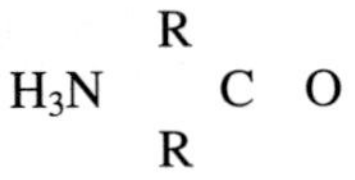

2. Put in the bonds and electron pairs that do not change.

$$H_3N \quad \begin{matrix} R \\ \diagdown \\ \diagup \\ R \end{matrix} C-\ddot{O}:$$

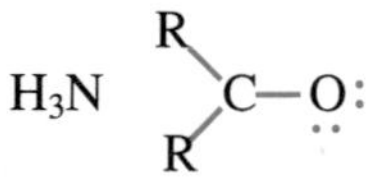

3. Draw the new bonds or electron pairs indicated by the arrow formalism.

new bond new electron pair

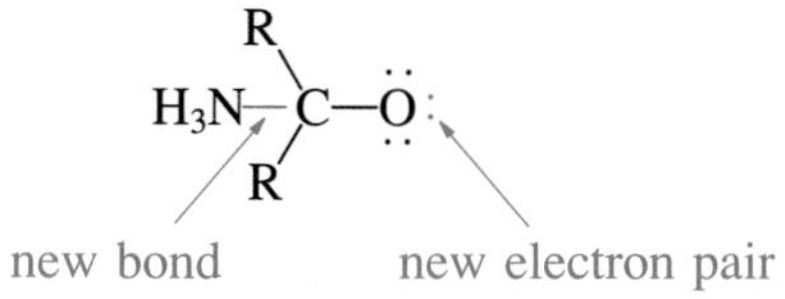

4. Be sure that the formal charges are correct.

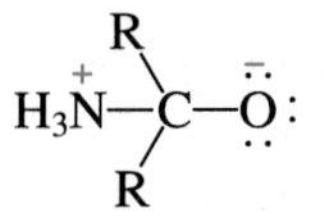

$$H_3\overset{+}{N}-\overset{R}{\underset{R}{C}}-\overset{..}{\underset{..}{O}}: \qquad \text{(Sum of formal charges is zero, as in reactants.)}$$

The last structure represents the product of this transformation.

The arrow formalism is a language that must be used precisely. It is crucial to remember that it represents the formal *flow of electrons,* not the flow of atoms. One of the more common errors that students make with this formalism is to confuse the flow of electrons with the flow of atoms. For example, the proton transfer from HCl to OH^- is sometimes *incorrectly* depicted as follows:

$$HO:^- \quad (H)-Cl: \longrightarrow\!\!\!\!\times \quad HO-H \quad :Cl:^- \qquad \text{Incorrect!}$$

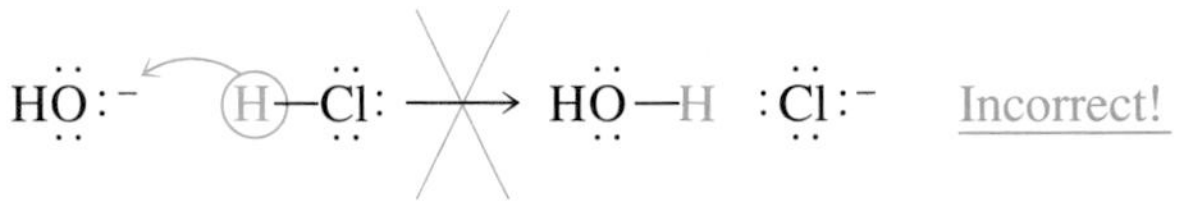

This is not correct because it shows the movement of the proton rather than the movement of electrons. Someone accustomed to using the formalism correctly would take this to imply the transfer of H^- to OH^-, an absurd reaction! In the *correct* use of the arrow formalism only the flow of *electrons* is shown.

$$HO:^- \quad H-Cl: \longrightarrow \quad HO-H \quad :Cl:^- \qquad \text{Correct!}$$

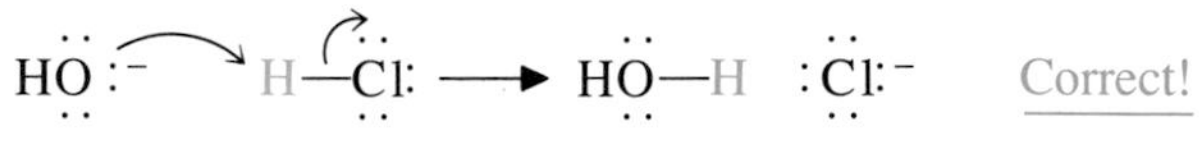

Remember: arrows are always drawn from the source of *electrons* to their destination.

Problems

13 For each of the following cases, give the product(s) of the transformation indicated by the arrow formalism:

(a) $HO:^-$ $CH_2-Cl:$ with CH_3 →

(b) H_2C, H_2C with O, O, O: + →

(c) H_3Al-H , $CH_3-Br:$ →

(d) CH_2 , C , H_3C CH_3 , $H-Br:$ →

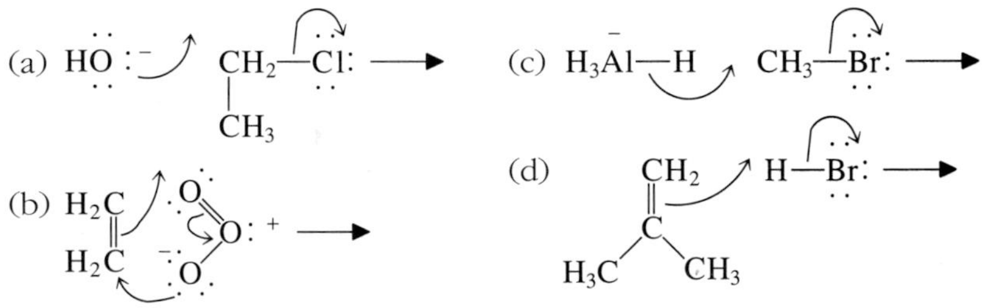

14 Suggest an arrow formalism for the following reaction, which occurs in a single step.

$$H_3N: + CH_3-Br: \longrightarrow H_3\overset{+}{N}-CH_3 \quad :Br:^-$$

1.6 STRUCTURES OF COVALENT COMPOUNDS

Most of us realize from our earlier chemical training that each covalent chemical compound has a **structure**—that is, a definite arrangement of its constituent atoms in space. The concept of covalent compounds as three-dimensional objects was developed in the latter part of the last century. Chemists who lived before that time regarded covalent compounds as shapeless groups of atoms held together in a rather undefined way by poorly understood electrical forces. Although the currently accepted structural characteristics of organic compounds were first suggested in 1874, these postulates were based on indirect chemical and physical evidence. Until the early twentieth century no one knew whether they had any physical reality, since scientists had no techniques for viewing molecules at the atomic level of resolution. Thus, as recently as the second decade of the twentieth century investigators could ask two questions: (1) Do organic molecules have specific geometries and, if so, what are they? (2) Can we develop simple principles to predict molecular geometry?

A. Methods for Determining Molecular Structure

Among the greatest developments of chemical physics in the early twentieth century were the discoveries of ways to peer into molecules and deduce the arrangement in space of their constituent atoms. Most information of this type today comes from three sources: X-ray crystallography, electron diffraction, and microwave spectroscopy. The arrangement of atoms in the crystalline solid state can be determined by **X-ray crystallography.** This technique, discovered in 1915 and revolutionized in recent years by the availability of high-speed computers, uses the fact that X rays are diffracted from the atoms of a crystal in precise patterns that can be translated into a molecular structure. In 1930 another technique, **electron diffraction,** was developed from the observation that electrons are scattered by the atoms in molecules of gaseous substances. The diffraction patterns resulting from this scattering can also be used to deduce the arrangements of atoms in molecules. Following the development of radar in World War II came **microwave spectroscopy,** in which the absorption of microwave radiation by gaseous molecules provides structural information.

Most of the details of atomic structure in this book are derived from gas-phase methods—electron diffraction and microwave spectroscopy. For molecules that are not readily studied in the gas phase, X-ray crystallography is the most important source of structural information. There is no comparable method that allows the study of structures in solution, a fact that is unfortunate because most chemical reactions take place in solution. The consistency of gas-phase and crystalline structures suggests, however, that molecular structures in solution probably differ little from those of molecules in the solid or gaseous state.

B. Prediction of Molecular Geometry

The geometry of a simple covalent molecule is defined by two quantities, bond length and bond angle. The **bond length** is defined as the distance between the centers of bonded nuclei. Bond length is usually measured in angstroms; 1 Å = 10^{-10} meter = 10^{-8} centimeter. **Bond angle** is the angle between two bonds to the same atom. Consider, for example, the compound methane, CH_4. When the C—H bond length and H—C—H bond angles are known, we know the structure of methane.

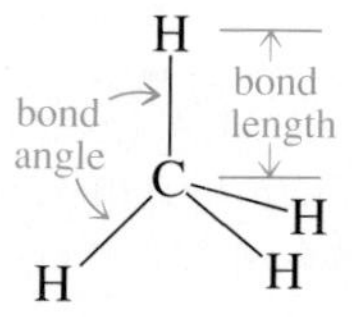

Molecular structure is important because the way a molecule reacts chemically is closely linked to its structure. From the many structures that have been determined, we can now make several generalizations about the structures of covalent compounds.

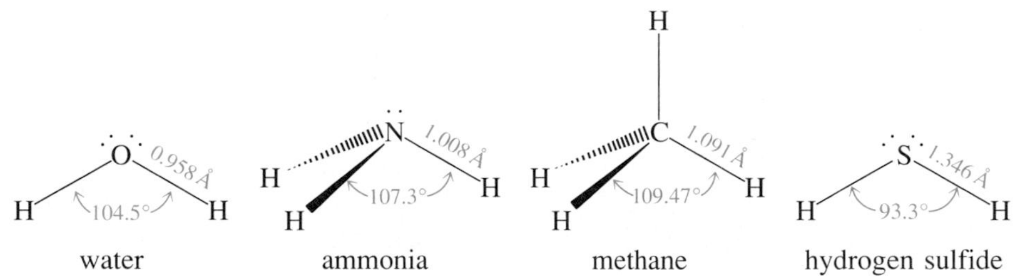

Figure 1.2 Structures of simple molecules. Within each of these molecules, all bonds to hydrogen are equivalent.

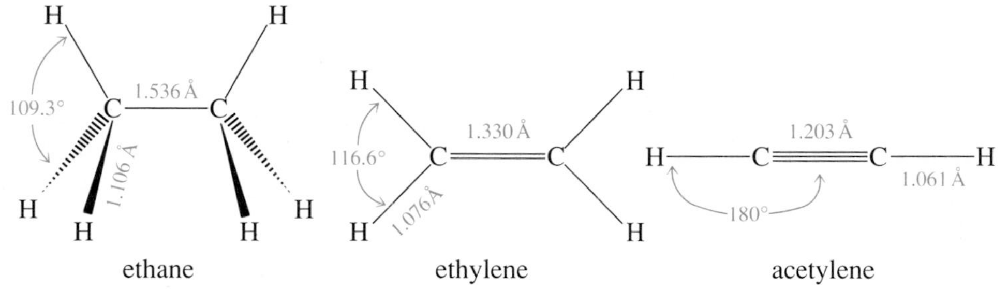

Figure 1.3 Structures of hydrocarbons with different types of carbon–carbon bonds.

Bond Length The following generalizations can be made about bond length:

1. *Bond lengths between atoms of a given type decrease with the amount of multiple bonding.* Thus, bond lengths for carbon–carbon bonds are in the order C—C > C═C > C≡C.

2. *Bond lengths tend to increase with the size of the bonded atoms.* This effect is most dramatic as we proceed down the periodic table. Thus, a C—H bond is shorter than a C—F bond, which is shorter than a C—Cl bond. Since bond length is the distance between the centers of bonded atoms, it is reasonable that larger atoms should form longer bonds.

3. When we make comparisons within a given row of the periodic table, *bonds of a certain type (single, double, or triple) between a given atom and a series of other atoms become shorter with increasing electronegativity.* Thus, the C—F bond in H_3C—F is shorter than the C—C bond in H_3C—CH_3. This effect occurs because a more electronegative atom has a greater attraction for the electrons of the bonding partner, and therefore "pulls it closer," than a less electronegative atom.

Figures 1.2 and 1.3 show the structures of some simple covalent organic molecules. You should show how each of the three previous generalizations is illustrated by the examples in these figures.

Bond Angle The bond angles within a molecule determine its *shape*—whether it is bent or linear. Two generalizations allow us to predict the approximate bond angles, and therefore the general shapes, of many simple molecules. The first is that *the*

Figure 1.4 Structure of methane. The tetrahedral shape is emphasized with shading.

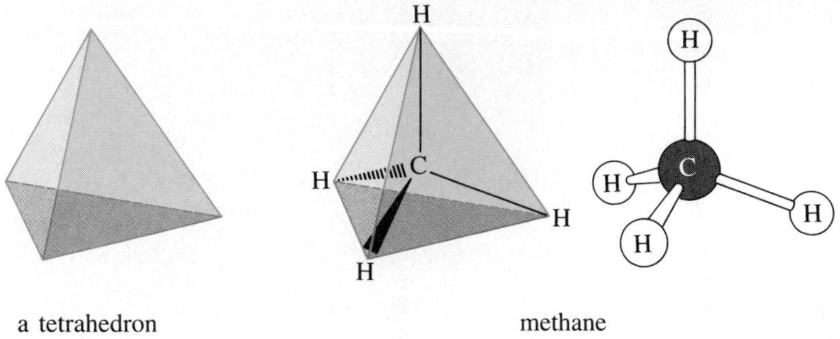

groups bound to a central atom are arranged so that they are as far apart as possible. For methane, CH_4, the central atom is carbon and the groups are the four hydrogens. The hydrogens of methane are farthest apart when they occupy the vertices of a *tetrahedron* centered on the carbon atom. The tetrahedral structure of methane is shown in Fig. 1.4. Because the four C—H bonds of methane are identical, the hydrogen atoms of methane lie at the vertices of a *regular tetrahedron* (a tetrahedron with equal sides). The tetrahedral shape of methane requires an H—C—H bond angle of 109.5° (Fig. 1.2).

In applying this rule for the purpose of predicting bond angles, we regard all groups as identical. Thus the groups that surround carbon in H_3CCl (methyl chloride) are treated as identical, even though in reality the C—Cl bond is considerably longer than the C—H bonds. Although the bond angles show minor deviations from the exact tetrahedral bond angle of 109.5°, methyl chloride in fact has the general tetrahedral shape.

The tetrahedral structure, then, is assumed by molecules when four groups are arranged about a central atom. Since carbon is tetravalent, this is an extremely important geometry for many organic compounds. Since we shall see this geometry repeatedly, it is worth the effort to become familiar with it. Tetrahedral carbon is often represented, as shown below in the structure of methylene chloride, CH_2Cl_2, with two of its bound groups in the plane of the page. One of the remaining groups, indicated with a dashed line, is behind the page, and the other, indicated with a wedge-shaped heavy line, is in front of the page.

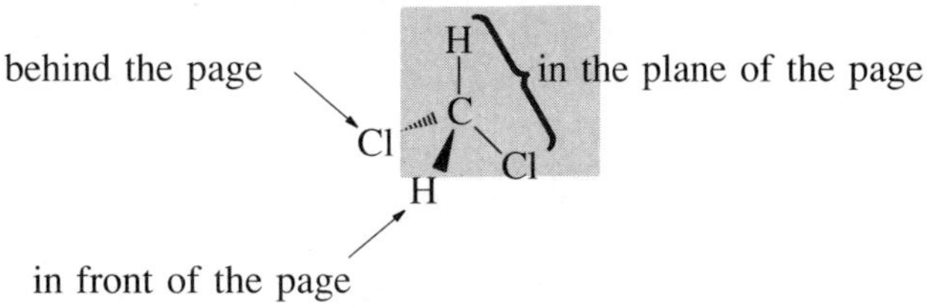

You should make a *molecular model* of a simple tetrahedral molecule such as CH_2Cl_2 and relate the three-dimensional model to its two-dimensional representation above.

Chemists often use **molecular models** to study questions of molecular geometry. It is a rare person who can grasp the three-dimensional geometrical concepts in organic chemistry without recourse to models. Some of the types of models available are shown in Fig. 1.5. We strongly recommend that you procure an inexpensive set of molecular models and use them frequently when a question of geometry is discussed.

Figure 1.5 Molecular models of methane, **CH_4.** Models (a) and (b) are of the ball-and-stick type, often inexpensive and suitable for use by beginning students. The Dreiding model (c) is a model of the framework type, which shows only the chemical bonds. In the space-filling model (d), the size of each atom is proportional to its covalent or atomic radius.

When *three* groups surround a central atom, the groups are as far apart as possible when all bonds lie in the same plane with bond angles of 120°. This is, for example, the geometry of boron trifluoride:

In such a situation the central atom (in this case boron) is said to have **trigonal** geometry.

When a central atom is surrounded by *two* groups, maximum separation of the groups demands a bond angle of 180°. This is the situation with each carbon in acetylene, H—C≡C—H. Each carbon is surrounded by two groups, a hydrogen and another carbon. (It makes no difference that the carbon has a triple bond.) Because of the 180° bond angle at each carbon, acetylene is a linear molecule.

acetylene

The second generalization about molecular structure applies to molecules with unshared electron pairs. In predicting the geometry of a molecule, *an unshared electron pair can be considered as a bond without a nucleus at one end.* This rule allows us to handle, for example, the geometry of ammonia. In view of this rule, ammonia, $:NH_3$, has four groups about the central nitrogen, three hydrogens and an electron pair. To a first approximation, these groups adopt the tetrahedral geometry with the electron pair occupying one corner of the tetrahedron. (This geometry is sometimes called **pyramidal,** or pyramid-like, geometry.) We can refine our prediction of geometry even more if we recognize that an electron pair without a nucleus at one end has an especially repulsive interaction with electrons in adjacent bonds. As a result, the bond angle between the electron pair and the N—H bond is a little larger than tetrahedral, leaving the H—N—H angle a little smaller than tetrahedral; in fact, the H—N—H angle is 107.3°.

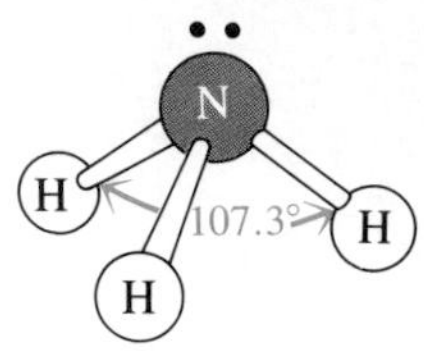

ammonia

Problem

15 Predict the approximate geometry in each of the following molecules:

(a) water

(b) the BF_4^- anion

(c) $H_2C{=}\ddot{O}$ (formaldehyde)

(d) $CH_3{-}C{\equiv}N{:}$ (acetonitrile)

If we examine the structure of a molecule that is somewhat more complex than those we have just discussed, we discover one further aspect of structure. Consider, for example, the structure of ethylene, $H_2C{=}CH_2$. Each carbon of ethylene is bound to three groups: two hydrogens and the other carbon. Our rules for predicting bond angle require, then, that the geometry at each carbon should be trigonal. The structure of ethylene in Figure 1.3 shows that the H—C—H bond angle, 117°, indeed approaches the idealized trigonal value of 120° quite closely, and that the three groups bonded to either of the carbons lie in the same plane. However, ethylene poses a new problem of geometry. Imagine sighting along the carbon–carbon double bond from one end of the molecule, as shown in Figure 1.6. The resulting angle between the C—H bonds on *adjacent* carbons is called the **dihedral angle.** The value of the dihedral angle determines the **conformation** of ethylene: the spatial relationship of the groups on one carbon to those on the other. Two extreme conformations are shown in Fig. 1.6. In the top drawing, the dihedral angle between nearest hydrogens is 0°; in the bottom drawing, this angle is 90°. Both conformations satisfy the trigonal geometry of carbon. (Be sure you understand this point!) At this stage of our study we cannot predict which conformation is the preferred one, but it turns out that ethylene exists in only one of these two conformations. In Chapter 4 we shall learn which conformation is preferred and why.

In summary, then, the structure of a molecule is completely determined by three elements: its bond lengths, its bond angles, and its conformation. There are simple molecules whose structures are completely determined by bond lengths and bond angles. We have learned what trends in bond lengths to expect and how to predict the shapes of molecules from bond angles. Conformation enters the picture for more complex molecules. We shall learn at various stages of our study some of the principles that determine molecular conformation.

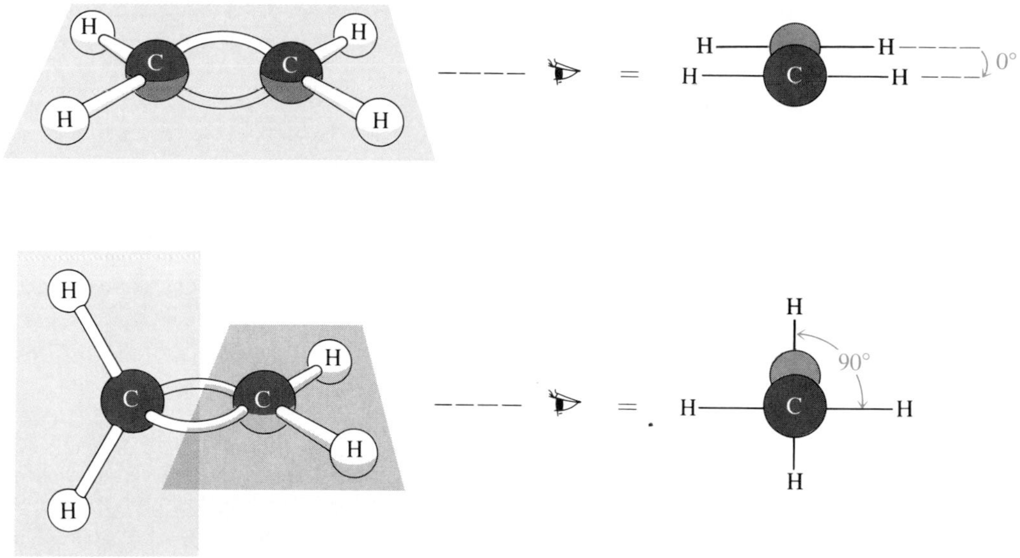

Figure 1.6 Definition of the possible conformations and dihedral angles in ethylene. The **H—C—C—H** dihedral angle is the angle between the **C—H** bonds on adjacent carbons when one sights along the **C=C** bond from one end of the molecule. In planar ethylene the dihedral angle between nearest hydrogens is 0°. In the nonplanar structure, the dihedral angle is 90°.

Problem

16 Using molecular models, examine the conformations of ethane, CH_3—CH_3. What are the dihedral angles between nearest hydrogens when (a) they are as far apart as possible? (b) they are as close together as possible? Draw a diagram like the one in Figure 1.6 to illustrate these two conformations.

1.7 RESONANCE STRUCTURES

Many compounds are not accurately described by a single Lewis structure. Consider, for example, nitromethane, H_3C—NO_2.

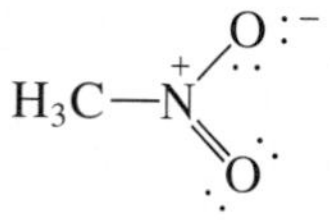

This Lewis structure shows an N—O single bond and an N=O double bond. On the basis of the last section, we expect double bonds to be shorter than single bonds. However, it is found experimentally that the two nitrogen–oxygen bonds of nitromethane have the same length, intermediate between the length expected for a single bond and that for a double bond. In order to convey this idea, nitromethane can be written as follows:

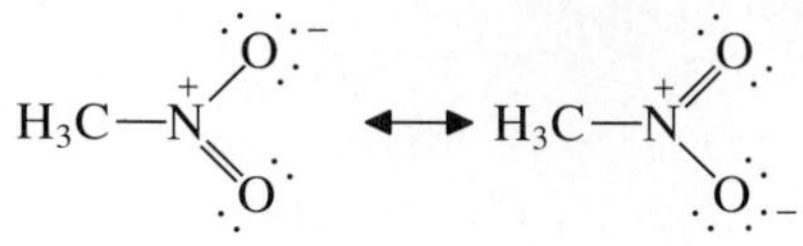

The double-headed arrow means that nitromethane is a single compound that has characteristics of both structures; nitromethane is said to be a **resonance hybrid** of these two structures. The double-headed arrow does *not* mean that the two structures are in rapid equilibrium; rather, nitromethane is *one* compound with *one* structure.

Suppose Joe lives in a purple house. Suppose also that his friend E. T. asks him, "What color is your house?" Joe finds that E. T. has never heard of the color purple, and Joe does not have a purple crayon or pen handy. Assuming that E. T. is familiar with the colors red and blue, Joe might express the color of his house as a resonance hybrid.

red house ◄──► blue house

The arrow formalism can be used to convert one resonance structure into another. The displacement arrow formalism, like the one used in Brønsted acid–base reactions, can be used to derive the resonance structures for nitromethane.

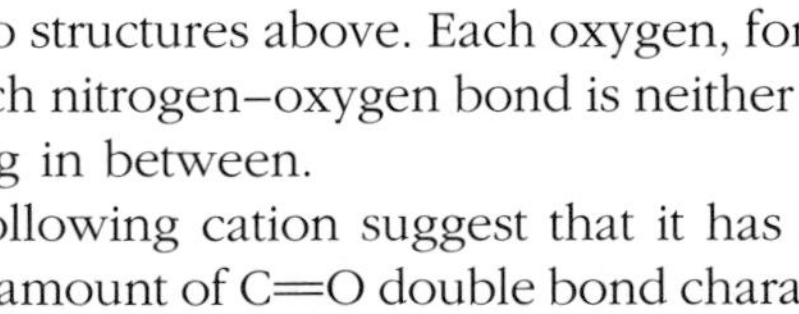

When two resonance structures are identical, as they are for nitromethane, they are equally important in describing the molecule. We can think of nitromethane as a 50 : 50 average of the two structures above. Each oxygen, for example, bears half of a negative charge, and each nitrogen–oxygen bond is neither a single bond nor a double bond, but something in between.

Properties of the following cation suggest that it has both positive charge on carbon and a significant amount of C=O double bond character; these characteristics can be shown with resonance structures.

$^+CH_2$—$\ddot{O}$—CH_3 ◄──► CH_2=$\overset{+}{O}$—CH_3 methoxymethyl cation

We should notice in this case that the structure on the right can be derived from the one on the left by donation of an oxygen unshared electron pair to the carbon that is two electrons short of an octet. The arrow formalism resembles the one used in Lewis acid–base reactions.

These two structures for the methoxymethyl cation are not identical, and they are not equally important in describing the ion. When two resonance structures of a molecule differ, the molecule is described as a *weighted average* of its structures. For example, the methoxymethyl cation is more accurately described by the structure on the right than by the one on the left. Later in our study, when we know more about the structures of organic molecules, we shall learn how to decide on the relative importance of resonance structures.

How do we know when to draw resonance structures? Before we can answer this question we shall have to develop a greater familiarity with organic compounds and their properties. In Chapter 15 we shall return to a study of resonance in more detail. In the meantime you can rely on the text to tell you when to use resonance structures. From this information you will develop an intuition that will prove helpful when we study resonance again. For now it will be helpful to remember the following points about the use of resonance structures:

1. Resonance structures are written for compounds whose properties are not adequately described by a single Lewis structure.
2. Resonance structures are *not* in equilibrium; they are alternate descriptions of one species.
3. The structure of a molecule is the *weighted average* of its resonance structures. When resonance structures are identical, they are equally important descriptions of the molecule.
4. Resonance structures may be derived by use of the arrow formalism.

Problem

17 The compound benzene has only one type of carbon–carbon bond, and this bond is found to have a length intermediate between that of a single bond and a double bond. Draw a resonance structure of benzene that, taken with the structure below, accounts for the carbon–carbon bond length. Use the arrow formalism to derive your structure.

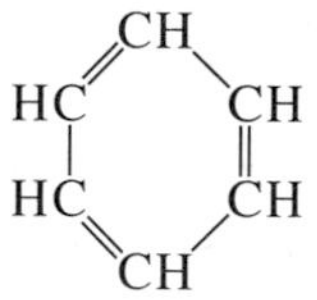

benzene

KEY IDEAS IN CHAPTER 1

- Chemical compounds contain two types of bonds: ionic and covalent. In ionic compounds ions are held together by electrostatic attraction (the attraction of opposite charges). In covalent compounds atoms are held together by the sharing of electrons.

- Both the formation of ions and the reactions of covalent compounds of the main-group elements generally follow the octet rule: there is a tendency for each atom to be surrounded by eight valence electrons.
- In covalent compounds, electrons may be shared unequally between bonding partners. This unequal sharing imparts to the bond a bond dipole. The vector sum of all bond dipoles in a molecule is its dipole moment.
- The formal charge convention allows us to assign the total charge of a given species to its constituent atoms.
- The curved-arrow formalism is used to diagram bonding changes in chemical reactions by showing the formal flow of electron pairs. Reactions that employ the arrow formalism are of two types. In one, Lewis acid–base reactions, an electron pair on one atom forms a bond to another atom that lacks an electron octet. In the other, an electron pair on one atom displaces an electron-pair bond on another. This is illustrated by proton transfer in the Brønsted–Lowry acid–base reaction.
- The structure of a molecule is determined by its bond lengths, bond angles, and conformation. Bond angles, which determine shape, may be predicted for simple molecules by assuming that the groups bound to a central atom are as far apart as possible.
- Molecules that are not adequately described by a single Lewis structure are represented by resonance structures. A molecule is the weighted average of its resonance structures. Resonance structures can be derived from each other by use of the curved-arrow formalism.

ADDITIONAL PROBLEMS

18 Of the following compounds, which one is most likely to exist as free ions in its liquid state?
(a) CS_2 (b) CCl_4 (c) CsF (d) HF (e) XeF_4

19 Which of the atoms in the following species have a complete octet? What is the formal charge on each? Assume all valence electrons are shown.
(a) CH_3 (b) $:NH_3$ (c) $:CH_3$ (d) BH_3 (e) $:\ddot{I}:$

20 Draw one Lewis structure for each of the following compounds; show all unshared electron pairs. None of the compounds bears a net charge, and all atoms except hydrogen have octets.
(a) C_3H_6O (b) C_2H_3Cl (c) C_5H_5N
(d) ketene, C_2H_2O, which has a carbon–carbon double bond
(e) a compound other than allene (Problem 5a), C_3H_4

21 Which of the following compounds are Lewis acids? Explain.

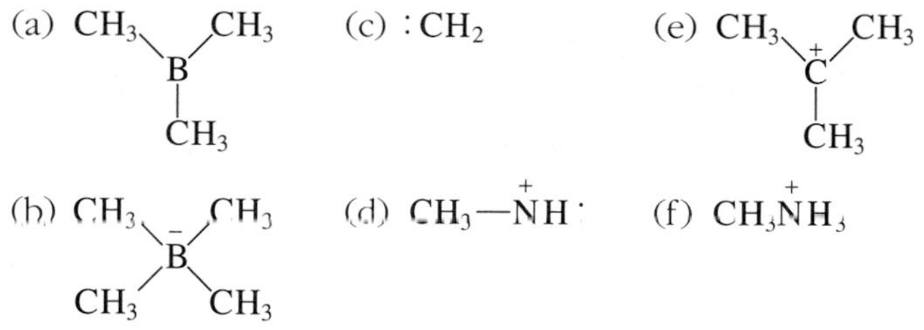

22 Give the formal charge on each atom and the net charge on each species in each of the following structures. All unshared valence electrons are shown.

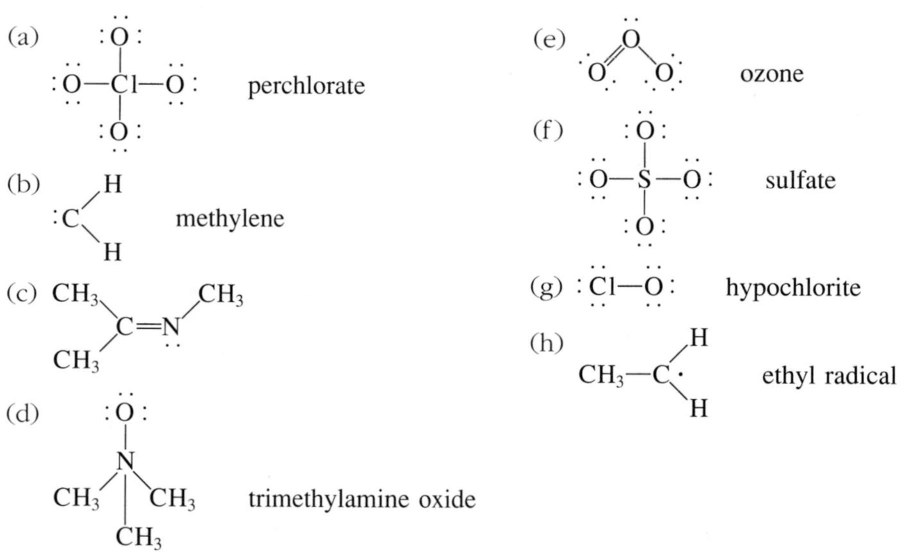

23 Which of the following Lewis acid–base reactions are reasonable, and which are not? Explain.

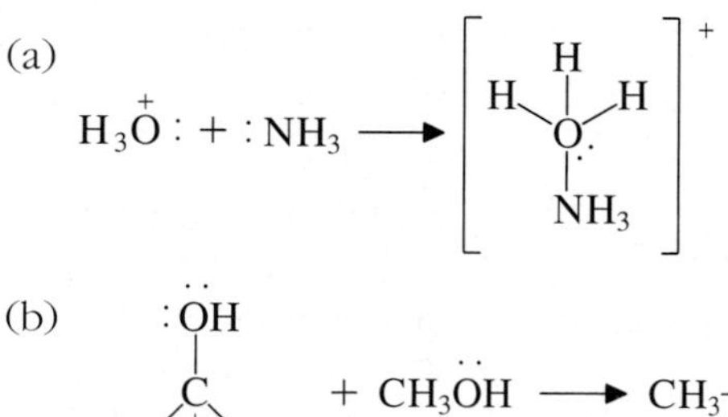

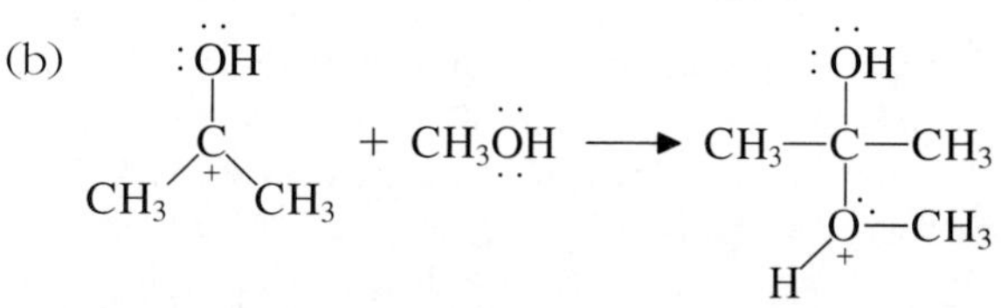

Problems (Cont.) (c)

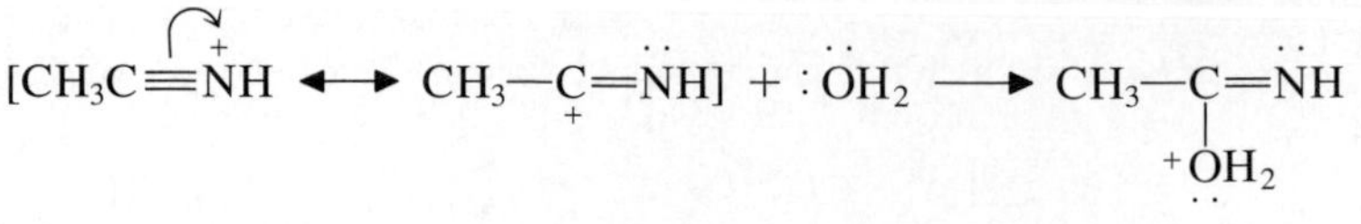

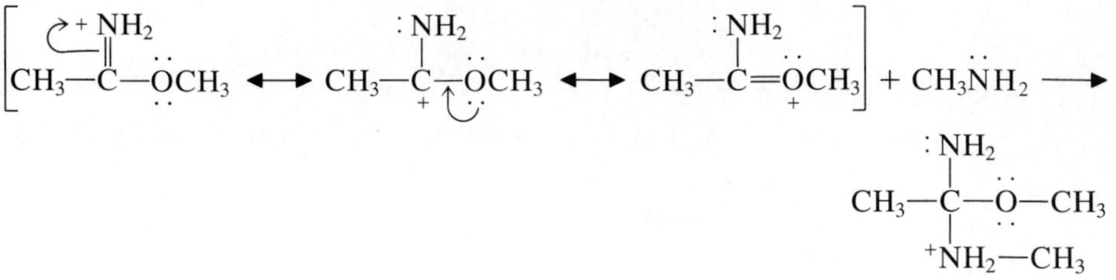

(e) The starting ion in (d)

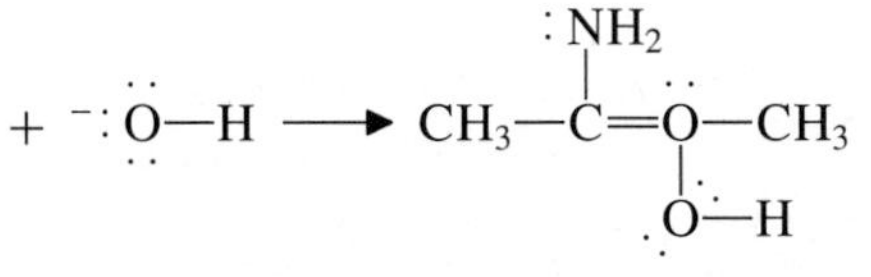

24 Give the arrow formalism for and predict the immediate products of each of the following Lewis acid–base reactions:

(a) $H_3C—\overset{..}{\underset{..}{O}}—CH_3 + BF_3 \longrightarrow$

(b)
$$CH_3—\overset{CH_3}{\underset{+}{\overset{|}{C}}}—CH_3 + :\overset{..}{\underset{..}{Cl}}:^- \longrightarrow$$

(c)
$$H_3C—\overset{:\overset{..}{O}H}{\underset{+}{\overset{|}{C}}}—CH_3 + CH_3—\overset{..}{N}H_2 \longrightarrow$$

(d)
$$CH_3—\overset{..}{\underset{..}{Mg}}—\overset{..}{\underset{..}{Cl}}: + 2 \underset{CH_2}{\overset{CH_2}{\underset{\underset{CH_2}{|}}{\overset{CH_2}{\overset{|}{}}}}}\overset{..}{\underset{..}{O}}: \longrightarrow$$

(e) $HO—CH_2$... $\longrightarrow$

(f)
$$:C\overset{H}{\underset{H}{\diagup}} + CH_3\overset{..}{N}H_2 \longrightarrow$$

(g) $(H_3C)_3 B + {}^-:C{\equiv}O:^+ \longrightarrow$

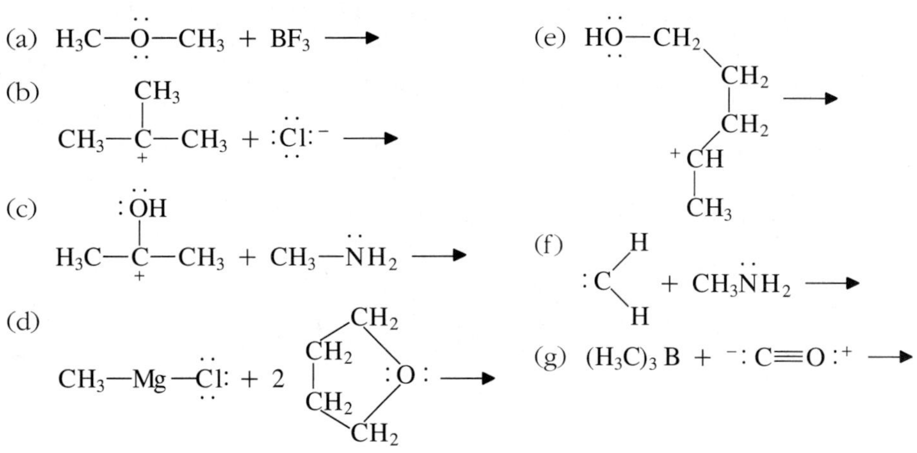

25 The reaction of $^-BH_4$ and water proceeds to completion as follows:

$$\bar{B}H_4 + 4H_2O \longrightarrow \bar{B}(OH)_4 + 4H_2$$

The reaction occurs in several steps, the first of which is shown in Eq. 1.17. Notice that BH_3 is a Lewis acid and OH^- is a Lewis base. Show the complete sequence of individual steps leading to the products shown above, and give the arrow formalism for each step.

26 Predict the approximate bond angles in each of the following molecules:

(a) BeH_2

(b) $:CH_2$

(c) $:\overset{..}{\underset{..}{Cl}}—Si(CH_3)_3$ (Give Cl—Si—C angle.)

(d) $\overset{+}{C}H_3$

(e) $\overset{..}{\underset{..}{O}}=\overset{+}{\underset{..}{O}}—\overset{..}{\underset{..}{O}}:^{-}$ ozone

(f) $CH_2=C=CH_2$ allene (Give H—C—C and C—C—C angles.)

(g)
$$H_3C—\overset{+}{N} \begin{smallmatrix} \nearrow \overset{..}{\underset{..}{O}}:^{-} \\ \searrow \underset{.}{\overset{..}{O}} \end{smallmatrix}$$

(h) $(H_3C)_3\overset{+}{N}—\overset{..}{\underset{..}{O}}:^{-}$

(i) $H_3\overset{+}{O}:$

(j)
$$\begin{smallmatrix} ^{-}:\overset{..}{O} \searrow \\ \\ ^{-}:\underset{..}{O} \nearrow \end{smallmatrix} C=\overset{..}{\underset{..}{O}}$$ carbonate ion

(k) $Mn(H_2O)_6^{2+}$, a coordination compound of six waters with the Mn^{2+} ion. (*Hint*: How would you arrange six equivalent ligands (the water molecules) about a central atom (the manganese) so that the ligands are as far apart as possible?)

27 In the following ion identify
(a) the longest bond(s) (c) the two bonds that are equivalent by resonance
(b) the shortest bond(s) (d) the most polar bond(s)

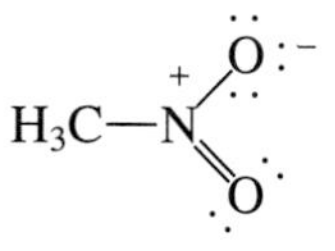

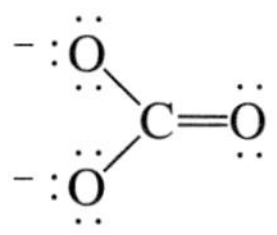

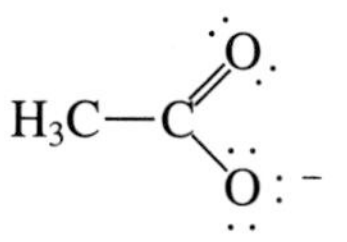

acetate ion

28 Explain why CH_3Cl (dipole moment 1.90 D) and CH_3F (dipole moment 1.85 D) have almost identical dipole moments although fluorine is more electronegative than chlorine.

29 The dipole moment of HF is 1.75 D, and the HF bond length is 0.92 Å. What is the apparent charge separation in HF, expressed in terms of the charge of an electron?

Problems (Cont.)

30 In each of the following Brønsted acid–base reactions, label the conjugate acid–base pairs. Then give the arrow formalism for each reaction

(a)

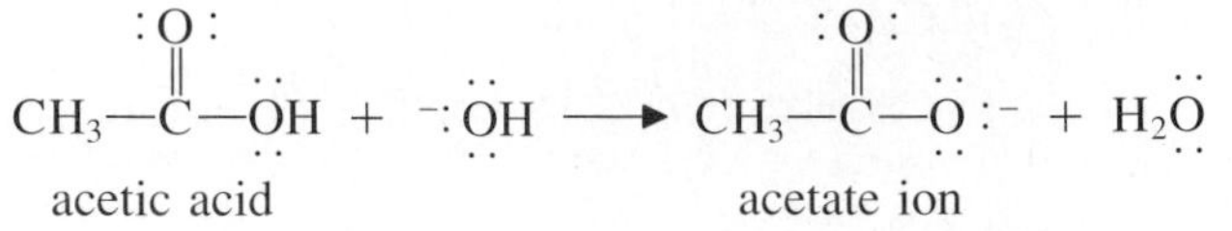

(b)

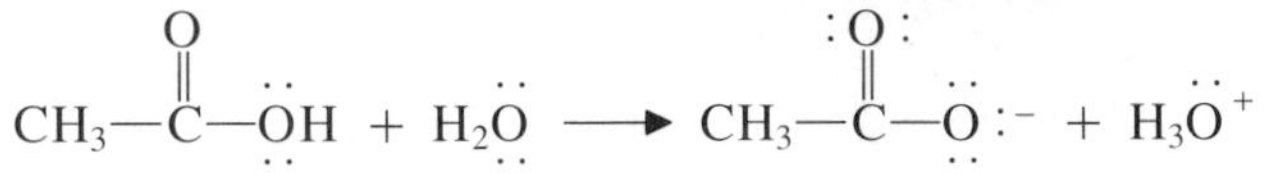

(c) $(CH_3)_3\overset{+}{N}H + CH_3\ddot{\underset{..}{O}}:^- \longrightarrow (CH_3)_3N: + CH_3\ddot{\underset{..}{O}}{-}H$

(d)

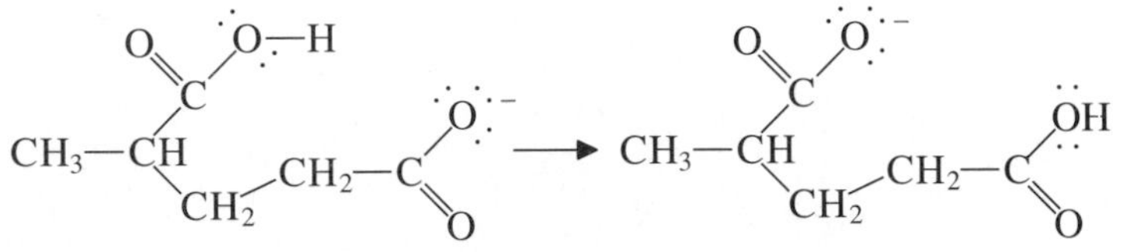

(e)

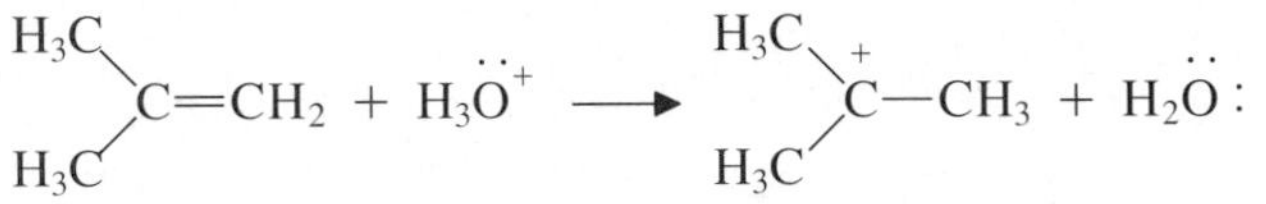

(f)

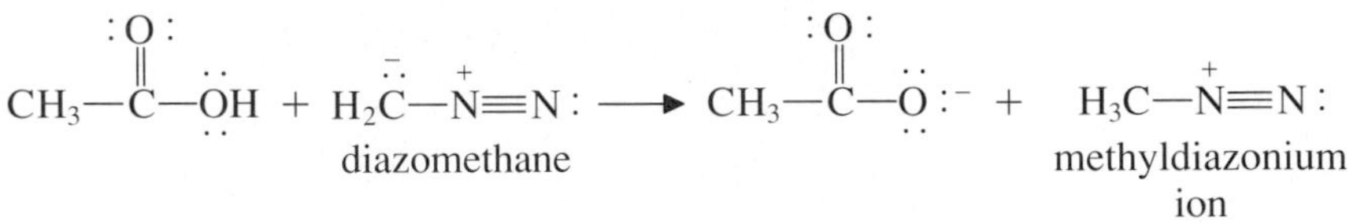

31 A well-known chemist, Havno Szents, has heard you apply the rules for predicting molecular geometry to water; you have proposed (Problem 15a) a bent geometry for this compound. Dr. Szents is unconvinced by your arguments and continues to propose that water is a linear molecule. He demands that you debate the issue with him before a learned academy. You must therefore come up with *experimental* data that will prove to an objective body of scientists that water indeed has bent geometry. Explain why the dipole moment of water, 1.84 D, could be used to support your case.

32 When a solvent dissolves an ionic compound, it brings about the separation of the ions that were very close together in the crystal. What type of solvent would best dissolve ions: one with a low dielectric constant or one with a high dielectric constant? Explain. (*Hint*: See Eq. 1.2.)

33 In Sec. 1.6B we noted that the principles for predicting bond angles do not permit a distinction between the following two conceivable forms of ethylene:

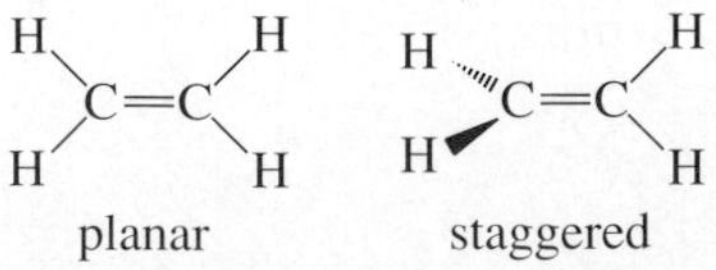

The dipole moment of ethylene is zero. Does this experimental fact provide a clue to the conformation of ethylene? Why or why not?

---

34 The C—O bond dipole is large, typically between 1.5 and 2.5 D. On this basis we would expect carbon monoxide, $:C{=}\ddot{O}:$, to have a substantial dipole moment. Yet it is found experimentally that the dipole moment is 0.1 D—almost zero. Draw a resonance structure for carbon monoxide that, taken together with the structure described previously, explains why this dipole moment is very small.

---

35 Draw resonance structure(s) for each compound below to rationalize the given fact.

(a) All C—O bonds in the carbonate ion are of equal length.

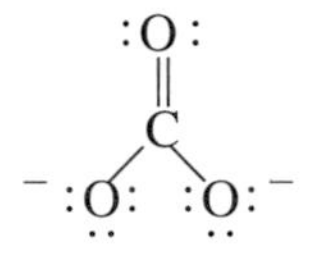

carbonate ion

(b) The two outer oxygens of ozone, $:\ddot{O}{=}\overset{+}{O}{-}\ddot{O}:^-$, have an equal amount of negative charge.

(c) The conjugate acid of formaldehyde, $H_2C{=}\overset{+}{O}{-}H$, has substantial positive charge on carbon.

(d) The inner carbon of acetonitrile oxide, $H_3C{-}C{\equiv}\overset{+}{N}{-}\ddot{O}:^-$, can act as a Lewis acid.

---

36 In the carbonate ion (Problem 35a), what is the average negative charge on each oxygen?

---

37 (a) What do the resonance structures for the allyl cation imply about its structure?

$$^+CH_2{-}CH{=}CH_2 \longleftrightarrow CH_2{=}CH{-}\overset{+}{C}H_2 \qquad \text{allyl cation}$$

(b) Although the structures in part (a) are reasonable descriptions of the allyl cation, the following cation *cannot* be described by analogous resonance structures. Explain why the structure on the right is not a reasonable resonance structure.

$$CH_2{=}CH{-}\overset{+}{N}H_3 \;\not\longleftrightarrow\; {}^+CH_2{-}CH{=}NH_3$$

---

38 (a) Consider the hypothetical compound CH_2Cl_2 in which there is a *planar* arrangement of the four groups about the central carbon atom. In such a compound, two distinct structures are possible. Draw them.

(b) Suppose you have in hand a sample of each of the hypothetical compounds in part (a), but you do not know which is which. Explain how a dipole moment measurement could help you decide.

Problems (Cont.)

(c) Now tell why the *failure* to isolate more than one compound with the formula CH_2Cl_2 in all of history is consistent with the tetrahedral, rather than the planar structure, for this compound. Does this failure *prove* a tetrahedral structure? Why or why not?

39 Using models, perform a vector construction to show that CCl_4 (carbon tetrachloride) should have zero dipole moment. It will help to observe from your models that the plane defined by the carbon and any two chlorines is perpendicular to the plane defined by the carbon and the other two chlorines. To perform the construction, first determine the direction of the resultant from vector addition of any two C—Cl bond dipoles. Then determine the direction of the resultant of the remaining two C—Cl bond dipoles. Finally, perform a vector addition of the two resultants. This problem illustrates the point that *all* tetrahedral molecules of the form CX_4 have zero dipole moment.

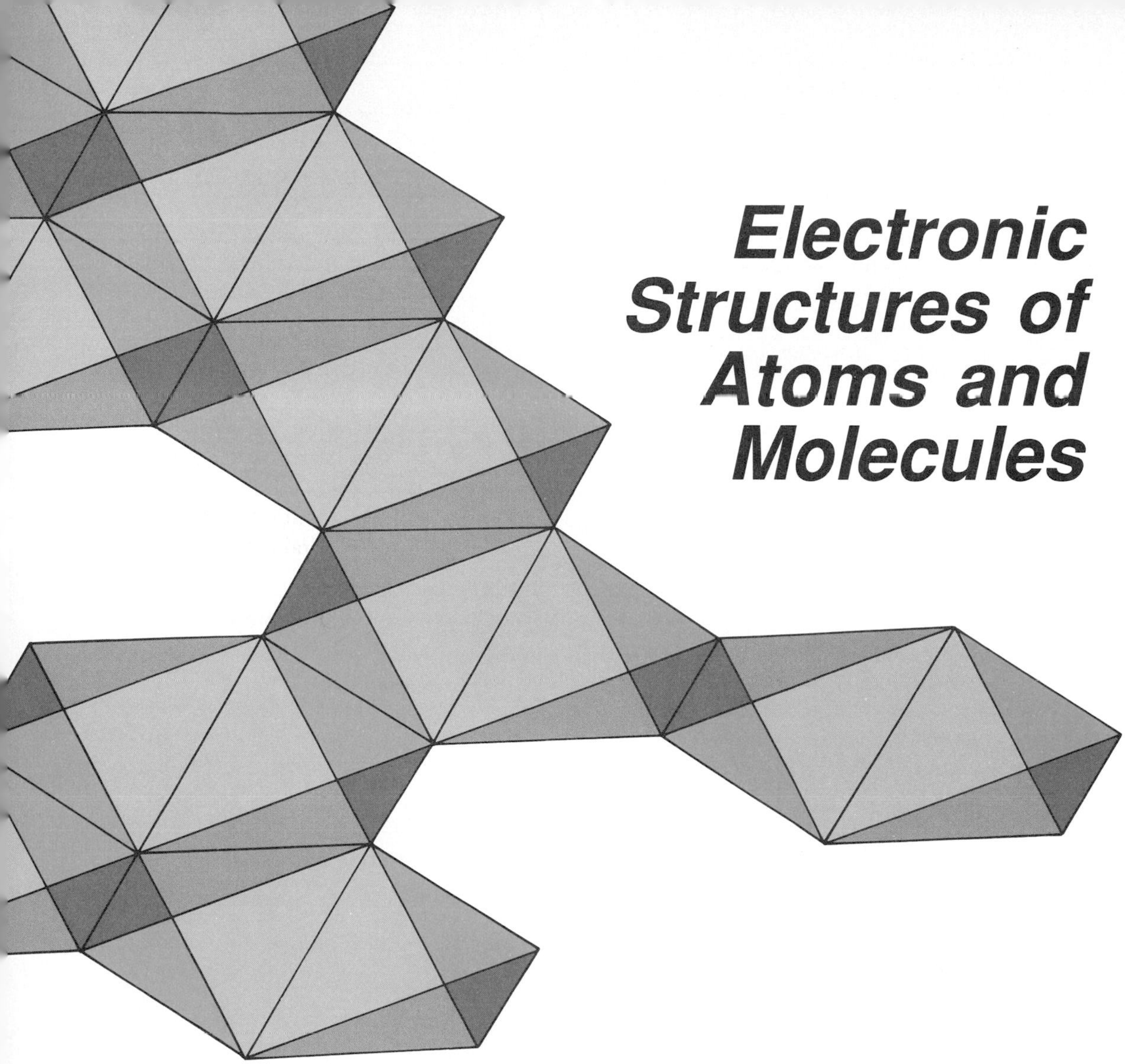

2 Electronic Structures of Atoms and Molecules

In Chapter 1 we learned that the covalent chemical bond can be viewed as the sharing of one or more electron pairs between two atoms. Although this simple model of the chemical bond, the *Lewis electron-pair bond,* will prove to be very useful in much of the organic chemistry we shall study, there are some situations in which it will not be adequate. In order to delve more deeply into the nature of the chemical bond we must turn to an area of physics called *quantum mechanics.* Quantum mechanics is a theory that deals in detail with, among other things, the behavior of electrons in atoms and molecules. Although this theory involves some sophisticated mathematics, we need not explore the mathematical detail in order to appreciate the general conclusions of the theory. We shall begin by considering the general ideas of quantum mechanics. Then we shall consider the quantum-mechanical picture of the electron in the simplest atom, hydrogen. This will be followed by an electronic description of more complex atoms and, finally, by a quantum-mechanical picture of chemical bonding.

2.1 THE WAVE NATURE OF THE ELECTRON

As the twentieth century opened, it became clear that certain things about the behavior of electrons could not be explained by conventional theories. There seemed to be no doubt that the electron was a particle; after all, both its charge and mass had been measured. However, electrons could also be diffracted like light, and diffraction phenomena were associated with waves, not particles. The traditional views of the physical world treated particles and waves as unrelated phenomena. In the mid-1920s, this mode of thinking was changed by the advent of *quantum mechanics.* This theory holds that in the submicroscopic world of the electron and other small particles, there is no real distinction between particles and waves. The behavior of small particles such as the electron can be described by the physics of waves. In other words, matter can be regarded as a *wave–particle duality.*

How does this wave–particle duality require us to alter our thinking about the electron? In our everyday life we are accustomed to a deterministic world. That is, the position of any familiar object can be measured precisely, and its velocity can be determined for all practical purposes to any desired degree of accuracy. For example, we can point to a ball resting on a table and state with confidence, "That ball is at rest (its velocity is zero), and it is located exactly one foot from the edge of the table." Nothing in our experience indicates that we could not make similar measurements for an electron. The problem is that humans, chemistry books, and baseballs are of a certain scale. Electrons and other tiny objects are of a much smaller scale. A central principle of quantum mechanics, the **Heisenberg uncertainty principle,** tells us that *the precision with which we can determine the position and velocity of a particle depends on its scale.* According to this principle, it is impossible to define exactly the position and velocity of an electron. Rather, we are limited to stating the *probability* with which we might expect to find an electron in any given region of space.

In summary:

1. Electrons have wavelike properties.
2. The exact position of an electron cannot be specified; only the probability that it occupies a certain region of space can be specified.

2.2 ELECTRONIC STRUCTURE OF THE HYDROGEN ATOM

Let us consider some of the important results that emerge from application of quantum theory to the simplest atom, hydrogen. The electronic structure of the hydrogen atom is important because it serves as the model on which the electronic structures of more complex atoms are based.

1. *The electron can exist only in certain states, called* **atomic orbitals.**

In the old view of the atom, an electron was thought to circle the nucleus in a well-defined orbit, much as the earth circles the sun. Quantum mechanics replaces the orbit with the *orbital,* which, despite the similar name, is something quite different. An **atomic orbital** is a mathematical description of the wave properties of the electron in an atom. It is sometimes called a **wavefunction** and abbreviated with the Greek letter *psi* (ψ). The physical meaning of the wavefunction is that its square (ψ^2) evaluated over a region of space gives the *relative probability of finding an electron* in that region. This relative electron probability is sometimes called the **electron density.**

There are many different orbitals, or wave motions, available to the electron in a hydrogen atom. How does one orbital differ from another? There are two ways that are particularly important for us: the orbitals may differ in *energy;* and they may differ in the *region of space* occupied by the electron. Let us focus on each of these aspects of atomic orbitals in turn.

2. *The energy of an electron in a particular orbital has a very precise value.*

When an electron is in one orbital—let us say ψ_1—it has a particular energy E_1. If an electron occupies a different orbital, for example ψ_2, it has a different energy E_2. When an electron gains or loses energy, it jumps between orbitals. This means that it ceases to have one wave motion and takes on another. The electron cannot have energy values between E_1 and E_2 if there are no orbitals available between ψ_1 and ψ_2.

The fact that an electron can have only certain allowed values of energy and can exist only in the corresponding allowed states, or orbitals, is one of the central ideas of quantum theory. The energy of the electron is said to be *quantized,* or limited to certain values. This feature of the electron is a direct consequence of the wave properties of the electron. An analogy to this may be familiar. If you have ever blown across a soda-pop bottle, you know that only certain sounds can be produced by a bottle of a given size. As you blow harder across the bottle, the pitch of the note you hear does not rise continuously; but, if you blow hard enough, it suddenly jumps to a note of higher pitch. We might say that the pitch is quantized; only certain sound frequencies (pitches) are allowed. Such phenomena are observed because sound is wave motion of the air. Only sound waves of certain frequencies can exist in the bottle without cancelling themselves out. The progressively higher pitches you hear as you blow harder (called *overtones* of the lowest pitch) are analogous to the progressively higher energy states (orbitals) of the electron in the atom. One might say that the various orbitals of progressively higher energy are an "atomic overtone series" for the electron.

Each orbital in the hydrogen atom is described by four **quantum numbers.** The first quantum number, n, is called the **principal quantum number.** This quantum number can assume any integral value from one to infinity; that is, $n = 1,2,3, \ldots$ The principal quantum number of an orbital tells us something about the energy of an electron occupying the orbital.

3. *Electrons occupying orbitals of higher principal quantum number have higher energy.*

Thus, an electron in an orbital of the hydrogen atom with $n = 2$ has a higher energy than an electron in an orbital with $n = 1$.

The second quantum number, l, called the **azimuthal quantum number,** can take on values that depend on n. The quantum number l can assume values from zero up to $n - 1$; that is, $0, 1, 2, \ldots, n - 1$. Thus, if an electron is in an orbital with $n = 1$, the only possible value of l is 0. For the $n = 2$ orbital, however, l may be 0 or 1. In order not to confuse the numerical designations of n and l, the l quantum numbers are "encoded" as letters. To $l = 0$ is assigned the letter *s*; to $l = 1$ the letter *p*; to $l = 2$

TABLE 2.1 Relationship Among All Four Quantum Numbers

n	l	m	s	n	l	m	s	n	l	m	s
1	0 (1s)	0	$\pm\frac{1}{2}$*	2	0 (2s)	0	$\pm\frac{1}{2}$	3	0 (3s)	0	$\pm\frac{1}{2}$
					1 (2p)	−1	$\pm\frac{1}{2}$		1 (3p)	−1	$\pm\frac{1}{2}$
						0	$\pm\frac{1}{2}$			0	$\pm\frac{1}{2}$
						+1	$\pm\frac{1}{2}$			+1	$\pm\frac{1}{2}$
									2 (3d)	−2	$\pm\frac{1}{2}$
										−1	$\pm\frac{1}{2}$
										0	$\pm\frac{1}{2}$
										+1	$\pm\frac{1}{2}$
										+2	$\pm\frac{1}{2}$

*$\pm\frac{1}{2}$ means that the spin quantum number s may assume either the value $+\frac{1}{2}$ or $-\frac{1}{2}$

the letter d; and to $l = 3$ the letter f. (We shall not be concerned with l values higher than 3.) The allowed values of l are summarized in Table 2.1.

The third quantum number, m, the **magnetic quantum number,** depends on the value of the azimuthal quantum number. For a given l, m may assume the values 0, ± 1, ± 2, . . . , $\pm l$. Thus, for $l = 0$ (an s orbital), m can assume only the value 0. For $l = 1$ (a p orbital), m can have the values -1, 0, $+1$. In other words, there is one s orbital with a given principal quantum number, but (for $n > 1$) there are three p orbitals with a given principal quantum number. It follows that there are more orbitals of higher principal quantum number than there are of lower quantum number (Table 2.1).

The fourth quantum number, s, the **spin quantum number,** can assume either of two values: $+\frac{1}{2}$ or $-\frac{1}{2}$. An electron may have either of these two allowed values of s for every value of m. The allowed quantum numbers of an electron for $n \leq 3$ in the hydrogen atom are given in Table 2.1.

The exact numerical values of the quantum numbers have mathematical significance in quantum theory, but for us they simply serve as convenient labels for the various orbitals available to the electron.

Problem

1 Continue Table 2.1 for $n = 4$.

In a hydrogen atom the only quantum number that affects the energy of the electron is the principal quantum number. The one electron of the hydrogen atom is in its lowest energy state when it occupies the orbital with $n = 1$. Since l must equal 0 when $n = 1$, an electron in the lowest allowed energy state of the hydrogen atom occupies a 1s orbital, and has $m = 0$. It can have either value ($\pm\frac{1}{2}$) of its spin quantum number. Upon absorption of energy (from light, for example) equal to the difference in energy $E_2 - E_1$ between the two states $n = 2$ and $n = 1$, the electron immediately transfers to an orbital with $n = 2$. This electron in principal quantum level 2 can subsequently emit energy equal to $E_2 - E_1$ (as light, for example) and return to the 1s state. Such energy absorption or emission phenomena gave the earliest clues to the energy-level structure of the atom (see Problem 13 at the end of the chapter).

Now let us consider how orbitals differ in the regions of space occupied by the electron. According to the uncertainty principle, we cannot pinpoint the exact location of an electron, but we can define with reasonable certainty a region of space within which an electron in a given orbital may be found with high probability. This region of space defines the approximate shape and dimensions of the orbital. As we shall learn shortly, the geometrical aspects of orbitals prove to be very important in chemical bonding.

4. *Each orbital is characterized by a three-dimensional region of space in which an electron in that orbital is most likely to exist.*

This point is best illustrated by example. When an electron occupies a 1*s* orbital, it is most likely to be found in a sphere with the atomic nucleus at its center (Fig. 2.1). We cannot say where the electron is within such a sphere because of the uncertainty principle. Again, locating the electron is strictly a matter of probability. The mathematics of quantum theory tells us that there is a greater than 90% probability that this electron will be found within a sphere of radius 1 Å about the nucleus. Thus, we depict an electron in a 1*s* orbital as a smear of *electron density,* or electron probability, with the densest part of the smear (or highest electron density) near the nucleus.

The orbital of next higher energy in the hydrogen atom is a 2*s* orbital ($n = 2$, $l = 0$). The wave motion of an electron in this orbital is illustrated in Fig. 2.2. The electron density in this orbital, like that in a 1*s* orbital, is described as a spherical smear. A new feature of this orbital is present, the *node,* that the 1*s* orbital does not have. The node is another consequence of the wave nature of the electron. What is a node?

Most of us are familiar with waves in a vibrating string or waves in a pool of water. We know that the waves have *peaks* and *troughs,* regions where the waves are, respectively, at their maximum and minimum heights (Fig. 2.3). Suppose we say that the

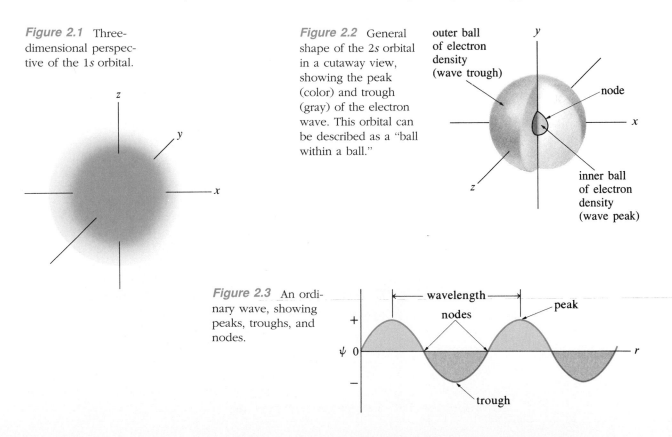

Figure 2.1 Three-dimensional perspective of the 1*s* orbital.

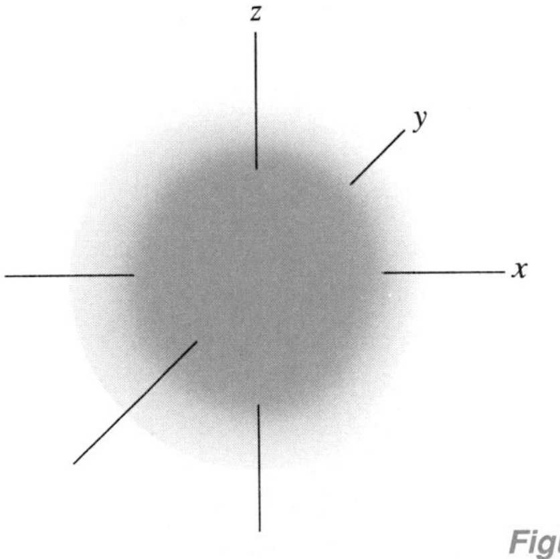

Figure 2.2 General shape of the 2*s* orbital in a cutaway view, showing the peak (color) and trough (gray) of the electron wave. This orbital can be described as a "ball within a ball."

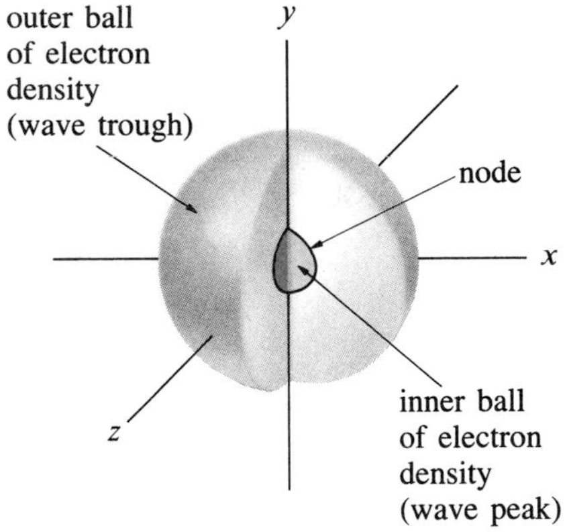

Figure 2.3 An ordinary wave, showing peaks, troughs, and nodes.

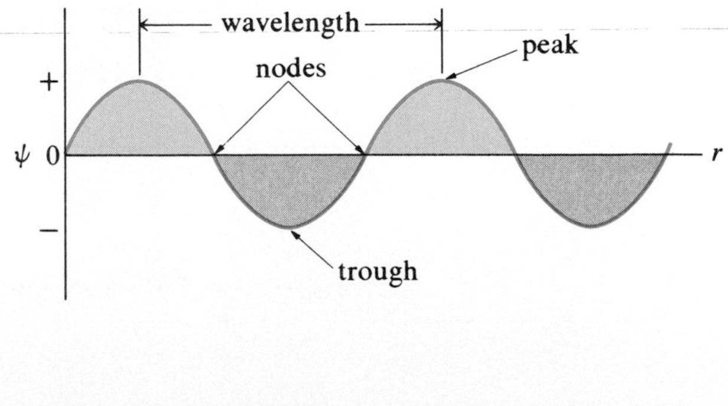

wave is positive at a wave peak and negative at a wave trough. When we trace the wave from a peak to a trough, somewhere between the two we pass through a point at which the wave is zero. This point is called a **node.**

The electron in a hydrogen 2*s* orbital has a wave peak at the nucleus (color), a node, and a wave trough far away from the nucleus. Now the wave in Fig. 2.3 is a simple wave confined to the plane of the paper, and the nodes in this wave are points. Because the electron waves (orbitals) are three-dimensional, the nodes are *surfaces*. The nodal surface in the 2*s* orbital is an infinitely thin sphere. Thus the 2*s* orbital takes on the characteristics of a "ball within a ball." Physically the wave peaks correspond to a positive value for the wavefunction ψ, the wave troughs to a negative value, and the nodes to a zero value. Since the electron density is the *square* of the wavefunction, the electron has a high probability of being found in either a peak or a trough; but it has *zero probability* of being found on a node.

5. *Some orbitals contain wave peaks and wave troughs separated by nodes.*

Some students ask, "If the electron cannot exist at the node, how does it cross the node?" The answer is that the electron *is* a wave, and the node is part of its wave motion, just as a node is part of a wave in a vibrating string.

Problem

2 Use the trends in orbital shapes you have just learned to outline the general features of a 3*s* orbital. Indicate where you expect to find peaks, troughs, and nodes. (*Hint:* All orbitals of the *s* type are spherical.)

The 2*p* orbital ($n = 2$, $l = 1$) is the simplest example of an orbital for which $l \neq 0$. This orbital is not spherical but has a dumbbell shape and is directed in space (Fig. 2.4a). One lobe of the 2*p* orbital corresponds to a wave peak, and the other to a wave trough; the electron density is identical in corresponding parts of each lobe. There is a node at the nucleus; this node is a plane rather than a spherical shell. There are three equivalent and mutually perpendicular 2*p* orbitals corresponding to $m = -1$, 0, and +1 (Fig. 2.4b).

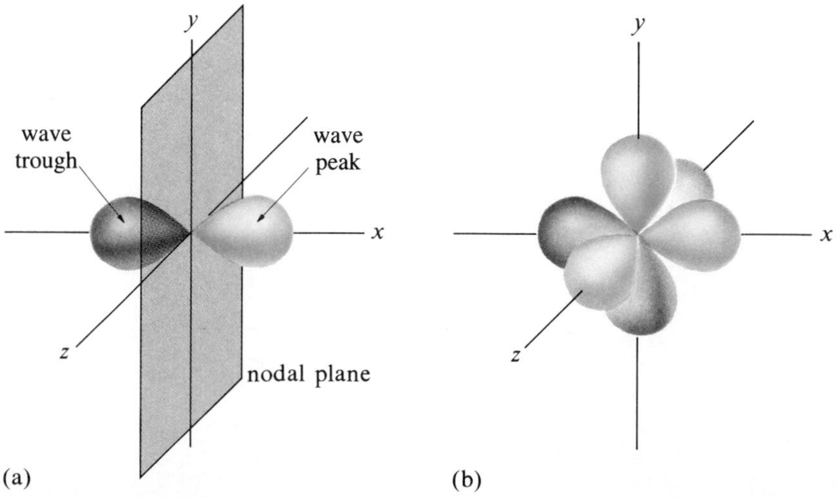

Figure 2.4 (a) General shape of a 2*p* orbital. (b) The three 2*p* orbitals shown together.

The $3p$ orbital has lobes much like those of the $2p$ orbital, and there are three $3p$ orbitals, one directed along each of the Cartesian (x, y, and z) axes. There is, however, one spherical node in each $3p$ orbital in addition to the nodal plane at the nucleus (Fig. 2.5).

If we now look at the pictures of the various orbitals, we discover another wave-like feature. The $1s$ orbital of lowest energy has no nodes; the $2s$ and $2p$ orbitals each have one node; and the $3s$ (Problem 2) and $3p$ orbitals each have two nodes. The following principle emerges from this comparison:

6. *Orbitals of higher principal quantum number (and therefore of higher energy) have more nodes.*

Specifically, an orbital with principal quantum number n has $n - 1$ nodes. The analogy to sound waves in a soda-pop bottle is again very striking: overtones of higher pitch have greater numbers of nodes.

If we compare the dimensions of the $1s$ and $2s$ orbitals, or of the $2p$ and $3p$ orbitals, we find that orbitals of higher principal quantum number are larger and more diffuse (that is, more spread out). This observation is generalized as follows:

7. *The electron density, on the average, is concentrated farther from the nucleus in orbitals of higher principal quantum number.*

This makes sense when we realize that the electron in an orbital of higher principal quantum number has greater energy than one in an orbital of lower quantum number. An electron of higher energy can overcome the attraction of the positive nucleus and exist at a greater distance from it.

2.3 ELECTRONIC STRUCTURES OF MORE COMPLEX ATOMS

The orbitals available to electrons in atoms with atomic number greater than one are, to a useful approximation, essentially like those of the hydrogen atom. There is, how-

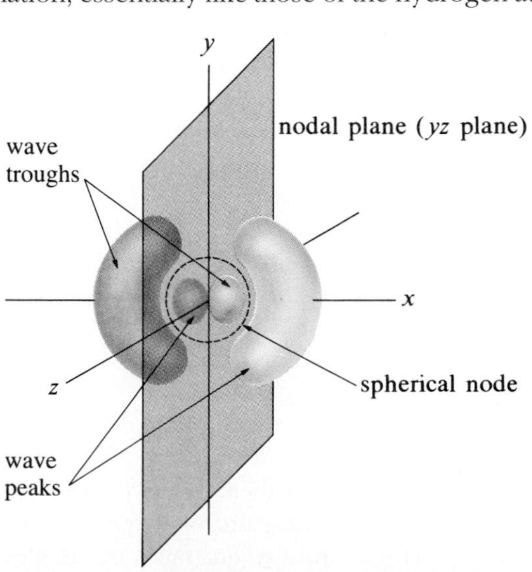

Figure 2.5 The general shape of a $3p$ orbital. There are three such orbitals, each mutually perpendicular.

Figure 2.6 Schematic representation of the relative energies of different orbitals in a many-electron atom. The exact scale varies from atom to atom, but the energy levels tend to be closer together as the principal quantum number increases.

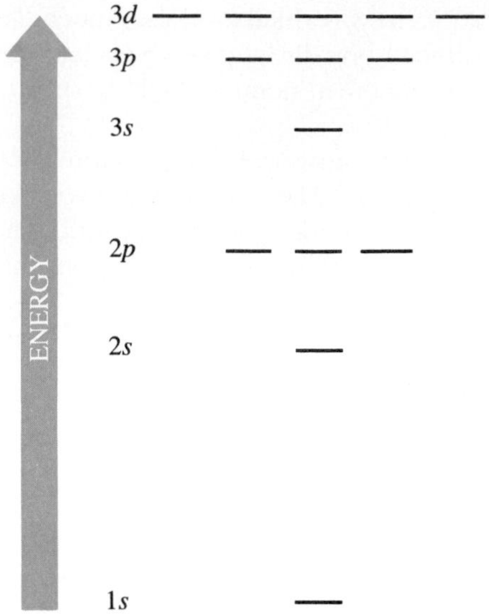

ever, one important difference: In atoms other than hydrogen, electrons with the same principal quantum number n but with different values of l have different energies. Thus helium, carbon, and oxygen, like carbon, have $2s$ and $2p$ orbitals, but, unlike hydrogen, electrons in these orbitals differ in energy. The ordering of energy levels for atoms with more than one electron is illustrated schematically in Fig. 2.6. As this figure shows, the gaps between energy levels become progressively smaller as the principal quantum number increases. Furthermore, the energy gap between orbitals that differ in principal quantum number is greater than the gap between two orbitals within the same principal quantum level. Thus, the difference in energy between $2s$ and $3s$ orbitals is greater than the difference in energy between $3s$ and $3p$ orbitals.

Atoms beyond hydrogen, of course, have more than one electron. How are the electrons of such atoms arranged in their atomic orbitals? The distribution of electrons in the orbitals of more complex atoms is governed by two principles, the Pauli exclusion principle and Hund's rule. The **Pauli exclusion principle** states that any two electrons within an atom must differ in at least one of their quantum numbers. Electrons occupy the orbitals of lowest possible energy as long as this principle is not violated. Hence a maximum of two electrons are present in any given orbital, and they must have opposite spins. Thus in helium, two electrons of opposite spin can be accommodated in the $1s$ orbital; the quantum numbers of these electrons are $(n, l, m, s) = (1, 0, 0, +\frac{1}{2})$ and $(1, 0, 0, -\frac{1}{2})$.

The **electronic configuration** of an atom specifies how its electrons are distributed among its atomic orbitals. The electronic configuration of helium can be indicated as follows:

$$\text{helium, He: } 1s^2$$

This means that there are two electrons in the $1s$ orbital of the helium atom. It is assumed that they are of differing spin, as required by the Pauli principle.

Hund's rule is illustrated in the electronic configuration of carbon, obviously a very important element in organic chemistry. Carbon has six electrons per atom. Two each occupy the 1*s* and 2*s* orbitals. The remaining two occupy 2*p* orbitals. Since the three 2*p* orbitals have equal energy, a second principle is required to indicate how the electrons are distributed within these orbitals. **Hund's rule** states that when distributing electrons among orbitals of equal energy, single electrons of identical spin are placed in each equivalent orbital. Only when each of these orbitals contains one electron are electrons of opposite spin introduced. Thus, the electronic configuration of carbon can be written as follows:

carbon, C: ($1s^2 2s^2 2p^2$)

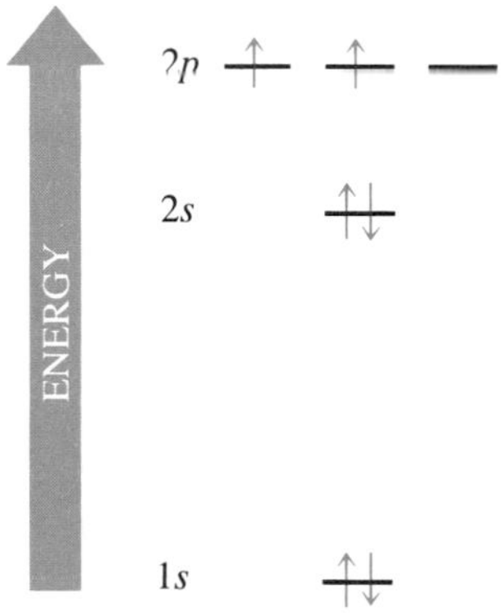

The colored arrows correspond to electrons, and their relative direction indicates the electron's relative spin. Thus, the electrons in the 2*p* orbitals of carbon are unpaired with identical spin. This configuration ensures that repulsions between electrons are minimized since electrons in different *p* orbitals move in different regions of space. (The different *p* orbitals are at right angles to each other; Fig. 2.4b.) When we write the carbon electronic configuration as $1s^2 2s^2 2p^2$, it is understood that the 2*p* electrons are distributed in accordance with Hund's rule.

The electronic configuration of oxygen shows what happens when more than three electrons occupy the 2*p* orbitals. In this case two electrons each occupy the 1*s* and 2*s* orbitals, and three electrons of identical spin occupy the 2*p* orbitals in accordance with Hund's rule. The one remaining electron is then placed with opposite spin into a 2*p* orbital.

oxygen, O: ($1s^2 2s^2 2p^4$)

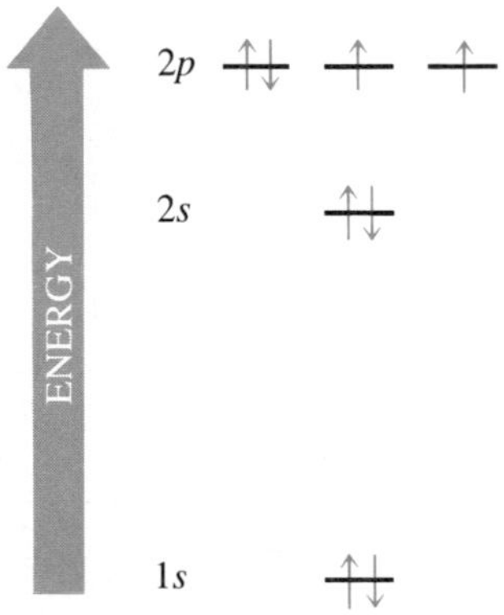

Because successive principal quantum levels contain more available orbitals (Table 2.1), each can accommodate more electrons than the last. Principal level 1 accommodates 2 electrons; level 2, 8 electrons; level 3, 18 electrons, and so on. The numbers 2, 8, 18, . . . correspond to lengths of successive rows of the periodic table. Elements in the same column of the periodic table have related chemical properties and also have related electronic configurations. Thus, each of the alkali metals has a single electron in an *s* orbital; the halogens each lack one electron of a completely filled *p* orbital; and the noble gases each have completely filled *s* and *p* levels.

Problem

3 Give the electronic configurations of the following atoms and ions:

(a) the lithium atom
(b) the nitrogen atom
(c) the sodium atom
(d) the fluoride ion, F^-
(e) the potassium ion, K^+

2.4 ELECTRONIC STRUCTURES OF MOLECULES. MOLECULAR ORBITALS

Having examined atomic structure in the light of quantum mechanics, we now turn to the quantum mechanical view of the chemical bond. We shall find that the electronic structures of atoms and molecules can be understood with many of the same concepts.

We have found that atomic structure involves a system of orbitals. Conceptually, we can specify the electronic configuration of an atom from a knowledge of its orbitals by assigning the available electrons to the atomic orbitals in accordance with the Pauli principle and Hund's rule. Similarly, quantum mechanics shows that we can construct a system of orbitals for *molecules* as well. Such orbitals are called **molecular orbitals.** The electronic description of a molecule is developed by assigning electrons to these molecular orbitals in a manner also consistent with the Pauli principle and Hund's rule. These ideas can be illustrated with the simplest of molecules: the hydrogen molecule, H—H.

The first step in deriving the electronic configuration of the hydrogen molecule is to construct its molecular orbitals. Quantum mechanics tells us that the molecular orbitals of a molecule can be constructed by combining the atomic orbitals of its constituent atoms in a certain way. Moreover, if we combine n atomic orbitals we must get n molecular orbitals. To see how these ideas work in practice, let us imagine that two hydrogen atoms are brought together to the distance by which they are separated in the hydrogen molecule. At this distance there is substantial overlap in the $1s$ orbitals of the two atoms. This interaction causes the $1s$ orbitals to combine, yielding two new orbitals—the two molecular orbitals of the hydrogen molecule. The way that the atomic orbitals combine is very simple. One molecular orbital is generated from the *addition* of atomic orbitals—that is, by the additive overlap of the $1s$ wavefunctions. The second molecular orbital is formed by *subtraction* of atomic orbitals—the subtractive overlap of $1s$ wavefunctions.

Let us first examine the molecular orbital formed by the addition of the two atomic $1s$ orbitals. In the region in which the waves overlap they reinforce because a wave peak is added to a wave peak (Fig. 2.7a). The resulting electron wave is one of the molecular orbitals of the hydrogen molecule, called the **bonding molecular orbital** (Fig. 2.8).

Subtraction of two atomic $1s$ orbitals is the same as the overlap of a wave peak with a wave trough (Fig. 2.7b). In the region of overlap the waves cancel, and the resulting electron wave contains a node between the two nuclei. The orbital resulting from this subtractive overlap is the second molecular orbital of the hydrogen molecule, called the **antibonding molecular orbital** (Fig. 2.8).

The energy of an electron in the bonding molecular orbital is lower than that of an electron in an isolated hydrogen atom. On the other hand, the energy of an electron in the antibonding molecular orbital is higher than its energy in the hydrogen

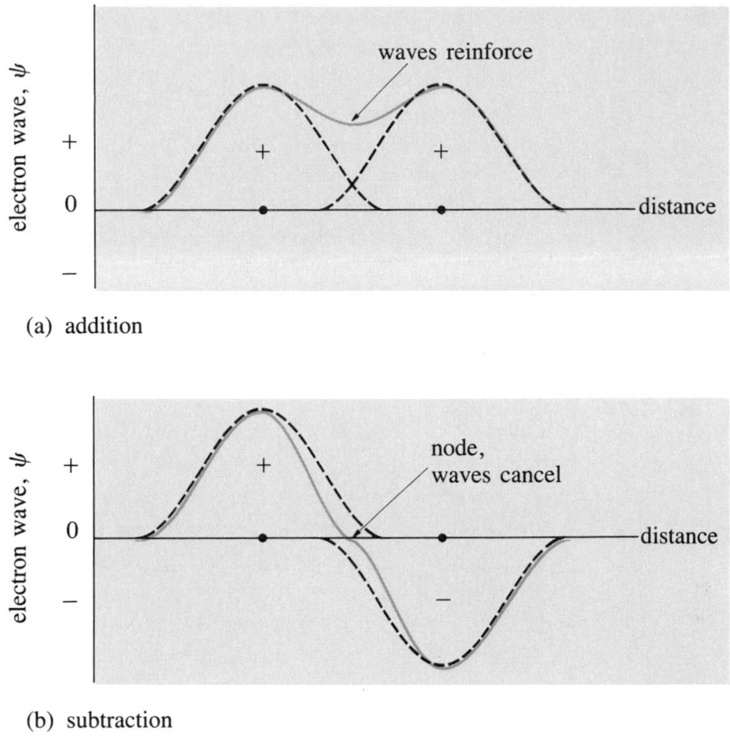

Figure 2.7 The addition and subtraction of two waves (ψ) such as the 1*s* (orbitals) of two hydrogen atoms. The dots represent the hydrogen nuclei; the dashed lines are the (orbitals) of the individual nuclei; and the colored line is the wave resulting from the combination.

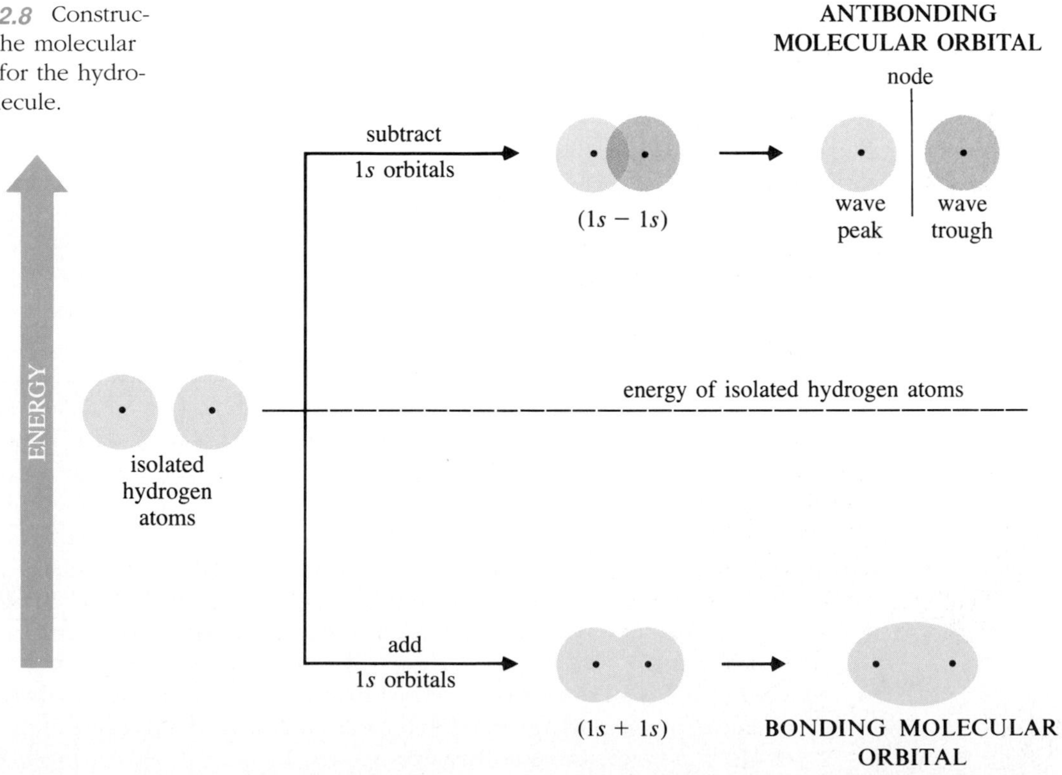

Figure 2.8 Construction of the molecular orbitals for the hydrogen molecule.

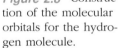

atom. Notice that the antibonding molecular orbital has a node; like atomic orbitals, molecular orbitals of higher energy have more nodes.

The Pauli exclusion principle operates for molecular orbitals just as it does for atomic orbitals. Hence, the two electrons of the hydrogen molecule occupy the bonding molecular orbital; the antibonding molecular orbital is unoccupied. Since electrons in the bonding orbital of H_2 have lower energy than electrons in two isolated hydrogen atoms, it is evident that *chemical bonding is an energetically favorable process.* In fact, formation of a hydrogen molecule from two hydrogen atoms releases 104 kcal/mol of energy. This is a large amount of energy on a chemical scale—more than enough to raise the temperature of one kilogram of water from freezing to boiling.

According to the picture we have just developed, the chemical bond in a hydrogen molecule results from the occupation of a bonding molecular orbital by two electrons. Some students wonder why we concern ourselves with the antibonding molecular orbital if it is not occupied. One way to answer this is by analogy. The hydrogen atom has one electron in a 1*s* orbital; however, other orbitals (2*s*, 3*p*, etc.) are available to this electron but are normally not occupied. Similarly, a hydrogen molecule also has the unoccupied antibonding orbital. When the hydrogen atom absorbs enough energy, the electron jumps to an atomic orbital of higher energy. Similarly, when the hydrogen molecule absorbs enough energy, an electron jumps from the bonding to the antibonding molecular orbital. This means that by gaining energy, this electron adopts the wave motion characteristic of the antibonding orbital, including the node between the two nuclei. Another demonstration of the importance of the antibonding orbital is evident from an attempt to construct the molecule He_2, diatomic helium. The molecular orbitals for this molecule are conceptually identical to those for H_2 (Fig. 2.8)—that is, one bonding and one antibonding molecular orbital, formed by addition and subtraction of the 1*s* orbitals of the isolated helium atoms. However, diatomic helium would have *four* electrons (two from each helium atom) to distribute between the molecular orbitals. The Pauli principle dictates that two electrons go into the bonding molecular orbital, and the remaining two go into the antibonding molecular orbital. The lower energy of the electrons in the bonding molecular orbital is offset by the higher energy of the electrons in the antibonding molecular orbital. Since there is no energetic advantage to bonding in the He_2 molecule, helium is monatomic.

In the bonding molecular orbital of the hydrogen molecule the electrons occupy an ellipsoidal region of space. No matter how we turn the hydrogen molecule about a line joining the two nuclei, its electron density looks the same. This is another way of saying that the bond in the hydrogen atom has **cylindrical symmetry.** Other cylindrically symmetric objects are shown in Fig. 2.9. Bonds that are cylindrically symmetric about the internuclear axis are called **sigma bonds** (abbreviated σ-bonds). The bond in the hydrogen molecule is thus a σ-bond.

Problem

4 (a) Draw an orbital diagram corresponding to Figure 2.8 for (1) the He_2^+ ion; (2) the H_2^- ion; (3) the H_2^{2-} ion; (4) the H_2^+ ion. Which of these species are likely to exist as diatomic species, and which would dissociate into monatomic fragments? Explain.

(b) The bond dissociation energy of H_2 is 104 kcal/mol; that is, it takes this amount of energy to dissociate H_2 into its atoms. Estimate the bond dissociation energy of He_2^+ and explain your answer.

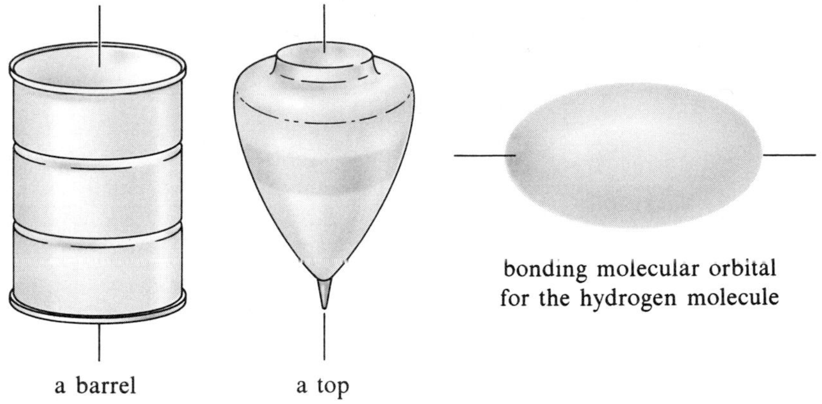

Figure 2.9 Some cylindrically symmetric objects. Objects are cylindrically symmetric when they appear the same no matter how they are rotated about their cylindrical axis (black line).

2.5 MOLECULAR ORBITAL THEORY AND LEWIS STRUCTURES

Let us now relate the quantum mechanical view of the chemical bond to the concept of the Lewis electron-pair bond. Lewis structures are widely used and are extremely important to modern organic chemists. Quantum mechanics does not require that we discard the idea of the electron-pair bond; rather, quantum theory has enriched our view of the chemical bond. Although Lewis structures suffice for describing the vast majority of the chemical situations we shall encounter, at times we shall have to supplement Lewis structures with some quantum mechanical refinements in order to understand certain experimental facts.

In the Lewis picture, each covalent bond is represented by at least one electron pair shared between two nuclei. In the quantum-mechanical description, the σ-bond exists because of the presence of electrons in a bonding molecular orbital and the resulting electron density between the two nuclei. The electrons are attracted to each nucleus and therefore act as the cement that holds the nuclei together. Since a bonding molecular orbital can hold two electrons, *the Lewis view of the electron-pair bond is approximately equivalent to the quantum-mechanical idea of a bonding molecular orbital occupied by a pair of electrons.* The Lewis picture places the electrons squarely between the nuclei. Quantum theory says that although the electrons have a high probability of being between the bound nuclei, they can also occupy other regions of space.

Molecular orbital theory also shows that a chemical bond need not be an electron *pair*. For example, H_2^+ (the hydrogen molecule cation, which we might represent in the Lewis sense as $H\overset{+}{\cdot}H$) is a stable species in the gas phase (Problem 4). It is not as stable as the hydrogen molecule because the ion has only one electron in the bonding molecular orbital, rather than the two in a neutral hydrogen molecule. The He_2^+ and H_2^- ions, also known species, might each be considered to have a three-electron bond—two bonding electrons and one antibonding electron in each case. The electron in the antibonding orbital is also shared by the two nuclei, but in a way that reduces the energetic advantage of bonding. These examples show that the sharing of electrons between nuclei need not contribute to bonding. Thus, these species are also not as stable as neutral H_2. It is not surprising, then, that the most common bonding situations occur when bonding molecular orbitals contain electron pairs and antibonding molecular orbitals are empty. This is why ordinary chemical bonds can be represented as electron pairs.

2.6 HYBRID ORBITALS

A. Bonding in Methane

When quantum theory is applied to methane, CH_4, we find that the bonding to four hydrogen atoms alters the simple orbital picture derived for carbon in Sec. 2.3. That is, *the carbon in methane has an arrangement of orbitals that is different from the orbitals in atomic carbon.* The orbital arrangement for carbon in methane can, however, be derived very simply from that of free carbon. For carbon in methane, we imagine that the 2*s* orbital and the three 2*p* orbitals are mixed to give four *equivalent* orbitals, each with a character intermediate between pure *s* and pure *p*. Since each carbon orbital in methane is one part *s* and three parts *p*, we call it an sp^3 orbital (pronounced "s-p-three," not "s-p-cubed"). This means that the six carbon electrons are distributed between one 1*s* orbital and four equivalent sp^3 orbitals. Recalling Hund's rule, this mental transformation can be summarized as follows:

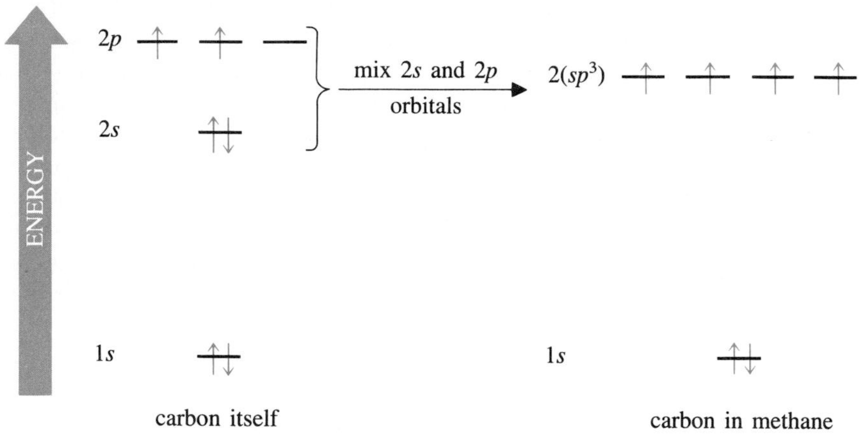

The sp^3 orbitals are examples of **hybrid orbitals,** because they come from the mixing of pure orbitals, just as a hybrid rose is formed by the mixing of pure strains. The shape of an sp^3 orbital is shown in Fig. 2.10a. The orbital consists of two lobes separated by a node, much like a *p* orbital. However, one of the lobes is very small, and the other is very large. In other words, the electron density in an sp^3 hybrid orbital is highly *directed* in space.

The number of hybrid orbitals (four in this case) is the same as the number of orbitals that are mixed to obtain them. (One *s* orbital + three *p* orbitals = four sp^3 orbitals.) The large lobes of the four carbon sp^3 orbitals are directed to the corners of a regular tetrahedron as shown in Fig. 2.10b. Four hydrogen atoms, each with a single 1*s* electron, overlap with the four carbon sp^3 orbitals, each also with a single electron, to give the four bonding C—H molecular orbitals containing two electrons apiece (Fig. 2.10c). (As a result of this overlap four unoccupied antibonding orbitals also arise, which we shall not consider further.) The four C—H bonds thus produced are σ-bonds.

Why are hybrid orbitals formed? If the hydrogens in methane are to be as far apart as possible, the tetrahedral geometry is required. The pure *s* and *p* orbitals

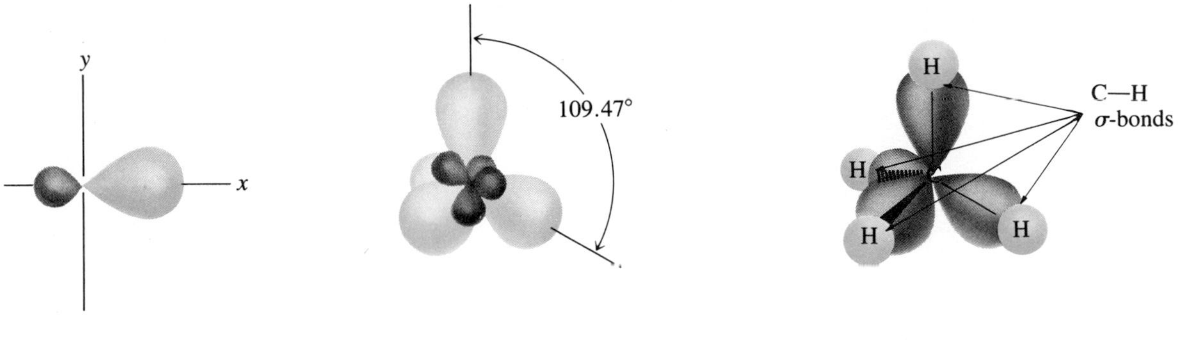

(a) (b) (c)

Figure 2.10 (a) General shape of a carbon sp^3 hybrid orbital. (b) The four sp^3 orbitals of carbon shown together. (c) An orbital picture of tetrahedral methane showing the four equivalent σ-bonds formed from overlap of carbon sp^3 and hydrogen $1s$ orbitals.

available on nonhybridized carbon are not directed tetrahedrally. Rehybridization provides orbitals that have the bulk of their electron density directed toward the hydrogen nuclei. This directional character provides more electron "cement" between the nuclei and gives stronger (that is, more stable) bonds.

B. Bonding in Ammonia

Ammonia, $:NH_3$, is an example of a compound with an unshared electron pair. The electronic configuration of nitrogen in ammonia is, like carbon in methane, hybridized to yield four sp^3 hybrid orbitals; one of these orbitals contains an electron pair.

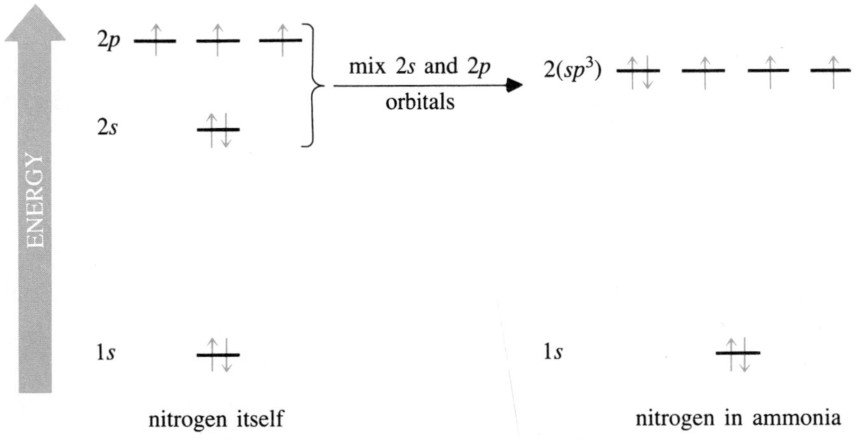

Each sp^3 orbital on nitrogen containing one electron can overlap with the $1s$ orbital of a hydrogen atom, also containing one electron, to give the three N—H σ-bonds of ammonia. The sp^3 orbital containing the pair of electrons is filled. The electrons in this orbital become the nitrogen unshared pair. The unshared pair and the three N—H

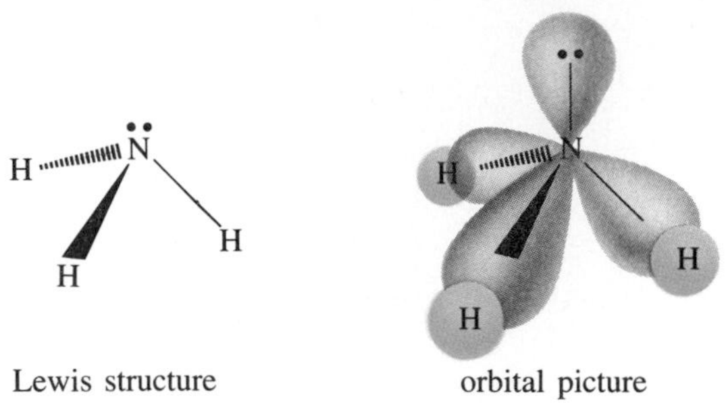

Figure 2.11 Bonding molecular orbital picture for ammonia, NH_3.

bonds, because they are made up of sp^3 hybrid orbitals, are directed to the corners of a regular tetrahedron (Fig. 2.11). The advantage of orbital hybridization in ammonia is the same as that in carbon: hybridization accommodates the maximum separation of the unshared pair and the three hydrogens and, at the same time, provides strong, directed N—H bonds.

We are now in a position to appreciate the quantum-mechanical reason for the wide adherence to the octet rule in covalent compounds. The atoms in the first row of the periodic table have partially filled orbitals in principal quantum level 2. These are the valence orbitals. There are four of these orbitals in all, one 2*s* orbital and three 2*p* orbitals. (Other atoms in the A groups of the periodic table have a similar valence orbital arrangement in higher principal quantum levels.) When these four orbitals (or hybrid orbitals derived from them) can be completely filled with electron pairs, the atoms of the first row achieve the noble-gas electron configuration. This can happen by the formation of bonds and sharing of electrons with other atoms. When this occurs, the valence orbitals contain eight electrons—an octet.

Problem

5 Construct a hybrid orbital picture for the water molecule using oxygen sp^3 hybrid orbitals.

KEY IDEAS IN CHAPTER 2

- As a consequence of their wave properties, electrons in atoms and molecules can exist only in certain allowed energy states. These states, called orbitals, are described mathematically by wavefunctions. The square of the wavefunction gives the electron density, or the probability of finding the electron.
- Electrons in orbitals are characterized by quantum numbers which, for atoms, are designated n, l, m, and s. These can assume only certain allowed values. The higher the principal quantum number n of an electron, the higher is its energy. In atoms more complex than hydrogen, the energy is also a function of the l quantum number.
- The wave motion of an electron can have nodes, which separate wave peaks from wave troughs. The number of nodes of an electron increases with its principal quantum number.

- The distribution of electron density in a given type of orbital has a characteristic arrangement in space: all *s* orbitals are spheres, all *p* orbitals have two lobes, etc.
- Bonds are formed when orbitals on different atoms overlap. Overlap of the 1*s* orbitals of two hydrogen atoms results in the formation of two molecular orbitals of the hydrogen molecule, one bonding and the other antibonding.
- Both atomic orbitals and molecular orbitals are populated with electrons according to Hund's rule and the Pauli exclusion principle.
- Hybrid orbitals, formed by mixing atomic orbitals, are involved in bonding because of their directional characteristics.

ADDITIONAL PROBLEMS

6 Give the electronic configuration of (a) the chlorine atom; (b) the chloride ion; (c) the argon atom; (d) the magnesium atom.

7 Which of the following orbitals is/are *not* permitted by the quantum mechanics of the hydrogen atom? Explain.
(a) 2*s* (b) 6*s* (c) 5*d* (d) 2*d* (e) 3*p*

8 (a) Two types of nodes occur in atomic orbitals: spherical surfaces and planes. Examine the nodes in 2*s*, 2*p*, and 3*p* orbitals and show that they agree with the following statements:

1. An orbital of principal quantum number n has $n - 1$ nodes.
2. The value of l gives the number of planar nodes.

(b) How many spherical nodes are there in a 4*s* orbital? In a 3*d* orbital? How many nodes of all types are there in a 3*d* orbital?

9 The shape of one of the five equivalent 3*d* orbitals is shown below. From your answer to the previous question, sketch the nodes of this 3*d* orbital, and associate a wave peak or a wave trough with each lobe of the orbital.

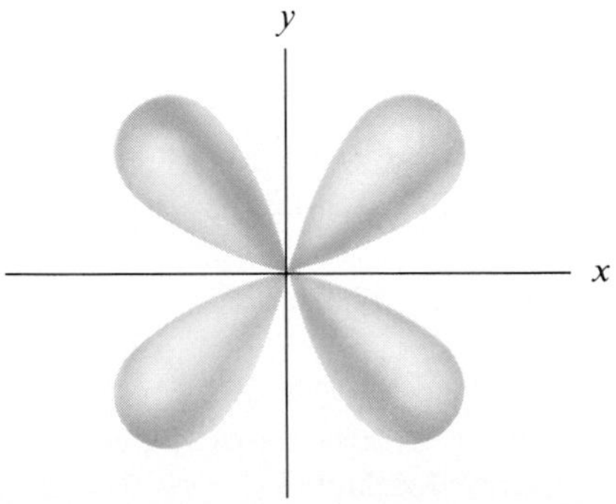

3*d* orbital

Problems (Cont.)

10 Orbitals with $l = 3$ are called f orbitals. (a) How many equivalent f orbitals are there? (b) In what principal quantum level do f orbitals first appear? (c) How many nodes would you expect for a $5f$ orbital? (d) How many spherical nodes would you expect for a $5f$ orbital? (See problem 8.)

11 It is common to visualize the electron as a tiny charged sphere. This idea would seem to require that the electron have a certain radius and volume—that is, a particular *size*. Why is it nonsense to ask, "How large is an electron?"

12 Sketch a picture of the $4p$ orbital. Show the nodes and the regions of wave peaks and wave troughs.

13 The difference in energy between an electron in the first quantum level of the hydrogen atom and one in the second is 1.635×10^{-11} erg. The energy available from a photon is

$$E = hc/\lambda$$

where h is Planck's constant, 6.625×10^{-27} erg-sec, c is the velocity of light (3×10^{10} cm/sec), and λ is the wavelength of the light. Calculate the wavelength of light (in angstroms) required to bring about the jump of an electron in a hydrogen atom from the $n = 1$ to $n = 2$ level. (The wavelength is defined in Fig. 2.3.) The hydrogen atom is found to absorb light at precisely this wavelength, which is in the far ultraviolet region of electromagnetic radiation. The elevation of an electron from the $n = 1$ to $n = 2$ level by absorption of light is analogous to the jump in sound pitch that occurs when we blow across a pop bottle (Sec. 2.2).

14 Consider two $2p$ orbitals, one on each of two atoms, oriented head to head as follows:

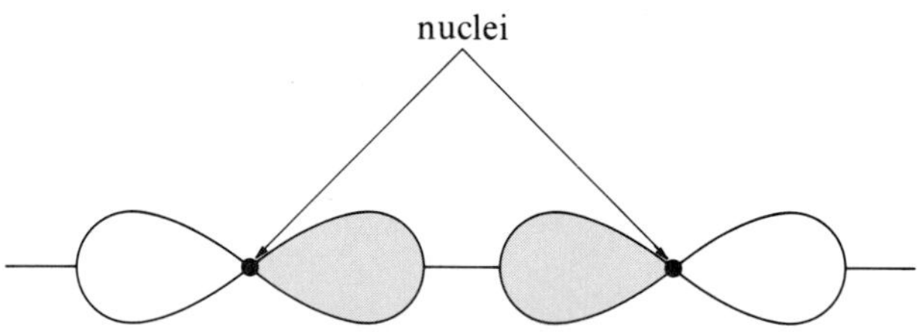

Imagine gradually bringing the nuclei closer together until the two wave peaks of the orbitals just overlap. A new system of molecular orbitals is formed by this overlap.

(a) Sketch the shape of the resulting bonding and antibonding molecular orbitals.
(b) Identify the nodes in each molecular orbital.
(c) If two electrons occupy the bonding molecular orbital, is the resulting bond a σ-bond? Explain.

15 Now consider two $2p$ orbitals, one on each of two different atoms, oriented side to side, as follows:

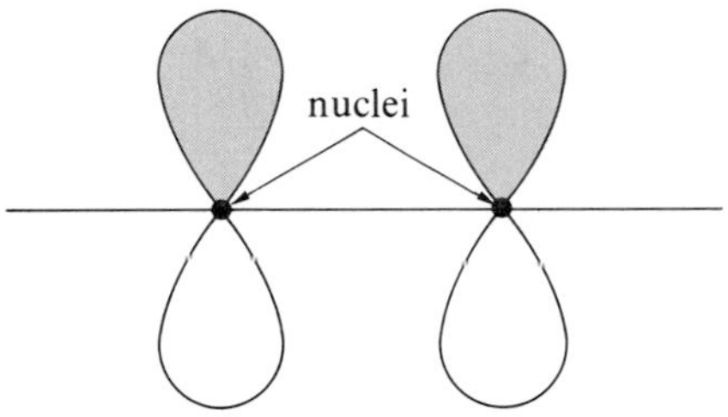

Imagine bringing these nuclei close enough together that overlap occurs between the wave peaks and between the wave troughs. This overlap results in a system of molecular orbitals.

(a) Sketch the shape of the resulting bonding and antibonding molecular orbitals.

(b) Identify the node(s) in each.

(c) When two electrons occupy the bonding molecular orbital, is the resulting bond a σ-bond? Explain.

16 When a hydrogen molecule absorbs light, an electron jumps from the bonding molecular orbital to the antibonding molecular orbital. Explain why this light absorption can lead to the dissociation of the hydrogen molecule into two hydrogen atoms. (This process, called *photodissociation,* can sometimes be used to initiate chemical reactions.)

17 Suppose you take a trip to a distant universe and find that the periodic table there is derived from an arrangement of quantum numbers different from the one on Earth. The rules there are:

principal quantum number $n = 1, 2, \ldots$ (as on Earth)
azimuthal quantum number $l = 0, 1, 2, \ldots, n - 1$ (as on Earth)
magnetic quantum number $m = 0, 1, 2, \ldots, l$ (that is, only *positive* integers up to and including l are allowed)
spin quantum number $s = -1, 0, +1$ (that is, *three* allowed values of spin)

(a) Write the electronic configuration of the element with atomic number 8 in the periodic table.

(b) What is the atomic number of the second noble gas?

(c) Assuming that the Pauli principle remains valid, how many electrons may populate a given orbital?

(d) What sort of rule would replace the octet rule?

18 Using both Lewis structures and molecular orbital arguments, explain why reaction (1) below is more likely to occur than reaction 2.

1. $:NH_3 + H^+ \longrightarrow \overset{+}{N}H_4$

2. $:NH_3 + H\cdot \longrightarrow \cdot NH_4$

Is reaction 2 strictly forbidden by molecular orbital arguments? Explain.

3 Alkanes and the Functional Groups

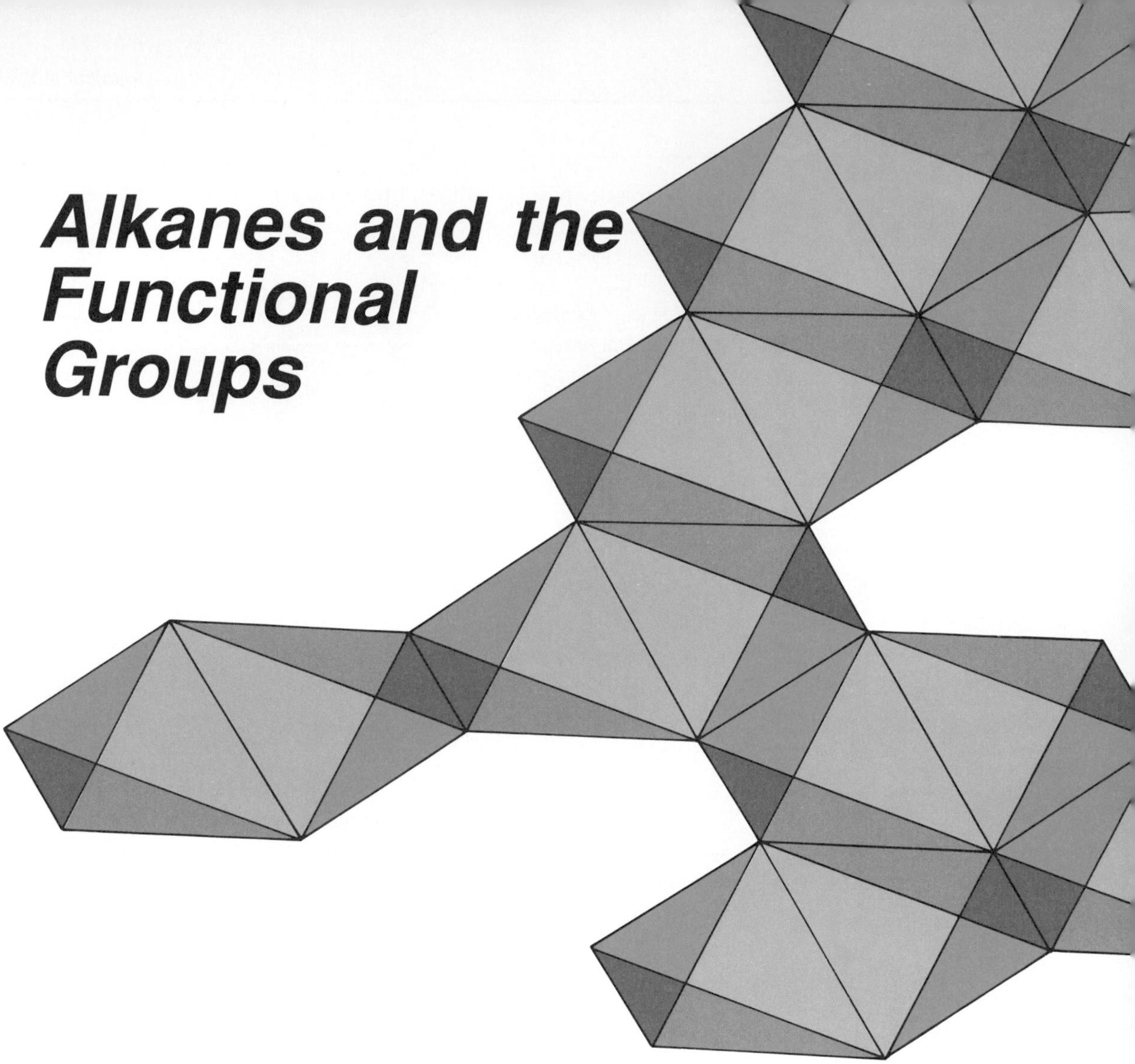

Our study of the various classes of organic compounds begins with the **hydrocarbons,** compounds that contain only carbon and hydrogen. Methane, CH_4, is the simplest hydrocarbon. In methane, the hydrogen atoms are all equivalent, occupying the corners of a regular tetrahedron. Imagine now the replacement of one of the hydrogen atoms of methane with another CH_3 group, as follows.

ethane:

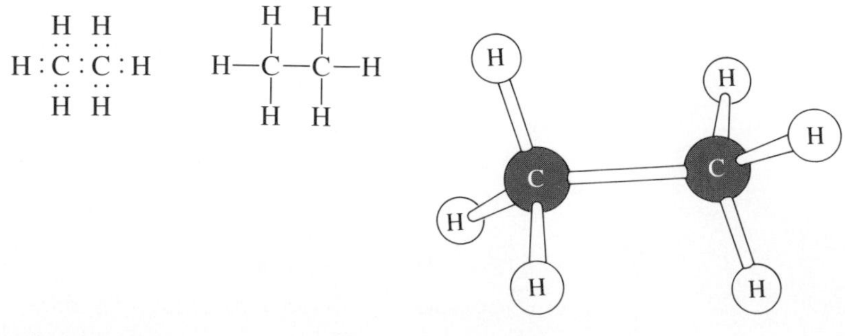

This compound is the hydrocarbon called *ethane*. The bond between the two carbon atoms is longer than a C—H bond, but it is an electron-pair bond in the Lewis sense much like the C—H bond. In terms of orbitals, the carbon–carbon bond in ethane consists of two electrons in a bonding molecular orbital formed by the overlap of two sp^3 hybrid orbitals, one from each carbon. Thus we say that the carbon–carbon bond in ethane is an sp^3-sp^3 single bond (Fig. 3.1). The C—H bonds in ethane are much like those of methane. They consist of electron pairs in bonding molecular orbitals, each of which is formed by the overlap of a carbon sp^3 orbital with a hydrogen 1*s* orbital. Both the H—C—C and H—C—H bond angles in ethane are approximately tetrahedral, because each carbon bears four groups.

We can go on to envision other hydrocarbons in which any number of carbons are bonded in this way to form chains of carbon atoms with their associated hydrogens. The idea of carbon chains, a revolutionary one in the early days of chemistry, was developed independently by the German chemist August Kekulé and the Scotsman Archibald Scott Couper in about 1858. Kekulé's account of his inspiration for this idea is amusing.

> During my stay in London I resided for a considerable time in Clapham Road in the neighborhood of Clapham Common . . . One fine summer evening I was returning by the last bus, "outside" as usual, through the deserted streets of the city that are at other times so full of life. I fell into a reverie, and lo, the atoms were gamboling before my eyes. Whenever, hitherto, these diminutive beings had appeared to me they had always been in motion. Now, however, I saw how, frequently, two smaller atoms united to form a pair . . . *I saw how the larger ones formed a chain,* dragging the smaller ones after them but only at the ends of the chain . . . The cry of the conductor, "Clapham Road," awakened me from my dreaming, but I spent a part of the night putting on paper at least sketches of these dream forms. This was the origin of the "Structure Theory."

We shall study two broad classes of hydrocarbons: **aliphatic hydrocarbons** and **aromatic hydrocarbons.** There are three types of aliphatic hydrocarbons: *alkanes, alkenes,* and *alkynes.* We shall begin our study of aliphatic hydrocarbons with the **alkanes,** also known as **paraffins.** These are hydrocarbons that contain only single bonds. Later we shall consider the **alkenes,** or **olefins,** hydrocarbons that contain carbon–carbon double bonds; and the **alkynes,** or **acetylenes,** hydrocarbons that

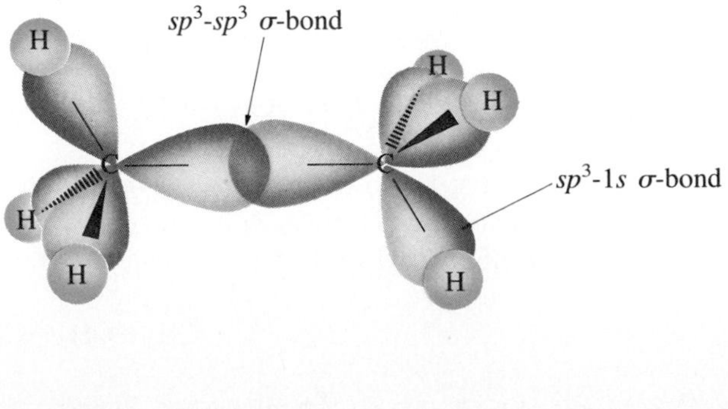

Figure 3.1 Bonding molecular orbitals in ethane.

contain carbon–carbon triple bonds. The class of **aromatic hydrocarbons** consists of benzene and its substituted derivatives.

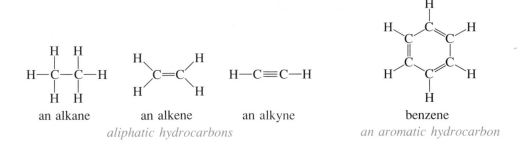

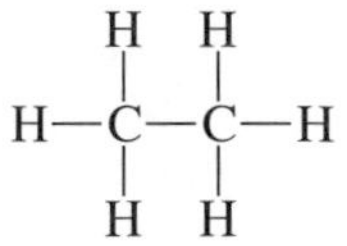

an alkane

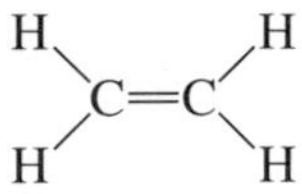

an alkene

$H—C \equiv C—H$

an alkyne

aliphatic hydrocarbons

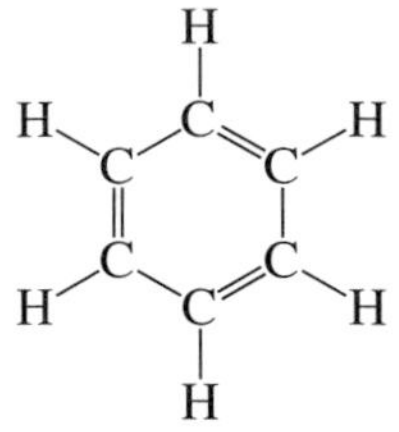

benzene

an aromatic hydrocarbon

3.1 NORMAL ALKANES

Carbon chains can take many forms in the alkanes; they may be unbranched or branched, and they can even exist in rings (cyclic alkanes). The alkanes with unbranched carbon chains are called **normal alkanes,** or ***n*-alkanes.** A few of the *n*-alkanes are shown in Table 3.1, along with some of their physical properties. The names of the first twelve *n*-alkanes should be learned, since they are the basis for naming many organic compounds. The names methane, ethane, propane, and butane have their origins in the early history of organic chemistry, but the names of the higher alkanes are derived from the corresponding Greek numerical names: *pent*ane (*cf.* pent- = five); *hex*ane (*cf.* hex- = six); etc.

Organic molecules are represented in different ways, which can be illustrated with the *n*-alkanes. The **molecular formula** of a compound, given in the second column of Table 3.1 for the *n*-alkanes, indicates its atomic composition. All noncyclic alkanes (alkanes without rings) have the general formula C_nH_{2n+2}, in which n is the number of carbons in the alkane. The **structural formula** of a molecule is its Lewis structure, which shows the **connectivity** of its atoms—that is, the order in which its atoms are connected. For example, a structural formula for *n*-hexane is the following:

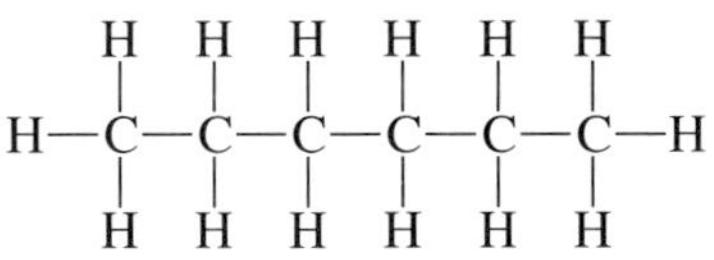

hexane

Writing each hydrogen atom in this way is very time-consuming, and a simpler representation of this molecule, called a **condensed structural formula,** conveys the same information.

$$CH_3—CH_2—CH_2—CH_2—CH_2—CH_3 \qquad \text{hexane}$$

In such a structure, the hydrogen atoms are understood to be connected to carbon atoms with single bonds. The structural formula may be further abbreviated using the device illustrated in Table 3.1. In this type of formula, for example, $(CH_2)_4$ means

$$—CH_2—CH_2—CH_2—CH_2—$$

TABLE 3.1 The n-Alkanes

Compound name	Molecular formula	Condensed structural formula	Melting point, °C	Boiling point, °C	Density,* g/mL
methane	CH_4	CH_4	−182.5	−161.7	—
ethane	C_2H_6	CH_3CH_3	−183.3	−88.6	—
propane	C_3H_8	$CH_3CH_2CH_3$	−187.7	−42.1	0.5005
butane	C_4H_{10}	$CH_3(CH_2)_2CH_3$	−138.3	−0.5	0.5788
pentane	C_5H_{12}	$CH_3(CH_2)_3CH_3$	−129.8	36.1	0.6262
hexane	C_6H_{14}	$CH_3(CH_2)_4CH_3$	−95.3	68.7	0.6603
heptane	C_7H_{16}	$CH_3(CH_2)_5CH_3$	−90.6	98.4	0.6837
octane	C_8H_{18}	$CH_3(CH_2)_6CH_3$	−56.8	125.7	0.7026
nonane	C_9H_{20}	$CH_3(CH_2)_7CH_3$	−53.5	150.8	0.7177
decane	$C_{10}H_{22}$	$CH_3(CH_2)_8CH_3$	−29.7	174.0	0.7299
undecane	$C_{11}H_{24}$	$CH_3(CH_2)_9CH_3$	−25.6	195.8	0.7402
dodecane	$C_{12}H_{26}$	$CH_3(CH_2)_{10}CH_3$	−9.6	216.3	0.7487
eicosane	$C_{20}H_{42}$	$CH_3(CH_2)_{18}CH_3$	+36.8	343.0	0.7886

*The densities tabulated in this text are at 20° unless otherwise noted.

The family of *n*-alkanes is one example of a **homologous series.** Successive members of a homologous series differ from each other by one —CH_2— group (**methylene group**) in the carbon chain. Generally, the physical properties of the members of a homologous series vary in a regular way down the series. An examination of Table 3.1, for example, reveals that the boiling points, melting points, and densities of the *n*-alkanes vary in a regular way with increasing number of carbon atoms. This variation can be useful for quickly estimating the properties of a member of the series.

Even more important, the members of a homologous series usually undergo the same chemical reactions. This observation greatly simplifies the learning of organic chemistry. For example, we can study the chemical reactions of propane with the confidence that ethane, butane, or dodecane will undergo analogous reactions.

The French chemist Charles Gerhardt (1816–1856) recognized in 1845 that many related compounds differ in their molecular formulas by multiples of (CH_2). He wrote, "These (related) substances undergo reactions according to the same equations, and it is only necessary to know the reactions of one in order to predict the reactions of the others."

Problems

1 (a) How many hydrogen atoms are in the *n*-alkane with 18 carbon atoms?
(b) How many carbon atoms are in the *n*-alkane with 30 hydrogen atoms?
(c) Is there an *n*-alkane containing 29 hydrogen atoms? If so, give its structural formula; if not, explain why.

2 Estimate the boiling point of tridecane, $C_{13}H_{28}$. Give a condensed structural formula for this compound.

3.2 STRUCTURAL ISOMERS. RULES FOR ALKANE NOMENCLATURE

When a carbon atom in an alkane is bound to more than two other carbon atoms, a branch in the carbon chain occurs at that position. The smallest branched alkane has four carbon atoms. As a result, there are two four-carbon alkanes; one is *butane,* and the other is *isobutane.*

$CH_3—CH_2—CH_2—CH_3$

butane

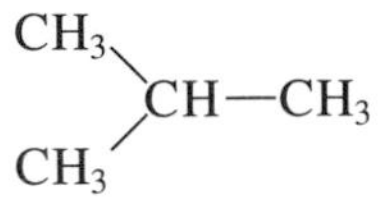

isobutane

Both compounds are butanes, and both have the molecular formula C_4H_{10}. Different compounds with the same molecular formula are called **isomers.** Isomers that differ in the connectivity of their atoms are called **structural isomers.** (Structural isomers are sometimes termed **constitutional isomers.**) Isobutane and butane are therefore structural isomers. These two compounds are the only structural isomers of butane.

More structural isomers are possible for alkanes with more carbon atoms. There are nine isomers of the heptanes (C_7H_{16}); 75 isomers of the decanes ($C_{10}H_{22}$); and 366,319 isomers of the eicosanes ($C_{20}H_{42}$)! The large number of isomers that are possible for organic compounds of even modest size creates a problem of nomenclature. How can we unambiguously designate each one of many isomeric compounds with a name that can be easily constructed and remembered?

The first proposals to standardize organic nomenclature, made at Geneva in 1892, have subsequently led to the modern rules for **systematic nomenclature** developed by the Commission on Nomenclature of the International Union of Pure and Applied Chemistry (IUPAC), an organization of chemists. Because the systematic, or IUPAC, nomenclature of alkanes forms the basis for the nomenclature of all other organic compounds, it is important to master the principles of systematic IUPAC nomenclature at an early stage.

A. Nomenclature of Alkanes

Alkanes are named by applying the following rules:

1. *The unbranched alkanes are named according to the number of carbons as shown in Table 3.1.*
2. *The following older names for branched-chain alkane isomers of four, five, and six carbons are still recognized.*

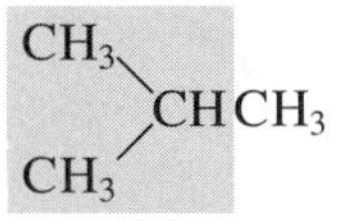

or $(CH_3)_2CHCH_3$ isobutane

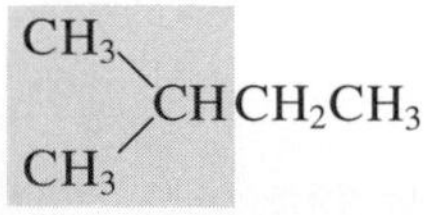

or $(CH_3)_2CHCH_2CH_3$ isopentane

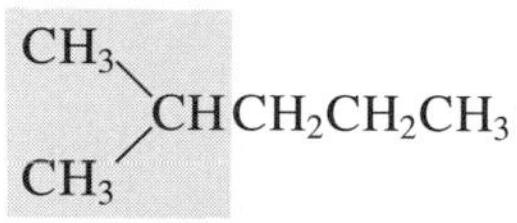 or $(CH_3)_2CHCH_2CH_2CH_3$ isohexane

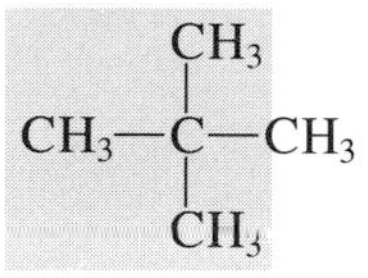 or $(CH_3)_4C$ neopentane

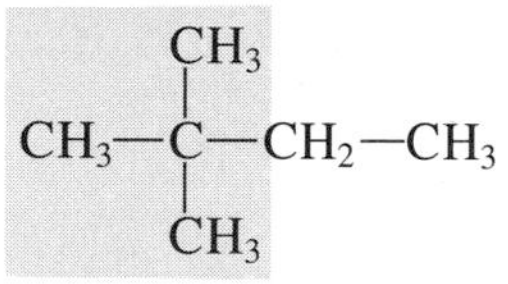 or $(CH_3)_3CCH_2CH_3$ neohexane

Notice that the iso prefix is used for structures with a one-carbon branch at the penultimate carbon of a chain (colored shading in structures above), and the neo prefix for structures with two one-carbon branches at the penultimate carbon of a chain (grey shading in structures above).

Other branched alkanes are named according to their **principal chains.** The principal chain of an alkane is the longest continuous carbon chain in the molecule. The name is derived by naming the principal chain and its branching groups, or **substituent groups.** In the following compound, for example, the principal chain contains six carbons; so the molecule is named as a *substituted derivative of hexane,* even though the compound is *an isomer of heptane.*

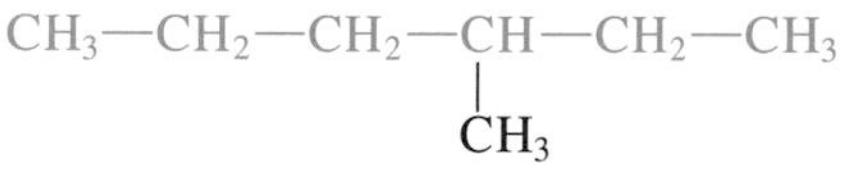 principal chain

3. *To name a substituted alkane, determine the principal chain and number its carbons consecutively from one end to the other in the direction that gives the substituent group the lower number. The name is constructed by writing the carbon number at which the branch occurs, a hyphen, the name of the substituent group at the branch, and the name of the parent alkane.*

In the previous example, the principal chain is numbered to give the lower number to the $—CH_3$ branch. The branch occurs at carbon-3 of the principal chain.

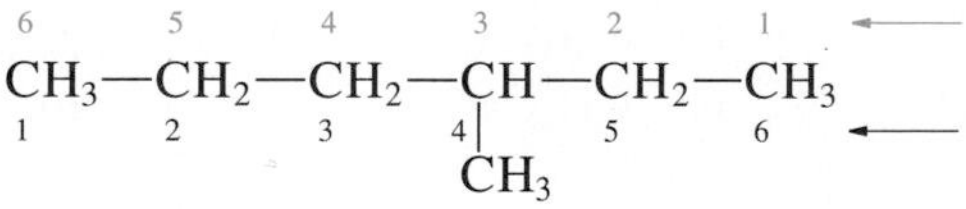
proper numbering
improper numbering

The $—CH_3$ group is called a *methyl* group. Following rule 3, then, the name of this compound is 3-methylhexane. (Notice that *methyl* and *hexane* are joined in one word.)

It is important to realize that the principal chain is the longest continuous carbon chain *no matter how the molecule is drawn*. Thus, the following structures represent the same molecule:

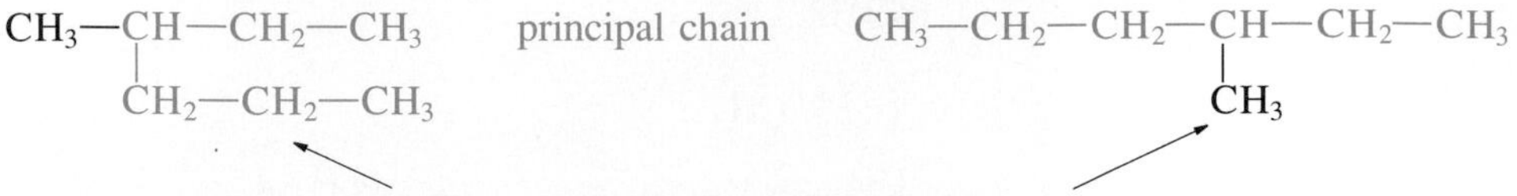

different ways to draw 3-methylhexane

4. *In the case of multiple substituents, each substituent receives its own number. The prefixes di, tri, tetra, etc., are used to indicate the number of identical groups.*

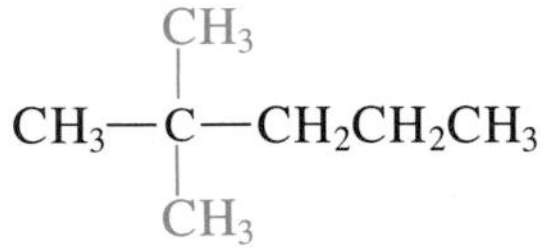

2,2-dimethylpentane

two methyl substituents

both methyl groups at carbon-2 of the principal chain

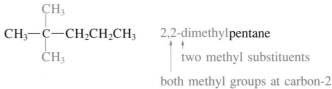

5. *When the branching groups occur on different carbons of the principal chain, alternate numbering schemes are compared number-by-number, and the one is chosen that gives the lowest number* at the first point of difference.

Consider, for example, the following compound:

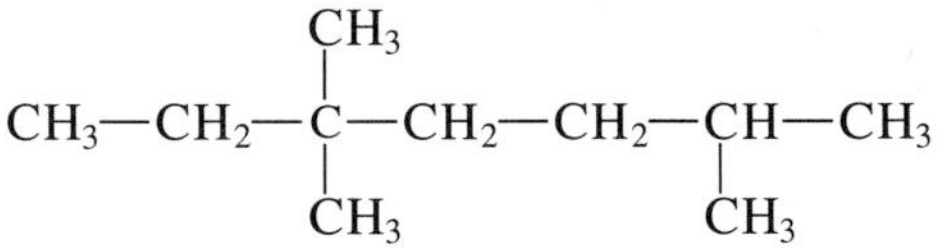

possible names:
2,5,5-trimethylheptane (correct)
3,3,6-trimethylheptane (incorrect)

To apply this rule, we write down the two possible names derived by numbering from either end of the chain. The numbering schemes are 2,5,5- and 3,3,6-. A decision between these is made by a pairwise comparison of the numbers—2 with 3, 5 with 3, and 5 with 6. Since the first point of difference in these schemes is in the initial numbers 2 and 3, the first scheme is chosen. The second point of difference, 5 *vs.* 3, does not enter in the choice.

The substituent groups may contain more than one carbon. When substituent groups are derived from alkanes, they are called **alkyl groups.** The names of unbranched alkyl groups are derived from the *n*-alkane with the same number of carbons by dropping the final *ane* and adding *yl.*

$—CH_3$	methyl (= meth~~ane~~ + yl)
$—CH_2CH_3$ or $—C_2H_5$	ethyl (= eth~~ane~~ + yl)
$—CH_2CH_2CH_3$	propyl

The substituents may also be branched. The most common branched alkyl groups have special names that must be learned. These are given in Table 3.2.

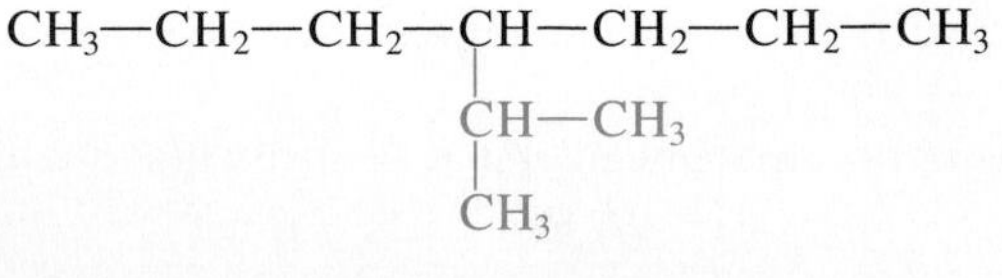

4-isopropylheptane

alkyl group name from Table 3.2

TABLE 3.2 Nomenclature of Short Branched-Chain Alkyl Groups

Group structure	Written name	Pronounced name
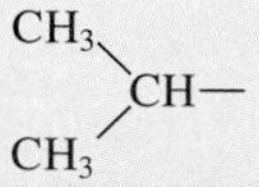	isopropyl	isopropyl
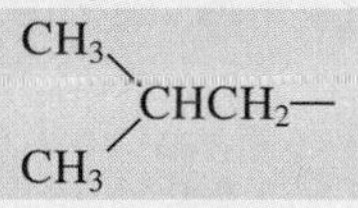	isobutyl	isobutyl
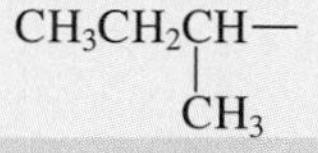	*sec*-butyl	secondary butyl
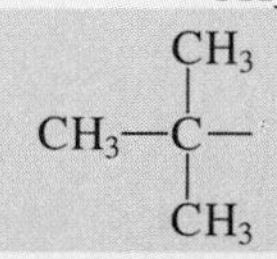	*tert*-butyl (or *t*-butyl)	tertiary butyl
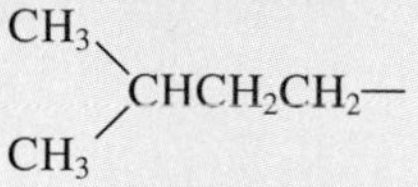	isopentyl	isopentyl
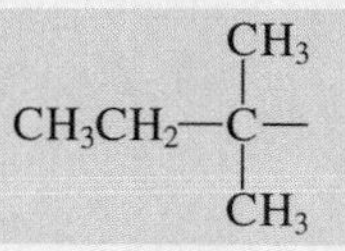	*tert*-pentyl (or *t*-pentyl)	tertiary pentyl
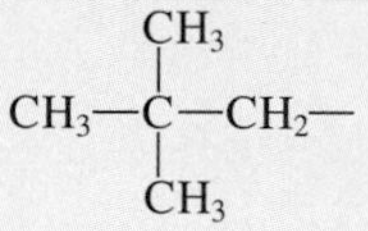	neopentyl	neopentyl

6. *Substituent groups are cited in alphabetical order regardless of their location in the principal chain.* The numerical prefixes di, tri, etc., as well as the prefixes *tert-* and *sec-* are ignored in alphabetizing, but the prefixes iso, neo, and cyclo are considered in alphabetizing substituent groups.

The following compounds illustrate the application of this rule:

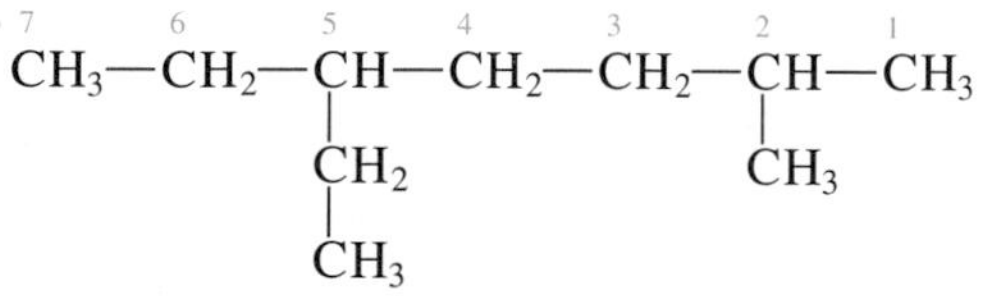

5-ethyl-2-methylheptane (*ethyl* is cited before *methyl* even though it has a higher number)

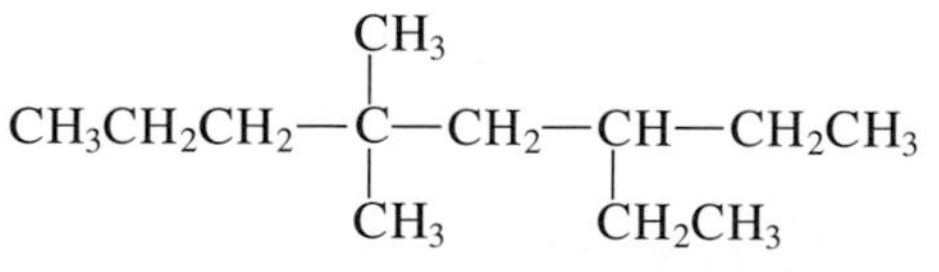

3-ethyl-5,5-dimethyloctane (note that *dimethyl* begins with the letter *m* for purposes of citation)

7. *When the numbering of different groups is not resolved by the other rules, the first-cited group receives the lowest number.*

In the following compound, rules 1–6 do not dictate a choice between the names 3-ethyl-5-methylheptane and 5-ethyl-3-methylheptane. Since the ethyl group is cited first in the name, it receives the lower number, by rule 7.

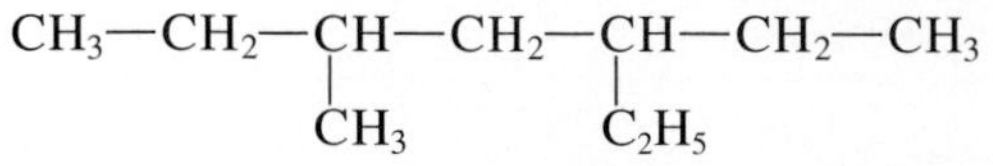

$$CH_3-CH_2-\underset{\underset{CH_3}{|}}{CH}-CH_2-\underset{\underset{C_2H_5}{|}}{CH}-CH_2-CH_3 \qquad \text{3-ethyl-5-methylheptane}$$

There are situations of greater complexity that are not covered by these seven rules; however, these rules will suffice for most of the cases that we are likely to encounter.

From the large number of alkane isomers alone, we can appreciate that there exist vast numbers of organic compounds. Many have not been synthesized—not because their syntheses would be unreasonable, but because there has simply been no need for them. As more organic compounds are added each year to the list of known compounds, the importance of automated information storage and retrieval systems becomes more obvious. Nomenclature is a central problem in any such system; even the IUPAC names are unwieldy for many compounds. For this reason new systems of chemical nomenclature that are especially suitable for computer storage are currently under development.

Problems

3 Name the following compounds. Be sure to designate the principal chain properly before constructing the name.

(a) $CH_3CH_2-\underset{\underset{\underset{\underset{CH_2-CH_3}{|}}{CH_2}}{|}}{CH}\underset{\underset{CH_3}{|}}{}\overset{}{CH}-CH_2CH_2CH_3$

(b) $CH_3CH_2CH_2\underset{\underset{CH_3}{|}}{CH}-\underset{\underset{CH_3}{|}}{\overset{\overset{CH_3}{|}}{C}}\overset{\overset{C_2H_5}{|}}{}CH-CH_2CH_2CH_3$

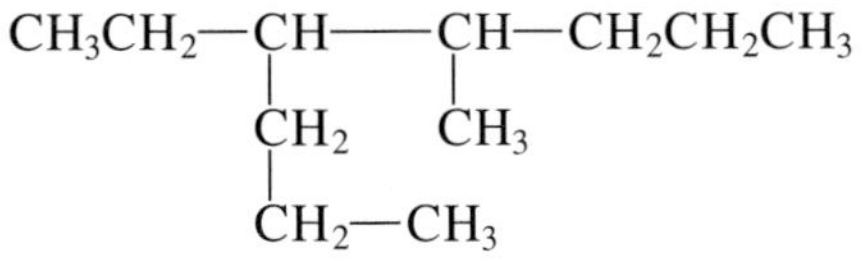

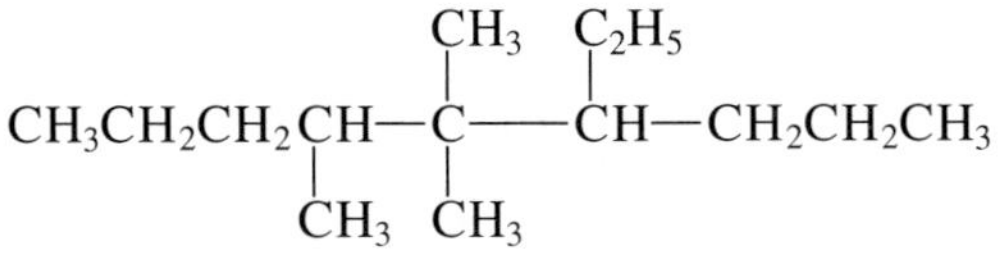

4 Draw condensed structures for all isomers of heptane and give their systematic names.

B. Classification of Carbon Substitution

It is important to be able to recognize different types of carbon substitution in branched compounds, because chemical reactivity often varies with the degree of substitution. A carbon is said to be **primary, secondary, tertiary,** or **quaternary** when it is bonded to one, two, three, or four other carbons, respectively.

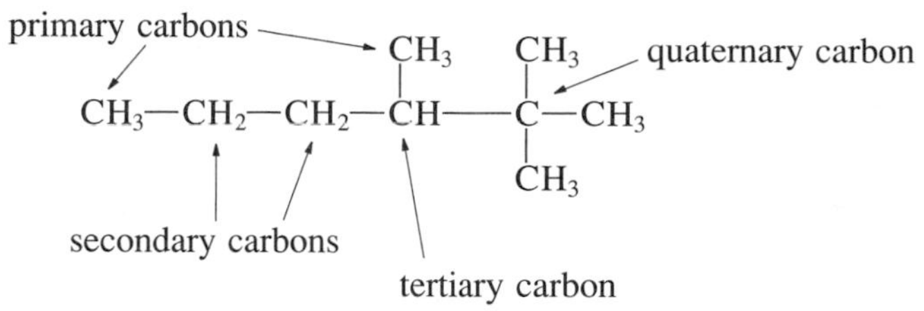

Likewise, the hydrogens bonded to each type of carbon are called primary, secondary, or tertiary hydrogens, respectively.

Problem

5 (a) Draw the structure of 4-isopropyl-2,4,5-trimethylheptane.
(b) In the structure of this compound identify the primary, secondary, and quaternary carbon atoms.
(c) In the structure of this compound, circle one example of each of the following groups: a methyl group; an ethyl group; an isopropyl group; a *sec*-butyl group; an isobutyl group.

3.3 CYCLOALKANES. SKELETAL STRUCTURES

Alkanes that contain carbon chains in closed loops, or rings, are called **cycloalkanes.** These compounds are named by adding the prefix *cyclo* to the name of the alkane. Thus, the six-membered cycloalkane is called *cyclohexane.*

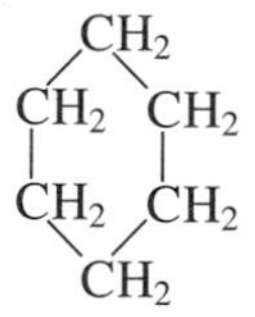
cyclohexane

The cycloalkanes form another homologous series. The names and some physical properties of the simple cycloalkanes are given in Table 3.3. Notice that the general formula for an alkane containing a single ring has two fewer hydrogens than that of

TABLE 3.3 Physical Properties of the Cycloalkanes

Compound	Boiling point, °C	Melting point, °C	Density, g/mL
cyclopropane	−32.7	−127.6	
cyclobutane	12.5	−50.0	
cyclopentane	49.3	−93.9	0.7457
cyclohexane	80.7	6.6	0.7786
cycloheptane	118.5	−12.0	0.8098
cyclooctane	150.0	14.3	0.8340

the open-chain alkane with the same number of carbon atoms. For example, cyclohexane has the formula C_6H_{12}, but hexane has the formula C_6H_{14}. The general formula for the cycloalkanes with one ring is C_nH_{2n}.

It is very common to write the structures of cycloalkanes using **skeletal structures**—structures that show only the carbon–carbon bonds. In this notation a cycloalkane is written as a closed geometrical figure. In a skeletal structure, it is understood that there is a carbon at each vertex of the figure, and that enough hydrogens are present on each carbon to fulfill its tetravalence. Thus, the skeletal structure of cyclohexane is written as follows:

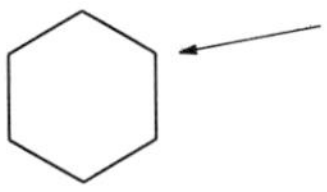

Skeletal structures may also be written for open-chain alkanes. For example, hexane might be indicated this way:

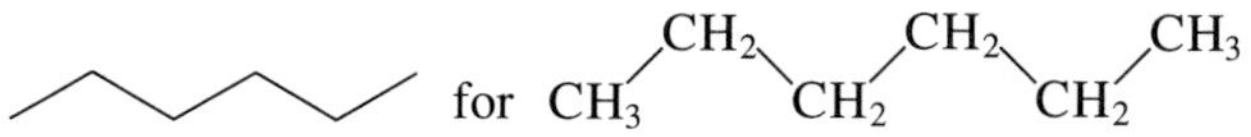

When a skeletal structure is written for an open-chain compound, it is important to remember that there are carbons not only at each vertex, but also at the ends of the structure. Thus, the six carbons of hexane above are indicated by the four vertices and two ends of the skeletal structure. Here are a few other examples of skeletal structures:

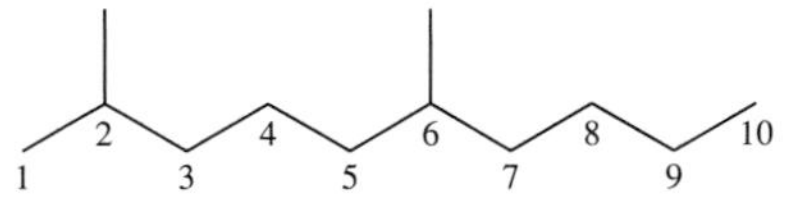

2,6-dimethyldecane

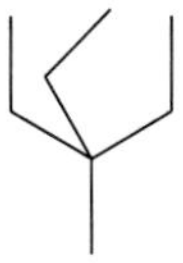

3-ethyl-3-methylpentane

The nomenclature of cycloalkanes follows essentially the same rules used for open-chain alkanes, as the following examples show:

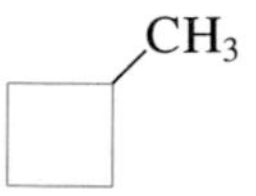

methylcyclobutane

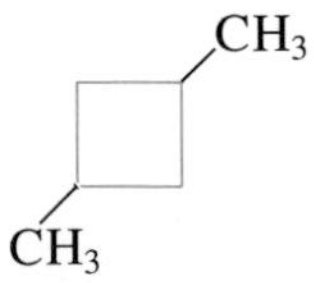

1,3-dimethylcyclobutane

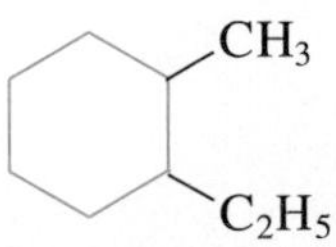

1-ethyl-2-methylcyclohexane
(Note alphabetical citation, rule 7.)

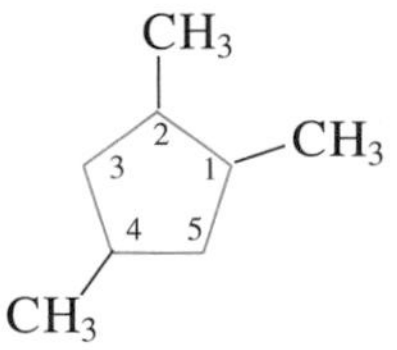

1,2,4-trimethylcyclopentane
(Note application of rule 5,
the lowest number at first
point of difference.)

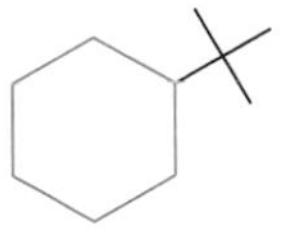

tert-butylcyclohexane

Notice that the numerical prefix 1- is not necessary for monosubstituted cycloalkanes. Thus, the first compound is methylcyclobutane, not 1-methylcyclobutane. When there are two or more substituents, however, they must be numbered to indicate their relative position. The lowest number is assigned in accordance with the usual rules. For example, there are several possible numbering schemes for the cyclopentane derivative above in which the number 1 could be assigned to each substituent. All of these are compared number-by-number according to rule 5 to arrive at the correct numbering. Finally, notice the use of the skeletal formula in the last example; the cross is a common way of representing the *t*-butyl group.

Problems

6 Represent each of the following compounds with a skeletal structure:

 (a) cyclopropane

 (b) ethylcyclopentane

 (c)

$$CH_3CH_2CH_2CH-\overset{\overset{\displaystyle CH_3}{|}}{CH}-C(CH_3)_3$$
$$\underset{\displaystyle CH_3}{|}$$

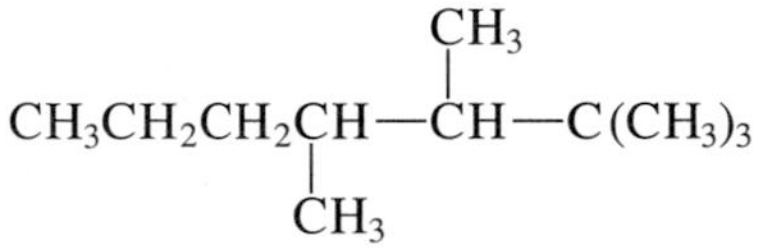

 (d) $CH_3(CH_2)_{15}CH_3$

7 Name the following compound:

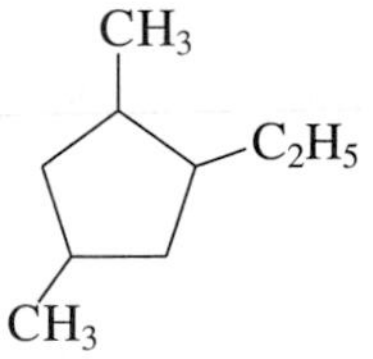

3.4 PHYSICAL PROPERTIES OF ALKANES

A. Boiling Points

The **boiling point** is the temperature at which a substance undergoes a transition from the liquid to the gaseous state. We discovered in Sec. 3.1 that there is a regular change in the boiling points of the *n*-alkanes with increasing number of carbons. This trend of boiling point within the homologous series of *n*-alkanes is particularly apparent in a

Figure 3.2 Boiling points of some *n*-alkanes plotted against number of carbon atoms.

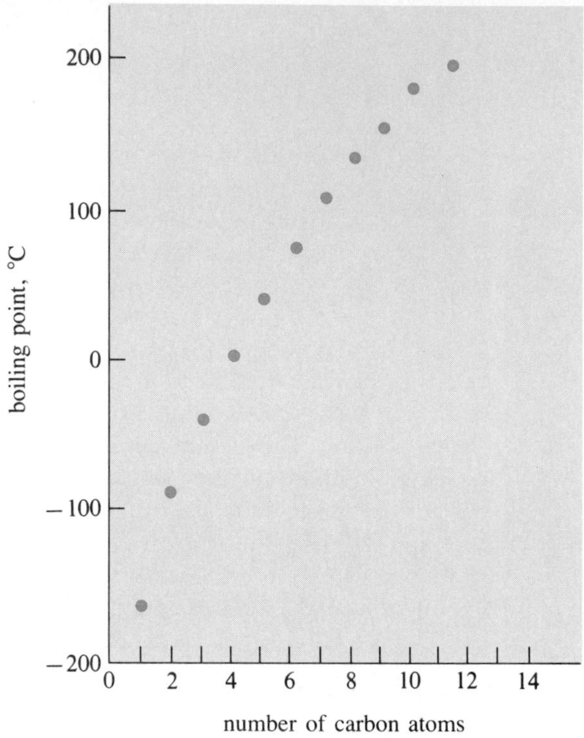

plot of boiling point against carbon number (Fig. 3.2). *The regular increase in boiling point of 20–30° per carbon atom within a homologous series is a general trend observed for many types of organic compounds.* What is the reason for this increase?

The boiling point is a crude measure of the attractive forces among molecules in the liquid state compared to those in the gaseous state. Because, to a useful approximation, the attractive forces among molecules in the gaseous state are negligible at atmospheric pressure, it follows that the boiling point measures the relative strengths of cohesive interactions among the molecules of a liquid. Yet no chemical bond is formed between separate molecules. What, then, is the origin of this attraction of one molecule for another?

In the previous chapter we learned that electrons in bonds are not confined between the nuclei, but rather reside in bonding molecular orbitals that surround the nuclei; we can think of these occupied molecular orbitals as "electron clouds." We might visualize an organic molecule such as an *n*-alkane as something akin to a cotton ball (the electron clouds) with embedded cotton seeds (the nuclei). The shapes of these electron clouds can be altered by external forces. One such external force is the electric field of the electrons in nearby molecules. When two molecules approach each other closely, as in a liquid, the electron clouds of one molecule repel the electron clouds of the other. As a result, both molecules *temporarily* acquire small localized separations of charge (Fig. 3.3) called *induced dipoles*. That is, the molecules take on a *temporary* dipole moment. The deficiency of electrons (positive charge) in part of one molecule is attracted by the excess of electrons (negative charge) in part of the other. This attraction, an example of a **van der Waals force** or **dispersion force,** is the cohesive force that must be overcome in order to vaporize a liquid hydrocarbon. Note that alkanes do *not* have appreciable permanent dipole moments. The dipole moment that causes the attraction between alkane molecules is induced *tempo-*

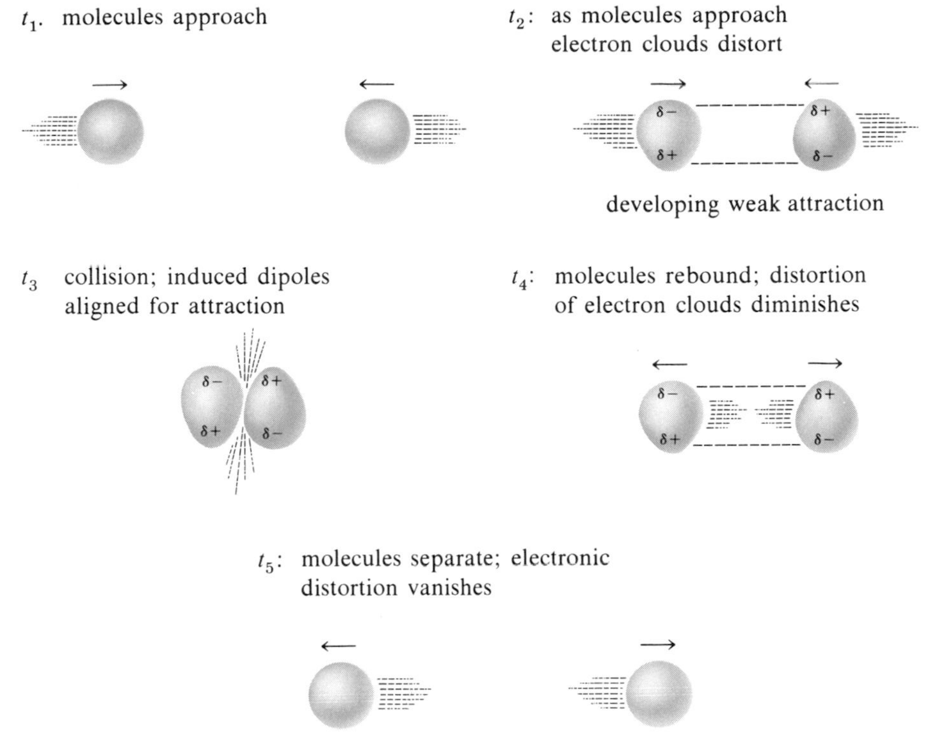
t_1. molecules approach
t_2: as molecules approach electron clouds distort
δ− δ+ δ+ δ−
developing weak attraction
t_3 collision; induced dipoles aligned for attraction
δ− δ+ δ+ δ−
t_4: molecules rebound; distortion of electron clouds diminishes
δ− δ+ δ+ δ−
t_5: molecules separate; electronic distortion vanishes

Figure 3.3 A stop-frame cartoon of the formation of induced dipoles as two hypothetical molecules collide and rebound. Notice that the distortion of the electron clouds is not permanent but varies with time. The electronic distortion in one species is induced by the proximity to the other. The frames are labeled t_1, t_2, etc., for successive times.

rarily in one molecule by the proximity of another. We might say that "nearness makes the molecules grow fonder."

We are now in a position to understand why larger molecules have higher boiling points. A larger molecule has a greater surface of electron clouds available for van der Waals interactions with other molecules. Since van der Waals attractions are greater between large molecules, large molecules have higher boiling points.

The *shape* of a molecule is also important in determining its boiling point. For example, a comparison of the boiling point of the highly branched alkane neopentane (9.4°) and its unbranched isomer pentane (36.1°) is particularly striking. As we can see from the following models, neopentane has four methyl groups disposed in a tetrahedral fashion about a central carbon. The molecule almost resembles a compact ball, and could fit comfortably within a sphere.

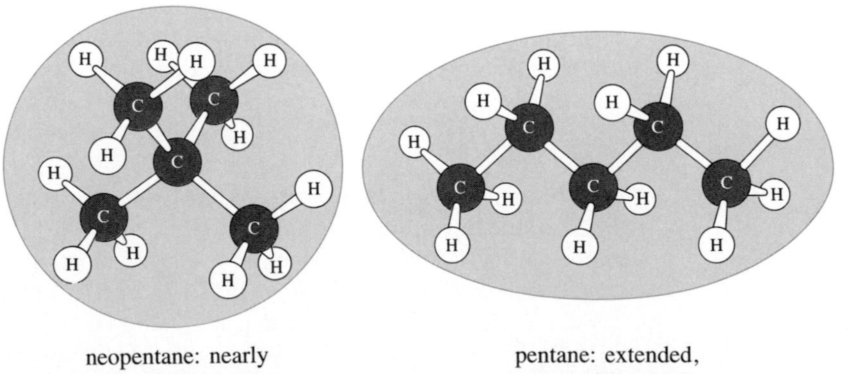

neopentane: nearly spherical

pentane: extended, more ellipsoidal

On the other hand, pentane is rather extended, ellipsoidal in shape, and would not fit within the same sphere. The more a molecule approaches spherical proportions, the less surface area it presents to other molecules, because a sphere is the three-dimensional object with the minimum surface-to-volume ratio. Since neopentane has less surface area at which van der Waals interactions with other neopentane molecules can occur, it has fewer cohesive interactions than pentane, and thus, a lower boiling point.

We now know two factors that can affect boiling point:

1. Boiling points increase with increasing molecular weight within a homologous series—typically 20–30° per carbon atom. This increase is due to the greater van der Waals attractions among larger molecules.
2. Boiling points tend to be lower for highly branched molecules that approach spherical proportions because they have less molecular surface available for van der Waals attractions.

B. Melting Points

The **melting point** of a substance is the temperature at which it undergoes the transition from the solid to the liquid state. The melting point is an especially important physical property in organic chemistry because it is used both to identify organic compounds and to make a general assessment of their purity. Melting points are usually depressed, or lowered, by impurities, and the melting range (the range of temperature over which a substance melts), usually quite narrow for a pure substance, is substantially broadened by impurities. The melting point is a measure of the forces stabilizing the solid state weighed against those stabilizing the liquid state. Figure 3.4, which is a plot of melting point against carbon number for the *n*-alkanes, shows that melting points tend to increase with number of carbons. This is a general trend within many homologous series of organic compounds.

Since both solid and liquid are *condensed states* of matter in which molecules are close together, it is not surprising to find that the same types of forces contribute to the cohesive interactions in both states. However, when we see unusual patterns in the melting point that are not present in the boiling point, we can be reasonably sure that they reflect some characteristic of the solid state. An example of such a pattern that is readily evident in Fig. 3.4 is the intriguing sawtooth alternation of melting points between compounds with odd and even numbers of carbon atoms. The even-carbon compounds appear to lie on a separate, higher curve from the odd-carbon compounds. The reason for this behavior is that the attractive van der Waals forces among molecules in the solid state of the odd-carbon alkanes are weaker than those in the even-carbon alkanes. We might say that the odd-carbon alkane molecules do not fit together as well in the crystal as even-carbon ones. Similar alternation of melting points is observed in other homologous series. Can you see a similar trend in Table 3.3?

Branched-chain hydrocarbons tend to have lower melting points than linear ones because the branching interferes with regular packing in the crystal. When a branched molecule has a substantial symmetry, however, its melting point is typically relatively high because of the ease with which symmetrical molecules fit together within the crystal. For example, the melting point of the very symmetrical molecule neopentane, −16.8°, is considerably higher than that of the less symmetrical pentane, −129.8°, which is higher still than that of isopentane, −159.9°.

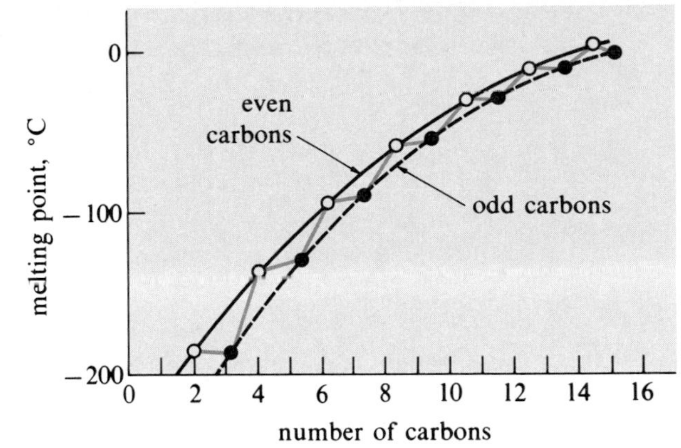

Figure 3.4 Plot of melting points of the *n*-alkanes against number of carbon atoms.

In summary, melting points show the following general trends:

1. Melting points tend to increase with increasing molecular weight within a homologous series.
2. Many highly symmetrical molecules have unusually high melting points.
3. A sawtooth pattern of melting point behavior (see Fig. 3.4) is observed within a number of homologous series.

Problem

8 Match each of the following compounds with the correct boiling points and melting points. Explain your choices.
Compounds: 2,2,3,3-tetramethylbutane and octane
Boiling points: 106.5°, 125.7°
Melting points: −118.3°, +100.7°

C. Other Physical Properties

The alkanes are, for all practical purposes, insoluble in water—thus the saying, "Oil and water don't mix." (Alkanes are a major constituent of crude oil.) Alkanes also have considerably lower densities than water. For these reasons a mixture of an alkane and water will separate into two distinct layers with the less dense alkane layer on top. An oil slick is an example of this behavior.

Problem

9 Explain why water is not usually very effective in extinguishing a gasoline fire. (Gasoline is mostly alkanes.)

3.5 CONFORMATIONS OF ALKANES. ROTATION ABOUT CARBON–CARBON SINGLE BONDS

A. Conformation of Ethane

The tetrahedral geometry of each carbon of ethane still leaves undefined the *conformation* of ethane: the angular relationship of the hydrogens on one carbon to those on the other. (We encountered the same problem with ethylene; Sec. 1.6B, Fig. 1.7.) An easy way to visualize this problem is as follows:

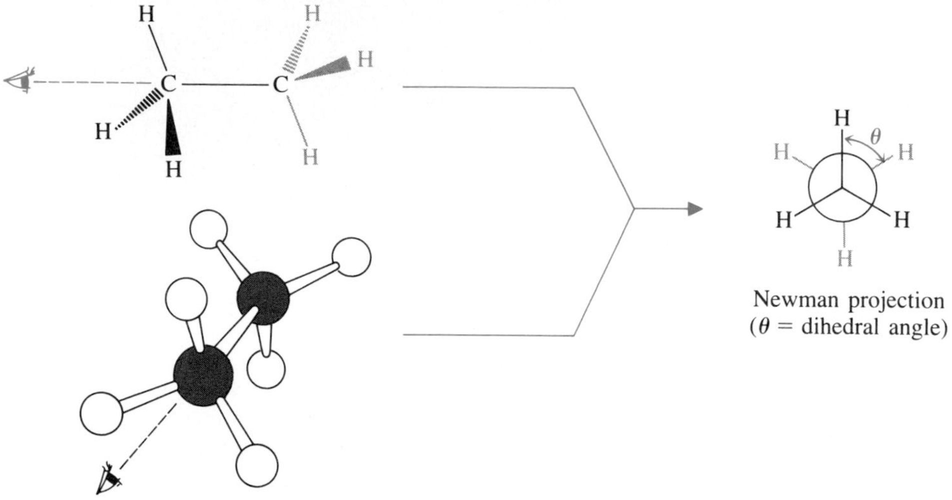

"sawhorse" representations of ethane

If we sight along the carbon–carbon bond of the ethane molecule (as shown in the sawhorse representation) and project what we see on the plane of the paper, we obtain what is called a **Newman projection.** In the Newman projection, a circle is used to represent the two carbon atoms at each end of the bond of interest. The bonds drawn to the center of the circle are attached to the carbon nearest the observer; the bonds drawn to the periphery of the circle are attached to the carbon farthest from the observer. The angle θ between the carbon–hydrogen bonds on different carbons of ethane is called the **dihedral angle.** There are two limiting possibilities for the conformation of ethane. In one, a C—H bond of one carbon bisects the angle between two C—H bonds of the other; this conformation of ethane is called the **staggered form,** ($\theta = 60°$). In the other conformation, the **eclipsed form,** the C—H bonds on the respective carbons are superimposed in the Newman projection ($\theta = 0°$).

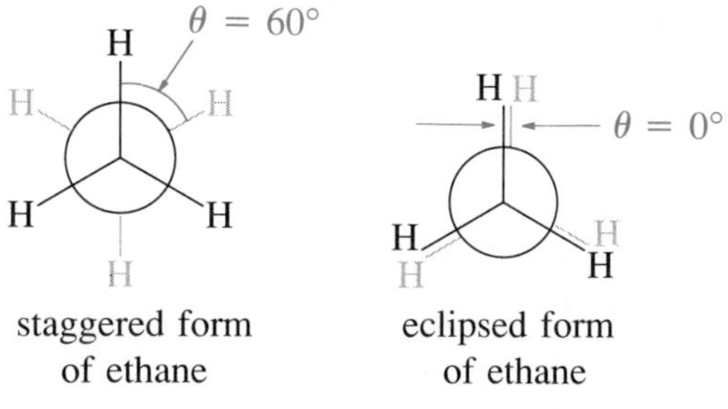

Of course, dihedral angles other than 60° and 0° are possible, but these two will prove to be of central importance. Which is the preferred conformation of ethane?

Physical chemists have investigated the relative energies of these forms of ethane. The results can be described by a plot of relative energy *vs.* angle of rotation about the C—C bond (dihedral angle), shown in Fig. 3.5. In this figure, 0° is defined by an eclipsed form, and a hydrogen on each carbon has been shown in color so that the rotation can be followed. If, using a model of ethane, we hold either carbon fixed and

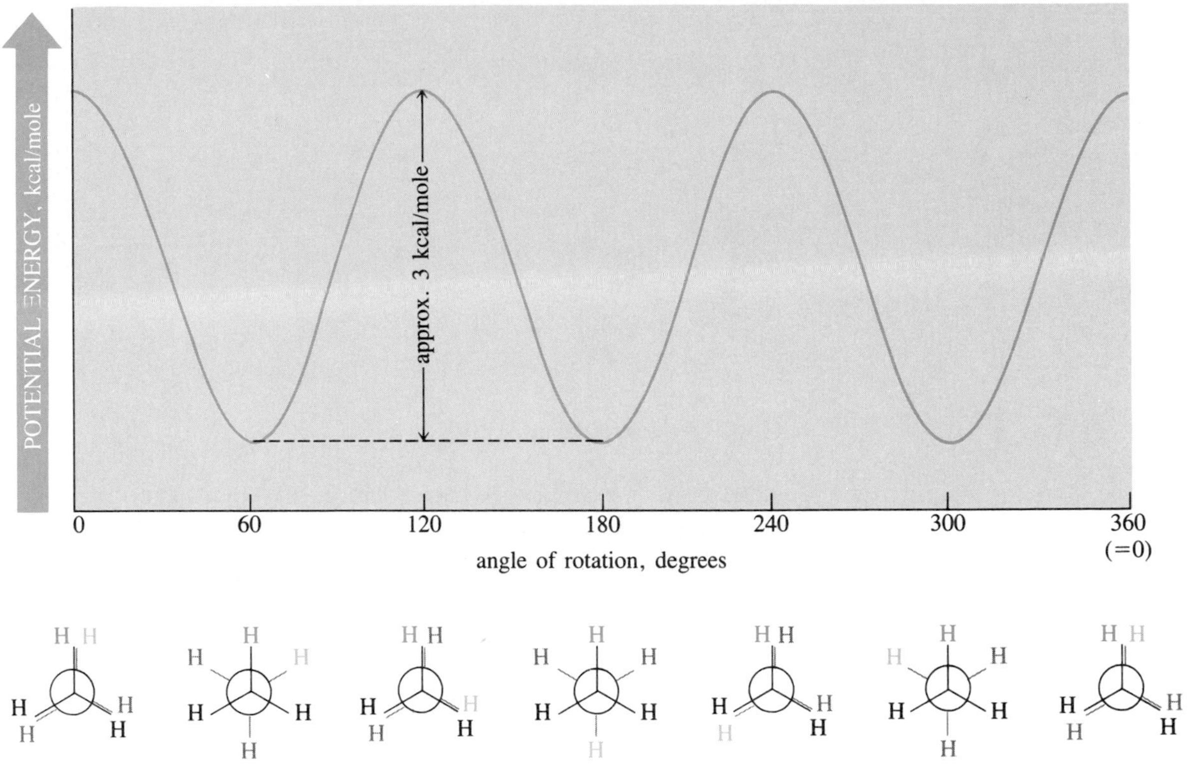

Figure 3.5 Variation of energy with rotation about the carbon–carbon bond of ethane.

turn the other about the C—C bond, the angle of rotation increases, and we pass alternately through three identical staggered and three identical eclipsed forms of ethane. (Demonstrate this to yourself with models.) The graph in Fig. 3.5 tells us first that identical forms have identical energy, as intuition dictates. The graph also shows that the eclipsed form of ethane is characterized by an energy *maximum,* and the staggered form is characterized by an energy *minimum*. The staggered form is thus the *stable* form of ethane. The graph shows that the staggered form is less energetic than the eclipsed form by about 3 kcal/mol. This means that it would take about 3 kcal of energy to convert one mole of staggered ethane to one mole of eclipsed ethane, if such a conversion were practical.

Chemists are not in complete agreement about the reasons why the staggered form of ethane is more stable, even though this problem has been studied for years. According to one argument, when ethane adopts the eclipsed form, the electrons in the eclipsed C—H bonds are closer together than they are in the staggered form, and therefore have a greater repulsive interaction. This interaction is manifested as a greater energy for the eclipsed ethane.

One staggered form of ethane can convert into another by rotation about the carbon–carbon bond. This rotation is called an **internal rotation** (to differentiate it from a rotation of the entire molecule). When an internal rotation occurs, an ethane molecule must briefly pass through the eclipsed form; thus, it must acquire the 3 kcal/mol additional energy of the eclipsed form and then lose it again. What is the source of this energy?

At temperatures above absolute zero, molecules are in constant motion. Heat is a manifestation of this molecular motion. Molecules in motion have velocity, v; since kinetic energy is given by $mv^2/2$ (m = mass of the molecule), molecules in motion

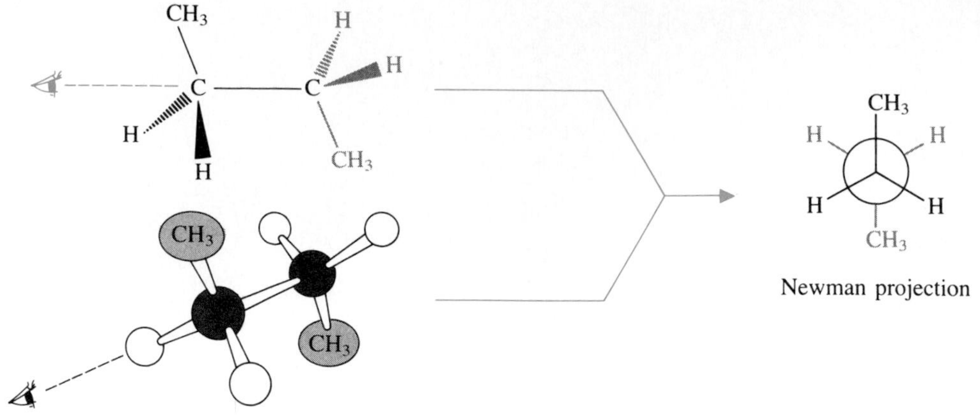

Newman projection

Figure 3.6 Derivation of the Newman projection between carbons 2 and 3 of butane. (Only one of the rotational forms is shown.)

also have kinetic energy. In a sample of ethane the molecules move about in a random manner, much as thousands of people might mill about in a crowd. These moving molecules frequently collide, and molecules can gain or lose energy in such collisions. (An analogy is the collision of a bat with a ball; some of the kinetic energy of the bat is lost to the ball.) When an ethane molecule gains sufficient energy from a collision, it can pass through the higher-energy eclipsed form into another staggered form—that is, it can undergo internal rotation. Whether a given ethane molecule acquires sufficient energy to undergo an internal rotation is strictly a matter of *probability* (random chance). However, there is a better chance for an internal rotation at higher temperature, because molecules have greater kinetic energy at higher temperature. The likelihood that ethane undergoes internal rotation is expressed as its rate of rotation. The additional 3 kcal/mol of energy required for such a rotation is sufficiently small that the internal rotation of ethane is very rapid even at room temperature. At 25 °C a typical ethane molecule undergoes a rotation from one staggered form to another at a rate of about 10^{11} times per second! This means that the interconversion between staggered forms takes place about once every 10^{-11} second. Despite this short lifetime for any one staggered form, an ethane molecule nevertheless spends most of its time in its staggered forms, passing only transiently through its eclipsed forms. Thus, an internal rotation is best characterized not as a continuous spinning, but a constant succession of abrupt jumps from one staggered form to another.

B. Conformations of Butane

Internal rotation about the central carbon–carbon bond of butane is a somewhat more complex case of rotation about a carbon–carbon single bond. The derivation of the Newman projections for this rotation is illustrated in Fig. 3.6, and the graph of energy as a function of angle of rotation is given in Fig. 3.7. Note once again that the various rotational possibilities are generated with a model by holding one carbon fixed (the carbon away from the observer in Fig. 3.7) and rotating the other one.

Figure 3.7 shows that the staggered forms of butane, like those of ethane, are at energy minima, and are thus the stable forms of butane. However, there are important differences between the butane and ethane situations. Not all of the staggered forms (nor the eclipsed forms) are alike. The different staggered forms have been given special names. The forms with a dihedral angle of 60° between the C—CH_3 bonds are

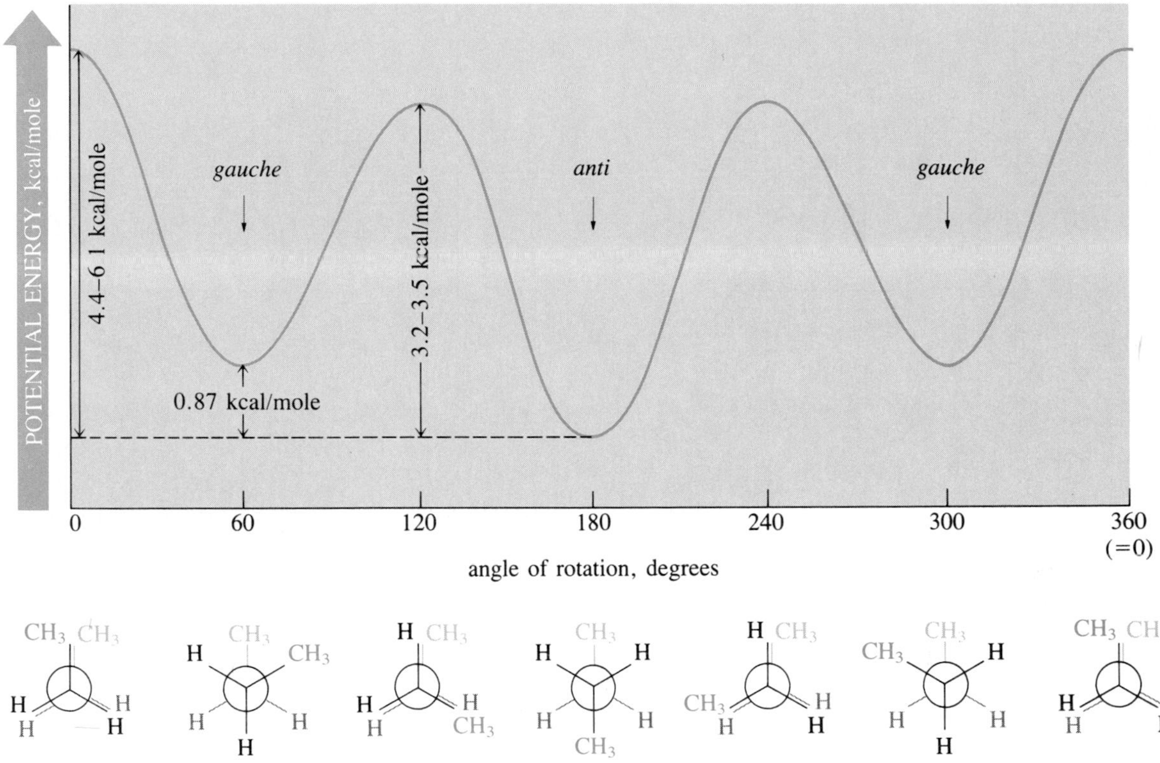

Figure 3.7 Internal rotation in butane.

called **gauche** forms; the form in which this dihedral angle is 180° is called the **anti** form. The *anti* and *gauche* forms of butane are examples of **rotational isomers** (also called **conformational isomers** or **torsional isomers**)—isomers that differ only by an internal rotation.

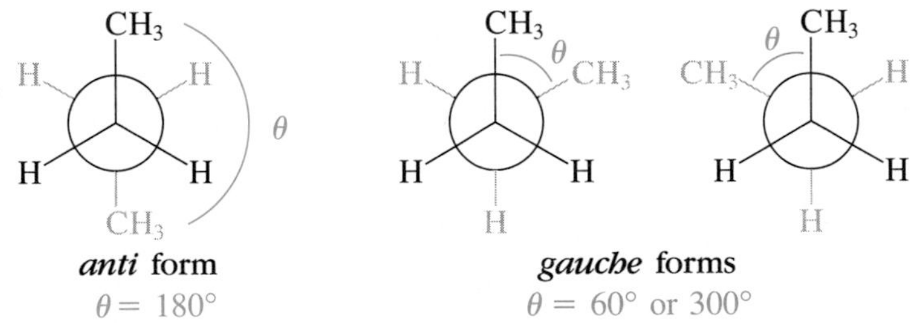

Figure 3.7 shows that the *gauche* and *anti* forms of butane have different energies. The *anti* form is the more stable of the two by about 0.87 kcal/mol. The difference in the two forms is that the methyl groups are closer together in the *gauche* form. The two kinds of eclipsed forms also have different energies. In the eclipsed form of higher energy, the methyl groups are very close together. From a model of this form we can see that the hydrogens of the two methyl groups almost touch as each methyl group rotates. It appears, then, that bringing the two methyl groups close together in butane causes an increase in the energy of the molecule. With the aid of Fig. 3.8, let us examine the physical laws that govern the interactions of the two methyl groups of butane.

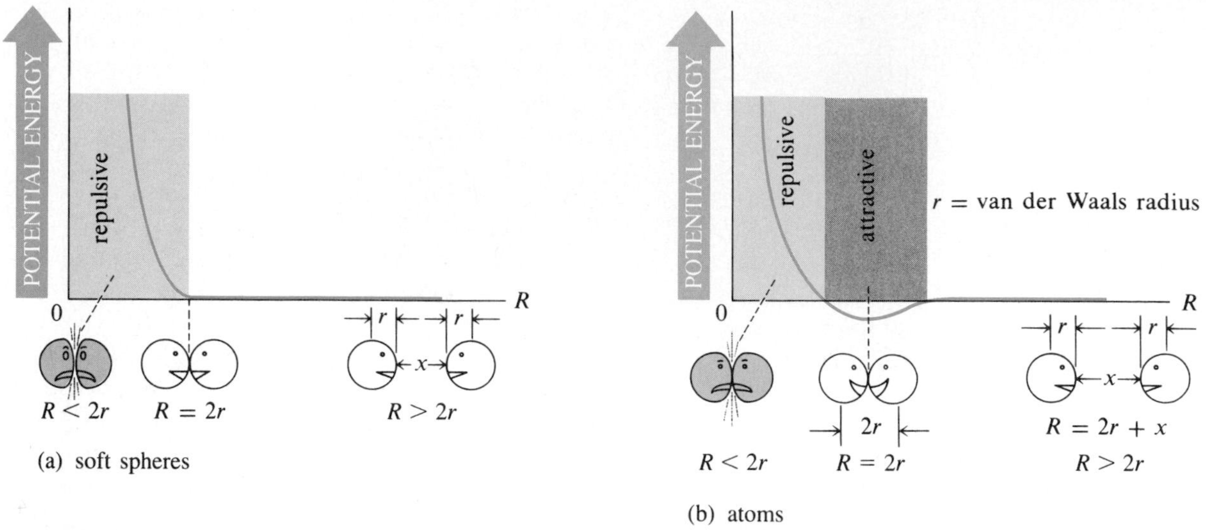

Figure 3.8 Energy curves for the approach of two spheres of radius *r*. Note that positive energies are defined as repulsive and negative energies as attractive.

Imagine bringing two soft spherical objects of radius *r*, such as balloons, together from an infinite separation, and consider the resulting energy of interaction. Because the balloons can be deformed, it is possible to bring their centers closer together than $2r$ with the expenditure of some energy. The closer we force them together, though, the more energy is required. Groups of atoms (alkanes, or alkyl groups, for example) behave a lot like the balloons in the previous example, except that these groups develop a weak attraction for each other as they approach each other. This is the attractive van der Waals, or dispersion, interaction discussed in Sec. 3.4A. In fact, one measure of the size of an atom or group of atoms is its **van der Waals radius,** which is defined in terms of this attraction: when two nonbonded groups are separated by the sum of their van der Waals radii their mutual attractive interaction is greatest. When the distance separating two groups of atoms—such as the methyl groups of butane—is less than the sum of their van der Waals radii, their attraction gives way to a **van der Waals repulsion.** At these smaller separations, the electron clouds of these groups cannot distort enough to get out of each other's way. The repulsion between these electron clouds on different atoms is the essence of the van der Waals repulsion. Thus, it is hard to force atoms too close together in the same sense that it is difficult to force spheres too close together—or in the same sense that it is difficult to get eight people into a car meant for four!

The repulsive part of the curve shown in Fig. 3.8b rises rapidly as the distance between two interacting groups decreases. Once two groups are brought within the repulsive part of the curve, forcing them together even a little more can require an enormous amount of energy. As we can see from Fig. 3.7, moving the methyl groups the small distance from their relative positions in *gauche*-butane to their positions in the eclipsed form of highest energy requires on the order of 3.5–5.2 kcal/mol of energy.

An energetically unfavorable effect on any chemical or physical process (such as internal rotation in butane) that results from van der Waals repulsions is termed a steric effect (from Greek *stereos,* solid). Thus, we say that the *gauche* or eclipsed forms of butane are destabilized by a steric effect, or unfavorable steric interaction. A molecule destabilized by the operation of a steric effect is said to be **strained.** Thus, some strain appears in a butane molecule when it is converted from the *anti* into the *gauche* form. When strain is introduced into a molecule by an internal rotation, the strain is termed **torsional strain.**

Why should we care about the relative energies of molecules? The reason is that these energies tell us something very fundamental about chemical equilibria. *When compounds are in equilibrium, the most stable compound—the compound of lowest energy—is present in greatest amount.* For example, the *anti* form of butane is in equilibrium with the *gauche* form at room temperature, and because the *anti* form is more stable, there is more of it (about twice as much) present in a sample of butane.

The *gauche* and *anti* forms of butane interconvert rapidly at room temperature— about as rapidly as the staggered forms of ethane. Because they are unstable, the eclipsed forms of butane do not exist to any measurable extent; they are characterized by energy maxima in Fig. 3.7.

Problems

10 (a) Draw Newman projections about the C2–C3 bond for each rotational isomer of isopentane; for each eclipsed form.

$$CH_3-\overset{2}{\underset{\underset{\displaystyle CH_3}{|}}{CH}}-\overset{3}{CH_2}-CH_3$$

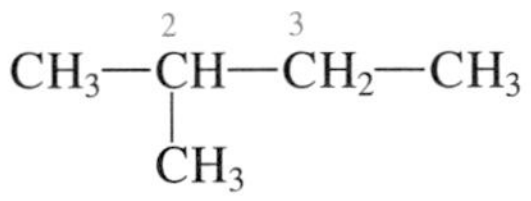

(b) Sketch a curve of relative potential energy *vs.* angle of rotation similar to that of butane in Fig. 3.7. Label each energy minimum and energy maximum with one of the forms you drew in (a). Your curve should clearly indicate the relative stabilities of the different forms.

(c) Which one form is likely to be present in the greatest amount at equilibrium? Explain.

11 Construct models for each *gauche* form of butane. These forms have the same energy, but are they truly identical? To answer this question, try to superimpose the two models, atom-for-atom. To do this systematically, superimpose one carbon and all three groups attached to it; do the remaining atoms superimpose?

3.6 COMBUSTION OF ALKANES. ELEMENTAL ANALYSIS

Although alkanes are among the least reactive organic compounds, one of the few important reactions of alkanes, *combustion,* is one of the most widely utilized chemical reactions on the earth. In complete combustion, an alkane combines with oxygen to produce carbon dioxide and water. For example, the burning of natural gas (methane) can be written as follows:

$$CH_4 + 2O_2 \longrightarrow CO_2 + 2H_2O \tag{3.1}$$

The general combustion reaction for a noncyclic alkane can be written:

$$C_nH_{2n+2} + \frac{3n+1}{2}O_2 \longrightarrow (n+1)H_2O + nCO_2 \tag{3.2}$$

Under conditions of oxygen deficiency, incomplete combustion may also occur with the formation of such by-products as carbon monoxide, CO.

The fact that one can carry a container of gasoline in the open air shows that simple mixing of alkanes and oxygen does not initiate combustion. However, once a small amount of heat is applied (in the form of a flame or a spark from a spark plug), the combustion reaction proceeds vigorously and spontaneously with the liberation of large amounts of energy.

Combustion is of tremendous commercial importance, since the energy it liberates can be used to keep us warm, generate electricity, or move motor vehicles. An important but less obvious use of the combustion reaction comes from the fact that the heat liberated during this reaction can be accurately measured. We shall learn in Sec. 4.4B how heats of combustion can be used to determine the relative stabilities of isomeric organic compounds.

Yet another use of combustion is in the quantitative determination of elemental compositions, called **elemental analysis.** The determination of the relative molar proportions of the elements in a compound is the first step in deriving its *molecular formula*. Combustion has been used for this purpose since the beginning of the modern era of chemistry. The results of combustion analyses led early chemists to the realization that most compounds contain their constituent elements in definite whole-number ratios.

Since most organic compounds contain both carbon and hydrogen, the analysis for these elements is particularly important in organic chemistry. The proportions of both carbon and hydrogen in a compound can be determined simultaneously by a technique that has changed little since its inception. A small sample (typically 5–10 mg) of the substance to be analyzed is completely burned, and the CO_2 and H_2O produced in the combustion are collected and weighed. From the weight of the carbon dioxide produced, the weight of carbon in the sample can be determined. Similarly, from the weight of water produced, the amount of hydrogen in the sample can be determined. Oxygen, if present, is usually not determined directly, but by difference. Let us see with an example how elemental analysis works.

Suppose 7.00 mg of a liquid hydrocarbon is burned and found to yield 21.58 mg of CO_2 and 9.94 mg of H_2O. First let us find the weight of carbon that is present in the CO_2; this same weight of carbon, then, must have been present in the sample. The weight of carbon is

$$\text{weight of carbon} = (\text{fraction of carbon in } CO_2)(\text{weight of } CO_2) \tag{3.3}$$

The fraction of carbon in CO_2 is simply the atomic weight of carbon over the molecular weight of CO_2. Thus, Eq. 3.3 becomes

$$\begin{aligned}\text{weight of carbon} &= (12.01/44.01)(21.58 \text{ mg of } CO_2) \\ &= 5.89 \text{ mg}\end{aligned} \tag{3.4}$$

Likewise, we calculate the weight of hydrogen in the water, noting that there are two hydrogens in every water molecule:

$$\text{weight of hydrogen} = (2\ \text{hydrogens/water})(1.008/18.02)(9.94\ \text{mg of water}) = 1.11\ \text{mg} \qquad (3.5)$$

Since the weight of hydrogen plus the weight of carbon equals the weight of the sample, 7.00 mg, the sample can contain no other elements. The weight percent of carbon is $(5.89/7.00) \times 100 = 84.14\%$, and the weight percent of hydrogen is $(1.11/7.00) \times 100 = 15.86\%$. The results of elemental analysis are commonly reported as weight percent of each element.

Let us now see how to derive a molecular formula from these data. Since the formula of a compound expresses the molar proportions of its elements, we must convert the weight percent data to mole ratios. The weight percentages of carbon and hydrogen mean that there are 84.14 g of carbon and 15.86 g of hydrogen in 100 g of the sample. Hence there are $(84.14/12.01) = 7.00$ moles of carbon and $(15.83/1.008) = 15.73$ moles of hydrogen in 100 g of the sample. A formula that expresses the relative molar proportions of carbon and hydrogen is therefore $C_{7.00}H_{15.73}$. Since organic compounds have whole numbers of elements, this must be converted into a formula in which both elements are present as whole numbers. The easiest way to do this is to divide through the formula above by the element present in least amount—in this case, carbon. This yields the formula $C_{1.00}H_{2.25}$. We now multiply the formula by successive integers (2, 3, 4, . . .) until we arrive at a whole number for both elements. Clearly, multiplication by 4 yields the smallest whole-number formula, C_4H_9. The formula that gives the smallest whole-number proportions of the elements is called the **empirical formula.** Sometimes the empirical formula is equal to the molecular formula, but more often it is not. In this case, the formula C_4H_9 cannot be the molecular formula, because all hydrocarbons have even numbers of hydrogens. The smallest molecular formula that has a carbon : hydrogen ratio of 4 : 9 is C_8H_{18}; this is an acceptable candidate for the molecular formula. The compound could be an octane.

Problem

12 We could also multiply the formula C_4H_9 by 4 to obtain the formula $C_{16}H_{36}$. Would this be an acceptable candidate for the molecular formula of the example above? Explain.

Let us consider another example in which the molecular formula cannot be derived from the empirical formula alone. Suppose combustion analysis of a liquid hydrocarbon yields the composition 85.63% carbon and 14.37% hydrogen. Using the previously described procedure, we obtain first $C_{7.13}H_{14.26}$ from which we derive $C_{1.00}H_{2.00}$, or CH_2, as an acceptable empirical formula. (Verify these calculations!) This empirical formula could correspond to an infinite variety of hydrocarbons: C_2H_4, C_3H_6, C_4H_8, and so on. In order to determine the molecular formula in this example, we need one additional piece of data: the molecular weight. Suppose that the molecular weight of the compound in this example is determined to be 84. The empirical formula C_nH_{2n} implies the following relationship between number of carbons n and molecular weight:

$$12 \cdot n \;+\; 1 \cdot \underline{(2n)} \;=\; 84 \qquad\qquad (3.6)$$

molecular weight of compound

number of hydrogens

atomic weight of hydrogen

number of carbons

atomic weight of carbon

Solving this equation gives $n = 6$; thus, our compound has the molecular formula C_6H_{12}. This example shows that the empirical formula must in some cases be combined with a molecular weight to obtain the molecular formula. Molecular weights are determined by special techniques such as mass spectrometry, which we shall consider later in the text.

Let us summarize the method of deriving the molecular formula from elemental analysis:

1. Find the weight percent of each element.

2. Convert the weight percents into mole ratios of the elements present.

3. Convert the mole ratios into whole numbers to obtain the empirical formula.

4. Use the molecular weight if necessary to find the molecular formula.

Problems

13 Calculate the weight percent of carbon, hydrogen, and oxygen in a compound with the molecular formula C_3H_6O.

14 In an abandoned laboratory you have found a compound in a bottle labeled "alkane X." In an attempt to determine its structure, you carry out an elemental analysis. Combustion of 6.50 mg of X in a stream of O_2 yields 20.92 mg of CO_2 and 7.04 mg of H_2O. Calculate a molecular formula for X. Assuming that the label on the bottle can be believed, what can you say about the structure of X?

3.7 OCCURRENCE AND USE OF ALKANES

Most alkanes come from **petroleum,** or crude oil. Crude oil is a dark, viscous mixture composed mostly of alkanes and aromatic hydrocarbons that are separated by a technique called **fractional distillation.** In fractional distillation, a mixture of compounds is slowly boiled; the vapor is then cooled and recondensed to a liquid. Because the compounds with the lowest boiling points vaporize most readily, the condensate from a fractional distillation is richest in the more volatile components of the mixture. As distillation continues, components of progressively higher boiling point appear in the condensate. A student who takes an organic laboratory course will almost certainly become acquainted with this technique on a laboratory scale. Indus-

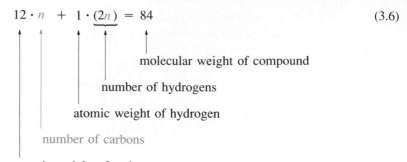

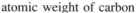

Figure 3.9 Industrial fractionating towers used to separate mixtures of compounds on the basis of their boiling points.

trial fractional distillations are carried out on a large scale in fractionating towers that are several stories tall (Fig. 3.9). The typical fractions obtained from distillation of crude oil are given in Table 3.4.

Another important alkane source is natural gas, much of which comes from gas wells of various types. Natural gas is mostly methane. Gas wells are also the major source of helium, a nonrenewable resource that is under strict government control. Methane is also produced by the action of certain anaerobic bacteria (bacteria that function without oxygen) on decaying organic matter (Fig. 3.10). Methane was known as "marsh gas" for many years before it was characterized by organic chemists. Although the production of methane by bacterial fermentation has not been widely

TABLE 3.4 Components of Crude Oil

Fraction name	Boiling range, °C	Number of carbon atoms
gas	<20	1–4
petroleum ether	30–60	5–6
ligroin (light naphtha)	60–90	6–7
gasoline (naphtha)	85–200	6–12
kerosene	200–300	12–15
heating oil	300–400	15–18
lubricating oil, asphalt	>400	16–24

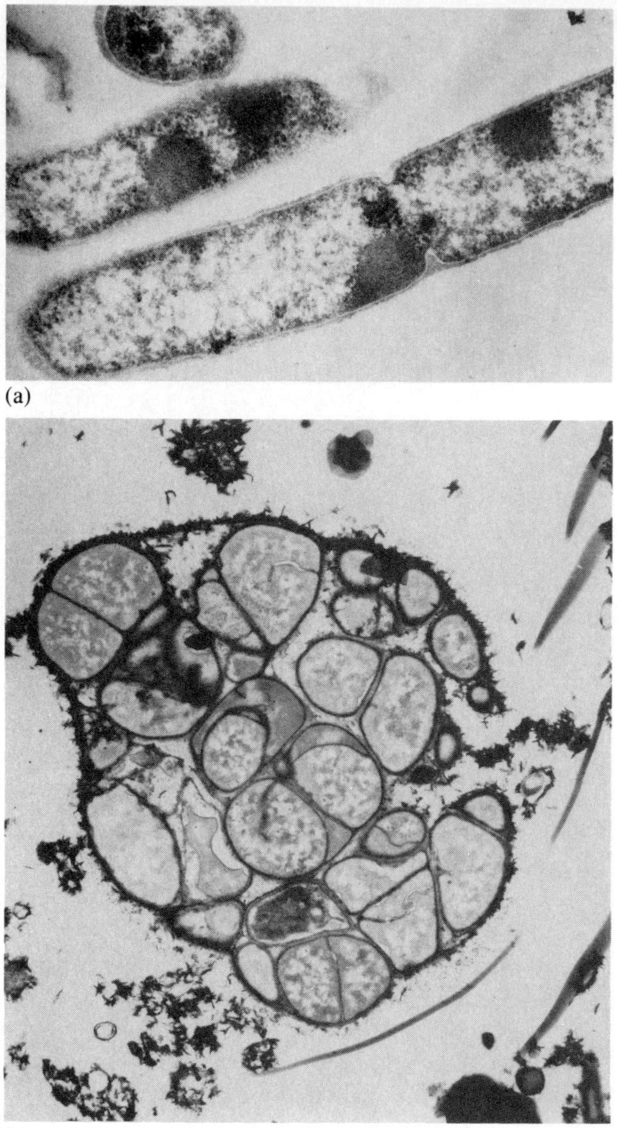

Figure 3.10
Methanogens. (a) *Methanobacterium* MOH and (b) *Methanosarcina* sp. are found in anaerobic sewage digesters and in anaerobic sediments of natural waters.

(a)

(b)

exploited in commerce, the biochemical reactions associated with methane production are becoming better understood, and they may become a hydrocarbon source of the future (Fig. 3.10).

Alkanes of low molecular weight are in great demand for a variety of purposes—especially motor fuels—and those available directly from wells do not satisfy the demand. The petroleum industry has developed methods for *cracking* high molecular weight alkanes into lower molecular weight alkanes and alkenes (Sec. 5.9B), and has also developed processes for *reforming* n-alkanes into branched-chain ones, which have superior ignition properties as motor fuels.

Alkanes as Motor Fuels Alkanes vary significantly in their quality as motor fuels. Branched-chain alkanes are better motor fuels than *n*-alkanes. The quality of a motor fuel relates to its rate of ignition in an internal combustion engine. Premature ignition results in engine knock, a condition that indicates poor engine performance. Severe engine knock can result in significant engine damage. The *octane number* is a measure of the quality of a motor fuel: the higher the octane number, the better the fuel. Octane numbers of 100 and 0 are assigned to 2,2,4-trimethylpentane and heptane, respectively. Mixtures of the two compounds are used to define octane numbers between 0 and 100. For example, a fuel that performs as well as a 1 : 1 mixture of 2,2,4-trimethylpentane and heptane has an octane number of 50. Good quality motor fuels used in modern automobiles have octane numbers in the 87–95 range. Various additives, such as tetraethyllead, $(C_2H_5)_4Pb$, and *t*-butyl methyl ether, $(CH_3)_3C—O—CH_3$, are used to boost the octane number of gasoline. The use of tetraethyllead is being curtailed for environmental reasons.

Recently, periods of relative scarcity of petroleum products have alternated with periods of relative abundance. There is no doubt, however, that eventually the world will exhaust its natural petroleum reserves. It is thus important that scientists develop new sources of energy, which may include new ways of producing petroleum.

At present the greatest use of alkanes is for fuel. Typically, motor fuels, fuel oils, and aviation fuels account for about 80% of all hydrocarbon consumption. An Arabian oil minister was once heard to remark, "Oil is too precious to burn." He was undoubtedly referring to the fact that there are important uses for petroleum besides fuels. Petroleum will remain for the foreseeable future the principal source of *carbon,* from which the organic starting materials for such diverse products as plastics and pharmaceuticals are made. Petroleum is thus the basis for organic chemical *feedstocks*—the basic organic compounds from which more complex chemical substances are fabricated.

3.8 FUNCTIONAL GROUPS. THE "R" NOTATION

The alkane family represents one of the many functional groups of organic chemistry. A **functional group** is a group of atoms and their associated chemical bonds that has about the same type of chemical reactivity whenever it occurs in different compounds. The common organic functional groups are listed inside the back cover. For example, the functional group characteristic of alkenes is the carbon–carbon double bond. Most alkenes undergo the same types of reactions, and these reactions occur at or near the double bond. Once we learn the reactions of one alkene, we have automatically learned the reactions of many others. Similarly, the characteristic functional group of alcohols is the $-\overset{|}{\underset{|}{C}}-OH$ group. The reactions of alcohols occur for the most part at or near the —OH group, and the functional group undergoes the same general chemical transformations regardless of the structure of the remainder of the molecule. The organization of this text is centered for the most part around the common functional groups. Although we shall study in detail each major functional group in subsequent chapters, it is important to recognize the common functional groups at an early stage in our study of organic chemistry.

Because the chemical properties of functional groups are general, it is often convenient to employ a general notation for organic compounds. When organic chemists wish to indicate a general structure, they use an R, much as mathematicians use x to indicate a general number. The "R," unless otherwise indicated, stands for an alkyl group. Thus, the general formula of an alkyl chloride, R—Cl, might stand for any of the following structures—or a host of others.

$$R-Cl \qquad CH_3-Cl \qquad (CH_3)_2CH-Cl$$

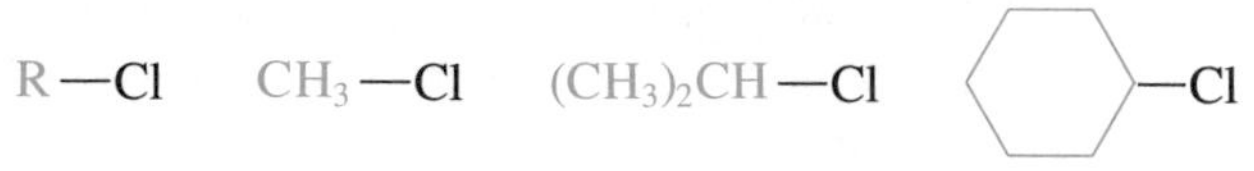

Just as alkyl groups are substituent groups derived from alkanes, **aryl groups** are substituent groups derived from benzene and its derivatives. The simplest aryl group is the **phenyl group,** abbreviated Ph—. This group is a monosubstituted benzene ring.

skeletal structure of benzene

—CH=CH$_2$ can be written Ph—CH=CH$_2$

phenyl group

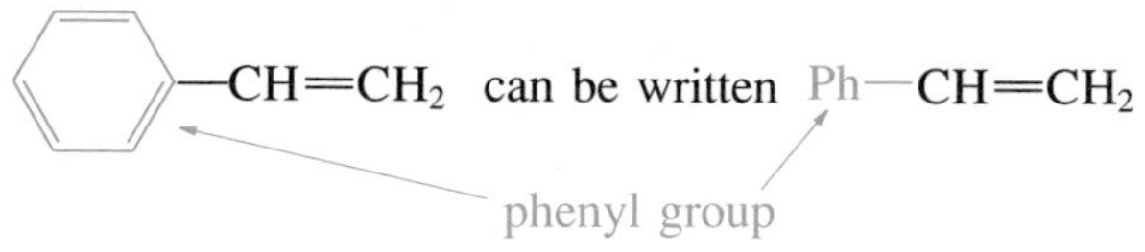

Other aryl groups are designated by Ar—. Thus Ar—OH could refer to any of the following compounds:

Ar—OH

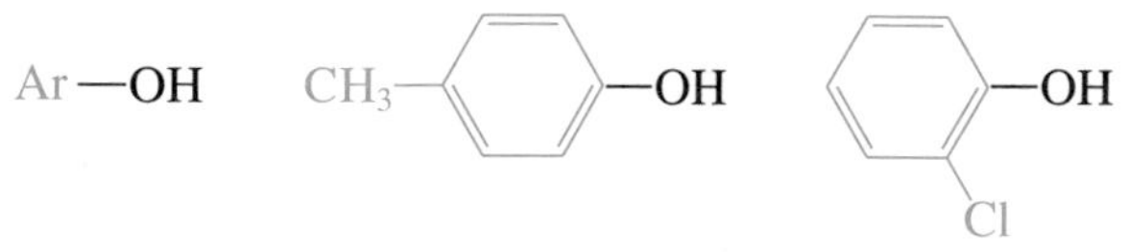

Although we shall not study benzene and its derivatives until Chapter 15, we shall see many examples in the meantime in which phenyl and aryl groups are used as substituent groups.

Problems

15 Draw a structural formula for each of the following compounds. (There may be several possible formulas.)
(a) an amine with molecular formula $C_4H_{11}N$
(b) an alcohol with molecular formula $C_5H_{10}O$
(c) a carboxylic acid with the elemental analysis 40.00% carbon, 6.71% hydrogen, and 53.29% oxygen

16 A certain compound was found to have the molecular formula $C_5H_{12}O_2$. Which of the following functional groups could be present in the compound? Give one example for each positive answer, and explain any negative responses.

(a) an amide
(b) an ether
(c) a carboxylic acid
(d) a phenol
(e) an alcohol
(f) an ester

KEY IDEAS IN CHAPTER 3

- Alkanes are hydrocarbons that contain only carbon–carbon single bonds; alkanes may exist with unbranched chains, branched chains, or rings.
- In general more than one alkane of a given molecular formula are known. Compounds that have the same molecular formula but differ in the way their atoms are connected are structural isomers.
- Alkanes are named systematically according to the IUPAC rules. The name is based on the longest continuous carbon chain in the molecule.
- The boiling point of an alkane is determined by the van der Waals attractions among alkane molecules, which in turn depend on molecular weight and shape. Large molecules have relatively high boiling points; molecules that fit into compact spheres have relatively low boiling points. The boiling points of compounds within a homologous series increase 20–30° per carbon atom.
- Melting points of alkanes increase with molecular weight and with molecular symmetry.
- Alkanes exist in staggered conformations, and the form that minimizes steric repulsions is preferred. In butane the form of lowest energy is the *anti* form.
- Combustion is the most important reaction of alkanes. It is used not only to provide energy, but also in determining empirical formulas of organic compounds by elemental analysis.
- Alkanes are derived from petroleum and are used mostly as fuels; however, they are also important as raw materials for the preparation of other organic compounds.
- Organic compounds are classified by their functional groups. Different organic compounds containing the same functional groups undergo similar types of reactions.

ADDITIONAL PROBLEMS

17 Given the boiling point of the first compound in each set, estimate the boiling point of the second.

(a)

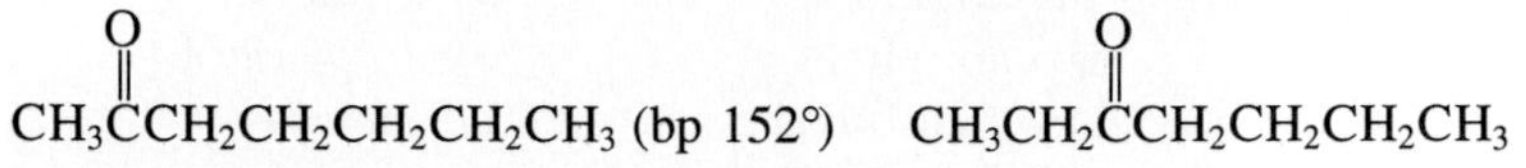

$$CH_3\overset{\overset{\displaystyle O}{\|}}{C}CH_2CH_2CH_2CH_2CH_3 \text{ (bp 152°)} \quad CH_3CH_2\overset{\overset{\displaystyle O}{\|}}{C}CH_2CH_2CH_2CH_3$$

(b)

$$CH_3\overset{\overset{\displaystyle O}{\|}}{C}CH_2CH_2CH_2CH_2CH_3 \text{ (bp 152°)} \quad CH_3\overset{\overset{\displaystyle O}{\|}}{C}CH_2CH_2CH_2CH_3$$

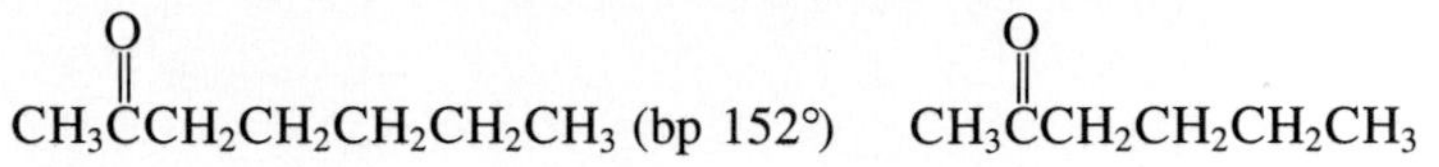

(c) $CH_3CH_2CH_2CH{=}CH_2$ (bp 30°) $CH_3CH_2CH_2CH_2CH{=}CH_2$

(d) $CH_3CH_2CH_2CH_2CH_2CH_2Br$ (bp 155°) $CH_3CH_2CH_2CH_2CH_2CH_2CH_2CH_2Br$

18 Draw the structures and give the systematic IUPAC names of all the isomeric octanes.

19 Label each carbon atom in the following molecule as primary, secondary, tertiary, or quaternary.

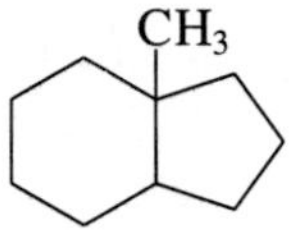

20 Draw the structure and give the name of an alkane
(a) that has more than three carbons and has only primary hydrogens.
(b) that has only five carbons and has only secondary hydrogens.
(c) that has a molecular weight of 84.2.
(d) that has only tertiary hydrogens.

21 Give the systematic IUPAC name for each of the following alkanes:

(a)

$$\begin{array}{c} CH_2{-}CH_2{-}CH_3 \\ | \\ CH_3{-}CH{-}CH{-}CH_2{-}CH_3 \\ | \\ CH_2{-}CH_2{-}CH_3 \end{array}$$

(b)

$$\begin{array}{c} CH_3 \quad CH_2{-}CH_2{-}CH_3 \\ | \qquad | \\ CH_3{-}C{-\!-\!-}CH{-}CH_2{-}CH_3 \\ | \\ CH_2{-}CH_2{-}CH_3 \end{array}$$

(c) $CH_3{-}CH{-}CH_2{-}CH_2{-}CH_3$
$$\begin{array}{c} \qquad\quad | \\ \qquad\quad CH \\ \qquad\; \diagup \;\; \diagdown \\ CH_3 \quad CH_3 \end{array}$$

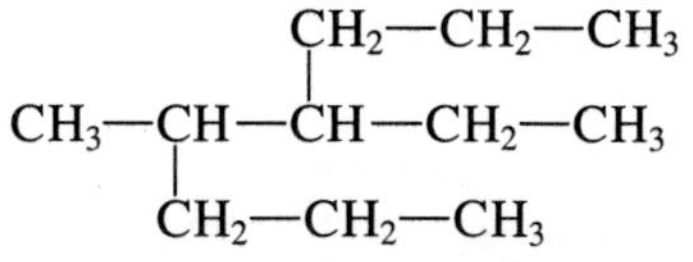

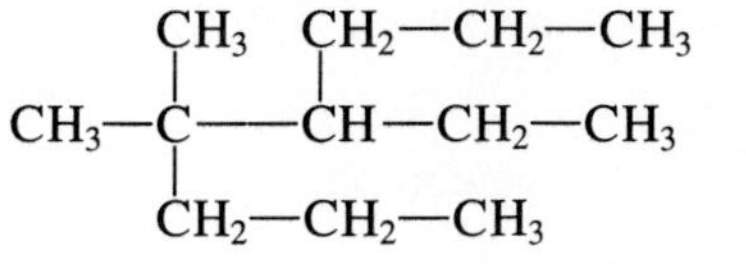

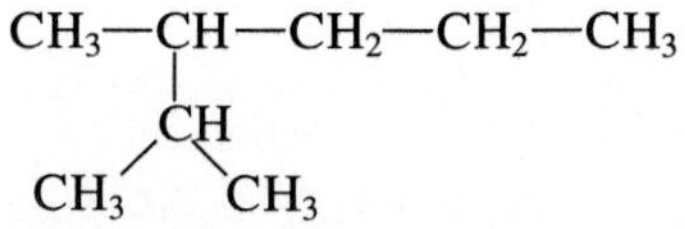

(d)

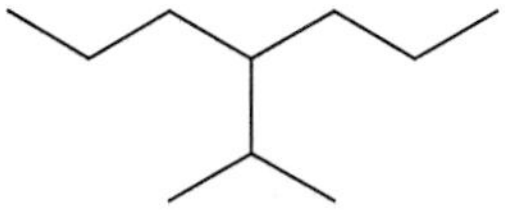

(e)

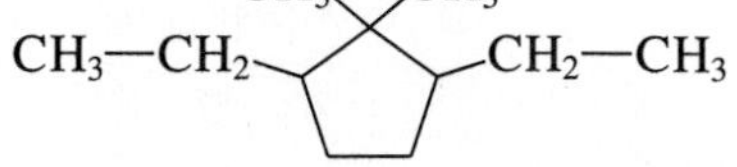

22 Draw structures that correspond to the following names:
(a) 2,2-dimethylpentane
(b) 2,3,5-trimethyl-4-propylheptane
(c) 4-isobutyl-2,5-dimethylheptane

23 The following labels were found on bottles of liquid hydrocarbons in the laboratory of Dr. Ima Turkey following his demise in a mysterious laboratory explosion. Although each defines a structure unambiguously, some are not correct IUPAC names. Give the correct names for the compounds that are named incorrectly.
(a) 3-ethyl-4-methylpentane
(b) 5-neopentyldecane
(c) 2-ethyl-2,4,6-trimethylheptane
(d) 1-cyclopropyl-3,4-dimethylcyclohexane
(e) 3-butyl-2,2-dimethylhexane

24 Sketch a diagram of potential energy *vs.* angle of rotation about the carbon–carbon bond of chloroethane, CH_3—CH_2—Cl. The energy barrier to internal rotation in this compound is 3.7 kcal/mol. Label this barrier on your diagram.

25 Explain how you would expect the diagram of potential energy *vs.* angle of rotation about the C2—C3 (central) carbon–carbon bond of 2,2,3,3-tetramethylbutane to differ from that for ethane.

26 The *anti* form of 1,2-dichloroethane, Cl—CH_2—CH_2—Cl, is 1.15 kcal/mol more stable than the *gauche* form. The two energy barriers (measured relative to the energy of the *gauche* form) for carbon–carbon bond rotation are 5.15 kcal/mol and 9.3 kcal/mol.
(a) Sketch a graph of potential energy *vs.* angle of rotation about the carbon–carbon bond. Show the energy differences on your graph and label each minimum and maximum with the appropriate conformation of 1,2-dichloroethane.
(b) Which conformation of this compound is present in greatest amount? Explain.

27 It has been argued that a covalently bound chlorine has about the same effective size as a methyl group. Assuming this is true, explain why the energy barrier for internal rotation of 1,2-dichloroethane (previous problem) is considerably larger than that in butane (Fig. 3.7). (*Hint:* the C—Cl bond is a polar bond; what is the relationship of the C—Cl bond dipoles when the two C—Cl bonds are eclipsed?)

28 When the structure of the compound on the left was determined in 1972, it was found to have an unusually long C—C bond and unusually large C—C—C bond angles, compared to the similar parameters for isobutane.

Problems (Cont.)

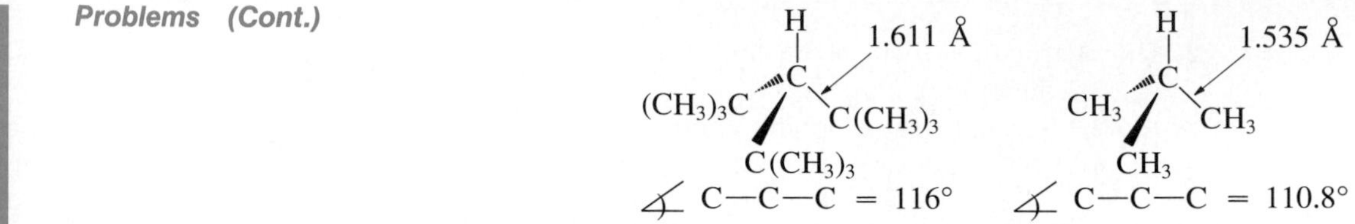

Explain why the indicated bond length and bond angle is larger for the compound on the left.

---

29 Which of the following compounds has the larger energy barrier to internal rotation about the central bond? Explain.

$(CH_3)_3C—C(CH_3)_3$ $(CH_3)_3Si—Si(CH_3)_3$

---

30 (a) What value is expected for the dipole moment of the *anti* form of 1,2-dibromoethane, $Br—CH_2—CH_2—Br$? Explain.

(b) The dipole moment $\boldsymbol{\mu}$ of any compound that undergoes internal rotation can be expressed as a weighted average of the dipole moments of each of its rotational isomers by the following equation:

$$\boldsymbol{\mu} = \boldsymbol{\mu}_1 N_1 + \boldsymbol{\mu}_2 N_2 + \boldsymbol{\mu}_3 N_3$$

in which $\boldsymbol{\mu}_i$ is the dipole moment of form i, and N_i is the mole fraction of form i. (The mole fraction of any form i is the number of moles of i divided by the total moles of all forms.) There are about 82 mole percent of *anti* form and about 9 mole percent of each *gauche* form present at equilibrium in 1,2-dibromoethane, and the observed dipole moment of 1,2-dibromoethane is 1.0 D. Using the above equation, and the answer to part (a), calculate the dipole moment of the *gauche* form of 1,2-dibromoethane.

---

31 Write a balanced equation for the combustion of a general cycloalkane C_nH_{2n}.

---

32 What is the weight percent carbon in the hydrocarbon decalin?

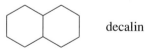

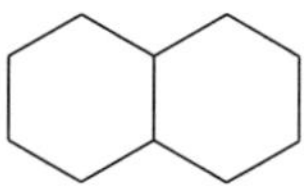
decalin

---

33 (a) A hydrocarbon is found to contain 87.42% carbon and 12.58% hydrogen by weight. Calculate the minimum molecular formula for this hydrocarbon.

(b) The compound in (a) has only carbon–carbon single bonds, no primary hydrogens, no tertiary carbon atoms, and one quaternary carbon atom. Draw one structure that is consistent with these facts and the minimum molecular formula.

34 A 7-mg sample of a hydrocarbon with a molecular weight of 140.3 is burned in a stream of oxygen to yield 21.96 mg of CO_2 and 8.99 mg of water.
(a) How many milliliters of oxygen (assume 25 °C, 1 atm pressure) are consumed in this experiment?
(b) What is the molecular formula of the hydrocarbon?

35 In an abandoned laboratory you have found a bottle marked "amide *X*, elemental analysis 55.14% carbon, 10.41% hydrogen, and 16.08% nitrogen, molecular weight 87." Evidence gathered by certain physical methods leads you to believe that the compound contains an isopropyl group. Suggest two structures for *X* that are consistent with all the data.

36 Identify the different functional groups (aside from the alkane carbons) present in acebutolol, a drug that blocks a certain part of the nervous system.

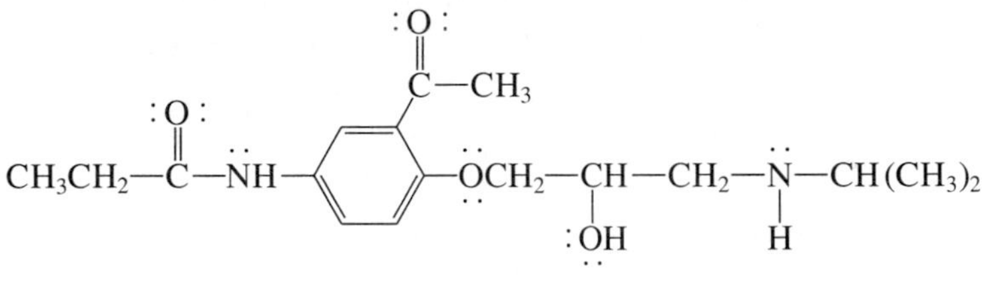

acebutolol

4 INTRODUCTION TO ALKENES. EQUILIBRIA AND REACTION RATES

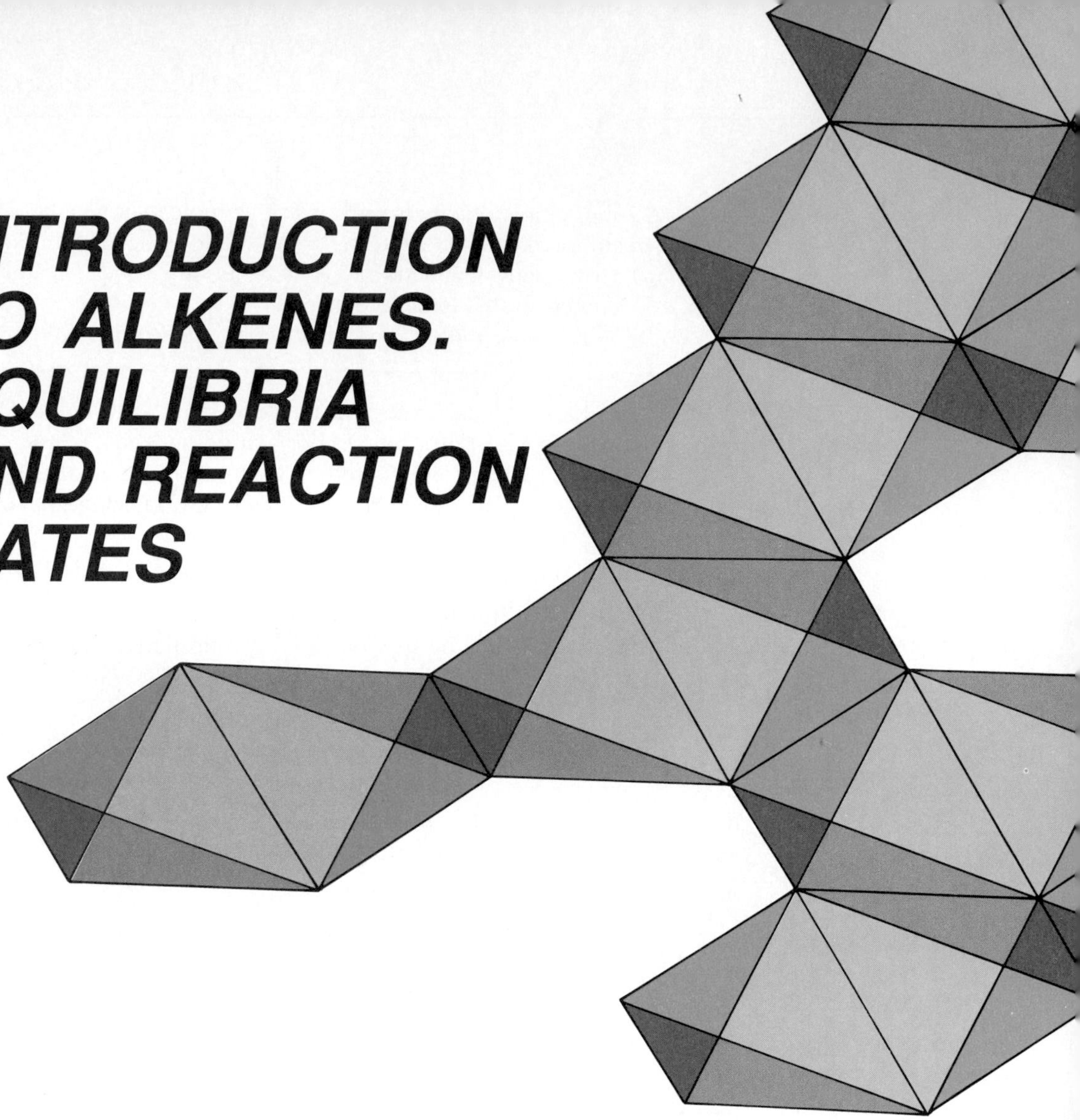

Alkenes, or **olefins,** are hydrocarbons that contain carbon–carbon double bonds. Ethylene and propylene are the two simplest alkenes.

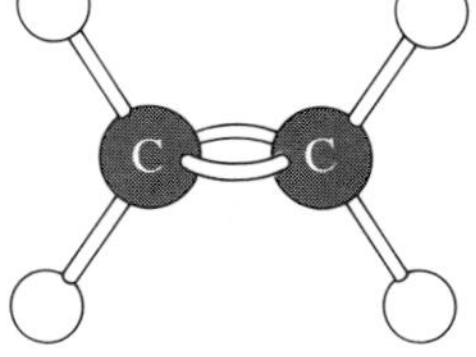

ethylene

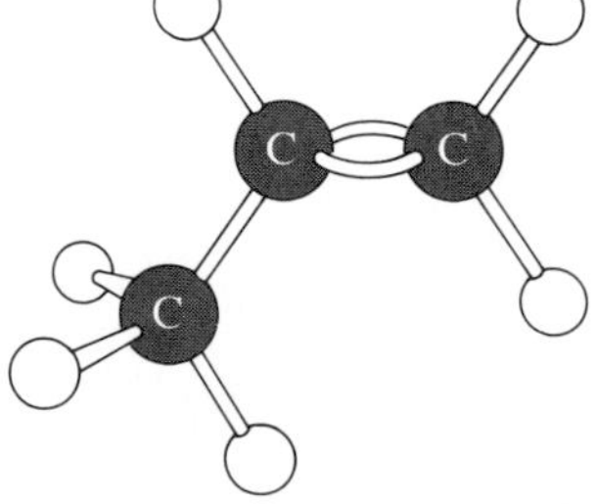

propylene, or propene

Because compounds containing double or triple bonds have fewer hydrogens than the corresponding alkanes, they are sometimes referred to as **unsaturated hydrocarbons,** in contrast to alkanes, which are **saturated hydrocarbons.**

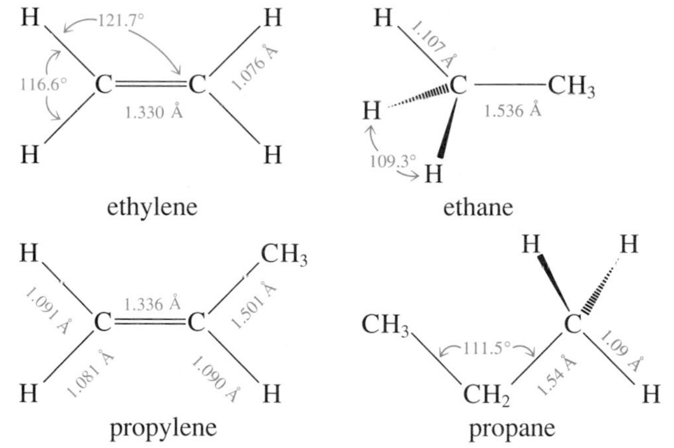

Figure 4.1 Structures of ethylene, ethane, propylene, and propane.

In this chapter, we shall consider the structure, bonding, nomenclature, and physical properties of alkenes. Then, using a few alkene reactions, we shall develop some of the physical principles that are important in understanding the reactions of organic compounds in general.

4.1 STRUCTURE AND BONDING IN ALKENES

The rules for predicting molecular geometry developed in Sec. 1.6B suggest that each carbon of ethylene should have trigonal geometry; that is, all the groups surrounding each carbon should lie in one plane, with bond angles approximating 120°. The structure of ethylene (Fig. 4.1) shows that these expectations are largely met; each carbon does indeed have approximately trigonal geometry, with H—C—H bond angles of about 117°.

As we found in Sec. 1.6B, the rules for predicting geometry do not extend to the *conformation* of ethylene. It turns out that *ethylene is a planar molecule.* For alkenes in general, the carbons of a double bond and the atoms attached to them all lie in the same plane. Another feature of alkene structure is also evident when we compare the structures of propylene and propane in Fig. 4.1. Although we know that a carbon–carbon double bond is shorter than a carbon–carbon single bond, it is also interesting that the CH_3—C and the H—C *single* bonds of propylene are somewhat shorter than the corresponding bonds in propane. The bonding picture developed for alkenes in Sec. A below will help us understand why these structural features of alkenes are reasonable.

A. Carbon Hybridization in Alkenes. The π-Bond

The bonding in ethylene will serve as our simple model for the bonding in other alkenes. The carbons of ethylene are hybridized differently from those of an alkane. In this hybridization (Fig. 4.2), the carbon 2*s* orbital is mixed, or hybridized, with only two of the three available 2*p* orbitals. Since three orbitals are mixed, the result is three hybrid orbitals. Each has one part *s* character and two parts *p* character. These hybrid orbitals are called sp^2 (pronounced s-p-two) orbitals, and the carbon is said to be sp^2

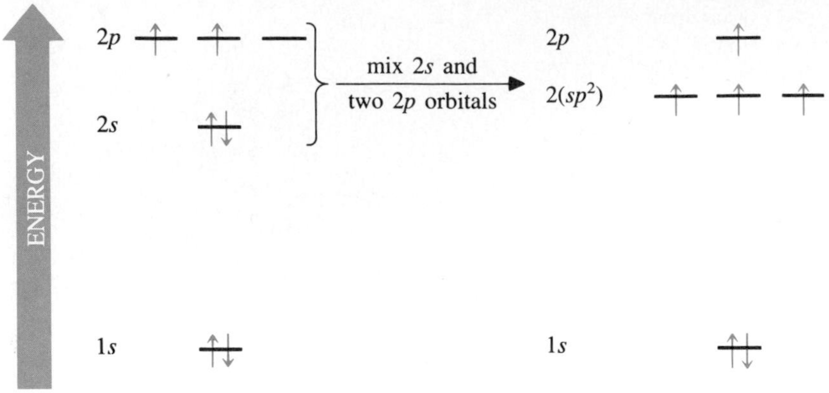

Figure 4.2 Orbitals of an sp^2-hybridized carbon.

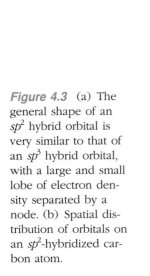

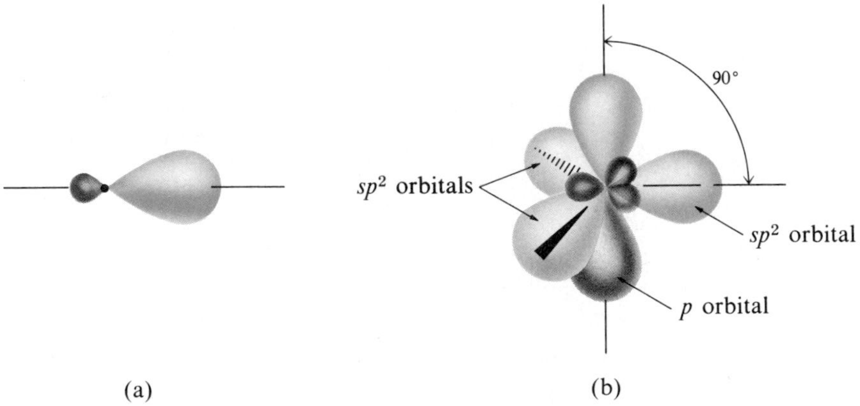

Figure 4.3 (a) The general shape of an sp^2 hybrid orbital is very similar to that of an sp^3 hybrid orbital, with a large and small lobe of electron density separated by a node. (b) Spatial distribution of orbitals on an sp^2-hybridized carbon atom.

hybridized. The shape of an sp^2 orbital (Fig. 4.3a) is much like that of an sp^3 orbital. A diagram of the arrangement of orbitals in sp^2-hybridized carbon (Fig. 4.3b) shows that the axes of the three sp^2 orbitals are disposed in the same plane at mutual angles of 120°. After mixing one *s* and two *p* orbitals to give the three sp^2 hybrid orbitals, a *p* orbital is left over. Thus, an sp^2-hybridized carbon atom bears three sp^2-hybrid orbitals and a *p* orbital. The axis of the *p* orbital is perpendicular to the plane containing the sp^2 orbitals. Whenever we see a *trigonal* carbon atom—a carbon bonded to three groups—we think of its electronic configuration in terms of sp^2 hybridization.

Like an sp^3 orbital, an sp^2 orbital is highly directed in space. The difference between the two types of hybrid orbitals is that the electron density of an electron in an sp^2 orbital is somewhat closer to the nucleus. This difference is a consequence of the larger amount of *s* character in an sp^2 orbital. An sp^3 orbital has 25% *s* character (one *s* orbital mixed with three *p* orbitals), but an sp^2 orbital has 33% *s* character. Since *s* electrons, on the average, are closer to the nucleus than *p* electrons, electrons in hybrid orbitals with more *s* character also have a greater probability of being found closer to the nucleus.

Conceptually ethylene can be formed by the bonding of two sp^2-hybridized carbon atoms and four hydrogen atoms. An sp^2 orbital on one carbon overlaps with an sp^2 orbital on another to form a bonding molecular orbital containing two electrons—a carbon–carbon σ-bond. (There are also corresponding antibonding orbitals that we

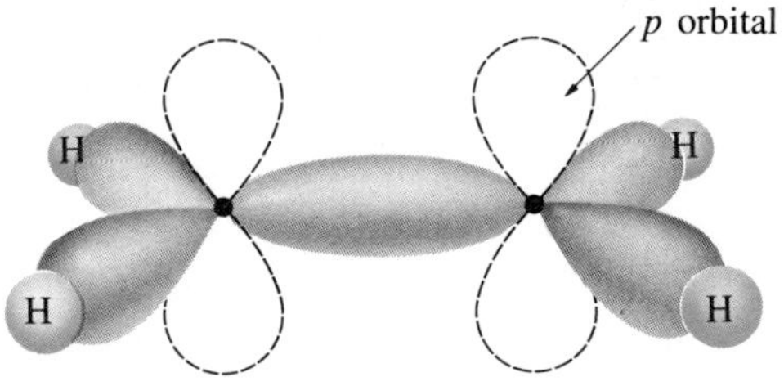

Figure 4.4 The σ-bonds in ethylene.

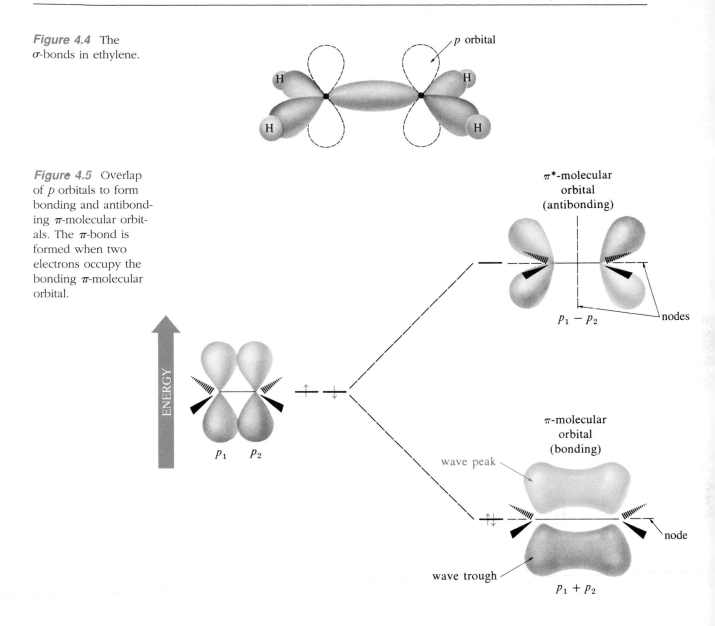

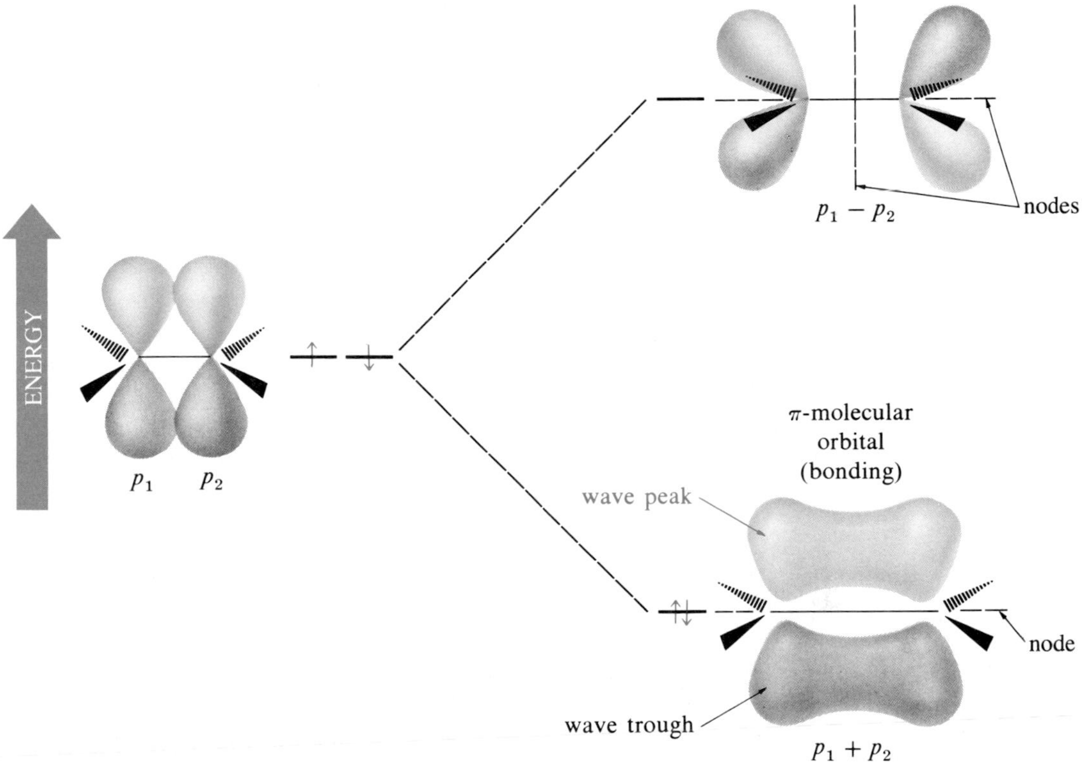

Figure 4.5 Overlap of p orbitals to form bonding and antibonding π-molecular orbitals. The π-bond is formed when two electrons occupy the bonding π-molecular orbital.

shall not consider.) Each of the two remaining sp^2 orbitals on each carbon overlaps with a hydrogen 1s orbital to form a C—H sp^2-1s σ-bond. These orbitals account for the four carbon-hydrogen bonds and *one* of the two carbon–carbon bonds of ethylene (Fig. 4.4); what about the second carbon–carbon bond?

Each carbon of ethylene bears a single 2p orbital; the second carbon–carbon bond comes from the overlap of the p orbitals from each carbon atom, forming bonding and antibonding molecular orbitals (Fig. 4.5; see also Problem 15, Chapter 2). These molecular orbitals are formed by additive and subtractive combinations of the p orbitals in much the same way that the molecular orbitals of the hydrogen molecule are formed from combinations of the hydrogen 1s orbitals (Sec. 2.4). However, the overlap of p orbitals, which is different from that in a σ-bond, in this case is side-to-side. The bonding molecular orbital, like the p orbitals from which it is formed, has a nodal plane; this plane coincides with the plane of the ethylene molecule. An electron in this bonding molecular orbital has lower energy than an electron in an isolated *p* orbital. The antibonding molecular orbital has two nodes: One node is the same as that in the bonding molecular orbital, and the other node is a plane between the two carbon atoms perpendicular to the plane of the molecule. An electron in this antibonding molecular orbital has higher energy than an electron in an isolated *p* orbital. Since the bonding molecular orbital has lower energy than the antibonding molecular orbital, the two 2*p* electrons (one from each carbon), by the Pauli principle, occupy this molecular orbital. The resulting bond is called a **π-bond.** The antibonding molecular orbital is unoccupied.

Unlike a σ-bond, a π-bond is not cylindrically symmetric about the line connecting the two nuclei. The π-bond has electron density both above and below the plane of the ethylene molecule, with a wave peak on one side of the molecule, a wave trough on the other, and a node in the plane of the molecule. The π-bond is *one bond* with two lobes, just as a *p* orbital is *one orbital* with two lobes. In this bonding picture, then, there are two types of carbon–carbon bonds: a σ-bond, with most of its electron density relatively concentrated between the carbon atoms, and a π-bond, with most of its electron density concentrated above and below the plane of the ethylene molecule.

This bonding picture accounts nicely for the planar conformation of ethylene. If the two CH_2 groups were twisted away from coplanarity, the *p* orbitals could not overlap to form the π-bond. Thus, the overlap of the *p* orbitals and, consequently, the existence of the π-bond, *require* the planarity of the ethylene molecule.

This bonding picture shows why the single bonds at an sp^2-hybridized carbon are somewhat shorter than those at an sp^3-hybridized carbon. The carbon–carbon single bond of propylene, for example, is derived from the overlap of a carbon sp^3 orbital of the CH_3 group with a carbon sp^2 orbital of the alkene carbon. A carbon–carbon bond of propane is derived from the overlap of two carbon sp^3 orbitals. Since the electron density of sp^2 orbitals is somewhat closer to the nucleus, it follows that the sp^2-sp^3 σ-bond of propylene is a little shorter than the sp^3-sp^3 σ-bond of propane. A similar analysis explains why the C—H bonds of an alkene are shorter than the C—H bonds of an alkane (see Fig. 4.1).

B. Cis–Trans *Isomerism*

The alkenes with four carbons (the butenes) illustrate an interesting and important consequence of bonding in alkenes. If we consider only the butenes with unbranched carbon chains, we see that the double bond may be located either at the end or the middle of the carbon chain.

$$\underset{\text{1-butene}}{CH_2{=}CH{-}CH_2{-}CH_3} \qquad \underset{\text{2-butene}}{CH_3{-}CH{=}CH{-}CH_3}$$

It is found, however, that there are not two, but *three* unbranched butene isomers. Moreover, two of these have a double bond in the middle of the carbon chain. These are separable compounds with different physical properties; one of these compounds has a boiling point of 3.7°, and the other a boiling point of 0.88°. The difference in these two compounds is the way the methyl groups are arranged about the double bond—that is, their *conformations.* In the compound with higher boiling point, *cis*-2-butene, the methyl groups are on the same side of the double bond. In the other, *trans*-2-butene, the methyl groups are on opposite sides of the double bond.

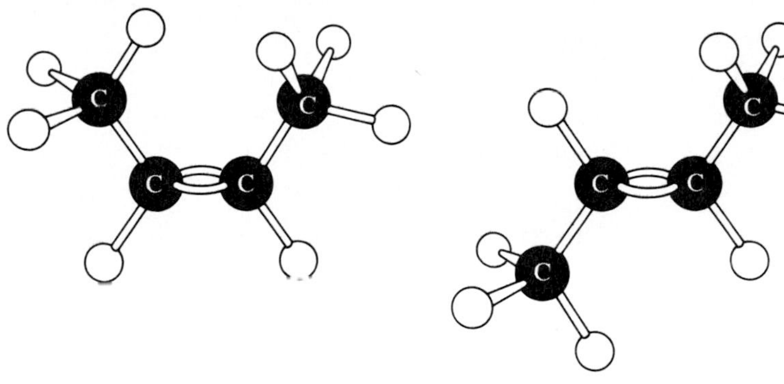

cis-2-butene *trans*-2-butene

These isomers have the same *connectivity* of their constituent atoms. That is, the order in which atoms are bound to each other is the same in both compounds. (A methyl group is connected to a CH═, the CH═ is connected to another CH, etc.) Despite their identical connectivity, these two compounds are different: they differ in the *way their constituent atoms are arranged in space.* Compounds that have the same atomic connectivity but differ in the spatial arrangement of constituent atoms are called **stereoisomers.** *Cis–trans* isomerism is but one type of stereoisomerism; other types will be considered in Chapter 6.

The interconversion of *cis*- and *trans*-2-butene requires a 180° rotation about the double bond.

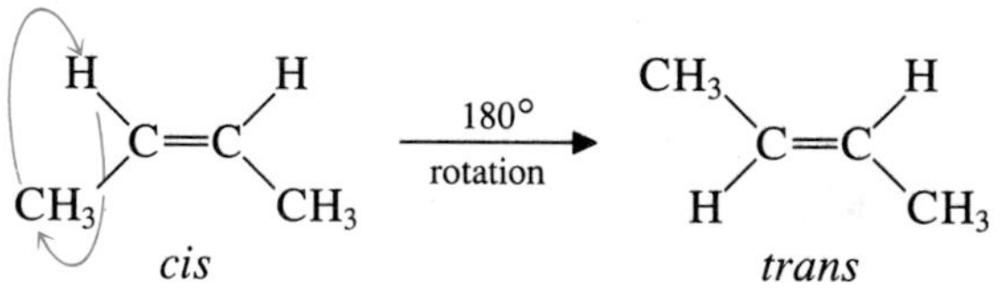

(4.1)

Since *cis*- and *trans*-2-butene do not interconvert at room temperature, it follows that this rotation must be very slow. Why is this so? In order for such an internal rotation to occur, the *p* orbitals on each carbon must be twisted away from coplanarity; that is, *the π-bond must be broken* (Fig. 4.6). Since bonding is energetically favorable, lack of it is energetically costly. It takes more energy to break the π-bond than is available under normal conditions; thus, the π-bond in alkenes remains intact and internal rotation about the double bond does not occur. In contrast, internal rotation about the carbon–carbon *single* bonds of ethane or butane is rapid (Sec. 3.5) because no chemical bond is broken in the process.

Figure 4.6 Rotation about the carbon–carbon double bond in an alkene requires breaking the π-bond.

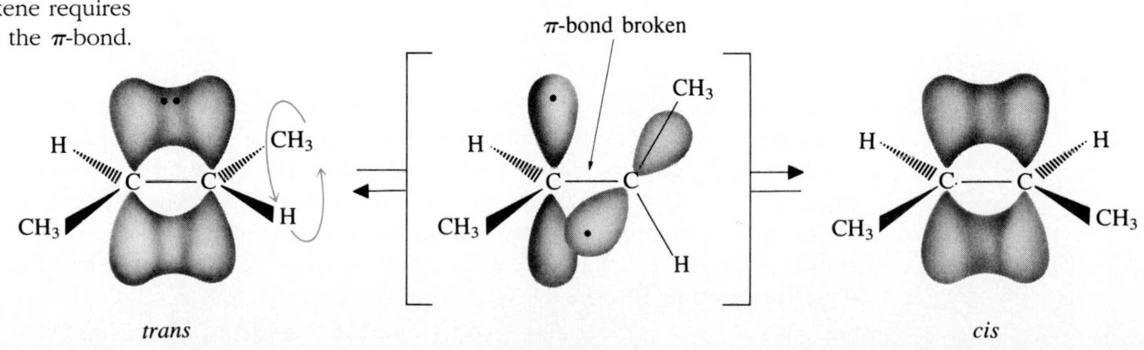

Problem

1 Which of the following alkenes can exist as *cis* and *trans* isomers? Explain. If necessary, make a model of each compound. After working this exercise, propose a general test for the existence of *cis* and *trans* alkene isomers.

(a) $CH_2{=}CH{-}CH_2CH_2CH_3$ (1-pentene)

(b) $CH_3CH_2{-}CH{=}CH{-}CH_2CH_3$ (3-hexene)

(c) $CH_3CH_2{-}CH{=}C(CH_3){-}CH_2CH_3$ (3-methyl-3-hexene)

(d) $CH_3{-}CH{=}CH{-}CH{=}CH{-}CH_3$ (2,4-hexadiene)

(e)  (cyclobutene)

(*Hint:* Try to build a model of both isomers, but don't break your models!)

4.2 NOMENCLATURE OF ALKENES

A. Common and Systematic Nomenclature

Although most alkenes are known by systematic IUPAC names, some are known by common names. A few that should be learned are the following:

$CH_2{=}CH_2$	$CH_3{-}CH{=}CH_2$	$(CH_3)_2C{=}CH_2$	$Ph{-}CH{=}CH_2$	$CH_2{=}C(CH_3){-}CH{=}CH_2$
ethylene	propylene	isobutylene	styrene	isoprene

(The Ph— notation in styrene was discussed in Sec. 3.8.)

The systematic IUPAC nomenclature of alkenes is derived by modifying alkane nomenclature in a simple way. An unbranched alkene is named by replacing the *ane* suffix in the name of the corresponding alkane with the ending *ene*. The chain is numbered from one end to the other so that the double bond receives the lowest number.

$$\overset{1}{C}H_2{=}\overset{2}{C}H{-}\overset{3}{C}H_2\overset{4}{C}H_2\overset{5}{C}H_2\overset{6}{C}H_3 \qquad \text{1-hexene}$$

hex~~ane~~ + ene = hexene

↑ position of double bond

The names of alkenes with branched chains are, like those of alkanes, derived from their *principal chains.* In an alkene, the principal chain is defined as *the chain containing the greatest number of double bonds.* If more than one candidate for the principal chain have equal numbers of double bonds, the principal chain is the longest of these. The principal chain is numbered to give the lowest numbers to double bonds at the first point of difference. (The meaning of the "first point of difference" was discussed in Sec. 3.2A.)

The following example illustrates the numbering of the principal chain and the details of the construction of the name. Notice that the position of the double bond is specified *after* the name of the substituent group.

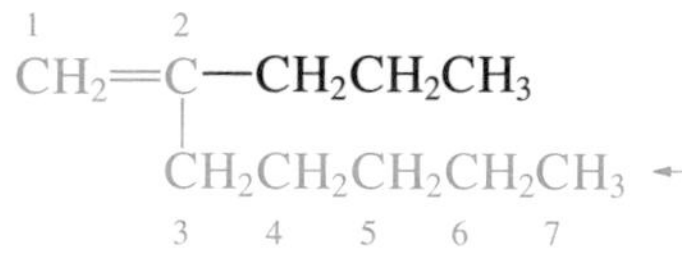

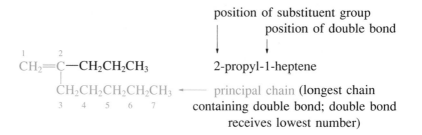

position of substituent group
position of double bond
2-propyl-1-heptene
principal chain (longest chain containing double bond; double bond receives lowest number)

The next example illustrates the requirement that the principal chain be the chain containing the greater number of double bonds.

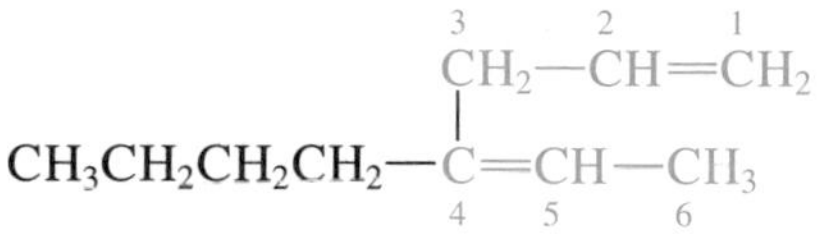

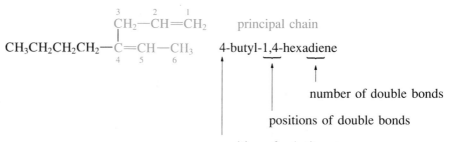

principal chain
4-butyl-1,4-hexadiene
number of double bonds
positions of double bonds
position of substituent

This example also shows how to name compounds with more than one double bond. The *ane* ending of the corresponding alkane is replaced with *adiene* (two double bonds), *atriene* (three double bonds), and so on. The location of each double bond is given in the name.

In the following example the position of substituent groups dictates the correct name because, in either scheme, the double bond has the number 1:

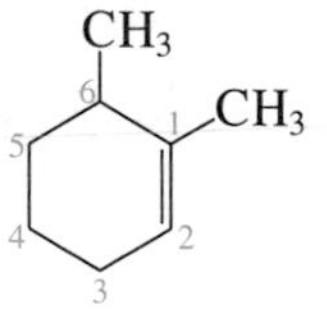

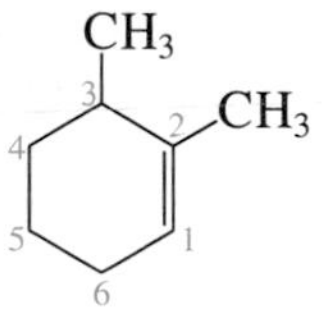

possible names:	1,6-dimethyl-1-cyclohexene (correct)	2,3-dimethyl-1-cyclohexene (incorrect)

In this situation we choose the numbering scheme with the *lowest number at the first point of difference*. In comparing (1,6) with (2,3), the former is correct because 1 is lower than 2.

Substituent groups may also contain double bonds. Some have special names that must be learned:

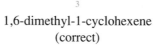

$CH_2{=}CH-$ vinyl

$CH_2{=}CH-CH_2-$ allyl

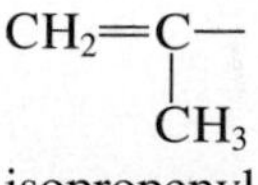

isopropenyl

Other substituent groups are numbered *from the point of attachment to the principal chain.*

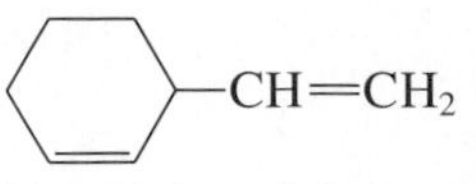

3-vinyl-1-cyclohexene

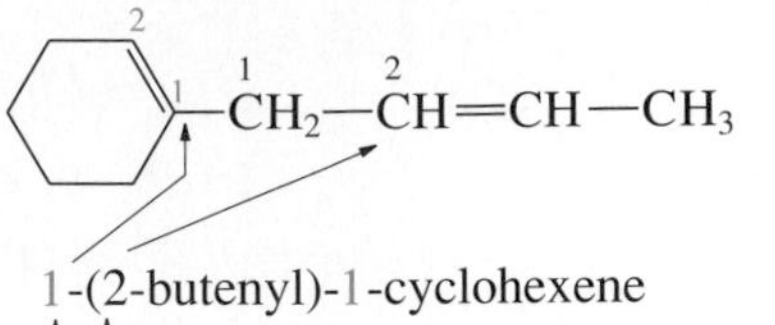

1-(2-butenyl)-1-cyclohexene

position of double bond within the substituent

position of the substituent group on the principal chain

The names of these groups, like those of ordinary alkyl groups, are constructed from the name of the parent hydrocarbon by dropping the final *e* and replacing it with *yl*. Thus, the substituent in the second example above is butene̸ + yl = butenyl. Notice the use of parentheses to set off the names of substituents with internal numbering.

Problems

2 Give the structure of each of the following:
(a) 1-isopropenyl-1-cyclopentene
(b) 3-ethyl-1,4-cyclohexadiene
(c) 5-(3-pentenyl)-1,3,6,8-decatetraene

---

3 Name the following compounds:
(a) $CH_3CH_2—CH{=}CH—CH_2CH_2CH_3$
(b)

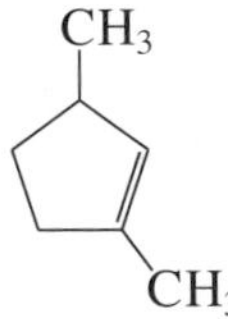

(c) $CH_3—CH{=}CH—CH_2—CH—CH{=}CH—CH_2—CH_3$, with $CH_2—CH{=}CH_2$ on the central CH

B. Nomenclature of Stereoisomers: the Cahn–Ingold–Prelog *E,Z* System

In Sec. 4.1A we learned about the existence of *cis–trans* stereoisomers in alkenes. How do we name these stereoisomers? In some cases the designation *cis* or *trans* can be used in the name.

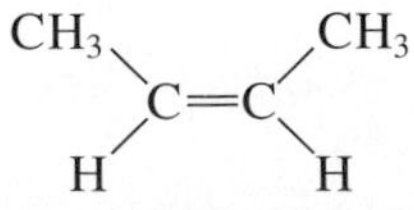

cis-2-butene

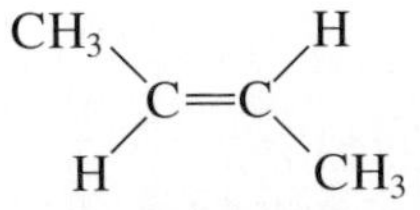

trans-2-butene

However, in some important situations use of the terms *cis* and *trans* is ambiguous. An example is the following compound:

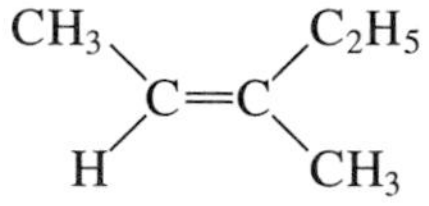

This compound is 3-methyl-2-pentene, but is it *cis* or *trans*? One person might label it *trans,* because the two identical groups are on opposite sides of the double bond. Another might label it *cis,* because the larger groups are on the same side of the double bond. Exactly this sort of ambiguity—and the use of both conventions simultaneously in the chemical literature—brought about the adoption of an unambiguous system, called the **Cahn–Ingold–Prelog system** after its inventors, for the nomenclature of stereoisomers. As this system is applied to alkenes, we assign a *priority* to the two groups on an alkene carbon according to a set of rules given in the list that follows. We then compare the relative locations of these groups on each alkene carbon. If the groups of higher priority are on the same side of the double bond, the compound is said to have the *Z* conformation (*Z* = *zusammen,* German, together). If the groups of higher priority are on opposite sides of the double bond, the compound is said to have the *E* conformation (*E* = *entgegen,* German, across). For a compound with more than one double bond, the conformation of each double bond is specified independently.

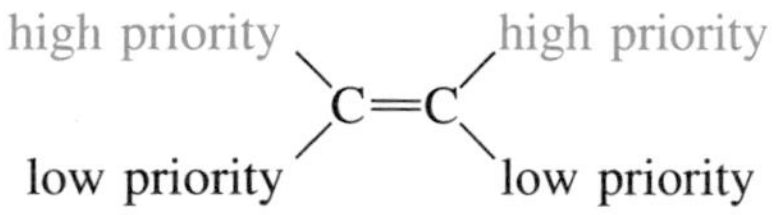

(*Z*)

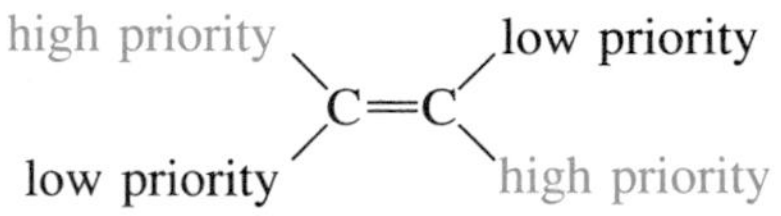

(*E*)

To assign relative group priority we examine the atoms directly attached to the carbons of the double bond, and follow these rules:

1. *Atoms of higher atomic number have higher priority.*

For the substitution

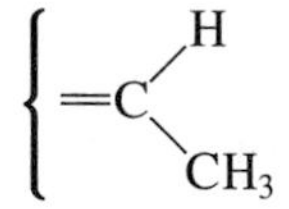

the atoms attached to the alkene carbon are H and C. Because C has higher atomic number than H, the methyl group has the higher priority. As a consequence of this rule, the following alkene has the *Z* conformation, because at each alkene carbon groups of higher priority are on the same side of the double bond.

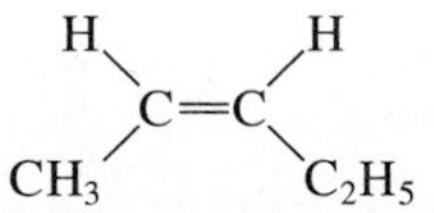

(*Z*)-2-pentene

2. *An isotope of higher atomic mass receives higher priority.*

For example, D (deuterium, or ^2H, an isotope of hydrogen) receives higher priority than ^1H.

When the atoms attached directly to the alkene carbons are identical, we then move out to the next atom in each group; when more than one atom is attached, the atom used is the one of highest atomic number. This process is continued *along the path of highest priority* until a point of difference in the attached atoms is reached. This difference can be due to either the *identity* of attached atoms or the *number* of attached atoms. The priority of the two groups is decided by applying each of the following rules *in order:*

3a. *The higher priority is assigned to the group with the atom of higher atomic number (or atomic mass, in the case of isotopes) at the first point of difference.*

3b. *If the difference between two groups is due to the* number *of otherwise identical atoms, the higher priority is assigned to the group with the greater number of atoms of higher atomic number (or mass).*

The following situation illustrates the application of rule 3a:

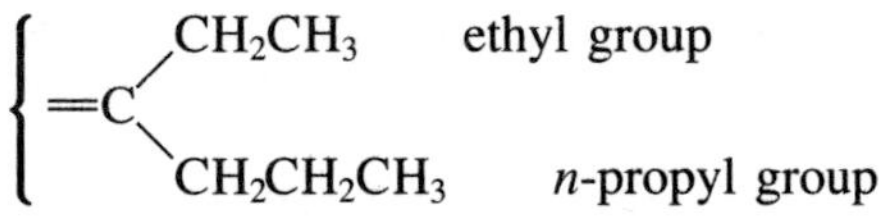

We first begin at the alkene carbon; the attached atoms are both carbon, and no decision can be made. We then move within each group to the next atom of highest priority, which is a carbon in each case (carbon-1). The atoms attached to carbon-1 in each group are (C,H,H). Again, no decision can be made. We move out in each group once more to the next atom of highest priority, again a carbon in each case (carbon-2). At this point, there is a difference: Carbon-2 of the ethyl group has atoms (H,H,H) attached; carbon-2 of the propyl group has atoms (C,H,H) attached. Thus, the propyl group has the higher priority. This procedure might be represented as follows:

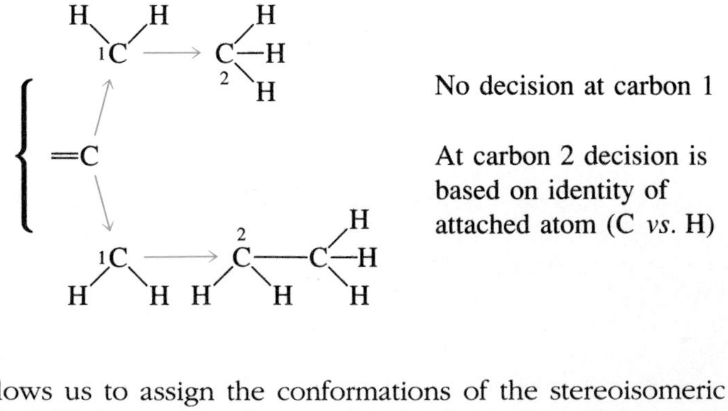

No decision at carbon 1
At carbon 2 decision is based on identity of attached atom (C vs. H)

This rule allows us to assign the conformations of the stereoisomeric 3-methyl-2-pentenes, the ambiguous case discussed at the beginning of this section.

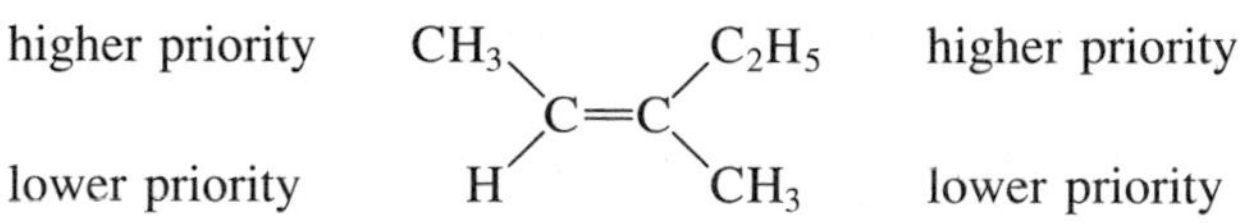

$$(Z)\text{-3-methyl-2-pentene}$$

The use of rule 3b is illustrated by the following situation:

$$=C\begin{cases} CH(CH_3)_2 & \text{higher priority (isopropyl group)} \\ CH_2CH_2CH_2CH_3 & \text{lower priority (} n\text{-butyl group)} \end{cases}$$

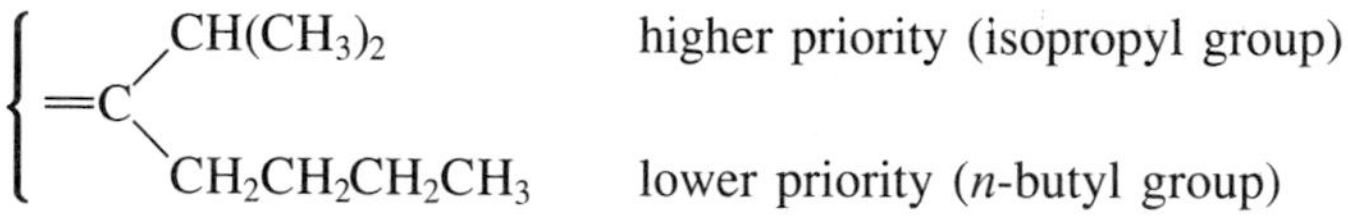

The first point of difference in these two groups occurs at the carbon attached to the double bond. In the isopropyl group the atoms attached to this carbon are (C,C,H); in the *n*-butyl group the attached atoms are (C,H,H). Carbon is attached to this atom in both groups, so rule 3a does not apply. Hence, rule 3b is used: The group of higher priority is the one with *more* carbons at the point of difference.

One last rule involves the treatment of double and triple bonds.

4. *For purposes of assigning priority, doubly and triply bonded atoms are replicated.*

By "replicated" we mean that the atoms in the double or triple bond are written out in duplicate or triplicate, respectively. A vinyl group is written this way:

$$-CH=CH_2 \quad \text{is taken as} \quad -\underset{\underset{C}{|}}{CH}-\underset{\underset{C}{|}}{CH_2}$$

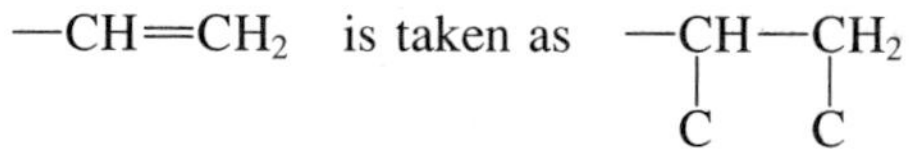

This convention recognizes that a doubly bonded carbon is really two carbon–carbon bonds. A triple bond is similarly recognized as three carbon–carbon bonds.

$$-C\equiv CH \quad \text{is taken as} \quad -\overset{\overset{C}{|}}{\underset{\underset{C}{|}}{C}}-\overset{\overset{C}{|}}{\underset{\underset{C}{|}}{CH}}$$

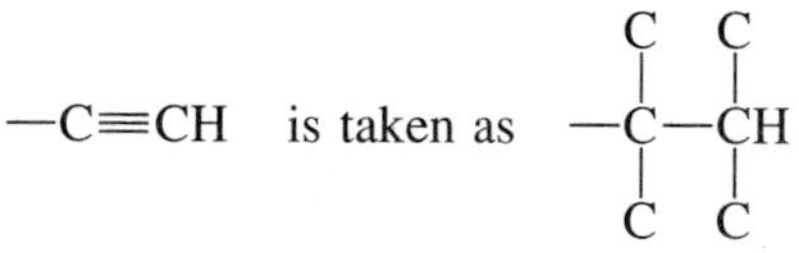

This rule allows us, for example, to establish the relative priority of the isopropyl and vinyl groups. According to rule 4, the groups should be written as follows.

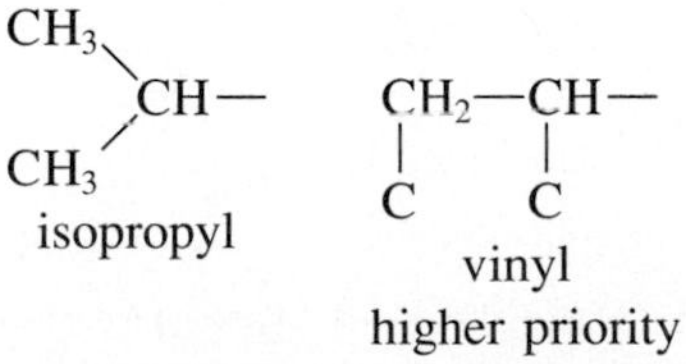

Application of rule 3a yields the path C(H,C,C) → C(H,H,H) for isopropyl and C(H,C,C) → C(H,H,C) for vinyl; thus, the priority decision in favor of vinyl is made at the second carbon in the path.

Additional rules deal with more complex situations, but we need not be concerned with these here. The *E,Z* nomenclature of compounds with more than one double bond is illustrated by the following example:

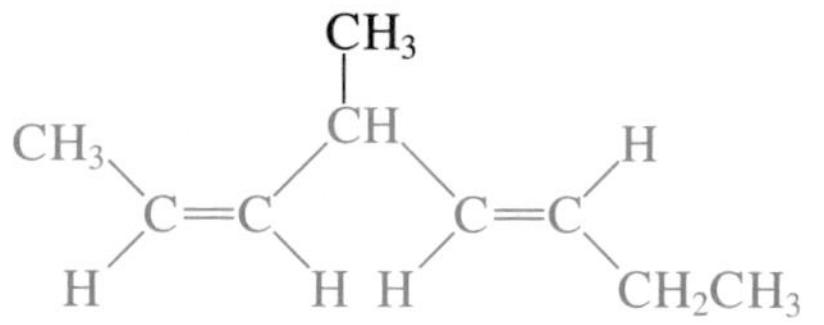

(2*Z*,5*E*)-4-methyl-2,5-octadiene

We can still use the *cis* or *trans* designation when each carbon of a double bond bears a hydrogen, because there is no ambiguity in such cases. Use of the *E,Z* system must be used in other cases, and is always appropriate in any situation.

Problems

4 Name each of the following compounds, including the proper designation of double-bond stereochemistry:

(a) 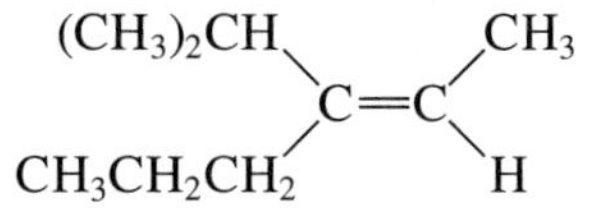

(b) 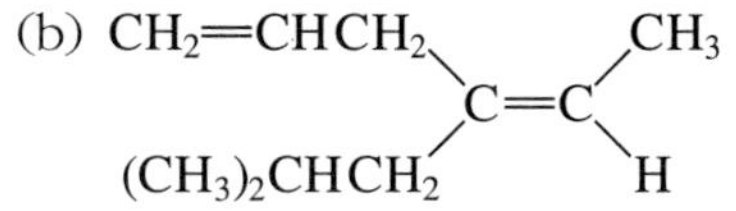

5 Give the structure of 5-[(*E*)-1-propenyl]-(2*E*,7*Z*)-2,7-nonadiene.

6 In each case, which group receives the higher priority?

(a) 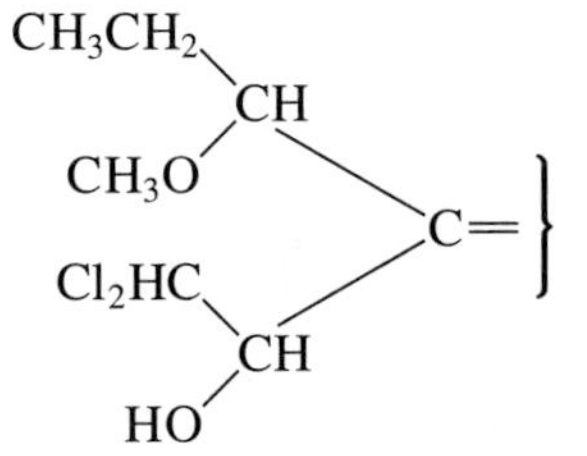

(b) 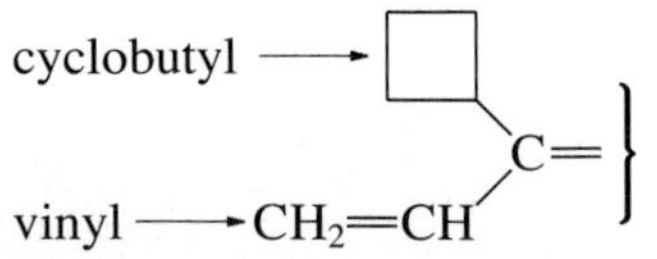

(c) 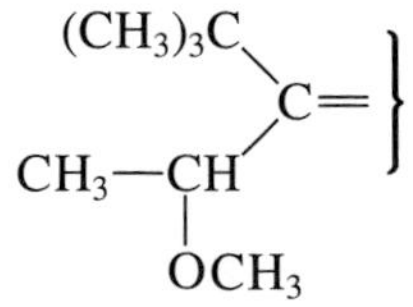

(d) 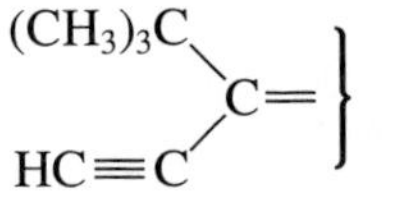

4.3 PHYSICAL PROPERTIES OF ALKENES

Except for their melting points and dipole moments, many alkenes differ little in their physical properties from the corresponding alkane.

	$CH_2{=}CH{-}(CH_2)_3{-}CH_3$ 1-hexene	$CH_3{-}(CH_2)_4{-}CH_3$ hexane
boiling point	63.4°	68.7°
melting point	−139.8°	−95.3°
density	0.673 g/mL	0.660 g/mL
water solubility	negligible	negligible
dipole moment	0.46 D	0.085 D

Like alkanes, alkenes are flammable, nonpolar compounds that are less dense than, and insoluble in, water. The lower molecular weight alkenes are gases.

The dipole moments of some alkenes, though small, are greater than those of the corresponding alkanes.

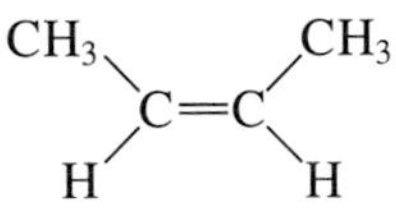

$\mu = 0.25$ D

$CH_3{-}CH_2{-}CH_2{-}CH_3$

$\mu = 0$ D

How can we account for the dipole moments of alkenes? In Sec. 4.1A, it was observed that electron density lies closer to the nucleus in sp^2 orbitals than it does in sp^3 orbitals. As a result, in alkenes with alkyl groups attached to double bonds, there is polarization of electrons *away* from the alkyl group, toward the trigonal carbon atom.

This polarization results in a bond dipole. The dipole moment of *cis*-2-butene is the vector sum of the $CH_3{-}C{=}$ and $H{-}C{=}$ bond dipoles. Although both types of bond dipoles are probably oriented toward the alkene carbon, there is good evidence (Problem 35) that the polarization of the $CH_3{-}C$ bond is greater. This is why *cis*-2-butene has a net dipole moment.

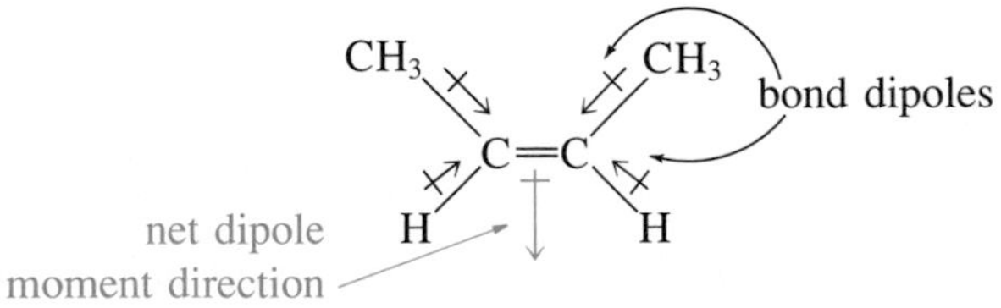

In summary: Bonds from alkyl groups to trigonal carbon are polarized so that *electrons are drawn away from alkyl groups toward trigonal carbon*. An equivalent statement is that *trigonal carbon withdraws electrons from alkyl groups.*

Problem

7 (a) Which compound should have the larger dipole moment, *cis*-2-butene or *trans*-2-butene? Explain.

(b) Interaction between the permanent dipole moments of molecules is one of the cohesive forces within the liquid state. Show one way in which two *cis*-2-butene molecules might orient themselves within a liquid to interact in an attractive manner. Use this result with your answer to (a) to explain why the boiling point of *cis*-2-butene is somewhat higher than that of its *trans*-stereoisomer.

4.4 RELATIVE STABILITIES OF ALKENE ISOMERS

Throughout our study of organic chemistry we shall talk about the relative stabilities of different molecules. When we ask, "Which compound is more stable?" we are really asking, "Which compound has lower energy?" We care about relative energies for at least two reasons. First, as we shall see in Sec. 4.4A, a knowledge of the relative *standard free energies* (ΔG^0) of two compounds tells us immediately which of the two is present in greater amount at equilibrium. An equally important reason for inquiring about relative stabilities is that they tell us something fundamental about bonding in chemical compounds. *A more stable compound has a more stable arrangement of bonds.* By examining how the relative *standard enthalpies,* or *heat contents* (ΔH^0) of molecules vary with structure, we can discern some general structural features that give rise to stability (or instability) in organic molecules. We shall consider this important point in Sec. 4.4C after learning about enthalpy calculations in Sec. 4.4B.

A. Free Energy and Chemical Equilibrium

Although *cis*- and *trans*-2-butene do not interconvert under normal conditions, special conditions, such as strong acid or high temperature, bring these two compounds to equilibrium.

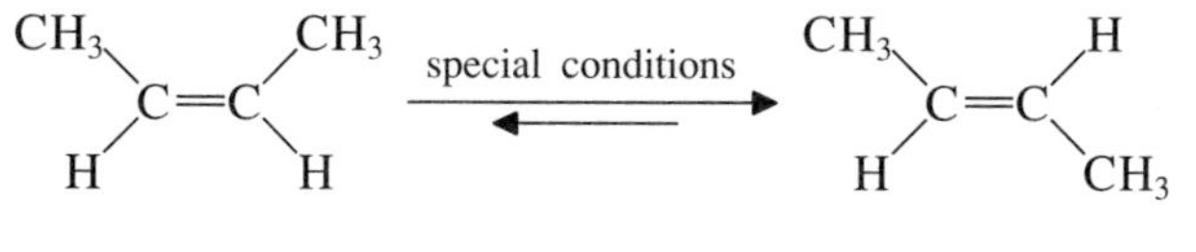

(4.2)

At equilibrium we find that the *trans* isomer is 76.2% of the equilibrium mixture, and the *cis* isomer is 23.8%. Thus, the equilibrium constant for reaction 4.2 is

$$K_{eq} = \frac{[trans\text{-2-butene}]}{[cis\text{-2-butene}]} = \frac{76.2}{23.8} = 3.20 \qquad (4.3)$$

The equilibrium constant is fundamentally related to the relative standard free energies of the two isomers. Let us define ΔG^0 as the difference between the standard free energies of the *trans* and *cis* isomers; that is,

$$\Delta G^0 = (\text{standard free energy of } trans) - (\text{standard free energy of } cis) \qquad (4.4)$$

The equilibrium constant is related to this standard free energy difference by the following equation:

$$\Delta G^0 = -RT\ln K_{eq} \tag{4.5a}$$

or, in common logarithms,

$$\Delta G^0 = -2.303RT\log K_{eq} \tag{4.5b}$$

in which R is the gas constant (1.987×10^{-3} kcal/degree-mol) and T is the absolute temperature in kelvins (degrees K). Solving this equation for ΔG^0, we find that at room temperature (25 °C or 298 K) the standard free energy difference is -0.69 kcal/mol. This means that *trans*-2-butene is more stable than *cis*-2-butene by 0.69 kcal under standard conditions, usually taken as 1 atm pressure (in the gas phase) or 1*M* concentration (in solution) at 25 °C.

Let us generalize this result for a reaction in which the reactant is *A* and the product is *B*. Suppose the standard free-energy difference ($G_B^0 - G_A^0$) between *B* and *A* is positive. This means that *B* has a higher free energy than *A*, or *B* is less stable than *A*. If *A* and *B* are brought to equilibrium, *A* will be present in greater amount. We can see this by rearranging Eq. 4.5b.

$$\log K_{eq} = \frac{-\Delta G^0}{2.303RT} \tag{4.6}$$

When ΔG^0 is positive (*A* more stable), $\log K_{eq}$ is a negative number, and consequently $K_{eq} = [B]/[A]$ is less than unity—in other words, the concentration of *A* is higher than that of *B* at equilibrium. On the other hand, suppose that the free energy difference is negative; that is, *B* has a lower energy than *A*, or *B* is more stable than *A*. In this case $\log K_{eq}$ is positive and K_{eq} is greater than unity—in other words, the concentration of *B* is higher than that of *A* at equilibrium. These examples show that *chemical equilibrium favors the more stable compound—the compound of lower free energy* (Fig. 4.7).

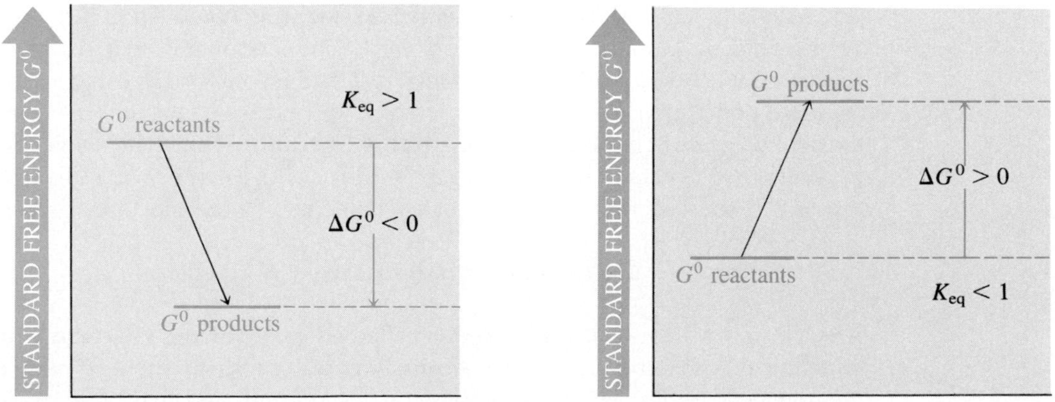

Figure 4.7 Relationship of free energies and equilibrium constants. The more stable compound is favored in a chemical equilibrium.

TABLE 4.1 Relationship Between Equilibrium Constant and Standard Free-Energy Changes for Reactions at 25 °C

A. ΔG^0 (kcal/mol)	means K_{eq}	B. K_{eq}	means ΔG^0 (kcal/mol)
+10	4.7×10^{-8}	10^6	−8.18
+6	4.0×10^{-5}	10^4	−5.46
+3	0.0063	10^2	−2.73
+1	0.19	50	−2.32
0	1.0	30	−2.01
−1	5.4	10	−1.36
−3	158	5	−0.95
−6	2.5×10^4	3	−0.65
−10	2.1×10^7	1	0.0

The more two compounds differ in stability, the greater is the difference in their concentrations at equilibrium. Because of the *logarithmic* relationship of free energy and equilibrium constant, rather small differences in energy are reflected in rather large changes in K_{eq}, as we can see in Table 4.1.

Problems

8 A reaction has a standard free energy change of −3.5 kcal/mol. Calculate the equilibrium constant of the reaction at 25 °C.

9 A reaction $A + B \rightleftarrows C$ has a standard free-energy change of −0.7 kcal/mol at 25 °C. What are the concentrations of *A, B,* and *C at equilibrium* if, at the beginning of the reaction, their concentrations are 0.1*M*, 0.2*M*, and 0*M*, respectively?

B. Heats of Formation

Relative enthalpies, or heat contents, of molecules can be obtained from the combustion reaction (Sec. 3.6). When the energy released from combustion is not used to do work, it is all liberated in the form of heat. The amount of heat released can be measured in a device called a *calorimeter*. When the combustion is carried out under standard conditions of constant temperature and pressure, this energy is called the **standard heat of combustion** or **enthalpy of combustion.** The conventions used in dealing with heats of reactions are the same as with free energies: The heat of any reaction is the *difference* in the enthalpy contents of products and reactants.

$$\Delta H^0(\text{reaction}) = H^0(\text{products}) - H^0(\text{reactants}) \tag{4.7}$$

A reaction in which heat is liberated is called an **exothermic reaction,** and one in which heat is absorbed is called an **endothermic reaction.** The ΔH^0 of an exothermic reaction, by Eq. 4.7, has a negative algebraic sign, and that of an endothermic reaction has a positive sign. All combustion reactions, as we know from experience, are exothermic. The combustion of *trans*-2-butene is an example.

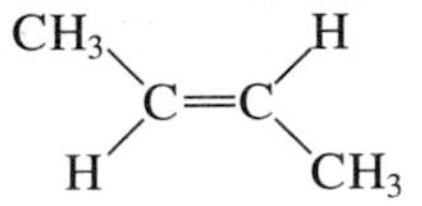 $+ 6O_2 \longrightarrow 4CO_2 + 4H_2O \qquad \Delta H^0 = -604.73$ kcal/mol (4.8)

The standard enthalpy of this reaction, ΔH^0, of −604.73 kcal/mol, means that burning one mole of *trans*-2-butene under standard conditions (298 K, 1 atm pressure) liberates 604.73 kcal of energy as heat.

Enthalpy data are generally tabulated not as heats of combustion, but instead as *heats of formation*. The standard **heat of formation** of a compound, abbreviated ΔH_f^0, is simply the heat of the reaction in which a compound is formed from its elements in their natural state under standard conditions of temperature and pressure. For example, the standard heat of formation of *trans*-2-butene is −2.67 kcal/mol. This means that 2.67 kcal/mol of heat would be liberated when 2-butene is formed from H_2 gas (the natural state of hydrogen) and carbon:

$$4H_2 + 4C \longrightarrow \textit{trans}\text{-2-butene} \qquad \Delta H^0 = -2.67 \text{ kcal/mol} \tag{4.9}$$

Heats of formation can be derived from heats of combustion by applying **Hess's law,** which states that *chemical reactions and their associated energies may be added like algebraic equations*. For example, the combustion of hydrogen, carbon, and *trans*-2-butene can be combined in the following equations. Items that appear on both sides of the equation are cancelled.

Equations:	ΔH^0 *(kcal/mol:)*	
$4H_2 + 2O_2 \longrightarrow 4H_2O$	$4(-57.80) = -231.20$	(4.10a)
$4C + 4O_2 \longrightarrow 4CO_2$	$4(-94.05) = -376.20$	(4.10b)
$4CO_2 + 4H_2O \longrightarrow 6O_2 +$ *trans*-2-butene	$+604.73$	(4.10c)
Sum: $4C + 4H_2 \longrightarrow$ *trans*-2-butene	$\Delta H^0 = -2.67$	(4.10d)

In Eq. 4.10a, the −57.80 kcal/mol quantity is the heat of combustion for hydrogen; the heat of combustion for four moles of hydrogen is 4(−57.80). To this is added Eq. 4.10b, the heat of combustion of four carbons; and Eq. 4.10c (the reverse of Eq. 4.8), the negative heat of combustion of *trans*-2-butene. (Since heats of combustion are negative, reverse heats of combustion are positive.) Algebraic addition of these equations gives the desired equation for formation of *trans*-2-butene, and addition of the associated heats gives the heat of formation. Fortunately, we do not have to go through a calculation like this every time we need heats of formation, because these quantities are available in standard tables.

Heats of formation can be used to derive relative energies of molecules by applying Hess's law. Suppose we wish to calculate ΔH^0 for the reaction

$$\begin{array}{ccccc} & \textit{cis}\text{-2-butene} & \longrightarrow & \textit{trans}\text{-2-butene} \\ \Delta H_f^0 & -1.67 & & -2.67 \end{array} \tag{4.11}$$

We simply subtract the heat of formation of the reactant, the *cis* isomer, from that of the product, the *trans* isomer. The ΔH^0 for this reaction is therefore −2.67 − (−1.67) or −1.00 kcal/mol. This means that *trans*-2-butene is more stable than *cis*-2-butene by

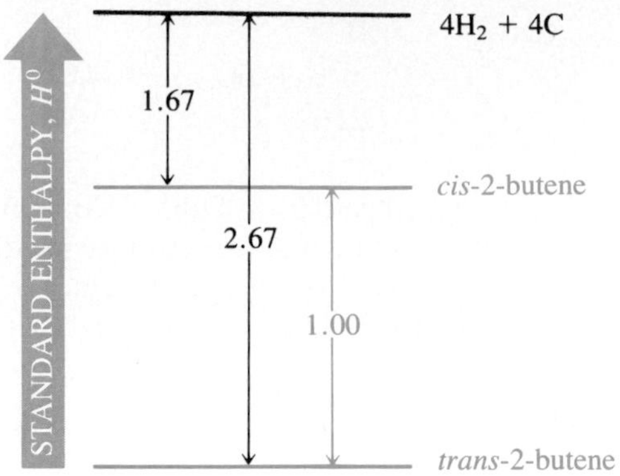

Figure 4.8 Use of heats of formation to derive relative energies of two isomeric compounds.

1.00 kcal/mol. What we have really done in Eq. 4.11 is to subtract the two formation reactions and their associated energies:

Equations:			ΔH^0 (*kcal/mol:*)
$\cancel{4C} + \cancel{4H_2}$	⟶	*trans*-2-butene	−2.67
cis-2-butene	⟶	$\cancel{4C} + \cancel{4H_2}$	+1.67
Sum: *cis*-2-butene	⟶	*trans*-2-butene	−1.00

(4.12)

Since *cis*- and *trans*-2-butene are isomers, the elements from which they are formed are the same and *cancel in the comparison*. This is shown diagramatically in Fig. 4.8.

It is important to notice that this procedure works because *cis*- and *trans*-2-butene are isomers. Were we to attempt to compare the enthalpies of compounds that are not isomers, there would be different quantities of carbon and hydrogen in the two formation equations, and the sum would contain leftover C and H_2. This sum would be valid, but it would not be the direct comparison we are seeking.

Notice that the *enthalpy difference* (ΔH^0) between *cis*- and *trans*-2-butene, 1.00 kcal/mol, is not the same as the *free-energy difference* (ΔG^0), 0.69 kcal/mol, although the numbers are of similar magnitude. The reason is that enthalpy and free energy measure somewhat different energy quantities. In a mechanical sense, ΔH^0 measures the total energy liberated in a reaction; ΔG^0 measures the part of this energy that could in principle be harnessed to do work (as in a battery). From a theoretical standpoint, ΔG^0 is related to *the position of chemical equilibrium,* as we learned in Sec. 4.4A. On the other hand, ΔH^0 data are most appropriate for examining the *energy differences associated with different arrangements of chemical bonds,* as we shall now see.

Problem

10 (a) Calculate the heat of formation of 2-methylpropene from its heat of combustion, −603.36 kcal/mol, using the data in Eq. 4.10a and 4.10b.
(b) Calculate the heat of combustion of 1-butene from its heat of formation, −0.03 kcal/mol.

C. Relative Stabilities of Alkene Isomers

We have found that *trans*-2-butene has a lower enthalpy than the *cis*-isomer by 1 kcal/mol (Eq. 4.12). In fact, it is rather general that *trans*-alkenes are more stable than their *cis* isomers. Why is this so? If we examine a model of *cis*-2-butene, we find that the methyl groups are forced to occupy the same plane on the same side of the double bond (Fig. 4.9). The methyl groups slightly violate each other's van der Waals radius.

Figure 4.9 The steric interaction between the methyl groups in *cis*-2-butene is absent in the *trans* isomer.

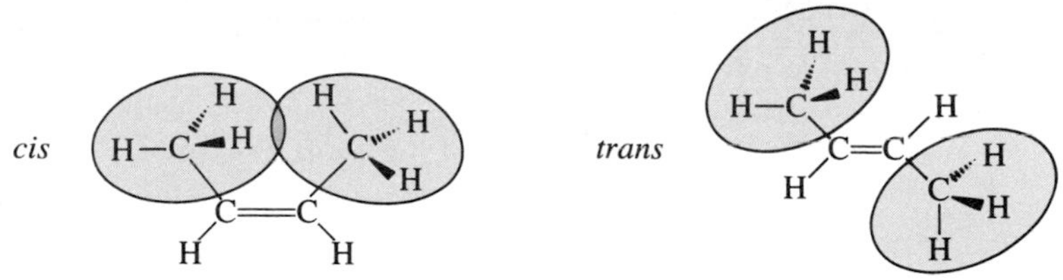

This situation does not occur in the *trans* isomer. This is another example of a *steric effect* (Sec. 3.5B); two groups are close enough together that their energy of interaction is repulsive. Not only do the enthalpies of *cis*- and *trans*-2-butene suggest the importance of strain in the *cis* isomer, but they also give us quantitative information about the magnitude of this effect.

Problem

11 (a) Construct models of the *cis*- and *trans*-isomers of 2,2,5,5-tetramethyl-3-hexene. Which isomer would you expect to have the lower enthalpy? Why?

(b) Assuming that the difference between the enthalpies of the two isomers in part (a) is about the same as the difference between their free energies, tell which reaction would have the greater equilibrium constant: the formation of *trans*- from *cis*-2-butene or the formation of *trans*- from *cis*-2,2,5,5-tetramethyl-3-hexene? Explain your reasoning.

How does branching at the double bond affect the relative energies of alkenes? First, let us compare the heats of formation of the following two isomers, of which one has alkene carbons bearing a single alkyl group (a *monosubstituted double bond,* and the other has alkene carbons with two alkyl groups (a *disubstituted double bond*):

monosubstituted: CH_2=CH—$CH(CH_3)_2$ $\Delta H_f^0 = -6.92$ kcal/mol

disubstituted: CH_2=$\overset{\overset{\displaystyle CH_3}{|}}{C}$—$C_2H_5$ $\Delta H_f^0 = -8.68$ kcal/mol

Notice that both compounds have equal numbers of branches; they differ only in the *position* of the branch. In this case, the isomer with the branch at the double bond is more stable. The following data for isomeric hexenes show that this trend continues for alkenes with increasing numbers of branches at the double bond:

monosubstituted: CH_2=CH—CH_2—$CH(CH_3)_2$ $\Delta H_f^0 = -10.54$ kcal/mol

disubstituted: CH_2=$C\begin{smallmatrix}\diagup C_2H_5 \\ \diagdown C_2H_5\end{smallmatrix}$ $\Delta H_f^0 = -12.32$ kcal/mol

trisubstituted: $\begin{smallmatrix}CH_3 \diagdown \\ \qquad H \diagup\end{smallmatrix}C$=$C\begin{smallmatrix}\diagup CH_3 \\ \diagdown C_2H_5\end{smallmatrix}$ $\Delta H_f^0 = -14.02$ kcal/mol

tetrasubstituted: $\begin{smallmatrix}CH_3 \diagdown \\ CH_3 \diagup\end{smallmatrix}C$=$C\begin{smallmatrix}\diagup CH_3 \\ \diagdown CH_3\end{smallmatrix}$ $\Delta H_f^0 = -14.15$ kcal/mol

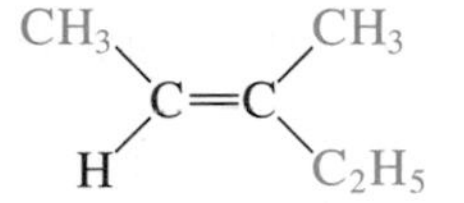

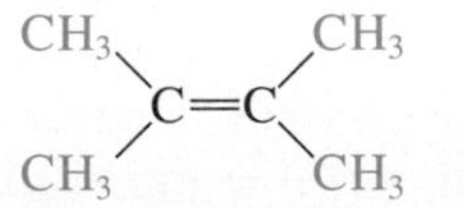

Evidently, *an alkene is stabilized by substitution of alkyl groups on the double bond.* When we compare the stability of alkene isomers we find that *the alkene with the greatest number of alkyl groups on the double bond is usually the most stable one.*

Notice that, to a useful approximation, it is the *number* of alkyl groups on the double bond more than their *identities* that governs the stability of an alkene. In other words, a molecule with two small alkyl groups on the double bond is more stable than its isomer with one large group on the double bond.

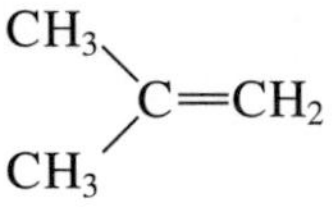

$$
\begin{array}{c}
CH_3 \\
\diagdown \\
C{=}CH_2 \\
\diagup \\
CH_3
\end{array}
\qquad \Delta H_f^0 = -4.04 \text{ kcal/mol}
$$

$$CH_3CH_2CH{=}CH_2 \qquad \Delta H_f^0 = -0.03 \text{ kcal/mol}$$

Heats of formation have given us considerable information about how alkene stabilities vary with structure. Let us summarize:

increasing stability:

$$R{-}CH{=}CH_2 < R{-}CH{=}CH{-}R \approx$$

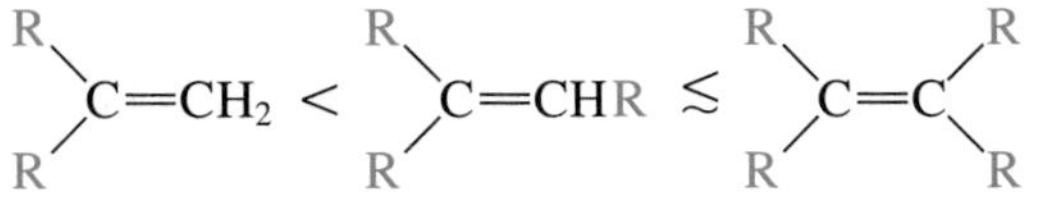

$$
\begin{array}{ccc}
R\diagdown \quad \diagup & R\diagdown \quad \diagup & R\diagdown \quad \diagup R \\
C{=}CH_2 < C{=}CHR \lesssim C{=}C \\
R\diagup & R\diagup & R\diagup \quad \diagdown R
\end{array}
\qquad (4.13a)
$$

and

$$
\begin{array}{cc}
R\diagdown \quad \diagup R & R\diagdown \quad \diagup H \\
C{=}C < C{=}C \\
H\diagup \quad \diagdown H & H\diagup \quad \diagdown R
\end{array}
\qquad (4.13b)
$$

Problem

12 There is very little difference (0.13 kcal/mol) between the heats of formation of 3-methyl-2-pentene and 2,3-dimethyl-2-butene. Suggest one reason why the latter compound, despite its greater degree of branching at the double bond, is not much more stable than the former.

4.5 ADDITION OF HYDROGEN HALIDES TO ALKENES

We turn now from a study of relative alkene energies to a study of alkene reactions. Two reactions will concern us for the duration of this chapter: the reaction with hydrogen halides, and the reaction with water (hydration). We shall use these two reactions to establish some important principles of chemical reactivity that we shall put to use throughout our study of organic chemistry. In Chapter 5 we shall consider a number of other alkene reactions.

The most characteristic reaction of the carbon–carbon double bond is **addition,** which can be represented generally as follows:

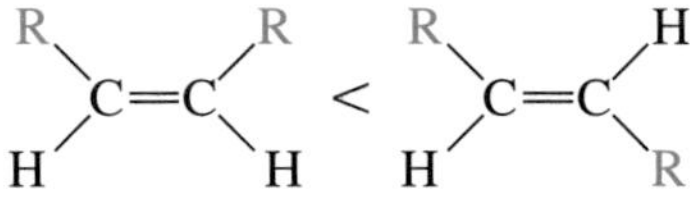

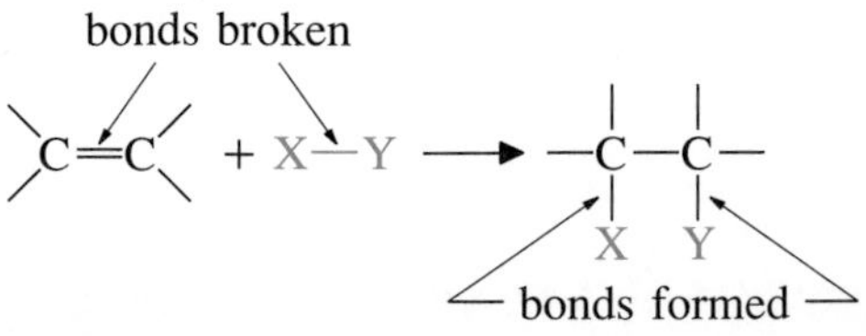

$$(4.14)$$

In such a reaction, a carbon–carbon bond and the X—Y bond are broken, and new C—X and C—Y bonds are formed. An example of such a reaction is the addition of a hydrogen halide to an alkene.

$$\underset{}{HBr} + \underset{(\textit{cis}\ \text{or}\ \textit{trans})}{CH_3{-}CH{=}CH{-}CH_3} \longrightarrow CH_3{-}\underset{\underset{Br}{|}}{CH}{-}\underset{\underset{H}{|}}{CH}{-}CH_3 \qquad \underset{(\textit{sec}\text{-butyl bromide})}{\text{2-bromobutane}} \tag{4.15}$$

When the alkene has an unsymmetrically located double bond (one that is not in the center of the molecule), two isomeric products are possible.

$$CH_2{=}CH{-}(CH_2)_3CH_3 + HI \longrightarrow$$

$$\underset{\substack{\text{2-iodohexane}\\ \text{(observed)}}}{CH_3{-}\underset{\underset{I}{|}}{CH}{-}(CH_2)_3CH_3} \quad \text{or} \quad \underset{\substack{\text{1-iodohexane}\\ \text{(not observed)}}}{I{-}CH_2{-}CH_2{-}(CH_2)_3CH_3} \tag{4.16}$$

As shown in Eq. 4.16, *only one of the two possible products is formed* from a 1-alkene in significant amount. Generally, *the main product is that isomer in which the halogen is bonded to the more highly substituted* (more highly branched) *carbon of the double bond.*

$$CH_2{=}CH{-}(CH_2)_3CH_3$$

H goes here (to CH_2) I goes here (to CH)

A reaction such as hydrogen halide addition that gives only one of several possible structural isomers is said to be **regioselective.** The regioselectivity of hydrogen halide addition to alkenes is said to follow **Markovnikov's rule.** The original statement of this rule was that *addition of a dipolar molecule H—X to an alkene proceeds so that the hydrogen goes to the less substituted carbon of the double bond.*

> Vladimir Markovnikov (1838–1904), director of the Chemical Institute of Moscow, observed the regioselectivity of hydrogen halide addition to alkenes in 1869. Because Markovnikov refused to publish his observation in a language other than Russian, his observations remained largely unknown until about 1899.

When there are equal amounts of branching at the two alkene carbons, little or no regioselectivity is observed in hydrogen halide addition, even though one branch may be a larger alkyl group than another.

$$HBr + CH_3{-}CH{=}CH{-}C_2H_5 \longrightarrow$$

$$CH_3{-}\underset{\underset{Br}{|}}{CH}{-}CH_2{-}C_2H_5 + CH_3{-}CH_2{-}\underset{\underset{Br}{|}}{CH}{-}C_2H_5 \tag{4.17}$$

(nearly equal amounts)

For many years the Markovnikov rule had only an *empirical* (experimental) basis. Chemists now understand why the rule is followed, as we shall see in the next section. The reasoning behind the Markovnikov rule will provide a broad framework that we can use to understand a wide variety of organic reactions.

Problem

13 Assuming operation of the Markovnikov rule, predict the product of addition of HBr to 1-methyl-1-cyclohexene.

A. Carbocations as Intermediates in the Addition of Hydrogen Halides to Alkenes

Electrons in the π-bond of an alkene can attack the proton of an acid. When isobutylene, for example, reacts with HBr, the carbon–carbon double bond is protonated by the acid.

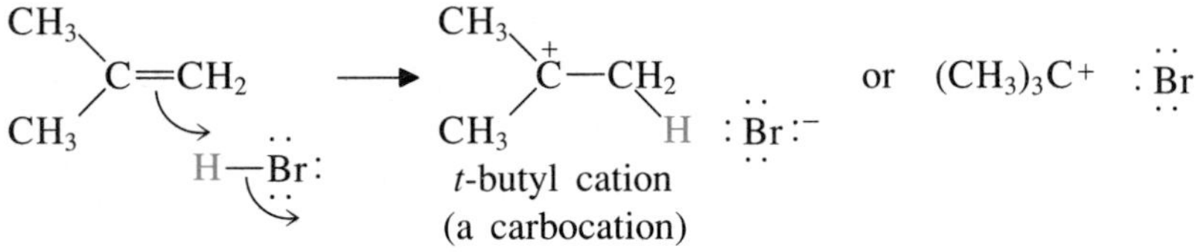

(4.18a)

The π-bond is thus acting as a Brønsted base toward the Brønsted acid H—Br. The π-bond is not an ordinary base in the sense that we consider $:NH_3$, $:\ddot{O}H^-$, or even water as bases. Rather, it is a very weak base. It can, however, be protonated to a small extent by the strong acid HBr. The product of such a reaction is an ionic species called a **carbocation** (pronounced CAR-bo-CAT-ion), or **carbonium ion** in earlier literature. A carbocation has an electron-deficient carbon atom; that is, it has a carbon that is short of an octet by one electron pair. The carbocation is therefore a Lewis acid, or *electrophile,* and it can be attacked by the Lewis base, or *nucleophile,* Br^- to complete the addition of HBr.

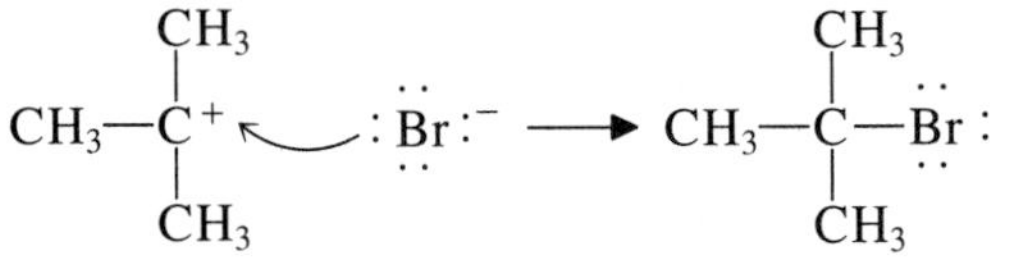

(4.18b)

Notice that the addition of HBr to isobutylene is really the result of *two consecutive reactions*:

1. Protonation of the π-bond
2. Attack of bromide ion on the resulting carbocation

The carbocation is an example of a *reactive intermediate:* a species that is present in very low concentration because it reacts as quickly as it is formed. The *t*-butyl cation cannot be isolated from the reaction of HBr and isobutylene because it reacts with Br^- as soon as it is produced. Indeed, most carbocations are so reactive that they cannot be isolated except under special circumstances.

Why does HBr react with the π-bond of an alkene rather than with the σ-bonds? In π-bonds the electron density is farther from the carbon nuclei than it is in σ-bonds. Because π-electrons are held less tightly than σ-electrons, they are more readily pulled away from their parent carbon atoms to form a bond with electron-seeking reagents such as protons. For this reason, *reactions of the π-bond are the typical reactions of alkenes.*

In the addition of HBr to isobutylene, the *t*-butyl cation is not the only carbocation that might form by reaction of the π-bond with HBr. Were the proton to react at the other carbon of the double bond, the isobutyl cation would form instead.

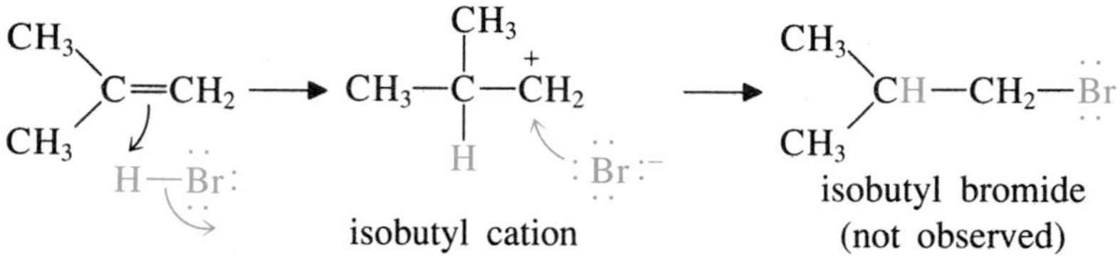

(4.19)

If the isobutyl cation were formed, Br^- would also attack it to yield isobutyl bromide as a product; yet *t*-butyl bromide is the only product observed. The reaction of HBr with isobutylene gives only *t*-butyl bromide because only the *t*-butyl cation is formed. What is the reason for the formation of only one of the two possible carbocations? The answer lies in the relative stability of these two species, which we shall now examine.

B. Structure and Stability of Carbocations

Carbocations are classified by the degree of alkyl substitution at the electron-deficient carbon atom.

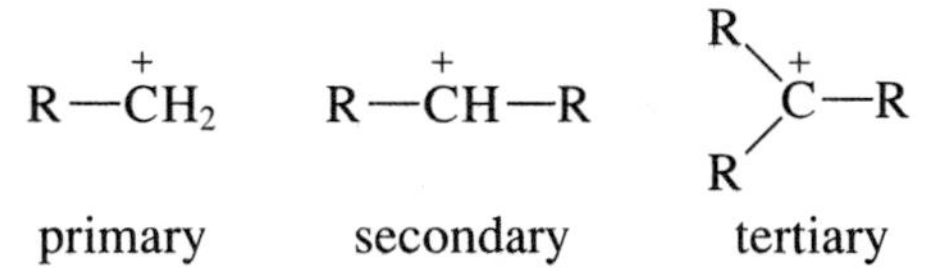

For example, the isobutyl cation in Eq. 4.19 is a primary carbocation, and the *t*-butyl cation is a tertiary carbocation. The relative stability of the isomeric carbocations is in the following order:

Stability of carbocations:

tertiary > secondary > primary (4.20)

When isobutylene reacts with HBr, two possible reaction pathways are in competition. One pathway involves a more stable tertiary carbocation; the other involves a less stable primary carbocation. Since the product derived from the tertiary carbocation is observed, it follows that many more molecules have reacted by the pathway involving this ion. In other words, *the reaction pathway involving the more stable carbocation is the one that is observed.*

What is the reason for the stability order in Eq. 4.20? To answer this question, we first must consider the geometry and electronic structure of carbocations. The rules for predicting molecular geometry (Sec. 1.6B) suggest that the three groups surrounding the electron-deficient carbon should all lie in the same plane with an angular separation of 120°. Thus, the electron-deficient carbon of a carbocation is a *trigonal*

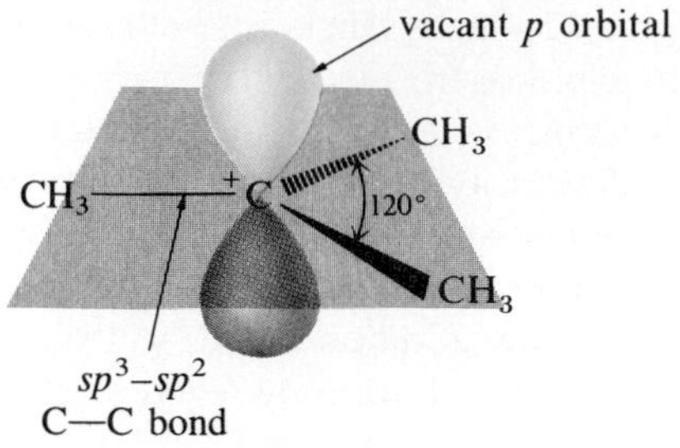

Figure 4.10 Hybridization and geometry of the *t*-butyl cation.

carbon. Trigonal carbons are sp^2 hybridized (Fig. 4.10). The $2p$ orbital on the trigonal carbon of the carbocation contains no electrons.

Because alkenes are stabilized by alkyl substituents on the trigonal carbon (Sec. 4.4C), it is reasonable to expect that the same alkyl substituents on the trigonal carbon should also stabilize a carbocation. Although this is true, the stabilization of carbocations is considerably greater (by tens of kilocalories per mole) than the stabilization of alkenes (2–3 kcal/mol). (See Problem 38 at the end of the chapter.) The origin of this greater stabilization for carbocations is the fact that the electrons in adjacent C—H bonds overlap, or spill over, into the empty p orbital.

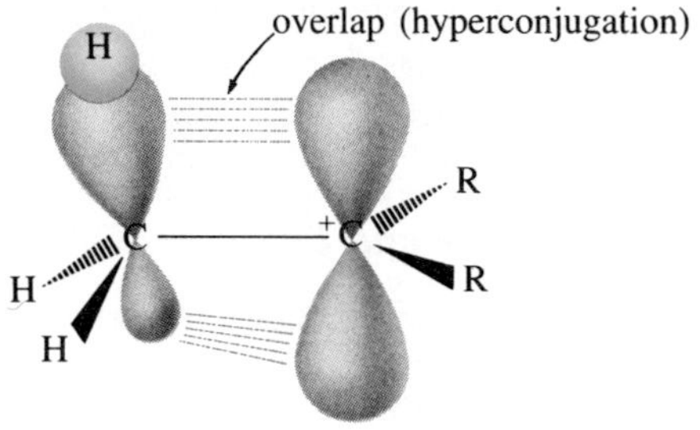

This overlap, called **hyperconjugation,** produces some additional bonding between the electron-deficient carbon and the adjacent carbon atom. Hyperconjugation allows the electron pair in the adjacent C—H bond to move over a larger region of space. These electrons are able to move out of each other's way and reduce their repulsive interactions. The greater the substitution at the trigonal carbon, the greater the stabilization by hyperconjugation.

C. Summary: Mechanism of Hydrogen-Halide Addition to Alkenes

The complete description of a reaction pathway, including any reactive intermediates such as carbocations, is called the **mechanism** of the reaction. We now understand the mechanism of hydrogen-halide addition to alkenes, which can be represented with Lewis structures and the arrow formalism as follows:

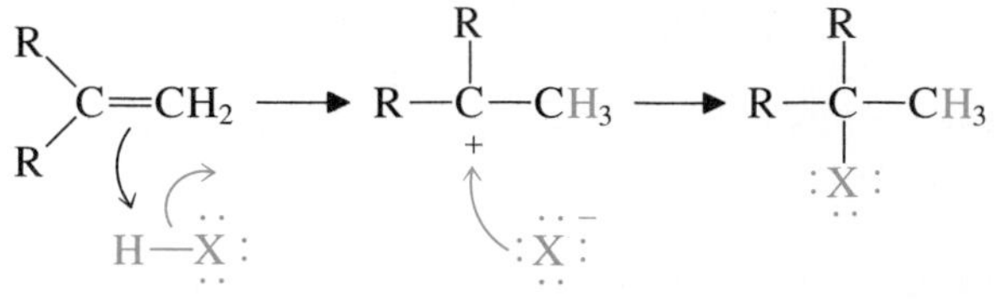

(X = halogen) (4.21)

We can see now why the Markovnikov rule is followed: in order for the reaction to proceed through the more stable carbocation, the proton of the hydrogen halide must form a bond to the *less substituted carbon* of the double bond, leaving the *more substituted carbon* electron-deficient. This is the carbon that then forms a bond with the halide. In fact, we can reformulate the Markovnikov rule in modern terms as follows: *In the addition of a dipolar molecule to an alkene, the positive atom of the dipole becomes attached at the less substituted carbon, and the negative atom at the more substituted carbon, of the alkene double bond.* This statement suggests that any suitable Lewis acid (*electrophile*)-Lewis base (*nucleophile*) pair will react at the π-bond of an alkene in a similar manner. Indeed, addition of hydrogen halides to alkenes is but one member of a broad class of reactions called **electrophilic addition reactions,** many of which we shall study in this and the following chapter. All electrophilic additions to alkenes are initiated by the interaction of π-electrons with a Brønsted or Lewis acid.

Problem **14** Predict the product of the reaction of I—Cl with isobutylene. (*Hint*: Consult Table 1.1.)

We have been writing our addition reactions using Lewis structures; what is happening to the orbitals involved in the reactions? These changes are shown in Fig. 4.11. The reaction begins when the electrophilic (electron-seeking, or positive) end of the H—X dipole reacts with the two π-electrons of the alkene. The carbon to which the proton bonds becomes sp^3 hybridized, and the other alkene carbon becomes the electron-deficient carbon of the carbocation—still sp^2 hybridized, but with its $2p$ orbital empty. When the p orbital on this carbon is attacked by the electron pair on the halide ion, this carbon also becomes sp^3 hybridized, completing the reaction.

4.6 REARRANGEMENT OF CARBOCATIONS

In the following example of HCl addition to an alkene, something unusual happens:

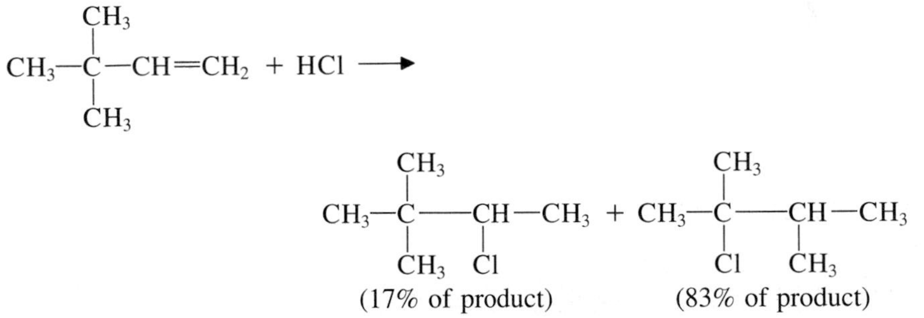

$$\underset{\underset{\text{CH}_3}{|}}{\overset{\overset{\text{CH}_3}{|}}{\text{CH}_3-\text{C}-\text{CH}=\text{CH}_2}} + \text{HCl} \longrightarrow$$

$$\underset{\underset{\text{CH}_3 \quad \text{Cl}}{|\qquad|}}{\overset{\overset{\text{CH}_3}{|}}{\text{CH}_3-\text{C}----\text{CH}-\text{CH}_3}} + \underset{\underset{\text{Cl} \quad \text{CH}_3}{|\qquad|}}{\overset{\overset{\text{CH}_3}{|}}{\text{CH}_3-\text{C}----\text{CH}-\text{CH}_3}} \quad (4.22)$$

(17% of product) (83% of product)

The minor product is clearly the result of ordinary Markovnikov addition of HCl across the double bond. The origin of the major product, however, is not so obvious. If we examine the carbon skeleton of the major product, we find that a **rearrangement** has occurred: The carbons of the alkyl halide product are connected differently than those of the alkene starting material. One of the methyl groups has apparently shifted its position during the transformation of the alkene into the alkyl halide. Although the rearrangement leading to the second product seems strange at first sight, it is readily understood if we consider the fate of the carbocation intermediate in the reaction.

The reaction begins like a normal addition of HCl—by protonation of the double bond to yield the more substituted carbocation.

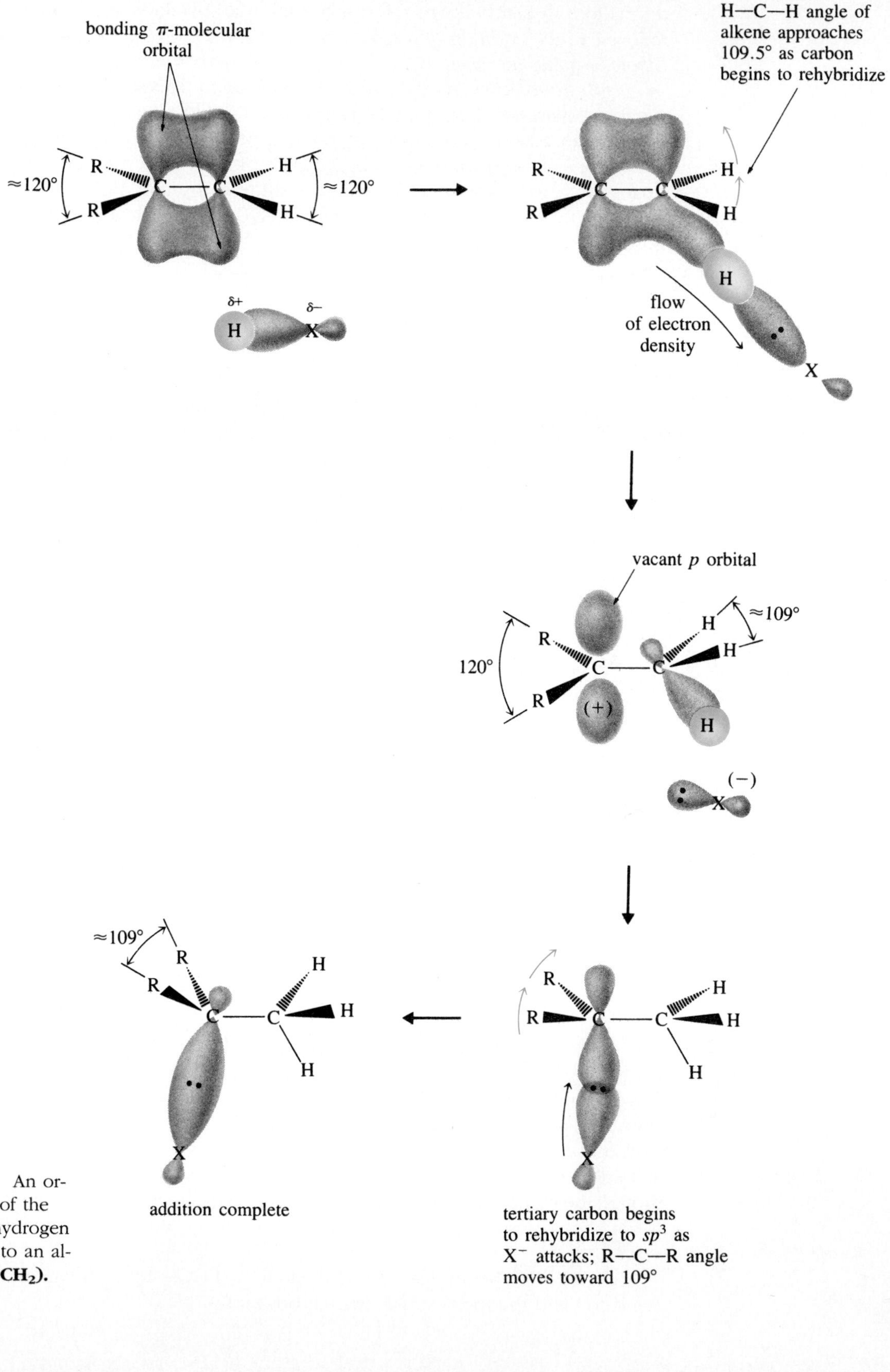

Figure 4.11 An orbital picture of the addition of hydrogen halide **(HX)** to an alkene **($R_2C{=}CH_2$).**

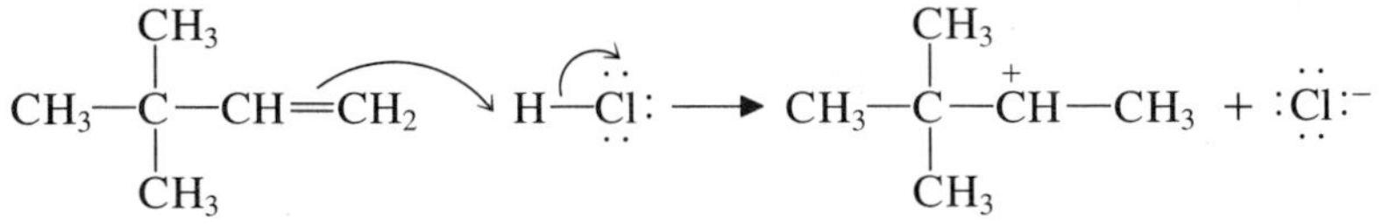

$$\underset{\substack{\text{CH}_3}}{\overset{\substack{\text{CH}_3 \\ |}}{\text{CH}_3-\overset{|}{\underset{|}{\text{C}}}-\text{CH}=\text{CH}_2}} \quad \text{H}-\ddot{\text{Cl}}: \longrightarrow \underset{\substack{\text{CH}_3}}{\overset{\substack{\text{CH}_3 \\ |}}{\text{CH}_3-\overset{|}{\underset{|}{\text{C}}}-\overset{+}{\text{CH}}-\text{CH}_3}} + :\ddot{\text{Cl}}:^- \quad (4.23)$$

Reaction of this carbocation with Cl⁻ completes a normal addition to yield the minor product of Eq. 4.22. However, the carbocation can also undergo a second type of reaction: it can *rearrange*.

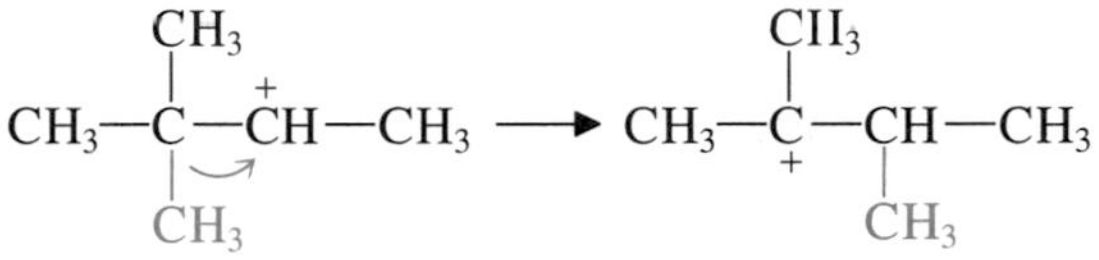

$$\underset{\substack{\text{CH}_3}}{\overset{\substack{\text{CH}_3 \\ |}}{\text{CH}_3-\overset{|}{\underset{|}{\text{C}}}-\overset{+}{\text{CH}}-\text{CH}_3}} \longrightarrow \underset{\substack{\text{CH}_3}}{\overset{\substack{\text{CII}_3 \\ |}}{\text{CH}_3-\overset{|}{\underset{+}{\text{C}}}-\text{CH}-\text{CH}_3}} \quad (4.24a)$$

In this reaction, the methyl group moves *with its pair of bonding electrons* from the carbon adjacent to the electron-deficient carbon. The carbon from which this group departs, as a result, becomes electron-deficient and positively charged. That is, the rearrangement converts one carbocation into another. The major product of Eq. 4.22 is formed by the attack of Cl⁻ on the new carbocation.

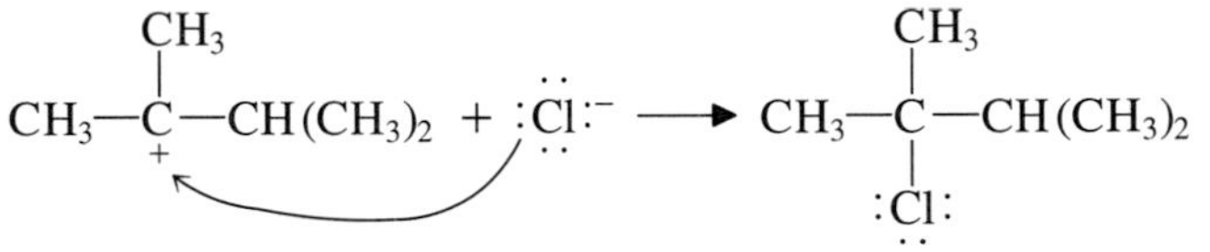

$$\underset{\substack{+}}{\overset{\substack{\text{CH}_3 \\ |}}{\text{CH}_3-\overset{|}{\text{C}}-\text{CH}(\text{CH}_3)_2}} + :\ddot{\text{Cl}}:^- \longrightarrow \underset{\substack{:\ddot{\text{Cl}}:}}{\overset{\substack{\text{CH}_3 \\ |}}{\text{CH}_3-\overset{|}{\underset{|}{\text{C}}}-\text{CH}(\text{CH}_3)_2}} \quad (4.24b)$$

Why does rearrangement of the carbocation occur? In the case of reaction 4.24a, a more stable tertiary carbocation is formed from a less stable secondary one. Therefore, *rearrangement is favored by the increased stability of the rearranged ion.*

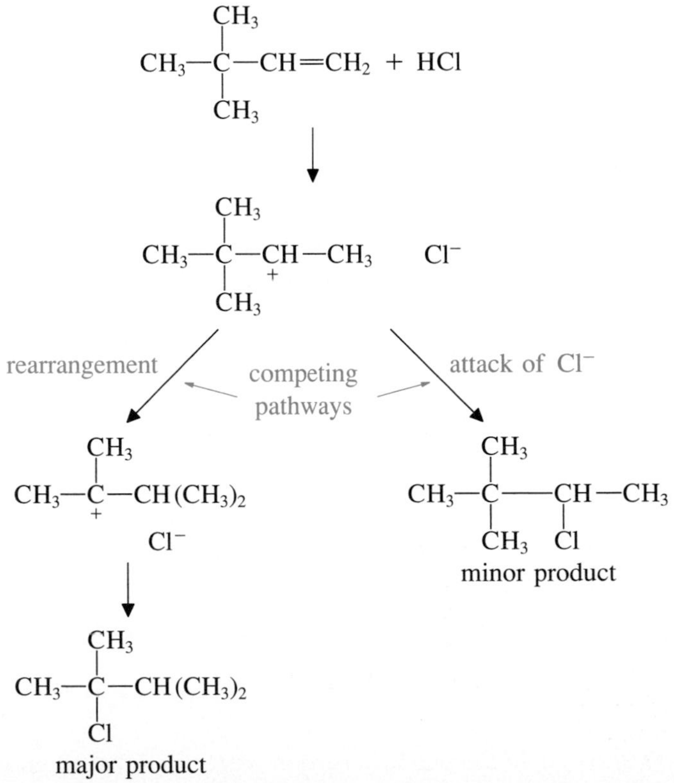

We have now learned two pathways by which carbocations can react. They can (1) react with a nucleophile and (2) rearrange to more stable carbocations. The outcome of Eq. 4.25 represents a competition between these two pathways. In any particular case, it is difficult to predict exactly how much of each different product will be obtained. Nevertheless, the reactions of carbocation intermediates show why both products are reasonable. Rearrangements do not occur in the reactions of some alkenes (for example, Eqs. 4.15 and 4.16) because rearrangement would not give a more stable carbocation.

The first clear formulation of the involvement of carbocations in molecular rearrangements was proposed by Frank C. Whitmore (1887–1947) of Pennsylvania State University. Whitmore said that carbocation rearrangements result when "an atom in an electron-hungry condition seeks its missing electron pair from the next atom in the molecule." We can see from Whitmore's description that a carbocation rearrangement is simply another Lewis acid–base reaction: The Lewis acid is the electron-deficient carbon of the carbocation, and the Lewis base is the electron pair in the bond to the group that rearranges.

Carbocation rearrangements are not limited to the migration of alkyl groups. In the following reaction, the major product is also derived from the rearrangement of a carbocation intermediate. This rearrangement involves the migration of a hydrogen with its two electrons, called a **hydride shift.**

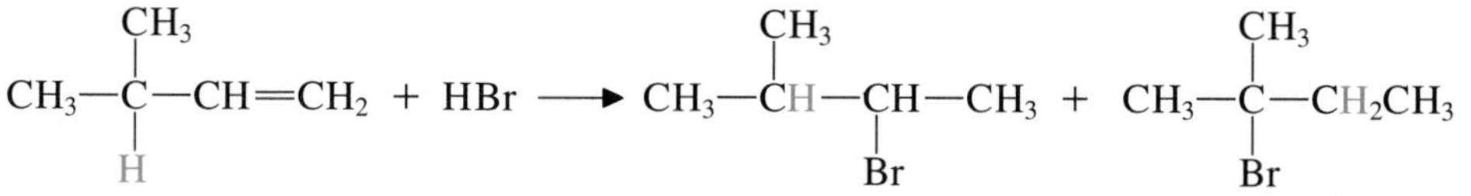

$$CH_3\!-\!\underset{\underset{H}{|}}{\overset{\overset{CH_3}{|}}{C}}\!-\!CH\!=\!CH_2 + HBr \longrightarrow CH_3\!-\!\overset{\overset{CH_3}{|}}{CH}\!-\!\underset{\underset{Br}{|}}{CH}\!-\!CH_3 + CH_3\!-\!\underset{\underset{Br}{|}}{\overset{\overset{CH_3}{|}}{C}}\!-\!CH_2CH_3 \quad (4.26)$$

(39–49% of product) (51–61% of product)

Problems

15 Give a mechanism for reaction in Eq. 4.26 that accounts for the formation of both products. Use the arrow formalism.

16 Which of the following alkyl halides could be prepared relatively free of isomers by the Markovnikov addition of HBr to an alkene? Which alkene would be used in each case?

(a) $CH_3\!-\!\underset{\underset{Br}{|}}{CH}\!-\!CH_2CH_2CH_3$

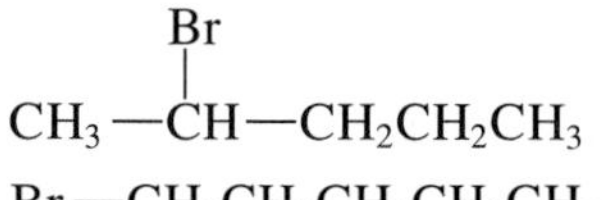

(b) $Br\!-\!CH_2CH_2CH_2CH_2CH_3$

(c) $CH_3\!-\!\underset{\underset{Br}{|}}{CH}\!-\!\underset{\underset{CH_3}{|}}{\overset{\overset{CH_3}{|}}{C}}\!-\!C_2H_5$

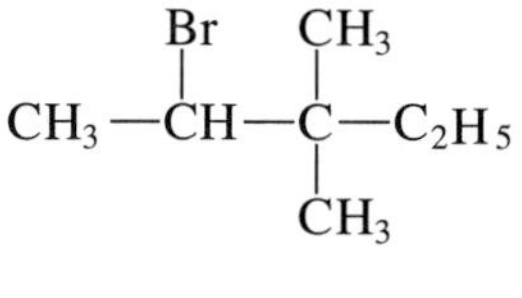

4.7 REACTION RATES. TRANSITION-STATE THEORY AND HAMMOND'S POSTULATE

In the previous section we learned that addition of hydrogen halides to alkenes such as isobutylene gives only one of two possible products.

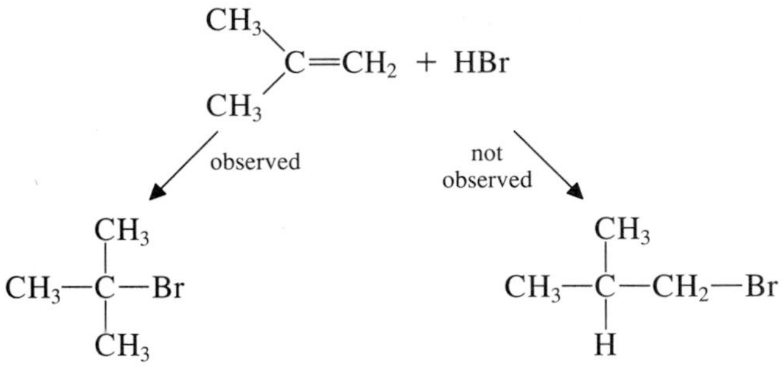

(4.27)

Let us reexamine this reaction and ask why this should be so. It probably seems reasonable that the observed reaction should be the one that involves the more stable carbocation intermediate. However, there is more to it than that. In order to understand what is happening here, we need to recognize an important point: The Markovnikov addition of HBr is observed because it is much *faster* than the opposite mode of addition. Why should Markovnikov addition be faster? What does the stability of the carbocation intermediate have to do with the reaction rate? To answer these questions, we need to understand some of the factors that affect the *rates* of chemical reactions.

A. Transition-State Theory

The **rate** of a chemical reaction can be defined for our purposes as the number of reactant molecules converted to product in a given time. Before we learn what factors *do* determine reaction rates, let us first establish what *does not:* the equilibrium constant for a reaction tells us *absolutely nothing* about the rate of the reaction. Some reactions with very large equilibrium constants are in fact very slow. For example, the equilibrium constant for the combustion of gasoline (alkanes) is very large; yet gasoline can be handled in the open air because its reaction with oxygen, in the absence of heat, is immeasurably slow. On the other hand, some very unfavorable reactions come to equilibrium very rapidly. In this category are a number of Brønsted acid–base reactions.

The approach used by many organic chemists to predict reaction rates assumes that as the reactants change into products, they pass through an unstable state of maximum free energy, called the **transition state.** The transition state has a higher energy than either the reactants or products and therefore represents an **energy barrier** to their interconversion. This energy barrier is shown graphically in a **reaction free-energy diagram** (Fig. 4.12). This is a diagram of the standard free energy of a reacting system as old bonds break and new ones form along the reaction pathway. In this diagram the pathway of the reaction from reactants to products is called the *reaction coordinate.* The energy barrier, $\Delta G^{0\ddagger}$, is called the **standard free energy of activation,** and is equal to the difference between the standard free energies of the transition state and reactants. (The double dagger, ‡, is the symbol associated with transition states.) The size of the energy barrier $\Delta G^{0\ddagger}$ determines the rate of a reaction: *the higher the barrier, the lower the rate.* Thus, the reaction shown in Fig. 4.12a is slower than the one in Fig. 4.12b because it has a larger energy barrier. In the same sense that relative free energies of reactants and products determine the equilibrium constant, the relative free energies of transition state and reactants determine the reaction rate.

Where do reactant molecules get the energy to cross the energy barrier and form products? The basic ideas needed to answer this question were developed in Sec. 3.5,

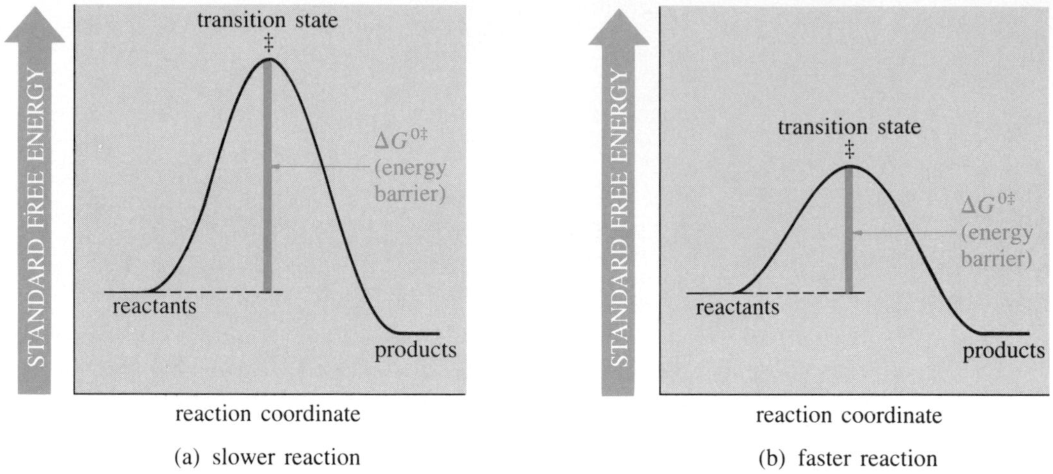

Figure 4.12 Reaction free-energy diagrams for two hypothetical reactions. The reaction in (a) is slower than the one in (b) because it has a larger $\Delta G^{0\ddagger}$.

where we learned how a butane molecule gets the energy to pass between *gauche* and *anti* forms through the eclipsed forms. (The eclipsed forms can be thought of as the transition states for the internal rotation "reaction.") In general, molecules obtain the energy from their thermal motions. The higher the energy barrier, the less probable it is that a molecule will have sufficient energy to achieve the transition state and form products. On the other hand, if the energy barrier for a reaction is low, it is much more probable that a molecule will have sufficient energy to form the transition state and hence, the products. A greater *probability* of reaction means a faster *rate* of reaction. It follows from this discussion that higher temperature (greater thermal energy) increases the number of molecules with sufficient energy to form the transition state, and therefore also increases the rate of reaction.

Let us summarize. Two factors that govern the reaction rate are:

1. The size of the energy barrier, or standard free energy of activation: reactions with larger $\Delta G^{0\ddagger}$ are slower.

2. The temperature: reactions are faster at higher temperature

An analogy that can help us to understand these concepts is shown in Fig. 4.13. Water in the upper cup would flow into the pan below if it could somehow gain enough energy to surmount the wall of the cup. The wall of the cup is a potential-energy barrier to the downhill flow of water. Likewise, molecules have to achieve a transient state of high energy—the *transition state*—in order to break stable chemical bonds and undergo reaction. The effect of thermal motion can be simulated by shaking the cup. If the cup is shallow (low barrier), there is a good probability that the shaking will cause water molecules to slosh over the sides of the cup and drop to the pan. If the cup is very deep (high barrier), it is less probable that water will flow from cup to pan in a given length of time. The effect of increasing temperature can be simulated by shaking the cup more violently. The more violent motion increases the probability,

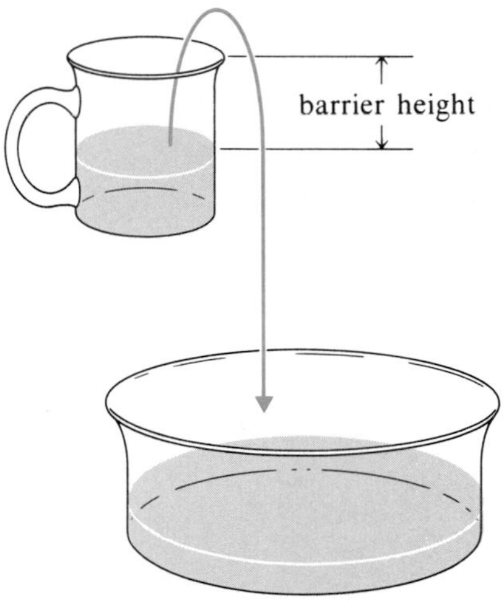

Figure 4.13 A difference of potential energy is not enough to cause the water in the cup to drop to the bowl below. The water must first climb the barrier imposed by the walls of the cup.

and hence the *rate,* that water will surmount the wall of the cup and fall to the pan below. Likewise, higher temperature increases the probabilities, and hence the rates, of chemical reactions.

B. Multistep Reactions

Many chemical reactions take place with the formation of intermediates. For example, we have already studied reactions that involve carbocation intermediates. If intermediates exist in a chemical reaction, then what we commonly express as one reaction—for example, addition of HBr to isobutylene

$$(CH_3)_2C{=}CH_2 + HBr \longrightarrow (CH_3)_3C{-}Br \qquad (4.28)$$

is really a sequence of reactions:

$$(CH_3)_2C{=}CH_2 + HBr \longrightarrow (CH_3)_3C^+ + Br^- \qquad (4.29a)$$

$$(CH_3)_3C^+ + Br^- \longrightarrow (CH_3)_3C{-}Br \qquad (4.29b)$$

Each step of a multistep reaction has its own characteristic rate, and therefore its own transition state. The energy changes in such a reaction can also be depicted in a reaction free-energy diagram. Such a diagram for the addition of HBr to isobutylene is shown in Fig. 4.14. Each free-energy minimum between reactants and products represents an intermediate, and each maximum a transition state.

The rate of the overall reaction depends in detail on the rates of its various steps. However, it quite often happens that one step of a multistep reaction is considerably slower than any of the others. This slowest step in a multistep chemical reaction is called the **rate-determining step,** or **rate-limiting step,** of the reaction. In such a case *the rate of the overall reaction is equal to the rate of the rate-determining step.* In terms of the reaction free-energy diagram in Fig. 4.13, *the rate-determining step is the step with the transition state of highest free energy.* This diagram indicates that in the addition of HBr to isobutylene, the rate-determining step is the first step of the reaction—the protonation of the alkene to give the carbocation. The overall rate of addi-

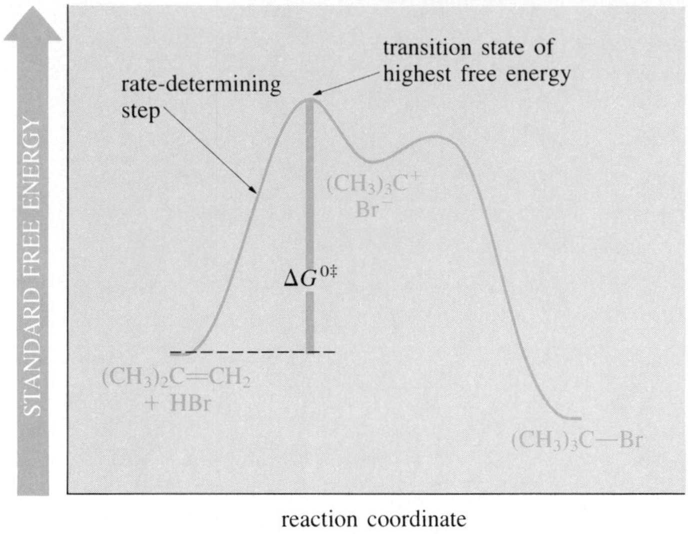

Figure 4.14 Reaction free-energy diagram for a multistep reaction. The rate-determining step of a multistep reaction is the step that has the transition state of highest standard free energy. In the addition of **HBr** to isobutylene, the rate-determining step is protonation of the double bond to give the carbocation intermediate.

tion of HBr to isobutylene is equal simply to the rate of this first step.

The rate-determining step of a reaction has a special importance. Anything we do to increase the rate of this step increases the overall reaction rate. Conversely, if we find that a change in the reaction conditions (for example, a change in temperature) affects the rate of the reaction, it is the effect on the rate-determining step that is being observed.

The idea of a rate-determining step is analogous to a toll station on a modern freeway at rush hour. We can divide the rate of passage of cars through a toll plaza into three steps: (1) entry of the cars into the toll area; (2) taking of the toll by the collector; and (3) exit of cars from the toll area. Let us suppose that our toll station has a very slow, lackadaisical toll collector. He takes the toll so slowly that the rate of passage of cars through the plaza is determined strictly by how fast he works. Toll-taking—the second step—becomes the rate-determining process for passage of cars through the toll booth. Cars can line up more or less frequently, but as long as there is a line of cars, the rate of passage through the toll plaza is the same. If the collector takes one toll per minute, then cars exit at one per minute.

Imagine now a different situation: a super-fast toll collector. He is so fast, in fact, that he can keep pace with any number of cars likely to pass through the toll booth in a given time. The rate at which cars exit from the toll plaza is determined strictly by how fast they arrive. In this case, step (1), the entry of cars into the toll plaza, can be considered the rate-determining step.

Let us imagine that an efficiency expert has been retained to increase the rate of passage of cars through the toll plaza with the slow toll collector. His first job must be to locate the bottleneck. Only if he affects the rate-determining step by replacing the slow toll collector will he improve the rate of passage of cars through the toll plaza. Likewise, if we want to increase the rate of reaction, we must do something to increase the rate of its slowest step.

Because the rate-determining step of a reaction has special importance, its identification receives particular emphasis when chemists seek to understand the mechanism of a reaction.

Problem

17 Draw a reaction free-energy diagram for a reaction $A \rightleftarrows B \rightleftarrows C$ that meets the following criteria: The standard free energies are in the order $C < A < B$; and the rate-determining step of the reaction is $B \rightleftarrows C$.

C. Use of Transition-State Theory. Hammond's Postulate

Transition states possess a maximum of free energy with respect to reactants and products, and therefore are not stable and cannot be isolated. How, then, do organic chemists use transition-state theory to predict reaction rates? In order to answer this question, we first need to understand what factors are responsible for the occurrence of energy barriers to chemical reactions.

Again consider as a concrete example the addition of HBr to isobutylene. The rate-determining step of this reaction is the first step: protonation of the alkene to form the carbocation. The rate of the overall reaction therefore corresponds to the rate of this step. The free energy of the transition state for this reaction governs the rate at which this protonation occurs.

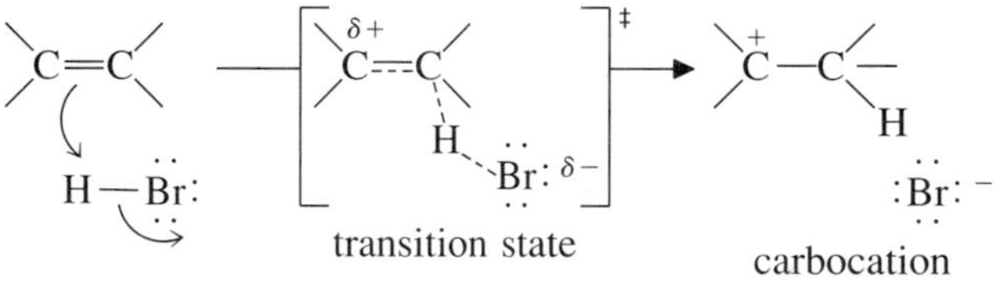

(4.30)

The advantage of transition-state theory is that the transition state can be visualized as a *structure*. Usually, we think of the transition state as a structure on the way from reactants to products. We know some of the factors that contribute to the stability or instability of ordinary structures, and we can apply this same sort of reasoning to structures of transition states. For example, in the protonation of an alkene, we see that several bonds are partially broken. Bond breaking requires energy that manifests itself as part of the energy barrier for the reaction.

Another factor contributing to the high energy of a transition state can be its resemblance to unstable intermediates. For example, the transition state for alkene protonation, except for the partial bonds, looks a lot like the carbocation. It has many of the same features that make the carbocation a rather high-energy species, such as the separation of positive and negative charge and the formation of an electron-deficient carbon atom. Therefore, the factors that contribute to the relatively high energy of the carbocation also contribute to the instability of the transition state. In fact, organic chemists are in the habit of *assuming* that the transition states for the formation of reactive intermediates of relatively high energy (such as carbocations) from stable starting materials closely resemble the intermediates themselves. This assumption is called **Hammond's postulate,** and was first applied to organic reactions in 1955 by George F. Hammond, then a Professor at Iowa State University. According to Hammond's postulate, the structure of the transition state for alkene protonation ought to approximate the structure of the carbocation intermediate formed in the reaction. If the structures are similar, it follows that their free energies are also similar; that is, the free energy of the transition state does not differ much from the free energy of the carbocation intermediate. Although there is no reason that Hammond's postulate *must* be true, it is often adopted as a working hypothesis in attempting to predict reaction rates.

Let us see how Hammond's postulate predicts the Markovnikov addition of HBr to isobutylene. As we indicated in the discussion of Eq. 4.27, *Markovnikov addition is observed because it is the faster of the two competing reactions.* Since the rate-determining step in this reaction is the protonation of isobutylene to give a carbocation, the rate of addition of HBr can be understood by looking only at the rate of this step. This protonation can occur in two ways. One way gives a tertiary carbocation, and the other way gives a primary carbocation. Applying Hammond's postulate, we assume that the transition states for the different modes of addition resemble the carbocations themselves.

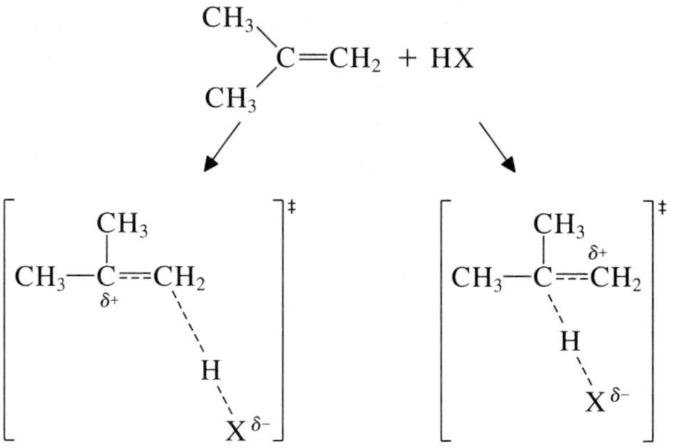

(4.31)

transition state resembles *t*-butyl cation (more stable)

transition state resembles isobutyl cation (less stable)

Since the more substituted carbocation is more stable, *the transition state leading to this ion should also be more stable.* As a result, protonation of isobutylene to give the tertiary carbocation has the more stable transition state, the lower $\Delta G^{0\ddagger}$, and thus should be the faster of the two reactions (Fig. 4.15). The more substituted carbocation, formed more rapidly, is then attacked by Br^- to complete the reaction. *Markovnikov addition is observed because the protonation of the alkene to give the tertiary carbocation is faster than the protonation of the alkene to give the primary carbocation.* Note that it is not the stabilities of the carbocations themselves that determine which reaction path is faster; it is the relative stabilities of the *transition states for carbocation formation* that determine the relative rates of the two processes. It is only because of the validity of Hammond's postulate that the relative energies of the two possible transition states parallel the relative energies of their respective carbocation intermediates.

Problem

18 Apply Hammond's postulate to decide which reaction is faster: addition of HBr to isobutylene or addition of HBr to *trans*-2-butene. Assume that the energy difference between the starting alkenes can be ignored. Why is this assumption necessary?

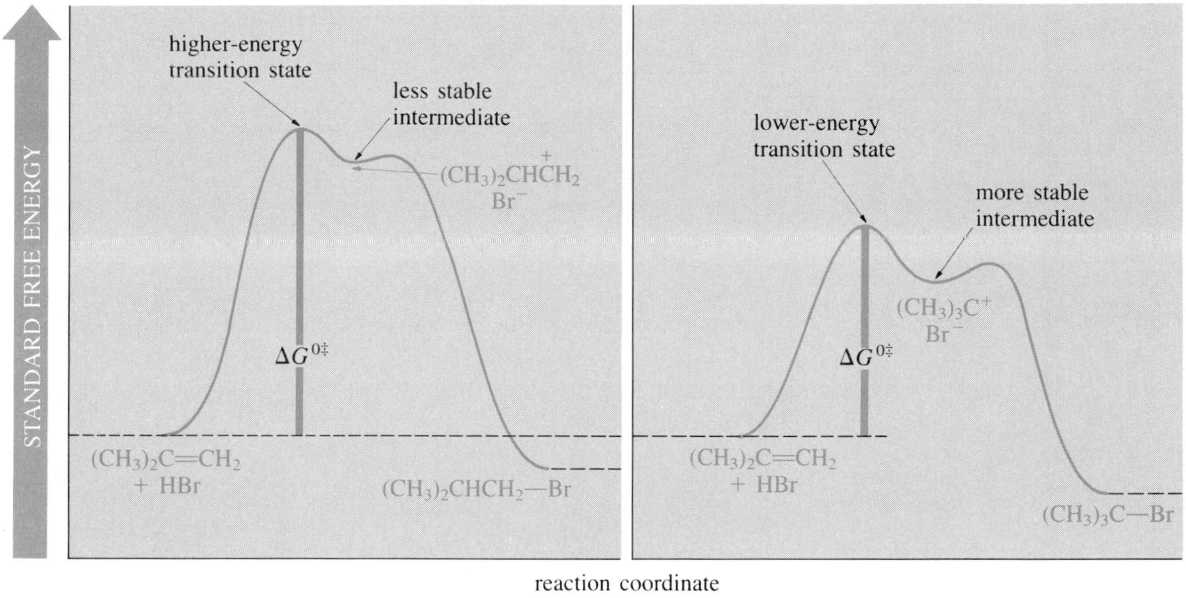

Figure 4.15 A reaction free-energy diagram for the two possible modes of addition of **HBr** to isobutylene. The reaction that gives the more stable intermediate is faster, and is therefore the one observed.

4.8 HYDRATION OF ALKENES. CATALYSIS

In the presence of moderately concentrated mineral acids (such as $HClO_4$, H_2SO_4, HNO_3), water undergoes an addition to the alkene double bond. The addition of the elements of water is called **hydration.**

$$\underset{CH_3}{\overset{CH_3}{>}}C{=}CH_2 \ + \ H{-}OH \ \xrightarrow{H_3O^+} \ CH_3{-}\underset{OH}{\overset{CH_3}{\underset{|}{\overset{|}{C}}}}{-}CH_3 \tag{4.32}$$

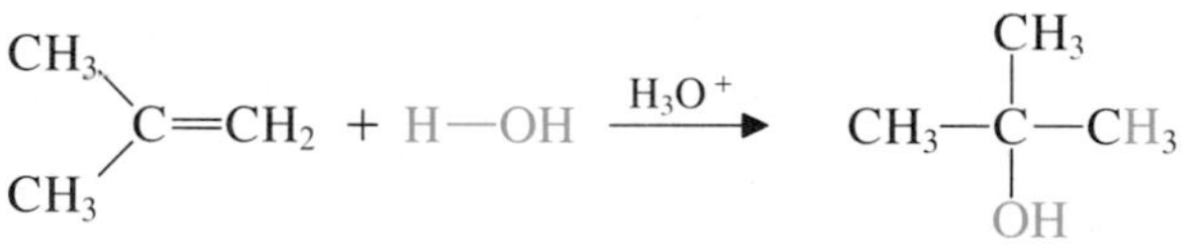

2-methyl-2-propanol
(t-butyl alcohol)

This addition also follows the Markovnikov rule: the hydrogen of the water molecule adds to the less substituted alkene carbon atom, and the —OH group adds to the more substituted carbon.

The reaction follows a pattern with which we are now familiar. The alkene is first protonated to give a carbocation.

$$\underset{CH_3}{\overset{CH_3}{>}}C{=}CH_2 \ \longrightarrow \ \underset{CH_3}{\overset{CH_3}{>}}\overset{+}{C}{-}CH_3 \ + \ H_2\overset{..}{\underset{..}{O}} \tag{4.33a}$$
$$H{-}\overset{+}{\underset{..}{O}H_2}$$

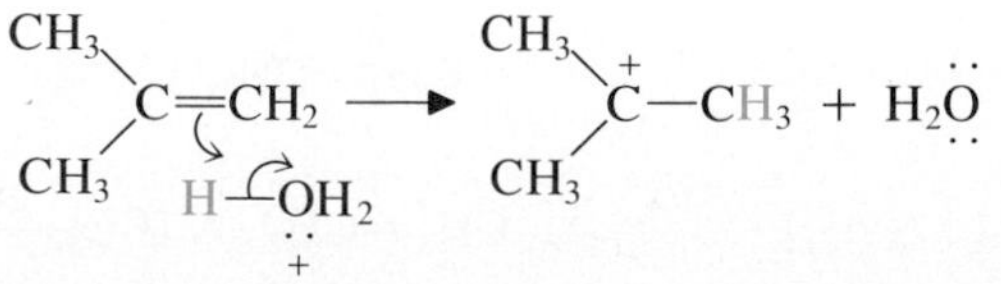

Attack of the nucleophile (Lewis base) water on the carbocation and loss of a proton from the resulting adduct completes the reaction.

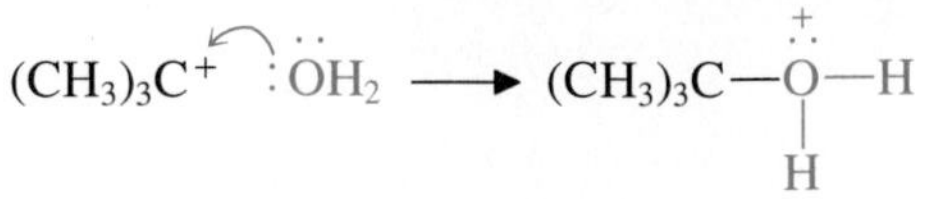

$$(CH_3)_3C^+ \quad :\ddot{O}H_2 \longrightarrow (CH_3)_3C\!-\!\overset{+}{\underset{\underset{H}{|}}{\ddot{O}}}\!-\!H \tag{4.33b}$$

$$(CH_3)_3C\!-\!\overset{+\cdot}{\underset{\underset{H}{|}}{\ddot{O}}}H \; \rightleftharpoons \; (CH_3)_3C\!-\!\ddot{O}H \; + \; H_3\overset{+}{O} \tag{4.33c}$$
$$H_2O:$$

Because carbocation intermediates are involved, the hydration of some alkenes is accompanied by rearrangement.

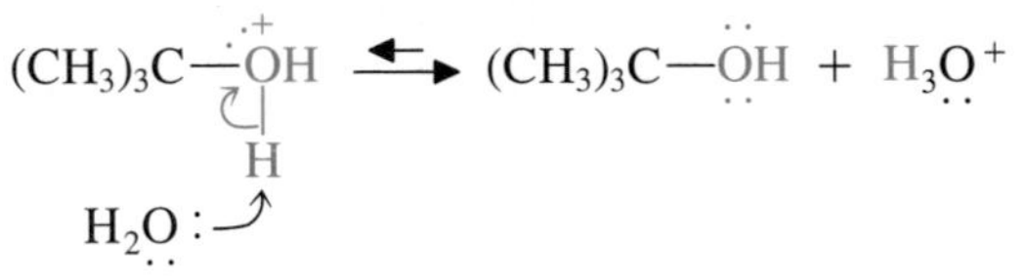

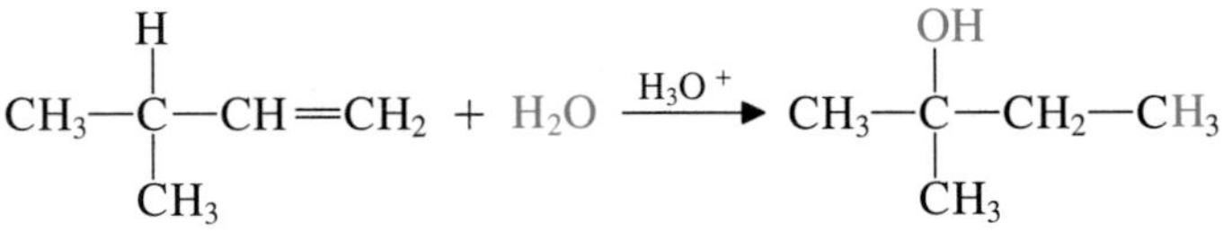

$$CH_3\!-\!\overset{\overset{\displaystyle H}{|}}{\underset{\underset{\displaystyle CH_3}{|}}{C}}\!-\!CH\!=\!CH_2 \; + \; H_2O \; \xrightarrow{H_3O^+} \; CH_3\!-\!\overset{\overset{\displaystyle OH}{|}}{\underset{\underset{\displaystyle CH_3}{|}}{C}}\!-\!CH_2\!-\!CH_3 \tag{4.34}$$

Problems

19 Give the mechanism for the reaction in Eq. 4.34. Show each step of the reaction separately, and use the arrow formalism.

20 Give the hydration product(s) expected from each of the following compounds:

(a) ⬦=CH₂ (methylenecyclobutane)

(b) $PhCH\!=\!CH_2$ (styrene)

(c) 3,3-dimethyl-1-butene

Alkene hydration is *not* a useful laboratory method for the preparation of most alcohols because of the rearrangements and other side reactions that can occur. Nevertheless it is important in our study of organic chemistry for two reasons. First, it is a widely used industrial method for the preparation of two organic compounds of economic importance, ethanol (ethyl alcohol, C_2H_5OH), and isopropyl alcohol $[(CH_3)_2CH\!-\!OH]$. Ethanol was at one time produced mostly by **fermentation,** which is the microbiological degradation of sugars to ethanol and CO_2. Today, however, the hydration of ethylene accounts for most of the ethanol produced industrially. (Alcoholic beverages account for most of the alcohol produced by fermentation.) For example, in 1986 almost 315,000,000 gallons of ethanol were produced in the United States, most of it by the hydration of ethylene.

$$CH_2\!=\!CH_2 \; + \; H_2O \; \xrightarrow{H_2SO_4,\, 100°} \; CH_3CH_2OH \tag{4.35}$$

$$CH_2\!=\!CH_2 \; + \; H_2O \; \xrightarrow[H_3PO_4]{300°,\text{ vapor phase}} \; CH_3CH_2OH \tag{4.36}$$

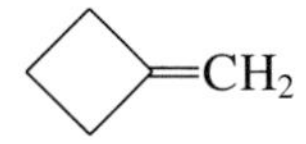

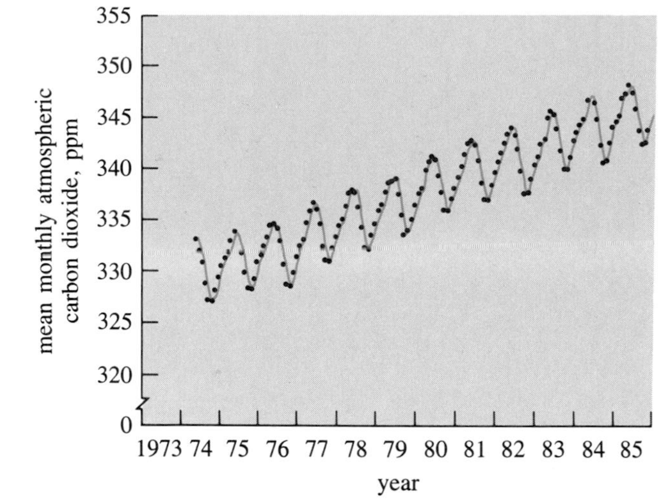

Figure 4.16 Carbon dioxide concentration in the atmosphere as monitored by a National Oceanic and Atmospheric Administration station at Mauna Loa, Hawaii. Although there are seasonal fluctuations (small peaks), a general trend toward higher CO_2 concentrations is also apparent.

There has been some interest in a return to the industrial production of ethanol by fermentation. This interest was especially strong during times of petroleum shortages in the United States, and it has remained important in the developing nations where plant sources of sugar are readily available and indigenous sources of petroleum are in short supply. Brazil, for example, currently produces more than 2.5 billion gallons of ethanol annually by fermentation of sugars from plants. The idea is that ethanol, rather than ethylene, can then serve as a raw material for the chemical industry as well as for fuel.

A very real problem in the combustion of hydrocarbons for fuel is the formation of CO_2 and the increase in the CO_2 content in the atmosphere, which has risen in the last 100 years from 290 to 350 parts per million, with 20% of this increase occurring in the last ten years (Fig. 4.16). Although the exact consequences of this increase cannot be predicted with certainty, some scientists have argued that it might cause a gradual warming of the earth followed by return to another ice age. At the very least, we appear to be conducting a gigantic chemical experiment of uncertain outcome on our environment. From this point of view, the production of fuels from fermentation-derived ethanol looks particularly attractive. Ethanol is fermented from sugars such as glucose. When plants synthesize glucose, they also fix CO_2; that is, the carbons of glucose ultimately come from atmospheric CO_2. The energy for the plant synthesis of glucose is derived from the sun (photosynthesis). If we sum up the reactions for the plant synthesis of glucose, its conversion into ethanol, and the combustion of ethanol (or anything derived from it) as fuel, we see that no net change is effected in the environment.

$$6CO_2 + 6H_2O \xrightarrow{\text{light}} C_6H_{12}O_6 + 6O_2 \quad \text{(biosynthesis of glucose, } C_6H_{12}O_6\text{)}$$

$$C_6H_{12}O_6 \longrightarrow 2C_2H_5OH + 2CO_2 \quad \text{(anaerobic fermentation)} \qquad (4.37)$$

$$2C_2H_5OH + 6O_2 \longrightarrow 4CO_2 + 6H_2O \quad \text{(combustion of ethanol)}$$

Sum: No Net Change

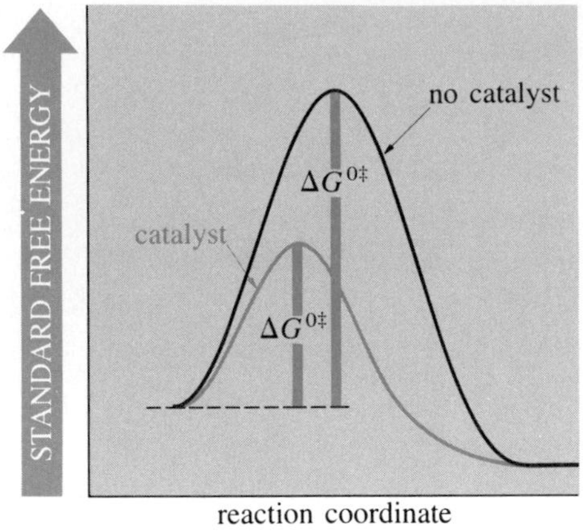

Figure 4.17 A catalyst lowers the standard free energy of activation of a chemical reaction, and therefore increases the reaction rate.

The second reason for our study of the hydration reaction is that it illustrates the phenomenon of *catalysis.* The hydration reaction proceeds at a useful rate only in the presence of acid. Furthermore, the acid is not consumed in the reaction; only the elements of water are added to the double bond. The proton added to the alkene in the first step of the reaction (Eq. 4.33a) is regenerated in the last step (Eq. 4.33c). Anything that increases the rate of a reaction without itself being consumed is called a **catalyst.** Hydration of alkenes is therefore an *acid-catalyzed reaction.*

Let us look at catalysis with the aid of a reaction free-energy diagram, as shown in Fig. 4.17. Since catalysts increase the reaction rate, they also lower the standard free energy of activation for the reaction. (Figure 4.17 is a generalized drawing and does not show any intermediates that might be formed in the reactions involved.)

How do catalysts increase reaction rates? There is no general answer, since there are many different mechanisms of catalysis. However, in the acid-catalyzed hydration of alkenes, the rate-determining step is the protonation of the alkene to form a carbocation. The rate of the hydration reaction is therefore increased when the rate of alkene protonation is increased. A strong mineral acid is more effective in protonating an alkene than is water alone; that is, a stronger acid is more effective in protonating a given base (the alkene) than a weaker acid.

Catalysis is not limited to the laboratory or chemical industry. The biological processes of nature involve thousands of chemical reactions, and the rates at which these reactions occur within any organism must be carefully coordinated. Almost every biological reaction of importance has its own unique naturally-occurring catalyst. These biological catalysts are called **enzymes.** Under physiological conditions, most important biological reactions would be too slow to be useful in the absence of their enzyme catalysts.

An example of an important enzyme-catalyzed addition to an alkene is the hydration of fumarate ion to malate ion.

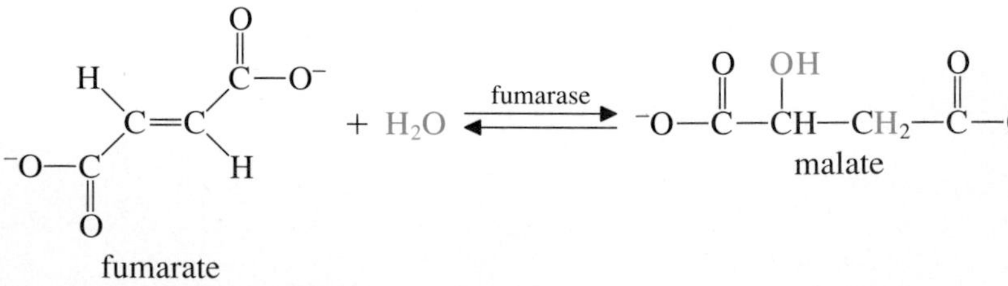

(4.38)

This reaction is catalyzed by the enzyme *fumarase*. It is one reaction in the Krebs cycle, or citric acid cycle, a series of reactions that plays a central role in the generation of energy in biological systems. The effectiveness of fumarase catalysis can be appreciated by the following comparison: At physiological pH and temperature (pH = 7, 37°), the enzyme-catalyzed reaction is 10^4 times faster than the same reaction in the absence of enzyme at 175°! Compared at a common temperature, the enzyme-catalyzed reaction is many orders of magnitude faster. (At 37° the reaction rate in the absence of enzyme is too slow to measure.) We shall learn more about enzymes in Chapters 10 and 27.

KEY IDEAS IN CHAPTER 4

- Alkenes are compounds containing carbon–carbon double bonds. Alkene carbon atoms, as well as other trigonal carbon atoms, are sp^2 hybridized.
- The carbon–carbon double bond consists of a σ-bond and a π-bond. In electrophilic addition reactions of alkenes the π-electrons react with Brønsted acids or Lewis acids (electrophiles).
- Because rotation about the alkene double bond does not occur under normal conditions, alkenes exist as *cis–trans* isomers. These can be named using the *E,Z* priority system.
- Relative free energies of molecules determine the position of chemical equilibrium; the most stable molecule is present in greatest amount in an equilibrium mixture. Relative enthalpies, determined from heats of formation by applying Hess's law, can be used to examine the relative stabilities of different arrangements of chemical bonds.
- Alkene isomers with highly substituted double bonds are more stable than those with hydrogens at the alkene carbons; *trans*-alkenes are more stable than *cis*-alkenes.
- Reactants are converted into products through unstable species called transition states. Reactions with smaller standard free energy differences between transition states and reactants are faster.
- The rates of many multistep reactions are governed by the rate of the slowest step, called the rate-determining step; this step has the transition state of highest standard free energy.
- The electrophilic addition reactions of hydrogen halides and water to alkenes involve carbocation intermediates and follow the Markovnikov rule. This mode of addition is a consequence of several facts: (a) the rate-determining transition state of each reaction resembles a carbocation; (b) the relative stability of carbocations is in the order tertiary > secondary > primary; and (c) Hammond's postulate applies to the rates of the two competing modes of addition. Hammond's postu-

late says that the structure and energy of the transition state resemble the structure and energy of the carbocation intermediate.

- Reactions involving carbocation intermediates show rearrangements in some cases.
- A catalyst increases the rate of a reaction without being consumed in the reaction. Acid catalyzes the hydration of alkenes; enzymes catalyze biological reactions.

ADDITIONAL PROBLEMS

21 Give the structures and systematic names of all the isomeric hexenes (compounds with molecular formula C_6H_{12}).

22 Which of the hexenes in Problem 21 gives predominately a single structural isomer when treated with HBr? Which gives a mixture of isomers? Explain.

23 Which of the hexenes in Problem 21 has the smallest heat of formation? Which has the largest heat of formation? Explain.

24 Give the structural formula for each of the following compounds:
(a) isobutylene
(b) styrene
(c) cyclopropene
(d) 3-methyl-1-octene
(e) (*Z*)-3-methyl-2-octene
(f) 5,5-dimethyl-1,3-cycloheptadiene
(g) 3-allyl-1-cyclopentene

25 Name the following compounds:

(a) $CH_2{=}CHCH_2CH_2CH_3$

(b) 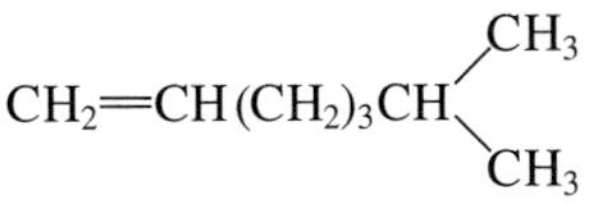

(c) 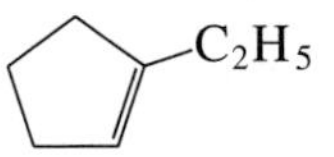

(d) 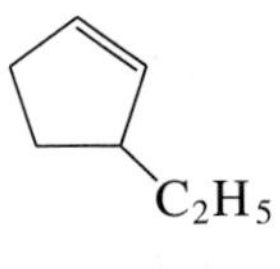

(e) 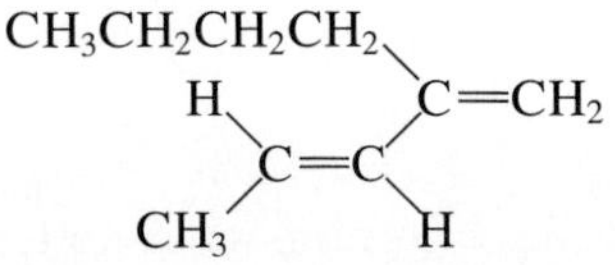

(f) 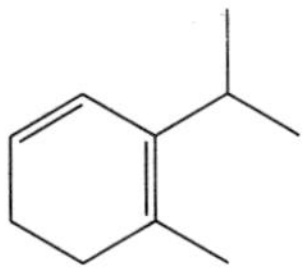

(g) 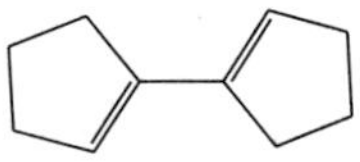

(h) 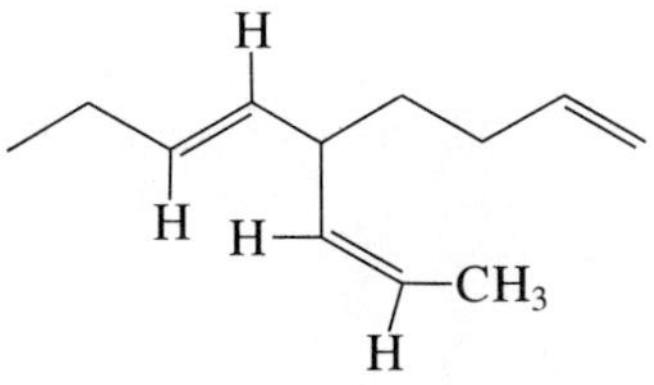

26 A confused chemist used the following names in a paper dealing with the properties of alkenes. Although each name specifies a structure, in some cases the name is not correct. Correct the names that are incorrect.

(a) 3-butene
(b) 2-hexene
(c) *trans*-1-*tert*-butyl-1-propene
(d) 6-methylcycloheptene
(e) 2-methyl-1,3-butadiene
(f) (*Z*)-3-methyl-3-pentene

27 Give the conformation (*E* or *Z*) of each of the following alkenes. (D = deuterium, or ^{2}H, is the isotope of hydrogen with atomic mass 2.)

(a)
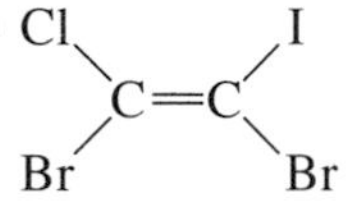

(b)
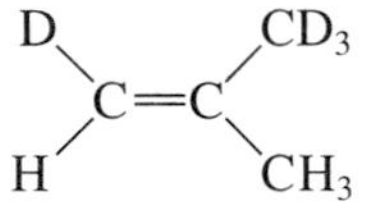

(c)
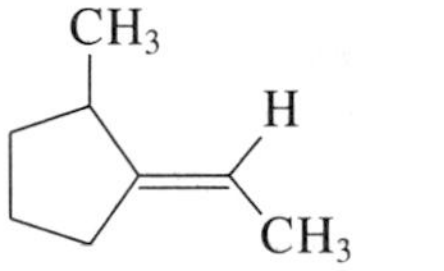

(d)
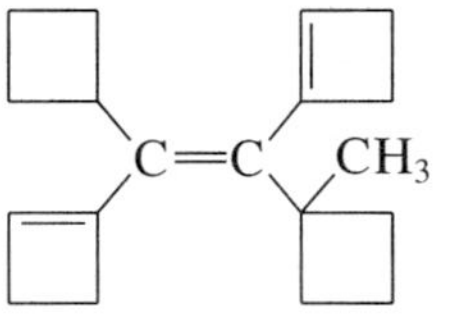

28 The preparation of isopropyl alcohol [$(CH_3)_2CH$—OH] by hydration of an alkene is an important industrial reaction.

(a) What alkene is used in this reaction?
(b) Could the isomeric alcohol *n*-propyl alcohol ($CH_3CH_2CH_2OH$) also be made by alkene hydration? Explain.

29 (a) The following compound can be prepared by the addition of HBr to either of two alkenes; give their structures.

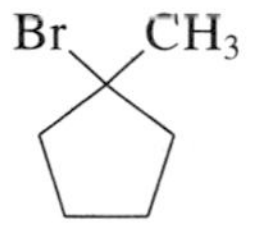

(b) Starting with the same two alkenes, would the products be different if DBr were used? Explain. (See note about deuterium in Problem 27.)

30 Supply the correct curved-arrow formalism for the following acid-catalyzed isomerization.

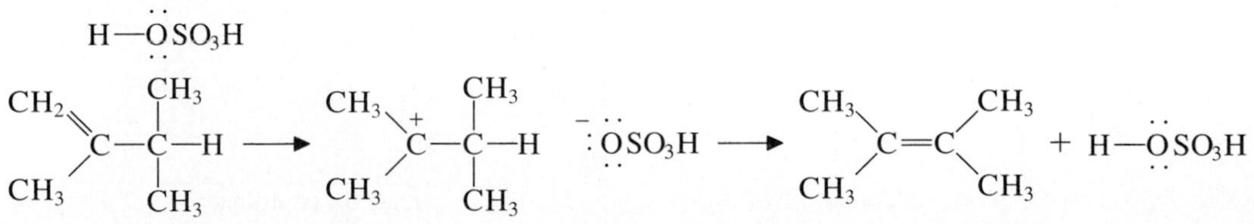

Problems (Cont.)

31 Predict the geometry of BF_3. What hybridization of boron is suggested by this geometry? Draw an orbital diagram for boron similar to that for the carbons in ethylene shown in Fig. 4.2.

32 Small- and medium-ring cycloalkenes are an exception to the generalization that *trans*-alkenes are more stable than *cis*-alkenes. Try to build a model of *trans*-cyclohexene. Explain why this *trans*-cycloalkene is much less stable than its *cis* isomer. (In fact, *trans*-cyclohexene cannot be prepared under normal conditions as a stable compound.)

33 A certain compound *A* is converted into a compound *B* in a reaction without intermediates. The reaction has an equilibrium constant $K_{eq} = [B]/[A] = 150$ and, with the free energy of *A* as a reference point, a standard free energy of activation of 23 kcal/mol.

(a) Draw a reaction free-energy diagram for this process, showing the relative free energies of *A, B,* and the transition state for the reaction.

(b) What is the standard free energy of activation for the reverse reaction $B \rightarrow A$? How do you know?

34 A reaction $A \rightleftarrows B \rightleftarrows C \rightleftarrows D$ has the reaction free-energy diagram shown in Fig. 4.18.

(a) Which compound is present in greatest amount when the reaction comes to equilibrium? In least amount?

(b) What is the rate-determining step of the reaction?

(c) Using a vertical arrow, label the standard free energy of activation for the overall $A \rightleftarrows D$ reaction.

(d) Which reaction of compound *C* is faster: $C \rightarrow B$ or $C \rightarrow D$? How do you know?

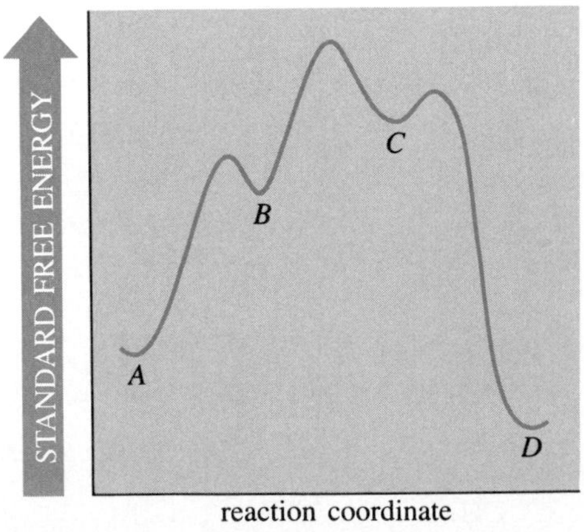

Figure 4.18 Reaction free-energy diagram for Problem 34.

35 Consider the following compounds and their dipole moments:

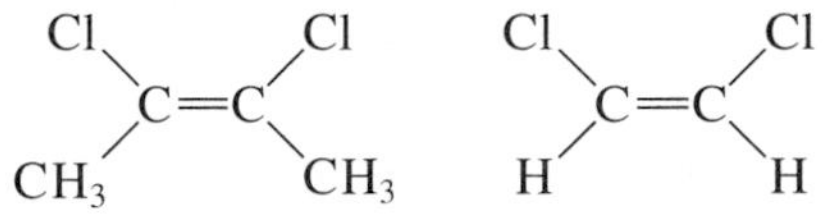

2.4 D   1.9 D

Because of the large difference in electronegativities between carbon and chlorine, it is safe to assume that the C—Cl bond dipole is oriented as follows in each of these compounds.

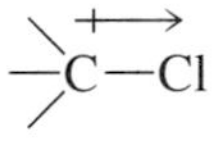

(a) According to the dipole moments above, which is *more* electron-donating toward a double bond, methyl or hydrogen? Explain.
(b) Which of the compounds below should have the greater dipole moment? Explain.

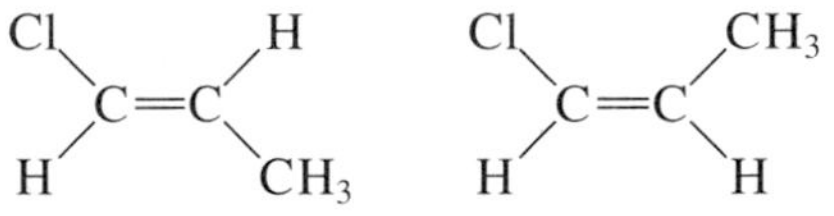

36 The ΔH_f^0 of (2*Z*,4*Z*)-2,4-octadiene is about 4 kcal/mol less than that for (2*Z*,5*Z*)-2,5-octadiene.
(a) Which compound is more stable?
(b) Let us try to deduce a structural reason for the relative stability you gave in (a). Assume for both alkenes that the alkene carbons (the carbons involved in the double bonds) and the atoms attached directly to them lie in the same plane. Sketch the *p* orbitals in each compound. Use the relative positions of the *p* orbitals in each compound to suggest a reason for your answer to (a). (*Hint*: More bonding implies greater stability.)

37 Treatment of isobutylene with methanol (CH_3OH) in the presence of mineral acid gives *t*-butyl methyl ether. (This ether is replacing tetraethyllead as an antiknock additive in motor fuels and, for this reason, has become one of the most important industrial organic chemicals in the United States.)

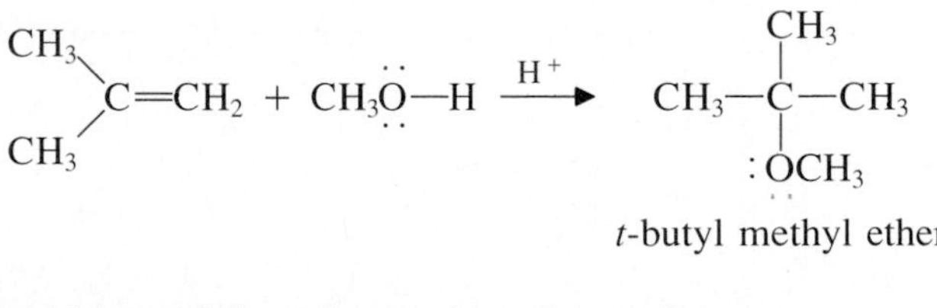

t-butyl methyl ether

(a) Does this addition follow the Markovnikov rule?
(b) Use the arrow formalism to write out the mechanism of this reaction in detail. (See following box)

Problems (Cont.)

WORKING MECHANISM PROBLEMS WITH THE ARROW FORMALISM

Problems that ask for mechanisms using the arrow formalism test your understanding of why chemical reactions proceed as they do. One of our goals in this book is to be able to predict reactions without having to memorize them. This type of problem, in which we rationalize the outcome of a known reaction, will help in developing predictive skills. To work this type of problem, always use the following approach:

1. Compare the reactants and products. From which atoms of the reactants do the atoms of the products originate?

2. Do the conditions of this reaction resemble any reaction that you have seen before? (In this case, the answer is yes. In which reaction(s) have we seen protons reacting with alkenes?) Use one of these earlier reactions as a model, or *analogy*, to begin writing the mechanism of this one. (In this case, show the reaction of the proton with the double bond.)

3. Write only one mechanistic step per structure! Trying to cram too many steps into one set of structures is confusing.

4. Once you begin the mechanism as in step 2, the reactive intermediate formed should suggest a way to complete the mechanism, given your answer to step 1 above. Complete it using the arrow formalism; if necessary review Secs. 1.3–1.5.

38 Consider the ΔH_f^0 values (in kcal/mol) of the following compounds: butane, -30.15; isobutane, -32.15; 2-methylpropene, -4.04; 1-butene, -0.03.

(a) On the basis of these results, which is stabilized more by branching, alkanes or alkenes?

The *t*-butyl cation, $(CH_3)_3C^+$, is 108 kcal/mol more stable than the *sec*-butyl cation, $CH_3\overset{+}{-}CH-CH_2-CH_3$, which is 23 kcal/mol more stable than the isobutyl cation, $(CH_3)_2CH\overset{+}{C}H_2$. (These results are from the gas phase. The differences in stability in solution are less pronounced.)

(b) Which is stabilized more by branching, carbocations or alkenes?

(c) Use your answer to (b) to determine which of the three following reactions has the smallest ΔH^0:

(1) $CH_2{=}CHCH_2CH_3 + H^+ \longrightarrow CH_3\overset{+}{C}HCH_2CH_3$

(2) $\begin{array}{c}CH_3\\ \diagdown\\ {/}\\ CH_3\end{array}C{=}CH_2 + H^+ \longrightarrow (CH_3)_3C^+$

(3) $\begin{array}{c}CH_3\\ \diagdown\\ {/}\\ CH_3\end{array}C{=}CH_2 + H^+ \longrightarrow \begin{array}{c}CH_3\\ \diagdown\\ {/}\\ CH_3\end{array}CH{-}\overset{+}{C}H_2$

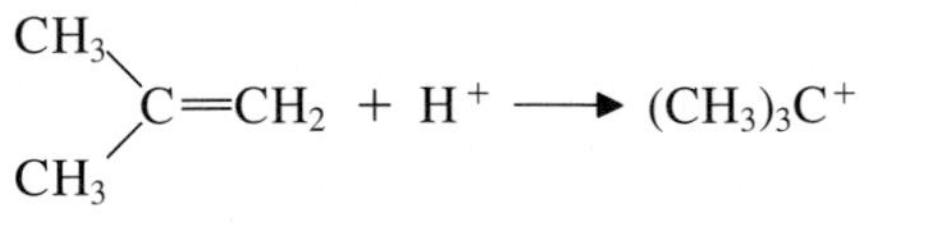

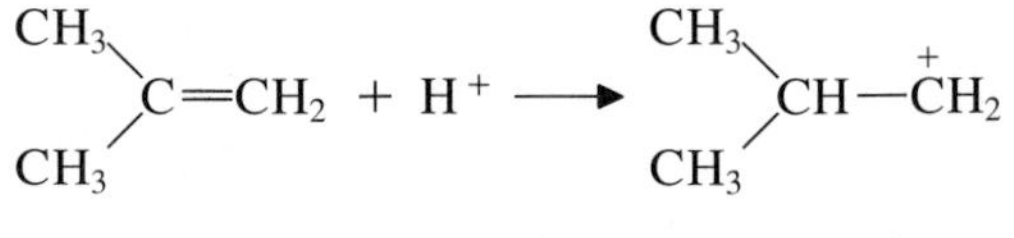

(d) Taking the enthalpy of the proton arbitrarily as zero, calculate the relative ΔH^0 values for these reactions and verify your answer to (c).

39 The standard free energy of formation, ΔG_f^0, is the free-energy change for the formation of a substance at 25 °C and 1 atm pressure from its elements in their natural states under the same conditions.

(a) Calculate the equilibrium constant for the interconversion of the alkenes below, given the standard free energy of formation of each. Indicate which compound is favored at equilibrium.

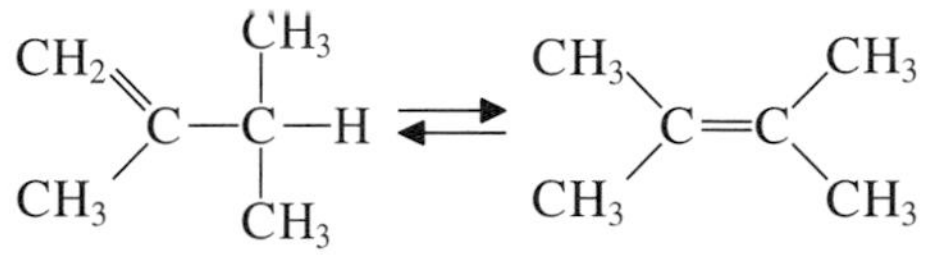

ΔG_f^0 18.89 kcal/mol 18.13 kcal/mol

(b) What does the equilibrium constant tell us about the rate with which this interconversion takes place?

40 The difference in the standard free energies of formation (see Problem 39) for 1-butene and 2-methylpropene is 3.2 kcal/mol.

(a) Which compound is more stable? Why?

The standard free energy of activation for the hydration of 2-methylpropene is 5.46 kcal/mol less than that for the hydration of 1-butene.

(b) Which reaction is faster?

(c) Draw reaction free-energy diagrams on the same scale for the hydration reactions of these two alkenes, showing the relative free energies of both starting materials and rate-determining transition states.

(d) By how many kcal/mol is the transition state for the hydration of 2-methylpropene more stable than that for hydration of 1-butene? Using the mechanism of the reaction, suggest a reason why it is more stable.

41 Consider the rearrangement leading to product A in the following reaction:

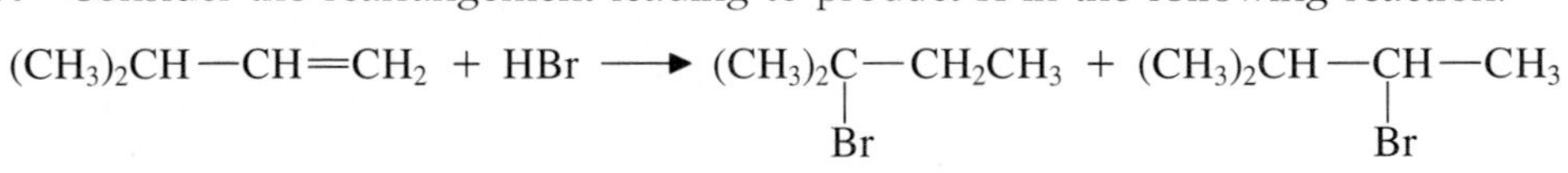

$$(CH_3)_2CH-CH=CH_2 + HBr \longrightarrow (CH_3)_2\underset{\underset{A}{\overset{|}{Br}}}{C}-CH_2CH_3 + (CH_3)_2CH-\underset{\underset{B}{\overset{|}{Br}}}{CH}-CH_3$$

This rearrangement could occur by two mechanisms. In the first mechanism, the carbocation intermediate rearranges simply by a shift of hydrogen and its bonding electrons from one carbon to the other (see Problem 15, Sec. 4.6).

mechanism 1:

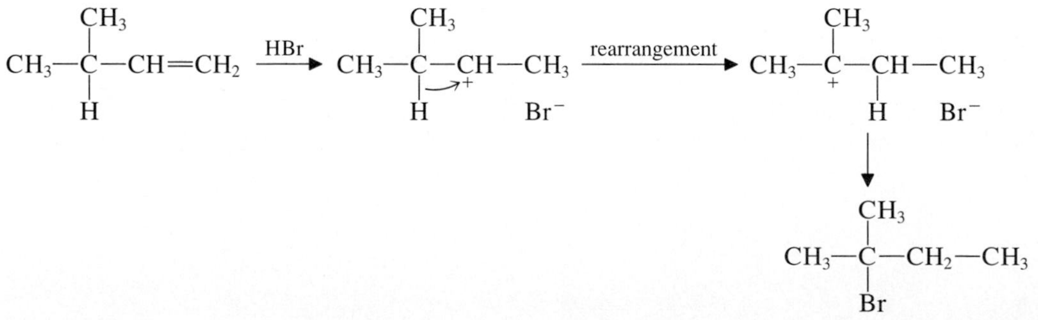

Problems (Cont.)

In the second mechanism, the carbocation formed initially loses a proton to give an isomer of the starting alkene. This alkene is then reprotonated by HBr to give the rearranged carbocation.

mechanism 2:

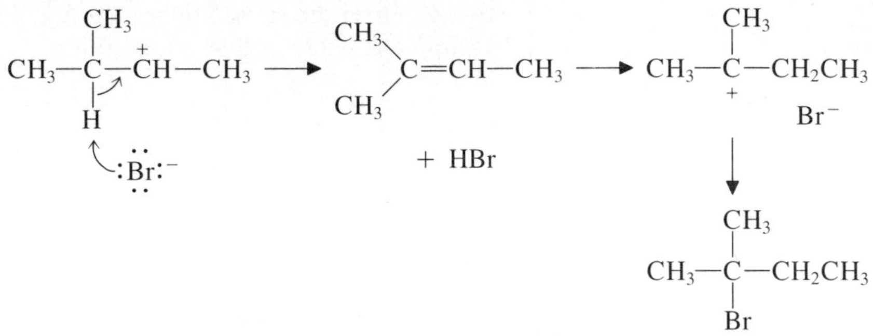

When the following isotopically labeled compound containing 1.0 atom of D per molecule is allowed to react with HBr, the only products isolated are the following:

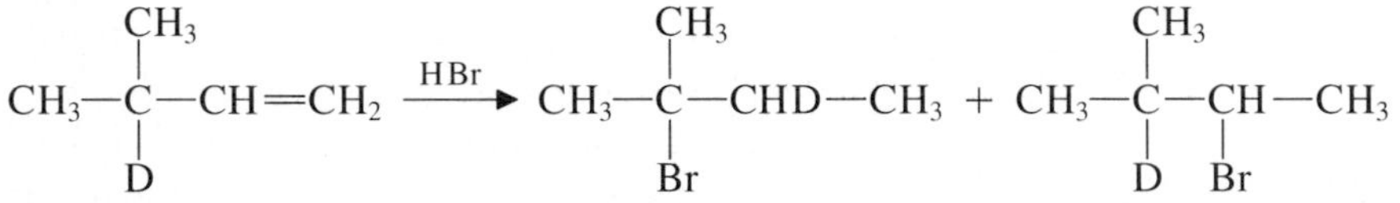

Each also contains 1.0 atom of D per molecule. With which mechanism is this result in agreement? Explain. (See note about deuterium in Problem 27.)

5 Addition Reactions of Alkenes

In Chapter 4 we studied two *addition reactions* of alkenes: addition of hydrogen halides and addition of water. We learned that carbocations are important reactive intermediates in these additions. In this chapter we shall survey several other addition reactions of alkenes, and we shall learn about another important type of reactive intermediate, the *free radical.*

5.1 HALOGENATION OF ALKENES

A. Addition of Chlorine and Bromine to Alkenes

The halogens undergo electrophilic addition to alkenes.

$$\underset{(\textit{cis}\ \text{or}\ \textit{trans})}{CH_3{-}CH{=}CH{-}CH_3} + Br_2 \xrightarrow[\text{(solvent)}]{CCl_4} \underset{\text{2,3-dibromobutane}}{CH_3{-}\underset{\substack{|\\ Br}}{CH}{-}\underset{\substack{|\\ Br}}{CH}{-}CH_3} \qquad (5.1)$$

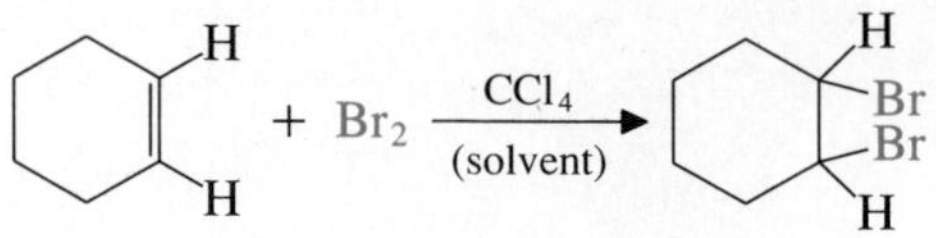

(5.2)

The products of these reactions are examples of **vicinal dihalides.** "Vicinal" (L. *vicinus,* neighborhood) means "on adjacent sites." Thus, vicinal dihalides are compounds with halogens on adjacent carbons.

Bromine and chlorine are the two halogens used most frequently in halogenation. Fluorine is so reactive that it not only adds to the double bond but also rapidly replaces all the hydrogens with fluorines, often with considerable violence. Iodine adds to alkenes at low temperature, but most diiodides are unstable and decompose to the corresponding alkenes and I_2 at room temperature. Because bromine is a liquid that is more easily handled than chlorine gas, many halogenations are carried out with bromine. Typically these reactions are run in an inert solvent that readily dissolves the halogen, such as carbon tetrachloride, CCl_4, or methylene chloride, CH_2Cl_2. The bromination of most alkenes is so fast that when bromine is added dropwise to a solution of the alkene the red bromine color disappears almost immediately. In fact this discharge of color is a useful test for alkenes.

The mechanism of bromination involves a reactive intermediate called a **bromonium ion.**

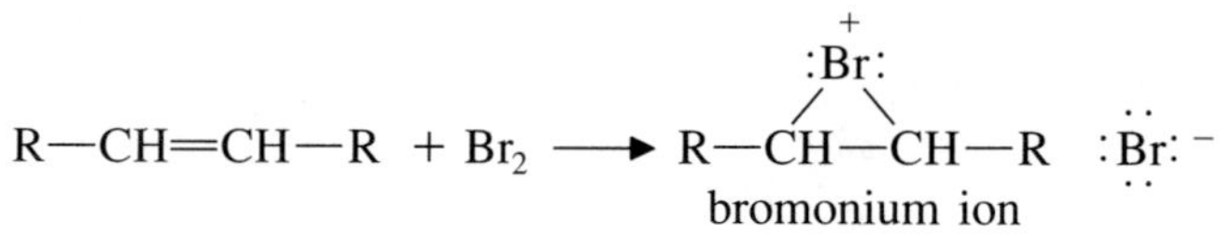

(5.3)

Although the bromonium ion probably forms in a single step, it is helpful if we envision its formation as follows:

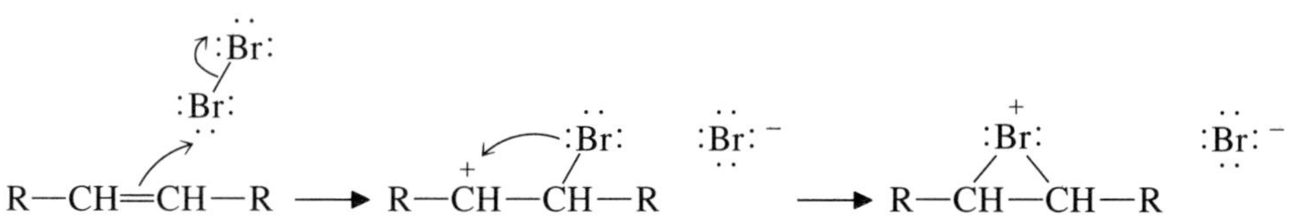

(5.4)

The bromonium ion has resonance structures in which positive charge is shared by carbon atoms:

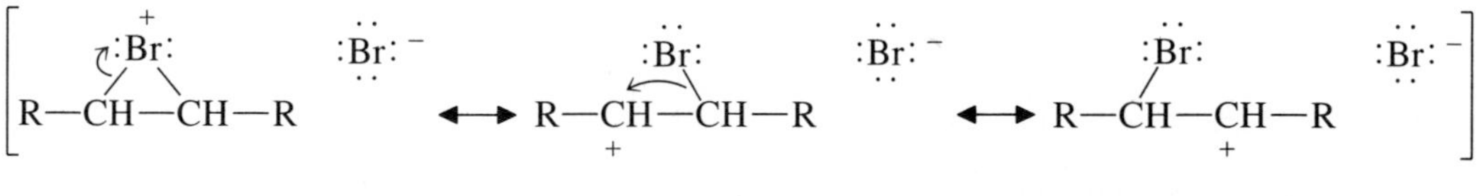

(5.5)

Attack of the bromide ion on either of these carbon atoms completes the addition.

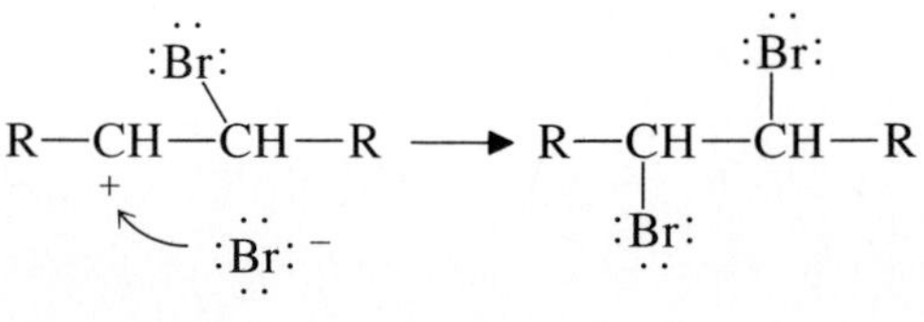

(5.6)

One reason for postulating a bromonium ion intermediate rather than a carbocation is that the rearrangements expected from carbocations are not observed in many halogenation reactions. (Other evidence for the bromonium ion will be considered in Chapter 7.) The greater amount of bonding in a bromonium ion gives it an energetic advantage over an isomeric carbocation. Furthermore, every atom in a bromonium ion has an octet, but in a carbocation there is an electron-deficient carbon. Analogous cyclic ions are intermediates in the attack of chlorine and iodine on alkene double bonds.

B. Halohydrin Formation

In Eq. 5.6 the dibromide product is formed because bromide is the only nucleophile (Lewis base) present in solution to complete the reaction. If other nucleophiles are present, they can compete with bromide for the bromonium ion intermediate. For example, if bromination is carried out in the presence of water, the nucleophilic water molecule can attack the bromonium ion to yield a *bromohydrin*.

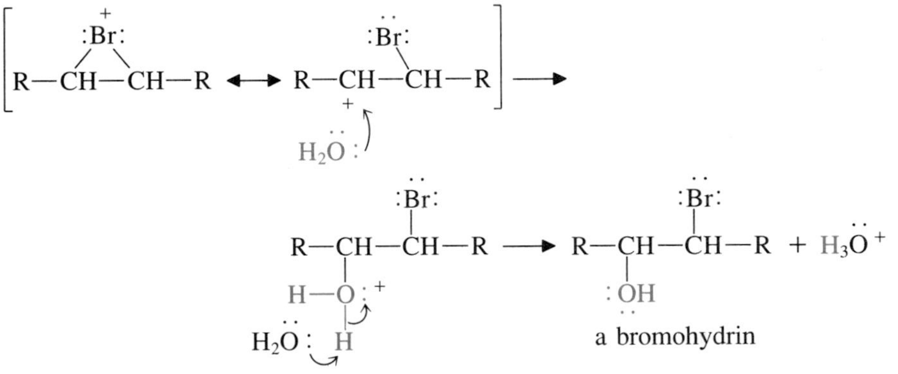

(5.7)

A bromohydrin is an example of a class of compounds called **halohydrins.** A halohydrin is an organic compound containing both a hydroxy (—OH) group and a halogen. The most common type of halohydrin is one in which the two groups occupy vicinal positions.

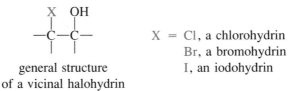

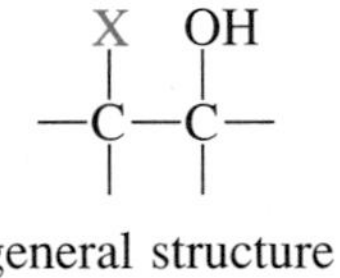

general structure of a vicinal halohydrin

X = Cl, a chlorohydrin
Br, a bromohydrin
I, an iodohydrin

Halohydrin formation involves the net addition of the elements of HO—Br (hypobromous acid) or HO—Cl (hypochlorous acid) across the alkene double bond. In fact, solutions of bromine or chlorine in water contain appreciable amounts of the respective hypohalous acids. (Household chlorine bleach is an aqueous solution of sodium hypochlorite, the sodium salt of hypochlorous acid.)

The addition of hypohalous acid follows our modern version of the Markovnikov rule (Sec. 4.5D). In the following example, the negative end of the Br—OH dipole, the OH group, adds to the more substituted alkene carbon atom.

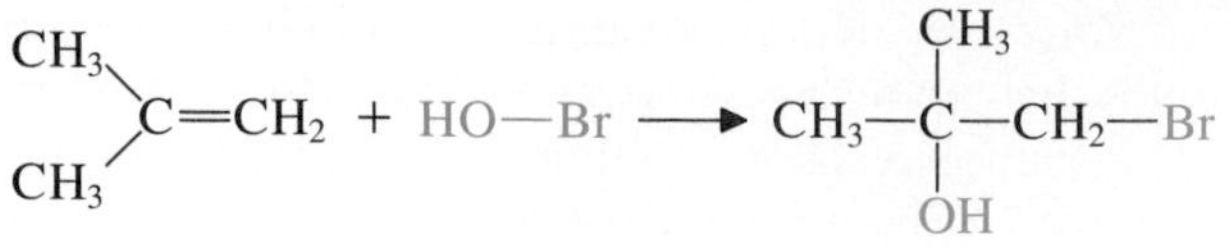

$$\underset{CH_3}{\overset{CH_3}{\diagdown}}C=CH_2 \ + \ HO-Br \ \longrightarrow \ CH_3-\underset{\underset{OH}{|}}{\overset{\overset{CH_3}{|}}{C}}-CH_2-Br \qquad (77\% \text{ yield}) \quad (5.8)$$

Why does this addition follow the Markovnikov rule? The answer lies in the nature of the bromonium ion intermediate. As we have already seen (Eq. 5.5), this ion has resonance structures in which the positive charge is shared between the bromine atom and the adjacent carbon atoms.

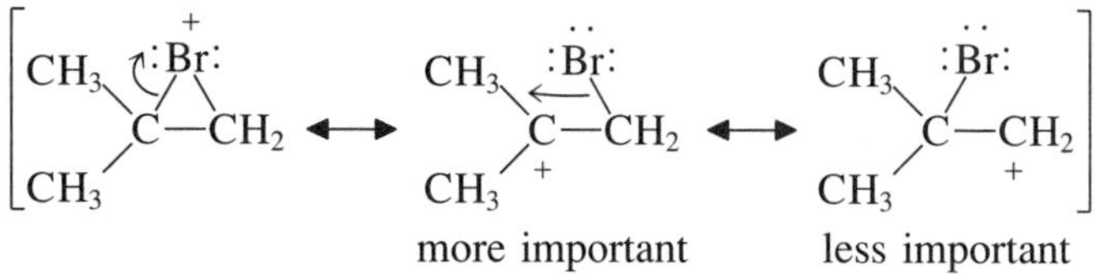

$$(5.9)$$

more important less important

As we have already learned (Sec. 4.5B), electron-deficient carbon is stabilized by alkyl substituents. Hence, the resonance structure in Eq 5.9 with the *tertiary* electron-deficient carbon is more important than the structure with the *primary* electron-deficient carbon. The Markovnikov rule is followed because water attacks the carbon with the greater amount of positive charge—the tertiary carbon.

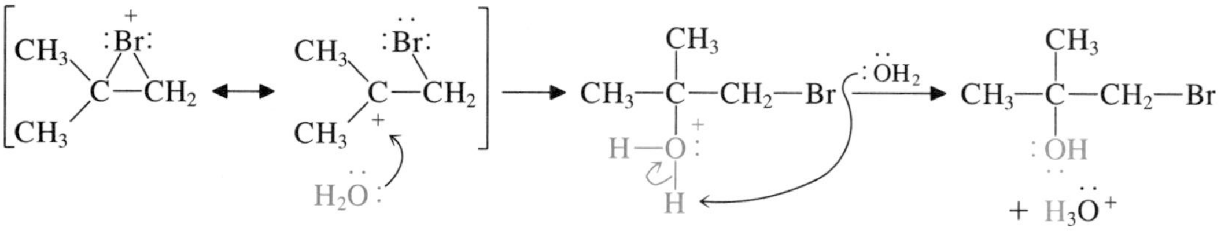

$$(5.10)$$

Problems

1 Give the structure of an alkene that would give the following product when it reacts with bromine water:

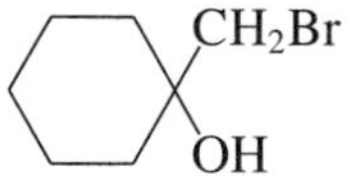

2 What is the product, and the mechanism for its formation, when iodine azide, I—N₃, reacts with isobutylene? (*Hint:* The azide ion, N₃⁻, behaves much like a halide ion.)

5.2 WRITING ORGANIC REACTIONS

In the previous section we saw several examples of organic reactions. Before we go any further, let us examine some of the conventions used by organic chemists in writing reactions. Of course, the most thorough way to write a reaction is to use a complete, balanced equation. Equations 5.1 and 5.8 are examples of such equations. Other information, such as the reaction conditions, is sometimes included in equations. For example, in Eq. 5.1 the solvent is written over the arrow, even though the solvent is not an actual reactant. In the following equation, the H^+ written over the arrow indicates that an acid catalyst (Sec. 4.8) is required.

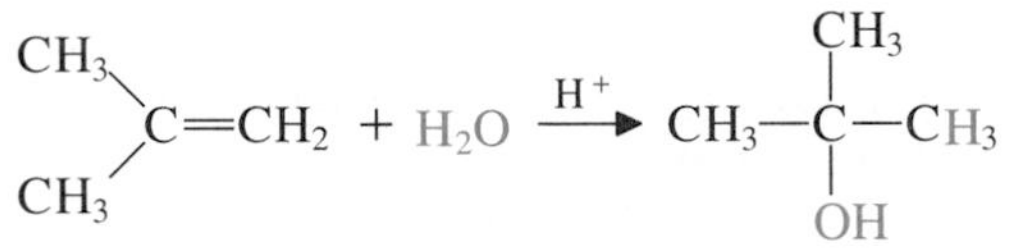

$$
\begin{array}{c}
CH_3 \\
\diagdown \\
\diagup \quad C=CH_2 \ + \ H_2O \ \xrightarrow{\ H^+\ } \ CH_3-\overset{\displaystyle CH_3}{\underset{\displaystyle OH}{C}}-CH_3 \\
CH_3
\end{array}
\qquad (5.11)
$$

Equation 5.8 includes a *percentage yield* figure. A **percentage yield** is the percentage of the theoretical amount of product formed that has actually been isolated from the reaction mixture by a chemist in the laboratory. Although different chemists might get different yields in the same reaction, the percentage yield gives some idea of how free the reaction is from contaminating by-products and/or how easy it is to isolate the product from the reaction mixture. Thus a reaction $2A + B \rightarrow 3C + D$ should give three moles of C for every one of B or two of A used (assuming that one of these reactants is not present in excess). A 90% yield of C means that 2.7 moles of C were *actually isolated* under these conditions. The 10% loss may have been due to handling problems, small amounts of by-products, or other reasons. Virtually all of the reactions given in this book are actual laboratory examples; the percentage yield figures included in many of these reactions are not meant to be learned, but are simply given to indicate how successful a reaction actually is in practice.

In many cases it is convenient to abbreviate reactions by showing only the *organic starting materials* and the *major organic product(s)*. The other reactants and conditions are written over the arrow. Thus, Eq. 5.1 might have been written

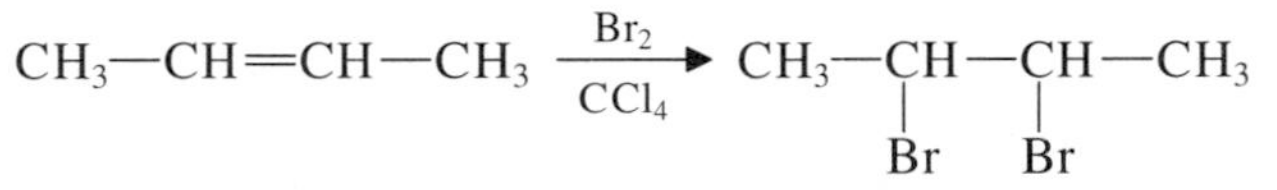

$$
CH_3-CH{=}CH-CH_3 \ \xrightarrow[CCl_4]{Br_2} \ CH_3-\underset{\underset{\displaystyle Br}{|}}{CH}-\underset{\underset{\displaystyle Br}{|}}{CH}-CH_3
\qquad (5.12)
$$

When equations are written this way, by-products are not given and, in many cases, the equation is not balanced. We can tell reactants from catalysts because reactants are consumed and catalysts are not. Thus in Eq. 5.12 we know that Br_2 is a reactant rather than a catalyst because it is consumed in the reaction. This "shorthand" way of writing organic reactions is frequently used because it saves space and time.

5.3 CONVERSION OF ALKENES INTO ALCOHOLS

A. Hydroboration–Oxidation of Alkenes

Conversion of Alkenes into Organoboranes When an alkene reacts with diborane, B_2H_6, an addition to the double bond takes place to yield an *organoborane*—a compound with a carbon–boron bond.

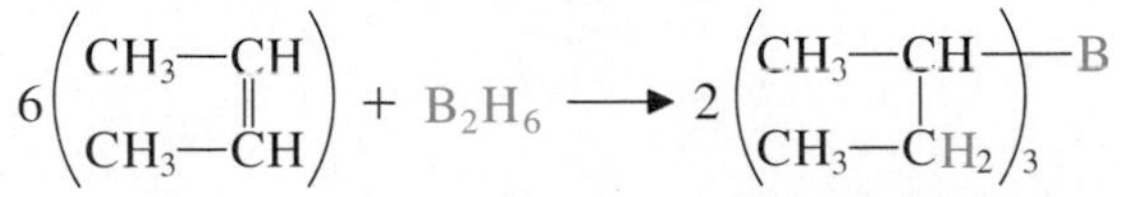

$$
6\left(\begin{array}{c} CH_3-CH \\ \| \\ CH_3-CH \end{array}\right) \ + \ B_2H_6 \ \longrightarrow \ 2\left(\begin{array}{c} CH_3-CH \\ | \\ CH_3-CH_2 \end{array}\!\!\!-B\right)_{\!3}
\qquad (5.13)
$$

tri-*sec*-butylboron
(an organoborane)

This reaction, called **hydroboration,** is another electrophilic addition.

Before considering the mechanism of hydroboration, let us examine the nature of the reactant diborane. This toxic, colorless gas is a *dimer* of borane, BH_3; this means diborane is a combination of two borane units. As a Lewis acid, borane has a tendency to acquire additional electrons. In the structure of diborane, this is accomplished to a certain degree by an unusual sharing of hydrogen atoms in "half-bonds" between two boron atoms. We can represent this bonding using resonance structures.

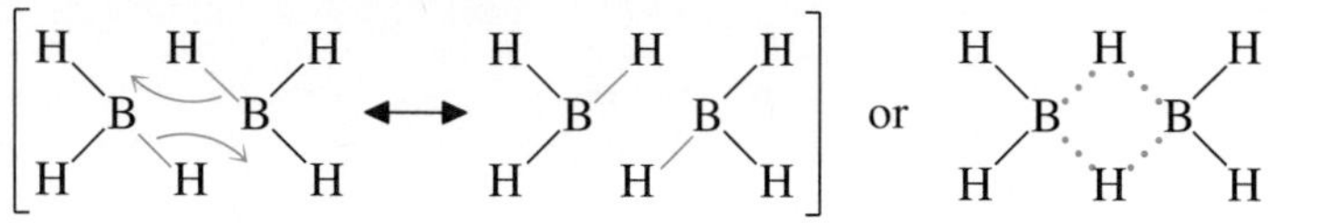

(5.14)

(The dotted lines in the structure on the right imply partial bonding.) When dissolved in an ether, diborane dissociates to form a borane–ether complex. Because it is a Lewis base, the ether can satisfy the electron deficiency of the boron atom in borane.

$$B_2H_6 + 2R-\ddot{\underset{..}{O}}-R \rightleftharpoons 2H_3\overset{-}{B}:\overset{+}{O}(R)_2 \qquad (5.15)$$

an ether; borane-ether complex

The following ethers are commonly used as solvents in the hydroboration reaction.

$C_2H_5-\ddot{O}-C_2H_5$ diethyl ether

tetrahydrofuran (THF)

$CH_3\ddot{O}-CH_2CH_2-\ddot{O}-CH_2CH_2-\ddot{O}CH_3$ diethylene glycol dimethyl ether (diglyme)

Borane–ether complexes are the actual reagents involved in hydroboration reactions.

Hydroboration begins as the addition of H—BH_2 across the alkene π-bond.

$$CH_3-CH=CH-CH_3 + H-BH_2 \longrightarrow CH_3-\underset{H}{CH}-\underset{BH_2}{CH}-CH_3 \qquad (5.16)$$

an alkylborane

Because the alkylborane product of this reaction has B—H bonds, it can react with another alkene molecule.

$$CH_3CH_2\underset{H-BH}{CH}CH_3 + CH_3-CH=CH-CH_3 \longrightarrow (CH_3CH_2CHCH_3)_2B-H \qquad (5.17)$$

a dialkylborane

Finally, the dialkylborane can add to a third alkene molecule, yielding the trialkylborane product of Eq. 5.13.

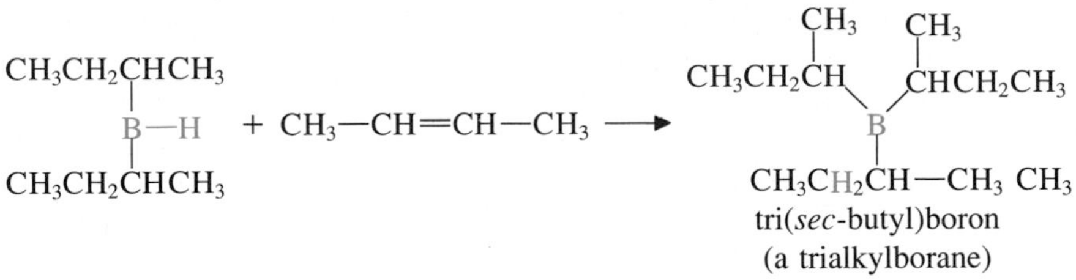

tri(*sec*-butyl)boron
(a trialkylborane)

Thus, hydroboration is a sequence of three additions that culminates in the formation of a trialkylborane.

The addition of borane to an alkene conforms to our modern view of the Markovnikov rule: the more electronegative atom of BH_3 (hydrogen) adds to the more substituted alkene carbon. (Hydrogen is more electronegative than boron; Table 1.1.)

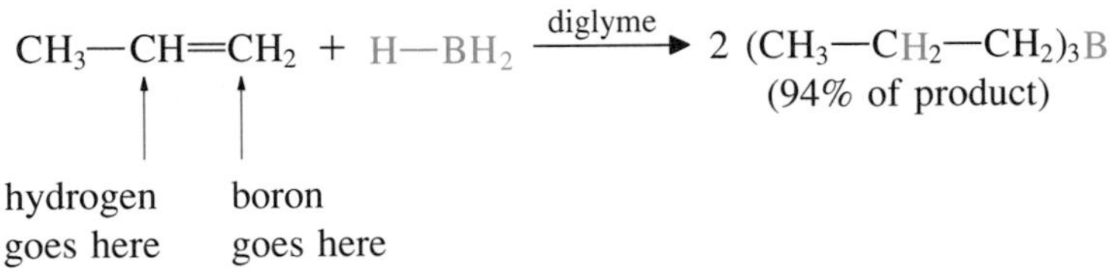

$$CH_3-CH{=}CH_2 \ + \ H-BH_2 \ \xrightarrow{\text{diglyme}} \ 2 \ (CH_3-CH_2-CH_2)_3B \qquad \text{(5.19)}$$
(94% of product)

hydrogen boron
goes here goes here

The addition of BH_3 to alkenes is believed to occur in a single bond-forming step.

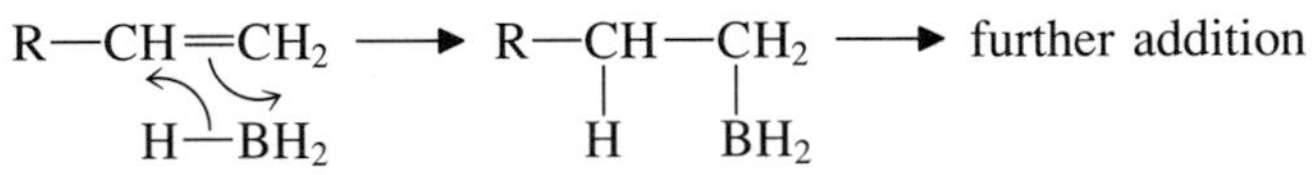

$$R-CH{=}CH_2 \longrightarrow R-\underset{\underset{H}{|}}{C}H-\underset{\underset{BH_2}{|}}{C}H_2 \longrightarrow \text{ further addition} \qquad \text{(5.20)}$$
$$H-BH_2$$

Reactions in which the bonds involved are broken or formed in a single step are classified as **concerted reactions.** Hydroboration is thus a *concerted addition reaction*. By definition, there are no intermediates in concerted reactions. Hydroboration can be further classified as a **pericyclic reaction:** a concerted reaction that occurs by the cyclic shift of electrons. By this we mean that the curved arrows used to describe the mechanism (Eq. 5.20) move in a closed loop. (The word *pericyclic* means "around the circle.")

Given that hydroboration is a concerted reaction, it cannot involve carbocation intermediates. One piece of experimental evidence that supports this conclusion is that rearrangements are *not* observed.

$$B_2H_6 \ + \ 6CH_3-\underset{\underset{CH_3}{|}}{\overset{\overset{CH_3}{|}}{C}}-CH{=}CH_2 \longrightarrow 2\left(CH_3-\underset{\underset{CH_3}{|}}{\overset{\overset{CH_3}{|}}{C}}-CH_2-CH_2\right)_{\!3}\!\!-B \qquad \text{(5.21a)}$$

$$B_2H_6 \ + \ 6CH_3-\underset{\underset{CH_3}{|}}{\overset{\overset{CH_3}{|}}{C}}-CH{=}CH_2 \ \ {\times}\!\!\longrightarrow \left(CH_3-CH-\underset{\underset{CH_3}{|}}{C}H-CH_2\right)_{\!3}\!\!-B \qquad \text{(5.21b)}$$

rearranged product;
not observed

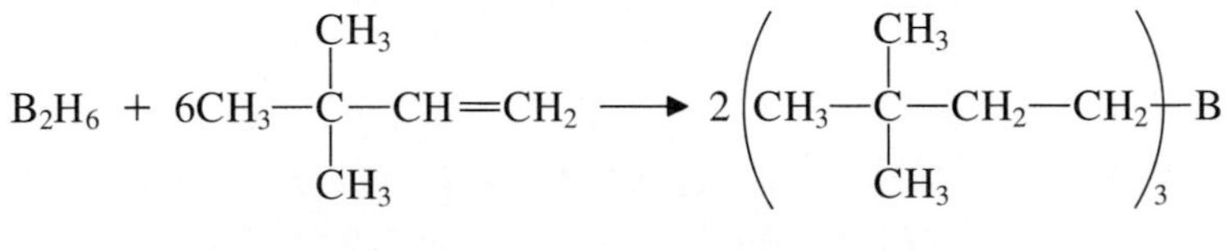

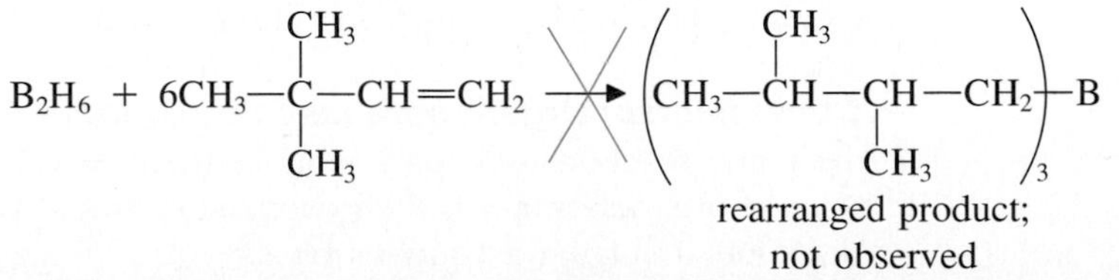

Recall that addition of HCl to this same alkene—a reaction that does involve carbocation intermediates—occurs with rearrangement (Eq. 4.25). Other evidence that we shall consider in Chapter 7 also supports a concerted mechanism for hydroboration.

Since ionic intermediates are not involved in hydroboration, why does it follow the Markovnikov rule? The regioselectivity of the reaction is attributable to two factors. One is the electron distribution in the transition state, which we can *approximate* as a complex of borane with the alkene π-electrons.

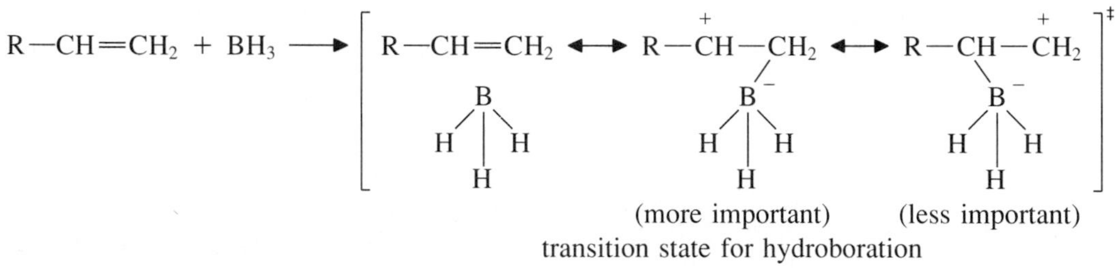

(5.22)

In this transition state, borane satisfies its Lewis-acid character by complex formation with the alkene π-electrons, and there is some partial positive charge on the alkene carbons. The hydrogen from borane is directed to the more highly substituted carbon because that carbon bears the greater amount of positive charge. In other words, the transition state for hydroboration has some carbocation character, even though carbocations are not involved in the reaction.

The second reason for the regioselectivity of hydroboration is that in the transition state the R_2B— group avoids van der Waals repulsions with alkyl substituents on the double bond by forming a bond at the less substituted alkene carbon.

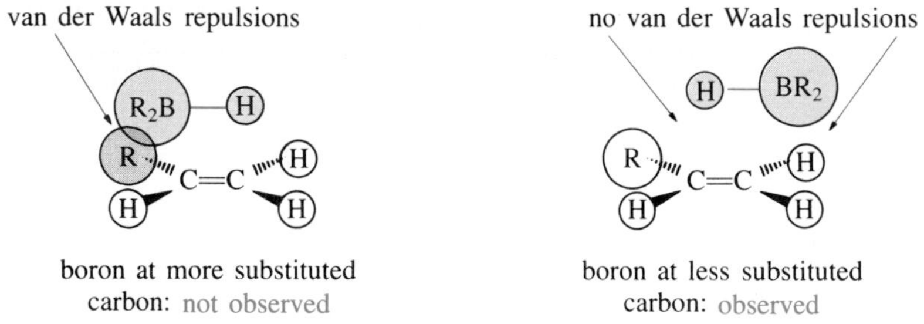

These van der Waals repulsions are conceptually much like those between the eclipsed methyl groups in the most unstable form of butane (Sec. 3.5B). In other words, the course of the hydroboration reaction is determined in part by a *steric effect*.

Conversion of Organoboranes into Alcohols The utility of hydroboration lies in the many reactions of organoboranes themselves. One of the most important reactions of organoboranes is their conversion into alcohols by oxidation with hydrogen peroxide (H_2O_2) and aqueous base.

$$(CH_3CH_2CH_2CH_2CH_2CH_2)_3B + 3H_2O_2 + OH^- \longrightarrow$$

$$\underset{\substack{\text{1-hexanol} \\ (n\text{-hexyl alcohol;} \\ \text{94–97\% yield)}}}{3CH_3CH_2CH_2CH_2CH_2CH_2{-}OH} + \underset{\substack{\text{borate} \\ \text{anion}}}{B(OH)_4^-} \qquad (5.23)$$

Although we shall not consider the mechanism of this process, it is important to notice that its end result is *replacement of the boron by an —OH group at each carbon atom.* Tracing the fate of an alkene through the entire hydroboration–peroxidation sequence, we find that the *net* result is addition of the elements of water (H and OH) across the double bond; the —OH group ends up *at the less highly substituted carbon of the alkene double bond* (color in Eq. 5.24).

$$6R{-}CH{=}CH_2 + B_2H_6 \longrightarrow 2\left(\underset{\substack{|\\H}}{RCH}{-}CH_2\right)_3\!B \xrightarrow[OH^-]{H_2O_2} 6R{-}\underset{\substack{|\\H}}{CH}{-}\underset{\substack{|\\OH}}{CH_2} \qquad (5.24)$$

Hydroboration–oxidation is therefore a useful way to synthesize certain alcohols from alkenes in high yield. Because carbocations are not involved in hydroboration, the alcohol products are not contaminated by structural isomers arising from rearrangements.

The following example shows that the benzene ring does not react with BH_3 even though the ring contains formal double bonds:

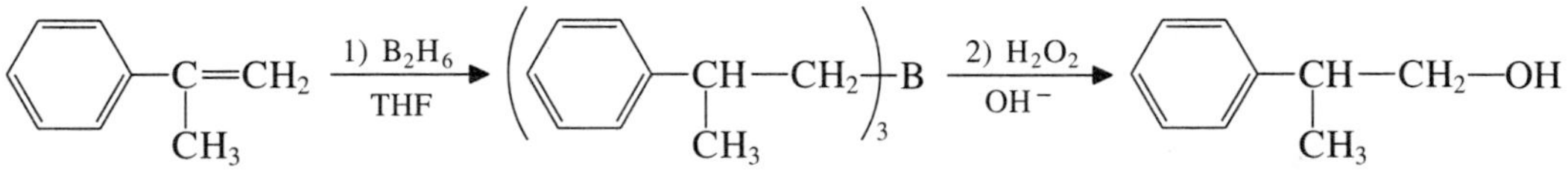

(5.25)

2-phenyl-1-propanol
(95% yield)

This is a general result: benzene rings are unreactive in most alkene addition reactions. Indeed, benzene has a special chemistry that we shall study in Chapter 16.

> Hydroboration was discovered accidentally in 1955 by Professor H. C. Brown (1912–   ) and his colleagues of Purdue University. Brown quickly realized its significance and in subsequent years has carried out research demonstrating the versatility of organoboranes as intermediates in organic synthesis. Brown, now an Emeritus Professor at Purdue, still conducts research in this area, which he calls "a vast unexplored continent." In 1979 his research was recognized with the Nobel Prize in Chemistry.

Problems

3 Give the principal product(s) expected from the hydroboration-oxidation of each of the following alkenes:
(a) ethylene (c) cyclohexene
(b) 2-methyl-2-pentene (d) *trans*-4-methyl-2-pentene

4 From which alkene could each of the following alcohols be prepared by hydroboration–oxidation?

(a) C_2H_5—CH—CH_2OH
 |
 CH_3

(b) ⬡—CH_2OH

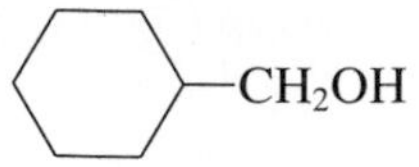

B. Oxymercuration–Reduction of Alkenes

Oxymercuration of Alkenes Compounds of mercury(II) add to double bonds. A mercury(II) salt that is particularly useful for this purpose is mercuric acetate, a water-soluble white crystalline solid.

$$\text{AcO—Hg—OAc}$$
$$\text{mercuric acetate}$$

The acetate group is abbreviated —OAc or AcO—.

$$-\text{OAc} = -\text{O}-\overset{\overset{\displaystyle O}{\|}}{C}-CH_3 \tag{5.26}$$
$$\text{acetate group}$$

When mercuric acetate reacts with an alkene in the presence of water, an addition to the double bond, called **oxymercuration,** takes place. The product of this addition has an acetoxymercuri group (—HgOAc) at the less substituted alkene carbon and a hydroxy (—OH) group at the more substituted alkene carbon.

$$CH_3CH_2CH_2CH_2—CH=CH_2 + Hg(OAc)_2 + H_2O \xrightarrow{\text{THF, H}_2\text{O}}$$

$$CH_3CH_2CH_2CH_2—\underset{\underset{\displaystyle OH}{|}}{CH}—CH_2—Hg—OAc + HOAc \tag{5.27}$$
$$\text{(95\% yield)}$$

Tetrahydrofuran (THF, Sec. 5.3A) is used as a cosolvent to dissolve the water-insoluble alkene.

Oxymercuration is yet another example of an electrophilic addition, and the mechanism closely resembles that of halohydrin formation (Sec. 5.1B). A cyclic ion called a *mercurinium ion* is an intermediate in this reaction. This ion is rather analogous to the bromonium ion intermediate in bromination (Sec. 5.1A).

$$R—CH=CH_2 + Hg(OAc)_2 \longrightarrow R—\underset{\underset{\displaystyle \underset{\displaystyle OAc}{|}}{\overset{\displaystyle +}{Hg}}}{CH—CH_2} \quad {}^-OAc \tag{5.28a}$$
$$\text{mercurinium ion}$$

The mercurinium ion, like the bromonium ion, has resonance structures in which the positive charge is shared by the carbon atoms attached to mercury. Water attacks the mercurinium ion at the more substituted of these carbon atoms.

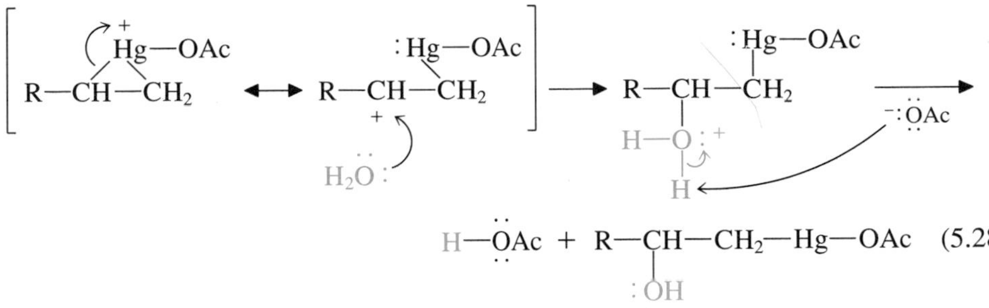

$$H\!-\!\ddot{O}Ac \;+\; R\!-\!\underset{\underset{\ddot{\,}}{\overset{|}{\ddot{O}H}}}{\overset{|}{C}}H\!-\!CH_2\!-\!Hg\!-\!OAc \qquad (5.28b)$$

Conversion of Oxymercuration Adducts into Alcohols The utility of oxymercuration lies in the fact that oxymercuration adducts are easily converted into alcohols by treatment with the reducing agent sodium borohydride (NaBH$_4$) in base.

$$4CH_3CH_2CH_2CH_2\!-\!\underset{\overset{|}{OH}}{\overset{|}{C}}H\!-\!CH_2\!-\!Hg\!-\!OAc \;+\; 4OH^- \;+\; \underset{\substack{\text{sodium}\\\text{borohydride}}}{NaBH_4} \longrightarrow$$

$$4CH_3CH_2CH_2CH_2\!-\!\underset{\overset{|}{OH}}{\overset{|}{C}}H\!-\!CH_3 \;+\; Na^+\,B(OH)_4^- \;+\; 4Hg^0\!\downarrow \;+\; 4AcO^- \qquad (5.29)$$

Although we shall not discuss the mechanism of this reaction, we should notice that the mercury is replaced by a hydrogen of NaBH$_4$ in this transformation. Thus, alkenes taken through the entire oxymercuration/NaBH$_4$ reduction undergo a net addition of the elements of water (H and OH) so that the —OH group is added to the *more substituted carbon of the double bond*.

$$CH_3CH_2CH_2CH_2\!-\!CH\!=\!CH_2 \xrightarrow[\text{2) NaBH}_4/\text{OH}^-]{\text{1) Hg(OAc)}_2/\text{H}_2\text{O}} CH_3CH_2CH_2CH_2\!-\!\underset{\overset{|}{OH}}{\overset{|}{C}}H\!-\!CH_3 \qquad (96\% \text{ yield}) \quad (5.30)$$

Because rearrangements are not observed in this reaction, it usually gives good yields of single products free of isomers.

$$(CH_3)_3C\!-\!CH\!=\!CH_2 \xrightarrow[\text{2) NaBH}_4/\text{OH}^-]{\text{1) Hg(OAc)}_2/\text{H}_2\text{O}} (CH_3)_3C\!-\!\underset{\overset{|}{OH}}{\overset{|}{C}}H\!-\!CH_3 \qquad (5.31)$$

(94% yield; note lack
of rearrangement)

The absence of rearrangement in alkenes such as this—alkenes that do give rearrangements in the addition of hydrogen halides—is one of the reasons that carbocations are thought not to be involved as reactive intermediates.

C. Comparison of Methods of Alcohol Synthesis

We have now studied several ways of preparing alcohols from alkenes. In Sec. 4.8 we learned that *hydration* is a useful industrial method for the preparation of a few alcohols, but that it is not a good laboratory method. Indeed, many industrial methods for the preparation of organic compounds are not *general*. That is, an industrial method typically works well in the specific case for which it was designed, but cannot be applied to other related cases. The reason is that the chemical industry has gone to great effort to work out conditions that are optimal for the preparation of *particular compounds* of great commercial importance (such as ethanol) using reagents that are readily available and cheap (such as water and mineral acid). We could, of course, duplicate the industrial ethanol synthesis in the laboratory, but there is no need to do so precisely because ethanol *is* cheap and readily available from industrial sources. For laboratory work it is not practical for chemists to design a specific procedure for each new compound. Thus the development of general methods that work with a wide variety of compounds is important. Because laboratory synthesis is generally carried out on a relatively small scale, the expense of reagents is less of a concern.

Hydroboration–oxidation and oxymercuration–reduction are both *general laboratory methods* for the preparation of alcohols from alkenes. That is, they can be applied successfully to a large variety of alkene starting materials. Which method is better for preparation of a given alcohol? A choice between the two methods usually hinges on the difference in their regioselectivities. As shown in the following equation, hydroboration–oxidation gives an alcohol in which the —OH group has been added to the *less substituted alkene carbon*. Oxymercuration–reduction gives an alcohol in which the —OH group has been added to the *more substituted alkene carbon*.

$$R\text{—}CH\text{=}CH_2 \quad \xrightarrow[\text{of } H\text{—}OH]{\text{net addition}}$$

$$R\text{—}\underset{OH}{CH}\text{—}\underset{H}{CH_2} \qquad \text{oxymercuration–reduction}$$

$$R\text{—}\underset{H}{CH}\text{—}\underset{OH}{CH_2} \qquad \text{hydroboration–oxidation}$$

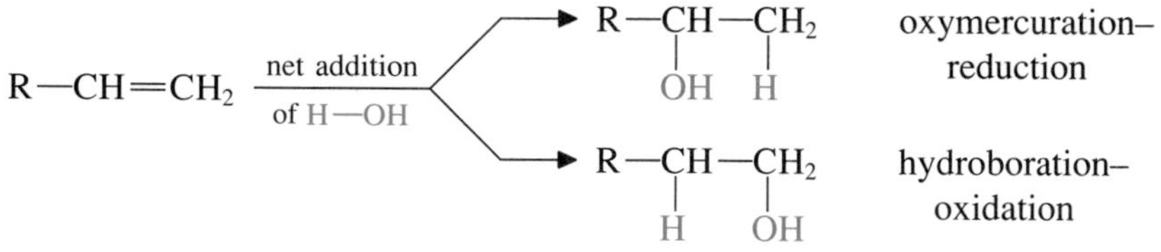

(5.32)

For alkenes that yield the same alcohol by either method, the choice between the two is in principle arbitrary.

Problem

5 Give the products expected from oxymercuration-$NaBH_4$ reduction of each of the following alkenes.

(a) ethylene (c) 1-methyl-1-cyclopentene

(b) 1-decene (d)

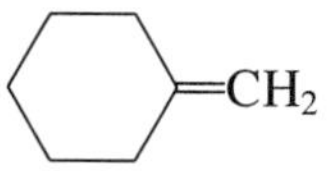

$$\text{⬡}\text{=}CH_2 \qquad \text{(methylenecyclohexane)}$$

Contrast the results of these reactions with the results from hydroboration–oxidation of the same alkenes.

5.4 CONVERSION OF ALKENES INTO GLYCOLS

Alkenes are readily converted into *vicinal glycols* using either alkaline potassium permanganate ($KMnO_4$) or osmium tetroxide (OsO_4). A **glycol,** or **diol,** is an alcohol containing two —OH groups; in a **vicinal glycol** these —OH groups are on adjacent carbons.

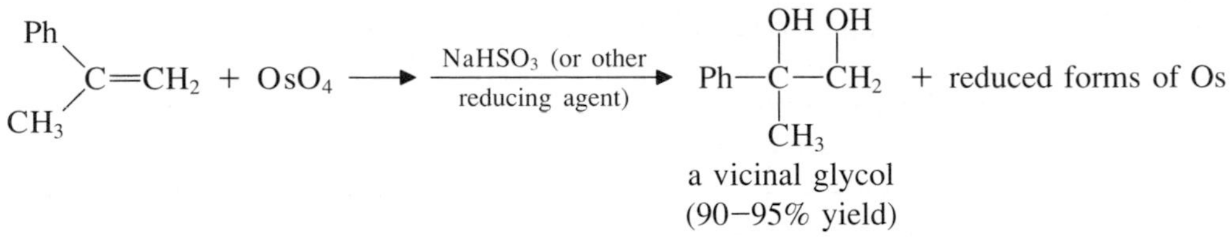

(5.33)

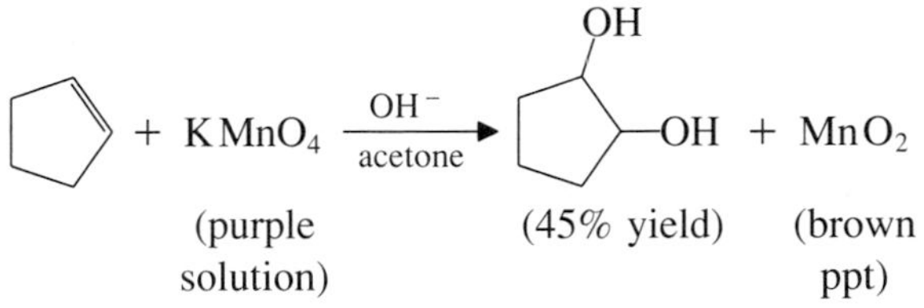

(5.34)

The reaction with $KMnO_4$ is the basis for a well-known qualitative test for alkenes called the *Baeyer test for unsaturation*. When an alkene is added to a solution containing aqueous permanganate, the brilliant purple color of the permanganate is replaced by a murky brown precipitate of MnO_2 (manganese dioxide). (Other compounds such as aldehydes and some types of alcohols give a positive test as well.)

The mechanisms of these two reactions are similar. The OsO_4 reaction, for example, involves a cyclic intermediate called an *osmate ester*. Although the mechanism by which it is formed is still debated, we can think of it in the following way:

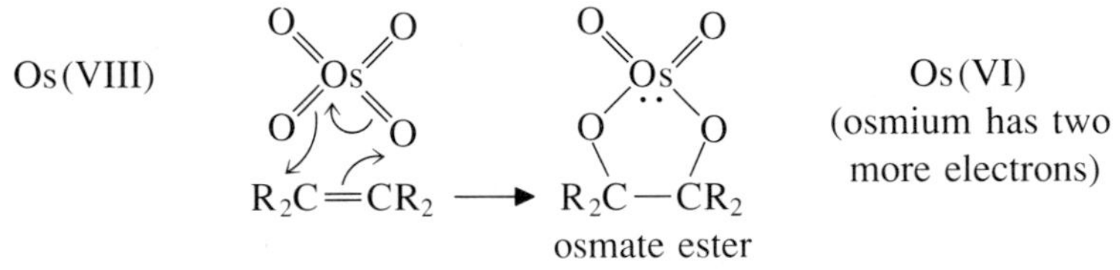

(5.35a)

The reaction occurs because osmium in its +8 oxidation state is rather electronegative and therefore attracts electrons. (Notice in Eq. 5.35a that osmium gains a pair of electrons in this reaction.) Thus, this reaction is yet another *electrophilic addition*. This reaction can also be classified as a **cycloaddition:** an addition that creates a ring. Finally, we can also recognize the mechanism of the reaction as another example of a *pericyclic reaction* (Sec. 5.3a): a reaction that occurs in one step by a cyclic flow of electrons. The mechanism of the $KMnO_4$ reaction is probably similar.

A glycol is formed when the cyclic osmate ester is treated with water and a mild reducing agent such as sodium bisulfite, $NaHSO_3$. The water converts the osmate ester into a glycol, and the sodium bisulfite converts the Os(VI) by-products into reduced forms of osmium that are easy to separate by filtration.

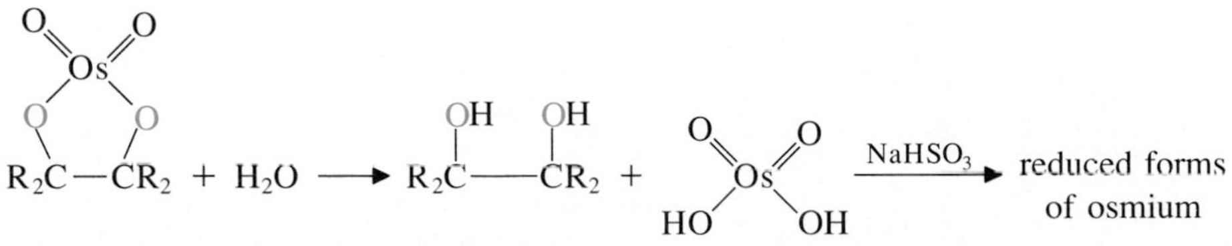

(5.35b)

In the $KMnO_4$ reaction, the glycol product forms spontaneously, and the MnO_2 by-product precipitates; no other reagents need to be added.

In many cases the $KMnO_4$ reaction gives lower yields than the OsO_4 reaction, partly because the cyclic intermediate in the permanganate reaction can undergo side reactions. (The alkaline conditions of the permanganate reaction keep these side reactions to a minimum.) However, because compounds of osmium are both toxic and expensive, the permanganate reaction is preferred for large-scale operations. Since chemists have developed ways to reoxidize the osmium(VI) by-product back to OsO_4, which can then be reused, the expense of the osmium reagent is not a serious problem on a small scale.

Problems

6 Give the products expected when each of the following alkenes is allowed to react with either OsO_4 followed by aqueous $NaHSO_3$, or alkaline $KMnO_4$:

(a) isobutylene

(b) 1-methyl-1-cyclopentene

(c)

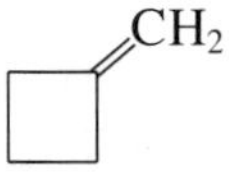

7 From what alkene could each of the following glycols be prepared by the methods described in this section?

(a) HO—CH(CH₃)—CH(OH)—CH₃

(b) HO—CH₂—CH(OH)—CH(OH)—CH₂—OH

8 Suggest a structure for the cyclic intermediate in the reaction of alkenes with $KMnO_4$. Show the formation of this intermediate with the arrow formalism. (*Hint:* The Lewis structure of MnO_4^- can be written as follows.)

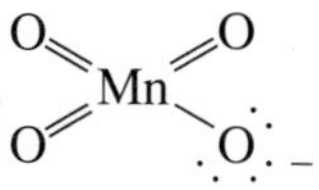

5.5 OZONOLYSIS OF ALKENES

Ozone, O_3, adds to alkenes at low temperature to yield an unstable compound called a *molozonide*. The molozonide is rapidly transformed into a second adduct, called simply an *ozonide*.

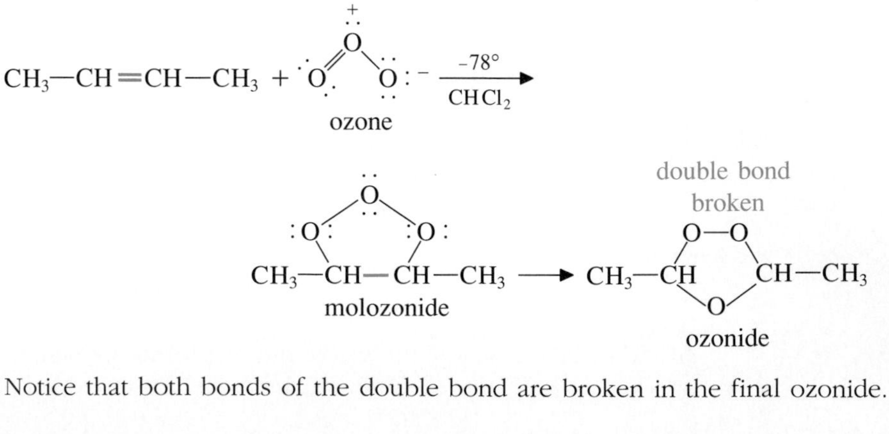

(5.36)

Notice that both bonds of the double bond are broken in the final ozonide.

How does formation of the molozonide occur? To begin with, the resonance structures of ozone show that the outer oxygens are electron-deficient, and thus have electrophilic (Lewis acid) character.

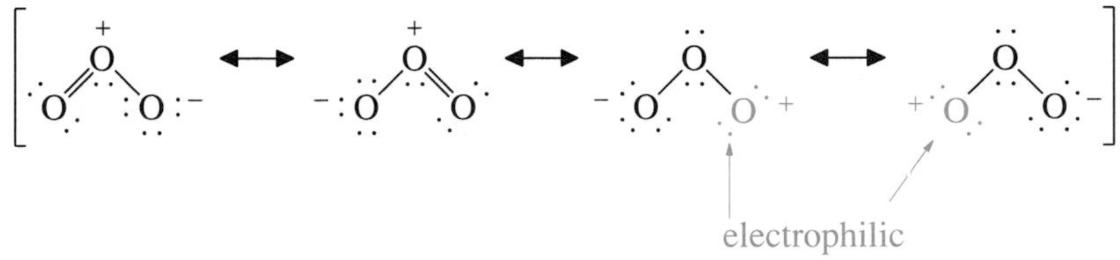

(5.37)

This property, as well as the general electronegativity of oxygen, generates an attraction of the ozone molecule for the alkene π-electrons. The ozone molecule therefore reacts with an alkene in a *pericyclic reaction* (Sec. 5.3A), which, depending on the resonance form of ozone that we use, can be written in either of two ways:

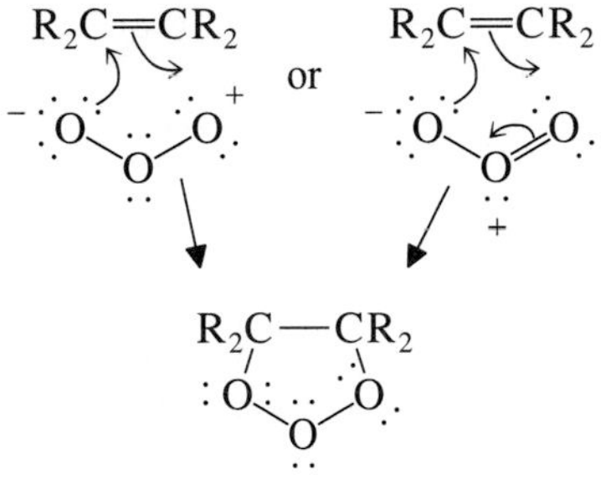

(5.38)

This reaction is also a *cycloaddition reaction* (Sec. 5.4), because a ring is created in the addition. In a series of subsequent reactions, which we shall not discuss, the molozonide is converted into the ozonide.

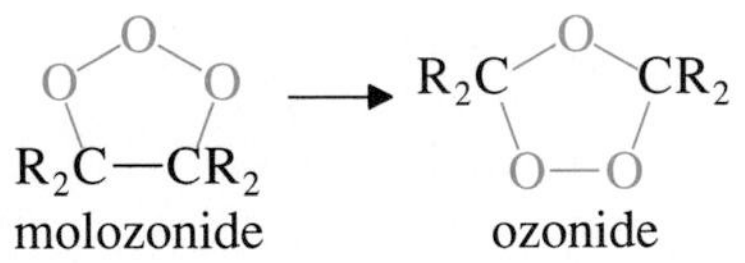

(5.39)

Ozonolysis is useful because of the reactions that ozonides can undergo. For example, when an alkene reacts with ozone and the resulting ozonide is treated with a mild reducing agent such as dimethyl sulfide [$(CH_3)_2S$], the alkene is "split" to give aldehydes and/or ketones derived from the fragments at each end of the double bond. (We shall not consider the mechanism of this conversion.)

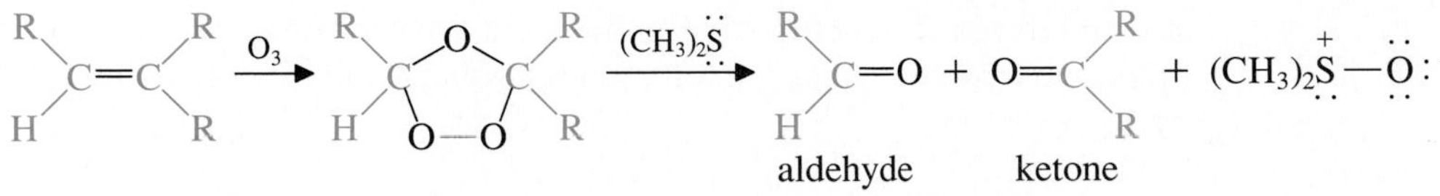

(5.40)

In effect, >C=C< is replaced by >C=O + O=C<

If the two ends of the alkene are identical, then two moles of the same compound are obtained from each mole of the alkene.

$$CH_3{-}CH{=}CH{-}CH_3 \xrightarrow{O_3} \xrightarrow{(CH_3)_2S} 2CH_3CH{=}O \qquad (5.41)$$

When the two ends of the alkene double bond bear different substituents, two different compounds are obtained.

$$CH_3(CH_2)_5CH{=}CH_2 \xrightarrow[CH_2Cl_2]{O_3} \xrightarrow{(CH_3)_2S} \underset{\text{hexanal (75\% yield)}}{CH_3(CH_2)_5CH{=}O} + \underset{\text{formaldehyde}}{CH_2{=}O} \qquad (5.42)$$

If the ozonide is treated with hydrogen peroxide (H_2O_2), a carboxylic acid instead of an aldehyde forms from each end of the double bond that bears a hydrogen; ketones are obtained if there is no hydrogen on the alkene carbon, as in Eq. 5.40.

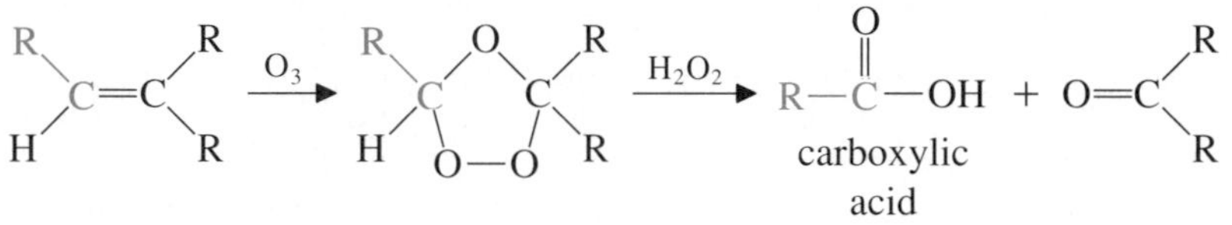

(5.43)

Returning, for example, to the alkene used in Eq. 5.42, the products are different if the ozonide is treated with hydrogen peroxide instead of dimethyl sulfide.

$$CH_3{-}(CH_2)_5{-}CH{=}CH_2 \xrightarrow{O_3} \xrightarrow{H_2O_2} CH_3{-}(CH_2)_5{-}\overset{O}{\overset{\|}{C}}{-}OH + HO{-}\overset{O}{\overset{\|}{C}}{-}H \qquad (5.44)$$

Problem

9 Give the products (if any) expected from treatment of each of the following compounds with ozone followed by (1) dimethyl sulfide; (2) hydrogen peroxide.

(a) 3-methyl-2-pentene

(b)

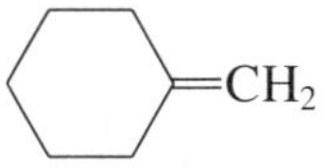

(c) cyclooctene

(d) 2-methylpentane

One use of ozonolysis is to determine the structure of unknown alkenes. Thus, if ozonolysis of an alkene, followed by one of the usual treatments, yields known aldehydes, ketones, or carboxylic acids, the structure of the alkene can be mentally reconstructed from the structures of the fragments. To illustrate, suppose that an alkene of unknown structure gives the following products after treatment with ozone followed by H_2O_2:

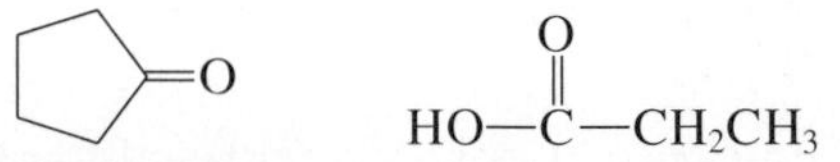

cyclopentanone propionic acid

The structure of the alkene can be deduced by *mentally reversing* the ozonolysis reaction.

ozonolysis products imply alkene starting material (5.45)

Problem

10 In each case, give the structure of an eight-carbon alkene that would yield each of the following compounds (and no others) after treatment with ozone followed by dimethyl sulfide:

(a) $CH_3CH_2CH_2CH{=}O$

(b) $O{=}CH—(CH_2)_5—\overset{\overset{\large O}{\|}}{C}—CH_3$

(c)

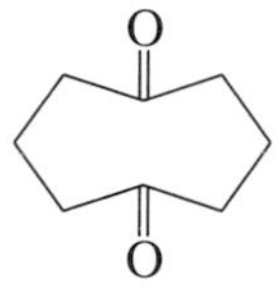

What aspect of alkene structure cannot be determined by ozonolysis?

5.6 SUMMARY OF ELECTROPHILIC ADDITION REACTIONS

We have now surveyed a number of electrophilic addition reactions of alkenes. Some involve carbocation intermediates (addition of hydrogen halides, acid-catalyzed hydration). Through a study of these reactions, we have learned some of the chemical properties of carbocations. Because they are Lewis acids, carbocations can react with Lewis bases (nucleophiles), and they can rearrange to other, more stable, carbocations. Other addition reactions involve *cyclic ions* as intermediates (oxymercuration, halogenation). These reactions are very similar to those involving carbocation intermediates; but because the cyclic-ion intermediates have only partial positive charge on carbon, and are not fully developed carbocations, they have a decreased tendency toward rearrangements. Other electrophilic addition reactions occur as *concerted reactions* (hydroboration, glycol formation, and ozonolysis).

All of these reactions have in common the fact that they are initiated by the interaction of an electrophile (Lewis acid), or the electron-deficient atom of a polar bond, with the π-electron cloud of the alkene. We shall study other electrophilic addition reactions of alkenes in subsequent chapters, but we shall usually find that their mechanisms fit into one of these categories. We shall also find that much alkyne chemistry—that is, chemistry of the carbon–carbon triple bond—is closely related to that of alkenes. Thus if we master the principles discussed in this and the preceding chapter we shall have mastered much of the chemistry to be encountered later.

As important as electrophilic additions are, alkenes also undergo other types of addition reactions. Some of these will occupy our attention in the next two sections.

5.7 CATALYTIC HYDROGENATION OF ALKENES

A. Conversion of Alkenes into Alkanes

Hydrogen can be added to the alkene double bond in the presence of certain insoluble, finely divided metal catalysts, such as platinum, palladium, or nickel. These catalysts are often adsorbed on support materials such as alumina (Al_2O_3), barium sulfate ($BaSO_4$), or activated charcoal (finely divided carbon). For example, in the equations below, Pt/C means "platinum catalyst adsorbed on a carbon support."

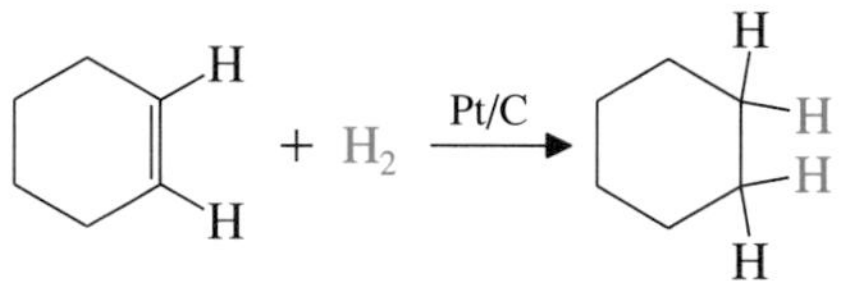

(5.46)

$$CH_3(CH_2)_5CH{=}CH_2 + H_2 \xrightarrow{Pt/C} CH_3(CH_2)_5CH_2{-}CH_3 \quad (5.47)$$

These reactions are examples of **catalytic hydrogenation,** which is one of the best ways to convert alkenes into alkanes. Catalytic hydrogenation is an important reaction in both industry and the laboratory. The fact that a special apparatus must be used for the handling of a flammable gas (hydrogen) is more than offset by the yields in this reaction, which typically approach 100%.

In the absence of appropriate catalysts, hydrogenation does not take place. The effect of metal catalysts on the rate of the hydrogenation reaction is one of the most dramatic and important examples of catalysis known in chemistry. Because hydrogenation catalysts are not consumed in the hydrogenation reaction, they can be used in small quantities.

The typical metal catalysts used in catalytic hydrogenation are insoluble in the reaction medium, and for this reason are classified as **heterogeneous catalysts.** In contrast, soluble catalysts, such as the acid catalysts used in alkene hydration (Sec. 4.8), are classified as **homogeneous catalysts.** The insolubility of heterogeneous hydrogenation catalysts means that they are easy to separate from a solution of the reaction products by filtration. This is important for two reasons. First, purification of the products is simple. Second, the catalyst materials, which are often rather expensive precious or semiprecious metals, can be easily recycled and reused.

The hydrogenation reaction takes place on the metal surface of the catalyst, at the interface between solution and metal. However, very few details of the mechanism of catalytic hydrogenation are known. There has been considerable recent research into this and related areas of catalysis by physicists who study the properties of surfaces, and by inorganic chemists who synthesize metal clusters of defined structure that emulate some of the catalytic properties of metal surfaces.

The energy crisis of 1974 stimulated a great deal of research in many aspects of noble-metal catalysis. For example, interest has increased in new methods of catalyzing the synthesis of methane or other gaseous fuels from readily available materials other than petroleum. Noble-metal catalysts also have a role in pollution control. The catalytic converter used on automobiles contains a platinum catalyst that promotes the complete oxidation of hydrocarbon exhaust emissions. Although catalysts should in theory function indefinitely, they slowly adsorb impurities (called *poisons*) and lose their ability to function effectively. The lead in leaded gasoline, for example, serves as a very potent poison of the catalyst in the catalytic converter. (Lead is also an effective poison of hydrogenation catalysts.) This is why leaded gasoline cannot be used in cars equipped with the catalytic converter.

The following example illustrates again (see Eq. 5.25) the resistance of benzene rings to conditions under which double bonds readily react:

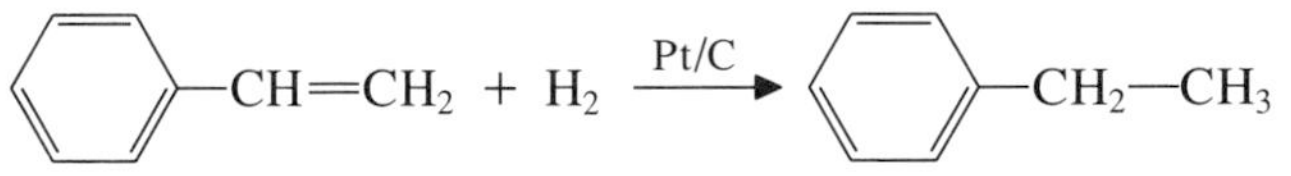

$$\text{styrene} \qquad\qquad \text{ethylbenzene} \qquad\qquad (5.48)$$

(Benzene rings can be hydrogenated under much more vigorous conditions.)

Problem **11** Give the structure of five alkenes with the formula C_6H_{12} that would give hexane as the product of catalytic hydrogenation.

B. Heats of Hydrogenation

Hydrogenation, like combustion, is a source of thermochemical data that can be used for the comparison of relative molecular stabilities. **Heats of hydrogenation** were at one time the most important source of such data. For example, the heats of hydrogenation of *cis*- and *trans*-2-butene are -28.6 and -27.6 kcal/mol, respectively.

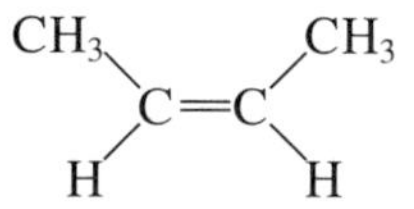

$$\begin{array}{c}\text{CH}_3 \qquad \text{CH}_3 \\ \diagdown\text{C}{=}\text{C}\diagup \\ \text{H} \qquad \text{H}\end{array} + \text{H}_2 \xrightarrow{\text{catalyst}} \text{CH}_3{-}\text{CH}_2{-}\text{CH}_2{-}\text{CH}_3 \qquad \Delta H^0 = -28.6 \text{ kcal/mol} \quad (5.49a)$$

$$\begin{array}{c}\text{CH}_3 \qquad \text{H} \\ \diagdown\text{C}{=}\text{C}\diagup \\ \text{H} \qquad \text{CH}_3\end{array} + \text{H}_2 \xrightarrow{\text{catalyst}} \text{CH}_3{-}\text{CH}_2{-}\text{CH}_2{-}\text{CH}_3 \qquad \Delta H^0 = -27.6 \text{ kcal/mol} \quad (5.49b)$$

Notice that the difference in the heats of hydrogenation of these two compounds is precisely the same as the difference in their heats of formation (1.0 kcal/mol; Sec. 4.4B). Whether we subtract their heats of hydrogenation, heats of combustion, or heats of formation, application of Hess's Law (Sec. 4.4B) shows that each of these differences is equal to the relative ΔH^0 of the two compounds.

Problems **12** From the heat of hydrogenation of 2-methylpropene (-28.1 kcal/mol) and its heat of formation (-4.04 kcal/mol), calculate the heat of formation of 2-methylpropane.

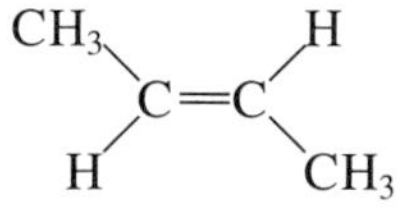

Problems (Cont.)

13 (a) Use Hess's Law to show that the difference in the heats of hydrogenation for *cis*- and *trans*-2-butene must equal the difference in their heats of formation.

(b) The heats of hydrogenation of 1-butene and 2-methylpropene are −30.3 and −28.1 kcal/mol, respectively. Explain why the difference between the heats of hydrogenation of these two alkenes is *not* expected to be the same as the difference between their heats of formation.

5.8 ADDITION OF HYDROGEN BROMIDE TO ALKENES: THE PEROXIDE EFFECT. FREE-RADICAL CHAIN REACTIONS

The regioselectivity of HBr addition to alkenes is dictated by the Markovnikov rule (Sec. 4.5A). We have learned that the basis of this rule for HBr addition can be understood in terms of carbocation intermediates. However, in organic chemistry prior to 1930, not only were the explanations of the Markovnikov rule not clear, but the observations themselves were not always reproducible. As an example, consider the following reaction:

$$\underset{}{HBr} + \underset{\text{allyl bromide}}{CH_2{=}CH{-}CH_2Br} \longrightarrow \underset{\text{1,2-dibromopropane}}{CH_3{-}\underset{\underset{Br}{|}}{C}H{-}CH_2Br} + \underset{\text{1,3-dibromopropane}}{Br{-}CH_2CH_2CH_2{-}Br} \quad (5.50)$$

(Which of the two products is predicted to predominate by the Markovnikov Rule?) When this reaction was carried out on different samples of allyl bromide from the same laboratory, sometimes 1,2-dibromopropane would predominate in the product mixture, but at other times 1,3-dibromopropane would predominate.

One of the precepts on which scientific inquiry is based is that different experiments carried out under identical conditions must yield the same results. Why, then, were the results of these seemingly identical reactions so different? A careful investigation (by Morris Kharasch of the University of Chicago) revealed that some of the samples of allyl bromide were contaminated by **peroxides,** compounds with the general structure R—O—O—R. Indeed, when peroxides were added to the reaction, addition of HBr to allyl bromide occurred rapidly, and the *anti-Markovnikov* product was the major product formed.

$$CH_2{=}CH{-}CH_2Br + HBr \xrightarrow[\substack{\text{(benzoyl peroxide)}\\ \text{trace; 18 hr}}]{C_6H_5\overset{\overset{O}{\|}}{C}{-}O{-}O{-}\overset{\overset{O}{\|}}{C}{-}C_6H_5}$$

$$\underset{\substack{\text{anti-Markovnikov product}\\ \text{(82\% of product)}}}{Br{-}CH_2CH_2CH_2{-}Br} + \underset{\text{(18\% of product)}}{CH_3{-}\underset{\underset{Br}{|}}{C}H{-}CH_2{-}Br} \quad (5.51)$$

It was also found that light promotes the anti-Markovnikov addition of HBr in the presence of peroxides. When peroxides and light were scrupulously excluded from the reaction, HBr addition to allyl bromide took place slowly and conformed to the Markovnikov rule.

$$CH_2{=}CH{-}CH_2Br + HBr \xrightarrow{\text{8 days}} \underset{\substack{\text{Markovnikov product}\\ \text{(90\% of product)}}}{CH_3{-}\overset{\displaystyle Br}{\overset{|}{C}}H{-}CH_2Br} + \underset{\text{(10\% of product)}}{Br{-}CH_2CH_2CH_2{-}Br} \qquad (5.52)$$

We now know that addition of peroxides reverses the regioselectivity of HBr addition to alkenes. In other words, *in the presence of peroxides, the addition of HBr to alkenes occurs such that the hydrogen is bound to the alkene carbon bearing the greater number of alkyl groups.* Very small amounts of peroxides are required in order to bring about this effect.

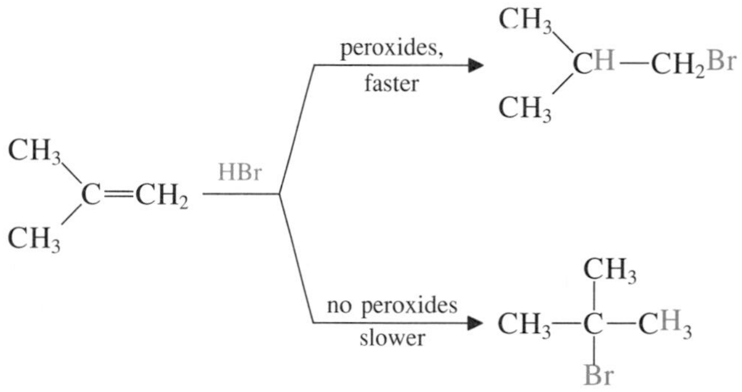

(5.53)

Further, the regioselectivity of HI or HCl addition to alkenes is much less sensitive to the presence of peroxides: Markovnikov addition predominates with HCl and HI whether peroxides are present or not!

This reversal of the Markovnikov rule suggests a different mechanism for HBr addition in the presence of peroxides than for normal HBr addition. This mechanism must explain not only why the Markovnikov rule is violated, but also why HCl and HI additions are relatively unaffected by peroxides. It turns out that this mechanism involves a type of reactive intermediate that we have not previously studied: the *free radical.* Before continuing with our study of the peroxide-promoted addition of HBr to alkenes, let us learn some basic facts about free radicals.

A. Homolysis vs. Heterolysis. Free Radicals

In all reactions considered previously we used an arrow formalism that indicates the movement of electrons in pairs. The dissociation of HBr, for example, is written

$$H{-}\ddot{\underset{..}{Br}}{:} \longrightarrow H^+ + {:}\ddot{\underset{..}{Br}}{:}^- \qquad (5.54)$$

In this reaction, both electrons of the H—Br covalent bond move to the bromine to give a bromide ion, and the hydrogen is left as the electron-deficient proton. This type of bond breaking is called **heterolytic cleavage,** or **heterolysis.**

However, this is not the only type of bond rupture that can occur. Electron-pair bonds may also break so that each bonding partner retains one electron of the chemical bond.

$$H \!-\! \ddot{B}\ddot{r}: \longrightarrow H\cdot + \cdot\ddot{B}\ddot{r}: \qquad (5.55)$$

In this process, a hydrogen *atom* and a bromine *atom* are produced. As we can readily verify, these atoms are uncharged. This type of bond breaking is called **homolytic cleavage,** or **homolysis.**

A different arrow formalism is used for homolysis. In this formalism *electrons are moved individually rather than in pairs.* This type of electron flow is represented with singly-barbed arrows, or fishhooks; one fishhook is used for each electron:

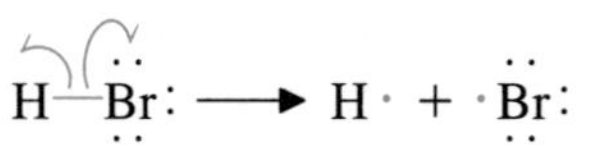

$$H \!-\! \ddot{B}\ddot{r}: \longrightarrow H\cdot + \cdot\ddot{B}\ddot{r}: \qquad (5.56)$$

Homolytic bond cleavage is not restricted to diatomic molecules. For example, peroxides readily undergo homolytic cleavage at the O—O bond.

$$(CH_3)_3C \!-\! \ddot{O} \!-\! \ddot{O} \!-\! C(CH_3)_3 \longrightarrow 2(CH_3)_3C \!-\! \ddot{O}\cdot \qquad (5.57)$$

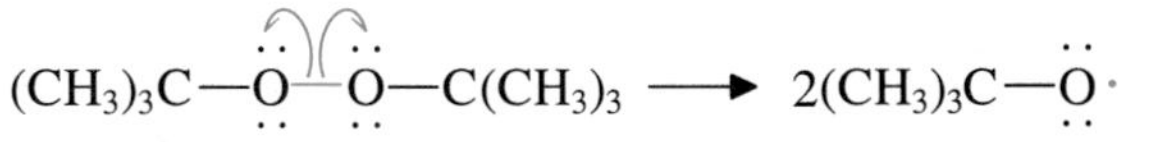

di-t-butyl peroxide t-butoxy radical

The fragments on the right side of this equation possess unpaired electrons. Any species with at least one unpaired electron is called a **free radical.**

Problems

14 Draw the products of each of the following transformations shown by the arrow formalism:

(a) $(CH_3)_2C\!=\!CH_2$ $\longrightarrow$ (c) $R\!-\!CH_2\!-\!CH_2\cdot \longrightarrow$

(b) $CH_2\!=\!CH\!-\!CH\!=\!CH\!-\!CH_2\cdot \longleftrightarrow$

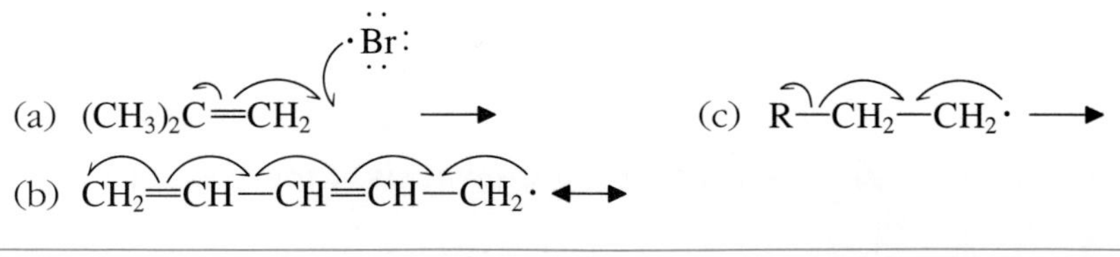

15 Indicate whether each of the following reactions is homolytic or heterolytic. Write the appropriate arrow formalism for each.

(a) $H_2 \longrightarrow 2H\cdot$ (d) $CH_2\!=\!CH_2 + HBr \longrightarrow {}^+CH_2\!-\!CH_3 + Br^-$

(b) $H_2\ddot{O} \longrightarrow H^+ + :\ddot{O}H^-$ (e) $CH_2\!=\!CH_2 + HBr \longrightarrow \cdot CH_2\!-\!CH_3 + Br\cdot$

(c) $H_2\ddot{O} \longrightarrow H\cdot + \cdot\ddot{O}H$

B. Free-Radical Addition of HBr to Alkenes. Free-Radical Chain Reactions

Although a few stable free radicals are known, most free radicals are very reactive and cannot be isolated. Nevertheless, free radicals are reactive intermediates in a number of organic reactions. For example, the peroxide-initiated addition of HBr to alkenes involves free-radical intermediates.

Let us assume that a peroxide such as the one in Eq. 5.57, di-*t*-butyl peroxide, is present in reaction mixture during the addition of HBr to a general alkene. Homolytic cleavage of the peroxide to give free radicals is promoted when the peroxide absorbs energy from heat or light. A radical formed in this way then reacts with HBr by removing a hydrogen *atom* to form a neutral alcohol and a new free radical, the bromine *atom*.

$$(CH_3)_3C\!-\!\ddot{O}\!:\overset{\frown}{}H\!\overset{\frown}{\underset{\smile}{}}\!\ddot{B}\ddot{r}: \longrightarrow (CH_3)_3C\!-\!\ddot{O}\!-\!H + \cdot\ddot{B}\ddot{r}: \qquad (5.58a)$$

This step illustrates a very common reaction of free radicals: *atom abstraction,* or the removal of an atom from another molecule. The bromine atom thus formed reacts with the π-bond of an alkene.

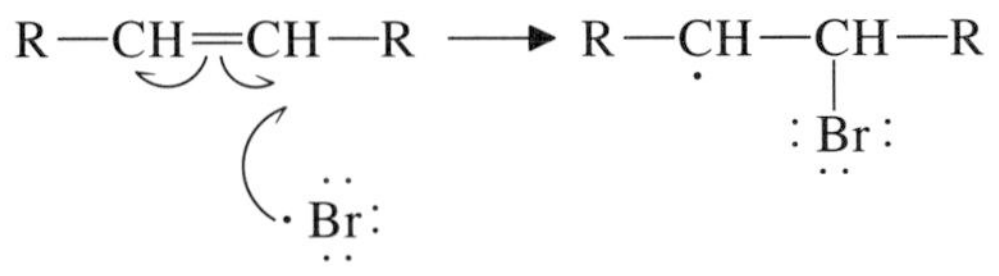

$$(5.58b)$$

This step illustrates a second common reaction of free radicals: *addition to a double bond.* Finally, the new radical generated in this step removes a hydrogen atom from HBr to generate another bromine atom.

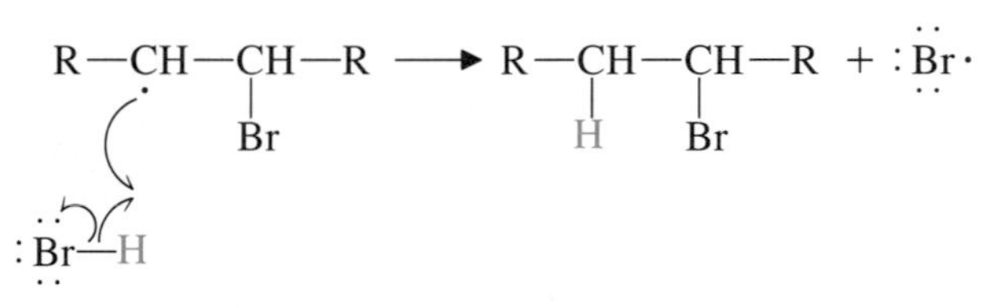

$$(5.58c)$$

The bromine atom, in turn, may react with another molecule of alkene (Eq. 5.58b), and this can be followed again by the generation of another bromine atom (Eq. 5.58c). This process can be repeated many times in a chain-like fashion: each reaction of an alkene with a bromine atom ultimately leads to the generation of a new bromine atom. The bromine atom concentration is never large, because for each radical formed in this process, one is also consumed. Such a reaction is an example of a **free-radical chain reaction.** In a free-radical chain reaction, the radical product of one reaction becomes the starting material for another. The radical product of the last reaction in the chain (Br · in this example) is used as a starting material for the first reaction.

The radical that is consumed in one step of a chain reaction and formed in another is said to *propagate the chain*. In the reactions above, the bromine atom and the alkyl radical are the chain-propagating radicals. For each "link," or cycle, of the free-radical chain reaction above, one molecule of alkyl bromide is produced.

An analogy for a chain reaction can be found in the world of business. A businessperson uses a little seed money, or capital, to purchase a small business. In time, this business produces profit that is used to buy another business. This business produces profit that can be used to buy yet another business, and so on. All this time, the businessperson is accumulating business property (instead of alkyl bromides) although the total amount of cash on hand is, by analogy to the chain-propagating radicals in the reaction sequence above, small compared with the total amount of the investments. Each expenditure of capital is ultimately replaced by profit.

While the chain reaction in Eqs. 5.58b,c is taking place, a competing reaction is occurring. Some bromine atoms react with other bromine atoms to form Br_2.

$$: \overset{..}{\underset{..}{Br}} \cdot \; + \; : \overset{..}{\underset{..}{Br}} \cdot \; \longrightarrow \; Br_2 \tag{5.58d}$$

Similarly, the radical products of Eq. 5.58b may also recombine.

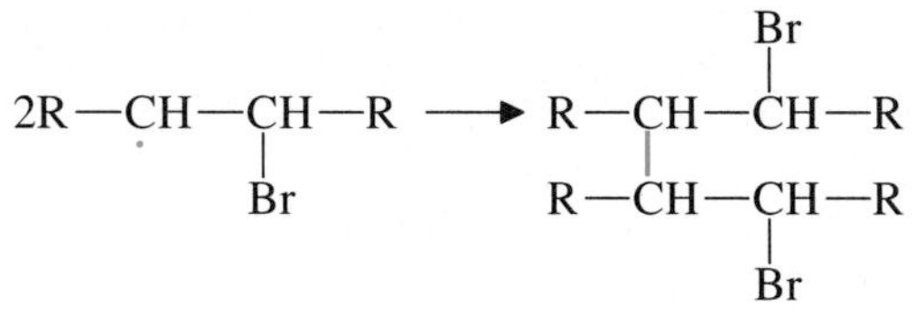

$$2R-CH-CH-R \longrightarrow \begin{array}{c} Br \\ | \\ R-CH-CH-R \\ R-CH-CH-R \\ | \\ Br \end{array} \tag{5.58e}$$

Such *recombination* reactions represent a third type of free-radical reaction.

Since breaking chemical bonds requires substantial energy, formation of chemical bonds liberates substantial energy. Hence, recombination of radicals is, energetically, a very favorable process. Most radical-recombination reactions are very fast. In view of this fact, we might ask why radicals do not recombine before they propagate any chains. The answer is simply a matter of the relative concentrations of the various species involved. Free radicals are present, as we have noted above, in very low concentration, but the other reactants in a chain reaction are present in much higher concentration. Thus, it is much more probable for a bromine atom in the chain reaction sequence of Eq. 5.58 to collide and react with an alkene molecule than with a bromine atom.

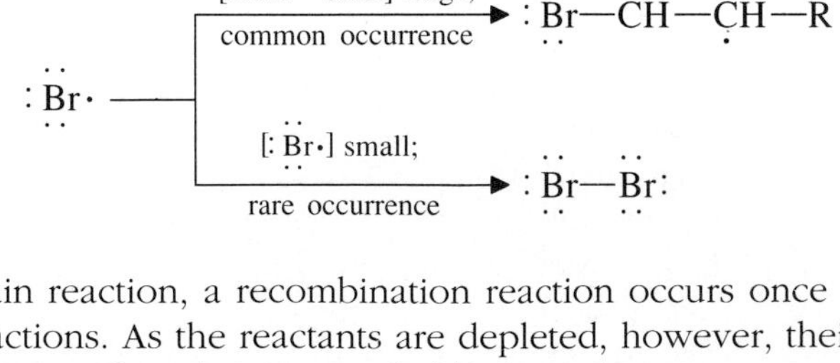

$$: \overset{..}{\underset{..}{Br}} \cdot \; \begin{cases} \xrightarrow[\text{common occurrence}]{\text{[RCH=CHR] large;}} \; : \overset{..}{\underset{..}{Br}}-CH-\overset{\cdot}{C}H-R \\ \\ \xrightarrow[\text{rare occurrence}]{\text{[: } \overset{..}{Br} \cdot \text{] small;}} \; : \overset{..}{\underset{..}{Br}}-\overset{..}{\underset{..}{Br}}: \end{cases} \tag{5.59}$$

In a typical chain reaction, a recombination reaction occurs once for every 10,000 propagation reactions. As the reactants are depleted, however, their concentrations are low, and it is significantly more probable that a bromine atom will find another radical with which to recombine.

Problem

16 Suggest a reason why small amounts of dihalide R—CHBr—CHBr—R are formed during the peroxide-initiated addition of HBr to an alkene R—CH=CH—R.

Free-radical chain reactions can be divided into three types of mechanistic steps:

1. Initiation
2. Propagation
3. Termination

In the **initiation** steps, free radicals are formed from a molecule that easily undergoes homolysis, called a **free-radical initiator.** In the reactions we have been discussing, peroxides serve as the free-radical initiators. The initiation steps are complete when a chain-propagating radical is formed.

initiation steps:

$$(CH_3)_3C{-}\ddot{\underset{..}{O}}{-}\ddot{\underset{..}{O}}{-}C(CH_3)_3 \longrightarrow 2(CH_3)_3C{-}\ddot{\underset{..}{O}}\cdot \qquad (5.60a)$$

$$(CH_3)_3C{-}\ddot{\underset{..}{O}}\cdot + H\ddot{\underset{..}{Br}}: \longrightarrow (CH_3)_3C{-}OH + :\ddot{\underset{..}{Br}}\cdot \qquad (5.60b)$$

Many compounds besides peroxides can serve as free-radical initiators. One is azoisobutyronitrile, known to chemists by the acronym AIBN. This substance readily forms free radicals because the very stable nitrogen molecule is liberated as a result of homolytic cleavage:

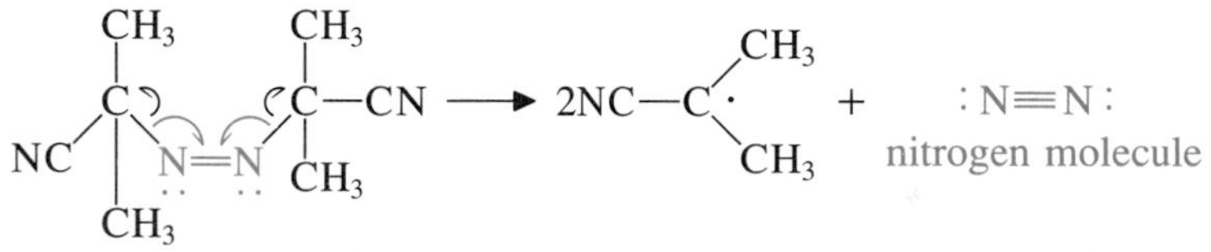

azoisobutyronitrile
(AIBN)

(5.61)

Ordinarily much less initiator than reactants is required because only a small concentration of free radicals is needed to propagate the radical chain. Typically the concentration of peroxide initiator might be 2% of the alkene concentration. Heat or light sometimes accelerates free-radical reactions because initiators or, in some cases, the reactants themselves, absorb energy from heat or light to form radicals.

In the **propagation** steps, the chain-carrying radicals are alternately consumed and produced. The algebraic sum of the propagation steps is equal to the overall chemical transformation.

propagation steps:

$$Br\cdot + R{-}CH{=}CH{-}R \longrightarrow R{-}\dot{C}H{-}CH(Br){-}R \qquad (5.62a)$$

$$R{-}\dot{C}H{-}CH(Br){-}R + H{-}Br \longrightarrow R{-}CH(H){-}CH(Br){-}R + Br\cdot \qquad (5.62b)$$

sum:

$$R{-}CH{=}CH{-}R + HBr \longrightarrow R{-}CH_2{-}CHBr{-}R \qquad (5.62c)$$

In the **termination** steps, radicals are destroyed by recombination reactions and other processes. Relatively minor amounts of by-products are produced by termination reactions, because there are only small amounts of free radicals present at any time.

a termination step:

$$2\ :\!\ddot{\underset{..}{Br}}\cdot \longrightarrow Br_2 \qquad (5.63)$$

How do we know when to write a free-radical mechanism for a given reaction? One telltale clue is that free-radical reactions are promoted by initiators such as peroxides and AIBN. When a reaction occurs in the presence of an initiator but does not occur in its absence, we can be fairly sure that free-radical intermediates are involved.

A second clue is that other molecules, called **free-radical inhibitors,** have exactly the opposite effect. When these compounds are present, free-radical chain reactions cannot occur. Examples of inhibitors are iodine, amines, molecular oxygen, elemental sulfur, and phenols. For example, hydroquinone and resorcinol are commonly used as free-radical inhibitors.

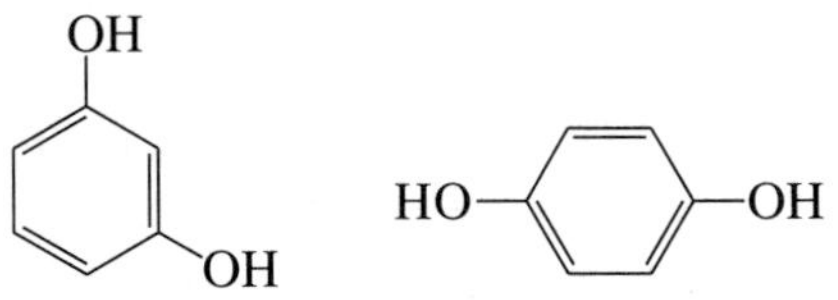

resorcinol hydroquinone

If a reaction that takes place readily does not occur in the presence of an inhibitor, we can reasonably suspect that the reaction occurs by a free-radical chain mechanism.

Inhibitors function by a variety of mechanisms. Some inhibitors react to form stable free radicals that are not reactive enough to propagate a radical chain. Other inhibitors form free radicals that scavenge other free radicals by recombination or other termination reactions. In some cases the mechanism of inhibition is not known.

C. Structure and Stability of Free Radicals. Explanation of the Peroxide Effect

Let us now try to understand why peroxides and other free-radical initiators cause HBr addition to take place with anti-Markovnikov regioselectivity, as in the following example:

$$(CH_3)_2C{=}CH_2 \xrightarrow{\text{HBr, peroxides}} (CH_3)_2CH{-}CH_2{-}Br \qquad (5.64)$$

The reaction is initiated by the formation of a bromine atom from HBr (Eq. 5.58a). When the bromine atom adds to the π-bond of an alkene, it has a choice: it can react at either of the two alkene carbons of the double bond to give different free-radical intermediates.

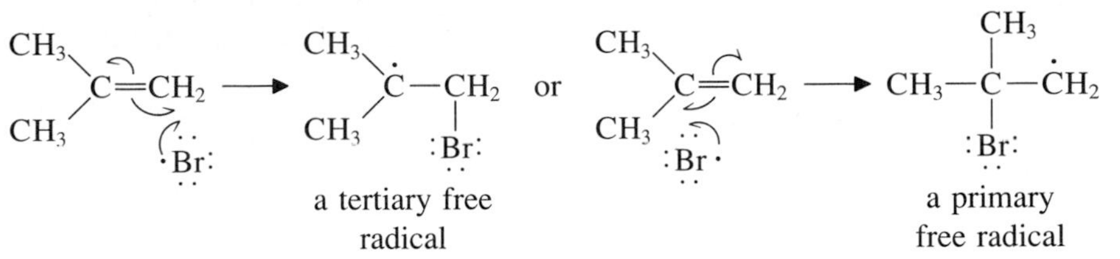

(5.65)

Free radicals, like carbocations, can be classified as primary, secondary, or tertiary.

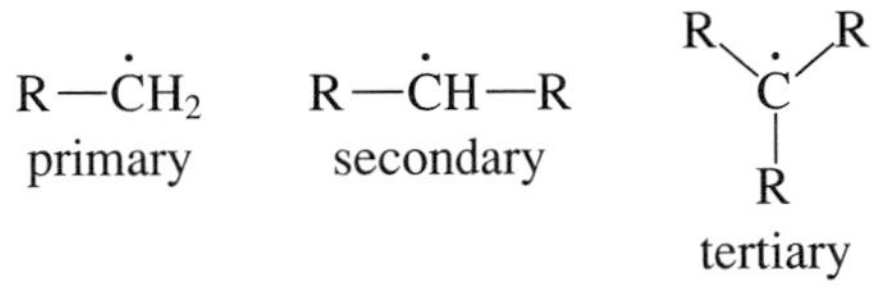

Equation 5.65 thus involves a competition between the formation of a primary and a tertiary free radical. *The formation of the tertiary radical is preferred,* and its reaction with HBr accounts for the product of Eq. 5.64. Why is the tertiary free radical formed rather than the primary one?

The first reason is that when the rather large bromine atom reacts with the alkene, it experiences fewer van der Waals repulsions when approaching the less substituted alkene carbon. That is, the reaction is governed in part by a *steric effect.*

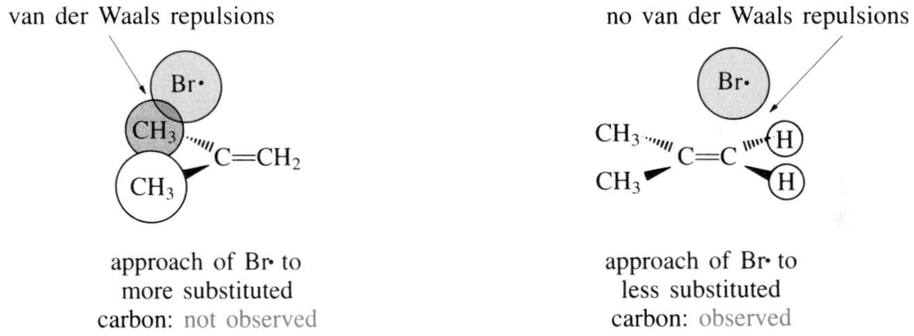

The second reason that the tertiary radical is formed has to do with its relative stability. The heats of formation of several free radicals are given in Table 5.1. By comparing the ΔH_f^0 values for *n*-propyl and isopropyl radicals, we find that the secondary radical is more stable than the primary one by 3.6 kcal/mol. Similarly, the tertiary *t*-butyl radical is more stable than the secondary *sec*-butyl radical by 4.5 kcal/mol.

stability of free radicals:

$$\text{tertiary} > \text{secondary} > \text{primary} \qquad (5.66)$$

Notice that free radicals have the same stability order as carbocations. However, the energy differences between isomeric radicals are considerably smaller than the differences between carbocations.

TABLE 5.1 Heats of Formation of Some Free Radicals (27 °C)

Radical	Structure	ΔH_f°, kcal/mol
methyl	$\cdot CH_3$	34.4
ethyl	$\cdot CH_2CH_3$	28.0
n-propyl	$\cdot CH_2CH_2CH_3$	22.8
isopropyl	$(CH_3)_2\dot{C}H$	19.2
sec-butyl	$CH_3\dot{C}HCH_2CH_3$	13.9
tert-butyl	$(CH_3)_3C\cdot$	9.4

We can see why this free-radical stability order is reasonable if we examine the geometry and hybridization of a typical carbon radical (Fig. 5.1). We have not developed rules for predicting the geometry of molecules with unpaired electrons. Although the subject continues to be debated, it appears that alkyl radicals have a trigonal planar, or nearly planar, geometry. As we learned in Sec. 4.1, trigonal carbons are sp^2 hybridized. Hence, alkyl radicals are sp^2 hybridized; the unpaired electron is in a carbon $2p$ orbital. The stability order in Eq. 5.66 implies, then, that *free radicals, like alkenes, are stabilized by alkyl-group substitution at sp^2-hybridized carbons.*

Now that we understand how the stability of free radicals varies with alkyl substitution, we can understand why HBr addition to alkenes in the presence of peroxides occurs with anti-Markovnikov regioselectivity. *Hammond's postulate* (Sec. 4.7C) states that the more stable of two reactive intermediates should be formed more rapidly. Thus, when a bromine atom adds to the π-bond of an alkene, the *more stable,* and hence, *more substituted,* radical is formed more rapidly than the less substituted one. The radical intermediate thus formed abstracts a hydrogen atom from HBr to give the product of anti-Markovnikov addition.

Let us be sure we also understand why the regioselectivity of HBr addition to alkenes differs in the presence and absence of peroxides. Both reactions begin by attachment of an atom to the less substituted alkene carbon. However, in the free-radical addition, Br · is the first group that attacks the alkene π-bond, and as a result ends up on the *less* substituted alkene carbon atom. In electrophilic addition (absence of peroxides), the proton of HBr adds first, and the bromine ends up on the *more*

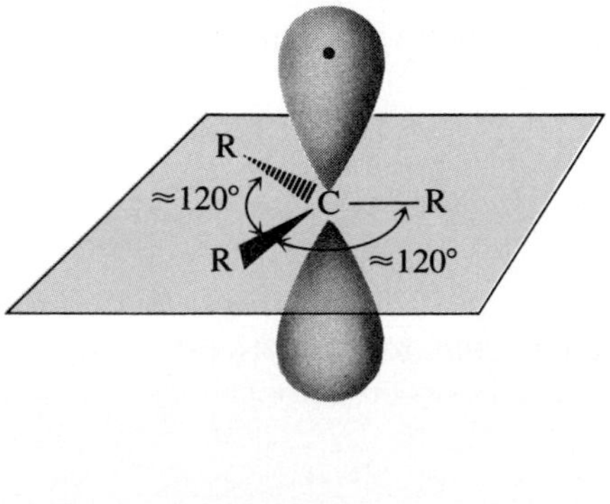

Figure 5.1 Carbon radicals are sp^2 hybridized and have trigonal geometry.

substituted alkene carbon. It should also be clear that the peroxide effect is not limited to peroxides: any good free-radical initiator will bring about the same effect.

Problem

17 Predict the major organic product(s) formed when HBr reacts with each of the following alkenes in the presence of peroxides:

(a) 2-methyl-2-butene (c) *trans*-4-methyl-2-hexene

(b) 1-ethyl-1-cyclopentene

Write out the entire mechanism for the reaction of (b); identify initiation, propagation, and termination steps of the reaction.

D. Bond Dissociation Energies. Absence of the Peroxide Effect in Addition of HCl and HI to Alkenes

How easily does a chemical bond break homolytically to form free radicals? This question can be answered by examining the **bond dissociation energy.** For a bond between two atoms A—B, the bond dissociation energy is defined as the enthalpy ΔH^0 of the reaction

$$\mathrm{A{-}B} \longrightarrow \mathrm{A}\cdot + \mathrm{B}\cdot \qquad (5.67)$$

Notice that a bond dissociation energy always corresponds to the enthalpy required to break a bond *homolytically*. Thus, the bond energy of H—Br refers to the process

$$\mathrm{H{-}\ddot{\underset{..}{Br}}\!:} \longrightarrow \mathrm{H}\cdot + \cdot\mathrm{\ddot{\underset{..}{Br}}\!:}$$

and *not* to the heterolytic process

$$\mathrm{H{-}\ddot{\underset{..}{Br}}\!:} \longrightarrow \mathrm{H^+} + {}^{-}\!:\mathrm{\ddot{\underset{..}{Br}}\!:}$$

Some bond dissociation energies are collected in Table 5.2. *A bond dissociation energy measures the intrinsic strength of a chemical bond.* For example, breaking the H—H bond requires 104 kcal/mol. It then follows that forming the hydrogen molecule from two hydrogen atoms liberates 104 kcal/mol of energy. Table 5.2 shows that different bonds exhibit significant differences in bond strength; even bonds of the same general type—for example, the various C—H bonds—can differ in bond strength by many kilocalories per mole.

From these data we can see why peroxides are excellent sources of free radicals. The homolysis of di-*tert*-butyl peroxide (the last entry in the table) requires only 35 kcal/mol. This energy is much lower than that required to break most carbon–carbon or carbon–hydrogen bonds.

Problems

18 Use the definition of the heat of formation (Sec. 4.4B) to calculate ΔH_f^0 for (a) the hydrogen atom; (b) the bromine atom.

19 Use your answer from Problem 18 and data from Tables 5.1 and 5.2 to calculate the heat of formation of ethane. (*Hint:* $C_2H_6 \rightarrow \cdot CH_2CH_3 + \cdot H$)

TABLE 5.2 Bond Dissociation Energies (kcal/mol)

X—X bonds	
H—H	104
F—F	38
Cl—Cl	58
Br—Br	46
I—I	36
C—H bonds	
CH_3—H	104
CH_3CH_2—H	100
$(CH_3)_2CH$—H	96
$(CH_3)_3C$—H	94
$PhCH_2$—H	88
CH_2=$CHCH_2$—H	87
RCH=CH—H	106
Ph—H	103
RC≡C—H	121
H—CN	120
C—halogen bonds	
CH_3—F	108
CH_3CH_2—F	106
CH_3—Cl	84
CH_3CH_2—Cl	82
CH_3—Br	70
CH_3CH_2—Br	69
CH_3—I	56
CH_3CH_2—I	54
Ph—F	116
Ph—Cl	97
Ph—Br	72
Ph—I	65
C—C bonds	
CH_3—CH_3	89
Ph—CH_3	93
$PhCH_2$—CH_3	72
CH_2=CH_2 (both)	163–173
CH_2=CH_2 (one)	63
HC≡CH	≈230
CH_3—CN	105
C—O bonds	
CH_3—OH	92
CH_3—OCH_3	80
Ph—OH	103
CH_2=O (both)	175
CH_2=O (one)	90
C—N bonds	
CH_3—NH_2	79
Ph—NH_2	91
CH_2=NH	≈154
HC≡N	224
H—X bonds	
H—OH	119
CH_3COO—H	112
PhO—H	103
CH_3O—H	102
H—F	135
H—Cl	103
H—Br	88
H—I	71
H—NH_2	103
H—SH	90
H—CN	120
Other	
HO—OH	51
$(CH_3)_3CO$—$OC(CH_3)_3$	35

Problems (Cont.)

20 Suggest a reason why the bond dissociation energy of the tertiary C—H bond of 2-methylpropane (96 kcal/mol) is considerably smaller than that of its primary C—H bonds (about 100 kcal/mol). (*Hint:* Look at the products of bond dissociation.)

As Problem 19 illustrates, bond dissociation energies may be used to estimate heats of reaction. This type of calculation is particularly useful for reactions in which the heats of formation of some reactants or products are not known. Let us apply this technique to the addition of hydrogen halides to alkenes. A puzzling feature of this reaction is that the peroxide effect is usually observed only for HBr addition; HCl and HI additions to alkenes follow the Markovnikov rule whether peroxides are present or not. The use of bond dissociation energies can help us to make sense of these observations.

The overall free-radical addition of a hydrogen halide (HX) to an alkene takes place in two propagation steps (Eqs. 5.58b and 5.58c). Let us use bond dissociation energies to calculate the standard enthalpy change for each of these propagation steps. We shall use the simplest alkene, ethylene. Hess's Law (Sec. 4.4B) allows us to calculate the ΔH^0 of each propagation step by breaking it down into simpler (fictitious) reactions. The first propagation step, addition of the halogen radical to the double bond, is given in Eq. 5.68c as the sum of two other reactions.

	ΔH^0 X = Cl	Br	I	
$CH_2{=}CH_2 \longrightarrow \cdot CH_2{-}CH_2\cdot$	+63	+63	+63	(5.68a)
$X\cdot + \cdot CH_2{-}CH_2\cdot \longrightarrow \cdot CH_2{-}CH_2{-}X$	−82	−69	−54	(5.68b)
Sum: $X\cdot + CH_2{=}CH_2 \longrightarrow X{-}CH_2{-}CH_2\cdot$	−19	−6	+9	(5.68c)

The first of these equations (Eq. 5.68a), breaking the π-bond, requires 63 kcal/mol (Table 5.2). The second equation represents formation of a primary carbon–halogen bond. The energy liberated in this reaction is the negative of the energy required for breaking this bond. The values used for this carbon–halogen bond energy are those in Table 5.2 for the $CH_3CH_2{-}X$ bonds. What we really need are the bond energies for carbon–halogen bonds with an unpaired electron on the adjacent carbon ($\cdot CH_2CH_2{-}X$). Since this value is not available in Table 5.2, we *assume* that the nearby unpaired electron has no effect on the carbon–halogen bond energy. Of course, this cannot be exactly right, but since we are making the same approximation in comparing the different halogens, any error introduced tends to cancel in the comparison.

Similarly, the ΔH^0 of the second propagation step, reaction 5.69c, can be estimated from another Hess's Law calculation.

	ΔH^0 X = Cl	Br	I	
$HX \longrightarrow H\cdot + X\cdot$	+103	+88	+71	(5.69a)
$H\cdot + \cdot CH_2CH_2X \longrightarrow CH_3CH_2X$	−100	−100	−100	(5.69b)
Sum: $HX + \cdot CH_2CH_2X \longrightarrow CH_3CH_2X + X\cdot$	+3	−12	−29	(5.69c)

Again, we use an approximation: the bond energy of the carbon-hydrogen bond in $H{-}CH_2CH_2X$ is the same for all halogens, and is equal to the bond energy of the $H{-}CH_2CH_3$ bond.

If we look at the ΔH^0 for each propagation step in the free-radical mechanism, we find that reaction 5.68c is energetically unfavorable (has a positive ΔH^0) when X = I, and reaction 5.69c is energetically unfavorable when X = Cl. Only when X = Br are *both* propagation steps energetically favorable. Now, in any free-radical reaction, the steps that propagate the radical chain are always in competition with recombination reactions (Eqs. 5.58d and e) that terminate the radical chain. These recombination reactions are so fast that only the exothermic propagation steps (steps with negative ΔH^0 values) are fast enough to compete with them. In the HBr addition both propagation steps are exothermic and fast enough to prevent the termination of radical chains. In addition of the other hydrogen halides, one or the other of the propagation steps is endothermic. In other words, there is an *energy barrier* to these reactions that makes them relatively slow. In addition of HCl or HI, therefore, the free-radical chain reaction is terminated by competing recombination reactions, and Markovnikov addition by the ionic pathway is observed instead.

Problem

21 Using bond energies, estimate the ΔH^0 values for each of the following halogenation reactions. (*Hint:* Break each reaction up into several free-radical reaction steps, calculate the ΔH^0 for each step, and apply Hess's Law.)

(a) $CH_2{=}CH_2 + Cl_2 \longrightarrow Cl{-}CH_2{-}CH_2{-}Cl$

(b) $CH_2{=}CH_2 + Br_2 \longrightarrow Br{-}CH_2{-}CH_2{-}Br$

(c) $CH_2{=}CH_2 + I_2 \longrightarrow I{-}CH_2{-}CH_2{-}I$

According to your calculation, which reaction is most likely to be reversible? Explain. (See Sec. 5.1A for the experimental facts.)

5.9 INDUSTRIAL USE AND PREPARATION OF ALKENES. OTHER TYPES OF FREE-RADICAL REACTIONS

A. Free-Radical Polymerization of Alkenes

In the presence of free-radical initiators such as peroxides or AIBN, many alkenes undergo a reaction called **polymerization.** In this reaction alkene molecules react to give a **polymer.** Polymers are very large molecules composed of repeating units derived from small molecules, in the same sense that a wall is composed of bricks. In polymer chemistry the small molecule from which the polymer is derived is called the **monomer.** For example, ethylene can be used as a monomer and polymerized under free-radical conditions to give its polymer, an industrially important compound called **polyethylene.**

$$n\ H_2C{=}CH_2 \xrightarrow{\text{initiator}} \left[CH_2{-}CH_2 \right]_n \qquad (5.70)$$

ethylene — polyethylene, the polymer of ethylene

In this formula for polyethylene, the subscript n means that each polyethylene molecule contains a very large and not very precisely defined number of repeating units, $-CH_2-CH_2-$. Typically n might be in the range of 3,000 to 40,000. Polyethylene is an example of an **addition polymer**—that is, a polymer in which no atoms of the monomer unit have been lost as a result of the polymerization reaction. (We shall examine other types of polymers later in this text.)

The polymerization of ethylene shown in Eq. 5.70 occurs by a free-radical mechanism, and is therefore an example of **free-radical polymerization.** The reaction is initiated by radicals derived from peroxides or other initiators.

In the mechanism of this reaction, an initiating radical R· adds to the double bond of ethylene to form a new radical.

$$\underset{\text{initiating radical}}{R\cdot}\ CH_2{=}CH_2 \longrightarrow R{-}CH_2{-}\dot{C}H_2 \qquad (5.71a)$$

The propagation phase of the reaction begins when the new radical adds to another molecule of ethylene.

$$R{-}CH_2{-}\dot{C}H_2\ \ CH_2{=}CH_2 \longrightarrow R{-}CH_2{-}CH_2{-}CH_2{-}\dot{C}H_2 \qquad (5.71b)$$

Notice that Eqs. 5.71a and b are further examples of a typical free-radical reaction: addition to a double bond. (Compare with Eq. 5.68b.) This process continues indefinitely until the ethylene supply is exhausted.

$$R{+}CH_2{-}CH_2{+}_n CH_2{-}\dot{C}H_2 + CH_2{=}CH_2 \longrightarrow R{+}CH_2{-}CH_2{+}_{n+1} CH_2{-}\dot{C}H_2 \qquad (5.71c)$$

The reaction terminates when the radical is eventually consumed by one of several termination reactions. Typically, polymers with molecular weights of 10^5 to 10^7 daltons are formed in free-radical polymerizations. The polymer chain is so long that the groups at its ends represent an insignificant part of the total structure. Hence, these terminating groups are ignored when the structure is written as $+CH_2{-}CH_2+_n$.

Free-radical polymerization of alkenes is very important commercially. In 1985 more than 14 billion pounds of polyethylene were manufactured in the United States, of which more than half was made by free-radical polymerization. The free-radical process yields a very transparent polymer, called *low-density polyethylene,* used in films and packaging. (Freezer bags and sandwich bags are usually made of low-density polyethylene.) Another method of polyethylene manufacture, called the *Ziegler process,* employs a titanium catalyst and does not involve free-radical intermediates. This process yields a *high-density polyethylene* used in blow-molded plastic containers.

There are many other commercially important polymers produced from alkene monomers by free-radical polymerization. Some of these are listed in Table 5.3 (see also Fig. 5.2).

Problem

22 Draw the structures of (a) polyvinyl chloride (PVC); (b) polypropylene, the polymer of propylene (propene). Show the free-radical mechanism for the formation of PVC initiated by AIBN.

An amusing story is associated with the discovery of polytetrafluoroethylene. One day in the early 1940s a scientist at one of the laboratories of the DuPont company opened the valve on a cylinder of tetrafluoroethylene, a gas, but no gas emerged from the cylinder. Since the weight of the empty cylinder was known, the scientist determined that the cylinder indeed had the weight expected for a full container of the alkene. The

TABLE 5.3 Some Polymers Produced by Free-Radical Polymerization

Polymer name (Trade name)	Structure of monomer	Properties of the plastic	Uses
Polyethylene	$CH_2{=}CH_2$	Flexible, semiopaque, generally inert	Containers, film
Polystyrene	$Ph{-}CH{=}CH_2$	Clear, rigid; can be foamed with air	Containers, toys, packing material and insulation
Polyvinyl chloride (PVC)	$CH_2{=}CH{-}Cl$	Rigid, but can be plasticized with certain additives	Plumbing, leatherette, hoses. Monomer has been implicated as a carcinogen
Polychlorotri-fluoroethylene (Kel-F)	$CF_2{=}CF{-}Cl$	Inert	Chemically inert apparatus, fittings, and gaskets
Polytetrafluoro-ethylene (Teflon)	$CF_2{=}CF_2$	Very high melting point; chemically inert	Gaskets; chemically resistant apparatus and parts
Poly(methyl methacrylate) (Plexiglas, Lucite)	$CH_2{=}C(CH_3){-}CO_2CH_3$	Clear and semiflexible	Lenses and windows; fiber optics
Polyacrylonitrile (Orlon, Acrilan)	$CH_2{=}CH{-}CN$	Crystalline, strong, high luster	Fibers

cylinder was cut open and a polymeric material was found on the inside of the tank. This substance felt slippery to the touch, could not be melted with extreme heat, and was chemically inert to almost everything. Evidently the formation of this polymer had been initiated by impurities in the tank. Thus was made the accidental discovery of the polymer that we know today as Teflon. This polymer was investigated further and became valued because of its unique properties. The consumer is perhaps most familiar with Teflon as the nonstick coating on cookware.

B. Industrial Source of Alkenes. Thermal Cracking of Alkanes

Simple alkenes are considered to be petroleum products, but they are not obtained directly from oil wells. Rather, they are produced industrially from saturated hydrocarbons in a process called **cracking.** Cracking breaks larger saturated hydrocarbons into a mixture of smaller hydrocarbons, some of which are alkenes. The chemistry of the cracking process is of interest, not only because of its commercial significance, but also because it illustrates more important chemistry of free radicals.

Ethylene, the alkene of greatest commercial importance, is produced by a process called *thermal cracking*. In this process, a hydrocarbon feedstock (Sec. 3.7) is mixed with steam and heated in a furnace at 750–900 °C for a fraction of a second, and is then quenched (rapidly cooled) to prevent secondary reactions. The products of cracking are then separated by fractional distillation.

Let us see how ethylene is formed from the cracking of a larger alkane such as nonane. The cracking temperatures are sufficiently high that initiating radicals are provided by the spontaneous rupture of carbon–carbon bonds. Although any of the carbon–carbon bonds might break, let us consider specifically what happens as a result of bond rupture in the middle of the carbon chain.

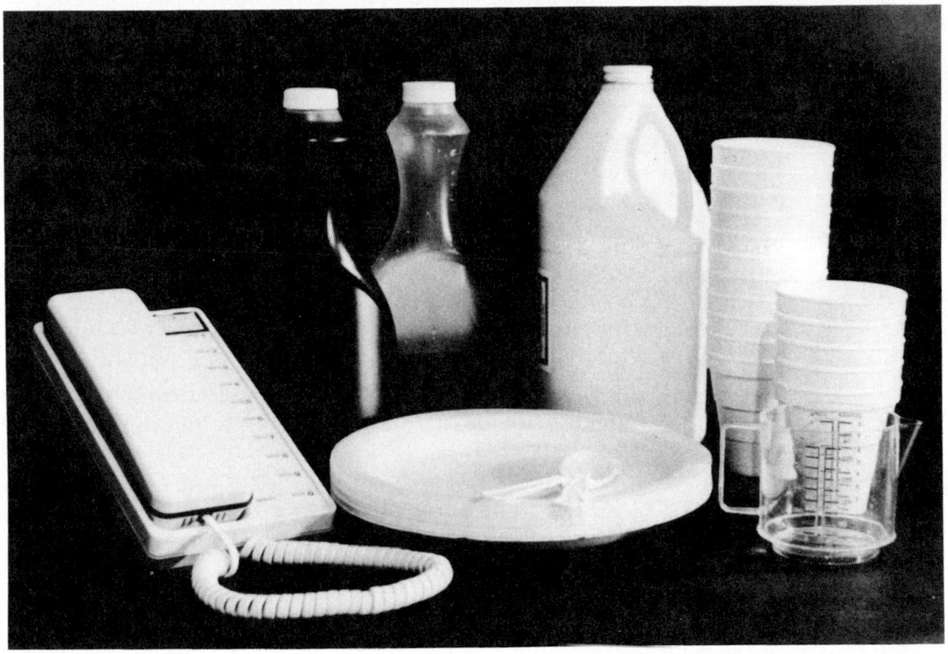

Figure 5.2 Plastics produced by free-radical polymerization are a part of everyday life.

$$CH_3CH_2CH_2CH_2CH_2\text{—}CH_2CH_2CH_2CH_3 \longrightarrow \underset{n\text{-pentyl radical}}{CH_3CH_2CH_2CH_2\dot{C}H_2} + \underset{n\text{-butyl radical}}{\dot{C}H_2CH_2CH_2CH_3} \quad (5.72)$$

The *n*-pentyl radical, for example, can "unzip" to ethylene by a process called ***β*-scission.**

$$CH_3CH_2CH_2\text{—}CH_2\text{—}\dot{C}H_2 \xrightarrow{\beta\text{-scission}} CH_3\text{—}CH_2\text{—}\dot{C}H_2 + CH_2{=}CH_2 \quad (5.73)$$

$$\downarrow \beta\text{-scission}$$

$$\dot{C}H_3 + CH_2{=}CH_2$$

β-Scission is another typical reaction of free radicals. The word "scission" means "cleavage" (it is derived from the same root as "scissors"). The Greek letter beta (*β*) refers to the fact that the cleavage occurs one carbon away from the radical site. (The Greek letters *α*, alpha; *β*, beta; *γ*, gamma; and so on are sometimes used to indicate the relative positions of groups on carbon chains.)

cleavage occurs here

$$R\text{—}\underset{\gamma}{CH_2}\text{—}\underset{\beta}{CH_2}\text{—}\underset{\alpha}{\dot{C}H_2}$$

Although *β*-scission might at first look like a new reaction, in fact it is nothing more than the *reverse* of free-radical addition to a double bond. (Examine Eq. 5.73 in reverse.)

The methyl radical, $\cdot CH_3$, that is produced in Eq. 5.73 can propagate the radical chain by hydrogen abstraction from other hydrocarbons. Among the possible hydrogen-abstraction reactions of the methyl radical with nonane is the following:

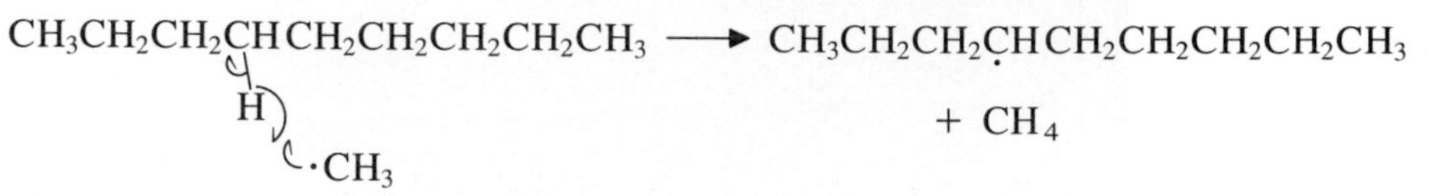

$$CH_3CH_2CH_2CHCH_2CH_2CH_2CH_2CH_3 \longrightarrow CH_3CH_2CH_2\overset{\cdot}{C}HCH_2CH_2CH_2CH_2CH_3 \quad (5.74)$$

$$\underset{\underset{\cdot CH_3}{H}}{\big|} \qquad\qquad\qquad\qquad + CH_4$$

The resulting radical can fragment by β-scission to give, for example, 1-pentene and the *n*-butyl radical, which can undergo further fragmentation.

Ethylene is the most important product of alkane cracking, although, as we can see from Eq. 5.74, other compounds are obtained as well. Among these are hydrogen, methane and other low molecular weight alkanes, propene, gasoline, fuel oil, and acetylene ($HC\equiv CH$).

In 1985 approximately 31 billion pounds of ethylene, valued at about $5 billion, were produced in the United States.

5.10 ELEMENTAL ANALYSIS OF ALKENES. UNSATURATION NUMBER

The empirical formula of an alkene, like that of an alkane (Sec. 3.9), can be determined by combustion analysis. An alkene with one double bond has two fewer hydrogens than the alkane with the same carbon skeleton. Likewise, a compound containing a ring also has two fewer hydrogens in its molecular formula than the corresponding open-chain compound. (Compare cyclohexane, C_6H_{12}, with hexane, C_6H_{14}.) As we can see from this simple example, the molecular formula of an organic compound contains information about the number of rings and double (or triple) bonds in the compound.

To explore this idea further, let us define a quantity called the **unsaturation number,** or **degree of unsaturation,** U. *The unsaturation number of a molecule is equal to the total number of its rings and multiple bonds.* The unsaturation number of a hydrocarbon is readily calculated from the molecular formula as follows. The maximum number of hydrogens possible in a hydrocarbon with C carbon atoms is $2C + 2$. Since *every ring or double bond reduces the number of hydrogens from this maximum by 2,* the unsaturation number is equal to half the difference between the maximum number of hydrogens and the actual number H:

$$U = \frac{2C + 2 - H}{2} = \text{number of rings + multiple bonds} \quad (5.75)$$

For example, cyclohexene, C_6H_{10}, has $U = [2(6) + 2 - 10]/2 = 2$. Cyclohexene has two degrees of unsaturation: one ring and one double bond.

How does the presence of other elements affect the calculation of the unsaturation number? We can readily convince ourselves from common examples (for instance, ethanol, C_2H_5OH) that Eq. 5.75 remains valid when *oxygen* is present in an organic compound. Since a *halogen* is monovalent, each halogen atom in an organic compound always reduces the maximum possible number of hydrogens by one. Thus, if the number of carbons is C, the maximum number of *halogens plus hydrogens* is $2C + 2$. If X equals the actual number of halogens present, the formula for unsaturation number therefore becomes

$$U = \frac{2C + 2 - (X + H)}{2} \tag{5.76}$$

Finally, when *nitrogen* is present, the number of hydrogens in a saturated compound increases by one for each nitrogen. (For example, the saturated compound methylamine, CH_3—NH_2, has $2C + 3$ hydrogens.) Therefore, if N is the number of nitrogens, the formula for unsaturation number becomes

$$U = \frac{2C + 2 + N - (H + X)}{2} \tag{5.77}$$

The utility of the unsaturation number is that it gives us structural information about an unknown compound from the molecular formula alone. This idea is illustrated in Problem 24.

Problems

23 Calculate the unsaturation number for each of the following compounds:
(a) $C_3H_4Cl_4$ (b) $C_5H_8N_2$

24 An unknown compound contains 85.60% carbon and 14.40% hydrogen. How many rings and/or double bonds does it have?

25 What degree of unsaturation is introduced into a compound by a carbon–carbon triple bond?

KEY IDEAS IN CHAPTER 5

- The characteristic reaction of alkenes is addition to the double bond.
- Addition to the alkene double bond occurs by a variety of mechanisms:

    1. By mechanisms involving carbocation intermediates (addition of hydrogen halides, hydration)
    2. By mechanisms involving cyclic ion intermediates (oxymercuration, halogenation)
    3. By pericyclic mechanisms (hydroboration, glycol formation, ozonolysis)
    4. By mechanisms involving free-radical intermediates (free-radical addition of HBr, polymerization)

    The mechanism of catalytic hydrogenation is not known with certainty.

- In some cases, useful products are obtained by the treatment of initially formed addition products with other reagents. Thus, organoboranes are treated with alkaline H_2O_2, and oxymercuration adducts with $NaBH_4$, to yield alcohols; cyclic compounds from OsO_4 or $KMnO_4$ oxidation of alkenes are converted by water into glycols; and ozonides from ozonolysis are converted into aldehydes, ketones, or acids, by treatment with $(CH_3)_2S$ or H_2O_2.
- Typical reactions of free radicals are:
  1. Addition to a double bond
  2. Atom abstraction
  3. β-Scission (the reverse of addition to a double bond)
  4. Recombination (the reverse of bond rupture)
- Reactions that occur by free-radical pathways are in many cases promoted by free-radical initiators (peroxides, AIBN), heat, or light, and are retarded by free-radical inhibitors.
- The stability of free radicals is in the order tertiary > secondary > primary, but the effect of substitution on free-radical stability is considerably smaller than the effect of substitution on carbocation stability.
- Anti-Markovnikov addition of HBr to alkenes in the presence of peroxides is a consequence of the free-radical mechanism of the reaction. The bromine atom attacks the double bond at the less substituted carbon in order to avoid steric repulsions and to give the more highly substituted, and hence more stable, radical. This radical then reacts with HBr to complete the addition.
- The bond dissociation energy measures the energy required to rupture a covalent bond and form free radicals. The ΔH^0 of a reaction can be approximately calculated by breaking up the reaction into separate (even fictitious) free-radical reactions, calculating the ΔH^0 of each, and applying Hess's Law.
- The unsaturation number measures the number of rings plus multiple bonds, and is determined from the molecular formula by Eq. 5.77.

ADDITIONAL PROBLEMS

26 Give the principal organic products expected when 1-butene reacts with each of the following reagents:

(a) Br_2 in CCl_4 solvent
(b) O_3, $-78°$
(c) product of (b) with $(CH_3)_2S$
(d) product of (b) with H_2O_2
(e) O_2, flame
(f) HBr
(g) OsO_4, then H_2O, $NaHSO_3$
(h) I_2, H_2O
(i) H_2, Pt/C
(j) HBr, peroxides
(k) BH_3 in tetrahydrofuran (THF)
(l) product of (k) with NaOH, H_2O_2
(m) $Hg(OAc)_2$, H_2O
(n) product of (m) with $NaBH_4$
(o) HI
(p) HI, AIBN

27 Repeat Problem 26 for 1-ethyl-1-cyclopentene.

28 Draw the structure of

(a) a six-carbon alkene that would give the same product from reaction with HBr whether peroxides are present or not.

(b) a compound C_5H_{10} that would *not* react with alkaline $KMnO_4$.

(c) four compounds of formula $C_{10}H_{16}$ that would undergo catalytic hydrogenation to give decalin.

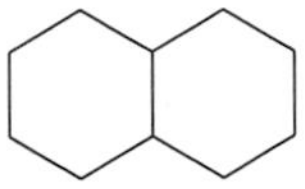

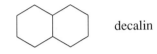

decalin

(d) two alkenes that would yield 1-methyl-1-cyclohexanol when treated with $Hg(OAc)_2$ in water, then $NaBH_4$.

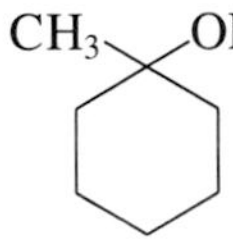

1-methyl-1-cyclohexanol

(e) an alkene that would give the following compound as the *only* product after ozonolysis followed by H_2O_2:

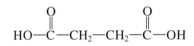

$$HO-\overset{\overset{\displaystyle O}{\|}}{C}-CH_2-CH_2-\overset{\overset{\displaystyle O}{\|}}{C}-OH$$

(f) two alkenes that would give the following alcohol as the major product of hydroboration followed by treatment with alkaline H_2O_2:

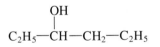

$$C_2H_5-\overset{\overset{\displaystyle OH}{|}}{C}H-CH_2-C_2H_5$$

(g) an alkene of five carbons that would give the same product as a result of treatment with *either* $Hg(OAc)_2$ in water, then $NaBH_4$; *or* B_2H_6, then alkaline H_2O_2.

29 Tell whether each of the following formulas could be that of a neutral organic compound containing no unpaired electrons. Explain.

(a) C_3H_7N (b) $C_3H_7N_2$ (c) C_3H_7NO

30 Give the missing reactant or product in each of the following equations:

(a) 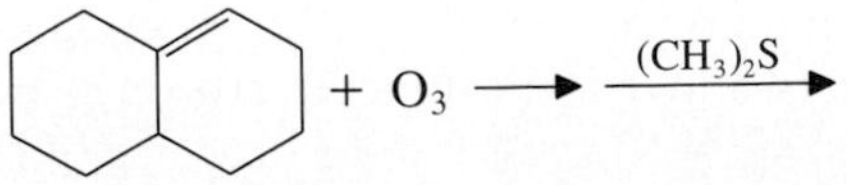

Problems (Cont.)

(b) 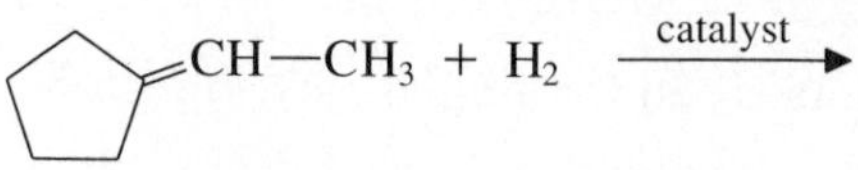

(c) 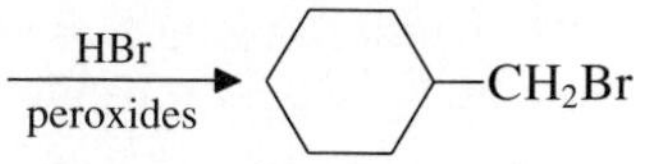

(d) 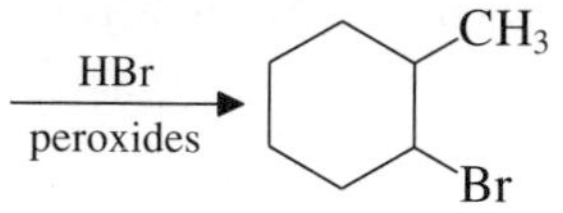 (only product)

---

31 Outline a laboratory preparation of each of the following compounds. Each should be prepared from an alkene with the same number of carbon atoms. All compounds should be virtually free of contaminating structural isomers. (See note about deuterium, D, in Chapter 4, Problem 27.)

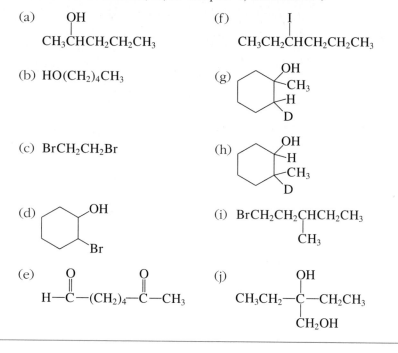

(a) $CH_3CH(OH)CH_2CH_2CH_3$

(b) $HO(CH_2)_4CH_3$

(c) $BrCH_2CH_2Br$

(d) 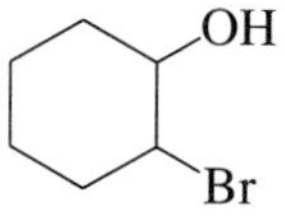

(e) $H-\overset{O}{\overset{\|}{C}}-(CH_2)_4-\overset{O}{\overset{\|}{C}}-CH_3$

(f) $CH_3CH_2CH(I)CH_2CH_2CH_3$

(g) 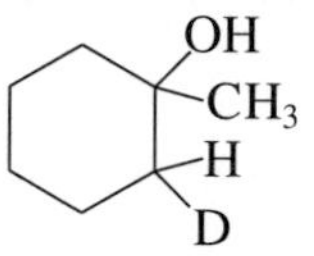

(h) 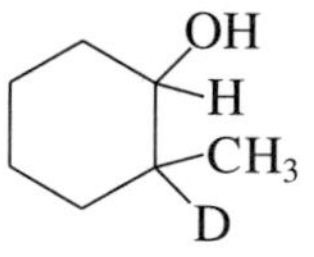

(i) $BrCH_2CH_2CH(CH_3)CH_2CH_3$

(j) $CH_3CH_2-C(OH)(CH_2OH)-CH_2CH_3$

---

32 From what you know about the mechanism of alkene halogenation, predict the product(s) that would be obtained when 2-methyl-1-butene is subjected to each of the following conditions. Explain your answers.

(a) Br_2 in CCl_4 (an inert solvent)

(b) Br_2 in H_2O

(c) Br_2 in CH_3OH solvent

(d) Br_2 in CH_3OH solvent containing concentrated Li^+ Br^-

---

33 Using the mechanism of the oxymercuration reaction, predict the product(s) that would be obtained when the 1-hexene is treated with mercuric acetate in each of the following solvents, and the resulting products are treated with $NaBH_4$. Explain your answers, and tell what functional groups are present in each of the products.

(a) CH_3OH (methanol)

(b) $(CH_3)_2CH\!-\!OH$ (isopropyl alcohol)

(c)

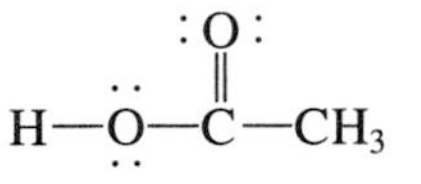

$$H\!-\!\overset{..}{\underset{..}{O}}\!-\!\overset{\displaystyle :O:}{\overset{\|}{C}}\!-\!CH_3 \quad\text{(acetic acid; see Eq. 5.26)}$$

34 In the addition of HBr to 3,3-dimethyl-1-butene, the following results are observed:

$$CH_3\!-\!\underset{\underset{CH_3}{|}}{\overset{\overset{CH_3}{|}}{C}}\!-\!CH\!=\!CH_2 + HBr \longrightarrow$$

$$CH_3\!-\!\underset{\underset{Br}{|}}{\overset{\overset{CH_3}{|}}{C}}\!-\!CH(CH_3)_2 + CH_3\!-\!\underset{\underset{CH_3}{|}}{\overset{\overset{CH_3}{|}}{C}}\!\!-\!\!\underset{\underset{Br}{|}}{CH}\!-\!CH_3 + CH_3\!-\!\underset{\underset{CH_3}{|}}{\overset{\overset{CH_3}{|}}{C}}\!-\!CH_2CH_2Br$$

| no peroxides: 71% | 29% | none |
| with peroxides: trace | trace | 100% |

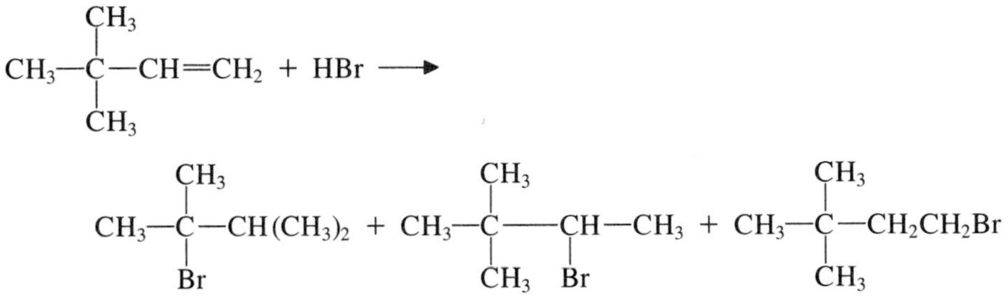

(a) Explain why there is a different product distribution under the different sets of conditions.

(b) Write a detailed mechanism for each reaction that explains the origin of all products.

(c) Which set of conditions gives the faster reaction? Explain.

35 Trifluoroiodomethane undergoes addition to alkenes in the presence of light:

$$CH_3CH_2CH_2CH\!=\!CH_2 + CF_3I \xrightarrow{\text{light}} CH_3CH_2CH_2\underset{\underset{I}{|}}{CH}\!-\!CH_2\!-\!CF_3$$

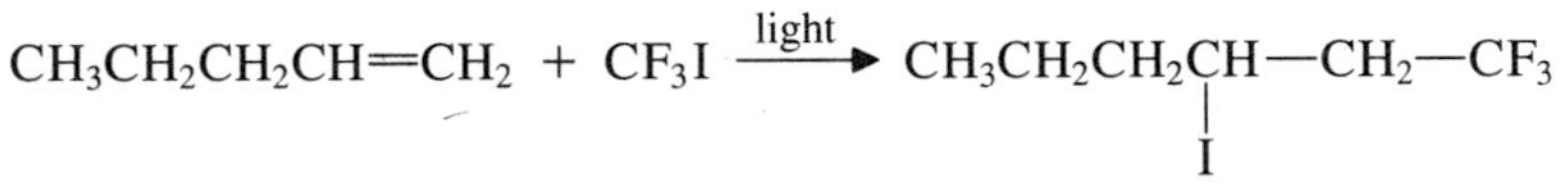

(a) Suggest a mechanism for this reaction.

(b) To which alkene would CF_3I add more rapidly: 2-methyl-1-pentene or 3-methyl-1-pentene? Explain.

36 A bottle containing the liquid styrene (Ph—CH=CH$_2$) was observed after several months to change first into a viscous syrup, then later into a solid mass. Another bottle of styrene bearing the label, "Contains 0.01% hydroquinone," did not change appearance during the same time period. Explain these observations.

37 (a) In the thermal cracking of 2,2,3,3-tetramethylbutane, which carbon–carbon bond would be most likely to break? Explain.

(b) Which compound, 2,2,3,3-tetramethylbutane or ethane, undergoes thermal cracking at the lower temperature? Explain.

(c) Check your answer to (b) by calculating the ΔH^0 for the initial carbon–

Problems (Cont.)

carbon bond breaking. Use the ΔH_f^0 values in Table 5.1 as well as those of 2,2,3,3-tetramethylbutane (−53.99 kcal/mol) and ethane (−20.24 kcal/mol).

38 Natural rubber is a polymer with the following structure.

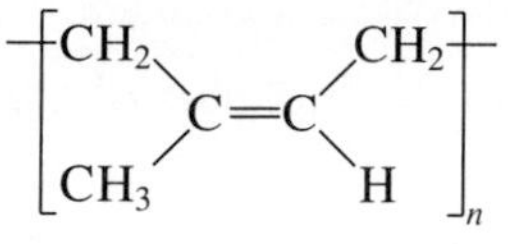

(a) What product would be obtained from ozonolysis of natural rubber, followed by reaction with H_2O_2?

(b) *Gutta-percha* is another type of natural polymer that gives the same ozonolysis product as natural rubber. Suggest a structure for gutta-percha.

39 We could envision the free-radical addition of hydrogen cyanide, H—CN, to alkenes by analogy to free-radical addition of HBr. Yet, in the presence of free-radical initiators, the addition of HCN to alkenes does not occur. Explain.

40 (a) Write the structure of *polystyrene,* the polymer obtained from the free-radical polymerization of styrene.

(b) How would the structure of the polymer product differ from the one in (a) if a few percent of 1,4-divinylbenzene were included in the reaction mixture?

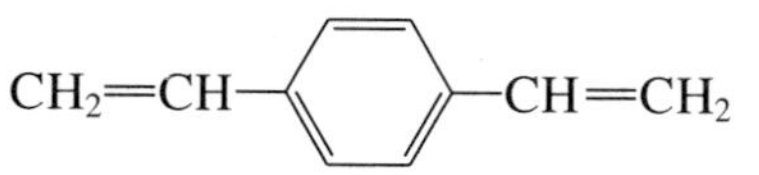

1,4-divinylbenzene

41 Using the corresponding reactions of alkenes as analogies, what products would you expect when phenylacetylene (Ph—C≡CH) is treated with (a) Br_2 in CCl_4; and (b) H_2 over a catalyst?

42 In an abandoned laboratory a bottle was found containing a clear liquid *A*. The bottle was labeled, "Isolated from a pine tree." You have been offered a substantial reward by the Department of Agriculture to identify this substance. Elemental analysis reveals that *A* contains 88.16% C and 11.84% H. Compound *A* decolorizes Br_2 in CCl_4, and gives a brown precipitate with alkaline $KMnO_4$. When 100 mg of *A* is hydrogenated at 0 °C over a catalyst, 33 mL of H_2 is consumed and the product is found to be 4-isopropyl-1-methylcyclohexane. (Recall that one mole of H_2 at 0° has a volume of about 22.4 L.) Ozonolysis of *A* followed by treatment of the reaction mixture with H_2O_2 gives the following compound as a major product:

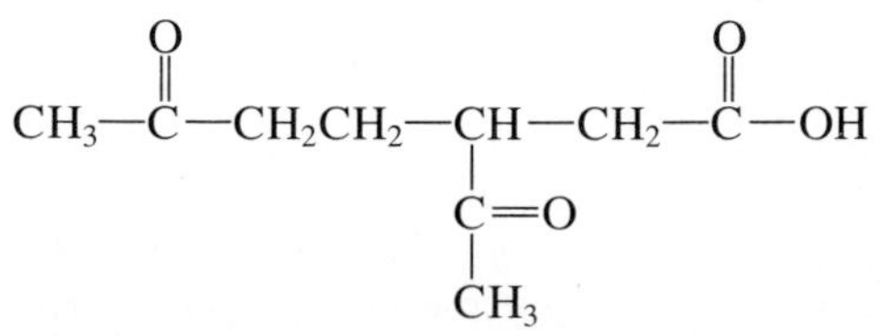

Suggest a structure for *A* and explain all observations. (See following box.)

WORKING PROBLEMS IN STRUCTURE DETERMINATION

Many students tend to attack a structural problem like Problem 42 from the top down. This is not always the most efficient way to solve this type of problem. Instead, read the problem through, calculate empirical formulas, molecular weights, and unsaturation numbers, and write down the general implications of each piece of information. Then focus as quickly as possible on any data that directly give a structure, a carbon skeleton, or other partial structure for *any* compound in the problem (not necessarily the first compound mentioned). Use this structural information and your knowledge of the reactions involved to work toward detailed structures of unknown compounds.

In Problem 42, for example, try the following procedure. (1) Calculate the empirical formula, molecular weight, and unsaturation number of the unknown. How many rings and double bonds are there? (2) Focus on the first structure mentioned, the product of catalytic hydrogenation. Compare this structure with the information from step (1). Are all the carbons of the unknown accounted for? What does this product tell you about the structure of the unknown? Do the same for the product of ozonolysis. (3) Write out *all* the structures that could have given the ozonolysis product. Use the structure of the hydrogenation product to decide which of these is the correct structure of the unknown.

43 A compound *A* was found by elemental analysis to contain 85.60% carbon and 14.40% hydrogen. The compound decolorized Br_2 in CCl_4 and gave a brown precipitate when treated with $KMnO_4$. Catalytic hydrogenation of 1 g of *A* at 0 °C consumed 200 mL of H_2, and gave octane as the product. Treatment of the unknown with O_3, followed by H_2O_2, gave butanoic acid, *B*. Suggest a structure for *A*. What part of the structure is not determined by the data?

$$CH_3CH_2CH_2\overset{\overset{\displaystyle O}{\|}}{C}—OH \qquad B$$

44 Knowing your expertise with the arrow formalism, a student in a more advanced organic chemistry course has come to you with the following reaction mechanism. The student does not understand what is happening in the reaction. Add the arrow formalism to show what is happening at each step as well as the relationship of the resonance structures. For each step of the reaction suggest an analogy from this or the preceding chapter (that is, a related transformation).

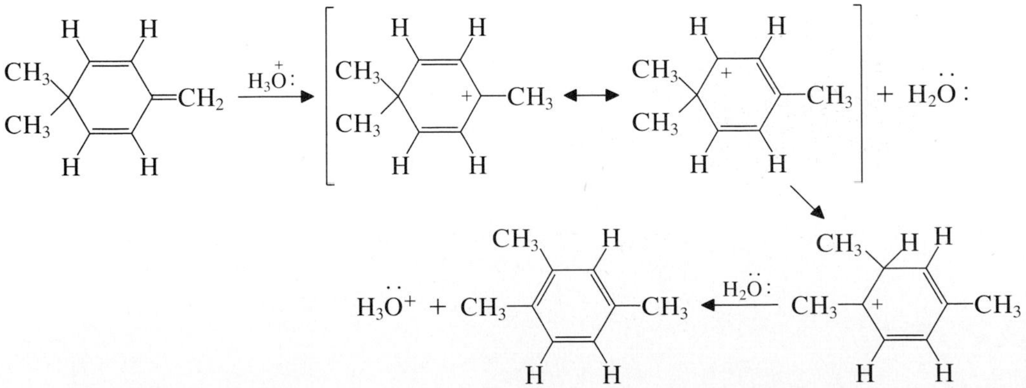

Problems (Cont.)

45 Using the curved-arrow formalism, suggest mechanisms for each of the following reactions.

(a) 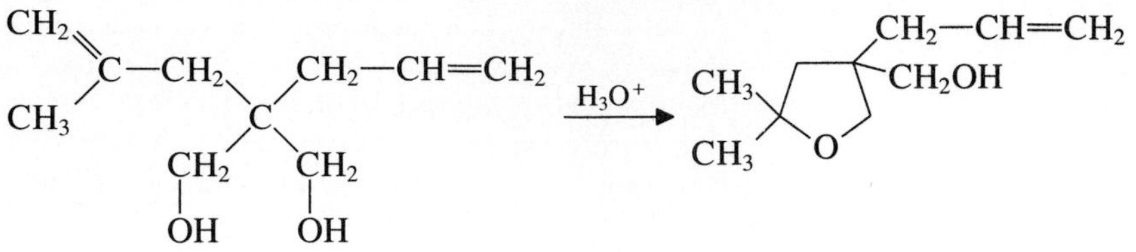

(*Hint:* Read again the box on p. 177. Now, write out the ring structure of the product using carbons and hydrogens. Indicate the correspondence between atoms in the product and those in the starting material. What happens to double bonds in the presence of acid? What species is formed? How can it lead to the observed bond connections? Use the *correct* arrow formalism and indicate only *one step* per structure. Use this same approach for the following problems.)

(b) 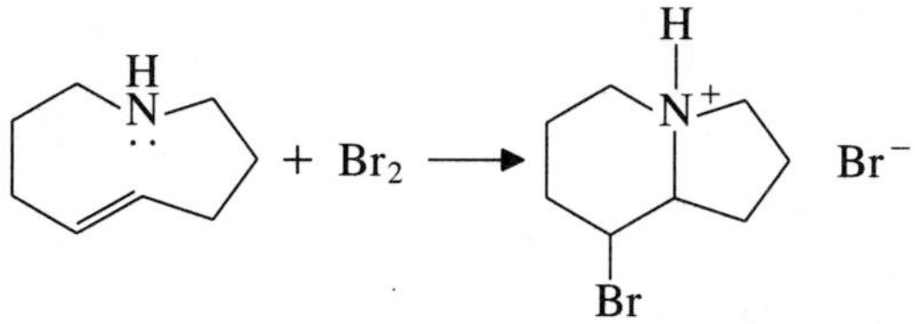

(c) 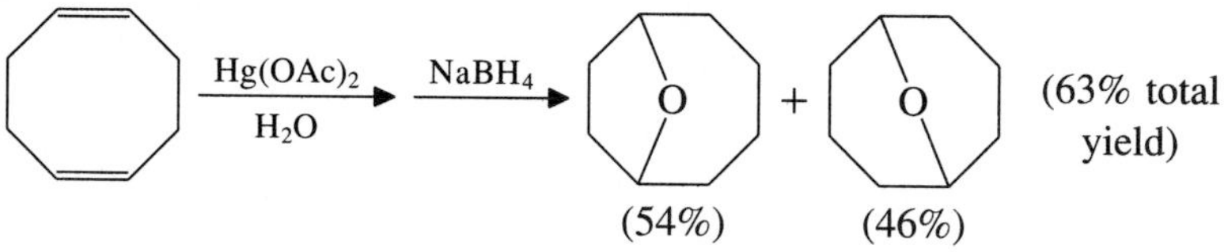

(d) 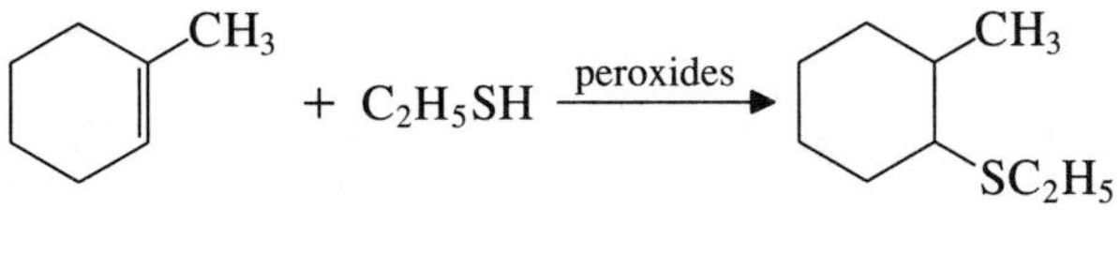

(e) $CH_2{=}CH(CH_2)_5CH_3 + CBr_4 \xrightarrow[\text{light}]{\text{peroxides}} Br_3C{-}CH_2{-}CH(Br)(CH_2)_5CH_3$

(96% yield)

(f) 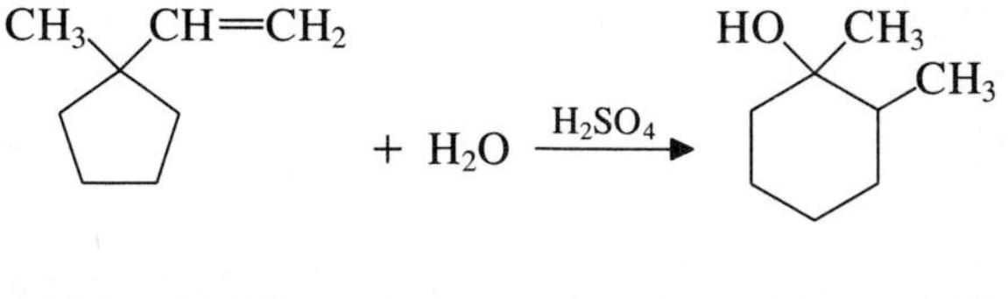

(*Hint:* Notice that this is a rearrangement reaction. When dealing with rearrangements involving ring carbon atoms, it helps to *draw out the carbon atoms of the ring*.)

46 Both parts of the following problem say something about the *reversibility* of hydroboration.

(a) When 1-ethyl-1-cyclohexene reacts with diborane and the resulting organoborane is heated at 160° for several hours, a mixture of organoboranes is formed in which the following compound predominates. Show how this compound is formed.

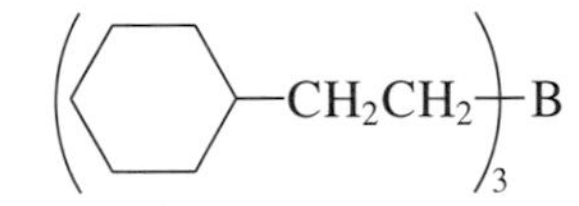

(b) After cyclopentene is treated with diborane and the diborane is completely consumed, 1-decene is added and the reaction mixture is heated to 160°. The mixture is then fractionally distilled, and cyclopentene, having a lower boiling point, is selectively removed. Treatment of the undistilled residue with alkaline H_2O_2 gives mostly 1-decanol, $CH_3(CH_2)_8CH_2OH$. Explain these observations.

47 In an abandoned laboratory a flammable liquid *A* has been found in a bottle bearing only the label "Compound *A*: 87.42% C, 12.58% H." Government agents have offered you a considerable sum to determine the structure of this compound. After verifying the elemental analysis, you find that compound *A* decolorizes Br_2 in CCl_4, and after ozonolysis followed by treatment with H_2O_2, gives α-methyladipic acid (structure below). Another bottle from the same laboratory is labeled "Compound *B* (isomer of *A*)." Compound *B* also decolorizes Br_2 in CCl_4, but yields cyclohexanone (structure below) as one of the products after ozonolysis/H_2O_2. Suggest structures for *A* and *B*.

α-methyladipic acid

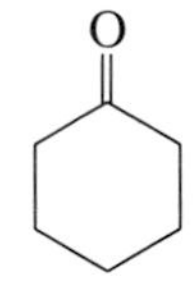

cyclohexanone

48 (a) Using O—Cl homolysis as an initiation step, give a free-radical chain mechanism for the following reaction:

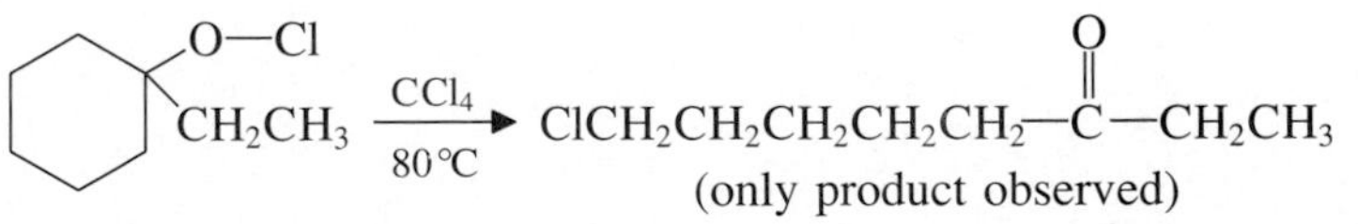

(only product observed)

(b) Account for all three products in the following reaction. Suggest a reason why ring opening, observed in the reaction of part (a), is only a minor pathway in the reaction on the next page.

Problems (Cont.)

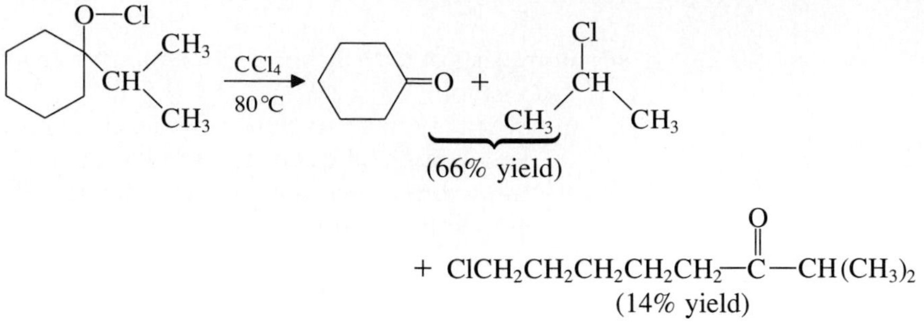

(*Hint:* Look at the free radicals formed in the two competing reactions.)

---

49 In the following sequence, the second reaction is unfamiliar. Nevertheless, identify compounds *A* and *B* from the information provided.

$$(CH_3)_3C{-}CH{=}CH_2 + HBr \longrightarrow \underset{\text{(mostly)}}{A} + \text{other compounds(s)}$$

$$A + \underset{\text{(a strong base)}}{(CH_3)_3C{-}\ddot{\underset{..}{O}}{:}^-\ K^+} \longrightarrow (CH_3)_3C{-}\ddot{\underset{..}{O}}H + KBr + B$$

$$B + O_3 \longrightarrow \xrightarrow{(CH_3)_2S} (CH_3)_2C{=}O \qquad \text{(only compound formed)}$$

Compound *B* contains 85.60% C and 14.40% H, decolorizes Br_2 in CCl_4, and 0.500 g of *B* takes up 130–135 mL of H_2 over a Pt/C catalyst. Once you have identified *B*, try to give an arrow formalism for its formation from *A* in one step.

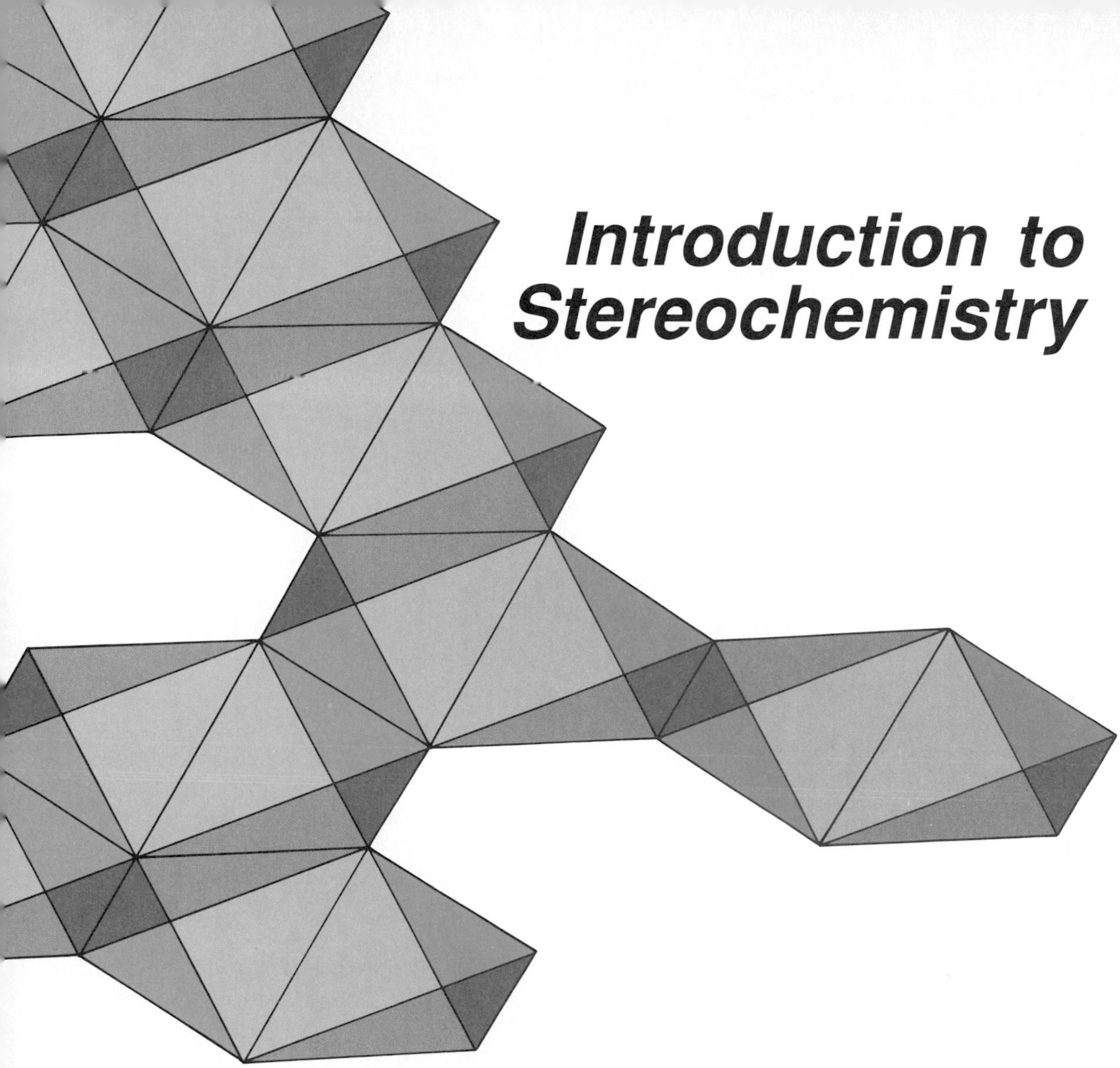

6 Introduction to Stereochemistry

Stereoisomers are compounds that have the same atomic connectivity but a different arrangement of their atoms in space. Our first encounter with stereoisomers was in Sec. 4.1B, where we studied *cis–trans* isomers of alkenes. In this and the following chapter we shall consider the phenomenon of stereoisomerism in more detail. The study of stereoisomers and their properties is called **stereochemistry.** First we shall concentrate on some of the basic definitions and principles of stereochemistry. Then we shall see how stereochemistry played a key role in the determination of the geometry of tetravalent carbon. In Chapter 7, we shall consider how the spatial arrangement of atoms influences chemical behavior.

We strongly recommend the use of molecular models during the study of this chapter. Eventually you will learn how to visualize three-dimensional structures on a two-dimensional surface (blackboard or paper) without recourse to models, but most students attain this skill only after a great deal of work with models.

6.1 ENANTIOMERS AND CHIRALITY

Any molecule—indeed, any object—has a mirror image (Fig. 6.1). Some molecules are superimposable on their mirror images, atom for atom. An example of such a molecule is ethanol or ethyl alcohol, CH_3—CH_2—OH (Fig. 6.2). Try to construct a model of ethanol and another model of its mirror image and show that these are superimposable. For simplicity, use a single colored ball to represent the methyl group, and a single ball of another color for the hydroxy (—OH) group. To show that these objects are superimposable, place the two central carbons side-by-side and align any two of the attached groups, as shown in Figure 6.2. The remaining two groups should then superimpose as well. *The fact that ethanol and its mirror image can be superimposed shows that these two objects are identical.*

There are some molecules, however, that cannot be superimposed on their mirror images. One such molecule is 2-butanol (Fig. 6.3).

$$\begin{array}{c} \quad OH \\ \quad | \\ CH_3\text{—}CH\text{—}C_2H_5 \end{array} \qquad \text{2-butanol}$$

If we try to match up the central carbon and any other two attached groups, the remaining groups do not align. (Verify this statement with models.) *The fact that a 2-butanol molecule and its mirror image cannot be superimposed shows that they are different molecules.* Since these two molecules have the same connectivity, then by definition they are stereoisomers. Molecules that are nonsuperimposable mirror images are called **enantiomers.** Thus, the two 2-butanol stereoisomers are enantiomers of each other; they have an **enantiomeric** relationship.

Notice that enantiomers must not only be mirror images; they must also be *nonsuperimposable* mirror images. Thus, ethanol (Fig. 6.2) has no enantiomer because an ethanol molecule and its mirror image are superimposable.

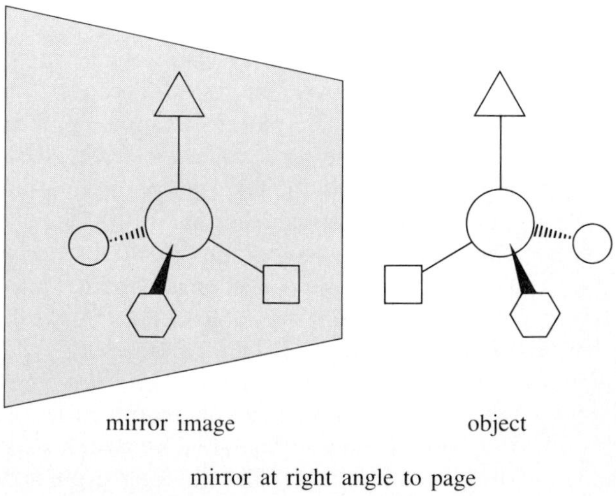

Figure 6.1 A general molecule and its mirror image.

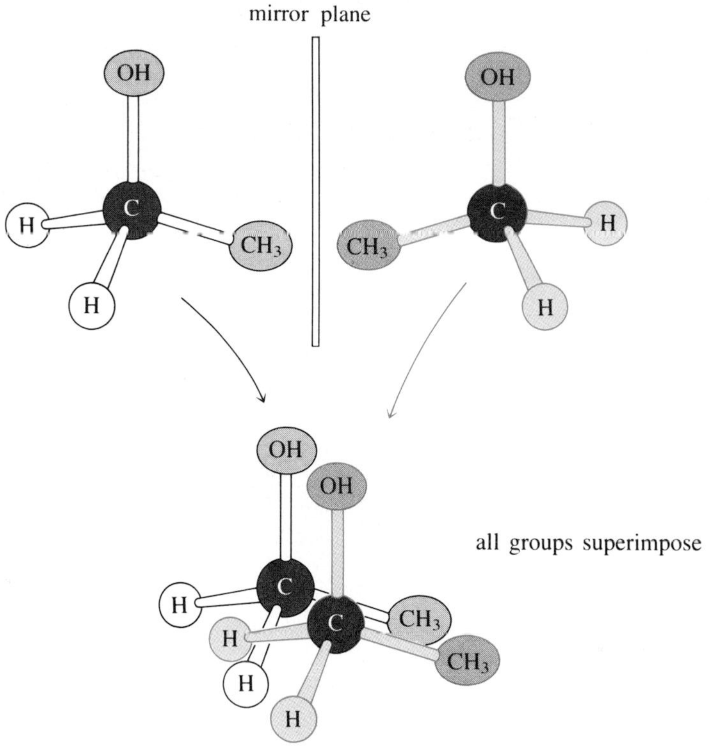

Figure 6.2 Testing mirror-image ethanol molecules for superimposability.

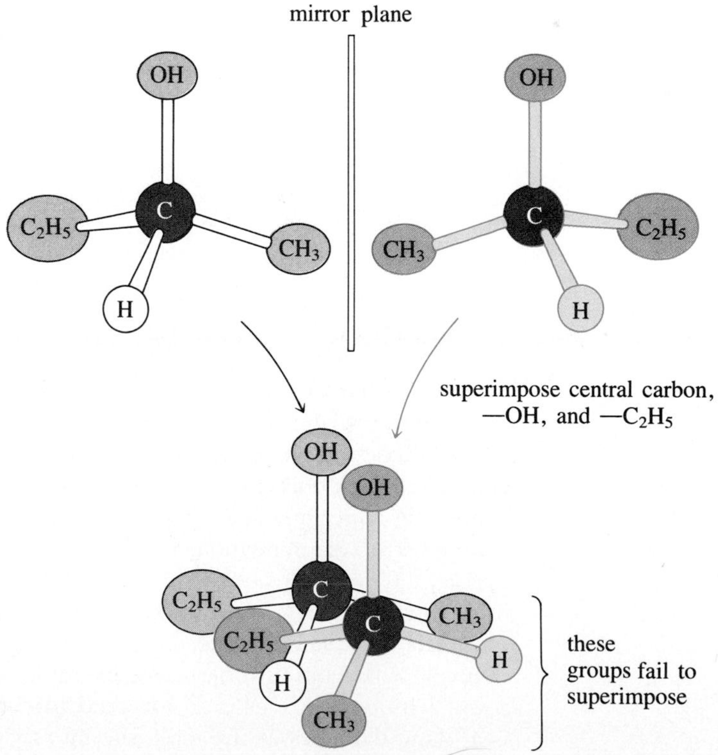

Figure 6.3 Testing mirror-image 2-butanol molecules for superimposability.

Figure 6.4 Concept of chirality, as conceived by the Swiss artist Hans Erni.

Molecules (or other objects) that can exist as enantiomers are said to be **chiral** (pronounced kī rəl); they possess the property of **chirality,** or handedness (Fig. 6.4). (*Chiral* comes from the Greek word for hand.) Enantiomeric molecules have the same relationship as the right and left hands—the relationship of object and nonsuperimposable mirror image. Thus, 2-butanol is a chiral molecule. Molecules (or other objects) that are not chiral are said to be **achiral,** without chirality. Ethanol is an achiral molecule. Both chiral and achiral objects are matters of everyday acquaintance. A foot or a hand is chiral; the helical thread of a screw gives it chirality. Achiral objects include a ball and an acorn.

What is it about the structure of 2-butanol or other chiral molecules that makes them chiral? 2-Butanol has one carbon bearing four *different* groups: —CH_3, —C_2H_5, —H, and —OH. Ethanol, however, has no such carbon atom. A carbon with four different attached groups is called an **asymmetric carbon.** When a molecule contains only one asymmetric carbon, it is a chiral molecule. Some molecules with more than one asymmetric carbon are chiral, but others are not; we shall see why in Sec. 6.7. There is nothing particularly unique about carbon as an asymmetric atom; other atoms can also be asymmetric. Most of our discussion of chirality, however, will center on asymmetric carbon.

Is there a general test for chirality that works for every molecule? The most general test for chirality is embodied in the definition of enantiomers: if an object cannot be superimposed on its mirror image, it is chiral. However, there is also a simple way to examine molecules for the *absence* of chirality. This method recognizes that achiral molecules possess certain kinds of symmetry that chiral molecules do not; this symmetry can be described by one or more *symmetry elements.* A **symmetry element** is a plane, a line, or a point used to describe the relationship between equivalent parts of an object. For example, a common symmetry element found in some achiral molecules is an **internal mirror plane,** or **plane of symmetry.** This is a plane that divides the molecule into two halves that are mirror images of each

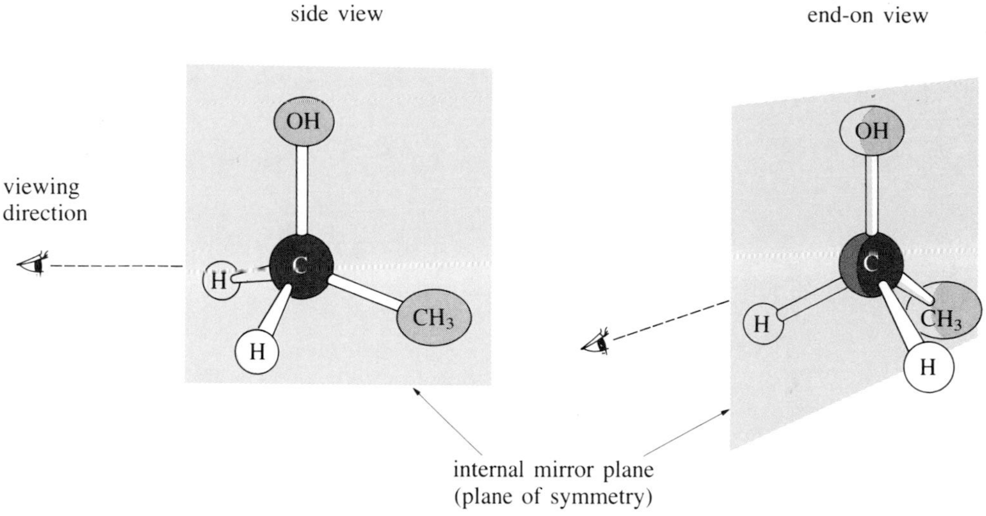

Figure 6.5 Structure of ethanol (an achiral molecule) showing the internal mirror plane.

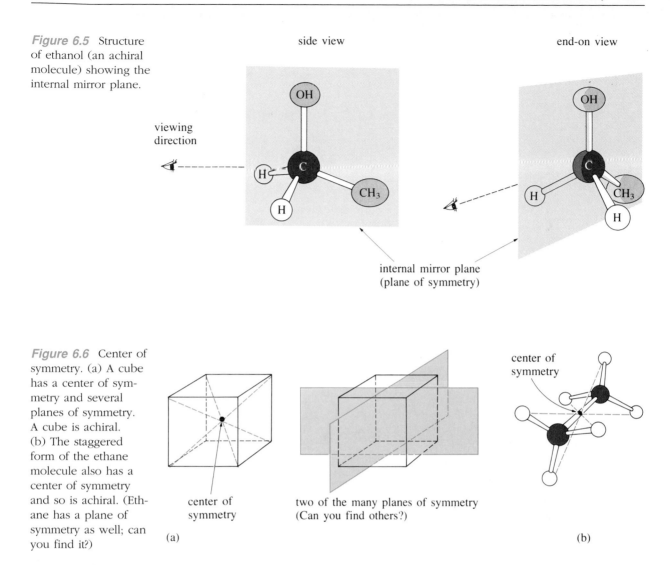

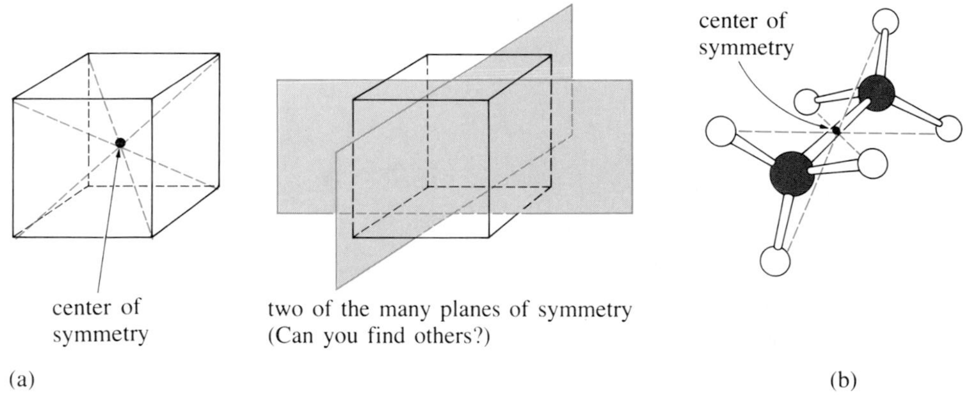

Figure 6.6 Center of symmetry. (a) A cube has a center of symmetry and several planes of symmetry. A cube is achiral. (b) The staggered form of the ethane molecule also has a center of symmetry and so is achiral. (Ethane has a plane of symmetry as well; can you find it?)

other. The ethanol molecule has a plane of symmetry (Fig. 6.5). This plane passes through the central carbon, the —CH_3 group, and the —OH group, and bisects the H—C—H angle.

Another common symmetry element found in other achiral molecules is a **center of symmetry.** A center of symmetry is a point at the center of an object. More precisely, a center of symmetry is a unique point such that all straight lines drawn through it touch equivalent parts of the object at equal distances. Thus, a cube has a center of symmetry (Fig. 6.6a) as well as many planes of symmetry. If a line drawn through the center of symmetry touches one corner of the cube at a certain distance from the center, it touches another corner at the same distance in the opposite direction. The molecule ethane (Fig. 6.6b) has a center of symmetry.

We can sometimes show that a molecule is achiral by identifying its symmetry elements. Thus, if a molecule contains either a plane of symmetry or a center of symmetry, it is not chiral. Some achiral molecules have other, less common symmetry elements. However, since these are rarely encountered, we shall not consider them here. If no plane of symmetry or center of symmetry can be found, then the only reliable test for chirality (or lack of it) of a molecule is to construct its mirror image and test it for superimposability.

Problems

1 Which of the following simple molecules are chiral?

(a) $\overset{\displaystyle Cl}{\underset{\displaystyle F}{H-C-Br}}$ (b) methane (c) water (d) $CH_3 - \overset{\displaystyle H}{\underset{\displaystyle C_3H_7}{N^+} }- C_2H_5$ Cl^-

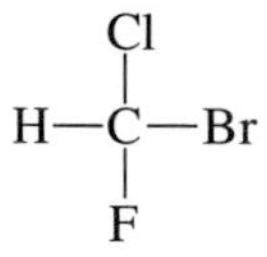

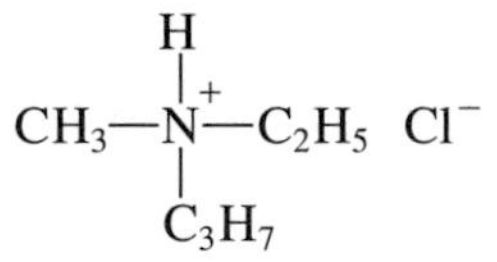

2 Indicate whether the following objects are chiral or achiral. (State any assumptions.)
(a) a shoe (d) a man
(b) this book (e) a pair of shoes (consider the pair as one object)
(c) a pencil (f) a pair of scissors

3 Show the plane of symmetry in each of the following achiral objects. (Some have more than one.) Which of these objects has a center of symmetry?
(a) the molecule methane (d) a regular pyramid
(b) a spherical ball (e) the molecule dichloromethane, CH_2Cl_2
(c) a cone

6.2 NOMENCLATURE OF ENANTIOMERS: THE R,S SYSTEM

The existence of enantiomers poses special problems of nomenclature. For example, suppose we are holding a model of one of the two enantiomers of 2-butanol. Is there a way to indicate which 2-butanol stereoisomer it is without actually using a model? This can be done quite easily with the same Cahn-Ingold-Prelog priority rules used to assign *E* and *Z* conformations to alkene stereoisomers (Sec. 4.2B). When we deal with compounds that have asymmetric carbons, we assign a *stereochemical configuration*, or arrangement of atoms, at each asymmetric carbon in the molecule using the following steps:

1. Identify an asymmetric carbon and the four different groups bound to it.

2. Assign priorities to the four different groups according to the rules given in Sec. 4.2B.

3. View the molecule along the bond *from the asymmetric carbon to the group of lowest priority*—that is, with the asymmetric carbon nearer and the low-priority group farther away (Fig. 6.7).

4. Consider the clockwise or counterclockwise order of the remaining group priorities. If the priorities of these groups decrease in the *clockwise* direction the asymmetric carbon is said to have the *R* configuration (*R* from *rectus,* Latin for right). If the priorities of these groups decrease in the *counterclockwise* direction, the asymmetric carbon is said to have the *S* configuration (*S* from *sinister,* Latin for left).

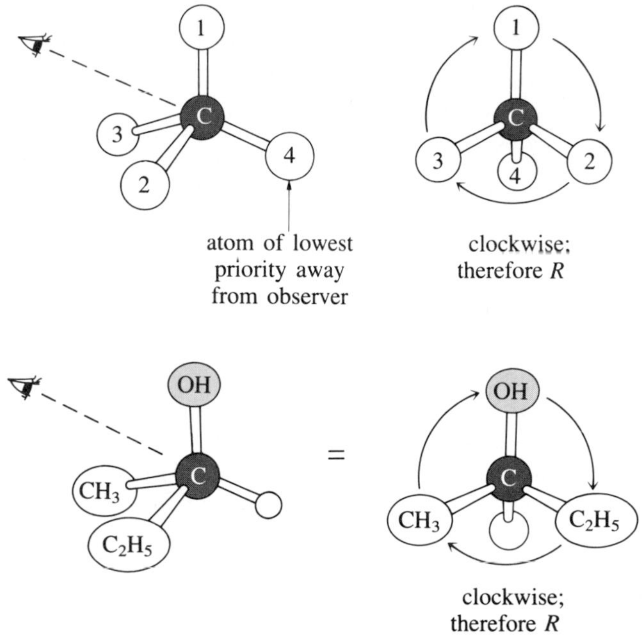

Figure 6.7 Use of the Cahn-Ingold-Prelog system to designate stereochemistry (a) of a general chiral carbon atom, and (b) of (*R*)-2-butanol.

A stereoisomer is named by indicating the configuration of each asymmetric carbon before the systematic name of the compound, as in the following examples:

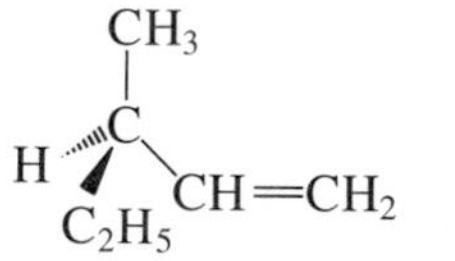

(*R*)-3-methyl-1-pentene

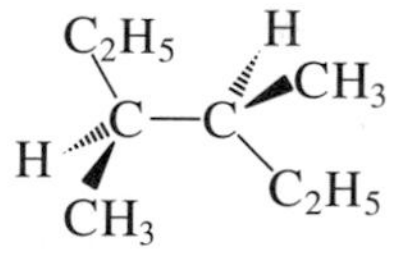

(3*S*,4*S*)-3,4-dimethylhexane

As illustrated by the second example, numbers are used with the *R*,*S* designations when there are more than one asymmetric carbon.

When we have assigned an *R* or *S* configuration to every asymmetric carbon in a molecule, we have specified the **absolute stereochemical configuration,** or **absolute stereochemistry,** of the molecule. There is another, older system for specifying absolute stereochemistry, the D,L system, which is still used in amino acid and carbohydrate chemistry, and we shall study that system in Chapters 26 and 27. With this exception, the *R*,*S* system has gained virtually complete acceptance.

Problems

4 Draw perspective representations of the indicated enantiomer of the following chiral molecules. Use models if necessary.

(a) (*S*)-CH_3—CHD—OH

(b) (*S*)-CH_3—CH—NH_2 (with C=O and OH below the CH, i.e. CH bearing a —C(=O)—OH group)

(c) (*R*)-4-methyl-(*Z*)-2-hexene

Problems (Cont.)

5 Indicate whether the asymmetric atom in each of the following compounds has the *R* or *S* configuration.

(a) malic acid

(b) alanine

(c)

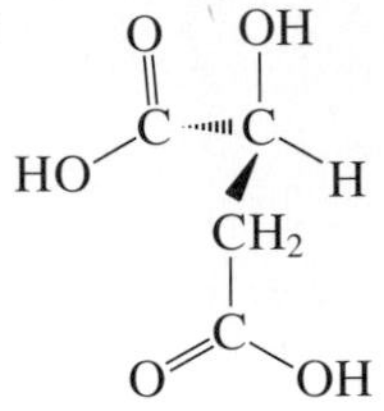

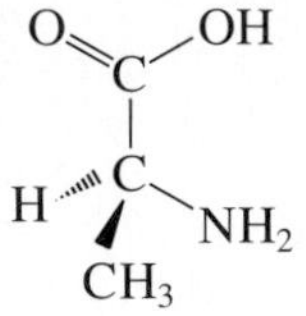

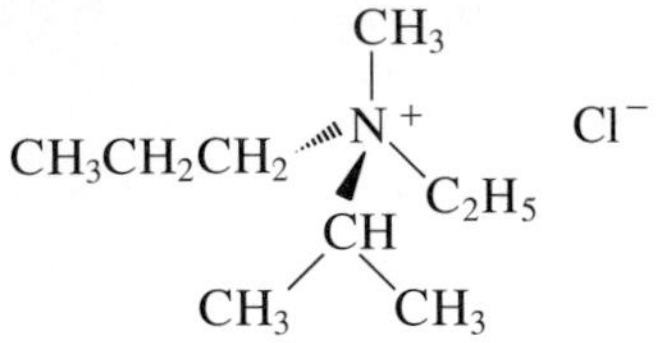

6.3 PHYSICAL PROPERTIES OF ENANTIOMERS. OPTICAL ACTIVITY

Many of the physical properties of two enantiomeric compounds have identical values. For example, both (*R*)- and (*S*)-2-butanol have the same boiling point, 99.5°. Likewise, both (*R*)- and (*S*)-lactic acid have the same melting point, 53°.

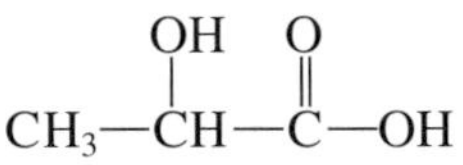

$$CH_3\!-\!\overset{\displaystyle OH}{\underset{\displaystyle |}{CH}}\!-\!\overset{\displaystyle O}{\underset{\displaystyle \|}{C}}\!-\!OH \qquad \text{lactic acid}$$

A pair of enantiomers also have identical densities, indices of refraction, heats of formation, standard free energies, and many other properties.

If a pair of enantiomers in fact have so many identical properties, how do we tell one enantiomer from the other? That is, how do we *distinguish* between enantiomers? The answer is that *we can distinguish a compound from its enantiomer by its behavior toward polarized light.* In order to understand this behavior, we first have to understand something about the nature of polarized light.

A. Polarized Light

Light is a wave motion that contains oscillating electric and magnetic fields. The electric field of ordinary light oscillates in all planes. However, it is possible to obtain light with an electric field that oscillates in only one plane. Such light is called **plane-polarized light,** or simply, polarized light (Fig. 6.8).

Polarized light is obtained by passing ordinary light through a polarizer, such as a Nicol prism. A polarizer transmits only the light waves that oscillate in a single plane. The orientation of the polarizer's axis of polarization determines the plane of the resulting polarized light. If we pass plane-polarized light through a second polarizer whose axis of polarization is perpendicular to that of the first, it follows that no light will pass the second polarizer (Fig. 6.9a). This same effect can be observed with two pairs of Polaroid® sunglasses (Fig. 6.9b). When the lenses are oriented in the same direction, light will pass. When the lenses are turned at right angles, their axes of polarization are crossed, no light is transmitted, and the lenses appear dark.

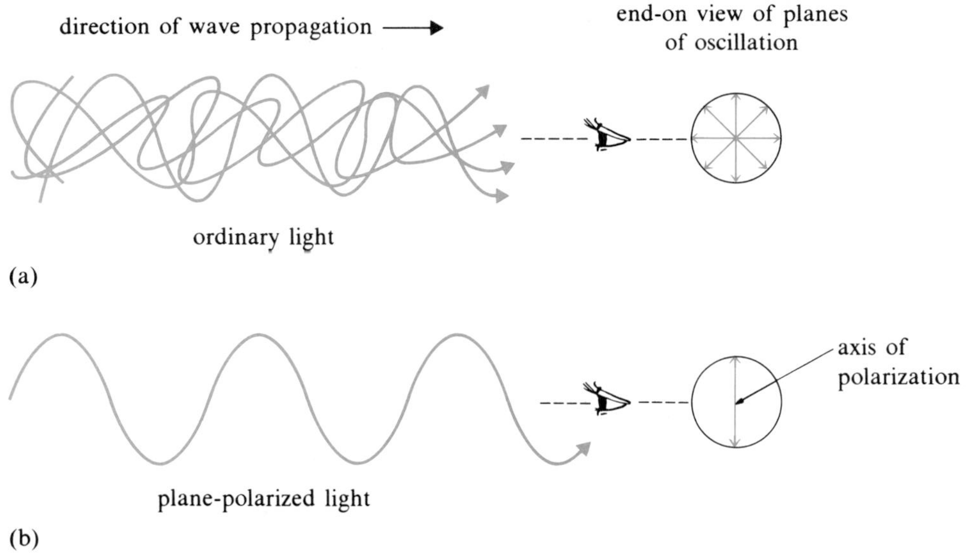

Figure 6.8 (a) Ordinary light has electromagnetic fields oscillating in all directions. (b) In plane-polarized light the oscillation is only in the direction of the polarization axis.

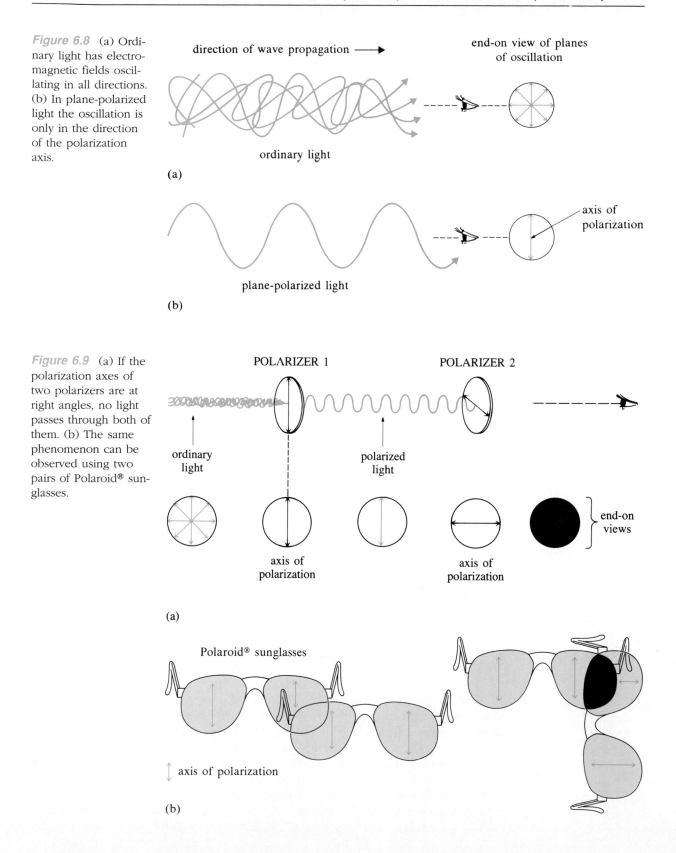

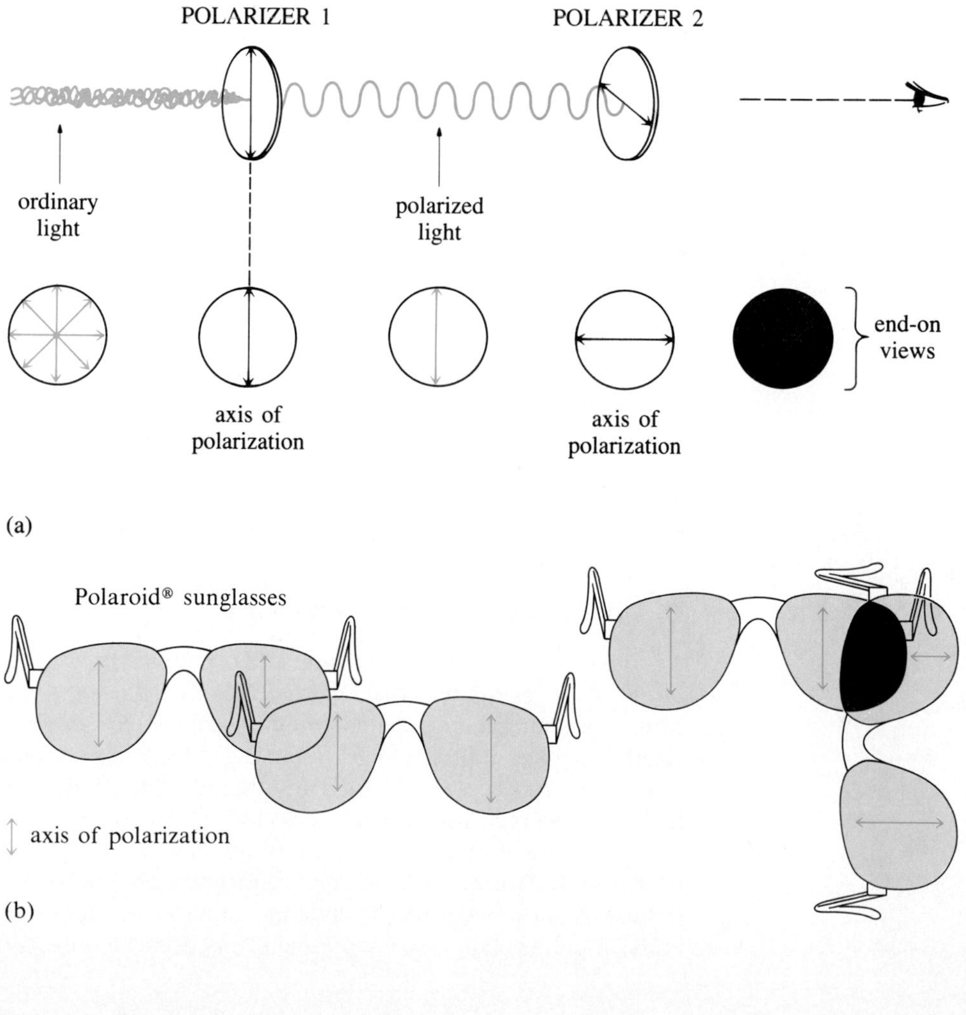

Figure 6.9 (a) If the polarization axes of two polarizers are at right angles, no light passes through both of them. (b) The same phenomenon can be observed using two pairs of Polaroid® sunglasses.

B. Optical Activity

If we pass plane-polarized light through one enantiomer of a chiral substance (either the pure enantiomer or a solution of it), we find that *the plane of polarization of the emergent light has been rotated*. A substance that rotates the plane of polarized light is said to be **optically active.** In general, *individual enantiomers of chiral substances are optically active.*

Optical activity is measured in a device called a **polarimeter** (Fig. 6.10). This is basically nothing more than the system of two polarizers shown in Fig. 6.9. The sample to be studied is placed in the light beam between the two polarizers. Because optical activity changes with the wavelength (color) of the light, it is necessary to use monochromatic light—light of a single color—to measure optical activity. The yellow light from a sodium arc (the sodium D-line with a wavelength of 5893 Å) is often used in this type of experiment. First, an optically inactive sample (such as air or solvent) is placed in the light beam. The light, polarized by the first polarizer, passes through the sample, and the analyzer is turned to establish a dark field. This setting of the analyzer defines the zero of optical rotation. Next, the sample whose optical activity is to be measured is placed in the light beam. The number of degrees α that the analyzer must be turned to reestablish the dark field corresponds to the optical activity, or **optical rotation,** of the sample. If the sample rotates the plane of polarized light in the clockwise direction, the optical rotation is given a plus sign. Such a sample is said to be **dextrorotatory** (*dextro* = right). If the sample rotates the plane of polarized light in the counterclockwise direction, the optical rotation is given a minus sign, and the sample is said to be **levorotatory** (*levo* = left).

The observed optical rotation α is proportional to the number of optically active molecules present in the light beam. Thus, α is proportional to both the concentration c of the optically active compound in the sample as well as the length l of the sample cell:

$$\alpha = [\alpha]cl \tag{6.1}$$

The constant of proportionality, $[\alpha]$, is called the **specific rotation.** The concentration of the sample is given in g/mL, and the path length in dm (decimeters). (For a pure liquid, c is taken as the density.) Thus, the specific rotation is the observed rotation at a concentration of 1 g/mL and a path length of 1 dm. The specific rotation is conventionally reported with a subscript that indicates the wavelength of light used and a superscript that indicates the temperature. Thus, a specific rotation reported as $[\alpha]_D^{20}$ has been determined at 20 °C using the sodium D-line.

C. Optical Activities of Enantiomers

A pair of enantiomers are distinguished by their optical activities because *a pair of enantiomers rotate the plane of polarized light by equal amounts in opposite directions.* For example, the specific rotation $[\alpha]_D^{20}$ of (*S*)-2-butanol is +13.9°. The specific rotation of its enantiomer (*R*)-2-butanol is −13.9°. If a solution of (*S*)-2-butanol has an observed rotation of +3.5°, then an otherwise identical solution of (*R*)-2-butanol will have an observed rotation of −3.5°.

A sample of a pure chiral compound uncontaminated by its enantiomer is said to be **optically pure.** In a mixture of the two enantiomers, each contributes to the optical rotation according to its concentration. It follows that a sample containing equal amounts of the two enantiomers must have an observed optical rotation of zero.

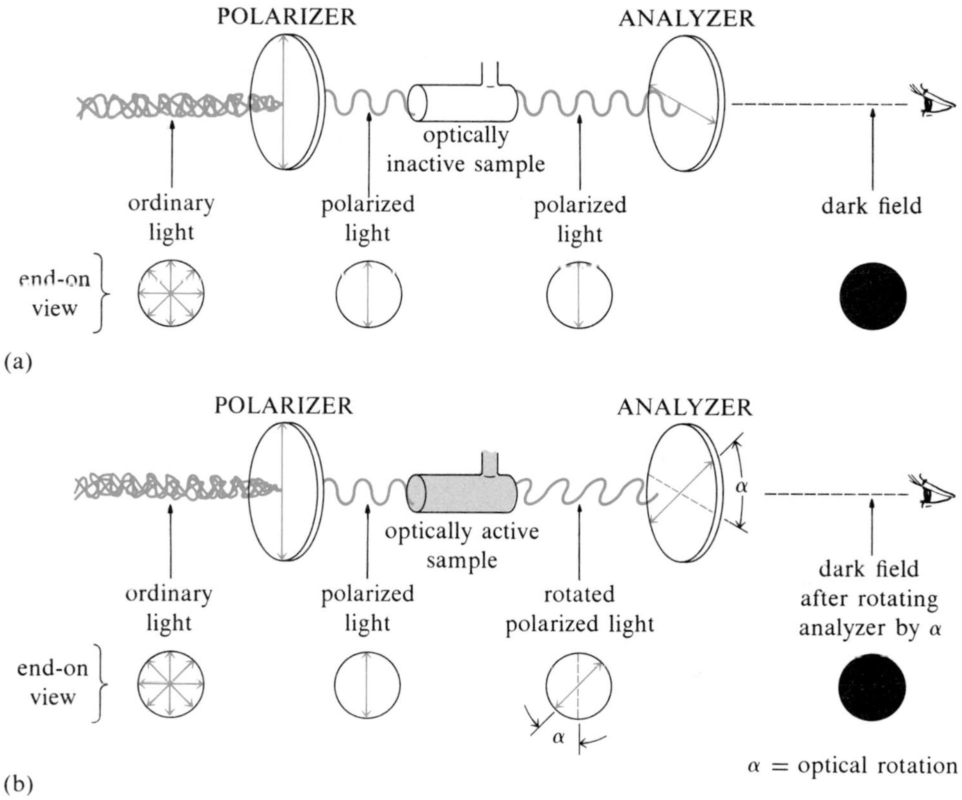

Figure 6.10 Determination of optical rotation. In (a), the reference condition of zero rotation is established as a dark field. In (b), the polarized light has passed through an optically active sample with observed rotation α. The analyzer must therefore be rotated through the same angle α to establish the dark-field condition again. The rotation α is read from the calibrated scale on the analyzer.

Sometimes a plus or minus sign is used with the name of a chiral compound to indicate the sign of its optical rotation. Thus, (*S*)-2-butanol is sometimes called (*S*)-(+)-2-butanol because it has a positive optical rotation. Similarly, (*R*)-2-butanol would be termed (*R*)-(−)-2-butanol.

There is no simple relationship between the sign of optical rotation and absolute configuration. Thus, some compounds with the *S* configuration have positive rotations, and others have negative rotations. We have seen that (*S*)-2-butanol is dextrorotatory ($[\alpha]_D^{20} = +13.9°$). However, the *R* enantiomer of glyceraldehyde is dextrorotatory:

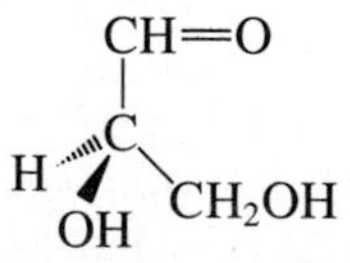

(*R*)-(+)-glyceraldehyde
$[\alpha]_D^{20} = +13.5°$

Why don't achiral molecules show optical activity? When polarized light passes through a sample of a compound, each molecule makes its own tiny contribution to optical activity, because the electrons in the molecule interact with the electric field of the light wave. For example, when polarized light passes through a sample of ethanol (Fig. 6.11), the light may "see" one ethanol molecule in a certain orientation with respect to the light beam. This molecule may actually make a nonzero contribution to

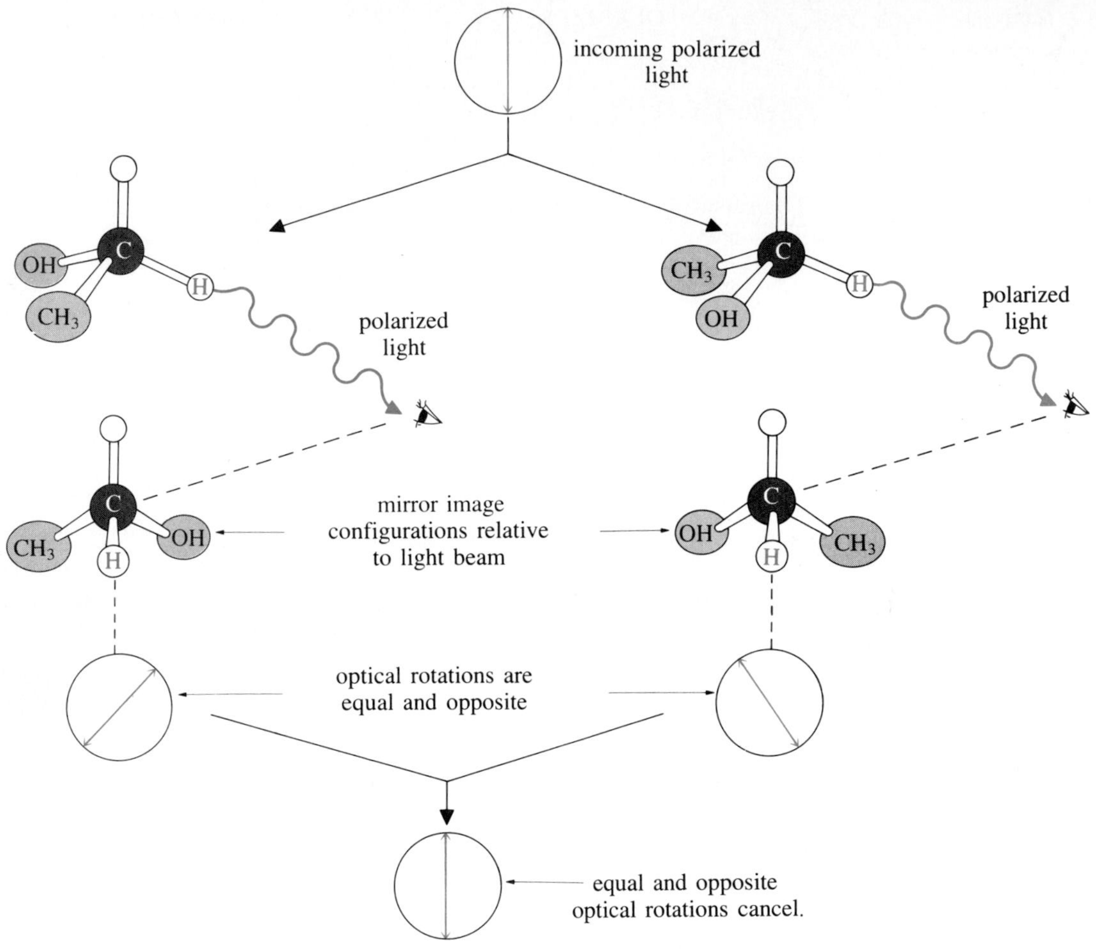

Figure 6.11 . Light "sees" mirror-image configurations of an achiral molecule within the same sample.

the optical activity. However, it is equally likely that the light will "see" another ethanol molecule in the mirror-image orientation. This orientation is equally likely because ethanol is achiral. The optical rotation of the second orientation will be equal and opposite to that of the first, and the two will cancel. In an optically active sample of chiral molecules, molecules cannot assume mirror-image orientations with respect to the light beam precisely because they lack the appropriate symmetry—they are chiral. Thus, samples of these molecules show a net optical activity.

Problems

6 A $0.1M$ solution of a pure chiral compound D has an observed rotation in a 1 dm cell of $+0.20°$. The molecular weight of the compound is 150.

(a) What is the specific rotation of D?

(b) What is the observed rotation if this solution is mixed with an equal volume of a solution that is 0.1 *M* in *L*, the enantiomer of *D*?

(c) What is the observed rotation if the solution of *D* is diluted with an equal volume of solvent?

(d) What is the specific rotation of *D* after the dilution described in (c)?

(e) What is the specific rotation of *L*, the enantiomer of *D*?

(f) What is the observed rotation of 100 mL of a solution that contains 0.01 mole of *D* and 0.005 mole of *L*? (Assume a 1 dm path length.)

7 Suppose a sample of an optically active substance has an observed rotation of +10°. The scale on the analyzer of a polarimeter is circular; +10° is the same as −350°. How would you determine whether the observed rotation is +10°, −350°, or some other value?

6.4 RACEMATES

A mixture containing equal amounts of each enantiomer of a pair is called a **racemate,** a **racemic mixture,** or a **racemic modification.** There are two ways to refer to a racemate. The racemate of 2-butanol, for example, could be called racemic 2-butanol or (±)-2-butanol.

Racemates typically have values for their physical properties that are different from those of the pure enantiomers. For example, the melting point of either enantiomer of lactic acid (Sec. 6.3) is 53°, but the melting point of racemic lactic acid is 18°. The optical rotation of the racemate is zero, because a racemate contains equal amounts of two enantiomers whose equal optical rotations of opposite sign exactly cancel each other.

It is important here to notice one point of terminology. It is sometimes said, incorrectly, that racemates are achiral. This statement is inappropriate because chirality is a property of individual molecules (or other objects). A chiral molecule has this property whether it is mixed with its enantiomer or not. Thus, 2-butanol is a chiral molecule because it exists in two enantiomeric forms. The fact that it might also occur as a mixture of two enantiomers does not alter the fact that it is chiral. What *is* true about racemates is that they are *optically inactive.* Optical activity is a physical property; chirality is a structural attribute. Optical activity requires a chiral sample; however, a sample that contains chiral molecules (a racemate, for example) need not be optically active. Although optical activity and chirality are terms that are closely associated, they should not be confused.

The process of forming a racemate from a pure enantiomer is called **racemization.** The simplest method of racemization is to mix equal amounts of each member of an enantiomeric pair. As we shall also learn, racemization can occur as a result of conformational changes or chemical reactions.

Problem

8 What is the observed rotation when a 0.1*M* solution of (*R*)-2-butanol is mixed with an equal volume of a 0.05*M* solution of racemic 2-butanol, and the resulting solution is analyzed in a cell that is 5 cm long? [The specific rotation of (*R*)-2-butanol is +13.9°.]

6.5 DETERMINATION OF ABSOLUTE CONFIGURATION. STEREOCHEMICAL CORRELATION

It is a simple matter to assign the *R* or *S* configuration to compounds with known three-dimensional structures. But how is a three-dimensional structure determined in the first place? [Recall again (Sec. 6.3B) that the sign of optical rotation *cannot* be used to assign an *R* or *S* configuration.] One way to determine the absolute stereochemistry of a compound would be to use a variation of X-ray crystallography, called *anomalous dispersion*. This technique, however, has been applied to relatively few compounds. The absolute configurations of most organic compounds are determined by using chemical reactions to correlate them with other compounds of known absolute configurations.

To illustrate, suppose we have in hand a pure sample of the following chiral alcohol, but we do *not* know its absolute stereochemical configuration.

$$\text{Ph—}\underset{\text{CH}_3}{\underset{|}{\text{CH}}}\text{—CH}_2\text{CH}_2\text{OH} \qquad [\alpha]_D^{25} = -16.82°$$

3-phenyl-1-butanol

Suppose further that we search the chemical literature and find that the absolute configuration of the following alkene is known.

$$\text{Ph—}\underset{\text{CH}_3}{\underset{|}{\text{CH}}}\text{—CH=CH}_2 \qquad R \text{ enantiomer has } [\alpha]_D^{22} = -6.39°$$

3-phenyl-1-butene

We then go to the laboratory and find that the alcohol of *unknown* configuration can be prepared from the alkene of *known* configuration by the following reaction:

$$(R)\text{-}(-)\text{-Ph—}\underset{\text{CH}_3}{\underset{|}{\text{CH}}}\text{—CH=CH}_2 \xrightarrow{B_2H_6} \xrightarrow{H_2O_2/OH^-} (-)\text{-Ph—}\underset{\text{CH}_3}{\underset{|}{\text{CH}}}\text{—CH}_2\text{CH}_2\text{OH} \qquad (6.2)$$

Notice that this reaction does not break any of the bonds to the asymmetric carbon. Thus, the way that corresponding groups are arranged about the asymmetric carbon must be the same in both reactant and product. Therefore, since the configuration of the alkene is known, that of the alcohol is also known. Application of the *R,S* system to the alcohol product shows that it is *R*.

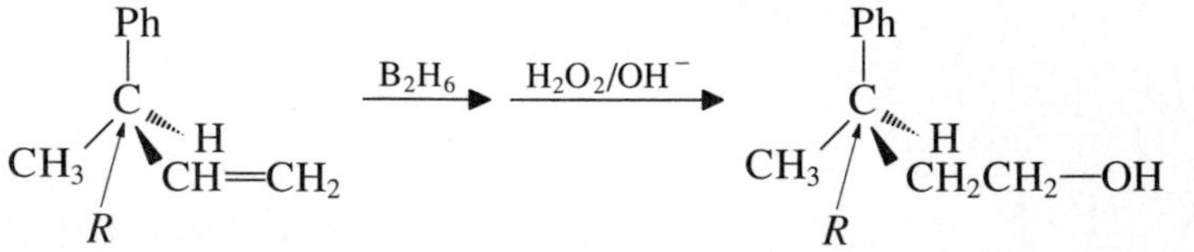

(6.3)

Although both reactant and product in this example have the same *R,S* designations, this does not have to be true in general. It is possible for the *R,S* designations of reactant and product to differ if the reaction results in a change in the relative priority of groups at the asymmetric carbon, as in Problem 9.

> The most secure way of relating absolute configurations is to use reactions that do not break the bonds at asymmetric carbon, as shown above. A reaction that breaks these bonds, however, may also be used provided that the stereochemical outcome of such a reaction has been established previously on a number of related compounds.

Problem

9 From the outcome of the following transformation, indicated whether the levorotatory enantiomer of the product has the *R* or *S* configuration. Draw the absolute configuration of the product. (*Hint*: The phenyl group has a higher priority than the vinyl group in the *R,S* system.)

$$(S)\text{-}(+)\text{-Ph}\underset{\underset{\displaystyle \text{CH}_2\text{CH}_2\text{CH}_3}{|}}{\text{CH}}\text{—CH}=\text{CH}_2 + \text{H}_2 \xrightarrow{\text{Pd/C}} (-)\text{-Ph}\underset{\underset{\displaystyle \text{CH}_2\text{CH}_2\text{CH}_3}{|}}{\text{CH}}\text{—CH}_2\text{CH}_3$$

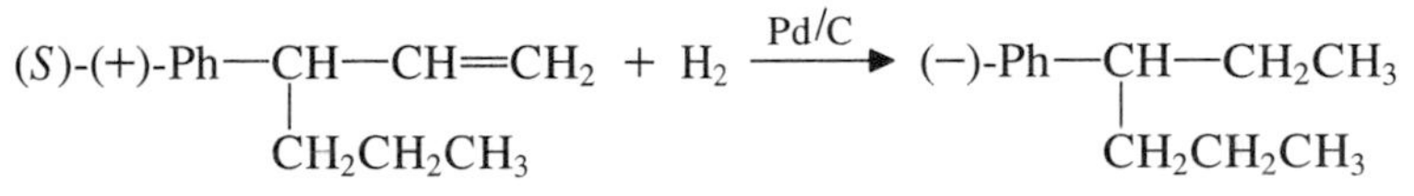

6.6 DIASTEREOMERS. ANALYSIS OF ISOMERISM

Up to this point, we have restricted most of our attention to molecules with single asymmetric carbons. What happens when two or more asymmetric carbons are present in the same molecule? This situation is illustrated by 2,3-pentanediol.

$$\underset{\text{carbon number} \qquad 1 \qquad 2 \qquad 3 \qquad 4 \qquad 5}{\overset{\overset{\displaystyle \text{OH} \quad \text{OH}}{|\qquad |}}{\text{CH}_3\text{—CH—CH—CH}_2\text{CH}_3}}$$

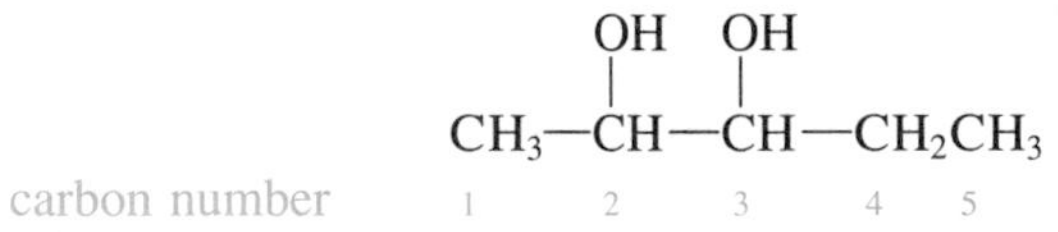

This compound has two asymmetric carbons: carbons 2 and 3. Each might have the *R* or *S* configuration. With two possible configurations at each carbon, there are four possible stereoisomers:

(2*R*,3*R*) (2*R*,3*S*)

(2*S*,3*R*) (2*S*,3*S*)

These possibilities are shown as sawhorse structures in Fig. 6.12. What are the relationships among these stereoisomers?

The 2*R*,3*R* and 2*S*,3*S* forms are a pair of enantiomers, because they are nonsuperimposable mirror images; the 2*S*,3*R* and 2*R*,3*S* forms are also an enantiomeric pair. (Demonstrate this point to yourself!) These structures illustrate the following generalization: *In order for a pair of molecules with more than one asymmetric carbon to be enantiomers, they must have different configurations at every asymmetric carbon.*

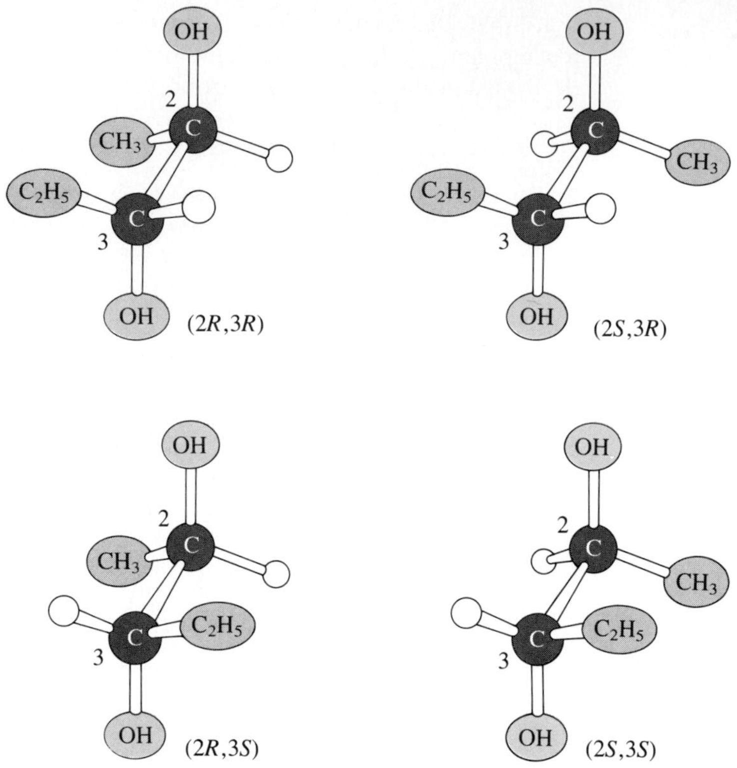

Figure 6.12 Stereoisomers of 2,3-pentanediol.

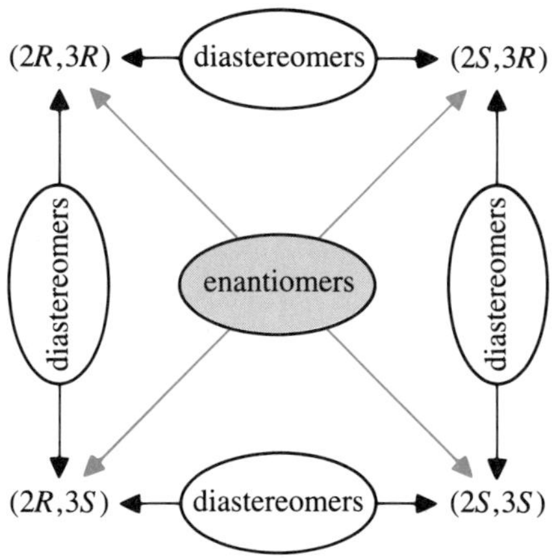

Figure 6.13 Relationships among the stereoisomers of 2,3-pentanediol.

Since neither the 2*R*,3*R* and 2*R*,3*S* pair, nor the 2*S*,3*S* and 2*S*,3*R* pair, are enantiomers, they must have a different stereochemical relationship. Stereoisomers such as these that are not enantiomers are called **diastereoisomers,** or, more simply, **diastereomers.** They have a **diastereomeric** relationship. All of the relationships among the stereoisomeric 2,3-pentanediols are shown in Fig. 6.13.

TABLE 6.1 *Properties of Four Chiral Stereoisomers*

$$CH_3-\overset{\overset{\displaystyle O}{\|}}{C}-NH-\overset{\underset{\displaystyle \overset{3}{C}H-CH_3}{|}}{\overset{2}{C}H}-\overset{\overset{\displaystyle O}{\|}}{C}-OH$$

$$\underset{\displaystyle C_2H_5}{|}$$

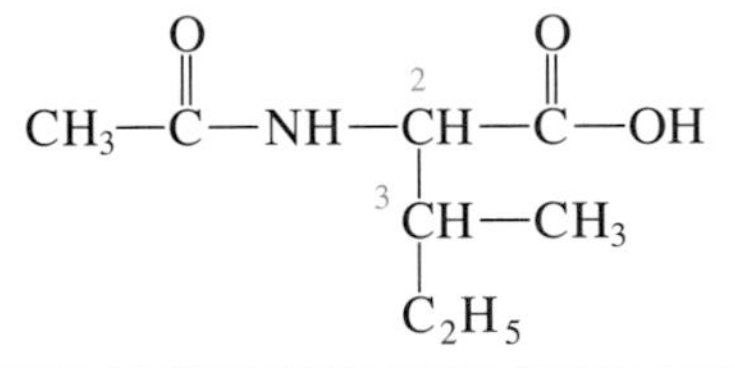

Configuration	Specific rotation (25°, ethanol)	Melting point, °C
(2S,3S)	+15°	150–151°
(2R,3R)	−15°	150–151°
(2S,3R)	+21.5°	155–156°
(2R,3S)	−21.5°	155–156°
racemate of (2S,3S) and (2R,3R)	0°	117–123°
racemate of (2S,3R) and (2R,3S)	0°	165–166°

Diastereomers have different values of their physical properties: different melting points, boiling points, indices of refraction, heats of formation, standard free energies, and, if a pair of diastereomers happen to be chiral, different optical rotations. Furthermore, the optical rotations of chiral diastereomers have no general relationship to one another. The physical properties of four chiral stereoisomers and their racemates are illustrated in Table 6.1. Because diastereomers differ in their physical properties, they can in principle be separated by conventional means: for example, by fractional distillation or crystallization.

We have now seen an example of every common type of isomerism. **Isomers** are compounds with the same molecular formula. **Structural isomers** or **constitutional isomers** have *different* atomic connectivities. **Stereoisomers** have the *same* atomic connectivities. Stereoisomers are either **enantiomers** or **diastereoisomers.**

If we work systematically, it is easy to determine whether a pair of molecules are isomers and, if so, what their isomeric relationship is. (We always consider the relationship of molecules *by pairs.*) Given a pair of molecules, we work our way down the flow chart in Fig. 6.14 asking each question in turn and following the appropriate branch. When we get to a box labeled "END," the isomeric relationship has been determined. Let us consider two examples to illustrate the use of Fig. 6.14.

Example 1: Relationship of (*R*)- and (*S*)-2-butanol (Fig. 6.3). The molecules have the same molecular formula and atomic connectivity, and they are mirror images. This leads us through all the "Yes" branches to the last question. The answer to this question is "No": the molecules are not superimposable. Thus they are enantiomers.

Example 2: Relationship of *cis*- and *trans*-2-butene. The first two questions are answered "Yes." However, at the third question—Are the molecules mirror images?—the answer is "No." The conclusion: *cis*- and *trans*-2-butene are diastereomers.

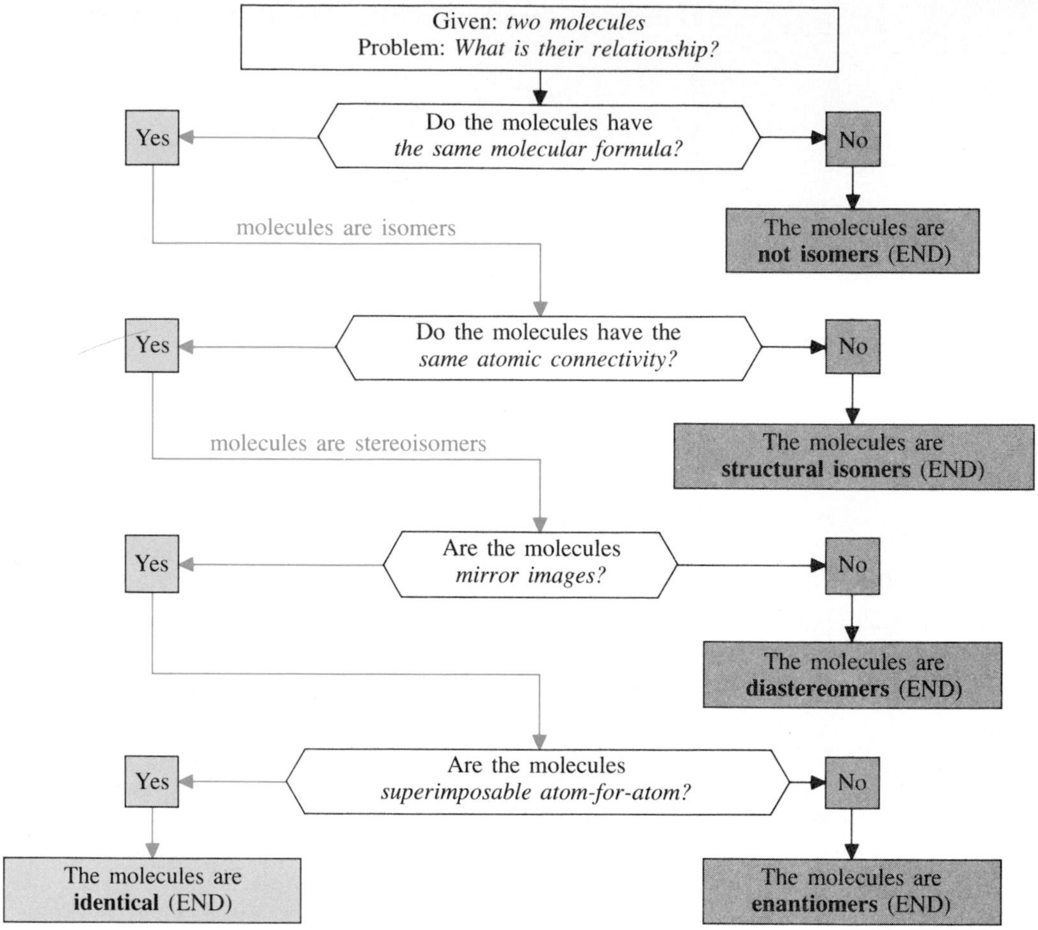

Figure 6.14 A systematic way to analyze the relationship between any two molecules.

From Example 2 we see that *cis–trans* isomerism is one type of diastereomeric relationship. Since neither *cis-* nor *trans*-2-butene is chiral, we see that diastereomers need *not* be chiral compounds. On the other hand, diastereomers *may* be chiral, as in the case of the diastereomeric 2,3-pentanediols (Fig. 6.12).

Problem

10 Give the relationship between the compounds in each of the following pairs:
(a) (*R*)-4-methyl-(*Z*)-2-hexene and (*R*)-4-methyl-(*E*)-2-hexene
(b) (*R*)-4-methyl-(*Z*)-2-hexene and (*S*)-4-methyl-(*Z*)-2-hexene
(c) (*R*)-4-methyl-(*Z*)-2-hexene and (*S*)-4-methyl-(*E*)-2-hexene

6.7 MESO *COMPOUNDS*

Another aspect of stereoisomerism is illustrated with 2,3-butanediol.

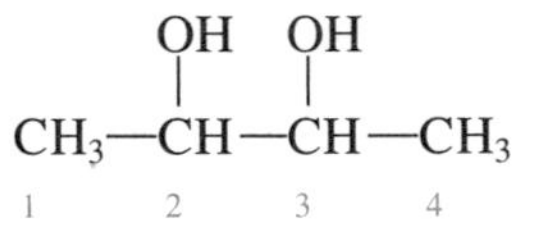

2,3-butanediol

carbon numbering

As with 2,3-pentanediol in the previous section, there appear to be four stereochemical possibilities:

(2*R*,3*R*) (2*R*,3*S*)
(2*S*,3*R*) (2*S*,3*S*)

Let us consider the relationships among these compounds. As shown in the sawhorse structures below, the 2*R*,3*R* and 2*S*,3*S* compounds are an enantiomeric pair.

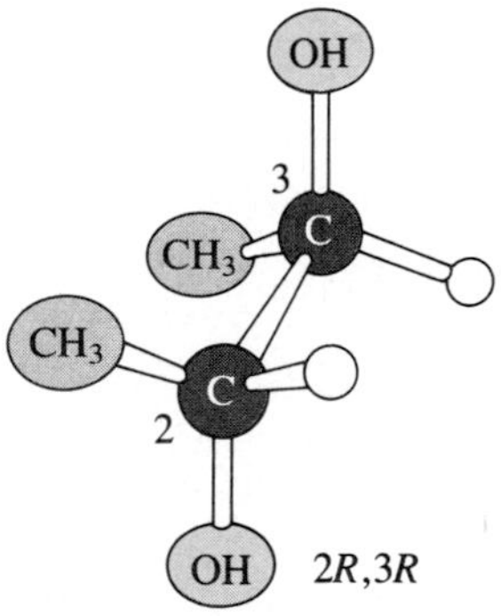

2*R*,3*R*

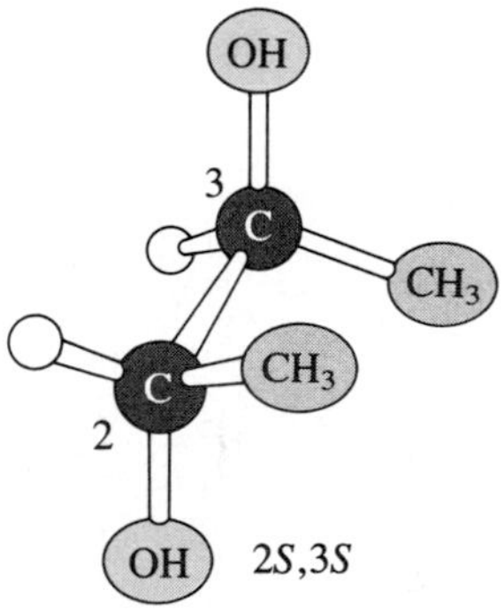

2*S*,3*S*

enantiomers

What is the relationship of the other two compounds—the 2*S*,3*R* and 2*R*,3*S* pair? These compounds, as they are drawn below, are mirror images. But are they *nonsuperimposable* mirror images?

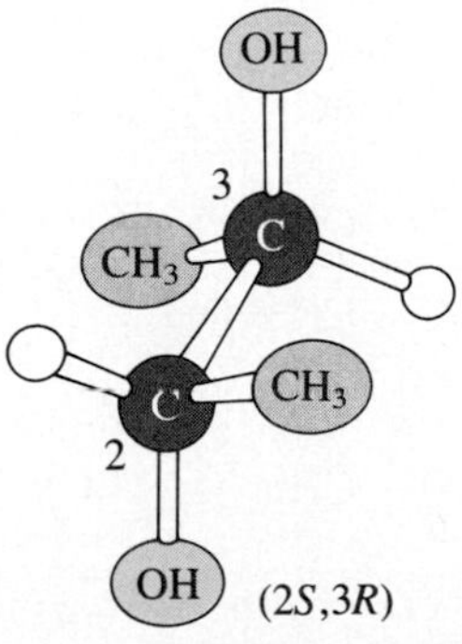

(2*S*,3*R*)

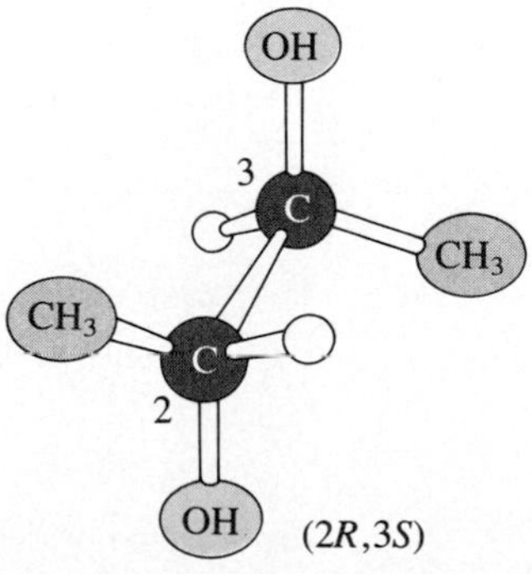

(2*R*,3*S*)

The answer is no. In fact, these two structures are not isomers at all; *they are the same molecule!* This can be seen by rotating either structure 180° about an axis perpendicular to the C2–C3 bond, as follows:

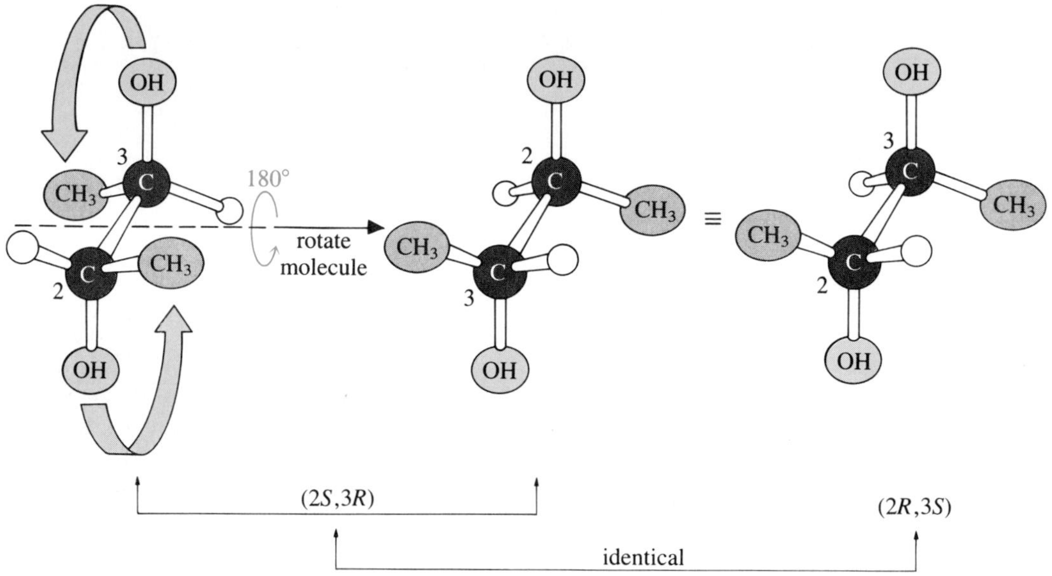

The 2*S*,3*R* compound is an example of a *meso* compound. In fact, this compound is properly called *meso*-2,3-butanediol. A ***meso* compound** is an achiral compound with asymmetric atoms (in this case, asymmetric carbons). Because the *meso* compound is superimposable on its mirror image, it is not chiral. Because it is not chiral, it is also not optically active. Notice carefully the difference between a racemate and a *meso* compound. Although both are optically inactive, a racemate is a *mixture* of compounds, but a *meso* compound is a *single* compound. *Meso*-2,3-butanediol is a diastereomer of the (2*S*,3*S*)- and (2*R*,3*R*)-butanediols; a summary of the relationships between the isomers of 2,3-butanediol is given in Fig. 6.15.

The existence of *meso* compounds shows that *an achiral compound may have asymmetric carbons.* Thus, the existence of asymmetric carbons in a molecule is *not* a sufficient condition for the molecule to be chiral, unless there is only *one* asymmetric carbon. If a molecule contains n asymmetric carbons, then it has 2^n stereoisomers unless there are *meso* compounds. If there are *meso* compounds, then there are fewer than 2^n stereoisomers. We should be wary of the existence of *meso* compounds when a molecule can be divided into two structurally identical halves. (The word *meso* means "middle," or "in between.")

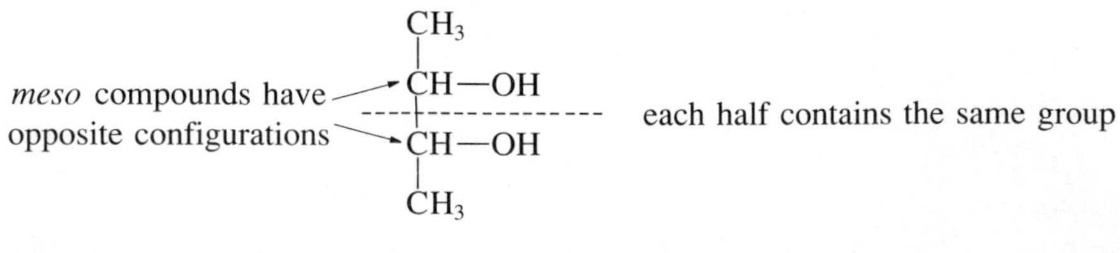

In a *meso* compound, the corresponding asymmetric carbons on each side of this dividing line must have opposite stereochemical configurations. This is why *meso*

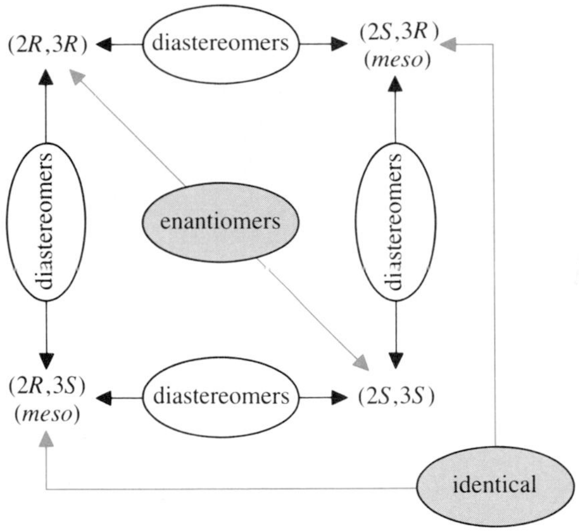

Figure 6.15 Relationships among the stereoisomers of 2,3-butanediol. Note that the (2*S*,3*R*) and (2*R*,3*S*) "isomers" are actually identical molecules and are therefore not enantiomers.

compounds have no optical activity: the two halves of the molecule contribute equal and opposite amounts to the optical rotation.

Problem

11 Explain why there are two *meso* stereoisomers of the following compound:

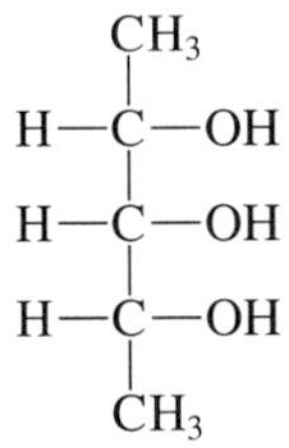

2,3,4-pentanetriol

(*Hint*: The dividing line that separates the molecule into two structurally identical groups can go *through* an atom.)

6.8 CONFORMATIONAL STEREOISOMERS. AMINE INVERSION

A. Chiral Molecules without Asymmetric Atoms

The existence of *meso* compounds shows that the presence of asymmetric carbons is not a *sufficient* condition for chirality of a molecule. We can still inquire whether the existence of an asymmetric atom is a *necessary* condition for the chirality of a molecule. The rotational isomers of butane provide an interesting example (see Problem 11 in Chapter 3). In Fig. 6.16 we see that the two *gauche* conformations of butane are nonsuperimposable mirror images, and are therefore a pair of enantiomers. The *anti* conformation has a center of symmetry (find it) and is therefore superimposable on

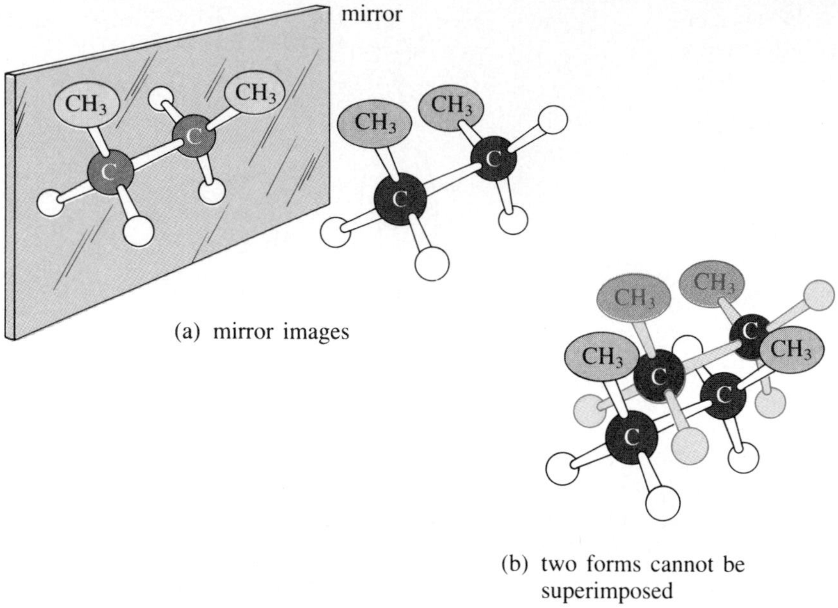

Figure 6.16 The two *gauche* forms of butane are (a) mirror images, that are (b) nonsuperimposable.

its mirror image; it is achiral. It follows, then, that butane is a mixture of a pair of chiral *gauche* enantiomers and one achiral *anti* diastereomer. The two *gauche* forms of butane are **conformational enantiomers;** *gauche-* and *anti-*butane are **conformational diastereomers.**

The compound butane is not optically active because the two *gauche* forms are present in equal amounts. The optical activity of one *gauche* enantiomer thus cancels the optical activity of the other. (The *anti* form, because it is achiral, would not be optically active even if it were present alone.) However, we might imagine an amusing experiment in which we separate the two *gauche* forms of butane (by an as yet undisclosed method!) at such a low temperature that the interconversion between the *gauche* and *anti* forms of butane is very slow. We would find that each *gauche* butane isomer is optically active! The two *gauche* isomers would have equal specific rotations of opposite signs, but many of their other properties would be the same. *Anti-*butane would have zero optical rotation, and many of its properties would be different from those of *gauche-*butane. Of course, at room temperature, this separation would be impossible, because the butane isomers come to equilibrium within 10^{-9} second by rotation about the central carbon–carbon bond. (This is another example of *racemization*; Sec. 6.5.) It is conceivable, though, that on some planet with a temperature near absolute zero, *gauche-* and *anti-*butanes exist as separate compounds. (It would also be interesting to meet any inhabitants of such a planet capable of appreciating this fact!)

Is butane considered to be a chiral molecule? Although butane contains chiral conformations, this compound *behaves as if* it is achiral, because it has zero optical activity and it cannot be separated into enantiomers on any reasonable time scale. We shall agree that if a molecule is made up of enantiomeric pairs that are rapidly interconverting under ordinary conditions, the molecule is considered to be achiral. That is, when we consider whether a molecule is chiral, we do not concern ourselves with the existence of conformational stereoisomers; instead, we shall consider the chirality of a molecule only as it is averaged out over time. Thus, butane is said to be achiral.

Although the conformations of butane interconvert rapidly at room temperature, some molecules have chiral conformations that interconvert slowly and can be separately isolated. The compound 2,3-pentadiene is an example.

enantiomers of 2,3-pentadiene

The carbons on each end of the diene and their two attached atoms lie in perpendicular planes. (We shall study the reason for this geometry in Sec. 15.1B.) Because of this geometry, 2,3-pentadiene and related compounds are chiral, even though they have no asymmetric carbon. (Be sure to demonstrate to yourself that the two enantiomers of 2,3-pentadiene are nonsuperimposable.) These enantiomers would be interconverted by internal rotation about a carbon–carbon double bond; but because such a rotation does not occur (Sec. 4.1B), the enantiomers of 2,3-pentadiene can be separately isolated.

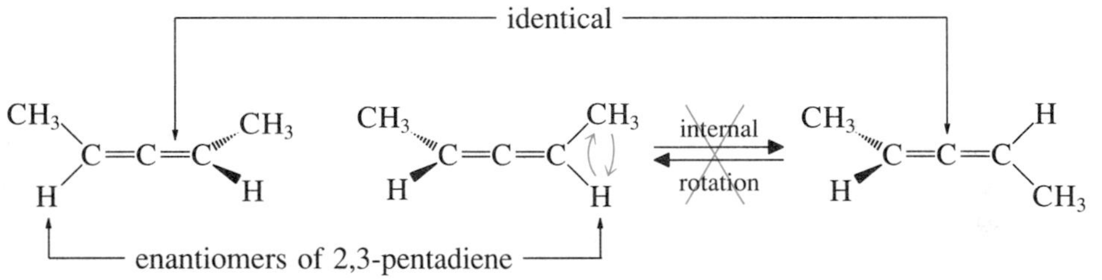

(6.4)

Problem

12 What are the relationships among the three rotational isomers of *meso*-2,3-butanediol, the compound discussed in Sec. 6.7? Point out any planes or centers of symmetry in the achiral isomer(s). Explain why this compound is achiral even though it has chiral rotational isomers.

B. Asymmetric Nitrogen: Amine Inversion

Amines provide another type of situation in which the rapid interconversion of stereoisomers takes place. *Amines* are derivatives of ammonia in which one or more of the hydrogen atoms have been replaced by an organic group. One example of an amine is ethylmethylamine.

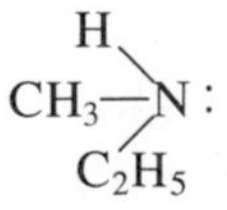

ethylmethylamine

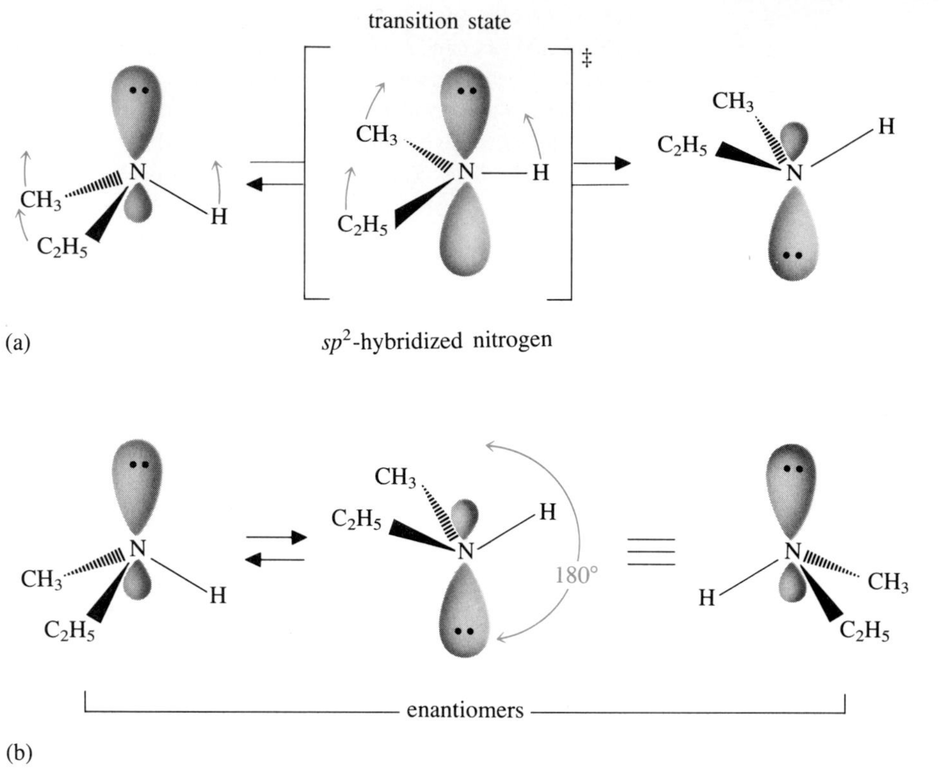

Figure 6.17 Inversion of amines. (a) As the inversion takes place, the large lobe of the electron pair appears to push through the nitrogen to the other side. As this occurs, the three other groups move, first, into a common plane with the nitrogen, then to the other side (colored arrows). (b) The enantiomeric relationship of the inverted amines.

Ethylmethylamine has four different groups around the nitrogen: a hydrogen, an ethyl group, a methyl group, and an electron pair. Since the geometry of this molecule is essentially tetrahedral, ethylmethylamine should be a chiral molecule—it should exist as two enantiomers. The asymmetric atom in this molecule is a nitrogen.

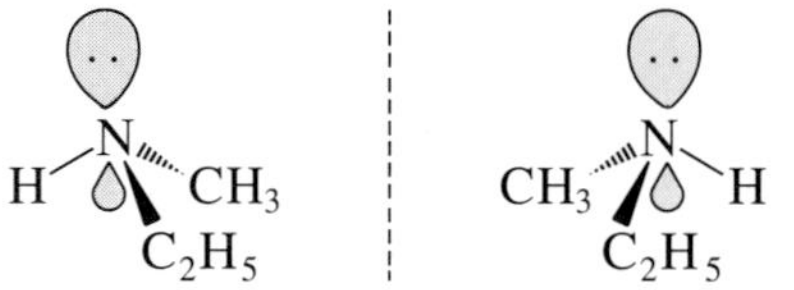

enantiomers of ethylmethylamine

In fact, the two enantiomers of amines such as ethylmethylamine cannot be isolated separately, because they rapidly interconvert by a process called **amine inversion,** shown in Fig. 6.17. In this process, the larger lobe of the electron pair seems to push through the nucleus to emerge on the other side. Notice that the molecule is not simply turning over; it is actually turning itself inside out! This is something akin to what happens when an umbrella turns inside out in the wind. This process occurs through a transition state in which the amine nitrogen adopts an sp^2 hybridization. Figure 6.17b shows that this inversion process interconverts the enantiomeric forms of the amine. Since this process is rapid at room temperature, it is impossible to isolate the separate enantiomers. Therefore, ethylmethylamine is a mixture of rapidly interconverting enantiomers. Amine inversion is yet another example of *racemization* (Sec. 6.4).

Problem

13 What is the isomeric relationship between the two forms of the following compound that are interconverted by amine inversion? Assume that the compound has the *S* configuration at its asymmetric carbon. Could this compound be isolated in optically active form?

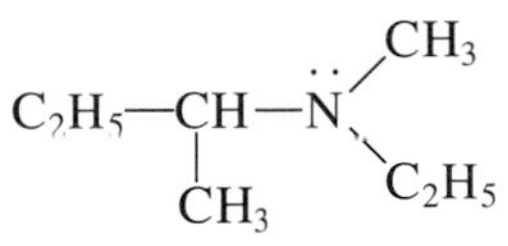
CH3
C2H5—CH—N
CH3
C2H5

6.9 FISCHER PROJECTIONS

We have learned how we can represent the three-dimensional structures of molecules on a two-dimensional page using perspective drawings that employ lines and wedges. It is simple enough to draw a perspective formula for a compound with a single asymmetric carbon, but when a molecule contains several asymmetric carbons, the drawing of perspective structures can become very tedious. For this reason, chemists have adopted another way of writing three-dimensional structures on a two-dimensional surface (paper or blackboard). The resulting structures are called **Fischer projections,** after the great German chemist Emil Fischer.

To represent a molecule in a Fischer projection, we view each asymmetric carbon in such a way that two of the bonds to this carbon are vertical and pointing away from us, and two are horizontal and pointing toward us. The Fischer projection is the structure obtained when this view is projected on a plane (Fig. 6.18a). The asymmetric carbons themselves are not drawn, but are assumed to be located at the intersections of vertical and horizontal bonds. Such a projection is, in effect, a flattened-out picture of the molecule. (In fact, one of the author's students pointed out that the Fischer projection is the way that the molecule would look after we put it on the floor and step on it!)

Fischer projections of molecules containing more than one asymmetric carbon are most useful when the asymmetric carbons are part of the same continuous carbon chain. In this case, the molecule is first placed (or imagined) in an eclipsed conformation (Figs. 6.18b,c) and the projection is derived by viewing each carbon in the way described above. For a compound with many asymmetric carbons, the carbon backbone can be imagined to be written on a curved, convex surface (for example, a drum or barrel). Projecting all horizontal bonds onto this surface and then mentally slicing the barrel open and flattening it gives the Fischer projection.

Although the Fischer projection is derived from an eclipsed conformation, this does *not* mean that the molecule actually has such a conformation. As we have learned, most molecules assume staggered conformations (Sec. 3.5). The use of an eclipsed conformation to draw the Fischer projection is a convention for showing the *stereochemical configuration* of each asymmetric carbon; it is not meant to convey the actual *conformation* of the molecule.

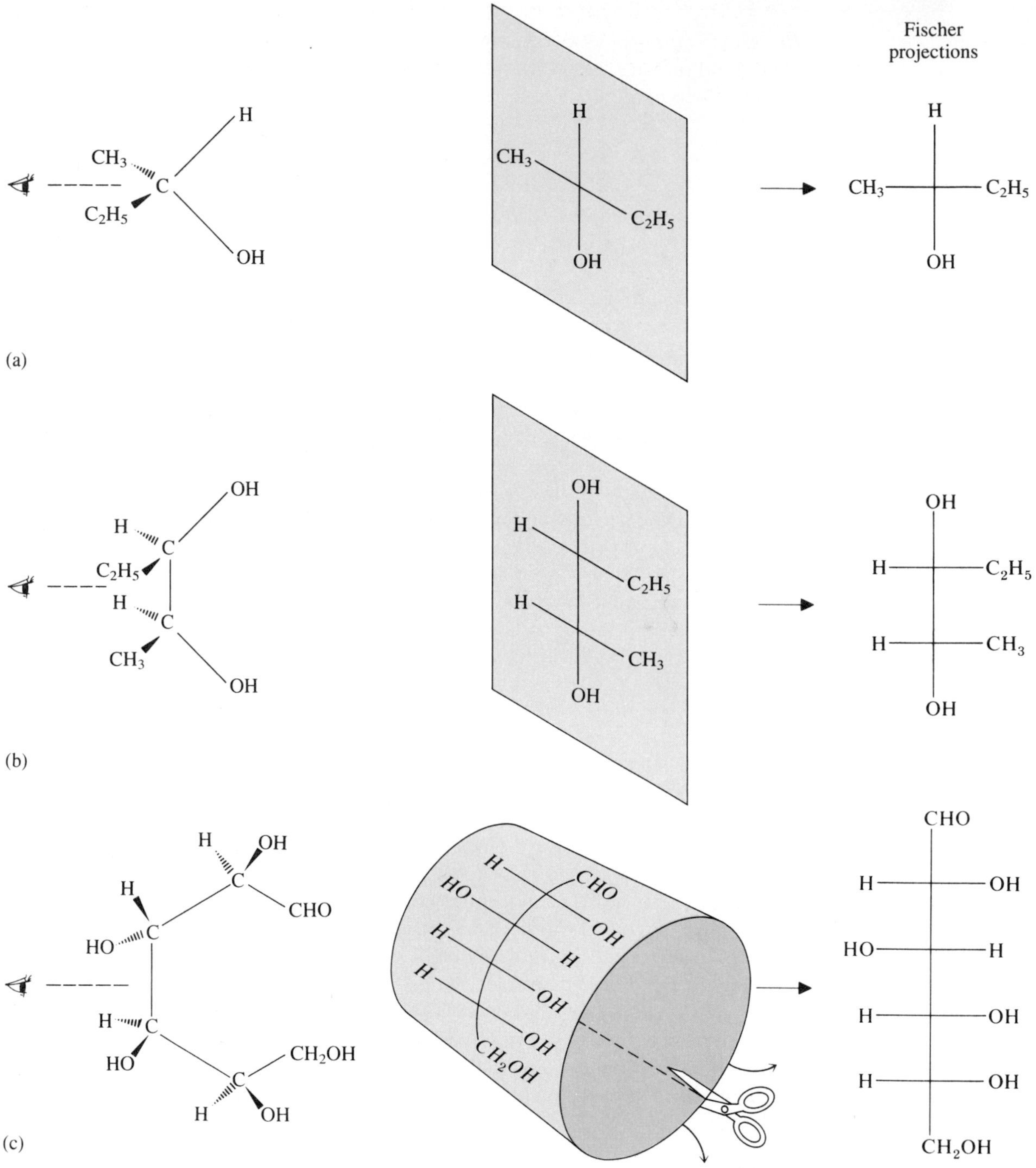

Figure 6.18 Formation of Fischer projections for stereoisomers of (a) 2-butanol, a compound with one asymmetric carbon; (b) 2,3-pentanediol, a compound with two asymmetric carbons; and (c) glucose, a compound with four asymmetric carbons.

To derive a three-dimensional model of a molecule from its Fischer projection, we simply reverse the process we have just described. Always remember that the vertical bonds in the Fischer projection are *away from* us, and the horizontal bonds are *toward* us.

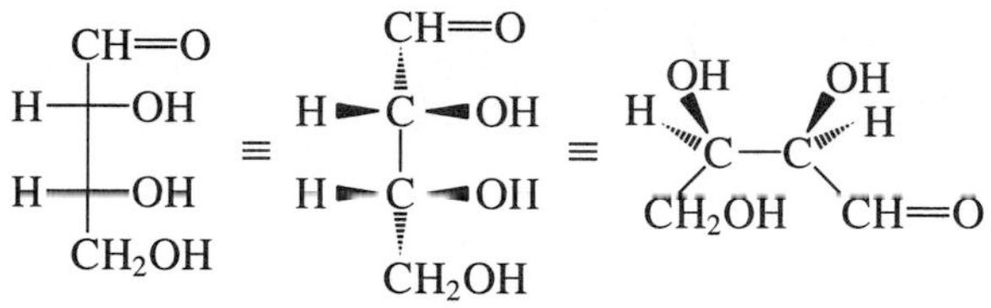

Because more than one group may be placed in a given vertical position, *several different Fischer projections can be written for the same molecule.* Because of this possibility, it is useful to be able to draw different Fischer projections of the same molecule without going back and forth to a three-dimensional model. Several rules for manipulation of Fischer projections are helpful. You should use models to convince yourself of the validity of these rules.

1. A Fischer projection may be turned 180° in the plane of the paper.

By this rule, the following two Fischer projections represent the same stereoisomer.

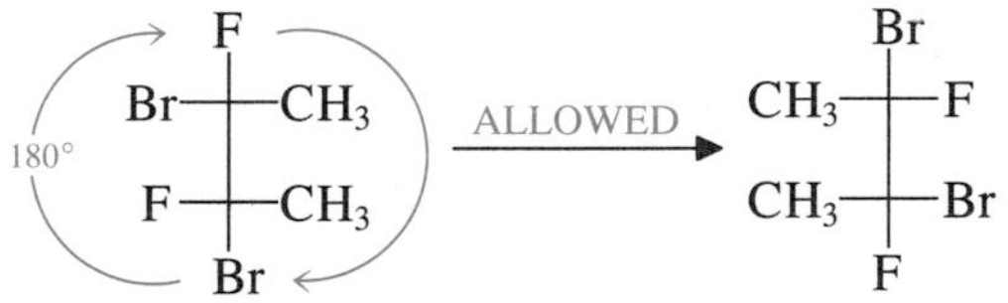

This manipulation is allowed because it leaves horizontal bonds horizontal and vertical bonds vertical; therefore, it does not alter the meaning of the Fischer projection.

2. A Fischer projection may *not* be turned 90°.

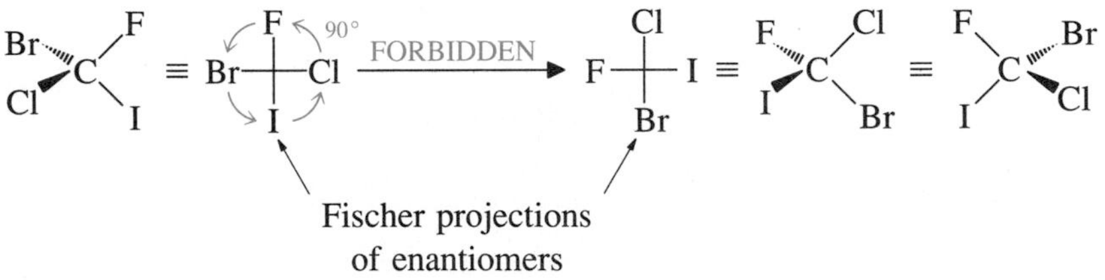

The reason this operation is forbidden is that it causes "out" bonds to be portrayed as "back" and vice versa. In other words, the original structure is converted into its enantiomer by this operation. This is disastrous, since the whole idea of Fischer projections is to convey stereochemical information. The following rule has a similar rationale.

3. A Fischer projection may *not* be lifted from the plane of the paper and turned over.

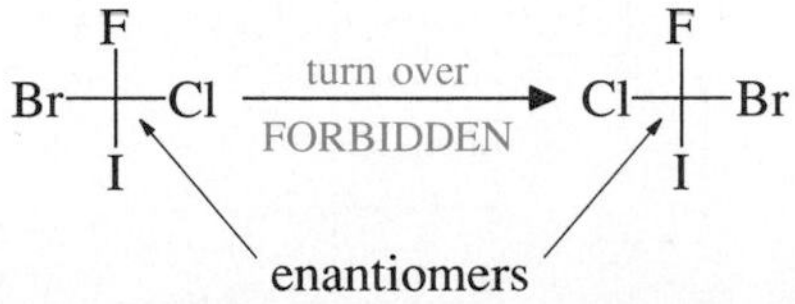

4. The three groups at either end of a Fischer projection may be interchanged in a *cyclic permutation*. That is, all three groups can be moved at the same time in a closed loop so that each occupies an adjacent position.

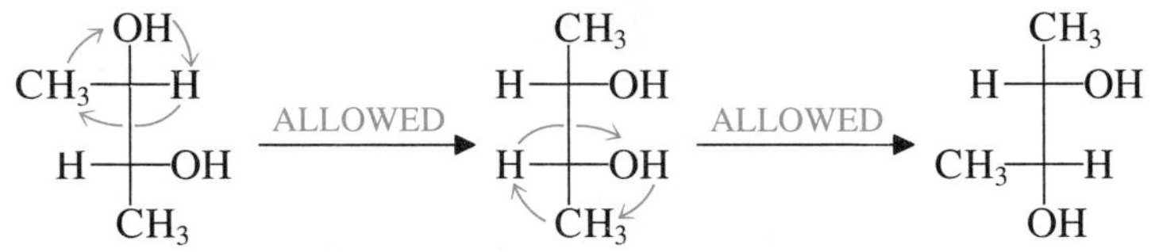

Fischer projections of the same molecule

This operation is equivalent to an internal bond rotation. This becomes clear if we build a model of any one of the compounds above and rotate about any carbon–carbon bond in the chain. We can see that each 120° bond rotation is equivalent to the cyclic permutation described in rule 4, and each rotational isomer can be written as a different Fischer projection of the same molecule.

It is particularly easy to recognize enantiomers and *meso* compounds from the appropriate Fischer projections, because planes of symmetry in the actual molecules reduce to lines of symmetry in their projections.

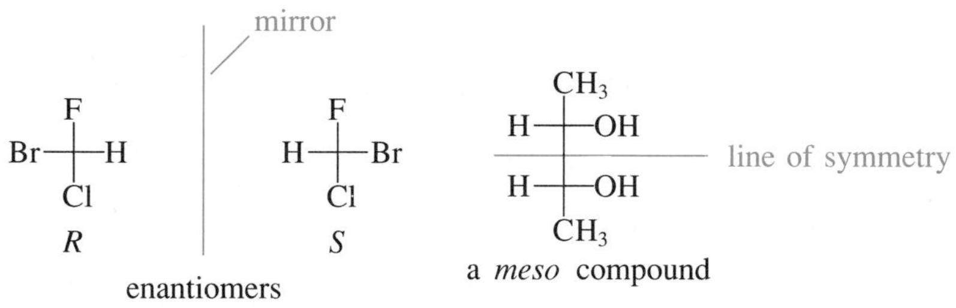

It is also easy to apply the *R,S* system to a Fischer projection. To do this, we first draw an equivalent Fischer projection (if necessary) in which the group of lowest priority is in either of the two vertical positions. We then simply apply the priority rules to the remaining three groups.

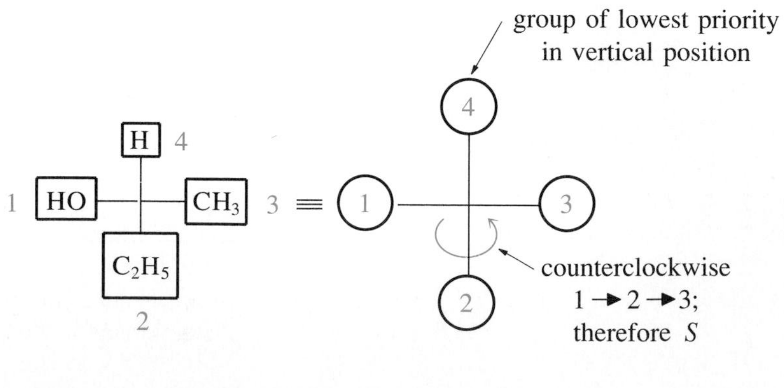

This method works because, if the lowest-priority group is in a vertical position in the Fischer projection, it is oriented away from us as required for application of the priority rules.

Problems

14 Draw at least two Fischer projections for each of the following molecules:

(a) (*S*)-2-butanol (see Fig. 6.3)

(b) $\overset{R}{\text{HO}_2\text{C}}—\overset{S}{\text{CH}}—\overset{S}{\text{CH}}—\text{CH}—\text{CH}_2\text{OH}$
 | | |
 Cl Cl Br

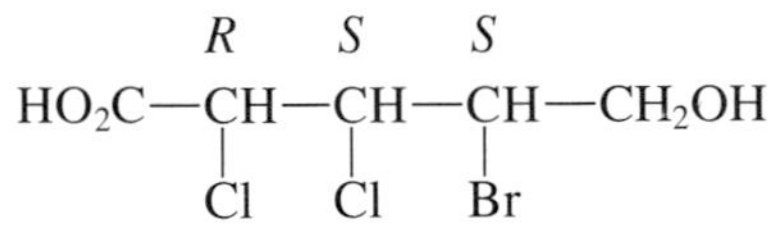

15 Indicate whether the structures in each of the following pairs are enantiomers, diastereomers, or identical molecules.

(a)

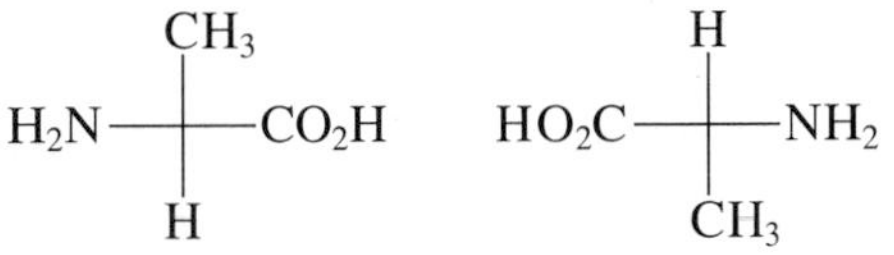

(b)

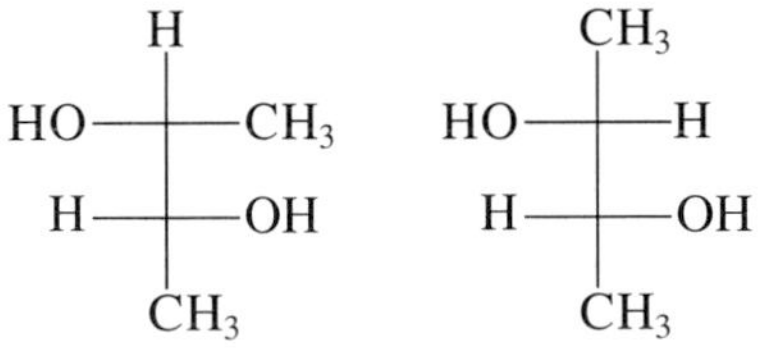

(c)

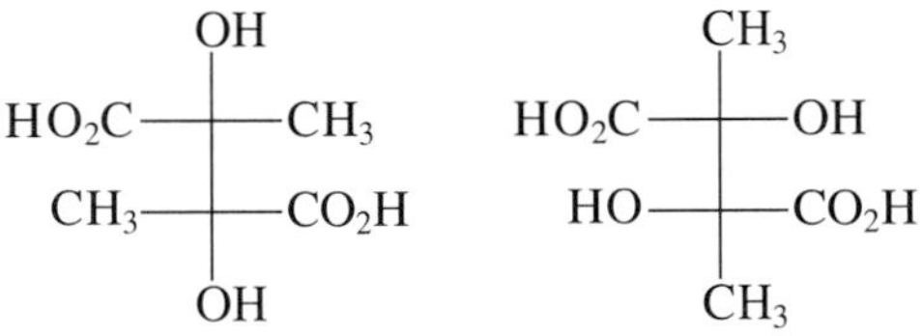

16 Specify the configuration of each asymmetric carbon atom in each of the following Fischer projections:

(a) (b) (c)

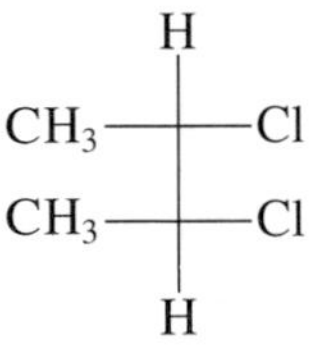

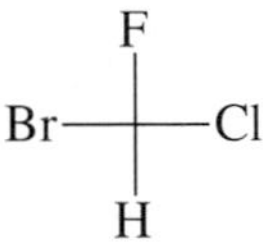

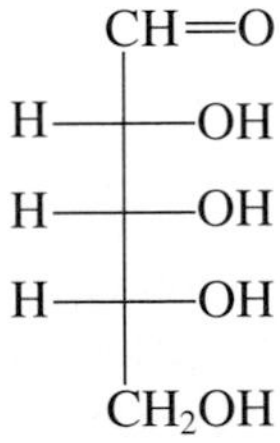

6.10 CHIRALITY, OPTICAL ACTIVITY, AND THE POSTULATION OF TETRAHEDRAL CARBON

Chemists recognized the tetrahedral configuration of tetracoordinate carbon almost one-half century before physical methods confirmed the idea with direct evidence. The phenomena of optical activity and chirality played a key role in this development, one of the most important in the history of organic chemistry. Let us consider the chain of events and follow some of the reasoning that led to the postulation of tetrahedral carbon.

The first chemical substance in which optical activity was observed was quartz. It was discovered that when a quartz crystal is cut in a certain way and exposed to polarized light along a particular axis, the plane of polarization of the light is rotated. In 1815, the French chemist Jean-Baptiste Biot (1774–1862) showed that there are both levorotatory and dextrorotatory quartz crystals. The Abbé Rene Just Haüy (1743–1822), a French crystallographer, had earlier shown that there are two kinds of quartz crystals related as object and nonsuperimposable mirror image. Sir John E. W. Herschel (1792–1871), a British astronomer, found a correlation between these crystal forms and their optical activity: one of these forms of quartz is dextrorotatory and the other levorotatory.

During the period 1815–1838, Biot examined several organic substances, both pure and in solution, for optical activity. He found that some (for example, oil of terpentine) show optical activity, and others do not. He recognized that since optical activity can be displayed by compounds in solution, *it must be a property of the molecules themselves.* (The dependence of optical activity on concentration, Eq. 6.1, is sometimes called *Biot's Law.*) What Biot did *not* observe is that some organic molecules exist in both dextrorotatory and levorotatory forms. The reason Biot never made this observation is undoubtedly that many optically active compounds are obtained from a given natural source as single enantiomers.

The next substance to figure prominently in the history of stereochemistry was tartaric acid.

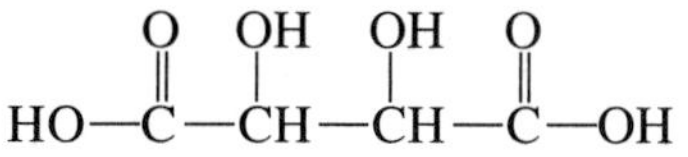

$$\underset{\text{HO}}{}-\overset{\overset{\text{O}}{\|}}{\text{C}}-\overset{\overset{\text{OH}}{|}}{\text{CH}}-\overset{\overset{\text{OH}}{|}}{\text{CH}}-\overset{\overset{\text{O}}{\|}}{\text{C}}-\text{OH} \qquad \text{tartaric acid}$$

This substance had been known by the ancient Romans as its monopotassium salt, *tartar,* which deposits from fermenting grape juice. Tartaric acid derived from tartar was one of the compounds examined by Biot for optical activity; he found that it has a positive rotation. An isomer of tartaric acid discovered in crude tartar, called *racemic acid* (Latin *racemus,* a bunch of grapes), was also studied by Biot and found to be optically inactive. The exact structural relationship of (+)-tartaric acid and its isomer racemic acid remained obscure.

All of these observations were known to Louis Pasteur (1822–1895), the French chemist and biologist. One day the young Pasteur was viewing crystals of the sodium ammonium double salts of (+)-tartaric acid and racemic acid under the microscope. Pasteur noted that the crystals of the salt derived from (+)-tartaric acid were hemihedral (chiral). But he noted that the racemic acid salt was not a single type of crystal, but actually was a mixture of hemihedral crystals: some crystals were "right-handed," like those in the corresponding salt of (+)-tartaric acid, and some were "left-handed" (Fig. 6.19). Pasteur meticulously separated the two types of crystals with a pair of tweezers; he found that the right-handed crystals were identical in every way to the crystals of the salt of (+)-tartaric acid. When equally concentrated solutions of the two types of crystals were prepared, he found that the optical rotations of the left- and right-handed crystals were equal in magnitude, but opposite in sign. Pasteur had thus performed the first separation of enantiomers by human hands! Racemic acid, then, was the first organic compound shown to exist as enantiomers—object and nonsuperimposable

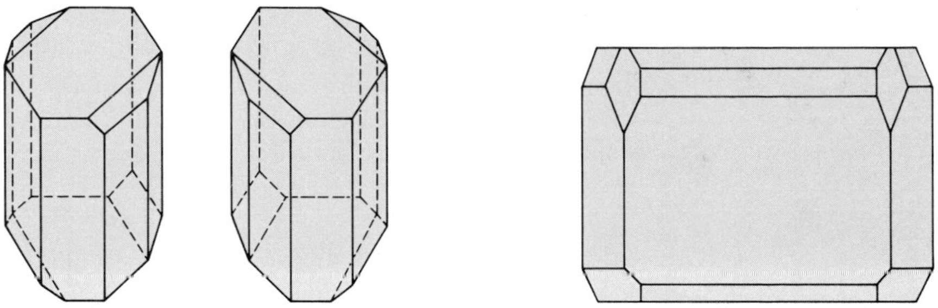

Figure 6.19 Diagrams of the crystals of the tartaric acid isomers that figured prominently in the history of stereochemistry. (a) The hemihedral crystals of sodium ammonium tartrate separated by Pasteur. (b) The holohedral crystal of sodium ammonium racemate that crystallizes at a higher temperature.

mirror image. One of these mirror-image molecules was identical to (+)-tartaric acid, but the other was previously unknown. Pasteur's own words tell us what then took place.

> The announcement of the above facts naturally placed me in communication with Biot, who was not without doubts concerning their accuracy. Being charged with giving an account of them to the Academy, he made me come to him and repeat before his very eyes the decisive experiment. He handed over to me some racemic acid that he himself had studied with particular care, and that he found to be perfectly indifferent to polarized light. I prepared the double salt in his presence with soda and ammonia that he also desired to provide. The liquid was set aside for slow evaporation in one of his rooms. When it had furnished about thirty to forty grams of crystals, he asked me to call at the *College de France* in order to collect them and isolate, before his very eyes, by recognition of their crystallographic character, the right and left crystals, requesting me to state once more whether I really affirmed that the crystals that I should place at his right would really deviate [the plane of polarized light] to the right and the others to the left. This done, he told me that he would undertake the rest. He prepared the solution with carefully measured quantities, and when ready to examine them in the polarizing apparatus, he once more invited me to come into his room. He first placed in the apparatus the more interesting solution, that which should deviate to the left [previously unknown]. Without even making a measurement, he saw by the tints of the images . . . in the analyzer that there was a strong deviation to the left. Then, very visibly affected, the illustrious old man took me by the arm and said, "My dear child, I have loved science so much all my life that this makes my heart throb!"

Pasteur's discovery of the two types of crystals of racemic acid salt was serendipitous. It is now known that the sodium ammonium salt of racemic acid forms separate right- and left-handed crystals only at temperatures below 26 °C. Had Pasteur's laboratories been warmer, he would not have made the discovery. Above this temperature, this salt forms only one type of crystal: a holohedral (achiral) crystal of the racemate (Fig. 6.19b)! From his discovery, Pasteur recognized that some molecules could, like the quartz crystals, have an enantiomeric relationship, but he was never able to deduce a structural basis for this relationship.

Problem

17 As described in the previous account, Pasteur discovered two stereoisomers of tartaric acid. Draw their structures [you cannot of course tell which is (+) and which is (−)]. What stereoisomer of tartaric acid was yet to be discovered? What can you say about its optical activity? (It was discovered in 1906.)

In 1874, Jacobus Hendricus van't Hoff (1852–1911), a Professor at the veterinary college at Utrecht, The Netherlands, and Achille Le Bel (1847–1930), a French chemist, independently arrived at the idea that if a molecule contains a carbon atom bearing four different groups, these groups can be arranged in different ways to give enantiomers. Van't Hoff suggested a tetrahedral arrangement of groups about the central carbon, but Le Bel was less specific. Van't Hoff's conclusions, published in a treatise of eleven pages entitled "*La chemie dans l'espace,*" were not immediately accepted. A caustic reply, reproduced in considerably censored form below, came from the famous German chemist Hermann Kolbe:

> A Dr. van't Hoff of the veterinary college, Utrecht, appears to have no taste for exact chemical research. Instead, he finds it a less arduous task to mount his Pegasus (evidently borrowed from the stables of the College) and soar to his chemical Parnassus, there to reveal in his "*La chemie dans l'espace*" how he finds atoms situated in universal space. This paper is fanciful nonsense! What times are these, that an unknown chemist should be given such attention!"

Kolbe's reply notwithstanding, van't Hoff's ideas prevailed to become a cornerstone of organic chemistry.

How can the existence of enantiomers be used to deduce a tetrahedral arrangement of groups around carbon? Let us examine some other possible carbon geometries to see the sort of reasoning that was used by van't Hoff and Le Bel. Consider a general molecule in which the carbon and its four groups lie in a single plane.

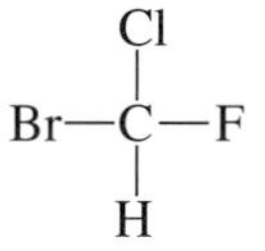

all atoms in the same plane

Because the mirror image of such a *planar* molecule is superimposable (show this!), enantiomeric forms are not possible. The existence of enantiomers thus *rules out* this planar geometry.

However, other, nonplanar nontetrahedral structures could exist as enantiomers. One structure has a pyramidal geometry:

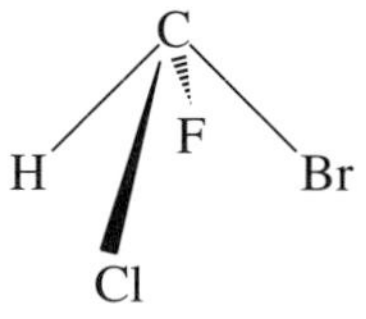

(Convince yourself that such a structure can have an enantiomer.) This geometry could not, however, account for other facts. Consider, for example, the compound methylene chloride (CH_2Cl_2). In the pyramidal geometry, two *diastereomers* would be known. In one, the chlorines are on opposite corners of the pyramid; in the other the chlorines are adjacent. (Why are these diastereomers?)

pyramidal CH_2Cl_2

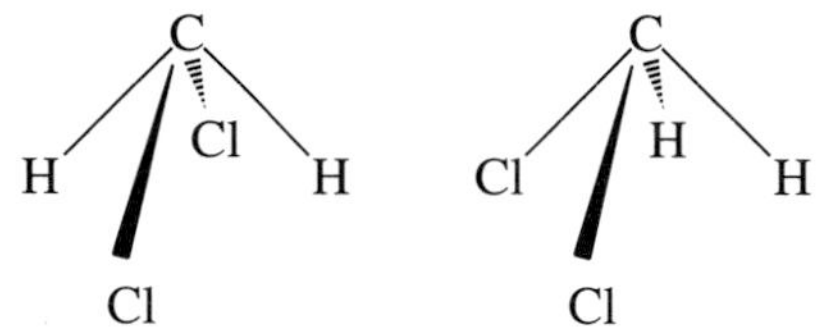

opposite adjacent

These molecules should be separable because diastereomers have different properties. Yet in the entire history of chemistry, only one isomer of CH_2Cl_2, CH_2Br_2, or any similar molecule, has ever been found. Now, we should recognize that this is *negative* evidence. To take this evidence as conclusive would be like saying to the Wright brothers in 1902, "No one has ever seen an airplane fly; therefore airplanes can't fly." Yet this evidence is certainly suggestive, and other types of experiments (Problems 18 and 38) were later carried out that could only be interpreted in terms of tetrahedral carbon. Indeed, modern methods of structure determination have shown that van't Hoff's original proposal—tetrahedral geometry—is essentially correct.

Problem

18 Show how a dipole moment measurement could readily distinguish between the following two structures for carbon tetrachloride, CCl_4.

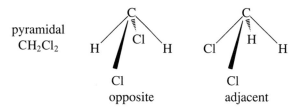

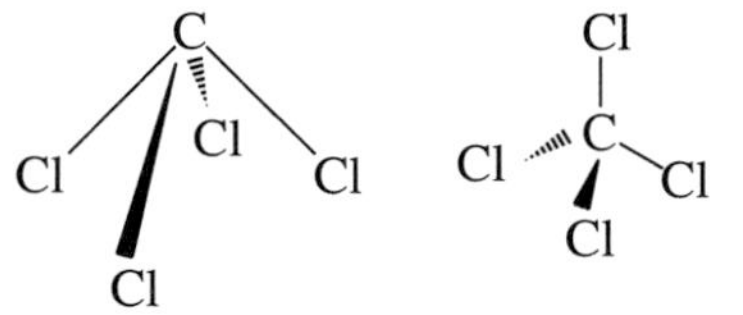

pyramidal tetrahedral

KEY IDEAS IN CHAPTER 6

- Stereoisomers are molecules that have the same atomic connectivity but differ in the arrangement of their atoms in space.
- There are two types of stereoisomers:
  1. Enantiomers—compounds that are related as object and nonsuperimposable mirror image
  2. Diastereomers—stereoisomers that are not enantiomers
- Internal rotation in some cases interconverts stereoisomeric conformations of molecules. Amine inversion at an appropriately substituted nitrogen interconverts stereoisomeric amines.

- A carbon atom bonded to four different groups is called an asymmetric carbon. A compound with a single asymmetric carbon is chiral. Compounds with more than one asymmetric carbon may be chiral, or they may be *meso* compounds, which are achiral compounds with asymmetric atoms. *Meso* isomers can occur when molecules containing asymmetric carbons can be divided into two structurally equivalent parts.
- The compounds of an enantiomeric pair have the same melting points and boiling points, but rotate the plane of polarized light in equal and opposite directions. The rotation of the plane of polarized light by a chiral compound is called optical activity.
- Diastereomers in general differ in their physical behavior: they have different melting points, boiling points, etc.
- Chiral compounds can be drawn in planar representations called Fischer projections. In a Fischer projection, all vertical bonds are assumed to be away from the observer, and all horizontal bonds are toward the observer. There are several valid Fischer projections for any chiral compound. These are derived by the rules in Sec. 6.9.
- Historically, optical activity and chirality formed the logical foundation for the postulate of the tetrahedral carbon atom.

ADDITIONAL PROBLEMS

19 Give the structures of all alkenes with the molecular formula C_6H_{12}. Using Fig. 6.14, give the isomeric relationship of every pair. Indicate which isomers are chiral.

20 Point out all of the asymmetric carbon atoms in each of the following structures:

(a)
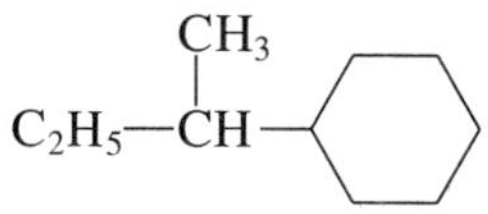

(b)
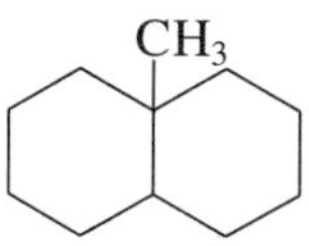

(c)
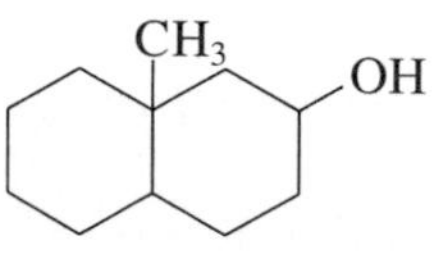

(d)
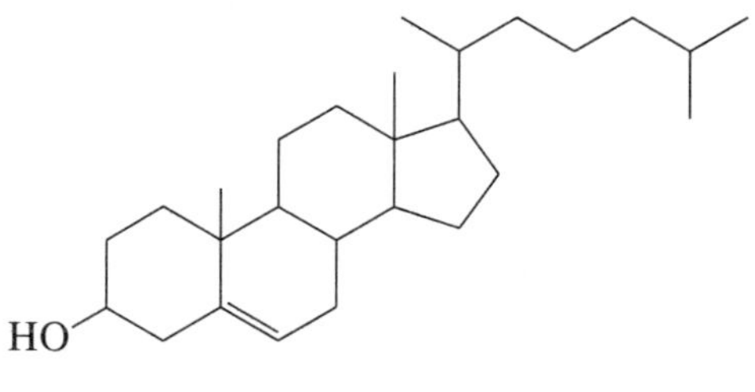

(e) $CH_3—CH(NH_2)—CH_2OH$

21 Give the configuration of each asymmetric atom in the following compounds. [Compounds (a) and (b) are drawn in Fischer projection.]

(a) 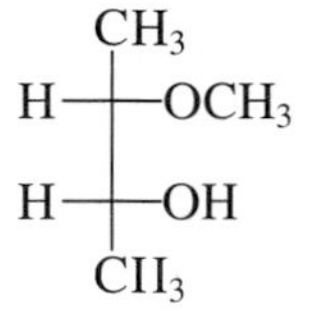

(c) 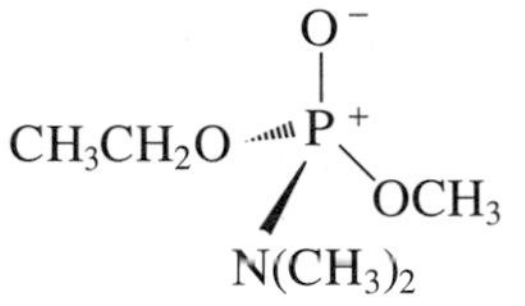

(b) 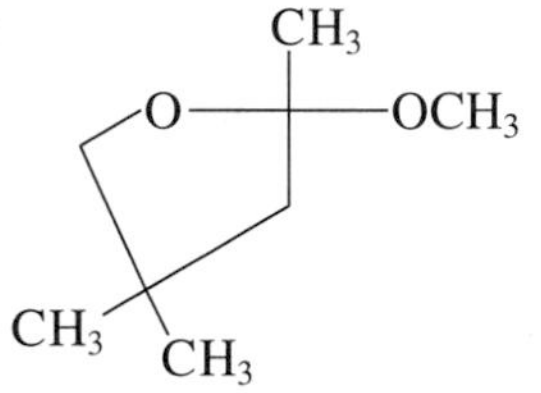

(d) *meso*-3,4-dimethylhexane

22 Draw all the allowed Fischer projections of (2*S*,3*R*)-2,3-pentanediol (see Fig. 6.12).

23 Draw a Fischer projection for each stereoisomer of 2,3-dibromobutane. Using Fig. 6.14, indicate the isomeric relationship between each pair of stereoisomers.

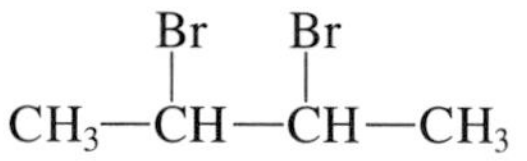 2,3-dibromobutane

24 Indicate whether each of the following statements is true or false. If false, explain why.

(a) Two structural isomers may be chiral.
(b) A pair of enantiomers always have a mirror-image relationship.
(c) Mirror-image molecules are in all cases enantiomers.
(d) All chiral compounds have diastereomers.
(e) If a compound has an enantiomer it must be chiral.
(f) If a compound has a diastereomer it must be chiral.
(g) All compounds with asymmetric carbons are chiral.
(h) Some chiral compounds may be optically inactive.
(i) Some diastereomers can have a mirror-image relationship.
(j) Any chiral compound with a single asymmetric carbon of the *R* configuration must have a positive optical rotation.
(k) If a compound does not have a plane of symmetry it must be chiral.

25 Imagine substituting each hydrogen atom of 3-methylpentane, in turn, with a chlorine atom to give a series of isomers with molecular formula $C_6H_{13}Cl$. Give the structure of each of these isomers. Which of these are chiral? Classify the relationship within every pair of these stereoisomers.

Problems (Cont.)

26 Construct models or draw Newman projections of the three rotational isomers of 2-methylbutane (isopentane) that result from rotation about the C2–C3 bond.
(a) Identify the rotational isomers that are chiral.
(b) Explain why 2-methylbutane is not considered a chiral compound, even though it has chiral rotational isomers.
(c) Suppose we could isolate each of the three rotational isomers in part (a) and determine their heats of formation. Rank these isomers in order of increasing heat of formation (that is, smallest first). Explain your choice. Indicate whether the ΔH_f^0 values for any of the isomers are equal and why.

27 Explain why compound *A* can be resolved into enantiomers, but compound *B* cannot.

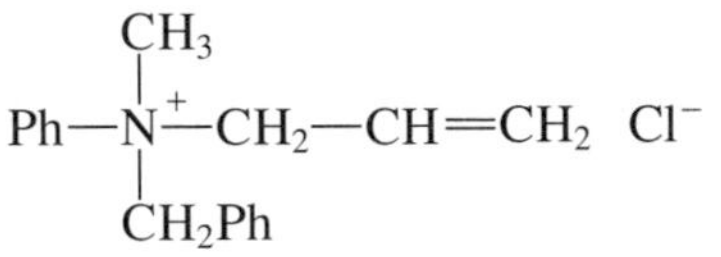

$$\underset{A}{\overset{\displaystyle CH_3}{Ph-\overset{+}{\underset{CH_2Ph}{N}}-CH_2-CH=CH_2}} \quad Cl^-$$

$$\underset{B}{\overset{\displaystyle CH_3}{Ph-\underset{..}{N}-CH_2-CH=CH_2}}$$

28 The specific rotation of the *R* enantiomer of the following alkene is $[\alpha]_D^{25} = +76°$. What is the observed rotation of a 0.5*M* solution of the same compound in a 5-cm sample cell?

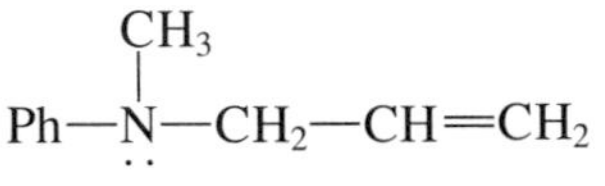

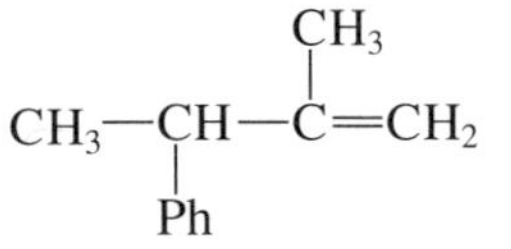

$$\underset{Ph}{\overset{\displaystyle CH_3}{CH_3-CH-C=CH_2}} \qquad \text{2-methyl-3-phenyl-1-butene}$$
$$\text{(MW = 146.2)}$$

29 The *percent optical purity* of a compound contaminated with its enantiomer is given by

$$\text{percent optical purity} = \left(\frac{\alpha}{\alpha^*}\right) \times 100$$

where α is the observed rotation of the mixture, and α^* is the observed rotation of the pure enantiomer under the same conditions.
(a) What is the percent optical purity of a racemate?
(b) Calculate the percent optical purity of a mixture of a pair of enantiomers that has an observed optical rotation of +5°, given that the pure dextrorotatory enantiomer of this compound has an optical rotation under the same conditions of +53°.
(c) Show that percent optical purity is also given by

$$\text{percent optical purity} = \left(\frac{n^* - n}{n^* + n}\right) \times 100$$

where n^* is the number of moles of the predominant enantiomer present in the mixture, and n is the number of moles of the minor enantiomer present.

(d) What is the mole fraction of each enantiomer present in the mixture described in (b)?

30 Polypropylene, the polymer of propylene, can be produced in either of two stereoisomeric forms, each with a regular, repeating stereochemistry.

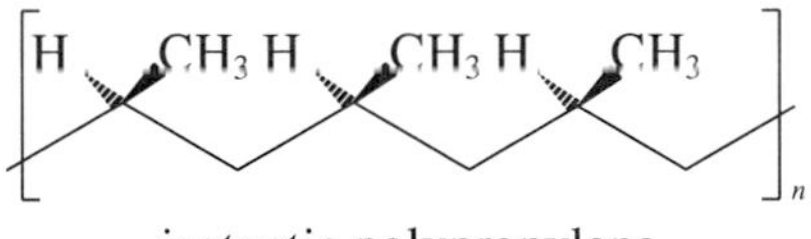

isotactic polypropylene

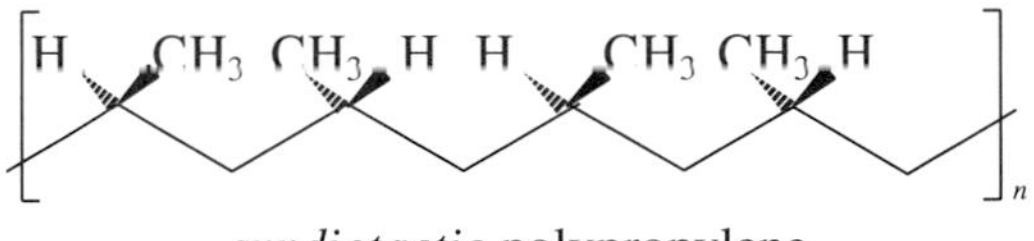

syndiotactic polypropylene

In addition, there exists a third form, *atactic* polypropylene, which has random rather than regular stereochemistry. (Isotactic and syndiotactic polymers tend to be high-melting, crystalline compounds, but atactic polymers tend to be low-melting, amorphous glasses.)

(a) Draw Fischer projections that show several repeating units of isotactic and syndiotactic polypropylene.
(b) Are either of these polymers chiral? Explain.
(c) Plexiglas® is an atactic form of poly(methyl methacrylate); see Table 5.3. Draw a Fischer projection for several repeating units of this polymer.

31 Draw a Fischer projection for each stereoisomer of 2,3,4-trichloropentane.

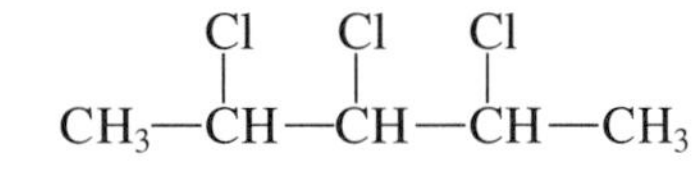

2,3,4,-trichloropentane

(a) Assign an *R* or *S* configuration to carbons 2 and 4 in each stereoisomer.
(b) Indicate which of the stereoisomers are chiral. Explain.
(c) Indicate whether carbon 3 is an asymmetric carbon in each of the stereoisomers. Explain.

32 When (−)-3-methyl-1-pentene is treated with hydrogen and a catalyst, the product is optically inactive. Explain.

33 When (3*S*,4*S*)-4-methoxy-3-methyl-1-pentene (structure below) is treated with mercuric acetate in methanol (CH_3OH) solvent, then with $NaBH_4$, two isomeric compounds with the formula $C_8H_{18}O_2$ are isolated. One, compound *A*, is optically inactive, and the other, compound *B*, is optically active. Give the structures and absolute configurations of both compounds. (*Hint*: See Problem 33a, Chapter 5.)

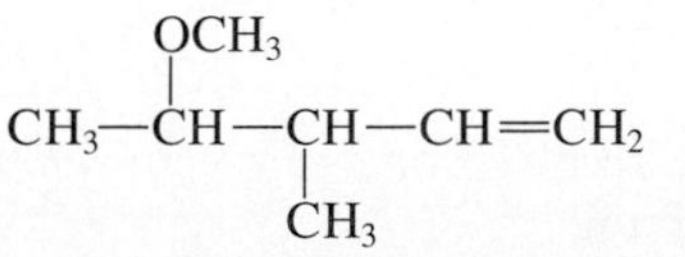

4-methoxy-3-methyl-1-pentene

Problems (Cont.)

34 What is the relationship between every pair of compounds in the following set (given in Fischer projections)?

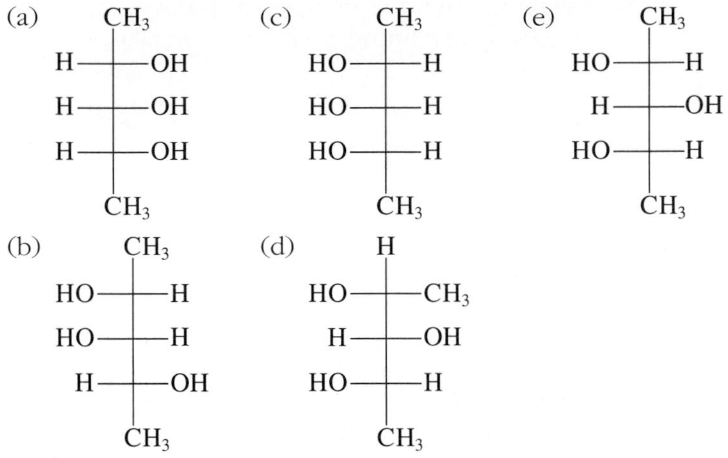

35 Two isomers of the compound $(H_3N)_2Pt(Cl)_2$ with different properties are known. Show that this fact allows a choice between the tetrahedral and square planar arrangements of the groups around the platinum.

36 Draw the structures of the different stereoisomers that are possible for the compound below, given (a) tetrahedral, (b) square planar, and (c) pyramidal geometries at the carbon atom. For each of these geometries, what is the relationship between each pair of stereoisomers?

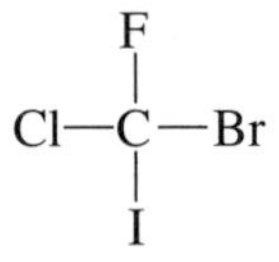

37 In 1914, the German Chemist Emil Fischer showed that he could carry out the following conversion in which the optically active starting material was transformed into the product with an identical melting point and an optical rotation of equal magnitude and opposite sign. No bonds to the asymmetric carbon were broken in the process.

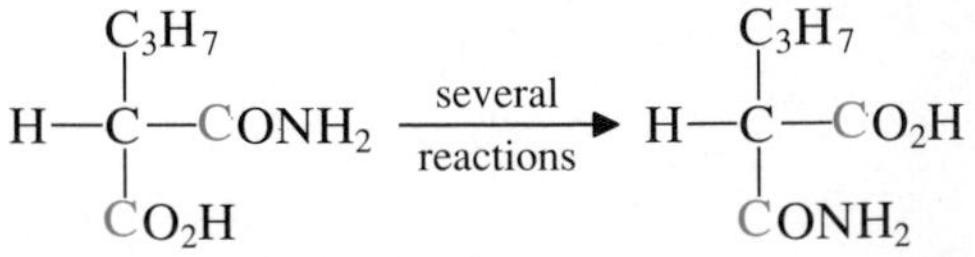

Show that this result is consistent with *either* tetrahedral or pyramidal geometry at the chiral carbon.

38 Fischer (see previous problem) also carried out the following pair of conversions. Again, no bonds to the asymmetric carbon were broken.

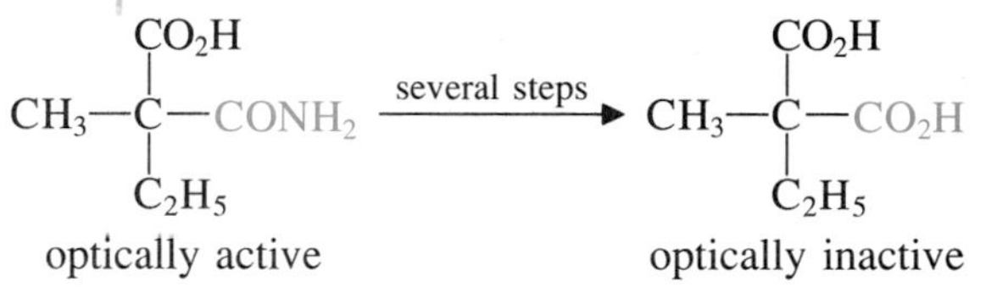

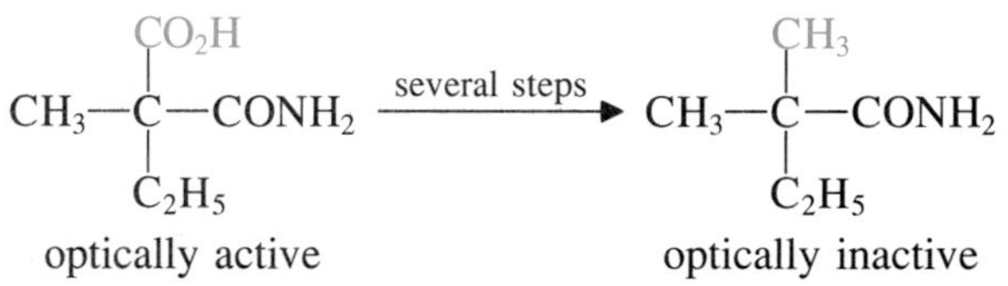

Explain why this *pair* of conversions (but not either one alone) and the associated optical activities rule out pyramidal geometry at the asymmetric carbon but are consistent with tetrahedral geometry.

7 Cyclic Compounds. Stereochemistry and Chemical Reactions

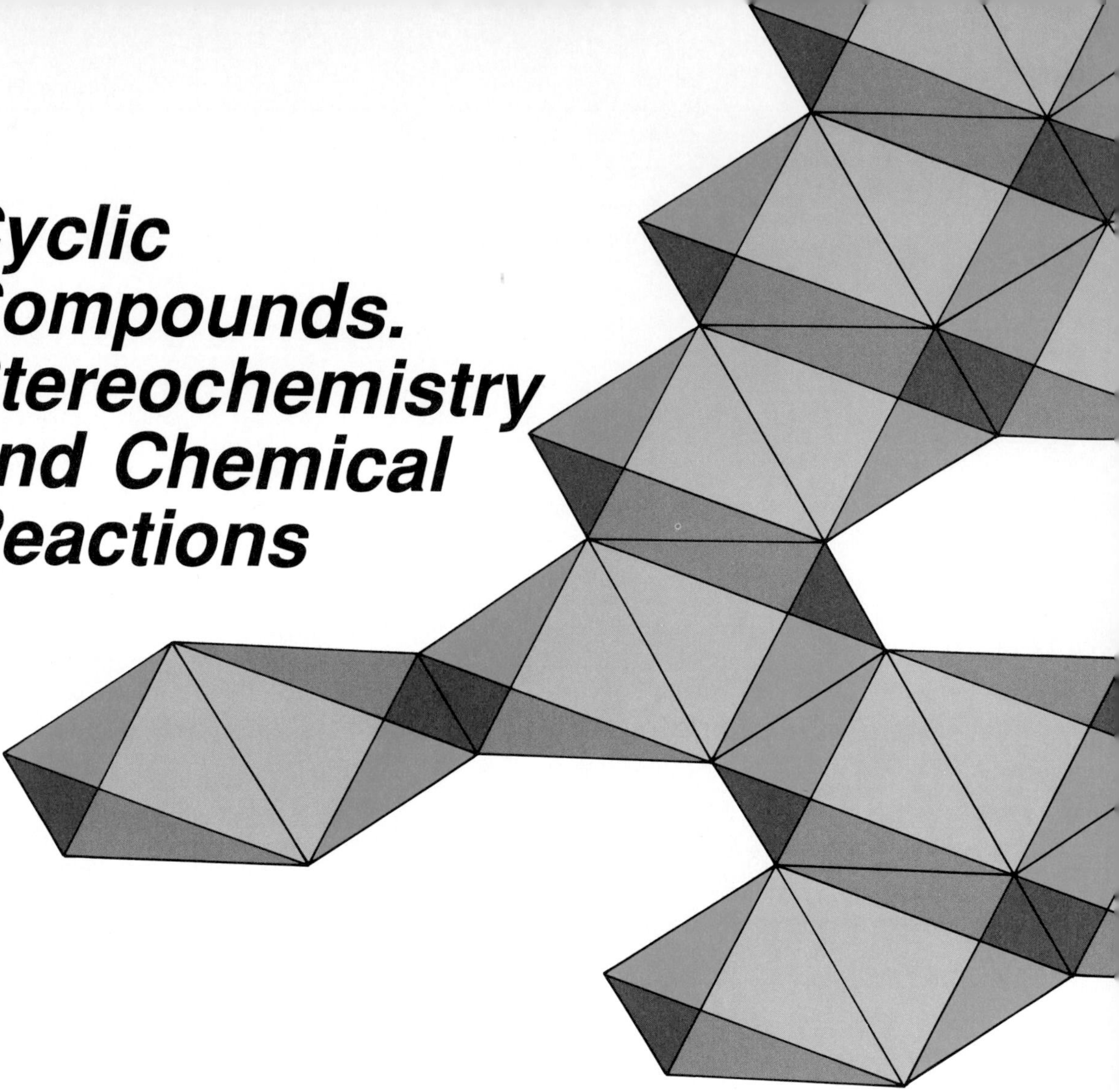

Although cyclic compounds have a great deal in common with their open-chain counterparts, they also present unique problems of stereochemistry and conformation that we shall consider in this chapter. The nomenclature of cyclic alkanes and alkenes, discussed in Secs. 3.2A and 4.2, provides the basis for naming simple cyclic compounds and should be reviewed.

Our introduction to cyclic compounds will focus on cyclic hydrocarbons and their derivatives. Then we shall consider the role of stereochemistry in chemical reactions. We have already learned about several reactions that yield one *structural* isomer in preference to another (for example, addition of HBr to alkenes). Many reactions also yield only one of several possible *stereoisomers*. We shall examine several such reactions and try to understand some of the general principles that govern the formation of stereoisomers. We shall also find that the mechanism of a reaction and its stereochemistry are intimately related. In particular, we shall see the crucial role played by stereochemical studies in determining the mechanisms of several reactions with which we have become familiar.

7.1 RELATIVE STABILITIES OF THE MONOCYCLIC ALKANES

A compound that contains a single ring is called a **monocyclic** compound. Cyclohexane, cyclopentane, cyclohexene, and methylcyclohexane are examples of monocyclic alkanes.

An important problem in the early days of organic chemistry was the conformations of the simple monocyclic hydrocarbons. One of the first chemists to consider this problem was Adolf von Baeyer (1835–1917). In 1885 von Baeyer proposed (erroneously, as we shall see) that the carbon skeletons of the cycloalkanes have the shapes of the corresponding planar polygons, with all carbon–carbon bonds of equal length. Cognizant of the writings of van't Hoff and the notion of tetrahedral carbon, von Baeyer suggested that the stability of the cycloalkanes would decrease as their C—C—C bond angles departed from the ideal value of 109.5°. This type of instability is called **angle strain,** and the carbon rings with substantial angle strain are called **strained rings.** According to von Baeyer's proposal, the internal bond angles in the cycloalkanes should be the same as the internal angles of the corresponding regular polygons:

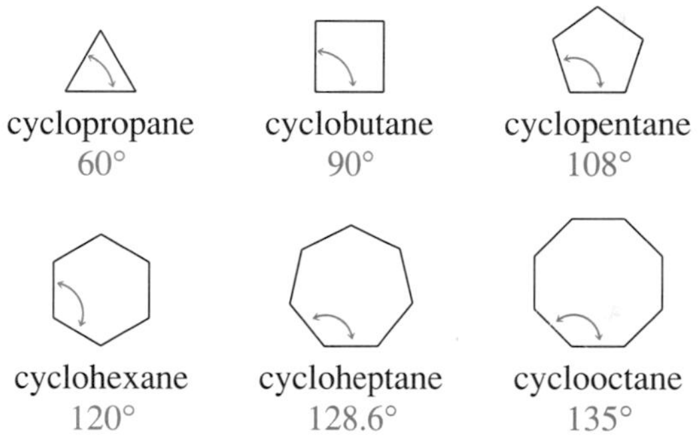

If this theory were true, then the cycloalkanes from cyclopropane to cyclopentane should become progressively more stable, and the cycloalkanes beyond cyclopentane should become progressively less stable. Cyclopentane should be the most stable cycloalkane because it should have the least angle strain.

Let us see whether the experimental facts support the von Baeyer theory. We have learned (Sec. 4.4B) that one way to determine the relative stability of two compounds is to examine their heats of formation. Previously we have used heats of formation to examine the relative stabilities of *isomers*. The cyclic alkanes are not isomers, but they do have the same empirical formula: CH_2. Hence, in order to compare their heats of formation, we can divide the heat of formation of each compound by its number of carbons to obtain a *heat of formation per CH_2 group*. The compound with the smallest heat of formation per CH_2 is the most stable cycloalkane. From the heats of formation given in Table 7.1, we see that *cyclohexane, not cyclopentane, is the most stable cycloalkane*. Furthermore, cyclic hydrocarbons larger than cyclodecane do not differ very much in stability. Clearly, von Baeyer's prediction, that cyclopentane should be the most stable monocyclic alkane, is not supported by the data.

It is also interesting to see how the stability of cyclohexane compares with that of a typical *noncyclic* alkane. The heats of formation of butane, pentane, and hexane are

TABLE 7.1 Heats of Formation Per —CH_2— for Some Cycloalkanes (n = number of carbon atoms)

n	Compound	$\Delta H_f^0/n$, kcal/mol	n	Compound	$\Delta H_f^0/n$, kcal/mol
3	cyclopropane	+4.25	9	cyclononane	−3.95
4	cyclobutane	+1.65	10	cyclodecane	−3.75
5	cyclopentane	−3.65	11	cycloundecane	−4.95
6	cyclohexane	−4.95	12	cyclododecane	−4.65
7	cycloheptane	−4.05	13	cyclotridecane	−4.55
8	cyclooctane	−3.75	14	cyclotetradecane	−4.95

−30.15, −35.00, and −39.96 kcal/mol, respectively. These data show that heats of formation, like other properties, change regularly down a homologous series; each CH_2 group in the carbon chain of an *n*-alkane contributes −4.9 kcal/mol to the heat of formation. This is the same value as the heat of formation per CH_2 of cyclohexane. Hence, we see that *cyclohexane has the same stability as a typical n-alkane.*

The von Baeyer theory fails because of the assumption that the carbon skeletons of the rings are planar polygons. Of course, cyclopropane has a planar carbon skeleton, since three carbons must lie in a single plane, but rings larger than cyclopropane are puckered. In cyclohexane this puckering allows the C—C—C bond angles to have the tetrahedral value of 109.5°, and the C—C bond length to be 1.54 Å, the same as in an open-chain alkane. Puckering also accounts for the relatively small ΔH_f^0 values of the larger cycloalkanes.

What conformation does this puckering impart to the cyclohexane ring? The answer to this and related questions will occupy our attention for the next few sections. A study of the conformational properties of cyclohexane is important, not only because of the wide occurrence of cyclohexane rings in nature, but also because much of what we learn here can be applied to the analysis of other cyclic compounds.

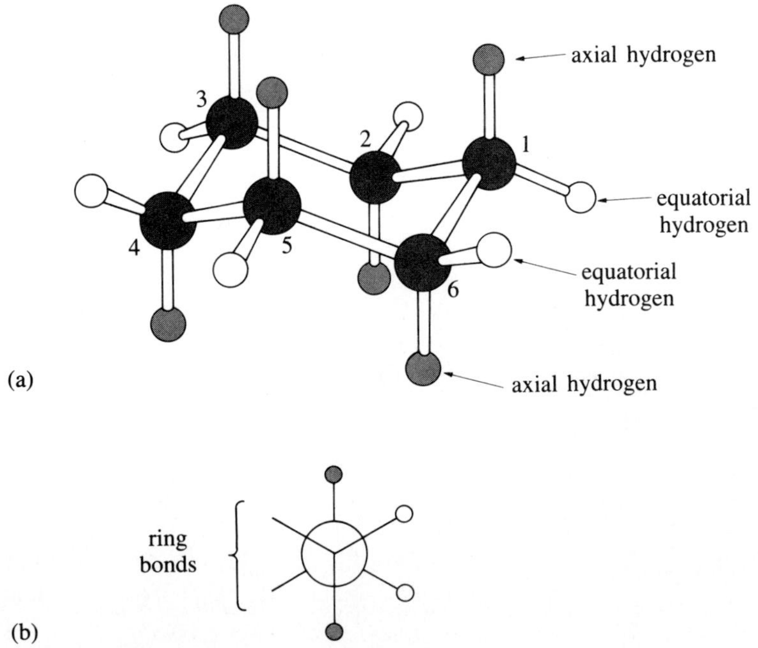

Figure 7.1 (a) Chair conformation of cyclohexane. The axial hydrogens are shown in gray and the equatorial hydrogens in white. (b) Newman projection from C6 to C1. Notice that all bonds are staggered. (The same projection would be obtained by viewing from C2 to C3, or C4 to C5.)

7.2 CONFORMATIONS OF CYCLOHEXANE

The most stable conformation of cyclohexane, shown in Fig. 7.1, is called the **chair form** because of its resemblance to a lawn chair. (You should construct a model of cyclohexane *now* and use it to follow the subsequent discussion.) Notice first that opposite sides of the chair are parallel. That is, if we number the carbons around the ring 1 through 6, the C1–C2 bond is parallel to the C4–C5 bond; the C2–C3 bond is parallel to the C5–C6 bond; and so on. If we look down any carbon–carbon bond, we see that both the ring carbons and the hydrogens are in a staggered arrangement.

Let us now look carefully at the hydrogens on the cyclohexane ring. If we place the cyclohexane model on a tabletop, we find that six C—H bonds are perpendicular to the plane of the table. These hydrogens, shown in black in Fig. 7.1, are called **axial** hydrogens. The remaining six C—H bonds point outward from the ring along its periphery. These hydrogens, shown in white in Fig. 7.1, are called **equatorial** hydrogens. Notice that each bond to an equatorial hydrogen is parallel to two carbon–carbon bonds of the ring. For example, the C1–H bond is parallel to the C2–C3 and C5–C6 bonds.

(a)

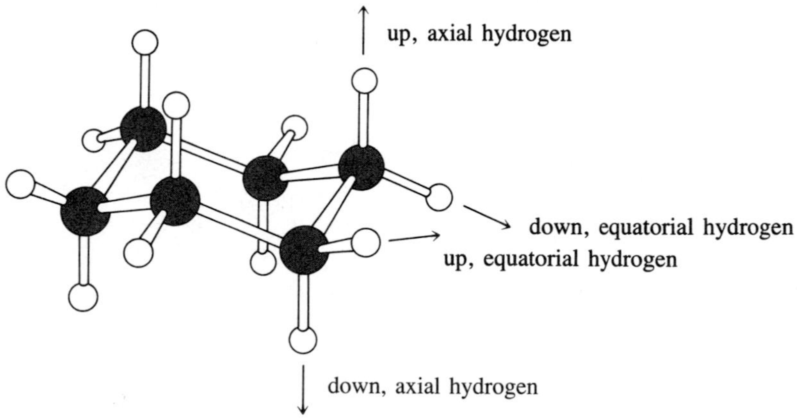

(b)

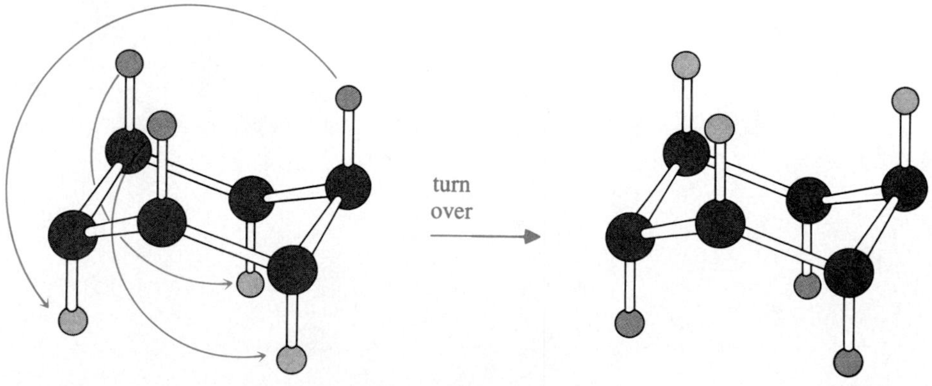

Figure 7.2 (a) Up and down equatorial and axial hydrogens. (b) The up and down axial hydrogens are equivalent. This can be demonstrated by turning the ring (colored arrows). The up axial hydrogens (color) become equivalent to the down axial hydrogens (gray). The same operation shows the equivalence of the up and down equatorial hydrogens.

Notice that three of the axial hydrogens are down (these support the model on the tabletop); the other three are up (Fig. 7.2a). Likewise, three of the equatorial hydrogens are down and three are up. The up and down hydrogens of a given type are completely equivalent. If we turn the ring over as shown in Fig. 7.2b, the up and down hydrogens simply exchange places and the ring looks the same. Another observation that will prove useful is that if a hydrogen of a given type (that is, axial or equatorial) is up on one carbon, it is down on the adjacent carbons.

Cycloalkanes, like alkanes, undergo internal rotations (Sec. 3.5) but, because the carbon atoms are tied together in a ring, several internal rotations must occur at the same time. (There is an analogy here to men on a chain gang: one cannot move very far unless they all move.) When a cyclohexane molecule undergoes internal rotations, the result is *a change in the conformation of the ring*. The various conformations of a cyclohexane molecule are shown in Fig. 7.3. The colored arrows in the figure show how one conformation is converted into the next. In the following discussion we shall examine these conformations individually.

If we hold the carbon at the upper right tip (the head) of the chair and its two neighboring carbons so that they cannot move and raise the carbon at the lower left tip (the foot) of the chair halfway up, we form a **half-chair** cyclohexane. The half-chair form of cyclohexane is not stable because the hydrogens at the foot are eclipsed

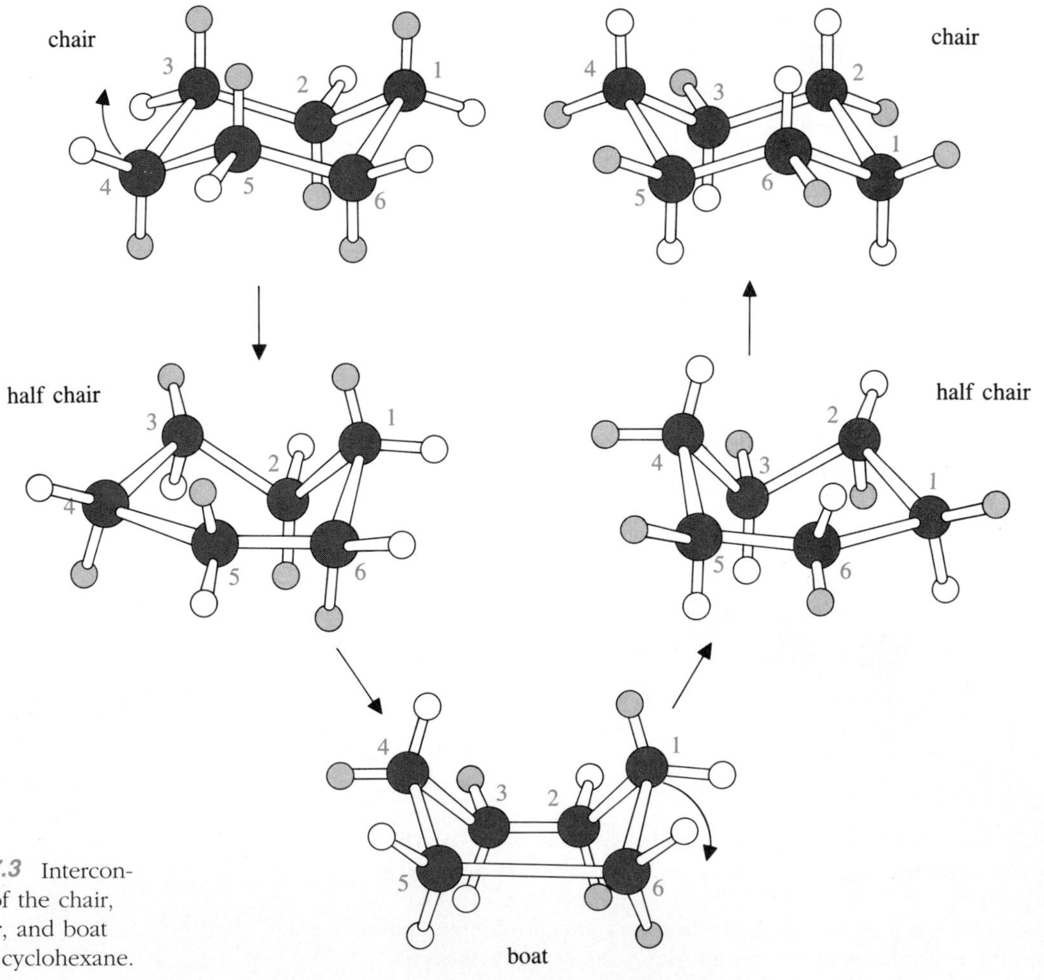

Figure 7.3 Interconversion of the chair, half-chair, and boat forms of cyclohexane.

Figure 7.4 Relative free energies of cyclohexane conformations.

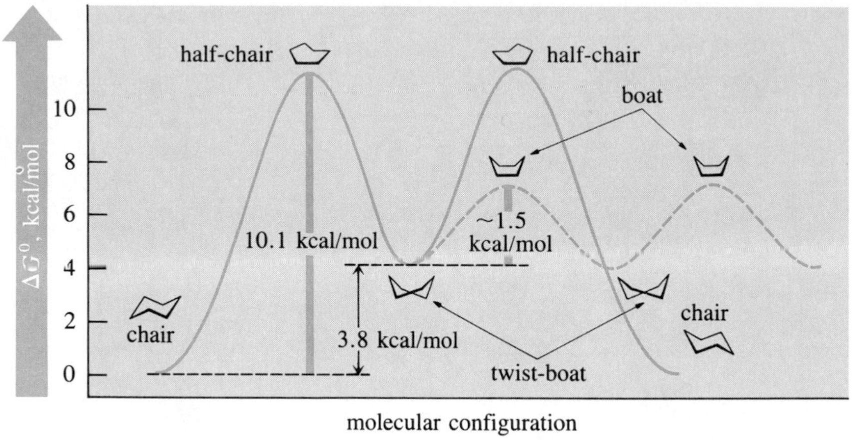

with those on the adjacent carbons, and because there is some angle strain. Recall that eclipsed rotational isomers in *n*-alkanes are also not stable (Sec. 3.5). Angle strain occurs because the molecule is more nearly planar, and (as von Baeyer predicted) when cyclohexane approaches planarity, it should become unstable. (Your model may actually spring back into a chair if you release it.) The half-chair conformation lies at an energy maximum, 10.1 kcal/mol higher in standard free energy than the chair conformation (Fig. 7.4).

If we continue to raise the leftmost carbon on the cyclohexane ring while still holding the right-hand carbons rigidly, we come to the **boat** form of cyclohexane. The boat form, shown in more detail in Fig. 7.5, is also unstable. Although there is no angle strain, several of the C—H bonds are eclipsed. In addition, the two hydrogens at the bow and stern of the boat (flagpole hydrogens) experience some van der Waals repulsion.

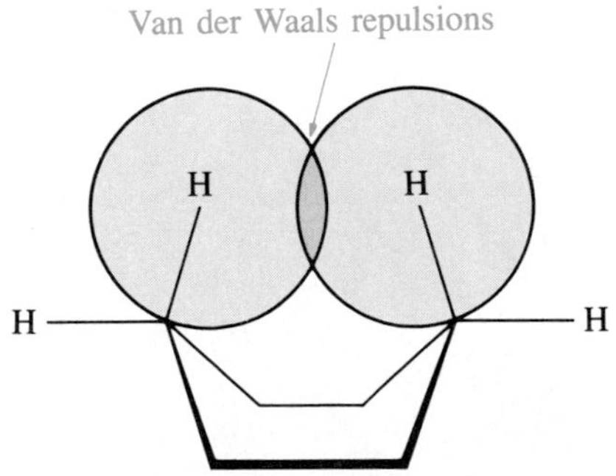

If the bow of the boat is moved across the molecule toward the observer, and the stern away (or vice versa), the resulting internal rotations give a **twist-boat** form of cyclohexane, also shown in Fig. 7.5. In this conformation the flagpole interaction and the eclipsing interactions are somewhat relieved. The twist-boat conformation lies at an energy minimum and is another stable form of cyclohexane. However, since it is about 3.8 kcal/mol less stable than the chair form, it is present in only a very small concentration in any sample of cyclohexane (Problem 1).

Returning to Fig. 7.3, if we now grasp the leftmost carbon of the boat and its neighbors so that they cannot move, we can move the rightmost carbon down. As we do this, we alternately pass through another half-chair form and, finally, into another chair conformation. What we have done is this: by raising the leftmost carbon of the

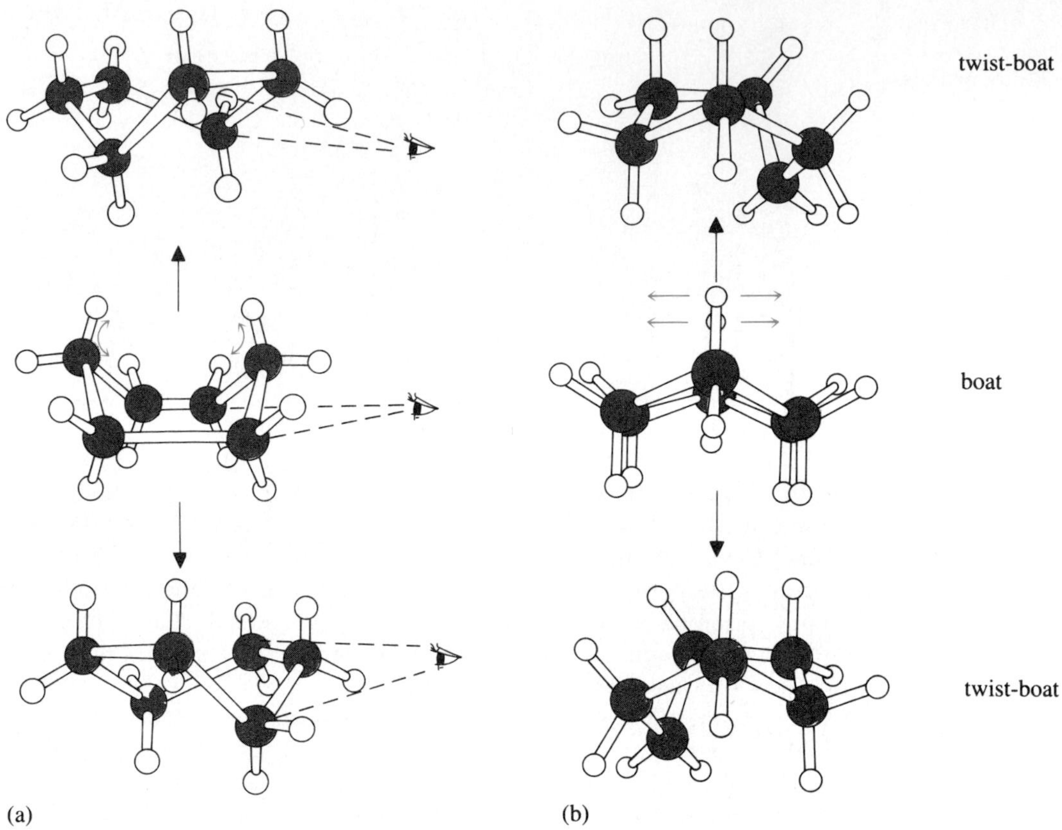

Figure 7.5 Boat and twist-boat forms of cyclohexane. (a) Side view (b) end-on view as seen by the eyeball.

cyclohexane up and the rightmost carbon down, we have changed one chair conformation into another completely equivalent chair conformation. In the process, *the equatorial hydrogens have become axial, and the axial hydrogens have become equatorial.*

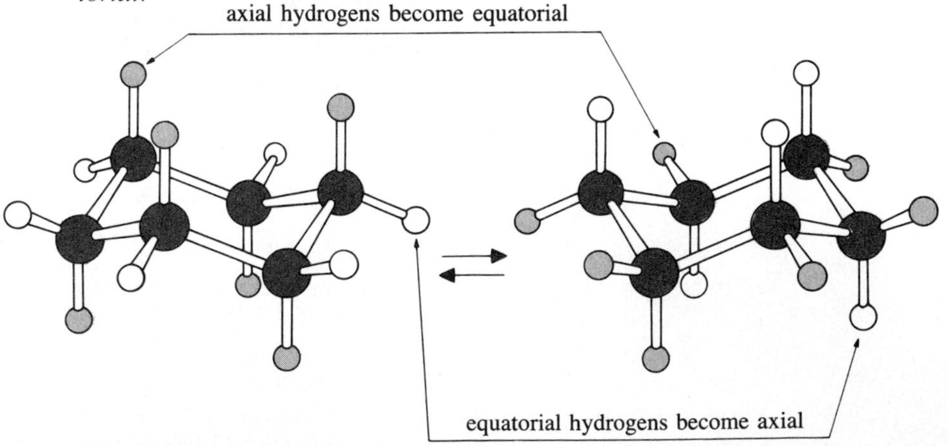

(7.1)

(You should confirm this observation with a model, using groups of different color for equatorial and axial hydrogens.) The standard free energy of activation for this process is the energy difference between the chair and half-chair forms (Fig. 7.4): 10.1 kcal/mol. Although this barrier is larger than that for internal rotation in butane, the chair–chair interconversion is nevertheless very rapid at room temperature. The axial and equatorial protons are different and distinguishable in any one chair form, but the high rate of the chair–chair interconversion causes the axial and equatorial protons over time to average out to be equivalent and indistinguishable.

Now we can see why cyclohexane is a nonplanar molecule. The answer is quite obvious merely from handling a model; it is virtually impossible to force the carbon atoms of a cyclohexane model into a single plane. If von Baeyer knew about tetrahedral carbon, why, then, did he assume that the cycloalkanes have planar conformations? It has been suggested that von Baeyer had a set of molecular models that had unusually floppy joints between the carbon atoms; von Baeyer's cyclohexane models evidently assumed a planar arrangement without difficulty!

Problem

1 Using the information in Fig. 7.4, determine the relative amounts of the chair and twist-boat forms of cyclohexane in equilibrium at 25 °C.

7.3 MONOSUBSTITUTED CYCLOHEXANES. CONFORMATIONAL ANALYSIS

A substituent group in a substituted cyclohexane, such as the methyl group in methylcyclohexane, can be in either an axial or an equatorial position.

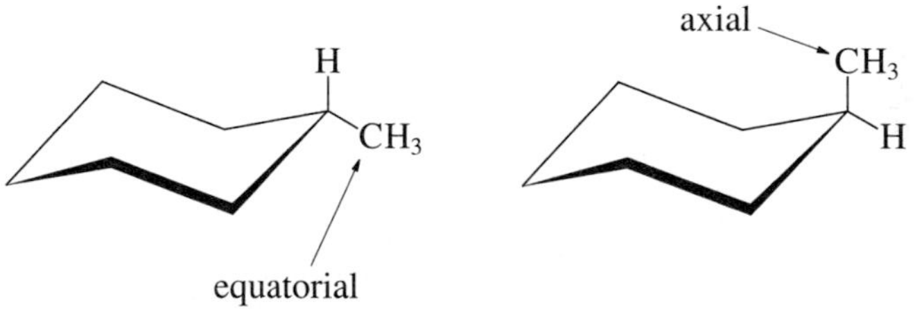

These two compounds are obviously stereoisomers, and since they are not enantiomers, they must be diastereomers. Substituted cyclohexanes such as methylcyclohexane also undergo chair–chair interconversion. As we can see in Fig. 7.6, axial methylcyclohexane and equatorial methylcyclohexane are interconverted by this process. (Demonstrate this to yourself with models!) Since this process is rapid at room temperature, methylcyclohexane is a mixture of two *conformational diastereomers* (Sec. 6.8A). Since diastereomers have different energies, one form is more stable than the other, and the more stable form is present in greater amount. Which form is more stable?

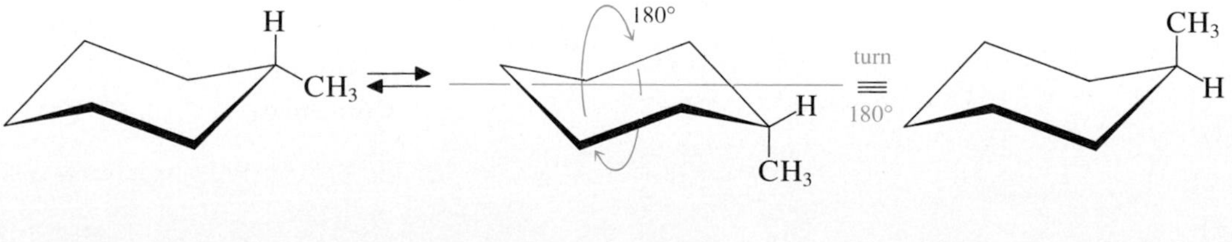

Figure 7.6 The chair–chair interconversion interconverts axial and equatorial methylcyclohexane.

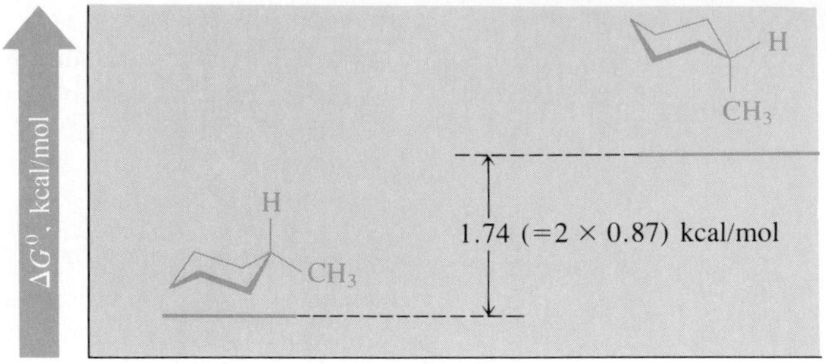

Figure 7.7 Relative free energies of axial and equatorial methylcyclohexane.

Equatorial methylcyclohexane is more stable than axial methylcyclohexane by 1.74 kcal/mol (Fig. 7.7). The reason for this difference can be understood if we think of methylcyclohexane in terms of what we know about butane. We learned that the *gauche* rotational isomer of butane is less stable than the *anti* isomer because of the unfavorable steric interaction of the methyl groups in the *gauche* form (Sec. 3.5B). We call these interactions ***gauche** interactions*. The more *gauche* interactions a molecule has, the more unstable it should be. As we shall now see, the relative energy of the methylcyclohexane conformations can also be understood in terms of *gauche* interactions.

First, let us realize that certain *gauche* interactions are common to both conformations of methylcyclohexane: these are the interactions between the ring carbons.

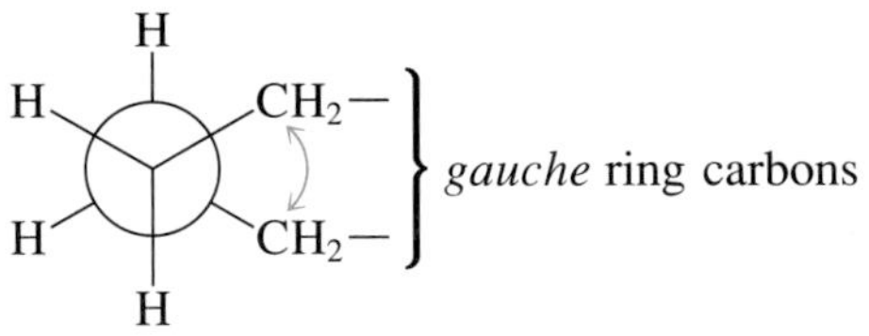

When we *compare* the energies of the two forms of methylcyclohexane, we can ignore these interactions, because there are equal numbers of them in both forms of methylcyclohexane. However, we do need to be concerned with the *gauche* interactions between the methyl group and the —CH_2— groups of the ring. There are two of these interactions in axial methylcyclohexane. These can be seen by sighting from the carbon bearing the methyl group (C1) down the C1–C2 and C1–C6 bonds of the ring, as shown.

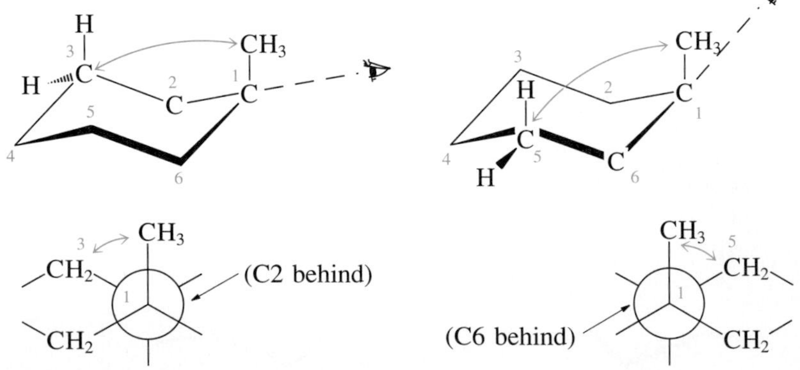

There are *gauche* interactions between the methyl group and both C1 and C3, very much like the *gauche* interactions between methyl groups in *gauche*-butane. In contrast, equatorial methylcyclohexane has no *gauche* interactions; the methyl group is *anti* to the ring carbons.

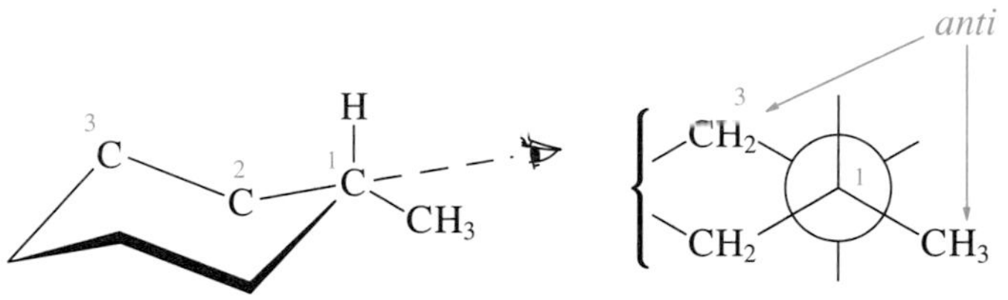

Since the *gauche* form of butane is destabilized relative to the *anti* form by 0.87 kcal/mol (Fig. 3.7), we should expect axial methylcyclohexane to be destabilized by the same amount for *each gauche interaction*. Because there are two *gauche* interactions in the axial conformation that do not exist in the equatorial conformation, axial methylcyclohexane is destabilized relative to equatorial cyclohexane by (2 × 0.87) = 1.74 kcal/mol (Fig. 7.7).

The *gauche* interaction just discussed is an example of what is sometimes called a **1,3-diaxial interaction,** because it is an interaction between the hydrogens of the axial methyl group and the axial hydrogens three carbons away.

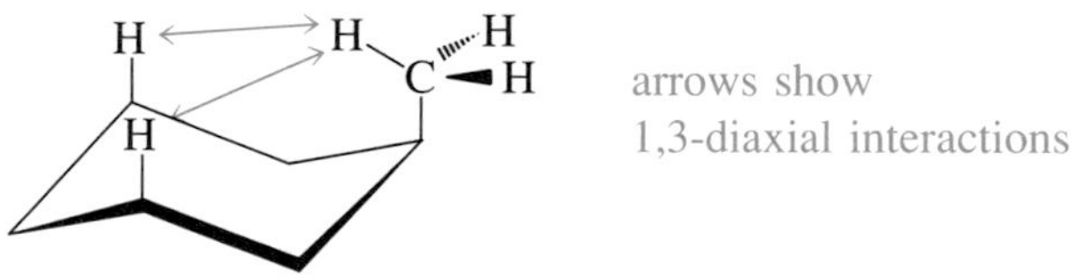

The analysis of molecular conformations and their relative energies is called **conformational analysis.** We have just seen that it is possible to do a conformational analysis of methylcyclohexane if we simply apply the principles that we learned in studying the internal rotations of butane. The conformational equilibria of many substituted cyclohexanes have been investigated experimentally. As we might expect, larger groups have more severe *gauche* interactions. Thus, *the larger the substituent group, the more the equilibrium favors the equatorial conformation*. For example, the equatorial conformation of isopropylcyclohexane is favored by 2.2 kcal/mol, and that of *t*-butylcyclohexane is favored by 5–6 kcal/mol.

Problems

2 Calculate the relative amounts of axial and equatorial forms present at equilibrium in a sample of methylcyclohexane.

3 Since we expect cyclohexanes with larger substituent groups to contain more of the equatorial conformation, it comes as a surprise that the energy difference between the conformations of ethylcyclohexane is about the same as that for methylcyclohexane. Suggest an explanation for this fact.

Since the conformational equilibrium of a monosubstituted cyclohexane interconverts diastereomers, we must admit that if we could separate the two chair forms and study them independently they would have different physical properties. In the late 1960s, C. Hackett Bushweller, a graduate student at the University of California, Berkeley (presently Professor of Chemistry at the University of Vermont) cooled a solution of chlorocyclohexane in an inert solvent to −150 °C. Crystals suddenly appeared in the solution. He filtered the crystals at low temperature; subsequent investigations showed that he had selectively crystallized the equatorial form of chlorocyclohexane!

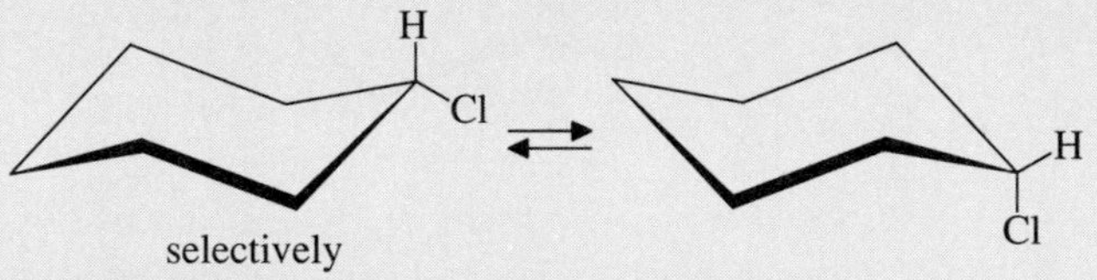

selectively crystallizes at low temperature

Similar types of experiments have been carried out with other monosubstituted cyclohexanes.

7.4 DISUBSTITUTED CYCLOHEXANES. CIS–TRANS ISOMERISM IN CYCLIC COMPOUNDS

An example of a disubstituted cyclohexane is 1-chloro-2-methylcyclohexane.

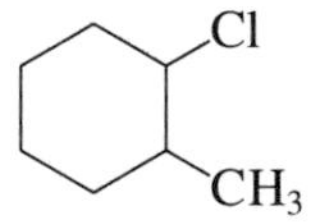

1-chloro-2-methylcyclohexane

In one stereoisomer of this compound, both the chloro and methyl groups assume equatorial positions. This compound is in rapid equilibrium with a conformational diastereomer in which both the chloro and methyl groups assume axial positions.

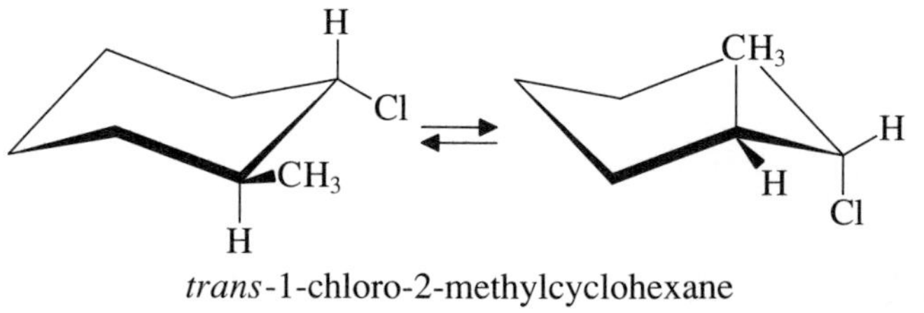

(7.2)

trans-1-chloro-2-methylcyclohexane

Either conformational isomer (or the mixture of them) is called *trans*-1-chloro-2-methylcyclohexane. A *trans*-disubstituted cyclohexane is one in which the substituent groups have an up–down relationship.

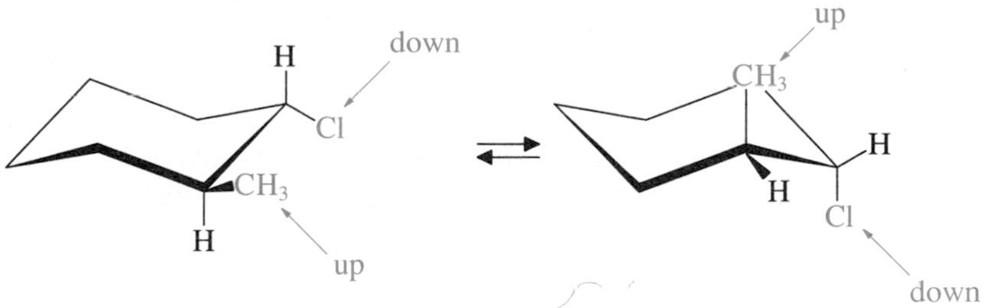

(7.3)

Notice that the up–down relationship is not affected by the conformational equilibrium.

In another stereoisomer of this compound, the chloro and methyl groups occupy adjacent equatorial and axial positions.

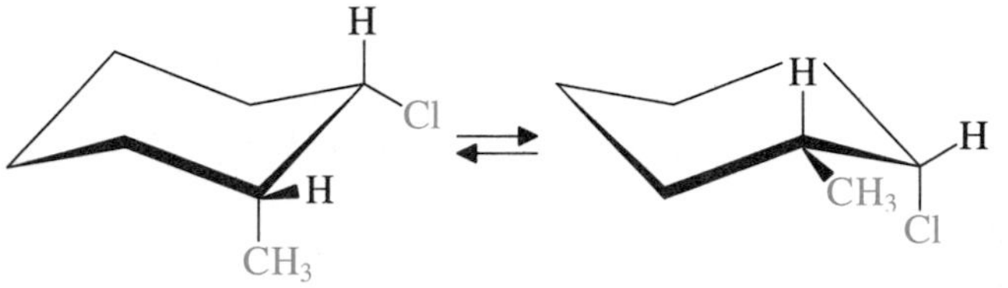

cis-1-chloro-2-methylcyclohexane

(7.4)

This compound, called *cis*-1-chloro-2-methylcyclohexane, is also a rapidly equilibrating mixture of conformational diastereomers. A *cis*-disubstituted cyclohexane is one in which the substituents have an up–up or a down–down relationship.

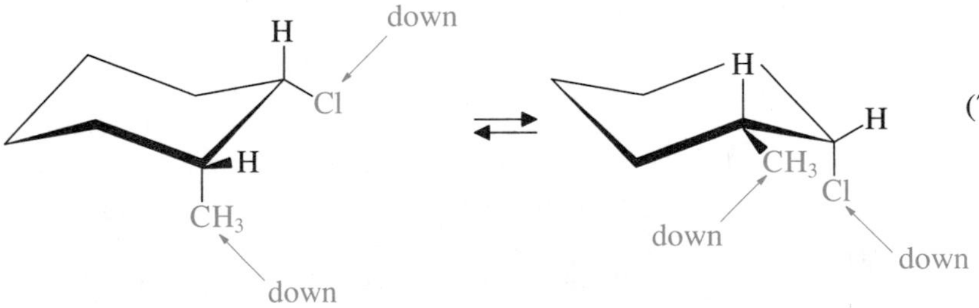

(7.5)

Again, this relationship is not altered by the conformational equilibrium.

The same definition of *cis* and *trans* disubstitution can be applied to substituents in other positions of a cyclohexane ring.

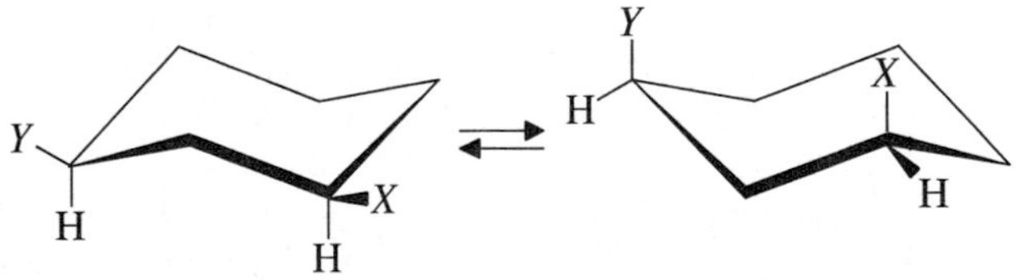

cis-1,3-disubstitution

(7.6)

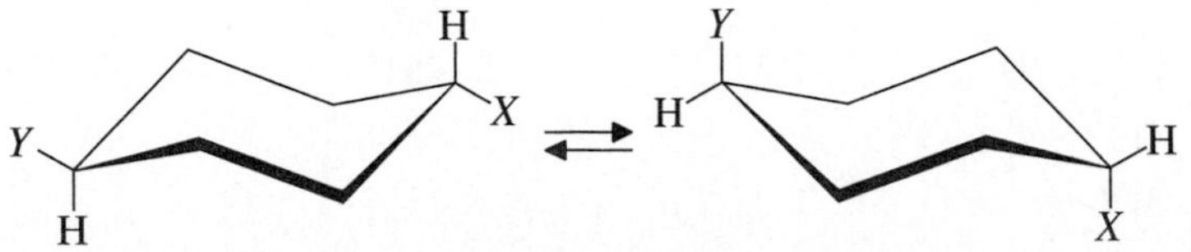

trans-1,4-disubstitution

Problems

4 Give the structure of
(a) *trans*-1,3-dimethylcyclohexane
(b) *cis*-1-ethyl-4-isopropylcyclohexane
For each compound, draw the two chair isomers that are in conformational equilibrium. For compound (b), draw a boat conformation.

5 In the following compounds identify the *one* pair of conformational isomers:

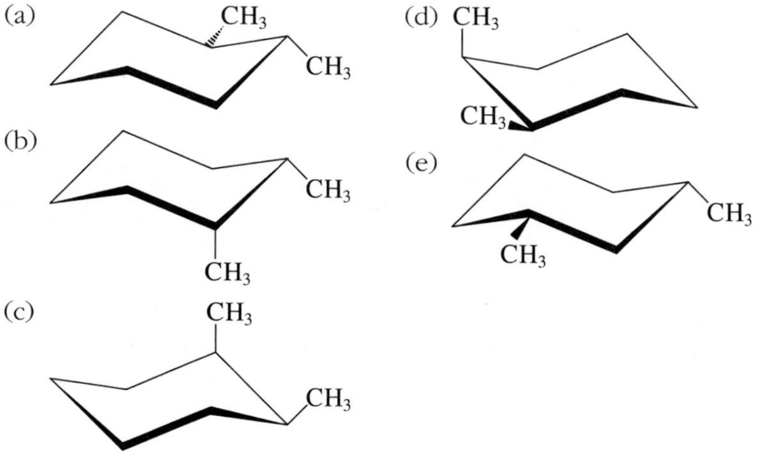

As we have just seen, *cis*- and *trans*-1-chloro-2-methylcyclohexane are both mixtures of *conformational diastereomers*. Some substituted cyclohexanes, however, are mixtures of *conformational enantiomers*, in the same sense that butane is a mixture of two rapidly equilibrating enantiomeric *gauche* forms (Sec. 6.8A). An example of such a compound is *cis*-1,2-dimethylcyclohexane:

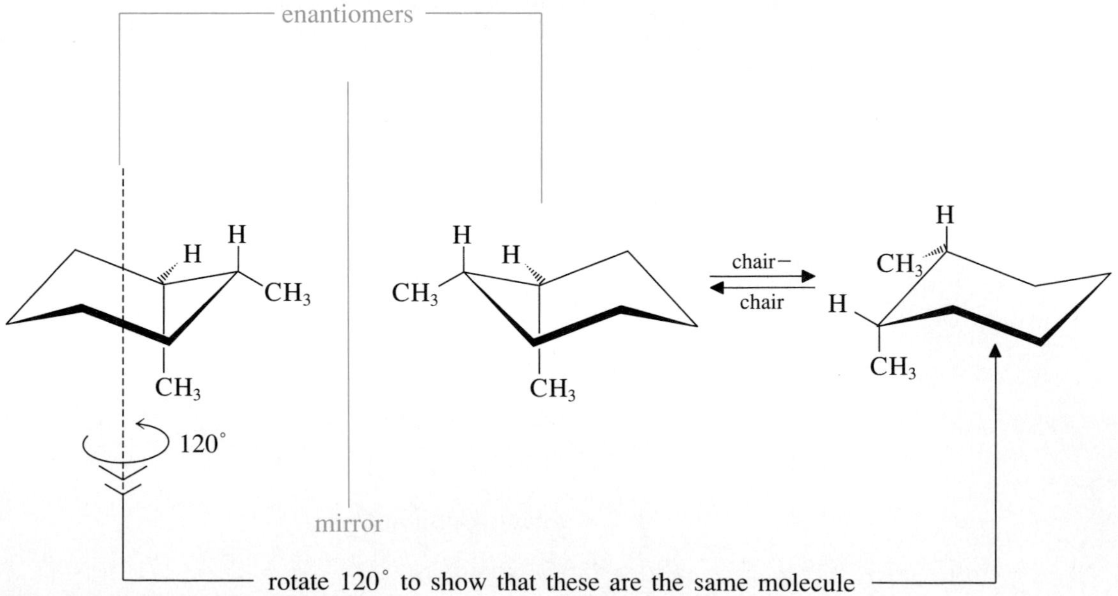

(7.7)

Because the enantiomers of *cis*-1,2-dimethylcyclohexane are interconverted rapidly, they cannot be separately isolated. Therefore, *cis*-1,2-dimethylcyclohexane cannot be optically active under normal conditions. On the other hand, the enantiomers of *cis*- or *trans*-1-chloro-2-methylcyclohexane *can* be separately isolated, because these enantiomers are not interconverted; therefore, these compounds can show optical activity. Each of these enantiomers, however, is a mixture of conformational diastereomers (Eq. 7.2, 7.3) that cannot be separately isolated under normal circumstances.

Problem

6 For each of the following compounds, indicate whether the two chair conformations are identical molecules, enantiomers, or diastereomers. Then indicate which of these compounds, if any, could be optically active under normal conditions.

(a) *trans*-1,3-dimethylcyclohexane
(b) *cis*-1-chloro-3-methylcyclohexane
(c) *cis*-1,3-dimethylcyclohexane
(d) *cis*-1,4-dimethylcyclohexane
(e) *cis*-1-[(*R*)-*sec*-butyl]-2-[(*R*)-*sec*-butyl]cyclohexane

Disubstituted cyclohexanes, like monosubstituted cyclohexanes, can be subjected to conformational analysis. The relative stabilities of the chair conformations of the *trans*-1,2-dimethylcyclohexanes, for example, can be determined from the *gauche*-butane interactions, as shown in Fig. 7.8. In the diequatorial form, there are no *gauche* interactions of the methyl groups with ring carbons; however, there is a *gauche* interaction between the two methyl groups themselves. In the diaxial form, there are two *gauche* interactions of *each* methyl group with ring carbons, for a total of four. Since the diaxial form has three more *gauche* interactions than the diequatorial form, it is $(3 \times 0.87) = 2.61$ kcal/mol less stable.

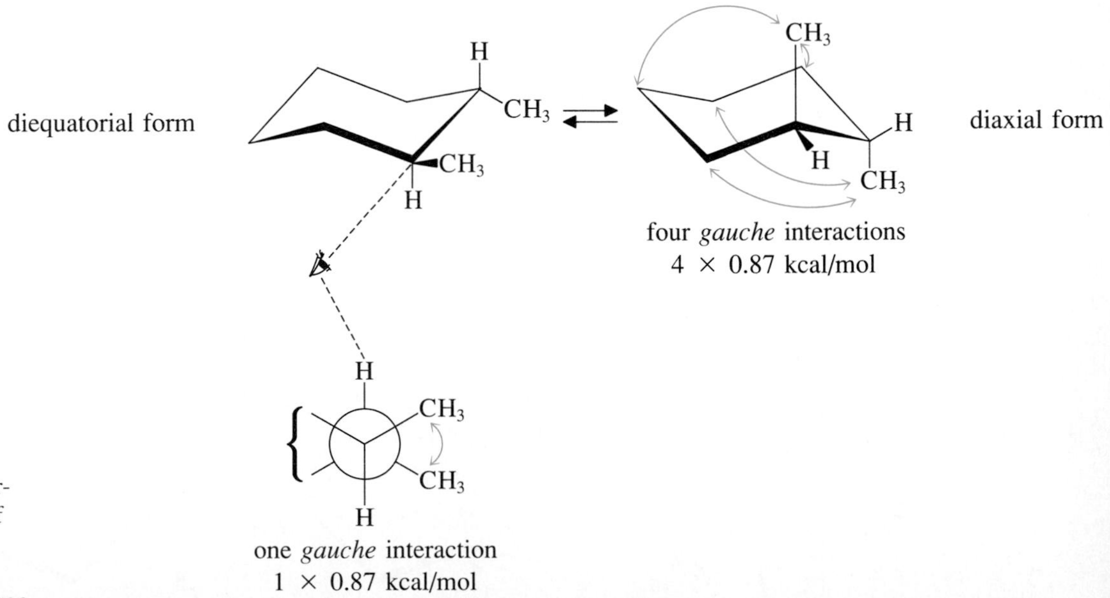

$$\text{energy (diaxial)} - \text{energy (diequatorial)} = (4 \times 0.87) - (1 \times 0.87) = 2.61 \text{ kcal/mol}$$

Figure 7.8 Conformational analysis of *trans*-1,2-dimethylcyclohexane using *gauche*-butane interactions.

In some cases effects other than *gauche*-butane interactions enter the picture in conformational analysis. One example of such an effect occurs in the diaxial form of *cis*-1,3-dimethylcyclohexane. The 1,3-diaxial interaction between the two methyl groups is considerably more severe than a *gauche*-butane interaction. (Recall that a *gauche*-butane interaction is the same as a 1,3-diaxial interaction between a methyl and a *hydrogen*.) It turns out (Problem 50) that the diaxial form of this compound is 5.5 kcal/mol less stable than the diequatorial form.

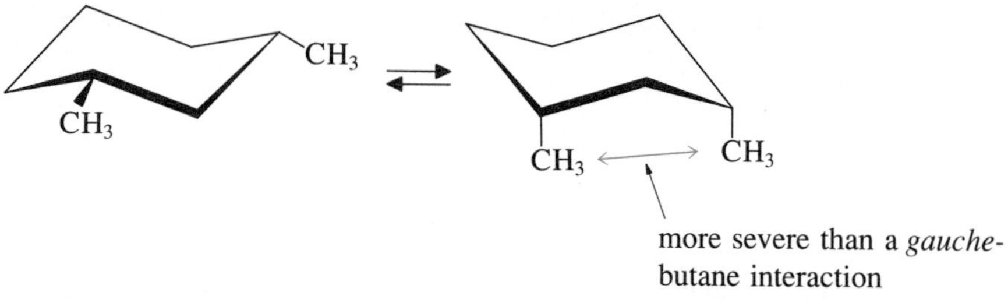

(7.8)

When two groups on a substituted cyclohexane conflict in their preference for the equatorial position, the result can usually be predicted from the relative conformational preferences of the two groups. An extreme example of this situation occurs in a cyclohexane substituted with a *t*-butyl group—for example, *cis*-4-*t*-butyl-1-methylcyclohexane.

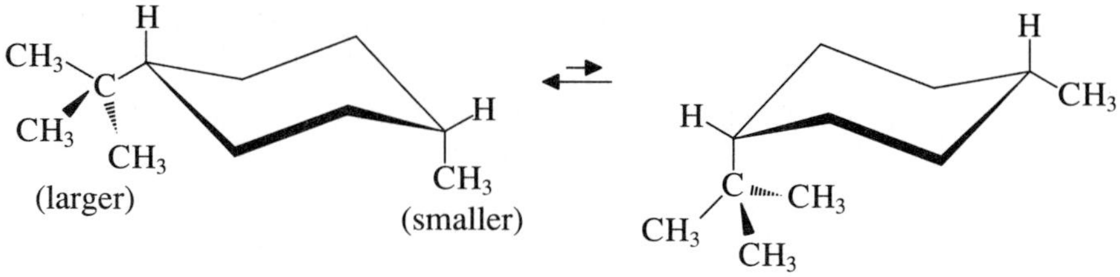

(7.9)

The *t*-butyl group is so large that its conformational preference dominates the conformational equilibrium. Hence, the conformational equilibrium strongly favors the chair form in which the *t*-butyl group assumes the equatorial position. The methyl group has no choice, then, but to take up the axial position. In fact, there is so little of the form with an *axial t*-butyl group that chemists say that the conformational equilibrium is "locked."

Problem

7 The principles of conformational analysis can also be applied to two compounds that are not conformational isomers. Calculate the relative energy of the following forms of *cis*- and *trans*-1,3-dimethylcyclohexane:

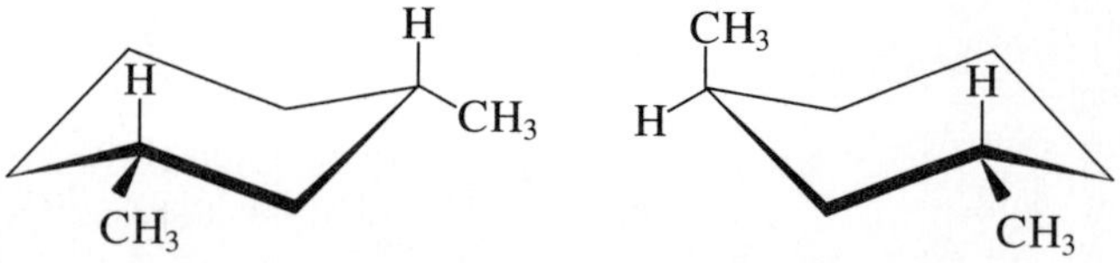

7.5 PLANAR REPRESENTATION OF CYCLIC COMPOUNDS

Although it is important for us to understand that many cyclic compounds have nonplanar conformations, it is not always necessary to draw the nonplanar structures; sometimes planar projections of cyclic compounds convey adequate information for the task at hand. We can represent the structures of nonplanar cyclic compounds by using planar polygons with the stereochemistry of substituents indicated by dashed lines or wedges. In this type of structure, we imagine that we are viewing the ring from above or below a face. If a substituent is up, it is represented with a wedge; if it is down, it is represented with a dashed line. In this convention *trans*-1,2-dimethylcyclohexane is written as follows:

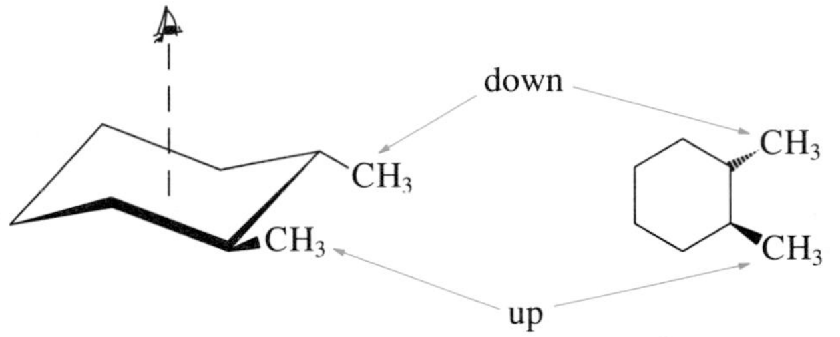

Cis-1,2-dimethylcyclohexane is drawn as

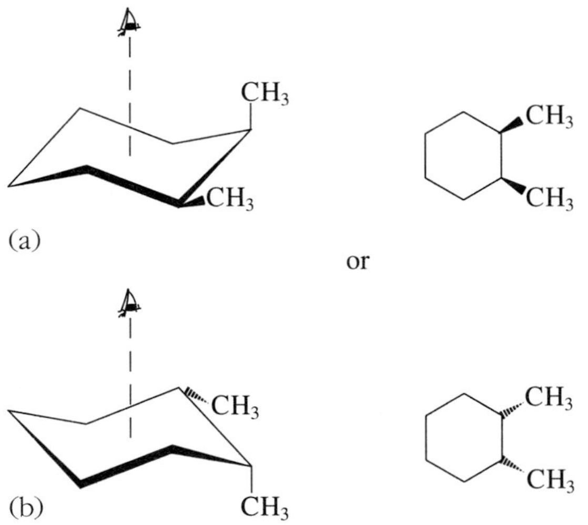

The two methyl groups in *cis*-1,2-dimethylcyclohexane can be represented with either two wedges or two dashed lines. If we look at the ring as shown in (a) above, the methyl groups are both up; if we turn the ring over, as shown in (b), they are both down.

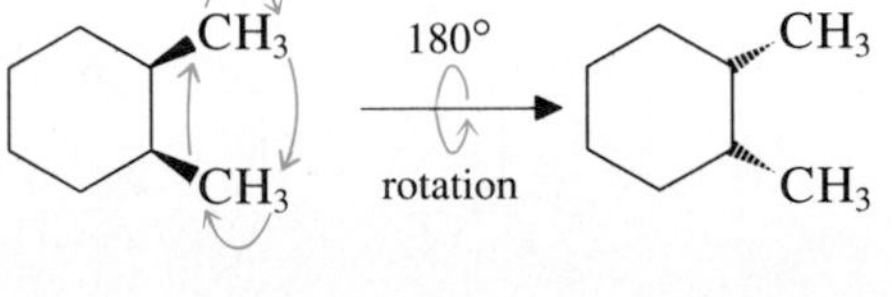

(Use models to convince yourself that these statements are true.)

The same planar structure describes any conformation of the ring provided that all conformations are viewed from the same face of the molecule. That is, the chair–chair conformational change does not interchange the dashed lines and wedges because it does not change the up or down relationship of groups on the ring.

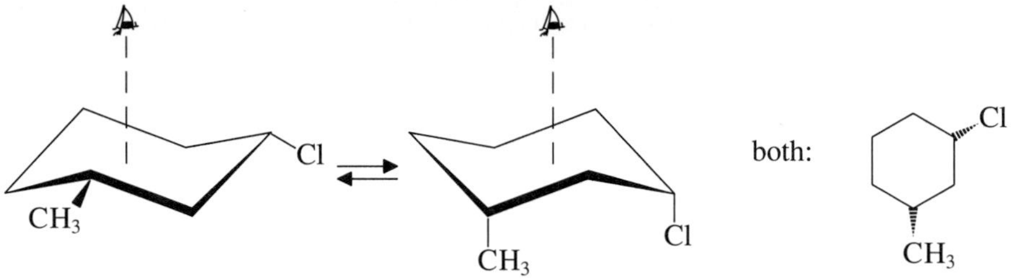

Notice also that these projections are *not* the same as Fischer projections. We can turn these projections in any direction, or lift them out of the paper and turn them over, provided that the dashed line or wedge to each substituent group is changed accordingly. The convention used in drawing Fischer projections is that all horizontal bonds are wedges; the rules for manipulating Fischer projections are necessary precisely because these wedges are not drawn in explicitly (Sec. 6.9).

Problems

8 Draw a planar representation for each of the following compounds using dashed lines and wedges to show the stereochemistry of the substituent groups.
(a) *cis*-1,3-dimethylcyclohexane
(b) *trans*-1,3-dimethylcyclohexane
(c) (1*R*,2*S*,3*R*)-2-chloro-1-ethyl-3-methylcyclohexane
(d) *trans*-1,2-dimethylcyclopentane

9 Draw a chair conformation for each of the following compounds:

(a)
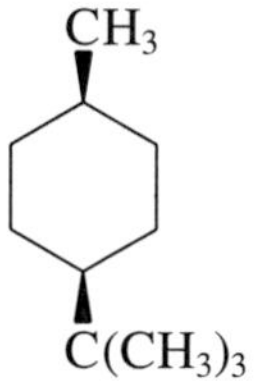

(b)
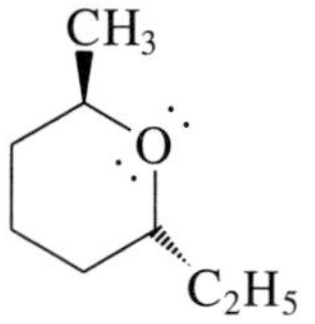

10 What is the relationship of the two structures in each set (identical molecules, enantiomers, or diastereomers)?

(a)
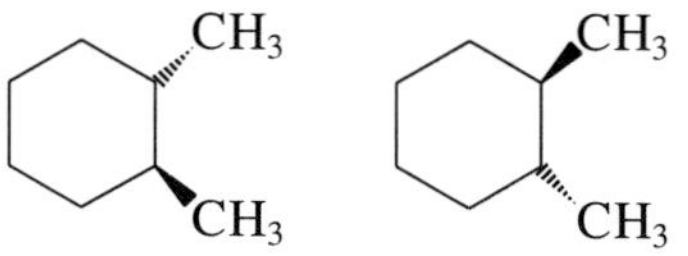

(c)
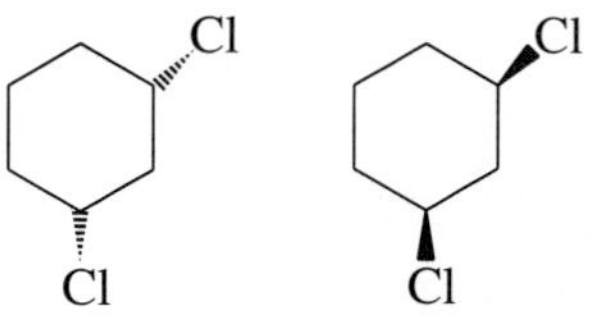

(b)
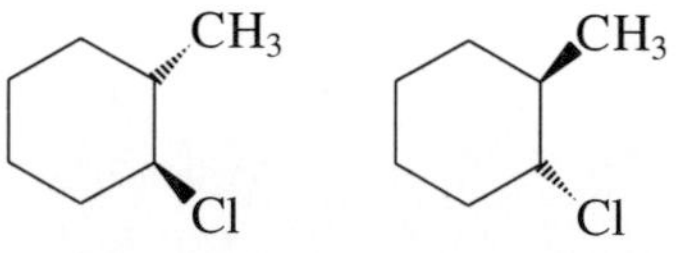

7.6 CYCLOPENTANE, CYCLOBUTANE, AND CYCLOPROPANE

A. Conformation of Cyclopentane

Cyclopentane, like cyclohexane, exists in a nonplanar conformation. This puckered form of the five-membered ring, called the **envelope conformation** (Fig. 7.9), relieves some of the eclipsed-hydrogen interactions that would be present in fully planar cyclopentane. In this conformation, the hydrogens at three of the five carbon atoms assume axial or equatorial positions. The envelope conformation of cyclopentane undergoes rapid conformational changes in which each carbon successively alternates as the point of the envelope.

Substituted cyclopentanes also exist in envelope conformations in which the substituents adopt positions that minimize van der Waals repulsions with neighboring groups. For example, in methylcyclopentane, the methyl group assumes an equatorial position at the point of the envelope.

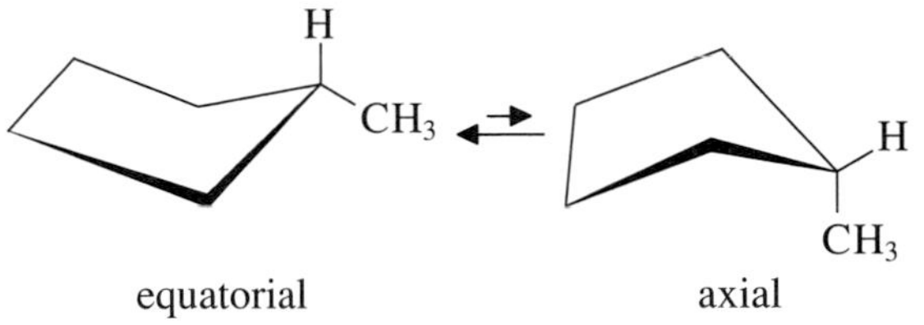

(7.10)

Cis and *trans* substitution patterns are possible in cyclopentanes just as in cyclohexanes. For example, the structures of the 1,3-dimethylcyclopentanes are shown in Fig. 7.10.

B. Cyclobutane and Cyclopropane

The data in Table 7.1 show that cyclobutane and cyclopropane are the most highly strained monocyclic hydrocarbons. Cyclobutane exists in a puckered conformation (Fig. 7.11) that relieves eclipsing interactions between the hydrogens on adjacent carbons; two equivalent puckered conformations are in rapid equilibrium.

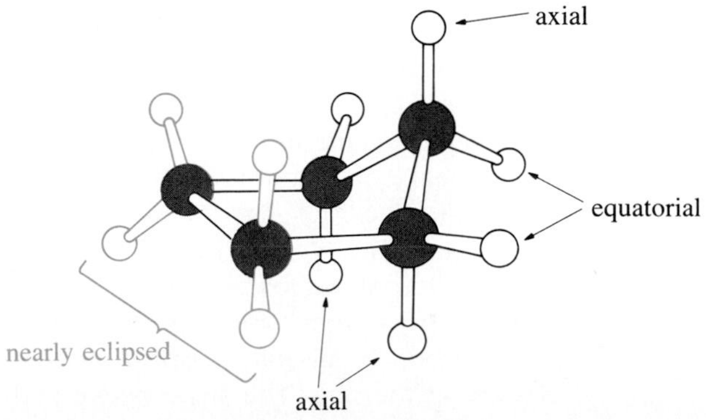

Figure 7.9 Envelope conformation of cyclopentane.

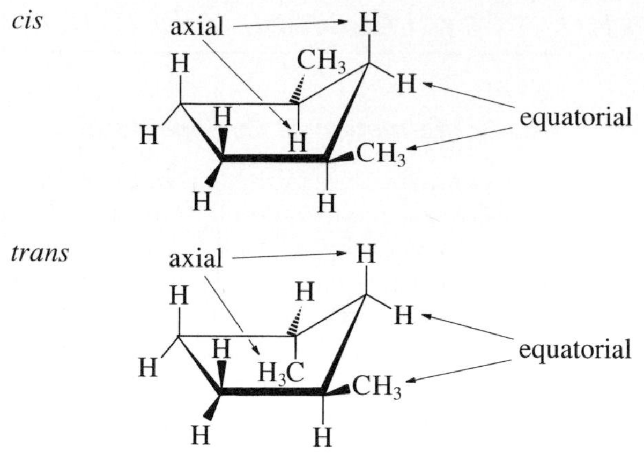

Figure 7.10 Structures of *cis*- and *trans*-1,3-dimethylcyclopentane.

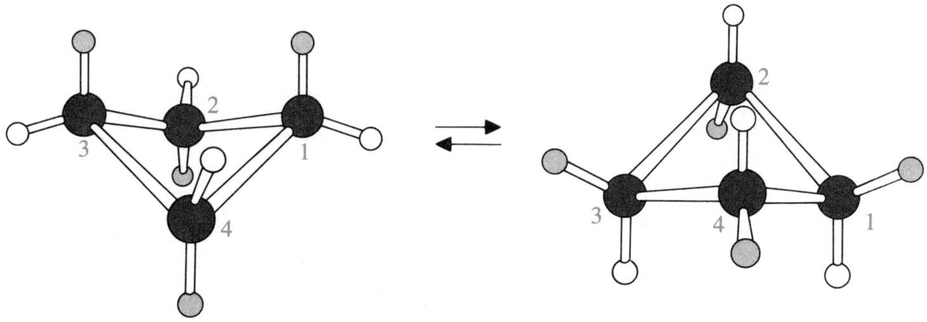

Figure 7.11 Structure of cyclobutane.

Problem

11 Explain why the dipole moment of *trans*-1,3-dibromocyclobutane, 1.10 D, is more in accord with the puckered structure than the planar structure. (What dipole moment would you expect for this compound if it were planar?) Draw a structure for this compound that shows its conformation.

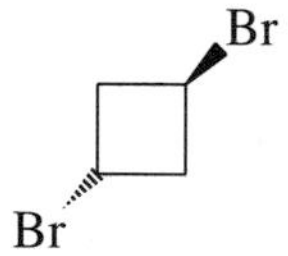

trans-1,3-dibromocyclobutane

Because three points determine a plane, cyclopropane must have a planar carbon skeleton, and the C—H bonds on adjacent carbon atoms are eclipsed. The internal angles of cyclobutane and cyclopropane, as von Baeyer predicted, are greatly distorted from the ideal tetrahedral value of 109.5°; these compounds therefore have significant *angle strain* (Sec. 7.1). This strain has interesting consequences for the bonding in cyclopıopane. The preferred C—C—C bond angle is 109.5°, but the actual C—C—C angle of cyclopropane, by plane geometry, must be 60°. The bonding in cyclopropane appears to be a compromise of these two angles: it is generally believed that the carbon–carbon bonds of cyclopropane are bent in a "banana" shape around the periphery of the ring. The angle between the *orbitals* is about 104° (Fig. 7.12).

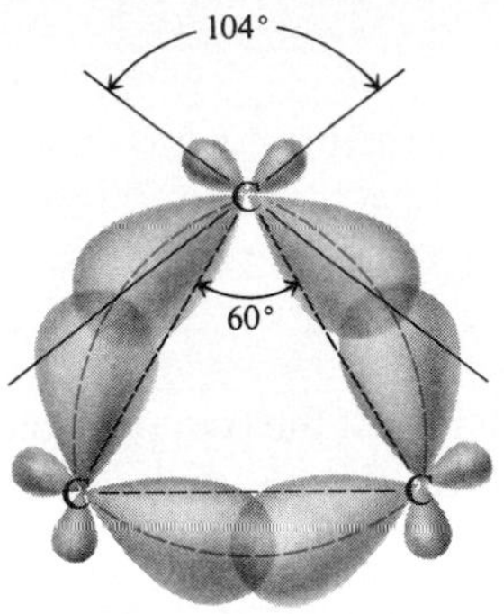

Figure 7.12 The orbitals that overlap to form the **C—C** bonds in cyclopropane do not lie along the straight line between the carbon atoms. These carbon–carbon bonds are sometimes called "bent" or "banana" bonds.

The ring strain of cyclobutanes and cyclopropanes is reflected to some degree in their reactivity. As ring size decreases, cyclic alkanes become increasingly like alkenes in their chemical behavior. For example, cyclopropanes can be hydrogenated:

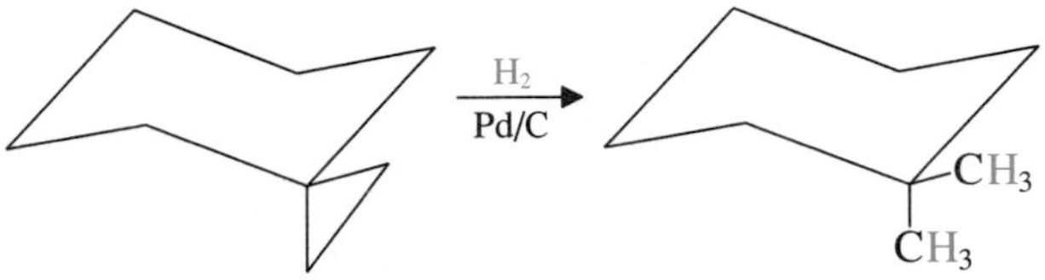

$$(7.11)$$

Hydrogenation of cyclopropanes, however, usually occurs with more difficulty than hydrogenation of alkenes:

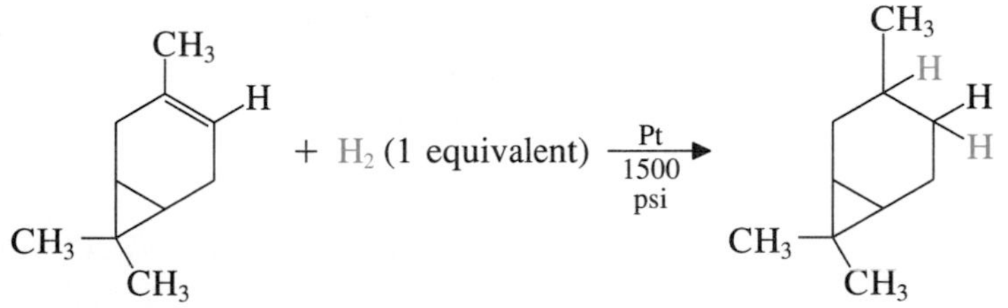

$$(7.12)$$

Even some cyclobutanes can be hydrogenated under extreme conditions:

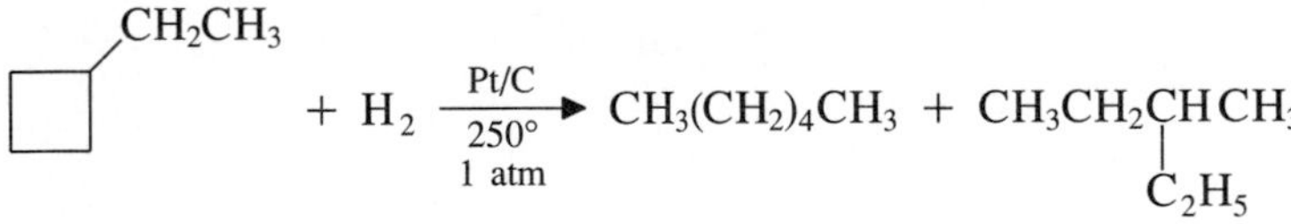

$$\square\text{—CH}_2\text{CH}_3 \; + \; H_2 \; \xrightarrow[\substack{250° \\ 1\ atm}]{Pt/C} \; CH_3(CH_2)_4CH_3 \; + \; CH_3CH_2\underset{\underset{C_2H_5}{|}}{C}HCH_3 \qquad (7.13)$$

Although we might say that small-ring cycloalkanes behave like alkenes, we could just as easily say that an alkene behaves like a two-membered cycloalkane: a "cycloethane." We would expect a "two-membered ring" to be very strained, and indeed ethylene has a higher heat of formation per —CH$_2$— than any other cyclo-alkane.

	cyclobutane	cyclopropane	"cycloethane" (ethylene)
ΔH_f^0 per CH$_2$: (kcal/mol)	+1.65	+4.25	+6.25

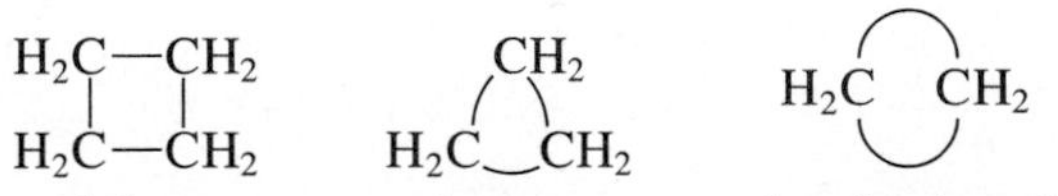

—— increasing ΔH_f^0 ⟶
—— increasing reactivity ⟶

Problem

12 Give the product(s) expected when cyclopropane and methylcyclopropane react with each of the following reagents under vigorous conditions:
(a) H_2, catalyst
(b) Br_2
(c) HBr
Which of the two possible products of the reaction (c) with methylcyclopropane would be favored, and why?

7.7 BICYCLIC AND POLYCYCLIC COMPOUNDS

A. Classification and Nomenclature

Cyclic compounds may, of course, contain more than one ring. An important and interesting situation arises when two rings have common atoms. If two rings share two common atoms, the compound is called a **bicyclic** compound. If two rings have a single common atom, the compound is called a **spirocyclic** compound.

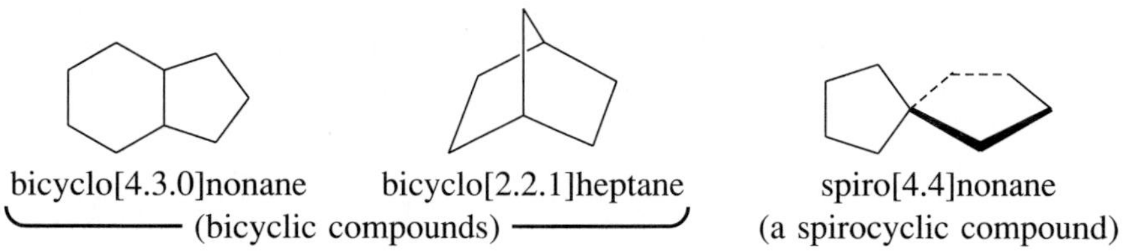

When a bicyclic compound contains two rings joined at adjacent carbons, as in bicyclo[4.3.0]nonane above, it is classified as a **fused** bicyclic compound. When the two rings are joined at nonadjacent carbons, as in bicyclo[2.2.1]heptane, the compound is classified as a **bridged** bicyclic compound.

The nomenclature of bicyclic hydrocarbons is best illustrated by example. We first identify the carbons at which the rings are joined; these are called the *bridgehead* carbons.

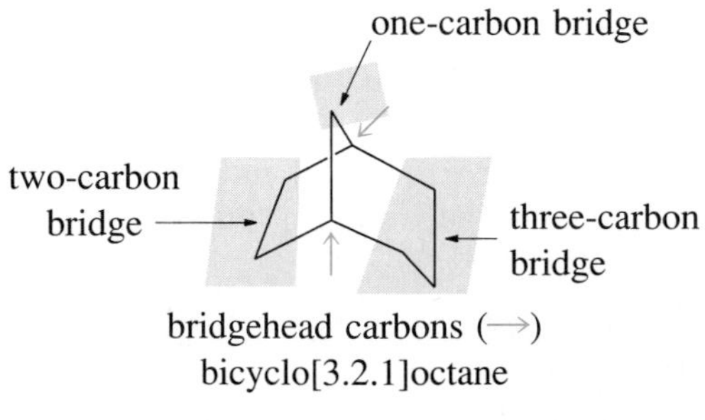

This compound is named as an octane because it has eight carbon atoms. It is called a bicyclooctane because it has two bridgehead carbon atoms. The numbers in brackets represent the number of the carbon atoms in the respective bridges, in order of decreasing size.

Notice that in the following example there is a bridge containing no carbons; this fact is reflected in the name by the number zero within the brackets.

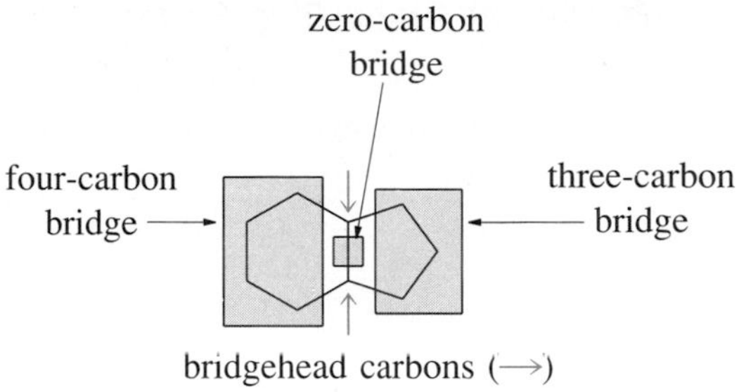

bicyclo[4.3.0]nonane

Although spirocyclic compounds are well known, they are less important than bicyclic compounds, and we shall not be concerned with their nomenclature.

Problem

13 Name the following compounds:

(a)

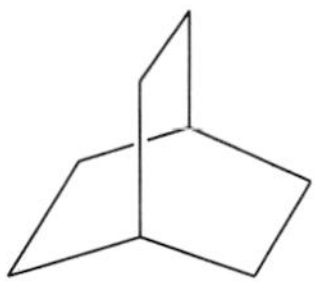

(b)

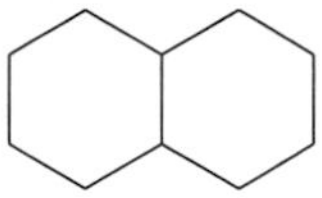

Some organic compounds contain many rings joined at common atoms; these compounds are called **polycyclic compounds.** Among the more intriguing polycyclic compounds are those that have the shapes of regular geometric solids. Three of the more spectacular examples are cubane, dodecahedrane, and tetrahedrane.

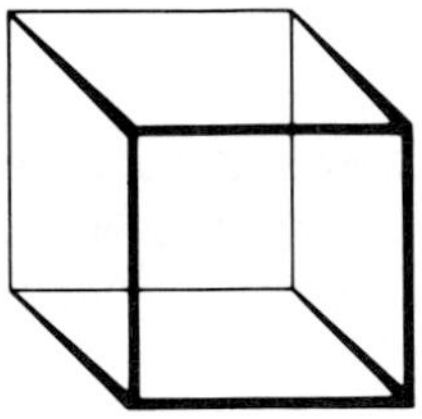

cubane

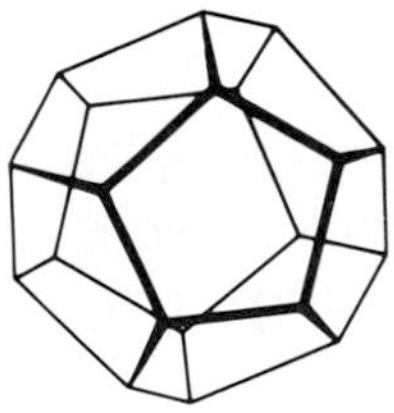

dodecahedrane

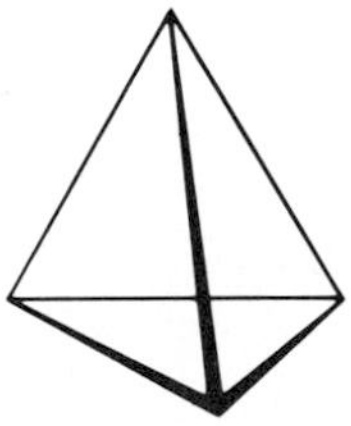

tetrahedrane

Cubane, which contains eight —CH— groups at the corners of a cube, was first synthesized in 1964 by Professor Philip Eaton and his associate Thomas W. Cole at the University of Chicago. Dodecahedrane, in which twenty —CH— groups occupy the corners of a dodecahedron, was synthesized in 1982 by a team of organic chemists led by Leo Paquette of Ohio State University. Tetrahedrane itself has not yielded to synthesis, although a derivative containing *t*-butyl substituent groups at each corner has been prepared. Chemists tackle the syntheses of these very pretty molecules not only because they represent interesting problems in chemical bonding, but also because of the sheer challenge of the endeavor.

B. Cis *and* Trans *Ring Fusion*

Two rings in a bicyclic compound can be fused in more than one way. Consider, for example, the *decalins* (the bicyclo[4.4.0]decanes):

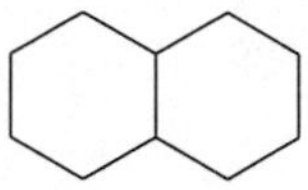

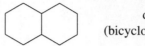

decalin
(bicyclo[4.4.0]decane)

There are two stereoisomers of decalin. In *cis*-decalin, the two —CH_2— groups on ring *B* (circles) are *cis* substituents on ring *A*; likewise, two —CH_2— groups on ring *A* (squares) are *cis* substituents on ring *B*.

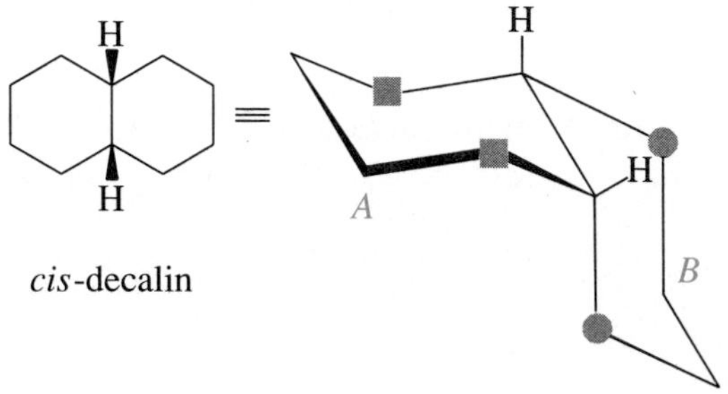

cis-decalin

In *trans*-decalin, the analogous —CH_2— groups are in a *trans*-diequatorial arrangement.

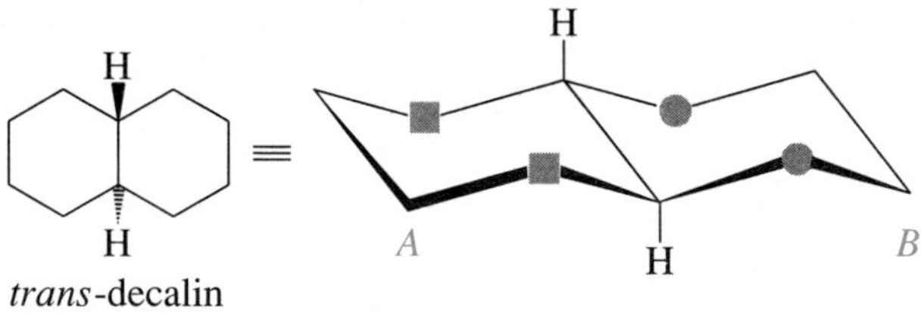

trans-decalin

In other words, the relationship between *cis*- and *trans*-decalin is the same as the relationship between *cis*- and *trans*-1,2-dimethylcyclohexane.

The planar representations of *cis*- and *trans*-decalin are as follows:

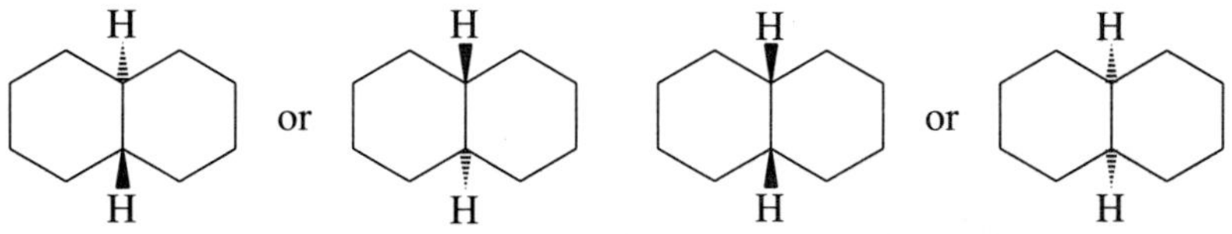

trans-decalin   *cis*-decalin

Each cyclohexane ring in *cis*-decalin can undergo the chair–chair interconversion, like cyclohexane itself. However, in *trans*-decalin, the six-membered rings can assume twist-boat conformations, but they cannot flip into their alternate chair conformations. You should try the chair–chair interconversion with models of *cis*- and *trans*-decalin to see for yourself the validity of these statements. Why is it that *trans*-decalin cannot undergo the chair–chair interconversion? To answer this question, focus on ring *B* in the structure above. Notice that the two circles represent carbons that are in effect *equatorial* substituents on ring *A*. If ring *A* were to flip into the alternate chair form, these two carbons in ring *B* would have to assume *axial* positions, since the chair–chair interconversion converts equatorial groups into axial groups, and vice versa. When these two carbons are in axial positions they are much further apart than they are in equatorial positions; the distance between them is simply too great to be spanned easily by the remaining two carbons of ring *B*. As a result, the chair–chair interconversion introduces so much ring strain into ring *B* that it cannot occur. Of course, exactly the same problem occurs with ring *A* when ring *B* tries to convert into its alternate chair conformation.

Problem

14 How many different *gauche* interactions are there in *cis*-decalin? How many in *trans*-decalin? Which of the two decalins is more stable, and by how many kcal/mol? (*Hint:* Use your models, and don't count the same interaction twice.)

Cis-decalin is less stable than *trans*-decalin because of its greater number of *gauche* interactions (Problem 14). *Trans* ring fusion is not, however, the more stable way of joining rings in all bicyclic molecules. In fact, if both of the rings are small, a *trans* ring fusion is virtually impossible. For example, only the *cis*-fused isomers of the following two compounds are known:

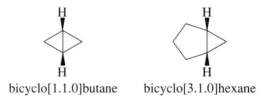

bicyclo[1.1.0]butane bicyclo[3.1.0]hexane

Attempting to join two small rings with a *trans* ring junction simply introduces too much ring strain. The best way to see this is with models, using the following exercise as your guide.

Problem

15 Construct models of cyclopentane and cyclononane. In each model, add *trans* bonds at two adjacent carbons; the hydrogen atoms need not be added.

(a) Compare the difficulty of making models of the *cis* and *trans* isomers of bicyclo[3.1.0]hexane. (Don't break your models!) Which is easier to make? Why?

(b) Compare the difficulty of making models of *trans*-bicyclo[7.1.0]decane and *trans*-bicyclo[3.1.0]hexane as in part (a). Which is easier to make? Explain.

Let us summarize what we have learned about compounds containing fused rings.

1. Two rings may be fused in a *cis* or *trans* arrangement.
2. When the rings are small, only *cis* fusion is observed because *trans* fusion introduces too much ring strain.
3. In larger rings, both *cis*- and *trans*-fused isomers are well known, but the *trans*-fused ones are more stable because *gauche*-butane interactions are minimized (as in the decalins).

Effects (2) and (3) are about equally balanced in the *hydrindanes* (bicyclo[4.3.0]nonanes); the *trans* isomer is only 0.66 kcal/mol more stable than the *cis* isomer.

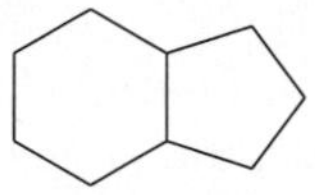

hydrindane
(bicyclo[4.3.0]nonane)

C. Trans-*Cycloalkenes. Bredt's Rule*

Cyclohexene and other cycloalkenes with small rings clearly have *cis* (or *Z*) stereochemistry at the double bond. Is there a *trans*-cyclohexene? The answer is that the *trans*-cycloalkenes with six or fewer carbons have never been observed. The reason is obvious if we try to build a model of *trans*-cyclohexene. In this molecule the carbons attached to the double bond are so far apart that it is difficult to connect them with only two other atoms. To do so either introduces a great amount of ring strain, or requires twisting the molecule about the double bond, thus weakening the overlap of *p* orbitals involved in the π-bond. *Trans*-cyclooctene is the smallest *trans*-cycloalkene that can be isolated under ordinary conditions; however, it is 9–11 kcal/mol less stable than its *cis* isomer.

It has long been known that a small bridged bicyclic compound is unstable if it has a double bond at a bridgehead carbon. This empirical observation is called **Bredt's rule.** The following compound, for example, is very unstable and has never been isolated in pure form:

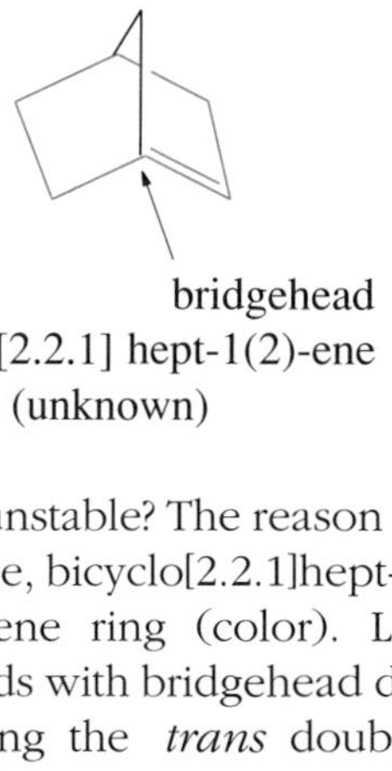

bicyclo[2.2.1] hept-1(2)-ene
(unknown)

Why are such compounds unstable? The reason is that they contain *trans* double bonds in small rings. For example, bicyclo[2.2.1]hept-1(2)-ene, the compound above, incorporates a *trans*-cyclohexene ring (color). Like the corresponding *trans*-cycloalkenes, bicyclic compounds with bridgehead double bonds are too unstable to isolate when the ring containing the *trans* double bond has fewer than eight members.

Bicyclic compounds that incorporate bridgehead double bonds in larger rings are correspondingly more stable. The following two compounds, for example, can both be prepared in pure form. The one on the left, in which the double bond is incorporated in an eight-membered ring (color), is stable, but reactive. (Recall that *trans*-cyclooctene can be isolated, but is considerably less stable than its *cis* isomer.) The one on the right is quite stable and behaves like an ordinary alkene; *trans*-cyclodecene, by analogy, is a stable *trans*-cycloalkene.

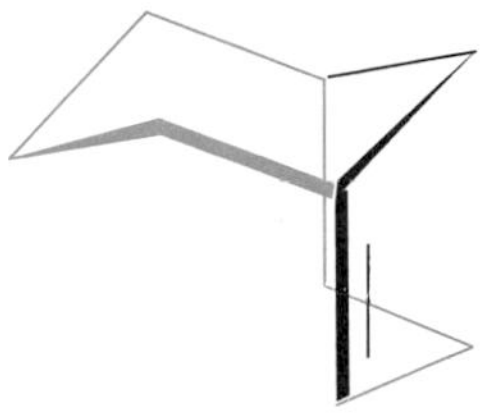

bicyclo[3.3.1]non-1(2)-ene (incorporates the *trans* cyclo octene shown in color)

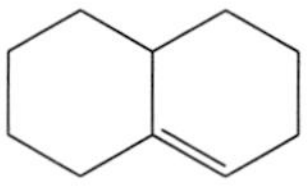

bicyclo[4.4.0]dec-1(2)-ene

Problems

16 (a) Using models, explain why enantiomeric forms of *trans*-cyclooctene can be isolated, even though the compound does not contain an asymmetric carbon.

(b) Optically active *trans*-cyclooctene is racemized on heating to about 200 °C, but the racemization is very slow at room temperature. Using your models, show the conformational change required for racemization, and explain why it is so slow.

17 Which of the following compounds should have the higher heat of formation? Explain.

(a)

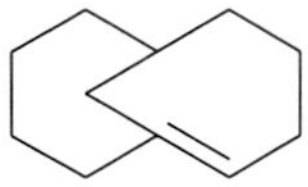

(b)

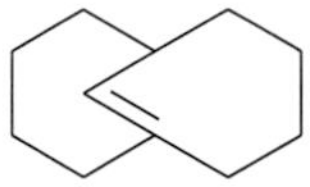

D. Steroids

Of the many naturally occurring compounds with fused rings, the class of compounds called the **steroids** is particularly important. The basic ring system of the steroids, with its special numbering, is as follows:

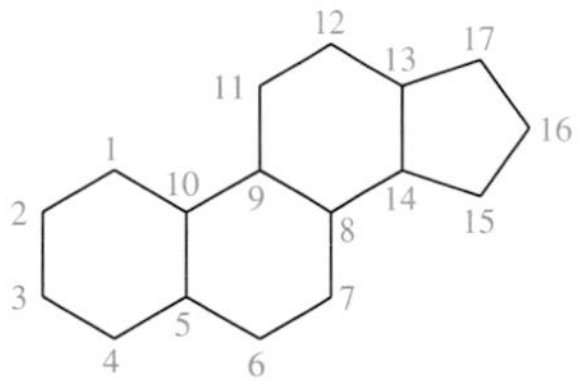

The various steroids differ in the types of functional groups that are present on this carbon skeleton.

It is common to find two particular structural features in naturally occurring steroids. The first is that in many cases all ring fusions are *trans*. Since *trans*-fused cyclohexane rings cannot undergo the chair–chair interconversion, all-*trans* ring fusion causes a steroid to be conformationally rigid and relatively flat. Second, many steroids have methyl groups, called *angular methyls*, at carbons 10 and 13.

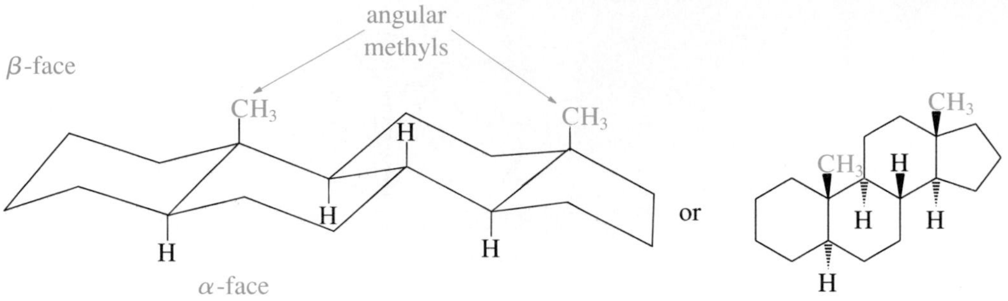

α-face behind the page
β-face in front of the page

These methyl groups occupy axial positions over one face, called the β-face, of the steroid molecule. The other face is called the α-face.

Many important hormones and other natural products are steroids. Cholesterol occurs widely and was the first steroid to be discovered (1775). The corticosteroids and the sex hormones represent two medicinally important classes of steroid hormones.

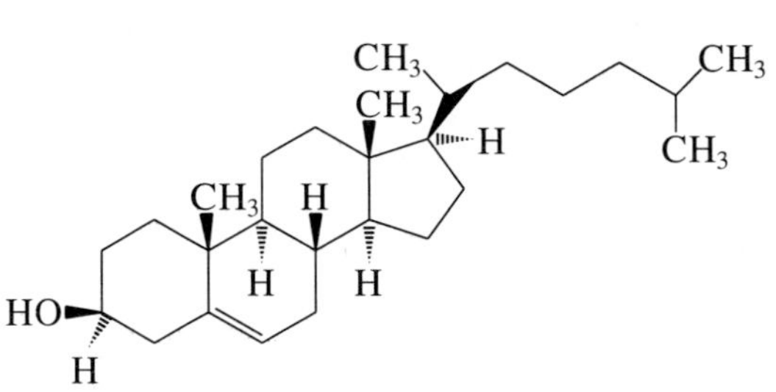

cholesterol
(important component of membranes: principal component of gallstones)

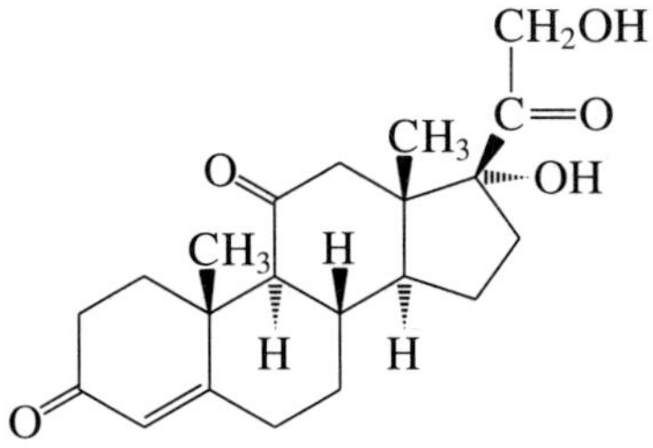

cortisone
(anti-inflammatory hormone)

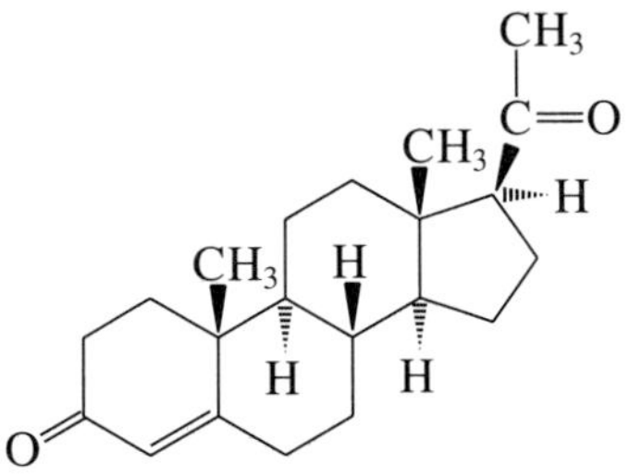

progesterone
(human female sex hormone)

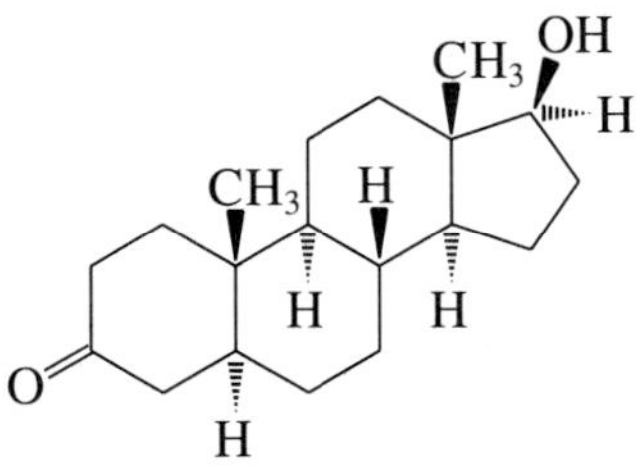

testosterone
(human male sex hormone)

Prior to 1940, steroids were isolated from such inconvenient sources as sows' ovaries or the urine of pregnant mares, and were very scarce and expensive. In the 1940s, however, a Pennsylvania State University chemist, Russell Marker (1902–) developed a process that could bring about the conversion of a naturally occurring compound called *diosgenin* into progesterone.

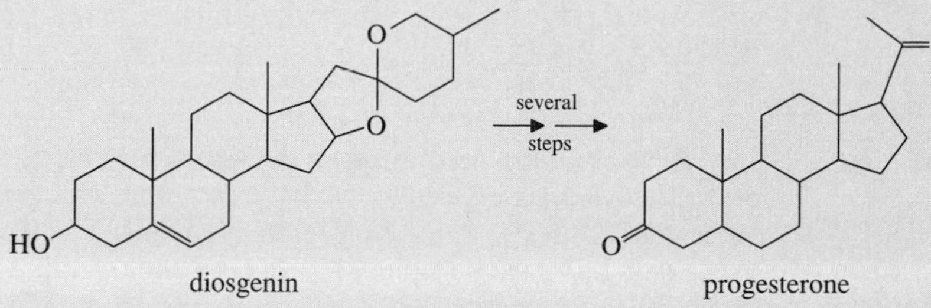

(7.14)

The natural source of diosgenin is the root of a vine, *cabeza de negro,* genus *Dioscorea,* indigenous to Mexico. Because Marker had shown (Eq. 7.14) that these vines are a potential new source of valuable steroids, he was eager to find a larger supply of them. Unable to convince any pharmaceutical company to finance an expedition, he quit his academic post and trekked off into the wild hill country of southern Mexico, where he rented a cottage and found a profuse growth of *Dioscorea.* In his cottage, working in secrecy with native assistants and coded reagent bottles, he carried out the conversion of diosgenin into progesterone. Marker carried his notes with him everywhere—even to bed! Perhaps his caution was warranted, for the European cartel that had hitherto dominated the steroid market was disturbed at the news of Marker's discovery, which threatened to disrupt their monopoly. Soon Marker had synthesized enough progesterone to fill two large jars—worth about $240,000 on the day's market.

Subsequently, Marker cofounded Syntex, S.A., where the industrial production of steroids from *Dioscorea* was developed. (The United States firm of Syntex Laboratories is the modern successor to this company.) The price of steroid hormones dropped from $80 to $2/gram. Complex licensing arrangements and competition have resulted in the entry of many companies into the steroid field, and there are now reasonably adequate supplies of these important drugs. *Dioscorea,* which takes three years to mature, is now grown on plantations, and the Mexican government justifiably regards this plant as a valuable natural resource. Russell Marker feuded with the management of Syntex and left the company, retiring from all research and production activities in 1949. This colorful and visionary scientist was honored in 1975 at a joint Chemical Congress of the United States and Mexico for his contributions in the field of steroids.

7.8 STEREOCHEMISTRY AND CHEMICAL REACTIONS

We have now learned the basic definitions of stereochemistry, and we have considered the conformational and stereochemical aspects of cyclic compounds. Now we are ready to ask: What is the importance of stereochemistry to the reactions of organic compounds?

A. Reactions of Achiral Compounds that Give Enantiomeric Pairs of Products

Suppose we carry out a chemical reaction with achiral starting materials that yields a chiral product. An example of such a reaction is the addition of HBr to styrene.

$$\underset{\text{styrene}}{Ph{-}CH{=}CH_2} + HBr \longrightarrow \underset{\text{1-bromo-1-phenylethane}}{Ph{-}\underset{\underset{Br}{|}}{C}H{-}CH_3} \qquad (7.15)$$

Neither of the reactants—styrene nor HBr—is chiral. However, the product of the reaction, 1-bromo-1-phenylethane, *is* chiral. This product could be either of two enantiomers. Thus, a question of stereochemistry arises in this reaction: Which of these stereoisomers is formed? A very general principle applies to this situation: *When chiral products are formed from achiral starting materials, both enantiomers of a pair are always formed at the same rate.* That is, the product is always the *racemate*. Thus, in the example of Eq. 7.15, equal amounts of (*R*)- and (*S*)-1-bromo-1-phenylethane are formed.

Since enantiomers have equal energies, there is no reason to expect that one enantiomer should be formed in preference to another. For each chiral transition state in the reaction pathway, there is an enantiomeric transition state of equal energy. Since the enantiomeric transition states are formed from the same starting material, the two enantiomers of the product are formed with the same free energy of activation. Hence, they are formed at the same rate, and therefore in the same amounts. As a result, the product is a 1 : 1 mixture of *R* and *S* enantiomers—the racemate.

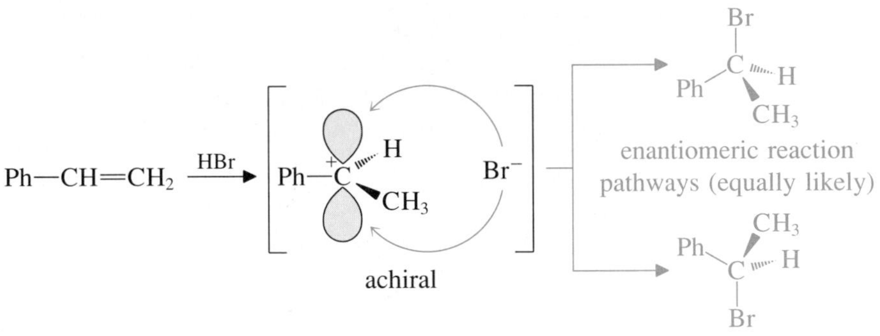

(7.16)

We can summarize by saying that *optical activity never arises spontaneously in the reactions of achiral compounds.*

B. Reactions that Give Diastereomeric Products

Some reactions can in principle give pairs of diastereomeric products, as in the following example:

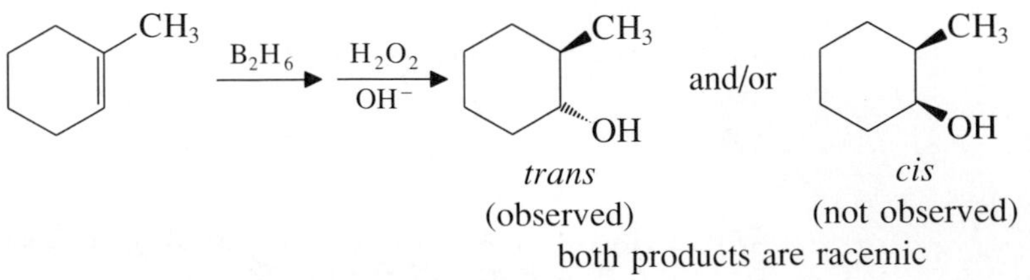

(7.17)

In this case either the *cis* or *trans* diastereomer of the product might be formed. Another principle applies to this situation: *When diastereomeric products can be* formed in a reaction, they are always formed at different rates. This means that *different amounts of each product are formed.*

We might not be able to predict, without knowing more about the reaction, *which* diastereomer is the major one; the point is that, in general, one product will be formed in greater amount than the other. (In Eq. 7.17, the *trans* isomer is formed exclusively; we shall see why in Sec. 7.10.)

Reactions that give a pair of diastereomeric products occur through diastereomeric transition states. Since diastereomers have different energies, one transition state has lower standard free energy than its diastereomer. The diastereomeric reaction pathways thus have different standard free energies of activation and therefore different rates, and their respective products are formed in different amounts.

Two points about what we have just learned are worth special note. First, although diastereomers are formed in different amounts, the differences in their amounts might be beyond detection: for example, one might be formed as 49.99% and the other as 50.01% of the product mixture. The important point here is that *in principle* the two are formed in different amounts. The second point worth noting is that when the starting materials are achiral, each diastereomer of the product will be formed as a pair of enantiomers (the racemate), by the principle of Sec. 7.8A. This is the situation, for example, in Eq. 7.17. For convenience we sometimes draw just one enantiomer of each product formed, as in this equation, but in situations like this it is understood that each of these diastereomers *must* be racemic.

C. Reactions of Enantiomers

In the first two parts of this section we have been concerned with the stereochemical results expected when one compound can react to give more than one stereoisomer. Let us now turn the question around: Are there any differences in reactivity when stereoisomers react with a particular reagent? Let us begin by considering how a pair of enantiomers might differ in their reactivity. Imagine, for example, the reaction of the following chiral alkene with diborane:

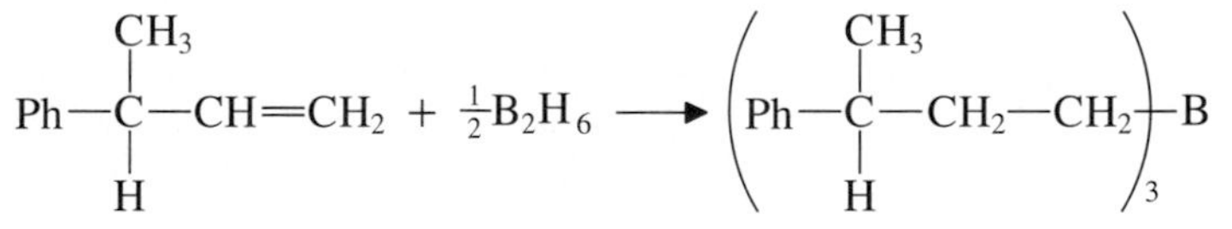

$$
\text{Ph}-\overset{\overset{\displaystyle CH_3}{\mid}}{\underset{\underset{\displaystyle H}{\mid}}{C}}-CH{=}CH_2 \;+\; \tfrac{1}{2}B_2H_6 \;\longrightarrow\; \left(\text{Ph}-\overset{\overset{\displaystyle CH_3}{\mid}}{\underset{\underset{\displaystyle H}{\mid}}{C}}-CH_2{-}CH_2\right)_{\!3}\!\!-B \qquad (7.18)
$$

(a chiral alkene)

The question here is: Do the *R* and *S* enantiomers of this alkene differ in their reactivities? Again, a general principle applies. *A pair of enantiomers react at the same rates with an achiral reagent.* This means that the enantiomers of the alkene in Eq. 7.18 react with diborane, an achiral reagent, with exactly the same rates to give their respective products in exactly the same yield.

An analogy from common experience can help us to understand why this should be so. Consider your feet, an enantiomeric pair of objects. Imagine placing first your right foot, then your left, in a perfectly square box—an achiral object. (We shall assume that your two feet have exactly the same dimensions.) Each foot will fit this box in exactly the same way. If the box pinches the big toe on your right foot, it will also pinch the big toe on your left foot in the same way. Just as your feet interact in the same way with the achiral box, so enantiomeric molecules react in exactly the same way with achiral reagents. Since diborane is an achiral reagent, the two alkene enantiomers in Eq. 7.18 react with diborane in exactly the same way.

Suppose, though, that we allow each enantiomer of the alkene in Eq. 7.18 to react in turn with one enantiomer of a *chiral* reagent, such as the following borane:

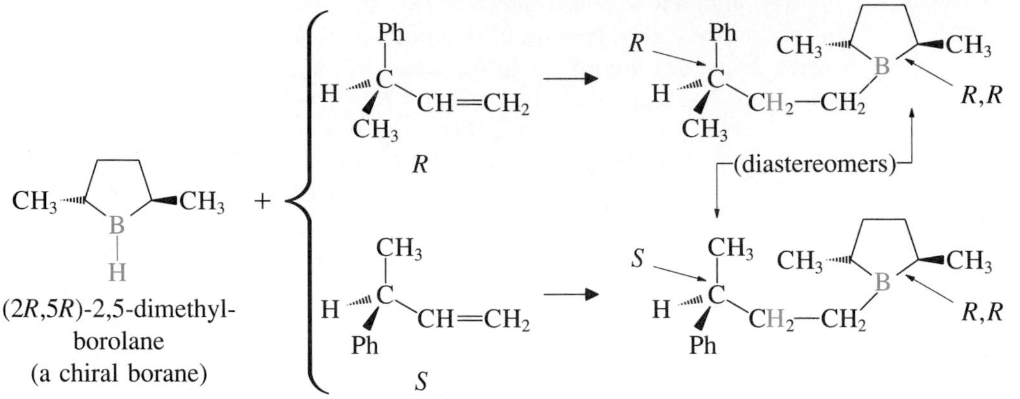

(7.19a)

(7.19b)

The following general principle applies to this situation: *The two enantiomers of a pair react at different rates with a chiral reagent.* This means that if we were to allow one mole of the alkene racemate (one-half mole of each enantiomer) to react with one-half mole of the chiral borane, one of the diastereomeric products in Eq. 7.19 would be formed in a greater amount than the other; the alkene remaining after the borane is consumed would be enriched in the less reactive enantiomer.

Let us again use an analogy to see why this principle is reasonable. Imagine placing first your right foot, then your left, in your right shoe—a chiral object. Your right and left feet interact differently with your right shoe; the shoe fits one foot and not the other. The enantiomeric objects—feet—interact differently with the chiral shoe. Likewise, a pair of chiral molecules interact differently with a chiral reagent. The interaction of one alkene enantiomer in Eq. 7.19 with the chiral borane is more favorable than the interaction of the other. As a result, one transition state has lower energy than the other, and the products derived from this transition state form more rapidly.

The ideas we have just discussed can be restated in a more general way: *Enantiomers differ in their chemical or physical behavior only when they interact with other chiral objects or forces.*

An important example of this idea is the fact that a pair of enantiomers have different optical activities. If, as we have just learned, enantiomers differ in their behavior only when they interact with chiral forces, then it follows that some aspect of plane-polarized light must be chiral. In fact, plane-polarized light is actually a mixture of two light forms, called respectively *right and left circularly polarized light* (Fig. 7.13). The electric fields of these light forms propagate through space respectively as right- and left-handed helices. Since a helix is chiral, these two forms of light are a pair of enantiomers! It is the vector addition of these enantiomeric light forms that gives plane-polarized light; plane-polarized light can therefore be thought of as "racemic" light. Let us call the circularly polarized light components *R* and *S*. When these two light forms pass through a pure enantiomer with, say, the *S* configuration, the interaction of the molecules with each light form in turn creates *S,R* and *S,S* interactions, respectively. Because these are diastereomeric interactions, they occur to different extents: one is stronger than the other. Although we shall not be concerned with precisely how this interaction takes place, it is the difference between these two interactions that gives rise to optical activity. Since an achiral compound has the same interaction with right and left circularly polarized light, such a compound has no optical activity.

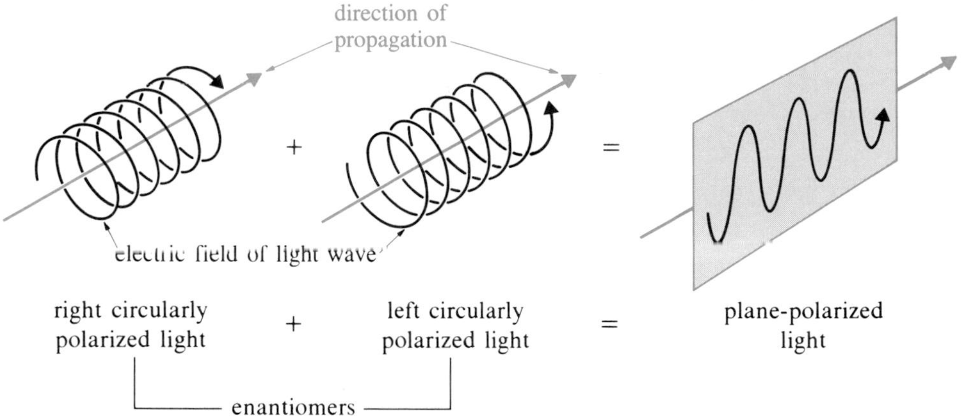

Figure 7.13 The electric fields of right and left circularly polarized light propagate through space as helices, and the two add vectorially to give plane-polarized light.

Problems

18 Assuming equal strength in both hands, would your right and left hands differ in their ability to drive a nail? To drive a screw with a screwdriver?

19 Imagine that a certain Mr. D. has been visited by a certain Mr. L. from elsewhere in the universe. Mr. D. and Mr. L. are alike in every way, except that they are nonsuperimposable mirror images! You have to introduce each of them at an international press conference, but neither of them will agree to give their names. How would you tell them apart? (There may be several ways.)

D. Reactions of Diastereomers

How do diastereomers differ in their chemical reactivity? The answer to this question is easy. Since a pair of diastereomers have different properties, they behave differently toward any reagent, whether it is chiral or achiral. Thus the following two diastereomeric alkenes would be expected to differ in their reactivities with *any* reagent, chiral or achiral, and consequently, the products from each would be formed at different rates.

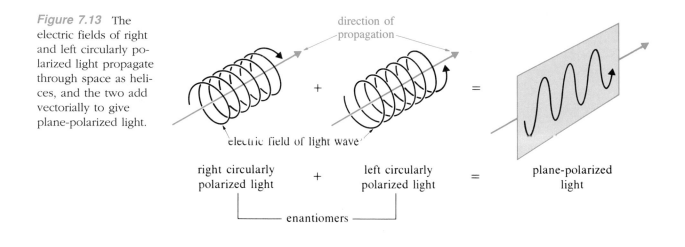

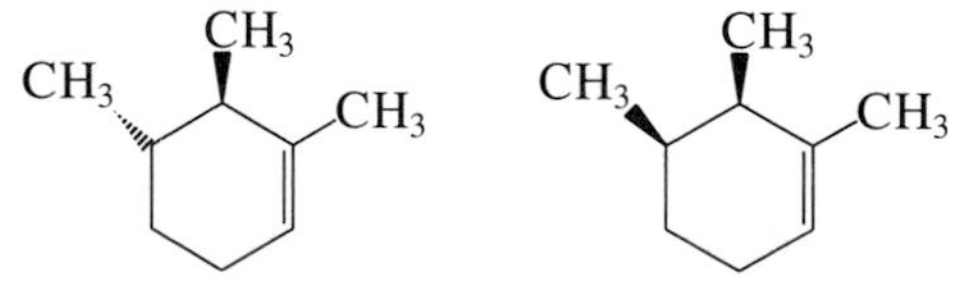

We may not be able to predict *which* alkene is more reactive; the point is simply that the two alkenes will not have the same reactivity.

Problem

20 Write all the possible products that might form when racemic 3-methylcyclohexene reacts with Br_2. What is the relationship of each pair? Which compounds should in principle be formed in the same amounts, and which in different amounts? Explain.

7.9 CHIRALITY IN NATURE

When a chiral compound occurs in nature, typically only one of its two enantiomers is found. That is, *nature is a source of optically active compounds.* For example, the sugar glucose occurs only as the dextrorotatory form shown below; the naturally occurring amino acid leucine is the levorotatory enantiomer.

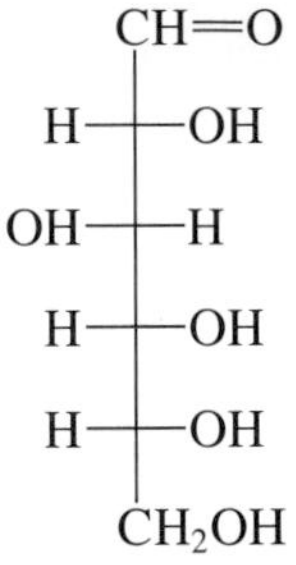

(+)-glucose

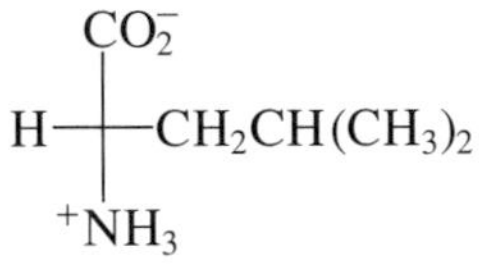

(−)-leucine

Many scientists believe that eons ago the first chiral compounds were formed from simple, achiral starting materials such as methane, water, and HCN. This theory presents a problem. As we learned in Sec. 7.8A, reactions that give chiral products from achiral starting materials always give the racemate; net optical activity cannot be generated in the reactions of achiral molecules. If these biological starting materials are all achiral, why do we have a world full of optically active compounds? We should in principle have instead a world full of racemates! (This means that somewhere in the world your nonsuperimposable mirror image should exist!) The only way out of this dilemma is to postulate that at some point in geologic time an **optical resolution** (separation of enantiomers) must have occurred. How could this have happened?

There has been much speculation about the answer to this question, some of it with metaphysical overtones. However, many people believe that the first optical resolution occurred purely by chance. Although a spontaneous optical resolution is *highly improbable,* it is not strictly *impossible.* In fact, such chance optical resolutions have been known to occur in the laboratory. It has been observed, for example, that one enantiomer of Pasteur's sodium ammonium tartrate sometimes crystallizes from a solution of the racemate, particularly if the solution is seeded by a crystal of the enantiomer. Perhaps the spontaneous crystallization of a pure enantiomer took place on the prebiotic earth, seeded by a speck of dust with just the right shape. The question is an intriguing one, and no one really knows the answer.

Given that one or more optical resolutions occurred by chance at some time during the course of natural history, we might ask how nature continues to manufacture one enantiomer and not the other of a given natural product. Consider, for example, the conversion of fumarate and water, both achiral reagents, into malate, a chiral compound (Sec. 4.8).

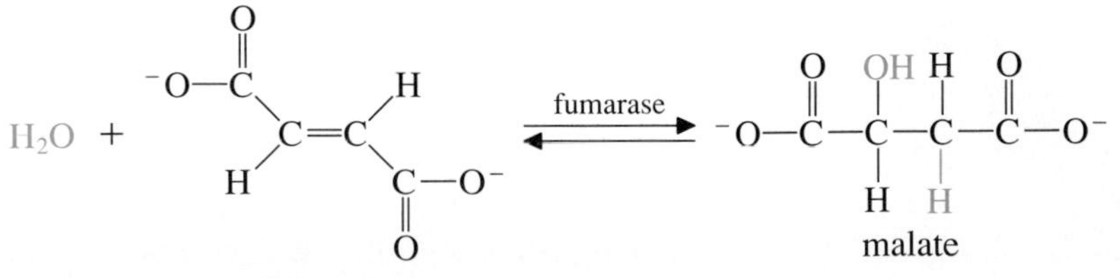

(7.20)

If we carry out this reaction in the laboratory at 175°, we find, as we expect from Sec. 7.8A, that the malate produced in this reaction is the racemate. In biological systems, however, this reaction is catalyzed at much lower temperature by the enzyme fumarase (Sec. 4.8), and the product malate is the optically pure *S* enantiomer. This does not violate any of our principles because *all enzymes are optically pure chiral compounds.* Water and fumarate react while they are in contact with the chiral catalyst fumarase; the chiral transition state leading to (*S*)-malate "fits" the chiral enzyme better than the transition state leading to (*R*)-malate. In other words, the two transition states, which are enantiomers in the *absence* of enzyme, are diastereomers in the *presence* of the chiral enzyme because *the enzyme itself is part of the transition state.*

enzyme + H_2O + fumarate → [(*R*)-transition state + chiral enzyme]

diastereomers (7.21)

enzyme + H_2O + fumarate → [(*S*)-transition state + chiral enzyme] → (*S*)-malate + enzyme

Because the two transition states are diastereomers, they differ in energy. It happens that the transition state leading to (*S*)-malate has lower energy, and therefore (*S*)-malate is formed more rapidly—so much more rapidly, in fact, that (*R*)-malate is not formed at all. Thus, enzymes not only catalyze biological reactions; they also catalyze *selectively* the formation of specific enantiomers. We shall learn more about enzymes in Chapter 26.

7.10 STEREOCHEMISTRY OF ALKENE ADDITION REACTIONS

At this point it may seem that stereochemistry adds a complicated new dimension to the study and practice of chemistry. To some extent this is true. No chemical structure is complete without stereochemical detail, and no chemical synthesis can be planned without considering problems of stereochemistry that might arise. In this section we shall reexamine some of the alkene addition reactions we have studied, with particular attention to their stereochemistry. We shall also learn how stereochemistry can be used as a tool to study reaction mechanisms.

A. Syn *and* Anti *Addition. Stereochemistry of Hydroboration*

The addition of any reagent to an alkene π-bond can occur in two stereochemically different ways. To illustrate, let us consider the addition of borane, or any of its substituted derivatives (which we shall represent as >B—H) to 1-methyl-1-cyclohexene. The first type of addition that can occur is called **syn addition.** In a *syn* addition *both groups add to the double bond from the same face;* the groups added in a *syn* addition to a cycloalkene have a *cis* relationship in the product. We can imagine *syn* addition occurring from either face of the alkene, as follows:

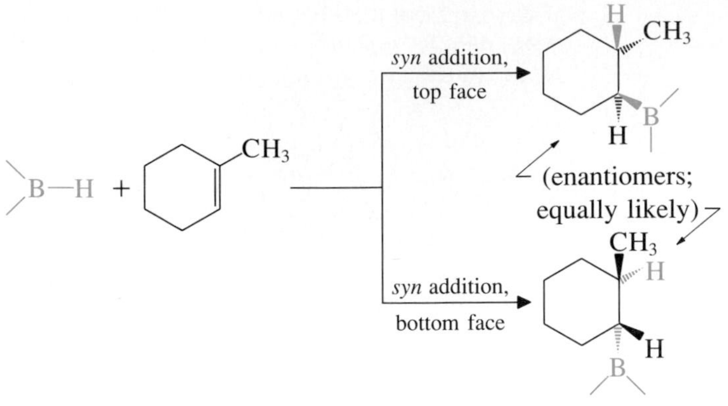

(7.22)

Since the product of addition from one face is the enantiomer of the product of addition from the other, addition from either face is equally likely (Sec. 7.8A).

The second type of addition that could occur is called ***anti* addition.** In *anti* addition, *the two groups add to the double bond from opposite faces*; in *anti* addition to a cycloalkene, these groups have a *trans* relationship in the product. As in *syn* addition, there are two enantiomeric, equally likely ways in which *anti* addition can occur.

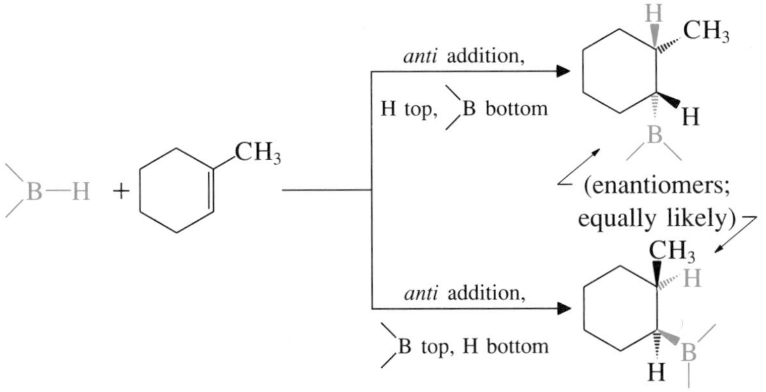

(7.23)

The products of *syn* addition are diastereomers of the products of *anti* addition. Therefore, by Sec. 7.8B, we expect different amounts of *syn* and *anti* addition. What are the experimental results? Which set of diastereomers predominates, and why? It is found experimentally that *hydroboration is a syn addition*. Thus, Eq. 7.22 is observed; Eq. 7.23 is *not* observed.

The stereochemistry of addition is determined by the *mechanism* of the reaction. In Sec. 5.3A we learned that the addition of borane to an alkene double bond is a one-step pericyclic reaction. We can see that the stereochemistry of the reaction is consistent with this mode of addition. It would be very difficult for an *anti* addition to occur in a single step; it would require an abnormally long B—H bond length to bridge opposite faces of the alkene π-bond.

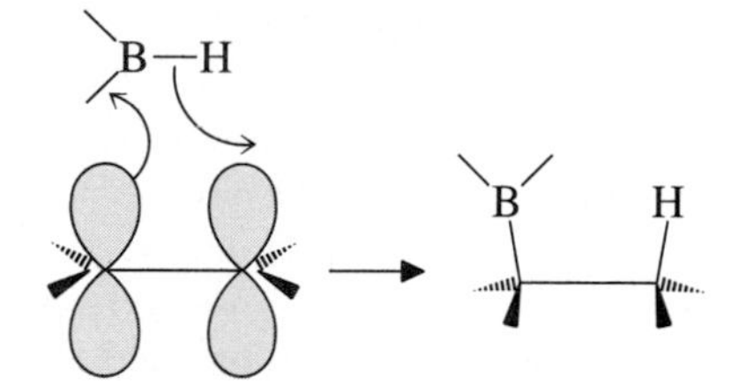

pericyclic
syn addition

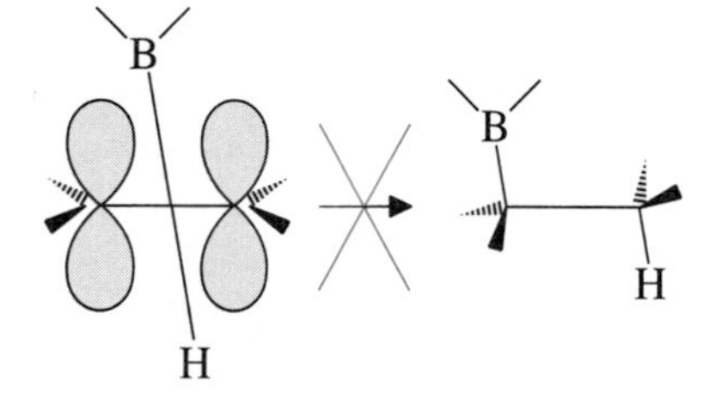

pericyclic *anti*
addition; requires
an unrealistic B—H
bond length

B. Reactions at Asymmetric Carbons. Oxidation of Organoboranes

We learned in Sec. 5.3A that when boranes are treated with alkaline H_2O_2, they are converted into alcohols. This reaction presents a new stereochemical question. In this reaction *a bond is broken, and a new bond is formed, at an asymmetric carbon.* From a stereochemical viewpoint, the reaction might take place in two distinct ways. In the first way, the —OH could directly replace the boron with no change in stereochemistry. In the following example, since boron is *trans* to the methyl group, the —OH would also be *trans*. In this case, we say that the reaction has occurred with **retention of configuration** at the asymmetric carbon.

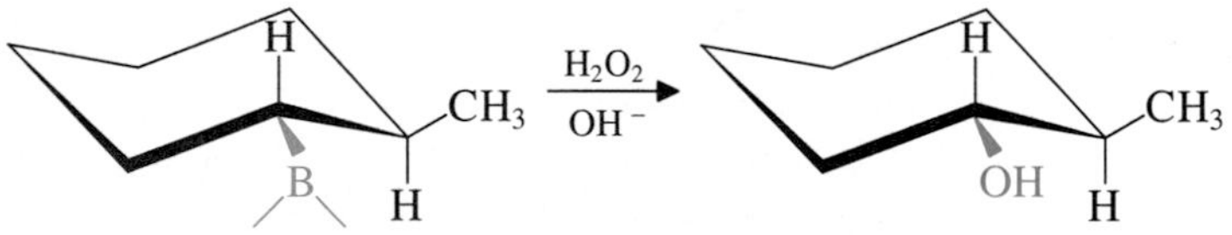

(7.24)

retention of configuration

The second way that the reaction could occur is that the —OH could replace the boron with a change in stereochemistry. This means that if the boron is *trans* to the methyl group, the —OH would be *cis*. In this case, we say that the reaction has occurred with **inversion of configuration** at the asymmetric carbon.

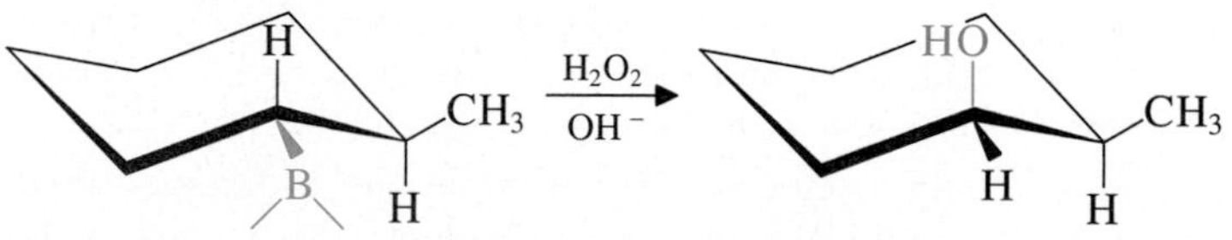

(7.25)

inversion of configuration

Of course, it is also possible that a mixture of products from both retention and inversion could be observed. Such a result is called **loss of configuration** at the asymmetric carbon.

Let us once again turn to experiment for the result. It is found that replacement of the boron by the —OH group occurs with *retention of configuration* as shown in Eq. 7.24.

Although the cleavage of organoboranes occurs with retention of configuration, we shall find that other reactions at asymmetric carbon occur with inversion of configuration, and still others are accompanied by loss of configuration. It is the *mechanism* of a reaction that determines its stereochemical outcome. Although we shall not be concerned here with the mechanism of organoborane cleavage, we shall shortly study other reactions in which the stereochemistry at an asymmetric carbon plays a crucial role in understanding their reaction mechanisms.

If we combine the results of this section with those of Sec. 7.10A, we see that *hydroboration–oxidation of an alkene brings about the net syn addition of the elements of H—OH to the double bond.*

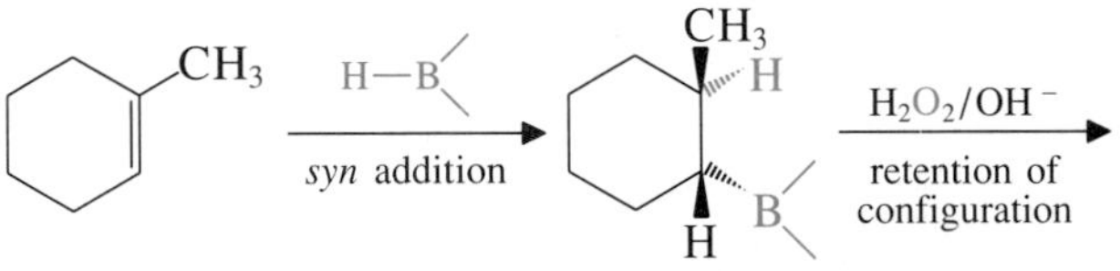

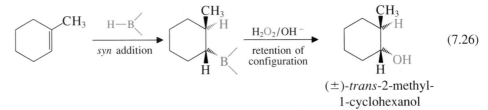

(7.26)

(±)-*trans*-2-methyl-1-cyclohexanol

As far as is known, *all* hydroboration–oxidation reactions of alkenes are *syn* additions.

Notice carefully that it is the —H and —OH that have added in a *syn* manner. The *trans* designation in the name of the product of Eq. 7.26 has nothing to do with the groups that have added—it refers to the relationship of the methyl group, which was part of the alkene starting material, and the —OH group.

A reaction that gives a large predominance of one stereoisomer over another is called a **stereoselective reaction.** Thus, hydroboration–oxidation is our first example of a stereoselective reaction; it is a stereoselective *syn* addition. Recall (Sec. 5.3A) that hydroboration–oxidation is also a *regioselective* reaction: the —OH group ends up on the less substituted carbon of the double bond. Reactions such as hydroboration that are both regioselective and stereoselective are very important to the chemist involved in preparing new compounds. These reactions can be used to introduce functional groups (such as the —OH group in this example) in structurally and stereochemically precise ways. Such reactions are not only efficient in providing pure compounds, but also allow the chemist to avoid time-wasting and tedious purifications to remove unwanted isomers.

We should make one further observation about reaction stereochemistry: sometimes the stereochemistry does not matter. When *syn* and *anti* addition to an alkene give the same products, then we do not have to be concerned whether a reaction is stereoselective. For example, the addition of anything to isobutylene, $(CH_3)_2C{=}CH_2$, would give the same product whether the addition is *syn* or *anti.* We need to be concerned about the stereoselectivity of a reaction only when it can give stereoisomers—as in Eq. 7.26.

C. Use of Stereochemistry to Investigate Reaction Mechanism. Stereochemistry of Alkene Bromination

The bromination of alkenes (Sec. 5.1A) is in many cases also a stereoselective reaction. Let us study the bromination of *cis*- and *trans*-2-butene with two goals in mind. First, we can use this reaction to illustrate certain aspects of reaction stereochemistry on noncyclic alkenes. Second, we shall see how the stereochemistry of a reaction can be used to understand its mechanism.

As we have already learned, when *cis*-2-butene reacts with Br_2, the product is 2,3-dibromobutane.

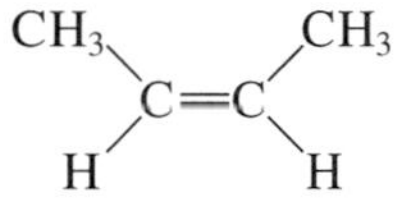

$+ Br_2 \longrightarrow CH_3—CHBr—CHBr—CH_3$ (7.27)

2,3-dibromobutane

We now know that several stereoisomers of this product are possible: a pair of enantiomers and the *meso* compound. In general, we expect the *meso* compound and the enantiomeric pair to be formed in different amounts (Sec. 7.8B). If the enantiomers are formed, we expect them to be present as the racemate because the starting materials are achiral (Sec. 7.8A).

If we carry out the bromination of *cis*-2-butene in the laboratory, we find that the racemate makes up virtually all (99%) of the product mixture. If we carry out the bromination of the *trans*-alkene, we find, in contrast, that the *meso* compound is the major stereoisomer formed. In summary,

experimental facts: $CH_3—CH{=}CH—CH_3 \xrightarrow{Br_2} CH_3—CHBr—CHBr—CH_3$ (7.28)

cis ⟶ racemate

trans ⟶ *meso*

This information tells us that addition of bromine to either *cis*- or *trans*-2-butene is an *anti* addition. Let us be sure we understand this point by analyzing the addition to *cis*-2-butene in detail.

If bromination were a *syn* addition, the Br_2 could add to either face of the double bond. (In the following structures we are viewing the alkene edge-on.) Addition from either direction gives the *meso* diastereomer.

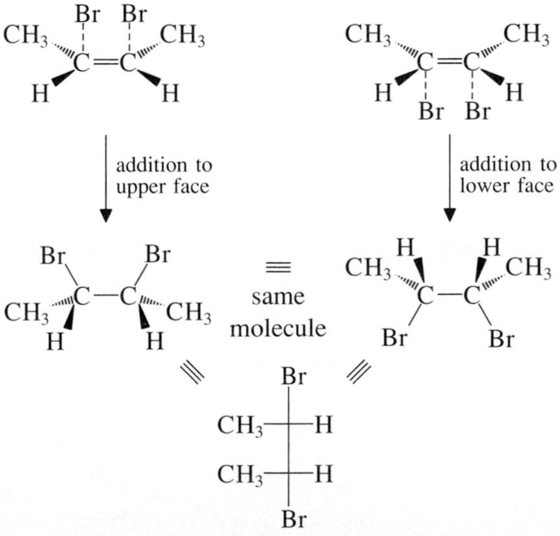

(7.29)

meso-2,3-dibromobutane

Since the experimental facts (Eq. 7.28) show that *cis*-2-butene does *not* give the *meso* isomer, it follows that the two bromine atoms *cannot* be adding from the same face of the molecule. Therefore *syn* addition does not occur.

Let us now look at the *anti* addition of the two bromines to *cis*-2-butene. This addition, too, can occur in two equally probable ways. Each mode of addition gives the enantiomer of the other; that is, the two modes of *anti* addition operating at the same time should give the racemate.

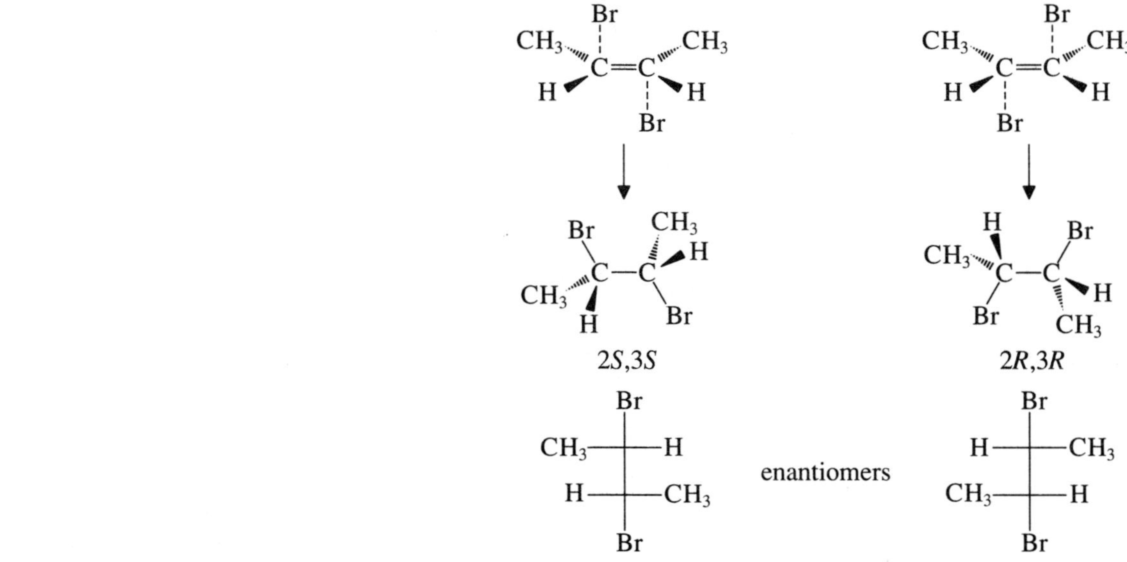

(7.30)

(±)-2,3-dibromobutane

The experimental facts of Eq. 7.28 show that bromination of *cis*-2-butene indeed gives the racemate; therefore, bromination of *cis*-2-butene is an *anti* addition.

Problem

21 Show by a similar analysis that *syn* addition of Br_2 to *trans*-2-butene should yield the racemate and *anti* addition should yield the *meso* compound. What mode of addition is actually observed?

By a similar analysis (Problem 21), we can show that the *anti* addition of Br_2 to *trans*-2-butene should yield the *meso* stereoisomer of 2,3-dibromobutane, and the experimental results (Eq. 7.28) indicate that this is the observed mode of addition. In fact, the bromination of most simple alkenes occurs with *anti* stereochemistry. Bromination is therefore another example of a *stereoselective reaction*.

The stereochemistry of bromination is one of the main reasons that the bromonium-ion mechanism, shown in Eqs. 5.3–5.6, was postulated. Let us see how this mechanism can account for the observed results. First, the bromonium ion can form at either face of the alkene. (Attack at one face is shown in the following equation; you should show attack at the other face and take your structures through the subsequent discussion.)

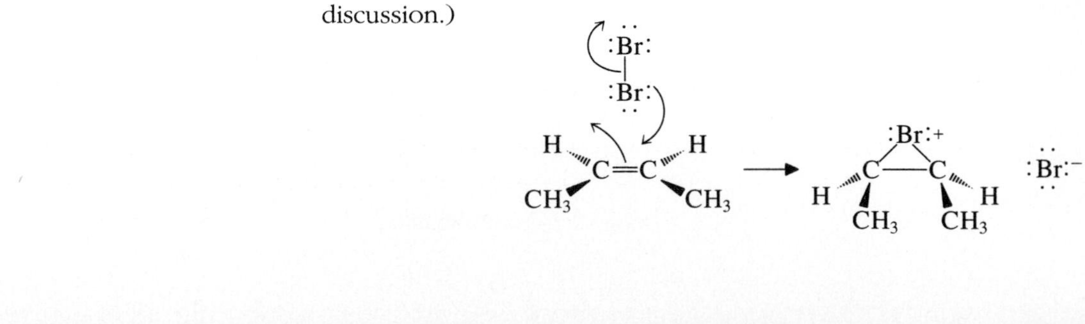

(7.31)

If the bromide ion then attacks the carbon at the *opposite* face of the molecule, the bromonium ion is opened to give the observed product. Notice that the attack of bromide ion occurs with *inversion of configuration* (Sec. 7.10B). As this reaction takes place, the methyl and the hydrogen move up (colored arrows) to maintain the tetrahedral configuration of carbon. Attack of the bromide ion at one carbon yields one enantiomer; attack at the other carbon yields the other enantiomer.

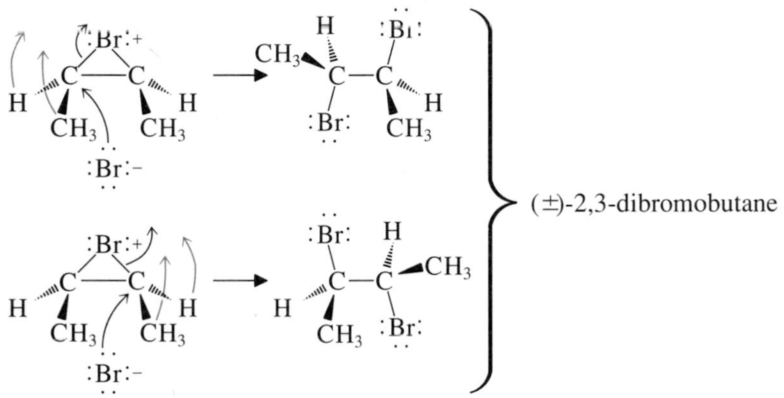

(7.32)

Thus, formation of a bromonium ion followed by *backside attack* of bromide is a mechanism that accounts for the observed *anti* addition of Br_2 to alkenes. We shall learn in Sec. 9.3B that when nucleophiles attack a saturated carbon atom, backside attack is always observed.

Might there be other mechanisms that could predict the *anti* stereochemistry of bromine addition? Let us see what sort of prediction a carbocation mechanism makes about the stereochemistry of the reaction.

Imagine the addition of Br_2 to either face of *cis*-2-butene to give a carbocation intermediate.

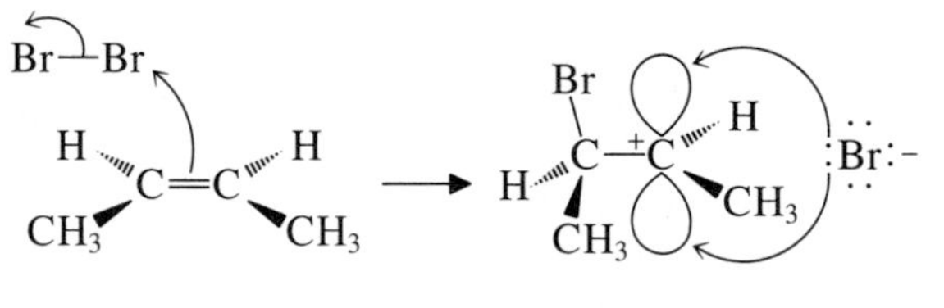
(7.33a)

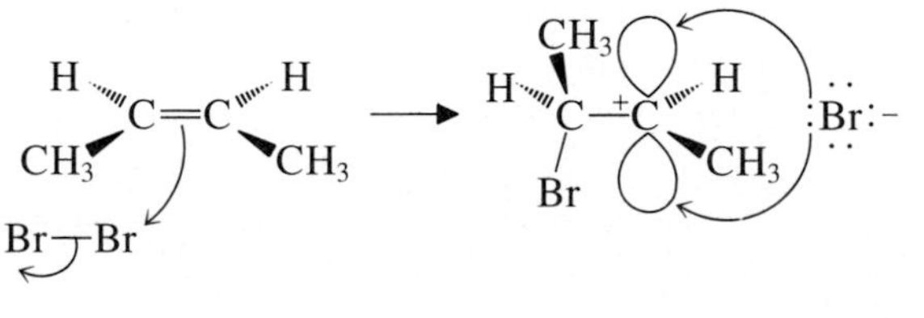
(7.33b)

Attack of Br^- could occur at either face of this carbocation; this would give a mixture of *meso* and racemic diastereomers. Yet such a mixture is not observed (Eq. 7.28). Hence, this carbocation mechanism is not in accord with the data. This mechanism also is not in accord with the observed absence of rearrangements when the reaction is run on alkenes that are known to give rearrangements in other reactions involving carbocations. The bromonium-ion mechanism, however, accounts for the results in a direct and simple way. Adding to the credibility of this mechanism is the fact that bromonium ions have been directly observed under special conditions and, in 1985, the structure of a bromonium ion was proved by X-ray crystallography.

Does the observation of *anti* stereochemistry *prove* the bromonium-ion mechanism? The answer is no. *No mechanism is ever proved.* Chemists deduce a mechanism by gathering as much information as possible about a reaction, such as its stereochemistry, presence and absence of rearrangements, etc., and ruling out all mechanisms that do not fit the experimental facts. If someone can think of another mechanism that explains the facts, then that mechanism is just as good until someone finds a way to decide between the two by a new experiment.

Problem

22 (a) Assuming the operation of the bromonium-ion mechanism, give the structure of the product(s) (including all stereoisomers) expected from the bromination of cyclohexene.

(b) In view of the brominium-ion mechanism, which of the products in your answer to Problem 20 are likely to be the major ones?

D. Stereochemistry of Other Addition Reactions

Let us now consider some of the other stereoselective alkene addition reactions we have studied.

Catalytic hydrogenation of most alkenes (Sec. 5.7A) is a *syn* addition. The following example is illustrative:

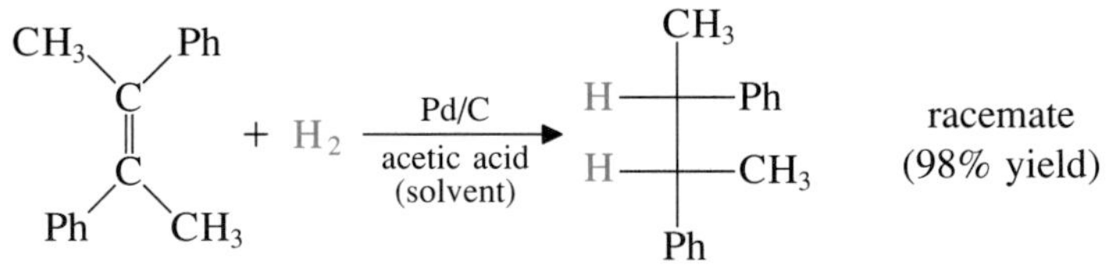

(7.34a)

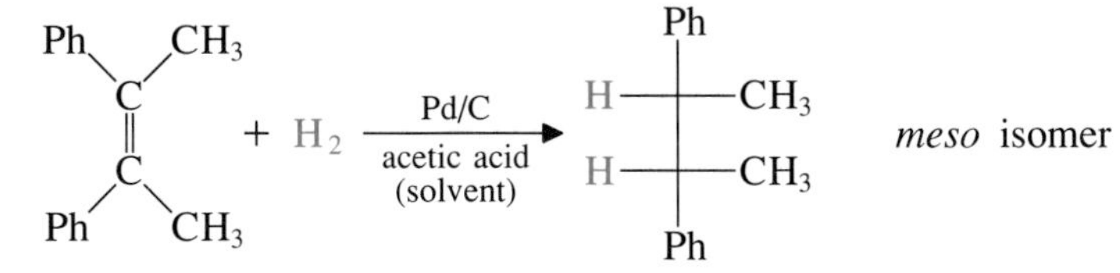

(7.34b)

Results like this show that the two hydrogen atoms are delivered from the catalyst to the same face of the alkene double bond. (Be sure to analyze this case to see why this is so.) Its stereoselectivity is another reason that hydrogenation is such an important reaction in organic chemistry.

Oxymercuration of alkenes (Sec. 5.3B) is typically a stereoselective *anti* addition.

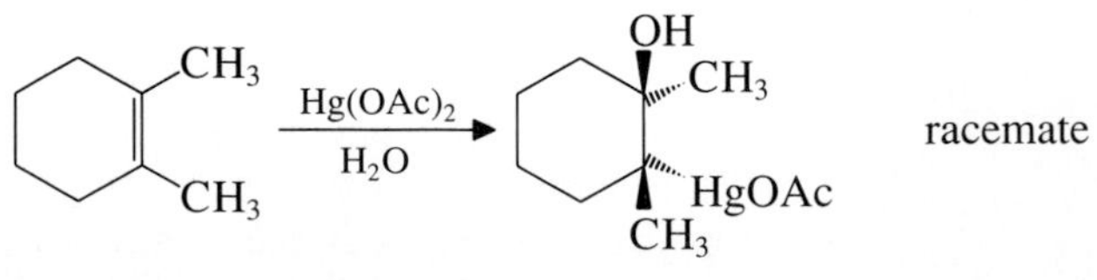

(7.35)

In the reduction of the mercury-containing product with $NaBH_4$, however, the mercury is replaced by hydrogen with *loss of stereochemical configuration.*

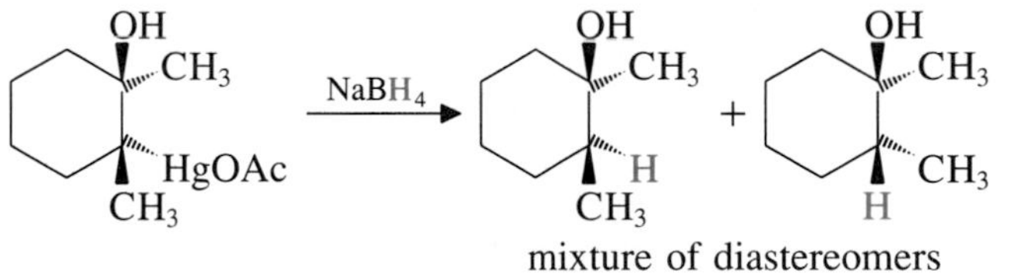

(7.36)

Although oxymercuration–reduction gives a mixture of diastereomers in cases like this, there are many alkenes for which both *syn* and *anti* addition give the same product. Even though oxymercuration–reduction is not *stereoselective,* it is nevertheless a useful reaction because it is *regioselective:* it gives an overall Markovnikov addition of H—OH across the double bond.

Problem

23 Show that the stereoselectivity of oxymercuration is in accord with the mercurinium-ion mechanism for this reaction given in Sec. 5.3B.

The formation of glycols from alkenes with either alkaline $KMnO_4$ or OsO_4 is a stereoselective *syn* addition.

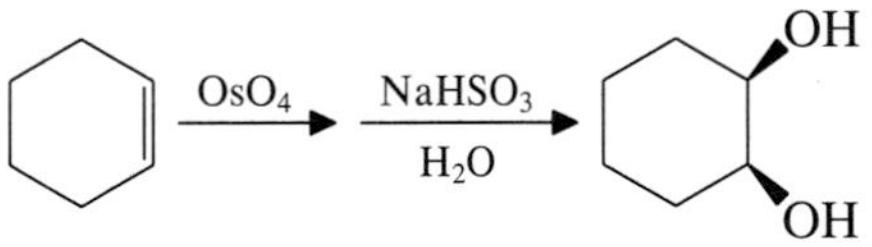

(7.37)

It is easy to see why this should be so from the mechanism of the addition, which we learned is a pericyclic reaction (Eq. 5.31a). The five-membered osmate ester ring is easily formed when the two oxygens are added from the same face of the double bond.

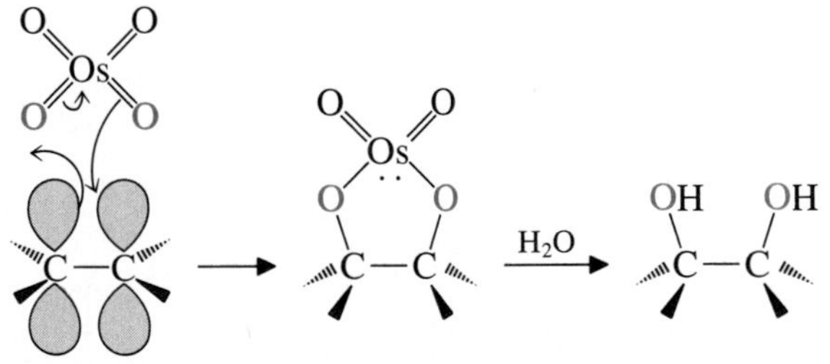

(7.38)

On the other hand, an *anti* addition by this mechanism would be very difficult: the two oxygens cannot simultaneously bridge the upper and lower faces of the π-bond. *Syn* addition of the two —OH groups with $KMnO_4$ occurs for similar reasons.

Problem

24 What products do you expect (including their stereochemistry) when alkaline $KMnO_4$ reacts with each of the following alkenes:
(a) 1-methyl-1-cyclopentene (c) *trans*-2-butene
(b) *cis*-2-butene
Contrast these results with those from bromination of the same alkenes, and explain the difference in stereochemistry.

KEY IDEAS IN CHAPTER 7

- Except for cyclopropane, the cycloalkanes have nonplanar carbon skeletons. The nonplanar conformations of the cycloalkanes avoid eclipsed-hydrogen interactions and angle strain that would be present in the planar forms.
- The most stable conformation of cyclohexane is the chair conformation. In this conformation, the hydrogens or substituent groups can assume axial and equatorial positions. Cyclohexane undergoes a chair–chair interconversion in which equatorial groups become axial, and vice versa. In some substituted cyclohexanes, this conformational change interconverts enantiomeric conformations; in others, diastereomeric conformations; and in others, identical conformations.
- Cyclohexane conformations with axial substituents are less stable than those with equatorial substituents because groups in axial positions have unfavorable steric (*gauche*) interactions with the ring three carbons away.
- Bicyclic compounds contain two rings that share two common atoms. If the common atoms are adjacent, the compound is a fused bicyclic compound; if the common atoms are not adjacent, the compound is a bridged bicyclic compound. Both *cis* or *trans* ring fusion are possible. *Trans* fusion, which avoids *gauche* interactions, is the most stable way to connect larger rings; *cis* fusion, which minimizes angle strain, is the most stable way to connect smaller rings. Polycyclic compounds contain many fused or bridged rings.
- *Trans* cycloalkenes containing rings with fewer than eight members are too unstable to exist. The instability of bridged bicyclic compounds containing bridgehead double bonds (Bredt's rule) is attributable to the fact that these compounds incorporate a *trans*-cycloalkene.
- Certain fundamental principles govern the outcome of reactions that give stereoisomeric products.

    1. Chiral products that result from chemical reactions of achiral starting materials are always formed as their racemates; that is, optical activity is not spontaneously generated.
    2. Diastereomeric products of chemical reactions are in general formed at different rates and in unequal amounts.
    3. A pair of enantiomers react at the same rates with achiral reagents, but at different rates with chiral reagents.
    4. Diastereomers react at different rates with any reagent.

- Additions to alkenes can occur as *syn* additions (glycol formation, hydroboration, catalytic hydrogenation), *anti* additions (bromination), or as a mixture of the two.
- Reactions that break chemical bonds to asymmetric carbons can occur with inversion of configuration (as attack of bromide ion on the bromonium ion in bromination), retention of configuration (as in the oxidation following hydroboration), or loss of configuration, which is a mixture of retention and inversion (as in the reduction following oxymercuration).

ADDITIONAL PROBLEMS

25 Draw the structures of
(a) two bicyclic alkanes with six carbon atoms.
(b) (S)-4-cyclobutyl-1-cyclohexene.
Name the compounds you drew in (a).

26 Which of the following would distinguish (in principle) between methylcyclohexane and (E)-4-methyl-2-hexene?
(a) molecular weight determination
(b) uptake of H_2 in the presence of a catalyst
(c) treatment with alkaline $KMnO_4$
(d) determination of the empirical formula
(e) determination of the heat of formation
(f) optical resolution (separation into enantiomers)
(g) bromination with Br_2 in CCl_4

27 Chlorocyclohexane contains 2.07 times as much of the equatorial form as the axial form at equilibrium. What is the standard free-energy difference between the two forms? Which is more stable?

28 Which of the following compounds can in principle be resolved into enantiomers at room temperature? Explain.
(a) cis-1,3-dimethylcyclopentane
(b) trans-1,3-dimethylcyclopentane
(c) 1,2,2-trimethylcyclopropane
(d) trans-1,3-dimethylcyclobutane

29 For each of the following reactions, provide the following information:
(a) Give the structures of all products (including stereoisomers).
(b) If more than one product is formed, give the stereochemical relationship (if any) of each pair of products.
(c) If more than one product is formed, indicate which products are formed in identical amounts and which in different amounts.
(d) If more than one product is formed, indicate which products have different physical properties (melting point or boiling point).

(1) C_3H_7—C=CH_2 + HBr $\longrightarrow$
　　　　　|
　　　　C_2H_5

(2) 　C_2H_5CH—C=CH_2 + HBr $\longrightarrow$
　　　／　　|　　|
　　S　　CH_3　C_2H_5

(3) CH_3CH_2—C=CH_2 + HBr $\xrightarrow{\text{peroxides}}$
　　　　　　　|
　　　　　　CH_3

(4) CH_3—CH—CH=CH_2 + Br_2 $\xrightarrow{H_2O}$
　　　　　|
　　　　Ph ＼R

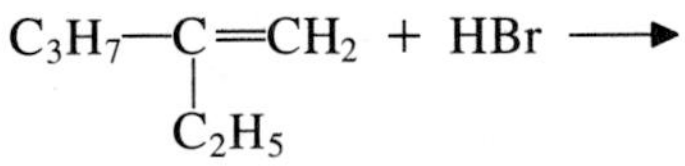

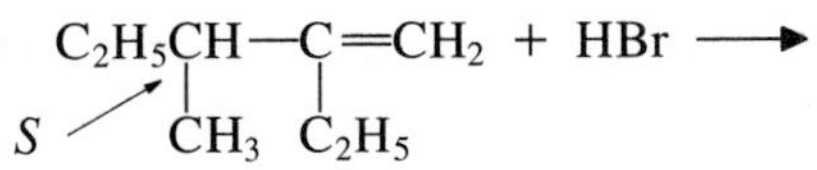

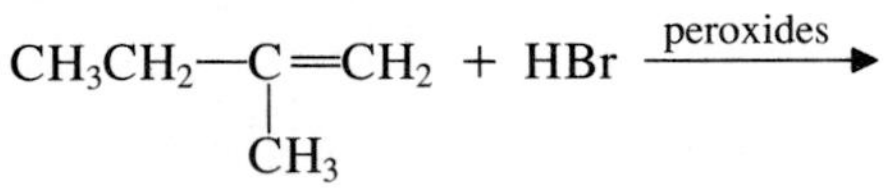

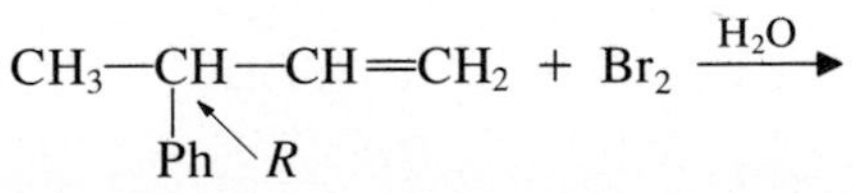

Problems (Cont.)

(5) 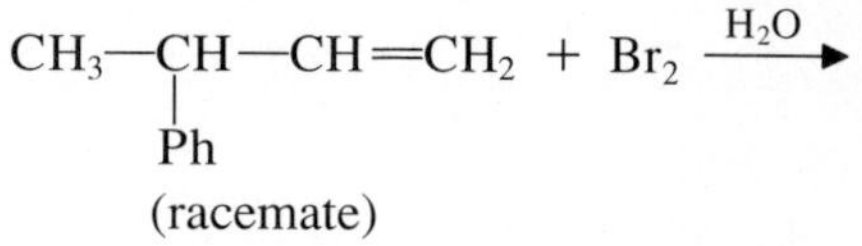

(6) 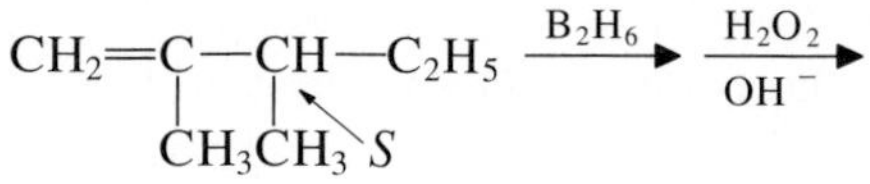

(7) 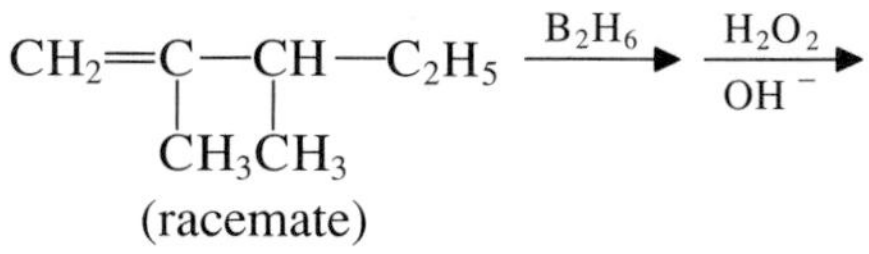

(8) 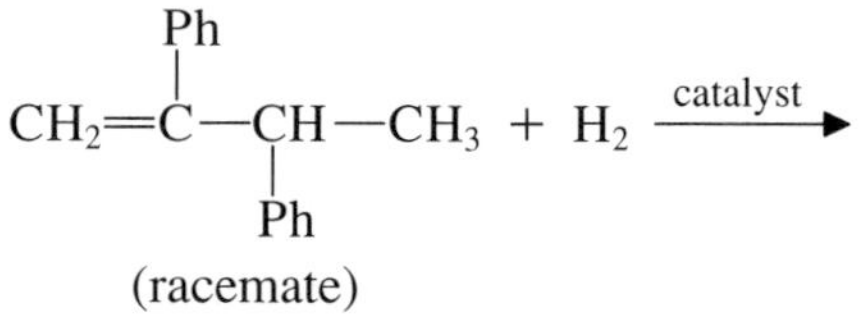

---

30 Which of the following alcohols can be synthesized free of structural isomers and diastereomers by hydroboration–oxidation of an alkene? Explain.

(a) 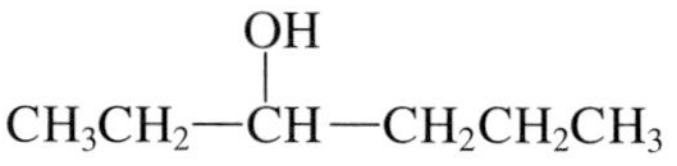

(b) 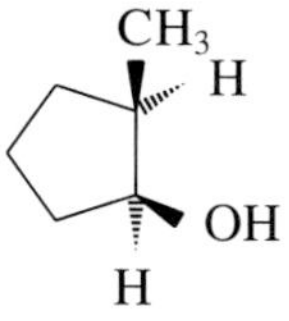

(c) 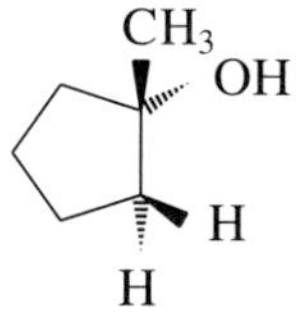

(d) 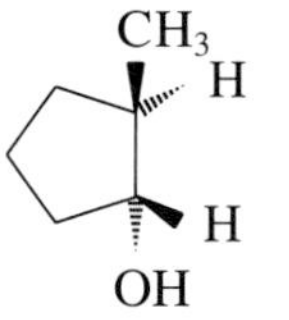

---

31 Draw the structures of

(a) a dimethylcyclohexane with two identical chair forms.

(b) a dimethylcyclohexane with two chair forms that are achiral diastereomers.

(c) a dimethylcyclohexane with chair forms that are a pair of enantiomers.

(There may be more than one answer for each question.)

---

32 (a) Draw the two chair forms of the sugar α-(+)-glucopyranose, one form of the sugar glucose. Which of these two forms is the major one at equilibrium?

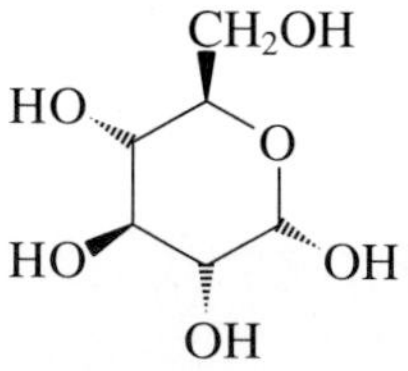

α-(+)-glucopyranose

(b) Draw a conformational representation of the following steroid. Show the α- and β-faces of the steroid, and label the angular methyl groups.

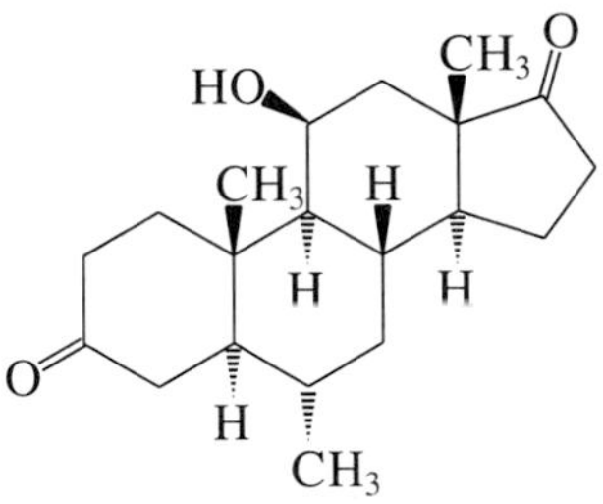

---

33 State whether you would expect each of the following properties to be identical or different for the two enantiomers of 2-pentanol. Explain.

$$CH_3—\underset{}{\overset{OH}{\overset{|}{C}}}H—CH_2CH_2CH_3 \qquad \text{2-pentanol}$$

(a) boiling point
(b) melting point
(c) rotation of the plane of polarized light
(d) solubility in hexane
(e) density
(f) solubility in (*S*)-3-methylhexane
(g) dipole moment
(h) taste (*Hint:* Your taste buds are chiral.)

---

34 From your knowledge of the mechanism of bromination, give the structure and stereochemistry of the product(s) when
(a) cyclopentene reacts with bromine in the presence of water. (*Hint:* See Sec. 5.1B.)
(b) (3*R*,5*R*)-3,5-dimethyl-1-cyclopentene reacts with bromine in the inert solvent CCl_4.

---

35 The bromination of the following bicyclic alkene in CCl_4 gives two separable dibromides. Suggest structures for each. (Remember that *trans*-decalin derivatives cannot undergo the chair–chair interconversion.)

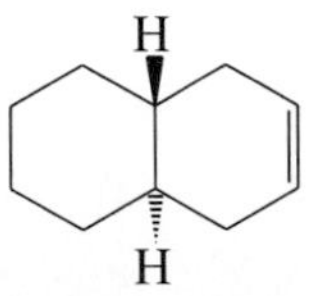

Problems (Cont.)

36 Draw a chair conformation for 3-methylpiperidine showing the orbital that contains the nitrogen unshared electron pair.

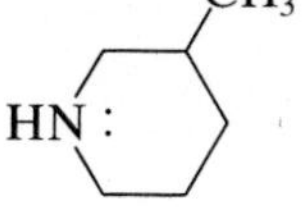

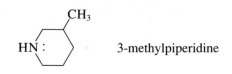

3-methylpiperidine

How many conformational isomers (chair forms only) of this compound are in rapid equilibrium? Explain. (*Hint:* See Sec. 6.8B.)

37 An alkene *A* of unknown stereochemistry was oxidized with alkaline KMnO$_4$ and the *meso* stereoisomer of the product was obtained:

$$CH_3(CH_2)_7CH=CH(CH_2)_7CH_3 \xrightarrow[OH^-]{KMnO_4} CH_3(CH_2)_7\overset{\overset{\displaystyle OH}{|}}{C}H-\overset{\overset{\displaystyle OH}{|}}{C}H(CH_2)_7CH_3$$

A *meso*

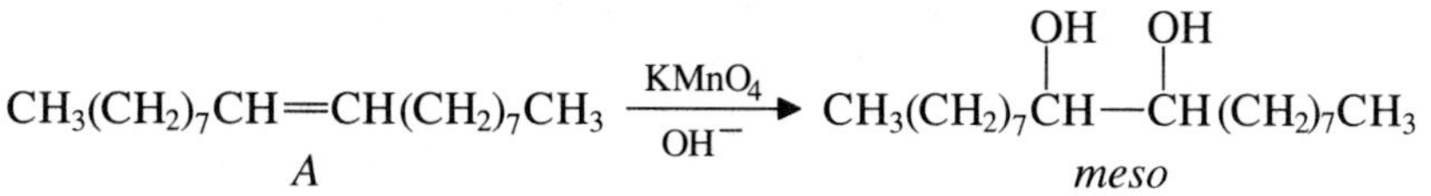

What stereoisomer of *A* was used in the reaction?

38 When fumarate reacts with D$_2$O in the presence of the enzyme *fumarase* (Sec. 4.8 and 7.9), only one stereoisomer of deuterated malate is formed.

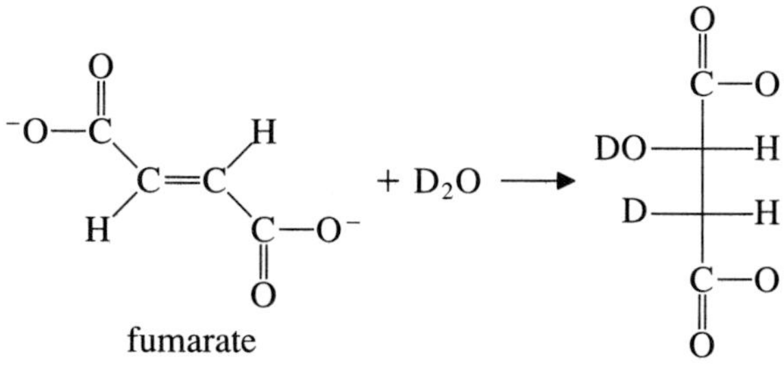

Is this a *syn* or an *anti* addition? Explain.

39 Give the structure of every stereoisomer of 1,2,3-trimethylcyclohexane; label the chiral pairs. Show a plane of symmetry in each of the achiral stereoisomers.

40 Which *one* of the following compounds would be most likely to exist with one of its cyclohexane rings in a boat conformation? Explain.

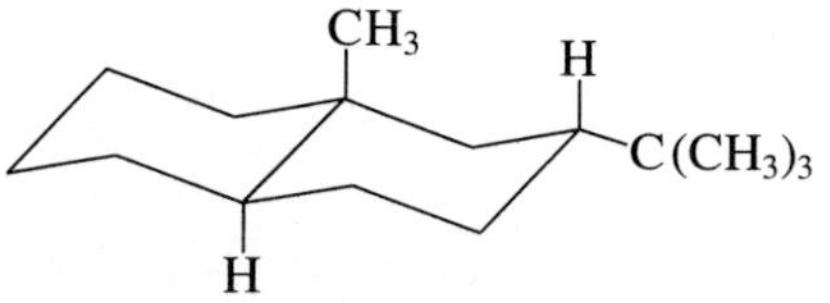

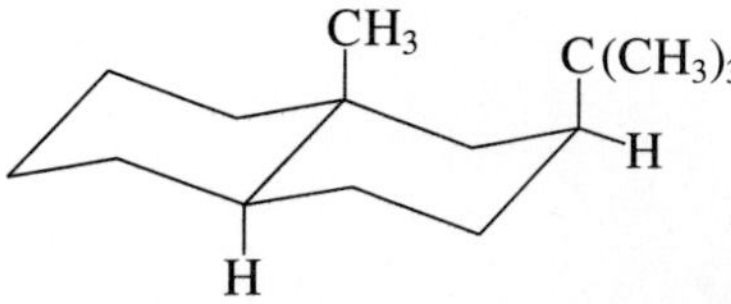

(c)

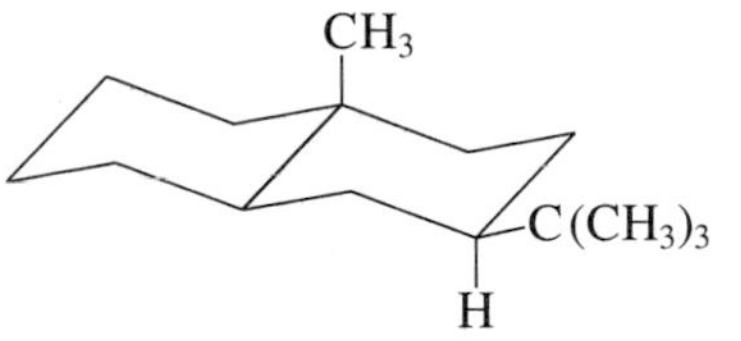

(d)

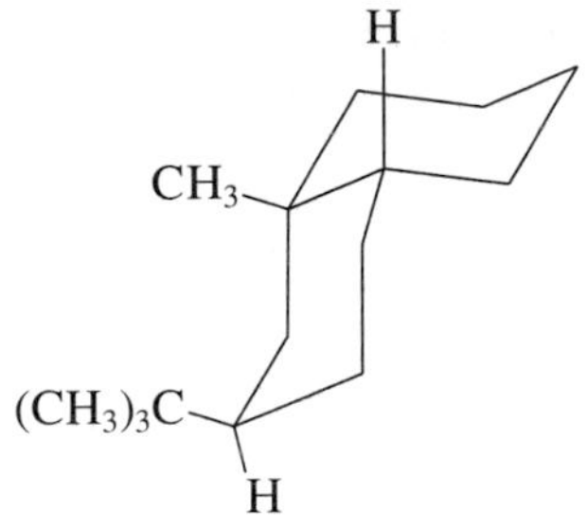

41 Which of the following statements about *cis*- and *trans*-decalin are true? Explain your answers.

(a) They are conformational isomers.
(b) They are structural isomers.
(c) They are diastereomers.
(d) At least one chemical bond would have to be broken in order to convert one into the other.
(e) They are enantiomers.
(f) They interconvert rapidly.

42 Which *one* of the following molecules can be resolved into enantiomers? Explain.

(a) $C_2H_5-\ddot{N}-C(CH_3)_3$ with $CH_2CH(CH_3)_2$ on N

(b)

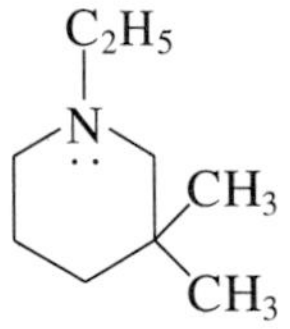

(c)

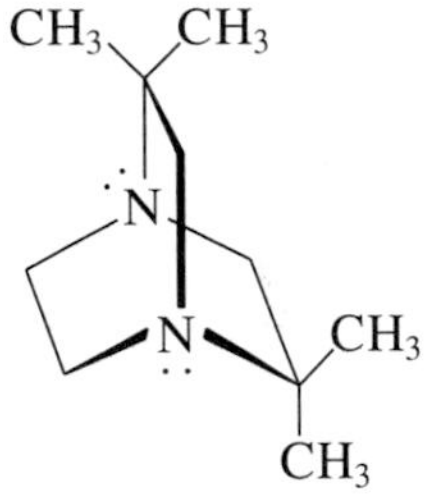

43 Indicate the relationship within each pair of structures below by using one of the following terms:

CE = conformational enantiomers
CD = conformational diastereomers
E = enantiomers, that are not conformational enantiomers
D = separable diastereomers
N = none of the above (explain)

(a)

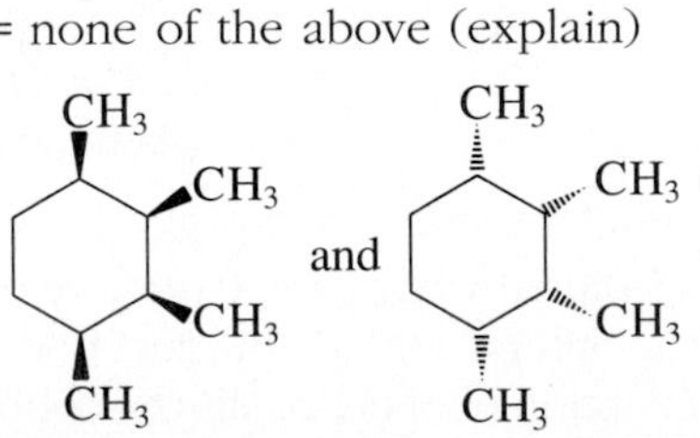

Problems (Cont.)

(b)

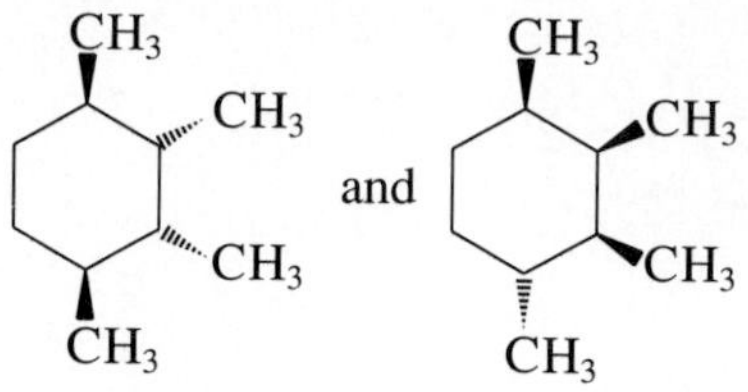

(c)

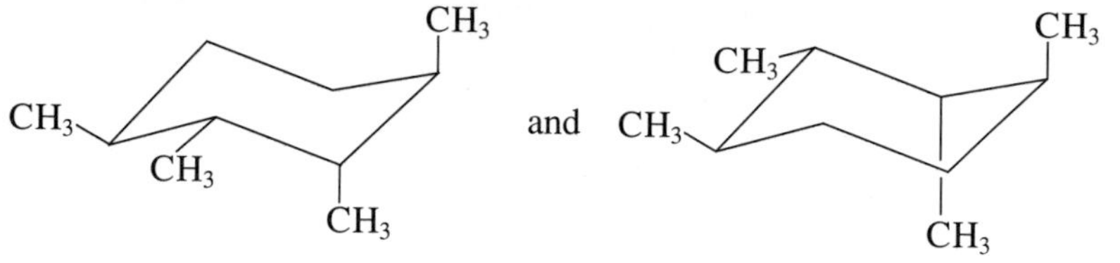

(d)

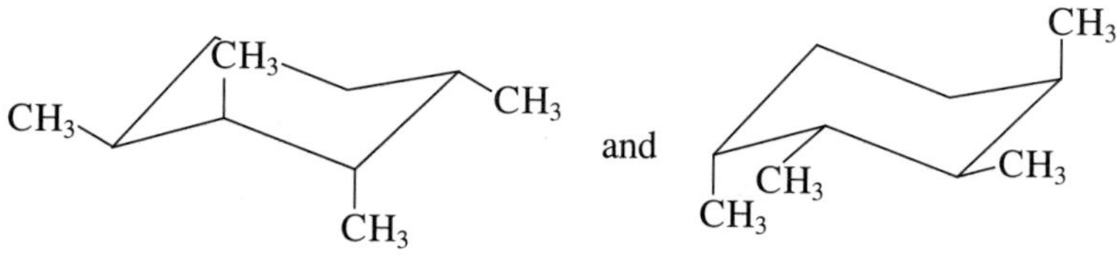

(e)

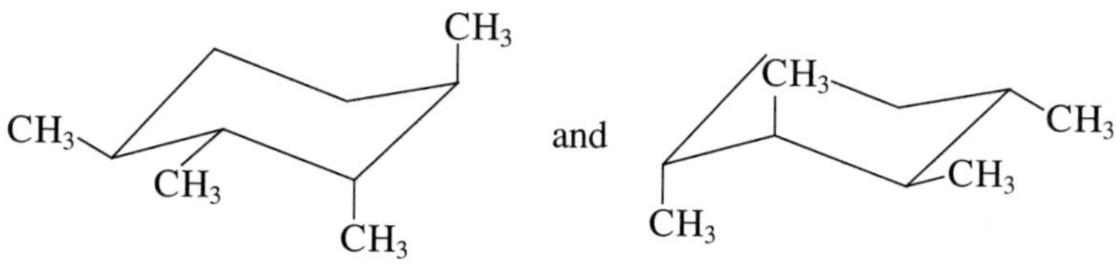

---

44 Explain why 1-methylaziridine undergoes amine inversion much more slowly than 1-methylpyrrolidine. (*Hint:* What are the hybridization and bond angles at nitrogen in the transition state for inversion?)

1-methylaziridine 1-methylpyrrolidine

---

45 When 1-methyl-1-cyclohexene is hydrated in D_2O, the product is a mixture of diastereomers; the hydration is thus *not* a stereoselective reaction. Show why the accepted mechanism for this reaction is consistent with these stereochemical results.

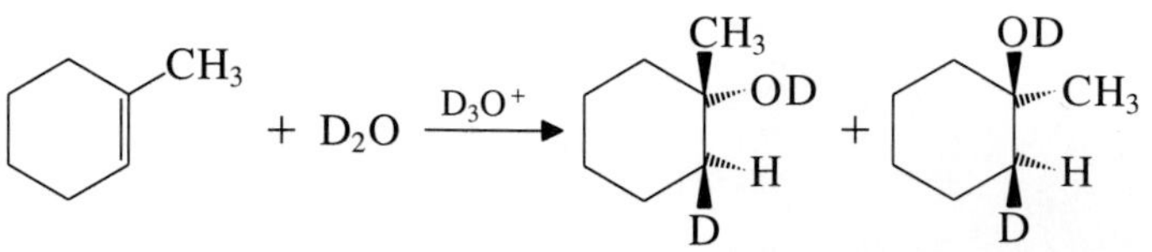

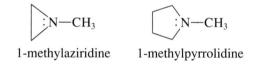

both compounds are racemates

---

46 When (2*R*,5*R*)-2,5-dimethylborolane (structure below) is used to hydroborate either *cis*- or *trans*-2-butene, and the product borane is treated with alkaline H_2O_2, mostly a single enantiomer of the product alcohol is formed. What is the absolute configuration of this alcohol? Explain why the other enantiomer is not formed.

(*Hint:* It will be easiest to work this problem if you build models of the borane and the alkenes. Let the borane model approach the alkene model from one face of the π-bond, then the other. Which reaction is preferred, and why? Then complete the analysis.)

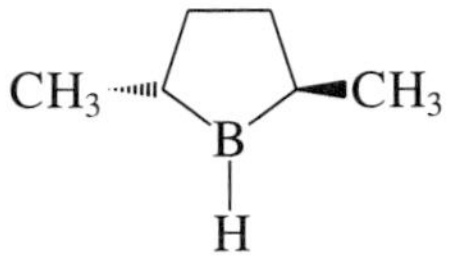

(2*R*,5*R*)-2,5-dimethylborolane

---

47 In compounds like the one below, the methyl group blocks the approach of reagents to one face of the π-bond. With this in mind, predict the product of hydroboration–oxidation of (a) the alkene below; (b) cholesterol (Sec. 7.7D).

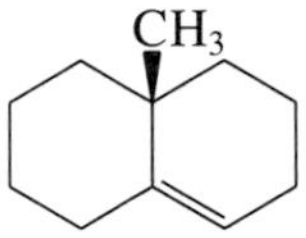

---

48 Rank the compounds within each of the following sets according to their heats of formation, lowest first. Explain.

(a)
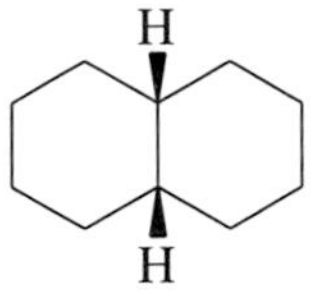

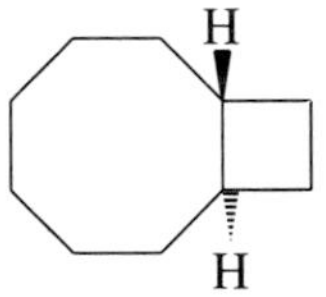

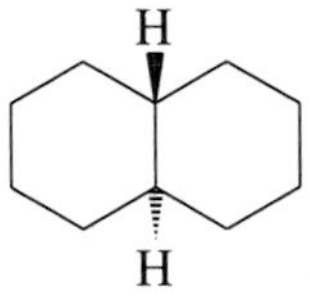

(b) methylcyclohexane, *trans*-1,2-dimethylcyclopentane, *cis*-1,2-dimethylcyclopentane

(c)
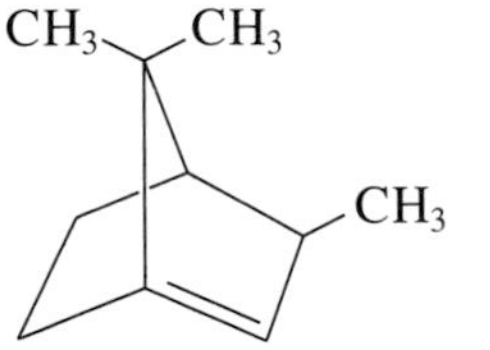

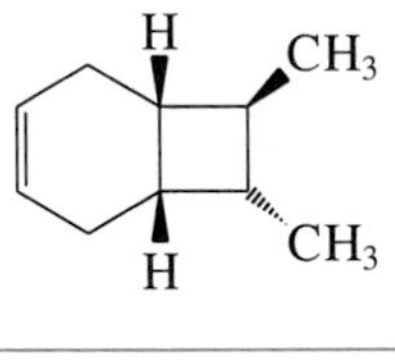

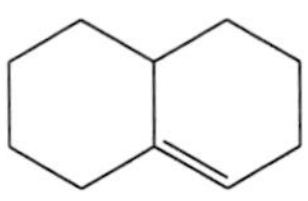

---

49 Consider the following compound:

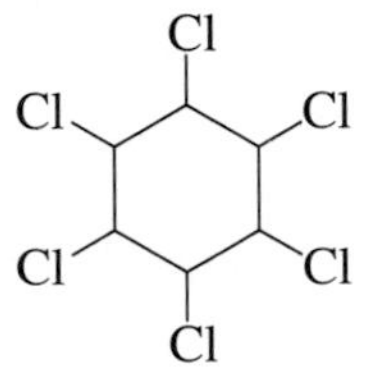

1,2,3,4,5,6-hexachlorocyclohexane

(a) Of the nine stereoisomers of this compound, only two are chiral. Give the structures of this enantiomeric pair.

(b) Give the structures of the two stereoisomers that each have two identical chair forms.

Problems (Cont.)

50 The energy difference between the two chair forms of *cis*-1,3-dimethylcyclohexane is 5.5 kcal/mol. This is a sum of both *gauche*-butane interactions and the severe 1,3-diaxial interaction shown below.

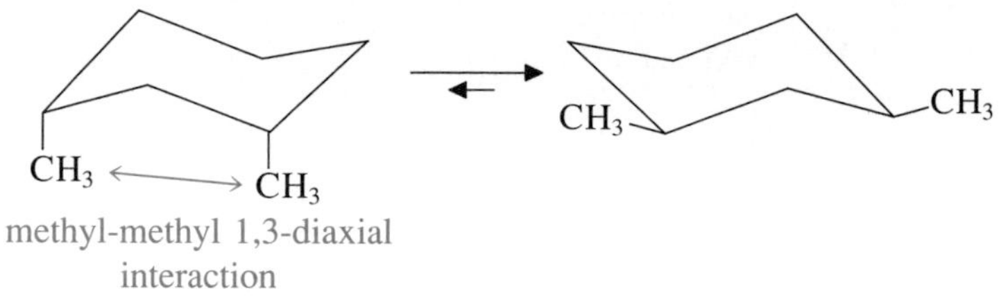

$\Delta G° = -5.5$ kcal/mol

Using the energy difference given, analyze the *gauche* interactions and derive an estimate (in kcal/mol) for the energy of the methyl-methyl 1,3-diaxial interaction.

51 When 1,4-cyclohexadiene reacts with two equivalents of Br_2, a solid that melts at 174–178° is formed. The wide melting range indicates a mixture of compounds. If this solid is dissolved in $CHCl_3$ and the solvent is slowly evaporated, two different types of crystals separate. Each type melts sharply, one at 188° and the other at 218°. If the crystals are mixed and melted together, then cooled, the solid that melts at 174–178° is formed again. Account for these observations.

52 It has been argued that the energy difference between *cis*- and *trans*-1,3-di-*tert*-butylcyclohexane is a good approximation for the energy difference between the chair and twist-boat forms of cyclohexane. Using models to assist you, explain why this view is reasonable.

53 It is possible to synthesize certain monosubstituted derivatives of borane; "thexylborane" is such a derivative:

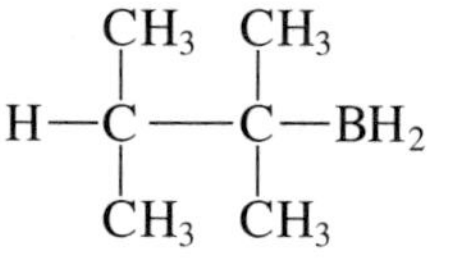

"thexylborane"

"Thexylborane" reacts with 1-vinyl-1-cyclohexene to give an organoborane *A*. When *A* reacts with H_2O_2/NaOH, a diol *B* (a compound with two —OH groups) is formed. The molecular formula of *B* is $C_8H_{16}O_2$. Propose structures for both *A* and *B*, including stereochemistry.

54 The ΔG^0 for the following equilibrium is +1.13 kcal/mol:

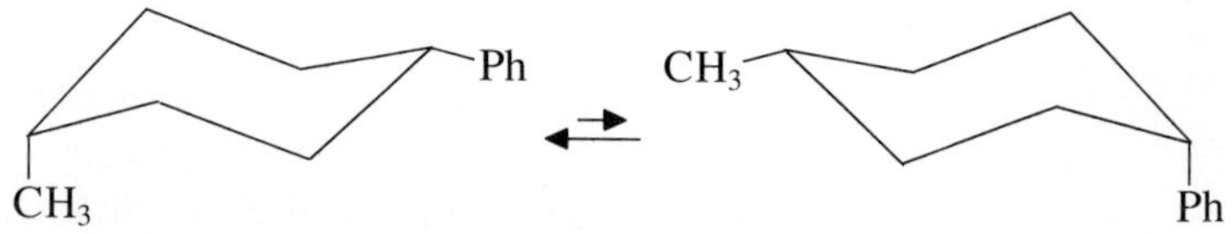

(a) Which behaves as if it is larger, methyl or phenyl (Ph)? Why is this reasonable?
(b) Use the ΔG^0 given above along with any other appropriate data to estimate ΔG^0 for the following two equilibria:

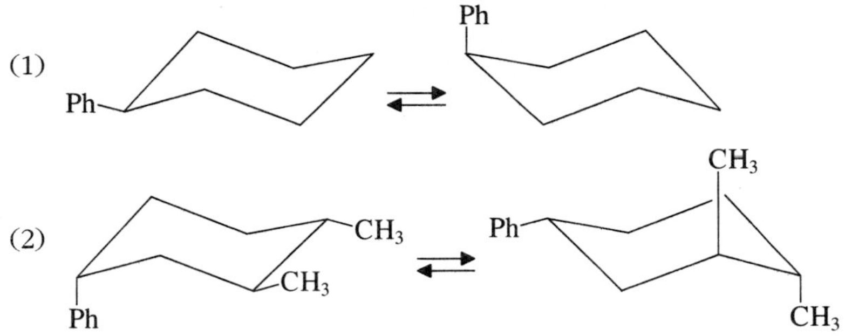

55 The claim has been made that the following compound prefers to exist in form *B*, in which one ring assumes a twist-boat conformation. Using models if necessary, explain why such a claim is reasonable.

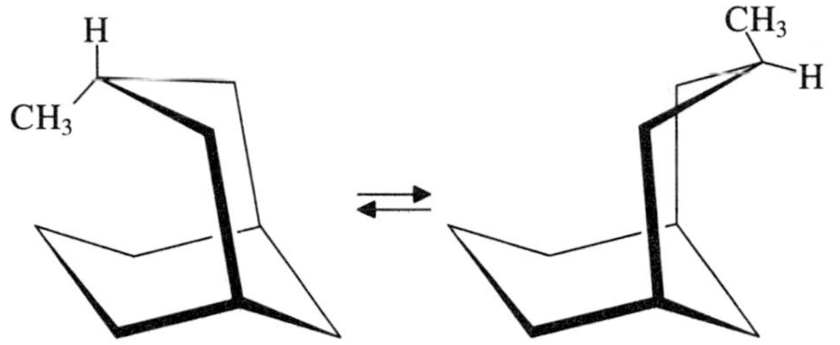

A *B*

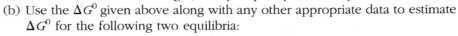

56 (a) Does the spirocyclic compound below have any asymmetric carbons?
(b) Is this compound chiral? Explain.

8 Introduction to Alkyl Halides, Alcohols, Ethers, Thiols, and Sulfides

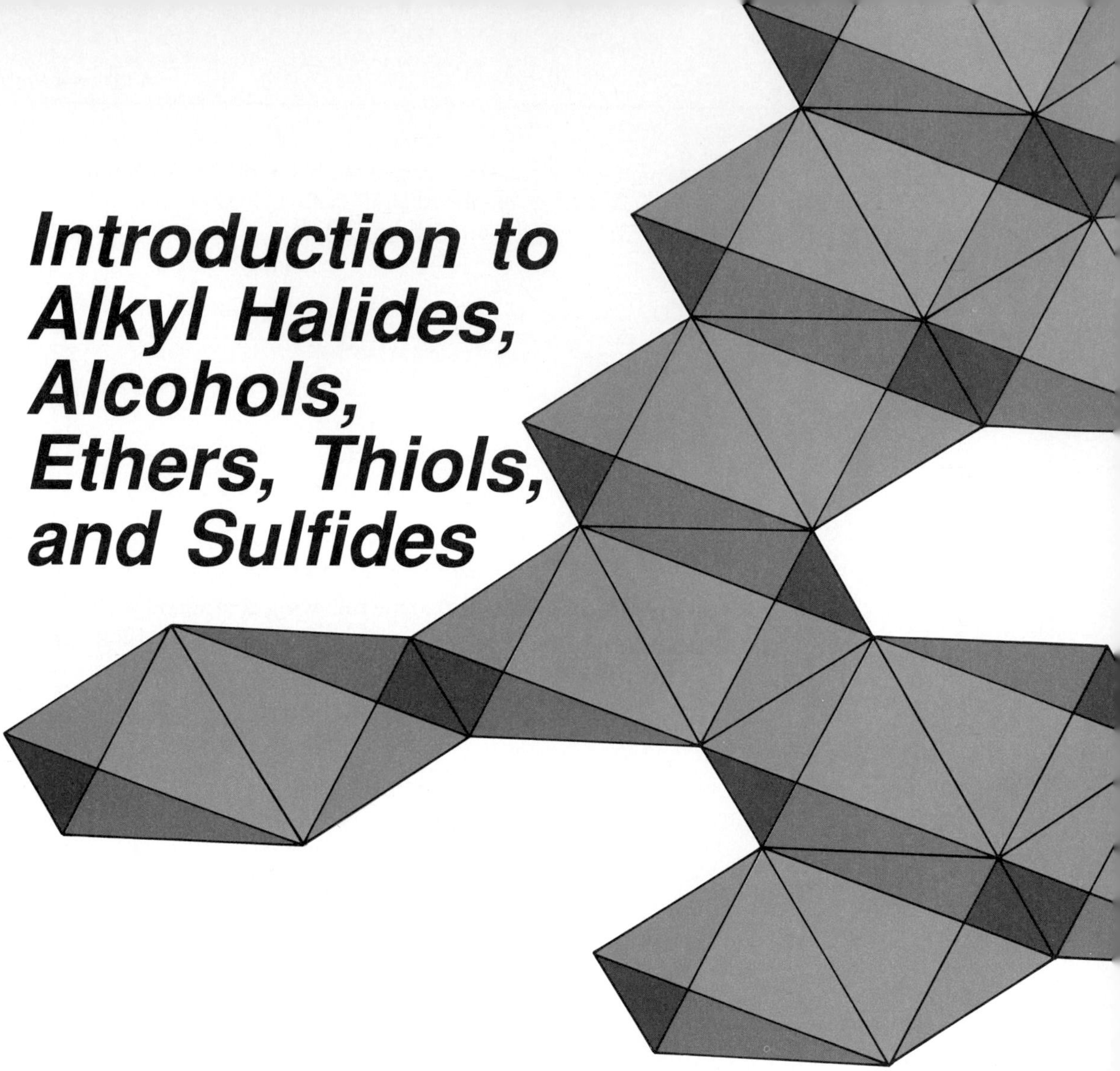

In this chapter we shall study the nomenclature and properties of several classes of compounds. In the **alkyl halides,** a halogen atom is bonded to the carbon of an alkyl group. Alkyl halides are classified as methyl, primary, secondary, or tertiary, depending on the number of alkyl groups attached to the carbon bearing the halogen (color in the structures below). A methyl halide has no alkyl groups, a primary halide has one, a secondary halide has two, and a tertiary halide has three.

CH_3—Br	CH_3—CH_2—Br	CH_3—CH(CH_3)—Br	CH_3—C(CH_3)(CH_3)—Br
methyl bromide (bromomethane) a methyl halide	ethyl bromide (bromoethane) a primary alkyl halide	isopropyl bromide (2-bromopropane) a secondary alkyl halide	*t*-butyl bromide (2-bromo-2-methyl propane) a tertiary alkyl halide

In the **alcohols,** a **hydroxy group,** —OH, is bonded to the carbon of an alkyl group. Alcohols, too, are classified as methyl, primary, secondary, or tertiary.

CH_3—OH	CH_3—CH_2—OH	CH_3—CH(CH_3)—OH
methyl alcohol (methanol)	ethyl alcohol (ethanol) a primary alcohol	isopropyl alcohol (2-propanol) a secondary alcohol

CH_3—C(CH_3)(CH_3)—OH *t*-butyl alcohol (2-methyl-2-propanol) a tertiary alcohol

Alcohols that contain two hydroxy groups are called **glycols.** If the hydroxy groups are on adjacent carbons, the compound is a **vicinal glycol,** the most important type of glycol.

HO—CH_2—CH_2—OH ethylene glycol (a vicinal glycol)

Thiols, sometimes called **mercaptans,** are the sulfur analogs of alcohols. In a thiol, a **sulfhydryl group,** —SH, also called a **mercapto group,** is bonded to an alkyl group. An example of a thiol is ethyl mercaptan (ethanethiol), CH_3CH_2—SH.

In an **ether,** an oxygen is bonded to two carbon groups, which may or may not be the same. A **thioether,** or **sulfide,** is simply the sulfur analog of an ether.

CH_3CH_2—O—CH_2CH_3	CH_3CH_2—O—$CH(CH_3)_2$	CH_3—S—CH_2CH_3
diethyl ether	ethyl isopropyl ether	ethyl methyl sulfide

Our introduction to the functional groups in this chapter will be followed in subsequent chapters by a study of the chemistry of each group in turn. We shall find that there is a very good reason for considering these groups together: much of their chemistry is related.

8.1 NOMENCLATURE

Several systems are recognized by the IUPAC for the nomenclature of organic compounds. Of these systems, two are the most widely used; we shall call these **common nomenclature** and **systematic nomenclature.** (The IUPAC calls these systems "radicofunctional nomenclature" and "substitutive nomenclature," respectively.) Common nomenclature is typically used for simple unbranched compounds. Systematic nomenclature is more often used for more complex branched structures. We learned the basics of systematic nomenclature when we studied alkanes (Sec. 3.2A) and alkenes (Sec. 4.2A). Although it might seem desirable to adopt only one system of nomenclature, historical usage and other factors have dictated that many organic compounds are widely known by both types of names. Fortunately, both systems are fairly easy to learn.

A. Nomenclature of Alkyl Halides

The common name of an alkyl halide is constructed from the name of the alkyl group (see Table 3.2) followed by the name of the halide as a separate word.

$CH_3CH_2—Cl$	CH_2Cl_2
ethyl chloride	methylene chloride
	($—CH_2—$ group = methylene group)

$CH_3—Br$	$(CH_3)_2CH—I$
methyl bromide	isopropyl iodide

Special common names for certain alkyl halides should be learned.

$CH_2{=}CH—CH_2—Cl$	$Ph—CH_2—Br$	$CH_2{=}CH—Cl$	CCl_4
allyl chloride	benzyl bromide	vinyl chloride	carbon tetrachloride

(Compounds with halogens attached to alkene carbons, such as vinyl chloride, are not alkyl halides, but it is convenient to discuss their nomenclature here.)

The **allyl group,** as the name above implies, is the $CH_2{=}CH—CH_2—$ group. This should not be confused with the **vinyl group,** $CH_2{=}CH—$, which lacks the additional $—CH_2—$. Similarly, the **benzyl group,** $Ph—CH_2—$, should not be confused with the *phenyl* group.

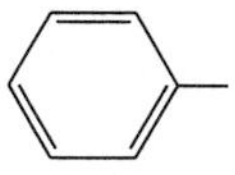 or Ph—

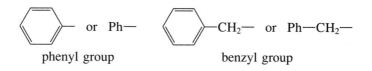

phenyl group

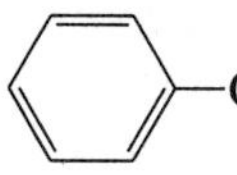 $—CH_2—$ or $Ph—CH_2—$

benzyl group

The **haloforms** are the methyl trihalides.

$HCCl_3$	$HCBr_3$	HCI_3
chloroform	bromoform	iodoform

Compounds in which all hydrogen atoms are formally replaced by halogen are called **perhalo compounds.**

$CF_3—CF_2—CF_3$
perfluoropropane
(a perfluoro compound)

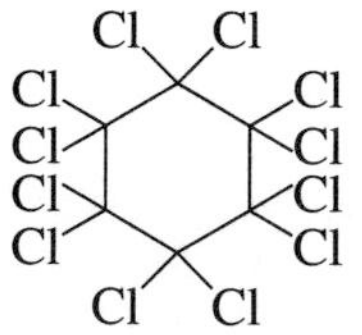

perchlorocyclohexane
(a perchloro
compound)

The systematic name of an alkyl halide is constructed by applying the rules of alkane and alkene nomenclature (Sec. 3.2A, 4.2A). We first determine the *principal chain,* which is the longest continuous *carbon* chain containing the greatest number of double bonds. Simple alkyl halides are named as alkanes in such a way that the substituents receive the lowest possible number. The halogen substituents are named fluoro, chloro, bromo, or iodo.

$CH_3CH_2—Cl$
chloroethane

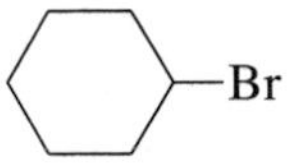
bromocyclohexane

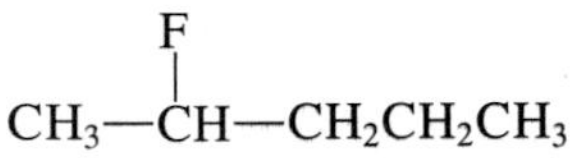
2-fluoropentane

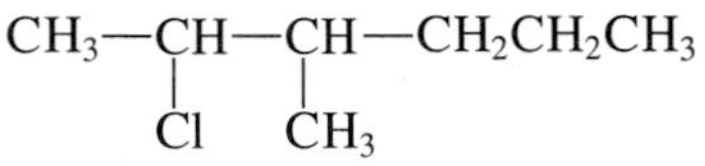
2-chloro-3-methylhexane

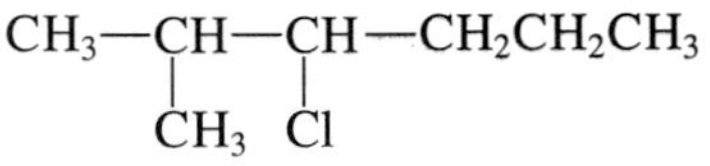
3-chloro-2-methylhexane

If a double bond is present, it receives the lowest number, as in hydrocarbon nomenclature.

$CH_3—CH{=}CH—CH_2—CH_2—Cl$ 5-chloro-2-pentene

Problems

1 Give the common name for each of the following compounds, and tell whether (a) and (c) are primary, secondary or tertiary alkyl halides.

(a) $(CH_3)_2CHCH_2—F$
(b) $CH_3CH_2CH_2CH_2CH_2CH_2—I$
(c) 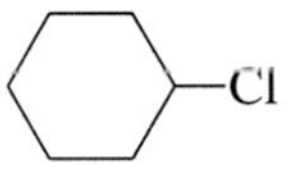

2 Give the structure of
(a) chlorocyclopropane
(b) 2,2-dichloro-5-methylhexane
(c) 6-bromo-1-chloro-3-methyl-1-cyclohexene
(d) methylene iodide

3 Give the systematic name for each of the following compounds:

(a) 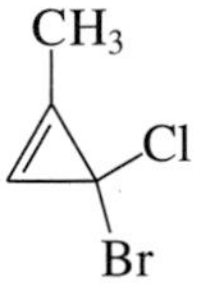

(b) 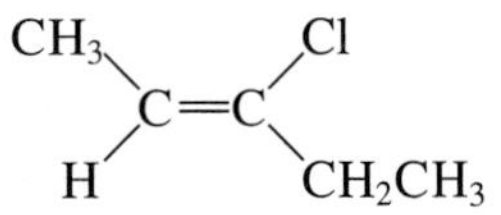

(c) 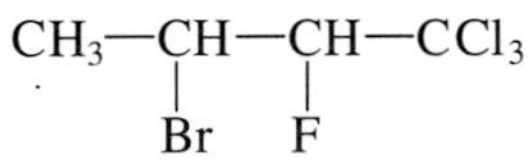

(d) chloroform

(e) 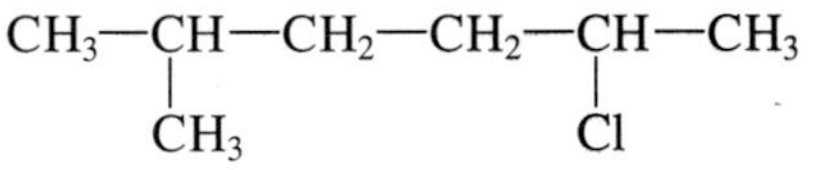

(f) neopentyl bromide (See Table 3.2)

B. Nomenclature of Alcohols and Thiols

The common name of an alcohol is derived by specifying the alkyl group to which the —OH group is attached, followed by the word *alcohol*.

$$CH_3-OH \qquad (CH_3)_2CH-OH \qquad \bigcirc-OH \qquad CH_3CH_2CH_2-OH$$

methyl alcohol isopropyl alcohol cyclohexyl alcohol propyl alcohol

$$CH_2=CH-CH_2-OH \qquad Ph-CH_2-OH$$
allyl alcohol benzyl alcohol

A few glycols have important traditional names.

$$HO-CH_2CH_2-OH$$
ethylene glycol

$$\underset{\underset{OH}{|} \quad \underset{OH}{|}}{CH_2-CH-CH_3}$$
propylene glycol

$$\underset{\underset{OH}{|} \quad \underset{OH}{|} \quad \underset{OH}{|}}{CH_2-CH-CH_2}$$
glycerol (glycerin)

Thiols are named in the common system as *mercaptans;* this name, which means "captures mercury," comes from the fact that thiols readily form heavy-metal derivatives (Sec. 8.6A).

$$CH_3CH_2-SH \qquad \text{ethyl mercaptan}$$

The systematic name of an alcohol is constructed by dropping the final *e* from the name of the parent alkane and adding the suffix *ol*. The —OH group receives the lowest number in the carbon chain.

$$CH_3-CH_2-OH$$
ethan*e* + ol = ethanol

$$\overset{4}{CH_3}-\overset{3}{CH_2}-\underset{\underset{OH}{|}}{\overset{2}{CH}}-\overset{1}{CH_3}$$
2-butanol

To name an alcohol containing more than one —OH group, the suffixes *diol, triol,* etc., are added to the name of the appropriate alkane *without* dropping the final *e*.

$$\overset{1}{CH_3}-\underset{\underset{OH}{|}}{\overset{2}{CH}}-\underset{\underset{OH}{|}}{\overset{3}{CH}}-\overset{4}{CH_2}-\overset{5}{CH_3}$$
2,3-pentanediol

The systematic nomenclature of thiols is similar to that of alcohols: the name of the parent alkane is followed by the suffix *thiol.* However, the final *e* in the name of the alkane is *not* dropped.

$$CH_3CH_2SH$$
ethanethiol

$$(CH_3)_2CH\underset{\underset{SH}{|}}{CH}CH_3$$
3-methyl-2-butanethiol

Suppose we want to name a compound that has both an —OH group and another functional group in the same molecule. In such a case we have to know which group receives precedence in the name. For example, if a compound contains both a hydroxy group and a mercapto group, do we name the compound as an alcohol or as a thiol?

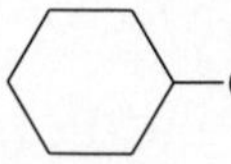

The IUPAC has established rules for designating the functional group on which the name is based; this group is called the **principal group.** Of the functional groups we have studied so far, only hydroxy groups and mercapto groups can be candidates for the principal group. An —OH group receives precedence over an —SH group.

$$\textit{priority as principal group:}\quad \text{—OH} > \text{—SH} \tag{8.1}$$

Once we have identified the principal group, we then identify the *principal chain,* the carbon chain on which the name is based (Sec. 3.2A); this is *the longest continuous carbon chain that contains the greatest number of principal groups.* (As we learned in Sec. 4.2A, if there are no principal groups, the principal chain is the chain containing the greatest number of double bonds; and if there are no double bonds, the principal chain is simply the longest continuous carbon chain.) The principal group and principal chain determine the name.

Let us apply these ideas by constructing the systematic name of the following compound:

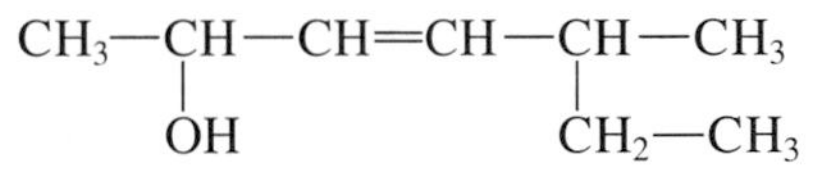

1. *Identify the principal group.* The principal group is the —OH group. For this reason the compound is named as an alcohol. Its name ends with the suffix *ol.*
2. *Identify the principal carbon chain.* The principal carbon chain is the longest carbon chain bearing the —OH group.
3. *Number the carbon chain consecutively from one end so that the principal group receives the lowest possible number.* In this example the —OH group receives the number 2. This numbering automatically determines the position of the methyl group and the double bond.

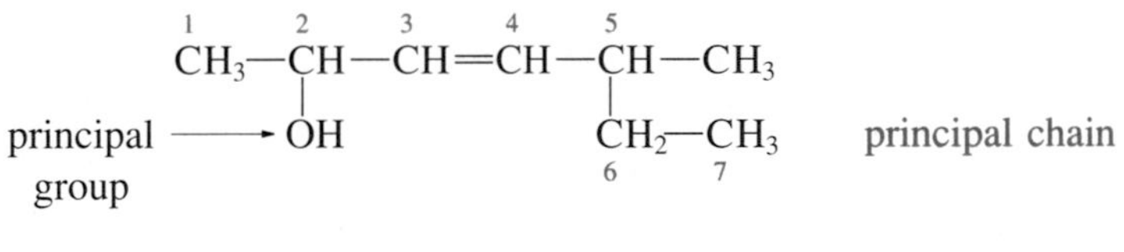

4. *Construct the name using the appropriate numbers for the other groups.*

5-methyl-3-hepten-2-ol

- 2: number of the principal group
- 3: number of the double bond
- 5: number of the methyl substituent

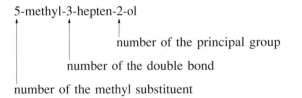

The following compound contains two groups that are eligible for citation as principal groups. According to the priority sequence (8.1), the compound is named as an alcohol, and the —SH (mercapto) group is named as a substituent.

$HS—CH_2—CH_2—OH$ 2-mercapto-1-ethanol

Alkyl substituents and halogens are *never* cited as principal groups. They are simply numbered as substituents according to the rules of alkane nomenclature.

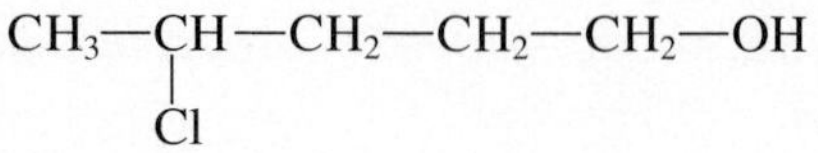

$$CH_3-CH-CH_2-CH_2-CH_2-OH \qquad \text{4-chloro-1-pentanol}$$
$$|$$
$$Cl$$

In the following cyclic compounds an —OH is the principal group and receives the number 1. However, we might wonder which way to number around the ring. The rule is that first we give precedence to the principal groups. Thus, the —OH groups receive the lowest numbers. Next, we give precedence to double bonds. These two rules determine the direction of numbering in these examples.

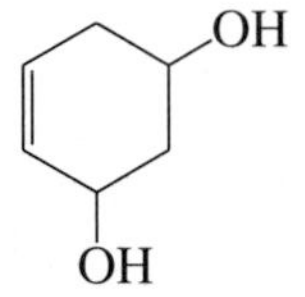

4-cyclohexene-1,3-diol 6-methyl-2-cyclohexen-1-ol

If there are no double bonds, we then give the lowest number to other substituents.

Common and systematic nomenclature should never be mixed. The following compounds are frequently named incorrectly:

$$CH_3$$
$$|$$
$$CH_3-C-OH$$
$$|$$
$$CH_3$$

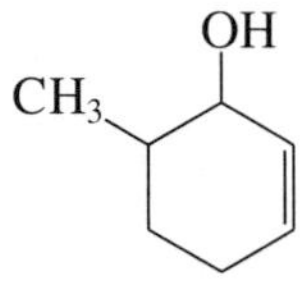

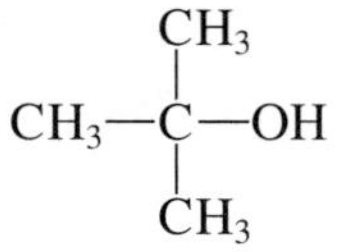

common: *tert*-butyl (or *t*-butyl) alcohol
systematic: 2-methyl-2-propanol
incorrect: *t*-butanol

$$CH_3-CH-CH_3$$
$$|$$
$$OH$$

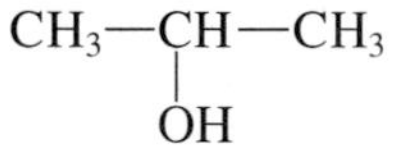

common: isopropyl alcohol
systematic: 2-propanol
incorrect: isopropanol

Problems

4 Draw the structure of
(a) 3-methyl-2-pentanol
(b) 2-cyclohexen-1-ol
(c) 3-ethyl-1-cyclopentanol
(d) (*E*)-6-chloro-4-hepten-2-ol
(e) 3-cyclopentene-*cis*-1,2-diol
(f) isobutyl alcohol

5 Give the systematic name for each of the following compounds:
(a) $CH_3CH_2CH_2CH_2OH$
(b) $CH_3CHCH_2CH_2OH$
$|$
Br

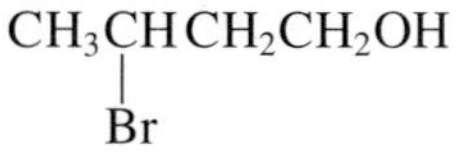

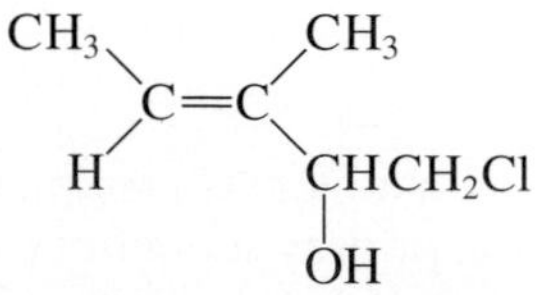

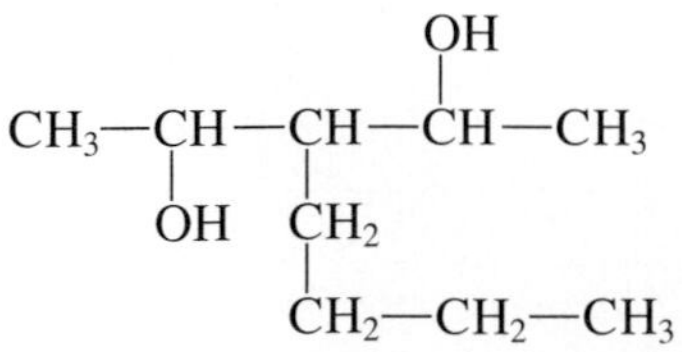

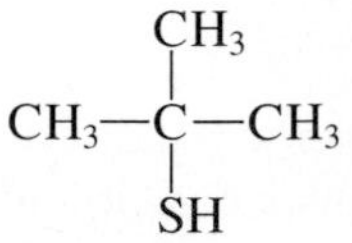

(f) 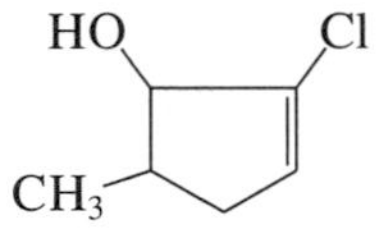

(g) 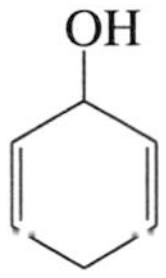

(h) $HO{-}CH_2{-}CH(OH){-}CH_2{-}OH$ (OH on the central CH)

(i) $CH_3CH_2CH_2CH(OH)CH_2SH$ (OH on the CH)

C. Nomenclature of Ethers and Sulfides

The common name of an ether is constructed by naming the two groups on the ether oxygen, followed by the word *ether.*

$CH_3CH_2{-}O{-}CH_2CH_3$ diethyl ether (also called ethyl ether or simply ether)	$CH_3{-}O{-}C_2H_5$ ethyl methyl ether

A sulfide is named in a similar manner, using the word *sulfide.* (In older literature, the word *thioether* was also used.)

$CH_3CH_2{-}S{-}CH_3$ ethyl methyl sulfide (also ethyl methyl thioether)	$(CH_3)_2CH{-}S{-}CH(CH_3)_2$ diisopropyl sulfide

In systematic nomenclature, there are no suffixes for *alkoxy* groups (RO—) or *alkylthio* groups (RS—). Hence, these groups are always cited as substituents.

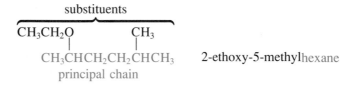

substituents: CH_3CH_2O (on C-2), CH_3 (on C-5)

$CH_3CH(OCH_2CH_3)CH_2CH_2CH(CH_3)CH_3$ — principal chain: $CH_3CHCH_2CH_2CHCH_3$ — 2-ethoxy-5-methylhexane

In this example, the principal chain is a six-carbon chain. Hence, the compound is named as a hexane, and the C_2H_5O— group and the methyl group are treated as substituents. The C_2H_5O— group is named by dropping the final *yl* from the name of the alkyl group and adding the suffix *oxy.* Thus the C_2H_5O— group is the (eth~~yl~~ + oxy) = ethoxy group. The numbering follows from alkane nomenclature, rule 7, Sec. 3.2A.

If double bonds are present, the rules of alkene nomenclature apply; the principal chain is the one that contains the greatest number of double bonds.

$$CH_3CH_2CH_2CH_2CH_2{-}O{-}CH_2CH_2CH{=}CH_2 \qquad \text{4-pentoxy-1-butene}$$

If an ether or sulfide contains a principal group cited as a suffix, such as an —OH or —SH, this group receives precedence in the name:

$$CH_3CH_2CH_2CH_2{-}O{-}\underbrace{\overset{3}{C}H_2\overset{2}{C}H_2\overset{1}{C}H_2}_{\text{principal chain}}{-}OH \qquad \text{3-butoxy-1-propanol}$$

(contains principal group —OH)

Nomenclature of sulfides is similar. An RS— group is named by adding the suffix *thio* to the name of the R group; the final *yl* is not dropped.

substituent: SCH_3

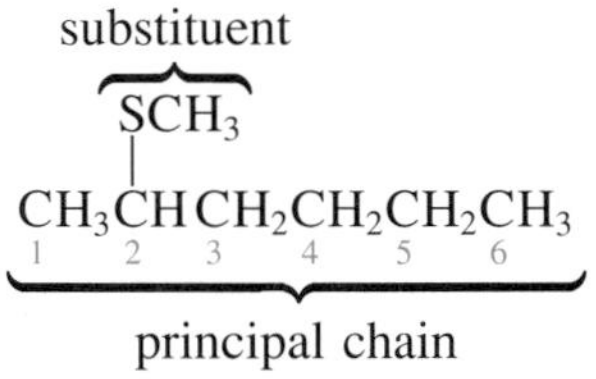

$$\underbrace{\overset{1}{C}H_3\overset{2}{C}H(SCH_3)\overset{3}{C}H_2\overset{4}{C}H_2\overset{5}{C}H_2\overset{6}{C}H_3}_{\text{principal chain}} \qquad \text{2-(methylthio)hexane}$$

The parentheses in the name indicate that "thio" is associated with "methyl" rather than with "hexane."

In a number of important ethers and sulfides, the oxygen or sulfur atom is part of a ring. Cyclic compounds with rings that contain more than one type of atom are called **heterocyclic compounds.** We shall study heterocyclic compounds in more detail in Chapter 24. However, there are a few heterocyclic ethers and sulfides whose names should be learned now.

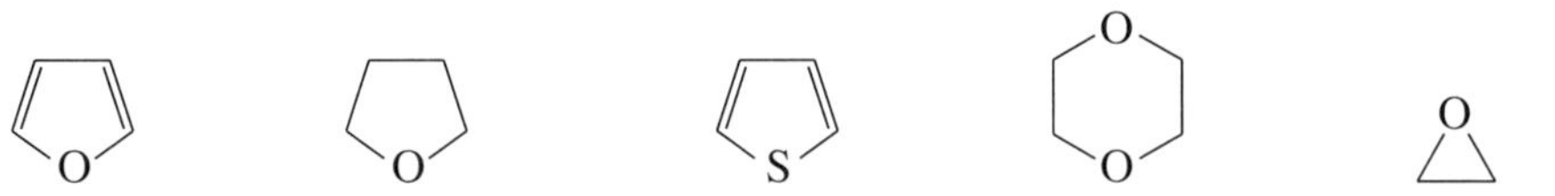

furan — tetrahydrofuran (often called THF) — thiophene — 1,4-dioxane (often called simply dioxane) — oxirane (ethylene oxide)

Oxirane is the parent compound of a special class of heterocyclic ethers, called **epoxides,** that contain three-membered rings. A few epoxides are named as oxides of the corresponding alkenes:

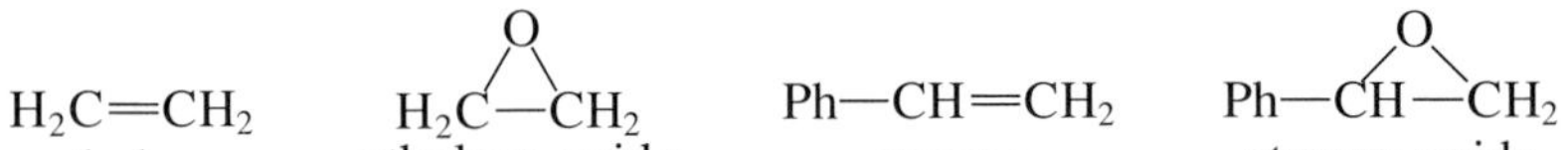

ethylene — ethylene oxide — styrene — styrene oxide

However, most epoxides are named systematically as derivatives of oxirane. The atoms of the epoxide ring are numbered consecutively with the oxygen receiving the number 1 *regardless of the substituents present.*

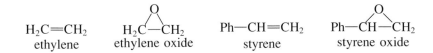

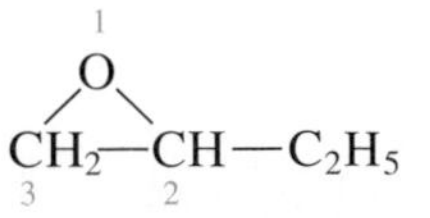

2-ethyloxirane

Problems

6 Draw the structure of each of the following compounds:

(a) ethyl *n*-propyl ether
(b) dicyclohexyl sulfide
(c) diisobutyl ether
(d) (2*R*,3*R*)-2,3-dimethyloxirane
(e) 5-(ethylthio)-2-methylheptane
(f) allyl benzyl ether

7 Give a systematic name for each of the following compounds:

(a) $(CH_3)_3C—O—CH_3$
(b) 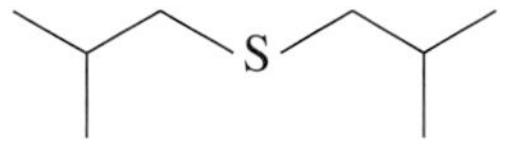
(c) $CH_3CH_2—O—CH_2CH_2—OH$
(d) 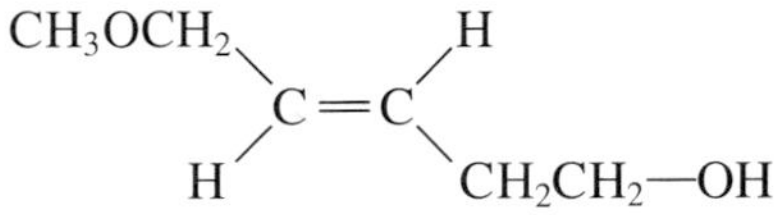

8.2 STRUCTURES

In all the compounds we are studying in this chapter, the bond angles about carbon are very nearly tetrahedral. For example, in the simple methyl derivatives—the methyl halides, methanol, methanethiol, dimethyl ether, and dimethyl sulfide—the H—C—H bond angle in the methyl group does not deviate more than a degree or so from 109.5°. In an alcohol, thiol, ether, or sulfide the bond angle at oxygen or sulfur further defines the shape of the molecule. We learned in Sec. 1.6B that we can predict the shapes of such molecules by thinking of an unshared electron pair as a bond without an atom at the end. This means that the oxygen or sulfur has four "groups": two electron pairs and two alkyl groups or hydrogens. These molecules are therefore bent at oxygen and sulfur, as we can see from the structures in Fig. 8.1. The angle at sulfur is generally found to be closer to 90° than the angle at oxygen. One reason for this trend is that the unshared electron pairs on sulfur occupy more diffuse orbitals that take up more space. The repulsion between these unshared pairs and the electrons in the chemical bonds forces the bonds closer together than they are on oxygen.

The lengths of bonds between carbon and other atoms follow the trends that we discussed in Sec. 1.6B. Bonds are longer as we go down the periodic table (larger atoms) or, within a given row, to the left (less electronegative atoms). These trends are clear in the methyl derivatives shown in Table 8.1.

TABLE 8.1 Bond Lengths (in Angstroms) in Some Methyl Derivatives

$CH_3—CH_3$ 1.536	$CH_3—NH_2$ 1.474	$CH_3—OH$ 1.426	$CH_3—F$ 1.391
		$CH_3—SH$ 1.82	$CH_3—Cl$ 1.781
	— Increasing electronegativity →		$CH_3—Br$ 1.939
	Increasing atomic radius ↓		$CH_3—I$ 2.129

Figure 8.1 Bond lengths and bond angles in a simple alcohol, thiol, ether, and sulfide.

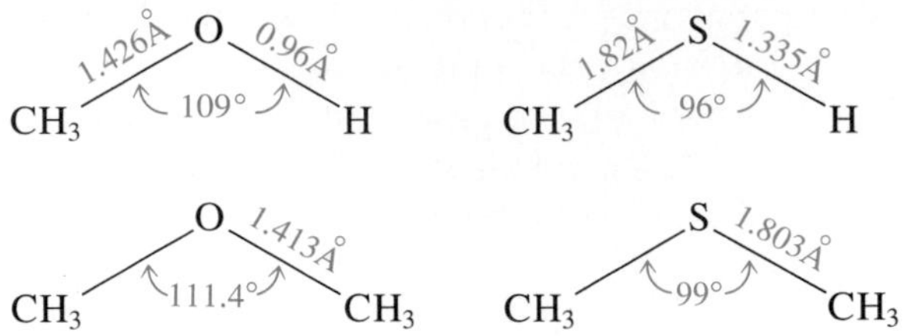

Problem

8 Predict the approximate carbon–selenium bond length in CH_3—Se—CH_3.

8.3 PHYSICAL PROPERTIES. POLARITY AND HYDROGEN BONDING

A. Boiling Points of Ethers and Alkyl Halides: Effect of Polarity

Most alkyl halides, alcohols, and ethers are *polar molecules*—that is, they have permanent dipole moments (Sec. 1.2C). The following examples are typical.

	CH_3—F	CH_3—Cl	CH_3—O—CH_3	CH_3—OH
dipole moment	1.82 D	1.94 D	1.31 D	1.7 D

The polarity of a compound affects its boiling point. When we compare two molecules of the same shape and size, in many cases the more polar molecule has the higher boiling point.

	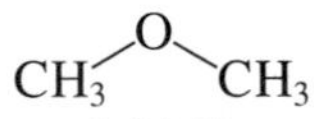	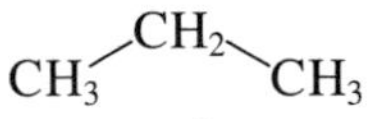
dipole moment	1.31 D	≈0
boiling point	−23.7°	−42.1°

dipole moment	1.7 D	0 D
boiling point	64.5°	49.3°

What is the reason for this effect? A higher boiling point, as we learned in Sec. 3.4A, results from greater attraction between molecules in the liquid state. Polar molecules are attracted to each other because they can align in such a way that the negative end of one dipole is attracted to the positive end of the other.

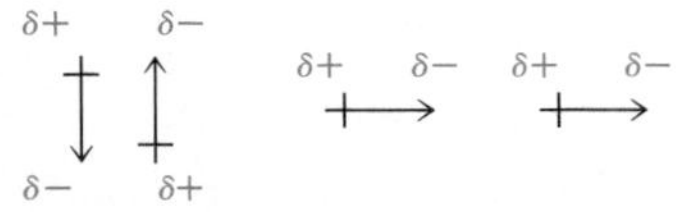

two ways in which dipoles can align attractively

Of course, since molecules in the liquid state are in constant motion, their relative positions are changing all the time; however, on the average, this attraction exists and serves to raise the boiling point of a polar compound.

When a polar molecule contains a hydrocarbon portion of even moderate size its polarity has little effect on its physical properties; it is sufficiently alkanelike that its properties resemble those of an alkane.

	$CH_3OCH_2CH_2CH_2CH_2CH_3$	$CH_3CH_2CH_2CH_2CH_2CH_2CH_3$
boiling point	99°	98°

When we compare the boiling points of alkyl halides to those of alkanes, we might expect from what we have just learned that the more polar alkyl halides should have higher boiling points. However, this is not so. Alkyl chlorides have about the same boiling points as alkanes of the same molecular weight, and alkyl bromides and iodides have lower boiling points than the alkanes of the same molecular weight.

	$CH_3CH_2CH_2CH_2Cl$	$CH_3CH_2CH_2CH_2CH_2CH_3$
molecular weight	92.6	86.2
boiling point	78.4°	68.7°
density	0.886 g/mL	0.660 g/mL
	CH_3CH_2Br	$CH_3CH_2CH_2CH_2CH_2CH_2CH_3$
molecular weight	109	100.2
boiling point	38.4°	98.4°
density	1.46 g/mL	0.684 g/mL
	CH_3I	$CH_3CH_2CH_2CH_2CH_2CH_2CH_2CH_2CH_2CH_3$
molecular weight	142	142
boiling point	42.5°	174°
density	2.28 g/mL	0.73 g/mL

The key to understanding these facts is to realize that we are comparing molecules of the same molecular weight, but *not the same size*. Alkyl halide molecules have a relatively large mass concentrated in a given volume; we can see this from their relatively high densities. Thus, *for a given molecular weight,* alkyl halides are smaller molecules than alkanes. In Sec. 3.4A we learned that attractive van der Waals (dispersion) forces act to increase boiling point, and that these forces are greater for larger molecules. Since alkanes of a given molecular weight are larger in size than alkyl halides, van der Waals forces between alkane molecules are greater than those between alkyl halide molecules. This analysis suggests that alkanes should have *higher* boiling points than alkyl halides of similar molecular weight. The polarity of alkyl halides, on the other hand, works in the opposite direction: if polarity were the only effect, alkanes should have *lower* boiling points than alkyl halides of similar molecular weights. The two effects thus oppose each other; they essentially cancel in the case of alkyl chlorides, which have about the same boiling points as alkanes of the same molecular weight. Alkanes, however, are so much larger than alkyl bromides and iodides of the same molecular weight that alkanes have considerably higher boiling points.

The boiling points of perfluoro compounds are particularly interesting; they are quite low for compounds of their molecular weight.

	$CF_3CF_2CF_2CF_2CF_2CF_3$	$CH_3CH_2CH_2CH_2CH_2F$
boiling point	57.1°	91.5°

In trying to understand this phenomenon, let us first realize that, although such compounds contain many polar C—F *bonds,* perfluoro compounds are not polar *molecules* for the same reason that CCl_4 is not a polar molecule: their *overall* dipole moments are about zero (see Problem 39, Chapter 1). Hence, there are no dipolar attractions acting to increase the boiling points of perfluoro compounds. In addition, the electron clouds around the electronegative fluorine atoms are held very tightly and are much more difficult to distort than the electron clouds in an alkane. (We can think of the electron cloud around fluorine as a "hard rubber ball" and the electron cloud around an alkane carbon as a "balloon.") The ease with which the electron clouds of a molecule can distort, or *polarize,* determines how effectively that molecule can take part in attractive van der Waals interactions (see Fig. 3.3). It is so difficult to polarize the tightly held electron clouds in a perfluoro compound that the dispersion forces among its molecules are relatively weak. Since these forces of attraction determine the boiling point, boiling points of perfluoro compounds are relatively low.

B. Boiling Points of Alcohols: Effect of Hydrogen Bonding

The boiling points of alcohols, especially alcohols of lower molecular weight, are unusually high in comparison to those of other organic compounds. For example, ethanol has a much higher boiling point than other organic compounds of about the same shape and molecular weight.

	CH_3CH_2—OH	$CH_3CH_2CH_3$	CH_3—O—CH_3	CH_3CH_2—F
compound	ethanol	propane	dimethyl ether	ethyl fluoride
boiling point	78°	−42°	−23°	−38°
dipole moment	1.7 D	0 D	1.3 D	1.8 D

The contrast between ethanol and the last two compounds is particularly striking: all have similar dipole moments, and yet the boiling point of ethanol is much higher. The fact that something is unusual about the boiling points of alcohols is also apparent from a comparison of the boiling points of ethanol, methanol, and the simplest "alcohol," water.

	C_2H_5—OH	CH_3—OH	H—OH
boiling point	78°	65°	100°

We learned in Sec. 3.4A that each additional —CH_2— group results in a 20–30° difference in boiling point between successive compounds in a homologous series. Yet the difference in boiling point between methanol and ethanol is only 13°, and water, although the "alcohol" of lowest molecular weight, has the highest boiling point of the three compounds.

These unusual trends in the boiling points of alcohols are the result of a phenomenon called **hydrogen bonding.** In the case of alcohols, hydrogen bonding is a weak association of the O—H proton of one molecule with the oxygen of another.

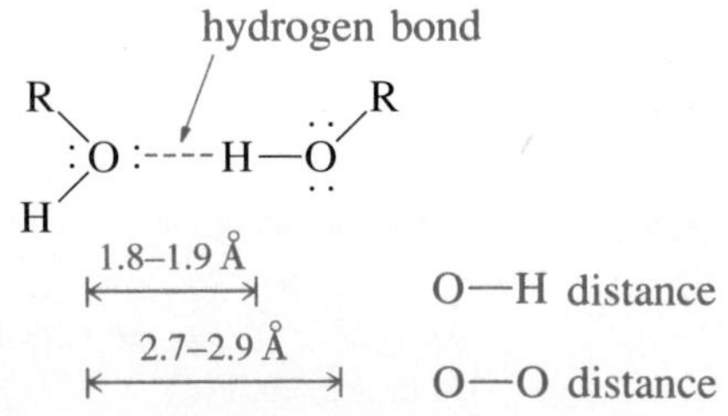

Formation of a hydrogen bond requires two partners, the **hydrogen-bond donor** (the atom to which the hydrogen is fully bonded) and the **hydrogen-bond acceptor** (the atom to which the hydrogen is partially bonded).

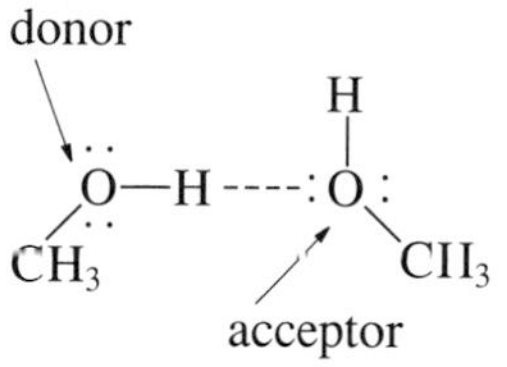

In a classical Lewis sense, a proton can only share two electrons. Thus, a hydrogen bond is difficult to describe with conventional Lewis structures. We can think of the hydrogen bond as the combination of two factors: first, a weak covalent interaction between a hydrogen on the donor atom and unshared electron pairs on the acceptor atom of another molecule; and second, an electrostatic attraction between oppositely charged ends of two dipoles.

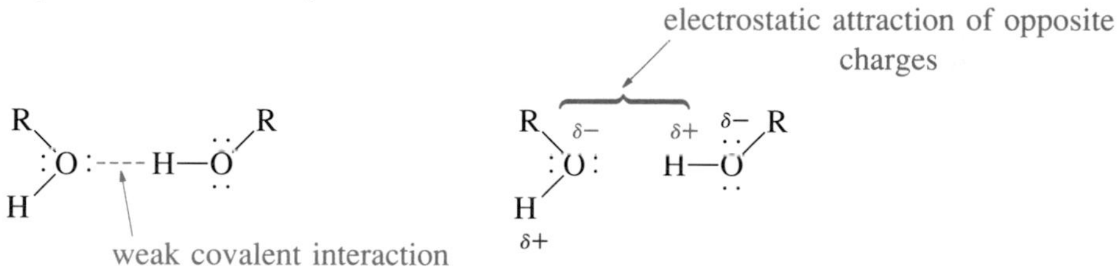

We can also think of the hydrogen bond in terms of a Brønsted acid–base reaction. For example, the hydrogen bond between two water molecules resembles the same two water molecules poised to undergo a Brønsted acid–base reaction:

hydrogen bonding:

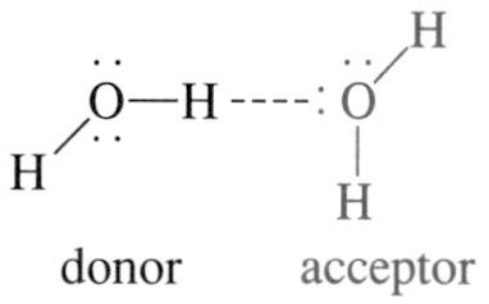

acid–base reaction:

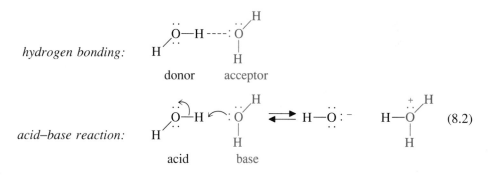

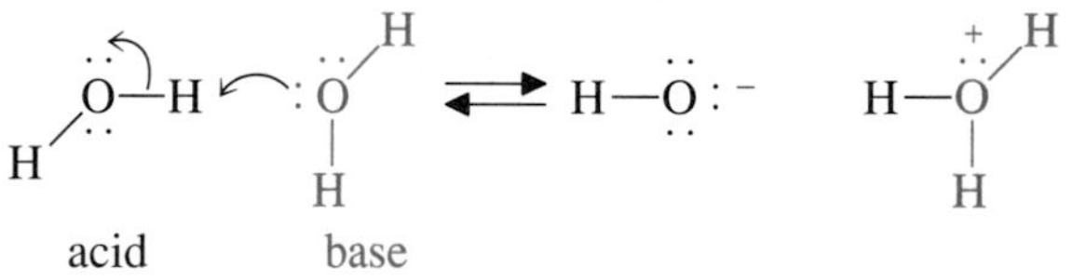

(8.2)

The hydrogen-bond donor is analogous to the Brønsted acid in Eq. 8.2, and the acceptor is analogous to the Brønsted base. In fact, it is not a bad analogy to think of the hydrogen bond as an acid–base reaction that has not quite started! In an acid–base reaction, the proton is fully transferred from the acid to the base; in a hydrogen bond, the proton interacts weakly with the acceptor but remains covalently bound to the donor.

The best hydrogen-bond donor atoms in neutral molecules are oxygens, nitrogens, and halogens. In addition, as we might expect from the similarity between hydrogen-bond interactions and Brønsted acid–base reactions, all strong proton acids are also good hydrogen-bond donors. The best hydrogen-bond acceptors in neutral molecules are the electronegative first-row atoms oxygen, nitrogen, and fluorine. All Brønsted bases are also good hydrogen-bond acceptors.

Sometimes an atom can act as both a donor and an acceptor of hydrogen bonds. For example, because the oxygen atoms in water or alcohols can act as both donors and acceptors, some of the molecules in liquid water and alcohols exist in hydrogen-bonded chains.

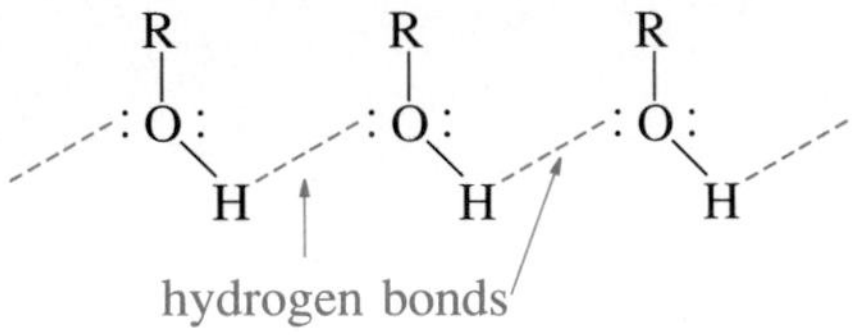

In contrast, the oxygen atom of an ether is a hydrogen-bond acceptor, but it is not a donor because it has no hydrogen to donate. Finally, some atoms are donors but not acceptors. The ammonium ion, $^{+}NH_4$, is a good hydrogen-bond donor; but because the nitrogen has no unshared electron pair, it is not a hydrogen-bond acceptor.

Although we have focused on the effects of hydrogen bonding *between* molecules (*intermolecular* hydrogen bonding), hydrogen bonding can also occur within the same molecule (*intramolecular* hydrogen bonding).

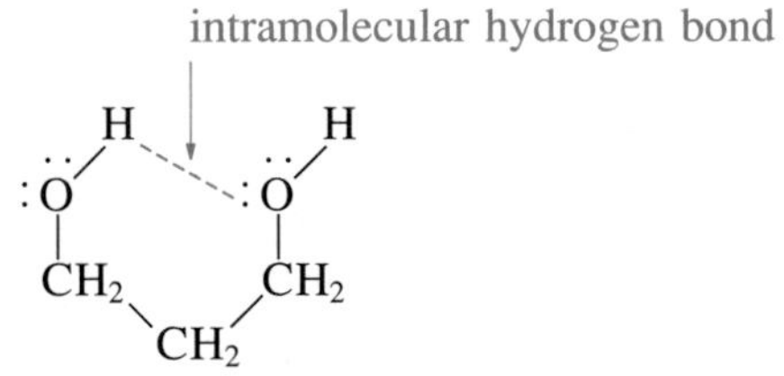

How can hydrogen bonding account for the unusually high boiling points of alcohols? We have learned that in the liquid state, hydrogen bonding is a force that holds molecules together. In the gas phase, hydrogen bonding is much less important and, at low pressures, does not exist. In order to vaporize a hydrogen-bonded compound, hydrogen bonds between molecules must be broken, and this requires energy. The energy required to break hydrogen bonds is manifested as an unusually high boiling point.

Hydrogen bonding is also important in other ways. As we shall see in Sec. 8.4, it can affect the solubility of organic compounds. It is also a very important phenomenon in biology. We shall see in Chapters 26 and 27 that hydrogen bonds have critical roles in maintaining the structures of proteins and nucleic acids (for example, enzymes and genes). Without hydrogen bonds, these materials would not function properly, and life as we know it would not exist.

Let us summarize the intermolecular forces that increase the boiling points of organic compounds. These are:

1. Attractive van der Waals forces, which are influenced by
   (a) Molecular size
   (b) Molecular shape
   (c) Polarizability of electrons
2. Attractive interactions between permanent dipoles
3. Association of molecules by hydrogen bonding

Problems

9 Label each of the following species as a hydrogen-bond acceptor, donor, or both. Indicate the hydrogen that is donated and/or the atom that serves as the hydrogen-bond acceptor.

(a) H—$\overset{..}{\underset{..}{Br}}$:

(b) H—$\overset{..}{\underset{..}{F}}$:

(c)

$$CH_3-\overset{\overset{\displaystyle :O:}{\|}}{C}-\overset{..}{\underset{..}{O}}H \quad \text{acetic acid}$$

(d)

$$CH_3-\overset{\overset{\displaystyle :O:}{\|}}{C}-\overset{..}{N}H-CH_3 \quad \textit{N}\text{-methylacetamide}$$

(e)

$\overset{..}{\underset{..}{O}}H$ (phenol ring)

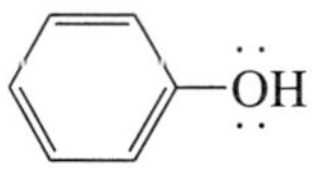

(f) $CH_3-CH_2-\overset{+}{N}H_3$

10 Ethyl alcohol in CCl_4 solvent forms a hydrogen-bonded dimer with an equilibrium constant $K_{eq} = 11$.

$$2C_2H_5OH \rightleftharpoons C_2H_5-\overset{..}{\underset{\underset{H}{..}}{O}}\cdots H-\overset{\overset{..}{..}}{\underset{C_2H_5}{O}}$$

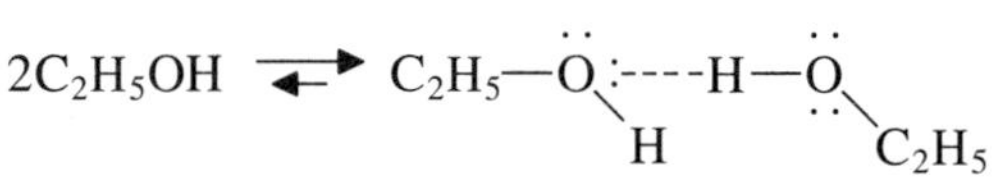

(a) What happens to the concentration of the dimer as the concentration of ethanol is increased?

(b) What is the standard free energy change for this reaction at 25 °C?

(c) If one mole of ethanol is dissolved in one liter of CCl_4, what are the concentrations of free ethanol and dimer?

(d) The equilibrium constant for the analogous dimerization of ethanethiol is 0.004. Answer questions (b) and (c) for this compound. Which has the stronger hydrogen bond, ethanethiol or ethanol?

8.4 SOLVENTS IN ORGANIC CHEMISTRY

Certain alkyl halides, alcohols, and ethers are numbered among the most important organic solvents. It is thus appropriate that we study the solvent properties of these and other organic compounds. In the past our knowledge about solvents was mostly empirical (that is, based solely on experience). More recently, however, chemists have begun to understand some of the underlying reasons for solvent properties.

Among the solvent effects we shall learn about are how and why solvents affect acidity and basicity, a topic that we shall consider first in Sec. 8.6C; and why solvents have dramatic effects on the rates of some chemical reactions, a topic that we shall address in Secs. 9.3D and 9.5D. To prepare ourselves for these sections, we first must learn to *classify* solvents. Identifying the category in which a solvent belongs will give us some idea what properties we can expect it to have. Then we shall examine some of the principles on which solubility is based. The effects of solvents on chemical reactions are closely tied to the principles governing solubility.

A. Classification of Solvents

Solvents can be classified according to three properties:

1. Proticity

2. Polarity

3. Electron-donor ability

Protic solvents consist of molecules that can act as hydrogen-bond donors. Water, alcohols, and carboxylic acids are examples of protic solvents. Solvents that cannot act as hydrogen-bond donors are termed **aprotic solvents.** Ether, carbon tetrachloride, and hexane are examples of aprotic solvents.

Polar solvents consist of molecules that have relatively high dielectric constants ϵ (see Eq. 1.2). Rather arbitrarily, we can say that if a solvent has a dielectric constant of about 20 or greater, it is polar. Water ($\epsilon = 78$), methanol ($\epsilon = 33$), and formic acid ($\epsilon = 58$) are polar solvents. Hexane ($\epsilon = 2$), ether ($\epsilon = 6$), and acetic acid ($\epsilon = 6$) are **apolar** solvents. It is important to understand that the word *polar* has a rather unfortunate double usage in organic chemistry. When we say that a *molecule* is polar, we mean that it has a significant dipole moment, μ (Sec. 1.2C). When we say that a *solvent* is polar, we mean that it has a high dielectric constant. In other words, solvent polarity, or dielectric constant, is a property of many molecules acting together, but molecular polarity, or dipole moment, is a property of individual molecules. The two types of polarity are not necessarily correlated. The contrast between acetic acid and formic acid is particularly striking:

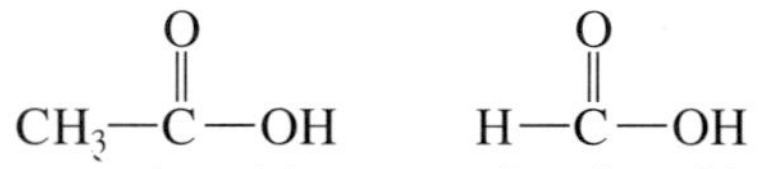

$$
\begin{array}{cc}
\overset{\displaystyle O}{\overset{\displaystyle \|}{CH_3-C-OH}} & \overset{\displaystyle O}{\overset{\displaystyle \|}{H-C-OH}} \\
\text{acetic acid} & \text{formic acid} \\
\mu = 1.5\text{--}1.7\ D & \mu = 1.6\text{--}1.8\ D \\
\epsilon = 6.15 & \epsilon = 58.5
\end{array}
$$

These two compounds contain identical functional groups and have very similar structures and dipole moments. Both are polar *molecules*. Yet they differ substantially in their dielectric constants *and in their solvent properties*! Formic acid is a polar solvent; acetic acid is not.

A third solvent classification is derived from the ability of solvent molecules to donate electron pairs. A **donor solvent** consists of molecules that can act as Lewis bases. Ether, THF, and methanol are donor solvents; pentane and benzene are not.

In Table 8.2 are listed some common solvents used in organic chemistry and their classifications. As this table shows, different solvents have different combinations of properties. For example, some polar solvents are protic (for example, water and methanol), but other polar solvents are aprotic (for example, acetone).

B. Solubility

One role of a solvent is simply to dissolve compounds of interest. Although finding a suitable solvent sometimes involves trial and error, certain principles can help us choose a solvent rationally. We shall divide our discussion of solubility into two parts: the solubility of covalent compounds, and the solubility of ionic compounds.

Solubility of Covalent Compounds In determining a solvent for a covalent compound, the best rule of thumb is *like dissolves like*. That is, a good solvent usually has at

TABLE 8.2 Properties of Some Common Organic Solvents (Listed in order of increasing dielectric constant)

Solvent	Structure	Common abbreviation	Boiling point, °C	Dielectric constant, ϵ*	Class: Polar	Class: Protic	Class: Donor
hexane	$CH_3(CH_2)_4CH_3$	—	68.7	1.9			
1,4-dioxane**	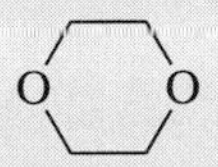	—	101.3	2.2			x
carbon tetrachloride	CCl_4	—	76.8	2.2			
benzene**	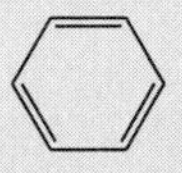	—	80.1	2.3			
diethyl ether	$(C_2H_5)_2O$	Et_2O	34.6	4.3			x
chloroform	$CHCl_3$	—	61.2	4.8			
ethyl acetate	$CH_3\overset{\overset{O}{\|}}{C}OC_2H_5$	EtOAc	77.1	6.0			x
acetic acid	$CH_3\overset{\overset{O}{\|}}{C}OH$	HOAc	117.9	6.1		x	x
tetrahydrofuran		THF	66	7.6			x
methylene chloride	CH_2Cl_2	—	39.8	8.9			
acetone	$CH_3\overset{\overset{O}{\|}}{C}CH_3$	Me_2CO, DMK	56.3	21	x		x
ethanol	C_2H_5OH	EtOH	78.3	25	x	x	x
hexamethylphosphoric triamide**	$[(CH_3)_2N]_3\overset{+}{P}—\overset{-}{O}$	HMPA, HMPT	233	30	x		x
methanol	CH_3OH	MeOH	64.7	33	x	x	x
nitromethane	CH_3NO_2	$MeNO_2$	101.2	36	x		
N,N-dimethylformamide	$H\overset{\overset{O}{\|}}{C}N(CH_3)_2$	DMF	153.0	37	x		x
acetonitrile	$CH_3C{\equiv}N$	MeCN	81.6	38	x		
sulfolane	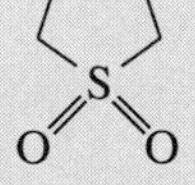	—	287 (dec)	43	x		x
dimethylsulfoxide	$CH_3\overset{\overset{O^-}{|}}{\underset{+}{S}}CH_3$	DMSO	189	47	x		x
formic acid	$H\overset{\overset{O}{\|}}{C}OH$	—	100.6	59	x	x	x
water	H_2O	—	100.0	78	x	x	x
formamide	$H\overset{\overset{O}{\|}}{C}NH_2$	—	211 (dec)	111	x	x	x

*Most values are at or near 25° **Known carcinogen

least some of the molecular characteristics of the compound we wish to dissolve. To illustrate this point, let us consider the water solubility of the following compounds of comparable size and molecular weight:

	$CH_3CH_2CH_2CH_3$	CH_3CH_2Cl	$CH_3CH_2—O—CH_3$	$CH_3CH_2CH_2—OH$
water solubility:	virtually insoluble		soluble	miscible

Of these compounds, the alcohol is most soluble; in fact, it is *miscible* with water. This means that a solution is obtained when the alcohol is mixed with water in any proportions.

Let us try to understand these observations using what we know about water and alcohols. In order for a substance to dissolve in water, some of the hydrogen bonds between water molecules have to be broken to make room for the invading solute. Breaking hydrogen bonds requires energy. Therefore water resists forming solutions with other compounds unless this energy is liberated again in interaction with the solute. An alcohol can provide favorable hydrogen-bonding interactions with water because its —OH group can both donate and accept hydrogen bonds. Since an ether can only accept hydrogen bonds, dissolving an ether costs more energy; hence, ethers are less soluble in water than alcohols. Finally, an alkane and an alkyl chloride can neither donate nor accept hydrogen bonds. Because it costs a great deal of energy for water to dissolve these compounds, they are very insoluble. We can say that an alcohol molecule is most like water, and therefore dissolves to a greater extent; an alkane molecule is least like water, and is insoluble.

We can see the same effect in the following series:

	CH_3OH	C_2H_5OH	$CH_3CH_2CH_2OH$	$CH_3CH_2CH_2CH_2OH$	$CH_3CH_2CH_2CH_2CH_2CH_2OH$
water solubility:	miscible			7.7 wt %	0.58 wt %

Alcohols with longer hydrocarbon tails are more like alkanes—and alkanes are insoluble in water. More hydrogen bonds in water must be broken in order to accommodate the "greasy" hydrocarbon tails of these alcohols, and the energy required for this process is not recovered from interactions of the water with the dissolved molecules. Hence, alcohols (or any other organic compounds) with long hydrocarbon tails are relatively insoluble.

In contrast, hydrocarbons dissolve well in other hydrocarbons. Since the attractive interactions between molecules of a hydrocarbon are mainly van der Waals attractions, the energy cost for substituting a solvent molecule with another one of the same general type is small.

Solvents consisting of polar molecules lie in between the extremes of water on one hand and hydrocarbons on the other. For example, consider the popular solvent diethyl ether. Because ether can accept hydrogen bonds, it dissolves many alcohols. Because its dipole moment can interact attractively with the dipoles of other polar molecules, it also dissolves these types of compounds (for example, alkyl halides). Finally, because its hydrocarbon portion provides attractive van der Waals interactions, it also dissolves hydrocarbons.

The solvent tetrahydrofuran (THF) is even more versatile. It dissolves everything ether dissolves, but it also dissolves more water than does ether. As a solvent for the reaction of a water-insoluble compound with water, THF is typically an excellent choice, since it dissolves both compounds. For example, THF is the solvent of choice in oxymercuration of alkenes; it dissolves both water and alkenes (Sec. 5.3B).

Problems

11 In which of the following solvents would you expect hexane to be *least* soluble: diethyl ether, methylene chloride (CH_2Cl_2), ethanol, or 1-octanol? Explain.

12 A popular undergraduate experiment is the recrystallization of acetanilide from water. Acetanilide is moderately soluble in hot water, but much less soluble in cold water. Identify one structural feature of the acetanilide molecule that would be expected to contribute positively to its solubility in water, and one that would be expected to contribute negatively.

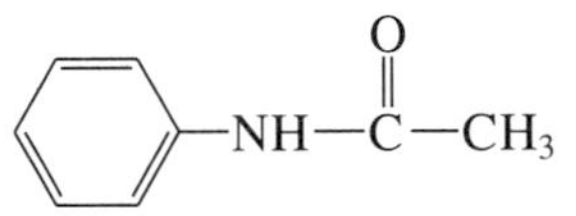

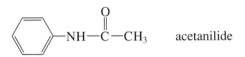

acetanilide

Solubility of Ionic Compounds Ionic compounds in solution can exist in several forms, two of which, **ion pairs** and **dissociated ions,** are shown in Fig. 8.2. In an ion pair, each ion is closely associated with an ion of opposite charge. When ions are dissociated, however, each ion moves more or less independently of every other in solution, and is surrounded by several solvent molecules, called collectively the **solvent shell** or **solvent cage** of the ion. These solvent molecules are said to **solvate** the ion. They interact with the ion in very specific ways and are actively involved in keeping the ion in solution. A solvent dissolves an ionic compound, first, by *separating* the ions of opposite charge, and second, by *solvating* the separated ions.

The ability of a solvent to *separate ions* is measured quantitatively by its dielectric constant, ϵ. We can see why this is so from the electrostatic law, Eq. 1.2. If two ions have energy of attraction E, this energy is reduced in solvents with high dielectric constants—that is, in polar solvents. Therefore ions in such solvents have less tendency to associate. Hence, all things being equal, ionic compounds have greater solubility in polar solvents.

Solvents *interact with ions to form solvation shells* in several ways. Many anions are solvated by *hydrogen bonding*. Solvation of cations takes place through what are called *donor interactions*. There are two types of donor interactions. In the first, a *covalent* interaction, atoms of solvent with unshared electron pairs (such as the oxy-

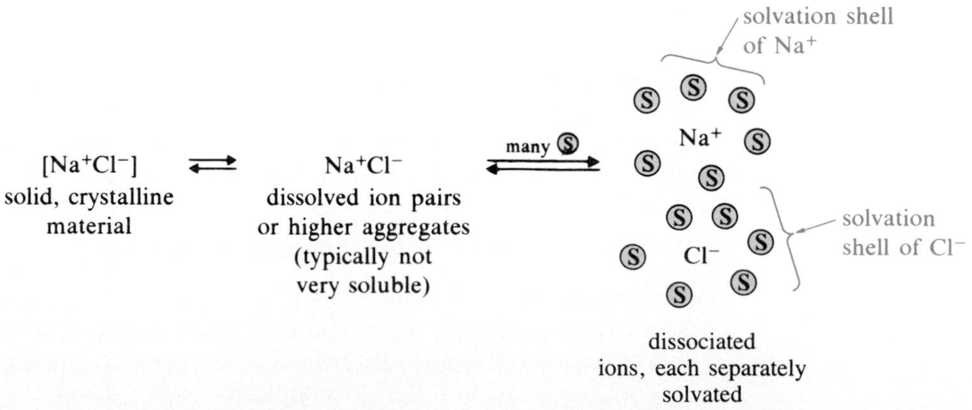

Figure 8.2 Ions in solution can exist as ion pairs and dissociated ions. The solubility of an ionic compound depends on the ability of the solvent to break the electrostatic attractions between ions and form separate solvation shells around the dissociated ions. (The colored circles are solvent molecules.)

gens of water molecules) act as Lewis bases toward an electron-deficient cation, such as the sodium ion in NaCl. In the second type of interaction, a *noncovalent* interaction, the dipole moments of the solvent molecules align themselves around a solute cation so that the negative end of the solvent dipole points toward the positive ion. This provides attractive interactions that are termed *ion–dipole* attractions.

The interactions between sodium chloride and water in aqueous solution illustrate the ion–solvent interactions we have been discussing. Dissolved sodium and chloride ions might be represented as follows:

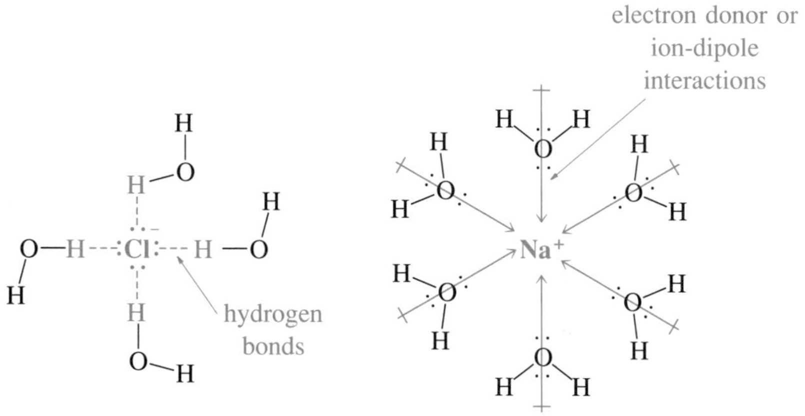

Water molecules solvate the chloride anion by hydrogen bonding, and they solvate the the sodium cation by donor interactions.

Let us summarize: three solvent properties contribute to the solubility of ionic compounds: *polarity* (dielectric constant), by which solvent molecules separate ions of opposite charge; *proticity* (hydrogen-bond donor capability), by which solvent molecules solvate anions; and *electron-donor ability,* by which solvent molecules solvate cations through Lewis-base and ion–dipole interactions. It follows that *the best solvents for dissolving ionic compounds are polar, protic, donor solvents.*

Thus, water is the ideal solvent for ionic compounds, something we perhaps know from experience. First, because it is polar—it has a very large dielectric constant—it is effective in separating ions of opposite charge. Second, because it is a donor solvent—a good Lewis base—it readily solvates cations. Finally, because it is protic—a good hydrogen-bond donor—it readily solvates anions. In contrast, hydrocarbons such as hexane do not dissolve ordinary ionic compounds because such solvents are apolar, aprotic, and nondonor solvents. Some ionic compounds, however, have appreciable solubilities in *polar aprotic* solvents such as acetone or DMSO. Although these solvents lack the proticity that solvates anions, their donor capacity solvates cations and their polarity separates ions of opposite charge. However, it is not surprising that because polar aprotic solvents lack the proticity that stabilizes anions, most salts are less soluble in these solvents than in water, and salts dissolved in polar aprotic solvents exist to a greater extent as ion pairs (Fig. 8.2).

C. Crown Ethers and Ionophorous Antibiotics

Crown Ethers Some metal ions form stable complexes with an interesting class of synthetic organic compounds known as **crown ethers,** which were first prepared in 1967. These compounds are heterocyclic ethers containing a number of regularly spaced oxygen atoms. Some examples of crown ethers are the following:

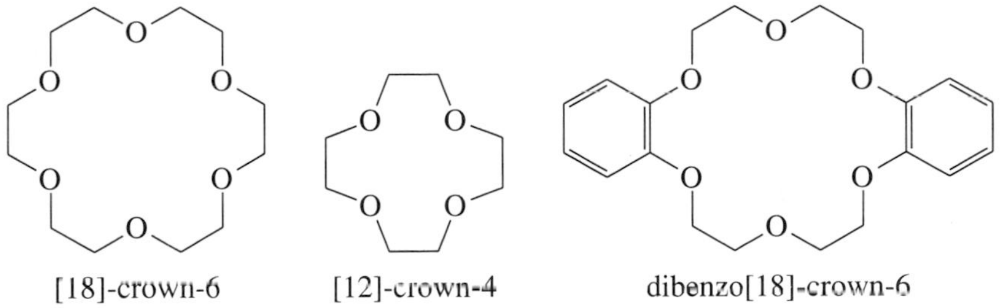

The number in brackets indicates the number of atoms in the ring containing the oxygens, and the number following the hyphen indicates the number of oxygens. The term "crown" was suggested by the three-dimensional shape of these molecules, shown in Fig. 8.3 for the rubidium ion (Rb^+) complex of dibenzo[18]-crown-6. The oxygens of the host crown ether wrap around the guest metal cation, chelating it within the cavity of the ether using the donor interactions discussed above: a Lewis-base-type electron donation and ion–dipole interactions. In fact, we can think of a crown ether molecule as a "synthetic solvation shell" for a cation. Because the metal ion must fit within the cavity, the crown ethers have some selectivity for metal ions according to size. For example, dibenzo[18]-crown-6 forms the strongest complexes with potassium ion, somewhat weaker complexes with sodium, cesium, and rubidium ions, and does not chelate lithium or ammonium ions appreciably. On the other hand, [12]-crown-4, with its smaller cavity, specifically chelates the lithium ion.

Because their structures contain hydrocarbon groups, crown ethers have significant solubilities in hydrocarbon solvents such as hexane or benzene. The remarkable thing about the crown ethers is that they can cause inorganic salts to dissolve in hydrocarbons—solvents in which these salts otherwise have no solubility whatsoever. For example, when potassium permanganate is added by itself to the hydrocarbon benzene, the $KMnO_4$ remains suspended, undissolved. Upon addition to the benzene of a little dibenzo[18]-crown-6, which complexes potassium ion, the benzene takes on the purple color of a $KMnO_4$ solution and this solution (nicknamed "purple ben-

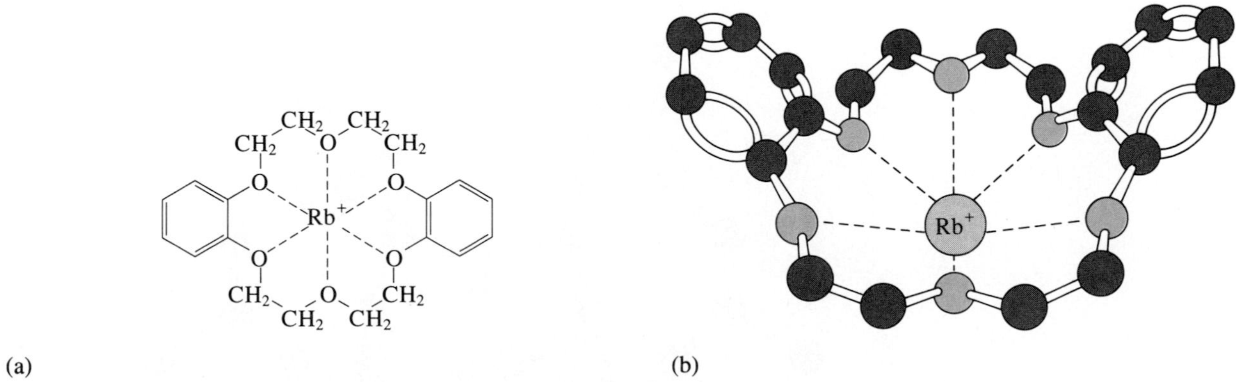

Figure 8.3 Structure of the dibenzo[18]-crown-6 complex of rubidium ion in (a) planar projection and (b) perspective drawing based on the three-dimensional structure determined by X-ray crystallography. In (b), the metal ion is shaded and the oxygens are shown in color. None of the hydrogen atoms are shown.

zene") acquires the oxidizing power typical of $KMnO_4$. What happens is that the crown ether complexes the potassium cation and dissolves it in benzene; electrical neutrality demands that the permanganate ion accompany the complexed potassium ion into solution. The stabilization of the potassium ion by the crown ether compensates for the fact that the permanganate anion is essentially unsolvated, or "naked." Other potassium salts can be dissolved in hydrocarbon solvents in a similar manner. For example, KCl and KBr can be dissolved in hydrocarbons in the presence of crown ethers to give solutions of "naked chloride" and "naked bromide," respectively.

Problem

13 Explain the fact that the complex of [18]-crown-6 with potassium ion has a much larger dissociation constant in water than it does in benzene.

Ionophorous Antibiotics Closely related to the crown ethers is a group of compounds called *ionophorous antibiotics*. An *antibiotic* is a compound found in nature (or a synthetically prepared analog) that interferes with the growth or survival of one or more microorganisms. An *ionophore* is a compound that binds ions very tightly. (The name "ionophore" means "ion-bearing"; the crown ethers are ionophores.) The ionophorous antibiotics form strong complexes with metal ions in much the same way as crown ethers. Nonactin is an example of an ionophorous antibiotic.

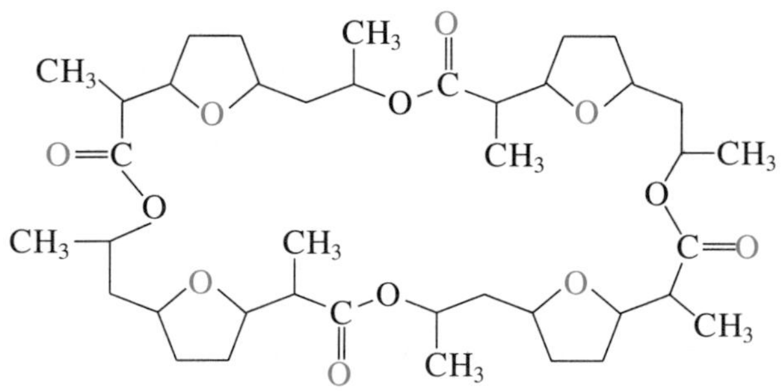

nonactin

Nonactin has a strong affinity for potassium ion; the molecule contains a cavity in which the colored oxygens in the structure above chelate the ion.

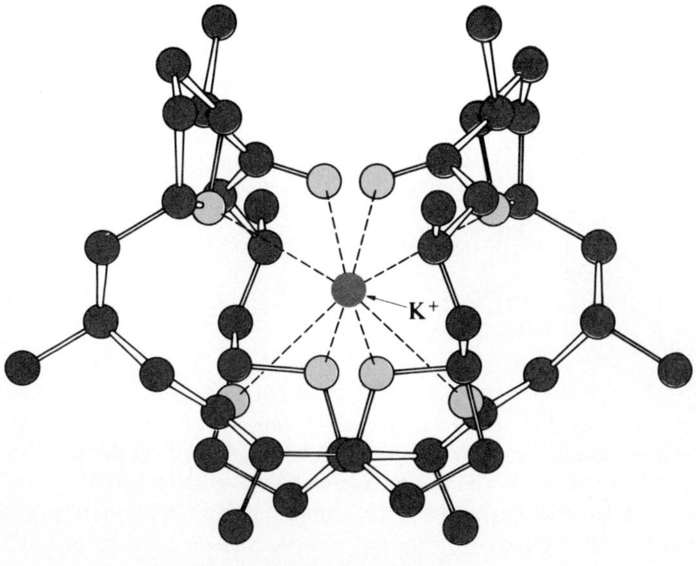

The ion-binding properties of nonactin and related compounds are the basis of their action as antibiotics.

8.5 REVIEW OF BRØNSTED ACIDITY AND BASICITY

A. Dissociation Constants and pK_a

Brønsted acid–base reactions are among the most important chemical reactions of some organic compounds. Alcohols and thiols can act as both Brønsted acids and bases; ethers and sulfides can act as Brønsted bases. In order to understand the acid–base behavior of these compounds (Sec. 8.6), we must first consider some of the quantitative aspects of Brønsted acidity and basicity. The fundamental definitions associated with Brønsted acid–base concepts are given in Sec. 1.4, and should be reviewed.

The relative strengths of acids are determined by how well they transfer a proton to a standard base. The standard base traditionally used for comparison is water. The equilibrium constant for transfer of a proton from an acid HA to water is called the **dissociation constant.** That is, for the equilibrium

$$\mathrm{HA} + \mathrm{H_2O} \rightleftharpoons \mathrm{A\!:^-} + \mathrm{H_3O^+} \qquad (8.3)$$

the dissociation constant K_a is defined as

$$K_a = \frac{[\mathrm{H_3O^+}][\mathrm{A\!:^-}]}{[\mathrm{HA}]} \qquad (8.4)$$

(The quantities in brackets are the molar concentrations at equilibrium.) Each acid has its own unique dissociation constant. The larger the dissociation constant of an acid, the more H_3O^+ ions are formed when the acid is dissolved in water at a given concentration. Thus, *the strength of an acid is measured by the magnitude of its dissociation constant.*

Because the dissociation constants of different Brønsted acids cover a range of many powers of ten, it is useful to express acid strength in a logarithmic manner. Using p as an abbreviation for negative logarithm, we can write the following definitions:

$$\mathrm{p}K_a = -\log K_a \qquad (8.5)$$

$$\mathrm{pH} = -\log\,[\mathrm{H_3O^+}] \qquad (8.6)$$

The approximate pK_a values of several Brønsted acids are given in Table 8.3 in order of increasing acidity. Since stronger acids have larger K_a values, it follows from Eq. 8.5 that stronger acids also have *smaller* pK_a values. Thus, HCN (pK_a = 9.4) is a stronger acid than water (pK_a = 15.7). In other words, acid strength in the first column of Table 8.3 increases from the top to the bottom.

Base strength is conveniently expressed for a base $\mathrm{A\!:^-}$ in terms of the pK_a of its *conjugate acid* HA. That is, when we say that a base is weak, we could just as easily say that its conjugate acid is strong. Thus, it is easy to tell which of two bases is stronger by looking at the pK_a values of their conjugate acids: *the stronger base has the conjugate*

TABLE 8.3 Relative Strengths of Some Acids and Bases

Conjugate acid	Conjugate base	pK_a
NH_3 (ammonia)	$^-:NH_2$ (amide ion)	~35
ROH (alcohol)	$RO:^-$ (alkoxide ion)	15–18*
HOH (water)	$HO:^-$ (hydroxide)	15.7
RSH (thiol)	$RS:^-$ (thiolate)	10–12*
R_3NH^+ (trialkylammonium ion)	$R_3N:$ (trialkylamine)	9–11*
NH_4^+ (ammonium ion)	$H_3N:$ (ammonia)	9.25
HCN: (hydrocyanic acid)	$^-:CN:$ (cyanide)	9.40
H_2S (hydrogen sulfide)	$HS:^-$ (hydrosulfide)	7.0
R—C(=O)—OH (carboxylic acid)	R—C(=O)—$O:^-$ (carboxylate)	4–5*
HF: (hydrofluoric acid)	$:F:^-$ (fluoride)	3.2
CH_3—C_6H_4—SO_3H (*p*-toluenesulfonic acid)	CH_3—C_6H_4—SO_3^- (*p*-toluenesulfonate, or "tosylate")	−1
H_3O^+ (hydronium ion)	H_2O (water)	−1.7
H_2SO_4 (sulfuric acid)	HSO_4^- (bisulfate)	−5
HCl: (hydrochloric acid)	$:Cl:^-$ (chloride)	−7
HBr: (hydrobromic acid)	$:Br:^-$ (bromide)	−9
HI: (hydroiodic acid)	$:I:^-$ (iodide)	−9.5
$HClO_4$ (perchloric acid)	ClO_4^- (perchlorate)	−10

* Precise value varies with the structure of R.

acid with the higher pK_a. For example, ^-CN, the conjugate base of HCN, is a weaker base than ^-OH, the conjugate base of water, because the pK_a of HCN is lower than the pK_a of water. Thus, base strength, in the second column of Table 8.3, increases from the bottom to the top.

When a Brønsted acid and base react, we can tell immediately whether the equilibrium lies to the right or left by looking at the pK_a values of the two acids involved. *The equilibrium in the reaction of an acid and a base always favors the side with the weaker acid and weaker base.* For example, in the following acid–base reaction, the equilibrium lies well to the right, because H_2O is the weaker acid and ^-CN the weaker base.

$$\underset{\substack{pK_a = 9.4 \\ \text{(stronger acid)}}}{HCN} + \underset{}{\overset{\substack{\text{(stronger} \\ \text{base)}}}{OH^-}} \rightleftharpoons \overset{\substack{\text{(weaker} \\ \text{base)}}}{^-CN} + \underset{\substack{pK_a = 15.7 \\ \text{(weaker acid)}}}{H_2O} \qquad (8.7)$$

The equilibrium constant for an acid–base reaction can be calculated in a straightforward way from the pK_a values of the acids involved. Let us illustrate by calculating the equilibrium constant for reaction 8.7. We first subtract the pK_a of the acid on the left from the pK_a of the acid on the right. This difference, $(15.7 - 9.4) = 6.3$, is the logarithm of the equilibrium constant for this reaction. The equilibrium constant itself is found by taking the antilog of this number. Hence, the equilibrium constant K_{eq} for reaction 8.7 is $10^{6.3}$, or 2×10^6, a very large number. (The basis for this procedure is explored in Problem 40.) The large value of K_{eq} means that the equilibrium in Eq. 8.7 lies *far to the right*. That is, if we dissolve HCN in a solution of NaOH, a reaction occurs to give a solution in which there is a lot of ^-CN and almost no HCN. On the other hand, if we dissolve NaCN in water, only a very small amount of it reacts with the water to give HCN and ^-OH.

Problems

14 What is the (a) strongest base and (b) the weakest base listed in the second column of Table 8.3?

15 Using the pK_a values in Table 8.3, calculate the equilibrium constant for the reaction of
(a) ammonia with HCN
(b) fluoride ion with HCN

B. Factors that Determine Acid Strength

At various points in our study of organic chemistry, we shall ask how the strengths of organic acids and bases vary with their structures. In order to lay the groundwork for these discussions, let us ask: Why are some acids strong and others weak? Although there are several factors that govern acidity, let us focus on two. These are evident when we write the dissociation of a general acid H—A as the following sequence of steps.

$$\text{H—A} \longrightarrow \text{H}\cdot + \text{A}\cdot \qquad (8.8a)$$

$$\text{A}\cdot + e^- \longrightarrow \text{A}:^- \qquad (8.8b)$$

$$\text{H}\cdot \longrightarrow \text{H}^+ + e^- \qquad (8.8c)$$

$$\textit{Sum:}\quad \text{H—A} \longrightarrow \text{H}^+ + \text{A}:^- \qquad (8.8d)$$

Even though acid dissociation does not occur this way, Hess's law allows us to think of the enthalpy or free energy for acid dissociation in terms of these steps. The overall enthalpy or free-energy change for acid dissociation is the sum of the enthalpy or free-energy changes for the individual steps. The smaller is each of these energy changes, the more favorable is acid dissociation.

The first step, Eq. 8.8a, is the homolytic dissociation of the H—A bond—that is, its dissociation to give radicals. The energy of this step is simply the bond dissociation energy (Table 5.2). This step is most favorable when the bond dissociation energy is small. We can conclude that *a weaker H—A bond contributes to stronger acidity*. Even though acid dissociation is *not* a free-radical reaction, the ease of homolytic dissociation is a factor that contributes to the strength of an acid.

The second step, Eq. 8.8b, is the ability of the radical A· to attract an electron. Because the energy of this process is related to the *electronegativity* of the group A, we can conclude that *electronegative groups in an acid contribute to enhanced acidity*.

The third step, Eq. 8.8c, is the loss of an electron from H· to give a proton. This step is the same for all acids, and thus does not contribute to the *difference* in acidity between acids.

Let us see the influence of these effects within the periodic table. As we go down a column in the periodic table, the bond-strength effect dominates. As we can see from Table 5.2, hydrogen–halogen bonds are much weaker as we proceed down the halogen group; from Table 8.3, the strength of the halogen acids is in the order HI > HBr > HCl >> HF. Hydrogen iodide is the strongest acid *even though it contains the least electronegative halogen,* because the HI bond is very weak. In other words, the greater electronegativity of fluorine, which would tend to make HF the strongest acid, is outweighed by the relative weakness of the HI bond.

On the other hand, as we go across a row in the periodic table, bond strengths change less drastically, and electronegativity differences determine the trend in acidity. Since electronegativity increases to the right in the periodic table (Table 1.1), the relative acid strength along the first full row is in the order

$$\begin{array}{lcccc} & \text{H—F} > & \text{H—OH} > & \text{H—NH}_2 > & \text{H—CH}_3 \\ \text{p}K_a & 3.2 & 15.7 & \approx 35 & >55 \end{array} \qquad (8.9)$$

We can see from this order, for example, that both ammonia and methane are very weak acids. In fact, methane is such a weak acid that its pK_a value cannot be measured accurately.

In summary: for the simple acids HA, stronger Brønsted acids are derived from the elements A nearest the *bottom* and *right side* of the periodic table.

Problem

16 Which acid in each set should be stronger? Why?
(a) CH_3O—H or CH_3S—H
(b) H—Cl or CH_3S—H
(c) $(CH_3)_2P$—H or $(CH_3)_2N$—H

8.6 ACIDITY OF ALCOHOLS AND THIOLS

If we notice the similarity in the structures of water and alcohols, it may come as no surprise that alcohols and water have about the same acidity—they are both weak acids.

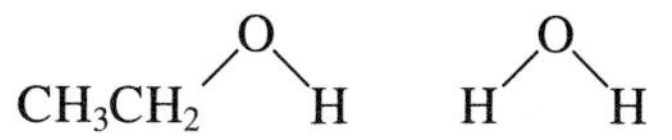

pK_a 15.9 15.7

The conjugate bases of alcohols are generally called *alkoxides*. The common name of an alkoxide is constructed by deleting the final *yl* from the name of the alkyl group and adding the suffix *oxide*. To name an alkoxide systematically, the suffix *ate* is simply added to the name of the alcohol.

$CH_3CH_2\ddot{\underset{\cdot\cdot}{O}}:^-\ Na^+$ *common:* sodium ethoxide
systematic: sodium ethanolate

Just as H_2S is more acidic than water, thiols are more acidic than alcohols; typically, thiols have pK_a values near 10. For example, the pK_a of ethanethiol, CH_3CH_2SH, is 10.5. That the acidity of thiols is greater than the acidity of alcohols reflects the fact that the S—H bond is weaker than the O—H bond. (Recall that acidity increases as we go down the periodic table.)

The conjugate bases of thiols are called *thiolates* or *mercaptides.*

$$CH_3\ddot{S}:^- Na^+ \qquad \textit{common:}\ \text{sodium methyl mercaptide} \quad \textit{systematic:}\ \text{sodium methanethiolate}$$

Problems

17 Give the structure of
(a) potassium *t*-butoxide
(b) lithium methanolate
(c) sodium isopropoxide
(d) sodium 2,2-dimethyl-1-butanolate

18 Name the following compounds
(a) $Ca(OCH_3)_2$ (b) $Cu—S—C_2H_5$

A. Formation of Alkoxides and Mercaptides

Since the acidity of a typical alcohol is about the same as that of water, an alcohol *cannot* be converted completely into its alkoxide conjugate base in an aqueous NaOH solution.

$$\underset{pK_a\,=\,15.9}{C_2H_5—\ddot{O}H} + {}^-:\ddot{O}H \rightleftharpoons C_2H_5\ddot{O}:^- + \underset{pK_a\,=\,15.7}{H_2\ddot{O}:} \qquad (8.10)$$

We can see why this is true from the relative pK_a values: since these are nearly the same for ethanol and water, both sides of the equation contribute significantly at equilibrium. In other words, *hydroxide is not a sufficiently strong base to convert an alcohol completely into its conjugate base alkoxide.*

We *can* form alkoxides from alcohols if we use stronger bases. One convenient base is sodium hydride, NaH, which is a source of the *hydride ion,* $H:^-$. Because hydride ion is a very strong base—the pK_a of its conjugate acid, H_2, is estimated to be 42—its reaction with alcohols goes essentially to completion. In addition, when NaH reacts with an alcohol, the reaction cannot be reversed because the by-product, hydrogen gas, simply bubbles out of solution.

$$Na^+\ H:^- + H—\ddot{O}—\underset{\displaystyle CH_3}{\underset{|}{C}}HCH_2CH_3 \longrightarrow \underset{\text{quantitative yield}}{Na^+\ {}^-:\ddot{O}—\underset{\displaystyle CH_3}{\underset{|}{C}}HCH_2CH_3} + H_2\uparrow \qquad (8.11)$$

Potassium hydride and sodium hydride are supplied as dispersions in mineral oil to protect them from reaction with moisture. When these compounds are used to convert an alcohol into an alkoxide, the mineral oil is rinsed away with pentane, a solvent such as ether or THF is added, and the alcohol is introduced cautiously with stirring. Hydrogen is evolved vigorously and a solution or suspension of the pure sodium or potassium alkoxide is formed.

Solution of alkoxides in their conjugate-acid alcohols are often useful in organic chemistry. To prepare such solutions, we can take a cue from a familiar reaction of water. Many of us have observed the violent reaction of sodium metal with water to give an aqueous sodium hydroxide solution. The analogous reaction occurs with many alcohols. Thus, sodium reacts with an alcohol to afford a solution of the sodium alkoxide.

$$2\text{H}{-}\ddot{\underset{\cdot\cdot}{\text{O}}}\text{H} + 2\text{Na} \longrightarrow 2\text{Na}^+ \;{}^-\!:\!\ddot{\underset{\cdot\cdot}{\text{O}}}\text{H} + \text{H}_2\uparrow \qquad (8.12)$$

$$2\text{R}{-}\ddot{\underset{\cdot\cdot}{\text{O}}}\text{H} + 2\text{Na} \longrightarrow \underset{\text{sodium alkoxide}}{2\text{Na}^+ \;{}^-\!:\!\ddot{\underset{\cdot\cdot}{\text{O}}}\text{R}} + \text{H}_2\uparrow \qquad (8.13)$$

The rate of this reaction depends strongly on the alcohol. The reactions of sodium with anhydrous (water-free) ethanol and methanol are vigorous, but not violent. However, the reactions of sodium with some alcohols, such as *t*-butyl alcohol, are rather slow. The alkoxides of such alcohols can be formed more rapidly with the more reactive potassium metal.

Because thiols are much more acidic than water or alcohols, they can be converted completely into their conjugate-base mercaptide anions by reaction with one equivalent of hydroxide or alkoxide. In fact, a common method of forming alkali-metal mercaptides is to dissolve them in ethanol containing one equivalent of sodium ethoxide:

$$\underset{\text{p}K_a\,=\,10.5}{\text{C}_2\text{H}_5\ddot{\underset{\cdot\cdot}{\text{S}}}\text{H}} + \text{C}_2\text{H}_5\ddot{\underset{\cdot\cdot}{\text{O}}}\!:^- \rightleftharpoons \text{C}_2\text{H}_5\ddot{\underset{\cdot\cdot}{\text{S}}}\!:^- + \underset{\text{p}K_a\,=\,15.9}{\text{C}_2\text{H}_5\ddot{\underset{\cdot\cdot}{\text{O}}}\text{H}} \qquad (8.14)$$

Because the equilibrium constant for this reaction is $>10^5$ (how do we know this?), the reaction goes essentially to completion.

Although alkali-metal mercaptides are soluble in water and alcohols, thiols form insoluble mercaptides with many heavy-metal ions, such as Hg^{2+}, Cu^{2+}, and Pb^{2+}

$$2\,n\text{-C}_{10}\text{H}_{21}{-}\text{SH} + \text{PbCl}_2 \xrightarrow{\text{C}_2\text{H}_5\text{OH}} \underset{(87\%\text{ yield})}{(n\text{-C}_{10}\text{H}_{21}\text{S})_2\text{Pb}} + 2\text{HCl} \qquad (8.15)$$

$$2\text{PhSH} + \text{HgCl}_2 \longrightarrow \underset{(98\%\text{ yield})}{(\text{PhS})_2\text{Hg}} + 2\text{HCl} \qquad (8.16)$$

The insolubility of heavy-metal mercaptides is analogous to the insolubility of heavy-metal sulfides (for example, PbS), which are among the most insoluble inorganic compounds known. One reason for the toxicity of heavy-metal salts is that they form tight mercaptide complexes with important biomolecules that normally contain free thiol groups.

B. Inductive Effects on Alcohol Acidity

Although many simple alcohols have pK_a values in the 15–16 range, *the exact acidity of an alcohol depends on its structure*. For example, alcohols containing electronegative substituent groups have enhanced acidity. Thus, 2,2,2-trifluoro-1-ethanol is more than three pK_a units more acidic than ethanol itself.

	$CF_3—CH_2—OH$	$CH_3—CH_2—OH$
pK_a	12.4	15.9

What is the reason for this effect? To begin with, let us consider the free energy of the ionization process. Since the dissociation constant for ionization is an equilibrium constant, it is directly related to the standard free energies of the species involved in the ionization equilibrium by the familiar equation

$$\Delta G^0 = -2.3RT \log K_a \tag{8.17a}$$

(See Sec. 4.4A.) Since $-\log K_a = pK_a$, then

$$\Delta G^0 = 2.3RT\, pK_a \tag{8.17b}$$

That is, for the ionization of an alcohol with dissociation constant K_a, the pK_a is directly proportional to ΔG^0, the standard free-energy difference between reactants and products. This is shown diagramatically in Fig. 8.4. Thus, when an alkoxide has a relatively high energy, its alcohol is a relatively weak acid (large pK_a). In contrast, when an alkoxide has a relatively low energy, its alcohol is a relatively strong acid (small pK_a). In other words, *the acidity of an alcohol is increased by stabilizing its conjugate-base alkoxide.*

Electronegative substituent groups increase the acidities of alcohols by stabilizing their corresponding alkoxides. This stabilization results from an electrostatic interaction of the negatively charged oxygen with nearby C—F bond dipoles.

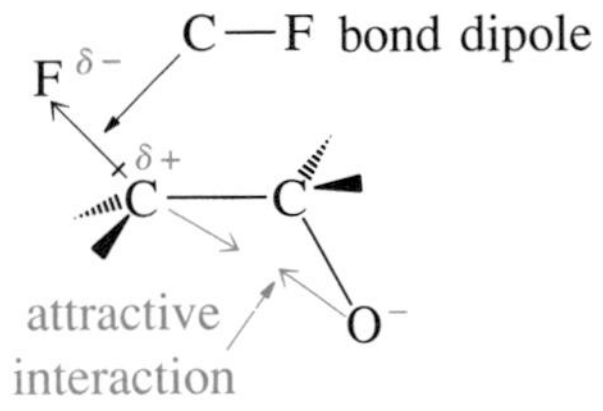

Since the negative charge of the alkoxide is closer to the positive end of the bond dipole (carbon) than it is to the negative end (fluorine), there is a net attraction, or

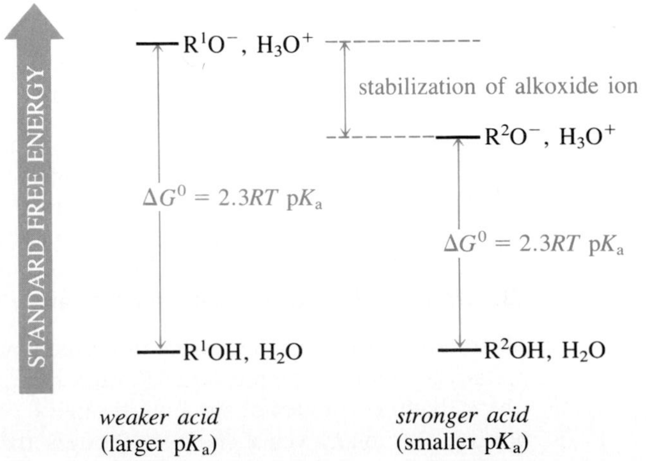

Figure 8.4 Alcohols are more acidic when their alkoxide conjugate bases are stabilized.

stabilization. Recall that the attraction between positive and negative charges lowers the energy of the system (the electrostatic law, Eq. 1.2).

This type of interaction is called an **inductive effect** or **polar effect.** The inductive effect stabilizes the conjugate base of 2,2,2-trifluoro-1-ethanol, but is not present in the conjugate base of ethanol itself. As a result, less energy is required to form the conjugate base of 2,2,2-trifluoro-1-ethanol, which is thus the more acidic compound. Because of the inductive effect it is quite general that *alcohols with electronegative substituents are more acidic than unsubstituted alcohols.*

> We can also think of the inductive effect in terms of Eq. 8.8b. Electronegative substituents increase the ability of oxygen to accept negative charge. This makes Eq. 8.8b more favorable and consequently, increases acidity.

As we have seen, the inductive effect is simply a manifestation of the electrostatic law, Eq. 1.2. One consequence of this law is that the attractive interaction between two charges diminishes with increasing distance between them. Thus, the inductive effect on alcohol acidity diminishes as the distance between the —OH and the electronegative substituent group is increased, as the following series shows:

$$CF_3—CH_2—OH < CF_3—CH_2—CH_2—OH < CF_3—CH_2—CH_2—CH_2—OH \quad (8.18)$$
$$\text{p}K_a \qquad 12.4 \qquad\qquad 14.6 \qquad\qquad\qquad 15.4$$

The fluorines have a negligible effect on acidity when they are separated from the —OH group by four or more carbons.

Problems

19 In each of the following sets, arrange the compounds in order of increasing acidity (decreasing pK_a). Explain your choices.
(a) $ClCH_2CH_2OH$, Cl_2CHCH_2OH, $Cl(CH_2)_3OH$
(b) HCF_2CH_2OH, $CF_3CF_2CH_2OH$, $HCF_2CH_2CH_2OH$
(c) $ClCH_2CH_2SH$, $ClCH_2CH_2OH$, CH_3CH_2OH
(d) $CH_3CH_2CH_2CH_2OH$, $CH_3OCH_2CH_2OH$

20 Calculate the amount of $CF_3CH_2—\overset{..}{\underset{..}{O}}:^-$ in a solution, initially $0.1M$ in CF_3CH_2OH, that has a pH of 11.0. (See Eq. 8.18.)

21 Calculate the standard free energy of ionization at 25 °C of (a) 2,2,2-trifluoro-1-ethanol; and (b) 4,4,4-trifluoro-1-butanol. (See Eq. 8.18.) Explain why it takes more energy to ionize one alcohol than another.

C. Effect of Branching on Alcohol Acidity. Role of the Solvent in Acidity

There are significant differences in the acidities of methyl, primary, secondary, and tertiary alcohols; some relevant pK_a values are given in Table 8.4. The data in this table show that the acidities of alcohols are in the order methyl > primary > secondary > tertiary. For many years chemists thought that this order was due to some sort of

TABLE 8.4 Acidities of Alcohols in Aqueous Solution

Alcohol	pK_a	Alcohol	pK_a
CH_3OH	15.1	$(CH_3)_2CHOH$	17.1
CH_3CH_2OH	15.9	$(CH_3)_3COH$	19.2

inductive effect (Sec. 8.6B) of the alkyl groups around the alcohol oxygen. However, in relatively recent times, chemists were fascinated to learn that in the *gas phase*—in the absence of solvent—the order of acidity of alcohols is exactly reversed.

relative gas-phase acidity:

$$(CH_3)_3COH > (CH_3)_2CHOH > CH_3CH_2OH > CH_3OH \qquad (8.19)$$

Notice carefully what is being stated here. It is *not* true that alcohols are more acidic in the gas phase than they are in solution; all alcohols are more acidic in solution than they are in the gas phase. What *is* true is that the *relative order* of acidity of different types of alcohols is reversed in solution compared to the *relative order* of acidity in the gas phase.

As we learned in Sec. 8.6B, the relative acidities of alcohols are determined largely by the stabilities of their conjugate base alkoxides. *In the gas phase,* the negatively charged oxygen of an alkoxide anion interacts with nearby alkyl groups in the same molecule by a polarization mechanism. That is, the electron clouds of the alkyl groups distort so that electron density moves away from the negative charge on the alkoxide oxygen. The resulting positive charge interacts attractively with the negative oxygen and stabilizes the anion.

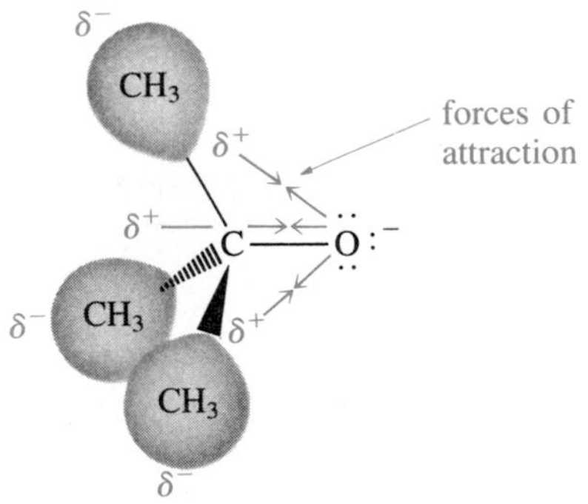

Since there are more alkyl groups in a tertiary alcohol, a tertiary alkoxide is stabilized by this effect more than a primary alkoxide. As a result, tertiary alcohols are more acidic.

In solution the same polarization effect is present, but another effect is much more important in stabilizing the alkoxide: hydrogen bonding of the alkoxide oxygen with the surrounding solvent molecules. Since hydrogen bonding is relatively unimportant in the gas phase, this interaction is a very important difference between gas phase and solution. Anything that interferes with hydrogen bonding makes the alkoxide less stable. It is thought that the alkyl groups of a tertiary alkoxide interfere with this hydrogen bonding by a *steric effect*—that is, the alkyl groups simply get in the way

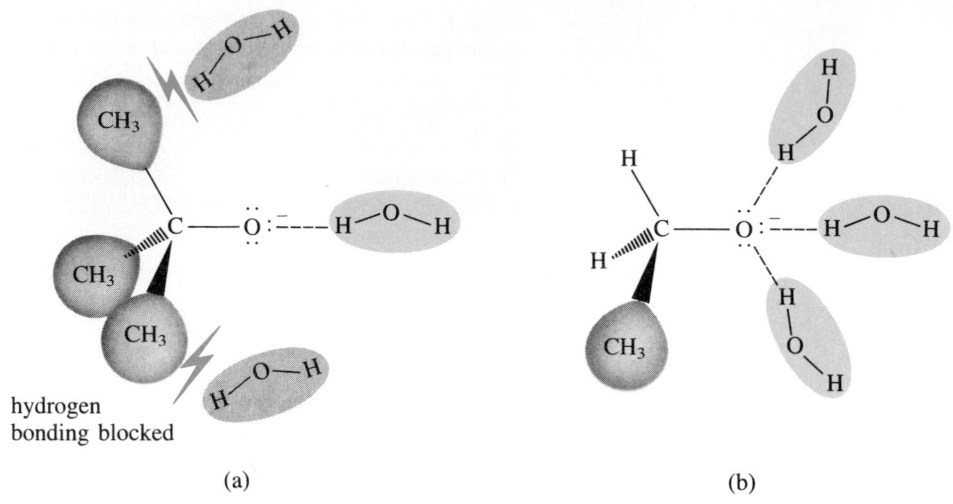

Figure 8.5 (a) The formation of hydrogen bonds to a tertiary alkoxide is sterically blocked, (b) but that to a primary alkoxide is not.

of hydrogen-bonding solvent molecules, as shown in Fig. 8.5a. Primary alkoxides do not have as many alkyl groups around the alkoxide oxygen, and hydrogen bonding from solvent to the anion can occur without difficulty (Fig. 8.5b). The greater stabilization of primary alkoxides by hydrogen bonding means that *alcohols with unbranched carbon chains near the —OH group are stronger acids in solution*. An equivalent conclusion is that *highly branched alkoxides are more basic than unbranched ones in solution*.

This discussion shows that the solvent is not an idle bystander in the acid–base reaction. It takes an active role in stabilizing the molecules involved. The nature of the solvent can dramatically affect the strength of a dissolved acid or base (Problem 22). We shall learn subsequently that solvents also play active roles in other chemical reactions.

Problem

22 In which of the following solvents is Na^+ ^-OH the strongest base? Explain.
(a) water (b) C_2H_5OH (c) 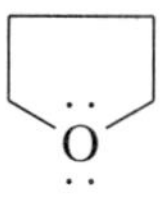

8.7 BASICITY OF ALCOHOLS AND ETHERS

Just as water can accept a proton to form the hydronium ion, alcohols, ethers, thiols, and sulfides can also be protonated to form positively charged conjugate acids. Alcohols and ethers do not differ greatly from water in their basicities.

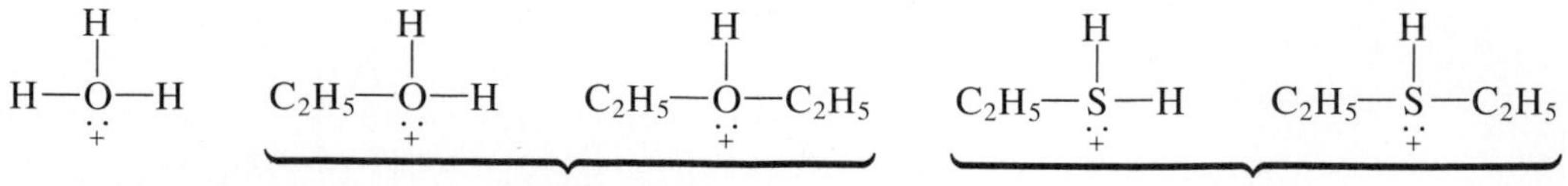

pK_a	−1.74	−2 to −3	−7

The negative pK_a values mean that these protonated species are very strong acids, and that their neutral conjugate bases are rather weak. Nevertheless, the ability of alcohols, ethers, and their sulfur analogs to accept a proton plays a very important role in many of their reactions, as we shall see.

Be sure to understand that the pK_a values above refer to the conjugate acid of the *neutral* base. Like water, alcohols can both gain and lose a proton. Thus, there are two acid–base equilibria associated with water or alcohols:

loss of a proton:

$$C_2H_5\!-\!\overset{..}{\underset{..}{O}}\!-\!H \; + \; {}^-\!:\!\overset{..}{\underset{..}{O}}H \; \rightleftharpoons \; C_2H_5\!-\!\overset{..}{\underset{..}{O}}:^- \; + \; H\!-\!\overset{..}{\underset{..}{O}}H \qquad (8.20a)$$

$$pK_a = 15.9$$

gain of a proton:

$$C_2H_5\!-\!\overset{..}{\underset{..}{O}}\!-\!H \; + \; H_3\overset{..}{O}{}^+ \; \rightleftharpoons \; C_2H_5\!-\!\overset{\overset{\textstyle H}{\textstyle |}}{\underset{\overset{..}{+}}{O}}\!-\!H \; + \; H_2\overset{..}{O}: \qquad (8.20b)$$

$$pK_a = -2 \text{ to } -3$$

When we speak about the acidity of alcohols—the loss of a proton—we are speaking about the reaction in Eq. 8.20a. Because alcohols are weak acids, this reaction is usually significant only in the presence of good bases. When we discuss the basicity of alcohols—the gain of a proton—we are speaking about the reaction in Eq. 8.20b. Because alcohols are weak bases, this reaction is usually significant only in the presence of strong acids.

Ethers are also important Lewis bases. For example, the Lewis acid–Lewis base complex of boron trifluoride and diethyl ether is stable enough that it can be distilled (bp 126°). This complex provides a convenient way to handle BF_3.

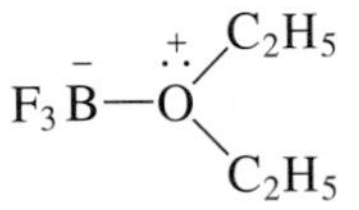

$$F_3\overset{-}{B}\!-\!\overset{..}{\underset{}{\overset{+}{O}}}\!\!\begin{array}{l}\diagup C_2H_5 \\ \diagdown C_2H_5\end{array} \qquad \text{boron trifluoride etherate}$$

Another example of the Lewis basicity of ethers is the complexation of BH_3 by tetrahydrofuran (THF), the solvent used in hydroboration (Sec. 5.3A).

Water and alcohols are also excellent Lewis bases, but in a practical sense they often cannot be used for forming complexes with Lewis acids. The reason is that the protons on water and alcohols can react further and, as a result, the complex is destroyed.

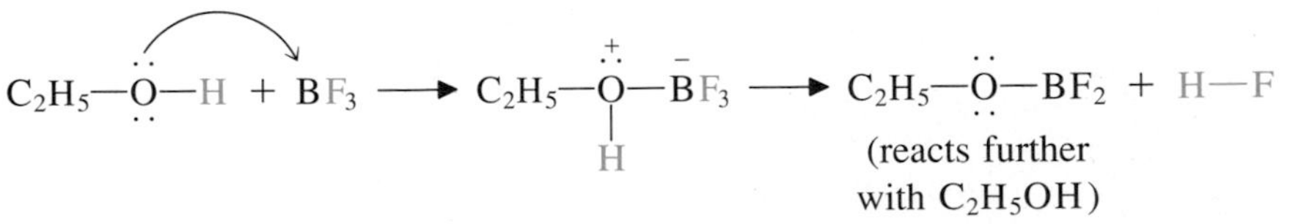

$$C_2H_5\!-\!\overset{..}{\underset{..}{O}}\!-\!H \; + \; B\,F_3 \; \longrightarrow \; C_2H_5\!-\!\overset{\overset{\textstyle +}{}}{\underset{\underset{\textstyle H}{}}{O}}\!-\!\overset{-}{B}F_3 \; \longrightarrow \; C_2H_5\!-\!\overset{..}{\underset{..}{O}}\!-\!BF_2 \; + \; H\!-\!F \qquad (8.21)$$

$$\text{(reacts further} \\ \text{with } C_2H_5OH)$$

Because ethers lack these protons, their complexes with Lewis acids do not react further.

8.8 ORGANOMETALLIC COMPOUNDS. GRIGNARD AND ORGANOLITHIUM REAGENTS

A. Formation of Grignard and Organolithium Reagents

When an alkyl halide is added to a rapidly stirred suspension of magnesium turnings in an ether solvent, such as diethyl ether or tetrahydrofuran (THF; Sec. 8.1C), an alkylmagnesium halide is formed. Alkylmagnesium halides are known as **Grignard reagents.**

$$CH_3CH_2—Br + Mg \xrightarrow{(C_2H_5)_2O} CH_3CH_2—Mg—Br \qquad (8.22)$$

ethylmagnesium bromide
(a Grignard reagent)

$$\bigcirc—Cl + Mg \xrightarrow{THF} \bigcirc—Mg—Cl \qquad (8.23)$$

cyclohexylmagnesium
chloride
(a Grignard reagent)

A Grignard reagent contains a carbon–magnesium bond. Compounds that contain carbon–metal bonds are called generally **organometallic compounds.** Grignard reagents are among the most important organometallic compounds.

Well known for many years, Grignard reagents are among the most versatile reagents in organic chemistry, and are extremely important, as we shall see in subsequent chapters. The utility of these reagents was originally investigated by Professor Francois Phillipe Antoine Barbier (1848–1922), of the University of Lyon in France. However, it was Barbier's successor at Lyon, Victor Grignard (1871–1935), who developed many applications of organomagnesium halides during the early part of the twentieth century. For this work, Grignard received the Nobel Prize in 1912.

The same kind of reaction can be used to prepare another type of organometallic compound, an **organolithium reagent.**

$$CH_3CH_2CH_2CH_2—Cl + 2Li \xrightarrow{hexane} CH_3CH_2CH_2CH_2—Li + LiCl \qquad (8.24)$$

n-butyllithium
(an organolithium reagent)

The reactions that form Grignard and organolithium reagents occur on the surface of the insoluble metals; thus, they are examples of *heterogeneous reactions* (see Sec. 5.7A). Formation of Grignard reagents has been shown to involve radical intermediates.

$$R—\overset{..}{\underset{..}{Br}} : \quad \cdot Mg \longrightarrow R \quad {}^{\frown\frown} Mg\overset{..}{\underset{..}{Br}} : \longrightarrow R—Mg\overset{..}{\underset{..}{Br}} : \quad \text{(on magnesium surface)} \qquad (8.25)$$

Formation of organolithium reagents may occur in a similar manner.

The solubility of Grignard reagents in ether solvents plays a crucial role in the formation of the reagents. As they form, Grignard reagents are dissolved from the metal surface by the ether. If this solubilization did not occur, the metal would gradually be covered by a coating of the Grignard reagent, and formation of the reagent would cease. Grignard reagents are soluble in ether solvents because the magnesium is coordinated to the ether oxygen in a *Lewis acid–base interaction.*

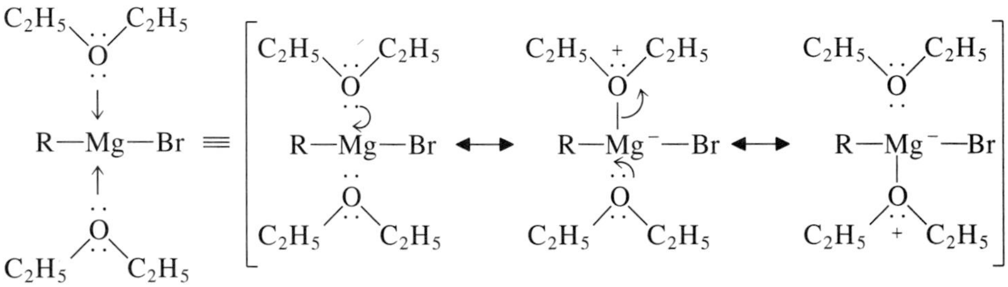
(8.26)

The magnesium of the Grignard reagent is two electron pairs short of an octet, and the ether can donate these electron pairs to the metal. (This interaction is very similar to the donor interactions that stabilize cations in solution; Sec. 8.4B.) In fact, ether solvents are generally required for the formation of Grignard reagents because of the importance of this coordination to the solubility of the reagent. Many organolithium reagents are sufficiently soluble in hexane or other hydrocarbons that ether need not be used.

Grignard and organolithium reagents react violently with oxygen and (as we shall see subsequently) with water. For this reason these reagents must be prepared under rigorous oxygen- and moisture-free conditions. In the case of Grignard reagents, exclusion of oxygen is easily assured by the low boiling points of the ether solvents that are normally used. As the Grignard reagent begins to form, heat is liberated and the ether boils. Since the reaction flask is filled with ether vapor, oxygen is virtually excluded.

Problem

23 Although Grignard reagents are normally insoluble in hydrocarbons, they can be dissolved in hydrocarbon solvents if a *tertiary amine,* $R_3N:$, is added. Explain.

B. Protonolysis of Grignard and Organolithium Reagents

Much of the reactivity of Grignard and organolithium reagents can be understood in terms of the polarity of the carbon–metal bond. Because the metal is the more electropositive partner in this bond, the carbon atom has a partial negative charge. In virtually all reactions of the Grignard reagent, the carbon reacts as if it were a carbon anion, or **carbanion.**

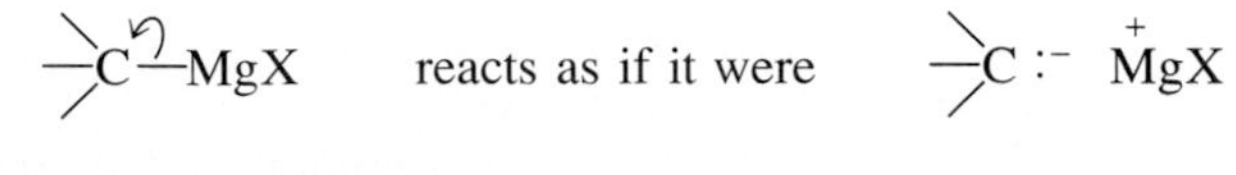

Organolithium and Grignard reagents are not *true* carbanions because they have covalent carbon–metal bonds. However, their reactions are so much like those we would expect of carbanions, that we can think of Grignard and organolithium reagents conceptually as carbanions.

We would expect simple carbanions to be very strong bases. For example, the *methyl anion,* $^{-}\!:CH_3$, is a strong base because the pK_a of methane is about 55. Hence, it is reasonable to expect the carbon in the carbon–metal bond of a Grignard or organolithium reagent to behave as a strong Brønsted base. One of the simplest reactions of a Brønsted base is its reaction with acids. In fact, any Grignard or organolithium reagent reacts vigorously with even relatively weak acids such as water and alcohols to give the conjugate-base hydroxide or alkoxides and the conjugate acid of the carbanion, a hydrocarbon.

$$CH_3CH_2{-}MgBr + H{-}OH \longrightarrow CH_3CH_2{-}H + HO{-}MgBr \qquad (8.27)$$

$$(CH_3)_3C{-}Li + H_2O \longrightarrow (CH_3)_3C{-}H + LiOH \qquad (8.28)$$

$$CH_3CH_2CH_2{-}MgBr + CH_3OH \longrightarrow CH_3CH_2CH_3 + CH_3O{-}MgBr \qquad (8.29)$$

Each of these reactions can be viewed as the reaction of the carbanion base with the proton of water or alcohol.

$$CH_3\ddot{C}H_2^{-}\ \ \overset{+}{M}gX\ \ H{-}\ddot{O}R \longrightarrow \underset{\text{conjugate acid of } CH_3CH_2:^-}{CH_3CH_2{-}H} + \underset{\text{conjugate base of RO—H}}{R\ddot{O}:^-\ {}^+MgX} \qquad (8.30)$$

This reaction is termed **protonolysis.** A protonolysis is a reaction with the proton of an acid that breaks chemical bonds. In this case, the carbon–metal bond of the Grignard reagent is broken. The protonolysis reaction can be an annoyance, since, because of it, Grignard and organolithium reagents must be prepared in the absence of moisture. However, the protonolysis reaction also has its uses. It provides a *method for the preparation of hydrocarbons from alkyl halides.* Notice, for example, in Eq. 8.27 that ethane (a hydrocarbon) is produced from ethylmagnesium bromide, which, in turn, comes from ethyl bromide (an alkyl halide). A particularly useful variation of this reaction is the preparation of hydrocarbons labeled with the hydrogen isotopes deuterium (D, or ^2H) or tritium (T, or ^3H) by reaction of a Grignard reagent with the corresponding isotopically labeled water.

$$(CH_3)_3CCH_2{-}Br \xrightarrow[\text{ether}]{Mg} (CH_3)_3CCH_2{-}MgBr \xrightarrow{D_2O} (CH_3)_3CCH_2{-}D \qquad (8.31)$$

Problems

24 Give the products of the following reactions. Draw the arrow formalism for reaction (b).
(a) isobutylmagnesium chloride + $H_2O \longrightarrow$
(b) methyllithium + $CH_3OH \longrightarrow$

25 Give the structures of two alkylmagnesium bromides that would react with water to give propane. What compounds would be formed from the protonolysis of each Grignard reagent in D_2O?

8.9 INDUSTRIAL PREPARATION AND USE OF ALKYL HALIDES, ALCOHOLS, AND ETHERS

A. Industrial Preparation of Alkyl Halides. Free-Radical Halogenation

Among the methods used in industry to produce simple alkyl halides is direct halogenation of alkanes. When an alkane such as methane is treated with Cl_2 or Br_2 in the presence of heat or light, a mixture of alkyl halides is formed.

$$CH_4 + Cl_2 \xrightarrow[\text{light}]{\text{heat or}} CH_3Cl + CH_2Cl_2 + CHCl_3 + CCl_4 + HCl \qquad (8.32)$$

The relative amounts of the various products can be controlled by varying the reaction conditions. This reaction is used particularly for the preparation of extensively halogenated compounds such as carbon tetrachloride.

The conditions of this reaction (initiation by heat or light) suggest the involvement of free-radical intermediates (Sec. 5.8C). The reaction of alkanes with halogens is a type of free-radical reaction called a **free-radical substitution** reaction. (This contrasts with the peroxide-mediated addition of HBr to alkenes, which is a free-radical *addition* reaction.) It is a substitution because a halogen atom is substituted for an alkane hydrogen. This becomes clear if we focus on the simplest product of Eq. 8.32, methyl chloride:

$$\text{H—}\underset{\text{H}}{\overset{\text{H}}{\text{C}}}\text{—H} + Cl_2 \xrightarrow{\text{heat or light}} \text{H—}\underset{\text{H}}{\overset{\text{H}}{\text{C}}}\text{—Cl} + HCl \qquad (8.33)$$

The mechanism of this reaction follows the typical pattern of other *free-radical chain reactions*; it has initiation, propagation, and termination steps. The reaction is initiated when a halogen molecule absorbs energy from the heat or light and dissociates homolytically into halogen atoms:

$$:\ddot{\underset{..}{Cl}}\text{—}\ddot{\underset{..}{Cl}}: \xrightarrow{\text{light}} :\ddot{\underset{..}{Cl}}\cdot + \cdot\ddot{\underset{..}{Cl}}: \qquad (8.34a)$$

The ensuing chain reaction has the following propagation steps:

$$:\ddot{\underset{..}{Cl}}\cdot + \text{H—CH}_3 \longrightarrow :\ddot{\underset{..}{Cl}}\text{—H} + \cdot CH_3 \qquad (8.34b)$$

methyl radical

$$:\ddot{\underset{..}{Cl}}\text{—}\ddot{\underset{..}{Cl}}: \; \cdot CH_3 \longrightarrow :\ddot{\underset{..}{Cl}}\text{—CH}_3 + :\ddot{\underset{..}{Cl}}\cdot \qquad (8.34c)$$

Termination steps result from the recombination of radical species (Problem 27).

It is found experimentally that free-radical halogenations with chlorine and bromine proceed smoothly, halogenation with fluorine is violent, and halogenation with iodine does not occur. The reason for these observations follows from the ΔH^0 of the halogenation reaction. Halogenation with fluorine is so strongly exothermic that the reaction easily gets out of control: fluorination produces so much heat that the temperature of the reaction mixture rises faster than the heat can be dissipated. Iodination, on the other hand, is endothermic; the reaction is so unfavorable energetically that it does not take place to a useful extent. Chlorination and bromination are mildly exothermic; these reactions proceed to completion, but they are not so exothermic that they get out of control.

Problems

26 Give the mechanism for the formation of methylene chloride from methane and chlorine in the presence of heat or light.

27 Explain why ethane is formed as a minor by-product in the free-radical chlorination of methane.

B. Uses of Halogen-Containing Compounds

There are very few naturally occurring halogen-containing organic compounds. Those that do occur originate in marine organisms that inhabit salt water, in which the concentration of halide ions is relatively high.

Alkyl halides and other halogen-containing organic compounds have many practical uses. Methylene chloride, chloroform, and carbon tetrachloride are important solvents (Table 8.2) that do not pose the flammability hazard of ethers. In fact, CCl_4 was used to extinguish fires before its liver toxicity was fully appreciated. Tetrachloroethylene, trichlorofluoroethane, and trichloroethylene are used industrially as dry-cleaning solvents.

Some insecticides are organohalogen compounds:

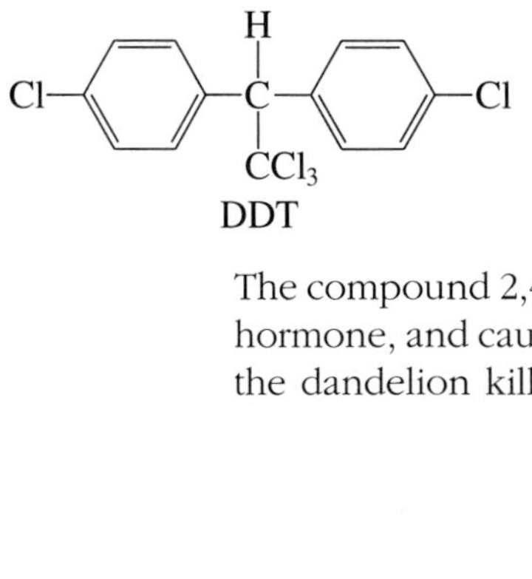

DDT

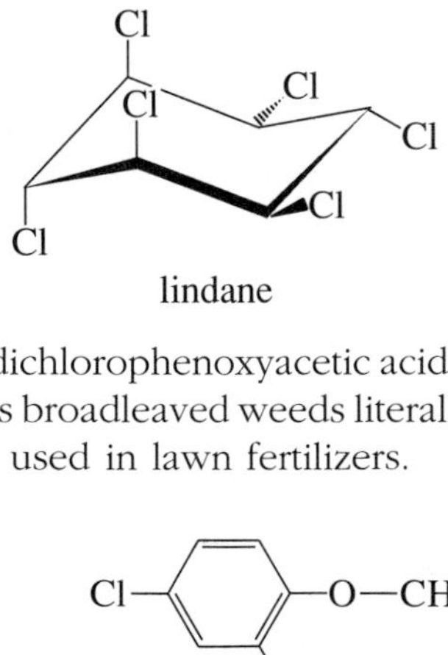

lindane

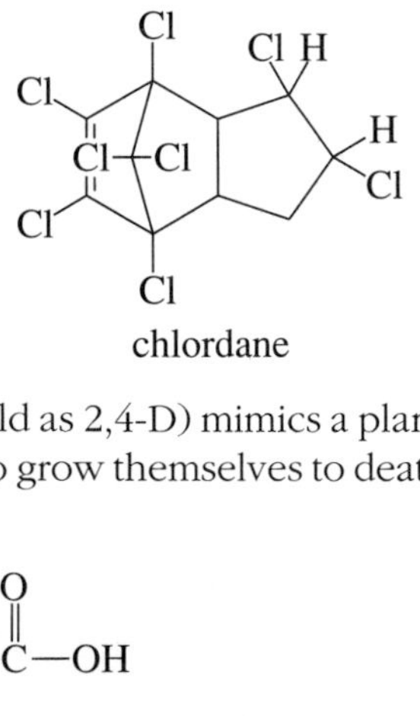

chlordane

The compound 2,4-dichlorophenoxyacetic acid (sold as 2,4-D) mimics a plant growth hormone, and causes broadleaved weeds literally to grow themselves to death. This is the dandelion killer used in lawn fertilizers.

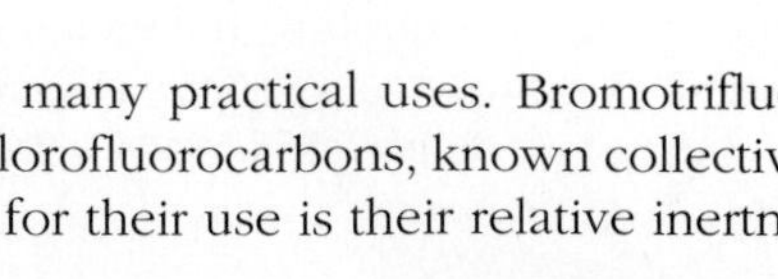

2,4-D

We have previously seen that a number of halogen-containing alkenes serve as monomers for the synthesis of useful polymers, for example, PVC, Teflon, and Kel-F (Table 5.3).

Fluorocarbons have many practical uses. Bromotrifluoromethane is used as a flame retardant. Many chlorofluorocarbons, known collectively as Freons, find use as refrigerants. One reason for their use is their relative inertness and safety. The compounds used as refrigerants include CCl_2F_2 (Freon 12), $HCClF_2$ (Freon 22), and $HCCl_2F$ (Freon 21). These materials, which are gases at room temperature, liquefy when compressed by the compressor in a refrigeration system. After being cooled to ambient temperature, the compressed liquid is circulated into the cooling coils of a refrigerator or air conditioner, where it is allowed to vaporize. The expanding liquid absorbs heat, thus cooling the surroundings.

Alkyl halides also have medical uses. Halothane, $ClBrCH—CF_3$, and methoxyflurane, $Cl_2CH—CF_2—OCH_3$, are safe and inert general anesthetics. These compounds do not pose the flammability hazards of cyclopropane and ether, other popular general anesthetics. Certain fluorocarbons dissolve substantial amounts of oxygen, and there have been reports that these compounds have been used as artificial blood in surgical applications.

Since alkyl halides are rarely found in nature, and since many are not biologically degraded, it is perhaps not too surprising that the large amounts of them released into the environment have become the focus of concern. At one time, a major use of chlorofluorocarbons was for propellants in sprays (hair sprays, deodorants, and the like). A complex series of reactions that may lead to the destruction of stratospheric ozone caused some environmentalists to question the use of chlorofluorocarbons as aerosol propellants. In October 1978, the United States government banned the use of virtually all aerosol products containing chlorofluorocarbons. The effects of the accumulation of the insecticide DDT in the fatty tissues of birds and fish have resulted in the curtailment of the use of this substance for pest control. Chlordane, too, has been essentially banned except for certain specialized applications. Vinyl chloride, the monomer used for the production of polyvinyl chloride (PVC) products such as plastic plumbing fixtures has been recognized as a potent carcinogen.

The conflict between the use of chemistry to better humanity's standard of living and the release of chemical hazards into the environment finds real focus in the controversies surrounding the use of many organohalogen compounds. Ultimately, the potential benefits of any chemical product must be weighed against its hazards. As one official has said, those who shoulder the responsibility for risk–benefit analysis must ultimately make a subjective judgement over which reasonable persons may differ.

Pollution by organohalogen compounds is neither new, nor totally without amusing aspects. Perhaps the first recorded case of pollution by alkyl halides (or their byproducts) involved the great French chemist Jean-Baptiste André Dumas (1800–1884), who was asked to investigate an unusual occurrence during the reign of Charles X. During a ball given at the Tuileries, the candles sputtered and gave off noxious fumes, driving the guests from the ballroom. Dumas found that the beeswax (which contains large numbers of double bonds) used to make the candles had been bleached with chlorine gas. (What chemical reaction took place?) Under the influence of heat from the candle flame, the chlorinated beeswax decomposed to liberate HCl gas—the noxious fumes.

C. Production and Use of Alcohols and Ethers

Ethanol A number of alcohols are important articles of commerce. Most industrial ethanol is made by the hydration of ethylene in the presence of phosphoric acid and alumina (Sec. 4.8).

$$CH_2{=}CH_2 + H_2O \xrightarrow[300°]{H_3PO_4} CH_3CH_2OH \tag{8.35}$$

Ethanol obtained from this reaction, called 95% ethanol, is 95.6 weight percent pure; the remainder is water. Anhydrous ethanol, or *absolute ethanol,* is obtained by further drying.

Industrial alcohol is chemically pure ethanol used as a starting material for the preparation of other compounds. It is subject to extensive federal controls to ensure that it is not diverted for illicit use in beverages. *Denatured alcohol,* used as solvents for inks, fragrances, and the like, is ethanol that has been made unfit for human consumption by the addition of certain toxic additives, such as methanol. About 55% of the ethanol produced synthetically finds use in the formulation of solvents; about 35% is used for other industrial processes.

Beverage alcohol is produced by the fermentation of malt, barley, grape juice, corn mash, or other sources of natural sugar. Beverage alcohol is not isolated; rather, alcoholic beverages are the actual mixtures of ethanol, water, and the natural colors and flavorings produced in the fermentation process and purified by sedimentation (as in wine) or distillation (as in brandy or whiskey). Industrial alcohol is not used to alter the alcoholic composition of beverages.

Most chemical industry relies on hydrocarbons as the principal raw materials, and ethylene is the single most important hydrocarbon in the chemical industry. However, when the price of crude oil rises, as it did in the late 1970s, production of ethanol by fermentation becomes increasingly attractive as a source of carbon, as discussed in Sec. 4.8. The production of gasohol, a mixture of ethanol and gasoline, has been taking place in the United States for about a decade, particularly in the corn belt. The ethanol used in gasohol is produced by the fermentation of the sugars in corn. This process provides farmers with an additional market for their surplus corn crop, and offers an alternative to imported oil. Although gasohol production is currently small, it seems likely to increase in importance.

Ethanol is a drug. Like many useful drugs, ethanol consumed in excess is toxic. Despite concern over the abuse of other chemical substances, ethanol is the single most abused drug in the world.

Methanol and t-Butyl Methyl Ether Methanol is formed from a mixture of carbon monoxide and hydrogen, called *synthesis gas,* at high temperature over special catalysts.

$$\underbrace{CO + H_2}_{\text{synthesis gas}} \xrightarrow[\substack{\text{Cr-Zn catalyst} \\ \text{100–600 atm}}]{250\text{–}400°} CH_3OH \tag{8.36}$$

Synthesis gas comes from the partial oxidation of methane, which is, in turn, derived from the cracking of hydrocarbons (Sec. 5.9B), or from the gasification of coal. If petroleum prices rise sharply, synthesis gas from coal may become another important source of carbon.

In 1985 1.2 billion gallons of methanol valued at $480 million were produced in the United States. Traditional, but nevertheless important, uses of methanol include oxidation to formaldehyde ($CH_2{=}O$) and reaction with carbon monoxide over special catalysts to give acetic acid, CH_3CO_2H. Newer uses of methanol include its reaction with isobutylene to give *t*-butyl methyl ether (MTBE), which is used as an antiknock additive in gasoline (see Problem 37, Chapter 4). Consumption of this ether is growing at a rate of about 75% per year; 1.5 billion pounds were produced in 1984. Methanol, which has an octane rating of 116, also has a largely unrealized potential use as a motor fuel. (It has been used in racing engines for years.) In fact, California has pioneered the introduction of automobile engines that burn only methanol.

Ethylene Oxide and Ethylene Glycol Ethylene oxide, produced by oxidation of ethylene over a silver catalyst, is one of the most important industrial derivatives of ethylene.

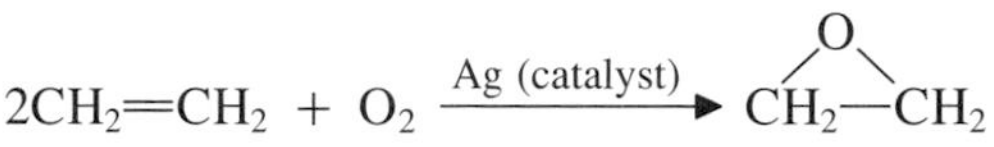

$$2CH_2{=}CH_2 \ + \ O_2 \ \xrightarrow{\text{Ag (catalyst)}} \ \overset{\displaystyle O}{\overset{\displaystyle /\backslash}{CH_2{-}CH_2}} \tag{8.37}$$

In 1986, 5.5 billion pounds of ethylene oxide were produced in the United States. The most important single use of ethylene oxide is its reaction with water to give ethylene glycol.

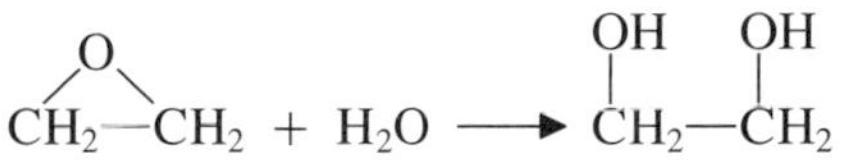

$$\overset{\displaystyle O}{\overset{\displaystyle /\backslash}{CH_2{-}CH_2}} \ + \ H_2O \ \longrightarrow \ \overset{\displaystyle OH \quad OH}{\underset{}{CH_2{-}CH_2}} \tag{8.38}$$

Ethylene glycol, 4.7 billion pounds of which were produced in the United States in 1985, is used both for automotive antifreeze and as a starting material for polyester fibers and films. When automotive sales in the United States are depressed, the ethylene glycol and ethylene oxide markets are also depressed by the reduced demand for antifreeze.

D. Safety Hazards of Ethers

There are two safety hazards that are generally associated with the use of ethers. The first is peroxide formation. On standing in air, ethers slowly undergo a type of reaction called **autoxidation,** which leads to dangerously explosive peroxides and hydroperoxides.

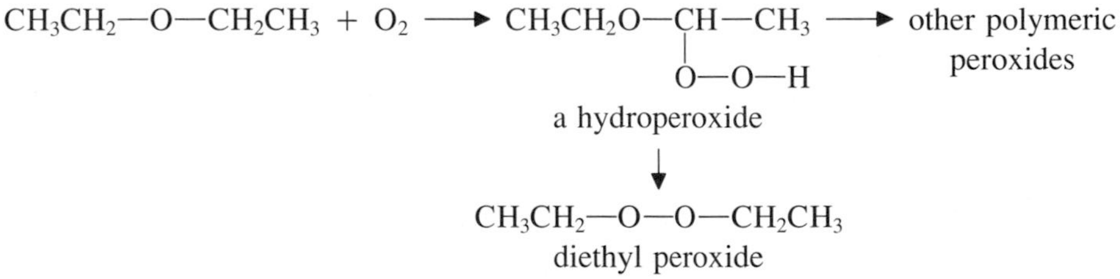

$$CH_3CH_2{-}O{-}CH_2CH_3 \ + \ O_2 \ \longrightarrow \ \underset{\underset{O{-}O{-}H}{|}}{CH_3CH_2O{-}CH{-}CH_3} \ \longrightarrow \ \begin{matrix}\text{other polymeric} \\ \text{peroxides}\end{matrix} \tag{8.39}$$

$$\text{a hydroperoxide}$$

$$\downarrow$$

$$CH_3CH_2{-}O{-}O{-}CH_2CH_3$$

$$\text{diethyl peroxide}$$

These peroxides can form by free-radical processes in samples of anhydrous diethyl ether or tetrahydrofuran (THF) within less than two weeks. For this reason, some ethers are sold with small amounts of free-radical inhibitors such as hydroquinone, which can be removed by distilling the ether. Because peroxides are particularly explosive when heated, it is a good practice not to distill ethers to dryness. Peroxides in an ether can be detected by shaking the ether with 10% aqueous potassium iodide solution. If peroxides are present, they oxidize the iodide to iodine, which imparts a yellow tinge to the solution. Small amounts of peroxides can be removed by distillation of ethers from lithium aluminum hydride ($LiAlH_4$), which both reduces the peroxides and removes contaminating water and alcohols.

The second ether hazard is the high flammability of diethyl ether, the ether most commonly used in the laboratory. Its flammability is indicated by its very low flash point of −45 °C. The **flash point** of a material is the minimum temperature at which it is ignited by a small flame under certain standard conditions. In contrast, the flash point of THF is −14 °C. Compounding the flammability hazard of ether is the fact that its vapor is 2.6 times as dense as air. This means that ether vapors from an open vessel will accumulate in a heavy layer along a laboratory floor or benchtop. For this reason flames can ignite ether vapors that have spread from a remote source. Good safety practice demands that open flames or sparks are not permitted anywhere in a laboratory in which ether is in active use. Even the spark from an electric switch (such as that on a hot plate) can ignite ether vapors. A steam bath is therefore one of the safest ways to heat ether.

The flammability of ether, however, has its uses. Ether is one of the active ingredients in the starter fluid used to start automotive engines on very cold mornings.

KEY IDEAS IN CHAPTER 8

- Organic compounds can be named by both common and systematic nomenclature. In systematic nomenclature, the name of a compound is based on its principal group and principal chain. Alcohols and thiols are among the principal groups that can be designated by suffixes; halogen and ether groups are always named as prefixes.
- Boiling points of organic compounds are increased by
  1. Attractive van der Waals forces between molecules, which are greater for molecules with large size, extended shapes, and polarizable electron clouds
  2. Attraction between permanent dipoles
  3. Association through hydrogen bonding
- The strength of a Brønsted acid is measured as its dissociation constant. The dissociation of a strong acid gives a weak base; the dissociation of a weak acid gives a strong base.
- For a Brønsted acid H—A, acid strength tends to increase with
  1. Increasing electronegativity of A (thus H—F is more acidic than H—OH)
  2. Decreasing H—A bond strength (thus thiols are more acidic than alcohols)
- A Brønsted acid–base equilibrium favors the side with the weaker acid and weaker base.
- Alcohols react with bases to give alkoxides; thiols react with bases to give mercaptides (thiolates).
- Typical primary alcohols have pK_a values near 15–16 in aqueous solution. The acidity of alcohols is reduced by chain branching; thus, tertiary alkoxides are more

basic than primary alkoxides. The acidity of alcohols is increased by the inductive effect of electronegative substituents; thus, 2,2,2-trichloroethanol is more acidic than ethanol.

- Alcohols, thiols, and ethers are weak Brønsted bases, and react with strong acids to form positively charged conjugate acid cations that have negative pK_a values. The Lewis basicity of ethers is useful in forming stable complexes with Lewis acids such as boron compounds and Grignard reagents.
- A solvent is classified as protic or aprotic, depending on its ability to donate hydrogen bonds; polar or apolar, depending on the size of its dielectric constant; donor or nondonor, depending on its ability to act as a Lewis base.
- The solubility of covalent compounds follows the "like-dissolves-like" rule. The solubility of ionic compounds tends to be greatest in solvents that have high dielectric constants, and in solvents that can solvate anions by hydrogen bonding and cations by donor interactions. Crown ethers and other ionophores chelate cations by creating artificial solvation shells for them.
- Reaction of alkyl halides with magnesium metal gives Grignard reagents; reaction with lithium gives organolithium reagents. Both types of reagents behave as strong Brønsted bases and react readily with acids, including water and alcohols, to give alkanes.
- Alkanes react with bromine and chlorine in the presence of heat or light in free-radical substitution reactions to give alkyl halides.

ADDITIONAL PROBLEMS

28 Give the structures of all alcohols with molecular formula $C_5H_{11}OH$. Which ones are chiral? Name each compound using systematic nomenclature, and classify each as a primary, secondary, or tertiary alcohol.

29 Give the systematic name for the anesthetic *halothane.*

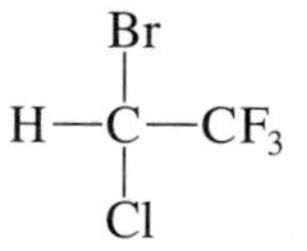

halothane

30 Without consulting tables, arrange the compounds of each set in order of increasing boiling point and give reasons for your answers.
(a) 1-hexanol, 2-pentanol, *t*-butyl alcohol
(b) 1-hexanol, 1-hexene, 1-chloropentane
(c) diethyl ether, propane, 1,2-propanediol
(d) 1-chlorobutane, 1-chloropropane, 1-pentanol
(e) 4-ethylheptane, 2-bromopropane
(f) cyclooctane, chlorocyclobutane, perfluorocyclobutane
(g) *n*-butyl alcohol, *t*-butyl alcohol

Problems (Cont.)

31 Give the structures of the following compounds:
(a) a chiral ether $C_5H_{10}O$ that has no double or triple bonds
(b) a chiral alcohol C_4H_6O
(c) an achiral vicinal glycol $C_6H_{10}O_2$
(d) the four stereoisomeric 4-*t*-butyl-*trans*-cyclohexane-1,2-diols
(e) a diol $C_4H_{10}O_2$ that exists in three and only three stereoisomeric forms
(f) a diol $C_4H_{10}O_2$ that exists in two and only two stereoisomeric forms
(g) the *six* epoxides with molecular formula C_4H_8O (*Hint:* Don't forget stereoisomers.)

32 In each case, give the structure of the hydrocarbon that would react with Br_2 and light to give the indicated products:
(a) C_5H_{12}, that gives only one monobromination product
(b) C_4H_{10}, that gives three monobromination products, two of which are enantiomers
(c) C_4H_{10}, that gives two and only two monobromination products, both achiral

33 A student, Flick Flaskflinger, in his twelfth year of graduate work, needed to prepare ethylmagnesium bromide from ethyl bromide and magnesium, but found that his laboratory was out of diethyl ether. From his years of accumulated knowledge he recalled that Grignard reagents will form in other ether solvents. He therefore attempted to form ethylmagnesium bromide in the ether C_2H_5—O—CH_2CH_2OH and was shocked to find that no Grignard reagent was present after several hours' stirring. Explain why Flick's reaction failed.

34 Identify the unknown compound in each case.
(a) A compound believed to be either diethyl ether or *n*-propyl alcohol is miscible in water.
(b) A compound believed to be either cyclohexyl methyl ether or 2-methylcyclohexanol gives off a gas when it is treated with NaH.
(c) A compound believed to be either *n*-propyl alcohol or allyl methyl ether decolorizes a solution of Br_2 in CCl_4.
(d) An optically active compound C_4H_8O decolorizes Br_2 in CCl_4 and evolves a gas when treated with CH_3MgI.

35 Arrange the compounds within each set in order of increasing acidity in solution. Explain your reasoning.
(a) *n*-propyl alcohol, isopropyl alcohol, *t*-butyl alcohol
(b) cyclohexyl mercaptan, cyclohexanol
(c) 2-chloropropanethiol, 2-chloroethanol, 3-chloropropanethiol
(d) CH_3NH—CH_2CH_2—OH, H_2N—$CH_2CH_2CH_2$—OH, $(CH_3)_3\overset{+}{N}$—CH_2CH_2—OH

36 One of the following compounds is an unusual example of a salt that is soluble in hydrocarbon solvents. Which one is it?

$$A \qquad n\text{-}C_{16}H_{33}\!-\!\overset{\overset{\displaystyle CH_2CH_2CH_2CH_3}{\displaystyle |}}{\underset{\displaystyle CH_2CH_2CH_2CH_3}{\underset{\displaystyle |}{N}}}\!\!\overset{+}{{}}\!-\!CH_2CH_2CH_2CH_3 \quad Br^- \qquad \overset{+}{N}H_4\ Cl^- \qquad B$$

Which would be present in greater amount in a hexane solution of this compound: separately solvated ions or ion pairs and aggregates? Explain your reasoning.

37 The pK_a of water is 15.7. However, titration of an aqueous solution containing Cu^{2+} ion suggests the presence of a species that acts as a Brønsted acid with pK_a = 8.30. Explain. (*Hint:* Cu^{2+} is a Lewis acid.)

38 Normally, di-*n*-butyl ether can be extracted from water into benzene. Explain why this ether can be extracted from benzene into water if the aqueous solution contains moderately concentrated nitric acid.

39 In an abandoned laboratory three alkyl halides are found with different boiling points, each with the formula $C_7H_{15}Br$. One of the compounds is optically active. Following reaction with Mg in ether, then with water, each compound gives 2,4-dimethylpentane. After the same reaction with D_2O, a different product is obtained from each compound. Suggest a structure for each of the three alkyl halides.

40 Let the equilibrium constant for the following acid–base reaction be K_{eq}.

$$HA + B^- \;\rightleftharpoons\; HB + A^-$$

Let K_{HA} be the dissociation constant of HA, and K_{HB} be the dissociation constant of HB. Justify the procedure used to calculate the equilibrium constant for Eq. 8.7 by showing that

$$K_{eq} = 10^{(pK_{HB} - pK_{HA})}$$

(*Hint:* Show first that $K_{eq} = K_{HA}/K_{HB}$.)

41 Offer a rational explanation for each of the following observations:
(a) Compound *A* exists preferentially in a chair conformation with an equatorial —OH group, but compound *B* prefers a chair conformation with an axial —OH group.

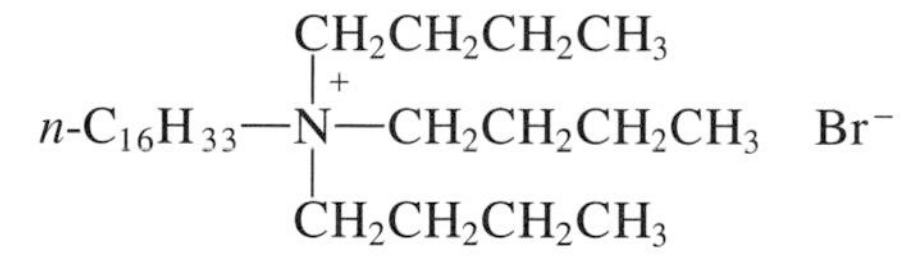

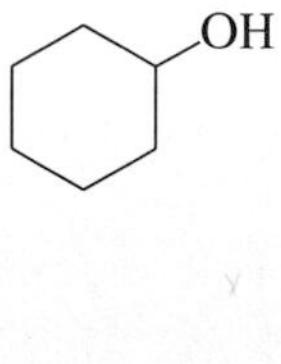

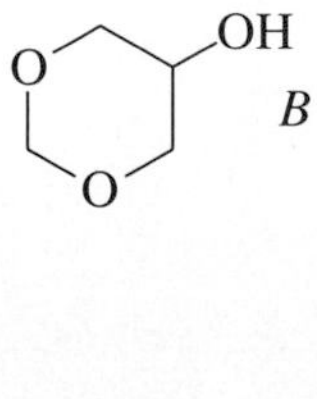

Problems (Cont.)

(b) Ethylene glycol contains a greater fraction of *gauche* conformer than butane in carbon tetrachloride solution.

(c) The racemate of 2,2,5,5-tetramethyl-3,4-hexanediol exists with a strong intramolecular hydrogen bond, but the *meso* compound has no intramolecular hydrogen bond.

---

42 (a) Show that the dipole moment of 1,4-dioxane (Sec. 8.1C) should be zero if we assume that the molecule exists in a chair conformation.

(b) What other conformation of dioxane might account for the fact that the dipole moment of 1,4-dioxane is not zero (0.38 D)?

---

43 Although tetrahydrofuran (*A*) is an excellent solvent for Grignard reagents, 2,2,5,5-tetramethyltetrahydrofuran (*B*) is a very poor solvent. Explain.

A  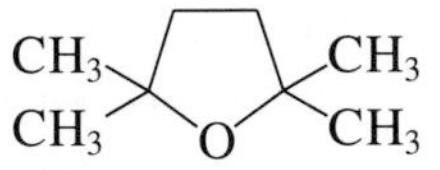 *B*

---

44 Ethylene glycol, HO—CH_2CH_2—OH, is a diprotic acid (has two ionizable protons). Suggest one or more reasons why ethylene glycol is *more* acidic than ethanol, but $^-$O—CH_2CH_2—OH is *less* acidic than ethanol.

---

45 (a) The bromination of 2-methylpropane in the presence of light can give two monobromination products; give their structures.

(b) In fact, only *one* of these compounds is formed in appreciable amount. Write the mechanism for formation of each compound, and use what you know about the stabilities of free radicals to predict which compound should be the major one formed.

---

46 (a) The acid HI is considerably stronger than HCl (see Table 8.3). Why, then, do 10^{-3} *M* aqueous solutions of either acid in water give the same pH reading of about 3?

(b) Potassium amide, $K^+\ {}^-{:}\ddot{N}H_2$, whose conjugate acid has a pK_a of 35, is a much stronger base than potassium *t*-butoxide, $K^+\ (CH_3)_3C{-}\ddot{\underset{..}{O}}{:}^-$, whose conjugate acid has a pK_a of 19. Yet a 10^{-3} *M* solution of either compound in water has an identical pH reading of about 11. Explain.

---

47 Give a mechanism for the following reaction, which takes place in several steps. Use the curved-arrow formalism.

$$3C_2H_5\ddot{\underset{..}{O}}H + BF_3 \longrightarrow C_2H_5\ddot{\underset{..}{O}}{-}B(\ddot{\underset{..}{O}}C_2H_5)_2 + 3H{-}F$$

9 Substitution and Elimination Reactions of Alkyl Halides

In this chapter we are going to examine carefully two very important types of alkyl halide reactions. One is the **nucleophilic substitution reaction.** An example of such a reaction is the reaction of hydroxide ion with methyl iodide.

$$CH_3{-}\ddot{\underset{..}{I}}{:} + {}^{-}{:}\ddot{\underset{..}{O}}H \longrightarrow CH_3{-}\ddot{\underset{..}{O}}H + {:}\ddot{\underset{..}{I}}{:}^{-} \qquad (9.1)$$

This reaction is a *substitution* because one group, in this case hydroxide, is substituted for another, iodide. It is a *nucleophilic* substitution because the incoming hydroxide uses a pair of electrons to form a bond to carbon, much as a Lewis base (nucleophile) uses a pair of electrons to form a bond to a Lewis acid. Nucleophilic substitution reactions are also called **nucleophilic displacement reactions.** (In the reaction above iodide is displaced by hydroxide.)

The second type of alkyl halide reaction we shall study is the **elimination reaction.** An example of such a reaction is the formation of isobutylene from *t*-butyl bromide and the base sodium ethoxide.

$$Na^+\ C_2H_5O^- + H{-}CH_2{-}\underset{\displaystyle CH_3}{\overset{\displaystyle Br}{C}}{-}CH_3 \longrightarrow C_2H_5O{-}H + CH_2{=}C(CH_3)_2 + Na^+\ Br^- \qquad (9.2)$$

This reaction is an *elimination* because the elements of H—Br (shown in color in Eq. 9.2) are lost from the alkyl halide when the alkene is formed.

Although our knowledge of nucleophilic substitution and elimination reactions will be derived from a study of alkyl halides, we shall find that such reactions occur widely with other functional groups. Therefore, the principles developed in this chapter apply to a great deal of the chemistry that we shall study later.

9.1 NUCLEOPHILIC SUBSTITUTION REACTIONS: INTRODUCTION

Let us use the reaction of methyl iodide and hydroxide in Eq. 9.1 to learn some of the terminology used in discussing nucleophilic substitution reactions. The group that forms the new bond to carbon, in this case hydroxide, is called the **nucleophile.** Recall that the word *nucleophile* is just another word for *Lewis base* (Sec. 1.3). When we study the mechanism of this reaction, we shall see that the hydroxide ion indeed behaves as a Lewis base. The group that loses its bond to carbon and takes on an additional pair of electrons, in this case the iodine, is called the **leaving group.** (Notice that the roles of the nucleophile and leaving group would be reversed if the reaction were thought of in the reverse sense.)

$$\underset{\text{leaving group: I}}{CH_3{-}\ddot{\underset{\cdot\cdot}{I}}:} + \underset{\text{nucleophile}}{{}^{-}:\ddot{\underset{\cdot\cdot}{O}}H} \longrightarrow H\ddot{\underset{\cdot\cdot}{O}}{-}CH_3 + {}^{-}:\ddot{\underset{\cdot\cdot}{I}}: \qquad (9.3)$$

The nucleophilic atom of the nucleophile may be anionic, neutral, or in a few cases, even positively charged. Some nucleophiles have a proton that is lost following the substitution:

$$H{-}\underset{\displaystyle H}{\ddot{O}}: + (CH_3)_3C{-}Br \longrightarrow (CH_3)_3C{-}\underset{\displaystyle H}{\overset{+}{\ddot{O}}}{-}H\ :\ddot{\underset{\cdot\cdot}{Br}}:^- \rightleftharpoons (CH_3)_3C{-}\ddot{\underset{\cdot\cdot}{O}}H + H\ddot{\underset{\cdot\cdot}{Br}}: \qquad (9.4)$$

This equation also illustrates an important special case of a nucleophilic substitution that occurs when the solvent (water in this case) is also the nucleophile. The reaction is then called a **solvolysis**—literally, a bond cleavage by solvent.

Another special situation arises when the nucleophile and the leaving group are situated in the same molecule. In this case, the nucleophilic substitution is *intramolecular* and a ring is formed, as in the following example:

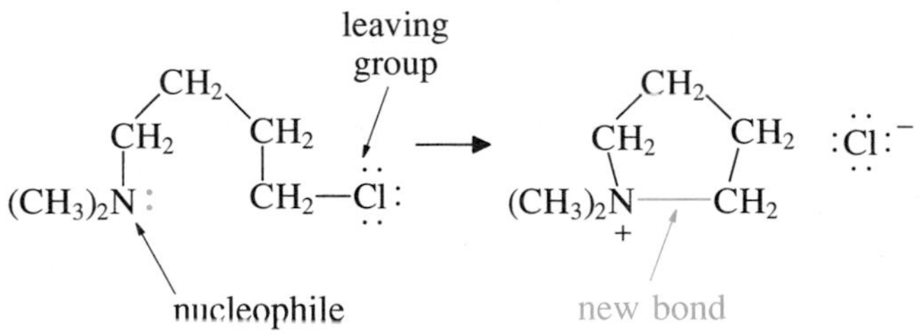

(9.5)

The nucleophilic substitution reaction is so important and general that we shall spend a good deal of time discussing it. A few nucleophilic substitution reactions are summarized in Table 9.1. It should be clear from this table that nucleophilic substitution reactions can be used to transform alkyl halides into a wide variety of other organic compounds.

9.2 EQUILIBRIUM IN NUCLEOPHILIC SUBSTITUTION REACTIONS

The first question we are going to ask about nucleophilic substitution reactions concerns their *equilibria*: how do we know when the equilibrium for a substitution is favorable? This problem is illustrated by the reaction of cyanide ion with methyl iodide, which has an equilibrium constant that favors the product acetonitrile by many powers of ten.

$$^{-}\!:C{\equiv}N: + CH_3\ddot{I}: \longrightarrow \underset{\text{acetonitrile}}{CH_3C{\equiv}N:} + :\ddot{I}:^{-} \tag{9.6}$$

TABLE 9.1 Some Nucleophilic Substitution Reactions (X = halogen or other leaving group; R,R′ = alkyl groups)

R—X: + Nucleophile (name)		⟶ :X:⁻ + Product (name)	
R—X + :Y:⁻	(another halide)	⟶ :X:⁻ + R—Y:	(another alkyl halide)
+ ⁻:C≡N:	(cyanide)	⟶ + R—C≡N:	(nitrile)
+ ⁻:OH	(hydroxide)	⟶ + R—OH	(alcohol)
+ ⁻:OR′	(alkoxide)	⟶ + R—O—R′	(ether)
+ ⁻N₃	(azide = :N̈⁻=N⁺=N̈⁻:)	⟶ + R—N₃	(alkyl azide)
+ ⁻:SR′	(alkanethiolate)	⟶ + R—S—R′	(thioether)
+ :NR′₃	(amine)	⟶ R—N⁺R′₃ :X:⁻	(alkylammonium salt)
+ :OH₂	(water)	⟶ R—O⁺(H)—H :X:⁻ ⇌ R—O—H + HX:	(alcohol)
+ :O(H)—R′	(alcohol)	⟶ R—O⁺(H)—R′ :X:⁻ ⇌ R—O—R′ + HX:	(ether)

Other substitution reactions, however, are reversible or even unfavorable.

$$:\ddot{\mathrm{I}}:^- + \mathrm{CH_3{-}\ddot{Br}}: \rightleftharpoons \mathrm{CH_3{-}\ddot{I}}: + :\ddot{\mathrm{Br}}:^- \qquad (9.7)$$

$$\mathrm{H_2\ddot{O}} + \mathrm{CH_3{-}\ddot{I}}: \rightleftharpoons \mathrm{CH_3{-}\overset{+}{\ddot{O}}H_2} \quad :\ddot{\mathrm{I}}:^- \qquad (9.8)$$

$$:\ddot{\mathrm{I}}:^- + \mathrm{CH_3{-}\ddot{O}H} \longleftarrow \mathrm{CH_3{-}\ddot{I}}: + {}^-\!:\ddot{\mathrm{O}}\mathrm{H} \quad \text{(does not proceed to the right)} \qquad (9.9)$$

We can predict results such as these by recognizing that each nucleophilic substitution reaction is formally very similar to a Brønsted acid–base reaction. That is, if we simply replace the alkyl group of the alkyl halide with a hydrogen, the substitution looks like an acid–base reaction.

$$\mathrm{^-OH + HI \longrightarrow H_2O + I^-} \quad \text{(acid–base neutralization)} \qquad (9.10)$$

$$\mathrm{^-OH + CH_3I \longrightarrow CH_3OH + I^-} \quad \text{(displacement reaction)} \qquad (9.11)$$

In the acid–base reaction, $^-$OH displaces I$^-$ from the *proton*; in the substitution reaction, $^-$OH displaces I$^-$ from *carbon*. What makes this analogy particularly useful is that *the position of equilibrium in a nucleophilic substitution reaction is very similar to that of its acid–base counterpart.* Thus, if the acid–base reaction

$$\mathrm{H{-}X + Y:^- \rightleftharpoons Y{-}H + X:^-} \qquad (9.12a)$$

strongly favors the right side of the equation, then the analogous substitution reaction

$$\mathrm{R{-}X + Y:^- \rightleftharpoons Y{-}R + X:^-} \qquad (9.12b)$$

likewise favors the right side of the equation. This means that *the equilibrium in any nucleophilic substitution reaction, as in an acid–base reaction, favors release of the weaker base.* Thus, we can see immediately why I$^-$ will *not* displace $^-$OH from CH_3OH: I$^-$ is a much weaker base than $^-$OH (Table 8.3). In fact, just the opposite reaction occurs: $^-$OH readily displaces I$^-$ from CH_3I. It should be clear from this discussion that a mastery of the acid–base principles in Sec. 8.5 is important for an understanding of nucleophilic substitution reactions.

Some equilibria that are not too unfavorable can be driven to completion by applying **LeChatelier's principle,** a central principle of chemical equilibrium. LeChatelier's principle states that if an equilibrium is disturbed, the components of the equilibrium will react so as to offset the effect of the disturbance. For example, alkyl chlorides normally do not react to completion with iodide ion because iodide is a weaker base than chloride; the equilibrium favors the weaker base, iodide. However, in the solvent acetone, it happens that potassium iodide is relatively soluble and potassium chloride is relatively insoluble. Thus, when an alkyl chloride reacts with KI in acetone, KCl precipitates, and the equilibrium compensates for this disturbance—the loss of KCl—by forming more of it (along with, of course, more alkyl iodide).

$$\mathrm{R{-}Cl + KI \longrightarrow R{-}I + KCl} \text{ (precipitates in acetone)} \qquad (9.13)$$

Problem

1 Which of the following reactions most strongly favors the products at equilibrium? (*Hint:* Use Table 8.3.)

(a) $CH_3Cl + I^- \longrightarrow CH_3I + Cl^-$
(b) $CH_3Cl + F^- \longrightarrow CH_3F + Cl^-$
(c) $CH_3Cl + CH_3O^- \longrightarrow CH_3OCH_3 + Cl^-$

9.3 THE S$_N$2 REACTION

A. Reaction Rates and Mechanism

The second question we are going to ask about nucleophilic substitution reactions concerns their *rates*. Some substitution reactions with favorable equilibria proceed rapidly; others proceed slowly. For example, the reaction of methyl iodide with cyanide is a relatively fast reaction, whereas the reaction of cyanide with neopentyl iodide is so slow that it is virtually useless:

$$^-:C \equiv N: + \; CH_3\!-\!\ddot{I}: \xrightarrow[\text{(rapid)}]{} \; CH_3\!-\!C \equiv N: + \; :\ddot{I}:^- \tag{9.14}$$

$$^-:C \equiv N: + \; CH_3\!-\!\underset{\underset{CH_3}{|}}{\overset{\overset{CH_3}{|}}{C}}\!-\!CH_2\!-\!\ddot{I}: \xrightarrow[\text{(very slow)}]{}$$

$$CH_3\!-\!\underset{\underset{CH_3}{|}}{\overset{\overset{CH_3}{|}}{C}}\!-\!CH_2\!-\!C \equiv N: + \; :\ddot{I}:^- \tag{9.15}$$

Why do reactions that are so similar in a formal sense differ so drastically in their rates? In order to answer this question, we shall need to define a reaction rate precisely and understand how reaction rates are determined experimentally. Then we shall be in a position to see how reaction rates can help us understand the mechanisms of nucleophilic substitution reactions. These mechanisms will provide us with the tools we need to understand effects like the ones in these equations.

Definition of Reaction Rate The term *rate* implies that something is changing with time. For example, when we speak of the rate of travel, or velocity, of a car, the "something" that is changing is the car's position:

$$\text{velocity} = v = \frac{\text{change in position}}{\text{corresponding change in time}} \tag{9.16}$$

The quantities that change with time in a chemical reaction are the *concentrations of the reactants and products.*

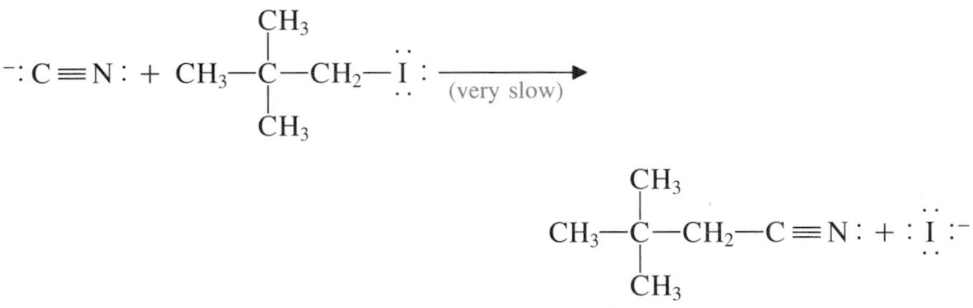

$$\text{reaction rate} = \frac{\text{change in product concentration}}{\text{corresponding change in time}} \tag{9.17a}$$

$$= -\frac{\text{change in reactant concentration}}{\text{corresponding change in time}} \tag{9.17b}$$

The reason for the different signs is that the concentrations of the reactants decrease with time, and the concentrations of the products increase.

A mechanical rate has the dimensions of length per unit time—for example, meters per second. A reaction rate, by analogy, has the dimensions of concentration per unit time. When the concentration unit is the mole per liter (*M*), and if time is measured in seconds, then the unit of reaction rate is

$$\frac{\text{concentration}}{\text{time}} = \frac{\text{mol/L}}{\text{sec}} = \frac{\text{mol}}{\text{L} \cdot \text{sec}} = M/\text{sec} = M\text{-sec}^{-1} \tag{9.18}$$

The Rate Law In order for molecules to react, they must "get together," or collide. Since molecules at high concentration are more likely to collide than molecules at low concentration, we expect that the rate of a reaction will be some function of the concentrations of the reactants. The mathematical statement of how the reaction rate depends on concentration is called the **rate law.** This rate law is determined experimentally by varying the concentration of each reactant (including any catalysts) independently. Each reaction has its own characteristic rate law. For example, suppose we find that for the reaction $A + B \rightarrow C$ the reaction rate doubles if *either* $[A]$ or $[B]$ is doubled, and increases by a factor of four if *both* $[A]$ and $[B]$ are doubled. The rate law for this reaction is then

$$\text{rate} = k[A][B] \tag{9.19}$$

The concentrations in the rate law are the concentrations of reactants at any time during the reaction, and the rate is the velocity of the reaction at that same time. The constant of proportionality, k, is called the **rate constant.** In general the rate constant is different for every reaction, and is a *fundamental physical constant* for a given reaction under particular conditions of temperature, pressure, solvent, and so on. As we can see from Eq. 9.19, the rate constant is numerically equal to the rate of the reaction when all reactants are present at $1M$ concentration; that is, the rate constant is the rate of the reaction under standard conditions of unit concentration. The rates of two reactions are compared using their rate constants.

An important aspect of a reaction is its **kinetic order.** The **overall kinetic order** for a reaction is the sum of the powers of all the concentrations in the rate law. For the rate law in Eq. 9.19, the overall kinetic order is two; the reaction described by this rate law is said to be a *second-order reaction.* The **kinetic order in each reactant** is the power to which its concentration is raised in the rate law. Thus, the reaction described by the rate law in Eq. 9.19 is said to be *first order in each reagent.*

The *dimensions* of the rate constant depend on the kinetic order of the reaction. With concentrations in moles/liter, and time in seconds, the rate of any reaction has the dimensions of M/sec (Eq. 9.18). For a second-order reaction, then, dimensional consistency requires that the rate constant have the dimensions of $M^{-1}\text{sec}^{-1}$.

$$\begin{aligned} \text{rate} &= k[A][B] \\ &= M^{-1}\text{sec}^{-1} \cdot M \cdot M = M/\text{sec} \end{aligned} \qquad (9.20)$$

Similarly, the rate constant for a first-order reaction has dimensions of sec^{-1}.

Problems

2 The bromination of an alkene under certain conditions follows the rate law

$$\text{rate} = k[\text{alkene}][Br_2]^2 \qquad (9.21)$$

(a) What is the overall kinetic order of this reaction?
(b) What is the order in each reagent?
(c) What are the dimensions of the rate constant?

3 What prediction does the rate law in Eq. 9.19 make about how the rate of the reaction changes as the reactants *A* and *B* are converted into *C* over time? Does the rate increase, decrease, or stay the same? Explain. Use your answer to sketch a plot of the concentrations of starting materials and products against time.

Let us relate what we have learned here about reaction rates to what we learned in Sec. 4.7. There we found that the standard free energy of activation, or energy barrier, determines the rate of a reaction under standard conditions. Since the rate constant *is* the rate of a reaction under standard conditions, it follows that the rate constant is related to the standard free energy of activation $\Delta G^{0\ddagger}$. If $\Delta G^{0\ddagger}$ is large for a reaction, the rate constant is small; if $\Delta G^{0\ddagger}$ is small, the rate constant is large.

Rate Law and Mechanism. The S_N2 Reaction The rate law gives us fundamental information about the mechanism of a reaction. This idea is best illustrated by example. Let us consider a simple nucleophilic substitution reaction: the reaction of ethoxide ion with methyl iodide in ethanol at 25°.

$$CH_3{-}I + C_2H_5O^- \longrightarrow C_2H_5O{-}CH_3 + I^- \qquad (9.22)$$

The rate law for this reaction was experimentally determined by measuring the reaction rate as the concentrations of reactants were varied. The rate law is

$$\text{rate} = k[CH_3I][C_2H_5O^-] \qquad (9.23)$$

with $k = 6.0 \times 10^{-4}\ M^{-1}\text{sec}^{-1}$. That is, this is a *second-order reaction, first order in each reagent.* This type of rate law is general for nucleophilic substitution reactions of *primary alkyl halides*: all such reactions are second order, first order in the alkyl halide and first order in the nucleophile.

The rate law tells us *what elements are present in the transition state of the rate-determining step.* Hence, the transition state of reaction 9.22 consists of the elements of one methyl iodide molecule and one ethoxide ion. The rate law does *not* tell us how these elements are arranged, or whether the reactant molecules are even still intact.

Let us now write a mechanism that is consistent with the rate law. The simplest possible mechanism is one in which the nucleophile ethoxide *directly displaces* iodide from the methyl carbon:

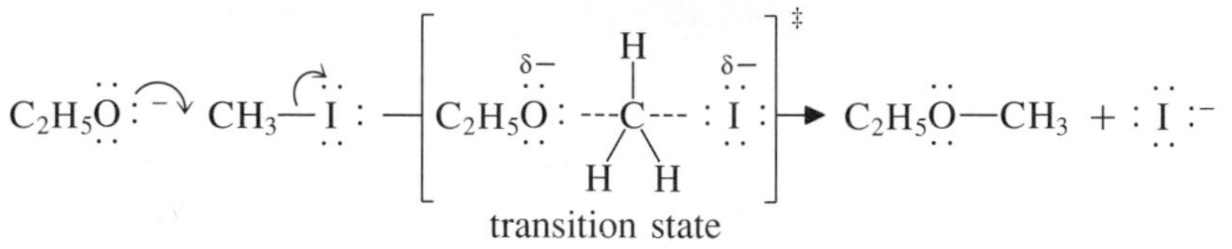

$$C_2H_5\ddot{O}: \curvearrowright CH_3 \overset{\curvearrowright}{\longrightarrow} \ddot{I}: \longrightarrow \left[\overset{\delta-}{C_2H_5\ddot{O}} : \cdots \overset{H}{\underset{H}{\overset{|}{C}}} \cdots : \overset{\delta-}{\ddot{I}} : \right]^{\ddagger} \longrightarrow C_2H_5\ddot{O}{-}CH_3 \; + \; : \ddot{I} :^- \quad (9.24)$$

transition state

The same type of mechanism accounts for the reactions of all methyl and primary alkyl halides with nucleophiles. Because this mechanism is general, it is given a special name, the **S$_N$2 reaction.**

substitution $\longrightarrow$ S$_N$2 $\longleftarrow$ bimolecular

nucleophilic

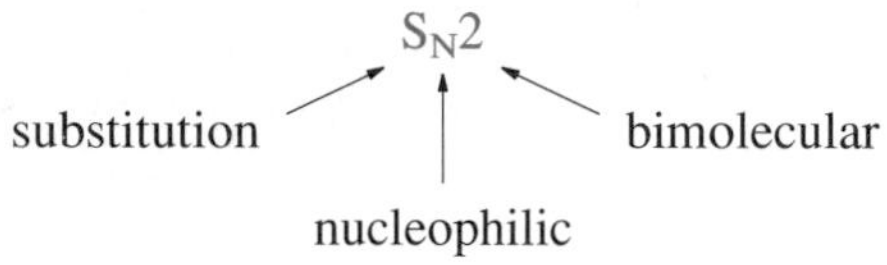

(The word *bimolecular* means that the transition state of the reaction involves the collision of two species—in this case, one methyl iodide and one ethoxide.) To summarize: nucleophilic substitution reactions of methyl and primary alkyl halides occur by the S$_N$2 mechanism.

How did we know to write *exactly* this mechanism? In fact, the rate law does not tell us exactly what mechanism to write. To reiterate what we said earlier: *the rate law tells us what atoms are present in the transition state, but not how they are arranged.* For example, the rate law does not tell us whether the nucleophile in the S$_N$2 reaction comes in from the front or back side of the carbon relative to the departing halide ion:

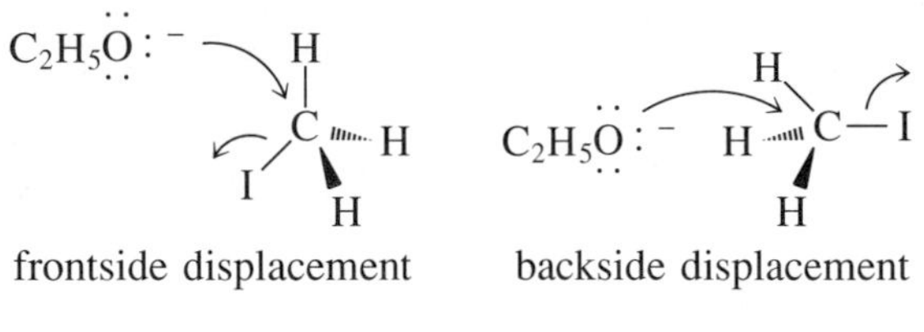

$$\text{(9.25)}$$

frontside displacement backside displacement

As far as the rate law is concerned, either mechanism is acceptable. To decide between these two possibilities, we have to do other types of experiments (see Sec. 9.3B below). However, the rate law does tell us what mechanisms *not* to write! For example, a transition state that contains two molecules of the nucleophile is *ruled out* by the rate law, because the rate law for such a mechanism would have to be second order in ethoxide.

Let us summarize the connection between the rate law and the mechanism of a reaction.

1. The concentration terms of the rate law tell us what species are involved in the rate-determining transition state.

2. Mechanisms not consistent with the rate law are ruled out.

3. Of the chemically reasonable mechanisms consistent with the rate law, the simplest one is provisionally adopted.
4. The mechanism of a reaction is modified or refined if required by subsequent experiments.

Point (4) may seem disturbing; after all, it means that a mechanism can be changed at a later time. It perhaps seems that there should be an "absolutely true" mechanism for every reaction. In fact, a mechanism is nothing more than a conceptual framework, or theory, for generalizing the results of many experiments and predicting the outcomes of new experiments. It is our way of looking at and rationalizing nature. A mechanism is of value precisely because it allows us to forecast the results of reactions that we have not yet observed in the laboratory. However, when someone observes an experimental result different from that predicted by a mechanism, then the mechanism must be modified to accommodate both previously known facts *and* the new facts. Knowledge in chemistry—indeed, in all of science—is dynamic: theories predict the results of experiments; a test of these predictions by experiment leads to new theories (sometimes); and so on.

Problems

4 Consider the reaction of acetic acid with ammonia.

$$\underset{\text{acetic acid}}{CH_3-\overset{\overset{\displaystyle O}{\|}}{C}-\ddot{\underset{..}{O}}-H} + :NH_3 \rightleftharpoons CH_3-\overset{\overset{\displaystyle O}{\|}}{C}-\ddot{\underset{..}{O}}:^- + \overset{+}{N}H_4$$

This reaction is very rapid and follows the simple rate law

$$\text{rate} = k[CH_3-\overset{\overset{\displaystyle O}{\|}}{C}-OH][NH_3]$$

Propose a mechanism that is consistent with this rate law.

5 What rate law would you expect for the reaction of cyanide ion ($^-$:CN) with ethyl bromide, assuming the reaction occurs by the S_N2 mechanism?

B. Stereochemistry of the S_N2 Reaction

The mechanism of the S_N2 reaction can be specified in more detail by considering what happens when substitution occurs at an asymmetric carbon. We learned in Sec. 7.10B that when a bond is broken at an asymmetric carbon, three results are possible: the reaction can occur with (1) *retention of configuration,* (2) *inversion of configuration,* or (3) *loss of configuration* (a mixture of retention and inversion).

If attack of the nucleophile Nuc:$^-$ and departure of the leaving group X:$^-$ occur from more or less the same direction (frontside attack), we see that the reaction would give the product with *retention of configuration.*

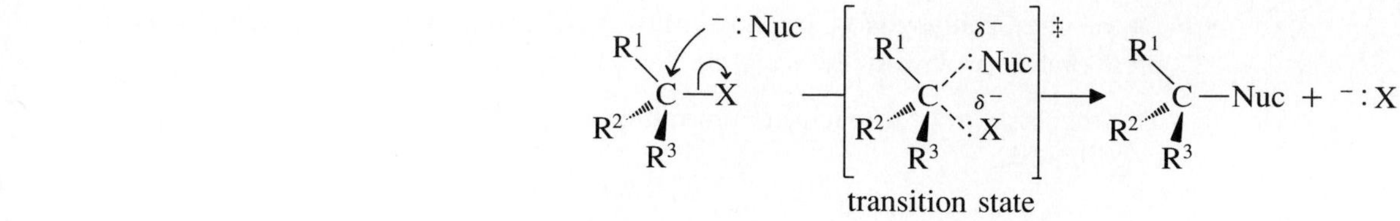

(9.26a)

If attack of the nucleophile and loss of the leaving group occur from opposite directions (backside attack), the other three groups on carbon must invert, or "turn inside out," in order to maintain the tetrahedral bond angle. This mechanism would lead to a product with *inversion of configuration* at the asymmetric carbon.

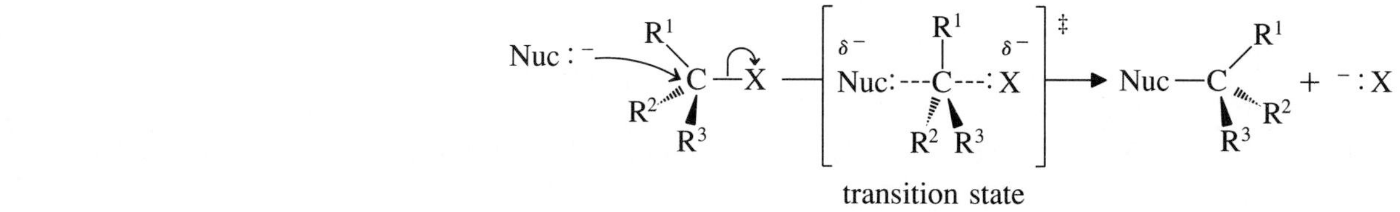

(9.26b)

If we compare the products of Eqs. 9.26a and 9.26b, we see that they are *enantiomers*. Thus, we can tell which type of attack is occurring by subjecting one enantiomer of a chiral alkyl halide to the S_N2 reaction and determining which enantiomer of the product is formed. Of course, if both paths occur at equal rates, then we expect to obtain the racemate.

What are the experimental results? Hydroxide ion reacts with 2-bromooctane, a chiral alkyl halide, to give 2-octanol. The reaction shows second-order kinetics, first order in ^{-}OH and first order in the alkyl halide. When (*R*)-2-bromooctane is used in the reaction, the product is (*S*)-2-octanol, the product enantiomer of inverted configuration.

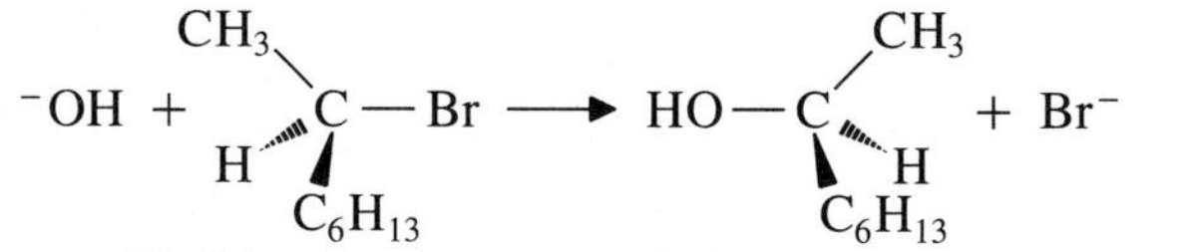

(*R*)-2-bromooctane (*S*)-2-octanol

(9.27)

The stereochemistry of this S_N2 reaction shows that it proceeds with *inversion of configuration*. Thus, the reaction occurs by ***backside attack*** of hydroxide ion on the alkyl halide.

We recall that backside attack is also observed for the attack of bromide ion and other nucleophiles on the bromonium ion intermediate in alkene bromination (Sec. 7.10C). We can now recognize that this reaction too is an S_N2 reaction.

The stereochemistry of the S_N2 reaction calls to mind the *inversion of amines* (Sec. 6.8B). In both processes, the central atom is turned "inside out," and is approximately sp^2 hybridized at the transition state. In the transition state for amine inversion, the *p* orbital on the nitrogen contains an unshared electron pair. In the transition state for an S_N2 reaction on carbon, the nucleophile and the leaving group are partially bonded to opposite lobes of the carbon *p* orbital. The stereochemistry of the S_N2 reaction is summarized in Fig. 9.1.

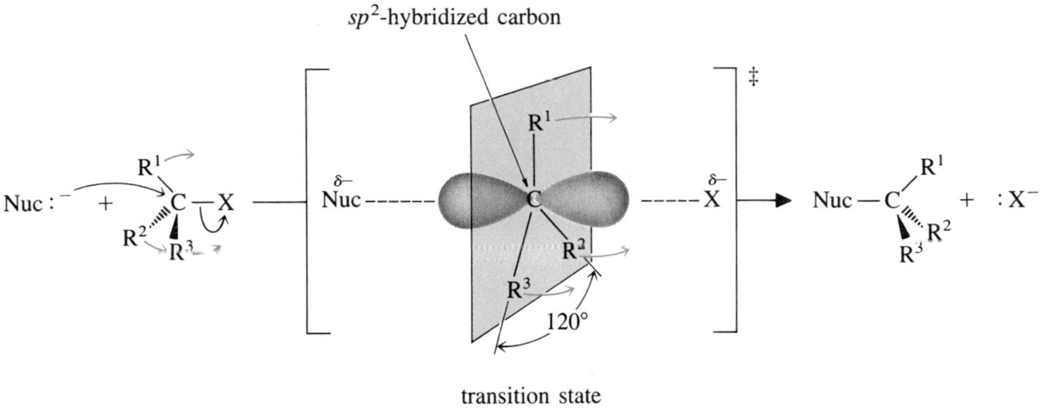

Figure 9.1 Stereochemistry of the S_N2 reaction. The small colored arrows show how the various groups change position during the reaction. (Nuc :⁻ = a general nucleophile.)

Problem

6 What is the expected product (including its stereochemical configuration) in the reaction of sodium hydroxide with (*R*)-CH_3—CHD—I?

C. Effect of Alkyl Halide Structure on the S_N2 Reaction

We now have a fairly detailed picture of the S_N2 reaction mechanism. Let us see how this mechanism accounts for some of the other experimental facts known about the reaction. First let us consider how the structure of an alkyl halide affects the rate of its S_N2 reactions. Alkyl halides differ, in some cases by many orders of magnitude, in the rates with which they undergo the S_N2 reaction. Table 9.2 shows how the rate of an S_N2 reaction varies as the structure of the alkyl halide is changed. This table shows that

TABLE 9.2 Effect of Branching in the Alkyl Halide on the Rate of an S_N2 Reaction:

$$R{-}Br + I^- \xrightarrow{\text{acetone, 25 °C}} R{-}I + Br^-$$

R—	Name of R	Relative rate*
CH_3—	methyl	145
CH_3CH_2—	ethyl	1.0
$CH_3CH_2CH_2$—	propyl	0.82
$(CH_3)_2CHCH_2$—	isobutyl	0.036
$(CH_3)_3CCH_2$—	neopentyl	0.000012

*All rates are relative to that of ethyl bromide.

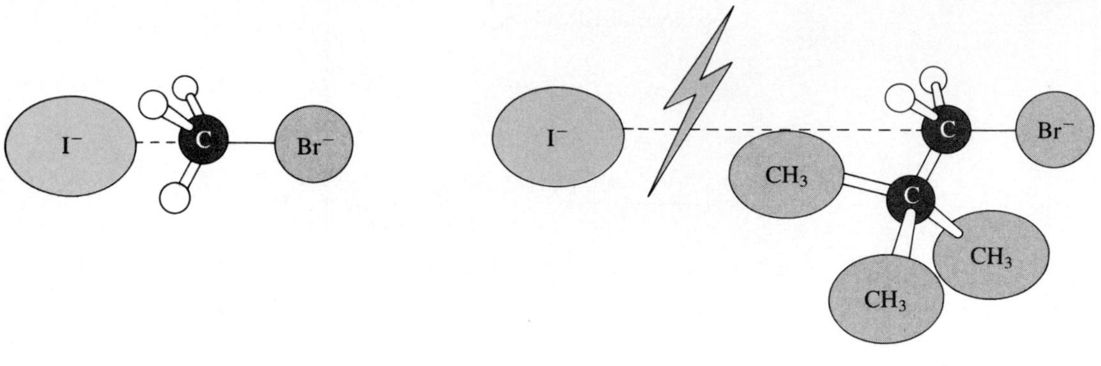

(a) $I^- + CH_3Br$ (b) $I^- + (CH_3)_3CCH_2Br$

Figure 9.2 (a) Backside displacement on methyl bromide is relatively unhindered. (b) Backside displacement on neopentyl bromide is virtually impossible because of the van der Waals repulsions between the attacking nucleophile and the methyl groups.

increased branching near the site of substitution retards the S_N2 reaction. As Fig. 9.2 shows, this observation is consistent with the backside displacement mechanism. When a methyl halide undergoes substitution, access of the nucleophile to the back side of the carbon atom is relatively unrestricted. However, when a neopentyl halide is attacked by a nucleophile, access to the back side of the carbon is almost totally blocked by the methyl branches. The likelihood that the nucleophile completes a successful substitution is about the same as the probability that a football player will score a touchdown by running unassisted through a wall of tacklers. In terms of transition state theory, the steric repulsions—that is, repulsive van der Waals forces—present in the transition state raise its energy and therefore reduce the reaction rate.

The cases cited in Table 9.2 involve either methyl or primary alkyl halides. What about secondary or tertiary alkyl halides? These compounds are branched at the carbon bearing the halogen. As we might expect, branching at this carbon also retards the reaction for the same steric reasons. Secondary alkyl halides undergo S_N2 reactions much more slowly than primary alkyl halides, and tertiary alkyl halides—for example, *t*-butyl bromide—do *not* undergo S_N2 reactions. These compounds *do* undergo nucleophilic substitution reactions, but, as we shall see in Sec. 9.5, they react by a different mechanism.

D. Effect of Leaving Group and Nucleophile Basicities on the S_N2 Reaction. Solvent Effects

When we keep the alkyl group the same and vary the leaving halide in an S_N2 reaction, we find that alkyl bromides and alkyl iodides are more reactive than alkyl chlorides, which in turn are more reactive than alkyl fluorides.

alkyl halide	R—F <	R—Cl <	R—Br <	R—I
relative S_N2 reaction rate:	0.005	1	50	150

(The relative reactivities above are typical, but vary from case to case.) In general, *the best leaving groups in S_N2 reactions are the weakest bases.*

TABLE 9.3 Rate Constants for Reaction of Various Nucleophiles with Methyl Bromide Under Different Conditions at 25°*

$$:\ddot{\underset{..}{X}}:^- + CH_3Br \longrightarrow X{-}CH_3 + :\ddot{\underset{..}{Br}}:^-$$

Nucleophile, $:\ddot{\underset{..}{X}}:^-$	Solvent: H_2O	Solvent: CH_3OH†	Gas Phase
$:\ddot{\underset{..}{F}}:^-$	3×10^{-7}	5×10^{-8}	$4 \times 10^{+11}$
$:\ddot{\underset{..}{Cl}}:^-$	5×10^{-6}	3×10^{-6}	$8 \times 10^{+9}$
$:\ddot{\underset{..}{Br}}:^-$	$\sim 10^{-4}$	8×10^{-5}	$<6 \times 10^{+9}$

*Rate constants are in $M^{-1}sec^{-1}$.
†Displacement on CH_3I.

This trend is exactly what we expect from the S_N2 reaction mechanism. In the transition state, the leaving group is accepting a pair of electrons. The more readily a leaving group accepts electrons, the lower is the energy of the transition state. The best electron acceptors, as we might expect, are those groups that give the weakest bases.

We expect the opposite trend as we vary the nucleophile. Since the nucleophile donates a pair of electrons in the S_N2 reaction, we expect the best electron donors—the best bases—to react most rapidly. *Within a row of the periodic table* this trend is indeed observed. For example, the rate for attack of cyanide ion ($^-$: CN) on methyl bromide in aqueous ethanol solvent is about one thousand times greater than the rate of the weaker base fluoride (F^-). Examination of Table 8.3 shows that cyanide is considerably more basic than fluoride.

When we compare nucleophiles from different rows of the periodic table, however, we find that *the relative reaction rates of different nucleophiles depends on the solvent* (Table 9.3). In *protic solvents,* such as methanol or water, the reaction rates of halide ions with a given alkyl halide do *not* correspond to their relative basicities. For example, the weaker base bromide ion is three thousand times more reactive as a nucleophile in the S_N2 reaction than the stronger base fluoride. That this has something to do with the solvent is apparent when we compare the rates for the reaction of the same halide ions with methyl bromide in the *gas phase* (Table 9.3), in which, of course, there is no solvent. In the gas phase, the relative reactivities of the halide ions correlate with their relative basicities; fluoride ion is more reactive than bromide ion.

The origin of this effect is the *hydrogen bonding* that occurs between a nucleophile and the molecules of a protic solvent (Fig. 9.3). Nucleophiles act as hydrogen-bond acceptors, and anions near the top of the periodic table—the more basic anions—form stronger hydrogen bonds than anions near the bottom of the periodic table. Thus, fluoride ion, the stronger base, forms stronger hydrogen bonds than iodide ion, the weaker base. In order for a halide ion to react as a nucleophile in the S_N2 reaction, at least one of its hydrogen bonds must be broken. It requires more energy to break a relatively strong hydrogen bond to fluoride than it does to break a relatively weak hydrogen bond to bromide. This extra energy is reflected in the free energy of activation—the energy barrier—and, as a result, the reaction of fluoride ion is slower.

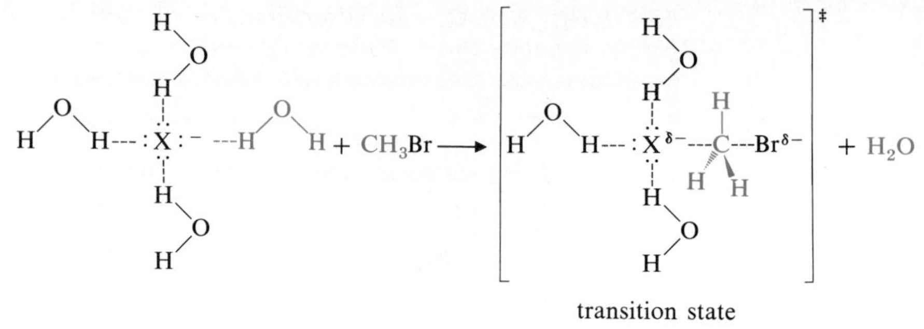

Figure 9.3 An S_N2 reaction by a halide nucleophile ($:\ddot{X}:^-$) in a protic solvent requires breaking a hydrogen bond to the nucleophile. When $X^- = F^-$, the hydrogen bond is strong, but when $X^- = I^-$, it is weaker. For this reason more energy is required to reach the transition state when $X^- = F^-$, and, as a result, the reaction is slower.

This analysis suggests two predictions. First, in solvents that cannot donate hydrogen bonds, the relative reactivities of the halide anions ought to parallel their relative basicities. In Table 9.3 we can see, for example, that in the gas phase—certainly a "solvent" that cannot hydrogen bond—the halide ions follow the expected reactivity order $F^- > Cl^- > Br^-$. We also find that in *polar aprotic solvents* such as dimethylsulfoxide, acetone, and *N,N*-dimethylformamide—solvents that solvate cations well but not anions (Sec. 8.4B)—the relative reactivities of nucleophiles correlate more closely with their relative basicities. The second prediction is that the rate of reaction of a given nucleophile and alkyl halide ought to be much greater in aprotic solvents, because there are no hydrogen bonds that have to be broken before the nucleophile can attack the alkyl halide. Indeed, we find this to be so: for example, the rate of reaction of chloride ion with CH_3Br is about a million times greater in the polar aprotic solvent *N,N*-dimethylformamide, and about 10^{15} times greater in the gas phase, than it is in methanol.

What solvents are best in a practical sense for the S_N2 reaction? First, the solvent must be able to dissolve the nucleophile, which is usually an ionic compound. For this reason water, alcohols, and mixtures of water with other solvents are frequently used, in spite of the fact that S_N2 reactions are relatively slow in these solvents. As we have seen, S_N2 reactions in aprotic solvents are faster. Unfortunately, aprotic solvents do not dissolve a number of ionic compounds. The very thing that makes the S_N2 reaction faster—poor solvation of the nucleophile—also makes the nucleophile less soluble! However, *polar aprotic solvents* do dissolve some salts, and such solvents are superb for many S_N2 reactions. These solvents dissolve many ionic compounds by solvating the *cation,* but they do not hydrogen bond to the anion. The unsolvated anions are very nucleophilic (basic), and as a result, S_N2 reactions in polar aprotic solvents are very fast. Widely used solvents of this type include acetone, dimethyl sulfoxide (DMSO), and *N,N*-dimethylformamide (DMF); structures of these compounds are given in Table 8.2. The reactions of KBr and KI with alkyl chlorides in acetone (Eq. 9.13), for example, take advantage of not only the insolubility of KCl, but also the enhanced nucleophilicity of bromide and iodide ions in acetone, a polar aprotic solvent.

Problem

7 In each pair, choose the species that reacts more rapidly with CH_3I. Explain your choices.

(a) ^-OH and H_2O (b) F^- and ^-OH (c) Cl^- and CH_3S^-

E. Summary of the S_N2 Reaction

Primary and some secondary alkyl halides react with nucleophiles by the S_N2 mechanism. Let us summarize the characteristic features of this mechanism.

1. The reaction rate is second order, first order in the nucleophile and first order in the alkyl halide.
2. Backside attack of the nucleophile on the alkyl halide occurs, and inversion of configuration is observed.
3. The reaction rate is retarded by branching near the site of reaction.
4. The fastest reactions are observed with nucleophiles that are the strongest bases. In protic solvents, this statement is valid as long as the comparison is made within the same row of the periodic table. In protic solvents, nucleophiles nearest the bottom of the periodic table give the fastest reactions because these nucleophiles are poorly solvated.
5. The fastest reactions are observed with leaving groups that give the weakest bases.
6. Of the solvents that dissolve salts, polar aprotic solvents give the fastest reactions.

9.4 THE E2 REACTION

The effect of chain branching on the rates of S_N2 reactions (Sec. 9.3C) leads us to believe that S_N2 reactions of tertiary alkyl halides are extremely slow. Indeed, when a strong base reacts with a tertiary alkyl halide, a substitution reaction does not take place; instead an *elimination* occurs.

$$CH_3-\underset{\displaystyle CH_3}{\overset{\displaystyle CH_3}{\overset{|}{\underset{|}{C}}}}-Br + Na^+C_2H_5O^- \xrightarrow[\text{ethanol}]{25\ ^\circ\text{C}} CH_2{=}C\begin{matrix} CH_3 \\ CH_3 \end{matrix} + C_2H_5OH + Na^+Br^- \quad (9.28)$$

If we call the carbon bearing the halogen the ***α*-carbon,** and the adjacent carbon the ***β*-carbon,** we see that the halide on the α-carbon and a hydrogen atom on the β-carbon are lost and an alkene is formed.

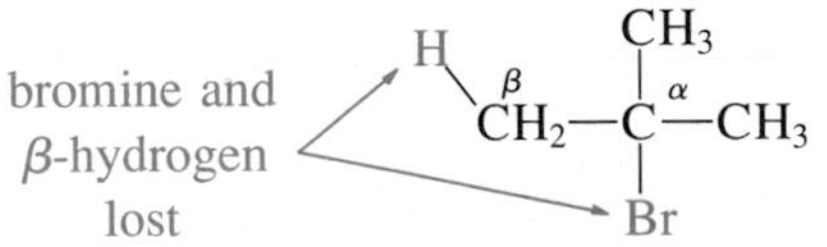

This reaction is an example of a ***β*-elimination.** In a β-elimination reaction two groups are eliminated from adjacent atoms. It is by far the most common type of elimination reaction in organic chemistry. Notice that the reaction is *formally* the reverse of H—Br addition.

Problem

8 List all the alkene products that could be formed in a β-elimination reaction of 3-bromo-3-methylpentane.

Among the bases commonly used to promote β-elimination reactions of alkyl halides are the alkoxide salts (Sec. 8.6A), such as sodium ethoxide (Na$^+$ C$_2$H$_5$O$^-$) and potassium *t*-butoxide (K$^+$ (CH$_3$)$_3$C—O$^-$). Often these bases are used in solutions of their conjugate acid alcohols: sodium ethoxide in ethanol, and potassium *t*-butoxide in *t*-butyl alcohol.

A. The E2 Reaction: Rate Law and Mechanism

Base-promoted elimination reactions of alkyl halides typically follow a rate law that is second order overall, and first order in each reactant.

$$\text{rate} = k[(CH_3)_3C\!-\!Br][C_2H_5O^-] \tag{9.29}$$

A mechanism consistent with this rate law is the following one, in which a proton and the halide are removed simultaneously to give the alkene:

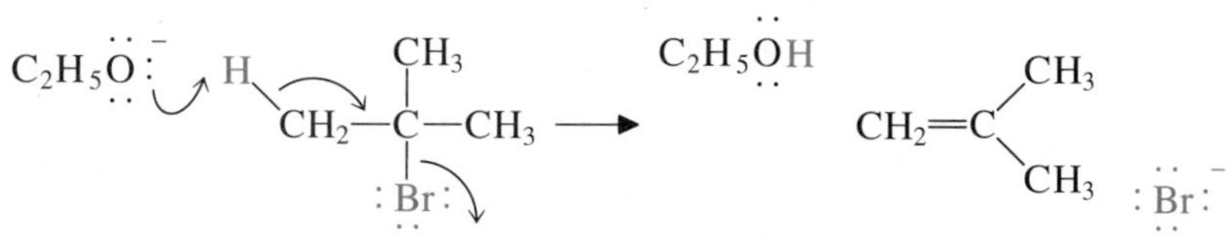

$$\tag{9.30}$$

This reaction also has a special name: the **E2 reaction.**

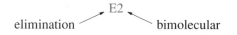

B. Effect of Leaving Group on the E2 Reaction. Isotope Effects

In the mechanism of the E2 reaction, the leaving halide takes on a negative charge, just as it does in the S$_N$2 reaction. Consequently, it should be no surprise to find that the rates of the E2 and S$_N$2 reactions are affected in very similar ways by changing the halide leaving group. Thus, in the E2 reactions of the 2-halo-2-methylbutanes to give 2-methyl-2-butene,

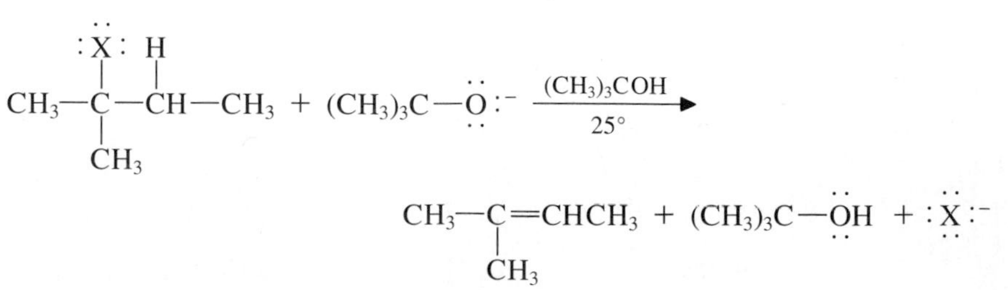

$$CH_3\!-\!\underset{\underset{CH_3}{|}}{\overset{\overset{:\ddot{X}:\ \ H}{|\ \ \ |}}{C}}\!-\!CH\!-\!CH_3 + (CH_3)_3C\!-\!\ddot{O}:^- \xrightarrow[25°]{(CH_3)_3COH}$$

$$CH_3\!-\!\underset{\underset{CH_3}{|}}{C}\!=\!CHCH_3 + (CH_3)_3C\!-\!\ddot{O}H + :\ddot{X}:^- \tag{9.31}$$

the rate of the reaction is greater when the leaving group is a weaker base:

halide leaving group X: Cl < Br < I
relative rate: 1 58 401

The mechanism in Eq. 9.30 implies that a proton is removed in the transition state of the elimination reaction. There is an interesting way of testing this aspect of the mechanism. When a proton is transferred in the rate-determining step of a reaction, a compound in which that proton is replaced by its isotope deuterium will react more slowly in the same reaction. This effect of isotopic substitution on reaction rates is called a **primary deuterium isotope effect.** For example, suppose we call the rate constant for the following E2 reaction of 2-phenyl-1-bromoethane k_H, and the rate constant for the reaction of its β-deuterium analog k_D:

$$\text{Ph—CH}_2\text{—CH}_2\text{—Br} + \text{C}_2\text{H}_5\text{O}^- \xrightarrow[\text{rate constant } k_H]{\text{C}_2\text{H}_5\text{OH}} \text{Ph—CH=CH}_2 + \text{Br}^- + \text{C}_2\text{H}_5\text{OH} \quad (9.32a)$$

$$\text{Ph—CD}_2\text{—CH}_2\text{—Br} + \text{C}_2\text{H}_5\text{O}^- \xrightarrow[\text{rate constant } k_D]{\text{C}_2\text{H}_5\text{OH}} \text{Ph—CD=CH}_2 + \text{Br}^- + \text{C}_2\text{H}_5\text{OD} \quad (9.32b)$$

The primary deuterium isotope effect is the ratio k_H/k_D; typically such isotope effects are in the range 2.5–8. In fact, k_H/k_D for the reaction in Eq. 9.32 is 7.1. The observation of a primary isotope effect of this magnitude shows that the bond to a β-hydrogen is broken in the rate-determining step of this reaction.

The theoretical basis for the primary isotope effect lies in the comparative strengths of C—H and C—D bonds. In the starting material, the bond to the heavier isotope D is stronger (and thus requires more energy to break; Sec. 5.8E) than the bond to the lighter isotope H. However, in the transition states for both reactions, the bond from H or D to carbon is partly broken, and the bond from H or D to the attacking group is partly formed. To a crude approximation, the isotope undergoing transfer is not bonded to anything—it is "in flight." Since there is no bond, there is no bond-energy difference between the two isotopes. Therefore, the compound with the C—D bond starts out at a lower energy than the compound with the C—H bond, and requires more energy to get to the transition state (Fig. 9.4). In other words, the energy barrier, or free energy of activation, for the compound with the C—D bond is greater; as a result, its rate of reaction is smaller.

Be sure you understand that a primary deuterium isotope effect is observed only when the *hydrogen that is transferred* in the rate-determining step is substituted by

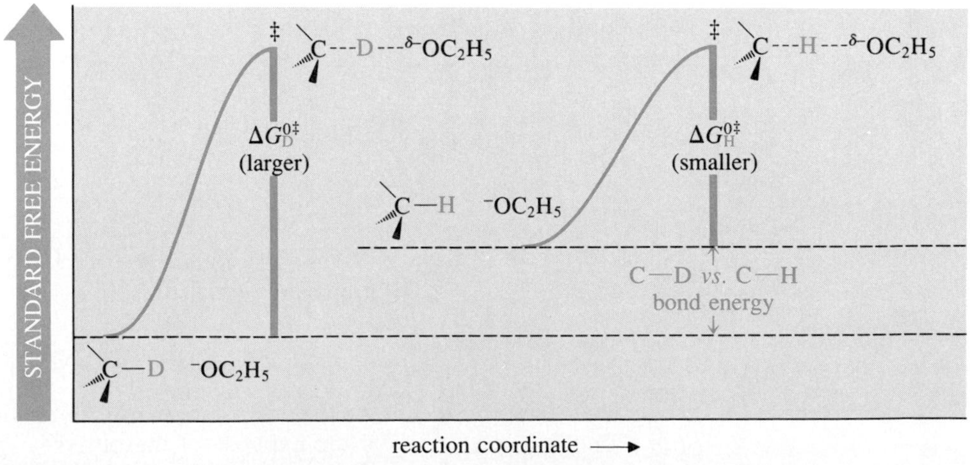

Figure 9.4 The source of the primary deuterium isotope effect is the stronger carbon–deuterium bond.

deuterium. Substitution of other hydrogens with deuterium usually has little or no effect on the rate of the reaction.

An ingenious practical application of the isotope effect was used in the design of the experimental antibiotic 3-fluoroalanine. It was suspected that the drug is destroyed in the body by a β-elimination reaction:

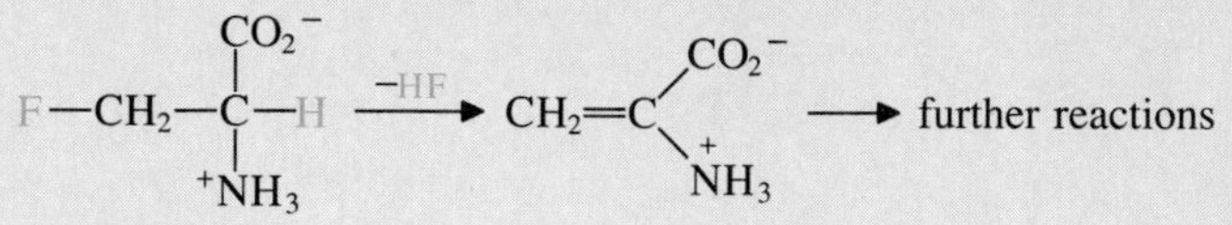

$$\underset{\text{3-fluoroalanine}}{\text{F}-\text{CH}_2-\overset{\displaystyle \overset{\text{CO}_2^-}{|}}{\underset{\underset{^+\text{NH}_3}{|}}{\text{C}}}-\text{H}} \xrightarrow{-\text{HF}} \text{CH}_2{=}\text{C}\overset{\text{CO}_2^-}{\underset{^+\text{NH}_3}{\diagdown}} \longrightarrow \text{further reactions} \qquad (9.33)$$

On the possibility that such an elimination reaction might be retarded by a deuterium isotope effect (in the same sense that the E2 reaction of alkyl halides is retarded), a new drug was synthesized in which the hydrogen shown in color was replaced by deuterium. It was found that the isotopically substituted drug is much more effective as an antibiotic, presumably because its metabolic destruction is slower and it lasts longer in the body.

Problem

9 (a) The rate-determining step in the hydration of styrene ($\text{Ph}-\text{CH}{=}\text{CH}_2$) is the initial transfer of the proton from H_3O^+ to the alkene. (See Sec. 4.8.) How would you expect the rate of the reaction to change if the reaction were run in $\text{D}_2\text{O}/\text{D}_3\text{O}^+$ instead of $\text{H}_2\text{O}/\text{H}_3\text{O}^+$? Would the product be the same?

 (b) How would the rate of styrene hydration in $\text{H}_2\text{O}/\text{H}_3\text{O}^+$ differ for an isotopically substituted styrene $\text{Ph}-\text{CH}{=}\text{CD}_2$? Explain.

C. Competition between the E2 and S_N2 Reactions

Many primary alkyl halides undergo the S_N2 reaction when they react with nucleophilic bases, as we have seen. On the other hand, we have just learned that tertiary alkyl halides react with bases to give the E2 reaction. What about secondary alkyl halides?

Many secondary alkyl halides react to give a mixture of products resulting from both substitution and elimination reactions. The following example is typical:

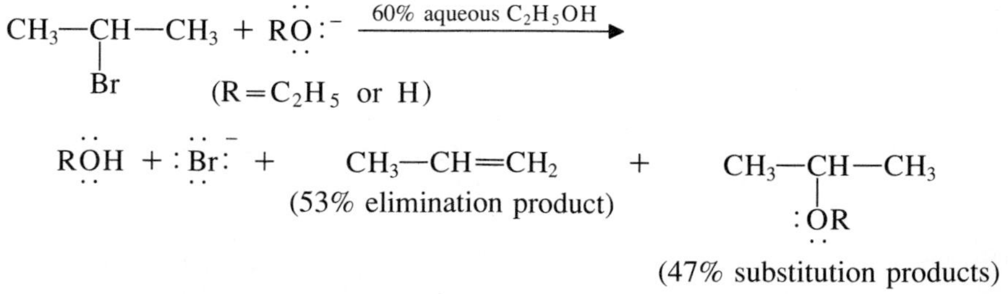

$$\text{CH}_3-\underset{\underset{\text{Br}}{|}}{\text{CH}}-\text{CH}_3 + \text{R}\ddot{\text{O}}\!:^- \xrightarrow{\;60\%\ \text{aqueous C}_2\text{H}_5\text{OH}\;}$$

$$(\text{R}=\text{C}_2\text{H}_5 \text{ or } \text{H})$$

$$\text{R}\ddot{\text{O}}\text{H} + :\!\ddot{\text{Br}}\!:^- + \underset{(53\%\ \text{elimination product})}{\text{CH}_3-\text{CH}{=}\text{CH}_2} + \underset{\underset{:\ddot{\text{O}}\text{R}}{|}}{\text{CH}_3-\text{CH}-\text{CH}_3} \qquad (9.34)$$

$$(47\%\ \text{substitution products})$$

From this example, we can see that substitution and elimination are *competitive reactions*: one reaction occurs at the expense of the other.

$$\text{alkyl halide with } \beta\text{-hydrogens} + \text{a base} \begin{cases} S_N2 \text{ reaction (substitution)} \\ \\ E2 \text{ reaction (elimination)} \end{cases} \qquad (9.35)$$

We should recognize that this competition is a matter of *relative rates*: the reaction pathway that occurs most rapidly is the one that predominates.

What determines whether the S_N2 reaction or the E2 reaction is faster? Two variables determine the outcome of this competition: (1) *the structure of the alkyl halide,* and (2) *the structure of the base.* Let us examine each effect in turn.

In order for the S_N2 reaction to occur, the base (nucleophile) must attack a carbon atom. When attack at carbon is retarded by steric hindrance, as in a tertiary alkyl halide, attack at relatively unhindered hydrogen—that is, elimination—predominates. This is one reason that tertiary halides react with strong bases to give elimination products exclusively.

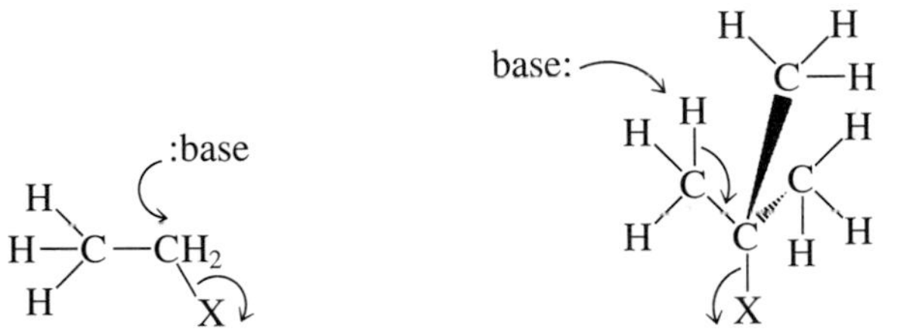

$$(9.36)$$

attack at carbon is unhindered; substitution occurs

attack at carbon is blocked; elimination occurs

A second reason has to do with the effect of branching on alkene stability. The transition state in the E2 reaction can be thought of as a structure that is something between an alkyl halide and an alkene. To the extent that the transition state resembles alkene, this transition state is stabilized by the same things that stabilize an alkene—such as alkyl branches at the double bond. Because branching stabilizes the transition state, branching increases the rate of the E2 reaction.

To summarize the effect of alkyl halide structure: branching *reduces* the rate of substitution and *increases* the rate of elimination. Since elimination and substitution are competing reactions (Eq. 9.35), branching increases the proportion of elimination.

These effects can be seen not only in tertiary alkyl halides, but in secondary and even primary alkyl halides as well.

secondary alkyl halides:

$$\beta\text{-carbons} \longrightarrow \underset{\text{CH}_3}{\text{CH}_3-\text{CH}-\text{Br}} + \text{C}_2\text{H}_5\text{O}^- \longrightarrow \underset{\text{(50–60\% elimination)}}{\text{CH}_2=\text{CH}-\text{CH}_3} + \underset{\substack{\text{CH}_3 \\ \text{(40–50\% substitution)}}}{\text{CH}_3-\text{CH}-\text{OC}_2\text{H}_5} \qquad (9.37a)$$

$$\text{one } \beta\text{-branch} \longrightarrow \underset{\text{CH}_3}{\text{CH}_3-\text{CH}_2-\text{CH}-\text{Br}} + \text{C}_2\text{H}_5\text{O}^- \longrightarrow$$

$$\underbrace{\text{CH}_3-\text{CH}=\text{CH}-\text{CH}_3 + \text{CH}_3-\text{CH}_2-\text{CH}=\text{CH}_2}_{\text{(82\% elimination)}} + \underset{\substack{\text{CH}_3 \\ \text{(18\% substitution)}}}{\text{CH}_3-\text{CH}_2-\text{CH}-\text{OC}_2\text{H}_5} \qquad (9.37b)$$

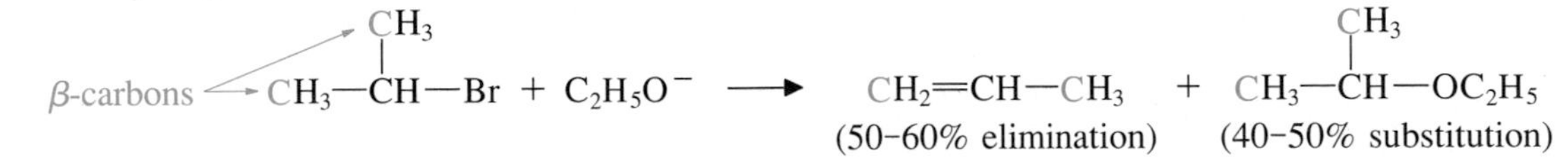

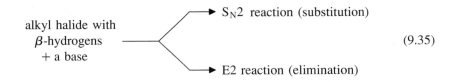

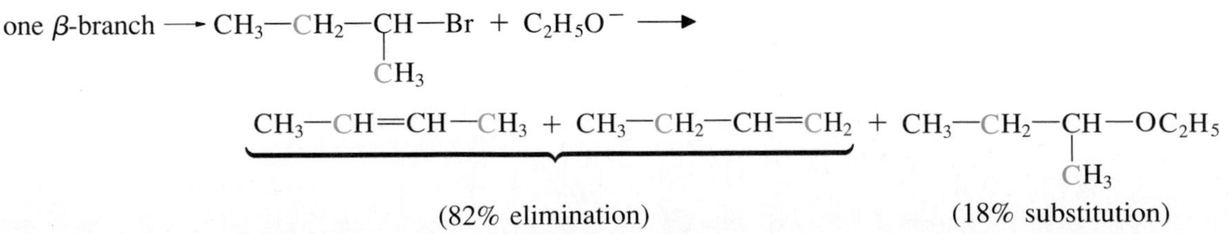

primary alkyl halides:

β-carbon → $CH_3—CH_2—Br + C_2H_5O^- \longrightarrow$

$$CH_2{=}CH_2 \; (1\% \text{ elimination}) + CH_3—CH_2—OC_2H_5 \; (99\% \text{ substitution}) \quad (9.37c)$$

one β-branch → $CH_3—CH_2—CH_2—Br + C_2H_5O^- \longrightarrow$

$$CH_3—CH{=}CH_2 \; (10\% \text{ elimination}) + CH_3—CH_2—CH_2—OC_2H_5 \; (90\% \text{ substitution}) \quad (9.37d)$$

two β-branches → $(CH_3)_2CH—CH_2—Br + C_2H_5O^- \longrightarrow$

$$(CH_3)_2C{=}CH_2 \; (62\% \text{ elimination}) + (CH_3)_2CH—CH_2—OC_2H_5 \; (38\% \text{ substitution}) \quad (9.37e)$$

The second variable affecting the ratio of elimination to substitution is the structure of the base. A highly branched base such as *t*-butoxide increases the proportion of the E2 product at the expense of the S_N2 product.

$$(CH_3)_2CHCH_2—Br + {}^-OCH_2CH_3 \xrightarrow{C_2H_5OH} (CH_3)_2C{=}CH_2 + (CH_3)_2CHCH_2—OCH_2CH_3 \quad (9.38a)$$

ethoxide (a primary, unbranched alkoxide base) — (62% elimination) — (38% substitution)

$$(CH_3)_2CHCH_2—Br + {}^-O—C(CH_3)_3 \xrightarrow{(CH_3)_3COH} (CH_3)_2C{=}CH_2 + (CH_3)_2CHCH_2—O—C(CH_3)_3 \quad (9.38b)$$

t-butoxide (a tertiary, branched alkoxide base) — (92% elimination) — (8% substitution)

When a highly branched base attacks a *carbon* to give the substitution product, the alkyl branches of the base suffer steric interference with the alkyl branches of the alkyl halide. These steric repulsions raise the energy of the transition state for substitution. This effect is more pronounced for a bulky, highly branched base like *t*-butoxide than it is for a smaller base like ethoxide. When the base attacks a *β-proton* to give the elimination product, the base is further removed from the site of branching in the alkyl halide, and steric repulsions are less severe, as shown in Eq. 9.36. The net effect of these two factors is that branching in the base increases the ratio of elimination to substitution.

Problems

10 Arrange the following four alkyl halides in descending order with respect to the ratio of E2 elimination to S_N2 substitution products observed in their reactions with sodium ethoxide in ethyl alcohol:

(a) CH_3I

(b) $(CH_3)_2CHCH_2—Br$

(c) $(CH_3)_3CCH_2CH_2CH_2—Br$

(d) $(CH_3)_2CHCH(CH_3)—Br$

11 Arrange the following three alkoxide bases in descending order with respect to the ratio of E2 elimination to S_N2 substitution products observed when they react with isobutyl bromide:
 (a) $(CH_3)_2CH—O^-$ (b) CH_3O^- (c) $(C_2H_5)_3C—O^-$

D. Double Bond Position (Regiochemistry) in E2 Reaction Products

When an alkyl halide has more than one type of β-hydrogen, its E2 reactions can in principle give several different alkenes (see Problem 8).

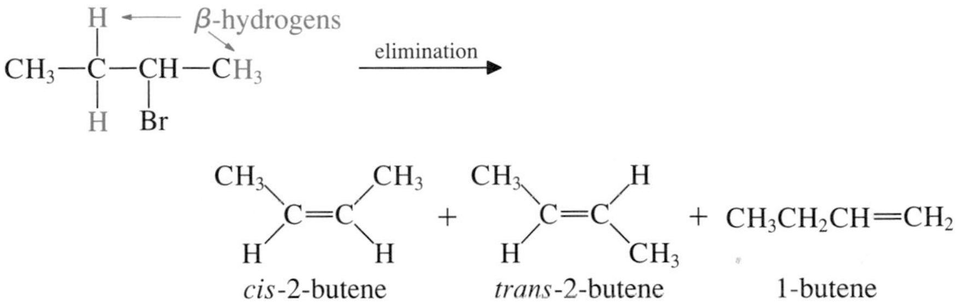

$$cis\text{-2-butene} \qquad trans\text{-2-butene} \qquad \text{1-butene} \qquad (9.39)$$

Is one of these alkenes formed in preference to another?

The outcome of the reaction again depends on several factors. One of the most important is the structure of the base. When relatively unbranched bases are used, such as primary alkoxides (for example, sodium ethoxide), we usually observe that *the predominant product is the most stable alkene isomer.* We learned in Sec. 4.4C that the most stable alkenes are generally those with the most alkyl groups on the alkene carbons. It follows, then, that these isomers are usually the ones formed in greatest amount.

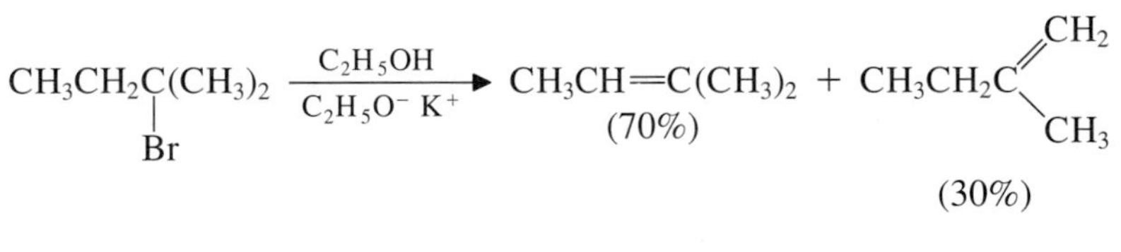

$$CH_3CH_2C(CH_3)_2 \xrightarrow[\text{C}_2\text{H}_5\text{O}^-\text{ K}^+]{\text{C}_2\text{H}_5\text{OH}} CH_3CH=C(CH_3)_2 + CH_3CH_2C\overset{CH_2}{\underset{CH_3}{\diagup\diagdown}}$$
$$\underset{\text{Br}}{|} \qquad\qquad (70\%)$$
$$(30\%) \qquad (9.40)$$

Notice that in this reaction, the minor alkene isomer is favored on statistical grounds: there are six equivalent hydrogens that can be lost to give the minor product, but only two that can be lost to give the major product. In the absence of a structural effect on the product distribution, three times as much of the minor isomer would have been formed. The greater stability of the transition state leading to the major isomer overrides this statistical factor.

The formation of the most highly branched alkene isomers is sometimes called **Saytzeff elimination,** after Alexander Saytzeff, the chemist who observed this behavior in 1875. The reason for Saytzeff elimination in the E2 reaction can be found in transition-state theory. As we discussed in Sec. 9.4C, branching at the developing double bond stabilizes the transition state for elimination. Because the transition state with the most alkyl branches at its developing double bond is the one of least energy, the reaction path with this transition state has the greatest rate. The result is Saytzeff elimination.

TABLE 9.4 Effect of Alkoxide Base Structure on the Regiochemistry of an E2 Elimination

$$(CH_3)_2CH\underset{\underset{Br}{|}}{C}(CH_3)_2 \longrightarrow (CH_3)_2C{=}C(CH_3)_2 + (CH_3)_2CH\underset{\underset{CH_3}{|}}{C}{=}CH_2$$

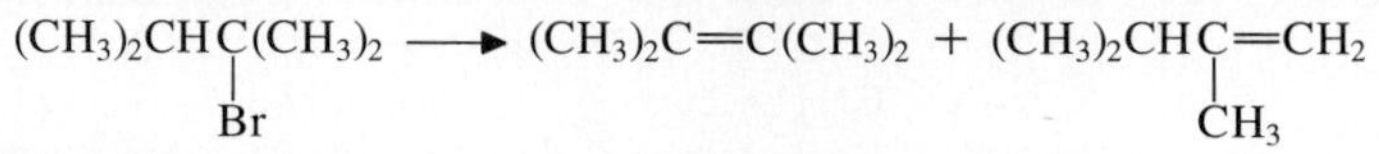

Base	Percentage of 1-alkene		
$CH_3CH_2{-}O^-$	21		
$CH_3{-}\overset{\overset{\displaystyle CH_3}{	}}{\underset{\underset{\displaystyle CH_3}{	}}{C}}{-}O^-$	73
$CH_3{-}\overset{\overset{\displaystyle CH_3}{	}}{\underset{\underset{\displaystyle CH_2CH_3}{	}}{C}}{-}O^-$	81
$CH_3CH_2{-}\overset{\overset{\displaystyle CH_2CH_3}{	}}{\underset{\underset{\displaystyle CH_2CH_3}{	}}{C}}{-}O^-$	92

Problem

12 (a) What are the relative rates of formation of the two alkenes in Eq. 9.40?

(b) What are the relative rates in (a) on a "per hydrogen" basis—that is, after correction for the number of equivalent hydrogens?

(c) Sketch a reaction free-energy diagram that shows the formation of both alkenes in Eq. 9.40 from the starting alkyl halide.

When a highly branched base such as K^+ $(CH_3)_3C{-}O^-$ is used, a smaller percentage of Saytzeff elimination is observed; that is, more of the alkene with the *least* amount of branching at the double bond is observed. This point is illustrated by the data in Table 9.4. Notice that the change from the relatively unbranched ethoxide to the highly branched $(C_2H_5)_3C{-}O^-$ results in formation of more 1-alkene (the least stable alkene isomer), and virtual exclusion of the more stable alkene isomer. The basis for these observations is probably a steric effect. A large, highly branched base preferentially attacks not only hydrogen rather than carbon, but also the *least sterically hindered* hydrogens; these are typically the hydrogens on the carbon with the least number of alkyl branches (Fig. 9.5).

We see, then, that branching in the base used in the E2 reaction has two consequences. First, it increases the relative proportion of E2 to S_N2 products; and second, it increases the proportion of alkenes with less branching at the double bonds.

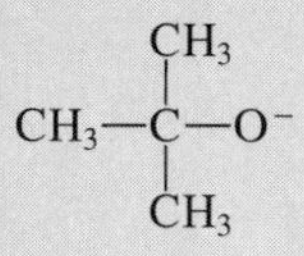

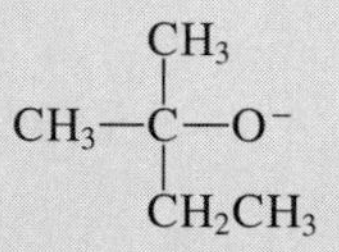

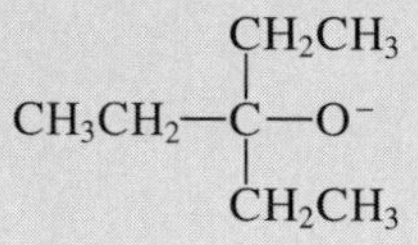

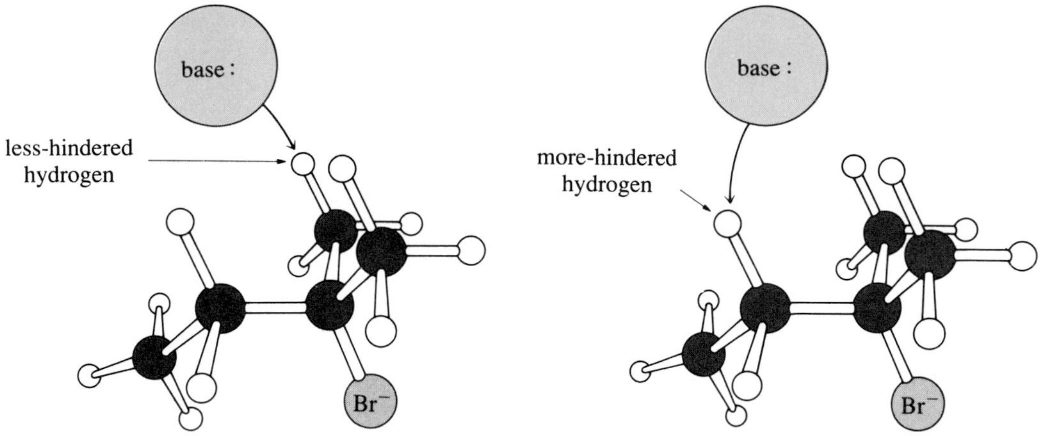

Figure 9.5 A comparison of the steric requirements for attack of a large base on two hydrogens in the same molecule; (a) a primary hydrogen and (b) a secondary hydrogen surrounded by branching.

Problem

13 (a) Predict the predominant elimination product when the following alkyl halide reacts with sodium ethoxide in ethanol:

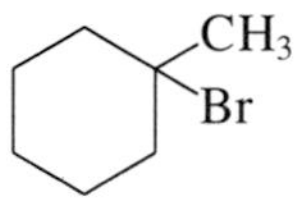

(b) How would your answer to (a) change if the base were changed to $K^+ (C_2H_5)_3C—O^-$?

We have just examined two types of competition: the competition between the E2 and S_N2 reactions, and the competition among E2 reactions themselves in the formation of different alkenes. The occurrence of competing reaction pathways has an undesirable practical consequence: the formation of mixtures. When an alkyl halide reacts to give a mixture of products, the yield of any one component of the mixture is low and, in some cases, the individual components of the mixture can be difficult to separate and purify. The separation of the components of a mixture is particularly tedious when, as in Eq. 9.39, they are isomers with very similar physical properties. On the other hand, when a reaction favors mostly one product, difficult separations are avoided. Therefore, the S_N2 and E2 reactions are most useful when they give largely a single product. The discussion in this and the previous sections tell us when to expect this situation. To summarize:

1. The S_N2 reaction can be used to prepare substitution products in good yield from primary and, in some cases, secondary halides that are largely unbranched near the α-carbon.

2. The E2 reaction can be used to prepare alkenes in good yield from tertiary alkyl halides and from branched secondary alkyl halides. However, mixtures can be expected from alkyl halides with more than one type of β-proton (as in Eq. 9.39).

3. The E2 reaction can be used to prepare alkenes from primary and unbranched secondary alkyl halides if a tertiary alkoxide base, such as *t*-butoxide, is used.

E. Stereochemistry of the E2 Reaction

An elimination reaction can occur in two stereochemically different ways, illustrated as follows for the elimination of H—X from a general alkyl halide:

syn:

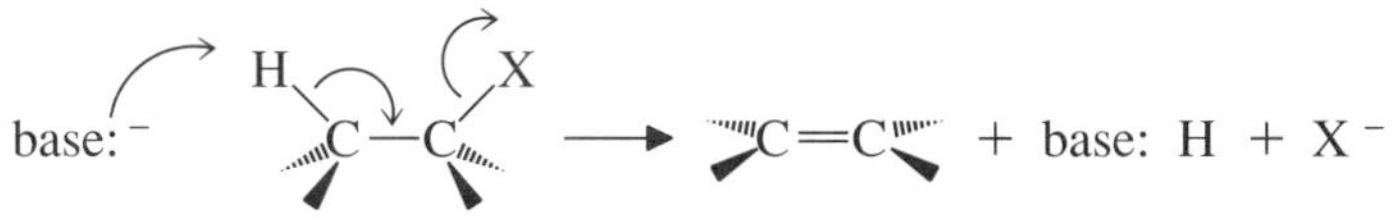

(9.41a)

anti:

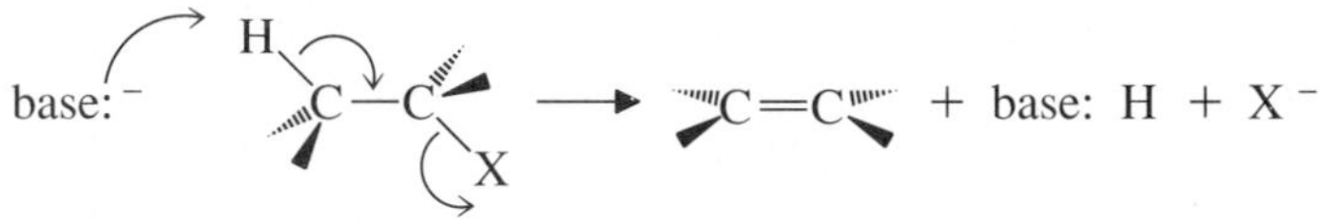

(9.41b)

In a *syn* elimination, H and X⁻ leave the alkyl halide molecule from the same side; in an *anti* elimination, H and X⁻ leave from opposite sides.

> We have seen the terms *syn* and *anti* before, when we discussed the stereochemistry of additions to double bonds (Sec. 7.10A). Notice that *syn* elimination is the formal reverse of a *syn* addition, and *anti* elimination is the formal reverse of an *anti* addition.

It is found experimentally that most E2 elimination reactions are stereoselective *anti* eliminations.

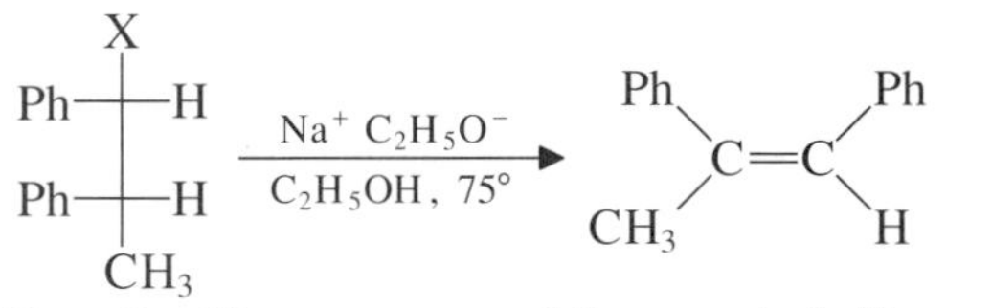

(9.42a)

X = Br, Cl
(Fischer projection)

(*Z*)-α-methylstilbene
(100%)

That is, if we put the alkyl halide in a conformation in which the H and halogen are on opposite sides of the molecule, the Ph groups end up on the same side, and elimination must then give the (*Z*)-alkene, as observed.

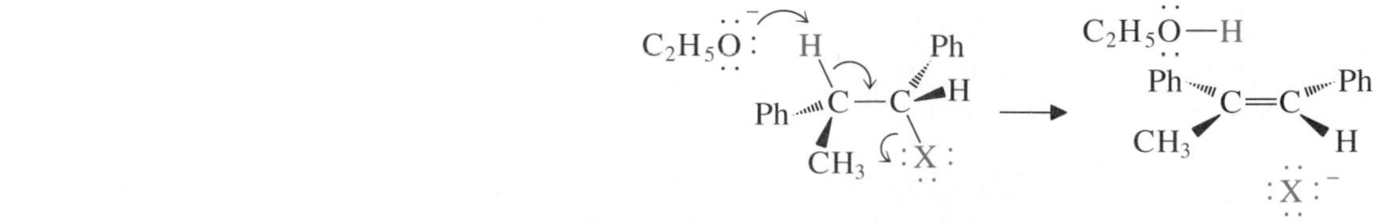

(9.42b)

(You should convince yourself that *syn* elimination, in contrast, would give the *E* isomer of the alkene.)

Both *syn* and *anti* eliminations allow the alkene product to form with its *p* orbit-

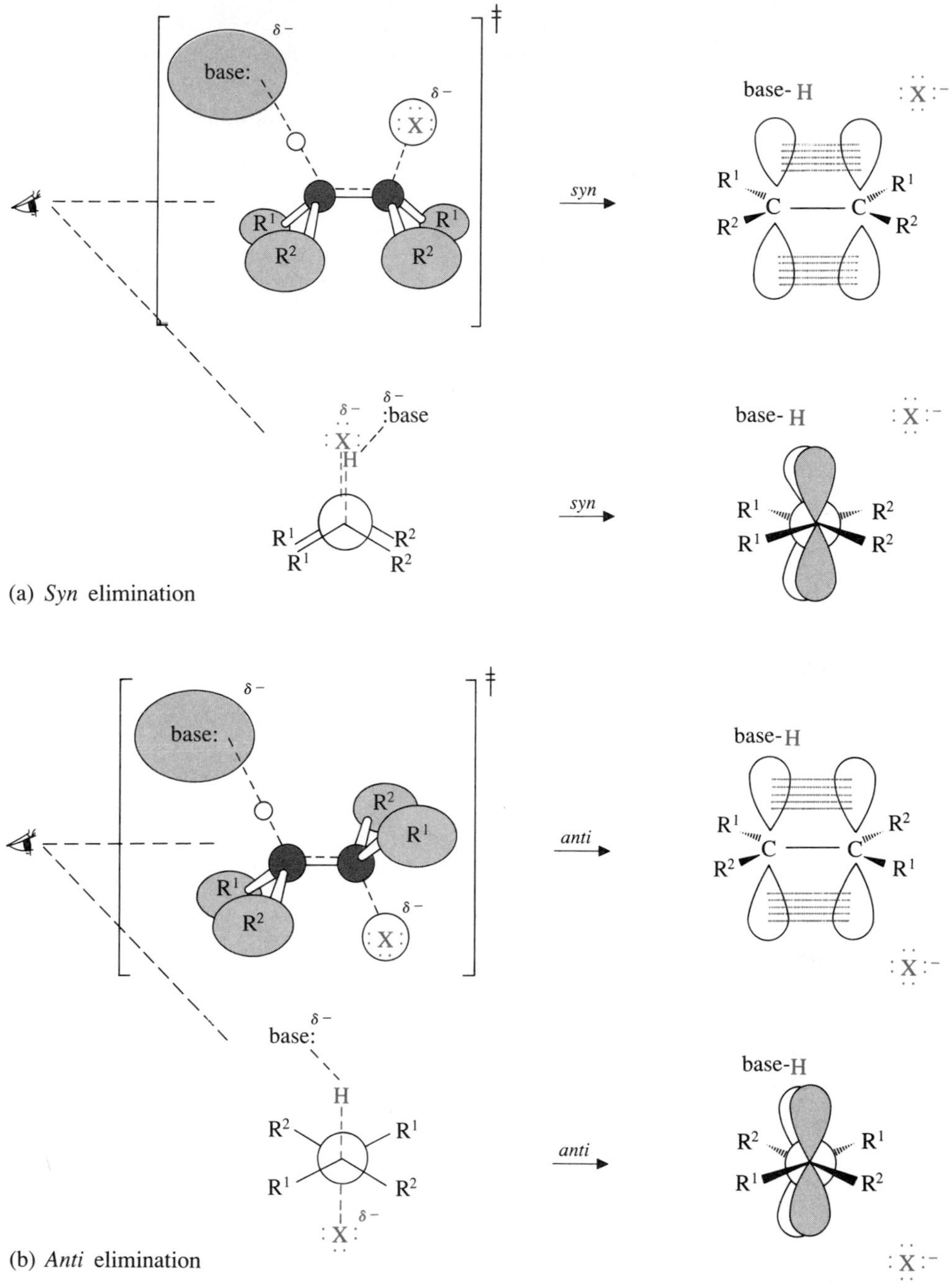

Figure 9.6 *Syn* and *anti* eliminations with corresponding Newman projections.

als in the coplanar arrangement required for π-bond formation, as we can see from the orbital diagrams in Fig. 9.6. Why, then, is *anti* elimination preferred? There are three reasons. First, *anti* elimination occurs through a transition state in which the molecule assumes a staggered conformation (Fig. 9.6b). However, *syn* elimination occurs through a transition state that has an eclipsed conformation (Fig. 9.6a). Since eclipsed conformations are unstable, the transition state for *syn* elimination is less stable than that for *anti* elimination. As a consequence, *anti* elimination is faster. The second reason that *anti* elimination is preferred is that the base and leaving group are on opposite sides of the molecule, out of each other's way. In *syn* elimination, they are on the same side of the molecule, and can interfere sterically with each other. Finally, there are quantum-mechanical reasons for believing that the *anti* elimination is preferable; these are related to the fact that *anti* elimination involves all-backside electronic displacements, whereas *syn* elimination requires a formal frontside electronic displacement on the carbon–halogen bond. (Recall that direct displacements occur by backside attack.)

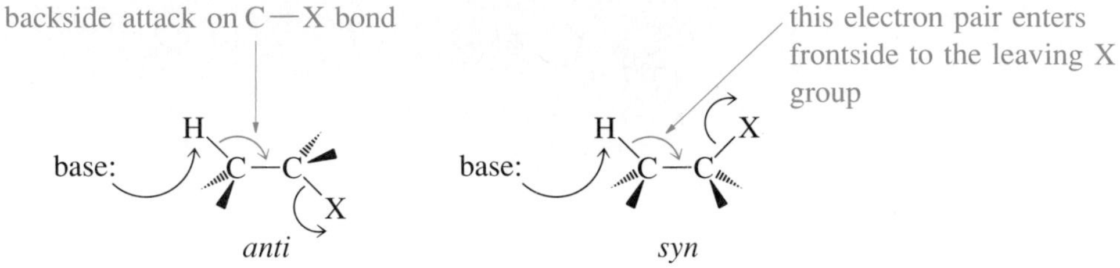

Problems

14 What starting material would give the *E* isomer of the alkene in the E2 reaction of Eq. 9.42a?

15 Predict the products, including their stereochemistry, from the reactions of (a) *meso*-dibromostilbene (Ph—CHBr—CHBr—Ph) and (b) (±)-dibromostilbene with sodium ethoxide in ethanol. (Assume that one equivalent of HBr is eliminated from each compound.)

F. Summary of the E2 Reaction

The E2 reaction is an elimination reaction of tertiary and other branched alkyl halides containing β-hydrogens, and is promoted by strong base. Some key points about this reaction are:

1. The rates of E2 reactions are second order, first order in base and first order in the alkyl halide.
2. E2 reactions usually occur with *anti* stereochemistry.
3. The weakest bases make the best leaving groups.
4. The rates of E2 reactions show substantial primary deuterium isotope effects.
5. E2 reactions compete with S_N2 reactions. Elimination is favored by branching in the alkyl halide at the α- or β-carbons, and by branching in the base.
6. The major products of E2 reactions are the alkenes with the most alkyl branches at their double bonds (the Saytzeff products). Branching in the base increases the percentage of other alkenes.

9.5 SUBSTITUTION REACTIONS OF TERTIARY ALKYL HALIDES. THE S_N1 REACTION

A. Rate Law and Mechanism of the S_N1 Reaction

If ethyl bromide is dissolved in ethanol in the absence of ethoxide ion, a very slow solvolysis reaction takes place by the S_N2 mechanism.

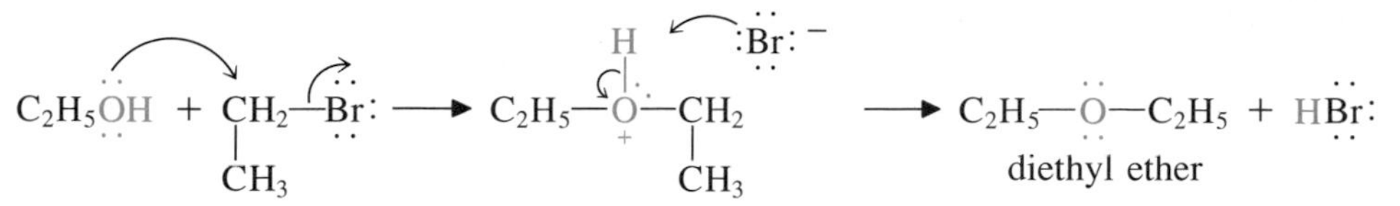

$$C_2H_5\overset{..}{O}H + CH_2\!-\!\overset{..}{Br}\!: \longrightarrow C_2H_5\!-\!\overset{H}{\underset{+}{\overset{..}{O}}}\!-\!CH_2 \qquad \longrightarrow C_2H_5\!-\!\overset{..}{O}\!-\!C_2H_5 + H\overset{..}{Br}\!: \qquad (9.43)$$

with CH_3 groups and $:Br:^-$ labeling; diethyl ether

The reaction is slow because ethyl alcohol is a weak base and a poor nucleophile. Because chain branching at the α-carbon severely retards the S_N2 reaction, we expect the solvolysis of t-butyl bromide to be even slower still—immeasurably slow—if it occurs by the same mechanism. But in fact, when t-butyl bromide is dissolved in ethanol, a reaction occurs that is about 800 times faster than the reaction of ethyl bromide! This reaction gives a mixture of substitution and elimination products.

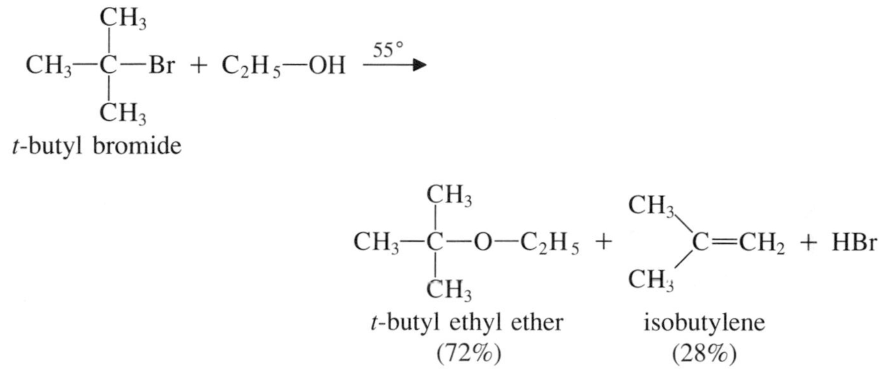

$$CH_3\!-\!\underset{\underset{CH_3}{|}}{\overset{\overset{CH_3}{|}}{C}}\!-\!Br + C_2H_5\!-\!OH \xrightarrow{55°}$$

t-butyl bromide

$$CH_3\!-\!\underset{\underset{CH_3}{|}}{\overset{\overset{CH_3}{|}}{C}}\!-\!O\!-\!C_2H_5 + \underset{CH_3}{\overset{CH_3}{}}\!\!C\!=\!CH_2 + HBr \qquad (9.44)$$

t-butyl ethyl ether isobutylene
(72%) (28%)

The substitution reaction of t-butyl bromide is faster because it occurs by a *different mechanism*, one that is characteristic of tertiary alkyl halides in the absence of strong base.

One clue to this new mechanism comes from a comparison of the rate laws for the reactions in Eqs. 9.43 and 9.44 when they are run in the presence of a small amount of sodium ethoxide. The reaction in Eq. 9.43 is accelerated by the addition of ethoxide, because, like other S_N2 reactions, its rate law contains a term in the nucleophile concentration.

$$rate = k[CH_3CH_2Br][C_2H_5O^-] \qquad (9.45)$$

However, the rate of the reaction in Eq. 9.44 is unaffected by small amounts of ethoxide because it follows a *first-order rate law* that contains no term in the nucleophile concentration.

$$rate = k[(CH_3)_3C\!-\!Br] \qquad (9.46)$$

This means that *neither* the substitution *nor* the elimination component of the reaction in Eq. 9.44 is affected by base concentration. Hence, the substitution reaction cannot be an S_N2 reaction, nor can the elimination reaction be an E2 reaction, because S_N2 and E2 reactions follow second-order rate laws. (At higher ethoxide concentrations, an E2 reaction of the tertiary halide *is* observed, and alkene is the major product formed.)

A mechanism that is consistent with this rate law must contain only the alkyl halide, *but not the nucleophile,* in the rate-determining transition state. The following substitution mechanism fits this rate law:

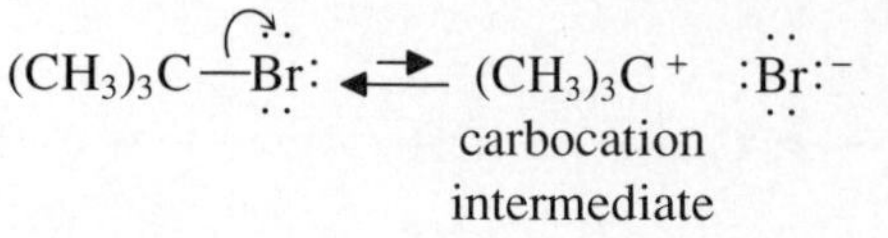

$$(CH_3)_3C \overset{\frown}{\longrightarrow} \ddot{B}r\!: \;\; \underset{\longleftarrow}{\rightleftharpoons} \;\; (CH_3)_3C^+ \;\; :\ddot{B}r\!:^- \qquad \text{(rate-determining step)} \qquad (9.47a)$$

carbocation
intermediate

$$(CH_3)_3C^+ + C_2H_5\ddot{O}H \; \rightleftharpoons \; (CH_3)_3C\overset{+}{-}\overset{}{\underset{H}{\ddot{O}C_2H_5}} \;\; :\ddot{B}r\!:^- \qquad (9.47b)$$
$$:\ddot{B}r\!:^-$$

$$(CH_3)_3C\overset{+}{-}\underset{H}{\ddot{O}C_2H_5} \qquad :\ddot{B}r\!:^- \; \rightleftharpoons \; (CH_3)_3C-\ddot{O}C_2H_5 + H\ddot{B}r\!: \qquad (9.47c)$$

The first step in this mechanism—ionization of the alkyl halide to a carbocation—is rate determining. This ionization is a simple heterolytic bond cleavage. When a tertiary alkyl halide is dissolved in a solvent such as ethanol, it reacts by dissociating slowly into a carbocation and a halide anion. Since the nucleophile concentration does not appear in the rate law, it enters the mechanism *after* the rate-determining step. Thus, once the carbocation is formed (slowly), it is rapidly consumed by reaction with any nucleophiles in solution.

Because the rate-determining step of this reaction is unimolecular (involves a single molecule), it is named the **S_N1 reaction.**

substitution S_N1 unimolecular

nucleophilic

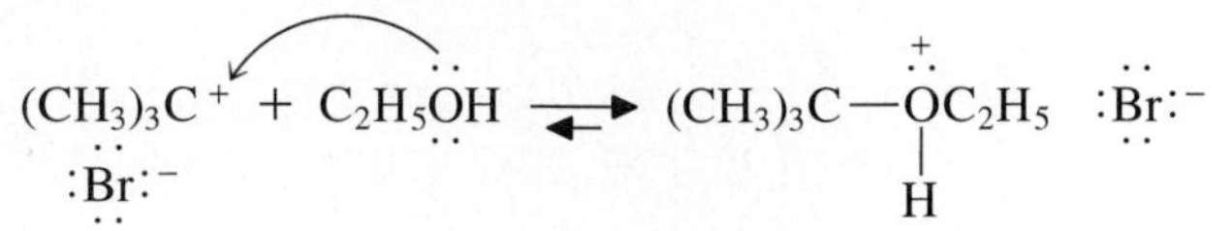

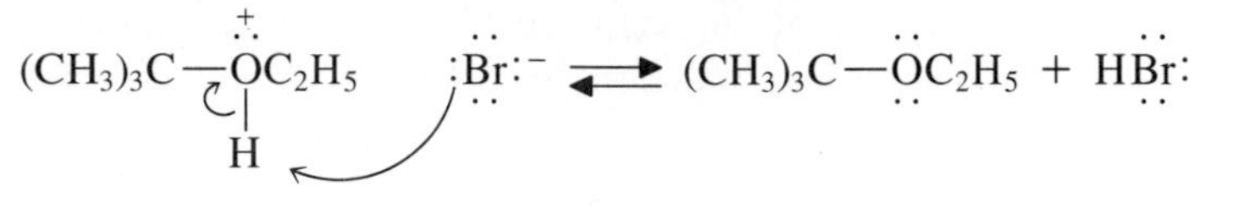

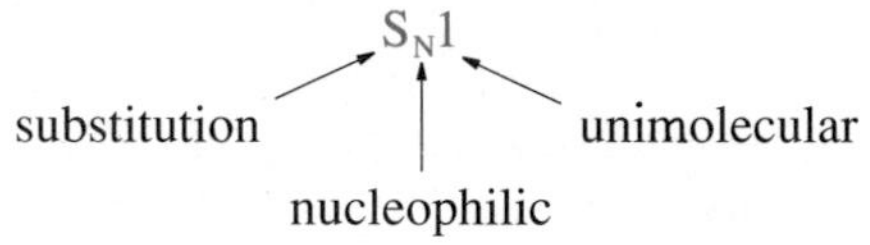

B. Rate-Determining and Product-Determining Steps. The E1 Reaction

The carbocation intermediate in Eq. 9.47a should be familiar to us because we have learned that carbocations are reactive intermediates in other reactions (Secs. 4.5B and 4.8). We have also seen that carbocations undergo certain typical reactions. One is, of course, that they react with nucleophiles. Indeed, the second step of the S_N1 mechanism (Eq. 9.47b) is exactly this reaction. Another reaction of carbocations is that they can lose a *β-proton* (a proton from the adjacent carbon) to give alkenes.

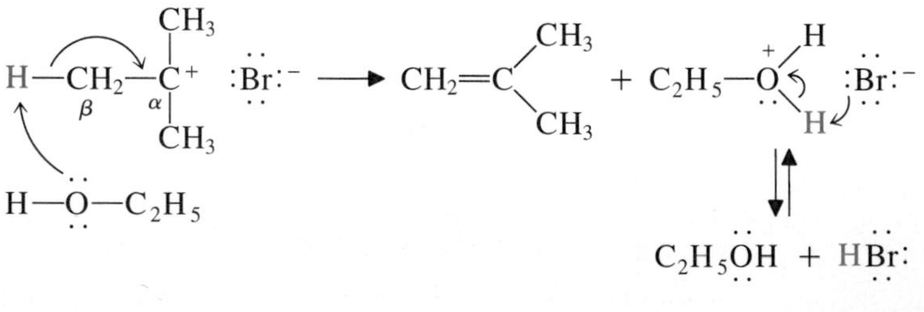

$$C_2H_5\ddot{O}H + H\ddot{B}r\!:$$

Thus, under the conditions of the S_N1 reaction, substitution products are accompanied by alkenes, which are formed by loss of a *β*-proton from the carbocation. That is, the S_N1 reaction is accompanied by some products of *elimination* (provided that the starting alkyl halide has *β*-hydrogens). The base that removes the proton from the carbocation is typically either the solvent or the halide ion (Eq. 9.48); notice that a strong base such as ethoxide is *not* present. (If substantial ethoxide ion were present, the E2 reaction would be the major reaction.)

Let us consider how this elimination reaction fits within the mechanism and rate law for the S_N1 reaction. First, a rate law like Eq. 9.46 describes the overall reaction rate regardless of the products that are formed from the carbocation. In other words, *the rate at which the alkyl halide disappears is determined by its rate of ionization*—the rate-determining step. Second, the fate of the carbocation intermediate, and thus the types and amounts of products formed, are determined by the steps that *follow* the rate-determining step: the **product-determining steps.** The relative amounts of the various products are determined by the relative rates of the product-determining steps, but *the rates of these steps have nothing to do with the rate at which the starting material disappears.*

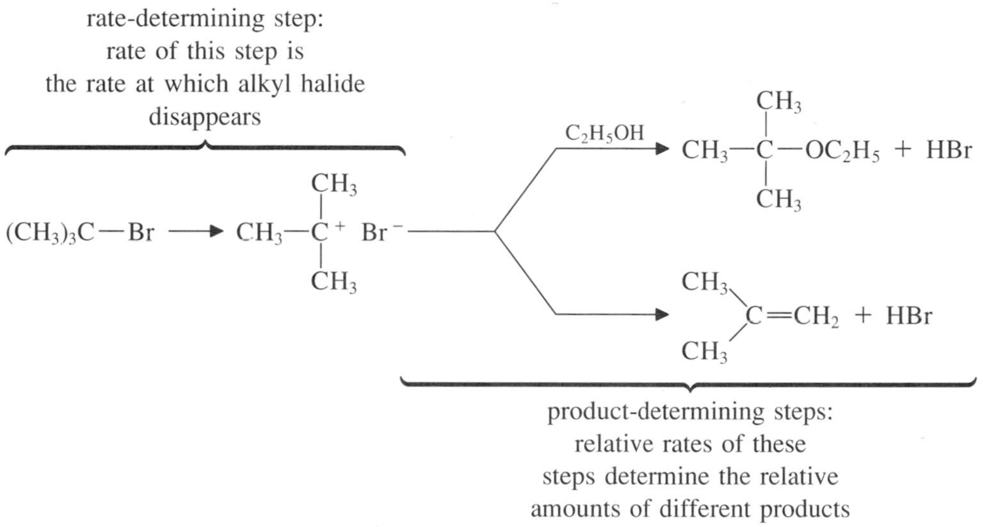

(9.49)

It happens that in this reaction, more substitution than elimination product is observed (see Eq. 9.44). This means that the rate of formation of substitution product from the carbocation is greater than the rate of formation of elimination product from the same ion. The reaction free-energy diagram in Fig. 9.7 summarizes these ideas. The first step, ionization of the alkyl halide to a carbocation, is the rate-determining step and therefore has the transition state of highest free energy. The rate of this step is the rate at which the alkyl halide reacts. The relative heights of the barriers for the product-determining steps determine the relative amounts of products formed.

Returning to the toll-taker analogy in Sec. 4.7B, ionization of an alkyl halide to a carbocation is analogous to the operation of a slow toll collector. Just as the rate of exit of automobiles from a toll booth with a slow toll collector is determined solely by the rate at which the collector works, the rate of reaction of a tertiary alkyl halide is determined by how fast it can form a carbocation.

Once cars leave the toll booth, the number of cars that take the turnoff for New York and the number that take the turnoff for Boston determine the relative number of cars that arrive at the two destinations. However, the rates at which angry drivers at the toll booth subsequently exit for Boston or New York has no effect on the rates at which tolls are taken. Likewise, the rates at which the carbocation is converted into either substitution or elimination products determine the ratio of the two types of products, but these rates have no effect on how fast the alkyl halide disappears.

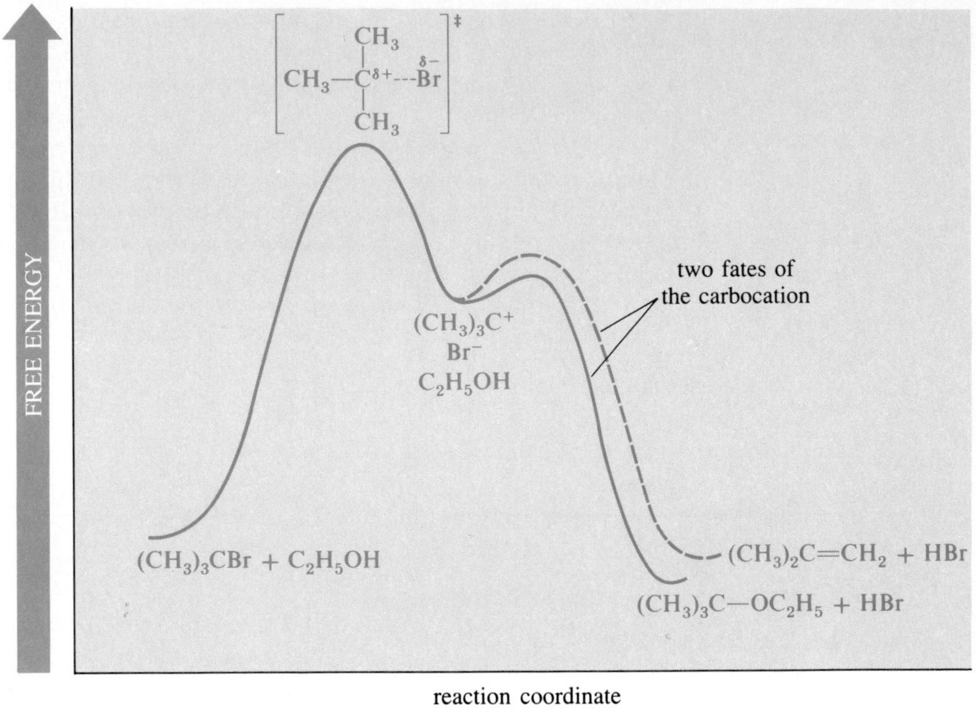

Figure 9.7 Energetics of the reaction of **$(CH_3)_3C{-}Br$** with ethanol. Notice that the rate-determining step—the step with the transition state of highest free energy—is ionization of the alkyl halide.

Problem

16 (a) From the data in Eq. 9.44, calculate the ratio of the rates for formation of substitution and elimination products.
(b) Explain why the calculation in (a) is valid only if it is shown experimentally that one product is not converted into the other under the conditions of the reaction.

From the previous discussion, it follows that the elimination and substitution pathways in Eq. 9.49 are competing reactions; that is, elimination product is formed at the expense of substitution product. Just as the substitution pathway in this reaction has been called the S_N1 reaction, the elimination component of this reaction has also been given a special name: the **E1 reaction.**

The competition between the S_N1 and E1 reactions is somewhat different from the competition between the S_N2 and E2 reactions. The latter two reactions share nothing in common but starting materials; they follow completely separate reaction pathways with no common intermediates.

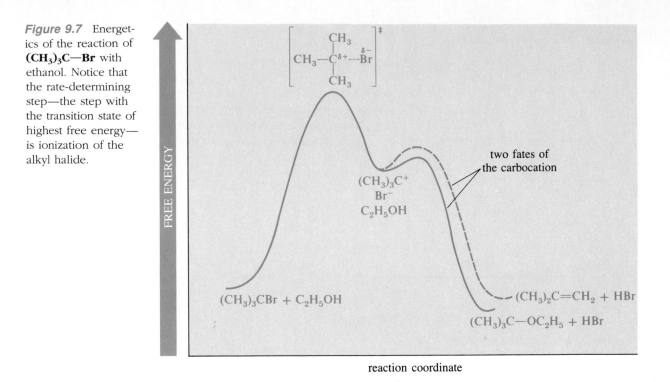

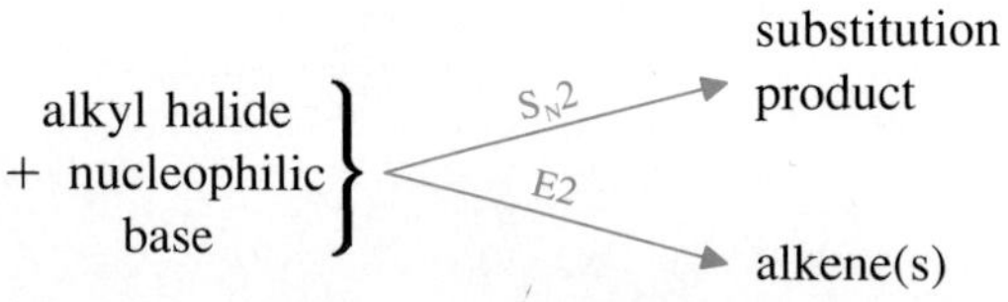

(9.50)

In contrast, the S_N1 and El reactions of an alkyl halide share not only common starting materials, but also a common rate-determining step, and hence *a common intermediate*—the carbocation.

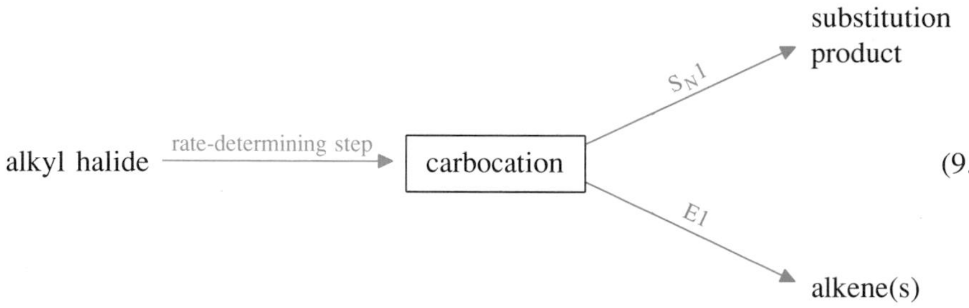

(9.51)

In the E1 reaction, the proton is not removed from the alkyl halide, as it is in the E2 reaction, but from the carbocation. Because the carbocation is a strong acid, a strong base is not required for this reaction (see Eq. 9.47).

What determines the ratio of substitution to elimination products in the S_N1–E1 reactions? The most important variable is the structure of the alkyl halide. The effect is similar to what is observed in the S_N2–E2 competition: the ratio of alkene to substitution product is usually large when there is substantial chain branching at the β-carbons of the alkyl halide. Such chain branching no doubt interferes sterically with the attack of the nucleophile on the carbocation, and it also stabilizes the transition states leading to alkene products (Sec. 4.4C). The following examples are illustrative:

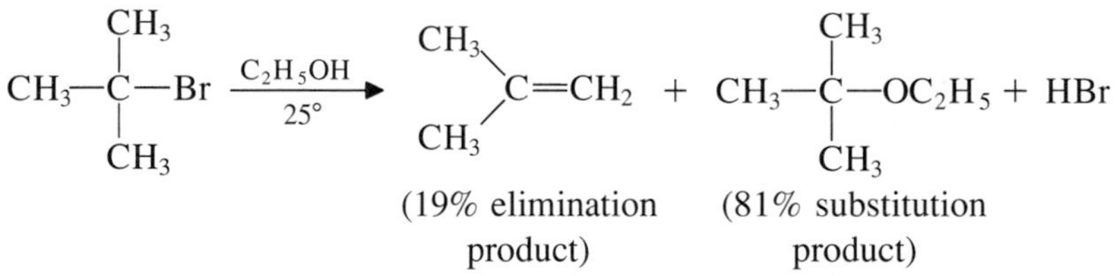

(9.52a)

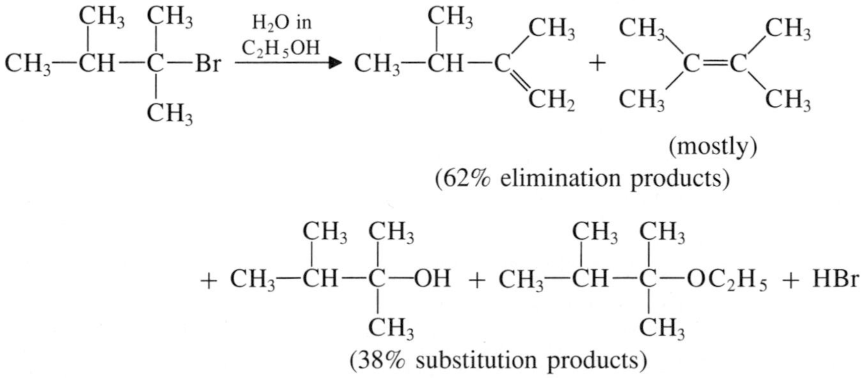

(9.52b)

In Eq. 9.52a, relatively little alkene is formed. In the reaction of the highly branched alkyl halide, Eq. 9.52b, more alkene is formed. The reaction in Eq. 9.52b shows in addition that the E1 reaction is a *Saytzeff elimination*: that is, the alkene with the most branching at the alkene carbons is formed in greater amount, just as in the E2 reaction.

Problem

17 Give all the products that might be formed when 2-bromo-2-methylbutane reacts in aqueous ethanol containing no added base. Of the alkenes formed, which should predominate? Why?

We have just considered the formation of substitution and elimination products from the carbocation intermediate in an S_N1 reaction. Other reactions of carbocations can also act as product-determining steps that compete with the S_N1 and E1 pathways. For example, some carbocations undergo *rearrangement* reactions (Sec. 4.6). As a case in point, a carbocation rearrangement occurs during the solvolysis of 3-chloro-2,2-dimethylbutane. Among the products of the reaction is 2-ethoxy-2,3-dimethyl-butane, which is formed by a rearrangement of the carbon skeleton (see Problem 18).

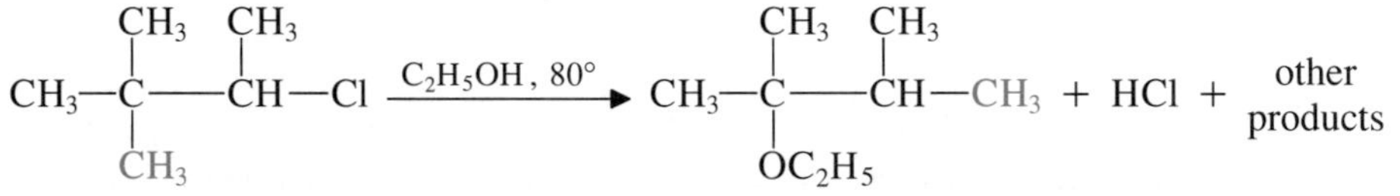

$$CH_3-\underset{\underset{CH_3}{|}}{\overset{\overset{CH_3}{|}}{C}}-\overset{\overset{CH_3}{|}}{CH}-Cl \xrightarrow{C_2H_5OH,\ 80°} CH_3-\underset{\underset{OC_2H_5}{|}}{\overset{\overset{CH_3}{|}}{C}}-\overset{\overset{CH_3}{|}}{CH}-CH_3 + HCl + \begin{matrix} \text{other} \\ \text{products} \end{matrix} \quad (9.53)$$

The different products in the S_N1/E1 reaction reflect three reactions of carbocation intermediates that we have learned about so far:

1. Reaction with a nucleophile

2. Loss of a *β*-proton

3. Rearrangement to a new carbocation followed by (1) or (2)

In any reactions involving carbocations as reactive intermediates—including the S_N1 reaction—we can expect to observe products that reflect the occurrence of one or more of these processes.

C. Effect of Leaving Group on the S_N1–E1 Reaction. Lewis-Acid Catalysis

In the rate-determining step of the S_N1–E1 reaction, the halogen leaving group takes on negative charge in much the same way that it does in the S_N2 reaction. Thus we find roughly the same type of leaving group effects on the rates of S_N1–E1 reactions that we found for S_N2 reactions. That is, alkyl halides with the least basic leaving groups react most rapidly in the S_N1–E1 reaction. For example, the relative rates of solvolysis of the *t*-butyl halides are in the order $F \ll Cl < Br < I$.

Alkyl halides undergo S_N1–E1 reactions much more rapidly if a Lewis acid, such as silver ion (Ag^+) or mercuric ion (Hg^{2+}) is present during the reaction.

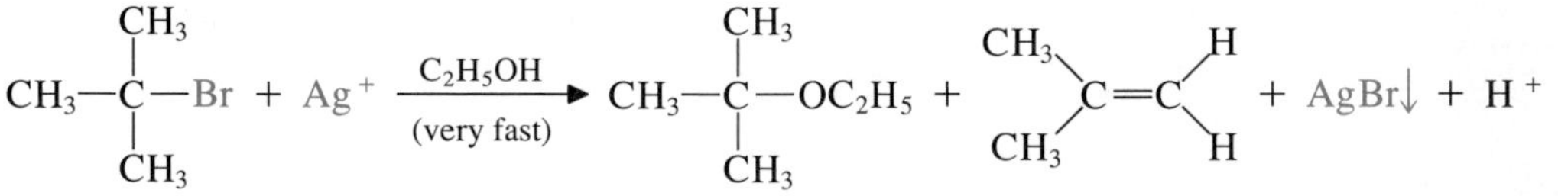

$$CH_3-\underset{\underset{CH_3}{|}}{\overset{\overset{CH_3}{|}}{C}}-Br + Ag^+ \xrightarrow[\text{(very fast)}]{C_2H_5OH} CH_3-\underset{\underset{CH_3}{|}}{\overset{\overset{CH_3}{|}}{C}}-OC_2H_5 + \underset{CH_3}{\overset{CH_3}{}}C{=}C\overset{H}{\underset{H}{}} + AgBr{\downarrow} + H^+ \quad (9.54)$$

This effect is due to the fact that the halide leaving group acts as a weak Lewis base and coordinates with the metal ion.

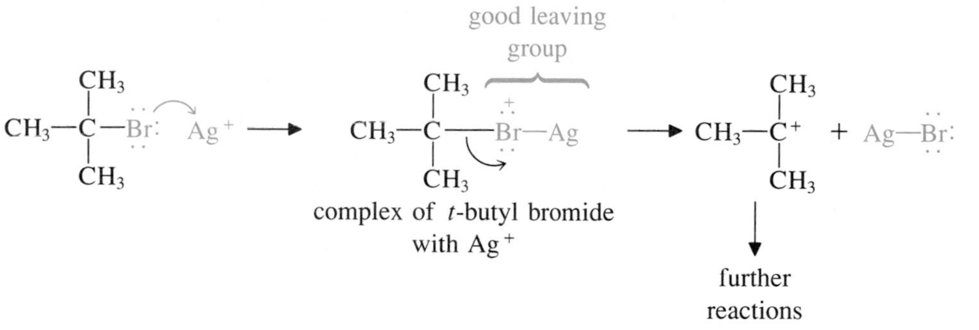

As a result of this coordination, the leaving group in effect becomes AgBr instead of Br⁻. Since a neutral AgBr is much less basic than Br⁻, it is a much better leaving group.

In the S_N2 reaction, we think of the nucleophile as something that provides an electronic "push" from the back side of the carbon, forcing the halide leaving group away from the other side. We can think of the effect of a Lewis acid like Ag^+ in the opposite sense: it provides an electron-deficient "pull" for the leaving group that causes the alkyl halide to ionize more rapidly.

Alkyl halides that ordinarily react by the S_N1–El mechanism react so rapidly in the presence of $AgNO_3$ that a precipitate of silver halide forms almost instantaneously. Alkyl halides that do not ordinarily react by an ionization pathway, such as primary alkyl halides, yield such a precipitate only with heating and/or prolonged reaction times. The rapid formation of a silver halide precipitate is a useful diagnostic test for an alkyl halide that forms a relatively stable carbocation. Notice that this silver nitrate test is similar to that used for ionic halide salts. For example, the combination of sodium bromide and silver nitrate rapidly gives a precipitate of silver bromide. The tertiary alkyl halide t-butyl bromide in the presence of silver ion rapidly becomes an ionic halide—the carbocation salt—and therefore also gives a precipitate of AgBr.

Problems

18 Give a mechanism for the reaction in Eq. 9.53 that accounts for the product shown. Use the arrow formalism for each step.

19 Rank the following alkyl halides in order of increasing rate at which they yield a silver halide precipitate with $AgNO_3$.

(a) $CH_3-\underset{\underset{CH_3}{|}}{\overset{\overset{CH_3}{|}}{C}}-\underset{\underset{}{}}{\overset{\overset{CH_3}{|}}{CH}}-Br$ (b) $C_2H_5-\underset{\underset{C_2H_5}{|}}{\overset{\overset{C_2H_5}{|}}{C}}-Br$ (c) CH_3-Br

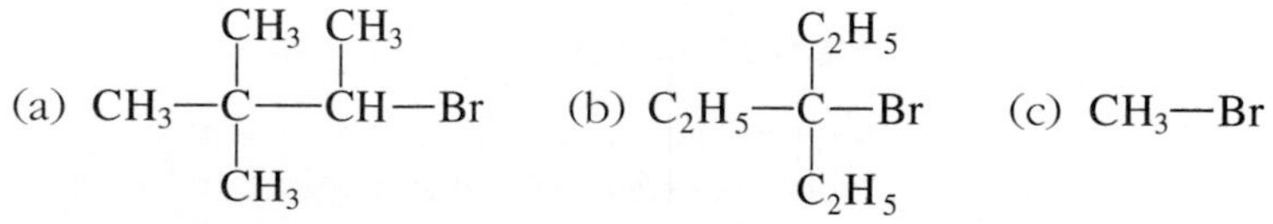

D. Solvent Effects on the S_N1 Reaction

Since the rate-determining step in the S_N1 reaction of an alkyl halide is ionization of the alkyl halide to a carbocation, solvents that promote this ionization accelerate the

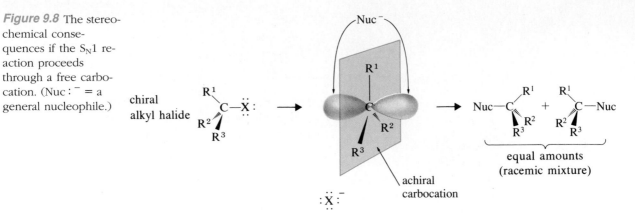

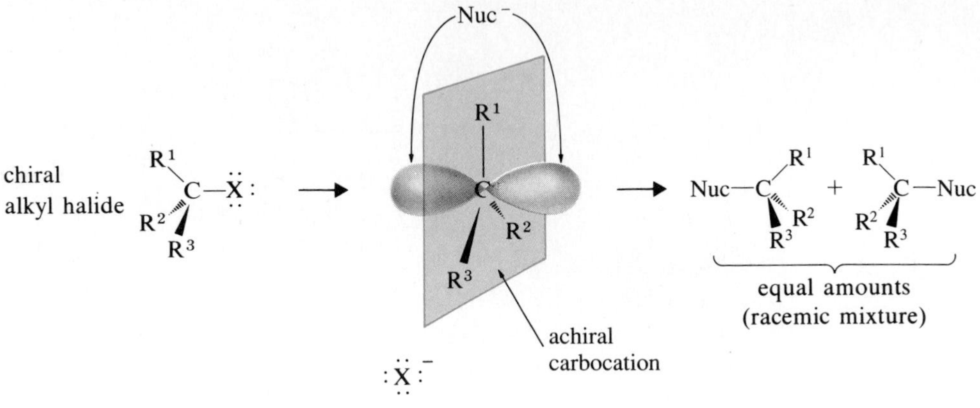

Figure 9.8 The stereochemical consequences if the S_N1 reaction proceeds through a free carbocation. (Nuc:$^-$ = a general nucleophile.)

reaction. Conceptually, ionization of an alkyl halide has a great deal in common with dissolution of an ionic compound (see Fig. 8.2). In both processes, ions of opposite charge must be separated from each other. Thus, to a useful approximation, solvents that dissolve ions well—polar, protic, donor solvents—also accelerate the S_N1 reaction. This is why many S_N1 reactions are solvolysis reactions run in water, aqueous acetone, or alcohol solvents. The molecules in these solvents promote the S_N1 reaction by donor interactions with the carbocation; by hydrogen bonding to the leaving-group anion; and, by virtue of their high dielectric constants, by separating oppositely charged ions. This last effect is dramatically illustrated by a comparison of the relative S_N1 solvolysis rates of *t*-butyl chloride in formic and acetic acid.

	relative rate	
$(CH_3)_3C{-}Cl + H{-}O\overset{O}{\overset{\Vert}{C}}H \longrightarrow (CH_3)_3C{-}O\overset{O}{\overset{\Vert}{C}}H + HCl$ formic acid ($\epsilon = 58$)	5000	(9.56a)
$(CH_3)_3C{-}Cl + H{-}O\overset{O}{\overset{\Vert}{C}}CH_3 \longrightarrow (CH_3)_3CO\overset{O}{\overset{\Vert}{C}}CH_3 + HCl$ acetic acid ($\epsilon = 6$)	1	(9.56b)

Formic acid and acetic acid have similar structures and dipole moments; both are protic. However, they have one important difference: their dielectric constants, ϵ. The significantly greater dielectric constant (polarity) of formic acid promotes separation of the *t*-butyl cation and the chloride anion, the very process that occurs in the rate-determining step. As a result, the S_N1 reaction is significantly accelerated in formic acid.

E. Stereochemistry of the S_N1 Reaction

Let us use the mechanism of the S_N1 reaction to predict whether it proceeds with retention, inversion, or loss of configuration. In a carbocation, the positively charged carbon and the three groups attached to it lie in a common plane (Fig. 4.10). That is, the carbocation is *achiral*. Since achiral compounds must react to give optically inactive products (Sec. 7.8A), this cation can, in principle, be attacked by a nucleophile at either face with equal probability (Fig. 9.8). Thus the S_N1 reaction of an optically active alkyl halide should give *racemic* substitution products.

What are the experimental facts? When 6-chloro-2,6-dimethyloctane, a chiral tertiary alkyl halide, undergoes solvolysis, the substitution products are only *partially* racemic; and they contain significantly more *inversion* product than *retention* product.

$$\underset{100\%\ R}{(CH_3)_2CHCH_2CH_2CH_2-\overset{\displaystyle CH_3}{\underset{\displaystyle Cl}{\overset{|}{\underset{|}{C}}}}-C_2H_5} \xrightarrow{H_2O \text{ in acetone, } 60°}$$

$$\underset{\left.\begin{matrix}60.5\%\ S \\ 39.5\%\ R\end{matrix}\right\}\ 21\%\text{ inversion}}{(CH_3)_2CHCH_2CH_2CH_2-\overset{\displaystyle CH_3}{\underset{\displaystyle OH}{\overset{|}{\underset{|}{C}}}}-C_2H_5} + HCl + \text{elimination products} \qquad (9.57)$$

We might be tempted to argue that substitution with inversion takes place by an S_N2 mechanism. However, this cannot be correct, because this reaction is much faster than S_N2 reactions of *primary* alkyl halides in the same solvents; yet the branching in this tertiary halide should retard, not accelerate, the S_N2 reaction (Sec. 9.3C). Results such as this require that we modify our S_N1 reaction mechanism as shown in Fig. 9.9.

Figure 9.9 Stereochemical consequences of ion-pair formation in the S_N1 reaction. (Nuc:$^-$ = a general nucleophile.)

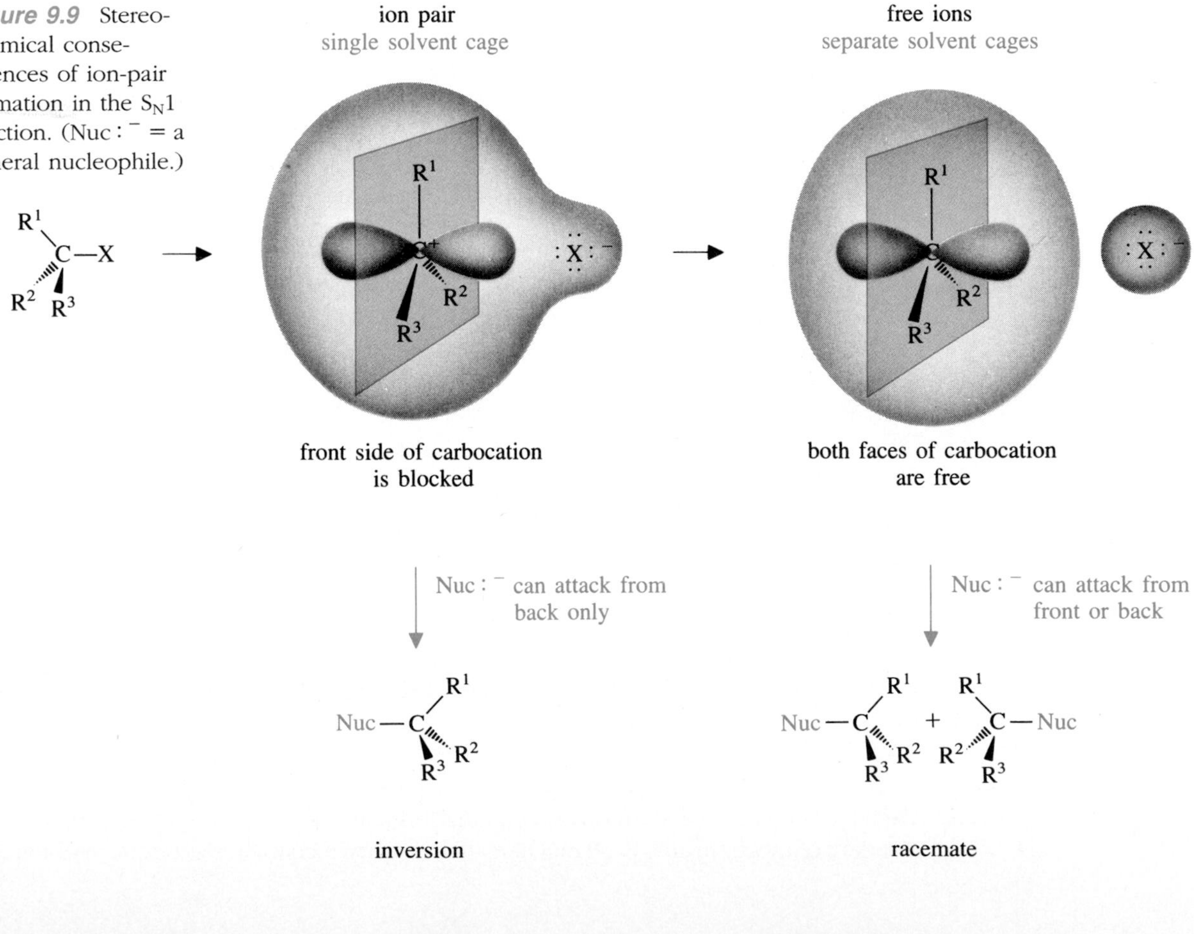

Let us imagine that the first intermediate in the S_N1 reaction is a carbocation intimately associated with a halide anion—an **ion pair** (compare with Fig. 8.2). If the halide ion blocks access of the nucleophile (which, in Eq. 9.57, is also the solvent) to the front face of the carbocation, attack of the nucleophile on the ion pair can occur more easily at the back side, giving the product with inverted configuration. However, the ion pair can dissociate to a carbocation that is separated from the halide anion by solvent. Attack of the nucleophile on this ion can occur from either the front or the back side, giving the racemic product. The mixture of racemization and inversion in Eq. 9.57 suggests that both types of carbocations—ion pairs and free ions—are important intermediates in the S_N1 reaction.

Problem

20 Give the structure of the expected product(s), including their stereochemistry, when the following alkyl halide undergoes solvolysis in aqueous acetone:

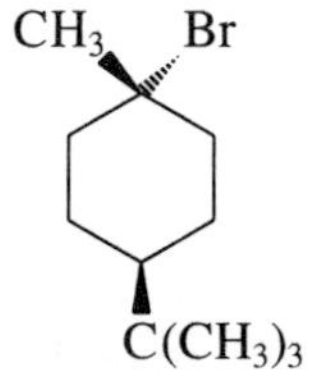

F. Summary of the S_N1 and E1 Reactions

Tertiary and some secondary alkyl halides undergo nucleophilic substitution reactions with weak bases such as water and alcohols by the S_N1 mechanism. If an alkyl halide contains β-hydrogens, elimination products formed by the E1 mechanism generally accompany substitution products. Let us summarize what we have learned about the S_N1 and E1 reactions:

1. Both reactions have the same rate-determining step: formation of a carbocation.
2. The reaction rate is first order in the alkyl halide.
3. The reactions differ in their product-determining steps. The product-determining step in the S_N1 reaction is attack of a nucleophile, and in the E1 reaction it is loss of a β-proton.
4. Carbocation rearrangements occur in appropriate cases.
5. The best leaving groups are those that form the weakest bases.
6. The reactions are accelerated by polar, protic, donor solvents.
7. The S_N1 reactions of chiral alkyl halides are accompanied by some racemization.

9.6 SUBSTITUTION AND ELIMINATION REACTIONS OF ALKYL HALIDES: SUMMARY

We have seen that substitution and elimination reactions of alkyl halides can occur by a variety of mechanisms. Each mechanism has its own characteristic pattern of reactivity that depends on the structure of the alkyl halide and the reaction conditions. Up to now, we have considered each mechanism independently, examining cases in which one mechanism predominates. A more practical question, however, is to ask how a particular alkyl halide will react under a given set of conditions. By what mechanism(s) will its reactions be likely to occur, and what products should predominate? Let us summarize what we have learned:

In the absence of a strong Brønsted base:

1. In a protic solvent, an unbranched primary alkyl halide will react very slowly unless heat is applied; any reaction observed will be a substitution that occurs by the S_N2 mechanism. In a polar aprotic solvent, the S_N2 reaction is faster, and some relatively weak bases (such as halide ions) can be used as nucleophiles. Neither substitution by the S_N1 mechanism nor elimination by any mechanism will be observed.
2. In a protic solvent, a tertiary alkyl halide will react by the S_N1 mechanism; if β-hydrogens are present, some elimination by the E1 mechanism can generally be expected. The proportion of elimination is greater when there is more branching at the β-carbons, and the most stable alkene product will form in greatest amount.
3. A secondary alkyl halide will have intermediate reactivity. In a protic solvent, reaction will generally occur by the S_N1–E1 mechanism. Elimination and rearrangement products can be expected in appropriate cases.

In the presence of a strong Brønsted base:

1. An unbranched primary alkyl halide will undergo substitution by the S_N2 mechanism. (Many strong Brønsted bases are excellent nucleophiles.) Elimination will be a minor reaction if an unbranched base is used. Increased branching at the β-carbon atom of the primary alkyl halide and use of a more highly branched base will give more elimination product by the E2 mechanism. The rates of both the S_N2 and E2 reactions are increased in polar aprotic solvents.
2. A tertiary alkyl halide invariably reacts by the E2 mechanism if it has β-hydrogens. The most stable alkene is usually the major product of an E2 reaction. However, if a highly branched base is used, the alkene with the least amount of branching at the double bond will be the predominant product.
3. A secondary alkyl halide will show intermediate reactivity and product distributions. Typically, a secondary alkyl halide will give a mixture of substitution and elimination products formed by the S_N2 and E2 mechanisms.

An instructive case study that illustrates these points is provided by the reaction of 2-bromo-2-methylbutane in ethanol solvent in the presence and absence of sodium ethoxide (Table 9.5). Let us examine the trends in this table with the aid of the rate law for the reaction. This rate law contains contributions from all possible mechanisms:

$$\text{rate} = \underbrace{k_1[\text{alkyl halide}]}_{S_N1\,+\,E1} + \underbrace{k_2[\text{alkyl halide}][C_2H_5O^-]}_{S_N2\,+\,E2} \tag{9.58}$$

(S_N2 negligible for tertiary alkyl halide)

TABLE 9.5 Effect of Base Concentration on the Proportion of Elimination and Substitution in the Reaction of a Tertiary Alkyl Halide

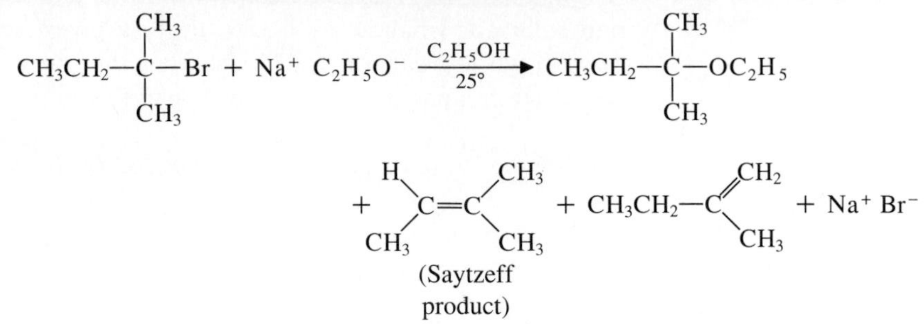

$[C_2H_5O^-]$, *M*	Percentage of product that is alkene	Percentage of alkene formed that is Saytzeff product
0	36.3	82
0.015	46.3	
0.05	56.0	
1.00	98.0	72
2.00	99.0	72

Because the alkyl halide is tertiary, an S_N2 reaction does not occur; hence, the second term in the rate law arises only from the E2 reaction. At zero base concentration, this term disappears. The polar, protic solvent ethanol promotes the S_N1 and E1 mechanisms; hence, the products in the absence of added base are derived solely from these pathways. Of the alkenes formed, the Saytzeff product predominates, as we expect from the E1 mechanism.

As base concentration is increased, the second term in Eq. 9.58, the rate of the E2 reaction, becomes larger; because the first term is not a function of the base concentration, it is essentially unchanged by the increase. At high base concentrations, the E2 mechanism is the dominant process, and consequently alkene is the only product observed. The predominant alkene is still the Saytzeff product; thus very little difference in the relative amounts of the two alkenes is observed in the presence and absence of base.

9.7 DIVALENT CARBON: CARBENES

We have seen that *β*-elimination is one of the possible reactions that can occur when certain alkyl halides containing *β*-hydrogens are treated with base.

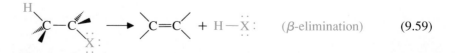

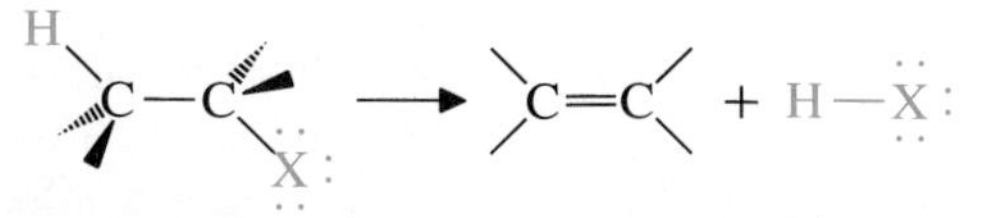

(*β*-elimination) (9.59)

When an alkyl halide contains no β-hydrogens, but has an α-hydrogen, a different sort of base-promoted elimination is sometimes observed. Chloroform is an alkyl halide that undergoes such a reaction. When chloroform, a weak acid with $pK_a = 25$, is treated with an alkoxide base such as potassium t-butoxide, small amounts of its conjugate-base anion are formed.

$$(CH_3)_3C\!-\!\ddot{O}\!:^- \quad H\!-\!CCl_3 \;\rightleftharpoons\; (CH_3)_3C\!-\!OH \;+\; {}^-\!:CCl_3 \qquad (9.60a)$$

This anion can lose chloride ion to give a neutral species called *dichloromethylene.*

$$^-\!:\!C\!\!\begin{array}{c}\ddot{C}l: \\ \diagdown Cl \\ Cl\end{array} \quad\rightleftharpoons\quad :\!C\!\!\begin{array}{c}Cl \\ \diagdown Cl\end{array} \;+\; :\!\ddot{C}l:^- \qquad (9.60b)$$

dichloromethylene

Dichloromethylene is an example of a **carbene**—a species with a divalent carbon atom. Carbenes are unstable and highly reactive species, as we shall see.

The formation of dichloromethylene shown in Eqs. 9.60a and 9.60b involves an elimination of the elements of HCl from the *same* carbon atom. An elimination of two groups from the same atom is termed an **α-elimination.**

$$\begin{array}{c}R^1 \\ \diagdown \\ R^2\end{array}\!\!C\!\!\begin{array}{c}H \\ \diagup \\ \ddot{X}:\end{array} \quad\longrightarrow\quad \begin{array}{c}R^1 \\ \diagdown \\ R^2\end{array}\!\!C: \;+\; H\!-\!\ddot{X}: \qquad (\alpha\text{-elimination}) \qquad (9.61)$$

Chloroform cannot undergo a β-elimination because it has no β-hydrogens. When an alkyl halide has β-hydrogens, β-elimination occurs in preference to α-elimination because alkenes, the products of β-elimination, are much more stable than carbenes, the products of α-elimination. For example, CH_3CHCl_2 reacts with base to form the alkene $CH_2\!\!=\!\!CHCl$ rather than the carbene $CH_3\!-\!\ddot{C}\!-\!Cl$.

The reactivity of dichloromethylene follows from its electronic structure. If we include the unshared pair of electrons, the carbon atom of dichloromethylene bears three groups (two chlorines and the lone pair), and therefore has approximately trigonal geometry. Because trigonal carbon atoms are sp^2 hybridized, the Cl—C—Cl bond angle is bent rather than linear, the unshared pair of electrons occupies an sp^2 orbital, and the p orbital is empty:

empty p orbital

unshared electron pair in an sp^2 orbital

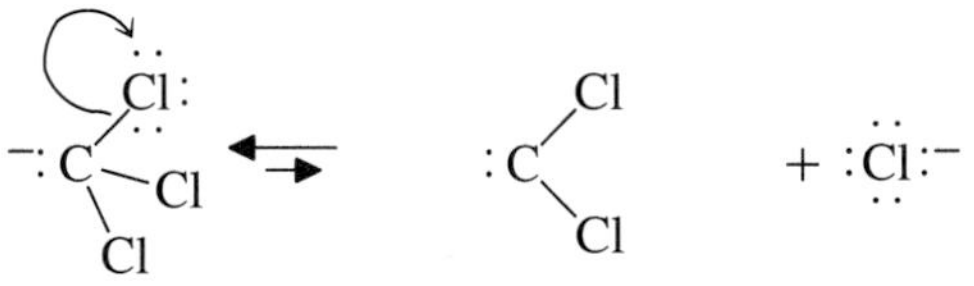

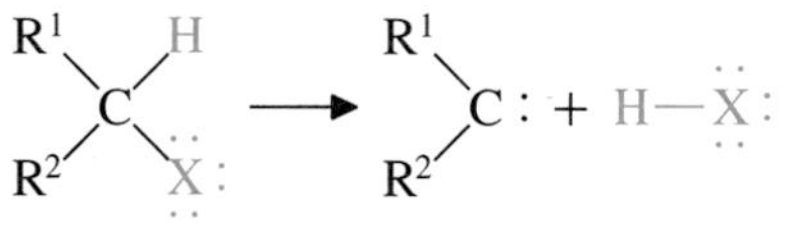

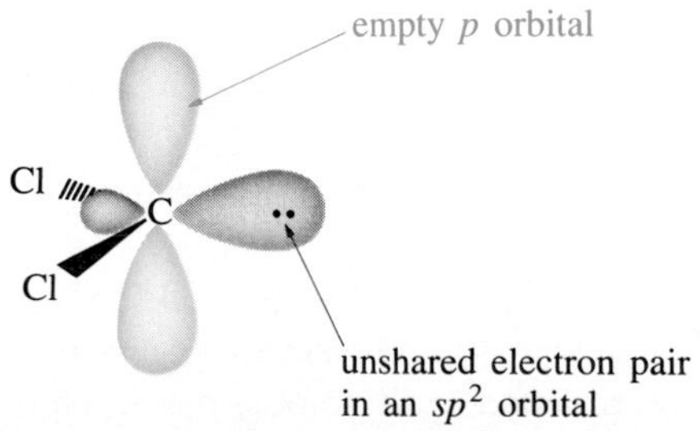

Since dichloromethylene lacks an electronic octet, it can accept an electron pair; that is, dichloromethylene is a Lewis acid, or electrophile, in the same sense that a carbocation is a Lewis acid. On the other hand, we expect an atom with an unshared electron pair to react as a Lewis base, or nucleophile. The divalent carbon of dichloromethylene, with its unshared electron pair, appears to fit into this category as well. Indeed, it seems that the divalent carbon of a carbene could act as both a nucleophile and an electrophile at the same time!

An important reaction of carbenes that fits the foregoing analysis is cyclopropane formation. When dichloromethylene is generated in the presence of an alkene, a cyclopropane is formed.

$$HCCl_3 + (CH_3)_3C\text{—}\ddot{\underset{..}{O}}:^- \, K^+ + (CH_3)_2C\text{=}CH_2 \longrightarrow$$

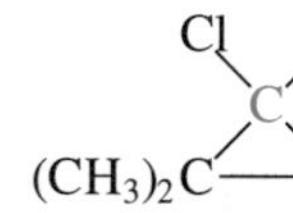

$$+ \; KCl + (CH_3)_3C\text{—}OH \qquad (9.62)$$

In general, reaction of a haloform with base in the presence of an alkene yields a 1,1-dihalocyclopropane. In the arrow formalism for cyclopropane formation, the π-electrons of the alkene attack the empty orbital of the carbene (carbene acts as a Lewis acid) while the unshared electron pair of the carbene attacks an alkene carbon (carbene acts as a Lewis base.)

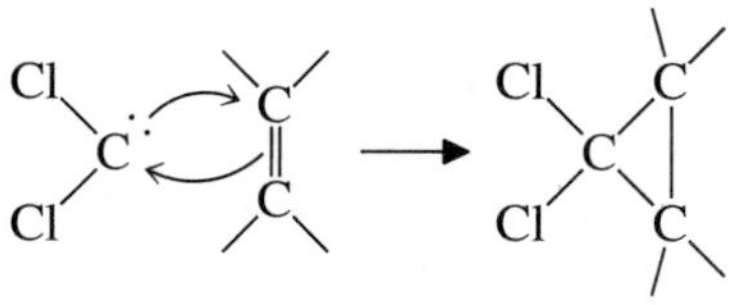

(9.63)

As the formalism in Eq. 9.63 suggests, the addition of dichloromethylene to an alkene is a *concerted pericyclic reaction* (Sec. 5.3A). The stereochemistry of the reaction is what we expect for this type of mechanism: dichloromethylene addition to an alkene is a *syn* addition. That is, groups that are *cis* in the reacting alkene are also *cis* in the cyclopropane product.

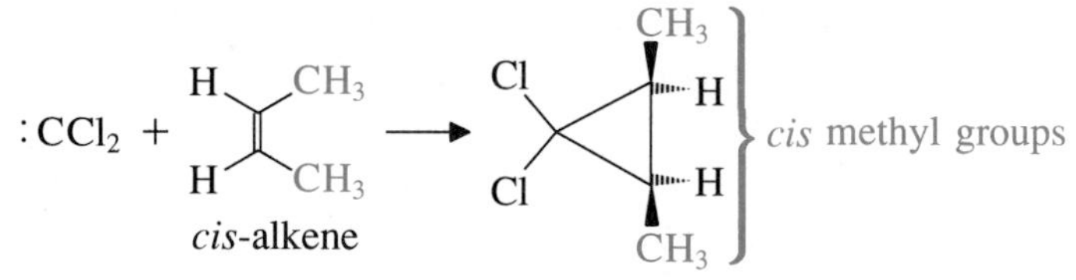

(9.64a)

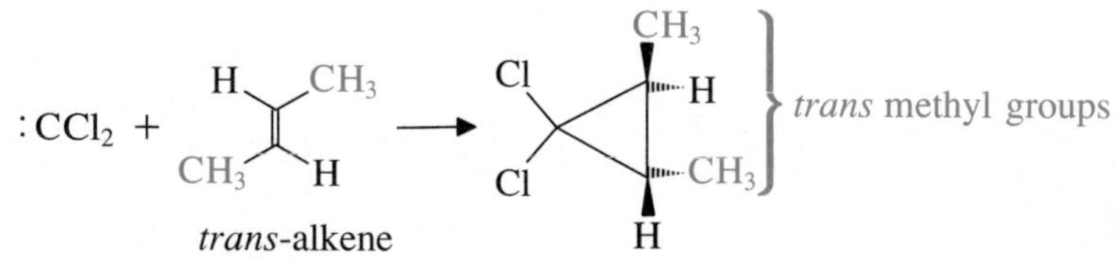

(9.64b)

A concerted *anti* addition would require that the empty orbital of the divalent carbon and its unshared electron pair react simultaneously at opposite faces of the alkene, a stereochemically impossible situation.

Problems

21 Predict the products when each of the following alkenes reacts with chloroform and potassium *t*-butoxide. Give the structures of all product stereoisomers, and, if more than one stereoisomer are formed, indicate whether they are formed in the same or different amounts.

(a) cyclopentene (b) *cis*-2-pentene (c) (*R*)-3-methyl-1-cyclohexene

22 What alkyl halide and alkene would yield each of the following cyclopropane derivatives in the presence of a suitable strong base?

(a)

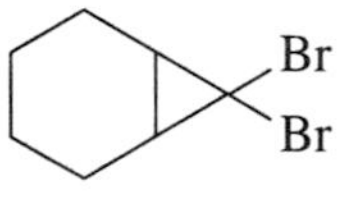

(b)

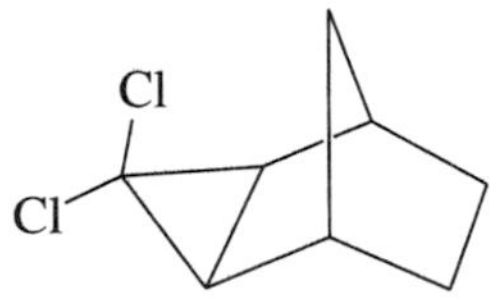

(c)

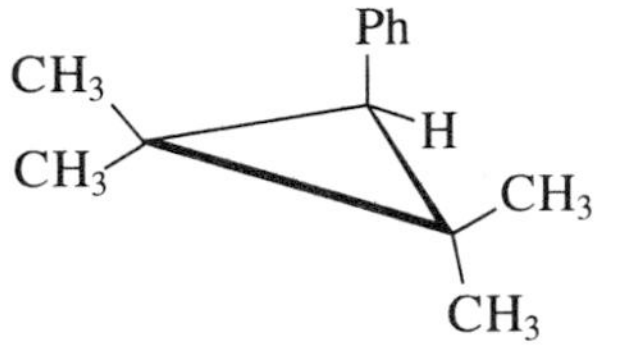

KEY IDEAS IN CHAPTER 9

- Two of the most important types of alkyl halide reactions are nucleophilic substitution and elimination.
- Nucleophilic substitution occurs by two mechanisms.

1. Substitution in methyl, primary, and some secondary alkyl halides occurs by the S_N2 mechanism. This type of reaction occurs in a single step with inversion of stereochemical configuration, and is characterized by a second-order rate law. It occurs most readily when the nucleophile is a relatively strong base. Branching in the alkyl halide retards the reaction and favors competing E2 elimination. Although the reaction is frequently run in polar, protic solvents for reasons of solubility, it is faster in polar, aprotic solvents; even weak bases like halide ions are good nucleophiles in these solvents.

2. Substitution in tertiary and some secondary alkyl halides occurs by the S_N1 mechanism. This type of reaction is characterized by a first-order rate law that contains only a term in alkyl halide concentration. It occurs mostly in the absence of a strong base, and is accelerated by Lewis acids such as Ag^+ and by polar, protic solvents. In fact, the most common S_N1 reactions are solvolysis reactions. The rate-determining step of the reaction is ionization of the alkyl halide to a carbocation intermediate.

- β-Elimination also occurs by two mechanisms.
  1. The E2 mechanism competes with the S_N2 mechanism, has a second-order rate law, occurs with *anti* stereochemistry, and is favored both by strong base and by branching in both the alkyl halide and the base. It is the major reaction of tertiary and many secondary alkyl halides in the presence of strong base. When an alkyl halide has more than one type of β-hydrogen, several alkene products are obtained. The formation of the more stable (Saytzeff) alkene isomers is observed with unbranched bases, whereas highly branched bases favor the formation of less stable alkene isomers.
  2. The E1 mechanism is an alternate product-determining step of the S_N1 mechanism in which the carbocation intermediate loses a β-proton to form alkene. Alkene formation is favored over substitution by branching at the double bond.

- The rate law indicates what species are involved in the rate-determining transition state of a reaction.

- A primary deuterium isotope effect indicates that proton transfer takes place in the rate-determining step of a reaction.

- Chloroform reacts with base in an α-elimination to give dichloromethylene, a carbene. Carbenes are unstable species containing divalent carbon. Dichloromethylene reacts with alkenes to give dichlorocyclopropanes.

ADDITIONAL PROBLEMS

23 Choose the alkyl halide(s) from the following list of $C_6H_{13}Br$ isomers that meet each criterion below.

(1) 1-bromohexane
(2) 3-bromo-3-methylpentane
(3) 1-bromo-2,2-dimethylbutane
(4) 3-bromo-2-methylpentane
(5) 2-bromo-3-methylpentane

(a) the compound(s) that can exist as enantiomers
(b) the compound(s) that can exist as diastereomers
(c) the compound that gives the fastest S_N2 reaction with sodium methoxide
(d) the compound that is least reactive to sodium methoxide in methanol
(e) the compound(s) that give an E2 reaction but not an S_N2 reaction with sodium methoxide in methanol
(f) the compound that gives the fastest S_N1 reaction
(g) the compound(s) that give an S_N1 reaction with rearrangement
(h) the compound(s) that give only one alkene in the E2 reaction

24 Give the products expected when isopentyl bromide or other substances indicated react with the following reagents:
(a) KI in aqueous acetone
(b) KOH in aqueous ethanol
(c) K^+ $(CH_3)_3C—O^-$ in $(CH_3)_3C—OH$
(d) product of (c) + HBr
(e) product of (c) + chloroform + potassium *t*-butoxide

(f) Mg and ether, then water
(g) sodium methoxide in methanol

25 Give the products expected when 2-bromo-2-methylhexane or other substances indicated react with the following reagents:
(a) 50% aqueous ethanol
(b) sodium ethoxide in ethanol
(c) product(s) of (b) + HBr, peroxides

26 Rank the following compounds in order of increasing rate of the S_N2 reaction with KI in acetone:
(a) $(CH_3)_3CCl$ (b) $(CH_3)_2CHCl$ (c) $(CH_3)_2CHCH_2Cl$ (d) $CH_3CH_2CH_2CH_2Br$ (e) $(CH_3)_2CHBr$

27 Give all the product(s) expected, including pertinent stereochemistry, when (*R*)-2-bromopentane reacts with sodium ethoxide in ethanol.

28 The pK_a of HCN is 9.40, and that of hydrazoic acid, HN_3, is 4.72. Suppose that CH_3I is dissolved in aqueous acetone with one equivalent each of potassium azide ($K^+\ N_3^-$) and potassium cyanide ($K^+\ ^-{:}CN$). What product would you expect to isolate in greatest amount from the reaction that occurs? Why?

29 In the *Williamson ether synthesis,* an alkoxide reacts with an alkyl halide to give an ether.

$$R{-}\ddot{\underset{\cdot\cdot}{O}}{:}^- + R'{-}\ddot{\underset{\cdot\cdot}{X}}{:} \longrightarrow R{-}\ddot{\underset{\cdot\cdot}{O}}{-}R' + {:}\ddot{\underset{\cdot\cdot}{X}}{:}^-$$

You are in charge of a research group for a large company, Ethers Unlimited, and you have been assigned the task of synthesizing *t*-butyl methyl ether, $(CH_3)_3C{-}O{-}CH_3$. You have decided to delegate this task to two of your staff chemists. One chemist, Ima Smart, allows $(CH_3)_3C{-}\ddot{\underset{\cdot\cdot}{O}}{:}^-$ to react with $CH_3{-}I$ and indeed obtains a good yield of the desired ether. The other chemist, Notso Bright, allows $CH_3\ddot{\underset{\cdot\cdot}{O}}{:}^-$ to react with $(CH_3)_3C{-}Br$. To his surprise, no ether was obtained. Before terminating Notso Bright's employment, explain to him why his reaction failed.

30 The insecticide chlordane is reported to lose some of its chlorine content when exposed to alkaline conditions. Explain.

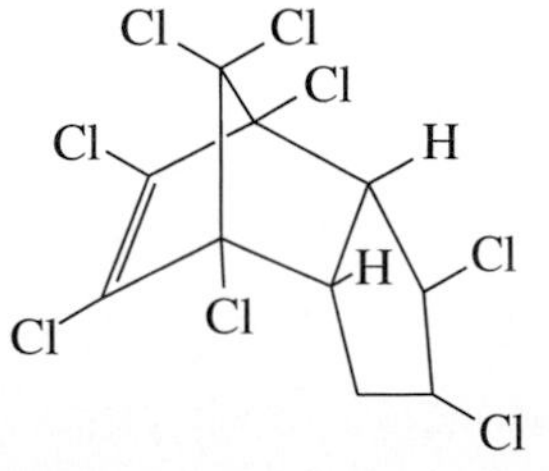

principal component of chlordane

Problems (Cont.)

31 Consider the following experiments with trityl chloride, $Ph_3C—Cl$, a very reactive tertiary alkyl halide:

1. In aqueous acetone the reaction of trityl chloride follows a rate law that is first order in the alkyl halide, and the product is trityl alcohol, $Ph_3C—OH$.
2. In another reaction, when a small amount of sodium azide ($Na^+ N_3^-$) is added to a solution that is otherwise identical to that used in experiment (1), the reaction rate is virtually the same as in (1); however, the product isolated in good yield is trityl azide, $Ph_3C—N_3$.
3. In a reaction mixture in which both sodium azide and sodium hydroxide are present in equal concentrations, trityl alcohol is the product, but the reaction rate is again unchanged.

Explain why the reaction rate is the same but the products are different in these three experiments. (See Problem 28 for the pK_a of HN_3, the conjugate acid of azide ion.)

32 When benzyl bromide ($Ph—CH_2—Br$) is added to a suspension of potassium fluoride in benzene, no reaction occurs. However, when a *catalytic* amount of the crown ether [18]-crown-6 (Sec. 8.4C) is added to the solution, benzyl fluoride can be isolated in high yield. If lithium fluoride is substituted for potassium fluoride, there is no reaction even in the presence of the crown ether. Explain these observations.

33 In the following two cases, explain (a) why different products are obtained under different sets of conditions, and (b) why so little reaction takes place in the first case even under the harsh conditions.

$$(CH_3)_3C—CH_2—I + OH^- \xrightarrow[100°,\ 10\ hr]{H_2O} \underset{\text{(9\%; plus 91\% starting material)}}{(CH_3)_3C—CH_2—OH} + I^-$$

$$(CH_3)_3C—CH_2—I + AgNO_3 \xrightarrow[\text{several minutes}]{H_2O} \underset{(97\%)}{(CH_3)_2\underset{|}{\overset{}{C}}(OH)—C_2H_5} + AgI + HNO_3$$

34 It is found that 2-bromo-3-methylbutane, on standing, is converted into 2-bromo-2-methylbutane. Suggest a mechanism for this transformation.

35 Which of the following compounds reacts faster with ^-CN in the S_N2 reaction? (*Hint*: Consider the hybridization and geometry of the S_N2 transition state.)

(a) 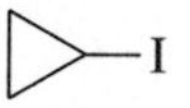 (b) $(CH_3)_2CH—I$

36 Outline a sequence of reactions that would allow you to carry out the following transformation in good yield:

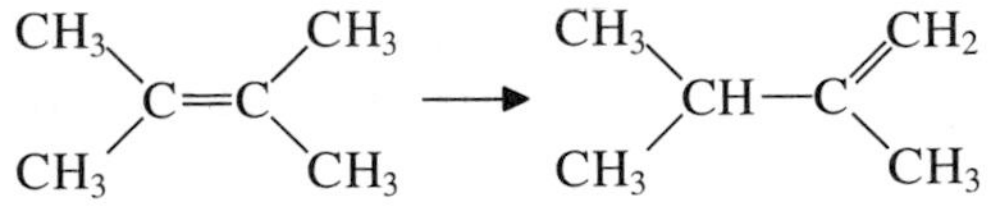

37 When menthyl chloride (see following structure) is treated with sodium ethoxide in ethanol, 2-menthene is the only alkene product observed. When neomenthyl chloride is subjected to the same conditions, the alkene products are mostly 3-menthene (78%) along with some 2-menthene (22%). Explain why different alkene products are formed from the different alkyl halides, and why 3-menthene is the major product in the second reaction. (*Hint*: Remember the stereochemistry of the E2 reaction, and don't forget about the chair–chair interconversion of cyclohexanes.)

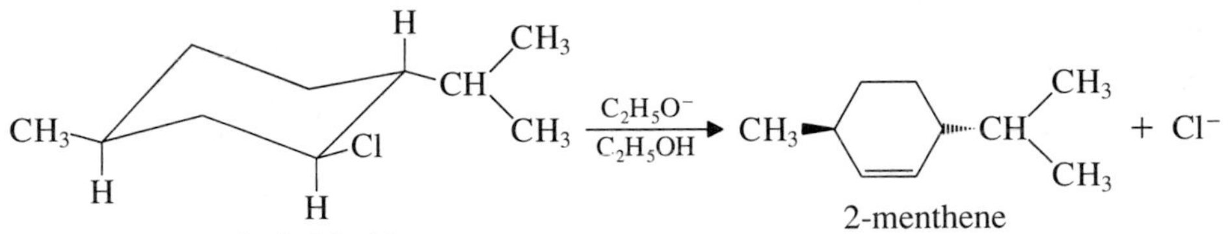

menthyl chloride — 2-menthene (100%)

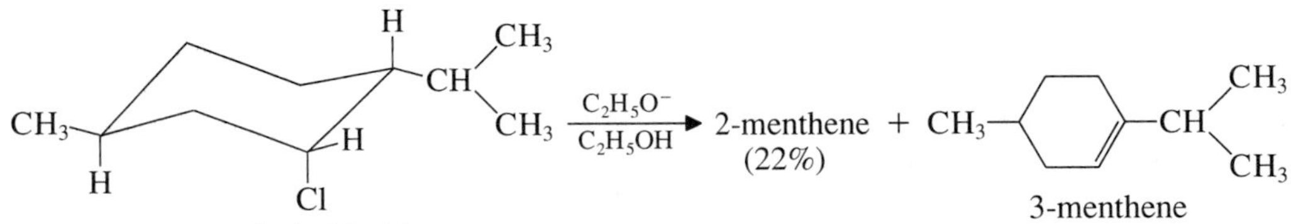

neomenthyl chloride — 3-menthene (78%)

38 Propose a reaction sequence for the conversion of (bromomethyl)cyclohexane into each of the following compounds. The products in each case should be relatively free of contaminating isomers.

(a) 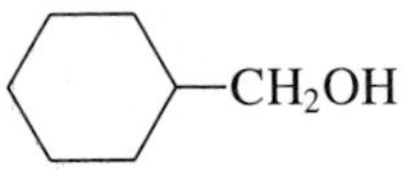 (b) 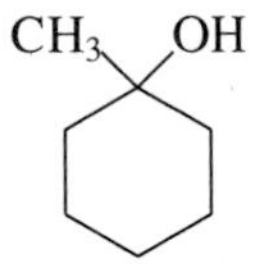 (c) 1-bromo-1-methylcyclohexane

39 (a) The reagent *tri-n-butyltin hydride*, $(C_4H_9)_3Sn—H$, brings about the rapid conversion of 1-bromo-1-methylcyclohexane into methylcyclohexane. The reaction is particularly fast in the presence of AIBN (Sec. 5.8C). Suggest a mechanism for this reaction.

(b) Suggest two other reaction sequences that would bring about the same transformation.

40 In an abandoned laboratory has been found an optically active compound *A* that gives the following elemental analysis: C, 50.81%; H, 6.93%; Br, 42.26%. Compound *A* gives no reaction with Br_2 in the dark, but it reacts with $K^+(CH_3)_3C$—O^- to give a single new compound *B* in good yield that decolorizes Br_2 in CCl_4 and takes up hydrogen over a catalyst. When compound *B* is treated with O_3, followed by workup with H_2O_2, the following compound is isolated in excellent yield; notice its stereochemistry.

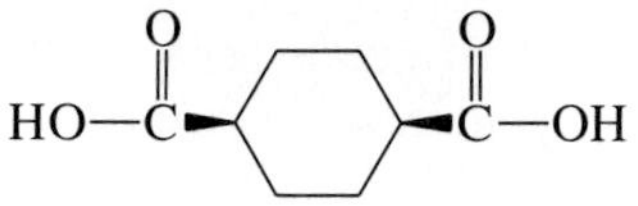

Identify compound *A* and account for all observations. (See the box on p. 177).

41 The reaction of *n*-butylamine, *n*-C_4H_9—NH_2, with 1-bromobutane in 60% aqueous ethanol follows the rate law

$$\text{rate} = k[n\text{-}C_4H_9NH_2][C_4H_9\text{—Br}]$$

The product of this reaction is $(n\text{-}C_4H_9)_2\overset{+}{N}H_2\ Br^-$. The following very similar reaction, however, has a first-order rate law:

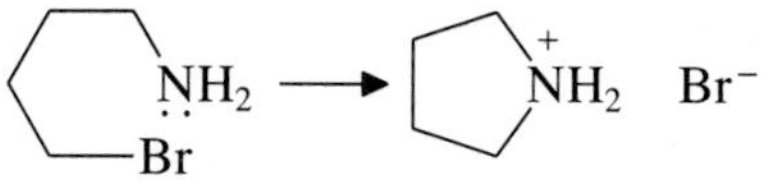

⬡—$\overset{..}{N}H_2$ ⟶ ⬠—$\overset{+}{N}H_2$ Br^- rate $= k[A]$
 —Br

A

(a) Give a mechanism for each reaction that is consistent with their rate laws and the other facts you know about nucleophilic substitution reactions. Use the arrow formalism.

When the compound *B* was subjected to the same conditions, two products were obtained.

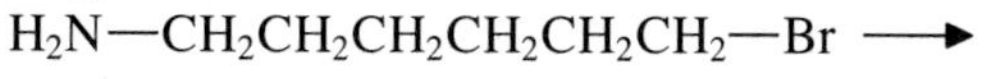

$$\overset{..}{H_2N}\text{—CH}_2CH_2CH_2CH_2CH_2CH_2\text{—Br} \longrightarrow$$

B

⬡—$\overset{+}{N}H_2$ Br^- + $H_2\overset{+}{N}(CH_2)_5CH_2Br$
 |
 $CH_2(CH_2)_4CH_2NH_2$ Br^-

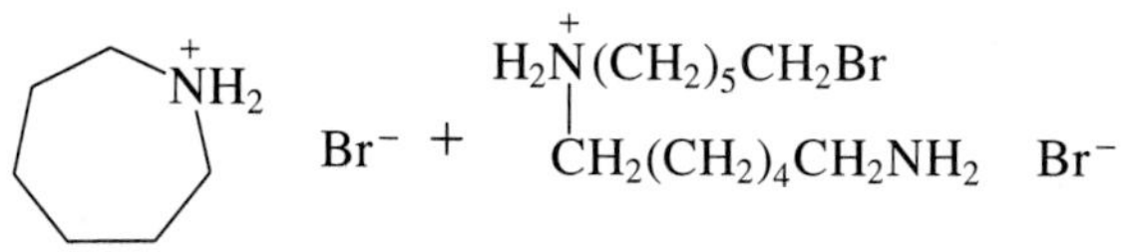

C *D*

(b) Show how each compound is formed.
(c) Assuming the rate laws for formation of each product are analogous to those above, indicate what could be done experimentally to maximize the ratio of *C* to *D* formed in the reaction.

42 Suggest a mechanism for the bromination of an alkene that follows the rate law in Eq. 9.21, Problem 2. (Several reasonable mechanisms are possible.)

43 (a) Tell whether each of the following eliminations is *syn* or *anti*.

(1)

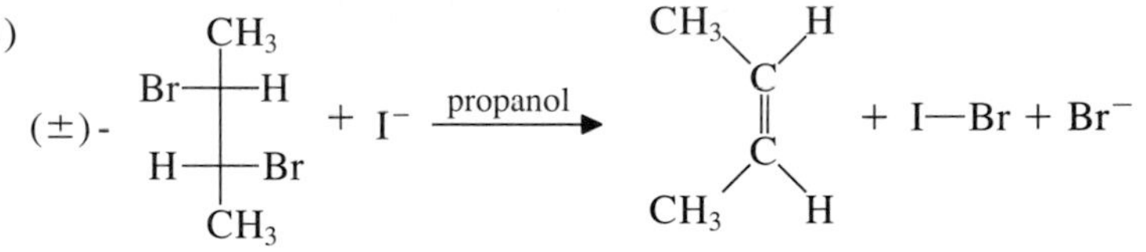

(2)

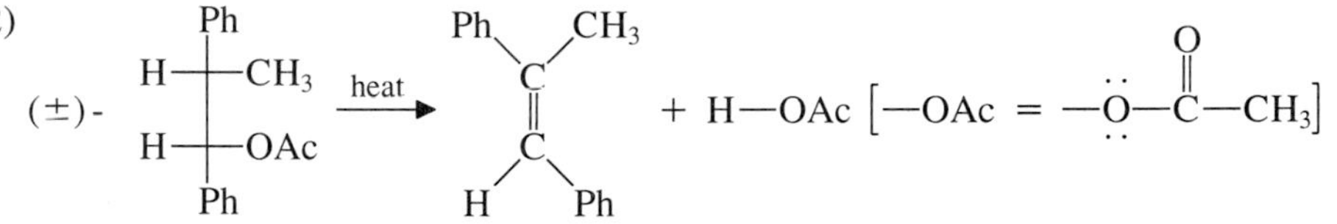

(b) Suggest a mechanism for reaction (1) consistent with its stereoselectivity and the fact that it follows a second order rate law that is first order in alkyl halide and first order in I^-.

(c) Suggest a mechanism for reaction (2) consistent with its stereoselectivity and the fact that it follows a rate law that is first order in the reactant only.

44 Given the following results, explain why diastereomer *A* reacts to give a mixture of alkene stereoisomers in which only the *Z* isomer contains deuterium, and diastereomer *B* reacts to give a mixture in which the *E* isomer contains deuterium.

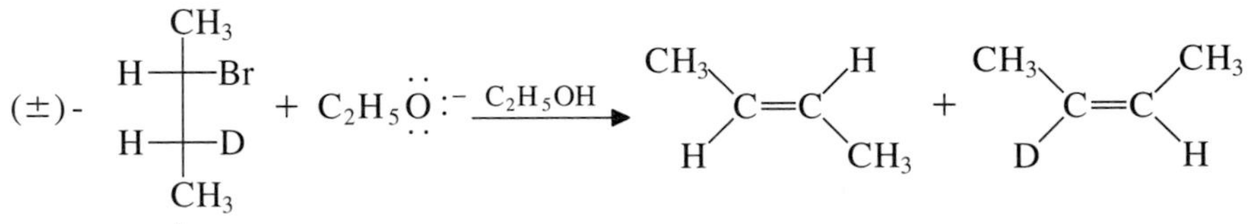

A

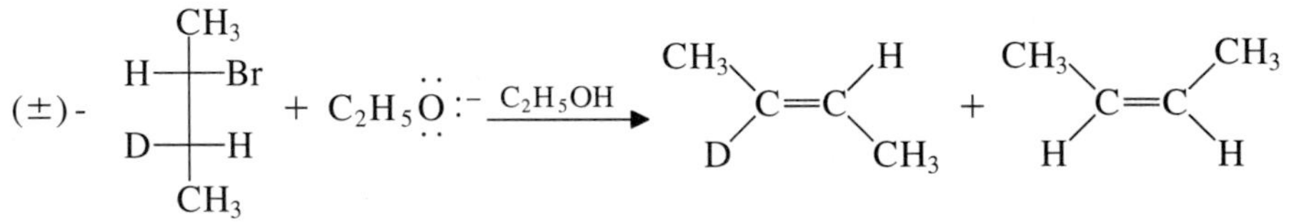

B

45 In an abandoned laboratory two liquids, *A* and *B*, were found in a box labeled only "isomeric alkyl halides $C_5H_{11}Br$." You have been employed to deduce the structures of these compounds from the following data left in an accompanying laboratory notebook. Reaction of each compound with Mg in ether, then water, gives the same hydrocarbon. Compound *A* reacts rapidly at room temperature with $AgNO_3$ in ethanol to give a light yellow precipitate; compound *B* reacts more slowly but gives the same precipitate. Both compounds give the same product, an ethyl ether, in this reaction. Reaction of compound *B* with sodium ethoxide in ethanol gives an alkene that reacts with O_3, then H_2O_2, to give acetone $(CH_3)_2C{=}O$ as one product. Give the structures for *A* and *B* and explain your reasoning.

Problems (Cont.)

46 Explain why each alkyl halide stereoisomer gives a different alkene in the E2 reactions shown. It will probably help to build models or draw conformational structures of the two starting materials.

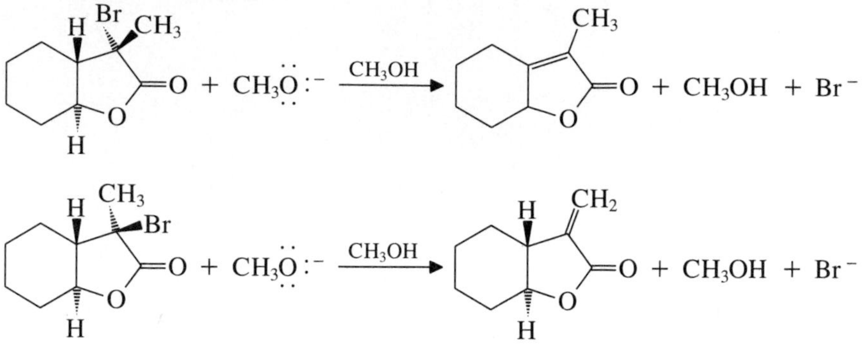

47 The *cis* and *trans* stereoisomers of 4-chlorocyclohexanol give different products when they react with OH^-.

trans-4-chlorocyclohexanol + OH^- →

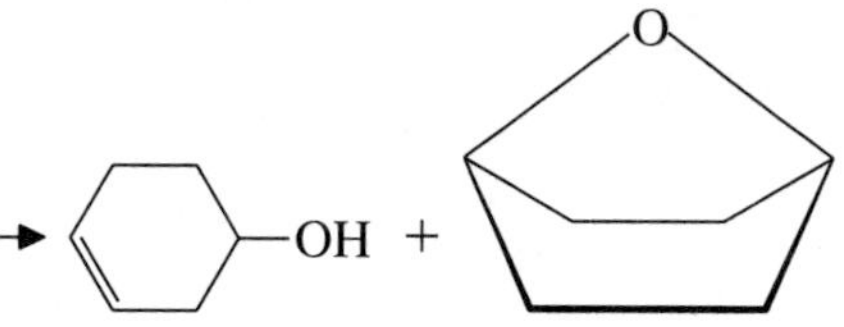

A *B*

cis-4-chlorocyclohexanol + OH^- →

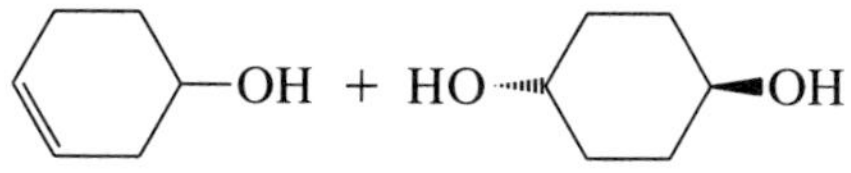

A *C*

(a) Give a mechanism for the formation of each product.
(b) Explain why the bicyclic material *B* is observed in the reaction of the *trans* isomer, but not in that of the *cis* isomer.

48 Consider the following equilibrium:

$$(CH_3)_2\ddot{S}: + CH_3\text{—}\ddot{Br}: \rightleftharpoons (CH_3)_2\overset{+}{S}\text{—}CH_3 + :\ddot{Br}:^-$$

In each case (a) and (b), choose the solvent in which the equilibrium would lie further *to the right*. Explain. (Assume that the products are soluble in all solvents considered.)
(a) ethanol or diethyl ether
(b) dimethylacetamide (a polar, aprotic solvent, $\epsilon = 38$) or a mixture of water and methanol that has the same dielectric constant

49 When Br_2 is added to the alkene *A*, the expected dibromide is formed.

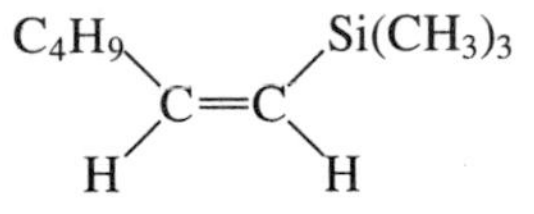

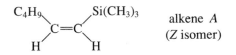

alkene *A*
(*Z* isomer)

When the dibromide product is treated with CH_3O^-, the compound $(CH_3)_3Si—OCH_3$, bromide ion, and the alkene (*E*)-1-bromo-1-hexene are formed.

(a) Is this a *syn* or an *anti* elimination? (*Hint*: Consider first the stereochemistry bromine addition. Then answer this question.)

(b) What would happen if the *E* isomer of the starting alkene *A* above were carried through this same sequence of reactions?

50 Account for each of the following results with a mechanism:

(a) $$HCCl_3 + Na^+ \, I^- \xrightarrow[35^\circ]{NaOH/H_2O} HCCl_2I + Na^+Cl^-$$

(This reaction is not observed in the absence of NaOH.)

(b)

$Ph—CH_2—Cl + CH_3CH_2CH_2CH_2—Li +$ 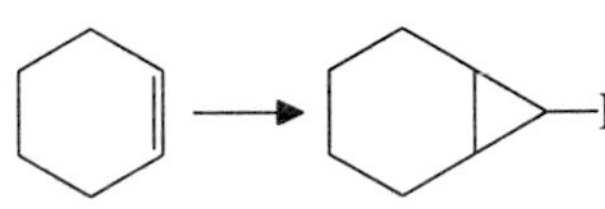 $—Ph + LiCl$

(*Hint*: Organolithium reagents are strong bases.)

51 In an abandoned government laboratory has been found a compound *A* ($C_{10}H_{17}Br$). All available evidence suggests that compound *A* is achiral. The Department of Agriculture thinks that the compound might be useful as a pesticide but needs to know its structure. You are called in as a consultant at a handsome fee. Compound *A*, when treated with KOH in ethanol, yields two compounds, *B* and *C*, each with the molecular formula $C_{10}H_{16}$. Compound *A* rapidly gives a precipitate with $AgNO_3$. Ozonolysis of *A* followed by treatment with $(CH_3)_2S$ affords as one of the products acetone $(CH_3)_2C=O$ plus halogen-containing material. Hydrogenation of either *B* or *C* gives the same compound, *cis*-4-isopropyl-1-methylcyclohexane. Compound *A* reacts with one equivalent of Br_2 to give a mixture of two separable compounds, *D* and *E*, neither of which can be resolved into enantiomers. Propose structures for compounds *A* through *E* that best fit the data.

10 Chemistry of Alcohols, Glycols, and Thiols

In this chapter we focus on the reactions of alcohols, thiols, and glycols. We shall find that alcohols, like alkyl halides, undergo substitution and elimination reactions. However, alcohols and thiols, unlike alkyl halides, undergo *oxidation* reactions. We shall learn to recognize oxidations and examine some of the ways oxidations are carried out in the laboratory and in nature. We shall also begin to discuss the strategies used in organic synthesis, a subject to which we shall return repeatedly in subsequent chapters as we learn more synthetically useful reactions.

10.1 DEHYDRATION OF ALCOHOLS

A. Conversion of Alcohols into Alkenes

Mineral acids such as H_2SO_4 and H_3PO_4 catalyze a β-elimination of water from alcohols to give alkenes.

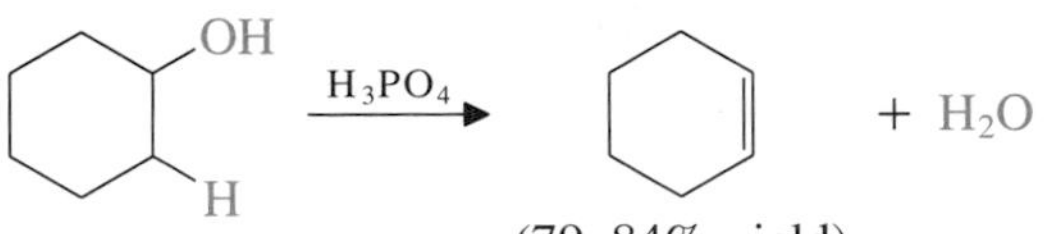

(79–84% yield)

(10.1)

A reaction such as this, in which the elements of water are lost from the starting material, is called a **dehydration.** Lewis acids such as alumina (aluminum oxide, Al_2O_3) and/or heat can also be used to catalyze dehydration reactions.

Most acid-catalyzed dehydrations of alcohols are reversible reactions. However, these reactions can easily be driven toward alkene by applying LeChatelier's principle (Sec. 9.2). For example, in Eq. 10.1, the equilibrium is driven toward the alkene product because the water that is formed complexes tightly with the mineral acid H_3PO_4, and the cyclohexene product is distilled from the reaction mixture. The dehydration of alcohols to alkenes is easily carried out in the laboratory and is an important procedure for the preparation of some alkenes.

The most common mechanism for the dehydration of alcohols is simply the reverse of the mechanism for hydration of alkenes (Sec. 4.8).

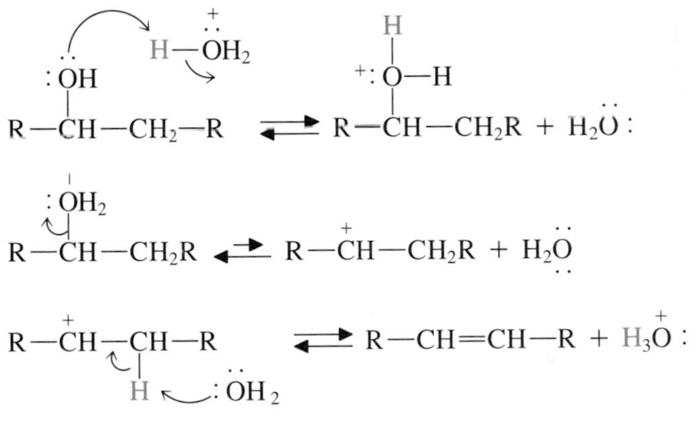

(10.2a)

(10.2b)

(10.2c)

Notice, for example, that the reverse of the sequence in Eq. 10.2 is the same as the hydration mechanism given in Eq. 4.33, and a carbocation intermediate is involved, as it is in hydration. In general, *a reaction and its reverse proceed by the forward and reverse of the same mechanism.* This idea is known as the **principle of microscopic reversibility.** It follows from this principle that forward and reverse reactions must have the same intermediates and the same rate-determining transition states. Thus, since protonation of the alkene is the rate-determining step in alkene hydration, the reverse of this step—loss of the proton from the carbocation intermediate (Eq. 10.2c)—is the rate-determining step in alcohol dehydration. This principle also requires that any reaction catalyzed in one direction is also catalyzed in the other. Thus, both hydration of alkenes to alcohols and dehydration of alcohols to alkenes are catalyzed by acids.

The first step in the dehydration of an alcohol (Eq. 10.2a) is protonation of the alcohol oxygen. Thus, the basicity of the alcohol (Sec. 8.7) is important in this reaction. Protonation of the oxygen converts the —OH group of the alcohol into an excellent leaving group. Loss of this group as water gives a carbocation, which then loses a β-proton to give the alkene. Notice that once the alcohol is protonated, the mechanism of dehydration is essentially identical to that of the E1 reaction of an alkyl halide, except that water, rather than a halide ion, is the leaving group.

alcohol dehydration: E1 *reaction of alkyl halide:*

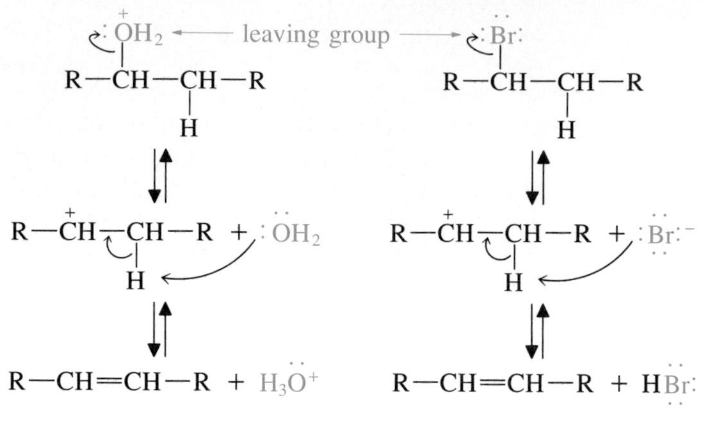

(10.3)

The involvement of carbocation intermediates in this reaction helps us understand several experimental facts about alcohol dehydration. First, the relative rates of alcohol dehydration are in the order tertiary > secondary >> primary. Since tertiary carbocations are more stable, Hammond's postulate (Sec. 4.7C) suggests that reactions involving these ions are faster. In fact, dehydration of primary alcohols is generally not a useful laboratory procedure. Second, if more than one type of β-proton is present in the alcohol, then a mixture of alkene products can be expected. As in the E1 reaction, the more stable alkenes usually predominate in the mixture (Saytzeff elimination; Sec. 9.4D).

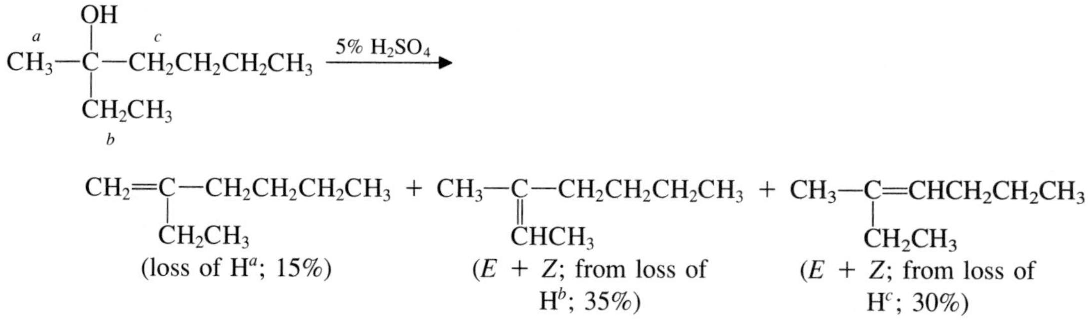

$$CH_2{=}C{-}CH_2CH_2CH_2CH_3 \ + \ CH_3{-}C{-}CH_2CH_2CH_2CH_3 \ + \ CH_3{-}C{=}CHCH_2CH_2CH_3 \quad (10.4)$$

（loss of H^a; 15%） （$E + Z$; from loss of H^b; 35%） （$E + Z$; from loss of H^c; 30%）

Finally, alcohols that react to give rearrangement-prone carbocation intermediates yield rearranged alkenes.

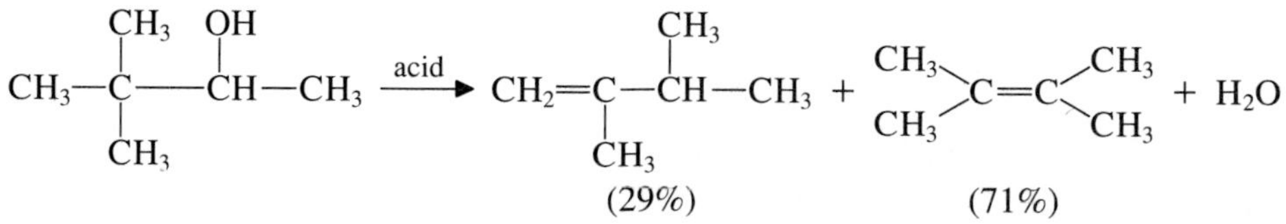

(10.5)

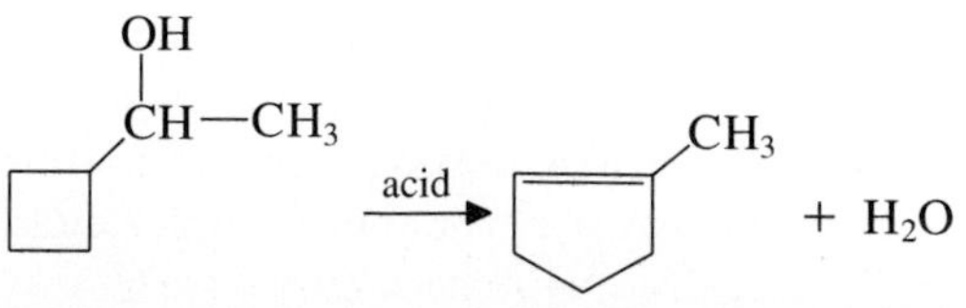

(10.6)

Let us digress briefly to consider the reaction in Eq. 10.6. The first step in the reaction is formation of the carbocation from the alcohol (show this). Once this carbocation is formed, it rearranges. Rearrangements involving rings may seem difficult to follow, but they are like any other rearrangement if we analyze carefully what is happening by *drawing out the individual carbon atoms implied by the skeletal formula.*

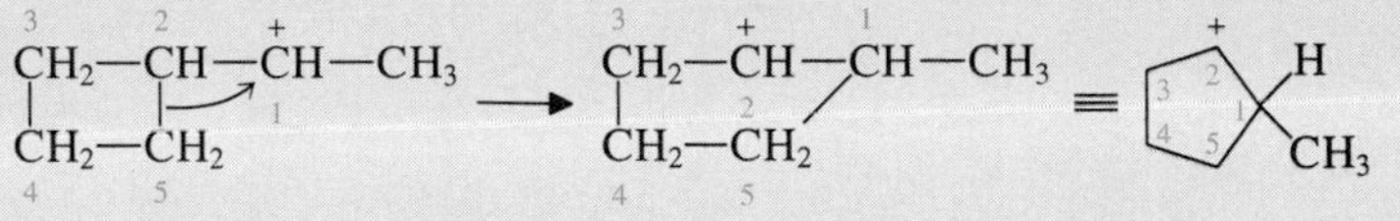

(10.7a)

The rearrangement is simply the migration of carbon-5 from carbon-2 to carbon-1. This migration leaves carbon-2 electron-deficient, carbon-1 becomes a part of the ring, and the ring size is expanded by one carbon. Conceptually, this ring expansion is analogous to the expansion of a noose in a rope:

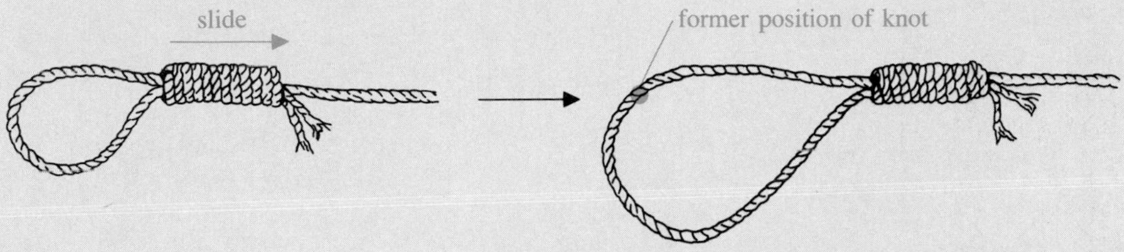

The rearranged carbocation forms the product in Eq. 10.6 by loss of a proton.

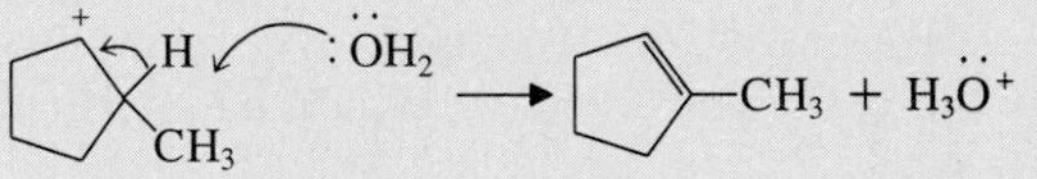

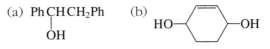

(10.7b)

Problems

1 (a) Give the mechanism for the reaction in Eq. 10.5.
(b) Explain why the rearrangement in Eq. 10.6 is energetically favorable even though both carbocations are secondary.

2 What alkene(s) are produced in the acid-catalyzed dehydration of each of the following alcohols?

(a) PhCHCH$_2$Ph (with OH on the CH) (b)

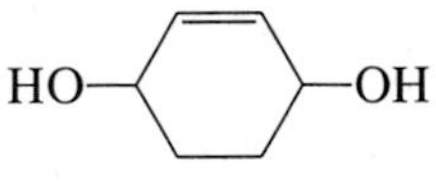

3 Give the structures of three alcohols that could in principle give 1-methyl-1-cyclohexene as the major acid-catalyzed dehydration product. Indicate which of the alcohols undergoes dehydration most rapidly.

B. Pinacol Rearrangement

Vicinal glycols also undergo dehydration in concentrated mineral acid. However, this reaction of glycols is always accompanied by rearrangement.

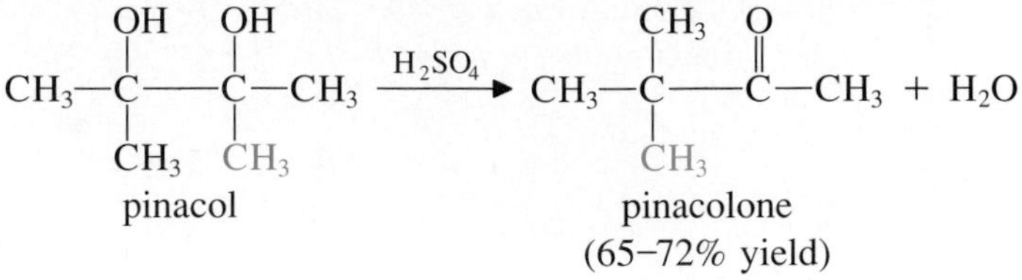

(10.8)

Although this is a general reaction of glycols, its name, the **pinacol rearrangement,** is derived from the glycol used as the starting material in Eq. 10.8.

The first steps of the reaction mechanism are like those of other alcohol dehydrations: protonation of the —OH group is followed by loss of water to give a carbocation.

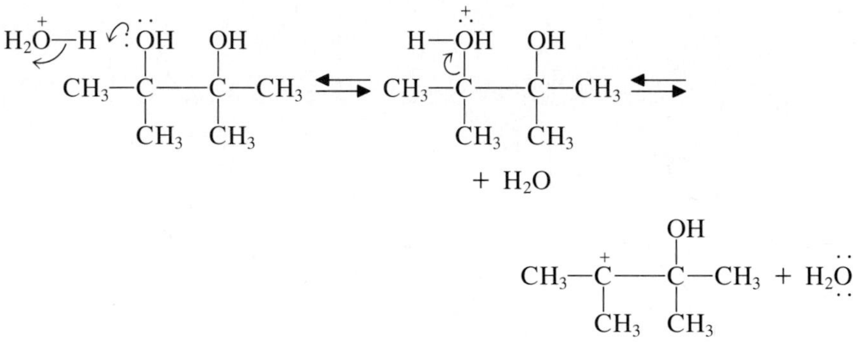

(10.9a)

However, the resulting carbocation does not lose a proton; instead, it rearranges to another carbocation, which has resonance structures.

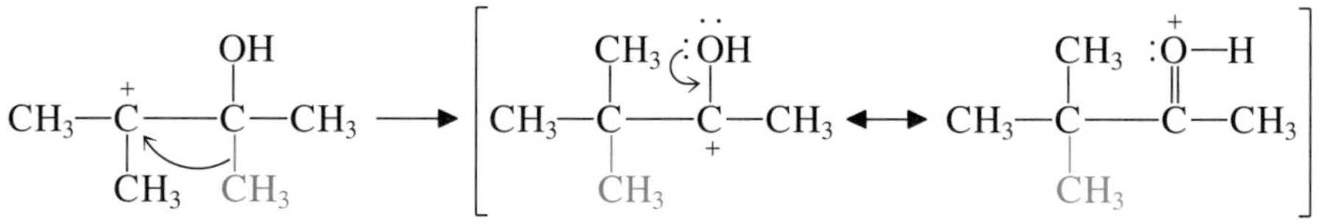

(10.9b)

(As we shall learn in Sec. 15.5, carbocations in which the positive charge is delocalized by resonance are particularly stable.) The resonance structures of this carbocation show that positive charge is shared by the neighboring oxygen. Loss of a proton from this oxygen yields the product.

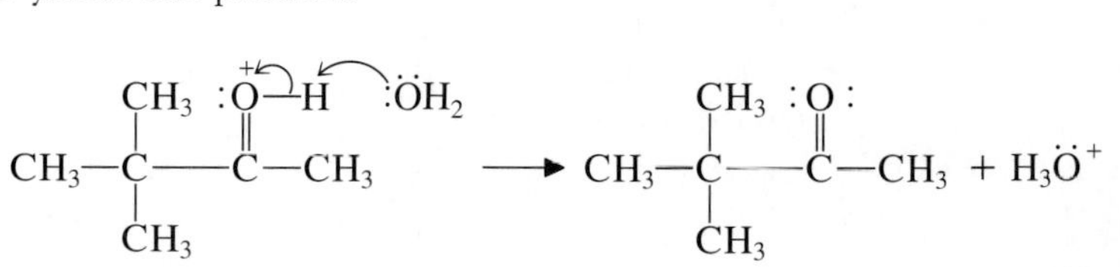

$$CH_3-\overset{\overset{\displaystyle CH_3}{|}}{\underset{\underset{\displaystyle CH_3}{|}}{C}}-\overset{\displaystyle :\overset{+}{O}-H}{\overset{\|}{C}}-CH_3 \longrightarrow CH_3-\overset{\overset{\displaystyle CH_3}{|}}{\underset{\underset{\displaystyle CH_3}{|}}{C}}-\overset{\displaystyle :O:}{\overset{\|}{C}}-CH_3 + H_3\overset{..}{O}{}^+$$ (10.9c)

Since the pinacol rearrangement involves carbocation intermediates, it follows that the most rapid reactions are those that involve relatively stable carbocations. Thus, pinacol rearrangements of tertiary glycols are rapid, but the rearrangement of ethylene glycol, which would involve a primary carbocation, is not a useful reaction under usual laboratory conditions.

More than one pinacol rearrangement are possible for some glycols.

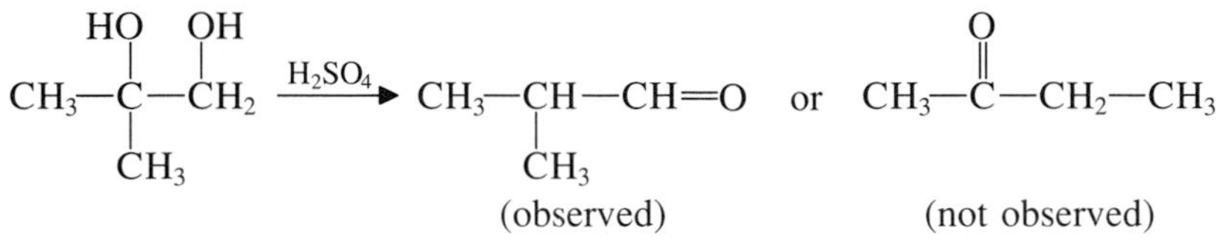

(10.10)

The course of this reaction can be understood in terms of carbocation theory. There are two —OH groups in the glycol that could, after protonation, be lost as water. Loss of water occurs so as to give the more stable carbocation.

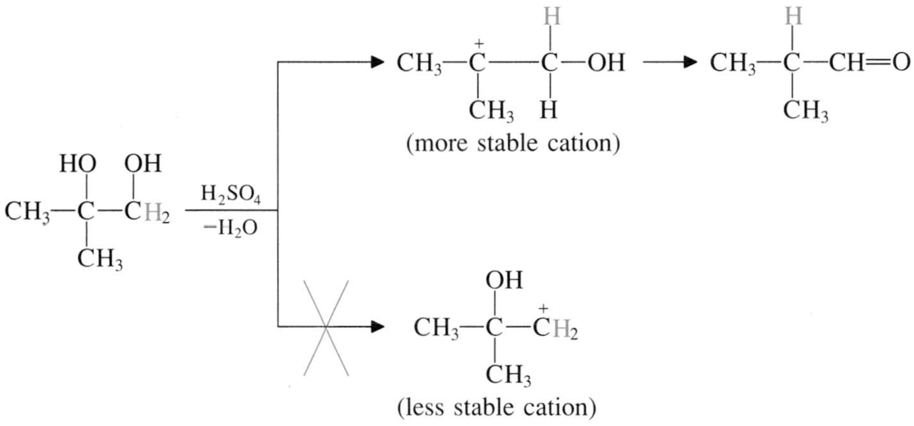

(10.11)

Problem

4 Predict the product of each of the following reactions. Explain your answers.

(a)

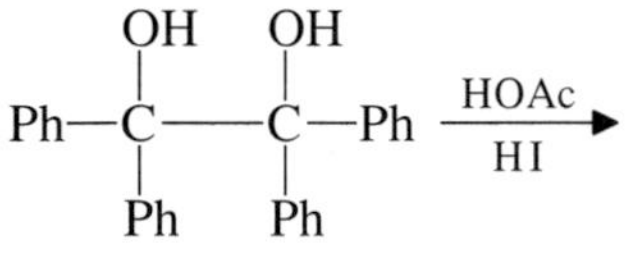

(b)

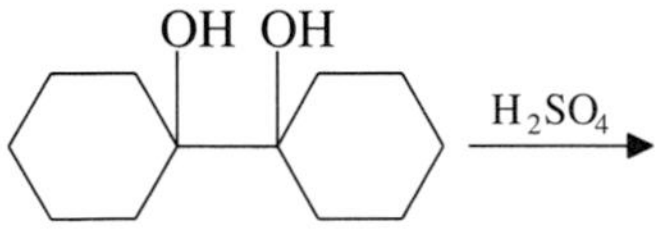

(*Hint*: For (b), re-read the discussion associated with Eq. 10.6.)

10.2 REACTION OF ALCOHOLS WITH HYDROGEN HALIDES

Alcohols react with hydrogen halides to give alkyl halides. (See Eqs. 10.12 and 10.13.)

$$(CH_3)_3C{-}OH + HCl \xrightarrow[25°,\ 20\ min]{H_2O} (CH_3)_3C{-}Cl + H_2O \quad (10.12)$$

(almost quantitative)

$$(CH_3)_2CHCH_2CH_2-OH \ + \ HBr \ \xrightarrow[\text{heat, 5-6 hr}]{H_2SO_4} \ (CH_3)_2CHCH_2CH_2-Br \ + \ H_2O \quad (10.13)$$
$$\text{(93\% yield)}$$

In Chapter 9 we learned that alkyl halides react with water to give alcohols and hydrogen halides. Many such reactions (for example, the reaction of *t*-butyl chloride with water) are quite fast. Why, then, can we make alkyl halides from alcohols? Like the dehydration of alcohols to alkenes, the success of the reaction generally depends on the application of Le Chatelier's principle (Sec. 9.2). For example, in the synthesis of *t*-butyl chloride from the reaction of *t*-butyl alcohol with aqueous HCl (Eq. 10.12), the starting alcohol is soluble in aqueous solution, but the product *t*-butyl chloride is not. Hence, the *t*-butyl chloride comes out of solution as a separate layer, thereby driving the equilibrium toward the alkyl halide. The excess of HCl used also drives the reaction. In some procedures, water is removed by using an H_2SO_4 catalyst along with the halogen acid. (H_2SO_4 is a strong dehydrating acid; that is, it has a great affinity for water.)

The mechanism of formation of alkyl halides from alcohols depends on the type of alcohol used as the starting material. Carbocation intermediates are involved in the reactions of tertiary alcohols.

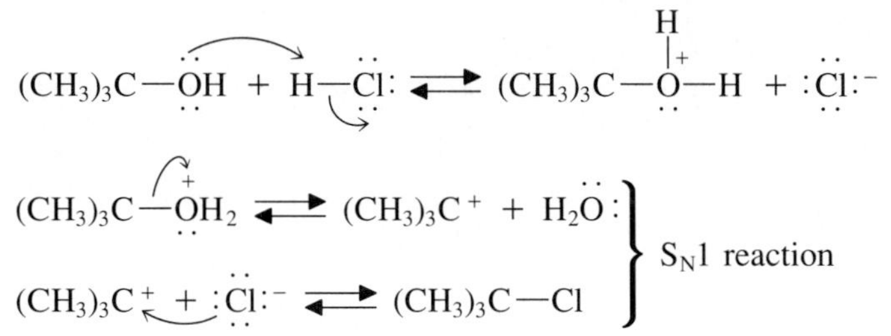

$$(CH_3)_3C-\overset{..}{\underset{..}{O}}H \ + \ H-\overset{..}{\underset{..}{Cl}}: \ \rightleftharpoons \ (CH_3)_3C-\overset{\overset{H}{|}+}{\underset{..}{O}}-H \ + \ :\overset{..}{\underset{..}{Cl}}:^- \quad (10.14a)$$

$$(CH_3)_3C-\overset{+}{\underset{..}{O}}H_2 \ \rightleftharpoons \ (CH_3)_3C^+ \ + \ H_2\overset{..}{O}: \quad \left.\begin{array}{c} \\ \\ \end{array}\right\} \quad (10.14b)$$

$$(CH_3)_3C^+ \ + \ :\overset{..}{\underset{..}{Cl}}:^- \ \rightleftharpoons \ (CH_3)_3C-Cl \quad \left.\begin{array}{c} \\ \end{array}\right. \quad S_N1 \text{ reaction} \quad (10.14c)$$

Notice that once the alcohol is protonated, *the reaction is essentially an S_N1 reaction with H_2O as the leaving group.* When a primary alcohol is the starting material, the reaction occurs as a direct displacement of water from the protonated alcohol.

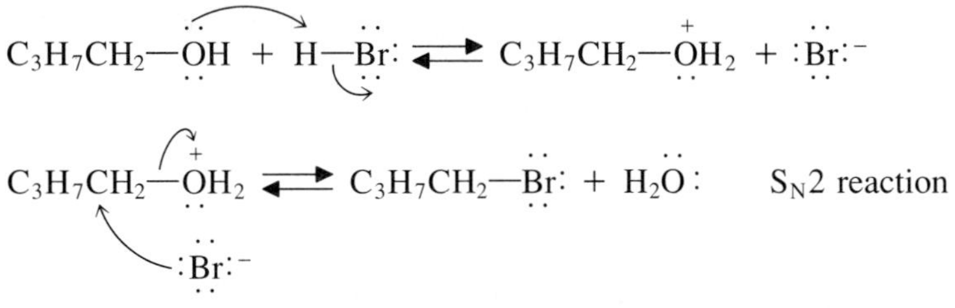

$$C_3H_7CH_2-\overset{..}{\underset{..}{O}}H \ + \ H-\overset{..}{\underset{..}{Br}}: \ \rightleftharpoons \ C_3H_7CH_2-\overset{+}{\underset{..}{O}}H_2 \ + \ :\overset{..}{\underset{..}{Br}}:^- \quad (10.15a)$$

$$C_3H_7CH_2-\overset{+}{\underset{..}{O}}H_2 \ \rightleftharpoons \ C_3H_7CH_2-\overset{..}{\underset{..}{Br}}: \ + \ H_2\overset{..}{O}: \quad S_N2 \text{ reaction} \quad (10.15b)$$
$$:\overset{..}{\underset{..}{Br}}:^-$$

In other words, *the reaction is an S_N2 reaction in which water is the leaving group.* Secondary alcohols can react by either the S_N1 or S_N2 mechanism, or both.

The reactions of tertiary alcohols with hydrogen halides are much faster than the reactions of primary alcohols. Typically, tertiary alcohols react with hydrogen halides rapidly at room temperature, whereas the reactions of primary alcohols require heating for several hours. In fact, the reactions of primary and many secondary alcohols with HCl are so slow that both heat and a Lewis-acid catalyst are required.

$$(CH_3)_2CH-OH \ + \ HCl \ \xrightarrow{ZnCl_2} \ (CH_3)_2CH-Cl \ + \ H_2O \quad (10.16)$$
$$\text{(75\% yield)}$$

In this example, Zn^{2+} is the Lewis-acid catalyst. The zinc ion coordinates to the oxygen of the alcohol and assists its departure in exactly the same way that electrophilic catalysts assist the ionization of alkyl halides (Sec. 9.5C).

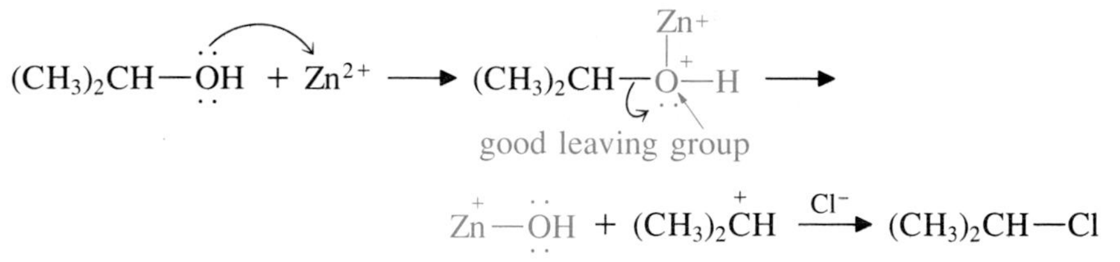

$$(CH_3)_2CH-\overset{..}{\underset{..}{O}}H + Zn^{2+} \longrightarrow (CH_3)_2CH-\underset{\underset{\text{good leaving group}}{}}{\overset{\overset{Zn+}{\overset{|}{\overset{+}{O}}}}{}}-H \longrightarrow$$

$$\overset{+}{Zn}-\overset{..}{\underset{..}{O}}H + (CH_3)_2\overset{+}{CH} \xrightarrow{Cl^-} (CH_3)_2CH-Cl \quad (10.17)$$

When carbocation intermediates are involved in the reactions of alcohols, we should expect the telltale signs of these intermediates—rearrangements—in appropriate cases.

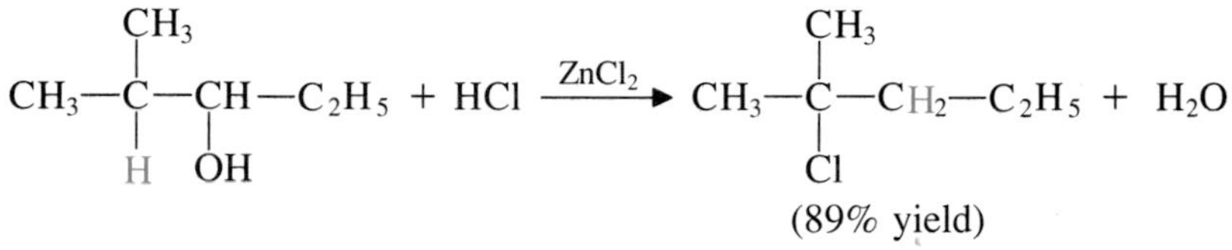

$$\underset{\underset{H}{|}\;\underset{OH}{|}}{CH_3-\overset{\overset{CH_3}{|}}{C}-CH-C_2H_5} + HCl \xrightarrow{ZnCl_2} \underset{\underset{Cl}{|}}{CH_3-\overset{\overset{CH_3}{|}}{C}-CH_2-C_2H_5} + H_2O \quad (10.18)$$

$$(89\% \text{ yield})$$

Problems

5 Suggest a synthesis for each of the following alkyl halides from an alcohol:

(a) (b) $I-CH_2CH_2CH_2CH_2CH_2-I$

6 Give a mechanism for the reaction in Eq. 10.18.

7 Give the products expected in each of the following conversions:

(a) $HO-CH_2CH_2CH_2-OH + HBr \xrightarrow{H_2SO_4}$

(b) HOCH₂ CH₂OH

$+ HCl \longrightarrow$

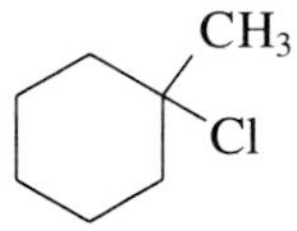

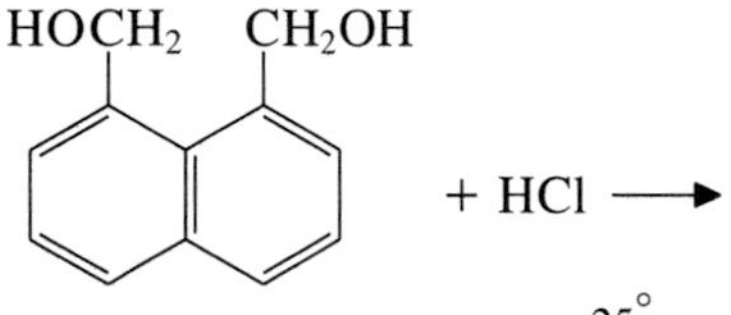

(c) $(CH_3)_3CCH_2OH + HCl \xrightarrow{25°}$
(*Hint:* See Sec. 9.3C, Fig. 9.2.)

(d)

$$\underset{\underset{CH_3}{|}}{CH_3-CH-\overset{\overset{OH}{|}}{CH}-CH_3} + HBr \longrightarrow$$

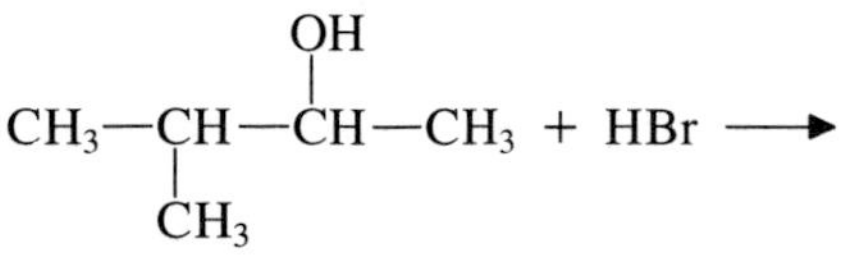

The dehydration of alcohols to alkenes and the reaction of alcohols with hydrogen halides are very similar reactions. Both reactions take place in acidic solution; in both reactions the acid converts the —OH group into a good leaving group. If acid were not present, the halide ion would have to displace $^-$OH in order to form the alkyl halide. This reaction would not take place because $^-$OH is a much stronger base than any halide ion (Table 8.3).

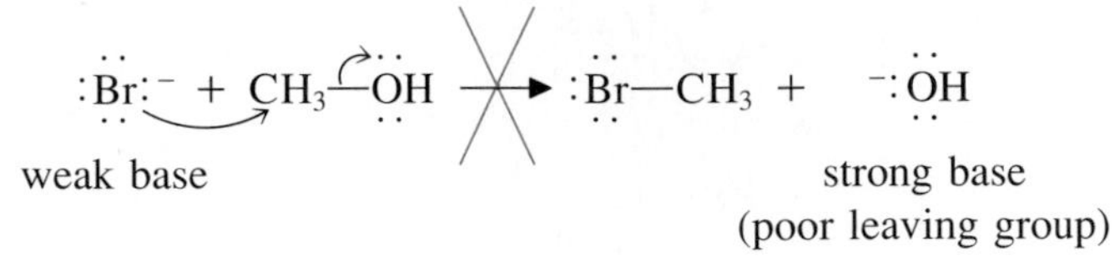

This analysis shows that *substitution and elimination reactions of alcohols are possible if the —OH group is first converted into a better leaving group.* This point is important because we are about to see other examples of this idea in the following sections.

Notice that formation of secondary and tertiary alkyl halides and dehydration of secondary and tertiary alcohols have the same initial steps, protonation of the alcohol oxygen and formation of a carbocation.

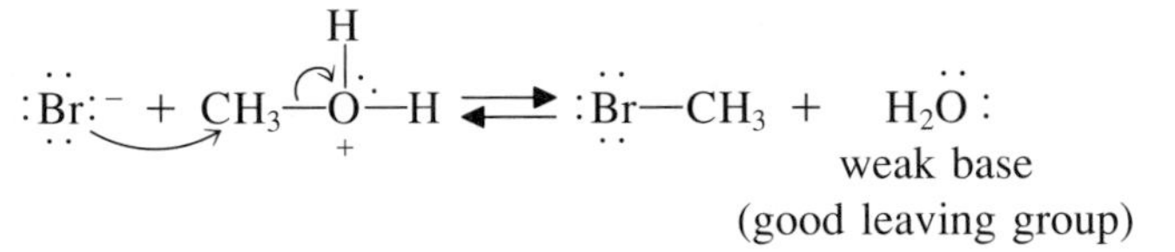

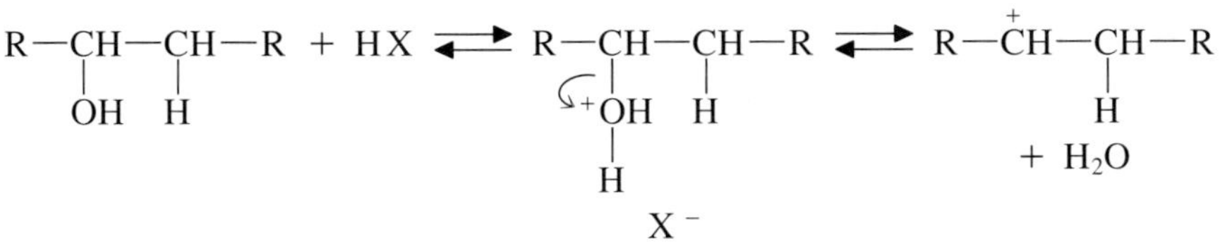

$$R-\underset{\underset{OH}{|}}{CH}-\underset{\underset{H}{|}}{CH}-R + HX \;\rightleftharpoons\; R-\underset{\underset{\underset{H}{|}}{+OH}}{CH}-\underset{\underset{H}{|}}{CH}-R \;\rightleftharpoons\; R-\overset{+}{\underset{\underset{H}{|}}{CH}}-\underset{\underset{H}{|}}{CH}-R \quad (10.20a)$$
$$X^-$$
$$+ H_2O$$

The two reactions differ in the fate of this carbocation. In the presence of hydrogen halide, the halide ion attacks the carbocation, and the excess of hydrogen halide used drives the equilibrium toward alkyl halide. Any alkene that does form reacts with hydrogen halide to give alkyl halide (Sec. 4.5). In dehydration, the carbocation loses a proton from a β-carbon. Because dehydration reactions are usually run with nonvolatile mineral acids such as H_2SO_4 or H_3PO_4, the dehydration equilibrium can be driven by distilling the alkene and water away from the acid catalyst as they are formed.

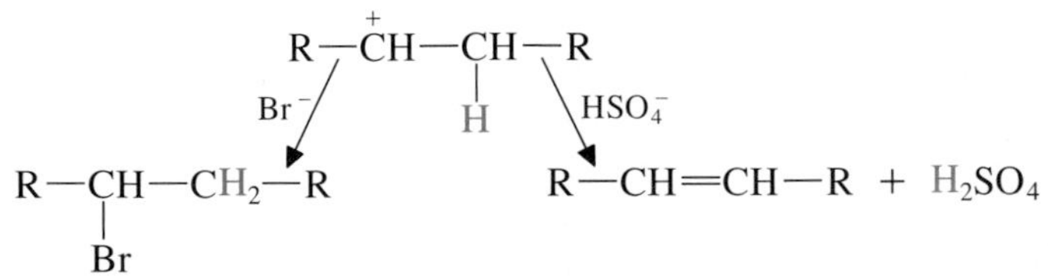

$$(10.20b)$$

It follows, then, that *alkyl halide formation and dehydration to alkenes are alternate branches of a common mechanism.* Notice that the principles developed for substitutions and eliminations of alkyl halides are also valid for other functional groups—in this case, alcohols.

10.3 SULFONATE AND INORGANIC ESTER DERIVATIVES OF ALCOHOLS

When alkyl halides are prepared from alcohols with hydrogen halides, protons or Lewis acids are used to convert the —OH groups of the alcohols into good leaving groups. However, the strongly acidic conditions required in these reactions are sometimes too harsh to be useful. In this section we shall study a different way of transforming the hydroxy group into a good leaving group.

A. Sulfonate Ester Derivatives of Alcohols

A **sulfonic acid** is a compound of the form R—SO_3H. Some typical sulfonic acids are the following:

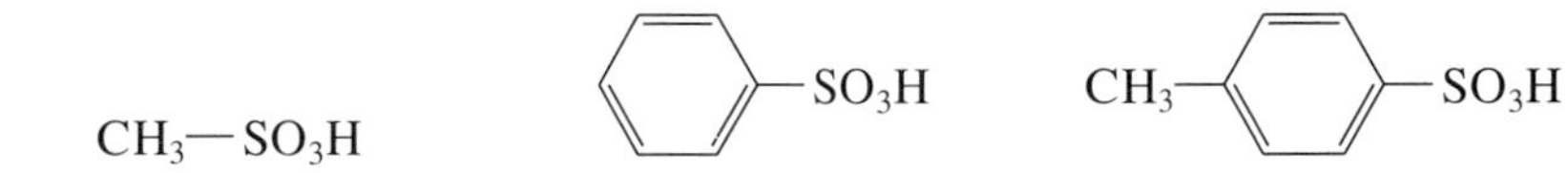

CH3—SO3H
SO3H
CH3
SO3H

methanesulfonic acid   benzenesulfonic acid   *p*-toluenesulfonic acid

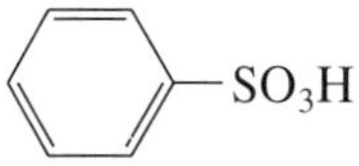

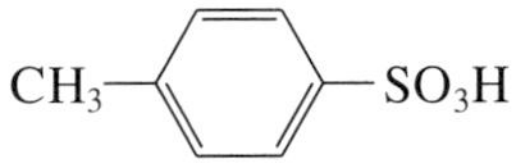

A compound in which the acidic proton (color) of the sulfonic acid is formally replaced with an alkyl group is called a **sulfonate ester.**

O
S—O—H
O
acidic proton
O
S—O—C2H5
O

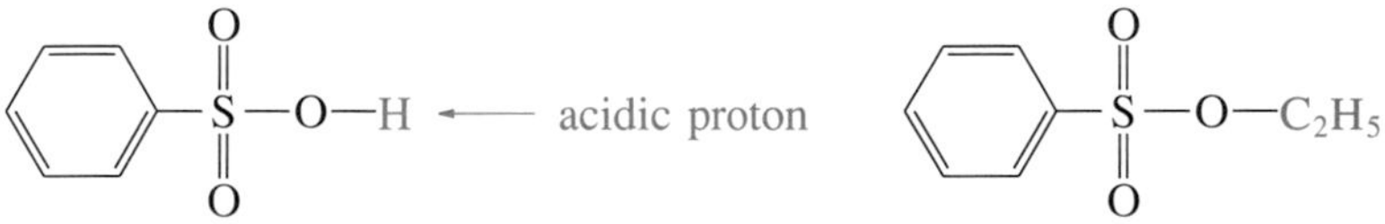

benzenesulfonic acid   ethyl benzenesulfonate (a sulfonate ester)

Esters are important compounds that we shall study in detail in Chapter 21. Many esters are prepared from alcohols, and are important alcohol derivatives. The chemistry of sulfonate esters is so closely related to the other alcohol chemistry we have discussed that we shall consider the reactions of sulfonate esters here.

Organic chemists use certain abbreviations for sulfonate esters. Esters of methanesulfonic acid are called *mesylates* (abbreviated R—OMs), and esters of *p*-toluenesulfonic acid are called *tosylates* (abbreviated R—OTs).

O
C2H5—O—S—CH3
O
is the same as
C2H5—OMs

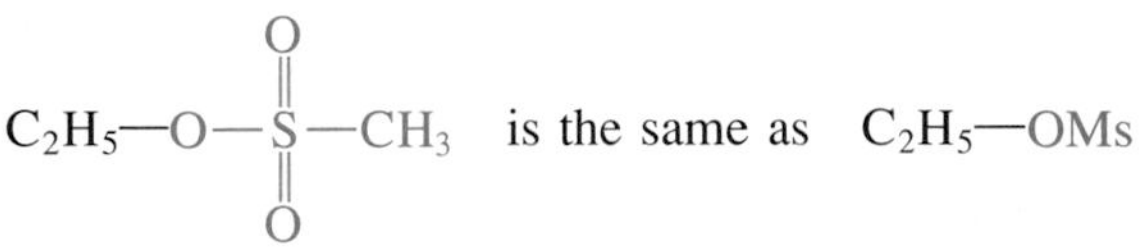

ethyl methanesulfonate   ethyl mesylate

O
CH3—CH—O—S
C2H5
O
CH3
is the same as
CH3—CH—OTs
C2H5

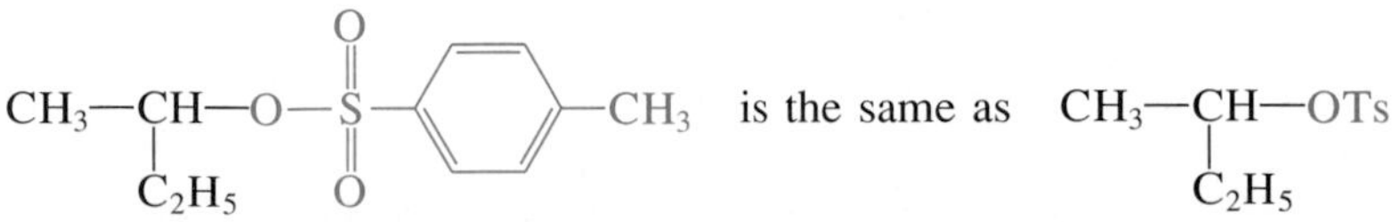

sec-butyl *p*-toluenesulfonate   *sec*-butyl tosylate

Problem

8 Draw a structure of
(a) isopropyl methanesulfonate
(b) methyl *p*-toluenesulfonate
(c) phenyl tosylate
(d) cyclohexyl mesylate

Sulfonate esters are prepared from alcohols and sulfonic acid derivatives called sulfonyl chlorides. For example, *p*-toluenesulfonyl chloride, often known as *tosyl chloride,* and abbreviated TsCl, is the sulfonyl chloride used to prepare tosylate esters.

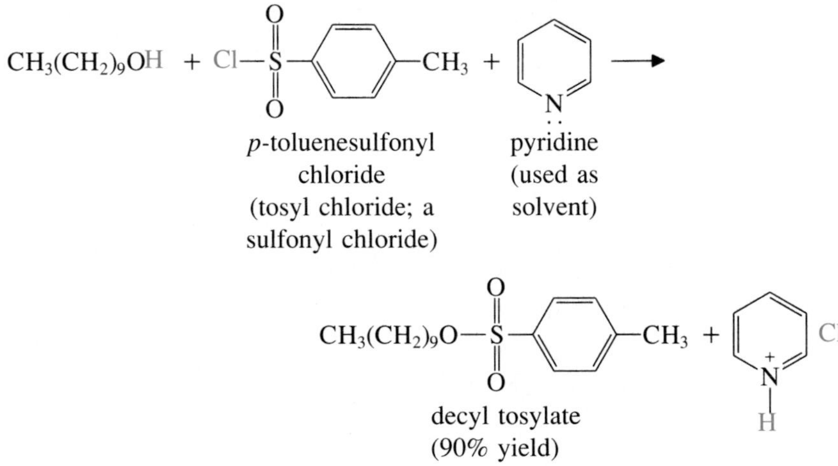

(10.21)

The pyridine used as the solvent is a base. Besides catalyzing the reaction (by a mechanism we shall not consider here), pyridine also neutralizes the HCl formed in the reaction (color in Eq. 10.21).

Problem

9 Suggest preparations for each of the following compounds from alcohols:
(a) isobutyl tosylate
(b) cyclohexyl mesylate

The utility of alkyl sulfonates hinges on the fact that, to a useful approximation, *they react like alkyl bromides or alkyl iodides in substitution and elimination reactions.* The reason for this similarity is that, like halide ions, *sulfonate anions are excellent leaving groups.* Sulfonate anions are good leaving groups because they are weak bases (Table 8.3, Sec. 9.3D); in other words, their conjugate acids, sulfonic acids, are strong acids. Thus, in the presence of a strong base, primary alkyl sulfonate esters, like primary alkyl halides, undergo the S_N2 reaction; tertiary alkyl sulfonate esters, like tertiary alkyl halides, undergo the E2 reaction.

For example, suppose we wish to displace the —OH group of an alcohol with bromide ion. Since hydroxide is a poor leaving group, the reaction cannot be carried out on the alcohol itself (Eq. 10.19a). Instead, the alcohol can be converted into a good leaving group—a sulfonate ester—and the sulfonate anion can be displaced by bromide ion.

$$CH_3CH_2\underset{}{\overset{OH}{\overset{|}{C}}}HCH_2CH_3 \xrightarrow{\text{TsCl, pyridine}} CH_3CH_2\overset{OTs}{\overset{|}{C}}HCH_2CH_3 \qquad (10.22)$$

$$CH_3CH_2CHCH_2CH_3 + Na^+Br^- \xrightarrow{\text{DMSO}} CH_3CH_2CHCH_2CH_3 + Na^+ {}^-OTs \quad (10.23)$$

with OTs below the left structure and Br below the right structure

(85% yield)

(Notice the use of the polar aprotic solvent DMSO in the S_N2 reaction; Sec. 9.3D.)
The E2 reaction of sulfonate esters, like the analogous reaction of alkyl halides, can be used to prepare alkenes.

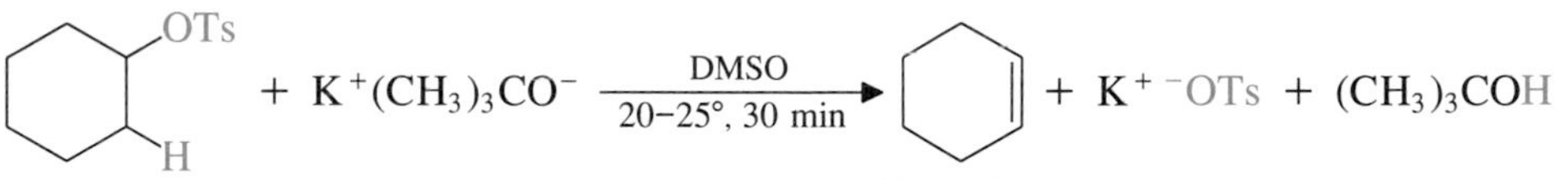

$$\text{(cyclohexyl-OTs with H)} + K^+(CH_3)_3CO^- \xrightarrow[20-25°, 30 \text{ min}]{\text{DMSO}} \text{(cyclohexene)} + K^+ {}^-OTs + (CH_3)_3COH \quad (10.24)$$

(83% yield)

Problem

10 Give the product of the following sequence of reactions:

$$(CH_3)_2CHCH_2CH_2OH \xrightarrow[\text{pyridine}]{\text{TsCl}} \xrightarrow[\text{DMSO}]{\text{NaCN}}$$

B. Esters Derived from Alcohols and Strong Inorganic Acids

Esters of strong inorganic acids such as sulfuric and phosphoric acids are well-known compounds and, like sulfonate esters, are formally derived by replacing the acidic protons of the acids with alkyl groups.

acidic protons

sulfuric acid dimethyl sulfate

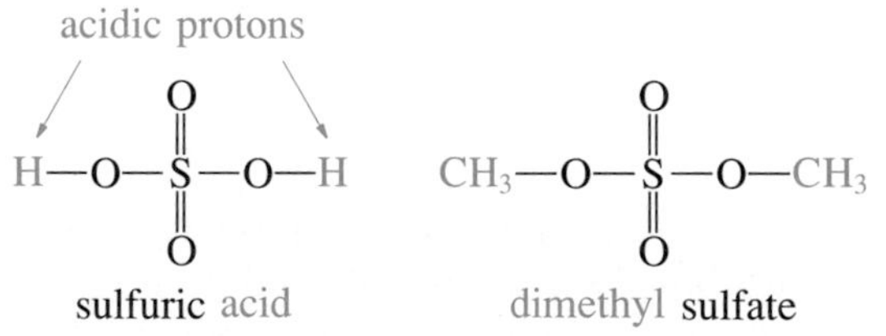

Dimethyl sulfate, for example, is a derivative of sulfuric acid in which the acidic protons of sulfuric acid (color) are replaced by methyl groups.
Esters derived from inorganic acids are also derivatives of alcohols. For example, dimethyl sulfate is a derivative of methanol, since it is prepared commercially from methanol and sulfuric acid:

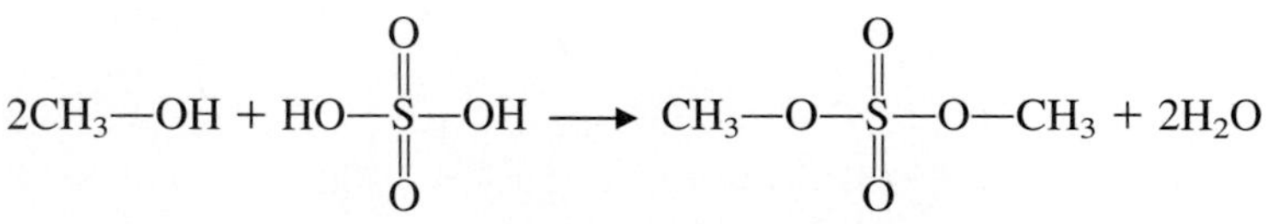

$$2CH_3{-}OH + HO{-}\underset{O}{\overset{O}{S}}{-}OH \longrightarrow CH_3{-}O{-}\underset{O}{\overset{O}{S}}{-}O{-}CH_3 + 2H_2O \quad (10.25)$$

Dimethyl sulfate is a very important methylating agent; that is, it readily transfers a methyl group to nucleophiles, as in the following example:

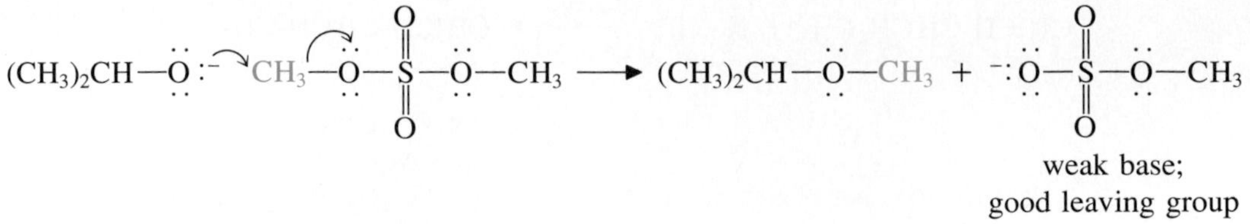

$$(CH_3)_2CH-\ddot{O}: \curvearrowright CH_3-\ddot{O}-\overset{\overset{O}{\parallel}}{\underset{\underset{O}{\parallel}}{S}}-\ddot{O}-CH_3 \longrightarrow (CH_3)_2CH-\ddot{O}-CH_3 + {}^-:\ddot{O}-\overset{\overset{O}{\parallel}}{\underset{\underset{O}{\parallel}}{S}}-\ddot{O}-CH_3 \quad (10.26)$$

weak base;
good leaving group

Dimethyl sulfate and diethyl sulfate are available commercially.

Along the same line, alkyl halides can be thought of as alkyl esters of the halogen acids. Methyl bromide, for example, is formally derived by replacing the acidic proton of HBr with a methyl group. As we have seen (Chapter 9), these "esters" also react readily with nucleophiles.

Problems

11 Phosphoric acid, H_3PO_4, has the following structure. Draw the structure of triethyl phosphate.

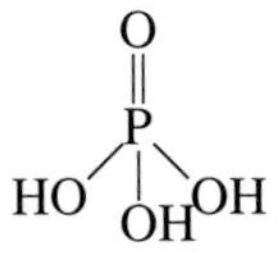

12 Predict the products in the reaction of dimethyl sulfate with each of the following species:
(a) water
(c) sodium 1-propanethiolate
(b) sodium ethoxide
(d) $CH_3-\ddot{N}H_2$ (methylamine)

10.4 REACTION OF ALCOHOLS WITH THIONYL CHLORIDE AND PHOSPHORUS TRIBROMIDE

The synthesis of alkyl halides from alcohols is one of the most important synthetic conversions of alcohols. We have already studied two ways of effecting this conversion: the reactions of alcohols with hydrogen halides, and the conversions of alcohols to sulfonate esters, followed by the S_N2 reactions of these esters with halide ions. Thionyl chloride is another reagent that is useful for the preparation of alkyl chlorides from alcohols.

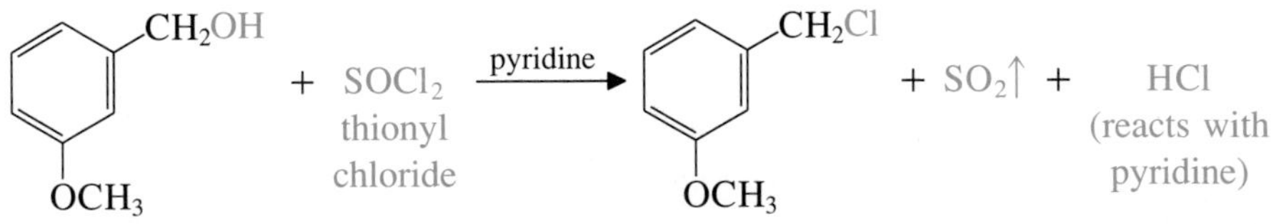

Thionyl chloride is a dense, fuming liquid (bp 75–76°). One advantage of using thionyl chloride for the preparation of alkyl chlorides is that the by-products of the reaction are gases; consequently, there are no separation problems in the purification of the product alkyl chlorides.

The preparation of alkyl chlorides from alcohols with thionyl chloride, like the use of sulfonate esters, involves the conversion of the alcohol—OH group into a good leaving group. When an alcohol reacts with thionyl chloride, a *chlorosulfite ester* intermediate is formed. (This reaction is analogous to that in Eq. 10.21.)

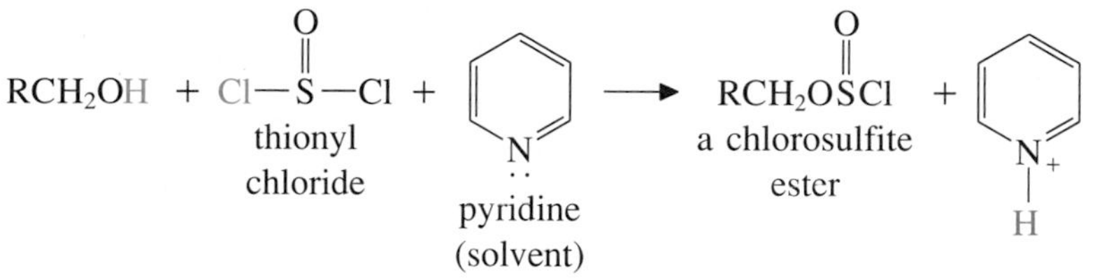

(10.28)

The chlorosulfite ester reacts readily with nucleophiles because the chlorosulfite group, —O—SO—Cl, is a very good leaving group. The chlorosulfite ester is usually not isolated, but reacts with the chloride ion formed in Eq. 10.28 to give the alkyl chloride. The displaced $^-$O—SO—Cl ion is unstable, and decomposes to SO_2 and Cl^-.

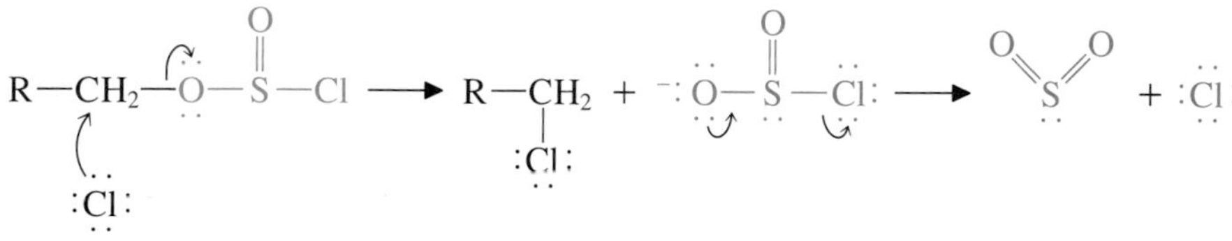

(10.29)

The thionyl chloride method is a considerably milder technique for synthesis of alkyl chlorides than the HCl method. The reaction of primary alcohols with HCl is very slow and requires heat and/or Lewis acids to proceed at a reasonable rate. With some secondary alcohols the thionyl chloride method seems to give somewhat less rearrangement than the HCl method, although rearrangements do occur with thionyl chloride.

If we are interested in obtaining alkyl bromides instead of alkyl chlorides, we can synthesize the bromides from alcohols with phosphorus tribromide, PBr_3.

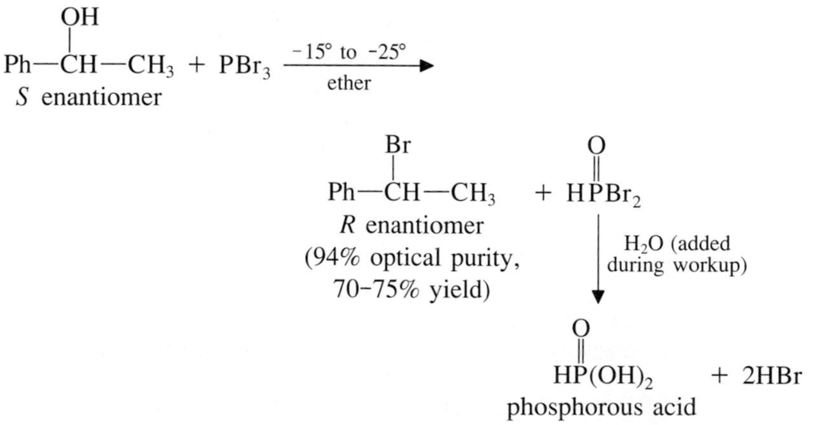

(10.30)

This reaction follows the same general mechanistic pattern observed for thionyl chloride: the —OH is converted into a good leaving group, in this case a bromophosphite ester, which is displaced by bromide.

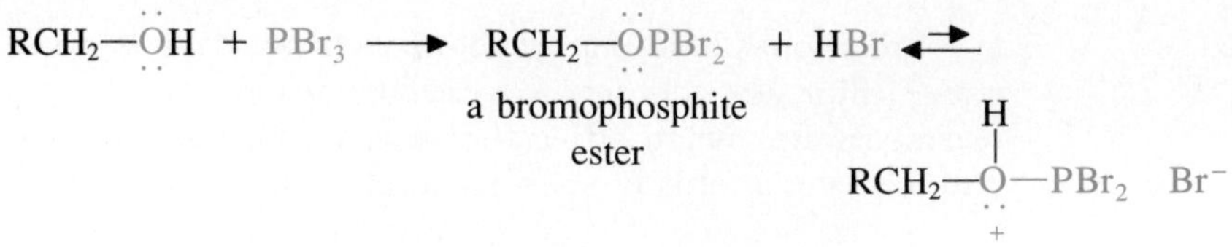

$$RCH_2{-}\overset{..}{\underset{..}{O}}H + PBr_3 \longrightarrow RCH_2{-}\overset{..}{\underset{..}{O}}PBr_2 + HBr \; \rightleftharpoons$$

a bromophosphite
ester

$$RCH_2{-}\overset{\displaystyle H}{\underset{..}{\overset{|}{O}}}{-}PBr_2 \quad Br^- \quad (10.31a)$$

$$RCH_2{-}\overset{\displaystyle H}{\underset{\overset{+}{..}}{\overset{|}{O}}}{-}PBr_2 \longrightarrow RCH_2Br + H\overset{..}{\underset{..}{O}}PBr_2 \qquad (10.31b)$$

$$:\!\overset{..}{\underset{..}{Br}}\!:^-$$

Because the mechanisms of both the thionyl chloride and phosphorus tribro-mide reactions involve S_N2 displacement steps, predominant stereochemical inversion at the site of substitution is observed in most cases (Eq. 10.30).

Problem **13** Give three methods that can be used to convert 1-butanol into 1-bromobutane.

10.5 CONVERSION OF ALCOHOLS INTO ALKYL HALIDES: SUMMARY

We have now studied a variety of reactions for the conversion of alcohols into alkyl halides. These are:

1. Reaction with hydrogen halides

2. Formation of sulfonate esters and displacement with halide ions

3. Reaction with $SOCl_2$ or PBr_3

How do we know which method to use in a given situation? The answer to this question depends on the structure of the alcohol.

Primary Alcohols: Alkyl halides are prepared from simple primary alcohols (example: 1-hexanol) by reaction of the alcohol with concentrated solutions of HI, HBr, or HCl. Heating is usually required, and, in the preparation of alkyl chlorides with HCl, a Lewis acid such as $ZnCl_2$ is usually added. Primary alcohols can also be converted into alkyl halides under milder conditions with $SOCl_2$ or PBr_3, or by conversion of the alcohol into a sulfonate ester such as a tosylate and reaction of the ester with halide ion in a polar aprotic solvent. In other words, just about any method can be used to prepare primary alkyl halides, because rearrangements are not a problem. Of course, neopentyl alcohol $(CH_3)_3C{-}CH_2OH$ and other highly branched primary alcohols do not react readily with any of the common reagents (why?). If forced to react with heat and/or Lewis acids, rearrangements are observed.

$$CH_3{-}\overset{\displaystyle CH_3}{\underset{\displaystyle CH_3}{\overset{|}{\underset{|}{C}}}}{-}CH_2OH \xrightarrow{\;SOCl_2\;} CH_3{-}\overset{\displaystyle CH_3}{\underset{\displaystyle CH_3}{\overset{|}{\underset{|}{C}}}}{-}CH_2Cl + CH_3{-}\overset{\displaystyle Cl}{\underset{\displaystyle CH_3}{\overset{|}{\underset{|}{C}}}}{-}CH_2CH_3 \quad (10.32)$$

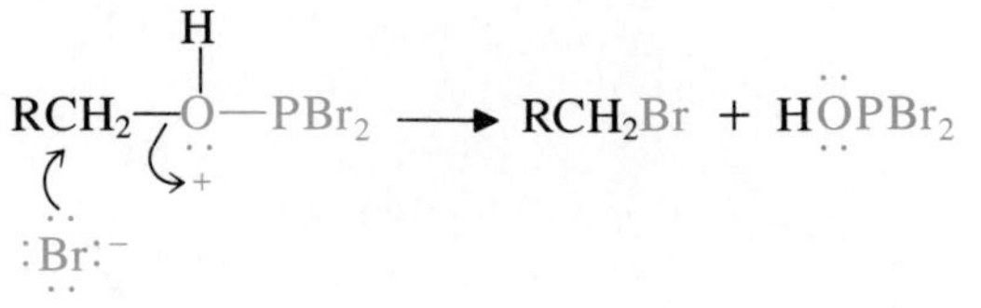

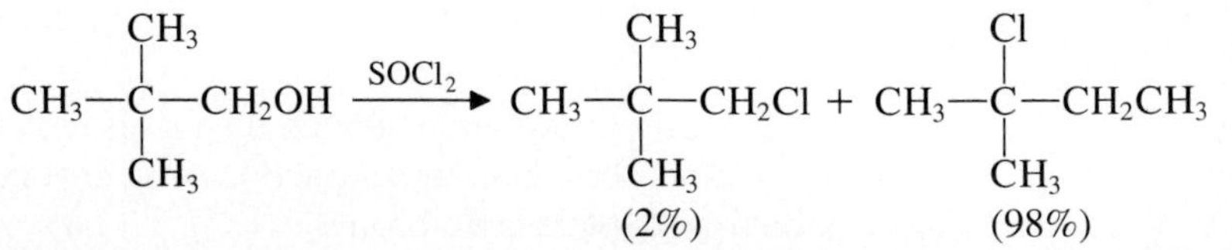

Tertiary Alcohols: Simple tertiary alcohols (example: *t*-butyl alcohol) react rapidly with HCl or HBr under mild conditions to give the corresponding alkyl halide. The sulfonate ester method does not work because tertiary systems do not undergo S_N2 reactions.

Secondary Alcohols: The synthesis of alkyl halides from simple secondary alcohols (example: 2-pentanol) with hydrogen halides in some cases is accompanied by rearrangements. Rearrangements are avoided by conversion of the alcohol into a sulfonate ester followed by S_N2 displacement (Eq. 10.23) if the alcohol is not too highly branched. In such a reaction we use a polar aprotic solvent and avoid protic solvents, which promote carbocation formation. Similarly, some unbranched secondary alcohols also react with $SOCl_2$ or PBr_3 to give the corresponding alkyl halide without rearrangement. For more highly branched secondary alcohols, or especially rearrangement-prone cases (example: 3-methyl-2-butanol), none of these methods is very good. Although some reagents do seem to work in these cases, we have not learned any general methods for the synthesis of alkyl halides from these alcohols.

Problem

14 Suggest conditions for carrying out each of the following conversions to yield a product as free of isomers as possible.

(a) $(CH_3)_2CHCH_2CH_2CH_2CH_2OH \longrightarrow (CH_3)_2CHCH_2CH_2CH_2CH_2Cl$

(b)

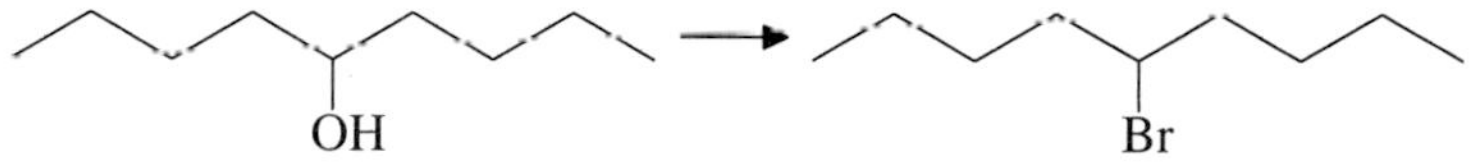

(c)

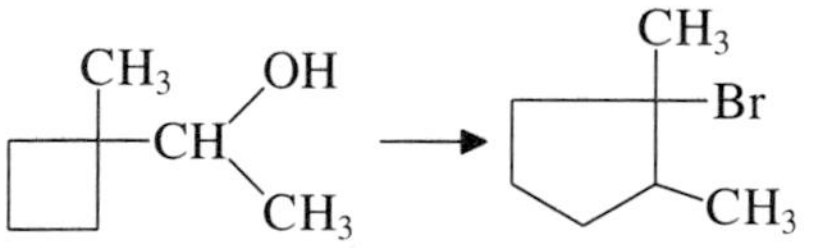

We have learned in this chapter that alcohols can undergo substitution and elimination reactions. However, the —OH group itself cannot act as a leaving group; it is far too basic. In order to break the carbon–oxygen bond, the —OH group must first be converted into a good leaving group. In this chapter we have seen several strategies by which this can be done:

1. Protonation: protonated alcohols are intermediates in both dehydration to alkenes and substitution to give alkyl halides.
2. Conversion into sulfonate esters or inorganic esters: these esters, to a useful approximation, react just like alkyl halides.
3. Reaction with thionyl chloride or phosphorus tribromide: these reagents effect the conversion of alcohols into chlorosulfite or bromophosphite esters, respectively. These intermediates are converted in the reaction mixture into alkyl halides.

10.6 OXIDATION OF ALCOHOLS. OXIDATION AND REDUCTION IN ORGANIC CHEMISTRY

In the previous sections we have been studying substitution and elimination reactions of alcohols. These reactions have much in common with the analogous reactions of alkyl halides. Now we turn to a different class of reactions: oxidations. Oxidation is a type of alcohol reaction that has no simple analogy in the alkyl halide chemistry we have studied. We first examine a common example of alcohol oxidation—oxidation with chromium(VI) reagents. Then we shall learn more generally how to recognize oxidation and reduction reactions.

A. Conversion of Alcohols into Carbonyl Compounds with Chromium(VI)

Primary and secondary alcohols react with reagents containing chromium in the +6 oxidation state to give *carbonyl compounds* (compounds containing the carbonyl group, $>$C=O). For example, secondary alcohols react to give ketones.

$$CH_3\underset{}{\overset{OH}{\overset{|}{C}}}HCH_2CH_2CH_2CH_2CH_2CH_3 \xrightarrow[\text{aqueous } H_2SO_4,\ 2\text{ hr}]{Na_2Cr_2O_7} CH_3\overset{O}{\overset{\|}{C}}CH_2CH_2CH_2CH_2CH_2CH_3 \quad (10.33)$$

(94% yield)

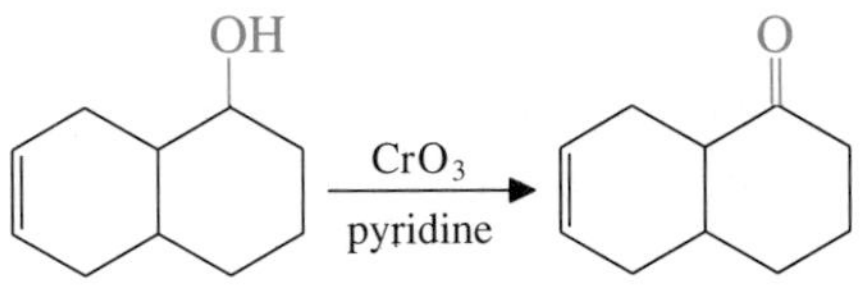

(89% yield) (10.34)

Several forms of Cr(VI) can be used to convert alcohols into carbonyl compounds. Three of these are chromate (CrO_4^{2-}), dichromate ($Cr_2O_7^{2-}$), and chromic anhydride or chromium trioxide (CrO_3). The first two reagents are customarily used under strongly acidic conditions; the last is often used in pyridine.

Primary alcohols react with these reagents to give aldehydes.

$$CH_3CH_2\underset{CH_3}{\underset{|}{C}}HCH_2OH \xrightarrow[\text{aqueous } H_2SO_4]{K_2Cr_2O_7} CH_3CH_2\underset{CH_3}{\underset{|}{C}}HCH{=}O \quad (10.35)$$

(52% yield; distilled from the reaction mixture)

One problem with this reaction of primary alcohols is that aldehydes react further with Cr(VI) reagents, particularly in aqueous solution, to give carboxylic acids.

$$CH_3CH_2\underset{CH_3}{\underset{|}{C}}H{-}\overset{O}{\overset{\|}{C}}{-}H \xrightarrow[H_2SO_4]{K_2Cr_2O_7} CH_3CH_2\underset{CH_3}{\underset{|}{C}}H{-}\overset{O}{\overset{\|}{C}}{-}OH \quad (10.36)$$

The aldehyde may be isolated if it is volatile and can be distilled from the reaction mixture as it is formed, as in Eq. 10.35. Alternatively, the reaction can be stopped at the aldehyde by excluding water from the reaction mixture. This is conveniently accomplished by the use of CrO_3 in pyridine, a reagent that contains no water.

$$CH_3(CH_2)_5CH_2OH \xrightarrow[\text{pyridine}]{CrO_3} CH_3(CH_2)_5CH{=}O \qquad (10.37)$$
$$(93\% \text{ yield})$$

Water promotes the transformation of aldehydes into carboxylic acids because, in water, aldehydes are in equilibrium with hydrates formed by addition of water across the C=O double bond.

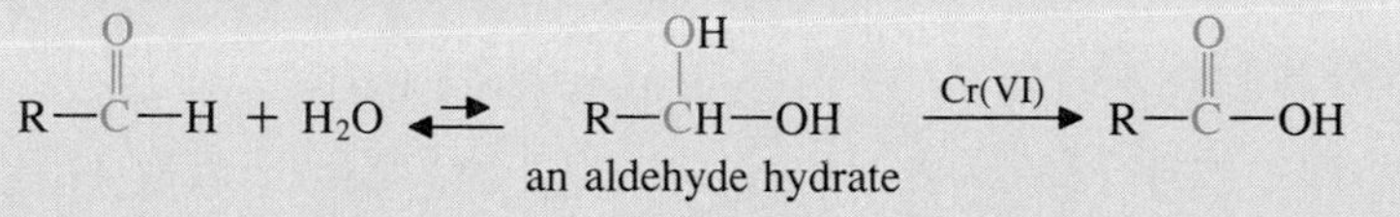

$$\underset{\text{an aldehyde hydrate}}{R{-}\overset{\overset{\displaystyle O}{\|}}{C}{-}H + H_2O \rightleftharpoons R{-}\overset{\overset{\displaystyle OH}{|}}{C}H{-}OH} \xrightarrow{Cr(VI)} R{-}\overset{\overset{\displaystyle O}{\|}}{C}{-}OH \qquad (10.38)$$

These compounds are really alcohols, and therefore oxidize just like secondary alcohols. Because of the absence of water in the anhydrous reagents, no 1,1-diol forms, and the reaction stops at the aldehyde.

Tertiary alcohols do not react under the usual conditions with Cr(VI).

The mechanism of alcohol oxidation by Cr(VI) involves several steps that have close analogies in other reactions that we have studied. Let us consider as a specific example the oxidation of isopropyl alcohol to the ketone acetone by chromic acid (H_2CrO_4).

The first steps of the reaction involve an acid-catalyzed displacement of water from chromic acid by the alcohol to form a *chromate ester*. (This ester is analogous to ester derivatives of other strong acids; see Sec. 10.3B.)

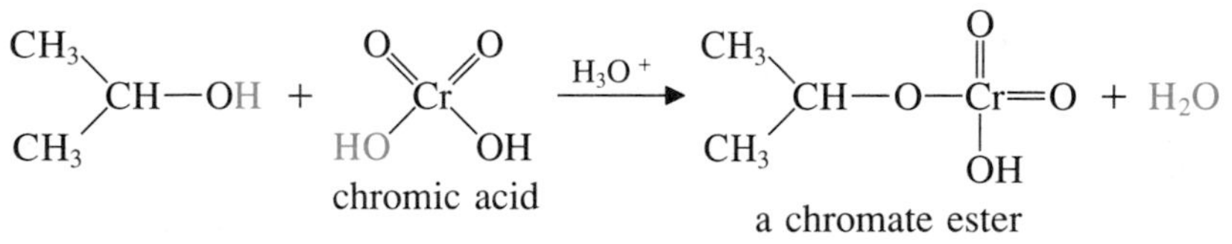

$$(10.39a)$$

After protonation of the chromate ester (Eq. 10.39b), it decomposes in a β-elimination reaction (Eq. 10.39c).

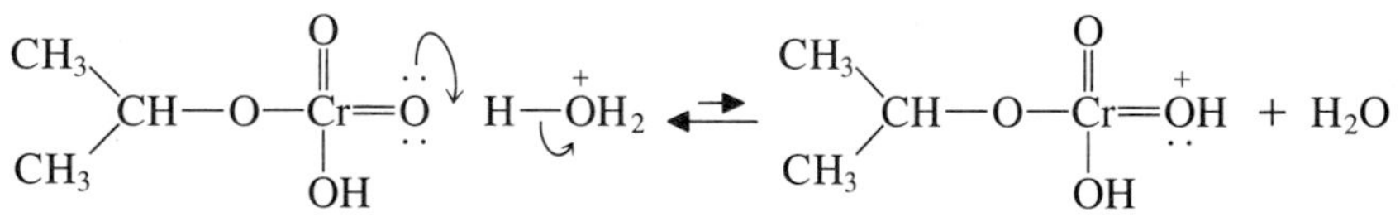

$$(10.39b)$$

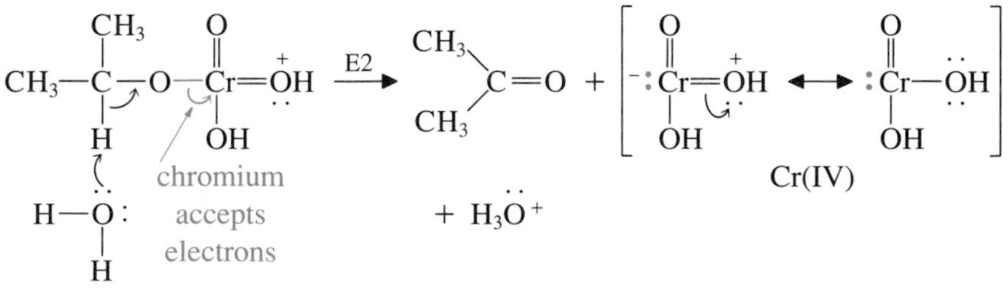

$$(10.39c)$$

This last step is like an E2 reaction, except that it does not involve a strong base. In this step chromium takes on an additional electron pair; the resulting H_2CrO_3 product contains chromium in a +4 oxidation state. Since the chromium began the reaction sequence in the +6 oxidation state, it has been *reduced.* Whenever something is reduced, something else is oxidized; that something is the alcohol, which is oxidized to a ketone. In subsequent reactions (the details of which we need not consider here), Cr(IV) and Cr(VI) react to give two equivalents of a Cr(V) species, which then oxidizes an additional molecule of alcohol. Because of these additional reactions, Cr(III) is the ultimate reduction product.

$$\text{Cr(IV)} + \text{Cr(VI)} \longrightarrow 2\text{Cr(V)} \qquad (10.40a)$$

$$\text{Cr(V)} + (\text{CH}_3)_2\text{CH—OH} \longrightarrow \text{Cr(III)} + (\text{CH}_3)_2\text{C{=}O} + 2\text{H}^+ \qquad (10.40b)$$

B. Oxidation and Reduction in Organic Chemistry

The conversion of alcohols into carbonyl compounds is an important reaction of primary and secondary alcohols, and is one of many examples in organic chemistry of **oxidation.** How do we know when an organic compound has been oxidized? In the last section, we recognized that conversion of an alcohol to a ketone is an oxidation because it is brought about by the reduction of Cr(VI). But there are other oxidations in which the oxidizing agent is less obvious. For example, the following reaction, which we studied in Sec. 5.1, is also an oxidation:

$$\text{R—CH{=}CH—R} + \text{Br}_2 \longrightarrow \text{R—}\underset{}{\overset{\text{Br}}{\overset{|}{\text{CH}}}}\text{—}\overset{\text{Br}}{\overset{|}{\text{CH}}}\text{—R} \qquad (10.41)$$

Our goal in this section is to be able to recognize an oxidation or reduction merely by examining the transformation of the organic compound itself. The procedure for doing this involves three steps:

Step 1. Assign an **oxidation level** to each carbon atom in reactant and product. (It is only necessary to assign an oxidation level to carbons that undergo some chemical change during the transformation; other carbons may be ignored.) The oxidation level of a particular carbon is assigned by considering the relative electronegativities of the groups bound to the carbon, as follows. (Table 1.1 is useful in this procedure.)

(a) For every bond to an element less electronegative than carbon (including hydrogen), and for every negative charge on the carbon, assign a −1.

(b) For every bond to another carbon atom, and for every unpaired electron on the carbon, assign a zero.

(c) For every bond to an element more electronegative than carbon, and for every positive charge on the carbon, assign a +1.

(d) Add the numbers assigned in (a), (b), and (c) to obtain the oxidation level for the carbon atom under consideration.

Let us apply this first step to the transformation of isopropyl alcohol to acetone.

$$\underset{\text{isopropyl alcohol}}{\text{CH}_3\text{—}\overset{\text{OH}}{\overset{|}{\text{CH}}}\text{—CH}_3} \xrightarrow{\text{Cr(VI)}} \underset{\text{acetone}}{\text{CH}_3\text{—}\overset{\text{O}}{\overset{\|}{\text{C}}}\text{—CH}_3} \qquad (10.42)$$

Since the carbon atoms of the two methyl groups do not change, we do not need to assign oxidation levels to these carbons. Notice in the treatment of acetone that the

C=O double bond is counted as two bonds; +1 for each bond gives a total of +2 for the double bond.

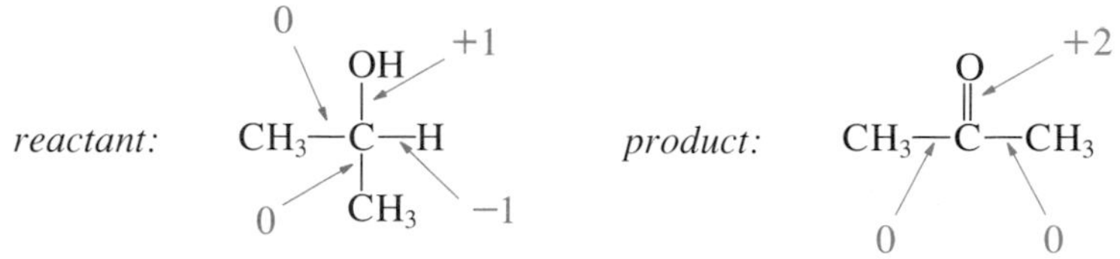

Sum: 0 + 0 + (+1) + (−1) = 0   Sum: 0 + 0 + (+2) = +2

Step 2. The oxidation number N_{ox} for each compound is computed by adding the oxidation levels of all carbons. In the structures above, only one carbon has changed its oxidation level, so the N_{ox} values of the reactant and product are simply equal to the respective oxidation levels of this carbon. Therefore, the oxidation level of the reactant is 0 and that of the product is +2. In other reactions involving more than one carbon atom, N_{ox} is computed by summing the oxidation levels of all carbon atoms that undergo a chemical change.

Step 3. Compute the difference.

$$N_{ox}(\text{product}) - N_{ox}(\text{reactant})$$

If this difference is positive, the transformation is an *oxidation*. If this difference is negative, the transformation is a *reduction*. If the difference is zero, neither an oxidation nor a reduction has taken place. For the reaction of Eq. 10.42, this difference is +2 − 0 = +2. This transformation is thus an oxidation.

Problem

15 Verify that the acid-catalyzed addition of water to isobutylene is neither an oxidation nor a reduction.

Although the oxidation-number formalism is very useful, we should not lose sight of the following two general characteristics of organic oxidations and reductions. These two points can enable us to spot an oxidation or reduction at a glance.

1. In most oxidations of organic compounds, either hydrogen in a C—H bond or carbon in a C—C bond is replaced by a more electronegative element, such as halogen or oxygen. The converse is true for reductions.
2. The oxidation state of a molecule is determined from the oxidation states of its individual carbon atoms.

The oxidation number concept can be simply related to a definition of oxidation that is often used in inorganic chemistry. According to this definition, oxidation is the *loss of electrons* and reduction is *the gain of electrons*. To see how this definition applies to organic compounds, let us consider as an example the oxidation of ethanol to acetic acid:

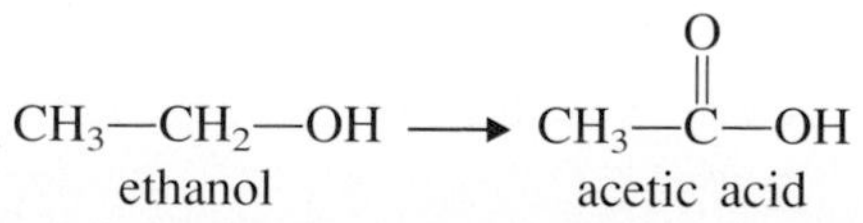

(10.43)

We can write this oxidation as a balanced half-reaction using H_2O to balance missing oxygens, protons to balance missing hydrogens, and "dummy electrons" to balance charges.

$$CH_3{-}CH_2{-}OH + H_2O \longrightarrow CH_3{-}\overset{\overset{\displaystyle O}{\|}}{C}{-}OH + 4H^+ + 4e^- \qquad (10.44)$$

According to this half-reaction, four electrons are lost from the ethanol molecule when acetic acid is formed. (Since this is only a half-reaction, a corresponding number of electrons must be gained by the species that brings about the oxidation.) It can be said that the oxidation of ethanol to acetic acid is a *four-electron oxidation*. This type of terminology, which is frequently used in biochemistry, comes from the half-reaction formalism.

If we compute the oxidation numbers of ethanol and acetic acid, we can see that the change in oxidation number for Eq. 10.43 is +4 (verify this statement). This example illustrates the following point: *the change in oxidation number is equal to the number of electrons lost*. If the change in oxidation number is negative, the reaction is a reduction, and the number corresponds to electrons gained.

Problem

16 Classify each of the following transformations, some of which may be unfamiliar, as an oxidation, reduction, or neither. For those that are oxidations or reductions, tell how many electrons are gained or lost.

(a) $CH_4 \xrightarrow[\text{light}]{Br_2} CH_3Br$

(b) $Ph{-}CH_3 \xrightarrow[H_2O]{Cr^{6+}} Ph{-}\overset{\overset{\displaystyle O}{\|}}{C}{-}OH$

(c) $CH_3CH_2CH_2I \xrightarrow{LiAlH_4} CH_3CH_2CH_3 + I^-$

(d)

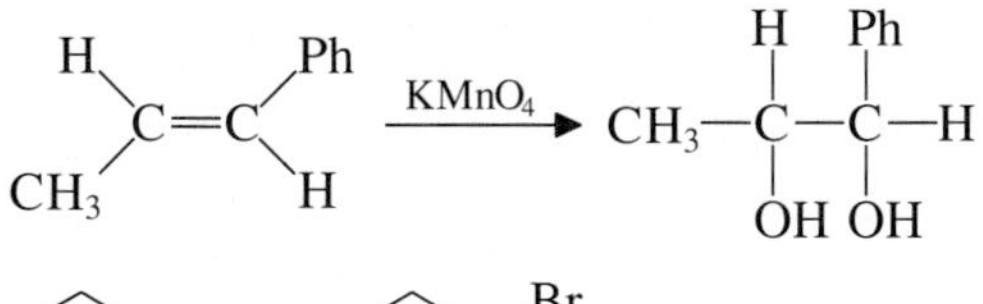

(e)

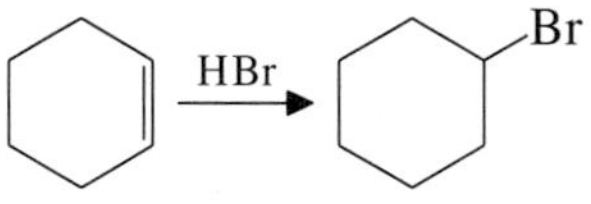

(f)

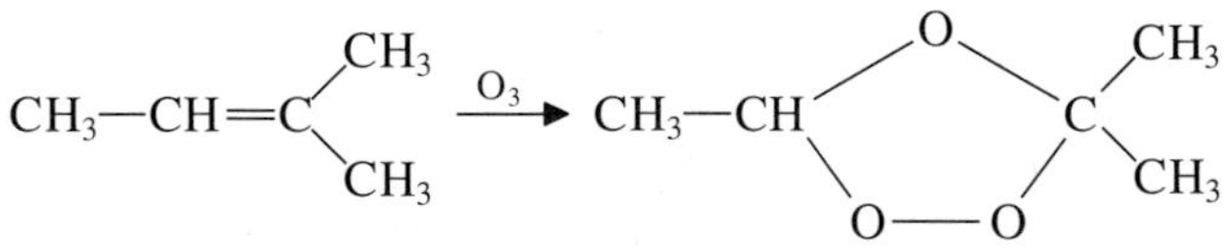

(g)

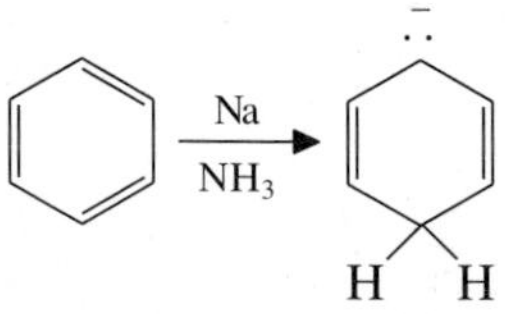

Oxidations and reductions, like acid–base reactions, always occur in pairs. Therefore, *whenever something is oxidized, something else is reduced.* When an organic compound is oxidized, the reagent that brings about the transformation is called an **oxidizing agent.** Likewise, when an organic compound is reduced, the reagent that effects the transformation is called a **reducing agent.** For example, suppose that chromate ion (CrO_4^{2-}) is used to bring about the oxidation of ethanol to acetic acid in Eq. 10.43; in this reaction, chromate ion is reduced to Cr^{3+}.

$$8H^+ + 3e^- + CrO_4^{2-} \longrightarrow Cr^{3+} + 4H_2O \qquad (10.45)$$

Three electrons are gained in the reduction of chromate to Cr^{3+}. Since four electrons are lost in the oxidation (Eq. 10.44), stoichiometry requires that for every *three* ethanol molecules oxidized to acetic acid (twelve electrons lost), *four* CrO_4^{2-} are reduced (twelve electrons gained).

By considering the change in oxidation number for a transformation, we can tell whether an oxidizing or reducing agent is required to bring about the reaction. For example, the following transformation is neither an oxidation nor a reduction (verify this statement):

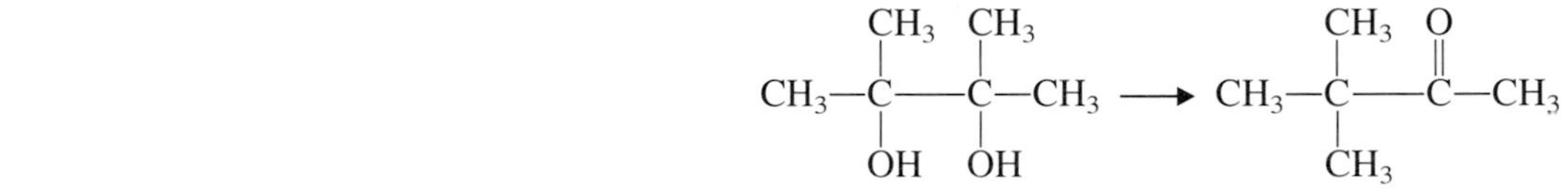
CH3 CH3
CH3—C—C—CH3
OH OH
CH3 O
CH3—C—C—CH3
CH3

(10.46)

Although one carbon is oxidized, another is reduced. Even though we might know nothing else about the reaction, it is clear that an oxidizing or reducing agent alone would not effect this transformation. (In fact, the reaction is the pinacol rearrangement (Sec. 10.1B), which is brought about by mineral acid.)

The oxidation-number concept can be used to organize organic compounds into functional groups with the same oxidation level, as shown in Table 10.1. Compounds within a given box are generally interconverted by reagents that are neither oxidizing nor reducing agents. For example, we know that alcohols can be converted into alkyl halides with HBr, which is neither an oxidizing nor a reducing agent. On the other hand, conversion of an alcohol into a carboxylic acid involves a change in oxidation level, and indeed this transformation requires an oxidizing agent. We can also see from Table 10.1 that there are a greater number of possible oxidation states for carbons with larger numbers of hydrogens. Thus, a tertiary alcohol cannot be oxidized at the α-carbon (without breaking carbon–carbon bonds) because this carbon bears no hydrogens. Methane, on the other hand, can be oxidized to CO_2. (Of course, any hydrocarbon can be oxidized to CO_2 if carbon–carbon bonds are broken; Sec. 3.6.)

Problems

17 Indicate which of the following balanced reactions are oxidation–reduction reactions and which are not. For those involving oxidation–reduction, indicate which compound(s) are oxidized and which are reduced. (*Hint*: Consider the organic compounds in each reaction first.)

(a) $CH_2{=}CH_2 + H_2 \xrightarrow{Pd/C} CH_3{-}CH_3$

(b) $CH_3CH_2Br + Li^+\ {}^-AlH_4 \longrightarrow CH_3CH_3 + Li^+Br^- + AlH_3$

(Problem continues on page 391.)

TABLE 10.1 Comparison of Oxidation States of Various Functional Groups
All molecules in the same box are at the same oxidation number.
X = an electronegative group such as halogen, etc.

METHANE ———— **increasing oxidation number** ————→

CH_4	$CH_3{-}OH$ $CH_3{-}X$	$H_2C{=}O$ H_2CX_2	$H{-}C({=}O){-}OH$ $H{-}CX_3$	$O{=}C{=}O$ CX_4

PRIMARY CARBON ———— **increasing oxidation number** ————→

$RCH_2{-}CH_3$	$RCH_2{-}CH_2{-}OH$ $RCH_2{-}CH_2{-}X$ $RCH{=}CH_2$	$R{-}CH{=}O$ $R{-}CHX_2$ $R{-}C{\equiv}C{-}H$	$R{-}C({=}O){-}OH$ $R{-}CX_3$

SECONDARY CARBON ———— **increasing oxidation number** ————→

$RCH_2{-}CH_2{-}R$	$RCH_2{-}CH(R){-}OH$ $RCH_2{-}CH(R){-}X$ $R{-}CH{=}CH{-}R$	$RCH_2{-}C(R){=}O$ $RCH_2{-}CX_2{-}R$ $R{-}C{\equiv}C{-}R$

TERTIARY CARBON ———— **increasing oxidation number** ————→

$R{-}CH(CH_2R){-}R$	$R{-}C(CH_2R)(R){-}OH$ $R{-}C(CH_2R)(R){-}X$ $R{-}C(R){=}CH{-}R$

Problems (Cont.)

(c) $Ph-\overset{\overset{\displaystyle CH_3}{|}}{CH}-CH_3 + O_2 \longrightarrow Ph-\overset{\overset{\displaystyle CH_3}{|}}{\underset{\underset{\displaystyle O-O-H}{|}}{C}}-CH_3$

(d) $CH_3-CH{=}CH_2 + Br_2 \longrightarrow CH_3-\overset{}{\underset{\underset{\displaystyle Br}{|}}{CH}}-\overset{}{\underset{\underset{\displaystyle Br}{|}}{CH_2}}$

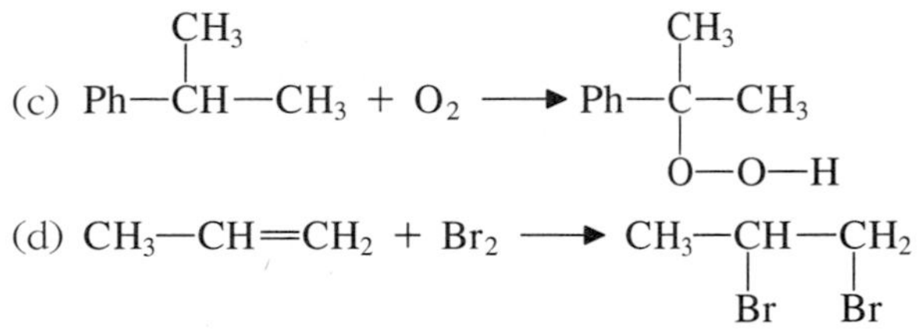

18 How many moles of permanganate are required to oxidize one mole of toluene to benzoic acid?

$$Ph-CH_3 + MnO_4^- \longrightarrow Ph-CO_2H + MnO_2$$
toluene permanganate benzoic acid manganese dioxide

(*Hint:* Balance each half-reaction; then reconcile dummy electrons.)

C. Other Methods for Oxidizing Alcohols

Besides Cr(VI) reagents, a number of other reagents can be used to oxidize alcohols. Potassium permanganate, $KMnO_4$, oxidizes primary alcohols to carboxylic acids. It is not possible to stop this oxidation at the aldehyde stage.

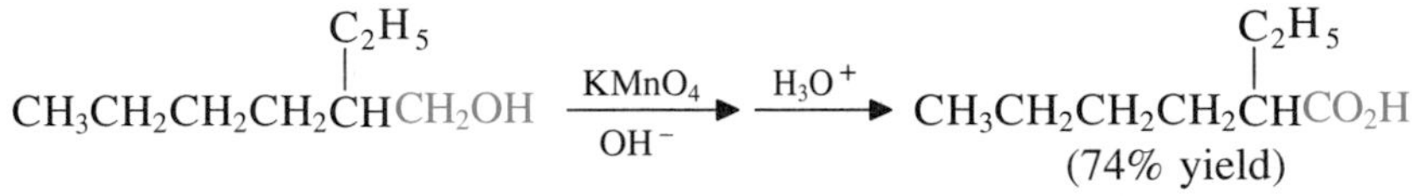

$$CH_3CH_2CH_2CH_2\overset{\overset{\displaystyle C_2H_5}{|}}{CH}CH_2OH \xrightarrow[OH^-]{KMnO_4} \xrightarrow{H_3O^+} CH_3CH_2CH_2CH_2\overset{\overset{\displaystyle C_2H_5}{|}}{CH}CO_2H \quad (10.47)$$
$$(74\% \text{ yield})$$

Manganese in $KMnO_4$ is in the Mn(VII) oxidation state; in this reaction, it is reduced to MnO_2, a common form of Mn(IV). Because $KMnO_4$ reacts with alkene double bonds (Sec. 5.4), Cr(VI) is preferred for the oxidation of alcohols that contain double or triple bonds (see Eq. 10.34).

Potassium permanganate is not used for the oxidation of secondary alcohols to ketones because many ketones react further with the alkaline permanganate reagent.

Nitric acid is sometimes useful as an oxidizing agent for saturated primary alcohols.

$$BrCH_2CH_2CH_2CH_2CH_2CH_2OH \xrightarrow{HNO_3} BrCH_2CH_2CH_2CH_2CH_2CO_2H \quad (10.48)$$
$$(91\% \text{ yield})$$

Under mild conditions, nitric acid selectively oxidizes both primary alcohols and aldehydes, but secondary alcohols remain unaffected. However, under more vigorous conditions (heat, more concentrated acid, and/or special catalysts), nitric acid will oxidize a secondary alcohol and cleave an adjacent carbon–carbon bond. For example, in the oxidation of cyclohexanol, a carbon–carbon bond at the CH—OH group is broken; a carboxylic acid group is formed at the carbons on each side of the break.

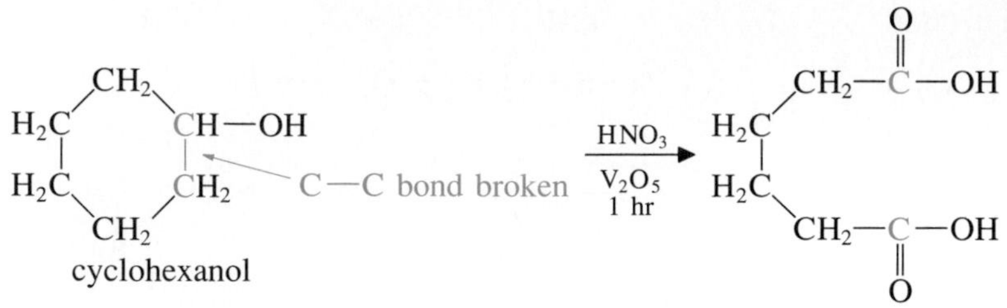

cyclohexanol → adipic acid (70% yield) (10.49)

We should not be overly concerned with exactly what constitutes "mild" or "vigorous" conditions for an oxidation; the best conditions for any oxidation—indeed, any reaction—must be determined by experiment. The important point is to understand in a general way that the outcome of an oxidation reaction can depend on the conditions under which it is carried out.

Problems

19 From what alcohol, and by what method, would each of the following compounds best be prepared by an oxidation?

(a) $(CH_3)_2CHCH_2CH_2CH_2CO_2H$

(b) $CH_3CH_2\overset{O}{\overset{\|}{C}}CH_2CH_3$

(c) 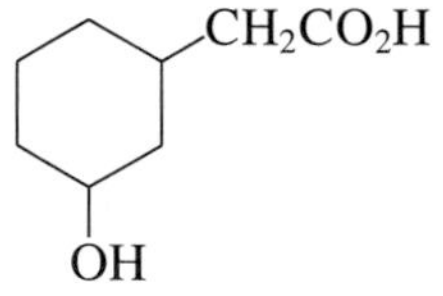

(d) 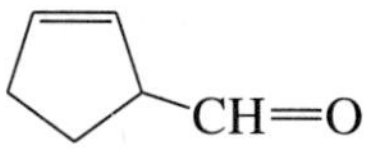

(e) $CH_3-C(OH)(CH_3)-CH_2CH_2CH_2CH{=}O$

---

20 Give the structures of the *two* compounds that are formed when *trans*-3-methyl-1-cyclohexanol is oxidized by HNO_3 under vigorous conditions.

D. Oxidative Cleavage of Glycols

The carbon–carbon bond between the —OH groups of a vicinal glycol can be cleaved with periodic acid to give two carbonyl compounds.

$$\underset{\text{periodic acid}}{H_5IO_6} + Ph-\overset{OH}{CH}-\overset{OH}{\underset{CH_3}{C}}-CH_3 \xrightarrow{\text{dil HOAc}}$$

$$\underset{(77\text{–}83\%\ \text{yield})}{Ph-\overset{O}{\overset{\|}{C}}-H} + CH_3-\overset{O}{\overset{\|}{C}}-CH_3 + 2H_2O + HIO_3 \cdot H_2O \qquad (10.50)$$

Periodic acid is the iodine analog of perchloric acid.

$HClO_4$	HIO_4
perchloric acid	periodic acid

Periodic acid is commercially available as the dihydrate, $HIO_4 \cdot 2H_2O$, often abbreviated, as in the above example, as H_5IO_6 (sometimes called *para-periodic acid*). Its sodium salt, $NaIO_4$ (sodium metaperiodate) is sometimes also used. Periodic acid is a fairly strong acid ($pK_a = -1.64$). Since periodate can be determined by titration, the periodate cleavage reaction has been used as a test for glycols as well as for synthesis. We shall use HIO_4 or H_5IO_6 interchangeably as formulas for periodic acid.

The cleavage of glycols with periodic acid takes place through a cyclic periodate ester intermediate (Sec. 10.3B) that forms when the glycol displaces water from H_5IO_6. The cyclic ester then breaks down in a pericyclic reaction.

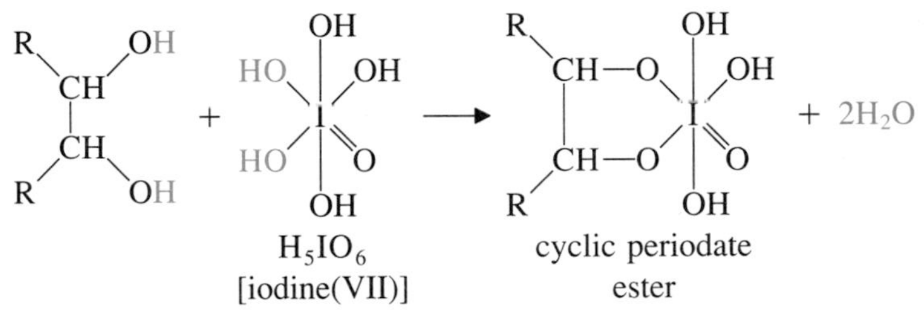

(10.51a)

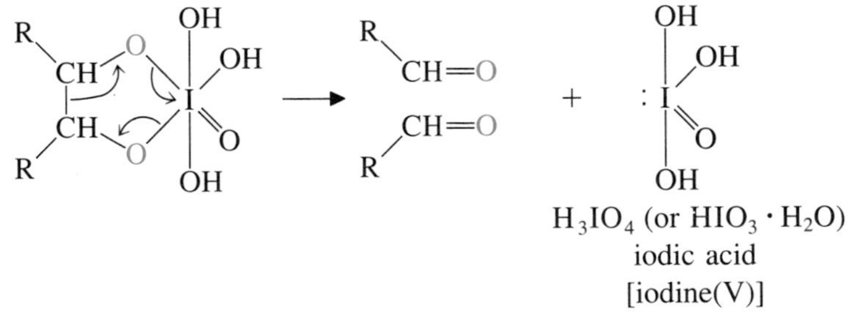

(10.51b)

Any glycol that cannot form a cyclic ester intermediate is not cleaved by periodic acid. For example, the following compound cannot form such an ester (can you see why?), and is therefore inert to periodate:

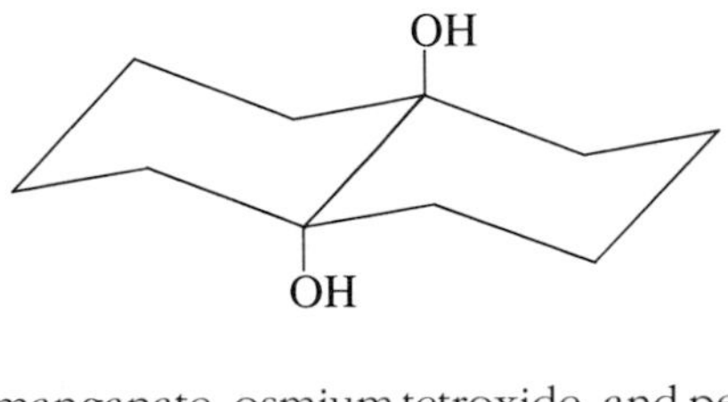

(10.52)

Do not confuse permanganate, osmium tetroxide, and periodate oxidations, all of which occur through cyclic ester intermediates (see Sec. 5.4). Periodate oxidizes *gly-*

cols, but the other reagents oxidize *alkenes* to give glycols. In all of these reactions, oxidation occurs because an atom in a highly positive oxidation state can accept an additional pair of electrons. In the periodate oxidation, the reduction of the iodine occurs during the *breakdown* of the cyclic ester; in the permanganate and osmium tetroxide oxidations, the metals are reduced during the *formation* of the cyclic ester.

Problems

21 Give the product(s) expected when each of the following compounds is treated with periodic acid:

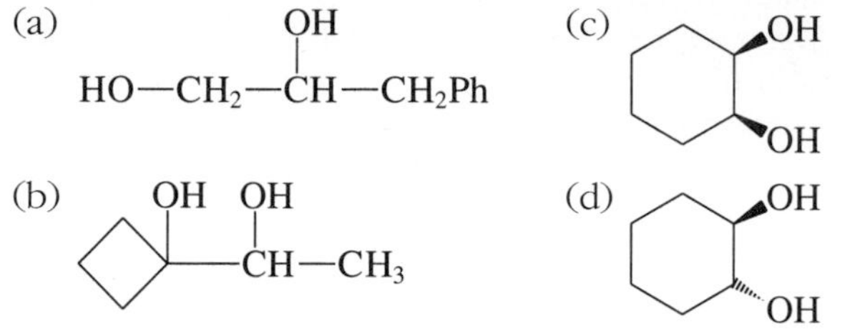

22 In each pair below, *one* of the glycols is virtually inert to periodate oxidation. Which glycol is inert? Explain why.

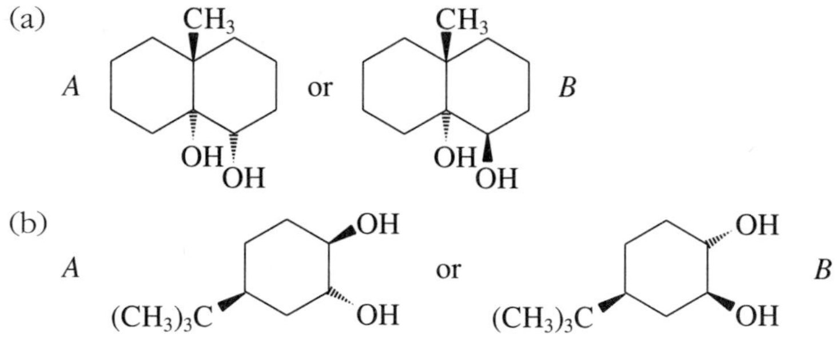

10.7 BIOLOGICAL OXIDATION OF ETHANOL

Oxidation and reduction reactions are extremely important in living systems. A typical biological oxidation is the conversion of ethanol into acetaldehyde, the principal reaction by which ethanol is removed from the bloodstream.

$$\underset{\text{ethanol}}{CH_3CH_2OH} \xrightarrow{\text{biological oxidation}} \underset{\text{acetaldehyde}}{CH_3CH{=}O} \qquad (10.53)$$

The reaction is carried out in the liver and is catalyzed by an enzyme called *alcohol dehydrogenase.* (Recall from Sec. 4.8 that enzymes are biological catalysts.) The oxidizing agent is not the enzyme, but a complex-looking molecule called *nicotinamide adenine dinucleotide,* abbreviated NAD^+ by biochemists; the structure of NAD^+ and a convenient abbreviated structure for it are shown in Fig. 10.1. When ethanol is oxidized, the NAD^+ is reduced to a product called NADH. The hydrogen removed from carbon-1 of the ethanol ends up in the NADH; the —OH hydrogen is lost as a proton.

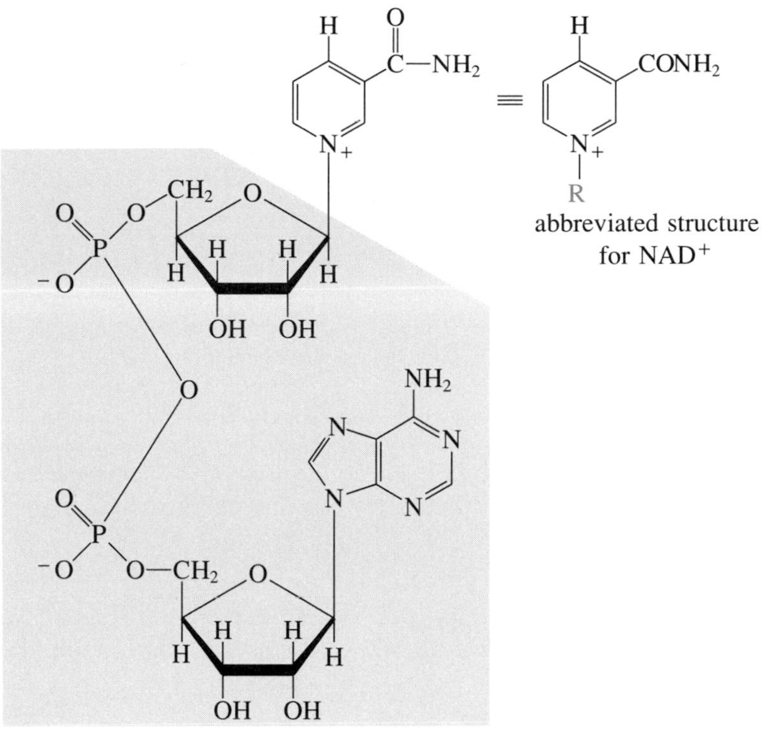

NAD$^+$

Figure 10.1 Structure of **NAD$^+$.** The portion of the structure in the box is abbreviated R.

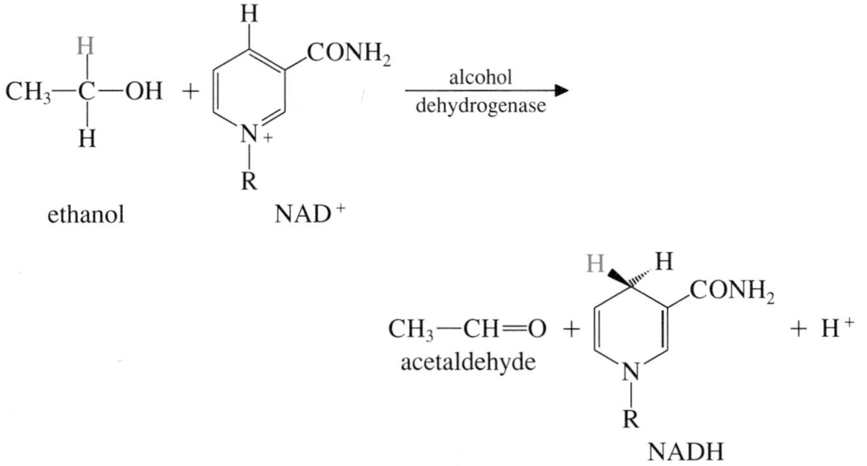

(10.54)

The compound NAD$^+$ is an example of a **coenzyme,** and is one of nature's important oxidizing agents. (We might call it "nature's substitute for Cr(VI).") An ethanol molecule and an NAD$^+$ molecule are brought together when they bind noncovalently to alcohol dehydrogenase, the enzyme catalyst. It is on the enzyme that the alcohol is oxidized to acetaldehyde and NAD$^+$ is reduced to NADH. The NADH is oxidized back to NAD$^+$ in other biological reactions.

The coenzymes NAD^+ and NADH are derived from the vitamin *niacin,* a deficiency of which is associated with the disease pellagra (black tongue). Many biochemical oxidation–reduction reactions employ the $NAD^+ \rightleftarrows$ NADH interconversion.

The human body uses the ethanol-to-acetaldehyde reaction of Eq. 10.54 to remove ethanol, but yeast cells use the reaction in reverse as the last step in the production of ethanol. Thus, yeast added to dough produces ethanol by the reduction of acetaldehyde, which, in turn, is produced in other reactions from sugars in the dough. Ethanol vapors, wafted away from the bread by CO_2 produced in other reactions, give rising bread its pleasant odor. Special strains of yeast ferment the sugars in corn syrup, grape juice, or malt and barley to whiskey, wine, or beer. In order for the fermentation reaction to take place, the reaction must be run in the absence of oxygen. Otherwise, acetic acid, CH_3CO_2H (vinegar, or "spoiled wine"), is formed instead by other reactions. In winemaking, air is excluded by trapping the CO_2 formed during fermentation as a blanket in the fermentation vessel. Because the production of alcohol by yeast occurs in the absence of air, it is called *anaerobic fermentation*. This is one of the oldest chemical reactions known to civilization.

How does NAD^+ work as an oxidizing agent? The resonance structure of NAD^+ shows that it has the character of a carbocation.

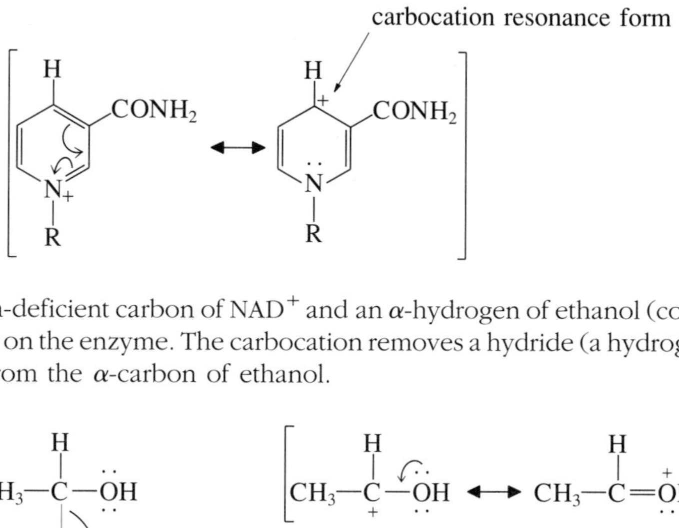

(10.55)

The electron-deficient carbon of NAD^+ and an α-hydrogen of ethanol (color) are held in proximity on the enzyme. The carbocation removes a hydride (a hydrogen with two electrons) from the α-carbon of ethanol.

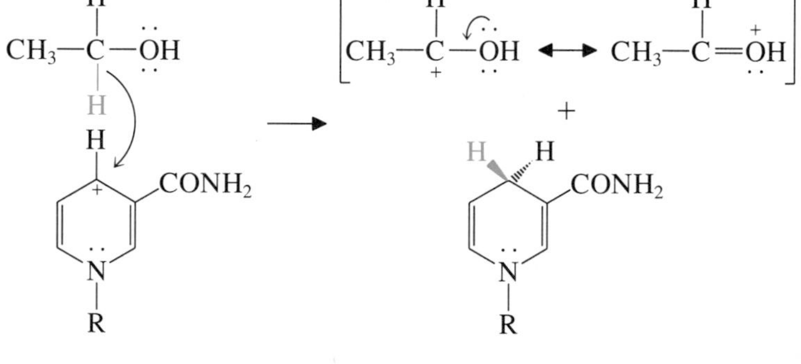

(10.56a)

Although this reaction may initially look strange, it is really just like a carbocation rearrangement involving the migration of a hydride, except that in this case the hydride moves to a different molecule. As a result, NADH and a new carbocation are formed. By loss of the proton bound to oxygen, this ion forms acetaldehyde.

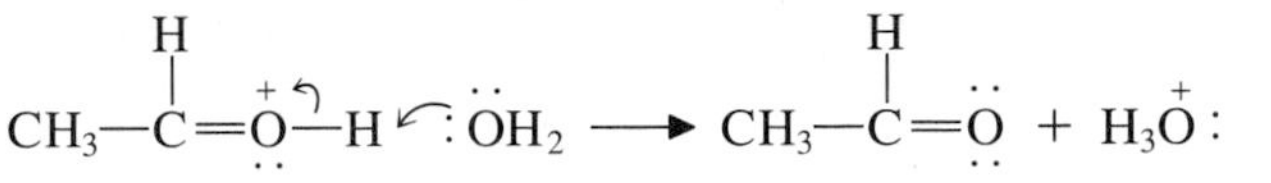

$$CH_3-\overset{\overset{\displaystyle H}{|}}{C}=\overset{+}{\underset{..}{O}}-H \; \overset{\frown}{:\underset{..}{O}}H_2 \longrightarrow CH_3-\overset{\overset{\displaystyle H}{|}}{C}=\underset{..}{\overset{..}{O}} + H_3\overset{+}{O}: \qquad (10.56b)$$

The acetaldehyde and NADH then leave the enzyme, which is then ready for another round of catalysis.

Despite the fact that the NAD^+ molecule looks large and complicated, the chemical changes that occur when it serves as an oxidizing agent take place in a relatively small part of the molecule. There are a number of other coenzymes that also have complex structures but undergo simple reactions. The part of the molecule abbreviated by "R" in Fig. 10.1 provides the groups that have particular affinity for the enzyme catalyst, but this part of the molecule remains unchanged in the oxidation reaction. The chemical changes that occur in NAD^+-promoted oxidations find analogy in laboratory reactions with which we are already familiar. This example illustrates that an appreciation of the fundamental types of organic reactions and their mechanisms is useful in the study of biochemical processes.

Problem

23 Suggest a reasonable mechanism for the following oxidation of 2-heptanol, which proceeds in 82% yield:

$$\underset{\begin{array}{c}\text{a stable}\\\text{carbocation}\end{array}}{CH_3\overset{\overset{\displaystyle OH}{|}}{C}HCH_2CH_2CH_2CH_2CH_3 + Ph_3C^+\ BF_4^-} \longrightarrow$$

$$CH_3\overset{\overset{\displaystyle O}{\|}}{C}CH_2CH_2CH_2CH_2CH_3 + HBF_4 + Ph_3CH \quad (10.57)$$

10.8 CHEMICAL AND STEREOCHEMICAL GROUP EQUIVALENCE

A. Stereochemistry of the Alcohol Dehydrogenase Reaction

The alcohol dehydrogenase-catalyzed oxidation of ethanol provides an interesting example of a subtle, but important, stereochemical phenomenon. Suppose we replace, in turn, each of the α-hydrogens of ethanol with the isotope deuterium (2H, or D). Replacing one hydrogen, called the pro-(R)-hydrogen, with deuterium, gives (R)-1-deuterioethanol; replacing the other, called the pro-(S)-hydrogen, gives (S)-1-deuterioethanol. Although ethanol itself is *not* chiral, the deuterium-substituted analogs *are* chiral; they are a pair of enantiomers.

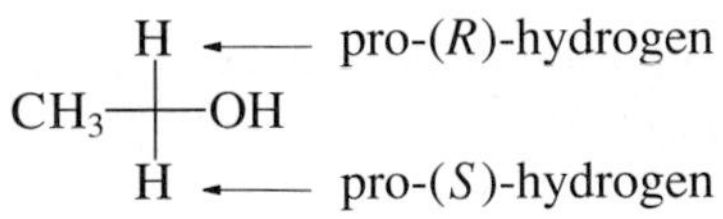

H ⟵ pro-(R)-hydrogen
CH$_3$ ┼ OH ethanol (Fischer projection)
H ⟵ pro-(S)-hydrogen

D
CH$_3$ ┼ OH
H
(R)-($+$)-1-deuterioethanol

H
CH$_3$ ┼ OH
D
(S)-($-$)-1-deuterioethanol

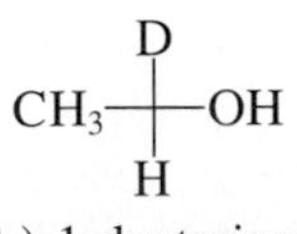

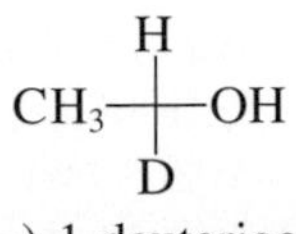

If the alcohol dehydrogenase reaction is carried out with (*R*)-1-deuterioethanol, *only* the deuterium is transferred from the alcohol to the NAD^+.

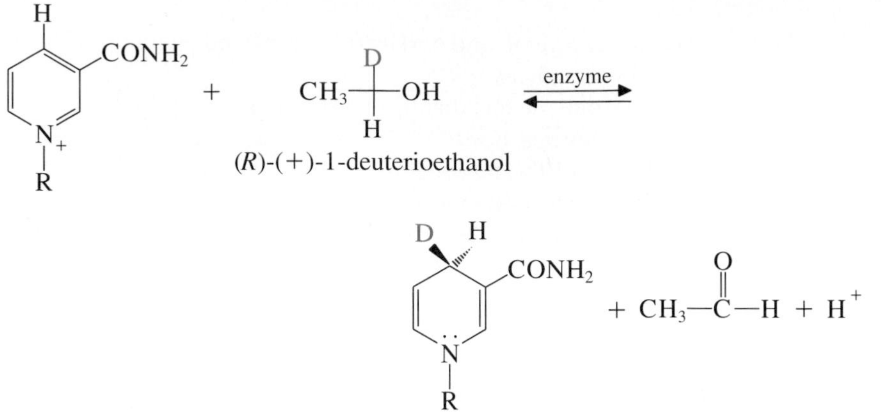

(10.58)

However, if the alcohol dehydrogenase reaction is carried out on (*S*)-1-deuterioethanol, *only* the hydrogen is transferred.

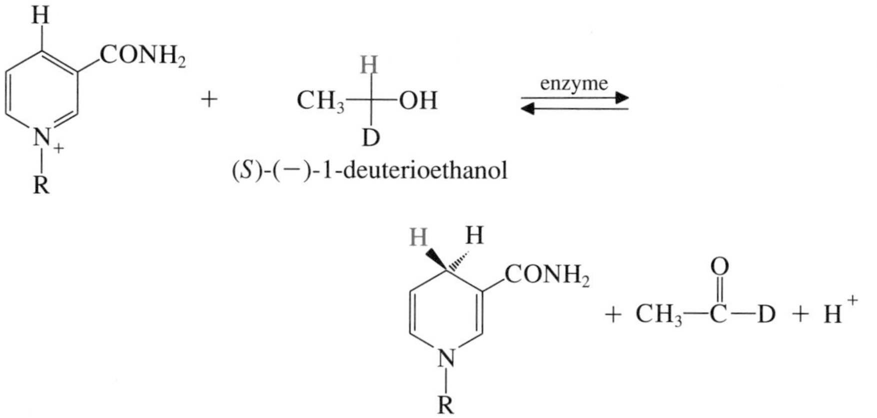

(10.59)

These two experiments show that *the enzyme distinguishes between the two α-hydrogens of ethanol.* These results cannot be attributed to a primary deuterium isotope effect (Sec. 9.4B), because an isotope effect would cause the enzyme to transfer the hydrogen in preference to the deuterium in both cases. Although we have to use an isotope to see the preference for transfer of one hydrogen and not the other, this experiment requires that even in the absence of the isotope the enzyme prefers to transfer the pro-(*R*)-hydrogen of ethanol.

Another interesting stereochemical point about the alcohol dehydrogenase reaction is also shown in Eq. 10.58: the hydrogen (or deuterium) that is removed from the isotopically substituted ethanol molecule is transferred specifically to one particular face, or side, of the NAD^+ molecule. That is, the deuterium in the product NADH (color) occupies the position above the plane of the page. This result and the principle of microscopic reversibility (Sec. 10.1A) require that if we were to start with the products acetaldehyde and isotopically substituted NADH shown on the right of Eq. 10.58 and run the reaction in reverse, only the deuterium in this position should be transferred to the acetaldehyde, and (*R*)-1-deuterioethanol should be formed. Indeed,

Eq. 10.58 *can* be run in reverse, and the experimental result is as predicted. No matter how many times the reaction runs back and forth, the H and the D on both the ethanol and the NADH molecules are never scrambled; they maintain their identity.

How can the enzyme distinguish between the two α-hydrogens of ethanol, and between the two ring hydrogens of NADH? This distinction is possible because the two hydrogens in each case are *stereochemically different*. The nature of this stereochemical difference is the subject of the next section.

Problem

24 In each of the following cases, imagine that the two reactants shown are allowed to react in the presence of alcohol dehydrogenase. Tell whether the ethanol formed is chiral. If the ethanol is chiral, draw a Fischer projection of the enantiomer that is formed.

(a)

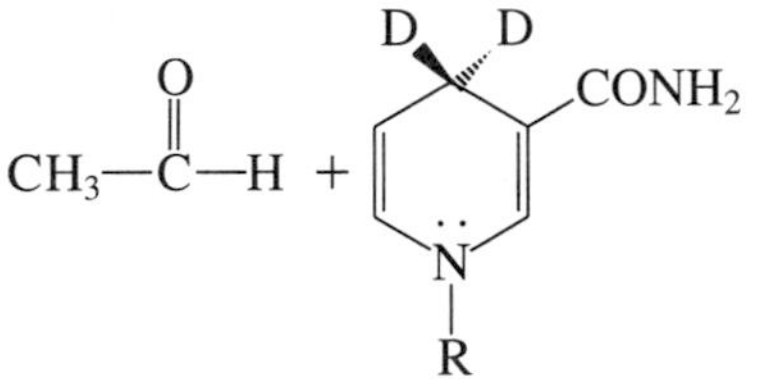

(c)

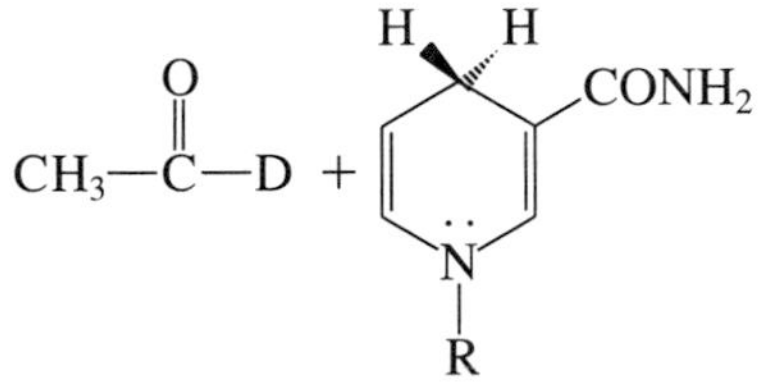

(b)

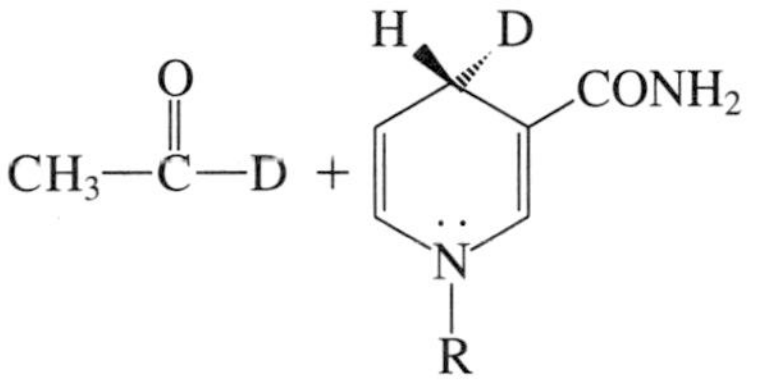

(d)

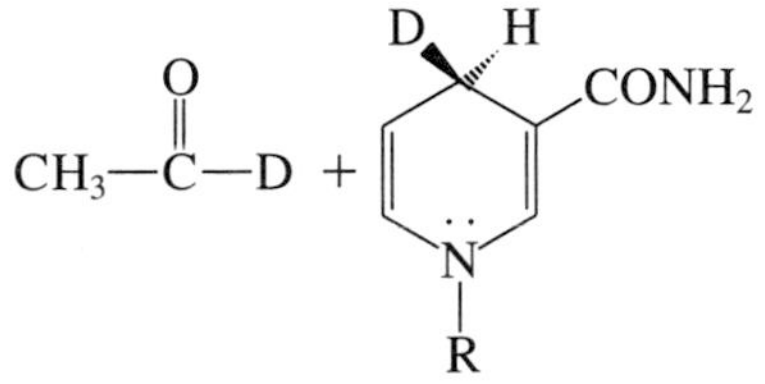

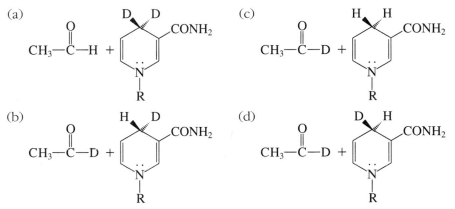

B. Chemical Equivalence and Nonequivalence

To understand why the enzyme alcohol dehydrogenase catalyzes the transfer of *a particular hydrogen* of ethanol to NAD^+, we have to answer a broader question: Under what conditions are two or more groups in a molecule *chemically equivalent?* Chemically equivalent groups cannot be distinguished; for example, they behave the same toward a chemical reagent, and one group does not react in preference to the other. Evidently, the two α-hydrogens of ethanol *are not* chemically equivalent toward the enzyme alcohol dehydrogenase, because the enzyme can tell these hydrogens apart; one reacts in preference to the other.

We can understand chemical equivalence by looking systematically at the *stereochemical relationships of groups within molecules*. In order to understand these relationships, we first have to discuss the concept of *connective equivalence*. Groups within a molecule are **connectively equivalent** when they have the same connectivity relationship to all other atoms in the molecule. Thus, within each of the following molecules, the protons shown in color are connectively equivalent:

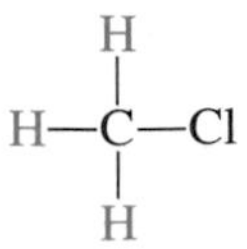

A

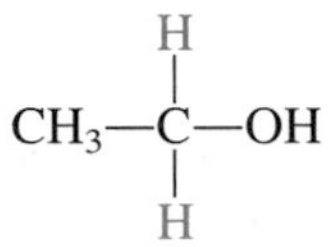

B

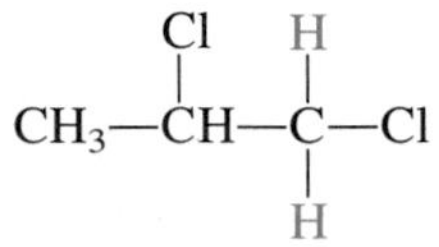

C

For example, in compound *C*, each of the colored hydrogens is connected to a carbon that is connected to a chlorine and a (CH_3CHCl—) group. Each of these protons has the same connectivity relationship to the other atoms in the molecule. On the other hand, the CH_3— and —CH_2— protons in ethanol are connectively nonequivalent. The CH_3— protons are connected to a carbon that is connected to a —CH_2OH, but the —CH_2— protons are connected to a carbon that is connected to an —OH and a —CH_3. However, the two —CH_2— protons of ethanol are connectively equivalent to each other, as are the three CH_3— protons.

Now we are ready for our first generalization about chemical equivalence and nonequivalence. In general, *connectively nonequivalent groups are chemically nonequivalent.* By this we mean that connectively nonequivalent groups have different chemical behavior. Thus, the CH_3— and —CH_2— protons of ethyl alcohol—which, as we have just seen, are connectively nonequivalent—have different reactivities with chemical reagents. A reagent that reacts with one of these sets of protons will have a different reactivity (or perhaps none at all) with the other set. We have already seen an example of this idea: oxidation of ethanol with Cr(VI) reagents causes the loss of one of the —CH_2— hydrogens, but the CH_3— hydrogens are unaffected. The oxidizing agent can tell the —CH_2— and CH_3— hydrogens apart. The converse of the statement above, however, is not true: some connectively equivalent groups are chemically equivalent, whereas others are not. *Whether two connectively equivalent groups are chemically equivalent depends on their stereochemical relationship.* The stereochemical relationship between connectively equivalent groups becomes very easy to see if we make a simple *substitution test*: Substitute each connectively equivalent group in turn with a fictitious circled group. The stereochemical relationship of the resulting molecules then determines the stereochemical relationship of the circled groups. This test is best illustrated by example, using compounds *A, B,* and *C* above.

Let us substitute each hydrogen of *A* in turn with a circled hydrogen.

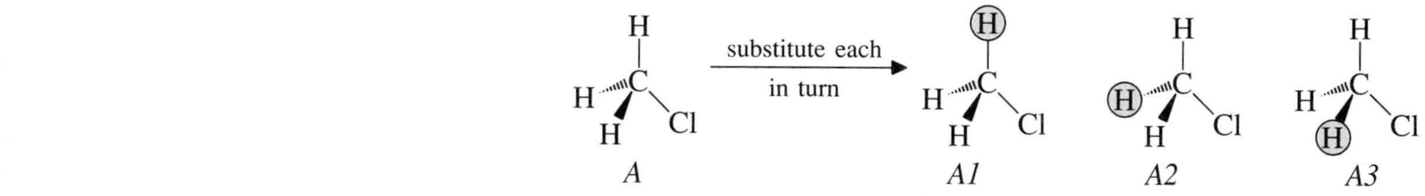

Each of these "new" molecules can be superimposed on the other, atom-for-atom. For example,

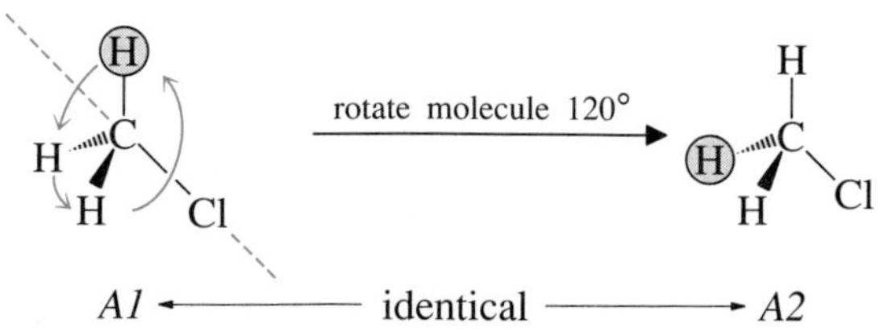

When the substitution test gives identical molecules, as in this example, the connectively equivalent groups are said to be **homotopic.** *Homotopic groups are chemically equivalent and indistinguishable under all circumstances.* Thus, the homotopic protons of methyl chloride all have the same reactivity toward any chemical reagent; there is no way to tell these protons apart.

Substitution of each connectively equivalent hydrogen in *B* (ethanol) gives *enantiomers.*

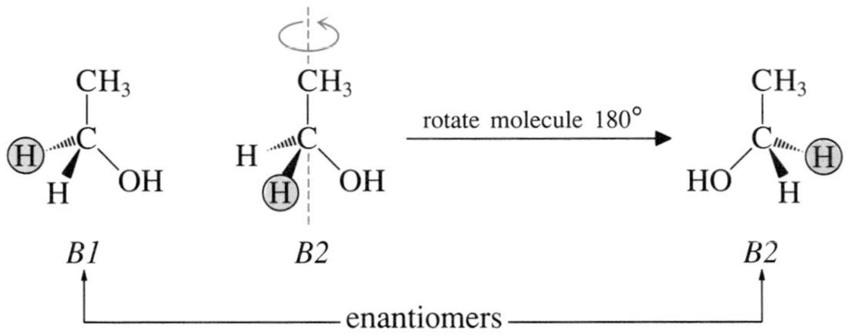

When the substitution test gives enantiomers, the connectively equivalent groups are said to be **enantiotopic.** This case is particularly relevant to the alcohol dehydrogenase-catalyzed oxidation of ethanol. *Enantiotopic groups are chemically nonequivalent toward chiral reagents, but are chemically equivalent toward achiral reagents.* The enzyme alcohol dehydrogenase, like all enzymes, is a chiral compound, and can therefore distinguish between the two enantiotopic hydrogens of ethanol. These two hydrogens, as we have seen in the last section, indeed react differently in the presence of the enzyme. In fact, one reacts and the other is completely unreactive. An achiral reagent, however, cannot distinguish between enantiotopic groups. Thus, chromic acid, which is an achiral reagent, does not distinguish between the α-hydrogens of ethanol; these hydrogens are removed indiscriminately when ethanol is oxidized to acetaldehyde.

Finally, substitution of each connectively equivalent hydrogen in *C* gives *diastereomers.*

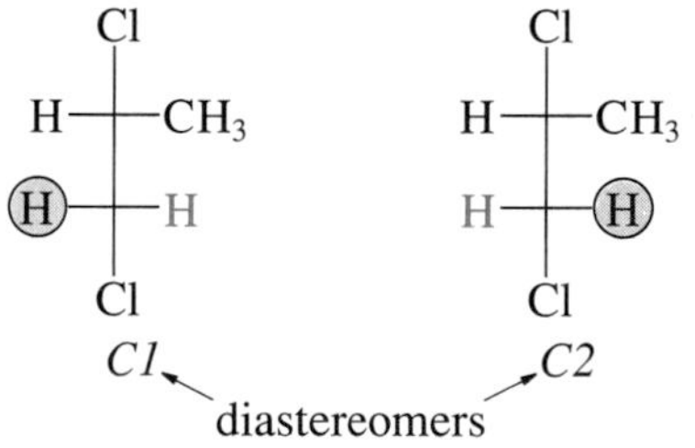

When the substitution test gives diastereomers, the connectively equivalent groups are said to be **diastereotopic.** *Diastereotopic groups are chemically nonequivalent under all conditions.* For example, the protons labeled H^a and H^b in 2-bromobutane (see Eq. 10.60) are diastereotopic. In the E2 reaction of this compound, *anti* elimination of H^b and Br gives *cis*-2-butene, and *anti* elimination of H^a and Br gives *trans*-2-butene. (Verify these statements using models if necessary.)

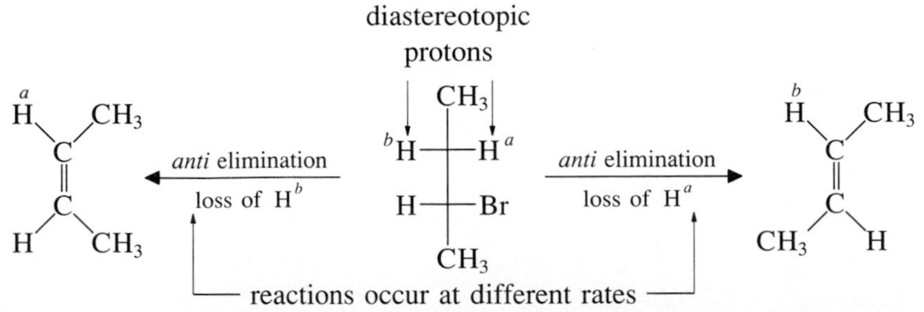

(10.60)

Figure 10.2 A flow chart for classifying groups within molecules.

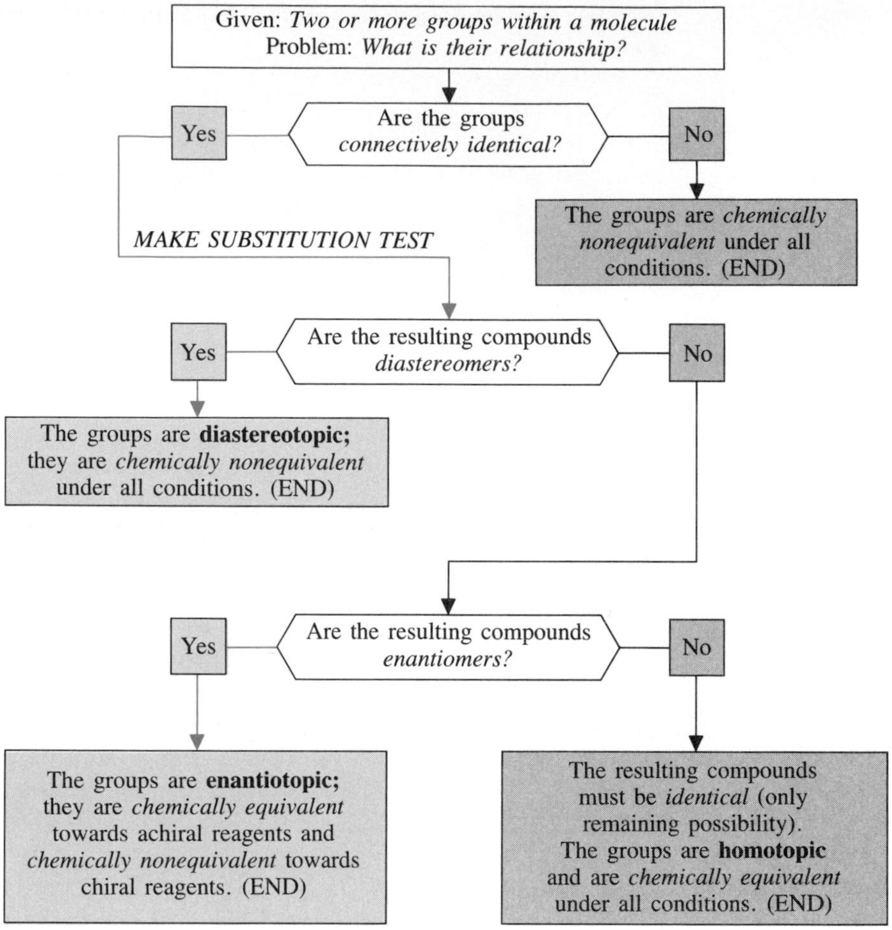

Different amounts of these two alkenes are formed in the reaction precisely because the two diastereotopic protons are removed at different rates—that is, these protons are distinguished by the molecule of base that promotes the elimination.

There are two situations in which diastereotopic groups are easily recognized at a glance. The first is when two connectively equivalent groups are present in a molecule that contains an asymmetric carbon.

The second situation occurs when two groups on one carbon of a double bond are the same and the two groups on the other carbon are different.

As you should readily verify, the substitution test on the colored (geminal) hydrogens gives *E,Z* isomers, which are diastereomers.

In Chapter 6 we learned to classify the stereochemical relationships *between molecules*. Here we have learned to classify the stereochemical relationships *between groups within molecules*. Just as we used the flow chart in Fig. 6.14 to summarize the relationships between molecules, we can also use the flow chart in Fig. 10.2 to summarize the relationships between groups.

We have now answered the question we posed at the beginning of this section: How do we know when two groups are chemically equivalent? Connectively nonequivalent groups are *never* chemically equivalent. Of the connectively identical groups, homotopic groups are *always* chemically equivalent; diastereotopic groups are *never* chemically equivalent; and enantiotopic groups (such as the two α-hydrogens of ethanol) are chemically equivalent toward achiral reagents, but chemically nonequivalent toward chiral reagents.

Problems

25 For each of the following molecules state whether the groups indicated are connectively equivalent or nonequivalent. If they are connectively equivalent, classify them as homotopic, enantiotopic, or diastereotopic.

(a)
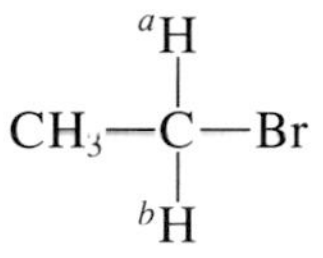

(d)
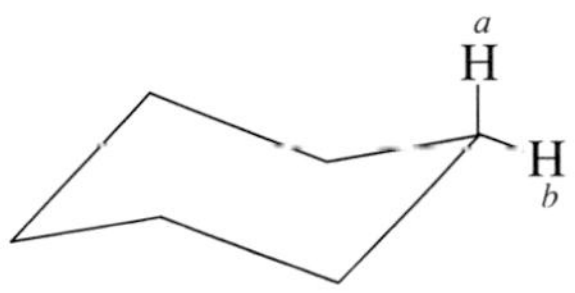

(b)
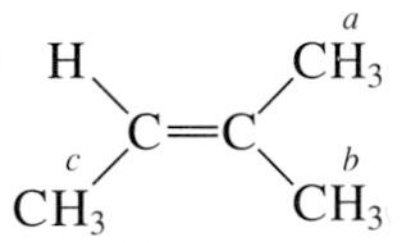
classify each pair of methyl groups

(e)
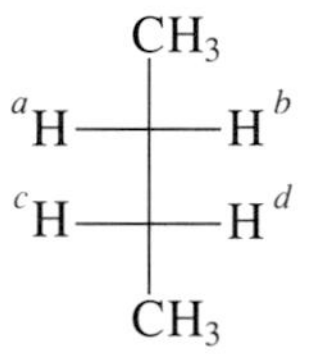

(f)
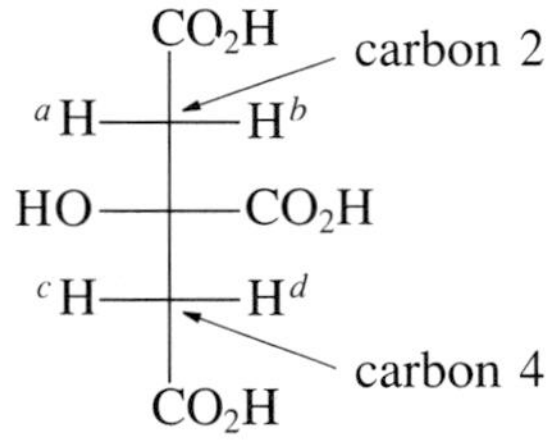

(citric acid)

classify both the hydrogen relationships and the relationship between carbons 2 and 4

(c)
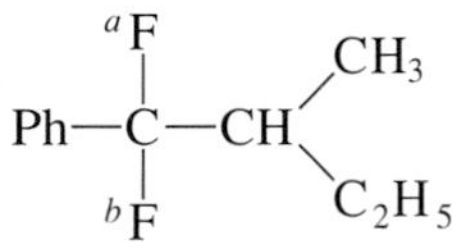

26 What is the relationship between the two ring hydrogens (shown in color) of NADH? Before answering, look at the structure of the R-group in Fig. 10.1. Why is it reasonable that the enzyme alcohol dehydrogenase can chemically distinguish these hydrogens? Would these hydrogens be distinguished in principle by achiral reagents? Explain.

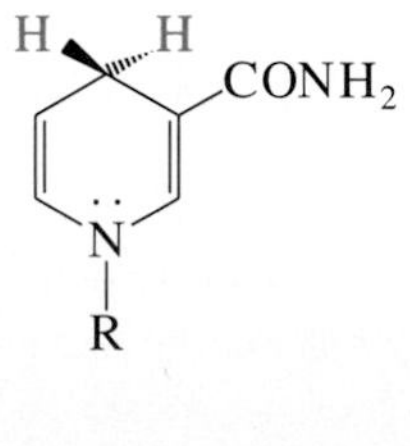
NADH

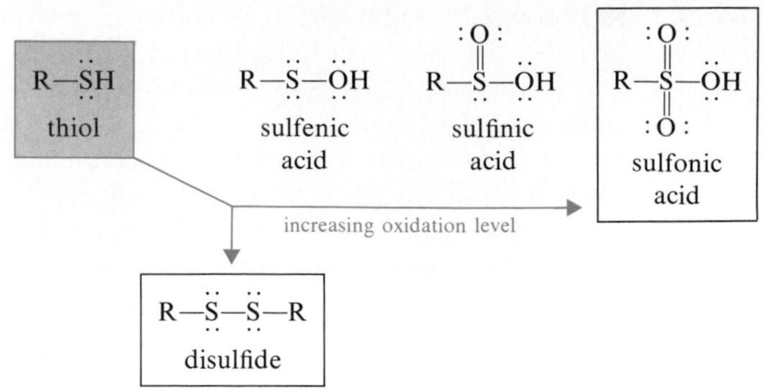

Figure 10.3 Oxidation of thiols gives several possible oxidation products. Of these, disulfides and sulfonic acids (boxed) are the most common.

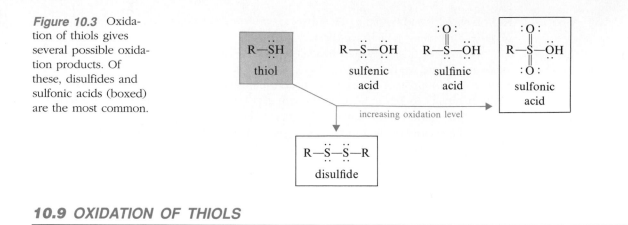

10.9 OXIDATION OF THIOLS

Much of the chemistry of thiols is closely analogous to the chemistry of alcohols, because sulfur and oxygen are in the same group of the periodic table. However, when it comes to oxidation reactions, this formal similarity disappears. Oxidation of an alcohol (Sec. 10.6) occurs at the *carbon* atom bearing the —OH group. However, oxidation of a thiol takes place not at the carbon, but at the *sulfur*. Although sulfur analogs of aldehydes, ketones, and acids are known, they are *not* obtained by simple oxidation of thiols.

$$RCH_2OH \xrightarrow{\text{oxidation}} RCH{=}O \xrightarrow{\text{oxidation}} RCO_2H \qquad (10.61a)$$

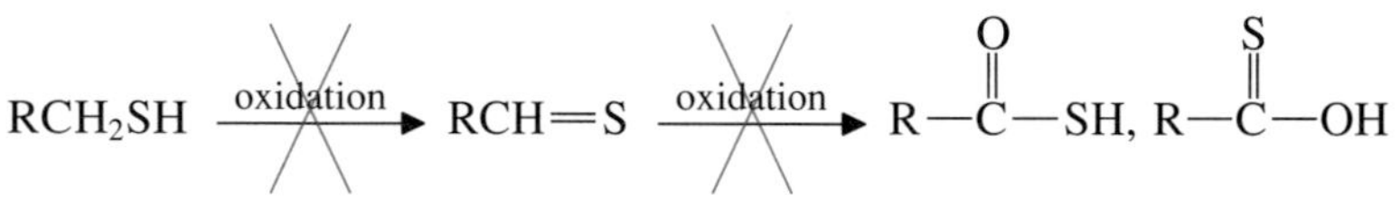

$$RCH_2SH \xrightarrow{\text{oxidation}} RCH{=}S \xrightarrow{\text{oxidation}} R{-}\overset{O}{\underset{}{C}}{-}SH, R{-}\overset{S}{\underset{}{C}}{-}OH \quad (10.61b)$$

Some oxidation products of thiols are given in Fig. 10.3. The most commonly occurring oxidation products of thiols are disulfides and sulfonic acids (boxed in the figure). The same formalism used for oxidation at carbon (Sec. 10.6B) can be applied to oxidation at sulfur.

Problem

27 How many electrons are involved in the oxidation of 1-propanethiol to 1-propanesulfonic acid, $CH_3CH_2CH_2SO_3H$?

In our first courses in chemistry we learn that multiple oxidation states are common for elements in rows of the periodic table beyond the first. The various oxidation products of thiols exemplify the multiple oxidation states available to sulfur. The Lewis structures of these derivatives require either violation of the octet rule or separation of formal charge.

$$(10.62)$$

octet structure
has charge separation

uncharged structure violates
octet rule:
12 electrons around sulfur

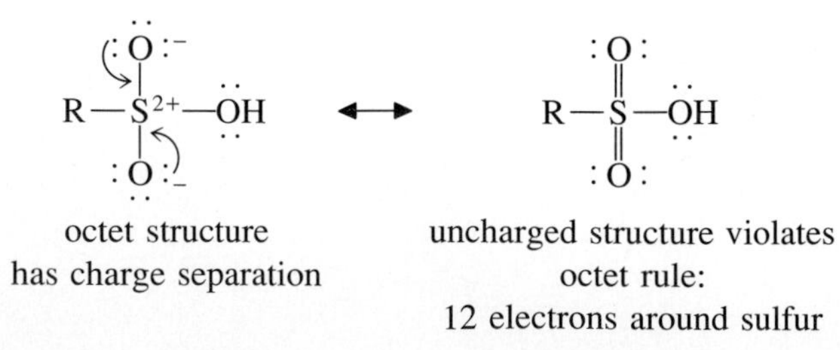

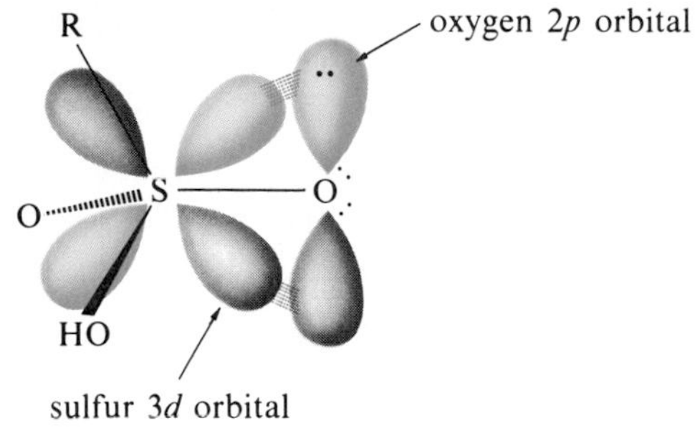

Figure 10.4 Bonding in the higher oxidation states of sulfur involves sulfur *d* orbitals. In a sulfonic acid **(RSO_3H)**, the oxygen electrons overlap with one of several sulfur 3*d* orbitals.

Sulfur can accommodate more than eight valence electrons because, in addition to its 3*s* and 3*p* orbitals, it has unfilled 3*d* orbitals of relatively low energy (Fig. 2.6). The electronic overlap between one of the oxygens in a sulfonic acid with a sulfur *d* orbital is shown in Fig. 10.4; this is essentially an orbital picture of the S=O double bond.

Sulfonic acids are formed by vigorous oxidation of thiols or disulfides with $KMnO_4$ or HNO_3.

$$\underset{\text{1-butanethiol}}{CH_3CH_2CH_2CH_2{-}SH} \xrightarrow{\text{concd } HNO_3} \underset{\substack{\text{1-butanesulfonic acid} \\ (72-96\% \text{ yield})}}{CH_3CH_2CH_2CH_2{-}SO_3H} \qquad (10.63)$$

We have already studied the chemistry of sulfonate esters (Sec. 10.3), and we shall consider other sulfonic acid chemistry in Chapter 20.

Thiols can be converted into disulfides by mild oxidants such as I_2 in base or Br_2 in CCl_4.

$$2C_5H_{11}SH + I_2 + 2NaOH \longrightarrow \underset{(70\% \text{ yield})}{C_5H_{11}S{-}SC_5H_{11}} + 2NaI + 2H_2O \qquad (10.64)$$

$$2C_2H_5SH + Br_2 \longrightarrow \underset{\text{(nearly quantitative yield)}}{C_2H_5S{-}SC_2H_5} + 2HBr \qquad (10.65)$$

These reactions can be viewed simply as a series of S_N2 reactions on halogen and sulfur:

$$R{-}\ddot{\underset{..}{S}}{-}H + {}^{-}{:}\ddot{\underset{..}{O}}H \longrightarrow R{-}\ddot{\underset{..}{S}}{:}^{-} + H_2\ddot{O}{:} \qquad (10.66a)$$

$$R{-}\ddot{\underset{..}{S}}{:}^{-} \quad {:}\ddot{\underset{..}{I}}{-}\ddot{\underset{..}{I}}{:} \longrightarrow R{-}\ddot{\underset{..}{S}}{-}\ddot{\underset{..}{I}}{:} + {:}\ddot{\underset{..}{I}}{:}^{-} \qquad (10.66b)$$

$$R{-}\ddot{\underset{..}{S}}{:}^{-} \quad R{-}\ddot{\underset{..}{S}}{-}\ddot{\underset{..}{I}}{:} \longrightarrow R{-}\ddot{\underset{..}{S}}{-}\ddot{\underset{..}{S}}{-}R + {:}\ddot{\underset{..}{I}}{:}^{-} \qquad (10.66c)$$

Many thiols spontaneously oxidize to disulfides merely on standing in air (O_2).

When thiols and disulfides are present together in the same solution, an equilibrium among them is rapidly established. For example, if ethanethiol and *n*-propyl disulfide are combined, they react to give a mixture of all possible thiols and disulfides:

$$C_2H_5SH + C_3H_7S{-}SC_3H_7 \rightleftharpoons C_2H_5S{-}SC_3H_7 + C_3H_7SH \qquad (10.67a)$$

$$C_2H_5SH + C_2H_5S{-}SC_3H_7 \rightleftharpoons C_2H_5S{-}SC_2H_5 + C_3H_7SH \qquad (10.67b)$$

Because the compounds on both sides of these equilibria have very similar structures, it is not surprising that the equilibrium constants for these reactions are near 1.0.

Problem

28 The rates of the reactions in Eq. 10.67 are increased when the thiol is ionized—that is, in the presence of a base such as sodium ethoxide. Suggest a mechanism for Eq. 10.67a that is consistent with this fact.

An interesting example of a biochemically important thiol–disulfide equilibrium is the reduction of *lipoic acid,* another coenzyme.

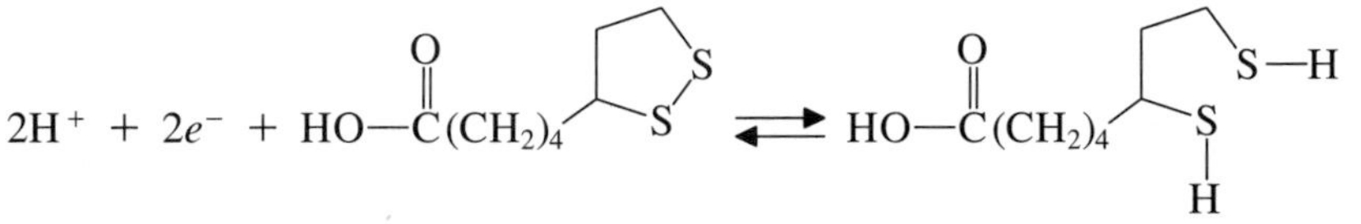
(10.68)

lipoic acid — dihydrolipoic acid

(Lipoic acid occurs naturally as a covalent adduct with proteins.) When lipoic acid is reduced, the corresponding disulfide, dihydrolipoic acid, is formed. Dihydrolipoic acid, in turn, is then used as a reducing agent for other molecules. When it acts as a reducing agent, dihydrolipoic acid is reoxidized back to lipoic acid.

10.10 SYNTHESIS OF ALCOHOLS AND GLYCOLS. DESIGN OF ORGANIC SYNTHESIS

A. Synthesis of Alcohols and Glycols

Let us review the methods we have learned for the synthesis of alcohols and glycols. The methods we have studied so far involve alkenes as starting materials:

1. Hydroboration–oxidation of alkenes (Sec. 5.3A).

$$6RCH{=}CH_2 + B_2H_6 \longrightarrow 2(RCH_2CH_2)_3B \xrightarrow[OH^-]{H_2O_2} 6RCH_2CH_2OH \qquad (10.69)$$

2. Oxymercuration–reduction of alkenes (Sec. 5.3B).

$$R{-}CH{=}CH_2 + Hg(OAc)_2 \xrightarrow{H_2O} R{-}\underset{}{\overset{OH}{\overset{|}{C}}}H{-}CH_2{-}HgOAc \xrightarrow{NaBH_4} R{-}\overset{OH}{\overset{|}{C}}H{-}CH_3 \qquad (10.70)$$

Acid-catalyzed hydration of alkenes is used industrially to prepare certain alcohols, but is not an important laboratory method. In principle, the S_N2 reaction of ^-OH with an alkyl halide can also be used to prepare alcohols from alkyl halides. However, since alkyl halides are generally prepared from alcohols themselves, this method is of little practical importance.

Some of the most important methods for the synthesis of alcohols involve the reduction of carbonyl compounds (aldehydes, ketones, or carboxylic acids and their derivatives), as well as the reactions of carbonyl compounds with Grignard or organolithium reagents. We shall study these methods in Chapters 19, 20, and 21. A summary of methods used to prepare alcohols is found in Appendix IV.

We have learned only one method for preparing glycols, namely, oxidation of alkenes with either OsO_4 or $KMnO_4$ (Sec. 5.4):

$$2KMnO_4 + 4H_2O + 3RCH{=}CHR \longrightarrow 3\underset{\displaystyle OH}{RCH}{-}\underset{\displaystyle OH}{CHR} + 2MnO_2 + 2KOH \qquad (10.71)$$

$$RCH{=}CHR \xrightarrow{OsO_4} \xrightarrow{NaHSO_3} \underset{\displaystyle OH}{RCH}{-}\underset{\displaystyle OH}{CHR} \qquad (10.72)$$

We shall learn in Sec. 11.4A that glycols can also be prepared from epoxides.

B. Design of Organic Synthesis

We have now studied a number of reactions that can transform one functional group into another. These reactions can be used to prepare organic compounds from readily available starting materials. The preparation of an organic compound by the use of one or more reactions is called an **organic synthesis.**

Although many useful syntheses consist of only one reaction, more typically it is necessary to use several reactions to complete the synthesis of an organic compound. Let us examine the logic used in planning such a multistep synthesis.

We shall call the molecule we desire to synthesize the **target molecule.** In order to assess the best route to the target molecule from the starting material, we take the same approach that a military officer might take in planning his assault on an objective: namely, we *work backward from the target towards the starting material.* Just as the officer considers secondary objectives—a hill here, a tree there—from which the final assault on the target can be launched, in planning a synthesis we first assess what compound can be used as the immediate precursor of the target. We then continue to work backward from this precursor step-by-step until the route from the starting material becomes clear. Sometimes more than one synthetic route will be possible. In such a case, we evaluate each synthesis in terms of yield, limitations, expense, and so on.

Let us illustrate the strategy we have just discussed with a sample problem. Suppose we are asked to prepare hexanal from 1-hexene:

$$\underset{\text{1-hexene}}{CH_3CH_2CH_2CH_2CH{=}CH_2} \longrightarrow \underset{\text{hexanal}}{CH_3CH_2CH_2CH_2CH_2CH{=}O} \qquad (10.73)$$

The first question we ask is whether we know any ways to prepare aldehydes directly from alkenes. The answer is that we do. Ozonolysis (Sec. 5.5) can be used to transform alkenes into aldehydes and ketones. However, ozonolysis breaks a carbon–carbon double bond, and certainly would not work for preparing an aldehyde from an alkene *with the same number of carbon atoms,* because at least one carbon is lost when the double bond is broken. We have studied no other ways to prepare aldehydes directly from alkenes. The next step is to ask what ways we know to prepare aldehydes from other starting materials. The answer is that we know two: cleavage of glycols (Sec. 10.6D), and oxidation of primary alcohols (Sec. 10.6A). The cleavage of glycols, like ozonolysis, breaks a carbon–carbon bond, and requires that we lose at least one carbon atom. However, the oxidation of a primary alcohol would be a satisfactory last step in our synthesis.

$$\underset{\text{1-hexanol}}{CH_3CH_2CH_2CH_2CH_2CH_2{-}OH} \xrightarrow[\text{pyridine}]{CrO_3} CH_3CH_2CH_2CH_2CH_2CH{=}O \quad (10.74)$$

We now ask whether it is possible to prepare 1-hexanol from 1-hexene; the answer is yes. Hydroboration–oxidation will convert 1-hexene into 1-hexanol. Our synthesis is now complete.

$$\underset{\text{1-hexene}}{CH_3CH_2CH_2CH_2CH{=}CH_2} \xrightarrow{B_2H_6} \xrightarrow{H_2O_2/OH^-} \underset{\text{1-hexanol}}{CH_3CH_2CH_2CH_2CH_2CH_2{-}OH}$$

$$\xrightarrow{CrO_3/\text{pyridine}} \underset{\text{hexanal}}{CH_3CH_2CH_2CH_2CH_2CH{=}O} \quad (10.75)$$

Notice how we worked backward from the target molecule one step at a time.

Working problems in organic synthesis is one of the best ways to master organic chemistry. It is akin to mastering a language: it is rather easy to learn to read a language (be it a foreign language, English, or even a computer language), but we have to understand it more thoroughly in order to write it. Similarly, it is relatively easy to follow individual organic reactions, but to integrate them and use them out of context requires more understanding. One way to bring together organic reactions and study them systematically is to go back through the text and write a representative reaction for each of the methods we have studied for preparing each functional group. For example, which reactions can be used to prepare alkanes? Carboxylic acids? Then jot down some notes describing the stereochemistry of each reaction (if known) as well as its limitations—that is, the situations in which the reaction would not be expected to work. For example, dehydration of tertiary and secondary alcohols is a good laboratory method for preparing alkenes, but dehydration of primary alcohols is not. (Do you understand the reason for this limitation?) This type of study can be continued throughout future chapters.

We have begun this process for you by summarizing in Appendix IV the reactions discussed in this text that can be used to prepare compounds containing each functional group. It is now time to begin to integrate what you have learned about organic chemistry, and problems in organic synthesis will help you achieve that goal.

Problems

29 Outline syntheses of each of the following compounds from the indicated starting materials:

(a) hexane from 1-hexanol

(b) 2-methyl-3-pentanol from 2-methyl-2-pentanol

(c) 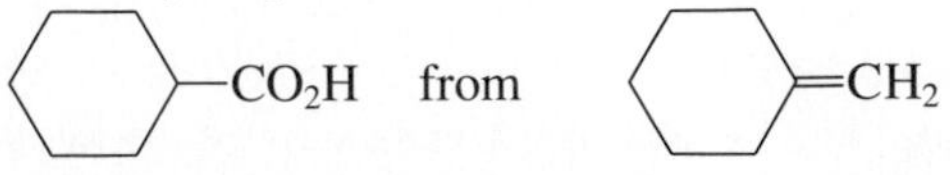

30 The reaction with KI in acetone is one way to prepare alkyl iodides from alkyl chlorides. What is the stereochemistry of this reaction? What is one limitation of this reaction? Explain the reasons for your answers to both of these questions.

KEY IDEAS IN CHAPTER 10

- Several reactions of alcohols involve breaking the C—O bond.

  1. In dehydration and reaction with hydrogen halides, the —OH group of an alcohol is converted by protonation into a good leaving group. The protonated —OH is eliminated as water (in dehydration) to give an alkene, or displaced by halide (in reaction with hydrogen halides) to give an alkyl halide. Many glycols react with acid to undergo the pinacol rearrangement, in which water also serves as a leaving group.
  2. In the reaction of an alcohol with $SOCl_2$ and PBr_3, the reagents themselves convert the —OH into a good leaving group, which is displaced in a subsequent step to form an alkyl halide.
  3. In the reaction with a sulfonyl chloride, the alcohol is converted into a sulfonate ester, such as a tosylate or a mesylate. The sulfonate group is an excellent leaving group in substitution or elimination reactions.

- Many oxidations in organic chemistry occur at carbon, and involve the replacement of the H or C in a C—H bond or a C—C bond, respectively, with a bond to a more electronegative element, such as oxygen or halogen. Reductions usually involve the converse replacements. The oxidation number formalism can be used to determine when an oxidation or reduction has occurred. This formalism also gives the number of electrons lost or gained in an oxidation or reduction.

- Alcohols can be oxidized to carbonyl compounds with Cr(VI). Primary alcohols are oxidized to aldehydes (in the absence of water) or carboxylic acids (in the presence of water), secondary alcohols to ketones, and tertiary alcohols are not oxidized. Primary alcohols are oxidized to carboxylic acids with $KMnO_4$ or HNO_3. Under more vigorous conditions, secondary alcohols react with HNO_3 to yield dicarboxylic acids resulting from cleavage of carbon–carbon bonds. Vicinal glycols react with periodic acid to give aldehydes or ketones derived from cleavage of the carbon–carbon bond between the —OH groups.

- Thiols, unlike alcohols, are oxidized at sulfur. Disulfides and sulfonic acids are two common thiol oxidation products.

- Oxidation is important in nature. An example of a naturally occurring oxidation is the conversion of ethanol into acetaldehyde by NAD^+, catalyzed by the enzyme alcohol dehydrogenase.

- Groups within molecules are either connectively equivalent or connectively nonequivalent. In general, connectively nonequivalent groups are chemically distinguishable. Connectively equivalent groups are of three types.

  1. Homotopic groups are equivalent in every way, and chemically indistinguishable.
  2. Enantiotopic groups are chemically distinguishable to a chiral reagent, such as an enzyme, but chemically indistinguishable to an achiral reagent.
  3. Diastereotopic groups are chemically distinguishable to all reagents.

- A useful strategy for the synthesis of a target compound is to work backward from the target systematically one step at a time.

ADDITIONAL PROBLEMS

31 Give the product expected when 1-butanol reacts with each of the following reagents:
(a) aqueous HBr, heat
(b) dilute HNO_3
(c) aqueous H_2SO_4, cold
(d) CrO_3 in pyridine
(e) NaH
(f) *p*-toluenesulfonyl chloride in pyridine
(g) propylmagnesium bromide
(h) $SOCl_2$ in pyridine
(i) product of (a) + Mg in ether
(j) product of (f) + K^+ $(CH_3)_3C—O^-$ in $(CH_3)_3COH$
(k) product of (j) + OsO_4, then $NaHSO_3$

32 Give the product expected when 2-methyl-2-propanol reacts with each of the following reagents:
(a) aqueous HBr
(b) CrO_3 in pyridine
(c) cold aqueous H_2SO_4
(d) Br_2 in CCl_4 (dark)
(e) potassium metal
(f) methanesulfonyl chloride in pyridine
(g) product of (f) + NaOH in DMSO
(h) product of (e) + product of (a)
(i) product of (g) + alkaline $KMnO_4$
(j) product of (i) + concd H_2SO_4
(k) product of (i) + periodic acid

33 Give the structure of a compound that satisfies the criterion given in each case. (There may be more than one correct answer.)

(a) a seven-carbon tertiary alcohol that yields a *single* alkene after acid-catalyzed dehydration

(b) an eight-carbon secondary alcohol that yields a *single* alkene after acid-catalyzed dehydration

(c) an alcohol, which, after acid-catalyzed dehydration, yields an alkene that in turn, upon ozonolysis and treatment with $(CH_3)_2S$, gives only benzaldehyde, Ph—CH=O.

(d) an optically active glycol that yields two equivalents of acetaldehyde on treatment with periodic acid

(e) two methylcyclohexanols that are oxidized to the same single dicarboxylic acid (a compound containing two carboxylic acid groups) by vigorous treatment with HNO_3

(f) an alcohol that gives the same product when it reacts with $KMnO_4$ that is obtained from the ozonolysis of 3,6-dimethyl-4-octene followed by treatment with H_2O_2

(g) a glycol that gives the following ketone in the pinacol rearrangement:

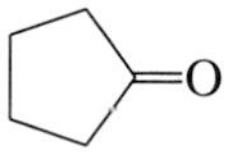

34 The following ester is a powerful explosive, but is also a medication for angina pectoris (chest pain). From what inorganic acid and what alcohol is it derived?

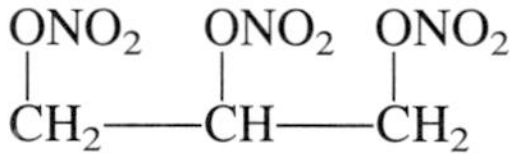

35 In each compound, identify (1) the diastereotopic fluorines, (2) the enantiotopic fluorines, (3) the homotopic fluorines, and (4) the connectively equivalent fluorines.

(a)
$$\begin{array}{c} F \quad F \\ | \quad\; | \\ F—C—C—H \\ | \quad\; | \\ F \quad F \end{array}$$

(b)
$$\begin{array}{c} F \qquad\quad F \\ \;\;\diagdown \quad\quad \diagup \\ \quad C{=}C \\ \;\;\diagup \quad\quad \diagdown \\ F \qquad\quad CH_3 \end{array}$$

(c)
$$\begin{array}{c} F \quad F \\ | \quad\; | \\ H—C—C—H \\ | \quad\;\; | \\ F \quad CH_3 \end{array}$$

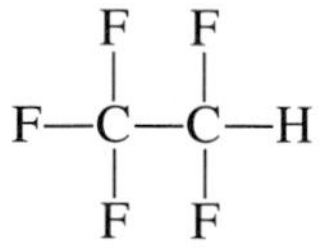

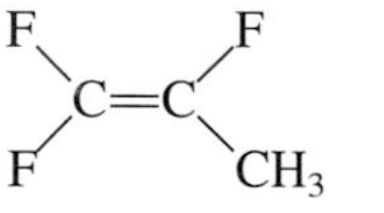

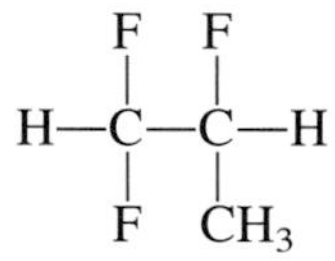

36 When *t*-butyl alcohol is treated with $H_2{}^{18}O$ (water derived from the heavy oxygen isotope ^{18}O) in the presence of acid, and the *t*-butyl alcohol is reisolated, it is found to contain ^{18}O. Explain how the isotope is incorporated into the alcohol.

Problems (Cont.)

37 Indicate whether each of the following transformations is an oxidation, a reduction, or neither, and how many electrons are involved in each oxidation or reduction process:

(a)

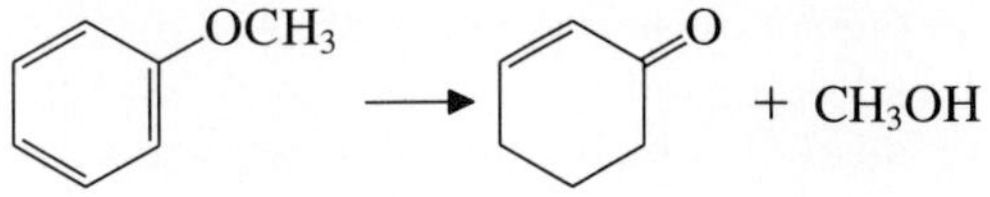

(b)

$$CH_3CH_2\overset{OH}{\underset{|}{C}}HCH=O \longrightarrow CH_3CH_2\overset{O}{\underset{||}{C}}CH_2OH$$

(c)

$$CH_3-\overset{O}{\underset{||}{C}}-NH_2 \longrightarrow CH_3NH_2 + O=C=O$$

(d) $(CH_3)_3C-Cl \longrightarrow (CH_3)_3C^+ + Cl^-$

(e)

$$2\ Ph-\overset{CH_3}{\underset{CH_3}{\overset{|}{\underset{|}{C}}}}-H \longrightarrow Ph-\overset{CH_3}{\underset{CH_3}{\overset{|}{\underset{|}{C}}}}-\overset{CH_3}{\underset{CH_3}{\overset{|}{\underset{|}{C}}}}-Ph$$

38 Outline a synthesis for each of the following compounds from the indicated starting material and any other reagents:

(a) $CH_3-\underset{\underset{D}{|}}{C}H-CH_2CH_2CH_3$ from 1-pentanol

(b) $D-CH_2CH_2CH_2CH_2CH_3$ from 1-pentanol

(c) CH_3 ⟍ ,OH
 $\overset{OH}{}$ OH from 1-methyl-1-cyclohexene

(d) ⬠—CH_2CO_2H from ⬠—$CH=CH_2$ (vinylcyclopentane)

(e) ⬠—CO_2H from vinylcyclopentane

(f) ⬠—$CH_2CH=O$ from vinylcyclopentane

(g) ethylcyclopentane from vinylcyclopentane

(h) $CH_3-\overset{O}{\underset{||}{C}}-C(CH_3)_3$ from 2,3-dimethyl-2-butene

(i) $CH_3CH_2CH_2CH_2-CN$ from 1-butene

(*Hint:* See Table 9.1.)

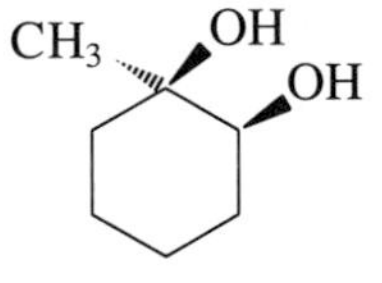

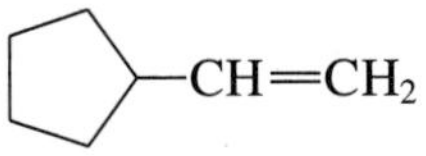

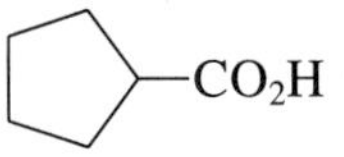

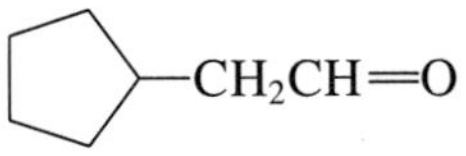

39 Which of the following sulfonate esters reacts more rapidly with sodium methoxide in methanol? Explain.

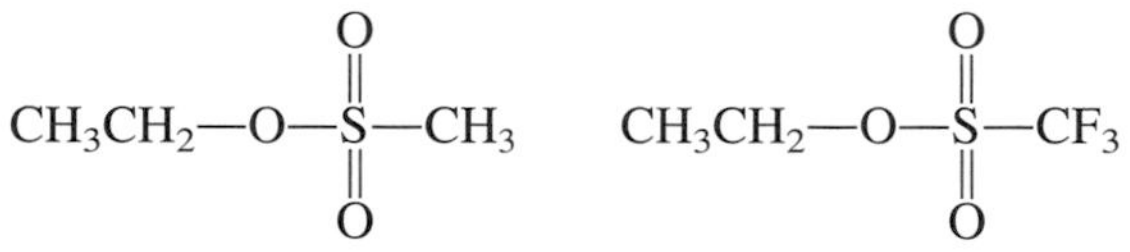

ethyl mesylate ethyl triflate

40 How many grams of $K_2Cr_2O_7$ are required to oxidize 10 g of 2-heptanol to the corresponding ketone?

41 Chemist Stench Thiall, intending to prepare the disulfide *A*, has mixed one mole each of *n*-butyl mercaptan and 2-octanethiol with I_2 and base. Stench is surprised at the low yield of the desired compound and has come to you for an explanation. Explain why Stench should not have expected a good yield in this reaction.

$$A \qquad CH_3CH_2CH_2CH_2-S-S-\overset{\displaystyle |}{\underset{\displaystyle CH_3}{CH}}(CH_2)_5CH_3$$

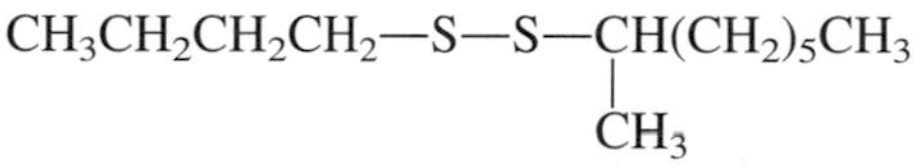

42 In an abandoned laboratory four bottles *A, B, C,* and *D* have been found, each containing a different compound. Near the bottles, scattered by adventitious breezes, are four labels, each bearing one of the structures below. It is assumed that one label came from each bottle, but it is not known which bottle goes with which label.

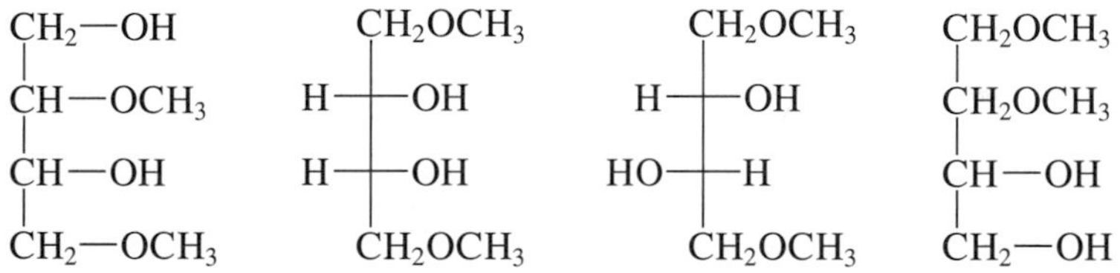

Compounds *A, B,* and *C* are optically active, but compound *D* is not. Compound *C,* on treatment with HIO_4, gives the same products as compound *D,* but *B* gives a different product. Compound *A* does not react with HIO_4. Because of your known expertise in the chemistry of glycols and their ethers, you have been called in as a consultant to identify the compounds. Which label goes with which bottle?

43 Compound *A,* C_7H_{14}, decolorizes Br_2 in CCl_4 and reacts with B_2H_6 followed by H_2O_2/OH^- to yield a compound *B*. When treated with $KMnO_4$, *B* is oxidized to a carboxylic acid *C* that can be resolved into enantiomers. Compound *A,* after ozonolysis and workup with H_2O_2, yields the same compound *D* formed by oxidation of 3-hexanol with chromic acid. Identify compounds *A, B, C,* and *D*.

44 Complete each of the following reactions by giving the principal organic product(s) formed in each case:

(a)

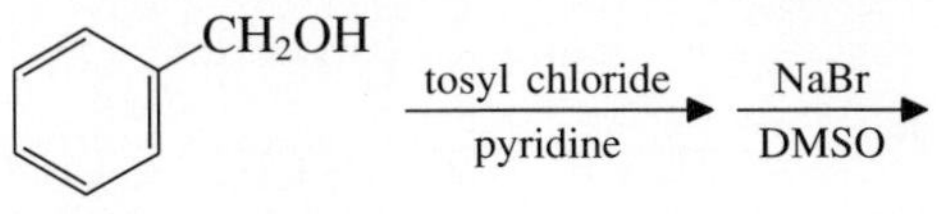

$\begin{array}{c}\text{CH}_2\text{OH}\end{array}$ $\xrightarrow[\text{pyridine}]{\text{tosyl chloride}}$ $\xrightarrow[\text{DMSO}]{\text{NaBr}}$

(b)

$\text{—CH}_2\text{CH}_2\text{CH}_2\text{OH} \ + \ \text{HI} \xrightarrow{\text{H}_3\text{PO}_4}$

(c) $(\text{CH}_3)_2\text{CH—SH} + \text{CH}_3\text{O}^- \xrightarrow{\text{CH}_3\text{OH}} \xrightarrow{\text{dimethyl sulfate}}$

(d) $\text{CH}_3\text{CHCH}_2\text{CH}_2\text{Ph} + \text{PBr}_3 \longrightarrow$
 |
 OH

(e) $\xrightarrow[\text{heat}]{\text{H}_3\text{PO}_4}$

(f) $\xrightarrow{\text{PBr}_3} \xrightarrow[(\text{CH}_3)_3\text{COH}]{\text{K}^+(\text{CH}_3)_3\text{CO}^-}$

(g) $(\text{CH}_3)_2\text{CH—CH=CH}_2 \xrightarrow{\text{OsO}_4, \text{ then NaHSO}_3} \xrightarrow{\text{dil HNO}_3}$

(h) $+ \ 2\text{NaIO}_4 \xrightarrow{\text{OsO}_4}$

(i)
$\begin{array}{cc}\text{OH} & \text{Cl}\end{array}$
$\text{CH}_3\text{—C—C—CH}_3 \xrightarrow{\text{H}_2\text{O, Ag}^+}$
$\begin{array}{cc}\text{CH}_3 & \text{CH}_3\end{array}$

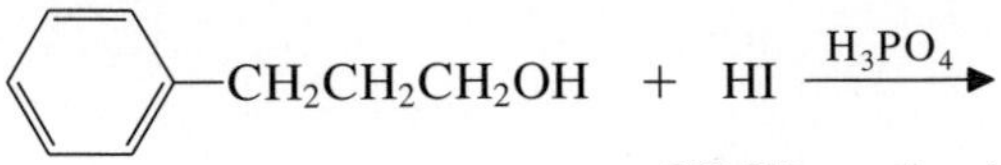

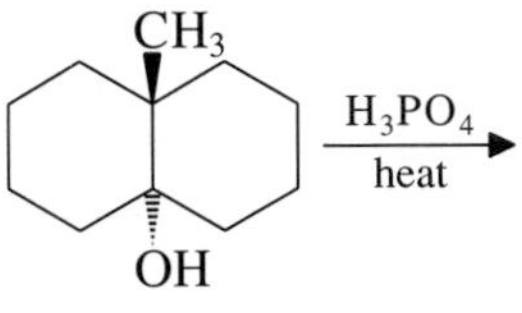

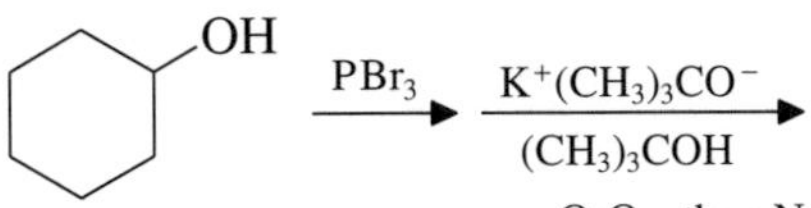

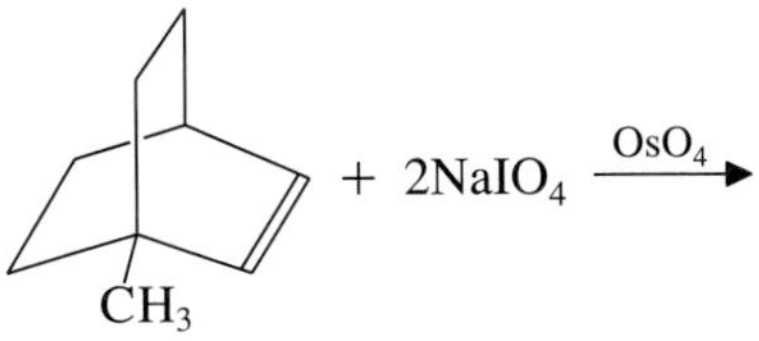

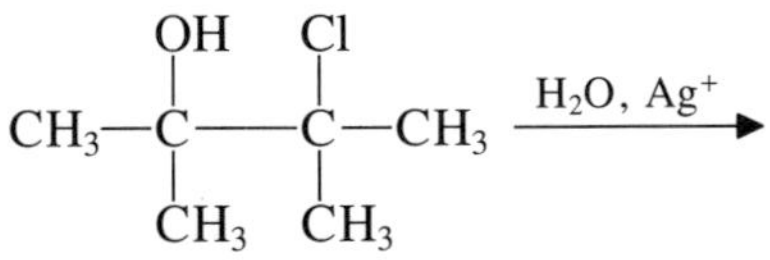

(j) isoamyl mercaptan (3-methyl-l-butanethiol) $+ \ \text{C}_2\text{H}_5\text{—S—S—C}_2\text{H}_5 \longrightarrow$

45 Compound A ($\text{C}_7\text{H}_{14}\text{O}$), which liberates a gas when treated with NaH in ether, can be resolved into enantiomers. Treatment of A with tosyl chloride in pyridine, and then with potassium *t*-butoxide, gives a mixture of alkenes B and C. When optically active A is subjected to the same treatment, *both B* and *C* are optically active. Treatment of the alkene mixture with H_2 and a catalyst gives methylcyclohexane. Give all structures for A, B, and C that fit the data, and explain your conclusions.

46 (a) When the rate of oxidation of isopropyl alcohol to acetone is compared with the rate of oxidation of a deuterated derivative, an isotope effect is observed.

$$\text{CH}_3\text{—C—OH} \xrightarrow{\text{H}_2\text{CrO}_4} \begin{array}{c}\text{CH}_3\\ \diagdown\\ \diagup\end{array}\text{C=O} \qquad \text{relative rate} = 6.6$$

with CH_3 above and H below the central carbon on the left, and CH_3 above and CH_3 below on the right.

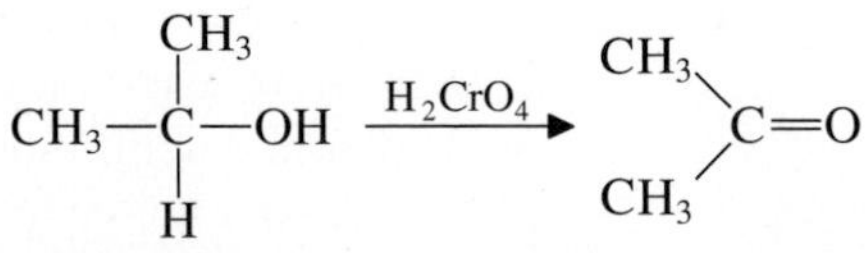

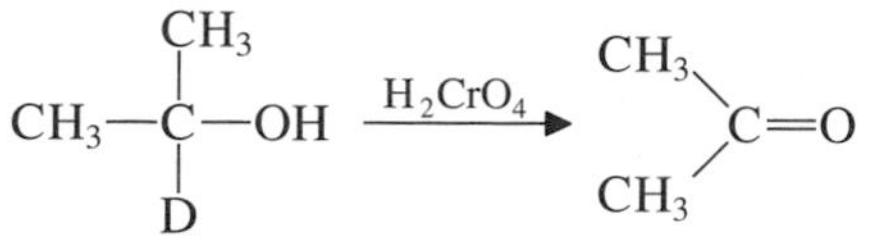

relative rate = 1.0

Which step in the mechanism of Eq. 10.39 is rate-determining?

(b) When either (*S*)- or (*R*)-1-deuterioethanol (see Eq. 10.58) is oxidized with CrO_3 in pyridine, the same product mixture results, and it contains significantly more $CH_3{-}CD{=}O$ than $CH_3{-}CH{=}O$. Contrast this result with the alcohol dehydrogenase-catalyzed oxidations of the same compounds. Explain why formation of the deuterium-containing product is observed in the CrO_3 oxidation of both enantiomers, while in the enzyme-catalyzed oxidation, only the *S* enantiomer gives the deuterium-containing product.

47 Explain why different products are obtained from the following alcohol under the different conditions:

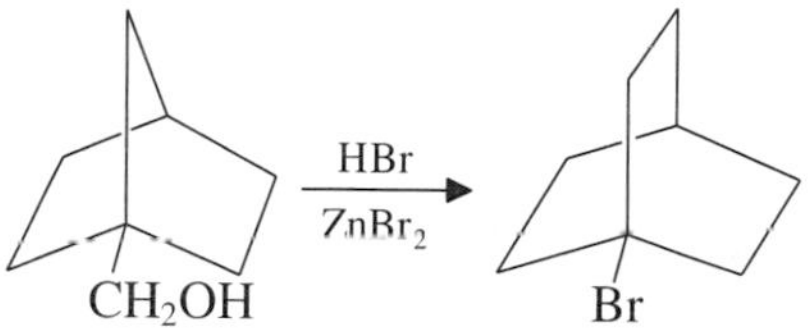

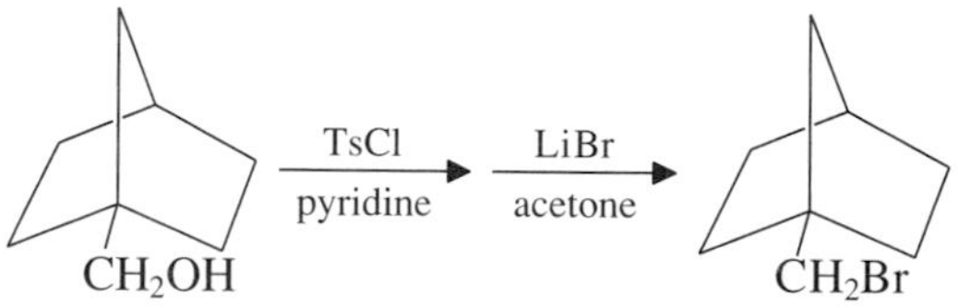

48 When the hydration of fumarate is catalyzed by the enzyme *fumarase* in D_2O, only (2*S*,3*R*)-3-deuteriomalate is formed as the product:

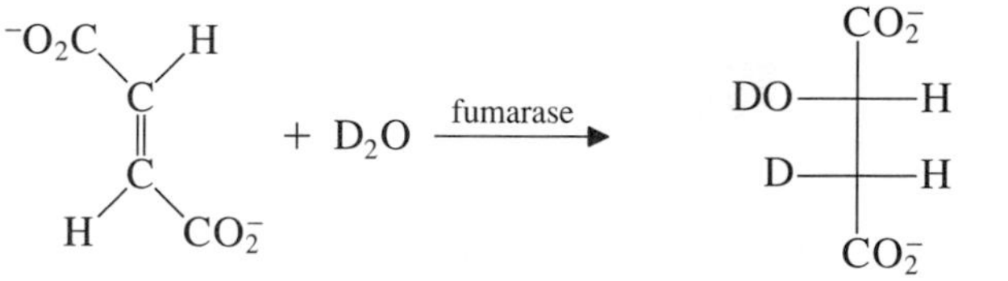

fumarate (2*S*,3*R*)-3-deuteriomalate

This reaction can also be run in reverse. By applying the principle of microscopic reversibility, predict the product (if any) when each of the following compounds is treated with fumarase in H_2O:

(a)

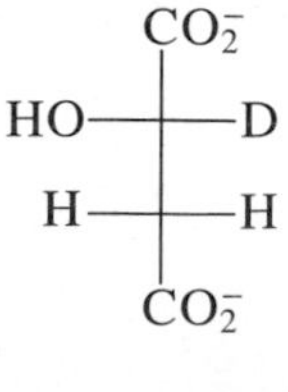

(b)

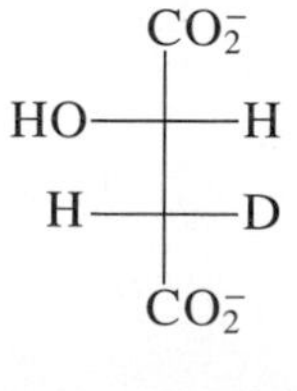

(c)

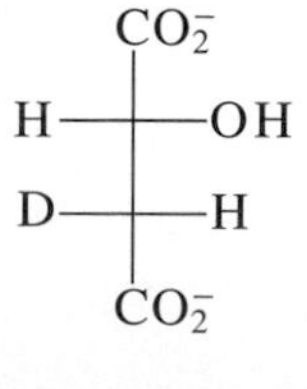

Problems (Cont.)

49 Buster Bluelip, a student repeating organic chemistry for the fifth time, has observed that alcohols can be converted into alkyl bromides by treatment with concentrated HBr, and has proposed that, by analogy, alcohols should be converted into nitriles (organic cyanides, R—CN) by treatment with concentrated HCN. Upon running the reaction, Bluelip finds that the alcohol does not react. Another student has suggested that the reason the reaction failed is the absence of an acid catalyst. Following this suggestion, Bluelip runs the reaction in the presence of H_2SO_4 and again observes no reaction of the alcohol. Explain the difference in the reaction of alcohols with HBr and HCN: why the latter fails but the former succeeds. (*Hint:* see Table 8.3.)

50 The following compound is unstable and breaks down rapidly to acetophenone and HCl. Suggest a mechanism for this reaction.

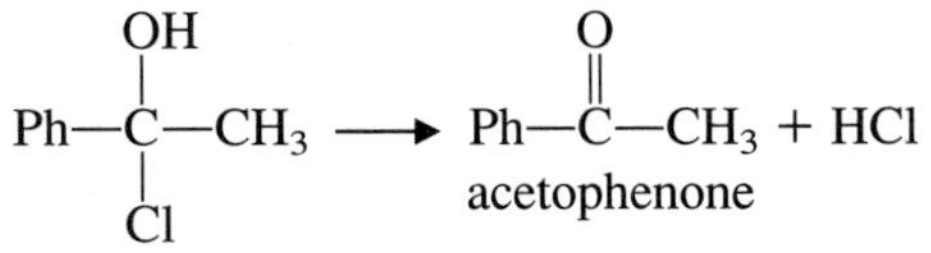

51 Suggest a mechanism for each of the following reactions. Use the arrow formalism. (*Hint:* Read the discussion associated with Eq. 10.7a).

(a)

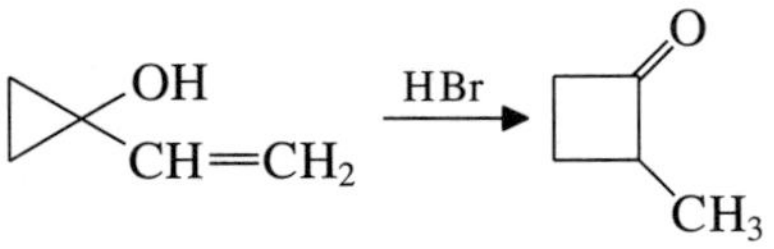

(b)

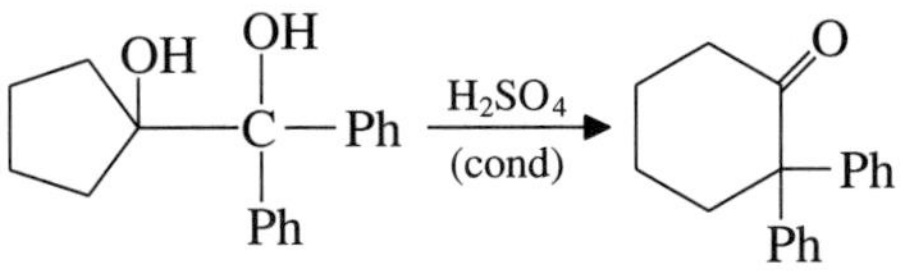

(99% yield)

(c)

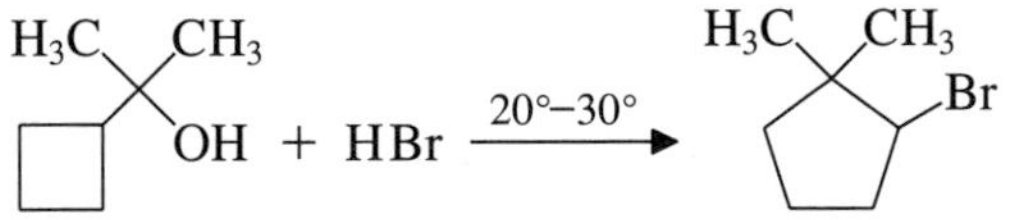

(74% yield)

(d)

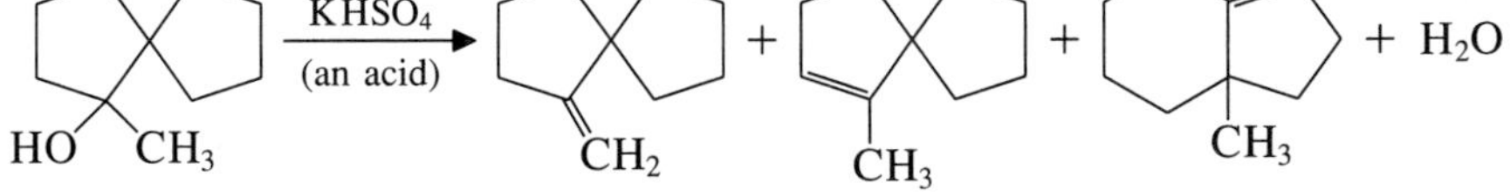

11 Chemistry of Ethers, Epoxides, and Sulfides

In our study of organic chemistry, ethers belong logically near alkyl halides and alcohols because, as we shall see, they can be readily prepared from, and transformed into, these types of compounds. We shall learn that the ether linkage is relatively unreactive, and is therefore rather difficult to break.

Epoxides are compounds in which the ether oxygen is part of a three-membered ring. In contrast to ethers, epoxides are quite reactive. In fact, we shall learn that breaking the ether linkage lies at the heart of epoxide chemistry.

Sulfides, or thioethers, are the sulfur analogs of ethers. We shall find that, just as thiols differ from alcohols in their oxidation reactions, so sulfides differ from ethers.

In this chapter we shall continue our discussion of organic synthesis with a detailed discussion of the strategy involved in planning a multistep synthesis.

11.1 SYNTHESIS OF ETHERS AND SULFIDES

A. Williamson Ether Synthesis

Some ethers and sulfides can be prepared by the reactions of alkoxides or thiolates with methyl halides, primary alkyl halides, or the corresponding sulfonate esters.

$$\underset{\text{an alkoxide}}{\overset{\overset{\displaystyle O^-\ Na^+}{|}}{Ph-CH-CH_3}} + CH_3-I \longrightarrow \underset{\text{(90\% yield)}}{\overset{\overset{\displaystyle O-CH_3}{|}}{Ph-CH-CH_3}} + Na^+\ I^- \qquad (11.1)$$

$$n\text{-}C_4H_9-SH \xrightarrow[CH_3OH]{NaOH} \underset{\text{a thiolate}}{n\text{-}C_4H_9-S^-} \xrightarrow{C_2H_5-OTs} \underset{\text{(78\% yield)}}{n\text{-}C_4H_9-S-C_2H_5} + Na^+\ {}^-OTs \quad (11.2)$$

Both of these reactions are examples of the **Williamson ether synthesis,** named for Alexander William Williamson (1824–1904), Professor of Chemistry at the University of London. (Williamson's syntheses of diethyl ether and ethyl methyl ether in 1850 settled a controversy over the structure and relationship of alcohols and ethers.) The Williamson ether synthesis is an important practical example of the S_N2 reaction (Table 9.1). In this reaction the conjugate base of an alcohol or thiol acts as a nucleophile; an ether is formed by displacement of a halide or other leaving group.

$$R-\ddot{O}:^- \quad CH_3-\ddot{I}: \longrightarrow R-\ddot{O}-CH_3 + :\ddot{I}:^- \qquad (11.3)$$

Tertiary and many secondary alkyl halides cannot be used in this reaction (why?).

In principle there are two different Williamson syntheses for any ether with two different alkyl groups.

$$\left.\begin{array}{l} R^1-\ddot{O}:^- + R^2-X \\ R^2-\ddot{O}:^- + R^1-X \end{array}\right\} \longrightarrow R^1-\ddot{O}-R^2 + X^- \qquad (11.4)$$

The preferred synthesis is usually the one that involves the alkyl halide with the greater S_N2 reactivity. If one of the alkyl groups is tertiary, for example, then it should be derived from the alkoxide rather than the alkyl halide.

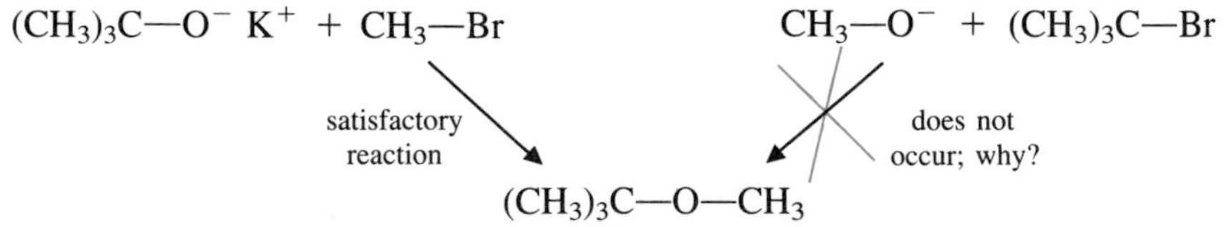

$$(CH_3)_3C-O^-\ K^+ + CH_3-Br \qquad\qquad CH_3-O^- + (CH_3)_3C-Br \quad (11.5)$$

satisfactory reaction / does not occur; why?

$$(CH_3)_3C-O-CH_3$$

Problems

1 Complete the following reactions:

(a) $(CH_3)_2CHOH + Na$, then $CH_3-I \xrightarrow{\text{isopropyl alcohol}}$

(b) $CH_3SH + CH_2{=}CH-CH_2-Cl \xrightarrow[CH_3OH]{NaOH}$

(c) $CH_3O^-\ Na^+ + (CH_3)_3C—Br \xrightarrow{CH_3OH}$ [Be careful!]

(d) $C_2H_5O^-\ K^+ + (CH_3)_3CCH_2—OTs \xrightarrow[25°]{C_2H_5OH}$

2 Suggest a Williamson ether synthesis, if one exists, for each of the following compounds. If a Williamson synthesis is not possible explain why.

(a) $\overset{\displaystyle OCH_3}{\underset{\displaystyle Ph—CH—CH_2—Ph}{|}}$ (c) $(CH_3)_3C—O—C(CH_3)_3$

(b) $PhCH_2CH_2—S—CH_2CH(CH_3)_2$

B. Alkoxymercuration–Reduction of Alkenes

Another ether synthesis is a variation on an alcohol synthesis we have already studied. We learned in Sec. 5.3B that alcohols can be prepared from alkenes by oxymercuration–reduction in *aqueous* solution. If the reaction is carried out with an *alcohol* solvent, an ether is formed instead. This process is called **alkoxymercuration-reduction.**

$$CH_2{=}CH—CH_2CH_2CH_2CH_3 + Hg(OAc)_2 + (CH_3)_2CHOH \longrightarrow$$
$$\text{(solvent)}$$

$$\underset{\displaystyle O—CH(CH_3)_2}{\overset{\displaystyle }{AcOHg—CH_2\underset{|}{C}H—CH_2CH_2CH_2CH_3}} + H—OAc \quad (11.6a)$$

$$\underset{\displaystyle O—CH(CH_3)_2}{AcOHg—CH_2\underset{|}{C}H—CH_2CH_2CH_2CH_3} + NaBH_4 \longrightarrow$$

$$\underset{\displaystyle \underset{\displaystyle (91\% \text{ yield})}{O—CH(CH_3)_2}}{CH_3\underset{|}{C}H—CH_2CH_2CH_2CH_3} + Hg + \text{borates} \quad (11.6b)$$

The mechanism of reaction 11.6a is completely analogous to the mechanism of oxymercuration, except that an alcohol instead of water is the nucleophile. Notice that the ether product of Eq. 11.6b could not have been made satisfactorily by the Williamson synthesis (why?).

Problems

3 Give the mechanism of Eq. 11.6a. Explain what would happen in an attempt to synthesize the product of Eq. 11.6b by the Williamson ether synthesis.

4 Complete the following reactions:

(a) $(CH_3)_2CH—CH{=}CH_2 + C_2H_5OH + Hg(OAc)_2 \longrightarrow \xrightarrow{NaBH_4}$

(b) $CH_3CH{=}CHC_2H_5 + CH_3OH + Hg(OAc)_2 \longrightarrow \xrightarrow{NaBH_4}$

5 Show how alkoxymercuration could be used to synthesize *t*-butyl isobutyl ether.

11.2 SYNTHESIS OF EPOXIDES

A. Oxidation of Alkenes with Peroxyacids

One of the best laboratory preparations of epoxides involves the direct oxidation of alkenes.

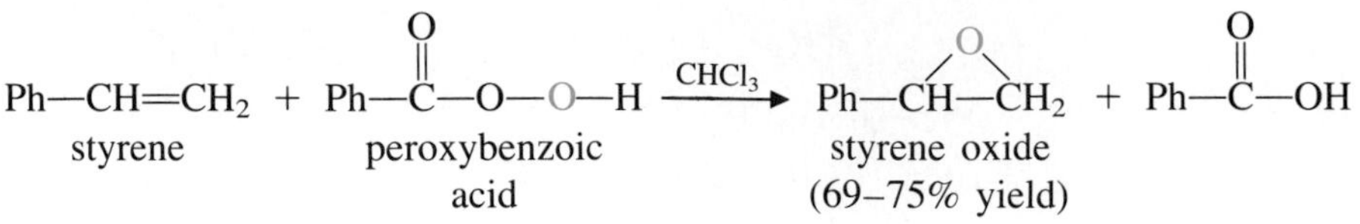

$$Ph—CH{=}CH_2 \;+\; Ph—\overset{\overset{O}{\|}}{C}—O—O—H \xrightarrow{\text{CHCl}_3} Ph—\overset{\overset{O}{\triangle}}{CH}—CH_2 \;+\; Ph—\overset{\overset{O}{\|}}{C}—OH \quad (11.7)$$

styrene peroxybenzoic acid styrene oxide (69–75% yield)

The use of alkenes as starting materials for epoxide synthesis is one reason that certain epoxides are named as oxidation products of their parent alkenes (for example, styrene oxide).

The oxidizing agent in Eq. 11.7, peroxybenzoic acid, is an example of a **peroxycarboxylic acid,** or **peroxyacid.** A peroxyacid is a peroxide analog of a carboxylic acid (that is, a carboxylic acid containing the peroxide group, —O—OH, rather than an —OH group).

a carboxylic acid a peroxyacid

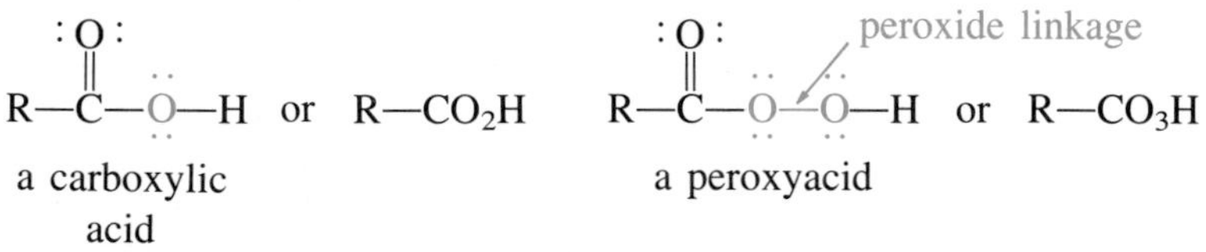

Many peroxyacids are unstable, but they can be formed just prior to use by mixing a carboxylic acid and hydrogen peroxide. However, one peroxyacid, *m*-chloroperbenzoic acid (MCPBA), is widely used for the synthesis of epoxides because it is a relatively stable crystalline solid that can be shipped commercially and stored in the laboratory.

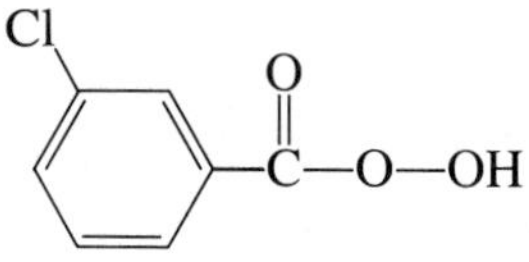

m-chloroperbenzoic acid (MCPBA)

The formation of an epoxide with a peroxyacid is another example of *electrophilic addition* to alkenes (Secs. 4.5D, 5.6). The electrophile in this reaction is the —OH group of the peroxyacid. Because the group attached to it is electronegative, this —OH group has some partially positive, electron-deficient character.

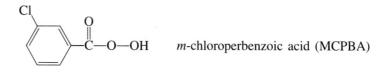

Furthermore, the O—O bond is very weak and easily broken.

In the mechanism of this reaction, the electron-deficient —OH of the peroxyacid interacts with the electron-rich π-bond of the alkene. This oxygen is then transferred to the alkene in a concerted fashion.

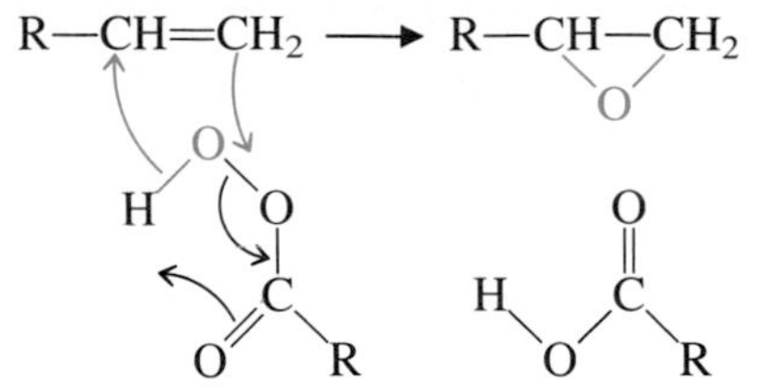

(11.8)

This reaction is quite similar mechanistically to the formation of the bromonium ion in the bromination of alkenes (Sec. 5.1):

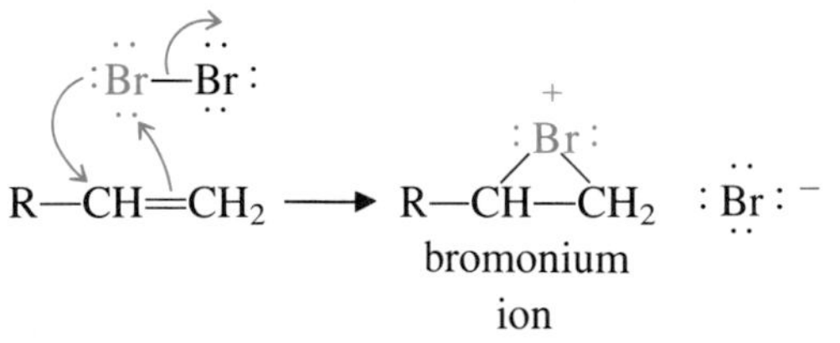

(11.9)

In fact, it is not a bad analogy to think of the bromonium ion as the bromine analog of an epoxide. Bromonium ions, however, are reactive intermediates; epoxides are isolable compounds.

The formation of epoxides with peroxyacids is a *stereoselective* reaction; it takes place with complete retention of the alkene stereochemistry. That is, if we begin with a *cis*-alkene, we get a *cis*-substituted epoxide; a *trans*-alkene gives a *trans*-substituted epoxide.

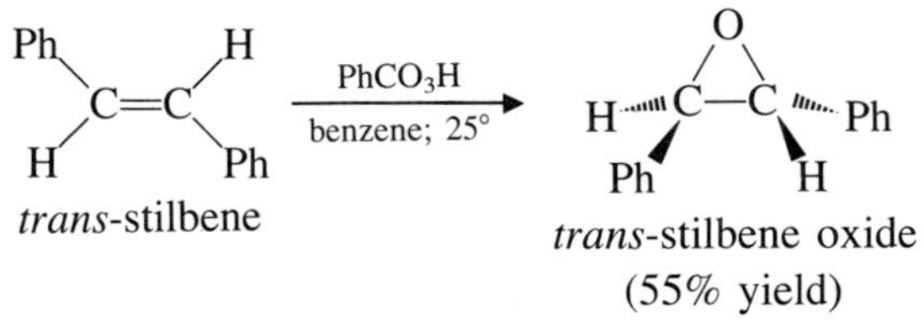

(11.10a)

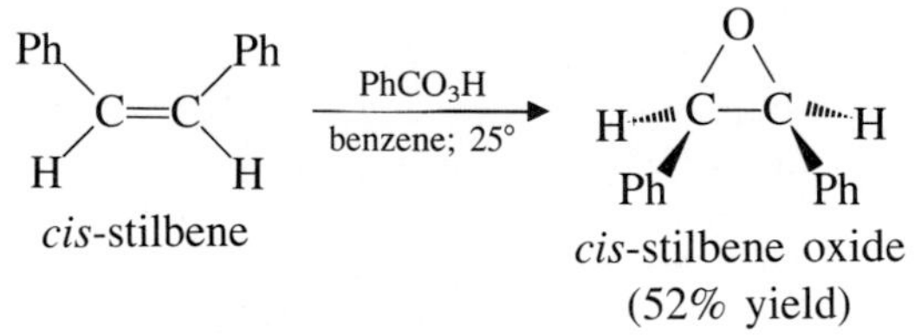

(11.10b)

This is precisely the result we expect from a concerted reaction. The oxygen from the peroxyacid must bond to both alkene carbons at the same face, since it cannot bridge both upper and lower faces simultaneously.

Problems

6 Give the structure of the alkene that would react with *m*-chloroperbenzoic acid to give each of the following epoxides:

(a) 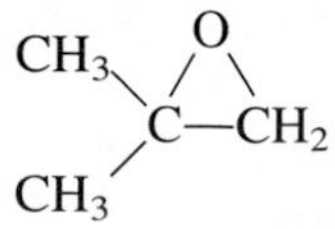

(b) 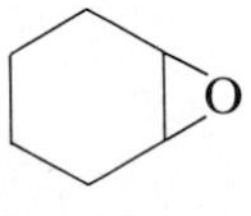

(c) 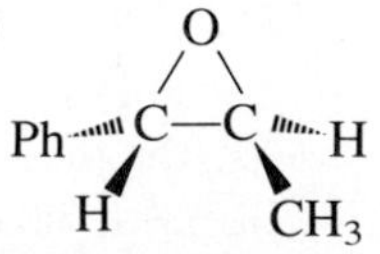

Problems (Cont.)

7 Give the structures of all epoxides that could in principle be formed when each of the following alkenes reacts with *m*-chloroperbenzoic acid. Which epoxide should predominate in each case? Why?

(a) *cis*-4,5-dimethyl-1-cyclohexene (b)

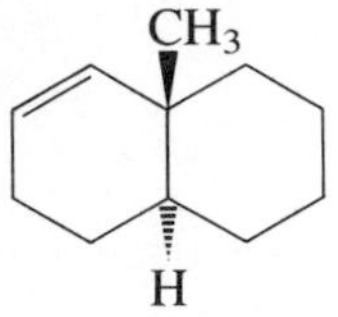

B. Cyclization of Halohydrins

Epoxides can also be synthesized by the treatment of halohydrins (Sec. 5.1B) with base.

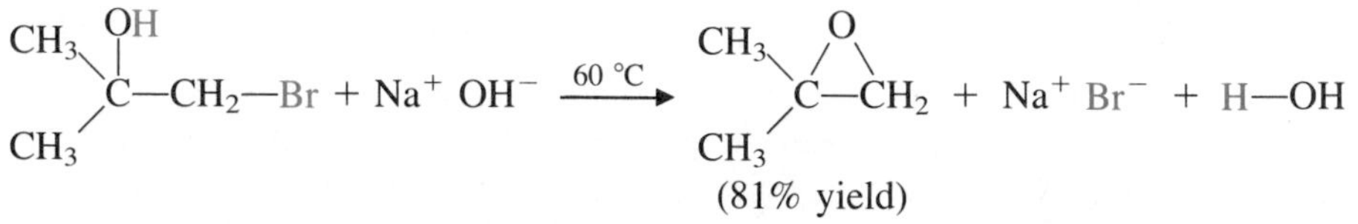

(11.11)

This reaction is an intramolecular variation of the Williamson ether synthesis (Sec. 11.1A); in this case, the alcohol and the alkyl halide are part of the same molecule. The alkoxide anion, formed reversibly by reaction of the alcohol with NaOH, displaces halide ion from the neighboring carbon.

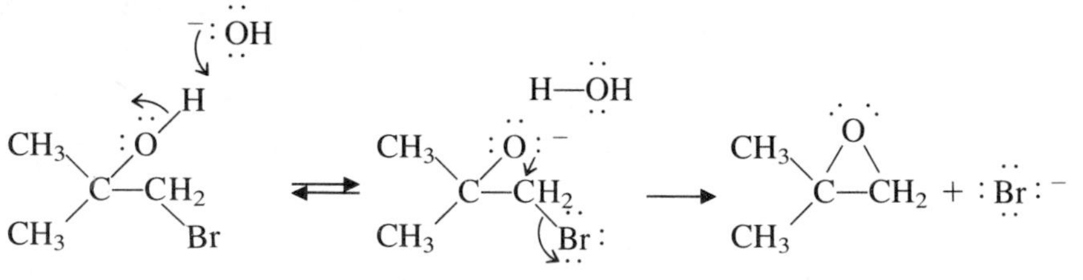

(11.12)

Like other S_N2 reactions, this reaction takes place by *backside attack* of the nucleophilic oxygen anion at the halide-bearing carbon. Such a backside attack requires that the attacking oxygen and the leaving halide assume an *anti* relationship in the transition state of the reaction. For noncyclic halohydrins, this relationship can generally be achieved through a simple bond rotation.

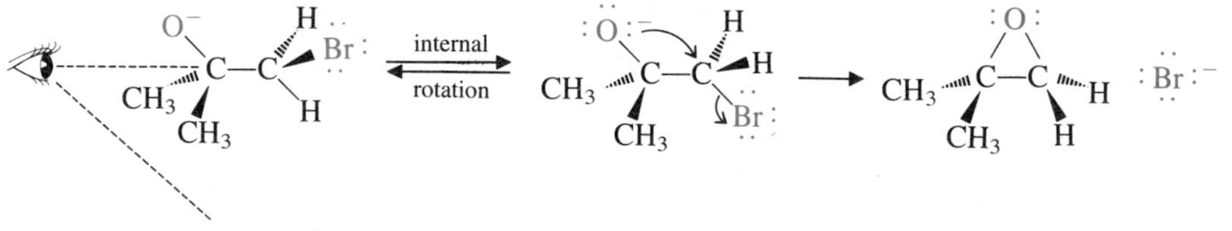

(11.13)

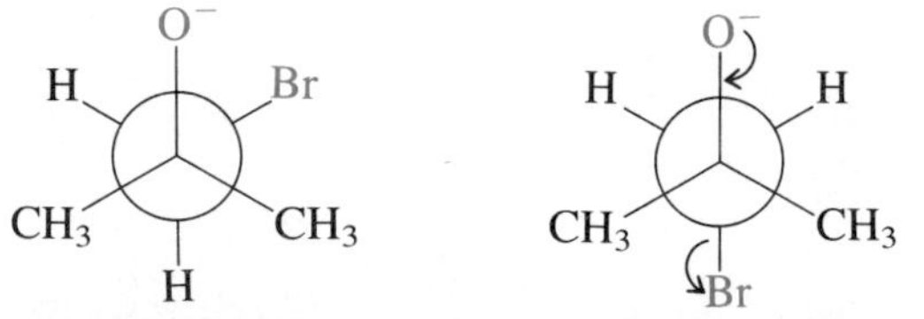

O, Br *gauche;* backside attack not possible

O, Br *anti;* backside attack possible

Halohydrins derived from cyclic compounds must be able to assume the required *anti* relationship through a conformational change if epoxide formation is to succeed. The cyclohexane derivative below, for example, must undergo the chair–chair interconversion before epoxide formation can occur.

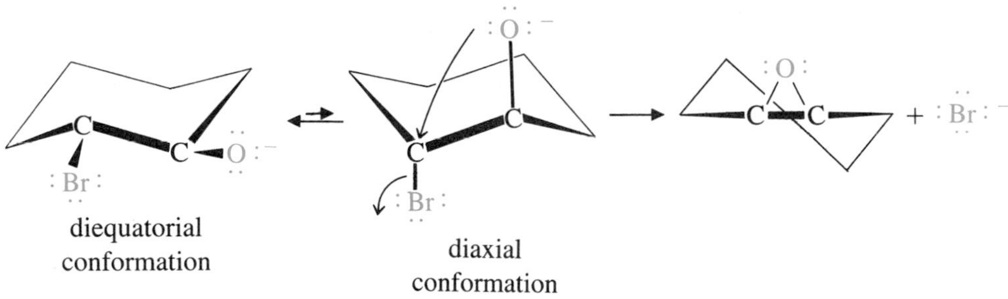

(11.14)

Even though the diaxial form is less stable than the diequatorial form, the two conformations are in rapid equilibrium. As the diaxial form reacts to give epoxide, it is replenished by the conformational equilibrium.

In this section we have learned two methods that can be used to synthesize epoxides from alkenes. Epoxides can be formed directly from alkenes with peroxyacids; and they can also be prepared from halohydrins, which in turn are also prepared from alkenes (Sec. 5.1B). We have already seen that oxidation of alkenes with peroxyacids occurs with retention of the alkene stereochemistry. Let us see how this compares with the stereochemistry of epoxide formation from alkenes by way of bromohydrins. Bromohydrin formation, like bromination (Sec. 7.10C), is an *anti* addition and therefore involves one stereochemical inversion of configuration. Epoxide formation from the bromohydrin involves a second inversion of configuration. Two inversions have the same overall effect as retention:

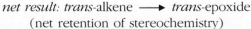

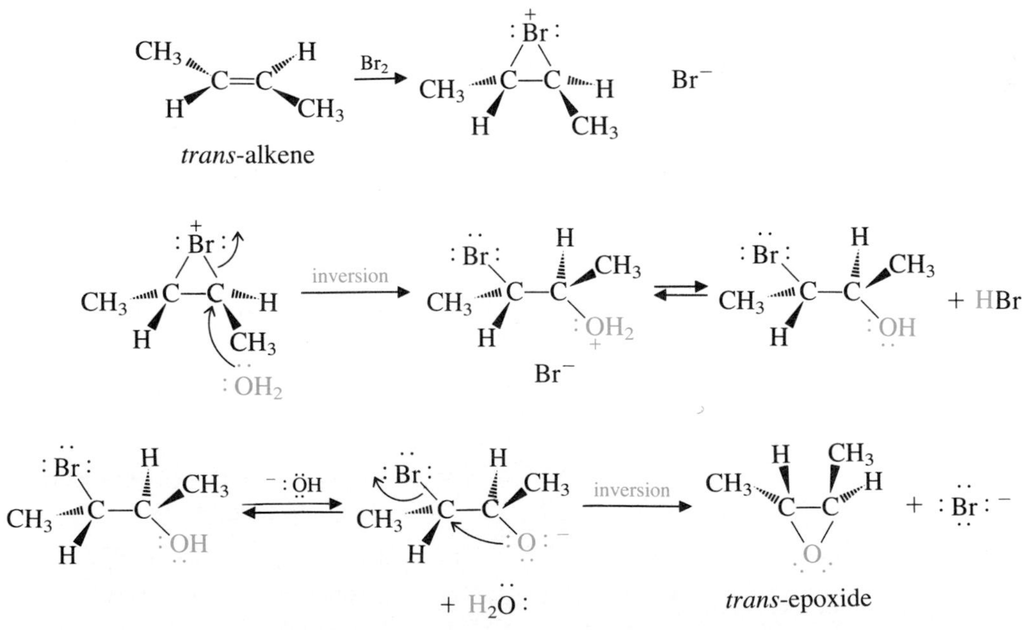

(11.15)

net result: trans-alkene ⟶ *trans*-epoxide
(net retention of stereochemistry)

Thus, epoxidation of alkenes by peroxyacids and epoxidation by bromohydrin formation/ring closure have the same overall stereochemistry.

Since both reactions have the same outcome, why would we choose one over the other? The answer really hinges on convenience and safety. The use of peroxyacids is a convenient reaction on a small laboratory scale, but on a large scale, the use of peroxyacids is hazardous (many are explosive), and the halohydrin approach is preferred.

Problems

8 From models of the transition states for their reactions, predict which of the two following stereoisomers (shown in Fischer projection) should form epoxide at the greater rate when treated with base. Explain.

A 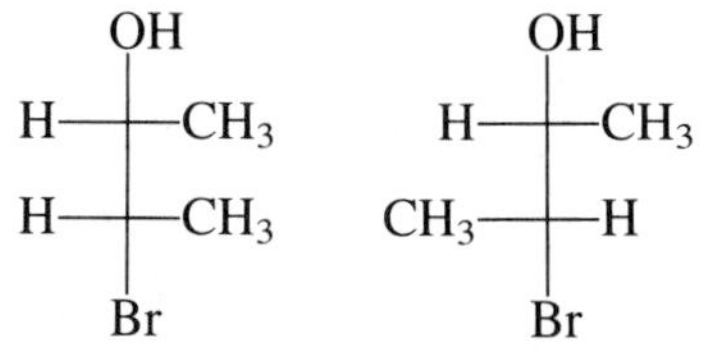 *B*

9 The chlorohydrin *trans*-2-chloro-1-cyclohexanol reacts rapidly in base to form an epoxide. The *cis* stereoisomer, however, is relatively unreactive and does not give an epoxide. Explain why the two stereoisomers behave so differently.

11.3 CLEAVAGE OF ETHERS AND SULFIDES

A. Cleavage of Ethers by Acids

The ether linkage is relatively inert to many reagents. This is one reason ethers are popular as solvents: a great many reactions can be carried out in ether solvents without affecting the ether linkage. However, there are a few important reactions of ethers. High concentrations of strong acids, particularly the hydrogen halides HI and HBr, promote the *cleavage* of the ether linkage to give alcohols and alkyl halides.

$$C_2H_5—O—C_2H_5 + H—I \longrightarrow C_2H_5—OH + C_2H_5—I \qquad (11.16)$$

$$CH_3CH_2CH_2—O—CH_2CH_3 + H—I \longrightarrow CH_3CH_2CH_2—OH + I—CH_2CH_3$$
(both pathways observed) (11.17)
$$\longrightarrow CH_3CH_2CH_2—I + HO—CH_2CH_3$$

$$CH_3—O—CH_2CH_2CH_3 + H—I \not\longrightarrow CH_3—OH + I—CH_2CH_2CH_3$$
(11.18)
$$\longrightarrow CH_3—I + HO—CH_2CH_2CH_3$$
(only products observed)

$$CH_3—O—C(CH_3)_3 + H—I \longrightarrow CH_3—OH + I—C(CH_3)_3$$
(only products observed) (11.19)
$$\not\longrightarrow CH_3—I + HO—C(CH_3)_3$$

The mechanisms of these reactions involve protonation of the ether oxygen.

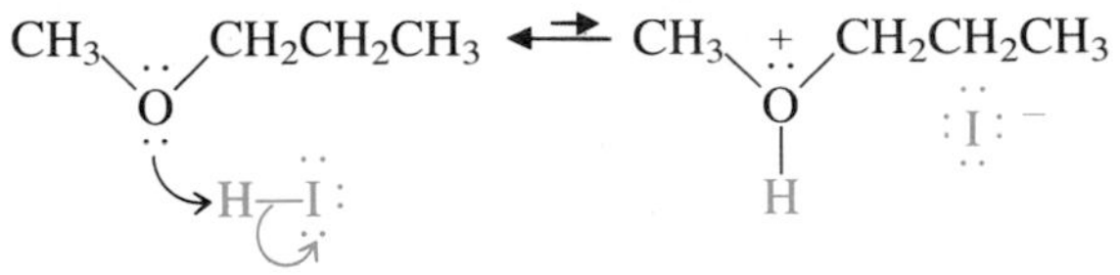

$$CH_3 \cdots O \cdots CH_2CH_2CH_3 \; \rightleftharpoons \; CH_3 \cdots \overset{+}{O} \cdots CH_2CH_2CH_3 \qquad (11.20a)$$
$$\underset{H}{\overset{\displaystyle O}{|}} \qquad \qquad \overset{\displaystyle O}{|} \quad \ddot{:}\ddot{I}\ddot{:}^-$$
$$H{-}\ddot{I}\ddot{:} \qquad \qquad H$$

If the ether is derived solely from methyl groups or primary alkyl groups, attack of halide ion occurs by the S_N2 mechanism on the more reactive alkyl group.

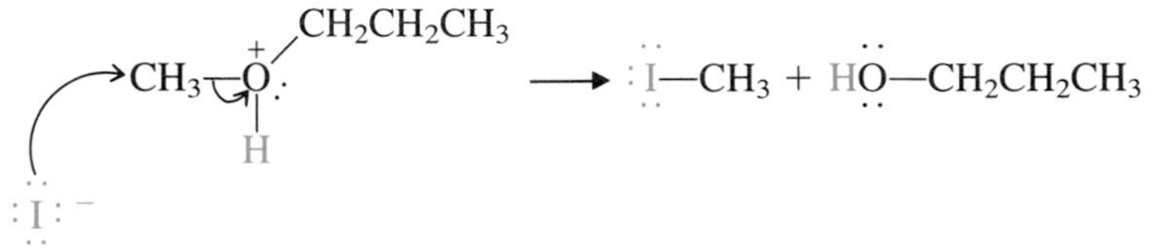

$$CH_3\overset{+}{\underset{H}{\overset{CH_2CH_2CH_3}{O}}} \longrightarrow \ddot{:}\ddot{I}{-}CH_3 + H\ddot{O}{-}CH_2CH_2CH_3 \quad (11.20b)$$
$$\ddot{:}\ddot{I}\ddot{:}^-$$

(Why does the halide attack the methyl group instead of the propyl group in Eq. 11.20b?) If the protonated ether can cleave to give a relatively stable carbocation, then cleavage occurs instead by the S_N1 pathway.

$$H{-}I + (CH_3)_3C{-}\ddot{O}{-}CH_3 \; \rightleftharpoons \; (CH_3)_3C{-}\overset{+}{\underset{H}{\overset{\displaystyle O}{|}}}{-}CH_3 \quad I^- \qquad (11.21a)$$

$$(CH_3)_3C{-}\overset{+}{\underset{H}{\overset{\displaystyle O}{|}}}{-}CH_3 \; \rightleftharpoons \; CH_3\ddot{O}{-}H + (CH_3)_3\overset{+}{C} \overset{\ddot{:}\ddot{I}\ddot{:}^-}{\longrightarrow} (CH_3)_3C{-}\ddot{I}\ddot{:} \quad (11.21b)$$

Because tertiary carbocations are relatively stable, cleavage reactions of tertiary ethers are generally faster, and occur at lower acid concentration, than cleavage reactions of primary or secondary ethers.

Notice the great similarity in the reactions of ethers and alcohols (Sec. 10.2) with halogen acids. In the reaction of an alcohol, *water* acts as a leaving group. In the reaction of an ether, an *alcohol* acts as a leaving group. Otherwise the reactions are essentially the same.

methyl or primary alcohol: $CH_3{-}\ddot{O}H \overset{HI}{\rightleftharpoons} CH_3{-}\overset{+}{\ddot{O}}H_2 \longrightarrow CH_3 + \ddot{O}H_2 \quad (11.22a)$
$$\underset{I^-}{} \qquad \qquad \underset{I}{|}$$

ether: $CH_3{-}\ddot{O}{-}CH_3 \overset{HI}{\rightleftharpoons} CH_3{-}\overset{\overset{\displaystyle H}{|}}{\overset{+}{O}}{-}CH_3 \longrightarrow CH_3 + CH_3\ddot{O}H \quad (11.22b)$
$$\underset{I^-}{} \qquad \qquad \underset{I}{|}$$

tertiary alcohol: $(CH_3)_3C{-}\ddot{O}H \overset{HBr}{\rightleftharpoons} (CH_3)_3C{-}\overset{+}{\ddot{O}}H_2 \rightleftharpoons (CH_3)_3\overset{+}{C} + \ddot{O}H_2 \quad (11.23a)$
$$\underset{Br^-}{\longrightarrow} (CH_3)_3C{-}Br$$

ether: $(CH_3)_3C{-}\ddot{O}{-}CH_3 \overset{HBr}{\rightleftharpoons} (CH_3)_3C{-}\overset{+}{\underset{H}{\overset{\displaystyle O}{|}}}{-}CH_3 \rightleftharpoons (CH_3)_3\overset{+}{C} + H\ddot{O}{-}CH_3 \quad (11.23b)$
$$\underset{Br^-}{\longrightarrow} (CH_3)_3C{-}Br$$

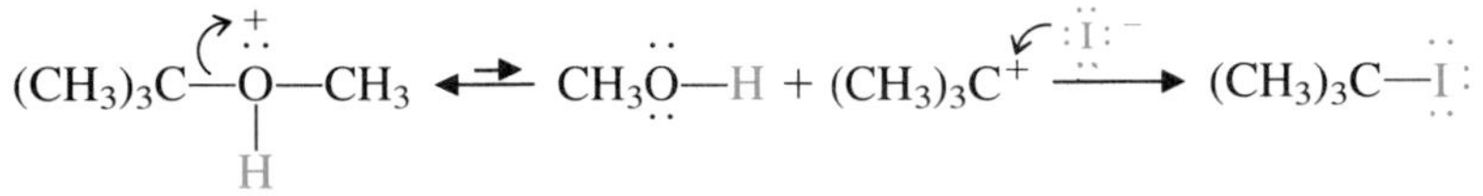

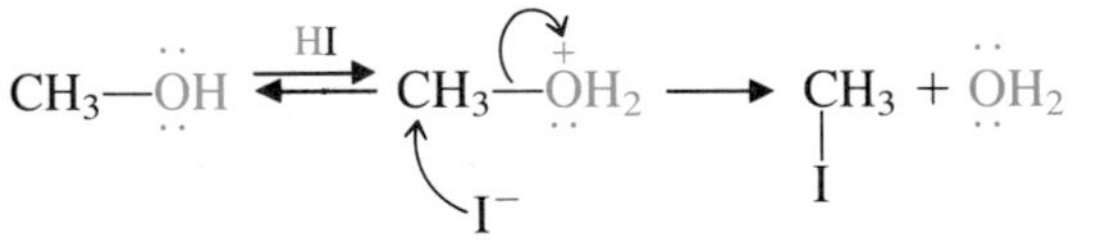

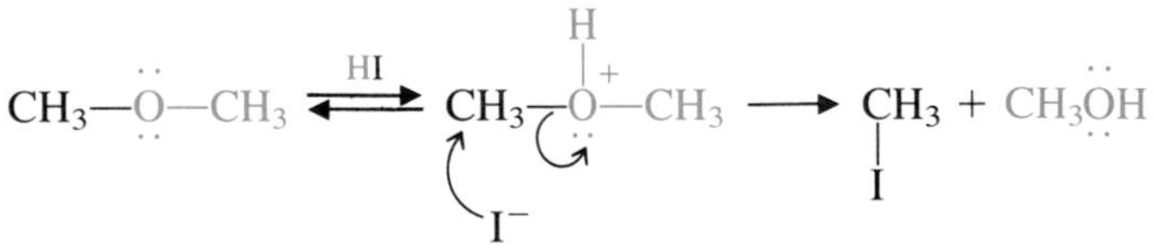

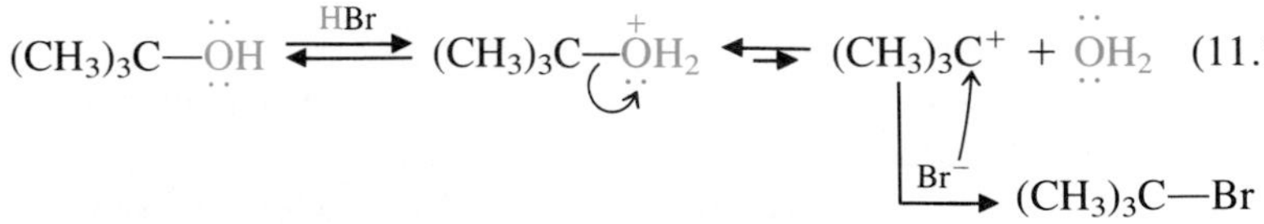

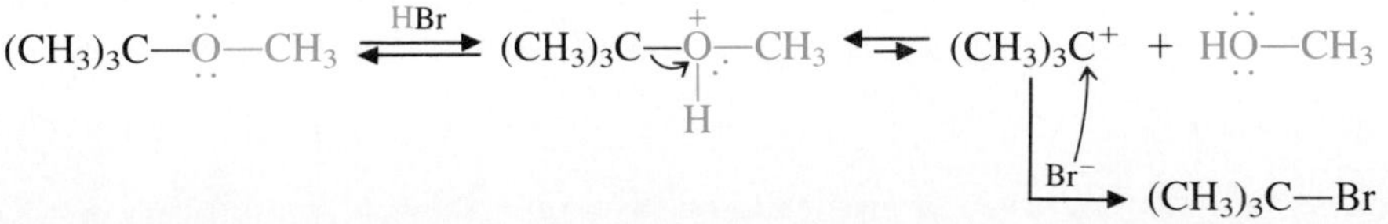

Because alcohols also react with concentrated HI or HBr, the alcohols initially formed in the ether cleavage reaction can react further to give alkyl halides, particularly if the acid concentration is high and the reaction mixture is heated.

$$CH_3-O-C_2H_5 + HI \longrightarrow CH_3I + C_2H_5OH \xrightarrow{HI} C_2H_5I + H_2O \quad (11.24)$$

Although cleavage of alkyl ethers gives alkyl halides and alcohols as products, this reaction is rarely used to prepare these compounds because ethers themselves are most often made from alkyl halides or alcohols.

Problems

10 Explain the following facts using a rational mechanistic argument. When butyl methyl ether is treated with HI, the initially formed products are mainly methyl iodide and butyl alcohol; little or no butyl iodide or methanol is formed. However, when *t*-butyl methyl ether is treated in the same way, the products are mainly methanol and *tert*-butyl iodide.

11 What products are formed when each of the following ethers reacts with concentrated aqueous HI?
(a) diisopropyl ether (b) 2-cyclopentyl-2-ethoxypropane

12 *t*-Butyl methyl ether cleaves much faster in HBr than its sulfur analog, *t*-butyl methyl sulfide. Suggest a reason for this observation. (*Hint:* See Sec. 8.7.)

B. Cleavage of Sulfides by Raney Nickel

Sulfides, disulfides, and even some thiols are readily cleaved by hydrogenation over *Raney nickel* (abbreviated RaNi), a finely divided nickel catalyst prepared from a nickel–aluminum alloy. In this reaction, the sulfur is completely removed from the molecule and is replaced by two hydrogen atoms derived from the hydrogen adsorbed on the metal.

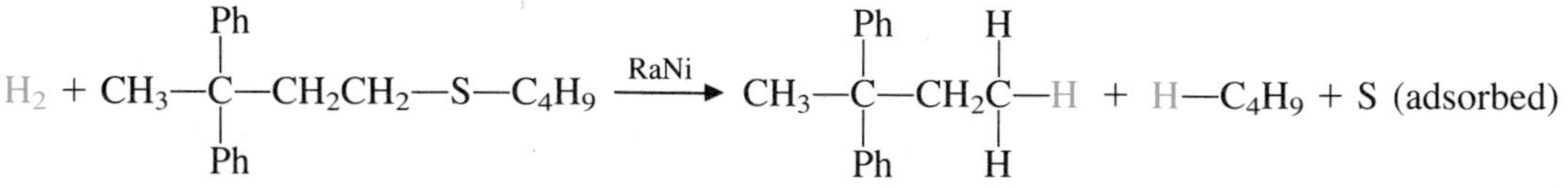

$$H_2 + CH_3-\overset{\overset{\displaystyle Ph}{|}}{\underset{\underset{\displaystyle Ph}{|}}{C}}-CH_2CH_2-S-C_4H_9 \xrightarrow{RaNi} CH_3-\overset{\overset{\displaystyle Ph}{|}}{\underset{\underset{\displaystyle Ph}{|}}{C}}-CH_2\overset{\overset{\displaystyle H}{|}}{\underset{\underset{\displaystyle H}{|}}{C}}-H + H-C_4H_9 + S \text{ (adsorbed)} \quad (11.25)$$

(quantitative)

The mechanism of this curious but useful reaction is not known definitively, but it is thought that adsorption of the sulfide on the nickel leads to the formation of radicals that abstract hydrogen atoms from the surface of the metal.

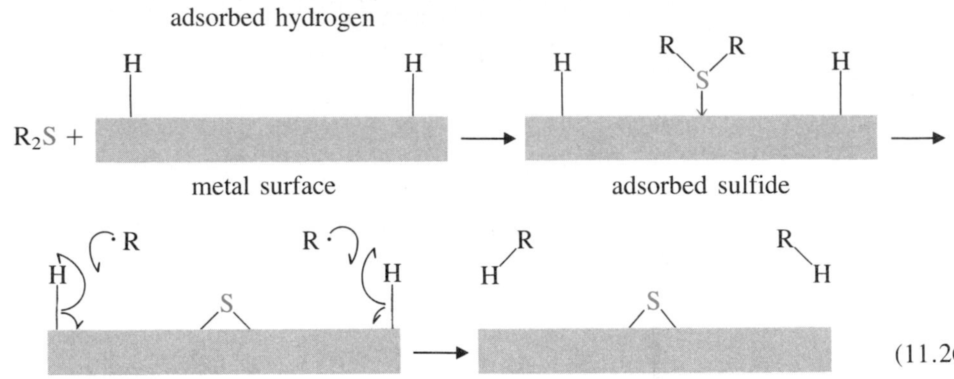

(11.26)

Problems

13 Explain how the radical mechanism of Eq. 11.26 can account for each of the products in the following reaction:

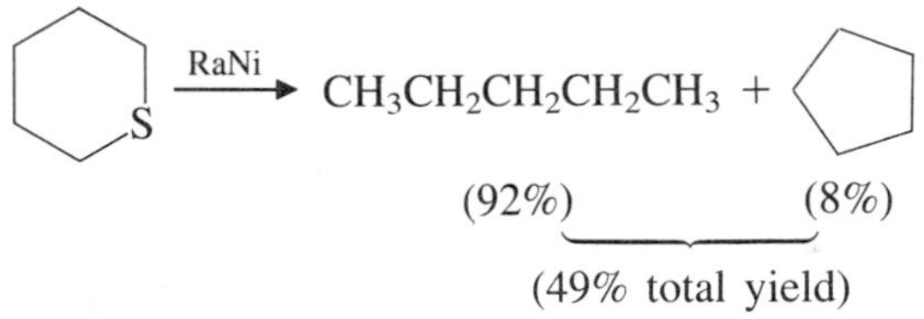

14 Give the product expected when each of the following compounds is treated with Raney nickel:

(a)

(b)

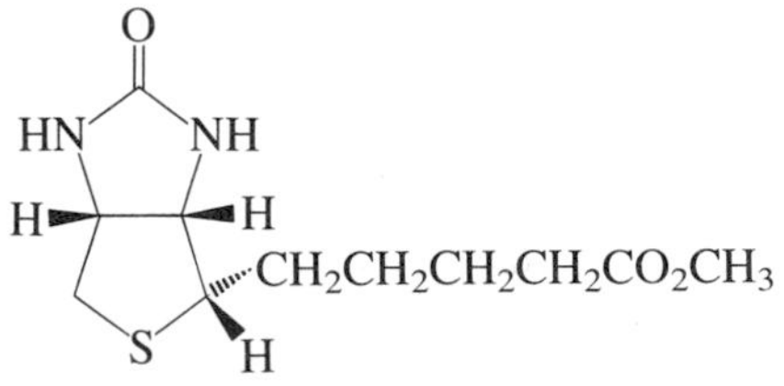

biotin methyl ester
(biotin is a vitamin)

11.4 NUCLEOPHILIC SUBSTITUTION REACTIONS OF EPOXIDES

A. Hydrolysis of Epoxides. Conversion of Epoxides into Glycols

Epoxides undergo a variety of nucleophilic substitution reactions that involve the opening, or cleavage, of the epoxide ring at one of the carbon–oxygen bonds. The net result of an epoxide cleavage is addition across one of the epoxide C—O bonds. The formation of a glycol by addition of water to an epoxide is a typical example of this type of reaction.

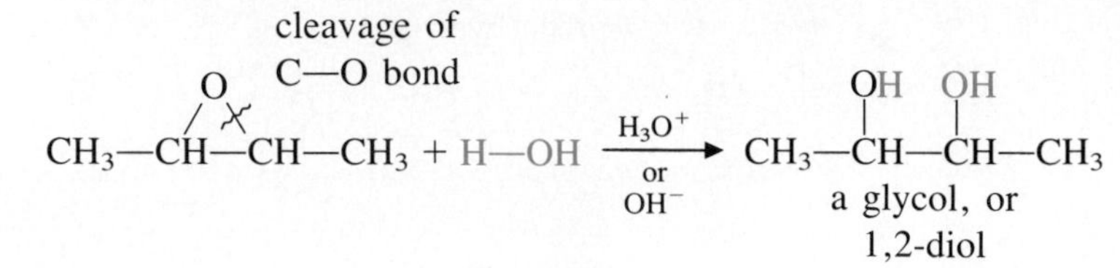
cleavage of
C—O bond
a glycol, or
1,2-diol
(11.27)

This reaction, which is catalyzed by either dilute acid or base, is an example of **hydrolysis**—a reaction in which a chemical bond is broken by reaction with water.

If we think of an epoxide as a type of ether, we can see that a reaction such as Eq. 11.27 is really an ether cleavage. The analogous cleavage of an ordinary ether, however, does *not* occur in dilute acid or base.

$$CH_3CH_2{-}O{-}CH_3 + H_2O \xrightarrow{\text{dil } H^+ \text{ or } OH^-} CH_3CH_2OH + HOCH_3 \qquad (11.28)$$

(not observed)

Rather, as we learned in the previous section, ether cleavage requires heat and strong acid.

What is the reason for the unusually high reactivity of epoxides? Epoxides, like their carbon analogs the cyclopropanes, are very strained molecules (Sec. 7.6). Opening an epoxide relieves the strain in the three-membered ring just as the snapping of a twig relieves the strain of its bending. Thus, epoxide cleavage occurs under mild conditions because epoxide opening is such a rapid reaction.

An unsymmetrical epoxide contains two C—O bonds that can break when ring opening occurs. Which of these bonds is broken? The answer depends on whether the reaction is carried out in acid or base. Let us first consider the reactions of epoxides under basic conditions. When the hydrolysis of isobutylene oxide (2,2-dimethyloxirane) is carried out with isotopically labeled hydroxide ($^{18}OH^-$), it is found that most of the ring opening occurs by attack of the hydroxide at the *less substituted carbon* of the epoxide.

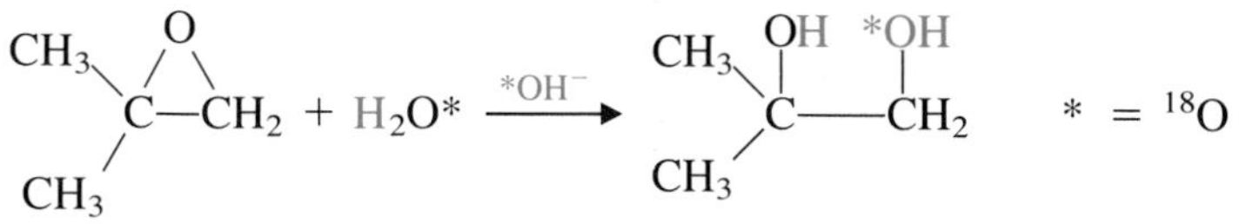
* = 18O
(11.29)

This makes sense if we think of the reaction as an S_N2 process that occurs by displacement of the epoxide oxygen with hydroxide. In this reaction, the epoxide oxygen acts as a leaving group, and forms an alkoxide. (Of course, this oxygen does not leave the molecule; nevertheless, it is still considered to be a leaving group because it leaves the carbon to which it was attached.) As we learned in Sec. 9.3C, the S_N2 reaction is retarded by branching at the site of substitution. Hence, we should not be surprised that hydroxide attacks the epoxide at the carbon with the smaller amount of branching.

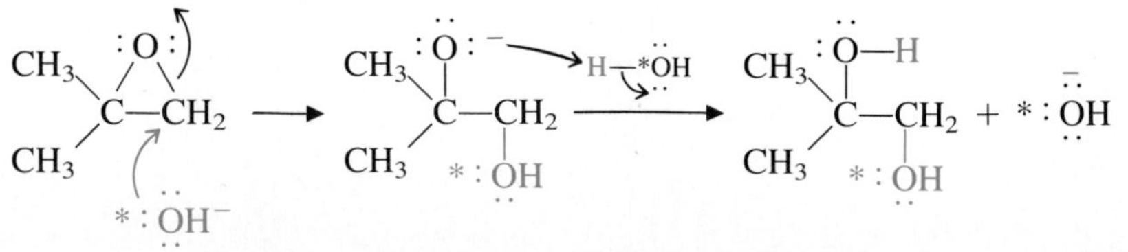
(11.30)

Epoxides undergo ring opening with a wide variety of bases (Problem 17, and Secs. 11.4B and C). It is found that generally, *basic reagents attack an epoxide at the less substituted carbon atom.*

Because the alkoxide leaving group in Eq. 11.30 is very basic, it is a very poor leaving group (Sec. 9.3D); alkoxides generally do not serve as leaving groups in ordinary S_N2 reactions. Indeed, this is the main reason that base-catalyzed cleavage of ordinary ethers is not observed (Eq. 11.28). Yet there is so much strain in the three-membered epoxide ring that ring opening is energetically favorable even at the cost of expelling such a poor leaving group.

Like other S_N2 reactions, ring opening of epoxides by bases involves *backside attack* of the nucleophile on the epoxide carbon, and consequently, inversion of configuration (Sec. 9.3B). Thus, *meso*-2,3-dimethyloxirane reacts with $^-$OH to give *racemic* 2,3-butanediol.

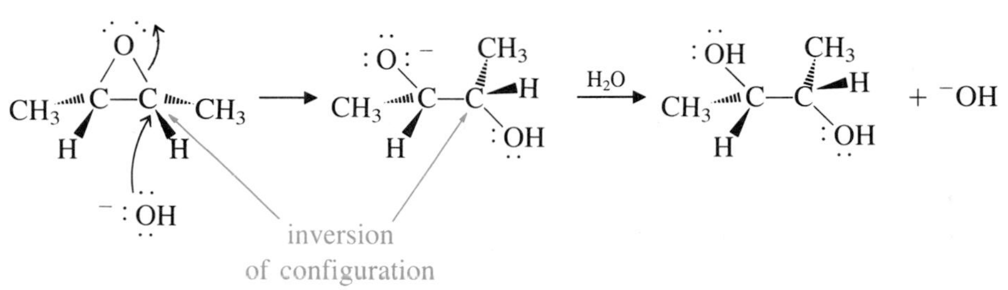

The reaction in Eq. 11.31 shows attack of $^-$OH at one epoxide carbon to give (2*S*,3*S*)-2,3-butanediol. Attack of $^-$OH at the other epoxide carbon gives the enantiomeric product, (2*R*,3*R*)-2,3-butanediol. (Write out this reaction!) Because attack on either carbon is equally likely, the product is the racemate. (Recall from Sec. 7.8A that optically active compounds cannot be formed from achiral reagents.)

Problems

15 What stereoisomer is formed from the reaction of (2*S*,3*S*)-2,3-dimethyloxirane with $^-$OH? Is this product optically active? Explain.

16 Predict the products of the following epoxide ring opening reactions:

(a)
$$\begin{array}{c} CH_3 \\ \diagdown \\ \diagup \\ C_2H_5 \end{array} \!\!\! \overset{O}{\underset{\diagdown}{C}} \!\!\! -CH_2 + Na^+\ N_3^- \xrightarrow[\text{(solvent)}]{C_2H_5OH/H_2O}$$

(b)
$$\begin{array}{c} CH_3 \\ \diagdown \\ \diagup \\ CH_3 \end{array} \!\!\! \overset{O}{\underset{\diagdown}{C}} \!\!\! -CH-CH_3 + Na^+\ CH_3\overset{..}{\underset{..}{O}}:^- \xrightarrow{CH_3OH}$$

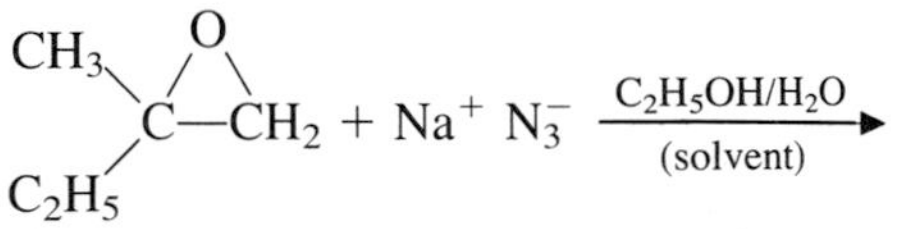

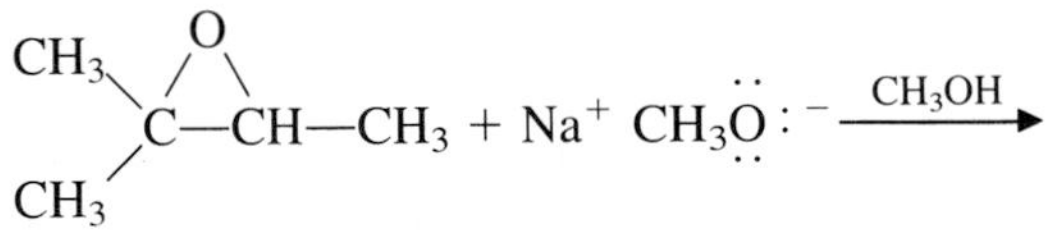

Acid-catalyzed hydrolysis of epoxides, like base-catalyzed hydrolysis, gives glycols. However, when acid-catalyzed hydrolysis of an epoxide is carried out in isotopically labeled water, it is found that attack of water occurs at a tertiary carbon in preference to a primary carbon—that is, at the *more* substituted carbon.

$$(CH_3)_2\overset{O}{\overset{\diagup\ \ \diagdown}{C}}\!\!-CH_2 + H_2O^* \xrightarrow{H_3O^+} (CH_3)_2\overset{\overset{*OH}{|}}{C}\!\!-CH_2-OH \qquad * = {}^{18}O \quad (11.32)$$
$$\text{(99\% of product)}$$

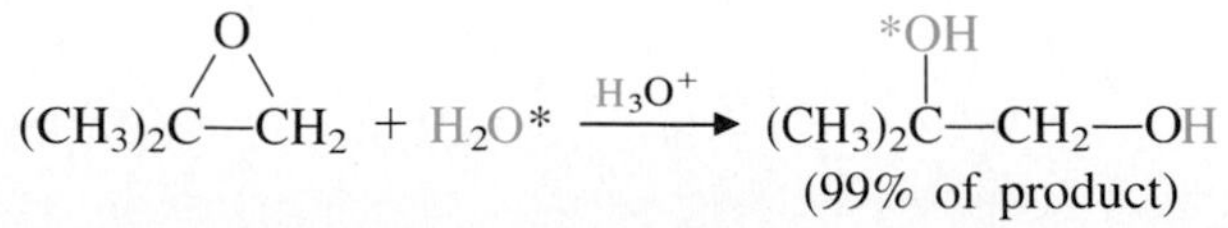

This reaction is analogous to the ring-opening reaction of bromonium ions (Eq. 5.10) or mercurinium ions (Eq. 5.28b). Thus, we can think of the protonated epoxide as a resonance hybrid that has some carbocation character:

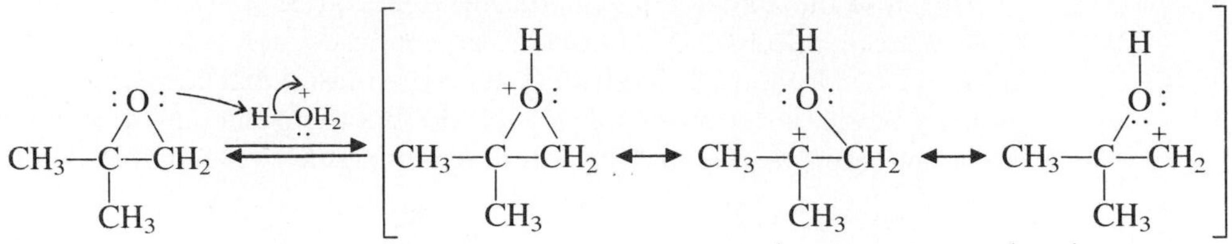

more important less important (11.33a)

Of the two resonance forms that have carbocation character, the one in which the tertiary carbon is electron deficient is more important. Hence, there is more positive charge at the tertiary carbon than at the primary carbon, and water attacks the site of greatest charge—the tertiary carbon.

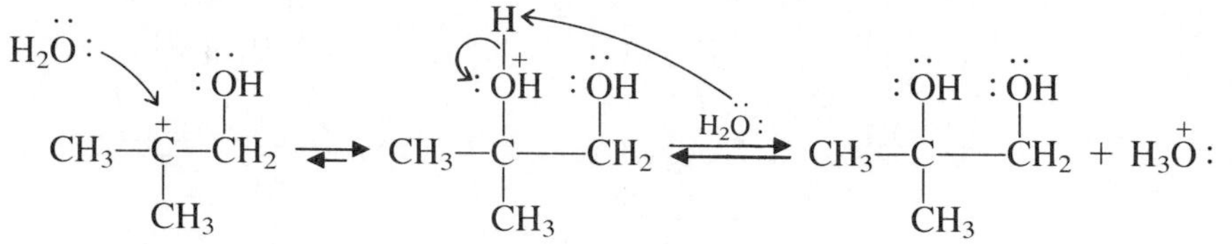

$$CH_3-\overset{+}{\underset{CH_3}{C}}-CH_2 \rightleftharpoons CH_3-\underset{CH_3}{C}-CH_2 \overset{H_2O:}{\rightleftharpoons} CH_3-\underset{CH_3}{C}-CH_2 + H_3\overset{+}{O}: \quad (11.33b)$$

The results of epoxide hydrolysis under acidic conditions can be generalized to the reactions of epoxides with a variety of nucleophiles under acidic conditions (see Problem 17). In general, nucleophiles attack protonated epoxides preferentially at a tertiary carbon. Mixtures of products are obtained under acidic conditions when the epoxide contains one primary and one secondary carbon.

Although both acid- and base-catalyzed hydrolyses of epoxides give glycols, the base-catalyzed reaction is sometimes inferior as a method of glycol synthesis because of side reactions that occur in some cases (see Problem 38); the acid-catalyzed method is preferred.

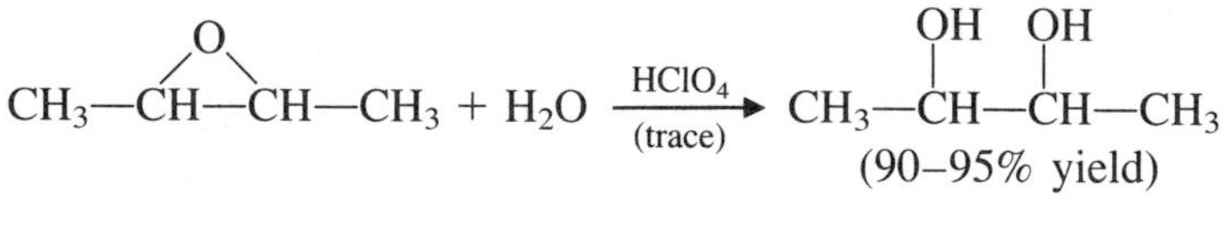

$$CH_3-\overset{O}{\overset{\diagup\diagdown}{CH-CH}}-CH_3 + H_2O \xrightarrow[\text{(trace)}]{HClO_4} CH_3-\underset{OH}{CH}-\underset{OH}{CH}-CH_3 \quad (11.34)$$
(90–95% yield)

From what we have learned, we can now devise two methods to prepare glycols from alkenes: epoxidation of alkenes (Sec. 11.2), followed by hydrolysis of the resulting epoxides; and the direct oxidation of alkenes with KMnO$_4$ or OsO$_4$ (Sec. 5.4). An important difference in the two methods is in the *stereochemistry* of the glycols that are obtained. We can see this difference clearly in the preparation of 1,2-cyclo-hexanediol by each method. Because the direct oxidation of alkenes by KMnO$_4$ or OsO$_4$ is a *syn* addition (Sec. 7.10D), oxidation of cyclohexene gives the *cis*-glycol.

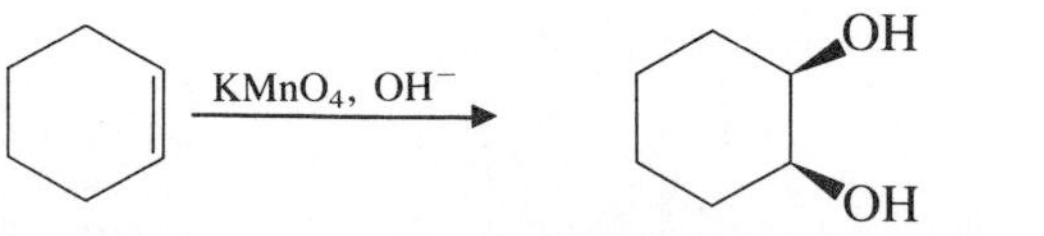

(11.35)

cis-1,2-cyclohexanediol

However, the epoxidation of cyclohexene, followed by hydrolysis of the epoxide, gives *trans*-1,2-cyclohexanediol.

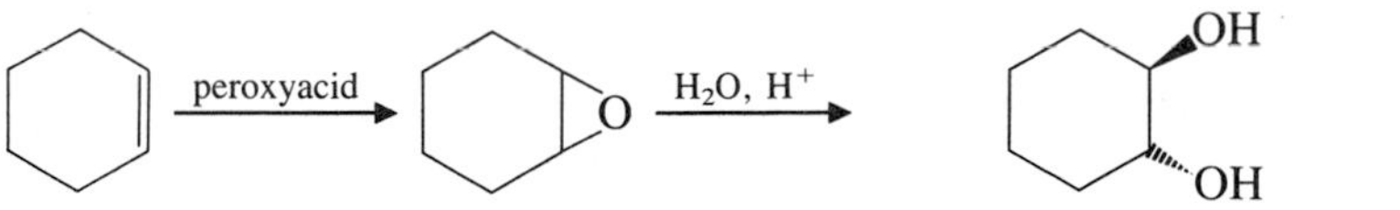

trans-1,2-cyclohexanediol (11.36)

This stereochemistry is consistent with a mechanism in which the protonated epoxide is opened by direct backside attack of water.

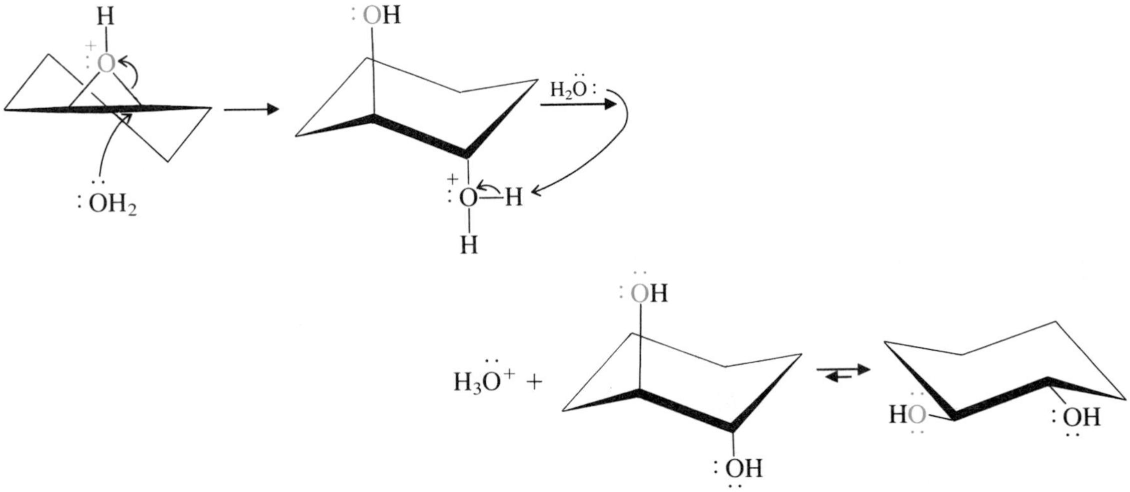

trans-1,2-cyclohexanediol (11.37)

(This is analogous to the stereochemistry of bromonium ion opening with nucleophiles; Eq. 7.32.) Hence, the two methods of glycol synthesis are complementary in a stereochemical sense, and in many cases can be used to prepare respective glycol diastereomers.

Problems

17 Predict the products of the following reactions:

(a)

(optically active)

$+ \; CH_3OH \; \xrightarrow[\text{(trace)}]{H_2SO_4}$

(b)

$(CH_3)_3C$... $CH_2 \; + \; H_2O \; \xrightarrow[\text{(trace)}]{H_2SO_4}$

(c)

$(CH_3)_3C$... $CH_2 \; + \; H_2O \; \xrightarrow{^-OH}$

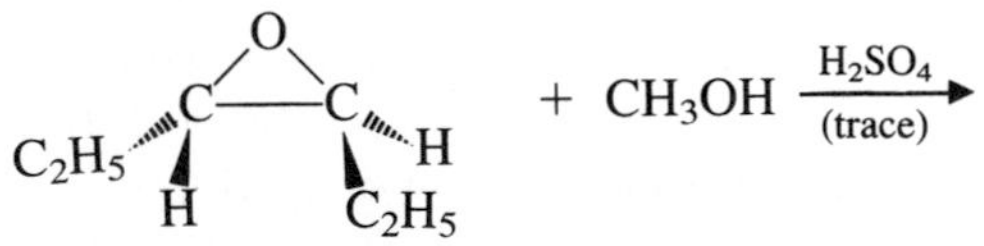

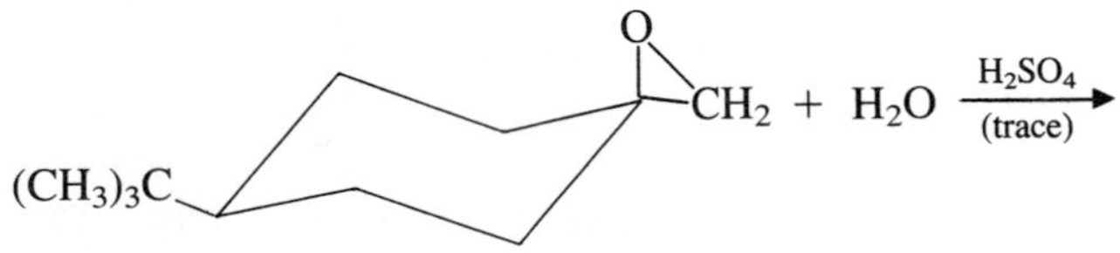

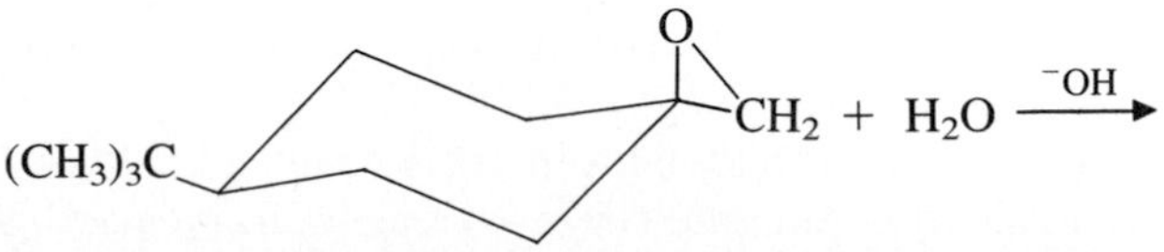

Problems (Cont.)

(d)

$$(CH_3)_2CH-C\underset{CH_3}{\overset{O}{\diagup}}CH_2 + :NH_3 \longrightarrow$$

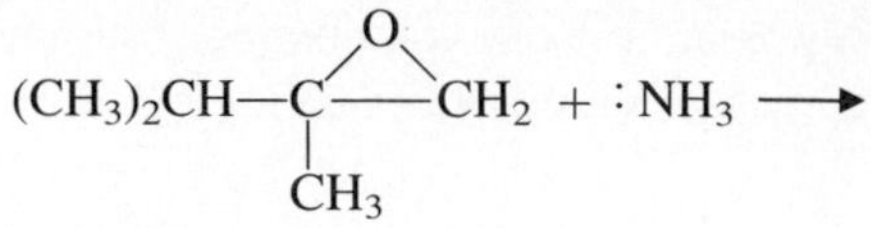

(e)

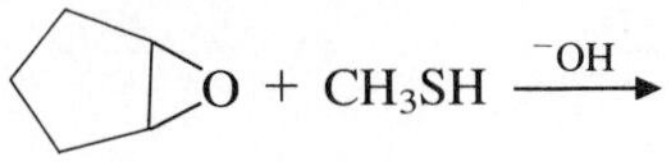

18 (a) In the following reaction, explain why products of opposite stereochemical configurations (shown in Fischer projection) are obtained in acid and base:

$$CH_3-CH-CH_2 \quad \begin{array}{c} \xrightarrow{\;H_3O^+\;} \\ \\ \xrightarrow[H_2O,\; ^-OH]{} \end{array}$$

optically active;
(+)-isomer

HO——H CH₂OH ... CH₃ (S)-(+) (mostly)

H——OH CH₂OH ... CH₃ (R)-(−)

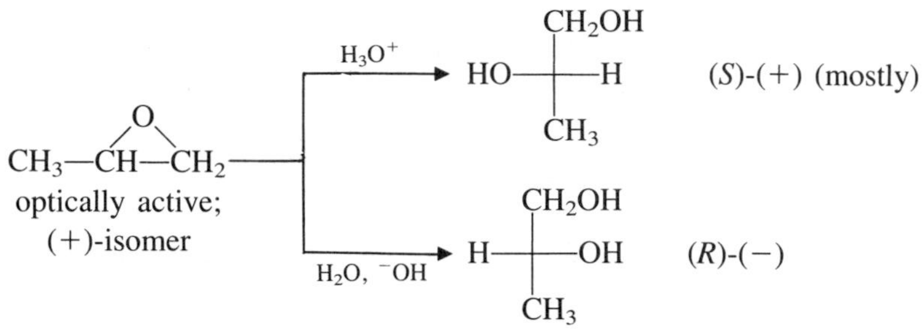

(b) Use these results to assign the absolute configuration of the epoxide.

B. Synthesis of Primary Alcohols: Reaction of Ethylene Oxide with Grignard Reagents

Grignard reagents react with ethylene oxide to give, after a protonation step, primary alcohols.

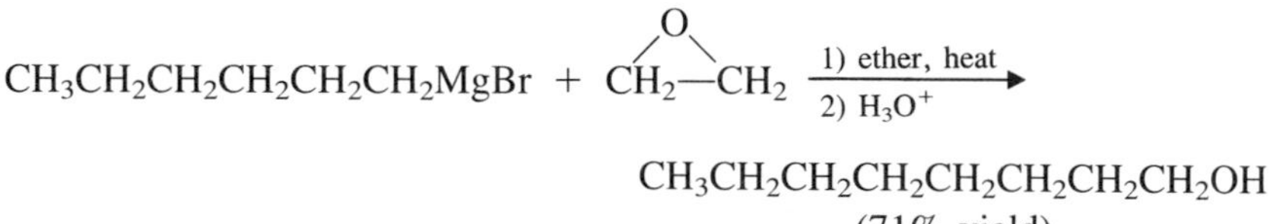

$$CH_3CH_2CH_2CH_2CH_2CH_2MgBr \; + \; CH_2\overset{O}{\diagup}CH_2 \xrightarrow[\text{2) }H_3O^+]{\text{1) ether, heat}}$$

$$CH_3CH_2CH_2CH_2CH_2CH_2CH_2CH_2OH \quad (11.40)$$
$$(71\% \text{ yield})$$

This reaction is another epoxide ring-opening reaction. To understand this reaction, let us recall (Sec. 8.8B) that the carbon in the C—Mg bond of the Grignard reagent has *carbanion* character and is therefore a very *basic* carbon. This carbon attacks the epoxide as a nucleophile. The —MgBr group of the Grignard reagent has Lewis-acid character and assists the reaction by coordinating with epoxide oxygen. (Recall from Eq. 8.26 that Grignard reagents coordinate strongly with ethers.)

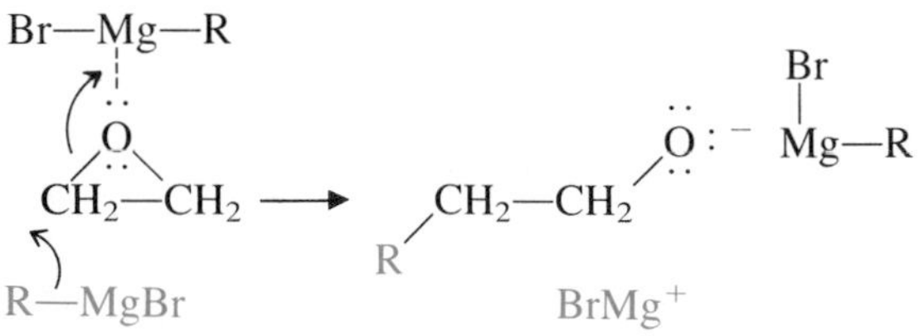

$$(11.41)$$

This reaction yields a bromomagnesium alkoxide—essentially, the magnesium salt of an alcohol. After the Grignard reagent has reacted, the alkoxide is converted into the alcohol product in a separate step by addition of water or dilute acid.

$$R{-}CH_2CH_2{-}\ddot{O}{:}^{-}\ \overset{+}{MgBr} \quad H{-}\overset{+}{O}H_2 \longrightarrow R{-}CH_2CH_2{-}\ddot{O}H + H_2\ddot{O} + Mg^{2+} + Br^{-} \qquad (11.42)$$

We might imagine that a Grignard reagent would react with other epoxides in the same way that it reacts with ethylene oxide. However, although other epoxides react, their reactions are unsatisfactory because they yield troublesome mixtures of products caused by rearrangements and other side reactions.

The reaction of Grignard reagents with ethylene oxide gives us another method for the synthesis of alcohols that we can add to our list in Sec. 10.10A. What are the limitations on the types of alcohols that can be prepared by each method?

Problem

19 (a) Outline a synthesis of 3-methyl-1-pentanol from ethylene oxide and any other reagents.

(b) Outline another synthesis of 3-methyl-1-pentanol from a six-carbon alkene.

11.5 OXONIUM AND SULFONIUM SALTS

A. Reactions of Oxonium and Sulfonium Salts

If we formally replace the proton of a protonated ether with an alkyl group, the resulting compound is called an **oxonium salt.** The sulfur analog of an oxonium salt is a **sulfonium salt.**

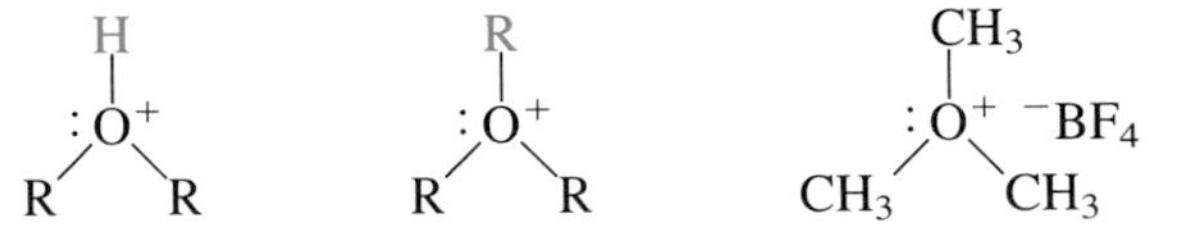

protonated ether — trialkyloxonium ion — trimethyloxonium fluoroborate (an oxonium salt)

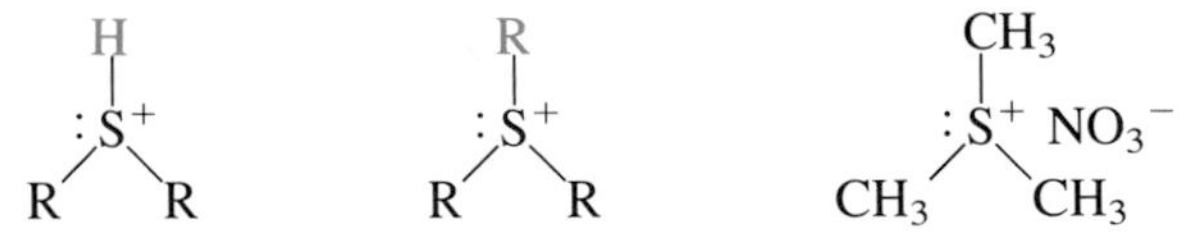

protonated sulfide — trialkylsulfonium ion — trimethylsulfonium nitrate (a sulfonium salt)

Oxonium and sulfonium salts react with nucleophiles in S_N2 reactions.

$$H\ddot{O}{:}^{-} + CH_3{-}\overset{+}{O}(CH_3)_2\ BF_4^{-} \longrightarrow HOCH_3 + (CH_3)_2\ddot{O}{:} + BF_4^{-} \qquad (11.43)$$

(89% yield)

$$(CH_3)_3N{:} + (CH_3)_3S^{+}\ NO_3^{-} \longrightarrow (CH_3)_4N^{+}\ NO_3^{-} + (CH_3)_2\ddot{S}{:} \qquad (11.44)$$

$$Ph{-}CH(CH_3){-}\overset{+}{S}(CH_3){-}CH_3 \quad {:}\ddot{Br}{:}^{-} \xrightarrow{heat} CH_3Br + Ph{-}CH(CH_3){-}\ddot{S}{-}CH_3 \qquad (11.45)$$

Oxonium salts react very rapidly with most nucleophiles. This reactivity reflects the great electronegativity of a positively charged oxygen—its ability to accept a pair of electrons and depart as a leaving group. Because of their reactivity, oxonium salts must be stored in the absence of moisture. For the same reason, these salts are stable only when they contain nonnucleophilic counter-ions, such as fluoroborate ($^-BF_4$). (Fluoroborate ion is not nucleophilic because the boron has no unshared electron pairs.) *Sulfonium salts* are considerably less reactive and therefore are handled more easily. Sulfonium salts are somewhat less reactive than the corresponding alkyl chlorides in S_N2 reactions.

Problems

20 Explain why all attempts to isolate trimethyloxonium iodide lead instead to methyl iodide and dimethyl ether.

21 Complete the following reactions:

(a)

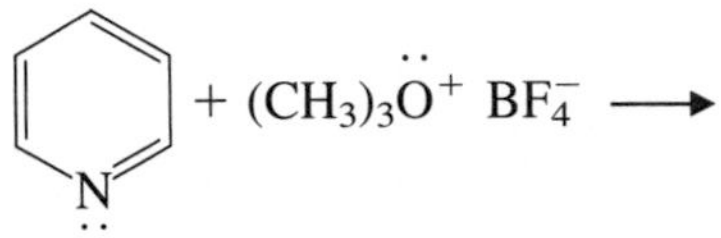

(b) $(CH_3)_2\ddot{S} + (CH_3)_3\overset{+}{\ddot{O}}\ BF_4^- \longrightarrow$

B. S-Adenosylmethionine: Nature's Methyl Donor

Nature provides an interesting example of a sulfonium salt that undergoes nucleophilic substitution reactions. The sulfonium salt *S*-adenosylmethionine (SAM) is used in nature to deliver methyl groups to biological nucleophiles. The structure of SAM is shown in Fig. 11.1. Although this structure is rather complex, we can understand the chemistry of SAM by focusing solely on its sulfonium salt functional group. Like other sulfonium salts, SAM reacts with nucleophiles at the methyl carbon, liberating a sulfide leaving group.

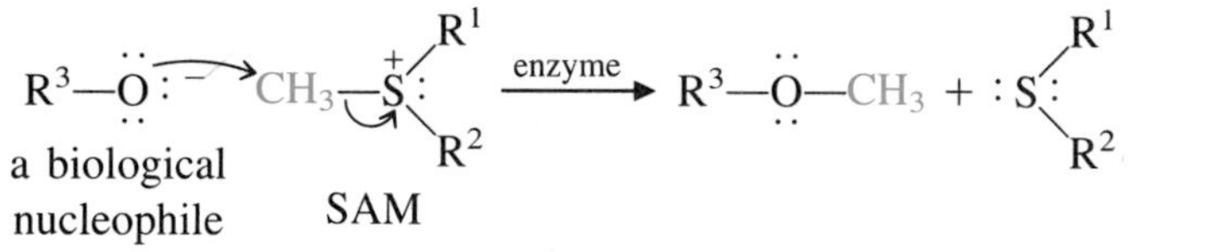

$$R^3 - \ddot{O} : \overset{-}{\longrightarrow} CH_3 - \overset{+}{\overset{R^1}{\underset{R^2}{\ddot{S}}}} \xrightarrow{\text{enzyme}} R^3 - \ddot{O} - CH_3 + : \overset{R^1}{\underset{R^2}{\ddot{S}}} \quad (11.46)$$

a biological nucleophile SAM

SAM is stable enough to survive in aqueous solution, but is reactive enough to undergo enzyme-catalyzed S_N2 reactions. Evidence for an S_N2 mechanism in methylation reactions involving SAM has been obtained by a very elegant type of experiment. The methyl carbon of SAM was made asymmetric solely by isotopic substitution. It was found that displacements on this methyl group proceed with inversion of configuration, exactly as expected for the S_N2 mechanism.

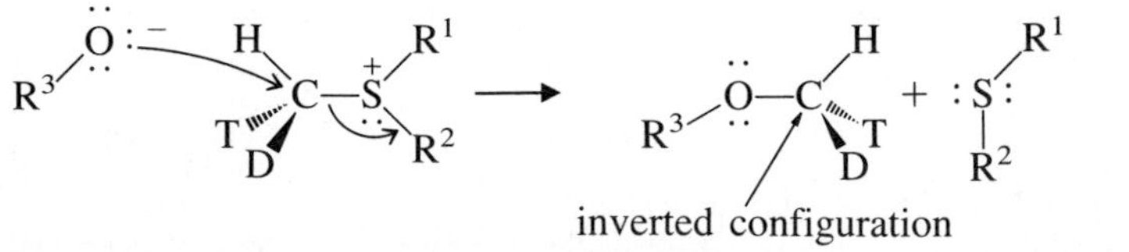

$$(11.47)$$

inverted configuration

The compound *S*-adenosylmethionine, like NAD$^+$ (Sec. 10.7), is another example

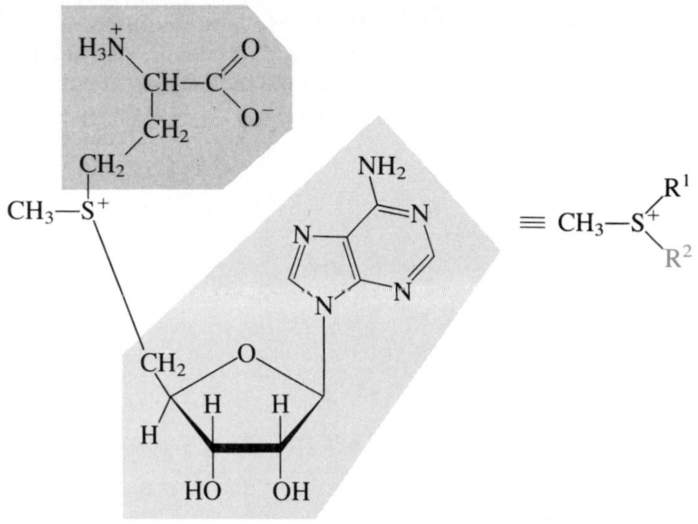

Figure 11.1 *S*-adenosylmethionine (**SAM**). The boxed parts of the structure are abbreviated $\mathbf{R^1}$ and $\mathbf{R^2}$ in the text.

of a very complex biological molecule that undergoes transformations that are readily understood in terms of simple analogies from organic chemistry.

11.6 NEIGHBORING-GROUP PARTICIPATION

An interesting phenomenon is the remarkable difference in the rates of the following two substitution reactions, which are formally very similar.

$$CH_3CH_2CH_2CH_2CH_2CH_2\text{—}Cl + H_2O \xrightarrow[100°]{dioxane}$$
hexyl chloride

relative rate:

$$CH_3CH_2CH_2CH_2CH_2CH_2\text{—}OH + HCl \qquad 1 \qquad (11.48a)$$

$$CH_3CH_2\overset{..}{\underset{..}{S}}CH_2CH_2\text{—}Cl + H_2O \xrightarrow[100°]{dioxane} CH_3CH_2\overset{..}{\underset{..}{S}}CH_2CH_2\text{—}OH + HCl \ \ 3200 \qquad (11.48b)$$
β-chloroethyl
ethyl sulfide

At first sight, both reactions appear to be simple S_N2 reactions in which chloride is displaced as a leaving group by water. In fact, this *is* the mechanism by which hexyl chloride reacts:

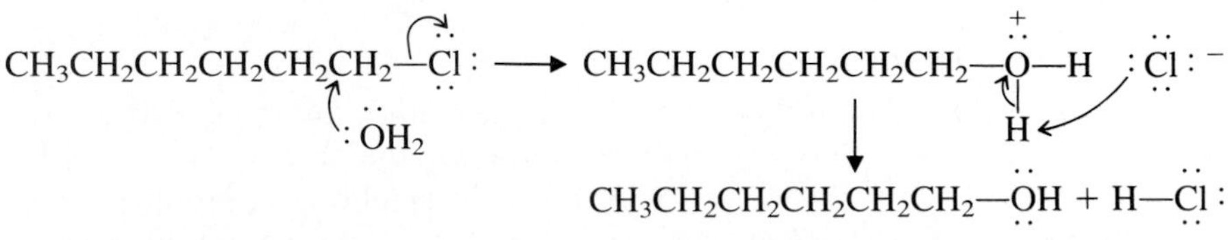

$$CH_3CH_2CH_2CH_2CH_2CH_2\text{—}\overset{..}{\underset{..}{Cl}}: \longrightarrow CH_3CH_2CH_2CH_2CH_2CH_2\text{—}\overset{+}{\underset{H}{O}}\text{—}H \quad :\overset{..}{\underset{..}{Cl}}:^-$$
$$:\overset{..}{O}H_2$$

$$CH_3CH_2CH_2CH_2CH_2CH_2\text{—}\overset{..}{O}H + H\text{—}\overset{..}{\underset{..}{Cl}}: \quad (11.49)$$

The presence of sulfur in an alkyl halide molecule should have little effect on the rate of an S_N2 reaction, because the S_N2 mechanism is not very sensitive to the electronegativities of substituent groups. (In fact, electronegative substituents are known to *re-tard* S_N2 reactions slightly.) The rate of Eq. 11.48b is unusually large because a special mechanism exists that facilitates this reaction, a mechanism not available to hexyl chloride. In this mechanism, the nearby sulfur displaces the chloride *within the same molecule.*

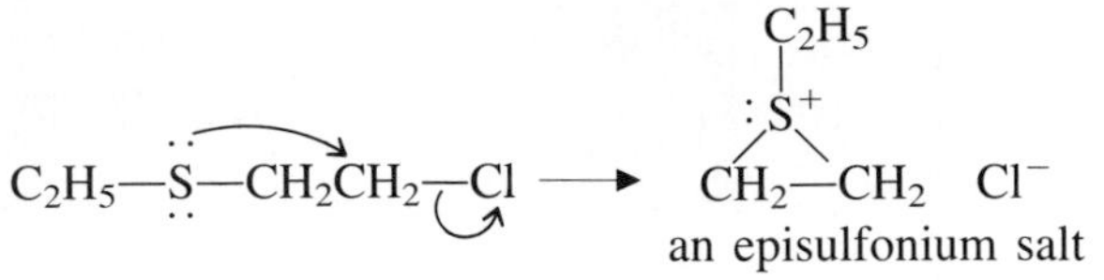

$$C_2H_5-\overset{..}{\underset{..}{S}}-CH_2CH_2-Cl \longrightarrow \overset{\displaystyle C_2H_5}{\underset{\displaystyle CH_2-CH_2}{\overset{|}{\underset{}{\overset{..}{S}^+}}}} \quad Cl^- \qquad (11.50a)$$
$$\text{an episulfonium salt}$$

$$\overset{\displaystyle C_2H_5}{\underset{\displaystyle CH_2-CH_2}{\overset{|}{\overset{..}{S}^+}}} \longrightarrow C_2H_5-\overset{..}{\underset{..}{S}}-CH_2CH_2-\overset{+}{\overset{..}{O}}H_2 \xrightarrow{\ Cl^-\ }$$
$$\underset{\displaystyle :\overset{..}{O}H_2}{}$$

$$C_2H_5-\overset{..}{\underset{..}{S}}-CH_2CH_2-\overset{..}{\underset{..}{O}}H + HCl \quad (11.50b)$$

The intermediate episulfonium ion that results from this internal displacement is relatively unstable because it contains a strained three-membered ring. It is structurally similar to a protonated epoxide or a bromonium ion (Sec. 5.1A, 7.10C). Water attacks this intermediate, reopening it to give the final substitution product. Notice that this product is identical to the one that would arise from direct displacement by water, but, as we have seen, is formed by a different mechanism.

The covalent involvement of neighboring groups in chemical reactions has been termed **neighboring-group participation** or **anchimeric assistance** (from the Greek *anchi,* near). We must recognize that the neighboring-group mechanism of Eq. 11.50 is in *competition* with an ordinary S_N2 mechanism in which water attacks the alkyl halide directly—a mechanism analogous to the one shown in Eq. 11.49. Because nature always seeks pathways of lowest energy, any compound reacts through the transition state of lowest energy—in other words, by the mechanism that gives the fastest reaction. Hence, in order for the neighboring-group mechanism to operate, it *must* give a faster reaction than other competing mechanisms. The rate of the reaction in Eq. 11.48a provides us with a basis of comparison—that is, a rough idea of what rate to expect for a reaction that occurs by direct substitution of water in the absence of neighboring-group participation. (This reaction cannot occur by a neighboring-group mechanism because there is no neighboring group.) A large rate acceleration, such as the one in Eq. 11.48b, is typical of the experimental evidence used to diagnose the involvement of a neighboring group in a chemical reaction. Professor Saul Winstein (1912–1969) of the University of California, Los Angeles, discovered numerous examples of neighboring-group participation and showed that all of these were associated with significant rate accelerations. Some other techniques used to investigate this phenomenon are explored in the following problems.

Problems

22 In the reaction of the following radioactively labeled compound with water, what labeling pattern do you expect in the product (a) if neighboring-group participation does not occur and (b) if neighboring-group participation does occur?

$$C_2H_5-S-CH_2\overset{*}{C}H_2-Cl \quad\quad * = {}^{14}C$$

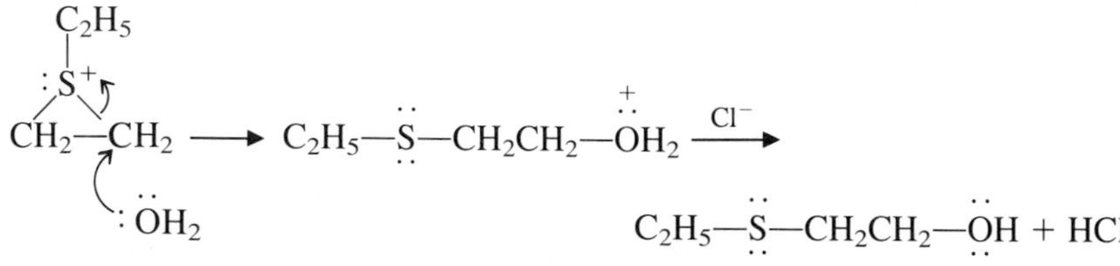

Use your answers to show how a labeling experiment of this sort might be used to argue for or against neighboring-group participation.

23 Explain why the following two alcohols each react with HCl to give the same alkyl chloride:

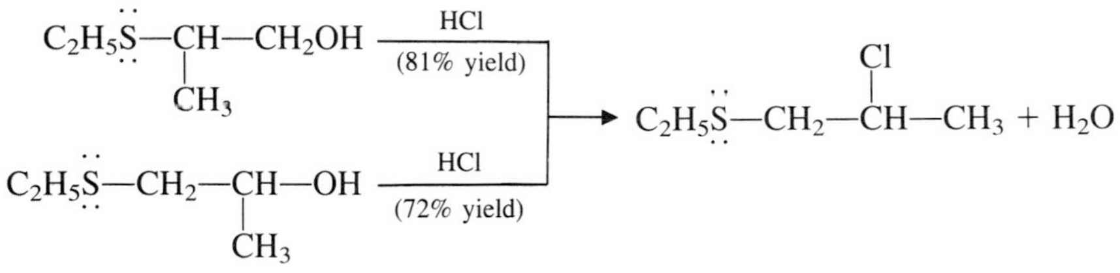

Why are reactions involving neighboring-group mechanisms faster? Of course, the nucleophiles in Eqs. 11.49 and 11.50a are different—an oxygen atom in one case, and a sulfur in the other. But careful studies have shown that the large difference in the rates of these reactions cannot be attributed solely to the difference in the nucleophilic groups. Rather, the difference has to do with the fact that the neighboring-group mechanism, Eq. 11.50a, is an **intramolecular reaction**—a reaction of groups within the same molecule—but the direct displacement mechanism, such as the one shown in Eq. 11.49, is an **intermolecular reaction**—a reaction between different molecules. Let us, then, rephrase our question: Why should an intramolecular reaction be faster than an intermolecular reaction?

The answer has to do with the *probability* that a reaction will occur. Other things being equal, reactions that occur with greater probability have a larger rate. What governs this probability? In order for two groups to react, they must "get together," or collide. When the reacting groups are in different molecules, such as water and an alkyl halide, they must find each other in solution by random diffusion. It is *relatively improbable* that two groups will diffuse together and collide in just the right way for a reaction to occur. However, when the two groups are present in the same molecule, they are *already together*; the only prerequisite for the reaction in Eq. 11.50a is that the C—S bond bend toward the back side of the C—Br bond. Thus the intramolecular reaction of the sulfide group with the alkyl halide group has a *greater probability* of occurring, and thus a *larger rate,* than the reaction of water with the same alkyl halide:

$$C_2H_5—\overset{..}{\underset{..}{S}}—CH_2CH_2—\overset{..}{\underset{..}{Cl}}: \quad \text{more probable than} \quad C_2H_5—S—CH_2CH_2—\overset{..}{\underset{..}{Cl}}: $$
$$: \overset{..}{O}H_2$$

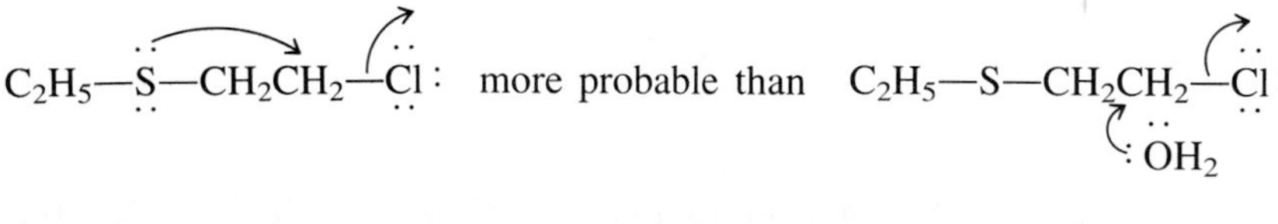

Consider the following analogy to reaction probability: Suppose you are left in a crowded airport and told to find a particular person and shake hands. This would take you a very long time if you had to search at random throughout the terminal. However, this "reaction" would be very rapid if the person you are looking for were tied to you by a very short rope!

The reaction probability is not the only factor that determines the reaction rate. The probability of an intramolecular reaction is balanced against the instability of the cyclic intermediate that is formed. Thus, ring strain raises the energy of the transition state for an intramolecular process that forms a three- or four-membered ring. Nevertheless, the reaction in Eq. 11.50a is so probable that *it occurs despite the strain in the intermediate episulfonium salt*. In fact, the strain in this three-membered ring is the reason that this salt does not survive, but reacts rapidly with water. However, when a four-membered ring is formed, the reaction is less probable (the intramolecular nucleophile is further away—on a longer tether), and the four-membered product contains a significant amount of ring strain. Although there are exceptions, instances of neighboring-group participation involving four-membered rings are rare. Reactions involving five- and six-membered rings are still less probable, but the rings thus formed are so stable that they have relatively low energy. Cases of neighboring-group participation involving five- and six-membered rings are quite common. (See, for example, Eq. 9.5.) Reactions involving rings larger than six members are so improbable that neighboring-group participation is generally not observed in these cases.

In summary: neighboring-group participation in nucleophilic substitution reactions is common for cases involving three-, five-, and six-membered rings.

Problem

24 Give the structure of an *intramolecular* substitution product and an *intermolecular* substitution product that would be obtained from each of the following compounds in the presence of water and NaOH. Which one of the three compounds will give the highest percentage of intermolecular reaction product? Why?

(a) HO—CH$_2$CH$_2$—Br (*Hint:* See Sec. 11.2B.)
(b) HO—CH$_2$CH$_2$CH$_2$CH$_2$—Br
(c) HO—CH$_2$CH$_2$CH$_2$CH$_2$CH$_2$CH$_2$CH$_2$—Br

11.7 OXIDATION OF ETHERS AND SULFIDES

Ethers are relatively inert toward many of the common oxidants used in organic chemistry if the reaction conditions are not too vigorous. For example, diethyl ether can be used as a solvent for oxidations with Cr(VI). On standing in air, however, ethers undergo the slow *autoxidation* discussed in Sec. 8.9D that leads to the formation of dangerous peroxide contaminants.

The sulfur analog of a peroxide is a disulfide, R—S—S—R. As we learned in Sec. 10.9, disulfides are more appropriately considered as oxidation products of thiols. Disulfides are not explosive.

Like thiols, sulfides oxidize at *sulfur* rather than carbon when they react with common oxidizing agents. Sulfides can be oxidized to **sulfoxides** and **sulfones.**

$$R—S—R \xrightarrow{\text{oxidize}} \left[R—\overset{\overset{\displaystyle :O:}{\|}}{\underset{..}{S}}—R \longleftrightarrow R—\overset{\overset{\displaystyle :O:^-}{|}}{\underset{+}{S}}—R \right] \xrightarrow{\text{oxidize}} \left[R—\overset{\overset{\displaystyle :O:}{\|}}{\underset{\underset{\displaystyle :O:}{\|}}{S}}—R \longleftrightarrow R—\overset{\overset{\displaystyle :O:^-}{|}}{\underset{\underset{\displaystyle :O:^-}{|}}{S^{2+}}}—R \right] \quad (11.51)$$

a sulfoxide a sulfone

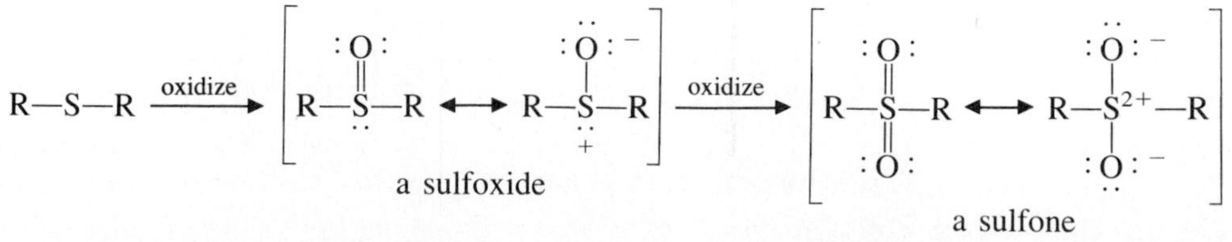

Dimethyl sulfoxide (DMSO) and sulfolane are well-known examples of a sulfoxide and a sulfone, respectively. (Both compounds are excellent dipolar aprotic solvents; see Table 8.2.)

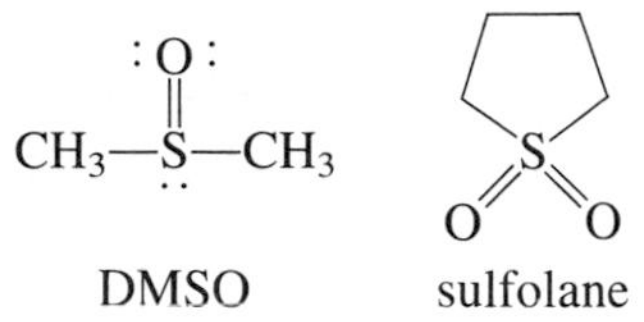

Notice that nonionic Lewis structures for sulfoxides and sulfones cannot be written without violating the octet rule (Sec. 10.9).

Sulfoxides and sulfones can be prepared by the direct oxidation of sulfides with one and two equivalents, respectively, of hydrogen peroxide, H_2O_2.

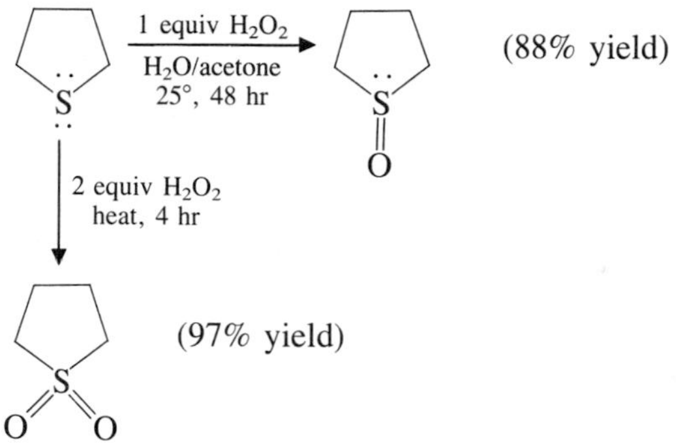

(11.52)

Other common oxidizing agents such as $KMnO_4$ or HNO_3 also oxidize sulfides to sulfones.

11.8 ORGANIC SYNTHESIS: THE THREE FUNDAMENTAL OPERATIONS

In Sec. 10.10B, we began our study of organic synthesis by considering a rational approach to synthesis problems. Let us now continue by analyzing the types of operations embodied in a typical synthesis. Organic synthesis, in the final analysis, can be broken down into *three fundamental types of operations.* Most reactions used in organic synthesis involve one or more of these fundamental operations. These operations are:

1. Functional-group transformation
2. Control of stereochemistry
3. Formation of carbon–carbon bonds

Functional-group transformation—the conversion of one functional group into another—is the most common type of synthetic operation. Most of the reactions we have studied so far involve functional-group transformation. For example, hydrolysis of epoxides transforms epoxides into glycols; hydroboration–oxidation converts alkenes into alcohols.

We have studied several stereoselective reactions that give us *control of stereochemistry*. Whenever we have to prepare a compound that can exist as several stereoisomers we think in terms of these reactions. Among the stereoselective reactions we have studied are hydroboration–oxidation, which is a *syn* addition; epoxide hydrolysis, which gives *trans*-glycols; and the S_N2 reaction of alkyl halides, which occurs with inversion of configuration.

Because most organic compounds contain many carbon–carbon bonds, reactions that bring about *carbon–carbon bond formation* occupy a place of special importance in the synthetic repertoire of the organic chemist. There are only two reactions of this type that we have studied in detail:

1. Cyclopropane formation from dichloromethylene and alkenes (Sec. 9.7)
2. Reaction of Grignard reagents with ethylene oxide (Sec. 11.4B)

As we learn others, we shall want to make special note of them, since it will be these reactions that allow us to elaborate new carbon skeletons.

Most reactions represent combinations of the three fundamental operations above. For example, hydroboration–oxidation is a functional-group transformation that also allows control of stereochemistry; two operations can be accomplished at once with this reagent. The reaction of ethylene oxide with Grignard reagents effects both carbon–carbon bond formation and the transformation of an epoxide into an alcohol.

Let us examine two more multistep syntheses with particular attention to the three fundamental operations. First, suppose we are asked to outline a synthesis of 1-hexanol from 1-butanol and any other reagents.

$$CH_3CH_2CH_2CH_2{-}OH \xrightarrow{?} CH_3CH_2CH_2CH_2CH_2CH_2{-}OH \qquad (11.53)$$

(Try to work this problem before reading on.) The first thing we should ask is whether new carbon–carbon bonds have to be formed. The answer is yes. We have to add a chain of two carbon atoms to the existing chain of four carbons in the starting material. Let us work backward from the product. The reaction of a Grignard reagent with ethylene oxide can be used both to extend a carbon chain by two atoms, and to prepare primary alcohols. This reaction is a suitable last step in our synthesis:

$$CH_3CH_2CH_2CH_2{-}MgBr + \overset{O}{CH_2{-}CH_2} \longrightarrow \xrightarrow{H_3O^+} CH_3CH_2CH_2CH_2CH_2CH_2{-}OH \qquad (11.54)$$

Next we ask how we can prepare the Grignard reagent. We only know one way.

$$CH_3CH_2CH_2CH_2{-}Br + Mg \xrightarrow{\text{ether}} CH_3CH_2CH_2CH_2{-}MgBr \qquad (11.55)$$

Since the alkyl halide has the same number of carbons as the starting alcohol, and since it cannot exist as stereoisomers, we only need to concern ourselves with a functional-group transformation to complete the synthesis. There are several ways that we can use to convert a primary alcohol into a primary alkyl halide. Any of these ways completes the synthesis.

$$CH_3CH_2CH_2CH_2{-}OH \xrightarrow{\text{HBr, heat}} CH_3CH_2CH_2CH_2{-}Br \xrightarrow{\text{Mg, ether}}$$

$$CH_3CH_2CH_2CH_2{-}MgBr \xrightarrow{\overset{O}{CH_2{-}CH_2}} \xrightarrow{H_3O^+} CH_3CH_2CH_2CH_2CH_2CH_2{-}OH \quad (11.56)$$

It is worth noting that the sequence used here is a general one for the net *two-carbon chain extension* of a primary alcohol.

$$R{-}OH \longrightarrow R{-}Br \longrightarrow RMgBr \xrightarrow{\overset{O}{CH_2{-}CH_2}} \xrightarrow{H_2O} R{-}CH_2CH_2{-}OH \quad (11.57)$$

net chain extension by two carbons

A second example illustrates a general use of epoxides—to place two functional groups at adjacent carbons. Suppose we are asked to prepare *trans*-2-methoxy-1-cyclohexanol from cyclohexene.

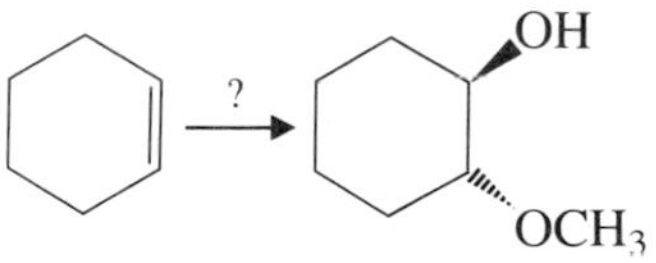

(11.58)

Immediately we see that no new carbon–carbon bonds are joined to the cyclohexene ring, so reactions that form carbon–carbon bonds are not likely to be useful. The next thing we should notice is that there is a stereochemical problem: the two oxygens must be introduced in a *trans* arrangement. Finally, we notice that there has been a net addition to the carbon–carbon double bond. Now we know no way of adding CH_3O— and HO— to a double bond in one step. However, we do know that in the opening of epoxides, the epoxide oxygen becomes an —OH group, and that this occurs with inversion of stereochemistry at the broken bond so that the resulting groups end up *trans*. Opening an epoxide with CH_3O^- in CH_3OH, or with CH_3OH and acid, is a good last step to the synthesis.

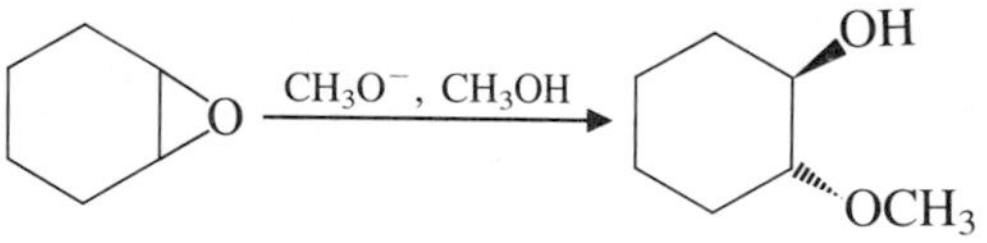

(11.59)

Completion of the synthesis requires only that we prepare the epoxide from cyclohexene (how is this accomplished?).

Problem

25 Outline a synthesis for each of the following compounds from the indicated starting materials and any other reagents:
(a) $(CH_3)_2CHCH_2CH_2CO_2H$ from $(CH_3)_2C{=}CH_2$ (isobutylene)
(b) $(CH_3)_2CHCO_2H$ from isobutylene
(c) *trans*-1-ethoxy-2-methoxycyclopentane from cyclopentene
(d) dibutylsulfone from 1-butanethiol

KEY IDEAS IN CHAPTER 11

- Ethers are relatively unreactive compounds. The major reaction of ethers is cleavage, which occurs under strongly acidic conditions. The cleavage of tertiary ethers occurs more readily than cleavage of primary or methyl ethers because carbocation intermediates can be formed.
- Epoxides, because of their ring strain, undergo ring-opening reactions with ease. For example, epoxides react with water to give glycols, and ethylene oxide reacts with Grignard reagents to give primary alcohols. In acid, a protonated epoxide is preferentially attacked at the tertiary carbon by nucleophiles. Bases attack an epoxide at the less substituted carbon.
- Ring-opening reactions of epoxides in either acid or base occur with inversion of configuration.
- Sulfides, disulfides, and thiols can be cleaved by Raney nickel to give the corresponding hydrocarbons.
- Ethers can be synthesized by the Williamson ether synthesis (an S_N2 reaction), and by alkoxymercuration (a variation of oxymercuration) followed by $NaBH_4$ reduction.
- Epoxides can be synthesized by direct oxidation of alkenes with peroxyacids, or by cyclization of halohydrins.
- Intramolecular nucleophilic substitution reactions involving the formation of three-, five-, and six-membered rings are in many cases considerably faster than competing intermolecular counterparts.
- Reactions involving the intramolecular covalent participation of neighboring groups are said to occur with neighboring-group participation. Reactions that occur by neighboring-group mechanisms are generally faster than analogous reactions that do not involve neighboring-group participation.
- Oxonium and sulfonium salts react with nucleophiles in substitution and elimination reactions; oxonium salts are more reactive than sulfonium salts. *S*-Adenosylmethionine (SAM) is a sulfonium salt used in nature to deliver methyl groups to nucleophiles in living organisms.
- Except for peroxide formation, which occurs on standing in air, ethers are relatively inert toward oxidizing conditions. Sulfides, however, are readily oxidized to sulfoxides and sulfones.
- The three fundamental operations of organic synthesis are (1) functional-group transformation, (2) control of stereochemistry, and (3) carbon–carbon bond formation. The reactions of dichloromethylene with alkenes, and Grignard reagents with ethylene oxide, are examples of reactions used for carbon–carbon bond formation.

ADDITIONAL PROBLEMS

26 Draw a structure of
(a) a nine-carbon ether that cannot be prepared by a Williamson synthesis
(b) an ether that would react with HBr to give propyl bromide as the only alkyl halide
(c) a four-carbon ether that would yield 1,4-diiodobutane after heating with an excess of K_3PO_4 and HI
(d) a four-carbon alkene that would give the same glycol after treatment either with alkaline $KMnO_4$ or with *m*-chloroperbenzoic acid, then dilute aqueous acid
(e) a four-carbon alkene that would give different glycols as a result of the different reaction conditions in (d)
(f) an alkene C_6H_8 that can form only one mono-epoxide and two di-epoxides (counting stereoisomers)

27 Give the products of the reaction of 2-ethyl-2-methyloxirane with each of the following reagents:
(a) water, H_3O^+
(b) Na^+ CH_3O^- in methanol
(c) product of (b) + HBr, 25°
(d) product of (c) + Mg in ether
(e) product of (d) + ethylene oxide, then H_3O^+
(f) product of (b) + NaH, then CH_3I
(g) product of (a) + periodic acid
(h) product of (a) + concd H_2SO_4

28 A student, Alise Analogy, noting that ethylene oxide is readily opened in aqueous NaOH, has theorized that because cyclopropane has a large amount of ring strain, it too should open in base. She has come to you puzzled because she is unable to find any examples of such a reaction in the chemical literature. Explain why the ring opening of epoxides in base occurs readily, whereas the ring opening of cyclopropanes in base does not occur.

29 Explain how you would differentiate between the compounds in each of the following sets by using simple chemical or physical tests that give readily observable results, such as solubility differences, color changes, evolution of gases, or formation of precipitates.
(a) 1-pentanol and 1-methoxybutane
(b) 3-ethoxypropene and 1-ethoxypropane
(c) 1-methoxy-2-methylpropane and 1-methoxy-2-chloro-2-methylpropane

30 Unwanted HCl is usually removed from reaction mixtures by neutralization with aqueous base. There are times, however, when the use of base is not compatible with the conditions of a reaction. It has been found that propylene oxide (2-methyloxirane) can be used to remove HCl quantitatively. Explain why this procedure works.

Problems (Cont.)

31 A student, I. Q. Low, has run each of the following reactions and is disappointed to find that each has given none of the desired product. Low has come to you for advice. Explain why each reaction failed.

(a) $(CH_3)_2CHCH_2CH_2Br \xrightarrow{Na^+ \; C_2H_5O^- \text{ in } H_2O} (CH_3)_2CHCH_2CH_2{-}O{-}C_2H_5$

(b) $HOCH_2CH_2\overset{O}{CH{-}CH_2}$ (epoxide) $\xrightarrow[\text{ether}]{1 \text{ equiv } CH_3MgI} \xrightarrow{H_3O^+} HOCH_2CH_2\underset{}{\overset{OH}{|}}{CH}CH_2CH_3$

32 Keeping in mind that many intramolecular reactions that form six-membered rings are faster than competing intermolecular reactions, predict the product of the following reaction:

$$HOCH_2CH_2C(CH_3)_2CH_2CH{=}CH_2 \xrightarrow[\text{THF/water}]{Hg(OAc)_2} \xrightarrow{NaBH_4} (C_8H_{16}O)$$

33 Compound *A*, $C_7H_{14}O_2$, cannot be resolved into enantiomers and liberates a gas on treatment with either metallic sodium or C_2H_5MgBr. Reaction of *A* with *p*-toluenesulfonyl chloride in pyridine yields a di-tosylate derivative, which, when allowed to react with Na_2S, gives compound *B*, $C_7H_{12}S$. When *B* is treated with Raney nickel, two products *C* and *D* are formed. What are the structures of *A* and *B*? (*Hint*: Think of Na_2S as $:\ddot{\underset{..}{S}}:^{2-}$.)

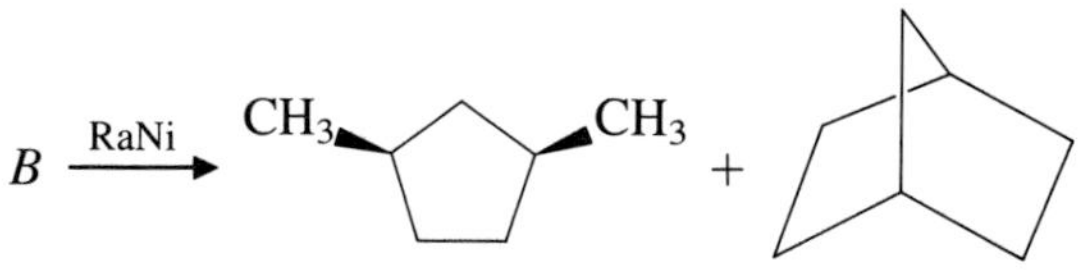

C (mostly) *D* (some)

34 Complete the following reactions by giving the principal organic products:

(a) $CH_3CH_2CH_2{-}Br + Na^+ \; C_2H_5O^- \xrightarrow{C_2H_5OH}$

(b) $CH_3{-}C(CH_3)(C_2H_5){-}Br + (CH_3)_3C{-}O^- \; K^+ \xrightarrow{(CH_3)_3C{-}OH}$

(c) $(CH_3)_2C{=}CH{-}CH_3 + (CH_3)_2CH{-}OH \xrightarrow{Hg(OAc)_2} \xrightarrow{NaBH_4}$

(d)

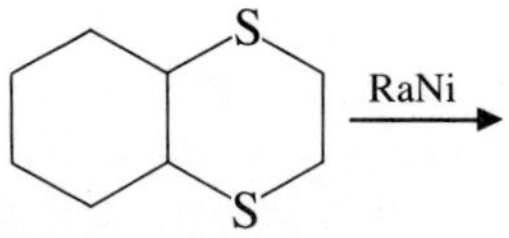

(e)

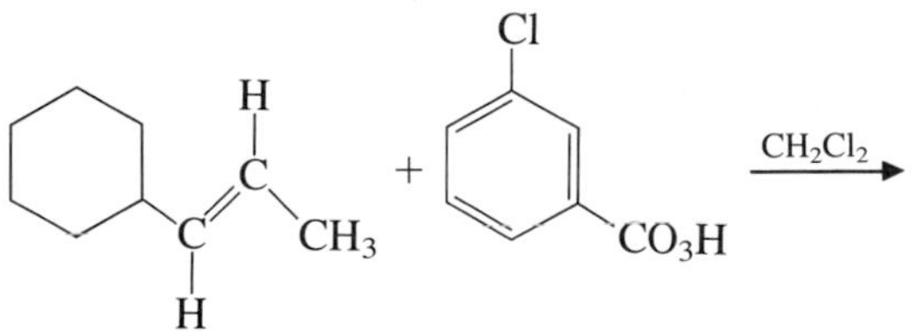

(f) $CH_3-CH=CH_2 + H_2O + Br_2 \longrightarrow \xrightarrow[\text{pyridine}]{CrO_3}$

(g) $BrCH_2CH_2CH_2-CH-CH_2 + HIO_4 \xrightarrow{H_2O}$

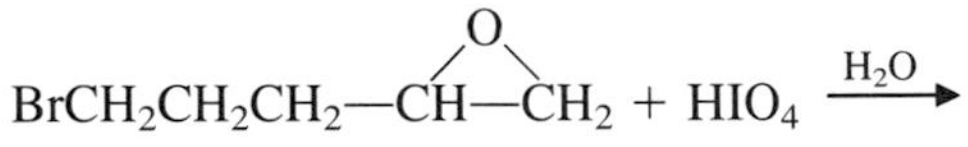

(*Hint:* HIO$_4$ is a fairly strong acid.)

(h) $O + NaN_3 \xrightarrow{H_2O}$

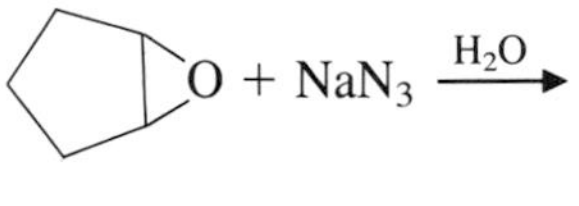

(i) $ClCH_2CH_2CH_2CH_2Cl + Na_2S \xrightarrow[\text{(a polar aprotic solvent)}]{DMF} (C_4H_8S)$

(*Hint:* Think of Na$_2$S as $:\!\ddot{S}\!:^{2-}$.)

(j) $CH_3-\underset{OH}{CH}-\underset{OH}{CH}-CH_3 + CH_3-\!\!\underset{\text{(1 equiv)}}{\!\!\!\!\!\!}\!\!\!\bigcirc\!\!-SO_2Cl \xrightarrow{\text{pyridine}} \xrightarrow{\text{NaOH}}$

35 When $CH_3CH_2-\ddot{S}-CH_2CH_2-\ddot{S}-CH_2CH_3$ reacts with two equivalents of CH_3I, the following double sulfonium salt precipitates:

$$CH_3CH_2-\overset{\overset{CH_3}{|}}{\underset{+}{\ddot{S}}}-CH_2CH_2-\overset{\overset{CH_3}{|}}{\underset{+}{\ddot{S}}}-CH_2CH_3 \quad 2I^-$$

(a) Give a mechanism for this reaction.
(b) Upon closer examination, this compound is found to be a mixture of two isomers with melting points of 123–124° and 154°, respectively. Explain why *two* compounds of this structure are formed. What is the relationship between these isomers?

36 The drug *mechloramine* is used in antitumor therapy.

$$Cl-CH_2CH_2-\overset{\overset{CH_3}{|}}{\ddot{N}}-CH_2CH_2-Cl \qquad \text{mechloramine}$$

It is one of a family of compounds called *nitrogen mustards,* which includes the antitumor drugs cyclophosphamide and chlorambucil.
(a) Mechloramine reacts with water several thousand times faster than 1,5-

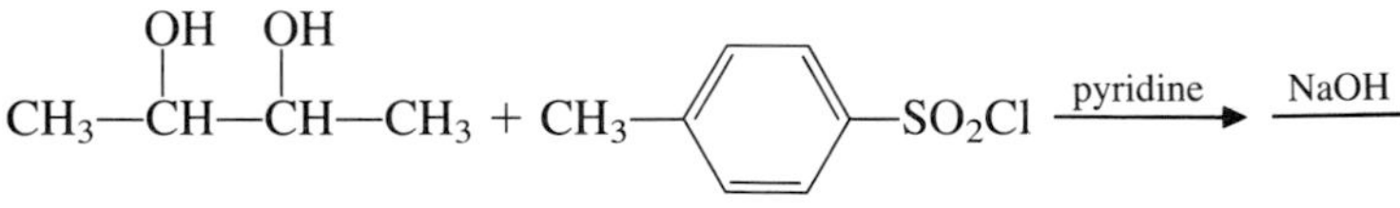

Problems (Cont.)

dichloropentane. Give the product of this reaction and the mechanism for its formation.

(b) It is theorized that the antitumor effects of mechloramine are due to its reaction with certain nucleophiles in the body. What product would you expect from the reaction of mechloramine and a general amine $R_3N:$?

37 Outline a synthesis for each of the following compounds from the indicated starting material and any other reagents. (All chiral compounds are prepared as racemates.)

(a) 2-ethoxy-3-methylbutane from 3-methyl-1-butene

(b) $C_2H_5SCH_2CH_2SC_2H_5$ from ethylene

(c)

$$C_2H_5{-}\overset{\overset{\displaystyle O}{\|}}{S}{-}CH_2CH_2CH_2CH_3 \text{ from compounds containing } \leq 2 \text{ carbons}$$

(d) from an alkene

The starting material for (e) and (f) is cyclohexene.

(e) (f)

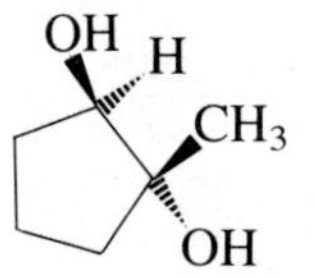

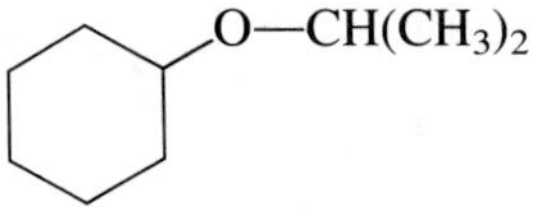

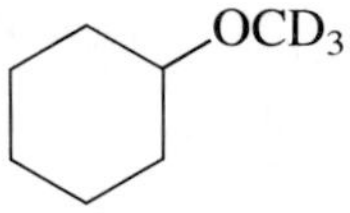

(Don't use CD_3OH as a solvent; it is expensive!)

(g) $(CH_3)_2CHCH_2CH_2CH_2CH{=}O$ from 3-methyl-1-butene

(h)
$$\overset{\displaystyle OCH_3}{\underset{\displaystyle |}{Ph{-}CH{-}CH_2Br}} \text{ from styrene } (Ph{-}CH{=}CH_2) \text{ by a single reaction}$$

(*Hint:* See Problem 32c, Chapter 5.)

(i) $\underset{\displaystyle CH_3}{\overset{\displaystyle C_2H_5}{\diagdown \diagup}}CH\overset{\overset{\displaystyle O}{\|}}{C}CH_2OCH_3$ from 3-methyl-1-pentene

38 One of the side reactions that take place when epoxides react with ^-OH is the formation of dimers or trimers, as in the following example:

$$CH_3{-}\overset{O}{\overset{\diagup \diagdown}{CH{-}CH_2}} \xrightarrow{\ ^-OH,\ H_2O\ } HO{-}CH_2{-}\underset{\underset{\displaystyle CH_3}{\displaystyle |}}{CH}{-}O{-}CH_2{-}\underset{\underset{\displaystyle CH_3}{\displaystyle |}}{CH}{-}O{-}CH_2{-}\underset{\underset{\displaystyle CH_3}{\displaystyle |}}{CH}{-}OH$$

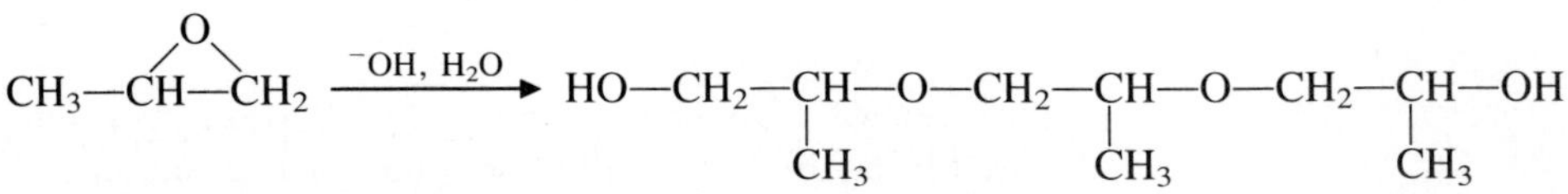

Give the mechanism of this reaction.

39 Compound *A*, C_8H_{16}, undergoes catalytic hydrogenation to give octane. When treated with *m*-chloroperbenzoic acid, *A* gives an epoxide *B*, which, when treated with aqueous acid, gives a compound *C*, $C_8H_{18}O_2$, that can be resolved into enantiomers. When *A* is treated with OsO_4, an achiral compound *D*, an isomer of *C*, forms. Identify all compounds, including stereochemistry where appropriate.

40 In an abandoned laboratory three bottles, *A*, *B*, and *C*, have been discovered, each containing a different substance. Scattered labels near the bottles suggest that each bottle contains one of the following three compounds:

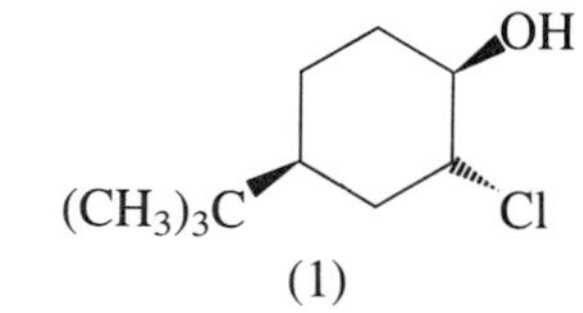
(1)

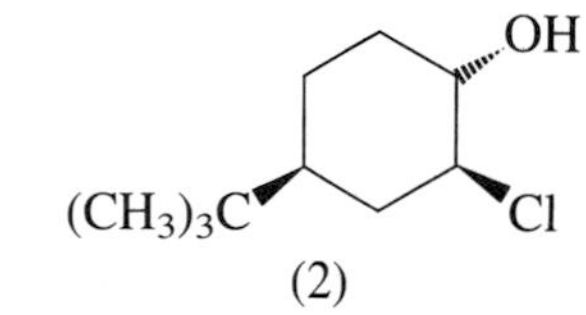
(2)

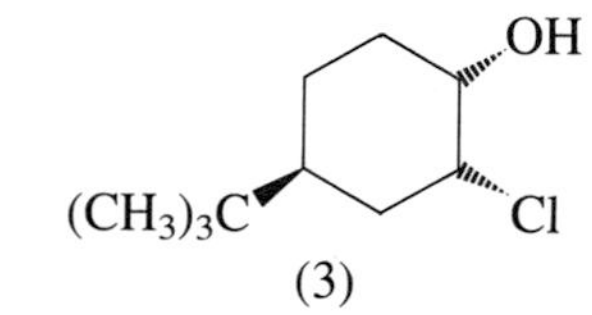
(3)

On treatment with KOH in methanol compound *A* gives no epoxide; compound *B* gives epoxide *D* and compound *C* gives epoxide *E*. Epoxides *D* and *E* are stereoisomers. Under identical conditions, *C* gives *E* much more slowly than *B* gives *D*. Identify *A*, *B*, and *C*, and explain all observations.

41 Suggest a reasonable mechanism for each of the following known conversions:

(a)

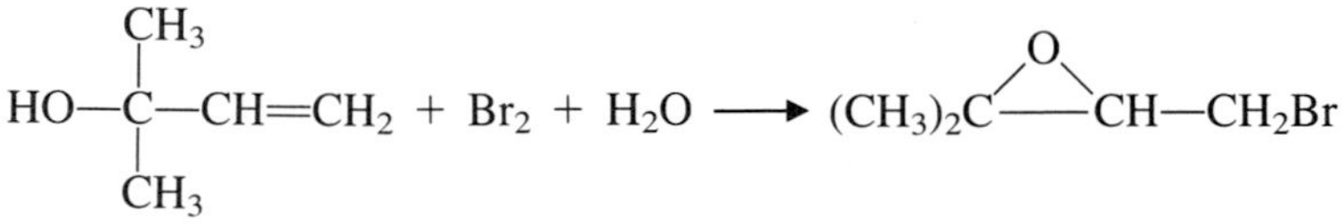

(b)

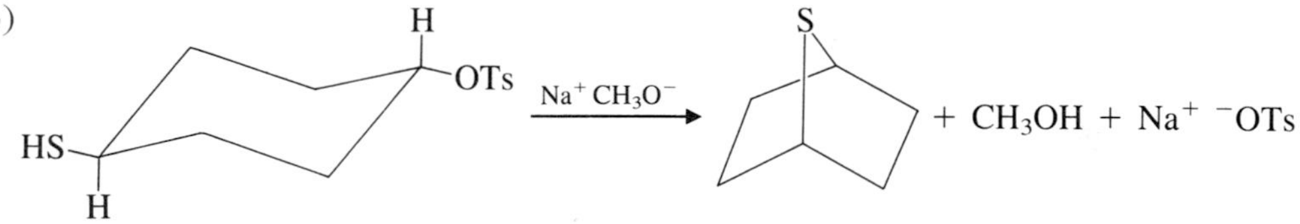

Hint: The structure of the product can be redrawn as follows:

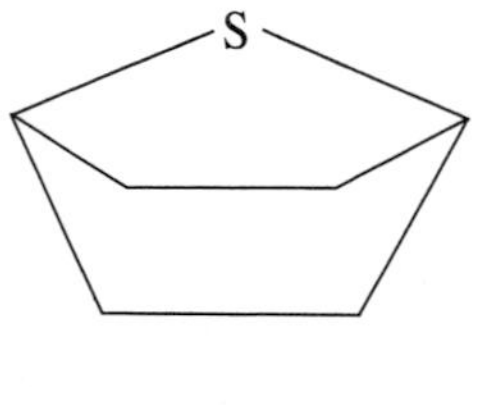

(c)

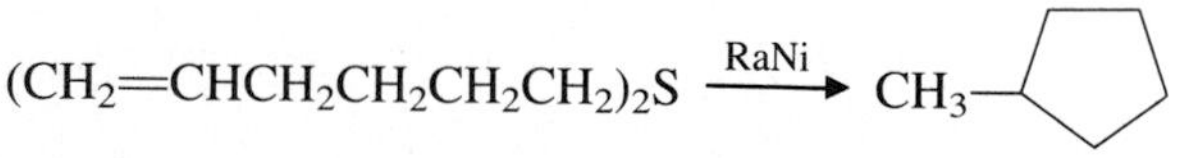

(d)

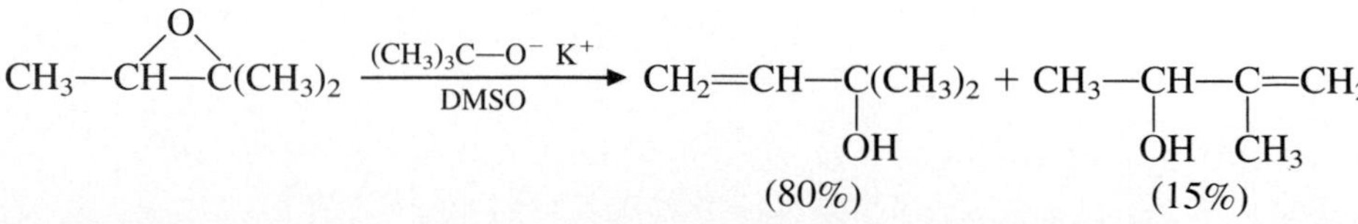

Problems (Cont.)

(e)

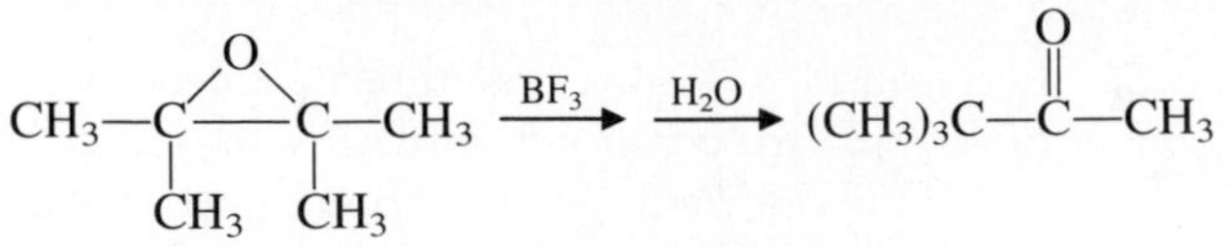

(f)

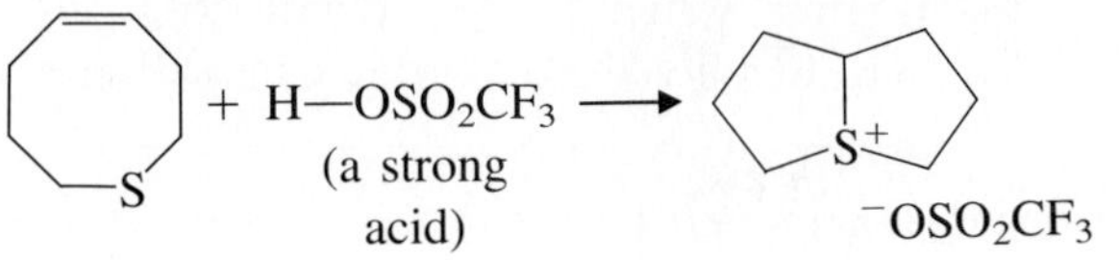

(g)

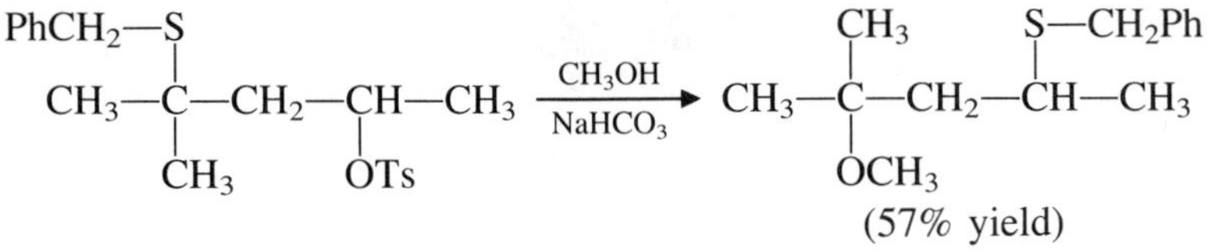

(h)

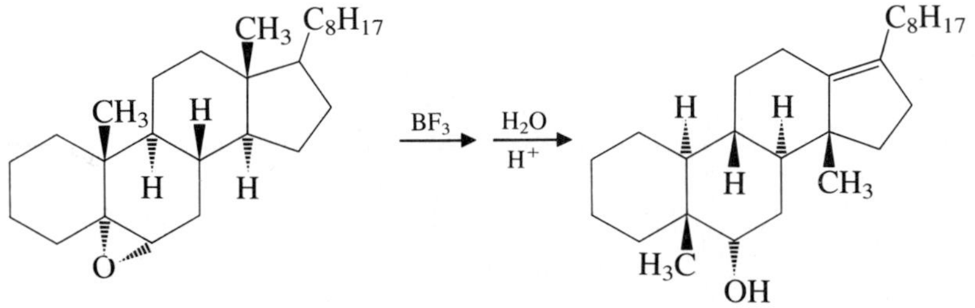

---

42 One of the following reactions is about 2,000 times faster in pure water than in pure ethanol. Another is about 20,000 times faster in pure ethanol than in pure water. The rate of the third changes very little as the solvent composition is changed from ethanol to water. Which of the reactions is faster in ethanol, which is faster in water, and which has a rate that is solvent-invariant? Explain. (*Hint:* Notice the difference in dielectric constants for ethanol and water in Table 8.2.)

(a) $(CH_3)_3S^+ + {}^-OH \longrightarrow CH_3{-}OH + (CH_3)_2S$ (S_N2 reaction)

(b) $(CH_3)_3C{-}Cl + \underset{\text{(solvent)}}{R{-}OH} \longrightarrow \left[\begin{matrix}(CH_3)_3C^+\\ Cl^-\end{matrix}\right] \longrightarrow (CH_3)_3C{-}OR + HCl$ (S_N1 reaction)

(c) $(CH_3)_3C{-}\overset{+}{S}(CH_3)_2 + \underset{\text{(solvent)}}{ROH} \longrightarrow [(CH_3)_3C^+] + S(CH_3)_2 \longrightarrow (CH_3)_3C{-}OR + H^+$ (S_N1 reaction)

---

43 Account for the following observations with a mechanism:

(1) In 80% aqueous ethanol compound *A* reacts to give compound *B*. Notice that *trans-B* is the only stereoisomer of this compound that is formed.

(2) Optically active *A* gives completely racemic *B*.

(3) The reaction of *A* is about 10^5 times faster than the analogous substitution reactions of both its stereoisomer *C* and chlorocyclohexane.

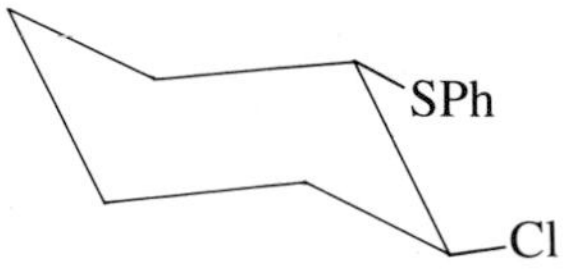

A

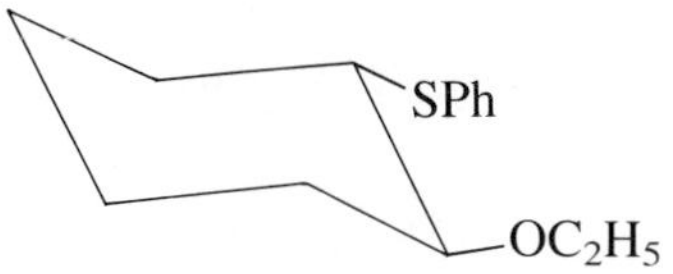

B

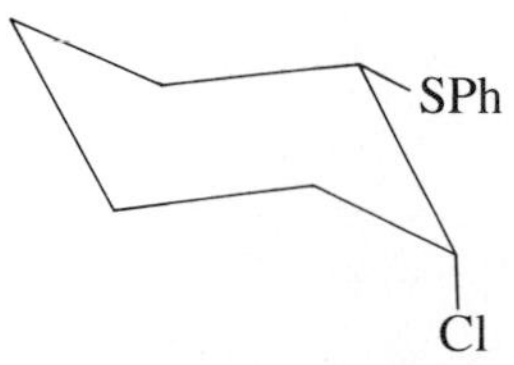

C

44 Outline a synthesis for each of the following compounds from (2*R*,3*R*)-2,3-dimethyloxirane:

(a) (2*R*,3*S*)-3-methoxy-2-butanol

(b)

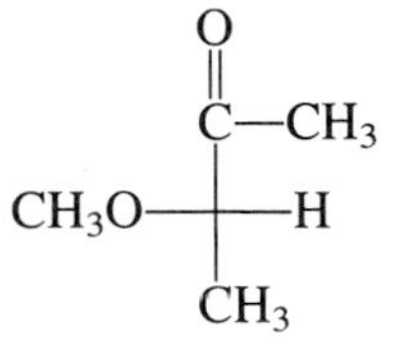

(c)

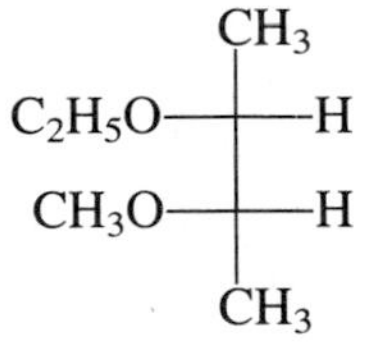

(d)

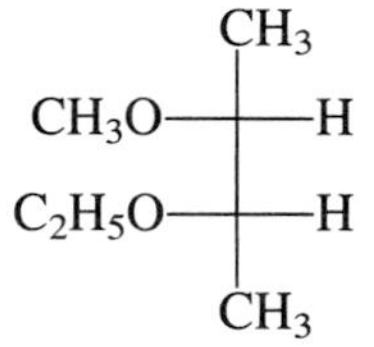

45 Account for the following stereochemical results in the reaction of 3-bromo-2-butanol with HBr. (*Hint:* See Sec. 11.6.)

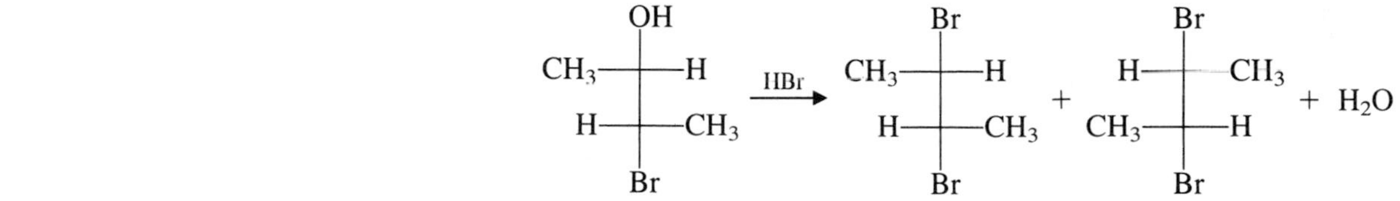

(optically active) (racemate)

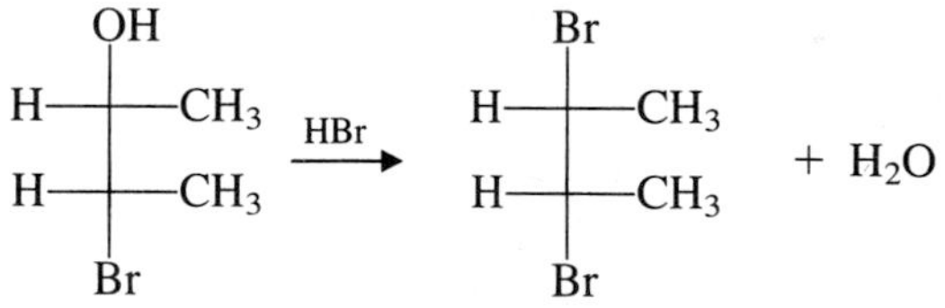

(*meso* compound)

46 Although the reaction of *cis*-4-*t*-butyl-1,2-epoxycyclohexane (compound *A*) with KOH could in principle yield two *trans*-glycols *B* and *C*, only *B* is formed. Explain.

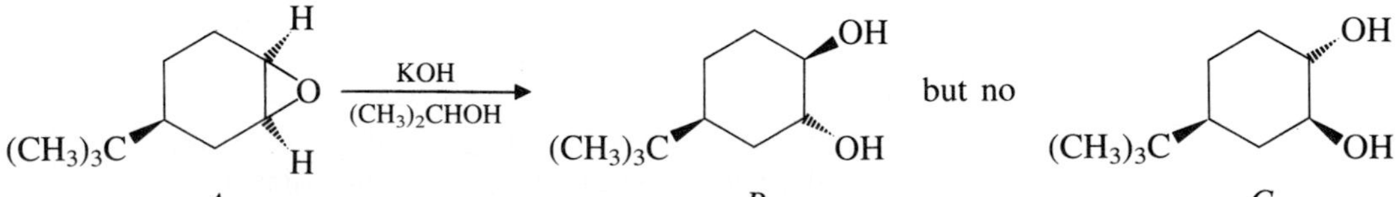

A *B* *C*

(*Hint:* Draw out the chair structures of *B* and *C*. Don't forget the effect of the *t*-butyl group (Sec. 7.4). Use the geometrical requirements of epoxide opening to determine which structure is more likely.)

12 Infrared Spectroscopy and Mass Spectrometry

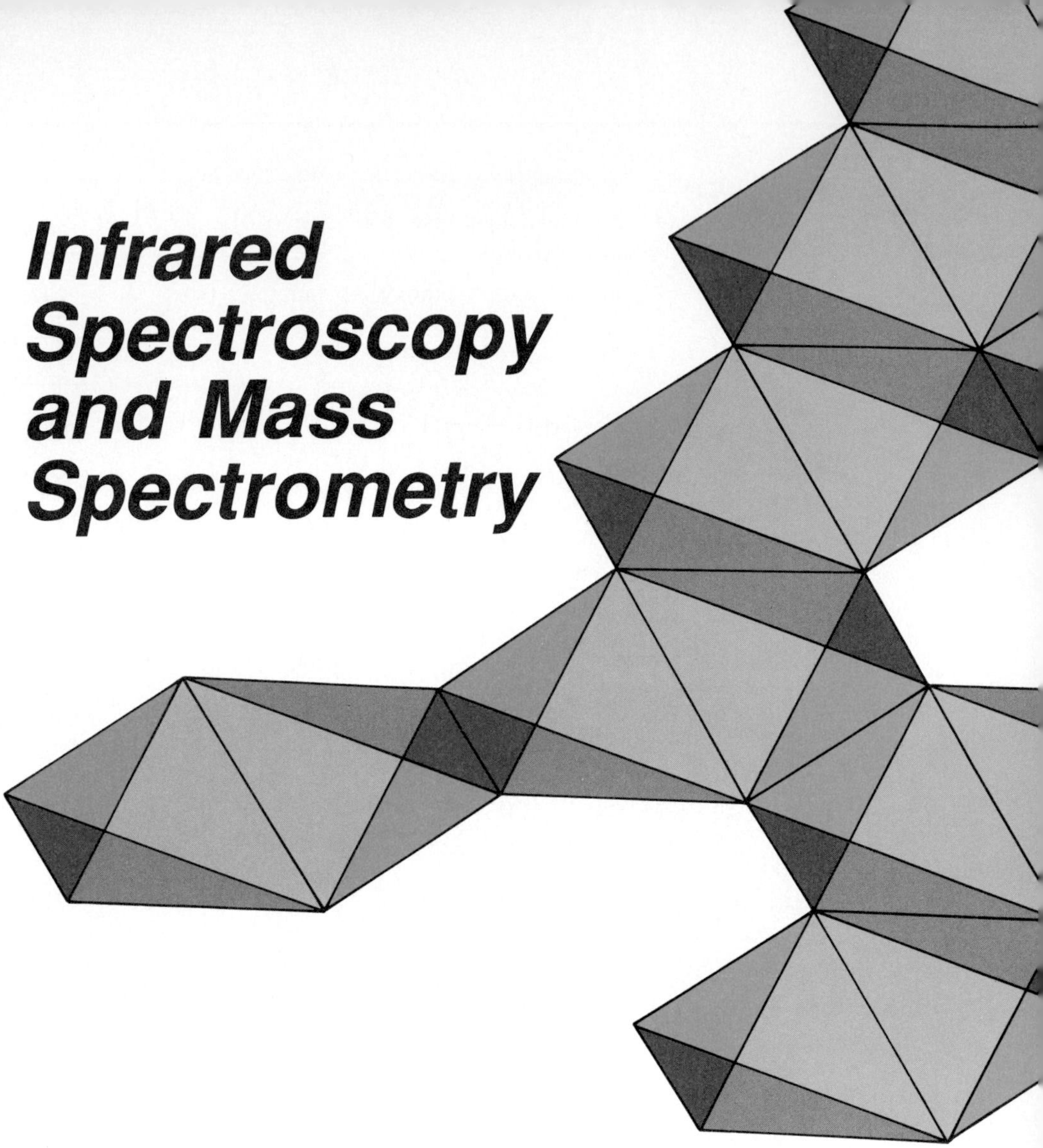

Up to this point in our study of organic chemistry we have taken it for granted that when a product of unknown structure is isolated from a reaction, it is possible somehow to determine its structure. At one time the structure determinations of many organic compounds required elaborate chemical degradation studies. Although many of these proofs were ingenious, they were also very time consuming, required relatively large amounts of compounds, and were subject to a variety of errors. In relatively recent years, however, physical methods have become available that allow chemists to determine molecular structures rapidly and nondestructively using very small quantities of material. With these methods it is not unusual for a chemist to do in thirty minutes or less a proof of structure that once took a year or more to perform! This and the following chapter will be devoted to a study of some of these methods.

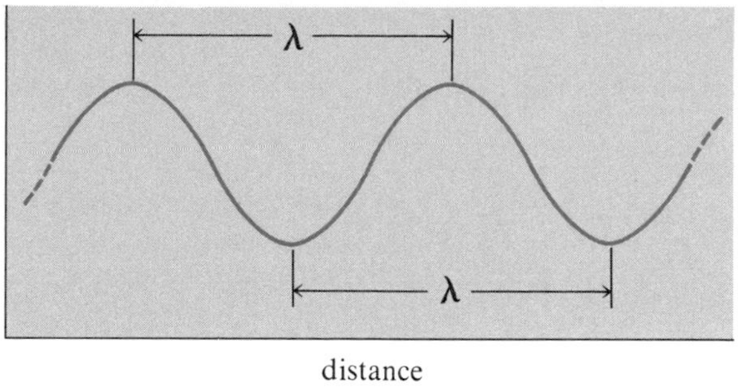

Figure 12.1 Definition of wavelength λ for a wave.

12.1 INTRODUCTION TO SPECTROSCOPY

Fundamental to modern techniques of structure determination is the field of **spectroscopy:** the study of the interaction of matter and light (or other electromagnetic radiation). Spectroscopy has been immensely important to many areas of chemistry and physics. For example, much of what we know about orbitals and bonding comes from spectroscopy. But spectroscopy is also important to the laboratory organic chemist because *it can be used to determine unknown molecular structures.* Although our study of spectroscopy will focus largely on its applications, we need to begin by considering some simple fundamentals of spectroscopy theory.

A. Electromagnetic Radiation

Visible light is one form of energy generally known as **electromagnetic radiation.** Other common forms of electromagnetic radiation are X-rays; ultraviolet radiation (UV, the radiation from a sun lamp); infrared radiation (IR, the radiation from a heat lamp); microwaves (used in radar and in the microwave oven); and radio waves (used to carry AM and FM radio and television signals). All these forms of electromagnetic radiation are different manifestations of the same basic phenomenon.

Because electromagnetic radiation has wave characteristics, some definitions associated with waves are important to our understanding of spectroscopy (Fig. 12.1). Electromagnetic radiation can be characterized by its **wavelength,** λ. The wavelength of a conventional sine or cosine wave is the distance between successive peaks or successive troughs in the wave. For example, red light has $\lambda = 6800$ Å and blue light has $\lambda = 4800$ Å. The **frequency** ν of a wave is defined by the equation

$$\nu = c/\lambda \tag{12.1}$$

in which c = the velocity of light = 3×10^{10} cm/sec. Since λ has the dimensions of length, ν has the dimensions of sec^{-1}, a unit more often called *cycles per second* (cps), or *hertz* (Hz). For example, the frequency of red light is

$$\begin{aligned} \nu &= \left(\frac{3 \times 10^{10}\ \cancel{\text{cm}}/\text{sec}}{6800\ \cancel{\text{Å}}}\right)\left(10^{8}\ \frac{\cancel{\text{Å}}}{\cancel{\text{cm}}}\right) \qquad (12.2) \\ &= 4.41 \times 10^{14}\ \text{sec}^{-1} = 4.41 \times 10^{14}\ \text{Hz} \end{aligned}$$

The physical meaning of the frequency is that if we stand in one place relative to a light source, successive peaks (or troughs) of the light wave will pass us with the frequency ν defined in Eq. 12.1.

Problem

1 Calculate the frequency of blue light with $\lambda = 4800$ Å; of infrared light with $\lambda = 9 \times 10^{-6}$ meter.

Although light can be described as a wave, it also shows particlelike behavior. The light particle is called a **photon.** The relationship between the energy of a photon and the wavelength or frequency of light is a fundamental law of physics:

$$E = h\nu = \frac{hc}{\lambda} \tag{12.3}$$

In this equation, h is **Planck's constant,** a universal constant that has the value

$$h = 6.625 \times 10^{-27} \text{ erg-sec} \tag{12.4}$$

This value of Planck's constant is used when energy is expressed in ergs. When we deal with energy in units of kcal/mol, Planck's constant has the value

$$h = 9.532 \times 10^{-14} \frac{\text{kcal-sec}}{\text{mol}} \text{ (or kcal-sec-mol}^{-1}\text{)} \tag{12.5}$$

Equation 12.3 shows that the *energy, frequency, and wavelength of electromagnetic radiation are simply related.* Thus, when the frequency or wavelength of electromagnetic radiation is known, its energy is also known.

Problems

2 Calculate the energy (in kcal/mol) of the two types of light described in Problem 1.

3 Given that X-rays are much more energetic than blue light, how do the wavelengths of X-rays compare to the wavelength of blue light?

The total range of electromagnetic radiation is called the *electromagnetic spectrum.* The types of radiation within the electromagnetic spectrum are shown in Fig. 12.2. Note that the frequency increases as the wavelength decreases, in accordance with Eq. 12.1. *All electromagnetic radiation is fundamentally the same; the various forms differ only in energy.*

B. Absorption Spectroscopy

The most common type of spectroscopy used for structure determination is **absorption spectroscopy.** The basis of absorption spectroscopy is that *matter can absorb energy from certain wavelengths of electromagnetic radiation.* In an absorption spectroscopy experiment, this absorption is determined as a function of wavelength, frequency, or energy in an instrument called a **spectrophotometer,** or **spectrometer.** The rudiments of an absorption spectroscopy experiment are shown schematically in

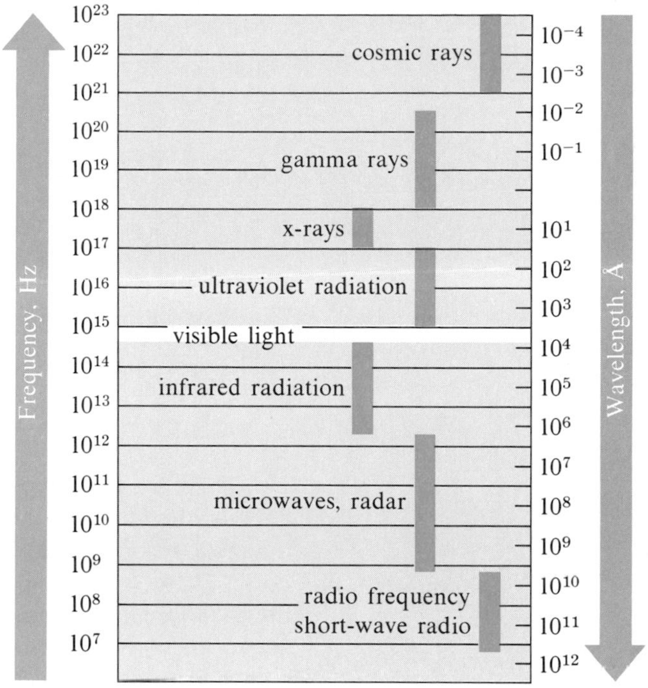

Figure 12.2 The electromagnetic spectrum.

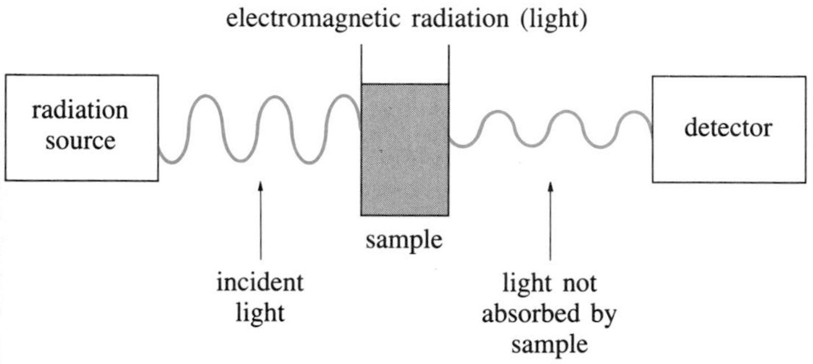

Figure 12.3 An absorption spectroscopy experiment.

Fig. 12.3. The experiment requires, first, a *source* of electromagnetic radiation. (If the experiment measures the absorption of visible light, the source could be a common light bulb.) We place the material to be examined, the *sample,* in the radiation beam. A *detector* measures the intensity of the radiation that passes through the sample unabsorbed; when this intensity is subtracted from the intensity of the source, the amount of radiation absorbed by the sample is known. We then vary the wavelength of the radiation falling on the sample, and the radiation absorbed at each wavelength is recorded as a graph of either radiation transmitted or radiation absorbed *vs.* wavelength or frequency. This graph is commonly called a **spectrum** of the sample.

The infrared spectrum of nonane shown in Figure 12.4 is an example of such a graph. It shows the radiation transmitted by a sample of nonane over a range of wavelengths in the infrared. (We shall learn how to read such a spectrum in more detail in the next section.)

We do not need experience with a spectrophotometer to appreciate exactly what is involved in a spectroscopy experiment. Imagine holding a piece of green glass up to the white light of the sun. The sun is the source, the glass is the sample, our eyes are the detector, and our brain provides the spectrum. White light is a mixture of all wavelengths. The glass appears green because only green light is transmitted through the glass; the other colors (wavelengths) in white light are absorbed. If we hold the same green glass up to red light, no light is transmitted to our eyes—the glass looks black—because the glass absorbs the red light.

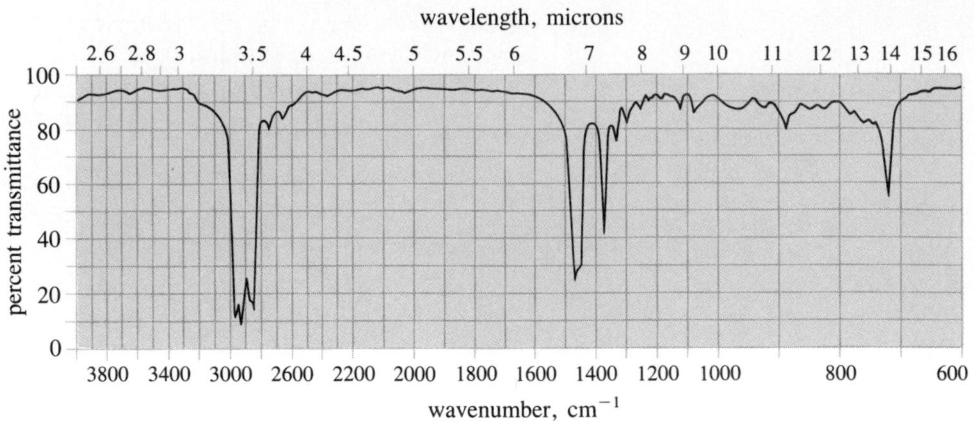

Figure 12.4 Infrared spectrum of nonane.

There are many types of spectroscopy. The three types of greatest use to organic chemists for structure determination, in descending order of importance, are nuclear magnetic resonance (NMR), infrared (IR), and ultraviolet–visible (UV–VIS) spectroscopy. These types of spectroscopy differ conceptually only in the type of radiation used, although the kinds of spectrometers required for each are quite different. Mass spectrometry is a fourth physical technique used for structure determination; it is not a type of absorption spectroscopy and is thus fundamentally different from NMR, IR, and UV spectroscopy. From the point of view of the chemist interested in the structure of an unknown material, each of these methods yields a different kind of information about molecular structure. The remainder of this chapter will be devoted to a study of IR spectroscopy and mass spectrometry. We shall study NMR spectroscopy in Chapter 13 and UV–VIS spectroscopy in Chapter 15.

12.2 INFRARED SPECTROSCOPY

A. The Infrared Spectrum

An infrared spectrum, like any absorption spectrum, is a record of the light absorbed by a substance as a function of wavelength. The IR spectrum is measured in an instrument called an **infrared spectrophotometer,** which we shall describe briefly in Sec. 12.4. In practice, the absorption of infrared radiation with wavelengths between 2.5×10^{-6} to 20×10^{-6} meters is of greatest interest to organic chemists. Let us consider the details of an IR spectrum by returning to the spectrum of nonane in Fig. 12.4.

The quantity plotted on the lower horizontal axis is the **wavenumber** $\tilde{\nu}$ of the light. The wavenumber, in units of *reciprocal centimeters* (cm^{-1}), is simply another way to express the wavelength of the radiation. Physically, the wavenumber can be envisioned as the number of wavelengths contained in one centimeter. Wavenumber is inversely proportional to the wavelength λ.

$$\tilde{\nu} = \frac{10^4}{\lambda} \qquad (\lambda \text{ in microns}) \tag{12.6a}$$

$$\lambda = \frac{10^4}{\tilde{\nu}} \tag{12.6b}$$

In these equations the wavelength is in **micrometers,** or **microns.** A micron is 10^{-6} meter, and is abbreviated with the Greek letter *mu* (μ) or, by more recent convention, μm. Thus, according to Eq. 12.6a, a wavelength of 10 μ corresponds to a wavenumber of 1000 cm^{-1}. In Fig. 12.4, wavenumber, across the bottom of the spectrum, increases to the left; wavelength, across the top of the spectrum, increases to the right. Notice that the wavenumber scale is linear, but is divided into three distinct regions in which the linear scale is different.

Scientists differ in the convention they prefer for citing the positions of IR absorptions. Many prefer to use the wavenumber convention because wavenumber is directly proportional to energy and frequency.

$$E = \frac{hc}{\lambda} = hc\tilde{\nu} \qquad \text{(for } \lambda \text{ in cm)} \tag{12.7a}$$

$$\nu = c\tilde{\nu} \tag{12.7b}$$

In fact, the wavenumber of a peak is sometimes loosely referred to as a frequency because of the direct proportion in Eq. 12.7b. Others prefer the wavelength convention because some common inexpensive IR spectrometers give spectra that are linear in λ. Since the spectra in this book are linear in wavenumber, we shall cite the positions of IR absorptions in cm^{-1}. However, it is a simple matter to convert from wavenumber to wavelength using Eq. 12.6b.

Referring again to the IR spectrum in Fig. 12.4, we see that on its vertical axis is plotted *percent transmittance.* This is the percent of the radiation falling on the sample that is transmitted to the detector. If the sample absorbs all the radiation, then none is transmitted, and the sample has 0% transmittance. If the sample does not absorb any radiation, then all of it is transmitted, and the sample has 100% transmittance. Thus, absorptions in the IR spectrum are registered as downward deflections—that is, "upside-down peaks." For example, absorptions in the spectrum of nonane occur at 2925, 1467, 1378, and 722 cm^{-1}.

Sometimes an IR spectrum is not presented in graphical form, but is summarized completely or in part using descriptions of peak *positions.* Intensities are often expressed qualitatively using the designations vs (very strong), s (strong), m (moderate), or w (weak). Some peaks are sharp (narrow), whereas others are broad (wide). The spectrum of nonane can be summarized as follows:

$\tilde{\nu}$ (cm^{-1}): 2925 (vs); 1467 (s); 1378 (m); 722 (m)

Problems

4 (a) What is the wavenumber of light with a wavelength of 6.0 μ?
(b) What is the wavelength of light with a wavenumber of 1720 cm^{-1}?

5 Show how Eq. 12.7b follows from Eqs. 12.7a and 12.1.

B. Physical Basis of IR Spectroscopy

Now that we know how to read an IR spectrum, our next task is to learn to *interpret* the spectrum. What does the spectrum tell us about molecular structure? In order to answer this question, we have to understand why molecules absorb infrared radiation.

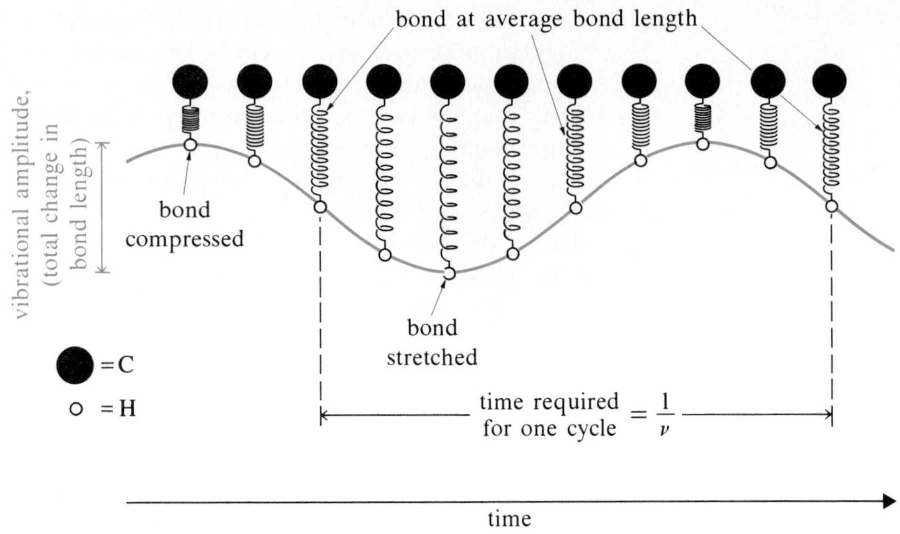

Figure 12.5 Chemical bonds undergo a variety of vibrations. The one illustrated here is a stretching vibration. The bond, represented as a spring, is shown at various times. The bond stretches and compresses about its average length over time. The time required for one complete cycle of vibration is the reciprocal of the vibrational frequency.

The absorptions we observe in an IR spectrum are the result of *vibrations* within a molecule. Atoms within a molecule are not stationary, but are constantly in motion. Consider, for example, the C—H bonds in a typical organic compound. These bonds undergo various stretching and bending motions. A useful analogy to the C—H stretching motion is the stretching and compression of a spring (Fig. 12.5). This vibration takes place with a certain frequency ν; that is, it occurs a certain number of times per second. Suppose that a C—H bond has a stretching frequency of 9×10^{13} times per second; this means that it undergoes a vibration every $1/(9 \times 10^{13})$ second or 1.1×10^{-14} second.

It is clear from Fig. 12.5 that the stretching of the C—H bond describes a wave motion. A wave of electromagnetic radiation can transfer its energy to the vibrational wave motion of the C—H bond only if the following very important condition is fulfilled: *There must be an exact match between the frequency of the radiation and the frequency of the vibration.* Thus, if our C—H vibration has a frequency of 9×10^{13} sec^{-1}, then it will absorb energy from radiation with the same frequency. From the relationship $\lambda = c/\nu$ (Eq. 12.1), we calculate that the radiation must have a wavelength of

$$\begin{aligned} \lambda &= (3 \times 10^{10})/(9 \times 10^{13}) \text{ cm} \\ &= 3.33 \times 10^{-4} \text{ cm} \\ &= 3.33 \times 10^{-6} \text{ m} = 3.33\ \mu \end{aligned}$$

The corresponding wavenumber of this radiation (Eq. 12.6a) is 3000 cm^{-1}. When radiation of this wavelength interacts with a vibrating C—H bond, radiant energy is absorbed and the bond vibration is excited. Mechanically, we can imagine that the bond, after energy absorption, vibrates with the same wavelength but with a larger *amplitude* (a larger stretch and tighter compression; Fig. 12.6). This absorption gives rise to the peak in the IR spectrum.

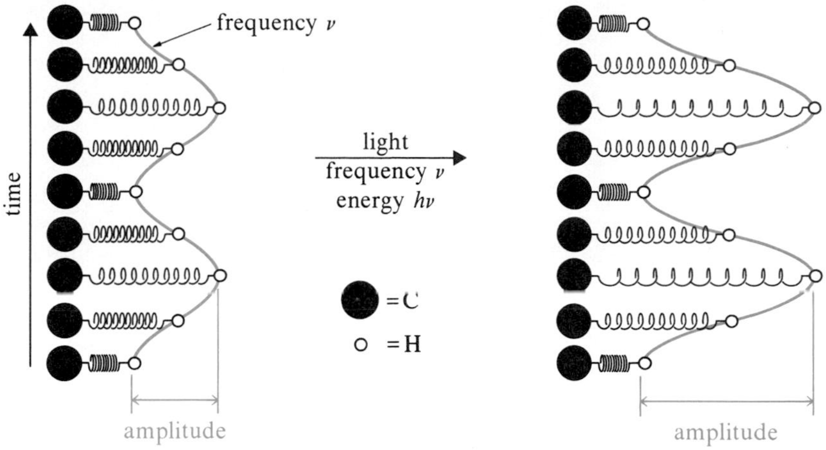

Figure 12.6 Light absorption by a bond vibration causes the bond (spring) to vibrate at the same wavelength but with a larger amplitude. The frequency of the light must exactly match the frequency of the bond vibration.

As a crude analogy, imagine a weight attached to a spring hanging from the ceiling; the weight is bouncing up and down. We grasp the weight, pull on it, and then release it—boin-n-ng! The weight now bounces up and down with larger "downs" and greater "ups"—a larger amplitude. We have imparted more energy to the spring by pulling on the weight. Similarly, infrared radiation transfers its radiant energy to the chemical bond, resulting in a more violent vibrational motion.

A major flaw in this analogy is that it does not explain why the wavelength of incident radiation has to match the wavelength of the bond vibration. This is a very important point, because it is this exact match that gives rise to the peaks in the spectrum. (If radiation of any wavelength could be absorbed with equal efficiency, the IR spectrum would be the same everywhere and would have no peaks and valleys.) Why can't a bond absorb energy from radiation of any wavelength? The answer lies in the fact that atoms are sufficiently small that they have the wavelike characteristics described by quantum theory. Although we shall not consider this theory here, we should realize that it is this wave character of atoms that requires equality between wavelength of the radiation and wavelength of the bond vibration in order for energy absorption to occur.

To summarize what we have learned so far:

1. Bonds vibrate with characteristic frequencies.
2. Absorption of energy from infrared radiation can occur only when there is a match between the wavelength of the radiation and the wavelength of the bond vibration.

Problem

6 The compound nonane contains many C—H bonds. Given that the stretching vibration of a typical C—H bond has a frequency of about 9×10^{13} sec^{-1}, which peak(s) in the IR spectrum of nonane (Fig. 12.4) would you assign to the C—H stretching vibration?

C. Infrared Absorption and Chemical Structure

What does an IR spectrum tell us about chemical structure? Each peak in the IR spectrum of a molecule corresponds to absorption of energy by the vibration of a bond or group of bonds. The utility of IR spectroscopy for the chemist is that *in all compounds, a given type of functional group absorbs in the same general region of the IR spectrum.* For example, in all organic compounds the stretching vibrations of C—H bonds show peaks in the 2700–3300 cm^{-1} region of the IR spectrum, and the stretching of C=C bonds gives rise to peaks in the 1640–1680 cm^{-1} region of the spectrum. We can diagnose the presence of these groups in a compound of unknown structure by looking for their characteristic absorptions in the IR spectrum. Furthermore, there are systematic variations for a given absorption that depend on the detailed structure of the absorbing group. For example, the double bond in a 1-alkene, R—CH=CH_2, usually absorbs at about 1640 cm^{-1}, but the disubstituted double bond in an internal alkene, R—CH=CH—R, absorbs at 1675 cm^{-1}. We can see from these few examples that *the IR spectrum serves as a fingerprint of the functional groups present in a molecule.*

D. Factors that Determine IR Absorption Position

One approach to the use of IR spectroscopy is simply to memorize the positions at which characteristic functional group absorbances appear, and look for peaks at these positions in the determination of unknown structures. However, we can use IR spectroscopy much more intelligently and learn the appropriate peak positions much more easily if we understand a little more about the physical basis of IR spectroscopy. Two aspects of IR absorption peaks are particularly important. First is the *position* of the peak—that is, the wavenumber or wavelength at which it occurs. Second is the *intensity* of the peak—that is, how strong it is. Let us consider each of these aspects in turn.

What factors govern the position of IR absorption? Three considerations are most important:

1. Strength of the bond
2. Masses of the atoms involved in the bond
3. The type of vibration being observed

Let us focus for the moment on the first two of these factors, and turn again to our mechanical spring analogy. How should bond strength affect the vibrational frequency? Intuitively, a stronger bond corresponds to a tighter spring. Balls connected by a tighter spring vibrate more rapidly—that is, with a higher frequency or wavenumber. Likewise, atoms connected by a stronger bond also vibrate at higher frequency.

Indeed, the physics of a vibrating spring gives us an equation that suggests exactly this result. Suppose we have two objects of mass m and M, respectively, attached by a spring. The tightness of the spring is measured by its **force constant,** κ: the tighter the spring, the greater its force constant. The vibrational frequency of the system in wavenumbers (cm^{-1}), and hence its absorption frequency, is given by

$$\tilde{\nu} = \frac{1}{2\pi c}\sqrt{\frac{\kappa(m + M)}{mM}} \tag{12.8}$$

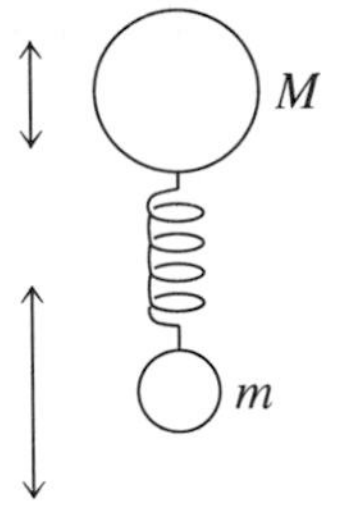

In this equation, c is the velocity of light (3×10^{10} cm/sec) and the units of the force constant are dynes/cm. Since the dyne, the cgs unit of force, has the units g-cm-sec^{-2}, then the force constant in dynes/cm has the units (g-cm-sec^{-2})(cm^{-1}), or g-sec^{-2}. If the objects are atoms, their masses are just their respective atomic weights divided by Avogadro's number. A particularly simple form of this equation is obtained if one mass, say M, is substantially greater than the other.

$$\tilde{\nu} = \frac{1}{2\pi c}\sqrt{\frac{\kappa}{m}} \tag{12.9}$$

(Show that this equation follows from Eq. 12.8.) Equations 12.8 and 12.9 are derived from the assumption that the chemical bond obeys the same laws of motion as a vibrating spring (Hooke's Law). Because this analogy is not perfect, this equation gives numbers that are only approximate. However, it does give a good semiquantitative idea of the trends to be expected in IR frequencies. Notice in Eq. 12.9 how the force constant affects the vibrational frequency: a larger force constant (tighter spring) increases the vibrational frequency, just as intuition suggests. In summary, *the vibration of a stronger bond occurs at higher wavenumber (lower wavelength) than the vibration of a weaker bond.*

How do we know whether one bond is stronger or weaker than another? A simple measure of bond strength is the difficulty with which it is broken. In other words, the *bond dissociation energies* in Table 5.2 give us information about bond strengths, and therefore information about force constants. *The higher the bond dissociation energy, the stronger the bond, and the greater is its force constant.* The effect of bond strength is particularly easy to see when we compare the vibration frequencies of single, double, and triple bonds between the same atoms. As intuition and the bond dissociation energies in Table 5.2 dictate, C≡C bonds are stronger than C=C bonds, which, in turn, are stronger than C—C bonds. Thus, the wavenumbers for the IR absorptions of these stretching motions are in the order C≡C > C=C > C—C.

What about the mass effect? We may have observed at one time or another that heavier balls connected to a spring of a given tightness vibrate more slowly (with lower frequency) than light balls connected to the same spring. This effect is also reflected in Eq. 12.9: larger m implies a smaller wavenumber. Thus, *vibrations of heavier atoms occur at lower frequency or wavenumber than vibrations of lighter atoms.* This effect is best illustrated by isotopic replacement. For example, the C—H and C—D bonds have essentially the same force constant. Thus, the difference in the IR absorption frequencies of these two bonds is due to the mass difference between H and D (Problem 7).

Problems

7 The C—H bond has a stretching frequency of about 3000 cm^{-1}. Calculate the stretching frequency of the C—D bond, assuming the same force constant. Note the effect of atomic mass on the vibrational frequency.

8 Which effect is most important in determining the absorption position in the following series, bond strength or atomic mass? How do you know?

bond:	H—C	H—O	H—F
wavenumber:	3000 cm^{-1}	3300 cm^{-1}	4000 cm^{-1}

The third factor that affects the absorption frequency is the *type of vibration*. So far, we have concentrated on **stretching** vibrations—vibrations along the line of the chemical bond. The other general type of vibration is the **bending** vibration—any vibration that does not occur along the line of the chemical bond. A bending vibration is easily imagined as a ball hanging on a spring and swinging side-to-side. In general, *bending vibrations occur at lower frequencies (higher wavelengths) than stretching vibrations of the same groups.*

Again, an analogy is helpful. If we imagine a ball attached to a tight spring hanging from the ceiling, we can tap the ball gently and make it swing back and forth. It takes considerably more energy to stretch the spring. Since the energy required to set the spring in motion is proportional to its frequency, then it follows that the swinging (bending) motion has a lower frequency than the stretching motion.

The only possible type of vibration in a diatomic molecule—for example, H—F—is the stretching vibration. However, when more than two atoms are present in a molecule, a variety of bending vibrations are possible. Consider, for example, a —CH_2— group. The types of C—H vibrations, called **normal vibrational modes,** that are characteristic of this group are shown in Fig. 12.7. They serve as models for the kinds of vibrations we can expect for other groups in organic molecules. The bending vibrations can be such that the hydrogens move *in the plane* of the —CH_2— group, or *out of the plane* of the —CH_2— group. Furthermore, stretching and bending vibrations can be *symmetrical* or *unsymmetrical* with respect to a symmetry plane bisecting the two vibrating hydrogens. The bending motions have been given very graphic names (scissoring, wagging, and so on) that describe the type of motion involved. Each of these motions occurs with a particular frequency and can have an associated peak in the IR spectrum (although some peaks are weak or absent for reasons that we shall consider later). A —CH_2— group in a typical organic molecule undergoes all of these motions simultaneously. That is, while the C—H bonds are stretching, they are also bending. The IR spectrum of nonane (Fig. 12.4) illustrates both C—H stretching and C—H bending vibrations. The peak at 2925 cm^{-1} is due to the C—H stretch; the peak at 722 cm^{-1} is due to the —CH_2— rocking vibration. As our analogy above suggests, the C—H stretching vibration absorbs at higher energy than the C—H bending vibration.

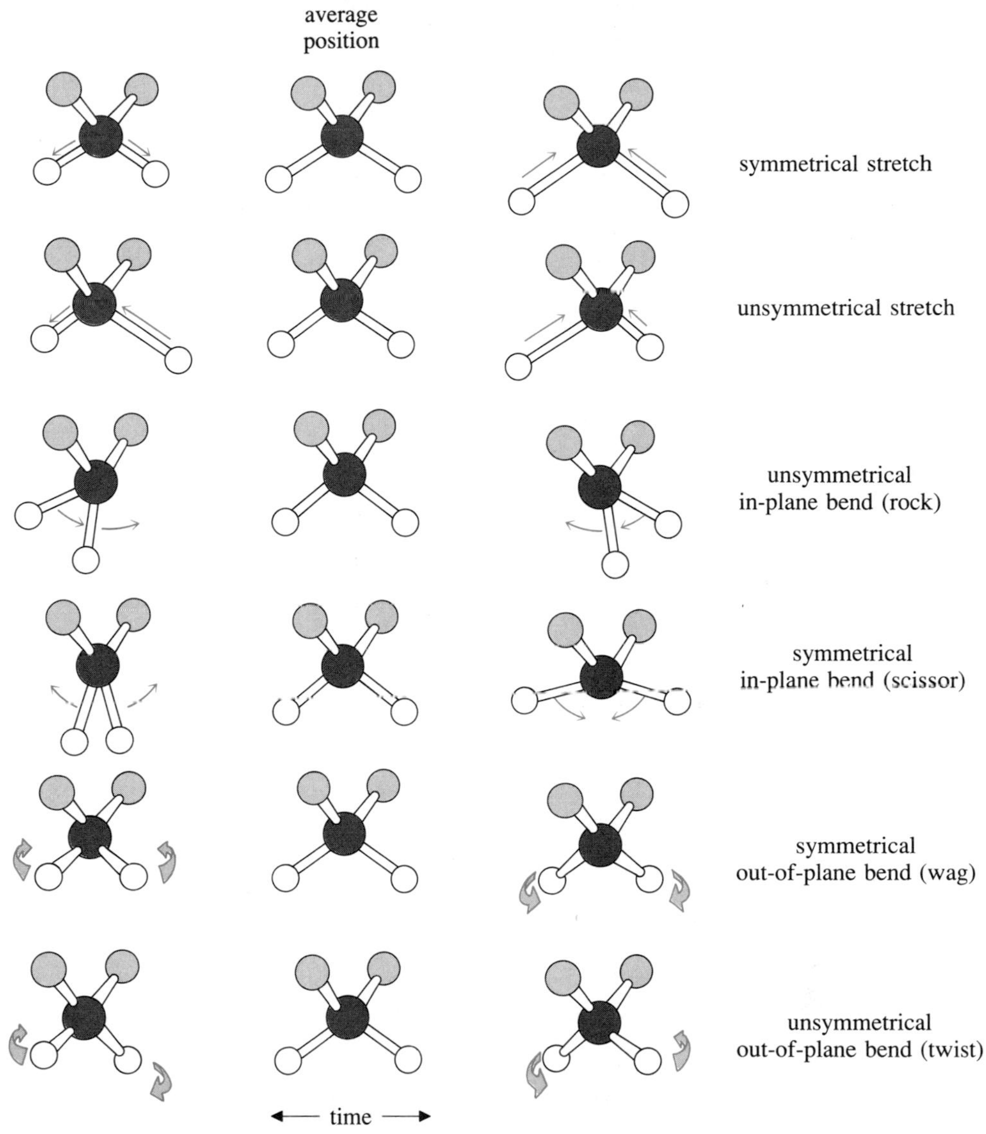

Figure 12.7 Typical vibrations of a —$\mathbf{CH_2}$— group in an organic compound. Start at the center figure in each case and move left and right to see how the bonds change with time. The white atoms are hydrogens, the black atoms are carbons, and the gray groups are the other groups attached to carbon.

E. Factors that Determine IR Absorption Intensity

The different peaks in an IR spectrum typically have very different intensities. It is important to understand in a general way the factors that control absorption intensity. It should seem reasonable that a greater number of molecules in the sample and more absorbing groups within a molecule give a more intense spectrum. Thus, a more concentrated sample gives a stronger spectrum than a less concentrated one, other things being equal. Similarly, a molecule such as nonane, which is rich in C—H bonds, has a stronger absorption for the C—H stretching vibration than a molecule of similar size with relatively few C—H bonds.

However, another important factor affects the intensity of an IR absorption. The theory that describes the interaction between infrared radiation and bond vibrations tells us that the vibration in question must result in a *dipole moment change* in the molecule if it is to be observed in the IR spectrum.

This point can be illustrated by comparing the C=C stretching vibrations of two alkenes. Because of the electron-donating characteristics of the methyl groups, isobutylene has a dipole moment along the C=C double bond (Sec. 4.3).

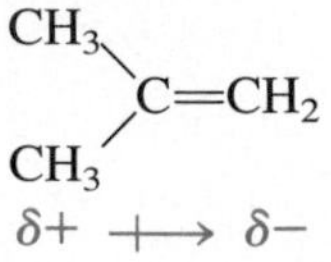

When this double bond stretches, the dipole moment increases, because dipole moment is the product of both charge separation and *distance*. Because it is associated with a changing dipole moment, the C=C stretching vibration in isobutylene, which occurs at 1640 cm^{-1}, is observed in the IR spectrum. A vibration like this one that gives an IR absorption is said to be **infrared active.** The alkene 2,3-dimethyl-2-butene, however, has no dipole moment, and stretching the double bond does not impart one to the molecule. Since the molecule and its stretched form have the same dipole moment—zero—the C=C stretching vibration of this alkene is not observed.

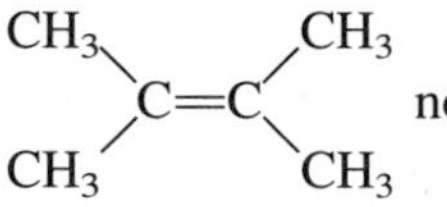

no IR absorption for the C=C stretching vibration

Indeed, this alkene has no IR absorption in the 1600–1700 cm^{-1} region of the IR spectrum. (This compound does have other IR absorptions.) Note carefully that *the C=C stretching vibration occurs;* it is simply *not observed in the IR spectrum.* Vibrations such as this one, which occur but are not observed, are said to be **infrared inactive.**

The intensity of an IR absorption depends on the size of the dipole moment change associated with the vibration. Thus, (*Z*)-3,4-dimethylheptene has a very small dipole moment that is affected by the C=C stretching motion; but, because the ethyl and propyl groups differ very little in their electronegativities, the dipole moment is so small that for all practical purposes the vibration is not infrared active.

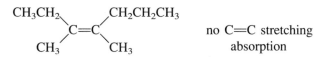

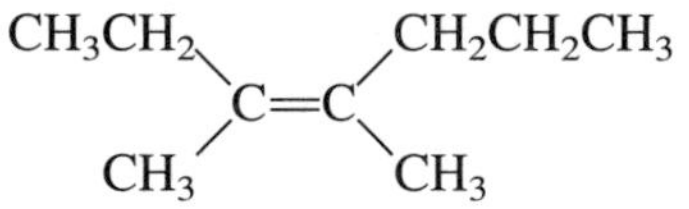

no C=C stretching absorption

To summarize, then, the intensity of an infrared absorption depends not only on how many groups of the type being observed are present, but also on the size of the dipole moment change associated with the vibration in question. We can now understand why certain IR peaks are absent in very symmetrical molecules (and extremely weak in almost symmetrical molecules) and yet are present in other compounds with the same functional groups.

Problem

9 Which of the following vibrations should be infrared active, and which should be infrared inactive (or nearly so)?

(a) $(CH_3)_2C{=}O$ C=O stretch

(b) O=C=O C=O unsymmetrical stretch (see Fig. 12.7)

(c) $CH_3CH_2CH_2CH_2—C{\equiv}C—H$ C≡C stretch

(d) $C_2H_5—C{\equiv}C—C_2H_5$ C≡C stretch

(e) 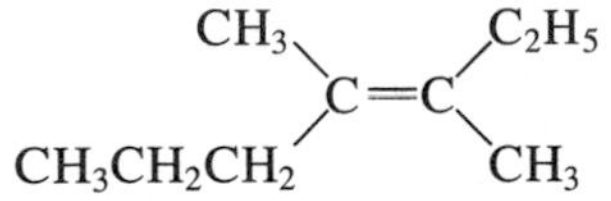

C=C stretch

(f) 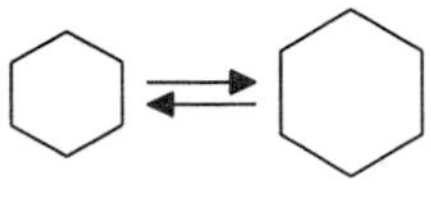
cyclohexane ring "breathing"
(simultaneous stretch of all C—C bonds)

(g) $(CH_3)_3C—Cl$ C—Cl stretch

(h) 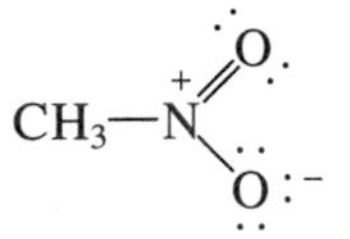

symmetrical N—O stretch
(note resonance structures; Sec. 1.7)

(i) *trans*-3-hexene (C=C stretch)

12.3 FUNCTIONAL-GROUP INFRARED ABSORPTIONS

A. IR Spectra of Alkanes, Alkyl Halides, and Alkenes

Let us now survey some of the important IR absorptions of the functional groups we have studied. First we consider alkanes. The obvious structural features of alkanes are the carbon–carbon and carbon–hydrogen single bonds. The stretching of the carbon–carbon single bond is infrared inactive (or nearly so) because this vibration is associated with little or no change of the dipole moment. The stretching absorptions of alkyl C—H bonds are typically observed in the 2850–2960 cm^{-1} region. We have already discussed this absorption in the IR spectrum of nonane (Fig. 12.4). Since a great many types of organic compounds contain alkyl groups, it is clear that this absorption is not unique to alkanes and hence, is not very useful.

The carbon–halogen stretching absorption of alkyl halides appears in the low wavenumber end of the spectrum, but there are many interfering absorptions in this region. NMR spectroscopy and mass spectrometry are more useful than IR spectroscopy for determining the structures of alkyl halides.

The infrared spectra of alkenes, in contrast to those of alkanes and alkyl halides, are very useful, and can help us determine not only whether a carbon–carbon double bond is present, but also the type of substitution pattern around the double bond. Typical alkene absorptions are given in Table 12.1. These fall into three categories: C=C stretching absorptions, =C—H stretching absorptions, and =C—H bending absorptions. The stretching vibration of the carbon–carbon double bond occurs in the 1640–1675 cm^{-1} range; the frequency of this absorption tends to increase with the degree of alkyl substitution on the double bond. As we learned in the last section, the C=C stretching absorption is infrared inactive in symmetrical or nearly symmetrical alkenes. Thus, the C=C stretching absorption is clearly evident in 1-octene at 1642 cm^{-1} (Fig. 12.8a), but is virtually absent in *trans*-3-hexene (Fig. 12.8b).

As we shall learn in Chapter 13, NMR spectroscopy is particularly useful for observing alkene hydrogens. Nevertheless, the =C—H stretching absorption that occurs

TABLE 12.1 Important Infrared Absorptions of Alkenes

Functional group	Absorption
$>C{=}C<$ stretching absorption	
$-CH{=}CH_2$ (terminal vinyl)	$1640\ cm^{-1}$ (m)
$>C{=}CH_2$ (terminal methylene)	$1655\ cm^{-1}$ (m)
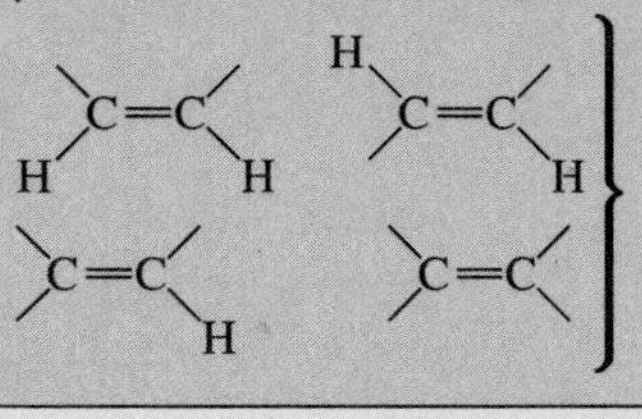	1660–$1675\ cm^{-1}$ (w) (absent in some compounds)
$=C-H$ stretching absorption	
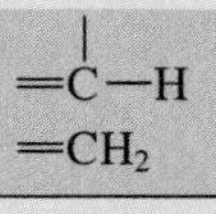	$3020\ cm^{-1}$ (m)
$=CH_2$	$3080\ cm^{-1}$ (m)
$=C-H$ bending absorption	
$-CH{=}CH_2$ (terminal vinyl)	$910, 990\ cm^{-1}$ (s) two absorptions
$>C{=}CH_2$ (terminal methylene)	$890\ cm^{-1}$ (s)
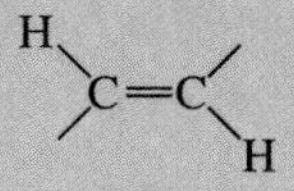 (*trans*-alkene)	960–$980\ cm^{-1}$ (s)
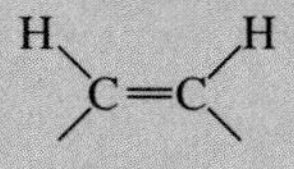 (*cis*-alkene)	675–$730\ cm^{-1}$ (br) (ambiguous and variable for different compounds)
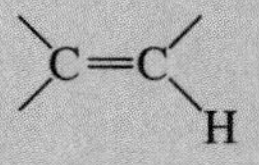 (trisubstituted)	800–$840\ cm^{-1}$ (s)

at somewhat higher wavenumber than the alkyl C—H stretching absorption can often be used as confirmation of the alkene functional group. This absorption is particularly easy to see in 1-alkenes; for example, it is present at $3080\ cm^{-1}$ in 1-octene (Fig. 12.8a) and less obvious, but still discernible, at $3030\ cm^{-1}$ in *trans*-3-hexene (Fig. 12.8b).

The alkene =C—H bending absorptions that appear in the low-wavenumber (high-wavelength) region of the IR spectrum are in many cases very strong and contain useful information about the substitution pattern at the alkene double bond. Of these absorptions, only the one for *cis*-alkenes tends to be somewhat variable and ambiguous. An example of useful =C—H bending absorptions are the peaks at 910 and $990\ cm^{-1}$ in the spectrum of 1-octene (Fig. 12.8a); these two peaks provide confirmation of the unbranched 1-alkene group ($-CH{=}CH_2$). One of the most useful =C—H bending absorptions is the vibration at 960–$980\ cm^{-1}$ characteristic of *trans*-alkenes. An example of this absorption is the peak at $965\ cm^{-1}$ in the spectrum of *trans*-3-hexene (Fig. 12.8b).

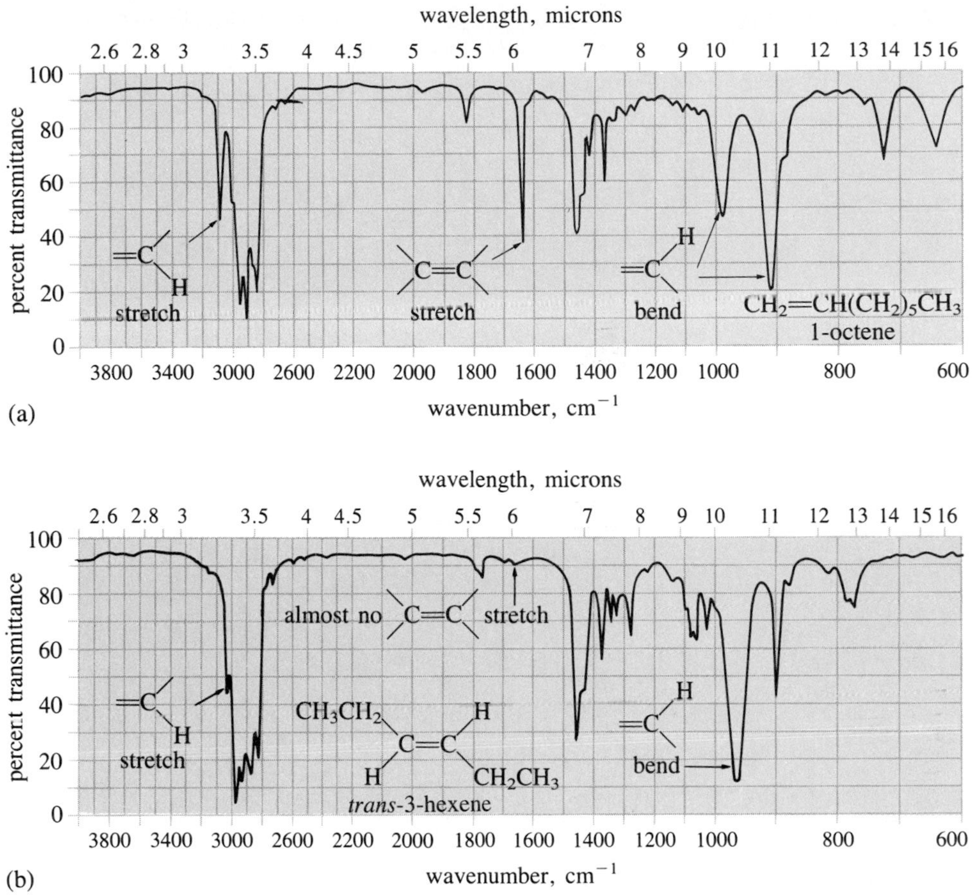

Figure 12.8 IR spectra of (a) 1-octene and (b) *trans*-3-hexene.

Problems

10 One of the spectra in Fig. 12.9 is that of *trans*-2-heptene, and the other is that of 2-methyl-1-hexene. Which is which? Explain.

11 What evidence can you cite from IR spectroscopy to support the view that the =C—H bond is a stronger bond than the alkyl C—H bond? Explain.

B. IR Spectra of Alcohols and Ethers

The O—H stretching absorption, which appears in the 3200–3400 cm^{-1} region of the IR spectrum, is a very important spectroscopic identifier for alcohols. The exact position of this absorption depends on whether the —OH group in the sample is free or hydrogen bonded. In typical samples the alcohol is concentrated enough that the hydrogen-bonded —OH absorption is observed. This peak, which is typically very broad and intense, is clearly evident in the IR spectrum of 1-hexanol (Fig. 12.10).

The other characteristic absorption of an alcohol is a strong C—O stretching peak that occurs in the 1050–1200 cm^{-1} region of the spectrum; primary alcohols absorb near the low end of this range and tertiary alcohols near the high end. For example, this absorption occurs at about 1060 cm^{-1} in the spectrum of 1-hexanol. There are

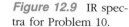

Figure 12.9 IR spectra for Problem 10.

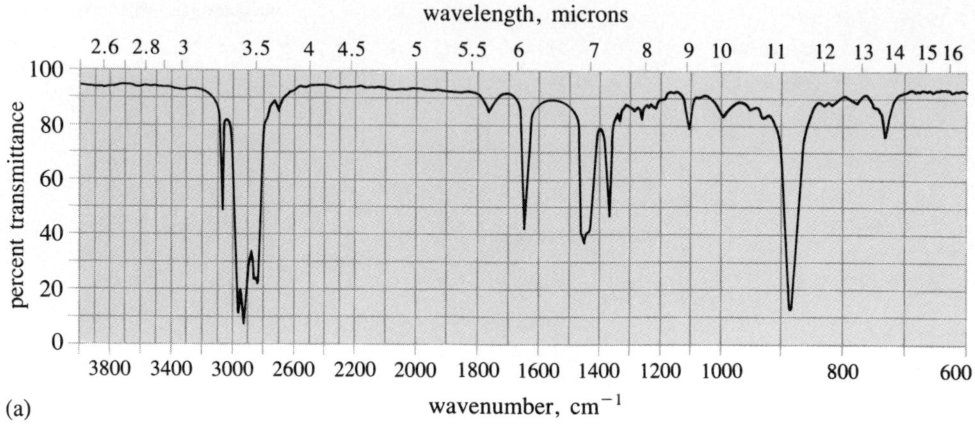

(a)

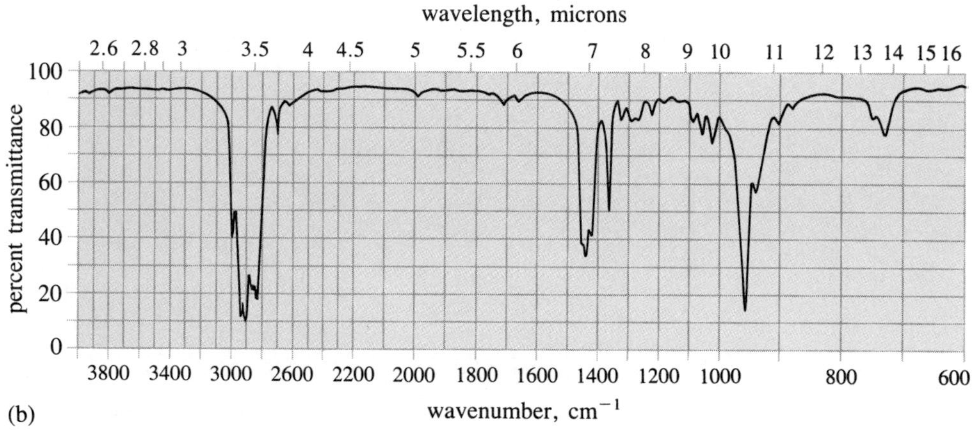

(b)

Figure 12.10 IR spectrum of 1-hexanol.

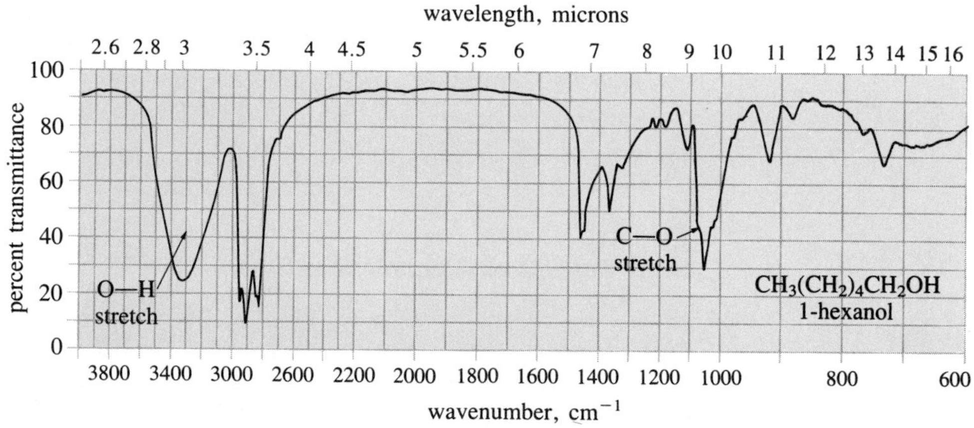

two reasons why this absorption is less useful for identification of alcohols. First, many other functional groups, such as ethers, esters, and carboxylic acids, have strong C—O stretching absorptions in the same general region of the spectrum. Second, the part of the spectrum from 900–1500 cm^{-1} contains IR absorptions of so many other functional groups that it is sometimes hard to make unique assignments using peaks in this region.

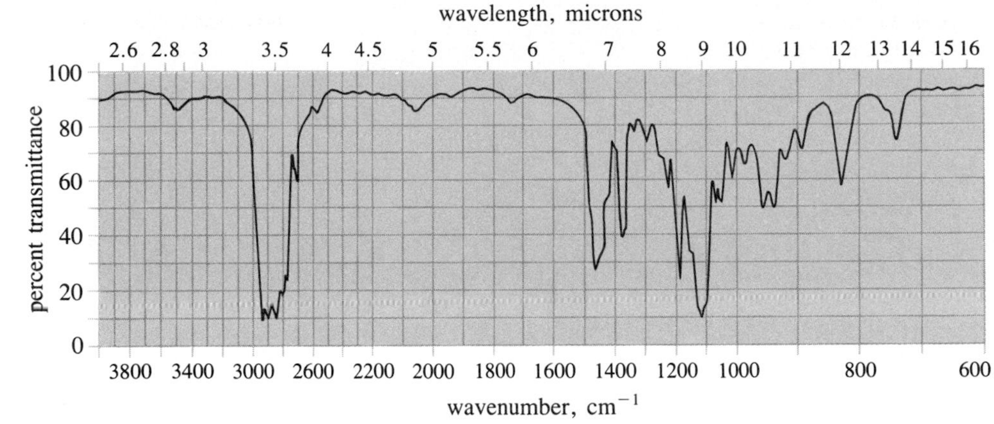

Figure 12.11 IR spectrum for Problem 12.

The most characteristic infrared absorption of ethers is the C—O stretching absorption, which, for the reasons we have just stated, is not very useful except for confirmation when an ether is already suspected from other data. For example, both dipropyl ether and its isomer 1-hexanol have strong C—O stretching absorptions near 1100 cm^{-1}.

The important IR absorptions of other functional groups are discussed in the respective chapters relating to these groups. In addition, a summary of key IR absorptions is given in Appendix II.

Problems

12 Match the IR spectrum in Fig. 12.11 to one of the following three compounds: 2-methyl-1-octene, butyl methyl ether, or 1-pentanol.

13 Explain why the IR spectra of some ethers have *two* C—O stretching absorptions. (*Hint:* See Fig. 12.7.)

14 Explain why the O—H stretching absorption of the hydrogen-bonded —OH group is observed in the IR spectrum of an alcohol run in concentrated solution, and why the O—H stretching absorption of the non-hydrogen-bonded —OH group is observed when the alcohol is in dilute solution.

12.4 EXPERIMENTAL ASPECTS OF INFRARED SPECTROSCOPY

Infrared spectrophotometers are available in most chemical laboratories. An infrared spectrophotometer (Fig. 12.12) consists essentially of a source of infrared radiation; a mirror system that splits the IR beam into a sample beam and a reference beam; a device for sorting the transmitted radiation into various wavelengths, called a monochromator; a detector, which converts the transmitted radiation into a voltage proportional to its intensity; and a recorder, on which the voltage is presented as a graph— the IR spectrum. The *x*-axis of the recorder and the monochromator are synchronized so that the wavelength or frequency presented on the graph corresponds to that transmitted by the monochromator. The double beam of IR radiation allows equal amounts of radiation to pass simultaneously through a cell containing sample plus

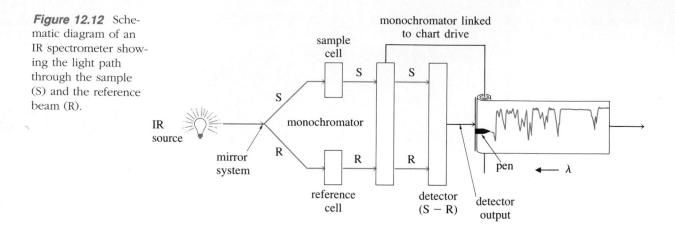

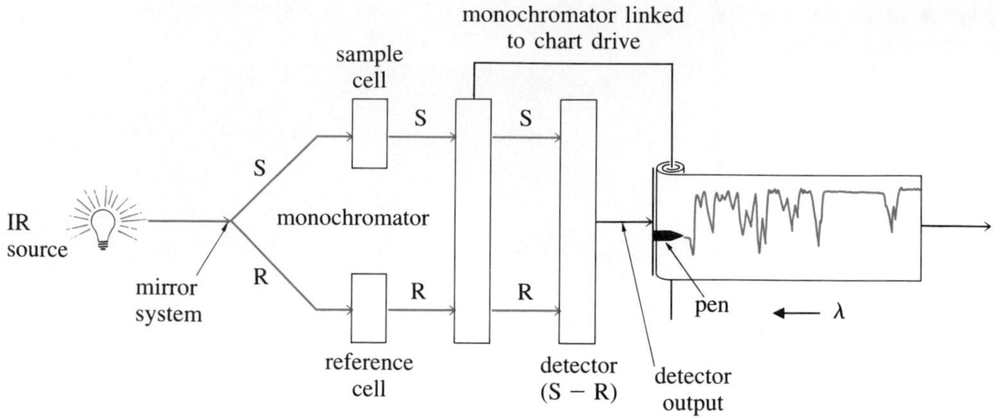

Figure 12.12 Schematic diagram of an IR spectrometer showing the light path through the sample (S) and the reference beam (R).

solvent and an identical reference cell containing only solvent. The reference beam is subtracted from the sample beam. In principle, this procedure removes interfering solvent absorptions.

IR spectra can be run on samples in the solid, liquid, or gaseous state. The spectrum of a sample in the solid state can be run as a dispersion either in solid anhydrous potassium bromide that has been fused (melted) to form a pellet, or in mineral oil. The spectrum of a pure liquid ("neat") sample can be run as a film between two small, optically polished plates of sodium chloride. (KBr and NaCl do not absorb in the infrared.) When a spectrum is run on a sample in solution, the solution is contained in a small sample holder, or cell, made of optically polished sodium chloride.

The only useful solvents for running solution spectra are those that have relatively few IR absorptions. Furthermore, solvents that dissolve IR sample cells are obviously of no use. Thus water, acetone, and alcohols cannot be used in IR cells made of sodium chloride. (There are cells available made of special polymers that permit IR spectra to be taken in aqueous solution.) As a practical matter, chloroform, carbon tetrachloride, and carbon disulfide are the three most widely used solvents for IR spectra. The solvent is chosen that has the fewest interfering absorptions at the wavelengths of interest.

12.5 INTRODUCTION TO MASS SPECTROMETRY

In contrast to other spectroscopic techniques, mass spectrometry does not involve the absorption of electromagnetic radiation, but operates on a completely different principle. As the name implies, mass spectrometry is used to determine molecular masses, and is one of the most important techniques for the determination of molecular weights; but its use is not limited to molecular weight determination. Our objective here will be to understand the fundamentals of mass spectrometry and how it can be applied to structure elucidation.

A. The Mass Spectrum

The mass spectrum is taken in a **mass spectrometer.** In this instrument a compound is vaporized in a vacuum and bombarded with an electron beam of high energy—

typically, 70 electron-volts (more than 1,600 kcal/mol). Since this energy is much greater than the bond energies of chemical bonds, some fairly drastic things happen when a molecule is subjected to such conditions. One thing that happens is that an electron is ejected from the molecule. For example, if methane is bombarded in this way, it loses an electron from one of the C—H bonds.

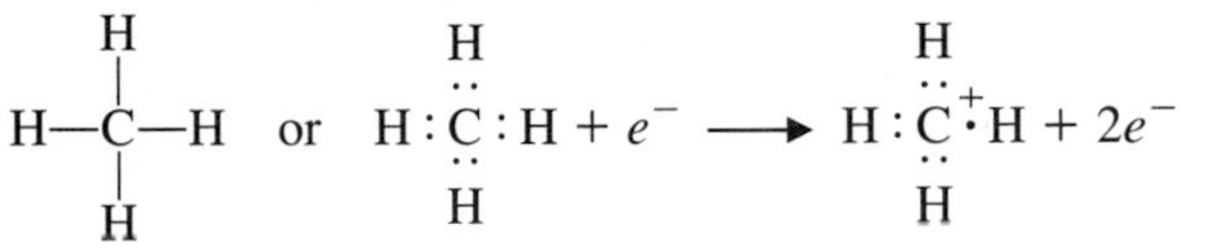

$$
\underset{H}{\overset{H}{H-C-H}} \quad \text{or} \quad H\!:\!\underset{H}{\overset{\cdot\cdot}{C}}\!:\!H + e^- \longrightarrow H\!:\!\underset{H}{\overset{\cdot\cdot}{C}}\!\!\overset{+}{\cdot}H + 2e^- \tag{12.10}
$$

The product of this reaction is sometimes abbreviated as follows:

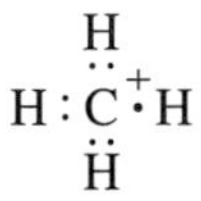

$$
H\!:\!\underset{H}{\overset{\overset{\textstyle H}{\cdot\cdot}}{C}}\!\!\overset{+}{\cdot}H \quad \text{abbreviated as} \quad CH_4\overset{+}{\cdot}
$$

The symbol $\overset{+}{\cdot}$ means that the molecule is both a radical and a cation—a **radical cation.** The species $CH_4\overset{+}{\cdot}$ is called the *methane radical cation.*

Following its formation, the methane radical cation decomposes in a series of reactions called **fragmentation reactions.** In one such reaction, it loses a hydrogen *atom* to generate the methyl cation, a carbocation.

$$
\underset{\text{mass} = 16}{CH_4\overset{+}{\cdot}} \longrightarrow \underset{\substack{\text{methyl}\\\text{cation}\\\text{mass} = 15}}{^+CH_3} + H\cdot \tag{12.11}
$$

The hydrogen atom carries the unpaired electron, and the methyl cation carries the charge. This process can be represented with the free-radical (fishhook) arrow formalism as follows:

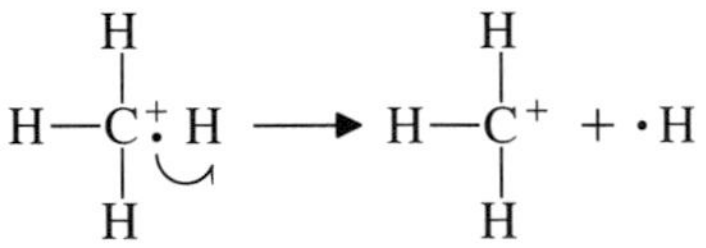

$$
\underset{H}{\overset{H}{H-\overset{+}{C}\cdot H}} \longrightarrow \underset{H}{\overset{H}{H-\overset{+}{C}}} + \cdot H \tag{12.12}
$$

Alternatively, the unpaired electron may remain associated with the carbon atom; in this case, the products of the fragmentation are a methyl radical and a proton.

$$
\underset{\text{mass} = 16}{CH_4\overset{+}{\cdot}} \longrightarrow \underset{\substack{\text{methyl}\\\text{radical}}}{\cdot CH_3} + \underset{\text{mass} = 1}{H^+} \tag{12.13}
$$

Further decomposition reactions give fragments of progressively lower mass. (Show how these occur by using the arrow formalism.)

$$
CH_3^+ \longrightarrow \underset{\text{mass} = 14}{CH_2\overset{+}{\cdot}} + H\cdot \tag{12.14a}
$$

$$
CH_2\overset{+}{\cdot} \longrightarrow \underset{\text{mass} = 13}{CH^+} + H\cdot \tag{12.14b}
$$

$$
CH^+ \longrightarrow \underset{\text{mass} = 12}{C\overset{+}{\cdot}} + H\cdot \tag{12.14c}
$$

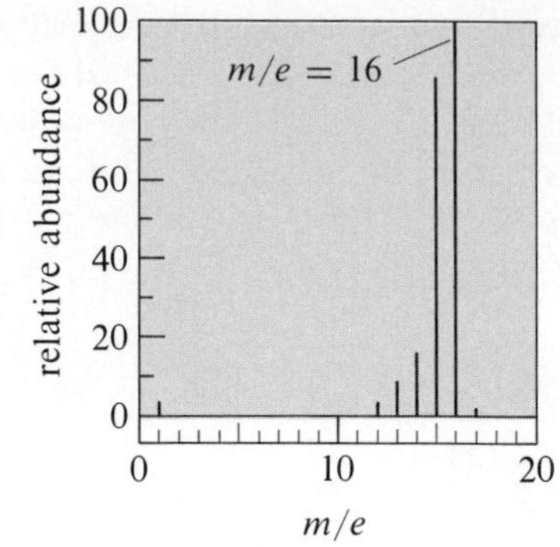

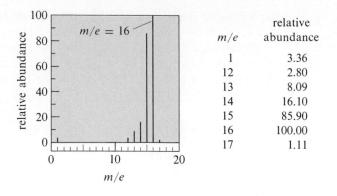

m/e	relative abundance
1	3.36
12	2.80
13	8.09
14	16.10
15	85.90
16	100.00
17	1.11

Figure 12.13 Mass spectrum of methane.

Thus methane undergoes fragmentation in the mass spectrometer to give several positively charged **fragment ions** of differing mass: $CH_4^{+\cdot}$, CH_3^+, $CH_2^{+\cdot}$, CH^+, and $C^{+\cdot}$. In the mass spectrometer, these fragment ions are separated according to their **mass-to-charge ratio,** *m/e* (m = mass, e = the charge of the electron). Since most ions formed in the mass spectrometer have unit charge, the *m/e* value can generally be taken as the mass of the ion. A **mass spectrum** is a graph of the relative amount of each ion (called the **relative abundance**) as a function of the ionic mass (or *m/e*). The mass spectrum of methane is shown in Fig. 12.13. It is important to understand that only *ions* are detected by the mass spectrometer—neutral molecules and radicals do not appear as peaks in the mass spectrum. The mass spectrum of methane shows peaks at *m/e* = 16, 15, 14, 13, 12, and 1, corresponding to the various ionic species that are produced from methane by electron ejection and fragmentation, as shown in Eqs. 12.12–12.14.

The mass spectrum can be determined for any molecule that can be vaporized in a high vacuum, and this includes most organic compounds. The utility of mass spectrometry is that (a) it can be used to determine the molecular weight of an unknown compound, and (b) it can be used to determine the structure of an unknown compound by an analysis of the fragment ions in the spectrum.

Let us examine some of the terminology used in mass spectrometry. The **parent ion** or **molecular ion,** sometimes abbreviated *p,* is the ion derived from electron ejection before any fragmentation takes place; hence, *the molecular ion occurs at an m/e value equal to the molecular weight of the sample molecule.* Thus, in the mass spectrum of methane, the molecular ion occurs at *m/e* = 16. In the mass spectrum of decane (Fig. 12.14), the molecular ion is at *m/e* = 142. Except for peaks due to isotopes, which we shall discuss in Sec. 12.5B, the molecular ion peak is the peak of highest *m/e* in any ordinary mass spectrum.

The **base peak** is the ion of greatest relative abundance. The base peak is arbitrarily assigned a relative abundance of 100%, and the other peaks in the mass spectrum are scaled relative to it. In the mass spectrum of methane, the base peak is the same as the molecular ion, but in the mass spectrum of decane the base peak occurs at *m/e* = 43. In this spectrum and most others, the molecular ion and the base peak are different.

B. Isotopic Peaks

The mass spectrum of methane (Fig. 12.13) shows a small but real peak at *m/e* = 17, a mass one unit higher than the molecular weight. This peak is termed a $p + 1$ peak, because it occurs one mass unit higher than the parent (p) peak. This ion occurs

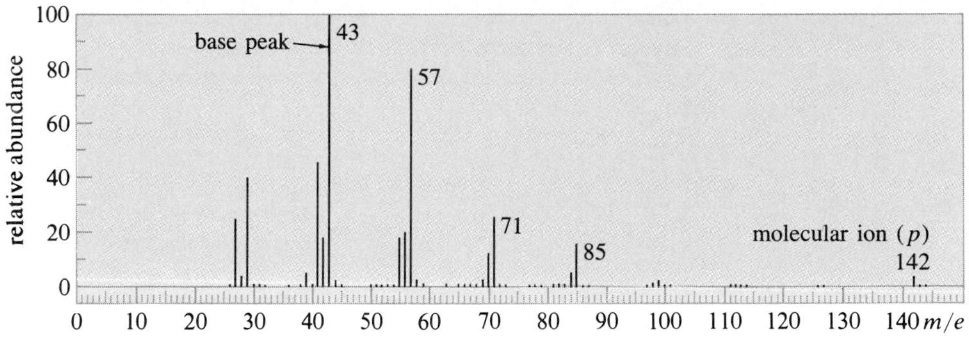

Figure 12.14 Mass spectrum of decane.

because chemically pure methane is really a mixture of compounds containing the various isotopes of carbon and hydrogen.

$$\text{methane} = {}^{12}CH_4,\ {}^{13}CH_4,\ {}^{12}CDH_3,\ \text{etc.}$$

$$m/e = 16 \quad 17 \quad 17$$

The isotopes of several elements and their natural abundances are given in Table 12.2.

Possible sources of the $m/e = 17$ peak for methane are ${}^{13}CH_4$ and ${}^{12}CDH_3$. *Each isotopic compound contributes a peak with a relative abundance in proportion to its amount.* In turn, the amount of each isotopic compound is directly related to the natural abundance of the isotope involved. The relative abundance of a peak due to ${}^{12}C$ methane and the peak due to the presence of a ${}^{13}C$ isotope is then given by the following equation:

$$\text{relative abundance} = \left(\frac{\text{abundance of } {}^{13}\text{C peak}}{\text{abundance of } {}^{12}\text{C peak}}\right) \tag{12.15a}$$

$$= (\text{number of carbons}) \times \left(\frac{\text{natural abundance of } {}^{13}\text{C}}{\text{natural abundance of } {}^{12}\text{C}}\right)$$

$$= (\text{number of carbons}) \times \left(\frac{0.0111}{0.9889}\right)$$

$$= (\text{number of carbons}) \times 0.0112 \tag{12.15b}$$

Since methane has only one carbon, the $m/e = 17$, or $p + 1$, peak is about 1.1% of the $m/e = 16$, or p, peak. We can make a similar calculation for deuterium.

$$\text{relative abundance} = (\text{number of hydrogens}) \times \left(\frac{\text{natural abundance of } {}^{2}\text{H}}{\text{natural abundance of } {}^{1}\text{H}}\right) \tag{12.16}$$

$$= (4) \times \left(\frac{0.00015}{0.99985}\right) = 0.0006$$

Thus, the CDH_3 naturally present in methane contributes 0.06% to the isotopic peak. Clearly, the contribution of deuterium is negligible; ${}^{13}C$ is the major isotopic contributor to the $p + 1$ peak.

TABLE 12.2 Exact Masses and Isotopic Abundances of Several Isotopes Important in Mass Spectrometry

Element	Isotope	Exact mass	Abundance, %
hydrogen	^{1}H	1.007825	99.985
deuterium	^{2}H	2.0140	0.015
carbon	^{12}C	12.0000	98.89
	^{13}C	13.00335	1.11
nitrogen	^{14}N	14.00307	99.63
	^{15}N	15.00011	0.37
oxygen	^{16}O	15.99491	99.759
	^{17}O	16.99991	0.037
	^{18}O	17.99992	0.204
fluorine	^{19}F	18.99984	100.
silicon	^{28}Si	27.97693	92.21
	^{29}Si	28.97649	4.70
	^{30}Si	29.97376	3.09
phosphorus	^{31}P	30.99376	100.
sulfur	^{32}S	31.97207	95.0
	^{33}S	32.97146	0.76
	^{34}S	33.96786	4.22
chlorine	^{35}Cl	34.96885	75.53
	^{37}Cl	36.9659	24.47
bromine	^{79}Br	78.9183	50.54
	^{81}Br	80.9163	49.46
iodine	^{127}I	126.90004	100.

In a compound containing more than one carbon, the $p + 1$ peak is larger relative to the p peak because there is a 1.1% probability that *each carbon* in the molecule will be present as ^{13}C. For example, cyclohexane has six carbons, and the abundance of its $p + 1$ ion relative to that of its parent ion should be $6(1.1) = 6.6\%$. In the mass spectrum of cyclohexane (Fig. 12.15), the parent ion has a relative abundance of about 70%; that of the $p + 1$ ion is calculated to be $(0.066)(70\%) = 4.6\%$, which corresponds closely to the value observed. (With careful measurement, it is possible to use these isotopic peaks to estimate the number of carbons in an unknown compound; see Problem 33.) Not only the parent peak, but every other peak in the mass spectrum has isotopic peaks.

Several elements of importance in organic chemistry have isotopes with significant natural abundances. Table 12.2 shows that silicon has significant $p + 1$ and $p + 2$ contributions; sulfur has a $p + 2$ contribution; and the halogens chlorine and bromine have very important $p + 2$ contributions. In fact, the naturally occurring form of the element bromine consists of about equal amounts of ^{79}Br and ^{81}Br. The mixture of isotopes leaves a characteristic trail in the mass spectrum that allows us to determine that the element is present even without an elemental analysis. Let us see how this determination is possible, using bromomethane as a simple example.

The mass spectrum of bromomethane is shown in Fig. 12.16. The peaks at $m/e =$ 94 and 96 result from the contributions of the two bromine isotopes to the parent ion.

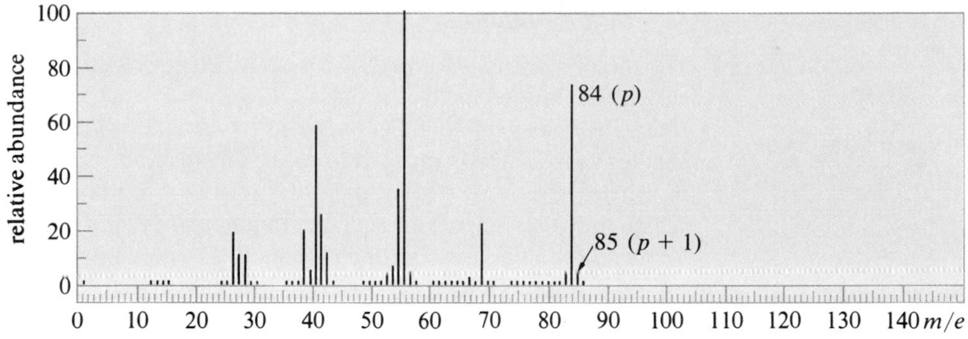

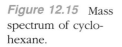

Figure 12.15 Mass spectrum of cyclohexane.

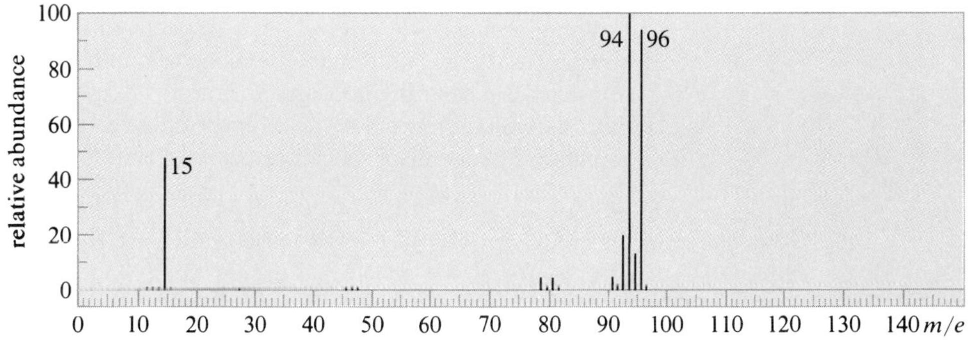

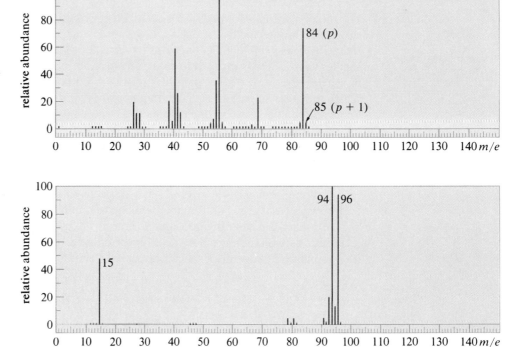

Figure 12.16 Mass spectrum of bromomethane.

They are in the relative abundance ratio 77.0/75.6 = 1.02, which is in good agreement with the ratio of the relative natural abundances of the bromine isotopes (Table 12.2). This double parent ion is a dead giveaway for a compound containing a single bromine. Notice that along with each major isotopic peak is a smaller isotopic peak one mass unit higher. These peaks are due to the natural abundance of ^{13}C present in bromomethane. For example, the $m/e = 95$ peak corresponds to bromomethane containing only ^{79}Br and one ^{13}C. The $m/e = 97$ peak arises from methyl bromide that contains only ^{81}Br and one ^{13}C.

Although isotopes such as ^{13}C and ^{18}O are normally present in small amounts in organic compounds, it is possible to synthesize compounds that are selectively enriched with these and other isotopes. Isotopes are especially useful because they provide a specific label at particular atoms without changing their chemical properties. One use of such compounds is to determine the fate of specific atoms in deciding between two mechanisms. (See, for example, Eq. 11.29 or Problem 22, Chapter 11.) Another use is to provide nonradioactive isotopes for biological metabolic studies (studies that deal with the fates of chemical compounds when they react in biological systems). When a compound has been isotopically enriched, isotopic peaks become much larger than they are normally. Quantitative measurements of the peak abundances permit us to calculate the amount of isotope present.

Problem

15 From the information in Table 12.2 predict the appearance of the molecular ion peak(s) in the mass spectrum of chloromethane.

C. Fragmentation

The molecular ion is formed by loss of an electron. If this ion is stable, it decomposes slowly and is detected by the mass spectrometer as a peak of large relative abundance. If this ion is less stable, it decomposes into smaller pieces, which are then detected as smaller ions, called **fragment ions.** The relative abundances of the various fragments in a mass spectrum depend on their relative lifetimes—that is, the relative rates at which they break apart into smaller fragments. Typically the most stable ions have the longest lifetimes, and are detected as the largest peaks in a mass spectrum.

The fragment ions produced in the mass spectrometer are literally pieces of the whole molecule. Just as a picture can be reconstructed from a jigsaw puzzle, the structure of a molecule can in principle be reconstructed from its mass spectrum. Let us now examine some examples of fragmentation.

The simplest way to analyze a mass spectrum is to think of fragment ions as coherent pieces that result from the breaking of chemical bonds in the molecular ion. This approach successfully accounts for the mass spectrum of methane, previously analyzed, as well as the mass spectrum of decane (Fig. 12.14). Decane undergoes fragmentation at several of its carbon–carbon bonds.

$$CH_3-CH_2 | CH_2 | CH_2 | CH_2 | CH_2 | CH_2-CH_2-CH_2-CH_3$$

$m/e = 29$
$m/e = 43$
$m/e = 57$
$m/e = 71$
$m/e = 85$

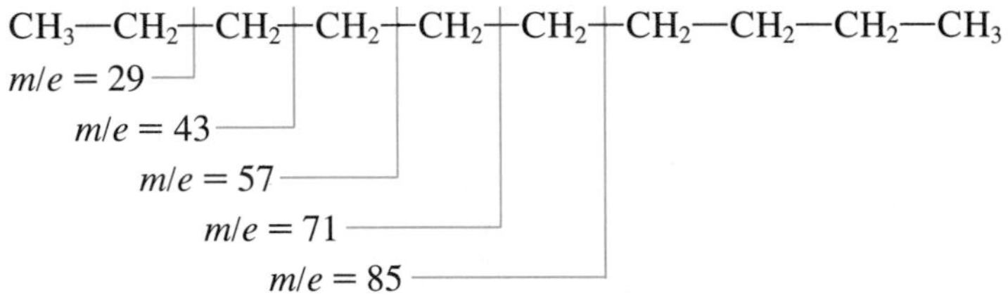

For example, the $m/e = 71$ fragment is formed in the following way:

electron ejection to yield parent ion:

$$C_4H_9-CH_2 : CH_2-C_4H_9 \xrightarrow{-e^-} C_4H_9-CH_2 \overset{+}{\cdot} CH_2-C_4H_9 \qquad (12.17a)$$

fragmentation:

$$C_4H_9-CH_2 \overset{+}{\cdot} CH_2-C_4H_9 \longrightarrow C_4H_9-CH_2^+ \ + \cdot CH_2-C_4H_9 \qquad (12.17b)$$
$$m/e = 71$$

(Write a mechanism for formation of one or two other fragments.) Notice that the different fragments are not formed in the same relative abundance. Furthermore, some possible fragment peaks are weak or missing. Thus, there is no $m/e = 15$ fragment, and the peaks corresponding to fragments at m/e greater than 85 are very weak. In some cases (see Problem 16) the reason for the predominance of certain fragments is clear from their relative stabilities, but in other cases, the reason is less well understood.

Problem

16 The most prominent peak (base peak) in the mass spectrum of 2,2,5,5-tetramethylhexane is at $m/e = 57$, which corresponds to a composition C_4H_9.
(a) Suggest a structure for the fragment ion that accounts for this peak.
(b) Offer a reason why this fragment ion is particularly stable.

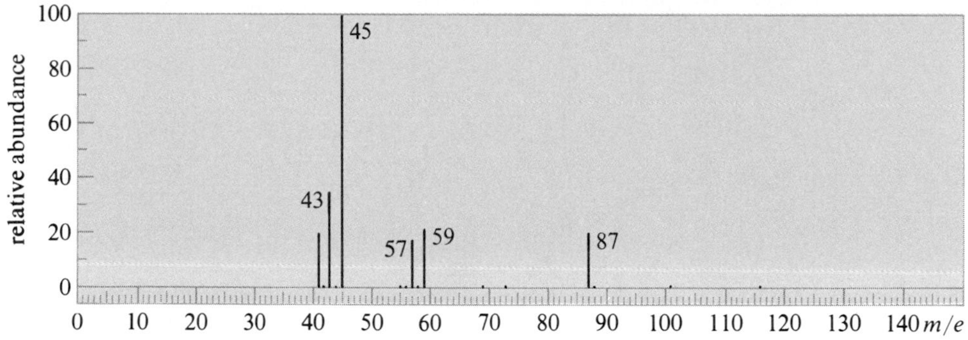

Figure 12.17 Mass spectrum of *sec*-butyl isopropyl ether.

The mass spectrum of *sec*-butyl isopropyl ether, shown in Fig. 12.17, illustrates two other important modes of fragmentation. The peaks at $m/e = 57$ and $m/e = 43$ come from obvious pieces of the molecule:

$$m/e = 57 \qquad m/e = 43$$
$$CH_3-CH|O|CH(CH_3)_2$$
$$\overset{|}{C_2H_5}$$

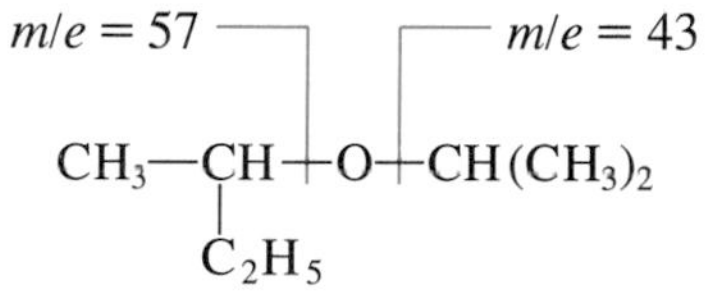

Notice that the fragmentations that give these peaks occur at bonds to the ether oxygen. The reason is that oxygen has unshared electron pairs; it takes less energy to eject an electron from an unshared electron pair than it does from a σ-bond, because unshared electrons are held less tightly than σ-electrons. Loss of an unshared electron from the ether oxygen leads to both fragments. For example, the $m/e = 57$ fragment is formed by the following mechanism:

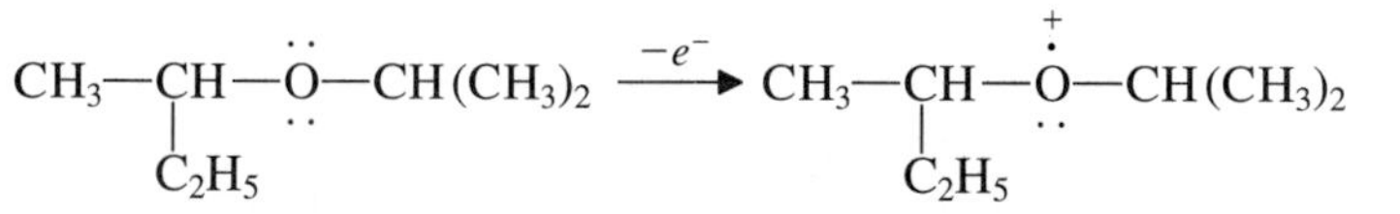

$$CH_3-\underset{\underset{C_2H_5}{|}}{CH}-\overset{..}{\underset{..}{O}}-CH(CH_3)_2 \xrightarrow{-e^-} CH_3-\underset{\underset{C_2H_5}{|}}{CH}-\overset{+}{\underset{..}{\overset{.}{O}}}-CH(CH_3)_2 \qquad (12.18a)$$

$$CH_3-\underset{\underset{C_2H_5}{|}}{CH}\overset{\overset{\curvearrowright}{+}}{-\underset{..}{O}}-CH(CH_3)_2 \longrightarrow CH_3-\underset{\underset{C_2H_5}{|}}{CH}{}^+ + \overset{.}{:}\underset{..}{O}-CH(CH_3)_2 \qquad (12.18b)$$
$$m/e = 57$$

This type of cleavage, called **inductive cleavage,** is very common in ethers, alkyl halides, and other compounds containing a very electronegative element.

Since the molecular weight of *sec*-butyl isopropyl ether is 116, the peak at $m/e = 87$ corresponds to a *loss* of 29 mass units; this corresponds to an ethyl radical, $\cdot CH_2CH_3$.

$$m/e = 87$$
$$CH_3-\underset{\underset{CH_2CH_3}{|}}{CH}-O-CH(CH_3)_2$$

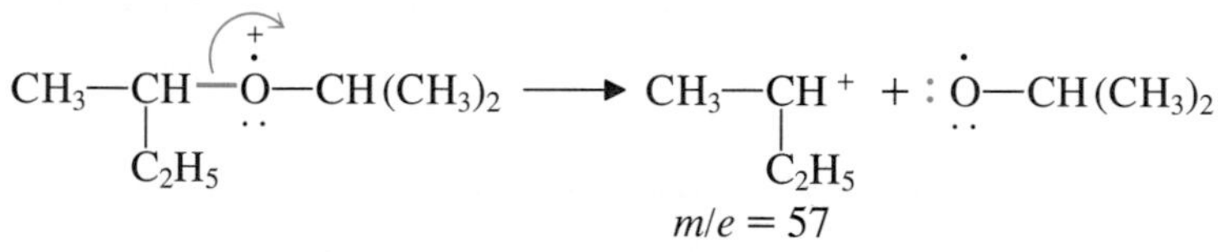

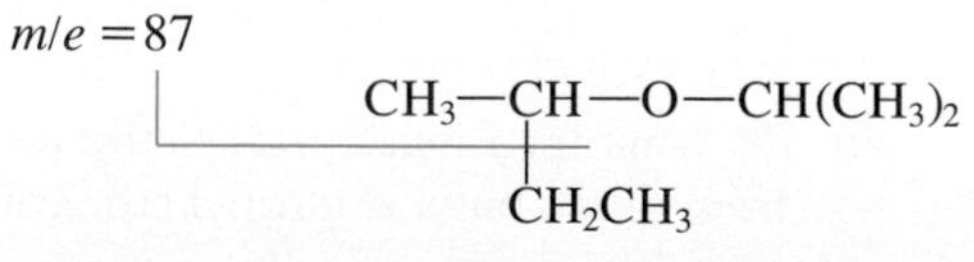

This peak is due to a fragmentation of the same parent ion by the following mechanism:

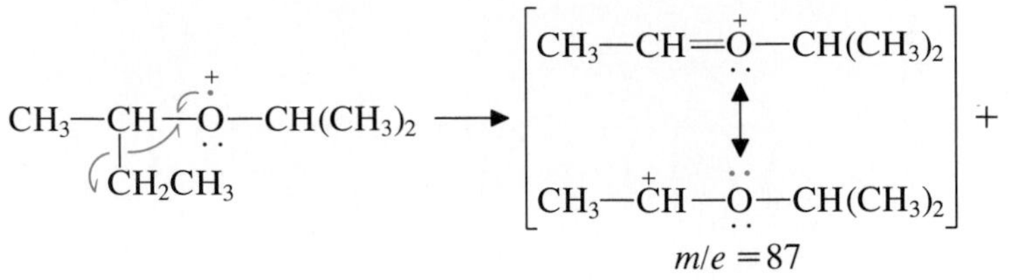

$$CH_3-CH\overset{+}{\underset{\overset{|}{CH_2CH_3}}{-}O}-CH(CH_3)_2 \longrightarrow \begin{bmatrix} CH_3-CH\overset{+}{=}O-CH(CH_3)_2 \\ \updownarrow \\ CH_3-\overset{+}{CH}-\overset{..}{\underset{..}{O}}-CH(CH_3)_2 \end{bmatrix} + \cdot CH_2CH_3 \quad (12.19)$$

$$m/e = 87$$

This type of cleavage, called **α-cleavage,** is important in secondary and tertiary ethers, alcohols, alkenes, and several other types of compounds. We have actually seen this type of cleavage before under a different name: the *β-scission* reaction discussed in Sec. 5.9B. Notice that another α-cleavage, giving a different cation, is also possible.

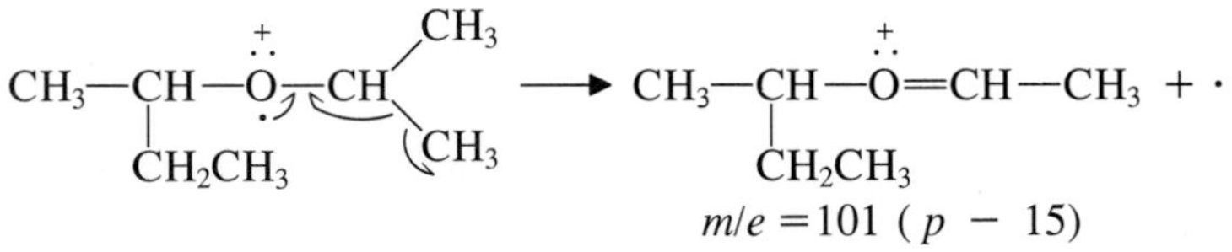

$$CH_3-CH\overset{+}{\underset{\overset{|}{CH_2CH_3}}{-}\overset{..}{O}}-CH\overset{CH_3}{\underset{CH_3}{\diagdown}} \longrightarrow CH_3-CH\overset{+}{\underset{\overset{|}{CH_2CH_3}}{-}\overset{..}{O}}=CH-CH_3 + \cdot CH_3 \quad (12.20)$$

$$m/e = 101 \; (p - 15)$$

Yet this ion is almost undetectable in the mass spectrum. This should not really surprise us, for the fragmentation shown in Eq. 12.19 results in the formation of the ethyl radical, and Eq. 12.20 results in the formation of the methyl radical. As we learned in Sec. 5.8D, more highly substituted free radicals are formed more easily.

Notice that we can use the mass of the fragments *lost* as well as the mass of the fragments *observed* to postulate the structure of fragment ions. The mass of the fragment lost (29 in the example above) is obtained merely by subtracting the mass of the observed fragment ion from the mass of the parent compound. We should be careful not to confuse each type of fragment, however. The fragment observed *must* be charged, because the mass spectrometer detects only charged particles; the fragment lost is uncharged. (It is a free radical in the example above.)

From these few examples, we can see that fragmentation is not random, but in many cases is a consequence of the rules of carbocation or free-radical stability that we learned from our study of chemical reactions. The important point here is that we should not expect to see every possible fragment in the mass spectrum of any compound; there are good reasons why some fragments predominate and others are missing.

Problems

17 The mass spectrum of 2-methyl-2-pentanol contains two prominent peaks at $m/e = 59$ and 87. To what ions do these peaks correspond? Suggest mechanisms for their formation.

18 Show how inductive cleavage accounts for the $m/e = 43$ peak in the mass spectrum of *sec*-butyl isopropyl ether (Fig. 12.17).

19 To what fragment ion does the peak at $m/e = 57$ correspond in the mass spectrum of *sec*-butyl isopropyl ether (Fig. 12.17)? Suggest a mechanism for its formation.

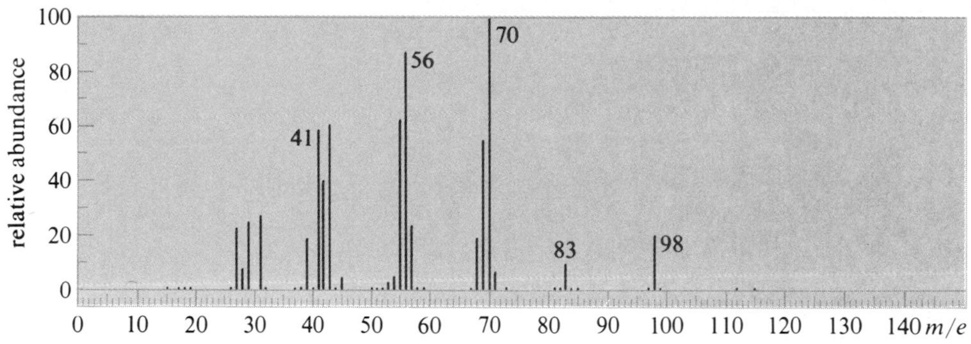

Figure 12.18 Mass spectrum of 1-heptanol.

An interesting thing to note about most of the fragment ions observed in the mass spectra of Figs. 12.14 and 12.17 is that they all have odd molecular masses. The reason for this is very simple. When an organic compound contains only C, H, and O (as the ones in these figures do) it has an even mass. When such a compound undergoes fragmentation to give separate radical and cation species, each of these must have odd mass.

$$CH_3-CH_2\overset{+}{\cdot}CH_3 \longrightarrow CH_3-\overset{+}{C}H_2 + \cdot CH_3 \qquad (12.21)$$
$$\text{parent ion} \qquad\qquad m/e = 29 \qquad \text{mass} = 15$$
$$m/e = 44 \text{ (even)} \qquad\qquad \text{(both odd)}$$

Furthermore, since the unpaired electron is carried off in the radical fragment, the carbocation fragment detected by the mass spectrometer must have no unpaired electrons. Ions with no unpaired electrons are called **even-electron ions.** On the other hand, molecular ions derived from compounds of even mass, as well as some fragment ions produced in other ways, have even mass. Such ions contain an unpaired electron, and therefore must be *radical cations,* or **odd-electron ions.** We see that the mass of the ion tells us something about its electronic structure. To summarize: *when a compound contains only C, H, and O, its fragment ions of odd mass must be even-electron ions and its fragment ions of even mass must be odd-electron ions.* (There are corresponding generalizations for compounds containing other elements, such as nitrogen; see Problem 38.)

An example of a mass spectrum containing odd-electron fragments is that of 1-heptanol (Fig. 12.18). Let us focus on the fragment at *m/e* = 98. This ion has even mass and is therefore an odd-electron ion. This corresponds to a loss of 18 mass units from the parent ion, and can be accounted for by the loss of water. A mechanism for this fragmentation begins with loss of an electron from one of the unshared pairs of oxygen. A hydrogen (color) then migrates from a β-carbon to this oxygen, and water leaves the molecule.

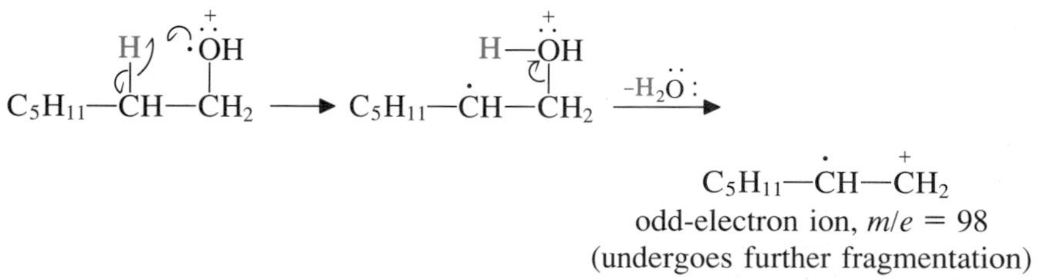

$$C_5H_{11}-\overset{|}{C}H-CH_2 \longrightarrow C_5H_{11}-\overset{|}{C}H-CH_2 \xrightarrow{-H_2\ddot{O}:}$$

$$\overset{\cdot}{C}_5H_{11}-\overset{+}{C}H-CH_2 \qquad (12.22)$$
$$\text{odd-electron ion, } m/e = 98$$
$$\text{(undergoes further fragmentation)}$$

Hydrogen-atom transfer followed by loss of a stable neutral molecule is a very common mechanism for the formation of odd-electron ions. This mode of fragmentation, for example, is common in primary alcohols, alkyl halides, and other compounds.

Problem

20 The mass spectrum of 2-chlorobutane shows large and almost equally intense peaks at $m/e = 57$ and $m/e = 56$.

(a) Which of these fragments is an even-electron ion? (The presence of chlorine in the parent ion does not affect the generalizations above.)

(b) What stable neutral molecule can be lost to give the odd-electron ion?

(c) Suggest a mechanism for the origin of each of these fragments.

D. Identifying the Molecular Ion

The molecular ion is the most important peak in the mass spectrum because it provides the molecular weight of the molecule under study and thus is the basis for calculating the losses involved in fragmentation. However, in some mass spectra, the abundance of the molecular ion is so small that it is undetectable. Thus there is a problem: suppose we are dealing with a compound of unknown molecular weight. How do we know whether the peak of highest mass is due to the molecular ion or a fragment ion? There is no general answer to this question, but there are some simple considerations that can help us decide.

All compounds containing only the elements C, H, and O have even molecular weights, since carbon and oxygen have even molecular weights and there must be an even number of hydrogen atoms. Such compounds must therefore have a molecular ion of even mass. (Analogous generalizations are possible for compounds containing other elements.) Suppose a compound known to contain only C, H, and O has a mass spectrum containing a peak of highest mass at an odd value of *m/e*. For example, the peak of highest mass in the mass spectrum of 3-methyl-3-pentanol occurs at $m/e = 87$.

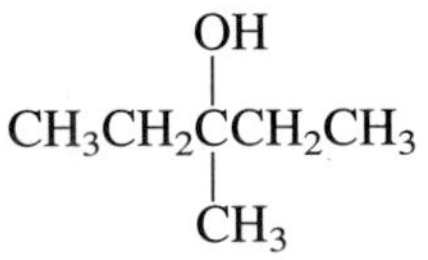

3-methyl-3-pentanol

Were this an unknown compound, we would know immediately that the peak at $m/e = 87$ cannot qualify as the molecular ion because of its odd mass.

Another test for the molecular ion involves the observed losses that are calculated from the candidate peak under the *assumption* that it is the molecular ion. Mass losses of 4–14, 21–25, 33, 37, and 38 from the molecular ion are relatively rare. The reason is that combinations of atoms that give these losses simply do not exist. For example, in the mass spectrum of 3-methyl-3-pentanol, the base peak occurs at $m/e = 73$, fourteen mass units below the peak of highest mass at $m/e = 87$. Even if we did not know that this compound, as an unknown, contains only C, H, and O, the loss of 14 mass units would cast serious doubt that the peak at $m/e = 87$ is the molecular ion.

Another way to determine the mass of the molecular ion is by chemical derivatization. For example, if the mass spectrum of an alcohol shows no molecular ion, a methyl ether that could be readily prepared from the alcohol might have one.

Yet another method of determining the mass of the molecular ion is to use a completely different ionization technique. Although we have focused on sample ionization by electron bombardment—*electron impact mass spectrometry* (EI)—another method, called *chemical-ionization mass spectrometry* (CI), is also used. In this technique, a sample in the gas phase is treated with protons, and the sample molecule is protonated at its most basic site. (Because a proton is attached, the molecular ion

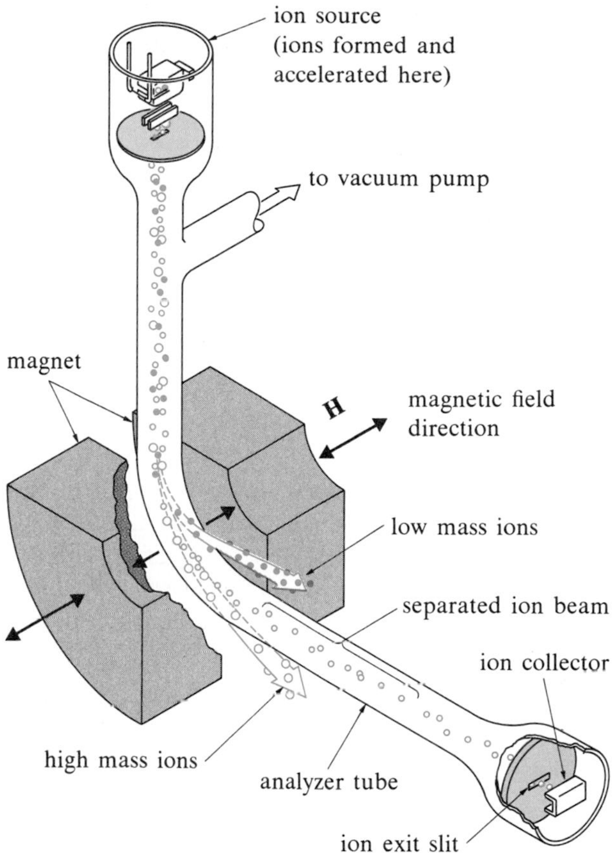

Figure 12.19 Diagram of a mass spectrometer.

appears at one mass unit higher than the molecular weight of the unprotonated molecule.) A much higher percentage of molecular ion is usually obtained by this technique, and different fragmentation patterns are observed containing fewer peaks.

E. The Mass Spectrometer

A diagram of a conventional mass spectrometer is shown in Fig. 12.19. Within the mass spectrometer the sample is ionized and fragment ions are formed, separated, and detected. After ionization of the sample by electron bombardment, the ions are accelerated with a voltage V (typically several kilovolts) and are passed into a magnetic field $\mathbf{H}$ along a path perpendicular to the field. Since a magnetic field exerts a force on any moving charge, the field causes the ions to bend in a circular path of radius r. The amount of bending depends on the kinetic energy of the ion, which is equal to $mv^2/2$ (m = mass, v = velocity). Since all ions have, to a good approximation, the same velocity as a result of the accelerating voltage, then it is the *ionic mass* that gives one ion a different kinetic energy from another. Ions of high kinetic energy—high mass—are least affected by the magnetic field and tend to "blast" through it with less bending. Ions of low mass are more easily bent by the magnetic field. The basic equation describing this behavior is

$$m/e = (4.82 \times 10^{-5}) \frac{\mathbf{H}^2 r^2}{V} \tag{12.23}$$

in which r is in cm, V is in volts, **H** is in gauss, m/e is the mass-to-charge ratio, and the constant (4.82×10^{-5}) is the product of several conversion factors. This equation shows that ions of high mass traverse a path of larger radius than those of low mass—in other words, the larger the mass, the harder it is for the field to bend the path of the ion. A collector is positioned to sense only the ions traversing a path of a certain radius r. By increasing the magnetic field in a regular way, ions of progressively increasing mass are brought into registration with the collector and the relative ionic intensities are recorded as an ion current.

A modern mass spectrometer is an extremely sensitive instrument, and can readily produce a mass spectrum from amounts of material in the range of micrograms (10^{-6} g) to nanograms (10^{-9} g). For this reason the instrument is very useful for the analysis of materials available in only trace quantities. It has played a key role in such projects as the analysis of drug levels in blood serum, and the elucidation of the structures of insect pheromones (Sec. 14.9) that are available only in minuscule amounts.

One of the operating characteristics of a mass spectrometer is its *resolution*—that is, how well it separates ions of different mass. A relatively simple mass spectrometer readily distinguishes, over a total m/e range of several hundred, ions that differ in mass by one unit. More complex mass spectrometers, called *high-resolution mass spectrometers,* can resolve ions that are separated in mass by only a few thousandths of a mass unit. Why would we want such high resolution? Suppose we have an unknown compound with a molecular ion at $m/e = 124$. Two possible formulas for this ion are $C_8H_{12}O$ and C_9H_{16}. Both formulas have the same **nominal mass** (mass to the nearest whole number). However, if we calculate the **exact mass** for each formula (using the values of the most abundant isotopes in Table 12.2), then different results are obtained:

$$C_8H_{16}O\text{, exact mass: } 124.0888$$

$$C_9H_{16}\text{, exact mass: } 124.1252$$

The difference of 0.0364 mass units is easily resolved by a high-resolution mass spectrometer. Such instruments, used in conjunction with computers, can work backwards from the exact mass and provide *an elemental analysis of the parent ion* (and therefore the compound of interest) *and that of each fragment in the mass spectrum*! The greater resolution can be achieved by a variety of techniques. In one method, the ions are sorted twice: once in an electric field, then in the magnetic field. Because a high-resolution mass spectrometer and its associated computer and other accessories cost nearly \$500,000, it is generally shared by a large number of researchers.

Before a compound can be analyzed by mass spectrometry, it must be vaporized. This presents a difficult problem in the case of a large molecule that has negligible vapor pressure. Research in mass spectrometry has focused on novel ways to produce ions in the gas phase from large molecules, many of which are of biological interest. In one technique, called *fast-atom bombardment* (FAB), compounds in solution are converted directly into gas-phase ions by subjecting them to a beam of heavy atoms (such as xenon, argon, or cesium) that have been accelerated to high velocities. At present it appears that ions with molecular weights of 5,000 to 10,000 can be produced by this technique.

KEY IDEAS IN CHAPTER 12

- Spectroscopy deals with the interaction of matter and electromagnetic radiation. Electromagnetic radiation is characterized by its energy, wavelength, and frequency, which are related by Eqs. 12.1, 12.3, and 12.7.
- Infrared spectroscopy deals with the absorption of infrared radiation by characteristic molecular vibrations. An infrared spectrum is a plot of the infrared radiation transmitted through a sample as a function of wavenumber or wavelength.
- The wavenumber of an absorption tends to be greatest for vibrations involving stronger bonds and smaller atomic masses.
- The intensity of an absorption increases with the number of absorbing groups in the sample and the size of the dipole moment change that occurs in the molecule when the vibration occurs. Absorptions that result in no dipole moment change are infrared inactive.
- The infrared spectrum provides information about the functional groups present in a molecule. The =C—H stretching and bending absorptions and the C=C stretching absorption are very useful for the identification of alkenes. The O—H stretching absorption is diagnostic for alcohols.
- In mass spectrometry a molecule loses an electron to form a radical cation called the molecular ion or parent ion, which then decomposes to fragment ions. The relative abundances of the fragment ions are recorded as a function of their mass-to-charge ratios *m/e,* which, for most ions, equal their masses. Both molecular weights and partial structures can be derived from the masses of these ionic fragments.
- Associated with each peak are other peaks at higher mass that arise from the presence of isotopes at their natural abundance. Such isotopic peaks are particularly useful for diagnosing the presence of certain elements, such as chlorine and bromine.
- Ionic fragments are of two types: even-electron ions, which contain no unpaired electrons; and odd-electron ions, which contain an unpaired electron. Even-electron ions are formed by such processes as α-cleavage, inductive cleavage, and direct fragmentation at a σ-bond. An important pathway for the formation of odd-electron ions is hydrogen transfer followed by loss of a small molecule such as water or hydrogen halide.

ADDITIONAL PROBLEMS

21 List the factors that determine the wavenumber of an infrared absorption band.

---

22 List two factors that determine the intensity of an infrared absorption band.

---

23 Indicate how you would carry out each of the following chemical transformations. What are some of the changes in the infrared spectrum that could be used to indicate whether the reaction has proceeded as indicated? (Your answer can include disappearance as well as appearance of IR bands.)

(a)

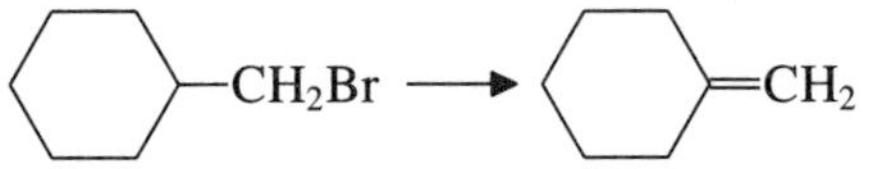

(b) 1-hexanol ⟶ hexyl methyl ether
(c) ethylene oxide ⟶ 1-butanol
(d) 1-methyl-1-cyclohexene ⟶ methylcyclohexane

---

24 Rationalize the indicated fragments in the mass spectrum of each of the following molecules by proposing a structure of the fragment and a mechanism by which it is produced.

(a) neopentane, $m/e = 57$
(b) 3-methyl-3-hexanol, $m/e = 73$
(c) 1-pentanol, $m/e = 70$
(d) $C_3H_7—C(CH_3)(C_2H_5)—\ddot{N}H_2$, $m/e = 72$

---

25 Suggest reasonable structures for the following *neutral* molecules commonly lost in mass spectral fragmentation:

(a) mass = 18 from a compound containing oxygen
(b) mass = 28 from a compound containing only C and H
(c) mass = 36 from a compound with a $p + 2$ peak about one-third as large as the parent peak

---

26 Which of the molecules in each of the following pairs should have identical IR spectra, and which should have different IR spectra (if only slightly different)? Explain your reasoning carefully.

(a) 3-pentanol and (±)-2-pentanol
(b) (*R*)-2-pentanol and (*S*)-2-pentanol
(c)

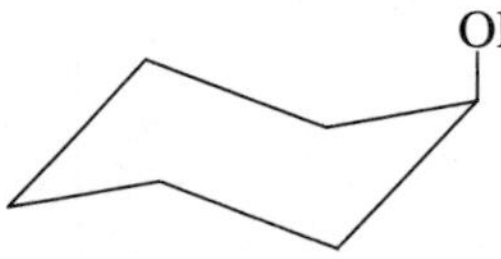

and

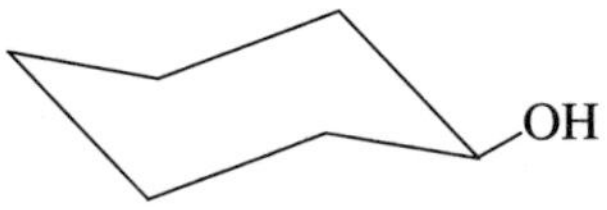

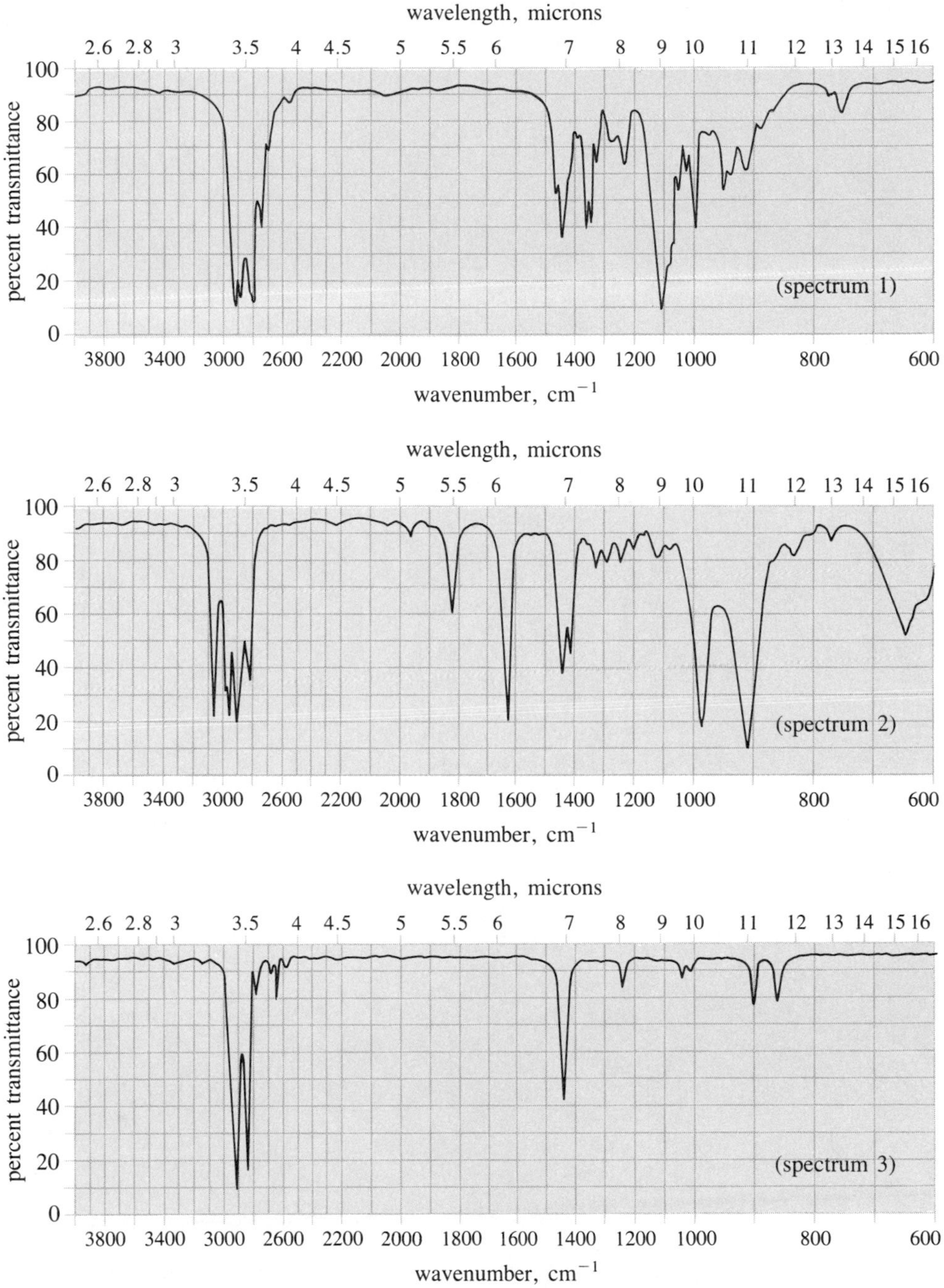

Figure 12.20 IR spectra for Problem 27.

(Figure continues on page 484.)

27 Match each of the IR spectra in Fig. 12.20 to one of the following compounds. (Notice that there is no spectrum for two of the compounds.)

(a) 1,5-hexadiene
(b) 1-methyl-1-cyclopentene
(c) 1-hexen-3-ol
(d) dipropyl ether
(e) *trans*-4-octene
(f) cyclohexane
(g) 3-hexanol

Figure 12.20 continued.

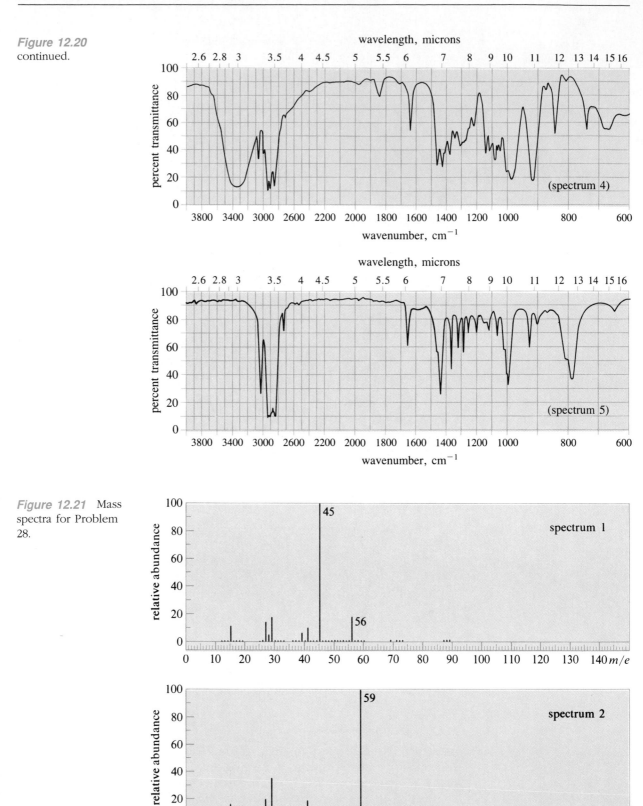

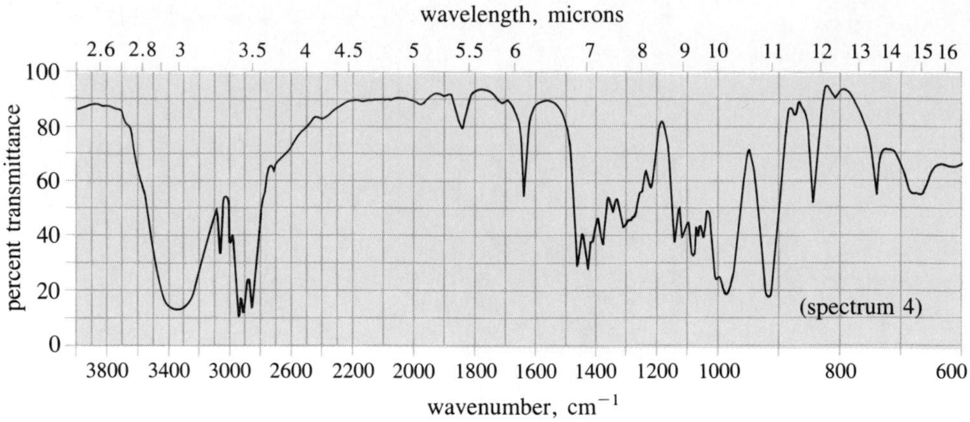

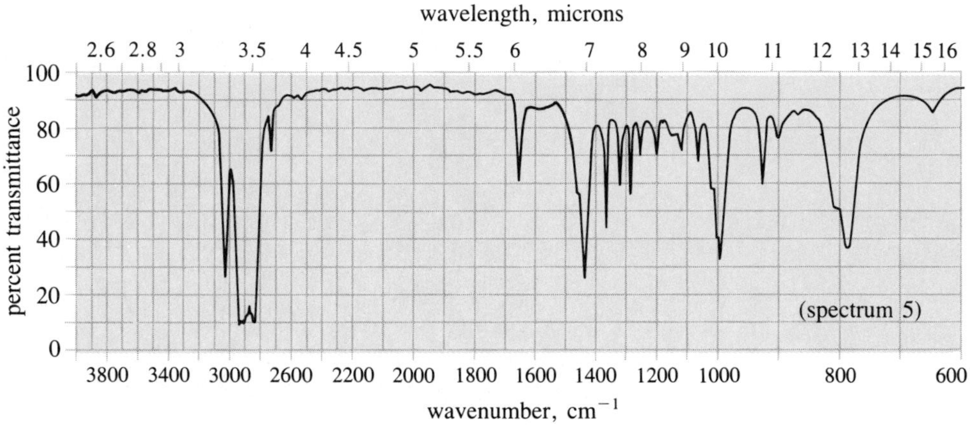

Figure 12.21 Mass spectra for Problem 28.

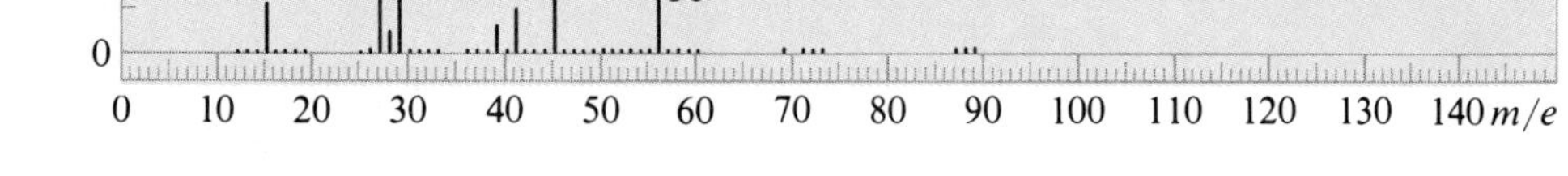

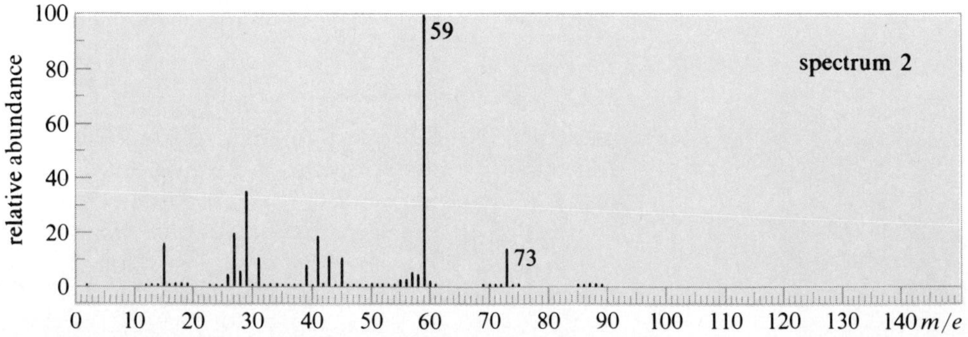

28 Fig. 12.21 shows the mass spectra of two isomeric ethers, 2-methoxybutane and 1-methoxybutane. Match each compound with its spectrum.

29 Explain why the mass spectrum of dibromomethane has three peaks at *m/e* = 172, 174, and 176 in the approximate relative abundances 1:2:1.

30 The mass spectrum of tetramethylsilane, $(CH_3)_4Si$, has a base peak at *m/e* = 73. Calculate the relative abundances of the isotopic peaks at *m/e* = 74 and 75.

31 A former theological student, Heavn Hardley, has turned to chemistry and, during his eighth year of graduate study, has carried out the following reaction:

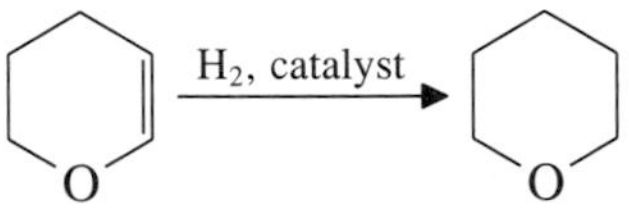

A *B*

Unfortunately, Hardley thinks he may have mislabeled his samples of *A* and *B*, but has wisely decided to take an IR spectrum of each sample. The spectra are reproduced in Fig. 12.22. Which sample goes with which spectrum? How do you know?

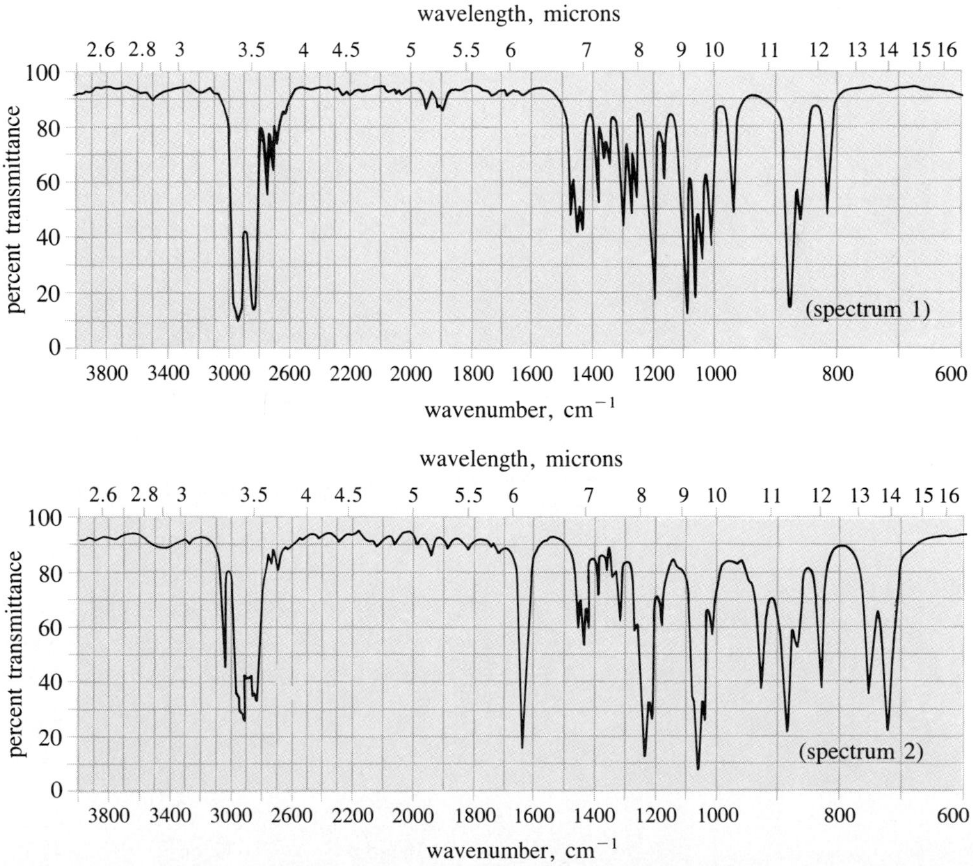

Figure 12.22 IR spectra for Problem 31.

Figure 12.23 Mass spectra for Problem 32.

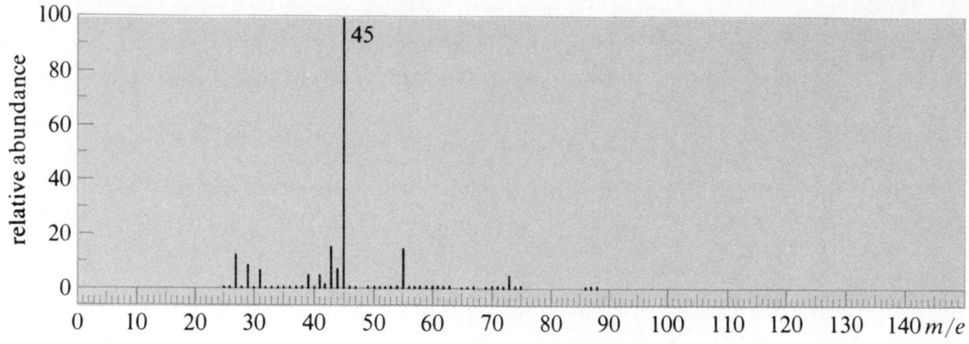

(a)

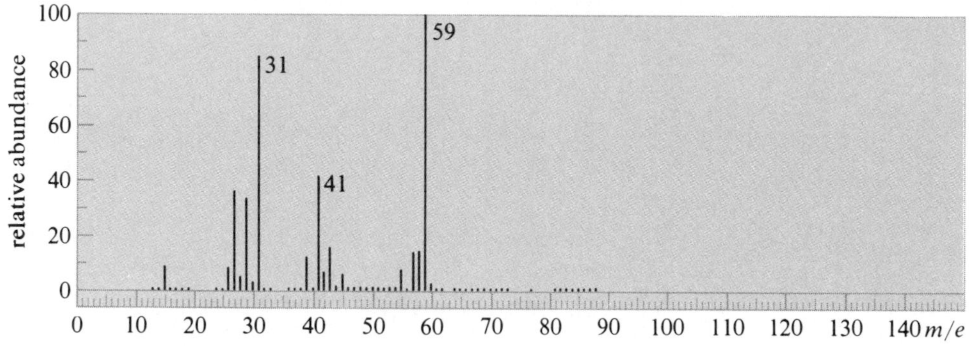

(b)

32 A chemist, Ilov Boron, carried out a reaction of *trans*-2-pentene with B_2H_6 followed by treatment with H_2O_2/OH^-. Two products were separated and isolated. Desperate to know their structures, Ilov took his compounds to the spectroscopy laboratory and found that only the mass spectrometer was operating. The mass spectra of the two products are given in Fig. 12.23. Suggest structures for the compounds, and indicate which mass spectrum goes with which compound.

33 From the molecular weights and relative intensities of their p and $p + 1$ peaks, suggest molecular formulas for the following compounds. (Assume the major isotopic contributor to the $p + 1$ peak is ^{13}C.)
(a) p at $m/e = 82$ (38%), $p + 1$ (2.5%)
(b) p at $m/e = 86$ (19%), $p + 1$ (1.06%)

34 Because peroxides (R—O—O—R) are very explosive, it would be very useful if they could be detected in small amounts by IR spectroscopy. Unfortunately, the O—O stretching frequency is very weak or absent in most peroxides. Explain.

35 Given the C=C stretching frequencies of the hydrocarbons below, which one has the strongest double bond? How do you know?

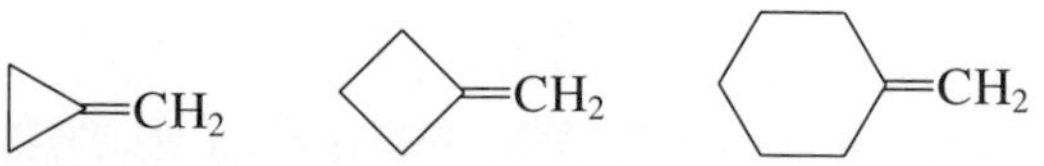

1781 cm^{-1} 1672 cm^{-1} 1650 cm^{-1}

36 The mass spectrum of ethyl bromide consists of the following peaks and relative abundances: *m/e* = 110 (98%), 108 (100%), 81 (5%), 79 (5%), 29 (61%), 28 (35%), and 27 (53%). Suggest a structure for each of the ions corresponding to these peaks. Give a mechanism for the formation of each ion.

37 (a) Explain why the S—H stretching absorption in the IR spectrum of a thiol is less intense and occurs at lower frequency (2550 cm^{-1}) than the O—H stretching absorption of an alcohol.

(b) A small entrepreneurial chemical company, EmuSkunk Inc., has been shut down by a group of irate citizens objecting to the stench emanating from the company's facilities. You have been called in as an expert to identify two of the compounds left in the company's laboratories. You find two bottles *A* and *B* containing liquids and near the bottles are labels that have fallen off the bottles. One label is marked "$(HSCH_2CH_2)_2O$" and the other is marked "$(HOCH_2CH_2)_2S$." The IR spectra of the two compounds are given in Fig. 12.24. Identify *A* and *B*.

(c) Suggest at least two chemical ways that could be used to confirm your diagnosis.

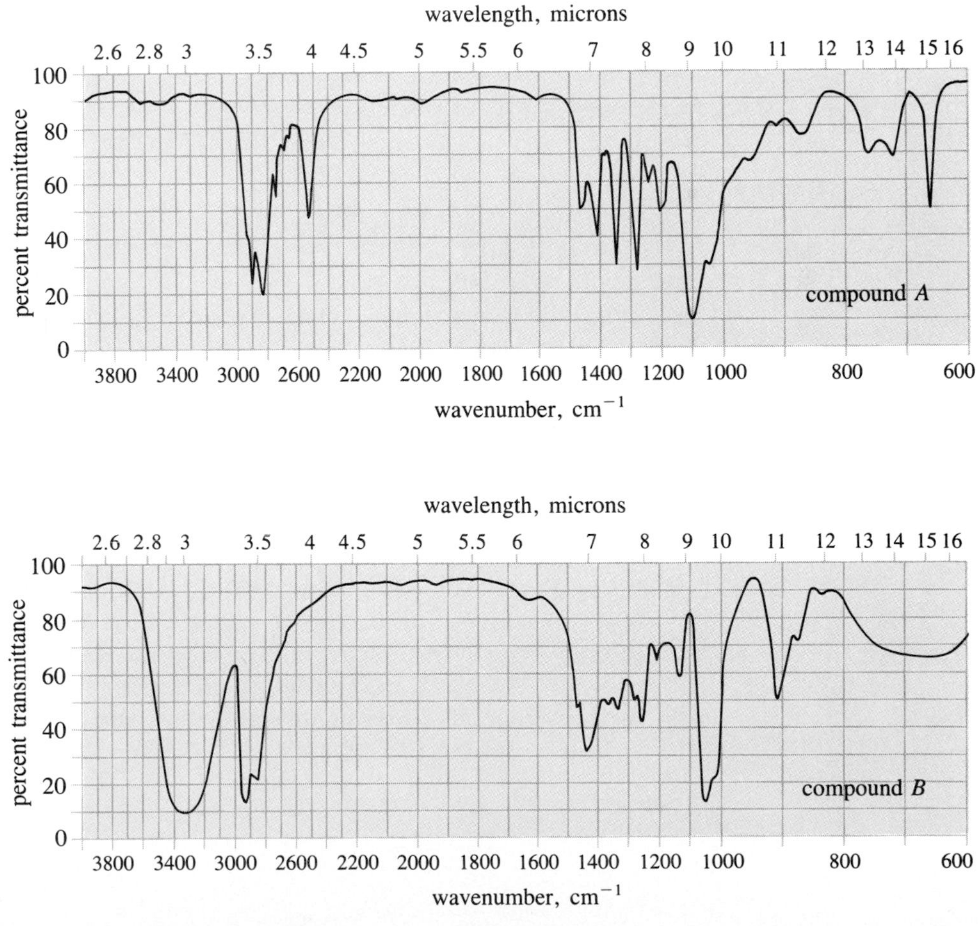

Figure 12.24 IR spectra for Problem 37.

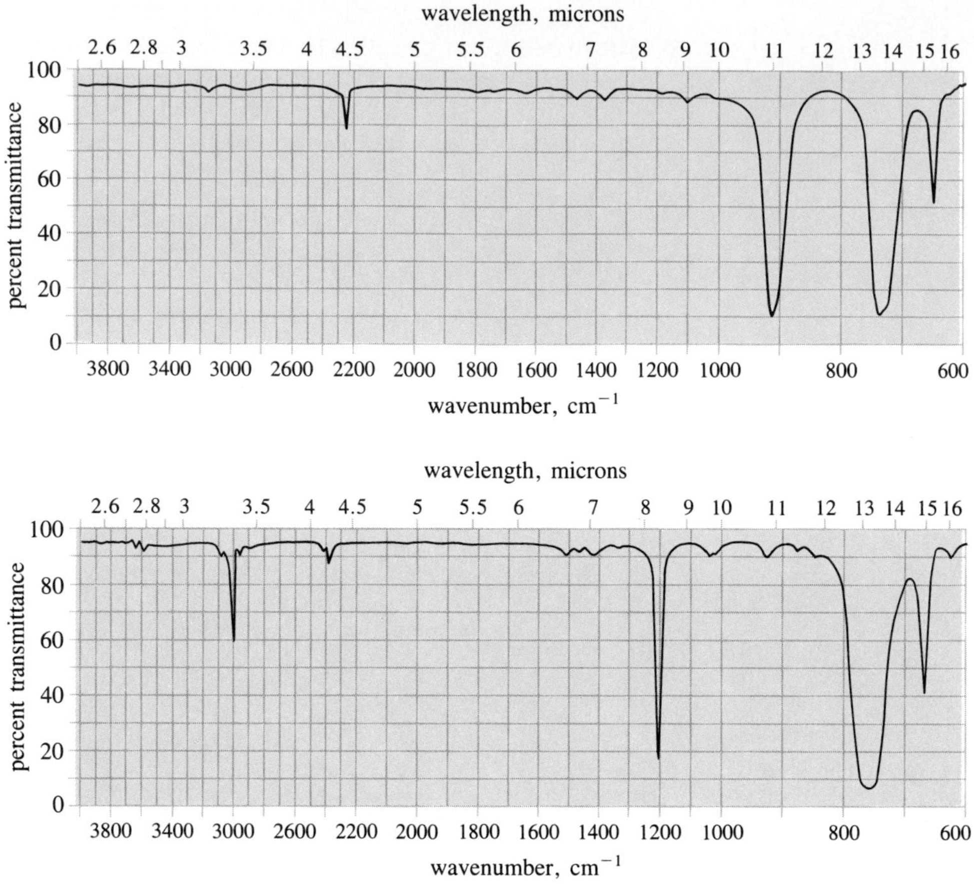

Figure 12.25 IR spectra for Problem 39.

---

38 A compound contains carbon, hydrogen, oxygen, and *one* nitrogen. Classify each of the following fragment ions derived from this compound as an odd-electron or an even-electron ion. Explain.

(a) the parent ion

(b) a fragment ion of even mass containing nitrogen

(c) a fragment ion of odd mass containing nitrogen

---

39 You have found in the laboratory two liquids in unlabeled bottles that both smell like chloroform, but you suspect that one is deuterated chloroform ($CDCl_3$) and the other is ordinary chloroform ($CHCl_3$). Unfortunately, the mass spectrometer is not operating because the same person who failed to label the bottles has been recently using the mass spectrometer! From the IR spectra of the two compounds, shown in Fig. 12.25, indicate which compound is which. Explain. How would these compounds be distinguished by mass spectrometry?

40 (a) Explain why ionization of a σ-electron requires less energy than ionization of a π-electron.

(b) Using the ionization of a π-electron to begin your fragmentation mechanism, account for the presence of a base peak at $m/e = 41$ in the mass spectrum of 1-heptene.

(c) Compare the parent ion of 1-heptene in (b) with the ion formed by the loss of water from 1-heptanol (Fig. 12.18, Eq. 12.22). Explain why the mass spectra of 1-heptanol and 1-heptene are nearly identical.

41 We take you back in our time machine to the earliest days of IR spectroscopy. You are the very first person to measure the IR spectrum of a 1-alkene—in this case, 2-methyl-1-pentene. From theory, you suspect that the sharp absorption peak at 3086 cm^{-1} is the =C—H stretching absorption, but you wish to gather experimental evidence to support your idea.

(a) Assuming you could prepare the necessary compounds, what isotopically substituted analog(s) of 2-methyl-1-pentene could be used to prove your point?

(b) Would you obtain a more definitive answer by replacing ^{12}C with ^{13}C or ^{1}H with ^{2}H (deuterium)? Explain.

42 The IR stretching absorption of the C—C bond is generally not observed. Estimate the frequency of this absorption from the absorption frequency of the C=C stretching vibration, stating any assumptions. Using the same logic, estimate the C≡C stretching frequency. (This absorption is an important one for many alkynes.)

13 Nuclear Magnetic Resonance Spectroscopy

In Chapter 12 we learned that infrared spectroscopy allows us to determine the functional groups present in a compound, and mass spectrometry permits us to determine the masses of a molecule and its coherent fragments. With rare exceptions, however, neither of these techniques allows us to determine a complete structure. In this chapter we shall study another form of spectroscopy, *nuclear magnetic resonance* (NMR) spectroscopy, which enables the chemist to probe molecular structure in much greater detail. Using NMR, sometimes in conjunction with other forms of spectroscopy, but often by itself, the chemist can usually define a complete molecular structure in a very short time. In the period since its introduction in the late 1950s, NMR spectroscopy has revolutionized organic chemistry. Our goal in this chapter is to learn the basic principles of NMR spectroscopy and see how it is used in structure determination.

13.1 INTRODUCTION TO NMR SPECTROSCOPY

One of the most important uses of NMR spectroscopy is to examine the *hydrogens* in organic compounds. The NMR experiment hinges on the *magnetic properties of the hydrogen nucleus.* Let us consider how these magnetic properties give rise to an NMR absorption.

We learned in Sec. 2.2 that electrons have two allowed spin states, designated by the quantum numbers $+\frac{1}{2}$ and $-\frac{1}{2}$. Some *nuclei* also have spin that is analogous to the electron spin. The hydrogen nucleus ^{1}H—that is, the proton—has a nuclear spin that also can assume either of two values, designated by quantum numbers $+\frac{1}{2}$ and $-\frac{1}{2}$.

The physical significance of nuclear spin is that the nucleus acts like a tiny magnet. This means that hydrogen nuclei, whether alone or in a compound, respond to a magnetic field. When a compound containing hydrogens is placed in a magnetic field, it becomes magnetized.

To see what causes this magnetization, let us represent the hydrogen nuclei in a chemical sample with arrows indicating their magnetic ("north–south") polarity. In the absence of a magnetic field, the nuclear magnetic poles are oriented randomly. After a magnetic field is applied, the magnetic poles of nuclei with spin of $+\frac{1}{2}$ are oriented parallel to the magnetic field, and those of nuclei with spin of $-\frac{1}{2}$ are oriented antiparallel to the field.

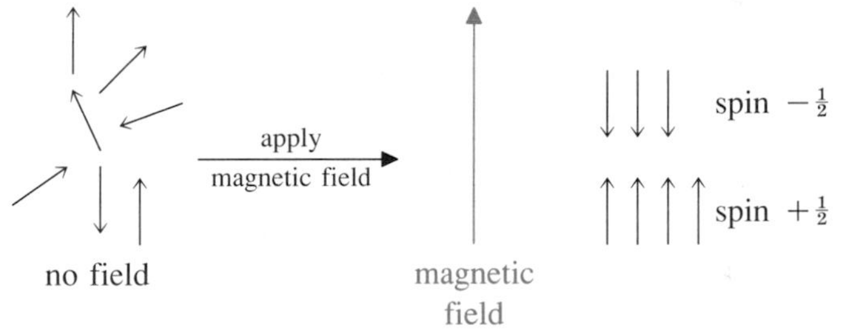

The presence of the magnetic field also causes the two spin states to have different energies: the $+\frac{1}{2}$ spin state has a lower energy than the $-\frac{1}{2}$ spin state. (In the absence of the field the two spin states have the same energy.) In addition, protons in the two spin states are in rapid equilibrium. The spin equilibrium, like a chemical equilibrium, favors the state with lower energy (Eq. 4.5). Thus, after the field is applied, there are a few more protons with spin $+\frac{1}{2}$ than there are with spin $-\frac{1}{2}$. For every *million* protons in a sample, the state with spin $+\frac{1}{2}$ typically contains an excess of only *ten to twenty* protons—certainly not a large difference, but enough to be detected. Since the nuclei with spin $+\frac{1}{2}$ have their magnetic poles aligned with the magnetic field, and there are more of these nuclei, the sample has a small net magnetization in the direction of the field.

For a given proton in our sample, the energy difference between its two spin states, ΔE, depends on the intensity of the magnetic field $\mathbf{H_p}$ that the proton itself "senses." The larger the field, the greater the difference in energy of the spin states. This idea, an important one for NMR spectroscopy, is illustrated in Fig. 13.1 and is expressed quantitatively by the following equation:

$$\Delta E = \frac{h\gamma_H}{2\pi}\mathbf{H_p} \tag{13.1}$$

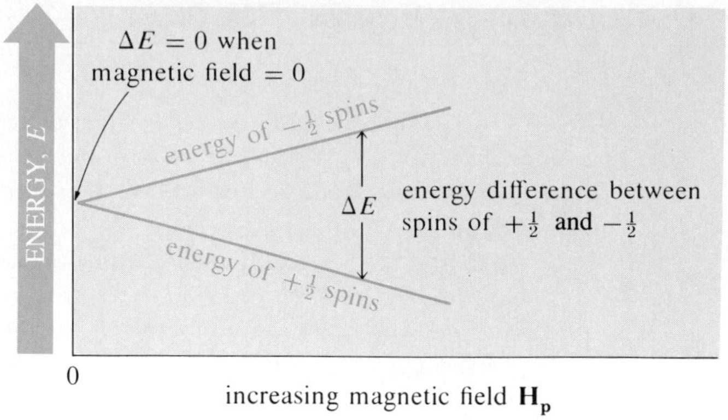

Figure 13.1 Effect of increasing magnetic field on the energy difference between $+\frac{1}{2}$ and $-\frac{1}{2}$ spin states of the proton.

In this equation, $\mathbf{H_p}$ is the intensity of the magnetic field *at the hydrogen nucleus*; the unit of magnetic field is the gauss (rhymes with *house*). The other quantities on the right side of this equation are constants. The **magnetogyric ratio,** γ_H, is a fundamental constant of the proton related to its magnetic properties, and has the value 26,753 radians gauss^{-1} sec^{-1}. (The subscript H in the symbol γ_H refers to hydrogen; do not confuse this with the symbol for the magnetic field.) Different nuclei (^{19}F, ^{31}P, etc.) have different magnetogyric ratios (γ_F, γ_P, etc.). The other constant in Eq. 13.1 is *h*, Planck's constant (Eq. 12.4 or 12.5).

Here is where things stand: Molecules of a sample are situated in a magnetic field; each nucleus is in one of two spin states that differ in energy by an amount ΔE; and a few more nuclei have spin $+\frac{1}{2}$. If our sample is now subjected to electromagnetic radiation of energy E_0 exactly equal to ΔE, this energy is *absorbed* by some of the nuclei in the $+\frac{1}{2}$ spin state. The absorbed energy causes these nuclei to invert or "flip" their spins and assume a more energetic state with spin $-\frac{1}{2}$.

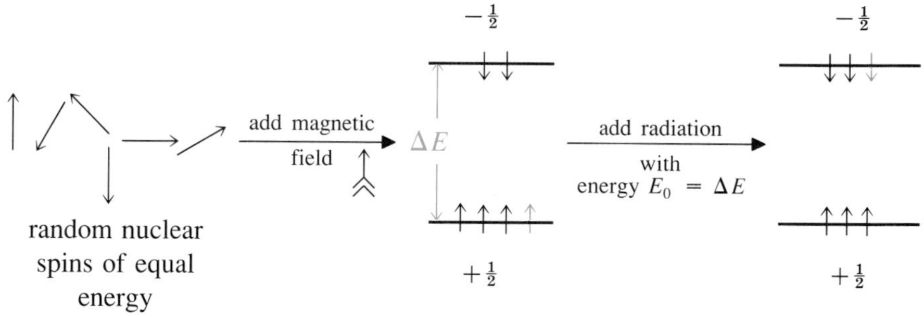

This energy absorption by nuclei in a magnetic field is termed **nuclear magnetic resonance,** and can be detected in a type of absorption spectrophotometer called a *nuclear magnetic resonance spectrometer,* or **NMR spectrometer** (Sec. 13.10). The study of this absorption is called **NMR spectroscopy.**

When nuclei situated in a magnetic field absorb energy, they are said to be "in resonance" with the electromagnetic radiation used in the NMR experiment. Do not confuse the term "resonance" used here with the "resonance" in Sec. 1.7; the two are unrelated phenomena.

To summarize: for nuclei to absorb energy, they must have a nuclear spin, and must be situated in a magnetic field. Once these two conditions are met, then the nuclei can be examined by an absorption spectroscopy experiment that is conceptually the same as the simple experiment shown in Fig. 12.3. The absorption of energy corresponds physically to the "flipping" of nuclear spins from the $+\frac{1}{2}$ to the $-\frac{1}{2}$ spin state.

What part of the electromagnetic spectrum is required for nuclear magnetic resonance? Suppose, for example, protons in our sample are subjected to a magnetic field of 14,092 gauss. (This is a large magnetic field, but one that is commonly used in obtaining NMR spectra, including most of those in this text.) Since the energy E_0 required for absorption must equal ΔE, application of Eq. 13.1 gives

$$\begin{aligned} E_0 &= \Delta E \\ &= \frac{h\gamma_H}{2\pi}\mathbf{H_p} \\ &= \frac{(9.5 \times 10^{-14} \text{ kcal-mol}^{-1}\text{-sec})(26{,}753 \text{ rad-gauss}^{-1}\text{-sec}^{-1})}{2(3.14) \text{ rad}} \times (14{,}092 \text{ gauss}) \\ &= 5.70 \times 10^{-6} \text{ kcal/mol} = 0.0057 \text{ cal/mol} \end{aligned} \qquad (13.2)$$

Notice that this is 0.0057 calories, not kilocalories! This shows that the energies of the two nuclear spin states, even in a large magnetic field of 14,092 gauss, differ very little. Corresponding to this energy E_0 is a frequency ν_0 given by Eq. 12.3:

$$\begin{aligned} \nu_0 = \frac{E_0}{h} = \frac{\Delta E}{h} = \frac{\gamma_H}{2\pi}\mathbf{H_p} &= \frac{(26{,}753)(14{,}092)}{2(3.14)} \text{ sec}^{-1} \\ &= 60 \times 10^6 \text{ sec}^{-1} \text{ (Hz)} \\ &= 60 \text{ megahertz (MHz)} \end{aligned} \qquad (13.3)$$

(Recall that the unit sec^{-1} is called the *hertz;* Sec. 12.1.) Thus, the frequency of the radiation required to bring about energy absorption by protons in a field of 14,092 gauss is sixty million hertz, or 60 megahertz (MHz). This is in the radiofrequency (rf) region of the electromagnetic spectrum (Fig. 12.2). Indeed, an ordinary radio receiver located near an NMR spectrometer and tuned to the appropriate frequency can transmit audible sounds associated with an NMR experiment!

Any nucleus with a spin can be detected by NMR. Since protons have nuclear spin, NMR can be used to detect and study the hydrogen nuclei in any compound. Carbon-12, however, has no nuclear spin, and therefore does not give an NMR signal. Other nuclei with spin are ^{19}F, ^{31}P, and the rare isotope of carbon, ^{13}C. These nuclei all have spins of $\pm\frac{1}{2}$ and can be observed with NMR. Some nuclei also have spin values other than $\pm\frac{1}{2}$. Nitrogen-14 and deuterium (^2H), for example, also can be detected by NMR, but these nuclei have three allowed spin states of ± 1 and 0. In this chapter, we shall restrict our attention to nuclei with spin of $\pm\frac{1}{2}$, and mostly we shall consider NMR spectroscopy of the hydrogen nucleus—the proton. Because almost all organic compounds contain protons, *proton NMR spectroscopy* is especially useful to the organic chemist. We shall return in Sec. 13.8 to a discussion of the NMR of other nuclei, particularly carbon-13. Despite the low abundance of this isotope, advances in instrumental techniques have made carbon-13 NMR a tool of major importance for the organic chemist.

Problem *1* (a) Calculate the magnetic field, in gauss, required for a proton to absorb energy at 100 MHz.
 (b) Calculate the frequency, in MHz, of the radiation from which a proton will absorb energy when it is exposed to a magnetic field of 51,671 gauss.

13.2 THE NMR SPECTRUM: CHEMICAL SHIFT AND INTEGRAL

A. The NMR Spectrum

Now we are ready to see how NMR absorption depends on chemical structure. Suppose we wish to take the proton NMR spectrum of dimethoxymethane, CH_3O—CH_2—OCH_3. In one type of NMR experiment, the radio frequency (rf) source on the NMR instrument is fixed at a convenient frequency—say, 60 MHz. This frequency, called the **operating frequency,** is ν_0 in Eq. 13.3. The magnetic field of the NMR spectrometer, H_0, called the **applied magnetic field,** is slowly and continuously changed over a very narrow range. When the protons in dimethoxymethane experience the magnetic field H_p that exactly satisfies Eq. 13.3, they absorb energy from the rf source. At this value of the field, the energy of the rf source and the energy separation of the two proton spin states exactly match. The resulting absorption, or resonance, is registered as a peak in the NMR spectrum.

The proton NMR spectrum of dimethoxymethane is shown in Fig. 13.2. The two peaks near the middle of the spectrum are the NMR absorptions of the ether; as indicated in the spectrum, one peak is the absorption of the —CH_2— protons, and the other is that of the —CH_3 protons. The small peak on the far right of the spectrum is the NMR absorption of another compound, tetramethylsilane (TMS), which is added to the sample as a reference, or internal standard. (This absorption peak is present in most of the NMR spectra in this text.)

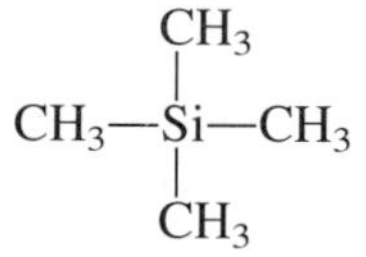

$$CH_3\text{—}\underset{\underset{\displaystyle CH_3}{|}}{\overset{\overset{\displaystyle CH_3}{|}}{Si}}\text{—}CH_3 \qquad \text{TMS (tetramethylsilane)}$$

Some of the terminology used in NMR spectroscopy is shown in this figure. The strength of the applied field H_0 is plotted on the horizontal axis, and increases to the right. The absorption positions of peaks, which are also called *lines* or *resonances,* can be designated by either of two scales, shown on the upper and lower horizontal axes of the spectrum. Both scales increase from *right to left* in the *downfield* direction—that is, toward *lower* field strength. Peak positions are always cited relative to the position of a standard, in this case TMS. The position of a peak relative to a standard is called its **chemical shift.** On the lower horizontal axis is the *parts-per-million* (ppm) scale. The symbol for parts per million is the small Greek letter delta, δ. Thus, the absorption of the —CH_3 protons occurs at δ 3.23, or 3.23 ppm downfield (to lower field) from TMS. The absorption of the —CH_2— protons occurs at δ 4.40, or 4.40 ppm downfield from TMS. On the upper horizontal axis is the chemical shift in hertz (Hz). Although the hertz is a unit of *frequency,* it can also be treated as a unit of magnetic field because of the proportionality between field and frequency in Eq. 13.3.

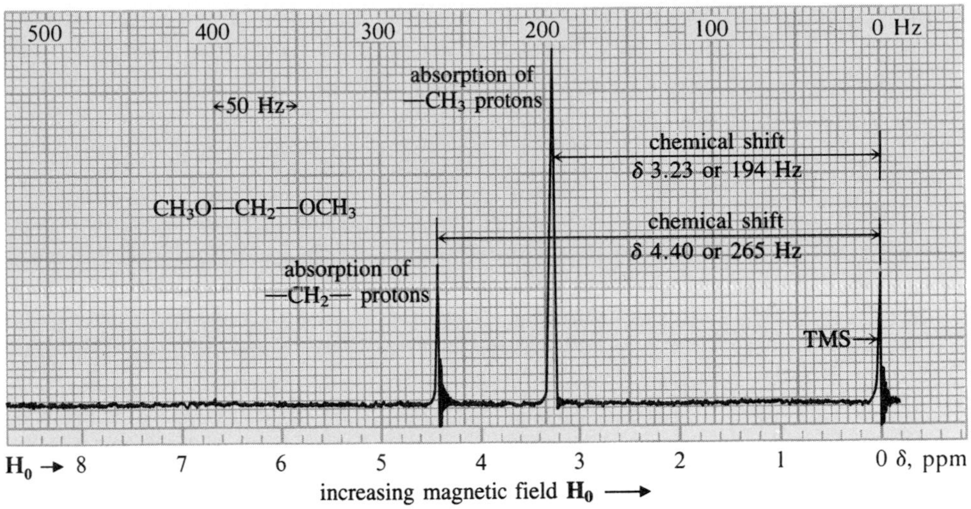

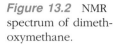

Figure 13.2 NMR spectrum of dimethoxymethane.

Each small space in the grid equals 5 Hz. Thus the chemical shift of the —CH_3 protons is 194 Hz downfield from TMS, and that of the —CH_2— protons is 265 Hz downfield from TMS.

The spectrum in Fig. 13.2 was taken with a ν_0 value of sixty million Hz (60 MHz). Although this operating frequency is used in many spectrometers, other operating frequencies are common. It turns out that *the chemical shift in Hz varies directly with the operating frequency.* Thus, at an operating frequency of 180 MHz, the chemical shift of the larger peak in Fig. 13.2 in Hz is three times as great and therefore occurs at (3×194), or 582 Hz, downfield from TMS. Because chemists needed a chemical shift scale that is independent of operating frequency, the ppm scale was devised. The chemical shift in ppm is defined by the ratio

$$\delta = \frac{\text{chemical shift in hertz}}{\text{operating frequency in megahertz}} = \frac{\nu_{\text{sample}} - \nu_{\text{TMS}}}{\nu_0} \qquad (13.4)$$

Since doubling (or tripling) the operating frequency also doubles (or triples) the chemical shift in Hz, then clearly *the chemical shift in ppm is independent of the operating frequency.* Thus, the chemical shift of the —CH_3 protons of dimethoxymethane is 194 Hz at $\nu_0 = 60$ MHz or 582 Hz at $\nu_0 = 180$ MHz, but it can be expressed as δ 3.23 at any value of ν_0.

In most organic molecules, protons give NMR absorptions over a chemical shift range of about 0–10 ppm downfield from TMS. Indeed, TMS was chosen as a standard not only because it is volatile and inert, but also because it absorbs at a higher field than most common organic compounds. (The rare absorptions that occur at higher field than TMS are assigned a negative δ value.)

Problems

2 What is the chemical shift, in Hz downfield from TMS, of the —CH_2— protons of dimethoxymethane (Fig. 13.2) in a spectrum taken at 90 MHz? At 360 MHz?

3 Two signals differ in chemical shift by 45 Hz at 60 MHz operating frequency. By how many ppm do the signals differ in chemical shift?

Problems (Cont.)

4 (a) What is the chemical shift of the —CH_3 protons of dimethoxymethane in gauss at $\nu_0 = 60$ MHz?
(b) Compare the magnitude of a typical chemical shift in gauss to the applied magnetic field and thus explain why the term parts per million is appropriate.

Obviously, the two NMR absorptions of dimethoxymethane in Fig. 13.2 are not the same size. The reason for this is quite simple: the size of an NMR absorption is governed almost entirely by the number of protons contributing to it. None of the complicating factors that govern, for example, IR absorption intensity are present in NMR spectroscopy. The exact intensity of an NMR absorption is given not by its peak height, but rather by the total area under the peak. This quantity can be determined by mathematical integration of the peak using more or less the same integration procedures that we use in calculus to determine the area under a curve. NMR instruments are equipped with an integrating device that can be used to display the integral on the spectrum. Such a spectrum integral is illustrated in Fig. 13.3 for the dimethoxymethane spectrum as the curve superimposed on each peak. *The relative height of the integral (in any convenient units, such as chart spaces) is proportional to the number of protons contributing to the peak.* Thus the heights of the integrals in Fig. 13.3 are 20.3 and 6.8 vertical chart spaces, in a ratio of 2.99 : 1 (that is, 3 : 1), and the relative number of the respective hydrogens are 6 and 2, also in the ratio 3 : 1.

It is important to remember that these values give the *relative* numbers of hydrogens under the NMR peaks, not the absolute numbers. Thus a sample giving a spectrum with three peaks having relative integrals 1 : 2 : 3 might contain any multiple of six protons.

B. Relationship of Chemical Shift and Integral to Structure

If all protons absorbed in the NMR at the same chemical shift, NMR spectroscopy would be useless to the organic chemist. The spectrum of dimethoxymethane in Fig. 13.2, however, illustrates that *protons in different chemical environments have different chemical shifts.* (The theory underlying this important point is discussed in Sec. 13.2D.) Thus, dimethoxymethane has two types of protons: the —CH_3 protons, all six of which are chemically identical, and the —CH_2— protons. As a result, there are two

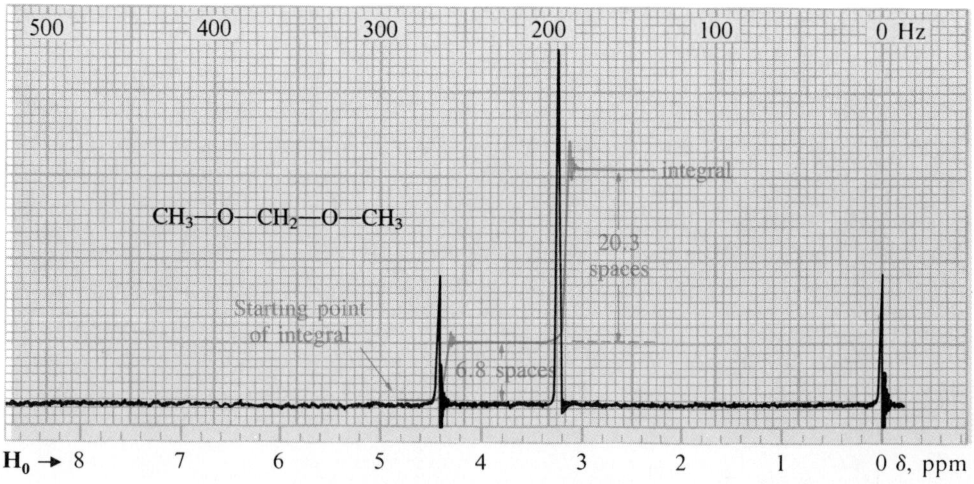

Figure 13.3 NMR spectrum of dimethoxymethane from Fig. 13.2 with superimposed integral.

NMR absorptions. The integral tells us *how many* protons of each type there are. Turning this statement around, *we can use the chemical shift and integral to determine for an unknown compound the types of chemically distinguishable protons as well as the number of protons within each type.*

The two methyl groups of dimethoxymethane are chemically equivalent because there is no way to tell one methyl group from the other; each would react the same with any chemical reagent because each is in exactly the same environment. Whether two groups are chemically equivalent is usually clear from inspection. However, recognizing chemical equivalence is occasionally less straightforward. The language of Sec. 10.8B can be used to define rigorously what we mean by "chemically equivalent protons." In order for two sets of protons to be **chemically equivalent,** they must be *connectively equivalent*—that is, they must have the same connectivity relationships to the rest of the molecule. In addition, they must be either *homotopic* or (assuming an achiral solvent is being used) *enantiotopic.* (The two methyl groups of dimethoxymethane are homotopic.) *Diastereotopic groups,* however, although connectively equivalent, are in principle chemically distinguishable; and indeed, we often find that diastereotopic protons have different chemical shifts. For example, protons *a* and *b* in the following compound are diastereotopic, and have slightly different chemical shifts:

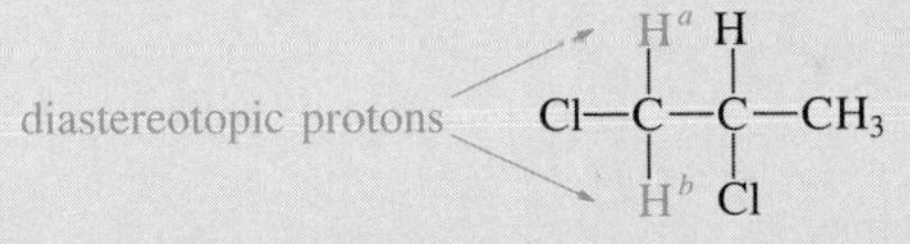

Just as diastereotopic groups have different *chemical behavior,* they also have different *spectroscopic behavior.*

Let us now try to understand how the chemical environment of a proton influences its chemical shift. One of the most important factors that affect proton chemical shifts is the *electronegativities of the groups in a molecule that are near the protons of interest.* Let us examine some data that illustrate this idea, using the following problem and Table 13.1 as our guide.

TABLE 13.1 Effect of Electronegativity on Proton Chemical Shift

Entry number	Compound	Chemical shift, δ
1	CH_3F	4.26
2	CH_3Cl	3.05
3	CH_3Br	2.68
4	CH_3I	2.16
5	CH_2Cl_2	5.30
6	$CHCl_3$	7.27
7	CH_3CCl_3	2.70
8	$(CH_3)_4C$	0.86
9	$(CH_3)_4Si$	0.00*

*By definition.

Problem

5 (a) Consider entries 1 through 4 of Table 13.1. How does the chemical shift of a proton vary with the electronegativity of the neighboring halogen?
(b) Compare entries 2, 5, and 6 of Table 13.1. How does chemical shift vary with number of neighboring halogens?
(c) Compare entries 6 and 7. How is the chemical shift of a proton affected by its distance from an electronegative group?
(d) Explain why $(CH_3)_4Si$ absorbs at higher field (lower chemical shift) than the other molecules in the table. Can you think of a molecule with protons that would have a smaller chemical shift than TMS (that is, a negative δ value)?

You should have concluded from your examination of Table 13.1 that the following factors *increase* the proton chemical shift:

1. Increasing *electronegativity* of nearby groups
2. Increasing *number* of nearby electronegative groups
3. Decreasing *distance* between the proton and nearby electronegative groups

The effect of electropositive groups is, as expected, opposite from the effect of electronegative groups.

In the spectrum of dimethoxymethane (Fig. 13.2) we can see these ideas in practice. The $—CH_3$ protons are adjacent to one oxygen, and therefore have a smaller chemical shift than the $—CH_2—$ protons, which are adjacent to two oxygens.

Chemical shift information for $—CH_2—$ and $—CH_3$ protons in different chemical environments is summarized in Table 13.2. (This table contains some data for unfamiliar functional groups that will be useful in later parts of the text.) The first column lists various groups G that we might find bound to an alkyl group in an organic compound. The second column lists the approximate chemical shift of a methyl group when bound to group G, that is, the shift of $CH_3—G$. For example, for G = F, we see that the chemical shift of $CH_3—F$ is δ 4.3. The third column contains effective shift contributions, σ_G, that can be used to calculate the approximate chemical shift for the protons of a $—CH_2—$ group bound to two groups G_1 and G_2. To carry out such a calculation, the following simple equation is used:

$$\delta\,(—CH_2—) = 0.2 + \sigma_{G_1} + \sigma_{G_2} \tag{13.5}$$

For example, suppose we want to calculate the chemical shift of the protons in CH_2Cl_2. This compound is of the form $G_1—CH_2—G_2$, in which $G_1 = G_2 = Cl$. From Table 13.2, $\sigma_{Cl} = 2.5$; application of Eq. 13.5 gives a predicted chemical shift of δ 5.2. The actual chemical shift is δ 5.3. As we can see from this example, calculations using Eq. 13.5 and Table 13.2 are not perfect, but their agreement with experiment is good enough to be useful. This example also illustrates that, to a good approximation, *the chemical shift contributions of different groups are additive.*

There are two other useful chemical shift correlations. The first is that a β-halogen or β-oxygen adds about 0.5 ppm to the chemical shift. For example, let us estimate the chemical shift of the colored protons in the following compound:

$$Br—\mathit{CH_2}—CH_2—Cl \longrightarrow \beta \text{ to colored protons}$$

TABLE 13.2 Contributions of Common Functional Groups to Chemical Shifts of CH_2 and CH_3 Groups

Group, G	δ for CH_3—G	Effective shift contribution σ_G*
—H	0.2	—
$—CR_3$ or $—CR_2—$ (R = H, alkyl)	0.9	0.5
—F	4.3	3.3
—Cl	3.0	2.5
—Br	2.7	2.3
—I	2.2	2.2
$—CR{=}CR_2$ (R = H, alkyl)	1.8	1.3
$—C{\equiv}C—R$ (R = H, alkyl)	2.0	1.4
—OH	3.5	2.6
—OR (R = alkyl)	3.3	2.4
—OR (R = aryl)	3.7	2.9
—SH	2.4	1.6
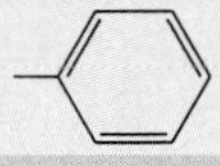 (phenyl)	2.3	1.8
—C(=O)—R	2.1 (R = alkyl) 2.6 (R = aryl)	1.5
—C(=O)—OR (R = alkyl, H)	2.1	1.5
—O—C(=O)—R	3.6 (R = alkyl) 3.8 (R = aryl)	3.0
$—C(=O)—NR_2$ (R = alkyl, H)	2.0	1.5
—NR—C(=O)—R	2.8	—
$—NR_2$ (R = alkyl, H)	2.2	1.6
—N(—)—R (R = aryl)	2.9	—
$—C{\equiv}N:$	2.0	1.6

* Used with Eq. 13.5 in text.

For the effects of the adjacent $—CH_2—$ and Br— groups, Eq. 13.5 and Table 13.2 give a contribution of 0.2 + 0.5 + 2.3 = 3.0, to which we add another 0.5 ppm for the β-Cl. The predicted chemical shift of the indicated hydrogens is δ 3.5; the observed value is δ 3.3.

The other useful correlation is that the chemical shifts of tertiary protons are greater than those of secondary protons, which are greater than those of primary protons. From Table 13.2 we can see that hydrogens at $—CH_2—G$ carbons usually have chemical shifts about 0.5 ppm higher than methyl hydrogens $CH_3—G$ at a carbon bearing the same group G. The chemical shift of an analogous tertiary hydrogen, although less predictable, is usually somewhat larger still.

Problems

6 In each of the following sets, indicate the proton that has the greatest chemical shift:

(a) CH_2Cl_2, CH_2I_2, CH_3I
(b) $(CH_3)_3C—C(CH_3)_3$, $Cl_2CH—CHCl_2$, $Cl_3C—CH_3$
(c) $(CH_3)_4Si$, $(CH_3)_4Sn$ (*Hint:* See Table 1.1.)

7 How many different NMR absorptions would we see in the spectrum of each of the following compounds?

(a) *t*-butyl bromide (b) $(CH_3)_3C—C(CH_3)_2$ with Cl on the second C (c) *cis*-2-butene

8 Estimate the chemical shifts of the italicized protons in the following molecules:

(a) $CH_3—\mathit{CH_2}—CH_3$
(b) $\mathit{CH_3}—Ph$
(c) $Br—\mathit{CH_2}—I$
(d) $CH_3—\mathit{CH_2}—O—\mathit{CH_2}—CH_3$
(e) $Br—CH_2—\mathit{CH_2}—CH{=}CH_2$
(f) $CH_3—CH_2—\mathit{CH_2}—Br$
(g) $CH_3—\mathit{CH_2}—CH_2—Br$
(h) $\mathit{CH_3}—CH_2—CH_2—Br$

C. Solving Unknown Structures with NMR (Part 1)

We now have the tools to determine some unknown structures with NMR. The approach to be used is best illustrated by the solution of a sample problem. You will profit most from this example if you will try to work the problem before reading the solution.

Problems

9 An unknown compound $C_5H_{11}Br$ has an NMR spectrum that consists of two lines, one at δ 1.01 (relative integral 4.5), and the other at δ 3.17 (relative integral 1.0). Give the structure of this compound.

Solution: We first learn what we can from the molecular formula. Since its unsaturation number is zero (Eq. 5.76), the compound has no rings or double bonds; it is a simple alkyl halide. The two signals in the NMR spectrum tell us immediately that there are only two chemically different types of hydrogens. Next we determine how many of each type there are. The total integral is (4.5 + 1.0), or 5.5 units, which, according to the molecular formula, must correspond to 11 protons. Thus, each unit of integration accounts for (5.5/11) or 0.5 units per proton. The larger peak therefore accounts for (4.5 units/0.5 units per proton) or 9 protons. By analogous reasoning, the smaller peak accounts for 2 protons. All protons are accounted for: there are two of one type and nine of another. The two hydrogens at δ 3.15 must be near an electronegative group. Since bromine is the only electronegative group in the molecule, the two-proton group is near the bromine. From Table 13.2 and Eq. 13.5 we see that if these two protons are on a carbon adjacent to both a carbon (shift contribution 0.5) and a —Br (shift contribution 2.3), the predicted shift is δ 3.0, very close to the δ 3.15 observed. A partial structure that accounts for these observations is

$$-\overset{|}{\underset{|}{C}}-CH_2-Br$$

The remaining nine protons are chemically equivalent. One useful thing to learn about chemically equivalent protons is that a methyl group often accounts for three equivalent protons and a *t*-butyl group accounts for nine. Indeed, completing the *t*-butyl group by adding three methyl groups to the partial structure illustrated gives the solution:

$(CH_3)_3C-CH_2-Br$ neopentyl bromide

δ 1.02 (9 hydrogens) δ 3.15 (2 hydrogens)

---

10 In each case give a single structure that fits the data provided.
(a) A compound $C_7H_{15}Cl$ has two NMR signals at δ 1.08 and δ 1.59, with relative integral 3 : 2.
(b) A compound $C_5H_9Cl_3$ has three NMR signals at δ 1.99, δ 4.31, and δ 6.55 with relative integral 6 : 2 : 1.

---

11 (a) Strange results in the undergraduate organic laboratory have led to the admission by a teaching assistant, Thumbs Throckmorton, that he has accidentally mixed some *t*-butyl bromide with the methyl iodide. The NMR spectrum of this mixture indeed contains two single lines at δ 2.20 and δ 1.8 with relative integrals of 5 : 1. What is the mole percent of each compound in the mixture? (*Hint:* be sure to assign each line to a compound before doing the analysis.)
(b) Which would be more easily detected by NMR: 1 mole percent CH_3I impurity in $(CH_3)_3C-Br$, or 1 mole percent $(CH_3)_3C-Br$ impurity in CH_3I? Explain.

D. Physical Basis of the Chemical Shift

Why do protons in different chemical environments have different chemical shifts? According to Eq. 13.3, for a given operating frequency ν_0, every proton absorbs rf radiation at the *same* value of the field $\mathbf{H_p}$. The fact that chemical shifts are observed, then, means that the field $\mathbf{H_p}$—the field "sensed" by a proton—is different for protons in different chemical environments. This is so because the field $\mathbf{H_p}$ felt by a proton is affected by the magnetic fields of its surrounding electrons in a way that is shown schematically in Fig. 13.4. The molecular orbitals containing the electrons shared by a hydrogen can be crudely envisioned as a loop of wire. According to the laws of electromagnetic induction, the applied field $\mathbf{H_0}$—the field provided by the magnet in the NMR spectrometer—induces a circulation of these electrons (analogous to the current induced by bringing a magnet up to a loop of wire), which, in turn, creates an induced magnetic field $\mathbf{H_e}$. Although this induced field takes the form of closed loops, its direction *at the hydrogen nucleus* is such that it opposes, or subtracts

Figure 13.4 Origin of the shielding phenomenon responsible for chemical shift.

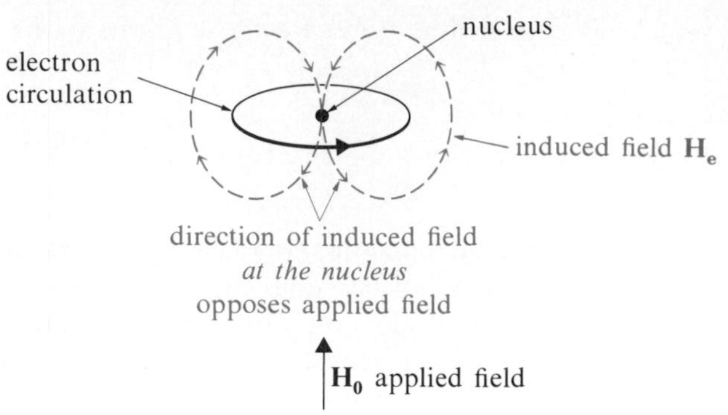

from, the applied field $\mathbf{H_0}$. The field $\mathbf{H_p}$ felt by a proton, then, is the applied field $\mathbf{H_0}$, less the opposing induced field $\mathbf{H_e}$.

$$\mathbf{H_p} = \mathbf{H_0} - \mathbf{H_e} \tag{13.6a}$$

(The different fields in this equation are taken as positive numbers; the opposing direction of $\mathbf{H_e}$ is taken into account by its minus sign in Eq. 13.6a.) Since the value of $\mathbf{H_p}$ at which absorption occurs is fixed by Eq. 13.3, the corresponding value of $\mathbf{H_0}$ required for absorption depends only on the magnitude of the induced field $\mathbf{H_e}$:

$$\mathbf{H_0} = \mathbf{H_p} + \mathbf{H_e} \tag{13.6b}$$

The greater the electron density about the proton, the greater will be the induced electron current and the greater will be the induced field, $\mathbf{H_e}$. The smaller this electron density, the smaller is $\mathbf{H_e}$. Nearby electronegative groups reduce the electron density at a proton; consequently, the same groups also reduce the induced electron current and the resulting field $\mathbf{H_e}$. Such a proton absorbs in the NMR spectrum at a lower value of $\mathbf{H_0}$, and is said to be **deshielded.** In the opposite sense, nearby electropositive groups increase the electron density at a proton; as a result, the induced electron current and the field $\mathbf{H_e}$ are increased. Such a proton absorbs at a higher value of $\mathbf{H_0}$, and is said to be **shielded.** Since chemical shift is defined so that smaller values of $\mathbf{H_0}$ correspond to greater chemical shift (Fig. 13.2), it follows that the NMR absorptions of protons near electronegative groups—the deshielded protons—occur at greater chemical shifts.

Problem

12 Which has the greater value of $\mathbf{H_e}$, and by how much (in gauss): TMS, or a proton that gives an NMR signal at δ 5.50? Assume an operating frequency of 60 MHz.

13.3 THE NMR SPECTRUM: SPIN–SPIN SPLITTING

Although substantial information can be gleaned from the chemical shift and integral, we are now going to learn about the aspect of NMR spectra that provides the most detailed information about chemical structures. Consider the compound ethyl bromide:

$$\overset{a}{C}H_3\text{—}\overset{b}{C}H_2\text{—}Br$$

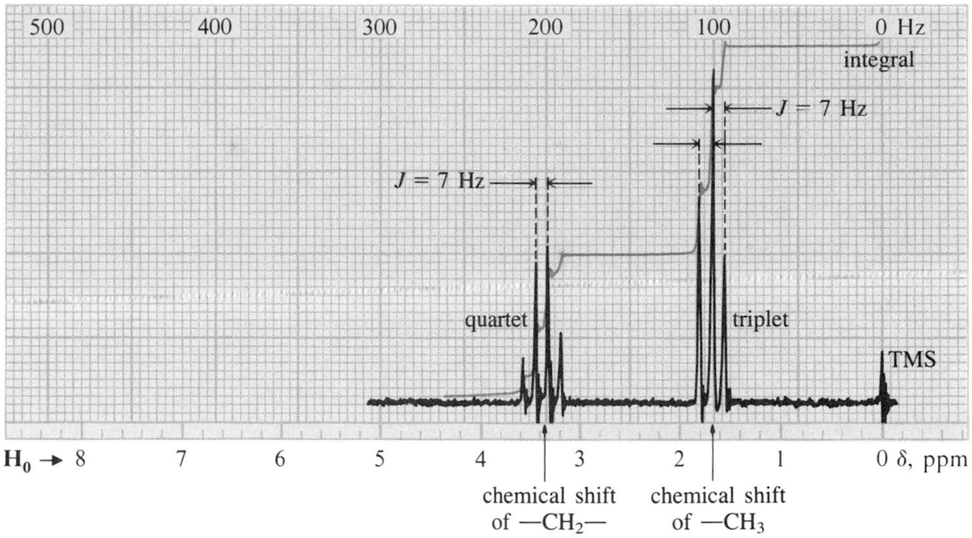

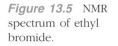

Figure 13.5 NMR spectrum of ethyl bromide.

This molecule has two chemically different sets of hydrogens, labeled *a* and *b*. We expect to find the NMR signals for these two protons in the integral ratio 3 : 2, respectively, with the absorption of protons *a* at higher field (smaller chemical shift). The NMR spectrum of ethyl bromide is shown in Fig. 13.5. The surprising feature of this spectrum, from what we have learned so far, is that it contains more lines than we might have expected—seven in all. Moreover, the lines fall into two distinct groups: a packet of three lines, or *triplet,* at high field; and a packet of four lines, or *quartet,* at low field. It turns out that all three lines of the triplet are the signal for the $—CH_3$ protons, and all four lines of the quartet are the signal for the $—CH_2—$ protons. The chemical shift of each packet of lines, taken at its center, is in agreement with the predictions of Eq. 13.5 and Table 13.2. The low-field quartet and the high-field triplet have total integrals, respectively, in the ratio 2 : 3.

The NMR signal for each group of protons is said to be *split.* **Splitting** *arises from the effect that one set of protons has on the NMR signal of neighboring protons.* We shall consider the physical reasons for splitting in Sec. 13.3B. First, let us focus on the appearance of the splitting pattern, and what it tells us about structure.

A. *The* n + 1 *Splitting Rule*

The number of lines in the NMR splitting pattern for a given set of equivalent protons depends on the number of *adjacent* protons according to the following rule: *If there are n equivalent protons in adjacent positions, a proton NMR signal is split into n + 1 lines.* This important rule is called the ***n* + 1 rule.**

Let us see how the $n + 1$ rule accounts for the splitting patterns in the spectrum of ethyl bromide. Since the carbon *adjacent* to the $—CH_3$ group has two protons, the signal for the $CH_3—$ protons of ethyl bromide is split into a pattern of $2 + 1 = 3$ lines, called a *triplet.* (The fact that there are also three methyl protons is a coincidence; the number of protons has *nothing* to do with their own splitting.) Since the carbon *adjacent* to the $—CH_2—$ group has three equivalent protons, the resonance for the $—CH_2—$ protons is split into a pattern of $3 + 1 = 4$ lines, called a *quartet.*

3 Hs on *adjacent* carbon; 3 + 1 = 4 lines

$CH_3—CH_2—Br$

2 Hs on *adjacent* carbon; 2 + 1 = 3 lines

The $—CH_3$ signal is split by the $—CH_2—$ protons, and the $—CH_2—$ signal is split by the $—CH_3$ protons. Because these two sets of protons split each other, they are **coupled protons.**

Why is no splitting observed in our previous examples of NMR spectra? The reason is that *splitting between protons is usually not observed if the protons are separated by more than two saturated atoms.* Thus since the protons in dimethoxymethane (Fig. 13.2) are on nonadjacent carbons, their splitting is negligible; the two signals in the NMR spectrum of this compound are unsplit singlets (single lines).

The spacing between adjacent peaks of a splitting pattern, measured in Hz, is called the **coupling constant** (abbreviated J). For ethyl bromide, this distance is about 7 Hz. *Two coupled protons must have the same value of J.* Thus, the coupling constants for both the $—CH_2—$ protons and the $—CH_3$ protons of ethyl bromide are the same because these protons split each other. If we call the $—CH_2—$ protons *a* and the $CH_3—$ protons *b,* then $J_{ab} = J_{ba}$.

Although chemical shifts in Hz increase in proportion to operating frequency ν_0 (Sec. 13.2), *the coupling constant between any two protons does not vary with operating frequency.* The importance of this point will become apparent in Sec. 13.4.

The chemical shift of a split line occurs at the midpoint of the splitting pattern (Fig. 13.5). (This is actually an approximation that we shall discuss further below, but it will suffice for most of the cases we shall study.) Thus, the chemical shift of the $—CH_2—$ protons occurs in the center of the quartet, and that of the $CH_3—$ protons occurs in the center of the triplet.

The relative intensities of the lines within a split signal follow a well-defined pattern. For example, the relative intensities of the triplet lines in ethyl bromide are approximately 1 : 2 : 1; the intensities within the quartet are in the approximate ratio 1 : 3 : 3 : 1. The intensity ratios of the lines in a splitting pattern can be read from the appropriate row of Table 13.3.

It turns out that the lines of the splitting pattern do not conform *exactly* to the intensity ratios in Table 13.3. This point is examined further in Fig. 13.6. In both the triplet and the quartet, the inner lines are a little taller than the outer lines, although,

TABLE 13.3 Relative Intensities of Lines Within Common NMR Splitting Patterns

Number of equivalent adjacent protons	Number of lines in splitting pattern (name)	Relative line intensity within splitting pattern
0	1 (singlet)	1
1	2 (doublet)	1 1
2	3 (triplet)	1 2 1
3	4 (quartet)	1 3 3 1
4	5 (quintet)	1 4 6 4 1
5	6 (sextet)	1 5 10 10 5 1
6	7 (septet)	1 6 15 20 15 6 1

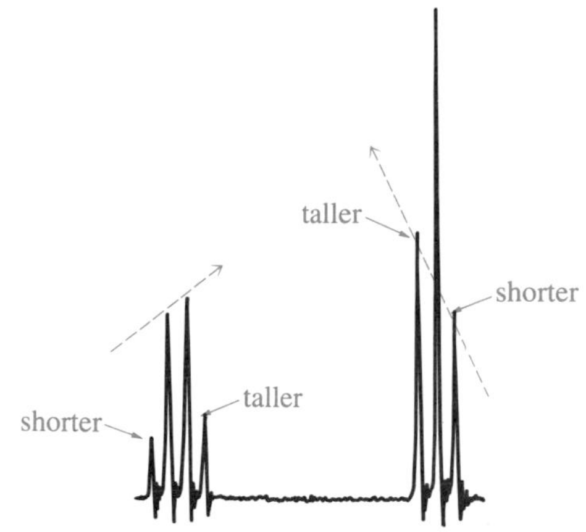

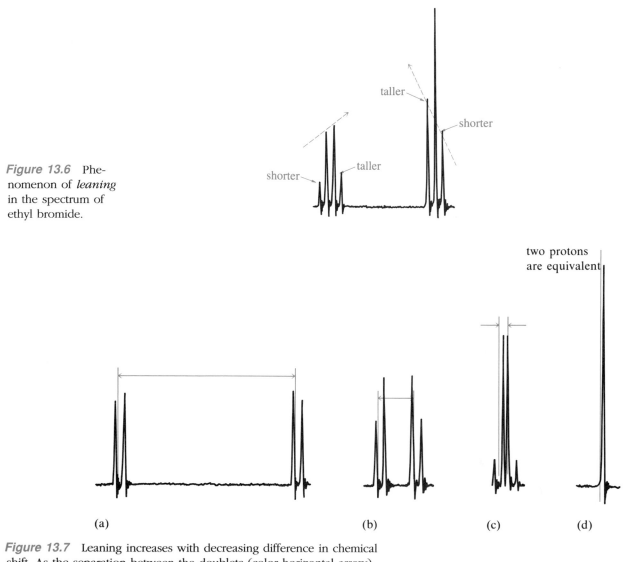

Figure 13.6 Phenomenon of *leaning* in the spectrum of ethyl bromide.

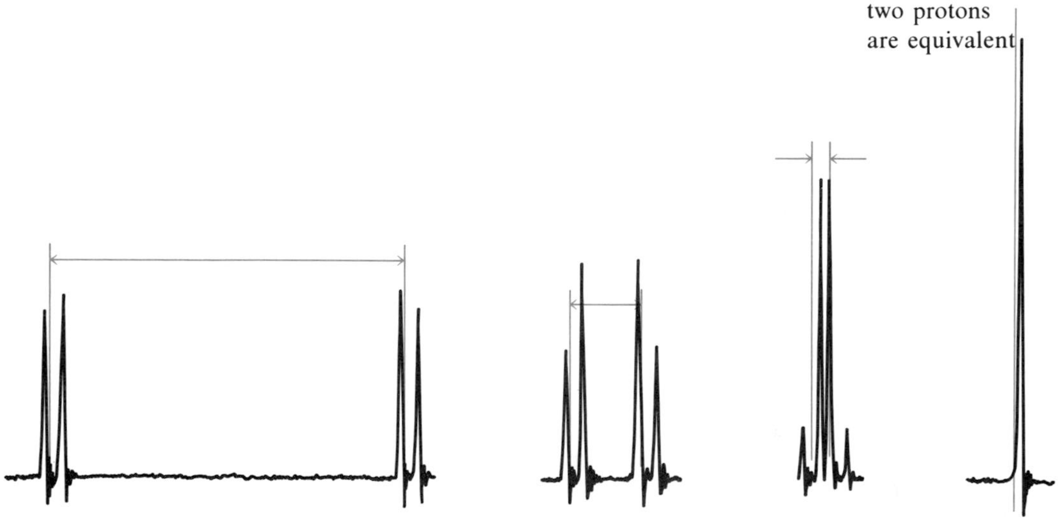

(a) (b) (c) (d)

Figure 13.7 Leaning increases with decreasing difference in chemical shift. As the separation between the doublets (color horizontal arrow) decreases, the leaning becomes more severe. The chemical shift position (color vertical line) of each doublet moves closer to the large line as the leaning increases. In (a), the doublets are well separated in chemical shift, and the relative intensities of the lines are close to 1 : 1 (Table 13.3). In (b), the chemical shifts are closer together than in (a), and substantial leaning is observed. In (c), the leaning is so severe that the outer lines are very small relative to the inner lines. In (d), the outer lines have disappeared altogether because the two protons are equivalent; that is, they appear as one line.

according to Table 13.3, they should be the same. This departure from the ratios in Table 13.3 is called *leaning*. Leaning is most severe when two signals that split each other are very close together in chemical shift; it does not occur when the two signals are very far apart. For example, Fig. 13.7 shows the NMR spectrum of a hypothetical compound XYCH—CHAB. According to the splitting rules, the spectrum of such a compound should consist of two doublets. In successive parts of the figure the coupling constant between the two protons is the same, but the chemical shift difference between them is made progressively smaller. According to Table 13.3, each line of each doublet should have the same intensity. As the chemical shifts of the two protons move closer together, the outer lines of each doublet become smaller relative to the inner lines. Furthermore, the chemical shift of each doublet moves away from the midpoint of each doublet and closer to the larger line. In Fig. 13.7c, for example, the outer line of each doublet is very small, and the chemical shift nearly coincides with the large inner line. In fact, when the chemical shifts of the two protons are identical, as in Fig. 13.7d, the outer lines of the doublets disappear altogether and the large inner lines merge to become a single line. A single line, of course, is exactly the signal observed for two protons that are chemically identical! This is why *no splitting is observed between protons with identical chemical shifts.* Thus, the nine methyl protons of *t*-butyl bromide, for example, appear as a single line because they all have the same chemical shift. These protons *are coupled,* but the coupling is invisible in the NMR. The same behavior can be observed even when two sets of protons are not chemically equivalent. For example, in the following molecule, the two sets of protons H^a and H^b are clearly in different chemical environments:

$$N{\equiv}C{-}\overset{a}{C}H_2{-}\overset{b}{C}H_2{-}\overset{\overset{\textstyle O}{\textstyle \|}}{C}{-}OCH_3$$

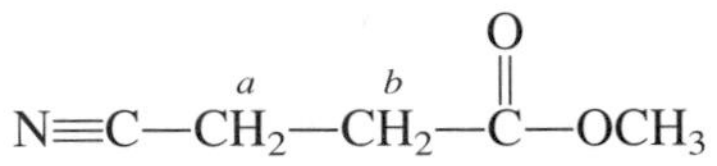

Yet *by chance* they happen to have the same chemical shift (δ 2.68). Therefore both sets of protons appear as a single line at δ 2.68 and there is no splitting observed. (We could not have predicted this equivalence in advance.)

Problem

13 (a) Predict the NMR spectrum of $CH_3{-}CHCl_2$, including the chemical shifts and the splitting. (Assume that the coupling constants are about the same as those for ethyl bromide.)

(b) What effect would leaning have on the appearance of this spectrum?

B. Physical Basis of Splitting

Splitting occurs because the magnetic field due to the spin of a neighboring proton adds to, or subtracts from, the applied magnetic field and affects the total field experienced by an observed proton. To see how this idea works, let us analyze the signal of a set of equivalent protons a (such as the methyl protons of ethyl bromide) adjacent to *two* equivalent neighboring protons b (Fig. 13.8). What are the spin possibilities for the neighboring protons b? In any one molecule, these protons could *both* have spin $+\frac{1}{2}$; they could *both* have spin $-\frac{1}{2}$; or they could have differing spins.

b protons			b protons			b protons		
1		2	1		2	1		2
			$+\frac{1}{2}$		$-\frac{1}{2}$			
$+\frac{1}{2}$		$+\frac{1}{2}$				$-\frac{1}{2}$		$-\frac{1}{2}$
			$-\frac{1}{2}$		$+\frac{1}{2}$			

(13.7)

(Notice that there are two ways in which the b protons can differ in spin: proton 1 can have spin $+\frac{1}{2}$ and proton 2 spin $-\frac{1}{2}$, or vice versa.) The spin of the neighboring b

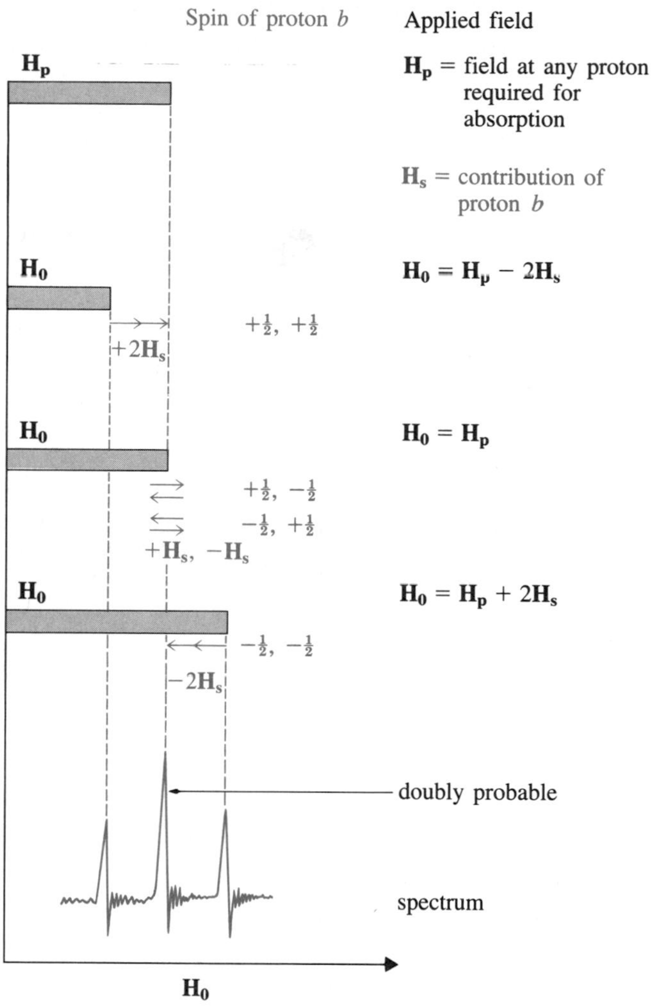

Figure 13.8 Analysis of the splitting of a proton a by two equivalent adjacent protons b.

protons exerts a magnetic field that influences the field felt by a proton a. When a neighboring b proton has spin $+\frac{1}{2}$, its magnetic field, which we shall call $\mathbf{H_s}$, is in the same direction as the applied field $\mathbf{H_0}$, and therefore adds to the applied field. When both b protons have spin $+\frac{1}{2}$, both add to the applied field, and the field felt by the set of a protons is

$$\mathbf{H_p} = \mathbf{H_0} + 2\mathbf{H_s} \tag{13.8a}$$

(The induced field $\mathbf{H_e}$ responsible for the chemical shift (Sec. 13.2) is also present, but is omitted here for simplicity.) When the protons b differ in spin, their fields cancel, and the field felt by the a protons is

$$\mathbf{H_p} = \mathbf{H_0} + \mathbf{H_s} - \mathbf{H_s} = \mathbf{H_0} \tag{13.8b}$$

Finally, when both of the b protons have spin $-\frac{1}{2}$, their magnetic fields subtract from the applied field. In this case the field felt by the set of a protons is

$$\mathbf{H_p} = \mathbf{H_0} - 2\mathbf{H_s} \tag{13.8c}$$

Rearranging Eqs. 13.8a–c, we see that there are three values of the applied field at which the protons *a* can absorb energy, depending on the spin of the neighboring *b* protons:

$$\mathbf{H_0} = \mathbf{H_p} - 2\mathbf{H_s} \quad \text{(both } b \text{ protons with spin } +\tfrac{1}{2}\text{)} \qquad (13.9a)$$

$$\mathbf{H_0} = \mathbf{H_p} \quad \text{(} b \text{ protons with opposite spin)} \qquad (13.9b)$$

$$\mathbf{H_0} = \mathbf{H_p} + 2\mathbf{H_s} \quad \text{(both } b \text{ protons with spin } -\tfrac{1}{2}\text{)} \qquad (13.9c)$$

These three situations correspond to the three lines of the triplet observed for the *a* protons. Furthermore, the situation described by Eqs. 13.8b and 13.9b is twice as probable, because, as shown in Eq. 13.7, there are two ways in which this situation can occur. Because this situation is twice as probable, the line at this value of $\mathbf{H_0}$ will be twice as intense—hence, a 1:2:1 triplet pattern is observed for the absorption of protons *a*.

The analogy here is to combinations that we can roll with a pair of dice. A 3 is twice as probable as a 2 because it can be rolled in two ways (2 + 1, 1 + 2), but a 2 can only be rolled in one way (1 + 1).

How does one proton "know" about the spin of adjacent protons? One of the most important ways that proton spins interact is through the electrons in the intervening chemical bonds. This interaction is weaker when the protons are separated by more chemical bonds. Thus, the coupling constant between hydrogens on adjacent saturated carbons is typically 5–8 Hz, but the coupling constant between more widely separated hydrogens is normally so small that it is not observed.

An analogy to this effect can be observed with an ordinary magnet and a few paper clips. If one paper clip is held to the magnet, it may be used to hold a second paper clip, and so on. The magnetic field of the magnet dies off with distance, so that typically the third or fourth paper clip is not magnetized.

Problem

14 Analyze the splitting pattern for a set of equivalent protons *a* in the presence of *three* equivalent adjacent protons *b*. Include an analysis of the relative intensity of each line of the splitting pattern. (This is the splitting pattern for the —CH_2— protons of ethyl bromide.)

C. Multiple Splitting. Splitting Diagrams

Several other important situations involve somewhat more complicated splitting patterns. Consider first the NMR spectrum of 1,3-dichloropropane, shown in Fig. 13.9.

$$\mathrm{Cl{-}\overset{b}{C}H_2{-}\overset{a}{C}H_2{-}\overset{b}{C}H_2{-}Cl}$$

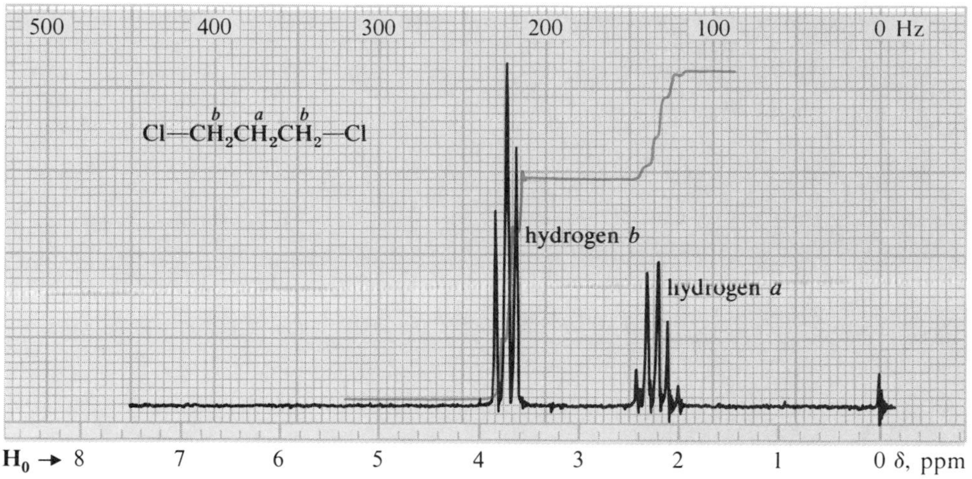

Figure 13.9 NMR spectrum of 1,3-dichloropropane.

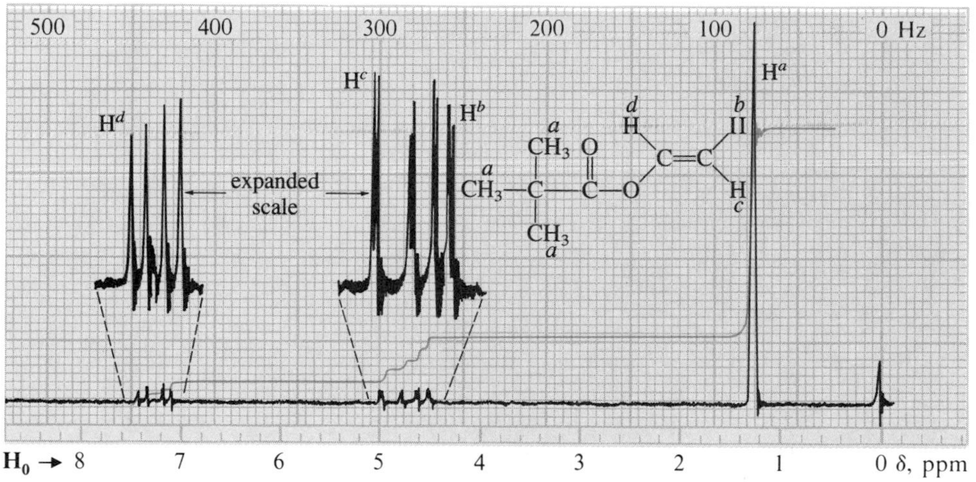

Figure 13.10 NMR spectrum of vinyl pivalate. The insets show the spectrum on a larger scale.

This molecule has two different types of protons, labeled *a* and *b*. The key to this spectrum is the recognition that all four protons H^b are chemically equivalent. The signal for H^a is therefore split into a quintet because there are four protons on adjacent carbons; the fact that two of the protons are on one carbon and two are on the other makes no difference. The signal for H^b is a triplet that appears at higher field.

A more complex situation occurs when a proton is split by more than one type of neighboring hydrogen. An instructive case of this sort is the NMR spectrum of vinyl pivalate, shown in Fig. 13.10.

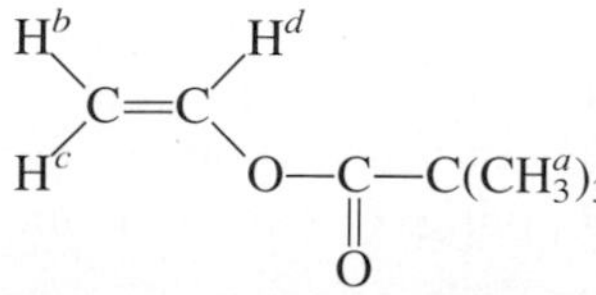

vinyl pivalate

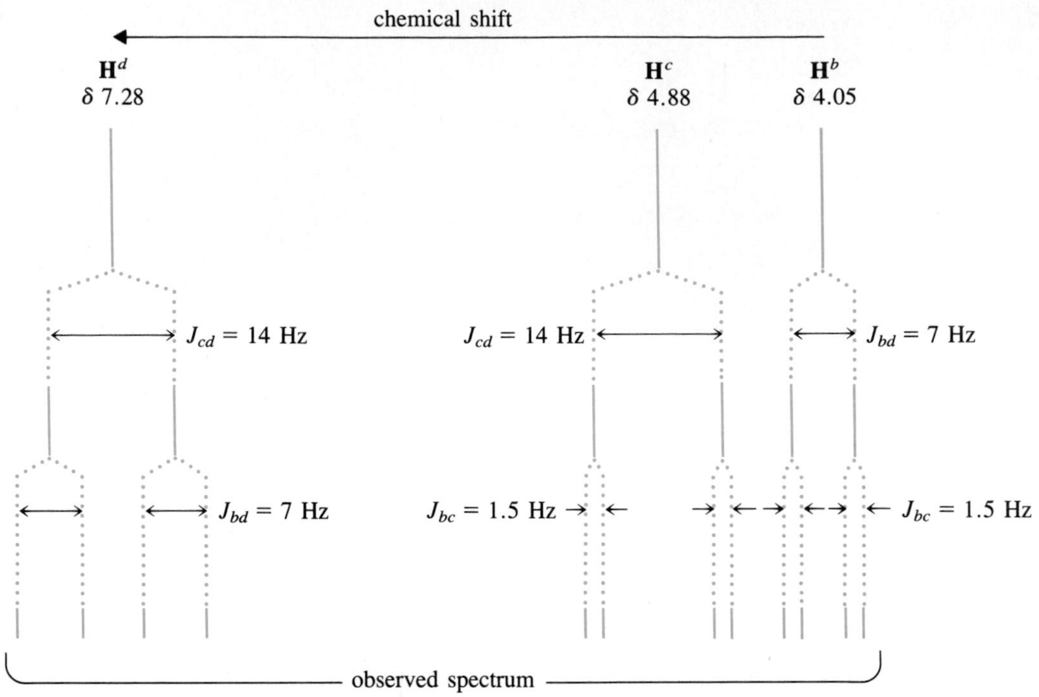

Figure 13.11 Splitting diagram for the NMR spectrum of the alkene protons in vinyl pivalate (Fig. 13.10).

(Do not be concerned that there is an unfamiliar functional group in this molecule; the principles are unchanged.) The nine equivalent *t*-butyl protons of vinyl pivalate (H^a) give the large singlet at δ 1.26. The interesting part of this spectrum is the region containing the resonances of the alkene protons. As we shall learn in Sec. 13.5A, alkene protons generally have chemical shifts greater than 4.5 ppm. The protons H^b and H^c are farthest from the electronegative oxygen and therefore have the smallest chemical shifts; the complex signals in the δ 4.5–5.0 region are from these protons. The four lines in the δ 7.0–7.5 region are all resonances of the one proton H^d.

The complexity in this spectrum is the result of two factors. First, each of these alkene protons is in a different chemical environment and has a different chemical shift from the others. Second, *each proton is split separately by the other two with a different coupling constant.* When a proton resonance is split separately by more than one other proton, *each splitting is applied successively.* Thus, the splitting of H^d by H^b gives a doublet; *each* line of this doublet is then split again by H^c. Thus H^d appears as a *doublet of doublets,* or four lines. For a similar reason, the signals of both H^b and H^c are also four-line patterns. The signal for all three alkene protons, then, consists of (3 × 4) or twelve lines, all of which are clearly discernible in the expanded part of the spectrum.

It is easiest to understand this situation with the aid of a **splitting diagram,** shown in Fig. 13.11. In a splitting diagram, NMR signals are approximated as vertical lines, and the effect of each splitting is shown separately. Consider, for example, the signal of H^d. The coupling constant of H^d and H^c, J_{cd}, is 14 Hz. We apply this splitting to the signal of H^d and get a doublet in which the intensity of each line is half the original. The coupling constant of H^d and H^b, J_{bd}, is 7 Hz. We next apply this splitting to

Figure 13.12
(a) Observed NMR signal of the H^b protons in 1-bromo-3-chloropropane. (b) A splitting diagram for the H^b protons for the case of equal coupling with H^a and H^c.

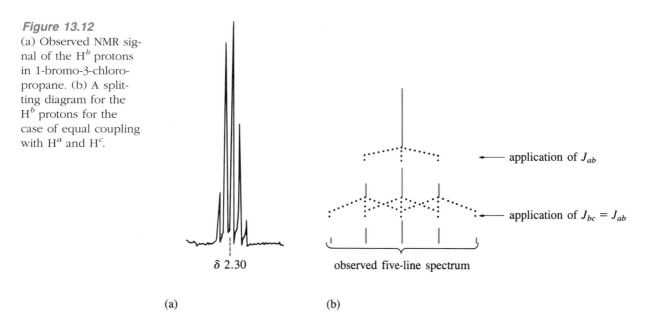

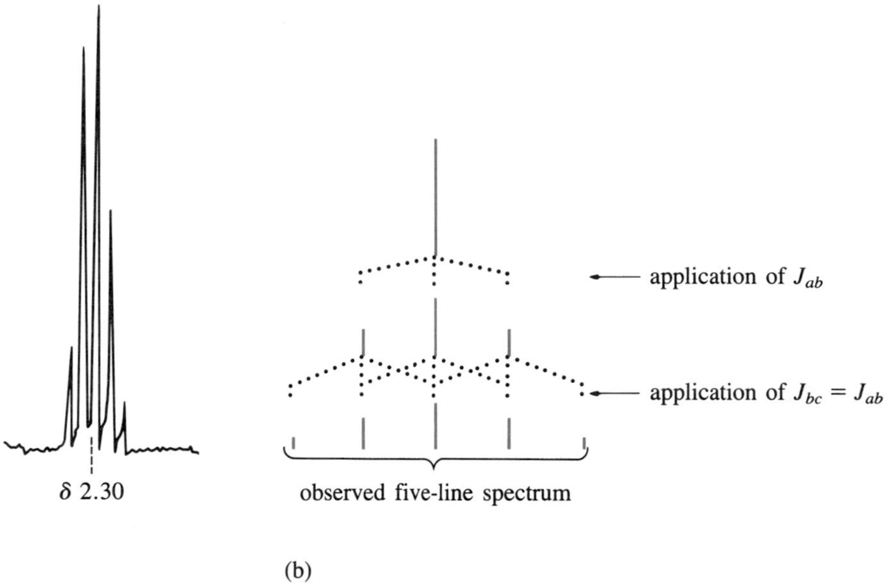

each line of the doublet to get two more doublets, in which the intensity of each line is halved again. Thus we end up with a doublet of doublets, or four lines, each with approximately one-fourth the intensity of the unsplit line. The same technique is illustrated in Fig. 13.11 for the other two alkene protons with $J_{bc} = 1.5$ Hz. (Recall that $J_{bc} = J_{cb}$.) In constructing this splitting diagram, it makes no difference which splitting is applied first. We could just as easily have applied the two splittings in the reverse order; the result is the same. (You should try this for yourself.) To summarize: observed splitting is the result of individual splittings applied successively.

Multiple splitting may occur any time one set of protons is split by two or more nonequivalent sets. However, such splitting is not always so complicated. Consider, for example, the spectrum of 1-bromo-3-chloropropane.

$$\mathrm{Br{-}\overset{a}{C}H_2{-}\overset{b}{C}H_2{-}\overset{c}{C}H_2{-}Cl}$$

The protons H^b are split by two nonequivalent sets, H^a and H^c; we might expect that the resonance of H^b would consist of (3 × 3) or 9 lines—a triplet of triplets. However, as shown in Fig. 13.12a, the signal for H^b in fact consists of five lines, exactly like the signal for H^b in 1,3-dichloropropane (Fig. 13.9). The reason for this unexpected simplicity is that, although there *are* indeed nine lines, some of them overlap because the coupling constant between H^a and H^b is the same as the coupling constant between H^b and H^c; that is, $J_{ab} = J_{bc}$. We can understand this point by again consulting a splitting diagram, this time for H^b, shown in Fig. 13.12b. The intensities of the overlapping lines add together, and as a result the splitting is exactly what we would have predicted if we had ignored the differences between H^a and H^c and applied the $n + 1$ rule. That is, there are four protons in positions adjacent to H^b, and the NMR signal for H^b consists of (4 + 1) or 5 lines with the intensity ratios shown in Table 13.3.

Summarizing: when we apply successive splittings to a proton signal and *each has about the same coupling constant,* the signal is the same as that predicted by the $n + 1$ rule. This situation turns out to be very common, particularly when the protons involved are on *adjacent saturated carbons.* In fact, it is usually a good guess to expect the $n + 1$ rule to hold in such cases; but we always must have in the back of our minds the possibility that the situation might be more complicated. (See, for example, Problem 15.)

Problems

15 The three absorptions in the NMR spectrum of 1,1,2-trichloropropane have the following characteristics:

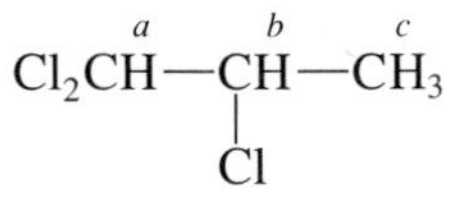

H^a: δ 5.82, $J_{ab} = 3.5$ Hz
H^b: δ 4.40, $J_{ab} = 3.5$ Hz, $J_{bc} = 6.0$ Hz
H^c: δ 1.78, $J_{bc} = 6.0$ Hz.

Using bars to represent lines in the spectrum, and a splitting diagram to determine the appearance of the H^b absorption, sketch the appearance of the spectrum. (Graph paper is especially useful in constructing splitting diagrams.)

16 Predict the NMR spectra of the following compounds. Consult Table 13.2 for chemical shift information when necessary.

(a)

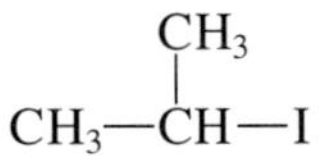

$$CH_3 - \overset{\displaystyle CH_3}{\underset{}{CH}} - I$$

(b) $Cl_2CH - CH_2 - Cl$

(c) $Cl - CH_2 - \overset{}{\underset{\displaystyle Cl}{C}}(CH_3)_2$

(d) [square with O—]

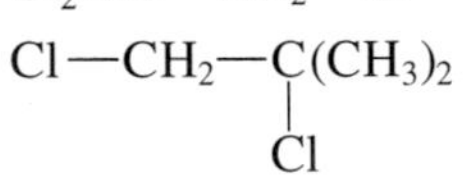

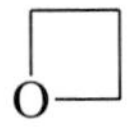

(e) $CH_3O - CH_2CH_2CH_2 - OCH_3$

D. Solving Unknown Structures with NMR (Part 2)

We have now learned all the basics of NMR. Let us summarize the type of information available from NMR spectra.

1. From the *chemical shift,* we obtain information about functional groups that are near an observed proton.

2. From the *integral,* we can determine the relative number of protons contributing to a given signal.

3. From the *splitting pattern,* we can ascertain the number of protons adjacent to an observed proton.

These three elements of an NMR spectrum can be put together like parts of a puzzle to deduce a great deal about chemical structure; it is not unusual for a complete structure to be determined from the NMR spectrum alone.

Because NMR spectra consume a large amount of space, it is common to see NMR spectra recorded in books and journals in an abbreviated form. In the form we shall use, the chemical shift of each resonance is followed by its integral, its splitting, and (if split) its coupling constants, if known. Abbreviations used to indicate splitting patterns are s (singlet), d (doublet), t (triplet), and q (quartet); complex patterns in which the nature of the splitting is not clear are simply designated m, for multiplet. It is assumed that the relative intensities of each splitting pattern, except for leaning, approximately match those in Table 13.3. For example, the spectrum of ethyl bromide (Fig. 13.5) would be summarized as follows:

$$\delta\ 1.67\ (3H,\ \mathrm{t},\ J = 7\ \mathrm{Hz});\ \delta\ 3.43\ (2H,\ \mathrm{q},\ J = 7\ \mathrm{Hz})$$

coupling constant

type of splitting pattern (triplet; intensities in Table 13.3)

integral

chemical shift (ppm downfield from TMS)

We are now ready to determine some structures using NMR spectra that contain splitting information. To work such problems, first use the information inherent in the molecular formula. What is the unsaturation number? Second, write out partial structures suggested by the integration and splitting pattern. When multiple possibilities exist, write them all! Use the chemical shift information to place functional groups at the appropriate carbons. Then write out all structures that seem to fit the data. Then decide what you would look for in the NMR spectrum to distinguish between them. Consult the spectrum to see whether such a feature is present. As in Sec. 13.2C, the general approach is best illustrated by an example. Try to work the problem before reading the solution.

Problems

17 Give the structure of a compound $C_4H_8Cl_2$ with the following NMR spectrum: $C_4H_8Cl_2$: $\delta\ 1.60\ (3H,\ \mathrm{d},\ J = 7.5\ \mathrm{Hz});\ \delta\ 2.15\ (2H,\ \mathrm{q},\ J = 7.5\ \mathrm{Hz});\ \delta\ 3.72\ (2H,\ \mathrm{t},\ J = 7.5\ \mathrm{Hz});\ \delta\ 4.27\ (1H,\ \mathrm{sextet},\ J = 7.5\ \mathrm{Hz})$

Solution: The formula $C_4H_8Cl_2$ requires an unsaturation number of zero; hence this compound can contain no rings or double bonds. The $\delta\ 4.27$ signal is a sextet, and integrates for one proton. Accordingly there must be five protons adjacent to the carbon bearing a single proton. Two partial structures satisfy this criterion.

$$\mathrm{CH_3-\overset{|}{C}H-CH_2-} \quad \text{or} \quad \mathrm{-CH_2-\overset{\overset{\displaystyle |}{C}H}{\underset{|}{C}H}-CH_2-}$$

(1) (2)

The chemical shift of $\delta\ 4.27$ requires that one of the chlorines be present on this carbon. Only partial structure (1) is capable of satisfying this requirement. This reasoning leads to the following:

$$\overset{\delta\ 4.27}{\mathrm{CH_3-\underset{\underset{\displaystyle Cl}{|}}{C}H-CH_2-}}$$

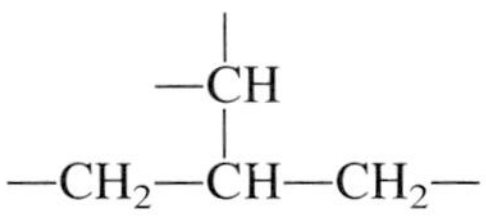

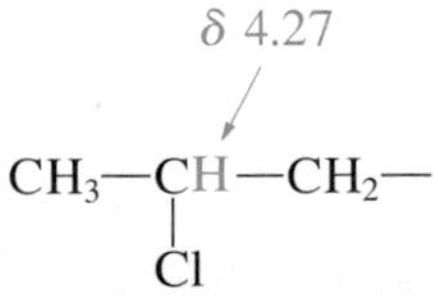

Problems (Cont.)

Notice that we have thus accounted also for the δ 1.60 doublet that integrates for three protons (why?). The δ 3.72 signal is a triplet, and the hydrogens giving this resonance must be on the same carbon as the remaining chlorine because of the chemical shift. This information suggests the partial structure

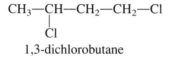

$$—CH_2—CH_2—Cl$$

δ 3.72

The one remaining signal is the δ 2.15 quartet for two protons. Since our two partial structures above have a total of five carbons, one of the carbons must be common to both. This carbon can only be that of the nonchlorinated $—CH_2—$ group. The unknown structure is therefore

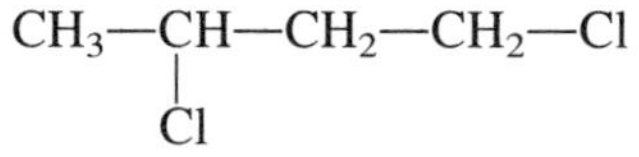

1,3-dichlorobutane

---

18 Give structures for each of the following compounds:

(a) C_3H_7Br: δ 1.03 (3*H*, t, *J* = 7 Hz); δ 1.88 (2*H*, sextet, *J* = 7 Hz); δ 3.40, (2*H*, t, *J* = 7 Hz)

(b) $C_2H_3Cl_3$: δ 3.98 (2*H*, d, *J* = 7 Hz); δ 5.87 (1*H*, t, *J* = 7 Hz)

(c) $C_5H_8Br_4$: δ 3.6 (s; only line in the spectrum)

---

19 When 3-bromopropene is allowed to react with HBr in the presence of peroxides, a compound *A* is formed that has the following NMR spectrum: δ 3.60 (4*H*, t, *J* = 6 Hz); δ 1.38 (2*H*, quintet, *J* = 6 Hz)

(a) From the reaction, what do you think *A* is?

(b) Use the NMR spectrum to confirm or refute your hypothesis. Identify *A*.

13.4 INTERPRETING COMPLEX NMR SPECTRA. HIGH-FIELD NMR

Mastery of the $n + 1$ splitting rule allows us to use NMR to solve a large number of structural problems. Spectra that conform to the $n + 1$ rule are called **first-order spectra.** However, many compounds have spectra that are not first order; they are more complicated. To say that a spectrum is *more complicated* means that there appear to be more lines in the spectrum than we would predict from the $n + 1$ splitting rule. Such spectra can be analyzed rigorously (in many cases) using special mathematical techniques. Even without such sophisticated methods, however, a great deal of information can be obtained from such complex spectra.

A good example of a spectrum that shows non-first-order behavior is that of 1-chlorobutane (Fig. 13.13). Consider, for example, the signal for the H^a protons.

$$\overset{a}{CH_3}—\overset{b}{CH_2}—\overset{c}{CH_2}—\overset{d}{CH_2}—Cl$$

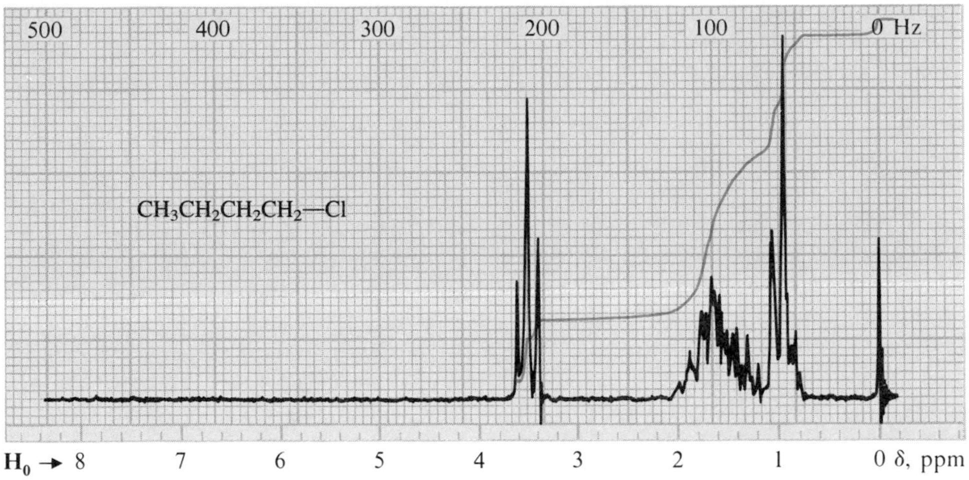

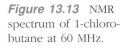

Figure 13.13 NMR spectrum of 1-chlorobutane at 60 MHz.

The resonance of this methyl group should occur at the smallest chemical shift (since this group is farthest from the —Cl), and the $n + 1$ rule predicts that it should be a triplet. We see near δ 1.0 a resonance vaguely resembling a triplet, but there are clearly more than just three lines in this signal; there is additional splitting that is not predicted by the $n + 1$ rule. The signals for H^b and H^c overlap in a chaotic jumble (called a "complex multiplet") between δ 1.1 and δ 2.1; no obvious first-order splitting pattern is evident here. Finally, we see that the signal for H^d is the expected triplet at δ 3.55.

Why is such a complex spectrum observed in this case? It turns out that first-order NMR spectra are generally observed when *the chemical shift difference, in Hz, between coupled protons is much greater than their coupling constant.* If we denote the difference in chemical shift of two signals A and B by $\Delta\nu_{AB}$ (in Hz) and their coupling constant J_{AB}, then this condition is simply expressed as follows:

condition for first-order behavior: $$\underset{\text{(in Hz)}}{\Delta\nu_{AB}} \gg J_{AB} \qquad (13.10)$$

As a practical matter, if $\Delta\nu_{AB}$ is greater than J_{AB} by a factor of about ten or more, a first-order spectrum can be expected. Since the coupling constants in many aliphatic compounds are on the order of 6–7 Hz, it follows that first-order behavior can be expected when the chemical shift difference between coupled protons is more than about 60 Hz, or 1 ppm at 60 MHz operating frequency. When the chemical shift difference of coupled protons is substantially less than this, non-first-order spectra usually can be anticipated. In the spectrum of 1-chlorobutane (Fig. 13.13), the differences in chemical shift between H^a, H^b, and H^c are small (all occur within 1 ppm of each other). Therefore, the $n + 1$ rule breaks down and a complex spectrum is observed. The signal for H^d, on the other hand, is well separated from the remainder of the spectrum, and this signal shows first-order splitting: It is a triplet, as predicted by the $n + 1$ rule.

Can we get any useful structural information from a complex spectrum? What we

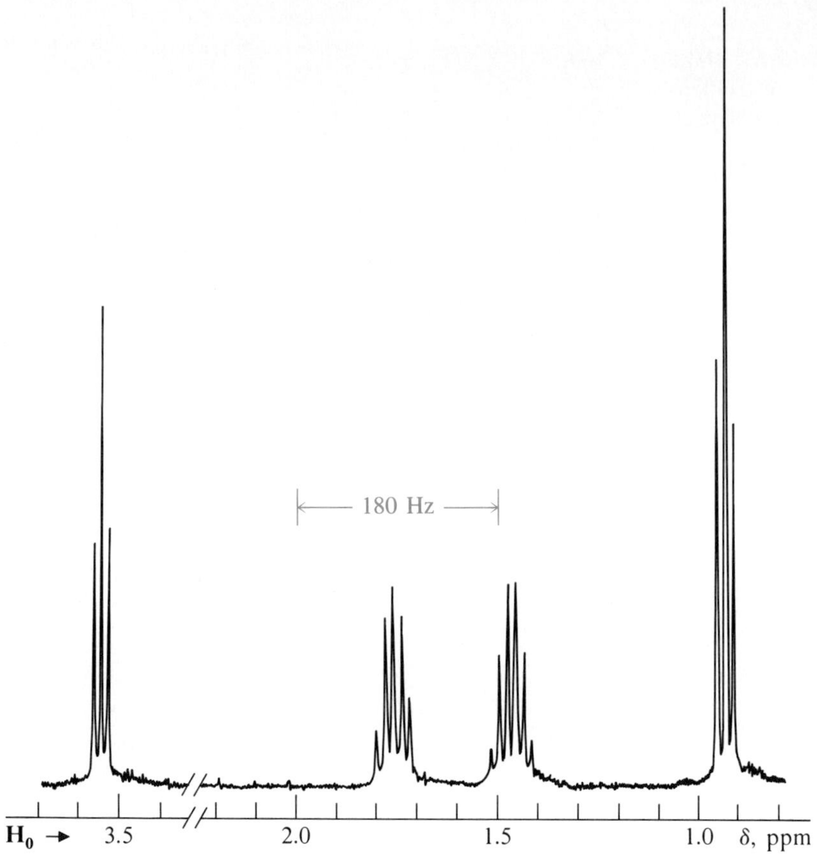

Figure 13.14 NMR spectrum of 1-chlorobutane taken at an operating frequency of 360 MHz.

have to understand is that the complexity in a spectrum such as that in Fig. 13.13 is in the splitting behavior; this complexity does not prevent us from extracting useful information from the chemical shift and integral. Furthermore, in Fig. 13.13, the low-field triplet provides partial splitting information. To illustrate the analysis of a complex spectrum, let us approach the spectrum of 1-chlorobutane in Fig. 13.13 as if it were that of an unknown structure. Given the molecular formula C_4H_9Cl, we can immediately deduce from the δ 3.55 resonance that the chlorine-bearing carbon has two hydrogens (from its integral) and is adjacent to a carbon bearing two hydrogens (from its splitting). This analysis immediately establishes the partial structure —CH_2—CH_2—Cl. A compound with this partial structure and the molecular formula C_4H_9Cl can be *only* 1-chlorobutane—structure solved!

With practice and experience we can learn to recognize some non-first-order patterns in NMR spectra. However, at this point our approach to such spectra should be to consider the chemical shift and integral of each set of peaks, and look for isolated first-order features (such as the low-field triplet in Fig. 13.13) that can reveal partial structures.

There is another approach to dealing with complex spectra—an *experimental* approach. It turns out that most complex spectra can be simplified by running them at a greater operating frequency ν_0. For example, the spectrum of 1-chlorobutane at ν_0 = 360 MHz is shown in Fig. 13.14. A greater operating frequency means that a greater magnetic field is required for resonance, by Eq. 13.3. Hence, an instrument with a much more powerful magnet is required to produce a spectrum like this. (Such spectra are sometimes called **high-field spectra.**) Notice that the spectrum of

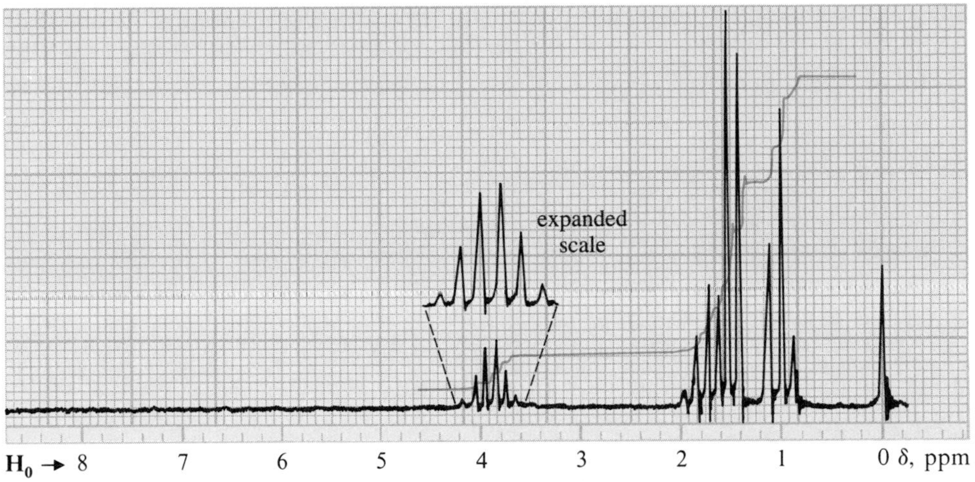

Figure 13.15 NMR spectrum for Problem 20.

1-chlorobutane in Fig. 13.14 is first order; it conforms completely to the $n + 1$ splitting rule. Compare this with the 60 MHz spectrum of the same compound in Fig. 13.13! This simplification occurs because chemical shifts (in Hz) increase with increasing ν_0 (Sec. 13.2A), but *coupling constants do not change* (Sec. 13.3A). The signals for H^b and H^c, for example, are separated by about 0.3 ppm. At 60 MHz, this separation corresponds to only 18 Hz, a number not much larger than the coupling constant J_{bc} (about 7 Hz). At 360 MHz a 0.3 ppm chemical shift difference equals 108 Hz, more than ten times greater than J_{bc}, which remains unchanged at the higher operating frequency. Since at the higher frequency the condition of Eq. 13.10 for first-order spectra is met, a simplified spectrum is observed.

Although it might seem that we should run all spectra at high field to avoid unnecessary complexity, in practice this is not a routine matter. The instruments required for such spectra are expensive to manufacture and maintain. The demand for time on them and the expense involved restricts their use to the most important samples whose structures cannot be determined by other more readily available means. We would not normally want to use a high-field instrument on a compound as simple as 1-chlorobutane; but the spectrum of this compound in Fig. 13.14 does illustrate the power of this technique.

Problem

20 A box labeled "C_4H_9Cl isomers" was found in an abandoned laboratory. In the box were two bottles, each containing a colorless liquid. The NMR of compound *A* is a singlet at δ 1.63, and the NMR of compound *B* is shown in Fig. 13.15. Because of your expertise in NMR, you have been hired as a consultant; identify each of the compounds.

13.5 FUNCTIONAL-GROUP NMR ABSORPTIONS

A. NMR Spectra of Alkenes

Let us now consider the important NMR absorptions of the major functional groups we have studied. Chemical shifts for the different types of protons found in alkenes are

TABLE 13.4 Chemical Shifts of Alkene Protons

Structural type and name		δ, ppm
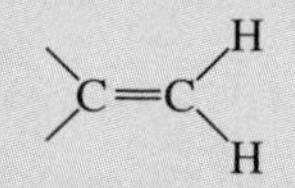	terminal vinylic	4.6–5.0
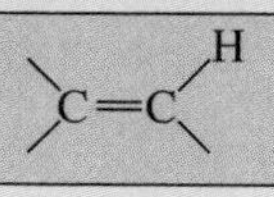	internal vinylic	5.0–5.7
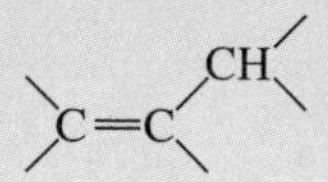	allylic	1.6–2.3

given in Tables 13.2 and 13.4, and are also illustrated by the following examples:

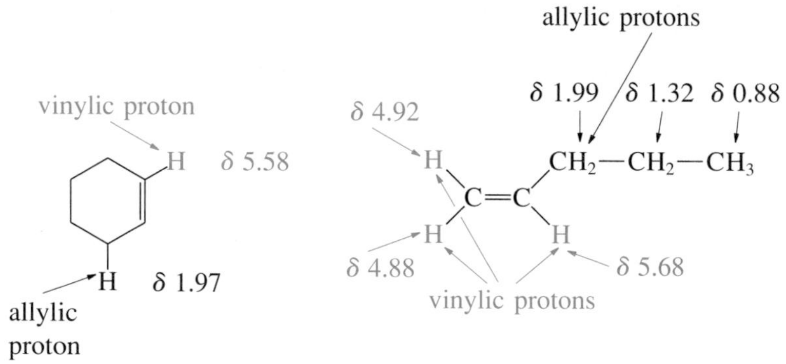

The protons attached to double bonds are called **vinylic protons** (color in the above structures). The chemical shifts of these protons are considerably greater than predicted from the electronegativity of the C=C functional group, and can be understood in the following way.

Imagine that an alkene molecule in an NMR spectrometer is oriented with respect to the applied field $\mathbf{H_0}$ as shown in Fig. 13.16. The applied field induces a circulation of the π-electrons in closed loops above and below the plane of the alkene. This electron circulation gives rise to an induced magnetic field $\mathbf{H_i}$ that *opposes* the applied field $\mathbf{H_0}$ at the center of the loop, very much like the induced field in Fig. 13.4. This induced field can be described as contours of closed circles. Although the induced field opposes the applied field $\mathbf{H_0}$ in the region of the π-bond, the curvature of the induced field causes it to lie in the same direction as $\mathbf{H_0}$ at the vinylic protons. The induced field therefore *augments* the applied field at the vinylic protons. These protons experience a net field $\mathbf{H_p}$ equal to the applied field $\mathbf{H_0}$ plus the induced field $\mathbf{H_i}$ less the usual shielding field $\mathbf{H_e}$ of Eq. 13.6a and Fig. 13.4.

$$\mathbf{H_p} = \mathbf{H_0} + \mathbf{H_i} - \mathbf{H_e} \qquad (13.11a)$$

When $\mathbf{H_p}$ meets the resonance condition of Eq. 13.3, $\mathbf{H_0}$ must be

$$\mathbf{H_0} = \mathbf{H_p} - \mathbf{H_i} + \mathbf{H_e} \qquad (13.11b)$$

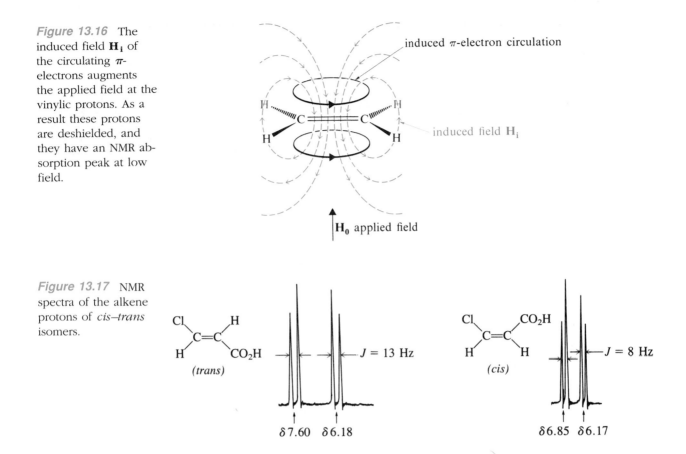

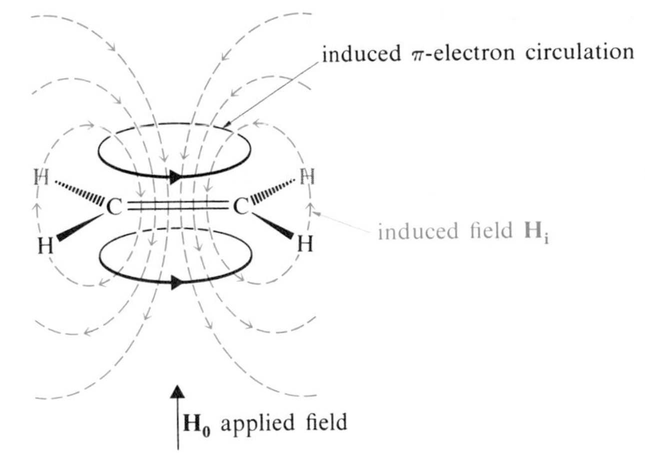

Figure 13.16 The induced field $\mathbf{H_i}$ of the circulating π-electrons augments the applied field at the vinylic protons. As a result these protons are deshielded, and they have an NMR absorption peak at low field.

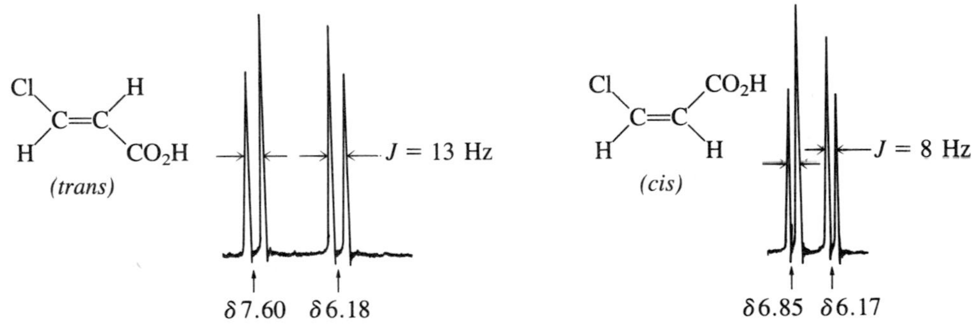

Figure 13.17 NMR spectra of the alkene protons of *cis–trans* isomers.

This means that the applied field necessary for resonance is reduced by the induced field $\mathbf{H_i}$. Thus vinylic protons are deshielded and absorb at lower field.

> Of course, molecules in solution are constantly in motion, tumbling wildly. At any given time, only a few alkene molecules are oriented with respect to the external field as shown in Fig. 13.16. The chemical shift of a vinylic proton is an average over all orientations of the molecule. However, this particular orientation makes such a large contribution that it dominates the chemical shift.

Protons on carbons adjacent to double bonds are called **allylic protons.** As shown in both Table 13.2 and the previous examples, these protons have greater chemical shifts than ordinary alkyl protons, but not so great as vinylic protons.

Splitting between vinylic protons in alkenes depends strongly on the geometrical relationship of the coupled protons. Typical coupling constants are given in Table 13.5. The spectra shown in Fig. 13.17 illustrate the very important observation that vinylic protons of a *cis*-alkene have smaller coupling constants than those of their *trans* isomers. (The same point is evident in the coupling constants of *cis* and *trans* protons shown in Figs. 13.10 and 13.11.) These coupling constants, along with the characteristic =C—H bending bands from IR spectroscopy (Sec. 12.3A), provide im-

TABLE 13.5 Coupling Constants for Proton Splitting in Alkenes

Relationship of protons		Coupling constant *J*, Hz
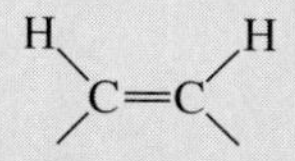	cis	6–14
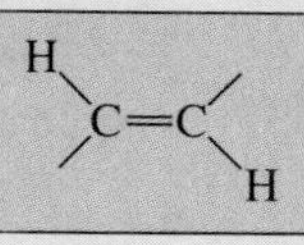	trans	11–18
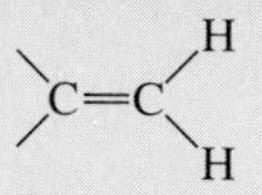	geminal	1–3.5
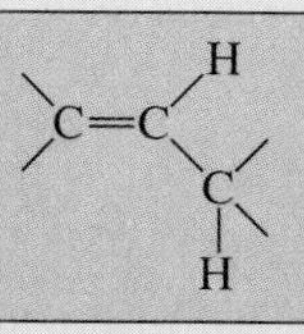	vicinal	4–10
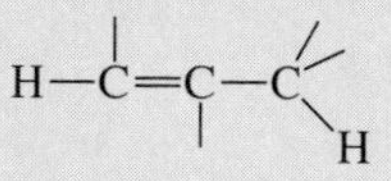	allylic (four bonds)	0.5–3.0
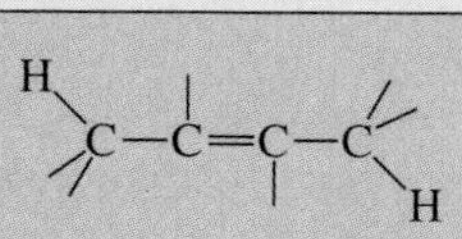	(five bonds)	0–1.5

portant ways to determine alkene stereochemistry. The very weak splitting (called *geminal splitting*) between vinylic protons on the *same* carbon stands in contrast to the much larger *cis* and *trans* splitting. Geminal splitting is also illustrated in Figs 13.10 and 13.11.

Splitting in alkenes may sometimes be observed between protons separated by more than three bonds. Such is the case in the last two examples shown in Table 13.5. Recall that splitting over these distances is usually *not* observed in saturated compounds. These long-distance interactions between protons are transmitted by the π-electrons.

Problem

21 Propose a structure for the compound C_7H_{14} with the NMR spectrum in Fig. 13.18. Explain in detail why your structure is consistent with the spectrum.

B. NMR Spectra of Alkanes and Cycloalkanes

Since all of the protons in a typical alkane are in very similar chemical environments, we should not be surprised to learn that the NMR spectra of alkanes and cycloalkanes cover a very narrow range of chemical shifts, typically δ 0.7 to δ 1.5. Because of this narrow range, the splitting in such spectra shows extensive non-first-order behavior.

An exception to these generalizations is the chemical shifts of protons on a cyclopropane ring, which are unusual for alkanes; they typically absorb at rather high field, typically δ 0 to δ 0.5. Some even have resonances at *higher* field than TMS (that is,

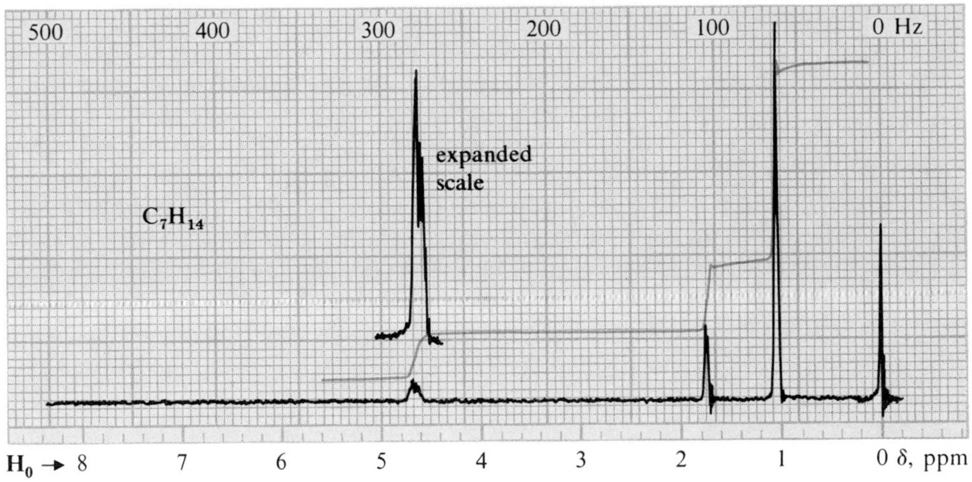

Figure 13.18 NMR spectrum for Problem 21.

negative δ values). For example, the chemical shifts of the ring protons of *cis*-1,2-dimethylcyclopropane shown in color below are δ (−0.11).

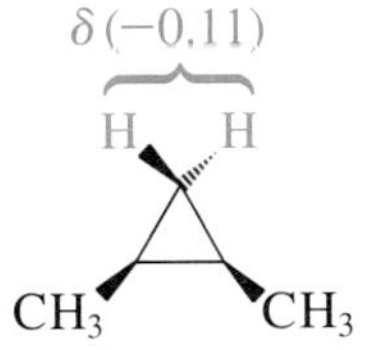

The reason for this unusual chemical shift is an induced electron current in the cyclopropane ring and resulting magnetic field, similar in principle to the one responsible for the large chemical shifts in alkenes (Fig. 13.16). In cyclopropanes, however, the induced field shields the ring protons, and their chemical shifts are decreased.

C. NMR Spectra of Alkyl Halides and Ethers

We are already familiar with the NMR spectra of alkyl halides and ethers from the many examples in the early part of this chapter. The chemical shifts caused by the halogens are usually in proportion to their electronegativity. For the most part, chloro groups and ether oxygens have about the same chemical-shift effect on neighboring protons (Table 13.2). However, epoxides, like cyclopropanes, have considerably smaller chemical shifts than their open-chain analogs.

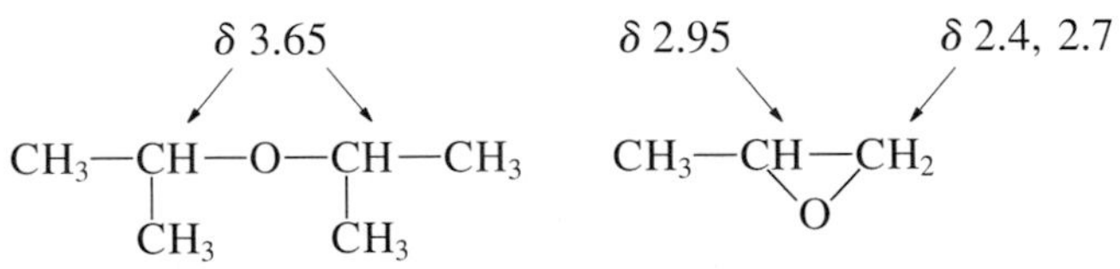

An interesting type of splitting is observed in the NMR spectra of compounds containing fluorine. The common isotope of fluorine (^{19}F) has a spin of $\pm\frac{1}{2}$. Proton signals are split by neighboring fluorine in the same general way that they are split by neighboring protons; the splitting rule is the same. For example, the proton in $HCCl_2F$ appears as a doublet centered at δ 7.43 with a large coupling constant J_{HF} of 54 Hz. Notice that this is *not* the NMR spectrum of the fluorine; it is the *splitting of the proton spectrum caused by the fluorine.* (It is also possible to do fluorine NMR (Sec. 13.8), but this requires, for the same magnetic field, a different operating frequency; the spectra of ^1H and ^{19}F do not overlap.) Values of H—F coupling constants are larger than H—H coupling constants. The J_{HF} value in $(CH_3)_3C$—F is 20 Hz; a typical J_{HH} value over the same number of bonds is 6–8 Hz. Because J_{HF} values are so large, coupling between protons and fluorines can sometimes be observed over as many as four single bonds. The common isotopes of chlorine, bromine, and iodine also have nuclear spins, but, as we have already observed in a number of spectra, they do not cause detectable proton splittings (for reasons that we shall not consider here).

Problems

22 Predict the NMR spectrum of ethyl fluoride. How would it differ from that of ethyl chloride?

23 Suggest structures for compounds with the following proton NMR spectra:
(a) $C_4H_{10}O$: δ 1.13 (3*H*, t, $J = 7$ Hz); δ 3.38 (2*H*, q, $J = 7$ Hz)
(b) $C_3H_5F_2Cl$: δ 1.75 (3*H*, t, $J = 17.5$ Hz); δ 3.63 (2*H*, t, $J = 13$ Hz)

D. NMR Spectra of Alcohols

As we might expect, protons on the α-carbons of primary and secondary alcohols have chemical shifts generally in the same range as ethers, from δ 3.0 to δ 4.1 (see Table 13.2). Since tertiary alcohols have no α-proton, the observation of an —OH stretching absorption in the IR spectrum accompanied by the *absence* of the —CH—O— absorption in the NMR is good evidence for a tertiary alcohol (or a phenol; see Sec. 16.3B).

CH_3—OH	CH_3—CH_2—OH	$(CH_3)_2CH$—OH	$(CH_3)_3C$—OH
δ 3.5	δ 3.6	δ 4.0	no proton in δ 3–δ 4 region

The chemical shift of the —OH proton in an alcohol depends on the degree to which the alcohol is involved in hydrogen bonding under the conditions that the spectrum is determined. For example, in pure ethanol, in which the alcohol molecules are extensively hydrogen bonded, the chemical shift of the —OH proton is δ 5.3. In CCl_4, the ethanol molecules are more dilute and less extensively hydrogen bonded, and the —OH absorption occurs at δ 2 to δ 3. In the gas phase, there is almost no hydrogen bonding, and the —OH resonance occurs at δ 0.8.

The chemical shift of the O—H proton in the gas phase is not as large as we might expect for a proton bound to an electronegative atom such as oxygen. The unusually high-field absorption of unassociated —OH protons is probably caused by the induced field of the lone-pair electrons on oxygen. This field shields the —OH proton from the applied field (Sec. 13.2D). Hydrogen-bonded protons, on the other hand, absorb at lower field because they bear less electron density and more positive charge (Sec. 8.3B).

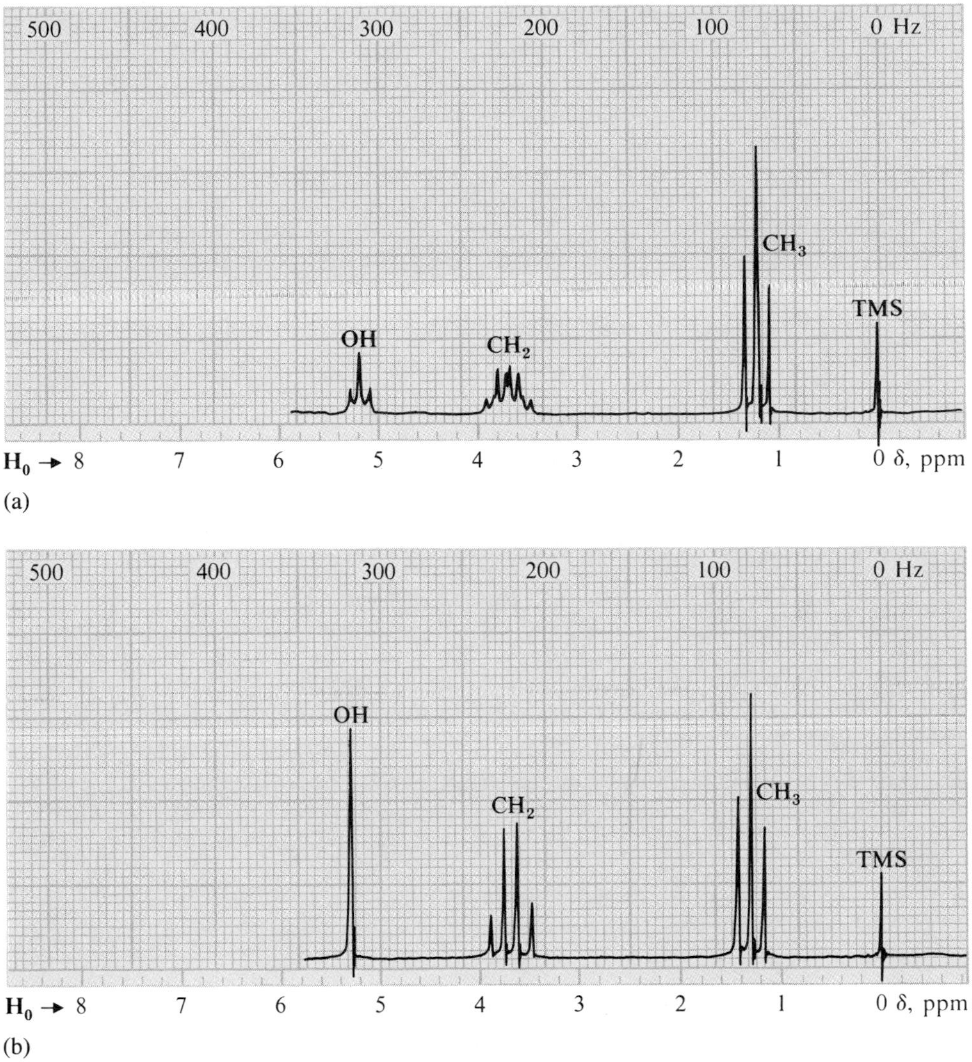

Figure 13.19 NMR spectra of ethanol. (a) Absolute, or dry, ethanol. Notice that the **—CH$_2$—** signal is split by both the **—OH** and the **—CH$_3$.** (b) Wet, acidified ethanol. Notice that the **—CH$_2$—** signal is now split only by the **—CH$_3$.**

The splitting between the —OH proton of an alcohol and neighboring protons is interesting. The $n + 1$ splitting rule predicts that the —OH resonance of ethanol should be a triplet, and the —CH$_2$— resonance should be split by both the adjacent —CH$_3$ and —OH protons, and should therefore appear as (4 × 2) or eight lines.

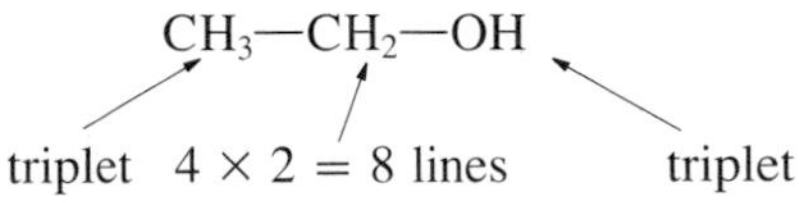

The NMR spectrum of very dry ethanol, shown in Fig. 13.19a, is as we expect. However, when a trace of water, acid, or base is added to the ethanol, the spectrum changes, as shown in Fig. 13.19b. *The presence of water, acid, or base causes collapse of the —OH resonance to a single line,* and obliterates all splitting associated with this proton. Thus, the —CH$_2$— proton resonance becomes a quartet, apparently split only by the CH_3— protons. *This type of behavior is quite general for alcohols, amines, and other compounds with a proton bonded to an electronegative atom.* In fact, most ordinary spectra of alcohols are taken on samples that contain a trace of moisture. In these samples the splitting of neighboring protons caused by the —OH proton is absent and the —OH resonance itself is somewhat broadened.

What is the reason for this effect? This phenomenon is a consequence of **chemical exchange:** an equilibrium involving chemical reactions that take place very rapidly as the NMR spectrum is being determined. In this case, the chemical reaction is proton exchange between the protons of the alcohol and those of water (or other alcohol molecules):

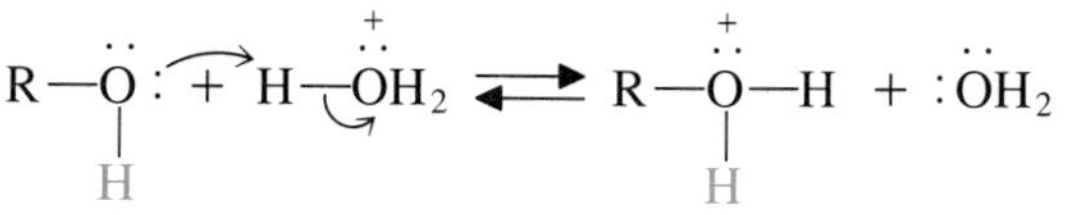

$$R—\overset{..}{\underset{|}{O}}: + H—\overset{+}{\underset{|}{O}}H_2 \rightleftharpoons R—\overset{..}{\underset{|}{O}}—H + :\overset{..}{O}H_2 \qquad (13.12a)$$
$$\quad H \qquad\qquad\qquad\qquad\quad H$$

$$R—\overset{+}{\underset{|}{O}}—H \qquad \rightleftharpoons R—\overset{..}{O}—H + H—\overset{+}{O}H_2 \qquad (13.12b)$$
$$\quad H \qquad :\overset{..}{O}H_2$$

This exchange is, of course, nothing more than two successive acid–base reactions. (Write the mechanism for $^-$OH-catalyzed exchange.) For reasons that we shall discuss in Sec. 13.7, *rapidly exchanging protons do not show spin–spin splitting with neighboring protons.* Acid and base catalyze this exchange reaction, accelerating it enough that splitting is no longer observed. In the absence of acid or base, this exchange is much slower, and splitting of the —OH proton and neighboring protons is observed.

Problems

24 In which solution does the —OH proton of ethanol have the greater chemical shift: $2M$ ethanol in CCl_4 or $0.2M$ ethanol in CCl_4? Explain.

25 Suggest structures for each of the following compounds:
 (a) $C_4H_{10}O$: δ 1.20 (9H, s); δ 2.40 (1H, broad s)
 (b) $C_5H_{10}O$: δ 1.71 (s); δ 1.78 (s), total integral of both resonances 6H; δ 2.31 (1H, broad s); δ 4.14 (2H, d, $J = 7$ Hz); δ 5.41 (1H, broad t, $J = 7$ Hz).

In this chapter we have studied the proton NMR spectra of alkanes, alkenes, alkyl halides, alcohols, and ethers. The proton NMR spectra of other classes of organic compounds will be considered in the chapters that cover the corresponding functional groups. A summary table of chemical shift information is given in Appendix III.

13.6 USE OF DEUTERIUM IN PROTON NMR

Deuterium (^{2}H, or D) finds special use in proton (^{1}H) NMR. Although deuterium has a nuclear spin, deuterium NMR and proton NMR require different operating frequencies. Consequently, deuterium NMR signals are not detected under the conditions used for proton NMR. Therefore deuterium is effectively "silent" in proton NMR. An important practical application of this fact is the use of deuterated solvents in

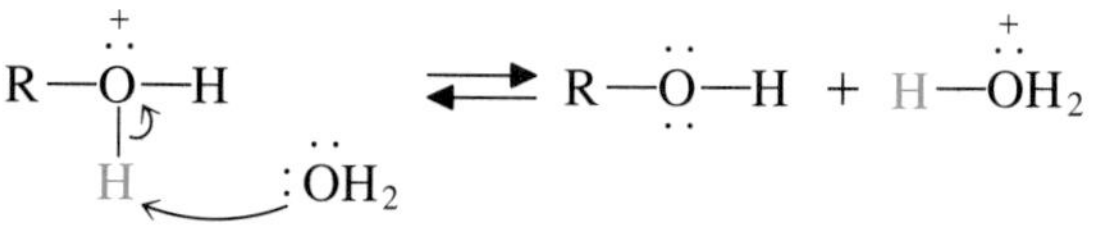

NMR experiments. (Solvents are needed for solid and viscous liquid samples because, in the usual type of proton NMR experiment, the sample must be in a free-flowing liquid state.) In order that the solvent not interfere with the NMR spectrum of the sample, it must either be devoid of protons, or its protons must not have NMR absorptions that obscure the sample absorptions. Carbon tetrachloride (CCl_4) is a useful solvent because it has no protons, and therefore has no NMR absorption. However, many important organic compounds do not dissolve in carbon tetrachloride. Chloroform, $CHCl_3$, is an excellent solvent for many organic compounds, but it has an interfering NMR absorption. However, the deuterium analog, $CDCl_3$ (chloroform-*d,* or deuterochloroform), has no proton-NMR absorption, but has all the desirable solvent properties of chloroform. This solvent is so widely used for NMR spectra that it is a relatively inexpensive article of commerce. Other deuterated solvents are also available.

Another useful observation about deuterium is that the coupling constants for proton–deuterium splitting are very small. Even when H and D are on adjacent carbons, the H–D coupling is negligible. For this reason, deuterium substitution can be used to simplify NMR spectra and assign resonances. Although this idea is normally most useful in more complex molecules, let us see how we might use it to assign the resonances of ethyl bromide (Fig. 13.5). If we were to synthesize $CH_3—CD_2—Br$ and record its NMR, we would find that the low-field quartet of ethyl bromide has disappeared from the NMR spectrum, and the high-field signal has become a singlet. This experiment would establish that the low-field quartet is the signal of the $—CH_2—$ group and that the high-field triplet is that of the $—CH_3$ group.

The same idea can be used to assign the —OH resonance of an alcohol. If we shake a solution of an alcohol in an NMR tube with a few drops of D_2O, the rapid exchange of the —OH proton for deuterium (Eq. 13.12) causes the disappearance of the —OH resonance. (The NMR signal of the DOH formed in the reaction is observed at a different chemical shift.)

Problem

26 The 60 MHz NMR spectrum of 1-chlorobutane, given in Fig. 13.13, is complex and not first order. Assuming you could synthesize the needed compounds, explain how you would use deuterium substitution to determine the chemical shift of each group of chemically equivalent protons in the molecule. Explain what you would see and how you would interpret the results.

13.7 NMR SPECTROSCOPY OF DYNAMIC SYSTEMS

Consider the NMR spectrum of cyclohexane, a compound that undergoes rapid conformational isomerizations. The spectrum of this compound is a singlet at δ 1.4. Yet we know from Eq. 7.1 and Sec. 7.2 that there are two distinguishable types of hydrogens in cyclohexane: the *axial* hydrogens and the *equatorial* hydrogens. It seems that there should be two signals in the NMR of cyclohexane: one for the axial protons and one for the equatorial ones. The reason that only one line is observed has to do with the *rate* of the chair–chair interconversion, which is so rapid that the NMR instrument detects only the average of the two conformations. Since the conformational equilibrium interchanges the positions of the axial and equatorial protons (Eq. 7.1), only one proton signal is observed. This is the signal of the "average" proton in cyclohexane—

one that is axial half the time and equatorial half the time. This example illustrates an important point about NMR spectroscopy: *when we take the spectrum of a compound that is involved in a rapid equilibrium, we observe a single spectrum that is the time-average of all species involved in the equilibrium.*

Although there are equations that describe this phenomenon exactly, we can begin to understand it by resorting to an analogy from common experience. Imagine looking at a three-blade fan or propeller that is rotating at a speed of about 100 times per second (Fig. 13.20). Our eyes do not see the individual blades, but only a blur. The appearance of the blur is a time average of the blades and the empty space between them. If we photograph the fan using a shutter speed of about 0.1 second, the fan appears as a blur in the resulting picture for the same reason: during the time the camera shutter is open (0.1 sec) the blades make ten full revolutions (Fig. 13.20a). Now imagine taking a picture with a shutter speed of 0.001 second. While the shutter is open, the fan blades make only 0.1 revolution—about 36°. We begin to see the fan blades; they are more distinct, but still somewhat blurred (Fig. 13.20b). Finally, imagine taking a picture with a shutter speed of 0.00001 second (very fast film). While the shutter is open, the fan blades traverse only $\frac{1}{1000}$ of a circle—about 0.36°. In our picture the individual blades will be visible and in relatively sharp focus (Fig. 13.20c). The rapid conformational flipping of cyclohexane is to the NMR spectrometer roughly what the rapidly rotating propeller is to the slow camera shutter. *The NMR spectrometer is intrinsically limited in its capacity to resolve events in time.*

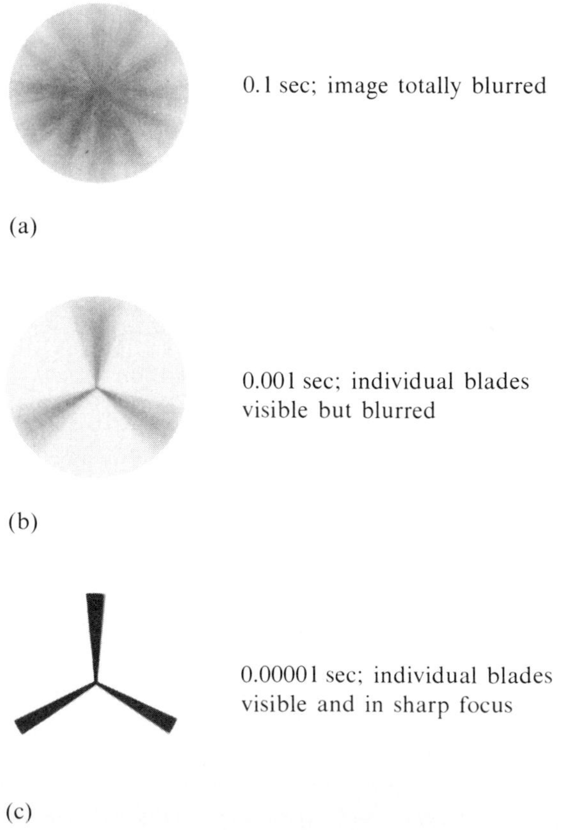

Figure 13.20 Imagine a three-blade propeller rotating at about 100 times per second (100 Hz). The diagrams show what we would see if we took snapshots using a shutter speed of (a) 0.1 sec, (b) 0.001 sec, and (c) 0.00001 sec.

We *can* observe both types of cyclohexane protons if we slow down the chair–chair interconversion by lowering the temperature. Let us imagine cooling a sample of cyclohexane in which all protons but one have been replaced by deuterium. (The use of deuterium eliminates splitting with neighboring protons; Sec. 13.6.)

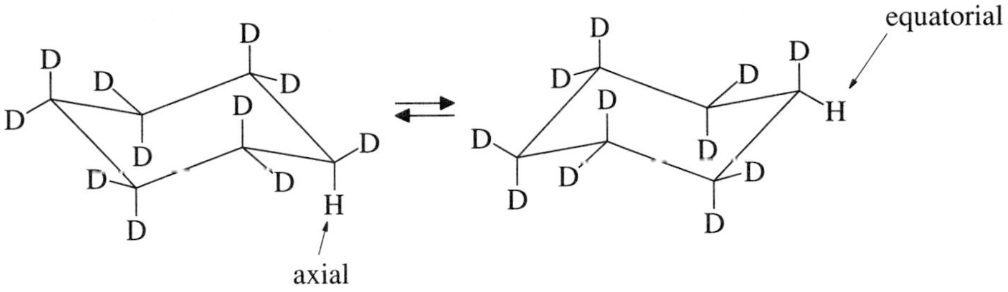

(13.13)

cyclohexane-d_{11}

In the chair–chair interconversion the single remaining proton alternates between axial and equatorial positions.

The NMR spectrum of cyclohexane-d_{11} at various temperatures is shown in Fig. 13.21. At room temperature, we see a single line, as we do with cyclohexane itself. As the temperature is lowered progressively, the resonance becomes broader until, just below −60°, the signal divides into two signals equally spaced about the original one. When the temperature is lowered still further, the spectrum becomes two sharp single lines. Lowering the temperature progressively retards the chair–chair interconversion until, at low temperature, NMR spectrometry can detect both chair forms indepen-

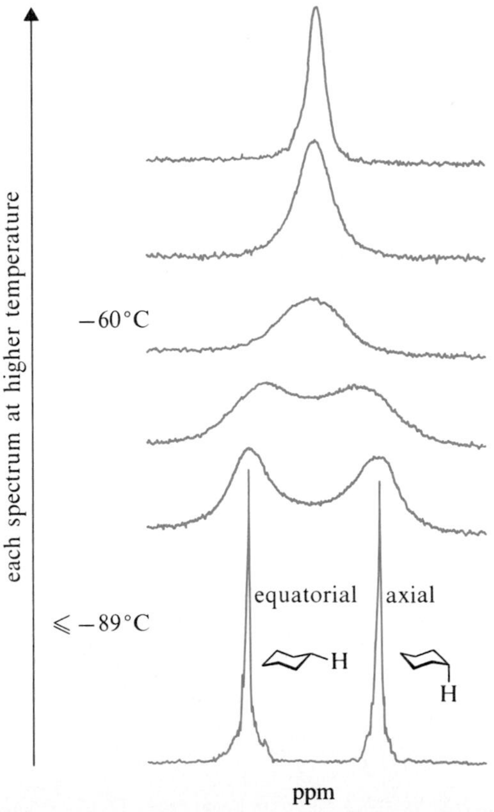

Figure 13.21 NMR spectrum of cyclohexane-d_{11} as a function of temperature.

dently. This is analogous to taking pictures of the propeller in Fig. 13.20 with a constant shutter speed, and slowing down the propeller until the blur disappears and the individual blades become clearly separated.

The information from these spectra at different temperatures can be used to calculate the rate of the chair–chair interconversion, although we shall not consider here how this is done. The energy barrier for the chair–chair interconversion shown in Fig. 7.4 was obtained from this type of experiment.

The time-averaging effect of NMR is not limited simply to conformational isomerizations. Molecules undergoing any rapid process, such as a chemical reaction, are also averaged by NMR spectroscopy. This is the reason, for example, that the splitting associated with the —OH protons of an alcohol is obliterated by chemical exchange (Sec. 13.5D). To see this more clearly, let us examine the effects of chemical exchange on the spectrum of the —CH_3 protons of methanol. In absolutely dry methanol, the resonance of these protons is split by the —OH proton into a doublet. Recall (Fig. 13.8) that this splitting occurs because the adjacent —OH proton can have either of two spins. If acid or base is added to the methanol, causing the —OH protons to exchange rapidly, protons of different spins jump quickly on and off the —OH. Thus, the —CH_3 protons *on any one molecule* are next to an —OH proton with spin $+\frac{1}{2}$ about half of the time, and an —OH proton with spin $-\frac{1}{2}$ about half of the time. (The very small difference between the numbers of protons in the two spin states can be ignored.) In other words, the —CH_3 protons "see" an adjacent —OH proton with a spin that averages to zero over time. Since a proton is not split by an adjacent nucleus with zero spin, rapid exchange eliminates splitting of the —CH_3 protons. Similar reasoning can be applied to the spectrum of the methanol —OH proton, which, in a dry sample, is a quartet, but is a singlet in a sample containing traces of moisture.

Problems

27 Describe in detail what changes you would expect to see in the NMR spectrum of the methyl group as 1-chloro-1-methylcyclohexane is cooled from room temperature to very low temperature.

28 Suppose you were able to cool a sample of 1-bromo-1,1,2-trichloroethane enough that rotation about the carbon–carbon bond becomes slow on the NMR time scale. What changes in the NMR spectrum would you anticipate? Be explicit.

13.8 NMR OF NUCLEI OTHER THAN HYDROGEN. CARBON-13 NMR

Although we have concentrated our attention on proton NMR, any nucleus with a nuclear spin can be studied by NMR spectroscopy. Table 13.6 lists a few nuclei with spin $+\frac{1}{2}$ and some of their properties. Splitting in the NMR spectra of these nuclei conforms to the same $n + 1$ rule that we learned for the proton, although the sizes of the coupling constants vary with the nuclei involved. (We have already seen an example of this point in the splitting of protons by neighboring ^{19}F, Sec. 13.5C.)

In a given magnetic field, different nuclei have different absorption frequencies. This follows from the basic equation for NMR (Eq. 13.14), generalized to any nucleus n.

TABLE 13.6 Properties of Some Nuclei with Spin $\pm\frac{1}{2}$

Isotope	Relative sensitivity	Natural abundance, %	Magnetogyric ratio*
^{1}H	(1.00)	99.98	26,753
^{13}C	0.0159	1.108	6,728
^{19}F	0.834	100	25,179
^{31}P	0.0665	100	10,840

*In radians-gauss^{-1}-sec^{-1}

$$\nu_0 = \frac{\gamma_n}{2\pi}\mathbf{H}_n \tag{13.14}$$

In this equation, ν_0 is the operating frequency, $\mathbf{H}_n$ is the magnetic field experienced by the nucleus n of interest, and γ_n is the magnetogyric ratio for the nucleus n. An absorption occurs when ν_0 and $\mathbf{H}_n$ meet the condition of this equation. For a given magnetic field $\mathbf{H}_n$, the value of ν_0 required for absorption clearly depends on the magnetogyric ratio γ_n, which is different for each nucleus. For example, the γ for phosphorus-31 is considerably less than that for the proton; therefore, at a given magnetic field strength, the NMR of ^{31}P requires an operating frequency that is also much smaller than the one used for proton NMR.

Problem

29 Given a magnetic field strength of 46,973 gauss, what operating frequency should be used to observe the magnetic resonance of (a) protons (b) ^{13}C (c) ^{31}P?

When we measure the *proton* NMR spectrum of a molecule containing another nucleus with spin—for example, ^{19}F—we do *not* see the NMR signals of the fluorines because ^{19}F requires a different operating frequency for observation. However, we *do* see the splitting of protons by neighboring fluorines, as we learned in Sec. 13.5C. Conversely, when we take a ^{19}F NMR spectrum, we *do not* see the NMR signals of protons; however, we *do* see the splitting of the fluorines by neighboring protons.

There are two special problems associated with NMR experiments on nuclei other than protons. The first problem is instrumentation. Because each nucleus requires a special operating frequency, instruments designed for the NMR of other nuclei must provide the proper operating frequency for each nucleus of interest. The second problem, that of detection, arises from the fact that *different nuclei give NMR signals with different intrinsic intensities.* It turns out that the relative signal intensities for nuclei with spin $\pm\frac{1}{2}$ are proportional to the cube of their magnetogyric ratios. This means that a proton gives a stronger NMR signal than ^{13}C by a factor of $(\gamma_H/\gamma_C)^3 = 62.9$; that is, ^{13}C NMR has $(1/62.9) = 0.0159$ times the sensitivity of proton NMR.

Aggravating the sensitivity problem is the problem of natural abundance. Some nuclei of great intrinsic interest to the organic chemist have very low natural abundance. This problem has particular focus in ^{13}C NMR. Since organic compounds are by definition carbon compounds, the NMR spectroscopy of carbon is very important. However, ^{12}C, the common isotope of carbon, has no nuclear spin, and therefore gives no NMR signal. If we wish to perform a carbon NMR experiment, either we have to synthesize molecules that are specifically enriched with (expensive) ^{13}C, or we have to detect somehow the tiny fraction (0.01108, or 1.11%) of ^{13}C present in a sample. And remember, even if ^{13}C were present in the same abundance as ^{1}H, it gives an NMR signal that is only about 1.5% as strong as a proton signal.

In spite of these difficulties, new instrumental methods developed in the past several years (Sec. 13.9) have allowed chemists to measure ^{13}C NMR spectra *at natural abundance* on a routine basis. As a result, carbon-13 NMR spectroscopy, generally called **carbon NMR** or **CMR,** has become very important to the organic chemist. Let us consider some of the basics of this important type of NMR spectroscopy.

Problem

30 (a) Verify the relative sensitivity numbers in Table 13.6 for ^{19}F and ^{31}P NMR.
(b) Tritium (^{3}H, the radioactive isotope of hydrogen) has a spin of $\pm\frac{1}{2}$ and a sensitivity relative to ^{1}H of 1.21. Calculate the magnetogyric ratio of tritium. Give at least two problems that are likely to be encountered in a tritium NMR experiment.

The principles of CMR and proton NMR are essentially the same. However, some points about CMR are unique and require special emphasis. First, coupling (splitting) between carbons is *not* generally observed. The reason for this is the low natural abundance of carbon-13. Recall that CMR measures the resonance of carbon-13, not the common isotope carbon-12. If the probability of finding a ^{13}C at a given carbon is 0.01108, then the probability of finding ^{13}C at any two carbons in the same molecule is $(0.01108)^2$, or 0.00012. In effect, this means that *two ^{13}C atoms almost never occur together within the same molecule.* (The two ^{13}C atoms would have to occur in the *same* molecule for coupling to be observed.) Of course, we can synthesize compounds that are isotopically enriched in ^{13}C, in which case the regular splitting rules apply (see Problem 59).

A second key point about CMR is that the range of chemical shifts is large compared to that in proton NMR. Typical carbon chemical shifts, shown in Table 13.7, cover a range of about 200 ppm. With a few exceptions, trends in carbon chemical shifts parallel those for proton chemical shifts; but the wider range of carbon chemical shifts means that chemical shifts in CMR are very sensitive to small changes in chemical environment. As a result we can often observe distinct signals from two carbons in very similar chemical environments.

A third point about CMR has to do with the splitting of carbon-13 signals by protons. It happens that ^{13}C—^{1}H splitting is large; typical coupling constants are 120–200 Hz for directly attached protons. Furthermore, carbon NMR signals are also split by protons further away. Although such splitting can sometimes be useful, more typically it presents a serious complication in the interpretation of CMR spectra. In the majority of CMR work, splitting is eliminated by a special technique called *proton spin decoupling*. Spectra in which proton coupling has been eliminated are called **proton-decoupled carbon NMR spectra.** In these spectra we observe a *single unsplit line* for each set of chemically equivalent carbon atoms.

These points are illustrated by the proton-decoupled CMR spectrum of 1-chlorobutane, shown in Fig. 13.22. (Contrast the simplicity of this spectrum with the complexity of the proton spectrum in Fig. 13.13.) The carbon spectrum consists of four single lines, one for each carbon of the molecule. As we can see from the assignment of the lines in Fig. 13.22, carbon chemical shifts, like proton chemical shifts, decrease

TABLE 13.7 Typical CMR Chemical Shift Ranges for Common Functional Groups

Functional group	δ, ppm*	Functional group	δ, ppm*
$-CH_3$, $-CH_2-$, $-\overset{|}{C}H-$, $-\overset{|}{\underset{|}{C}}-$	0–55	$-\overset{|}{\underset{|}{C}}-O-$	40–80
		$-\overset{|}{\underset{|}{C}}-\overset{|}{N}-$	25–70
$>C=C<$	100–145	$-\overset{O}{\overset{\|}{C}}-$	190–220
$-C\equiv C-$	70–85		
$-\overset{|}{\underset{|}{C}}-Cl$	20–60	$-\overset{O}{\overset{\|}{C}}-O-$	160–190
$-\overset{|}{\underset{|}{C}}-Br$	10–50	$-\overset{O}{\overset{\|}{C}}-N<$	160–180
$-\overset{|}{\underset{|}{C}}-I$	−15–25	$-C\equiv N:$	110–130

*Downfield from TMS.

with distance from the electronegative chlorine. Notice that each carbon gives a distinguishable signal—even the two carbons farthest removed from the chlorine. This type of spectrum is particularly useful in differentiating compounds on the basis of their molecular symmetry. For example, the proton-decoupled CMR spectrum of 1-chloropentane has five lines because there are five chemically nonequivalent carbon atoms. The spectrum of 3-chloropentane only has three lines.

$$\underset{a}{CH_3}-\underset{b}{CH_2}-\underset{c}{CH_2}-\underset{d}{CH_2}-\underset{e}{CH_2}-Cl$$

five lines in CMR

$$\underset{a}{CH_3}-\underset{b}{CH_2}-\underset{c}{\overset{\overset{Cl}{|}}{C}H}-\underset{b}{CH_2}-\underset{a}{CH_3}$$

three lines in CMR

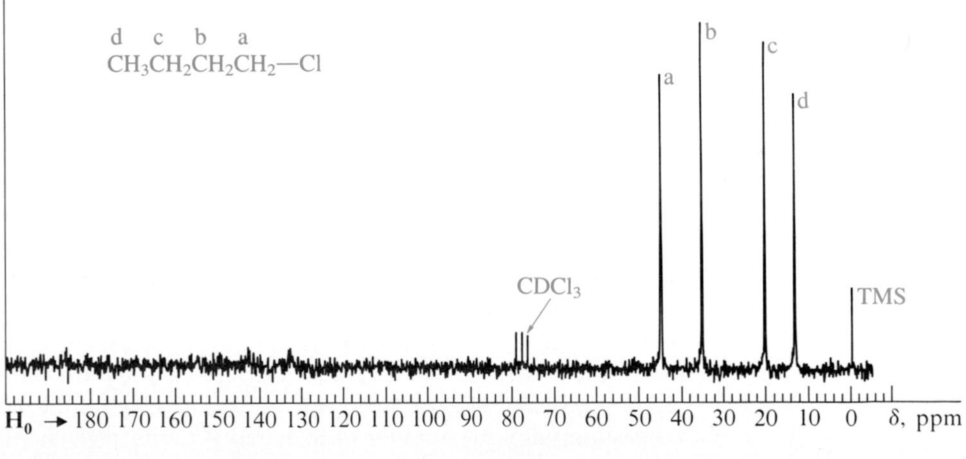

Figure 13.22 Proton-decoupled carbon NMR spectrum of 1-chlorobutane. (The reason for the $CDCl_3$ triplet is considered in Problem 50d.)

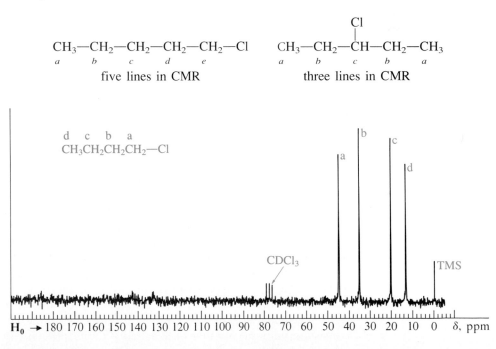

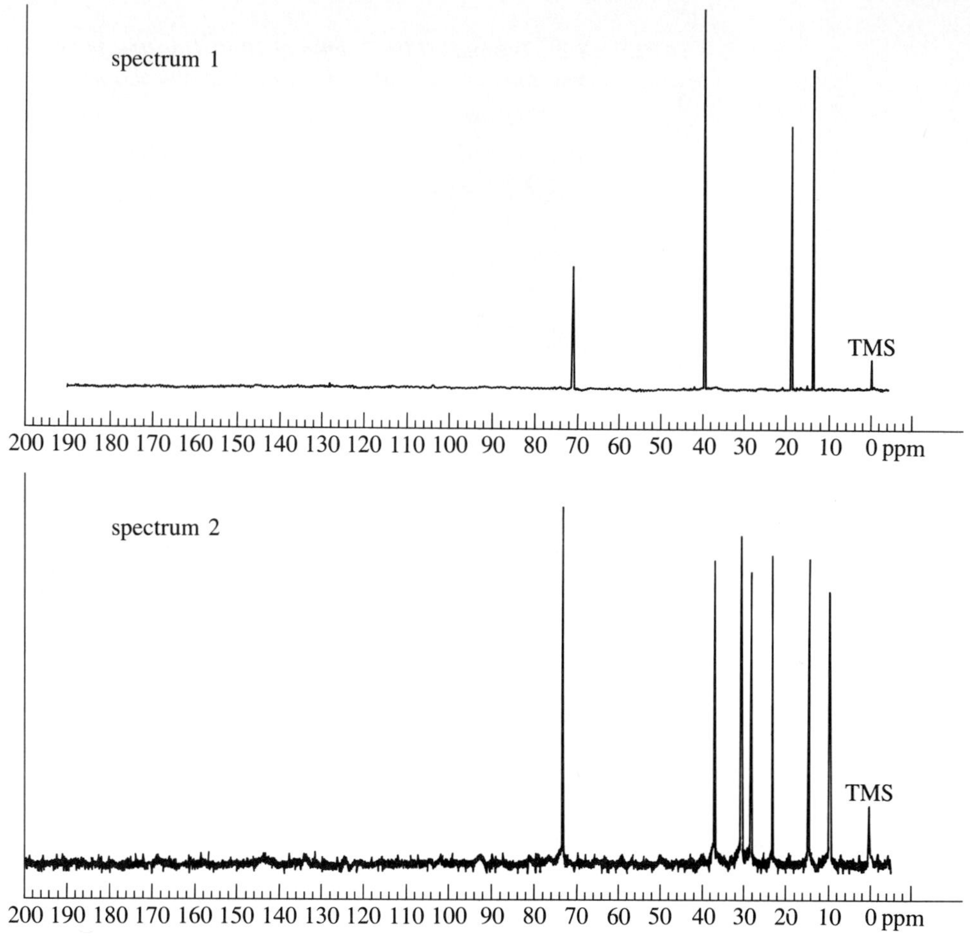

Figure 13.23
Proton-decoupled carbon-13 NMR spectra for Problem 31.

CMR spectra are generally not integrated, because the special technique used for taking the spectra (Sec. 13.10) gives relative peak integrals that are generally not accurate. Factors other than the number of carbons control the size of the CMR signal. For example, the decoupling technique enhances the peaks of carbons that bear hydrogens; hence, peaks for carbons that bear *no* hydrogens are usually smaller than those for other carbons.

Problem

31 The isomers 3-heptanol and 4-heptanol are difficult to distinguish by either their IR or proton NMR spectra. The proton-decoupled ^{13}C-NMR spectra of these compounds are given in Fig. 13.23. Indicate which compound goes with each spectrum, and explain your reasoning.

A special instrumental technique known as *off-resonance proton spin decoupling* can be used to decrease all ^{13}C—H coupling constants to the point that only the coupling of *directly attached protons* is observed. In such **off-resonance decoupled spectra** we can readily distinguish primary, secondary, tertiary, and quaternary carbons. For example, primary carbons (—CH_3) appear as quartets; secondary carbons (—CH_2—) as triplets; and so forth (n + 1 rule). Figure 13.24 shows the off-resonance decoupled CMR spectrum of 1-chlorobutane; compare this spectrum to the proton-decoupled CMR spectrum in Fig. 13.22. Notice that the splitting of each signal is consistent with the number of attached protons at each carbon.

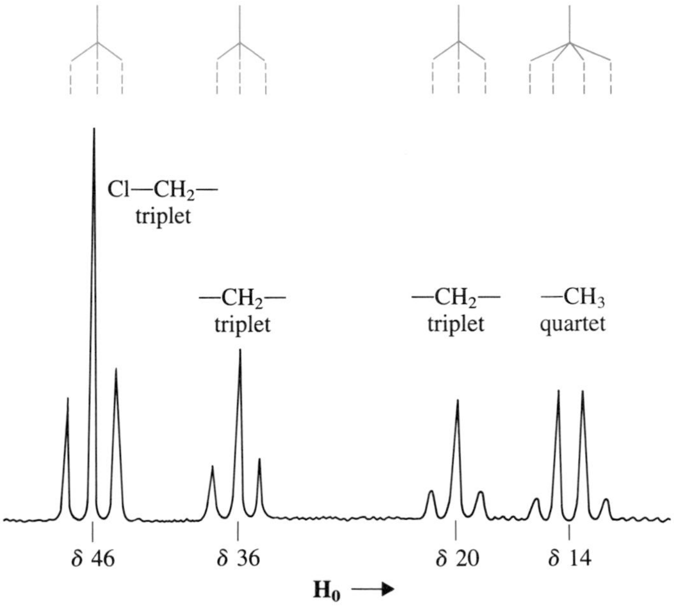

Figure 13.24 Off-resonance decoupled carbon-13 NMR spectrum of 1-chlorobutane. The splitting diagrams over the spectrum show the chemical shifts of each singlet in Fig. 13.22.

Problem

32 Indicate how you would use off-resonance decoupled CMR to distinguish between 1-chlorobutane and 2-chlorobutane.

13.9 SOLVING STRUCTURE PROBLEMS WITH SPECTROSCOPY

Let us now put what we know about IR, NMR, and mass spectrometry together and solve some problems that use more than one of these techniques at the same time. Two such problems are presented and solved below; as usual, you will gain most from them if you will attempt to work them before reading the solutions. Other problems of this type are given at the end of the chapter.

Problems

33 Propose a structure for the compound with the IR, NMR, and mass spectra given in Fig. 13.25.

Solution: The mass spectrum of this compound shows a *doublet* at $m/e = 90$ and 92, with the latter peak about one-third the size of the former. This pattern is very characteristic of a compound that contains chlorine. Furthermore, the base peak at $m/e = 55$ corresponds to a loss of Cl (35 and 37 mass units, respectively). Let us adopt the hypothesis that this is a chlorine-containing compound with molecular weight of 90 (for the chlorine-35 isotope). Next we turn to the IR spectrum. The peak at 1642 cm^{-1} suggests a C=C stretch, and, in the NMR spectrum, we see that there is a complex signal in the vinylic proton region. Evidently this compound is a chlorine-containing alkene. If the molecular weight is indeed 90, then chlorine and two alkene carbons account for 59 mass units; only 31 mass units remain to be accounted for.

In the NMR spectrum, the total integral is 43 spaces; the vinylic protons account for 17.5 spaces, or 40.7% of the integral. The doublet at δ 1.6 accounts for 19.5 integral spaces, or 45.3% of the total integral. The quintet at δ 4.4 ac-

Figure 13.25 Spectra for Problem 33.

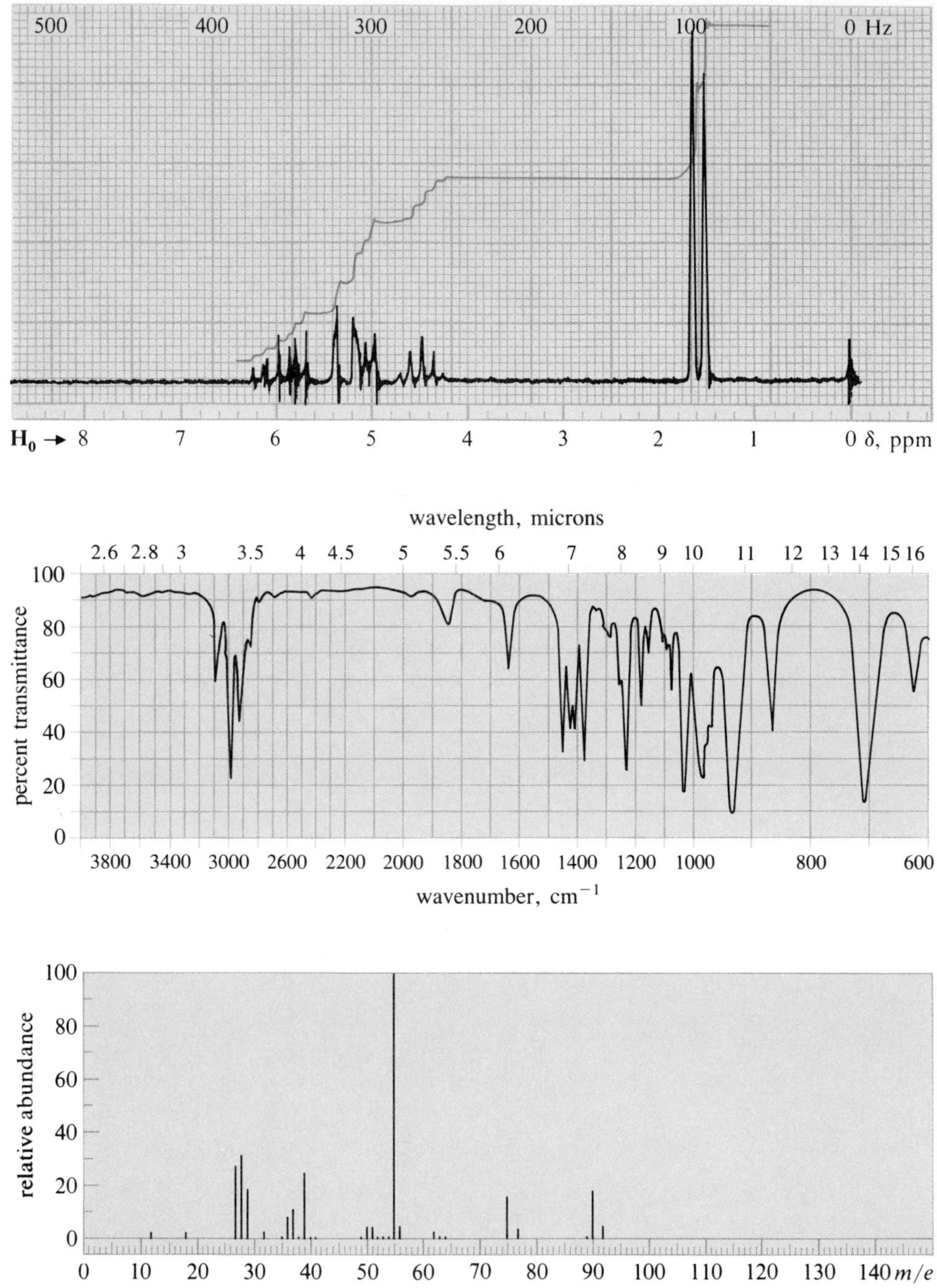

counts for the remaining 6 integral spaces, or 14% of the integral. The integrals of the three sets of signals are (in order from lowest field) in the ratio 17.5 : 6 : 19.5, or about 3 : 1 : 3, to nearest whole numbers. The integral suggests that there is some multiple of seven protons. If there are in fact seven protons (7 mass units) and one chlorine (35 mass units), then the remaining 48 mass units can be accounted for by four carbons. A possible molecular formula is then C_4H_7Cl. (Would fourteen protons be a likely possibility? Why or why not?)

The unsaturation number for this formula is one. Hence, there can be only one double bond in the molecule. Since the NMR integral suggests that there are three vinyl protons, then the molecule must contain a $-CH=CH_2$ group. In the IR spectrum the peaks at 930 and 990 cm^{-1} are consistent with such a group, although the former peak is at somewhat higher wavenumber than usual for this type of alkene. The three-proton doublet at δ 1.6 suggests a methyl group adjacent to a CH group.

$$CH_3-CH-$$

The δ 4.45 absorption accounts for one proton, and its coupling constant matches that of the absorption at δ 1.6. The splitting and chemical shift of the δ 4.45 absorption fit the partial structure

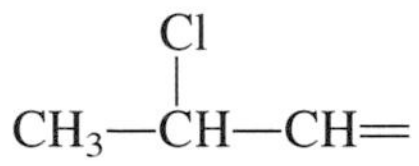

Given the molecular weight and the fact that there are three vinyl protons, the only possible complete structure is

$$\begin{array}{c} Cl \\ | \\ CH_3-CH-CH=CH_2 \end{array}$$

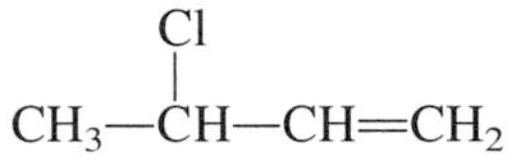

34 A compound $C_8H_{18}O_2$ with a strong, broad infrared absorption at 3293 cm^{-1} has the following NMR spectrum:

$$\delta \ 1.22 \ (12H, \ s); \ \delta \ 1.57 \ (4H, \ s); \ \delta \ 1.96 \ (2H, \ s)$$

(The resonance at δ 1.96 exchanges in D_2O; that is, it disappears when the sample is shaken with D_2O.) The proton-decoupled CMR of this compound consists of three lines, and the one at lowest field is considerably smaller than the others. Identify the compound.

Solution: The IR spectrum clearly indicates an alcohol, and the exchangeable NMR peak at δ 1.96 provides confirmation. Since the proton NMR spectrum has no peaks in the δ 3–4 region, the alcohol must be a tertiary alcohol. Since the molecular formula indicates *two* oxygens, there must either be two $-OH$ groups or an $-OH$ and an ether. The proton NMR also shows that there are only three chemically different sets of protons, and the CMR indicates that there are only three sets of chemically different carbons, two of which must therefore be α-carbons of tertiary alcohol groups (why?). The *only* structure that fits these data is

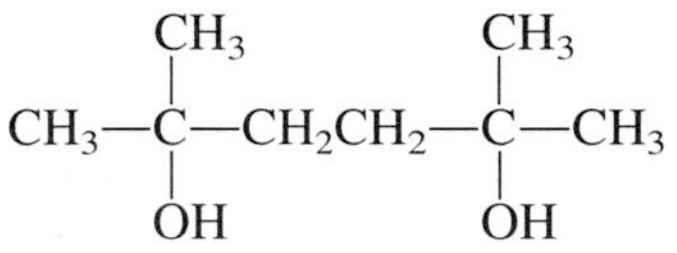

Notice that both NMR spectra require a molecule with a very high degree of symmetry.

Problems (Cont.)

35 Tell why the structures of each of the following compounds is not consistent with the data in Problem 33:
(a) *trans*-1-chloro-2-butene (b) 2-chloro-1-butene

36 Tell why each of the following structures is not consistent with the spectroscopic data in Problem 34.

(a)
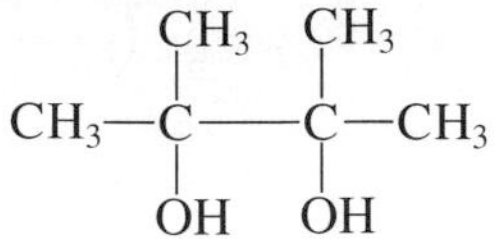

(b)
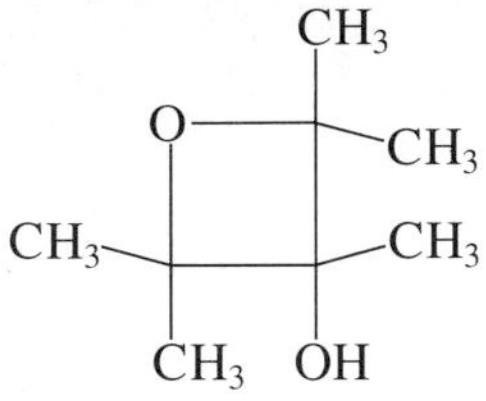

13.10 NMR INSTRUMENTATION. NEWER USES OF NMR

The basic components of an NMR instrument are shown in Fig. 13.26. The *magnetic field,* provided by an electromagnet, is required (Sec. 13.1) so that energy can be absorbed by the *sample* from the radio-frequency (rf) energy source, the *rf oscillator.* The *sweep coils* are used to vary the magnetic field across the range of chemical shift. The absorption of rf energy is detected by the *receiver coil,* amplified, and presented as a spectrum on a *recorder.*

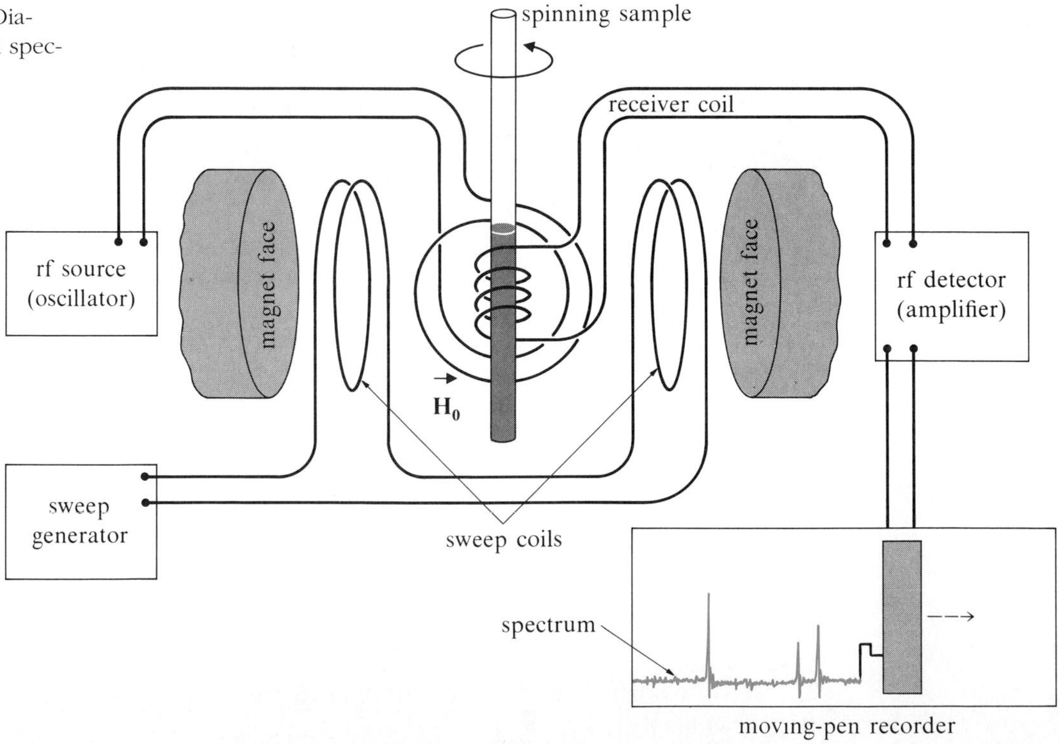

Figure 13.26 Diagram of an NMR spectrometer.

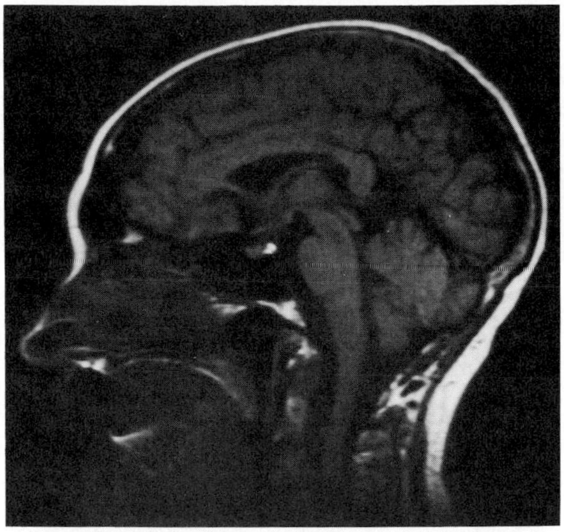

Figure 13.27 Magnetic-resonance imaging can show the details of soft tissue not visible in X-ray images, as in this brain scan of a 69-year old female.

The difficulty of detecting NMR signals from either very dilute samples or samples of rare isotopes like ^{13}C has been largely eliminated by a technique called **Fourier-transform NMR (FT-NMR).** In FT-NMR, an entire spectrum can be taken in less than a second. A large number of spectra (anywhere from 50 to 20,000, depending on the sample concentration and isotope) are recorded and stored in a computer. The computer mathematically adds all the spectra together. Since electronic noise is random, it sums to zero when averaged over many spectra, while the signals from the sample add together, and give a much stronger spectrum than could be obtained in a single experiment. The heart of FT-NMR is the method for rapid accumulation and storage of spectra, which we cannot discuss here. The cost of FT-NMR instruments has been reduced by the availability of relatively inexpensive dedicated small computers that are required for application of the FT-NMR technique.

The utility of NMR for structure determination should by now be fairly obvious. However, NMR also has many other uses. *Solid-state NMR* is being used to study the properties of important solid substances as diverse as drugs, coal, and industrial polymers. Phosphorus-31 NMR is being used to study biological processes, in some cases using intact cells or even whole organisms. One of the most exciting clinical applications of NMR is *NMR tomography,* or *magnetic-resonance imaging* (MRI). By monitoring the proton magnetic resonance signals from water in various parts of the body, clinical scientists can achieve organ imaging without using X-rays or other potentially harmful techniques. The brain image in Fig. 13.27 was obtained by this method.

KEY IDEAS IN CHAPTER 13

- The NMR spectrum measures the absorption of energy by nuclei from a radiofrequency (rf) source in the presence of a magnetic field.
- Only nuclei with spin give NMR spectra; protons have spin $\pm\frac{1}{2}$.

- The three elements of an NMR spectrum are the chemical shift, which provides information on the chemical environment of the observed nucleus; the integral, which tells the relative number of nuclei being observed; and the splitting, which gives information on the number of nuclei on adjacent atoms.
- The chemical shifts of protons in alkyl groups can be calculated with reasonable accuracy by Eq. 13.5 using the data in Table 13.2.
- Chemically equivalent nuclei have the same chemical shifts.
- The $n + 1$ splitting rule determines the splitting observed in most spectra. When a nucleus is split by more than one set of nonequivalent nuclei, the observed splitting is the result of successive applications of the splittings caused by each set.
- For nuclei of spin $\pm\frac{1}{2}$, such as the proton, the individual lines of a splitting pattern ideally have the relative intensities shown in Table 13.3. In practice, splitting patterns show leaning: they deviate from these intensities, and the deviation is greater as the chemical shifts of the mutually split protons are closer together.
- Splitting more complex than that predicted by the $n + 1$ rule is observed when the chemical shifts of two signals are separated by a frequency difference that is not much more than their mutual coupling constant. Because the chemical shift in Hz increases in proportion to the operating frequency, but coupling constants do not, many compounds that give complex (non-first-order) spectra at a lower frequency give simpler, first-order spectra at a higher frequency.
- Deuterium-NMR signals do not appear in the proton NMR spectrum. Thus, shaking the solution of an alcohol with D_2O removes the signal of the —OH proton because of its rapid exchange for deuterium.
- A time-averaged NMR spectrum is observed for species involved in rapid equilibria. It is possible to observe the individual species by retarding the reactions involved in the equilibria.
- The sensitivity of NMR spectroscopy depends on the natural abundance of the nucleus being observed and the cube of its magnetogyric ratio. Despite its low sensitivity, carbon-13 NMR, or CMR, is an important technique in organic chemistry.
- A proton-decoupled CMR spectrum is taken in such a way that carbon–proton couplings are removed; each chemically nonequivalent set of carbons appears as a single line. Carbon–proton couplings are partially restored in an off-resonance decoupled spectrum. Such a spectrum can be used to differentiate carbons on the basis of the number of attached protons.

ADDITIONAL PROBLEMS

(*Note*: in these problems, the term NMR refers to *proton* NMR unless otherwise indicated.)

37 What three pieces of information are available from an NMR spectrum? How is each used?

38 How would you distinguish among the compounds within each set below using their NMR spectra? Explain carefully and explicitly what features of the NMR spectrum you would use.
(a) cyclohexane and *trans*-2-hexene
(b) *trans*-3-hexene and 1-hexene
(c) 1,1,2,2-tetrabromoethane and 1,1,1,2-tetrabromoethane
(d) 1,1-dichlorohexane, 1,6-dichlorohexane, and 1,2-dichlorohexane
(e) *t*-butyl methyl ether and isopropyl methyl ether
(f) $Cl_3C—CH_2—CH_2—CHF_2$ and $CH_3—CH_2—CCl_2—CClF_2$
(g) *cis*-Cl—CH=CH—Br, *trans*-Cl—CH=CH—Br, and CH_2=CClBr

39 Answer each of the following questions as briefly as possible:
(a) What is the relationship between chemical shift in Hz and operating frequency ν_0?
(b) What is the relationship between coupling constant *J* and operating frequency?
(c) Why does the chemical shift in ppm not change with operating frequency?
(d) How does NMR spectroscopy differ conceptually from other forms of absorption spectroscopy?
(e) What condition must be met for an NMR spectrum to be first order?
(f) What happens physically when energy is absorbed in the NMR experiment?

40 Give the structure of
(a) a six-carbon hydrocarbon, not an alkene, whose proton NMR spectrum consists entirely of one singlet.
(b) a six-carbon alkene whose proton NMR spectrum consists of one singlet.
(c) a nine-carbon hydrocarbon whose proton NMR spectrum consists of two singlets.
(d) an eight-carbon ether whose proton NMR spectrum consists of one singlet and whose CMR spectrum consists of two singlets.
(e) a seven-carbon hydrocarbon whose proton NMR spectrum consists of two singlets at δ 0.23 and δ 1.21 (relative integral 1:6) and whose proton-decoupled CMR spectrum consists of three single lines.

41 Give the structure that corresponds to each of the following elemental analyses and NMR spectra:
(a) C_5H_{12}; δ 0.93, s
(b) C_5H_{10}; δ 1.5, s
The compounds in (c) and (d) can both be hydrogenated to give 2,2,4-trimethylpentane.
(c) C_8H_{16}: NMR spectrum in Fig. 13.28a
(d) C_8H_{16}: NMR spectrum in Fig. 13.28b
(e) $C_2H_2Br_2Cl_2$: δ 4.40, s
(f) $C_7H_{12}Cl_2$: δ 1.07 (s, 31); δ 2.28 (d, J = 6 Hz, 7.1); δ 5.77 (t, J = 6 Hz, 3.6). *Note*: The numbers in the spectrum of (f) are the actual values of the integral in chart spaces.
(g) $C_2H_2Br_2F_2$: δ 4.02 (t, J = 16 Hz)

Figure 13.28 NMR spectra for Problem 41.

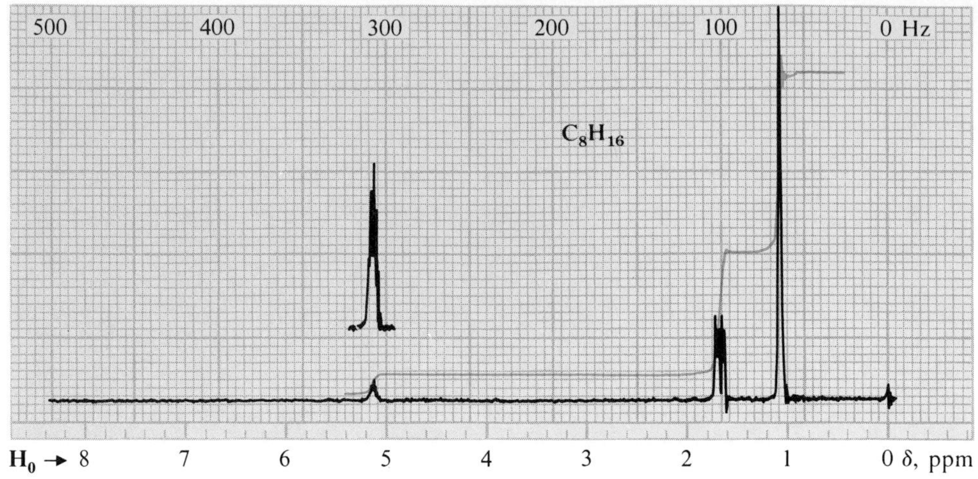

(a)

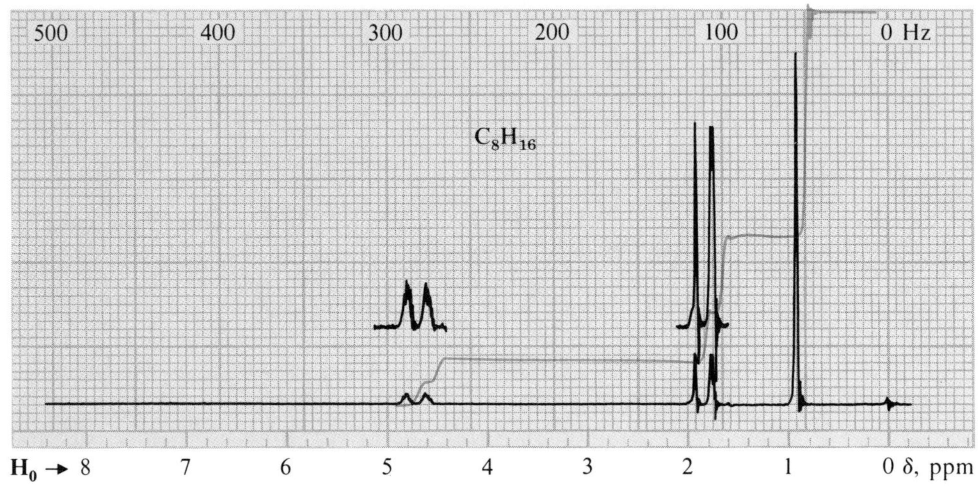

(b)

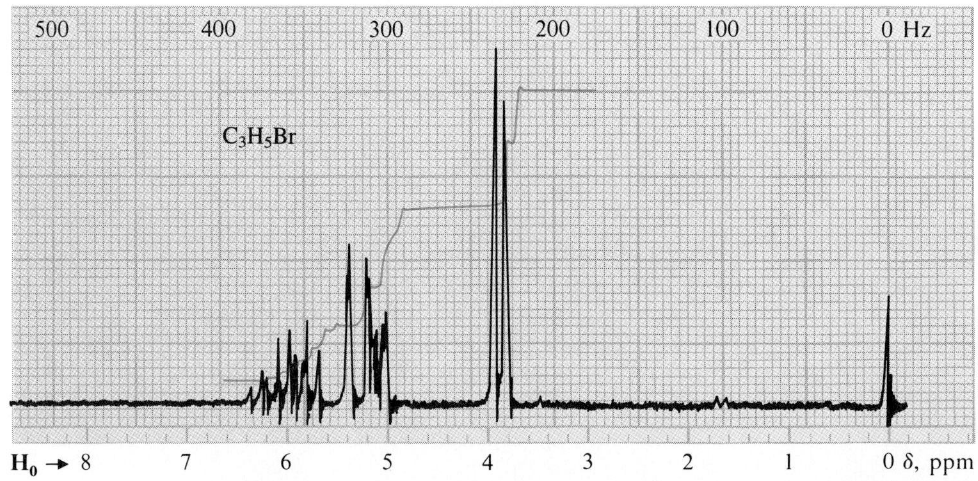

(c)

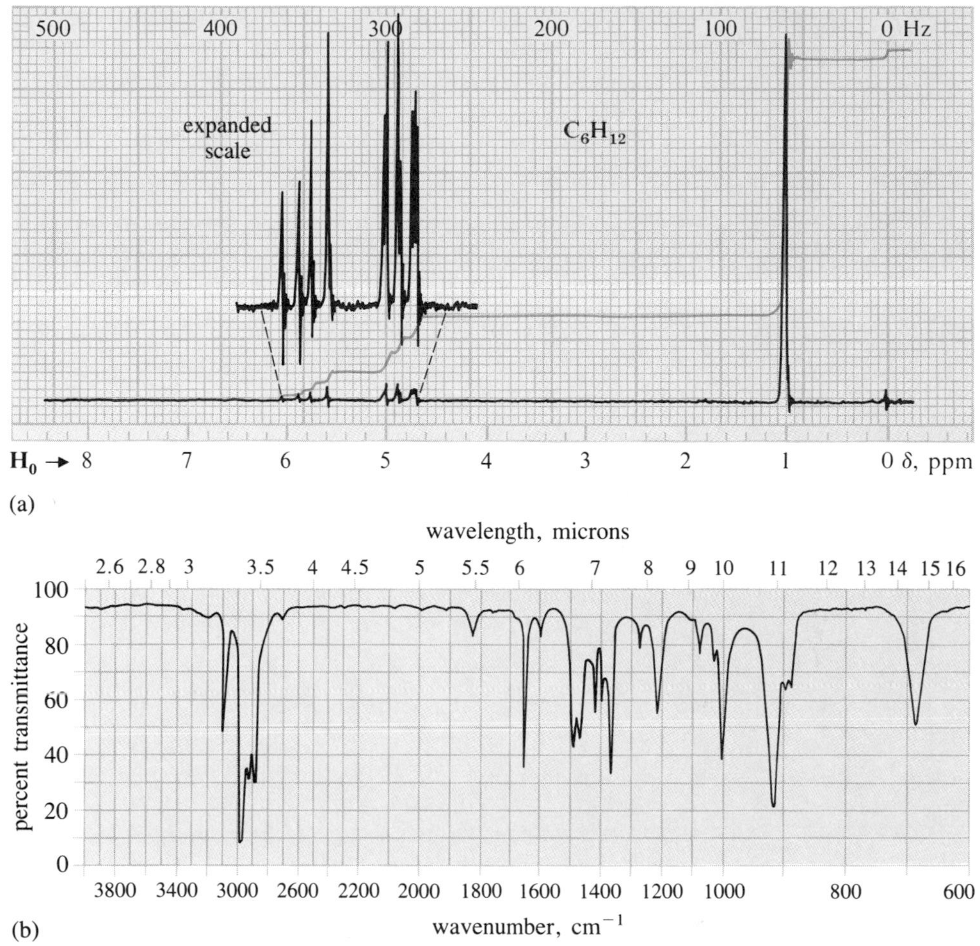

Figure 13.29 Spectra for compound *C* in Problem 42.

(h) C_3H_5Br: NMR spectrum in Fig. 13.28c
(i) $C_2H_4OCl_2$: δ 3.7 (3*H*, s); δ 7.33 (1*H*, s)
(j) $C_2H_2F_3I$: δ 3.56 (q, $J = 10$ Hz)
(k) $C_7H_{16}O_4$: δ 1.93 (t, $J = 6$ Hz); δ 3.35 (s); δ 4.49 (t, $J = 6$ Hz); relative integral 1 : 6 : 1

42 In an abandoned laboratory are found four bottles, *A, B, C,* and *D,* each labeled only "C_6H_{12}" and containing a colorless liquid. You have been called in as an expert to identify these compounds from their spectra:

Compound *A*:
NMR: δ 1.66 (s), decolorizes Br_2 in CCl_4; IR: no absorption in the range 1620–1700 cm^{-1}

Compound *B*:
NMR: δ 1.07 (6*H*, d, $J = 7$ Hz); δ 1.70 (3*H*, d, $J = 1.5$ Hz);
δ 2.20 (1*H*, sextet, $J = 7$ Hz); δ 4.60 (2*H*, d, $J = 1.5$ Hz)
IR: 1642, 891, 3080 cm^{-1}

Compound *C*:
NMR and IR spectra in Fig. 13.29

Compound *D*:
NMR: δ 1.40 (s); does not decolorize Br_2 in CCl_4.

Problems (Cont.)

43 At which one of the following operating frequencies is the spectrum of ethyl iodide likely to be more complex than predicted by the $n + 1$ splitting rule? Explain.
(a) 120 MHz (b) 60 MHz (c) 10 MHz

44 You wish to carry out the following reactions, and you have the NMR spectrum of each starting material. In each case explain how the NMR spectra of the product and starting material would be expected to differ.
(a) $(CH_3)_2C{=}CH_2 + HBr \longrightarrow (CH_3)_3CBr$
(b) $(CH_3)_2C{=}C(CH_3)_2 + HCl \longrightarrow (CH_3)_2CHC(Cl)(CH_3)_2$
(c) $(CH_3)_3C{-}OH + HCl \longrightarrow (CH_3)_3C{-}Cl$

45 When 3-methyl-1-butene reacts with HCl, two products are observed, a Markovnikov addition product and a rearrangement product. Assume that these two products have been separated. Explain what features you would look for in their NMR spectra to identify each of them.

46 A sample of 1-chlorobutane (NMR spectrum in Fig. 13.13) is believed to contain some *t*-butyl chloride. Although the *t*-butyl chloride singlet is visible in the spectrum at δ 1.63, it cannot be integrated independently because it is buried in the midst of the complex absorption in the 1-chlorobutane spectrum. The triplet of 1-chlorobutane centered at δ 3.55 integrates for 25 chart spaces. The *total* remaining signal integrates for 113 chart spaces. What mole percent of each alkyl halide is present in the mixture?

47 In the back of a reagent cabinet in an industrial laboratory, a compound has been found that has the following elemental analysis: %C, 87.42; %H, 12.58. The compound, whose NMR spectrum is shown in Fig. 13.30, reacts with H_2 over Pd/C to give methylcyclohexane. Al Keen, a chemist for the company, has declared that the compound must be either 1-methyl-1-cyclohexene or 3-methyl-1-cyclohexene. You have been called in as a consultant to help Keen decide between these two structures. Give reasons for your answer.

48 To which of the compounds below does the NMR spectrum shown in Fig. 13.31 belong? Explain your choice carefully.
(a) *cis*-3-hexene
(b) (*Z*)-1-ethoxy-1-butene
(c) 2-ethyl-1-butene
(d) Cl—CH(OCH$_2$CH$_3$)—CH(OCH$_2$CH$_3$)—Cl, i.e. $Cl{-}CH(O{-}CH_2CH_3){-}CH(O{-}CH_2CH_3){-}Cl$
(e) $Cl_2CH{-}CH(OCH_2CH_3)_2$

49 Explain how the proton NMR spectra of the compounds within each of the following sets would differ, if at all. (*Hint:* See Sec. 10.8.)

(Problem continues on page 543.)

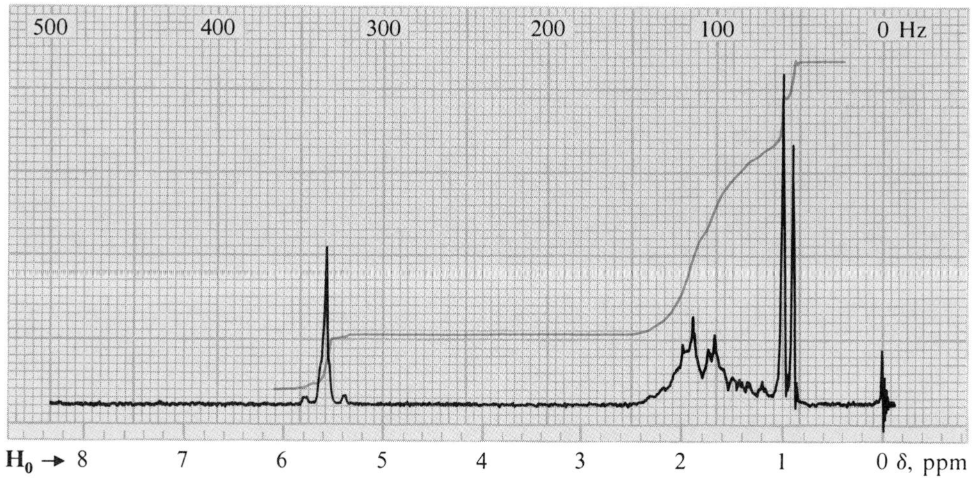

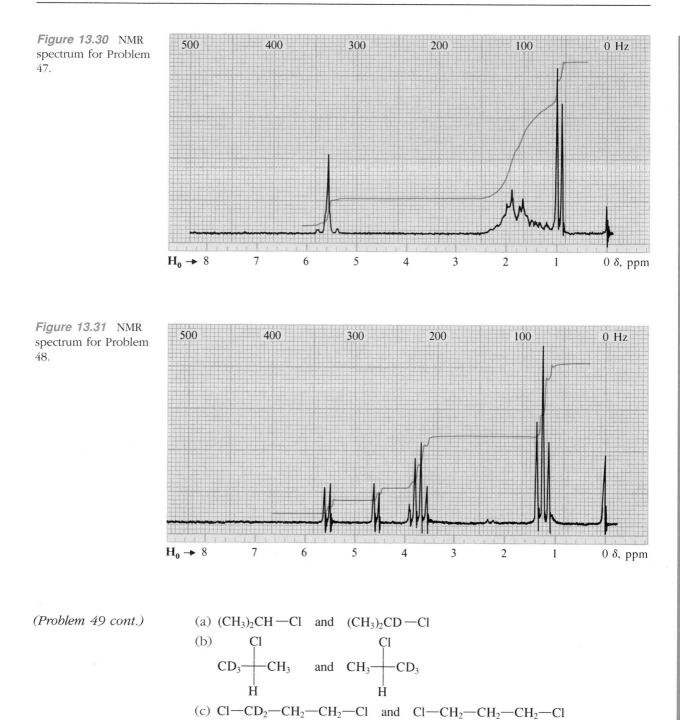

Figure 13.30 NMR spectrum for Problem 47.

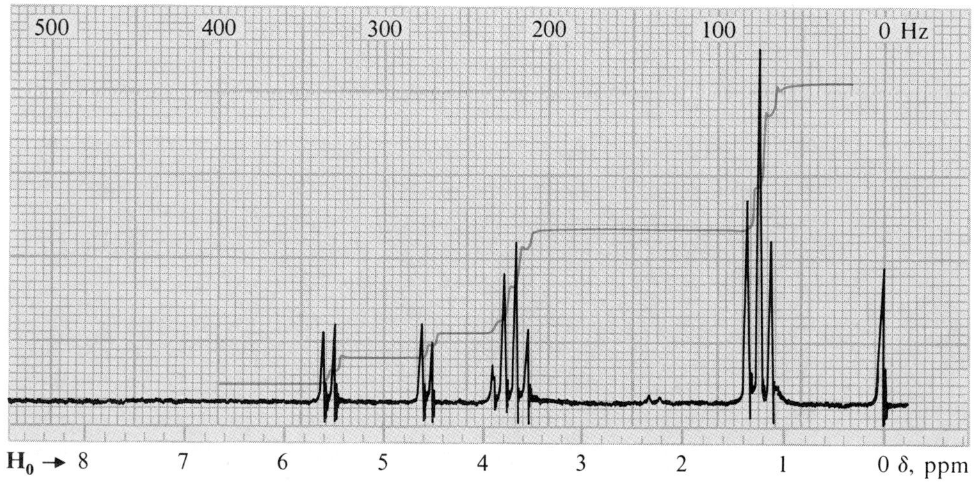

Figure 13.31 NMR spectrum for Problem 48.

(Problem 49 cont.)

(a) $(CH_3)_2CH—Cl$ and $(CH_3)_2CD—Cl$

(b)

$$CD_3{\overset{Cl}{\underset{H}{\rule{0pt}{0pt}}}}CH_3 \quad and \quad CH_3{\overset{Cl}{\underset{H}{\rule{0pt}{0pt}}}}CD_3$$

(c) $Cl—CD_2—CH_2—CH_2—Cl$ and $Cl—CH_2—CH_2—CH_2—Cl$

(d)

$$H{\overset{Cl}{\rule{0pt}{0pt}}}D \quad D{\overset{Cl}{\rule{0pt}{0pt}}}H \quad H{\overset{Cl}{\rule{0pt}{0pt}}}H$$
$$D{\underset{CH_3}{\rule{0pt}{0pt}}}Cl \quad D{\underset{CH_3}{\rule{0pt}{0pt}}}Cl \quad D{\underset{CH_3}{\rule{0pt}{0pt}}}Cl$$

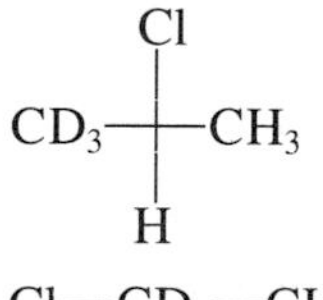

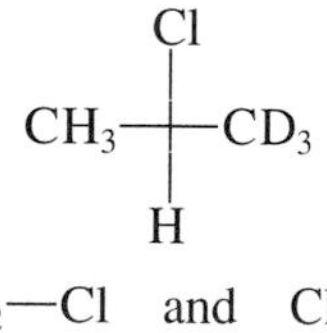

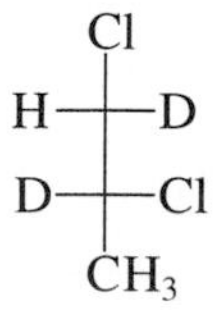

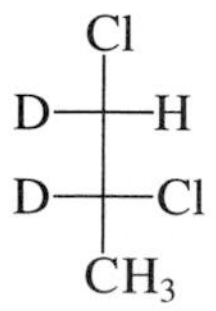

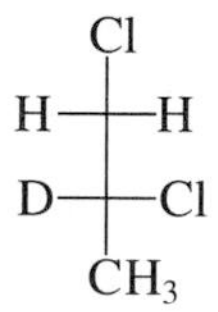

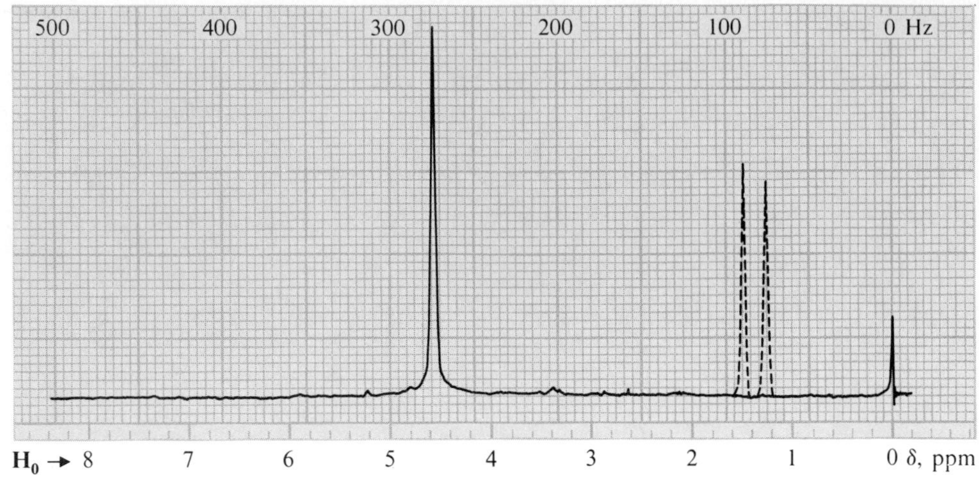

Figure 13.32 NMR spectrum for Problem 51. The dashed line is the spectrum of pure **$(CH_3)_3C—F$**; the solid line is the spectrum of the same compound after it reacts with the Lewis acid **SbF_5**.

50 Although we have been concerned with nuclei that have spin $\pm\frac{1}{2}$, several common nuclei such as nitrogen-14 and deuterium have a spin of 1. This means that the nuclear spin can be any of three values: +1, 0, and −1.

(a) How many lines would you expect to observe in the proton NMR of $^{+}NH_4$? What is the theoretical relative intensity of each line?

(b) The proton NMR spectrum of CHD_2—I is a quintet (five line pattern) with relative intensities 1 : 2 : 3 : 2 : 1. Explain why this pattern should be expected.

(c) How could you tell a sample of CH_2D—I apart from one of CHD_2—I by proton NMR? What other spectroscopic technique could be used for this determination?

(d) Explain why the carbon-13 NMR spectrum of $CDCl_3$ is a 1 : 1 : 1 triplet. (This triplet is sometimes observed in the spectra of compounds taken in $CDCl_3$ solvent; see, for example, Fig. 13.22.)

51 (a) When *t*-butyl fluoride (NMR spectrum in Fig. 13.32, dashed lines) reacts with the strong Lewis acid SbF_5, its NMR spectrum changes to the one shown with the solid lines. When the reaction mixture is added to ice water, *t*-butyl alcohol is formed in high yield. Explain these observations.

(b) Predict the spectrum of $(CH_3)_2CH$—F. What would you expect to observe in the NMR spectrum when this compound is treated as in (a)?

52 (a) Propose a structure for a compound $C_5H_{10}O$ that can be resolved into enantiomers and has the NMR and IR spectra shown in Fig. 13.33.

(b) Rationalize the base peak at $m/e = 45$ in the mass spectrum of this compound.

53 Review Sec. 10.8 if necessary, and identify the diastereotopic protons in the following molecules. Because diastereotopic protons are chemically nonequivalent, in many cases they have different chemical shifts. When diastereotopic protons are on the same carbon, they are split by each other as well as by adjacent protons. Assuming that the diastereotopic protons in the following molecules

Figure 13.33 Spectra for Problem 52. The NMR spectrum was taken at an operating frequency of 80 MHz.

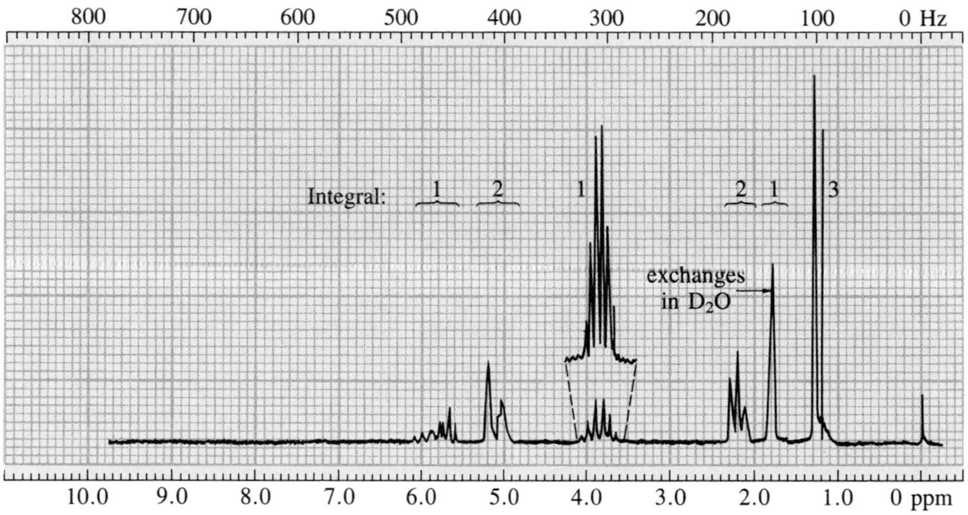

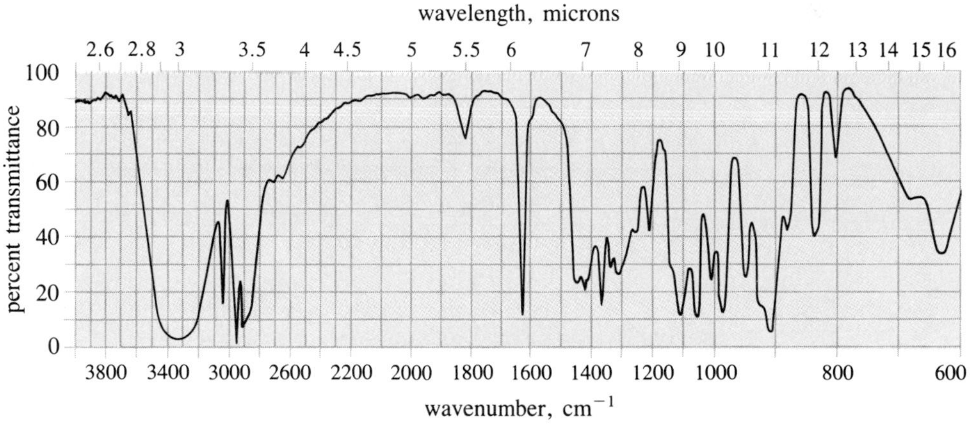

have different chemical shifts, indicate how many lines you would expect for these protons in the NMR spectrum.
(a) 4-methyl-1-penten-3-ol (b) 1,1-diethoxy-2-bromoethane

54 The proton-decoupled ^{13}C NMR spectrum of 4-methyl-2-pentanol below shows six lines; that of its isomer 4-methyl-1-pentanol shows five lines. Explain. (*Hint:* See the previous problem.)

55 In a dusty corner of a laboratory a bottle has been found containing a liquid *A*, which has attracted attention because of its penetrating, camphorlike odor. The IR, NMR, and mass spectra of *A* are given in Fig. 13.34; the proton-decoupled CMR spectrum of *A* consists of three lines at δ 8, δ 31, and δ 75. On treatment with concentrated H_3PO_4, compound *A* is converted into compound *B* that has the mass spectrum also shown in Fig. 13.34. Because of your widely known expertise in spectroscopic methods you have been asked to determine the structure of *A*.

Figure 13.34 Spectra for Problem 55.

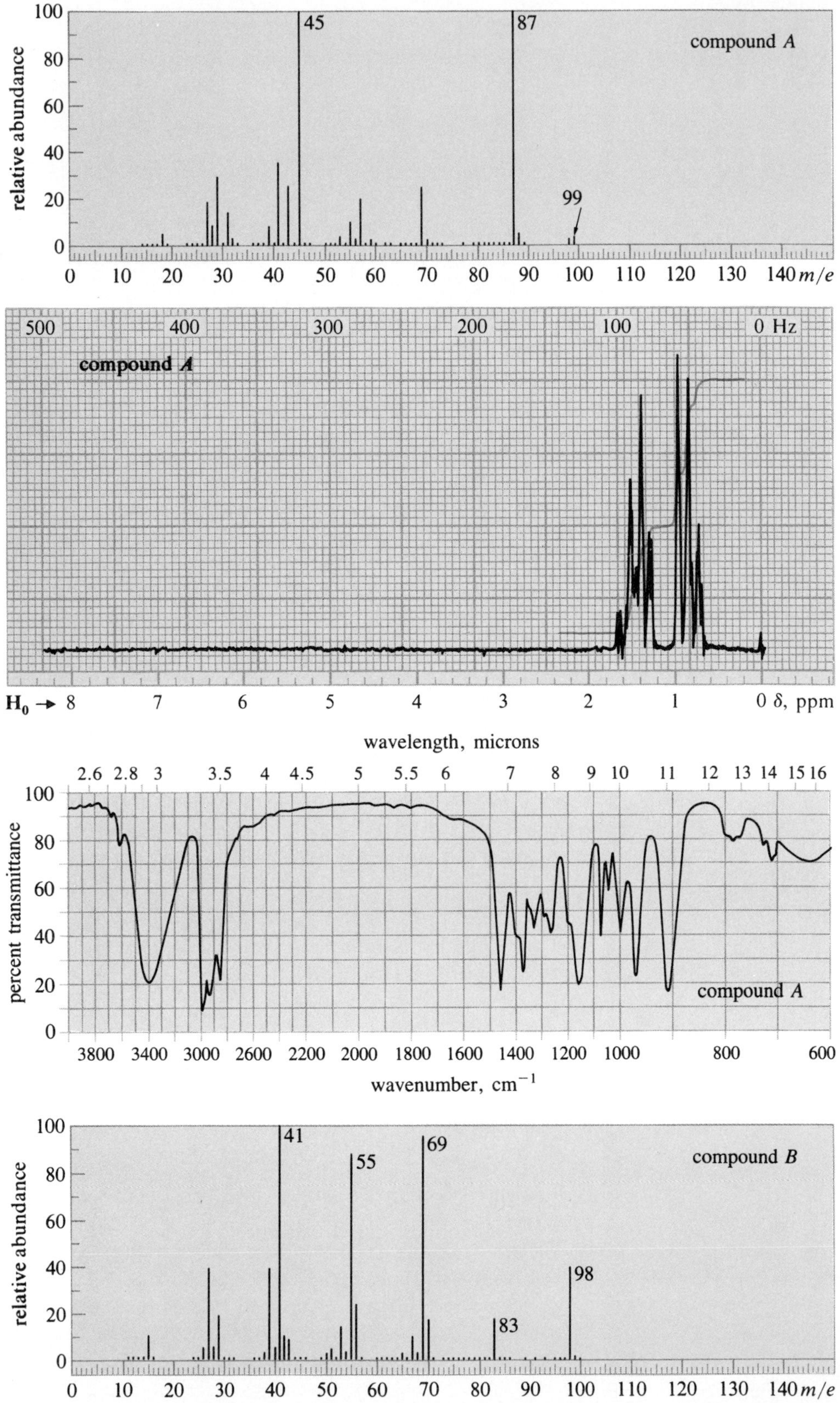

56 Propose a structure for the compound with the following spectra:
NMR: δ 1.28 (3H, t, J = 7 Hz); δ 3.91 (2H, q, J = 7 Hz); 5.0 (1H, d, J = 4 Hz); δ 6.49 (1H, d, J = 4 Hz)
IR: 3100, 1644 (strong), 1104, 1166, 694 cm^{-1} (strong); no IR absorptions in the range 700–1100 or above 3100 cm^{-1}
Mass spectrum: m/e = 152, 150 (equal intensity; double parent)

57 Oxygen-17 is a rare isotope that has a nuclear spin. The ^{17}O NMR of a small amount of water dissolved in CCl_4 is a triplet (intensity ratio 1 : 2 : 1). When water is dissolved in the strongly acidic HF-SbF_5 system (Problem 51), its ^{17}O NMR becomes a 1 : 3 : 3 : 1 quartet. Suggest a reason for these observations.

58 Indicate how the NMR spectrum of very dry 2-chloro-1-ethanol would differ from that of the same compound containing a trace of aqueous acid. Explain your answer.

59 Carbon–carbon splitting is not apparent in natural-abundance carbon-13 NMR spectra because of the rarity of the ^{13}C isotope. However, carbon–carbon splitting can be observed in compounds that are enriched in ^{13}C. A colleague, Buster Magnet, has just completed a synthesis of CH_3—CH_2—Br that contains 50% ^{13}C at each position. (What this really means is that some molecules contain no ^{13}C, some contain ^{13}C at one position, and some contain ^{13}C at both positions.) Buster does not know what to expect for the spectrum of this compound and has come to you for assistance. Describe the proton-decoupled ^{13}C NMR spectrum of this compound. (*Hint:* a spectrum of a mixture shows peaks for each compound in the mixture.)

60 **Electron spin resonance (ESR) spectroscopy,** a type of magnetic resonance spectroscopy, is used to study the unpaired electrons in free radicals. (ESR spectroscopy is to unpaired electrons as NMR spectroscopy is to protons.)
(a) The magnetogyric ratio of the electron is 658 times greater than that of the proton. What operating frequency in a field of 14,092 gauss would be required for the observation of the magnetic resonance of an electron? In what region of the electromagnetic spectrum does this frequency lie?
(b) Explain why the ESR spectrum of the unpaired electron in the methyl radical, $\cdot CH_3$, is a quartet of four lines in a 1 : 3 : 3 : 1 intensity ratio.

61 The 60-MHz proton NMR spectrum of 2,2,3,3-tetrachlorobutane is a sharp singlet at 25 °C, but at −45 °C is two singlets of different intensity separated by about 10 Hz.
(a) Explain the changes in the spectrum as a function of temperature.
(b) Explain why the two lines observed at low temperature have different intensities.

14 Chemistry of Alkynes

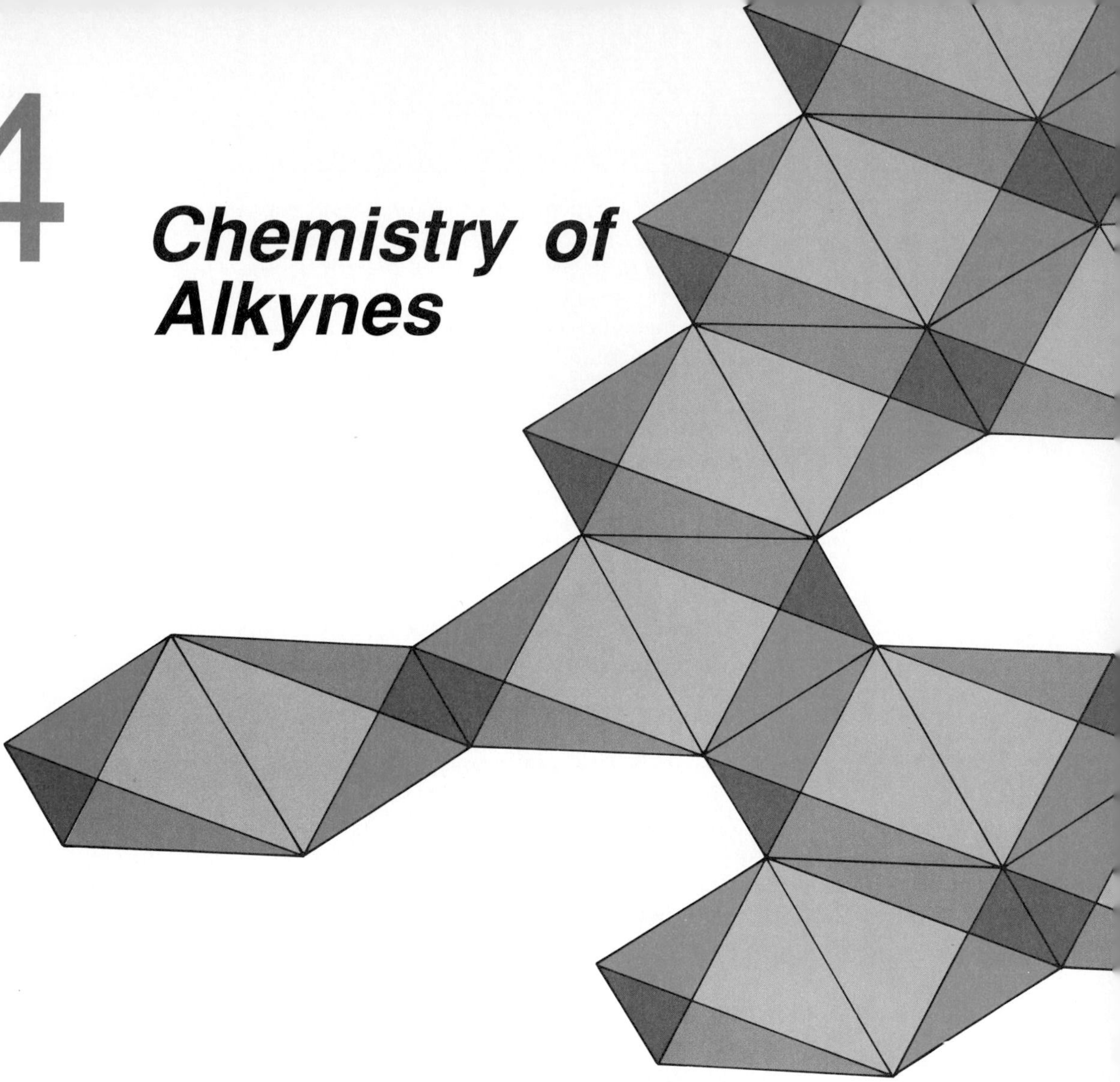

Alkynes, or **acetylenes,** are compounds with carbon–carbon triple bonds; the simplest member of this family is **acetylene,** H—C≡C—H. The chemistry of the carbon–carbon triple bond is similar in many respects to that of the carbon–carbon double bond. We shall find that alkynes and alkenes undergo many of the same addition reactions. Alkynes also have some unique chemistry, most of it associated with the bond between hydrogen and the triply bonded carbon, the ≡C—H bond.

With this chapter and the following two we return to an examination of hydrocarbon functional groups, a study that began with alkanes and alkenes in Chapters 3 and 4.

14.1 NOMENCLATURE OF ALKYNES

In common nomenclature, simple alkynes are named as derivatives of the parent compound acetylene.

$CH_3{-}C{\equiv}C{-}H$	methylacetylene
$CH_3{-}C{\equiv}C{-}CH_3$	dimethylacetylene
$C_2H_5{-}C{\equiv}C{-}CH_3$	ethylmethylacetylene

Compounds containing the **propargyl group,** $HC{\equiv}C{-}CH_2{-}$, are named as derivatives of this group in the common system. Notice that the propargyl group is the triple-bond analog of the allyl group.

$HC{\equiv}C{-}CH_2{-}Cl$	$CH_2{=}CH{-}CH_2{-}Cl$
propargyl chloride	allyl chloride

The systematic nomenclature of alkynes is very much like that of alkenes. The suffix *ane* in the name of the corresponding alkane is replaced by the suffix *yne,* and the triple bond is given the lowest possible number.

$CH_3{-}C{\equiv}C{-}H$	propyne
$CH_3{-}CH_2{-}C{\equiv}C{-}H$	1-butyne
$CH_3CH_2CH_2CH_2{-}C{\equiv}C{-}CH_3$	2-heptyne
$HC{\equiv}C{-}CH_2{-}CH_2{-}C{\equiv}C{-}CH_3$	1,5-heptadiyne
$CH_3{-}CH(CH_3){-}C{\equiv}C{-}CH_3$	4-methyl-2-pentyne

When double bonds and triple bonds are present in the same molecule, the principal chain is the carbon chain containing the greatest number of double and triple bonds. The numerical precedence of a double or triple bond within the principal chain is decided by the *first point of difference* rule (Sec. 3.2A, rule 5): precedence is given to the bond that gives the lowest number in the name of the compound, whether it is a double or triple bond. The double bond, however, is always *cited* first in the name by dropping the terminal *e* from the *ene* suffix, as in the following examples:

$\overset{1}{H}\overset{}{C}{\equiv}\overset{2}{C}{-}\overset{3}{C}H_2{-}\overset{4}{C}H{=}\overset{5}{C}H{-}\overset{6}{C}H_3$

possible names: 4-hexen-1-yne (correct) ← double bond cited first
2-hexen-5-yne (incorrect)

$\overset{1}{C}H_2{=}\overset{2}{C}H{-}\overset{3}{C}H_2{-}\overset{4}{C}{\equiv}\overset{5}{C}{-}\overset{6}{C}H_3$ 1-hexen-4-yne

$\overset{1}{C}H_2{=}\overset{2}{C}H{-}\overset{3}{C}H(CH{=}CH_2){-}\overset{4}{C}H{=}\overset{5}{C}H{-}\overset{6}{C}{\equiv}\overset{7}{C}{-}\overset{8}{C}H_3$ (the $CH{=}CH_2$ branch: vinyl group)

3-vinyl-1,4-octadien-6-yne (principal chain contains most double and triple bonds)

When the foregoing rules do not completely specify the name, the double bond takes precedence in numbering.

$CH_2{=}CH{-}CH_2{-}C{\equiv}CH$ 1-penten-4-yne

Substituent groups that contain a triple bond (called *alkynyl groups*) are named by replacing the final *e* in the name of the corresponding alkyne with the suffix *yl*. (This is exactly analogous to the nomenclature of substituent groups containing double bonds; see Sec. 4.2A.) The alkynyl group is numbered from its point of attachment to the main chain.

$HC \equiv C—$ ethynyl group (ethyne + yl)

$HC \equiv C—CH_2—$ 2-propynyl group

1-methyl-2-(2-propynyl)-1-cyclohexene

position of triple bond within the substituent

positions of substituents on the cyclohexane ring

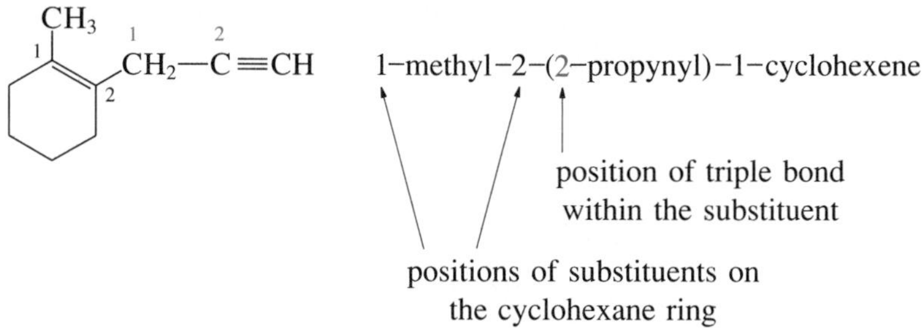

Groups that can be cited as principal groups are given numerical precedence over the triple bond. (See Appendix I.)

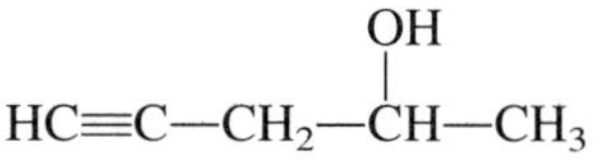

$HC \equiv C—CH_2—\overset{\underset{|}{OH}}{CH}—CH_3$ 4-pentyn-2-ol

Problems

1 Draw Lewis structures for the following alkynes:
(a) isopropylacetylene
(b) 4-methyl-1-pentyne
(c) cyclononyne
(d) 1-ethynyl-1-cyclohexanol
(e) 1,3-hexadiyne
(f) 3-pentoxy-5-decyne

2 Provide acceptable names for the following compounds. Use common nomenclature for (a).

(a) $CH_3CH_2CH_2CH_2C \equiv CCH_2CH_2CH_2CH_3$

(b)
$CH_2—CH—CH_2CH_2C \equiv CH$

(c)
$CH_3—\overset{\underset{|}{C}}{\underset{\underset{|}{CH_3}}{}}—C \equiv C—CH_3$ with OH

(d) $HC \equiv CCHCH_2CH_2CH_2CH_3$
$CH_2CH_2—CH=CH—CH_2OCH_3$

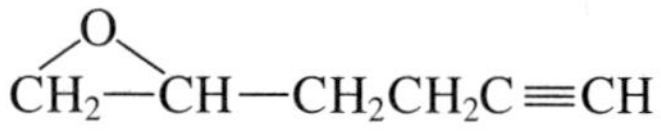

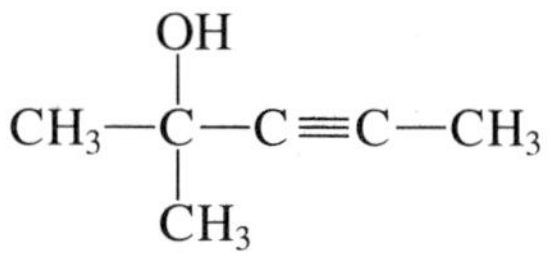

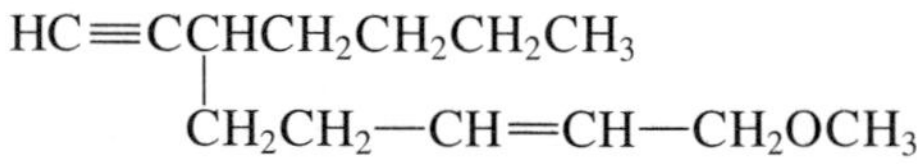

14.2 STRUCTURE AND BONDING IN ALKYNES

Since each carbon of acetylene is connected to two groups—a hydrogen and another carbon—our simple rules for predicting geometry (Sec. 1.6B) suggest that the

H—C≡C bond angle in acetylene should be 180°. Indeed, the acetylene molecule is *linear.*

Because of this linear shape, there is no *cis–trans* isomerism at a triple bond. Thus, although 2-butene can exist as *cis* and *trans* stereoisomers, there is only one isomer of 2-butyne. Another consequence of this linear geometry is that cycloalkynes smaller than cyclooctyne cannot be isolated under ordinary conditions (Problem 3).

We have studied two types of carbon hybridization so far: sp^3 (Sec. 2.6A) and sp^2 (Sec. 4.1A). The bonding in alkynes involves yet a third type of carbon hybridization. Imagine that the 2*s* orbital and *one* 2*p* orbital on carbon mix to form two new hybrid orbitals. Since these two new orbitals are each one part *s* and one part *p,* they are called ***sp* hybrid orbitals.**

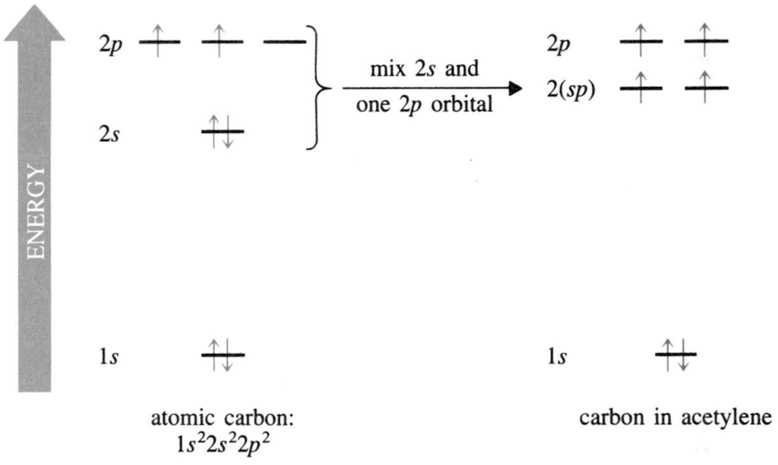

An *sp* orbital very much resembles an sp^2 or sp^3 orbital (Fig. 2.10a, 4.3a). However, electrons in an *sp* hybrid orbital are, on the average, somewhat closer to the carbon nucleus than they are in sp^2 or sp^3 hybrid orbitals. In other words, *sp* orbitals are more compact than sp^2 or sp^3 hybrid orbitals. An *sp*-hybridized carbon atom, shown in Fig. 14.1a, has two *sp* orbitals at a relative orientation of 180°. The two remaining unhybridized *p* orbitals lie along axes that are at right angles both to each other and to the *sp* orbitals.

The bonding in acetylene results from the combination of two *sp*-hybridized carbon atoms and two hydrogen atoms (Fig. 14.1b). One bond between the carbon atoms is a σ-bond resulting from the overlap of two *sp* hybrid orbitals. This bonding molecular orbital contains two electrons. Two π-bonds are formed by the side-to-side overlap of the *p* orbitals. These bonding π-molecular orbitals, like the *p* orbitals from which they are composed, are mutually perpendicular, and each contains two electrons. When the total electron density from the π-electrons is added mathematically, it is found to lie in a cylinder, or barrel, about the axis of the molecule (Fig. 14.2). The remaining *sp* orbital on each carbon overlaps with a hydrogen 1*s* orbital to form a carbon–hydrogen bond.

Figure 14.1 (a) An *sp*-hybridized carbon atom. (b) Bonding molecular orbitals in acetylene resulting from the combination of two *sp*-hybridized carbon atoms.

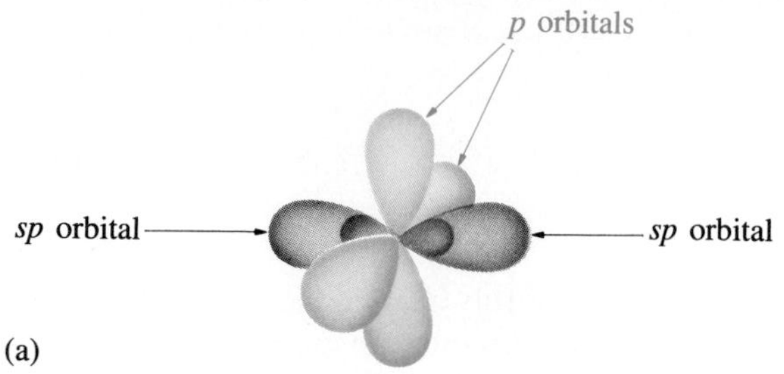

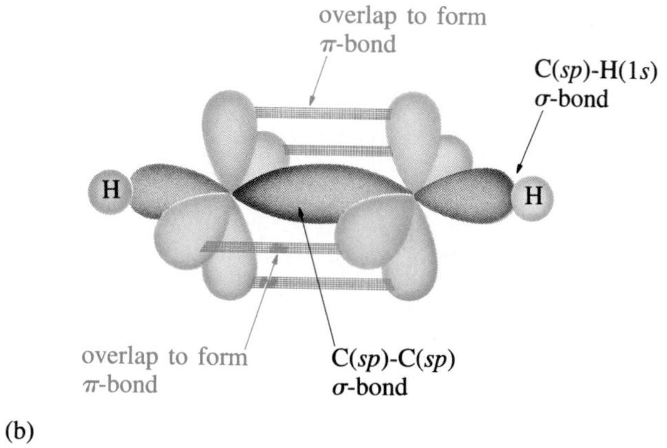

Figure 14.2 The π-electron density in acetylene lies in a cylinder about the axis of the molecule.

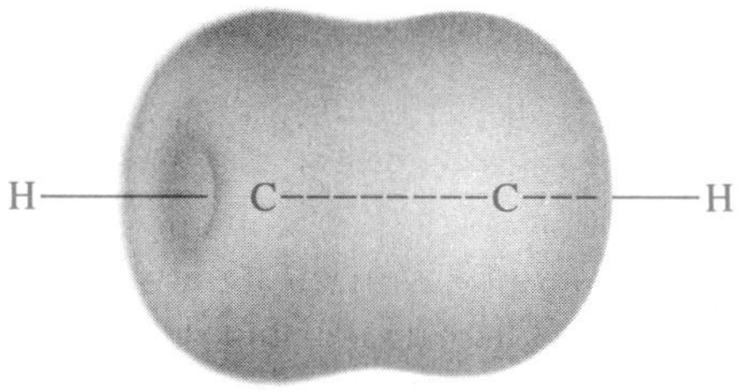

We already know (Sec. 1.6B) that carbon–carbon triple bonds are shorter than carbon–carbon double bonds or single bonds; the bond lengths are 1.20 Å, 1.33 Å, and 1.54 Å, respectively. The carbon–hydrogen bond lengths also decrease along the series: ethane (1.11 Å), ethylene (1.08 Å), and acetylene (1.06 Å). We can understand this trend from the hybridization of the carbon orbitals. The alkyne C—H bond is derived from the overlap of a carbon *sp* orbital with a hydrogen 1*s* orbital, whereas the alkane C—H bond is derived from the overlap of a carbon sp^3 orbital with a hydrogen 1*s* orbital. Since electrons in the *sp* orbital are closer to the positive carbon nucleus than those in an sp^3 orbital, it follows that the C—H bond of the alkyne should be shorter.

The following heats of formation show that alkynes are less stable than isomeric dienes:

	$H—C≡C—CH_2CH_2CH_3$	$CH_3—C≡C—CH_2CH_3$	$CH_2=CH—CH_2—CH=CH_2$
ΔH_f°	+34.50 kcal/mol	+30.80 kcal/mol	+25.42 kcal/mol

In other words, the *sp* hybridization state is inherently less stable than the sp^2 hybridization state, other things being equal. These heats of formation also show that a triple bond, like a double bond, prefers energetically to be on the interior of a carbon chain rather than at the end.

Problems

3 Try to build a model of cyclohexyne; explain why this compound is not stable.

4 Predict the relative order of carbon–carbon single bond lengths in propyne, propene, and propane. Explain.

14.3 PHYSICAL PROPERTIES OF ALKYNES

A. Boiling Points and Solubilities

The boiling points of most alkynes are not very different from those of analogous alkenes and alkanes.

	$HC≡C(CH_2)_3CH_3$	$CH_2=CH(CH_2)_3CH_3$	$CH_3—CH_2(CH_2)_3CH_3$
compound	1-hexyne	1-hexene	hexane
boiling point	71.3°	63.4°	68.7°
density	0.7155 g/mL	0.6731 g/mL	0.6603 g/mL

Like alkanes and alkenes, alkynes have much lower densities than water, and are also insoluble in water.

B. IR Spectroscopy of Alkynes

Many alkynes have a C≡C stretching absorption in the 2100–2200 cm^{-1} region of the infrared spectrum. This absorption is clearly evident, for example, at 2120 cm^{-1} in the IR spectrum of 1-octyne (Fig. 14.3). However, this absorption is absent in the IR spectra of many symmetrical, or nearly symmetrical, alkynes because of the dipole moment effect (Sec. 12.2E). For example, 4-octyne has no C≡C stretching absorption at all.

A very useful absorption of 1-alkynes is the ≡C—H stretching absorption, which occurs at about 3300 cm^{-1}. This absorption, also visible in the spectrum of 1-octyne (Fig. 14.3), is usually well-separated from other C—H absorptions. Of course, alkynes other than 1-alkynes lack the unique ≡C—H bond, and therefore do not show this absorption.

Problem

5 Use their infrared stretching absorptions to determine which bond is stronger: the alkane C—H bond, the alkene =C—H bond, or the alkyne ≡C—H bond. Explain. (You can check your answer in Table 5.2.)

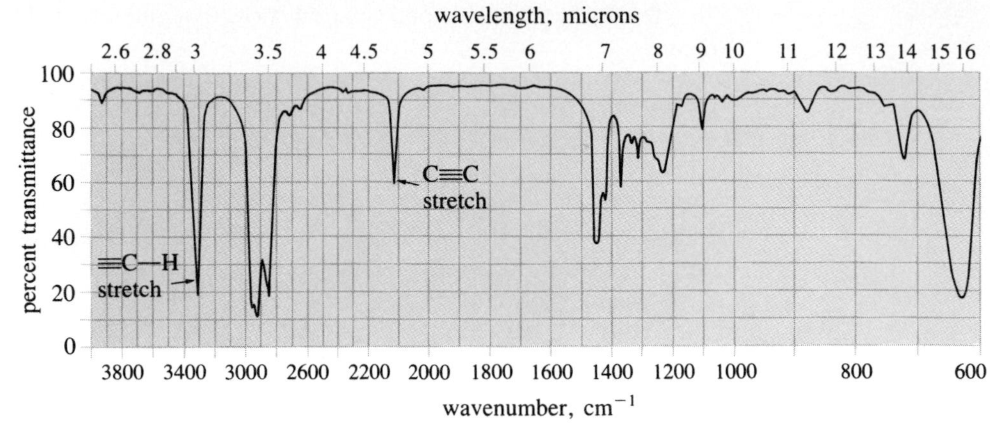

Figure 14.3 IR spectrum of 1-octyne. The two key absorptions shown are absent in the spectrum of 4-octyne.

C. NMR Spectroscopy of Alkynes

Typical chemical shifts observed in the NMR spectra of alkynes are contrasted below with the analogous shifts for alkenes.

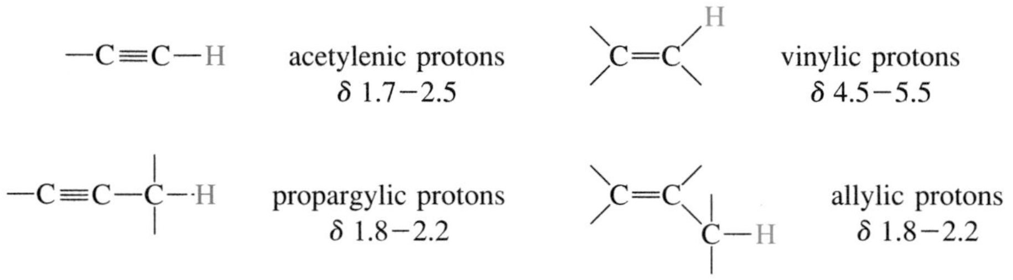

Although the chemical shifts of allylic and propargylic protons are very similar (as we might expect from the fact that both double and triple bonds involve π-electrons), the chemical shifts of acetylenic protons are much less than those of vinylic protons. What is the explanation for this difference?

The argument is closely related to the explanation for the chemical shifts of vinylic protons (Fig. 13.16), although the effect is in the opposite direction. If an alkyne is oriented in the applied field $\mathbf{H_0}$ as shown in Fig. 14.4, an induced electron circulation is set up in the cylinder of π-electrons that encircles the alkyne molecule. The resulting induced field $\mathbf{H_i}$ *opposes* the applied field along the axis of this cylinder. Since the acetylenic proton lies along this axis, the net field at this proton is reduced—that is, this proton is *shielded* by the induced field $\mathbf{H_i}$. Because of this shielding, a larger value of $\mathbf{H_0}$ is required for resonance. Consequently, acetylenic protons show NMR absorptions at higher field than they would if $\mathbf{H_i}$ were not present.

We might ask why there isn't a deshielding contribution like the one shown for alkenes in Fig. 13.16 if the alkyne is turned 90° from its orientation in Fig. 14.4. In fact, there *is* such a contribution, but it is weaker than the one described above. The chemical shift of any molecule represents an average of all orientations; the one shown in Fig. 14.4 is the most important contributor.

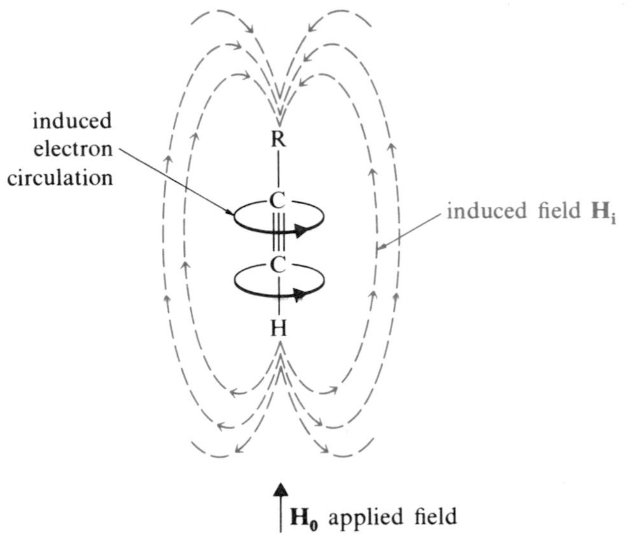

Figure 14.4 The acetylenic proton is shielded by the induced field $\mathbf{H_i}$ that results when the alkyne is oriented so that its axis is parallel to the applied field.

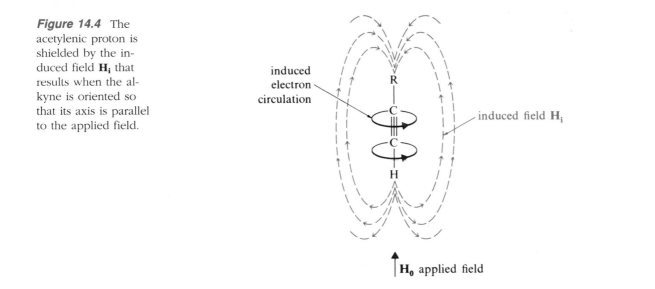

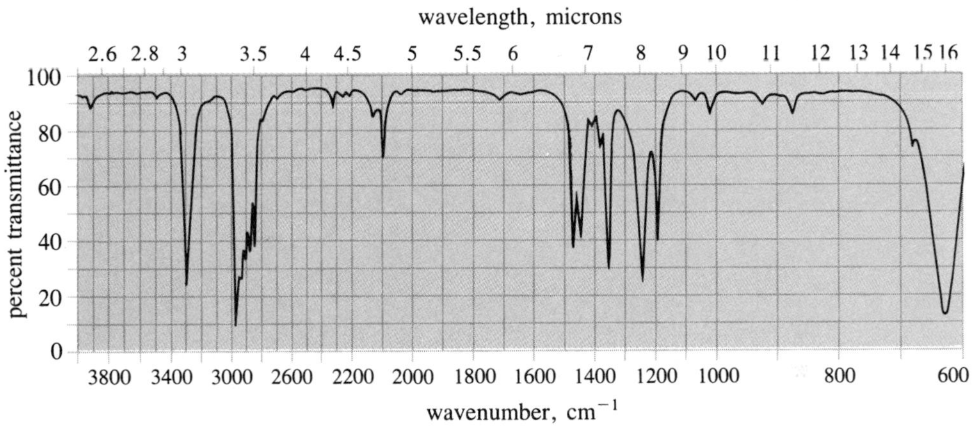

Figure 14.5 IR spectrum for Problem 7.

The absorptions of acetylenic protons lie in a very "busy" region of the NMR spectrum and in many cases are obscured by the absorptions of other protons (Problem 6). Fortunately the acetylenic proton is easily detected by its ≡C—H stretching absorption in the IR spectrum (Sec. 14.3B).

Problems

6 The NMR spectrum of propyne consists of a singlet at δ 1.8. Explain.

7 Identify the compound with a molecular weight of 82 that has the IR spectrum shown in Fig. 14.5 and the following NMR spectrum:
δ 1.90 (1*H*, s); δ 1.21 (9*H*, s)

14.4 ADDITIONS TO THE TRIPLE BOND: INTRODUCTION

In Chapters 4 and 5 we learned that electrophilic reagents and free radicals add to the π-bonds of alkenes. Alkynes undergo similar addition reactions at the carbon–carbon triple bond. For example, HCl and HBr add slowly to alkynes in accordance with the Markovnikov rule (Sec. 4.5A).

$$Ph—C\equiv C—H \;+\; HCl \;\xrightarrow{\;HOAc\;}\; Ph—\underset{\underset{Cl}{|}}{C}=CH_2 \tag{14.1}$$

phenylacetylene

α-chlorostyrene
(1-chloro-1-phenylethene)

In the presence of free-radical initiators such as peroxides, anti-Markovnikov addition of HBr is observed, as with alkenes (Sec. 5.8).

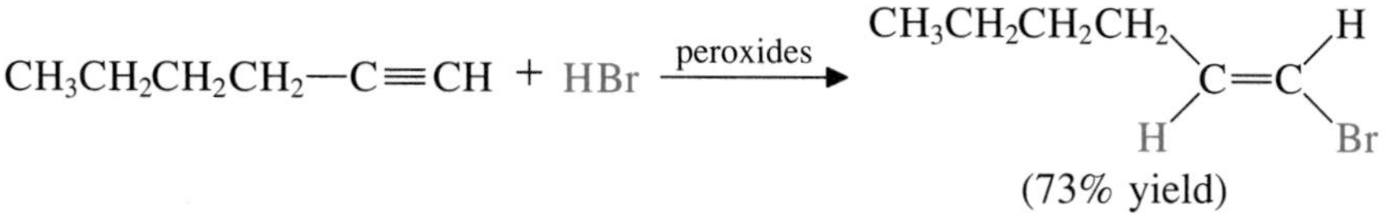

$$CH_3CH_2CH_2CH_2—C\equiv CH \;+\; HBr \;\xrightarrow{\;peroxides\;}\; \underset{H}{\overset{CH_3CH_2CH_2CH_2}{>}}C=C\underset{Br}{\overset{H}{<}} \tag{14.2}$$

(73% yield)

Certain complications arise in additions to the triple bond. The first is that the stereochemistry of addition, as with alkenes (Sec. 7.10), can be *syn* or *anti,* and can lead to *E* or *Z* adducts, respectively:

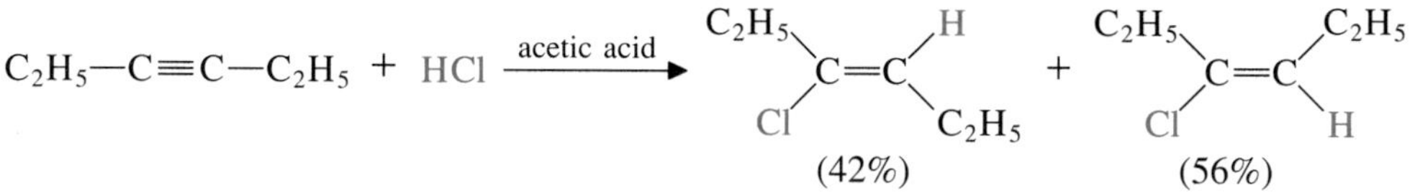

$$C_2H_5—C\equiv C—C_2H_5 \;+\; HCl \;\xrightarrow{\;acetic\ acid\;}\; \underset{Cl}{\overset{C_2H_5}{>}}C=C\underset{C_2H_5}{\overset{H}{<}} \;+\; \underset{Cl}{\overset{C_2H_5}{>}}C=C\underset{H}{\overset{C_2H_5}{<}} \tag{14.3}$$

(42%) (56%)

$$C_4H_9C\equiv CH \;+\; Br_2 \;\xrightarrow{\;acetic\ acid\;}\; \underset{Br}{\overset{C_4H_9}{>}}C=C\underset{Br}{\overset{H}{<}} \;+\; \underset{Br}{\overset{C_4H_9}{>}}C=C\underset{H}{\overset{Br}{<}} \tag{14.4}$$

(28%) (72%)

(80% yield)

As these examples illustrate, the addition of hydrogen halides or halogens to alkynes in some cases gives a mixture of stereoisomers.

A second complication that arises in alkyne chemistry has no counterpart in alkene chemistry: since an alkene is the product of a single addition to an alkyne, and since alkenes themselves undergo addition reactions, a second addition can, in principle, follow the first.

$$CH_3C\equiv C—H \;+\; HCl \;\xrightarrow{\;-70°\;}\; CH_3—\underset{\underset{Cl}{|}}{C}=CH_2 \;+\; CH_3—CCl_2—CH_3 \tag{14.5}$$

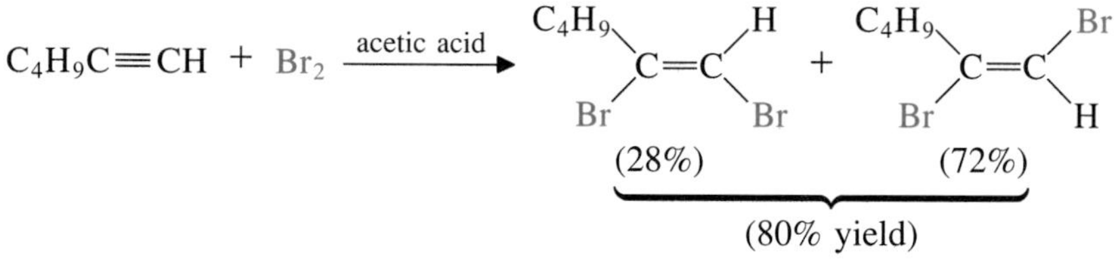

Both of these problems are difficult to avoid in additions of hydrogen halides and halogens to alkynes. For this and other reasons, these alkyne addition reactions are not very important in the laboratory, although they are used industrially (Sec. 14.9). There are, however, other important alkyne addition reactions that we shall discuss in the next two sections.

14.5 CONVERSION OF ALKYNES INTO ALDEHYDES AND KETONES

A. Hydration of Alkynes

Water adds to the triple bond of an alkyne in an acid-catalyzed addition reaction.

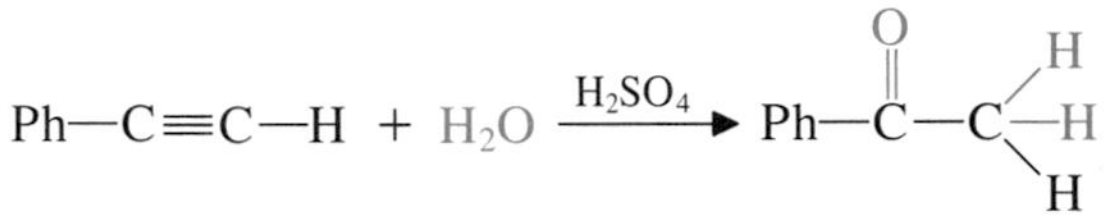

$$Ph-C\equiv C-H \ + \ H_2O \ \xrightarrow{H_2SO_4} \ Ph-\overset{O}{\overset{\|}{C}}-\overset{H}{\underset{H}{\overset{|}{C}}}-H \qquad (14.6)$$

Since the acid-catalyzed addition of water to an alkene gives an alcohol (Sec. 4.8), by analogy we might have expected the following reaction to occur in alkyne hydration:

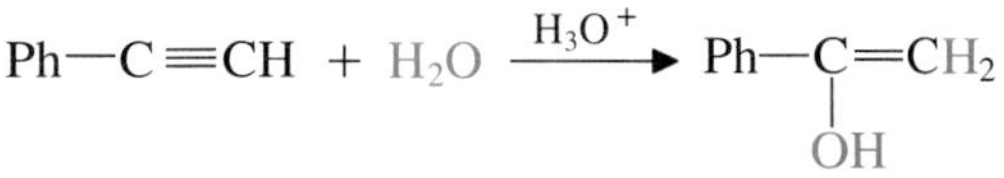

$$Ph-C\equiv CH \ + \ H_2O \ \xrightarrow{H_3O^+} \ Ph-\underset{OH}{\overset{|}{C}}=CH_2 \qquad \begin{array}{l} \text{a vinylic alcohol or } enol \quad (14.7) \\ \text{(reacts further)} \end{array}$$

An alcohol in which the —OH group is on the carbon of a double bond is called an **enol.** In fact, enols *are* formed in alkyne hydrations, but they are not isolated because they are unstable and react further to give aldehydes or ketones.

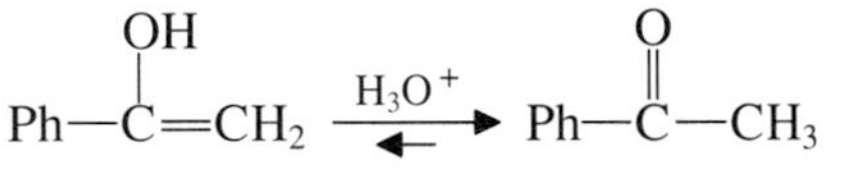

$$Ph-\underset{}{\overset{OH}{\overset{|}{C}}}=CH_2 \ \xrightarrow{H_3O^+} \ Ph-\overset{O}{\overset{\|}{C}}-CH_3 \qquad \text{acetophenone (a ketone)} \qquad (14.8)$$

$$CH_2=\underset{}{\overset{OH}{\overset{|}{C}}}-H \ \rightleftharpoons \ CH_3-\overset{O}{\overset{\|}{C}}-H \qquad \text{acetaldehyde (an aldehyde)} \qquad (14.9)$$

As we can see from these examples, the characteristic functional group in aldehydes and ketones is the **carbonyl group,** $\diagup \overset{}{C}=O$.

The mechanism of enol formation involves carbocation intermediates, and is quite analogous to the mechanism of alkene hydration (Sec. 4.8). The first step in the mechanism is protonation of the triple bond, giving a carbocation in which the electron-deficient carbon is part of a double bond:

$$Ph-C\equiv CH \ \ H-\overset{+}{\underset{}{O}}H_2 \ \longrightarrow \ Ph-\overset{+}{C}=C\overset{H}{\underset{H}{\diagdown}} \ + \ H_2\ddot{O}: \qquad (14.10a)$$

a vinylic cation

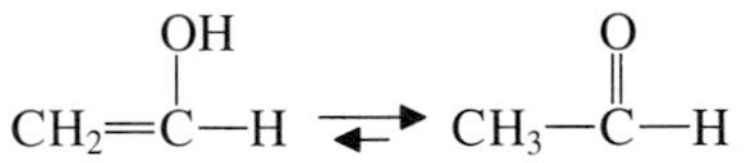

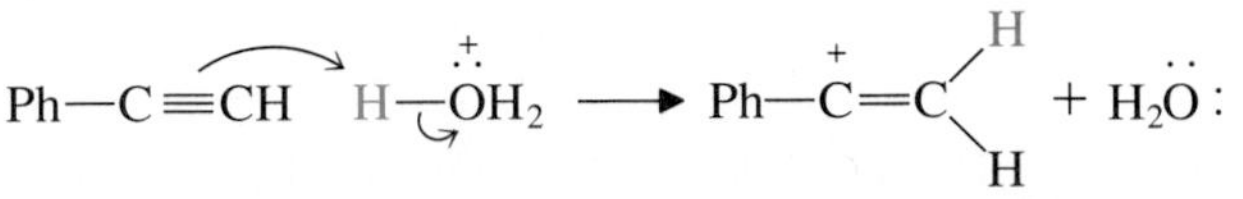

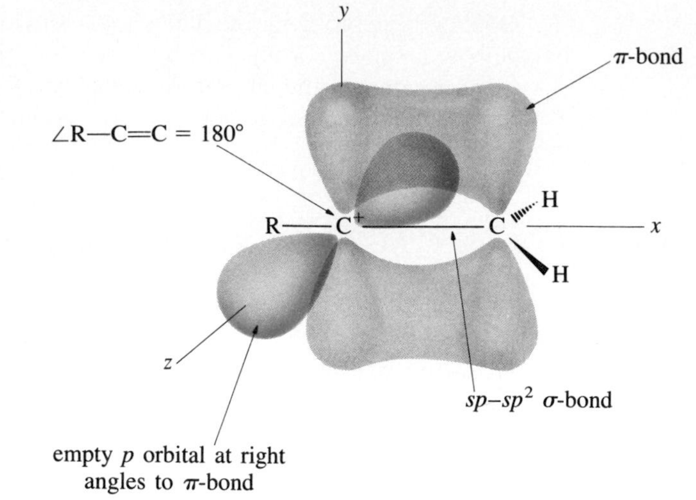

Figure 14.6 Geometry and electronic structure of a vinylic cation.

Such a cation is called a **vinylic cation.** Since the electron-deficient carbon of the vinylic cation is connected to two groups (Ph— and $=CH_2$ in Eq. 14.10a), the bond angle at this carbon is 180°, and this carbon is *sp* hybridized, as shown in Fig. 14.6. This carbon forms two σ-bonds, one to —Ph and one to $—CH_2$, using its *sp* hybrid orbitals. One of its remaining *p* orbitals is used to form the alkene π-bond; the other *p* orbital is vacant. Notice that this vacant *p* orbital lies at right angles to the alkene π-system, and therefore is not part of the π-bond.

The formation of the enol is completed by attack of water on the vinylic cation and subsequent loss of a proton:

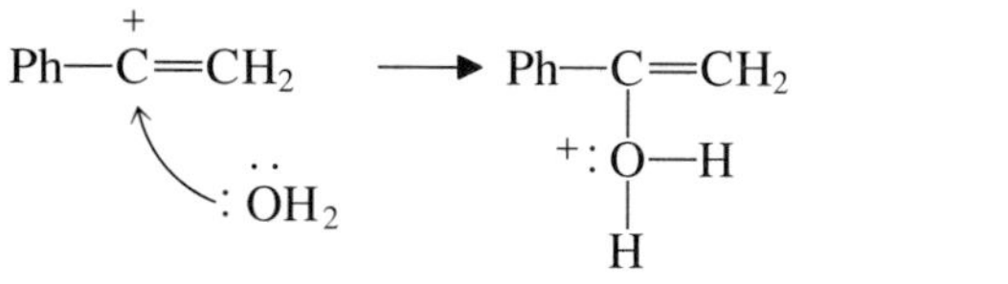

(14.10b)

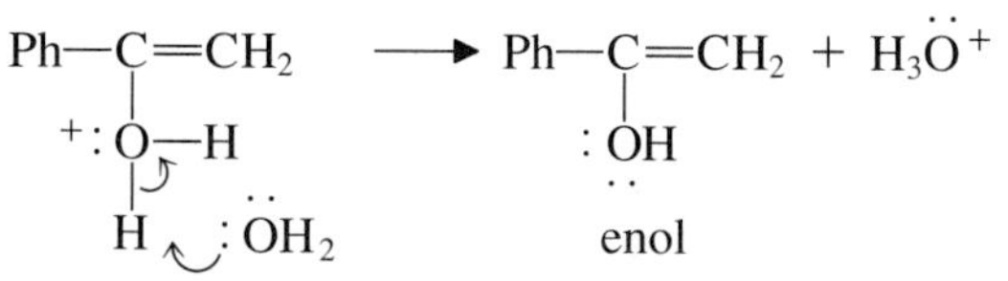

(14.10c)

Notice that the Markovnikov rule is obeyed, as we expect for a reaction involving carbocations.

Enols are converted into aldehydes or ketones by a mechanism that begins like hydration of the double bond (Sec. 4.8). The enol double bond is protonated to give another carbocation. The carbocation in this example, as its resonance structures show, is a protonated ketone. Instead of adding water, this ion loses a proton to give the ketone.

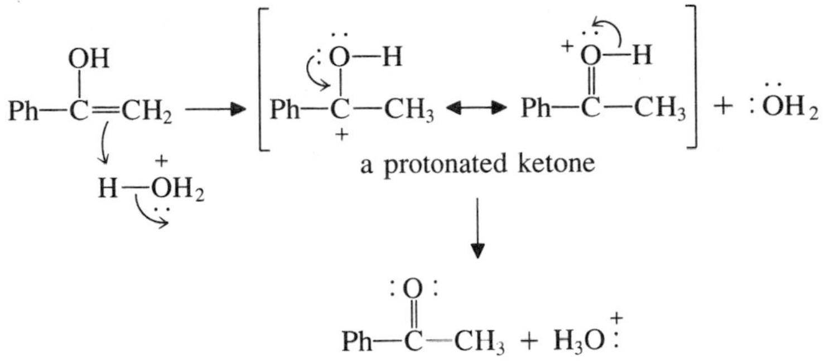

$$Ph-\overset{:O:}{\underset{}{C}}-CH_3 \; + \; H_3O^{\overset{+}{:}} \qquad (14.10d)$$

Why isn't the carbocation in Eq. 14.10d attacked by water to give another alcohol, as in the following equation?

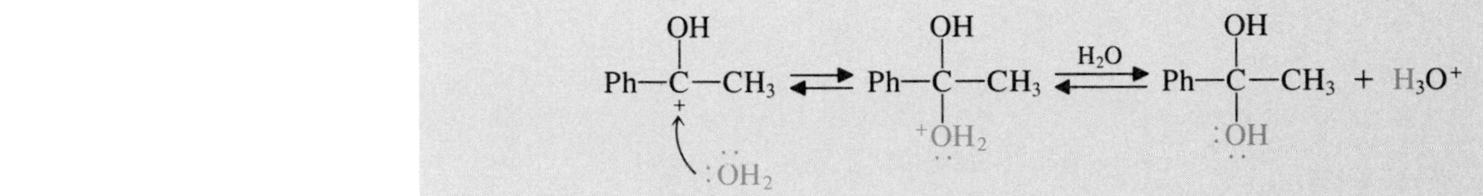

$$Ph-\overset{OH}{\underset{+}{C}}-CH_3 \; \rightleftharpoons \; Ph-\overset{OH}{\underset{+OH_2}{C}}-CH_3 \; \xrightarrow{H_2O} \; Ph-\overset{OH}{\underset{:OH}{C}}-CH_3 \; + \; H_3O^+ \quad (14.11)$$

The answer is that this reaction *does* occur; however, it is reversible, and the equilibrium between ketones and the corresponding diols in most cases favors the formation of the ketone.

All aldehydes or ketones with a hydrogen at the carbon *adjacent* to the carbonyl group are in equilibrium with their enols; however, in most cases, the fraction of enol present is very small.

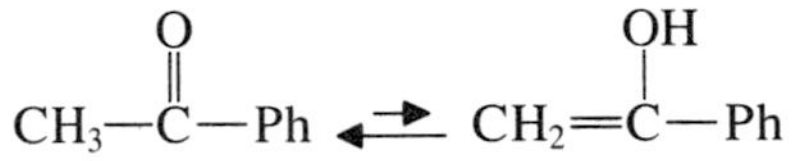

$$CH_3-\overset{O}{\overset{\|}{C}}-Ph \; \rightleftharpoons \; CH_2=\overset{OH}{\underset{}{C}}-Ph \quad (14.12)$$

relative proportions at equilibrium:

1	:	10^{-8}
acetophenone		(an enol)
(a ketone)		

We shall learn more about aldehydes, ketones, and enols in Chapters 19 and 22. The important point here, however, is that because most enols are unstable, *any synthesis of an enol will yield instead the corresponding carbonyl compound.*

Just as acid-catalyzed hydration of alkenes is not usually a good laboratory method for the preparation of alcohols, acid-catalyzed hydration of alkynes is not a good laboratory method for the preparation of ketones because the high acid concentration required for the reaction to proceed at a reasonable rate also promotes polymerization and other side reactions. However, the hydration of alkynes can be carried out under milder conditions when it is catalyzed by mercuric ion, Hg^{2+}.

$$\text{(91\% yield)}$$

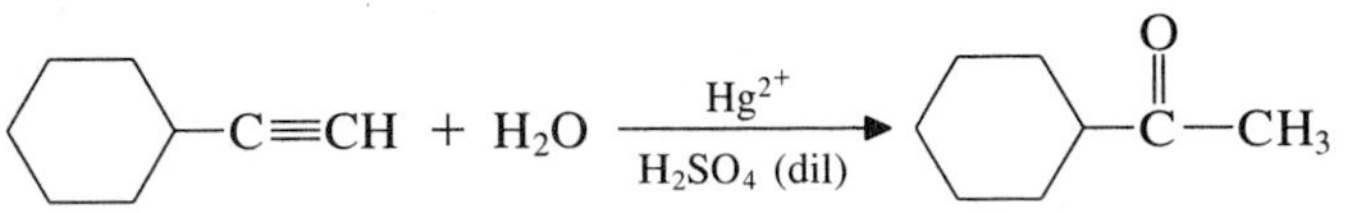

The first step in this reaction is probably oxymercuration of the triple bond (see Sec. 5.3B):

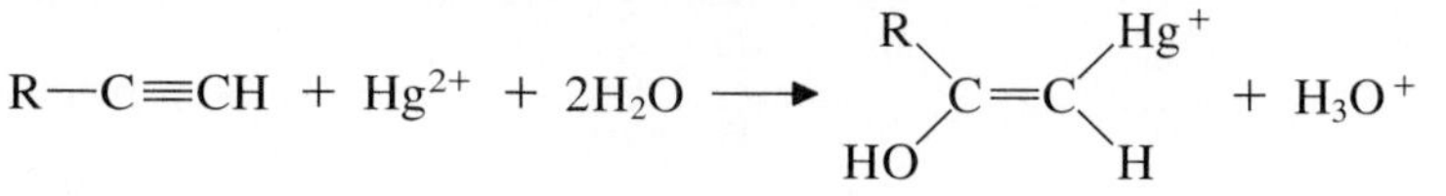

$$R-C\equiv CH + Hg^{2+} + 2H_2O \longrightarrow \underset{HO}{\overset{R}{>}}C=C\underset{H}{\overset{Hg^+}{<}} + H_3O^+ \quad (14.14a)$$

Protonation of the organomercury compound leads to loss of mercuric ion and formation of the enol, which then reacts to give the ketone (Eq. 14.8).

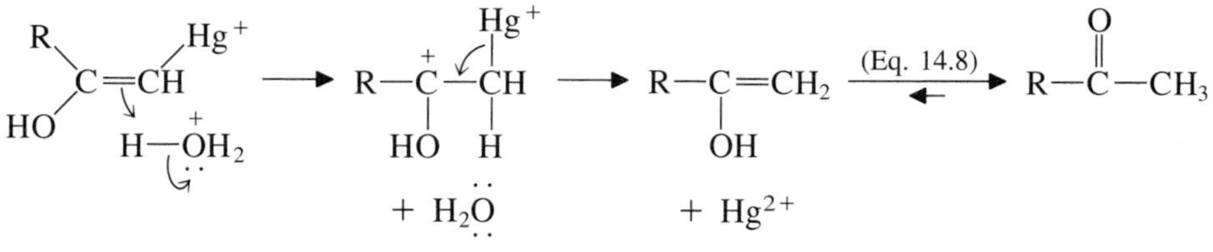

$$\underset{HO}{\overset{R}{>}}C=CH \atop H-\overset{+}{O}H_2 \longrightarrow R-\overset{+}{\underset{HO}{C}}{-}\overset{Hg^+}{\underset{H}{CH}} \longrightarrow R-\underset{OH}{C}=CH_2 \xrightarrow{(Eq.\ 14.8)} R-\overset{O}{\overset{\|}{C}}-CH_3 \quad (14.14b)$$

$$+\ H_2O \qquad +\ Hg^{2+}$$

This reaction is very similar to oxymercuration of alkenes, but with an important difference: the reaction of alkynes does *not* require $NaBH_4$ for the removal of mercury from the organomercury compound. Because the mercury is attached to a double bond, it can be removed by the acid in the reaction mixture. Since the Hg^{2+} is regenerated, it is a catalyst for the reaction.

The hydration of alkynes is a useful preparative method for ketones provided that the starting material is a 1-alkyne or a symmetrical alkyne (an alkyne with the same groups attached to each end of the triple bond). Other alkynes give difficult to separate mixtures of isomers (why?).

Problems

8 From which alkynes could the following compounds be prepared by acid-catalyzed hydration?

(a)
$$CH_3CH_2CH_2CH_2-\overset{O}{\overset{\|}{C}}-CH_2CH_2CH_2CH_2CH_3$$

(b)
$$(CH_3)_3C-\overset{O}{\overset{\|}{C}}-CH_3$$

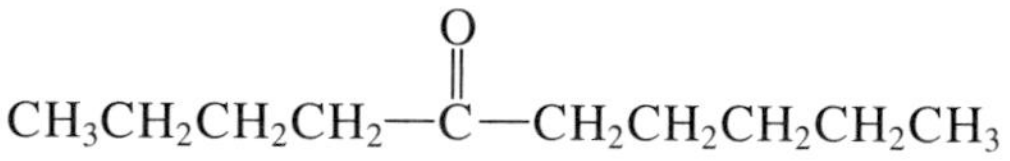

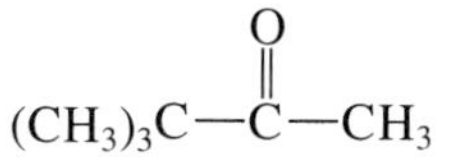

9 Hydration of an alkyne is *not* a reasonable preparative method for each of the following carbonyl compounds. Explain why.

(a)
$$(CH_3)_3C-\overset{O}{\overset{\|}{C}}-C(CH_3)_3$$

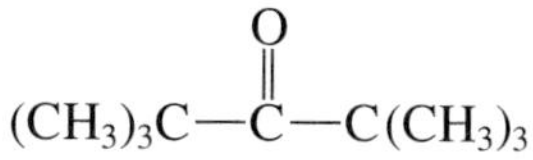

(b) $CH_3CH_2CH=O$

(c)

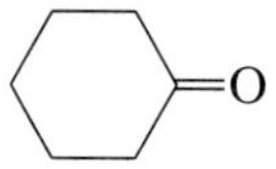

10 (a) Draw the structures of *all* enol forms of the following ketone.

$$CH_3CH_2-\overset{O}{\overset{\|}{C}}-CH_2CH_3$$

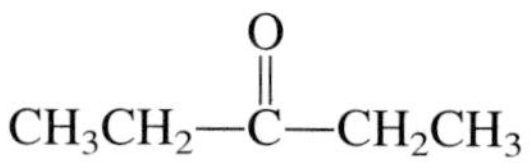

(b) Would alkyne hydration be a good preparative method for this compound? Explain.

B. Hydroboration of Alkynes

Alkynes react with diborane, B_2H_6, to add the elements of BH_3 to the triple bond, just as we would expect from the analogous reaction of alkenes.

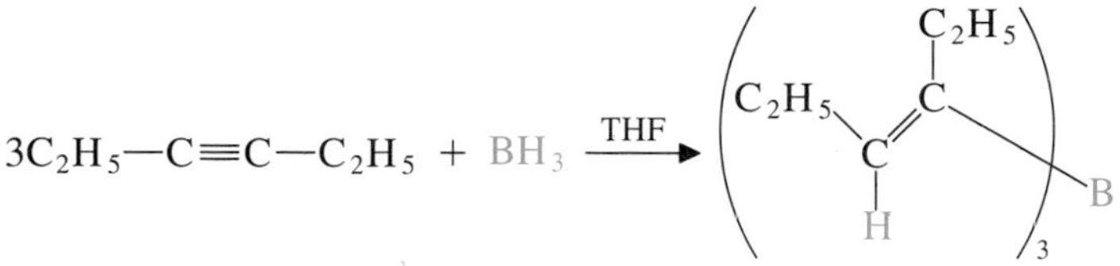

$$3C_2H_5—C\equiv C—C_2H_5 + BH_3 \xrightarrow{\text{THF}} \quad (14.15a)$$

By analogy to the similar reaction of alkenes, oxidation of the organoborane with alkaline hydrogen peroxide yields the corresponding "alcohol," which in this case is an *enol*. As we learned in Sec. 14.5A, enols react further to give the corresponding carbonyl compounds.

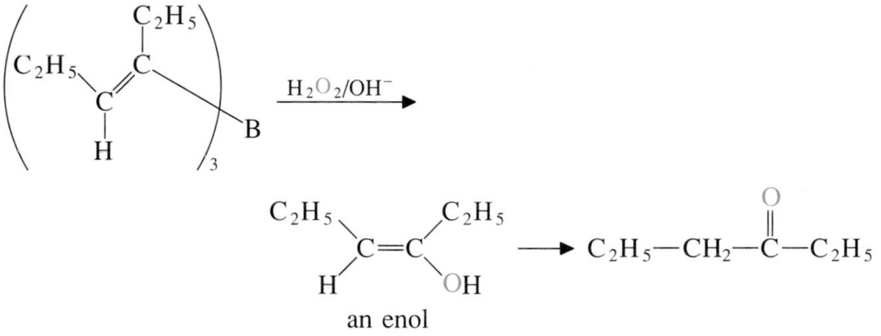

$$\xrightarrow{H_2O_2/OH^-}$$

$$\quad \longrightarrow \quad C_2H_5—CH_2—\overset{\overset{\displaystyle O}{\|}}{C}—C_2H_5 \quad (14.15b)$$

an enol

Since the organoborane product of Eq. 14.15a has a double bond, a second addition of BH_3 is possible, and in fact cannot be prevented in the reaction of 1-alkynes. However, it is possible to control the addition to internal alkynes so that addition stops after the addition of one equivalent of BH_3.

Even the hydroboration of 1-alkynes can be stopped after the first addition provided that a highly branched borane is used. One reagent developed for this purpose is called *disiamylborane*, abbreviated as shown in Eq. 14.16. (How would you synthesize disiamylborane?)

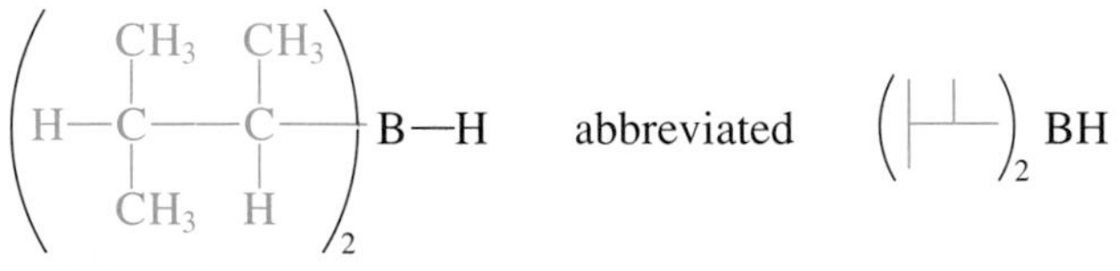

$$\left(H—\underset{\underset{\displaystyle CH_3}{|}}{\overset{\overset{\displaystyle CH_3}{|}}{C}}—\underset{\underset{\displaystyle H}{|}}{\overset{\overset{\displaystyle CH_3}{|}}{C}}—\right)_2 B—H \quad \text{abbreviated} \quad \left(\quad \right)_2 BH \quad (14.16)$$

disiamylborane

Since hydroboration is sensitive to steric effects (Sec. 5.3A), only one equivalent of the very large disiamylborane molecule adds to the triple bond of a 1-alkyne. Addition of a second equivalent of the reagent is retarded because the transition state for this addition is crowded and strained.

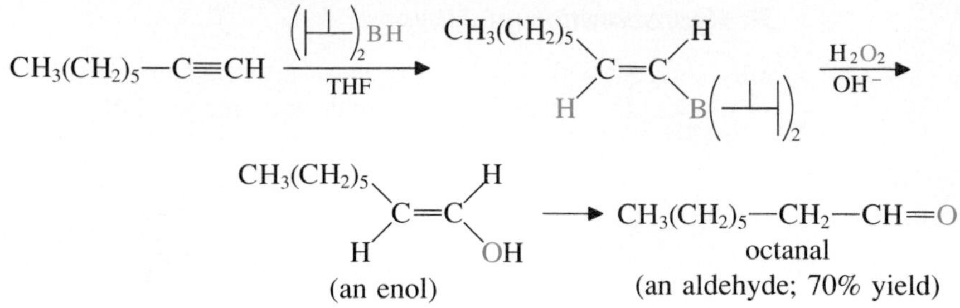

$$CH_3(CH_2)_5 \underset{H}{\overset{H}{>}}C=C\underset{OH}{\overset{H}{<}} \longrightarrow CH_3(CH_2)_5—CH_2—CH=O \quad (14.17)$$

(an enol) octanal

(an aldehyde; 70% yield)

Notice from this example that the regioselectivity of alkyne hydroboration is the same as observed in alkene hydroboration (Sec. 5.3A): boron adds to the least substituted carbon atom of the triple bond.

Because hydroboration–oxidation and mercury-catalyzed hydration give different products when a 1-alkyne is used as the starting material (why?), these are *complementary* methods for the preparation of carbonyl compounds in the same sense that hydroboration–oxidation and oxymercuration–reduction are complementary methods for the preparation of alcohols from alkenes.

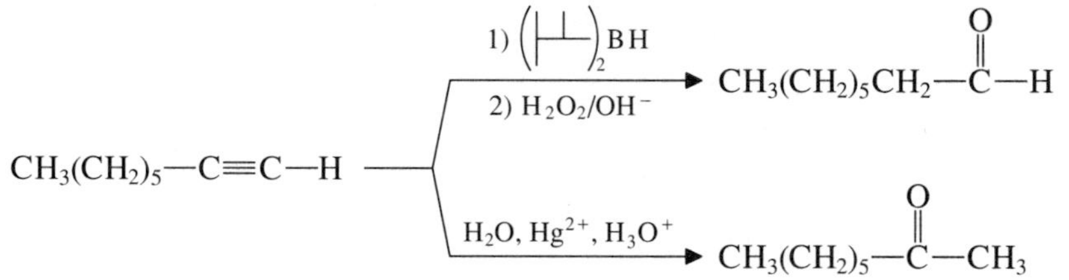

$$CH_3(CH_2)_5—C\equiv C—H \quad \begin{cases} \xrightarrow[\text{2) } H_2O_2/OH^-]{\text{1) } \left(\square\right)_2 BH} CH_3(CH_2)_5CH_2—\overset{O}{\overset{\|}{C}}—H \\[2em] \xrightarrow{H_2O,\ Hg^{2+},\ H_3O^+} CH_3(CH_2)_5—\overset{O}{\overset{\|}{C}}—CH_3 \end{cases} \quad (14.18)$$

Notice that hydroboration–oxidation of a 1-alkyne gives an *aldehyde;* hydration of any 1-alkyne (other than acetylene itself) gives a *ketone.*

Problem

11 Contrast the results of hydroboration–oxidation and mercuric ion-catalyzed hydration for (a) ethynylcyclohexane and (b) 2-butyne.

14.6 REDUCTION OF ALKYNES

A. Catalytic Hydrogenation of Alkynes

Alkynes, like alkenes (Sec. 5.7A), undergo catalytic hydrogenation. The first addition of hydrogen yields an alkene; since alkenes also undergo catalytic hydrogenation, a second addition of hydrogen to give an alkane is also possible.

$$R—C\equiv C—R \xrightarrow[\text{catalyst}]{H_2} R—CH=CH—R \xrightarrow[\text{catalyst}]{H_2} R—CH_2—CH_2—R \quad (14.19)$$

The utility of catalytic hydrogenation is enhanced considerably by the fact that hydrogenation of an alkyne may be stopped at the alkene stage if the reaction mixture contains a **catalyst poison**: a compound that disrupts the action of a catalyst. Among the useful catalyst poisons are salts of Pb^{2+}, and certain nitrogen compounds, such as pyridine, quinoline, or other amines. These compounds *selectively* block the hydrogenation of alkenes without stopping the hydrogenation of alkynes to alkenes.

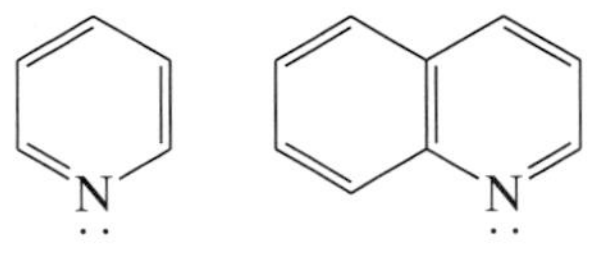

pyridine quinoline

For example, a $Pd/CaCO_3$ catalyst can be washed with $Pb(OAc)_2$ to give a poisoned catalyst known as a **Lindlar catalyst.** In the presence of Lindlar catalyst, an alkyne is hydrogenated to the corresponding alkene.

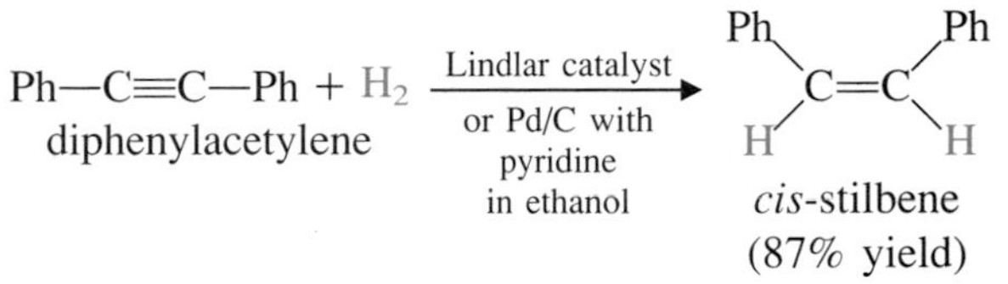

$$Ph—C \equiv C—Ph + H_2 \xrightarrow[\substack{\text{or Pd/C with} \\ \text{pyridine} \\ \text{in ethanol}}]{\text{Lindlar catalyst}} \underset{\substack{cis\text{-stilbene} \\ (87\% \text{ yield})}}{\overset{Ph \qquad Ph}{\underset{H \qquad H}{C=C}}} \qquad (14.20)$$

As Eq. 14.20 shows, hydrogenation of alkynes, like hydrogenation of alkenes (Sec. 7.10D), is a stereoselective *syn* addition. Thus, in the presence of a poisoned catalyst, hydrogenation of appropriate alkynes gives *cis*-alkenes. In fact, *catalytic hydrogenation of alkynes is one of the best ways to prepare cis-alkenes.*

In the absence of a catalyst poison, hydrogenation proceeds readily to the alkane.

$$Ph—C \equiv C—Ph + 2H_2 \xrightarrow[\text{no poison}]{Pd/C} \underset{(95\% \text{ yield})}{Ph—CH_2CH_2—Ph} \qquad (14.21)$$

Catalytic hydrogenation of alkynes may therefore be used to prepare alkenes or alkanes by either including or omitting the catalyst poison. How catalyst poisons exert their inhibitory effect on the hydrogenation of alkenes, while still permitting the hydrogenation of alkynes, is not well understood.

Problem

12 Give the principal organic product formed in each of the following reactions:

(a) $CH_3CH_2CH_2C \equiv C(CH_2)_4CH = CH_2 + H_2 \xrightarrow{\text{Lindlar catalyst}}$

(b) Same as (a), with no poison

(c)
$$\underset{N}{\bigcirc}\!\!-\!\!C \equiv CH \quad + H_2 \xrightarrow{Pd/C}$$

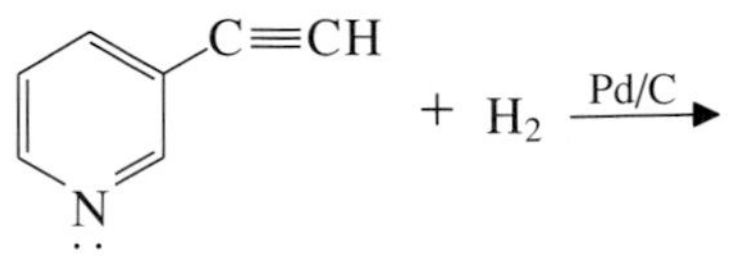

B. Reduction of Alkynes with Alkali Metals in Ammonia

Reaction of an alkyne with a solution of an alkali metal (usually sodium) in liquid ammonia gives a *trans*-alkene.

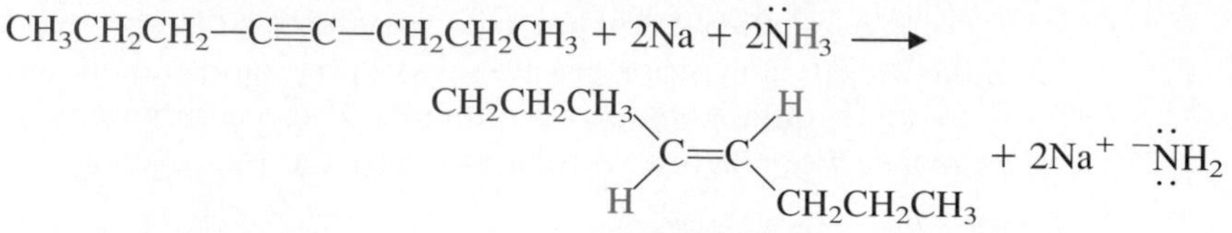

$$CH_3CH_2CH_2—C \equiv C—CH_2CH_2CH_3 + 2Na + 2\ddot{N}H_3 \longrightarrow$$

$$\underset{\underset{\text{(97\% yield)}}{}}{\begin{array}{c} CH_2CH_2CH_3 \\ \diagdown \\ \diagup \quad \diagdown \\ H \end{array} \overset{H}{\underset{CH_2CH_2CH_3}{C=C}}} \qquad + 2Na^+ \; {}^-\ddot{N}H_2 \quad (14.22)$$

Liquid ammonia (bp $-33°$) should not be confused with ammonium hydroxide, or "household ammonia." Liquid ammonia is formed by condensing pure gaseous ammonia at low temperatures. Household ammonia is formed by dissolving ammonia in water.

This reduction of alkynes with sodium in liquid ammonia is complementary to catalytic hydrogenation of alkynes, which, as we have just learned, is used to prepare *cis*-alkenes.

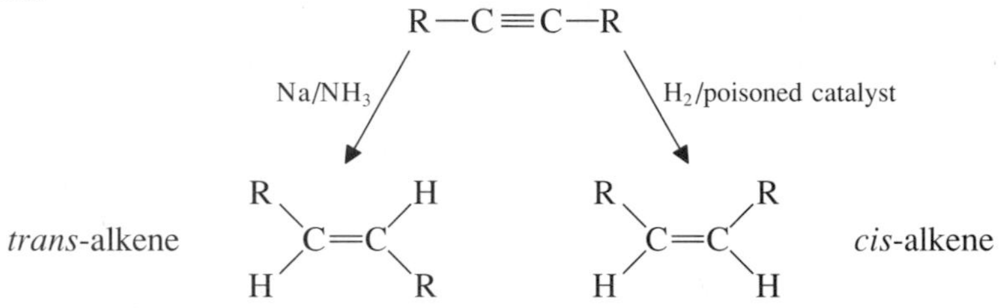

$$(14.23)$$

The reason for the stereochemistry of this reaction follows from its mechanism. If sodium or other alkali metals are dissolved in pure liquid ammonia, a deep blue solution forms that contains electrons complexed to ammonia (*solvated electrons*).

$$Na\cdot + nNH_3 \text{ (liq)} \longrightarrow Na^+ + \quad e^-(NH_3)_n \qquad (14.24)$$
$$\text{solvated electron}$$

The solvated electron can be thought of as the simplest free radical. We know that free radicals add to triple bonds (Eq. 14.2); the reaction of solvated electrons with alkynes begins with an analogous addition of the electron to the triple bond. The resulting species, called a **radical anion,** has both an unpaired electron and a negative charge.

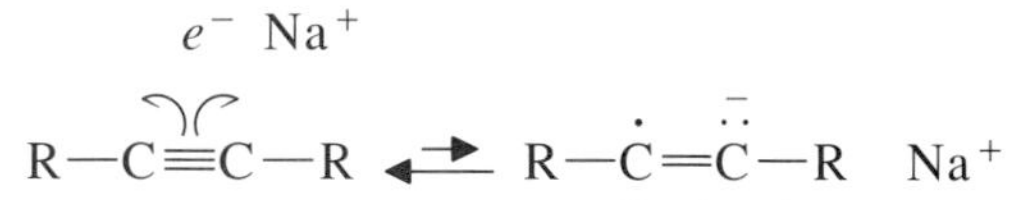

$$e^- \; Na^+$$

$$R—C \equiv C—R \; \rightleftharpoons \; R—\overset{\cdot}{C}=\overset{..}{C}—R \quad Na^+ \qquad (14.25a)$$
$$\text{a radical anion}$$

The formation of the radical anion has a very unfavorable equilibrium constant, but the radical anion is such a strong base that it readily removes a proton from ammonia to give a *vinylic radical*—a radical in which the unpaired electron is associated with one carbon of a double bond. The destruction of the radical anion in this manner pulls the equilibrium in Eq. 14.25a to the right.

$$H—\overset{..}{N}H_2$$

$$R—\overset{\cdot}{C}=\overset{..}{C}—R \qquad \longrightarrow \qquad R—\overset{\cdot}{C}=\overset{H}{\underset{R}{C}} \quad + \; {}^-\ddot{N}H_2 \; Na^+ \qquad (14.25b)$$
$$\qquad Na^+ \qquad\qquad\qquad \text{a vinylic radical}$$

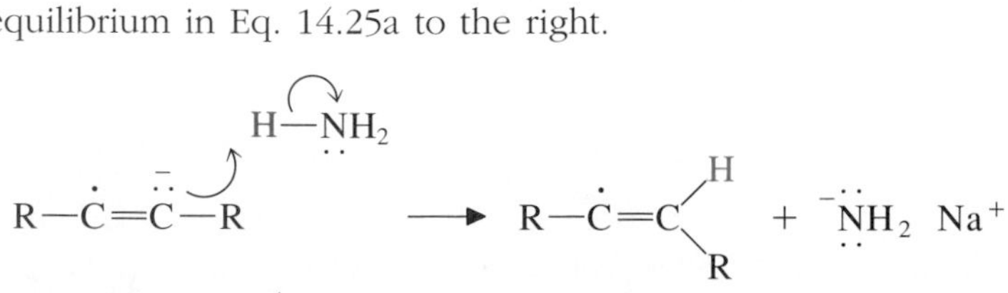

The vinylic radical, like the lone electron pair of an amine (Sec. 6.8B), rapidly undergoes inversion, and the equilibrium between the *cis* and *trans* radical favors the *trans* radical for the same reason that *trans*-alkenes are more stable than *cis*-alkenes: repulsions between the R groups are reduced.

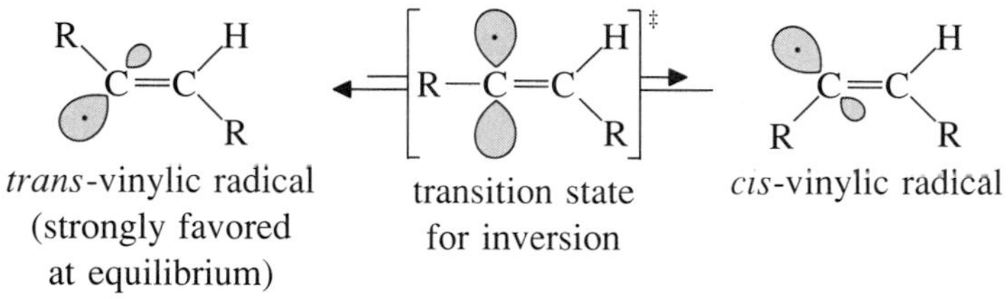

(14.25c)

The *cis* and *trans* stereoisomers of this radical probably react at about the same rate in the subsequent steps of the mechanism. However, since there is much more of the *trans* radical, the ultimate product of the reaction is the one derived from this radical—the *trans*-alkene.

Two more mechanistic steps give the alkene product. First, the vinylic radical accepts an electron to form an anion:

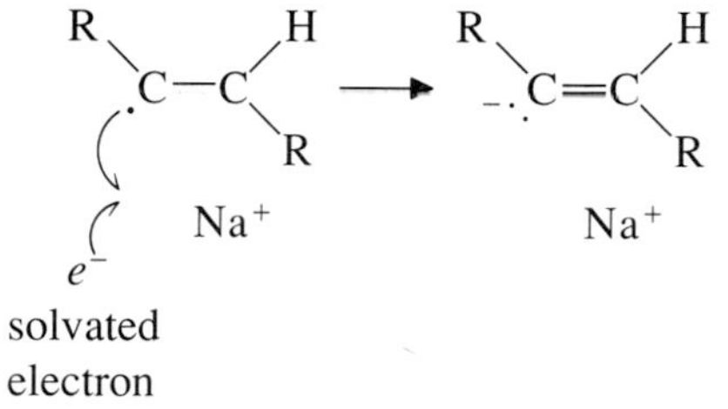

(14.25d)

This anion is also more basic than the solvent, and therefore removes a proton from ammonia to complete the addition.

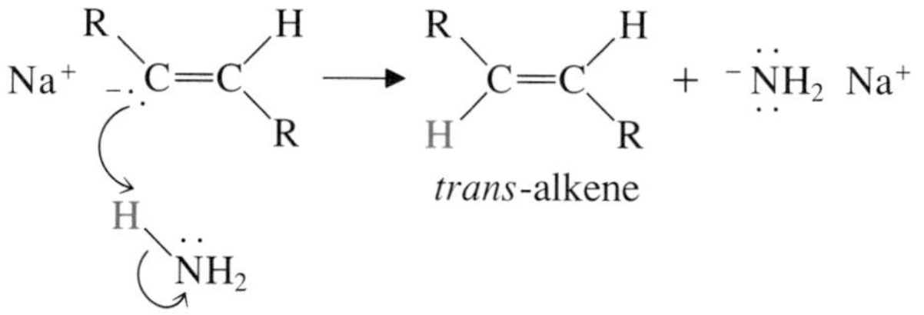

(14.25e)

Because ordinary alkenes do not react with the solvated electron (the initial equilibrium analogous to Eq. 14.25a is too unfavorable), the reaction stops at the *trans*-alkene stage.

Problem

13 What is the difference in the products obtained (if any) when 3-hexyne is treated in each of the following ways:

(a) with sodium in liquid ammonia, and the product of that reaction with D_2 over Pd/C

(b) with H_2 over Pd/C and quinoline, and the product of that reaction with D_2 over Pd/C

14.7 ACIDITY OF 1-ALKYNES

A. Carbanions. Acetylenic Anions

Hydrocarbons are not usually regarded as acids. Nevertheless we can imagine the removal of a proton from a hydrocarbon by a very strong base B:⁻.

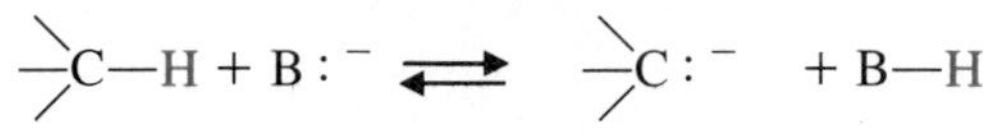

a carbanion (14.26)

In this equation the hydrocarbon acts as a Brønsted acid. Its conjugate base, a species with an electron pair and a negative charge on carbon, is a carbon anion, or *carbanion* (see Sec. 8.8B).

The conjugate base of an alkane, called generally an *alkyl anion,* has an electron pair in an sp^3 orbital. An example of such an ion is the 2-propanide anion:

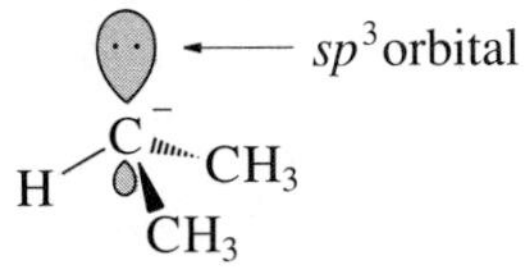

2-propanide anion
(an alkyl anion)

The conjugate base of an alkene, called generally a *vinylic anion,* has an electron pair in an sp^2 orbital. An example of this type of carbanion is the 1-propenide anion.

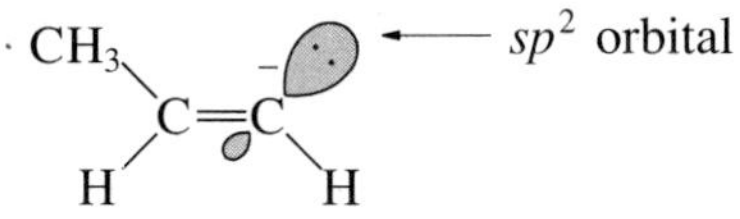

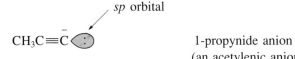

1-propenide anion
(a vinylic anion)

The anion derived from the ionization of a 1-alkyne, called generally an *acetylide anion,* has an electron pair in an *sp* orbital. An example of this type of anion is the 1-propynide anion.

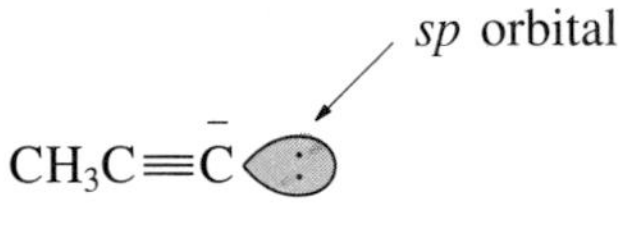

1-propynide anion
(an acetylenic anion)

The approximate acidities of the different types of aliphatic hydrocarbons have been measured or estimated.

	R_3C—H	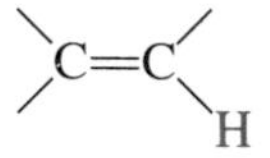	—C≡C—H	
type of hydrocarbon	alkane	alkene	alkyne	(14.27)
approximate pK_a	≥55	42	25	

From these data we can see, first, that carbanions are extremely strong bases (that is, hydrocarbons are very weak acids); and second, that alkynes are the most acidic of the aliphatic hydrocarbons.

Alkyl anions and vinylic anions are seldom if ever formed by proton removal from the corresponding hydrocarbons; the hydrocarbons are simply not acidic enough. However, alkynes are sufficiently acidic that their conjugate-base acetylide anions can be formed with strong bases. One base commonly used for this purpose is sodium amide, or sodamide, $Na^+ \ {}^-\!:\!\ddot{N}H_2$, dissolved in liquid ammonia. The amide ion, ${}^-\!:\!\ddot{N}H_2$, is the conjugate base of ammonia, which, as an *acid,* has a pK_a of about 35.

$$\underset{}{B:^-} + \underset{pK_a\,=\,35}{:NH_3} \rightleftharpoons \underset{\text{amide ion}}{{}^-\!:\!\ddot{N}H_2} + B\text{—}H \qquad (14.28)$$

Do not confuse this reaction of ammonia with the more common one in which ammonia acts as a *base*:

$$:NH_3 + H\text{—}\ddot{\underset{..}{O}}H \rightleftharpoons \underset{\substack{\text{ammonium}\\ \text{ion}\\ pK_a\,=\,9.2}}{{}^+NH_4} + {}^-\!:\!\ddot{\underset{..}{O}}H \qquad (14.29)$$

Also, do not confuse a solution of *sodium amide* in liquid ammonia (a strong base) with a solution of *sodium metal* in liquid ammonia (a source of solvated electrons; Sec. 14.6B).

Because the amide ion is a much stronger base than an acetylide ion, the following equilibrium lies well to the right:

$$R\text{—}C{\equiv}C\text{—}H \ \ {}^-\!:\!\ddot{N}H_2 \ Na^+ \xrightarrow[\longleftarrow]{NH_3\ (liq)} R\text{—}C{\equiv}C\!:^- \ Na^+ + \ddot{N}H_3 \qquad (14.30)$$

Thus, the sodium salt of an alkyne can be formed from a 1-alkyne quantitatively with $NaNH_2$. Since the amide ion is a much *weaker* base than either a vinylic anion or an alkyl anion, these ions *cannot* be prepared using sodium amide (Problem 15).

The relative acidity of alkynes is also reflected in the method usually used to prepare *acetylenic Grignard reagents*: the reaction of an alkylmagnesium halide with an alkyne.

$$CH_3CH_2CH_2CH_2\text{—}C{\equiv}C\text{—}H + C_2H_5\text{—}MgBr \longrightarrow$$
$$CH_3CH_2CH_2CH_2\text{—}C{\equiv}C\text{—}MgBr + C_2H_6 \qquad (14.31)$$

$$H\text{—}C{\equiv}C\text{—}H + C_2H_5\text{—}MgBr \xrightarrow{THF} H\text{—}C{\equiv}C\text{—}MgBr + C_2H_6 \qquad (14.32)$$

This reaction, called **transmetallation,** is really just another Brønsted acid–base reaction:

conjugate acid–base pair

$$BrMg^{+}\ ^{-}\!:CH_2CH_3 \quad H{-}C{\equiv}C{-}R \longrightarrow CH_3{-}CH_3 + BrMg^{+}\ ^{-}\!:C{\equiv}C{-}R \qquad (14.33)$$

conjugate base–acid pair

This reaction is similar in principle to the reaction of a Grignard reagent with water or alcohols (Eq. 8.30). Like all Brønsted acid–base equilibria, this one favors formation of the weaker base. The release of ethane gas in the reaction with ethylmagnesium bromide is a useful test for 1-alkynes. Alkynes with an internal triple bond do not react because they lack an acidic alkyne hydrogen.

What is the reason for the relative acidities of the hydrocarbons? We learned in Sec. 8.5B about two important factors that contribute to the acidity of an acid A—H: a relatively weak A—H bond, and the electronegativity of the group A. If we consult Table 5.2, we find that alkynes have the *strongest* C—H bond of all the aliphatic hydrocarbons:

$$\equiv C{-}H \ (121\ \text{kcal/mol}) > \ =C{-}H \ (106\ \text{kcal/mol}) > \ {-}C{-}H \ (94\text{–}100\ \text{kcal/mol}) \qquad (14.34)$$

acetylenic C—H   vinylic C—H   alkyl C—H

If relative bond strengths were the major factor controlling hydrocarbon acidity, then alkynes would be the *least* acidic hydrocarbons. Since they are in fact the *most* acidic hydrocarbons, it must be the electronegativity—the electron-attracting ability—of the RC≡C— group that is responsible for the enhanced acidity of alkynes. Thus, the electronegativities of carbon groups increase in the order alkyl < alkenyl < alkynyl.

The electronegativity of alkynyl groups can be understood in the following way. We learned that *sp* electrons are closer to the nucleus, on the average, than sp^2 electrons, which in turn are closer to the nucleus than sp^3 electrons (Sec. 4.1A and 14.2). Therefore *electrons are drawn closer to the nucleus when they occupy sp orbitals.* This is a stabilizing effect, because, by the electrostatic law (Eq. 1.2), the energy of interaction is more favorable as the positive nucleus and negative electrons are brought closer together. Thus the energy of electron pairs—including those in carbanions—is in the order $sp < sp^2 < sp^3$. In other words, *free electron pairs prefer to be in orbitals with the most s-character,* other things being equal.

Problems

14 Each of the following compounds protonates on nitrogen. Draw the conjugate acid of each. Rank the compounds in order of their basicity, least basic first, and explain your reasoning.

(a) $CH_3{-}CH{=}\ddot{N}H$ (b) $CH_3{-}C{\equiv}N{:}$ (c) $CH_3{-}\ddot{N}H_2$

15 Using the pK_a values of the hydrocarbons and ammonia, estimate the equilibrium constant for (a) reaction 14.30 and (b) the analogous reaction of an alkane with amide ion. (*Hint:* See Problem 40, Chapter 8.) Use your calculation to explain why sodium amide cannot be used to form alkyl anions from alkanes.

B. Acetylenic Anions as Nucleophiles

Although acetylenic anions are the weakest bases of the simple hydrocarbon anions, they are nevertheless strong bases—much stronger, for example, than hydroxide or alkoxides. They undergo many of the characteristic reactions of strong bases. One of the important reactions of bases we have learned about is the S_N2 reaction with an alkyl halide or alkyl sulfonate (Sec. 9.3, 10.3A). In fact, acetylenic anions may be used as nucleophiles in S_N2 reactions to prepare other alkynes.

$$Na^+ \ ^-\!:C{\equiv}CH + CH_3CH_2CH_2CH_2{-}Br \xrightarrow{NH_3\,(liq)} CH_3CH_2CH_2CH_2{-}C{\equiv}CH + Na^+\ Br^- \quad (14.35)$$

sodium acetylide; 1-hexyne (64% yield)

$$CH_3CH_2CH_2CH_2{-}C{\equiv}C:^-\ Na^+ + CH_3{-}Br \longrightarrow CH_3CH_2CH_2CH_2{-}C{\equiv}C{-}CH_3 + Na^+\ Br^- \quad (14.36)$$

The acetylenic anions in these reactions are formed by the reactions of the appropriate 1-alkynes with $NaNH_2$ in liquid ammonia (Sec. 14.7A). The alkyl halides and sulfonates, as in most other S_N2 reactions (Sec. 9.4C), must be primary or unbranched secondary compounds (why?).

The reaction of acetylenic anions with alkyl halides or sulfonates is important because *it is another method of carbon–carbon bond formation*. Let us review the methods we have learned so far:

1. Cyclopropane formation by addition of carbenes to alkenes (Sec. 9.7)
2. Reaction of Grignard reagents with ethylene oxide (Sec. 11.4B)
3. Reaction of acetylenic anions with alkyl halides or sulfonates (this section).

Problems

16 Predict the products of the following reactions.

(a) $CH_3C{\equiv}C:^-\ Na^+$ + ethyl iodide $\longrightarrow$

(b) $Br{-}(CH_2)_5{-}Br + Na^+\ HC{\equiv}C:^-$ (excess) $\longrightarrow$

(c) butyl tosylate + $Na^+\ Ph{-}C{\equiv}C:^-$ $\longrightarrow$

(d) $CH_3C{\equiv}C{-}MgBr$ + ethylene oxide $\longrightarrow \xrightarrow{H_3O^+}$

17 Explain why graduate student Choke Fumely, in attempting to synthesize 4,4-dimethyl-2-pentyne using the reaction of $Na^+\ CH_3{-}C{\equiv}C:^-$ with *t*-butyl bromide, obtained none of the desired product.

18 Suggest another pair of reagents that could be used to prepare the alkyne product of Eq. 14.36.

19 Outline a preparation of the following compounds from an alkyne and an alkyl halide:
(a) 3-heptyne (b) 4,4-dimethyl-2-pentyne (See Problem 17.)

14.8 ORGANIC SYNTHESIS USING ALKYNES

The reactions of alkynes illustrate all the fundamental operations of organic synthesis (Sec. 11.8): formation of carbon–carbon bonds, transformation of functional groups, and establishment of stereochemistry. Problem 20 provides an opportunity to use all of these operations; try to work it before looking at the solution, which follows.

Problem

20 Outline a synthesis of the following compound from acetylene and any other compounds containing no more than five carbons:

$$CH_3(CH_2)_6 \diagdown \qquad \diagup CH_2CH_2CH(CH_3)_2$$
$$C{=}C$$
$$H \diagup \qquad \diagdown H$$

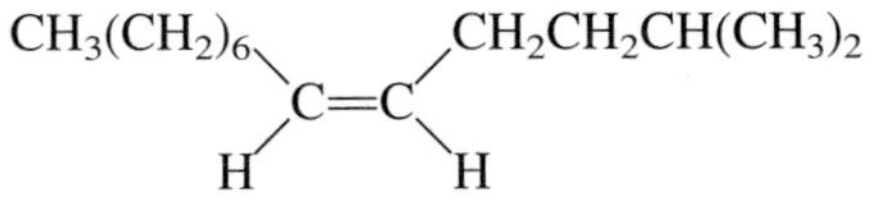

Solution: Adopting our usual strategy (Sec. 10.10B) of working backward from the target, we see immediately that there is a question of stereochemistry in the product: we are asked to prepare a *cis*-alkene. We only know *one* way to prepare *cis*-alkenes: by hydrogenation of alkynes (Sec. 14.6A). This immediately determines the last step of the synthesis:

$$CH_3(CH_2)_6{-}C{\equiv}C{-}CH_2CH_2CH(CH_3)_2 \xrightarrow[\text{Lindlar catalyst}]{H_2}$$

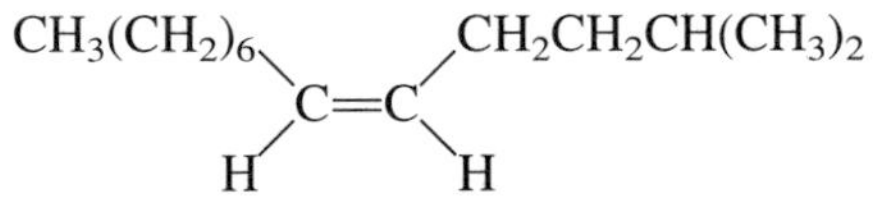

$$(14.37a)$$

(target molecule)

Our next task is to prepare the alkyne used as the starting material in Eq. 14.37a. According to the problem, we must use acetylene and other compounds of five or fewer carbons. Why should there be such a stipulation? The reason is that most simple compounds containing five or fewer carbons are readily available from commercial sources, and they are generally not very expensive.

Since the desired alkyne contains fourteen carbons, we shall have to use several reactions that form carbon–carbon bonds. There are two primary alkyl groups on the triple bond; the order in which we introduce them is arbitrary. Let us plan to introduce the five-carbon fragment on the right-hand side of this alkyne in the last step of the alkyne synthesis. This is easily done by forming its conjugate-base acetylenic anion and allowing it to react with the appropriate alkyl halide (Sec. 14.7B).

$$CH_3(CH_2)_6{-}C{\equiv}C{-}H \xrightarrow[\text{NH}_3 \text{ (liq)}]{\text{NaNH}_2} CH_3(CH_2)_6C{\equiv}C{:}^{-} \xrightarrow{\text{Br}{-}CH_2CH_2CH(CH_3)_2}$$
$$CH_3(CH_2)_6{-}C{\equiv}C{-}CH_2CH_2CH(CH_3)_2 \quad (14.37b)$$

To prepare 1-nonyne, the starting material for this reaction, we use the reaction of 1-bromoheptane with the sodium salt of acetylene itself.

$$H\!-\!C\!\equiv\!C\!-\!H \xrightarrow[\text{NH}_3 \text{ (liq)}]{\text{NaNH}_2} H\!-\!C\!\equiv\!C:^- Na^+ \xrightarrow{CH_3CH_2CH_2CH_2CH_2CH_2CH_2\!-\!Br}$$

(large excess
relative to
NaNH$_2$)

$$H\!-\!C\!\equiv\!C\!-\!CH_2CH_2CH_2CH_2CH_2CH_2CH_3 \quad (14.37c)$$

The large excess of acetylene relative to sodium amide is required to ensure formation of the *monoanion*. If there were more sodium amide than acetylene, some *dianion* $^-:C\!\equiv\!C:^-$ could form, and other reactions would occur (what are they?). Since acetylene is cheap, use of a large excess presents no practical problem.

Since the 1-bromoheptane used in Eq. 14.37c has more than five carbons, we have to prepare it as well. The following sequence of reactions will accomplish this objective. (There is also another way; see Problem 21a.)

$$CH_3CH_2CH_2CH_2CH_2\!-\!Br \xrightarrow[\text{ether}]{\text{Mg}} \overset{O}{\overset{\diagup\diagdown}{CH_2\!-\!CH_2}} \xrightarrow{H_3O^-} CH_3CH_2CH_2CH_2CH_2CH_2CH_2\!-\!OH$$

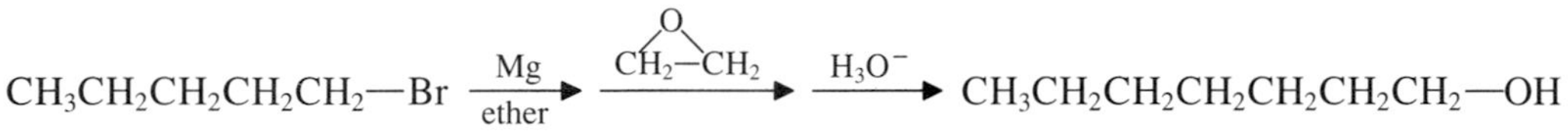

$$\xrightarrow{\text{concd HBr}} CH_3CH_2CH_2CH_2CH_2CH_2CH_2\!-\!Br \quad (14.37d)$$

Our synthesis is now complete. To summarize:

$$HC\!\equiv\!CH \text{ (excess)} \xrightarrow[\text{NH}_3 \text{ (liq)}]{\text{NaNH}_2} \xrightarrow{CH_3(CH_2)_6Br \text{ (Eq. 14.37d)}} CH_3(CH_2)_6C\!\equiv\!CH$$

$$\xrightarrow[\text{NH}_3 \text{ (liq)}]{\text{NaNH}_2} \xrightarrow{(CH_3)_2CHCH_2CH_2Br} CH_3(CH_2)_6C\!\equiv\!CCH_2CH_2CH(CH_3)_2 \xrightarrow[\text{Lindlar catalyst}]{\text{H}_2}$$

$$\underset{H}{\overset{CH_3(CH_2)_6}{C}}\!=\!\underset{H}{\overset{CH_2CH_2CH(CH_3)_2}{C}} \quad (14.37e)$$

21 Outline a synthesis for each of the following compounds from acetylene and any other compounds containing five or fewer carbons. Your synthesis for (a) should be different from that shown in Eq. 14.37d.

(a) 1-heptanol (b)

$$CH_3\!-\!\overset{\overset{\textstyle O}{\|}}{C}\!-\!(CH_2)_8CH_3$$

22 In the course of the synthesis of the sex attractant of the grape berry moth (see next section), both the *cis* and *trans* isomers of the following alkene were needed.

$$C_2H_5\!-\!CH\!=\!CH\!-\!CH_2CH_2CH_2CH_2CH_2CH_2CH_2CH_2\!-\!O\!\!\diagup$$

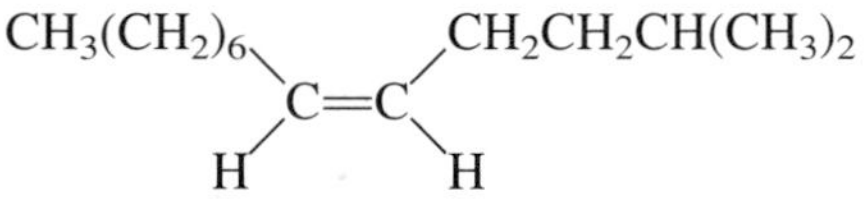

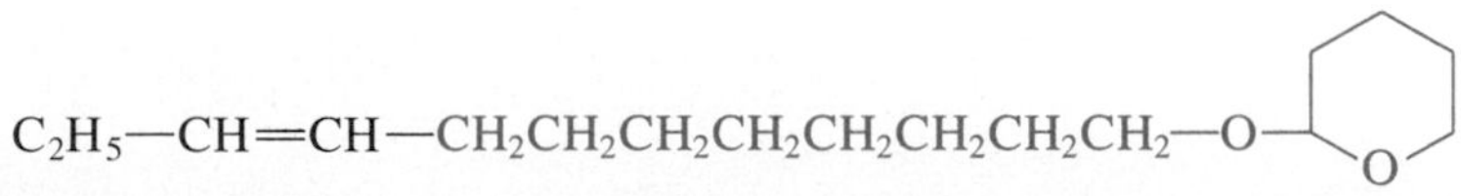

Problems (Cont.) Suggest a separate synthesis for each isomer of this alkene from the following alkyl halide and any other organic compounds.

$$Br—CH_2CH_2CH_2CH_2CH_2CH_2CH_2CH_2—O—\text{(tetrahydropyranyl)}$$

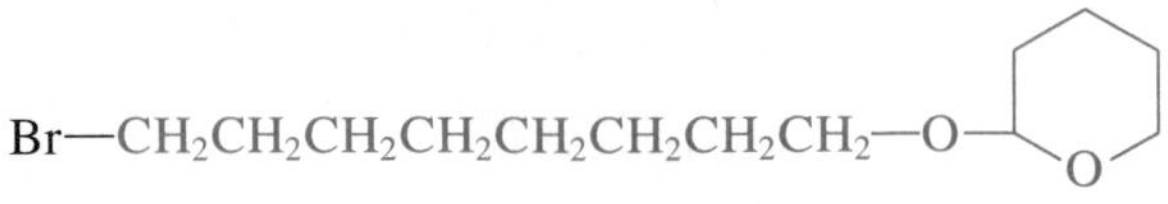

14.9 PHEROMONES

As Problem 22 illustrates, the chemistry of alkynes can be applied to the synthesis of a number of **pheromones**—chemical substances used in nature for communication, or signaling. For example, the female of an insect species might secrete one or more chemicals, called sex attractants, that signal her readiness for mating. The sex attractant of the grape berry moth (Problem 22) is an example of such a compound; it appears to be a mixture of *cis* and *trans* isomers of the following alkene:

$$C_2H_5—CH\!\!=\!\!CH—CH_2CH_2CH_2CH_2CH_2CH_2CH_2CH_2—O—\overset{\overset{\displaystyle O}{\|}}{C}—CH_3$$
(96% *cis*, 4% *trans*)

Pheromones are also used for defense (Fig. 14.7), to mark trails, and for many other purposes. It was recently discovered that the traditional use of sows in France to discover buried truffles owes its success to the fact that the truffles contain a steroid that happens to be identical to a sex attractant secreted in the saliva of boars during premating behavior!

Many scientists have become intrigued with the idea that pheromones might be used as a species-specific form of insect control. A sex attractant, for example, might be used to attract and trap the male of an insect species selectively without affecting other insect populations. Alternatively, the males of a species might become confused by a blanket of sex attractant and not be able to locate a suitable female. When used successfully, this strategy would break the reproductive cycle of the insect. The harmful environmental impact of the usual pesticides (for example, DDT) has stimulated interest in such highly specific methods.

Research along these lines has shown, unfortunately, that pheromones are not effective for broad control of insect populations. They are, however, very useful for trapping target insects, thereby providing an early warning for insect infestations. When this approach is used, conventional pesticides need be applied only when the target insects appear in the traps. This strategy has brought about reductions in the use of conventional pesticides by as much as 70% in many parts of the United States.

14.10 OCCURRENCE AND USE OF ALKYNES

Naturally occurring alkynes are relatively rare. Alkynes do not occur as constituents of petroleum, but instead are synthesized from other compounds.

Figure 14.7 Example of a pheromone used for defense. A whip scorpion is ejecting its spray toward an appendage pinched with forceps. The pattern of the spray is visible on acid sensitive indicator paper. The secretion is 84% acetic acid **(CH_3CO_2H)**, 5% octanoic acid **($CH_3(CH_2)_6CO_2H$)**, and 11% water.

There are two common sources of acetylene itself. Acetylene can be produced by heating coke (carbon from coal) with CaO in an electric furnace to yield calcium carbide, CaC_2.

$$CaO + 3C \xrightarrow{\text{heat}} \underset{\text{calcium carbide}}{CaC_2} + \underset{\text{carbon monoxide}}{CO} \tag{14.38}$$

Calcium carbide is formally an organometallic compound derived from the acetylene dianion.

$$Ca^{2+}\ {}^{-}\!:C\equiv C:^{-} \qquad \text{calcium carbide}$$

Like any acetylide ion, calcium carbide reacts vigorously with water to yield the hydrocarbon; the calcium oxide produced can be recycled in Eq. 14.38.

The second process for manufacture of acetylene is the thermal cracking (Sec. 5.9B) of ethylene at temperatures above 1200° to give acetylene and H_2. (This process is thermodynamically unfavorable at lower temperatures.) In the future, the choice between the carbide process and cracking as sources of acetylene will depend on the balance of energy costs (the carbide process requires more electricity), the availability of the starting materials (ethylene *vs.* coal), and capitalization costs necessary for plant conversions.

The most important general use of acetylene is for a chemical feedstock, as illustrated by the following examples.

$$HC{\equiv}CH + HCl \xrightarrow{HgCl_2} \underset{\text{vinyl chloride (a monomer used in the manufacture of polyvinyl chloride, PVC)}}{CH_2{=}CHCl} \tag{14.39}$$

$$2HC{\equiv}CH \xrightarrow{\text{catalyst}} \underset{\text{vinylacetylene}}{HC{\equiv}C{-}CH{=}CH_2} \xrightarrow{HCl} \underset{\text{chloroprene (polymerizes to give neoprene rubber; Sec. 15.5.)}}{CH_2{=}\underset{\displaystyle Cl}{\underset{|}{C}}{-}CH{=}CH_2} \tag{14.40}$$

Oxygen–acetylene welding is an important use of acetylene, although it accounts for a relatively small percentage of acetylene consumption. The acetylene used for this purpose is supplied in cylinders, but is hazardous because, at concentrations of 2.5–80% in air, it is explosive. Furthermore, since gaseous acetylene at even moderate pressures is unstable, this substance is not sold simply as a compressed gas. Acetylene cylinders contain a porous material saturated with a solvent such as acetone. Acetylene is so soluble in acetone that most of it actually dissolves. As acetylene gas is drawn off, more of the material escapes from solution as the gas is needed—another example of LeChatelier's principle in action!

KEY IDEAS IN CHAPTER 14

- Alkynes are compounds containing carbon–carbon triple bonds. The carbons of the triple bond are *sp* hybridized. Electrons in *sp* orbitals are held somewhat closer to the nucleus than those in sp^2 or sp^3 orbitals.
- The carbon–carbon triple bond in an alkyne consists of one σ-bond and two mutually perpendicular π-bonds. The electron density associated with the π-bonds resides in a cylinder surrounding the triple bond. The induced circulation of these π-electrons in a magnetic field causes the unique shielding of acetylenic protons observed in NMR spectra.
- The *sp* hybridization state is less stable than the sp^2 or sp^3 state. For this reason, alkynes have higher heats of formation than isomeric alkenes.
- Alkynes have two general types of reactivity:

  1. Addition to the triple bond
  2. Reactions at the acetylenic $\equiv$C—H bond

- Addition in some cases can occur twice, and is complicated by questions of stereochemistry. Nevertheless, there are some useful additions, such as Hg^{2+}-catalyzed hydration, hydroboration, catalytic hydrogenation, and reduction with sodium in liquid ammonia.
- Both hydration and hydroboration–oxidation of alkynes yield enols, which spontaneously form the isomeric aldehydes or ketones.
- Catalytic hydrogenation of alkynes can be stopped at the *cis* alkenes by the use of a catalyst poison. The reduction of alkynes with alkali metals in liquid ammonia, a reaction that involves radical-anion intermediates, provides a route to the corresponding *trans*-alkenes.
- 1-Alkynes, with pK_a values near 25, are the most acidic of the aliphatic hydrocarbons. Acetylenic anions are formed by the reactions of 1-alkynes with the strong base sodium amide. In a related reaction, acetylenic Grignard reagents can be formed in the reactions of 1-alkynes with alkylmagnesium halides.
- Acetylenic anions are good nucleophiles, and react with alkyl halides and sulfonates in S_N2 reactions to form new carbon–carbon bonds.

ADDITIONAL PROBLEMS

23 Give the principal product(s) expected when 4-octyne is treated with each of the following reagents:
(a) HBr
(b) H_2, Pd/C
(c) H_2, PtO_2, quinoline
(d) Product of (c) + O_3, then $(CH_3)_2S$
(e) Product of (c) + diborane, then H_2O_2/OH^-
(f) Br_2
(g) NaOH solution
(h) Hg^{2+}, H_2SO_4, H_2O
(i) Na/NH_3 (liq)
(j) Product of (i), then $Hg(OAc)_2$ in H_2O, then $NaBH_4$

24 In their latest catalog, Blarneystyne, Inc., a chemical company of dubious reputation specializing in alkynes, has offered some compounds for sale under the following names. Although each name unambiguously specifies a structure, all are incorrect. Propose a correct name for each compound.
(a) 2-hexyn-4-ol (c) 6-methoxy-1,5-hexadiyne
(b) 1-butyn-3-ene

25 Using only compounds containing carbon and hydrogen, draw the structure of
(a) a chiral alkyne of six carbon atoms
(b) an alkyne of six carbon atoms that gives the *same single* product in its reaction with either B_2H_6 followed by H_2O_2/OH^- or $H_2O/Hg^{2+}/H^+$
(c) an alkyne of six carbon atoms that gives the *same single* product in its reaction with either Na/NH_3 (liq) or H_2 over Pd/C in the presence of pyridine.
(d) a six-carbon alkyne that can exist as diastereomers.

Problems (Cont.)

26 Rank the anions within each series in order of increasing basicity, lowest first. Explain.

(a) $CH_3(CH_2)_3{-}C{\equiv}C{:}^-$, $CH_3(CH_2)_4\overset{-}{\ddot{C}}H_2$, $CH_3(CH_2)_3{-}CH{=}\overset{-}{\ddot{C}}H$

(b) $CH_3CH_2\ddot{\underset{\cdot\cdot}{O}}{:}^-$, $HC{\equiv}C{:}^-$, ${:}\ddot{\underset{\cdot\cdot}{F}}{:}^-$

---

27 Using simple observations or tests with readily observable results, show how you would distinguish between the compounds in each of the following pairs:
(a) *cis*-2-hexene and 1-hexyne
(b) 1-hexyne and 2-hexyne
(c) 4,4-dimethyl-2-hexyne and 3,3-dimethylhexane
(d) propyne and 1-decyne
(e) 5-chloro-5-methyl-2-hexyne and 6-chloro-5,5-dimethyl-2-hexyne

---

28 Outline a preparation of each of the following compounds from acetylene and any other reagents:

(a) 1-hexyne

(b) $CH_3CH_2\overset{\overset{\displaystyle OH}{|}}{C}HCH_2CH_2CH_3$

(c) $CH_3CH_2CD_2CD_2CH_2CH_3$

(d) 1-hexene

(e) $CH_3(CH_2)_7{-}\overset{\overset{\displaystyle O}{\|}}{C}{-}OH$

(f) *cis*-2-pentene
(g) *trans*-3-decene
(h) $(CH_3)_2CHCH_2CH_2CH_2CH{=}O$ (5-methylhexanal)
(i) *meso*-4,5-octanediol
(j) (±)-3,4-dibromohexane

---

29 Account with a detailed mechanism for the following known reaction of phenylacetylene:

$$Ph{-}C{\equiv}C{-}H \xrightarrow[\text{THF}]{\text{NaOD, } D_2O \text{ (large excess)}} Ph{-}C{\equiv}C{-}D$$

---

30 Using 1-butyne as the only source of carbon in the products, propose a synthesis for each of the following compounds:

(a) $CH_3CH_2C{\equiv}C{-}D$ (d) octane

(b) $CH_3CH_2CD_2CD_3$ (e) *cis*-3-octene

(c) $CH_3CH_2{-}\overset{\overset{\displaystyle O}{\|}}{C}{-}OH$

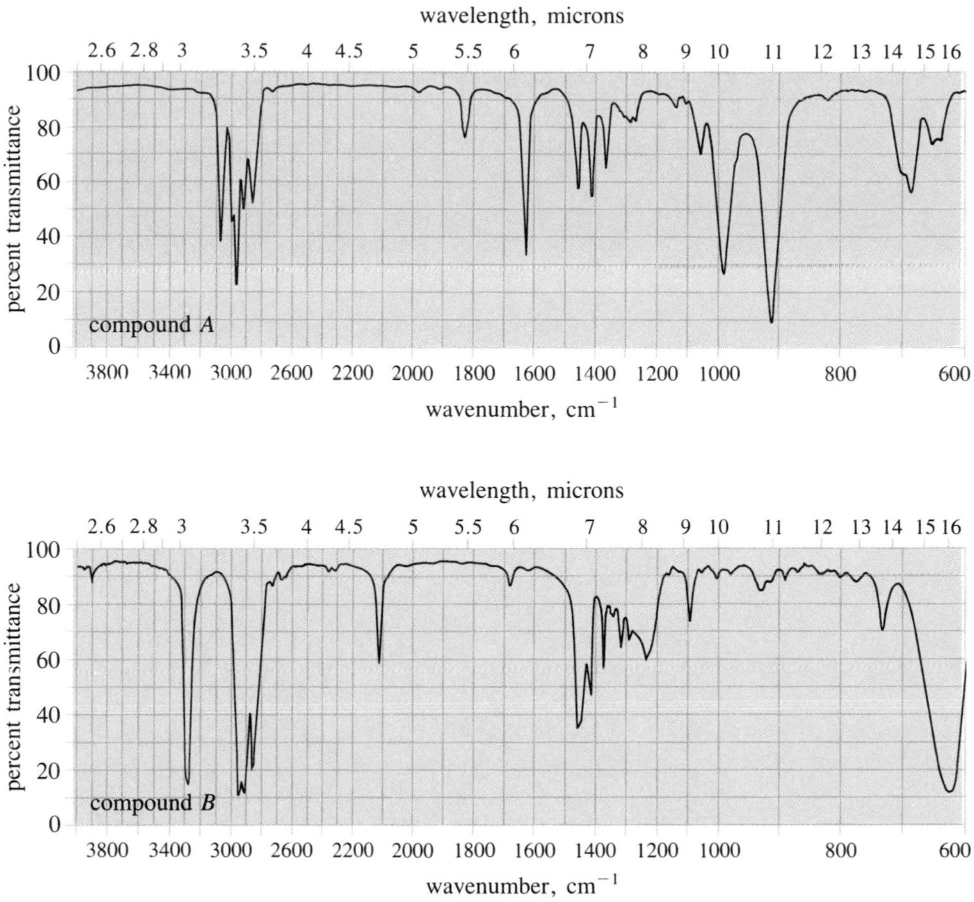

Figure 14.8 IR spectra for Problem 31.

31 In an abandoned laboratory has been found a box labeled "C_6H_{10} isomers" containing three compounds, *A, B,* and *C.* Along with the compounds are the IR spectra of *A* and *B,* shown in Fig. 14.8. Fragmentary data in a laboratory notebook suggest that the compounds are 1-hexyne, 2-hexyne, and 3-methyl-1,4-pentadiene. Identify the three compounds.

32 You have just been hired by Triple Bond, Inc., a company that specializes in the manufacture of alkynes containing five or fewer carbons. The President, Mr. Al Kyne, needs an outlet for the company's products. You have been asked to develop a synthesis of the housefly sex pheromone, *muscalure,* with the stipulation that all the carbon in the product must come only from the company's alkynes. The muscalure will then be used in an experimental fly trap.

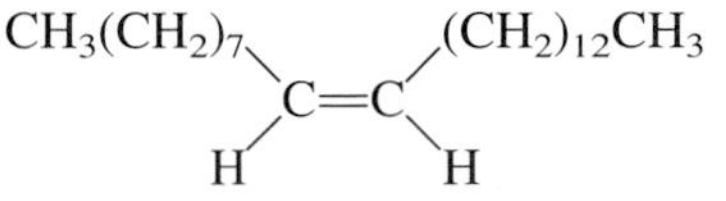

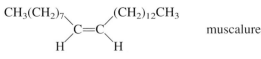

muscalure

(Problem continues on next page.)

Problems (Cont.)

You will be equipped with a laboratory containing all the company's alkynes, requisition forms for other reagents, and one gross of flyswatters in case you are successful. Outline a preparation of muscalure that meets the company's needs.

---

33 In the preparation of ethynylmagnesium bromide by Eq. 14.31, ethylmagnesium bromide is added to a large excess of acetylene in THF solution. Two side reactions that can occur in this procedure are the following:

$$H{-}C{\equiv}C{-}MgBr + C_2H_5{-}MgBr \longrightarrow C_2H_6 + BrMg{-}C{\equiv}C{-}MgBr \quad (1)$$

$$2H{-}C{\equiv}C{-}MgBr \rightleftharpoons H{-}C{\equiv}C{-}H + BrMg{-}C{\equiv}C{-}MgBr \quad (2)$$

(a) Suggest a mechanism for each of these side reactions.
(b) Explain why a large excess of acetylene minimizes the importance of these side reactions.
(c) Tetrahydrofuran (THF) is used as a solvent because the side product $BrMg{-}C{\equiv}C{-}MgBr$ is relatively soluble in THF. Explain why the solubility of this side product is important if both side reactions are to be avoided.

---

34 Identify the following compounds from their IR and NMR spectra:
(a) $C_6H_{10}O$:
NMR: δ 3.31 (3*H*, s); δ 2.41 (6*H*, s); δ 1.43 (1*H*, s)
IR: 2110, 3300 cm^{-1} (sharp)
(b) C_4H_6O:
liberates a gas when treated with C_2H_5MgBr
NMR: δ 2.43 (1*H*, t, $J = 2$ Hz); δ 3.41 (3*H*, s); δ 4.10 (2*H*, d, $J = 2$ Hz)
IR: 2125 cm^{-1}, 3300 cm^{-}
(c) C_4H_6O:
NMR in Fig. 14.9
IR: 2100, 3300 cm^{-1} (sharp), superimposed on a broad, strong band at 3350 cm^{-1}
(d) C_5H_6O (an alkyne):
NMR: δ 3.10 (1*H*, d, $J = 2$ Hz); δ 3.79 (3*H*, s); δ 4.52 (1*H*, doublet of doublets, $J = 6$ Hz and 2 Hz); δ 6.38 (1*H*, d, $J = 6$ Hz)

---

35 Outline a preparation of *disparlure,* a pheromone of the Gypsy moth, from acetylene and any other compounds containing not more than five carbon atoms.

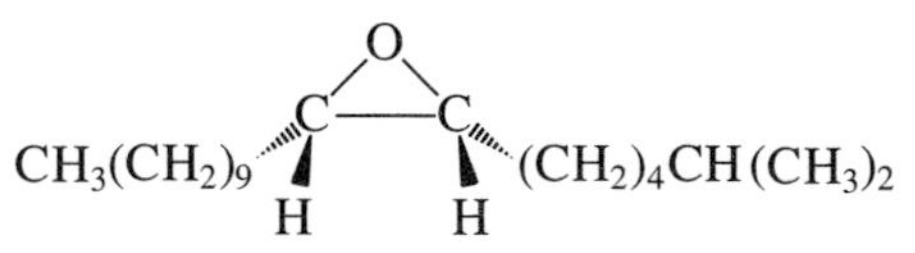

disparlure

---

36 Which of the following two secondary carbocations is more stable: the alkyl cation or the vinylic cation? (*Hint:* Which should have the greater heat of formation? See Sec. 14.2.)

$CH_3{-}CH{=}CH{-}CH_2{-}\overset{+}{C}H{-}CH_3$
secondary alkyl carbocation

$CH_3{-}CH{=}\overset{+}{C}{-}CH_2{-}CH_2{-}CH_3$
secondary vinylic carbocation

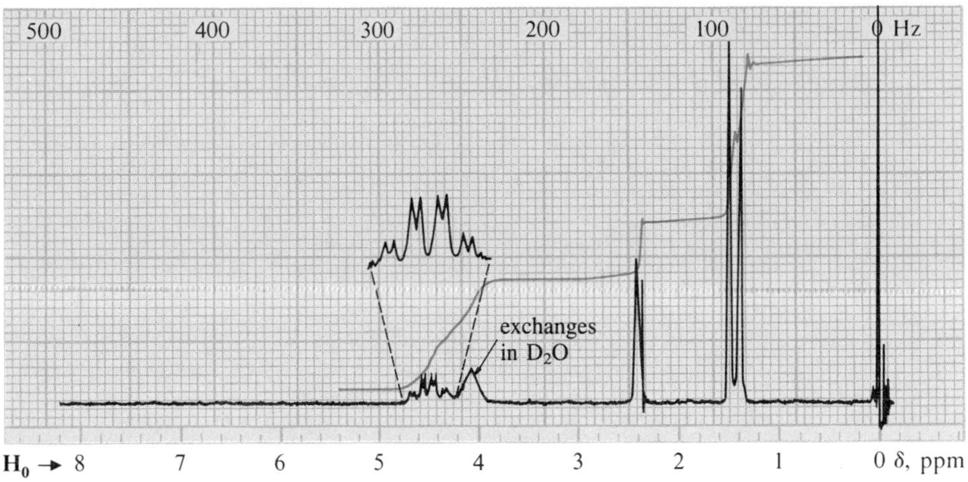

Figure 14.9 NMR spectrum for Problem 34c.

37 Complete the following reactions using intuition developed from this or previous chapters:

(a) $CH_3CH_2CH_2CH_2{-}C{\equiv}C{-}H + CH_3CH_2CH_2CH_2{-}Li \longrightarrow \xrightarrow{(CH_3)_3SiCl}$

(*Hint:* Tertiary silyl halides, unlike tertiary alkyl halides, undergo nucleophilic displacement reactions that are not complicated by competing elimination reactions.)

(b) $CH_3(CH_2)_6{-}C{\equiv}CH \xrightarrow{NaNH_2} \xrightarrow[\text{(diethyl sulfate)}]{CH_3CH_2{-}O{-}S(=O)_2{-}O{-}CH_2CH_3}$

(*Hint:* See Sec. 10.3B.)

(c) $Li{-}C{\equiv}CH + F{-}(CH_2)_5{-}Cl \longrightarrow$
(1 equiv)

(d) $Ph{-}CH{=}CH_2 + Br_2 \longrightarrow \xrightarrow{NaNH_2}$ (a hydrocarbon not containing bromine)

(*Hint:* See Sec. 9.4.)

(e) $Ph{-}C{\equiv}C{-}Ph + HCCl_3 \xrightarrow{K^+(CH_3)_3CO^-}$

(*Hint:* See Sec. 9.7.)

38 Propose mechanisms for each of the following known transformations; use the arrow formalism where possible:

(a) $HO{-}C(CH_3)_2{-}C{\equiv}C{-}C(CH_3)_2{-}OH \xrightarrow[Hg^{2+}]{H_2SO_4}$ 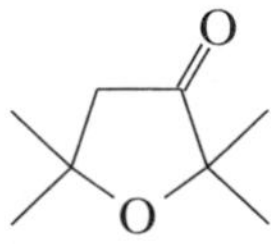

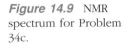

Problems *(Cont.)*

(b) $Ph-C\equiv CH + Br_2 \xrightarrow{H_2O} Ph-\overset{\overset{\displaystyle O}{\parallel}}{C}-CH_2Br$

(c) $Ph-C\equiv C-H \xrightarrow{C_2H_5MgBr} \xrightarrow{\overset{O}{\triangle}} \xrightarrow{H_3O^+}$

$Ph-C\equiv C-CH_2CH_2OH + Ph-\overset{}{\langle}\!\!\!\raisebox{-3pt}{O}$

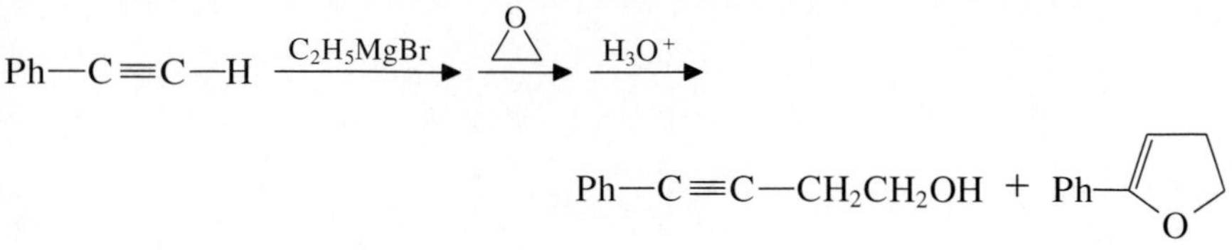

39 An optically active alkyne *A* has the following elemental analysis: 89.52% C; 10.48% H. Compound *A* can be catalytically hydrogenated to butylcyclohexane. Treatment of *A* with C_2H_5MgBr liberates no gas. Catalytic hydrogenation of *A* over Pd/C in the presence of quinoline, and treatment of the product with O_3, then H_2O_2, gives an *optically inactive* tricarboxylic acid $C_8H_{12}O_6$. (A tricarboxylic acid is a compound with three $-CO_2H$ groups.) Give the structure of *A* and account for all observations.

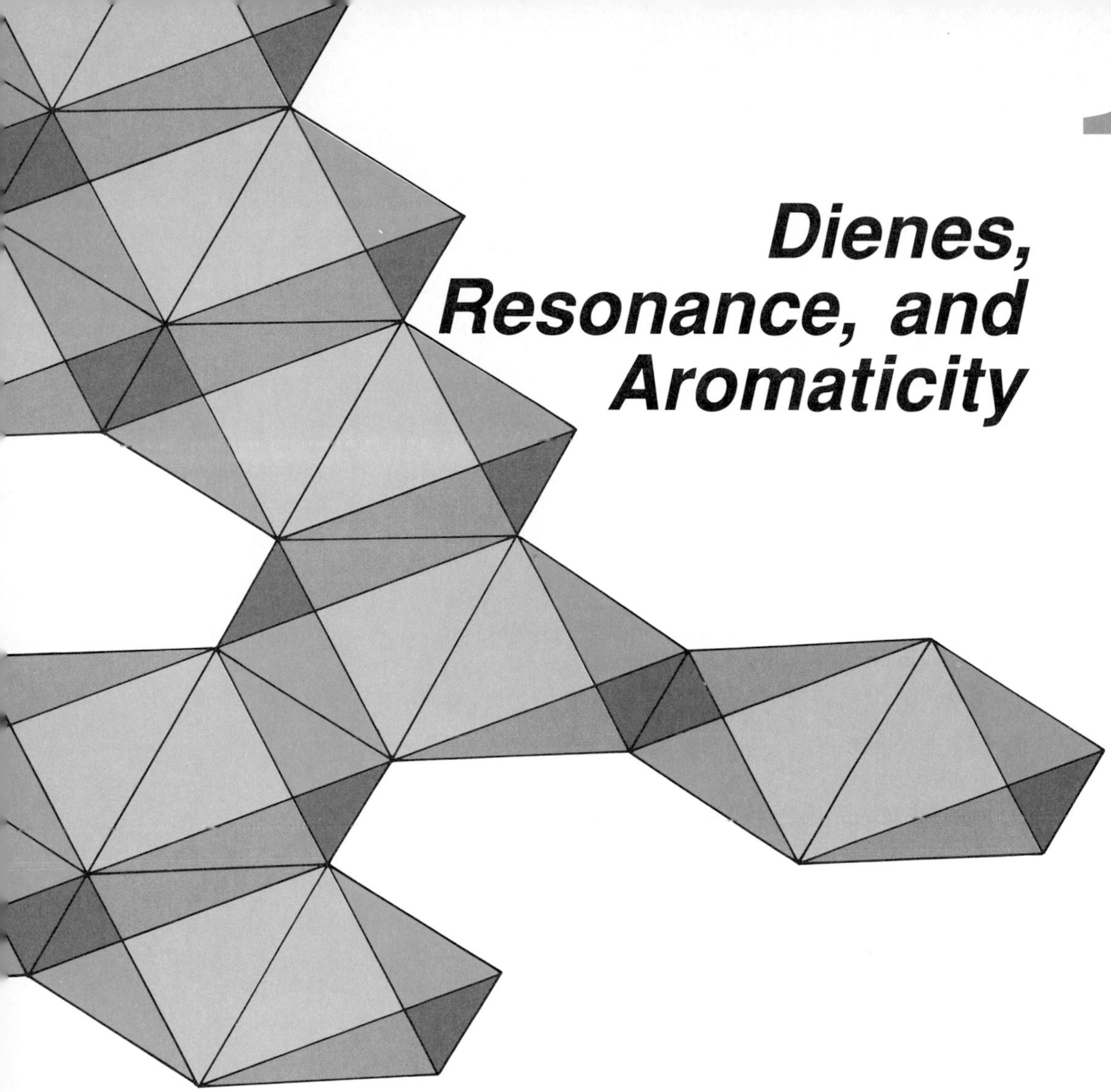

15 Dienes, Resonance, and Aromaticity

Dienes are compounds with two carbon–carbon double bonds. Their nomenclature was discussed along with the nomenclature of other alkenes (Sec. 4.2A). Dienes are classified according to the relationship of their double bonds. In **conjugated dienes,** two double bonds are separated by one single bond. These double bonds are called **conjugated double bonds.**

$$CH_2{=}CH{-}CH{=}CH_2 \qquad \text{1,3-butadiene (a conjugated diene)}$$

conjugated double bonds

Cumulenes are compounds in which one carbon participates in two carbon–carbon double bonds; these double bonds are called **cumulated double bonds.** Allene is the simplest cumulene. The term *allene* is also sometimes used as a family name for compounds containing only two cumulated double bonds.

$$CH_2{=}C{=}CH_2 \qquad \text{allene}$$

one carbon involved in two double bonds

Conjugated dienes and allenes have unique structures and chemical properties that are the basis for much of the discussion in this chapter. On the other hand, dienes in which the double bonds are separated by two or more saturated carbon atoms have structures and chemical properties more or less like those of simple alkenes, and do not require special discussion. We shall call these dienes "ordinary" dienes.

$$CH_2{=}CHCH_2CH_2CH_2CH{=}CH_2$$ 1,6-heptadiene (an ordinary diene)

In this chapter we shall have our first taste of what happens when two functional groups—in this case, two carbon–carbon double bonds—are close enough within a molecule that one functional group affects the properties of the other. In particular, we shall spend considerable time learning about the effects of *conjugation* on chemical stability and reactivity. This discussion will lead us into a consideration of benzene, a cyclic hydrocarbon in which the effects of conjugation are particularly dramatic. The chemistry of benzene derivatives and the effects of conjugation will continue as the central themes through Chapter 18.

15.1 STRUCTURE AND STABILITY OF DIENES

A. Structure and Stability of Conjugated Dienes

The heats of formation of several C_5H_8 isomers, tabulated in Table 15.1, provide information about the relative stabilities of dienes. The effect of conjugation on the stability of dienes can be deduced from a comparison of the numbers for 1,3-pentadiene and 1,4-pentadiene, provided that we first make a correction for the different degrees of alkyl substitution on the double bond in each compound (Sec. 4.4C).

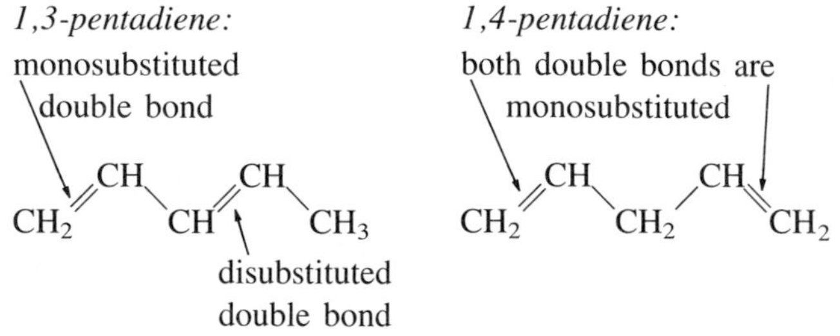

From Table 15.1, we see that *trans*-1,3-pentadiene is about 7 kcal/mol more stable than 1,4-pentadiene. The effect of conjugation on the relative stability of 1,3-pentadiene is determined by subtracting 2 kcal/mol, the effect due to the additional alkyl substitu-

TABLE 15.1 Heats of Formation of Some C_5H_8 Isomers

Compound	ΔH_f^0 (25°, gas phase), kcal/mol	Compound	ΔH_f^0 (25°, gas phase), kcal/mol
1,4-pentadiene	25.20	1-pentyne	34.50
1,3-pentadiene (*cis*)	19.10	2-pentyne	30.80
1,3-pentadiene (*trans*)	18.10	2,3-pentadiene	31.80
1,2-pentadiene	34.80	isoprene	18.06

tion on one double bond (Sec. 4.4C). The remaining 5 kcal/mol is the additional stability attributable to the conjugated arrangement of double bonds.

There are two reasons for this additional stability. The first is that the carbon–carbon single bond between the two double bonds in a conjugated diene is derived from the overlap of two carbon sp^2 orbitals; that is, it is an sp^2-sp^2 single bond. This is a stronger bond than the sp^2-sp^3 single bonds in an ordinary diene such as 1,4-pentadiene. The stronger bond gives a conjugated diene greater stability.

The second reason is that there is overlap of *p* orbitals across the carbon–carbon bond connecting the two alkene units. That is, not only is there π-bonding *within* the two alkene units, but *between* them as well. The additional bonding associated with this overlap provides additional stability to the molecule.

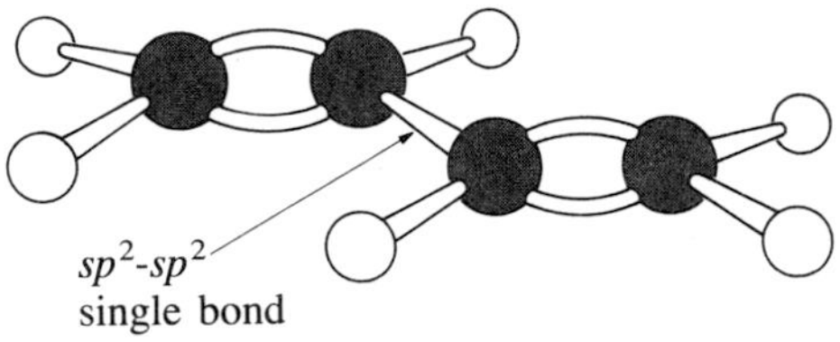

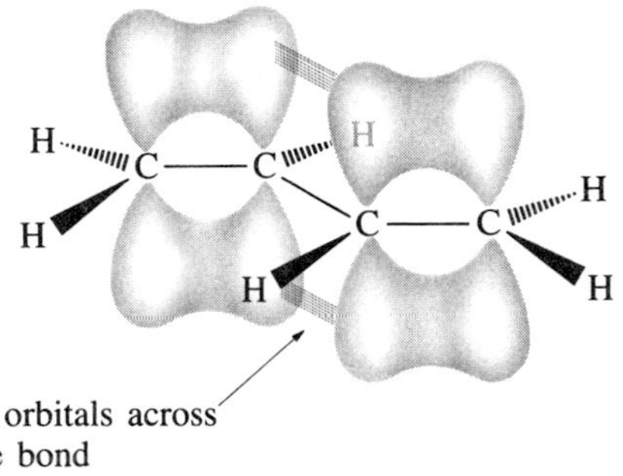

The length of the carbon–carbon single bond in 1,3-butadiene reflects the hybridization of the orbitals from which it is constructed. At 1.46 Å, this sp^2-sp^2 single bond is considerably shorter than the sp^2-sp^3 carbon–carbon single bond in propene (1.50 Å), or the sp^3-sp^3 carbon–carbon bond in ethane (1.54 Å).

$$\overset{1.34\ \text{Å}}{CH_2{=\!=}CH}-\!\!\!-\underset{1.46\ \text{Å}}{}\overset{1.34\ \text{Å}}{CH{=\!=}CH_2}$$

As the fraction *s* character in the component orbitals increases, the length of the bond decreases (Secs. 4.1A, 14.2).

There are two stable conformations of 1,3-butadiene: the transoid, or *s-trans* form, and the cisoid, or *s-cis* form. (These terms refer to conformation about the *single* bond—thus the *s* prefix.) These conformations are related by a 180° rotation about the central carbon–carbon single bond.

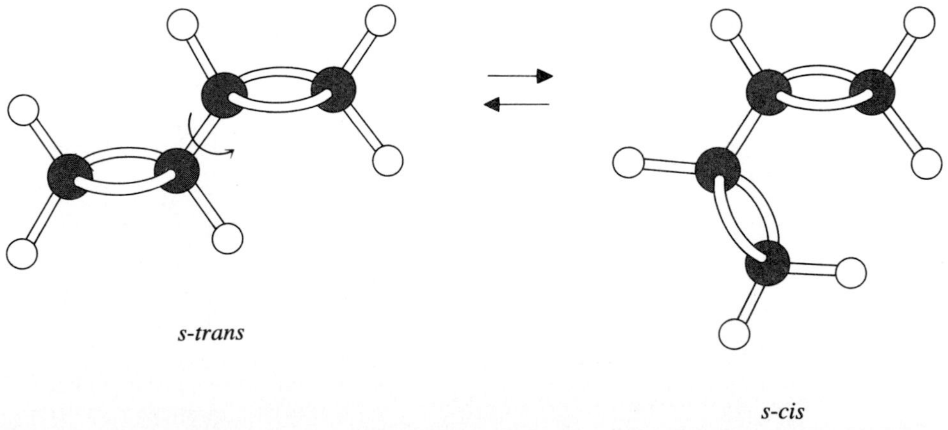

(15.1)

Both forms are planar, or nearly so. If 1,3-butadiene were not planar, its *p* orbitals could not overlap, and some of the extra stabilization associated with conjugation would be lost. The *s-trans* form of the molecule is about 3 kcal/mol more stable than the *s-cis* form. The internal rotation that interconverts these two forms is very rapid at room temperature.

Problems

1 Why does the *s-cis* form of 1,3-butadiene have higher energy than the *s-trans* form? Use models if necessary. It has been suggested that this form may be skewed away from planarity by about 15°. Why should this be energetically advantageous to the molecule?

2 Although the *s-cis* and *s-trans* forms of 1,3-butadiene interconvert rapidly, the energy barrier for this interconversion is about 3–4 kcal/mol higher than the barriers for internal rotation in butane. Suggest an explanation for this larger energy barrier.

B. Structure and Stability of Cumulated Dienes

The structure of allene (propadiene) is shown in Fig. 15.1. Since the central carbon of allene is bound to two groups, the carbon skeleton of this compound is linear (Sec. 1.6B). We have learned (Sec. 14.2) that a carbon atom with 180° bond angles is *sp* hybridized. Therefore, the central carbon of allene, like the carbons in an alkyne triple bond, is *sp* hybridized (Fig. 15.2). The two remaining carbons are sp^2 hybridized and have trigonal geometry. Notice that the bond length of each cumulated double bond is somewhat less than that of an ordinary alkene double bond.

As Fig. 15.2 shows, the two bonding π-molecular orbitals in allenes are mutually perpendicular, as required by the *sp* hybridization of the central carbon atom. Consequently, *the H—C—H plane at one end of the allene molecule is perpendicular to the H—C—H plane at the other end.* Note carefully the difference in the bonding arrange-

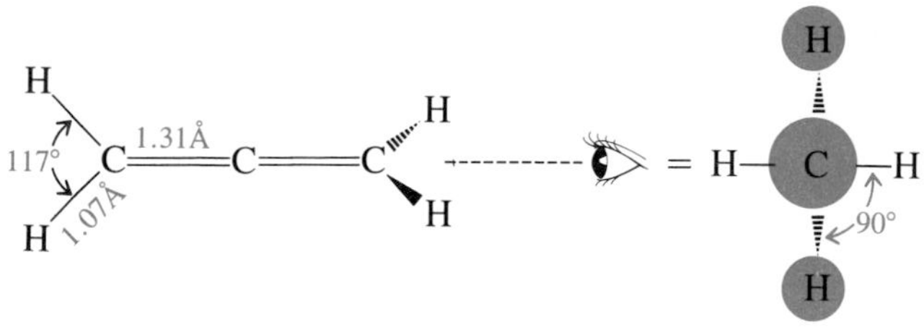

Figure 15.1 Structure of allene.

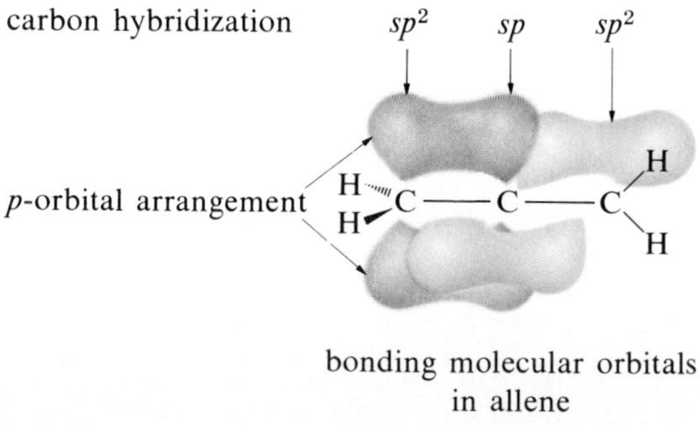

Figure 15.2 Electronic structure of allene.

ments in allene and the conjugated diene 1,3-butadiene. In the conjugated diene, the π-electron systems of the two double bonds are coplanar, and can overlap; all atoms are sp^2 hybridized. In allene, there are two mutually perpendicular π-systems, each spanning two carbons; the central carbon is part of both. Because these two π-systems are perpendicular, they do *not* overlap.

Because of their geometry, some allenes are chiral even though they do not possess an asymmetric carbon atom. The following molecule, 2,3-pentadiene, is an example of a chiral allene.

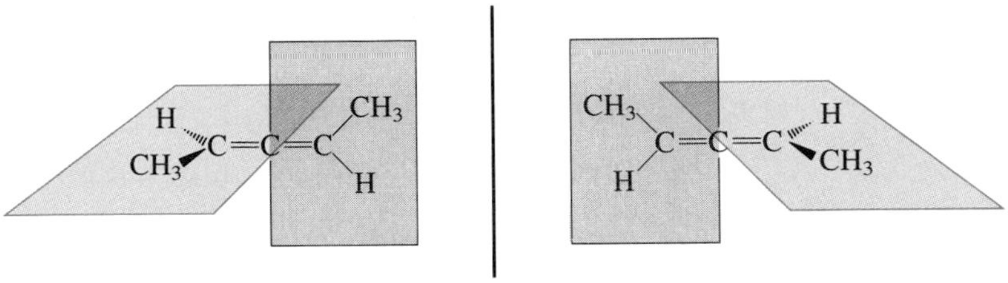

enantiomers

(Using models if necessary, verify the chirality of 2,3-pentadiene by attempting to superimpose these two structures.) These enantiomers differ by a rotation about either double bond. Since rotation about a double bond for all practical purposes does not occur, enantiomeric allenes can be isolated. (See Sec. 6.8A.)

The sp hybridization of allenes is reflected in their C=C stretching absorption in the infrared spectrum, which occurs near 1950 cm^{-1}, not far from the C≡C stretching absorption of alkynes.

An examination of Table 15.1 shows that allenes have greater heats of formation than other types of isomeric dienes. Thus, the cumulated arrangement is the least stable arrangement of two double bonds. Allenes are also somewhat less stable than isomeric alkynes. In fact, one of the common reactions of allenes is isomerization to alkynes.

Although there are a few naturally occurring allenes, generally allenes are relatively rare in nature.

Problems

3 Explain why the bond lengths of cumulated double bonds are somewhat shorter than those of an ordinary alkene.

4 (a) Draw the two enantiomers of the following allene:

$$CH_3-CH=C=C \begin{cases} CH_2CH_2CH_2CH_3 \\ CO_2H \end{cases}$$

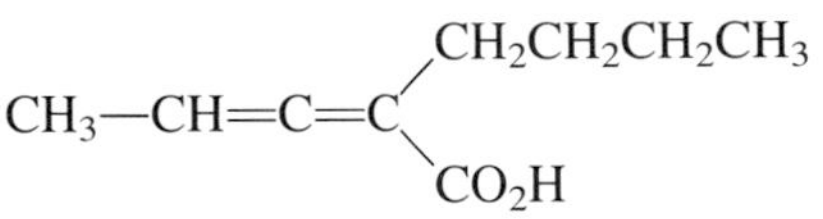

(b) One enantiomer of this compound has a specific rotation of $-30.7°$. What is the specific rotation of the other?

15.2 ULTRAVIOLET SPECTROSCOPY

The IR and NMR spectra of conjugated dienes show only minor differences from the spectra of ordinary alkenes. However, another type of spectroscopy can be used to identify organic compounds containing conjugated π-electron systems. In this type of spectroscopy, called **ultraviolet-visible spectroscopy,** the absorption of radiation in the ultraviolet or visible region of the spectrum is recorded as a function of wavelength. The part of the ultraviolet spectrum of greatest interest to organic chemists is the *near ultraviolet* (wavelength range 200×10^{-9} to 400×10^{-9} meter). Visible light, as the name implies, is electromagnetic radiation visible to the human eye (wavelengths from 400×10^{-9} to 750×10^{-9} meter). Since there is a common physical basis for the absorption of both ultraviolet and visible radiation by chemical compounds, both UV and visible spectroscopy are considered together as one type of spectroscopy, often called simply **UV spectroscopy.**

A. The UV Spectrum

Like any other absorption spectrum, the UV spectrum of a substance is the graph of radiation absorption by a substance *vs.* the wavelength of radiation that is passed through it. The instrument used to measure the UV spectrum is called a **UV spectrometer.** Except for the fact that it is designed to operate in a different part of the electromagnetic spectrum, it is conceptually much like an IR spectrometer (Fig. 12.12).

A typical UV spectrum, that of 2-methyl-1,3-butadiene (isoprene), is shown in Fig. 15.3. Since isoprene does not absorb visible light, only the ultraviolet region of the spectrum is shown. On the horizontal axis of the UV spectrum is plotted the wavelength λ of the ultraviolet radiation. In UV spectroscopy, the conventional unit of wavelength is the **nanometer** (abbreviated nm). One nanometer equals 10^{-9} meter. (In older literature, the term **millimicron,** abbreviated mμ, was used; one millimicron equals one nanometer.) The relationship between energy of the electromagnetic radiation and its frequency or wavelength, discussed in Sec. 12.1A, should be reviewed again.

The vertical axis of a UV spectrum shows the **absorbance.** (Absorbance is sometimes called **optical density,** abbreviated O. D.) The absorbance is a measure of the amount of radiant energy absorbed. Suppose the radiation entering a sample has intensity I_0, and the light emerging from the sample has intensity I. The absorbance A is defined as the logarithm of the ratio I_0/I:

$$A = \log (I_0/I) \tag{15.2}$$

The more radiant energy is absorbed, the larger is the ratio I_0/I, and the greater is the absorbance.

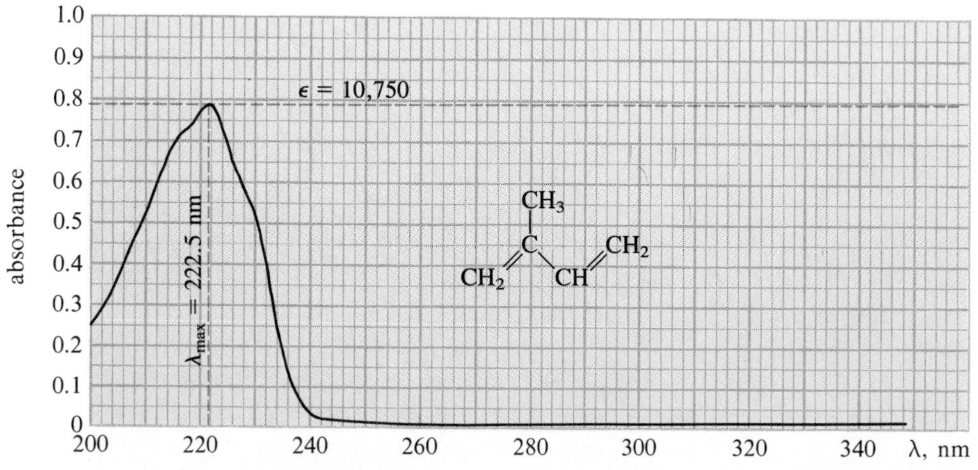

Figure 15.3 Ultraviolet spectrum of isoprene in methanol.

Problems

5 What is the energy, in kcal/mol, of light with a wavelength of 450 nm?

6 (a) What is the absorbance of a sample that transmits one-half of the incident radiation intensity?
(b) What percent of the incident radiation is transmitted by a sample when its absorbance is 1.0? Zero?

7 When a thin piece of red glass is held up to a white light, it appears brighter to the eye than a piece of the same glass that is twice as thick. Which piece has the higher absorbance?

In the UV spectra used in this text, absorbance increases from bottom to top of the spectrum. Therefore, absorption maxima occur as high points or peaks in the spectrum. Notice the difference in how UV and IR spectra are presented. (Absorption in IR spectra increases from top to bottom, since IR spectra are conventionally presented as plots of *transmittance*, or percentage of light transmitted.) In the UV spectrum shown in Fig. 15.3, the absorbance maximum occurs at a wavelength of 222.5 nm. The wavelength at the maximum of an absorption peak is called the $\boldsymbol{\lambda_{max}}$ (read "lambda-max"). Some compounds have several absorption peaks and a corresponding number of λ_{max} values. Absorption peaks in the UV spectra of compounds in solution are generally quite broad. That is, peak widths span a considerable range of wavelength, typically 50 nm or more.

UV spectroscopy is frequently used for quantitative analysis because the absorbance at a given wavelength depends on the number of molecules in the light path. If a sample is contained in a vessel with a thickness along the light path of l cm, and the absorbing compound is present at a concentration of c moles per liter, then the absorbance is proportional to the product lc.

$$A = \epsilon lc \tag{15.3}$$

This equation is called the *Beer–Lambert Law* or simply **Beer's Law.** The constant of proportionality ϵ is called the **molar extinction coefficient** or **molar absorptivity.** There is a unique extinction coefficient for each absorption of each compound, and it depends on wavelength, solvent, and temperature. The larger is ϵ, the greater is its light absorption at a given concentration c and path length. For example, isoprene (Fig. 15.3) has an extinction coefficient at 222.5 nm of 10,750 in methanol solvent at 25°; its extinction coefficient in alkane solvents is nearly twice as large.

Some UV spectra are presented in abbreviated form by citing the λ_{max} values of their principal peaks, the solvent used, and the extinction coefficients. For example, the spectrum in Fig. 15.3 is summarized as follows:

$$\lambda_{max}(CH_3OH) = 222.5 \text{ nm } (\epsilon = 10{,}750)$$

or

$$\lambda_{max}(CH_3OH) = 222.5 \text{ nm } (\log \epsilon = 4.03).$$

Problem

8 (a) From the extinction coefficient of isoprene and its observed absorbance at 222.5 nm (Fig. 15.3), calculate the concentration of isoprene in moles/liter (assume a 1-cm light path).

(b) From the results of part (a) and Fig. 15.3, calculate the extinction coefficient of isoprene at 235 nm.

B. Physical Basis of UV Spectroscopy

What determines whether an organic compound will absorb UV or visible radiation? Ultraviolet or visible radiation is absorbed by the π-electrons and, in some cases, by the unshared electron pairs in organic compounds. For this reason, UV and visible spectra are sometimes called *electronic spectra*. (The electrons of σ-bonds absorb at much lower wavelengths, in the far ultraviolet.) Absorption by compounds containing only unshared electron pairs is generally quite weak (that is, their extinction coefficients are small). However, intense absorption of UV radiation occurs when a compound contains π-electrons. Extinction coefficients in such cases are in the range 10^3 to 10^5. The simplest hydrocarbon containing π-electrons, ethylene, absorbs UV radiation at $\lambda_{max} = 165$ nm ($\epsilon = 15{,}000$). Although this is a strong absorption, the λ_{max} of ethylene and other simple alkenes is below the usual working range of most conventional UV spectrometers; the lower end of this range is about 200 nm. However, molecules with *conjugated* double or triple bonds (for example, isoprene, Fig. 15.3), have λ_{max} values greater than 200 nm. Therefore, *UV spectroscopy is especially useful for the diagnosis of conjugated double or triple bonds.*

$CH_2{=}CHCH_2CH_2CH{=}CH_2$ double bonds not conjugated; no UV spectrum above 200 nm

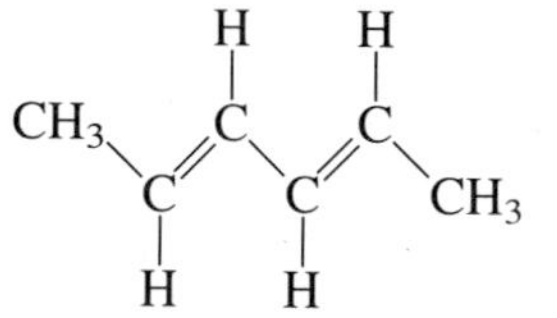

conjugated double bonds; this compound has UV absorption above 200 nm: $\lambda_{max} = 227$ nm ($\epsilon = 14{,}200$)

The structural feature of a molecule responsible for its UV or visible absorption is called a **chromophore.** Thus, the chromophore in isoprene (Fig. 15.3) is the system of conjugated double bonds. Because many important compounds do not contain conjugated double bonds or other chromophores, UV spectroscopy, compared to NMR and IR spectroscopy, has limited utility in structure determination. However, the technique is widely used for quantitative analysis in both chemistry and biology; and, when compounds do contain conjugated multiple bonds, the UV spectrum can be an important element in a structure proof.

What happens to the π-electrons when they absorb energy from electromagnetic radiation? Consider the π-electron structure of ethylene, shown in Fig. 4.5. The two π-electrons of ethylene occupy a *bonding* π-molecular orbital. When ethylene absorbs radiant energy, a π-electron is elevated from this bonding molecular orbital to the *antibonding* π^*-molecular orbital. This means that an electron assumes the wave motion characteristic of the π^*-orbital, which includes a node between the two carbon atoms. To a useful approximation, the radiant energy required for this absorption must match the difference in energy between the π- and π^*-orbitals (Fig. 15.4). The 165 nm-absorption of ethylene is called for this reason a **$\pi \rightarrow \pi^*$ transition** (read

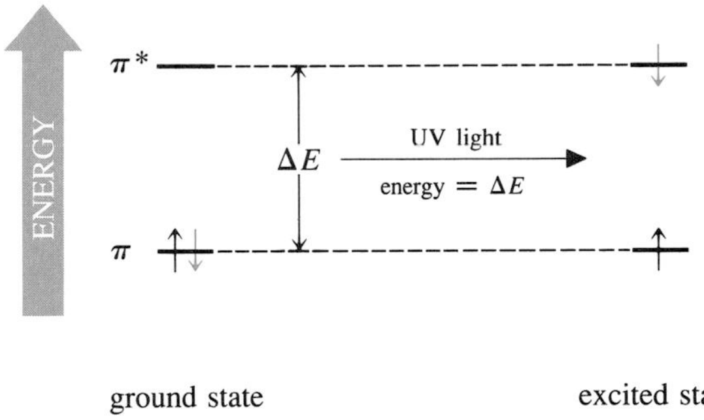

Figure 15.4 Absorption of UV light by ethylene causes an electron in the bonding π-molecular orbital to be promoted to the antibonding π*-molecular orbital.

"pi to pi star"). The UV absorptions of conjugated alkenes are also due to $\pi \rightarrow \pi^*$ transitions. However, their absorptions occur at higher wavelengths (lower energies) because the energy separation between the π- and π*-orbitals is smaller (for reasons that we need not consider here).

Problem

9 Estimate the energy difference between the π- and π*-molecular orbitals involved in the UV absorption of (a) ethylene; (b) isoprene (Fig. 15.3).

C. UV Spectroscopy of Conjugated Alkenes

When UV spectroscopy is used to determine chemical structure, the most important aspect of a spectrum is the λ_{max} values. The structural feature of a compound that is most important in determining the λ_{max} is the number of consecutive conjugated double (or triple) bonds. *The longer the conjugated system (that is, the more consecutive conjugated multiple bonds), the higher the wavelength of the absorption.* Table 15.2 gives the λ_{max} values for a homologous series of conjugated alkenes. (The skeletal structures used in this table were discussed in Sec. 3.3.) Notice how λ_{max} (as well as the extinction coefficient) increases with increasing number of conjugated double bonds. From these data we can see that each additional conjugated double bond increases λ_{max} by 30–50 nm. Molecules that contain many conjugated double bonds, such as the last one in Table 15.2, generally have several absorption peaks. The λ_{max} usually quoted for such compounds is the one at highest wavelength.

If a compound has enough double bonds in conjugation, its λ_{max} will be large enough to be in the visible region of the electromagnetic spectrum, and the compound will appear colored. An understanding of this idea is based on the concept of

TABLE 15.2 Ultraviolet Absorptions for Ethylene and Conjugated Alkenes

Alkene	λ_{max}, nm	ε
ethylene	165	15,000
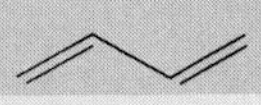	217	21,000
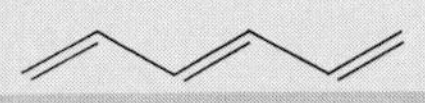	268	34,600
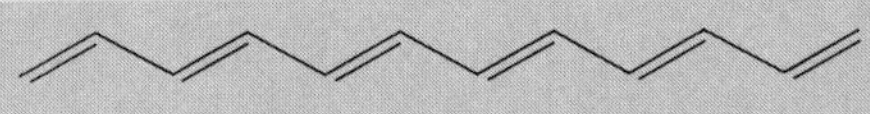	364	138,000

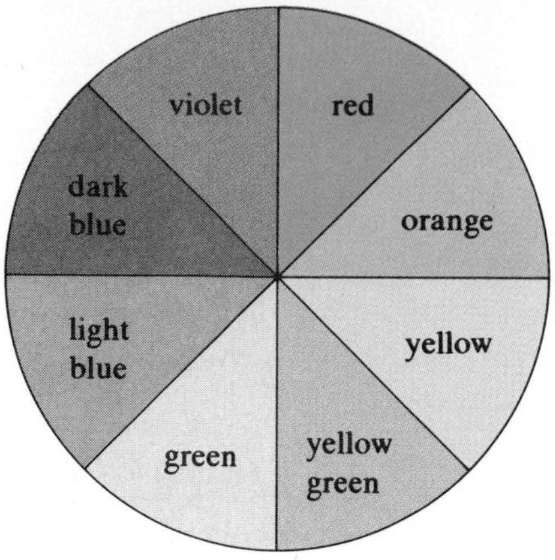

Figure 15.5 Colors in opposite wedges of the color rosette are complementary.

complementarity, illustrated by the color rosette shown in Fig. 15.5. Diametrically opposed segments of the color rosette are said to be *complementary*; for example, red and green are complementary colors. If we view a transparent object in white light (light that contains all wavelengths of the visible spectrum), its observed color is that of the light transmitted—that is, the part of the white light that the object *does not* absorb. In terms of the color rosette, if a particular color is absorbed, we see its complementary color. Thus if a compound absorbs red light (670 nm), it will appear green to the eye because green is the complement of red. Actually, this picture is too simple, because the eye has differing sensitivity to different colors, and because we rarely use pure white light as a source. Nevertheless, this discussion shows the general principles involved when color is produced by light absorption.

An example of a conjugated alkene with visible absorption is β-carotene, which is found in carrots and is known to be a biosynthetic precursor of vitamin A.

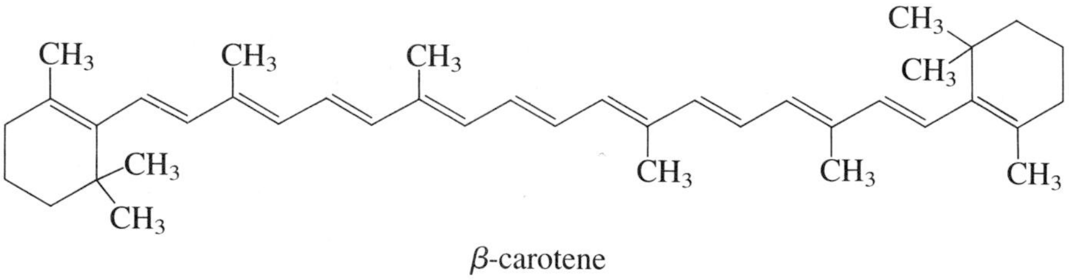

β-carotene

Because of the large number of double bonds in β-carotene, its absorption, which occurs between 400 and 500 nm, is in the visible (blue-green) part of the electromagnetic spectrum. From the color rosette (Fig. 15.5), we find that the complement of blue-green is red-orange. In fact, β-carotene is responsible for the orange color of carrots.

Since the human eye can detect visible light, it is not surprising that there are, within the eye, organic compounds that absorb light in the visible region of the spectrum. In fact, light absorption by a pigment, *rhodopsin,* in the rod cells of the eye (as well as a related pigment in the cone cells) is what triggers the series of physiological events associated with *vision*. The chromophore in rhodopsin is its group of six conjugated double bonds (color in the following structure):

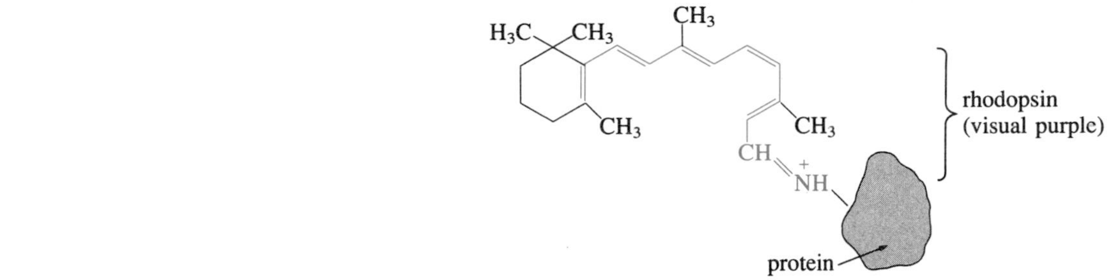

Although the number of double or triple bonds in conjugation is the most important thing that determines the λ_{max} of an organic compound, other factors are involved. One is *the conformation of a diene unit about its central single bond*—that is, whether the diene is in an *s-cis* or *s-trans* conformation (Sec. 15.1A). Noncyclic dienes generally assume the lower energy *s-trans* conformation. However, dienes that are locked into *s-cis* conformations have *higher* values of λ_{max} and *lower* extinction coefficients than comparably substituted *s-trans* compounds, as we see in the following examples:

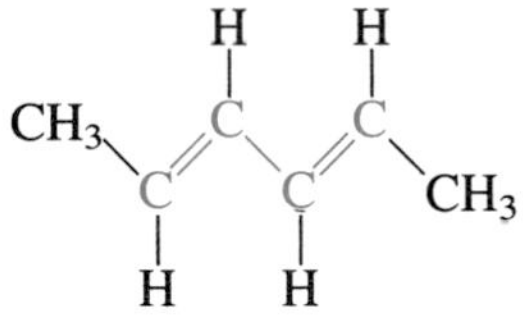

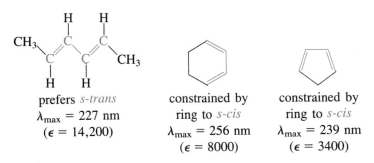

prefers *s-trans*
λ_{max} = 227 nm
(ϵ = 14,200)

constrained by ring to *s-cis*
λ_{max} = 256 nm
(ϵ = 8000)

constrained by ring to *s-cis*
λ_{max} = 239 nm
(ϵ = 3400)

A third variable that affects λ_{max} in a less dramatic but yet predictable way is the presence of *substituent groups on the double bond.* For example, each alkyl group on a double bond adds about 5 nm to the absorption position of a conjugated alkene. Thus, the two methyl groups of 2,3-dimethyl-1,3-butadiene add (2 × 5) = 10 nm to the λ_{max} of 1,3-butadiene, which is 217 nm (Table 15.2). The predicted λ_{max} is (217 + 10) = 227 nm; the observed value is 226 nm.

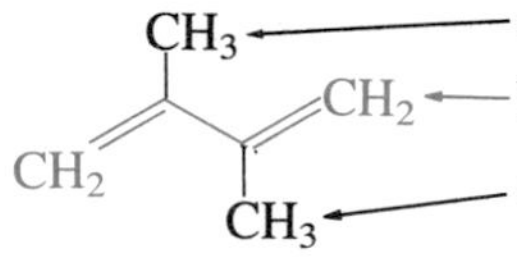

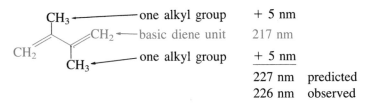

one alkyl group	+ 5 nm	
basic diene unit	217 nm	
one alkyl group	+ 5 nm	
	227 nm	predicted
	226 nm	observed

Although there are other structural features that affect the λ_{max} of a conjugated alkene, we do not need to learn them all here. There are really two important points to remember:

1. The λ_{max} is greater for compounds containing more conjugated double bonds.
2. The λ_{max} is affected by substituents, conformation, and other structural characteristics of the conjugated π-electron system.

Problem

10 Predict λ_{max} for the UV absorption of each of the following compounds:

(a)

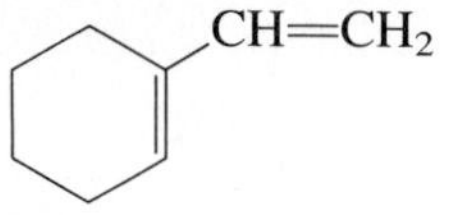

(c)

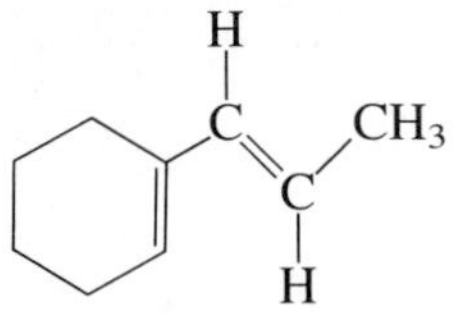

(b)

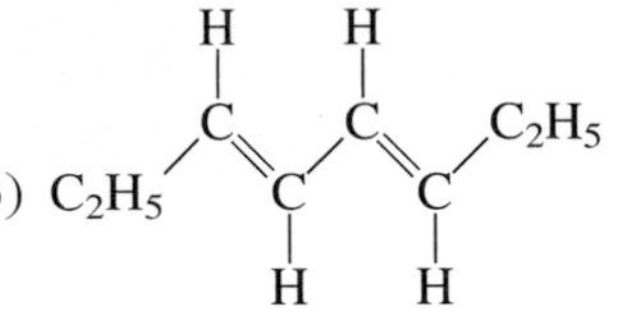

15.3 THE DIELS–ALDER REACTION

A. Reactions of Conjugated Dienes with Alkenes

Conjugated dienes undergo several unique reactions. One of these was discovered in 1928, when Otto Diels and Kurt Alder showed that many conjugated dienes undergo addition reactions with certain alkenes or alkynes. The following reaction is typical:

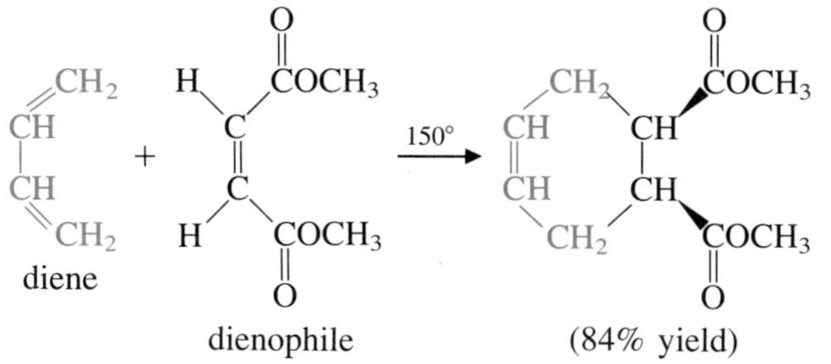

(15.4)

This type of reaction between a conjugated diene and an alkene is called the **Diels–Alder reaction.** For their extensive work on this reaction, Diels and Alder shared the 1950 Nobel Prize in chemistry. The conjugated diene component of this reaction is often referred to as the *diene,* and the other component as the *dienophile* (literally, "diene-lover"). Mechanistically, the Diels–Alder reaction resembles the OsO_4 oxidation and ozonolysis of alkenes (Secs. 5.4 and 5.5) in that it is both a *pericyclic reaction*—a reaction that occurs in one step by a cyclic flow of electrons—and a *cycloaddition*—an addition reaction that forms a ring.

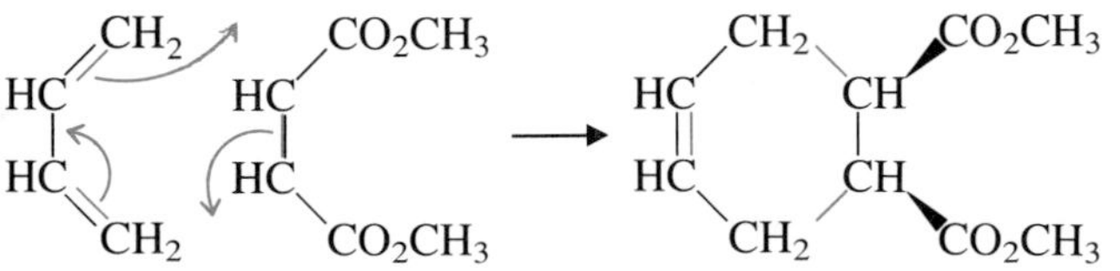

(15.5)

Some of the dienophiles that react most readily in the Diels–Alder reaction, as in the above example, bear substituent groups such as esters (—CO_2R), nitriles (—C≡N), or certain other unsaturated, electronegative groups. However, these substituents are not strictly necessary, as the reactions of many other dienophiles can be promoted by heat or pressure. Some alkynes also act as dienophiles:

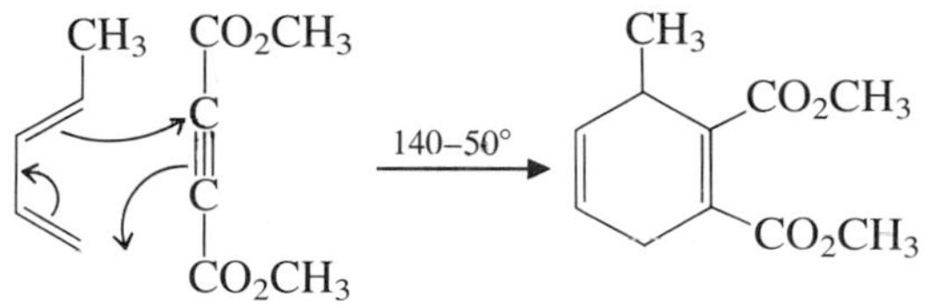

(15.6)

When the diene is cyclic, bicyclic products are obtained in the Diels–Alder reaction. In fact, this reaction is one of the best ways to prepare certain *bicyclic compounds* (Sec. 7.7).

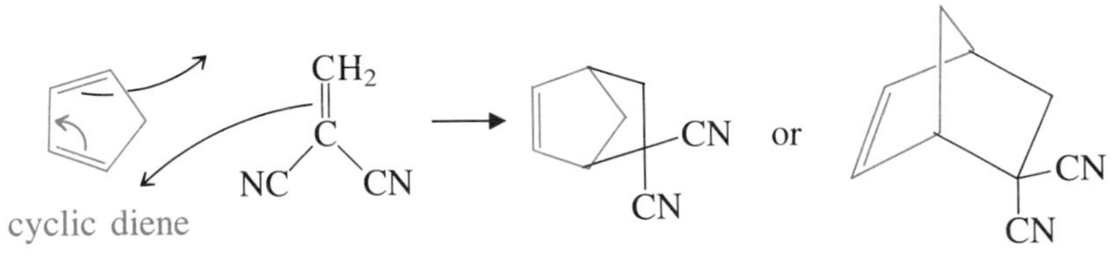

(15.7)

If we focus on what happens to the diene, we find that the Diels–Alder reaction is a **1,4-addition,** or **conjugate addition,** because the dienophile carbons add across the outer carbons of the diene. (Note that the 1,4-terminology indicates the relative locations of the carbons involved in the addition; it has nothing to do with the numbering of the diene used in its systematic nomenclature.)

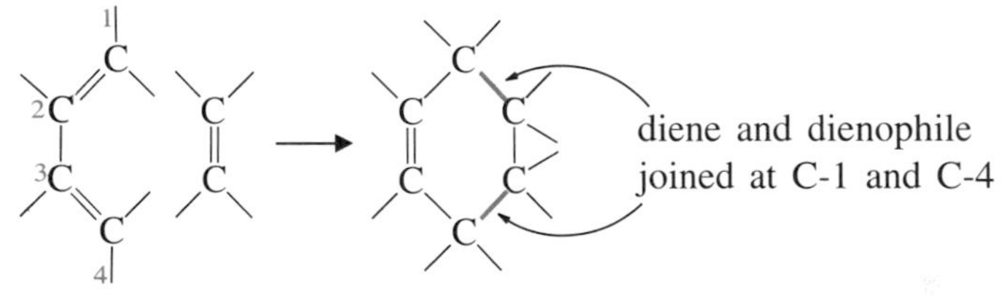

(15.8)

We shall learn in subsequent sections that conjugated dienes undergo a variety of other 1,4-addition additions.

Problems

11 What products are formed in the Diels–Alder reactions of the following dienes and dienophiles?

(a) 1,3-butadiene and ethylene

(b) 1,3-cyclohexadiene and $CH_2{=}C(CO_2CH_3)_2$

12 Give the diene and dienophile that would react in a Diels–Alder reaction to give each of the following products:

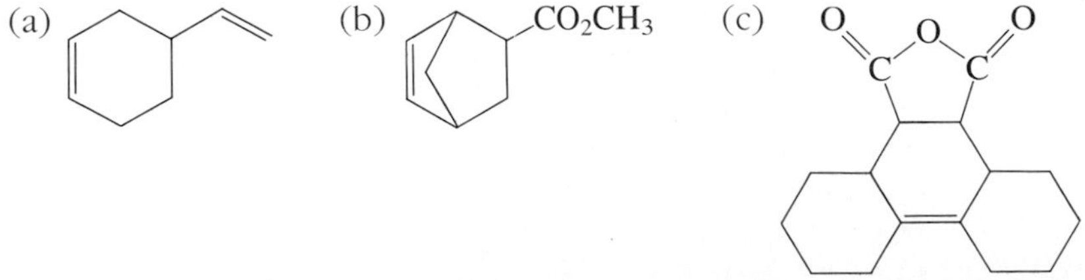

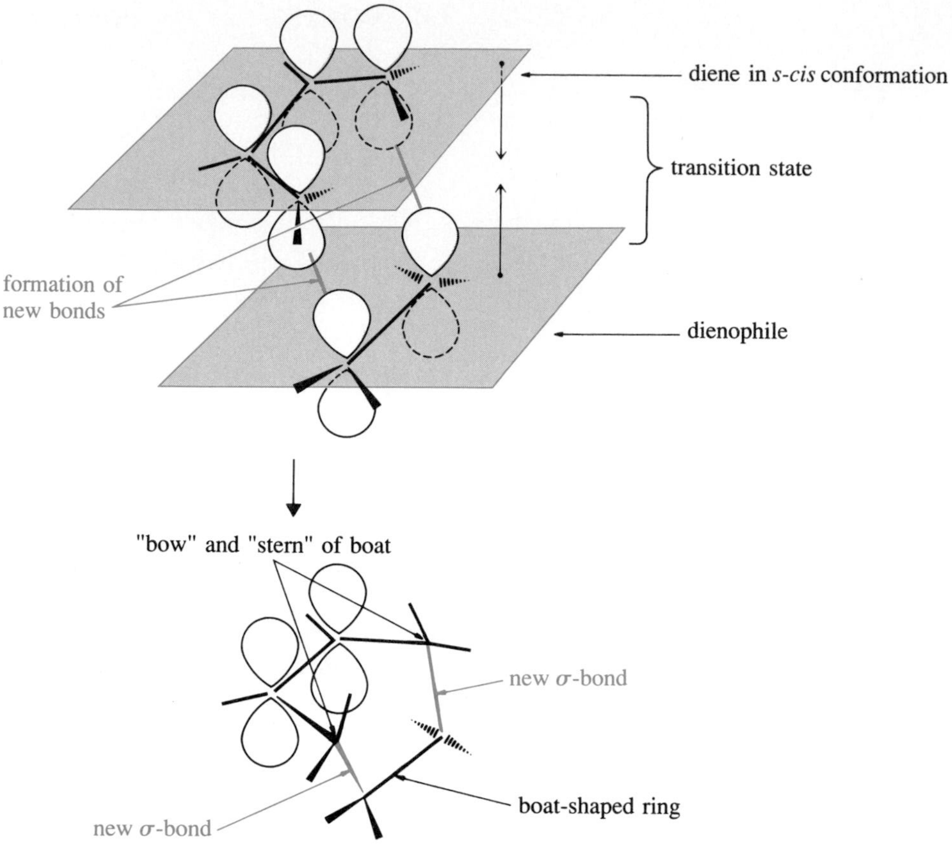

Figure 15.6 In the Diels–Alder reaction, the diene and dienophile approach in parallel planes. The diene is in the *s-cis* conformation. Bonds are formed by overlap of the π-electron clouds of both molecules. The product is formed first in a boat conformation.

Problems (Cont.)

13 Explain why two structural isomers are formed in the following Diels–Alder reaction:

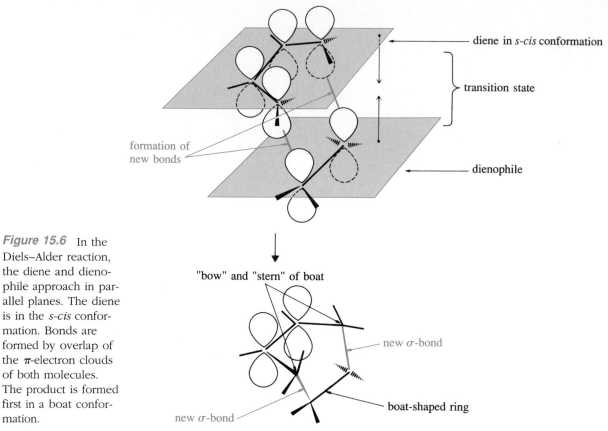

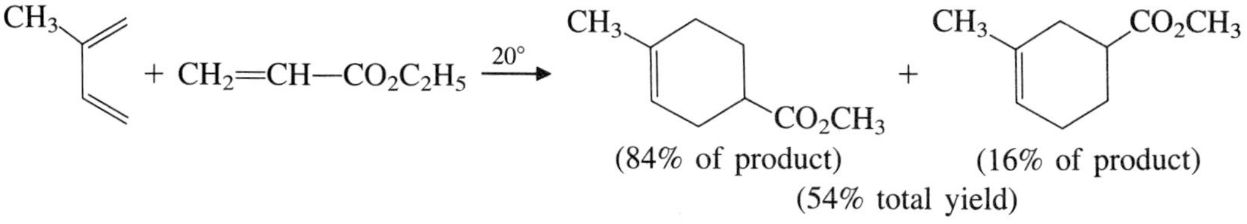

B. Effect of Diene Conformation on the Diels–Alder Reaction

Figure 15.6 shows a diagram of the Diels–transition state—that is, the way in which the diene and dienophile are believed to be oriented as they react. We can see from this diagram that the diene component is in the *s-cis* conformation. The experimental evidence supporting this view is that dienes "locked" into *s-trans* conformations are unreactive in Diels–Alder reactions.

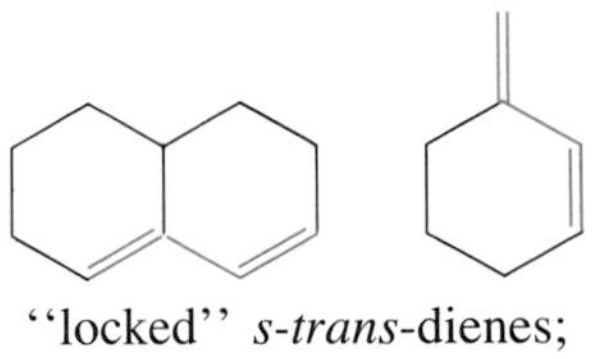

''locked'' *s-trans*-dienes;
unreactive in Diels-Alder reactions

Conversely, dienes "locked" into *s-cis* conformations are unusually reactive, and in many cases are much more reactive than corresponding open-chain dienes.

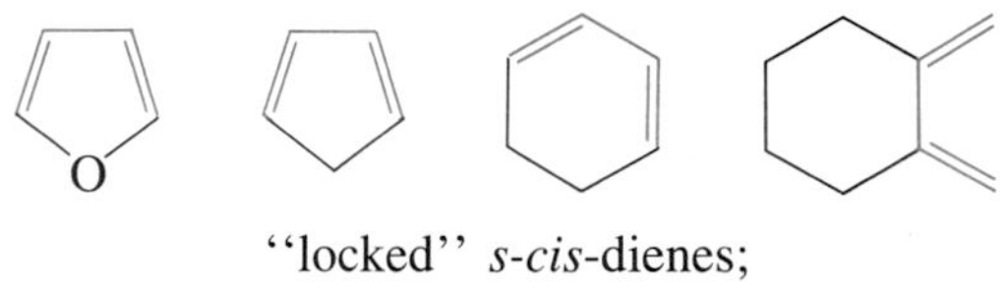

''locked'' *s-cis*-dienes;
all reactive in the Diels-Alder reaction

For example, 1,3-cyclopentadiene, which is locked in an *s-cis* conformation, reacts with maleic anhydride about one-thousand times faster than 1,3-butadiene, which prefers an *s-trans* conformation. On the other hand, conjugated dienes that have strained *s-cis* conformations are completely unreactive in the Diels–Alder reaction. Thus, the *trans* isomer of 1,3-pentadiene reacts with maleic anhydride, but the *cis* isomer is much less reactive under the same conditions.

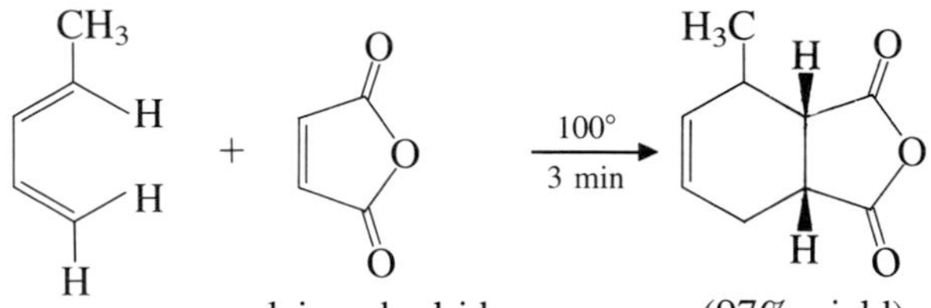

maleic anhydride (97% yield) (15.9)

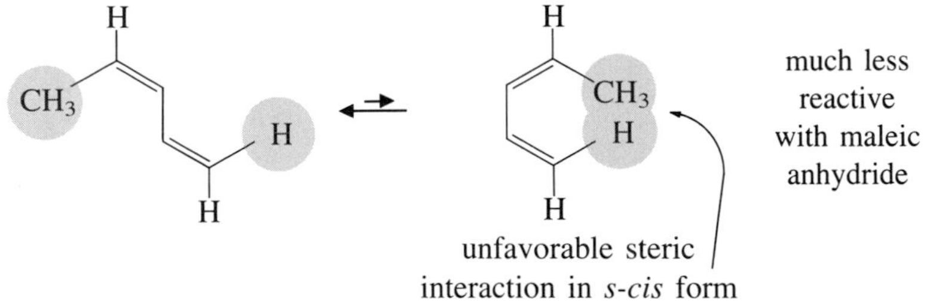

(15.10)

In other words, if a steric effect destabilizes the *s-cis* form of the diene, it also destabilizes the transition states of its Diels–Alder reactions. Because the transition state has high energy, the reaction is slow (Sec. 4.7A).

Problems

14 A mixture of 0.1 mole of *trans, trans*-2,4-hexadiene and 0.1 mole of *cis, trans*-2,4-hexadiene was allowed to react with 0.1 mole of maleic anhydride. After the reaction, the precipitated Diels–Alder adduct was filtered off, and the remaining unreacted diene was found to contain only one of the starting 2,4-hexadiene isomers. Which isomer did not react? Explain.

Problems (Cont.)

15 Complete the following Diels–Alder reaction:

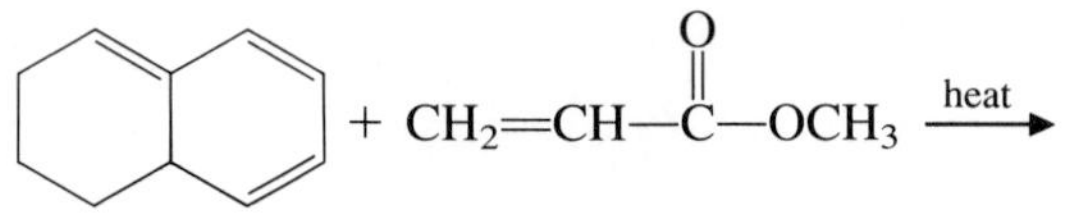

C. Stereochemistry of the Diels–Alder Reaction

When a diene and dienophile react in a Diels–Alder reaction, they approach each other in parallel planes (Fig. 15.6). This type of approach allows the π-electron clouds of the two components to overlap and form the bonds of the product. According to this picture, the diene undergoes a *syn* addition to the dienophile; likewise the dienophile undergoes a 1,4-*syn* addition to the diene. Each component adds to *one face* of the other. Observe also that the six-membered ring of the product is formed initially in a boat-like conformation; it subsequently relaxes to its most stable conformation. This point is particularly obvious when cyclic dienes are used, as in Eq. 15.7. In this case the initially formed boat conformation of the product (color) is "trapped" by the bicyclic structure of the product.

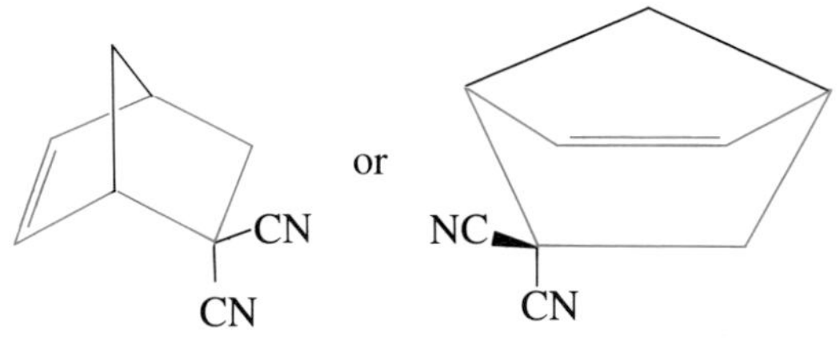

(boat structure in color)

The *syn* addition to the dienophile component is illustrated by the following reactions. Groups that are *cis* in the alkene starting material are also *cis* in the product.

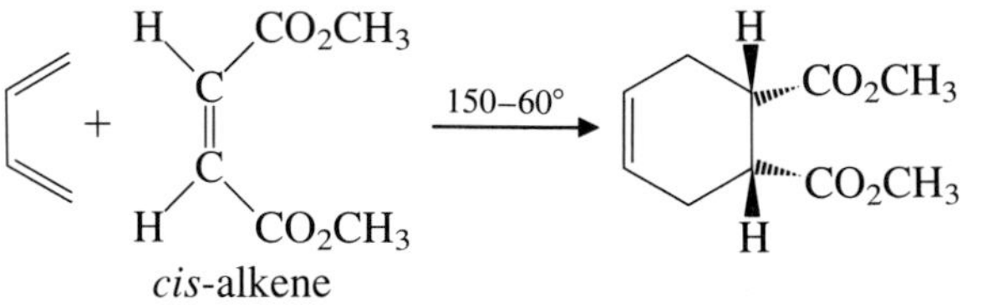

cis (15.11a)

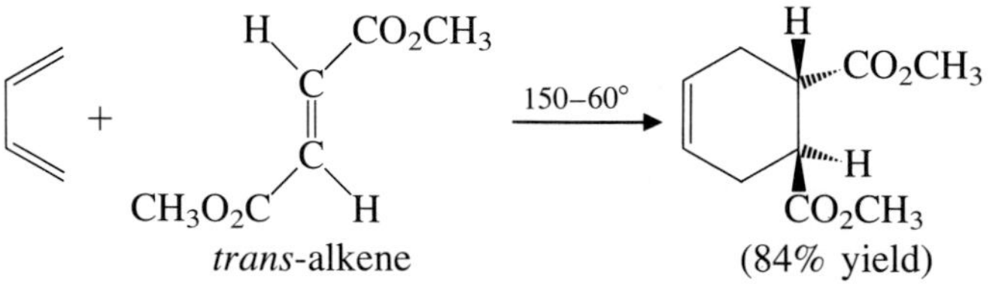

trans (15.11b)

This is, of course, the same stereochemical result observed in other *syn* additions, such as hydroboration and catalytic hydrogenation (Sec. 7.10).

Syn addition to the diene is apparent if either or both terminal carbons of the diene unit contain two different substituent groups. If we view the diene in its *s-cis* conformation, we can classify these groups as inner or outer groups.

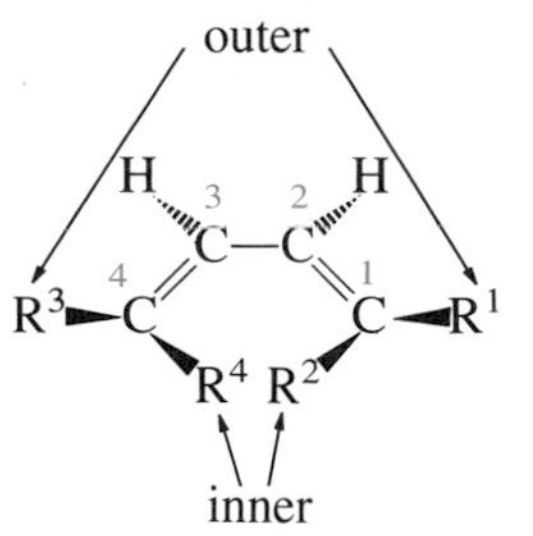

The two inner groups always end up in a *cis* relationship in the Diels–Alder product; similarly, the two outer groups also end up *cis*. Conversely, an inner group on one carbon always ends up *trans* to an outer group on the other in the Diels–Alder product. The inner groups become the "flagpole" groups (Sec. 7.2) on the boat-shaped ring.

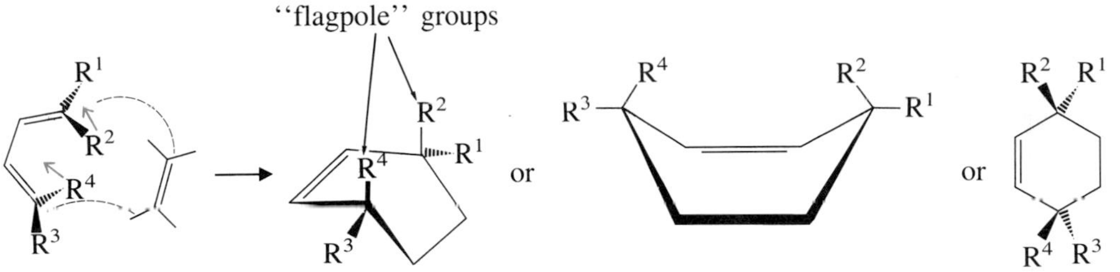

(15.12)

The following reactions illustrate these points:

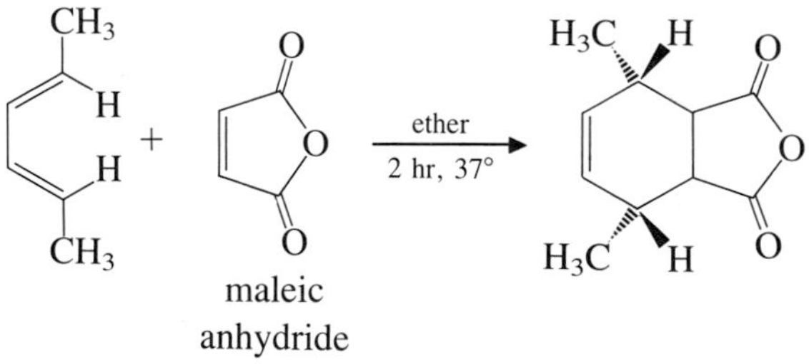

(15.13)

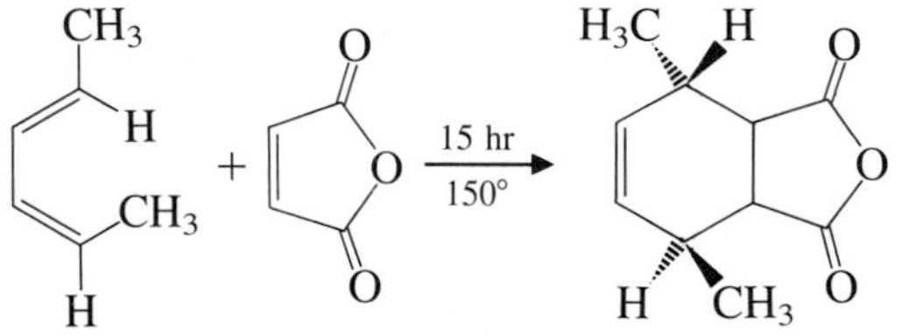

(15.14)

The same ideas are also illustrated by the reactions of cyclic dienes, such as the one in Eq. 15.7. In this case the same group is tied to both diene "inner" positions.

One further point of stereochemistry in the Diels–Alder reaction becomes important when both the diene and dienophile have substituent groups. In the Diels–Alder product, what is the stereochemical relationship between the diene substituents and those of the dienophile? There are two possibilities. One, called *endo* addition, involves a transition state in which the dienophile substituent groups (R^2 in Eq. 15.15a) are directly under (or over) the diene double bonds.

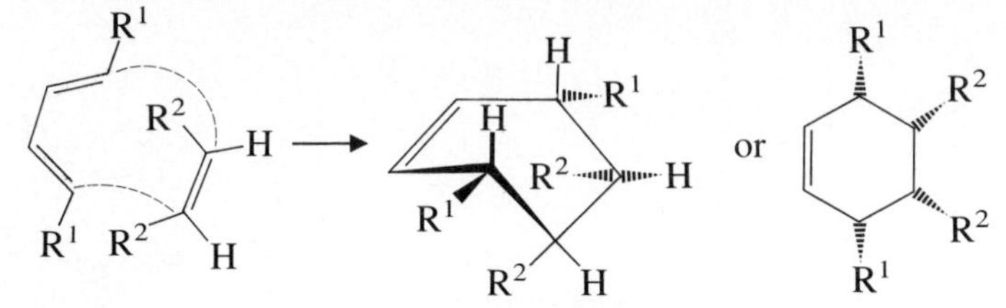

(15.15a)

endo product

In the other type of addition, called *exo* addition, the dienophile is "flipped" relative to the diene so that addition occurs to the opposite face of the dienophile; in this mode of addition the dienophile substituents project away from the diene.

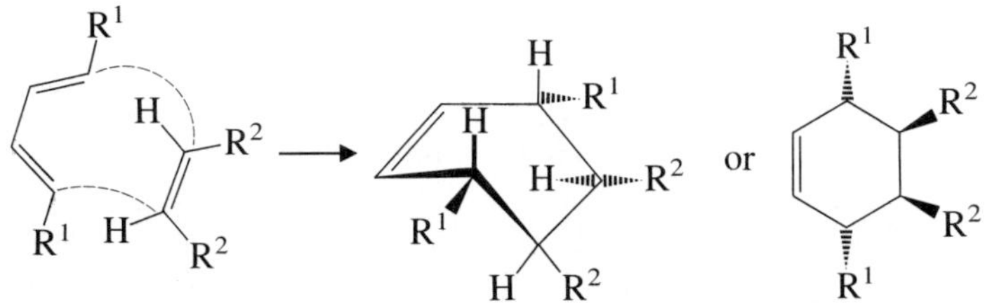

(15.15b)

exo product

As the following example illustrates, it is found in many Diels–Alder reactions that the *endo* product predominates, particularly when cyclic dienes are used:

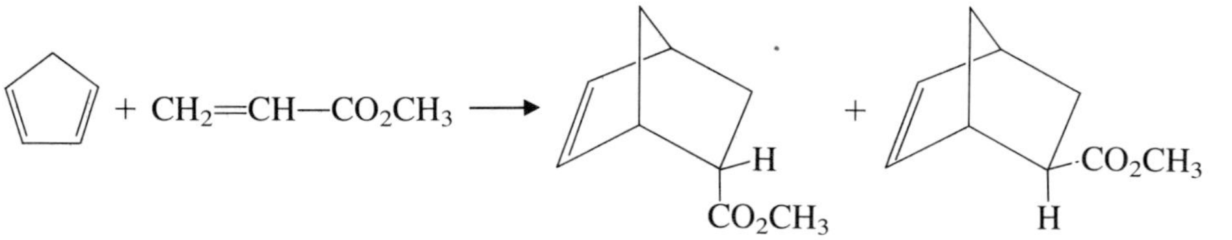

(15.16)

endo
(76%)

exo
(24%)

This observation is particularly intriguing because the *endo* product is the less stable one! (The reasons for this observation are beyond the scope of our discussion.)

Problems

16 Give the products of the following Diels–Alder reactions; show the relative stereochemistry of the substituent groups where appropriate:

(a)

$+ (CN)_2C{=}C(CN)_2 \longrightarrow$

OCH$_3$

(b) OAc

$+ CH_2{=}C(CO_2CH_3)_2 \longrightarrow$

OAc

$(-OAc = acetoxy = -O-\overset{\overset{\text{O}}{\|}}{C}-CH_3)$

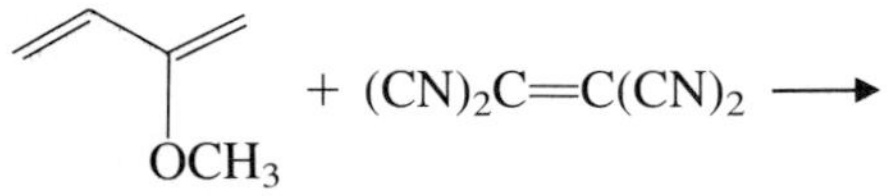

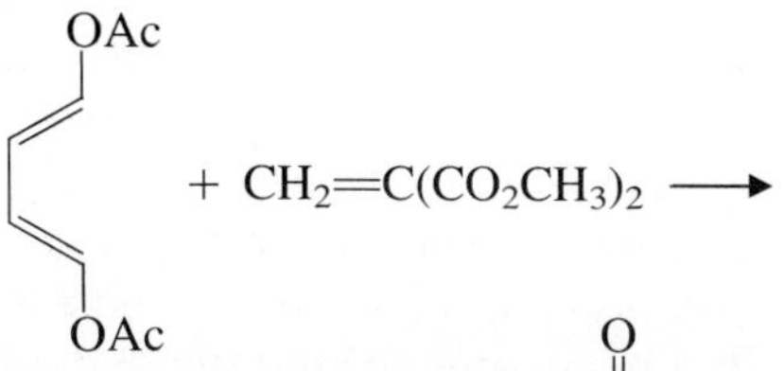

(c)

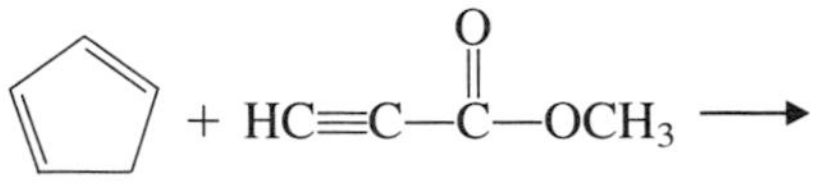

17 Give the structures of the starting materials that would yield each of the compounds below in Diels–Alder reactions. Pay careful attention to stereochemistry, where appropriate.

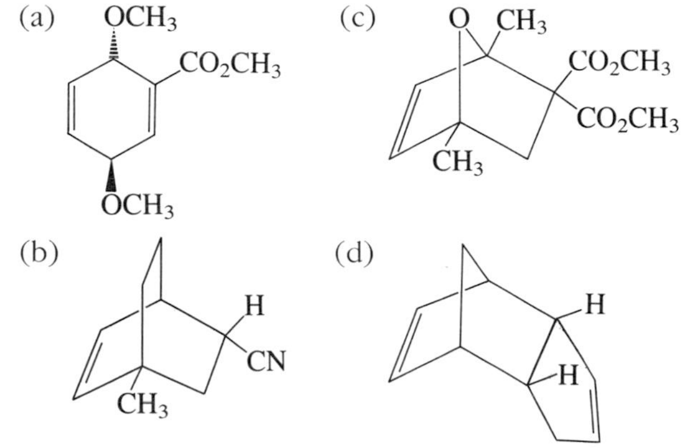

18 What aspect of the product stereochemistry is not shown in Eq. 15.13? Complete the stereochemistry of the product, assuming that it results from *endo* addition.

D. Uses of the Diels–Alder Reaction

The Diels–Alder reaction is a very useful organic reaction for several reasons. For one thing, *it is another method of forming carbon–carbon bonds*. Its most obvious application, as we can see from the examples in this section, is in the construction of carbon rings, including bicyclic compounds. Another reason for the utility of the reaction is its *stereoselectivity*: the reaction can be planned so that the diene and dienophile substituent groups end up in a well-defined stereochemical relationship in the Diels–Alder products. The Diels–Alder reaction has also been useful in proving the stereochemistry of double bonds in natural products (Problem 20).

Yet another use of the Diels–Alder reaction is the trapping of highly reactive dienophiles that cannot be isolated in pure form, but whose existence is nevertheless of theoretical interest. One diene used for this purpose is diphenylisobenzofuran. The diene unit in this molecule (shown below in color) is highly reactive.

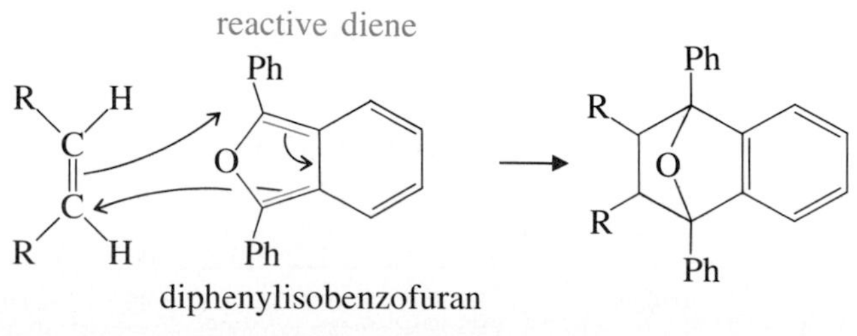

(15.17)

For example, when 1,2-dibromocyclopentene is treated with magnesium in the presence of diphenylisobenzofuran, a Diels–Alder adduct is formed.

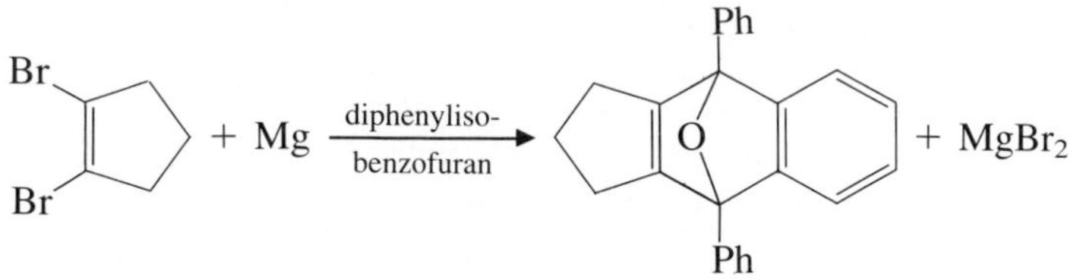

(15.18)

From the structure of the Diels–Alder adduct, it can be inferred that a cyclic alkyne, *cyclopentyne,* has been formed (Problem 19). This highly strained compound cannot be isolated in pure form under ordinary conditions (Problem 3, Chapter 14).

This type of experiment is called a *trapping* experiment. The philosophy of the chemist who carries out this type of experiment is something like that of the police detective who tries to identify a burglar: the next best thing to catching the thief is to have the culprit leave footprints in a bucket of wet plaster! Cyclopentyne cannot be "caught" and bottled, but the structure of the product in Eq. 15.18 is fairly convincing evidence that it is formed.

Problems

19 (a) Using the arrow formalism, show that cyclopentyne would be expected to give the Diels–Alder product shown in Eq. 15.18.

(b) The first step in the formation of cyclopentyne is the formation of a Grignard reagent, (2-bromo-1-cyclopentenyl)magnesium bromide. Using the arrow formalism, show how this Grignard reagent is converted into cyclopentyne.

20 The following natural product readily gives a Diels–Alder adduct with maleic anhydride. What is the most likely configuration of its two double bonds (*cis* or *trans*)? (*Hint*: See Eq. 15.10.)

$$CH_3(CH_2)_5CH{=}CH{-}CH{=}CH(CH_2)_7CO_2H$$

15.4 ADDITION OF HYDROGEN HALIDES TO CONJUGATED DIENES

A. 1,2- and 1,4-Additions. Allylic Cations

Conjugated dienes, like ordinary alkenes (Sec. 4.5), react with hydrogen halides; however, conjugated dienes give two types of addition products.

$$CH_3{-}CH{=}CH{-}CH{=}CH{-}CH_3 + HBr \xrightarrow{-20°}$$

$$CH_3{-}\underset{\displaystyle H}{\underset{|}{CH}}{-}\underset{\displaystyle Br}{\underset{|}{CH}}{-}CH{=}CH{-}CH_3 + CH_3{-}\underset{\displaystyle H}{\underset{|}{CH}}{-}CH{=}CH{-}\underset{\displaystyle Br}{\underset{|}{CH}}{-}CH_3 \qquad (15.19)$$

1,2-addition product (~65%)   1,4-addition product (~35%)

The major product is a **1,2-addition** product, so-called because it results from the addition of HBr at adjacent carbons. The minor product results from **1,4-addition,** or conjugate addition, because the hydrogen and bromine add to carbons that have a 1,4-relationship. (Note again that the terms 1,2-addition and 1,4-addition have nothing to do with systematic nomenclature.)

The 1,2-addition reaction is analogous to the reaction of HBr with an ordinary alkene. But how do we account for the conjugate-addition product? As in HBr addition to ordinary alkenes, the first mechanistic step is protonation of a double bond. Protonation of the diene in Eq. 15.19 at either of the equivalent carbons 2 or 5 gives the following carbocation:

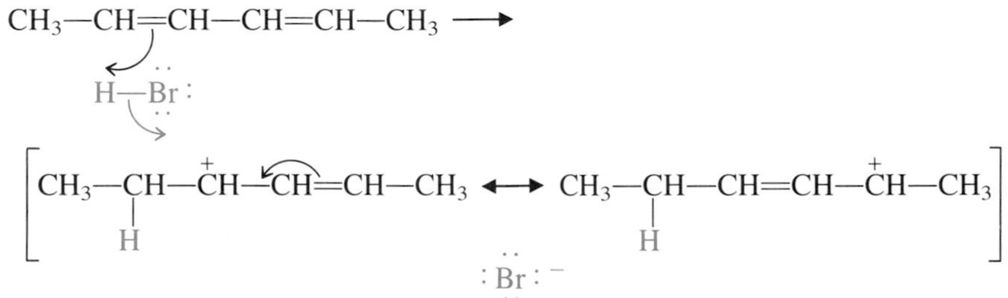

$$CH_3—CH=CH—CH=CH—CH_3 \longrightarrow$$

$$H—Br:$$

$$\left[CH_3—\underset{H}{CH}—\overset{+}{CH}—CH=CH—CH_3 \longleftrightarrow CH_3—\underset{H}{CH}—CH=CH—\overset{+}{CH}—CH_3 \right] \quad (15.20)$$

$$:Br:^-$$

The resonance structures for this carbocation show that the positive charge in this ion is not localized, but is instead *shared* by two different carbons. Thus, two products are formed in Eq. 15.19 because the bromide ion can attack at either of the electron-deficient carbons.

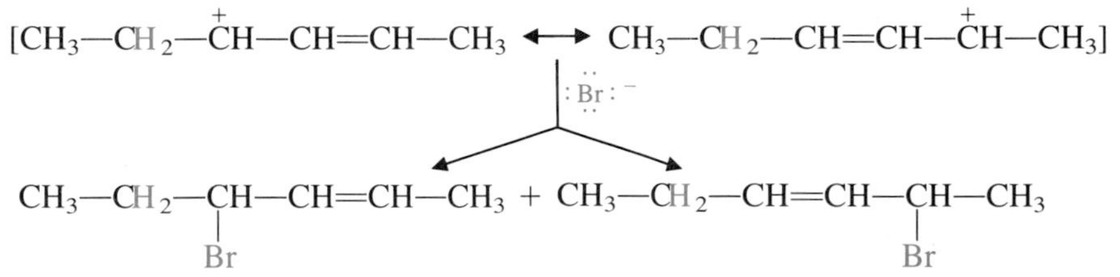

$$[CH_3—CH_2—\overset{+}{CH}—CH=CH—CH_3 \longleftrightarrow CH_3—CH_2—CH=CH—\overset{+}{CH}—CH_3]$$

$$:Br:^-$$

$$(15.21)$$

$$CH_3—CH_2—\underset{Br}{CH}—CH=CH—CH_3 + CH_3—CH_2—CH=CH—\underset{Br}{CH}—CH_3$$

What would happen if the starting diene were protonated at carbon-3? A different carbocation would be formed, and it would be attacked by bromide ion to give an alkyl halide that is different from those observed experimentally:

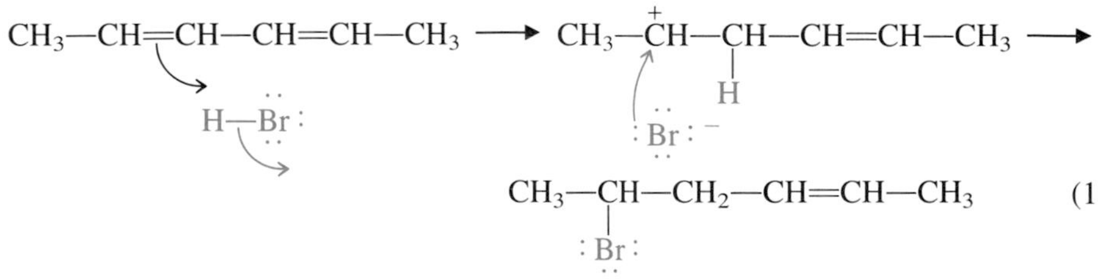

$$CH_3—CH=CH—CH=CH—CH_3 \longrightarrow CH_3—\overset{+}{CH}—\underset{H}{CH}—CH=CH—CH_3 \longrightarrow$$

$$H—Br:$$

$$:Br:^-$$

$$CH_3—\underset{:Br:}{CH}—CH_2—CH=CH—CH_3 \quad (15.22)$$

(not formed)

Because this product is not formed, we deduce that protonation at carbon-3 does not occur. What is the reason for the regioselectivity of this reaction?

In Sec. 4.7C we learned that the regioselectivity of HBr addition is determined by the stability of the carbocation intermediate. This means, then, that the carbocation in Eq. 15.21 is more stable than the carbocation in Eq. 15.22.

$$[CH_3—CH_2—\overset{+}{CH}—CH=CH—CH_3 \longleftrightarrow CH_3—CH_2—CH=CH—\overset{+}{CH}—CH_3] \qquad \text{more stable ion}$$

$$CH_3—\overset{+}{CH}—CH_2—CH=CH—CH_3 \qquad \text{less stable ion}$$

At first sight, it may seem surprising that these two carbocations have significantly different stabilities; after all, both are secondary. The difference in the two carbocations lies in the location of the electron-deficient carbon with respect to the double bond. In the more stable carbocation, the electron-deficient carbon is adjacent to the double bond; in the less stable cation, the electron-deficient carbon is further removed from the double bond. The more stable carbocation is an **allylic cation.**

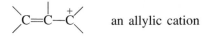

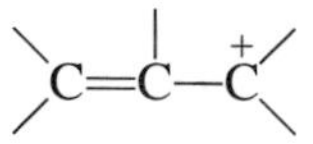
an allylic cation

The word *allylic* is a generic term applied to any functional group at a carbon adjacent to a double bond.

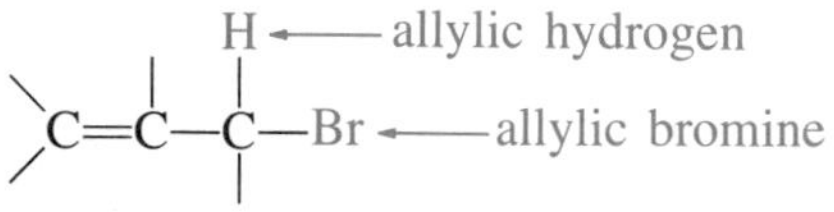

The parent group from which this name is derived is the **allyl group**:

$CH_2{=}CH{-}CH_2{-}$	$CH_2{=}CH{-}CH_2{-}Br$	$CH_2{=}CH{-}\overset{+}{C}H_2$
the allyl group	allyl bromide	the allyl cation

Be sure to distinguish between a **vinylic** group and an **allylic** group. A vinylic group is attached to a carbon of a double bond; an allylic group is attached to the carbon next to a double bond.

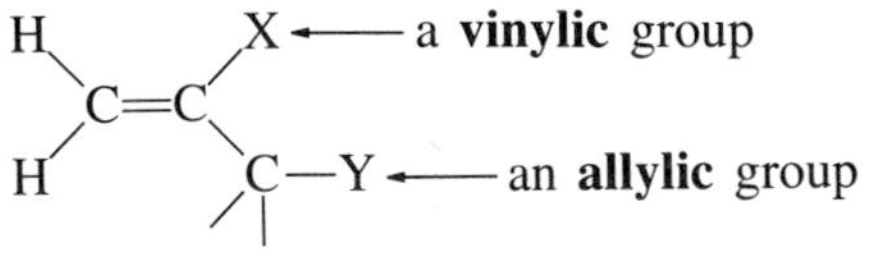

Allylic cations are more stable than comparably substituted nonallylic alkyl cations. Where does the stability of allylic cations fit into the overall stabilities of carbocations? Roughly speaking, an allylic cation is about as stable as a nonallylic alkyl cation with one additional alkyl substituent. Thus, a secondary allylic cation is about as stable as a tertiary nonallylic one. Summarizing:

stability of carbocations:

—increasing stability→

$$R{-}CH{=}\overset{+}{C}H < R{-}\overset{+}{C}{=}CH_2 \approx R{-}\overset{+}{C}H_2 < R{-}\overset{+}{C}H{-}R < R{-}CH{=}CH{-}\overset{+}{C}H{-}R \approx$$

primary vinyl; secondary vinyl; primary alkyl; secondary alkyl; secondary allylic

$R_3\overset{+}{C}$ (tertiary alkyl) < 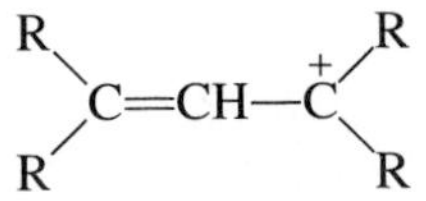 (tertiary allylic) (15.23)

Figure 15.7 Overlap of *p* orbitals in the allyl cation.

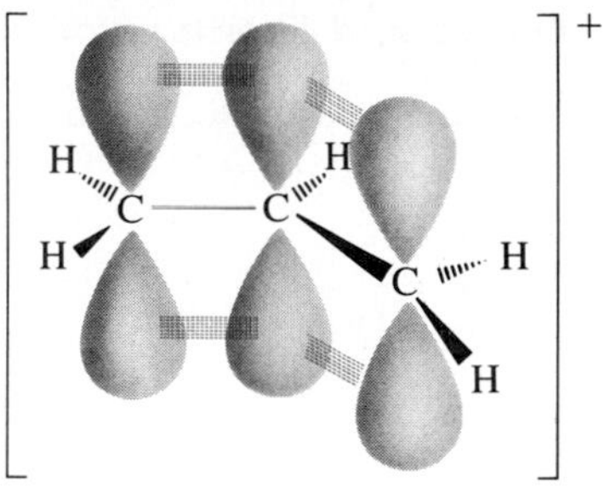

The reason for the unusual stability of allylic cations lies in their electronic structures, shown in Fig. 15.7. The electron-deficient carbon and the carbons of the double bond are all sp^2 hybridized; each carbon has a *p* orbital. The overlap of these *p* orbitals provides *additional bonding* in allylic cations, and hence, *additional stability*. This means that both the positive charge and the double bond character are *delocalized*. Since this delocalization is not adequately shown by a single Lewis structure, allylic cations are represented as resonance hybrids:

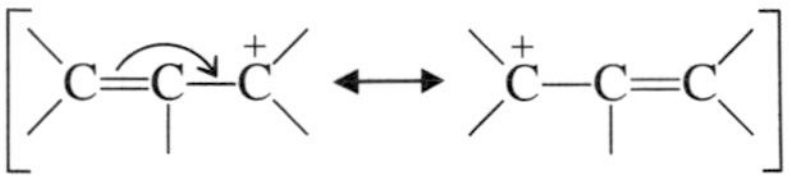
(15.24)

Because resonance structures show the additional bonding that gives allylic cations their additional stability, we see that there is a connection between resonance and stability: *species that have resonance structures are especially stable.* We shall examine this idea in more detail in Sec. 15.6.

Problems

21 Predict the products of each of the following reactions, and give the mechanisms for their formation:

(a) 1,3-butadiene + HCl $\longrightarrow$

(b) 3-chloro-1-methylcyclohexene + C_2H_5OH $\longrightarrow$

(c) $(CH_3)_2C{=}CH{-}CH{=}CH{-}\underset{\displaystyle CH_3}{\underset{|}{CH}}{-}Cl \xrightarrow{H_2O,\ Ag^+}$

22 Suggest a mechanism that accounts for both products of the following reaction:

$$CH_3{-}CH{=}CH{-}CH_2{-}OH + HBr\ (48\%) \xrightarrow[-15°]{H_2SO_4}$$

$$\underset{(84\%)}{CH_3{-}CH{=}CH{-}CH_2{-}Br} + \underset{(16\%)}{CH_3{-}\underset{\displaystyle Br}{\underset{|}{CH}}{-}CH{=}CH_2}$$

B. Kinetic and Thermodynamic Control

As we have just learned, conjugated dienes react with hydrogen halides to give mixtures of 1,2- and 1,4-addition products in which the 1,2-addition product predominates.

$$CH_2{=}CH{-}CH{=}CH_2 + HCl \longrightarrow \underset{(75\text{–}80\%)}{CH_3{-}\underset{|}{\overset{Cl}{CH}}{-}CH{=}CH_2} + \underset{(20\text{–}25\%)}{CH_3{-}CH{=}CH{-}CH_2{-}Cl} \qquad (15.25)$$

An interesting thing about this reaction is that *the relative proportions of products are not the same as those observed if the products are allowed to come to equilibrium.* We can establish an equilibrium between the 1,2- and 1,4-addition products by allowing the compounds to stand in solution for a long period of time, or by adding Lewis acids. When the 1,2- and 1,4-addition products come to equilibrium we find that the 1,4-isomer predominates.

$$\underset{(25\%\text{ at equilibrium})}{CH_3{-}\underset{Cl}{\underset{|}{CH}}{-}CH{=}CH_2} \xrightleftharpoons[\text{Lewis acid }(FeCl_3)]{\text{heat or}} \underset{(75\%\text{ at equilibrium})}{CH_3{-}CH{=}CH{-}CH_2{-}Cl} \qquad (15.26)$$

Since an equilibrium always favors the more stable isomer (Sec. 4.4A), the data in Eq. 15.26 show that *the 1,4-addition product is more stable than the 1,2-addition product.*

Problem

23 Suggest a reason why the 1,4-addition product is more stable than the 1,2-addition product. (*Hint:* See Sec. 4.4C.)

Evidently the products of Eq. 15.25 are isolated before they can come to equilibrium. Since the less stable isomer—the 1,2-addition product—is formed in greater amount, *the reaction that gives the less stable product is faster than the one that gives the more stable product.*

$$CH_2{=}CH{-}CH{=}CH_2 + HCl$$

1,2-addition / faster ↙ ↘ slower \ 1,4-addition

$$\underset{\text{less stable product}}{CH_3{-}\underset{Cl}{\underset{|}{CH}}{-}CH{=}CH_2} \qquad \underset{\text{more stable product}}{CH_3{-}CH{=}CH{-}CH_2{-}Cl} \qquad (15.27)$$

When the product distribution in a reaction differs substantially from the product distribution that would be observed if the products were at equilibrium, the reaction is said to be **kinetically controlled.** Thus, addition of hydrogen halides to conjugated dienes is a kinetically controlled reaction. On the other hand, if the products of

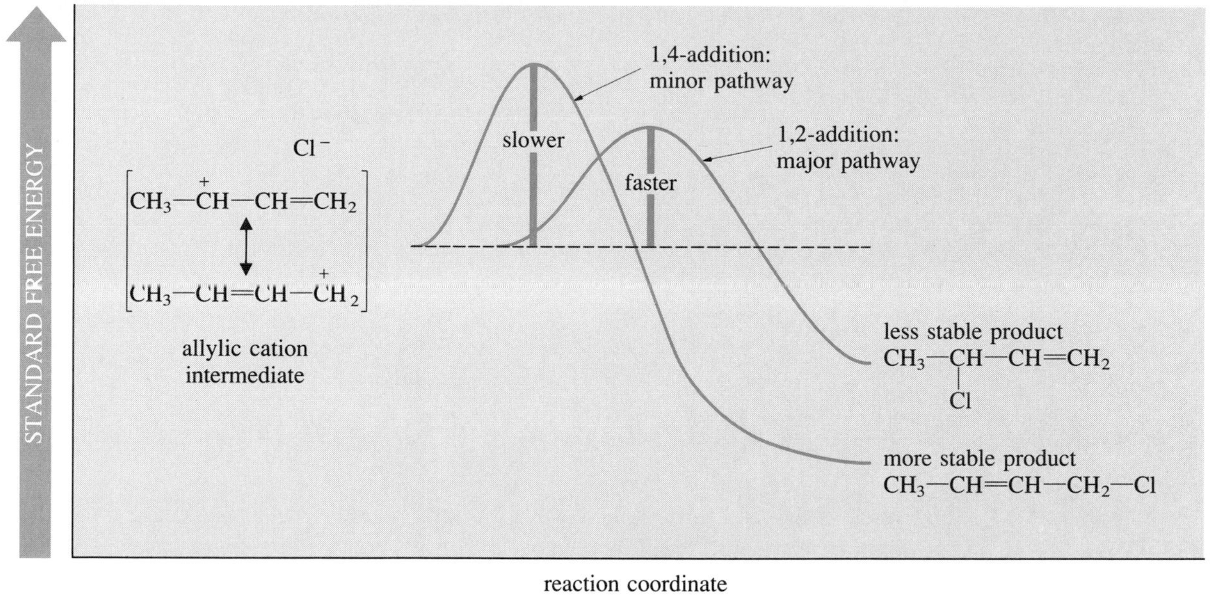

Figure 15.8 In hydrogen-halide addition, halide-ion attack on the allylic cation intermediate gives the less stable product more rapidly.

a reaction come to equilibrium under the reaction conditions, the product distribution is said to be **thermodynamically controlled.**

The product distribution in hydrogen halide addition to a conjugated diene is determined by the relative rates of the two *product-determining steps* (Sec. 9.5B): attack of the halide ion on one or the other of the electron-deficient carbons of the allylic cation intermediate (Fig. 15.8).

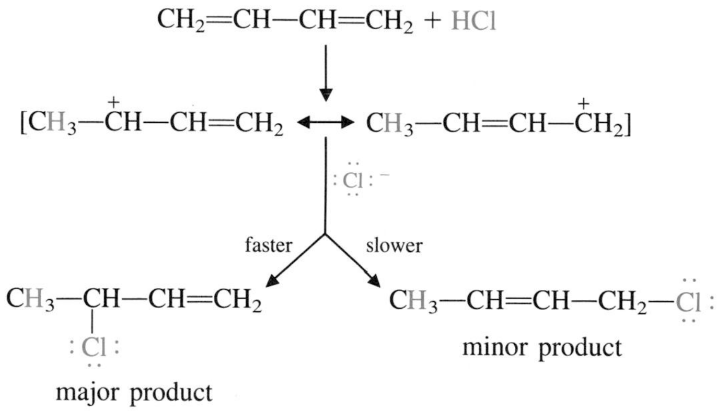

$$CH_2{=}CH{-}CH{=}CH_2 + HCl$$

$$[CH_3{-}\overset{+}{CH}{-}CH{=}CH_2 \longleftrightarrow CH_3{-}CH{=}CH{-}\overset{+}{CH_2}] \qquad (15.28)$$

When Lewis acids are added, or if the products stand in solution, the 1,2-addition product is converted into the 1,4-addition product by an S_N1 reaction that goes through the same allylic cation intermediate. This **allylic rearrangement** (interconversion of two allylic isomers) allows the two products eventually to come to equilibrium.

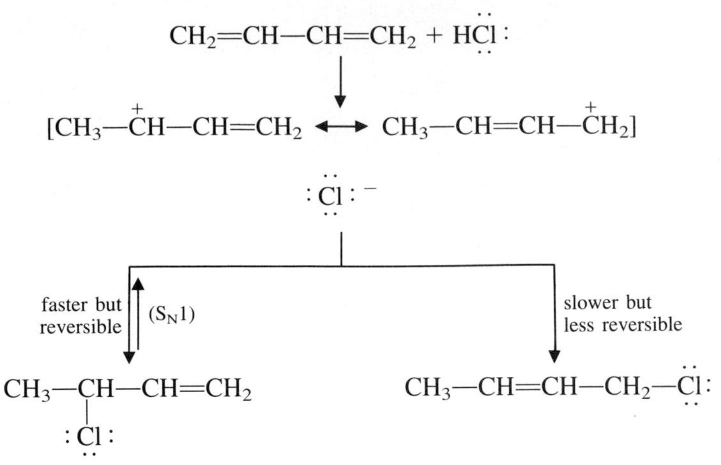

(15.29)

In other words, the formation of the 1,2-addition product is rapid but reversible; the formation of the 1,4-addition product is slower but almost irreversible.

An analogy may assist in understanding kinetically and thermodynamically controlled reactions. Imagine a very inebriated gentleman stumbling randomly around a pasture. Near each other in the pasture are a shallow watering hole and a deep well with a high fence around it. Our drunken friend is likely to fall in the hole several times, but because it is shallow, he can climb out of it and continue staggering around the pasture. He frequently tries to climb the high fence, but usually doesn't make it and falls back into the pasture. After a very long while, however, he makes it over the fence and falls into the well; once in the well, he is there to stay. If we now imagine Avogadro's number of people staggering around a (very large) pasture, we get a reasonably good picture of kinetic and thermodynamic control. Initially, a large number of people fall into the shallow hole. If we wait long enough, however, most of them will end up in the deep well. The frequent occurrence—falling in the shallow hole—is reversible, but the rare occurrence—climbing the fence and falling in the well—is irreversible.

Similarly, the allylic cation is rapidly attacked by halide to give the 1,2-addition product, which, however, is not very stable; this product is in a shallow "energy hole." This reaction reverses rapidly to give back the carbocation. Occasionally the cation is attacked to give the more stable 1,4-addition product. This product is in a deeper "energy well" and does not as readily give back the allylic cation. If the reaction mixture is analyzed before the products have a chance to reach equilibrium, then the distribution of products reflects kinetic control.

Why is the less stable 1,2-addition product formed more rapidly? A very clever experiment suggested the reason for the kinetic control observed in the addition of HCl to 1,3-butadiene (Eq. 15.25). In this reaction the allylic cation intermediate has two resonance forms that are not equivalent. In one form, the positive charge is on a secondary carbon, and in the other, on a primary carbon. It was hypothesized that the preference for 1,2-addition had something to do with this difference. In particular, if a greater fraction of the positive charge were on the secondary carbon, the chloride ion might be selectively drawn to this carbon, and the 1,2-addition product would predominate, as observed.

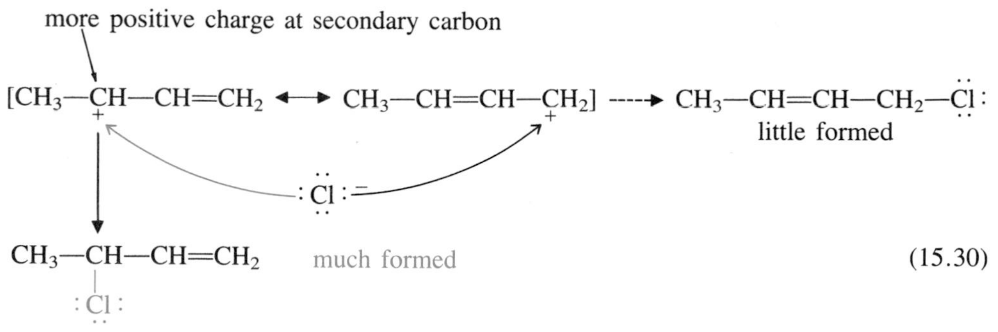

more positive charge at secondary carbon

$$[CH_3-\overset{+}{C}H-CH=CH_2 \longleftrightarrow CH_3-CH=CH-\overset{+}{C}H_2] \longrightarrow CH_3-CH=CH-CH_2-\overset{..}{\underset{..}{Cl}}:$$

little formed

:$\overset{..}{\underset{..}{Cl}}$:$^-$

$$CH_3-CH-CH=CH_2 \qquad \text{much formed} \tag{15.30}$$
$$\underset{:\overset{..}{\underset{..}{Cl}}:}{\mid}$$

According to this reasoning, a diene that reacts with HCl to give a carbocation with *identical* resonance forms should not show a preference for 1,2-addition. Thus, when 1,3-pentadiene is protonated by DCl at carbon-1, the carbocation intermediate, neglecting the difference between H and D, has two identical resonance forms. (The isotope was necessary to label the protonated carbon.)

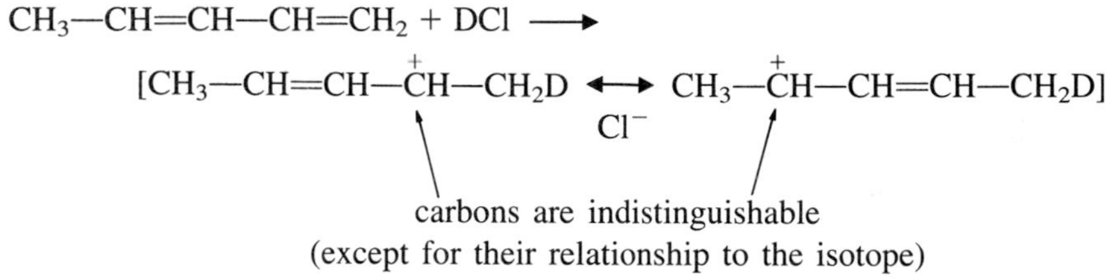

$$CH_3-CH=CH-CH=CH_2 + DCl \longrightarrow$$

$$[CH_3-CH=CH-\overset{+}{C}H-CH_2D \longleftrightarrow CH_3-\overset{+}{C}H-CH=CH-CH_2D] \tag{15.31}$$

$$Cl^-$$

carbons are indistinguishable
(except for their relationship to the isotope)

Yet even in this case, the 1,2-addition product is again the major one, even though the two electron-deficient carbons must have equal amounts of positive charge.

$$CH_3-CH=CH-CH=CH_2 + DCl \longrightarrow$$

$$\underset{\underset{Cl}{\mid}}{CH_3-CH=CH-CH-CH_2D} + \underset{\underset{Cl}{\mid}}{CH_3-CH-CH=CH-CH_2D} \tag{15.32}$$

$$(75\%) \qquad\qquad\qquad (25\%)$$

The reason for these observations appears to be that the carbocation and chloride ion are not really free, but exist as an *ion pair* (Sec. 9.5E, Fig. 9.9). When the alkene reacts with the proton of HCl, the chloride ion is therefore closely associated with the resulting cation. The chloride simply finds itself closer to one of the positively charged carbons than to the other. Addition is completed, therefore, at the nearest site of positive charge, giving the 1,2-product.

$$CH_2=CH-CH=CH_2 \longrightarrow [CH_2=CH-\overset{+}{C}H-CH_2 \longleftrightarrow \overset{+}{C}H_2-CH=CH-CH_2]$$

:$\overset{..}{\underset{..}{Cl}}$—H

:$\overset{..}{\underset{..}{Cl}}$:$^-$ H :$\overset{..}{\underset{..}{Cl}}$:$^-$ H

closer farther
away (15.33)

$$CH_2=CH-CH-CH_3 \qquad\qquad CH_2-CH=CH-CH_3$$
$$\underset{:\overset{..}{\underset{..}{Cl}}:}{\mid} \qquad\qquad\qquad \underset{:\overset{..}{\underset{..}{Cl}}:}{\mid}$$

(mostly) (little)

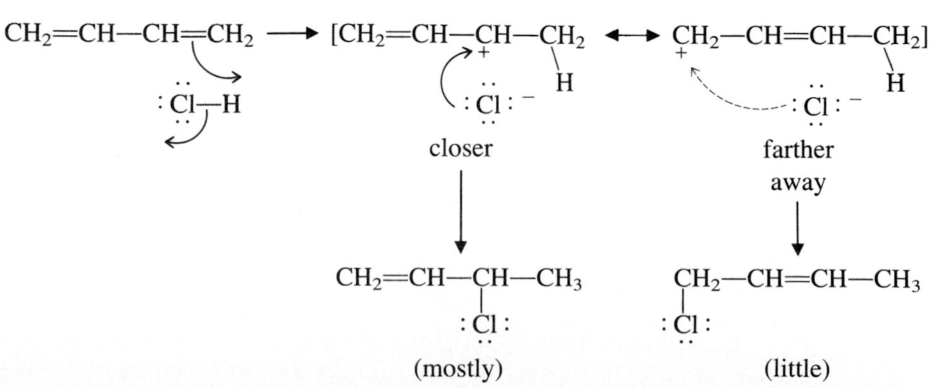

Many organic reactions give product mixtures that are kinetically controlled. Thus, *the most stable product is not always the one formed in greatest amount.* The precise reason for kinetic control varies from reaction to reaction. Whatever the reason, the relative amounts of products in a kinetically controlled reaction are determined by the relative energies of the transition states for each of the product-determining steps (Sec. 9.5B).

Problem	**24** Suggest structures for the two products formed when 1,3-butadiene reacts with one equivalent of chlorine. Which of these product(s) would predominate if they were allowed to come to equilibrium?

15.5 USES OF CONJUGATED DIENES: DIENE POLYMERS

1,3-Butadiene is one of the most important raw materials of the synthetic rubber industry. In 1985, butadiene production in the United States reached 2.5 billion pounds with a market value of about $750 million.

Polybutadiene, most of which is used for the manufacture of tires, is formed by the free-radical polymerization of 1,3-butadiene. The polymer results from a combination of mostly 1,4-addition and some 1,2-addition. The 1,4-polymerization of 1,3-butadiene can be represented as follows (R· is an initiating radical):

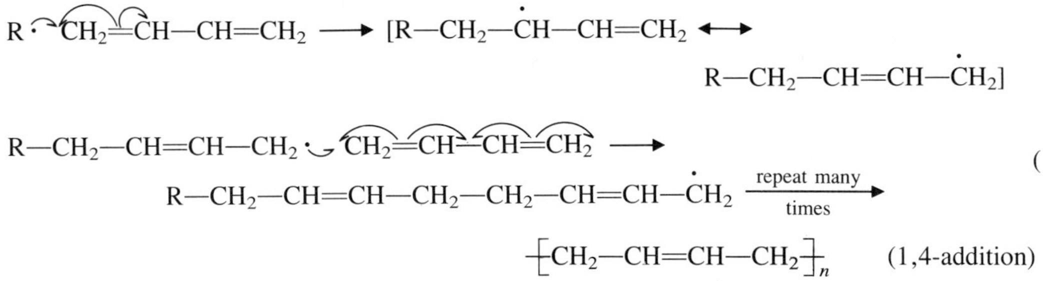

$$R \cdot CH_2 = CH - CH = CH_2 \longrightarrow [R - CH_2 - \overset{\cdot}{C}H - CH = CH_2 \longleftrightarrow$$

$$R - CH_2 - CH = CH - \overset{\cdot}{C}H_2]$$

$$R - CH_2 - CH = CH - CH_2 \curvearrowright CH_2 = CH - CH = CH_2 \longrightarrow$$

$$R - CH_2 - CH = CH - CH_2 - CH_2 - CH = CH - \overset{\cdot}{C}H_2 \xrightarrow[\text{times}]{\text{repeat many}}$$ (15.34)

$$\left[CH_2 - CH = CH - CH_2 \right]_n \quad \text{(1,4-addition)}$$

Butadiene can also be polymerized with styrene ($Ph - CH = CH_2$), usually in about a 3:1 ratio, to form *styrene–butadiene rubber* (SBR), most of which is used for tires and tread rubber. A simplified structure for SBR is as follows:

$$\left[(CH_2CH = CHCH_2)_3 CH_2\underset{\underset{Ph}{|}}{CH} \right]_n$$

SBR is an example of a **copolymer**: a polymer that is produced by the simultaneous polymerization of two or more monomers.

Natural rubber is (*Z*)-polyisoprene, another diene polymer.

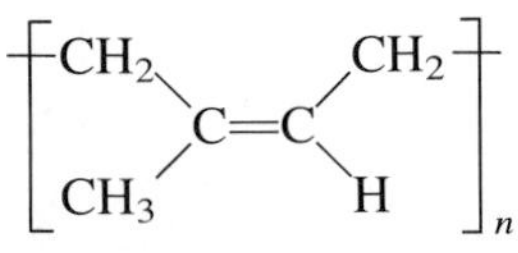

 natural rubber

Although it is formally a diene polymer, natural rubber is *not* made in nature from isoprene. (The biosynthesis of naturally occurring isoprene derivatives is discussed in

Sec. 17.6B.) Rubber hydrocarbon is obtained as a 40% aqueous emulsion from the rubber tree. After isolation, the polymer is subjected to a process called **vulcanization.** In this process, discovered in 1840 by Charles Goodyear, the rubber is kneaded and heated with sulfur. The sulfur forms crosslinks between the polymer chains, which can be represented schematically as follows:

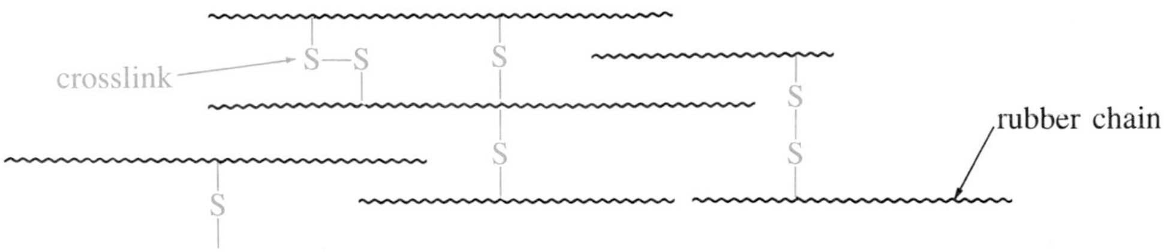

The crosslinks increase the rigidity and strength of the rubber at the cost of some flexibility. Although polyisoprene can be made synthetically, the natural material is generally preferred for economic reasons. Chemists are investigating the possibility of cultivating other hydrocarbon-producing plants that could become hydrocarbon sources of the future.

Problem

25 Write a mechanism for the free-radical copolymerization of butadiene and styrene to give styrene–butadiene rubber.

15.6 RESONANCE

In Sec. 1.7 we learned that resonance structures are used when a single Lewis structure is not an adequate representation of a molecular structure. In Sec. 15.4A we also learned that resonance structures have implications for the stability of a molecule. The ability to invoke and use resonance structures is therefore important for our understanding of *molecular structure* and *molecular stability.* In order to become proficient in the use of resonance structures, we must first understand *how* to write them—that is, how to derive one resonance structure from another. Then we must know *when* to write them; that is, we need to know when a resonance structure is important enough to be considered seriously as a contributor to a molecular structure.

A. Writing Resonance Structures

Resonance structures can be drawn when bonds, unshared electron pairs, or single electrons can be delocalized (moved) by the arrow formalism *without moving any atoms.* The following are all valid examples of resonance structures:

$CH_2{=}CH{-}\overset{+}{C}H_2 \longleftrightarrow \overset{+}{C}H_2{-}CH{=}CH_2$ movement of a bond using electron-pair arrow formalism

$CH_2{=}CH{-}\ddot{\underset{..}{O}}{:}^{-} \longleftrightarrow {}^{-}{:}CH_2{-}CH{=}\ddot{O}{:}$ movement of a bond and an unshared electron pair

$CH_2{=}CH{-}CH_2{\cdot} \longleftrightarrow {\cdot}CH_2{-}CH{=}CH_2$ movement of a single electron and a bond by the unpaired-electron arrow formalism (Sec. 5.8A)

As these examples illustrate, some of the most common situations in which resonance structures are used involve electron-deficient atoms, electron pairs, or unpaired electrons next to double or triple bonds.

The following structures, although reasonable Lewis structures, are *not* resonance structures, because the movement of an atom takes place; the location of the chlorine is different in the two structures.

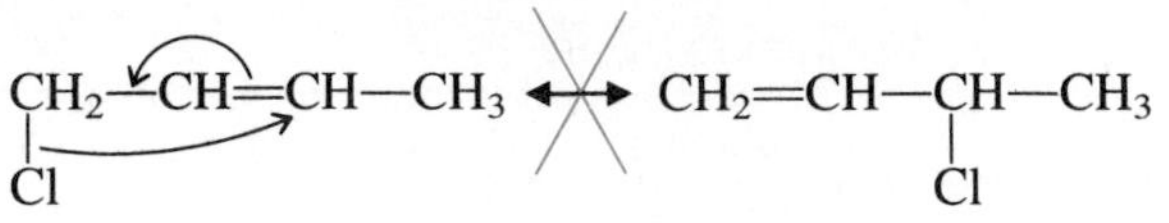

$$CH_2\!\!-\!\!CH\!\!=\!\!CH\!\!-\!\!CH_3 \longleftrightarrow CH_2\!\!=\!\!CH\!\!-\!\!CH\!\!-\!\!CH_3 \qquad (15.35)$$
$$\underset{Cl}{|} \qquad\qquad\qquad\qquad\qquad \underset{Cl}{|}$$

In fact, these are separate compounds:

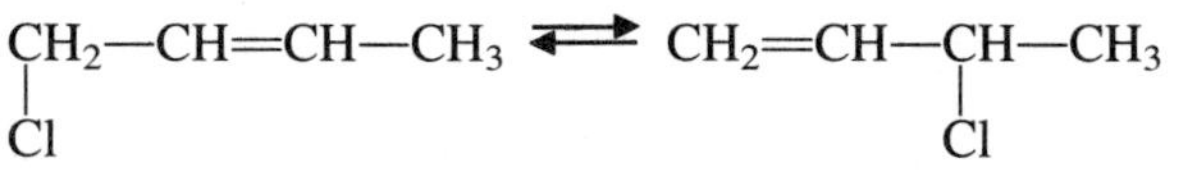

$$CH_2\!\!-\!\!CH\!\!=\!\!CH\!\!-\!\!CH_3 \rightleftharpoons CH_2\!\!=\!\!CH\!\!-\!\!CH\!\!-\!\!CH_3 \qquad (15.36)$$
$$\underset{Cl}{|} \qquad\qquad\qquad\qquad\qquad \underset{Cl}{|}$$

If two structures can exist as different compounds, they cannot be resonance structures. Resonance structures, on the other hand, are alternate electronic descriptions of a *single* molecule; the molecule is an average of its resonance structures. That is, resonance contributors are *fictitious* structures used to help us understand the structures of *real molecules* for which single Lewis structures are inadequate. Notice that the equilibrium double arrows $\rightleftharpoons$ and the resonance double-headed *single* arrow $\longleftrightarrow$ have very different implications; we must be careful not to use one symbol in a situation in which the other is appropriate.

The usual restrictions of Lewis structures (Sec. 1.2B) apply when we draw resonance structures. One of the most important of these restrictions is that when we deal with bonds to atoms in the first row of the periodic table, the octet rule is not violated. Thus, we can use a resonance structure to indicate the delocalization of electrons in the allyl cation because its positively charged carbons are electron deficient and can accommodate an additional pair of electrons:

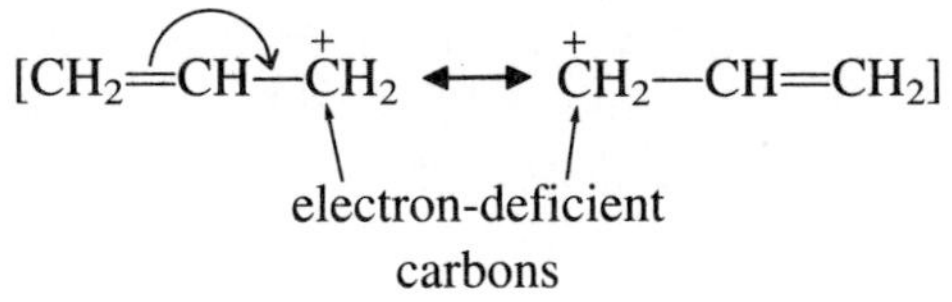

$$[CH_2\!\!=\!\!CH\!\!-\!\!\overset{+}{C}H_2 \longleftrightarrow \overset{+}{C}H_2\!\!-\!\!CH\!\!=\!\!CH_2] \qquad (15.37a)$$

electron-deficient
carbons

However, we cannot draw an analogous resonance structure for the following cation because the octet rule would be violated:

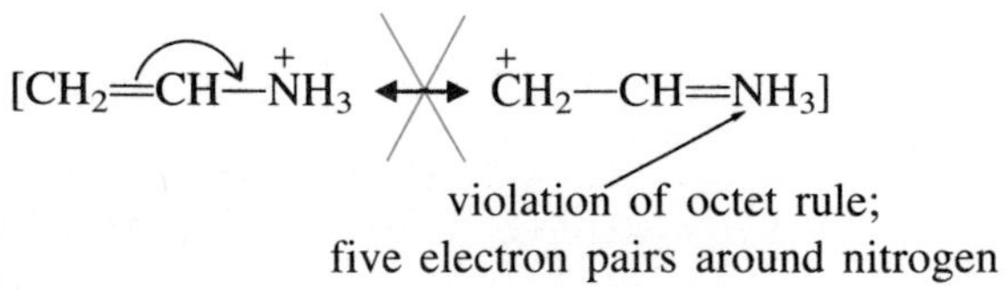

$$[CH_2\!\!=\!\!CH\!\!-\!\!\overset{+}{N}H_3 \longleftrightarrow \overset{+}{C}H_2\!\!-\!\!CH\!\!=\!\!NH_3] \qquad (15.37b)$$

violation of octet rule;
five electron pairs around nitrogen

B. Relative Importance of Resonance Structures

A molecule is an average of its contributing resonance structures. But things are a little more complicated than this simple statement suggests. In some cases a molecule resembles one of its resonance structures more than another. This means that a molecule is a *weighted* average of its resonance structures; some structures are *more important* than others. How do we know whether one resonance structure is more important than another?

To evaluate the relative importance of resonance structures, we adopt a very simple procedure: we compare the stabilities of all the resonance structures for a given molecule *as if each structure were a separate molecule*. That is, we *imagine* that each structure is real. Then we use what we know about relative stabilities to determine the relative importance of each resonance structure. *The most stable structures are the most important ones.* The following guidelines for evaluating resonance structures emerge from this type of analysis:

1. *Identical structures are equally important descriptions of a molecule.*

Example:

$$^{+}CH_2{-}CH{=}CH_2 \longleftrightarrow CH_2{=}CH{-}\overset{+}{C}H_2 \qquad (15.38)$$

Since these structures of the allyl cation are indistinguishable, they are both equally important in describing this species.

2. *Structures with the greater number of formal bonds are more important.*

Example:

$$CH_2{=}CH{-}CH{=}CH_2 \longleftrightarrow \cdot CH_2{-}CH{=}CH{-}\dot{C}H_2 \qquad (15.39)$$

Since bonding is energetically favorable, it follows that the more formal bonds a structure has, the more stable it is. Thus, the structure on the right is relatively unimportant.

3. *Structures that delocalize charge or unpaired electrons are important.*

Examples:

$$(CH_3)_2\overset{+}{C}{-}\ddot{\underset{..}{O}}{-}H \longleftrightarrow (CH_3)_2C{=}\overset{+}{\ddot{O}}{-}H \qquad \text{delocalizes positive charge}$$

$$\dot{C}H_2{=}CH{-}CH_2 \longleftrightarrow \dot{C}H_2{-}CH{=}CH_2 \qquad \text{delocalizes unpaired electron}$$

(The structures in Eq. 15.38 also delocalize charge.) Delocalization of charge, as we saw in Sec. 15.4A, implies overlap of orbitals and increased bonding. Since bonding is energetically favorable, structures that delocalize charge are important.

4. *Structures that require the separation of opposite formal charges are less important than those that do not.*

Example:

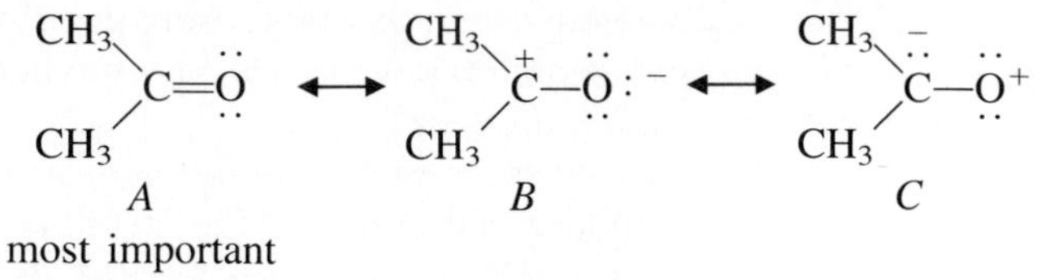

(15.40)

Since the electrostatic law (Eq. 1.2) tells us that energy is required to separate charge, structures *B* and *C* are less stable and thus less important. (They also have fewer formal bonds, and so are minor contributors on two counts.)

Be sure that you understand the difference between "delocalization of charge" and "separation of charge." When a charge is *delocalized* by resonance, charge of a given type is moved to different locations within a molecule. When charge is *separated,* two opposite charges are moved away from each other.

5. *Structures in which charges and electron deficiency are assigned to atoms of appropriate electronegativity are more important.*

Example: In Eq. 15.40, structure *B* is more important than structure *C* because structure *B* assigns an electron deficiency to the more electropositive atom (carbon) and negative charge to the more electronegative atom (oxygen).

In applying this rule, it is important not to confuse *electron deficiency* with *positive charge.* There are numerous situations in which positively charged atoms are not electron deficient. To illustrate, the oxygen atoms in both $H_3O:^+$ and in the following structure are positively charged, but these oxygens are *not* electron deficient, because they have complete electronic octets:

$$CH_2{=}\overset{+}{\ddot{O}}{-}H \qquad \text{oxygen has an electronic octet}$$

On the other hand, the oxygen in structure *C* of Eq. 15.40 is electron deficient because it is two electrons short of the octet. Because of the importance of the octet rule, an electron-deficient oxygen or nitrogen is a much worse situation energetically than a positively charged oxygen or nitrogen with an electronic octet. Thus, a resonance contributor that delocalizes charge and gives every atom an electronic octet is important even if it involves positive charge on an electronegative atom. A resonance structure that involves an electron-deficient electronegative atom, whether positively charged or not, is generally not very important.

6. *Highly strained structures are unimportant.*

Example:

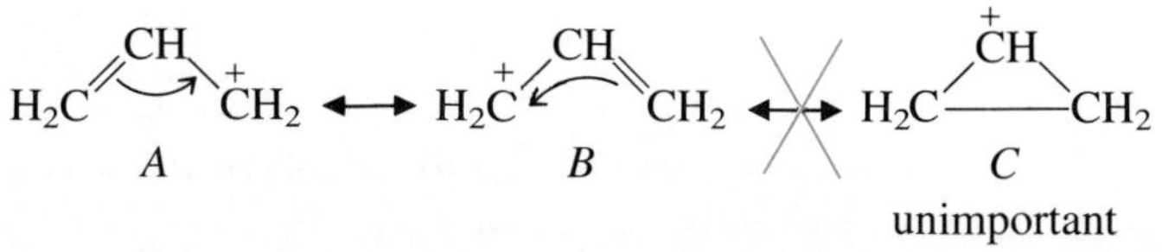

(15.41)

Structure *C* incorporates a strained three-membered ring and is therefore an unimportant structure for the allyl cation. Because this structure is unimportant, positive charge is shared only on the two terminal carbons of the allyl cation; there is no charge on the middle carbon, even though all three carbons have *p* orbitals (Fig. 15.7).

7. *If the orbital overlap symbolized by a resonance structure is not possible, the resonance structure is not important.*

Example:

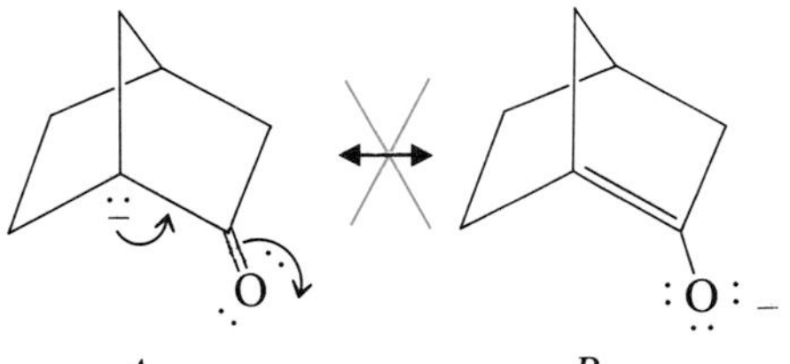

A *B* unimportant (15.42)

Structure *B* is unimportant because the orbital containing the unshared electron pair lies at a right angle to the π-orbital of the C═O group, and therefore cannot overlap with it.

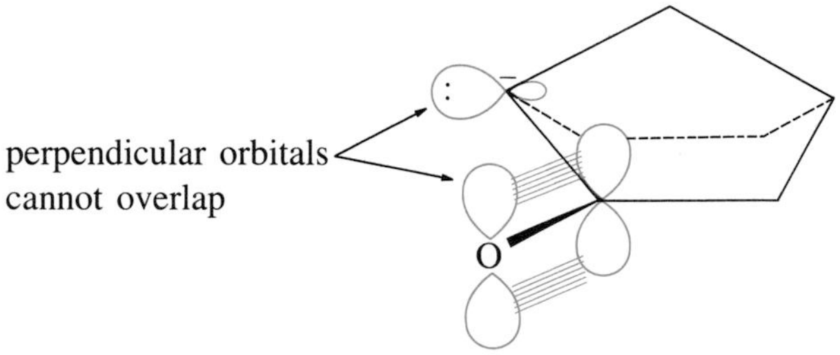

Only by an energetically costly distortion of the molecule could overlap of all the *p* orbitals occur. (Another way to look at this structure is that it violates Bredt's rule (Sec. 7.7C) and therefore is strained and violates guideline 6.) Notice that a formally similar type of resonance structure *is* important in other molecules.

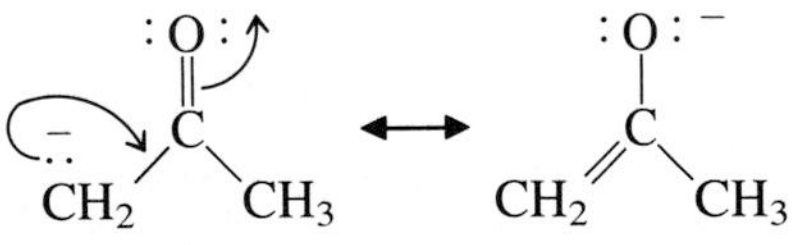

A *B* both important (15.43)

Both structures are important because they indicate delocalization of charge (guideline 3) and they have the same number of formal bonds (guideline 1). Structure *B* is particularly important because it places negative charge on the electronegative oxygen atom. In this example the carbon bearing the negative charge can easily rotate about its C—C bond so that its electron pair can overlap with the neighboring π-bond of the C═O group; this type of internal rotation is impossible in Eq. 15.42 because of the bicyclic structure.

C. Use of Resonance Structures

Although resonance does have a mathematical basis in quantum theory, organic chemists generally use resonance arguments in a qualitative way to compare the stabilities of molecules. To make this comparison, the following principle is applied: *Usually the compound with the greater number of important resonance structures is more stable.* (The reason for the qualifier "usually" will be apparent in Sec. 15.7.) The more stable cation of the two below, for example, is the allylic cation, and it has two resonance structures. The other carbocation has only one Lewis structure and is less stable.

$$[\text{CH}_3-\text{CH}_2-\overset{+}{\text{CH}}-\text{CH}=\text{CH}-\text{CH}_3 \quad \longleftrightarrow \quad \text{CH}_3-\text{CH}_2-\text{CH}=\text{CH}-\overset{+}{\text{CH}}-\text{CH}_3] \qquad \text{more stable ion}$$

$$\text{CH}_3-\overset{+}{\text{CH}}-\text{CH}_2-\text{CH}=\text{CH}-\text{CH}_3 \qquad \text{less stable ion}$$

The number of resonance structures for a compound is related to its stability because *resonance structures are a symbolic way of representing additional bonding associated with orbital overlap.* For example, allylic cations have additional orbital overlap and therefore additional bonding (Fig. 15.7) that is not present in isomeric nonallylic cations. The resonance structures for allylic cations are a symbolic way of showing this additional bonding and stability. Any species for which additional bonding and hence, additional stability, can be indicated by resonance structures is said to be **resonance stabilized.** Thus allylic cations are resonance-stabilized carbocations.

Let us consider one additional example. Suppose we want to decide which of the following cations is more stable:

$$\underset{A}{\text{CH}_3\overset{..}{\text{O}}-\text{CH}=\text{CH}-\overset{+}{\text{CH}}_2} \quad \text{or} \quad \underset{B}{\text{CH}_2=\overset{\overset{\displaystyle :\overset{..}{\text{O}}\text{CH}_3}{|}}{\text{C}}-\overset{+}{\text{CH}}_2}$$

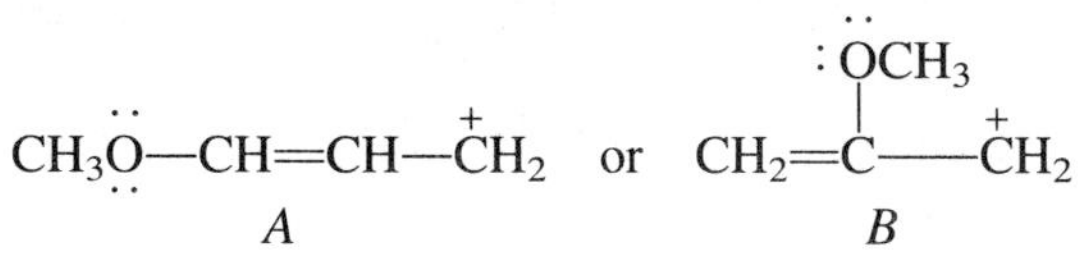

Carbocation *A* has the following important resonance structures:

$$\text{CH}_3\overset{..}{\text{O}}-\text{CH}=\overset{\frown}{\text{CH}}-\text{CH}_2^+ \quad \longleftrightarrow \quad \text{CH}_3\overset{.\,.}{\text{O}}\overset{+}{\text{CH}}-\text{CH}=\text{CH}_2 \quad \longleftrightarrow \quad \text{CH}_3\overset{+}{\text{O}}=\text{CH}-\text{CH}=\text{CH}_2 \quad (15.44)$$

Carbocation *B* has only two reasonable structures. In particular, the charge can be delocalized onto the oxygen in ion *A* but not in ion *B*:

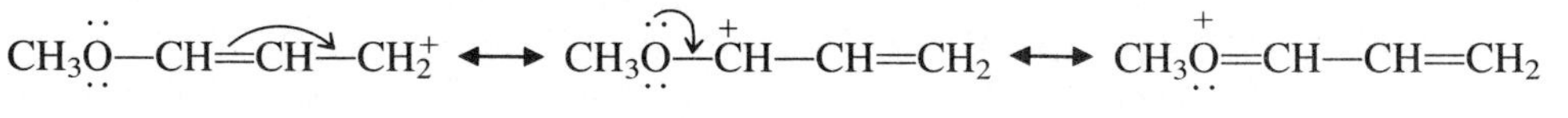

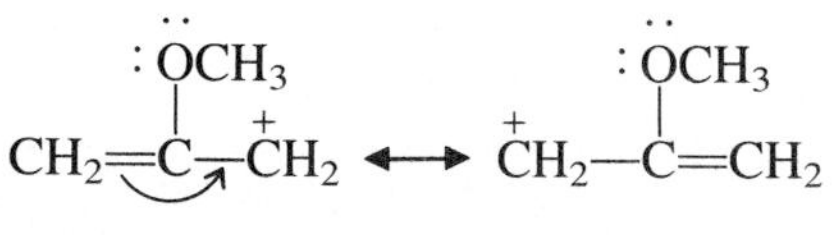

$$\underset{}{\text{CH}_2=\overset{\overset{\displaystyle :\overset{..}{\text{O}}\text{CH}_3}{|}}{\text{C}}\overset{+}{\text{CH}}_2} \quad \longleftrightarrow \quad \overset{+}{\text{CH}}_2-\overset{\overset{\displaystyle :\overset{..}{\text{O}}\text{CH}_3}{|}}{\text{C}}=\text{CH}_2 \qquad (15.45)$$

Thus, ion *A* is more stable because it has the larger number of important resonance structures.

Problems

26 In each of the following sets, show by the arrow formalism how each structure is derived from any other one, and indicate which structure(s) are most important and why:

(a) $\text{H}_3\text{C}-\text{CH}_3 \quad \longleftrightarrow \quad \text{H}_3\text{C}\cdot \quad \cdot\text{CH}_3 \quad \longleftrightarrow \quad \text{H}_3\text{C}:^- \; ^+\text{CH}_3 \quad \longleftrightarrow \quad \text{H}_3\text{C}^+ \; ^-:\text{CH}_3$

(b) $CH_3—\overset{+}{C}=\ddot{N}H \longleftrightarrow CH_3—C\equiv\overset{+}{N}H \longleftrightarrow CH_3—\overset{2+}{C}—\overset{-}{\ddot{N}}H$

(c)

(d) $CH_3—CH=CH—\overset{+}{C}H_2 \longleftrightarrow CH_3—\overset{+}{C}H—CH=CH_2$

27 Using resonance arguments, rank the molecules within each set in order of increasing stability, least stable first. Explain.

(a) $CH_3—CH(—\ddot{O}:^-)—CH=CH_2$ or $CH_3—C(—\ddot{O}:^-)=CH—CH_3$

(b) $O=CH—\overset{-}{\ddot{C}}H—C\equiv C—CH_3$ or $HC\equiv C—C(—\ddot{O}:^-)=CH—CH_3$

(c) $CH_2=CH—CH=CH—\dot{C}H_2$ or $\cdot CH_2—C(=CH_2)—CH=CH_2$

(d) $CH_2=C(CH_3)—\overset{+}{C}H_2$, $\overset{+}{C}H=C(CH_3)—CH_3$, or $CH_3—CH=CH—\overset{+}{C}H_2$

28 The following two compounds do not differ greatly in stability. Using Hammond's postulate, indicate which compound should react more rapidly in an S_N1 reaction in aqueous solution.

$CH_3\ddot{O}—CH=CH—CH_2—Cl$ $\quad$ $CH_3\ddot{O}—C(=CH_2)—CH_2—Cl$

15.7 AROMATICITY

A. Benzene: A Puzzling "Alkene"

We have been discussing conjugated alkenes and resonance. Let us now consider the structure and stability of benzene, a molecule whose Lewis structure is formally that of a cyclic conjugated triene. The chemistry of benzene, however, bears almost no resemblance to the chemistry of conjugated alkenes. Resonance arguments, as we shall learn, play a central role in understanding the structure and reactivity of benzene.

Benzene is the simplest of the **aromatic hydrocarbons.**

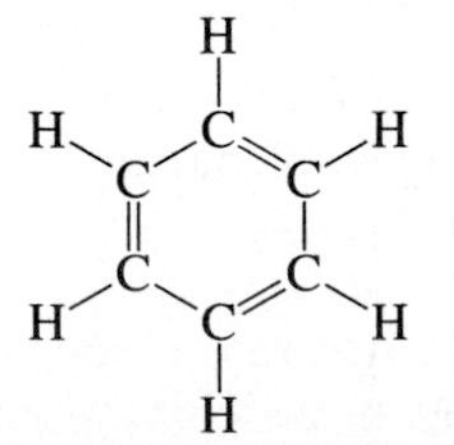

or

benzene

The reason for the term *aromatic* is historical: many fragrant compounds known from earliest times, such as the ones below, proved to be derivatives of benzene.

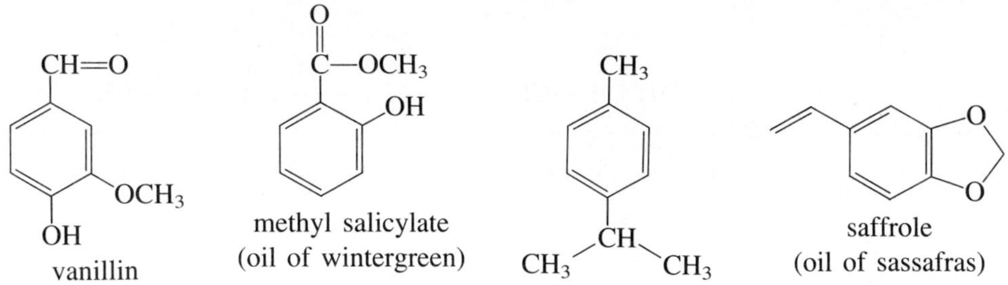

vanillin (vanilla) — methyl salicylate (oil of wintergreen) — *p*-cymene (oil of caraway) — saffrole (oil of sassafras)

Although today we know that many benzene derivatives are not distinguished by unique odors, the word *aromatic* continues to be used as a family name for all benzene derivatives and certain related compounds.

The structure that we use for benzene today was proposed in 1865 by August Kekulé, who claimed in 1890 that it came to him in a dream. (This claim has been disputed by modern historians.) To nineteenth-century chemists, the problem with the Kekulé structure was that it portrays benzene as a cyclic, conjugated triene. Yet benzene does not undergo any of the addition reactions that we have come to associate with either conjugated dienes or ordinary alkenes. Benzene itself, as well as benzene rings in other compounds, are inert to the usual conditions of halogenation, hydroboration, hydration, or ozonolysis. This property of the benzene ring is illustrated by the bromination of styrene, a compound that contains both a benzene ring and one additional double bond.

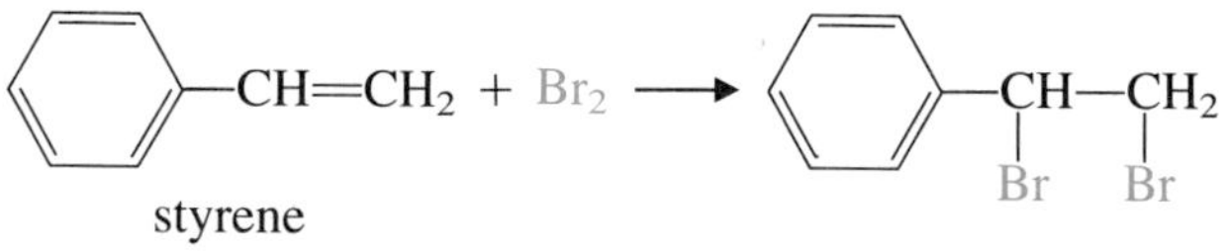

styrene (15.46)

The noncyclic double bond in styrene rapidly adds bromine, but the benzene ring remains unaffected.

We might speculate that benzene's lack of reactivity has something to do with its cyclic structure; yet cyclohexene also brominates readily. Perhaps, then, it is the cyclic structure and the conjugated double bonds that *together* account for the unusual behavior of benzene. However, 1,3,5,7-cyclooctatetraene (which we shall abbreviate as COT) brominates smoothly even at low temperature.

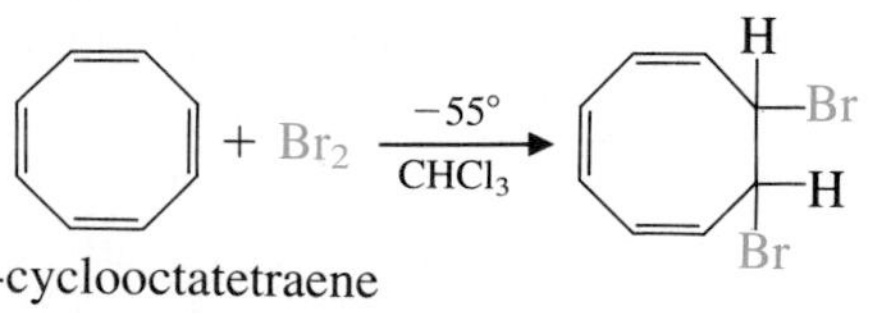

(100% yield) (15.47)

1,3,5,7-cyclooctatetraene (COT)

Thus, the Kekulé structure clearly had difficulties that could not be easily explained away. In 1869, A. Ladenburg proposed a structure for benzene, called both

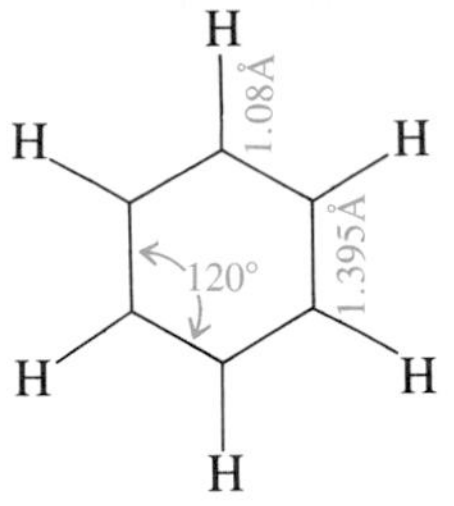

Figure 15.9 Structure of benzene.

Ladenburg benzene and *prismane*, that seemed to overcome these objections.

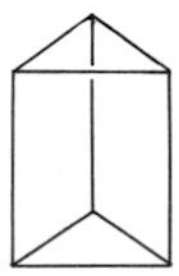

Ladenburg benzene, or prismane

Although we can recognize today that Ladenburg benzene is a highly strained molecule (it has been described as a "caged tiger"), to nineteenth-century chemists an attractive feature of this molecule was its lack of double bonds.

Several facts, however, ultimately led to adoption of the Kekulé structure. One of the most compelling arguments was that all efforts to prepare the alkene 1,3,5-cyclohexatriene using standard alkene syntheses led to benzene. The argument was, then, that benzene and 1,3,5-cyclohexatriene must be one and the same compound. Synthetic routes involving the same kinds of reactions were also used to prepare COT, which, as we have seen, has the reactivity of an ordinary alkene.

Although the Ladenburg-benzene structure had been discarded for all practical purposes decades earlier, its final refutation came in 1973 with its synthesis by Professor Thomas J. Katz and his colleagues at Columbia University. These chemists found that Ladenburg benzene is an explosive liquid with properties quite different from those of benzene.

How can we reconcile the Kekulé "cyclic triene" structure for benzene with the fact that benzene is inert to the usual reactions of alkenes? The answer to this question will occupy our attention in the remainder of this chapter.

B. Structure of Benzene

The structure of benzene (Fig. 15.9) shows that it has *one* type of carbon–carbon bond with a bond length (1.395 Å) intermediate between the lengths of single and double bonds. All atoms in the molecule lie in one plane. The Kekulé structure for benzene shows *two* types of carbon–carbon bonds: single bonds and double bonds. This inadequacy of the Kekulé structure can be remedied, however, by depicting benzene as *the hybrid of two equally contributing resonance structures.*

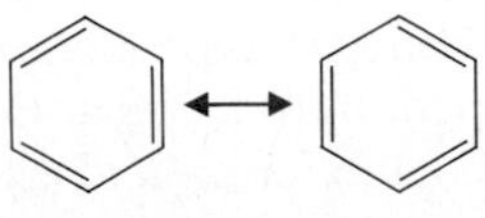

(15.48)

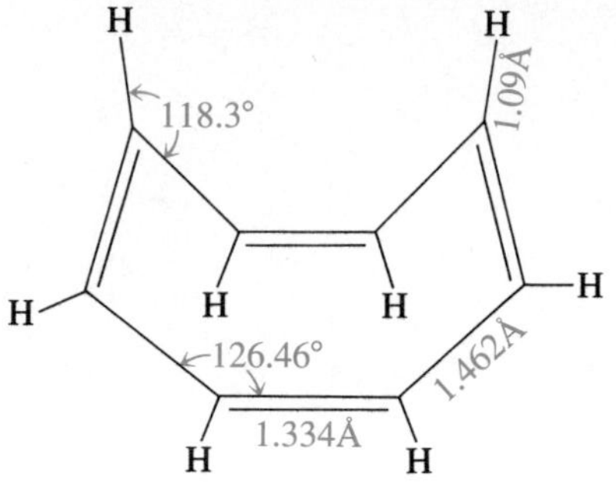

(a)

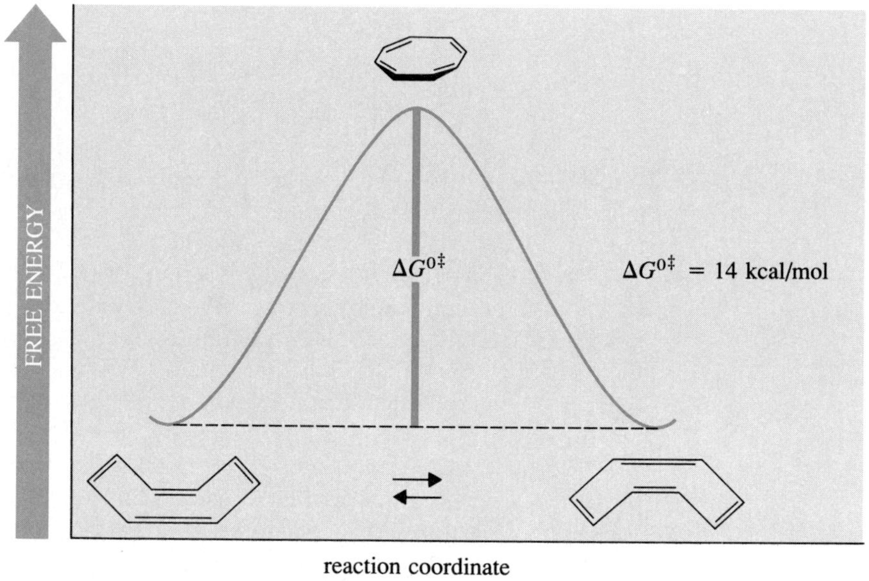

(b)

Figure 15.10 Structure of 1,3,5,7-cyclooctatetraene (COT). (a) Bond lengths and bond angles and (b) conformational equilibrium in COT. The planar conformation of the molecule is 14 kcal/mol less stable than the tub forms and is the transition state for the interconversion of the tub forms.

Benzene is an average of these two structures; it is *one* compound with *one* type of carbon–carbon bond that is neither a single bond nor a double bond, but something in between. Occasionally a benzene ring is represented with either of the following structures, which show the "smearing out" of double bond character.

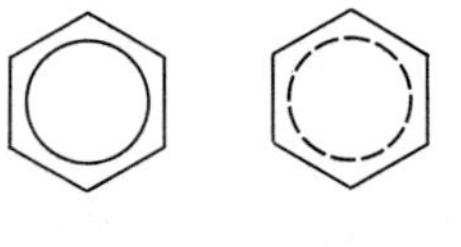

It is interesting to compare the structures of benzene and 1,3,5,7-cyclooctatetraene (COT), in view of their greatly different chemical reactivities (Eqs. 15.46 and 15.47). Their structures are remarkably different. First, COT has alternating single and double bonds (Fig. 15.10a). Second, COT is not planar like benzene, but instead is tub-shaped. In fact, planar COT is the *transition state* for the interconversion of two equivalent tub forms, and is less stable than the tub forms by about 14 kcal/mol (Fig. 15.10b).

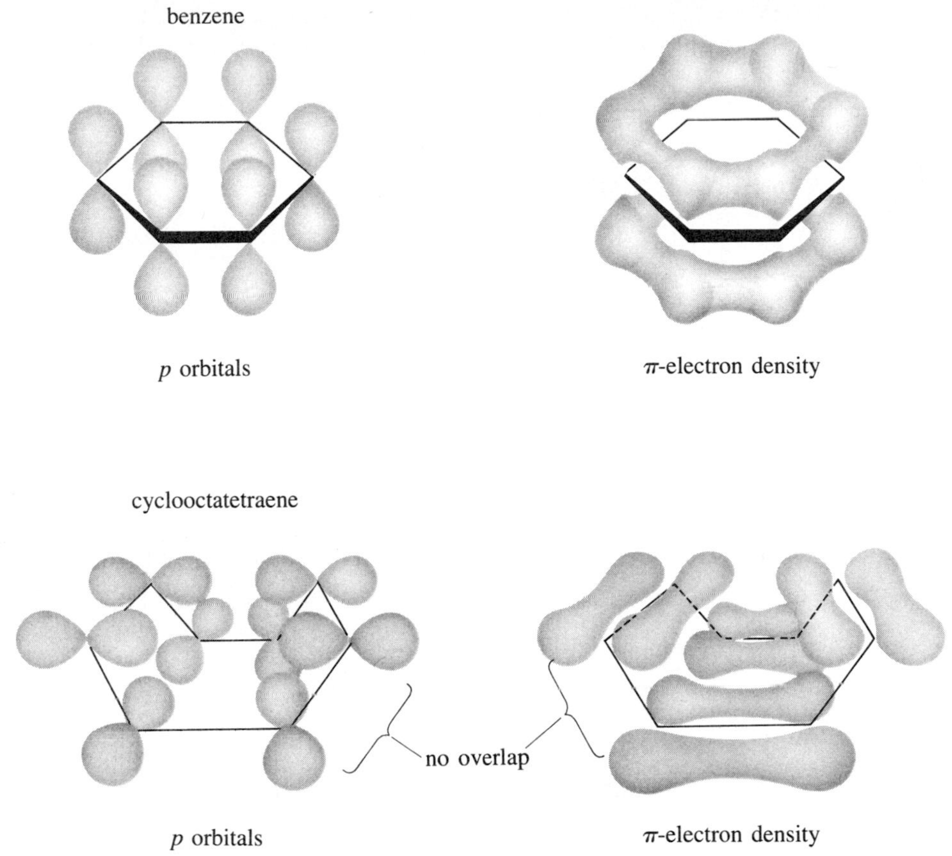

Figure 15.11 Comparison of the π-bonds in benzene and 1,3,5,7-cyclooctatetraene.

There are also differences in the π-bonds of benzene and COT (Fig. 15.11). The Kekulé structures for benzene suggest that each carbon atom should be trigonal, and therefore sp^2 hybridized. This means that there is a *p* orbital on each carbon atom. Since the molecule is planar, and the axes of all six *p* orbitals of benzene are parallel, these *p* orbitals overlap to form a continuous bonding π-molecular orbital (Fig. 15.11a). That is, the π-electron density in benzene lies in doughnut-shaped regions both above and below the plane of the ring. This overlap is symbolized by the resonance structures of benzene. In contrast, the carbon atoms of COT are not all coplanar, but they are nevertheless all trigonal. This means that there is a *p* orbital on each carbon atom of COT, but these orbitals do not overlap to form a continuous π-molecular orbital like the one in benzene. Instead (Fig. 15.11b) there are four π-electron systems of two carbons each; the π-bonds are mutually perpendicular, and therefore they do *not* overlap. As far as the π-electrons are concerned, COT looks like four isolated ethylene molecules. Because there is no electronic overlap between the π-orbitals of adjacent double bonds, *COT does not have resonance structures analogous to those of benzene* (Sec. 15.6B, guideline 7).

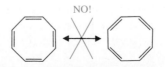

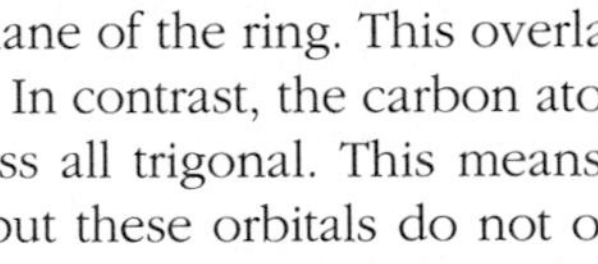

(15.49)

Let us summarize: we can write resonance structures for benzene, because the *p* orbitals of benzene overlap to provide additional bonding and additional stability. We *cannot* write resonance structures for COT because there is no overlap between *p* orbitals on adjacent double bonds.

C. Stability of Benzene

The chemists of the nineteenth century considered benzene to be unusually stable because it failed to react with reagents that attack ordinary alkenes (Sec. 15.7A). Let us examine the stability of benzene using heats of formation. Since benzene and COT have the same empirical formula (CH), their heats of formation per CH group can be compared. The ΔH_f^0 of benzene is +19.82 kcal/mol, or 19.82/6 = 3.3 kcal per CH group. The ΔH_f^0 of COT is +71.23 kcal/mol, or 71.23/8 = 8.9 kcal per CH group. This means that, per CH group, benzene is 8.9 − 3.3 = 5.6 kcal/mol more stable than COT. It follows, then, that benzene is 5.6 × 6 = 33.6 kcal/mol more stable than a hypothetical six-carbon cyclic triene with the same stability as COT.

This energy difference of about 34 kcal/mol is called the **empirical resonance energy** of benzene. We learned (Sec. 15.6C) that when a molecule can be represented with important resonance structures, it has enhanced stability. We also learned that resonance structures can be drawn for benzene, but not for COT. The 34 kcal/mol figure is an estimate of just how much special stability is implied by the resonance structures for benzene—thus the name "resonance energy."

Notice that the resonance energy is the energy by which benzene is *stabilized*; it is therefore an energy that benzene "doesn't have". The empirical resonance energy of benzene has been estimated in several different ways; these estimates range from 30–41 kcal/mol. The important point, however, is not the exact value of this number, but the fact that it is *large:* benzene is a very stable compound!

D. Aromaticity. The Hückel 4n + 2 Rule

It is clear that benzene is an especially stable molecule. A number of other compounds have a similar special stability, which is called **aromaticity.** (This term has *nothing* to do with the odor of the compound; it simply implies a special relationship to benzene.) To be aromatic, a compound must conform to *all* of the following criteria:

1. Aromatic compounds have a *cyclic* arrangement of *p* orbitals. Thus, aromaticity is a property of certain *cyclic* compounds.
2. There is a *p* orbital on *every* atom of an aromatic ring.
3. Aromatic rings are *planar;* this planarity allows the *p* orbitals on every atom to overlap.
4. The cyclic arrangement of *p* orbitals in an aromatic compound must contain $4n + 2$ π-electrons, where n is any integer (0, 1, 2, . . .). In other words, aromatic compounds have 2, 6, 10, . . . π-electrons.

These criteria for aromatic behavior were first recognized in 1931 by Erich Hückel, a German chemical physicist. They are often called collectively the **Hückel $4n + 2$ rule,** or simply the **$4n + 2$ rule.** The application of the $4n + 2$ rule is best illustrated by some examples.

Benzene conforms to the $4n + 2$ rule because it contains a planar, continuous ring of atoms with a total of six π-electrons; $4n + 2 = 6$ if $n = 1$. Notice that when we count the π-electrons, *each atom in a formal double bond contributes one π-electron.*

The alkene 1,3,5-hexatriene is *not* aromatic.

$$CH_2{=}CH{-}CH{=}CH{-}CH{=}CH_2$$ 1,3,5-hexatriene (*not* aromatic)

This compound has six π-electrons, and all p orbitals are parallel. However, this compound is not aromatic because the six π-electrons are not present in a *cyclic* arrangement; this compound does not meet the first criterion for aromaticity. (Of course, like all molecules containing conjugated double bonds, this compound does have a small amount of additional stability; Sec. 15.1A.)

Naphthalene *is* an aromatic compound.

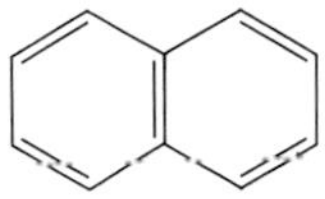

naphthalene (aromatic)

This compound has a planar cyclic array of ten carbon atoms; each carbon contributes one π-electron. Since $4n + 2 = 10$ for $n = 2$, naphthalene fulfills all the criteria for aromatic behavior. (For our purposes, it makes no difference that there is a bond across the center of the molecule.)

Toluene and biphenyl are both aromatic compounds.

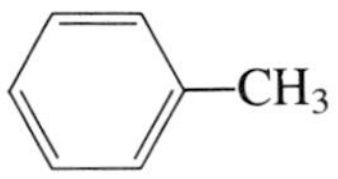

toluene (aromatic)

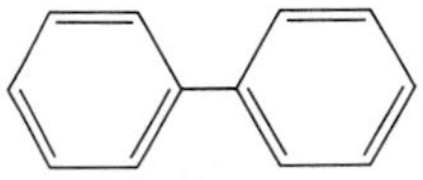

biphenyl (aromatic)

The benzene ring in toluene meets all of the criteria for aromaticity. The methyl group is a substituent group and is not part of the aromatic system of π-electrons. Each ring of biphenyl is a separate aromatic π-electron system. We do not count the number of π-electrons in both rings together because they are not all part of one continuous cycle.

The hydrocarbon 1,3,5-cycloheptatriene is not aromatic.

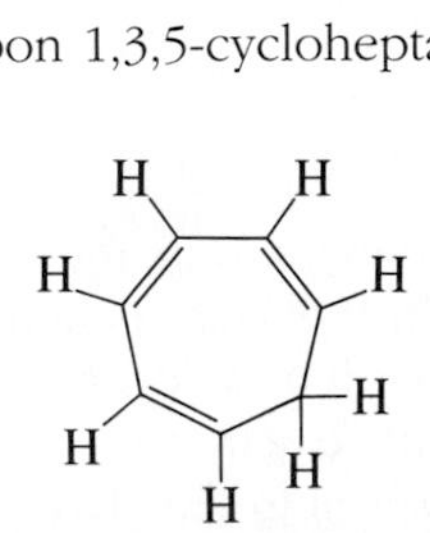

1,3,5-cycloheptatriene (*not* aromatic)

Even though this is a cyclic compound with $4n + 2$ π-electrons, it is not aromatic because it fails to meet the second criterion: one of the atoms in the ring does not have a *p* orbital. The π-electron system is not continuous, but is interrupted by the sp^3-hybridized —CH_2— carbon.

Neither of the following hydrocarbons are aromatic because they fail to meet the fourth criterion: they do not have $4n + 2$ π-electrons.

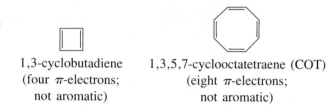

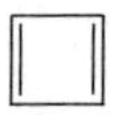

1,3-cyclobutadiene (four π-electrons; not aromatic)

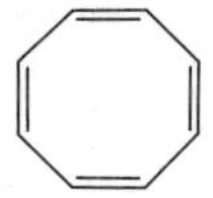

1,3,5,7-cyclooctatetraene (COT) (eight π-electrons; not aromatic)

There is no integer n such that $4n + 2 = 4$ or 8. These compounds have $4n$ π-electrons. In fact, compounds containing planar, continuous rings of $4n$ π-electrons are especially *unstable*; they are said to be **antiaromatic.** Cyclobutadiene is so unstable that it cannot be isolated, although it has been trapped at 4 K. *Planar* 1,3,5,7-cyclooctatetraene is also antiaromatic and unstable; this is one reason that this compound assumes a tub conformation (Fig. 15.10). Cyclobutadiene cannot adopt a nonplanar conformation, but chemists believe that this molecule seeks to avoid the continuous overlap of $4n$ π-electrons as much as possible by adopting a rectangular geometry with two short double bonds and two long single bonds:

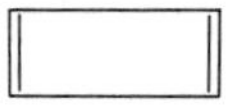

In other words, a molecule with $4n$ π-electrons is *destabilized* by resonance. This is why cyclobutadiene and COT adopt conformations that minimize the resonance interaction between double bonds. This destabilization of antiaromatic compounds by resonance is the reason we noted in Sec. 15.6C that resonance *usually* implies extra stability; here is a situation in which resonance implies special *instability*.

Aromatic behavior is not restricted to hydrocarbons. Several *heterocyclic compounds* (Sec. 8.1C) are aromatic; for example, pyridine and pyrrole are both aromatic compounds.

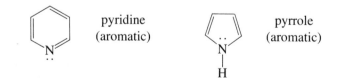

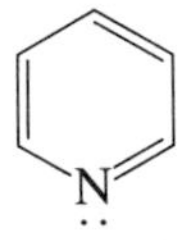

pyridine (aromatic)

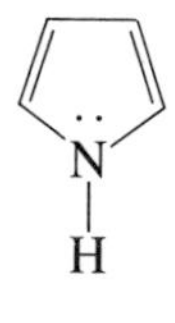

pyrrole (aromatic)

Except for the nitrogen in the ring, the structure of pyridine closely resembles that of benzene. Each atom in the ring, including the nitrogen, is part of a formal double bond and therefore contributes one π-electron. What about the extra electron pair on nitrogen? How does this electron pair figure in the π-electron count? This electron pair resides in an sp^2 orbital in the plane of the ring (Fig. 15.12a). (Formally, it has the same relationship to the ring as a hydrogen of benzene.) Because it does not overlap

Figure 15.12 The *p* orbitals in pyridine and pyrrole.

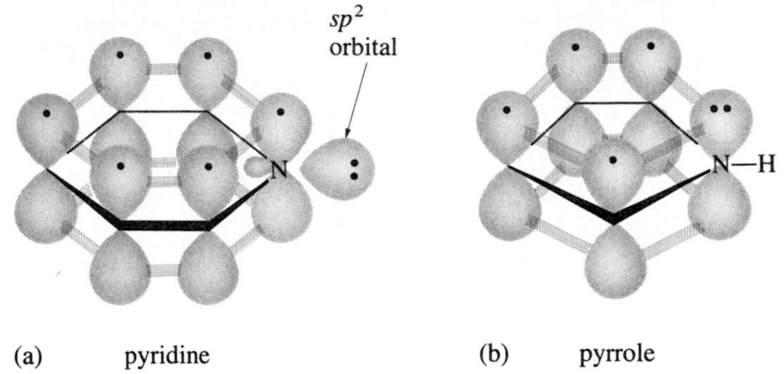

with the ring π-electron system, it is not included in the π-electron count. Thus *vinylic electrons (electrons on doubly bonded atoms) are not counted as π-electrons.*

In pyrrole, on the other hand, the electron pair on nitrogen is formally *allylic* (Fig. 15.12b). The nitrogen has a trigonal geometry and sp^2 hybridization that allow its electron pair to occupy a *p* orbital and contribute to the π-electron count. The N—H hydrogen lies in the plane of the ring. In general, *allylic electrons are counted as π-electrons when they reside in orbitals that are parallel to the other p orbitals in the molecule.* Therefore, pyrrole has six π-electrons—four from the double bonds and two from the nitrogen—and is aromatic.

Note carefully the different ways in which we handle the electron pairs on the nitrogens of pyridine and pyrrole. The nitrogen in pyridine is part of a formal double bond, and the electron pair *is not* part of the π-electron system. The nitrogen in pyrrole is formally allylic and its electron pair *is* part of the π-electron system.

Problem

29 Furan is an aromatic compound. Discuss the hybridization of its oxygen and the geometry of its two electron pairs.

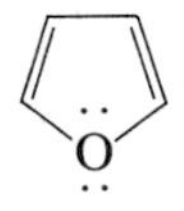

furan (aromatic)

Aromatic compounds need not be neutral species; a number of ions have special aromatic stability. One of the best characterized aromatic ions is the cyclopentadienyl anion:

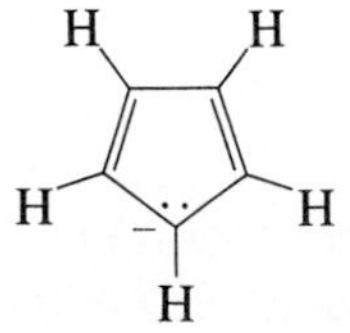

2,4-cyclopentadien-1-ide anion (cyclopentadienyl anion; aromatic)

This ion resembles pyrrole, except that since the atom bearing the allylic electron pair is carbon rather than nitrogen, its charge is −1. One way to form this ion is by the reaction of sodium with the conjugate acid hydrocarbon, 1,3-cyclopentadiene; notice the formal analogy to the reaction of Na with H_2O.

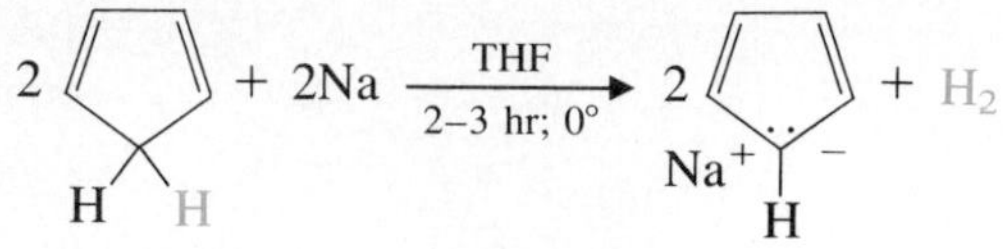

(15.50)

The cyclopentadienyl anion has five equivalent resonance structures; the negative charge can be delocalized to every carbon atom.

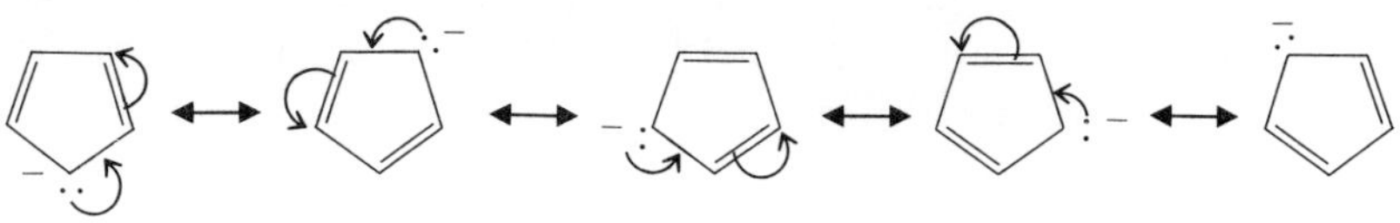

(15.51)

These structures show that all carbon atoms of the cyclopentadienyl anion are equivalent. Because of the stability of this anion, its conjugate acid, 1,3-cyclopentadiene, is an unusually strong acid. With a pK_a of 15, this compound is 10^{10} times as acidic as a 1-alkyne, and about as acidic as water!

A remarkable example of anion aromaticity is the aromatic *dianion* derived from 1,3,5,7-cyclooctatetraene (COT):

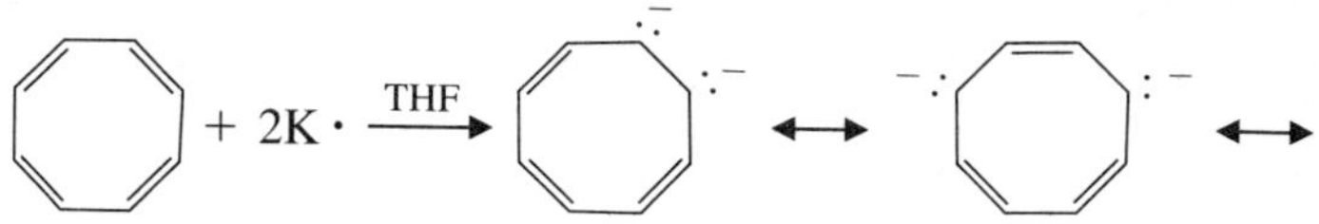

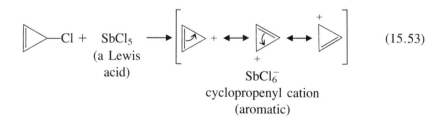

(draw the remaining resonance structures) (15.52)

COT

$2K^+$
cyclooctatetraenyl dianion

We have learned that COT is nonplanar and not aromatic, and that planar COT is antiaromatic. Yet when the hydrocarbon gains two electrons, it becomes planar in spite of the resulting angle strain, because the anion has ten π-electrons and is aromatic.

Cations, too, may be aromatic.

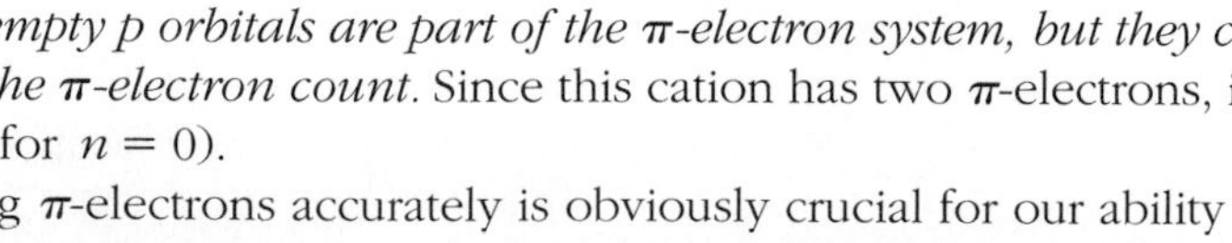

(15.53)

(a Lewis acid)

$SbCl_6^-$
cyclopropenyl cation
(aromatic)

Atoms with empty p orbitals are part of the π-electron system, but they contribute no electrons to the π-electron count. Since this cation has two π-electrons, it is aromatic ($4n + 2 = 2$ for $n = 0$).

Counting π-electrons accurately is obviously crucial for our ability to apply the $4n + 2$ rule. Let us summarize what we have learned about π-electron counting:

1. Each atom that is part of a formal double bond contributes one π-electron.
2. Vinylic electrons do not contribute to the π-electron count.
3. Allylic electrons contribute to the π-electron count if they occupy an orbital parallel to the other *p* orbitals in the molecule.

4. An atom with an empty *p* orbital can be part of a continuous aromatic π-electron system, but contributes no π-electrons.

Problem

30 Do you think it would be possible to have an aromatic free radical? If so, draw the structure of one. If not, explain why.

There are some remarkable organometallic compounds that demonstrate some effects of aromaticity. For example, the cyclopentadienyl anion forms stable complexes with a number of transition-metal ions. One of the best known of these complexes is *ferrocene,* which can be formally regarded as a salt of the ferrous ion (Fe^{2+}) and two cyclopentadienyl anions:

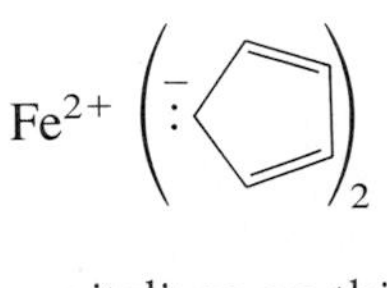

The synthesis of ferrocene in fact capitalizes on this relationship: two equivalents of cyclopentadienyl anion react with one of ferrous ion to give ferrocene.

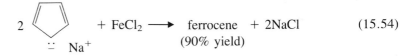

2 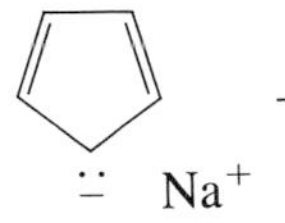 $+ FeCl_2 \longrightarrow$ ferrocene $+$ 2NaCl (90% yield) (15.54)

This synthesis is formally similar to a metathesis (exchange) reaction of inorganic salts. The actual structure of ferrocene, however, is not that of a salt, but is a remarkable "sandwich" structure in which the iron is embedded between two cyclopentadienyl anions:

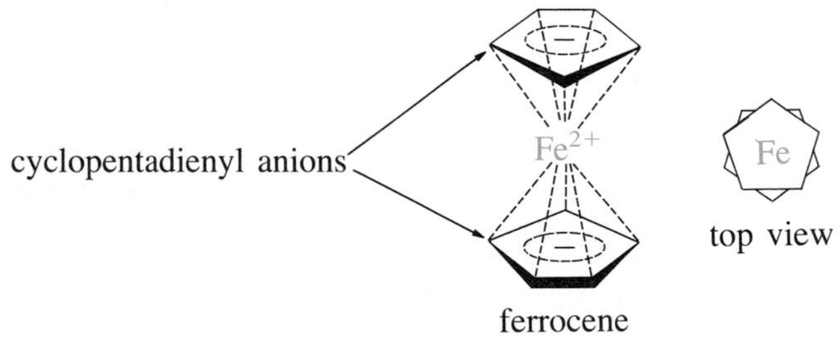

The dotted lines mean that the electrons are spread out around the ring. As we have learned (Eq. 15.51), all carbons in the cyclopentadienyl anion are equivalent, and they remain so in ferrocene; each carbon is bonded equally to the iron.

Despite the great instability of 1,3-cyclobutadiene, the following complex of this molecule is quite stable!

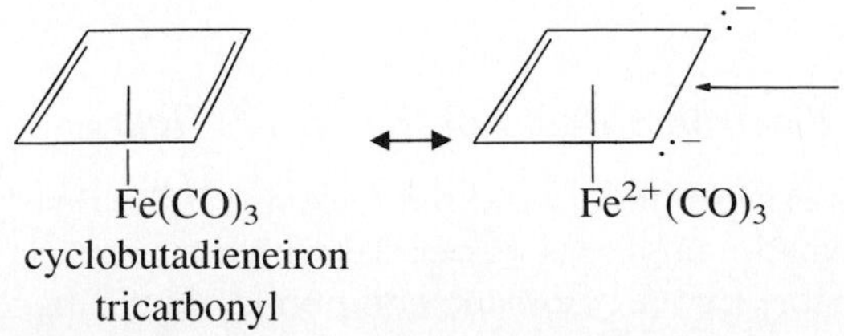

In this structure, the iron has a formal oxidation state of zero; the CO groups are neutral carbon monoxide ligands. A simplified way to view this compound, shown in the resonance structure above, is to realize that cyclobutadiene is short of a Hückel number of electrons (six) by two electrons, and that zero-valent iron, an electropositive metal, can lose two electrons to give a stable Fe^{2+} oxidation state. Thus, this complex has much of the character of a cyclobutadiene with two additional electrons—a cyclobutadienyl *dianion,* a six π-electron aromatic system—complexed to iron minus two of its electrons—that is, Fe^{2+}.

The $4n + 2$ rule has been very successful in predicting aromatic behavior. There are, however, some compounds that could be aromatic according to the rule, but are not. (For example, if the planar conformation of a molecule required for aromaticity is too strained to exist then the molecule cannot have aromatic stability; Problem 38.) However, no compound has ever been found to be aromatic that the $4n + 2$ rule says should *not* be aromatic.

Let us now return to the question we asked near the beginning of this section: what makes benzene inert to the usual reactions of alkenes? It is the *aromaticity* of benzene that is responsible for its unique chemical behavior. Were benzene to undergo the addition reactions typical of alkenes, its continuous cycle of $4n + 2$ π-electrons would be broken; it would lose its aromatic character, and thus much of its stability.

This is not to say, however, that benzene is unreactive under all conditions. Indeed, benzene and many other aromatic compounds undergo characteristic reactions that we shall study in the next chapter.

Problem

31 Which of the following species should be aromatic by the Hückel $4n + 2$ rule? Explain.

(a)

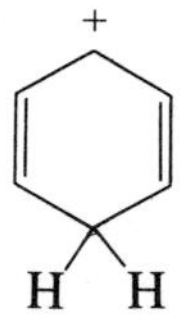

(b)

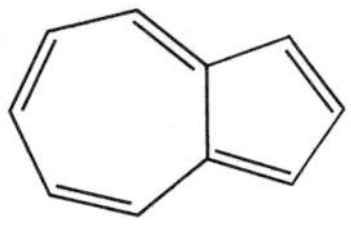

azulene

(c)

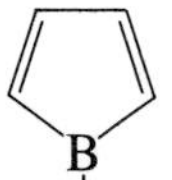

(d)

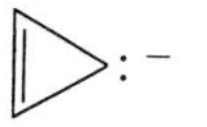

(e)

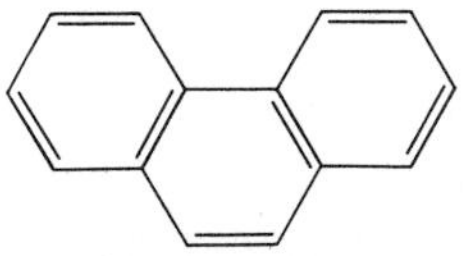

phenanthrene

(f)

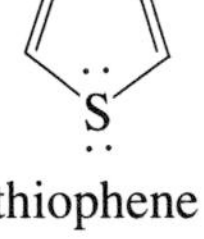

thiophene

(g)

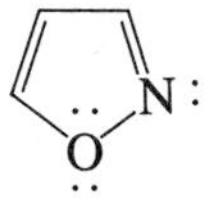

isoxazole

(h)

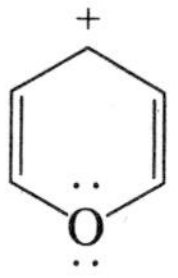

(i)

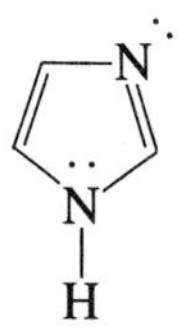

imidazole

E. Electronic Basis of the 4n + 2 Rule

It is easy to understand the reasons for the first three criteria for aromatic behavior. Aromatic stability is associated with the overlap of *p* orbitals; therefore it seems reasonable that an aromatic compound should have an uninterrupted planar arrange-

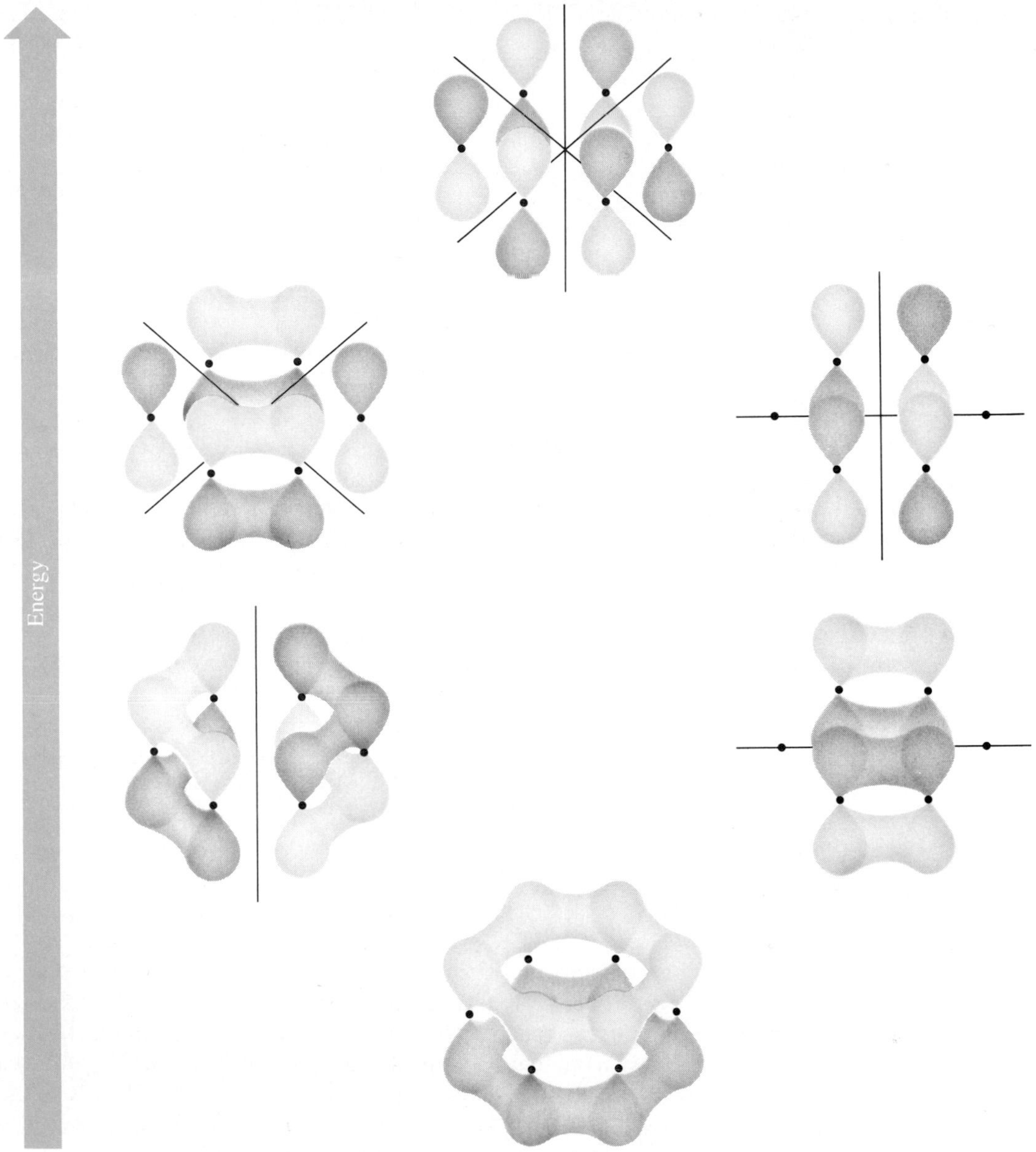

Figure 15.13 Benzene π-molecular orbitals. The dots represent the carbon nuclei; lines represent nodal planes; color represents wave peaks; and black represents wave troughs.

ment of *p* orbitals. What is more difficult to understand is the fourth criterion: the requirement for $4n + 2$ π-electrons.

The $4n + 2$ rule comes from application of the mathematics of quantum theory to cyclic π-electron systems—that is, from *molecular orbital theory* (MO theory). MO theory tells us that when a molecule contains a number n of *p* orbitals on adjacent atoms, these orbitals combine in very specific ways to give *n* π*-molecular orbitals.* Since benzene has six *p* orbitals, it therefore has six π-molecular orbitals, or MOs, which are shown in Fig. 15.13.

Figure 15.14 Construction of energy-level diagrams for the π-molecular orbitals of (a) benzene and (b) cyclobutadiene.

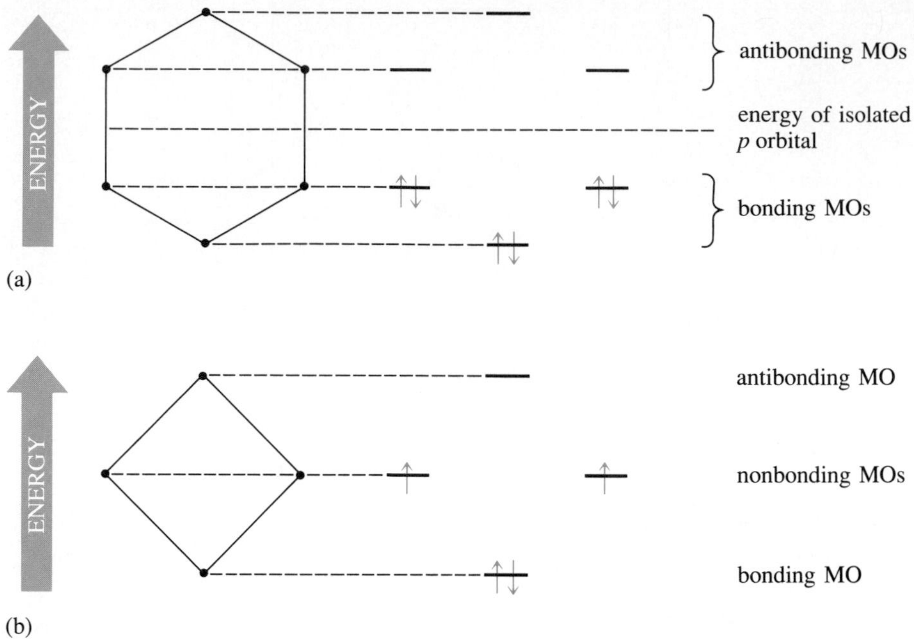

The shapes of these MOs come from the mathematics of quantum theory. Although how these shapes are determined is something that we cannot discuss here, they do have certain features that should seem reasonable from our discussion of molecular orbitals in Secs. 2.4 and 4.1A. First, the π-MOs have lobes on both faces of the benzene ring, like the *p* orbitals from which they are constructed. Second, MOs of progressively higher energy have more *nodes*. Hence, the MO of lowest energy has no nodes; the MOs of next higher energy have one node; and so on. (We can now understand that the molecular orbital shown in Fig. 15.11 is only one of the six benzene MOs.)

The most important feature of the benzene MOs in explaining the $4n + 2$ rule is the energies of the orbitals. MO theory gives us a very simple device, shown in Fig. 15.14, for determining the relative MO energies in cyclic conjugated hydrocarbons. We construct a geometrical figure identical to the carbon skeleton of the hydrocarbon under consideration; for benzene, this is a hexagon (Fig. 15.14a). The figure is drawn with a *vertex down*. If the energy of one MO is placed at this lowest vertex, the relative energies of the other MOs are proportional to the heights of the other vertices. The energy of an isolated *p* orbital lies across the middle of the figure.

This procedure shows that benzene has one MO at lowest energy; two MOs at next higher energy; two more at still higher energy; and one at highest energy. Different MOs of the same energy are said to be **degenerate MOs.** Hence, benzene has two pairs of degenerate MOs. The MOs at energies below that of the isolated *p* orbital are called **bonding molecular orbitals.** Those at higher energies are called **antibonding molecular orbitals.** Benzene has three bonding MOs and three antibonding MOs. Electrons that occupy bonding MOs contribute positively to π-bonding; electrons that occupy antibonding MOs contribute negatively to π-bonding.

Benzene has six π-electrons. These occupy the benzene MOs according to Hund's rule. As a result, the three benzene MOs of lowest energy—the bonding MOs—are exactly filled.

A similar procedure gives the energies of the cyclobutadiene MOs from the vertices of a square (Fig. 15.14b). Cyclobutadiene has one bonding MO, but the next two orbitals are degenerate orbitals that lie at the same energy as an isolated *p* orbital. These MOs are examples of **nonbonding molecular orbitals.** Electrons in nonbonding MOs contribute neither positively nor negatively to π-bonding.

Cyclobutadiene has four π-electrons. Two occupy the lowest MO, but the other two must be placed in the degenerate nonbonding MOs. According to Hund's rule, these electrons are unpaired; that is, *two MOs in cyclobutadiene are half-filled.*

Just as filled atomic orbitals are associated with stability of atoms such as the noble gases, filled *bonding molecular orbitals* are associated with stability in conjugated molecules. Hence, benzene, with its filled bonding MOs, is stable; cyclobutadiene is much less stable. In fact, it is so much less stable that it is termed *antiaromatic.*

Were we to construct MO energy diagrams for other molecules, we would find that *all molecules conforming to the $4n + 2$ rule have just enough electrons to fill exactly their bonding MOs.* Molecules with other numbers of π-electrons have partially filled MOs, and some of these are not bonding MOs. The $4n + 2$ rule, then, comes from the relationship between the number of bonding π-molecular orbitals in a molecule and the number of π-electrons available to fill them.

Problem

32 Construct a molecular-orbital energy-level diagram for each of the following species. Use the diagram, the number of π-electrons, and Hund's rule, to explain whether each is aromatic.

(a) 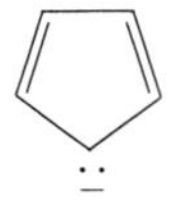 (b) 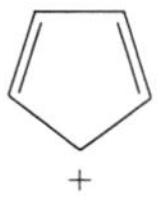 (c) 

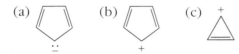

KEY IDEAS IN CHAPTER 15

- Molecules containing conjugated double or triple bonds have additional stability, relative to unconjugated isomers, that can be attributed to the continuous overlap of their *p* orbitals.
- A cumulene is a compound in which one or more carbon atoms are part of two double bonds. Adjacent π-bonds in a cumulene are mutually perpendicular; appropriately substituted cumulenes are chiral.
- Heats of formation are generally in the order: conjugated dienes < ordinary dienes < alkynes < cumulenes.
- Compounds with conjugated double or triple bonds have UV absorptions at $\lambda_{max} > 200$ nm.
- Each conjugated double or triple bond in a molecule contributes 30–50 nm to its λ_{max}. When a compound contains many conjugated double or triple bonds, it absorbs visible light and appears colored.

- The intensity of the UV absorption of a compound is proportional to its concentration (Beer's Law). The constant of proportionality ϵ, called the extinction coefficient, measures the intrinsic intensity of an absorption.
- The Diels–Alder reaction is a pericyclic reaction that involves the cycloaddition of a conjugated diene and a dienophile (usually an alkene). When the diene is cyclic, bicyclic products are produced.
- The diene assumes an *s-cis* conformation in the transition state of the Diels–Alder reaction; dienes that are locked into *s-trans* conformations are unreactive.
- Each component of the Diels–Alder reaction undergoes a *syn* addition to the other. In many cases the *endo* mode of addition is favored over the *exo* mode.
- Conjugated dienes react with hydrogen halides to give a mixture of 1,2- and 1,4-addition (conjugate addition) products. The mixture of products is accounted for by the formation of a resonance-stabilized allylic cation intermediate, which can be attacked by halide ion at either of two positively charged carbons. The 1,2-addition product is the product of kinetic control. On equilibration, the thermodynamically more stable 1,4-addition product predominates.
- Resonance structures are derived by the arrow formalism. A molecule is the weighted average of its resonance structures. That is, structures of greater importance make a greater contribution to the structure of the molecule.
- When comparing isomers, the one with the greatest number of important resonance structures is most stable, unless the resonance leads to antiaromatic character.
- Benzene has special stability that is termed *aromaticity*. All aromatic compounds contain $4n + 2$ π-electrons in a continuous, planar, cyclic array.
- The $4n + 2$ rule is a consequence of the stability of planar, cyclic species in which all bonding molecular orbitals (and only the bonding orbitals) are filled.

ADDITIONAL PROBLEMS

33 Give the principal product(s) expected when *trans*-1,3-pentadiene reacts under the following conditions. Assume one equivalent of each reagent reacts unless noted otherwise.

(a) Br_2 (dark), CCl_4
(b) HBr
(c) H_2 (2 equiv), Pd/C
(d) H_2O, H_3O^+
(e) Na^+ $C_2H_5O^-$ in C_2H_5OH
(f) maleic anhydride (Eq. 15.9) heat

34 Repeat Problem 33 for 3-methylene-1-cycloheptene.

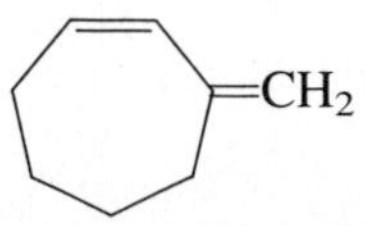

3-methylene-1-cycloheptene

35 What five-carbon conjugated diene would give the same *single* product from either 1,2- or 1,4-addition of HCl?

36 Explain each of the following observations:
(a) The allene 2,3-heptadiene can be resolved into enantiomers but the cumulene 2,3,4-heptatriene cannot.
(b) The cumulene in (a) can exist in diastereomeric forms, but the allene in (a) cannot.

37 Using the Hückel $4n + 2$ rule, determine whether each of the following compounds is likely to be aromatic. Explain how you arrived at the π-electron count in each case.

(a) 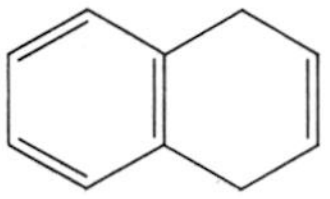

(d) 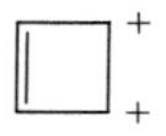

(g) 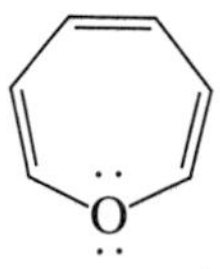

oxepin

(b) 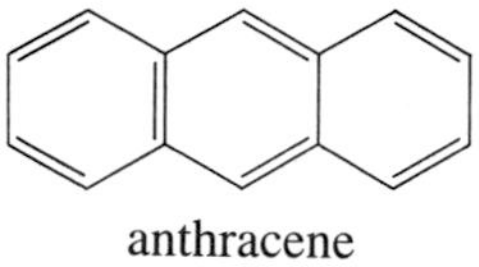

anthracene

(e) 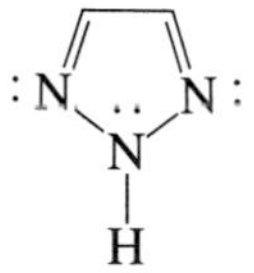

(h) 

(c) 

(f) 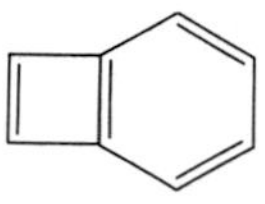

(i) 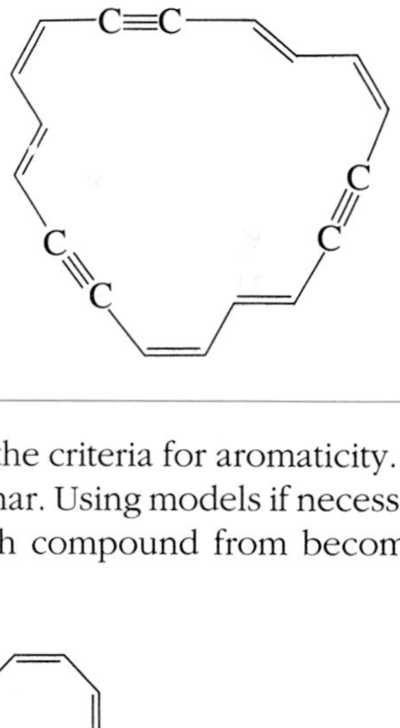

38 The following two compounds appear to meet all the criteria for aromaticity. Yet both are unstable because they cannot become planar. Using models if necessary, explain what it is in each case that prevents each compound from becoming planar and aromatic.

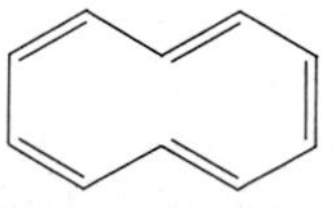

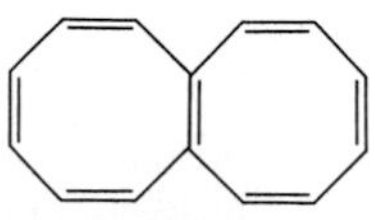

octalene

39 The hydrocarbon naphthacene absorbs blue light with $\lambda_{max} = 474$ nm. What color is naphthacene?

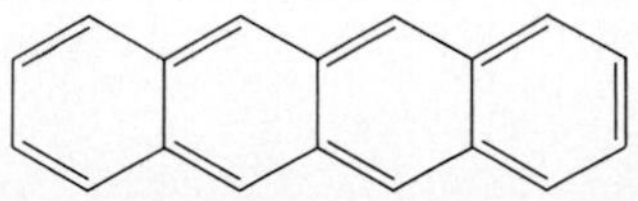

naphthacene

40 Assume you have unlabeled samples of the compounds within each set below. Explain how UV spectroscopy could be used to identify each compound.

(a) 1,4-cyclohexadiene and 1,3-cyclohexadiene

(b)

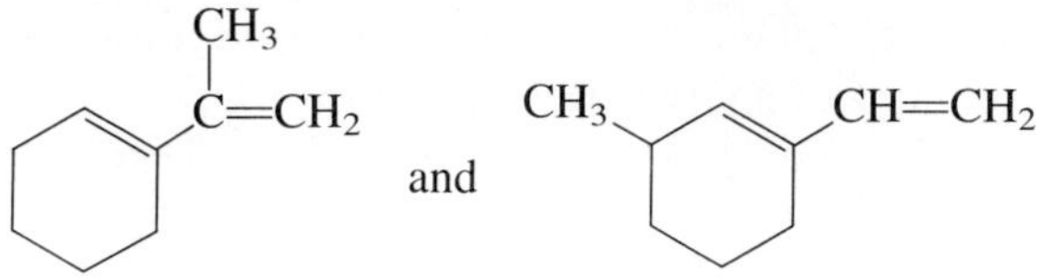

(c)

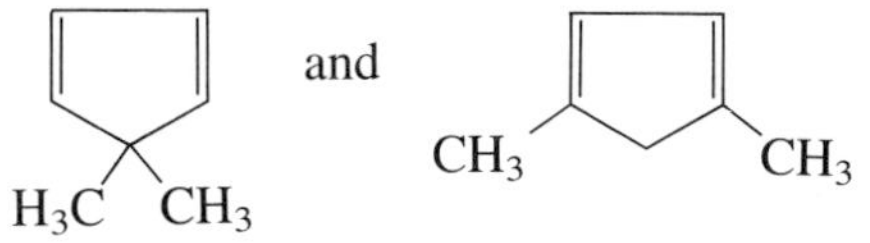

(d)

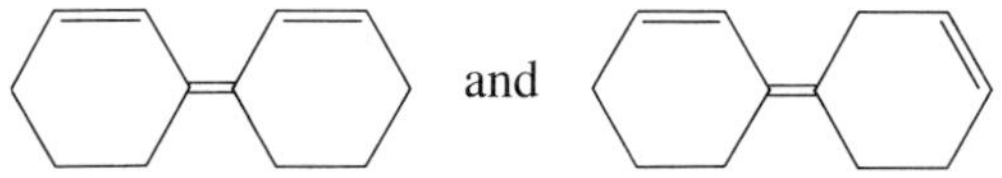

(e)

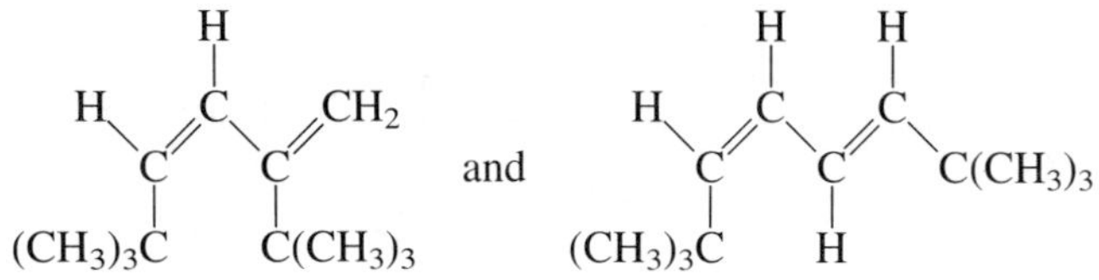

41 A colleague, Ima Hack, has subjected isoprene (Fig. 15.3) to catalytic hydrogenation to give isopentane. Hack has inadvertently stopped the hydrogenation prematurely and now wants to know how much unreacted isoprene remains in the sample. The mixture of isoprene and isopentane (75 mg total) is diluted to one liter with pure methanol and found to have an absorption at 222.5 nm and 1-cm path length of 0.356. Given an extinction coefficient of 10,750 at this wavelength, what weight percent of the sample is unreacted isoprene?

42 The 1,2-addition of one equivalent of HCl to the triple bond of vinylacetylene, HC≡C—CH=CH_2, gives a material called *chloroprene*. Chloroprene can be polymerized to give *neoprene*, valued for its resistance to oils, oxidative breakdown, and other deterioration. Give the structures of chloroprene and neoprene.

43 A chemist, I. M. Shoddy, has just purchased some compounds in a going-out-of-business sale from Pybond, Inc., a cut-rate chemical supply house. The company, whose motto is "You get what you pay for," has sent Shoddy a compound *A* at a bargain price in a bottle labeled only "C_6H_{10}." Unfortunately, Shoddy has lost his purchase order and cannot remember what he ordered, and he has come to ask your help in identifying the compound. Compound *A* is optically active, and has IR absorption at 2083 cm^{-1}. Partial hydrogenation of *A* with 0.2 equivalents of H_2 over a catalyst gives, in addition to recovered *A*, a mixture of *cis*-2-hexene and *cis*-3-hexene. Identify compound *A* and explain your reasoning.

44 The Diels–Alder reaction product below is consistent with the idea that the starting triene, 1,3,5-cyclooctatriene, is in equilibrium with another compound *A*.
(a) Deduce the structure of *A*.
(b) Using the arrow formalism, show how it is formed from the starting triene.

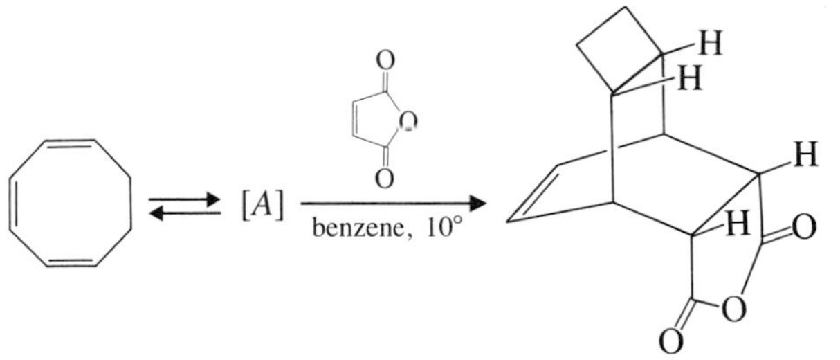

45 Each of the following problems can be answered by considering the aromaticity of the compound involved.
(a) Explain why borazole (sometimes called *inorganic benzene*) is a very stable compound.

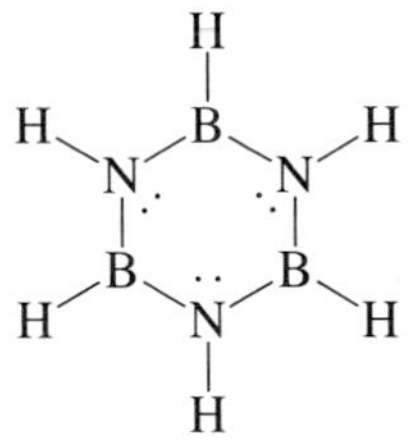

borazole

(b) When 1,3-cyclopentadiene, containing radioactive ^{14}C only at carbon-5, is treated with potassium metal, a gas is evolved. When the resulting mixture is added to water, radioactive 1,3-cyclopentadiene is again formed, but now 20% of the radioactive carbon is found at each position of the ring. Explain how the radioactive carbon is scrambled. (*Hint:* See Eq. 15.51.)
(c) When acenaphthylene is treated with chlorine gas in the dark, only one of its six double bonds reacts, and an addition product $C_{12}H_8Cl_2$ is formed. Propose a structure for this compound.

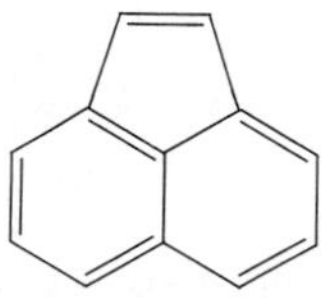
acenaphthylene

46 (a) Predict the products expected from the addition of one equivalent of HBr to (1) isoprene; (2) *trans*-1,3,5-hexatriene.
(b) In each case, explain which are likely to be the kinetically controlled products, and which the thermodynamically controlled ones.

47 Account for the fact that the antibiotic *mycomycin* is optically active.

$$HC{\equiv}C{-}C{\equiv}C{-}CH{=}C{=}CH{-}CH{=}CH{-}CH{=}CH{-}CH_2{-}CO_2H \quad \text{mycomycin}$$

48 Rank the compounds within each set in order of increasing heat of formation (lowest first).

(a)

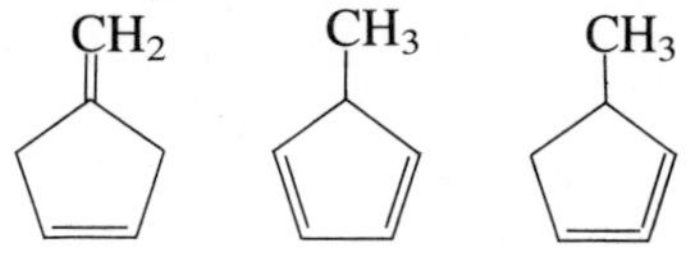

(b)

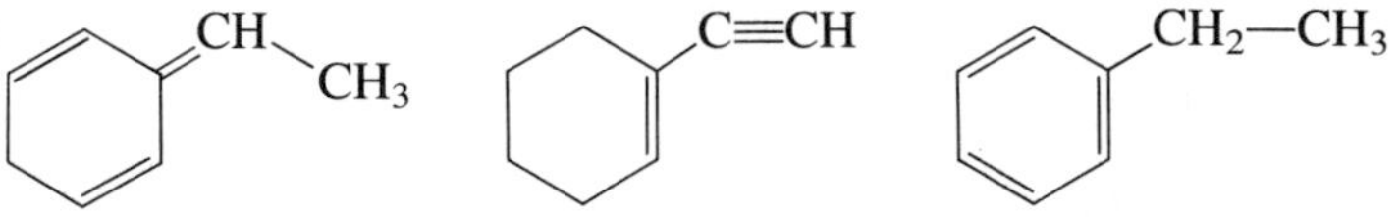

49 Explain the fact that 2,3-dimethyl-1,3-butadiene and maleic anhydride (Eq. 15.9) readily react to give a Diels–Alder adduct, but 2,3-di-*t*-butyl-1,3-butadiene and maleic anhydride do not.

50 What very unstable species is trapped as the Diels–Alder adduct in the following equation? Why is this species unstable, and by what mechanism is it formed?

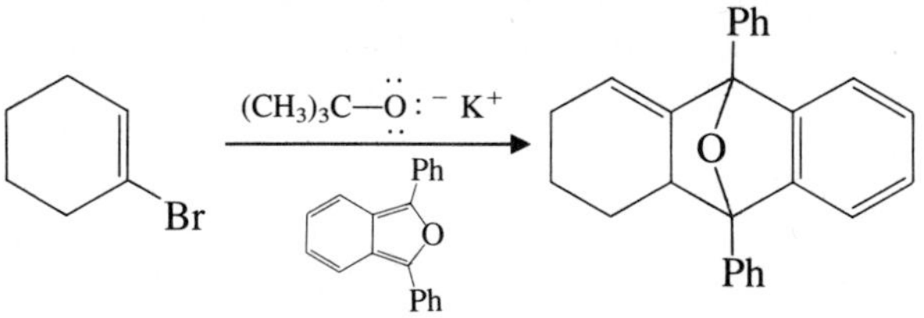

51 Knowing that conjugated dienes react in the Diels–Alder reaction, a student, M. T. Brainpan, has come to you, a noted authority on the Diels–Alder reaction, with an original research idea: to use conjugated alkynes as the diene component in the Diels–Alder reaction (such as the following one). Would Brainpan's idea work? Explain.

$$CH_3{-}C{\equiv}C{-}C{\equiv}C{-}CH_3 + \text{maleic anhydride (Eq. 15.9)} \longrightarrow$$

52 Complete the following reactions, giving the structures of all reasonable products and the reasoning used to obtain them:

(a) $Ph{-}CH{=}CH{-}CH{=}CH{-}Ph + CH_3O{-}\overset{O}{\overset{\|}{C}}{-}C{\equiv}C{-}\overset{O}{\overset{\|}{C}}{-}OCH_3 \xrightarrow{\text{heat}}$

(all *trans*)

(b)

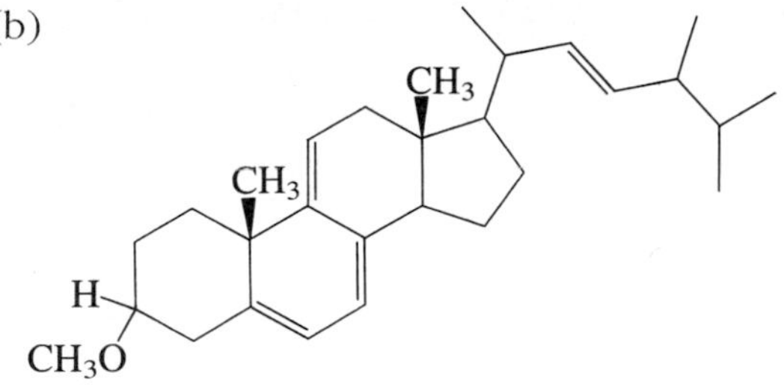

+ maleic anhydride ⟶
(Eq. 15.15)

(c) $CH_2{=}C{=}CH{-}CH{=}CH_2$ +
(2 equiv)

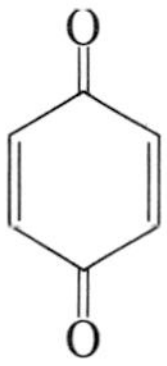

benzoquinone ⟶

(d)

$CH_2{=}CHC(=CH_2)CH_2CH_2CH_2C(=O)CH{=}CH_2 \xrightarrow{\text{heat}}$ (a compound with ten carbon atoms)

(e)

$NiCl_2$ + 2 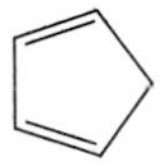 $:^- \; Na^+$ ⟶

53 Dextropimaric acid, isolated from an exudate resin of the cluster pine, is converted by acid and heat into abietic acid. Give a mechanism for this reaction using the arrow formalism.

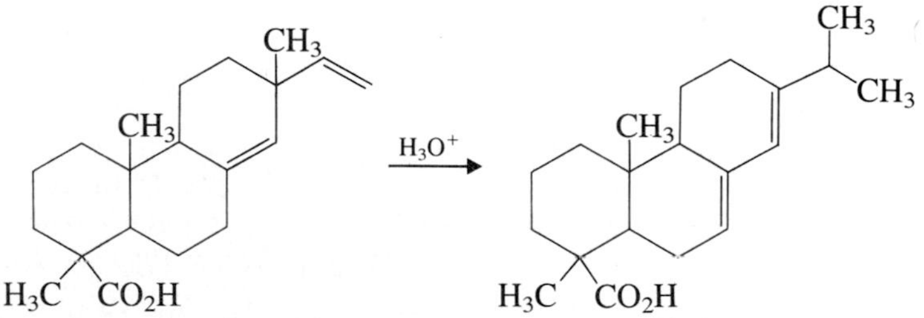

dextropimaric acid abietic acid

54 When the compound shown below is treated with acid a stable carbocation *A* is formed.

(a) Propose a structure for this carbocation and draw its resonance structures.

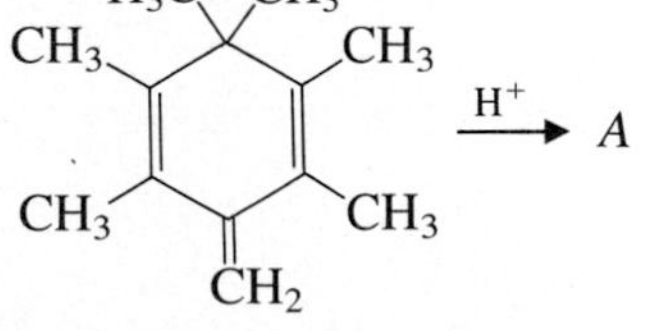

(b) The NMR spectrum of carbocation *A* at $-10°$ consists of four singlets at δ 1.54, δ 2.36, δ 2.63, and δ 2.82 (relative integral 2 : 2 : 2 : 1). Explain why the structure of *A* is consistent with this spectrum by assigning each resonance.

(c) Explain why the NMR spectrum of *A* becomes a single broad line when the temperature is raised to 113°. (*Hint:* See Sec. 13.7.)

55 The Diels–Alder reaction is an equilibrium that, in some cases, favors the decomposition of the Diels–Alder adduct:

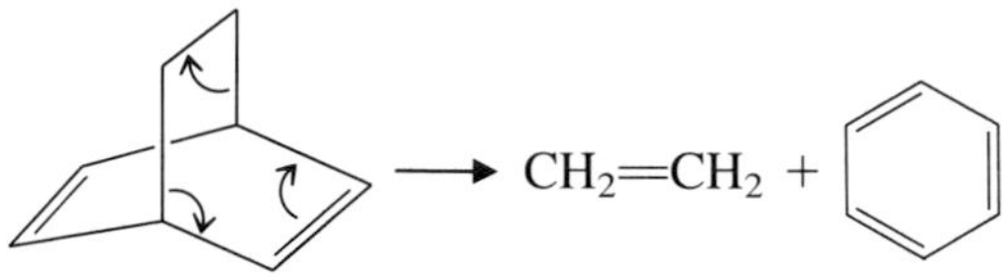

(a) Suggest two reasons why this reaction proceeds in the direction shown.

(b) The compound α-phellandrene ($C_{10}H_{16}$) adds H_2 in the presence of a catalyst to give 1-isopropyl-4-methylcyclohexane and undergoes the following reaction. Deduce the structure of α-phellandrene.

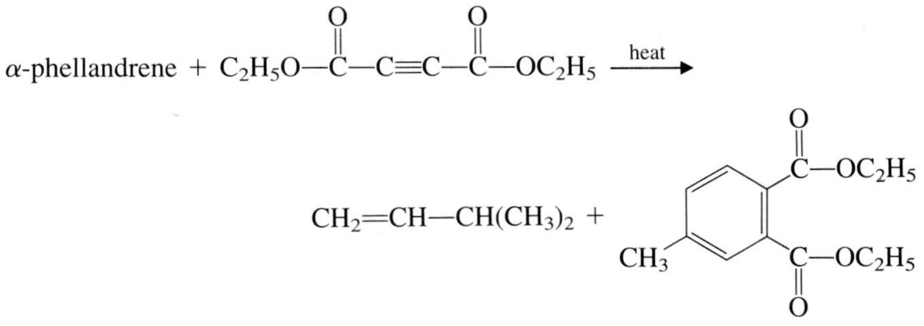

56 When 1,3-butadiene reacts with excess Cl_2 in $CHCl_3$, two dichlorides, *A* and *B,* as well as two tetrachlorides, *C* and *D,* are formed. When butadiene is in excess, only *A* and *B* are formed in a 2 : 1 ratio; if $ZnCl_2$ is added, the ratio of *A* to *B* in the mixture is 3 : 7. When compound *B* reacts with more Cl_2, mostly *D* is formed. After ozonolysis followed by treatment with H_2O_2, *A* gives acid *E,* and *B* gives acid *F.* Compound *D* is an achiral solid. Give the structures of *A, B, C,* and *D,* including relevant stereochemistry, and explain all observations.

E $\quad Cl—CH_2—CH(Cl)—C(=O)—OH \qquad Cl—CH_2—C(=O)—OH$ $\quad$ *F*

57 Which of the following molecules are chiral? Explain.

(a) 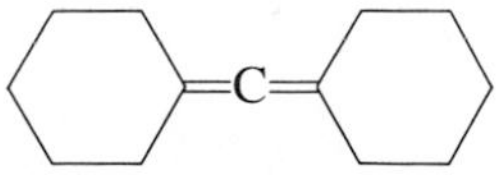

(b) $CH_3—CH(OH)—CH=C=CH_2$

(c) $CH_3CH_2CH_2CH_2—CH=C=CH—CH_3$

(d) 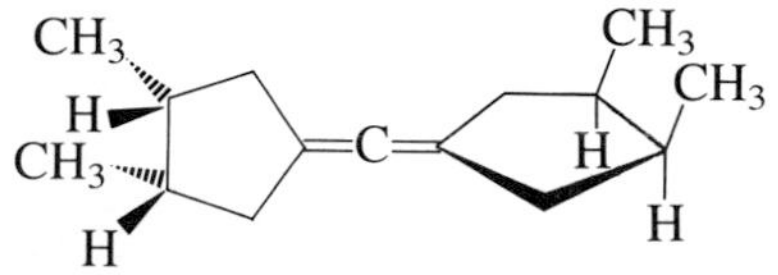

(e) 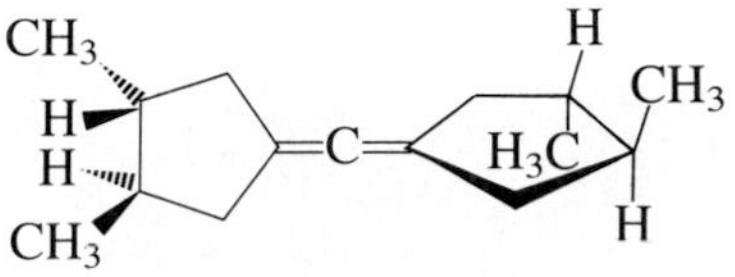

58 Each of the following compounds can be resolved into enantiomers. Explain why each is chiral, and why compounds (b) and (c) racemize when they are heated.

(a) 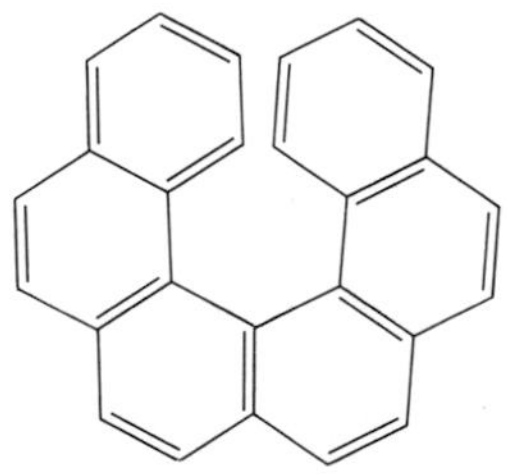

hexahelicene
$[\alpha]_D^{25} = 3700°$

(b) 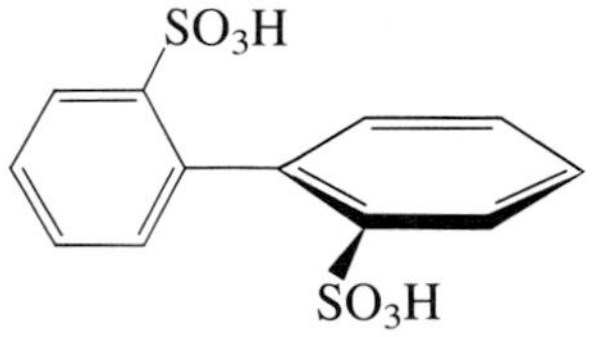

(c) 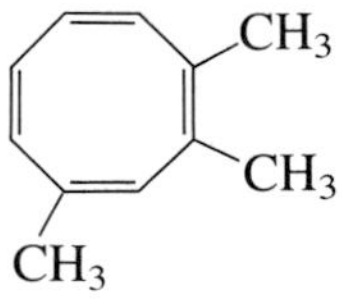

16

Chemistry of Benzene and Its Derivatives

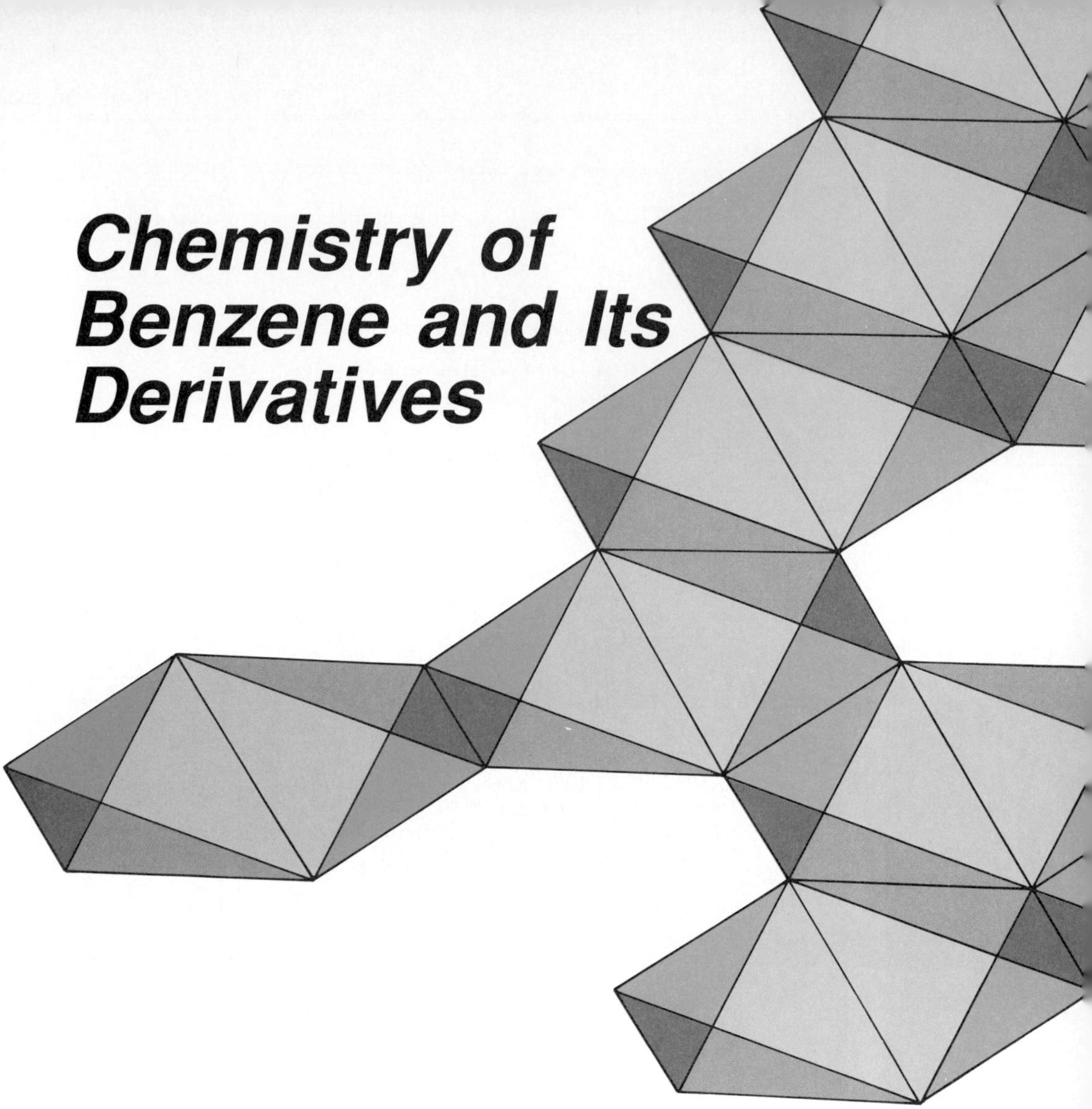

We learned in Chapter 15 that benzene and its derivatives are *aromatic* compounds. Because of their aromaticity, they do not undergo most of the usual addition reactions of alkenes. Instead, benzene and its derivatives undergo reactions in which a ring hydrogen is *substituted* by another group. As we shall learn, such substitution reactions allow us to prepare a variety of substituted benzenes from benzene itself.

These substitution reactions of benzene and its derivatives will occupy our attention for most of this chapter. In the following two chapters, we shall learn how the presence of a benzene ring in a molecule affects the reactivity of nearby functional groups.

16.1 NOMENCLATURE OF BENZENE DERIVATIVES

The nomenclature of many substituted derivatives of benzene follows the same rules used to name other substituted hydrocarbons.

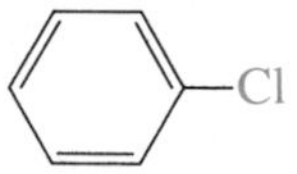

chlorobenzene

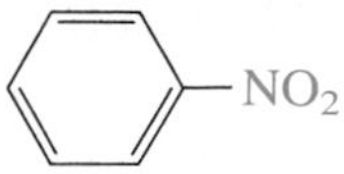

nitrobenzene

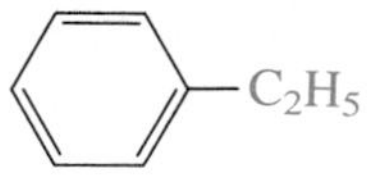

ethylbenzene

Other monosubstituted benzene derivatives have well-established common names that must be learned.

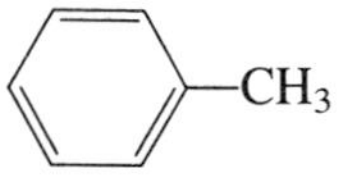

toluene

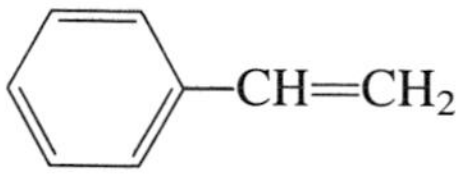

styrene

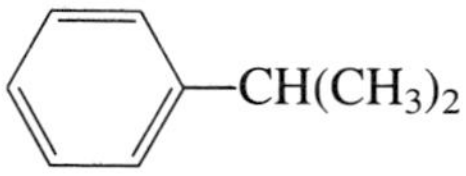

cumene

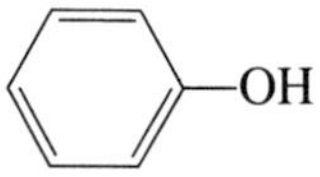

phenol

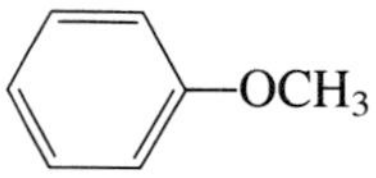

anisole

The substituent groups of disubstituted benzenes can be designated either by special prefixes or by number. The prefixes *o* (for *ortho*), *m* (for *meta*), or *p* (for *para*) can be used for disubstituted benzene derivatives when the substituent groups have a 1,2, 1,3, or 1,4 relationship, respectively.

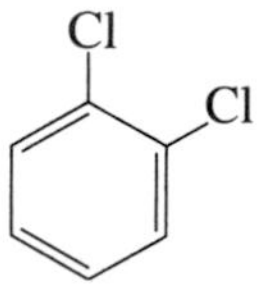

o-dichlorobenzene
1,2-dichlorobenzene

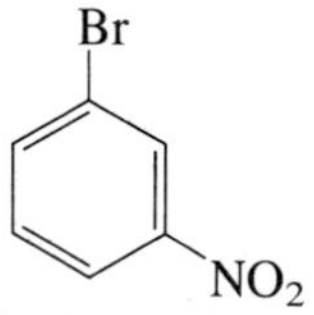

m-bromonitrobenzene
1-bromo-3-nitrobenzene

p-fluoroiodobenzene
1-fluoro-4-iodobenzene

As these examples illustrate, when none of the substituents qualify as principal groups, they are cited and numbered in alphabetical order. If a substituent is eligible for citation as a principal group, it is assumed to be at position-1 of the ring.

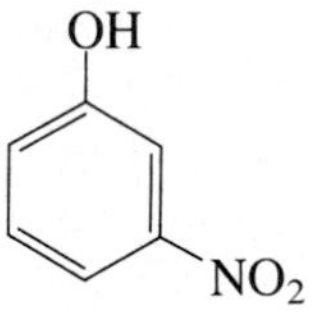

m-nitrophenol (3-nitrophenol)
—OH group is the principal group

Some substituted benzene derivatives also have time-honored common names. The dimethylbenzenes are called *xylenes*.

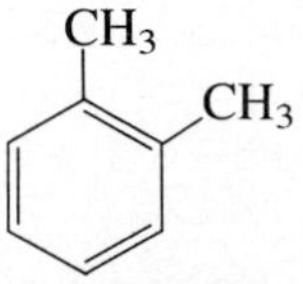

o-xylene

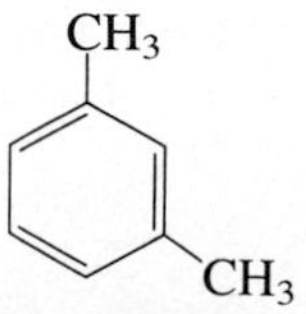

m-xylene

The methylphenols are called *cresols.*

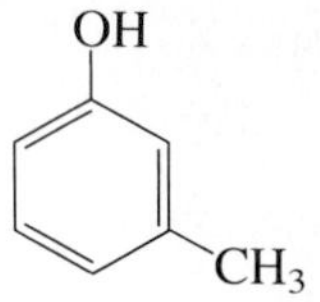

m-cresol

The hydroxyphenols also have important common names.

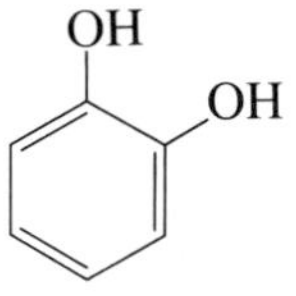

catechol

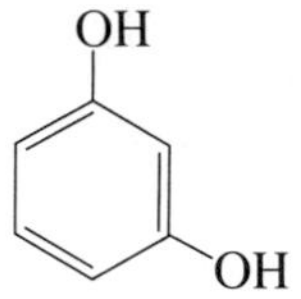

resorcinol

hydroquinone

Many other benzene derivatives have common names, but these can be learned in context as necessary.

When a benzene derivative contains more than two substituents on the ring, the *o, m,* and *p* designations are not appropriate; only numbers may be used to designate the positions of substituents. The usual nomenclature rules are followed (Sec. 3.2A, 4.2A, 8.1).

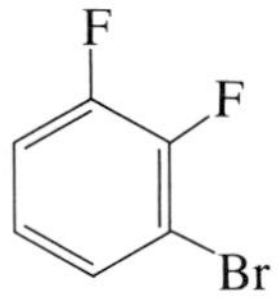

alphabetical citation: *b*romodi*f*luoro
numbering: 1,2,3
name: 1-bromo-2,3-difluorobenzene

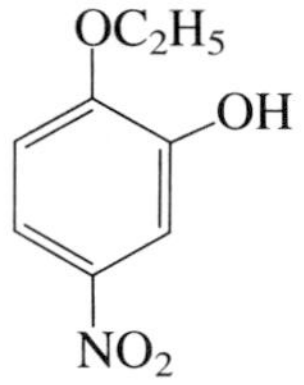

2-ethoxy-5-nitrophenol

Sometimes it is simpler to name the benzene ring as a substituent group. A benzene ring or substituted benzene ring used as a substituent is referred to generally as an **aryl group;** this term is analogous to *alkyl group* in nonaromatic compounds (Sec. 3.8). When an unsubstituted benzene ring is a substituent, it is called a **phenyl group.** This group, as we already know, can be abbreviated Ph—. It is also sometimes abbreviated by its group formula, C_6H_5—.

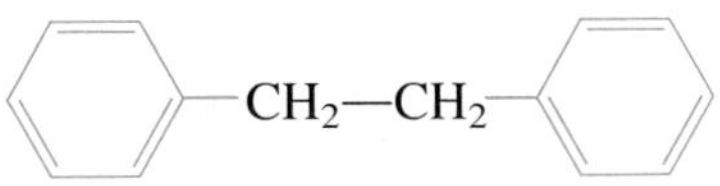

Ph—CH_2CH_2—Ph
C_6H_5—CH_2CH_2—C_6H_5
1,2-diphenylethane

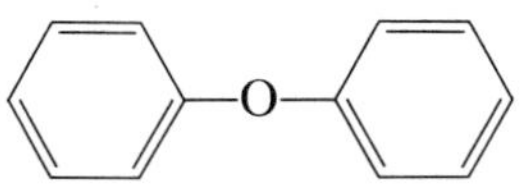

diphenyl ether
(phenoxybenzene)

The Ph—CH_2— group is called the **benzyl group.**

Ph—CH_2—Cl benzyl chloride
or (chloromethyl)benzene

Be sure to notice the difference between the *benzyl* group and the *phenyl* group.

Problems

1 Name the following compounds:

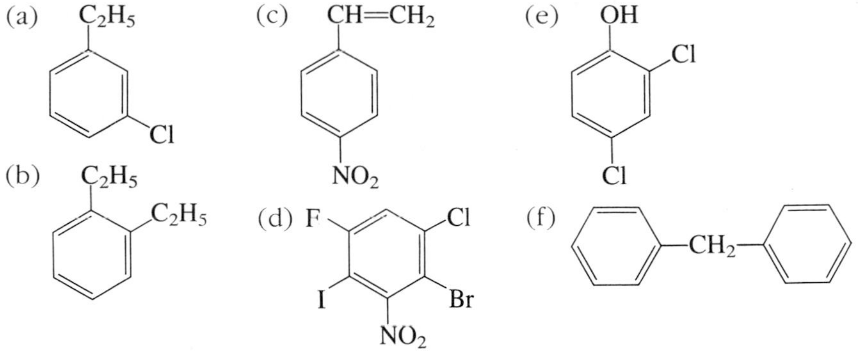

2 Give the structure for each of the following compounds:
(a) *p*-chloroanisole
(b) 3-nitrotoluene
(c) benzenethiol
(d) benzylcyclohexane
(e) *p*-cresol

16.2 PHYSICAL PROPERTIES OF BENZENE DERIVATIVES

The boiling points of benzene derivatives are similar to those of other hydrocarbons with similar shapes and molecular weights.

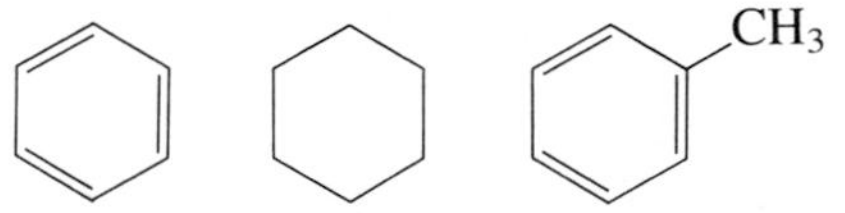

bp	80.1°	80.7°	110.6°
mp	5.5°	6.6°	−95°

The melting points of benzene and cyclohexane are unusually high because of their symmetry.

The melting points of *para*-disubstituted benzene derivatives are typically higher than the corresponding *ortho* or *meta* isomers.

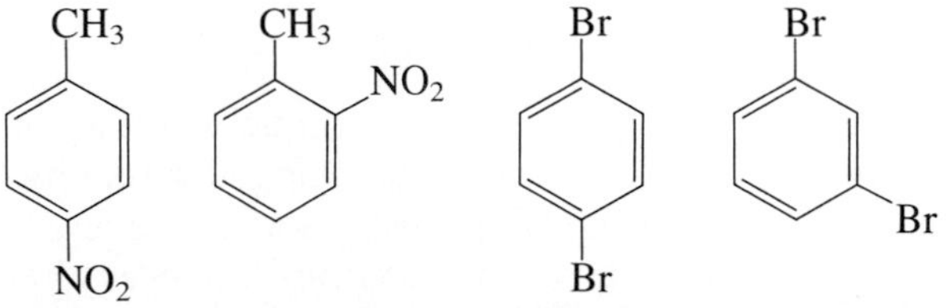

mp	54.5°	−9.6°	87.3°	−7°

This trend can be useful in purifying the *para* isomer of a benzene derivative from mixtures containing other isomers. Since the isomer with the highest melting point is usually the one that is most easily crystallized, many *para*-substituted compounds can be separated from their *ortho* and *meta* isomers by recrystallization.

Benzene and other aromatic hydrocarbons are less dense than water, but more dense than alkanes and alkenes of about the same molecular weight. Like other hydro-

Figure 16.1 (a) IR spectrum of toluene. (b) IR spectrum of phenol.

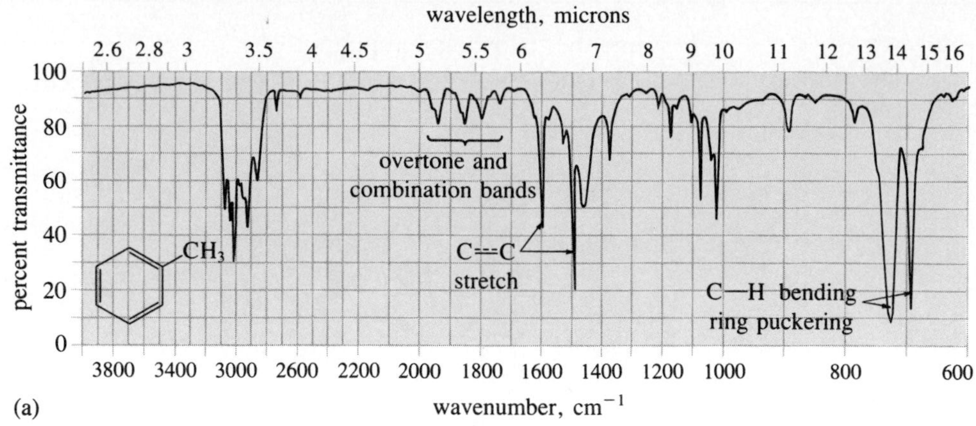

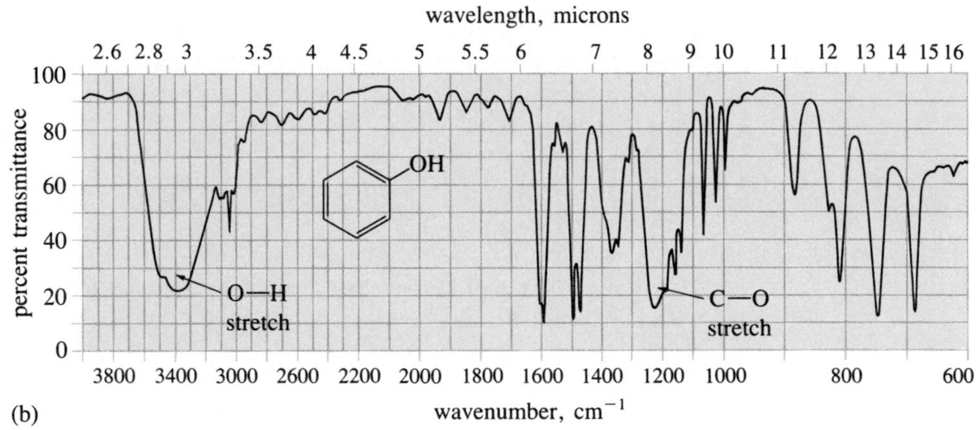

carbons, benzene and its hydrocarbon derivatives are insoluble in water. Benzene derivatives with substituents that form hydrogen bonds to water are more soluble, as we might expect. Thus, phenol has substantial water solubility; in fact, an 85:15 phenol–water solution is one of the commercially available forms of phenol.

16.3 SPECTROSCOPY OF BENZENE DERIVATIVES

A. IR Spectroscopy

The most useful absorptions in the infrared spectra of benzene derivatives are the carbon–carbon stretching absorptions of the ring, which occur at lower frequency than the C=C absorption of alkenes. Typically there are two such absorptions: one near 1600 cm^{-1} and the other near 1500 cm^{-1}; these are illustrated in the spectrum of toluene (Fig. 16.1a). Other absorptions that are sometimes useful are the very strong absorptions below 900 cm^{-1} characteristic of C—H bending and ring puckering. These are visible in the toluene spectrum at 728 cm^{-1} and 694 cm^{-1}, respectively. Notice that the C—H bending absorption is in the same region as the C—H bending absorption of a *cis*-alkene (Table 12.1); benzene is formally a *cis*-alkene. The IR spectra of aromatic compounds also contain a series of relatively weak absorptions in the 1660–2000 cm^{-1} region. These are called *overtone and combination bands,* for reasons that we need not discuss here. Such bands are very common in aromatic com-

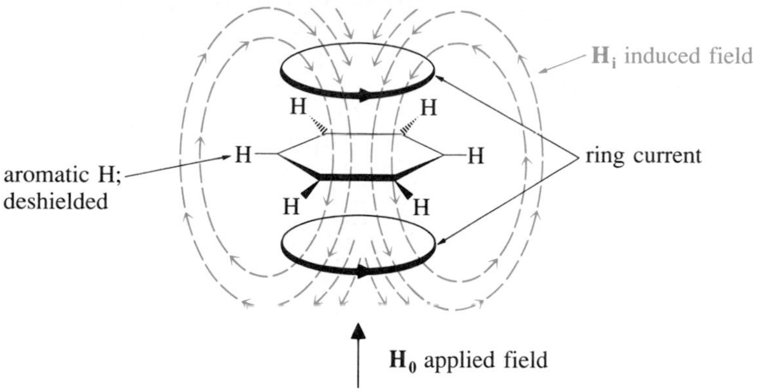

Figure 16.2 The induced field of the ring current deshields the benzene protons.

pounds, but generally do not appear in simple alkenes. These bands were once used to determine the substitution patterns of aromatic compounds. Although NMR spectroscopy is now a more reliable tool for this purpose, these IR absorptions are a good indication that a molecule contains a benzene ring.

Phenols have not only the characteristic aromatic ring absorptions, but also O—H and C—O stretching absorptions, which are very much like those of tertiary alcohols. The IR spectrum of phenol is shown in Fig. 16.1b.

B. NMR Spectroscopy

The NMR spectrum of benzene consists of a singlet at a chemical shift of δ 7.4. This chemical shift is greater than that of a typical alkene. For example, the NMR spectrum of 1,3,5,7-cyclooctatetraene (COT) is a singlet at δ 5.69, in the usual chemical-shift region for vinyl protons. Generally, *protons on aromatic rings show unusual downfield chemical shifts relative to alkenes of about 1.5 ppm.*

What is the reason for the unusual chemical shift of benzene? Recall that the π-electron density in benzene lies in two doughnut-shaped regions above and below the plane of the ring (Fig. 15.12). When benzene is oriented relative to the applied field as shown in Fig. 16.2, a circulation of π-electrons around the ring, called a **ring current,** is induced. The ring current, in turn, induces a magnetic field that forms closed loops through the ring. This induced field opposes the applied field along the axis of the ring, but augments the applied field outside of the ring, at the benzene protons. The result is a deshielding effect on the benzene protons and hence a larger downfield chemical shift. This chemical shift effect is analogous to the deshielding effect in alkenes (Fig. 13.15), except that it is larger. The ring current and the large chemical shift are characteristic of compounds that are aromatic by the Hückel $4n + 2$ rule (Sec. 15.7D). This is reasonable because the basis of both the ring current and aromaticity is the overlap of *p* orbitals in a continuous cyclic array. Many chemists believe that the existence of the ring current (detected by unusually low-field chemical shifts) is the best *experimental* evidence of aromatic behavior.

Problems

3 Which of the following two compounds has NMR signals at lower field? Explain.

(a)

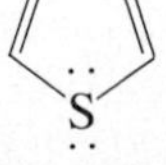

thiophene

(b)

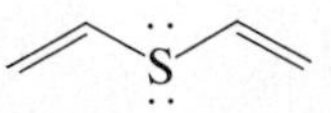

divinyl sulfide

Problems (Cont.)

4 The following compound is an example of an *annulene:*

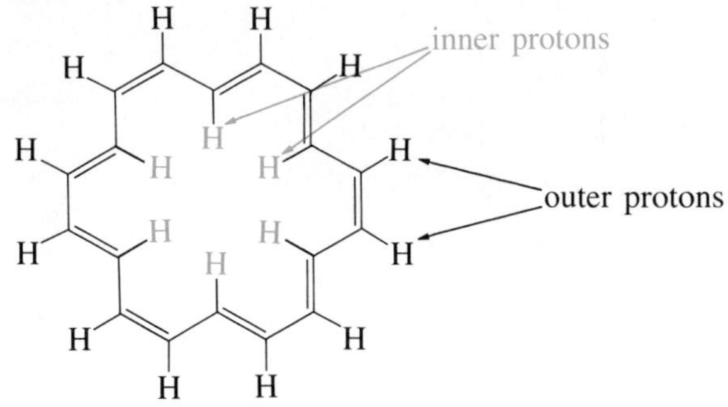

(a) Verify that this compound meets the criteria for aromaticity.

(b) The NMR spectrum of this compound consists of two sets of multiplets: one at δ 9.28, and the other at δ (−2.99)—*3 ppm higher field than TMS!* The relative integral of the two signals is 2 : 1, respectively. Assign the two sets of signals, and explain why the two chemical shifts are so different; in particular, explain the large upfield shift. (*Hint:* Look carefully at the direction of the induced field in Fig. 16.2.)

When the protons in a substituted benzene derivative are nonequivalent, they split each other, and their coupling constants depend on their positional relationships, as shown in Table 16.1. Because of this splitting, the NMR spectra of many monosubstituted benzene derivatives have complex absorptions in the aromatic re-

TABLE 16.1 Typical Coupling Constants of Aromatic Protons

Relationship of protons		Coupling constant
ortho	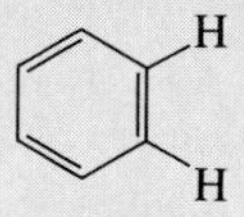	J_{ortho} = 6–10 Hz
meta	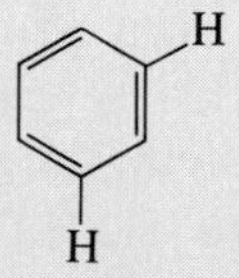	J_{meta} = 1–3 Hz
para	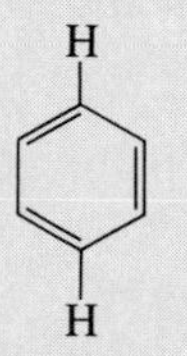	J_{para} = 0–1 Hz

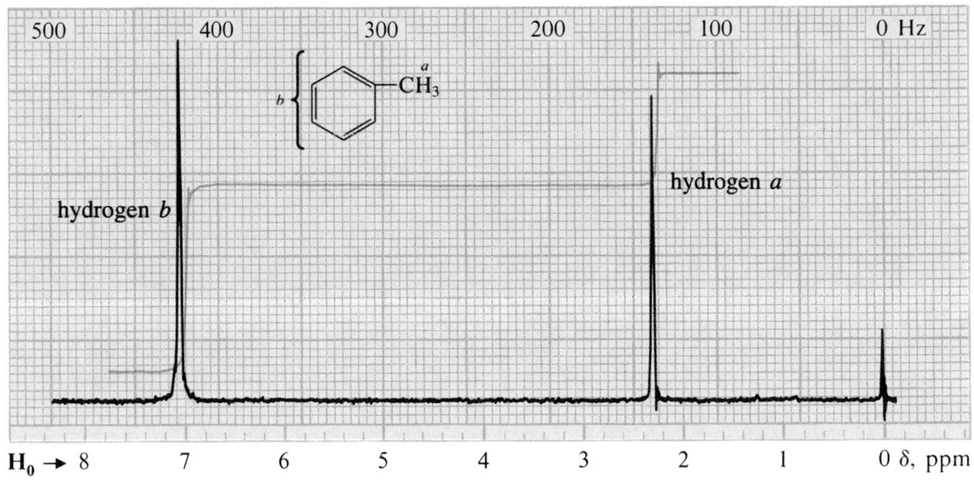

Figure 16.3 NMR spectrum of toluene.

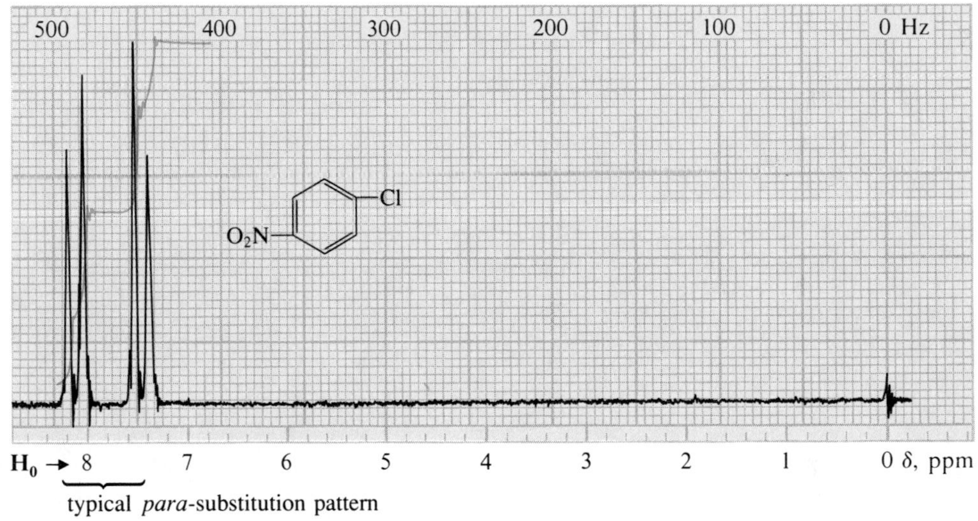

Figure 16.4 NMR spectrum of *p*-chloronitrobenzene.

gion. This is particularly true if the ring is substituted with a strongly electropositive or electronegative group. For example, the aromatic region of the nitrobenzene spectrum is very complex, with absorptions throughout the range δ 7.1–8.2. (The nitro group is very electronegative.) In contrast, an alkyl substituent has about the same effect on the chemical shifts of all ring protons; consequently, the ring protons of monosubstituted alkylbenzenes all have about the same chemical shift, and appear as a broadened singlet. For example, the ring protons in the spectrum of toluene appear as a broad singlet at δ 7.04 (Fig. 16.3).

Splitting patterns in many disubstituted benzenes are complex. However, one situation occurs often enough that it is worth remembering. Many *para*-disubstituted benzenes, such as *p*-chloronitrobenzene (NMR spectrum in Fig. 16.4), have a characteristic splitting pattern in the aromatic region consisting of two smaller outer lines and two larger inner lines—that is, a pair of doublets, in which each doublet leans toward the other. The splitting pattern in a *p*-disubstituted benzene is due to the fact that there are two sets of chemically equivalent protons. Since the largest splitting in a

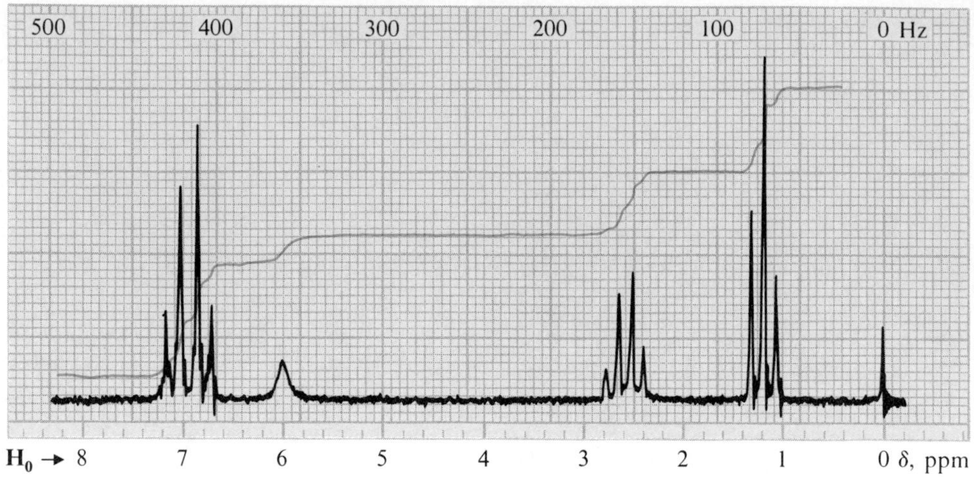

Figure 16.5 NMR spectrum for Problem 6b.

benzene ring is between *ortho* protons (Table 16.1), a given proton is split by the adjacent *ortho* proton into a doublet, with a typical *ortho*-coupling constant given in Table 16.1. Its nonequivalent neighbor shows a corresponding doublet.

The spectrum of *p*-chloronitrobenzene also shows how substituent groups affect the chemical shifts of ring protons. The protons *ortho* to the very electronegative nitro group have the resonance that is farthest downfield, at δ 8.1. The resonance of the protons *ortho* to the less electronegative chloro group is further upfield, at δ 7.4.

The chemical shifts of *benzylic protons*—protons on carbons adjacent to benzene rings—are in the δ 2 to δ 3 region, and are slightly greater than those of allylic protons (see Table 13.2). The chemical shifts of benzylic protons in toluene and ethylbenzene are typical.

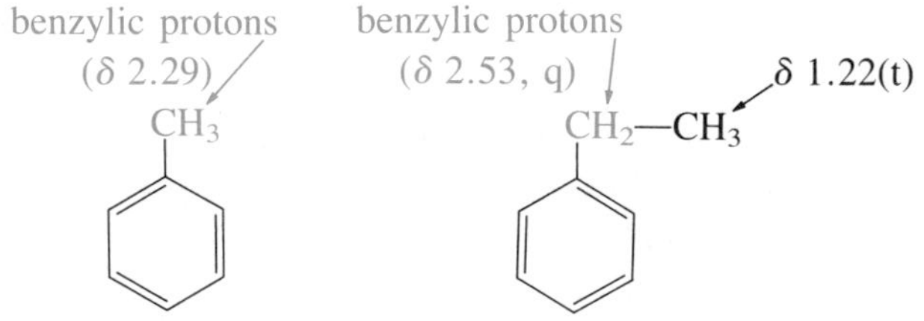

The O—H absorptions of phenols are typically at lower field (about δ 5–6) than those of alcohols (δ 2–3). The O—H protons of phenols, like those of alcohols, undergo exchange in D_2O.

Problems

5 Explain how you would differentiate the two C_9H_{12} isomers mesitylene (1,3,5-trimethylbenzene) and cumene (isopropylbenzene) by NMR. Explain in detail what you would expect in the NMR spectrum of each.

6 Give structures for each of the following compounds:

(a) $C_9H_{12}O$: δ 1.27 (3*H*, d, $J = 7$ Hz); δ 2.26 (3*H*, s); δ 3.76 (1*H*, broad s, exchanges with D_2O); δ 4.60 (1*H*, q, $J = 7$ Hz); δ 6.98 (4*H*, apparent pair of doublets)

(b) $C_8H_{10}O$: IR, 3150–3600 cm^{-1} (broad); NMR in Fig. 16.5.

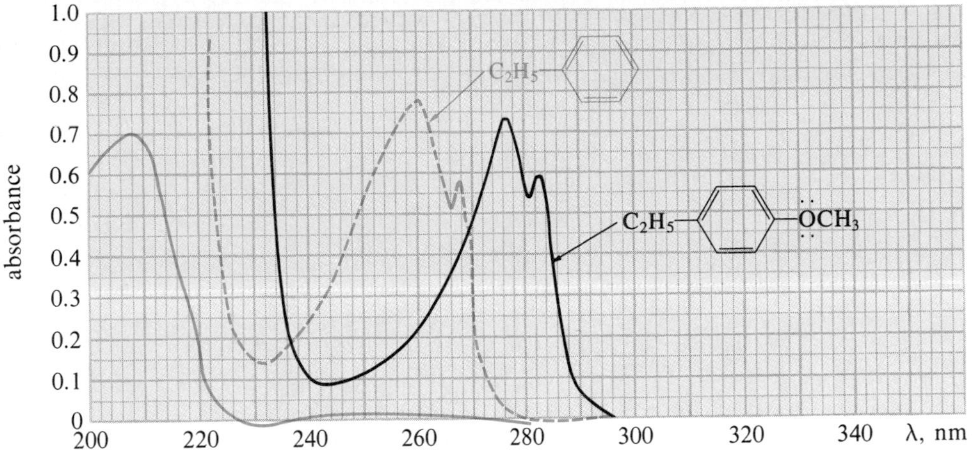

Figure 16.6 Comparison of the UV spectra of ethylbenzene (color) and *p*-ethylanisole (black). The solid lines are spectra taken at the same concentrations. Notice that the spectrum of *p*-ethylanisole is much more intense. The dashed line is the spectrum of ethylbenzene at higher concentration. Notice that the peak in the *p*-ethylanisole spectrum occurs at higher wavelength.

C. UV Spectroscopy

Simple aromatic hydrocarbons have two absorption bands in their UV spectra: a relatively strong band near 210 nm and a much weaker one near 260 nm. The spectrum of ethylbenzene in methanol solvent (Fig. 16.6) is typical: λ_{max} = 208 nm (ϵ = 7520); 261 nm (ϵ = 200). Substituent groups on the ring alter both the λ_{max} values and the intensities of both peaks, particularly if the substituent has an unshared electron pair or *p* orbitals that can overlap with the π-electron system of the aromatic ring. As is the case in alkenes, more extensive conjugation is associated with an increase in both λ_{max} and intensity. For example, 1-ethyl-4-methoxybenzene (*p*-ethylanisole) in methanol solvent has absorptions at λ_{max} = 224 nm (ϵ = 10,100) and 276 nm (ϵ = 1,930), at greater wavelengths and with greater intensities than the analogous absorptions of ethylbenzene (Fig. 16.6) because the —OCH_3 group has electron pairs in orbitals that overlap with the *p* orbitals of the benzene ring.

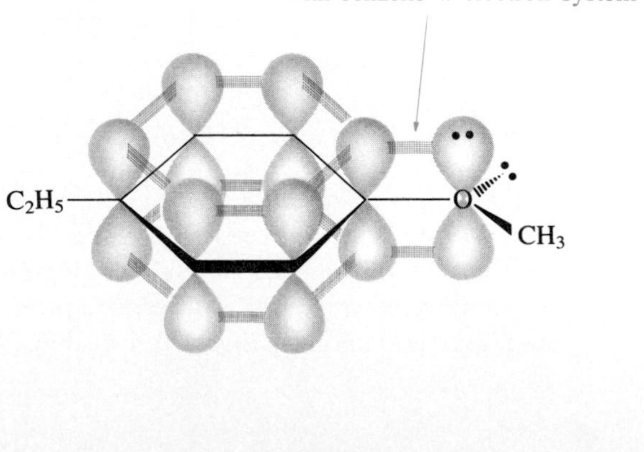

Problem **7** (a) Explain why compound A has a UV spectrum with considerably greater λ_{max} and intensity than that of ethylbenzene.

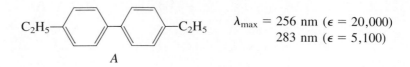

$$\lambda_{max} = 256 \text{ nm } (\epsilon = 20{,}000)$$
$$283 \text{ nm } (\epsilon = 5{,}100)$$

A

(b) In view of your answer to (a), explain why the UV spectra of compound B and C are virtually identical.

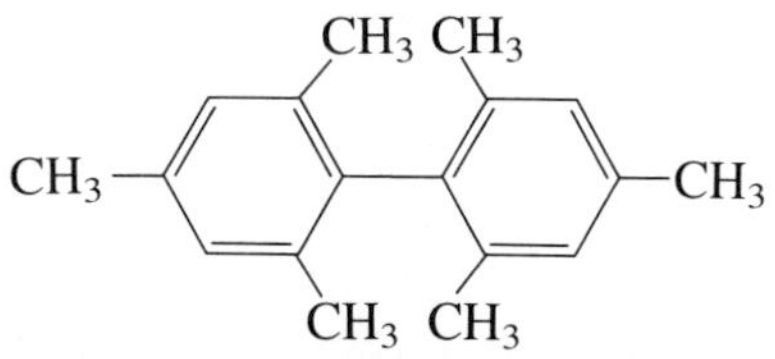

B
bimesityl
$\lambda_{max} = 266 \text{ nm } (\epsilon = 700)$

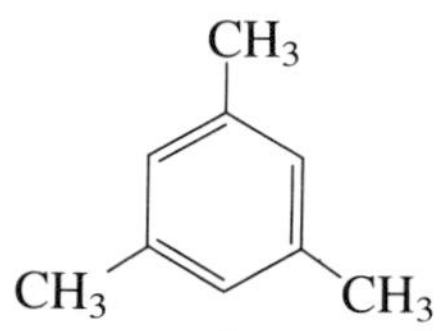

C
mesitylene
$\lambda_{max} = 266 \text{ nm } (\epsilon = 200)$

16.4 ELECTROPHILIC AROMATIC SUBSTITUTION REACTIONS OF BENZENE

Benzene undergoes a number of reactions, called *electrophilic aromatic substitution* reactions, that follow a common mechanistic pattern. In this section we shall examine several such reactions and try to understand the common thread that runs through all of them.

A. Halogenation of Benzene

When benzene reacts with bromine under harsh conditions—liquid bromine, no solvent, and the Lewis acid FeBr$_3$ as a catalyst—a reaction occurs in which *one* bromine is substituted for a ring hydrogen.

$$\text{C}_6\text{H}_5\text{—H} + \text{Br}_2 \xrightarrow[\text{(0.2 equiv)}]{\text{FeBr}_3 \text{ or Fe}} \text{C}_6\text{H}_5\text{—Br} + \text{HBr} \qquad (16.1)$$

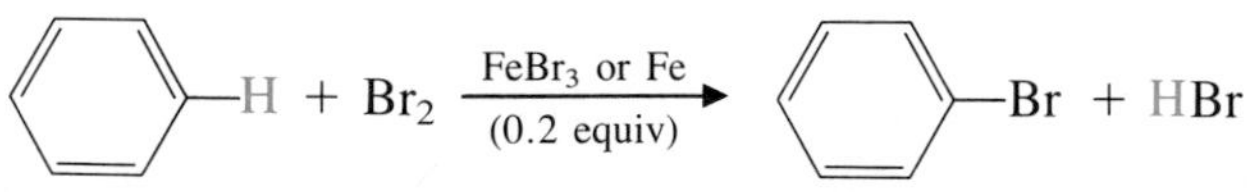

bromobenzene
(50% yield)

(Since iron reacts with Br$_2$ to give FeBr$_3$, iron filings can be used in place of FeBr$_3$.) An analogous chlorination reaction using Cl$_2$ and FeCl$_3$ gives chlorobenzene.

This reaction of benzene with halogens differs from the halogenation alkenes in two important ways. First is the type of product obtained. Alkenes react spontaneously with bromine and chlorine, even in dilute solution, to give *addition products*.

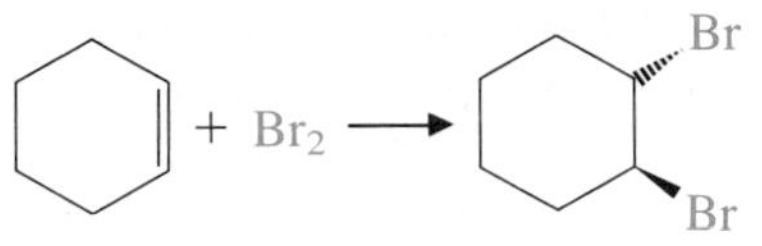
(16.2)

Halogenation of benzene, however, is a *substitution*; a ring hydrogen is *replaced* by a halogen. Second, the severity of the reaction conditions for benzene halogenation is much greater than for halogenation of an alkene.

The mechanism of the bromination reaction can help us to make sense of these observations. In the first step, bromine forms a complex with the Lewis acid $FeBr_3$.

$$:\ddot{Br}-\ddot{Br}: \;\; FeBr_3 \rightleftarrows :\ddot{Br}-\overset{+}{\ddot{Br}}-\overset{-}{Fe}Br_3 \qquad (16.3)$$

This complexation makes one of the bromines a better leaving group in the same sense that complexation of Ag^+ with an alkyl halide makes the halide a better leaving group (Sec. 9.5C). The other bromine in this complex takes on some electron-deficient character. In other words, the complex of Br_2 with $FeBr_3$ reacts *as if* it were the electron-deficient species $:\ddot{Br}:^+$:

$$:\ddot{Br}-\overset{+}{\ddot{Br}}-\overset{-}{Fe}Br_3 \quad \text{reacts as if it were} \quad :\ddot{Br}^+ \quad \overset{-}{Fe}Br_4 \qquad (16.4)$$

In the second step of the mechanism, the electron-deficient bromine in this complex is attacked by the π-electrons of the benzene ring.

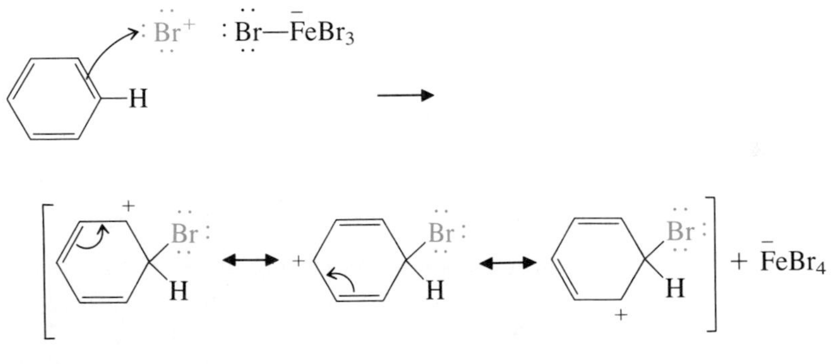
(16.5)

Bromination requires harsh conditions because this step disrupts the aromatic stability of the benzene ring. However, the carbocation intermediate, although not aromatic (why not?), is resonance-stabilized, as shown in Eq. 16.5.

The reaction is completed when a Br^- counter-ion (complexed to $FeBr_3$) acts as a base to remove the ring proton and give the products bromobenzene and HBr.

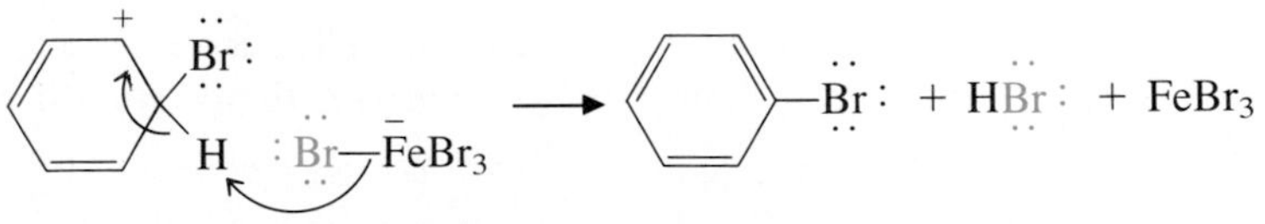
(16.6)

As we learned in Sec. 9.5B, loss of a β-proton is one of the characteristic reactions of carbocations. Another typical reaction of carbocations—attack of bromide ion at the electron-deficient carbon itself—doesn't occur because the resulting addition product would not be aromatic. By losing a β-proton instead, the carbocation can form a stable aromatic compound, bromobenzene.

Notice that $FeBr_3$ is a true catalyst for bromination of benzene. (For example, it is used in catalytic amount in Eq. 16.1.) Although $FeBr_3$ is consumed in Eq. 16.3, it is regenerated in Eq. 16.6.

Problem

8 A small amount of a by-product, *p*-dibromobenzene, is also formed in the bromination of benzene shown in Eq. 16.1. Suggest a mechanism for the formation of this compound.

B. Electrophilic Aromatic Substitution

Halogenation of benzene is one of many related reactions that are very typical of benzene and other aromatic compounds. These are called **electrophilic aromatic substitution** reactions. The bromination reaction, for example, is a *substitution* because hydrogen is replaced by another group (bromine). The reaction is *electrophilic* because it involves the reaction of an electrophilic (electron-deficient) species with the benzene π-electrons. In bromination the electrophilic species is "Br^+" in the complex of bromine and the Lewis acid catalyst.

We have studied two other types of substitution reactions: *nucleophilic substitution* (the S_N1 and S_N2 reactions, Secs. 9.3 and 9.5) and *free-radical substitution* (halogenation of alkanes, Sec. 8.9A). In a nucleophilic substitution reaction, the substituting group acts as a nucleophile, or Lewis base; and in free-radical substitution, free-radical intermediates are involved. In electrophilic substitution, the substituting group reacts as an electrophile, or Lewis acid.

Electrophilic aromatic substitution is the most typical reaction of benzene and its derivatives. We shall study other electrophilic substitution reactions in the following sections. These all have the following steps in common:

1. *Generation of an electrophile.* The electrophile in bromination is "Br^+" in the complex of bromine with $FeBr_3$, as shown in Eq. 16.4.
2. *Attack of the π-electrons of the aromatic ring on the electrophile and formation of a resonance-stabilized carbocation.* This step in the bromination reaction is shown in Eq. 16.5. As shown in Fig. 16.7, the electrophile approaches the π-electron cloud of the aromatic compound above or below the plane of the molecule, and the carbon at the point of attack becomes tetrahedral.
3. *Loss of a proton from the carbocation intermediate at the site of substitution to generate the substituted aromatic compound.* This step in the bromination reaction is shown in Eq. 16.6.

As we study other electrophilic substitution reactions, it is important for us to see that *all of them have in common these same three steps.*

Problem

9 Give a detailed mechanism for the following reaction:

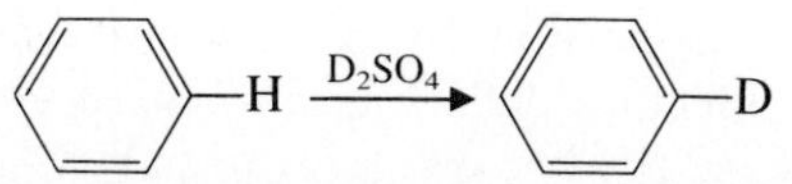

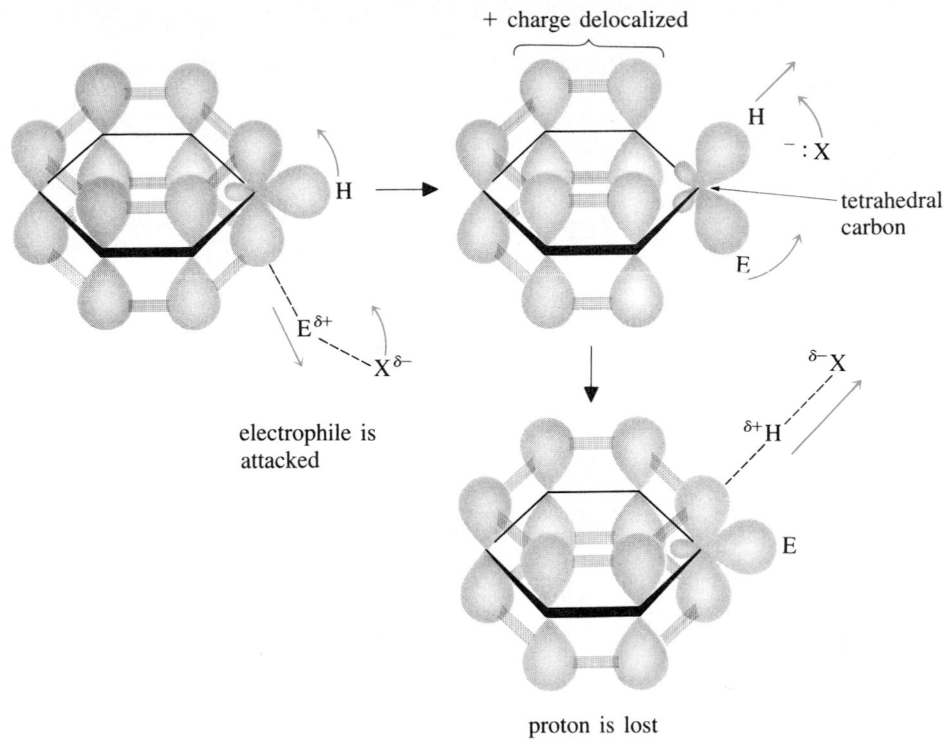

Figure 16.7 Electrophilic aromatic substitution. E is the electrophilic group, and X is a general Lewis base (leaving group). The colored arrows show the movement of atoms, not of electrons. Notice that attack on the electrophile and loss of the proton occur at opposite faces of the aromatic ring and that the carbocation intermediate has tetrahedral geometry at the site of substitution.

+ charge delocalized

tetrahedral carbon

electrophile is attacked

proton is lost

C. Nitration of Benzene

Benzene reacts with concentrated nitric acid, usually in the presence of a sulfuric acid catalyst, to form nitrobenzene. In this reaction, called *nitration,* the nitro group, —NO$_2$, is introduced into the benzene ring by electrophilic substitution.

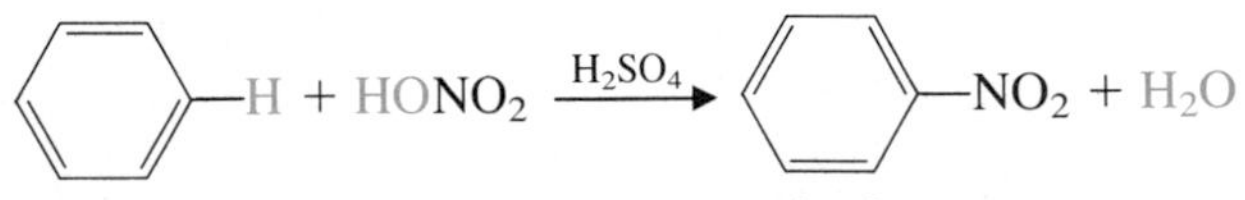

$$\text{—H} + \text{HONO}_2 \xrightarrow{\text{H}_2\text{SO}_4} \text{—NO}_2 + \text{H}_2\text{O} \qquad (16.7)$$

nitrobenzene
(81% yield)

This reaction perfectly fits the pattern of the electrophilic aromatic substitution mechanism outlined in the previous section:

1. *Generation of the electrophile:* In nitration, the electrophile is $^+\text{NO}_2$, the *nitronium ion*. This ion is formed by the acid-catalyzed removal of the elements of water from HNO$_3$.

$$\text{HO}_3\text{SO—H} \quad \text{HO—}\overset{+}{\text{N}}\overset{\text{O}^-}{\underset{\text{O}}{\Big|}} \longrightarrow \text{H—}\overset{}{\underset{\text{H}}{\text{O}}}\text{—}\overset{+}{\text{N}}\overset{\text{O}^-}{\underset{\text{O}}{\Big|}} \qquad {}^-\!:\text{O—SO}_3\text{H} \quad (16.8a)$$

$$\text{H—}\overset{+}{\underset{\text{H}}{\text{O}}}\text{—}\overset{+}{\text{N}}\overset{\text{O}^-}{\underset{\text{O}}{\Big|}} \longrightarrow \text{H}_2\text{O} + \quad \text{O}=\overset{+}{\text{N}}=\text{O} \qquad (16.8b)$$

nitronium ion

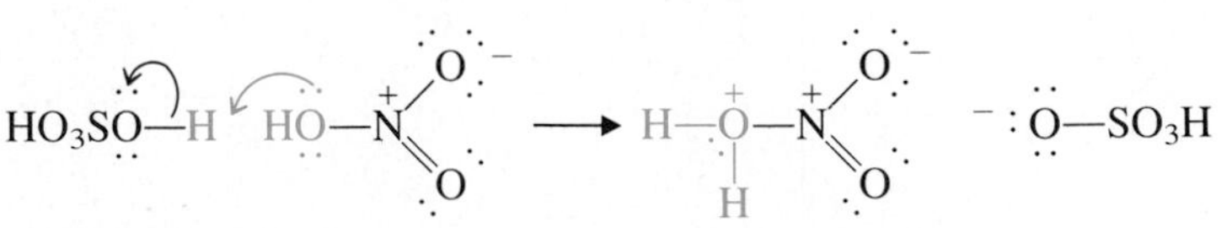

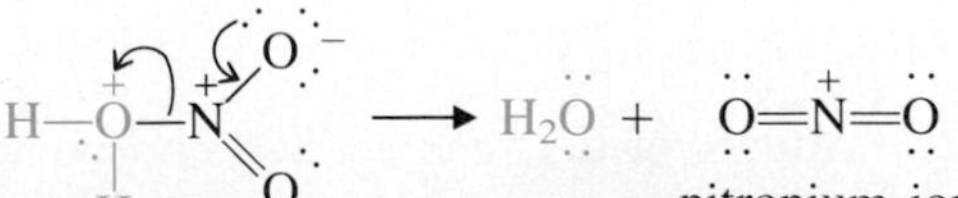

2. *Attack of the ring π-electrons on the electrophile.*
3. *Loss of a proton from the carbocation to give a new aromatic compound:*

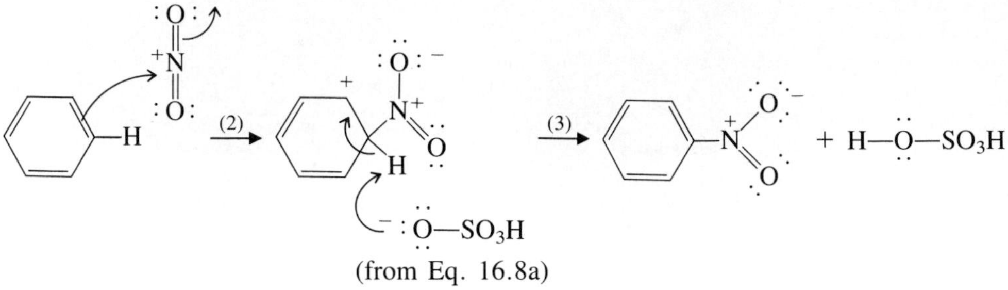

(16.9)

While we are discussing nitrobenzene, let us digress briefly to consider two important aspects of its structure. First, the Lewis structure of the nitro group cannot be written without a full formal charge on the nitrogen and one oxygen. Second, the nitro group has two equivalent resonance structures. (Recall the structures of nitromethane, Sec. 1.7.)

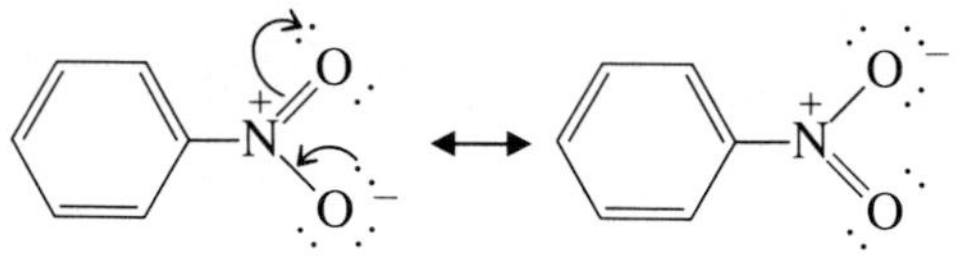
(16.10)

D. Sulfonation of Benzene

Another electrophilic substitution reaction of benzene is its conversion into benzenesulfonic acid by a solution of sulfur trioxide in H_2SO_4.

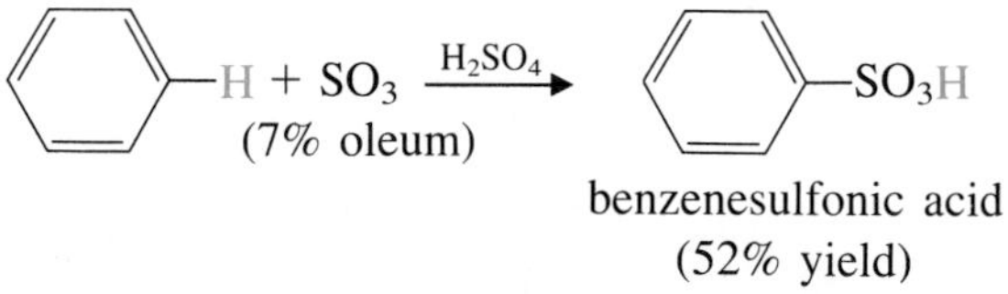

(16.11)

In this reaction, called *sulfonation,* the electrophile is the neutral compound sulfur trioxide, SO_3. Sulfur trioxide is a fuming liquid that reacts violently with water to give H_2SO_4. The source of SO_3 for sulfonation is usually a solution called *fuming sulfuric acid* or *oleum,* consisting of 10–30% SO_3 in concentrated H_2SO_4; this material is one of the most acidic Brønsted acids that can be purchased commercially. Although sulfur trioxide can be written without formal charge if we violate the octet rule for sulfur, its octet structure shows that the sulfur bears considerable positive charge.

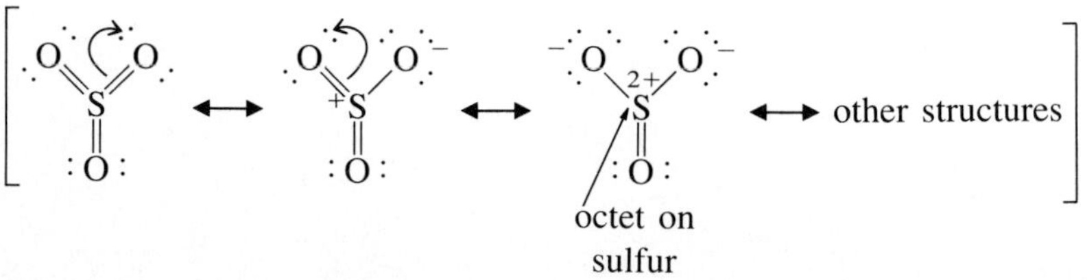

(16.12)

Attack of benzene on SO_3 and completion of the substitution reaction occur by a mechanism analogous to those discussed for bromination and nitration.

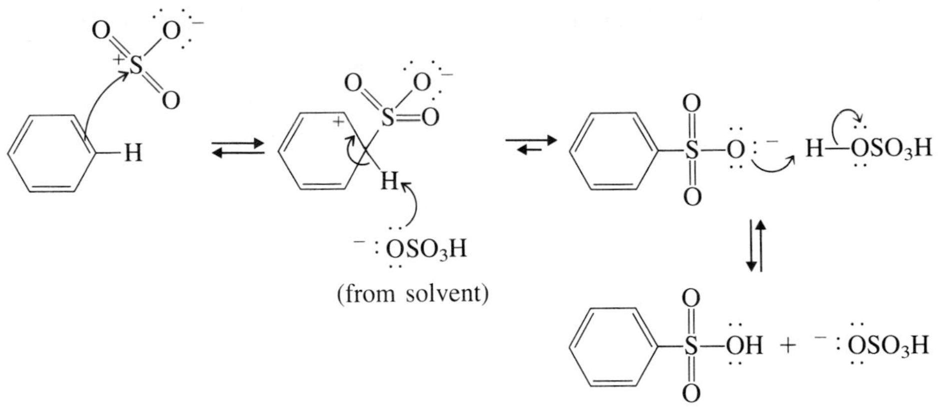

(16.13)

Sulfonic acids such as benzenesulfonic acid are rather strong acids. (Notice the last equilibrium in Eq. 16.13 and the formal resemblance of benzenesulfonic acid to sulfuric acid.) Many sulfonic acids are isolated from sulfonation reactions as their sodium salts.

Problem

10 A compound called *p*-toluenesulfonic acid is formed when toluene is sulfonated at the *para* position. Draw the structure of this material, and give the mechanism of its formation.

E. Friedel–Crafts Acylation of Benzene

When benzene reacts with an acid chloride in the presence of a Lewis acid catalyst such as $AlCl_3$, a ketone is formed.

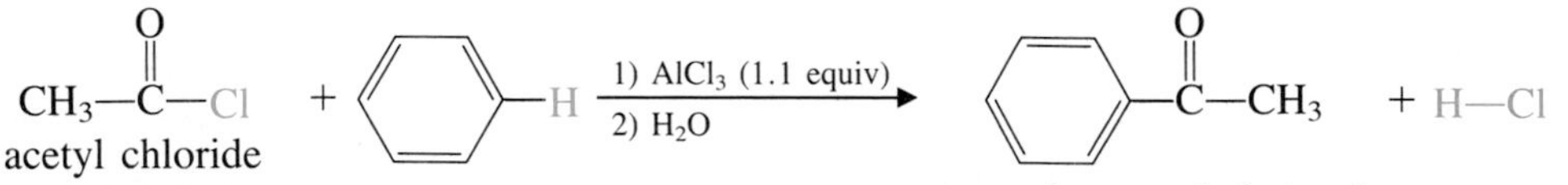

(16.14)

acetyl chloride (an acid chloride)

acetophenone (a ketone) (97% yield)

The group introduced into the benzene ring is called in general an *acyl* group, and this reaction is an example of a **Friedel–Crafts acylation.**

an acid chloride an acyl group

The Friedel–Crafts acylation is named for its discoverers, a Frenchman, Charles Friedel (1832–1899), and an American, James Mason Crafts (1839–1917). The two men met in the laboratory of the famous French chemist C. A. Wurtz, and in 1877 began their collaboration on the reactions that were to bear their names. Friedel became a very active figure in the development of chemistry in France, and Crafts served for a time as president of Massachusetts Institute of Technology.

The electrophile in the Friedel–Crafts acylation reaction is a carbocation called an *acylium ion*. This ion is formed when the acid chloride reacts with the Lewis acid $AlCl_3$.

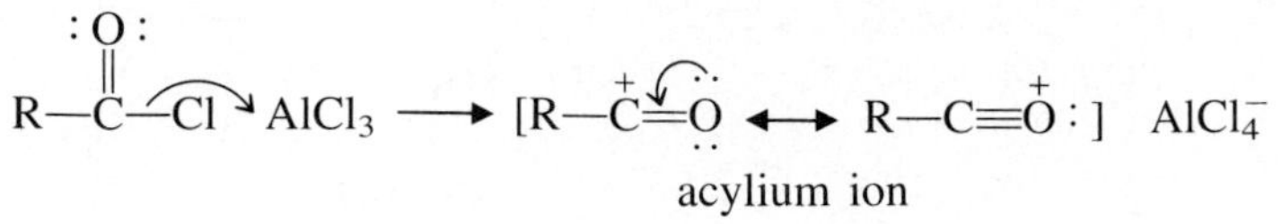

$$R-\overset{\displaystyle:O:}{\underset{\displaystyle\|}{C}}-Cl \;\; AlCl_3 \longrightarrow [R-\overset{+}{C}=\overset{..}{O} \longleftrightarrow R-C\equiv\overset{+}{O}:] \;\; AlCl_4^- \qquad (16.15)$$

acylium ion

Weaker Lewis acids, such as $FeCl_3$ and $ZnCl_2$, are also used in Friedel–Crafts acylations of aromatic compounds more reactive than benzene. The acylation reaction is completed by the usual steps of electrophilic aromatic substitution (Sec. 16.4B):

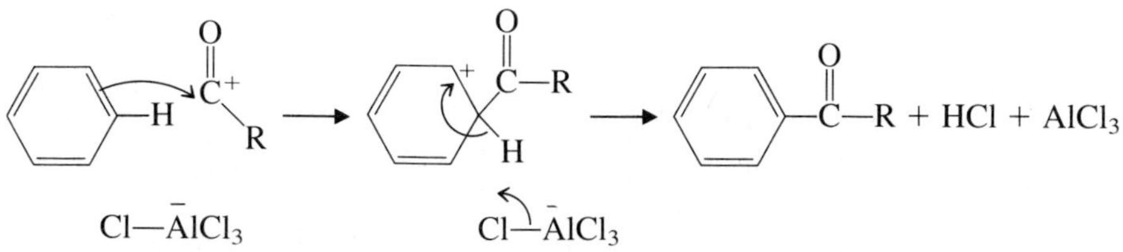

$$(16.16)$$

Attack of the benzene π-electrons on the acylium ion represents another in our list of carbocation reactions, which now includes the following:

1. Reaction with nucleophiles

2. Rearrangement to other carbocations

3. Loss of a β-proton to give an alkene

4. Reaction with the π-electrons of a double bond (or aromatic ring)

The acylation product is isolated by pouring the reaction mixture into ice water. This treatment is required to liberate the product ketone from a complex that it forms with the catalyst. Since complexed $AlCl_3$ is catalytically inactive, slightly more than one equivalent of the catalyst must be used in the Friedel–Crafts acylation.

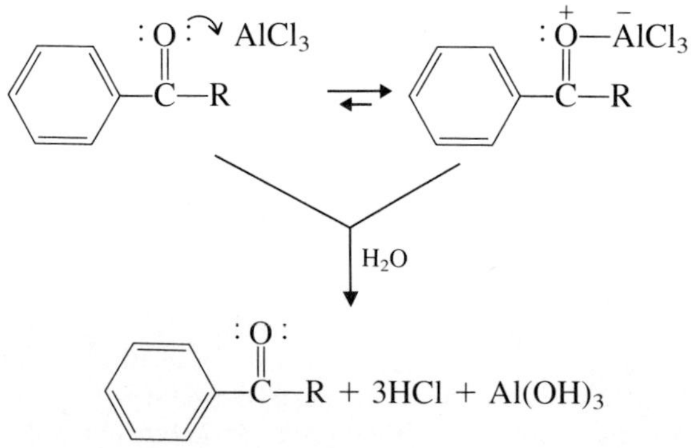

$$(16.17)$$

complex of $AlCl_3$ with ketone product

The Friedel–Crafts acylation occurs *intramolecularly* when the product contains a five- or six-membered ring.

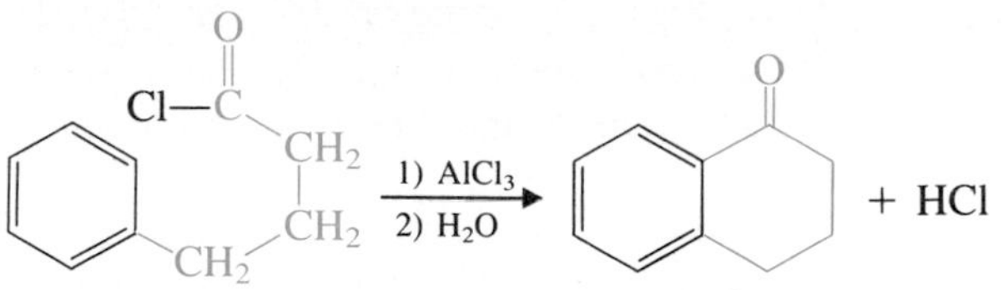

4-phenylbutanoyl chloride   α-tetralone (74–91% yield)   (16.18)

In this reaction, the acylium ion "bites back" on its own phenyl ring to form a bicyclic compound. When five- or six-membered rings are involved, this process is much faster than attack of the acylium ion on the phenyl ring of another molecule. (Sometimes this reaction can be used to form larger rings as well.) This type of reaction can only occur at an adjacent *ortho* position, since attack at other positions would produce highly strained products. (Use models to convince yourself that this is so.)

The Friedel–Crafts acylation reaction is important for two reasons. First, it is an excellent method for the synthesis of aromatic ketones. Second, it is another method for the formation of carbon–carbon bonds. Let us recapitulate our list of reactions that form carbon–carbon bonds:

1. Addition of carbenes to alkenes (Sec. 9.7)
2. Reaction of Grignard reagents with ethylene oxide (Sec. 11.4B)
3. Reaction of acetylide anions with alkyl halides or sulfonates (Sec. 14.7B)
4. Diels–Alder reactions (Sec. 15.3)
5. Friedel–Crafts reactions (Secs. 16.4E and 16.4F)

Problems

11 Give the structure of the product expected from the reaction of each of the following compounds with benzene in the presence of $AlCl_3$:

(a) $(CH_3)_2CH—\overset{O}{\overset{\|}{C}}—Cl$

isobutyryl chloride

(b) 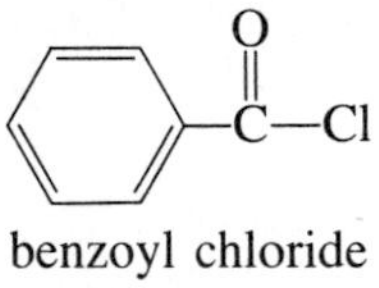

benzoyl chloride

12 Show two Friedel–Crafts acylation reactions that could be used to prepare the following compound:

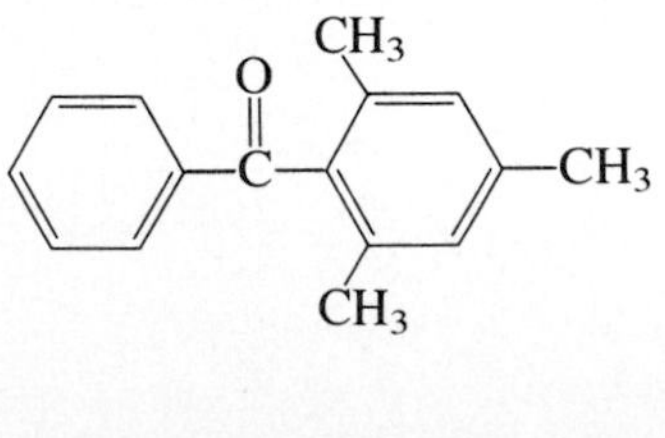

Problems (Cont.)

13 Give the product and the mechanism for its formation in the reaction of the following compound with aluminum chloride, followed by reaction with water.

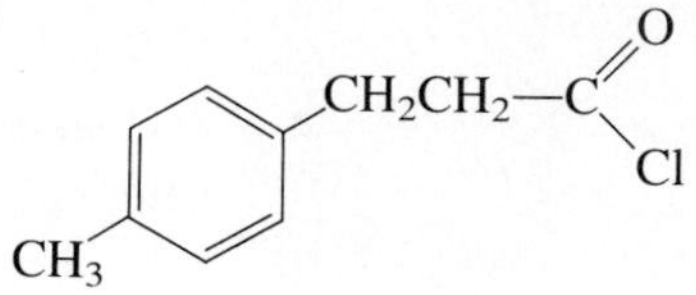

F. Friedel–Crafts Alkylation of Benzene

The reaction of an alkyl halide with benzene in the presence of a Lewis acid gives an alkylbenzene.

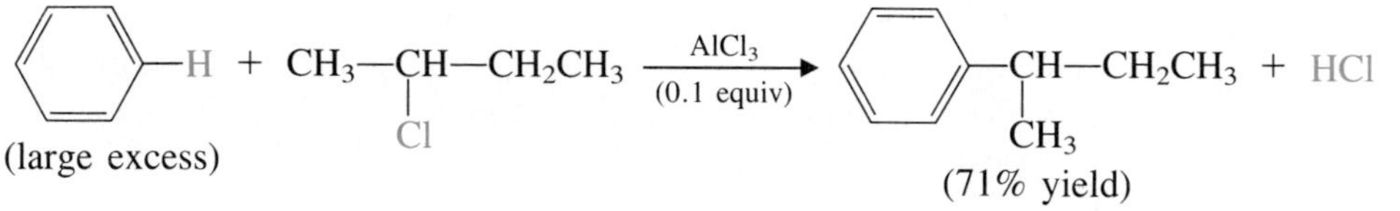

This reaction, called Friedel–Crafts *alkylation* (introduction of an alkyl group), has an obvious similarity to the Friedel–Crafts acylation reaction; both reactions involve the formation of new carbon–carbon bonds from compounds containing carbon–halogen bonds.

The electrophile in this reaction is formed by complexation of $AlCl_3$ with the alkyl halide. The resulting complex forms a carbocation, which is attacked by the benzene π-electrons.

$$R-\overset{..}{\underset{..}{Cl}}:\curvearrowright AlCl_3 \longrightarrow R-\overset{\overset{+}{..}}{\underset{\curvearrowleft}{Cl}}-\bar{A}lCl_3 \longrightarrow R^+ \quad \bar{A}lCl_4 \qquad (16.20a)$$

$$\text{(benzene)} \quad R^+ \ \bar{A}lCl_4 \longrightarrow \text{(complex)} \longrightarrow \text{(R)} + HCl + AlCl_3 \qquad (16.20b)$$

The promotion of carbocation formation by the Lewis acid, shown in Eq. 16.20a, is analogous to the S_N1 reaction of alkyl halides with Ag^+ (Eq. 9.55). In fact, it is not a bad analogy to think of the Friedel–Crafts alkylation reaction as an S_N1 reaction with benzene as the nucleophile. In contrast to the acylation reaction, much less than one equivalent of the $AlCl_3$ catalyst can be used in this reaction because the alkylbenzene product does not "tie up" the catalyst by formation of a strong complex.

Since some carbocations can rearrange, it is not surprising that rearrangements of alkyl groups are observed in some Friedel–Crafts alkylations.

$$\text{(benzene)}-H + (CH_3)_2CHCH_2CH_2Cl \xrightarrow{AlCl_3} \text{benzene}-\overset{\overset{CH_3}{|}}{\underset{\underset{CH_3}{|}}{C}}-C_2H_5 + HCl \qquad (16.21)$$

(70% yield)

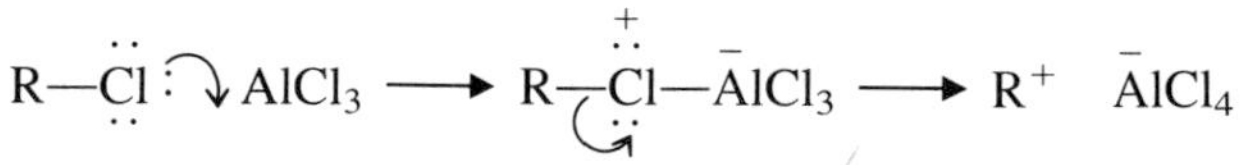

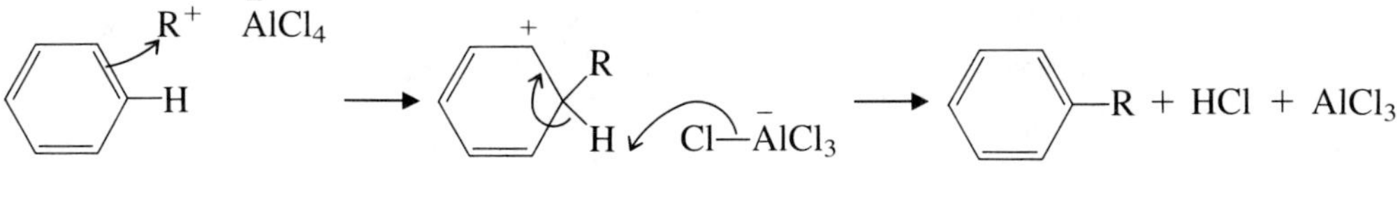

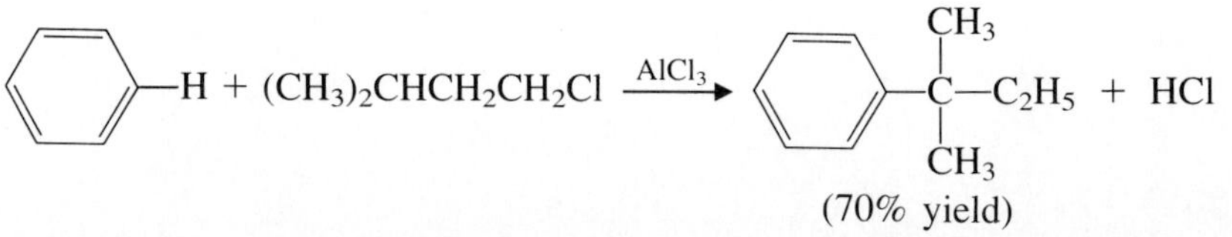

Of course, rearrangement of the alkyl group is not a problem if the carbocation intermediate is not prone to rearrangement.

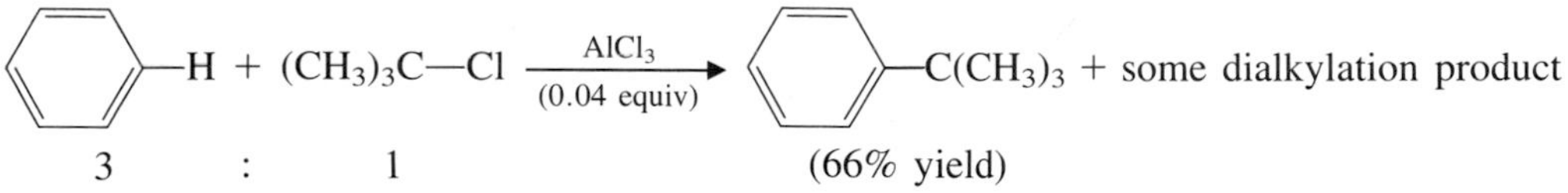

$$\text{benzene}-H + (CH_3)_3C-Cl \xrightarrow[\text{(0.04 equiv)}]{AlCl_3} \text{benzene}-C(CH_3)_3 + \text{some dialkylation product} \quad (16.22)$$

3 : 1 (66% yield)

Problem

14 Give a mechanism for the reaction shown in Eq. 16.21. Be sure to show how the rearrangement takes place.

Another complication in Friedel–Crafts alkylation is that the alkylbenzene products are more reactive than benzene itself (for reasons that we shall consider in Sec. 16.5B). This means that the product itself can undergo alkylation, and mixtures of products alkylated to different extents are observed.

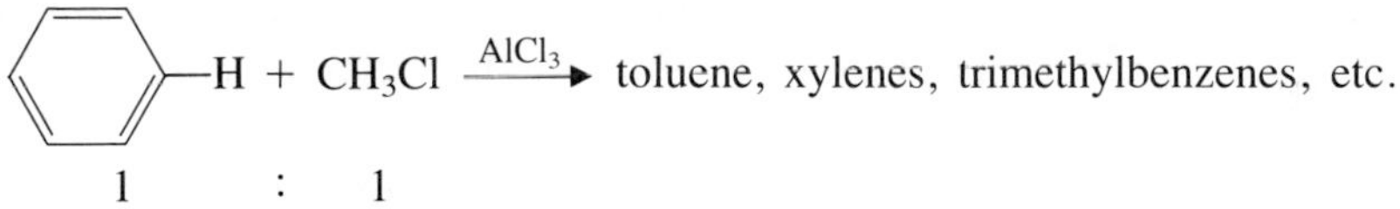

$$\text{benzene}-H + CH_3Cl \xrightarrow{AlCl_3} \text{toluene, xylenes, trimethylbenzenes, etc.} \quad (16.23)$$

1 : 1

(This is not a problem in acylation, since the ketone products of acylation are much less reactive than the benzene starting material.) However, a monoalkylation product can be obtained in good yield if a large excess of the starting material is used. When the starting material is in excess, the alkylating agent "sees"—and therefore reacts with—almost nothing but starting material.

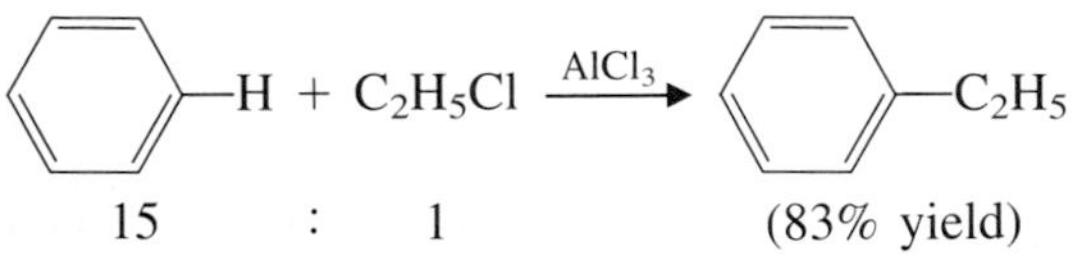

$$\text{benzene}-H + C_2H_5Cl \xrightarrow{AlCl_3} \text{benzene}-C_2H_5 \quad (16.24)$$

15 : 1 (83% yield)

(Notice also the use of excess starting material in Eqs. 16.19 and 16.22.) This strategy is practical only if the starting material is cheap, and if it can be readily separated from the product.

Alkenes can also be used in Friedel–Crafts alkylation reactions. The carbocation electrophiles in such reactions are generated from alkenes by protonation—another reaction that we have seen before (Sec. 4.5B).

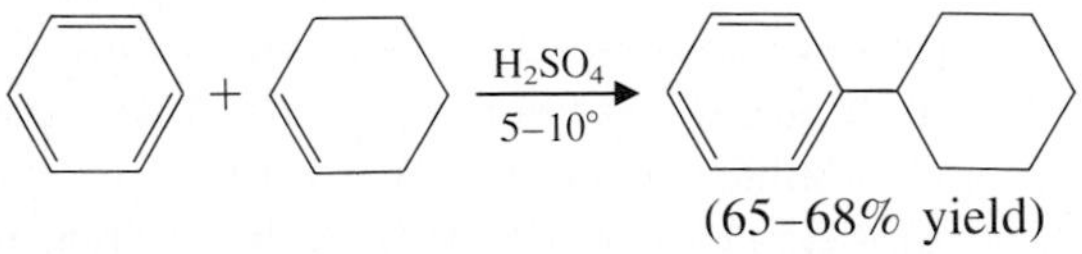

$$(16.25)$$

(65–68% yield)

Problems

15 Predict the product of the following reaction, and explain how it is formed:

$$\text{benzene}-(CH_2)_3-Cl \xrightarrow{AlCl_3} \text{a nine-carbon compound} + HCl$$

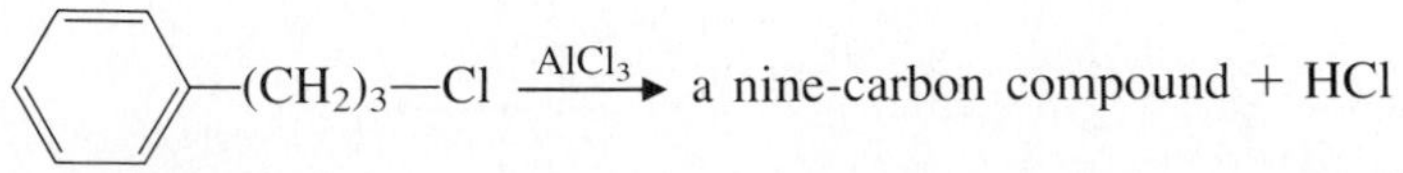

(*Hint:* See Eq. 16.18.)

Problems (Cont.)

16 (a) Suggest a mechanism for the formation of the product in Eq. 16.25.
(b) Explain why the same product is formed if cyclohexanol is substituted for cyclohexene in this reaction.

17 What product is formed when isobutylene is bubbled into a large excess of benzene containing HF and the Lewis acid boron trifluoride, BF_3?

16.5 ELECTROPHILIC AROMATIC SUBSTITUTION REACTIONS OF SUBSTITUTED BENZENES

A. Directing Effects of Substituents

When a monosubstituted benzene undergoes an electrophilic aromatic substitution reaction, three possible disubstitution products might be obtained. For example, nitration of bromobenzene could in principle give *ortho*-, *meta*-, or *para*-bromonitrobenzene. If substitution were totally random, we would expect to find an *ortho*:*meta*:*para* ratio of 2:2:1 (why?). It is found experimentally that a second substitution is *not* random, but is *regioselective*.

Further substitution on a substituted benzene ring occurs in one of two possible ways. Some substituted benzenes react to give mostly a mixture of *ortho*- and *para*-disubstitution products (in many cases with the *para* product predominating). Bromobenzene, for example, is nitrated to give mostly *o*- and *p*-bromonitrobenzene and almost none of the *meta* isomer.

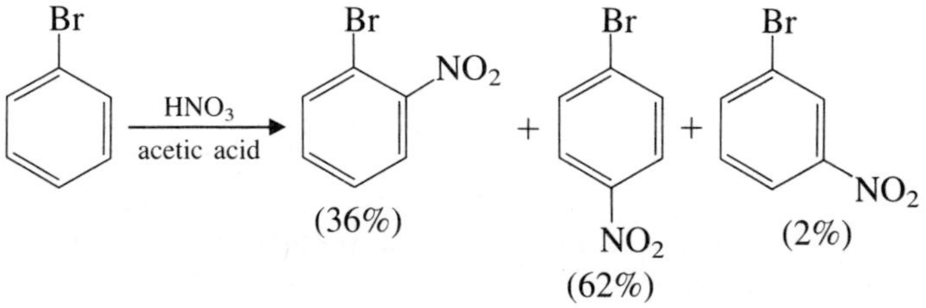

(16.26)

Other electrophilic substitution reactions of bromobenzene also give mostly *ortho* and *para* isomers. When a substituted benzene undergoes further substitution at the *ortho* and *para* positions, the original substituent is called an ***ortho, para*-directing group.** Thus, bromine is an *ortho, para*-directing group, because substitution in bromobenzene occurs at the *ortho* and *para* positions.

Some substituted benzenes react in electrophilic aromatic substitution to give mostly the *meta* disubstitution product. For example, the bromination of nitrobenzene gives only the *meta* isomer.

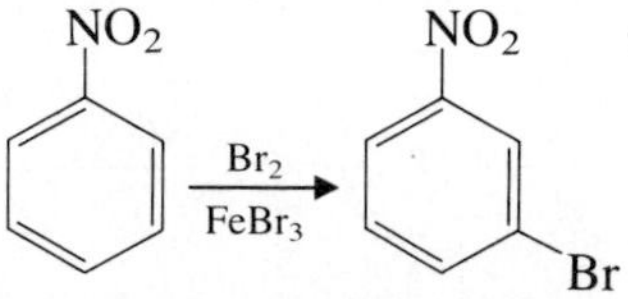

(only product observed) (16.27)

Other electrophilic substitution reactions of nitrobenzene also give mostly the *meta* isomers. If a substituted benzene undergoes further substitution at the *meta* position, the original substituent group is called a ***meta*-directing group.** Thus, the nitro group is a *meta*-directing group because substitution in nitrobenzene occurs at the *meta* position.

A substituent group is either an *ortho, para*-directing group or a *meta*-directing group in *all* electrophilic aromatic substitution reactions; no substituent is *ortho, para* directing in one reaction and *meta* directing in another. A summary of the directing effects of common substituent groups is given in the third column of Table 16.2.

TABLE 16.2 Summary of Directing and Activating or Deactivating Effects of Common Substituent Groups
(Groups are listed in decreasing order of activation)

Substituent group	Name of group	Directing effect	Activating (A) or deactivating (D)
$-\ddot{N}H_2$, $-\ddot{N}R_2$	amino	*ortho, para*	A
$-\ddot{\underset{..}{O}}H$	hydroxy	*ortho, para*	A
$-\ddot{\underset{..}{O}}R$	alkoxy	*ortho, para*	A
$-\ddot{N}H-C(=O)-R$	acylamino	*ortho, para*	A
$-R$	alkyl	*ortho, para*	A
$-\ddot{\underset{..}{F}}:$, $-\ddot{\underset{..}{Cl}}:$, $-\ddot{\underset{..}{Br}}:$, $-\ddot{\underset{..}{I}}:$	halogens	*ortho, para*	D
$-C(=O)-R$	acyl	*meta*	D
$-C(=O)-OH$, $-C(=O)-NH_2$, $-C(=O)-OR$	carboxy, carboxamido, carboalkoxy	*meta*	D
$-SO_3H$	sulfonic acid	*meta*	D
$-CN$	cyano	*meta*	D
$-NO_2$	nitro	*meta*	D

Problem

18 Using the information in Table 16.2, predict the product of Friedel–Crafts acylation of anisole with acetyl chloride (structure in Eq. 16.14).

What is the reason for these directing effects? These effects occur because *the rate of reaction at one position of a benzene derivative is much faster than the rate of reaction at another position.* For example, in Eq. 16.26, *o*- and *p*-bromonitrobenzenes are the major products because the rate of nitration is faster at the *ortho* and *para* positions of bromobenzene than it is at the *meta* position. In order to understand these effects, we must understand the factors that control the rates of aromatic substitution at each position.

Ortho, Para-*Directing Groups* If we examine Table 16.2, we find that *all ortho, para-directing substituents are either alkyl groups or groups that have atoms with unshared electron pairs directly attached to the benzene ring.* Let us try to understand the directing effect of these groups by considering the reactions of anisole (methoxybenzene) and toluene with a general electrophile E^+, first at the *para* position, then at the *meta* position.

Reaction of E^+ at the *para* position of anisole gives an intermediate carbocation with the following four important resonance structures:

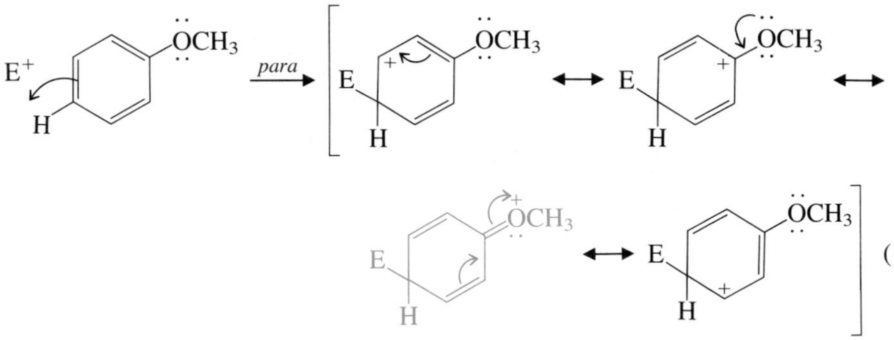

(16.28)

The colored structure shows that the unshared pair of the methoxy group can delocalize the positive charge on the carbocation. This is an especially important structure because it contains more formal bonds than the others, and every atom has an octet.

Problem

19 Draw the carbocation that results from the reaction of the electrophile at the *ortho* position of anisole; show that this ion also has four resonance structures.

If the electrophile reacts with anisole at the *meta* position, the carbocation intermediate that is formed has fewer resonance structures than the ion in Eq. 16.28. In particular, *the charge cannot be delocalized onto the* —*OCH_3 group when reaction occurs at the meta position.* There is no structure corresponding to the colored structure in Eq. 16.28.

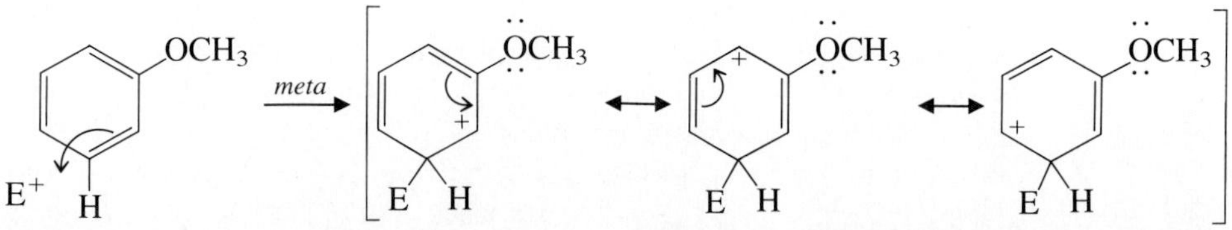

(16.29)

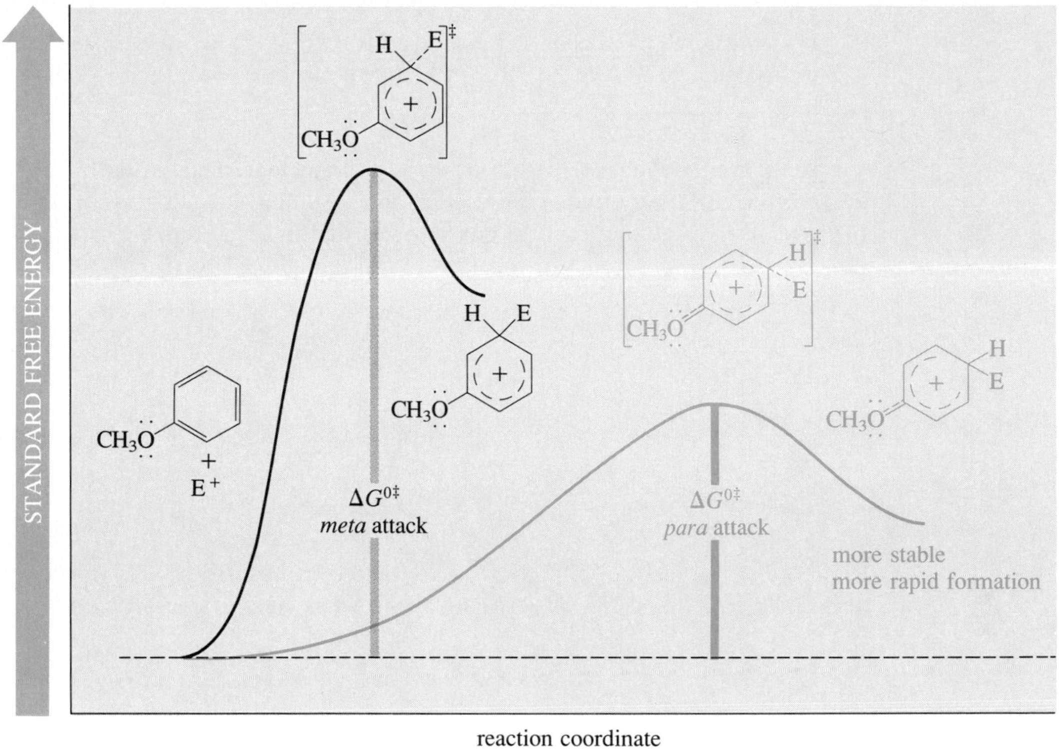

Figure 16.8 In electrophilic aromatic substitution reactions the more stable carbocation intermediate is formed more rapidly. The dashed lines symbolize the delocalization of electrons.

In order for the oxygen to delocalize the charge, it would have to be adjacent to an electron-deficient carbon, as in Eq. 16.28. The resonance formalism shows—and calculations of quantum theory verify—that the positive charge is shared on *alternate* carbons of the ring. When *meta* substitution occurs, the positive charge is not shared by the carbon adjacent to the oxygen.

A comparison of Eq. 16.28 and Problem 19 with Eq. 16.29 shows that the reaction of an electrophile at either the *ortho* or *para* positions of anisole gives a carbocation with more resonance structures—that is, a more stable carbocation. The rate-determining step in many electrophilic aromatic substitution reactions is *formation of the carbocation intermediate*. Hammond's postulate (Sec. 4.7C) leads to expect that the more stable carbocation is formed more rapidly. Hence, the products derived from the more rapidly formed cation—the more stable carbocation—are the ones observed. Since attack at the *ortho* and *para* positions of anisole gives more stable carbocations than attack at the *meta* positions, the products of *ortho, para* substitution are formed more rapidly, and are thus the products observed. This conclusion is shown with a reaction-free energy diagram in Fig. 16.8. This is why the —OCH_3 group is an *ortho, para*-directing group. An exactly analogous argument explains the *ortho, para*-directing effects of other groups with unshared electrons on atoms adjacent to the ring.

Problem

20 Using arguments similar to those used for attack of an electrophile on anisole, explain the results of Eq. 16.26 for the nitration of bromobenzene.

Alkyl groups such as a methyl group have no unshared electrons, but the explanation for the directing effects of these groups is similar. Reaction at a position that is *ortho* or *para* to an alkyl group gives an ion that has one tertiary carbocation resonance structure (colored structure in the following equation).

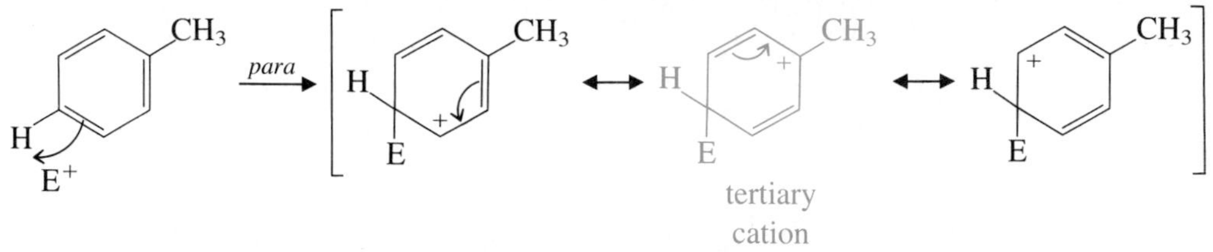

(16.30)

Reaction of the electrophile *meta* to the alkyl group also gives an ion with three resonance structures, but all resonance forms are secondary carbocations.

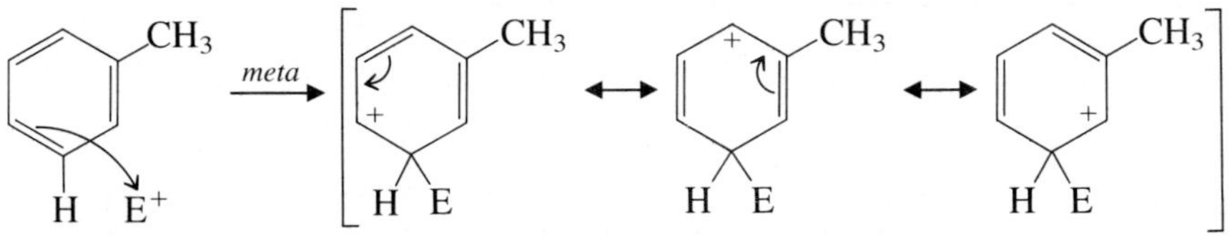

(16.31)

Since reaction at the *ortho* or *para* position gives the more stable carbocation, alkyl groups are *ortho, para*-directing groups.

Meta-*Directing Groups* The groups that act as *meta*-directing groups are all *electronegative groups without an unshared electron pair on an atom adjacent to the benzene ring*. Let us try to understand the directing effect of these groups by considering as an example the reactions of a general electrophile E^+ with nitrobenzene at the *meta* and *para* positions.

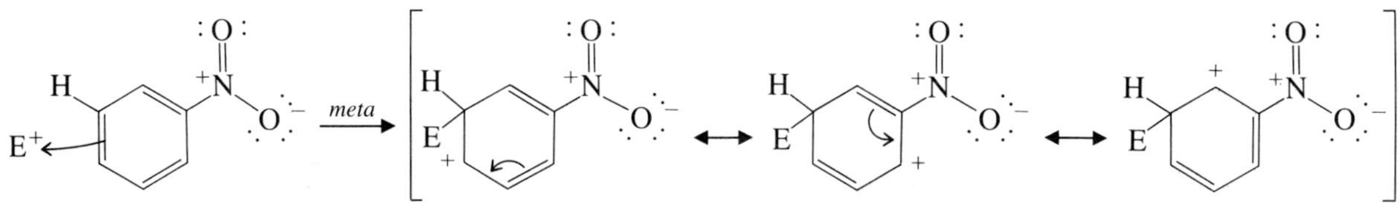

(16.32)

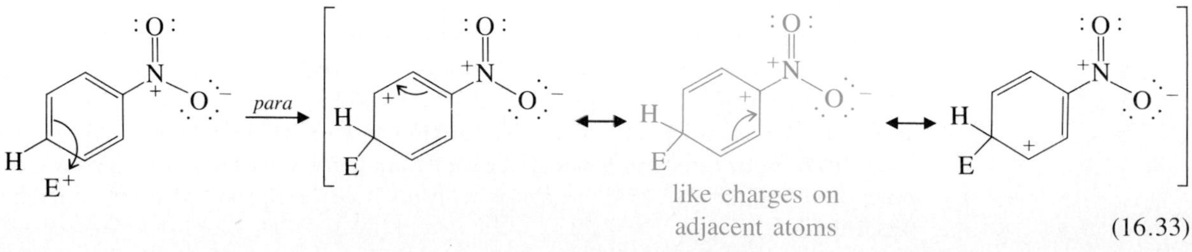

(16.33)

Both reactions give carbocations that have three resonance structures, but attack at the *para* position gives an ion with one particularly unfavorable structure (color). In this structure there are two positive charges on adjacent atoms. Since repulsion between two like charges, and consequently their energy of interaction, increases with decreasing distance between them (Eq. 1.2), the colored resonance structure in Eq. 16.33 is less important than the others. Reaction at the *meta* position is favored because it gives an ion in which the two like charges are farthest apart.

Problem

21 Show that reaction of an electrophile E^+ at the *ortho* position of nitrobenzene should also occur more slowly than reaction at the *meta* position.

Not all *meta*-directing groups have full formal positive charges like the nitro group, but all of them have bond dipoles that place a substantial degree of positive charge next to the benzene ring.

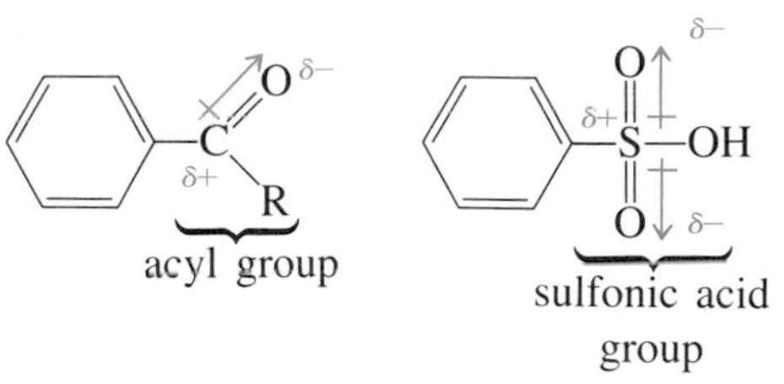

Problem

22 Predict the predominant products that would result from bromination of each of the following compounds. Classify each substituent group as an *ortho, para* director or a *meta* director and explain your reasoning.

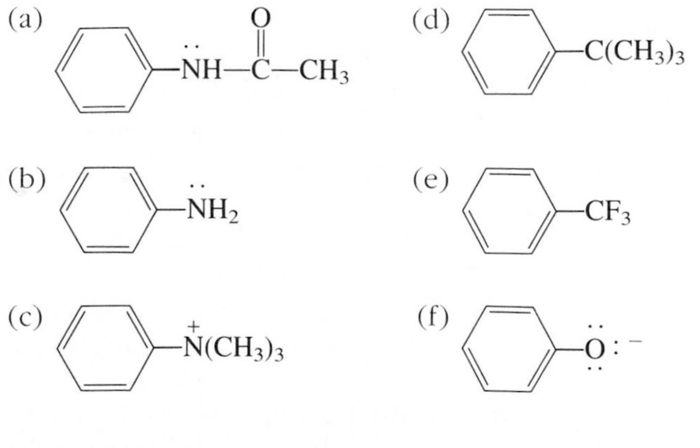

The* Ortho : Para *Ratio When a benzene derivative bears an *ortho, para*-directing group, most electrophilic aromatic substitution reactions give much more *para* than *ortho* product. This effect is illustrated by the greater amount of *para* isomer in the products of Eq. 16.26. An even more dramatic illustration is provided by the Friedel–Crafts acylation reactions of alkylbenzenes, which typically give almost all *para* products and only traces, if any, of the *ortho* products. In some cases, predominance of the *para* isomer can be explained as a *steric effect* (Sec. 3.5B, 5.3A). For example, when toluene reacts in a Friedel–Crafts acylation with the large complex of the acid chloride and $AlCl_3$, the complex reacts at the *para* position to avoid unfavorable van der Waals repulsions with the methyl group. On the other hand, although nitration of toluene yields about twice as much *ortho* as *para* substitution product—the ratio that we expect on a statistical (random) basis—nitration of fluorobenzene yields mostly *para* isomer. The latter result cannot be the consequence of steric effects because the fluoro group is considerably smaller than the methyl group. The *ortho*:*para* ratio, then, is a complex matter, and in many cases the reasons for the observed results are not well understood.

Whatever the reasons for the *ortho*:*para* ratio, we can see that if an electrophilic aromatic substitution reaction yields a mixture of *ortho* and *para* isomers, a problem of isomer separation arises that must be solved if the reaction is to be useful. Usually we try to avoid syntheses that give mixtures of isomers because, in many cases, isomers are difficult to separate. However, it happens that the *ortho* and *para* isomers obtained in many electrophilic aromatic substitution reactions have sufficiently different physical properties (Sec. 16.2) that they are readily separated. For example, the boiling points of *o*- and *p*-nitrotoluene, 220° and 238°, respectively, are sufficiently different that these isomers can be separated by careful fractional distillation. The melting points of *o*- and *p*-chloronitrobenzene, 34° and 84°, respectively, are so different that the *para* isomer selectively crystallizes. Most aromatic substitution reactions are so simple and inexpensive to run that when separation of isomeric products is not difficult, these reactions are useful for organic synthesis despite the product mixtures that are obtained.

Problem

23 Suggest an explanation for the trend in the *ortho*:*para* product ratios observed in nitration of the following compounds:

(a) toluene, 60:40
(b) ethylbenzene, 55:45
(c) cumene, 14:86
(d) *tert*-butylbenzene, 13:87

B. Activating and Deactivating Effects of Substituents

Different benzene derivatives have greatly different reactivities in electrophilic aromatic substitution reactions. If a substituted benzene derivative reacts more rapidly than benzene itself, then the substituent group is said to be an **activating group.** Friedel–Crafts acylation of anisole (methoxybenzene), for example, is 300,000 times as fast as the same reaction of benzene under comparable conditions. Furthermore, anisole shows a similar enhanced reactivity relative to benzene in all other electrophilic substitution reactions. Thus, the methoxy group is an activating group. On the other hand, if the derivative reacts more slowly than benzene itself, then the substituent is called a **deactivating group.** The bromination of nitrobenzene is more than 100,000 times slower than the bromination of benzene, and nitrobenzene reacts much more slowly than benzene in all other electrophilic substitution reactions as well. Thus, the nitro group is a deactivating group.

A given substituent group is either activating in all electrophilic aromatic substitution reactions or deactivating in all such reactions. Whether a substituent is activating or deactivating is shown in the last column of Table 16.2. In this table the most activating substituent groups are near the top of the table. Three generalizations emerge from examining this table:

1. All *meta*-directing groups are deactivating groups.
2. All *ortho, para*-directing groups except for the halogens are activating groups.
3. The halogens are deactivating groups.

Thus, except for the halogens, there appears to be a correlation between the activating and directing effects of substituents.

In view of this correlation, it is not surprising that the explanation of activating and deactivating effects is closely related to the explanation for directing effects. A key to understanding these effects is for us to realize that directing effects are concerned with the relative rates of substitution at different positions of the *same* compound, while activating or deactivating effects are concerned with the relative rates of substitution of *different* compounds—a substituted benzene compared with benzene itself. As in our discussion of directing effects, we consider the effect of the substituent on the stability of the intermediate carbocation, and assume that the stability of this carbocation is related to the stability of the transition state for its formation, as suggested by Hammond's postulate.

We must take into account two properties of substituents in order to understand activating and deactivating effects. First is the **resonance effect** of the substituent. This is the ability of the substituent to stabilize the carbocation intermediate in electrophilic substitution by delocalization of electrons from the substituent into the ring. The resonance effect is the same effect that is responsible for the *ortho, para*-directing effects of substituents with unshared electron pairs, such as $—OCH_3$ and halogen (colored structure in Eq. 16.28).

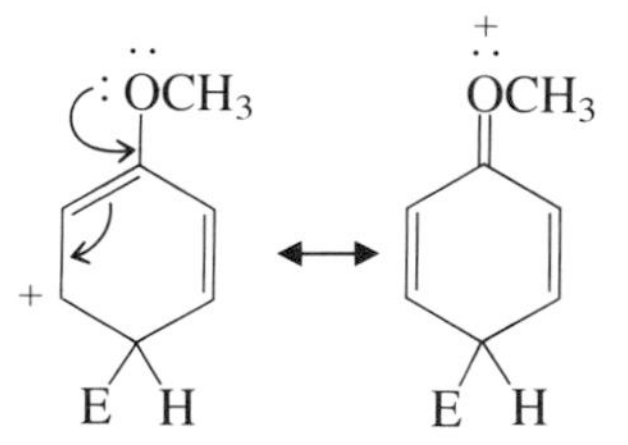

resonance effect of the methoxy group *stabilizes* the carbocation

The second property is the **polar effect** or **inductive effect** of the substituent. This is the tendency of the substituent group, by virtue of its electronegativity, to pull electrons away from the ring. This is the same effect we discussed in connection with substituent effects on alcohol acidity (Sec. 8.6B). When a ring substituent is electronegative, it pulls the electrons of the ring toward itself and creates a slight electron deficiency, or positive charge, in the ring. In the carbocation intermediate of an electrophilic substitution reaction, the positive end of the bond dipole interacts repulsively with the positive charge in the ring, raising the energy of the ion:

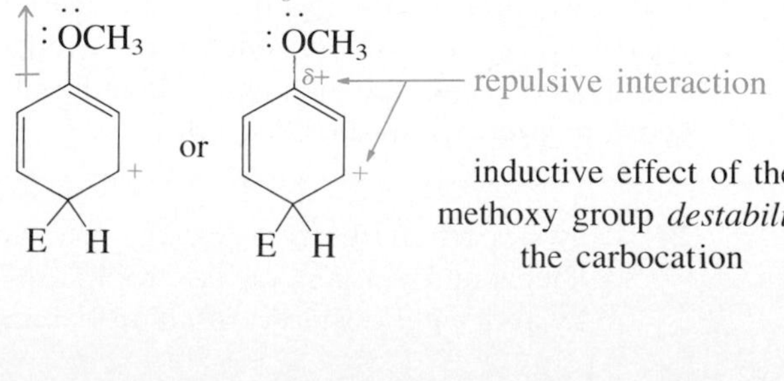

inductive effect of the methoxy group *destabilizes* the carbocation

Thus, the electron-donating resonance effect of a group with unshared electron pairs *stabilizes* positive charge and *activates* further substitution; the electron-withdrawing inductive effect of such an electronegative group *destabilizes* positive charge and *deactivates* further substitution. *Whether a substituted derivative of benzene is activated or deactivated toward further substitution depends on the balance of the resonance and inductive effects.*

Anisole (methoxybenzene) undergoes electrophilic substitution much more rapidly than benzene because the resonance effect of the methoxy group far outweighs its inductive effect. The benzene molecule, in contrast, has no substituent to help stabilize the carbocation intermediate by resonance. Hence, the carbocation intermediate (and the transition state) derived from substitution on anisole is more stable relative to starting materials than the carbocation (and transition state) derived from substitution on benzene. As a result, *para* substitution of anisole in a given reaction is faster than substitution on benzene—in other words, the methoxy group activates the benzene ring toward *para* substitution.

There is also an important subtlety here. Although the *ortho* and *para* positions of anisole are highly activated toward substitution, the *meta* position is deactivated. When substitution occurs in the *meta* position, the methoxy group cannot exert its resonance effect (Eq. 16.29), and only its rate-retarding inductive effect is operative. Thus, whether a group activates or deactivates further substitution really depends on the *position* on the ring that we are considering. Thus, the methoxy group activates *ortho, para* substitution and deactivates *meta* substitution. But this is just another way of saying that the methoxy group is an *ortho, para* director. Since *ortho, para* substitution is the *observed* mode of substitution, the methoxy group is considered to be an activating group. These ideas are summarized in the reaction free-energy diagram shown in Fig. 16.9.

The deactivating effects of halogen substituents reflect a different balance of resonance and inductive effects. Consider the chloro group, for example. Since chlorine and oxygen have similar electronegativities, the inductive effects of the chloro and methoxy groups are similar. However, the resonance interaction of chlorine electron pairs with the ring is much less effective than the interaction of oxygen electron pairs because the chlorine valence electrons reside in orbitals with higher quantum numbers. Because these orbitals and the carbon $2p$ orbitals in the benzene ring have different numbers of nodes, they do not overlap as well (Fig. 16.10). Since this overlap is the basis of the resonance effect, the resonance effect of chlorine is weak. With a weak rate-enhancing resonance effect and a strong rate-retarding inductive effect, chlorine is a deactivating group. Bromine and iodine exert weaker inductive effects than chlorine, but their resonance effects are also weaker (why?). Hence, these groups, too, are deactivating groups. Fluorine, as a first-row element, has a stronger resonance effect than the other halogens, but, as the most electronegative element, has a stronger inductive effect as well. Fluorine is also a deactivating group.

The deactivating, rate-retarding inductive effects of the halogens are offset somewhat by their resonance effects when substitution occurs *para* to the halogen. The resonance effect of a halogen cannot come into play when substitution occurs at the *meta* position of a halobenzene (why?). Hence, *meta* substitution in halobenzenes is deactivated even more than *para* substitution. This is another way of saying that halogens are *ortho, para*-directing groups.

Problem

24 Draw a reaction free-energy diagram analogous to Fig. 16.9 in which substitution on benzene by a general electrophile E^+ is compared with substitution at the *para* and *meta* positions of chlorobenzene.

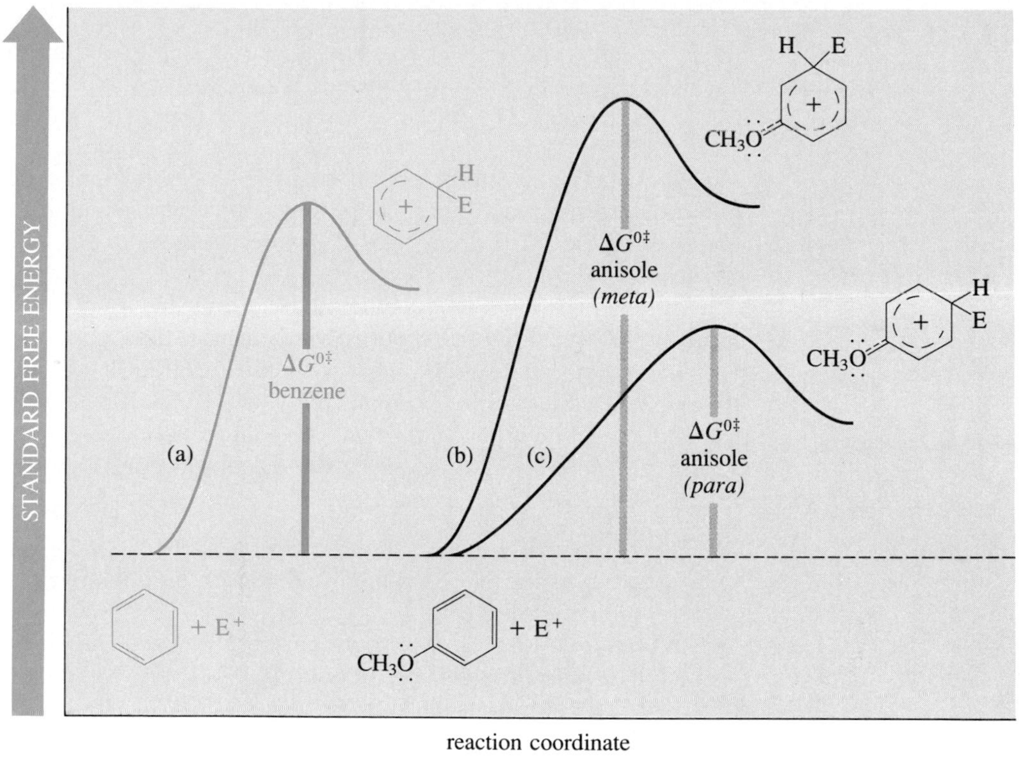

Figure 16.9 Reaction free-energy diagrams for the reactions of a general electrophile E^+ with (a) benzene, (b) anisole at the *para* position, and (c) anisole at the *meta* position. (The dashed lines represent delocalization of electrons in the carbocation intermediate.) The reaction of anisole at the *para* position is faster than the reaction of benzene; the reaction of anisole at the *meta* position is slower than the reaction of benzene.

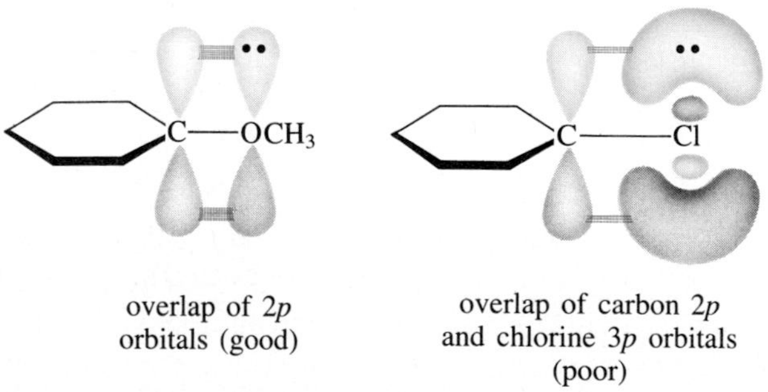

Figure 16.10 Chlorine–carbon orbital overlap is weaker than oxygen–carbon orbital overlap because the orbitals of chlorine and carbon have different quantum numbers. The colored and white parts of the orbitals represent wave peaks and wave troughs, respectively. Recall that bonding occurs only when peaks overlap with peaks and troughs overlap with troughs (Sec. 2.4).

Since a nitro group has no appreciable electron-donating resonance effect, the inductive effect of this electronegative group destabilizes the carbocation intermediate and retards electrophilic substitution at *all* positions of the ring. Since substitution is retarded more at the *ortho* and *para* positions than at the *meta* positions (Eqs. 16.32 and 16.33), the nitro group is a *meta*-directing group. In other words, the *meta*-directing effect of the nitro group is not due to selective activation of the *meta* positions, but rather to greater deactivation of the *ortho* and *para* positions. For this reason, we might term the nitro group and the other *meta*-directing groups *meta-allowing groups*.

Alkyl substituents such as the methyl group have no resonance effect, but the inductive effect of any alkyl group toward electron-deficient carbons, as we learned in Sec. 4.5C, is an electropositive, charge-stabilizing effect. It follows that alkyl groups on the benzene ring stabilize positive charge and, for this reason, activate electrophilic substitution. Alkyl groups are *ortho, para* directors because they activate *ortho, para* substitution more than they activate *meta* substitution (Eqs. 16.30 and 16.31).

Problem

25 Which should be faster:

(a) bromination of *N,N*-dimethylaniline or bromination of benzene under the same conditions?

(b) nitration of anisole or nitration of thioanisole under the same conditions? Explain your answers carefully.

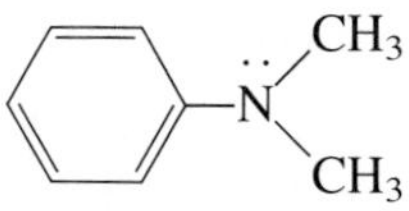

N,N-dimethylaniline

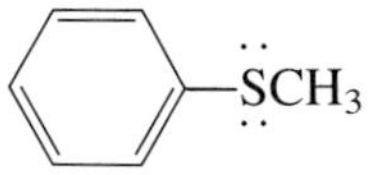

thioanisole

C. Use of Electrophilic Aromatic Substitution Reactions in Organic Synthesis

When we plan an organic synthesis that involves electrophilic aromatic substitution reactions, we must consider carefully the directing effects of substituents. For example, suppose we want to prepare *p*-bromonitrobenzene. We must introduce the bromine substituent first, and then use its directing effect to introduce the nitro group in the *para* position.

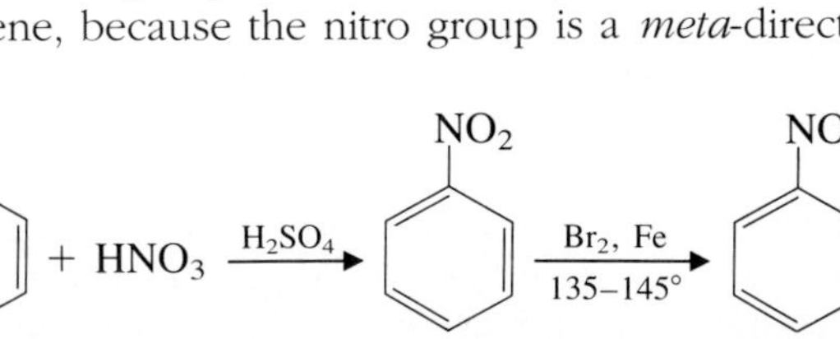

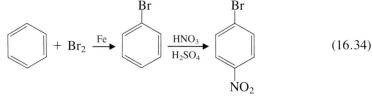

(16.34)

Introduction of the nitro group first followed by bromination would give instead *m*-bromonitrobenzene, because the nitro group is a *meta*-directing group.

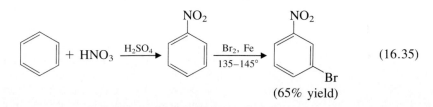

(16.35)

When an electrophilic substitution reaction is carried out on a benzene derivative with more than one substituent, the activating and directing effects are roughly the sum of the effects of the separate substituents. For example, the outcome of the Friedel–Crafts acylation of *m*-xylene is easy to predict because both methyl groups direct the substitution to the same positions.

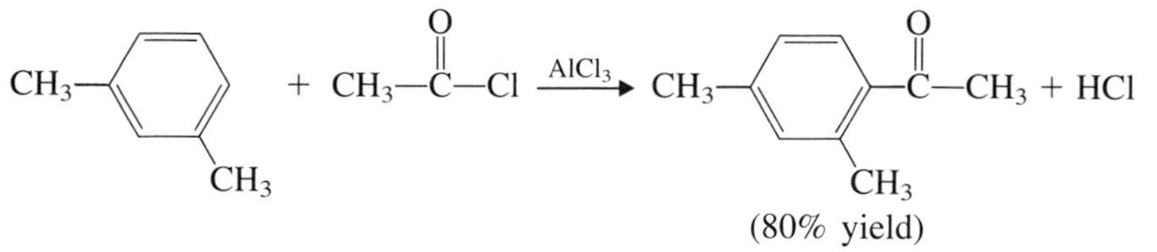

(16.36)

Methyl groups are *ortho, para* directors. Since the position *ortho* to both methyl groups is sterically hindered, substitution occurs *ortho* to one methyl group and *para* to the other.

Two *meta*-directing groups on a ring, such as the carboxylic acid (—CO_2H) groups in the following example, direct further substitution to the remaining open *meta* position:

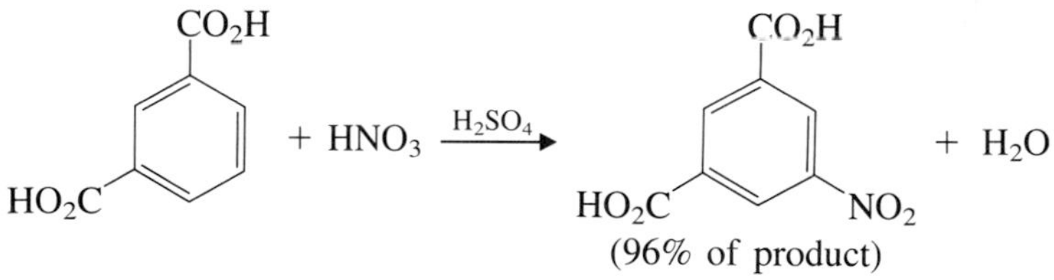

(16.37)

In each of the last two examples, both substituents direct the incoming group to the same position. What happens when the directing effects of the two groups are in conflict? If one group is much more strongly activating than the other, the directing effect of the more powerful activating group generally predominates. For example, the —OH group is such a powerful activating group that phenol can be brominated three times. (Notice that the —OH group is near the top of Table 16.2.)

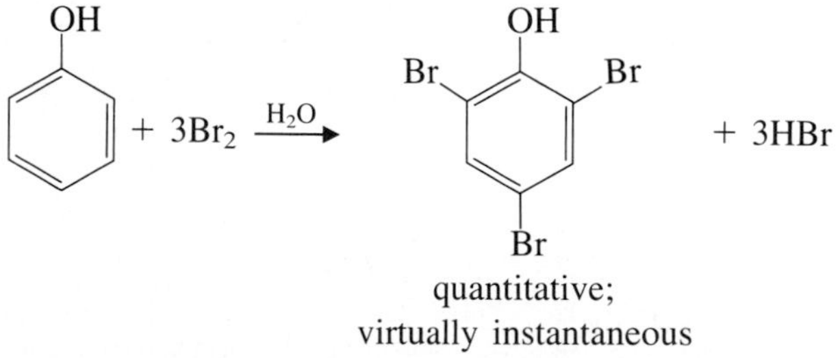

(16.38)

After the first bromination, the —OH and —Br groups direct subsequent brominations to different positions. The strong activating and directing effect of the —OH group at the *ortho* and *para* positions overrides the weaker directing effect of the —Br.

In other cases, mixtures of isomers are typically obtained.

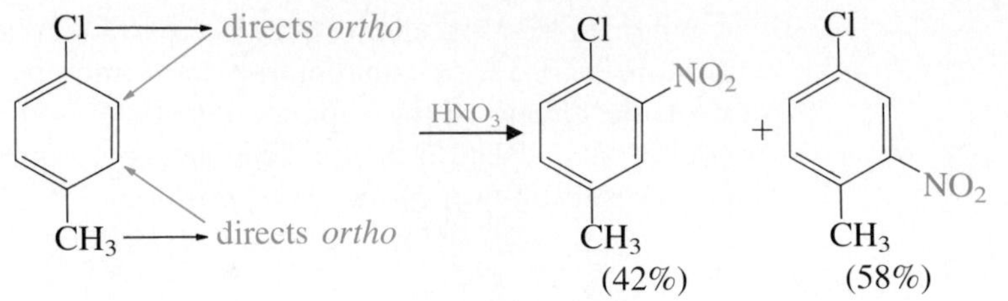

(16.39)

Problem

26 Predict the predominant product(s) from:
(a) monobromination of mesitylene (1,3,5-trimethylbenzene)
(b) monosulfonation of *m*-bromotoluene
(c) dinitration of toluene
(d) mononitration of *m*-bromoiodobenzene
Explain your answers.

Knowledge of the activating effects of substituents in an aromatic compound tells us what conditions we must use in an electrophilic substitution reaction. The bromination of nitrobenzene, for example (Eq. 16.27), requires heat and a Lewis acid catalyst because the nitro group deactivates the ring toward electrophilic substitution. In contrast, mesitylene can be brominated under very mild conditions, because the ring is activated by three methyl groups; a Lewis acid catalyst is unnecessary.

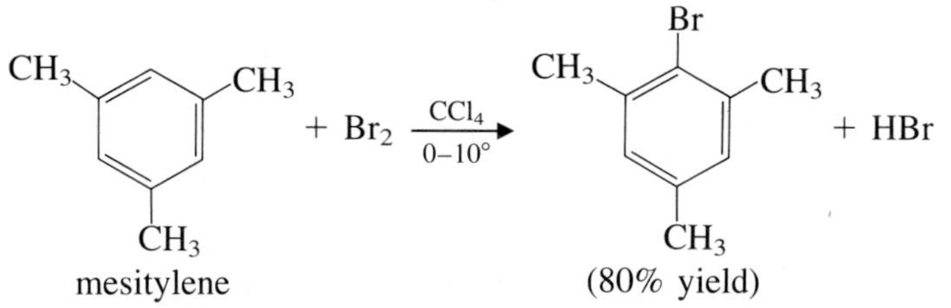

(16.40)

A similar contrast is apparent in the conditions required to sulfonate benzene and toluene. Sulfonation of benzene requires fuming sulfuric acid (Eq. 16.11). However, because toluene is more reactive than benzene, toluene can be sulfonated with concentrated sulfuric acid, a milder reagent than fuming sulfuric acid.

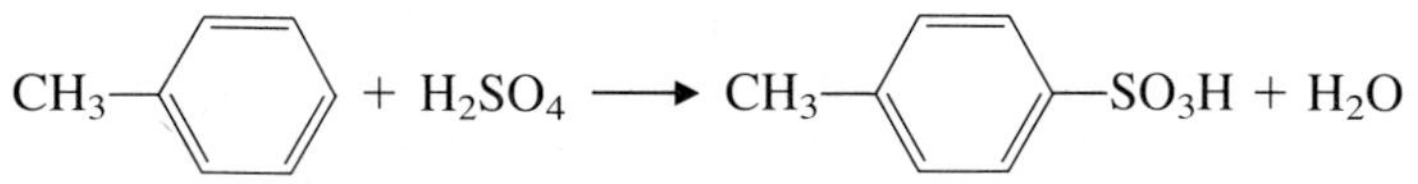

$$CH_3 - \bigcirc + H_2SO_4 \longrightarrow CH_3 - \bigcirc - SO_3H + H_2O \qquad (16.41)$$

Another important consequence of activating and deactivating effects is that when a deactivating group—for example, a nitro group—is being introduced by an electrophilic substitution reaction, it is easy to introduce one group at a time. Thus, it is easy to nitrate toluene only once because the nitro group that is introduced retards a second nitration on the same ring. Additional nitrations require increasingly harsh conditions.

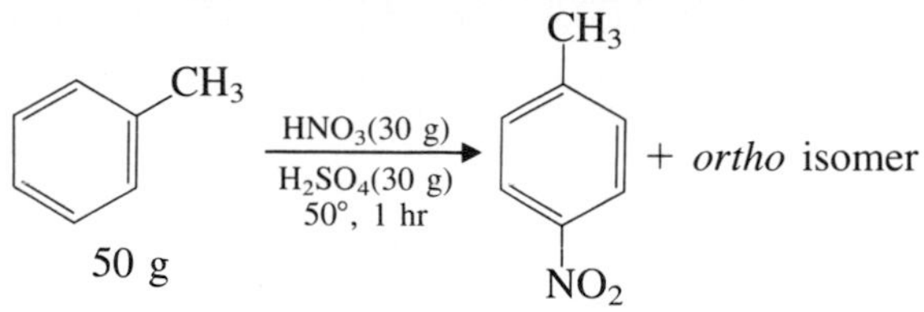

(16.42a)

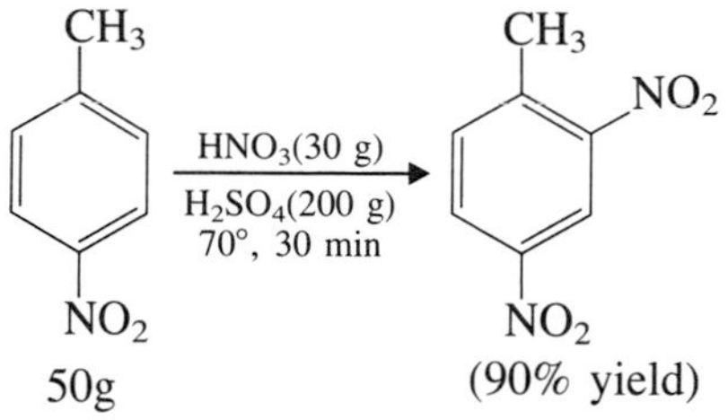

(16.42b)

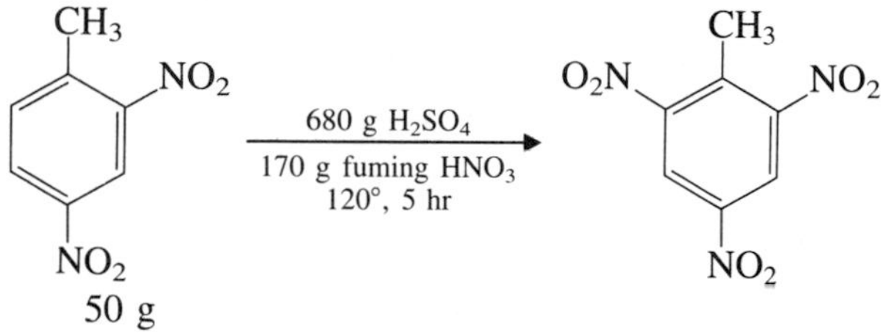

2,4,6-trinitrotoluene
"TNT"
(90% yield)

(16.42c)

Fuming nitric acid (Eq. 16.42c) is an especially concentrated form of nitric acid. Ordinary nitric acid contains 68% by weight of nitric acid; fuming nitric acid is 95% by weight nitric acid. It owes its name to the layer of colored fumes usually present in the bottle of the commercial product.

When an activating group is introduced by electrophilic substitution, additional substitutions can occur easily under the conditions of the first substitution, and mixtures of products are obtained. This is the situation in Friedel–Crafts alkylation. As we learned in the discussion of Eq. 16.23, one way to avoid multiple substitution is to use a large excess of the starting material.

Deactivating substituents retard some reactions to the point that they are not useful. For example, Friedel–Crafts *acylation* (Sec. 16.4E) does not occur on a benzene ring substituted *solely* with one or more *meta*-directing groups. In fact, nitrobenzene is so unreactive in the Friedel–Crafts acylation that it can be used as the solvent in the acylation of other aromatic compounds! Similarly, the Friedel–Crafts *alkylation* (Sec. 16.4F) is generally not useful on compounds that are more deactivated than benzene itself.

Problems

27 Rank the following compounds in order of increasing severity of the conditions required to effect monosulfonation: benzene, *m*-xylene, *p*-dichlorobenzene.

Problems (Cont.)

28 Outline a synthesis of *m*-nitroacetophenone from benzene; explain your reasoning.

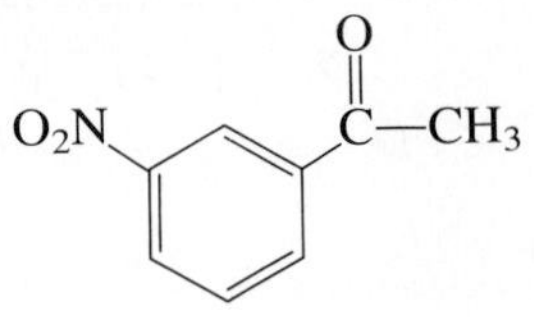

m-nitroacetophenone

16.6 HYDROGENATION OF BENZENE DERIVATIVES

Because of its aromatic stability, the benzene ring is resistant to conditions used to hydrogenate ordinary double bonds.

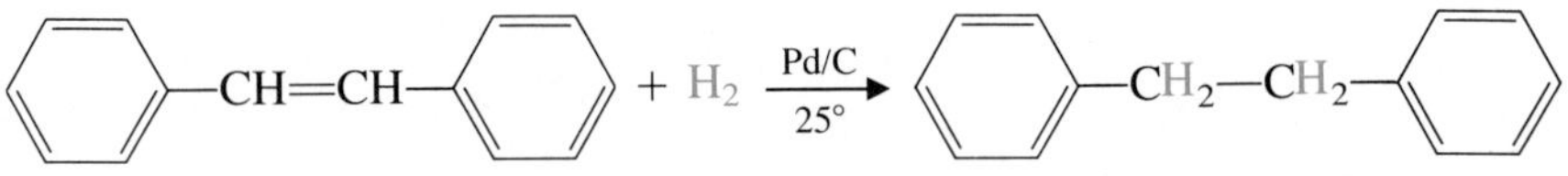

(16.43)

stilbene
(*cis* or *trans*)

1,2-diphenylethane
(bibenzyl)
(95% yield)

Nevertheless, aromatic rings can be hydrogenated under more extreme conditions of temperature and/or pressure. Typical conditions for carrying out the hydrogenation of benzene derivatives include Rh or Pt catalysts at 5–10 atm of hydrogen pressure and 50–100°, or Ni or Pd catalysts at 100–200 atm and 100–200°.

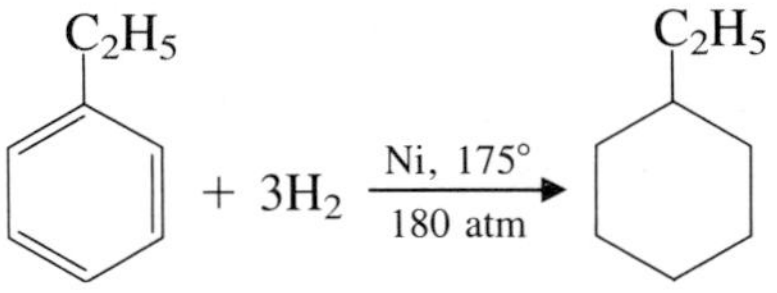

(93% yield) (16.44)

In some cases, the best way to prepare a substituted cyclohexane is to prepare the corresponding benzene derivative and then hydrogenate it.

Catalytic hydrogenation of benzene derivatives gives the corresponding cyclohexanes, and cannot be stopped at the cyclohexadiene or cyclohexene stage. We can understand why if we consider the heats of hydrogenation of benzene, 1,3-cyclohexadiene, and cyclohexene.

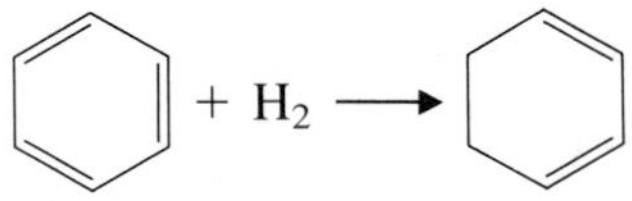

$\Delta H^0 = +5.8$ kcal/mol (16.45a)

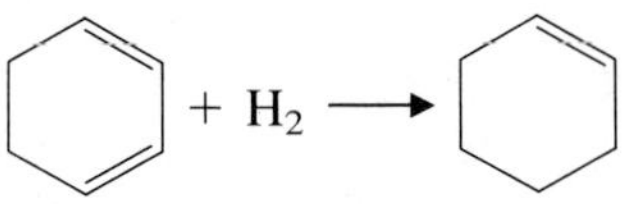

$\Delta H^0 = -26.5$ kcal/mol (16.45b)

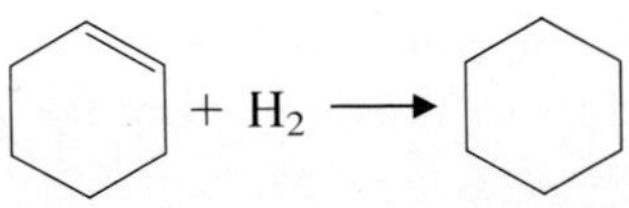

$\Delta H^0 = -28.3$ kcal/mol (16.45c)

The hydrogenation of most alkenes is *exothermic* by 27–30 kcal/mol; yet the reaction in Eq. 16.45a is *endothermic*. The unusual ΔH^0 of this reaction reflects the aromatic stability of benzene. Because it is thermodynamically unfavorable, energy must be added for this reaction to take place—thus the harsh conditions required for the hydrogenation of benzene derivatives. The hydrogenation of 1,3-cyclohexadiene and cyclohexene proceed so rapidly under these vigorous conditions that once these compounds are formed, they react instantaneously.

Problem **29** Outline a synthesis of *t*-butylcyclohexane from benzene and any other reagents.

16.7 SOURCE AND INDUSTRIAL USE OF AROMATIC HYDROCARBONS

The most common source of aromatic hydrocarbons is petroleum. Some petroleum sources are relatively rich in aromatic hydrocarbons, and aromatic hydrocarbons can be obtained by catalytic reforming of the hydrocarbons from other sources. Another potentially important, but relatively minor, source of aromatic hydrocarbons is *coal tar*, the tarry residue obtained when coal is heated in the absence of oxygen. Once a major source of aromatic hydrocarbons, coal tar may increase in importance as a source of aromatic compounds as the use of coal increases.

Benzene itself is obtained by separation from petroleum fractions, and by demethylation of toluene. In 1985, the production of benzene in the United States was about 1.3 billion gallons, valued at $1.5 billion. Benzene serves as a principal source of ethylbenzene, styrene, and cumene (see Eq. 16.46), and as one of the sources of cyclohexane. Since cyclohexane is an important intermediate in the production of nylon, benzene has substantial importance to the nylon industry.

Toluene is also obtained by separation from reformates. As previously noted, some toluene is used in the production of benzene. Toluene is also used as an octane booster for gasoline and as a starting material in the polyurethane industry.

Cumene and ethylbenzene are obtained by the alkylation of benzene with ethylene and propylene, respectively, in the presence of an acid catalyst (Friedel–Crafts alkylation).

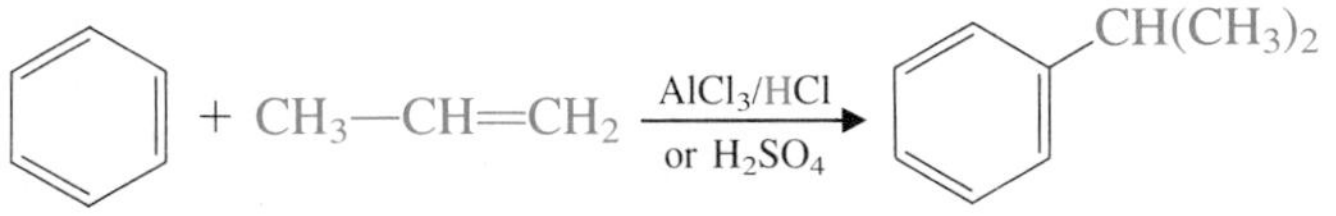

$$+ \ CH_3{-}CH{=}CH_2 \ \xrightarrow[\text{or } H_2SO_4]{\text{AlCl}_3/\text{HCl}} \qquad (16.46)$$

cumene

Cumene is an important intermediate in the manufacture of phenol and acetone (Sec. 18.7). The major use of ethylbenzene is dehydrogenation to styrene, one of the most commercially important of the aromatic hydrocarbons. Its principal uses are in the manufacture of polystyrene (Sec. 5.9A) and styrene–butadiene rubber (Sec. 15.5).

The xylenes are obtained by separation from petroleum and reforming C_8 petroleum fractions. Of the xylenes, *p*-xylene is the most important commercially. Virtually the entire production of *p*-xylene is used for oxidation to terephthalic acid (Eq. 16.47), an important intermediate in polyester synthesis (for example, Dacron; Sec. 21.12A). (Oxidation of alkylbenzenes is discussed in Sec. 17.5.)

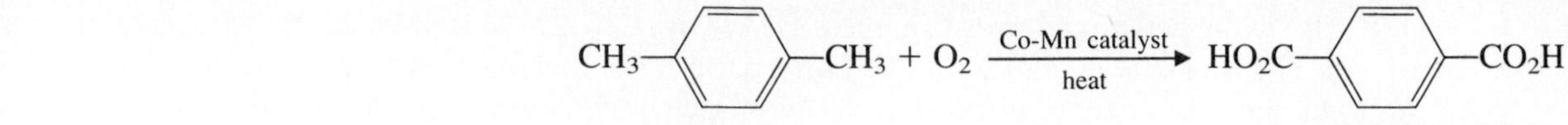

$$CH_3{-}\langle\rangle{-}CH_3 + O_2 \xrightarrow[\text{heat}]{\text{Co-Mn catalyst}} HO_2C{-}\langle\rangle{-}CO_2H \quad (16.47)$$

p-xylene terephthalic acid

16.8 BIOLOGICAL CONVERSION OF AROMATIC COMPOUNDS INTO EPOXIDES. THE CANCER CONNECTION

Most of us appreciate that certain chemicals are hazardous. Among the most worrisome chemical hazards is *carcinogenicity*—the proclivity of a substance to cause cancer. Certain aromatic compounds are **carcinogens,** or cancer-causing chemicals. Both the historical aspects of this finding and the reasons behind it are interesting.

After the great fire of London in 1666, Londoners began the practice of building homes with long and tortuous chimneys. The use of coal for heating resulted in deposits of black soot that had to be periodically removed from these chimneys, but the only people who could negotiate these narrow passages were small boys, called "sweeping boys." It was common for these boys to contract a disease that we now know is cancer of the scrotum. In 1775 Percivall Pott, a surgeon at London's St. Bartholomew's Hospital, identified coal dust as the source of "this noisome, painful, and fatal disease," and Pott's findings subsequently led to substantial reform in the child-labor statutes in England. In 1892 Henry T. Butlin, also of St. Bartholomew's, pointed out that the disease did not occur in countries in which the chimney sweeps washed thoroughly after each day's work.

In 1933, the compound benzo[*a*]pyrene was isolated from coal tar; this compound and 7,12-dimethylbenz[*a*]anthracene, both polycyclic aromatic hydrocarbons, are two of the most potent carcinogens known.

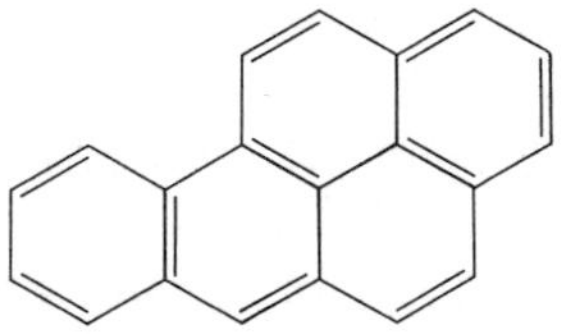

benzo[*a*]pyrene 7,12-dimethylbenz[*a*]anthracene
(7,12-DMBA)

These and related compounds are the active carcinogens in soot. Materials such as these are also found in cigarette smoke and in the exhaust of internal combustion engines (that is, automobile pollution). About three thousand tons of benzo[*a*]pyrene per year are released as particulates into the environment.

Benzene has also been found to be carcinogenic, and has been supplanted for many uses by toluene, which is not carcinogenic. (Not all aromatic compounds are carcinogens.)

Organic chemists and biochemists have learned that the *ultimate carcinogens* (true carcinogens) are not aromatic hydrocarbons themselves, but rather certain epoxide derivatives. The diol-epoxide shown in Eq. 16.48, formed when living cells attempt to metabolize benzo[*a*]pyrene, is the ultimate carcinogen derived from benzo[*a*]pyrene.

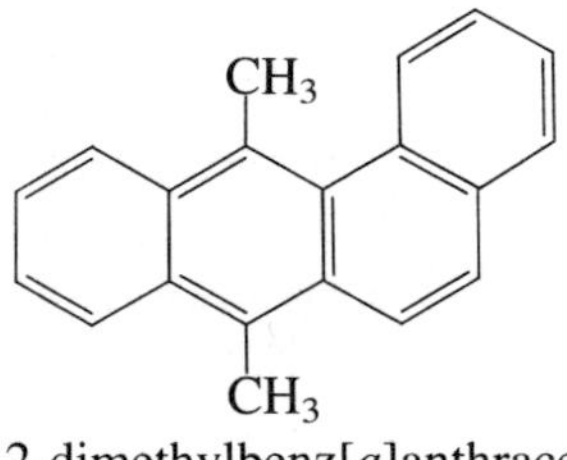

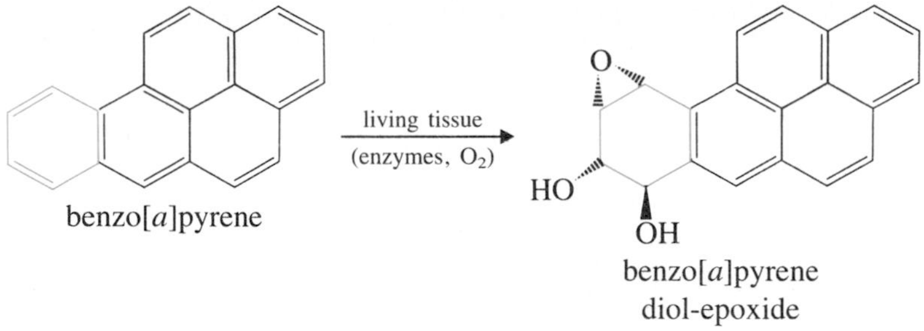

(16.48)

This transformation is particularly remarkable in view of the usual resistance of benzene rings to addition reactions. This type of oxidation is normally employed as a detoxification mechanism; the oxidized derivatives of many hydrocarbons, but not the hydrocarbons themselves, can be secreted and thus removed from the cell. Benzo[*a*]pyrene diol-epoxide, however, survives long enough to find its way into the nuclei of cells, where the epoxide group reacts with nucleophilic groups on DNA, the molecule that contains the genetic code. (We shall discuss this reaction in Sec. 27.11D.) It is strongly suspected, although not conclusively proven, that this is a key event in the transformation of cells to a cancerous state caused by benzo[*a*]pyrene.

KEY IDEAS IN CHAPTER 16

- Benzene derivatives are distinguished by the NMR absorptions of their ring protons, which occur at lower field than the absorptions of vinylic protons. The deshielding of protons on an aromatic ring is caused by the ring-current effect.
- The most characteristic reaction of aromatic compounds is electrophilic aromatic substitution. In this type of reaction, an electron-deficient species (electrophile) attacks the benzene ring to form a resonance-stabilized carbocation. Loss of a proton from this ion at the point of substitution gives a new aromatic compound.
- Examples of electrophilic aromatic substitution reactions discussed in this chapter are halogenation, used to prepare halobenzenes; nitration, used to prepare nitrobenzene derivatives; sulfonation, used to prepare benzenesulfonic acid derivatives; Friedel–Crafts acylation, used to prepare aryl ketones; and Friedel–Crafts alkylation, used to prepare alkylbenzenes.
- Derivatives containing substituted benzene rings undergo further substitution either at the *ortho* and *para* positions or at the *meta* position, depending on the ring substituent.
- Benzene rings with alkyl substituents or substituent groups that delocalize positive charge by resonance undergo further substitution at the *ortho* and *para* positions; these substituent groups are called *ortho, para*-directing groups.

- Benzene rings with substituents that cannot stabilize carbocations or delocalize positive charge by resonance undergo further substitution at the *meta* position. These substituents are called *meta*-directing groups.
- Whether a substituted benzene undergoes substitution more rapidly or more slowly than benzene itself is determined by the balance of resonance and inductive effects of the substituents. Benzene rings containing an *ortho, para*-directing group other than halogen react more rapidly in electrophilic aromatic substitution than benzene itself. Benzene rings containing a halogen substituent or any *meta*-directing group react more slowly in electrophilic aromatic substitution than benzene itself.
- The activating and directing effects of substituent groups must be taken into account when planning a synthesis.
- Alkene double bonds can be hydrogenated without affecting the benzene ring. Benzene derivatives, however, can be hydrogenated under relatively harsh conditions to cyclohexane derivatives.

ADDITIONAL PROBLEMS

30 Give the products expected (if any) when toluene reacts under the following conditions:

(a) Br_2 in CCl_4 (dark)
(b) HNO_3, H_2SO_4
(c) H_2SO_4
(d) $C_2H_5\text{—}\overset{\overset{\displaystyle O}{\|}}{C}\text{—}Cl$, $AlCl_3$ (1.1 equiv), then H_2O
(e) CH_3Br, $AlCl_3$
(f) Br_2, $FeBr_3$

31 Show how you would distinguish the compounds within each of the following pairs using only simple chemical or physical tests with readily observable results:

(a) ethylbenzene and styrene
(b) phenylacetylene and styrene
(c) benzene and ethylbenzene

32 Which of the following compounds *cannot* contain a benzene ring?

(a) $C_{10}H_{16}$ (b) $C_8H_6Cl_2$ (c) C_5H_4 (d) $C_{10}H_{15}N$

33 (a) Arrange the three isomeric dichlorobenzenes in order of increasing dipole moment (smallest first).

(b) Assuming that the dipole moment is the principal factor governing their relative boiling points, arrange the compounds in (a) in order of increasing boiling point (smallest first). Explain your reasoning.

34 Arrange the following compounds in order of increasing reactivity toward HNO_3 in H_2SO_4. (The references to equations will assist you with nomenclature.)
(a) chlorobenzene, benzene, nitrobenzene
(b) mesitylene (Eq. 16.40), toluene, 1,2,4-trimethylbenzene
(c) *m*-chloroanisole, *p*-chloroanisole, anisole
(d) acetophenone (Eq. 16.14), *p*-methoxyacetophenone, *p*-bromoacetophenone

35 Two alcohols, *A* and *B*, have the same molecular formula $C_9H_{10}O$ and react with sulfuric acid to give the same hydrocarbon *C*. Compound *A* is chiral and compound *B* is not. Catalytic hydrogenation of *C* gives a hydrocarbon *D*, C_9H_{10}, which gives two and only two products when nitrated once with HNO_3 in H_2SO_4. Give the structures of *A*, *B*, *C*, and *D*.

36 (a) In the following transformation, identify the reactions that comprise each of the three fundamental steps of electrophilic aromatic substitution.

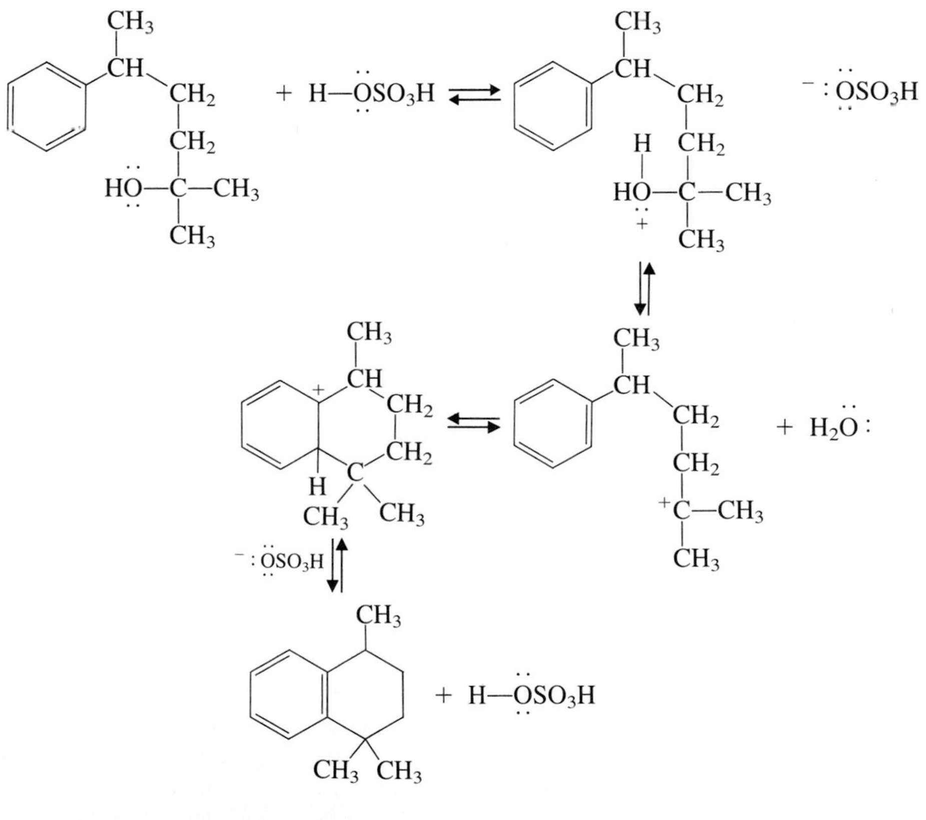

(b) Supply the curved arrows for each step.
(c) When the following compound is treated with H_2SO_4, the product of the resulting reaction has the same number of carbons as the starting material and does not decolorize Br_2 in CCl_4. By analogy with the reaction in (a), suggest a structure for this product and give the mechanism for its formation.

Problems (Cont.)

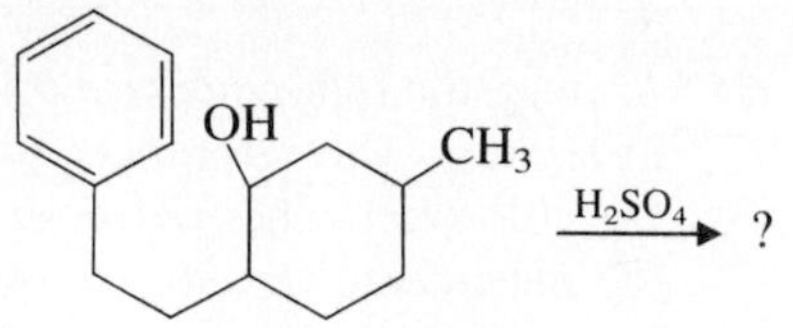

37 Outline laboratory syntheses of each of the following compounds, starting with benzene and any other reagents. (The references to equations will assist you with nomenclature.)

(a) *p*-nitrotoluene
(b) *p*-dibromobenzene
(c) *p*-chloroacetophenone (Eq. 16.14)
(d) *m*-nitrobenzenesulfonic acid (Eq. 16.11)
(e) *p*-chloronitrobenzene
(f) 1,3,5-trinitrobenzene
(g) 2,6-dibromo-4-nitrotoluene
(h) 2,4-dibromo-6-nitrotoluene
(i) 4-ethyl-3-nitroacetophenone (Eq. 16.14)
(j) methylcyclohexane
(k) cyclopentylbenzene

38 Explain how you would distinguish each of the following compounds from the others using NMR spectroscopy. Be explicit.

(a)
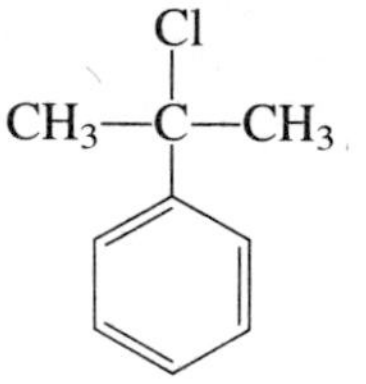

(c)
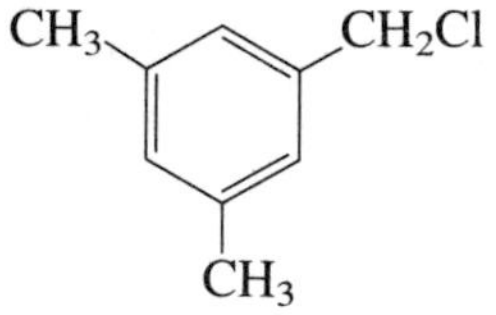

(b)
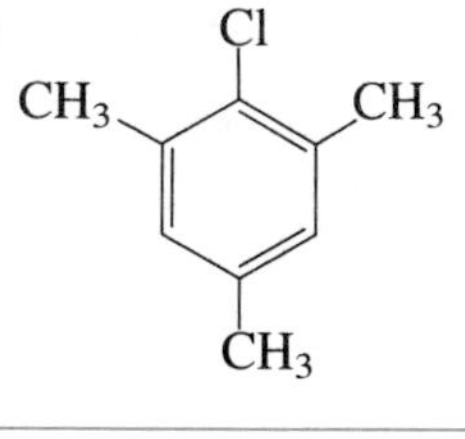

39 You have been assigned the task of cleaning out an abandoned student research laboratory and identifying unlabeled materials. You have found two bottles *A* and *B* containing colorless liquids; each bottle bears only the label C_8H_9Br. The NMR spectra of these compounds are given in Fig. 16.11. Give the structures of these compounds.

40 Indicate whether each of the following compounds should be nitrated more rapidly or more slowly than benzene, and give the structures of the principal mononitration products. Explain your reasoning.

(a)
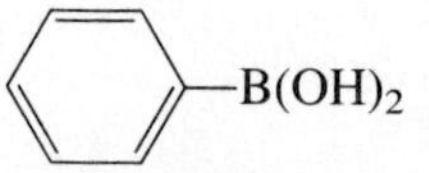

benzeneboronic acid

(b)
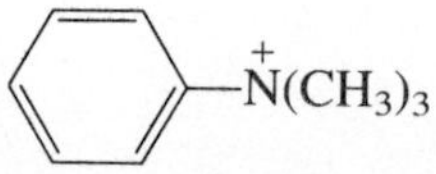

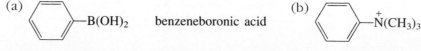

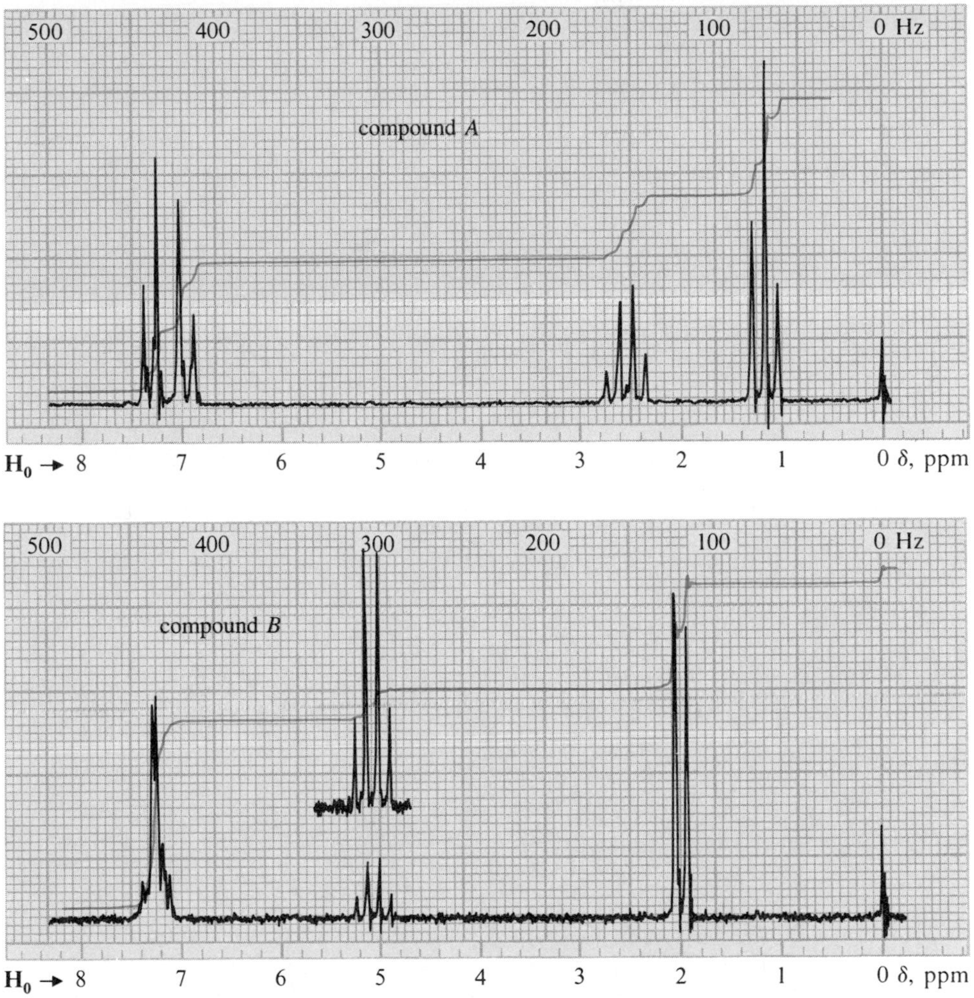

Figure 16.11 NMR spectra for Problem 39.

(c) 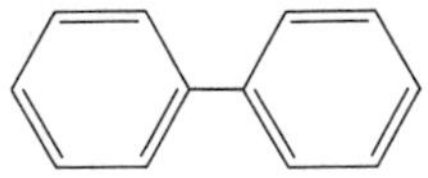 biphenyl

(d) 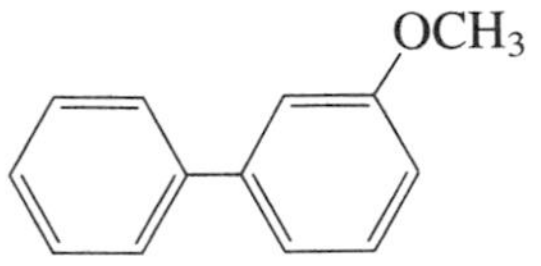

41 Give the structures of the principal organic product(s) of each of the following reactions:

(a) 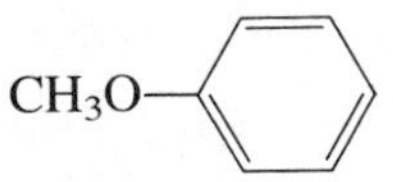 + dilute D_2SO_4 in D_2O ⟶ ($C_7H_5D_3O$)

anisole

(b) benzene (large excess) + 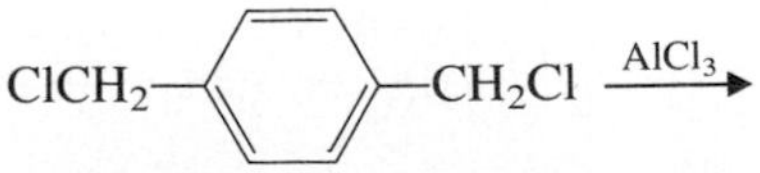

Figure 16.12 NMR spectrum for Problem 42.

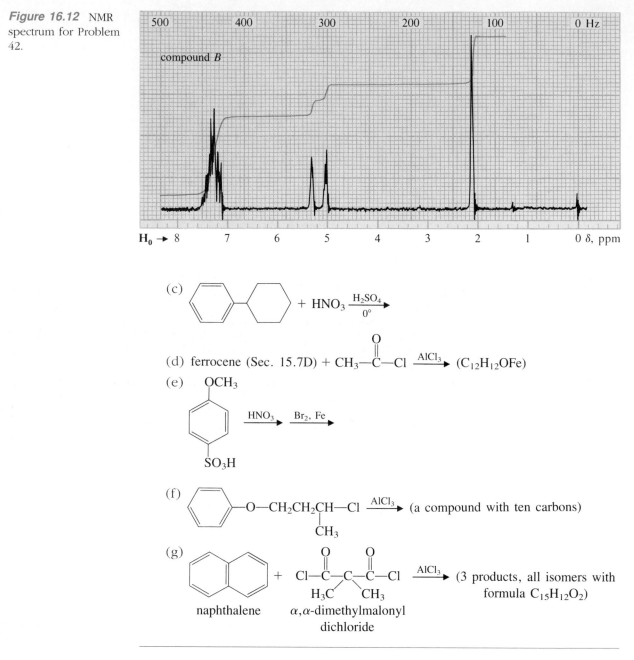

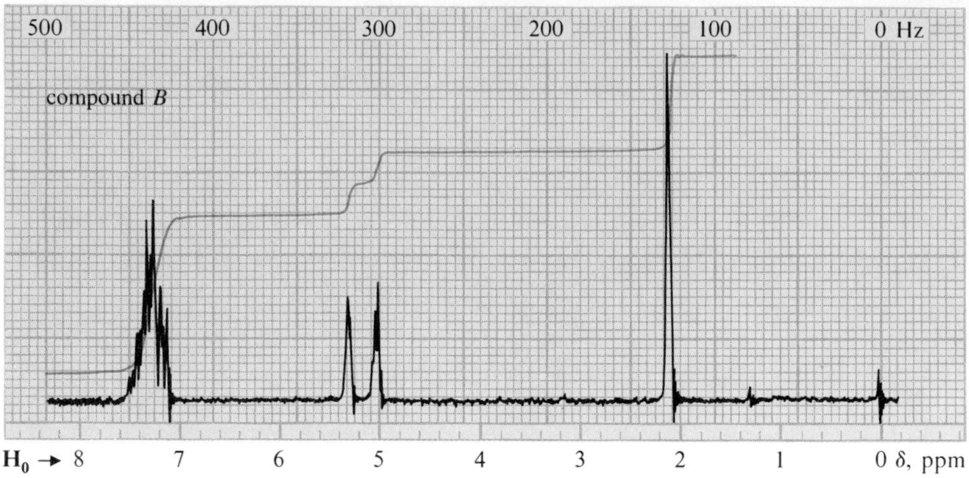

(c) + HNO$_3$ $\xrightarrow[0°]{\text{H}_2\text{SO}_4}$

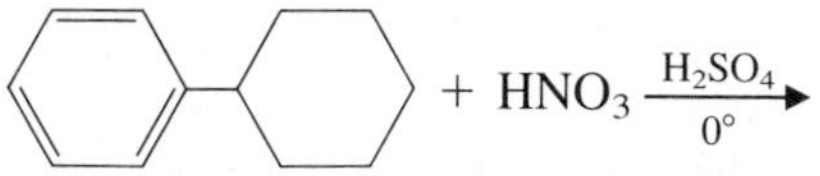

(d) ferrocene (Sec. 15.7D) + CH$_3$—C—Cl $\xrightarrow{\text{AlCl}_3}$ (C$_{12}$H$_{12}$OFe)

(e) OCH$_3$... $\xrightarrow{\text{HNO}_3}$ $\xrightarrow{\text{Br}_2, \text{Fe}}$ SO$_3$H

(f) —O—CH$_2$CH$_2$CH—Cl $\xrightarrow{\text{AlCl}_3}$ (a compound with ten carbons)
CH$_3$

(g) naphthalene + Cl—C—C—C—Cl $\xrightarrow{\text{AlCl}_3}$ (3 products, all isomers with formula C$_{15}$H$_{12}$O$_2$)
H$_3$C CH$_3$
α,α-dimethylmalonyl dichloride

42 An optically active compound A (C$_9$H$_{11}$Br) reacts with sodium ethoxide in ethanol to give an optically inactive hydrocarbon B (NMR spectrum in Fig. 16.12). Compound B undergoes hydrogenation over a Pd/C catalyst at room temperature to give a compound C. Give the structures of A, B, and C.

43 It is found experimentally that the heats of hydrogenation (Sec. 5.7B) of all alkenes with the same amount of branching at the double bond are nearly equal. Thus if benzene were an ordinary alkene, it should have about the same heat of hydrogenation as three cyclohexenes.

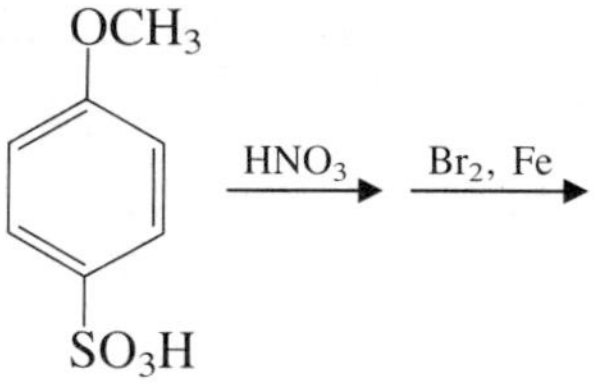

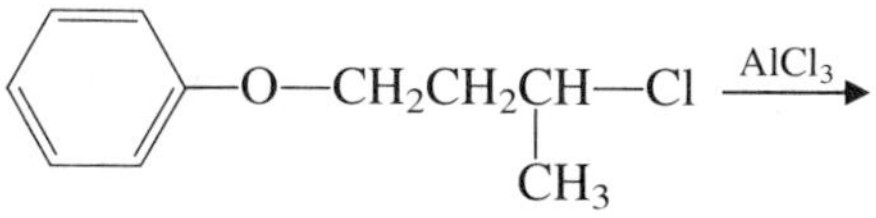

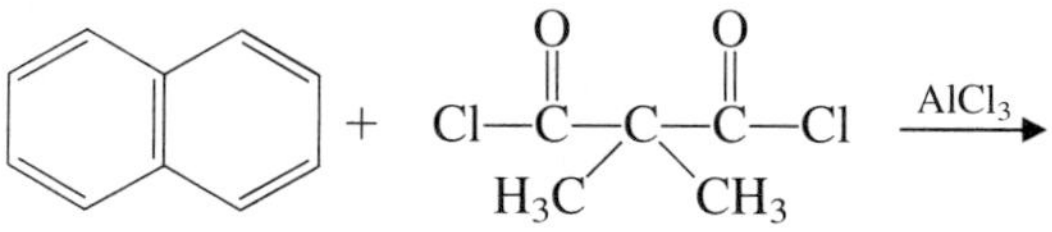

(a) Using the data in Eqs. 16.45a–c, calculate the heat liberated when benzene is hydrogenated to cyclohexane.
(b) Calculate the heat liberated in the hydrogenation of three cyclohexenes to cyclohexane.
(c) Calculate the discrepancy between the quantities calculated in (a) and (b). This number has been used as another estimate of the *empirical resonance energy* of benzene (Sec. 15.6C).

44 Outline a synthesis of each of the following compounds from anisole and any other reagents:

(a)

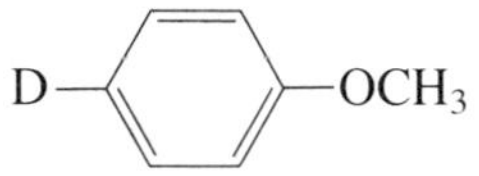

(*Hint:* Bromobenzene derivatives form Grignard reagents.)

(b)

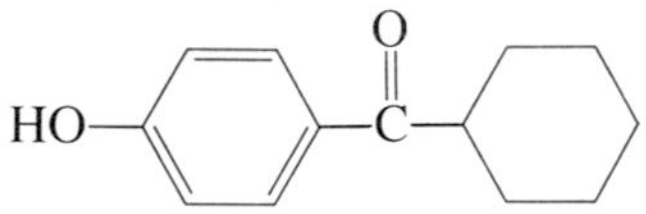

45 A method for determining the structures of disubstituted benzene derivatives was proposed in 1874 by Wilhelm Körner of the University of Milan. Körner had in hand three dibromobenzenes, *A*, *B*, and *C*, with melting points of 89°, 6.7°, and −6.5°, respectively. He nitrated each isomer in turn and meticulously isolated *all* of the mononitro derivatives of each. Compound *A* gave one mononitro derivative; compound *B* gave two mononitro derivatives; and compound *C* gave three. These experiments gave him enough information to assign the structures of *o*-, *m*-, and *p*-dibromobenzene.
(a) Assuming the correctness of the Kekulé structure for benzene, assign the structures of the dibromobenzene derivatives.
(b) Körner had no way of knowing whether the Kekulé or Ladenburg-benzene structure (Sec. 15.7A) was correct. Assuming the correctness of the Ladenburg-benzene structure, assign the structures of the dibromobenzene derivatives.
(c) It is a testament to Körner's experimental skill that he could isolate all the mononitration products. Of all the mononitration products that he isolated, which one(s) were formed in smallest amount? Explain.

46 Explain why
(a) the NMR spectrum of the sodium salt of cyclopentadiene consists of a singlet.

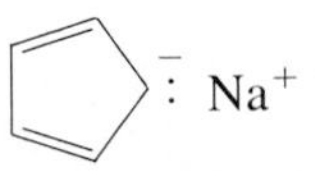

sodium salt of cyclopentadiene

(b) the methyl group in the following compound has an unusual chemical shift of δ (−1.67), about four ppm higher field than a typical allylic methyl group.

Problems (Cont.)

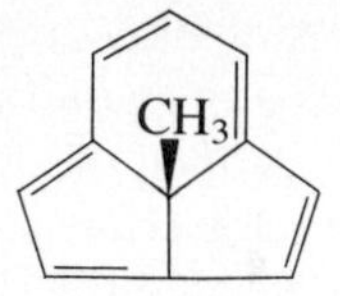

(c) the chemical shift of protons *a* in *p*-methoxytoluene is smaller than that of protons *b*, despite the fact that the electronegativity of oxygen is greater than that of the methyl group.

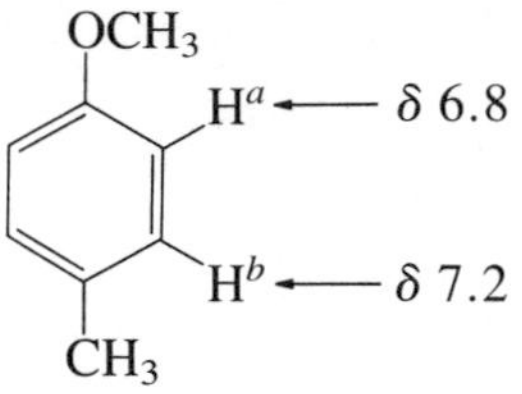

---

47 Given that anisole protonates primarily on oxygen in concentrated H_2SO_4, explain why 1,3,5-trimethoxybenzene protonates primarily on a carbon of the ring. Draw the structure of each conjugate acid.

---

48 Rank the following compounds in order of increasing reactivity in bromination. In each case, indicate whether the principal monobromination products will be the *o,p*-isomers or the *m*-isomer, and whether the compound will be more or less reactive than benzene. Explain carefully the points that cause any uncertainty.

(a) 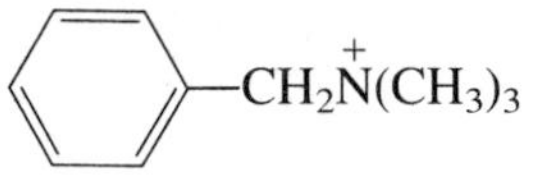

(c) 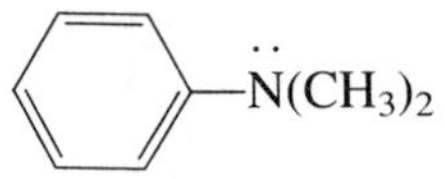

(b) 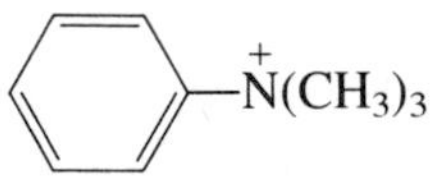

(d) 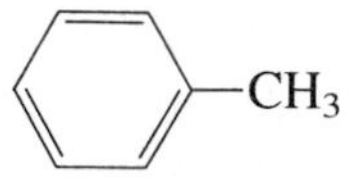

---

49 Diphenylsulfone is a by-product from the sulfonation of benzene. Explain how it is formed.

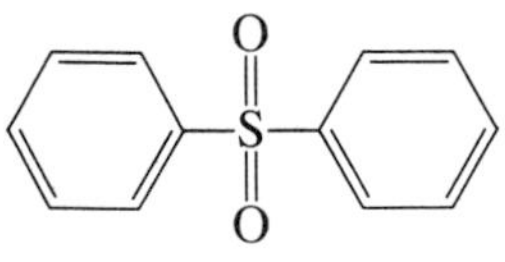

diphenylsulfone

---

50 (a) By drawing resonance structures for the carbocation intermediates, show that nitration of naphthalene should take place at carbon-1 and not carbon-2.

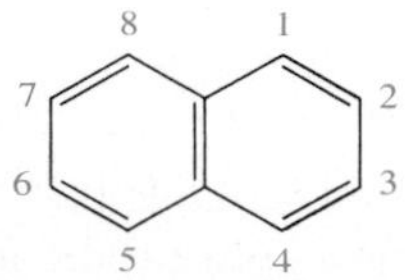

naphthalene

(b) Would a methoxy substituent at carbon-1 of naphthalene be expected to accelerate or retard nitration at carbon-5? Explain.

---

51 Using the arrow formalism when appropriate, give a detailed mechanism for each of the following reactions:

(a)

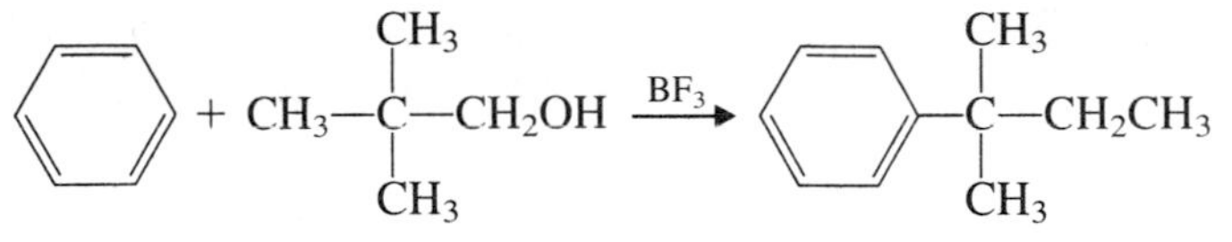

(b)

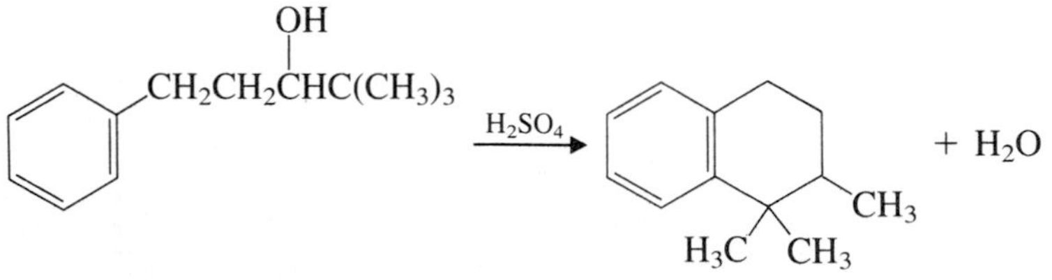

(*Hint:* See Problem 36.)

(c)

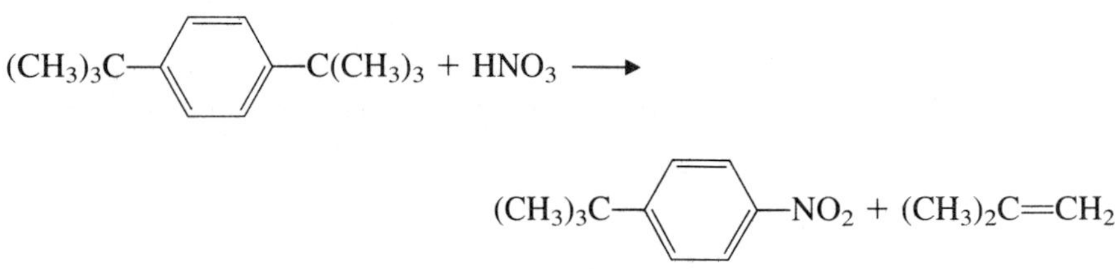

(*Hint:* Modify step 3 of the usual aromatic substitution mechanism in Sec. 16.4B.)

(d)

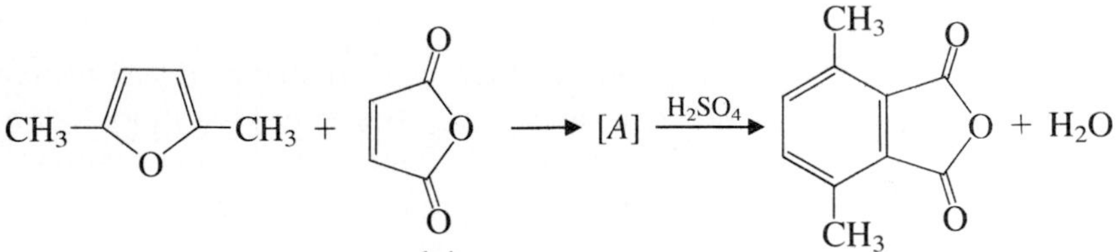

maleic anhydride

Your mechanism should include the identity of compound *A*.

---

52 Identify each of the following compounds:

(a) Compound *A*: IR 1605 cm^{-1}, no O—H stretch
NMR: δ 3.72 (3*H*, s); δ 6.72 (2*H*, apparent doublet, $J = 9$ Hz); δ 7.15 (2*H*, apparent doublet, $J = 9$ Hz).
Mass spectrum in Fig. 16.13a. (*Hint:* Notice the $p + 2$ peak.)

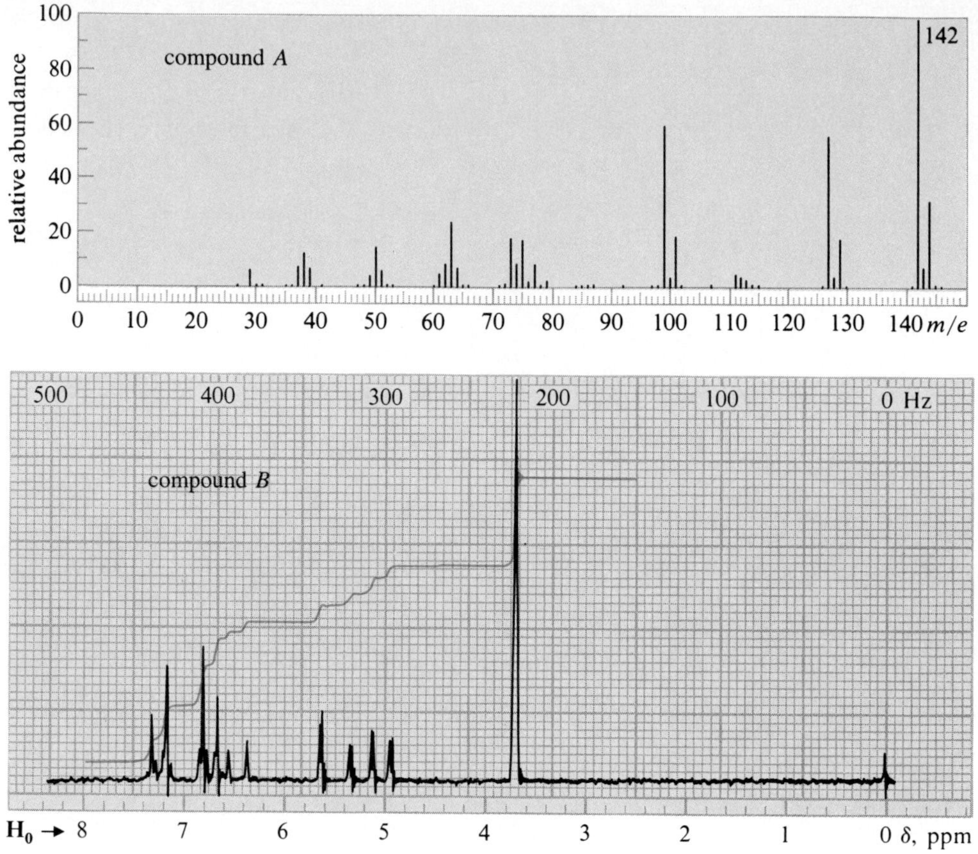

Figure 16.13 Spectra for Problem 52.

(b) Compound *B*: IR 1600, 1640 cm^{-1}, no O—H stretch
UV: λ_{max} = 207 nm (ϵ = 20,400), 258 nm (ϵ = 19,000), and 291 nm (ϵ = 4450).
NMR spectrum in Fig. 16.13b.

53 Explain why the dipole moment of *p*-chloronitrobenzene (2.69 D) is *less* than that of nitrobenzene (3.99 D), and the dipole moment of *p*-nitroanisole (4.92 D) is *greater* than that of nitrobenzene, even though the electronegativities of chlorine and oxygen are about the same.

54 At 36° the NMR signals for the ring methyl groups of "isopropylmesitylene" (protons *a* and *b* in the following structure) are two singlets at δ 2.25 and δ 2.13 with a 2 : 1 intensity ratio. When the spectrum is taken at −60°, however, it shows three signals of equal intensity for these groups at δ 2.25, δ 2.17, and δ 2.11. Explain these results.

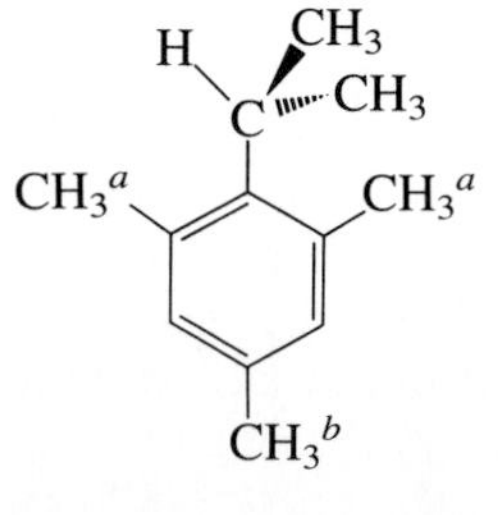

"isopropylmesitylene"

17

Allylic and Benzylic Reactivity

An **allylic group** is a group on a carbon adjacent to a double bond. A **benzylic group** is a group on a carbon adjacent to a benzene ring.

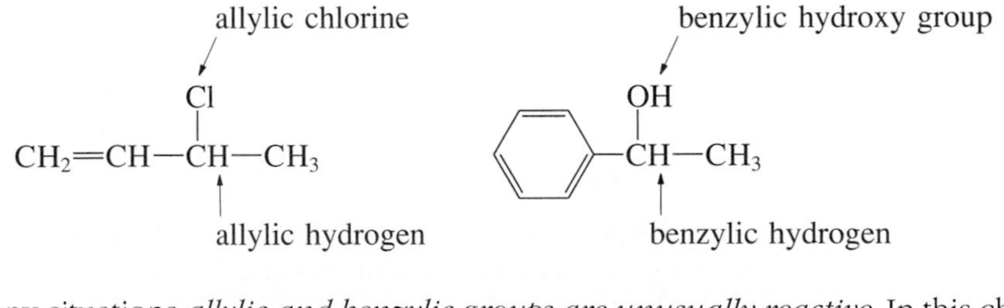

In many situations *allylic and benzylic groups are unusually reactive.* In this chapter we are going to examine what happens when some familiar reactions occur at allylic and benzylic positions, and we shall learn the reasons for allylic and benzylic reactivity. In Sec. 17.6 we shall also discover that allylic reactivity is brought to bear in some important chemistry that occurs in nature.

TABLE 17.1 Comparison of S_N1 Solvolysis Rates of Allylic and Nonallylic Alkyl Halides

$$R{-}Cl + C_2H_5OH + H_2O \xrightarrow{44.6°} R{-}OC_2H_5 + R{-}OH + HCl \text{ (50\% aqueous ethanol)}$$

R—Cl	Relative rate
$CH_2{=}CH{-}C(CH_3)_2{-}Cl$	162
$(CH_3)_2C{=}CH{-}CH_2{-}Cl$	38
$CH_3CH_2{-}C(CH_3)_2{-}Cl$	(1.00)
$(CH_3)_2CH{-}CH_2{-}CH_2{-}Cl$	0.00002

17.1 ALLYLIC AND BENZYLIC CARBOCATIONS. S_N1 REACTIONS

In Sec. 15.4A we learned that allylic carbocations are resonance stabilized.

$$\left[CH_2{=}CH{-}\overset{+}{C}H_2 \longleftrightarrow \overset{+}{C}H_2{-}CH{=}CH_2\right] \quad (17.1)$$

resonance structures of the allyl cation

These resonance structures symbolize the delocalization of electrons and positive charge by the overlap of p orbitals, as shown in Fig. 15.7.

Benzylic carbocations are also resonance stabilized.

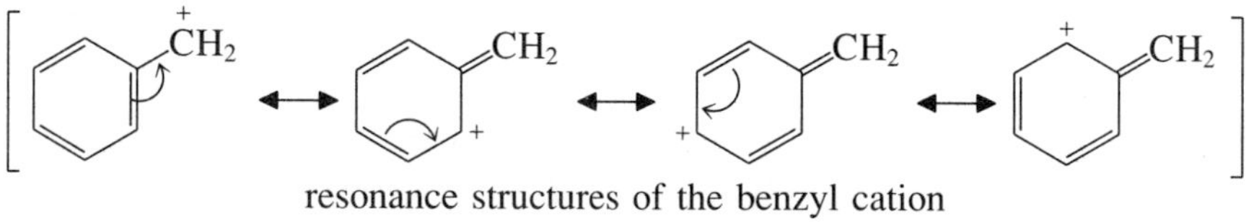
(17.2)

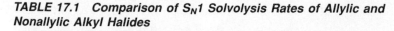

resonance structures of the benzyl cation

The charge on a benzylic carbocation is shared not only by the benzylic carbon, but also by alternate carbons of the ring.

The structures and stabilities of allylic and benzylic carbocations have important consequences for reactions in which they are involved. Consider, for example, the S_N1 reaction. *Allylic and benzylic alkyl halides are considerably more reactive in S_N1 reactions than their comparably substituted nonallylic or nonbenzylic counterparts.* Some representative cases are shown in Table 17.1. The tertiary allylic alkyl halide in

TABLE 17.2 Comparison of S$_N$1 Solvolysis Rates of Benzylic and Nonbenzylic Alkyl Halides

$$\text{R—Cl} + \text{H}_2\text{O} \xrightarrow{25°} \text{R—OH} + \text{HCl} \text{ (90\% aqueous acetone)}$$

Compound	Common name	Relative rate
(CH$_3$)$_3$C—Cl	*t*-butyl chloride	1.00
Ph—CH—Cl | CH$_3$	*α*-phenethyl chloride	1.0
CH$_3$ | Ph—C—Cl | CH$_3$	*t*-cumyl chloride	620
Ph$_2$CH—Cl	benzhydryl chloride	200*
Ph$_3$C—Cl	trityl chloride	>600,000

* In 80% aqueous ethanol.

entry 1 is more than one hundred times as reactive as the tertiary nonallylic compound in entry 3. Similar data for the solvolysis of some benzylic halides is given in Table 17.2. Comparing entries 1 and 3 of this table—both tertiary alkyl halides—we see that *t*-cumyl chloride, the benzylic compound, is more than six hundred times as reactive as *t*-butyl chloride.

The greater reactivity of allylic and benzylic halides is due to resonance stabilization of the carbocation intermediates that are formed when they react. For example, *t*-cumyl chloride ionizes to a carbocation with four important resonance structures:

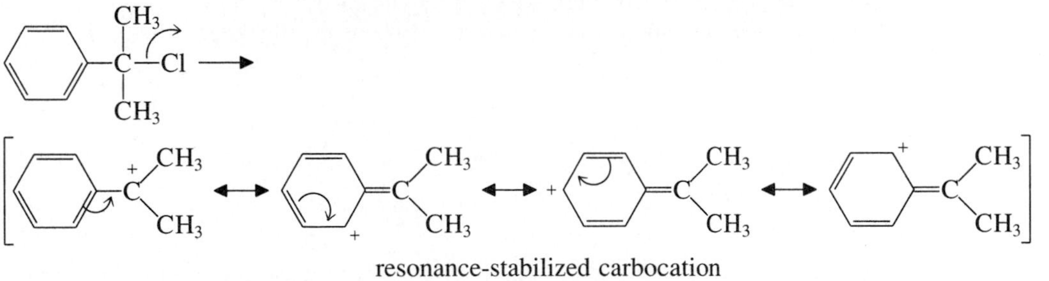

resonance-stabilized carbocation

(17.3)

Ionization of *t*-butyl chloride, on the other hand, gives the *t*-butyl cation, a carbocation with only one important contributing structure.

$$\text{CH}_3\text{—}\overset{\displaystyle \text{CH}_3}{\underset{\displaystyle \text{CH}_3}{\text{C}}}\text{—}\overset{..}{\underset{..}{\text{Cl}}}: \longrightarrow \text{CH}_3\text{—}\overset{\displaystyle \text{CH}_3}{\underset{\displaystyle \text{CH}_3}{\overset{+}{\text{C}}}} \quad :\overset{..}{\underset{..}{\text{Cl}}}:^{-} \qquad (17.4)$$

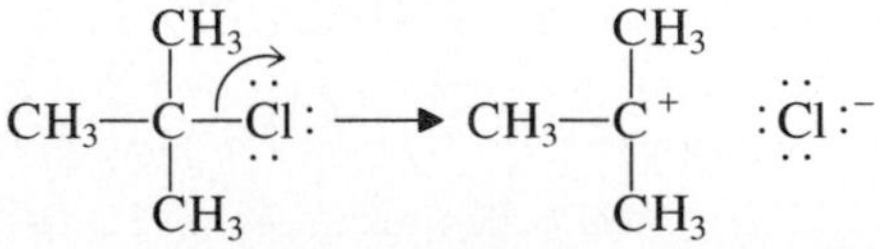

The benzylic cation is more stable relative to its alkyl halide starting material than the *t*-butyl cation is, and Hammond's postulate says that the more stable carbocation should be formed more rapidly. A similar analysis explains the reactivity of allylic alkyl halides.

Because of the possibility of resonance, *ortho* and *para* substituent groups on the benzene ring that activate electrophilic aromatic substitution accelerate S_N1 reactions at the benzylic position even further:

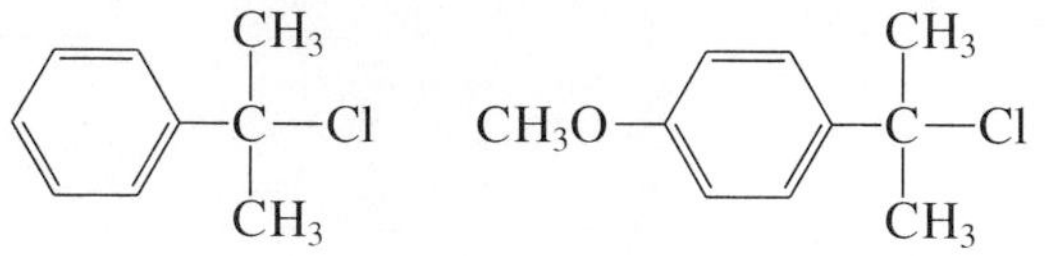

(17.5)

relative solvolysis rates (90% aqueous acetone, 25°): 1 | 3,400

The carbocation derived from the ionization of the *p*-methoxy derivative above not only has the same type of resonance structures as the unsubstituted compound, shown in Eq. 17.3, but also an additional structure (color) in which charge can be delocalized onto the substituent group itself:

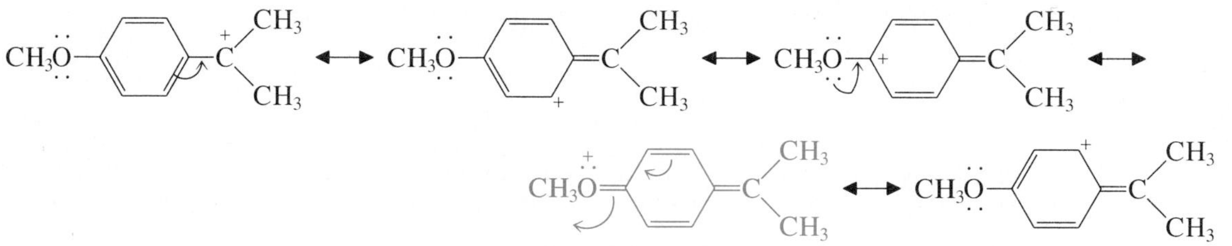

charge shared by oxygen

(17.6)

Other reactions that involve carbocation intermediates are accelerated when the carbocations are allylic or benzylic. Thus, dehydration of alcohols (Sec. 10.1A) and reaction of alcohols with hydrogen halides (Sec. 10.2) are also faster when the alcohol is allylic or benzylic. Similarly, in the addition of hydrogen halides to conjugated dienes, protonation of the diene gives an allylic carbocation rather than its nonallylic isomer because the allylic cation is formed more rapidly (Sec. 15.4A).

Problems

1 Suggest a reason why trityl chloride is much more reactive than the other alkyl halides in Table 17.2.

2 Predict the order of relative reactivities of the compounds below in an S_N1 solvolysis reaction, and explain your answer carefully.

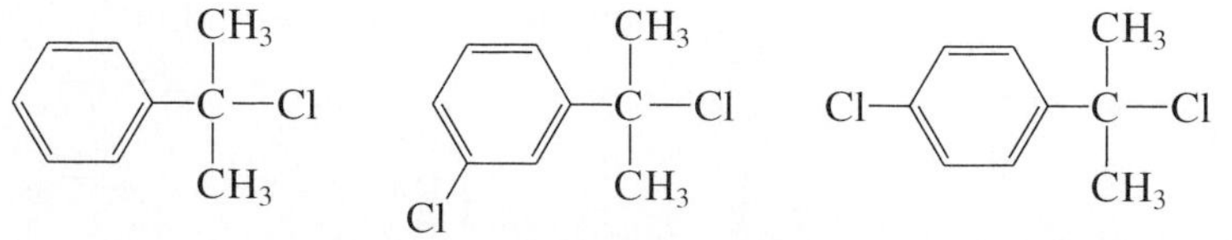

In an allylic cation, charge is shared between two carbons. Thus, two products are formed in the following S_N1 reaction because water can attack either of the two electron-deficient carbons in the allylic cation intermediate.

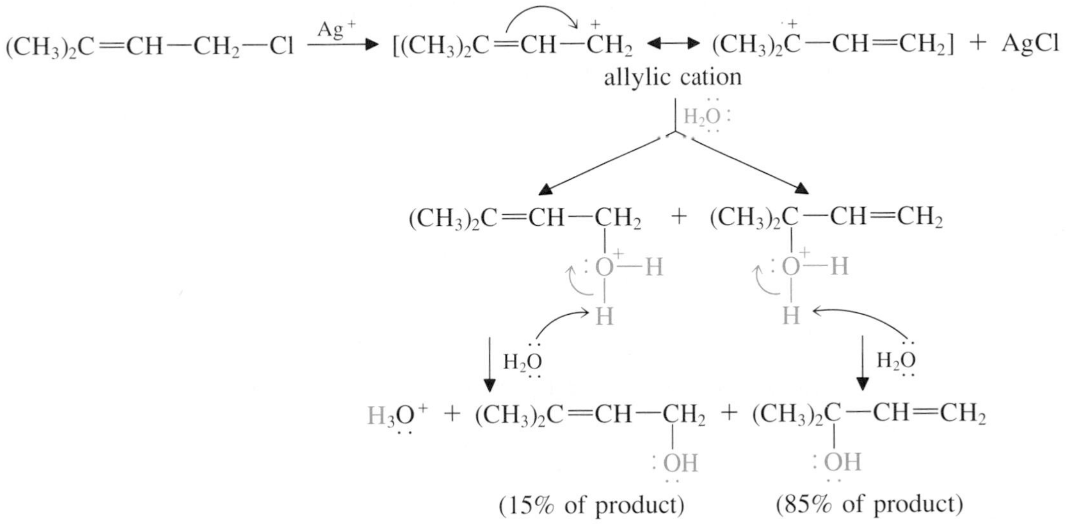

(17.7)

The two products are derived from *one* allylic cation that has two resonance forms. We learned in Sec. 15.4A that similar reasoning explains why a mixture of products (1,2- and 1,4-addition products) is obtained in the reactions of hydrogen halides with conjugated alkenes.

We might expect several substitution products in the S_N1 reactions of benzylic alkyl halides for the same reason.

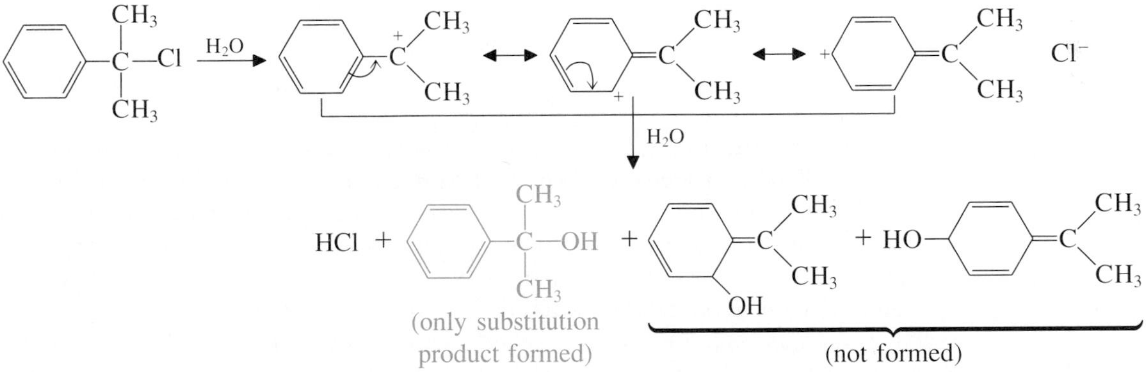

(17.8)

But as Eq. 17.8 shows, the products derived from attack of water on the ring are not formed. The reason is that these products are not aromatic, and thus lack the stability associated with the aromatic ring. Aromaticity is such an important stabilizing factor that only the aromatic product (color) is formed.

Problem

3 Give the structure of an isomer of the allylic halide in Eq. 17.7 that would react with water in an S_N1 reaction to give the same two products. Explain your reasoning.

17.2 ALLYLIC AND BENZYLIC RADICALS. ALLYLIC AND BENZYLIC BROMINATION

An **allylic radical** has an unpaired electron at an allylic position. Allylic radicals are resonance-stabilized, and are more stable than comparably substituted nonallylic radicals.

$$\left[CH_2=CH\!-\!\overset{\cdot}{C}H_2 \longleftrightarrow \overset{\cdot}{C}H_2\!-\!CH=CH_2 \right] \tag{17.9}$$

resonance structures of the allyl radical

Similarly, a **benzylic radical,** which has an unpaired electron at a benzylic position, is also resonance stabilized.

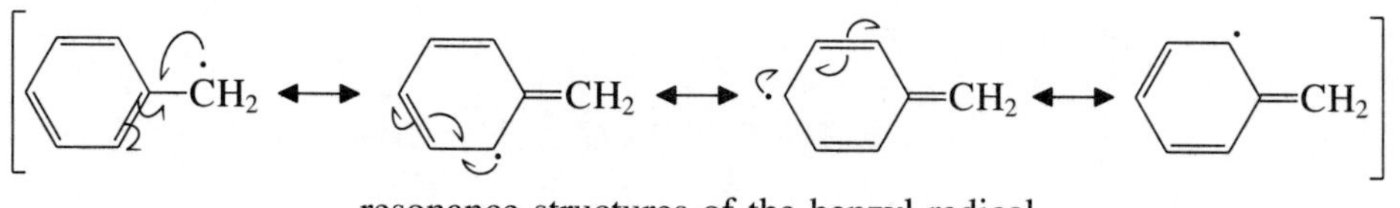

$$(17.10)$$

resonance structures of the benzyl radical

These resonance structures symbolize the delocalization (sharing) of the unpaired electron that results from overlap of p orbitals.

The enhanced stabilities of allylic and benzylic radicals can be experimentally demonstrated with bond dissociation energies. Let us compare the bond dissociation energies of the two types of $-CH_3$ hydrogens in 2-pentene:

$$H^a\!-\!CH_2\!-\!CH=CH\!-\!CH_2\!-\!CH_2\!-\!H^b$$

+87 kcal/mol +98 kcal/mol (17.11)

$$H\cdot + \cdot CH_2\!-\!CH=CH\!-\!CH_2\!-\!CH_3 \qquad\qquad CH_3\!-\!CH=CH\!-\!CH_2\!-\!\overset{\cdot}{C}H_2 + H\cdot$$

allylic radical alkyl radical

One set of hydrogens is allylic and the other is not. It takes 11 kcal/mol less energy to remove the allylic hydrogen than the nonallylic one. As Fig. 17.1 shows, the difference in bond dissociation energies is a direct measure of the relative energies of the two radicals. The allylic radical is stabilized by 11 kcal/mol relative to the nonallylic radical. A similar comparison suggests about the same relative stability for a benzylic radical.

Since allylic and benzylic radicals are especially stable, they are readily formed as reactive intermediates. One reaction that involves free-radical intermediates is the substitution of an allylic or benzylic hydrogen with a bromine or chlorine atom.

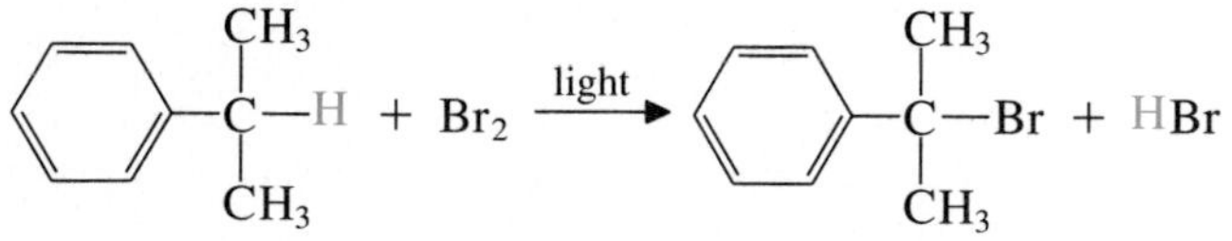

$$\text{Ph}\!\!\overset{CH_3}{\underset{CH_3}{\overset{|}{\underset{|}{C}}}}\!\!-\!H + Br_2 \xrightarrow{\text{light}} \text{Ph}\!\!\overset{CH_3}{\underset{CH_3}{\overset{|}{\underset{|}{C}}}}\!\!-\!Br + HBr \tag{17.12}$$

(nearly quantitative)

The initiation step in this reaction, dissociation of a bromine molecule into two bromine atoms, is promoted by heat or light:

Figure 17.1 Use of bond dissociation energies to determine energy relationships between allylic and nonallylic radicals.

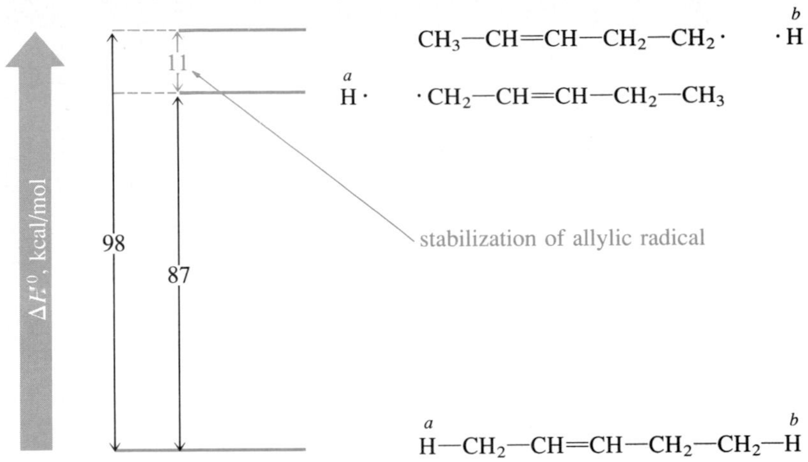

$$:\ddot{Br}-\ddot{Br}: \longrightarrow 2\ :\ddot{Br}\cdot \qquad (17.13a)$$

In the bromination reaction of cumene, shown in Eq. 17.12, a bromine atom abstracts the benzylic hydrogen in preference to the six nonbenzylic hydrogens.

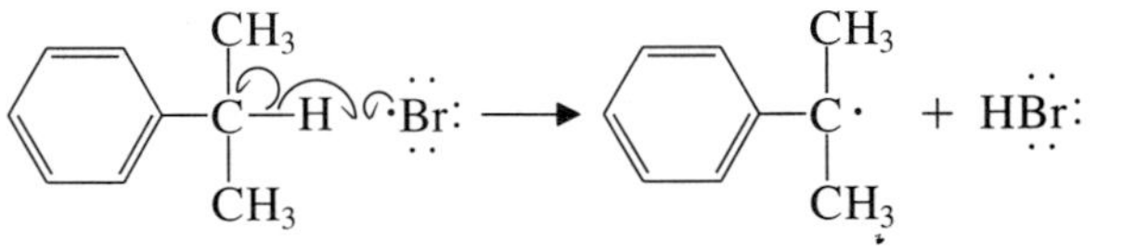
(17.13b)

The reason for this preference is the greater stability of the benzylic radical that is formed. The benzylic radical reacts with another molecule of bromine to generate a molecule of product as well as another bromine atom, which can react again in Eq. 17.13b.

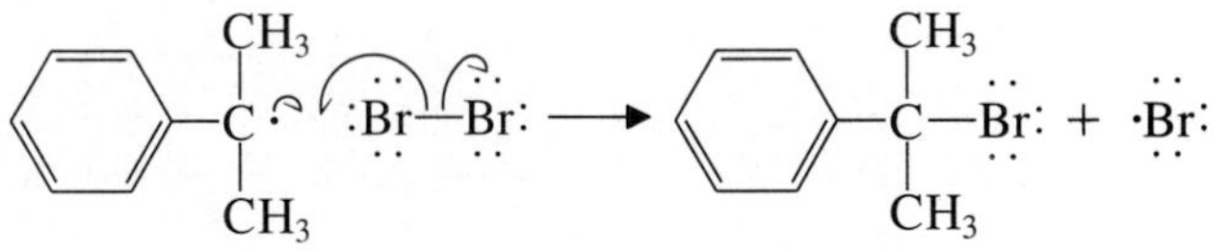
(17.13c)

Equations 17.13b and 17.13c together constitute the propagation steps of the reaction.

As we have already learned (Sec. 8.9A), free-radical halogenation is used to halogenate alkanes industrially. It is possible to replace just about any type of hydrogen with a halogen by using free-radical substitution. Since halogenation of most alkanes gives mixtures of products, this reaction is not very useful in the laboratory. (It can be used industrially because industry has developed elaborate and efficient fractional distillation methods that can separate mixtures of liquids.) But when a benzylic hydrogen is present, it reacts selectively, a single product is obtained, and the reaction can be used for the laboratory preparation of benzylic alkyl halides.

Because the allylic radical is also relatively stable, a similar substitution occurs preferentially at the allylic positions of an alkene. But there is a competing reaction in the case of an alkene that is not observed with benzylic substitution: addition of halogen to the alkene double bond.

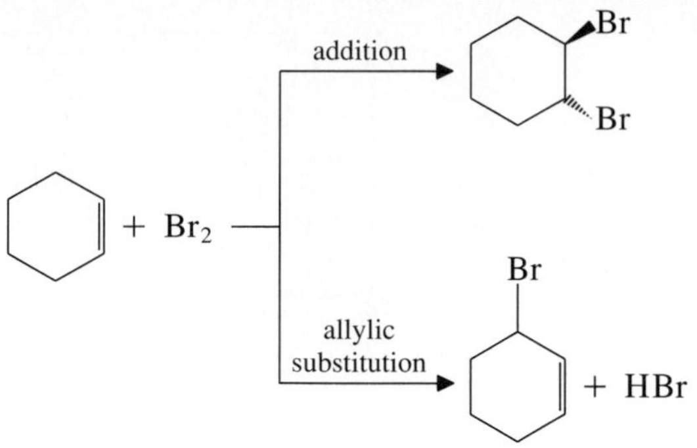

(17.14)

(Why isn't there an analogous competing reaction in Eq. 17.12?)

Can we promote one reaction over the other? The answer is yes—if we choose the reaction conditions carefully. We already know that free-radical substitution can be suppressed by avoiding conditions that promote most free-radical reactions: heat, light, or free-radical initiators. Thus, only addition occurs when an alkene reacts with bromine in the dark at room temperature. The substitution reaction occurs, however, in the presence of light and heat *at very low bromine concentrations.*

addition:

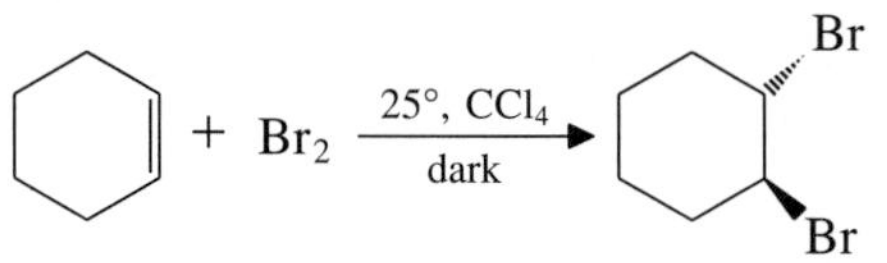

(17.15a)

substitution:

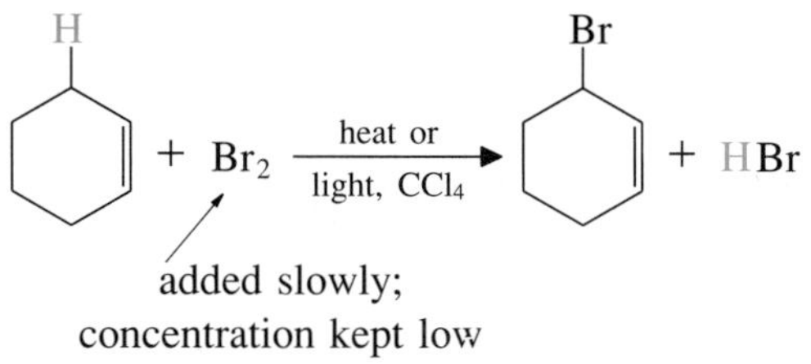

(17.15b)

It is not obvious why low bromine concentration should favor the free-radical substitution reaction. One reason is that the rate of bromine addition may be higher than first order in bromine.

$$\text{rate (ionic bromination)} = k[\text{alkene}][\text{Br}_2]^2 \quad (17.16)$$

When the bromine concentration is very small, then $[Br_2]^2$, and hence the rate of bromine addition, is negligible (the square of a small number is a very small number). When the addition reaction is slow enough, the free-radical substitution is observed instead.

Adding bromine to a reaction so slowly that it remains at very low concentration is experimentally inconvenient, but a very useful reagent can be used to accomplish the same objective: *N*-bromosuccinimide (often abbreviated NBS). When a compound that has allylic hydrogen atoms is treated with *N*-bromosuccinimide in CCl_4 under free-radical conditions (heat, light, and/or peroxides), allylic bromination takes place, and addition to the double bond is not observed.

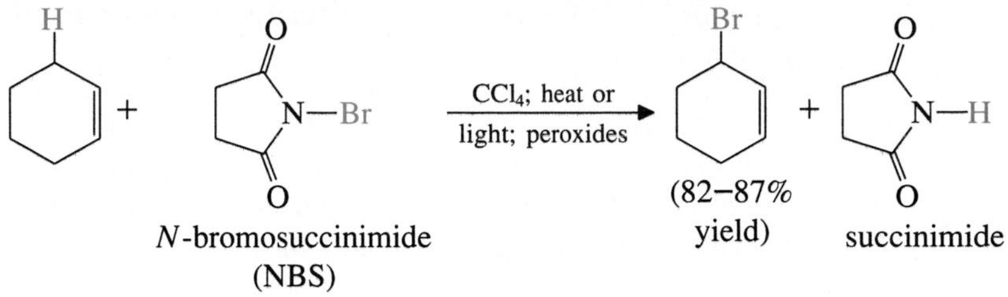

(17.17)

N-Bromosuccinimide can also be used to effect benzylic bromination.

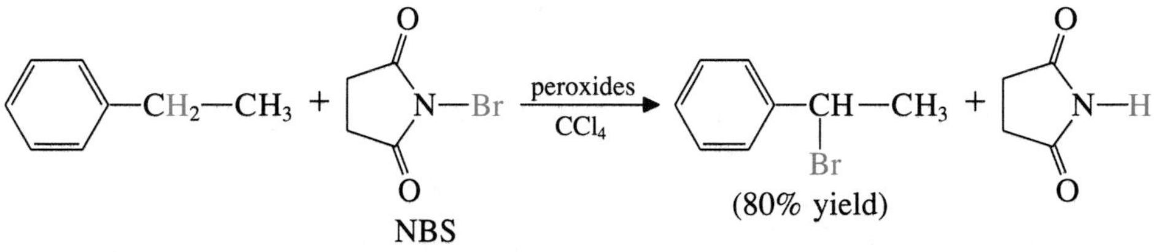

(17.18)

The initiation step in allylic and benzylic bromination with NBS is the formation of a bromine atom by the dissociation of the N—Br bond in NBS itself. The ensuing substitution reaction has three propagation steps. First, the bromine atom abstracts an allylic hydrogen from the alkene molecule:

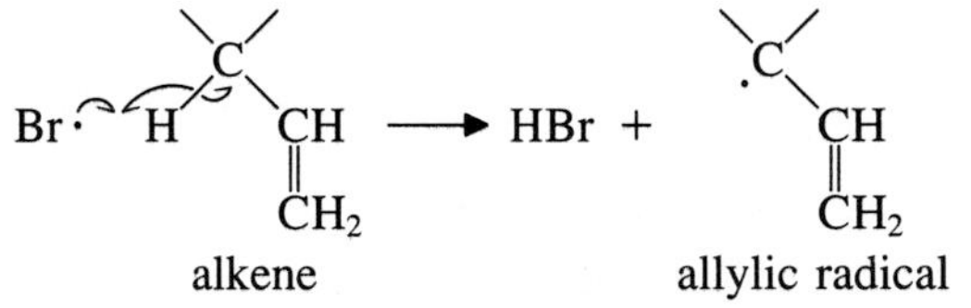

(17.19a)

The HBr thus formed reacts with the NBS in the second propagation step (by a mechanism that need not concern us here) to produce a Br_2 molecule.

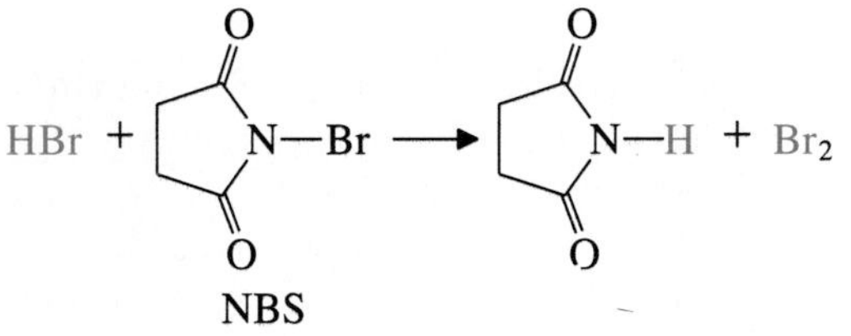

(17.19b)

The last propagation step is the reaction of this bromine molecule with the radical formed in Eq. 17.19a. A new bromine radical is produced that can begin the cycle anew.

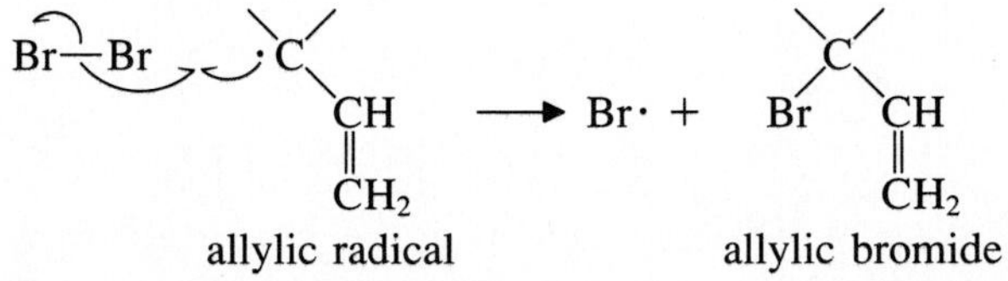

(17.19c)

The first and last propagation steps are identical to those for free-radical substitution with Br_2 itself (Eq. 17.13b,c). The unique role of NBS is to maintain the very low concentration of bromine by reacting with HBr in Eq. 17.19b. The Br_2 concentration remains low because it can be generated no faster than HBr and an allylic radical are generated in Eq. 17.19a. Thus every time a bromine molecule is formed, an allylic radical is also formed with which the bromine can react.

The low solubility of NBS in CCl_4 ($\leq 0.005M$) is crucial to the success of allylic bromination with NBS. When solvents that dissolve NBS are used, different reactions are observed! Hence CCl_4 *must* be used as the solvent in allylic or benzylic bromination with NBS. During the reaction, the insoluble NBS, which is more dense than CCl_4, disappears from the bottom of the flask and the less dense by-product succinimide (Eq. 17.17) forms a layer on the surface of the CCl_4. Equation 17.19b, and possibly other steps of the mechanism, occur at the surface of the insoluble NBS.

Because the unpaired electron of an allylic radical is shared between two carbons (Eq. 17.9), more than one product is obtained in some allylic brominations (Problem 4). This is not a problem in benzylic bromination reactions (why?).

Problems

4 Suggest a mechanism for the following reaction, and explain why more than one product are observed.

$$CH_3(CH_2)_4CH_2CH{=}CH_2 \xrightarrow[\text{peroxides}]{\text{NBS, CCl}_4} CH_3(CH_2)_4\underset{\displaystyle Br}{\underset{|}{C}}HCH{=}CH_2 + CH_3(CH_2)_4CH{=}CHCH_2Br$$

(*cis* and *trans*)

5 What product(s) are expected when each of the following compounds reacts with one equivalent of NBS in CCl_4 in the presence of light and/or peroxides?

(a) *trans*-2-pentene
(b) 3,3-dimethyl-1-cyclohexene
(c) *p*-cymene (1-isopropyl-4-methylbenzene)

17.3 ALLYLIC AND BENZYLIC GRIGNARD REAGENTS

Although allylic and benzylic Grignard reagents undergo the same reactions as other Grignard reagents, they also have some unique reactivity and properties that require special attention.

Allylic and benzylic Grignard reagents form easily, but an important side reaction occurs during their formation that does not occur with other Grignard reagents. In this reaction the Grignard reagent attacks the alkyl halide from which it is formed to give a coupling product.

$$CH_2{=}CH{-}CH_2{-}Cl + Mg \xrightarrow[25°,\ 5\ \text{hr}]{\text{ether}} CH_2{=}CH{-}CH_2{-}MgCl$$

(both combine) $\downarrow$ (17.20)

$$CH_2{=}CH{-}CH_2{-}CH_2{-}CH{=}CH_2 + MgCl_2$$

1,5-hexadiene
(biallyl; 55–65% yield)

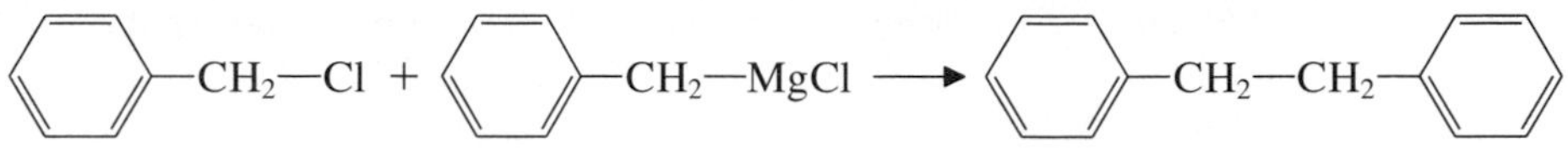

$$\langle \rangle-CH_2-Cl + \langle \rangle-CH_2-MgCl \longrightarrow \langle \rangle-CH_2-CH_2-\langle \rangle \quad (17.21)$$

1,2-diphenylethane
(bibenzyl; 68% yield)

As these examples show, the coupling reaction can be synthetically useful for preparing certain hydrocarbons. (Notice that it is another method for carbon–carbon bond formation.)

Why is this reaction unique to allylic and benzylic reagents? The answer lies in the nature of both the carbon–magnesium bond in the Grignard reagent and the reactivity of the allylic alkyl halide. As we learned in Sec. 8.8, Grignard reagents have many of the properties that we expect of *carbanions*. Allylic and benzylic Grignard reagents resemble allylic and benzylic carbanions, respectively.

$$\left[CH_2\!\!=\!\!CH\!\!-\!\!CH_2\!\!-\!\!MgBr \longleftrightarrow CH_2\!\!=\!\!CH\!\!-\!\!\overset{\bar{\cdot\cdot}}{C}H_2 \quad {}^+MgBr \right] \quad (17.22)$$

allyl anion

Allylic and benzylic anions are resonance-stabilized:

$$\left[CH_2\!\!=\!\!CH\!\!-\!\!\overset{\bar{\cdot\cdot}}{C}H_2 \longleftrightarrow \overset{\bar{\cdot\cdot}}{C}H_2\!\!-\!\!CH\!\!=\!\!CH_2 \right] \quad (17.23)$$

(Draw the resonance structures for the benzyl anion.) Both allylic and benzylic anions are about 14 kcal/mol more stable than comparably substituted, nonallylic, alkyl anions. A consequence of this greater stability is that the carbon–magnesium bonds of allylic and benzylic Grignard reagents have more ionic character than the carbon–magnesium bonds of other Grignard reagents. That is, in a Grignard reagent R—MgX, *there is more negative charge on R if it is allylic or benzylic than if it is an ordinary alkyl group.* As a result, allylic and benzylic Grignard reagents behave more like carbanions than ordinary alkylmagnesium halides. Now, one reaction of carbanions is reaction with alkyl halides in S_N2 reactions. (See, for example, Sec. 14.7B.) The coupling reaction of allylic and benzylic Grignard reagents is exactly this sort of reaction:

$$CH_2\!\!=\!\!CH\!\!-\!\!\overset{\bar{\cdot\cdot}}{C}H_2 \quad {}^+MgCl + Cl\!\!-\!\!CH_2\!\!-\!\!CH\!\!=\!\!CH_2 \longrightarrow$$

$$CH_2\!\!=\!\!CH\!\!-\!\!CH_2\!\!-\!\!CH_2\!\!-\!\!CH\!\!=\!\!CH_2 + MgCl_2 \quad (17.24)$$

An analogous coupling reaction occurs in the formation of allylic and benzylic lithium reagents.

The coupling reaction can be avoided if formation of the Grignard reagent is carried out in relatively dilute solution. (The reason has to do with the fact that the rates of the two competing reactions respond differently to changes in the alkyl halide concentration.) Once the allylic or benzylic Grignard reagent is formed, it can be used in the usual reactions of Grignard reagents.

Because of the relative stability of allylic anions, the carbon–magnesium bond of an allylic Grignard reagent dissociates readily. The magnesium moves rapidly back and forth between the two negatively charged carbon atoms at a rate of about 1000 times per second.

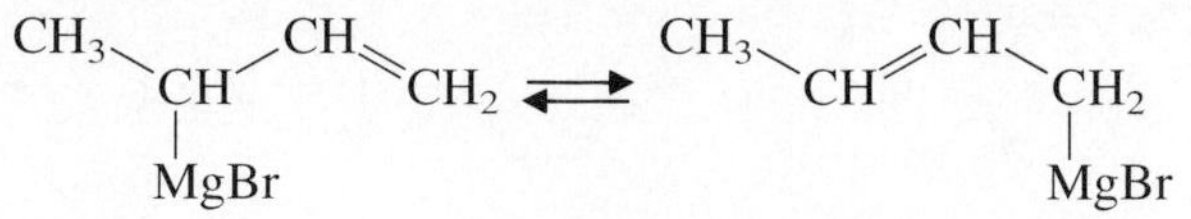

(17.25)

Thus, if the Grignard reagent is unsymmetrical, an allylic Grignard reagent is actually a rapidly equilibrating mixture of two different reagents. This means that the same mixture of reagents can be obtained from either of two allylically related alkyl halides.

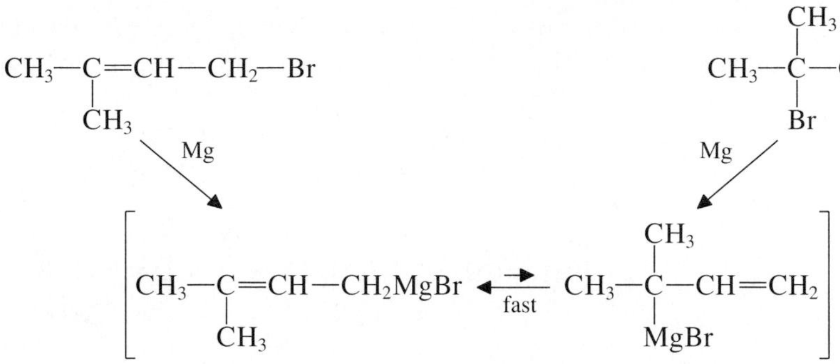

(17.26)

Reactions of the Grignard reagent obtained from either alkyl halide gives the same mixture of products. The following example illustrates this point for a protonolysis reaction:

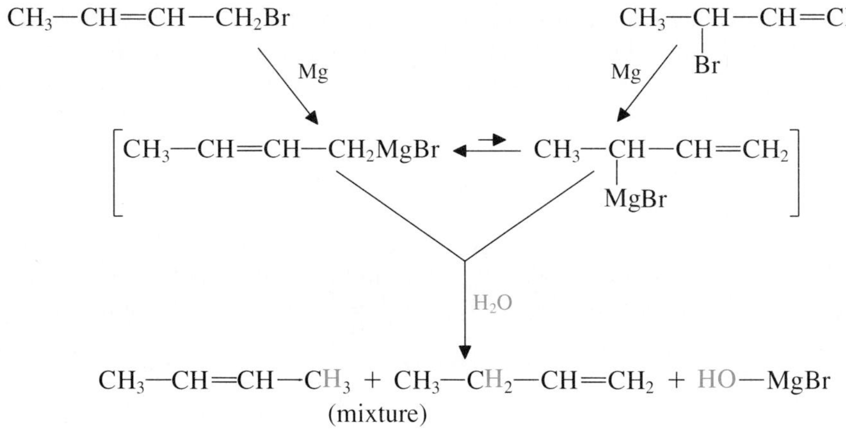

(17.27)

$$CH_3—CH=CH—CH_3 + CH_3—CH_2—CH=CH_2 + HO—MgBr$$
(mixture)

Problem

6 What products are formed when a Grignard reagent prepared from 1-(bromomethyl)-1-cyclohexene is treated with D_2O?

17.4 ALLYLIC AND BENZYLIC S_N2 REACTIONS

We have learned that allylic and benzylic cations, anions, and radicals are especially stable, and that reactions involving these species are relatively fast. S_N2 reactions of allylic and benzylic halides, although not involving any of these reactive intermediates, are also relatively fast. The following data for allyl chloride are typical:

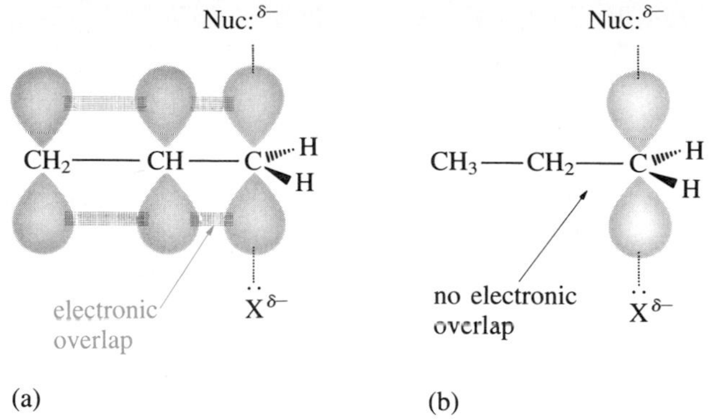

Figure 17.2 Transition states for S$_N$2 displacement reactions at (a) allylic and (b) non-allylic carbons. Nuc : $^-$ and X : $^-$ are the nucleophile and leaving group, respectively. The allylic displacement is faster because the transition state is stabilized by overlap of the p orbital at the site of substitution with the adjacent π-bond.

(a) (b)

$$CH_2{=}CH{-}CH_2{-}Cl + I^- \xrightarrow[50°]{\text{acetone}} CH_2{=}CH{-}CH_2{-}I + Cl^- \qquad\qquad 73 \qquad (17.28a)$$

relative rate

$$CH_3{-}CH_2{-}CH_2{-}Cl + I^- \xrightarrow[50°]{\text{acetone}} CH_3{-}CH_2{-}CH_2{-}I + Cl^- \qquad\qquad 1 \qquad (17.28b)$$

An even greater acceleration is observed for benzylic halides.

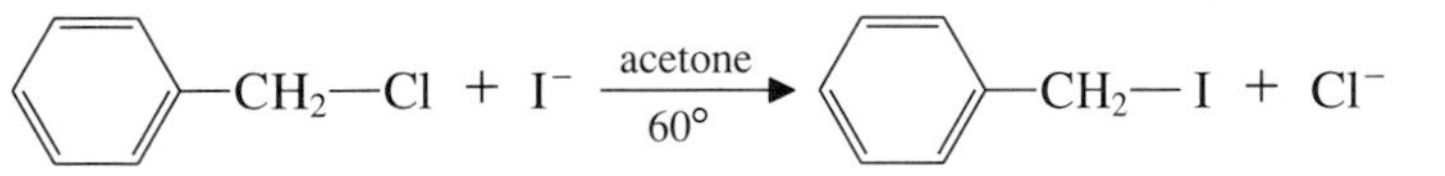

relative rate

$$\text{Ph}{-}CH_2{-}Cl + I^- \xrightarrow[60°]{\text{acetone}} \text{Ph}{-}CH_2{-}I + Cl^- \qquad \approx100{,}000 \quad (17.29a)$$

$$(CH_3)_2 CH{-}CH_2{-}Cl + I^- \xrightarrow[60°]{\text{acetone}} (CH_3)_2 CH{-}CH_2{-}I + Cl^- \qquad 1 \qquad (17.29b)$$

Allylic and benzylic S$_N$2 reactions are accelerated because the energies of their transition states are reduced by p-orbital overlap, shown in Fig. 17.2 for an allylic S$_N$2 reaction. We learned (Fig. 9.1) that in the transition state of the S$_N$2 reaction, the carbon at which substitution occurs is sp^2 hybridized; the incoming nucleophile and the departing leaving group are partially bonded to a p orbital on this carbon. Overlap of this p orbital with the p orbitals of an adjacent double bond or phenyl ring provides additional bonding in the transition state that lowers its energy and accelerates the reaction.

Problem

7 Explain how and why the product(s) would differ in the following reactions of *trans*-2-buten-1-ol:
(a) Reaction with concentrated aqueous HBr
(b) Conversion to the tosylate, then reaction with NaBr in acetone

17.5 SIDE-CHAIN OXIDATION OF ALKYLBENZENES

Treatment of alkylbenzene derivatives with strong oxidizing agents under vigorous conditions converts the alkyl side chain into a carboxylic acid group. Oxidants commonly used for this purpose are Cr^{6+} derivatives, such as $Na_2Cr_2O_7$ (sodium dichromate) or CrO_3; the Mn^{7+} reagent $KMnO_4$ (potassium permanganate); or O_2 and special catalysts, a procedure that is used industrially (Eq. 16.46).

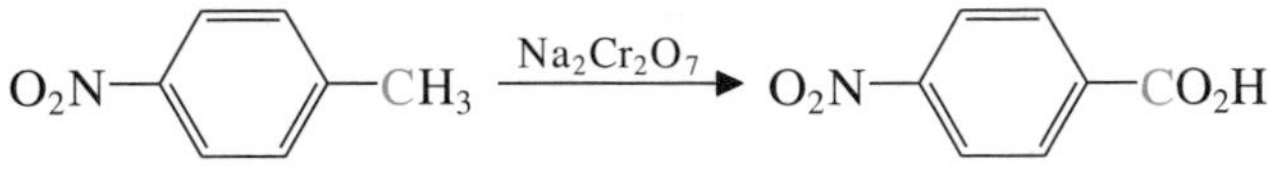

$$O_2N-\!\!\bigcirc\!\!-CH_3 \xrightarrow{Na_2Cr_2O_7} O_2N-\!\!\bigcirc\!\!-CO_2H \qquad (17.30)$$

$$\bigcirc\!\!-CH_2CH_2CH_2CH_3 \xrightarrow{KMnO_4} \bigcirc\!\!-CO_2H \qquad (17.31)$$
$$\text{benzoic acid}$$

$$\bigcirc\!\!-C_2H_5 \xrightarrow[\text{48 hr, 100°}]{CrO_3,\ 40\%\ H_2SO_4} \bigcirc\!\!-CO_2H \qquad (17.32)$$
$$\text{(80\% yield)}$$

Notice that the benzene ring is left intact, and the alkyl side-chain, *regardless of length,* is "burned away" to a single carboxylic acid group. This reaction is useful for preparing some carboxylic acids from alkylbenzenes.

We must recognize that the conditions for this side-chain oxidation are generally vigorous: heat, high concentrations of oxidant, or long reaction times. It is also possible to effect less extensive oxidations of side-chain groups. Thus 1-phenylethanol is readily oxidized to acetophenone under milder conditions—the normal oxidation of secondary alcohols to ketones (Sec. 10.6)—but it is converted into benzoic acid under more vigorous conditions.

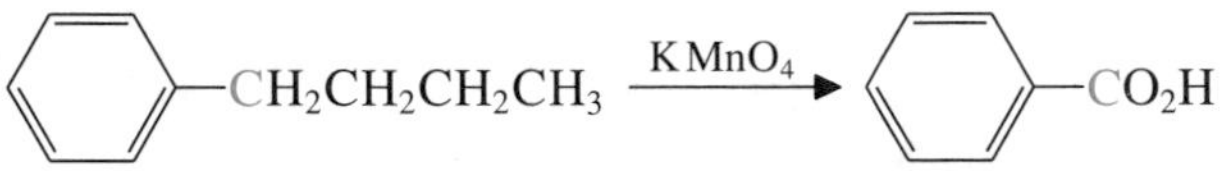

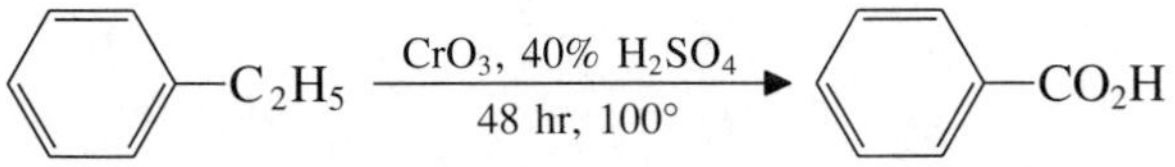

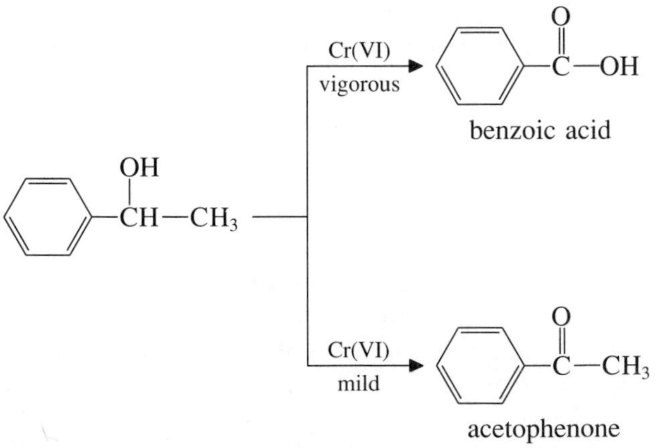

$$(17.33)$$

We do not need to concern ourselves with learning the exact conditions for these reactions; rather, it is important simply to be aware that it is usually possible to find appropriate conditions for each type of oxidation.

Although we shall not study the mechanism of this oxidation, it does require the presence of a benzylic hydrogen. Thus *t*-butylbenzene is resistant to side-chain oxidation.

Problems

8 Give the products of vigorous $KMnO_4$ oxidation of each of the following compounds:
(a) 1-butyl-4-*t*-butylbenzene
(b) *p*-nitrobenzyl bromide

9 Each of two compounds *A* and *B* has the formula C_8H_{10}. After vigorous oxidation, compound *A* yields phthalic acid; under the same conditions *B* yields benzoic acid (structure in Eq. 17.31). What are the structures of *A* and *B*?

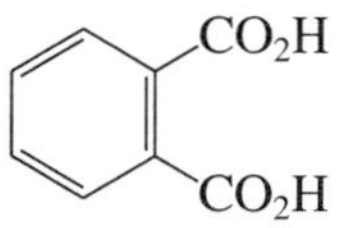

phthalic acid

17.6 THE ISOPRENE RULE. BIOSYNTHESIS OF TERPENES

A. Essential Oils and Terpenes

It has long been known that the flowers, leaves, and roots of many plants contain volatile, pleasant-smelling substances that have come to be called **essential oils.** Essential oils, particularly oil of turpentine, were known to the ancient Egyptians. However, not until early in the nineteenth century was an effort made to determine the chemical constitution of the essential oils. In 1818, it was found that the C : H ratio in oil of turpentine was 5 : 8. This same ratio was subsequently found for a wide variety of natural products. This collection of related natural products became known as the **terpenes,** a name coined by August Kekulé. The similarity in the atomic composition of the many terpenes led to the idea that they might possess some unifying structural element.

In 1887, Otto Wallach, a German chemist, pointed out the common structural feature of the terpenes: they all consist of repeating units that have the same carbon skeleton as the five-carbon diene isoprene. This generalization subsequently became known as the **isoprene rule.**

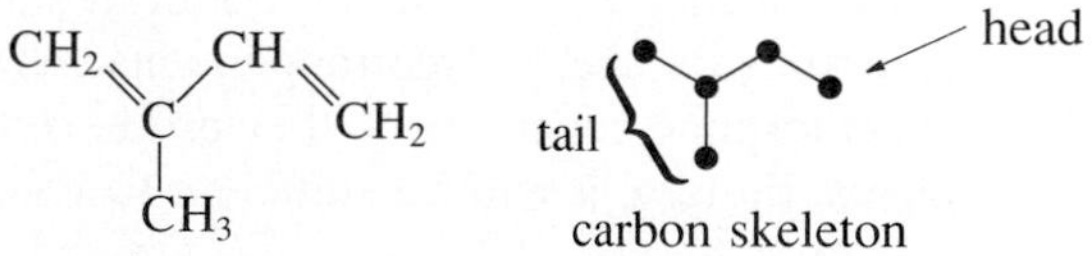

isoprene

carbon skeleton of isoprene

For example, citronellol (from oil of roses and other sources) incorporates two isoprene units:

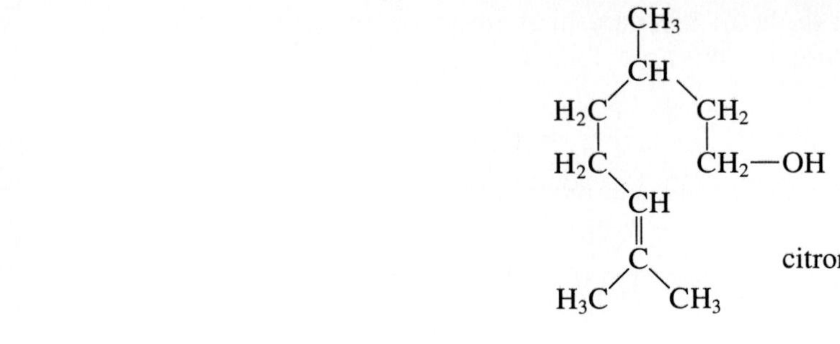

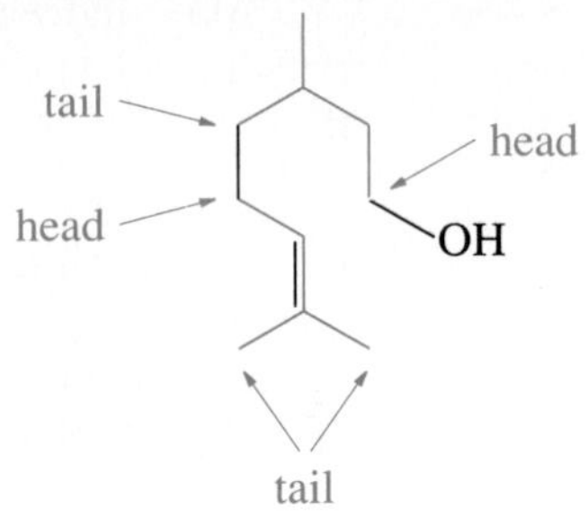

citronellol

(Because of this relationship to isoprene, terpenes are also called **isoprenoids.**) The basis of the terpene classification is the *connectivity of the carbon skeleton*. Different terpenes may have the same carbon skeleton functionalized in different ways: they may differ in the presence of double bonds, hydroxy groups, etc.

Because terpenes are assembled from five-carbon units, their carbon skeletons contain multiples of five carbon atoms (10, 15, 20, . . . , $5n$). Terpenes with ten carbon atoms in their carbon chains are classified as **monoterpenes,** those with fifteen carbons **sesquiterpenes,** those with twenty carbons **diterpenes,** and so on. Some examples of terpenes are given in Fig. 17.3. Many of these compounds are familiar natural flavorings or fragrances.

In this section we shall make frequent use of skeletal structures, which were explained in Sec. 3.3. Be especially careful to remember that there are carbon atoms at the ends of each structure as well as at each vertex.

In many terpenes, the isoprene units are connected in a "head-to-tail" arrangement.

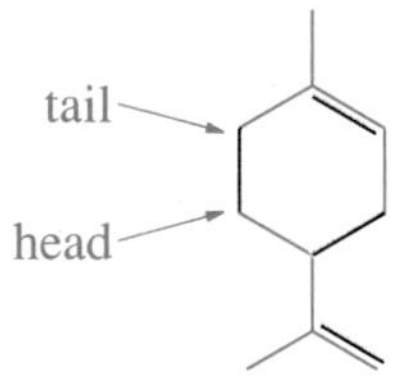

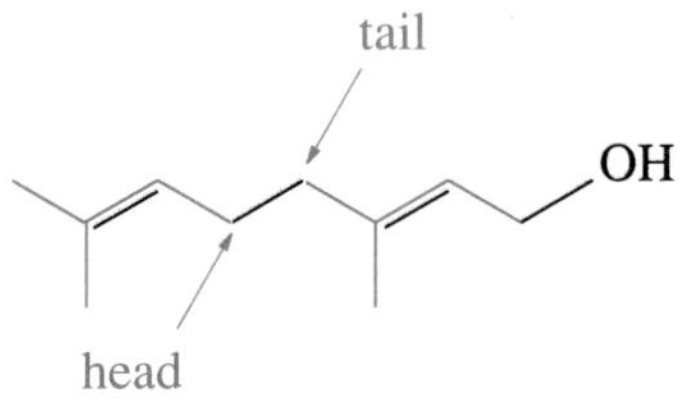

Because this arrangement is so common, Wallach assumed the generality of the head-to-tail connectivity in his original statement of the isoprene rule. However, many examples are now known in which the isoprene units have a "head-to-head" connectivity. Furthermore, some compounds are derived from the conventional terpene structures by skeletal rearrangements. Although these compounds do not have the exact terpene connectivity, they are nevertheless classified as terpenes. For our purposes, though, it will be sufficient to recognize terpenes by two criteria:

1. A multiple of five carbon atoms in the main carbon chain
2. The formal carbon connectivity of the isoprene carbon skeleton within each five-carbon unit

Figure 17.3 Examples of terpenes. In the monoterpenes, the isoprene units are shown in color.

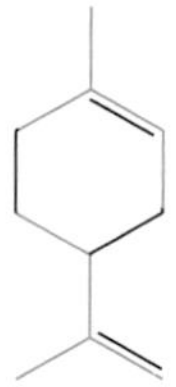

limonene
(oil of oranges and lemons)

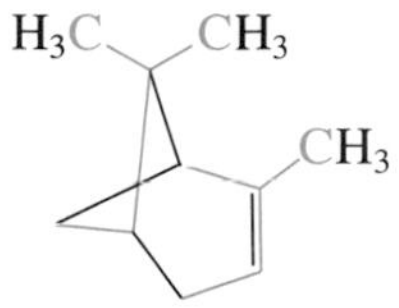

α-pinene
(ingredient of oil of turpentine

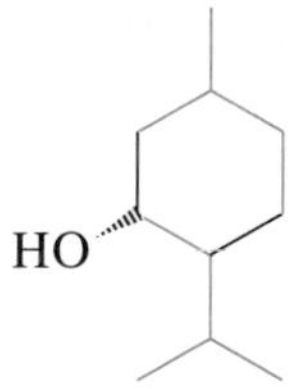

menthol
(oil of peppermint)

monoterpenes

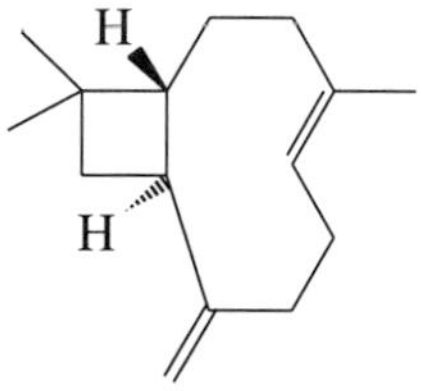

caryophyllene
(oil of cloves)
a sesquiterpene

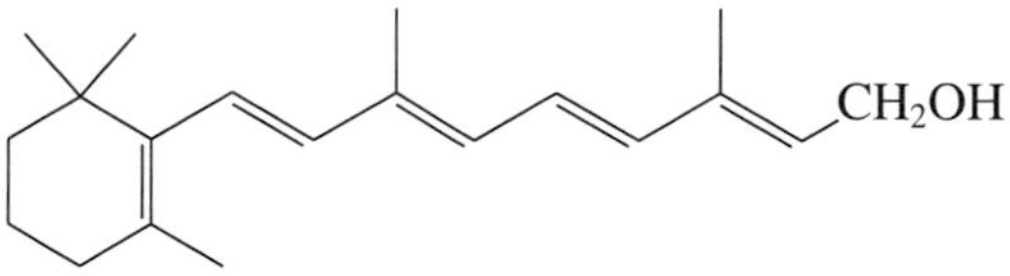

vitamin A
a diterpene

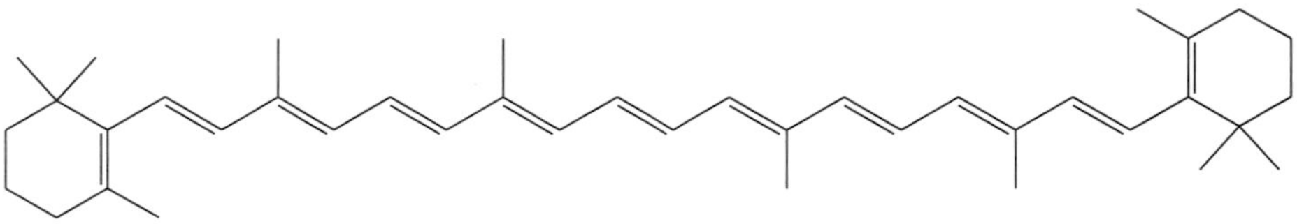

β-carotene (orange pigment in carrots; converted to vitamin A by human liver)

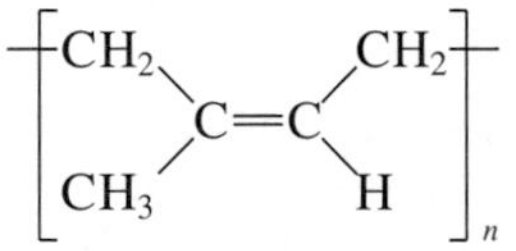

natural rubber (a polyterpene)

Problem

10 Show the isoprene units in the following compounds of Fig. 17.3:
(a) caryophyllene (b) vitamin A

B. Biosynthesis of Terpenes

How are terpenes synthesized in nature? What is responsible for the regular repetition of isoprene units? To answer this question, chemists have studied the biosynthesis of terpenes. **Biosynthesis** is the synthesis of chemical compounds by living organisms. The study of biosynthesis is an active area of research that lies at the interface of chemistry and biochemistry. A brief study of terpene biosynthesis is appropriate at this point, since it shows how nature takes advantage of allylic reactivity.

The repetitive isoprene unit in all terpenes has a common origin in two simple five-carbon compounds:

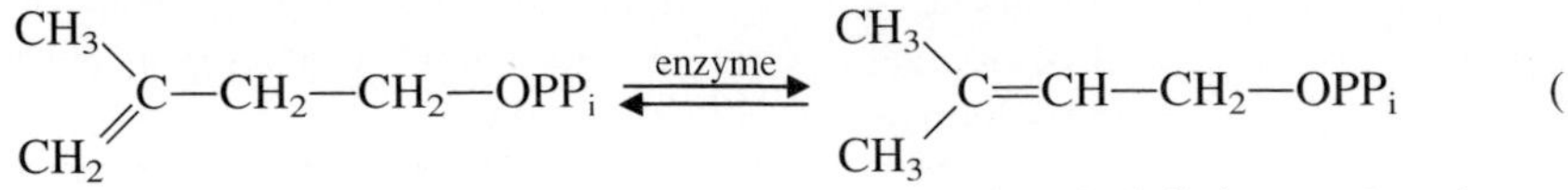

(17.34)

isopentenyl pyrophosphate (IPP) — γ,γ-dimethylallyl pyrophosphate (DMAP)

The $—OPP_i$ in these structures is an abbreviation for the *pyrophosphate group,* which, in nature, is usually found complexed to a metal ion like Mg^{2+} or Mn^{2+}.

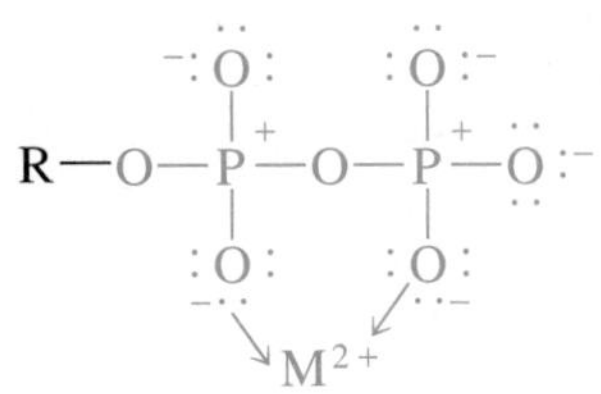

abbreviated $R—OPP_i$
$M^{2+} = Mn^{2+}$ or Mg^{2+}

a metal-complexed alkyl pyrophosphate

Alkyl pyrophosphates are esters of the inorganic acid *pyrophosphoric acid* (Sec. 10.3B).

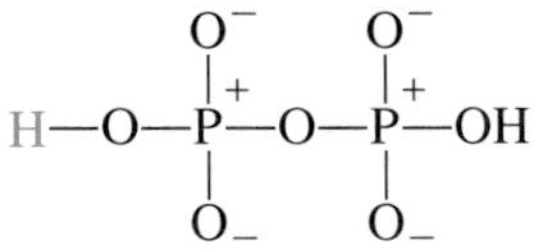

pyrophosphoric acid

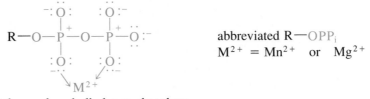

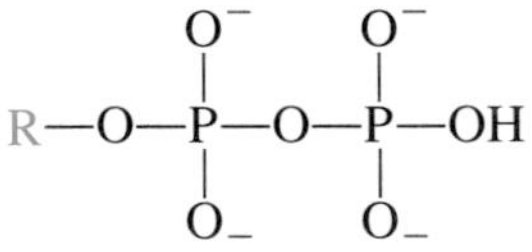

an alkyl pyrophosphate

Pyrophosphate and phosphate are "nature's leaving groups." Just as we use alkyl halides or alkyl tosylates in the laboratory as starting materials for nucleophilic substitution reactions, living organisms use alkyl pyrophosphates.

Because IPP and DMAP are readily interconverted in living systems by the reaction of Eq. 17.34, the presence of one ensures the presence of the other. Like all biochemical reactions, including the ones we are about to discuss, this reaction does not occur freely in solution, but is catalyzed by an *enzyme* (Secs. 4.8, 7.9). Roughly speaking, the role of the enzyme is to bind, or fix, reactants close to each other, and to provide acidic and basic groups (which we shall call BH^+ and B:, respectively) that carry out the proton transfers required in the reaction. The presence of enzyme catalysts does not alter the fact that the chemical reactions of living systems are reasonable and understandable in terms of other reactions with which we are familiar.

Let us illustrate the general pattern of terpene biosynthesis by examining how a simple monoterpene, *geraniol* (the fragrant compound in oil of geraniums), is assembled in nature. In the first step of geraniol biosynthesis, IPP and DMAP are bound to the enzyme *prenyl transferase.* The DMAP loses its pyrophosphate leaving group in an S_N1-like process.

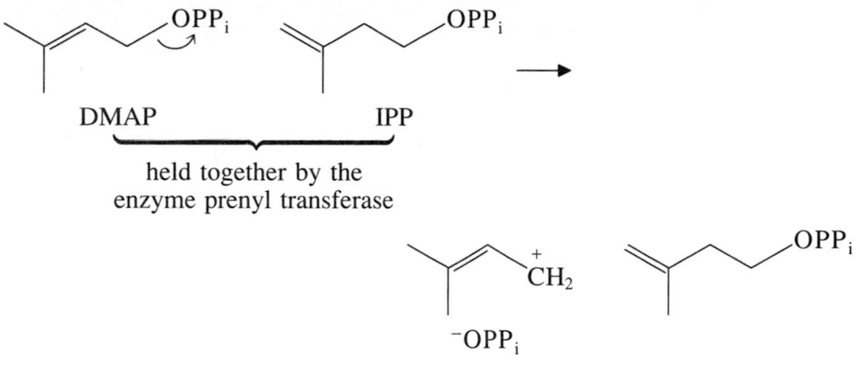

(17.35)

The carbocation formed in Eq. 17.35 is a relatively stable allylic cation (Sec. 17.1). Carbocations, like other electrophiles, can be attacked by the π-electrons of a double bond. (This same type of reaction is involved in Friedel–Crafts acylations and alkylations; see Secs. 16.7 and 16.8.) The reaction of this carbocation with the double bond of IPP gives a new carbocation. Loss of a proton from a β-carbon of this carbocation gives the monoterpene geranyl pyrophosphate.

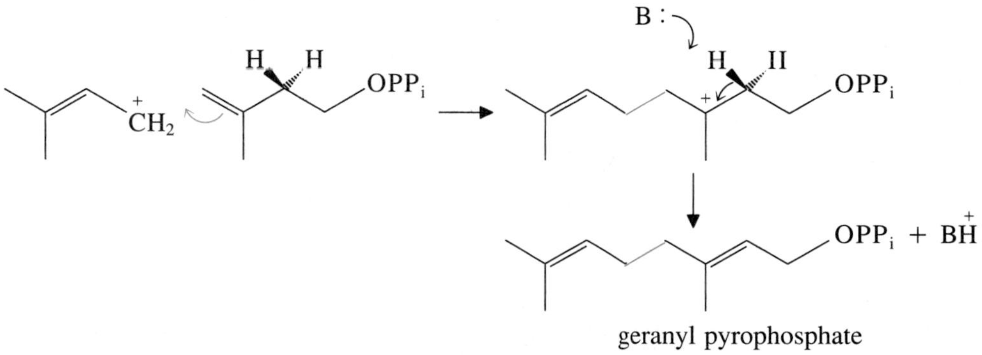

(17.36)

Geraniol is formed in a displacement of the pyrophosphate group by water.

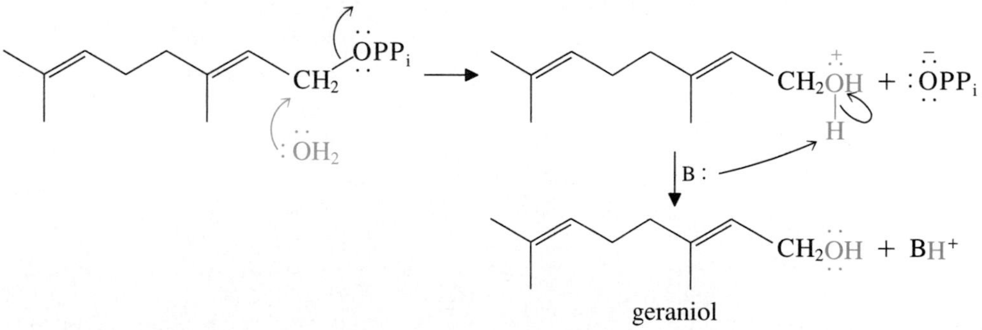

(17.37)

All head-to-tail terpenes are formed by reactions analogous to those shown above. For example, geranyl pyrophosphate, like DMAP, can itself react with IPP to give the sesquiterpene farnesyl pyrophosphate. This reaction is also catalyzed by the enzyme prenyl transferase.

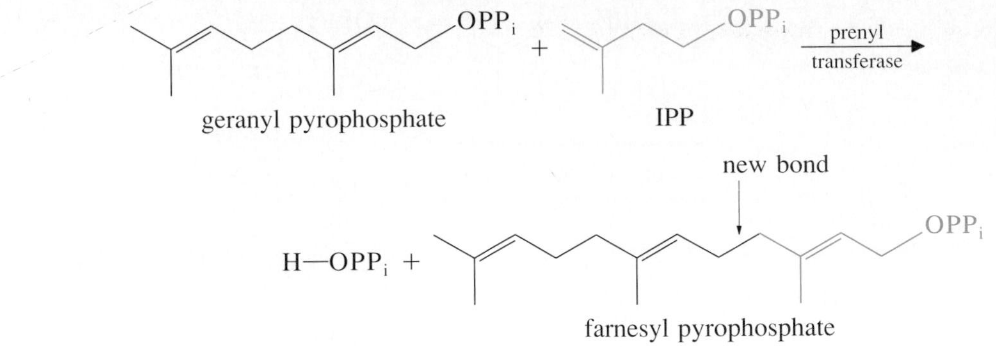

(17.38)

The formation of terpene rings occurs by the attack of double-bond π-electrons on carbocations within the same molecule rather than on another molecule of IPP. Consider, for example, the biosynthesis of limonene (structure in Fig. 17.3). Geranyl pyrophosphate first ionizes to an allylic carbocation that has the *E* configuration at one double bond. This cation reacts with pyrophosphate ion at the other electron-deficient carbon to form a tertiary pyrophosphate.

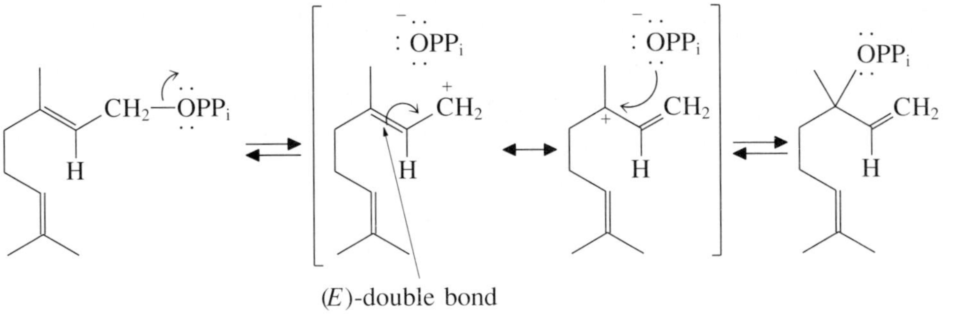

(17.39a)

This reaction is followed by an internal rotation, and then by another ionization that gives a carbocation with the *Z* configuration at the double bond.

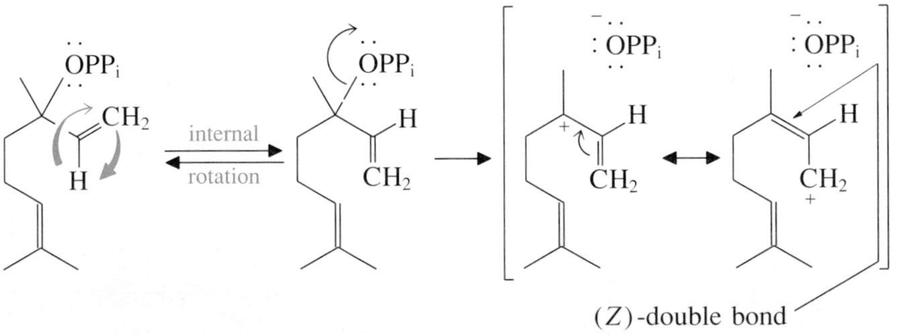

(17.39b)

The *Z* configuration places the electron-deficient carbon near the other double bond. Attack of the π-electrons of the double bond on the electron-deficient carbon gives a new carbocation, which loses a proton to complete the formation of limonene.

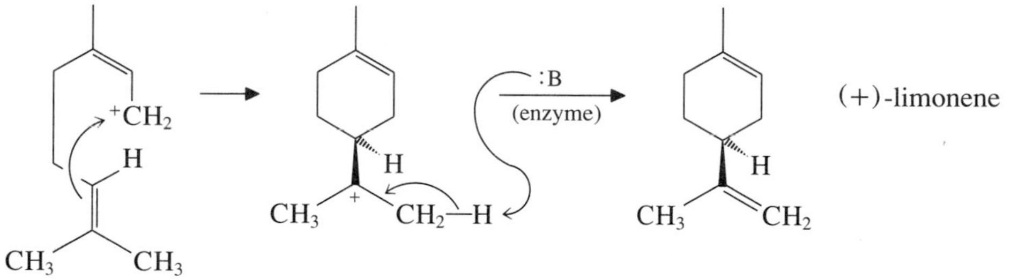

(17.40)

Introduction of oxygen substituents into terpene skeletons can occur by attack of water on carbocation intermediates, displacement of pyrophosphates by water (see Eq. 17.37), or by more obscure oxidation reactions.

As these examples illustrate, the biosynthesis of terpenes can be understood in terms of the same types of carbocation intermediates that we have encountered in laboratory chemistry. Once again (Secs. 10.7, 11.5B) we see that the organic chemistry of living systems is understandable in terms of laboratory analogies.

The biosynthesis of terpenes also illustrates the economy of nature: a truly remarkable array of substances is generated from a common starting material. This economy is evident also in other families of natural products. For example, terpenes also serve as the starting point for the biosynthesis of *steroids* (Sec. 7.7D). The isoprene rule is one of the unifying elements that underlie the chemical diversity of nature; we shall see others.

Problems

11 Suggest a mechanism for the reaction in Eq. 17.38. Assume that water, acids, and bases are present as necessary.

12 Propose a biosynthetic pathway for each of the following natural products. Assume that acids, bases, and water are available as needed.

(a) 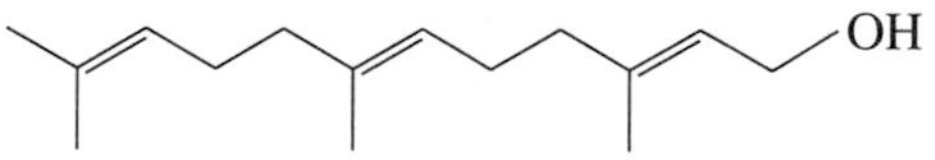

farnesol

(b) 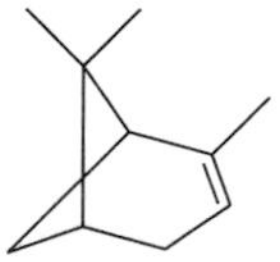

α-pinene

KEY IDEAS IN CHAPTER 17

- Functional groups at allylic and benzylic positions are in many cases unusually reactive.
- Addition of hydrogen halides to dienes, and solvolysis of allylic and benzylic alkyl halides, are reactions that involve allylic or benzylic carbocations. Typically these

reactions are accelerated relative to analogous reactions in which the carbocation intermediates are not allylic or benzylic.

- The acceleration of reactions that involve allylic or benzylic carbocations can be attributed to the resonance stabilization of these ions.
- Because charge is shared by more than one carbon of the ion, a mixture of products is typically obtained from reactions involving allylic cations. Reactions involving benzylic cations, on the other hand, usually give only the product derived from attack of the nucleophile at the benzylic position, since in this product the aromatic ring remains intact.
- Free-radical halogenation is selective for allylic and benzylic hydrogens because of the stability of the allylic or benzylic radical intermediate.
- *N*-Bromosuccinimide (NBS) in CCl_4 solution provides such a low concentration of bromine that addition to the double bond does not compete with allylic substitution. Since addition to the double bonds of a phenyl ring is not observed in benzylic substitution, ordinary halogenation conditions (for example, bromine and light) can be used for benzylic bromination, although NBS bromination also works.
- Because the unpaired electron in allylic radicals is shared on different carbons, some reactions that involve allylic radicals give more than one product.
- Allylic and benzylic Grignard and lithium reagents have more ionic character than ordinary reagents of this type. For this reason allylic Grignard and lithium reagents undergo rapid allylic rearrangement.
- S_N2 reactions at allylic and benzylic positions are accelerated because their transition states are stabilized by orbital overlap.
- In aromatic compounds, alkyl side chains that contain benzylic hydrogens can be oxidized to carboxylic acid groups under vigorous conditions.
- Terpenes, or isoprenoids, are natural products with carbon skeletons characterized by repetition of the five-carbon isoprene pattern.
- Terpenes are synthesized in nature by enzyme-catalyzed processes involving the reaction of allylic carbocations with double bonds. The biosynthetic precursor of all terpenes is isopentenyl pyrophosphate (IPP).

ADDITIONAL PROBLEMS

13 Give the principal products expected when *trans*-2-butene reacts under the following conditions. Assume one equivalent of each reagent reacts.

(a) Br_2, CCl_4, dark
(b) *N*-bromosuccinimide in CCl_4, light
(c) product(s) of (b) + Ag^+ + H_2O
(d) product(s) of (b) + Mg in ether
(e) product(s) of (d) + D_2O

14 Which of the following compounds, all known in nature, can be classified as terpenes? Show the isoprene units in each.

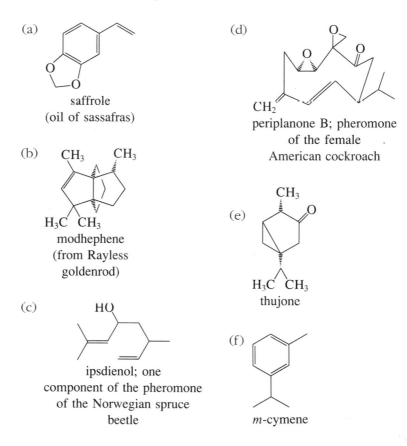

(a)

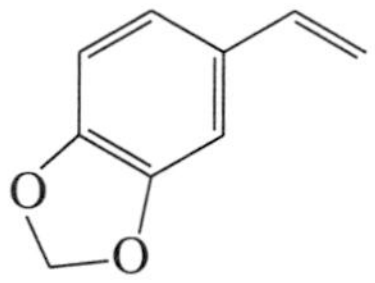

saffrole
(oil of sassafras)

(b)

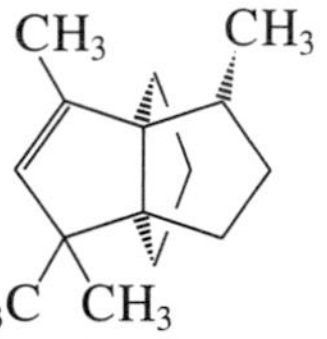

modhephene
(from Rayless
goldenrod)

(c)

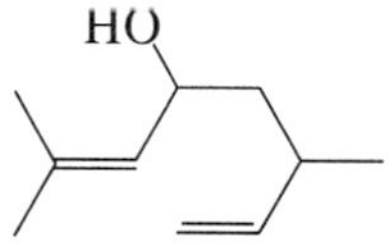

ipsdienol; one
component of the pheromone
of the Norwegian spruce
beetle

(d)

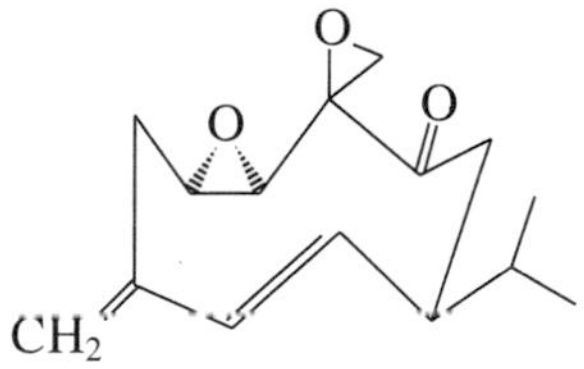

periplanone B; pheromone
of the female
American cockroach

(e)

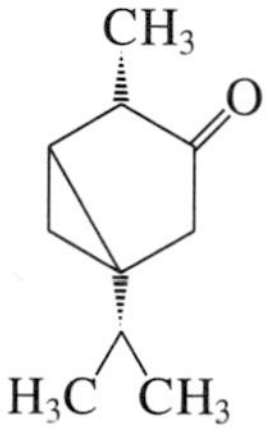

thujone

(f)

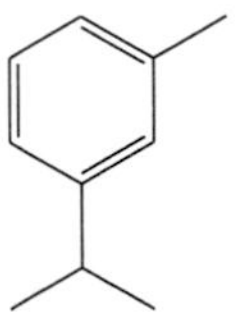

m-cymene

15 A graduate student, Al Lyllic, has prepared a pure sample of 3-bromo-1-butene, *A*. Several weeks later he finds that the sample is contaminated with an isomer *B*. Knowing your expertise in such matters, Lyllic has come to you for assistance.
(a) Postulate a structure for *B* and show how it is formed.
(b) Analysis shows that at equilibrium one isomer is present at 13% and the other at 87%. Which isomer is the major one? Why?
(c) Calculate the standard free energy difference between *A* and *B*.

16 How many grams of $KMnO_4$ would be required to oxidize 10 g of toluene to benzoic acid (structure in Eq. 17.31)? (*Hint:* See Sec. 10.6B.)

17 Outline a synthesis of each of the following compounds from the indicated starting materials and any other reagents.
(a) 3-ethoxy-1-cyclohexene from cyclohexene
(b) (4Z)-1,4-nonadiene from 1-hexyne
(c) 3-phenyl-1-propanol from toluene
(d) benzyl methyl ether from toluene

Problems (Cont.)

(e) CO_2H

from cumene

with NO_2 substituent

(f) CO_2H

from cumene

with NO_2 substituent

(g) O_2N—⟨ ⟩—$\overset{O}{\overset{\|}{C}}$—$CH_3$ from benzene

(h) cyclopentenone from cyclopentene

18 Rank the following compounds in order of increasing reactivity (least reactive first) in an S_N1 solvolysis reaction in aqueous acetone. (The systematic name of *t*-cumyl chloride is 2-chloro-2-phenylpropane.)

(a) *m*-nitro-*t*-cumyl chloride (c) *p*-fluoro-*t*-cumyl chloride
(b) *p*-methoxy-*t*-cumyl chloride (d) *p*-nitro-*t*-cumyl chloride

19 Complete the following reactions by proposing structures for the major organic products:

(a) CH_2=CH—CH=CH—CH_2MgBr + $\overset{O}{CH_2-CH_2}$ $\longrightarrow$ $\xrightarrow{H_3O^+}$

(b) Br—CH_2—CH=CH—$CH(CH_3)_2$ + Na^+ $C_2H_5O^-$ $\xrightarrow{\text{ethanol}}$

(c) same as (b), but with ethanol only; no Na^+ $C_2H_5O^-$

(d) ⟨Ph⟩—CH=CH—CH=CH_2 + HBr (1 equiv) $\longrightarrow$

(e) (tetrahydronaphthalene) + *N*-bromosuccinimide $\xrightarrow{CCl_4,\ AIBN}$

(f) Ph—$\overset{OH}{\underset{Ph}{C}}$—$\overset{OH}{\underset{CH_3}{C}}$—$CH_3$ $\xrightarrow{H_2SO_4}$

(g) (1-methyl tetrahydronaphthalene) $\xrightarrow{KMnO_4,\ \text{heat}}$

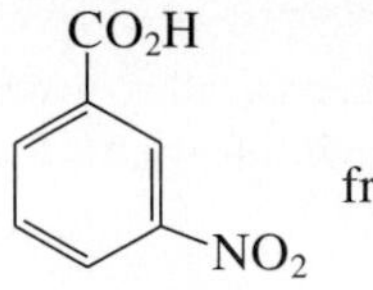

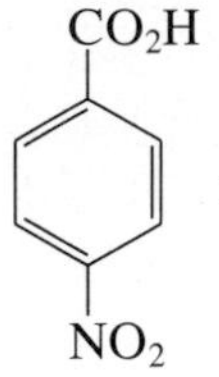

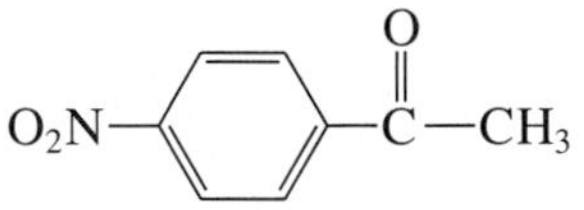

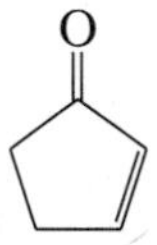

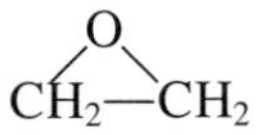

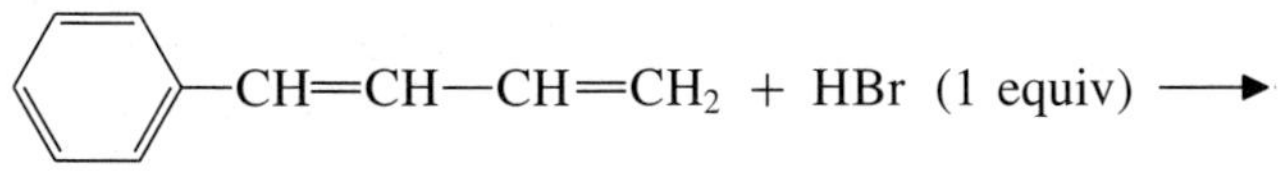

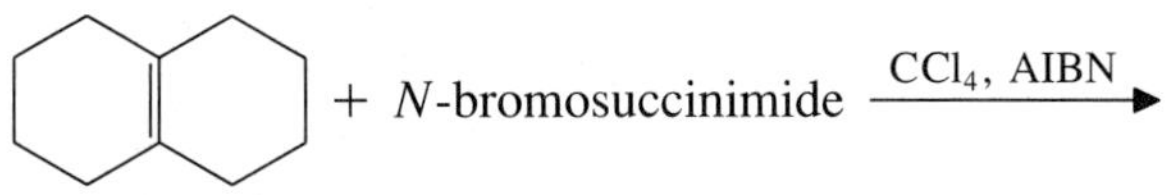

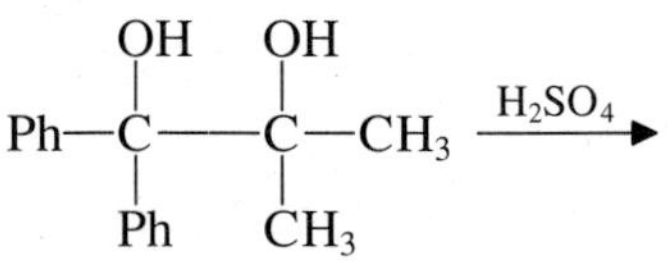

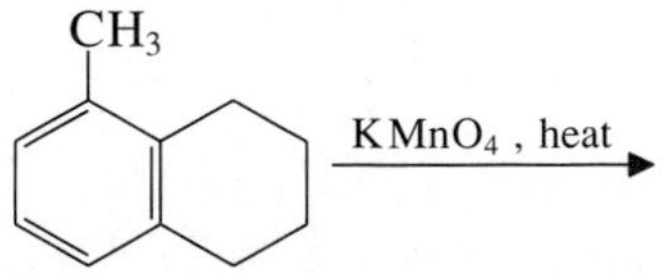

20 Arrange the following alcohols according to increasing rate of acid-catalyzed dehydration to alkene (smallest rate first), and explain your reasoning:

(a)

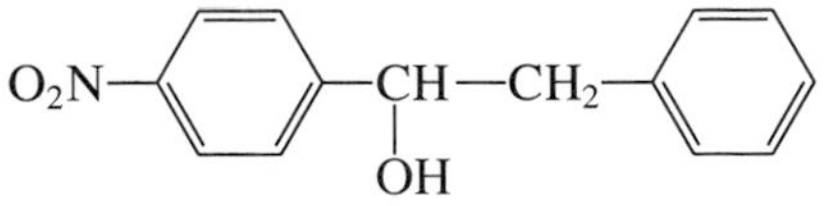

(b)

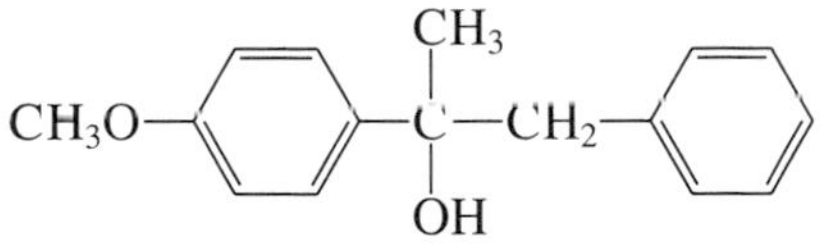

(c)

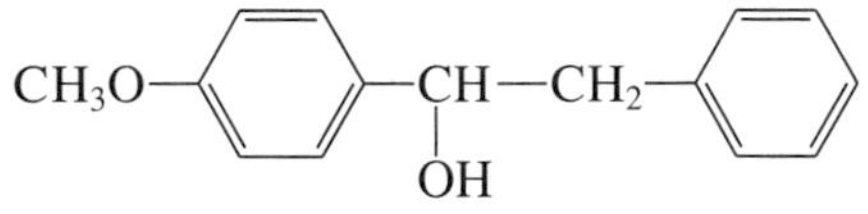

21 Propose an explanation for the distribution of the radioactive carbon in the following reaction (* = ^{14}C).

$$CH_2{=}CH{-}\overset{*}{C}H_2{-}OH \xrightarrow{PBr_3} CH_2{=}CH{-}\overset{*}{C}H_2{-}Br + \overset{*}{C}H_2{=}CH{-}CH_2{-}Br$$

(nearly equal amounts)

22 Commercial thionyl chloride, $SOCl_2$, is sometimes contaminated with sulfuryl chloride, SO_2Cl_2. Sulfuryl chloride is known to be a free-radical chlorinating reagent that reacts as follows:

$$SO_2Cl_2 + R{-}H \longrightarrow SO_2 + R{-}Cl + HCl$$

A good way to remove the sulfuryl chloride is to treat the impure thionyl chloride with limonene (structure in Fig. 17.3). What is the rationale of this procedure? Why is limonene better than an ordinary alkene for the removal of sulfuryl chloride?

23 Propose mechanisms for the biosynthesis of *A* and *B*. Compound *A* is eudesmol, from eucalyptus; *B* is zingiberene, from oil of ginger.

A

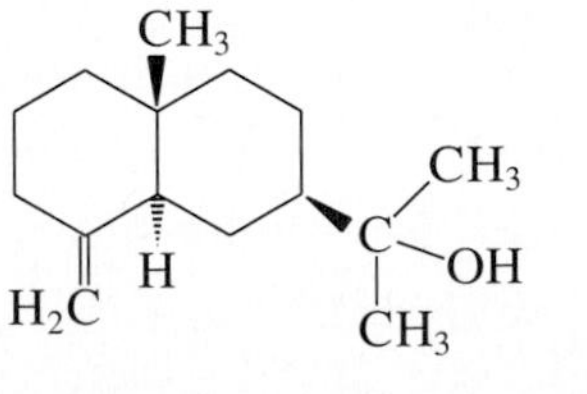

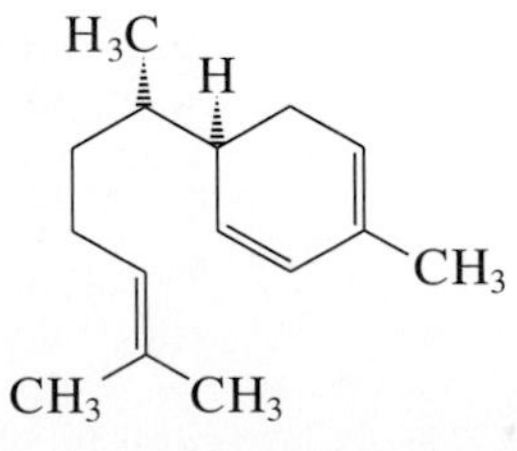

B

Problems (Cont.)

24 The following relative rate data were obtained for reactions catalyzed by the enzyme *prenyl transferase* (Sec. 17.6B). (Fluorine is not much larger than hydrogen, and therefore has no major steric effect on the reaction.)

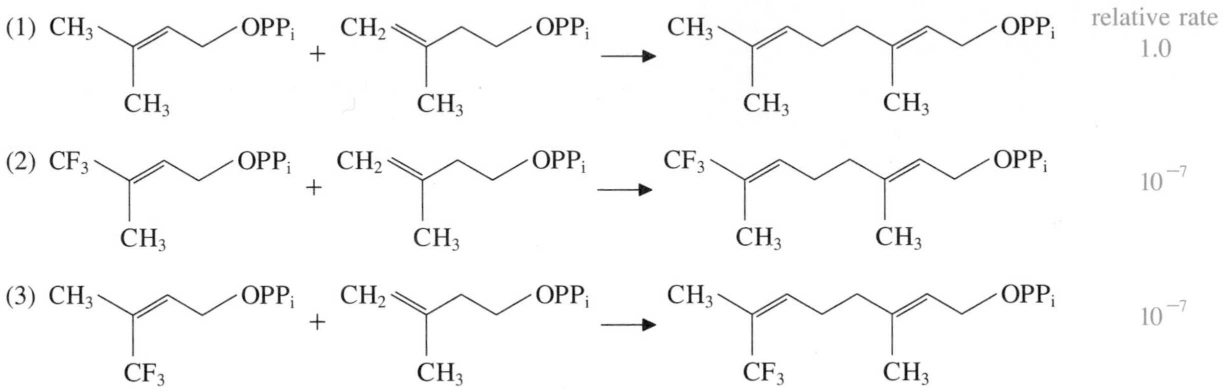

These data were used to argue that a carbocation intermediate is involved in the prenyl transferase-catalyzed biosynthesis of isoprenoids. Explain why this interpretation is reasonable. (*Hint:* What effect is fluorine likely to have on the stability of the carbocation intermediate?)

25 Explain why compound *A* reacts faster than compound *B* when each undergoes solvolysis in aqueous acetone.

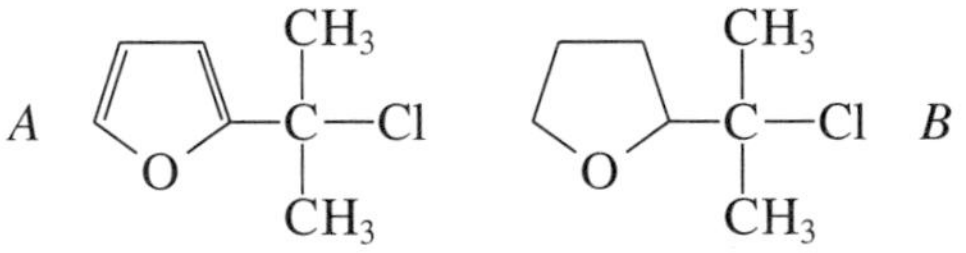

26 A hydrocarbon *A*, C_9H_{10}, is treated with *N*-bromosuccinimide to give a single monobromo compound *B*. When *B* is dissolved in aqueous acetone it reacts to give two compounds, *C* and *D*. Catalytic hydrogenation of *D* gives back *A*, and *C* can be separated into enantiomers. When optically active *C* is oxidized with CrO_3 and pyridine, an optically inactive ketone *E* is obtained. Vigorous oxidation of *A* with $KMnO_4$ affords phthalic acid (structure in Problem 9). Propose structures for compounds *A* through *E* and explain your reasoning.

27 When benzyl alcohol (λ_{max} = 258 nm, ϵ = 520) is dissolved in H_2SO_4, a colored solution is obtained that has a different UV spectrum: λ_{max} = 442 nm, ϵ = 53,000. When this solution is added to cold NaOH, the original spectrum of benzyl alcohol is restored. Account for these observations.

28 Using the arrow formalism when appropriate, suggest a mechanism for each of the following reactions:

(a) $CH_3(CH_2)_3-C\equiv C-CH_2-Br + Mg \xrightarrow{\text{ether}} \xrightarrow{H_2O}$

$\qquad CH_3(CH_2)_3-CH=C=CH_2 + CH_3(CH_2)_3-C\equiv C-CH_3$

(b) ... $CH-Cl + H_2O \xrightarrow{Ag^+}$

... $+$... $CH-OH + CH_3CH=CHCH_2CH_2OH + AgCl + H_3O^+$

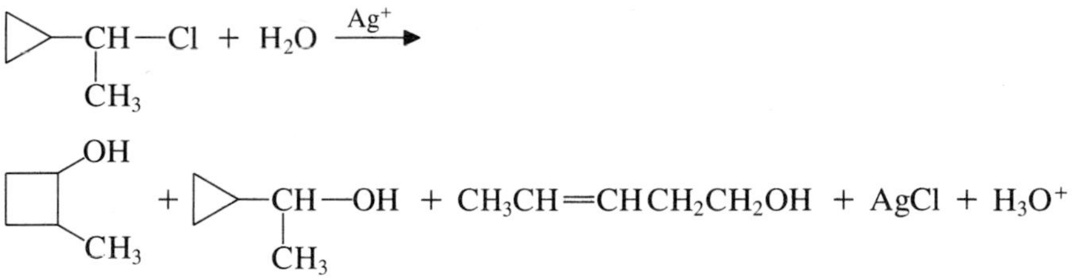

(c) 1-penten-4-yne $+ Na^+ \; ^-:C\equiv CH$; then allyl bromide $\longrightarrow$

$\qquad CH_2=CH-CH-C\equiv CH$

$\qquad\qquad\qquad CH_2-CH=CH_2$

(d) ... $\xrightarrow{H_3O^+}$...

(e) $CH_3-\underset{\underset{Cl}{|}}{\overset{\overset{CH_3}{|}}{C}}-C\equiv C-H + $... $\xrightarrow{(CH_3)_3C-O^- \; K^+}$...

(f) ... $+ C_2H_5OH \xrightarrow[\text{(dil)}]{H_2SO_4}$...

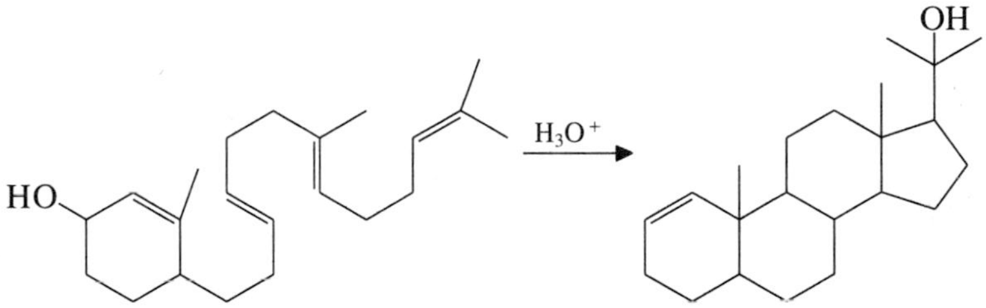

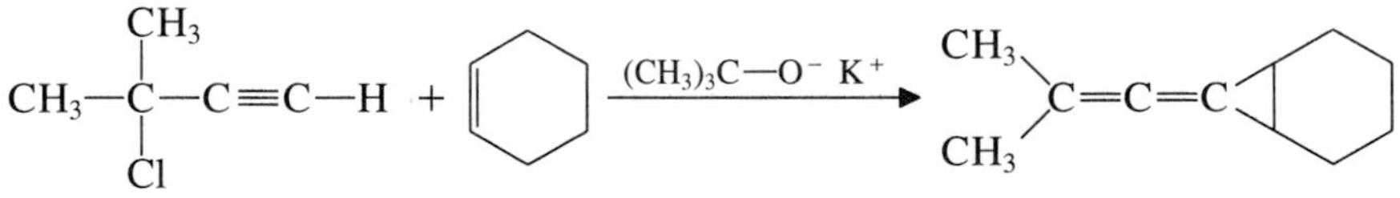

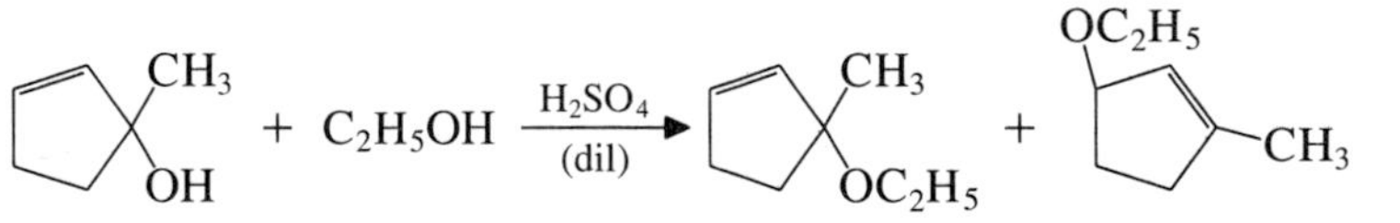

29 When 1-buten-3-yne reacts with aqueous HCl, two compounds, *A* and *B*, are formed in a ratio of 2.2 : 1. Only compound *A* reacts instantaneously with an acidic solution of $AgNO_3$ to give a white precipitate. Only compound *B* gives a Diels–Alder adduct with maleic anhydride. Neither compound liberates a gas when treated with methylmagnesium bromide. Suggest structures for both compounds and explain all observations.

30 Around 1900, Moses Gomberg, a pioneer in free-radical chemistry, prepared the triphenylmethyl radical, $Ph_3C\cdot$, sometimes called the trityl radical.
(a) Explain why the trityl radical is an unusually stable radical.
(b) The trityl radical is known to exist in equilibrium with a dimer which, for many years, was assumed to be hexaphenylethane. Show how hexaphenylethane could be formed from the trityl radical.
(c) In 1968 the structure of this dimer was investigated using modern methods and found not to be hexaphenylethane, but rather the following compound. Using the fishhook arrow formalism, show how this compound is formed from the trityl radical, and explain why this compound might form instead of

Problems (Cont.)

hexaphenylethane. (*Hint:* Can you think of any reason why hexaphenylethane might be unstable?)

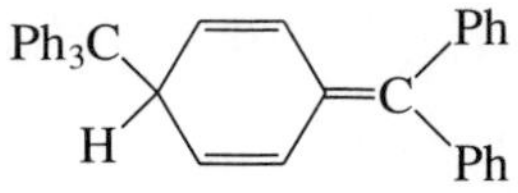

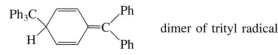

dimer of trityl radical

---

31 (a) Triphenylmethane (structure below) has a pK_a of 31.5 and, although an alkane, is almost as acidic as a 1-alkyne. (Most alkanes have p$K_a \geq 55$.) By considering the structure of its conjugate base, suggest a reason why triphenylmethane is such a strong hydrocarbon acid.

(b) Fluoradene is structurally very similar to triphenylmethane, except that the three aromatic rings are "tied together" with bonds. Fluoradene has a pK_a of 11! Suggest a reason why fluoradene is much more acidic than triphenylmethane.

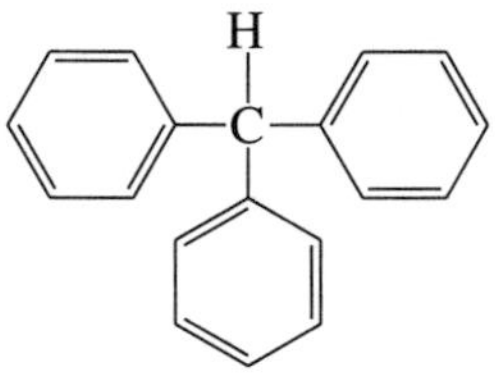

triphenylmethane

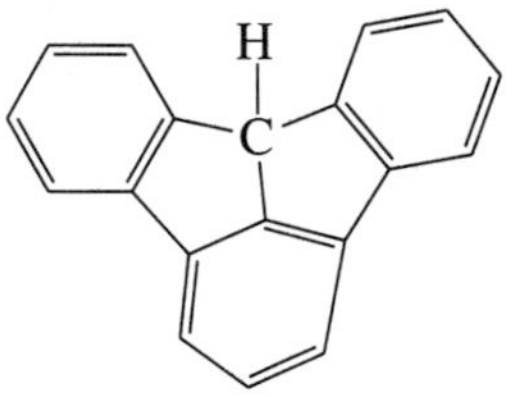

fluoradene

---

32 The amount of *anti* addition (see Sec. 7.10C) in the chlorination of alkenes varies with the structure of the alkene, as shown in the following table:

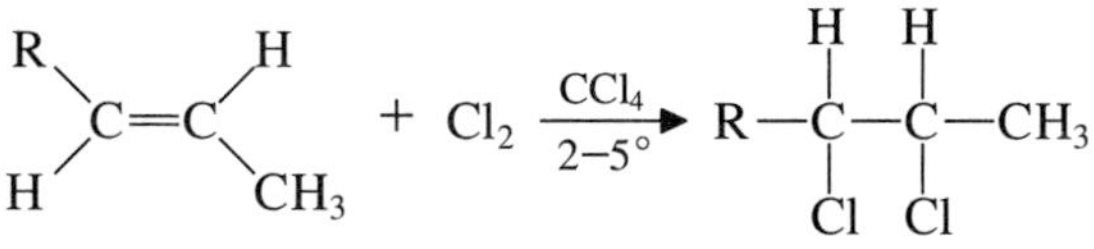

Structure of R	Percent *anti* addition
CH_3—	99
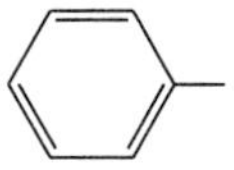	88
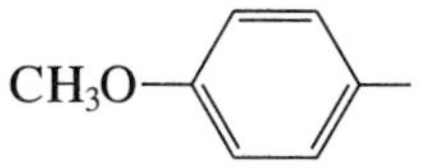	63

Suggest a reason for the variation in the stereochemistry of addition as the alkene structure is varied. (*Hint:* What types of reactive intermediate(s) could account for the stereochemical observations?)

---

33 For the following two reactions suggest mechanisms that are consistent with all of the experimental facts given:

reaction 1: $CH_3-CH=CH-CH_2-Cl + (C_2H_5)_2\overset{..}{N}H$ ——
 (trans)

reaction 2: $CH_3-\underset{\underset{Cl}{|}}{CH}-CH=CH_2 + (C_2H_5)_2\overset{..}{N}H$ ——

$CH_3-CH=CH-CH_2-\overset{+}{N}H(C_2H_5)_2\ Cl^-$
(82–85% yield)

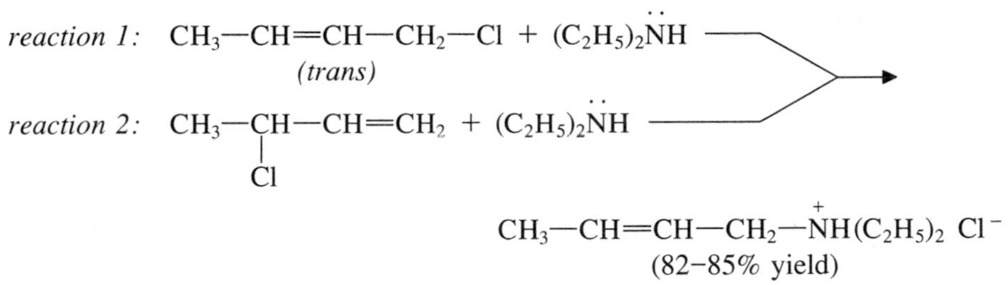

Experimental observations:
(a) Both reactions follow the rate law

$$\text{rate} = k[\text{alkyl halide}][(C_2H_5)_2\overset{..}{N}H]$$

with different rate constants.
(b) The alkyl chloride starting materials do *not* interconvert under the reaction conditions.
(c) The compound below, prepared separately, is *not* converted into the observed product under the reaction conditions.

$$CH_2=CH-\underset{\underset{CH_3}{|}}{CH}-\overset{+}{N}H(C_2H_5)_2\ Cl^-$$

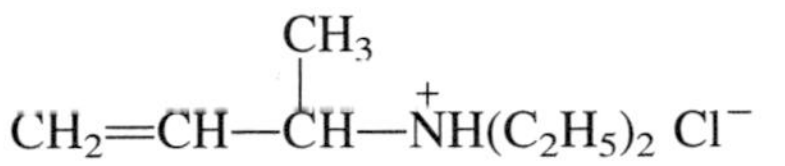

In particular, explain why facts (b) and (c) are crucial to the mechanistic discussion.

The mechanism of reaction (2) is called the S_N2' mechanism. Suggest a reason why this reaction occurs by the S_N2' mechanism and reaction (1) does not.

34 The stereochemistry of the S_N2' reaction discussed in Problem 33 was determined by the following experiment:

(95%) (5%)

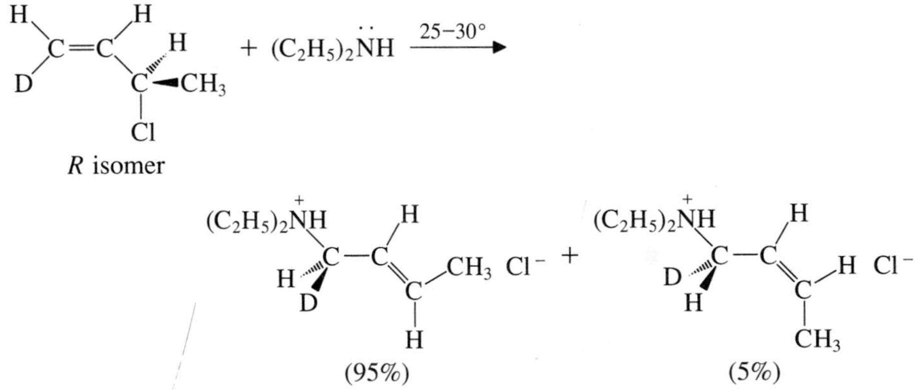

Is this reaction *syn* or *anti* with respect to the plane of the alkene? How does this result contrast with the stereochemistry of the S_N2 reaction?

Problems (Cont.)

35 In the late 1970s a graduate student at a major west-coast university began synthesizing new classes of drugs and testing them on himself. After being asked to leave the university he began making his living by illegally selling synthetic meperidine (Demerol) to heroin addicts. After shortening his synthetic procedure and injecting his product, he developed severe symptoms of Parkinson's disease, as did several of his young clients; one person died. Chemists found that his synthetic meperidine contained two by-products, alcohol *B* and another compound MPTP, which, when independently prepared and injected into animals, caused the same symptoms. (Ironically, this has been one of the most significant advances in parkinsonism research.)

The illicit chemistry is outlined below, and the elemental analysis of MPTP is 83.19 %C; 8.73 %H; 8.09 %N. Provide the structure of compound *A*, suggest a structure for MPTP, and show how all products are formed.

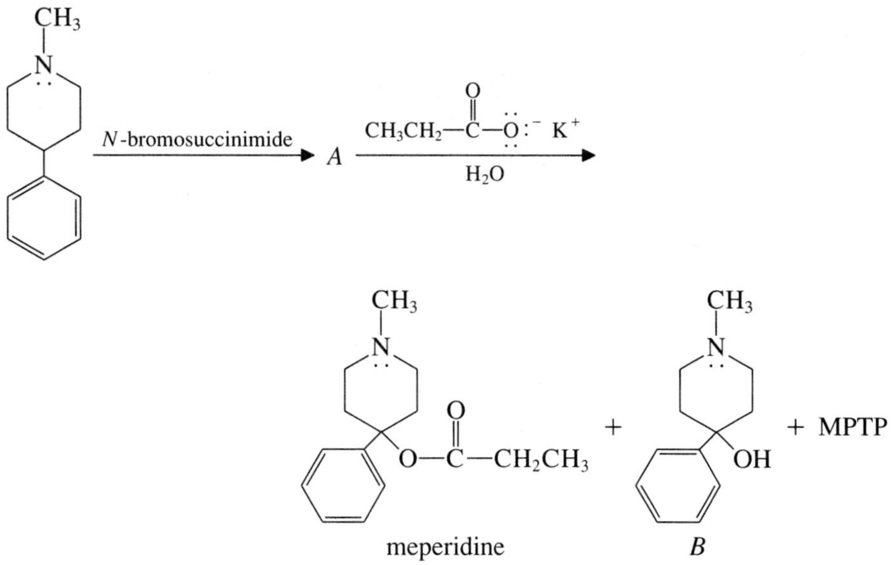

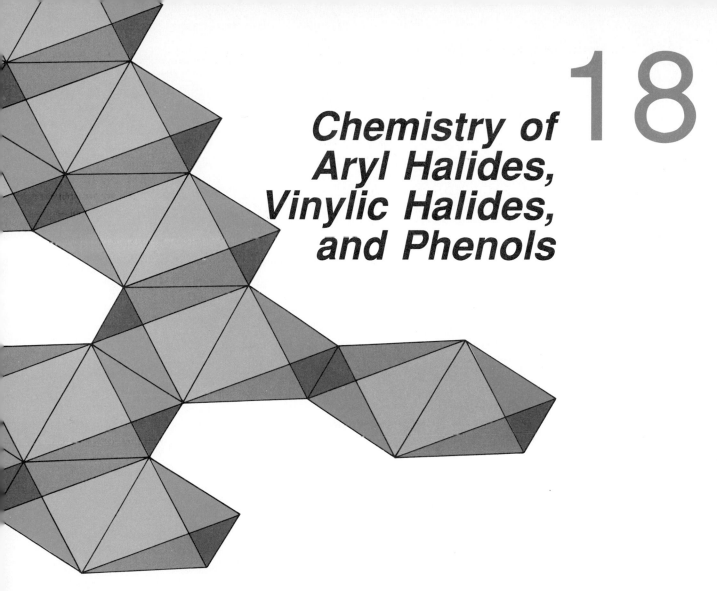

18 Chemistry of Aryl Halides, Vinylic Halides, and Phenols

An **aryl halide** is a compound in which a halogen is bound to the carbon of a benzene ring (or other aromatic ring).

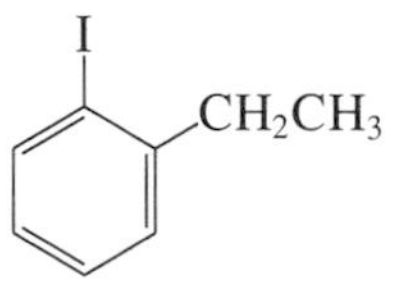

1-ethyl-2-iodobenzene
(an aryl iodide)

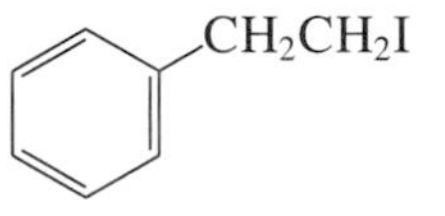

not an aryl halide; halogen not attached directly to benzene ring

A **vinylic halide** is a compound in which a halogen is bound directly to a carbon of a double bond.

$CH_2{=}CH{-}Cl$
vinyl chloride
(1-chloroethene)

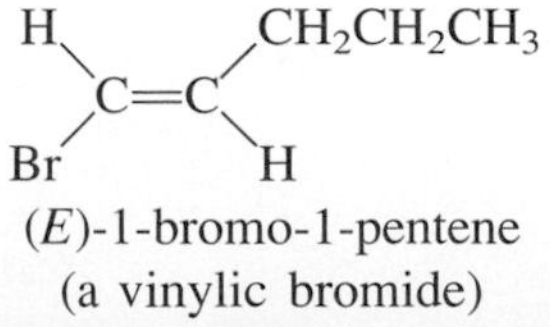

(*E*)-1-bromo-1-pentene
(a vinylic bromide)

Phenols are compounds in which a hydroxy (—OH) group is bound to an aromatic ring.

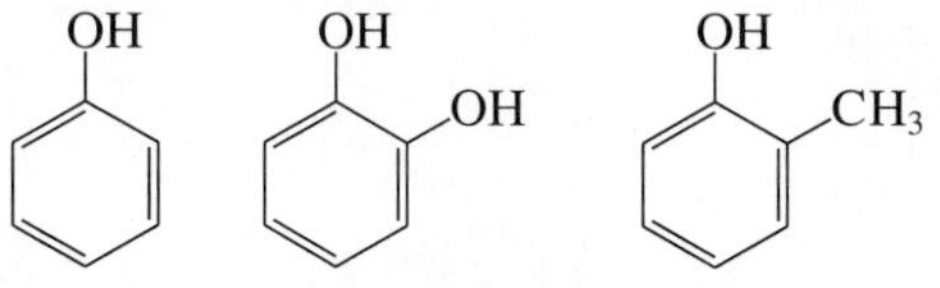

phenol catechol *o*-cresol

Be sure to differentiate carefully between *vinylic* and *allylic* halides (Chapter 17, Introduction). *Allylic* groups are on the carbon *adjacent* to the double bond. Likewise, be sure that the distinction between *aryl* and *benzylic* halides is clear. *Benzylic* groups are on the carbon *adjacent* to an aromatic ring.

The reactivity of aryl and vinylic halides is quite different from that of ordinary alkyl halides. We shall also find that, although phenols and alcohols have some reactions in common, there are important differences in the chemical behavior of these two functional groups.

The nomenclature and spectroscopy of aryl halides and phenols was discussed in Secs. 16.1 and 16.3, respectively. The nomenclature of vinylic halides follows the principles of alkene nomenclature (Sec. 4.2), and the spectroscopy of vinylic halides, except for minor differences due to the halogen, is also similar to that of alkenes.

18.1 LACK OF REACTIVITY OF VINYLIC AND ARYL HALIDES IN S_N2 REACTIONS

One of the most important differences between vinylic or aryl halides and alkyl halides is their reactivity in nucleophilic substitution reactions. The two most important mechanisms for nucleophilic substitution reactions of alkyl halides are the S_N2 (bimolecular backside attack) mechanism, and the S_N1 (unimolecular carbocation) mechanism (Secs. 9.3, 9.5). What happens to vinylic and aryl halides under the conditions used for S_N1 or S_N2 reactions of alkyl halides?

Let us consider first the S_N2 reaction. One of the most dramatic contrasts between vinylic or aryl halides and alkyl halides is that simple *vinylic and aryl halides are inert under S_N2 conditions.* For example, when ethyl bromide is allowed to react with Na^+ $C_2H_5O^-$ in C_2H_5OH solvent at 55°, the following S_N2 reaction proceeds to completion in about an hour with excellent yield:

$$CH_3CH_2\text{—}Br + Na^+\ CH_3CH_2\text{—}\ddot{\underset{..}{O}}:^- \xrightarrow[CH_3CH_2OH]{55°} CH_3CH_2\text{—}\ddot{\underset{..}{O}}\text{—}CH_2CH_3 + Na^+\ Br^- \quad (18.1)$$

Yet when vinyl bromide or bromobenzene is subjected to the same conditions, nothing happens!

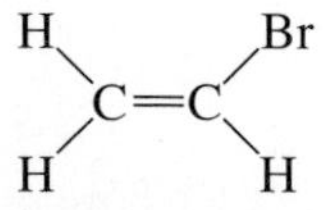

and

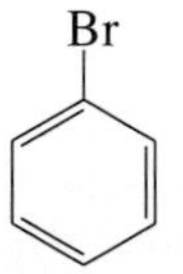

inert to $Na^+\ C_2H_5O^-$ in C_2H_5OH, 55°

Why don't aryl and vinylic halides undergo S_N2 reactions? Let us consider the consequences of the usual S_N2 mechanism applied to a vinylic halide.

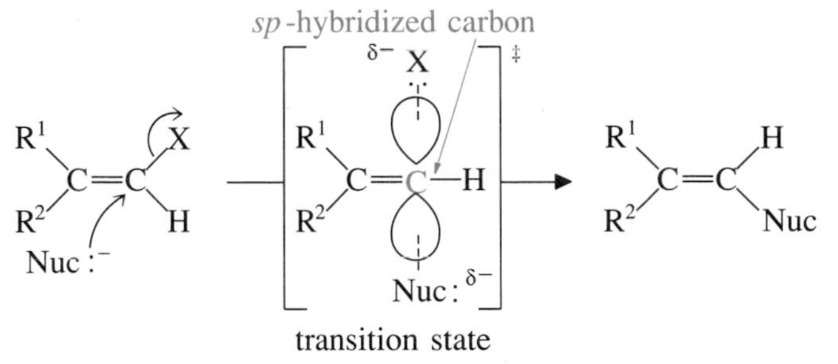

(18.2)

As we can see from the diagram of the transition state in Eq. 18.2, this mechanism has several unattractive features. First, the attacking nucleophile (Nuc : $^-$) must approach the vinyl halide at the backside of the halogen-bearing carbon *and in the plane of the alkene.* This places the nucleophile and the alkene group R^2 in the same plane and very close together. The resulting van der Waals (steric) repulsions between R^2 and Nuc : $^-$ raise the energy of the transition state and therefore lower the reaction rate.

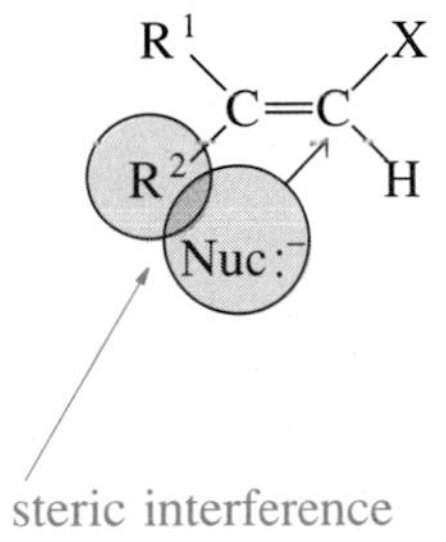

Similar repulsions in the transition state also occur between R^1 and the leaving group X : $^-$. Second, the carbon at which substitution occurs is *sp*-hybridized in the transition state. In Sec. 14.2 we learned that *sp* hybridization is the least favorable hybridization state of carbon. Since *sp* hybridization raises the energy of the transition state, it also lowers the reaction rate. Both effects are sufficiently important that the vinylic S_N2 reaction is too slow to be observed.

In addition to all of the unattractive features just discussed for the S_N2 reaction of a vinylic halide, the S_N2 reaction of an aryl halide has two others. First, backside attack requires that the attacking nucleophile would have to approach on a path that goes through the plane of the benzene ring—an obvious impossibility. Furthermore, since the carbon at the site of attack undergoes an inversion of configuration, the reaction would yield a benzene derivative containing a formal *trans*-double bond!

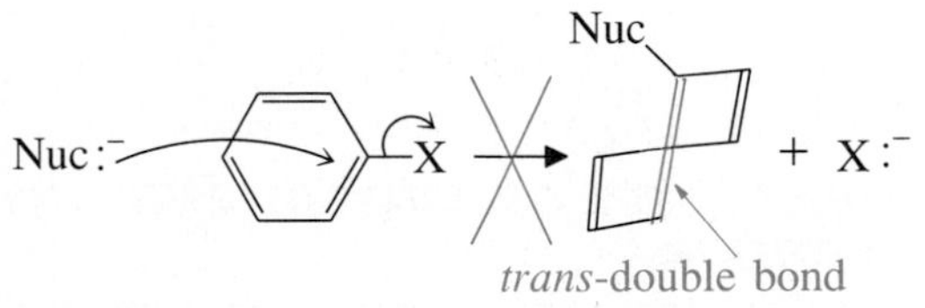

(18.3)

If the difficulty of this possibility is not obvious, try to make a model of the product—but don't break your models!

Problem

1 Rank the following compounds in order of increasing rate in an S_N2 reaction with NaCl in acetone. Explain your reasoning.
(a) benzyl bromide
(b) 1-bromo-3-phenylpropane
(c) *p*-bromotoluene

18.2 ELIMINATION REACTIONS OF VINYLIC HALIDES

Although S_N2 reactions of vinylic halides are completely unknown, base-promoted β-elimination reactions of vinyl halides do occur and can be useful in the synthesis of alkynes.

$$\text{Ph—CH=CH—Br} + \text{KOH} \xrightarrow{200^\circ} \underset{\substack{\text{(distills from}\\ \text{reaction mixture;}\\ \text{67\% yield)}}}{\text{Ph—C}\equiv\text{C—H}} + K^+\ Br^- \qquad (18.4)$$

$$\text{Ph—}\overset{\text{Br}}{\overset{|}{\text{CH}}}\text{—}\overset{\text{Br}}{\overset{|}{\text{CH}}}\text{—Ph} + 2\text{KOH} \xrightarrow{C_2H_5OH} \underset{\text{(67\% yield)}}{\text{Ph—C}\equiv\text{C—Ph}} + 2K^+\ Br^- + 2H_2O \qquad (18.5)$$

In Eq. 18.5, two successive eliminations take place. The first gives a vinylic halide and the second gives the alkyne.

Although these reactions involve the energetically unfavorable conversion of the sp^2-hybridized carbons of the starting material into the sp-hybridized carbons of the product, they do not have the steric problems associated with vinylic S_N2 reactions. Furthermore, the vinylic proton is more acidic than an alkyl proton (why? Sec. 14.7A), and is more easily removed by a base than an alkyl proton. As the examples in Eqs. 18.4 and 18.5 illustrate, some of the more useful examples of this reaction occur when the hydrogen to be removed is adjacent to an aromatic ring. Vinylic protons adjacent to aromatic rings, like other benzylic protons, are especially acidic (why?). Neverthe-less, the conditions for E2 reactions of vinylic halides are more severe than those for the analogous reactions of ordinary alkyl halides.

Problem

2 Arrange the following compounds according to increasing relative rate of elimina-tion with $NaOC_2H_5$ in C_2H_5OH. What is the product in each case?

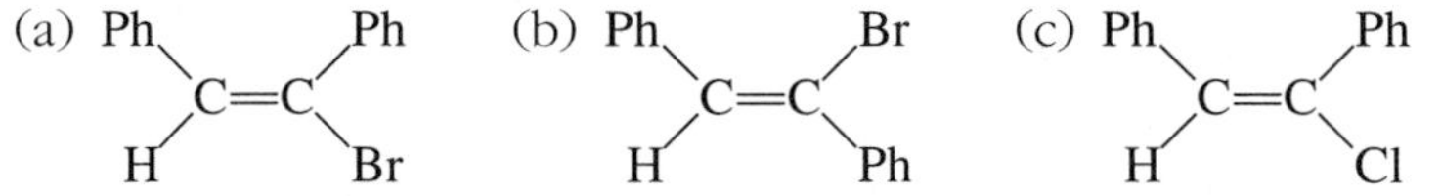

18.3 LACK OF REACTIVITY OF VINYLIC AND ARYL HALIDES IN S_N1 AND E1 REACTIONS

Tertiary and some secondary alkyl halides undergo nucleophilic substitution reactions by the S_N1 mechanism (Sec. 9.5). S_N1 reactions involve carbocation intermediates, and in many cases compete with the E1 elimination pathway. Thus, *t*-butyl bromide undergoes a rapid solvolysis in ethanol to give both substitution and elimination products.

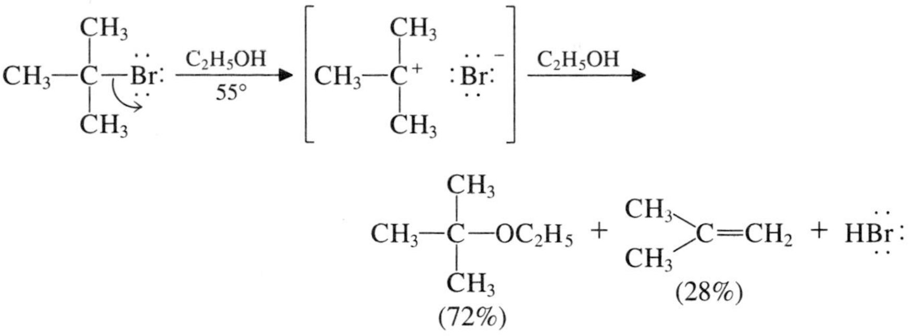

$$CH_3-\underset{\underset{CH_3}{|}}{\overset{\overset{CH_3}{|}}{C}}-\overset{..}{\underset{..}{Br}}: \xrightarrow[55°]{C_2H_5OH} \left[CH_3-\underset{\underset{CH_3}{|}}{\overset{\overset{CH_3}{|}}{C^+}} \quad :\overset{..}{\underset{..}{Br}}:^- \right] \xrightarrow{C_2H_5OH}$$

$$CH_3-\underset{\underset{CH_3}{|}}{\overset{\overset{CH_3}{|}}{C}}-OC_2H_5 \quad + \quad \underset{CH_3}{\overset{CH_3}{>}}C=CH_2 \quad + \quad H\overset{..}{\underset{..}{Br}}: \quad (18.6)$$
$$(72\%) \qquad\qquad (28\%)$$

Vinylic and aryl halides, however, are virtually inert to the conditions that promote S_N1 or E1 reactions of alkyl halides. Certain vinylic halides can be forced to react by the S_N1–E1 mechanism under extreme conditions, but such reactions are relatively uncommon and need not concern us.

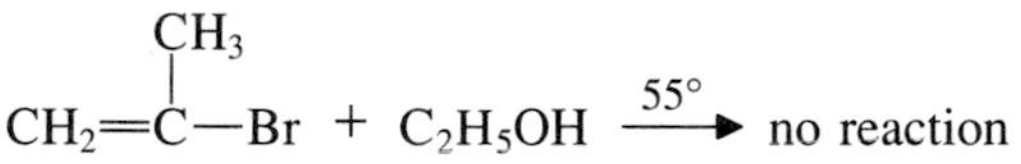

$$CH_2=\underset{\underset{CH_3}{|}}{C}-Br \quad + \quad C_2H_5OH \xrightarrow{55°} \text{ no reaction} \qquad (18.7)$$

$$\bigcirc\!\!-\!Br \quad + \quad C_2H_5OH \xrightarrow{55°} \text{ no reaction} \qquad (18.8)$$

To understand why vinylic and aryl halides are inert under S_N1 conditions, consider what would happen if they *were* to undergo the S_N1 reaction. If a vinylic halide undergoes the S_N1 reaction, it must ionize to form a *vinylic cation* (Sec. 14.5A).

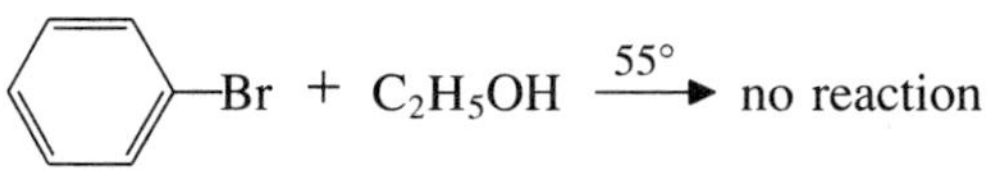

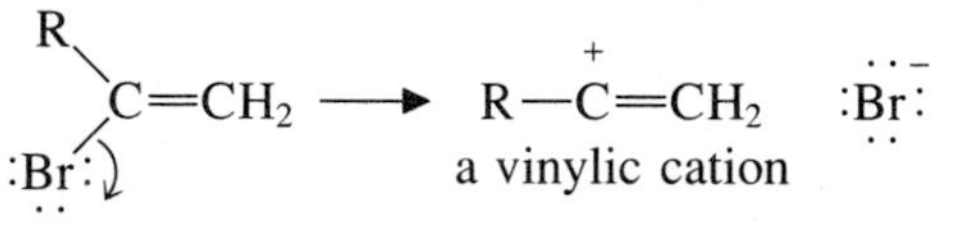

$$\underset{:\overset{..}{Br}:}{\overset{R}{>}}C=CH_2 \longrightarrow R-\overset{+}{C}=CH_2 \quad :\overset{..}{\underset{..}{Br}}:^- \qquad (18.9)$$
$$\text{a vinylic cation}$$

There are two reasons why this process is more unfavorable energetically than the corresponding ionization of an alkyl halide. First, the bond between a vinylic carbon and a halogen is stronger than an alkyl carbon–halogen bond because a carbon sp^2 orbital is involved; just as the strengths of bonds to hydrogen increase with increasing s character (Eq. 14.34), the strengths of bonds to halogen do also. Since it takes more energy to break a strong bond, the relative free energy of the S_N1 transition state is raised and the S_N1 reaction is slower for a vinylic halide. Second, vinylic cations are among the most unstable carbocations. (Be sure to distinguish between a *vinylic* cation and an *allylic* cation, which is relatively stable; Eq. 15.23.) The major source of instability in a vinylic cation is its sp hybridization (Fig. 14.6). Solvolysis of a vinylic halide changes the hybridization of one of its carbons from sp^2 to sp. This change requires a relatively large amount of energy, raises the energy of the transition state, and slows the reaction. Vinylic cations are formed more readily in the hydration of alkynes (Sec. 15.4A) because alkynes are already sp hybridized; an unfavorable hybridization change does not have to occur in order to form the cation.

S_N1 reactions of aryl halides presents all the same difficulties just discussed for vinylic halides. However, the cation derived from an aryl halide, called an **aryl**

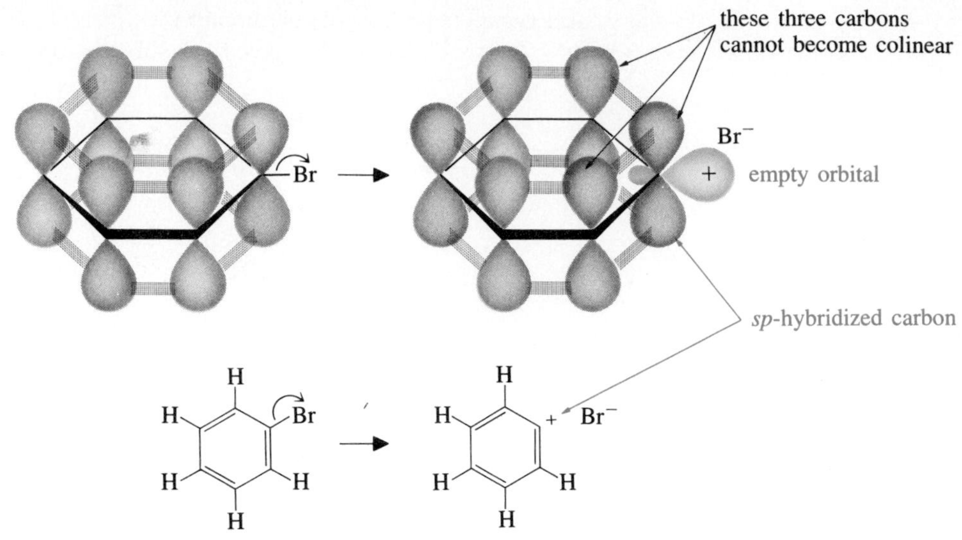

Figure 18.1 Aryl cations are unstable because an *sp*-hybridized carbon atom is incorporated in a six-membered ring.

cation, is formed with even more difficulty than the vinylic cation derived from a vinylic halide (Fig. 18.1). The electron-deficient carbon in the aryl cation is bonded to two groups (the two ring carbons) and therefore prefers a linear geometry; but this geometry is impossible because it introduces too much strain in a six-membered ring.

Note that the aryl cation is quite different from the cation formed in electrophilic aromatic substitution (Fig. 16.7), in which case the carbocation intermediate is stabilized by resonance. In the aryl cation, the empty orbital is not part of the ring π-electron system, but is orthogonal (at right angles) to it. Hence, this carbocation is *not* resonance stabilized. Although aryl cations are not completely unknown, they do not form from aryl halides.

Problem

3 When α-bromostyrene is heated in a solvent containing water, both acetophenone and phenylacetylene are formed. Show how the S_N1 or E1 mechanism accounts for each product. (*Hint:* See Sec. 14.5A.)

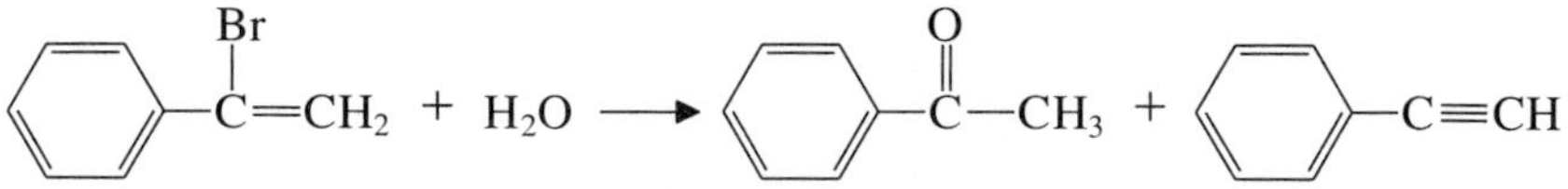

α-bromostyrene

acetophenone (80%)

phenylacetylene (20%)

18.4 NUCLEOPHILIC SUBSTITUTION REACTIONS OF ARYL HALIDES

We have seen that nucleophilic substitution reactions of aryl halides do not occur by the S_N1 and S_N2 mechanisms. However, under certain conditions, nucleophilic substitution reactions of aryl halides do occur by other mechanisms.

A. Nucleophilic Aromatic Substitution

Aryl halides that have one or more strongly electronegative groups (especially nitro groups) *ortho* or *para* to the halogen undergo nucleophilic substitution reactions.

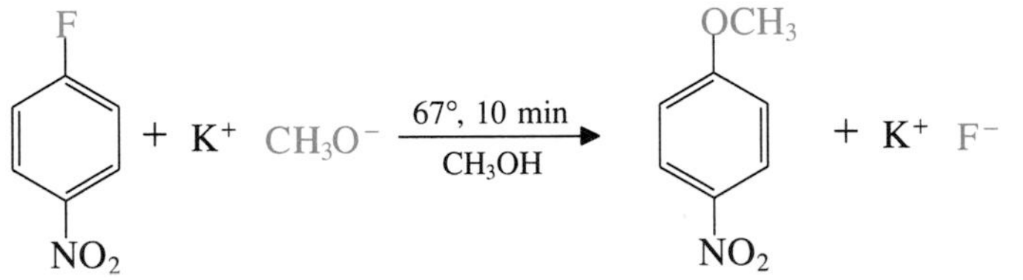

(93% yield) (18.10)

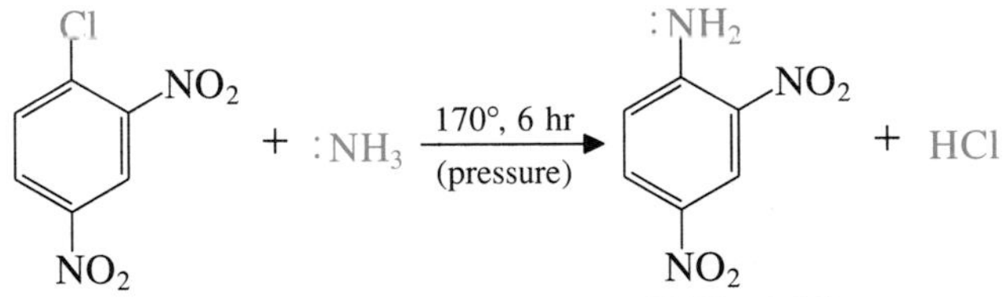

(70% yield) (18.11)

These reactions are examples of **nucleophilic aromatic substitution,** a type of substitution that occurs by a special mechanism.

Let us examine some of the characteristics of this mechanism. Like S_N2 reactions, nucleophilic aromatic substitution reactions involve nucleophiles and leaving groups, and they also give second-order kinetics.

$$\text{rate} = k[\text{aryl halide}][\text{nucleophile}] \quad (18.12)$$

However, nucleophilic aromatic substitution reactions do not involve a one-step backside attack for the reasons given in Sec. 18.1. Two clues about the reaction mechanism come from the reactivities of different aryl halides. First, the reaction goes faster when there are more nitro groups *ortho* and *para* to the halogen leaving group:

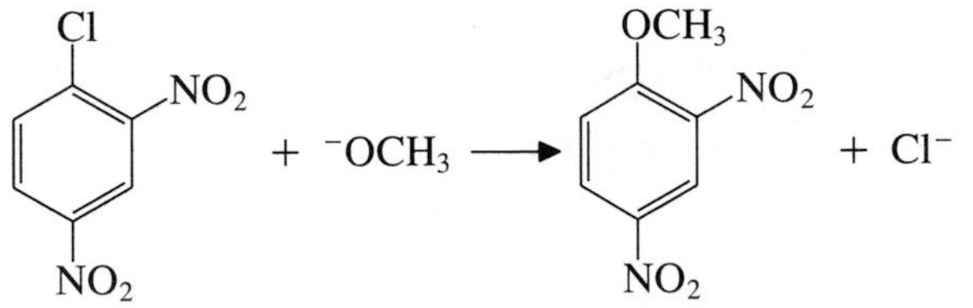

relative rate 10^5-10^6 (18.13a)

Compare with Equations 18.13b and c on the next page:

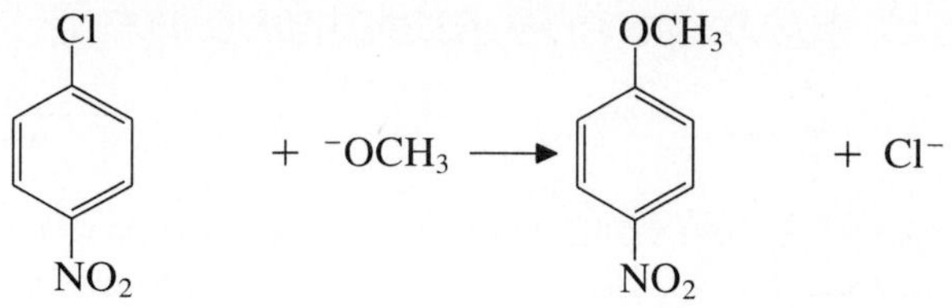

1 (18.13b)

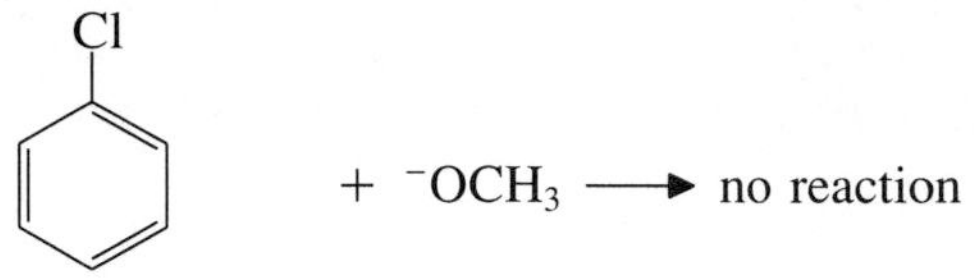

~0 (18.13c)

Second, the leaving-group effect in this reaction is quite different from that in the S_N1 or S_N2 reaction of alkyl halides. In nucleophilic aromatic substitution reactions, fluoride is the best halide leaving group.

$$\text{leaving group order: } F^- \gg Cl^- \approx Br^- \approx I^- \tag{18.14}$$

In S_N2 and S_N1 reactions of *alkyl* halides, the leaving-group order is exactly the reverse: fluoride is the worst of the halide leaving groups (Sec. 9.3D, 9.5C).

These data are consistent with a reaction mechanism in which the nucleophile attacks the π-electron system below (or above) the plane of the aromatic ring to yield a resonance-stabilized anion called a *Meisenheimer complex.* In this anion the negative charge is delocalized throughout the π-electron system of the ring. Formation of this anion is the rate-determining step in many nucleophilic aromatic substitution reactions.

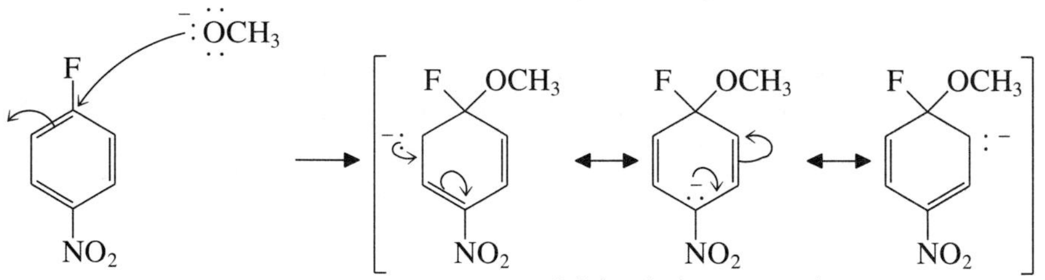

(18.15)

Meisenheimer complex

The negative charge in this complex is also delocalized into the nitro group.

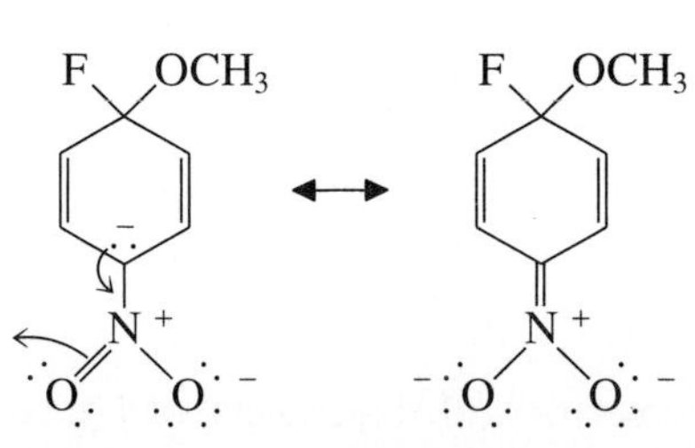

(18.16)

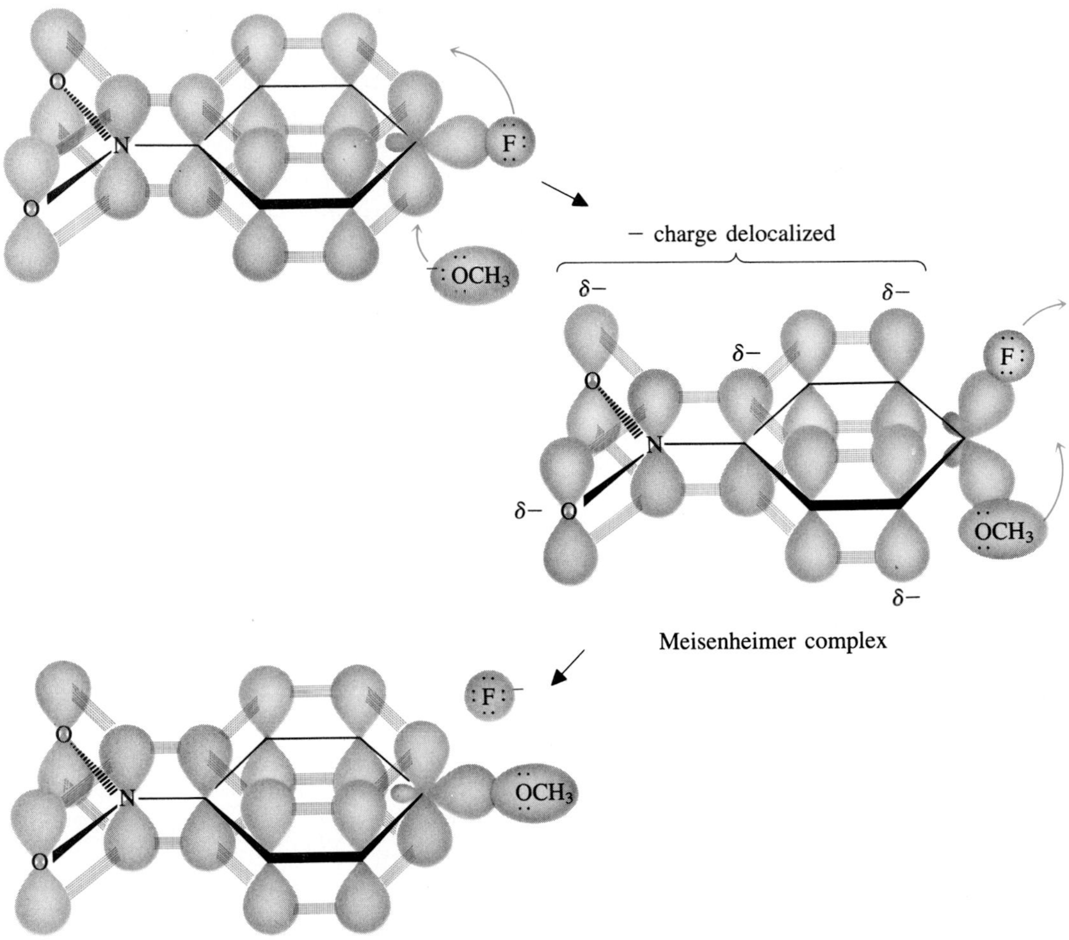

Figure 18.2 An orbital diagram for the nucleophilic aromatic substitution reaction of methoxide ion **($CH_3\ddot{O}:^-$)** with *p*-fluoronitrobenzene. The colored arrows show the movement of atoms. Notice that the nucleophile methoxide attacks at one face of the aromatic π-electron system and the leaving group fluoride departs from the opposite face.

The Meisenheimer complex breaks down to products by loss of the halide ion.

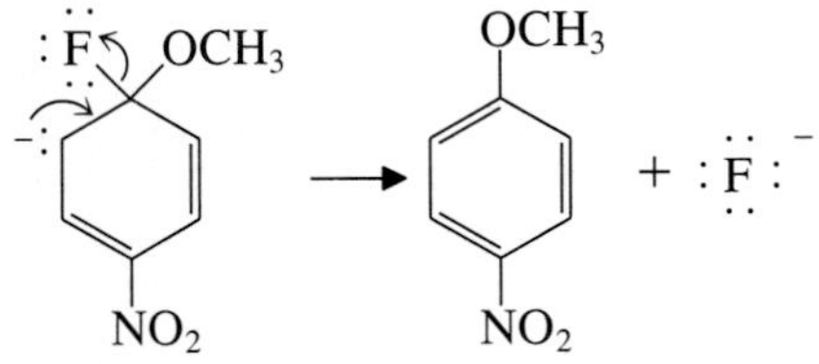

(18.17)

Figure 18.2 shows an orbital picture for this reaction.

Let us see how this mechanism fits the experimental facts. *Ortho* and *para* nitro groups accelerate the reaction because the transition state resembles the Meisenheimer complex, and the nitro groups stabilize this complex by resonance. Fluorine also stabilizes the negative charge by its electron-withdrawing *inductive effect,* which is greater than the inductive effect of the other halogens. Because the halide does not act as a leaving group *in the rate-determining step of the reaction,* the basicity of the halide is not important in determining the reaction rate.

Let us contrast the *nucleophilic aromatic substitution* mechanism with the *electrophilic aromatic substitution* mechanism that we studied in Chapter 16. Electrophilic aromatic substitution involves the attack of ring π-electrons on an electrophile (Lewis acid) to give a resonance-stabilized *carbocation* intermediate. The incoming electrophile is substituted for a ring *proton*. Nucleophilic aromatic substitution involves the attack of a nucleophile (Lewis base) on the ring π-electrons to give a resonance-stabilized *carbanion* intermediate. The incoming nucleophile is substituted for a leaving *halide.*

The effect of nitro substituent groups on the rate of each type of reaction is particularly striking: A nitro group *retards,* or deactivates, electrophilic aromatic substitution, because its electronegativity destabilizes carbocation intermediates; a nitro group *accelerates* nucleophilic aromatic substitution, because both its electronegativity and its electron-withdrawing resonance effect (when in the *ortho* or *para* position) stabilize carbanion intermediates.

Notice also how the nucleophilic aromatic substitution reaction differs from the S_N2 reaction of alkyl halides. First, there is an actual intermediate in the nucleophilic aromatic substitution reaction—the Meisenheimer complex. (In some cases this is sufficiently stable that it can be directly observed.) There is no evidence for an intermediate in any S_N2 reaction of alkyl halides. Second, the nucleophilic aromatic substitution reaction is an overall frontside displacement; it requires no inversion of configuration. The S_N2 reaction of an alkyl halide, in contrast, is a backside displacement with inversion of configuration. Finally, the effect of leaving group on the reaction rate (Eq. 18.14) is different in the two reactions. Aryl fluorides react most rapidly in nucleophilic aromatic substitution, whereas alkyl fluorides react most slowly in S_N2 reactions.

Problems

4 Complete the following reactions:

(a)

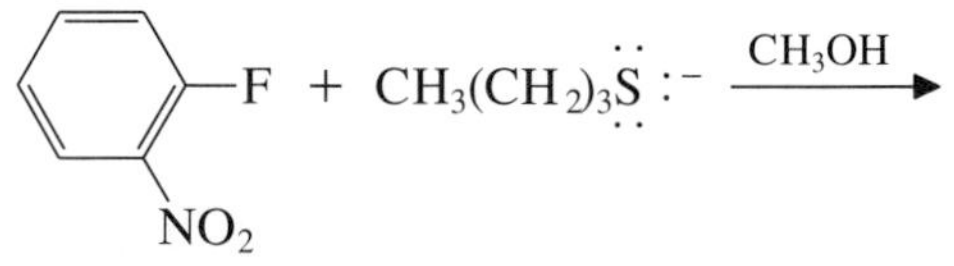

(b)

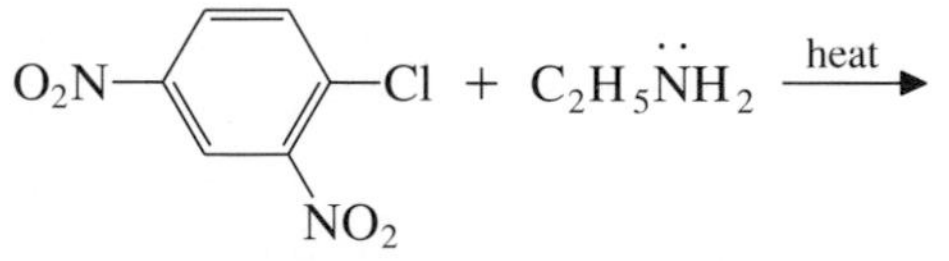

(c)

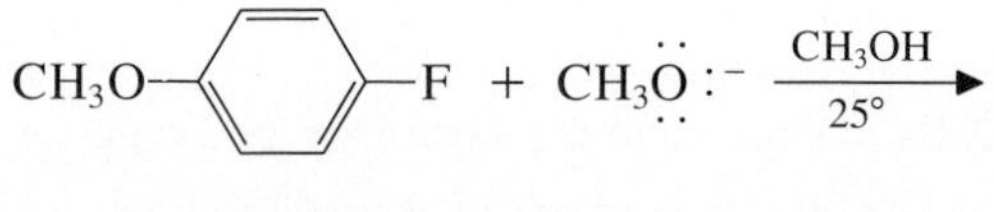

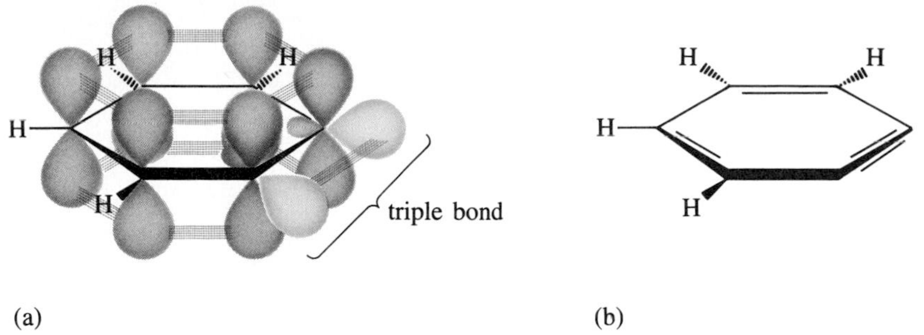

Figure 18.3 (a) Orbital picture of benzyne and (b) the corresponding Lewis structure. The orbitals of one π-bond (color) are perpendicular to the aromatic π-electron system of the ring.

5 Which of the two compounds in each set below should react more readily in a nucleophilic aromatic substitution reaction with CH_3O^- in CH_3OH? Explain your answers.

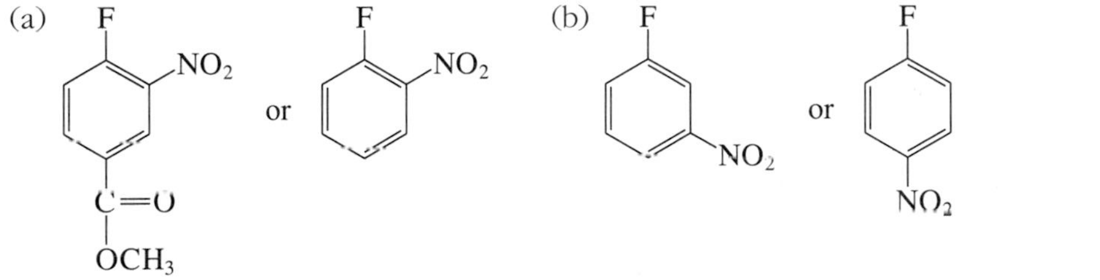

B. Substitution by Elimination–Addition: Benzyne

In Sec. 18.2 we learned that vinylic halides undergo β-elimination reactions under vigorous conditions. Let us contemplate what would happen if an aryl halide undergoes an analogous reaction. β-Elimination of an aryl halide would give an interesting "alkyne," called **benzyne.**

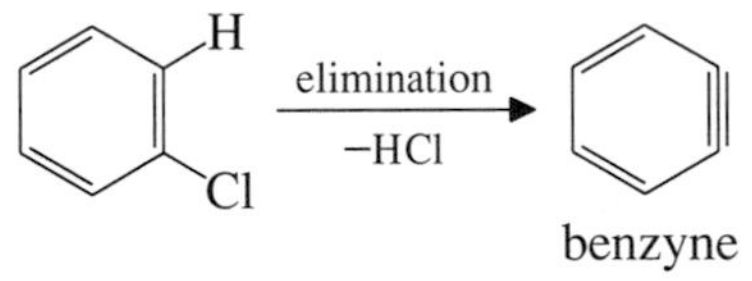

(18.18)

An orbital picture of benzyne is shown in Fig. 18.3. Notice that one of the two π-bonds in the triple bond of benzyne is perpendicular to the π-electron system of the aromatic ring. This extra π-bond is unusual, because the orbitals from which it is formed are more like sp^2 orbitals than p orbitals.

Because alkynes require linear geometry, it is difficult to incorporate them into six-membered rings. Therefore benzyne is highly *strained* (Problem 30) and, although it is a neutral molecule, it is very unstable. Indeed, benzyne has proven to be too reactive to isolate except at temperatures near absolute zero.

However, there is evidence that, despite its instability, benzyne is a reactive intermediate in certain reactions of aryl halides. For example, when chlorobenzene is treated with very strong bases such as potassium amide (KNH_2), a substitution reaction takes place.

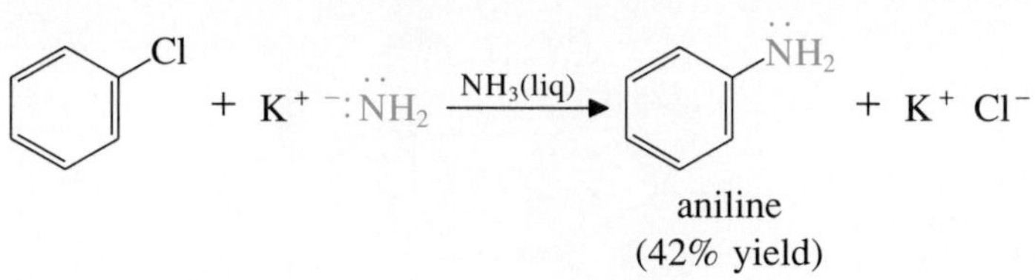

(18.19)

We know that this reaction cannot occur by the S_N1 or S_N2 mechanism for the reasons given in the previous section. Furthermore, since the aryl halide is not substituted with a nitro or other electron-withdrawing group, the structure of the aryl halide does not appear to be conducive to nucleophilic aromatic substitution.

Evidence for a mechanism involving a benzyne intermediate was obtained in a very elegant isotopic labeling experiment conducted by Professor John D. Roberts and his colleagues (of the California Institute of Technology). They carried out the reaction of Eq. 18.19 on a chlorobenzene sample in which the carbon bearing the chlorine (and *only* that carbon) was labeled with the radioactive isotope ^{14}C.

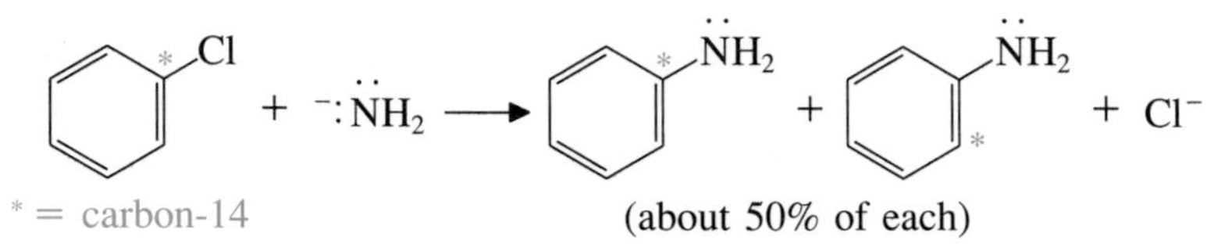

(18.20)

Analysis of the radioactivity in the product showed that in about half of the product, the radioactive carbon is *adjacent* to the substituted carbon, and in the other half, the radioactive carbon is the substituted carbon itself. Now, radioactivity cannot actually move from one carbon to the other. The only way to account for this result is that somehow *carbon-1 and carbon-2 have become equivalent and indistinguishable at some point in the reaction.* It is also important to notice that *only* carbon-1 and carbon-2 have become equivalent; the radioactivity is not found at any other carbons.

Let us see how a benzyne intermediate can account for these results. The first step in the mechanism of this reaction is formation of an anion at the *ortho* position. Because benzene derivatives are only weakly acidic (about as acidic as the corresponding alkenes; Sec. 14.7A), this requires a strong base.

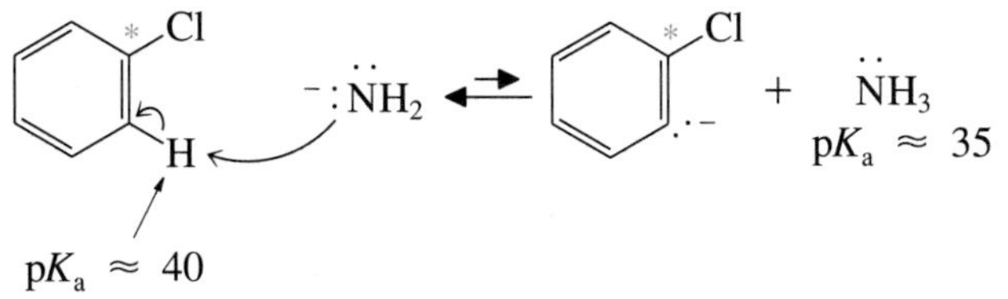

(18.21a)

This anion expels chloride ion, thus completing an elimination and forming benzyne.

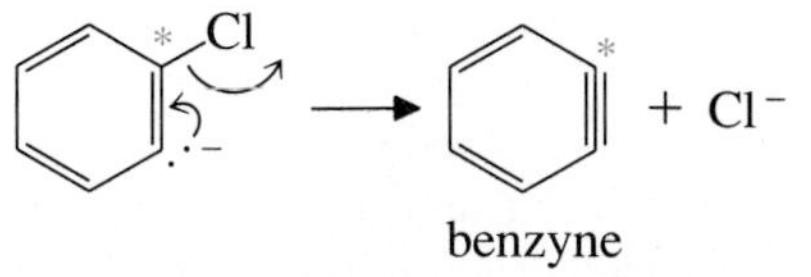

(18.21b)

Because benzyne is very unstable, it undergoes a reaction that is not typical of normal alkynes: *it is attacked by the amide ion to give a new anion.* Since benzyne is symmetric (except for its label), the carbons of the triple bond are indistinguishable. Hence, attack of $^{-}NH_2$ occurs with equal probability at each carbon, and the radioactive carbon ends up at two different positions relative to the $—NH_2$ group.

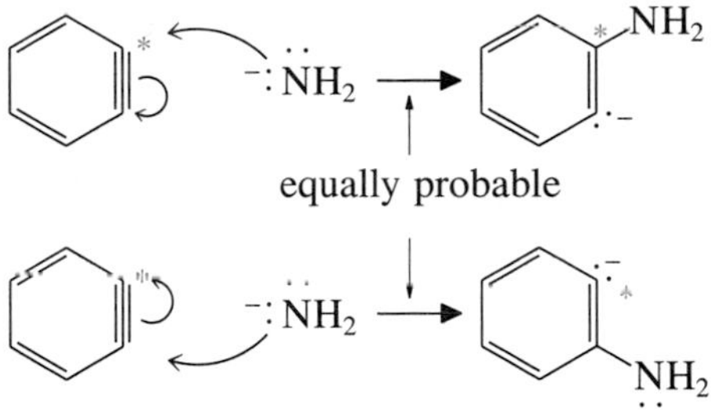

(18.21c)

Protonation of each anion gives the mixture shown in Eq. 18.20.

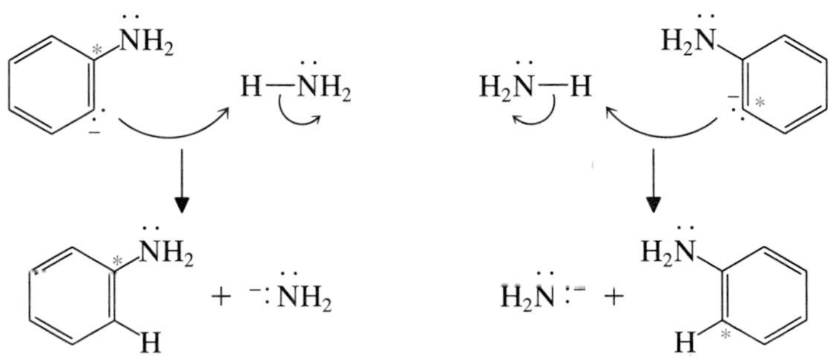

(18.21d)

The reaction that gives benzyne is a *β-elimination reaction*. It formally resembles an E2 elimination, except that the reaction occurs in two steps. (Recall that the E2 elimination takes place in a single step; Sec. 9.4A.) Benzyne is subsequently consumed in an *addition reaction*. Therefore, the overall displacement shown in Eq. 18.20 is really an *elimination–addition* process.

The benzyne mechanism accounts for the fact that, when an aryl halide bears ring substituents, more than one product is observed.

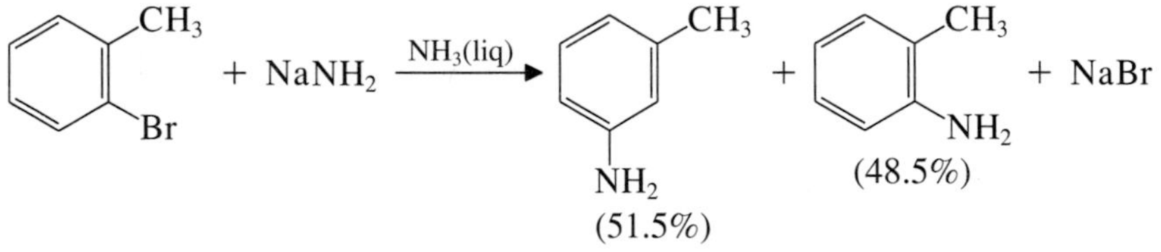

(18.22)

Problem

6 Explain why the benzyne mechanism predicts the formation of both products in Eq. 18.22.

We can see from these examples that when substitution occurs by the benzyne mechanism, some of the product results from attack of the nucleophile (for example, the amide ion) at the carbon *adjacent* to the one that bears the halide in the starting material. This type of substitution is called a **cine substitution** (*cine,* from Greek *kinēma,* movement). Thus the benzyne mechanism predicts a mixture of direct and *cine*-substitution products.

Although reactions involving benzyne intermediates are rarely important as preparative procedures, it is important to understand the benzyne mechanism because it explains some of the side reactions that occur when aryl halides are subjected to *strongly basic conditions*.

Problem

7 According to the benzyne mechanism, what product(s) are expected when each of the following compounds reacts with potassium amide in liquid ammonia?

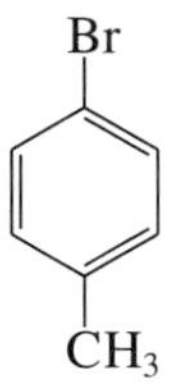

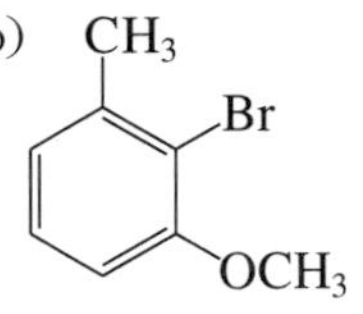

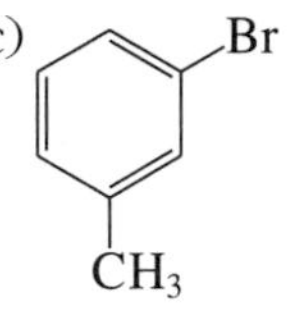

C. Substitution Reactions of Aryl Halides: Summary

We have seen that aryl halides undergo several different types of substitution reactions. Each reaction takes place under a specific set of conditions.

1. Aryl halides substituted with *ortho-* or *para*-nitro groups (or other conjugated, electron-withdrawing groups) react with Lewis bases in *nucleophilic aromatic substitution* reactions (Sec. 18.4A). This type of reaction involves resonance-stabilized anionic intermediates (Meisenheimer complexes) resulting from nucleophilic attack on the aromatic π-electron system, followed by loss of the halide from this ion.

2. Ordinary aryl halides undergo substitution by the *benzyne* mechanism (Sec. 18.4B) only in the presence of very strong bases such as alkali metal amides and organolithium reagents. This reaction involves β-elimination of the halide followed by addition of the nucleophile to benzyne. Except for the benzyne reaction, which requires a very strong base, ordinary alkyl halides do not undergo displacement reactions.

3. Aryl halides undergo *electrophilic aromatic substitution* (Secs. 16.4B, 16.5). However, this type of reaction involves the substitution of a *proton* of the aryl halide, not the halogen.

18.5 ARYL AND VINYLIC GRIGNARD REAGENTS

Preparation of Grignard and organolithium reagents from aryl and vinylic halides is analogous to the corresponding reactions of alkyl halides.

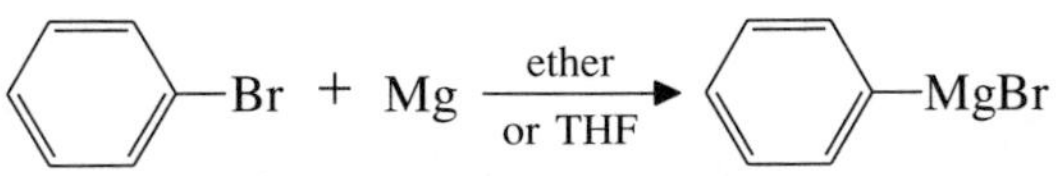

$$\text{C}_6\text{H}_5\text{—Br} + \text{Mg} \xrightarrow[\text{or THF}]{\text{ether}} \text{C}_6\text{H}_5\text{—MgBr} \qquad \text{(90–95\% yield)} \qquad (18.23)$$

Arylmagnesium bromides or iodides can be prepared from the corresponding bromo- or iodobenzenes in either tetrahydrofuran (THF) or diethyl ether solvent (see

Sec. 8.8A). However THF rather than ether *must* be used as the solvent when Grignard reagents are formed from chlorobenzenes.

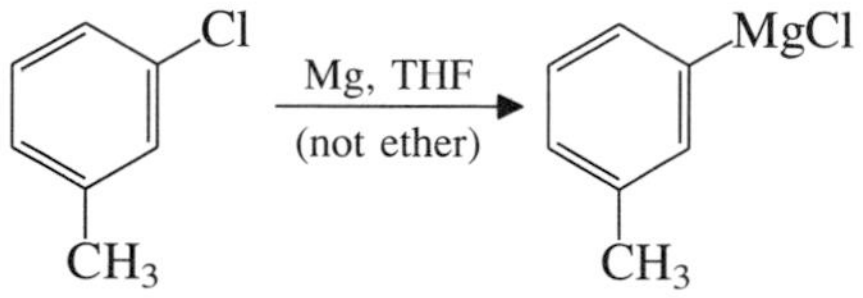

(96% yield) (18.24)

Because of this solvent effect, the following selective reaction is possible.

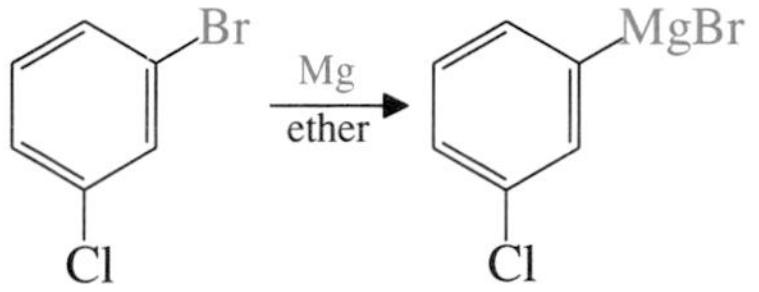

(18.25)

The preparation of vinylic Grignard reagents also requires THF.

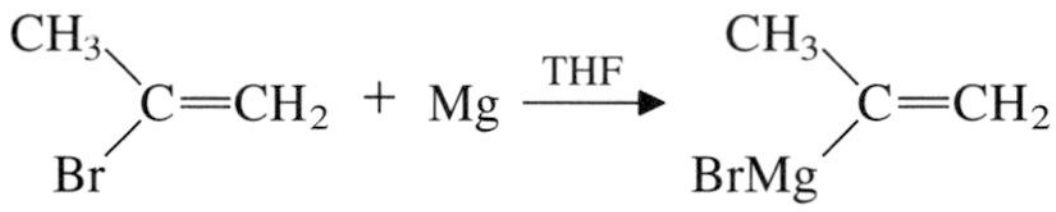

(18.26)

The reasons for these solvent effects are not clearly understood.

Problem

8 Outline a preparation of 2-phenyl-1-ethanol from benzene and any other reagents.

18.6 ACIDITY OF PHENOLS

A. Resonance Effects on Acidity

Phenols, like alcohols, can ionize.

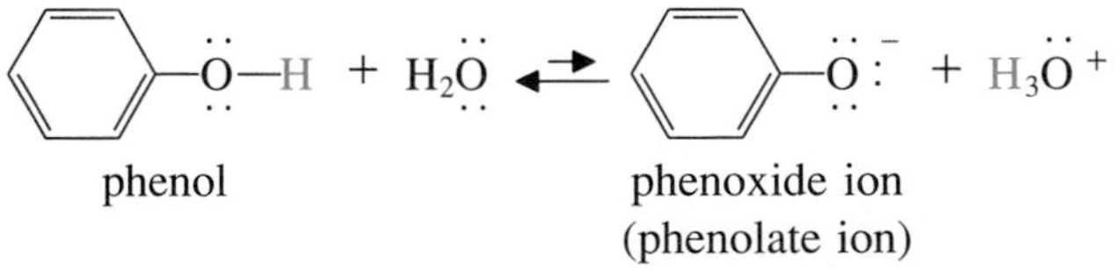

(18.27)

The conjugate base of a phenol is named, using common nomenclature, as a *phenoxide ion* or, using systematic nomenclature, as a *phenolate ion*. Thus, the sodium salt of phenol is called sodium phenoxide or sodium phenolate; the potassium salt of *p*-chlorophenol is called potassium *p*-chlorophenoxide or potassium 4-chlorophenolate.

Phenols are considerably more acidic than alcohols. For example, the pK_a of phenol is 9.95, but that of cyclohexanol is about 17. This means that phenol is approxi-

Figure 18.4 Stabilization of the phenoxide ion by resonance reduces the free energy of ionization of a phenol and therefore lowers its pK_a relative to that of an alcohol. (The free energies of cyclohexanol and phenol have been arbitrarily placed at the same level for comparison purposes.)

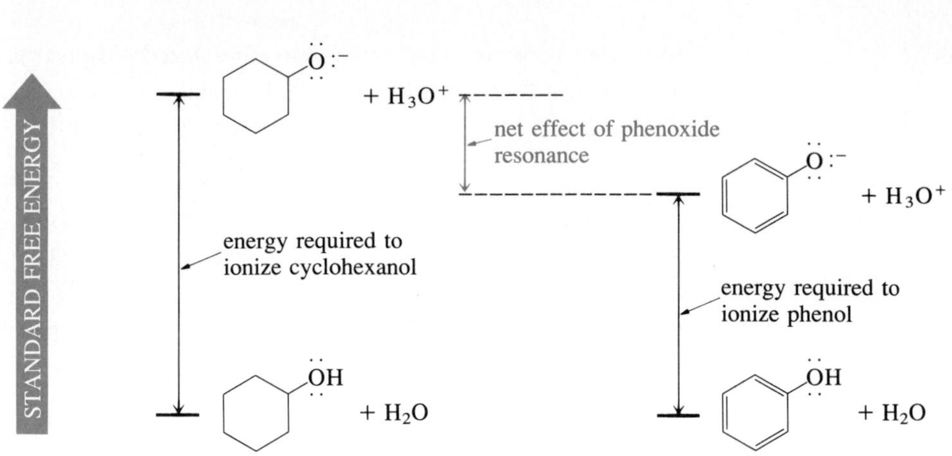

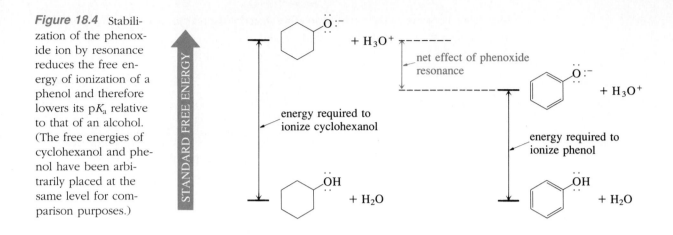

mately seven pK_a units, or 10^7 times, more acidic than an alcohol of similar size and shape.

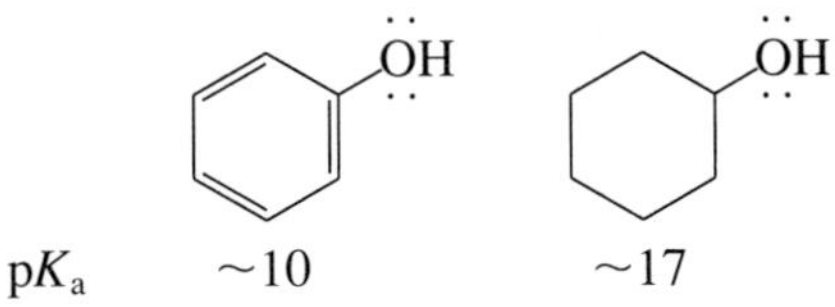

Recall from Fig. 8.4 and Eq. 8.17b that the pK_a of an acid is increased by stabilizing its conjugate base. *The enhanced acidity of phenol is due largely to stabilization of its conjugate-base anion by resonance.*

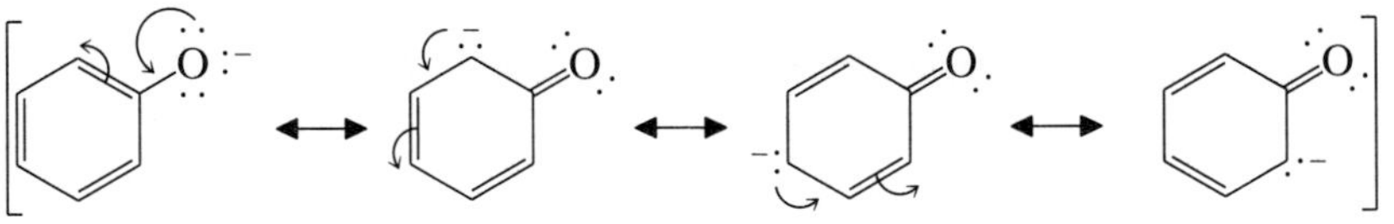

(18.28)

resonance structures for the phenoxide anion

There is no resonance stabilization of the alkoxide.

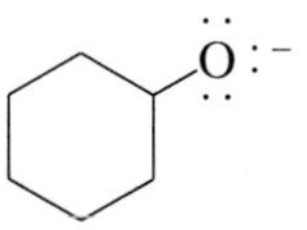

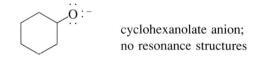

cyclohexanolate anion; no resonance structures

Because phenoxide ions are resonance stabilized, less energy is required to form phenoxides from phenols than is required to form alkoxides from alcohols. Therefore, phenols have lower free energies of ionization than alcohols. Since free energy of ionization is proportional to pK_a, phenols have lower pK_a values, and thus are more acidic, than alcohols (Fig. 18.4).

There is an important subtlety here. Neutral phenols are also stabilized by resonance. However, the resonance structures of a phenol involve *separation* of charge, but the structures of a phenoxide ion involve *dispersal* of charge.

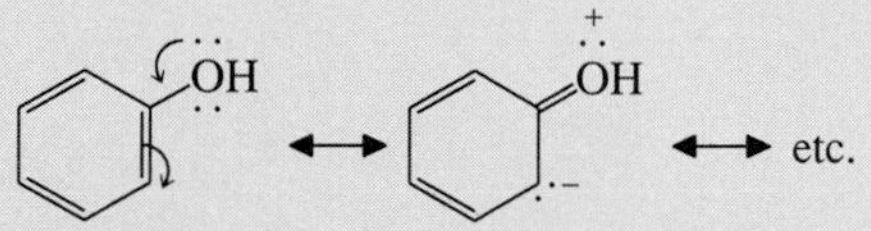

structures for phenol involve *separation* of charge

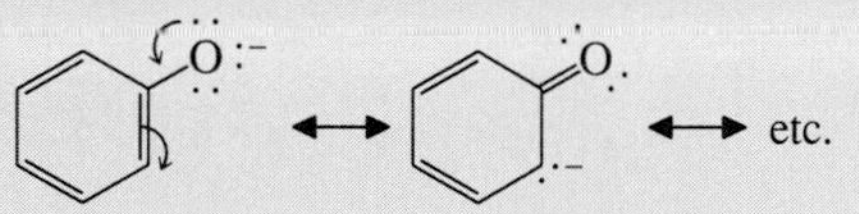

structures for phenolate ion involve *dispersal* of charge

Since structures that disperse charge are more important (Sec. 15.6B), a phenoxide ion is stabilized *more* by resonance than its conjugate-acid phenol. If the two were equally stabilized by resonance, the resonance effect would cancel when we compare the free energies of phenol and phenoxide and it would have no effect on the pK_a.

Substituent groups can affect the acidity of phenols by their *inductive effects,* as with alcohols. But a *resonance effect* also comes into play with some substituent groups. The relative acidities of phenol, *m*-nitrophenol and *p*-nitrophenol reflect the operation of both effects.

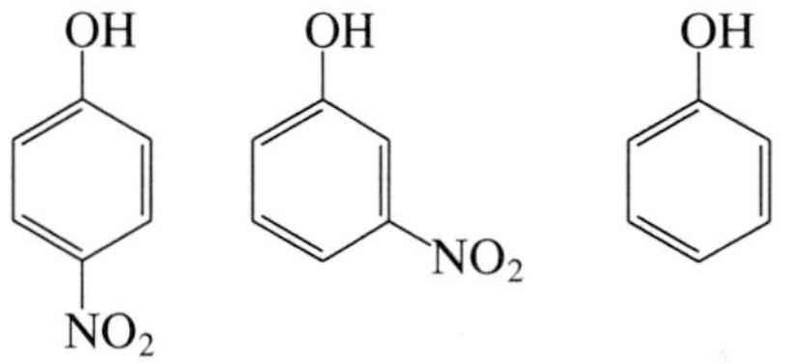

pK_a	7.21	8.35	9.95

The *m*-nitrophenol is more acidic than phenol because the nitro group is very electronegative. The inductive effect of the nitro substituent stabilizes the conjugate base anion for the same reason that it would stabilize the conjugate base of an alcohol (Sec. 8.6B). Yet *p*-nitrophenol is more acidic than *m*-nitrophenol by more than one pK_a unit, even though the *p*-nitro group is farther from the phenol oxygen. This cannot be an inductive effect, for inductive effects on acidity *decrease* in importance as the distance between the substituent and the acidic group increases. The reason for the increased acidity of *p*-nitrophenol is that the *p*-nitro group stabilizes the conjugate-base anion by resonance (colored structure).

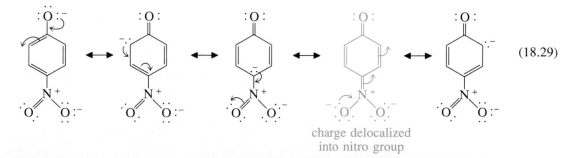

(18.29)

charge delocalized into nitro group

The colored structure is especially important because it places charge on the electronegative oxygen atom. In *m*-nitrophenol, however, it is not possible to draw resonance structures that delocalize the negative charge into the nitro group.

(18.30)

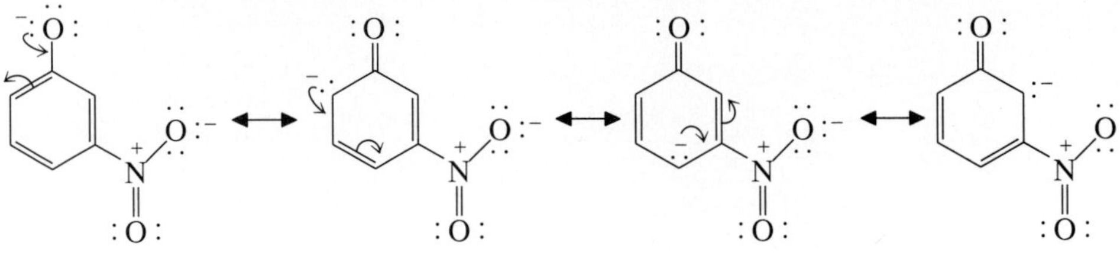

fewer important structures

Since *p*-nitrophenoxide has more resonance structures, it is more stable relative to its corresponding phenol than *m*-nitrophenoxide. Hence *p*-nitrophenol is the more acidic of the two nitrophenols. The acid-strengthening resonance effect of *ortho*- and *para*-nitro groups is so large that *picric acid* (2,4,6-trinitrophenol) is actually a strong acid.

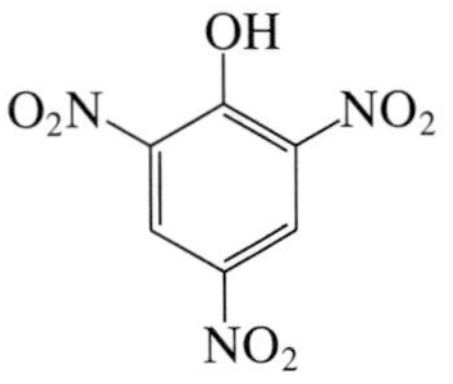

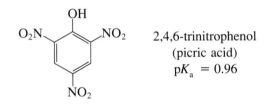

2,4,6-trinitrophenol
(picric acid)
$pK_a = 0.96$

Problem

9 Which of the two phenols in each set is the more acidic? Explain.
 (a) phenol or *m*-chlorophenol
 (b) 2,5-dinitrophenol or 2,4-dinitrophenol
 (c)

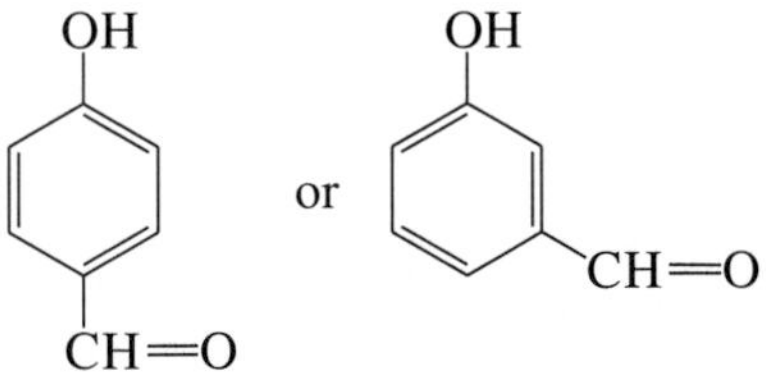

 (d) OH

(images showing structures with $\overset{+}{N}(CH_3)_3$ groups)

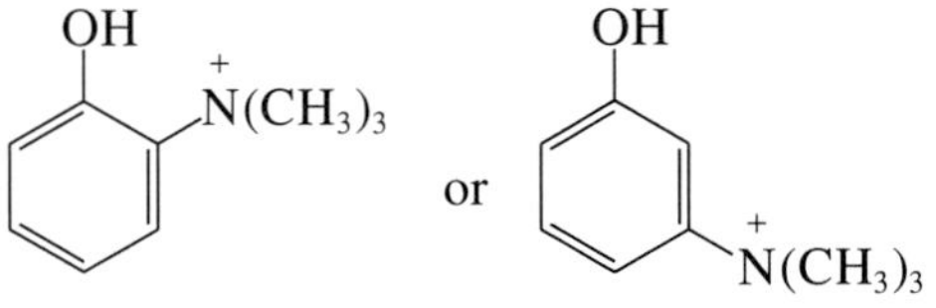

B. Formation and Use of Phenoxides

We learned in Sec. 8.6A that alcohols are not converted completely into alkoxides by aqueous NaOH solution because the pK_a values of water and alcohols are similar. In contrast, the equilibrium for the reaction of phenol and NaOH lies well to the right:

$$\text{C}_6\text{H}_5\text{—O—H} + \text{:}^-\text{OH} \;\rightleftharpoons\; \text{C}_6\text{H}_5\text{—O:}^- + \text{H}_2\text{O:} \qquad (18.31)$$

$pK_a = 9.95 \qquad\qquad\qquad\qquad pK_a = 15.7$

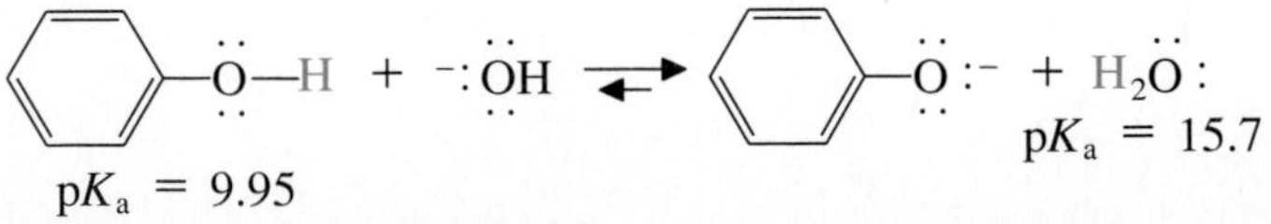

Because the difference in pK_a values of water and phenol is about 6, the equilibrium constant for this reaction is about 10^6. Thus, for all practical purposes, phenols are converted completely into their conjugate-base anions by NaOH solution. Although the stronger bases used to ionize alcohols (Sec. 8.6A) can also be used for phenols, hydroxide ion is perfectly adequate for the purpose. Thus when phenol is treated with one equivalent of sodium hydroxide solution, the phenol O—H proton is titrated completely to give a solution of sodium phenoxide.

The acidity of phenols can sometimes be used to separate them from mixtures with other organic compounds. For example, suppose that we wish to separate the water-insoluble phenol, 4-chlorophenol, from a water insoluble alcohol, 4-chlorocyclohexanol. Although the phenol itself is water insoluble, its sodium or potassium salt, like many other alkali metal salts, has considerable solubility in water because it is an *ionic compound.* (Recall from Sec. 8.4B that water is one of the best solvents for ionic compounds.)

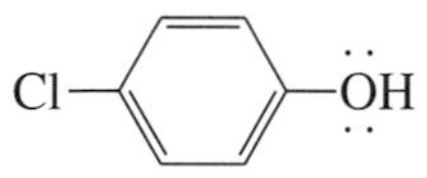

4-chlorophenol:
soluble in ether
insoluble in water

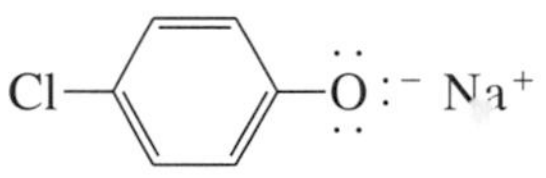

sodium *p*-chlorophenoxide
insoluble in ether
soluble in water
(an ionic compound)

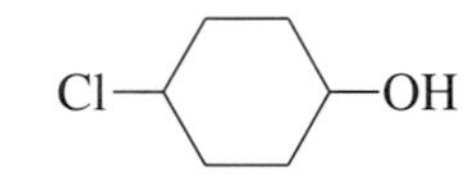

4-chlorocyclohexanol
soluble in ether
insoluble in water

The technique used to separate the phenol and the alcohol is summarized in Fig. 18.5. The mixture is dissolved in a solvent—for example, ether. Then aqueous NaOH solution is added to the ether solution and the mixture is vigorously stirred or shaken. Since ether is not very soluble in water, the resulting mixture contains two layers. The water layer contains sodium *p*-chlorophenoxide, because the NaOH has converted the phenol into its water-soluble conjugate base. The ether layer contains the 4-chlorocyclohexanol, because this alcohol is not acidic enough to be converted into its conjugate base by dilute NaOH solution. If we remove the water layer and acidify it with HCl, the *p*-chlorophenoxide is converted by protonation back into *p*-chlorophenol, which comes out of solution. The phenol may be re-extracted into fresh ether and isolated in pure form by drying and concentrating the ether solution. The original ether solution of 4-chlorocyclohexanol can also be dried and concentrated and the pure chlorocyclohexanol can be recovered.

It is usually said that phenols are "soluble in sodium hydroxide solution." What is really meant by this statement is that if sodium hydroxide is added to a phenol, the phenol is converted into its sodium salt, which is the species that actually dissolves. Solubility in 5% NaOH solution is an important qualitative test for phenols.

Problems

10 Would a separation scheme like the one just discussed be useful for separating phenol and ethanol? Explain. If not, suggest a method for separating these compounds.

11 The following compound, unlike most phenols, is *soluble* in acid, but *insoluble* in base. Explain this unusual behavior.

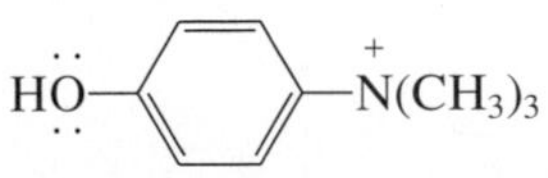

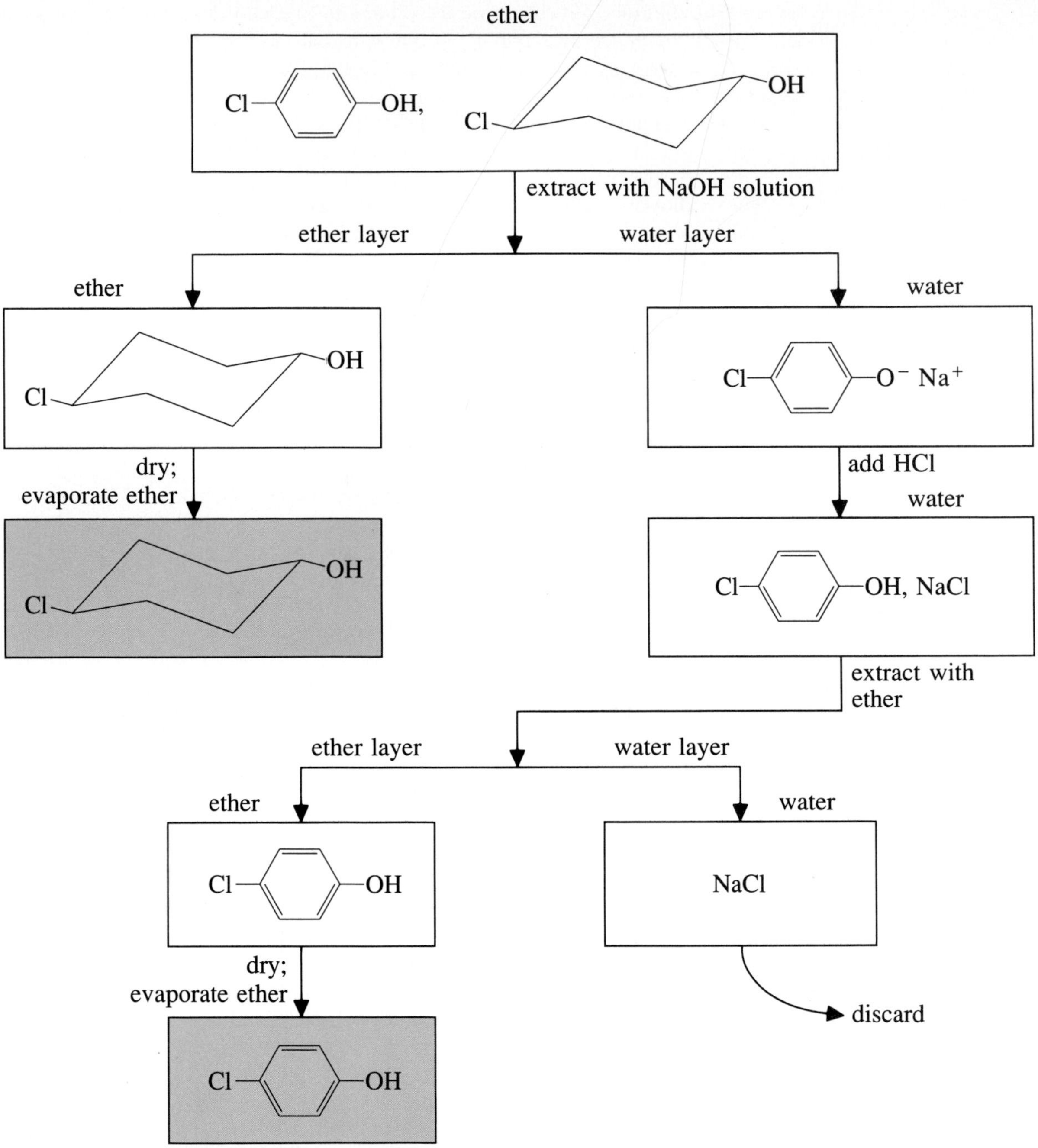

Figure 18.5 A flow sheet for the separation of a phenol and an alcohol. Boxes represent solutions, lines represent operations, and colored boxes represent pure materials.

Phenoxides, like alkoxides, can be used as nucleophiles. For example, aryl ethers can be prepared by the reaction of a phenoxide anion and an alkyl halide.

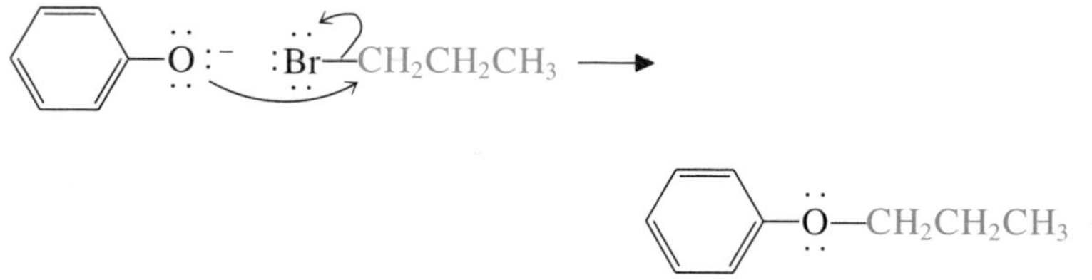

$$
\underset{\substack{\text{1-phenoxypropane}\\(63\%\ \text{yield})}}{\text{Ph}-\overset{..}{\underset{..}{O}}-CH_2CH_2CH_3} \; + \; :\overset{..}{\underset{..}{Br}}:^- \quad (18.32)
$$

This is another example of the Williamson ether synthesis (Sec. 11.1A).

Problem **12** Outline a preparation of 2-phenoxy-1-ethanol from phenol and any other reagents.

18.7 OTHER REACTIONS OF PHENOLS

In the previous section we learned about a similarity between phenols and alcohols: their ability to act as acids. In this section we shall study some of the reactions of phenols not shared by alcohols.

A. Oxidation of Phenols to Quinones

Since tertiary alcohols are not readily oxidized, we might also expect phenols to resist oxidation. Yet it is found that phenols are all oxidized to *quinones*.

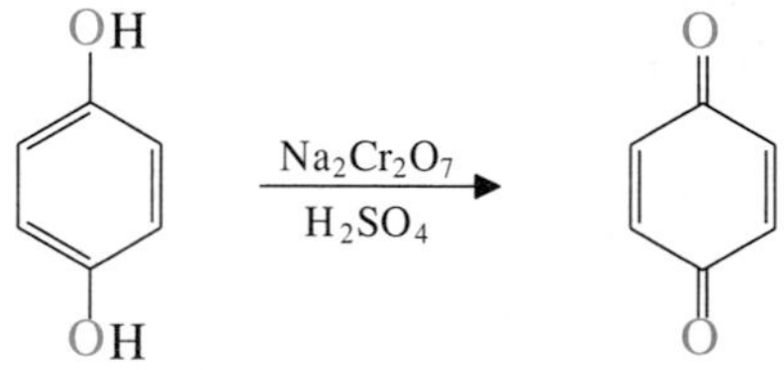

hydroquinone

p-benzoquinone
(86−92% yield)

(18.33)

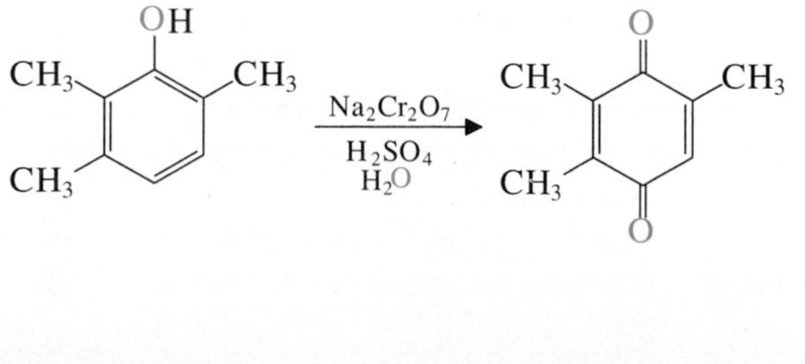

(50% yield) (18.34)

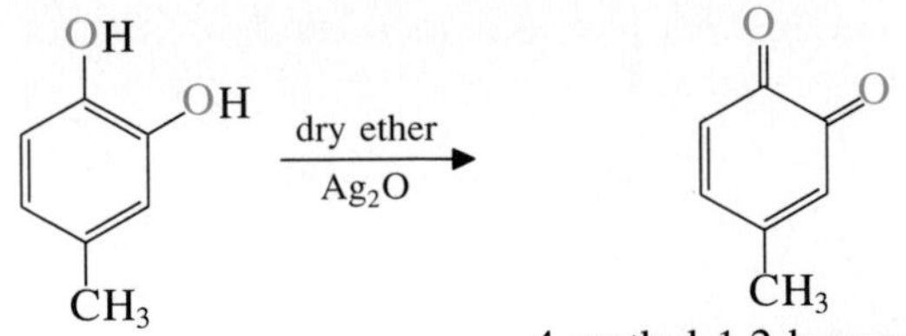

4-methyl-1,2-benzoquinone
(unstable red crystals)

(68% yield) (18.35)

Notice that *p*-hydroxyphenols (hydroquinones), *o*-hydroxyphenols (catechols), and phenols with an unsubstituted position *para* to the hydroxy group are oxidized to quinones.

Although the term **quinone** is sometimes used as a common name of *p*-benzoquinone (Eq. 18.33), it is a generic name for any compound containing either of the following structural units.

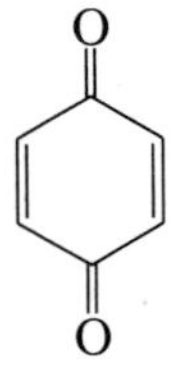

a *p*-quinone

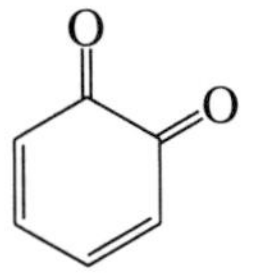

an *o*-quinone

If the quinone oxygens have a 1,4 (*para*) relationship, the quinone is called a *para*-quinone; if the oxygens are in a 1,2 (*ortho*) arrangement, the quinone is called an *ortho*-quinone. The following compounds are typical quinones.

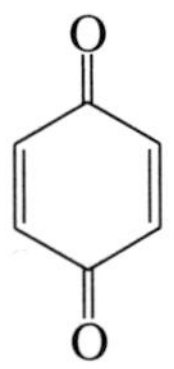

p-benzoquinone
(1,4-benzoquinone)

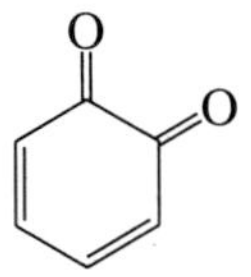

o-benzoquinone
(1,2-benzoquinone)

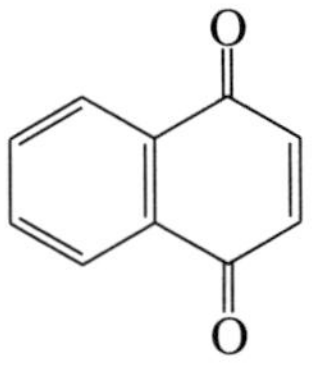

1,4-naphthoquinone

derived from

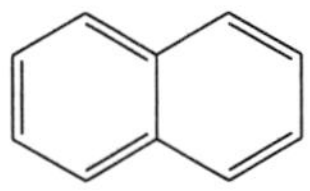

naphthalene

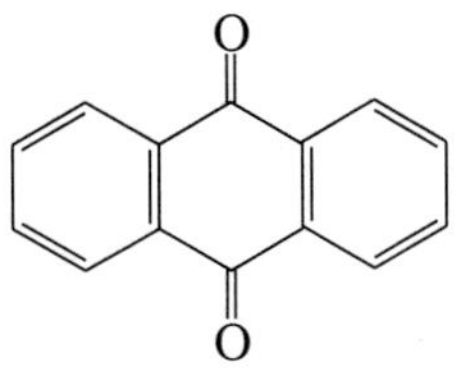

9,10-anthroquinone

derived from

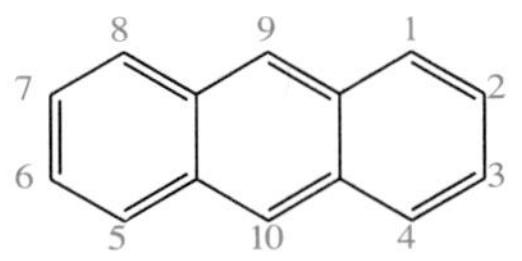

anthracene

The names of quinones are derived from the names of the corresponding aromatic hydrocarbons: *benzo*quinone is derived from benzene, *naphtho*quinone from naphthalene, and so on.

Ortho-quinones, particularly *ortho*-benzoquinones, are considerably less stable than the isomeric *para*-quinones. One reason for this difference is that in *ortho*-

quinones, the C=O bond dipoles are nearly aligned, and therefore have a repulsive, destabilizing interaction. In *p*-quinones these dipoles are further apart.

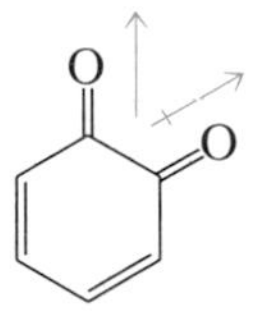

o-quinone:
bond dipoles nearly aligned

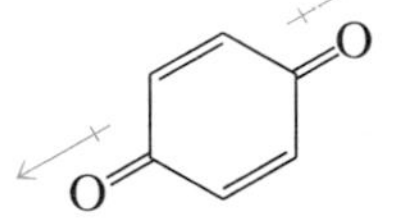

p-quinone
bond dipoles further apart

A number of quinones occur in nature. *Juglone* occurs in walnut shells. *Doxorubicin* (adriamycin), isolated from a microorganism, is an important antitumor drug.

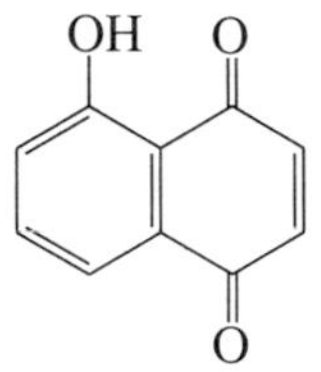

juglone

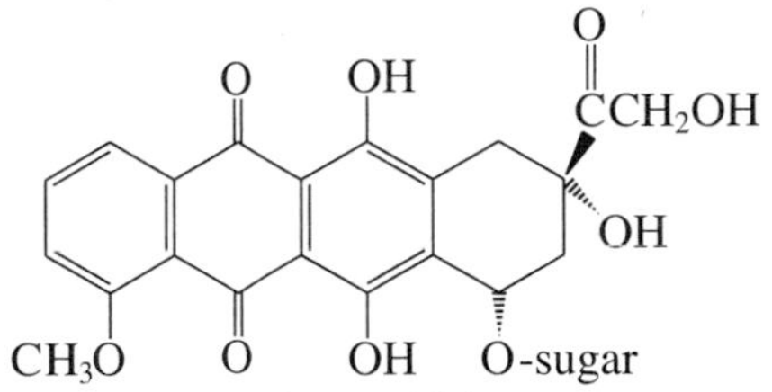

doxorubicin
(Adriamycin)

Problem

13 Draw the structure of 9,10-phenanthroquinone, which is formally derived from the hydrocarbon phenanthrene. Is 9,10-phenanthroquinone an *o*- or a *p*-quinone?

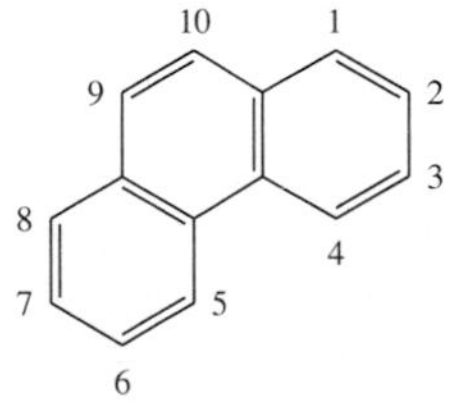

phenanthrene

The oxidation of phenols by air (O_2) to colored, quinone-containing products is responsible for the darkening that is observed when some phenols are stored for long periods of time. The oxidation of hydroquinone and its derivatives to the corresponding *p*-benzoquinones can also be carried out reversibly in an electrochemical cell. Oxidation potentials of a number of phenols with respect to standard electrodes are well known.

The oxidation of phenols has several practical applications. For example, hydroquinone and other phenols can be used as inhibitors of free-radical chain reactions (Sec. 5.8B). Many free radicals (R· in Eq. 18.36a) abstract a hydrogen from hydroquinone to form a very stable radical called a *semiquinone*.

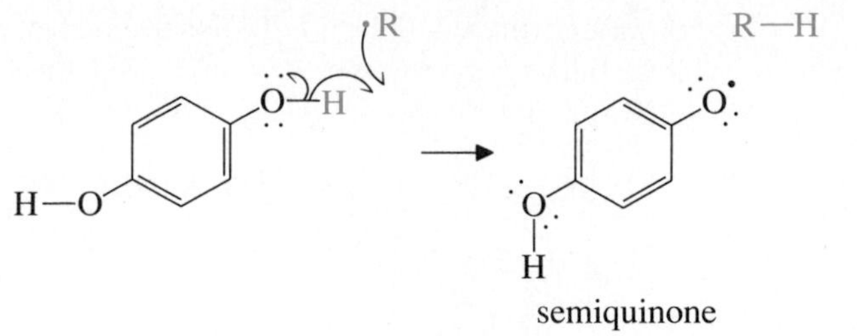

(18.36a)

(This radical is resonance stabilized; draw its resonance structures.) A second free radical can react with the semiquinone to complete its oxidation to quinone.

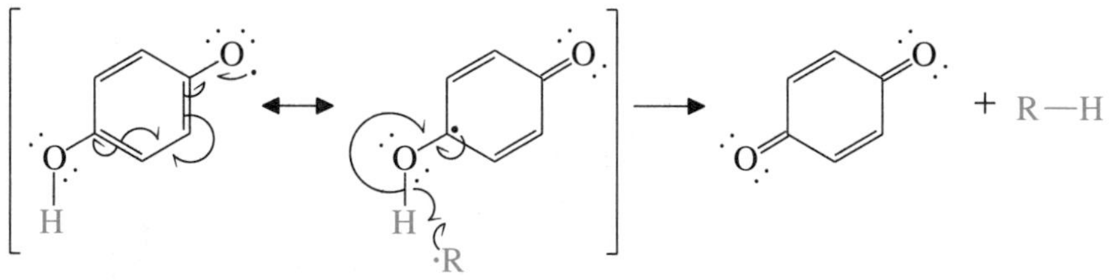

(18.36b)

Hydroquinone thus terminates free-radical chain reactions by intercepting free radical intermediates R· and reducing them to R—H.

Several widely used food preservatives are phenols. Two of these are "butylated hydroxytoluene" (BHT) and "butylated hydroxyanisole" (BHA).

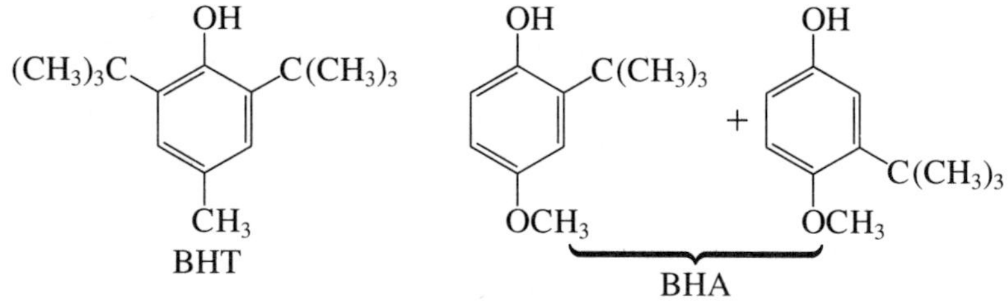

Oxidation involving free-radical processes is one way that foods discolor and spoil. A preservative such as BHT inhibits these processes by donating its O—H hydrogen atom to free radicals in the food (as in Eq. 18.36a). The BHT is thus transformed into a phenoxy radical, which is too stable and unreactive to propagate radical chain reactions. (See Problem 42.) Although the use of BHT and BHA as food additives has generated some controversy because of their potential side effects, without such additives foods could not be stored for reasonable periods of time and transported over long distances.

Recent research indicates that vitamin E, a phenol, is the major compound in the blood responsible for preventing oxidative damage by free radicals. Vitamin E acts by terminating radical chains in the manner shown in Eq. 18.36a.

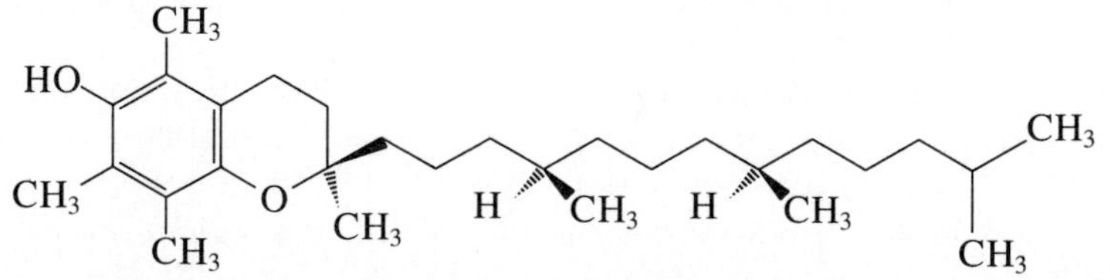

α-tocopherol, a major form of vitamin E

The oxidation of hydroquinone lies at the heart of the photographic process. When photographic film is exposed to light, grains of silver bromide in the photographic emulsion on the film absorb light and are activated or *sensitized.*

$$AgBr + \text{light} \longrightarrow \underset{\text{sensitized silver bromide}}{[AgBr]^*} \quad (18.37)$$

Because silver bromide is trapped in the photographic emulsion, it is immobile. Thus sensitized silver bromide molecules provide a faithful record of the positions on the film that have been struck by light. Now, sensitized silver bromide is a much better oxidizing agent than silver bromide that has not been exposed to light. When exposed film is treated with a solution of hydroquinone (a common photographic developer), [AgBr]* oxidizes the hydroquinone to quinone (which is subsequently washed away), and the silver(I) is reduced to finely divided silver metal, which remains trapped in the photographic emulsion.

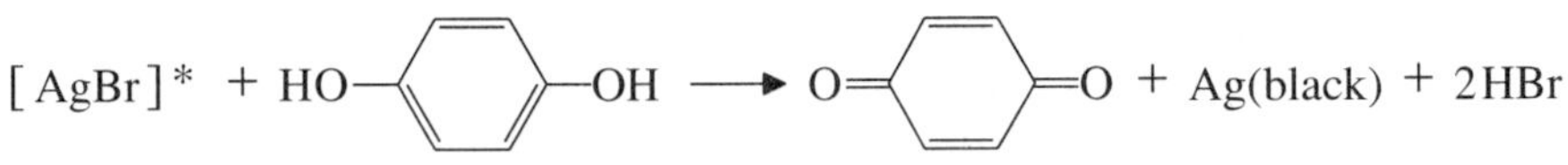
[AgBr]* + HO
OH
O
O + Ag(black) + 2HBr
(18.38)

Because unactivated AgBr oxidizes hydroquinone much more slowly, silver metal forms only where light has impinged on the film. This precipitated silver is the black part of a black-and-white negative.

Problem

14 Complete the following reactions:

(a)

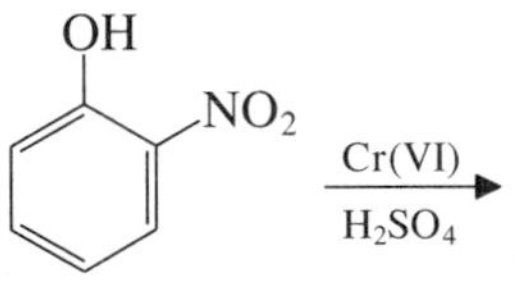
OH
NO2
Cr(VI)
H2SO4

(b)

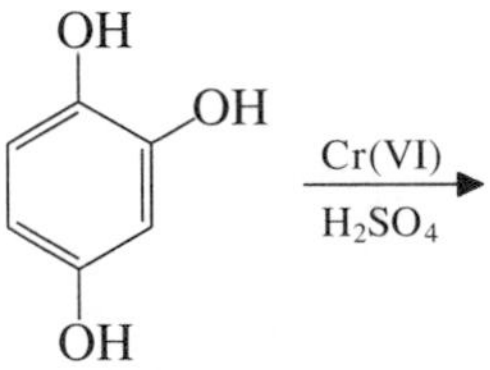
OH
OH
OH
Cr(VI)
H2SO4

B. Electrophilic Aromatic Substitution Reactions of Phenols

Phenols are aromatic compounds, and they undergo electrophilic aromatic substitution reactions like those described in Chapter 16. Some of these reactions are worth special attention because the —OH group has special effects that are not common to other substituent groups.

Because the —OH group is a strongly activating substituent, phenol may be halogenated once under mild conditions that do not affect benzene itself.

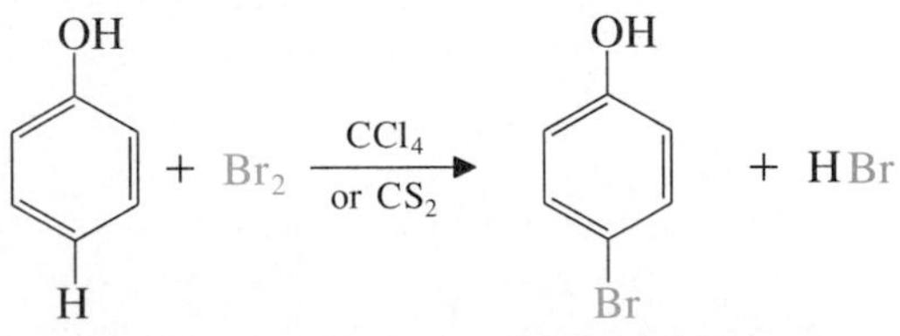
OH
H
+ Br2
CCl4
or CS2
OH
Br
+ HBr
(18.39)

(82% yield)

(A solution of Br_2 in CCl_4 is the reagent usually used for adding bromine to alkenes.) But when phenol reacts with Br_2 in H_2O (bromine water), more extensive bromination occurs and tribromophenol is obtained (Eq. 16.38).

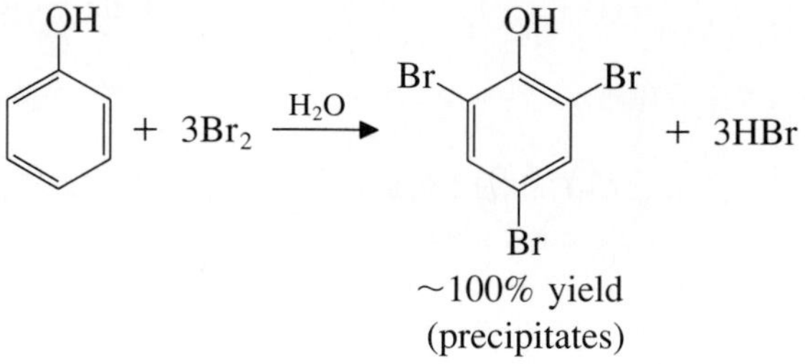

(18.40)

This more extensive bromination occurs for two reasons. First, bromine reacts with water to give protonated hypobromous acid, a more potent electrophile than bromine itself.

$$Br_2 + H_2\ddot{O}: \rightleftharpoons Br^- + \underset{\text{protonated hypobromous acid}}{Br\!-\!\overset{+}{\ddot{O}}H_2} \rightleftharpoons HBr + \underset{\text{hypobromous acid}}{Br\!-\!\ddot{\underset{..}{O}}H} \qquad (18.41)$$

Second, in aqueous solutions near neutrality, phenol partially ionizes to its conjugate-base phenoxide anion. Although only a small amount of this anion is present, it is very reactive and brominates instantly, thereby pulling the phenol–phenolate equilibrium to the right.

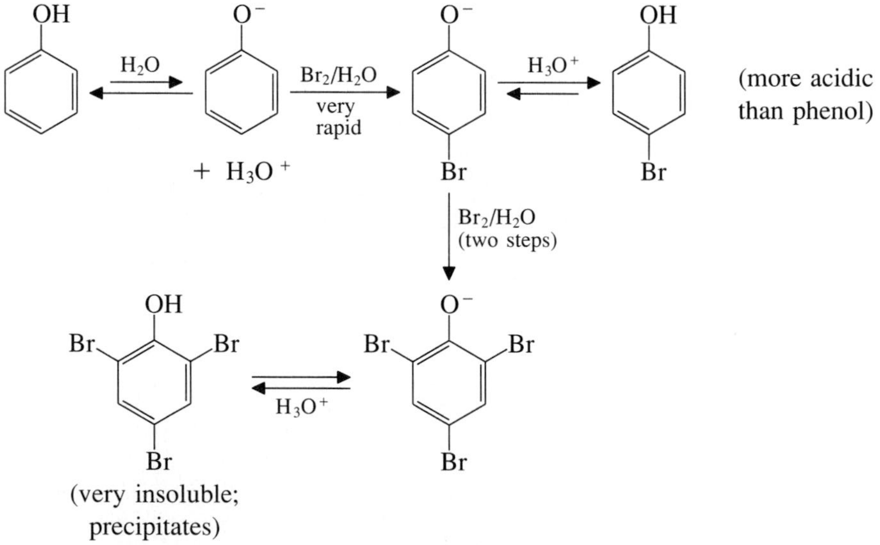

(18.42)

Phenoxide ion is much more reactive than phenol because the "carbocation" intermediate is not really an ion, but instead a more stable neutral molecule (colored structure).

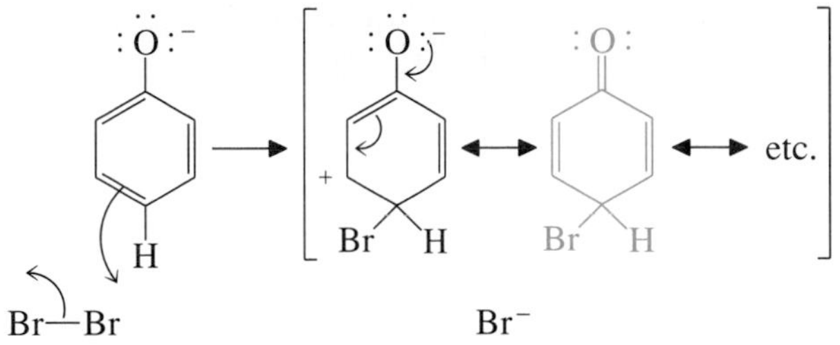

(18.43)

The *p*-bromophenol is also in equilibrium with its conjugate base *p*-bromophenoxide anion, which brominates again until all *ortho* and *para* positions have been substituted. Notice in Eq. 18.40 that in the second and third substitutions the powerful *ortho, para* directing and activating effects of the $—\ddot{\underset{..}{O}}:^-$ group override the weaker deactivating and directing effects of the bromine substituents. In strongly acidic solution, in which formation of the phenolate anion is suppressed, bromination can be stopped at the 2,4-dibromophenol stage.

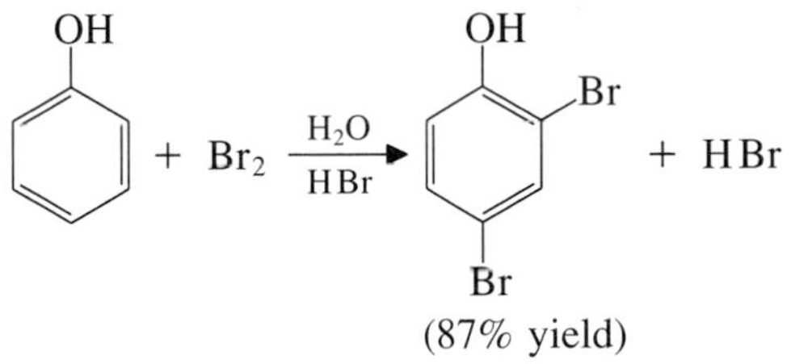

(18.44)

Phenol is also very reactive in other electrophilic substitution reactions, such as nitration. Phenol can be nitrated once under mild conditions. (Notice that H_2SO_4 is not present as it is in the nitration of benzene; Eq. 16.7.)

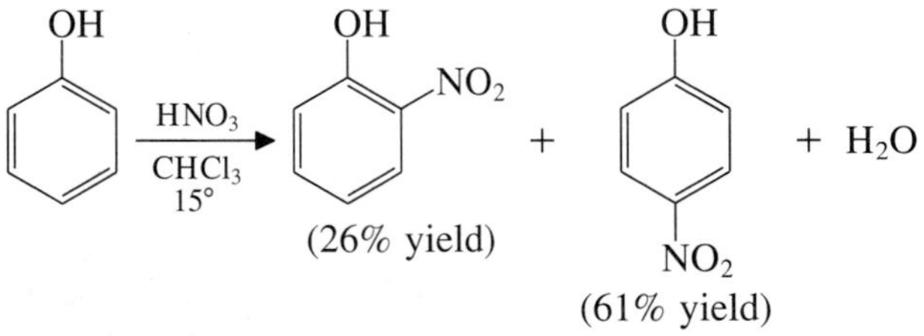

(18.45)

Because phenol is so activated toward electrophilic substitution, it is also possible to nitrate phenol two and three times. However, direct nitration is *not* the preferred method for synthesis of di- and trinitrophenol, because the concentrated HNO_3 required for nitration is an oxidizing agent (Sec. 10.6C) and phenols are easily oxidized (Sec. 18.7A). Thus 2,4-dinitrophenol is synthesized instead by the nucleophilic aromatic substitution reaction of 1-chloro-2,4-dinitrobenzene with OH^- (Sec. 18.4A).

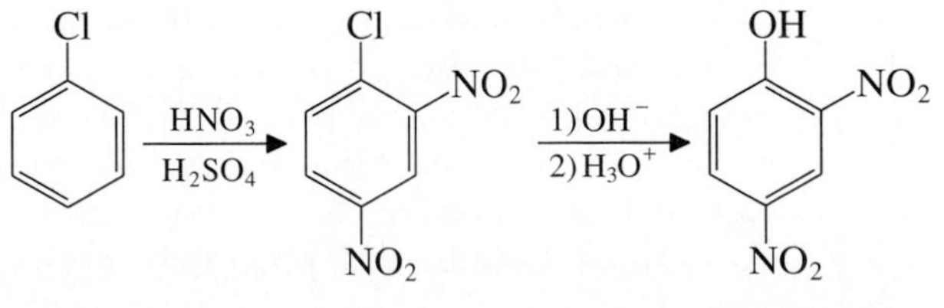

(18.46)

The great reactivity of phenol in electrophilic aromatic substitution does not extend to the Friedel–Crafts acylation reaction, because phenol reacts rapidly with the $AlCl_3$ catalyst.

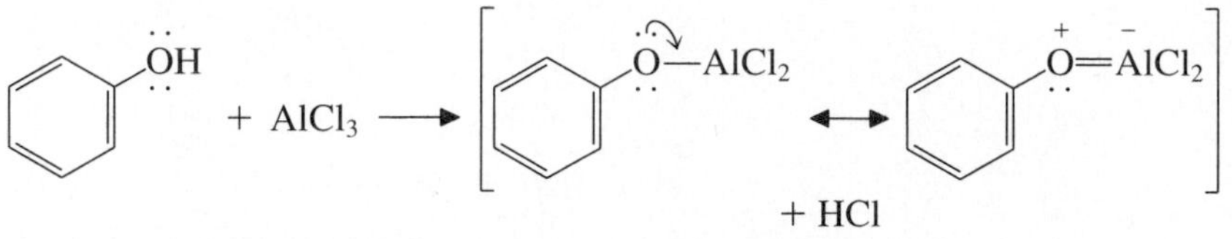

$$+ HCl \tag{18.47}$$

The adduct of phenol and $AlCl_3$ is much less reactive than phenol itself in electrophilic aromatic substitution reactions because, as shown in Eq. 18.47, the oxygen electrons are delocalized onto the electron-deficient aluminum. Because of their delocalization away from the benzene ring, these electrons are less available for resonance stabilization of the carbocation intermediate formed within the ring during Friedel–Crafts acylation (Eq. 16.16). Thus, Friedel–Crafts acylation of phenol occurs slowly, but may be carried out successfully at elevated temperatures. Because it is not highly activated, the ring is acylated only once.

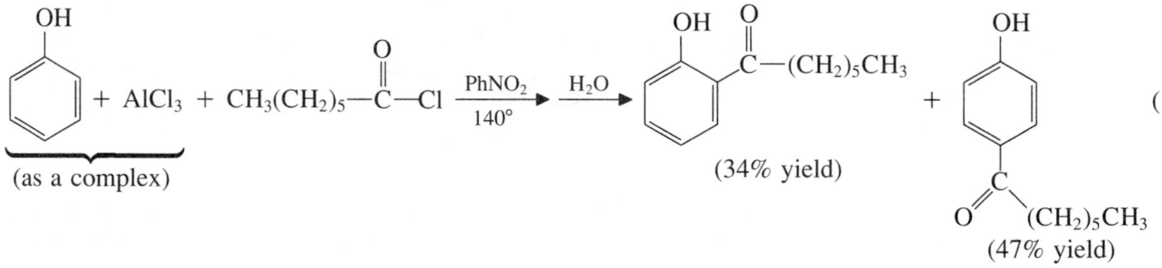

$$\tag{18.48}$$

Problem

15 Give the principal organic product(s) formed in each of the following reactions:

(a) *o*-cresol + Br_2 in CCl_4 ⟶

(b) *m*-chlorophenol + HNO_3, low temperature ⟶

(c) *p*-cresol + Br_2 in H_2O ⟶

(d)

$$\text{*p*-bromophenol} + CH_3-\overset{\overset{\displaystyle O}{\|}}{C}-Cl \xrightarrow{AlCl_3} \xrightarrow{H_3O^+}$$

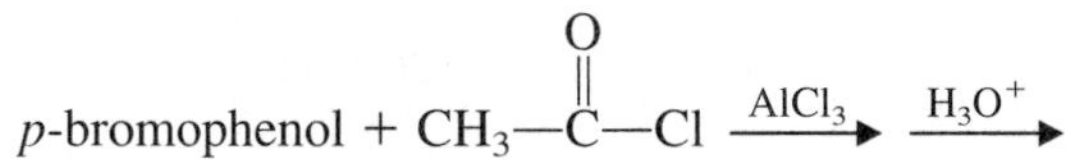

C. Lack of Reactivity of the Phenolic Carbon–Oxygen Bond

We know that alcohols react with hydrogen halides to form alkyl halides (Sec. 10.2), and with H_2SO_4 to form alkenes by dehydration (Sec. 10.1). Both of these reactions have in common the *breaking of the C—O bond*. What happens to phenols under the same conditions?

Just as the reactions of alcohols that break the carbon–oxygen bond have close analogy to the reactions of alkyl halides that break the carbon–halogen bond, we would expect that the carbon–oxygen reactivity of phenols should follow the carbon–halogen reactivity of *aryl* halides. (That is, we can think of phenols as "aryl alcohols.") We learned in Sec. 18.3 that aryl halides do not undergo S_N1 or E1 reactions; for the same reasons, phenols also do not react under conditions used for the S_N1 or E1 reactions of alcohols. Thus phenols do *not* form aryl bromides with concentrated HBr;

they do *not* dehydrate with concentrated H_2SO_4. The reasons for these observations parallel exactly those used to explain the lack of reactivity of aryl halides. (What are these reasons?)

We also learned in Sec. 18.1 that aryl halides do not undergo S_N2 reactions. Similarly, sulfonate esters of phenols (Sec. 10.3A) do not react in S_N2 reactions for the same reasons that aryl halides do not react.

Problem	**16** Although phenol does not dehydrate in the presence of concentrated H_2SO_4, it does react in another way. Give the products of this reaction. (*Hint:* See Sec. 16.4.)

18.8 INDUSTRIAL PREPARATION AND USE OF PHENOL

Historically, phenol has been made in a variety of ways, but the principal method used today is an elegant example of a process that gives two industrially important compounds, phenol and acetone, $(CH_3)_2C{=}O$, from a single starting material. The starting material for the manufacture of phenol is cumene, which comes from benzene and propylene, two compounds obtained from petroleum (Sec. 16.11). The production of phenol and acetone is a two-step process.

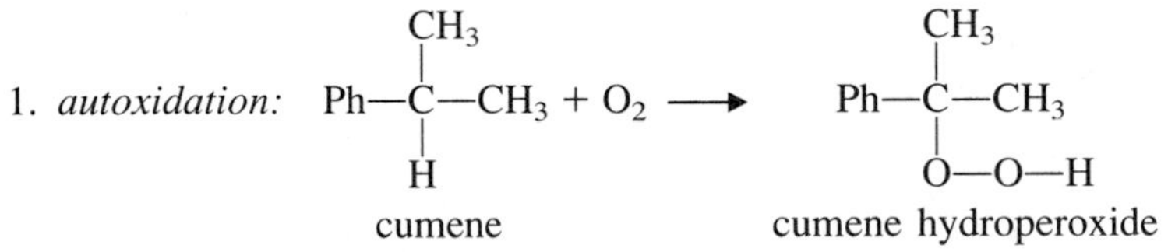

In the first step, cumene undergoes a reaction called an *autoxidation* with molecular oxygen. This is a free-radical chain reaction. Oxygen initiates this reaction because it is actually a "double free radical," or *diradical*.

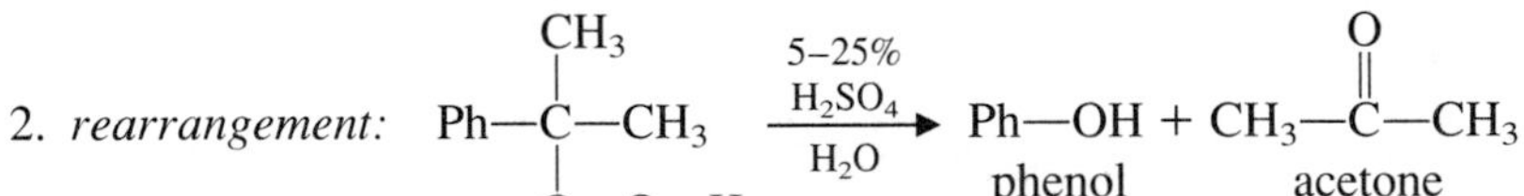

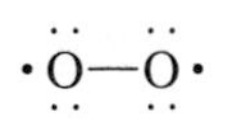

$$\cdot \ddot{O}{-}\ddot{O} \cdot \qquad \text{electronic structure of } O_2$$

In the initiation step, oxygen abstracts a hydrogen atom from cumene to give a resonance-stabilized benzylic free radical.

Initiation:

$$\underset{\substack{|\\H}}{\overset{\substack{CH_3\\|}}{Ph{-}C{-}CH_3}} \cdot \ddot{O}{-}\ddot{O}\cdot \longrightarrow \underset{\substack{|\\ \cdot}}{\overset{\substack{CH_3\\|}}{Ph{-}C{-}CH_3}} + H{-}\ddot{O}{-}\ddot{O}\cdot \qquad (18.50a)$$

a benzylic radical

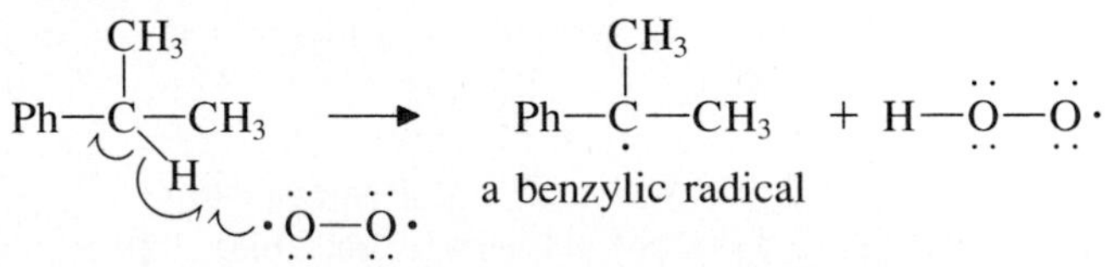

Cumene hydroperoxide is formed in the propagation steps of the reaction.

Propagation:

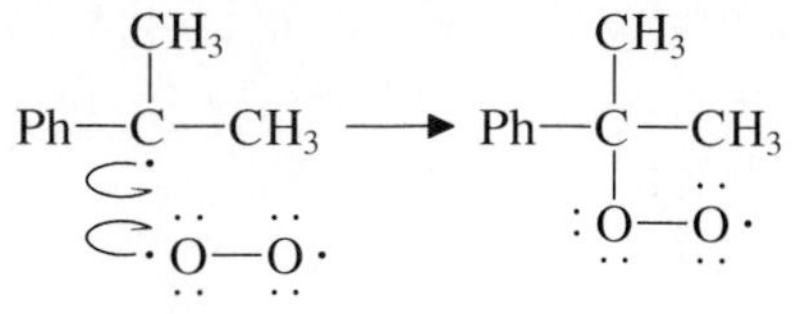

(18.50b)

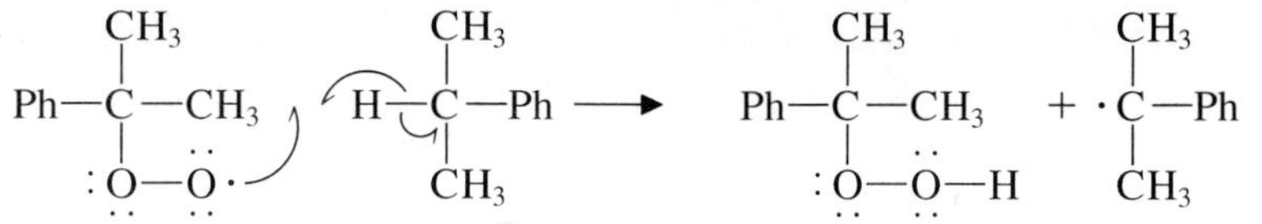

(18.50c)

cumene hydroperoxide

The rearrangement shown in Eq. 18.49b involves protonation of cumene hydroperoxide followed by loss of water to give an ion with an electron-deficient oxygen. This cation rearranges to a more stable carbocation.

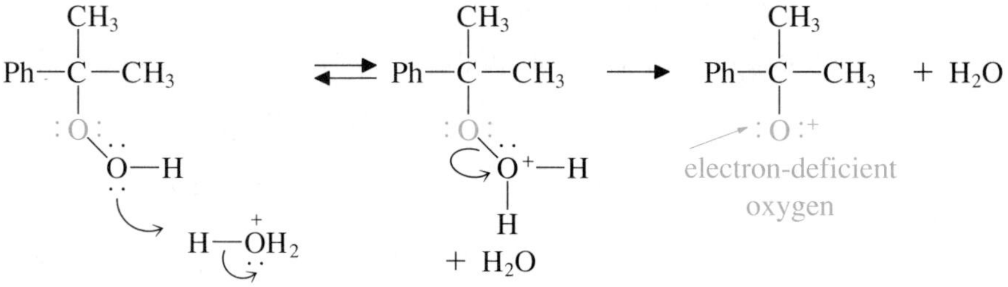

(18.51a)

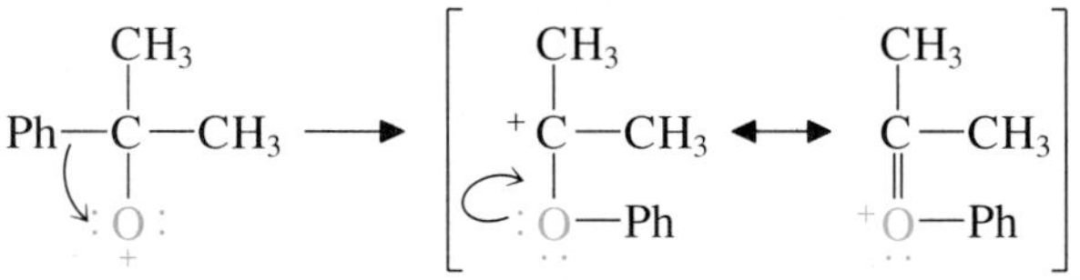

(18.51b)

(Actually, the ionic product of Eq. 18.51a is so unstable that both reactions probably occur as one step.) The carbocation formed in Eq. 18.51b reacts with water to give a type of alcohol (called a *hemiacetal;* Sec. 19.10A) that is unstable and breaks down spontaneously to phenol and acetone.

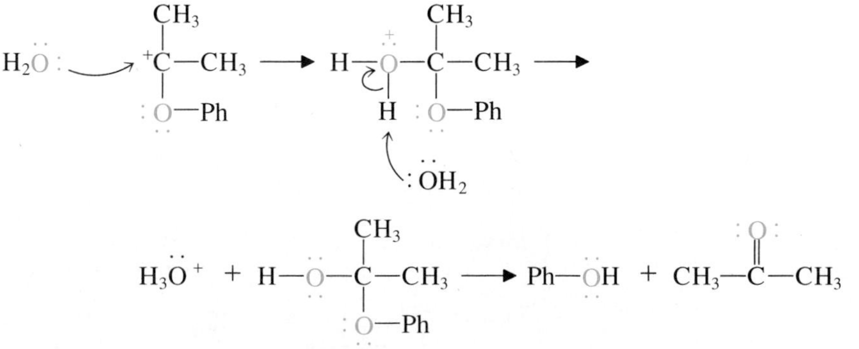

(18.51c)

a hemiacetal
(unstable)

Phenol is a very important article of commerce. In 1985 about 2.5 billion pounds, valued at $850 million, was produced. Phenol is a starting material for the production of phenol–formaldehyde resins (Sec. 19.15), which are polymers that have a variety of uses, including plywood adhesives, glass fiber (Fiberglass) insulation, molded phenolic plastics used in automobiles and appliances, and many others.

Now that we have seen how several organic compounds are synthesized and used industrially, it is appropriate to pause for an overview of some currently used industrial processes, shown in Fig. 18.6. From this diagram we can draw two very important conclusions. First, organic chemistry is a potent force in today's economy. It is difficult to imagine modern life without many of the finished products derived directly from the compounds in Fig. 18.6. Second, petroleum has overriding importance as a source of carbon in the chemical industry. Shortages of petroleum in the future will threaten not only our ability to heat our homes and workplaces, or our ability to move about in planes, trains, buses, and automobiles, but will also have a significant impact on a major economic force—the chemical industry.

Problem

17 Two by-products in the industrial synthesis of phenol and acetone are compounds *A* and *B*.

A 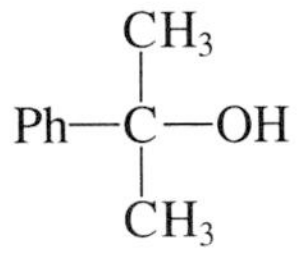 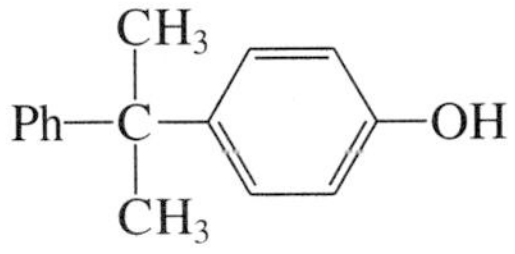 *B*

(a) Suggest a mechanism for the formation of compound *A* during the free-radical chain reaction of Eq. 18.49a. (*Hint*: Peroxides have weak O—O bonds.)

(b) Suggest a mechanism for the formation of compound *B* under the acidic conditions of the cumene hydroperoxide rearrangement (Eq. 18.49b). (*Hint:* Start with compound *A*.)

KEY IDEAS IN CHAPTER 18

- Neither aryl halides nor phenols undergo S_N1 reactions because of the instability of the aryl cation intermediates that would be involved in such reactions. Vinylic halides undergo S_N1 reactions only under extreme conditions.
- Neither aryl halides, vinylic halides, nor phenols undergo S_N2 reactions because of the *sp* hybridization of the S_N2 transition state and, primarily, because backside attack of nucleophiles is sterically blocked.

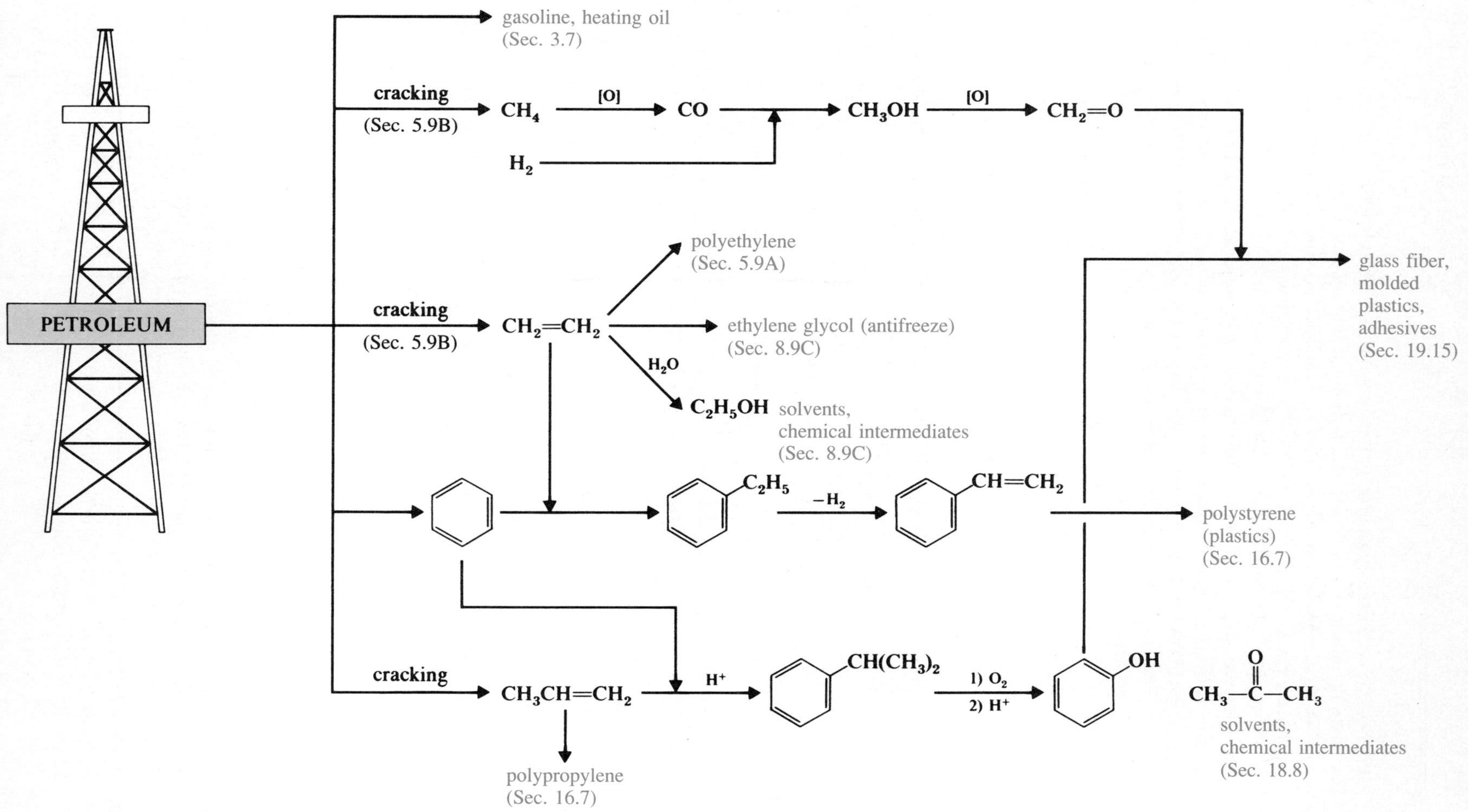

Figure 18.6 The organic chemical industry is a major economic force that depends strongly on the availability of petroleum. (In the reactions **[O]** indicates oxidation.)

- Vinylic halides undergo β-elimination reactions in base under vigorous conditions. Strong bases can also cause β-elimination in aryl halides to give benzyne intermediates, which are rapidly consumed by addition reactions with the bases.
- Aryl halides substituted with electron-withdrawing groups, particularly nitro groups, undergo nucleophilic aromatic substitution. This type of reaction involves attack of the nucleophile to form a resonance-stabilized anion (Meisenheimer complex), which then expels the halide leaving group.
- Aryl and vinyl halides form Grignard reagents. However, formation of Grignard reagents from chlorobenzene, and formation of most vinylic Grignard reagents, require tetrahydrofuran (THF) rather than ether as a solvent.
- Phenols are considerably more acidic than alcohols because phenoxide anions are resonance stabilized. Phenols with substituents that stabilize negative charge either by a resonance or inductive effect are even more acidic.
- Phenols dissolve in dilute hydroxide solution because they are converted almost entirely into their conjugate-base anions, which are water-soluble ionic compounds. This solubility can be used to separate phenols from water-insoluble organic compounds that are not acidic.
- Phenols are oxidized to quinones.
- Some electrophilic aromatic substitution reactions of phenols show unusual effects attributable to the —OH group. Thus phenol brominates three times in bromine water because the —OH group can ionize; and phenol is rather unreactive in the Friedel–Crafts acylation reaction because the —OH group reacts with the $AlCl_3$ catalyst.
- The preparation of phenol and acetone by the oxidation of cumene and rearrangement of the resulting hydroperoxide is an important industrial process.

ADDITIONAL PROBLEMS

18 What reaction (if any) takes place when *p*-iodotoluene is subjected to each of the following conditions?

(a) CH_3OH, 25°
(b) CH_3O^- in CH_3OH, 25°
(c) CH_3O^-, pressure, heat
(d) Mg in THF
(e) Li in hexane
(f) product of (d) + D_2O

19 Give the product(s) expected when *m*-cresol is subjected to each of the following conditions:

(a) concentrated H_2SO_4
(b) Br_2 in CCl_4 (dark)
(c) Br_2 (excess) in CCl_4, light
(d) dilute HCl
(e) 0.1*N* NaOH solution
(f) HNO_3, cold
(g) $C_2H_5{-}\overset{\overset{\displaystyle O}{\|}}{C}{-}Cl$, $AlCl_3$, heat; then H_2O
(h) $Na_2Cr_2O_7$ in H_2SO_4

Problems (Cont.)

20 Arrange the compounds within each set in order of increasing acidity, and explain your reasoning.

(a) cyclohexanol, phenol, benzyl alcohol

(b) *p*-nitrophenol, *p*-chlorophenol, 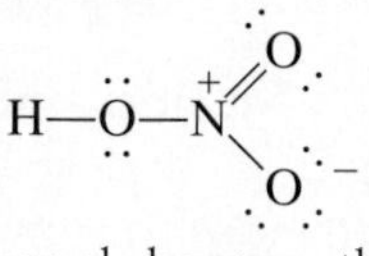

(c) cyclohexyl mercaptan, cyclohexanol, benzenethiol

(d) *p*-nitrobenzenethiol, *p*-nitrophenol, phenol

(e) 2,4,6-trimethylphenol, 2,6-di-*t*-butylphenol, 2,6-dichlorophenol

21 Identify compounds *A* and *B* from the following information:

(a) Compound *A*, $C_8H_{10}O$, is insoluble in water but soluble in aqueous NaOH solution, and yields 3,5-dimethylcyclohexanol when hydrogenated over a nickel catalyst at high pressure.

(b) Aromatic compound *B*, $C_8H_{10}O$, is insoluble in both water and aqueous NaOH solution. When treated successively with PBr_3, then Mg in THF, then water, it gives *p*-xylene.

22 Contrast the reactivities of cyclohexanol and phenol with each of the following reagents, and explain.

(a) aqueous NaOH solution

(b) *p*-toluenesulfonyl chloride in pyridine

(c) NaH in THF

(d) concentrated aqueous HBr, H_2SO_4

(e) Br_2 in CCl_4

(f) $Na_2Cr_2O_7$ in H_2SO_4

(g) H_2SO_4, heat

23 Choose the compound within each set that fits the criterion given, and explain your choices.

(a) The compound that reacts with alcoholic KOH to liberate fluoride ion:

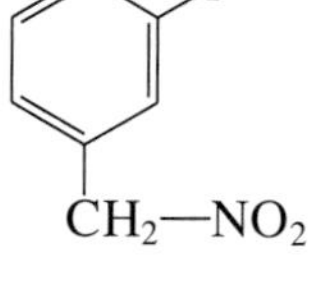

or

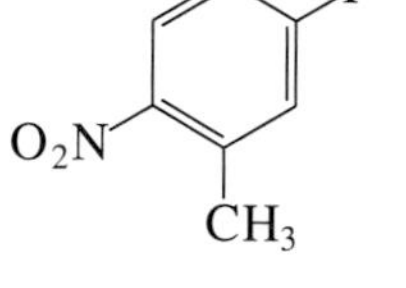

(b) The compound that *cannot* be prepared by a Williamson ether synthesis:

or

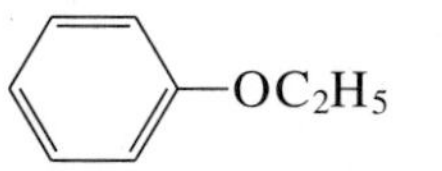

or

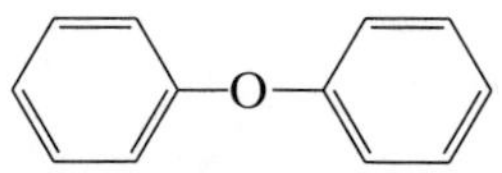

(c) The compound that gives an acidic solution when allowed to stand in aqueous ethanol:

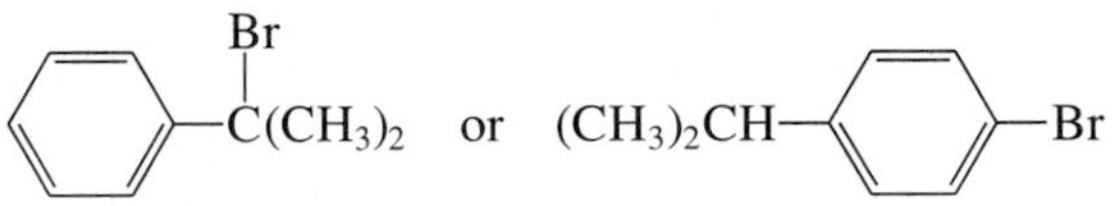

(d) The compound that reacts with magnesium in ether:

p-chlorotoluene or benzyl chloride

(e) The ether that would cleave more rapidly in HI:

phenyl cyclohexyl ether or diphenyl ether

(f) The compound that gives two products when it reacts with KNH_2 in liquid ammonia:

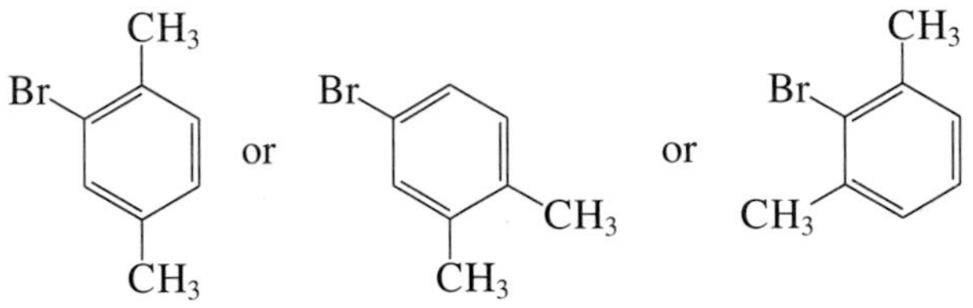

24 Phenols, like alcohols, are not only Brønsted acids, but are also Brønsted bases.
(a) Write the reaction in which phenol acts as a base toward H_2SO_4.
(b) Which is the weaker base: phenol or cyclohexanol? Why?

25 Give a rational explanation for the following observations, which were recorded in the chemical literature, concerning the reaction between *t*-butyl bromide and potassium benzenethiolate (the potassium salt of benzenethiol): "The attempts to prepare phenyl *t*-butyl sulfide by this route failed. If the reactants were kept at room temperature KBr was formed, but the benzenethiol was recovered unchanged."

26 (a) A water-insoluble phenol is most likely to dissolve in an aqueous solution with a pH that is:
(1) well below the pK_a of the phenol.
(2) equal to the pK_a of the phenol.
(3) well above the pK_a of the phenol.
Choose one and explain.
(b) A mixture of *p*-cresol, $pK_a = 10.2$, and 2,4-dinitrophenol, $pK_a = 4.11$, is dissolved in ether. The ether solution is then vigorously shaken with one of the following aqueous solutions. Which solution effects the best separation of the two phenols by dissolving one and leaving the other in solution? Explain.
(1) a $0.1M$ aqueous HCl solution
(2) a solution that contains a large excess of a pH = 4 buffer
(3) a solution that contains a large excess of a pH = 7 buffer
(4) $0.1M$ NaOH solution

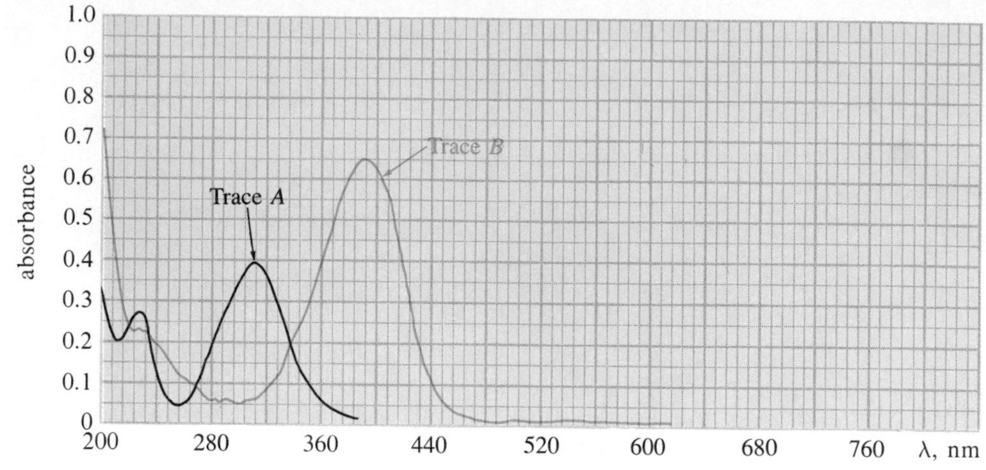

Figure 18.7 UV spectra for Problem 27.

27 The UV spectrum of *p*-nitrophenol in aqueous solution is shown in Fig. 18.7 (trace *A*). When a few drops of concentrated NaOH are added, the solution turns yellow and the spectrum changes (trace *B*). On addition of a few drops of concentrated acid, the color disappears and the spectrum shown in trace *A* is restored. Explain these observations.

28 When a suspension of 2,4,6-tribromophenol is treated with an excess of bromine water, the white precipitate of 2,4,6-tribromophenol disappears and is replaced by a precipitate of a yellow compound that has the following structure. Give a mechanism for the formation of this compound.

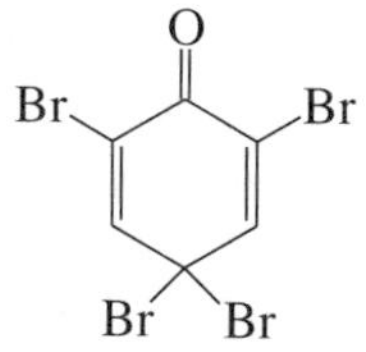

29 Complete each reaction by giving the major organic product(s). *No reaction* may be an appropriate response.

(a)

$$O_2N\text{—}\langle\text{—}\rangle\text{—}F \ + \ C_2H_5S^- \longrightarrow$$

(b)

$$\langle\text{—}\rangle\text{—}Cl \ + \ CH_3CH_2CH_2\text{—}NH_2 \xrightarrow{25°}$$

(c)

$$\begin{array}{c}\text{OCH}_3\\ \langle\text{—}\rangle\\ \text{Br}\end{array} \ + \ KNH_2 \xrightarrow{NH_3(\text{liq})}$$

(d) resorcinol + Br$_2$ $\xrightarrow{CCl_4}$

(e)

$$C_2H_5\text{—}\langle\text{—}\rangle\text{—}Br \ + \ K^+ (CH_3)_3C\text{—}O^- \xrightarrow{\text{heat}}$$

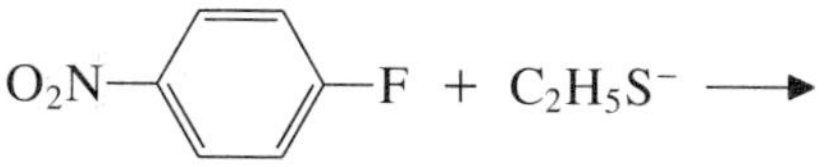

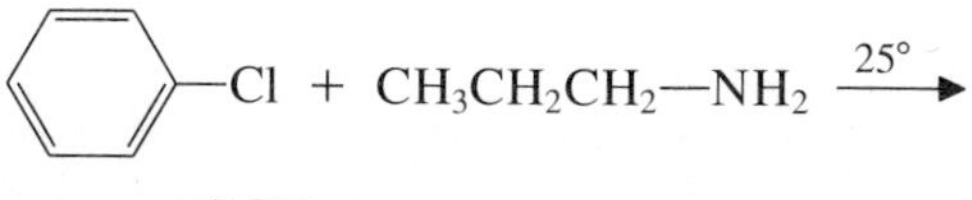

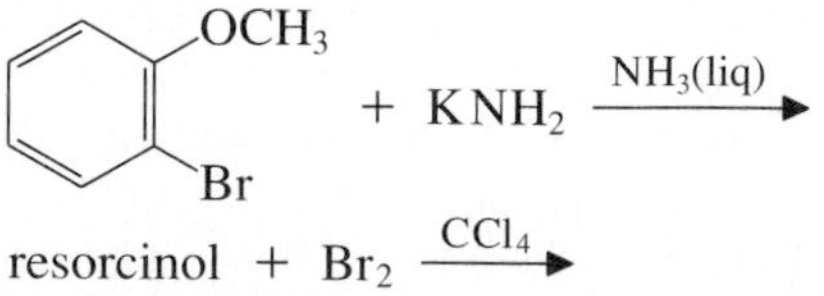

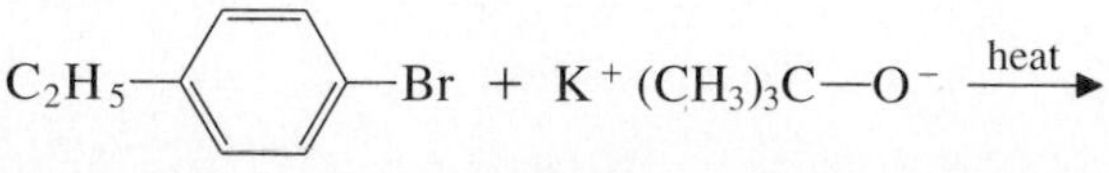

(f)

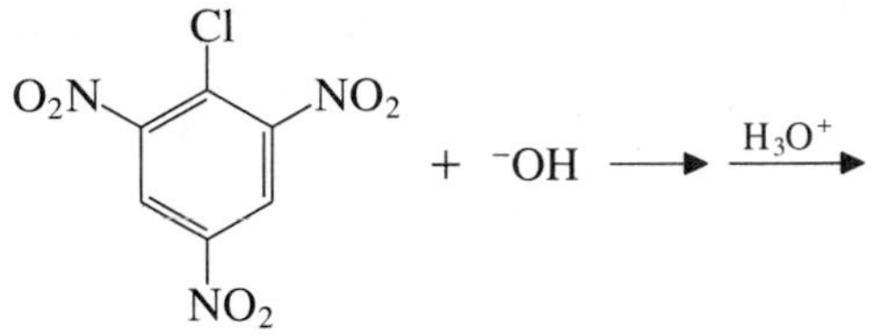

(g)

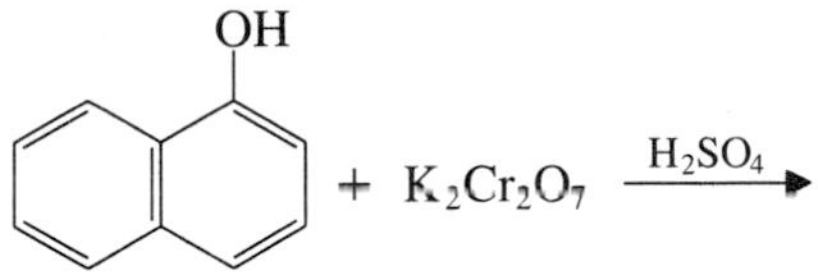

(h) 2,4-dinitrophenol $\xrightarrow[\text{pyridine}]{CH_3SO_2Cl}$ $\xrightarrow[CH_3OH]{CH_3O^-}$

(i) *m*-chlorophenol + $Na_2Cr_2O_7$ $\xrightarrow{H_2SO_4}$

(j)

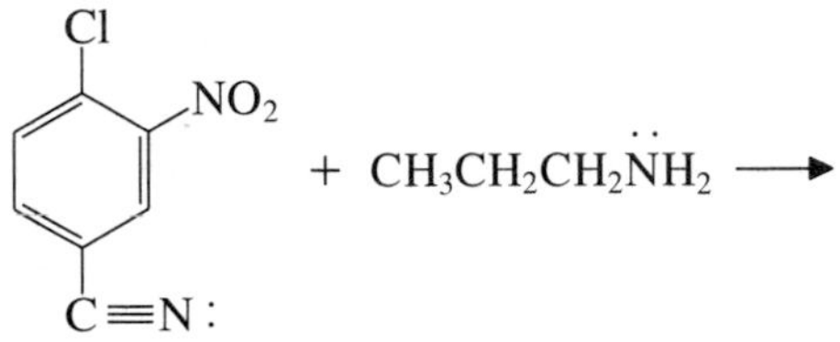

(k)

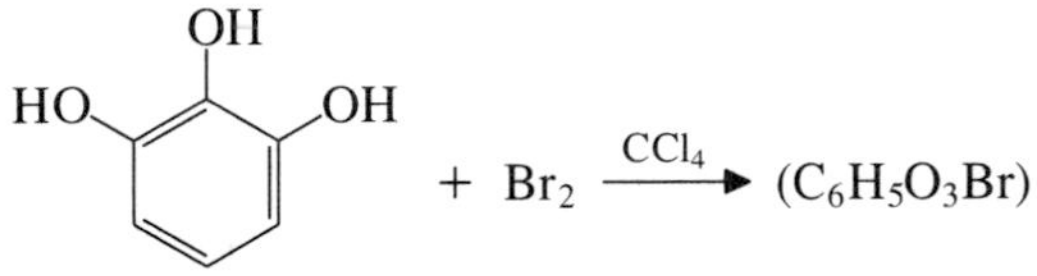

pyrogallol

(l)

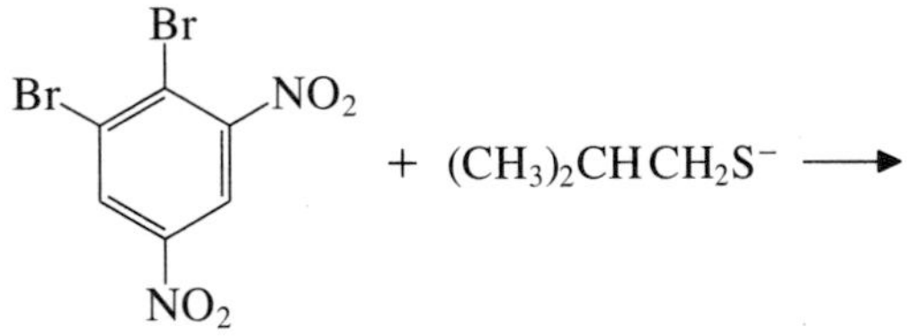

30 How unstable is benzyne? We can answer this question by comparing its heat of hydrogenation with the heats of hydrogenation of other alkynes. The justification for this procedure is that all ordinary disubstituted alkynes have almost the same heat of hydrogenation.

(a) Using the heats of formation below, calculate the approximate heat of hydrogenation of an ordinary alkyne.

Compound	ΔH_f^0 (kcal/mol)	Compound	ΔH_f^0 (kcal/mol)
2-butyne	35.0	benzyne	118.0
cis-2-butene	−1.7	benzene	19.8
2-pentyne	30.8		
cis-2-pentene	−6.7		

(Problem continues on next page.)

Problems (Cont.)

(b) Calculate the heat of hydrogenation of benzyne. By approximately how many kcal/mol is benzyne less stable than an ordinary alkyne?

---

31 Outline a synthesis for each of the following compounds from the indicated starting material and any other reagents:

(a) 1-chloro-2,4-dinitrobenzene from benzene

(b) 1-chloro-3,5-dinitrobenzene from benzene

(c) Ph—C≡C—Ph from

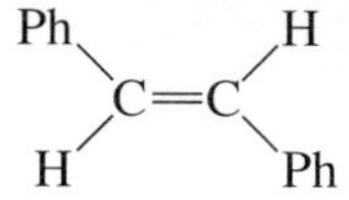

(d)

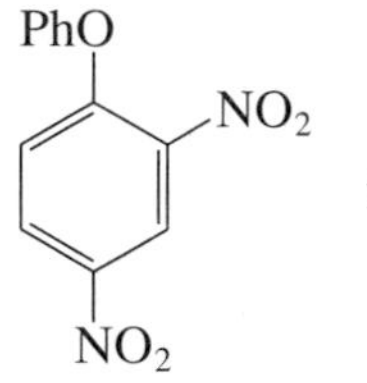

from benzene

(e) 2-chloro-4,6-dinitrophenol from chlorobenzene

(f)

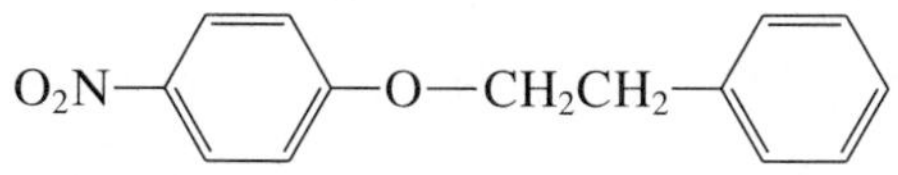

from fluorobenzene and bromobenzene

(g) "butylated hydroxytoluene" (BHT; Sec. 18.7A) from *p*-cresol

(h)

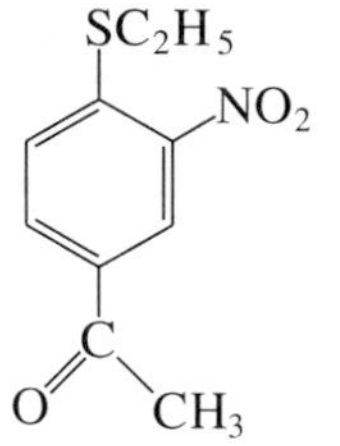

from chlorobenzene

(i)

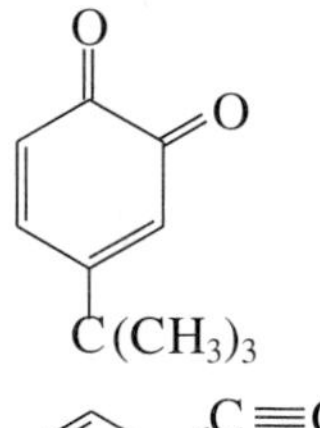

from catechol

(j)

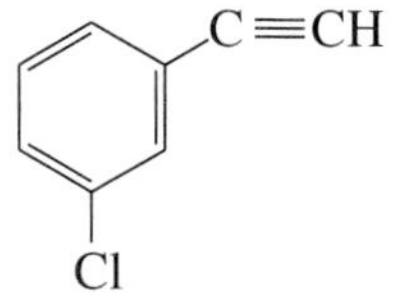

from *m*-bromochlorobenzene

---

32 Explain why, in the following reactions, different isomers of the starting material give different products:

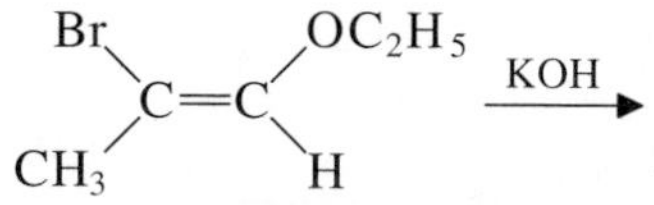

$CH_3—C≡C—OC_2H_5$ (only elimination product observed)

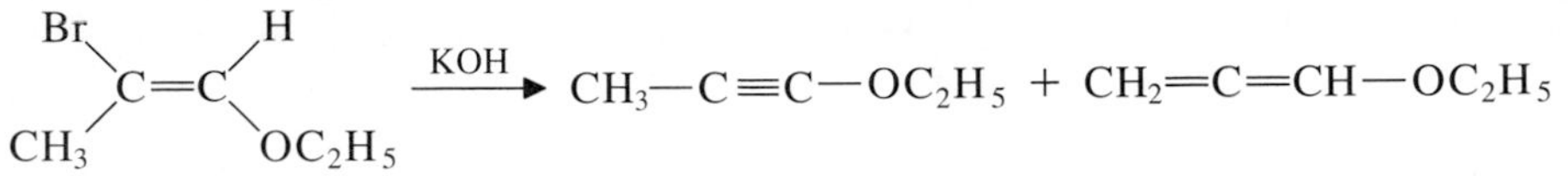

33 Arrange the following phenols in order of decreasing pK_a, and explain your reasoning:

(a) (b) (c) (d)

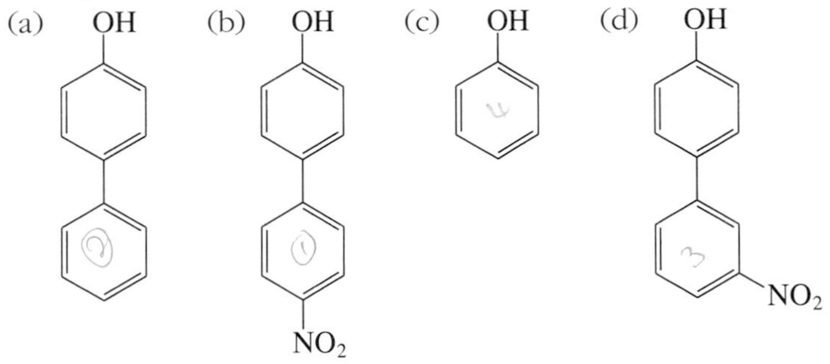

34 When 1,3,5-trinitrobenzene [NMR: δ 9.1(s)] is treated with Na$^+$ CH$_3$O$^-$, the ionic compound formed has the following NMR spectrum:

$$\delta\ 3.3\ (3H,\ s);\ \delta\ 6.3\ (1H,\ t,\ J = 1\ Hz);\ \delta\ 8.7\ (2H,\ d,\ J = 1\ Hz)$$

Suggest a structure for this compound.

35 (a) The following reaction was once used in a major commercial method for the preparation of phenol (Dow phenol process). Suggest a mechanism for this reaction. (Notice the temperature!)

$$\text{Ph—Cl + NaOH} \xrightarrow{300°} \text{Ph—O}^-\ \text{Na}^+ + \text{Cl}^-$$

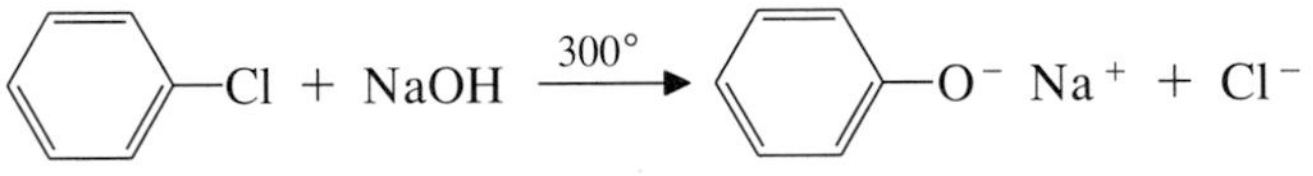

(b) A major by-product in the Dow phenol process is diphenyl ether. Use your mechanism from (a) to account for the formation of this by-product.

36 Although enols are unstable compounds (Sec. 14.5A), suppose that the acidity of an enol could be measured. Which would be more acidic: enol *A* or alcohol *B*? Why?

$$A\quad CH_3-\overset{\displaystyle OH}{\underset{\displaystyle |}{C}}=CH_2 \qquad CH_3-\overset{\displaystyle OH}{\underset{\displaystyle |}{CH}}-CH_3\quad B$$

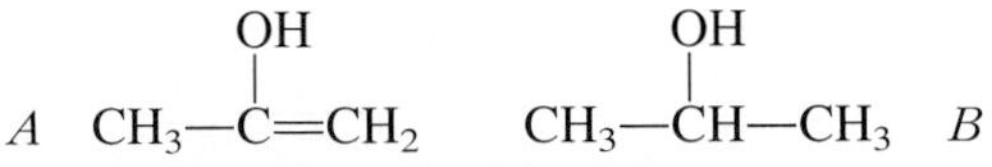

37 The following reaction occurs readily at 95° (X = halogen):

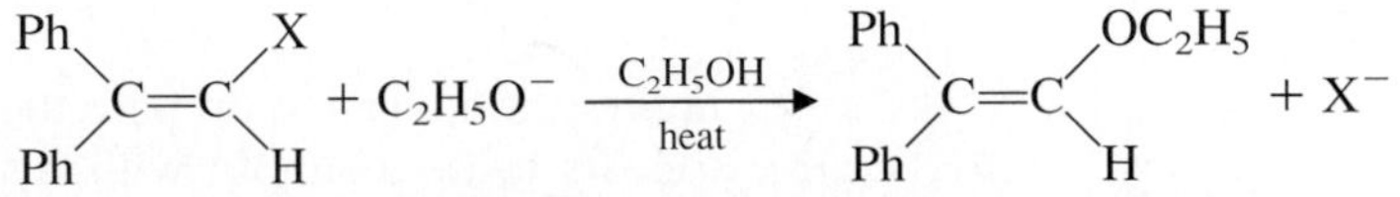

Problems (Cont.)

The relative rates for the various halogens are 290 (X = F); 1.4 (X = Cl); and 1.0 (X = Br). When there is a nitro group in the *para* position of each benzene ring, the reaction is substantially accelerated. Give a detailed mechanism for this reaction, and explain how it is consistent with the experimental facts.

---

38 The herbicide 2,4,5-T is synthesized by heating 2,4,5-trichlorophenol with chloroacetic acid and NaOH. A by-product of this process is the toxic compound dioxin. Dioxin accumulates in the environment because it is degraded very slowly. Give mechanisms for the formation of both 2,4,5-T and dioxin.

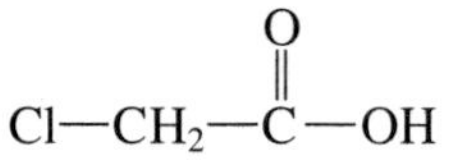

chloroacetic acid

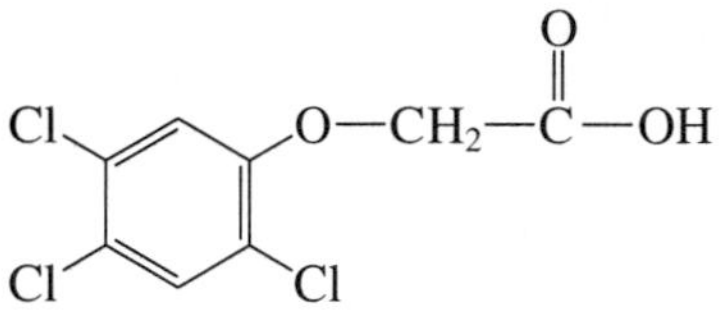

2,4,5-T

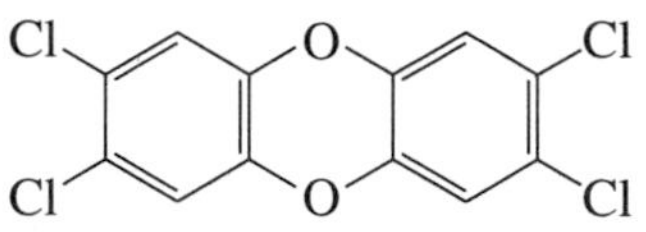

dioxin

---

39 Explain why biphenyl forms as a by-product when phenyllithium is synthesized by the reaction of bromobenzene and lithium metal. (*Hint:* Phenyllithium is a strong base.)

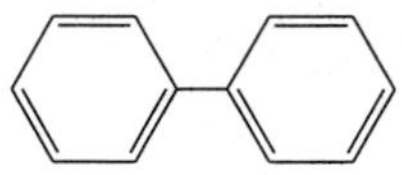

biphenyl

---

40 Using the arrow formalism where appropriate, give mechanisms for the following reactions:

(a)

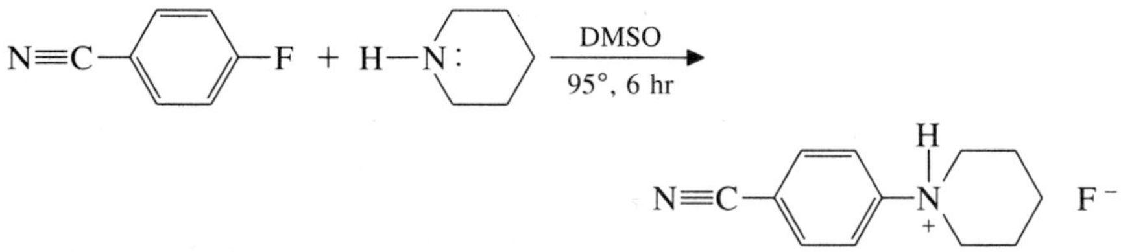

(b)

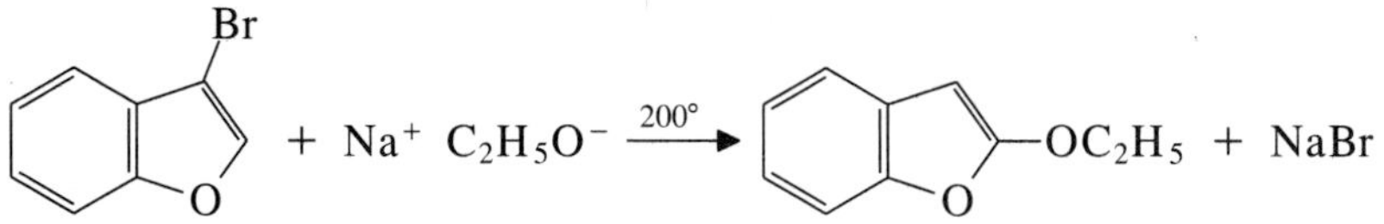

(c)

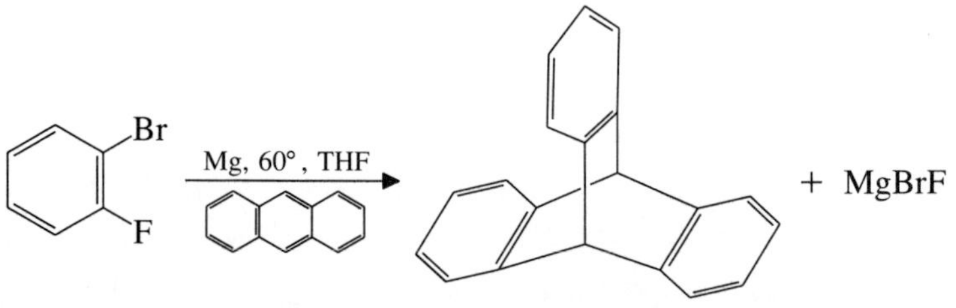

triptycene

---

41 The following reaction, used to prepare the drug Mephenesin (a skeletal muscle relaxant), appears to be a simple Williamson ether synthesis:

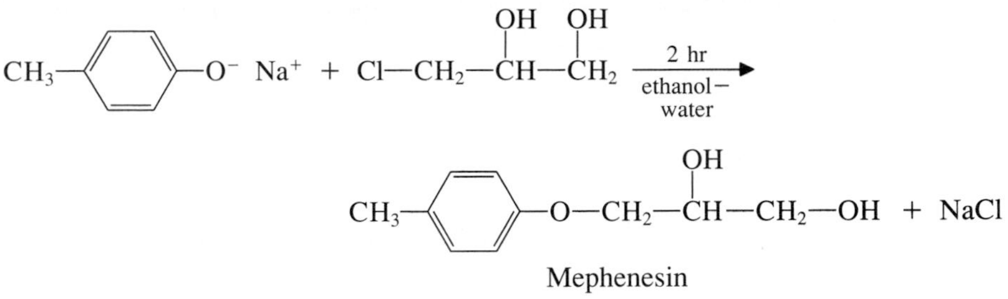

During this reaction a precipitate of NaCl forms after only about ten minutes, but a considerably longer reaction time (about two hours) is required to obtain a good yield of Mephenesin. Explain why NaCl formation occurs more rapidly than formation of Mephenesin. (*Hint:* See Sec. 11.6.)

42 You are a troubleshooter for Phenomenal Phenols, Inc., a company that specializes in the synthesis of phenol from cumene (Sec. 18.8). A batch of cumene has failed to undergo normal autoxidation with molecular oxygen and you have been called in to solve the problem. On questioning the staff you have found that a new employee (who formerly worked in a bakery) has added some BHT (Sec. 18.7A) to the cumene to "keep it from going bad." Examination of the reaction mixture reveals the presence of the following compound:

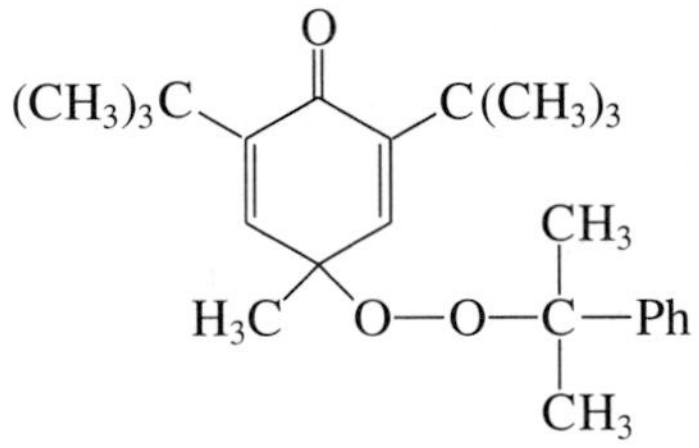

In the absence of BHT, the cumene autoxidizes normally and the new compound is not present in the reaction mixture. Explain these observations.

43 The following compound, 1,1-dichlorocyclohexane containing radioactive ^{14}C only at carbon-2, was subjected to an E2 reaction in base to give labeled 1-chloro-1-cyclohexene.

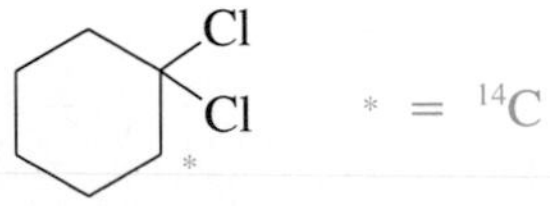

(a) What is the distribution of radioactive carbon in 1-chloro-1-cyclohexene?

When the labeled 1-chloro-1-cyclohexene formed in (a) was treated with phenyllithium (a strong base) at 150°, 1-phenyl-1-cyclohexene was formed. The radioactive carbon was found to be distributed in this product as follows:

Problems (Cont.)

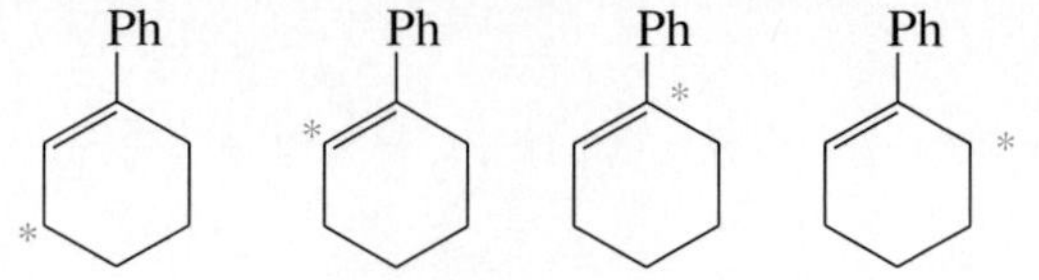

(25% of each)

(b) Propose a mechanism for 1-phenyl-1-cyclohexene formation that accounts for the observed distribution of ^{14}C.

44 Explain why it is that, in the reactions below, different products are obtained when different amounts of $AlBr_3$ catalyst are used. (*Hint:* See Sec. 18.7B.)

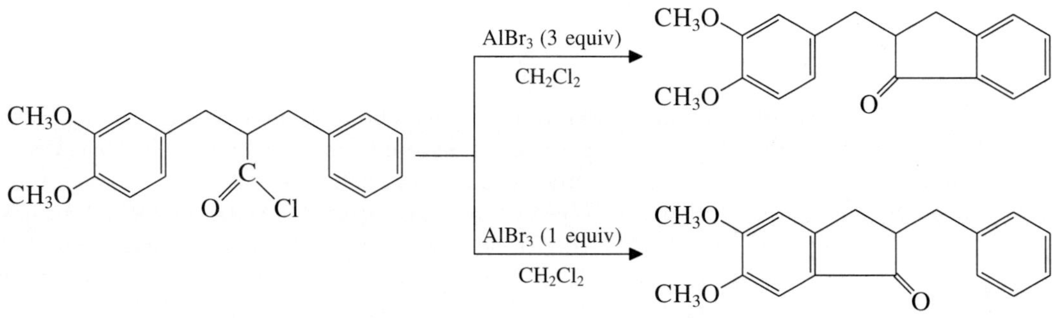

45 When the following unusual epoxide reacts with water, *p*-cresol is the only product formed, and it has the labeling pattern shown. This rearrangement is called the "NIH shift," because it was discovered by chemists at the National Institutes of Health.

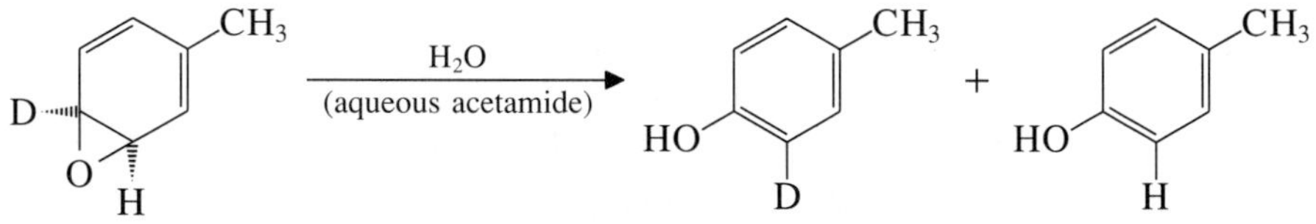

(75% of product) (25% of product)

(a) Suggest a mechanism for this reaction that accounts for the position of the label, and explain why some of the deuterium is lost.
(b) Explain why only *p*-cresol and no *m*-cresol is formed.

19 Chemistry of Aldehydes and Ketones. Carbonyl-Addition Reactions

In this chapter we begin a study of **carbonyl compounds**: compounds containing the **carbonyl group,** C=O. Aldehydes, ketones, carboxylic acids, and most derivatives of carboxylic acids—esters, amides, anhydrides, and acid chlorides—are all carbonyl compounds.

In this chapter we focus on the nomenclature, properties, and characteristic carbonyl-group reactions of aldehydes and ketones. In Chapters 20 and 21, we shall consider carboxylic acids and carboxylic-acid derivatives, respectively. In Chapter 22, we shall study ionization, enolization, and condensation reactions, which are common to the chemistry of all carbonyl compounds.

Aldehydes and ketones have the following general structures:

aldehyde ketone

Examples:

acetaldehyde (an aldehyde) acetone (a ketone)

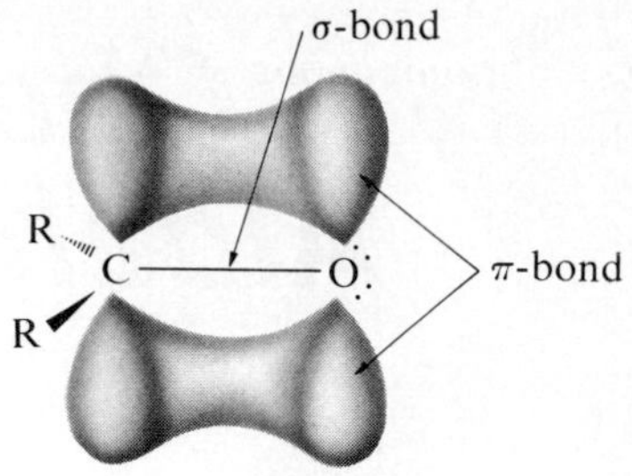

Figure 19.1 Bonding in a typical carbonyl compound. The carbonyl group and the two R-groups lie in the same plane, and there are both a σ-bond and a π-bond between the carbonyl carbon and oxygen.

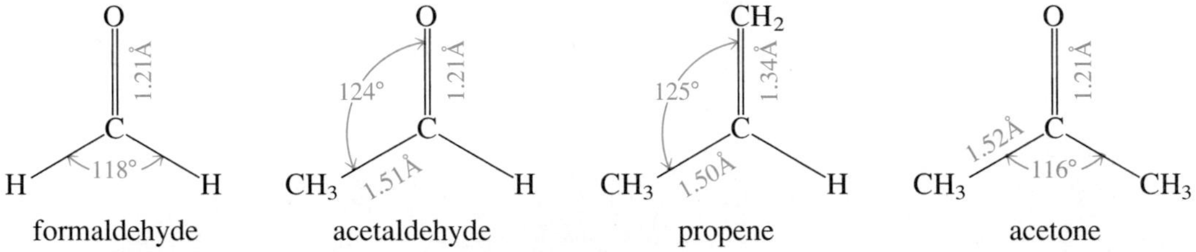

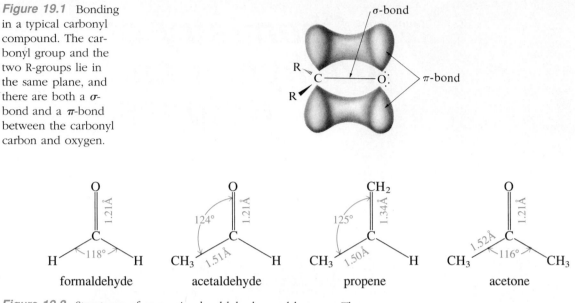

Figure 19.2 Structures of some simple aldehydes and ketones. The structure of propene is given for comparison.

In a **ketone,** the groups bound to the carbonyl carbon (R and R′ in the structures above) are alkyl or aryl groups. In an **aldehyde,** at least one of the groups at the carbonyl carbon atom is a hydrogen, and the other may be alkyl, aryl, or a second hydrogen.

The carbonyl carbon of a typical aldehyde or ketone is sp^2 hybridized with bond angles approximating 120°. The carbon–oxygen double bond consists of a σ-bond and a π-bond, much like the double bond of an alkene (Fig. 19.1). Just as C—O single bonds are shorter than C—C single bonds, C=O bonds are shorter than C=C bonds. The structures of some simple aldehydes and ketones are given in Fig. 19.2.

19.1 NOMENCLATURE OF ALDEHYDES AND KETONES

A. Common Nomenclature

In the common nomenclature of aldehydes the suffix *aldehyde* is added to a prefix that indicates the chain length. A list of prefixes is given in Table 19.1.

$$CH_3—CH_2—CH_2—CH{=}O \qquad CH_2{=}O$$

butyr + *aldehyde* = butyraldehyde formaldehyde

(The prefixes in Table 19.1 are also used in the nomenclature of carboxylic acids and their derivatives.) Common names are almost always used for the simplest aldehydes.

TABLE 19.1 Prefixes Used in Common Nomenclature of Carbonyl Compounds

Prefix	R— in R—CH=O or R—CO_2H	Prefix	R— in R—CH=O or R—CO_2H
form	H—	isobutyr	$(CH_3)_2CH$—
acet	CH_3—	valer	$CH_3CH_2CH_2CH_2$—
propion, propi*	CH_3CH_2—	isovaler	$(CH_3)_2CHCH_2$—
butyr	$CH_3CH_2CH_2$—	benz, benzo†	Ph—

*Used in phenone nomenclature as discussed in the text.
†Used in carboxylic acid nomenclature (Sec. 20.1A).

Acetone is the only nonaromatic ketone for which we shall use a common name.

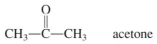

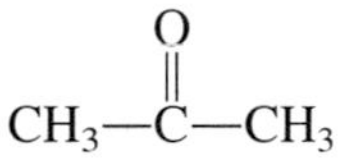
acetone

Aromatic aldehydes and ketones also have common names. Benzaldehyde is the simplest aromatic aldehyde.

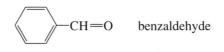

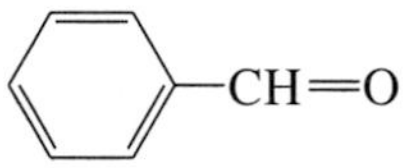
benzaldehyde

Certain aromatic ketones are named by attaching the suffix *ophenone* to the appropriate prefix from Table 19.1.

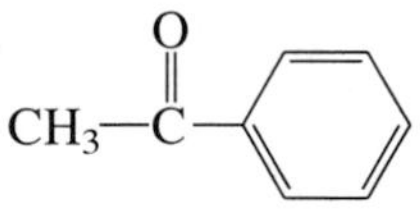
acet + *ophenone* = acetophenone

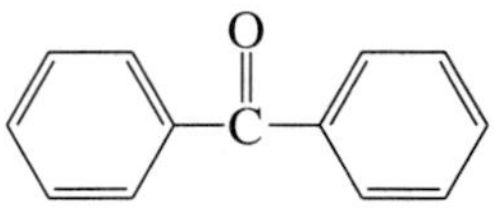
benzophenone

It is occasionally convenient to construct the common name of a ketone by citing the two groups on the carbonyl carbon followed by the word *ketone*.

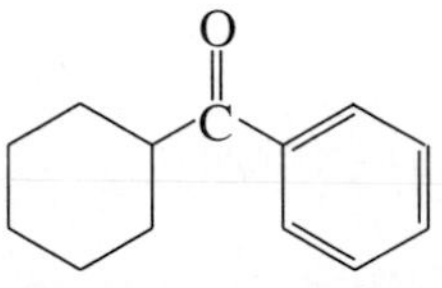
cyclohexyl phenyl ketone

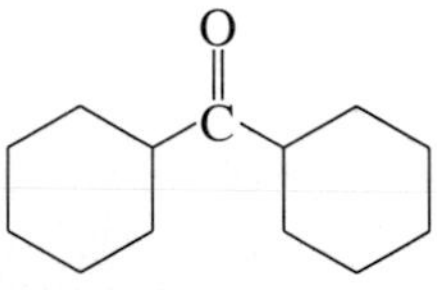
dicyclohexyl ketone

Simple substituted aldehydes and ketones can be named in the common system by designating the positions of substituents with Greek letters, beginning at the position *adjacent* to the carbonyl group.

$$\overset{\gamma}{-CH_2}-\overset{\beta}{CH_2}-\overset{\alpha}{CH_2}-\overset{\displaystyle O}{\overset{\|}{C}}- \qquad BrCH_2CH_2\overset{\displaystyle O}{\overset{\|}{C}}H$$

β-bromopropionaldehyde

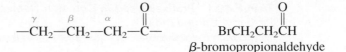

Many common carbonyl-containing substituent groups are named by a simple extension of the terminology in Table 19.1: the suffix *yl* is added to the appropriate prefix. The following names are examples:

$H-\overset{O}{\overset{\|}{C}}-$	$CH_3-\overset{O}{\overset{\|}{C}}-$	$CH_3CH_2-\overset{O}{\overset{\|}{C}}-$	$Ph-\overset{O}{\overset{\|}{C}}-$
formyl group	acetyl group	propionyl group	benzoyl group

Such groups are called in general **acyl groups.** (This is the source of the term *acylation,* used in Sec. 16.4E.) To be named as an acyl group, a substituent group must be connected to the remainder of the molecule *at its carbonyl carbon.*

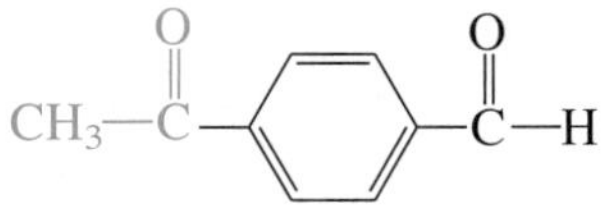

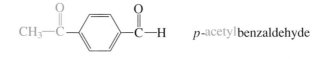

p-acetylbenzaldehyde

Be careful not to confuse the *benzoyl* group, an acyl group, with the *benzyl* group, an alkyl group.

$Ph-\overset{O}{\overset{\|}{C}}-$	$Ph-CH_2-$
benzoyl group	benzyl group

A great many aldehydes and ketones were well known long before any system of nomenclature existed. These are known by traditional names that are best learned as the need arises.

$Ph-CH=CH-CH=O$
cinnamaldehyde

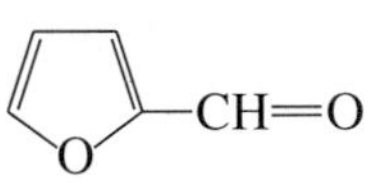

furfural

$CH_2=CH-CH=O$
acrolein

B. Systematic Nomenclature

The systematic name of an aldehyde is constructed from a prefix indicating the length of the carbon chain followed by the suffix *al.* The prefix is the name of the corresponding hydrocarbon without the final *e.*

$$CH_3CH_2CH_2CH=O \qquad \text{butan}\not{e} + al = \text{butanal}$$

In numbering the carbon chain of an aldehyde, the carbonyl carbon receives the number one.

$$\overset{4}{C}H_3\overset{3}{C}H_2\underset{\displaystyle CH_3}{\overset{2}{C}H}\overset{1}{C}H=O \qquad \text{2-methylbutanal}$$

Note carefully the difference in chain numbering of aldehydes in common and systematic nomenclature. In common nomenclature, numbering begins at the carbon *adjacent* to the carbonyl (the α-carbon); in systematic nomenclature, numbering begins at the carbonyl carbon itself.

As with diols, the final *e* is not dropped when there is more than one aldehyde group in the carbon chain.

$$O{=}CH{-}CH_2CH_2CH_2CH_2{-}CH{=}O \qquad \text{hexanedial}$$

When an aldehyde group is attached to a ring, the suffix *carbaldehyde* is appended to the name of the ring. (In older literature, the suffix *carboxaldehyde* was used.)

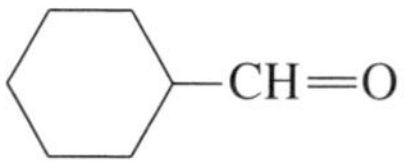

cyclohexanecarbaldehyde

In aldehydes of this type carbon-1 is not the carbonyl carbon, but rather the ring carbon attached to the carbonyl group.

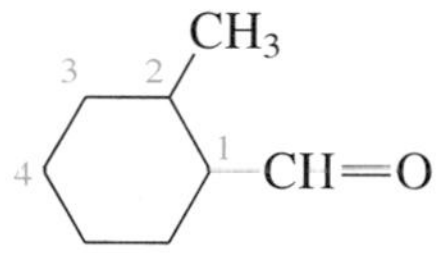

2-methylcylohexanecarbaldehyde

The name *benzaldehyde* (Sec. 19.1A) is used in both common and systematic nomenclature.

A ketone is named by giving the hydrocarbon name of the longest carbon chain containing the carbonyl group, dropping the final *e*, and adding the suffix *one*. The position of the carbonyl group is given the lowest possible number.

$$\underset{5}{CH_3}{-}\underset{4}{CH_2}{-}\underset{3}{CH_2}{-}\underset{2}{\overset{\overset{\displaystyle O}{\|}}{C}}{-}\underset{1}{CH_3}$$

five-carbon chain:
pentan~~e~~ + *one* = pentanone
position of carbonyl: 2-pentanone

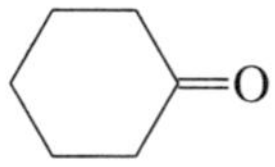

cyclohexan~~e~~ + *one* = cyclohexanone

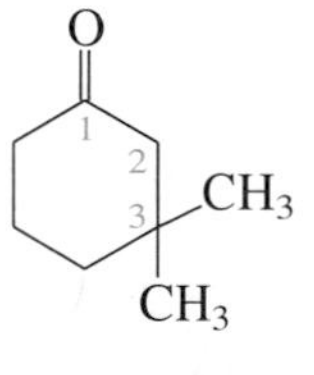

3,3-dimethyl-1-cyclohexanone

As with diols and dialdehydes, the final *e* of the hydrocarbon name is not dropped in the nomenclature of diones, triones, etc.

$$CH_3{-}\overset{\overset{\displaystyle O}{\|}}{C}{-}CH_2{-}\overset{\overset{\displaystyle O}{\|}}{C}{-}CH_2{-}CH_3$$

six-carbon chain:
hexane + *dione* = hexanedione
position of carbonyls: 2,4-hexanedione

Aldehyde and ketone carbonyl groups receive higher priority for citation as *principal groups* (Sec. 8.1B) than —OH or —SH groups:

Priority for citation as principal group:

$$-\overset{\overset{\displaystyle O}{\|}}{C}H \text{ (aldehyde)} > -\overset{\overset{\displaystyle O}{\|}}{C}- \text{ (ketone)} > -OH > -SH \qquad (19.1)$$

For example, the following compound is named as a ketone (that is, with the suffix *one*) because a ketone carbonyl group has higher priority than an —OH group; the —OH (hydroxy) group is treated as a substituent.

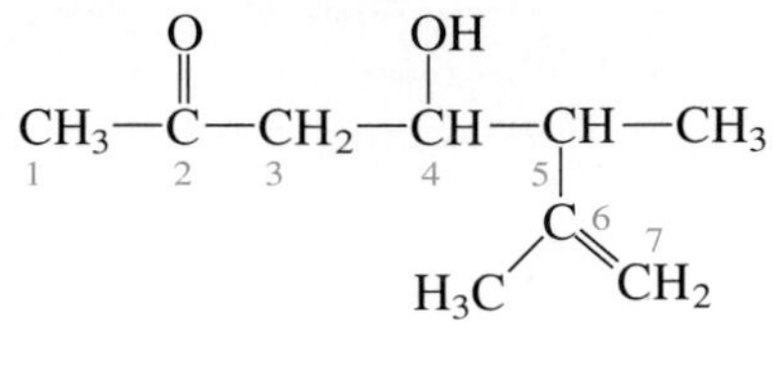

4-hydroxy-5,6-dimethyl-6-hepten-2-one

position of carbonyl carbon

position of double bond

positions of substituent groups

(A complete list of group priorities is given in Appendix I.)

When a ketone carbonyl group is treated as a substituent, its position is designated by the term *oxo*.

$$H-\overset{\overset{\displaystyle O}{\|}}{\underset{1}{C}}-\overset{}{\underset{2}{CH_2}}-\overset{}{\underset{3}{CH_2}}-\overset{\overset{\displaystyle O}{\|}}{\underset{4}{C}}-\overset{}{\underset{5}{CH_3}}$$

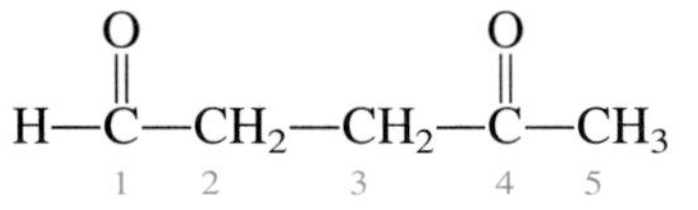

principal group: aldehyde carbonyl
name: 4-oxopentanal

Problems

1 Give the structure of:
(a) isobutyraldehyde
(b) valerophenone
(c) 3-cyclohexen-1-one
(d) γ-chlorobutyraldehyde
(e) 2-oxocyclopentanecarbaldehyde
(f) *o*-bromoacetophenone
(g) 3-hydroxy-2-butanone
(h) *m*-methoxypropiophenone
(i) 4-(2-chlorobutyryl)benzaldehyde

2 Give the systematic name for each of the following compounds:
(a) diisopropyl ketone
(b) acetone
(c)

$$CH_3-\overset{\overset{\displaystyle O}{\|}}{C}-\overset{\underset{\displaystyle |}{\displaystyle CH}}{CH}-\overset{\overset{\displaystyle O}{\|}}{C}-CH_3$$
$$CH-CH=CH_2$$

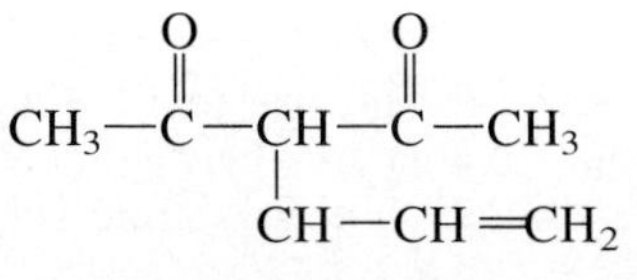

(d)

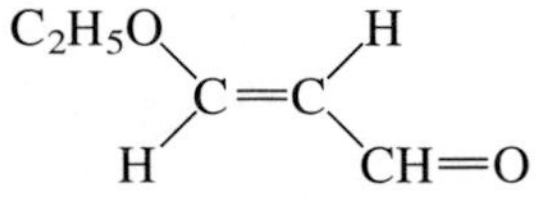

(e) 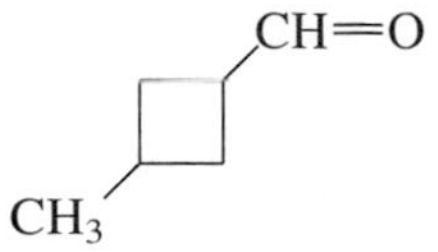

(f) 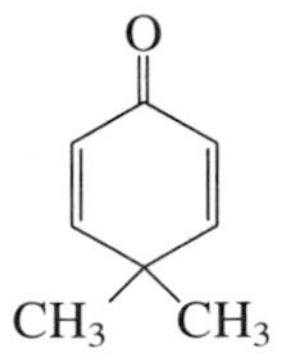

19.2 PHYSICAL PROPERTIES OF ALDEHYDES AND KETONES

Most simple aldehydes and ketones are liquids. However, formaldehyde is a gas, and acetaldehyde has a boiling point (20.8°) very near room temperature, although it is usually sold as a liquid.

Aldehydes and ketones are polar molecules because of the C=O bond dipole.

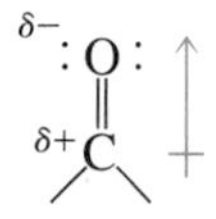

Because of their polarity, aldehydes and ketones have higher boiling points than alkenes or alkanes with similar molecular weights and shapes. But since aldehydes and ketones are not hydrogen-bond donors, their boiling points are considerably lower than those of the corresponding alcohols.

	$CH_3CH{=}CH_2$	$CH_3CH{=}O$	CH_3CH_2OH
boiling point	−47.4°	20.8°	78.3°
dipole moment	0.4 D	2.7 D	1.7 D

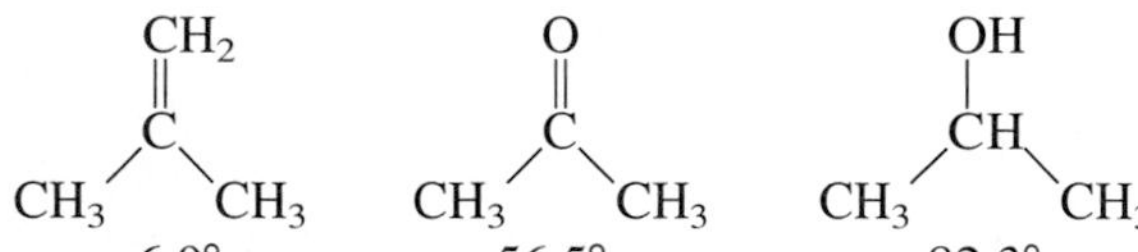

	$(CH_3)_2C{=}CH_2$	$(CH_3)_2C{=}O$	$(CH_3)_2CHOH$
boiling point	−6.9°	56.5°	82.3°
dipole moment	0.5 D	2.7 D	1.7 D

Aldehydes and ketones with four or fewer carbons have considerable solubility in water because they can accept hydrogen bonds from water at the carbonyl oxygen.

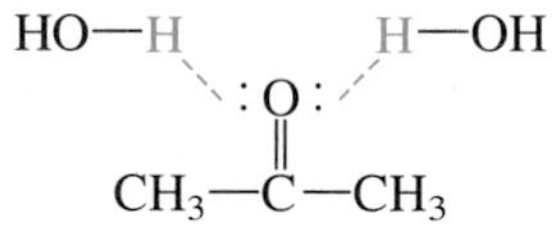

Acetaldehyde and acetone are miscible with water (soluble in all proportions). The water solubility of aldehydes and ketones along a homologous series diminishes rapidly as molecular weight increases.

Acetone and 2-butanone are especially valued as solvents because they dissolve not only water, but also a wide variety of organic compounds. These solvents have sufficiently low boiling points that they can be easily separated from other less volatile compounds. Acetone, with a dielectric constant of 20.7, is a polar solvent, and is often used as a solvent or co-solvent for nucleophilic substitution reactions.

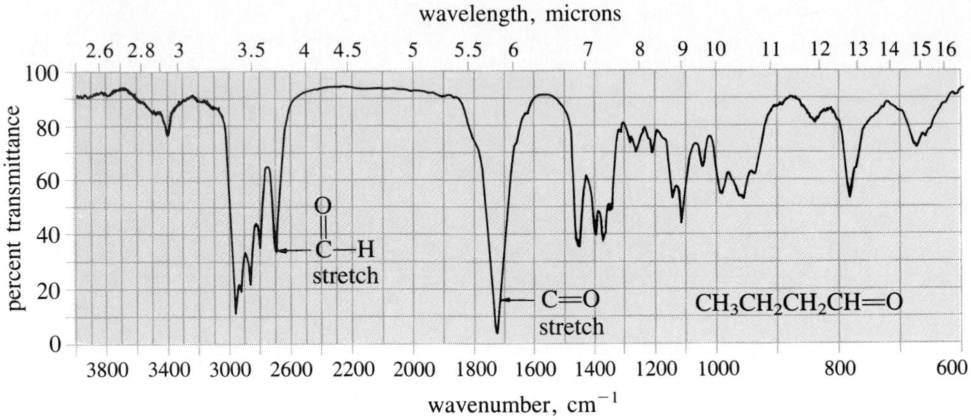

Figure 19.3 Infrared spectrum of butyraldehyde.

19.3 SPECTROSCOPY OF ALDEHYDES AND KETONES

A. IR Spectroscopy

The principal infrared absorption of aldehydes and ketones is the C=O stretching absorption, a strong absorption that occurs in the vicinity of 1700 cm^{-1}. In fact, this is one of the most important of all infrared absorptions. Because the C=O bond is stronger than the C=C bond, its stretching vibration has a larger force constant and, by Eq. 12.9, it absorbs at higher frequency.

The position of the C=O stretching absorption varies predictably for different types of carbonyl compounds. It generally occurs at 1710–1715 cm^{-1} for simple ketones and at 1720–1725 cm^{-1} for simple aldehydes. The carbonyl absorption is clearly evident, for example, in the IR spectrum of butyraldehyde (Fig. 19.3). The stretching absorption of the carbonyl–hydrogen bond of aldehydes near 2710 cm^{-1} is another characteristic absorption; however, as we shall see in Sec. 19.3B, NMR spectroscopy provides a more reliable way to diagnose the presence of this type of hydrogen.

Compounds in which the carbonyl group is conjugated with aromatic rings, double bonds, or triple bonds have lower carbonyl stretching frequencies than unconjugated carbonyl compounds.

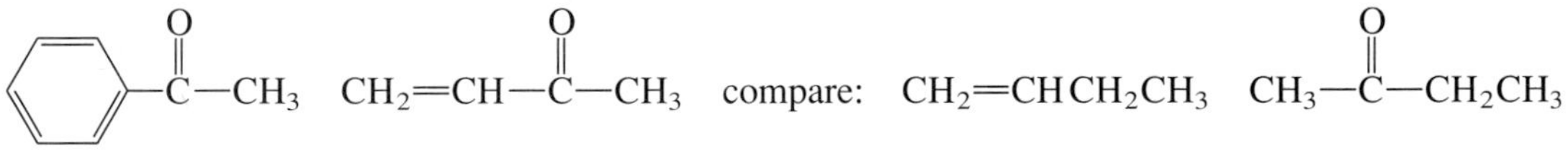

C=O	1685 cm^{-1}	1670 cm^{-1}	—	1715 cm^{-1}
C=C	1600 cm^{-1}	1613 cm^{-1}	1642 cm^{-1}	—

Note that the double-bond stretching frequencies are also lower in the conjugated molecules. These effects can be explained by the resonance structures for these compounds. Since the C=O and C=C bonds have some single bond character, as indicated by the following resonance structures, they are somewhat weaker than ordinary double bonds, and therefore absorb in the IR at lower frequency.

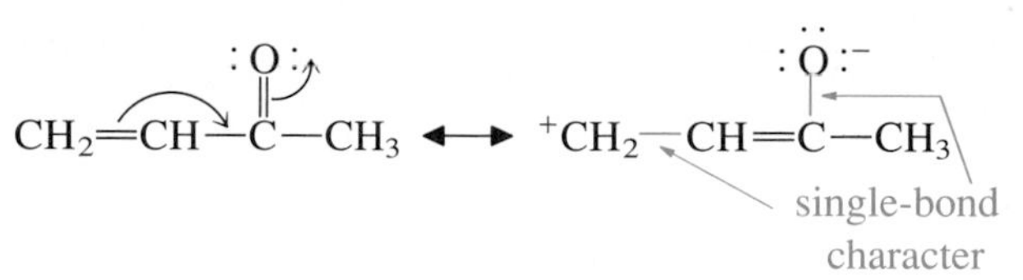

(19.2)

In cyclic ketones with rings containing fewer than six carbons, the carbonyl absorption frequency increases significantly as the ring size decreases.

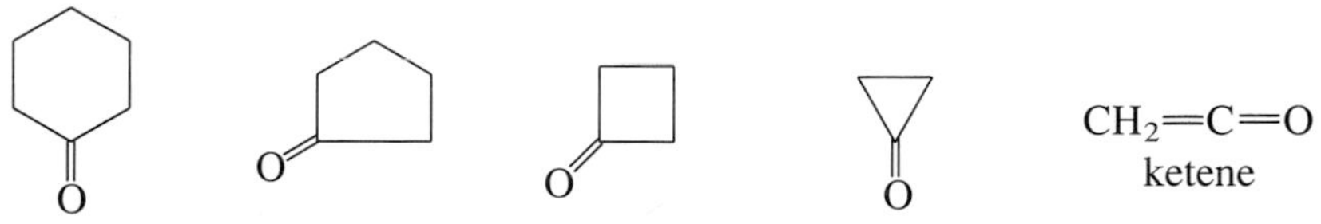

C=O	$1715\ cm^{-1}$ (normal)	$1745\ cm^{-1}$	$1780\ cm^{-1}$	$1850\ cm^{-1}$	$2150\ cm^{-1}$

This trend can be understood in terms of the relative strengths of the C=O bonds. The C=O bond in ketene involves an *sp*-hybridized carbonyl carbon, whereas that in cyclohexanone involves an sp^2-hybridized carbonyl carbon. The amount of *s* character in the C=O bond increases progressively across the series from cyclohexanone to ketene. As with C—H bonds (Eq. 14.34), the strengths of C=O bonds, and hence, their absorption frequencies, increase with degree of *s* character.

Problem

3 Explain how IR spectroscopy could be used to differentiate the compounds within each of the following pairs:
(a) cyclohexanone and hexanal
(b) 3-cyclohexen-1-one and 2-cyclohexen-1-one
(c) 2-butanone and 3-buten-2-ol

B. NMR Spectroscopy

The characteristic NMR absorption common to both aldehydes and ketones is that of the protons on the carbons *adjacent* to the carbonyl group—the α-protons—which is in the δ 2.0–2.5 region of the spectrum (see also Table 13.2). This absorption is slightly farther downfield than the absorptions of allylic protons; this makes sense because the α-protons of an aldehyde or ketone are "allylic" to the C=O bond. In addition, the absorption of the aldehydic proton is quite distinctive, occurring in the δ 9–10 region of the NMR spectrum, at lower field than most other NMR absorptions.

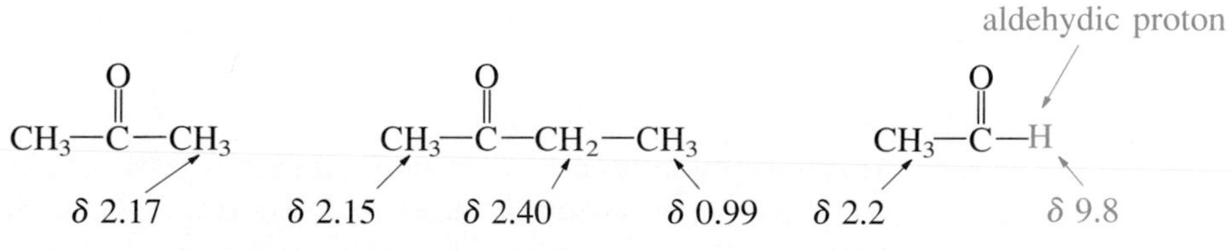

These NMR absorptions are illustrated by the spectrum of acetaldehyde (Fig. 19.4)

We can see from these examples that protons in the vicinity of the carbonyl group are significantly deshielded. The reason for this deshielding is the same as that for the

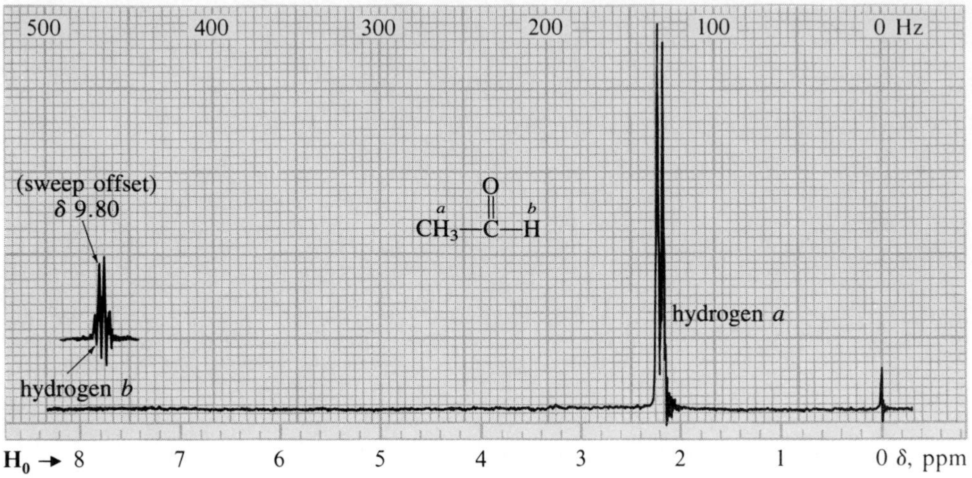

Figure 19.4 NMR spectrum of acetaldehyde.

deshielding effects of the carbon–carbon double bond (Sec. 13.5A). The carbonyl group has a greater deshielding effect because of the electronegativity of the carbonyl oxygen. As we have learned, electronegative groups induce significant downfield shifts in the absorptions of nearby protons.

Problem

4 Deduce the structures of the following compounds:

(a) C_4H_8O: IR 1720, 2710 cm^{-1}
NMR in Fig. 19.5a.

(b) C_4H_8O: IR 1717 cm^{-1}
NMR δ 0.95 (3*H*, t, $J = 8$ Hz); δ 2.03 (3*H*, s); δ 2.38 (2*H*, q, $J = 8$Hz)

(c) $C_{10}H_{12}O_2$: IR 1690 cm^{-1}, 1612 cm^{-1}
NMR in Fig. 19.5b.

C. UV Spectroscopy

The $\pi \rightarrow \pi^*$ absorptions of unconjugated aldehydes and ketones occur at about 150 nm, below the wavelength range of common UV spectrometers. Simple aldehydes and ketones also have another, much weaker, absorption at higher wavelength, in the 260–290 nm region. This absorption is caused by excitation of the unshared electrons on oxygen (sometimes called the *n*-electrons). This high-wavelength absorption is usually referred to as an ***n* → π* absorption.**

$$(CH_3)_2C{=}\ddot{\underset{\cdot\cdot}{O}} \qquad n \rightarrow \pi^* \qquad 271\text{ nm } (\epsilon = 16) \text{ (in ethanol)}$$

n-electrons

This absorption is easily distinguished from a $\pi \rightarrow \pi^*$ absorption because it is only 10^{-2}–10^{-3} times as strong. However, it is strong enough that aldehydes and ketones cannot be used as solvents for UV spectroscopy.

Like conjugated dienes, the π-electrons of compounds whose carbonyl groups are conjugated with double or triple bonds have strong absorption in the UV spectrum. The spectrum of 1-acetyl-1-cyclohexene (Fig. 19.6) is typical. The 232-nm peak is

Figure 19.5 (a) NMR spectrum for Problem 4(a). NMR spectrum for Problem 4(c).

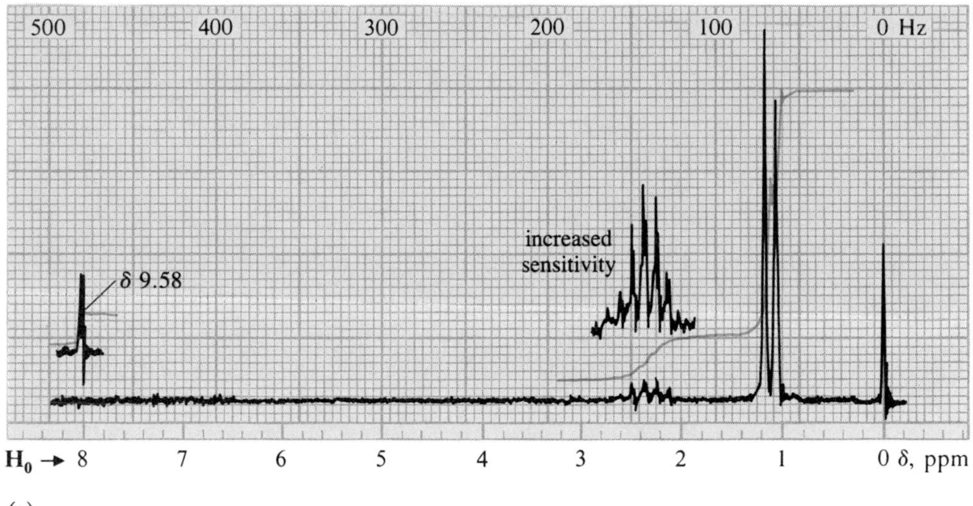

(a)

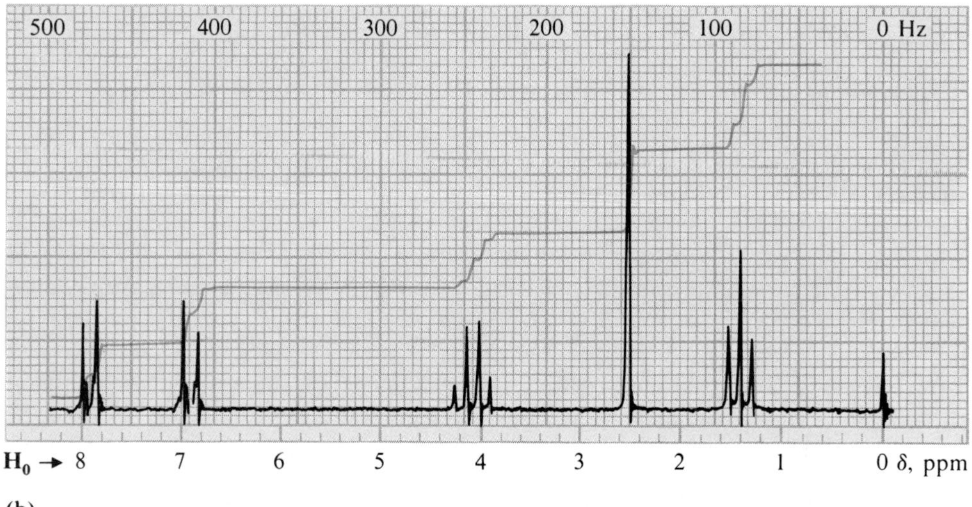

(b)

Figure 19.6 Ultraviolet spectrum of 1-acetyl-1-cyclohexene. The upper spectrum, which shows the weak $n \rightarrow \pi^*$ absorption, is run at eighty times the concentration of the lower.

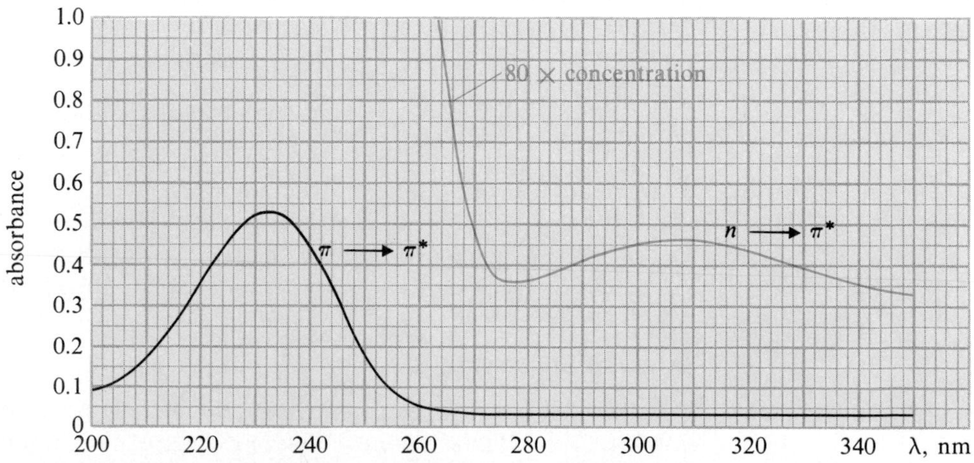

due to light absorption by the conjugated π-electron system and is thus a $\pi \rightarrow \pi^*$ absorption. It has a very large extinction coefficient, much like that of a conjugated diene. The weak 308 nm absorption is an $n \rightarrow \pi^*$ absorption.

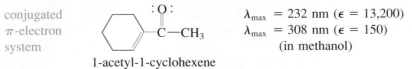

conjugated π-electron system

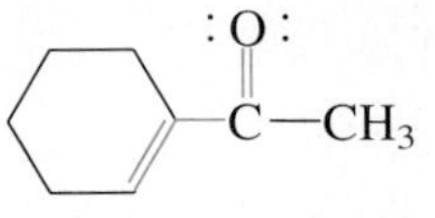

1-acetyl-1-cyclohexene

λ_{max} = 232 nm (ϵ = 13,200)
λ_{max} = 308 nm (ϵ = 150)
(in methanol)

The λ_{max} of a conjugated aldehyde or ketone is governed by the same variables that affect the λ_{max} of conjugated dienes: number of conjugated double bonds, substitution on the double bond, and so on. When an aromatic ring is conjugated with a carbonyl group the typical aromatic absorptions are more intense and shifted to higher wavelength than those of benzene.

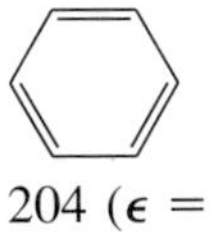

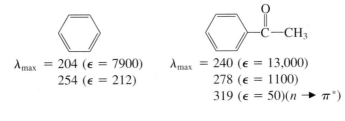

λ_{max} = 204 (ϵ = 7900)
254 (ϵ = 212)

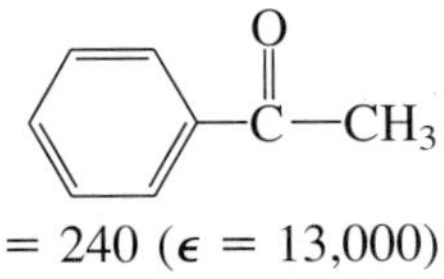

λ_{max} = 240 (ϵ = 13,000)
278 (ϵ = 1100)
319 (ϵ = 50)($n \rightarrow \pi^*$)

The $\pi \rightarrow \pi^*$ absorptions of conjugated carbonyl compounds, like those of conjugated alkenes, arise from the promotion of a π-electron from a bonding to an antibonding (π^*) molecular orbital (Sec. 15.2B). An $n \rightarrow \pi^*$ absorption arises from promotion of one of the n (unshared) electrons on a carbonyl oxygen to a π^* molecular orbital. As we learned above, $n \rightarrow \pi^*$ absorptions are weak. Physicists say that these absorptions are *forbidden*. This means that there are physical reasons (which we cannot discuss here) why these absorptions should have very low intensity. The 254 nm absorption of benzene, which has a very low extinction coefficient of 212, is another example of a forbidden absorption.

Problems

5 Explain how the compounds within each set can be distinguished using only UV spectroscopy.
(a) 2-cyclohexen-1-one and 3-cyclohexen-1-one
(b) 1-phenyl-2-propanone and *p*-methylacetophenone
(c) 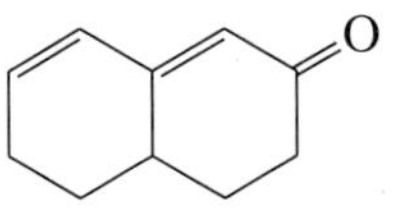 and 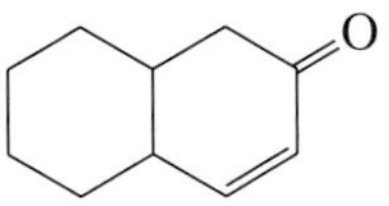

6 In neutral alcohol solution, the UV spectra of *p*-hydroxyacetophenone and *p*-methoxyacetophenone are virtually identical. When NaOH is added to the solution, the λ_{max} of *p*-hydroxyacetophenone increases by about 50 nm, but that of *p*-methoxyacetophenone is unaffected. Explain these observations.

D. Mass Spectrometry

Important fragmentations of aldehydes and ketones are illustrated by the mass spectrum of 5-methyl-2-hexanone, shown in Fig. 19.7. The three most important peaks

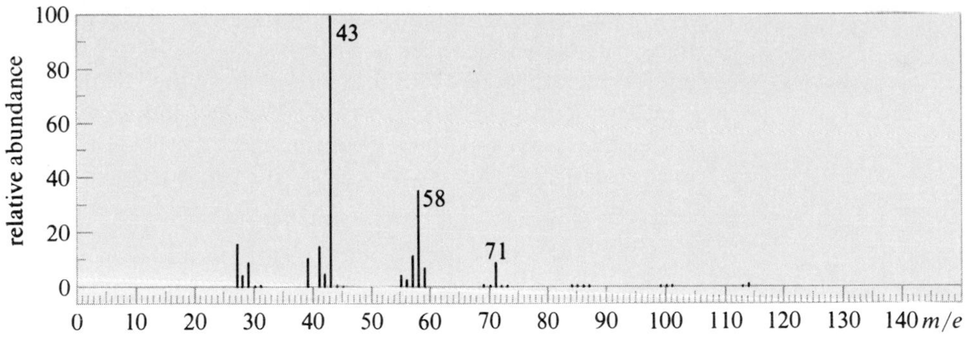

Figure 19.7 Mass spectrum of 5-methyl-2-hexanone.

occur at $m/e = 71$, 58, and 43. The peaks at $m/e = 71$ and $m/e = 43$ come from cleavage of the parent ion at the bond between the carbonyl group and the adjacent carbon atom. *Inductive cleavage* accounts for the $m/e = 71$ peak; in this cleavage the alkyl fragment carries the charge and the carbonyl fragment carries the unpaired electron.

$$\left[(CH_3)_2CHCH_2CH_2-\overset{\overset{\displaystyle +}{:\overset{..}{O}{\cdot}}}{\underset{\|}{C}}-CH_3 \longleftrightarrow (CH_3)_2CHCH_2CH_2-\overset{\overset{\displaystyle +}{:\overset{..}{O}:}}{\underset{\cdot}{C}}-CH_3 \right] \longrightarrow$$

parent ion from loss of unshared electron

$$(CH_3)_2CHCH_2\overset{+}{C}H_2 + {\cdot}\overset{:\overset{..}{O}:}{\underset{\|}{C}}-CH_3 \quad (19.3)$$
$$m/e = 71$$

α-Cleavage accounts for the $m/e = 43$ peak. In this case the same parent ion fragments in such a way that the carbonyl fragment carries the charge and the alkyl fragment carries the unpaired electron:

$$(CH_3)_2CHCH_2CH_2-\overset{\overset{\displaystyle +}{:\overset{..}{O}}}{\underset{\|}{C}}-CH_3 \longrightarrow (CH_3)_2CHCH_2\overset{\cdot}{C}H_2 + :\overset{..}{O}{\equiv}C-\overset{+}{C}H_3 \quad (19.4)$$
$$m/e = 43$$

An analogous cleavage at the carbon–hydrogen bond accounts for the fact that many aldehydes show a strong $p - 1$ peak.

How do we account for the $m/e = 58$ peak? This peak is an odd-electron ion. (How do we know?) As we learned in Sec. 12.5C, a common mechanism for formation of odd-electron ions is hydrogen transfer followed by loss of a stable neutral molecule; and indeed, exactly such a mechanism is responsible for the $m/e = 58$ peak. The oxygen radical in the parent ion abstracts a hydrogen atom from a carbon five atoms away, and the resulting radical then undergoes α-cleavage.

hydrogen transfer α-cleavage $m/e = 58$

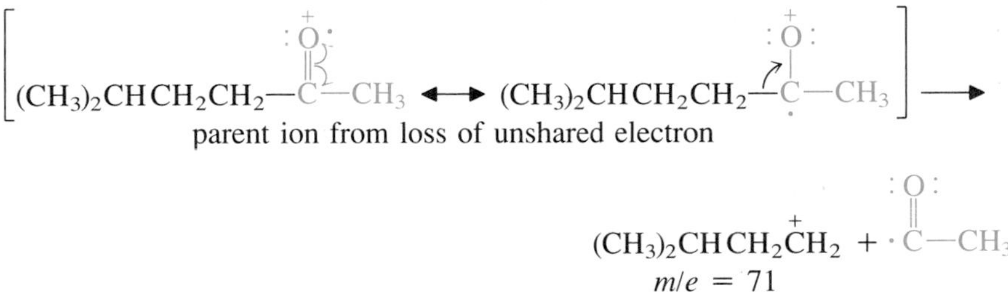

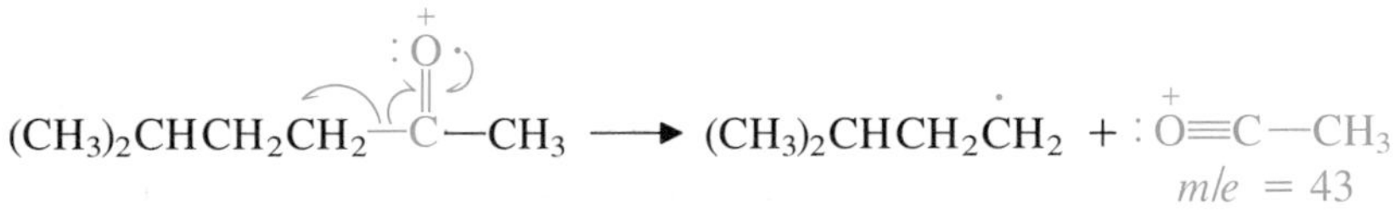

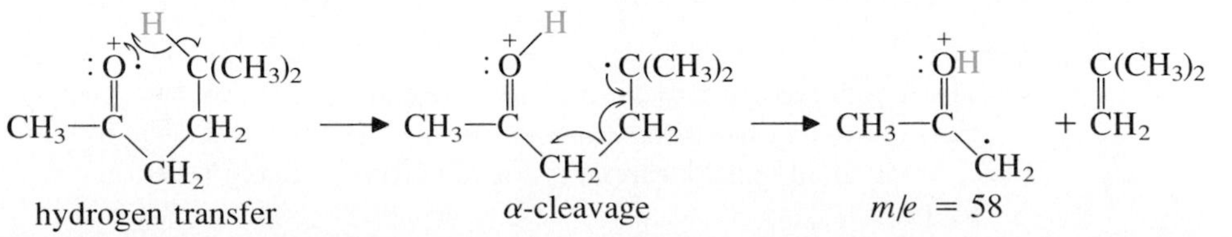

$$(19.5)$$

If we count the hydrogen that is transferred, the rearrangement occurs through a transient six-membered ring. This process is called a **McLafferty rearrangement,** after Professor Fred McLafferty, now of Cornell University, who investigated this type of fragmentation extensively. The McLafferty rearrangement and subsequent α-cleavage is a common mechanism for the production of odd-electron fragment ions in the mass spectrometry of carbonyl compounds.

Problem

7 Using only mass spectrometry, how would you distinguish 2-heptanone from 3-heptanone?

19.4 SYNTHESIS OF ALDEHYDES AND KETONES

We have learned several reactions that can be used for the preparation of aldehydes and ketones. The three most important of these are:

1. Oxidation of alcohols (Sec. 10.6). Primary alcohols can be oxidized to aldehydes, and secondary alcohols can be oxidized to ketones.
2. Friedel–Crafts acylation (Sec. 16.4E). This reaction provides a way to synthesize aryl ketones. It also involves the formation of a carbon–carbon bond—the bond between the aryl ring and the carbonyl group.
3. Hydration and hydroboration of alkynes (Sec. 14.5).

We have studied other reactions that give aldehydes or ketones as products, but are less important as synthetic methods. These are:

4. Ozonolysis of alkenes (Sec. 5.5).
5. Periodate cleavage of glycols (Sec. 10.6D).
6. Pinacol rearrangement of glycols (Sec. 10.1B).

Ozonolysis and periodate cleavage are reactions that break carbon–carbon bonds. Since an important aspect of organic synthesis is the *making* of carbon–carbon bonds, use of these reactions in effect wastes some of the effort that goes into making the alkene or glycol starting materials. Nevertheless, these reactions can be used synthetically in certain cases. The pinacol rearrangement is a useful reaction, but is less important than the first three methods because it is not as *general.*

In Chapter 21, we shall learn other important methods of preparing aldehydes and ketones from carboxylic acid derivatives.

19.5 INTRODUCTION TO ALDEHYDE AND KETONE REACTIONS

The reactions of aldehydes and ketones can be conveniently grouped into two categories: (1) reactions of the carbonyl group, and (2) reactions involving the α-carbon.

We shall consider three types of carbonyl-group reactions of aldehydes and ketones.

1. *Basicity of the carbonyl oxygen.* The carbonyl oxygen is weakly basic and thus reacts with Lewis and Brønsted acids. Letting E^+ be a general electrophile, we can represent this reaction as follows:

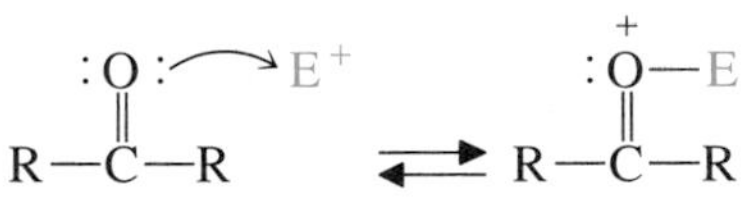

$$\overset{:O:}{\underset{R-C-R}{\|}} \quad \overset{\curvearrowright E^+}{} \quad \rightleftharpoons \quad \overset{:\overset{+}{O}-E}{\underset{R-C-R}{\|}} \tag{19.6}$$

Carbonyl basicity is important because it plays a role in several other carbonyl-group reactions.

2. *Addition reactions.* The most important carbonyl-group reaction is addition to the C=O double bond. Letting E—Y symbolize a general reagent, addition is represented in the following way:

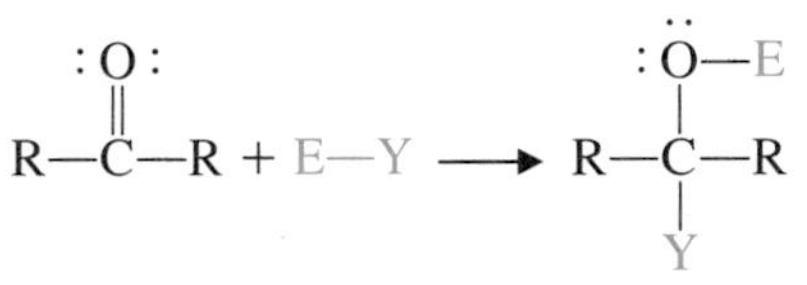

$$\overset{:O:}{\underset{R-C-R}{\|}} + E-Y \longrightarrow \overset{:\overset{..}{O}-E}{\underset{\underset{Y}{|}}{R-C-R}} \tag{19.7}$$

Superficially, carbonyl addition is analogous to alkene addition (Sec. 4.5A).

Many reactions of aldehydes and ketones are simple additions that conform exactly to the model in Eq. 19.7. Others are multistep processes in which addition is followed by other reactions.

3. *Oxidation of aldehydes.* Aldehydes can be oxidized to carboxylic acids:

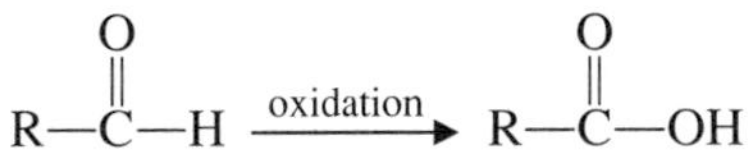

$$\overset{O}{\underset{R-C-H}{\|}} \xrightarrow{\text{oxidation}} \overset{O}{\underset{R-C-OH}{\|}} \tag{19.8}$$

The remainder of this chapter is devoted to carbonyl-group reactions of aldehydes and ketones. The following two chapters deal with the properties and carbonyl-group reactions of carboxylic acids and their derivatives. In Chapter 22, we shall study reactions at the α-carbon—reactions that are important not only in aldehyde and ketone chemistry, but also in the chemistry of other carbonyl compounds.

19.6 BASICITY OF ALDEHYDES AND KETONES

Aldehydes and ketones are weakly basic and react at the carbonyl oxygen with protons or Lewis acids.

$$H_3\overset{..}{O}{}^+ + \overset{:O:}{\underset{CH_3 \quad CH_3}{C}} \rightleftharpoons \left[\overset{+:\overset{\curvearrowleft}{O}-H}{\underset{CH_3 \quad CH_3}{C}} \longleftrightarrow \overset{:\overset{..}{O}-H}{\underset{CH_3 \quad \overset{+}{C}H_3}{C}} \right] + H_2\overset{..}{O}: \tag{19.9}$$

protonated ketone
$pK_a = -6 \text{ to } -7$

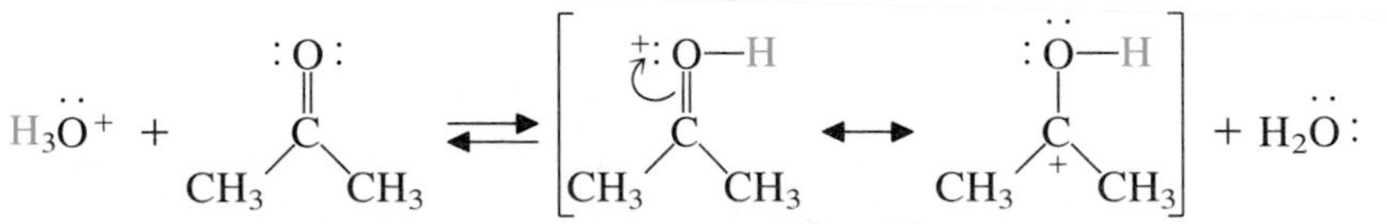

As Eq. 19.9 shows, the protonated form of an aldehyde or ketone is resonance stabilized. The resonance structure on the right shows that the protonated carbonyl compound has carbocation character. In fact, we shall find in some cases that the conjugate acids of aldehydes and ketones undergo typical carbocation reactions.

Closely related to protonated aldehydes and ketones are *α-alkoxycarbocations:* cations in which the acidic proton is replaced by an alkyl group.

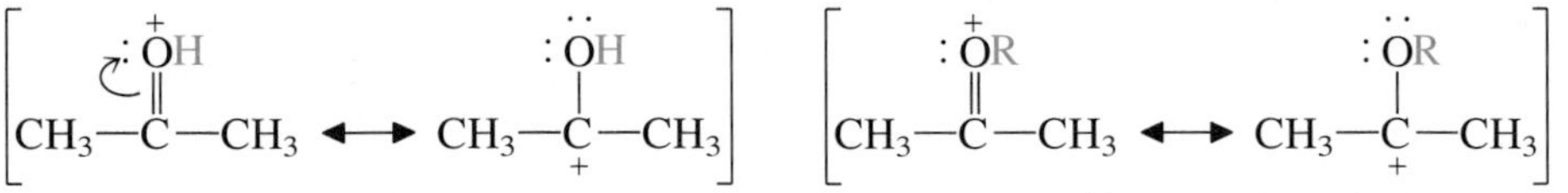

protonated acetone
(an *α*-hydroxy carbocation) (an *α*-alkoxy carbocation)

Such ions are considerably more stable than ordinary alkyl cations. For example, a secondary *α*-alkoxycarbocation is about 40 kcal/mol more stable in the gas phase than an ordinary secondary carbocation.

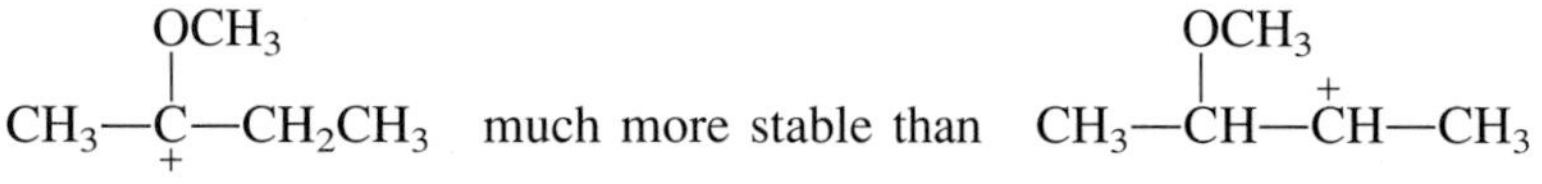

$$CH_3-\overset{+}{\underset{|}{C}}-CH_2CH_3 \quad \text{much more stable than} \quad CH_3-\underset{|}{CH}-\overset{+}{CH}-CH_3 \quad (19.10)$$
$$\overset{OCH_3}{} \qquad\qquad\qquad\qquad\qquad \overset{OCH_3}{}$$

These ions, like protonated aldehydes or ketones, owe their stabilities to the resonance interaction of the electron-deficient carbon with the neighboring oxygen. This resonance effect far outweighs the electron-attracting inductive effect of the oxygen, which, by itself, would destabilize the carbocation.

The stability of such ions explains, for example, the relative rates of the two nucleophilic substitution reactions in Eq. 19.11 (see Problem 8).

relative rate:

$$CH_3CH_2CH_2Cl + C_2H_5OH \xrightarrow{25°} CH_3CH_2CH_2-OC_2H_5 + HCl \qquad 1$$
$$CH_3-O-CH_2Cl + C_2H_5OH \xrightarrow{25°} CH_3-O-CH_2-OC_2H_5 + HCl \qquad 10^4$$

(19.11)

Problem

8 Using Hammond's postulate, explain why the second reaction in Eq. 19.11 is much faster than the first. Your explanation should include a mechanism for the two reactions.

Aldehydes and ketones in solution are considerably less basic than alcohols (Sec. 8.7). In other words, their conjugate acids are more acidic than those of alcohols.

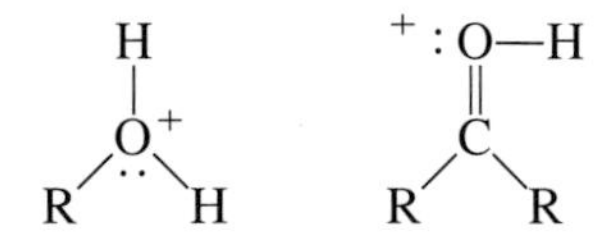

$pK_a \approx -2.5$ $pK_a \approx -7$

Since protonated aldehydes and ketones are resonance stabilized and protonated alcohols are not, we might have expected the protonated carbonyl compounds to be *more stable* relative to their conjugate bases and therefore *less acidic*. As the pK_a values above show, this is not the case. The relative acidity of protonated alcohols and carbonyl compounds has been found to be yet another example of a solvent effect. In the *gas phase,* aldehydes and ketones *are* indeed more basic than alcohols. One reason for the greater basicity of alcohols in solution is that protonated alcohols have more O—H protons to participate in hydrogen bonding to solvent than do protonated aldehydes or ketones.

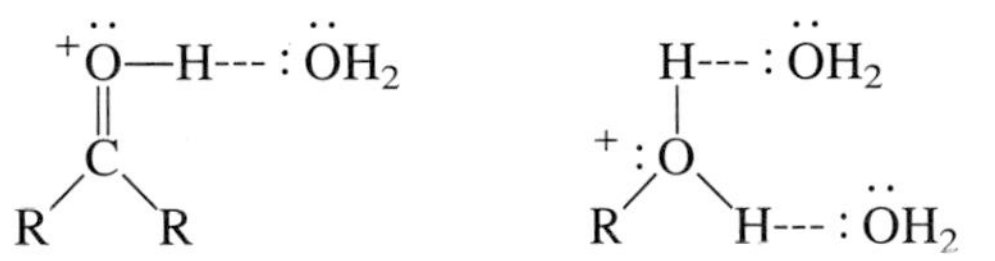

protonated ketone

protonated alcohol: better hydrogen bonding

Problem

9 Use resonance arguments to explain why 3-buten-2-one is more basic than 2-butanone.

19.7 REVERSIBLE ADDITION REACTIONS OF ALDEHYDES AND KETONES

One of the most typical reactions of aldehydes and ketones is *addition* to the carbon–oxygen double bond. To begin with, let us focus on two simple addition reactions, hydration and addition of hydrogen cyanide (HCN).

Hydration (addition of water):

$$CH_3-\overset{O}{\overset{\|}{C}}-H + H-OH \rightleftarrows CH_3-\underset{OH}{\overset{OH}{\underset{|}{\overset{|}{C}}}}-H \qquad (19.12)$$

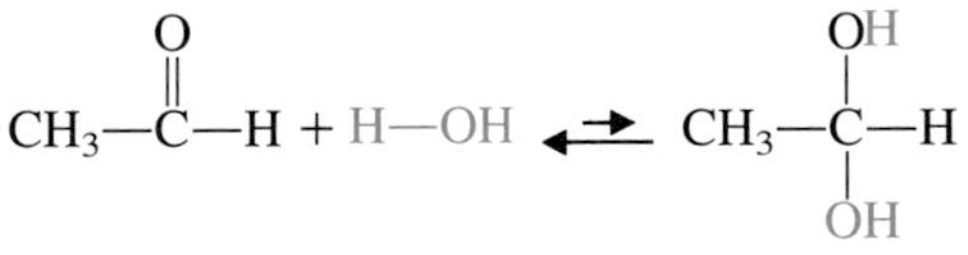

acetaldehyde
hydrate

The product of water addition is called a *hydrate,* or *gem*-diol. (The prefix *gem*- stands for *geminal,* from the Latin word for twin, and is used in chemistry when two identical groups are present on the same carbon.)

Addition of HCN:

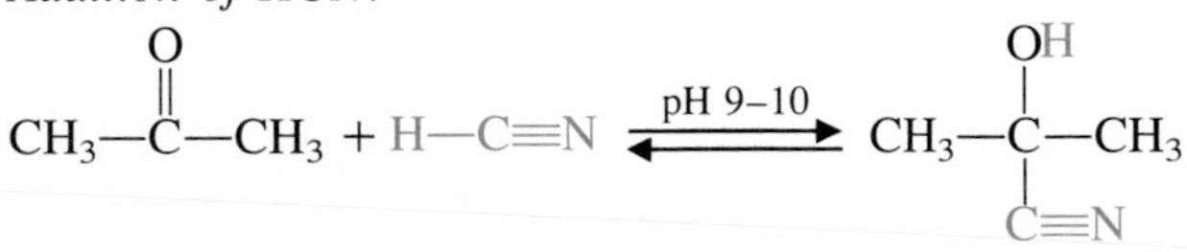

$$CH_3-\overset{O}{\overset{\|}{C}}-CH_3 + H-C\equiv N \underset{}{\overset{pH\ 9-10}{\rightleftarrows}} CH_3-\underset{C\equiv N}{\overset{OH}{\underset{|}{\overset{|}{C}}}}-CH_3 \quad \text{acetone cyanohydrin} \quad (19.13)$$

(77–78% yield)

The product of HCN addition is termed a *cyanohydrin.* Cyanohydrins constitute a special class of *nitriles* (organic cyanides). (We shall study the chemistry of nitriles in Chapter 21.) Notice that the preparation of cyanohydrins is another method of forming carbon–carbon bonds.

All carbonyl-addition reactions show a complementarity between the polarity of the addition reagent and the polarity of the carbonyl group. Thus, the electropositive end of the addition reagent (the proton) adds to the electronegative end of the carbonyl group (the oxygen), and the electronegative end of the reagent (—OH or —CN) adds to the electropositive end of the carbonyl group (the carbonyl carbon).

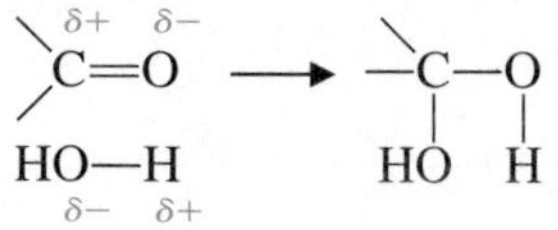

$$
\begin{array}{c}
\overset{\delta+}{}\overset{\delta-}{} \\
\text{C=O} \\
\text{HO—H} \\
\overset{\delta-}{}\overset{\delta+}{}
\end{array}
\quad\longrightarrow\quad
\begin{array}{c}
\text{—C—O} \\
\text{HO H}
\end{array}
\qquad (19.14)
$$

This generalization is analogous to the Markovnikov rule for alkene addition.

A. Mechanisms of Carbonyl-Addition Reactions

Carbonyl-addition reactions occur by two general types of mechanisms. The first general mechanism occurs under *basic conditions*. In this mechanism, a nucleophile attacks the carbonyl group at the carbonyl carbon, and the carbonyl oxygen becomes negatively charged. In cyanohydrin formation, for example, the cyanide ion, formed by ionization of HCN, is the nucleophile.

$$
\text{H—CN} + {}^-\text{OH} \rightleftharpoons {}^-:\text{CN} + H_2O \qquad (19.15a)
$$
$$
pK_a = 9.4
$$

$$
{}^-:\text{CN} + \underset{CH_3}{\overset{CH_3}{\diagdown}}\text{C=O} \rightleftharpoons CH_3-\underset{CN}{\overset{CH_3}{\underset{|}{\overset{|}{C}}}}-\ddot{O}:{}^- \qquad (19.15b)
$$

The negatively charged oxygen—essentially an alkoxide ion—is a relatively strong base, and is protonated by either water or HCN to complete the addition:

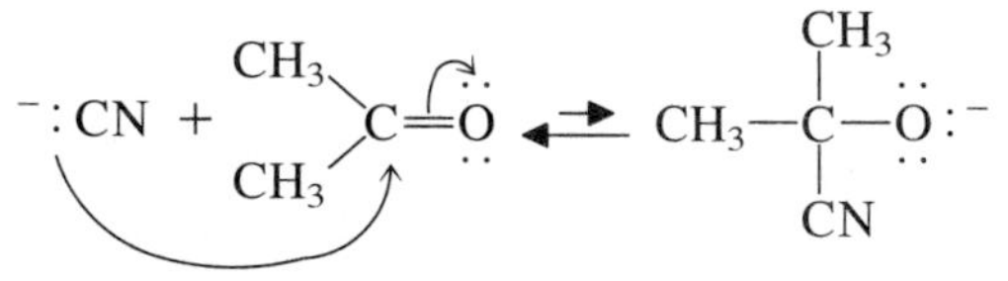

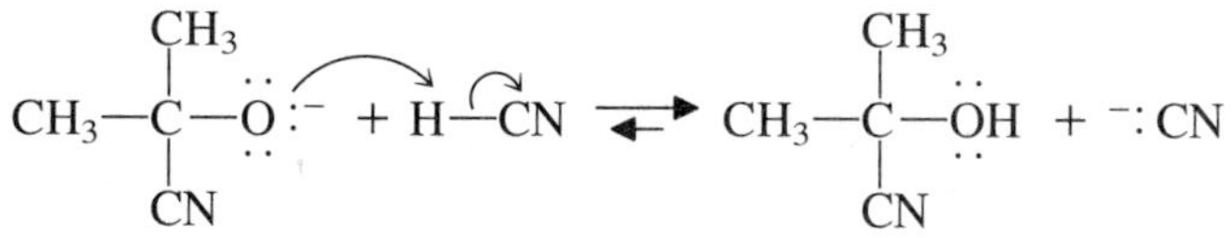

$$
CH_3-\underset{CN}{\overset{CH_3}{\underset{|}{\overset{|}{C}}}}-\ddot{O}:{}^- + H{-}CN \rightleftharpoons CH_3-\underset{CN}{\overset{CH_3}{\underset{|}{\overset{|}{C}}}}-\ddot{O}H + {}^-:\text{CN} \qquad (19.15c)
$$

This mechanism, called **nucleophilic addition,** has no analogy in the reactions of ordinary alkenes. This pathway occurs with aldehydes and ketones because, in the transition state, negative charge is placed on oxygen, an electronegative atom. The same reaction on an alkene would place negative charge on a relatively electropositive carbon atom.

negative charge on oxygen

negative charge on carbon

(observed)

(not observed with ordinary alkenes)

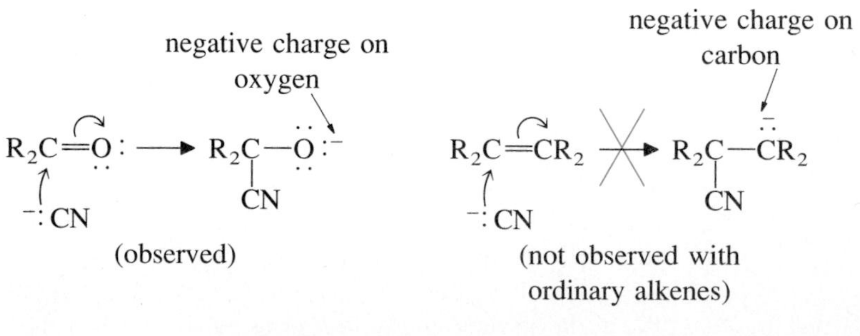

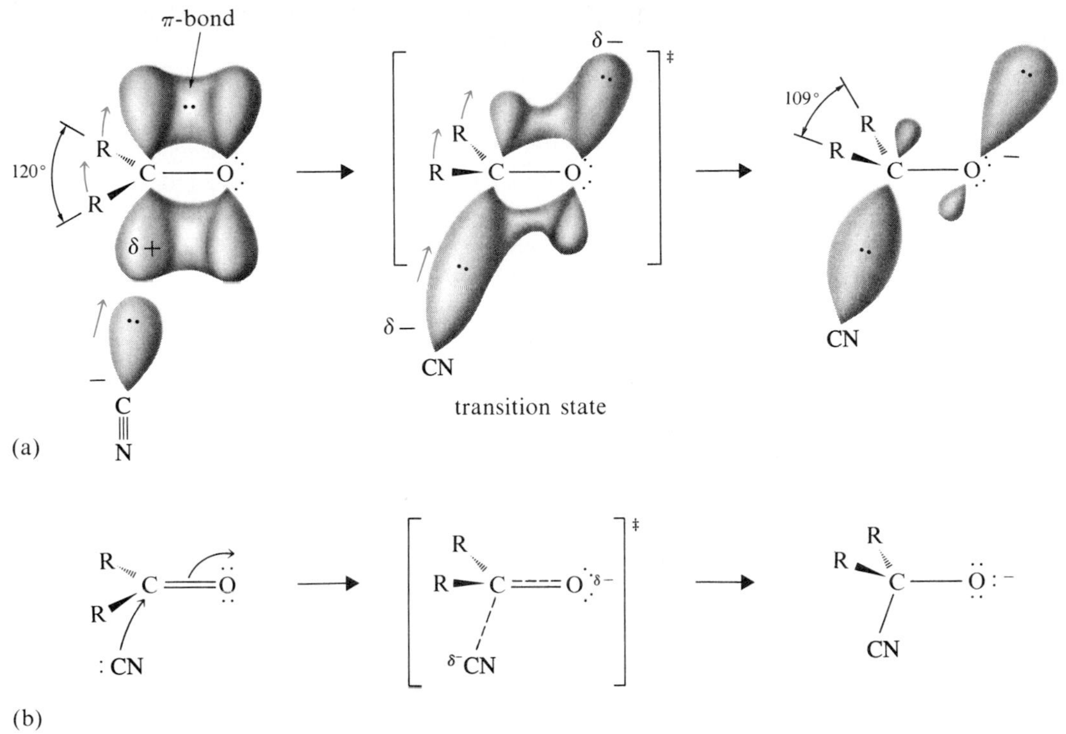

Figure 19.8 Nucleophilic attack on a carbonyl group. (a) Orbital picture; (b) the corresponding reaction in the curved-arrow formalism. The colored arrows show the movement of atoms.

Attack of the nucleophile occurs on the carbon of the carbonyl group rather than the oxygen for the same reason: negative charge is placed on the more electronegative atom—oxygen.

Among the reactions we have studied, the closest analogy to nucleophilic carbonyl addition is nucleophilic ring opening of epoxides (Sec. 11.4). Both mechanisms involve attack of a nucleophile at carbon, giving a negatively charged oxygen:

Attack at a carbonyl group: 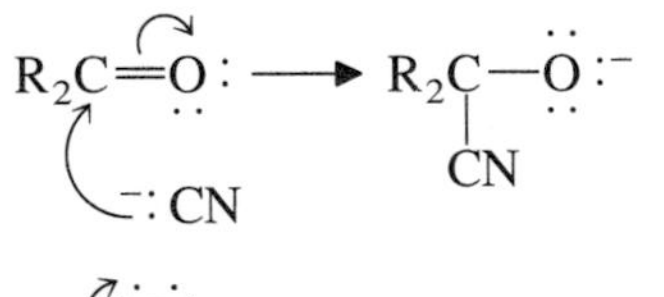 (19.16a)

Ring opening of an epoxide: 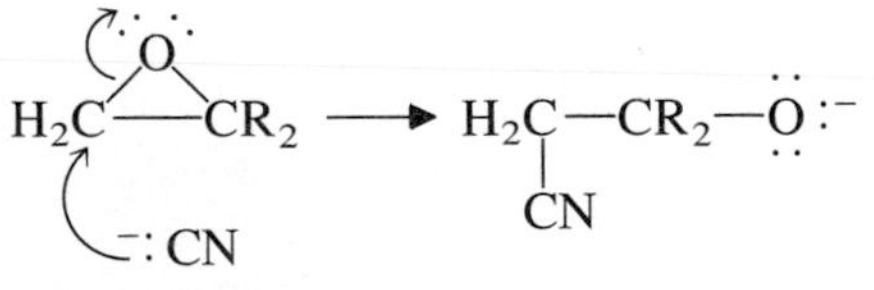 (19.16b)

An orbital picture of nucleophilic addition, along with the geometry of nucleophilic attack on the carbonyl group, is shown in Fig. 19.8. When a nucleophile attacks a carbonyl group, it attacks the π-bond from above or below the plane of the molecule, "pushing" an electron pair onto the carbonyl oxygen.

The second mechanism for carbonyl addition occurs under *acidic conditions*, and is closely analogous to electrophilic addition to alkenes (Sec. 4.5D). Acid-catalyzed hydration of carbonyl compounds is an example of this mechanism. The carbonyl group is first protonated by an acid in solution.

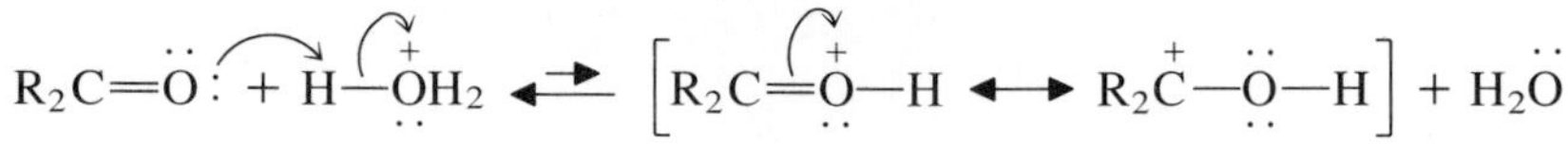

$$R_2C{=}\overset{..}{\underset{..}{O}}: + H{-}\overset{+}{O}H_2 \rightleftharpoons \left[R_2C{=}\overset{+}{O}{-}H \longleftrightarrow R_2\overset{+}{C}{-}\overset{..}{\underset{..}{O}}{-}H \right] + H_2\overset{..}{\underset{..}{O}}: \quad (19.17a)$$

The protonated carbonyl compound, as we learned in Sec. 19.6, has carbocation character. The electron-deficient carbon is attacked by a nucleophile, in this case water, which then loses a proton, completing the addition.

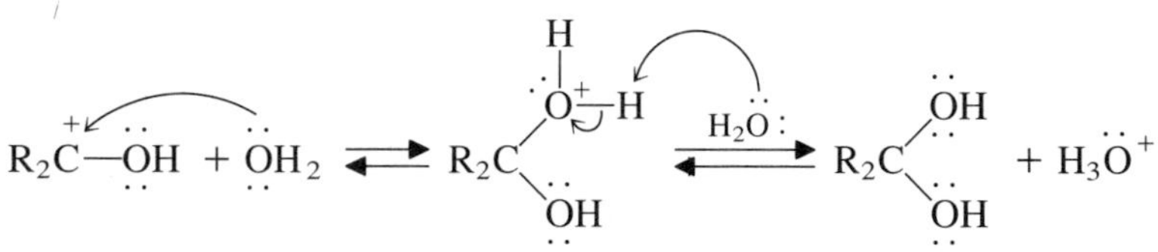

$$\overset{+}{R_2C}{-}\overset{..}{\underset{..}{O}}H + \overset{..}{\underset{..}{O}}H_2 \rightleftharpoons R_2C\underset{\overset{..}{\underset{..}{O}}H}{\overset{\overset{\textstyle H}{\underset{|}{\overset{+}{\underset{..}{O}}}{-}H}}{\diagdown}} \quad \underset{H_2\overset{..}{O}:}{\rightleftharpoons} \quad R_2C\underset{\overset{..}{\underset{..}{O}}H}{\overset{\overset{..}{O}H}{\diagdown}} + H_3\overset{+}{\underset{..}{O}}{}^+ \quad (19.17b)$$

This particular mechanism is very much like that for the hydration of alkenes (Sec. 4.8).

Notice that under basic conditions, the nucleophile is usually a fairly strong base, and the acid that protonates the negative oxygen (as in Eq. 19.15c) is usually a weak acid—in many cases the conjugate acid of the nucleophile. Under basic conditions strong acids are not available to complete the protonation, nor are they necessary, since the alkoxide is a fairly strong base. Conversely, under acidic conditions, the carbonyl is protonated by a relatively strong acid, and the nucleophile is usually a fairly weak base (water in Eq. 19.17b). Under acidic conditions, strong bases are not available, nor are they necessary, since a protonated carbonyl compound is very electrophilic—it is, after all, a carbocation.

Sometimes beginning students are tempted to write mechanisms involving strong acids and strong bases at the same time:

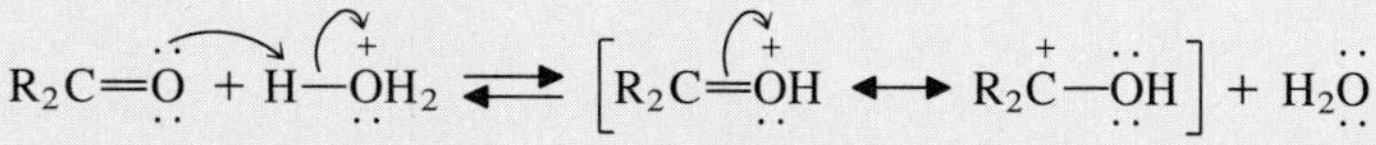

$$R_2C{=}\overset{..}{\underset{..}{O}} + H{-}\overset{+}{\underset{..}{O}}H_2 \rightleftharpoons \left[R_2C{=}\overset{+}{\underset{..}{O}}H \longleftrightarrow R_2\overset{+}{C}{-}\overset{..}{\underset{..}{O}}H \right] + H_2\overset{..}{\underset{..}{O}} \quad (19.18a)$$

$$R_2\overset{+}{C}{-}\overset{..}{\underset{..}{O}}H + {}^{-}:\overset{..}{\underset{..}{O}}H \rightleftharpoons R_2\underset{\overset{|}{:}\overset{..}{\underset{..}{O}}H}{C}{-}\overset{..}{\underset{..}{O}}H \quad (19.18b)$$

Mechanisms such as this are extremely rare, because high concentrations of H_3O^+ (a strong acid) and OH^- (a strong base) cannot exist at the same time in solution. The appropriate nucleophile for Eq. 19.18b is water, not hydroxide, because under acidic aqueous conditions, water usually acts as the base.

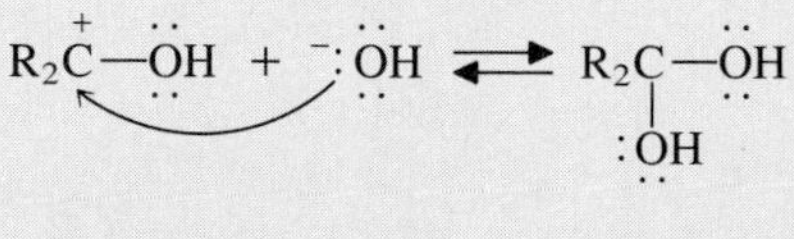

TABLE 19.2 Equilibrium Constants for Hydration of Aldehydes and Ketones

$$H_2O + R\!-\!\overset{\displaystyle O}{\overset{\|}{C}}\!-\!R \underset{}{\overset{K_{eq}}{\rightleftarrows}} R\!-\!\overset{\displaystyle OH}{\underset{\displaystyle OH}{\overset{|}{\underset{|}{C}}}}\!-\!R$$

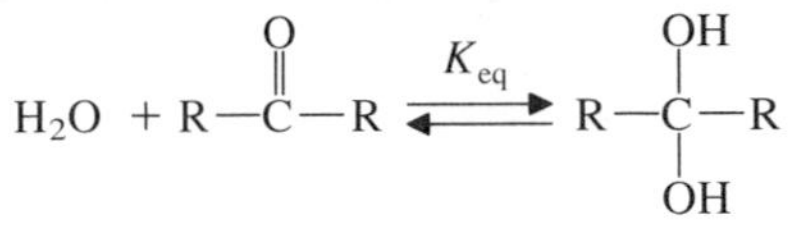

Compound	K_{eq}	Compound	K_{eq}
$CH_2\!=\!O$	2.2×10^3	$Cl_3C\!-\!CH\!=\!O$	2.8×10^4
$CH_3CH\!=\!O$	1.0	$(CH_3)_2C\!=\!O$	1.4×10^{-3}
$CH_3CH_2CH_2CH\!=\!O$	0.5	$Ph\!-\!\overset{\displaystyle O}{\overset{\|}{C}}\!-\!CH_3$	6.6×10^{-6}
$(CH_3)_2CHCH\!=\!O$	0.5–1.0	$Ph\!-\!\overset{\displaystyle O}{\overset{\|}{C}}\!-\!Ph$	1.2×10^{-7}
$Ph\!-\!CH\!=\!O$	8.3×10^{-3}	$(ClCH_2)_2C\!=\!O$	10
$ClCH_2CH\!=\!O$	37	$(CF_3)_2C\!=\!O$	too large to measure

Problems

10 Hydration of aldehydes and ketones is also catalyzed by bases. Write a mechanism for the base-catalyzed hydration of acetaldehyde.

11 Suggest a mechanism for the acid-catalyzed addition of methanol to benzaldehyde. (This is the first step in the formation of *acetals,* which we shall study in Sec. 19.10.)

B. Equilibrium in Carbonyl-Addition Reactions

Hydration and cyanohydrin formation have in common the fact that they are reversible. (Not all carbonyl additions are reversible.) Cyanohydrin formation, for example, favors the product in the case of aldehydes and methyl ketones, but not aromatic ketones. Hydration occurs more extensively with aldehydes than with ketones. What is the reason for these effects?

Let us answer this question by examining how the equilibrium constants for hydration of the carbonyl group vary with the structure of the aldehyde or ketone (Table 19.2). The data in Table 19.2 indicate the following trends:

1. Addition is more favorable for aldehydes than for ketones.

2. Electronegative groups near the carbonyl carbon make carbonyl addition more favorable.

3. Addition is less favorable when groups are present that donate electrons by resonance to the carbonyl carbon.

Each of these effects is understandable in terms of the relative stabilities of aldehydes and ketones. The stability of the carbonyl compound relative to the addition product

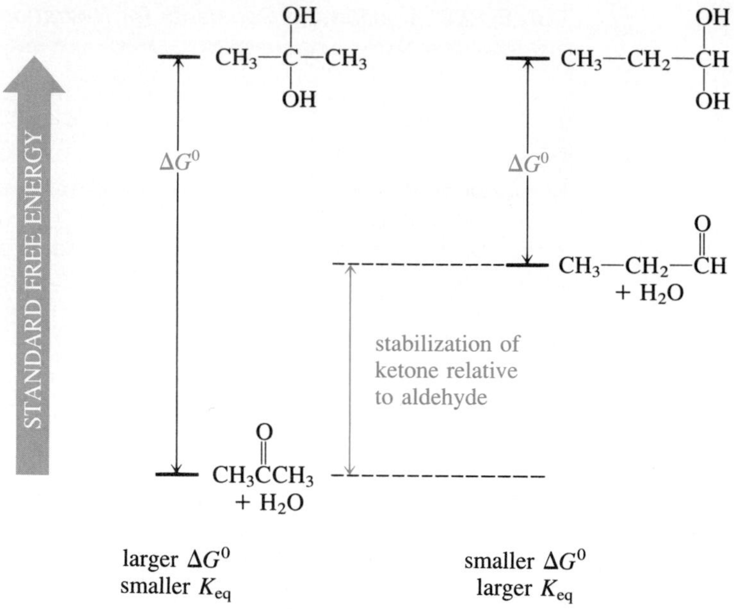

Figure 19.9 Stabilization of a ketone relative to an aldehyde increases its standard free energy of hydration and therefore lowers its equilibrium constant. (The two hydrates have been placed at the same energy level for comparison purposes.)

governs the ΔG^0 for addition. This point is illustrated in Fig. 19.9. The conclusion from this figure is that *added stability in the carbonyl compound increases the energy change (ΔG^0), and hence decreases the equilibrium constant, for formation of an addition product.*

What stabilizes carbonyl compounds? One way to answer this question is to consider the resonance structures of the carbonyl group:

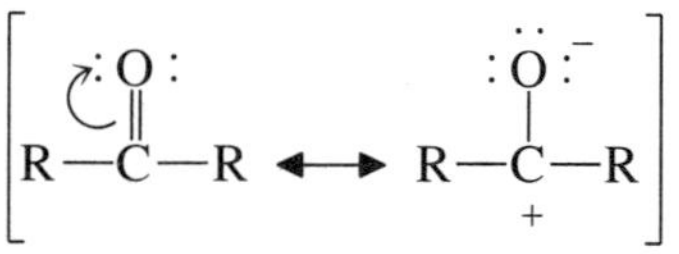

$$\left[\begin{array}{c} \overset{\displaystyle :O:}{\underset{\displaystyle R-C-R}{\parallel}} \end{array} \longleftrightarrow \begin{array}{c} \overset{\displaystyle :\overset{..}{O}:^-}{\underset{\displaystyle R-\overset{+}{C}-R}{\vert}} \end{array} \right] \tag{19.19}$$

The structure on the right, although not as important a contributor as the one on the left, reflects the polarity of the carbonyl group, and has the characteristics of a carbocation. Therefore anything that stabilizes carbocations also tends to stabilize carbonyl compounds. Since alkyl groups stabilize carbocations, ketones (R = alkyl) are more stable than aldehydes (R = H). We can see this stability reflected in the relative heats of formation of aldehydes and ketones. For example, acetone, with $\Delta H_f^0 = -52.0$ kcal/mol, is 6.1 kcal/mol more stable than its isomer propionaldehyde, which has $\Delta H_f^0 = -45.9$ kcal/mol. Since alkyl groups stabilize carbonyl compounds, the equilibria for additions to ketones are less favorable than those for additions to aldehydes (Trend 1). Formaldehyde, with *two* hydrogens and no alkyl groups bound to the carbonyl, has a very large equilibrium constant for hydration.

Electronegative groups such as halogens destabilize carbocations by an inductive effect, and also destabilize carbonyl compounds. Thus, halogens make the equilibria for addition more favorable (Trend 2). In fact, chloral hydrate (known in medicine as a hypnotic) is a stable crystalline compound.

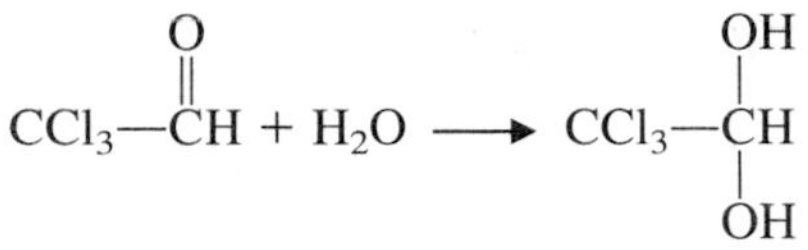

$$CCl_3-\overset{\overset{\displaystyle O}{\|}}{C}H + H_2O \longrightarrow CCl_3-\overset{\overset{\displaystyle OH}{|}}{\underset{\underset{\displaystyle OH}{|}}{C}}H \qquad \text{chloral hydrate} \qquad (19.20)$$

Groups that are conjugated with the carbonyl, such as the phenyl group of benzaldehyde, stabilize carbocations, and hence, carbonyl compounds, by resonance:

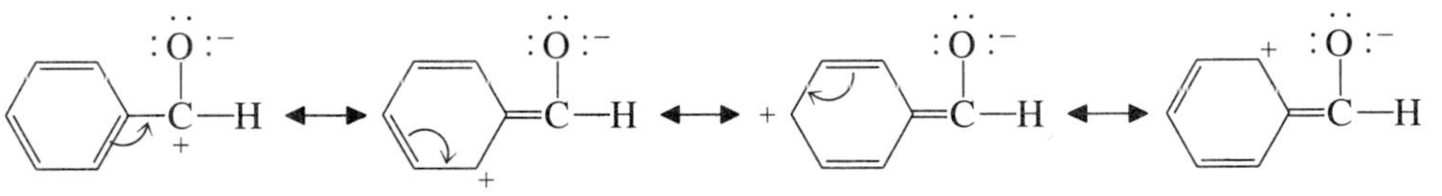

$$(19.21)$$

Such compounds have relatively unfavorable hydration equilibria (Trend 3).

The trends in relative rates of addition can be predicted from the trends in equilibrium constants. That is, *compounds with the most favorable addition equilibria tend to react most rapidly in addition reactions.* Thus, *aldehydes are generally more reactive than ketones in addition reactions; formaldehyde is more reactive than many other simple aldehydes.* We can see why this is reasonable if we imagine that the transition states for addition reactions resemble the addition products to some extent. Just as destabilization of aldehydes or ketones decreases the ΔG^0 for their addition reactions (Fig. 19.8), the same destabilization decreases the *free energy of activation* $DG^{0\ddagger}$ and thus increases the rates of their addition reactions.

Problems

12 The compound *ninhydrin* exists as a hydrate. Which carbonyl group is hydrated? Explain, and give the structure of the hydrate.

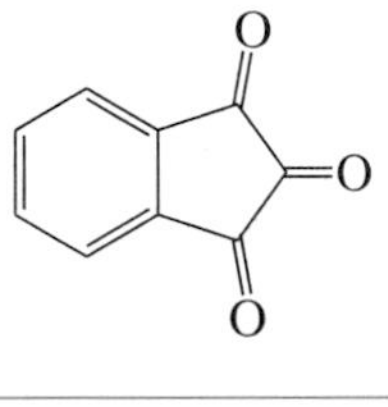

13 Within each set, which compound should be more reactive in carbonyl-addition reactions? Explain your choices.

(a)

$$CH_3-\overset{\overset{\displaystyle O}{\|}}{C}-CH_2CH_2Br \quad \text{or} \quad CH_3-\overset{\overset{\displaystyle O}{\|}}{C}-CH_2Br$$

(b)

$$CH_3-\overset{\overset{\displaystyle O}{\|}}{C}-\overset{\overset{\displaystyle O}{\|}}{C}-CH_3 \quad \text{or} \quad CH_3-\overset{\overset{\displaystyle O}{\|}}{C}-CH_2CH_3$$

(c)

$$O_2N-\!\!\!\!\bigcirc\!\!\!\!-CH{=}O \quad \text{or} \quad CH_3\overset{..}{O}-\!\!\!\!\bigcirc\!\!\!\!-CH{=}O$$

(d)

$$\bigcirc{=}O \quad \text{or} \quad \triangleright{=}O$$

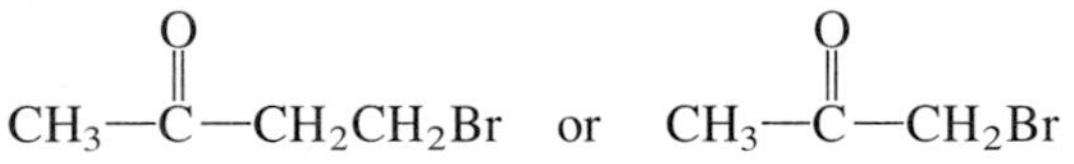

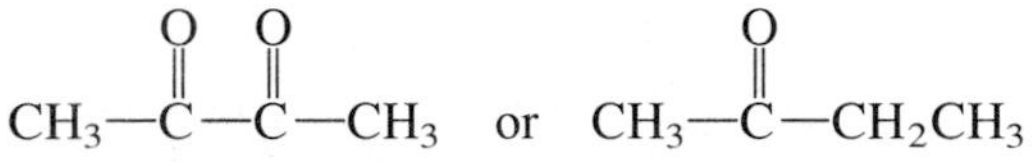

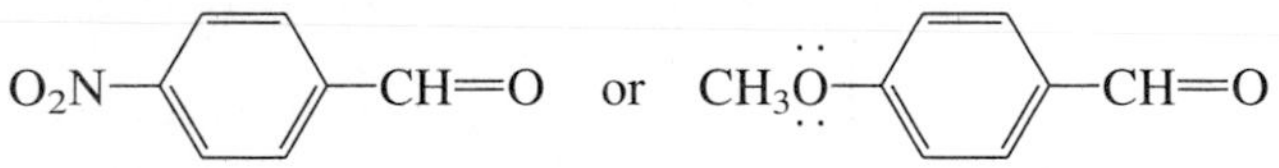

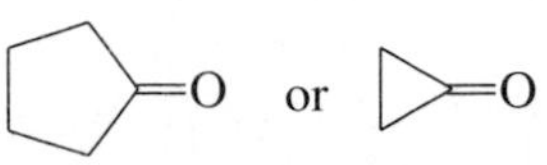

In this section we have studied two examples of addition to the carbonyl group. In subsequent sections we shall study other addition reactions, and then go on to consider more complex reactions that have mechanisms in which the initial steps are addition reactions. These addition reactions all have mechanisms similar to the ones discussed in this section. *Addition to the carbonyl group is a common thread that runs throughout most of aldehyde and ketone chemistry.*

19.8 REDUCTION OF ALDEHYDES AND KETONES TO ALCOHOLS

Aldehydes and ketones are reduced to alcohols with either lithium aluminum hydride, $LiAlH_4$, or sodium borohydride, $NaBH_4$. These reactions result in the net *addition* of the elements of H_2 across the C=O bond.

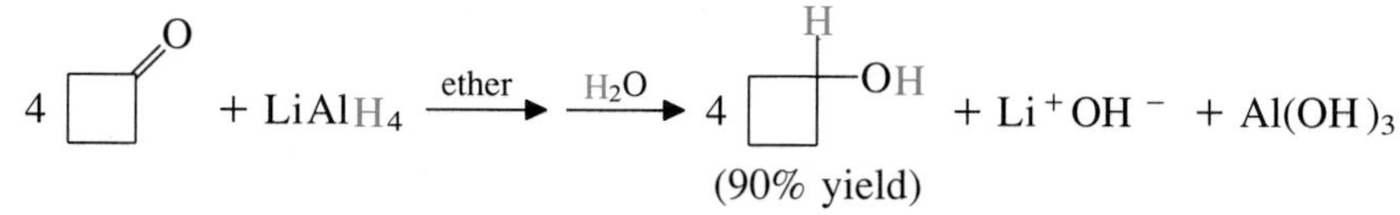

$$4\ \text{(cyclobutanone)} + LiAlH_4 \xrightarrow{\text{ether}} \xrightarrow{H_2O} 4\ \text{(cyclobutanol)} + Li^+OH^- + Al(OH)_3 \quad (19.22)$$
$$\text{(90\% yield)}$$

$$4CH_3O\text{—}\bigcirc\text{—}CH{=}O + 4CH_3OH + NaBH_4 \xrightarrow{CH_3OH}$$

$$4CH_3O\text{—}\bigcirc\text{—}CH_2OH + Na^+ CH_3O^- + B(OCH_3)_3 \quad (19.23)$$
$$\text{(96\% yield)}$$

As these examples illustrate, reduction of an aldehyde gives a primary alcohol, and reduction of a ketone gives a secondary alcohol.

Let us focus first on reduction with $LiAlH_4$. In order to understand this reaction, we need to know some of the characteristics of lithium aluminum hydride. This reagent, one of the most useful reducing agents in organic chemistry, serves generally as a source of $H:^-$, the *hydride ion*. This is reasonable because hydrogen is more electronegative than aluminum (Table 1.1). Thus, the Al—H bonds of the $^-AlH_4$ ion carry a substantial fraction of the negative charge. In other words,

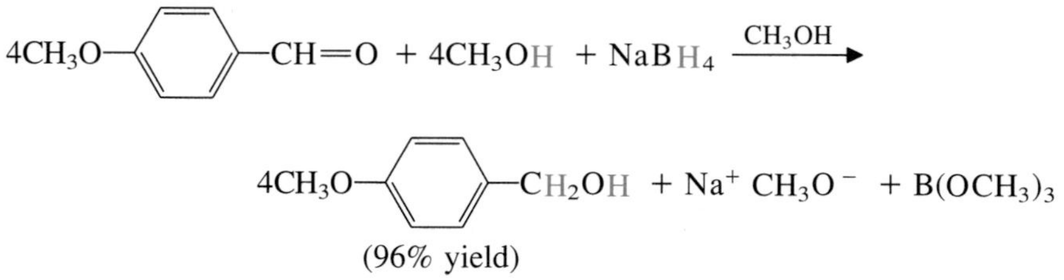

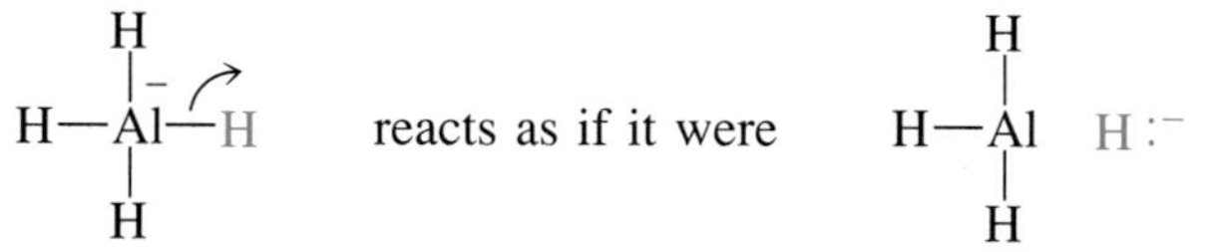

$$\text{H—}\overset{\overset{H}{|}}{\underset{\underset{H}{|}}{Al}}\text{—H} \quad \text{reacts as if it were} \quad \text{H—}\overset{\overset{H}{|}}{\underset{\underset{H}{|}}{Al}} \quad H:^- \quad (19.24)$$

The hydride ion in $LiAlH_4$ is very basic. For example, $LiAlH_4$ reacts violently with water, and therefore must be used in dry solvents such as anhydrous ether and THF.

$$Li^+ \text{ H—}\overset{\overset{H}{|}}{\underset{\underset{H}{|}}{Al}}\text{—H} \quad H\text{—}\ddot{O}H \longrightarrow \text{H—}\overset{\overset{H}{|}}{\underset{\underset{H}{|}}{Al}} + \quad H\text{—H} \quad + Li^+ \ ^-:\ddot{O}H \quad (19.25)$$
$$\text{(reacts further}\qquad\qquad\text{hydrogen gas}$$
$$\text{with water)}$$

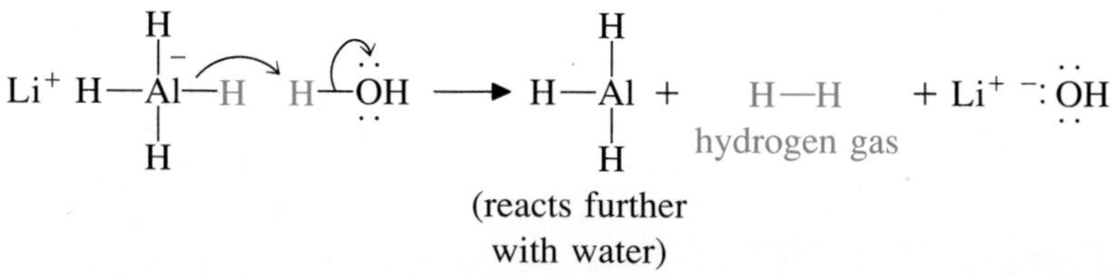

Like many good bases, the hydride ion in $LiAlH_4$ is a good nucleophile. The reaction of $LiAlH_4$ with aldehydes and ketones involves the nucleophilic attack of hydride (delivered from $^-AlH_4$) on the carbonyl carbon. A lithium ion coordinated to the carbonyl oxygen acts as a Lewis-acid catalyst.

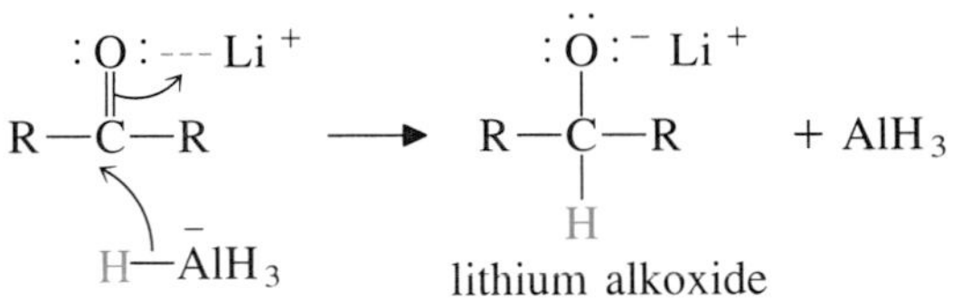

(19.26a)

The alkoxide salt (which actually exists as a complex with the Lewis acid AlH_3 or other trivalent-aluminum species present in solution) is converted by protonation into the alcohol product. The proton source is water (or an aqueous solution of a weak acid such as ammonium chloride), which is added in a separate step to the reaction mixture.

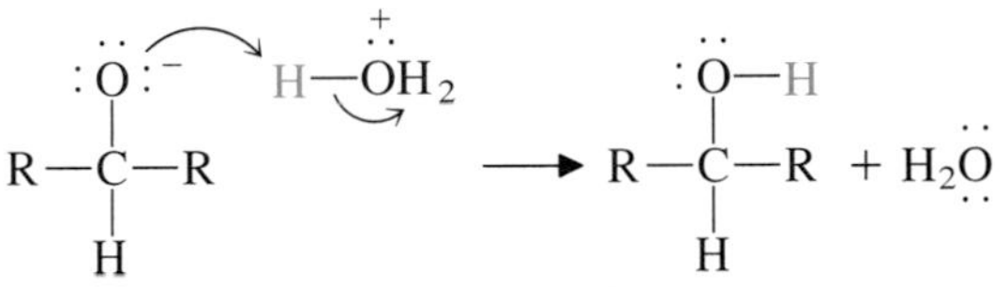

(19.26b)

As the stoichiometries of Eqs. 19.22 and 19.23 indicate, all four hydrides of $LiAlH_4$ are active, although we shall not consider here the detailed mechanisms for reduction by hydride equivalents beyond the first.

The reaction of sodium borohydride with aldehydes and ketones is conceptually similar to that of $LiAlH_4$. The sodium ion, however, does not coordinate to the carbonyl oxygen as well as the lithium ion does. For this reason, $NaBH_4$ reductions are carried out in protic solvents, such as alcohols. Hydrogen bonding between the alcohol solvent and the carbonyl group serves as a weak acid catalysis that activates the carbonyl group. $NaBH_4$ reacts only slowly with alcohols, and can even be used in water if the solution is not acidic.

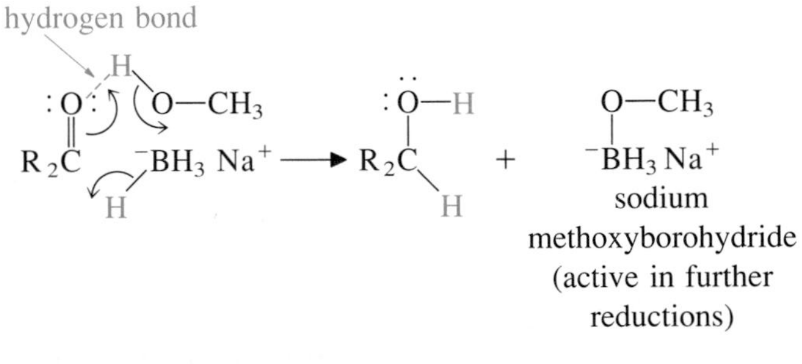

(19.27)

All four hydride equivalents of $NaBH_4$ are active in the reduction.

Because $LiAlH_4$ and $NaBH_4$ are hydride donors, reductions by these and related reagents are generally referred to as **hydride reductions.** The important mechanistic point about these reactions is that they are further examples of *nucleophilic addition.* Hydride ion from $LiAlH_4$ or $NaBH_4$ is the nucleophile, and the proton is delivered from acid added in a separate step (in the case of $LiAlH_4$ reductions) or solvent (in the case of $NaBH_4$ reductions).

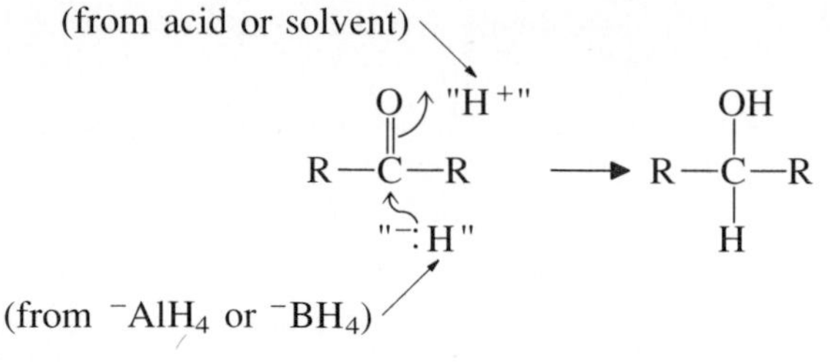

(19.28)

Unlike the additions discussed in Sec. 19.7, hydride reductions are *not* reversible. Reversal of carbonyl addition requires that the original attacking group—in this case, $H:^-$—be expelled as a leaving group. As in S_N1 or S_N2 reactions, the best leaving groups are the weakest bases. Hydride ion is such a strong base that it is not easily expelled as a leaving group. Hence, hydride reductions of all aldehydes and ketones go to completion.

Both $LiAlH_4$ and $NaBH_4$ are highly useful in the reduction of aldehydes and ketones. Lithium aluminum hydride is, however, a much more *reactive* reagent than sodium borohydride. There are a number of functional groups that react with $LiAlH_4$ but not $NaBH_4$—for example, alkyl halides, alkyl tosylates, and nitro groups. Sodium borohydride can be used as a reducing agent in the presence of these groups.

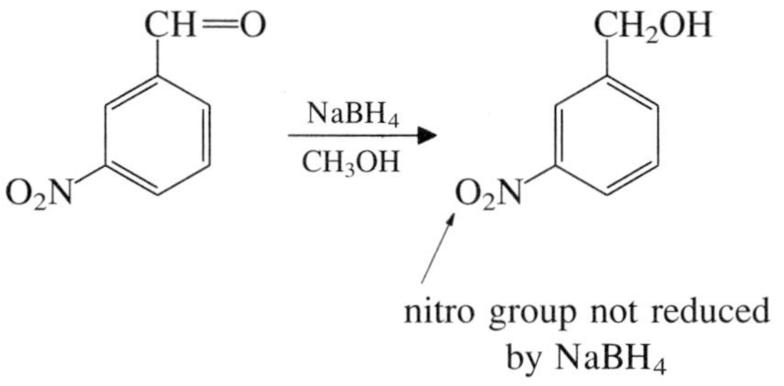

(82% yield) (19.29)

Sodium borohydride is also a much less hazardous reagent than lithium aluminum hydride. The greater selectivity and safety of $NaBH_4$ make it the preferred reagent in many applications, but either reagent can be used for the reduction of simple carbonyl compounds. Both are very important in organic chemistry.

The discovery of $NaBH_4$ illustrates that interesting research findings are sometimes obtained by accident. In the early 1940s, the U. S. Army Signal Corps became interested in methods for generation of hydrogen gas in the field. $NaBH_4$ was proposed as a relatively safe, portable source of hydrogen: addition of acidified water to $NaBH_4$ results in the evolution of hydrogen gas at a safe, moderate rate. In order to supply the required quantities of $NaBH_4$, a large-scale synthesis was necessary. The following reaction appeared to be suitable for this purpose.

$$4NaH + B(OCH_3)_3 \longrightarrow NaBH_4 + 3Na^+ \ CH_3O^- \qquad (19.30)$$

The problem with this process was that the sodium borohydride had to be separated from the sodium methoxide by-product. Several solvents were tried in the hope that a significant difference in solubilities could be found. In the course of this investigation, acetone was tried as a recrystallization solvent, and was found to react with the $NaBH_4$ to yield isopropyl alcohol. Thus was born the use of $NaBH_4$ as a reducing agent for carbonyl compounds.

These investigations, carried out by Herbert C. Brown (1912–), now Professor Emeritus of Chemistry at Purdue University, were part of what was to become a major research program in the boron hydrides, shortly thereafter leading to the discovery of hydroboration (Sec. 5.2A). Brown even describes his interest in the field of boron chemistry as something of an accident, because it sprung from his reading a book about boron and silicon hydrides that was given to him by his girl friend (now his wife) as a graduation present. Mrs. Brown observes that the choice of this particular book was dictated by the fact that it was among the least expensive chemical titles in the bookstore; in the depression era students had to be careful how they spent their money! For his work in organic chemistry Brown shared the Nobel Prize in Chemistry in 1979 with Georg Wittig (Sec. 19.12).

Aldehydes and ketones can also be reduced to alcohols by catalytic hydrogenation. This reaction is analogous to the hydrogenation of alkenes (Sec. 5.7).

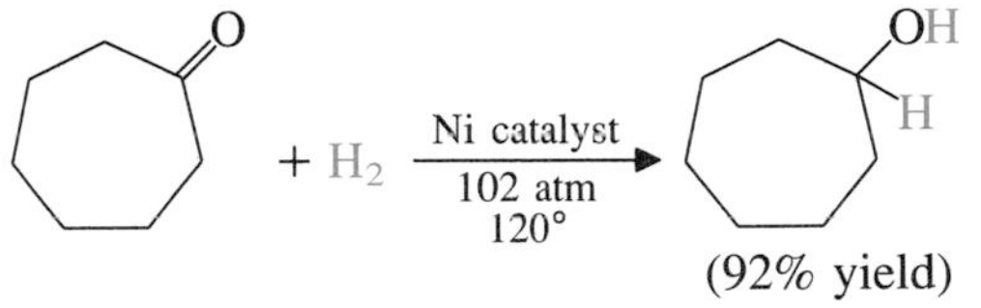

(19.31)

Catalytic hydrogenation is less important for the reduction of carbonyl groups than it once was because of the modern use of hydride reagents.

It is usually possible to use catalytic hydrogenation for the selective reduction of an alkene double bond in the presence of a carbonyl group. Palladium catalysts are particularly effective in this reduction.

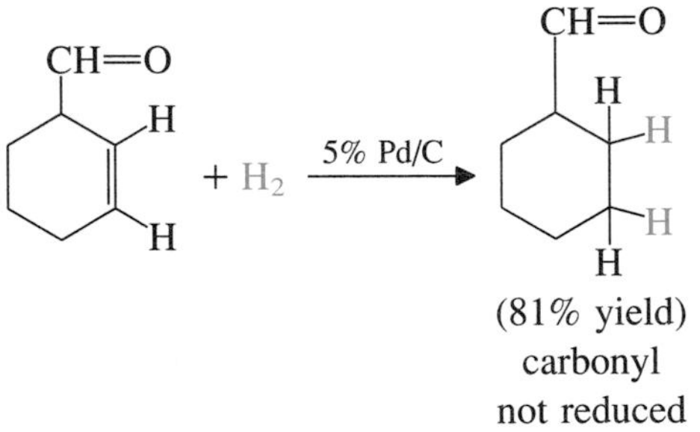

(19.32)

Problems

14 From what aldehydes or ketones could each of the following be synthesized by reduction with either $LiAlH_4$ or $NaBH_4$?

(a) 

(b) 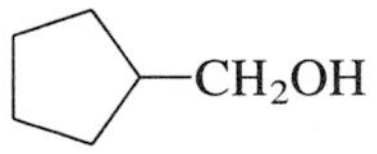

(c) 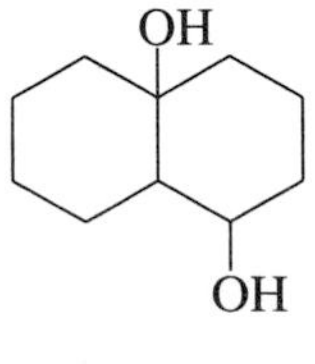

Problems (Cont.)

15 Which of the following alcohols could *not* be synthesized by a hydride reduction of an aldehyde or ketone? Explain.

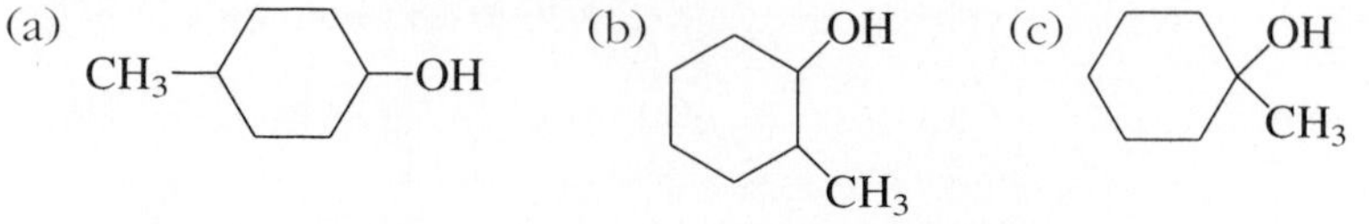

19.9 REACTIONS OF ALDEHYDES AND KETONES WITH GRIGNARD REAGENTS

Addition of Grignard reagents to aldehydes and ketones in an ether solvent, followed by protonolysis, gives alcohols. This is the most important single application of the Grignard reagent in organic chemistry.

$$(CH_3)_2CH\overset{\overset{\displaystyle O}{\|}}{C}H + C_2H_5MgBr \xrightarrow{\text{ether}} \xrightarrow{H_3O^+} (CH_3)_2CH\overset{\overset{\displaystyle OH}{|}}{C}HC_2H_5 \quad (68\%\ \text{yield}) \qquad (19.33)$$

$$CH_3—\overset{\overset{\displaystyle O}{\|}}{C}—CH_3 + CH_3CH_2CH_2—MgBr \xrightarrow{\text{ether}} \xrightarrow{H_3O^+} CH_3—\underset{\underset{\displaystyle CH_3CH_2CH_3}{|}}{\overset{\overset{\displaystyle OH}{|}}{C}}—CH_3 \quad (68\%\ \text{yield}) \qquad (19.34)$$

The reaction of Grignard reagents with aldehydes or ketones is another example of *carbonyl addition*. In this reaction, the magnesium of the Grignard reagent, a Lewis acid, coordinates with the carbonyl oxygen. This coordination, much like protonation in more conventional acid-catalyzed additions, increases the positive charge on the carbonyl carbon. The carbon group of the Grignard reagent attacks the carbonyl carbon. Recall that this group is a strong base that behaves much like a *carbanion* (Sec. 8.8B, 11.4B).

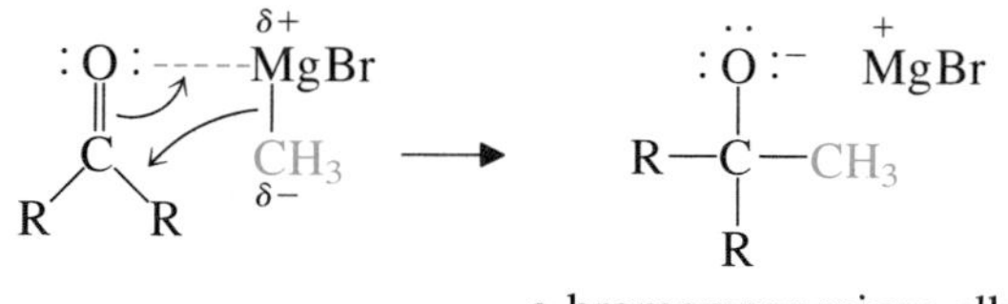

(19.35a)

a bromomagnesium alkoxide

The product of this addition, a bromomagnesium alkoxide, is essentially the magnesium salt of an alcohol. Addition of dilute acid to the reaction mixture gives an alcohol.

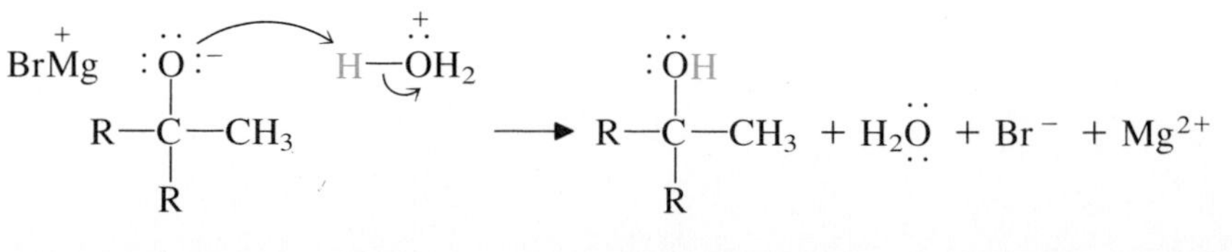

(19.35b)

Because of the great basicity of the Grignard reagent, this addition, like hydride reductions, is not reversible, and works with just about any aldehyde or ketone.

The reactions of organolithium and sodium acetylide reagents with aldehydes and ketones are fundamentally similar to the Grignard reaction.

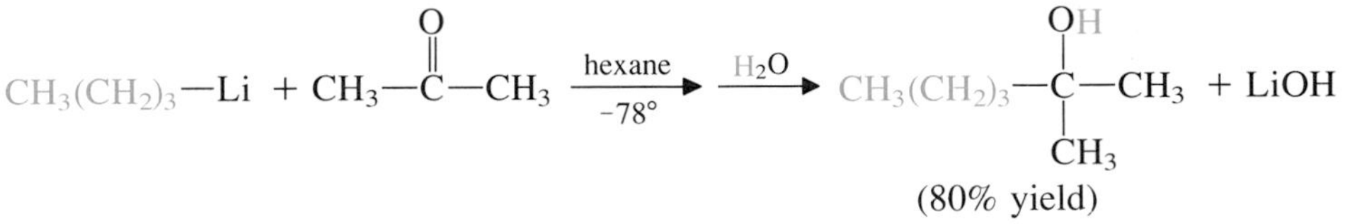

$CH_3(CH_2)_3$—Li + CH_3—C(=O)—CH_3
hexane
-78°
H_2O
$CH_3(CH_2)_3$—C(OH)(CH_3)—CH_3 + LiOH
(80% yield)

(19.36)

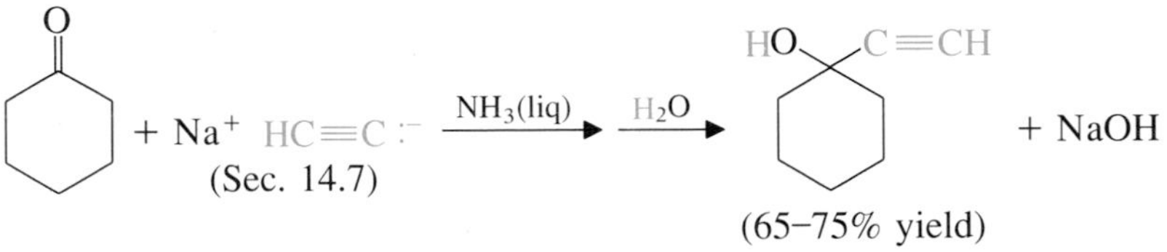

O
+ Na^+ HC≡C:$^-$
(Sec. 14.7)
NH_3(liq)
H_2O
HO
C≡CH
+ NaOH
(65–75% yield)

(19.37)

The reaction of Grignard and related reagents with aldehydes and ketones is important not only because it can be used to convert aldehydes or ketones into alcohols, but also because it is an excellent method of *carbon–carbon bond formation.*

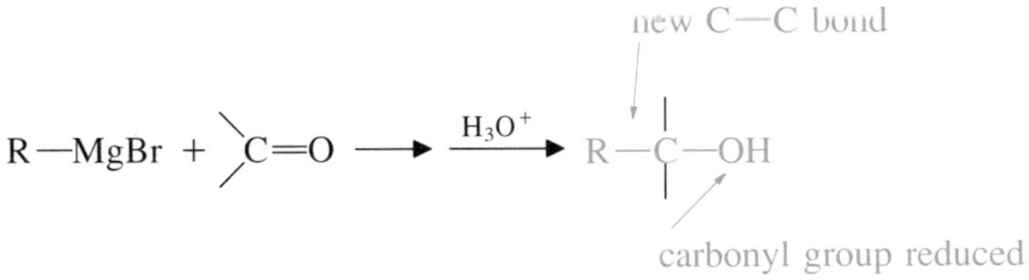

new C—C bond
R—MgBr + C=O
H_3O^+
R—C—OH
carbonyl group reduced

(19.38)

Let us review the methods of carbon–carbon bond formation that we have learned so far:

1. Cyclopropane formation with carbenes (Sec. 9.7)
2. Reaction of Grignard reagents with ethylene oxide (Sec. 11.4B)
3. Reaction of acetylide ions with alkyl halides (Sec. 14.7B)
4. Diels–Alder reactions (Sec. 15.3)
5. Friedel–Crafts reactions (Secs. 16.4E,F)
6. Cyanohydrin formation (Sec. 19.7)
7. Reaction of Grignard reagents with aldehydes and ketones (this section).

The possibilities for alcohol synthesis with the Grignard reaction are almost endless. Primary alcohols are synthesized by the addition of a Grignard reagent to formaldehyde.

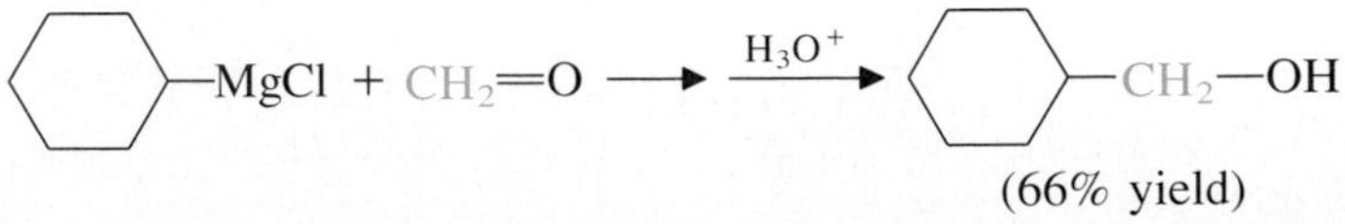

—MgCl + CH_2=O
H_3O^+
—CH_2—OH
(66% yield)

(19.39)

Since Grignard reagents are made from alkyl halides, which in many cases can be synthesized from alcohols, this reaction can be used as a net one-carbon chain extension of an alcohol:

$$\text{R—OH} \longrightarrow \text{R—Br} \xrightarrow[\text{ether}]{\text{Mg}} \text{R—MgBr} + \text{CH}_2\text{=O} \xrightarrow{\text{ether}} \xrightarrow{\text{H}_3\text{O}^+} \text{R—CH}_2\text{—OH} \quad (19.40)$$

(net one-carbon chain extension)

Addition of a Grignard reagent to an aldehyde other than formaldehyde gives a secondary alcohol (Eq. 19.33), and addition to a ketone gives a tertiary alcohol (Eq. 19.34). The Grignard synthesis of tertiary and some secondary alcohols can also be turned into an alkene synthesis by dehydration of the alcohol with strong acid during the protonolysis step.

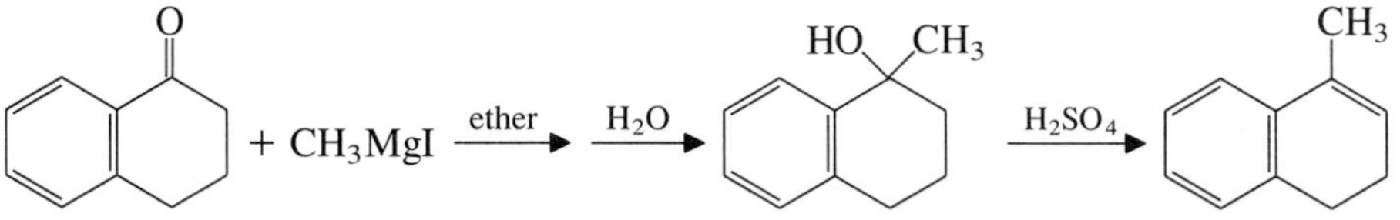

(68% yield) (19.41)

When we are asked to prepare an alcohol, it is a simple matter to determine whether it can be synthesized by the reaction of a Grignard reagent with an aldehyde or ketone. Let us first realize that the *net effect* of the Grignard reaction, followed by protonolysis, is addition of R—H (R = an alkyl or aryl group) across the C=O double bond:

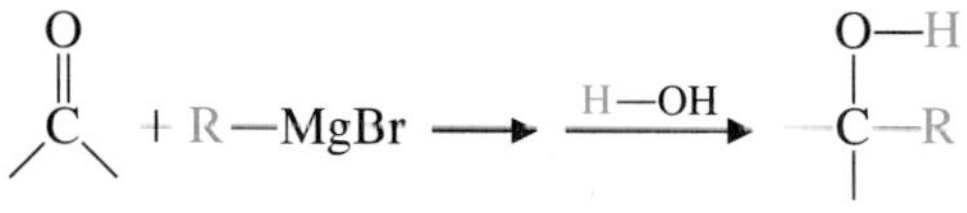

(19.42)

Once we see this relationship, we work backward, as usual, from the target compound by mentally subtracting R—H. For example, suppose we are asked to outline a synthesis of 2-butanol. One possible analysis is as follows:

$$\underset{\text{target compound}}{\text{CH}_3\text{—}\overset{\text{O—H}}{\overset{|}{\text{CH}}}\text{—CH}_2\text{—CH}_3} \xrightarrow[]{\text{subtract C}_2\text{H}_5\text{—H}} \text{CH}_3\text{—}\overset{\text{O}}{\overset{\|}{\text{CH}}} + \text{C}_2\text{H}_5\text{—MgBr, then H—OH} \quad (19.43)$$

(The double arrow is read, "Implies as starting materials.") There is also another possibility for a Grignard synthesis of 2-butanol that can be found by a similar analysis; what is it?

In this and the previous section we have learned two important methods for alcohol synthesis. Let us review the alcohol syntheses we have studied so far:

1. Hydroboration–oxidation of alkenes (Sec. 5.2A)
2. Oxymercuration–reduction of alkenes (Sec. 5.2B)
3. Reaction of ethylene oxide with Grignard reagents (Sec. 11.4B)

4. Hydride reduction of aldehydes and ketones (Sec. 19.8)

5. Addition of Grignard reagents to aldehydes or ketones (this section).

Of course, a great variety of compounds can be prepared from alcohols. Thus, these alcohol syntheses offer the further potential for the preparation of a wide variety of organic compounds.

Problems

16 Show how ethyl bromide can be used as a starting material in the preparation of each of the following compounds. (*Hint:* How are Grignard reagents prepared?)

(a) OH
 |
 PhCHC$_2$H$_5$

(b) OH
 |
 (C$_2$H$_5$)$_2$CCH$_3$

(c) OH
 |
 C$_2$H$_5$

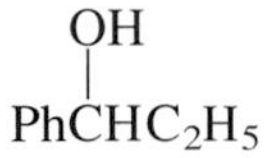

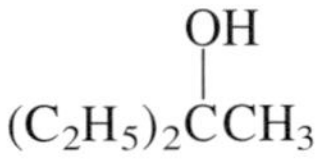

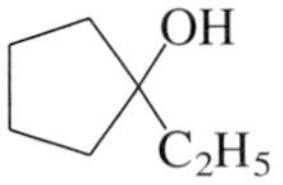

(d) 1-propanol

(e) 1-butanol

(f) CH$_3$CH$_2$CH=O

(g) Ph
 |
 Ph—C=CH—CH$_3$

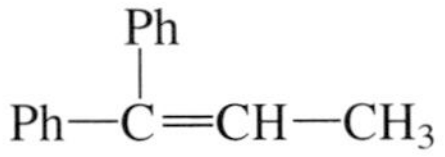

17 Outline two different Grignard syntheses for each of the following compounds:

(a) OH
 |
 CH—CH$_3$

(b) CH$_3$
 |
 PhCH$_2$COH
 |
 CH$_3$

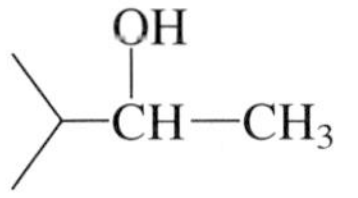

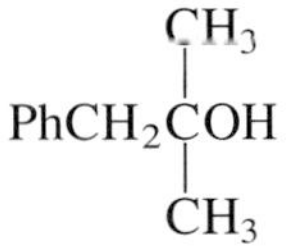

19.10 ACETALS. PROTECTING GROUPS

In the preceding sections, we discussed simple carbonyl-addition reactions—first, reversible additions (hydration and cyanohydrin formation); then, irreversible additions (hydride reduction and addition of Grignard reagents). In this and the following sections, we consider some reactions that begin as additions but incorporate other types of mechanistic steps.

A. Preparation and Hydrolysis of Acetals

When an aldehyde or ketone reacts with a large excess of an alcohol in the presence of a trace of mineral acid, an *acetal* is formed.

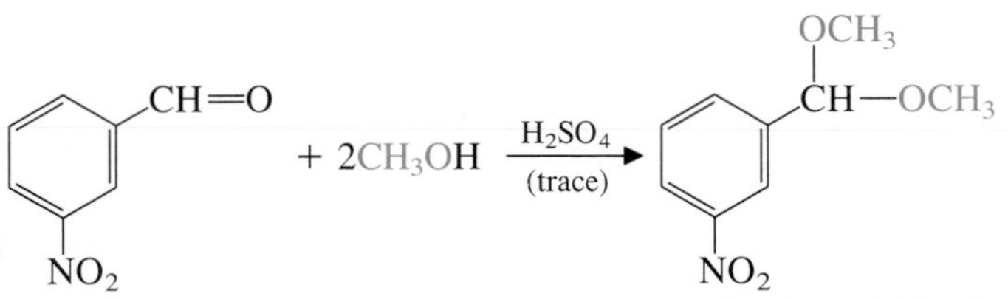

m-nitrobenzaldehyde
dimethyl acetal
(76–85% yield)

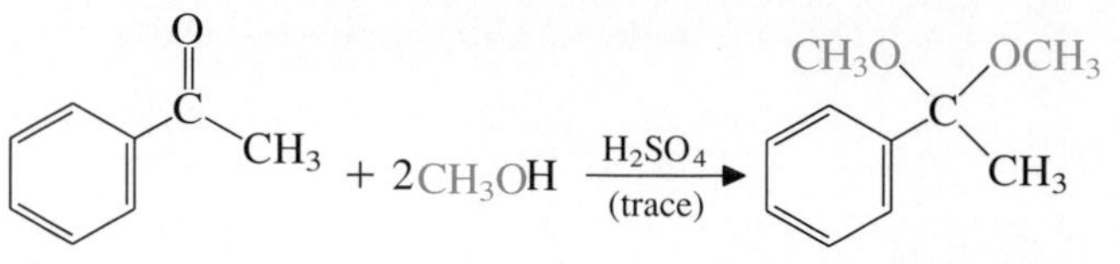

(19.45)

acetophenone dimethyl acetal (82% yield)

An **acetal** is simply a diether in which both ether oxygens are bound to the same carbon. In other words, acetals are the ethers of *gem*-diols (Sec. 19.7). (Acetals derived from ketones were once called *ketals,* but this name is no longer used.)

Notice that two moles of alcohol are required in the reactions above. However, 1,2- and 1,3-diols contain two —OH groups within the same molecule. Hence, one mole of a 1,2- or 1,3-diol reacts to form a cyclic acetal, in which the acetal group is part of a five- or six-membered ring.

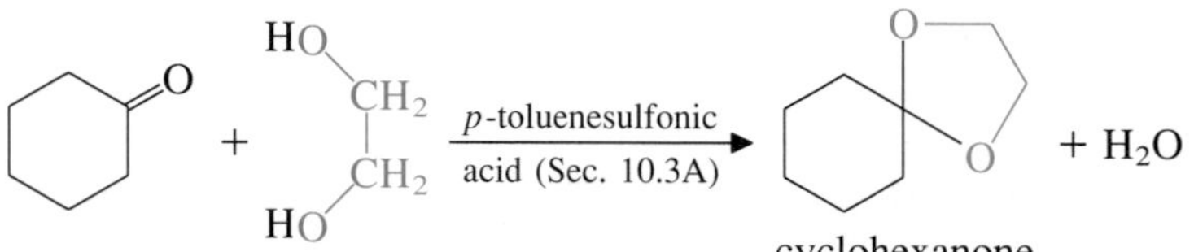

(19.46)

ethylene glycol

cyclohexanone ethylene acetal (85% yield)

The formation of acetals is reversible. The reaction is driven to the right by applying *LeChatelier's principle* (Sec. 9.2). In acetal formation this is accomplished either by the use of excess alcohol as the solvent, by removal of the water by-product, or both. In Eq. 19.46, for example, the water can be removed as an *azeotrope* with benzene. (The benzene–water azeotrope is a mixture of benzene and water that has a lower boiling point than either benzene or water alone.)

The first step in the mechanism of acetal formation is acid-catalyzed addition of the alcohol to the carbonyl group, a reaction completely analogous to acid-catalyzed hydration.

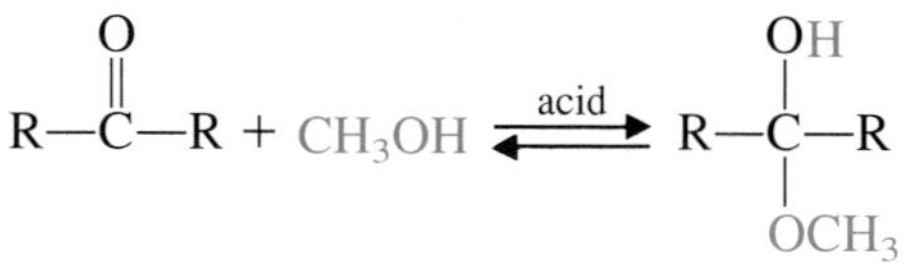

a hemiacetal (19.47a)

Problem

18 Give the detailed mechanism of acid-catalyzed hemiacetal formation from acetone and methanol.

The product of Eq. 19.47a is called a **hemiacetal.** (Hemi = half; hemiacetal = half acetal.) The hemiacetal reacts further by undergoing an S_N1 reaction: the —OH group is protonated and water is lost to give a relatively stable carbocation—an α-alkoxycarbocation (see Sec. 19.6).

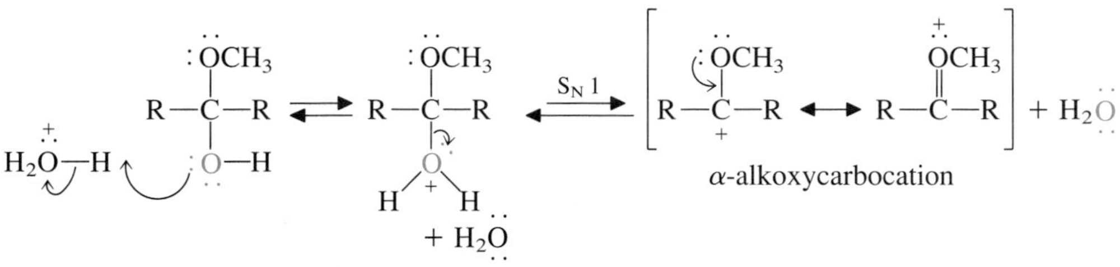

$$\alpha\text{-alkoxycarbocation} \qquad (19.47b)$$

Loss of water from the hemiacetal is analogous to the loss of water in the dehydration of an ordinary alcohol (Eq. 10.2b). Attack of an alcohol molecule on the cation and deprotonation of the attacking oxygen complete the mechanism.

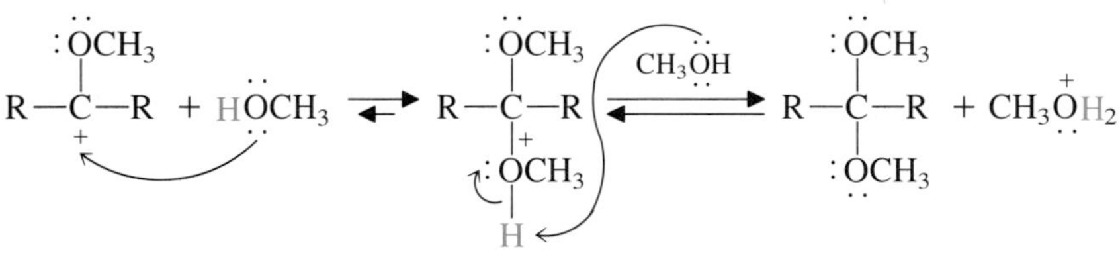

$$(19.47c)$$

Some students ask why the hemiacetal protonates on the —OH oxygen rather than on the —OCH$_3$ oxygen. Indeed, protonation can occur at either oxygen, since ethers and alcohols have similar basicities. However, protonation at the ether oxygen and loss of alcohol are the reverse of the first steps of hemiacetal formation and simply lead back to aldehyde or ketone. These reactions do occur, but under conditions of excess alcohol and removal of water, the equilibrium shifts to favor the acetal.

We can see that the mechanism for acetal formation is really a combination of other mechanisms we have studied already. It involves an *acid-catalyzed carbonyl addition* followed by a *substitution* that occurs by the S_N1 *mechanism*.

Because the formation of acetals is reversible, acetals in the presence of acid and excess water are transformed rapidly back into their corresponding carbonyl compounds; this process is called **acetal hydrolysis.** (A *hydrolysis* is a cleavage reaction involving water.) By the principle of microscopic reversibility, the mechanism of acetal hydrolysis is the reverse of the mechanism of acetal formation. Hence, acetal hydrolysis is acid catalyzed. Acetals are *stable in base* because protonation of the acetal is required in order to form the carbocation intermediate in the hydrolysis reaction, and the proton concentration in basic solution is too low for this protonation to occur to a significant extent.

Hemiacetals, the intermediates in acetal formation (Eq. 19.47a), in most cases cannot be isolated because they react further to yield acetals (in alcohol solution) or decompose to aldehydes or ketones (in water). Simple aldehydes, however, form appreciable amounts of hemiacetals in alcohol solution, just as they form appreciable amounts of hydrates in water.

$$CH_3\!-\!CH\!=\!O + C_2H_5OH \underset{solvent}{\rightleftarrows} CH_3\!-\!\overset{\overset{\displaystyle OH}{|}}{\underset{\underset{\displaystyle OC_2H_5}{|}}{CH}} \qquad (97\% \text{ at equilibrium}) \quad (19.48)$$

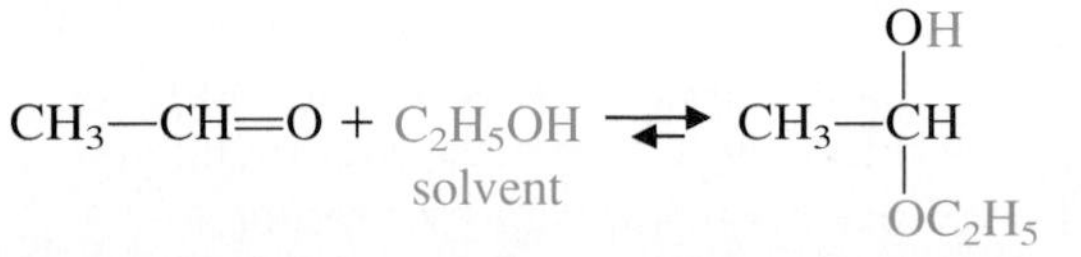

Five- and six-membered *cyclic* hemiacetals form spontaneously from the corresponding hydroxyaldehydes, and most are stable, isolable compounds.

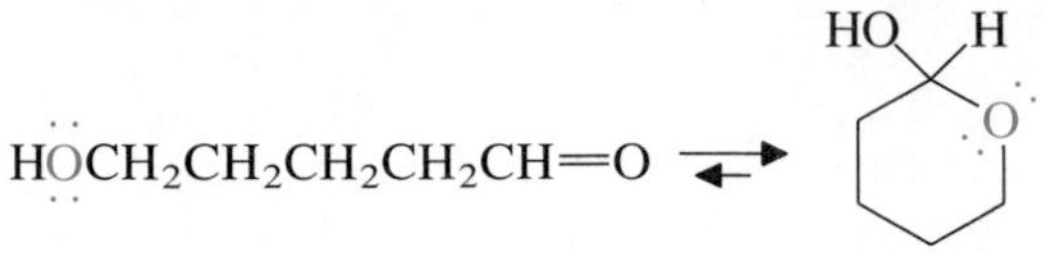

$$\text{HOCH}_2\text{CH}_2\text{CH}_2\text{CH}_2\text{CH}{=}\text{O} \;\rightleftharpoons\; \qquad\qquad \begin{array}{l}\text{a cyclic hemiacetal}\\ \text{(94\% at equilibrium)}\end{array} \qquad (19.49)$$

$$\text{HOCH}_2\text{CH}_2\text{CH}_2\text{CH}{=}\text{O} \;\rightleftharpoons\; \qquad\qquad \text{(89\% at equilibrium)} \qquad (19.50)$$

The five- and six-carbon sugars are important biological examples of cyclic hemiacetals.

$$(19.51)$$

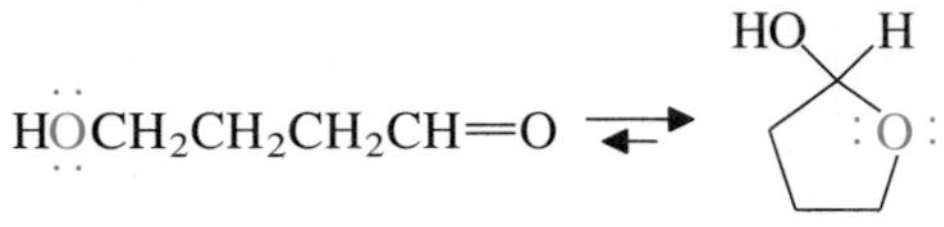

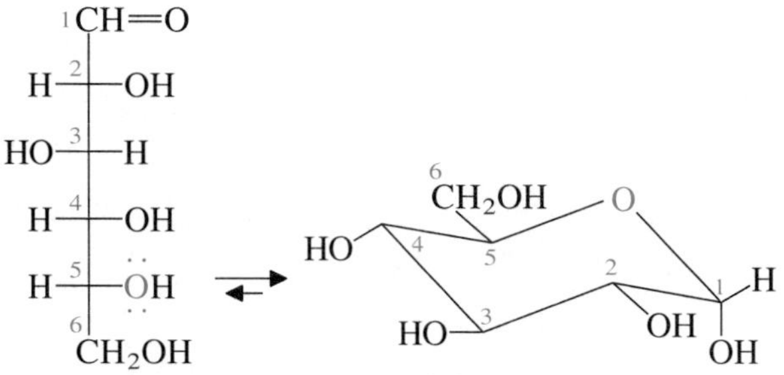

(+)-glucose α-(+)-glucopyranose,
(a cyclic form of glucose)

Some aldehydes can be stored as acetals. Acetaldehyde, when treated with a trace of acid, readily forms a cyclic acetal called *paraldehyde*. Each molecule of paraldehyde is formed from three molecules of acetaldehyde. (Notice that an alcohol is not required for formation of paraldehyde.) Paraldehyde, with a boiling point of 125°, is a particularly convenient way to store acetaldehyde, which boils near room temperature. Upon heating with a trace of acid, acetaldehyde can be distilled from a sample of paraldehyde.

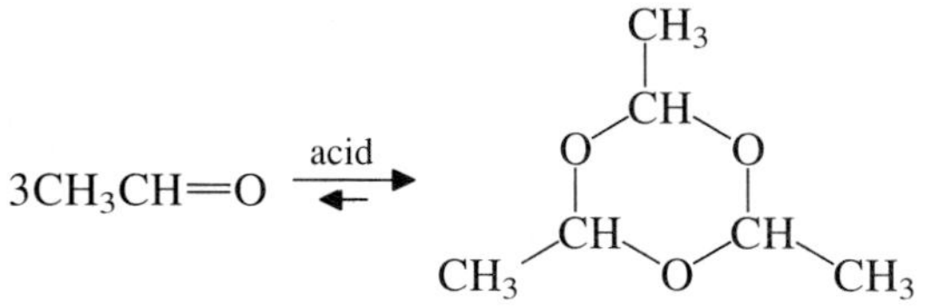

$$3\text{CH}_3\text{CH}{=}\text{O} \;\xrightleftharpoons[\;]{\text{acid}}\; \qquad\qquad\qquad \text{paraldehyde} \qquad (19.52)$$

Formaldehyde can be stored as the acetal polymer *paraformaldehyde,* which precipitates from concentrated formaldehyde solutions.

$$\text{HO}\!\!-\!\!\!\left[\text{CH}_2\!\!-\!\!\text{O}\right]_{\!\!n}\!\!\!-\!\!\text{H}$$
paraformaldehyde

(An alcohol is not involved in paraldehyde formation.) Because it is a solid, paraldehyde is a useful form in which to store formaldehyde, itself a gas. Formaldehyde is liberated from paraformaldehyde by heating.

Problems

19 Complete the following reactions.

(a) 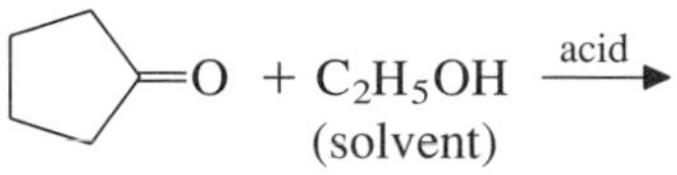

(b) 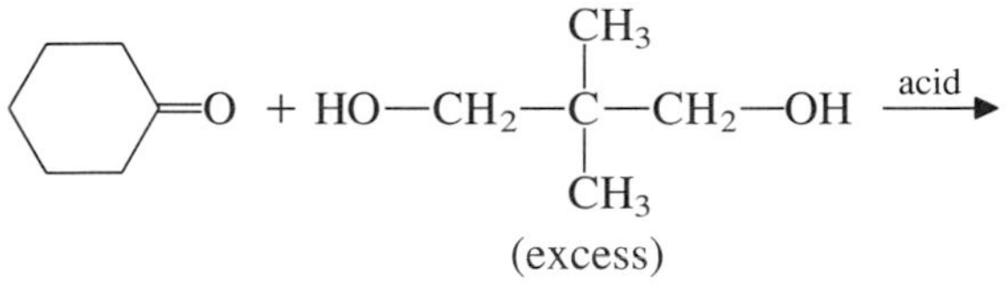

(c) 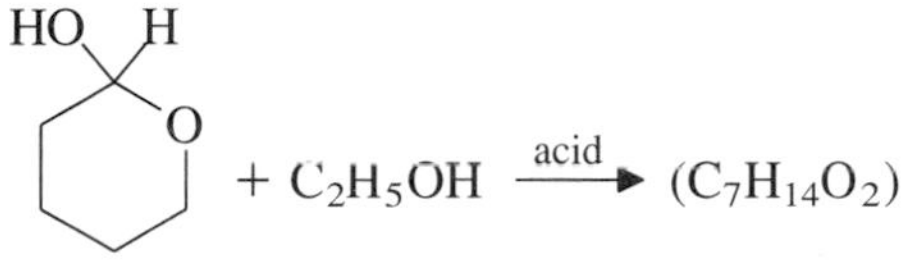

20 When hexanal reacts with glycerol (1,2,3-propanetriol), *four* separable isomers are formed. Give the structures of all four compounds.

B. Protecting Groups

A common tactic of organic synthesis is the use of *protecting groups*. The method is illustrated by the following analogy. Suppose you and a friend are both unwelcome at a party, but are determined to attend it anyway. To avoid recognition and confrontation you wear a disguise, which might be a wig, a false mustache, or even more drastic accoutrements. Your friend doesn't bother with such deception. The host recognizes your friend and throws him out of the party, but, because you are not recognized, you remain and enjoy the evening, removing your disguise only after the party is over. Now, suppose we have two groups in a molecule, *A* and *B*, that are both known to react with a certain reagent, but we want to let only group *A* react and leave group *B* unaffected. The solution to this problem is to *disguise*, or *protect*, group *B* in such a way that it cannot react. After group *A* is allowed to react, the disguise of group *B* is removed. The "chemical disguise" used with group *B* is called a **protecting group.**

The following synthesis illustrates the use of a protecting group.

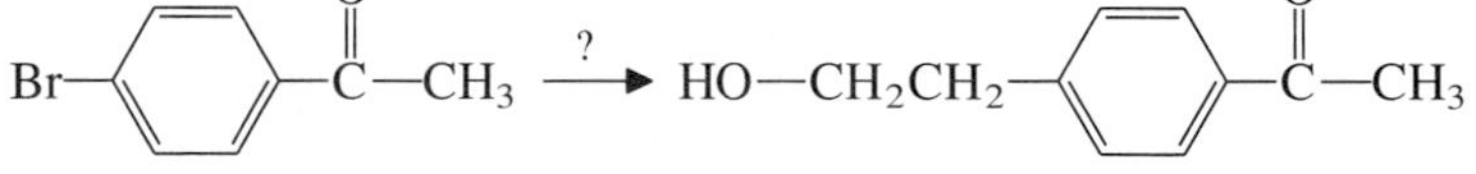
(19.53)

The obvious way to effect this conversion is to convert the starting material into the corresponding Grignard reagent, and then allow this reagent to react with ethylene oxide, followed by dilute aqueous acid (Sec. 11.4B). Unfortunately, as we learned in Sec. 19.9, Grignard reagents react with ketones. Hence, the Grignard reagent derived from one molecule of the starting material would react with the carbonyl group of another molecule. Hence, the ketone group would not survive this reaction. However, the ketone can be *protected* as an acetal, which does *not* react with Grignard reagents. (Acetals are really ethers, and ethers are unaffected by Grignard reagents.) The following synthesis incorporates this strategy.

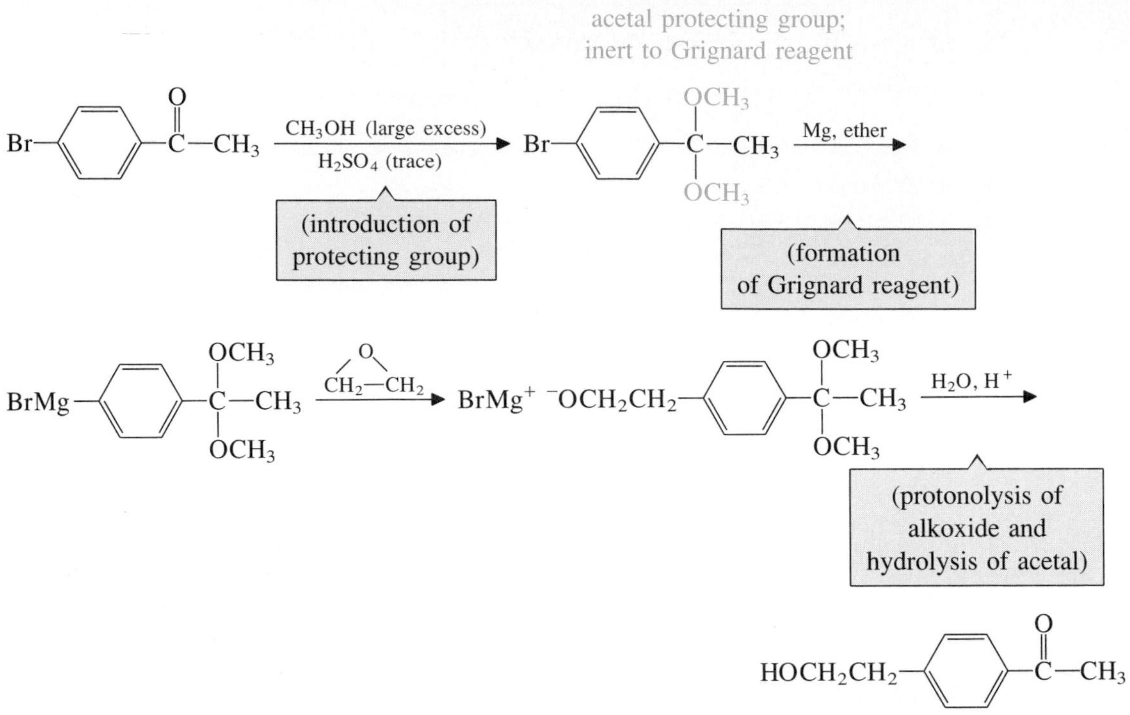

(19.54)

Carbonyl groups react with a number of reagents used with other functional groups. Acetals are commonly used to protect the carbonyl groups of aldehydes and ketones from basic, nucleophilic reagents. Once the protection is no longer needed, the protecting group is easily removed, and the carbonyl group re-exposed, by treatment with dilute aqueous acid. Because acetals are hydrolyzed in acid, they do *not* protect carbonyl groups under acidic conditions.

Problem

21 Outline a synthesis of the following compound from *p*-bromobenzaldehyde.

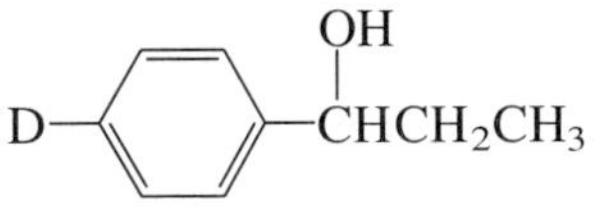

19.11 REACTIONS OF ALDEHYDES AND KETONES WITH AMINES

A. Imine Formation with Primary Amines

A **primary amine** is an organic derivative of ammonia in which only one ammonia hydrogen is replaced by an alkyl or aryl group. An **imine** is a nitrogen analog of an aldehyde or ketone in which the C=O group is replaced by a C=N—R group.

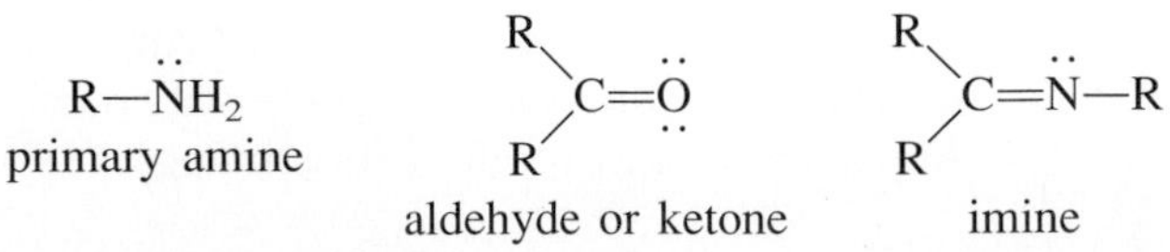

(Imines are sometimes called **Schiff bases.**) Imines are prepared by the reaction of aldehydes or ketones with primary amines.

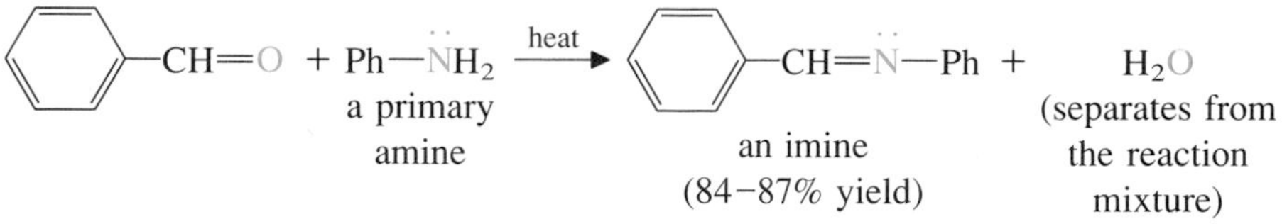

(19.55)

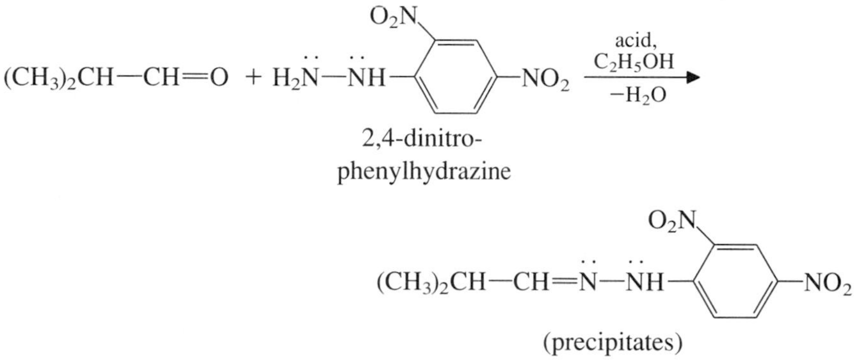

(19.56)

Formation of imines is reversible, and generally takes place with acid or base catalysis, or with heat. Imine formation is typically driven to completion by precipitation of the imine, removal of water, or both.

The mechanism of imine formation begins as a nucleophilic addition to the carbonyl group. In this case, the nucleophile is an amine. In the first step of the mechanism, the amine reacts with the aldehyde or ketone to give an unstable addition compound called a **carbinolamine.** The mechanism of the addition is analogous to the mechanisms of other reversible additions: attack of the nucleophile on the carbonyl carbon, followed by proton transfers to and from solvent.

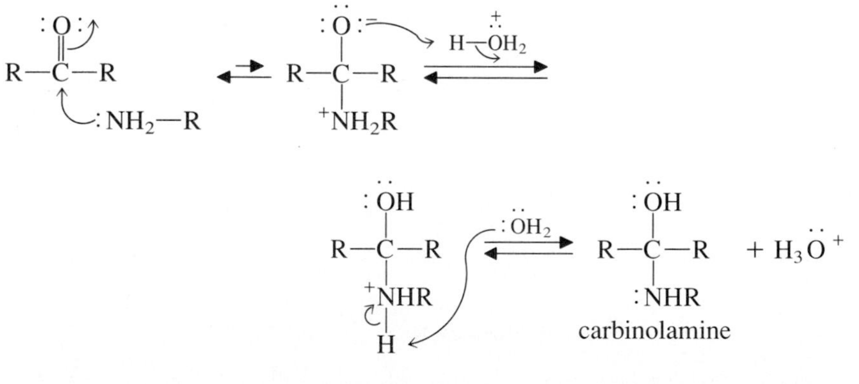

(19.57a)

Like other alcohols, carbinolamines undergo acid-catalyzed dehydration. (Write the mechanism of this reaction; see Sec. 10.1A.)

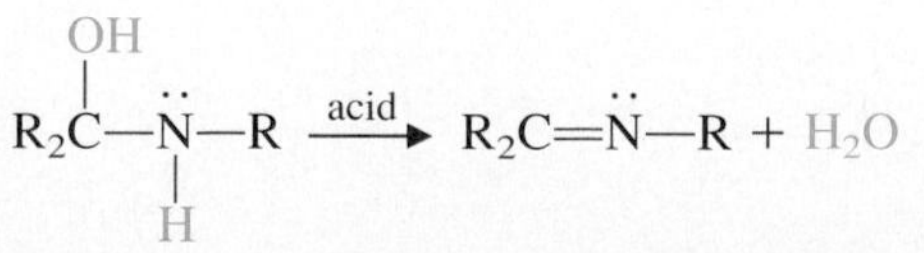

(acid-catalyzed dehydration) (19.57b)

Typically the dehydration of the carbinolamine is the rate-determining step of imine formation. This is why imine formation is catalyzed by acids. Yet the acid concentration cannot be too high because amines are basic compounds.

$$R\ddot{N}H_2 + H_3\ddot{O}^+ \rightleftarrows R\overset{+}{N}H_3 + H_2\ddot{O}: \tag{19.58}$$

Protonation of the amine pulls the equilibrium in Eq. 19.57a to the left and carbinolamine formation cannot occur. Therefore many imine syntheses are best carried out at mildly acidic pH values.

We can see that imine formation is really a sequence of two types of reactions with which we are already familiar: *carbonyl addition* followed by *elimination*.

Imines have uses in organic synthesis that we shall consider in Chapter 23. One use of imines, however, was especially important in organic chemistry before spectroscopy assumed a central role in structure elucidation. When a new compound was synthesized, it was typically characterized by conversion into at least two solid derivatives. These derivatives served as the basis for subsequent identification of the new compound when it was isolated from another source or from a different reaction. It is always possible that two different compounds might have similar melting points or boiling points. However, it almost never happens that two compounds with identical melting or boiling points also give two solid derivatives with the same melting points. Certain types of imine adducts found frequent use as derivatives of aldehydes and ketones because they are almost always solids with well-defined melting points. These adducts, and the amines from which they are derived, are summarized in Table 19.3.

To illustrate how such derivatives might be used in structure verification, let us suppose that a chemist has isolated a liquid that could be either 6-methyl-2-cyclohexen-1-one or 2-methyl-2-cyclohexen-1-one. The boiling points of these compounds are too similar for an unambiguous identification. Yet the melting point of either a 2,4-DNP derivative or a semicarbazone would quickly establish which compound has been isolated.

TABLE 19.3 *Some Common Imine Derivatives of Aldehydes and Ketones*

$$R_2C{=}O \; + \; H_2\ddot{N}{-}R' \longrightarrow R_2C{=}\ddot{N}{-}R' + H_2O$$

Amine	Name	Carbonyl Derivative	Name
$H_2\ddot{N}{-}\ddot{O}H$	hydroxylamine	$R_2C{=}\ddot{N}{-}\ddot{O}H$	oxime
$H_2\ddot{N}{-}\ddot{N}H_2$	hydrazine	$R_2C{=}\ddot{N}{-}\ddot{N}H_2$	hydrazone
$H_2\ddot{N}{-}\ddot{N}H{-}\bigcirc$	phenylhydrazine	$R_2C{=}\ddot{N}{-}\ddot{N}H{-}\bigcirc$	phenylhydrazone
$H_2\ddot{N}{-}\ddot{N}H{-}\bigcirc{-}NO_2$ (with O_2N)	2,4-dinitrophenyl-hydrazine	$R_2C{=}\ddot{N}{-}\ddot{N}H{-}\bigcirc{-}NO_2$ (with O_2N)	2,4-dinitrophenyl-hydrazone (2,4-DNP derivative)
$H_2\ddot{N}{-}\ddot{N}H{-}\overset{O}{\overset{\|}{C}}{-}\ddot{N}H_2$	semicarbazide	$R_2C{=}\ddot{N}{-}\ddot{N}H{-}\overset{O}{\overset{\|}{C}}{-}\ddot{N}H_2$	semicarbazone

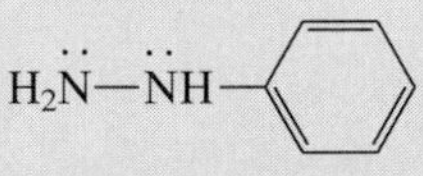

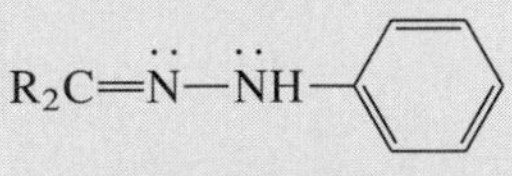

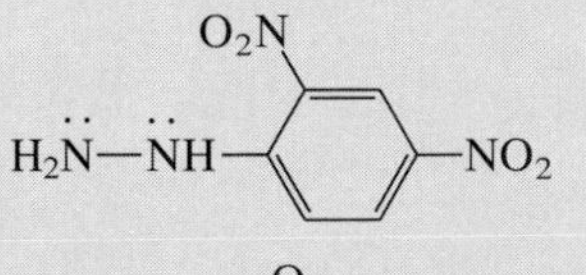

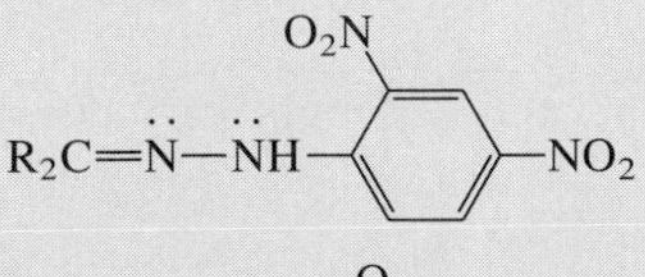

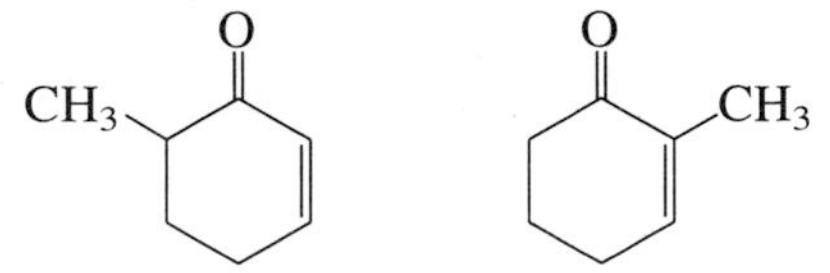

boiling point	69–71° (18 mm)	69–70° (16 mm)
semicarbazone, mp	177–178°	207–208°
2,4-DNP derivative, mp	162–164°	207–208°

Of course, the identity of the compound could be readily established today by spectroscopy (how?).

We can envision an imine derived from the reaction of an aldehyde or ketone with ammonia itself.

$$\overset{\overset{\displaystyle :O:}{\|}}{Ph-C-Ph} \qquad \overset{\overset{\displaystyle :NH}{\|}}{Ph-C-Ph}$$

benzophenone benzophenone imine

As a rule, imines derived from the reaction of aldehydes or ketones with ammonia are *not* stable, but they are believed to be intermediates in certain reactions. When aldehydes react with ammonia, cyclic derivatives analogous to paraldehyde (Eq. 19.52) can form; these are known as *aldehyde ammonias*.

$$3R-CH=O + 3NH_3 \longrightarrow \quad + 3H_2O \qquad (19.59)$$

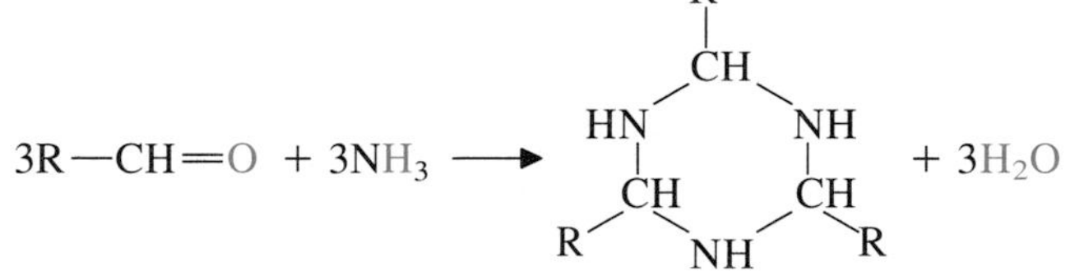

an aldehyde
ammonia

Formaldehyde reacts with ammonia to form a commercially important derivative called *hexamethylenetetramine*. This compound is used in the manufacture of phenol–formaldehyde resins (Sec. 19.15).

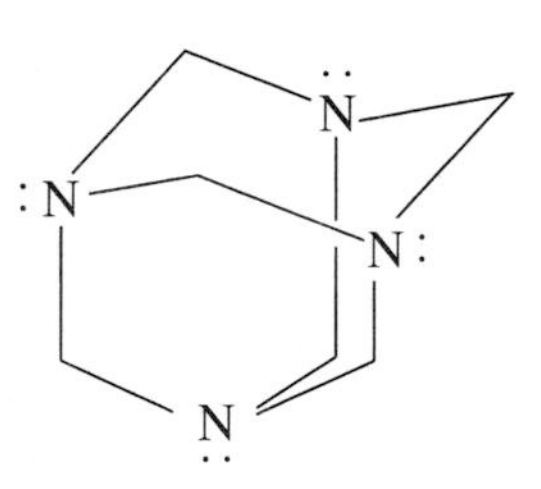

hexamethylenetetramine

Problems **22** Write a mechanism for the acid-catalyzed hydrolysis of the imine derived from benzaldehyde and $C_2H_5NH_2$ (ethylamine). Don't neglect what you learned in Sec. 10.1A about the principle of microscopic reversibility.

Problems (Cont.)

23 Draw the structure of
(a) the semicarbazone of cyclohexanone.
(b) the imine formed in the reaction between 2-methylhexanal and ethylamine ($C_2H_5NH_2$).
(c) the 2,4-DNP derivative of acetone.

24 There are actually *two* separable 2,4-dinitrophenylhydrazone derivatives of 2-butanone; give their structures. (*Hint:* See Sec. 4.1B.)

25 Semicarbazide (Table 19.3) has *two* —NH_2 groups that could in principle react with an aldehyde or ketone. Explain why semicarbazide reacts as it does.

B. Enamine Formation with Secondary Amines

An **enamine** has the following general structure:

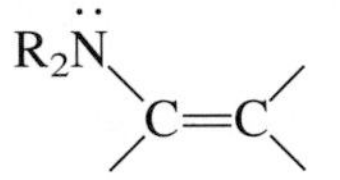

general enamine structure

A **secondary amine** has the general structure $R_2\ddot{N}$—H, in which two ammonia protons are formally replaced by alkyl or aryl groups; the nitrogen of a secondary amine may be part of a ring, as in Eq. 19.61. When an aldehyde or ketone with α-hydrogens reacts with a secondary amine an enamine is formed.

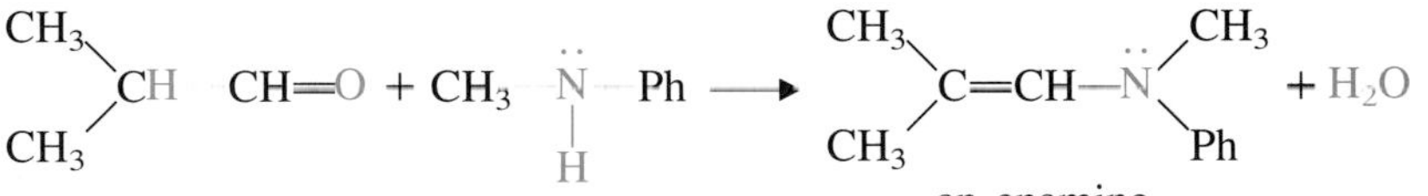

a secondary amine

an enamine (87% yield)

(19.60)

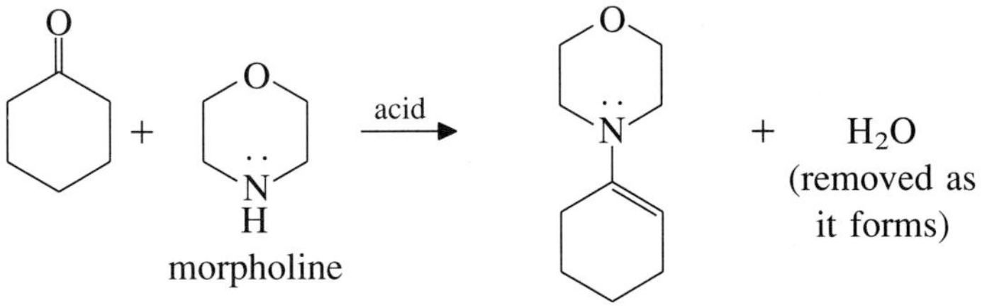

morpholine

(72–80% yield)

(19.61)

Like imine formation, enamine formation is reversible, and must be driven to completion by the removal of one of the reaction products (usually water; see Eq. 19.61). Enamines, like imines, are converted back into their corresponding carbonyl compounds and amines in aqueous acid.

The mechanism of enamine formation begins like the mechanism of imine formation, as a nucleophilic addition to give a carbinolamine intermediate. (Write the mechanism of this reaction; see Eq. 19.57a.)

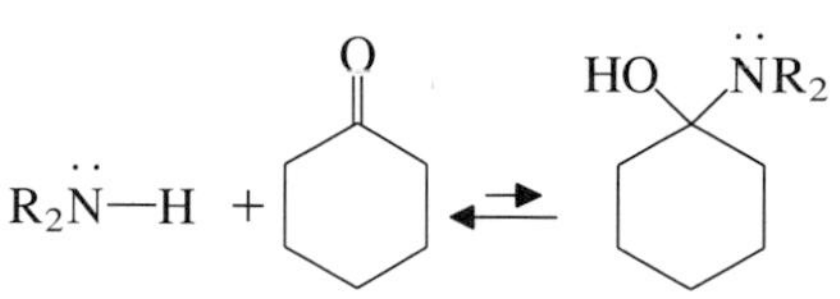

(19.62a)

Since there is no proton remaining on the nitrogen of this carbinolamine, imine formation cannot occur. Instead, dehydration of the carbinolamine involves proton loss from the adjacent *carbon*.

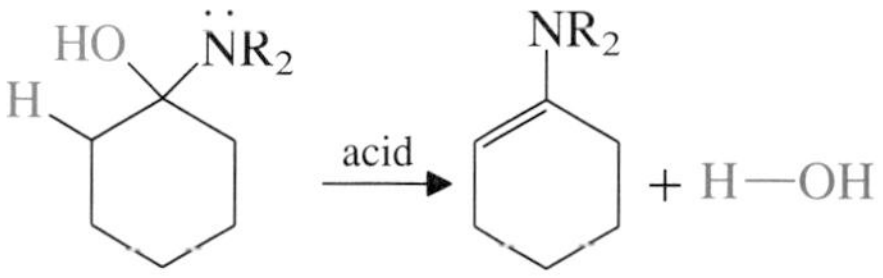

(19.62b)

We can see why secondary amines don't form imines. But why don't primary amines react with aldehydes or ketones to form enamines rather than imines? The answer is that enamines bear the same relationship to imines that *enols* bear to ketones.

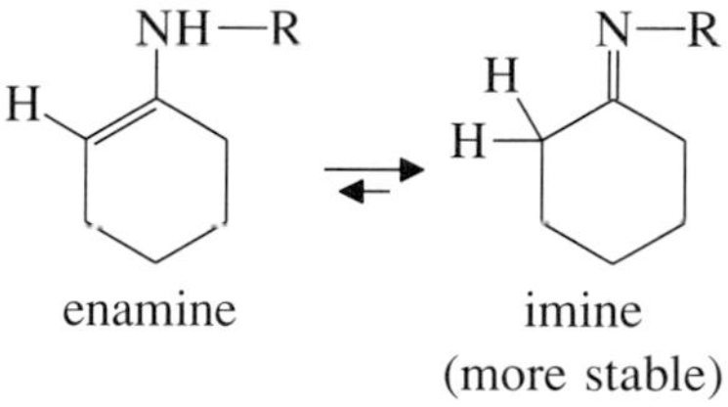

(19.63a)

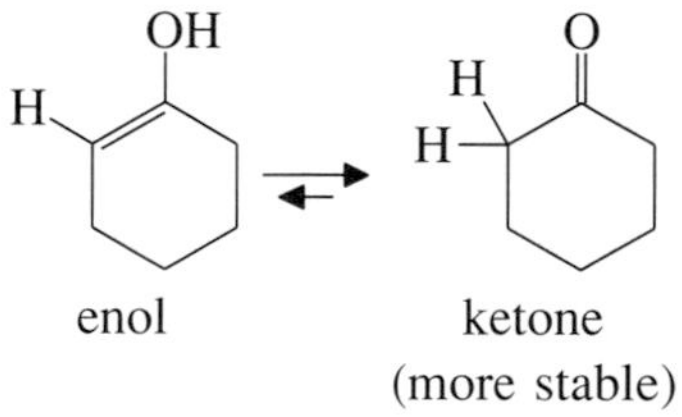

(19.63b)

Just as most aldehydes and ketones are more stable than their corresponding enols (Sec. 14.5A), most imines are more stable than their corresponding enamines. Since secondary amines *cannot* form imines, they have no choice but to form enamines instead.

Problem

26 Give the enamine product formed when the compounds in each of the following pairs react:

(a) acetone and 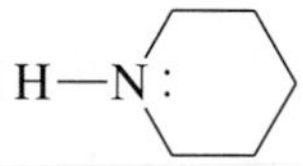

(b) $PhCH_2CH{=}O$, $(CH_3)_2\ddot{N}H$

We have seen that aldehydes and ketones react with primary amines ($R{-}NH_2$) to give imines, and with secondary amines (R_2NH) to give enamines. There is a third type of amine, a **tertiary amine** ($R_3N:$), in which all protons of ammonia are formally replaced by alkyl or aryl groups. *Tertiary amines do not react with aldehydes and* ketones to form stable derivatives. Although most tertiary amines are good nucleophiles, they have no N—H protons, and therefore cannot even form carbinolamines. Their adducts with aldehydes and ketones are unstable and can only break down to starting materials.

$$R_3\ddot{N}: + \; \text{C=O} \; \rightleftharpoons \; R_3\overset{+}{N}-\overset{|}{\underset{|}{C}}-\ddot{\underset{..}{O}}:^- \qquad (19.64)$$

19.12 REDUCTION OF CARBONYL GROUPS TO METHYLENE GROUPS

The most common reductive transformation of aldehydes or ketones is their conversion into alcohols (Sec. 19.8). But it is also possible to reduce the carbonyl group of an aldehyde or ketone completely to a methylene ($-CH_2-$) group. One procedure for effecting this transformation involves heating the aldehyde or ketone with hydrazine (NH_2-NH_2) and strong base.

$$\underset{\text{hydrazine}}{\underset{\text{(85\% aqueous}}{\underset{\text{solution)}}{Ph-\overset{O}{\overset{||}{C}}-CH_2CH_3 \; + \; NH_2-NH_2}}} \; \xrightarrow[\text{triethylene glycol}]{\text{KOH, heat, 1 hr}} \; \underset{\text{(82\% yield)}}{Ph-CH_2CH_2CH_3} \; + \; H_2O \; + \; N_2 \qquad (19.65)$$

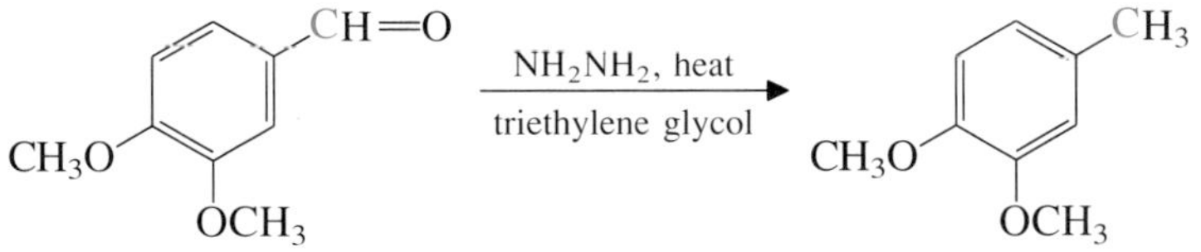

$$\underset{\text{OCH}_3}{\underset{CH_3O}{\bigcirc}}CH{=}O \; \xrightarrow[\text{triethylene glycol}]{NH_2NH_2, \text{ heat}} \; \underset{\underset{\text{(81\% yield)}}{OCH_3}}{\underset{CH_3O}{\bigcirc}}CH_3 \qquad (19.66)$$

This reaction, called the **Wolff–Kishner reduction,** typically utilizes ethylene glycol or similar compounds (triethylene glycol in the above cases) as co-solvents. The high boiling points of these solvents allow the reaction mixtures to reach the high temperatures required for the reduction to take place at a reasonable rate.

The Wolff–Kishner reduction is an extension of imine formation, discussed in the foregoing section. An intermediate in the reduction is a *hydrazone,* the imine of hydrazine (Table 19.3). A series of acid–base reactions leads to the expulsion of nitrogen and formation of a carbanion, which is instantaneously protonated by water to yield the product.

$$R-\overset{O}{\overset{||}{C}}-R \; + \; \ddot{N}H_2\ddot{N}H_2 \; \xrightarrow{-H_2O} \; \underset{\text{a hydrazone}}{R-\overset{N-N}{\overset{||}{C}}-R} \; \overset{H}{\underset{:OH}{\longrightarrow}} \; \left[R-\overset{N-NH}{\overset{||}{C}}-R \; \longleftrightarrow \; R-\overset{:N=\ddot{N}H}{\underset{..}{C}}-R \right] \qquad (19.67a)$$

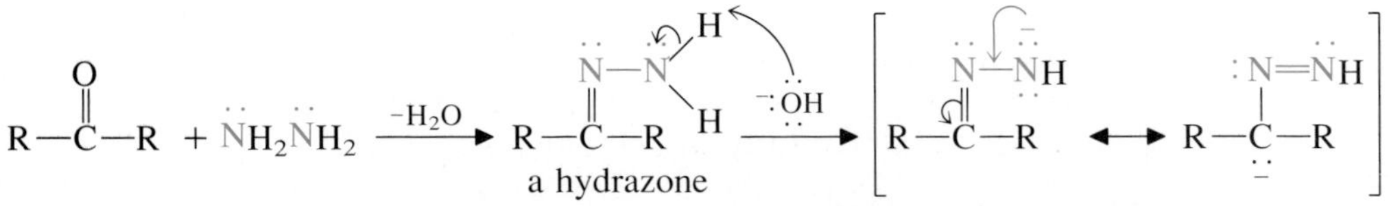

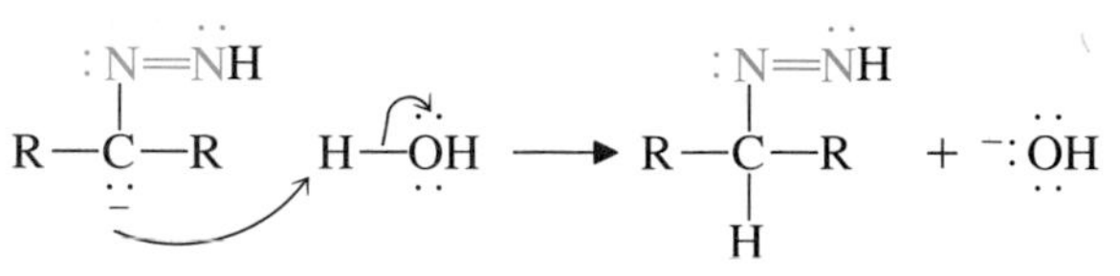

(19.67b)

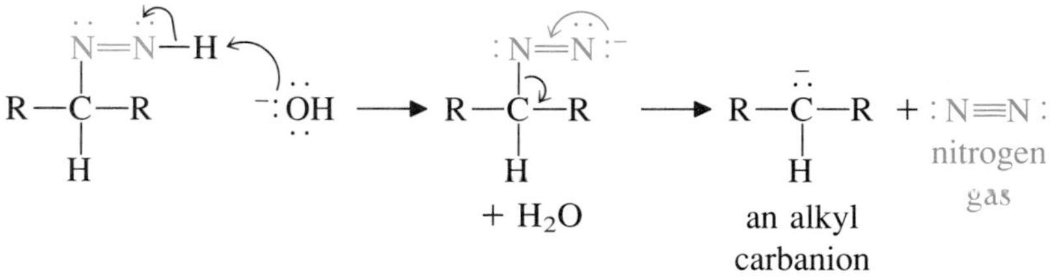

(19.67c)

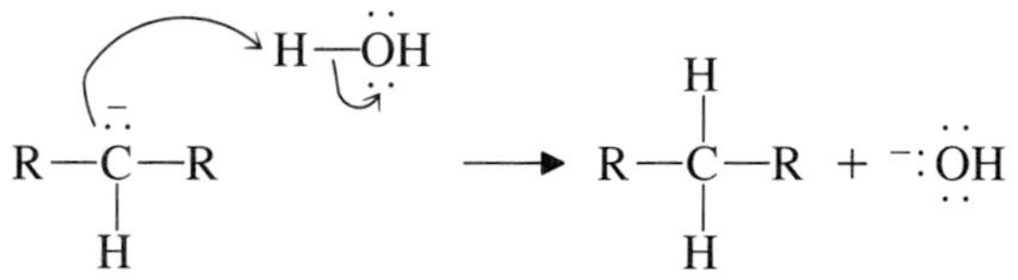

(19.67d)

As a rule, alkyl carbanions are too unstable to exist as reaction intermediates. In this reaction, however, the generation of the very stable nitrogen molecule helps to drive the reaction. In other words, the instability of the alkyl anion is offset by the stability of the nitrogen molecule.

The Wolff–Kishner reduction takes place under strongly basic conditions. The same overall transformation can be achieved under acidic conditions by a reaction called the **Clemmensen reduction.** In this reaction, an aldehyde or ketone is reduced with zinc amalgam (a solution of zinc metal in mercury) in the presence of HCl.

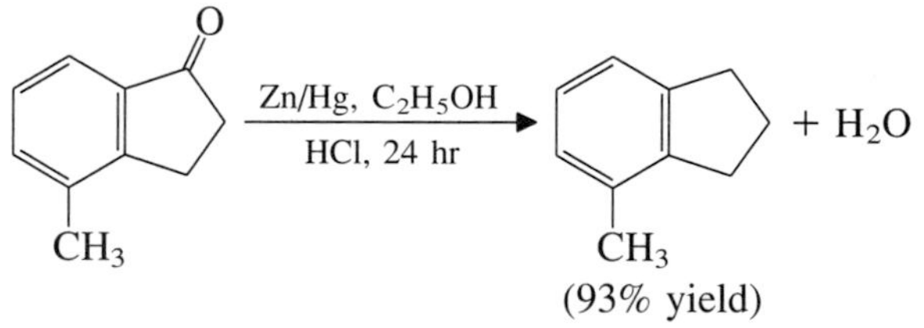

(19.68)

$$CH_3(CH_2)_5CH{=}O \xrightarrow[\text{25\% HCl}]{\text{Zn/Hg}} CH_3(CH_2)_5CH_3 \quad \text{(87\% yield)} \tag{19.69}$$

There is considerable uncertainty about the mechanism of the Clemmensen reduction.

When would we want to use the Wolff–Kishner or Clemmensen reductions in synthesis? These reactions are particularly useful for the introduction of alkyl groups into a benzene ring. For example, suppose we wish to prepare butylbenzene from benzene. The first reaction that comes to mind is the one used to introduce alkyl groups into benzene: Friedel–Crafts *alkylation* (Sec. 16.4F). The problem with Friedel–Crafts alkylation for introducing unbranched alkyl groups into benzene is that rearrangements of the alkyl groups are usually observed. This is precisely what happens when we attempt to prepare butylbenzene from benzene and 1-chlorobutane in the presence of $AlCl_3$:

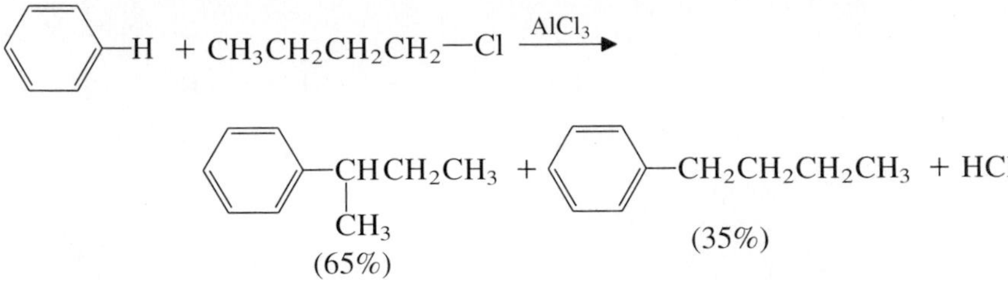

(19.70)

Butylbenzene can be easily prepared, however, by the Wolff–Kishner reduction of butyrophenone:

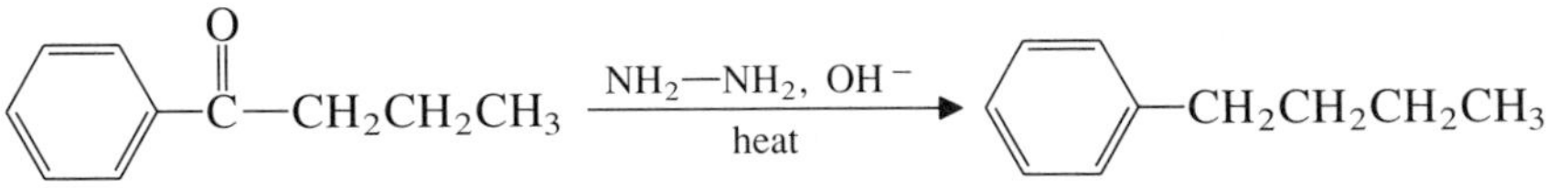

(19.71a)

In turn, butyrophenone is readily prepared by Friedel–Crafts *acylation* (Sec. 16.4E), which is not plagued by the rearrangement problems associated with the *alkylation*.

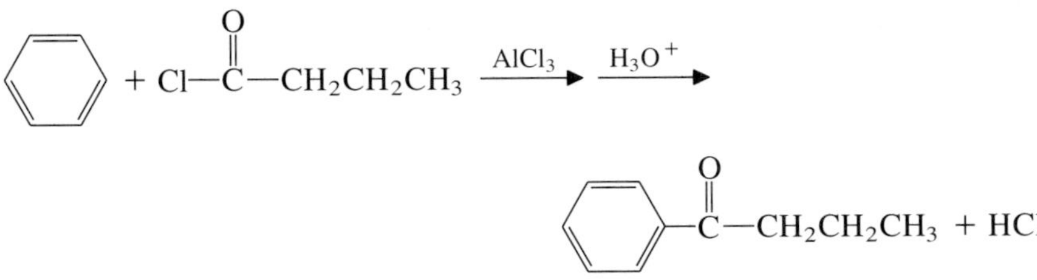

(19.71b)

Problems

27 Draw the structures of all aldehydes or ketones that could in principle give the following product after application of either the Wolff–Kishner or Clemmensen reduction.

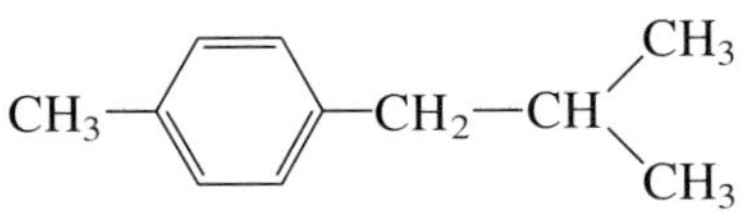

28 Outline a synthesis of 1,4-dimethoxy-2-propylbenzene from hydroquinone.

19.13 CONVERSION OF ALDEHYDES AND KETONES INTO ALKENES: THE WITTIG ALKENE SYNTHESIS

In our tour through aldehyde and ketone chemistry, we first discussed simple additions; then addition followed by substitution (acetal formation); then additions followed by elimination (imine and enamine formation). Another addition–elimination reaction, called the **Wittig reaction,** is an important method for preparing alkenes from aldehydes and ketones.

An example of the Wittig reaction is the preparation of methylenecyclohexane from cyclohexanone.

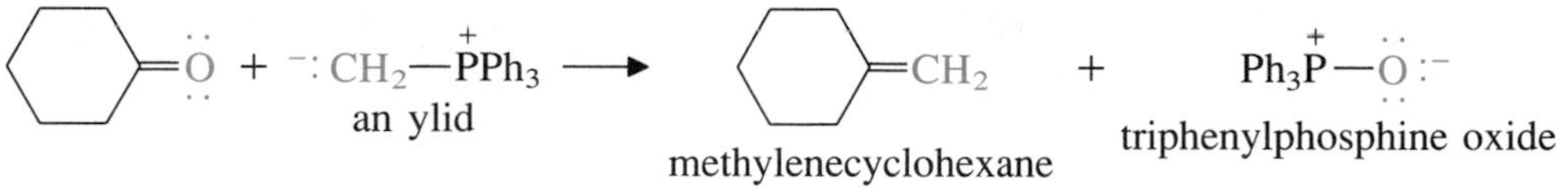

$$\text{[cyclohexylidene]}{=}\overset{\cdot\cdot}{\underset{\cdot\cdot}{O}} \; + \; {}^{-}{:}CH_2{-}\overset{+}{P}Ph_3 \longrightarrow \text{[cyclohexylidene]}{=}CH_2 \; + \; Ph_3\overset{+}{P}{-}\overset{\cdot\cdot}{\underset{\cdot\cdot}{O}}{:}^{-} \qquad (19.72)$$

an ylid methylenecyclohexane triphenylphosphine oxide

The nucleophile in the Wittig reaction is a type of **ylid.** An ylid is any compound with opposite charges on adjacent, covalently bound atoms, each of which has an electronic octet.

sextet of electrons — not an ylid

complete octet on each charged atom — an ylid

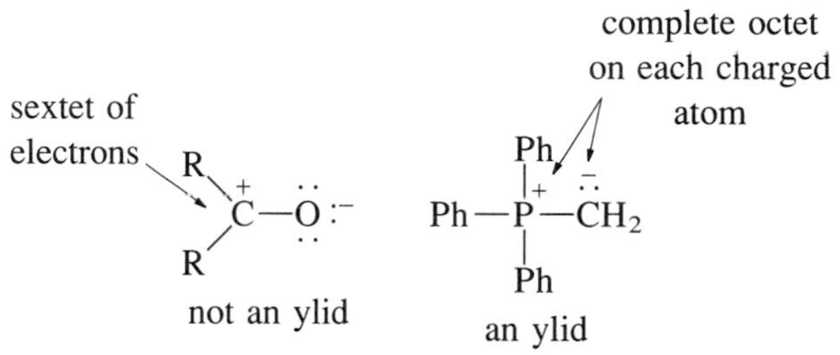

(We shall learn later in this section how ylids are prepared.)

Because phosphorus, like sulfur (Sec. 10.9), can accommodate more than eight valence electrons, a phosphorus ylid has an uncharged resonance structure.

$$Ph_3\overset{+}{P}{-}\overset{-}{\overset{\cdot\cdot}{C}}H_2 \longleftrightarrow Ph_3P{=}CH_2 \qquad \begin{array}{l}\text{ten electrons around} \\ \text{phosphorus}\end{array} \qquad (19.73)$$

Although the structures of phosphorus ylids are sometimes written with phosphorus–carbon double bonds, the charged structures, in which each atom has an octet of electrons, are very important contributors.

The mechanism of the Wittig reaction, like the other carbonyl reactions we have been studying, involves attack of a nucleophile on the carbonyl carbon. The nucleophile in the Wittig reaction is the anionic carbon of the ylid. The anionic oxygen in the resulting species reacts with phosphorus to form an *oxaphosphetane* intermediate. (An oxaphosphetane is a saturated four-membered ring containing both oxygen and phosphorus as ring atoms.)

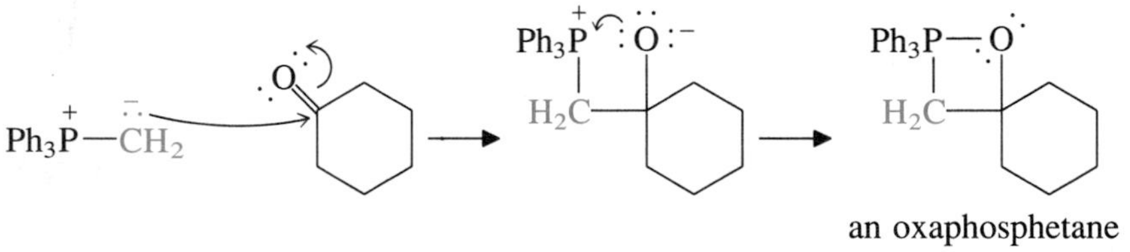

an oxaphosphetane

Under the usual reaction conditions, the oxaphosphetane spontaneously breaks down to the alkene and the by-product triphenylphosphine oxide.

triphenylphosphine oxide

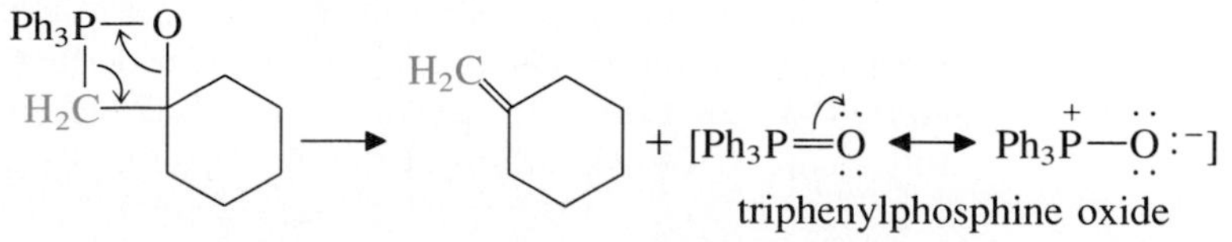

The Wittig reaction is important because it gives alkenes in which there is no ambiguity about the *position* of the alkene double bond; it is a completely *regioselective* reaction. In contrast, E2 elimination of alkyl halides and dehydration of alcohols in many cases give mixtures of products. The Wittig reaction is particularly useful for the synthesis of alkenes that are hard to obtain by other means. For example, an attempt to synthesize the methylenecyclohexane product of Eq. 19.72 by the following route gives 1-methyl-1-cyclohexene instead (why?):

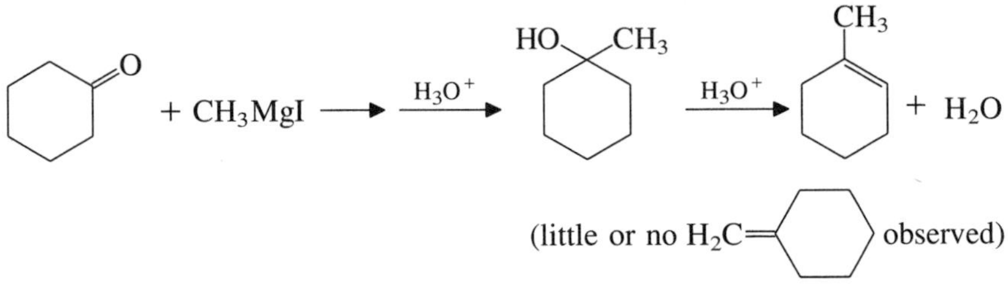

(little or no H$_2$C=⬡ observed)

Although we might be able to devise other routes to the desired product from cyclohexanone, the Wittig reaction is the most direct route because it requires the fewest synthetic operations.

How can we prepare the phosphorus ylids used in the Wittig reaction? First, an alkyl halide is allowed to react with triphenylphosphine (Ph$_3$P) to give a *phosphonium salt*.

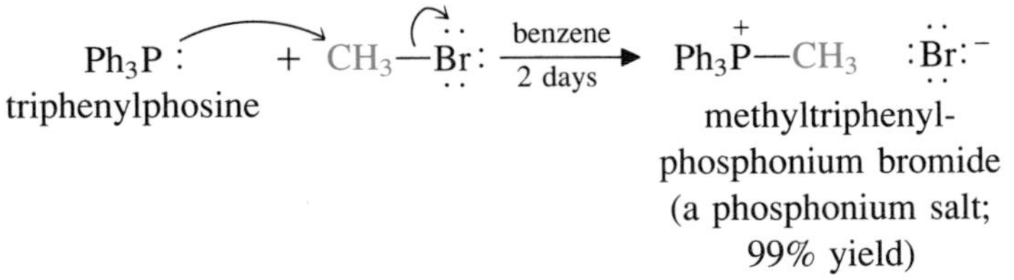

$$\text{Ph}_3\text{P}: \quad + \quad \text{CH}_3\!-\!\ddot{\text{Br}}: \xrightarrow[\text{2 days}]{\text{benzene}} \quad \text{Ph}_3\overset{+}{\text{P}}\!-\!\text{CH}_3 \quad :\!\ddot{\text{Br}}:^- \quad (19.76a)$$
triphenylphosine

methyltriphenyl-
phosphonium bromide
(a phosphonium salt;
99% yield)

This is a typical S$_N$2 reaction and is therefore limited for the most part to primary alkyl halides. The phosphonium salt can be converted into its conjugate base, the ylid, by reaction with a strong base such as a lithium reagent.

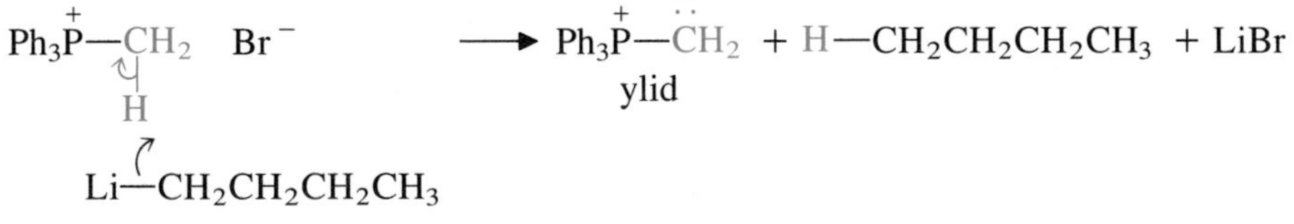

$$\text{Ph}_3\overset{+}{\text{P}}\!-\!\underset{\underset{\text{H}}{|}}{\text{CH}_2} \quad \text{Br}^- \qquad \longrightarrow \quad \text{Ph}_3\overset{+}{\text{P}}\!-\!\overset{..}{\text{CH}}_2 \;+\; \text{H}\!-\!\text{CH}_2\text{CH}_2\text{CH}_2\text{CH}_3 \;+\; \text{LiBr} \quad (19.76b)$$
ylid

$$\text{Li}\!-\!\text{CH}_2\text{CH}_2\text{CH}_2\text{CH}_3$$

To plan the synthesis of an alkene by the Wittig reaction, we consider the origin of each part of the product, and then reason deductively. Thus, one carbon of the alkene double bond originates from the alkyl halide used to prepare the ylid; the other is the carbonyl carbon of the aldehyde or ketone:

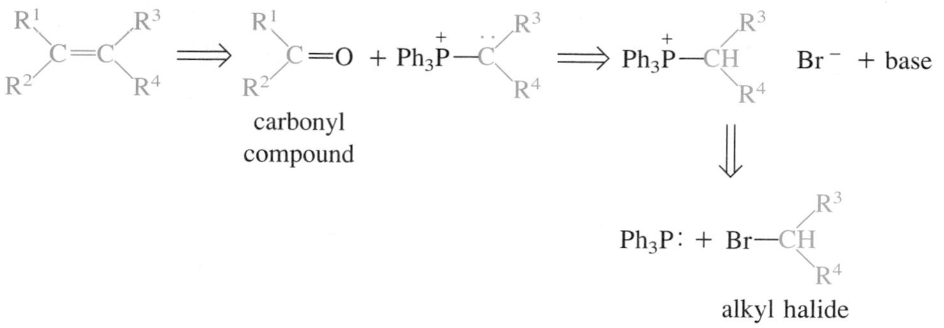

$$Ph_3P: + \; Br-\overset{R^3}{\underset{R^4}{CH}} \qquad (19.77)$$

alkyl halide

(Again, the double arrows used in this deductive analysis are read "implies as starting material.") This analysis also shows that there are, in principle, two possible Wittig syntheses for any given alkene; in the other possibility, the R^1 and R^2 groups could originate from the alkyl halide and the R^3 and R^4 groups from the aldehyde or ketone. However, we must remember the restrictions on the mechanism of phosphonium salt formation. Since this is an S_N2 reaction, it works best with primary alkyl halides. In other words, most Wittig syntheses are planned so that a primary alkyl halide can be used as one of the starting materials.

One problem with the Wittig reaction is that it gives mixtures of *E* and *Z* isomers.

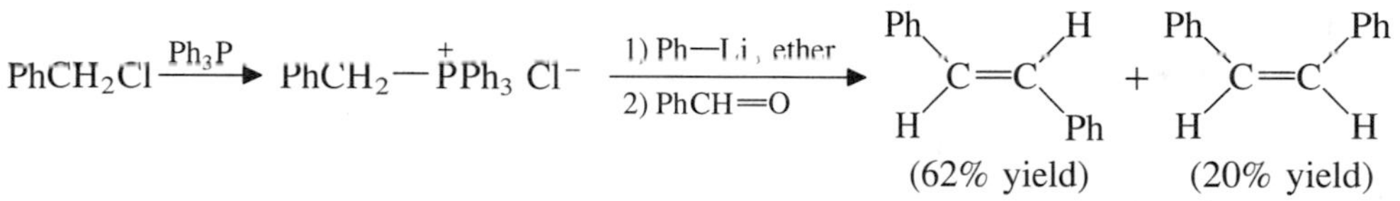

$$PhCH_2Cl \xrightarrow{Ph_3P} PhCH_2-\overset{+}{P}Ph_3 \; Cl^- \xrightarrow[\text{2) } PhCH=O]{\text{1) } Ph-Li, \text{ ether}} \underset{\text{(62\% yield)}}{\overset{Ph}{\underset{H}{}}C=C\overset{H}{\underset{Ph}{}}} + \underset{\text{(20\% yield)}}{\overset{Ph}{\underset{H}{}}C=C\overset{Ph}{\underset{H}{}}} \qquad (19.78)$$

Although certain modifications of the Wittig reaction that avoid this problem have been developed, these are outside the scope of our discussion.

Georg Wittig (1897–) was Professor of Chemistry at the University of Heidelberg. Wittig and his coworkers discovered the alkene synthesis in the course of other work in phosphorus chemistry; they had not set out to develop this reaction explicitly. Once the significance of the reaction was recognized, it was widely exploited. Wittig shared the Nobel Prize for Chemistry with H. C. Brown (Sec. 19.8) in 1979.

The Wittig reaction is not only important as a laboratory reaction; it has also been industrially useful. For example, it is an important reaction in the industrial synthesis of vitamin A derivatives.

Problems

29 Give the structure of the alkene formed in each of the following reactions:

(a) $CH_3CH_2I \xrightarrow{Ph_3P} \xrightarrow{\text{butyllithium}} \xrightarrow{\text{acetone}}$

(b) $CH_3Br \xrightarrow{Ph_3P} \xrightarrow{\text{butyllithium}} \xrightarrow{\text{benzaldehyde}}$

Problems (Cont.)

30 Outline a Wittig synthesis for each of the following alkenes; give two Wittig syntheses of (b):

(a)

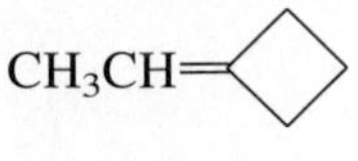

$CH_3CH=$

(c)

$$CH_2=\overset{\underset{\displaystyle |}{CH_3}}{C}CH_2CH_3$$

(b) CH_3O—⬡—$CH=CH$—⬡

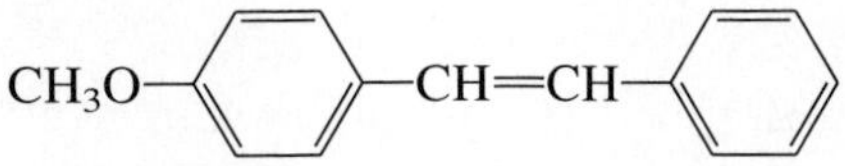

(mixture of *cis* and *trans*)

19.14 OXIDATION OF ALDEHYDES

Now we turn to our last carbonyl-group reaction of this chapter: oxidation of aldehydes to carboxylic acids.

$$CH_3CH_2CH_2CH_2\underset{\underset{\displaystyle C_2H_5}{|}}{CH}—CH=O \xrightarrow[H_2O,\ NaOH]{KMnO_4} \xrightarrow{H_3O^+} CH_3CH_2CH_2CH_2\underset{\underset{\displaystyle C_2H_5}{|}}{CH}—CO_2H \quad (19.79)$$

(78% yield)

Other common oxidants, such as Cr(VI) reagents and nitric acid, also work in this reaction. These oxidizing agents are also the ones used for oxidizing alcohols (Sec. 10.6A; Eq. 10.36).

Some aldehyde oxidations begin as addition reactions. For example, in the oxidation of aldehydes by Cr(VI) reagents, the *hydrate,* not the aldehyde, is actually the species oxidized. (See Eq. 10.38.)

$$\underset{\text{aldehyde}}{R—\overset{\overset{\displaystyle O}{\|}}{C}—H} + H_2O \rightleftharpoons \underset{\substack{\displaystyle | \\ \displaystyle OH \\ \text{aldehyde} \\ \text{hydrate}}}{R—\overset{\overset{\displaystyle OH}{|}}{C}—H} \xrightarrow{H_2Cr_2O_7} R—\overset{\overset{\displaystyle O}{\|}}{C}—OH \quad (19.80)$$

That is, the "aldehyde" oxidation is really an "alcohol" oxidation—the alcohol being the hydrate formed by addition of water to the aldehyde carbonyl group. For this reason, some water should be present in solution in order that aldehyde oxidations with Cr(VI) occur at a reasonable rate.

In the laboratory, aldehydes can be conveniently oxidized to carboxylic acids with Ag(I).

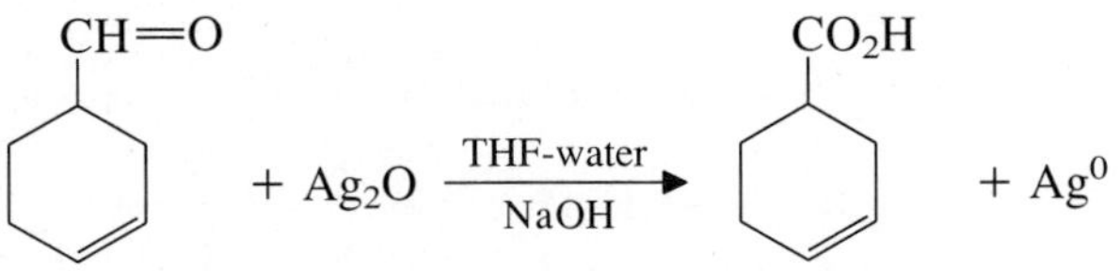

$$+ Ag_2O \xrightarrow[NaOH]{THF\text{-water}} + Ag^0 \quad (19.81)$$

(75% yield)

The expense of silver limits its use to small-scale reactions, as a rule. Sometimes, as in

Eq. 19.81, the silver is used as a slurry of brown Ag_2O, which changes to a black precipitate of silver metal as the reaction proceeds. If the silver ion is solubilized as its ammonia complex, $^{+}Ag(NH_3)_2$, oxidation of the aldehyde is accompanied by the deposition of a metallic silver mirror on the walls of the reaction vessel. This observation can be used as a convenient test for aldehydes, known as **Tollens' test.**

Many aldehydes are oxidized by the oxygen in air upon standing for long periods of time. This process, called *autoxidation,* is responsible for the contamination of some aldehyde samples with appreciable amounts of carboxylic acids.

Ketones cannot be oxidized without breaking carbon–carbon bonds (see Table 10.1). Ketones are resistant to mild oxidation with Cr(VI) reagents, and acetone can even be used as a solvent for oxidations with such reagents. Potassium permanganate, however, oxidizes ketones, and is therefore not a useful reagent for the preparation of ketones from secondary alcohols.

Problem

31 Give the structure of an aldehyde $C_8H_8O_2$ that would be oxidized to terephthalic acid by $KMnO_4$.

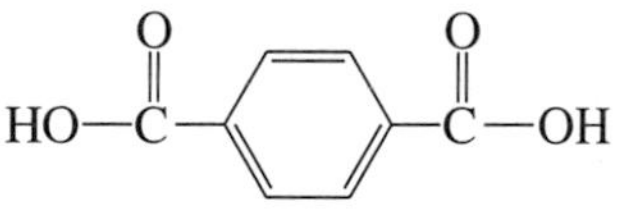

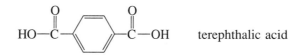

terephthalic acid

19.15 MANUFACTURE AND USE OF ALDEHYDES AND KETONES

The most important commercial aldehyde is formaldehyde, which is manufactured by the oxidation of methanol over a silver catalyst.

$$CH_3\text{—}OH \xrightarrow[O_2,\ 600\text{–}650°]{Ag(cat)} CH_2\text{=}O \qquad (19.82)$$

In 1985, 5.6 billion pounds of formaldehyde, valued at approximately $500 million, was produced in the United States.

The most important single use of formaldehyde is in the synthesis of a class of polymers known as *phenol–formaldehyde resins.* (A *resin* is a polymer with a rigid three-dimensional network of repeating units.) Although the exact structure and properties of a phenol–formaldehyde resin depend on the conditions of the reaction used to prepare it, a typical segment of such a resin can be represented schematically as follows:

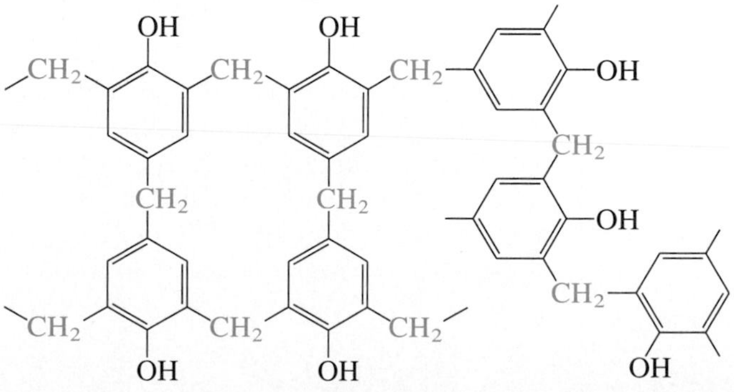

Phenol–formaldehyde resins are produced by heating phenol and formaldehyde with acidic or basic catalysts. The —CH_2— groups (color) are derived from the formaldehyde, which in some cases is supplied in the form of hexamethylenetetramine (Sec. 19.11A). Different formulations of these resins are used for telephones, adhesives in exterior-grade plywood, and heat-stable bondings for brake linings. A phenol–formaldehyde resin called *Bakelite,* patented in 1909 by the Belgian immigrant Leo H. Baekeland, was the first useful synthetic polymer.

The simplest ketone, acetone, is co-produced with phenol by the autoxidation–rearrangement of cumene (Sec. 18.8). About 1.7 billion pounds of acetone, valued at $425 million, was produced in the United States in 1985. Acetone is used both as a solvent and as a starting material for polymers.

KEY IDEAS IN CHAPTER 19

- The functional group in aldehydes and ketones is the carbonyl group.
- Aldehydes and ketones are polar molecules. Simple aldehydes and ketones have boiling points higher than those of hydrocarbons but lower than those of alcohols. The aldehydes and ketones of low molecular weight have substantial solubility in water.
- The carbonyl absorption is the most important infrared absorption of aldehydes and ketones. The NMR spectra of aldehydes have distinctive low-field absorptions for the aldehydic protons, and α-hydrogens of both aldehydes and ketones absorb near δ 2.5. Aldehydes and ketones have weak $n \rightarrow \pi^*$ UV absorptions, and compounds that contain double bonds conjugated with the carbonyl group have $\pi \rightarrow \pi^*$ absorptions. α-Cleavage, inductive cleavage, and the McLafferty rearrangement are the important fragmentation modes observed in the mass spectra of aldehydes and ketones.
- Aldehyde and ketone reactions can be subdivided into reactions at the carbonyl group, and reactions at the α-position.
- Aldehydes and ketones are weak bases, and are protonated on their carbonyl oxygens to give α-hydroxycarbocations. The interaction of a proton or Lewis acid with a carbonyl group activates it toward addition reactions.
- The most characteristic carbonyl-group reactions of aldehydes and ketones are carbonyl-addition reactions, which occur under both acidic and basic conditions. Hydration and cyanohydrin formation are examples of simple reversible carbonyl additions. Hydride reductions and Grignard reactions are examples of simple additions that are not reversible; in each of these cases the Lewis-acid fragment of the reagent (for example, ^+MgBr in the case of a Grignard reagent) is replaced by a proton when water is added to the reaction mixture.

- Acetal formation is an example of addition to the carbonyl group followed by substitution.
- Imine formation, enamine formation, and the Wittig reaction are examples of addition to the carbonyl group followed by elimination.
- The carbonyl group of an aldehyde or ketone can be reduced to a methylene group by either the Wolff–Kishner or Clemmensen reduction. Mechanistically, the Wolff–Kishner reduction begins with formation of an imine from hydrazine and the aldehyde or ketone. The nitrogen atoms of hydrazine are eliminated as N_2 in subsequent mechanistic steps.
- Aldehydes can be oxidized to carboxylic acids; ketones cannot be oxidized without breaking carbon–carbon bonds.
- Aldehydes and ketones can be converted into alcohols (Grignard reactions and hydride reductions); alkenes (Wittig reaction); and alkanes (Wolff–Kishner and Clemmensen reduction). Aldehydes can be converted into carboxylic acids by oxidation. These transformations illustrate why aldehydes and ketones are valued starting materials in organic synthesis.

ADDITIONAL PROBLEMS

32 Give the products expected when acetone reacts with each of the following reagents:

(a) H_3O^+
(b) $NaBH_4$ in CH_3OH, then H_2O
(c) CrO_3, pyridine
(d) NaCN, pH 10, H_2O
(e) CH_3OH (excess), H_2SO_4 (trace)
(f) 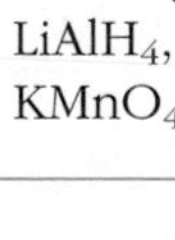, trace of acid
(g) 2,4-dinitrophenylhydrazine, H_2SO_4
(h) CH_3MgI, ether, then H_3O^+
(i) Product of (b) + $Na_2Cr_2O_7$ in H_2SO_4
(j) Product of (h) + H_2SO_4
(k) H_2, PtO_2
(l) $CH_2{=}PPh_3$
(m) Zn amalgam, HCl

33 Give the product expected when butyraldehyde reacts with each of the following reagents:

(a) Ph—MgBr, then H_3O^+
(b) $LiAlH_4$, then H_3O^+
(c) $KMnO_4$, OH^-
(d) $H_2Cr_2O_7$
(e) NH_2OH, pH = 5
(f) Ag_2O

34 Compound *A*, C_8H_8O, when treated with Zn amalgam and HCl, gives a xylene isomer that in turn gives one and only one ring monobromination product. What is *A*?

Problems (Cont.)

35 Sodium bisulfite adds reversibly to aldehydes and a few ketones to give *bisulfite addition products.*

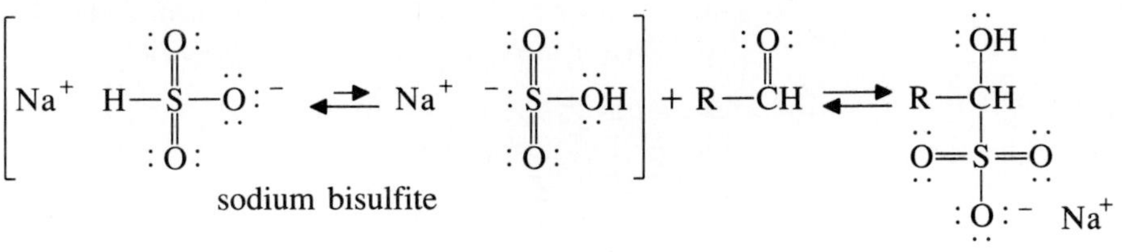

(a) Write a mechanism for this addition reaction; assume water is the solvent.
(b) The reaction can be reversed by adding H_3O^+ or ^-OH. Explain this observation using LeChatelier's principle and your knowledge of sodium bisulfite reactions from inorganic chemistry.

36 Each of the following reactions gives a mixture of two separable isomers. What are the two isomers formed in each case?

(a)

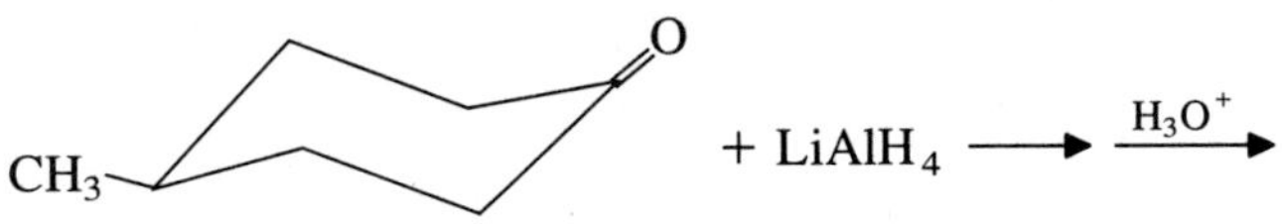

(b) $PhCH{=}O + Ph{-}CH{=}PPh_3 \longrightarrow$

(c)

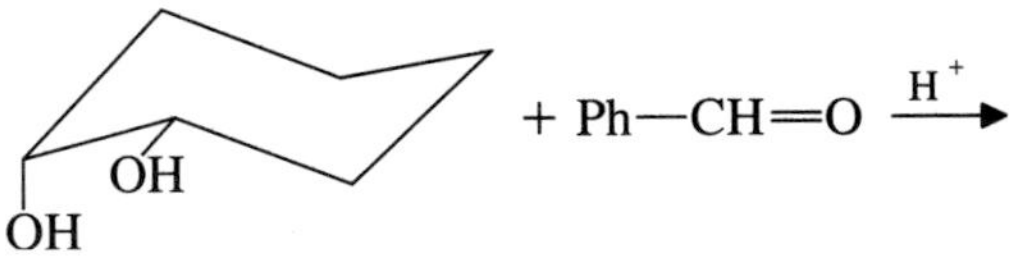

37 Complete the following reactions by giving the principal organic product(s):

(a)

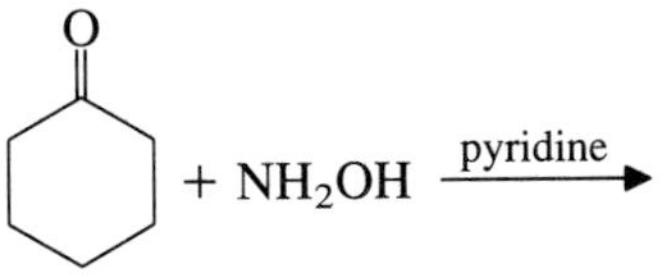

(b)

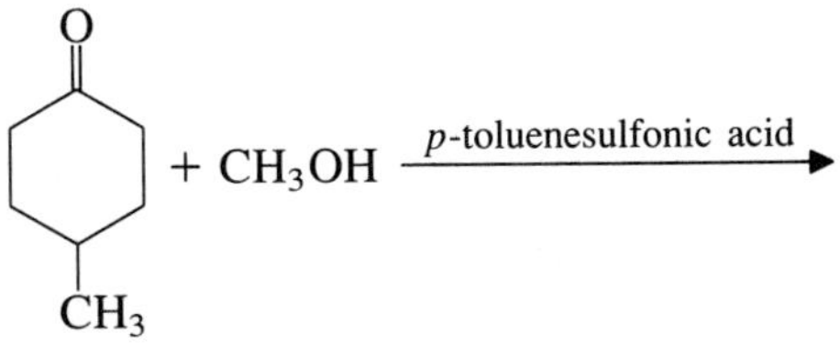

(c) propiophenone + Ph—MgBr $\longrightarrow \xrightarrow{H_2SO_4}$

(d) $CH_3I + Ph_3P \longrightarrow \xrightarrow[\text{benzene}]{CH_3CH_2CH_2CH_2{-}Li} \xrightarrow{Ph_2C{=}O}$

(e)

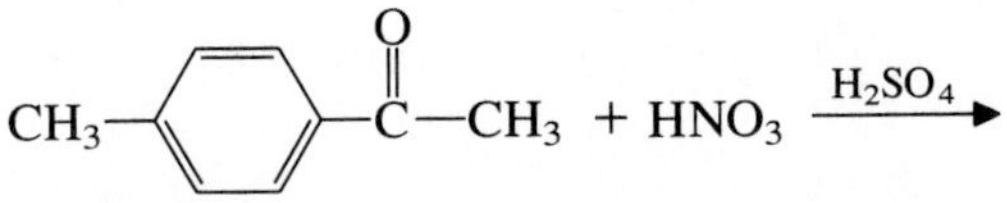

(f) 3-methylcyclopentanone + 1,3-propanediol $\xrightarrow{\text{H}^+}$

(g) [structure: 3-bromochlorobenzene] $\xrightarrow{\text{Mg, ether}}$ $\xrightarrow{\text{PhCH=O}}$ $\xrightarrow{\text{H}_3\text{O}^+}$

(h) $\text{CH}_3\text{OCH}_2\text{Cl} + \text{Ph}_3\text{P} \xrightarrow{\text{CH}_3\text{CH}_2\text{CH}_2\text{CH}_2\text{Li}} \xrightarrow{\text{cyclohexanone}}$

(i) $\text{CH}_3-\overset{\text{O}}{\underset{}{\text{C}}}-\overset{\text{O}}{\underset{}{\text{C}}}-\text{H} + \text{CH}_3\text{OH}$ (solvent) $\xrightarrow{\text{H}^+}$ ($\text{C}_5\text{H}_{10}\text{O}_3$)

(j) $\text{PhC}\equiv\text{CH} + \text{CH}_3\text{OH}$ (excess) $\xrightarrow{\text{H}_2\text{SO}_4 \text{ (cat)}}$

(k) $\underset{}{\text{HOCH}_2\overset{\text{CH}_3}{\underset{}{\text{CH}}}\text{CH}_2\text{CH}_2\overset{\text{O}}{\underset{}{\text{C}}}\text{CH}_2\text{CH}_2\overset{\text{CH}_3}{\underset{}{\text{CH}}}\text{CH}_2\text{OH}}$

$\xrightarrow[\text{benzene}]{\text{CH}_3-\langle\rangle-\text{SO}_3\text{H}}$ (a compound with eleven carbons)

(l) $\text{CH}_3\text{CH}_2\text{CH}_2\overset{\text{O}}{\underset{}{\text{C}}}\text{CH}_3 + \text{H}_2\ddot{\text{N}}-\text{Ph} \longrightarrow \underline{\quad} \xrightarrow{\text{NaBH}_4}$

(*Hint:* The C=N bond undergoes addition much like the C=O bond.)

(m) $\text{CH}_3\text{CH}_2\text{CH}_2\overset{\overset{\displaystyle\text{Ph}_2\text{P}\nearrow^{\text{O}}}{|}}{\text{CH}}-\overset{\text{O}}{\underset{}{\text{C}}}-\text{CH}_2\text{CH}_3 \xrightarrow[\text{MeOH}]{\text{NaBH}_4} \xrightarrow{\text{NaH}}$

(*Hint*: See Eq. 19.74.)

(n) [structure: biphenylene with two CH$_2$Br groups] $+ 2\text{Ph}_3\text{P} \longrightarrow \xrightarrow{\text{PhLi}} \xrightarrow{\text{H}-\overset{\text{O}}{\underset{}{\text{C}}}-\overset{\text{O}}{\underset{}{\text{C}}}-\text{H}}$ a compound with sixteen carbons

38 Suggest routes by which each of the following compounds could be synthesized from the indicated starting materials and any other reagents:
(a) 1-phenyl-1-butanone (butyrophenone) from butyraldehyde
(b) $(\text{CH}_3)_2\text{CHCH}=\text{CH}_2$ from isobutyraldehyde
(c) cyclohexyl methyl ether from cyclohexanone
(d) 2-cyclohexyl-2-propanol from cyclohexanone
(e) 2,3-dimethyl-2-hexene from 3-methyl-2-hexanone
(f) 2,3-dimethyl-1-hexene from 3-methyl-2-hexanone

(g) $\text{CH}_3\text{O}-\langle\rangle-\text{CH}_2\text{OH}$ from *p*-bromoanisole

(*Problem continued on next page.*)

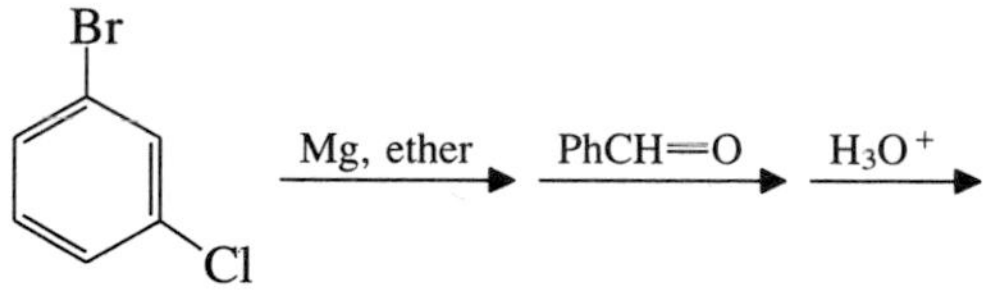

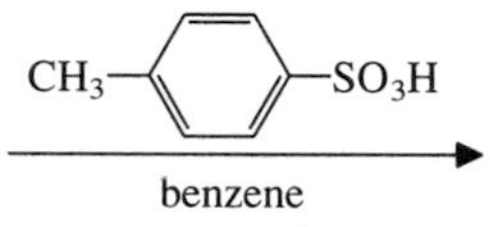

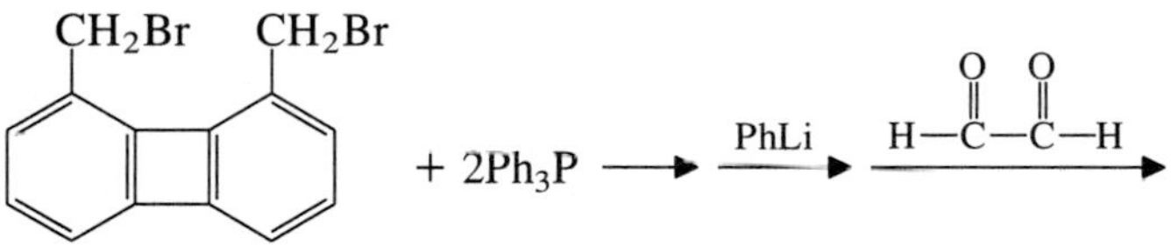

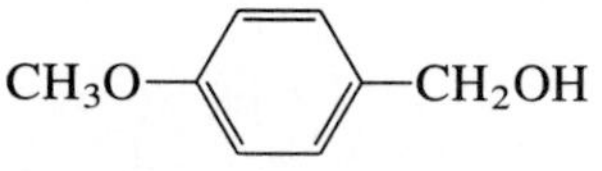

Problems (Cont.)

(h) 3-methyl-3-hexanol from propyl bromide

(i) 1,6-hexanediol from cyclohexene

(j) 1-butyl-4-propylbenzene from benzene

(k)

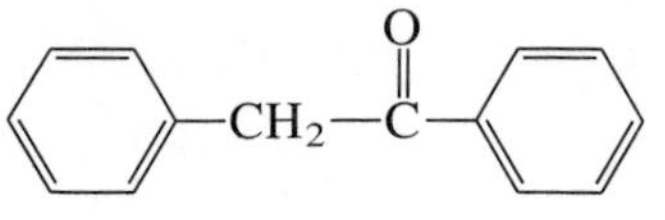

from benzaldehyde as the only source of carbon

(l)

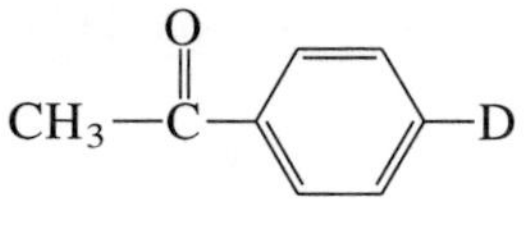

$$CH_3-\overset{\overset{\displaystyle O}{\|}}{C}-\!\!\!\raisebox{0pt}{⬡}\!\!\!-D \quad \text{from bromobenzene}$$

(m)

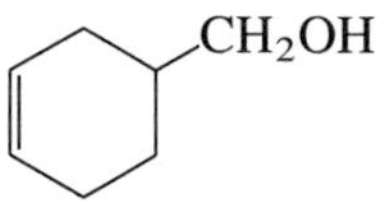

from $CH_2\!=\!CH\!-\!CH\!=\!O$ (*Hint:* See Sec. 15.3.)

(n)

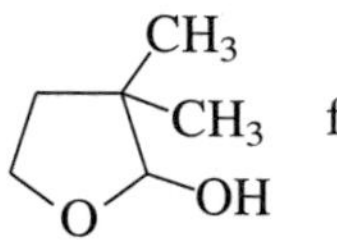

from $O\!=\!CH\!-\!\underset{\underset{\displaystyle CH_3}{|}}{\overset{\overset{\displaystyle CH_3}{|}}{C}}\!-\!CH\!=\!CH_2$

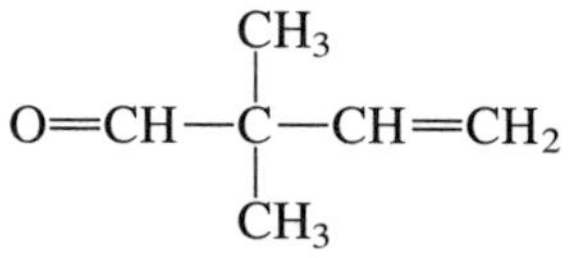

(*Hints:* 1. Diborane reduces aldehydes and ketones to alcohols.
2. Can you find an aldehyde lurking somewhere in the desired compound?)

39 What are the starting materials for synthesis of each of the following imines?

(a)
$$CH_3CH_2CH_2CH_2\!-\!CH\!=\!N\!-\!NH\!-\!\!\!\raisebox{0pt}{⬡}\!\!\!-OCH_3$$

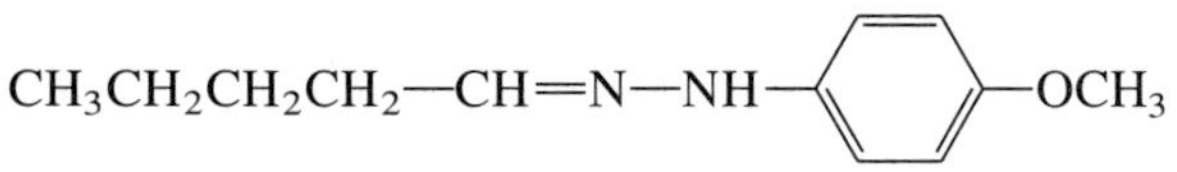

(b)

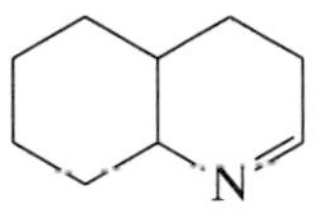

40 Compound *A*, $C_{11}H_{12}O$, that gave no Tollens' test, was found in an abandoned laboratory. On treatment with LiAlH$_4$, followed by dilute acid, compound *A* was converted into compound *B*, which could be resolved into enantiomers. When optically active *B* was treated with CrO$_3$ in pyridine, an optically inactive sample of *A* was obtained. Heating *A* with hydrazine in base gave hydrocarbon *C*, which, when heated with alkaline KMnO$_4$, gave carboxylic acid *D*. Identify all compounds and explain your reasoning.

D

$$\begin{array}{c} HO_2C\!\!\!\!\!\raisebox{0pt}{}\quad\quad CO_2H \\ \text{⬡} \\ HO_2C\quad\quad CO_2H \end{array}$$

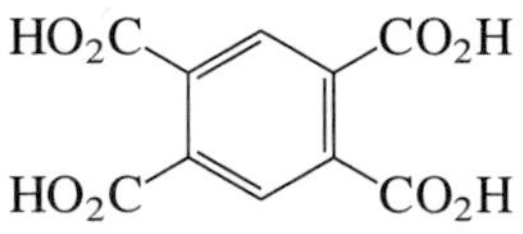

41 (a) The following compound is unstable and spontaneously decomposes to acetophenone and HBr. Suggest a mechanism for this transformation.

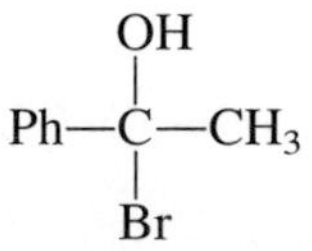

(b) Use the information in (a) to complete the following reaction:

$$CH_2{=}CH{-}Br + OsO_4 \xrightarrow{H_2O}$$

42 Protonated aldehydes and ketones can be used as electrophiles in electrophilic aromatic substitution. Give detailed mechanisms, including the arrow formalism, for the following reactions. The following sources can provide help if needed: Box, p. 132; Sec. 16.4B; and Problem 37, Chapter 16.

(a)

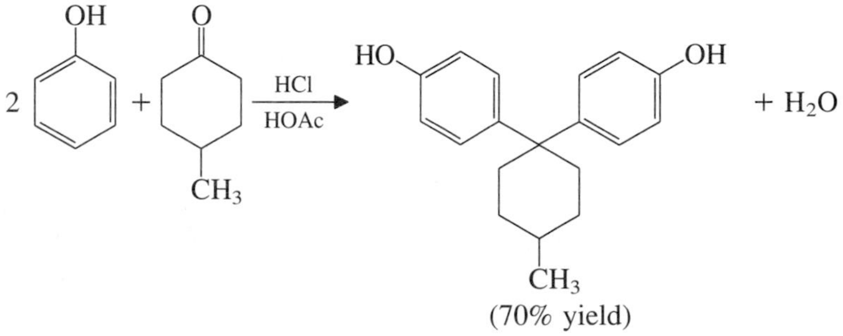

(b)

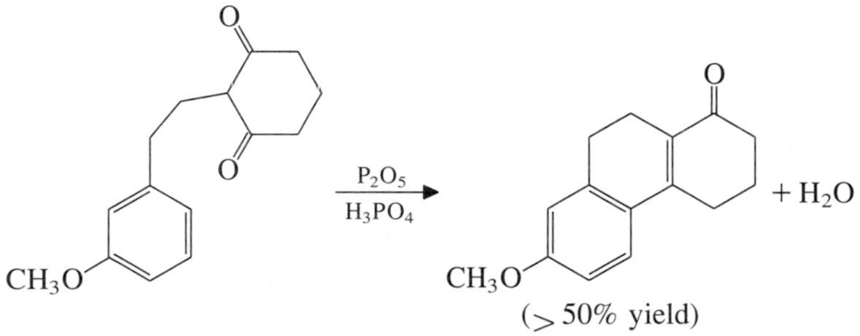

43 From your knowledge of the reactivity of $LiAlH_4$, as well as the reactivity of epoxides with nucleophiles, predict the product (including stereochemistry, if appropriate) in each of the following reactions:

(a)

(b)

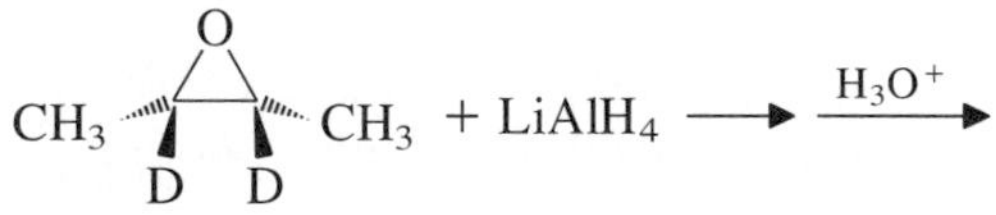

44 Give detailed mechanisms for the following reactions, using the arrow formalism when appropriate:

(a) $$Ph{-}\overset{S}{\overset{\|}{C}}{-}Ph \xrightarrow{H_2O} Ph{-}\overset{O}{\overset{\|}{C}}{-}Ph$$

(b) $$Ph{-}\underset{Cl}{\underset{|}{CH}}{-}OCH_3 \xrightarrow{H_2O} PhCH{=}O + CH_3OH + HCl$$

Problems (Cont.)

(c)

$(CH_3O)_3P:$ + CH_2—CH—CH_3 ⟶ $(CH_3O)_3P$=O + CH_3CH=CH_2

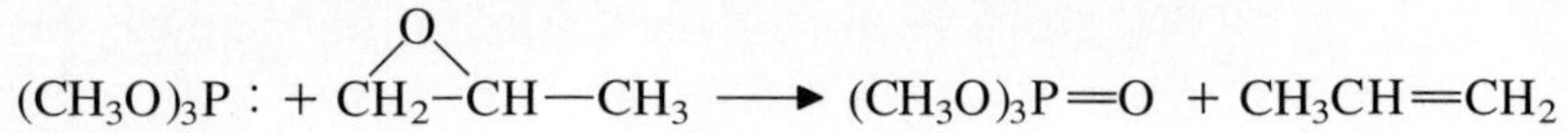

(d)

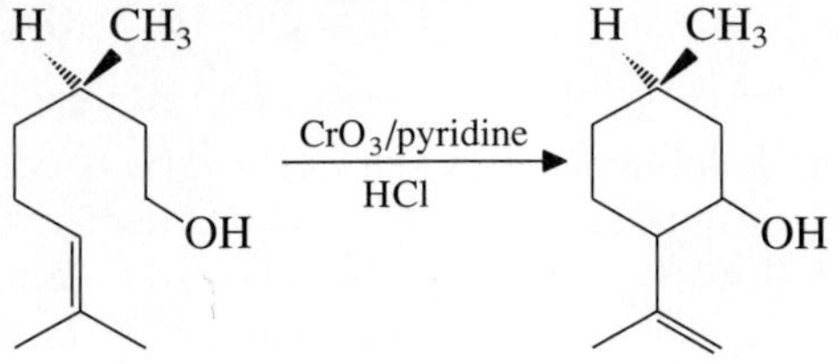

(e)

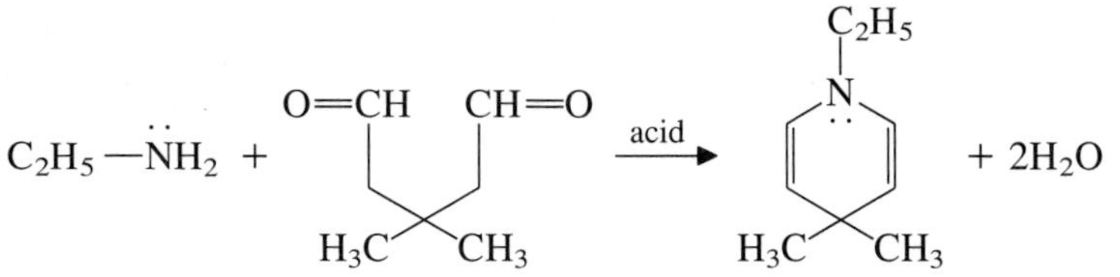

(f)

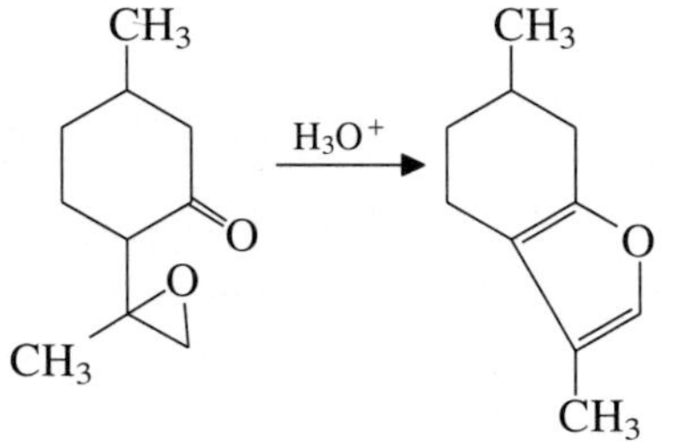

45 Thumbs Throckmorton, a graduate student in his twelfth year of study, has paused from his usual pursuit of testing the tensile strength of laboratory glassware long enough to design the following synthetic procedures. Comment on what problems (if any) each synthesis is likely to encounter.

(a)

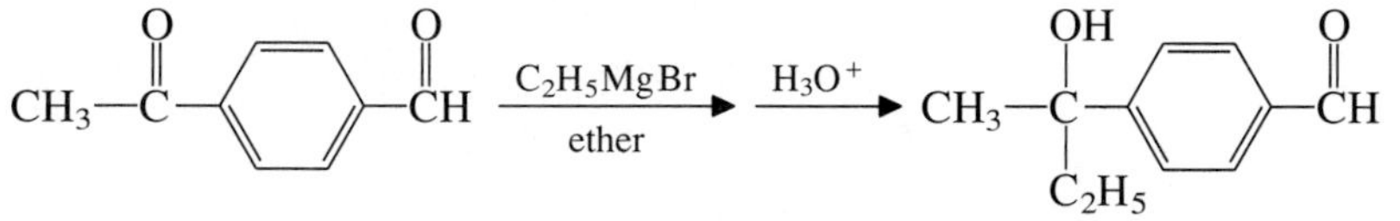

(b) Ph_3P + $(CH_3)_3C$—CH_2Br ⟶ $\xrightarrow{C_4H_9Li}$ $\xrightarrow{Ph—CH=O}$ $PhCH$=CH—$C(CH_3)_3$

46 Identify the compound with the mass spectrum and NMR spectrum given in Fig. 19.10. This compound has IR absorption at 1678 and 1600 cm^{-1}.

47 Identify the following compounds:
 (a) $C_5H_{10}O$ NMR: δ 9.8 (1*H*, s), δ 1.1 (9*H*, s)
 (b) $C_{10}H_{10}O_2$ NMR: δ 2.82 (6*H*, s), δ 8.13 (4*H*, s)
 IR: 1681 cm^{-1}, no O—H stretch
 (c) $C_6H_{10}O$ NMR in Fig. 19.11
 IR: 1701 cm^{-1}, 970 cm^{-1}
 UV: λ_{max} = 215 nm (ϵ = 17,400)
 329 nm (ϵ = 26)
 This compound gives a positive Tollens' test.

Figure 19.10 Spectra for Problem 46.

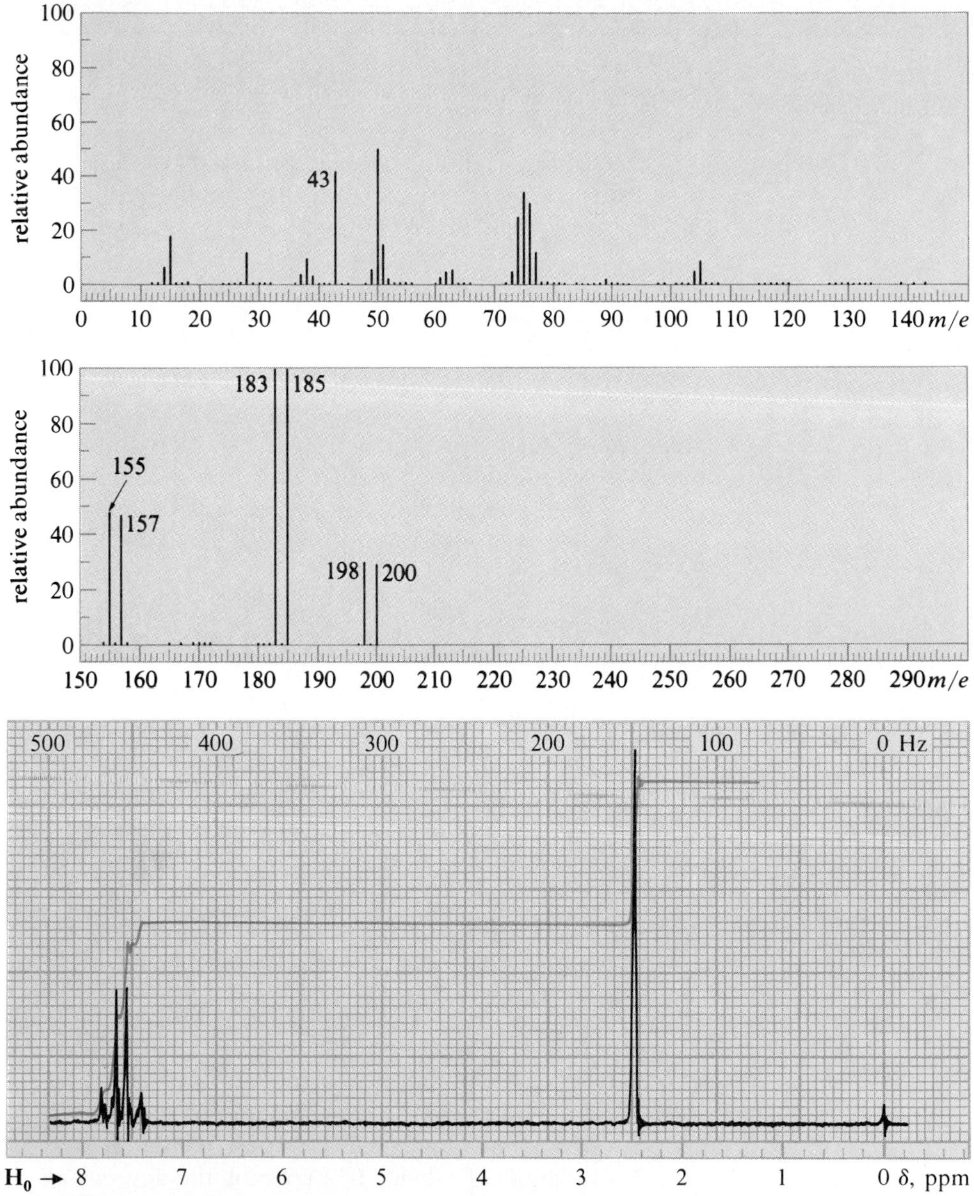

Figure 19.11 NMR spectrum for Problem 47(c).

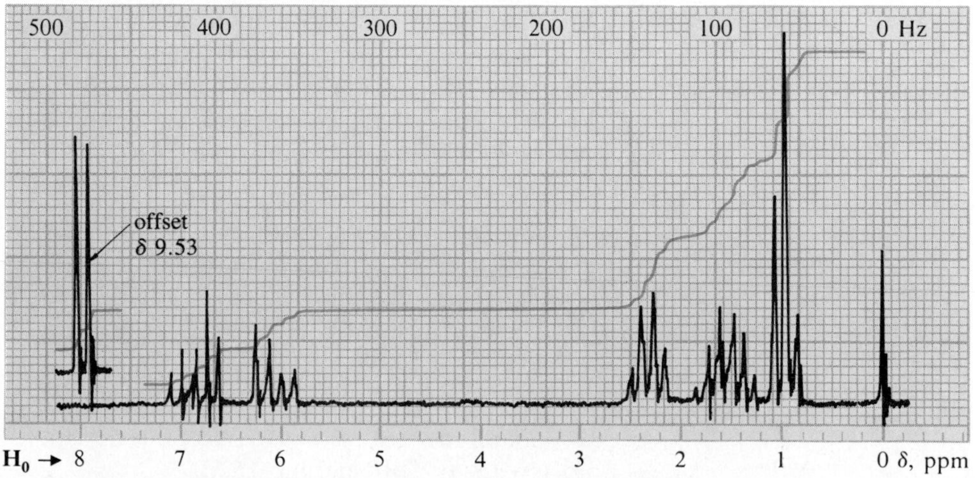

Problems (Cont.)

48 A graduate student needs the following isotopically labeled compound for a mechanism study and, knowing your expertise in organic synthesis, has come to you for assistance. Suggest a synthesis of the desired compound from compounds containing eight or fewer carbons.

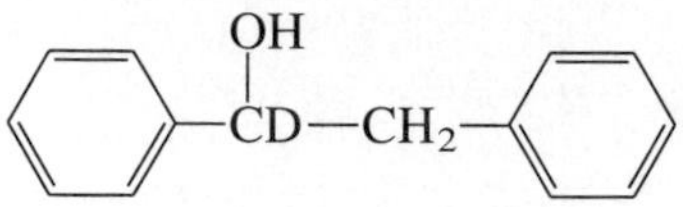

49 Offer a rational explanation for each of the following observations:

(a) Although biacetyl and 1,2-cyclopentanedione have the same types of functional groups, their dipole moments differ substantially.

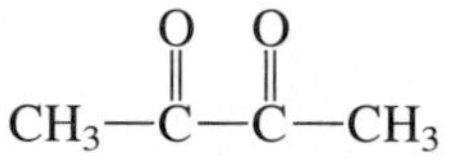

biacetyl
μ = 1.04 D

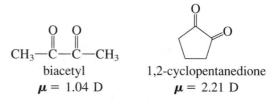

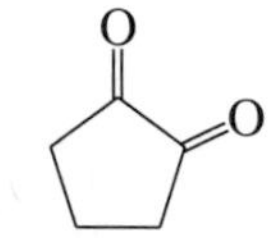

1,2-cyclopentanedione
μ = 2.21 D

(b) When acetaldehyde is mixed with a tenfold excess of ethanethiol, its $n \rightarrow \pi^*$ absorption at 280 nm is nearly eliminated.

(c) The following compound gives a Tollens' test much more slowly than benzaldehyde:

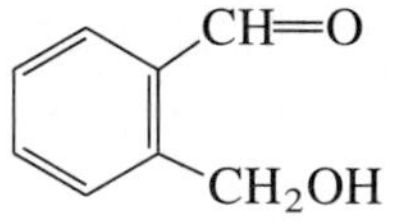

50 (a) You are the chief organic chemist for Bugs and Slugs, Inc., a firm that specializes in environmentally safe pest control. You have been asked to design a synthesis of 4-methyl-3-heptanol, the aggregation pheromone of the European elm beetle (the carrier of Dutch elm disease). Propose a synthesis of this compound from starting materials containing five or fewer carbons.

(b) After successfully completing the synthesis in part (a) and delivering your compound, you are advised that it appears to be a mixture of isomers. You are certain that you have prepared the correct compound, but your client wants a rational explanation. Provide one.

51 Starting with acetophenone, and using $H_2^{18}O$ as the source of the isotope, outline a synthesis of 1-phenyl-1-ethanol isotopically labeled with oxygen-18. (*Hint:* See Sec. 19.7.)

52 Trichloroacetaldehyde, CCl_3—CH═O, forms a cyclic trimer analogous to paraldehyde (Sec. 19.10A).

(a) Account for the fact that *two* forms of this trimer are known (α, bp 223° and mp 116°; β, bp 250° and mp 152°).

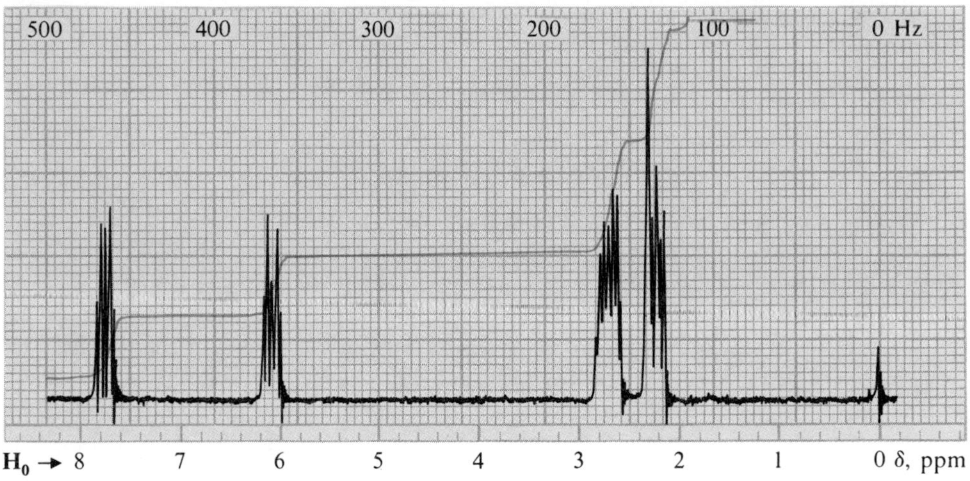

Figure 19.12 NMR spectrum for Problem 53.

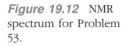

(b) Assume you have in hand samples of both the α- and β-forms, but you do not know which is which. Show how NMR spectroscopy could be used to distinguish one isomer from the other.

---

53 Identify the compound that has the following spectroscopic properties:
IR: 1686, 1600 cm^{-1}
Mass spectrum: Parent ion and base peak at $m/e = 82$
NMR spectrum in Fig. 19.12.

---

54 (a) What are the two structurally isomeric cyclic acetals that could in principle be formed in the acid-catalyzed reaction of acetone and glycerol (1,2,3-propanetriol)?

(b) Only one of the two compounds is actually observed to form. Given that it can be resolved into enantiomers, which isomer in (a) is the one that is observed?

---

55 Compound *A*, $C_6H_{12}O_2$, was found to be optically active. Compound *A* was slowly oxidized to an optically active carboxylic acid, *B*, $C_6H_{12}O_3$, by $^{+}Ag(NH_3)_2$. With HNO_3, compound *A* was oxidized to an achiral dicarboxylic acid $C_6H_{10}O_4$. Compound *A* was converted into an optically inactive compound with $NaBH_4$ in methanol. Give structures for *A* and *B* and explain all observations.

20 Chemistry of Carboxylic Acids

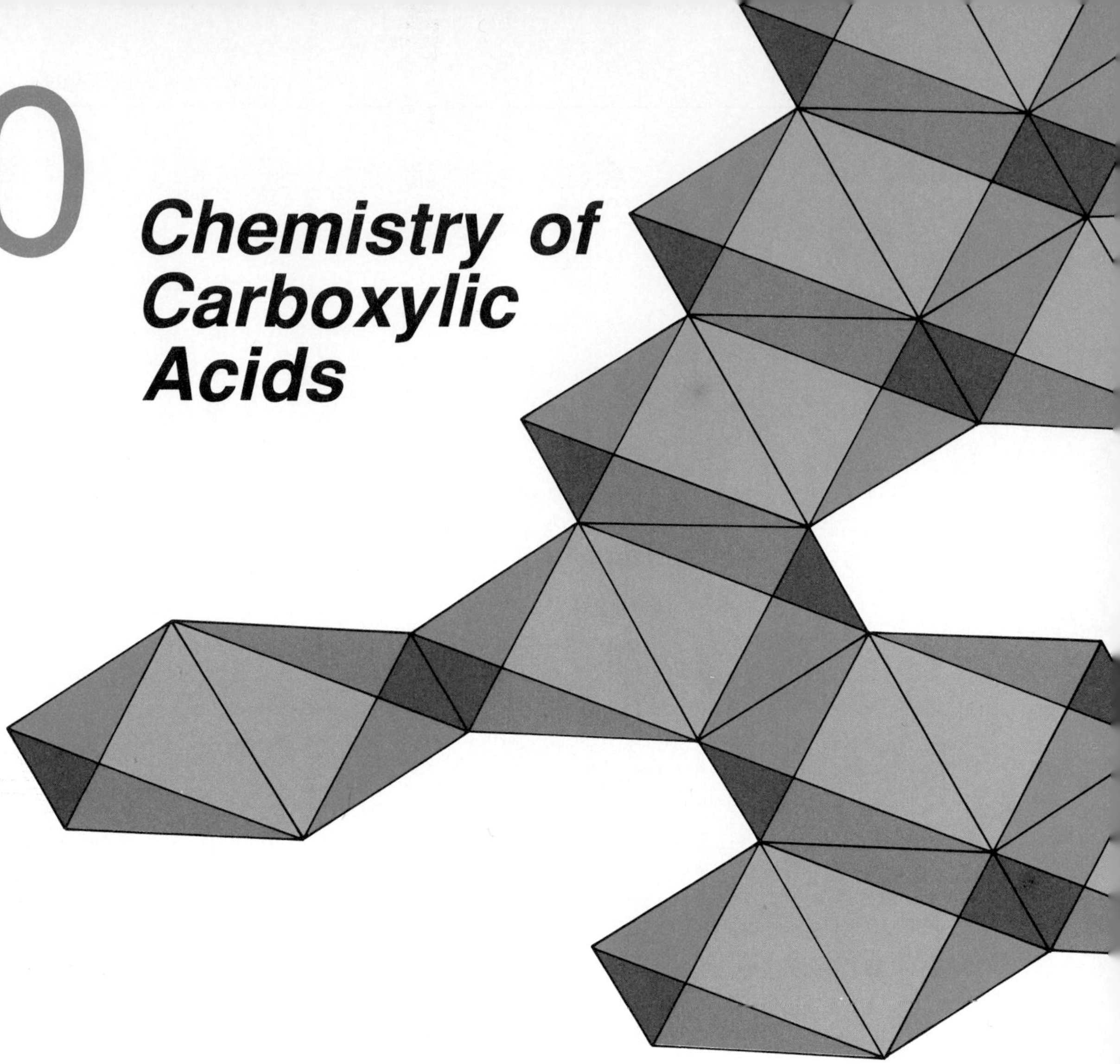

The characteristic functional group in a **carboxylic acid** is the **carboxy group.**

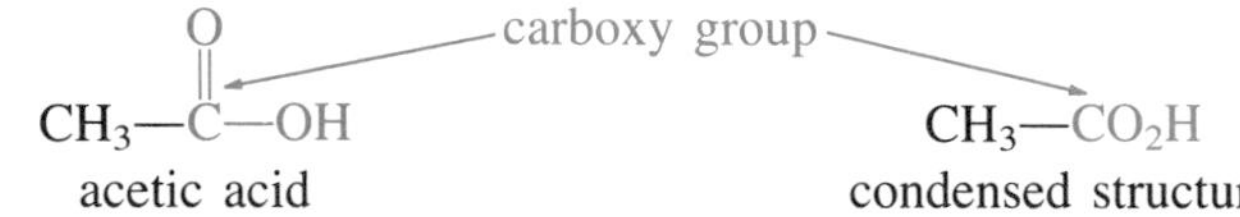

acetic acid
(a simple carboxylic acid)

condensed structure

Carboxylic acids and their derivatives rank with aldehydes and ketones among the most important organic compounds. Not only do they occur widely in nature, but thev also serve important roles in organic synthesis. This chapter is concerned with the structure, properties, acidity, and carbonyl-group reactions of carboxylic acids themselves. The following chapter is devoted to a study of carboxylic acid derivatives.

We shall also consider briefly in this chapter some of the chemistry of **sulfonic acids.**

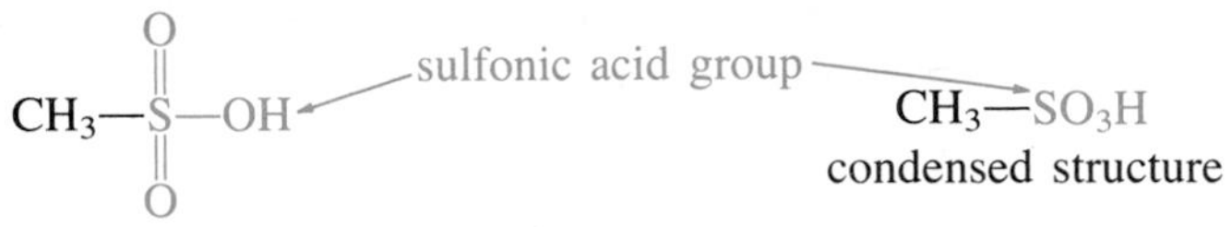

methanesulfonic acid
(a simple sulfonic acid)

20.1 NOMENCLATURE OF CARBOXYLIC ACIDS

A. Common Nomenclature

Common nomenclature is widely used for the simpler carboxylic acids. A carboxylic acid is named by adding the suffix *ic acid* to the prefix for the appropriate group given in Table 19.1.

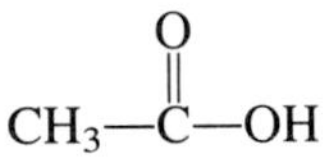
prefix (Table 19.1): acet + *ic acid* = acetic acid

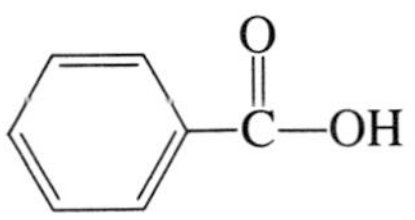
benzo + *ic acid* = benzoic acid

Some of these names owe their origin to the natural source of the acid. For example, formic acid occurs in the venom of the red ant (Latin, *formica* = ant); acetic acid is the acidic component of vinegar (Latin, *acetus* = vinegar); and butyric acid is the foul-smelling component of rancid butter (Latin, *butyrum* = butter). The common names of carboxylic acids, given in Table 20.1, are used as much or more than the systematic names.

As with aldehydes and ketones, substitution in the common system is denoted with Greek letters rather than numbers. The position *adjacent* to the carboxy group is designated as α.

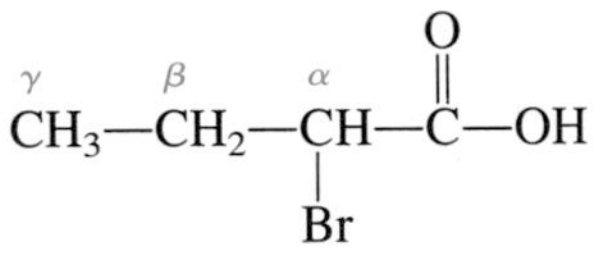
α-bromobutyric acid

Carboxylic acids with two carboxy groups are called **dicarboxylic acids.** The unbranched dicarboxylic acids are particularly important, and are invariably known by their common names. Some important dicarboxylic acids are also listed in Table 20.1.

$HO_2CCH_2CH_2CO_2H$ succinic acid

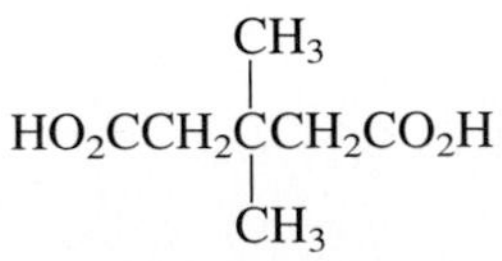
β,β-dimethylglutaric acid

TABLE 20.1 Names and Structures of Some Carboxylic Acids

Systematic name	Common name	Structure
methanoic* acid	formic acid	HCO_2H
ethanoic* acid	acetic acid	CH_3CO_2H
propanoic acid	propionic acid	$C_2H_5CO_2H$
butanoic acid	butyric acid	$n\text{-}C_3H_7CO_2H$
2-methylpropanoic acid	isobutyric acid	$(CH_3)_2CHCO_2H$
pentanoic acid	valeric acid	$n\text{-}C_4H_9CO_2H$
3-methylbutanoic acid	isovaleric acid	$(CH_3)_2CHCH_2CO_2H$
2-2-dimethylpropanoic acid	pivalic acid	$(CH_3)_3CCO_2H$
hexanoic acid	caproic acid	$n\text{-}C_5H_{11}CO_2H$
octanoic acid	caprylic acid	$n\text{-}C_7H_{15}CO_2H$
decanoic acid	capric acid	$n\text{-}C_9H_{19}CO_2H$
dodecanoic acid	lauric acid	$n\text{-}C_{11}H_{23}CO_2H$
tetradecanoic acid	myristic acid	$n\text{-}C_{13}H_{27}CO_2H$
hexadecanoic acid	palmitic acid	$n\text{-}C_{15}H_{31}CO_2H$
octadecanoic acid	stearic acid	$n\text{-}C_{17}H_{35}CO_2H$
2-propenoic* acid	acrylic acid	$CH_2{=}CHCO_2H$
2-butenoic* acid	crotonic acid	$CH_3CH{=}CHCO_2H$
benzoic acid	benzoic acid	$PhCO_2H$
	Dicarboxylic acids	
ethanedioic* acid	oxalic acid	$HO_2C{-}CO_2H$
propanedioic* acid	malonic acid	$HO_2CCH_2CO_2H$
butanedioic* acid	succinic acid	$HO_2C(CH_2)_2CO_2H$
pentanedioic* acid	glutaric acid	$HO_2C(CH_2)_3CO_2H$
hexanedioic* acid	adipic acid	$HO_2C(CH_2)_4CO_2H$
heptanedioic* acid	pimelic acid	$HO_2C(CH_2)_5CO_2H$
1,2-benzenedicarboxylic* acid	phthalic acid	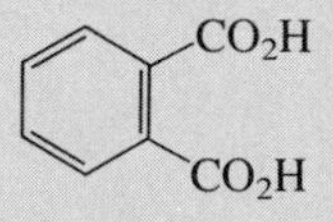
cis-2-butenedioic* acid	maleic acid	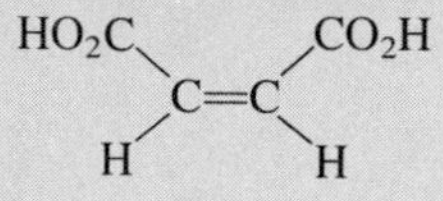
trans-2-butenedioic* acid	fumaric acid	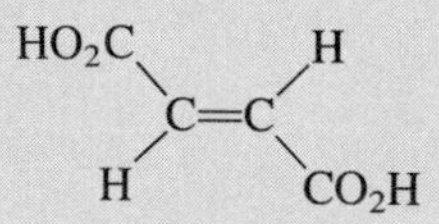

*Systematic name is almost never used.

A mnemonic device used by generations of organic chemistry students for remembering the names of the dicarboxylic acids is the phrase, "*O*h, *M*y, *S*uch *G*ood *A*pple *P*ie," in which the first letter of each word corresponds to the name of homologous dicarboxylic acids: oxalic, malonic, succinic, glutaric, adipic, and pimelic acids.

Phthalic acid is an important aromatic dicarboxylic acid.

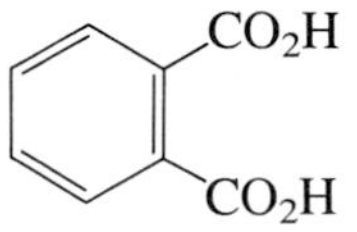

phthalic acid

Carbonic acid (H_2CO_3) has two —OH groups that share a single carbonyl.

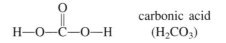

$$H{-}O{-}\overset{\overset{\large O}{\|}}{C}{-}O{-}H$$

carbonic acid (H_2CO_3)

As we shall discuss in Sec. 20.11, carbonic acid is unstable.

Many carboxylic acids were known long before any system of nomenclature existed, and their time-honored traditional names are widely used.

$$HO_2C{-}\underset{\underset{OH}{|}}{CH}{-}\underset{\underset{OH}{|}}{CH}{-}CO_2H$$

tartaric acid

$$Ph{-}CH{=}CH{-}CO_2H$$

cinnamic acid

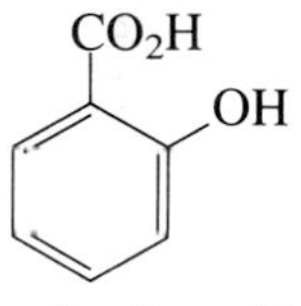

salicylic acid

B. Systematic Nomenclature

A carboxylic acid is named systematically by dropping the final *e* from the name of the hydrocarbon with the same number of carbon atoms and adding the suffix *oic acid*.

$$CH_3CH_2{-}\overset{\overset{\large O}{\|}}{C}{-}OH$$

propan~~e~~ + *oic acid* = propanoic acid

The final *e* is not dropped in the name of dicarboxylic acids.

$$HO_2C{-}CH_2CH_2CH_2CH_2CH_2CH_2{-}CO_2H$$

octanedioic acid

When the carboxylic acid is derived from a cyclic hydrocarbon, the suffix *carboxylic acid* is added to the name of the hydrocarbon. (This nomenclature is similar to that for the corresponding aldehydes; Sec. 19.1B.)

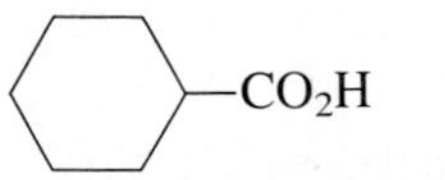

cyclohexanecarboxylic acid

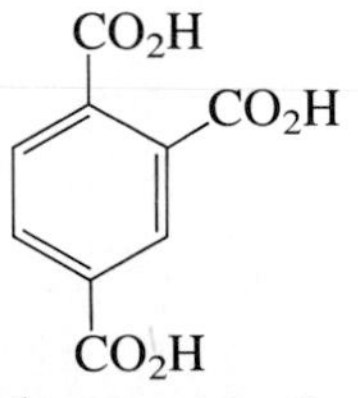

1,2,4-benzenetricarboxylic acid

One exception to this nomenclature is benzoic acid (Sec. 20.1A), for which the IUPAC recognizes the common name.

The numbering of the principal chain in substituted carboxylic acids, as in aldehydes, assigns the carbonyl carbon the number 1.

$$\underset{5}{CH_3}-\underset{4}{CH_2}-\underset{3}{\overset{\overset{CH_3}{|}}{CH}}-\underset{2}{CH_2}-\underset{1}{CO_2H}$$

3-methylpentanoic acid

This numbering scheme should be contrasted with that used in the common system, in which numbering begins with the Greek letter α at carbon-2.

In carboxylic acids derived from cyclic hydrocarbons, numbering begins at the ring carbon bearing the carboxy group.

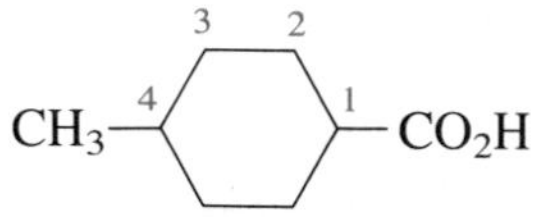

4-methylcyclohexanecarboxylic acid

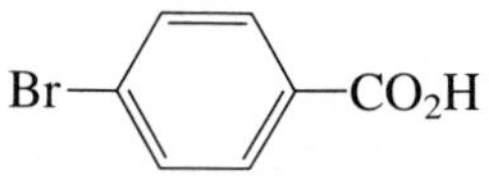

4-bromobenzoic acid
(or *p*-bromobenzoic acid)

When carboxylic acids contain other functional groups, the carboxy groups receive priority over aldehyde and ketone carbonyl groups, over hydroxy groups, and over mercapto groups for citation as the principal group.

priority for citation as principal group:

$$-\overset{\overset{O}{\|}}{C}-OH > -\overset{\overset{O}{\|}}{C}-H > -\overset{\overset{O}{\|}}{C}- > -OH > -SH \qquad (20.1)$$

$$H\overset{\overset{O}{\|}}{C}-CH{=}CH-\overset{\overset{OH}{|}}{CH}-CH_2-CH_2-\overset{\overset{O}{\|}}{C}-OH$$

4-hydroxy-7-oxo-5-heptenoic acid

(4 and 7: position of substituent groups; 5: position of double bond)

Notice in this example that the position of the carboxy group is not specified since it must by definition fall at the end of the principal chain.

The carboxy group is sometimes named as a substituent:

carboxymethyl group

$$\underset{1}{HO_2C}-\underset{2}{CH_2}\underset{3}{\overset{\overset{HO_2C-CH_2}{|}}{C}H}\underset{4}{CH_2}\underset{5}{CH_2}-\underset{6}{CO_2H}$$

3-(carboxymethyl)hexanedioic acid

A complete list of nomenclature priorities for all of the functional groups covered in this text is given in Appendix I.

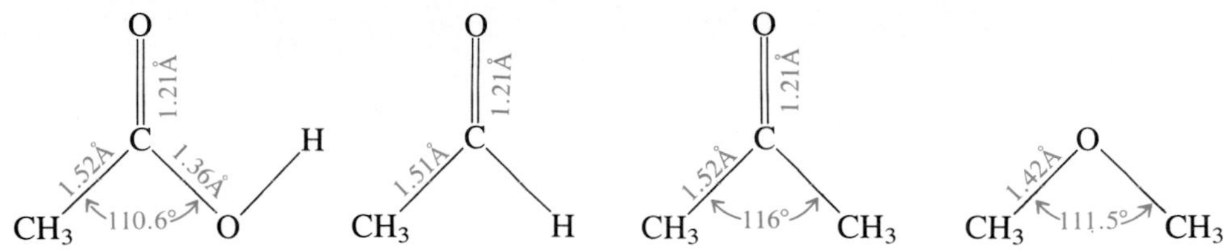

Figure 20.1 Comparison of the structures of acetic acid, acetone, acetaldehyde, and dimethyl ether.

acetic acid acetaldehyde acetone dimethyl ether

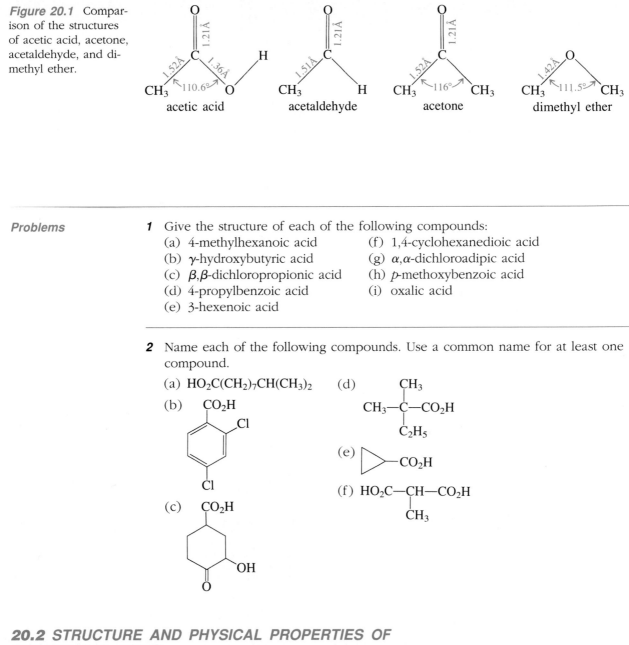

Problems

1 Give the structure of each of the following compounds:
 (a) 4-methylhexanoic acid (f) 1,4-cyclohexanedioic acid
 (b) γ-hydroxybutyric acid (g) α,α-dichloroadipic acid
 (c) β,β-dichloropropionic acid (h) p-methoxybenzoic acid
 (d) 4-propylbenzoic acid (i) oxalic acid
 (e) 3-hexenoic acid

2 Name each of the following compounds. Use a common name for at least one compound.

 (a) $HO_2C(CH_2)_7CH(CH_3)_2$

 (b) CO_2H

 (c) CO_2H

 (d) $CH_3-C(CH_3)(C_2H_5)-CO_2H$

 (e) $\triangleright-CO_2H$

 (f) $HO_2C-CH(CH_3)-CO_2H$

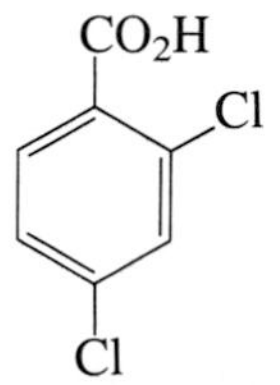

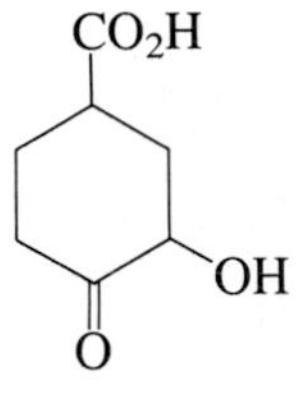

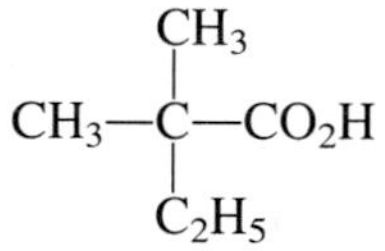

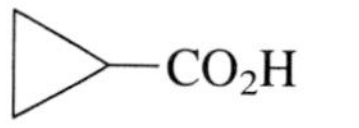

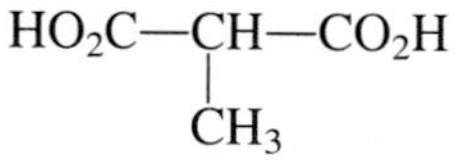

20.2 STRUCTURE AND PHYSICAL PROPERTIES OF CARBOXYLIC ACIDS

The data in Fig. 20.1 illustrate the fact that carbonyl groups in carboxylic acids have about the same bond lengths as carbonyl groups in aldehydes and ketones. We can also see from Fig. 20.1 that the C—O single-bond lengths of carboxylic acids are considerably smaller than those in alcohols or ethers (1.36 Å *vs.* 1.42 Å). One reason for this reduction in bond length is that the C—O bond in an acid is an sp^2-sp^3 single bond, whereas the C—O bond in an alcohol or ether is an sp^3-sp^3 single bond (Sec. 4.1A). Another reason is that carboxylic acids have a resonance structure in which this bond has some double bond character.

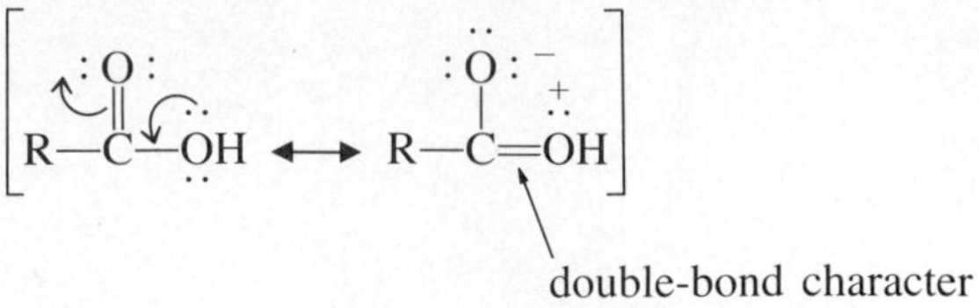

$$(20.2)$$

double-bond character

The carboxylic acids of lower molecular weight are high-boiling liquids with acrid, piercing odors. In fact, they have considerably higher boiling points than many other organic compounds of about the same molecular weight and shape:

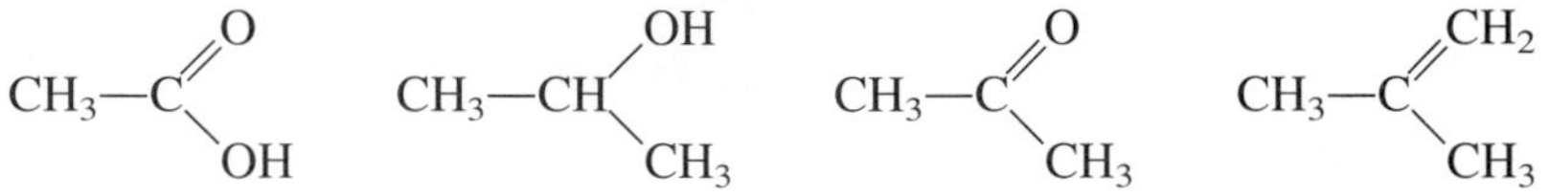

$CH_3-C(=O)OH$	$CH_3-CH(OH)CH_3$	$CH_3-C(=O)CH_3$	$CH_3-C(=CH_2)CH_3$
boiling point 117.9°	82.3°	56.5°	−6.9°

The high boiling points of carboxylic acids can be attributed not only to their polarity, but also to the fact that they form very strong hydrogen bonds. In the solid state, and under some conditions in both the gas phase and solution, carboxylic acids exist as hydrogen-bonded dimers.

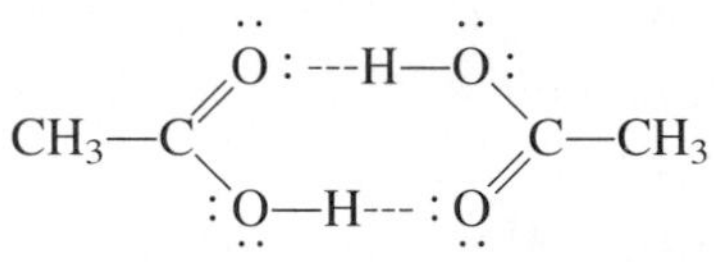

acetic acid dimer

The equilibrium constants for the formation of such dimers in solution are on the order of 10^6 to 10^7 M^{-1}. (The equilibrium constant for dimerization of ethanol, in contrast, is 11 M^{-1}.)

Many aromatic and dicarboxylic acids are solids. For example, the melting points of benzoic acid and succinic acid are 122° and 188°, respectively.

The simpler carboxylic acids have substantial solubilities in water, as we would expect from their hydrogen-bonding capabilities; the unbranched carboxylic acids below pentanoic acid are miscible with water. Many dicarboxylic acids also have significant water solubilities.

Problem

3 At a given concentration of acetic acid, in which solvent would you expect the amount of acetic acid dimer to be greater: CCl_4 or water? Explain.

20.3 SPECTROSCOPY OF CARBOXYLIC ACIDS

A. IR Spectroscopy

There are two important bands in the infrared spectrum of a typical carboxylic acid. One is the C=O stretching absorption, which occurs near 1710 cm^{-1} for carboxylic acid dimers. (The IR spectra of carboxylic acids are nearly always run under conditions such that they are in the dimer form. The carbonyl absorptions of carboxylic acid monomers occur near 1760 cm^{-1}, but are rarely observed.) The other important carboxylic acid absorption is the O—H stretching absorption. This absorption is much

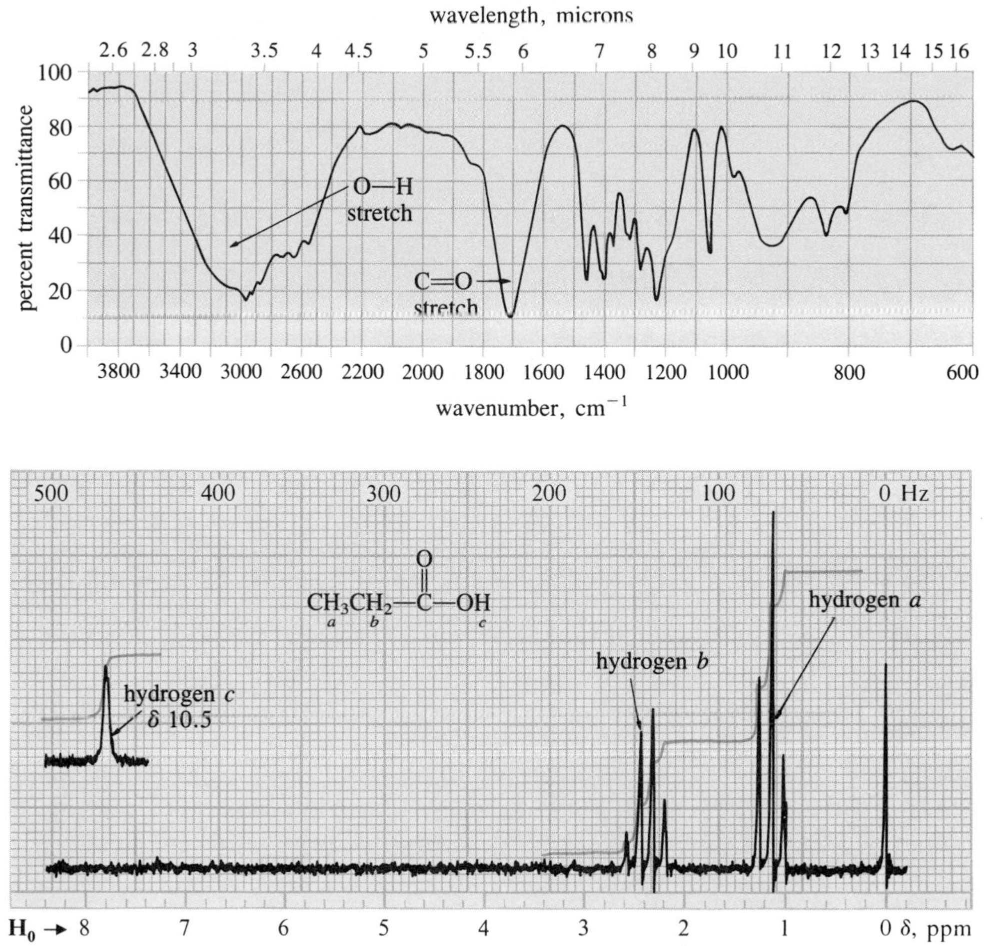

Figure 20.2 IR and NMR spectra of propanoic acid.

broader than the O—H stretching absorption of an alcohol or phenol, and covers a very wide region of the spectrum—typically 2400–3600 cm^{-1}. (In many cases this absorption obliterates the C—H stretching absorption of the acid.) A carbonyl band and this broad O—H stretching band, illustrated in the IR spectrum of propanoic acid (Fig. 20.2), are the hallmarks of a carboxylic acid. A conjugated carbon–carbon double bond affects the position of the carbonyl absorption much less in acids than it does in aldehydes and ketones. A substantial shift in the carbonyl absorption is observed, however, for acids in which the carboxy group is on an aromatic ring. Benzoic acid, for example, has a carbonyl absorption at 1680 cm^{-1}.

B. NMR Spectroscopy

The α-protons of carboxylic acids, like those of aldehydes and ketones, show NMR absorptions in the δ 2–2.5 chemical shift region. The O—H proton resonances of carboxylic acids occur at positions that depend on the acidity of the acid and its concentration. Typically, the carboxylic acid O—H proton signal is found far downfield, in the δ 9–13 region, and in many cases it is broad. It is readily distinguished

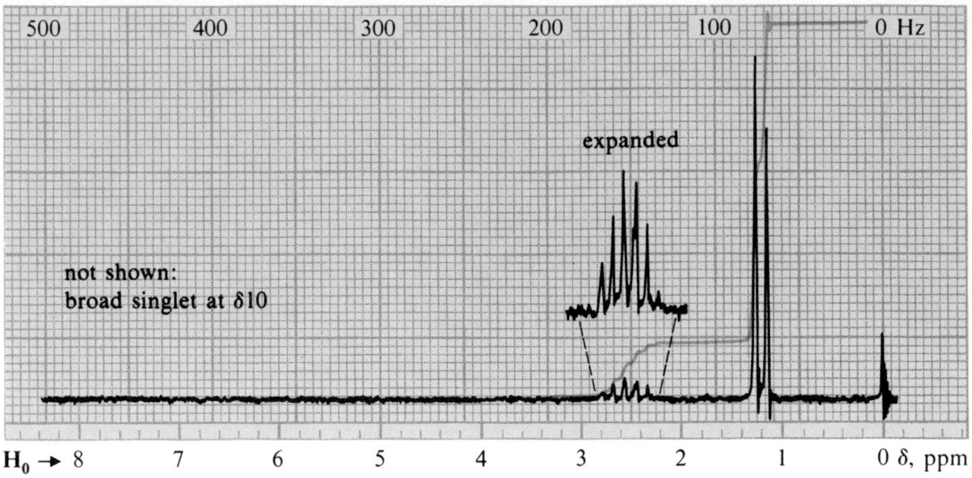

Figure 20.3 NMR spectrum for Problem 4.

from an aldehydic proton because the acid proton, like an alcohol O—H proton, rapidly exchanges with D_2O. The NMR spectrum of propanoic acid is also shown in Fig. 20.2.

Problem

4 Give the structure of the compound with the NMR spectrum shown in Fig. 20.3. The molecular weight of this compound is 88, and it has IR absorptions at 2600–3400 cm^{-1} (broad) and 1720 cm^{-1}.

20.4 ACIDITY AND BASICITY OF CARBOXYLIC ACIDS

A. Acidity of Carboxylic and Sulfonic Acids

The O—H proton of a carboxylic acid is acidic.

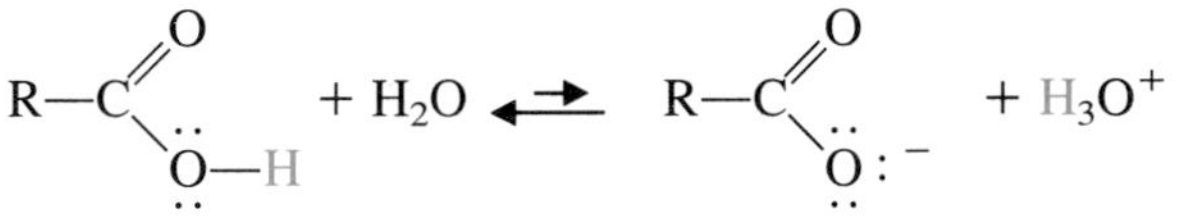

carboxylate ion

(20.3)

The conjugate bases of carboxylic acids are called generally **carboxylate ions.** Carboxylate salts are named by replacing the *ic* in the name of the acid (in any system of nomenclature) with the suffix *ate.*

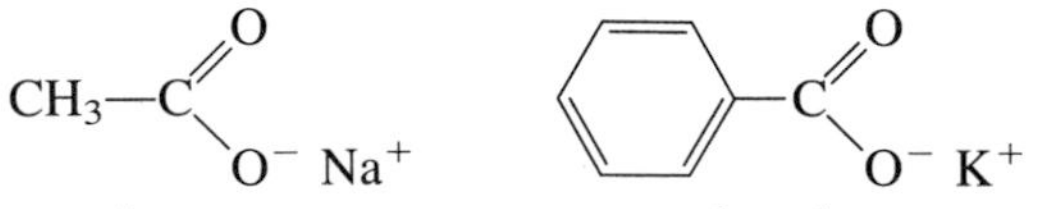

sodium acetate (acet~~ic~~ + *ate* = acetate)   potassium benzoate

Carboxylic acids are among the most acidic organic compounds; acetic acid, for example, has a pK_a of 4.67. This pK_a is low enough that an aqueous solution of acetic acid gives an acid reaction with litmus or pH paper.

Carboxylic acids are more acidic than alcohols or phenols, other compounds with O—H bonds.

$$\text{——— increasing acidity ———}\rightarrow$$

	C_2H_5—O—H	—O—H	CH_3—C—O—H
pK_a	15.9	9.95	4.67

The acidity of carboxylic acids is due to resonance stabilization of their conjugate-base carboxylate ions.

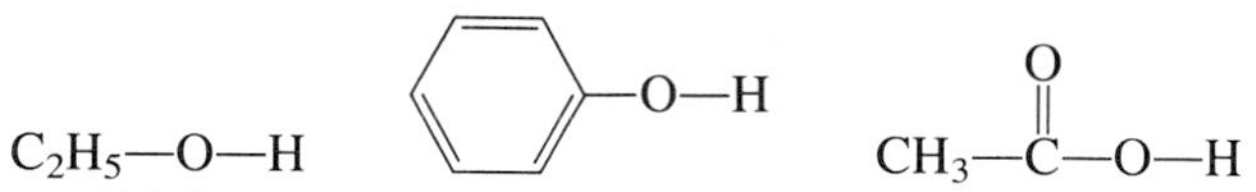

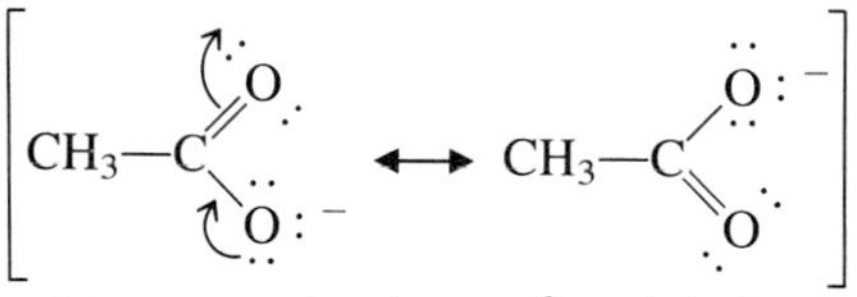

$$\left[CH_3-C\overset{\overset{\ddot{O}}{\|}}{\underset{\ddot{O}:^-}{}} \longleftrightarrow CH_3-C\overset{\ddot{O}:^-}{\underset{\ddot{O}}{}} \right] \qquad (20.4)$$

resonance structures of acetate ion

Since the conjugate bases of phenols (phenoxide anions) are also resonance stabilized (Sec. 18.6A), we might ask why carboxylic acids are much more acidic than phenols. The reason is that resonance stabilization of carboxylate anions is more important than resonance stabilization of phenoxide anions. In a carboxylate anion, charge is delocalized onto another oxygen, an electronegative atom. In contrast, charge in a phenoxide anion is delocalized onto carbon, a relatively electropositive atom.

Although typical carboxylic acids have pK_a values in the 4–5 range, the acidities of carboxylic acids vary with structure. As is the case with alcohols and phenols, the acidity of carboxylic acids is enhanced by the presence of electronegative groups near the carboxy group. This is another example of an *inductive effect* (Sec. 8.6B).

	CH_3—CO_2H	FCH_2—CO_2H	F_2CH—CO_2H	F_3C—CO_2H	(20.5)
pK_a	4.76	2.66	1.24	0.23	

Trifluoroacetic acid, the strongest acid in the examples above, is almost as strong as some mineral acids. Electronegative groups farther removed from the carboxy group have less of an effect on acidity.

$$CH_3CH_2\underset{\underset{Cl}{|}}{C}HCO_2H \qquad CH_3\underset{\underset{Cl}{|}}{C}HCH_2CO_2H \qquad \underset{\underset{Cl}{|}}{C}H_2CH_2CH_2CO_2H \qquad CH_3CH_2CH_2CO_2H \qquad (20.6)$$

pK_a	2.84	4.06	4.52	4.82

The pK_a values of some carboxylic acids are given in Table 20.2, and the pK_a values of the simple dicarboxylic acids in Table 20.3. The data in these tables give some idea of the range over which the acidities of carboxylic acids vary.

TABLE 20.2 pK_a Values of Some Carboxylic Acids

Acid	pK_a
formic	3.75
acetic	4.76
propionic	4.87
2,2-dimethylpropanoic (pivalic)	5.05
acrylic	4.26
chloroacetic	2.85
phenylacetic	4.31
benzoic	4.18
p-methylbenzoic (*p*-toluic)	4.37
p-nitrobenzoic	3.43
p-chlorobenzoic	3.98
p-methoxybenzoic (*p*-anisic)	4.47
2,4,6-trinitrobenzoic	0.65

TABLE 20.3 pK_a Values of Some Dicarboxylic Acids

Acid	First pK_a	Second pK_a
carbonic	3.58	6.35
oxalic	1.27	4.27
malonic	2.86	5.70
succinic	4.21	5.64
glutaric	4.34	5.27
adipic	4.41	5.28
phthalic	2.95	5.41

Sulfonic acids are much stronger than comparably substituted carboxylic acids.

p-toluenesulfonic acid
(TsOH, or tosic acid)
a strong acid; $pK_a \approx -1$

One reason that sulfonic acids are more acidic than carboxylic acids is the high oxidation state of sulfur. If we draw the octet structure for a sulfonate anion, we can see that sulfur has considerable positive charge. This positive charge stabilizes the negative charge on the oxygens.

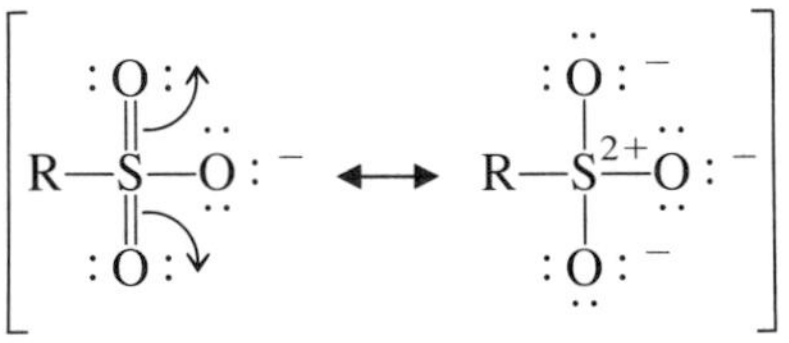

octet structure

Sulfonic acids are almost as strong as mineral acids, and are useful as acid catalysts in organic solvents because they are more soluble than most mineral acids. For example, *p*-toluenesulfonic acid is moderately soluble in benzene, and can be used as a strong acid catalyst in that solvent. (Sulfuric acid, in contrast, is completely insoluble in benzene.)

Many carboxylic acids of moderate molecular weight are not soluble in water. Their alkali metal salts, however, are ionic compounds, and in many cases are much more soluble in water. Therefore many water-insoluble carboxylic acids dissolve in solutions of alkali metal hydroxides (NaOH, KOH) because the insoluble acids are converted completely into their soluble salts.

$$\underset{}{R-\overset{O}{\overset{\|}{C}}-\ddot{O}H} + NaOH \longrightarrow R-\overset{O}{\overset{\|}{C}}-\ddot{O}:^- Na^+ + H_2O \qquad (20.7)$$

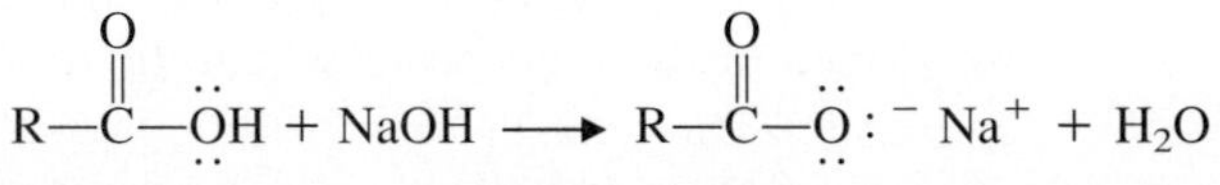

Even 5% sodium bicarbonate ($NaHCO_3$) solution is basic enough (pH ≈ 8.5) to dissolve a carboxylic acid. We can understand this point by examining the equilibrium expression for ionization of a carboxylic acid RCO_2H with a dissociation constant K_a.

$$K_a = \frac{[RCO_2^-][H_3O^+]}{[RCO_2H]} \tag{20.8a}$$

or

$$\frac{K_a}{[H_3O^+]} = \frac{[RCO_2^-]}{[RCO_2H]} \tag{20.8b}$$

In order for the carboxylic acid to dissolve in water, it has to be mostly ionized and in its conjugate base form RCO_2^-; that is, in Eq. 20.8b, the ratio $[RCO_2^-]/[RCO_2H]$ has to be *large*. As a practical matter, we can say that when this ratio is 100 or greater, the acid has been completely converted into its anion. (Of course, there is no pH at which the acid exists *completely* as its anion; but when this ratio is ≥100, the concentration of acid becomes negligible.) Since K_a is a constant, Eq. 20.8b shows that this ratio can be made large by making the hydrogen-ion concentration $[H_3O^+]$ small. In other words, $[H_3O^+]$ *must be small in comparison with the* K_a *of the acid.* If we take negative logarithms of Eq. 20.8b, we obtain the following result:

$$pH - pK_a = \log \frac{[RCO_2^-]}{[RCO_2H]} \tag{20.8c}$$

If $[RCO_2^-]/[RCO_2H]$ is ≥100, then the logarithm of this ratio is ≥2. Equation 20.8c shows that, for conversion of an acid into its anion, the pH of the solution must be two or more units greater than the pK_a of the acid. Since 5% sodium bicarbonate solution has a pH of about 8.5, and most carboxylic acids have pK_a values in the range 4–5, sodium bicarbonate solution is more than basic enough to dissolve a typical acid, provided of course that there is enough bicarbonate present that it is not completely consumed by reaction with the acid.

A typical carboxylic acid, then, can be separated from mixtures with other water-insoluble, nonacidic substances by extraction with NaOH, Na_2CO_3, or $NaHCO_3$ solution. The acid dissolves in the basic aqueous solution, but nonacidic compounds do not. After separating the basic aqueous solution, it can be acidified with mineral acid to yield the carboxylic acid, which may be isolated by filtration or extraction with organic solvents. (A similar idea was used in the separation of phenols; Sec. 18.6B.) Carboxylic acids may also be separated from phenols by extraction with 5% $NaHCO_3$ if the phenol is not one that is unusually acidic. Since the pK_a of a typical phenol is about 10, it remains largely un-ionized and thus insoluble in a solution with a pH of 8.5. This point follows from an equation for phenol ionization analogous to Eq. 20.8c. As we have just learned, carboxylic acids dissolve in this solution as their sodium salts.

Problems

5 (a) Write out the equations for the first and second ionizations of malonic acid and label each with the appropriate pK_a value from Table 20.3.

(b) Why is it that the first pK_a of malonic acid is much *lower* than that of acetic acid, but the second pK_a of malonic acid is much *higher* than that of acetic acid?

(c) Explain why the first and second pK_a values of the dicarboxylic acids become closer together as the lengths of their carbon chains increase (Table 20.3).

6 You have just carried out a conversion of *p*-bromotoluene into *p*-bromobenzoic acid and wish to separate the product from unreacted starting material. Design a separation of these two substances that would enable you to isolate the purified acid.

B. Basicity of Carboxylic Acids

The carbonyl group of a carboxylic acid, like that of an aldehyde or ketone, is weakly basic.

$$R-C\overset{O}{\underset{OH}{\diagdown}} + H_3O^+ \rightleftharpoons \left[R-C\overset{\overset{+}{O}-H}{\underset{O-H}{\diagup}} \longleftrightarrow R-\overset{+}{C}\overset{O-H}{\underset{O-H}{\diagdown}} \longleftrightarrow R-C\overset{O-H}{\underset{\overset{+}{O}-H}{\diagdown}} \right] + H_2O \quad (20.9)$$

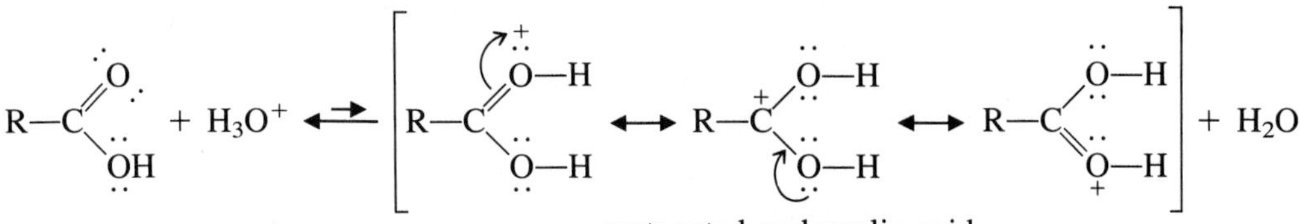

protonated carboxylic acid
$pK_a \approx -6$

The basicity of carboxylic acids plays a very important role in many of their reactions.
Protonation of an acid on the carbonyl oxygen occurs because, as Eq. 20.9 shows, a resonance-stabilized cation is formed. Protonation on the hydroxy oxygen is much less favorable because it does not give a resonance-stabilized cation.

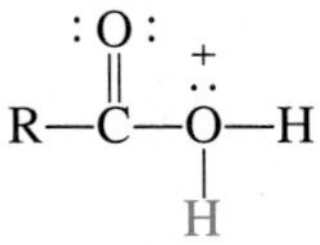

$$\underset{\substack{\overset{\displaystyle H}{|} \\ \text{not resonance-} \\ \text{stabilized}}}{R-\overset{\overset{\displaystyle :O:}{\|}}{C}-\overset{+}{O}-H} \qquad \underset{\substack{\text{resonance-stabilized} \\ \text{(Eq. 20.9)}}}{R-\overset{\overset{+}{O}-H}{\|}{C}-\ddot{O}H}$$

20.5 FATTY ACIDS, SOAPS, AND DETERGENTS

Carboxylic acids with long, unbranched carbon chains are called **fatty acids** because many of them are liberated from fats and oils by a hydrolytic process called *saponification* (Sec. 21.7). Some fatty acids contain carbon–carbon double bonds. Fatty acids with *cis*-double bonds occur widely in nature, but those with *trans*-double bonds are rare. The following compounds are examples of common fatty acids:

$$n\text{-}C_{15}H_{31}CO_2H \quad \text{or} \quad CH_3CH_2CH_2CH_2CH_2CH_2CH_2CH_2CH_2CH_2CH_2CH_2CH_2CH_2CH_2CO_2H$$
palmitic acid (from palm oil)

$$n\text{-}C_{17}H_{35}CO_2H$$
stearic acid
(Greek, *stear* = tallow, or beef fat)

$$\underset{H}{\overset{n\text{-}C_8H_{17}}{\diagdown}}C=C\underset{H}{\overset{(CH_2)_7CO_2H}{\diagup}}$$
oleic acid

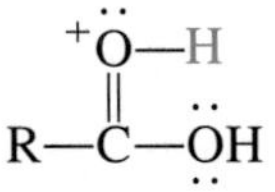

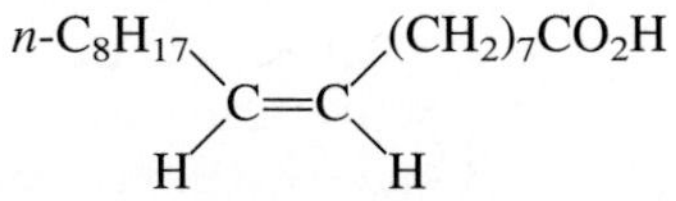

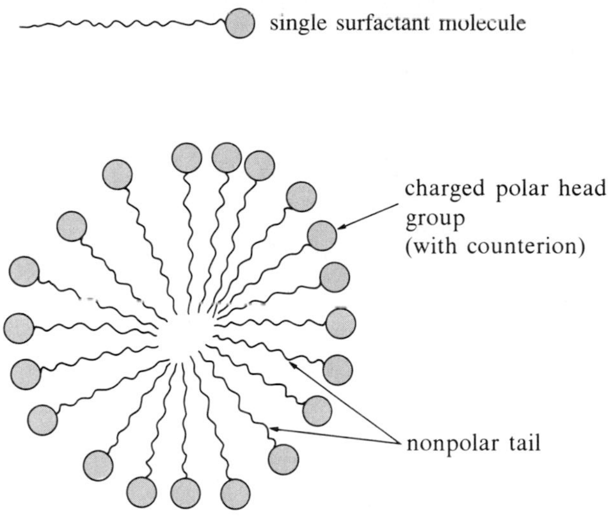

Figure 20.4
Schematic diagram of micelle structure. Each micelle contains 50 to 150 molecules and is approximately spherical.

single surfactant molecule

charged polar head group (with counterion)

nonpolar tail

The sodium and potassium salts of fatty acids, called **soaps,** are the major ingredients of commercial soap.

$$n\text{-}C_{17}H_{35}\overset{\overset{\displaystyle O}{\|}}{-C}-O^-\ Na^+ \qquad \text{sodium stearate: a soap}$$

Closely related to soaps are synthetic detergents. The following compound, the sodium salt of a sulfonic acid, is used in laundry detergents:

$$Na^+\ ^-O\overset{\overset{\displaystyle O}{\|}}{\underset{\underset{\displaystyle O}{\|}}{-S}}-\!\!\!\!\!\!\bigcirc\!\!\!\!\!\!-(CH_2)_{11}CH_3$$

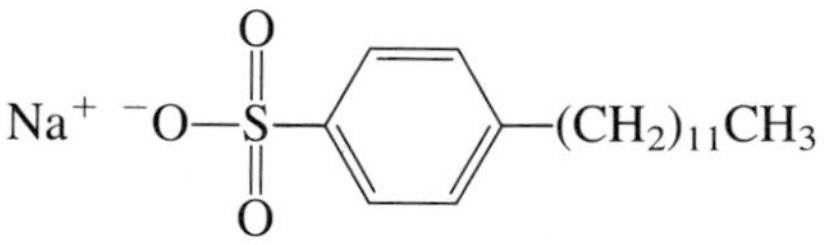

Many soaps and detergents have not only cleansing properties, but also germicidal characteristics.

Soaps and detergents are two examples of a larger class of molecules known as **surfactants.** Surfactants are molecules with two structural parts that interact with water in opposing ways: a *polar head group,* which energetically prefers to be solvated by water, and a *hydrocarbon tail,* which, like a long alkane, is not well solvated by water. In a soap, the polar head group is the carboxylate anion, and the hydrocarbon tail is obviously the carbon chain. The soap and the detergent shown above are examples of *anionic surfactants:* surfactants with an anionic polar head group. *Cationic surfactants* are also known:

$$PhCH_2\overset{\overset{\displaystyle CH_3}{|}}{\underset{\underset{\displaystyle CH_3}{|}}{-N^+}}-n\text{-}C_{16}H_{33}\quad Cl^-$$

benzylcetyldimethylammonium chloride (benzalkonium chloride, a cationic surfactant and germicide)

Surfactants in aqueous solution spontaneously form **micelles,** which are approximately spherical aggregates of 50–150 surfactant molecules (Fig. 20.4). We can think

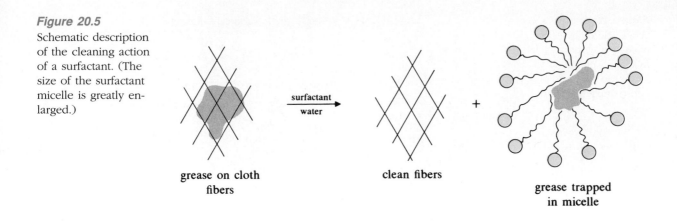

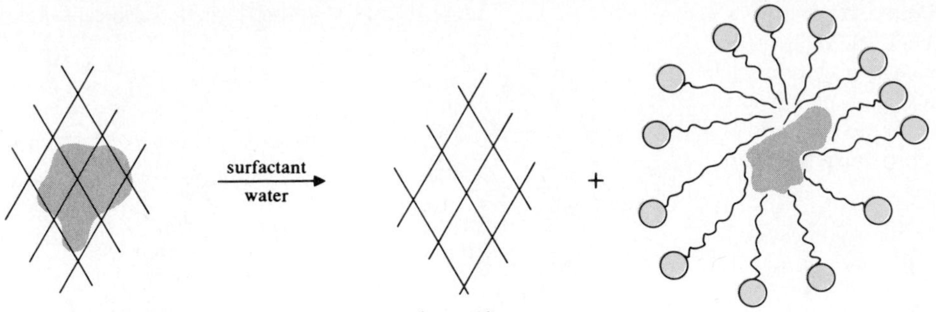

Figure 20.5
Schematic description of the cleaning action of a surfactant. (The size of the surfactant micelle is greatly enlarged.)

of a micelle as a large ball in which the polar head groups are exposed on the outside and the nonpolar tails are buried on the inside. The micellar structure satisfies the solvation requirements of both the polar head groups, which are close to water, and the "greasy groups"—the nonpolar tails—which associate with each other on the inside of the micelle. (Recall "like-dissolves-like;" Sec. 8.4B.)

The detergent properties of surfactants are easy to understand once the rationale for the formation of micelles is clear (Fig. 20.5). When a fabric with greasy dirt is exposed to an aqueous solution containing micelles of a soap or detergent, the dirt associates with the "greasy" hydrocarbon chains on the interior of the micelle and is incorporated into the micellar aggregate. Then the dirt is lifted away from the surface of the fabric and carried into solution. The antiseptic action of some surfactants owes its success to a similar phenomenon. A cell membrane—the envelope that surrounds the contents of the cell—is made up of molecules, called *phospholipids,* that are also surfactants (Sec. 21.12B). When the bacterial cell is exposed to a solution containing a surfactant, phospholipids of the cell membrane tend to associate with the surfactant. In some cases this disrupts the membrane enough that the cell can no longer function, and it dies.

The term *surfactant* means "surface-active compound." The name is derived from the fact that surfactants lower the surface tension of an aqueous solution. (This is why a soap solution feels slippery to the touch.)

So-called *hard water* disrupts the cleaning action of detergents, and causes the formation of a scum when mixed with soaps. Hard water contains Ca^{2+} and Mg^{2+} ions. Hard-water scum ("bathtub ring") is a precipitate of the calcium or magnesium salts of fatty acids, which (unlike the sodium salts) do not form micelles and are insoluble in water. These offending ions can be solubilized and removed by complexation with phosphates. Phosphates, however, have been found to cause excessive growth of algae in rivers and streams, and their use has been curtailed (thus, "low-phosphate detergents"). Unfortunately, no completely acceptable substitute for phosphates has yet been found.

20.6 SYNTHESIS OF CARBOXYLIC ACIDS

One useful synthesis of carboxylic acids is closely related to a reaction of aldehydes and ketones that we have already studied: addition of a Grignard reagent to a carbonyl group (Sec. 19.9). Reaction of a Grignard reagent with carbon dioxide, followed by protonolysis, gives a carboxylic acid.

$$
\underset{\overset{|}{CH_3}}{CH_3CH_2CHMgBr} + CO_2 \longrightarrow \xrightarrow{H_3O^+} \underset{\underset{(76-86\%\ \text{yield})}{}}{CH_3CH_2\overset{\overset{CH_3}{|}}{CH}-\overset{\overset{O}{\|}}{C}-OH} \qquad (20.10)
$$

Carbon dioxide, of course, is a carbonyl compound. *Addition* of the Grignard reagent to carbon dioxide gives the bromomagnesium salt of a carboxylic acid. When aqueous acid is added to the reaction mixture in a separate reaction step, the free carboxylic acid is formed.

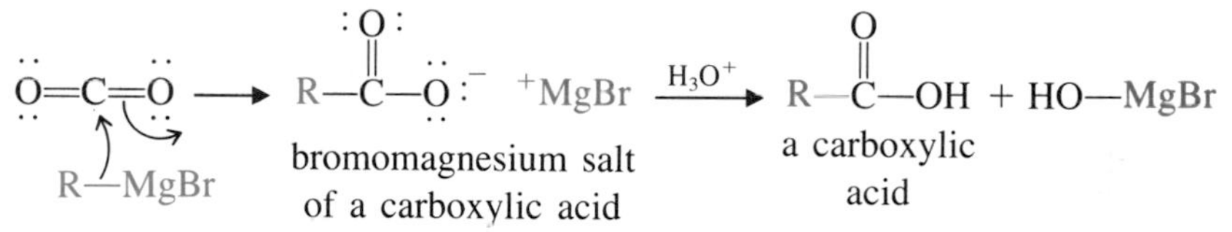

$$
\overset{..}{\underset{..}{O}}=C=\overset{..}{\underset{..}{O}} \longrightarrow R-\overset{\overset{:\overset{..}{O}:}{\|}}{C}-\overset{..}{\underset{..}{O}}:^- \; ^+MgBr \xrightarrow{H_3O^+} R-\overset{\overset{O}{\|}}{C}-OH + HO-MgBr \qquad (20.11)
$$

$$
\underset{R\!-\!MgBr}{\uparrow}
$$

bromomagnesium salt of a carboxylic acid · · · a carboxylic acid

The reaction of Grignard reagents with CO_2 is another method for the formation of carbon–carbon bonds. The methods we now know for carbon–carbon bond formation are:

1. Cyclopropane formation by carbene addition to alkenes (Sec. 9.7)

2. Reaction of Grignard reagents with ethylene oxide (Sec. 11.4B)

3. Reaction of acetylide ions with alkyl halides and sulfonates (Sec. 14.7B)

4. Diels–Alder reactions (Sec. 15.3)

5. Friedel–Crafts reactions (Secs. 16.4E and 16.4F)

6. Cyanohydrin formation (Sec. 19.7)

7. Reaction of Grignard reagents with aldehydes and ketones (Sec. 19.9)

8. The Wittig alkene synthesis (Sec. 19.13)

9. Reaction of Grignard reagents with carbon dioxide (this section)

Notice how many of these methods involve Grignard reagents. Perhaps it is becoming clear why Grignard reagents are very important in organic chemistry.

Let us also summarize the methods we have learned for preparing carboxylic acids:

1. Oxidation of primary alcohols and aldehydes (Secs. 10.6 and 19.14)

2. Side-chain oxidation of alkylbenzenes (Sec. 17.5)

3. Reaction of Grignard reagents with CO_2 (this section)

Ozonolysis of alkenes, followed by oxidative workup with H_2O_2 (Sec. 5.5), is a fourth method for preparing carboxylic acids, but it is less important than the others because it involves *breaking* the carbon–carbon double bond and therefore wastes the effort involved in assembling the carbon–carbon bonds of the alkene starting material. Nevertheless, the method is occasionally used for the preparation of carboxylic acids that are not easily accessible in other ways.

We shall learn a number of other important syntheses of carboxylic acids in Chapters 21 and 22.

Problem	**7** Outline a synthesis of cyclopentanecarboxylic acid from cyclopentanol.

20.7 INTRODUCTION TO CARBOXYLIC ACID REACTIONS

The reactions of carboxylic acids can be categorized into four types:

1. Carbonyl-group reactions

2. Reactions at the oxygen of the —OH group. (We shall refer to this oxygen as the *carboxylate oxygen,* to differentiate it from the carbonyl oxygen.)

3. Decarboxylation

4. Reactions involving the α-carbon

The most typical *carbonyl-group reaction* of carboxylic acids and their derivatives is *substitution at the carbonyl group.* For reaction of a carboxylic acid with a general reagent E—Y, we can represent this type of reaction generally in the following way:

$$R-\overset{\overset{\text{O}}{\|}}{C}-OH + E-Y \;\rightleftharpoons\; R-\overset{\overset{\text{O}}{\|}}{C}-Y + HO-E \qquad (20.12)$$

That is, the —OH of the carboxy group is substituted by a group Y.

Another carbonyl-group reaction of carboxylic acids and their derivatives is the reaction of the carbonyl oxygen with electrophiles (Lewis acids) or the protons of Brønsted acids—that is, the reaction of the carbonyl oxygen as a *base:*

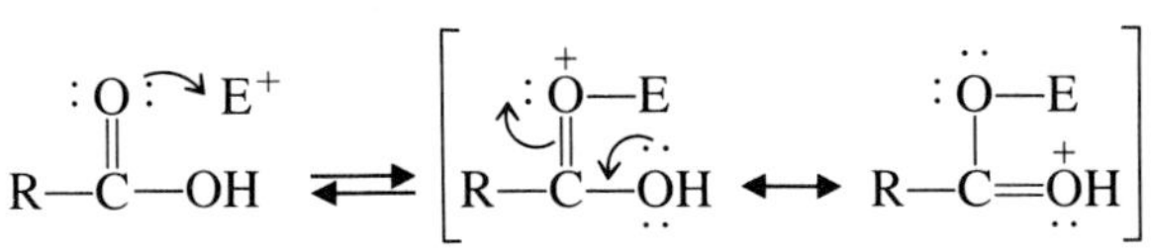

$$\overset{:\text{O}:\;E^+}{R-\overset{\|}{C}-OH} \;\rightleftharpoons\; \left[R-\overset{\overset{+}{\overset{..}{O}}-E}{\underset{..}{\overset{\|}{C}}\text{OH}} \;\longleftrightarrow\; R-\overset{:\overset{..}{O}-E}{\underset{..}{\overset{\|}{C}}=\overset{+}{O}H} \right] \qquad (E^+ = \text{electrophile}) \quad (20.13)$$

This reaction was discussed in Sec. 20.4B. We shall find that many substitution reactions at the carbonyl group are acid catalyzed, and that carbonyl basicity plays an important mechanistic role in these processes.

We have already studied one *reaction at the carboxylate oxygen:* the ionization of carboxylic acids (Sec. 20.4A):

$$R-\overset{\overset{:\text{O}:}{\|}}{C}-\overset{..}{O}-H + H_2\overset{..}{O}: \;\rightleftharpoons\; \left[R-\overset{\overset{:\text{O}:}{\|}}{C}-\overset{..}{O}:^- \;\longleftrightarrow\; R-\overset{:\overset{..}{O}:^-}{\underset{..}{\overset{|}{C}}=\overset{..}{O}:} \right] + H_3\overset{+}{O}: \quad (20.14)$$

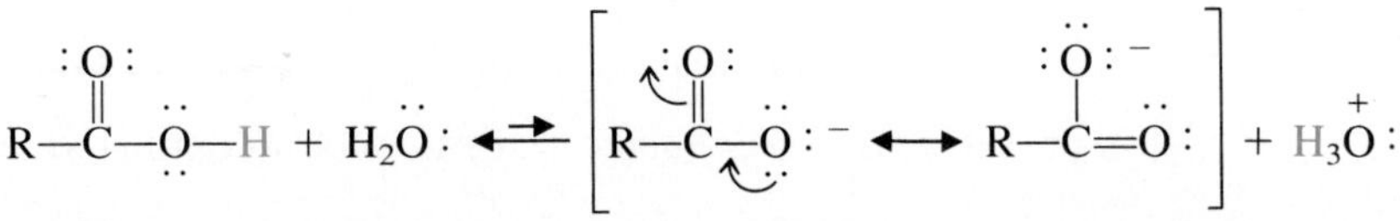

Another general reaction involves reaction of the carboxylate oxygen as a nucleophile (X:⁻ = halide, sulfonate ester, or other leaving group).

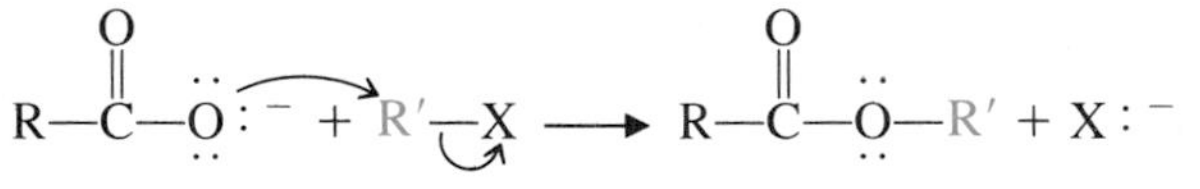

$$R-\overset{\overset{\textstyle O}{\|}}{C}-\overset{..}{\underset{..}{O}}:^- + R'-X \longrightarrow R-\overset{\overset{\textstyle O}{\|}}{C}-\overset{..}{\underset{..}{O}}-R' + X:^- \qquad (20.15)$$

Decarboxylation is loss of the carboxy group as CO_2:

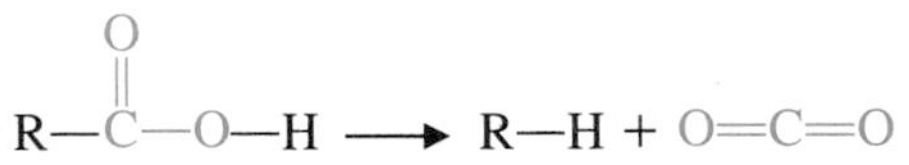

$$R-\overset{\overset{\textstyle O}{\|}}{C}-O-H \longrightarrow R-H + O{=}C{=}O \qquad (20.16)$$

This reaction is more important for some types of carboxylic acids than for others, as we shall learn in Sec. 20.11.

In this chapter, we shall concentrate on the first three types of reactions—carbonyl-group reactions, reactions at the carboxylate oxygen, and decarboxylation. In Chapter 21, we shall consider for the most part carbonyl-group reactions of carboxylic acid derivatives. In Chapter 22, we shall return not only to carboxylic acids, but also to aldehydes and ketones, when we take up the fourth type of reaction: reactions involving the α-carbon.

20.8 CONVERSION OF CARBOXYLIC ACIDS INTO ESTERS

A. Acid-Catalyzed Esterification

Esters are carboxylic acid derivatives with the following general structure:

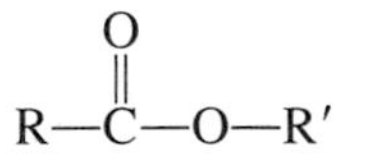

$$R-\overset{\overset{\textstyle O}{\|}}{C}-O-R' \qquad \text{(R and R' can be alkyl or aryl)}$$

When a carboxylic acid is treated with a large excess of an alcohol in the presence of a strong acid catalyst, an ester is formed.

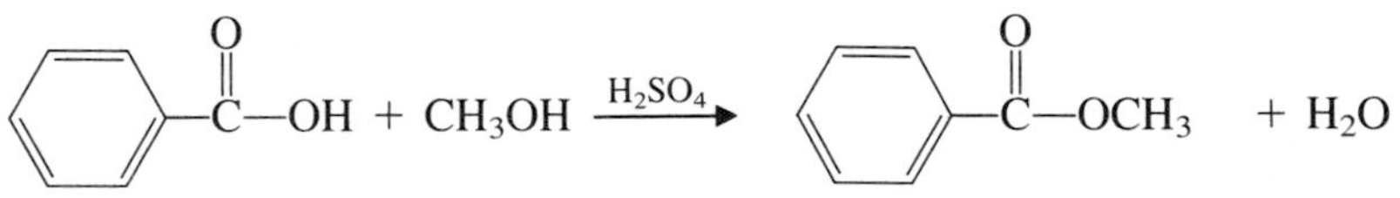

$$+ H_2O \quad (20.17)$$

methyl benzoate
(an ester; 85–95% yield)

This reaction is called **acid-catalyzed esterification,** or sometimes *Fischer esterification,* after the great German chemist Emil Fischer (1852–1919), about whom we shall learn more in Chapter 27.

The equilibrium constants for esterifications with most primary alcohols are near unity; for example, the equilibrium constant for the esterification of acetic acid with ethyl alcohol is 3.38. The reaction is driven to completion by applying *LeChatelier's principle.* The equilibrium is driven toward the ester by using a large excess of alcohol as the solvent and by using sulfuric acid as the catalyst. Besides its catalytic effect, sulfuric acid has strong dehydrating properties that help to remove the by-product water.

Acid-catalyzed esterification *cannot* be applied to the synthesis of esters from phenols or tertiary alcohols. Tertiary alcohols undergo other reactions under the acidic conditions of the reaction (can you think of one?), and the equilibrium constants for esterification of phenols are much less favorable than those for esterification of alcohols by a factor of about 10^4. Although it is possible in principle to drive the esterification of phenols to completion, there are simpler ways for preparing esters of both phenols and tertiary alcohols that we shall consider in Chapter 21.

Before writing a mechanism for acid-catalyzed esterification, it is essential that we know whether the oxygen of the water liberated in the reaction comes from the alcohol or from the acid.

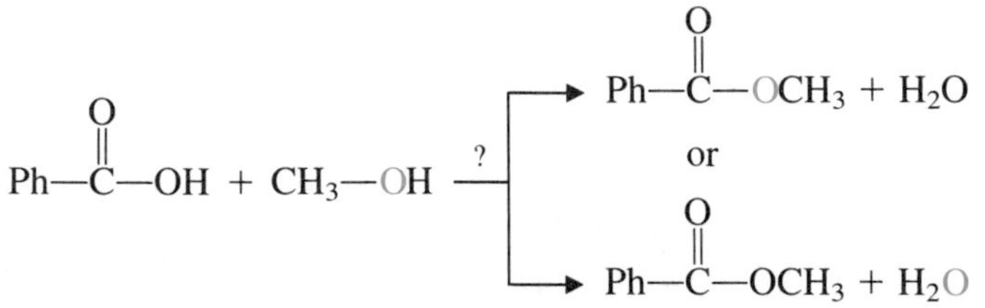

$$\text{Ph}\overset{\overset{\textstyle O}{\|}}{-\text{C}}-\text{OH} + \text{CH}_3-\text{OH} \xrightarrow{\ ?\ }
\begin{cases}
\text{Ph}\overset{\overset{\textstyle O}{\|}}{-\text{C}}-\text{OCH}_3 + \text{H}_2\text{O} & (20.18\text{a}) \\[2mm]
\text{or} \\[2mm]
\text{Ph}\overset{\overset{\textstyle O}{\|}}{-\text{C}}-\text{OCH}_3 + \text{H}_2\text{O} & (20.18\text{b})
\end{cases}$$

This question was answered in 1938, when it was found, using the ^{18}O isotope to label the alcohol oxygen, that the water produced comes exclusively from the carboxylic acid. Acid-catalyzed esterification is therefore a *substitution of —OH (as water) at the carbonyl group of the acid by the alcohol*(Eq. 20.18a). Thus, acid-catalyzed esterification is our first example of *substitution at a carbonyl group.*

The mechanism of acid-catalyzed esterification is very important, for it serves as a model for the mechanisms of other acid-catalyzed reactions of carboxylic acids and their derivatives. In this mechanism, the carboxylic acid is first protonated (Sec. 20.4B).

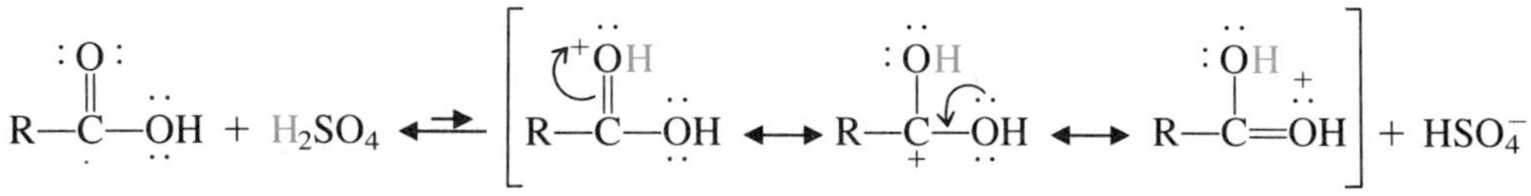

$$\text{R}\overset{\overset{\textstyle :O:}{\|}}{-\text{C}}-\overset{..}{\underset{..}{\text{O}}}\text{H} + \text{H}_2\text{SO}_4 \rightleftarrows \left[\ \text{R}\overset{\overset{\textstyle +\!\overset{..}{\text{O}}\text{H}}{\|}}{-\text{C}}-\overset{..}{\underset{..}{\text{O}}}\text{H} \longleftrightarrow \text{R}\overset{\overset{\textstyle :\overset{..}{\text{O}}\text{H}}{|}}{\underset{+}{-\text{C}}}-\overset{..}{\underset{..}{\text{O}}}\text{H} \longleftrightarrow \text{R}\overset{\overset{\textstyle :\overset{..}{\text{O}}\text{H}}{|}}{-\text{C}}=\overset{+}{\underset{..}{\text{O}}}\text{H}\ \right] + \text{HSO}_4^- \quad (20.19\text{a})$$

As the resonance structures show, the protonated intermediate has carbocation character. Alcohol attacks this "carbocation" to form, after loss of a proton, a **tetrahedral addition intermediate.** This reaction is essentially the same as the acid-catalyzed reaction of an alcohol with a protonated aldehyde or ketone to form a hemiacetal (Sec. 19.10A).

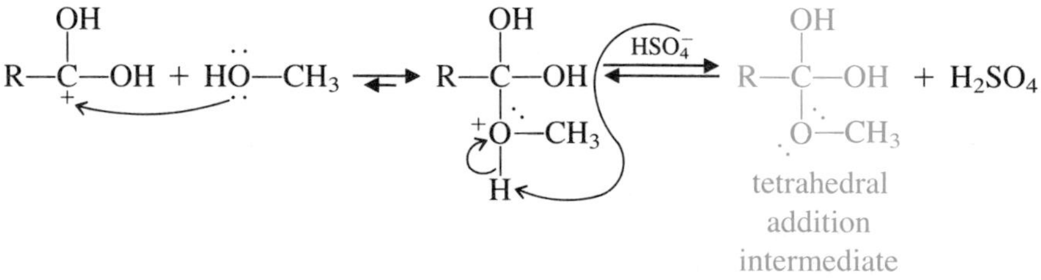

The tetrahedral addition intermediate reacts further by protonation, loss of water, and deprotonation to give the ester product.

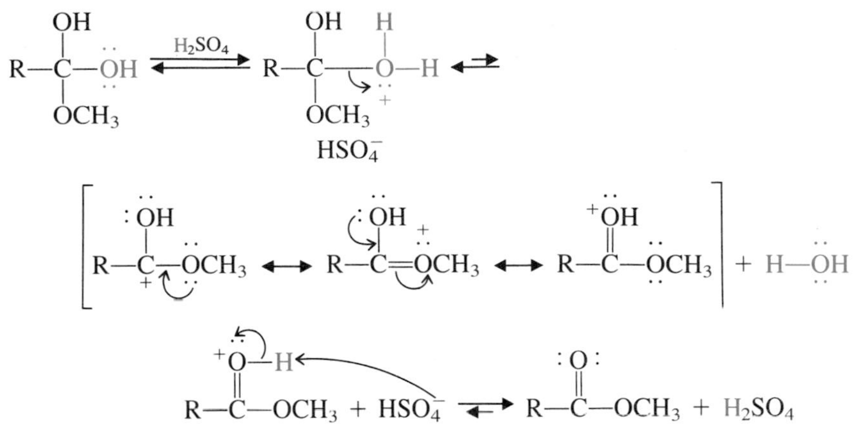

(20.19c)

(20.19d)

Tetrahedral intermediates in esterification reactions are in most cases too unstable to isolate and observe directly. How, then, do we know that such intermediates exist? Evidence for the formation of tetrahedral addition intermediates came from a very elegant isotopic labeling experiment carried out by Professor Myron Bender (now at Northwestern University) and his colleagues. This experiment actually involved a study of ester hydrolysis, the reverse of ester formation, using an ester labeled at the carbonyl oxygen atom with ^{18}O. The logic of the experiment uses the principle of microscopic reversibility (Sec. 10.1A): if a tetrahedral addition intermediate is involved in ester hydrolysis, it must also be involved in the reverse reaction, ester formation. The labeling experiment takes advantage of the fact that *the labeled oxygen of the ester and the unlabeled oxygen that originates from water become equivalent in the tetrahedral addition intermediate* ($O^* = {}^{18}O$):

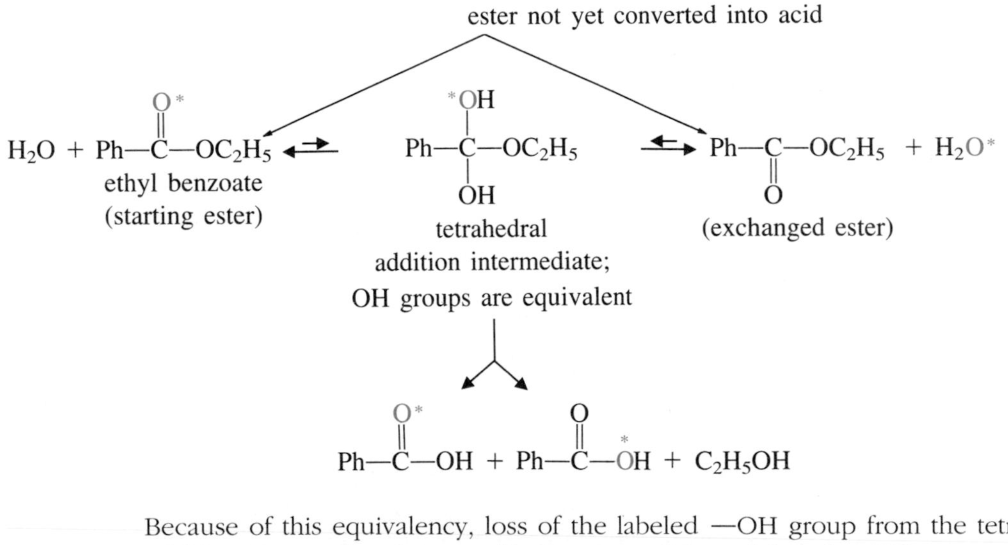

(20.20)

Because of this equivalency, loss of the labeled —OH group from the tetrahedral addition intermediate is just as likely as loss of the unlabeled —OH group when the intermediate reverts back to ester. Were the tetrahedral intermediate not formed, equivalency between the labeled and unlabeled —OH groups would not be established, and loss of ^{18}O from the starting ester could not occur. The experimental result was that unconverted ester recovered from the reaction mixture had lost ^{18}O, thus providing evidence for the existence of the tetrahedral addition intermediate.

Now we must recognize that the *existence* of this intermediate is a *necessary but not sufficient* condition for its involvement in the mechanism of esterification. Its existence proves its involvement in the mechanism no more than the presence of a man at the scene of a crime proves his responsibility for the crime; it is *circumstantial* evidence. Nevertheless, so much evidence consistent with the esterification mechanism in Eq. 20.19 has been gathered that this mechanism is generally accepted.

Esterification is one of many substitution reactions of carboxylic acids and their derivatives that involve tetrahedral addition intermediates. *The mechanisms of these reactions are extensions of the mechanisms of carbonyl addition.* In a carbonyl-substitution reaction, a nucleophile attacks the carbonyl carbon of a carboxylic acid derivative in exactly the same way that it attacks an aldehyde or ketone (see Fig. 19.8), and an addition compound is formed. Subsequently, a group —X from the original acid derivative (—X = —OH in a carboxylic acid itself) is expelled as a leaving group from the addition intermediate to give a new carboxylic acid derivative.

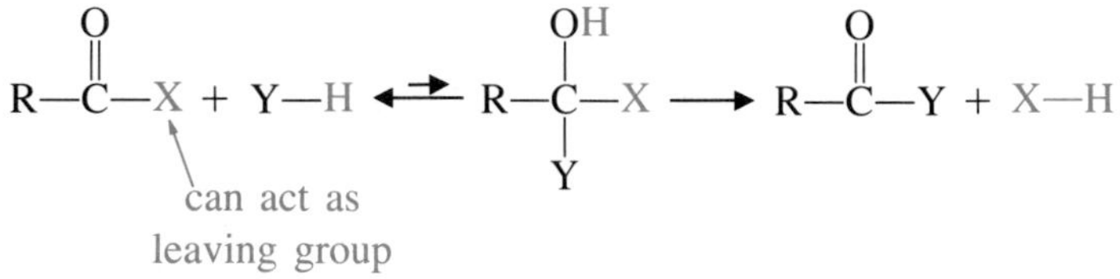

$$(20.21a)$$

In other words, substitution at a carbonyl group is really a sequence of two processes: *addition* to the carbonyl group followed by *elimination* to regenerate the carbonyl group.

Why don't aldehydes and ketones undergo substitution at their carbonyl groups? When a nucleophile attacks the carbonyl group of an aldehyde or ketone, *neither of the groups originally attached to the carbonyl can act as a leaving group.*

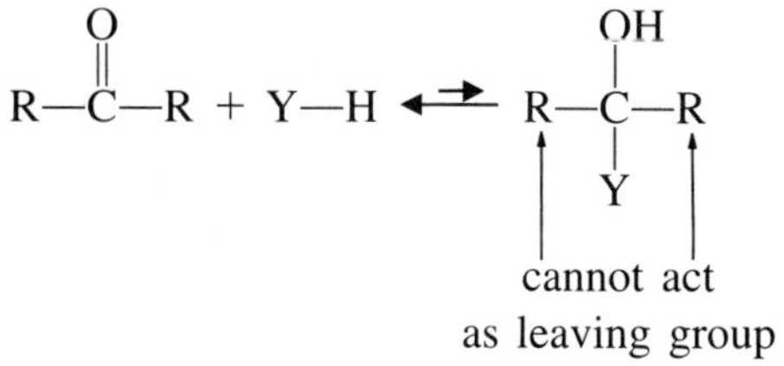

$$(20.21b)$$

The reason is that the best leaving groups are the weakest bases, as in the S_N1 and S_N2 reactions. The H— or R— groups of aldehydes and ketones, in order to act as leaving groups, would have to depart as the very basic anions H : ⁻ and R : ⁻, respectively. In carboxylic acids and their derivatives, however, one of the groups attached to the carbonyl carbon is a weak enough base that it can act as a leaving group in a substitution reaction. Thus, in acid-catalyzed esterification, the —OH group of a carboxylic acid, after protonation, leaves as H_2O.

We might ask whether esters might react further under the esterification conditions to form compounds analogous to acetals (Sec. 19.10A).

$$R-\overset{O}{\overset{\|}{C}}-OCH_3 + 2CH_3OH \xrightarrow{H^+} R-\overset{OCH_3}{\underset{OCH_3}{\overset{|}{\underset{|}{C}}}}-OCH_3 + H_2O \qquad \text{(not observed)} \quad (20.22)$$

an ortho-ester

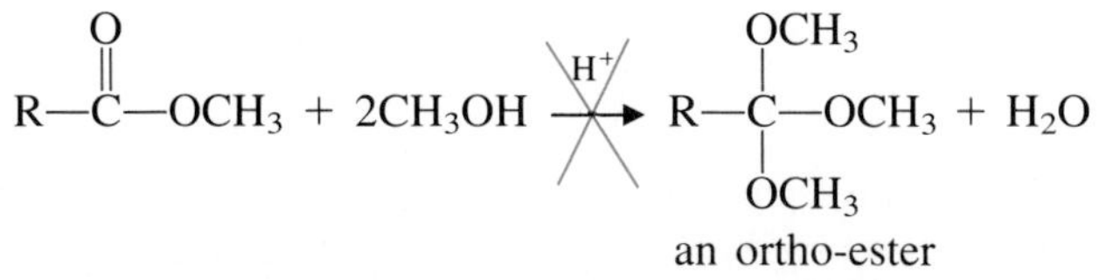

Such compounds, called ortho-esters, are indeed known. However, they *cannot* be formed in this manner because the equilibrium constants for addition to the carbonyl groups of acid derivatives are about 10^8 times smaller than those for formation of aldehyde addition compounds. These equilibria are simply too unfavorable to be driven to completion in the usual ways. The equilibrium constant for addition to a carboxylic acid derivative is unfavorable because the carboxylic acid derivative is stabilized by resonance interaction of the carboxylate oxygen with the carbonyl group:

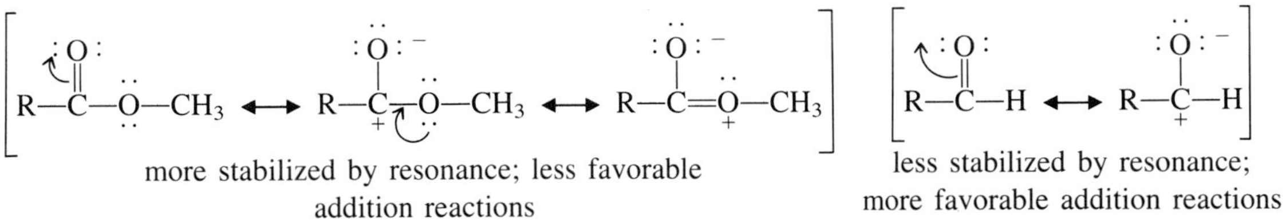

$$
\left[
\begin{array}{c}
\ddot{O} \\
\| \\
R-C-\ddot{O}-CH_3
\end{array}
\longleftrightarrow
\begin{array}{c}
:\ddot{O}:^- \\
| \\
R-\overset{+}{C}-\ddot{O}-CH_3
\end{array}
\longleftrightarrow
\begin{array}{c}
:\ddot{O}:^- \\
| \\
R-C=\overset{+}{O}-CH_3
\end{array}
\right]
\qquad
\left[
\begin{array}{c}
:\ddot{O}: \\
\| \\
R-C-H
\end{array}
\longleftrightarrow
\begin{array}{c}
:\ddot{O}:^- \\
| \\
R-\overset{+}{C}-H
\end{array}
\right]
\qquad (20.23)
$$

more stabilized by resonance; less favorable addition reactions less stabilized by resonance; more favorable addition reactions

As we learned in Sec. 19.7B, stabilization of a carbonyl compound reduces its equilibrium constant for addition.

Problems

8 What product is formed when adipic acid is heated in 1-propanol with H_2SO_4?

9 (a) Using the principle of microscopic reversibility, give a detailed mechanism for the acid-catalyzed hydrolysis of methyl benzoate (structure in Eq. 20.17) to benzoic acid.

(b) How would you change the reaction conditions of acid-catalyzed esterification to favor ester *hydrolysis* rather than ester *formation*?

(c) The hydrolysis of methyl benzoate is also promoted by ^-OH. Write a mechanism for the hydrolysis of methyl benzoate in NaOH solution.

(d) Methyl benzoate *cannot* be prepared from benzoic acid and methanol containing sodium methoxide ($Na^+ CH_3O^-$) because the acid reacts with methoxide in another way. What is this reaction?

B. Esterification by Alkylation

Acid-catalyzed esterification involves attack of a nucleophile at the carbonyl carbon. In this section, we consider different methods of forming esters that illustrate another mode of carboxylic acid reactivity: nucleophilic reactivity of the *carboxylate oxygen*.

When a carboxylic acid is treated with diazomethane in ether solution, it is rapidly converted into its methyl ester.

$$
CH_3(CH_2)_4CH=CH-\overset{\overset{\textstyle O}{\|}}{C}-OH + \,\,:\bar{C}H_2-\overset{+}{N}\equiv N: \xrightarrow{\text{ether}}
$$
diazomethane

$$
CH_3(CH_2)_4CH=CH-\overset{\overset{\textstyle O}{\|}}{C}-OCH_3 + \,:N\equiv N: \quad (20.24)
$$
(91% yield)

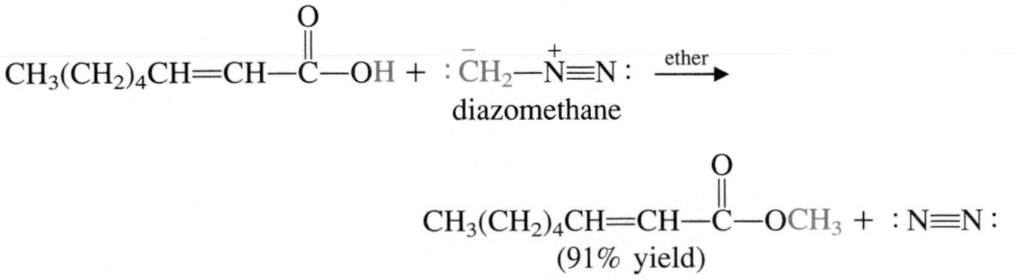

Diazomethane, a toxic yellow gas (bp = $-23°$), is usually generated chemically as it is needed and is codistilled with ether into a flask containing the carboxylic acid to be esterified. Diazomethane is both explosive and allergenic, and is therefore only used in small quantities. Nevertheless, esterification with diazomethane is so mild and free of side reactions that in many cases it is the method of choice for the synthesis of methyl esters.

The acidity of the carboxylic acid is important in the mechanism of this reaction. Protonation of diazomethane by the carboxylic acid gives the methyldiazonium ion.

$$R-\overset{\overset{\text{O}}{\|}}{C}-O-H \quad :CH_2-\overset{+}{N}\equiv N: \;\; \rightleftharpoons \;\; R-\overset{\overset{\text{O}}{\|}}{C}-O:^- \;+\; CH_3-\overset{+}{N}\equiv N: \qquad (20.25a)$$
$$\text{methyldiazonium ion}$$

This ion has one of the best leaving groups, molecular nitrogen. An S_N2 reaction of the methyldiazonium ion with the carboxylate oxygen results in the displacement of N_2 and formation of the ester.

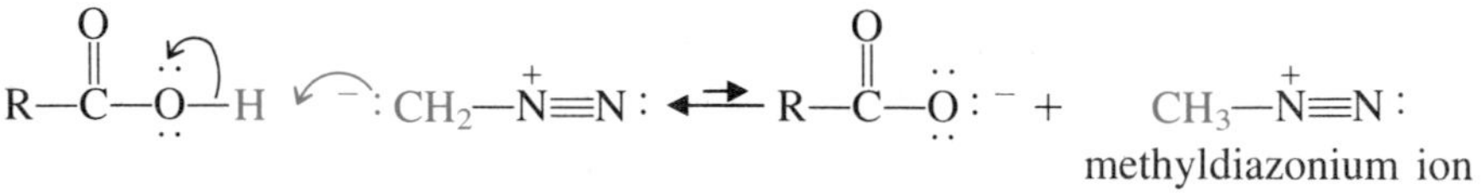

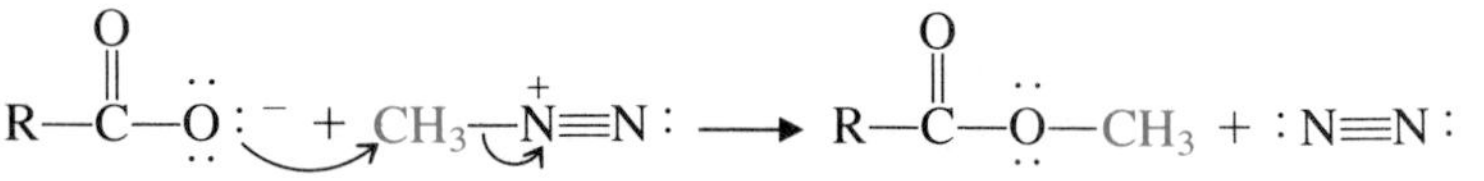

$$R-\overset{\overset{\text{O}}{\|}}{C}-O:^- \;+\; CH_3-\overset{+}{N}\equiv N: \;\longrightarrow\; R-\overset{\overset{\text{O}}{\|}}{C}-O-CH_3 \;+\; :N\equiv N: \quad (20.25b)$$

Carboxylate ions are less basic, and therefore less nucleophilic, than alkoxides or phenoxides, but they do react with especially reactive alkylating agents. The methyldiazonium ion formed by the protonation of diazomethane is one of the most reactive alkylating agents known. Notice, though, that it is the *acid* that reacts with diazomethane, because protonation must occur as the first step of the reaction.

The nucleophilic reactivity of carboxylates is also illustrated by the reaction of certain alkyl halides with carboxylate ions.

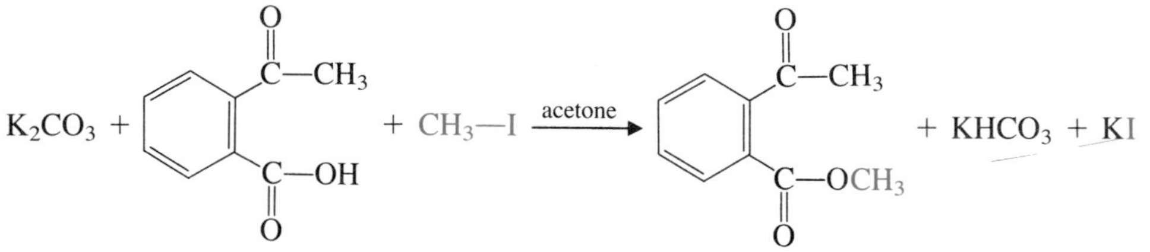

$$(65\% \text{ yield})$$

This is an S_N2 reaction in which the carboxylate ion, formed by the acid–base reaction of the acid and K_2CO_3, acts as the nucleophile that attacks the alkyl halide. Because carboxylate ions are such weak nucleophiles, this reaction works best on alkyl halides that are especially reactive in S_N2 reactions, such as methyl iodide and benzylic or allylic halides (Sec. 17.4).

Let us contrast the esterification reactions above with acid-catalyzed esterification in Sec. 20.8A. In all of the reactions in this section, *the carboxylate oxygen of the acid acts as a nucleophile.* This oxygen is *alkylated* by an alkyl halide or diazomethane. In acid-catalyzed esterification, *the carbonyl group, after protonation, acts as an electrophile* (a Lewis acid). The nucleophile in acid-catalyzed esterification is the oxygen atom of the solvent alcohol molecule. The two types of reactivity illustrated here—reaction of the carbonyl carbon as an electrophile (Lewis acid), and reaction of the —OH oxygen as a nucleophile (Lewis base)—illustrate two of the general ways in which carboxylic acids react. (See again Section 20.7.)

Problems

10 Give the structure of the ester formed when isobutyric acid reacts with
(a) diazomethane in ether.
(b) *n*-butyl alcohol (as solvent), H_2SO_4.
(c) benzyl bromide, K_2CO_3, acetone.

11 *tert*-Butyl esters can be prepared by the acid-catalyzed reaction of isobutylene with carboxylic acids.

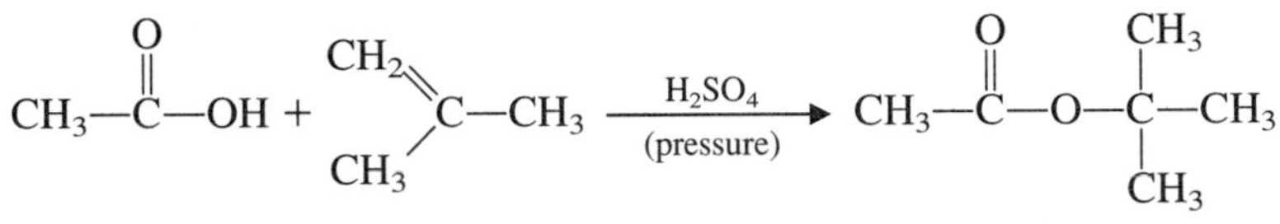

$$CH_3-\overset{\overset{\displaystyle O}{\|}}{C}-OH + \overset{\overset{\displaystyle CH_2}{\diagdown}}{\underset{\underset{\displaystyle CH_3}{\diagup}}{C}}-CH_3 \xrightarrow[\text{(pressure)}]{H_2SO_4} CH_3-\overset{\overset{\displaystyle O}{\|}}{C}-O-\overset{\overset{\displaystyle CH_3}{|}}{\underset{\underset{\displaystyle CH_3}{|}}{C}}-CH_3 \quad (20.27)$$

t-butyl acetate
(85% yield)

(a) Suggest a mechanism for this reaction. (*Hint:* See Sec. 4.5.) Be sure your mechanism accounts for the role of the acid catalyst.
(b) Would you classify this esterification reaction mechanistically as one that is related to acid-catalyzed esterification, or one that is related to the reactions in Sec. 20.8B? Explain.

20.9 CONVERSION OF CARBOXYLIC ACIDS INTO ACID CHLORIDES AND ANHYDRIDES

A. Synthesis of Acid Chlorides

Acid chlorides are carboxylic acid derivatives with the following general structure:

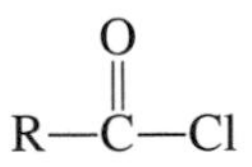

$$R-\overset{\overset{\displaystyle O}{\|}}{C}-Cl \quad (R = \text{alkyl or aryl})$$

Acid chlorides are invariably prepared from carboxylic acids. Two reagents used for this purpose are thionyl chloride, ($SOCl_2$), and phosphorus pentachloride (PCl_5).

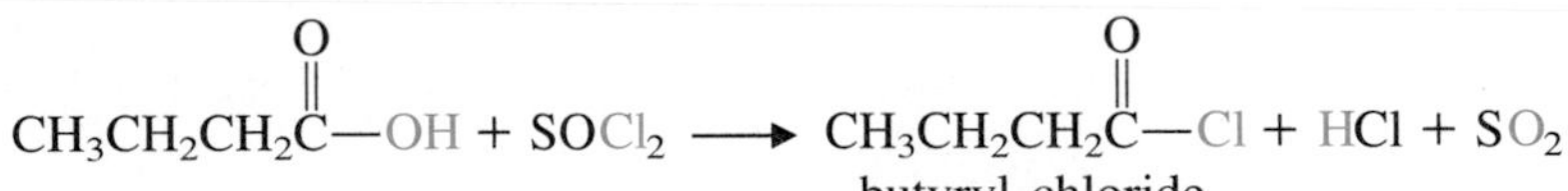

$$CH_3CH_2CH_2\overset{\overset{\displaystyle O}{\|}}{C}-OH + SOCl_2 \longrightarrow CH_3CH_2CH_2\overset{\overset{\displaystyle O}{\|}}{C}-Cl + HCl + SO_2 \quad (20.28)$$

butyryl chloride
(an acid chloride;
85% yield)

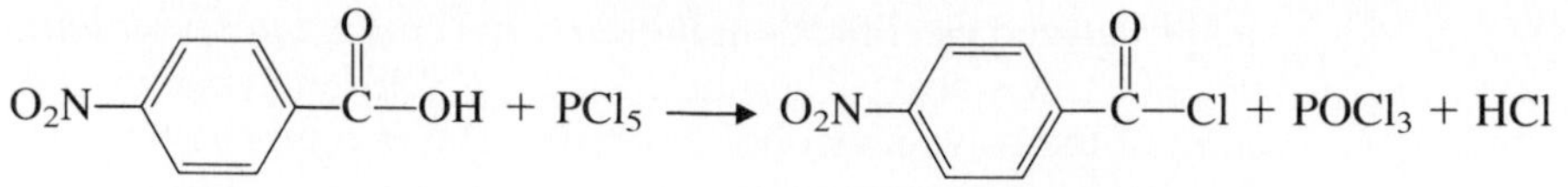

$$O_2N\!-\!\!\langle\ \rangle\!\!-\!\!\overset{\displaystyle O}{\overset{\|}{C}}\!-\!OH + PCl_5 \longrightarrow O_2N\!-\!\!\langle\ \rangle\!\!-\!\!\overset{\displaystyle O}{\overset{\|}{C}}\!-\!Cl + POCl_3 + HCl \quad (20.29)$$

p-nitrobenzoyl
chloride (90–96% yield)

Without concerning ourselves with the mechanistic details of these transformations, let us simply observe that acid chloride synthesis fits the general pattern of *substitution at a carbonyl group*, in this case, —OH is substituted by —Cl.

$$R\!-\!\overset{\displaystyle O}{\overset{\|}{C}}\!-\!OH \longrightarrow R\!-\!\overset{\displaystyle O}{\overset{\|}{C}}\!-\!Cl \quad (20.30)$$

Notice also that thionyl chloride is the same reagent used for making alkyl chlorides from alcohols (Sec. 10.4), a reaction in which —OH is replaced by —Cl at an *alkyl group.*

As we shall learn in Sec. 21.8A, acid chlorides are very reactive toward nucleophiles, and serve as useful reagents for the synthesis of other carboxylic acid derivatives. Recall that acid chlorides are also used as acylating reagents in Friedel–Crafts acylation (Sec. 16.4E). Hence, aromatic ketones can be prepared from carboxylic acids by conversion of an acid into an acid chloride, followed by a Friedel–Crafts reaction of the acid chloride with an aromatic compound.

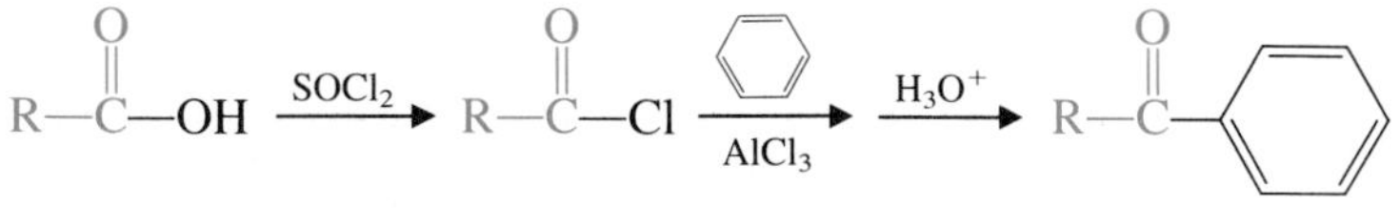

$$R\!-\!\overset{\displaystyle O}{\overset{\|}{C}}\!-\!OH \xrightarrow{\text{SOCl}_2} R\!-\!\overset{\displaystyle O}{\overset{\|}{C}}\!-\!Cl \xrightarrow[\text{AlCl}_3]{} \xrightarrow{\text{H}_3\text{O}^+} R\!-\!\overset{\displaystyle O}{\overset{\|}{C}}\!-\!\!\langle\ \rangle \quad (20.31)$$

Sulfonyl chlorides, the acid chlorides of sulfonic acids, can be prepared by treating sulfonic acids with PCl_5. Aromatic sulfonyl chlorides can be prepared directly by the reaction of aromatic compounds with chlorosulfonic acid.

$$\langle\ \rangle + 2Cl\!-\!\overset{\displaystyle O}{\underset{\displaystyle O}{\overset{\|}{\underset{\|}{S}}}}\!-\!OH \xrightarrow{20\text{–}25°} \langle\ \rangle\!\!-\!\!\overset{\displaystyle O}{\underset{\displaystyle O}{\overset{\|}{\underset{\|}{S}}}}\!-\!Cl + H_2SO_4 + HCl \quad (20.32)$$

chlorosulfonic
acid

benzenesulfonyl
chloride
(73–77% yield)

This reaction is a variation of aromatic sulfonation, an electrophilic aromatic substitution reaction (Sec. 16.4D). Chlorosulfonic acid, the acid chloride of sulfuric acid, acts as an electrophile in this reaction just as SO_3 does in sulfonation.

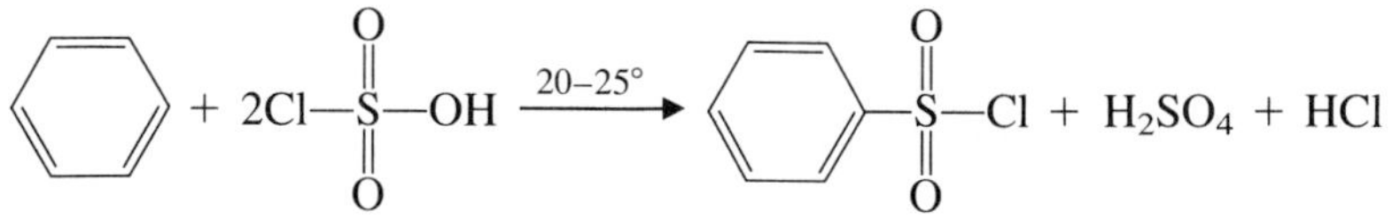

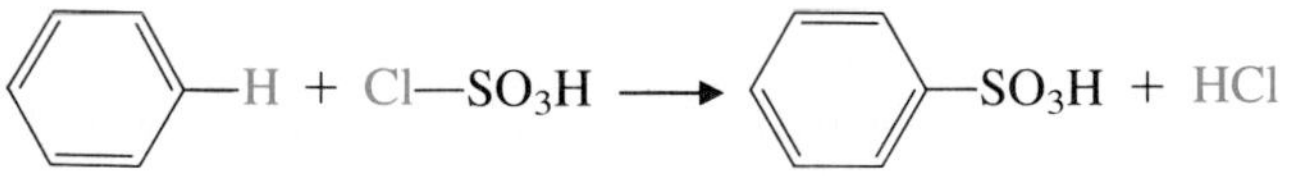

$$\langle\ \rangle\!\!-\!H + Cl\!-\!SO_3H \longrightarrow \langle\ \rangle\!\!-\!SO_3H + HCl \quad (20.33a)$$

The sulfonic acid produced in the reaction is converted into the sulfonyl chloride by reaction with another equivalent of chlorosulfonic acid.

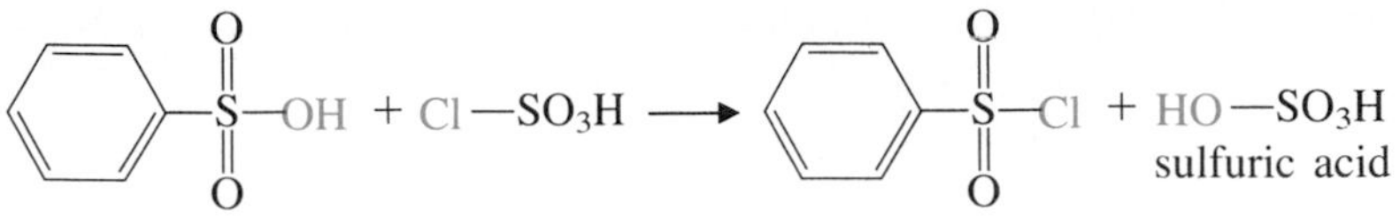

(20.33b)

This part of the reaction is analogous to the reaction of a carboxylic acid with thionyl chloride (Eq. 20.28).

Problem

12 Outline a synthesis of the following compound from benzoic acid and any other reagents:

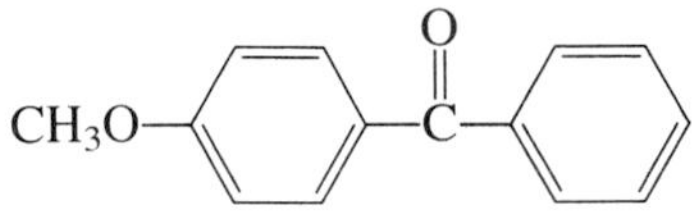

B. Synthesis of Anhydrides

Carboxylic acid *anhydrides* have the following general structure:

$$\mathrm{R{-}\overset{\overset{O}{\|}}{C}{-}O{-}\overset{\overset{O}{\|}}{C}{-}R} \qquad (\mathrm{R} = \text{alkyl or aryl})$$

The name *anhydride,* which means "without water," comes from the fact that an anhydride reacts with water to give two equivalents of a carboxylic acid.

$$\underset{\text{an anhydride}}{\mathrm{R{-}\overset{\overset{O}{\|}}{C}{-}O{-}\overset{\overset{O}{\|}}{C}{-}R}} + \mathrm{H_2O} \longrightarrow \mathrm{R{-}\overset{\overset{O}{\|}}{C}{-}OH} + \mathrm{HO{-}\overset{\overset{O}{\|}}{C}{-}R} \qquad (20.34)$$

The name anhydride also graphically describes the way that anhydrides are prepared: treatment of carboxylic acids with strong dehydrating agents.

$$2\mathrm{CF_3{-}\overset{\overset{O}{\|}}{C}{-}OH} + \underset{\substack{\text{phosphorus}\\ \text{pentoxide}}}{\mathrm{P_2O_5}} \longrightarrow \underset{\substack{\text{trifluoroacetic}\\ \text{anhydride}\\ \text{(74\% yield)}}}{\mathrm{CF_3{-}\overset{\overset{O}{\|}}{C}{-}O{-}\overset{\overset{O}{\|}}{C}{-}CF_3}} + \text{complex phosphates} \qquad (20.35)$$

Phosphorus pentoxide (actual formula P_4O_{10}) is an extremely hygroscopic white powder that reacts violently with water. It is also used as a potent desiccant. This compound is a complex anhydride of phosphoric acid, because it gives phosphoric acid when it reacts with water.

Most anhydrides may themselves be used to form other anhydrides. (As noted above, P_2O_5 is an inorganic anhydride that is used to form anhydrides of carboxylic acids.) In the following example, a dicarboxylic acid reacts with acetic anhydride to form a *cyclic anhydride*—a compound in which the anhydride group is part of a ring:

β-methylglutaric
acid

acetic
anhydride

β-methylglutaric
anhydride (>90% yield;
a cyclic anhydride)

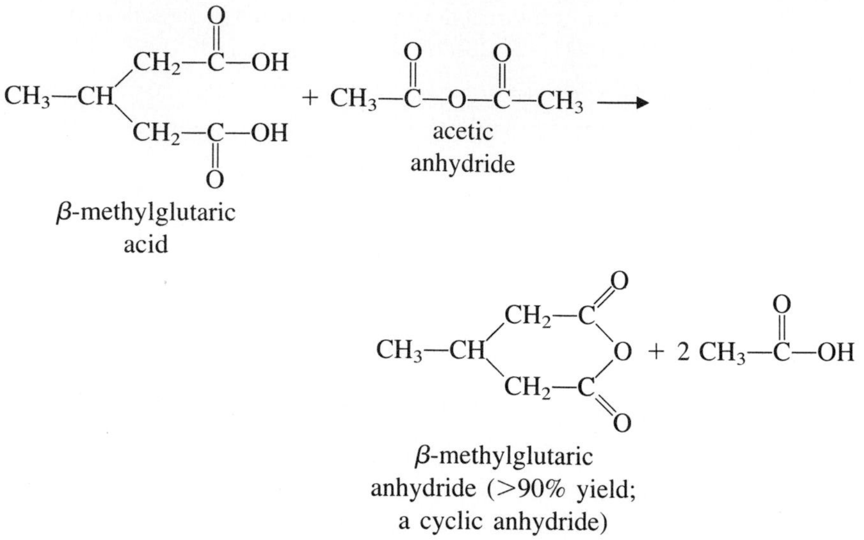

(20.36)

Phosphorus oxychloride ($POCl_3$) and P_2O_5 (Eq. 20.35) can also be used for the formation of cyclic anhydrides. Cyclic anhydrides containing five- and six-membered anhydride rings are readily prepared from their corresponding dicarboxylic acids. Compounds containing either larger or smaller anhydride rings generally cannot be prepared this way. Formation of cyclic anhydrides with five- and six-membered rings is so facile that in some cases it occurs on heating the dicarboxylic acid.

phthalic acid phthalic anhydride

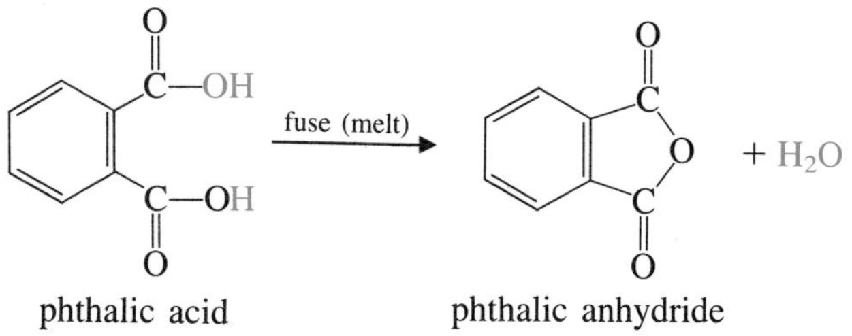

(20.37)

Although we shall not be concerned with the detailed mechanism of anhydride formation, it is important to observe that this reaction also fits the pattern of substitution at a carbonyl group: the —OH of one carboxylic acid molecule is substituted by the *acyloxy group* (color) of another.

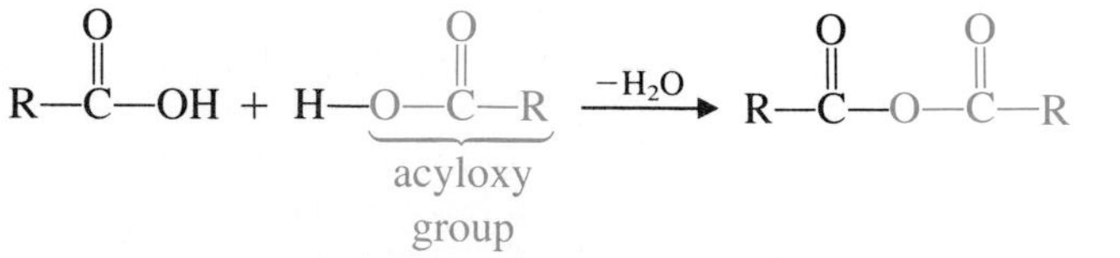

$$R-\overset{O}{\overset{\|}{C}}-OH + H\underbrace{-O-\overset{O}{\overset{\|}{C}}-R}_{\substack{\text{acyloxy} \\ \text{group}}} \xrightarrow{-H_2O} R-\overset{O}{\overset{\|}{C}}-O-\overset{O}{\overset{\|}{C}}-R \qquad (20.38)$$

Anhydrides, like acid chlorides, are used in the synthesis of other carboxylic acid derivatives (Sec. 21.8B).

Problems

13 Fumaric and maleic acids (Table 20.1) are *cis–trans* isomers. One forms a cyclic anhydride on heating and one does not. Which one forms the anhydride? Explain.

14 Give the product of each of the following reactions.
(a) benzoic acid + P_2O_5 $\longrightarrow$
(b) malonic acid + excess PCl_5 $\longrightarrow$
(c) toluene + excess chlorosulfonic acid $\longrightarrow$

20.10 REDUCTION OF CARBOXYLIC ACIDS TO ALCOHOLS

When a carboxylic acid is treated with lithium aluminum hydride, then dilute acid, it is reduced to a primary alcohol.

$$2CH_3CH_2\underset{\displaystyle CH_3}{\underset{|}{C}}H\overset{\displaystyle O}{\overset{\|}{C}}\text{—}OH + LiAlH_4 \xrightarrow{\text{ether}} \xrightarrow{H_3O^+} 2CH_3CH_2\underset{\displaystyle CH_3}{\underset{|}{C}}HCH_2\text{—}OH + LiAlO_2 \quad (20.39)$$

(83% yield)

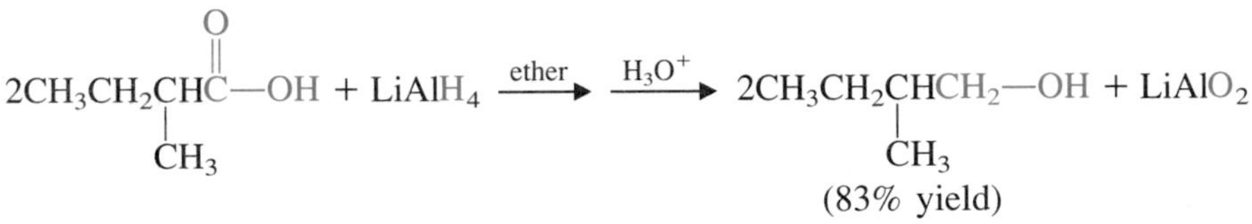

This reduction of carboxylic acids adds another alcohol synthesis to our repertoire (Sec. 19.8).

To understand how this reaction occurs, we recall once again that lithium aluminum hydride acts as a source of the strongly basic hydride ion, $H:^-$ (Sec. 19.8). The hydride from $^-AlH_4$ reacts with the acidic proton of the acid to give a lithium carboxylate and an equivalent of hydrogen.

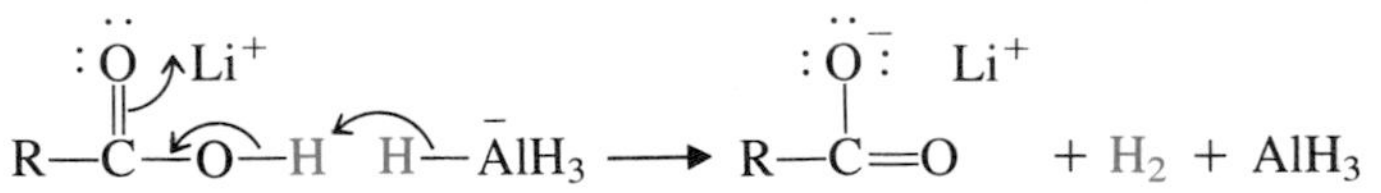

(20.40a)

Addition of AlH_3 to the carbonyl group of the lithium carboxylate—in much the same sense as addition of BH_3 to an alkene (Sec. 5.3A)—gives a tetrahedral addition intermediate.

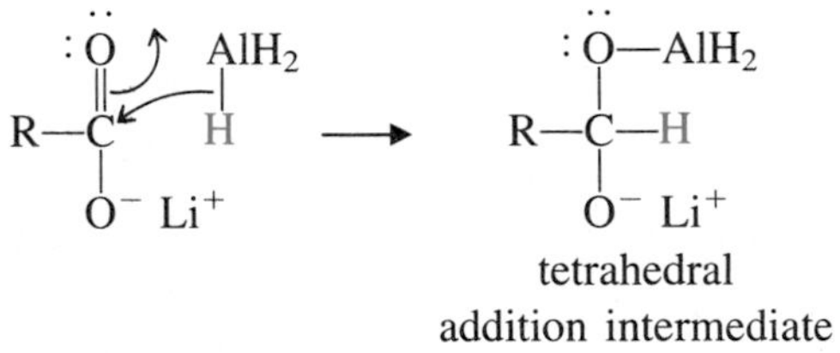

(20.40b)

This addition intermediate, like the one in esterification, breaks down by loss of a leaving group, in this case the $—OAlH_2$ group, and an aldehyde is formed.

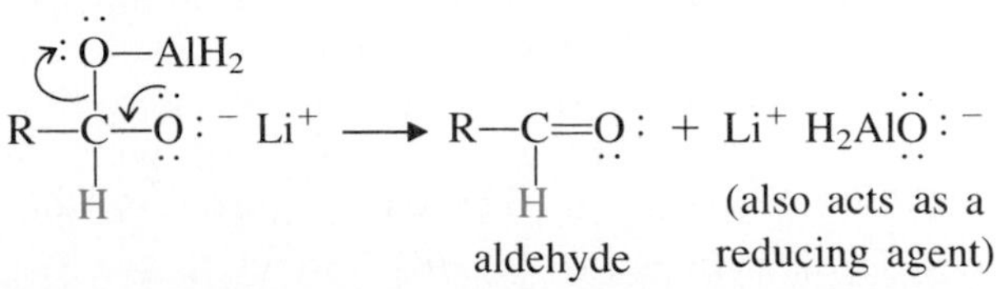

(20.40c)

(The Li^+ H_2AlO^- formed as a by-product is also active as a reducing agent.) However, the reaction does not stop with formation of the aldehyde. Aldehydes, as we have learned, are rapidly reduced to alcohols by $LiAlH_4$ (Sec. 19.8). Thus the aldehyde cannot be isolated, but reacts with another equivalent of $LiAlH_4$ to give the alkoxide of the primary alcohol. Addition of dilute acid in a separate step affords the alcohol.

$$\text{R—}\overset{\text{O}}{\overset{\|}{\text{C}}}\text{—H} \xrightarrow{\text{LiAlH}_4} \xrightarrow{\text{H}_3\text{O}^+} \text{R—}\overset{\text{OH}}{\overset{|}{\text{C}}}\text{H}_2 \qquad (20.40d)$$

Notice that the $LiAlH_4$ reduction of acids incorporates two different types of carbonyl reactions. The first, shown in Eqs. 20.40b,c, is a net *substitution* reaction at the carbonyl group to give the aldehyde intermediate. The second, shown in Eq. 20.40d, is an *addition* to the aldehyde thus formed.

$$\text{R—}\overset{\text{O}}{\overset{\|}{\text{C}}}\text{—OH} \xrightarrow{\text{(substitution)}} \text{R—}\overset{\text{O}}{\overset{\|}{\text{C}}}\text{—H} \xrightarrow{\text{(addition)}} \text{R—}\underset{\text{H}}{\underset{|}{\overset{\text{OH}}{\overset{|}{\text{C}}}}}\text{—H} \qquad (20.41)$$

In Chapter 21 we shall see other reactions that fit this pattern of reactivity.

Sodium borohydride ($NaBH_4$) does *not* reduce carboxylic acids, although it does react with the acidic protons of acids in a manner analogous to that shown in Eq. 20.40a.

Problem

15 Give the structure of a carboxylic acid $C_8H_6O_3$ that would give the product below in a $LiAlH_4$ reduction followed by hydrolysis.

20.11 DECARBOXYLATION OF CARBOXYLIC ACIDS

The loss of carbon dioxide from a carboxylic acid is called **decarboxylation.**

$$\text{R—}\overset{\text{O}}{\overset{\|}{\text{C}}}\text{—O—H} \longrightarrow \text{R—H} + \text{O=C=O} \qquad (20.42)$$

Although decarboxylation is not an important reaction for most ordinary carboxylic acids, certain types of carboxylic acids are readily decarboxylated. Among these are (1) β-ketoacids, (2) malonic acid derivatives, and (3) derivatives of carbonic acid.

β-Ketoacids—carboxylic acids with a keto group in the β-position—readily decarboxylate at room temperature in *acidic* solution.

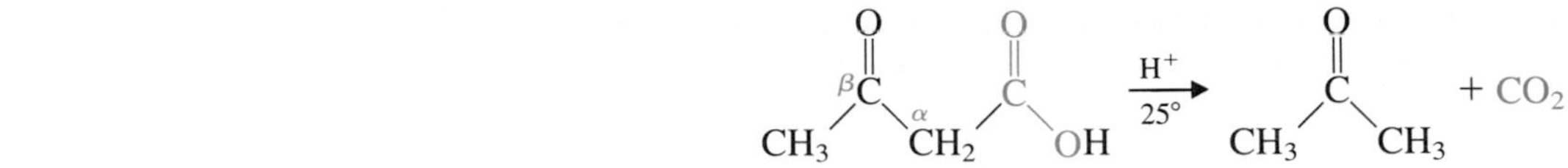
acetoacetic acid
(a β-ketoacid)

(20.43)

Decarboxylation of a β-ketoacid involves an *enol intermediate* that is formed by an internal proton transfer from the carboxylic acid group to the carbonyl oxygen atom of the ketone. As we learned in Sec. 14.5A, enols are transformed spontaneously into their corresponding ketones by the mechanism shown in Eq. 14.10d.

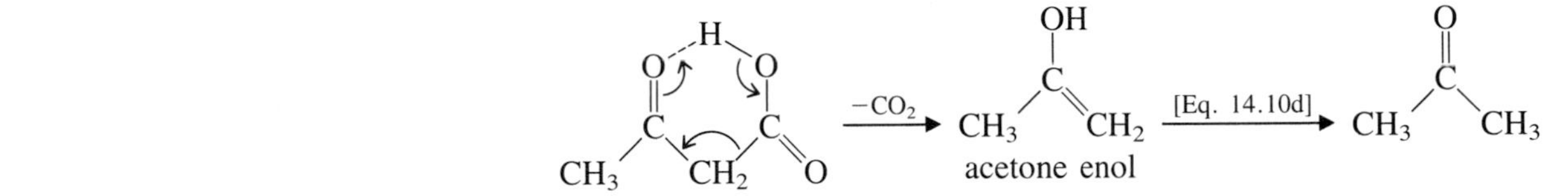

(20.44)

The *acid* form of the β-ketoacid decarboxylates more readily than the conjugate-base carboxylate form because the latter has no acidic proton that can be donated to the β-carbonyl oxygen. In effect, the carboxy group catalyzes its own removal!

Malonic acid and its derivatives readily decarboxylate upon heating in acidic solution.

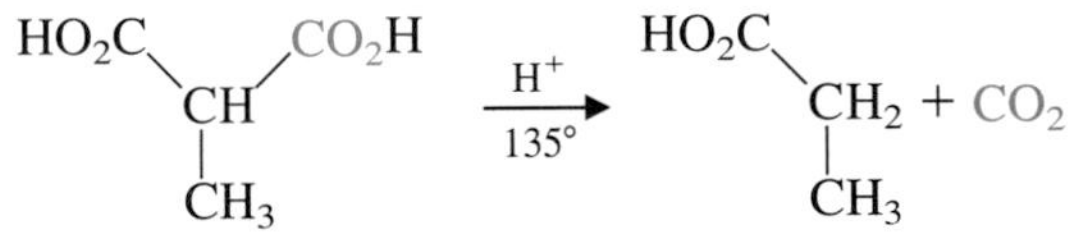
α-methylmalonic acid

(20.45)

This reaction, which also does not occur in base, bears a close resemblance to the decarboxylation of β-ketoacids, because both types of acids have a carbonyl group β to the carboxy group.

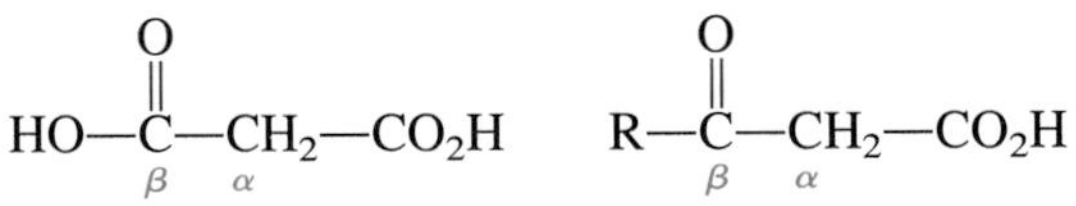
malonic acid　　　a β-keto acid

Because decarboxylation of malonic acid and its derivatives requires heating, the free acids can be isolated at room temperature.

Carbonic acid is unstable and decarboxylates spontaneously in acid solution to CO_2 and water. (Carbonic acid is formed reversibly when CO_2 is bubbled into water; carbonic acid gives carbonated beverages their acidity, and CO_2 gives them their "fizz.")

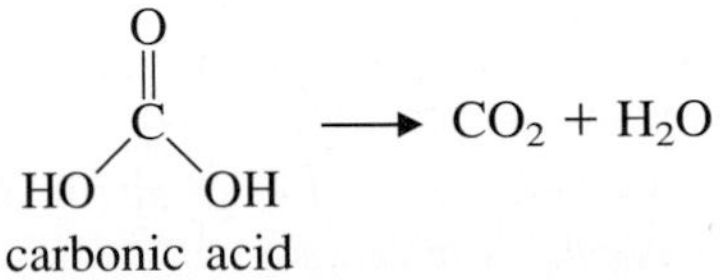
carbonic acid

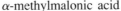

(20.46)

Similarly, any carbonic acid derivative with a free carboxylic acid group will also decarboxylate under acidic conditions.

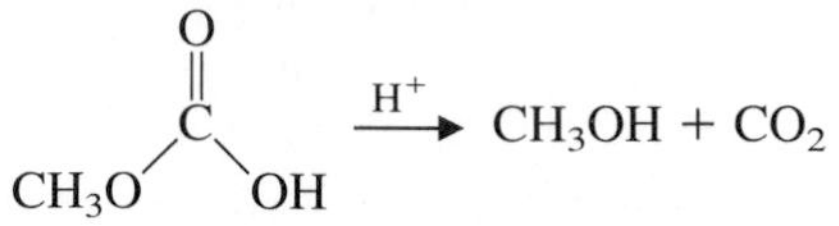

$$\underset{\substack{CH_3O \qquad OH}}{O \atop \|\atop C} \quad \xrightarrow{H^+} \quad CH_3OH + CO_2 \qquad (20.47)$$

$$\underset{\substack{H_2N \qquad OH}}{O \atop \|\atop C} \quad \xrightarrow{H^+} \quad CO_2 + NH_3 \quad \underset{\longleftarrow}{\overset{H_3O^+}{\longrightarrow}} \quad NH_4^+ \qquad (20.48)$$

Under basic conditions, carbonic acid is converted into its salts and does not decarboxylate. These salts—sodium bicarbonate ($NaHCO_3$) and sodium carbonate (Na_2CO_3)—are familiar stable compounds.

Carbonic acid derivatives in which *both* carboxylic acid groups are derivatized are stable. Dimethyl carbonate (a diester of carbonic acid) and urea (a diamide of carbonic acid) are examples of such compounds.

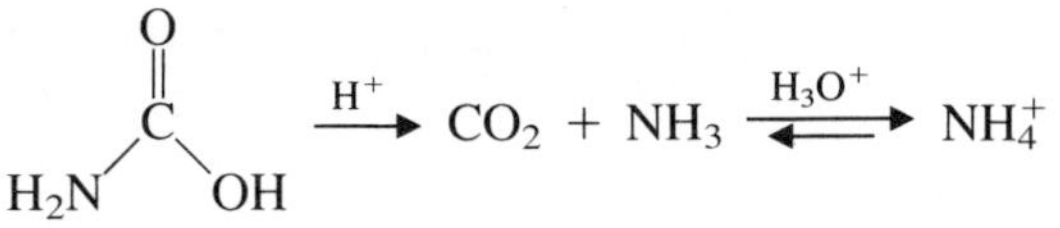

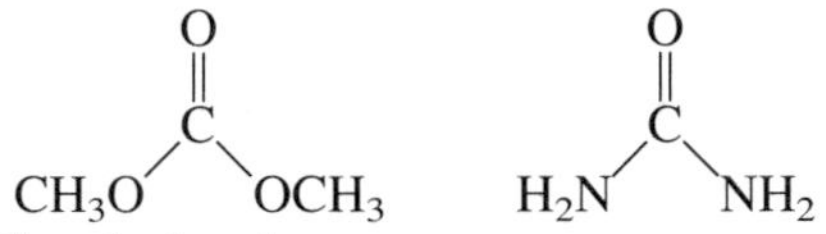

$$\underset{\substack{CH_3O \qquad OCH_3 \\ \text{dimethyl carbonate}}}{O \atop \|\atop C} \qquad\qquad \underset{\substack{H_2N \qquad NH_2 \\ \text{urea}}}{O \atop \|\atop C}$$

Problems

16 Give the product expected when each of the following compounds is treated with acid:

(a) (b) (+ heat) (c) $C_2H_5-NH-CO_2^-\ Na^+$

17 One piece of evidence supporting the enol mechanism in Eq. 20.44 is that β-ketoacids which *cannot* form enols are stable to decarboxylation. For example, the following β-ketoacid can be distilled at 310° without decomposition. Attempt to construct a model of the enol that would be formed when this compound decarboxylates. Use your model to explain why this β-ketoacid resists decarboxylation. (*Hint:* See Sec. 7.7C.)

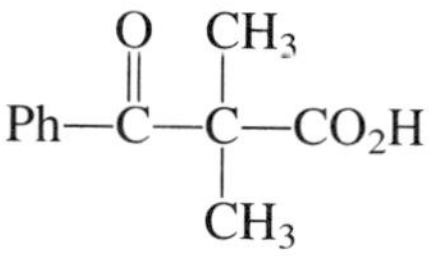

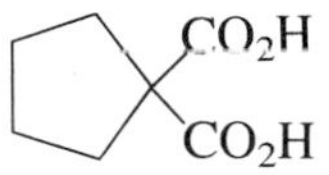

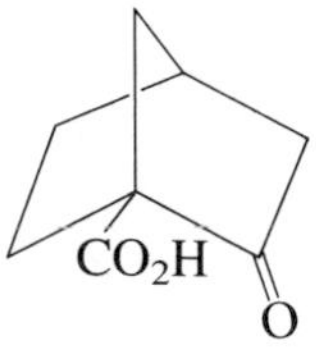

18 Give the structures of all the β-ketoacids that will decarboxylate to yield 2-methyl-1-cyclohexanone.

KEY IDEAS IN CHAPTER 20

- The carboxy group is the characteristic functional group of carboxylic acids.
- Carboxylic acids are high-boiling liquids or solids, and carboxylic acids of relatively low molecular weight have substantial solubility in water.
- Carbonyl and O—H absorptions are the most important infrared absorptions of carboxylic acids. In the NMR spectrum the α-hydrogens of carboxylic acids absorb in the δ 2–2.5 region, and the O—H proton, which can be exchanged with D_2O, absorbs in the δ 9–13 region.
- Typical carboxylic acids have pK_a values between 4 and 5, although pK_a values are influenced by inductive effects. Sulfonic acids are even more acidic. The conjugate base of a carboxylic acid is a carboxylate ion.
- Carboxylic acids with long unbranched carbon chains are called fatty acids. The alkali metal salts of fatty acids are soaps. Soaps and detergents are surfactants; they form micelles in aqueous solution.
- Because of their acidity, carboxylic acids dissolve not only in aqueous NaOH, but also in aqueous solutions of weaker bases such as sodium bicarbonate.
- The reaction of Grignard reagents with CO_2 is both a synthesis of carboxylic acids and a method of carbon–carbon bond formation.
- The reactivity of the carbonyl group toward nucleophiles plays an important role in many reactions of carboxylic acids. In many reactions of carboxylic acid derivatives a nucleophile attacks the carbonyl group to form a tetrahedral addition intermediate, which then breaks down by loss of a leaving group. The result is a net substitution reaction at the carbonyl group. Acid-catalyzed esterification and lithium aluminum hydride reduction are two examples of nucleophilic carbonyl substitution. In lithium aluminum hydride reduction the substitution product, an aldehyde, reacts further in an addition reaction.
- The nucleophilic reactivity of the carboxylate oxygen is important in some reactions of carboxylic acids, such as ester formation by alkylation with diazomethane or alkyl halides.
- β-Ketoacids and derivatives of malonic acid and carbonic acid decarboxylate in acidic solution; most malonic acid derivatives require heating. The decarboxylation of β-ketoacids, and probably malonic acids as well, involves enol intermediates.

ADDITIONAL PROBLEMS

19 Give the product expected when butyric acid reacts with each of the following reagents:
(a) ethanol, H_2SO_4
(b) NaOH solution
(c) $LiAlH_4$ (excess), then H_3O^+
(d) heat
(e) $SOCl_2$
(f) diazomethane
(g) product of (c) (excess) + $CH_3CH{=}O$, H^+
(h) product of (e), $AlCl_3$, benzene
(i) product of (h), NH_2NH_2, base, heat

20 Give the product expected when benzoic acid reacts with each of the following reagents:
(a) CH_3I, K_2CO_3
(b) concentrated HNO_3, H_2SO_4
(c) PCl_5
(d) P_2O_5, heat
(e) CH_3MgBr (1 equivalent), ether

21 Draw the structures and give the names of all dicarboxylic acids with the formula $C_6H_{10}O_4$. Indicate which are chiral; which would readily form cyclic anhydrides on heating; and which would decarboxylate on heating.

22 How many milliliters of 0.1*N* NaOH solution are required to neutralize 100 mg of (a) hexanoic acid; (b) succinic acid? How can this type of information be used to calculate the molecular weight of an unknown carboxylic acid? What additional piece of information is required if the molecular weight is to be determined this way?

23 Explain why all efforts to synthesize a carboxylic acid with oxygen-18 at *only* the carbonyl oxygen fail and yield instead an acid in which the labeled oxygen is distributed equally between both oxygens of the carbonyl group.

($^* = {}^{18}O$)

(50% of label on each oxygen)

24 Rank the following compounds in order of increasing acidity. Explain your answers.

(a)

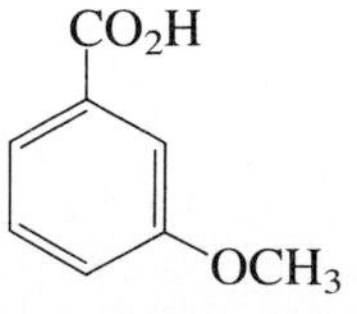

(b)

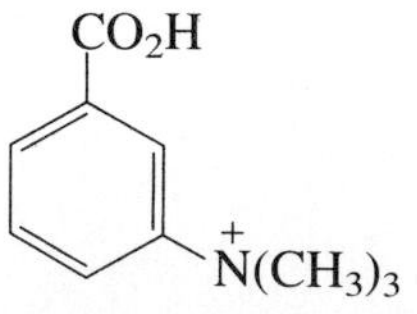

(c)

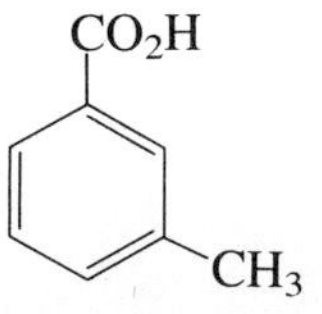

25 (a) Common pH indicator paper turns red at pH values below about 3. Perform calculations that explain why a $0.1M$ aqueous solution of acetic acid turns pH paper red but a $0.1M$ aqueous phenol solution does not.

(b) What is the pH of an acetic acid–sodium acetate buffer in which the actual [acetic acid]/[sodium acetate] ratio is 1/3? In which this ratio is 3? In which it is 1?

26 Outline a synthesis of each of the following compounds from isobutyric acid and any other reagents:

(a)
$$(CH_3)_2CH-\overset{\overset{\displaystyle O}{\|}}{C}-OCH_3$$

(b)
$$(CH_3)_2CHCH_2-\overset{\overset{\displaystyle O}{\|}}{C}-OCH_3$$

(c)
$$(CH_3)_2CHCH_2O-\overset{\overset{\displaystyle O}{\|}}{C}-CH_3$$

(d) isobutyrophenone

(e) 3-methyl-2-butanone

27 Fatty acids containing an even number of carbons are readily obtained from natural sources, but those containing an odd number of carbons are relatively rare. (This is a consequence of the way these compounds are synthesized in nature; Sec. 22.7.) Outline a synthesis of the rare tridecanoic acid ($n\text{-}C_{12}H_{25}-CO_2H$) from the readily available lauric acid (see Table 20.1).

28 A graduate student, Al Kane, has been given by his professor a very precious sample of $(-)$-3-methylhexane, along with optically active samples of both enantiomers of 4-methylhexanoic acid, each of known absolute configuration. Kane has been instructed to determine the absolute configuration of $(-)$-3-methylhexane. Kane has come to you for assistance. Show what he should do to deduce the configuration of the optically active hydrocarbon from the acids of known configuration. Be specific.

29 Explain why hydrogen bromide (HBr) is a much stronger acid when used in acetic acid solvent than when it is used in aqueous solvent.

30 The sodium salt of *valproic acid* is a drug that has been used in the treatment of epilepsy. (Valproic acid is a name used in medicine.)

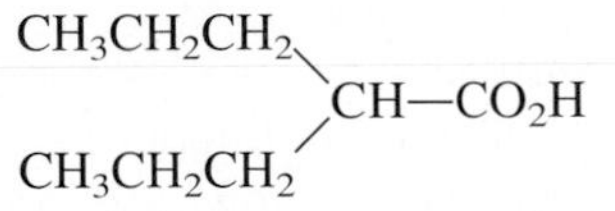

$$\begin{array}{c} CH_3CH_2CH_2 \\ \diagdown \\ \diagup \\ CH_3CH_2CH_2 \end{array}\!\!CH-CO_2H \qquad \text{valproic acid}$$

(a) Give the systematic name of valproic acid.

(b) Give the common name of valproic acid.

(c) Outline a synthesis of valproic acid from carbon sources containing fewer than five carbons, and any other reagents.

Problems (Cont.)

31 You have been employed by a biochemist, Fungus P. Gildersleeve, who has given you a very expensive sample of benzoic acid labeled equally in both oxygens with ^{18}O. He asks you to prepare methyl benzoate (structure in Eq. 20.17), preserving as much ^{18}O label in the ester as possible. Which method(s) of ester synthesis would you use to carry out this assignment? Why?

32 Peanut oil is a *polyunsaturated* fat because the oil contains esters of several unsaturated fatty acids as major constituents. When a fatty acid *A* isolated from the hydrolysis of peanut oil was treated with O_3, then H_2O_2, about equal amounts of malonic acid, nonanedioic acid, and hexanoic acid were obtained. When the acid was treated with diazomethane prior to the ozonolysis, malonic acid, hexanoic acid, and the *monomethyl ester* of nonanedioic acid were isolated. What is the structure of *A?*

33 Propose reasonable fragmentation mechanisms that explain why
(a) the mass spectrum of 2-methylpentanoic acid has a strong peak at $m/e = 74$.
(b) the mass spectrum of benzoic acid shows major peaks at $m/e = 105$ and $m/e = 77$.

34 You are a chemist for Chlorganics, Inc., a company specializing in chlorinated organic compounds. A process engineer, Turner Switchback, has accidentally mixed the contents of four vats containing, respectively, *p*-chlorophenol, 4-chlorocyclohexanol, *p*-chlorobenzoic acid, and chlorocyclohexane. The president of the company, Hal Ogen (green with anger) has ordered you to design an expeditious separation of these four compounds. Your job is on the line; accommodate him.

35 Give the structure of each of the following compounds:
(a) $C_9H_{10}O_3$: IR 2400–3200, 1700, 1630 cm^{-1}
NMR in Fig. 20.6a.
(b) $C_9H_{10}O_3$: IR 2300–3200, 1710, 1600 cm^{-1}
NMR in Fig. 20.6b.
(c) The compound with IR absorptions at 2300–3400, 1710, 1670, and 974 cm^{-1}, UV absorption at 208 nm ($\epsilon = 12{,}300$), peaks in the mass spectrum at $m/e =$ 86 (parent) and 41, and the NMR spectrum in Fig. 20.6c.

36 The radioactive isotope carbon-14 is used in tracer levels and cannot be determined by spectroscopy. It is generally detected by counting its radioactive decay in a device called a scintillation counter. The location of carbon-14 in a chemical compound must be determined by carrying out chemical degradations, isolating the resulting fragments that represent different carbons in the molecule, and counting them.

A well-known biologist, Fizzi O. Logical, has purchased a sample of α-phenylacetic acid advertised to be labeled with the radioactive isotope carbon-14 *only* at the carbonyl carbon. Before using this compound in feeding experiments designed to test a brilliant theory of biosynthesis, he has wisely decided that he should be sure the radiolabel is located only at the carbonyl carbon as

Figure 20.6 NMR spectra for Problem 35.

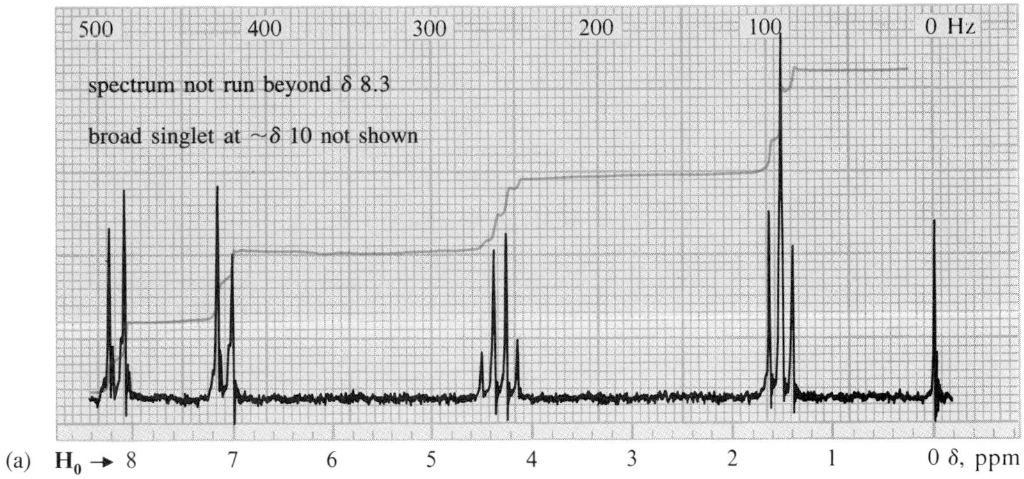

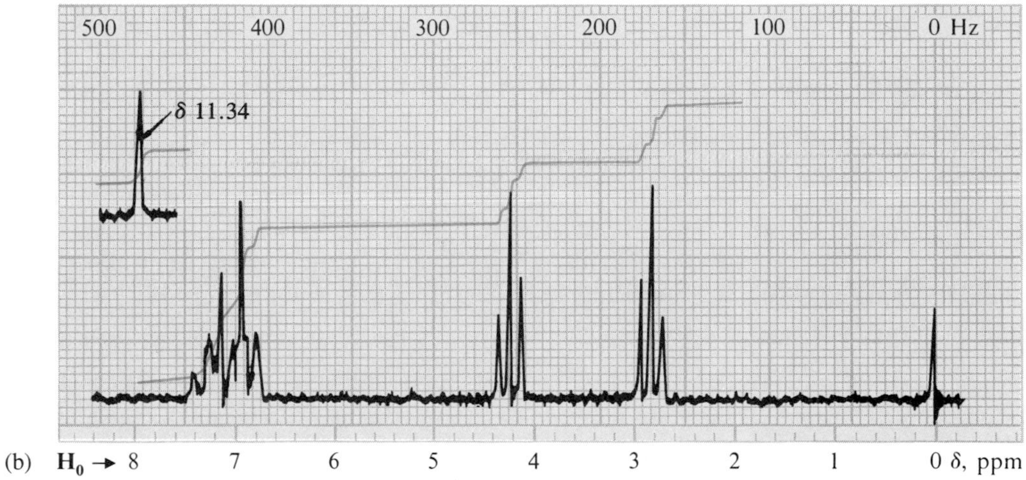

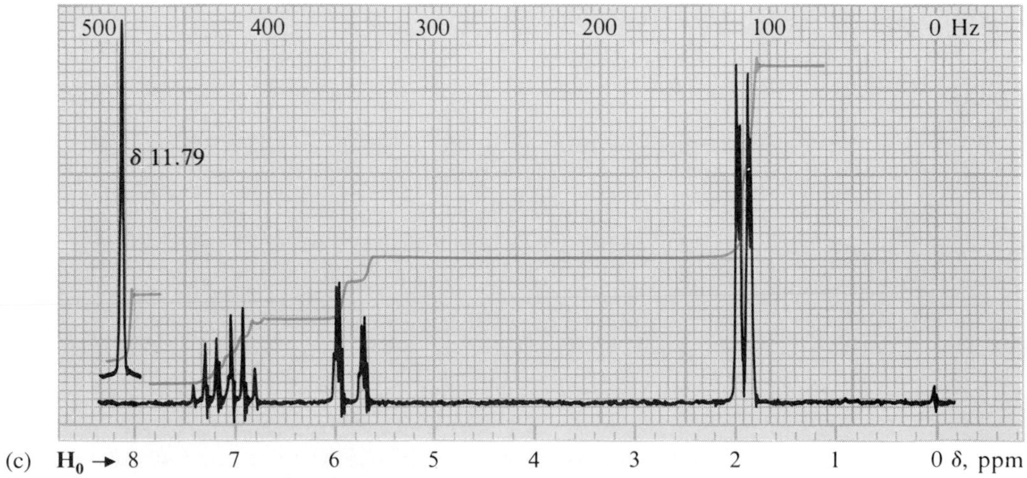

Problems (Cont.)

advertised. Knowing your expertise in organic chemistry, he has asked that you devise a way to determine what fraction of the ^{14}C is at the carbonyl carbon and what fraction is elsewhere in the molecule. Outline a reaction scheme that could be used to make this determination.

---

37 Which *one* of the following compounds would *not* readily form a cyclic anhydride on heating with acetic anhydride? Explain. Draw structures of the cyclic anhydrides of the other compounds. (*Hint:* Don't forget what you learned in Sec. 7.2.)

(a)

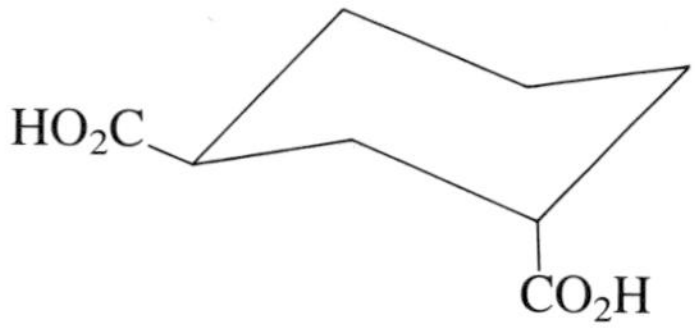

(d)

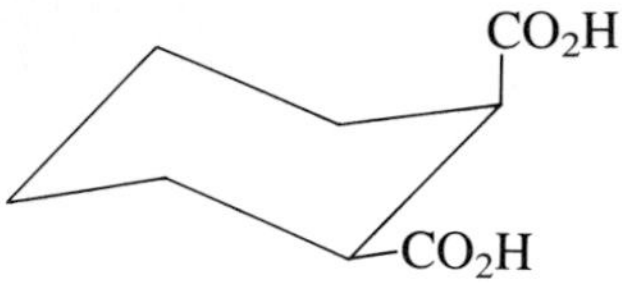

(e) α,β-dimethylsuccinic acid

(b)

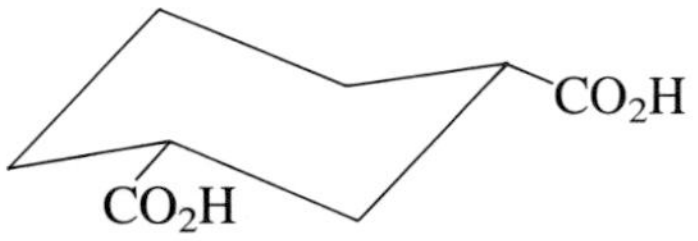

(c)

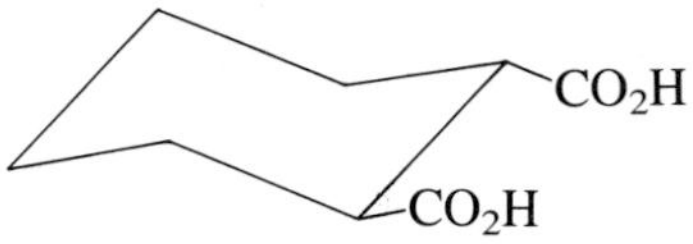

---

38 Propose a synthesis of each of the following compounds from the indicated starting material(s) and any other reagents:

(a) $CH_3CH_2-\overset{\overset{O}{\|}}{C}-O-CH_2CH=CH_2$ from allyl alcohol as the only carbon source

(b) 2-pentanol from propionic acid
(c) isohexane from 4-methylpentanoic acid
(d) 2-methylheptane from pentanoic acid
(e) *m*-nitrobenzoic acid from toluene
(f) *p*-nitrobenzoic acid from toluene
(g) 1,3-diphenylpropane from benzoic acid

(h) 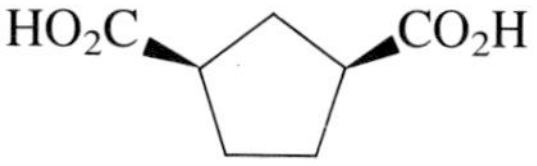 from 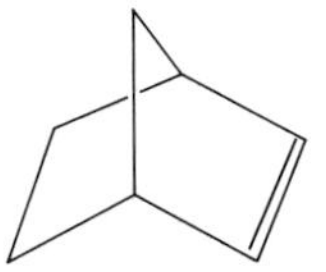 (norbornene)

(i) 5-oxohexanoic acid from 5-bromo-2-pentanone

---

39 Complete each of the following reactions by giving the principal organic product(s). Give the reasons for your answers.

(a)

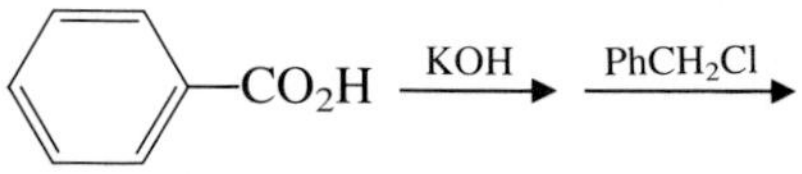

(b)

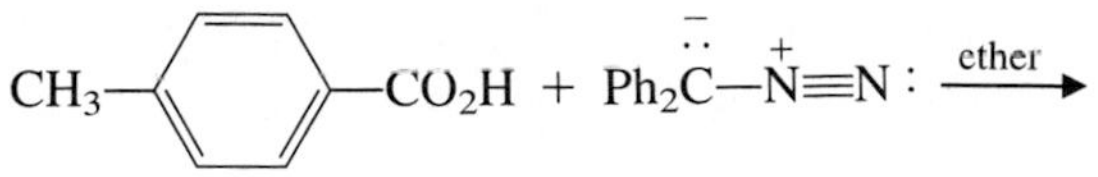

$CH_3\text{—}\langle\text{benzene}\rangle\text{—}CO_2H + Ph_2\ddot{C}\text{—}\overset{+}{N}{\equiv}N: \xrightarrow{\text{ether}}$

(c)

$Cl\text{—}\langle\text{cyclobutane}\rangle\overset{CO_2H}{\underset{CO_2H}{}} \xrightarrow{200°} \text{(two isomeric compounds)}$

(d) $Br\text{—}CH_2CH_2CH_2CH_2\text{—}CO_2H \xrightarrow[\text{acetone}]{K_2CO_3} (C_5H_8O_2)$

(e)

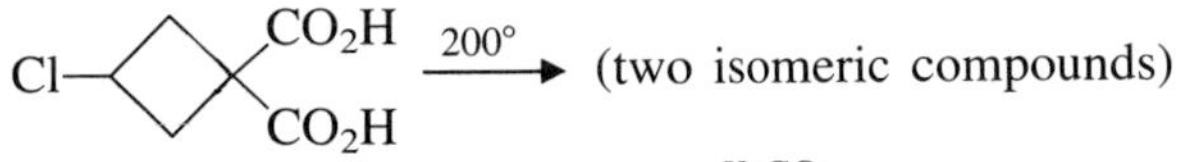

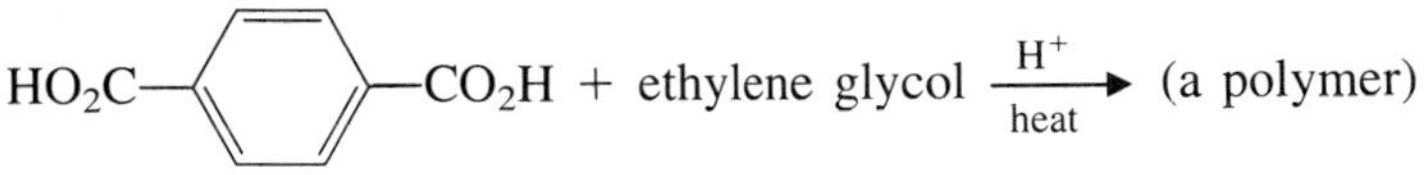

$HO_2C\text{—}\langle\text{benzene}\rangle\text{—}CO_2H + \text{ethylene glycol} \xrightarrow[\text{heat}]{H^+} \text{(a polymer)}$

(f)

$\langle\text{1-methylcyclohexene}\rangle + Hg(OAc)_2 + CH_3CO_2H \text{ (solvent)} \longrightarrow \xrightarrow{NaBH_4}$

(g)

$CH_3CH_2\text{—}\overset{O}{\overset{\|}{C}}\text{—}OH + \overset{O}{CH_2\text{—}CH\text{—}CH_3} \xrightarrow[\text{(1 equiv)}]{KOH}$

(h)

$\xrightarrow[\text{acetone}]{(CH_3)_2SO_4 \text{ (excess) } K_2CO_3 \text{ (excess)}}$

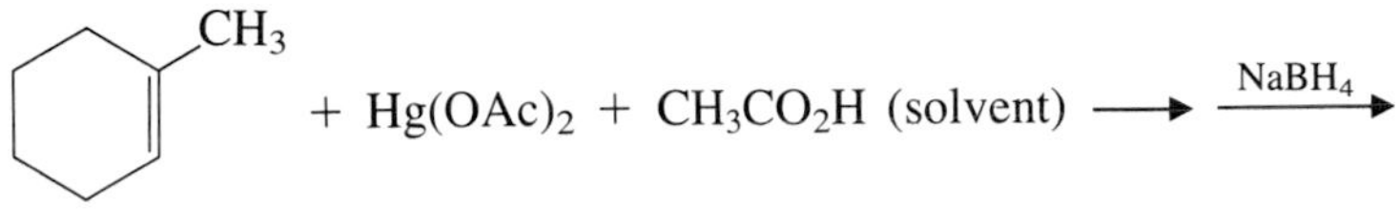

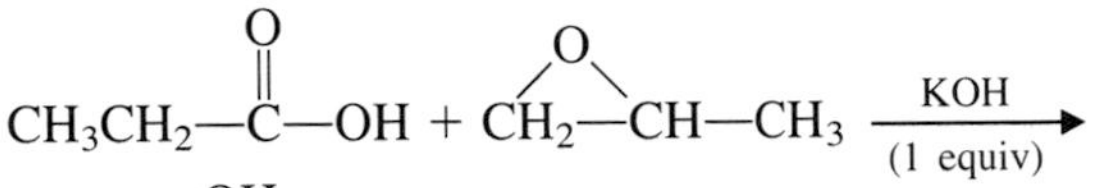

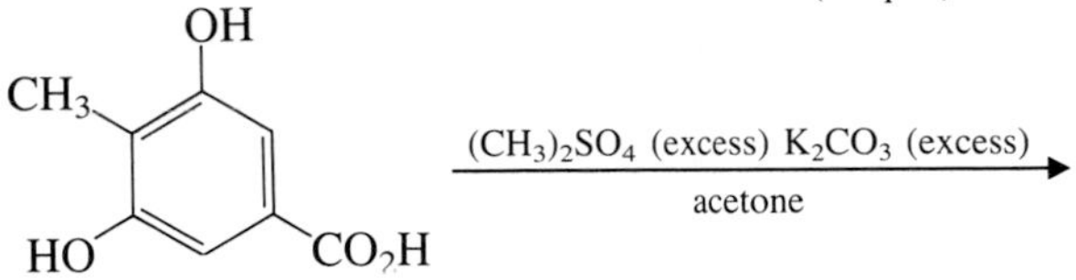

40 Penicillin-G is a widely used member of the penicillin family of drugs. In which fluid would you expect penicillin-G to be more soluble: stomach acid (pH = 2) or the bloodstream (pH 7.4)? Explain.

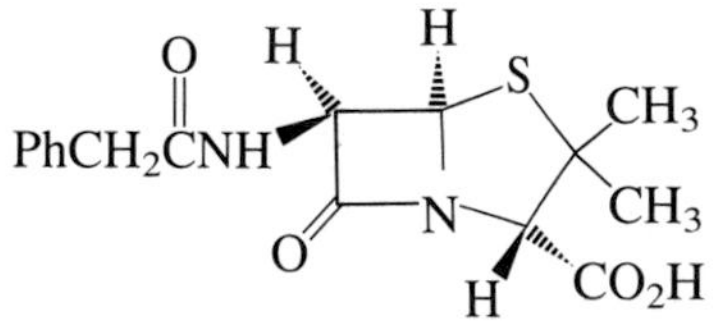

penicillin-G

41 Using the curved-arrow formalism, outline a reasonable mechanism for each of the following known reactions:

(a)

$CH_3\text{—}\overset{OC_2H_5}{\underset{OC_2H_5}{\overset{|}{\underset{|}{C}}}}\text{—}OC_2H_5 + H_2O \xrightarrow{H^+ \text{ (cat)}} CH_3\text{—}\overset{O}{\overset{\|}{C}}\text{—}OC_2H_5 + 2C_2H_5OH$

an ortho-ester

(b)

$\xrightarrow[\text{CH}_3\text{OH}]{H^+} \quad + H_2O$

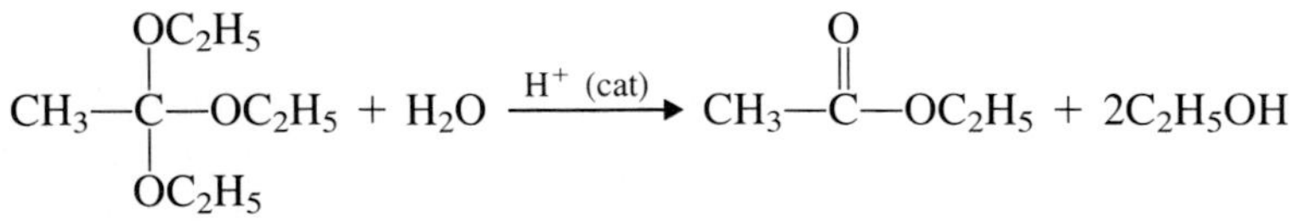

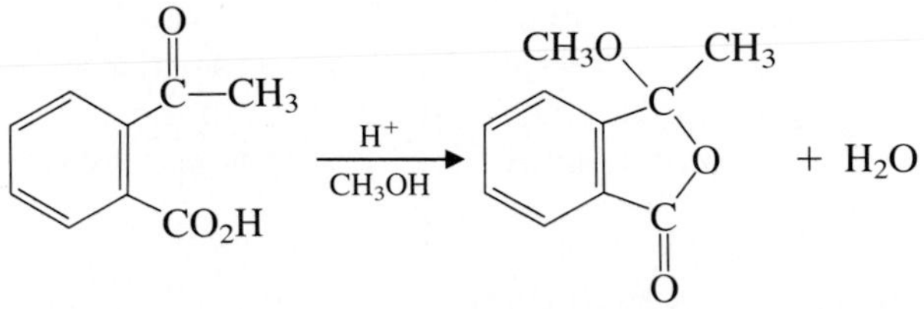

Problems (Cont.)

(c)

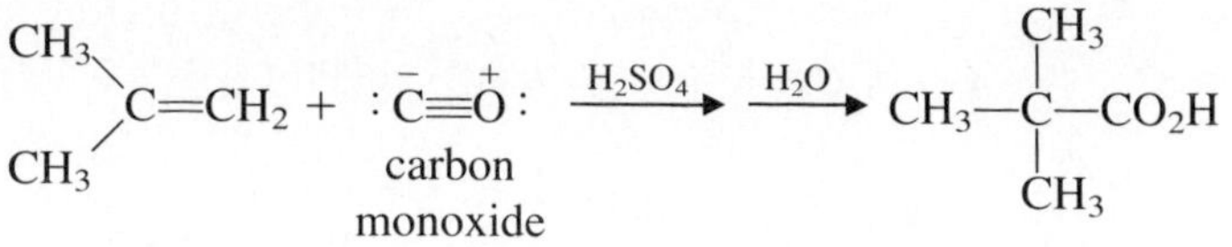

(d)

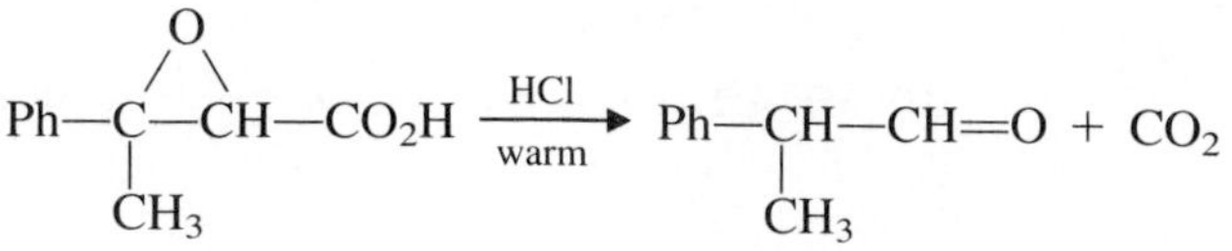

(e)

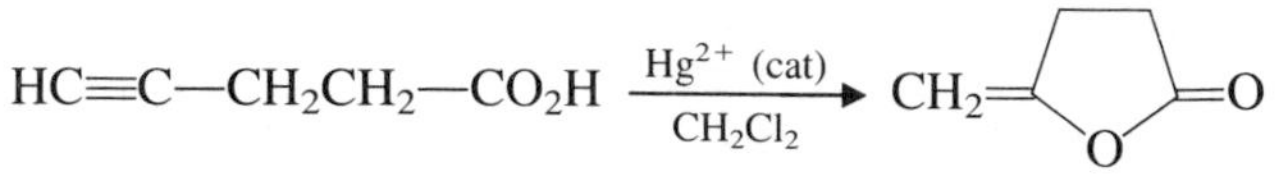

(f)

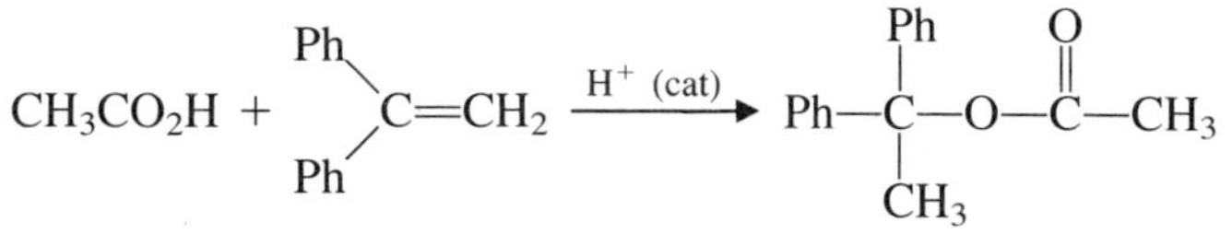

---

42 In an abandoned laboratory has been found a bottle labeled "isolated from natural sources." The bottle contains a solid *A*, mp 129–130°, which is soluble in base and hot water. The NMR and IR spectra of *A* are given in Fig. 20.7, and the mass spectrum of *A* has prominent peaks at $m/e = 166$ and 107. It is determined by titration that *A* has two acidic groups with pK_a values of 4.7 and 10.4, respectively. The UV spectrum of *A* is virtually unchanged as the pH is raised from 2 to 7, but the λ_{max} shifts to much higher wavelength when *A* is dissolved in 0.1*N* NaOH solution. Propose a structure of *A*. Rationalize the two peaks in its mass spectrum.

---

43 In an abandoned laboratory has been found an unlabeled bottle containing a flammable, water-insoluble substance *A* that decolorizes Br_2 in CCl_4 and has the following elemental analysis: 87.7% carbon, 12.3% hydrogen. The base peak in the mass spectrum of *A* occurs at $m/e = 67$. The NMR of *A* is complex, but integration shows that about 30% of the protons have chemical shifts in the δ 1.8–2.2 region of the spectrum. After ozonolysis of *A* and treatment with H_2O_2, a single dicarboxylic acid *B* is obtained that can be resolved into enantiomers. Neutralization of a solution containing 100 mg of *B* requires 13.7 mL of 0.1*N* NaOH solution (see Problem 22). Compound *B*, when treated with $POCl_3$, forms a cyclic anhydride. Give the structures of *A* and *B*.

---

44 An acidic compound *A*, $C_{10}H_{16}O_4$, on heating in acid, gives two isomers *B* and *C*, $C_9H_{16}O_2$, that have different melting points. An acidic optically active compound *D*, $C_{10}H_{16}O_4$, on heating in acid, gives a *single* optically active compound *E*, $C_9H_{16}O_2$. Compounds *B*, *C*, and *E* are individually treated with $LiAlH_4$, then H_3O^+, then with concentrated HBr, then with Mg in ether, and finally with H_2O. As a result of this sequence of reactions, *C* and *E* give the same optically *inactive* compound, which is different from the compound that *B* gives after the same treatment. Identify all compounds and explain your reasoning.

Figure 20.7 IR and NMR spectra for Problem 42.

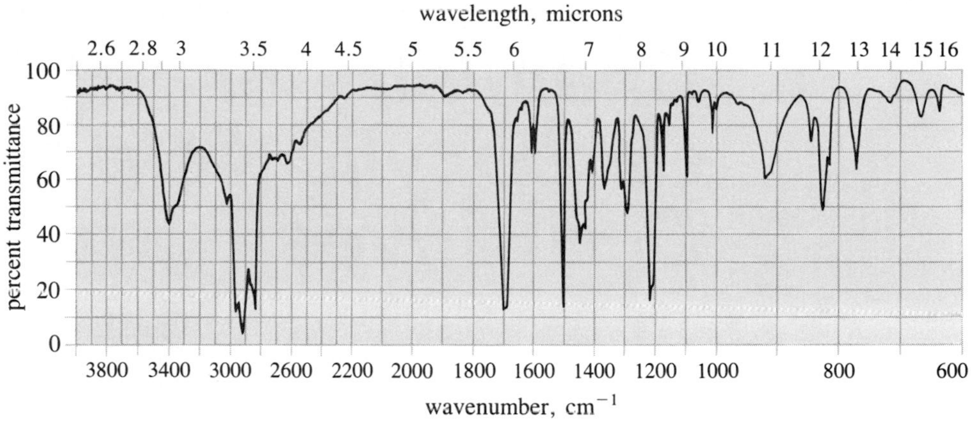

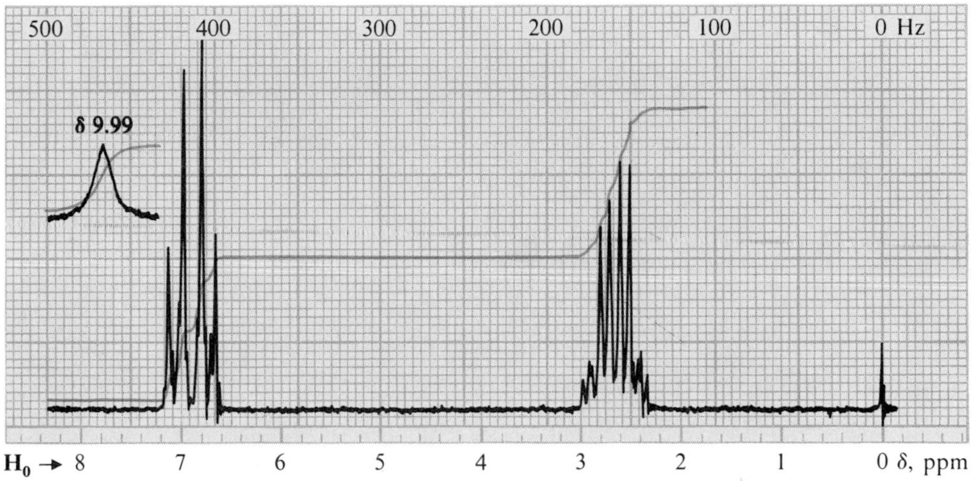

21 Chemistry of Carboxylic Acid Derivatives

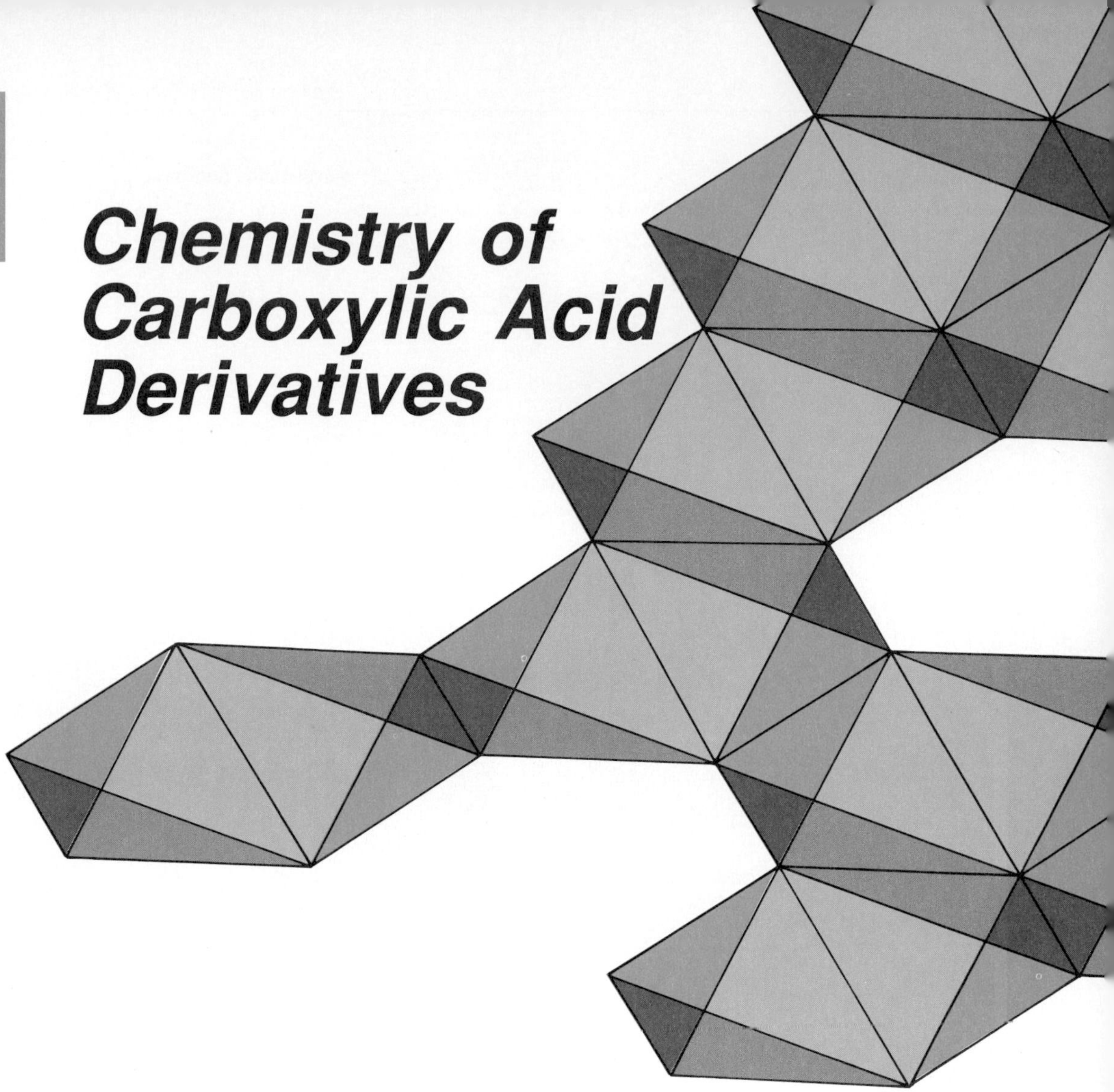

Carboxylic acid derivatives are compounds that can be hydrolyzed under acidic or basic conditions to give a related carboxylic acid. All of them can be formally derived by replacing a small part of the carboxylic acid structure with other groups, as shown in Table 21.1.

We shall find that there is not only a structural similarity among carboxylic acids and their derivatives, but also a close relationship in their chemistry. With the exception of nitriles, all carboxylic acid derivatives contain a *carbonyl group*. Many important reactions of these compounds occur at the carbonyl group, and the —C≡N (cyano) group of nitriles has reactivity that resembles that of a carbonyl group. Thus, the chemistry of carboxylic acid derivatives, like that of aldehydes, ketones, and carboxylic acids, involves the chemistry of the carbonyl group.

TABLE 21.1 Structures of Carboxylic Acid Derivatives

General structure, name of derivative	Condensed structure	Derived from acid by replacing	With	Example
R—C(=O)—O—R′ ester	$R—CO_2R'$	—H	—R′	$CH_3—C(=O)—O—C_2H_5$ ethyl acetate
R—C(=O)—O—C(=O)—R anhydride	$(R—C(=O))_2O$	—H	—C(=O)—R (acyl group)	$CH_3—C(=O)—O—C(=O)—CH_3$ acetic anhydride
R—C(=O)—X acid halide	R—CO—X	—OH	—X (halogen)	$CH_3—C(=O)—Cl$ acetyl chloride
R—C(=O)—NR′R′ amide	$R—CO—NR'_2$	—OH	—NR′R′	$CH_3—C(=O)—NH_2$ acetamide
R—C≡N nitrile	R—CN	$—CO_2H$	—CN (cyano group)	$CH_3—C≡N$ acetonitrile

21.1 NOMENCLATURE AND CLASSIFICATION OF CARBOXYLIC ACID DERIVATIVES

A. Esters and Lactones

Esters are named as derivatives of their parent carboxylic acids by applying a variation of the system used in naming carboxylate salts (Sec. 20.4A). The group attached to the carboxylate oxygen is named first as a simple alkyl or aryl group. This name is followed by the name of the parent carboxylate, which, as we learned, is constructed by dropping the final *ic* from the name of the acid and adding the suffix *ate*. This procedure is used in both common and systematic nomenclature.

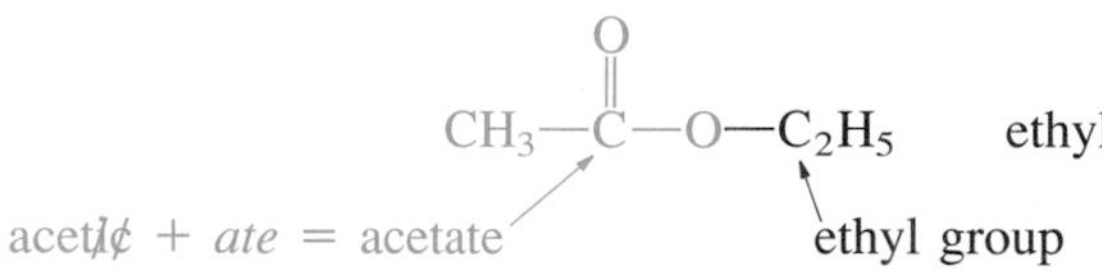

ethyl acetate (common)

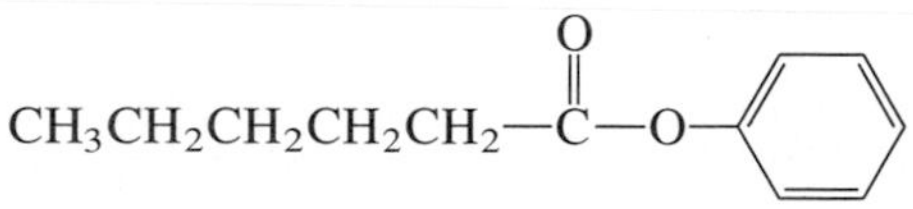

phenyl hexanoate (systematic)

Substitution is indicated by numbering the acid portion of the ester as in acid nomenclature, beginning with the carbonyl as carbon-1 (systematic nomenclature), or with the adjacent carbon as the α-position (common nomenclature). The alkyl or aryl group is numbered (using numbers in systematic nomenclature, Greek letters in common nomenclature) from the point of attachment to the carboxylate oxygen.

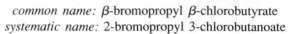

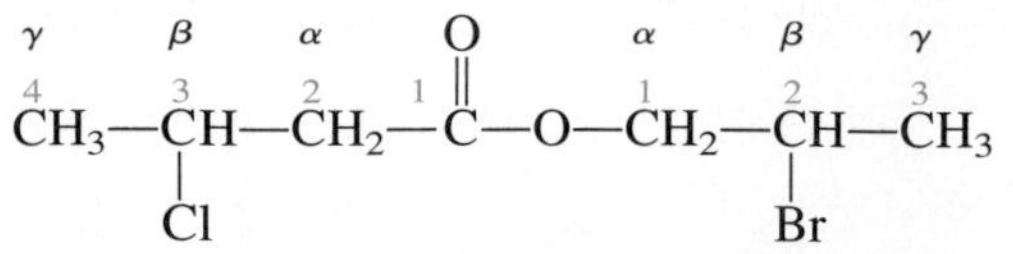

common name: β-bromopropyl β-chlorobutyrate
systematic name: 2-bromopropyl 3-chlorobutanoate

Esters of other acids are named by analogous extensions of acid nomenclature.

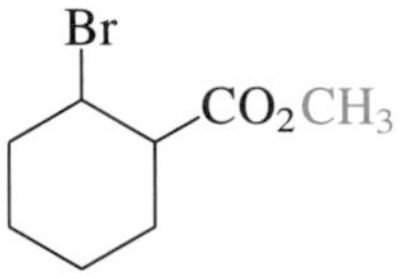

methyl 2-bromocyclohex-anecarboxylate

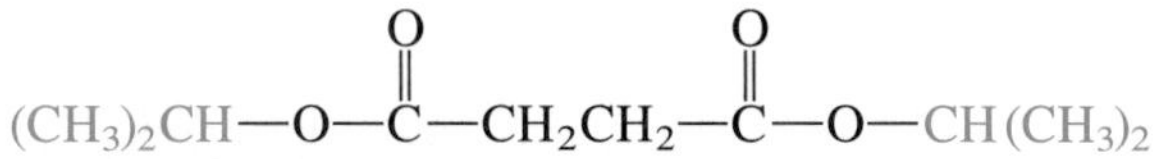

diisopropyl succinate

Cyclic esters are called **lactones.** We shall be concerned only with the common nomenclature of the simplest lactones, illustrated in the following examples:

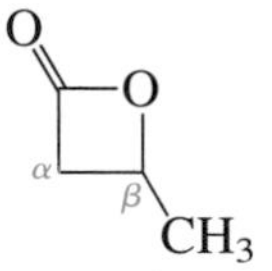

β-butyrolactone (a β-lactone)

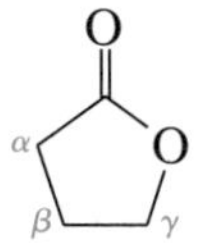

γ-butyrolactone (a γ-lactone)

As these examples illustrate, the *name* of a lactone is derived from the acid with the same number of carbons in its principal chain; the *ring size* is denoted by a Greek letter corresponding to the point of attachment of the lactone ring oxygen to the carbon chain. Thus, in a β-lactone, the ring oxygen is attached at the β-carbon to form a four-membered ring.

B. Acid Halides

Acid halides are named in any system of nomenclature by replacing the *ic* ending of the acid with the suffix *yl,* followed by the name of the halide.

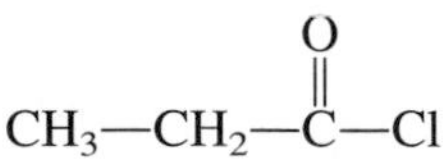

propion~~ic~~ + *yl* =
propionyl chloride

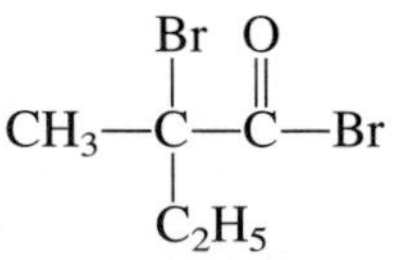

α-bromo-α-methylbutyryl bromide (common)
2-bromo-2-methylbutanoyl bromide (systematic)

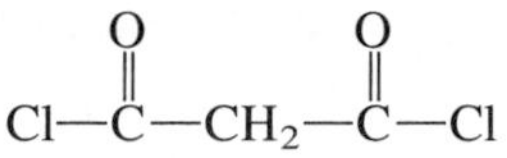

malonyl dichloride

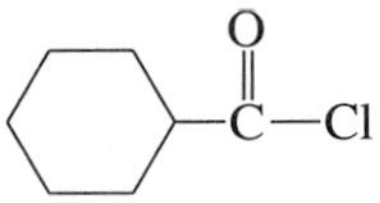

cyclohexanecarbonyl chloride

Notice in the last example the special nomenclature required when the acid halide group is attached to a ring: the compound is named as an alkanecarbonyl halide.

C. Anhydrides

To name an anhydride, the name of the parent acid is followed by the word *anhydride.*

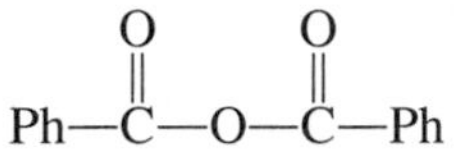

benzoic anhydride

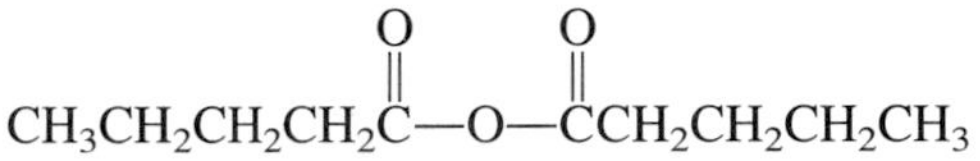

valeric anhydride (common)
pentanoic anhydride (systematic)

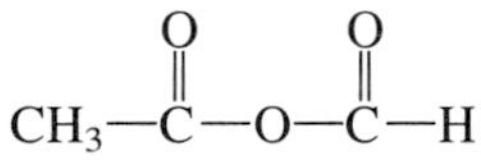

acetic formic anhydride
(a mixed anhydride)

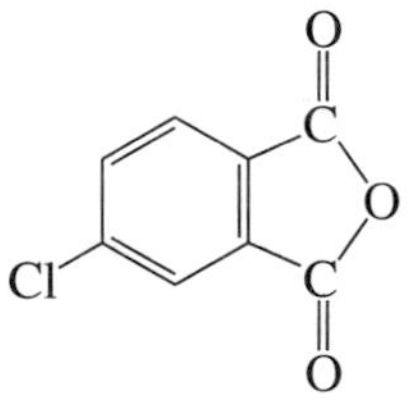

4-chlorophthalic anhydride
(a cyclic anhydride)

Acetic formic anhydride (above) is an example of a **mixed anhydride,** an anhydride derived from two different acids. Mixed anhydrides are named by citing the two parent acids in alphabetical order.

D. Nitriles

Nitriles are named in the common system by dropping the *ic* or *oic* from the name of the acid *with the same number of carbon atoms* (counting the nitrile carbon) and adding the suffix *onitrile.* In systematic nomenclature, the suffix *nitrile* is added to the name of the hydrocarbon with the same number of carbon atoms.

$Ph-C\equiv N:$ benzonitrile (benz~~oic~~ + *onitrile*)

$CH_3-C\equiv N:$ acetonitrile (acet~~ic~~ + *onitrile*)

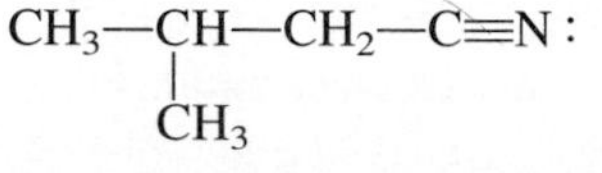

isovaleronitrile (common)
3-methylbutanenitrile (systematic)

$:N\equiv C-CH_2-CH_2-C\equiv N:$ succinonitrile

The name of the three-carbon nitrile is shortened in common nomenclature:

$$C_2H_5{-}CN \qquad \text{propionitrile (not propiononitrile)}$$

When the nitrile group is attached to a ring, a special *carbonitrile* nomenclature is used.

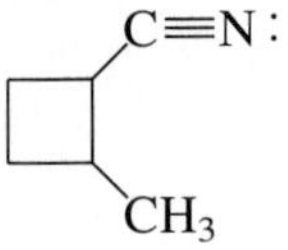

2-methylcyclobutanecarbonitrile

E. Amides, Lactams, and Imides

Simple amides are named in any system by replacing the *ic* or *oic* suffix of the acid name with the suffix *amide*.

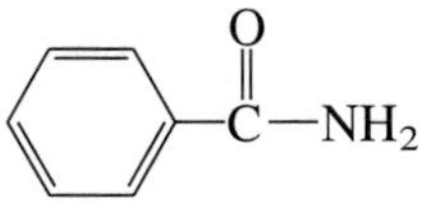

benzamide (benz~~oic~~ + amide)

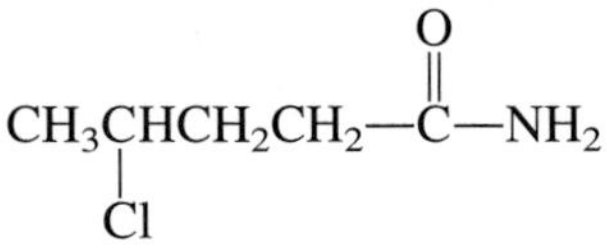

γ-chlorovaleramide (common)
4-chloropentanamide (systematic)

When the amide functional group is attached to a ring, the suffix *carboxamide* is used.

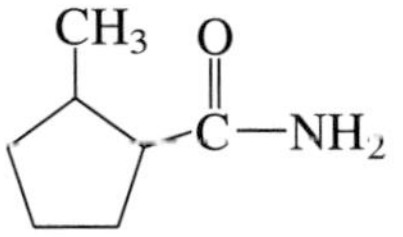

2-methylcyclopentanecarboxamide

Amides are classified as **primary, secondary,** or **tertiary** according to the number of hydrogens on the amide nitrogen.

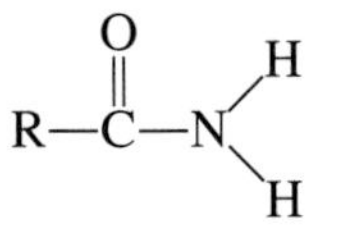

primary amide secondary amide tertiary amide

Notice that this classification, unlike that of alkyl halides and alcohols, refers to substitution *at nitrogen* rather than substitution at carbon. Thus, the following compound is a *secondary amide,* even though there is a tertiary alkyl group bound to nitrogen.

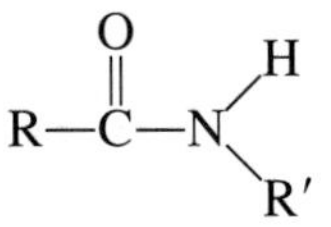

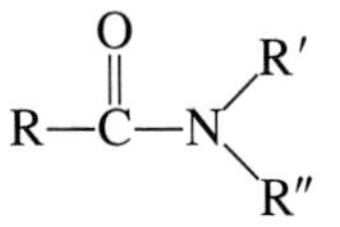

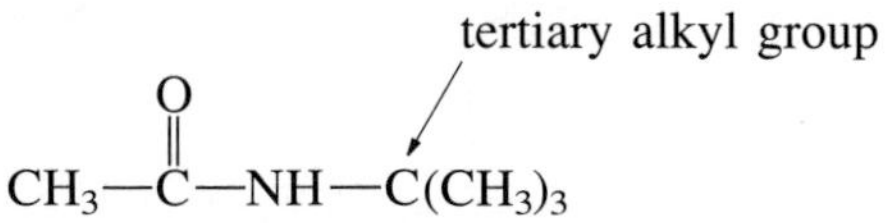

tertiary alkyl group

$$CH_3{-}\overset{\overset{\displaystyle O}{\|}}{C}{-}NH{-}C(CH_3)_3$$
a secondary amide

Substitution on nitrogen in secondary and tertiary amides is designated with the letter *N* (italicized or underlined).

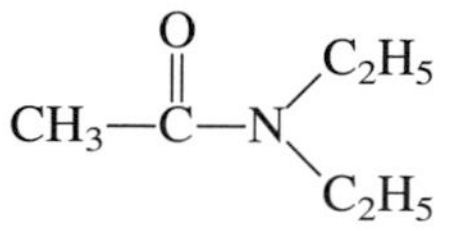

N,N-diethylacetamide (double *N* designation shows that both ethyl groups are on nitrogen)

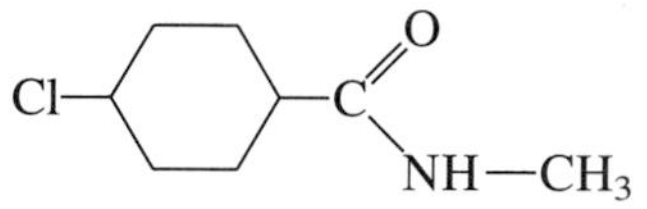

N-methyl-4-chlorocyclohexanecarboxamide

Cyclic amides are called **lactams,** and the common nomenclature of the simple lactams is analogous to that of lactones. Lactams, like lactones, are classified by ring size as γ-lactams (5-membered lactam ring), β-lactams (4-membered lactam ring), and so on.

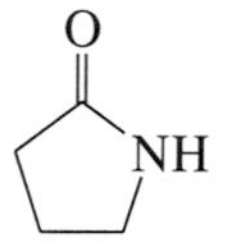

γ-butyrolactam (a γ-lactam)

Imides are formally the nitrogen analogs of anhydrides.

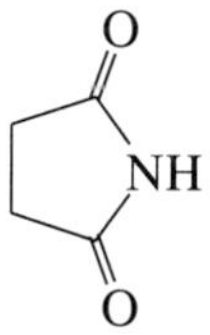

succinimide

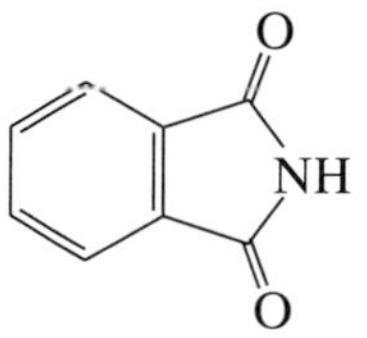

phthalimide

Cyclic imides, of which these two compounds are examples, are of greater importance than open-chain imides, although the latter are also known compounds.

F. Nomenclature of Substituent Groups

The priorities for citing principal groups in a carboxylic acid derivative are as follows:

$$\text{acid} > \text{anhydride} > \text{ester} > \text{acid halide} > \text{amide} > \text{nitrile} \qquad (21.1)$$

All of these groups have citation priority over aldehydes and ketones, as well as the other functional groups we have studied. (A complete list of group priorities is given in Appendix I.) The names used for citing these groups as substituents are given in Table 21.2. The following compounds illustrate the use of these names:

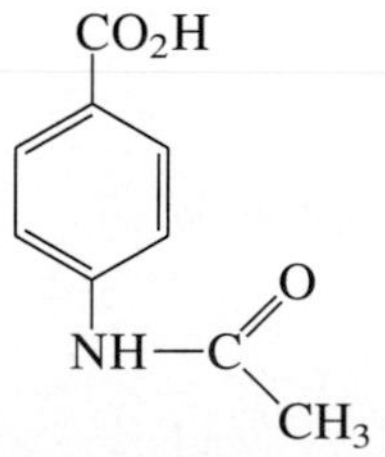

p-acetamidobenzoic acid

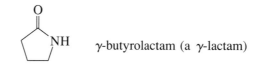

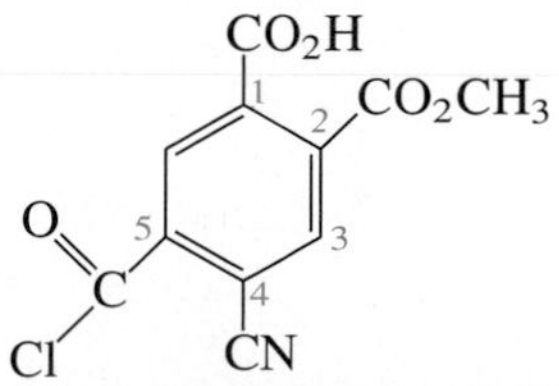

5-chloroformyl-4-cyano-2-meth-oxycarbonylbenzoic acid

TABLE 21.2 Names of Carboxylic Acid Derivatives When Used as Substituent Groups

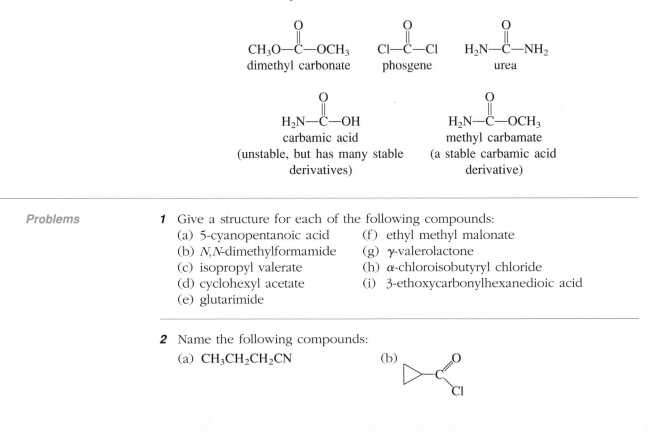

Group	Name	Group	Name
$-C(=O)-OH$	carboxy	$-C(=O)-Cl$	chloroformyl
$-C(=O)-OCH_3$	methoxycarbonyl	$-C(=O)-NH_2$	carbamoyl
$-C(=O)-OC_2H_5$	ethoxycarbonyl	$-NH-C(=O)-CH_3$	acetamido
$-CH_2-C(=O)-OH$	carboxymethyl	$-C\equiv N$	cyano
$-O-C(=O)-CH_3$	acetoxy		

G. Carbonic Acid Derivatives

Esters of carbonic acid are named like any other ester, but other important carbonic acid derivatives have special names that must be learned.

$CH_3O-C(=O)-OCH_3$
dimethyl carbonate

$Cl-C(=O)-Cl$
phosgene

$H_2N-C(=O)-NH_2$
urea

$H_2N-C(=O)-OH$
carbamic acid
(unstable, but has many stable derivatives)

$H_2N-C(=O)-OCH_3$
methyl carbamate
(a stable carbamic acid derivative)

Problems

1 Give a structure for each of the following compounds:

(a) 5-cyanopentanoic acid
(b) *N,N*-dimethylformamide
(c) isopropyl valerate
(d) cyclohexyl acetate
(e) glutarimide
(f) ethyl methyl malonate
(g) γ-valerolactone
(h) α-chloroisobutyryl chloride
(i) 3-ethoxycarbonylhexanedioic acid

2 Name the following compounds:

(a) $CH_3CH_2CH_2CN$

(b) 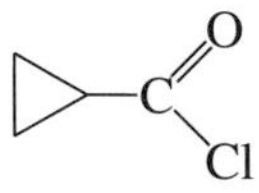

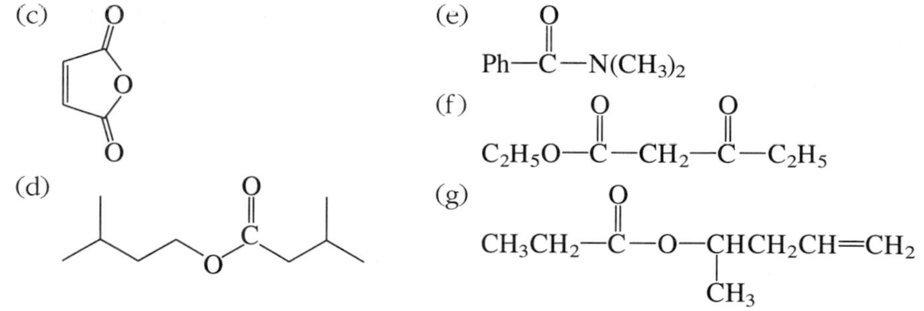

21.2 STRUCTURE OF CARBOXYLIC ACID DERIVATIVES

The structures of many carboxylic acid derivatives are very similar to what we would expect from the structures of other carbonyl compounds. The bond lengths in some carbonyl compounds are compared in Fig. 21.1.

The nitrile C≡N bond length, 1.16 Å, is significantly less than the acetylene C≡C bond length, 1.20 Å. This is another example of the shortening of bonds to electronegative atoms (Sec. 1.6B).

Secondary and tertiary amides can exist in both *E* and *Z* conformations about the carbonyl–nitrogen single bond; the *Z* form predominates in most cases for steric reasons.

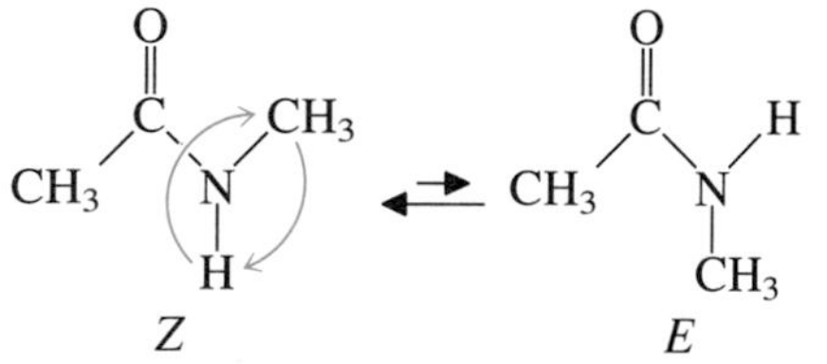

E and *Z* conformations of *N*-methylacetamide (21.2)

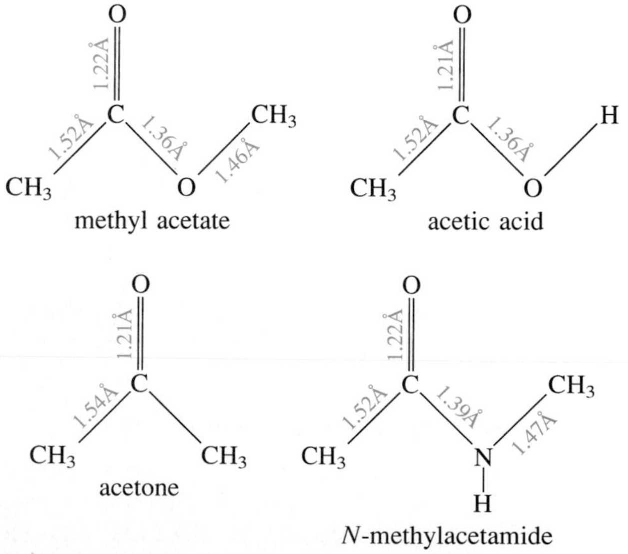

Figure 21.1 Comparison of bond lengths in some carbonyl compounds.

The interconversion of these conformations is too rapid at room temperature to permit their separate isolation, but it is very slow compared with rotation about ordinary single bonds. A typical energy barrier for rotation about the carbonyl–nitrogen bond of an amide is 17 kcal/mol; this means that the rotation occurs about ten times per second. (By contrast, internal rotation in butane occurs about 10^{11} times per second.) The relatively slow rotation in amides can be accounted for by the fact that an amide has an important resonance structure in which the carbonyl–nitrogen bond has substantial double-bond character. As we have learned, rotation about double bonds is much slower than rotation about single bonds.

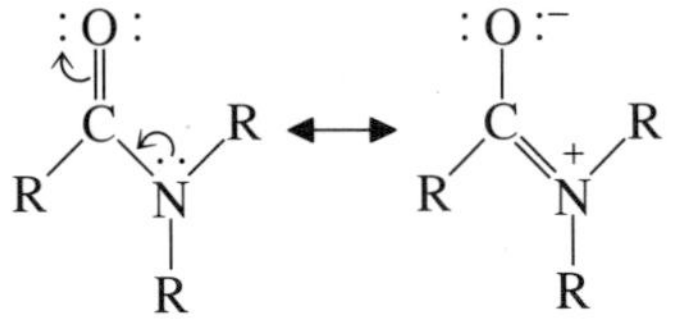
(21.3)

amide resonance structures

Problem

3 Draw the structure of an amide that *must* exist in an *E* conformation about the carbonyl–nitrogen bond.

21.3 PHYSICAL PROPERTIES OF CARBOXYLIC ACID DERIVATIVES

A. Esters

Esters are polar molecules, but they lack the capability of acids to donate hydrogen bonds. The lower esters are typically volatile, fragrant liquids with lower density than water. Most esters are not soluble in water. The low boiling point of a typical ester is illustrated by the following comparison:

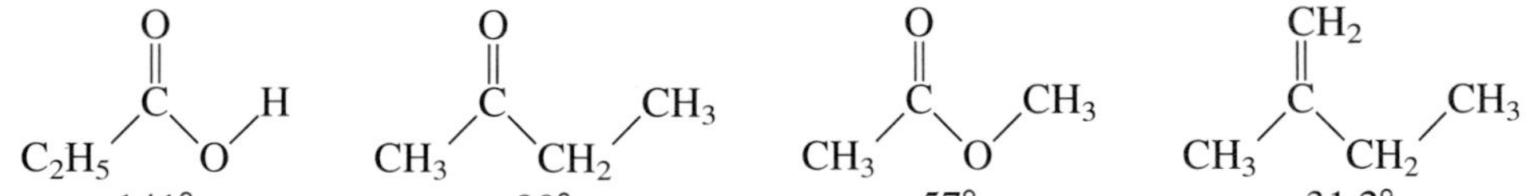

boiling point 141° 80° 57° 31.2°

Problem

4 Account for the fact that methyl acetate (μ = 1.8 D) has a lower dipole moment than 2-butanone (μ = 2.8 D), and hence a lower boiling point.

B. Anhydrides and Acid Chlorides

Most of the lower anhydrides and acid chlorides are dense, water-insoluble liquids with acrid, piercing odors. The boiling points are not very different from those of other polar molecules of about the same molecular weight and shape.

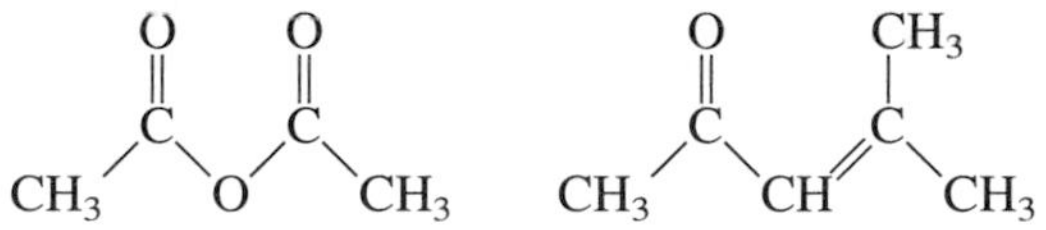

boiling point	139.6°	129.8°
density	1.082	

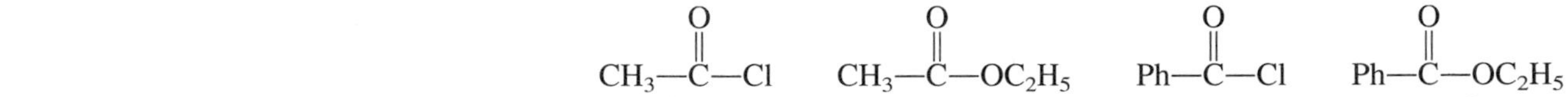

boiling point	50.9°	57°	197.2°	213°
density	1.051		1.212	

The simplest anhydride, formic anhydride, and the simplest acid chloride, formyl chloride, are unstable and cannot be isolated under normal conditions.

C. Nitriles

Nitriles are among the most polar organic compounds. Acetonitrile, for example, has a dipole moment of 3.4 D. The polarity of nitriles is reflected in their boiling points, which are rather high despite their lack of hydrogen bonding.

	$CH_3—C\equiv N:$	$C_2H_5—C\equiv N:$	$CH_3—C\equiv C—H$	$CH_2{=}C{=}O$
boiling point	81.6°	97.4°	−23.3°	−56°

Although nitriles are very poor hydrogen-bond acceptors (because they are very weak bases; see Sec. 21.5), acetonitrile is miscible with water and propionitrile has a moderate solubility in water. Higher nitriles are insoluble in water. Acetonitrile is a particularly valuable polar aprotic solvent because of its moderate boiling point and its relatively high dielectric constant of 38.

D. Amides

The lower amides are water-soluble, polar molecules with high boiling points. Primary and secondary amides, like carboxylic acids (Sec. 20.2), tend to associate into hydrogen-bonded dimers or higher aggregates in the solid state, in the pure liquid state, or in solvents that do not form hydrogen bonds. This association has a noticeable effect on the properties of amides, and is of substantial biological importance, as we shall learn in Chapter 26. For example, simple amides have very high boiling points; many are solids.

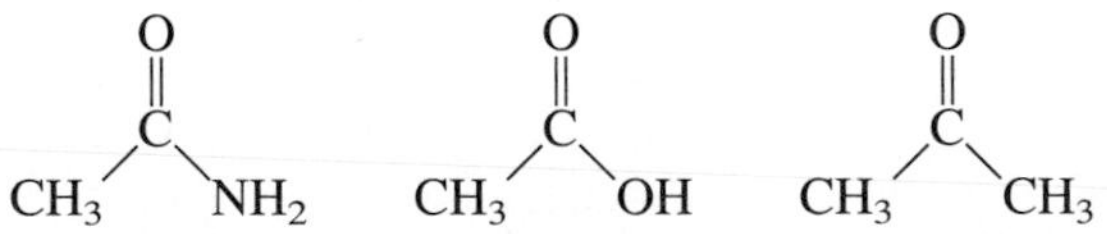

boiling point	221.2°	117.9°	56.5°
melting point	82.3°	16.7°	−94°

Primary amides have two protons on the amide nitrogen that can form hydrogen bonds. Along a series in which these protons are replaced by methyl groups, the capacity for hydrogen bonding is reduced, and boiling points decrease (in spite of the increase in molecular weight).

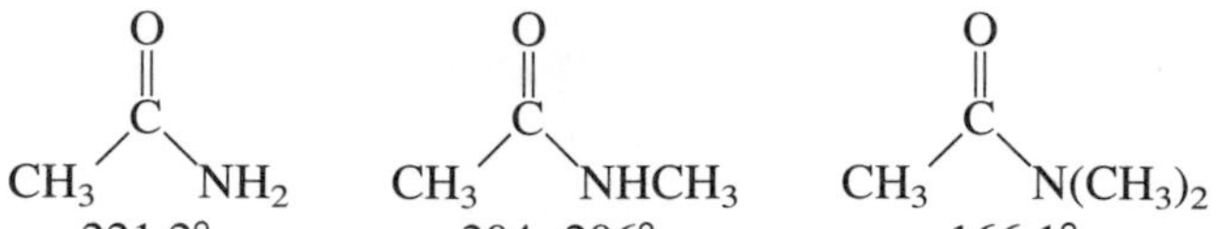

boiling point	221.2°	204–206°	166.1°
melting point	82.3°	28°	−20°

A number of amides have high dielectric constants (see Table 8.2). *N,N*-Dimethylformamide (DMF), for example, dissolves a number of inorganic salts, and is widely used as a polar aprotic solvent, despite its high boiling point.

21.4 SPECTROSCOPY OF CARBOXYLIC ACID DERIVATIVES

A. IR Spectroscopy

The most important feature in the IR spectra of most carboxylic acid derivatives is the C=O stretching absorption. For nitriles, the most important feature in the IR spectrum is the C≡N stretching absorption. These absorptions are summarized in Table 21.3, along with the absorptions of other carbonyl compounds. Some of the noteworthy trends in this table are the following:

1. Esters are readily differentiated from carboxylic acids, aldehydes, or ketones by the unique ester carbonyl absorption at 1735 cm^{-1}.

2. Lactones, lactams, and cyclic anhydrides, like cyclic ketones, have carbonyl absorption frequencies that increase dramatically as the ring size decreases.

3. Anhydrides and some acid chlorides have two carbonyl absorptions. The two carbonyl absorptions of anhydrides are due to the symmetrical and unsymmetrical stretching vibrations of the carbonyl group (Fig. 12.7). (The reason for the double absorption of acid chlorides is more obscure.)

4. The carbonyl absorption of amides occurs at much lower frequency than that of other carbonyl compounds.

5. The C≡N stretching absorption of nitriles generally occurs in the triple-bond region of the spectrum. This absorption is stronger, and occurs at a higher frequency, than the C≡C absorption of an acetylene (why?).

Examples of IR spectra of carboxylic acid derivatives are shown in Fig. 21.2a–c.

Other useful absorptions in the IR spectra of carboxylic acid derivatives are also summarized in Table 21.3. For example, primary and secondary amides show an N—H stretching absorption in the 3200–3400 cm^{-1} region of the spectrum. Many primary amides show a double N—H absorption, and secondary amides show a single strong N—H absorption. In addition, a strong N—H bending absorption occurs in the vicinity of 1640 cm^{-1}, typically appearing as a shoulder on the low-frequency side of the amide carbonyl band. Obviously, tertiary amides lack both of these N—H vibrations. The presence of these bands in a primary amide and their absence in a tertiary amide is evident in the comparison of the two spectra in Fig. 21.3.

TABLE 21.3 Important Infrared Absorptions of Carbonyl Compounds and Nitriles

Compound	Carbonyl absorption, cm^{-1}	Other absorption, cm^{-1}
ketone	1710–1715	
α,β-unsaturated ketone	1670–1680	
aryl ketone	1680–1690	
cyclopentanone	1745	
cyclobutanone	1780	
aldehyde	1720–1725	aldehydic C—H stretch at 2720
α,β-unsaturated aldehyde	1680–1690	
aryl aldehyde	1700	
carboxylic acid (dimer)	1710	OH stretch at 2400–3000 (strong, broad); C—O stretch at 1200–1300
aryl carboxylic acid	1680–1690	
ester or 6-membered lactone (δ-lactone)	1735	C—O stretch at 1000–1300
α,β-unsaturated ester	1720–1725	
5-membered lactone (γ-lactone)	1770	
4-membered lactone (β-lactone)	1840	
acid chloride	1800	second weaker band sometimes observed at 1700–1750
anhydride	1760, 1820 (two bands)	C—O stretch as in ester
6-membered cyclic anhydride	1750, 1800	
5-membered cyclic anhydride	1785, 1865	
amide	1650–1655	N—H bend at 1640 N—H stretch at 3200–3400; double absorption for primary amide
6-membered lactam (δ-lactam)	1670	
5-membered lactam (γ-lactam)	1700	
4-membered lactam (β-lactam)	1745	
nitrile		C≡N stretch at 2200–2250

Figure 21.2 Infrared spectra of some carboxylic acid derivatives: (a) ethyl acetate; (b) butyronitrile; (c) propionic anhydride.

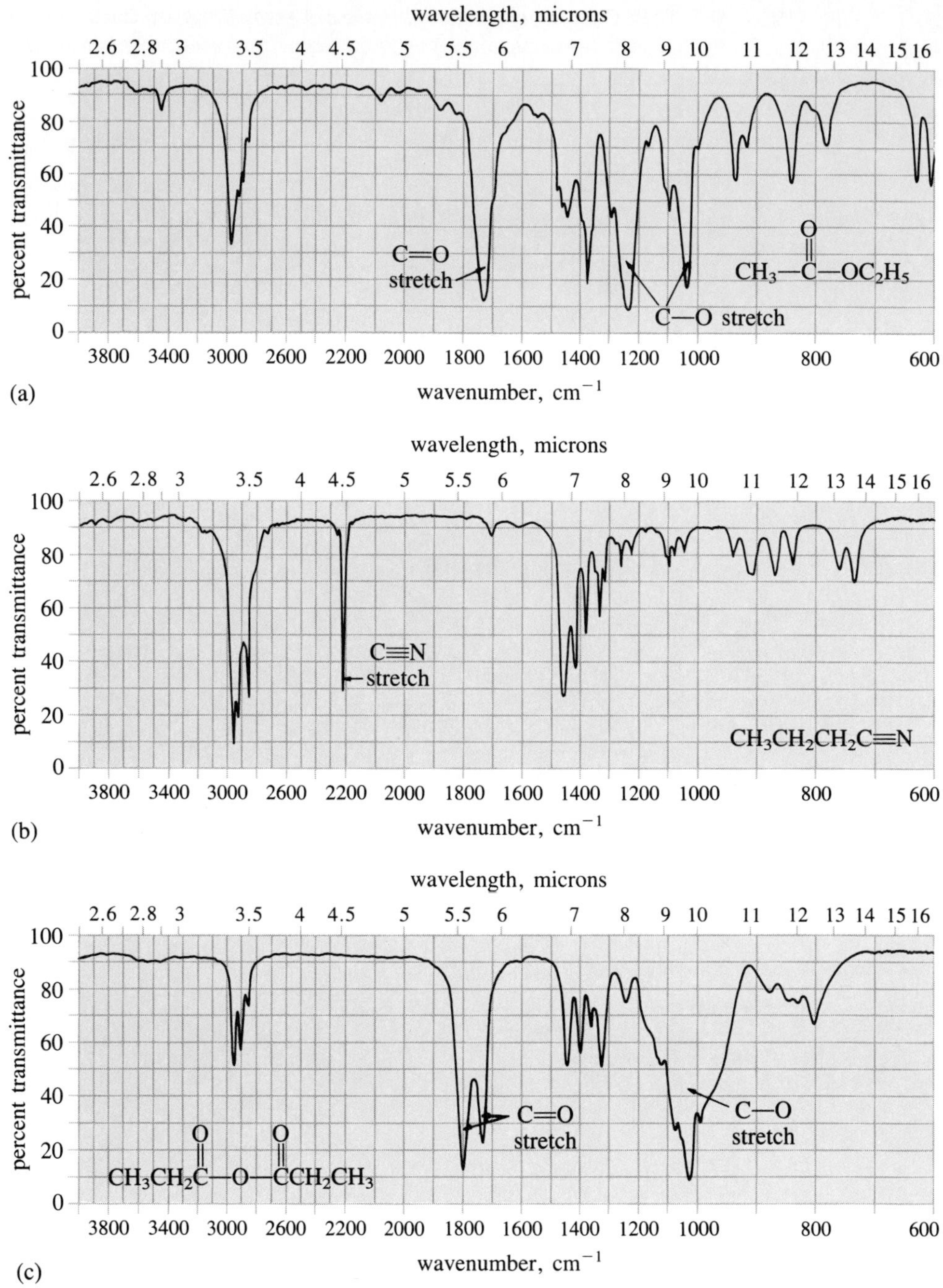

B. NMR Spectroscopy

The α-proton resonances of all carboxylic acid derivatives are observed in the δ 1.9–3 region of the NMR spectrum (see Table 13.2). In esters, the chemical shifts of protons on the carbon adjacent to oxygen occur at about 0.6 ppm lower field than the analogous protons in alcohols and ethers. This shift is attributable to the electronegativity of the carbonyl group.

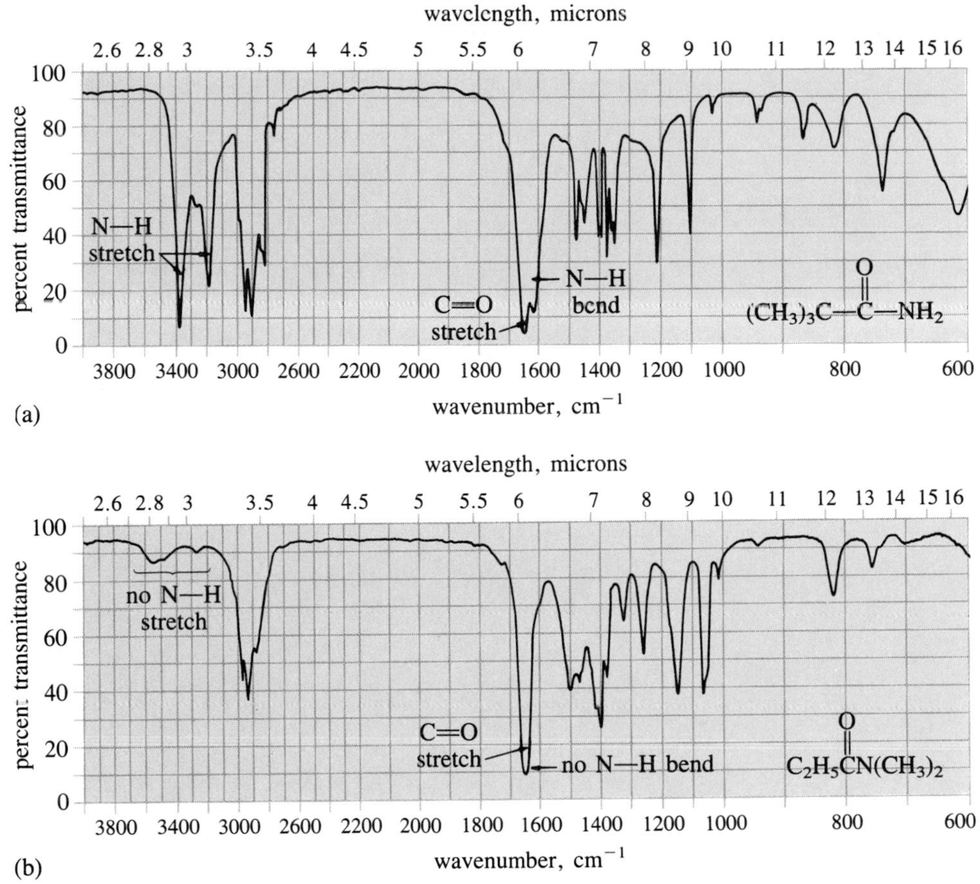

Figure 21.3 Infrared spectra of amides: (a) 2,2-dimethylpropanamide (pivalamide); (b) *N,N*-dimethylpropanamide.

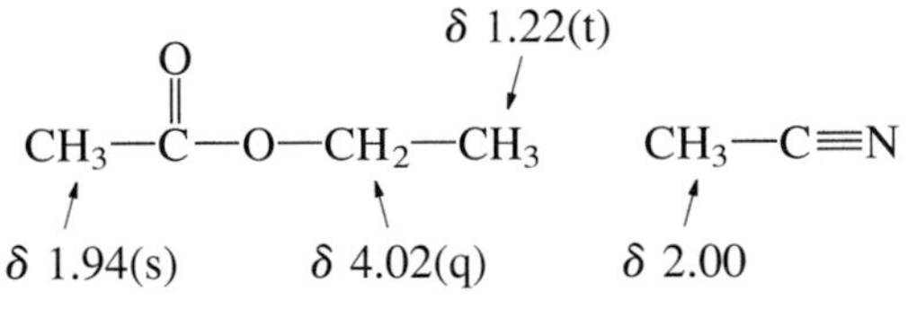

The *N*-alkyl protons of amides have chemical shifts in the δ 2.6–3 chemical shift region, and the N—H proton resonances of primary and secondary amides are observed in the δ 7.5–8.5 region. The signals for these protons, like those of carboxylic acid O—H protons, are broad. This broadening is caused by a slow chemical exchange with the protons of other protic substances (such as traces of moisture) and by unresolved splitting with ^{14}N, which has a nuclear spin. Amide N—H resonances, like the O—H signals of acids and alcohols, can be washed out by exchange with D_2O.

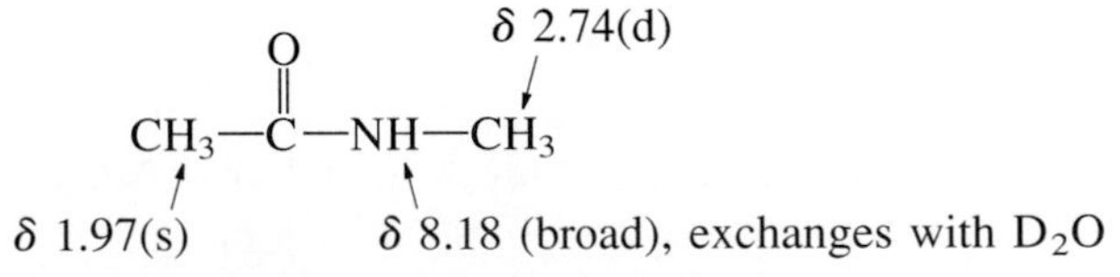

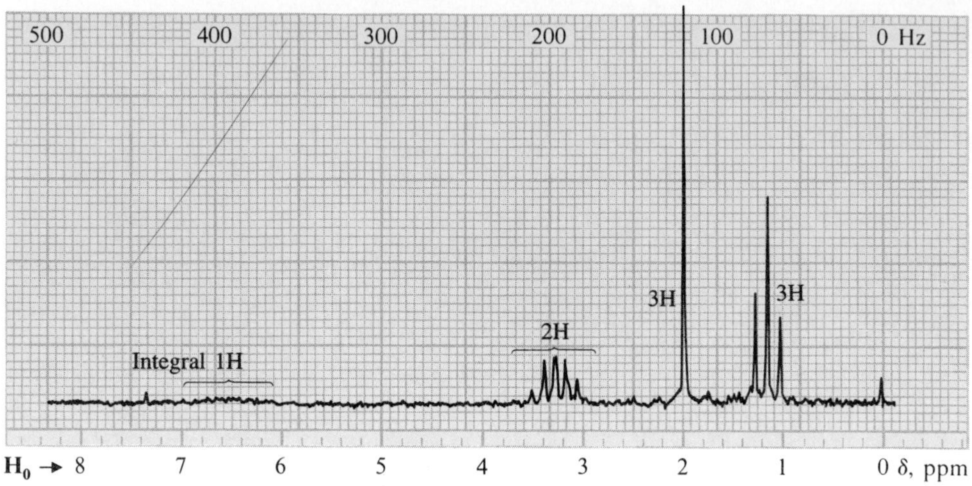

Figure 21.4 NMR spectrum for Problem 6.

C. Solving Structural Problems Involving Nitrogen-Containing Compounds

Amides and nitriles are the first nitrogen-containing functional groups we have studied. When we solve structural problems, there are certain things to remember about compounds that contain nitrogen. First, compounds containing an odd number of nitrogen atoms have *odd molecular weight*. This means that the parent ion in the mass spectrum of such a compound will occur at an odd *m/e* value. Second, there is a special formula, given in Eq. 5.77, for calculating the unsaturation number of a compound containing nitrogen. Finally, when a compound contains nitrogen, we cannot tell whether an ion in its mass spectrum is an odd-electron ion or even-electron ion simply by its mass, since the number of nitrogens present determines whether the mass is odd or even (see Problem 38, Chapter 12).

Problems

5 How would you differentiate between the compounds in each of the following pairs:
(a) *p*-ethylbenzoic acid and ethyl benzoate by IR spectroscopy?
(b) $CH_3—O—CH_2CH_2—O—CH_2CH{=}CH_2$ and ethyl butyrate by IR spectroscopy?
(c) 2,4-dimethylbenzonitrile and *N*-methylbenzamide by NMR spectroscopy?
(d) methyl propionate and ethyl acetate by NMR spectroscopy.

6 Identify the compound with the NMR spectrum given in Fig. 21.4. This compound has molecular weight = 87 and absorptions at 3300 and 1650 cm^{-1} in its IR spectrum.

21.5 BASICITY OF CARBOXYLIC ACID DERIVATIVES

Like carboxylic acids themselves, carboxylic acid derivatives are weakly basic at the carbonyl oxygen. This property is particularly important in some of the acid-catalyzed reactions of esters, amides, and nitriles.

The basicity of an ester is about the same as that of the corresponding carboxylic acid.

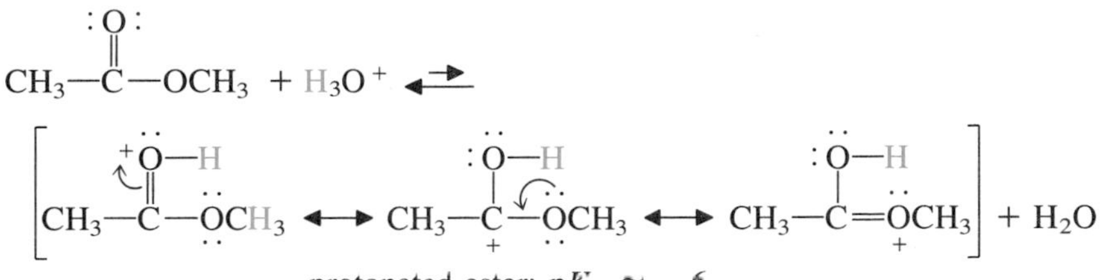

$$
CH_3-\overset{\overset{\displaystyle :O:}{\|}}{C}-OCH_3 \ + \ H_3O^+ \ \rightleftharpoons
$$

$$
\left[CH_3-\overset{\overset{\displaystyle \overset{+}{O}-H}{\|}}{C}-\overset{\displaystyle ..}{\underset{..}{O}}CH_3 \ \longleftrightarrow \ CH_3-\overset{\overset{\displaystyle :O-H}{\|}}{C}\underset{+}{\overset{..}{O}}CH_3 \ \longleftrightarrow \ CH_3-\overset{\overset{\displaystyle :O-H}{\|}}{C}=\underset{+}{\overset{..}{O}}CH_3 \right] \ + \ H_2O \quad (21.4a)
$$

protonated ester; $pK_a \approx -6$

Amides are considerably more basic than other acid derivatives. This basicity, relative to esters, is a reflection of the reduced electronegativity of nitrogen relative to oxygen. That is, the resonance structures in which positive charge is shared on nitrogen are particularly important for a protonated amide.

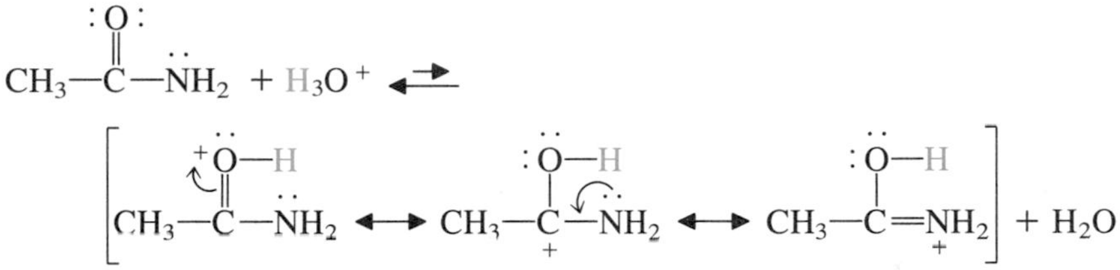

$$
CH_3-\overset{\overset{\displaystyle :O:}{\|}}{C}-\overset{..}{N}H_2 \ + \ H_3O^+ \ \rightleftharpoons
$$

$$
\left[CH_3-\overset{\overset{\displaystyle \overset{+}{O}-H}{\|}}{C}-\overset{..}{N}H_2 \ \longleftrightarrow \ CH_3-\overset{\overset{\displaystyle :O-H}{\|}}{C}\underset{+}{\overset{..}{N}}H_2 \ \longleftrightarrow \ CH_3-\overset{\overset{\displaystyle :O-H}{\|}}{C}=\underset{+}{N}H_2 \right] \ + \ H_2O \quad (21.4b)
$$

protonated amide;
$pK_a \approx -0.5$ to -1

Amides are not quite basic enough, however, to dissolve as their conjugate acids in *dilute* HCl solution.

Observe carefully that both esters and amides protonate on the *carbonyl oxygen;* protonation of esters on the carboxylate oxygen or amides on the nitrogen would give a cation that is not resonance stabilized (see Sec. 20.4B). The site of protonation of amides was for many years a subject of controversy, because ammonia and amines ($R_3N:$) are protonated on nitrogen. It has been estimated, however, that protonation on the amide nitrogen is less favorable than carbonyl protonation by a factor of about 10^8.

Nitriles are very weak bases; protonated nitriles have a pK_a of about -10.

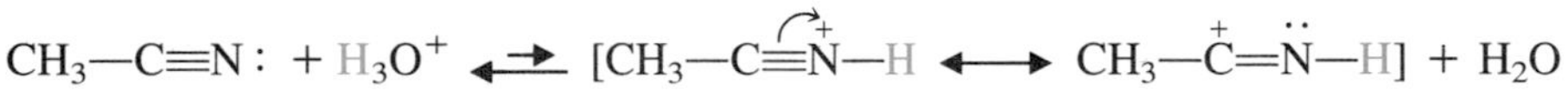

$$
CH_3-C\equiv N: \ + \ H_3O^+ \ \rightleftharpoons \ [CH_3-C\equiv\overset{+}{N}-H \ \longleftrightarrow \ CH_3-\overset{+}{C}=\overset{..}{N}-H] \ + \ H_2O \quad (21.5)
$$

protonated nitrile;
$pK_a \approx -10$

One reason for the poor basicity of nitriles is that the electron pair on the nitrogen is in an *sp* orbital. Just as acetylenic anions, which also have unshared electron pairs in *sp* orbitals, are the least basic of the simple carbanions (Eq. 14.27), nitriles are the least basic of the nitrogen analogs:

$$
\overset{}{\underset{}{C}}-\overset{+}{N}H_3 \qquad \overset{}{\underset{}{C}}=\overset{+}{N}H_2 \qquad -C\equiv\overset{+}{N}H \qquad\qquad (21.6)
$$

$$
pK_a \approx 9 \qquad pK_a \approx 3 \qquad pK_a \approx -10
$$

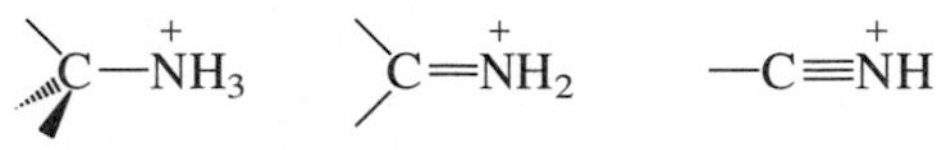

Problem

7 Which of the two compounds in each set below should have the greater basicity at the carbonyl oxygen? Explain.

(a) $CH_3—CH{=}CH—C({=}O)—OCH_3$ or $CH_3—C({=}O)—O—CH_2—CH{=}CH_2$

(b)

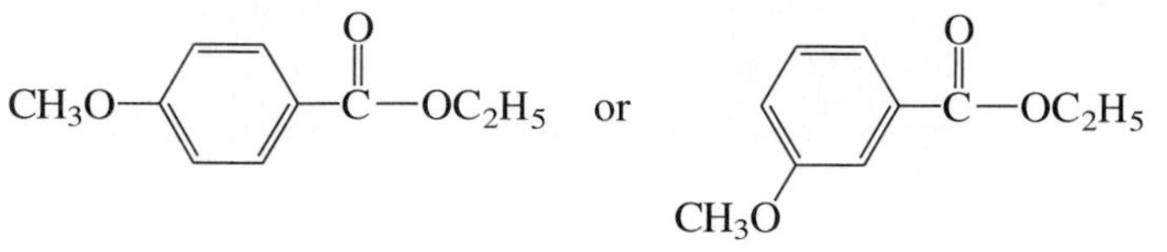

21.6 INTRODUCTION TO REACTIONS OF CARBOXYLIC ACID DERIVATIVES

The reactions of carboxylic acid derivatives can be categorized as follows:

1. Carbonyl group reactions
2. Addition to nitriles
3. Reactions involving the α-carbon
4. Reactions at the nitrogen of amides

One *carbonyl group reaction* of acid derivatives is the reaction of the carbonyl oxygen—and, by analogy, the nitrile nitrogen—as a base. This type of reaction was discussed in the previous section, and often serves as the first step in acid-catalyzed reactions of carboxylic acid derivatives.

As with carboxylic acids, the major carbonyl group reaction of carboxylic acid derivatives is *substitution at the carbonyl group,* which can be represented generally as follows:

$$\underset{\text{carboxylic acid derivative}}{R—\overset{O}{\overset{\|}{C}}—X} + E—Y \longrightarrow \underset{\text{another carboxylic acid derivative}}{R—\overset{O}{\overset{\|}{C}}—Y} + E—X \qquad (21.7)$$

The group —X might be the —Cl of an acid chloride, the —OR of an ester, and so on; this group is substituted by another group —Y. This is precisely the same type of reaction as esterification of carboxylic acids (—X = —OH, E—Y = H—OCH_3; Sec. 20.8A). Carbonyl substitution reactions of carboxylic acid derivatives are the major focus of this chapter.

Although nitriles lack a carbonyl group, the C≡N bond behaves chemically much like a carbonyl group. For example, a typical reaction of nitriles is *addition.*

$$R—C{\equiv}N\!: + E—Y \longrightarrow R—\overset{Y}{\overset{|}{C}}{=}N—E \qquad (21.8)$$

(Compare this reaction with addition to the carbonyl group of an aldehyde or ketone.) The resulting addition products are stable in some cases; but in most of the reactions we shall study, they react further.

Like aldehydes and ketones, carboxylic acid derivatives undergo certain reactions involving the α-carbon. The α-carbon reactions of all carbonyl compounds are grouped together in Chapter 22. The reactivity of amides at the nitrogen is discussed in Sec. 23.11C.

21.7 HYDROLYSIS OF CARBOXYLIC ACID DERIVATIVES

A. Hydrolysis and Cleavage of Esters

Saponification of Esters All carboxylic acid derivatives have in common the fact that they undergo *hydrolysis* (a cleavage reaction with water) to yield carboxylic acids. One of the most important of these reactions is the hydrolysis of esters in base to yield carboxylate salts and alcohols. The carboxylic acid itself is formed when mineral acid is subsequently added to the reaction mixture.

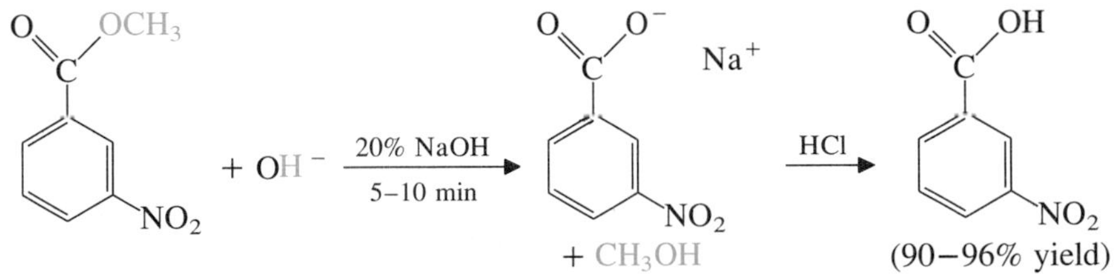

(21.9)

Ester hydrolysis in aqueous hydroxide is called **saponification** because it is used in the production of soaps from fats (Sec. 21.12B). Despite its association with fatty-acid chemistry, the term *saponification* is sometimes used to refer to hydrolysis in base of any acid derivative.

The mechanism of ester saponification involves attack by the nucleophilic hydroxide anion to give a tetrahedral addition intermediate from which an alkoxide ion is expelled.

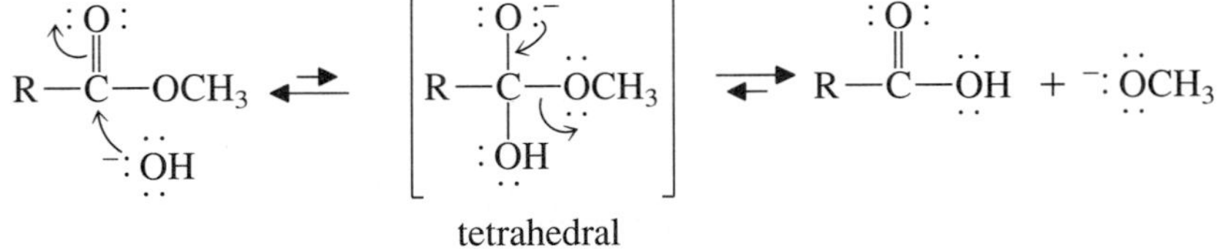

(21.10a)

tetrahedral
addition intermediate

The alkoxide ion thus formed (methoxide in the above example) reacts with the acid to give the carboxylate salt and the alcohol.

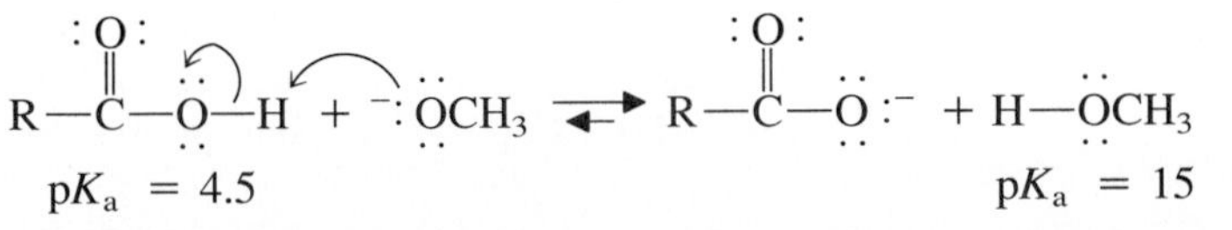

(21.10b)

The equilibrium in this reaction lies far to the right because the carboxylic acid is a much stronger acid than methanol. Once again we see LeChatelier's principle at work: the reaction in Eq. 21.10b removes the carboxylic acid from the equilibrium in Eq. 21.10a as its salt and thus drives the hydrolysis to completion. Hence, *saponification is effectively irreversible.* Although an excess of hydroxide ion is often used as a matter of convenience, many esters can be saponified with just one equivalent of ^{-}OH. The equilibrium for saponification is so favorable that it can be conducted in an alcohol solvent, even though alcohol is one of the saponification products.

Acid-Catalyzed Ester Hydrolysis Since esterification of an acid with an alcohol is a reversible reaction (Sec. 20.8A), esters can be hydrolyzed to carboxylic acids in aqueous mineral acid. In most cases this reaction is slow and must be carried out with an excess of water, in which most esters are insoluble. Saponification, followed by acidification, is a much more convenient method for hydrolysis of most esters because it is faster, it is irreversible, and it can be carried out not only in water, but also in a variety of solvents—even alcohols.

Let us contrast the mechanisms of ester hydrolysis in acid and base. By the principle of microscopic reversibility (Sec. 10.1A), the mechanism of acid-catalyzed hydrolysis is the exact reverse of acid-catalyzed esterification (Sec. 20.8A):

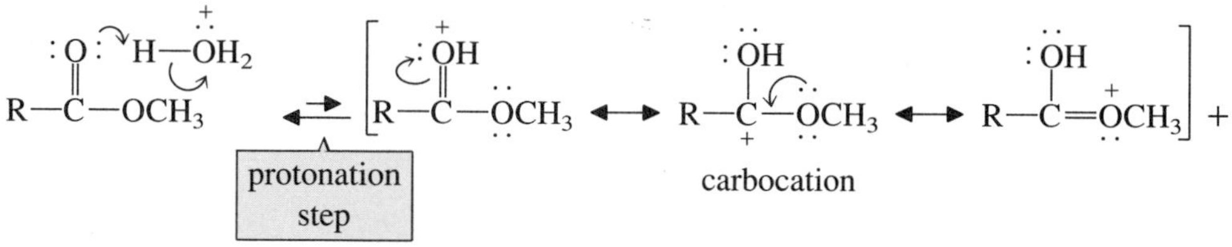

(21.11a)

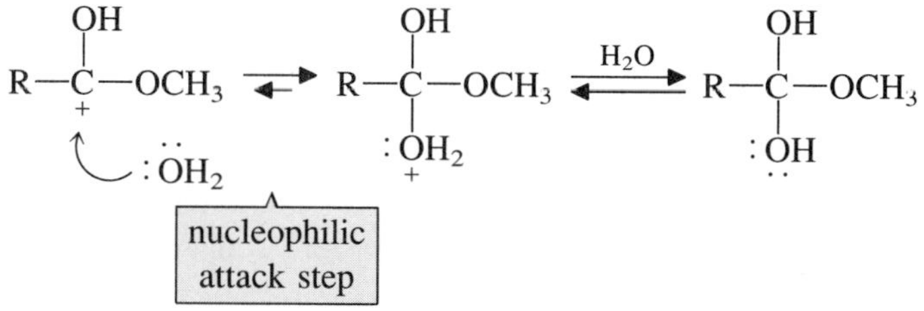

(21.11b)

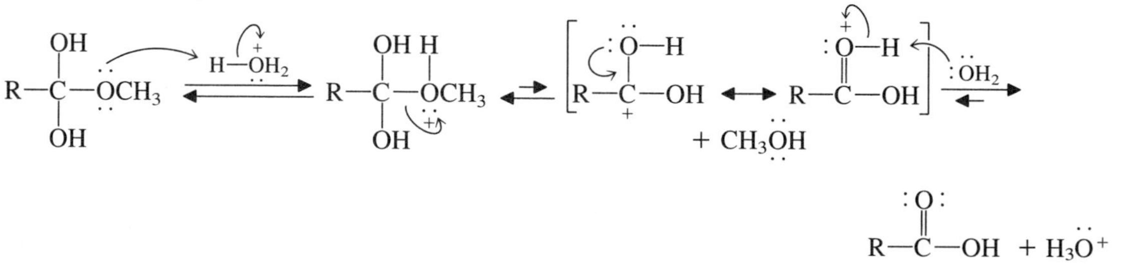

(21.11c)

When a strong acid is present, the carbonyl oxygen is protonated. This protonation gives the carbonyl carbon more carbocation character (Eq. 21.11a). Because there is considerable positive charge on the carbonyl carbon, it is readily attacked by the relatively weak Lewis base (nucleophile) water (Eq. 21.11b). From the saponification mechanism in Eq. 21.10, we see that in base, the carbonyl oxygen is *not* protonated, and the carbonyl carbon therefore has *less* carbocation character. Hence it takes a stronger base than water—hydroxide ion—to react with the carbonyl group (Eq. 21.10a).

Some students try to combine elements of the acidic and basic mechanisms by protonating the carbonyl group and then attacking it with hydroxide.

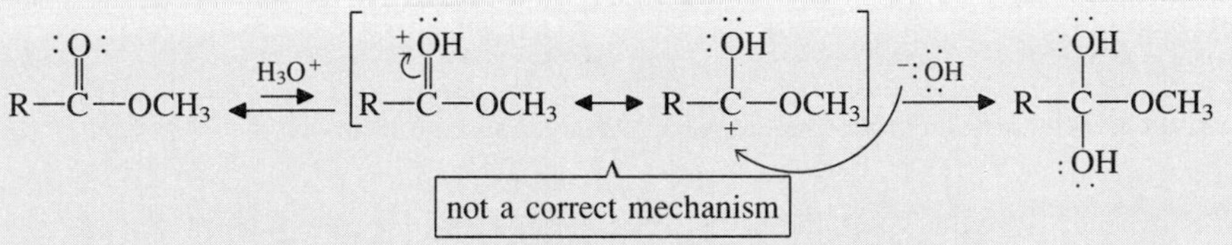

$$R-\overset{\overset{\displaystyle :O:}{\|}}{C}-OCH_3 \overset{H_3O^+}{\longleftrightarrow} \left[R-\overset{\overset{\displaystyle +OH}{\|}}{C}-OCH_3 \longleftrightarrow R-\overset{\overset{\displaystyle :OH}{}}{\underset{+}{C}}-OCH_3 \right] \overset{-:OH}{\longrightarrow} R-\overset{\overset{\displaystyle :OH}{}}{\underset{:OH}{C}}-OCH_3 \quad (21.12)$$

| not a correct mechanism |

Although this mechanism looks as if it ought to be especially attractive, it is not correct because *a solution cannot be acidic enough to protonate the carbonyl oxygen and basic enough to contain substantial hydroxide ion at the same time.* Reactions that require relatively strong acids (such as H_3O^+) typically involve relatively weak bases (such as H_2O); reactions that require relatively strong bases (such as OH^-) involve relatively weak acids (such as H_2O). This same point was made in Sec. 19.7A, where we discussed mechanisms of carbonyl-addition reactions.

Acidic Cleavage of Tertiary Esters As we have just seen, the hydrolysis of ordinary esters in aqueous acid involves attack of water at the carbonyl group. Esters derived from *tertiary alcohols,* such as *t*-butyl esters, are also converted rapidly by acid into carboxylic acids, but by a different mechanism. Consider, for example, cleavage of the following *t*-butyl ester by anhydrous HCl.

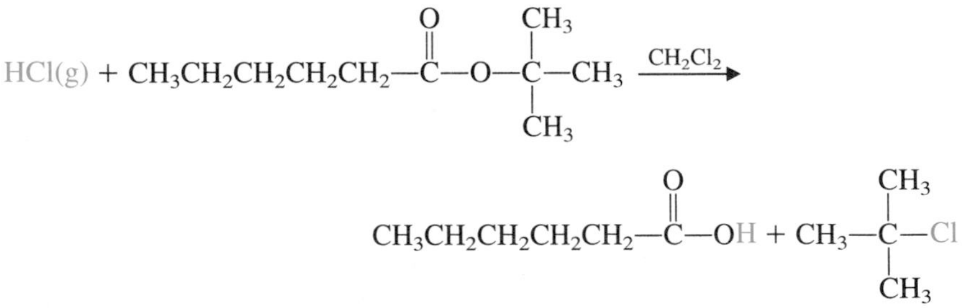

$$HCl(g) + CH_3CH_2CH_2CH_2CH_2-\overset{\overset{\displaystyle O}{\|}}{C}-O-\overset{\overset{\displaystyle CH_3}{|}}{\underset{\underset{\displaystyle CH_3}{|}}{C}}-CH_3 \xrightarrow{CH_2Cl_2}$$

$$CH_3CH_2CH_2CH_2CH_2-\overset{\overset{\displaystyle O}{\|}}{C}-OH + CH_3-\overset{\overset{\displaystyle CH_3}{|}}{\underset{\underset{\displaystyle CH_3}{|}}{C}}-Cl \quad (21.13)$$

As this example illustrates, conversion of tertiary esters into carboxylic acids does not even require water (although water *can* be present).

The first step in the mechanism of this reaction is the same as in mechanisms involving acid-catalyzed carbonyl substitution: protonation of the ester carbonyl oxygen, as in Eq. 21.11a. However, here the similarity to carbonyl substitution ends. The protonated ester undergoes a simple S_N1 reaction to give the carboxylic acid and a relatively stable tertiary carbocation:

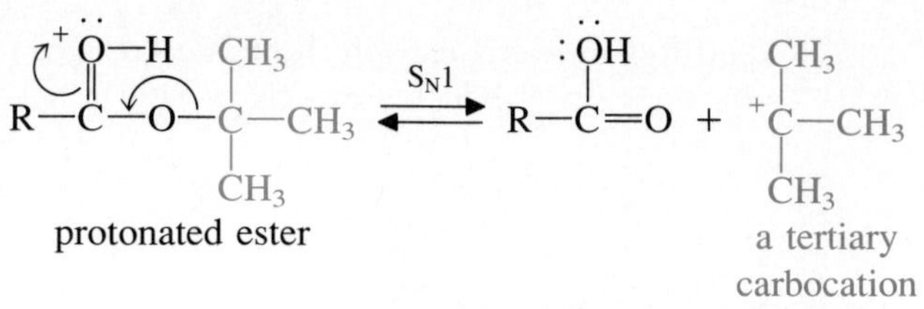

$$\text{protonated ester} \qquad\qquad\qquad \text{a tertiary} \atop \text{carbocation} \tag{21.14a}$$

This cation undergoes the usual carbocation reactions—for example, attack by a nucleophile.

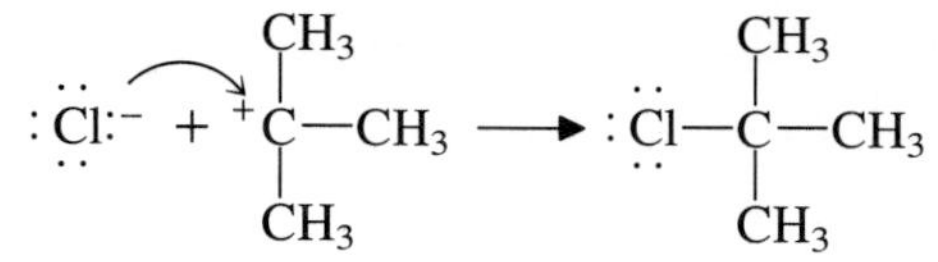

$$\tag{21.14b}$$

Notice that cleavage of tertiary esters by this mechanism involves breaking the *alkyl-oxygen bond* (Eq. 21.14a). In contrast, acid-catalyzed ester hydrolysis by the carbonyl substitution mechanism involves breaking the *carbonyl-oxygen bond* (Eq. 21.11c). Thus, the mechanism of tertiary ester cleavage does not conveniently fit into any of the categories discussed in Sec. 21.6. Rather, it is more like the S_N1 mechanism for substitution in a tertiary alcohol. (Compare Eqs. 21.14a,b with Eqs. 10.14a–c.) Esters of primary and secondary alcohols do not cleave by this mechanism, because the cations that would be formed are much less stable.

Problem

8 Give the products of the following reactions, including the fate of the ^{18}O isotope ($O^* = {}^{18}O$):

(a)

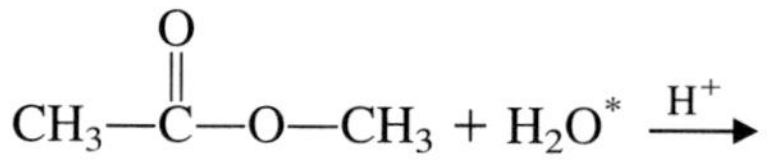

$$CH_3-\overset{\overset{\displaystyle O}{\|}}{C}-O-CH_3 + H_2O^* \xrightarrow{\ H^+\ }$$

(b)

$$CH_3-\overset{\overset{\displaystyle O}{\|}}{C}-O-C(CH_3)_3 + H_2O^* \xrightarrow{\ H^+\ }$$

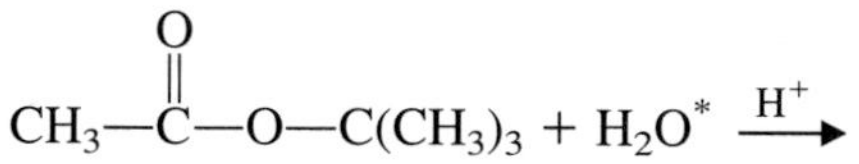

Hydrolysis and Formation of Lactones Because lactones are cyclic esters, they undergo many of the reactions of esters, including saponification. Saponification converts a lactone completely into the salt of the corresponding hydroxy acid.

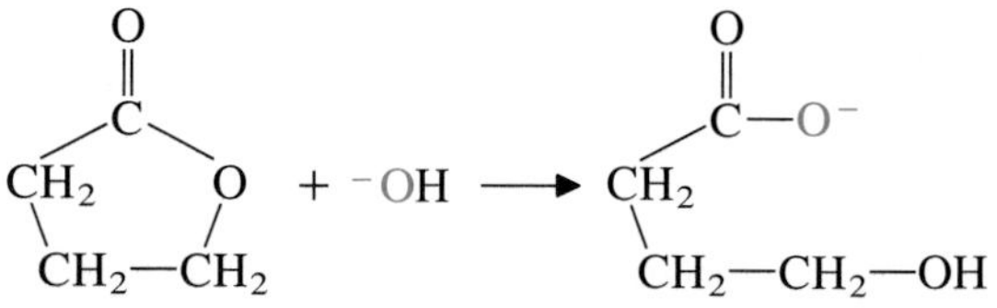

$$\tag{21.15}$$

Upon acidification, the hydroxy acid forms. However, *if a hydroxy acid is allowed to stand in acidic solution, it comes to equilibrium with the corresponding lactone.* The formation of a lactone from a hydroxy acid is nothing more than an *intramolecular* esterification (esterification within the same molecule) and, like esterification, the lactonization equilibrium is acid catalyzed.

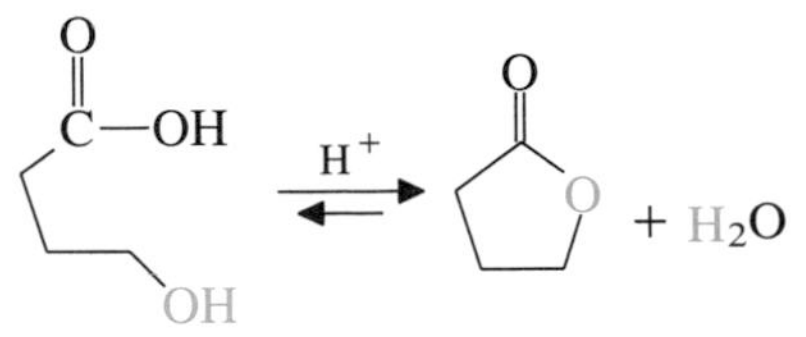

$$K_{eq} \approx 160 \qquad (21.16)$$

$$K_{eq} = 3.3 \qquad (21.17)$$

As the examples in Eqs. 21.16 and 21.17 illustrate, lactones containing five- and six-membered rings are favored at equilibrium over their corresponding hydroxy acids. Although lactones with ring sizes smaller than five or larger than six are well known, they are less stable than their corresponding hydroxy acids. Consequently, the lactonization equilibria for these compounds favor instead the hydroxy acids.

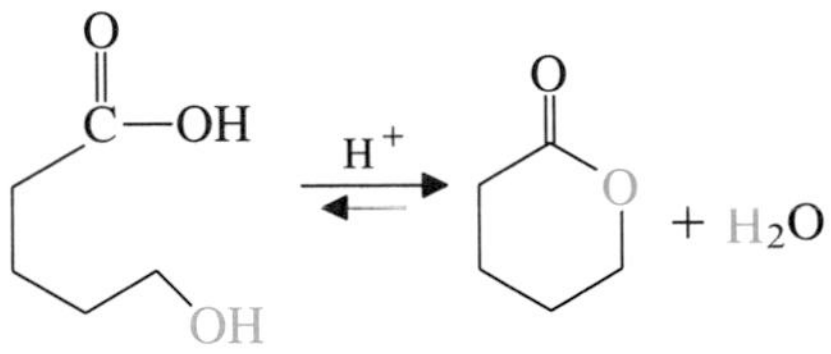

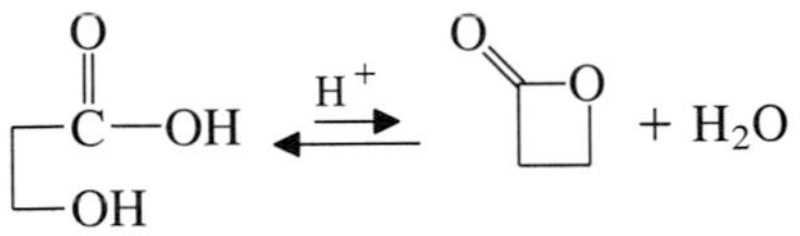

(almost no lactone present at equilibrium) (21.18)

B. Mechanisms of Substitution at the Carbonyl Group

The mechanisms of ester hydrolysis reactions in acid and base are worth special attention for a very important reason: *Most of the carbonyl-substitution reactions we shall study—not only hydrolysis reactions, but other substitution reactions as well— proceed by analogous mechanisms.* Hence, if we understand ester hydrolysis, we understand most of the other carbonyl-substitution reactions that we shall encounter.

Let us generalize: substitution reactions at the carbonyl group occur by two types of mechanisms. Under *basic* conditions, a nucleophile (Nuc : $^-$) attacks the carbonyl carbon to give a tetrahedral addition intermediate in which the carbonyl oxygen assumes a negative charge. The leaving group X : $^-$ is then expelled from the tetrahedral intermediate.

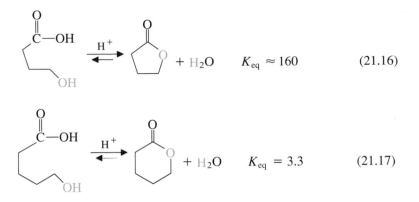

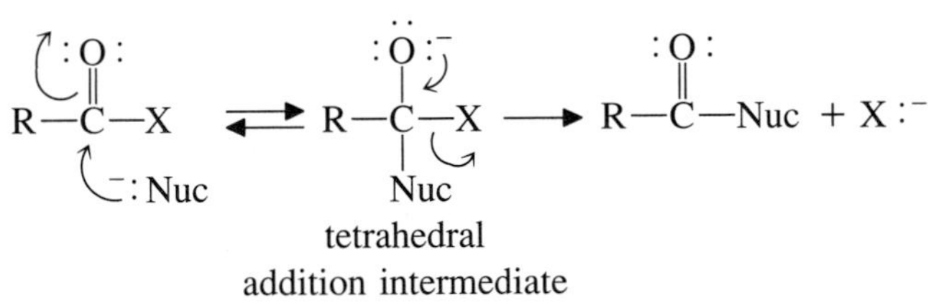

$$
\underset{}{R-\overset{\overset{\textstyle :O:}{\|}}{C}-X} \quad \rightleftharpoons \quad \underset{\text{Nuc}}{R-\overset{\overset{\textstyle :O:^-}{|}}{\underset{|}{C}}-X} \quad \longrightarrow \quad R-\overset{\overset{\textstyle :O:}{\|}}{C}-\text{Nuc} + X:^- \qquad (21.19)
$$

tetrahedral
addition intermediate

In ester saponification, for example, $^-$: Nuc is $^-$OH, and X : $^-$ is the $^-$OR group of the ester.

Under *acidic* conditions, the carbonyl is first protonated. This protonation imparts cationic character to the carbonyl carbon. This carbon is then attacked by a relatively weak base : Nuc—H to form a tetrahedral addition intermediate. After proton transfer to the leaving group —X, it is expelled as H—X.

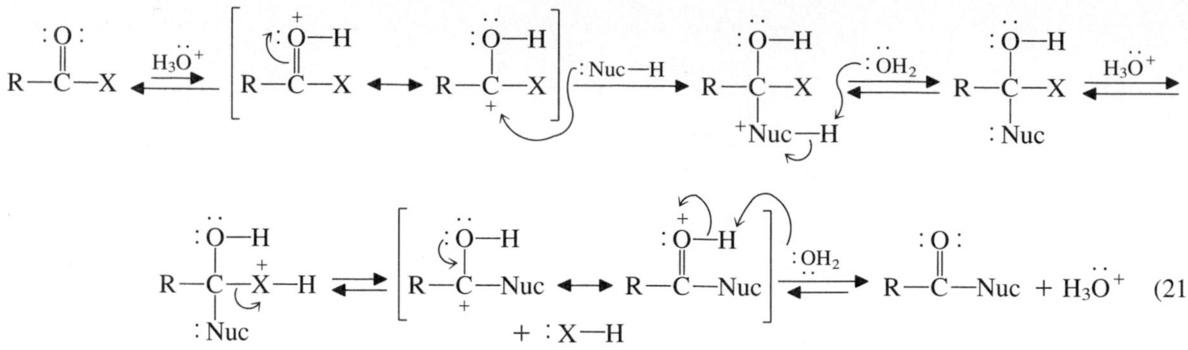

In acid-catalyzed ester hydrolysis, for example, :Nuc—H is HO—H, and X—H is the alcohol product RO—H.

Notice how each subsequent carbonyl substitution reaction that we consider falls into one of these categories.

C. Hydrolysis of Amides

Amides can be hydrolyzed to carboxylic acids and ammonia or amines by heating them in acidic or basic solution.

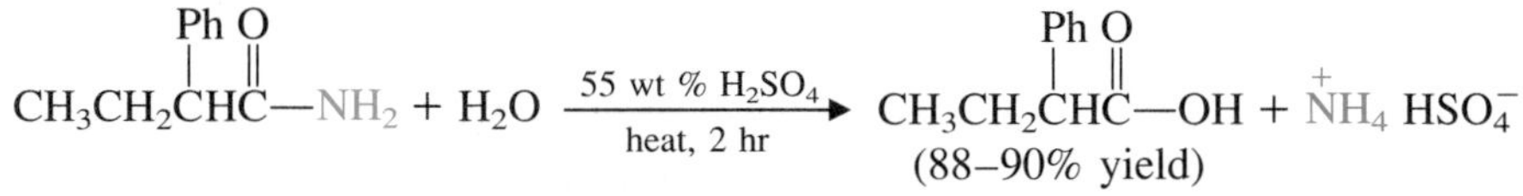

$$\underset{\substack{\text{Ph O} \\ | \; \| \\}}{CH_3CH_2CHC}\!-\!NH_2 + H_2O \xrightarrow[\text{heat, 2 hr}]{55 \text{ wt \% } H_2SO_4} \underset{\substack{\text{Ph O} \\ | \; \| \\}}{CH_3CH_2CHC}\!-\!OH + \overset{+}{N}H_4 \; HSO_4^- \quad (21.21)$$
$$(88\text{–}90\% \text{ yield})$$

In acid, protonation of the ammonia or amine by-product drives the hydrolysis equilibrium to completion. The amine can be isolated, if desired, by addition of base to the reaction mixture following hydrolysis.

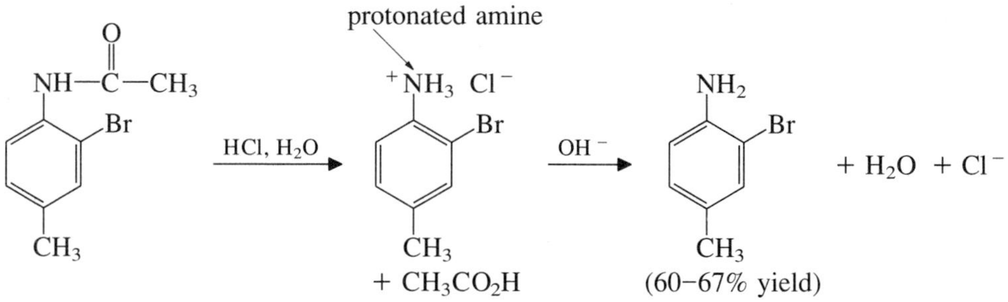

Hydrolysis of amides in base is analogous to saponification of esters. In base, the reaction is driven to completion by formation of the carboxylic acid salt.

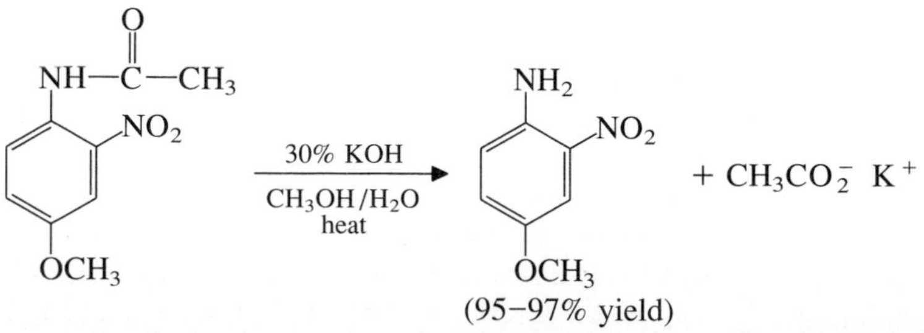

The mechanisms of amide hydrolysis are typical carbonyl substitution mechanisms, as you should explore in the following problem:

Problem

9 Show in detail the hydrolysis mechanism of *N*-methylbenzamide (a) in acid; (b) in base. Assume that each mechanism involves the formation of a tetrahedral addition intermediate.

D. Hydrolysis of Nitriles

Nitriles are hydrolyzed to carboxylic acids and ammonia by heating them in acidic or basic solution.

$$PhCH_2-C\equiv N + 2H_2O + \underset{(57\ wt\ \%)}{H_2SO_4} \xrightarrow[3\ hr]{heat} \underset{(78\%\ yield)}{PhCH_2-CO_2H + NH_4^+\ HSO_4^-} \quad (21.24)$$

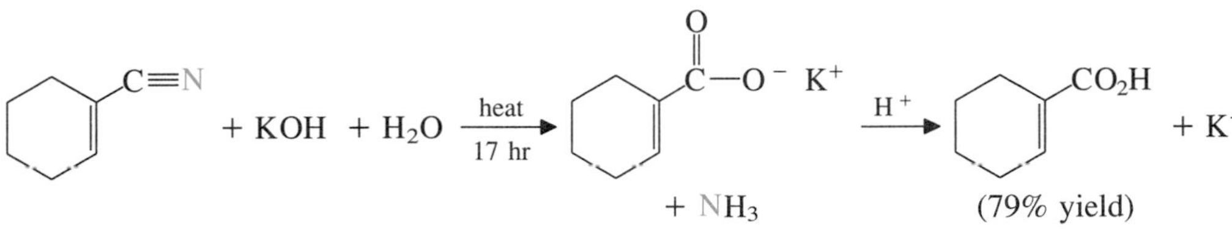

$$(21.25)$$

The mechanism of nitrile hydrolysis in acid begins as an *addition* to the C≡N bond:

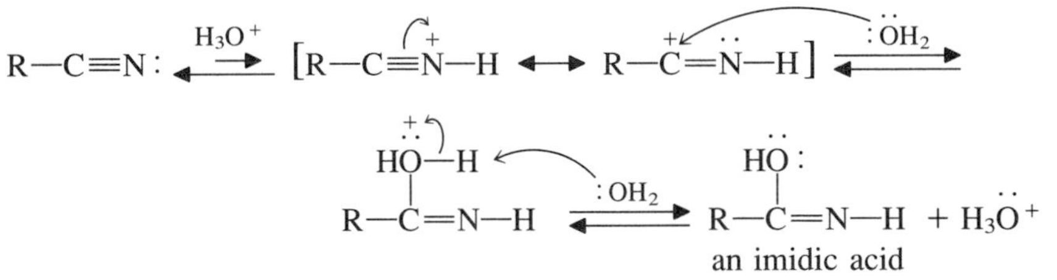

$$R-C\equiv N: \xrightarrow{H_3O^+} [R-C\equiv \overset{+}{N}-H \longleftrightarrow R-\overset{+}{C}=N-H] \xrightarrow{:\overset{..}{O}H_2}$$

$$R-\underset{HO-H}{C}=N-H \underset{}{\overset{:OH_2}{\rightleftarrows}} R-\underset{H\overset{..}{O}:}{C}=N-H + H_3O^+ \quad (21.26a)$$
an imidic acid

The addition product, an *imidic acid,* is unstable and is converted under the reaction conditions to an amide.

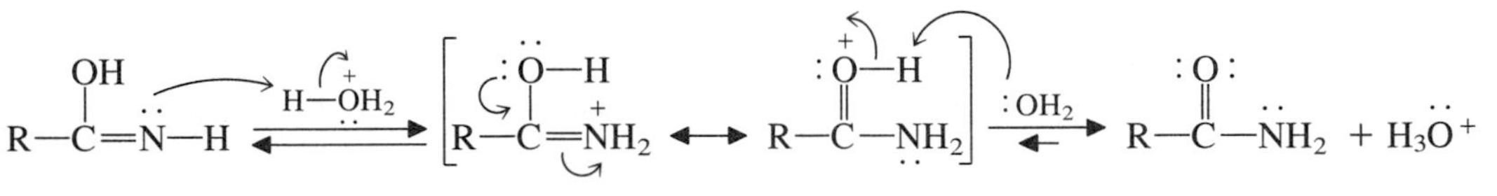

$$R-\underset{OH}{\overset{..}{C}}=N-H \underset{}{\overset{H-\overset{+}{O}H_2}{\rightleftarrows}} [R-\underset{\overset{..}{O}-H}{C}=\overset{+}{N}H_2 \longleftrightarrow R-\underset{:\overset{+}{O}-H}{C}-NH_2] \overset{:OH_2}{\rightleftarrows} R-\underset{:O:}{C}-NH_2 + H_3O^+ \quad (21.26b)$$

The amide thus formed does not survive under the vigorous conditions of nitrile hydrolysis, and is hydrolyzed to a carboxylic acid and ammonium ion, as we discussed in Sec. 21.7C. Thus the ultimate product of nitrile hydrolysis in acid is a carboxylic acid.

Notice that nitriles behave mechanistically much like carbonyl compounds. Compare, for example, the mechanism of acid-promoted nitrile hydrolysis in Eq. 21.26 with that for the acid-catalyzed hydration of an aldehyde or ketone (Sec. 19.7A). In both mechanisms, an electronegative atom is protonated (nitrogen of the C≡N bond, or oxygen of the C=O bond), and water attacks the carbon of the resulting cation.

The parallel between nitrile and carbonyl chemistry is further illustrated by the hydrolysis of nitriles in base. The nitrile group, like a carbonyl group, is attacked by basic nucleophiles and, as a result, the electronegative nitrogen assumes a negative charge. Proton transfer gives an imidic acid (which, like a carboxylic acid, ionizes in base).

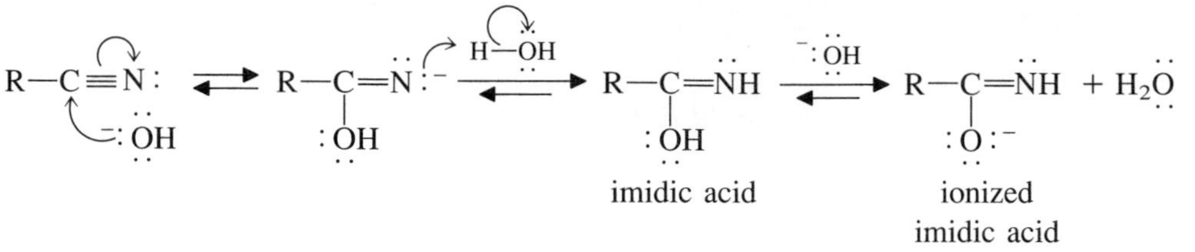

$$\text{(21.27a)}$$

imidic acid ionized
imidic acid

As in acid-promoted hydrolysis, the imidic acid reacts further to give the corresponding amide, which, in turn, hydrolyzes under the reaction conditions to the carboxylate salt of the corresponding carboxylic acid (Sec. 21.7C).

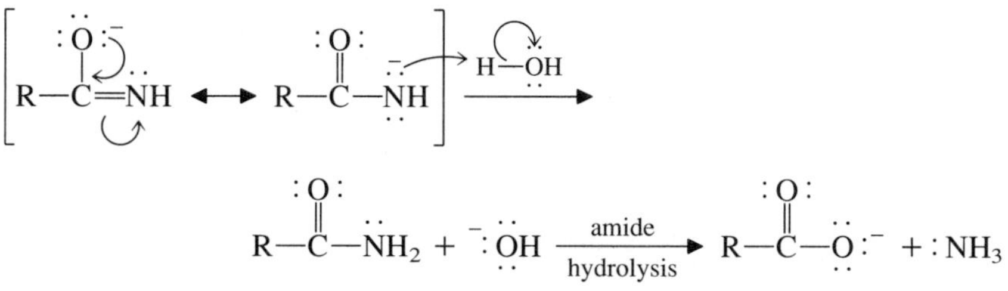

$$\text{(21.27b)}$$

One other useful analogy to nitrile hydrolysis is the hydrolysis of alkynes (Sec. 14.5A). Just as the acid-catalyzed hydrolysis of an alkyne gives a ketone by way of an enol intermediate, hydrolysis of a nitrile gives the nitrogen analog of a ketone—an amide—by way of an analogous *N*-enol intermediate, the imidic acid.

hydrolysis of an alkyne:

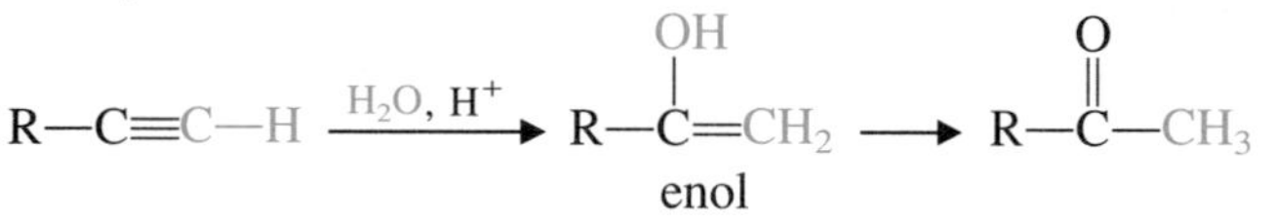

$$R-C\equiv C-H \xrightarrow{\text{H}_2\text{O, H}^+} R-\overset{\overset{\text{OH}}{|}}{C}=CH_2 \longrightarrow R-\overset{\overset{\text{O}}{\|}}{C}-CH_3 \qquad \text{(21.28a)}$$

enol

hydrolysis of a nitrile:

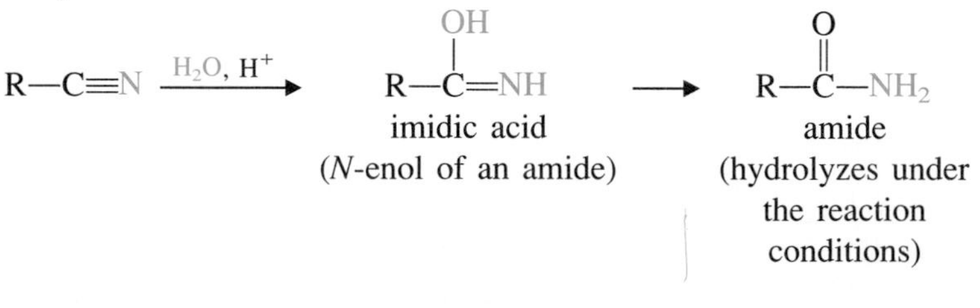

$$\text{(21.28b)}$$

imidic acid amide
(*N*-enol of an amide) (hydrolyzes under
 the reaction
 conditions)

The hydrolysis of nitriles is a useful way to prepare carboxylic acids because nitriles, unlike many other carboxylic acid derivatives, are generally synthesized from compounds other than the acids themselves. The preparation of nitriles is discussed in Sec. 21.11.

E. Hydrolysis of Acid Chlorides and Anhydrides

Acid chlorides and anhydrides react *rapidly* with water, even in the absence of acid or base catalysts.

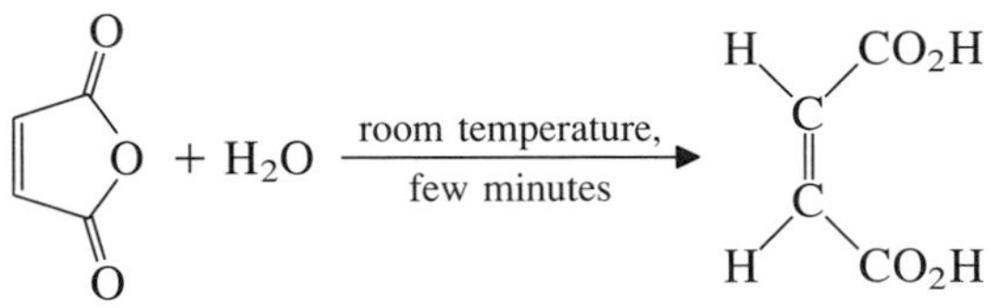

(94% yield) (21.29)

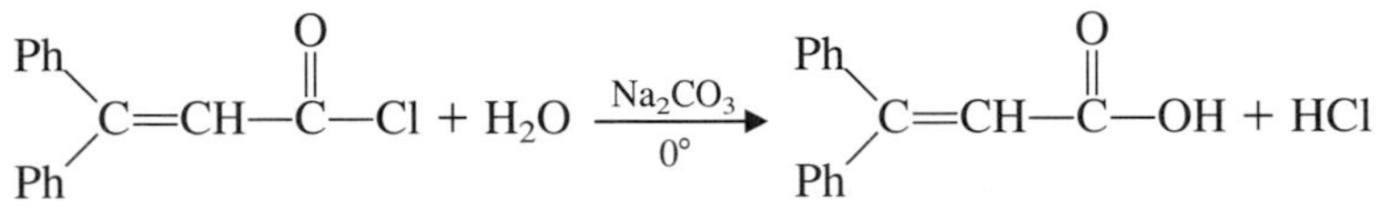

(21.30)

(>95% yield)

However, the hydrolysis reactions of acid chlorides and anhydrides are almost never used for the preparation of acids, because these derivatives are themselves usually prepared from acids (Sec. 20.9).

F. Relative Carbonyl Reactivity of Carboxylic Acid Derivatives

In the foregoing discussion we have seen that all carboxylic acid derivatives can be hydrolyzed to carboxylic acids. However, the *conditions* under which the different derivatives are hydrolyzed differ considerably. Hydrolysis of amides and nitriles requires heat as well as acid or base; hydrolysis of esters requires acid or base, but requires heating only briefly, if at all; and hydrolysis of acid chlorides and anhydrides occurs rapidly at room temperature even in the absence of acid or base. It turns out that the same trend in reactivity is observed in other reactions of carboxylic acid derivatives at the carbonyl group:

carbonyl reactivity of carboxylic acid derivatives:

$$\text{acid chlorides} > \text{anhydrides} \gg \text{esters, acids} > \text{amides} > \text{nitriles} \qquad (21.31)$$

What is the reason for this relative reactivity? We can use the same argument that we used to explain the relative reactivities of aldehydes and ketones (Sec. 19.7B). Anything that stabilizes a carboxylic acid derivative *relative to its addition intermediate* (or transition state) increases its free energy of activation and decreases its reactivity. This point is shown diagrammatically in Fig. 21.5. Substituent groups affect the stability of carbonyl compounds through the operation of the same two types of electronic effects that we have considered before: *inductive effects* and *resonance effects*. (Review again Sec. 16.5B.)

Let us first consider inductive effects. If we consider the polarity of the C=O bond, we are justified in thinking of a carbonyl compound, to a useful approximation, as a carbocation. Electronegative groups attached to the carbonyl carbon destabilize its positive charge.

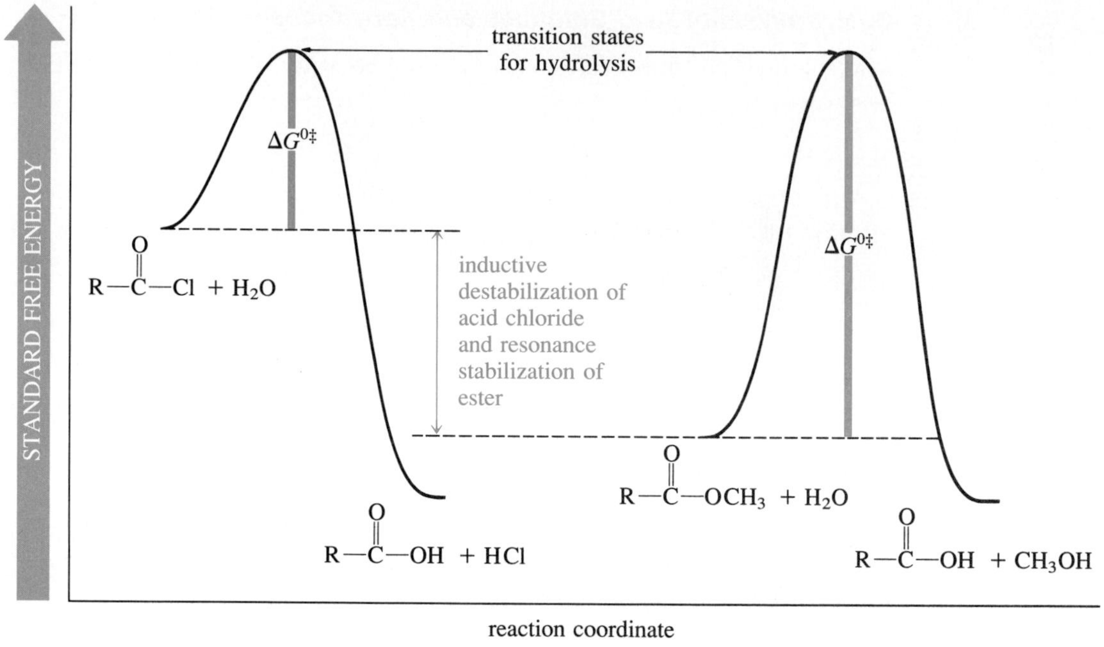

Figure 21.5 Inductive destabilization of an acid chloride and resonance stabilization of an ester, relative to the transition states for their hydrolysis reactions, lead to a smaller free energy of activation, and hence a larger rate, for hydrolysis of the acid chloride. (The two transition states have been placed at the same energy level for comparison purposes.)

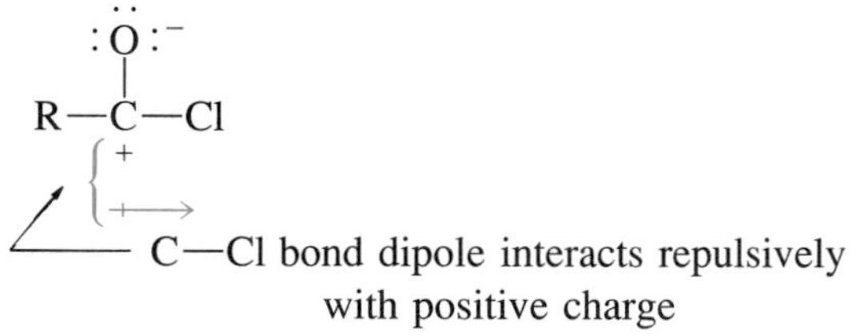

The tetrahedral addition intermediate (or transition state) has less positive charge. Hence, this repulsion is diminished in the tetrahedral intermediate. It follows that the inductive effect of an electron-withdrawing group raises the energy of a carbonyl compound more than it raises the energy of the transition state. This inductive effect thus *decreases* the free energy of activation and increases reactivity. As this analysis suggests, the electron-withdrawing inductive effects of substituent groups are in the same order as the relative reactivities of the corresponding carboxylic acid derivatives:

$$-\text{Cl} > -\text{O}-\overset{\text{O}}{\overset{\|}{\text{C}}}-\text{R} > -\text{OR} > -\text{NR}_2$$

(Notice that this order is also reflected in relative acid strengths of the corresponding acids: HCl is most acidic, then RCO_2H, then ROH, then R_2NH.)

But inductive effects are not the whole story. A resonance effect *stabilizes* carboxylic acid derivatives.

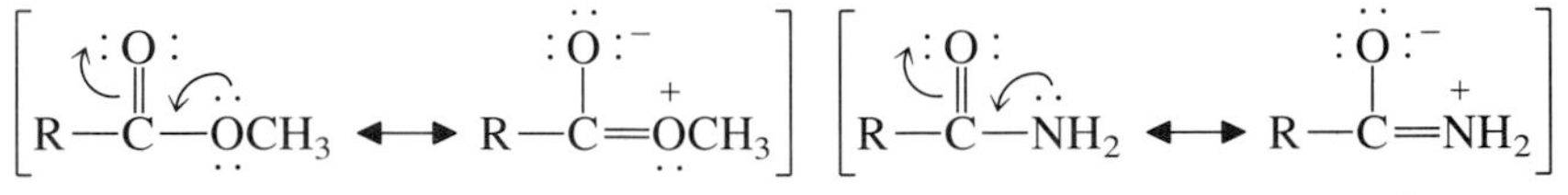

$$(21.32)$$

resonance structures of an ester resonance structures of an amide

Since this resonance stabilization does not exist in the tetrahedral intermediate, resonance *lowers* the energy of a carbonyl compound relative to that of its transition state. Consequently, the free energy of activation is increased and the reactivity is reduced. Now, the resonance effect of oxygen stabilizes esters and anhydrides more than the resonance effect of chlorine stabilizes acid chlorides. The reason is that resonance interaction of chlorine with a carbonyl group requires overlap of chlorine $3p$ orbitals with carbon $2p$ orbitals. As we learned in Fig. 16.10, this overlap is less effective than the overlap of p orbitals with the same quantum number. Furthermore, the resonance stabilization of an amide (Eq. 21.32) is more important than that of an ester because nitrogen is less electronegative than oxygen and accepts positive charge more readily.

Acid chlorides are the most reactive carboxylic acid derivatives because the *destabilizing* inductive effect of chlorine is more important than its stabilizing resonance effect. In contrast, esters and amides are less reactive because of strong resonance effects in these compounds, which override the inductive effects of oxygen and nitrogen. The balance of these effects in anhydrides is such that they are much more reactive than esters, but less reactive than acid chlorides.

The reason for the poor reactivity of nitriles depends on whether we are discussing reactions in base or reactions in acid. Reactions of nitriles in base are slower than those of other acid derivatives because nitrogen is less electronegative than oxygen, and therefore accepts negative charge less readily. The modest reactivity of nitriles in acid is due to their poor basicity. It is the protonated form of a nitrile that reacts with nucleophiles in acid solution, but there is simply so little of this form present (Sec. 21.5B) that the nitrile reacts very slowly.

Problem

10 Complete the following reactions:

(a)

F—⟨benzene ring⟩—CO_2CH_3 + H_2O $\xrightarrow{OH^-}$ $\xrightarrow{H^+}$

(b)

$N≡C-CH_2-\overset{\overset{\displaystyle O}{\|}}{C}-OCH_3$ + OH^- (1 equiv) $\xrightarrow[CH_3OH]{H_2O}$

(c)

$H_2N-\overset{\overset{\displaystyle O}{\|}}{C}-NH_2$ + H_2O $\xrightarrow[\text{heat}]{H^+}$

(d)

⟨cyclic structure with O and NH⟩ + H_2O $\xrightarrow[\text{heat}]{H^+}$

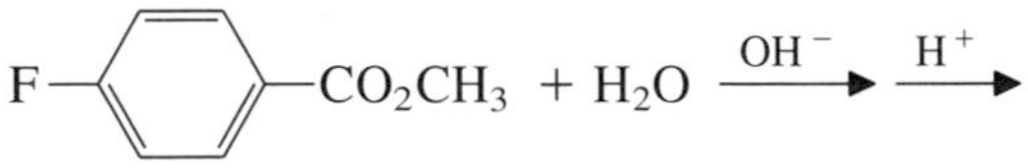

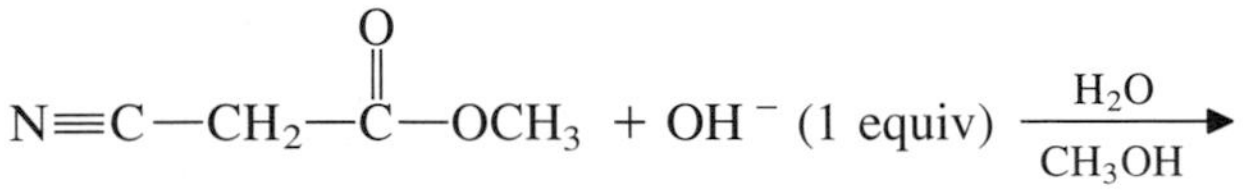

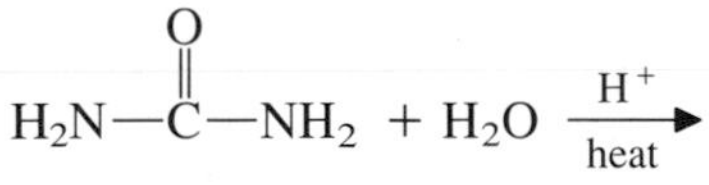

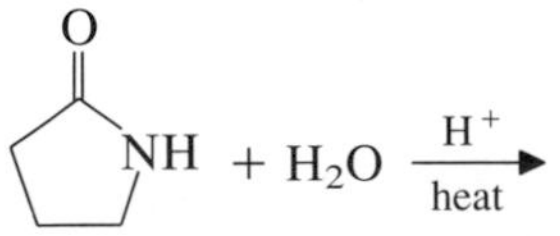

21.8 REACTIONS OF CARBOXYLIC ACID DERIVATIVES WITH NUCLEOPHILES

In the previous section, we learned that all carboxylic acid derivatives react with water to give carboxylic acids. Water is but one of the nucleophiles that react with carboxylic acid derivatives. In this section, we shall learn how the reactions of a variety of nucleophiles with carboxylic acid derivatives can be used to prepare other carboxylic acid derivatives.

A. Reactions of Acid Chlorides with Nucleophiles

One of the most useful ways of preparing carboxylic acid derivatives is the reaction of acid chlorides with various nucleophiles. The great reactivity of acid chlorides (Eq. 21.31) means that their reactions take place rapidly under mild conditions.

Reaction of Acid Chlorides with Ammonia and Amines The reaction of acid chlorides with ammonia or amines yields amides.

$$CH_3(CH_2)_8-\overset{O}{\overset{\|}{C}}-Cl + 2\ddot{N}H_3 \longrightarrow CH_3(CH_2)_8-\overset{O}{\overset{\|}{C}}-\ddot{N}H_2 + \overset{+}{N}H_4\ Cl^- \quad (21.33)$$
$$\text{(conc.} \qquad\qquad\qquad \text{(73\% yield)}$$
$$NH_4OH)$$

$$Ph-\overset{O}{\overset{\|}{C}}-Cl + PhCH_2CH_2\ddot{N}H_2 + \underset{\underset{\ddot{}}{N}}{\bigcirc} \longrightarrow Ph-\overset{O}{\overset{\|}{C}}-\ddot{N}HCH_2CH_2Ph + \underset{\underset{H\ Cl^-}{\overset{+}{N}}}{\bigcirc} \quad (21.34)$$
$$\text{pyridine} \qquad\qquad \text{(89--98\% yield)}$$

$$Ph-\overset{O}{\overset{\|}{C}}-Cl + \overset{}{\underset{}{\bigcirc}}{:}N-H + NaOH \longrightarrow Ph-\overset{O}{\overset{\|}{C}}-N{:}\overset{}{\underset{}{\bigcirc}} + H_2O + Na^+\ Cl^- \quad (21.35)$$
$$\text{(77--81\% yield)}$$

Reaction with *ammonia* yields a primary amide (Eq. 21.33); reaction with a *primary amine* (an amine of the form R—NH$_2$; Sec. 19.11A) yields a secondary amide (Eq. 21.34); and reaction with a *secondary amine* (an amine of the form R$_2$N—H; Sec. 19.11B) yields a tertiary amide (Eq. 21.35).

This reaction is another example of nucleophilic substitution at a carbonyl group. The amine attacks the carbonyl group to form a tetrahedral intermediate, which expels chloride ion; a proton-transfer step yields the amide.

$$R-\overset{\ddot{O}:}{\overset{\|}{C}}-\ddot{C}l: \underset{:\ddot{N}H_2-R}{\rightleftharpoons} R-\overset{:\ddot{O}:\bar{}}{\underset{+\ddot{N}H_2-R}{\overset{|}{C}}}-\ddot{C}l: \longrightarrow R-\overset{:O:}{\overset{\|}{\underset{+}{C}}}-\overset{H}{\overset{|}{N}H}-R \xrightarrow{H_2\ddot{N}R} R-\overset{:O:}{\overset{\|}{C}}-\ddot{N}H-R + H_3\overset{+}{N}R \quad (21.36)$$
$$+ :\ddot{C}l:\bar{}$$

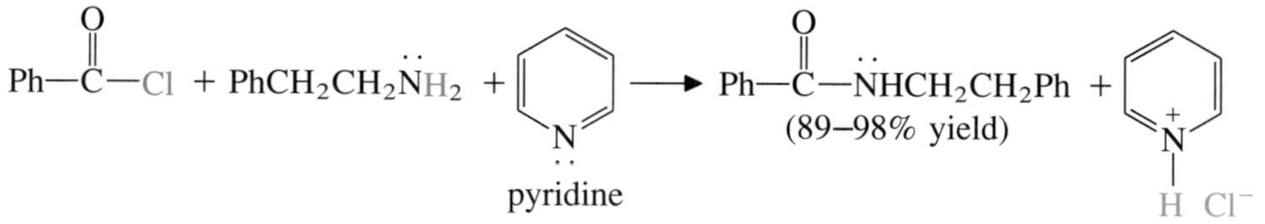

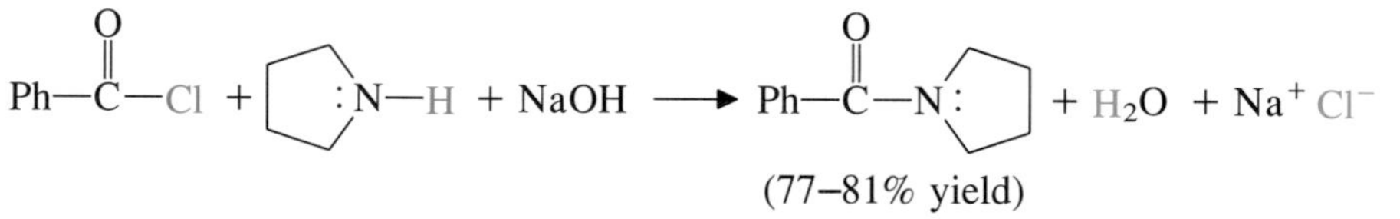

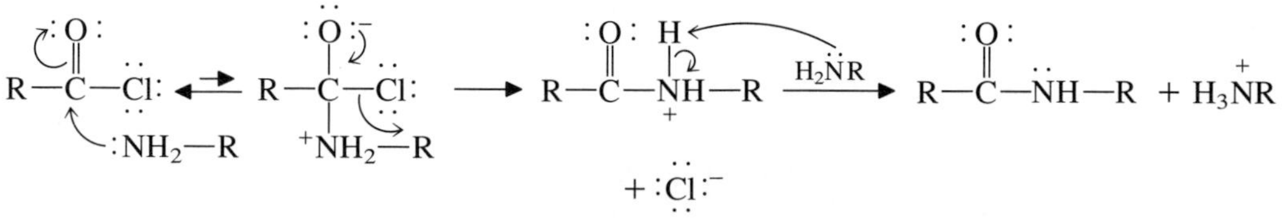

An important aspect of amide formation is the proton transfer in the last step of the mechanism. Unless another base is added to the reaction mixture, the starting amine acts as the base in this step. Hence, for each equivalent of amide that is formed, an equivalent of amine is protonated. When the amine is protonated, its electron pair is taken "out of action," and the amine is no longer nucleophilic.

$R—\ddot{N}H_2$	$R—\overset{+}{N}H_3$
strong base; good nucleophile	protonated amine; cannot act as a nucleophile

Hence, if the only base present is the amine nucleophile, then at least *two* equivalents must be used: one equivalent as the nucleophile, and one as the base in the final proton-transfer step.

The use of excess amine is practical when the amine is cheap and readily available (Eq. 21.33). Another alternative is to use one equivalent of amine and one equivalent of another base. In the *Schotten–Baumann* technique, the reaction is run with a water-insoluble acid chloride in a separate layer (either alone or in a solvent) over an aqueous solution of NaOH (Eq. 21.35). Hydrolysis of the acid chloride by NaOH is avoided because the water-insoluble acid chloride is not in contact with the water-soluble hydroxide ion. The amine, which is soluble in the acid chloride solution, reacts to yield an amide. The aqueous NaOH extracts and neutralizes the protonated amine that is formed.

occurs in organic layer

occurs in aqueous layer

$$\underset{\text{water-insoluble}}{R—\overset{O}{\overset{\|}{C}}—Cl} + R—NH_2 \longrightarrow R—\overset{O}{\overset{\|}{C}}—NH—R + \underset{\text{water-soluble}}{R—\overset{+}{N}H_3\ Cl^-} \xrightarrow{^-OH} R—NH_2 + H_2O \qquad (21.37)$$

In yet another procedure, a *tertiary amine* (an amine of the form $R_3N:$, such as triethylamine or pyridine) may be used as the base (Eq. 21.34).

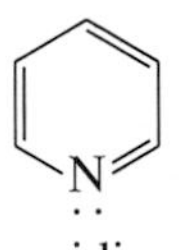
pyridine

$(C_2H_5)_3N:$
triethylamine

The presence of a tertiary amine does not interfere with amide formation by another amine because a tertiary amine itself cannot form an amide (why?). The use of a tertiary amine is particularly appropriate if the amine used to form the amide is expensive and cannot be used in excess.

Although they do not form amides, tertiary amines can form *acylammonium salts.*

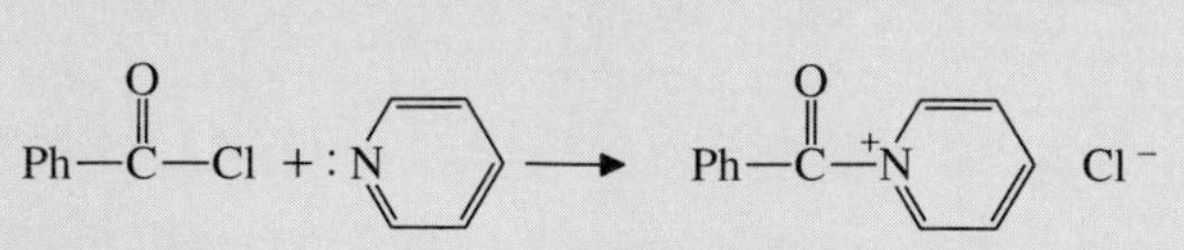

(21.38a)

benzoylpyridinium chloride
(an acylammonium salt)

Acylammonium salts are very reactive, and are converted into amides by reaction with primary or secondary amines.

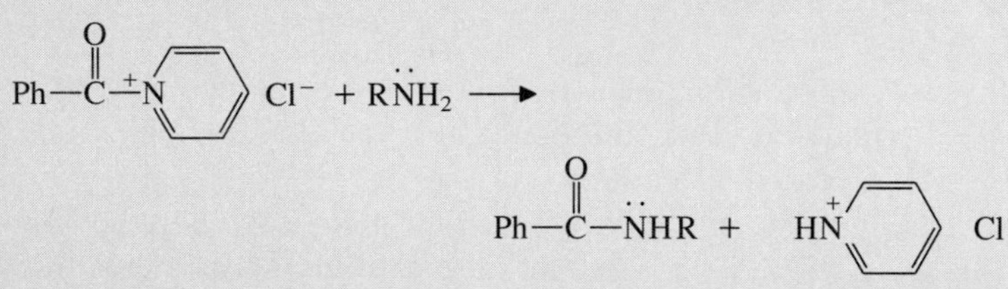

(21.38b)

pyridinium chloride
(HCl salt of pyridine)

Hence, acylammonium salt formation does not interfere with amide formation.

The important point about all the methods for preparing amides is that either two equivalents of amine must be used, or an equivalent of base must be added to effect the final neutralization.

Reaction of Acid Chlorides with Alcohols and Phenols Esters are formed rapidly when acid chlorides react with alcohols or phenols. In principle, the HCl liberated in the reaction need not be neutralized, since alcohols and phenols are not basic enough to be extensively protonated by the acid. However, some esters (such as *t*-butyl esters; Sec. 21.7A) and alcohols (such as tertiary alcohols; Secs. 10.1A., 10.2), are sensitive to acid. In practice, a tertiary amine like pyridine is added to the reaction mixture or is even used as solvent to neutralize the HCl.

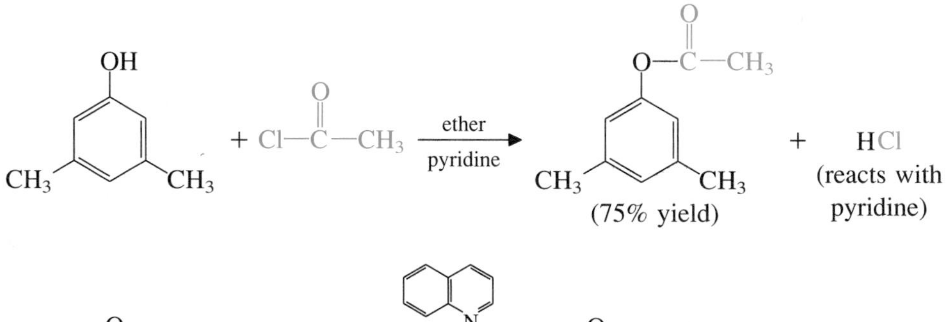

(21.39)

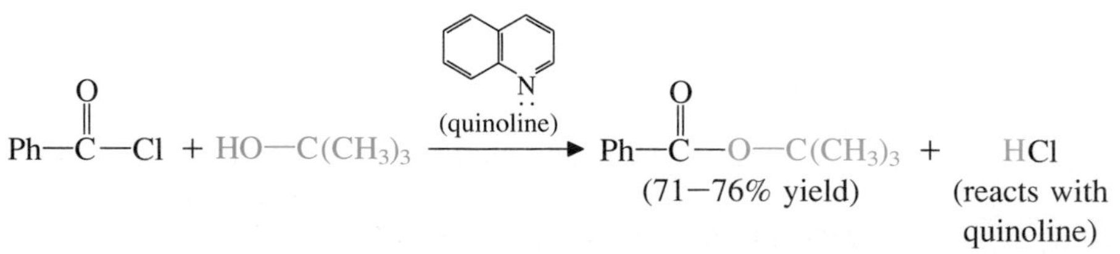

(21.40)

As these examples illustrate, esters of tertiary alcohols and phenols, which cannot be prepared by acid-catalyzed esterification, can be prepared by this method.

Sulfonate esters (the esters of sulfonic acids) are prepared by the analogous reactions of sulfonyl chlorides (the acid chlorides of sulfonic acids) with alcohols.

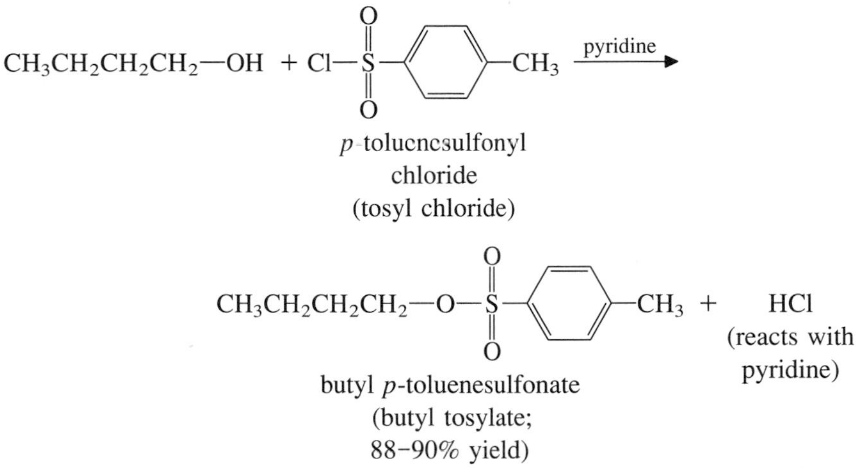
pyridine
p-toluenesulfonyl chloride
(tosyl chloride)
HCl
(reacts with pyridine)
butyl p-toluenesulfonate
(butyl tosylate;
88–90% yield)
(21.41)

We considered this reaction previously in Sec. 10.3A.

Reaction of Acid Chlorides with Carboxylate Salts Even though carboxylate salts are weak nucleophiles, acid chlorides are reactive enough to be attacked by carboxylate salts and afford anhydrides.

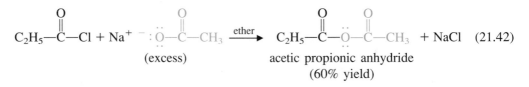

$$C_2H_5-\overset{O}{\overset{\|}{C}}-Cl + Na^{+}\ ^{-}:\ddot{O}-\overset{O}{\overset{\|}{C}}-CH_3 \xrightarrow{\text{ether}} C_2H_5-\overset{O}{\overset{\|}{C}}-\ddot{O}-\overset{O}{\overset{\|}{C}}-CH_3 + NaCl \quad (21.42)$$

(excess) — acetic propionic anhydride (60% yield)

This is our second general procedure for the synthesis of anhydrides. Unlike the anhydride synthesis we discussed in Sec. 20.9B, the reaction of acid chlorides with carboxylate salts can be used to prepare mixed anhydrides, as the example in Eq. 21.42 illustrates.

Friedel–Crafts Acylation In Sec. 16.4E we focused on Friedel–Crafts acylation as an *electrophilic substitution* reaction of benzene derivatives.

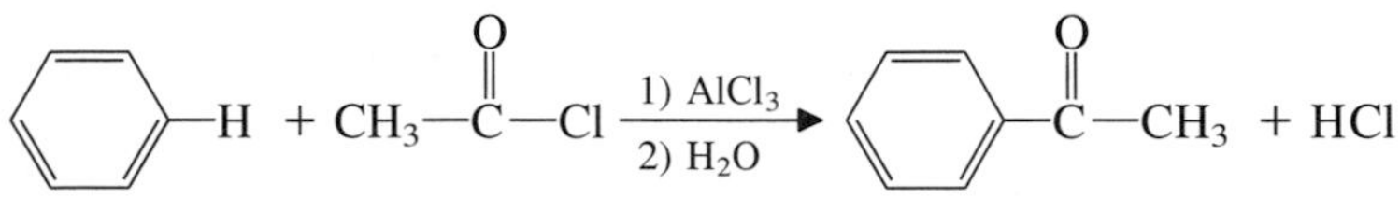
1) AlCl3
2) H2O
(21.43)

However, we can just as easily think of it as a *nucleophilic substitution* reaction of aromatic hydrocarbons with acid chlorides; the nucleophilic electrons are the π-electrons of the aromatic ring. These electrons are so weakly nucleophilic, however, that a strong Lewis acid such as $AlCl_3$ is required to activate the acid chloride by forming an acylium ion (Eq. 16.15). (Review again the mechanism of Friedel–Crafts acylation from this perspective.)

B. Reactions of Anhydrides with Nucleophiles

Anhydrides react with nucleophiles in much the same way as acid chlorides: reaction with amines yields amides, reaction with alcohols yields esters, and so on.

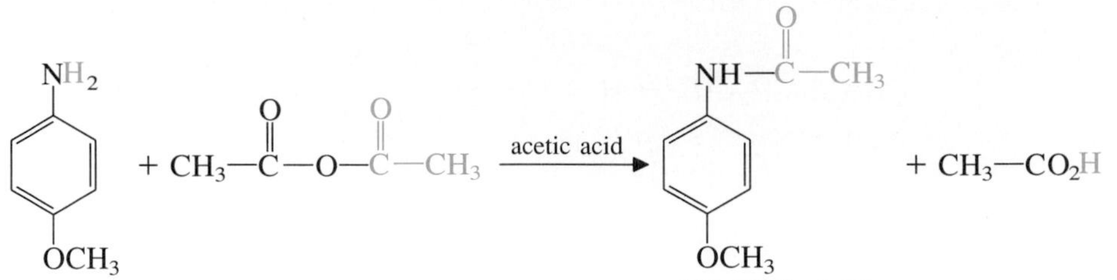

(75–79% yield) (21.44)

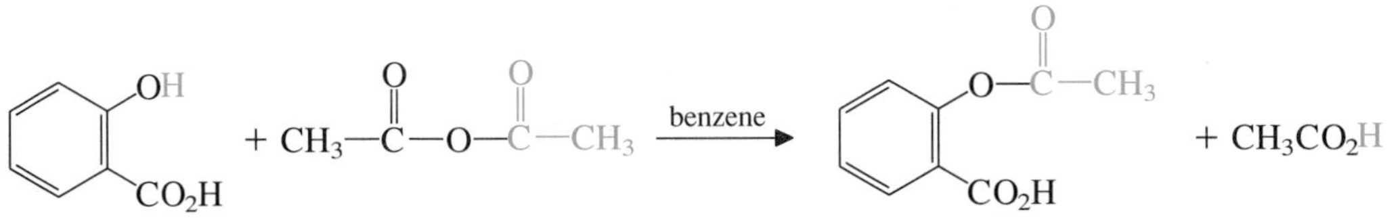

salicylic acid

O-acetylsalicylic acid (aspirin; 80–90% yield) (21.45)

Because most anhydrides are prepared from the corresponding carboxylic acids, the use of an anhydride to prepare an ester or amide wastes one equivalent of the parent acid as a leaving group. (Notice, for example, that acetic acid is a by-product in Eqs. 21.44 and 21.45.) Therefore, this reaction in practice is used only with inexpensive and readily available anhydrides, such as acetic anhydride.

One exception to this generalization is the formation of half-esters and -amides from cyclic anhydrides:

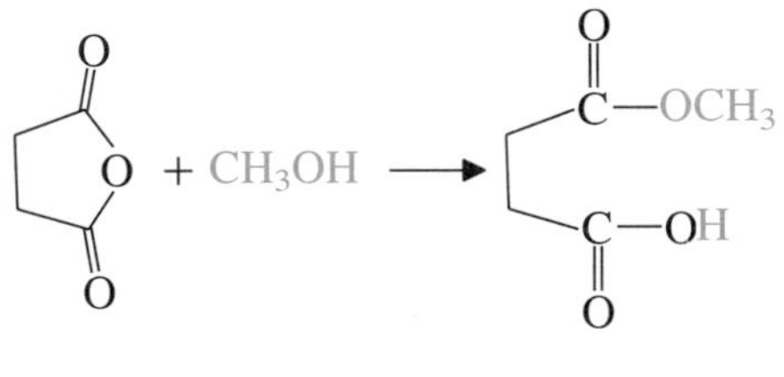

methyl hydrogen succinate (95–96% yield) (21.46)

Half-amides of dicarboxylic acids are produced in analogous reactions of amines and cyclic anhydrides. These compounds can be cyclized to imides by treatment with dehydrating agents, or in some cases just by heating. This reaction is the nitrogen analog of cyclic anhydride formation (Sec. 20.9B).

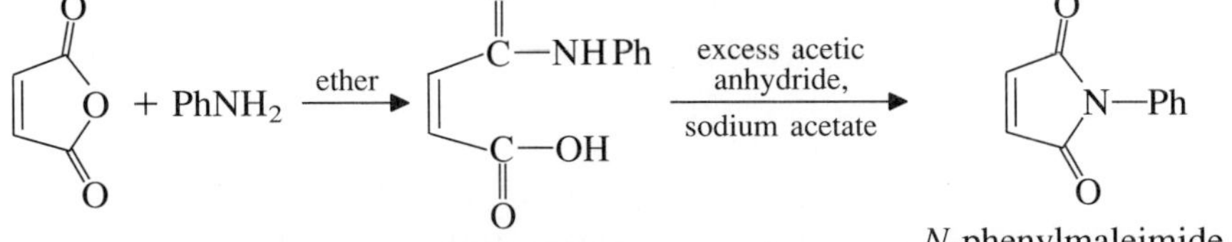

(97% yield)

N-phenylmaleimide (75–80% yield) (21.47)

C. Reactions of Esters with Nucleophiles

Reaction of Esters with Amines or Alcohols Just as esters are much less reactive than acid chlorides toward hydrolysis, they are also much less reactive toward amines and alcohols. Nevertheless, reactions of esters with these nucleophiles are sometimes useful. The reaction of esters with ammonia or amines yields amides.

$$N{\equiv}C{-}CH_2{-}\overset{\overset{\displaystyle O}{\|}}{C}{-}OC_2H_5 + NH_3 \xrightarrow{H_2O} \underset{\text{(86\% yield)}}{N{\equiv}C{-}CH_2{-}\overset{\overset{\displaystyle O}{\|}}{C}{-}NH_2} + C_2H_5OH \qquad (21.48)$$

The reaction of esters with *hydroxylamine* (:NH_2OH, Table 19.3) gives *N*-hydroxyamides; these compounds are known as **hydroxamic acids.**

$$R{-}\overset{\overset{\displaystyle O}{\|}}{C}{-}OC_2H_5 + \underset{\text{hydroxylamine}}{:NH_2OH} \longrightarrow \underset{\text{a hydroxamic acid}}{R{-}\overset{\overset{\displaystyle O}{\|}}{C}{-}NHOH} + C_2H_5OH \qquad (21.49)$$

(Acid chlorides and anhydrides also react with hydroxylamine to form hydroxamic acids.) This chemistry forms the basis for the *hydroxamate test,* used mostly for esters. The hydroxamic acid products are easily recognized because they form highly colored complexes with ferric ion.

When an ester reacts with an alcohol under acidic conditions, or with an alkoxide under basic conditions, a new ester is formed.

$$Ph{-}\overset{\overset{\displaystyle O}{\|}}{C}{-}OCH_3 + \underset{\text{(excess)}}{HO(CH_2)_3CH_3} \xrightleftharpoons{K^+\ CH_3(CH_2)_3\ddot{\underset{..}{O}}:^-} \underset{\text{(72\% yield)}}{Ph{-}\overset{\overset{\displaystyle O}{\|}}{C}{-}O(CH_2)_3CH_3} + CH_3OH \qquad (21.50)$$

This type of reaction, called **transesterification,** typically has an equilibrium constant near unity, because there is no reason for one ester to be strongly favored over the other at equilibrium. The reaction is driven to completion by the use of an excess of the displacing alcohol or by removal of a relatively volatile alcohol by-product as it is formed.

Problems

11 Using the arrow formalism, give detailed mechanisms for the reaction of acetyl chloride with
(a) methanol (b) sodium benzoate

12 Complete the following reactions by giving the major organic products:

(a) $CH_3CH_2CO_2H \xrightarrow[\text{(excess)}]{SOCl_2} \xrightarrow[\text{(excess)}]{(CH_3)_2NH}$

(b)
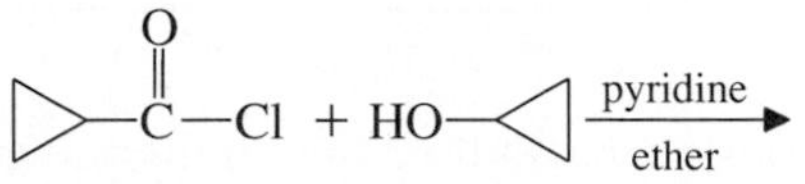

(Problem continues on page 890.)

Problems (Cont.)

(c)

$$CH_3CH_2CH_2-\overset{\overset{\displaystyle O}{\|}}{C}-Cl \; + \; Na^+ \; {}^-O-\overset{\overset{\displaystyle O}{\|}}{C}-CH_3 \longrightarrow$$

(d)

$$Cl-\overset{\overset{\displaystyle O}{\|}}{C}-Cl \;(\text{excess}) \; + \; CH_3OH \longrightarrow$$

(e)

$$Cl-\overset{\overset{\displaystyle O}{\|}}{C}-Cl \; + \; CH_3OH \;(\text{excess}) \longrightarrow$$

(f)

$$PhCH_2-\overset{\overset{\displaystyle O}{\|}}{C}-Cl \; + \; C_2H_5-SH \longrightarrow$$

(g) phthalic anhydride $+$ $CH_3OH \longrightarrow$

(h)

$$C_2H_5O-\overset{\overset{\displaystyle O}{\|}}{C}-OC_2H_5 \; + \; HO-CH_2CH_2-OH \; \xrightarrow[\text{heat}]{H^+} \; (C_3H_4O_3)$$

13 Contrast the location of ^{18}O in the products of the following two reactions, and explain:

(a)

$$PhCH_2-O-\overset{\overset{\displaystyle O}{\|}}{\underset{\underset{\displaystyle O}{\|}}{S}}-CH_3 \; + \; {}^{18}OH^- \longrightarrow$$

(b)

$$PhCH_2-O-\overset{\overset{\displaystyle O}{\|}}{C}-CH_3 \; + \; {}^{18}OH^- \longrightarrow$$

14 How would you synthesize each of the following compounds from an acid chloride?

(a)

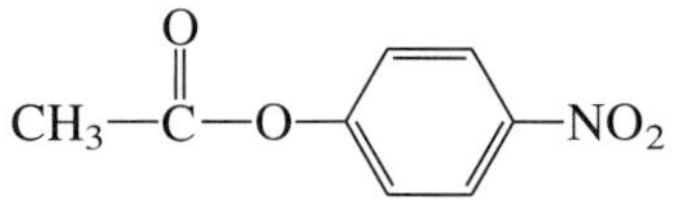

$$CH_3-\overset{\overset{\displaystyle O}{\|}}{C}-O-\!\!\left\langle\!\!\bigcirc\!\!\right\rangle\!\!-NO_2$$

(b)

$$\underset{\underset{\displaystyle CH_3\overset{\displaystyle |}{C}HOSO_2}{}}{\overset{\displaystyle Ph}{|}}-\!\!\left\langle\!\!\bigcirc\!\!\right\rangle\!\!-CH_3$$

(c)

(d)

$$(CH_3)_3C-O-\overset{\overset{\displaystyle O}{\|}}{C}-CH_2-\overset{\overset{\displaystyle O}{\|}}{C}-O-C(CH_3)_3$$

21.9 REDUCTION OF CARBOXYLIC ACID DERIVATIVES

Most of the reactions we have discussed in the previous sections have been straight-forward carbonyl substitution reactions. In this and the following section, we shall

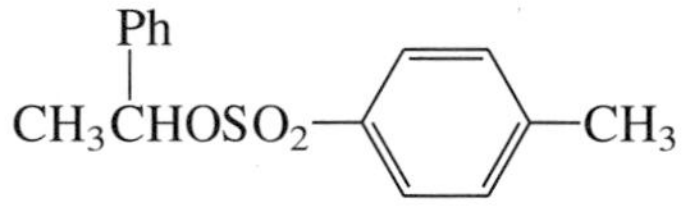

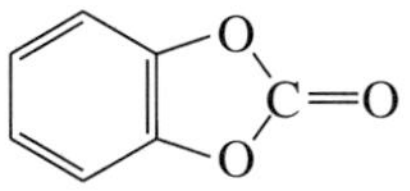

consider some reactions with mechanisms that involve carbonyl substitution steps followed by other reactions.

A. Reduction of Esters to Primary Alcohols

Lithium aluminum hydride reduces all carboxylic acid derivatives. Reduction of esters with this reagent, like reduction of carboxylic acids, gives primary alcohols.

$$2C_2H_5{-}\underset{\displaystyle CH_3}{\underset{|}{CH}}{-}\overset{\displaystyle O}{\overset{\|}{C}}{-}OC_2H_5 + LiAlH_4 \xrightarrow{\text{ether}} \xrightarrow{H_3O^+}$$

$$2C_2H_5{-}\underset{\displaystyle CH_3}{\underset{|}{CH}}{-}CH_2{-}OH + C_2H_5OH + LiAl(OH)_4 \quad (21.51)$$

(91% yield)

As we have already observed several times (Sec. 20.10), the active nucleophile in $LiAlH_4$ reductions is the *hydride ion* ($H:^-$) delivered from $^-AlH_4$, and this reduction is no exception. Hydride replaces alkoxide at the carbonyl group of the ester to give an aldehyde. (Write out a detailed mechanism for this reaction, another example of substitution at the carbonyl group.)

$$Li^+ \; ^-AlH_4 + R{-}\overset{\displaystyle O}{\overset{\|}{C}}{-}OC_2H_5 \longrightarrow R{-}\overset{\displaystyle O}{\overset{\|}{C}}{-}H + Li^+ \; C_2H_5O^- + AlH_3 \quad (21.52a)$$

As we learned in Sec. 19.11, the aldehyde reacts rapidly with $LiAlH_4$ to give, after protonolysis, the alcohol.

$$R{-}\overset{\displaystyle O}{\overset{\|}{C}}{-}H \xrightarrow{LiAlH_4} \xrightarrow{H_3O^+} R{-}\underset{\displaystyle H}{\underset{|}{\overset{\displaystyle OH}{\overset{|}{C}}}}{-}H \quad (21.52b)$$

The reduction of esters to alcohols thus involves a *carbonyl substitution* reaction followed by a *carbonyl addition* reaction.

Sodium borohydride, another useful hydride reducing agent, is much less reactive than lithium aluminum hydride. It reduces aldehydes and ketones, but reacts very sluggishly with most esters; in fact, $NaBH_4$ may be used to reduce aldehydes and ketones selectively in the presence of esters.

Acid chlorides and anhydrides also react with $LiAlH_4$ to give primary alcohols. However, since acid chlorides and anhydrides are usually prepared from carboxylic acids, and since carboxylic acids themselves can be reduced to alcohols with $LiAlH_4$, the reduction of acid chlorides and anhydrides is seldom used.

B. Reduction of Amides to Amines

Amines are formed when amides are reduced with $LiAlH_4$.

$$LiAlH_4 + Ph-\overset{\overset{O}{\|}}{C}-NH_2 \longrightarrow \xrightarrow{H_3O^+} Ph-CH_2-NH_2 + Li^+, Al^{3+} \text{ salts} + 2H_2 \quad (21.53)$$
$$\text{(80\% yield)}$$

Amide reduction can be used not only to prepare primary amines from primary amides, but also secondary and tertiary amines from secondary and tertiary amides, respectively.

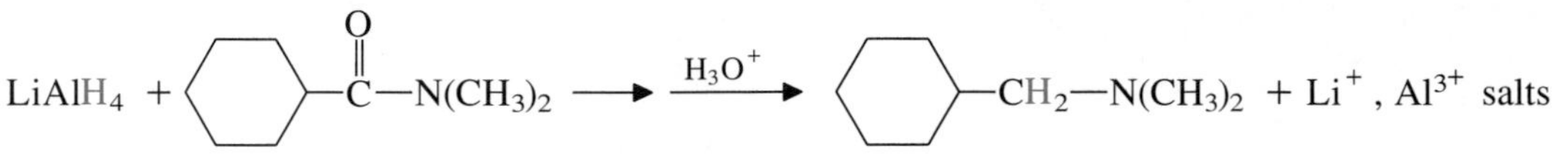

$$LiAlH_4 + \hspace{-0.3em}\text{(cyclohexyl)}\hspace{-0.3em}-\overset{\overset{O}{\|}}{C}-N(CH_3)_2 \longrightarrow \xrightarrow{H_3O^+} \text{(cyclohexyl)}-CH_2-N(CH_3)_2 + Li^+, Al^{3+} \text{ salts} \quad (21.54)$$
$$\text{(88\% yield)}$$

Notice carefully that the reaction of $LiAlH_4$ with an amide differs from its reaction with an ester. In the reduction of ester, the *carboxylate oxygen* is lost as a leaving group. If amide reduction were strictly analogous to ester reduction, the nitrogen would be lost, and a primary alcohol would be formed; clearly, this is not the case. Instead, it is the *carbonyl oxygen* that is lost in amide reduction.

ester reduction:

$$R-\overset{\overset{O}{\|}}{C}-OR \xrightarrow{LiAlH_4} \xrightarrow{H_3O^+} R-CH_2OH + ROH \quad (21.55a)$$
$$\text{(carbonyl oxygen retained)}$$

amide reduction:

$$R-\overset{\overset{O}{\|}}{C}-NR_2 \xrightarrow{LiAlH_4} \xrightarrow{H_3O^+} R-CH_2NR_2 \quad \text{(carbonyl oxygen lost)} \quad (21.55b)$$

Let us consider the reason for this difference, using as a case study the reduction of a secondary amide.

In the first step of the mechanism, the weakly acidic amide proton reacts with an equivalent of hydride, a strong base, to give hydrogen gas, AlH_3, and the lithium salt of the amide.

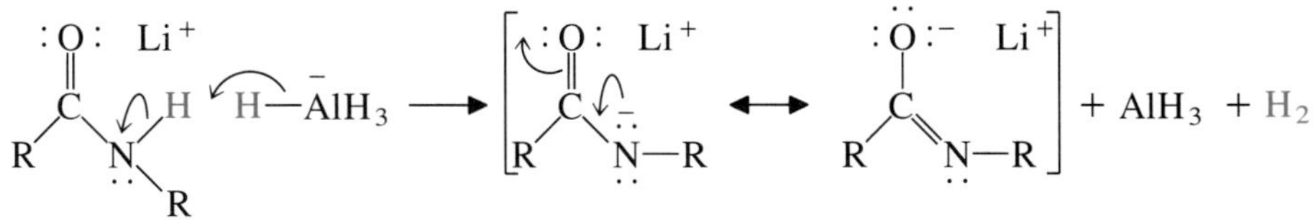

$$(21.56a)$$

The lithium salt of the amide, a Lewis base, reacts with the Lewis acid AlH_3.

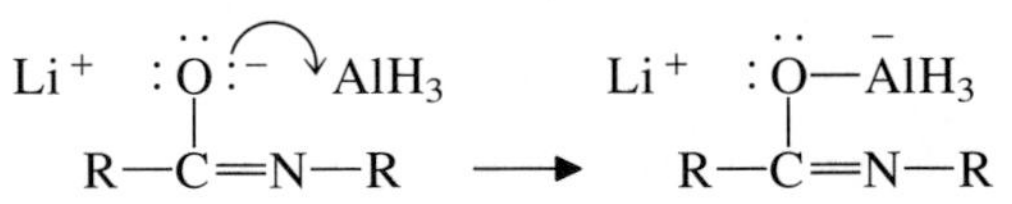

$$Li^+ \quad :\overset{..}{O}:^- \overset{\frown}{}AlH_3 \qquad Li^+ \quad :\overset{..}{O}-\overset{-}{A}lH_3$$
$$R-C{=}N-R \qquad \longrightarrow \qquad R-C{=}N-R \qquad (21.56b)$$

The resulting species is an active hydride reagent conceptually much like $LiAlH_4$, and can deliver hydride to the the C=N double bond.

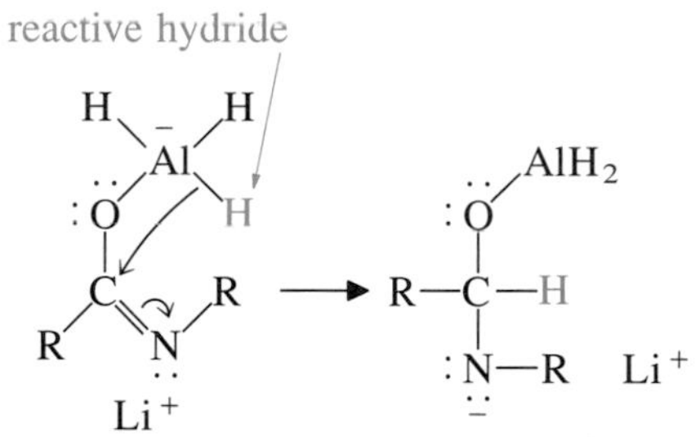

(21.56c)

Now we are prepared to see why oxygen rather than nitrogen is lost from the amide. If nitrogen were lost from the tetrahedral addition intermediate, it would have to assume a second negative charge. On the other hand, loss of oxygen requires expulsion of $^{-}OAlH_2$, which is considerably less basic than a nitrogen dianion, and which is actually a fairly good leaving group. Loss of this group gives an *imine* (Sec. 19.11).

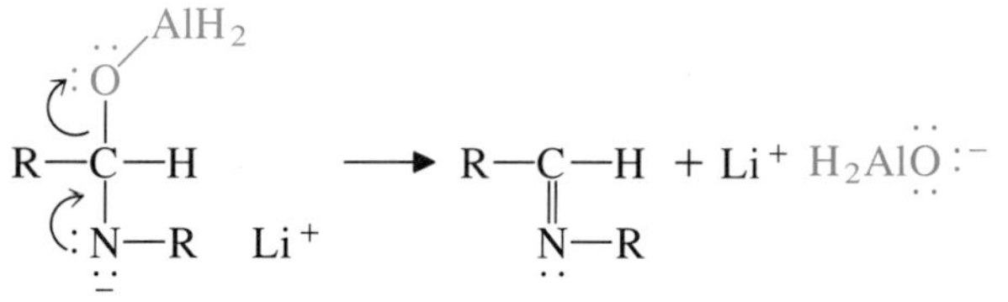

(21.56d)

Hence, we see that loss of the carbonyl oxygen occurs because it is converted into a better leaving group than the nitrogen.

The C=N of the imine, like the C=O of an aldehyde, undergoes nucleophilic addition with "H :$^{-}$" from $^{-}AlH_4$ or from one of the other hydride-containing species in the reaction mixture. Addition of water or acid to the reaction mixture converts the addition intermediate into an amine by protonolysis.

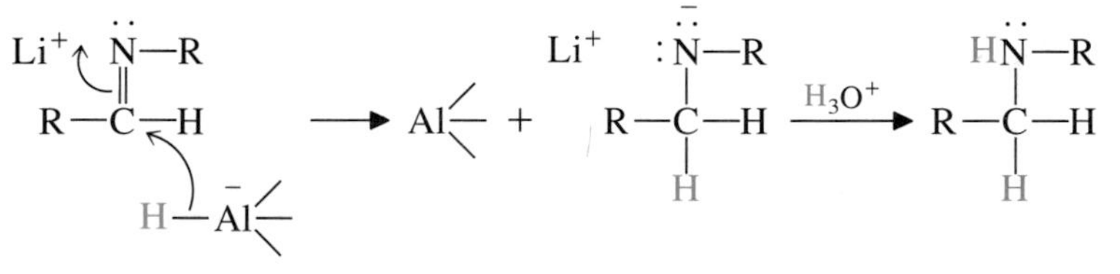

(21.56e)

The reductions of primary and tertiary amides involve somewhat different mechanisms, but, like the reduction of secondary amides, involve loss of oxygen rather than nitrogen as a leaving group.

C. Reduction of Nitriles to Primary Amines

Nitriles are reduced to primary amines by reaction with $LiAlH_4$, followed by the usual protonolysis step.

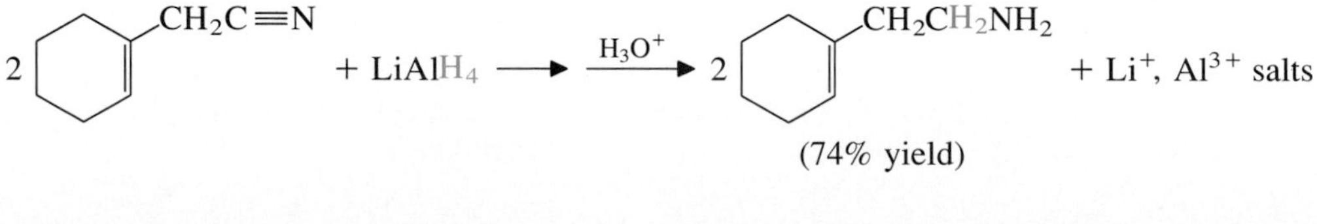

(21.57)

The mechanism of this reaction illustrates again how the C≡N and C=O bonds react in similar ways. This reaction probably occurs as two successive *nucleophilic additions*.

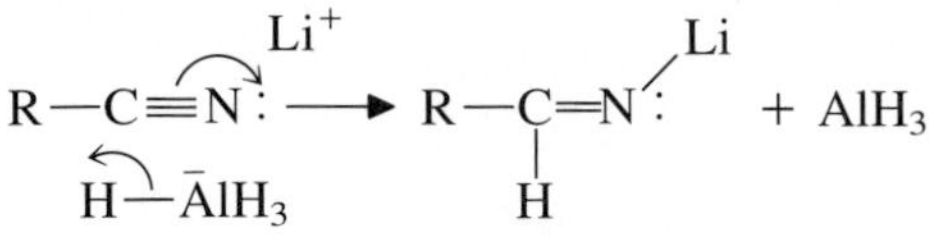

imine salt

(21.58a)

In the second addition, the imine salt reacts in a similar manner with AlH_3 (or another equivalent of AlH_4^-).

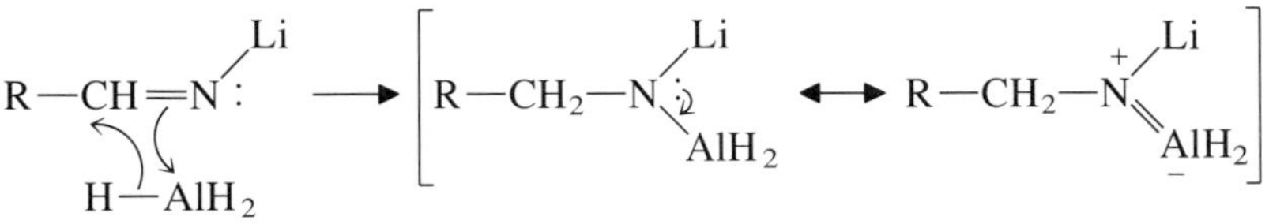

(21.58b)

In the resulting derivative, both the N—Li and the N—Al bonds are highly polar, and the nitrogen has a great deal of anionic character. Both bonds are susceptible to protonolysis. Hence, an amine is formed when water is added to the reaction mixture.

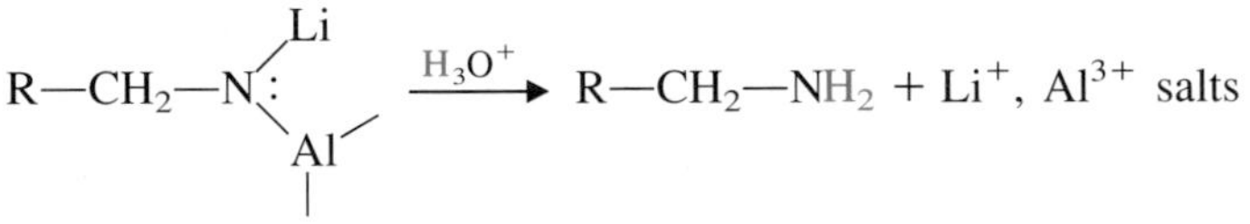

(21.58c)

Nitriles are also reduced to primary amines by catalytic hydrogenation.

$$CH_3(CH_2)_4C{\equiv}N + 2H_2 \xrightarrow[\substack{2000\text{ psi} \\ 120\text{–}130°}]{\text{Raney Ni}} CH_3(CH_2)_4CH_2NH_2 \qquad (21.59)$$

An intermediate in the reaction is the imine, which is not isolated but is hydrogenated to the amine product. (See also Problem 18.)

$$R{-}C{\equiv}N \xrightarrow{H_2,\text{ catalyst}} \underset{\text{imine}}{[R{-}CH{=}NH]} \xrightarrow{H_2,\text{ catalyst}} R{-}CH_2{-}NH_2 \qquad (21.60)$$

The reductions discussed in this and the previous section provide an entry into the *amine* functional group from amides and nitriles, the nitrogen-containing carboxylic acid derivatives. Hence, any synthesis of a carboxylic acid can be used as part of an amine synthesis, provided that the amine has the following form:

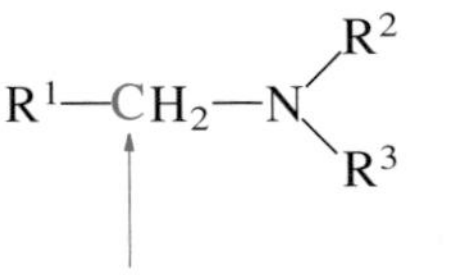

carbonyl or C≡N carbon of the carboxylic acid derivative

As indicated, the carbonyl or C≡N carbon of the carboxylic acid derivative ends up as the —CH_2— group adjacent to the amine nitrogen.

To illustrate: the following sequence of reactions could be used to convert an alkyl halide into an amine. (Fill in the reagents required; the bracketed numbers are the sections in the text in which the appropriate reactions are discussed.)

$$R-C≡N \longrightarrow R-CH_2NH_2$$

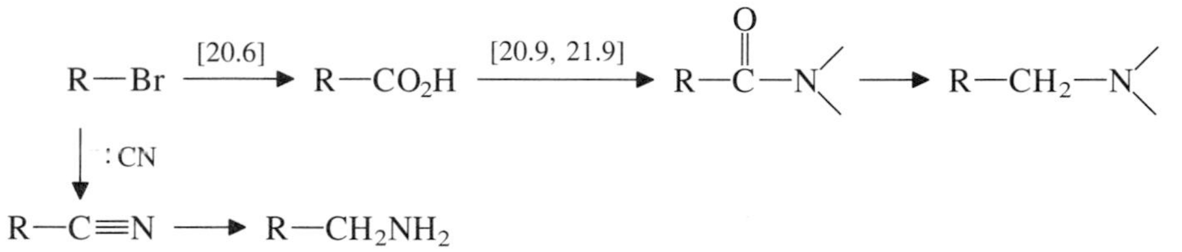

(21.61)

The synthesis of nitriles from alkyl halides, indicated by the vertical arrow, is discussed in Sec. 21.11.

D. Reduction of Acid Chlorides to Aldehydes

Acid chlorides can be reduced to aldehydes by either of two procedures. In the first, the acid chloride is hydrogenated over a catalyst that has been deactivated, or *poisoned,* with an amine, such as quinoline, that has been heated with sulfur. (Amines and sulfides are catalyst poisons.) This reaction is called the **Rosenmund reduction.**

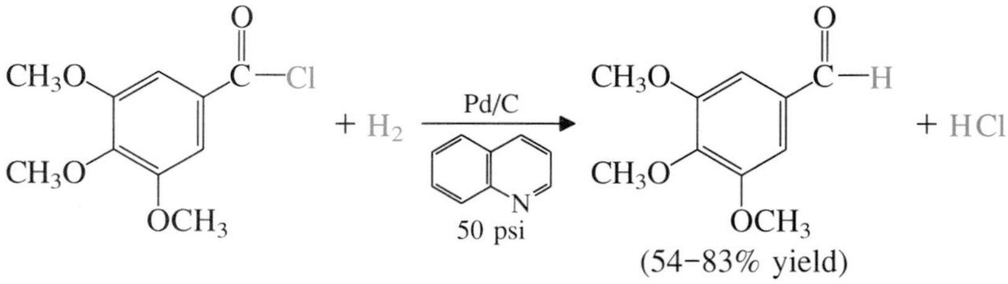

(54–83% yield) (21.62)

The poisoning of the catalyst prevents further reduction of the aldehyde product.

A second, more recent, method of converting acid chlorides into aldehydes is the reaction of an acid chloride at low temperature with a "cousin" of $LiAlH_4$, lithium tri(*t*-butoxy)aluminum hydride.

$$(CH_3)_3C-\overset{O}{\overset{\|}{C}}-Cl \;+\; Li^+\; H-\overset{-}{Al}\left(O-\overset{CH_3}{\underset{CH_3}{\overset{|}{\underset{|}{C}}}}-CH_3\right)_3 \xrightarrow[-78°]{\text{diglyme}} \xrightarrow{H_3O^+}$$

lithium tri(*t*-butoxy)aluminum hydride

$$(CH_3)_3C\overset{O}{\overset{\|}{C}}-H \;+\; 3CH_3-\overset{CH_3}{\underset{CH_3}{\overset{|}{\underset{|}{C}}}}-OH \;+\; LiCl \;+\; Al^{3+}\; \text{salts}$$ (21.63)

The hydride reagent used in this reduction is derived by the replacement of three hydrogens of lithium aluminum hydride by *t*-butoxy groups. As the hydrides of $LiAlH_4$ are replaced successively with alkoxy groups, less reactive reagents are obtained. (Can

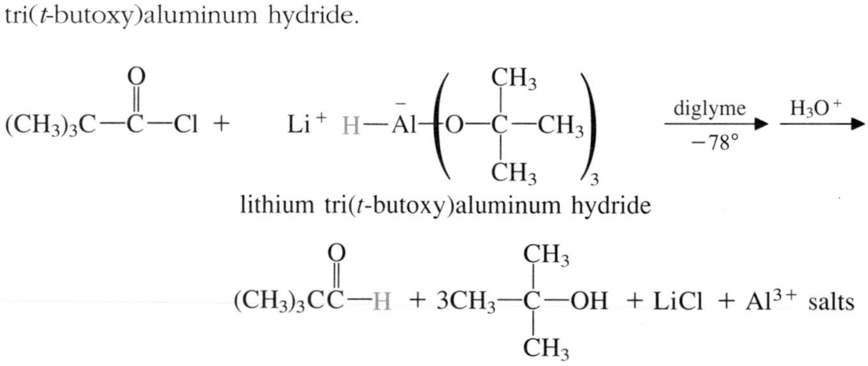

you think of a reason why this should be so?) In fact, the preparation of $LiAlH[O{-}C(CH_3)_3]_3$ owes its success to the poor reactivity of its hydride: the reaction of $LiAlH_4$ with *t*-butyl alcohol stops after three moles of alcohol have been consumed.

$$Li^+ \ {}^-AlH_4 + 3(CH_3)_3C{-}O{-}H \longrightarrow Li^+ \ H{-}\bar{Al}[O{-}C(CH_3)_3]_3 + 3H_2 \qquad (21.64)$$

The one remaining hydride reduces only the most reactive functional groups. Because *acid chlorides are more reactive than aldehydes toward nucleophiles,* the reagent reacts preferentially with the acid chloride reactant rather than with the product aldehyde. Lithium aluminum hydride, in contrast, is so reactive that it fails to discriminate to a useful degree between the aldehyde and acid chloride groups, and reduces acid chlorides, like esters, to primary alcohols.

The reduction of acid chlorides adds another synthesis of aldehydes and ketones to the list given in Sec. 19.4. Can you name the other methods?

Problems

15 Show how benzoyl chloride can be converted into each of the following compounds:

(a) benzaldehyde

(b) benzyl alcohol

(c) $PhCH_2{-}NH_2$ (benzylamine)

(d)

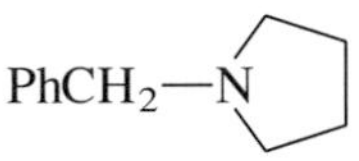

(*N*-benzylpyrrolidine)

16 Complete the following reactions by giving the principal organic product(s):

(a) $PhCH_2C{\equiv}N + H_2 \xrightarrow[\text{heat}]{\text{Raney Ni}}$

(b) $Ph{-}CH(O{-}\overset{O}{\overset{\|}{C}}{-}CH_3){-}CO_2C_2H_5 + LiAlH_4 \text{ (excess)} \longrightarrow \xrightarrow{H_3O^+}$

(c) $C_2H_5O{-}\overset{O}{\overset{\|}{C}}{-}CH_2{-}CN \xrightarrow[\text{(excess)}]{LiAlH_4} \xrightarrow{H_2O}$

17 Give the structures of two compounds that would give the following product after $LiAlH_4$ reduction:

$$(CH_3)_2CHCH_2CH_2CH_2NH_2$$

18 (a) In the catalytic hydrogenation of some nitriles, secondary amines are obtained as by-products:

$$R{-}C{\equiv}N \xrightarrow{H_2,\ cat} RCH_2NH_2 + \underset{\text{secondary amine}}{(RCH_2)_2NH}$$

Suggest a mechanism for the formation of this by-product. (*Hint:* What is the intermediate in the reduction? How can this intermediate react with an amine?)

(b) Explain why ammonia added to the reaction mixture prevents the formation of this by-product.

E. Relative Reactivity of Carbonyl Compounds with Nucleophiles

As we have learned, the reaction of lithium aluminum hydride with a carboxylic acid (Sec. 20.10) or ester (Sec. 21.9A) involves an aldehyde intermediate. But the product of such reactions is a primary alcohol, not an aldehyde, because *the aldehyde intermediate is more reactive than the acid or ester.* The instant a small amount of aldehyde is formed, it is in competition with the remaining acid or ester for the $LiAlH_4$ reagent. Since it is more reactive, the aldehyde reacts faster than the remaining ester does. Hence, we cannot expect to isolate the aldehyde under such circumstances. On the other hand, the lithium tri(*t*-butoxy)aluminum hydride reduction of acid chlorides can be stopped at the aldehyde because acid chlorides are more reactive than aldehydes. When the aldehyde is formed as a product, it is in competition with the remaining acid chloride for the hydride reagent. Because the acid chloride is more reactive, it is consumed before the aldehyde has a chance to react.

We can see from these examples that the outcome of many reactions of carboxylic acid derivatives is determined by the *relative reactivity of carbonyl compounds* toward nucleophilic reagents, which can be summarized as follows. (We also include nitriles, which we can consider to be "honorary carbonyl compounds.")

Relative reactivity toward nucleophiles:

acid chlorides > aldehydes > ketones >> esters, acids > amides > nitriles (21.65)

The explanation of this reactivity order is the same one used in Sec. 21.7F. Relative reactivity is determined by the stability of each type of compound relative to its transition state for addition or substitution. *The more a compound is stabilized, the less reactive it is* (Fig. 21.5). For example, esters are stabilized by resonance (Eq. 21.32) in a way that aldehydes and ketones are not. Hence esters are less reactive than aldehydes. Acid chlorides are destabilized by the inductive effect of the chlorine. Hence acid chlorides are more reactive than aldehydes.

In the next section we shall see other reactions in which the relative reactivity of carbonyl compounds plays an important role in determining what products are formed.

21.10 REACTIONS OF CARBOXYLIC ACID DERIVATIVES WITH ORGANOMETALLIC REAGENTS

A. Reaction of Esters with Grignard Reagents

Most carboxylic acid derivatives react with Grignard or organolithium reagents. One of the most important reactions of this type is the reaction of esters with Grignard reagents. In this reaction, a tertiary alcohol is formed after protonolysis. (Secondary alcohols are formed from esters of formic acid.)

$$(CH_3)_2CH-\overset{\overset{\displaystyle O}{\|}}{C}-OC_2H_5 + 2CH_3MgI \xrightarrow{\text{ether}} \xrightarrow{H_3O^+}$$

$$(CH_3)_2CH-\overset{\overset{\displaystyle OH}{|}}{\underset{\underset{\displaystyle CH_3}{|}}{C}}-CH_3 + C_2H_5OH + Mg^{2+} \text{ salts} \quad (21.66)$$

(92% yield)

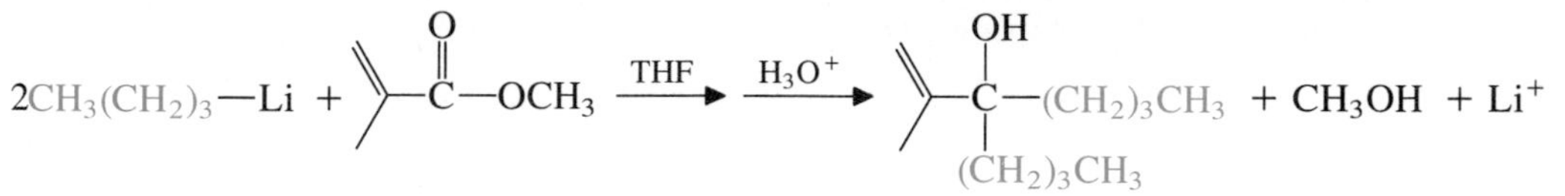

$$2CH_3(CH_2)_3-Li + \quad \overset{\overset{\displaystyle O}{\|}}{C}-OCH_3 \xrightarrow{\text{THF}} \xrightarrow{H_3O^+} \quad \overset{\overset{\displaystyle OH}{|}}{\underset{\underset{\displaystyle (CH_2)_3CH_3}{|}}{C}}-(CH_2)_3CH_3 + CH_3OH + Li^+ \quad (21.67)$$

Notice that two equivalents of organometallic reagent react per mole of ester.

Like the LiAlH$_4$ reduction of esters (Sec. 21.9A), this reaction is a substitution followed by an addition. A ketone is formed in the substitution step. (Fill in the details of the mechanism.)

$$R-\overset{\overset{\displaystyle O}{\|}}{C}-OC_2H_5 + CH_3-MgI \longrightarrow R-\overset{\overset{\displaystyle O}{\|}}{C}-CH_3 + I-Mg-OC_2H_5 \quad (21.68a)$$

The ketone intermediate is not isolated because *ketones are more reactive than esters toward nucleophilic reagents* (Eq. 21.65). The ketone therefore reacts with a second equivalent of the Grignard reagent to form a magnesium alkoxide, which, after hydrolysis, gives the alcohol (Sec. 19.9).

$$R-\overset{\overset{\displaystyle O}{\|}}{C}-CH_3 + CH_3-MgI \xrightarrow[\text{H}_2\text{O}]{\text{[Sec. 19.9]} \quad \text{H}^+} R-\overset{\overset{\displaystyle OH}{|}}{\underset{\underset{\displaystyle CH_3}{|}}{C}}-CH_3 \quad (21.68b)$$

This reaction is a very important method for the synthesis of alcohols in which at least two of the groups on the α-carbon of the alcohol product are identical. The reason for this stipulation is that two equivalents of the Grignard reagent are used in carbon–carbon bond formation (see color in Eqs. 21.68a and 21.68b). This reaction adds another alcohol synthesis to our list in Sec. 19.9, which we can update as follows:

1. Hydroboration–oxidation of alkenes (Sec. 5.3A)

2. Oxymercuration–reduction of alkenes (Sec. 5.3B)

3. Reaction of ethylene oxide with Grignard reagents (Sec. 11.4B)

4. Hydride reduction of aldehydes and ketones (Sec. 19.8)

5. Reaction of aldehydes and ketones with Grignard reagents (Sec. 19.9)

6. Hydride reduction of carboxylic acids and esters (Secs. 20.10, 21.9A)

7. Reaction of Grignard reagents with esters (this section)

B. Reaction of Acid Chlorides with Lithium Dialkylcuprates

Because acid chlorides are more reactive than ketones, the reaction of an acid chloride with a Grignard reagent can in principle be stopped at the ketone.

$$\mathrm{R{-}\overset{\overset{\displaystyle O}{\|}}{C}{-}Cl + R'{-}MgBr \longrightarrow R{-}\overset{\overset{\displaystyle O}{\|}}{C}{-}R' + Cl{-}Mg{-}Br} \qquad (21.69)$$

However, Grignard reagents are so reactive that this transformation is difficult to achieve in practice without careful control of the reaction conditions; that is, it is hard to prevent the reaction of the ketone with the Grignard reagent to give an alcohol.

Another type of organometallic reagent, called a **lithium dialkylcuprate,** can be used to effect this transformation cleanly. Lithium dialkylcuprate reagents are prepared by the reaction of two equivalents of an organolithium reagent with one equivalent of cuprous chloride, CuCl. The first equivalent forms an alkylcopper compound:

$$\mathrm{R{-}Li + Cu{-}Cl \longrightarrow R{-}Cu + LiCl} \qquad (21.70a)$$

The driving force for this reaction is the preference of lithium, the more electropositive metal, to exist as an ionic compound (LiCl). Because the copper of an alkylcopper reagent is a Lewis acid, it reacts accordingly with "alkyl anion" from a second equivalent of the organolithium reagent.

$$\mathrm{Li^{+}\ {}^{-}\!:\!R\ \ Cu{-}R \longrightarrow \underset{\text{lithium dialkylcuprate}}{R{-}\overset{-}{Cu}{-}R\ \ Li^{+}}} \qquad (21.70b)$$

The product of this reaction is a lithium dialkylcuprate. Although the copper bears a formal negative charge, it is electropositive relative to carbon, and the reagent conceptually can be considered as an alkyl anion complexed to copper:

$$\mathrm{R{-}Cu\cdots\cdots{}^{-}\!:\!R\ \ Li^{+}}$$

Lithium dialkylcuprates react much like Grignard or lithium reagents; however, since the "alkyl anion" is complexed by copper, a less electropositive element than lithium, these reagents are less reactive. They typically react with acid chlorides and aldehydes, very slowly with ketones, and not at all with esters. The reaction of lithium dialkylcuprates with acid chlorides gives ketones in excellent yield.

$$\mathrm{CH_3(CH_2)_4\overset{\overset{\displaystyle O}{\|}}{C}{-}Cl + \underset{\text{lithium dimethylcuprate}}{(CH_3)_2Cu^{-}\ Li^{+}} \xrightarrow[\text{THF}]{-78°,\ 15\ \text{min}} \xrightarrow{H_2O} \underset{(81\%\ \text{yield})}{CH_3(CH_2)_4\overset{\overset{\displaystyle O}{\|}}{C}{-}CH_3} + \underset{(\text{reacts with } H_2O)}{CH_3{-}Cu} + Li^{+}\ Cl^{-}} \qquad (21.71)$$

Because ketones are much less reactive than acid chlorides toward lithium dialkylcuprates, they do not react further.

> The description of lithium dialkylcuprates as "modified lithium reagents" is an oversimplification, albeit a useful one. We shall learn in Chapter 22 that these reagents react, at least in some cases, by a special mechanism, and that they have a particularly unusual type of reactivity with certain types of carbonyl compounds.

The reaction of acid chlorides with lithium dialkylcuprates adds another ketone synthesis to our repertoire. To review:

1. Oxidation of alcohols (Sec. 10.6)
2. Hydration and hydroboration of alkynes (Sec. 14.5)
3. Friedel–Crafts acylation (Sec. 16.4E)
4. Reaction of lithium dialkylcuprates with acid chlorides (this section)

Both the reaction of lithium dialkylcuprates with acid chlorides and the reaction of Grignard reagents with esters can be added to our list of reactions that form carbon–carbon bonds—a list that we shall update in the following section.

Problems

19 Suggest a sequence of reactions for carrying out each of the following conversions:
(a) benzoic acid to triphenylmethanol
(b) propionic acid to 3-pentanone
(c) butyric acid to 3-methyl-3-hexanol
(d) isobutyronitrile to 2,3-dimethyl-2-butanol (2 ways)

20 From what *ester* could 3-pentanol be synthesized using only a single Grignard reaction followed by protonolysis (and no other reaction)?

21 Predict the product when each of the following compounds reacts with one equivalent of lithium dimethylcuprate, followed by protonolysis. Explain.

(a) $N{\equiv}C(CH_2)_{10}{-}C(=O){-}Cl$

(b) $CH_3(CH_2)_3O{-}C(=O)(CH_2)_4C(=O){-}Cl$

21.11 SYNTHESIS OF CARBOXYLIC ACID DERIVATIVES

It should be clear from this and the previous chapter that many syntheses of carboxylic acid derivatives begin with other carboxylic acid derivatives. Let us review the methods we have learned:

Synthesis of Esters:

1. Acid-catalyzed esterification of carboxylic acids (Sec. 20.8A)
2. Alkylation of carboxylic acids or carboxylates (Sec. 20.8B)

3. Reaction of acid chlorides and anhydrides with alcohols or phenols (Sec. 21.8A)
4. Transesterification of other esters (Sec. 21.8C)

Synthesis of Acid Chlorides:

1. Reaction of carboxylic acids with $SOCl_2$ or PCl_5 (Sec. 20.9A)

Synthesis of Anhydrides:

1. Reaction of carboxylic acids with dehydrating agents (Sec. 20.9B)
2. Reaction of acid chlorides with carboxylate salts (Sec. 21.8A)

Synthesis of Amides:

1. Reaction of acid chlorides or anhydrides with amines (Sec. 21.8A)

Synthesis of Nitriles:

The synthesis of nitriles is an important exception to the generalization that carboxylic acid derivatives are usually prepared from other carboxylic acid derivatives. Two syntheses of nitriles are:

1. Cyanohydrin formation (Sec. 19.7)
2. S_N2 reaction of cyanide ion with alkyl halides or sulfonate esters

The S_N2 reaction of cyanide ion, like all S_N2 reactions, requires a primary or unbranched secondary alkyl halide or sulfonate ester.

$$PhCH_2Cl + Na^+ \; ^-{:}CN \xrightarrow{EtOH-H_2O} \underset{(80\text{–}90\%\ \text{yield})}{PhCH_2CN} + Na^+ \; Cl^- \qquad (21.72)$$

$$Br(CH_2)_3Br + 2Na^+ \; ^-CN \xrightarrow{EtOH-H_2O} \underset{(77\text{–}86\%\ \text{yield})}{NC(CH_2)_3CN} + 2Na^+ \; Br^- \qquad (21.73)$$

Both this reaction and the synthesis of cyanohydrins are noteworthy because they provide additional ways to form carbon–carbon bonds. Let us update our list of reactions that can be used for the construction of carbon–carbon bonds:

1. Cyclopropane formation by addition of carbenes to alkenes (Sec. 9.7)
2. Reaction of Grignard reagents with ethylene oxide (Sec. 11.4B)
3. Reaction of acetylide ions with alkyl halides or sulfonate esters (Sec. 14.7B)
4. Diels–Alder reactions (Sec. 15.3)
5. Friedel–Crafts acylation (Sec. 16.4E) and alkylation (Sec. 16.4F)
6. Cyanohydrin formation (Sec. 19.7)
7. Reaction of Grignard reagents with aldehydes and ketones (Sec. 19.9)

8. Reaction of Grignard reagents with carbon dioxide (Sec. 20.6)
9. Reaction of Grignard reagents with esters (Sec. 21.10A)
10. Reaction of lithium dialkylcuprates with acid chlorides (Sec. 21.10B)
11. Reaction of cyanide ion with alkyl halides or sulfonate esters (this section).

Since carboxylic acid derivatives can be hydrolyzed to carboxylic acids, any synthesis of a carboxylic acid derivative can be used as part of the synthesis of the corresponding carboxylic acid. Since nitriles are prepared from compounds other than carboxylic acid derivatives, the preparation of a nitrile can be particularly useful as an intermediate step in the preparation of a carboxylic acid. For example, if we wish to prepare pentanoic acid (valeric acid) from 1-butanol, the additional carbon atom required can be introduced as a nitrile:

$$CH_3CH_2CH_2CH_2OH \xrightarrow{HBr} CH_3CH_2CH_2CH_2Br \xrightarrow{^-CN} CH_3CH_2CH_2CH_2C{\equiv}N$$

$$\downarrow H_2O,\ \text{acid, heat}$$

$$CH_3CH_2CH_2CH_2{-}\overset{\overset{\displaystyle O}{\|}}{C}{-}OH \qquad (21.74)$$

Problems

22 Outline a preparation of valeric acid from 1-butanol different from the one shown in Eq. 21.74.

23 From what nitrile can 2-hydroxypropionic acid (lactic acid) be prepared? Show how this nitrile, in turn, can be prepared from acetaldehyde.

21.12 OCCURRENCE AND USE OF CARBOXYLIC ACIDS AND THEIR DERIVATIVES

A. Nylon and Polyesters

Two of the most important polymers produced on an industrial scale are **nylon** and **polyesters.** The chemistry of carboxylic acids and their derivatives plays an important role in the synthesis of these polymers.

Nylon is the general name given to a group of polymeric amides. The two most widely used are nylon-6,6 and nylon-6.

$$\cdots{-}\overset{\overset{\displaystyle O}{\|}}{C}{-}NH(CH_2)_6NH{-}\overset{\overset{\displaystyle O}{\|}}{C}{-}(CH_2)_4{-}\overset{\overset{\displaystyle O}{\|}}{C}{-}NH(CH_2)_6NH{-}\overset{\overset{\displaystyle O}{\|}}{C}{-}(CH_2)_4{-}\cdots$$

nylon-6,6

$$\text{or} \quad \left[NH(CH_2)_6NH\overset{\overset{\displaystyle O}{\|}}{C}(CH_2)_4\overset{\overset{\displaystyle O}{\|}}{C} \right]_n$$

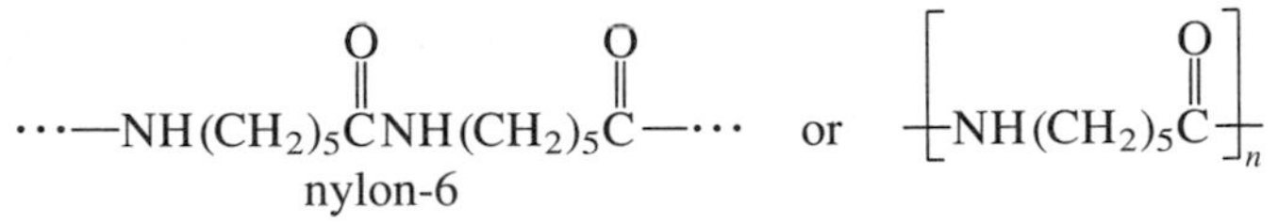

$$\cdots\!-\!NH(CH_2)_5\overset{O}{\overset{\|}{C}}NH(CH_2)_5\overset{O}{\overset{\|}{C}}\!-\!\cdots \quad \text{or} \quad \left[NH(CH_2)_5\overset{O}{\overset{\|}{C}}\right]_n$$

nylon-6

About 2.4 billion pounds of nylon was produced in the United States in 1986. Nylon is used in tire cord, carpet, and apparel.

The starting material for the industrial synthesis of nylon-6,6 is adipic acid. In one process, adipic acid is converted into its dinitrile and then into 1,6-diaminohexane (hexamethylenediamine).

$$\underset{\text{adipic acid}}{HO_2C\!-\!(CH_2)_4\!-\!CO_2H} \xrightarrow{\text{several steps}} N{\equiv}C\!-\!(CH_2)_4\!-\!C{\equiv}N \xrightarrow[\text{cat}]{H_2} \underset{\text{hexamethylenediamine}}{H_2N\!-\!(CH_2)_6\!-\!NH_2} \qquad (21.75)$$

When hexamethylenediamine and adipic acid are mixed, they form a salt. Heating the salt forms the polymeric amide.

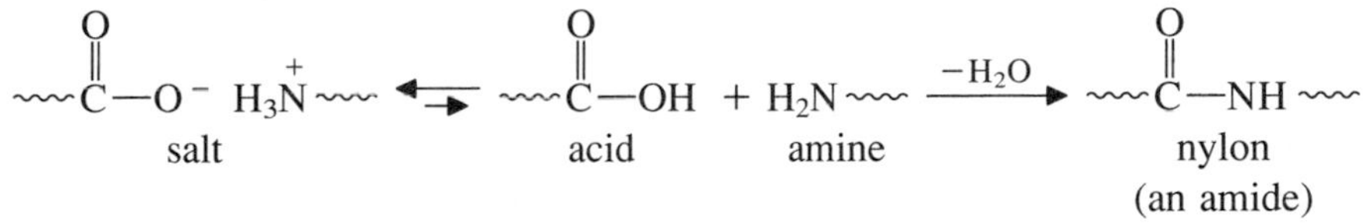

$$\underset{\text{salt}}{\sim\overset{O}{\overset{\|}{C}}\!-\!O^-\ \ H_3\overset{+}{N}\sim} \rightleftarrows \underset{\text{acid}}{\sim\overset{O}{\overset{\|}{C}}\!-\!OH} + \underset{\text{amine}}{H_2N\sim} \xrightarrow{-H_2O} \underset{\substack{\text{nylon}\\ \text{(an amide)}}}{\sim\overset{O}{\overset{\|}{C}}\!-\!NH\sim} \qquad (21.76)$$

The reaction of an amine with an acid to form an amide is analogous to the reaction of an amine with an ester (Sec. 21.8C). However, much more vigorous conditions are required because the equilibrium on the left of Eq. 21.76 strongly favors the salt, since the amine is basic and the carboxylic acid is acidic. In the salt, the amine is protonated and therefore not nucleophilic, and the carboxylate ion is very unreactive toward nucleophiles (why?). The small amount of amine and carboxylic acid in equilibrium with the salt react when the salt is heated, pulling the equilibrium to the right.

Nylon is an example of a **condensation polymer,** a polymer formed in a reaction that liberates a small molecule, in this case H_2O. This is in contrast to an *addition polymer,* which is formed without the liberation of a small molecule (compare Sec. 5.9A).

Adipic acid, from which nylon is made, is prepared industrially in several steps from petroleum. This is a classic example of the dependence of an important segment of the chemical economy on petroleum feedstocks. About 1.3 billion pounds of adipic acid was synthesized in the United States in 1985.

Polyesters are condensation polymers derived from the reaction of diols and diacids. One widely used polyester, polyethylene terephthalate, can be produced by the esterification of ethylene glycol and terephthalic acid.

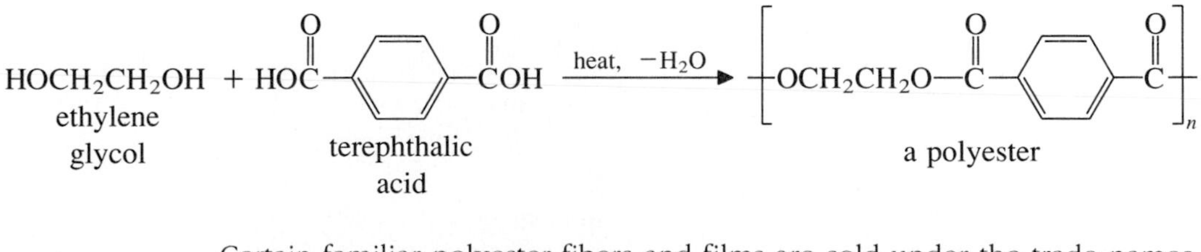

(21.77)

Certain familiar polyester fibers and films are sold under the trade names Dacron® and Mylar®, respectively. About 3.3 billion pounds of polyester was produced in the

United States in 1986. Polyester production also depends on raw materials produced from petroleum.

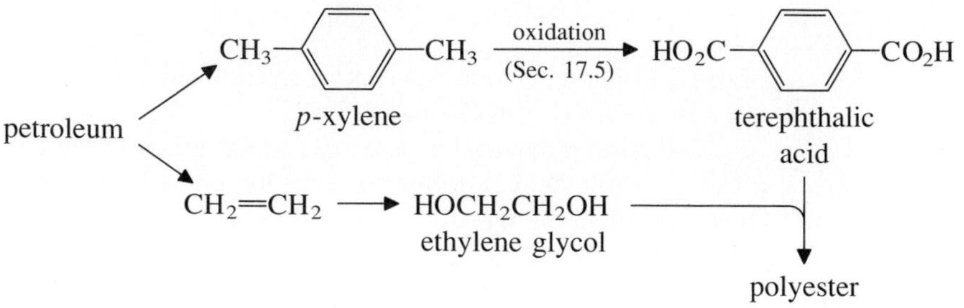

(21.78)

Problems

24 (a) Which polymer should be more resistant to strong base: nylon-6,6 or the polyester in Eq. 21.77? Explain.

(b) What would happen to the polyester in Eq. 21.77 if it were allowed to react with $LiAlH_4$, followed by dilute acid?

25 One interesting process for making nylon-6,6 demonstrates the potential of using biomass as a chemical feedstock. The raw material for this process, outlined below, is the aldehyde furfural, obtained from sugars found in oat hulls. Suggest conditions for carrying out each of the steps in this process indicated by an asterisk.

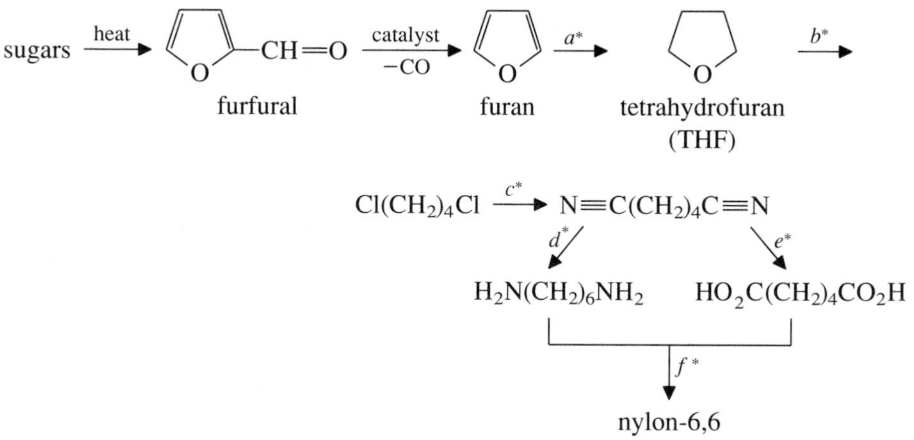

26 ε-Caprolactam is polymerized to nylon-6 when it is heated with a *catalytic amount* of water.

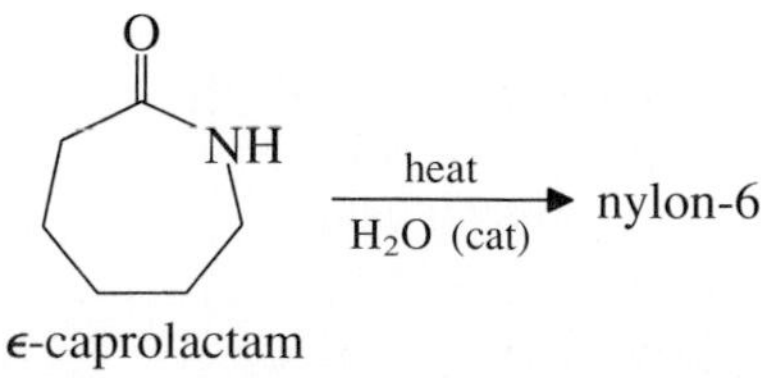

What is the role of the water? Give a mechanism for the polymerization.

B. Waxes, Fats, and Phospholipids

Waxes, fats, and phospholipids are all important naturally occurring ester derivatives of fatty acids. A **wax** is an ester of a fatty acid and a "fatty alcohol," an alcohol with a long unbranched carbon chain. For example, carnauba wax, obtained from the leaves of the Brazilian carnauba palm, and valued for its hard, brittle characteristics, consists of about 80% of esters derived from C_{24}, C_{26}, and C_{28} fatty acids and C_{30}, C_{32}, and C_{34} alcohols. The following compound is a typical constituent of carnauba wax.

$$CH_3(CH_2)_{24}—\overset{\overset{\displaystyle O}{\|}}{C}—O—(CH_2)_{31}CH_3 \qquad \text{constituent of carnauba wax}$$

A **fat** is an ester derived from a molecule of glycerol and three molecules of fatty acid.

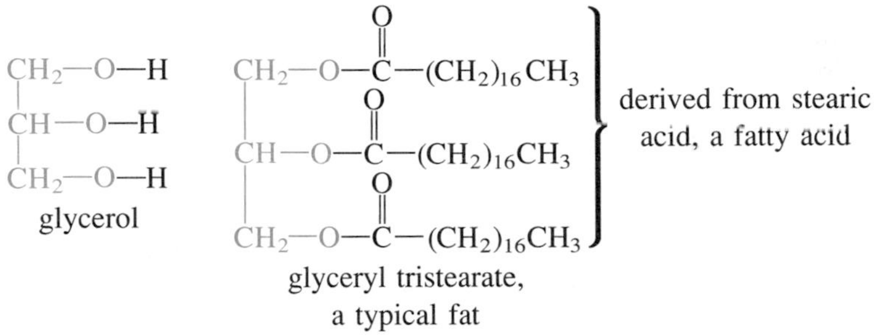

The three acyl groups in a fat may be the same, as in glyceryl tristearate, or different, and they may contain unsaturation, which is typically in the form of one or more *cis* double bonds. Treatment of a fat with NaOH yields glycerol and the sodium salts of the fatty acids (*soaps*; Sec. 20.5). This is the reaction from which *saponification* (Sec. 21.6A) gets its name.

$$\begin{array}{l} CH_2—O_2C—(CH_2)_{16}CH_3 \\ | \\ CH—O_2C—(CH_2)_{16}CH_3 \\ | \\ CH_2—O_2C—(CH_2)_{16}CH_3 \end{array} + 3NaOH \longrightarrow \begin{array}{l} CH_2OH \\ | \\ CHOH \\ | \\ CH_2OH \end{array} + \underset{\text{a soap}}{3CH_3(CH_2)_{16}—CO_2^- \; Na^+} \qquad (21.79)$$

Fats are stored in highly concentrated droplets in the body, and serve as a means of storing carbon until it is needed for other purposes.

Phospholipids are closely related to fats, since they too are esters of glycerol. Phospholipids differ from fats in that one of the terminal oxygens of glycerol in a phospholipid is esterified to a special type of organic phosphate derivative, forming a polar head group in the molecule. The remaining two oxygens of glycerol are esterified to fatty acids.

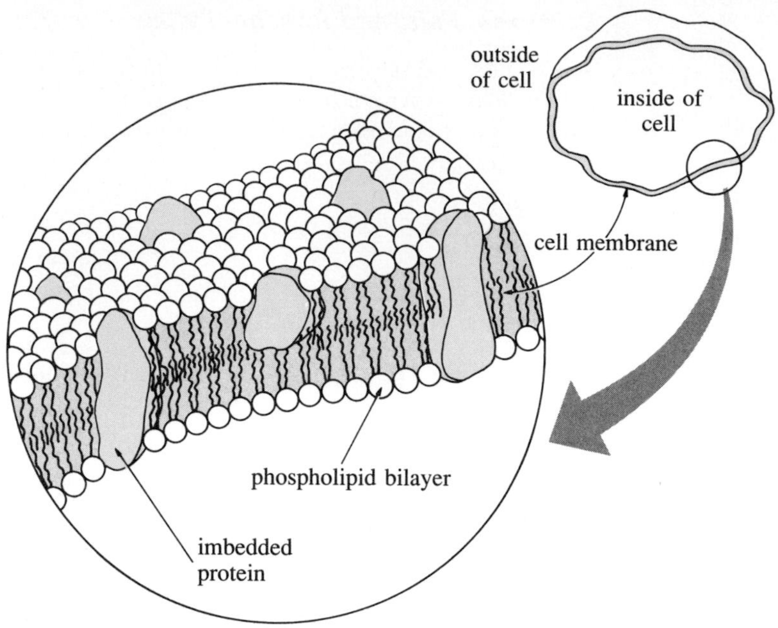

Figure 21.6 Schematic view of a cell membrane, with an enlargement showing the phospholipid bilayer containing imbedded proteins.

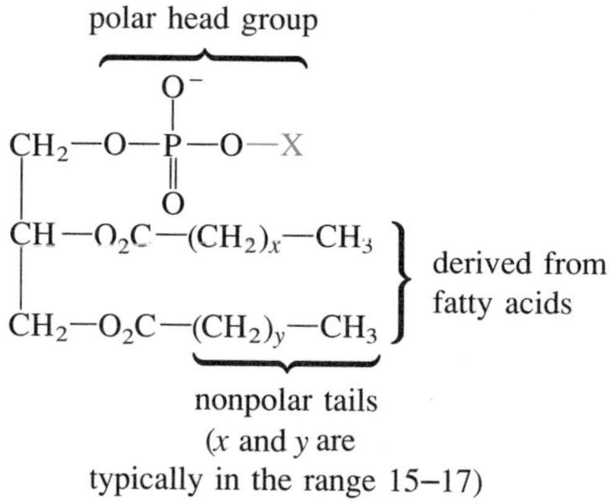

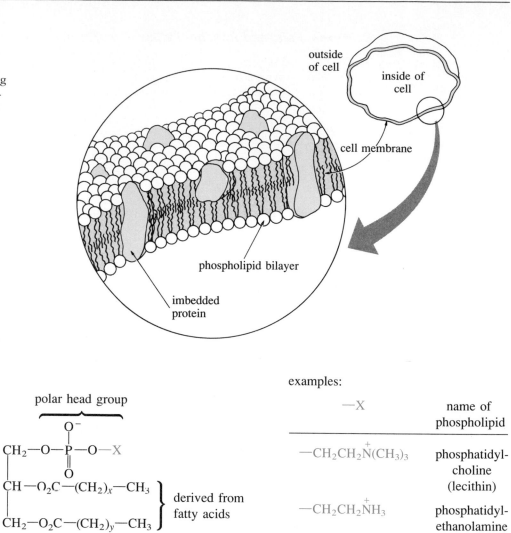

examples:

—X	name of phospholipid
$-CH_2CH_2\overset{+}{N}(CH_3)_3$	phosphatidylcholine (lecithin)
$-CH_2CH_2\overset{+}{N}H_3$	phosphatidylethanolamine

Phospholipids closely resemble detergents (Sec. 20.5), since each molecule contains a polar head group and nonpolar tails. Phospholipids are sometimes represented schematically as shown on the left; a space-filling model is shown on the right:

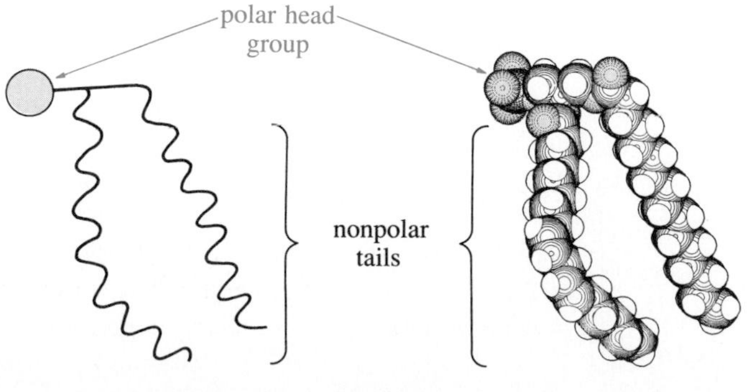

Phospholipids, along with cholesterol (a steroid) and imbedded proteins, are major constituents of *cell membranes,* the envelopes that surround living cells. The phospholipids give membranes a *bilayer* structure in which the polar head groups interact with aqueous solution, and the nonpolar tails interact with each other, away from aqueous solution (Fig. 21.6). This *phospholipid bilayer* forms a barrier that is impermeable to most ions and to many organic molecules, and allows the cell to control "what gets in" and "what gets out." (Look again at Sec. 20.5; can you see why detergents can disrupt cell membranes?)

KEY IDEAS IN CHAPTER 21

- Carboxylic acid derivatives are polar molecules; except for amides, they have small water solubility and low-to-moderate boiling points. Because of their capacity for hydrogen bonding, amides have high boiling points (many are solids) and moderate solubility in water.
- The carbonyl absorptions (and the C≡N absorption of nitriles) are the most important absorptions in the infrared spectra of carboxylic acid derivatives (Table 21.3). The NMR spectra show typical δ 2–3 absorptions for the α-protons, as well as other characteristic absorptions: δ 3.5–4.5 for the *O*-alkyl groups of esters, δ 2.6–3 for the *N*-alkyl groups and δ 7.5–8.5 for the N—H protons of amides.
- Carboxylic acid derivatives, like carboxylic acids themselves, are weak bases and are protonated on the carbonyl oxygen or nitrile nitrogen.
- The most characteristic reaction of carboxylic acid derivatives is substitution at the carbonyl group. Hydrolysis reactions and the reactions of acid chlorides, anhydrides, and esters with nucleophiles are examples of this type of reactivity. Substitution at the carbonyl group typically involves addition to form a tetrahedral intermediate, which then loses a leaving group to form the product.
- In some reactions of carboxylic acid derivatives, substitution is followed by addition. This is the case, for example, in the $LiAlH_4$ reductions and Grignard reactions of esters, in which the aldehyde and ketone intermediates, respectively, react further.
- The relative reactivity of carbonyl compounds and nitriles with nucleophiles is in the order acid chlorides > aldehydes > ketones ≫ esters > amides > nitriles. Thus the reduction of an acid chloride with lithium tri(*t*-butoxy)aluminum hydride can be stopped at the aldehyde because acid chlorides are more reactive than aldehydes; but the $LiAlH_4$ reduction of an ester cannot be stopped at the aldehyde, because aldehydes are more reactive than esters.
- Nitriles react by addition, much like aldehydes and ketones. In some reactions, the product of addition undergoes a second addition (as in reduction to primary amines); and in other cases it undergoes a substitution reaction (as in hydrolysis).

ADDITIONAL PROBLEMS

27 Give the principal organic product(s) expected when propionyl chloride reacts with each of the following reagents:
(a) H_2O
(b) ethanethiol, pyridine, 0°
(c) $(CH_3)_3COH$, pyridine
(d) $(CH_3)_2CuLi$, −78°, then H_2O
(e) H_2, Pd catalyst (quinoline/S poison)
(f) $AlCl_3$, toluene, then H_2O
(g) $(CH_3)_2CHNH_2$ (2 equiv)
(h) sodium benzoate
(i) *p*-cresol, pyridine

28 Give the principal organic product(s) expected when ethyl benzoate reacts with each of the following reagents.
(a) H^+, H_2O, heat
(b) NaOH, H_2O
(c) aqueous NH_3, heat
(d) $LiAlH_4$, then H_2O
(e) excess $CH_3CH_2CH_2MgBr$, then H_2O
(f) product of (e) + acetyl chloride, pyridine, 0°
(g) product of (e) + benzenesulfonyl chloride
(h) NaOEt, EtOH

29 Give the structure of a compound that satisfies each of the following criteria:
(a) A compound C_3H_7ON that liberates ammonia on treatment with hot aqueous KOH
(b) A compound C_3H_5N that liberates ammonia on treatment with hot aqueous KOH
(c) A compound that gives equal amounts of 1-hexanol and 2-hexanol on treatment with $LiAlH_4$ followed by protonolysis
(d) A compound that gives 1-butanol and 2-methyl-2-propanol on treatment with CH_3MgBr followed by protonolysis

30 Complete the following diagram by filling in all missing reagents or intermediates:

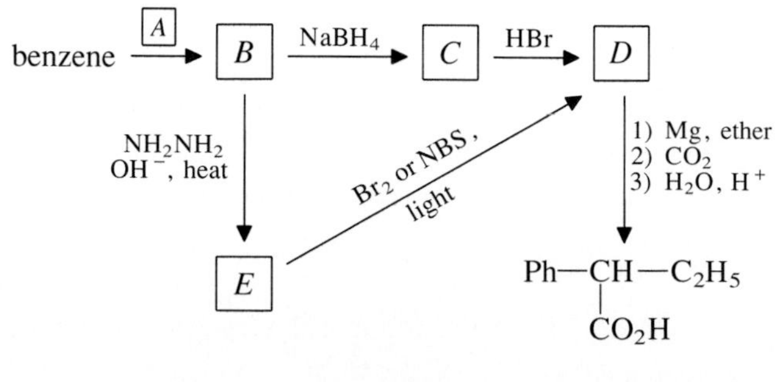

31 Propose a synthesis of the following compound, the active ingredient in some insect repellents, from 3-methylbenzaldehyde and any other reagents:

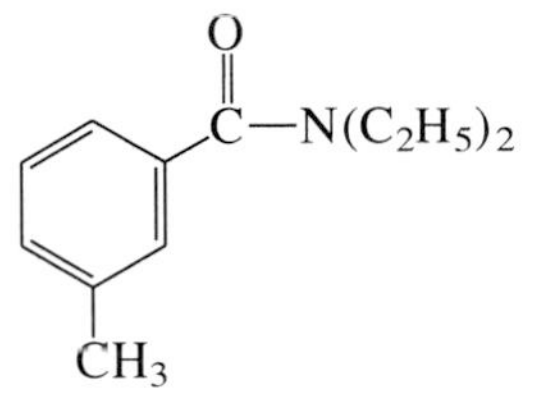

32 Propose a synthesis of each of the following compounds from butyric acid and any other reagents:
(a) 2-methyl-2-pentanol
(b) 4-methyl-4-heptanol
(c) 4-heptanol
(d) $CH_3CH_2CH_2CH_2NH_2$
(e) $CH_3CH_2CH_2CH_2CH_2NH_2$
(f) $CH_3CH_2CH_2CH_2CH_2CH_2NH_2$

33 Treatment of acetic propionic anhydride with ethanol gives a mixture of two esters consisting of 36% of the higher boiling one, *A*, and 64% of the lower boiling one, *B*. Identify *A* and *B*. Suggest a reason why *B* is formed in greater amount.

34 Amides associate to form dimers much like carboxylic acids.
(a) Draw the structure of the *N*-methylacetamide dimer.
(b) The dimerization equilibrium constant for *N*-methylacetamide is 4.7–5.4 M^{-1} in CCl_4, but is only 0.005 M^{-1} in water. Explain the effect of solvent on the dimerization constant.
(c) In CCl_4, the dimerization equilibrium constant for *N*-propylacetamide is 4.5 M^{-1}, but that of γ-butyrolactam is 288 M^{-1}. Suggest a reason why the lactam should have a much greater tendency to form dimers.

35 Explain why carboxylate salts are much less reactive than esters in carbonyl substitution reactions.

36 Predict the relative reactivity of thiol esters toward hydrolysis with aqueous NaOH:
(a) Much more reactive than acid chlorides
(b) Less reactive than acid chlorides, but more reactive than esters
(c) Less reactive than esters
Explain your choice.

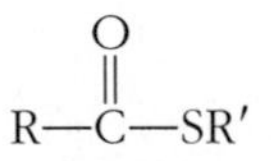
general structure of a thiol ester

Problems (Cont.)

37 A compound *A* has prominent infrared absorptions at 1050, 1786, and 1852 cm^{-1}, and shows a single absorption in the NMR at δ 3.00. When heated gently with methanol, compound *B*, $C_5H_8O_4$, is obtained. Compound *B* has IR absorption at 2500–3000 (broad), 1730, and 1701 cm^{-1}, and its NMR spectrum in D_2O consists of two singlets at δ 2.7 and δ 3.7 in the intensity ratio 4:3. Identify *A* and *B*, and explain your reasoning.

38 (a) Draw the structures of all the products that would be obtained when ($\pm$)-α-phenylglutaric anhydride is treated with ($\pm$)-1-phenyl-1-ethanol. Which products should be separable without optical resolution? Which products should be obtained in equal amounts? In different amounts?

(b) Answer the same question for the reaction of the *S* enantiomer of the same alcohol with ($\pm$)-α-phenylglutaric anhydride. (*Hint:* See Sec. 7.8B.)

39 Using the curved-arrow formalism, write a detailed mechanism for the acid-catalyzed hydration of carbon dioxide to give carbonic acid.

40 Klutz McFingers, a graduate student in his ninth year of study, has suggested the following synthetic procedures, and has come to you in the hope that you can explain why none of them works very well (or not at all).

(a) Noting the fact that primary alcohols + HBr give alkyl halides, Klutz has proposed by analogy the following nitrile synthesis:

$$\text{Ph—CH}_2\text{—OH} + \text{HCN} \xrightarrow{\text{H}_2\text{O}} \text{Ph—CH}_2\text{—CN} + \text{H}_2\text{O}$$

(b) Klutz has proposed the following synthesis for the half-ester of adipic acid:

$$\underset{\text{(1.0 mole)}}{\text{HO}_2\text{C—(CH}_2)_4\text{—CO}_2\text{H}} + \underset{\text{(1.0 mole)}}{\text{CH}_3\text{OH}} \xrightarrow{\text{H}_2\text{SO}_4} \text{HO}_2\text{C—(CH}_2)_4\text{—CO}_2\text{CH}_3 + \text{H}_2\text{O}$$

(c) Klutz has proposed the following synthesis of acetic benzoic anhydride:

$$\text{CH}_3\text{—}\overset{\overset{\displaystyle\text{O}}{\|}}{\text{C}}\text{—OH} + \text{Ph—}\overset{\overset{\displaystyle\text{O}}{\|}}{\text{C}}\text{—OH} \xrightarrow{\text{P}_2\text{O}_5} \text{CH}_3\text{—}\overset{\overset{\displaystyle\text{O}}{\|}}{\text{C}}\text{—O—}\overset{\overset{\displaystyle\text{O}}{\|}}{\text{C}}\text{—Ph}$$

(d) Noting correctly that methyl benzoate is completely saponified by one molar equivalent of NaOH, Klutz has suggested that methyl salicylate should also undergo saponification with one equivalent of NaOH.

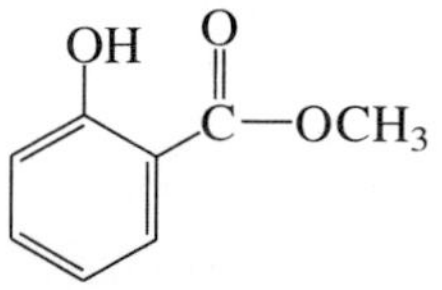

methyl salicylate

(e) Klutz, finally able to secure a position with a pharmaceutical company working on β-lactam antibiotics, has proposed the following reaction for deamidation of a cephalosporin derivative:

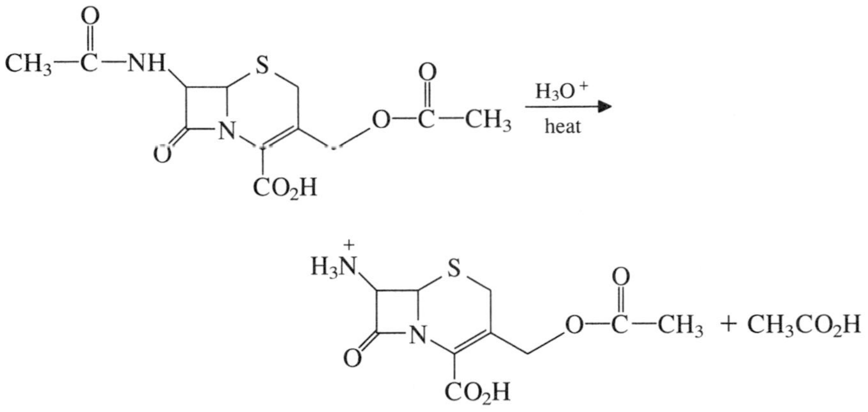

41 When (*R*)-(−)-mandelic acid (α-hydroxy-α-phenylacetic acid) is treated with CH_3OH and H_2SO_4, and the resulting compound is treated with excess $LiAlH_4$ in ether, then H_2O, a levorotatory product is obtained that reacts with sodium periodate. Give the structure, name, and absolute configuration of this product.

42 You are employed by Fibers Unlimited, a company specializing in the manufacture of specialty polymers. The vice-president for research, Strong Fishlein, has asked you to design laboratory preparations of the following polymers. You are to use as starting materials the company's extensive stock of dicarboxylic acids containing six or fewer carbon atoms. Accommodate him.

(a) 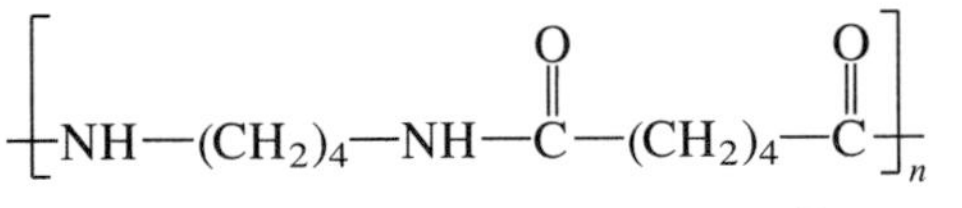 nylon-4,6

(b) 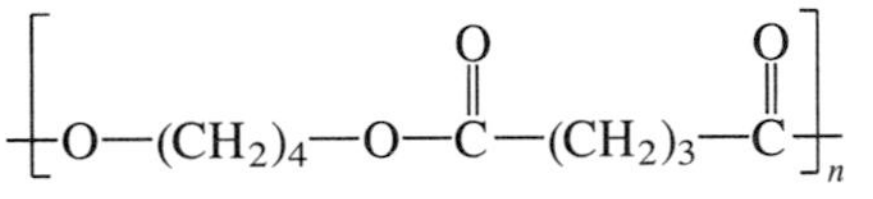

43 Lactic acid (α-hydroxypropionic acid), on standing, forms a cyclic compound $C_6H_8O_4$. Suggest a structure for this compound, and explain why lactic acid does not form an ordinary lactone.

44 You are a chemist working for the Ima Hot Pepper Company, and have been asked to provide some chemical information about *capsaicin,* the active ingredient of hot peppers.

(Problem continues on page 912.)

Problems (Cont.)

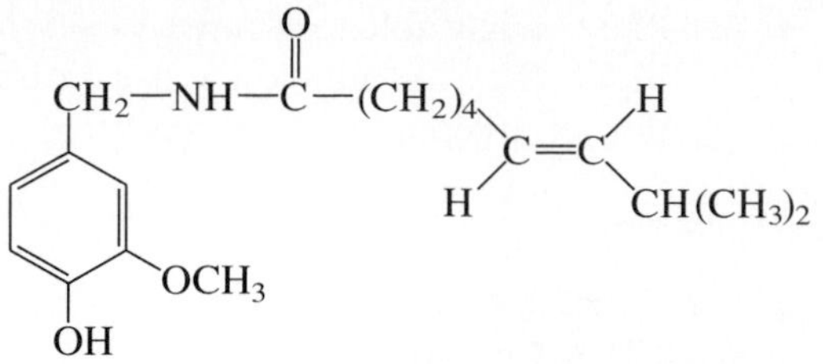

capsaicin

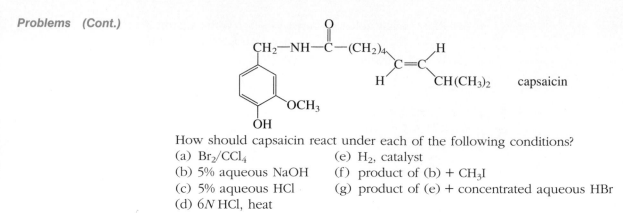

How should capsaicin react under each of the following conditions?

(a) Br_2/CCl_4
(b) 5% aqueous NaOH
(c) 5% aqueous HCl
(d) 6*N* HCl, heat
(e) H_2, catalyst
(f) product of (b) + CH_3I
(g) product of (e) + concentrated aqueous HBr

45 When compound *A* (below) is treated with ozone at −78° in CH_2Cl_2, compound *B* is formed. Compound *B* has IR absorption at 1820 cm^{-1} and a singlet NMR absorption at δ 1.5. Reaction of *B* with water gives α,α-dimethylmalonic acid, and reaction of *B* with methanol gives the monomethyl ester of the same acid. Suggest a structure for *B*. (This very interesting compound was unknown until this work was carried out in 1978 by chemists at the University of California, San Diego.)

A

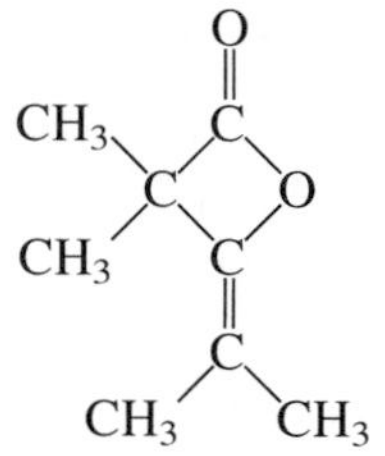

46 An optically active compound *A*, $C_6H_{10}O_2$, when dissolved in NaOH solution, consumed one equivalent of base. On acidification, compound *A* was slowly regenerated. Treatment of *A* with $LiAlH_4$ in ether followed by protonolysis gave an optically *inactive* compound *B* that reacted with acetic anhydride to give a diacetyl derivative *C*. Compound *B* was oxidized by dilute nitric acid to β-methylglutaric acid (3-methylpentanedioic acid). Identify compounds *A*, *B*, and *C* and explain your reasoning.

47 Exactly 2.00 g of an ester *A* containing only C, H, and O was saponified with 15.00 mL of a 1.00*N* NaOH solution. Following the saponification the solution required 5.30 mL of 1.00*N* HCl to titrate the unused NaOH. Ester *A*, as well as its acid and alcohol saponification products *B* and *C*, respectively, were all optically active. Compound *A* was not oxidized by $K_2Cr_2O_7$, nor did compound *A* decolorize Br_2 in CCl_4. Alcohol *C* was oxidized to acetophenone by $K_2Cr_2O_7$. When acetophenone was reduced with $NaBH_4$, a compound *D* was formed that reacted with the acid chloride derived from *B* to give two optically active compounds: *A* (identical with starting ester) and *E*. Propose a structure for each compound that is consistent with the data.

48 Complete the following reactions by giving the principal organic products. Explain your reasoning.

(a)

$$CH_3{-}\overset{\overset{\displaystyle O}{\|}}{C}{-}O{-}\underset{\underset{\displaystyle Ph}{|}}{C}{=}CH_2 + H_2O \xrightarrow{NaOH}$$

(b)

$$CH_3{-}\overset{\overset{\displaystyle O}{\|}}{C}{-}CH_2{-}O{-}\overset{\overset{\displaystyle O}{\|}}{C}H + CH_3OH \xrightarrow{CH_3O^-\ (trace)}$$

(c)

$$H_2N{-}NH_2 + Ph\overset{\overset{\displaystyle O}{\|}}{C}{-}Cl\ (excess) \xrightarrow{NaOH}$$

(d)

$$Ph{-}\overset{\overset{\displaystyle O}{\|}}{C}{-}NH(CH_2)_5CN \xrightarrow[heat]{H_3O^+}$$

(e)

$$CH_3{-}\overset{*}{O}H + Ph{-}\overset{\overset{\displaystyle O}{\|}}{\underset{\underset{\displaystyle O}{\|}}{S}}{-}Cl \xrightarrow{pyridine} \xrightarrow{OH^-} \qquad (\overset{*}{O} = {}^{18}O)$$

(f)

$$CH_3{-}\overset{*}{O}H + CH_3{-}\overset{\overset{\displaystyle O}{\|}}{C}{-}Cl \xrightarrow{pyridine} \xrightarrow{OH^-} \qquad (\overset{*}{O} = {}^{18}O)$$

(g)

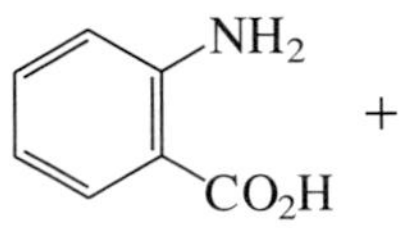

$$+ Cl{-}\overset{\overset{\displaystyle O}{\|}}{C}{-}Cl \longrightarrow (C_8H_5NO_3)$$

(h)

$$(CH_3)_2\underset{\underset{\displaystyle NH_2}{|}}{C}CH_2CH_2CO_2CH_3 \xrightarrow{stand\ in\ CH_3OH} (C_6H_{11}NO)$$

(i)

$$(CH_3)_3CNH{-}\overset{\overset{\displaystyle O}{\|}}{C}{-}NH_2 + H_2O \xrightarrow{H^+,\ heat}$$

(j)

$$\text{glyceryl tristearate} + CH_3OH \xrightarrow{CH_3O^-}$$

(k)

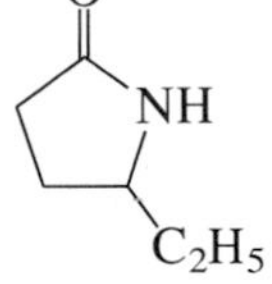

$$+ LiAlH_4\ (excess) \longrightarrow \xrightarrow{H_2O}$$

(l)

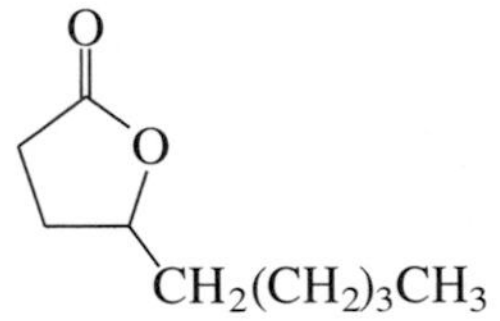

$$+ CH_3MgBr\ (excess) \longrightarrow \xrightarrow{H_3O^+}$$

(m)

$$Ph{-}MgBr\ (1\ equiv) + CH_3O{-}\overset{\overset{\displaystyle O}{\|}}{C}{-}(CH_2)_3{-}CH{=}O \longrightarrow \xrightarrow{H_3O^+}$$

(n)

$$C_2H_5O{-}\overset{\overset{\displaystyle O}{\|}}{C}{-}OC_2H_5 + EtMgBr\ (large\ excess) \longrightarrow \xrightarrow{H_3O^+}$$

(o)

$$(CH_3)_3C{-}\overset{\overset{\displaystyle O}{\|}}{C}{-}Cl + LiAlH_4\ (excess) \longrightarrow \xrightarrow{H_3O^+}$$

Figure 21.7 Spectra for Problem 49.

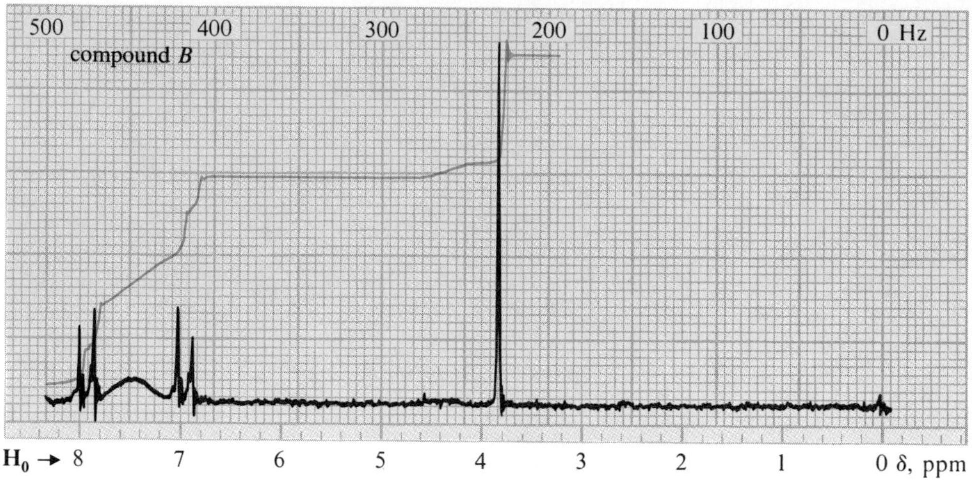

(a)

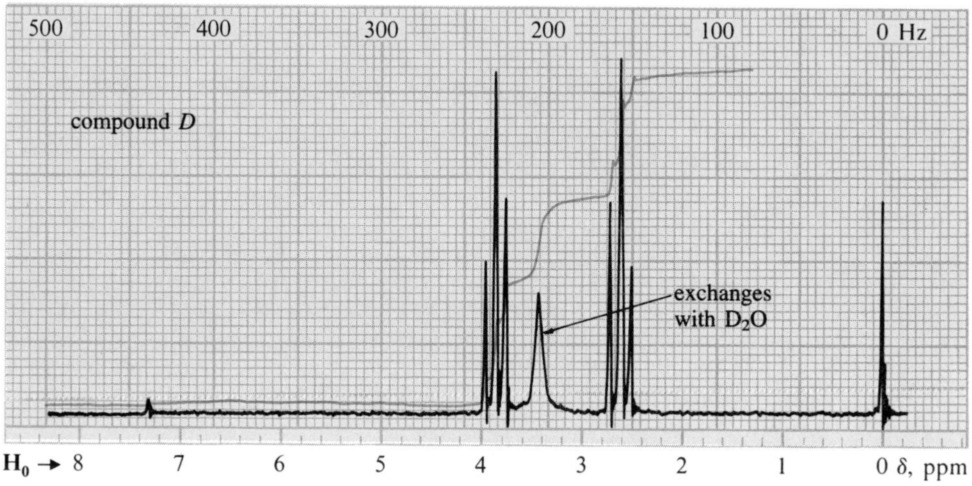

(b)

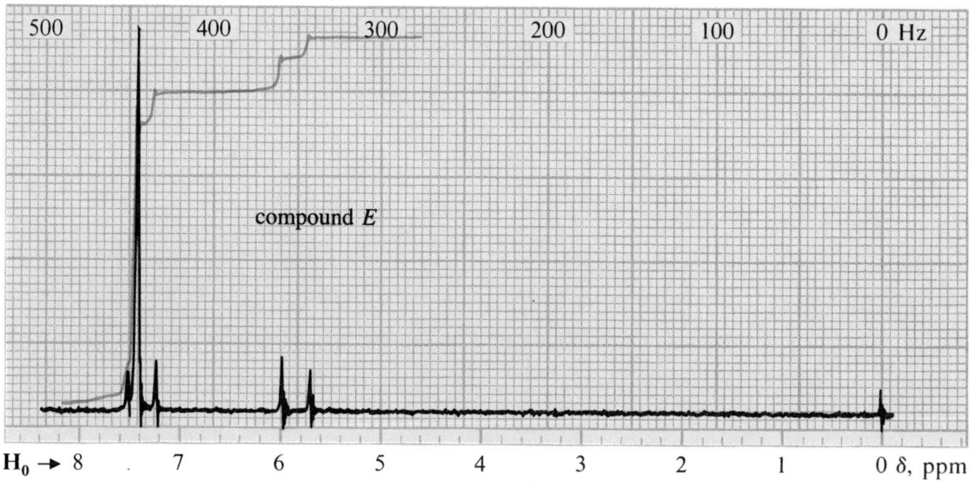

(c)

49 Identify each of the following compounds from their spectra:

(a) Compound *A:* MW 113; gives a positive hydroxamate test
IR: 2237, 1733, 1200 cm^{-1}
NMR: δ 1.33 (3*H*, t, $J = 7$ Hz), δ 3.45 (2*H*, s), δ 4.27 (2*H*, q, $J = 7$ Hz)

(b) Compound *B:* MW 151; IR 3400, 3200, 1658, 1618 cm^{-1}
NMR spectrum in Fig. 21.7a

(c) Compound *C:* slowly dissolves in NaOH solution
IR: 3278, 2120, 1721, 1242 cm^{-1}
NMR: δ 1.31 (3*H*, t, $J = 7$ Hz), δ 3.02 (1*H*, s), δ 4.21 (2*H*, q, $J = 7$ Hz)

(d) Compound *D:* MW 71; IR 3200 (strong, broad), 2250 cm^{-1}; no absorptions between 1500–2250 cm^{-1}
NMR spectrum in Fig. 21.7b

(e) Compound *E:* UV spectrum: $\lambda_{max} = 272$ ($\epsilon = 39{,}500$)
mass spectrum: m/e = 129 (parent ion and base peak)
IR: 2200 cm^{-1}, 970 cm^{-1}
NMR spectrum in Fig. 21.7c

(f) Compound *F:* IR 1724 cm^{-1}
NMR and mass spectra in Fig. 21.7d

Figure 21.7 continued

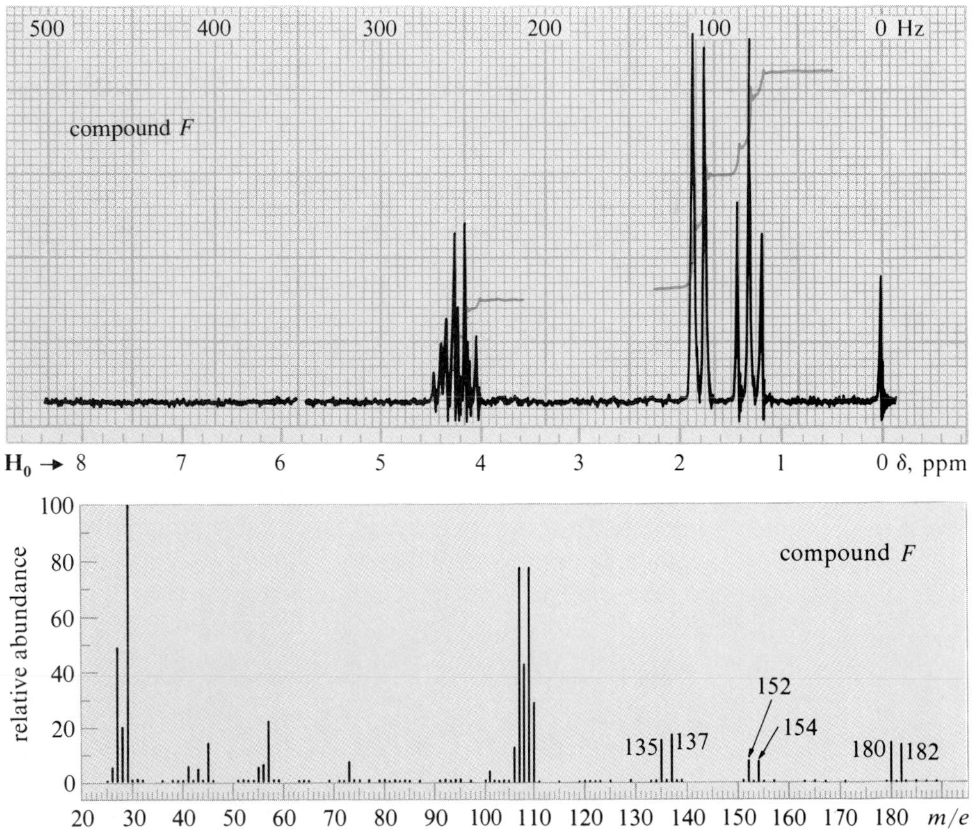

(d)

Problems (Cont.)

50 A sample of 3,7-dimethyl-3-octanol with a specific rotation of $-0.63°$ was treated with acetyl chloride/pyridine to give the acetate ester, which had a specific rotation of $-0.68°$. Saponification of this ester with 0.02–0.04*N* NaOH gave back the alcohol. The specific rotation of this sample of alcohol was $-0.62°$. When the acetate ester was treated with 0.02*N* aqueous HCl, the alcohol was also obtained, but its specific rotation was found to be $+0.06°$. Suggest an explanation for the different results in the acid- and base-promoted ester hydrolysis reactions.

51 Outline a synthesis of each of the following compounds from the indicated starting materials and any other reagents:

(a) *o*-methylbutyrophenone from *o*-bromotoluene

(b) $PhNHCH_2CH_2CH(CH_3)_2$ from isovaleric acid

(c) $HO^*—C(=O^*)—CH_2CH_2—C(=O^*)—O^*H$ from succinic acid; ($^*O = {}^{18}O$); each oxygen partially labeled

(d) 1-cyclohexyl-2-methyl-2-propanol from bromocyclohexane

(e) $PhO—C(=O)—OCH_3$ from phosgene (Sec. 21.1G)

(f) 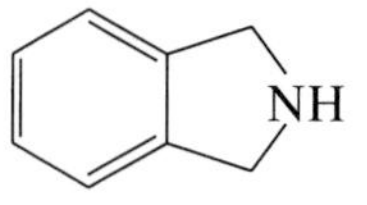 from phthalic acid

(g) 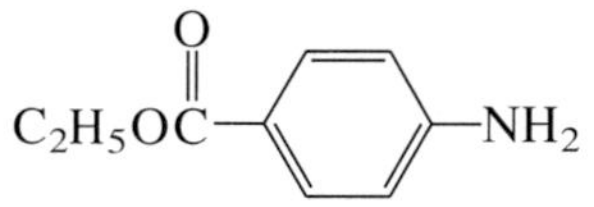 from 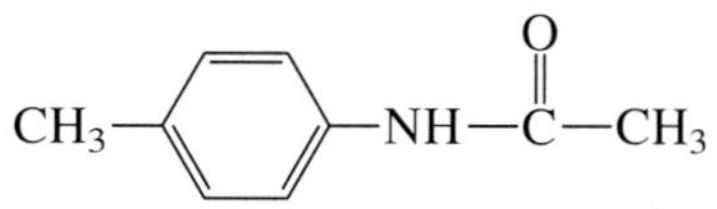

(benzocaine; a local anesthetic)

(h) 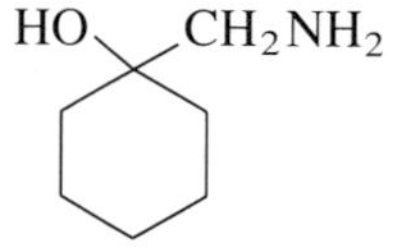 from cyclohexanone

(i) 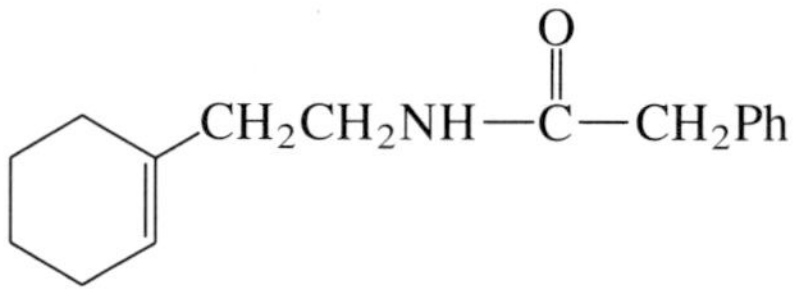 from 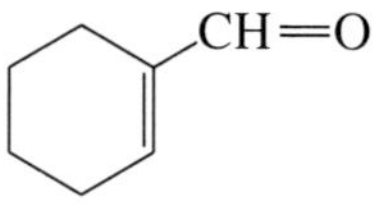

52 Rationalize each of the following known reactions with a reasonable mechanism, using the arrow formalism where possible:

(a) 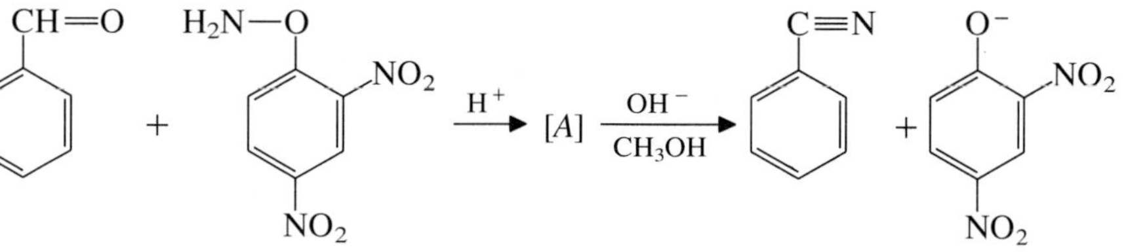

(This sequence is a method for making nitriles from aldehydes.)

(b)

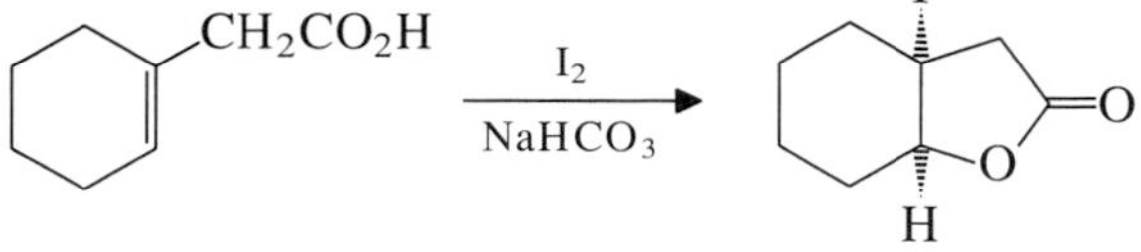

(Be sure to explain the observed stereochemistry.)

(c)

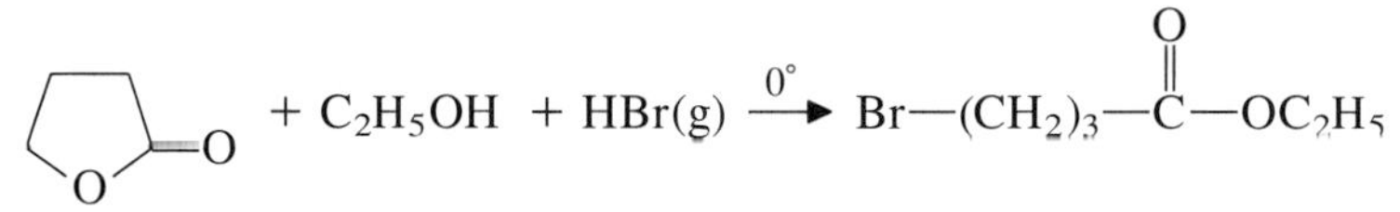

(d)

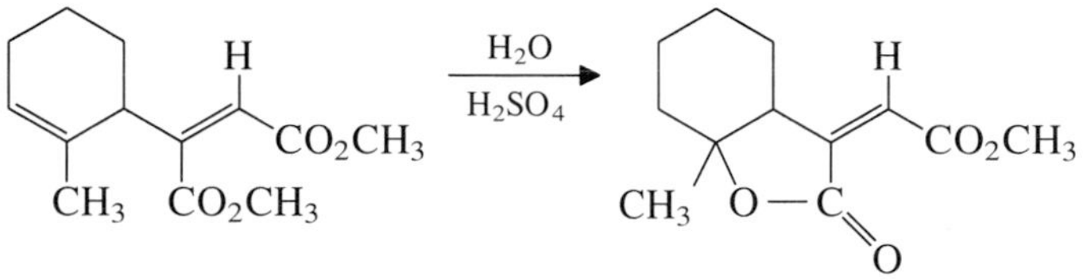

(e)

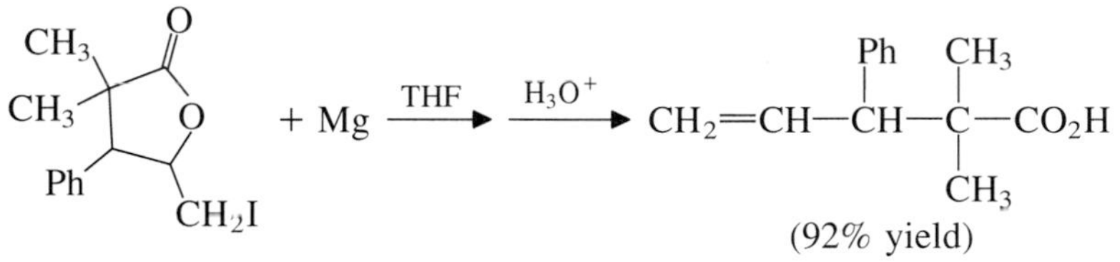

(f)

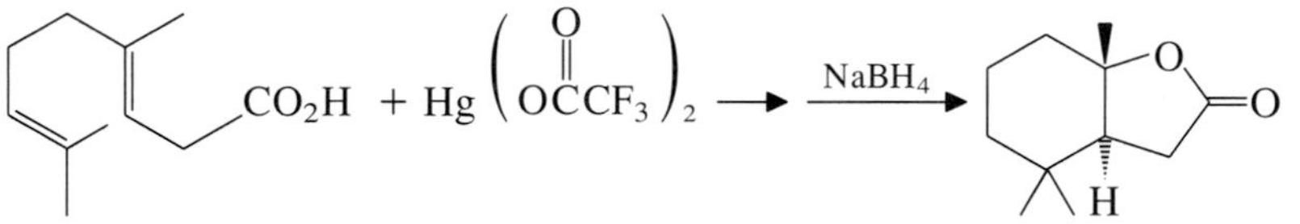

---

53 At room temperature the NMR spectrum of *N,N*-dimethylacetamide consists of *three* singlets of equal intensity at δ 2.1, δ 2.93, and δ 3.03. On warming to +95°, the two singlets at highest chemical shift coalesce into one singlet at δ 2.98 with twice the intensity of the resonance at δ 2.1, which is largely unaffected. Account for these results.

---

54 (*R*)-(+)-Ethyl α-hydroxypropionate, (+)-*A*, was treated with *p*-toluenesulfonyl chloride to yield the (+)-tosylate, which, when treated with KBr in acetone, gave (−)-ethyl α-bromopropionate, *B*. Saponification of (−)-*B* gave the (−)-α-bromoacid, (−)-*C*. When (−)-*C* reacted with 0.06*N* aqueous NaOH, a compound was obtained, which, when esterified with C_2H_5OH/H^+, gave the levorotatory ester (−)-*A*. When (−)-*C* reacted with aqueous 1*N* NaOH under the same conditions, the product, after esterification, was (+)-*A*.

(a) Indicate whether the reactions in dilute and concentrated NaOH, respectively, proceed with inversion or retention of configuration. (*Hint:* Determine the relative configurations of (+)-*A* and (−)-*C*.)

(b) Give a mechanism for the base-promoted hydrolysis reaction in each case that accounts for the stereochemical results. (*Hint:* See Sec. 11.6.)

Problems (Cont.)

55 Give an explanation for each of the following facts. The various explanations should be coherent and mutually consistent; that is, the explanation for one fact should contain nothing that is inconsistent with another fact.

(a) The esterification of mesitoic acid (2,4,6-trimethylbenzoic acid, below) does not take place in methanol/H_2SO_4, but takes place readily with diazomethane in ether.

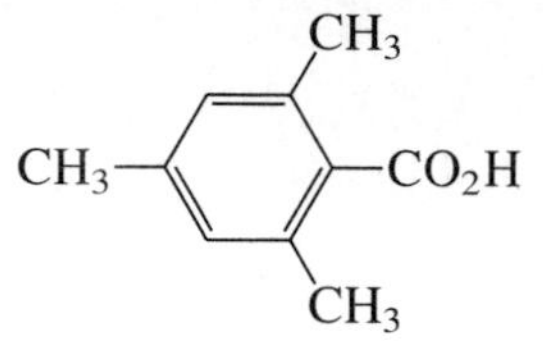

mesitoic acid

(b) The methyl ester of mesitoic acid, when treated with aqueous HCl and heat, does not hydrolyze, but the *t*-butyl ester hydrolyzes readily in aqueous HCl.

(c) Treatment of either the methyl ester or the *t*-butyl ester of mesitoic acid with aqueous NaOH and heat gives no reaction.

56 In acidic solution, the hydrolysis of phthalamic acid to phthalic acid (below) is 10^5 times faster than the hydrolysis of benzamide under the same conditions. Furthermore, an isotope *double-labeling experiment* gives the following results ($* = {}^{18}O$, $\# = {}^{13}C$).

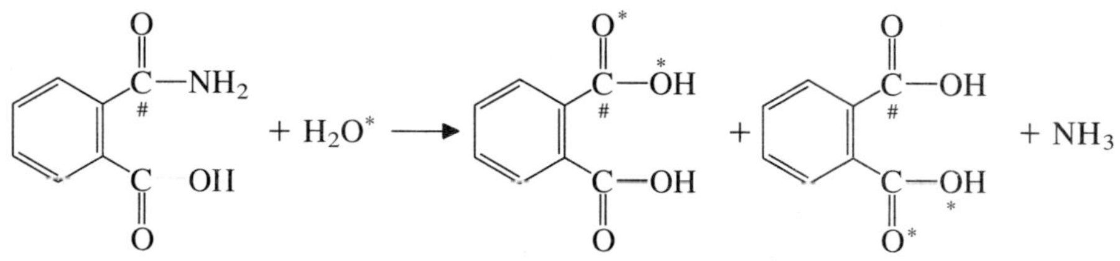

phthalamic acid

equal amounts of each

Suggest a mechanism that accounts for these two facts.

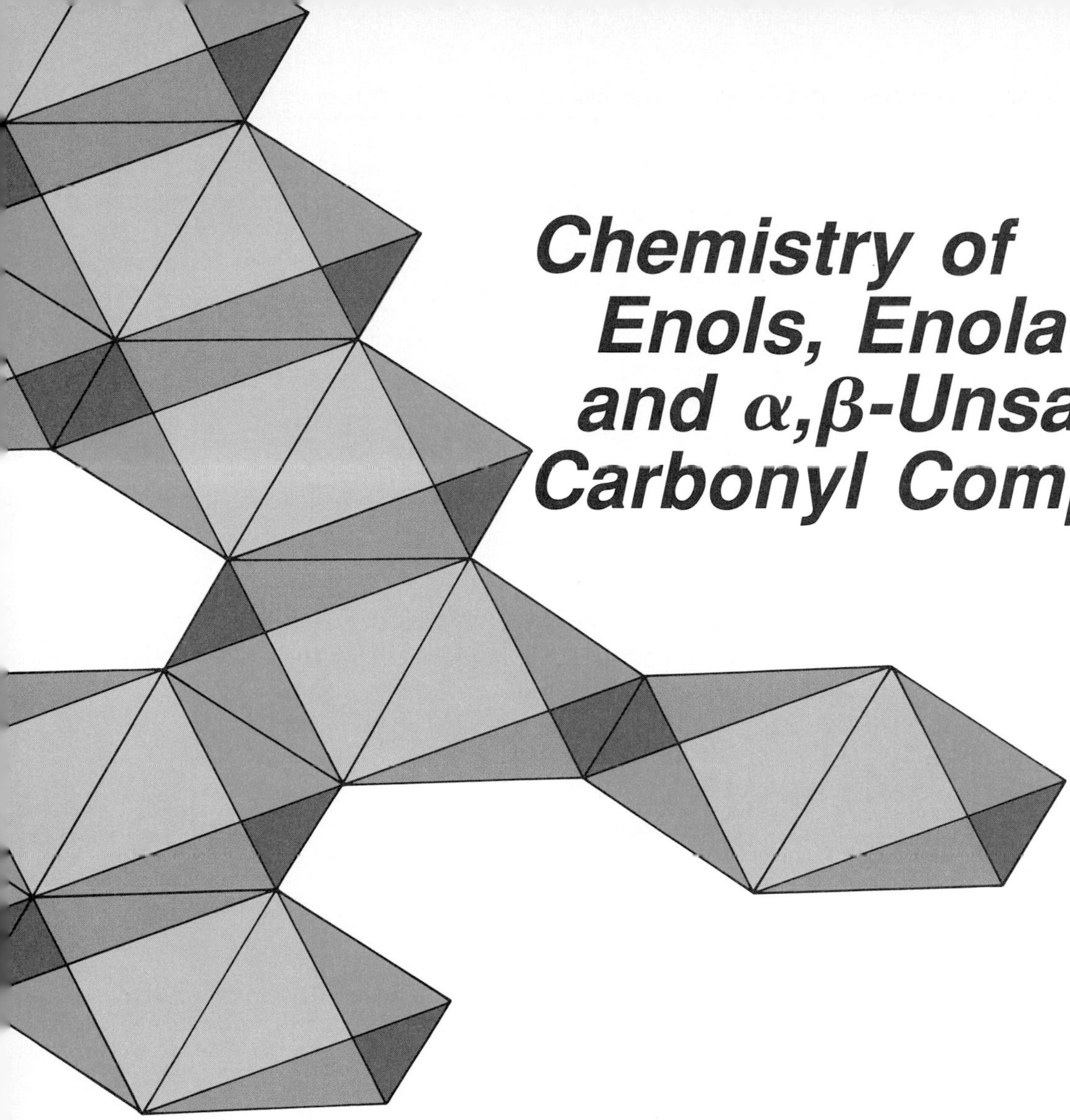

22 Chemistry of Enols, Enolate Ions, and α,β-Unsaturated Carbonyl Compounds

In the last three chapters we examined the chemistry of carbonyl compounds, concentrating largely on reactions at the carbonyl group. In this chapter, we complete our study of carbonyl compounds by considering the reactions of carbonyl compounds at the α-carbon.

We shall also study in this chapter the chemistry of **α,β-unsaturated carbonyl compounds**—compounds in which a carbonyl group is conjugated with a carbon-carbon double bond.

$$CH_3{-}CH{=}CH{-}\overset{\overset{\displaystyle O}{\|}}{C}{-}CH_3$$

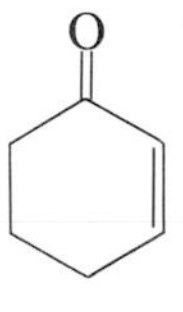

α,β-unsaturated ketones

$$CH_3{-}CH{=}CH{-}\overset{\overset{\displaystyle O}{\|}}{C}{-}OC_2H_5$$

an α,β-unsaturated ester

Just as the chemistry of conjugated alkenes is different from that of ordinary alkenes (Chapter 15), α,β-unsaturated carbonyl compounds have some unique chemistry that is also worth special attention.

Figure 22.1 Orbital picture of an enolate ion.

anionic carbon

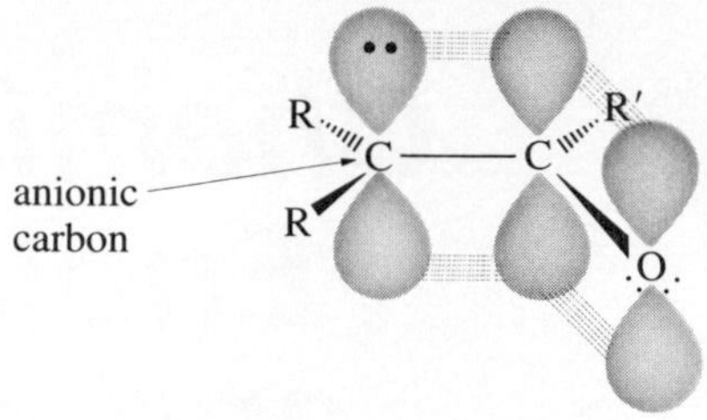

22.1 ACIDITY OF CARBONYL COMPOUNDS. ENOLATE IONS

A. Formation of Enolate Anions

The α-hydrogens of many carbonyl compounds, as well as those of nitriles, are weakly acidic. Ionization of an α-hydrogen gives the conjugate-base anion, called an **enolate ion.**

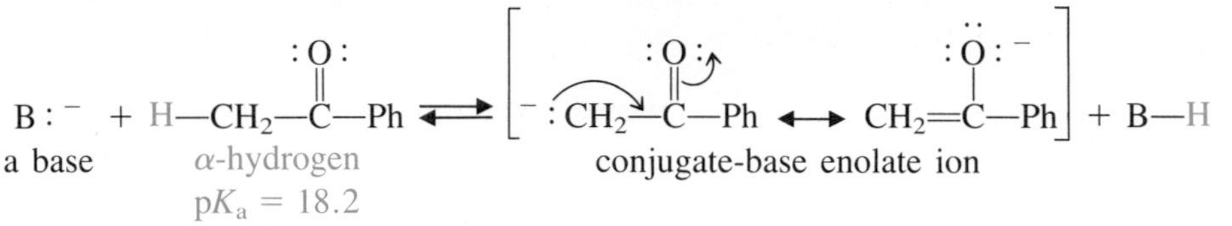

$$B:^- + H-CH_2-\overset{\overset{\displaystyle :O:}{\|}}{C}-Ph \rightleftharpoons \left[{}^-\!:CH_2-\overset{\overset{\displaystyle :O:}{\|}}{C}-Ph \longleftrightarrow CH_2{=}\overset{\overset{\displaystyle :\ddot{O}:^-}{|}}{C}-Ph \right] + B-H \quad (22.1)$$

a base α-hydrogen conjugate-base enolate ion
$pK_a = 18.2$

$$B:^- + H-CH_2-\overset{\overset{\displaystyle :O:}{\|}}{C}-OC_2H_5 \rightleftharpoons \left[{}^-\!:CH_2-\overset{\overset{\displaystyle :O:}{\|}}{C}-OC_2H_5 \longleftrightarrow CH_2{=}\overset{\overset{\displaystyle :\ddot{O}:^-}{|}}{C}-OC_2H_5 \right] + B-H \quad (22.2)$$

α-hydrogen conjugate-base enolate ion
$pK_a \approx 25$

The pK_a values of simple aldehydes or ketones are in the range 16–20, and the pK_a values of esters are about 25. Although we shall focus mostly on the acidity of aldehydes, ketones, and esters, the α-C—H bonds of nitriles and tertiary amides also have acidities similar to those of esters.

Although carbonyl compounds are classified as weak acids, their α-hydrogens are much more acidic than other types of hydrogens bound to carbon. For example, the dissociation constants of carbonyl compounds are greater than those of alkanes by about thirty powers of ten! What is the reason for the greater acidity of carbonyl compounds?

First, recall that stabilization of a base lowers the pK_a of its conjugate acid (Eq. 8.17b). Enolate ions are resonance stabilized, as shown in Eqs. 22.1 and 22.2. Hence, carbonyl compounds have lower pK_a—greater acidity—than carbon acids that lack this stabilization. Resonance is a symbolic way of depicting the orbital overlap in enolate ions, as shown in Fig. 22.1. The anionic carbon of an enolate ion is sp^2 hybridized. This hybridization allows the electron pair of an enolate anion to occupy a p orbital, which overlaps with the π-orbital of a carbonyl group. This additional overlap provides additional bonding and hence, additional stabilization.

The second reason for the acidity of α-hydrogens is that the negative charge in an enolate ion is delocalized onto oxygen, an electronegative atom. Thus, the α-hydrogens of carbonyl compounds are much more acidic than the allylic hydrogens

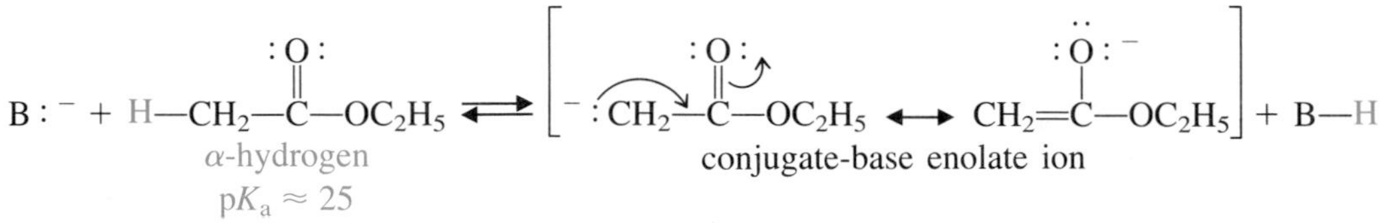

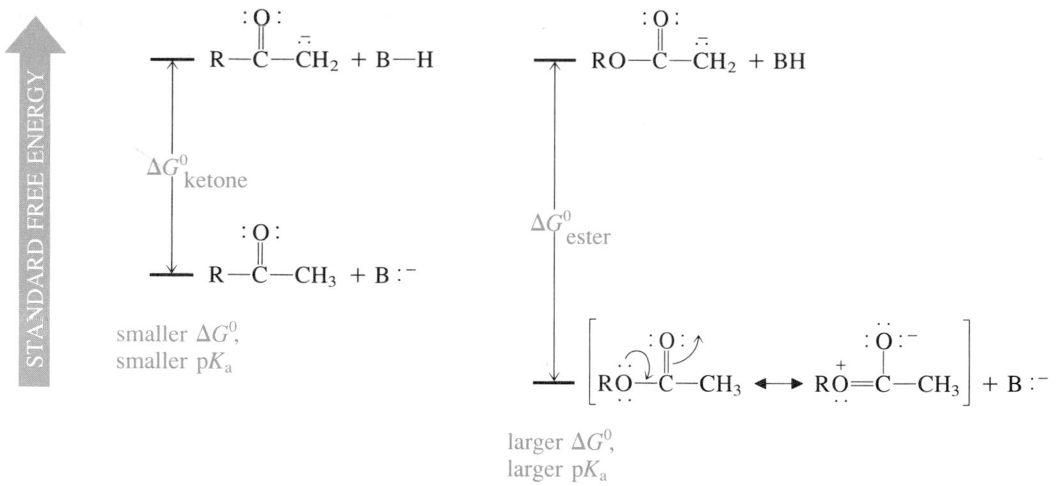

Figure 22.2 Resonance stabilization of an ester increases its standard free energy of ionization relative to that of a ketone and raises its pK_a. (The free energies of the conjugate-base enolate ions have been placed at the same level for comparison purposes.)

of alkenes, even though the conjugate-base anions of both types of compounds are resonance stabilized.

$$CH_3-CH{=}CH_2 \qquad CH_3-CH{=}\ddot{O}$$

allylic hydrogen α-hydrogen
$pK_a = 42$ $pK_a = 16.7$

Comparing the pK_a values in Eqs. 22.1 and 22.2, we see that aldehydes and ketones are about ten million times (seven pK_a units) as acidic as esters. What is the reason for this difference? Just as the pK_a of a carbonyl compound is *lowered* by resonance stabilization of its conjugate-base enolate ion, the pK_a is *raised* by resonance stabilization of the carbonyl compound itself (Fig. 22.2). Recall from Sec. 21.7F (Eq. 21.32) that esters have important resonance structures; aldehydes and ketones do not have analogous resonance contributors. The resonance stabilization of esters accounts for their lower acidity (higher pK_a). Notice that this resonance effect overrides the inductive effect of the carboxylate oxygen, which, in the absence of resonance, would increase the acidity of esters. These arguments are precisely the same ones that we used to explain the greater carbonyl reactivity of aldehydes and ketones (Sec. 21.9E), except that we are considering a different type of reaction.

Amide N—H protons are formally α-protons. That is, they are attached to an atom that is adjacent to a carbonyl group. The N—H protons are the most acidic protons in primary and secondary amides.

$$B{:}^- + CH_3-\overset{:O:}{\overset{\|}{C}}-\ddot{N}H_2 \rightleftharpoons \left[CH_3-\overset{:O:}{\overset{\|}{C}}\overset{..}{\underset{..}{N}}H \longleftrightarrow CH_3-\overset{:\ddot{O}:^-}{\overset{|}{C}}{=}\ddot{N}H \right] + BH \quad (22.3)$$

$pK_a = 15\text{--}17$

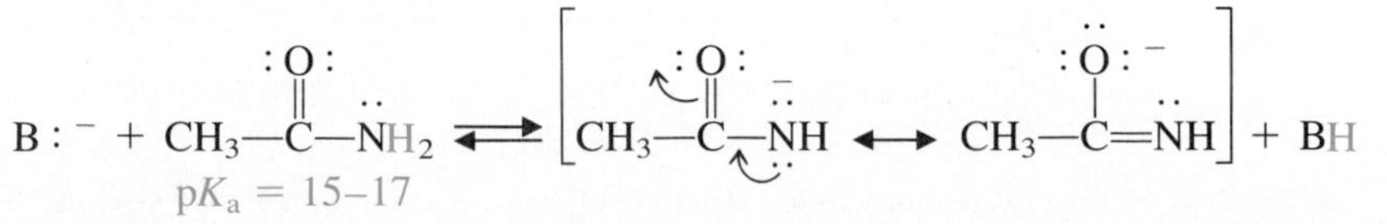

Similarly, carboxylic acid O—H protons ($pK_a \approx 4$–5) are also α-protons. We can think of amide conjugate-base anions as "nitrogen analogs" of enolate ions, and carboxylate anions as "oxygen analogs" of enolate anions. Notice that the acidity order carboxylic acids > amides > (aldehydes, ketones) corresponds to the relative electronegativities of the atoms to which the acidic hydrogens are bound—oxygen, nitrogen, and carbon, respectively.

Amides of sulfonic acids, called *sulfonamides,* are somewhat stronger acids than carboxylic acid amides, just as sulfonic acids are stronger acids than carboxylic acids. Thus, benzenesulfonamide has a pK_a of about 10, and it dissolves as its conjugate base in NaOH solution.

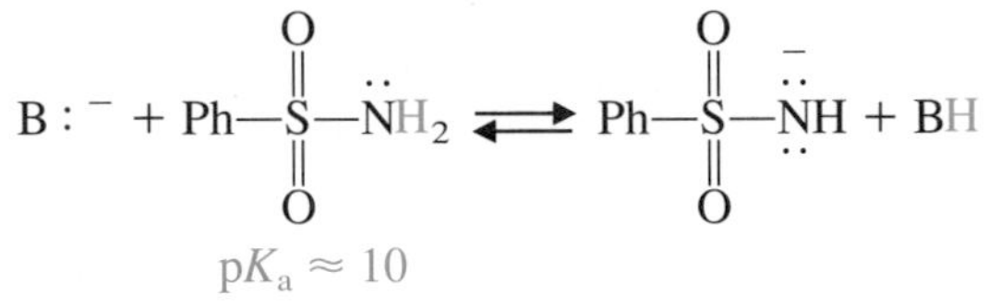

$$B:^- + Ph-\overset{\overset{O}{\|}}{\underset{\underset{O}{\|}}{S}}-\ddot{N}H_2 \;\rightleftharpoons\; Ph-\overset{\overset{O}{\|}}{\underset{\underset{O}{\|}}{S}}-\overset{-}{\ddot{N}}H + BH \qquad (22.4)$$

$$pK_a \approx 10$$

Problems

1 Explain why diethyl malonate ($pK_a = 12.9$) and ethyl acetoacetate (ethyl 3-oxobutanoate, $pK_a = 10.7$) are much more acidic than ordinary esters. (To answer this question, you first have to identify the acidic proton in each of these compounds.)

2 Which is more acidic: the diamide of succinic acid or the imide succinimide (Sec. 21.1E)? Why?

B. Introduction to Reactions of Enolate Ions

The acidity of aldehydes, ketones, and esters is particularly important because, as we shall learn later in this chapter, enolate ions are key reactive intermediates in many important reactions of carbonyl compounds. Let us consider the types of reactivity we can expect to observe with enolate ions.

First, enolate ions are Brønsted bases, and react with Brønsted acids. (This reaction is the reverse of Eq. 22.1 or 22.2.) The formation of enolate ions and their reaction with Brønsted acids have two simple but important consequences. First, the α-hydrogens of an aldehyde or ketone—and no others—can be exchanged for deuterium by treating the carbonyl compound with a base in D_2O.

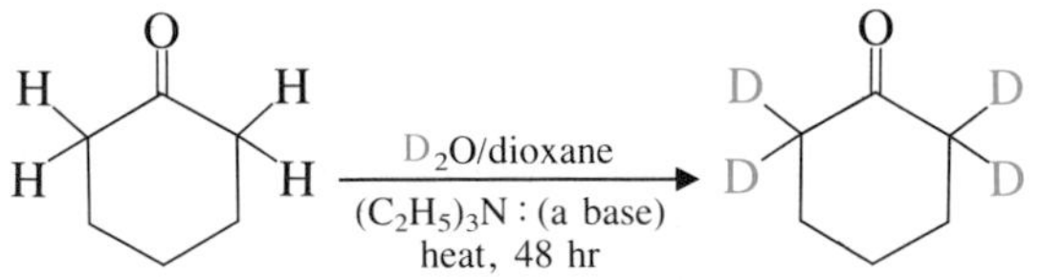

$$(22.5)$$

Problems

3 Write a mechanism for the reaction shown in Eq. 22.5. Explain why the α-hydrogens and *only* those hydrogens are replaced by deuterium.

4 Explain how the NMR spectrum of 2-butanone would change if the compound were treated with D_2O in base.

The second consequence of enolate-ion formation and protonation is that, if an optically active aldehyde or ketone owes its chirality solely to an asymmetric α-carbon, and if this carbon bears a hydrogen, the compound will be racemized by base.

optically active

racemate

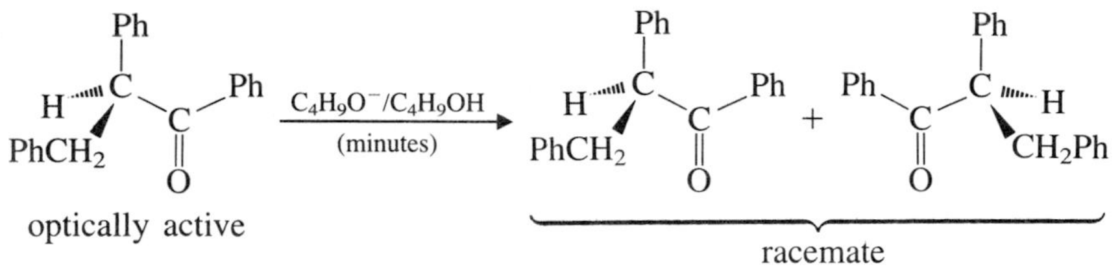

(22.6)

The reason that racemization occurs is that the enolate ion, which forms in base, is *achiral* because of its sp^2 hybridization at the anionic carbon (Fig. 22.1). That is, the ionic α-carbon and its attached groups lie in one plane. Thus, this carbon, which was an asymmetric carbon in the starting carbonyl compound, is no longer asymmetric in the enolate anion. The anion can be reprotonated at either face to give either enantiomer with equal probability. Although there is not very much enolate ion present at any one time, the reactions involved in the ionization equilibrium are relatively fast, and racemization occurs relatively quickly if the carbonyl compound is left in contact with base.

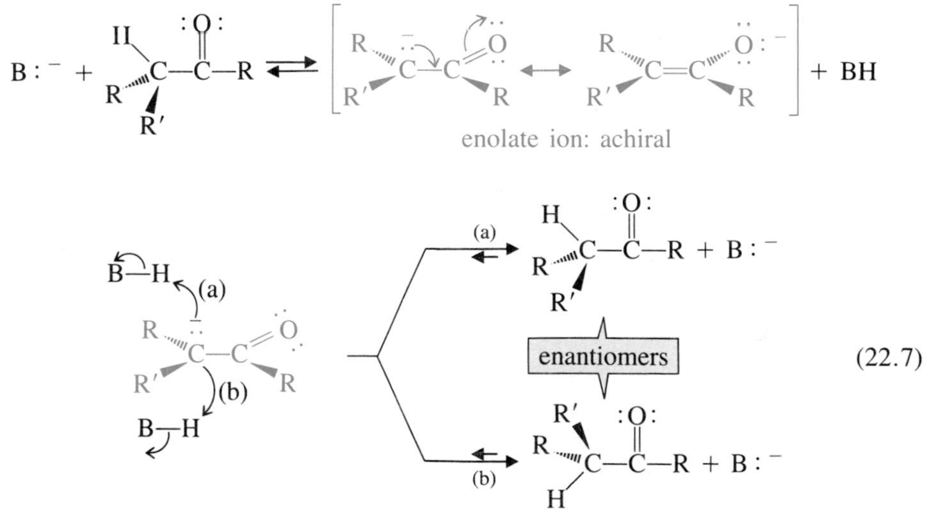

(22.7)

Chemical exchange and racemization occur much more readily on aldehydes and ketones than on esters, because aldehydes and ketones are more acidic, and therefore form enolate ions more rapidly and under milder conditions.

Enolate ions are not only Brønsted bases, but Lewis bases as well. Hence, enolate ions react as *nucleophiles*. Like other nucleophiles, enolate ions attack the carbon of carbonyl groups:

$$R\text{—}\overset{\text{O}}{\overset{\|}{C}}\text{—}\overset{..}{C}R_2 \quad \overset{:\text{O}:}{\underset{}{=}C\text{—}} \quad \rightleftharpoons \quad R\text{—}\overset{\text{O}}{\overset{\|}{C}}\text{—}CR_2\text{—}\overset{:\overset{..}{\text{O}}:^-}{\underset{|}{C}\text{—}} \quad \longrightarrow \quad \text{further reactions} \quad (22.8)$$

enolate ion carbonyl compound

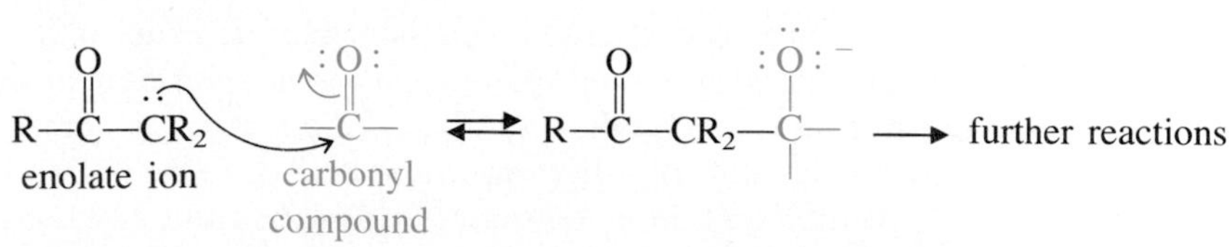

This type of process is the first step of *carbonyl addition* reactions and *carbonyl substitution* reactions of enolate ions. In subsequent sections, we shall study several such reactions in detail.

Enolate ions, like other nucleophiles, also react with alkyl halides and sulfonate esters:

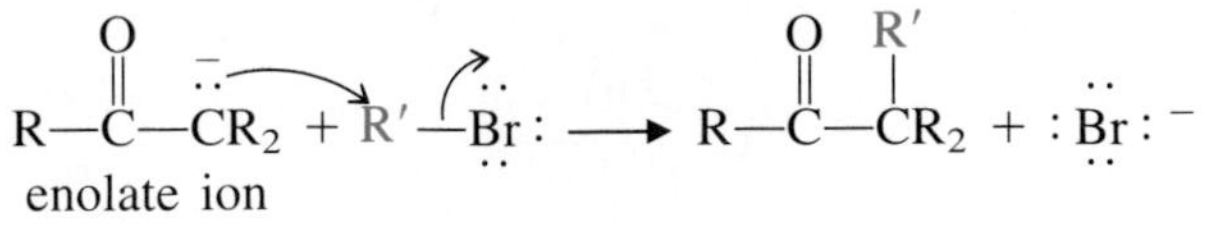

(22.9)

This type of reaction, too, is a very important part of the chemistry we are about to consider.

Problem

5 Explain why it is that the compound below does *not* undergo base-catalyzed exchange in D_2O even though it has an α-hydrogen. (*Hint*: See Sec. 7.7C.)

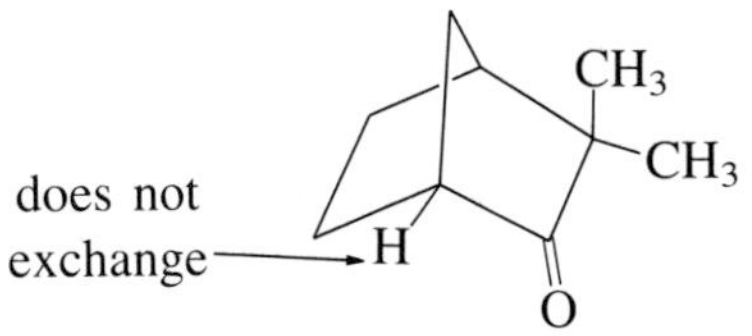

22.2 ENOLIZATION OF CARBONYL COMPOUNDS

Carbonyl compounds with α-hydrogens are in equilibrium with vinylic alcohol isomers called **enols.**

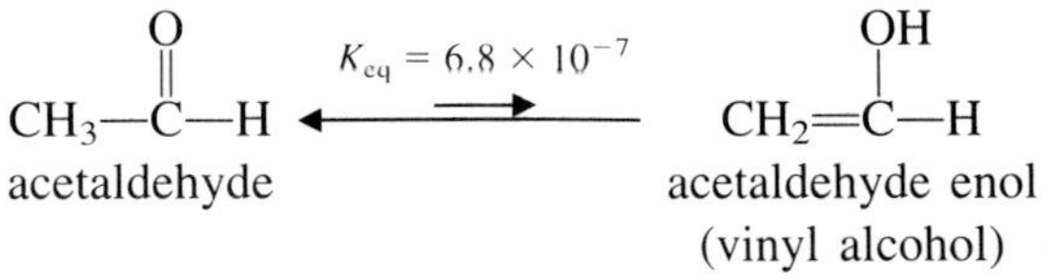

(22.10)

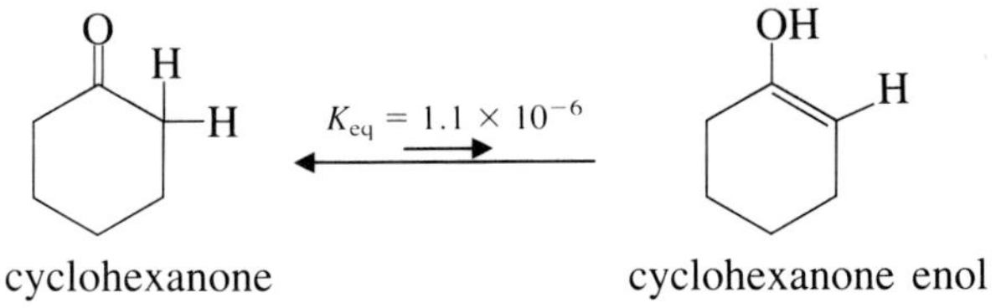

(22.11)

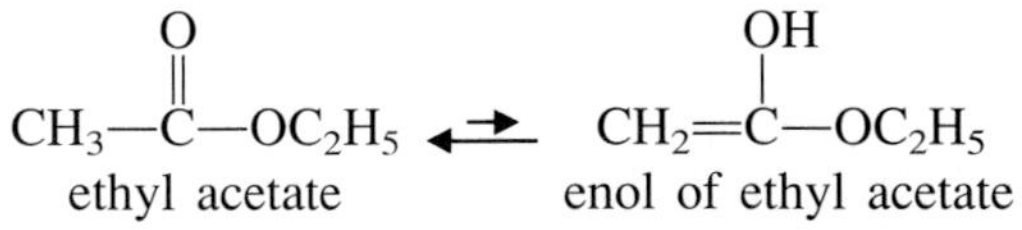

(22.12)

We have already encountered enols as intermediates in the hydration of alkynes (Sec. 14.5A). Enols and their parent carbonyl compounds are sometimes said to be **tautomers** of each other. Tautomers are structural isomers that are formally related only by the shift of a hydrogen and one or more π-bonds.

As the equilibrium constants in Eq. 22.10 and 22.11 suggest, most carbonyl compounds are considerably more stable than their corresponding enols. The reason is that the C=O double bond of a carbonyl group is a stronger bond than the C=C of an enol (Table 5.2). However, some enols are more stable than their corresponding carbonyl compounds. Phenol is formally an enol, but the enol form of phenol is more stable than its two keto tautomers because phenol is *aromatic*.

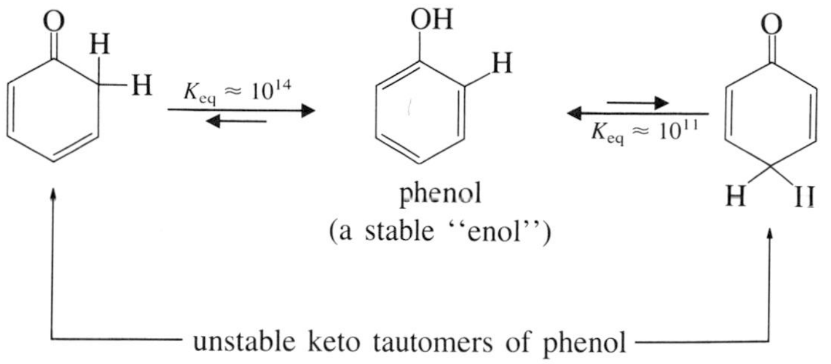

(22.13)

The enols of *β-dicarbonyl compounds* are also relatively stable. (β-Dicarbonyl compounds have two carbonyl groups separated by one carbon.)

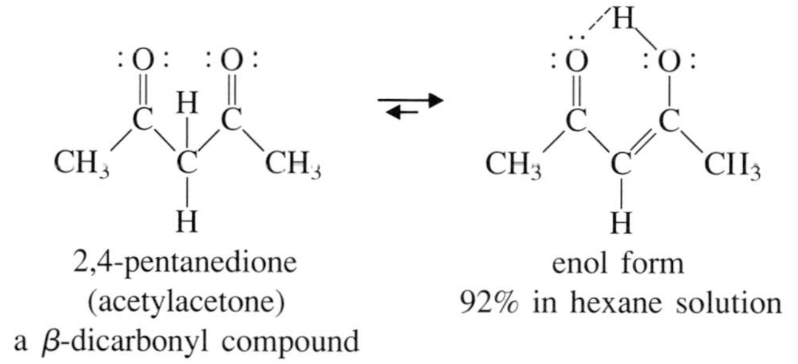

(22.14)

There are two reasons for the stability of these enols. The first is that they are conjugated, but their parent carbonyl compounds are not. The π-electron overlap associated with conjugation provides additional bonding that stabilizes the enol.

conjugation:

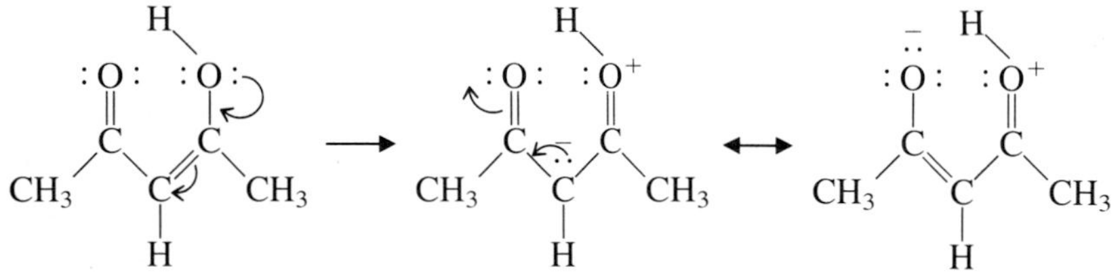

(22.15)

The second stabilizing effect is the intramolecular hydrogen bond present in each of these enols. This provides another source of increased bonding and hence, increased stabilization.

intramolecular hydrogen bond:

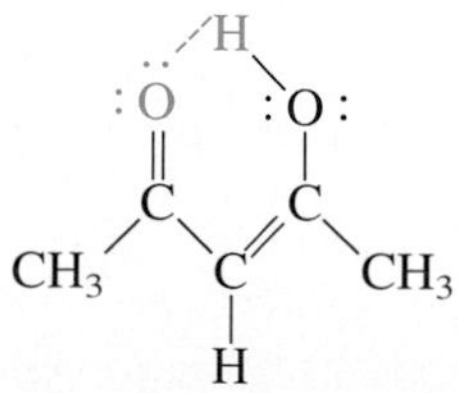

Problems

6 Draw all enol tautomers of the following compounds. If there are none, explain why.
(a) 2-methylcyclohexanone
(b) benzaldehyde
(c) *N,N*-dimethylacetamide
(d) 2-butanone
(e) 2-methylpentanoic acid
(f) propionitrile

7 (a) Explain why 2,4-pentanedione (Eq. 22.14) contains much less enol form in water (15%) than it does in hexane (92%).
(b) Explain why the same compound has a strong UV absorption in hexane solvent (λ_{max} = 272 nm, ϵ = 12,000), but a weaker absorption in water (λ_{max} = 274 nm, ϵ = 2050).

The formation of enols and the reverse reaction, conversion of enols into carbonyl compounds, are catalyzed by both acids and bases. The rapid conversion of enols into carbonyl compounds accounts for the fact that enols are difficult to isolate and observe under ordinary circumstances. Base-catalyzed enolization involves the intermediacy of the *enolate ion,* and is thus a consequence of the acidity of the α-hydrogen.

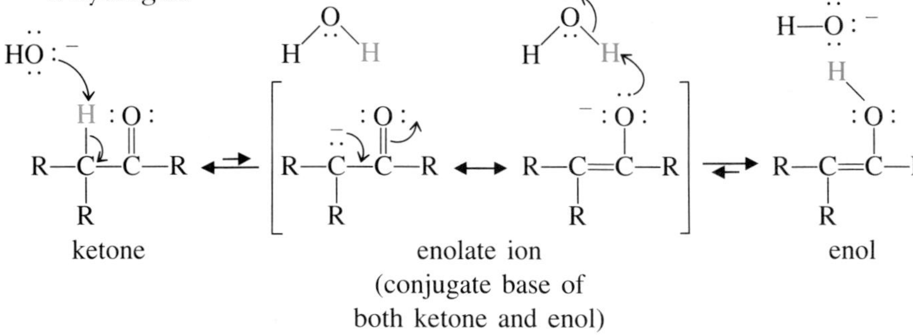

(22.16a)

Protonation of the enolate anion by water on the α-carbon gives back the carbonyl compound; protonation on oxygen gives the enol. Notice that the enolate ion is the conjugate base of not only the carbonyl compound, but also the enol—that is why it is called an *enolate ion*!

Acid-catalyzed enolization involves the conjugate acid of the carbonyl compound. As we learned in Sec. 19.6, this ion has carbocation characteristics; one of the reactions of carbocations is loss of a proton from an adjacent atom (Sec. 9.5B). Loss of the proton from oxygen gives back the starting carbonyl compound; loss of the proton from the α-carbon gives the enol. Notice that an enol and its carbonyl tautomer have the same conjugate acid.

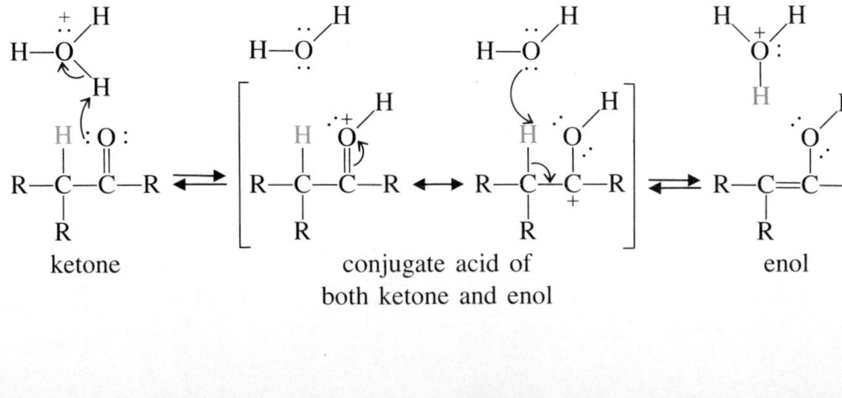

(22.16b)

Exchange of α-hydrogens for deuterium as well as racemization at the α-carbon are catalyzed not only by bases (Sec. 22.1B) but also by acids.

(22.17)

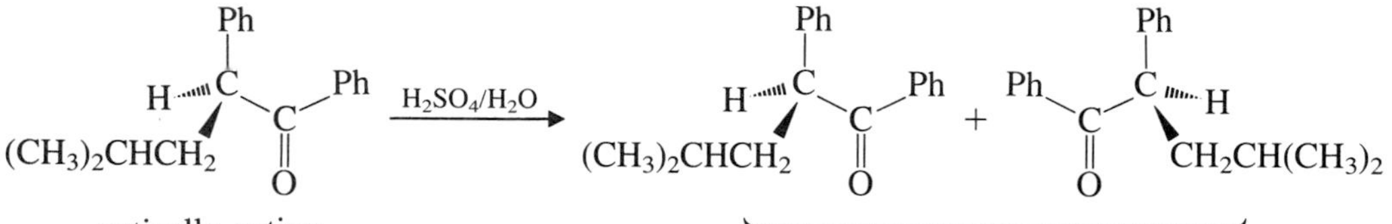

optically active

racemate

(22.18)

Both acid-catalyzed processes can be explained by the intermediacy of enols. As we can see from Eq. 22.16b, formation of a carbonyl compound from an enol reintroduces hydrogen from solution at the α-carbon; this fact accounts for the observed isotope exchange. The α-carbon of an enol, like that of an enolate ion, is not asymmetric. The absence of chirality in the enol accounts for the racemization observed in acid.

Problem

8 Using the arrow formalism, give the detailed mechanism for (a) the exchange shown in Eq. 22.17, and (b) the racemization shown in Eq. 22.18.

22.3 α-HALOGENATION OF CARBONYL COMPOUNDS

In this section we begin our study of reactions that involve enols and enolate ions as reactive intermediates. These reactions will occupy our attention through Sec. 22.7.

A. Acid-Catalyzed α-Halogenation of Aldehydes and Ketones

Halogenation of an aldehyde or ketone in *acidic* solution usually results in the replacement of one α-hydrogen by halogen.

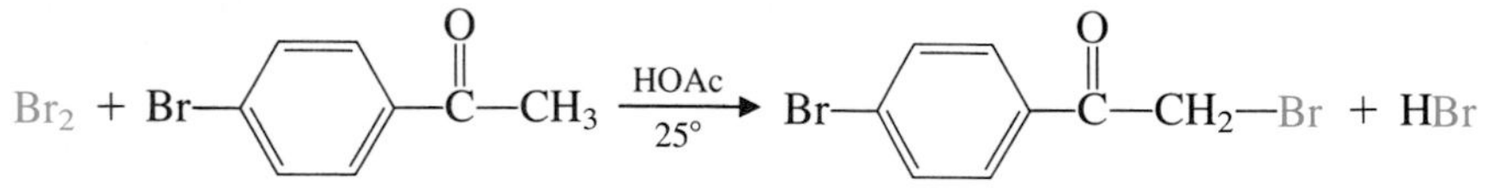

(69–72% yield)

(22.19)

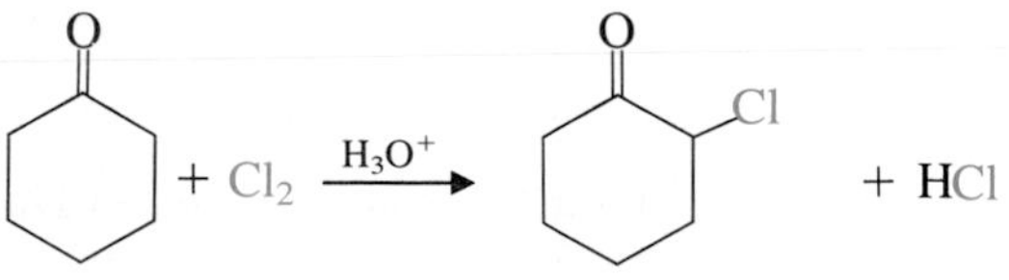

separates from solution (61–66% yield)

(22.20)

Enols are reactive intermediates in these reactions.

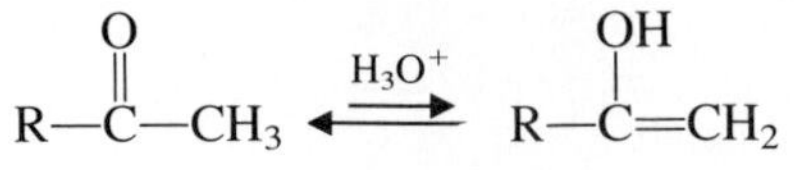

$$R\text{—}\overset{\overset{\textstyle O}{\|}}{C}\text{—}CH_3 \underset{\xrightarrow{\quad H_3O^+ \quad}}{\longleftrightarrow} R\text{—}\overset{\overset{\textstyle OH}{|}}{C}{=}CH_2 \tag{22.21a}$$

Like other "alkenes," enols react with halogens. Halogenation of an enol involves a resonance-stabilized carbocation intermediate, which loses a proton to give the α-halo ketone.

$$R\text{—}\overset{\overset{\textstyle OH}{|}}{C}{=}CH_2 \longrightarrow \left[R\text{—}\overset{\overset{\textstyle :\ddot{O}H}{|}}{\underset{+}{C}}\text{—}CH_2Br \longleftrightarrow R\text{—}\overset{\overset{\textstyle +\,:O\text{—}H}{\|}}{C}\text{—}CH_2Br \right] \longrightarrow R\text{—}\overset{\overset{\textstyle :O:}{\|}}{C}\text{—}CH_2Br + HBr \tag{22.21b}$$

$$\overset{\displaystyle Br\text{—}Br}{} \qquad \qquad \qquad \qquad Br^-$$

The rate law for acid-catalyzed halogenation proved to be very important in elucidating the reaction mechanism. Under the usual reaction conditions, the rate law for the reaction is found to be

$$\text{rate} = k[\text{ketone}][H_3O^+] \tag{22.22}$$

The rate law implies that even though the reaction is a halogenation, *the rate is independent of the halogen concentration.* This rate law means that halogens *cannot* be involved in the transition state for the rate-determining step of the reaction (Sec. 9.3A). From this observation and others, it was deduced that *enol formation (Eq. 22.21a) is the rate-determining step in acid-catalyzed halogenation of aldehydes and ketones.* Since halogen is not involved in enol formation, it does not appear in the rate law.

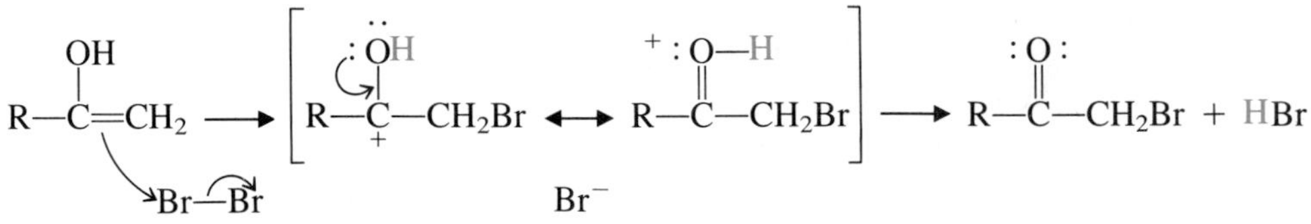

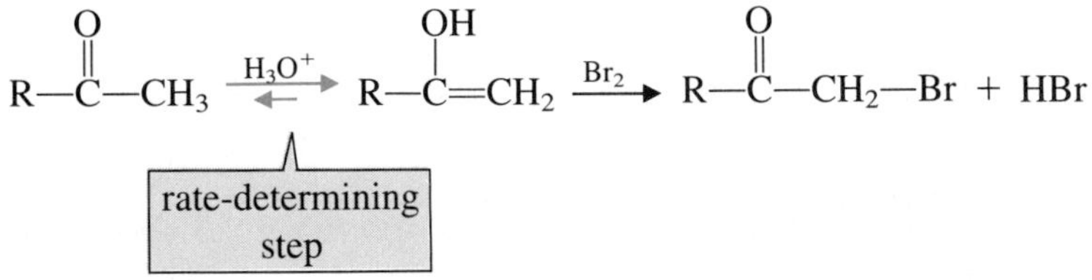

$$R\text{—}\overset{\overset{\textstyle O}{\|}}{C}\text{—}CH_3 \xrightarrow{\;H_3O^+\;} R\text{—}\overset{\overset{\textstyle OH}{|}}{C}{=}CH_2 \xrightarrow{\;Br_2\;} R\text{—}\overset{\overset{\textstyle O}{\|}}{C}\text{—}CH_2\text{—}Br + HBr \tag{22.23}$$

rate-determining step

Because halogenation in acid solution occurs only once, it follows that enolization of a halogenated ketone must be much slower than enolization of an unhalogenated ketone. This is exactly what we would expect from the mechanism of acid-catalyzed enolization (Eq. 22.16b). The inductive effect of a bromine at the α-carbon destabilizes the carbocation intermediate in enolization and, by Hammond's postulate, retards the reaction.

$$R\text{—}\overset{\overset{\textstyle :O:}{\|}}{C}\text{—}CH_2Br \underset{\xrightarrow{\; H_3O^+ \;}}{\longleftrightarrow} \left[R\text{—}\overset{\overset{\textstyle +\,\ddot{O}\text{—}H}{\|}}{C}\text{—}CH_2Br \longleftrightarrow R\text{—}\overset{\overset{\textstyle :\ddot{O}\text{—}H}{|}}{\underset{+}{C}}\text{—}CH_2Br \right] \underset{\xrightarrow{\;:\ddot{O}H_2\;}}{\longleftrightarrow} R\text{—}\overset{\overset{\textstyle :O\text{—}H}{|}}{C}{=}CHBr + H_3O^+ \tag{22.24}$$

destabilized by
inductive effect of bromine

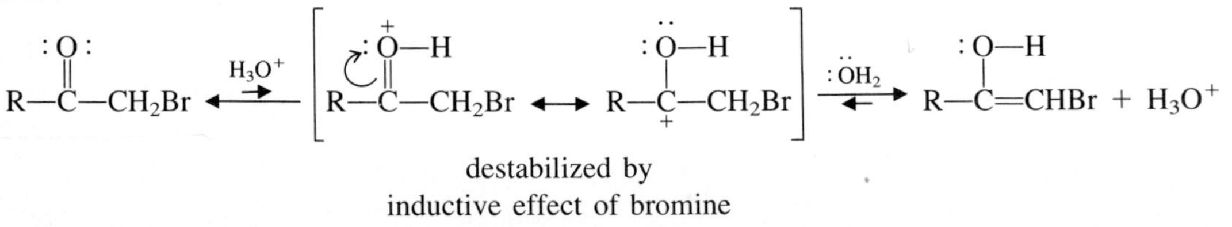

Problems

9 Sketch a reaction free-energy diagram (analogous to Fig. 4.14) for acid-catalyzed halogenation of a ketone, showing the enol as an intermediate and reflecting the fact that enol formation is rate determining. According to your diagram, which is faster: the reverse of enol formation or bromination of the enol?

10 Explain why:

(a) the rate of iodination of 1-phenyl-2-methyl-1-butanone in acetic acid/HNO_3 is essentially identical to the rate of racemization of the same optically active compound under the same conditions.

(b) The rates of bromination and iodination of acetophenone are identical at a given acid concentration.

B. Halogenation of Aldehydes and Ketones in Base: the Haloform Reaction

Halogenation of aldehydes and ketones with α-hydrogens also occurs in base. In this reaction, *all* α-protons are substituted by halogen.

$$\underset{\text{no }\alpha\text{-protons}}{(CH_3)_3C}\text{—}\overset{O}{\overset{\|}{C}}\text{—}\underset{\alpha\text{-protons}}{CH_3} + Br_2 \xrightarrow[\substack{H_2O/\text{dioxane}\\0^\circ}]{\text{NaOH}} (CH_3)_3C\text{—}\overset{O}{\overset{\|}{C}}\text{—}CBr_3 \qquad (22.25a)$$

When the aldehyde or ketone starting material is either acetaldehyde or a methyl ketone (as in Eq. 22.25a), the trihalo carbonyl compound is not stable under the reaction conditions, and it reacts further to give, after acidification of the reaction mixture, a carboxylic acid and a haloform. (Recall from Sec. 8.1A that a *haloform* is a trihalomethane—that is, a compound of the form HCX_3, where X = halogen.)

$$(CH_3)_3C\text{—}\overset{O}{\overset{\|}{C}}\text{—}CBr_3 \xrightarrow{OH^-} \xrightarrow{H_3O^+} \underset{(71\text{–}74\%\text{ yield})}{(CH_3)_3C\text{—}\overset{O}{\overset{\|}{C}}\text{—}OH} + \underset{\text{bromoform}}{HCBr_3} \qquad (22.25b)$$

This reaction is called the **haloform reaction.** Notice that, in the haloform reaction, a carbon–carbon bond is broken and the carbonyl compound is oxidized.

The mechanism of the haloform reaction begins with formation of an *enolate ion* as a reactive intermediate.

$$R\text{—}\overset{O}{\overset{\|}{C}}\text{—}CH_3 + OH^- \rightleftharpoons \underset{\text{enolate ion}}{R\text{—}\overset{O}{\overset{\|}{C}}\text{—}\ddot{C}H_2^-} + H_2O \qquad (22.26a)$$

The enolate ion reacts as a nucleophile with halogen to give an α-halo carbonyl compound.

$$R\text{—}\overset{O}{\overset{\|}{C}}\text{—}\ddot{C}H_2^- + :\ddot{Br}\text{—}\ddot{Br}: \longrightarrow R\text{—}\overset{O}{\overset{\|}{C}}\text{—}CH_2\ddot{Br}: + :\ddot{Br}:^- \qquad (22.26b)$$

However, halogenation does not stop here, because the enolate ion of the α-halo ketone is formed even more rapidly than the enolate ion of the starting ketone. The reason is that the inductive effect of the halogen stabilizes the enolate ion and, by Hammond's postulate, the transition state for its formation. Consequently, a second bromination occurs.

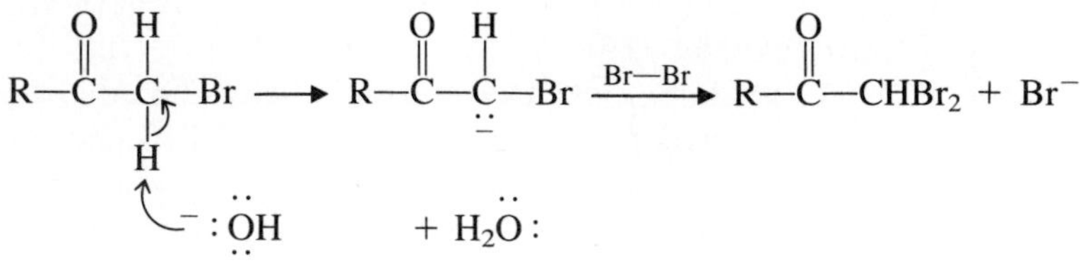

$$R-\overset{\overset{\displaystyle O}{\|}}{C}-\overset{\overset{\displaystyle H}{|}}{\underset{\underset{\displaystyle H}{|}}{C}}-Br \longrightarrow R-\overset{\overset{\displaystyle O}{\|}}{C}-\overset{\overset{\displaystyle H}{|}}{\underset{\displaystyle \cdot\cdot}{C}}-Br \xrightarrow{Br-Br} R-\overset{\overset{\displaystyle O}{\|}}{C}-CHBr_2 + Br^- \quad (22.26c)$$

$$^-:\overset{\cdot\cdot}{O}H \qquad + H_2\overset{\cdot\cdot}{O}:$$

The dihalo carbonyl compound brominates again (why?).

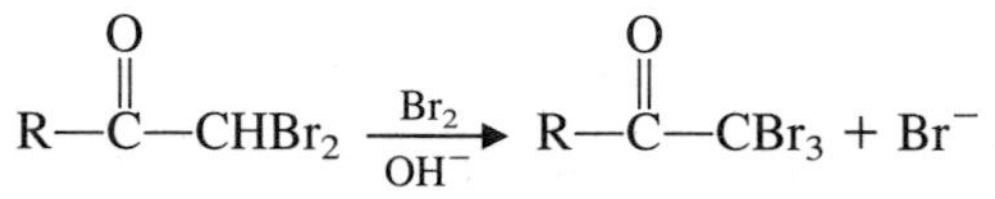

$$R-\overset{\overset{\displaystyle O}{\|}}{C}-CHBr_2 \xrightarrow[OH^-]{Br_2} R-\overset{\overset{\displaystyle O}{\|}}{C}-CBr_3 + Br^- \qquad (22.26d)$$

A carbon–carbon bond is broken when the trihalo carbonyl compound undergoes a *substitution reaction at the carbonyl group*.

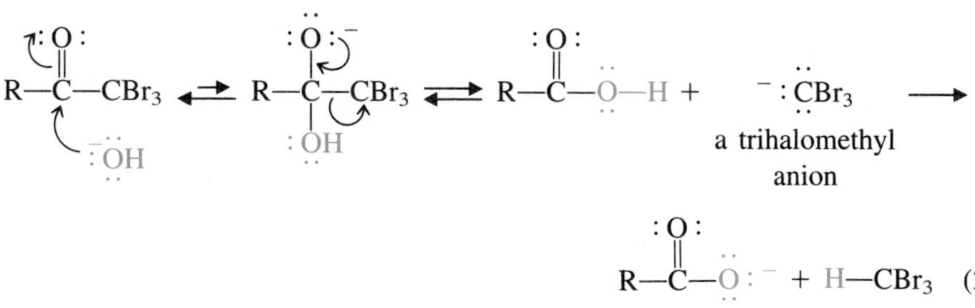

$$R-\overset{\overset{\displaystyle \cdot\cdot}{\overset{\displaystyle O:}{\|}}}{C}-CBr_3 \rightleftarrows R-\overset{\overset{\displaystyle :\overset{\cdot\cdot}{O}:^-}{|}}{C}-CBr_3 \rightleftarrows R-\overset{\overset{\displaystyle :O:}{\|}}{C}-\overset{\cdot\cdot}{O}-H + \quad ^-:\overset{\cdot\cdot}{C}Br_3 \longrightarrow$$

$$\overset{\cdot\cdot}{:}\overset{\cdot\cdot}{O}H \qquad\qquad :\overset{\cdot\cdot}{O}H \qquad\qquad\qquad\qquad \text{a trihalomethyl} \\ \text{anion}$$

$$R-\overset{\overset{\displaystyle :O:}{\|}}{C}-\overset{\cdot\cdot}{O}:^- + H-CBr_3 \quad (22.26e)$$

The leaving group in this reaction is a trihalomethyl anion. Usually, carbanions are too basic to serve as leaving groups; but trihalomethyl anions are much less basic than ordinary carbanions (why?). However, the basicity of trihalomethyl anions, while low enough for them to act as leaving groups, is high enough for them to react almost irreversibly with the carboxylic acid by-product. This acid–base reaction drives the overall haloform reaction to completion. (This is analogous to saponification, which is also driven to completion by ionization of the carboxylic acid product; Sec. 21.7A.) The carboxylic acid itself can be isolated by acidifying the reaction mixture.

Occasionally, the haloform reaction is used to prepare carboxylic acids from readily available methyl ketones. More often, however, this reaction is used as a qualitative test for methyl ketones, called the **iodoform test.** In this test, a compound of unknown structure is mixed with alkaline I_2. A yellow precipitate of iodoform is taken as evidence for a methyl ketone (or acetaldehyde, the "methyl aldehyde"). Alcohols of the form shown in Eq. 22.27 also give a positive iodoform test because they are oxidized to methyl ketones (or to acetaldehyde, in the case of ethanol) by the basic iodine solution.

$$R-\overset{\overset{\displaystyle OH}{|}}{C}H-CH_3 \xrightarrow[\text{base}]{I_2} R-\overset{\overset{\displaystyle O}{\|}}{C}-CH_3 \qquad (22.27)$$
$$\text{undergoes iodoform} \\ \text{reaction}$$

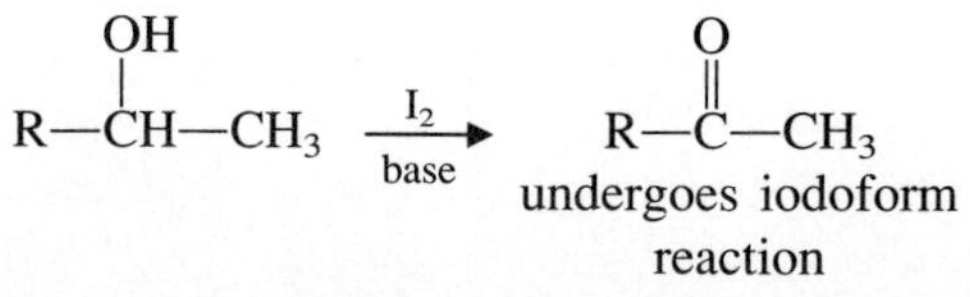

Problems

11 Give the products expected (if any) when each of the following compounds reacts with Br_2 in NaOH:

(a) 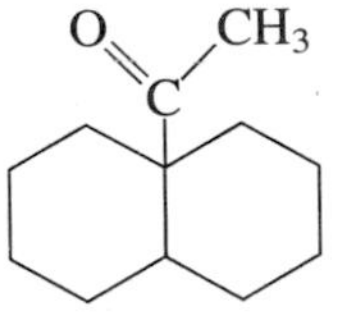 (b) 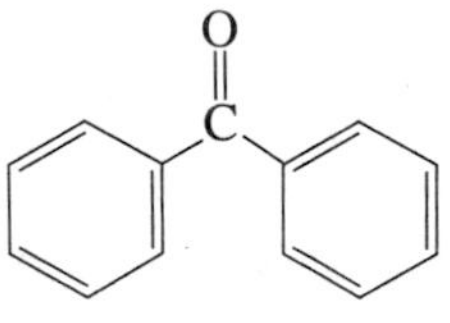 (c) $\text{Ph—}\overset{\text{OH}}{\overset{|}{\text{CH}}}\text{—CH}_3$

12 Give the structure of a compound $C_6H_{10}O_2$ that gives succinic acid and iodoform on treatment with an alkaline solution of iodine, followed by acidification.

C. α-Bromination of Carboxylic Acids: the Hell–Volhard–Zelinsky Reaction

Carboxylic acids can be brominated in the α-position. A bromine is substituted for an α-hydrogen when a carboxylic acid is treated with Br_2 and a catalytic amount of red phosphorus or PBr_3. (The actual catalyst is PBr_3; phosphorus can be used because it reacts with Br_2 to give PBr_3.)

$$CH_3CH_2CH_2CH_2CH_2CO_2H + Br_2 \xrightarrow{\text{P or PBr}_3} CH_3CH_2CH_2CH_2\underset{\underset{\text{Br}}{|}}{C}HCO_2H + HBr \qquad (22.28)$$

(83–89% yield)

This reaction is called the **Hell–Volhard–Zelinsky reaction** after its discoverers, and is sometimes nicknamed the *HVZ reaction.*

The first stage in the mechanism of this reaction is the conversion of the carboxylic acid into a small amount of acid bromide by the catalyst PBr_3 (Sec. 20.9A).

$$R\text{—}CH_2\text{—}\overset{\overset{\text{O}}{\|}}{C}\text{—}OH + \tfrac{1}{3}PBr_3 \longrightarrow \underset{\text{an acid bromide}}{R\text{—}CH_2\text{—}\overset{\overset{\text{O}}{\|}}{C}\text{—}Br} + \tfrac{1}{3}P(OH)_3 \qquad (22.29a)$$

From this point the mechanism closely resembles that for the acid-catalyzed bromination of ketones. The *enol* of the acid bromide is the species that actually brominates.

$$\underset{\text{acid bromide}}{R\text{—}CH_2\text{—}\overset{\overset{\text{O}}{\|}}{C}\text{—}Br} \rightleftarrows \underset{\text{enol form}}{R\text{—}CH\text{=}\overset{\overset{\text{OH}}{|}}{C}\text{—}Br} \xrightarrow{Br_2} \underset{\alpha\text{-bromo acid bromide}}{R\text{—}\underset{\underset{\text{Br}}{|}}{C}H\text{—}\overset{\overset{\text{O}}{\|}}{C}\text{—}Br} + HBr \qquad (22.29b)$$

If *one full equivalent* of PBr_3 catalyst is used, the α-bromo acid bromide is the reaction product; this can be used in many of the reactions of acid halides that we learned in Sec. 21.8A. For example, the reaction mixture can be quenched with an alcohol to give an α-bromo ester:

$$CH_3CH_2—\overset{O}{\overset{\|}{C}}—OH \xrightarrow[Br_2]{P(1\ equiv)} CH_3—\underset{Br}{\underset{|}{CH}}—\overset{O}{\overset{\|}{C}}—Br \xrightarrow[(CH_3)_2N—Ph]{(CH_3)_3COH} CH_3—\underset{Br}{\underset{|}{CH}}—\overset{O}{\overset{\|}{C}}—OC(CH_3)_3 \quad (22.30)$$

When *a small amount* of PBr_3 catalyst is used, the α-bromo acid bromide sets up an equilibrium with unreacted acid to form more acid bromide, which is then brominated as shown above.

$$R—CH_2—\overset{O}{\overset{\|}{C}}—OH + R—\underset{Br}{\underset{|}{CH}}—\overset{O}{\overset{\|}{C}}—Br \rightleftharpoons \underset{\text{enters bromination sequence at Eq. 22.29b}}{R—CH_2—\overset{O}{\overset{\|}{C}}—Br} + R—\underset{Br}{\underset{|}{CH}}—\overset{O}{\overset{\|}{C}}—OH \quad (22.31)$$

Thus, when a catalytic amount of PBr_3 is used, the reaction product is the α-bromo acid.

D. Reactions of α-Halo Carbonyl Compounds

In Secs. 22.3A and 22.3C we have learned two methods for preparing α-halo carbonyl compounds. The utility of these compounds is that most of them are very reactive in S_N2 reactions, and can be used to prepare other α-substituted carbonyl compounds.

$$(CH_3)_2\ddot{S}: \ :\ddot{Br}—CH_2—\overset{O}{\overset{\|}{C}}—Ph \xrightarrow[25°,\ 30\ min]{acetone/H_2O} \underset{\text{(85\% yield)}}{CH_3—\overset{CH_3}{\overset{|}{\underset{+}{\ddot{S}}}}—CH_2—\overset{O}{\overset{\|}{C}}—Ph} \ :\ddot{Br}:^- \quad (22.32)$$

In the case of α-halo ketones, nucleophiles used in these reactions must not be too basic. For example, dimethyl sulfide, used in Eq. 22.32, is a very weak base but a fairly good nucleophile. (Stronger bases promote enolate-ion formation; and the enolate ions of α-halo ketones undergo other reactions.) More basic nucleophiles can be used with α-halo acids since acids do not easily form enolate ions (why?).

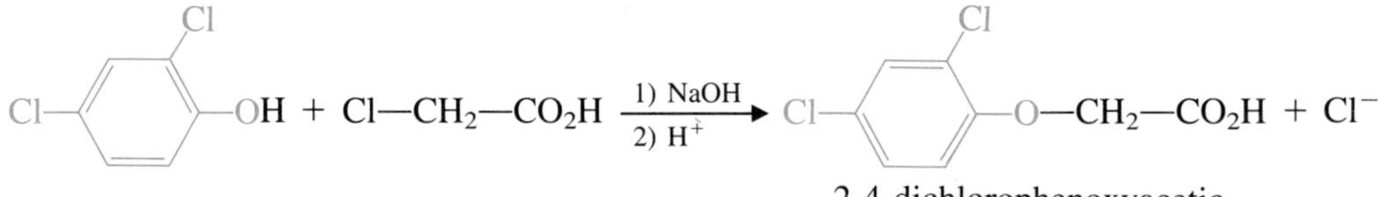
Cl
Cl
OH + Cl—CH₂—CO₂H
1) NaOH
2) H⁺
Cl
Cl
O—CH₂—CO₂H + Cl⁻
(22.33)

2,4-dichlorophenoxyacetic acid (2,4-D, a selective herbicide; 87% yield)

$$^-:C≡N: + Cl—CH_2—CO_2^- \longrightarrow \xrightarrow{H^+} \underset{\text{cyanoacetic acid (77–80\% yield)}}{:N≡C—CH_2—CO_2H} + Cl^- \quad (22.34)$$

The following comparison gives a quantitative measure of the S_N2 reactivity of α-halo carbonyl compounds:

relative rate:

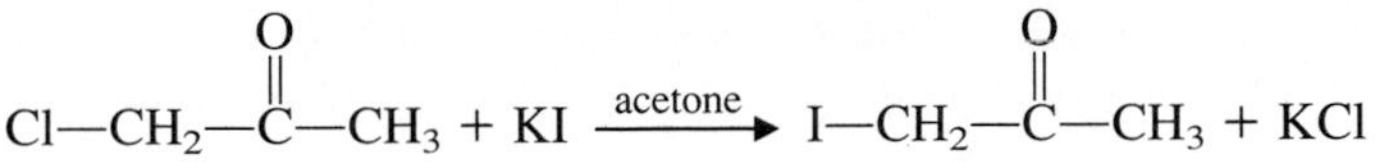

35,000 (22.35a)

$$\text{Cl—CH}_2\text{CH}_2\text{CH}_3 + \text{KI} \xrightarrow{\text{acetone}} \text{I—CH}_2\text{CH}_2\text{CH}_3 + \text{KCl} \qquad 1 \qquad (22.35b)$$

The explanation for the enhanced reactivity is probably similar to that for the increased reactivity of allylic alkyl halides in S_N2 displacements (Sec. 17.4).

In contrast, α-halo carbonyl compounds react so slowly by the S_N1 mechanism that this reaction is not useful.

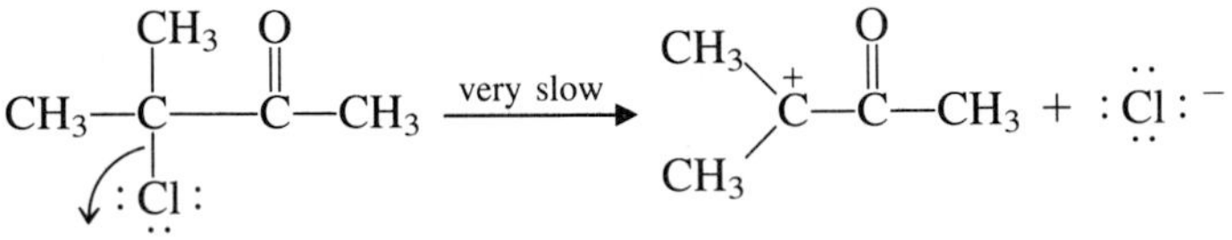

(22.36)

In fact, *reactions that require the formation of carbocations alpha to carbonyl groups generally do not occur.* Although an α-carbocation is formally resonance stabilized, the resonance structure is very unattractive (why?).

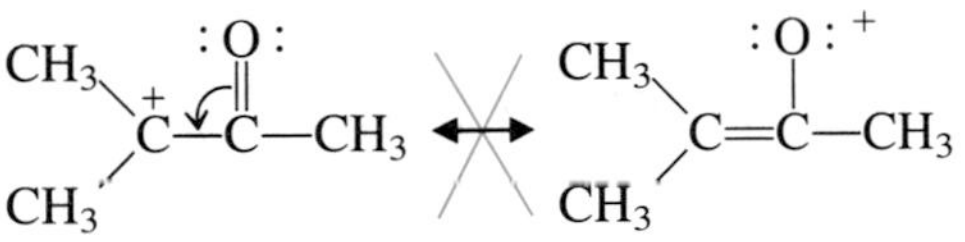

(very poor resonance structure) (22.37)

Problems

13 What product is formed when propionic acid is treated first with Br_2 and one equivalent of PBr_3, then with a large excess of ammonia?

14 Give the product of the reaction of 1-bromo-2-butanone with pyridine.

pyridine 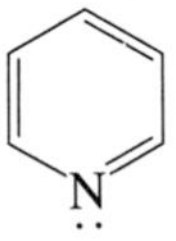

15 Give the mechanism of the reaction in Eq. 22.33. What is the purpose of the NaOH? What side reactions might be anticipated in this reaction? Explain.

22.4 ALDOL CONDENSATION

A. Base-Catalyzed Aldol Condensation

In aqueous base, acetaldehyde undergoes a reaction called the **aldol condensation.** In this reaction, two acetaldehyde molecules react to form a β-hydroxy aldehyde called *aldol.*

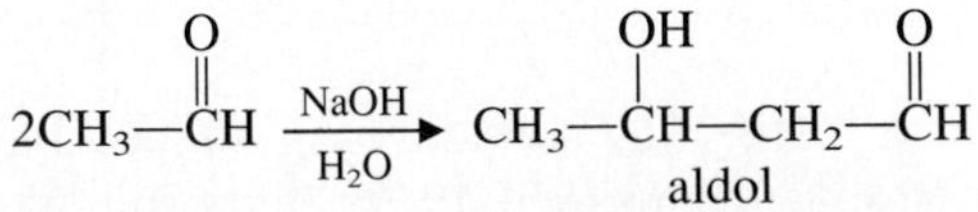

aldol
(50% yield)

(22.38)

The aldol condensation is a very important and general reaction of aldehydes and ketones having α-hydrogens. Notice that this reaction is another method of forming carbon–carbon bonds.

The base-catalyzed aldol condensation is another reaction that involves the *enolate ion* as an intermediate. In this reaction, an enolate ion, formed by reaction of acetaldehyde with aqueous NaOH, adds to a second molecule of acetaldehyde.

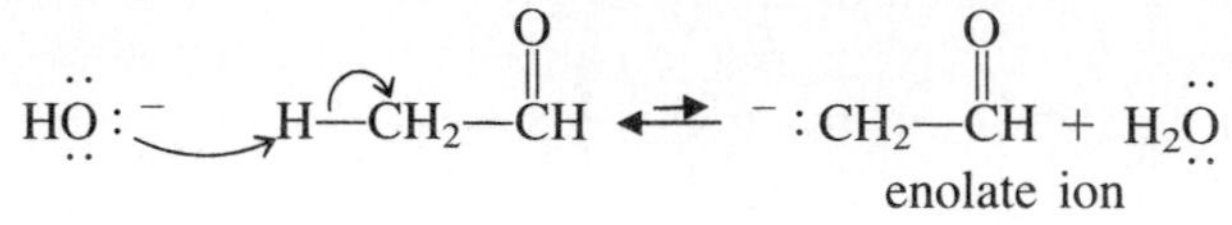

(22.39a)

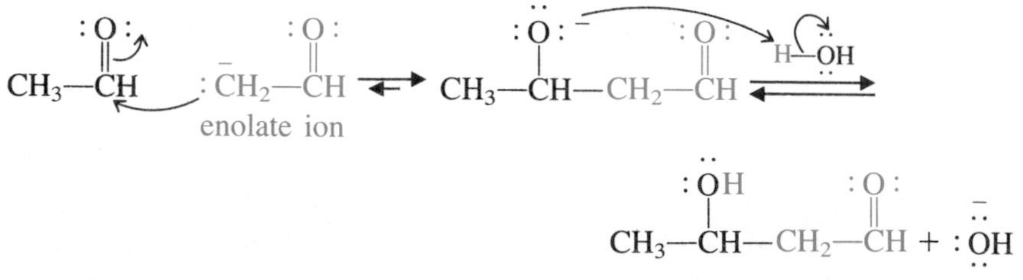

(22.39b)

We can see from this mechanism that the aldol condensation is simply another nucleophilic addition to a carbonyl group. In this reaction, the nucleophile is an enolate ion. The reaction may *look* more complicated than some additions because of the number of carbon atoms in the product. However, it is not conceptually different from other nucleophilic additions—for example, cyanohydrin formation.

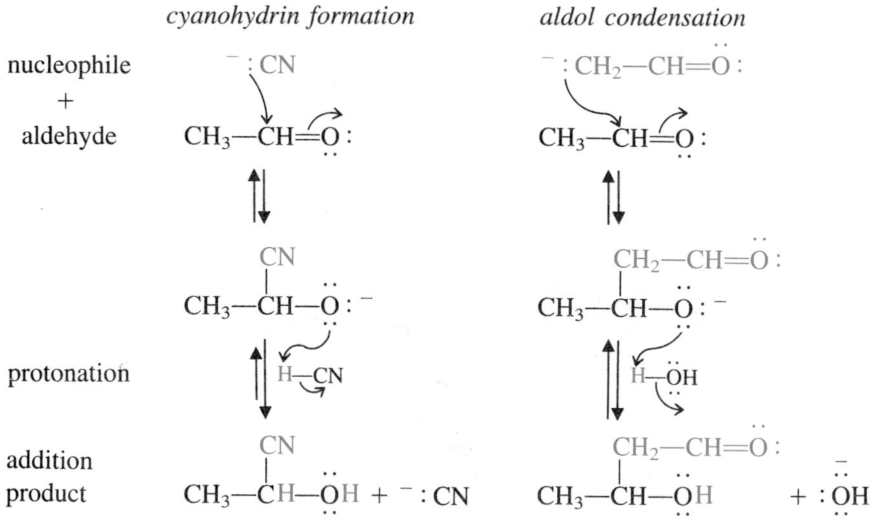

(22.40)

Problem

16 Use the mechanism of the reaction to deduce the product of the aldol condensation of propionaldehyde.

The aldol condensation is reversible. Like many other reversible carbonyl addition reactions (Sec. 19.7B), the equilibrium for the aldol condensation is much more favorable for aldehydes than for ketones.

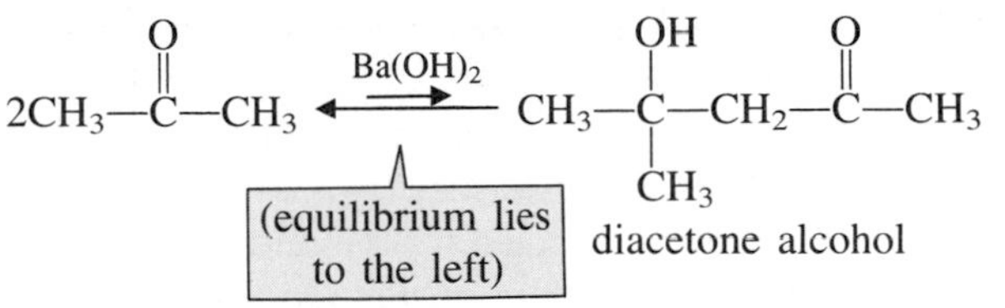

$$2CH_3-\overset{O}{\overset{\|}{C}}-CH_3 \rightleftharpoons CH_3-\overset{OH}{\overset{|}{C}}-CH_2-\overset{O}{\overset{\|}{C}}-CH_3 \quad (22.41)$$

(equilibrium lies to the left) diacetone alcohol

In this aldol condensation of acetone, the equilibrium favors acetone rather than the condensation product, diacetone alcohol. This reaction is useful for preparing diacetone alcohol only if the product is removed as it is formed.

In many cases, the conditions of an aldol condensation can be chosen to make the hydroxy aldehyde or hydroxy ketone addition product undergo *dehydration*. Moderately concentrated NaOH, for example, catalyzes this dehydration reaction.

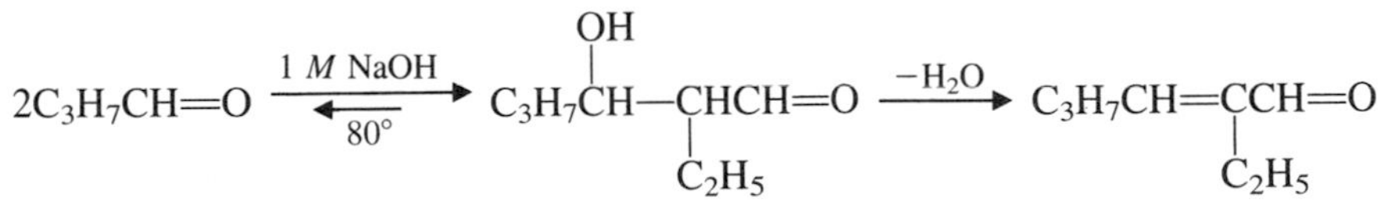

$$2C_3H_7CH{=}O \underset{80°}{\overset{1\ M\ \text{NaOH}}{\rightleftharpoons}} C_3H_7\overset{OH}{\overset{|}{CH}}-\overset{}{\underset{C_2H_5}{CHCH}}{=}O \xrightarrow{-H_2O} C_3H_7CH{=}\underset{C_2H_5}{CCH}{=}O \quad (22.42)$$

This reaction is a relatively rare example of a base-catalyzed dehydration; most alcohols do not dehydrate in base. However, β-hydroxy aldehydes and β-hydroxy ketones do, because (a) the α-hydrogen is relatively acidic, and (b) the product is conjugated and therefore particularly stable. Base-catalyzed dehydration occurs in two distinct steps through a *carbanion* intermediate.

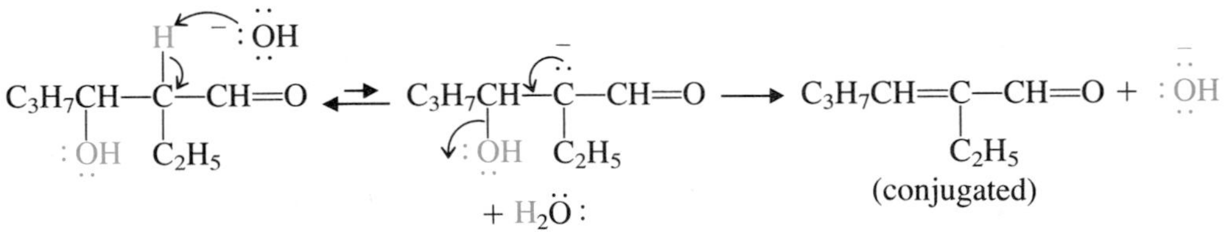

$$C_3H_7\underset{:\ddot{O}H}{\overset{H}{\overset{|}{CH}}}-\underset{C_2H_5}{\overset{|}{C}}-CH{=}O \rightleftharpoons C_3H_7CH-\underset{\underset{:\ddot{O}H}{\underset{+\ H_2\ddot{O}:}{}}}{\overset{\overset{:\ddot{O}H}{}}{\overset{-}{C}}}-CH{=}O \longrightarrow C_3H_7CH{=}\underset{C_2H_5}{\overset{-}{C}}-CH{=}O + :\ddot{O}H \quad (22.43)$$

(conjugated)

Although this type of dehydration is a β-elimination, notice that it is *not* concerted. In this respect it differs from the E2 reaction.

The product of an aldol condensation–dehydration sequence is an α,β-*unsaturated carbonyl compound*. In fact, *the major use of the aldol condensation is the preparation of α,β-unsaturated aldehydes and ketones.*

How do we know whether the aldol product or the dehydration product is formed? The answer depends on reaction conditions, and it must be determined on a case-by-case basis. The important thing to understand at this point is why the reaction takes place, and that α,β-unsaturated aldehydes and ketones can generally be isolated under appropriate conditions.

B. Acid-Catalyzed Aldol Condensation

Aldol condensations are also catalyzed by acid.

$$2CH_3-\overset{O}{\overset{\|}{C}}-CH_3 \xrightarrow{\text{acid}} \underset{CH_3}{\overset{CH_3}{}}C{=}CH-\overset{O}{\overset{\|}{C}}-CH_3 + H_2O \quad (22.44)$$

mesityl oxide
(79% yield)

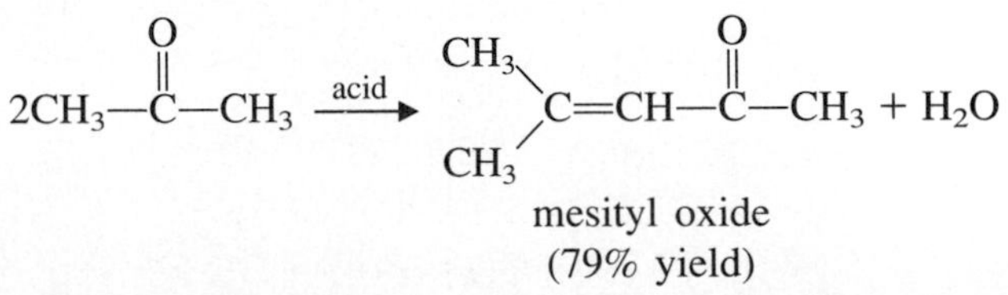

Acid-catalyzed aldol condensations, as in this example, generally give α,β-unsaturated carbonyl compounds as products.

In acid-catalyzed aldol condensations, *the reactive intermediates are enols,* not enolate ions. (Enolate ions are too basic to form in any appreciable amount in acidic solution.) Protonation of an aldehyde or ketone gives its conjugate acid, which behaves like a carbocation (Sec. 19.6).

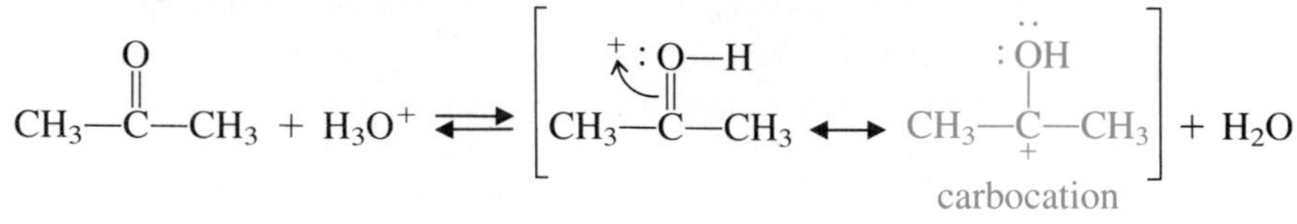

(22.45a)

Carbocations are electron-deficient species and, like other electron-deficient species (such as BH_3 or H^+), they add to double bonds. Thus a protonated aldehyde or ketone can add to the double bond of an enol derived from a second molecule of ketone starting material. This addition gives another protonated ketone, which upon loss of a proton gives an aldol condensation product.

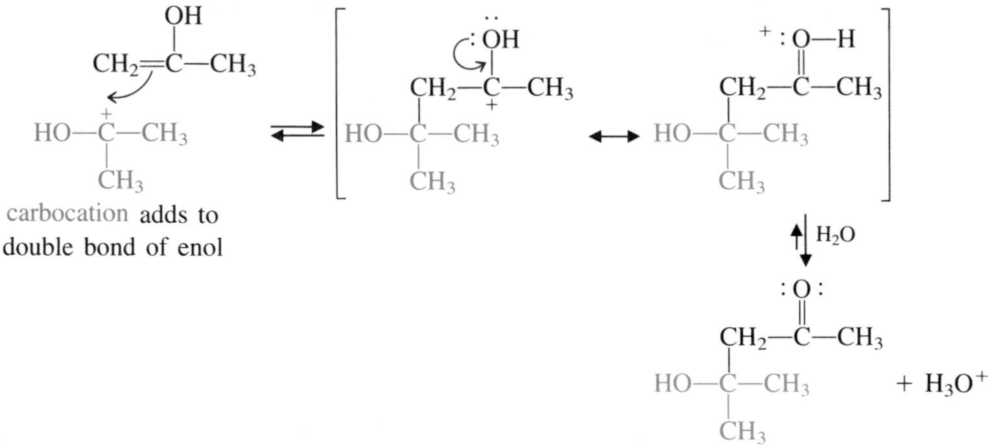

(22.45b)

Under the acidic conditions, the alcohol product dehydrates by the usual mechanism for alcohol dehydration (Sec. 10.1A) to give an α,β-unsaturated carbonyl compound.

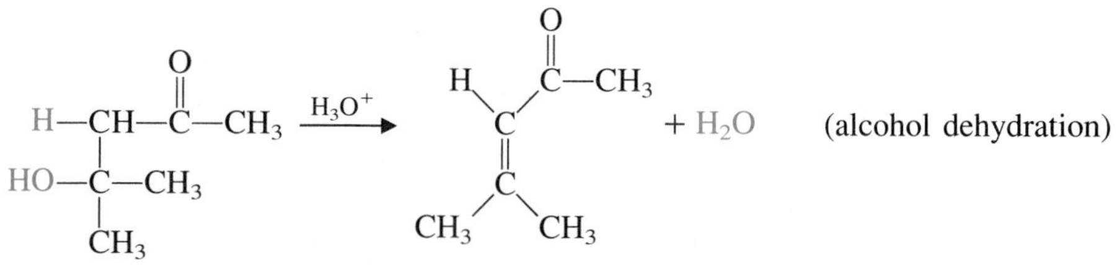

(22.45c)

Discovery of the aldol condensation is usually attributed solely to Charles Adolphe Wurtz, a French chemist who trained Friedel and Crafts. In fact, the reaction was first investigated during the period 1864–1873 by Aleksandr Borodin, a Russian chemist who was also a self-taught and proficient composer of music. (Borodin's musical themes were used as the basis of songs in the musical *Kismet.*) Borodin found it difficult to compete with Wurtz's large, modern, well-funded laboratory. Borodin also lamented that his professional duties so burdened him with "examinations and commissions" that he could only compose when he was at home ill. Knowing this, his friends used to greet him, "Aleksandr, I hope you are ill today!"

C. Special Types of Aldol Condensations

Crossed Aldol Condensations So far we have considered only aldol condensations between two molecules of the same carbonyl compound. What happens if two *different* carbonyl compounds are used? An example of this type of aldol condensation, called a **crossed aldol condensation,** is the reaction between acetaldehyde and propionaldehyde. In this case, four different aldol condensation products are obtained as a difficult-to-separate mixture. If we abbreviate our two aldehydes as *A* and *P*, respectively, then the enolate ion from *A* can react with the carbonyl of *P* and vice versa; and the enolate ion from *A* can react with another molecule of *A*, and the enolate ion from *P* with another molecule of *P*, giving four possible products:

$$\underset{A\,+\,A}{CH_3\overset{OH}{\overset{|}{C}}HCH_2CH{=}O} \qquad \underset{P\,+\,A}{CH_3CH_2\overset{OH}{\overset{|}{C}}HCH_2CH{=}O} \qquad \underset{A\,+\,P}{CH_3\overset{OH}{\overset{|}{C}}H\underset{CH_3}{\underset{|}{C}}HCH{=}O} \qquad \underset{P\,+\,P}{CH_3CH_2\overset{OH}{\overset{|}{C}}H\underset{CH_3}{\underset{|}{C}}HCH{=}O}$$

Problem

17 Give the structures of the four aldol condensation products (not counting stereoisomers) obtained in the crossed aldol condensation of butyraldehyde and propionaldehyde. (Assume that dehydration does not occur under the reaction conditions.)

Chemists have worked out some relatively sophisticated solutions to this mixture problem for a number of cases. However, under protic conditions (aqueous or alcoholic acid or base), useful crossed aldol condensations as a practical matter are limited to situations in which *a ketone with α-hydrogens is condensed with an aldehyde without α-hydrogens.* An important example of this type of crossed aldol condensation is the **Claisen–Schmidt condensation.** In this reaction, a ketone with α-hydrogens—acetone in the following example—is condensed with an aromatic aldehyde that has no α-hydrogens—benzaldehyde in this case.

$$PhCH{=}O + \underset{\text{(excess)}}{CH_3{-}\overset{O}{\overset{\|}{C}}{-}CH_3} \xrightarrow[\text{NaOH}]{\text{aqueous}} \underset{\substack{\text{benzalacetone}\\ \text{(65–78\% yield)}}}{Ph{-}CH{=}CH{-}\overset{O}{\overset{\|}{C}}{-}CH_3} + H_2O \qquad (22.46)$$

Why is only one product obtained from this aldol condensation? First, because the aldehyde in the Claisen–Schmidt reaction has no α-hydrogens, it cannot act as the enolate component of the aldol condensation. This eliminates two of the four products. The other possible side reaction is the condensation of the ketone with itself, as in Eq. 22.41; why doesn't this reaction occur? Once the enolate ion from acetone is formed, it has a choice: it can react with another molecule of acetone or it can react with benzaldehyde. As we have learned (Eq. 21.65), addition to a ketone occurs more slowly than addition to an aldehyde. Furthermore, even if addition to acetone does occur, the aldol condensation of two ketones is reversible (Eq. 22.41) and addition to an aldehyde has a more favorable equilibrium constant than addition to a ketone (Sec. 19.7B). Thus, in Eq. 22.46, both the rate and equilibrium for addition to benzaldehyde are more favorable than they are for addition to a second molecule of acetone. This leaves the product shown in Eq. 22.46 as the most likely possibility.

The Claisen–Schmidt condensation, like other aldol condensations, can also be catalyzed by acid.

$$\text{Ph—CH=O + Ph—}\overset{\displaystyle O}{\overset{\|}{\text{C}}}\text{—CH}_3 \xrightarrow[\text{CH}_3\text{CO}_2\text{H}]{\text{H}_2\text{SO}_4} \text{Ph—CH=CH—}\overset{\displaystyle O}{\overset{\|}{\text{C}}}\text{—Ph} \qquad (22.47)$$
$$\text{(95\% yield)}$$

Intramolecular Aldol Condensation When a molecule contains more than one aldehyde or ketone group, an *intramolecular* reaction (a reaction within the same molecule) is possible. In such a case the aldol condensation results in formation of a ring. Intramolecular aldol condensations are particularly favorable when unstrained five- and six-membered rings can be formed.

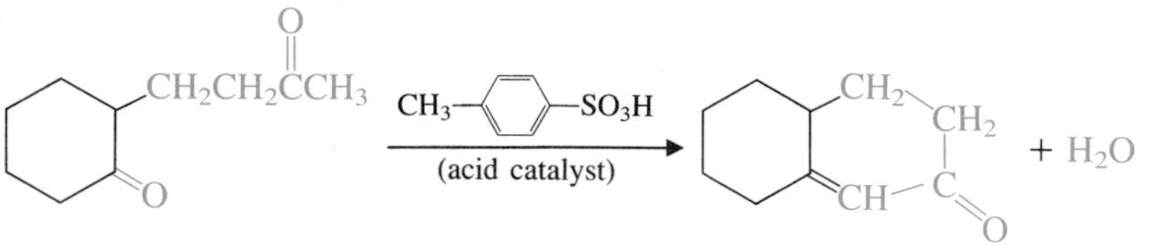

$$+ \text{ H}_2\text{O} \qquad (22.48)$$

(Write out the mechanism of this reaction.)

Problem

18 Predict the product(s) in each of the following aldol condensations:

(a)

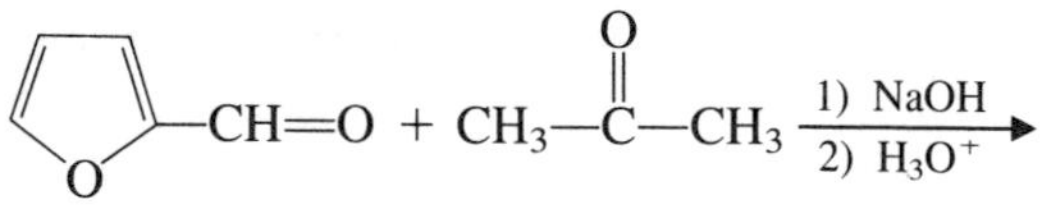

$$\text{(furan)—CH=O + CH}_3\text{—}\overset{\displaystyle O}{\overset{\|}{\text{C}}}\text{—CH}_3 \xrightarrow[\text{2) H}_3\text{O}^+]{\text{1) NaOH}}$$

(b)

$$\text{(cyclopentanone)—CH}_2\text{—}\overset{\displaystyle O}{\overset{\|}{\text{C}}}\text{—CH}_2\text{CH}_2\text{CH}_2\text{CH}_3 \xrightarrow{\text{KOH}}$$

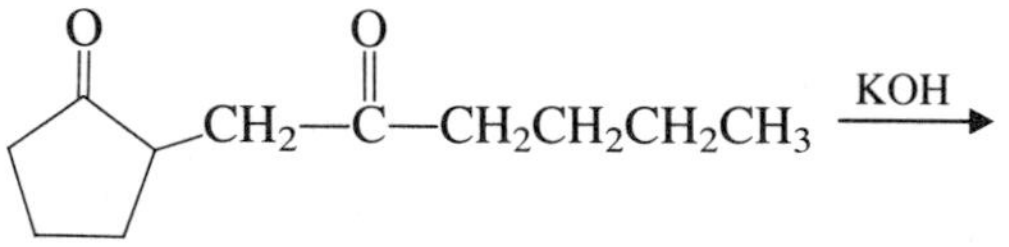

(c) acetophenone + hexanal $\xrightarrow{\text{NaOH}}$

D. Synthesis with the Aldol Condensation

Let us now focus on the use of the aldol condensation in organic synthesis. The previous discussion shows that the aldol condensation can be applied to the synthesis of α,β-unsaturated aldehydes and ketones. Notice that it also represents another method for the formation of carbon–carbon bonds. (See the list in Sec. 21.10.) Given an α,β-unsaturated aldehyde or ketone as our synthetic target, we must ask two questions: (1) What starting materials are required in the aldol condensation? (2) Knowing the required starting materials, is the aldol condensation of these compounds a feasible one?

The starting materials for an aldol condensation can be determined by mentally "splitting" the α,β-unsaturated carbonyl compound at the double bond:

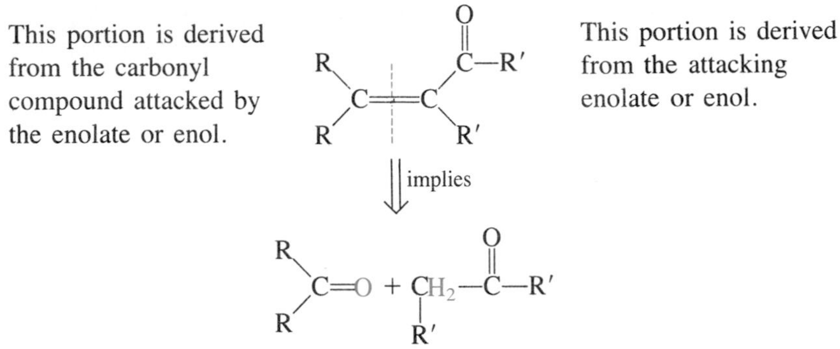
This portion is derived from the carbonyl compound attacked by the enolate or enol.
This portion is derived from the attacking enolate or enol.
implies

(22.49)

That is, we work backwards from the desired synthetic objective by replacing the double bond on the carbonyl side by two hydrogens and on the other side by a carbonyl oxygen (=O) to obtain the structures of the starting materials in the aldol condensation.

Once we know the starting materials, we still must ask whether the required aldol condensation is a feasible one. Consider, for example, the following case:

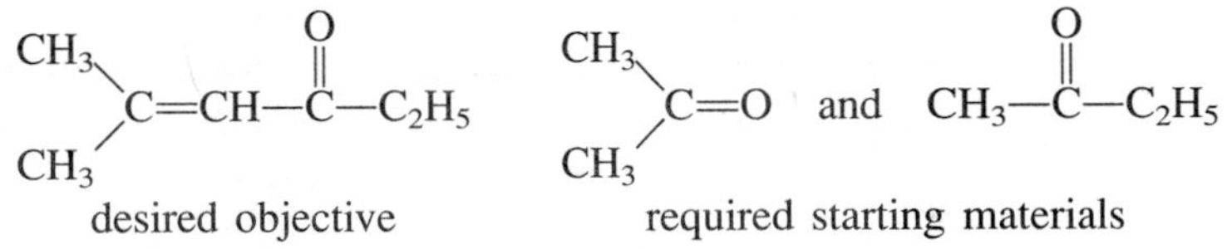
desired objective
required starting materials

In this example, the required aldol condensation is a crossed condensation between two similar ketones, and therefore will not succeed (why?). Direct synthesis of this compound by a simple aldol condensation is not feasible. In other words, a knowledge of the starting materials required for a given condensation is no guarantee that it will actually work.

Problems

19 Some of the following molecules can be synthesized in good yield using an aldol condensation. Identify these and give the structures of the required starting materials. Others cannot be synthesized in good yield by an aldol condensation. Identify these, and explain why the required aldol condensation would not be likely to succeed.

(a)

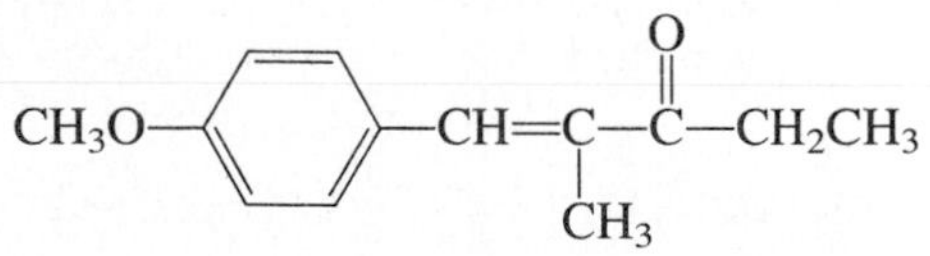

(c)

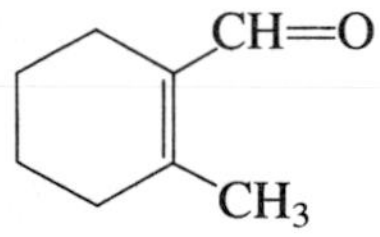

(b)

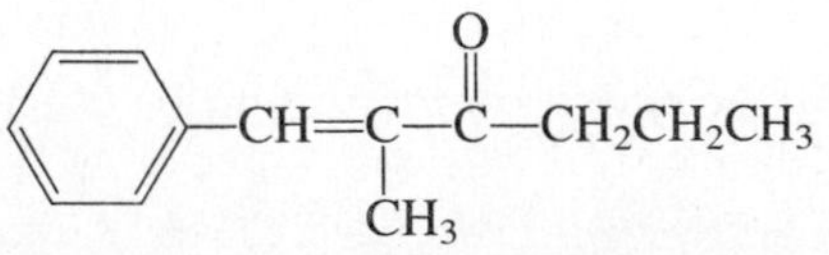

(Problem continues on p. 940.)

Problems (Cont.)

(d)

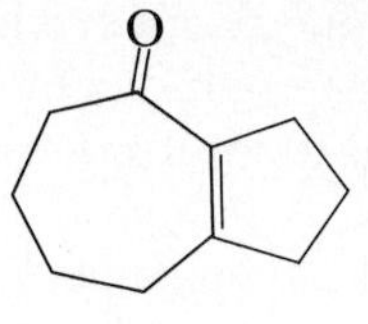

(f)

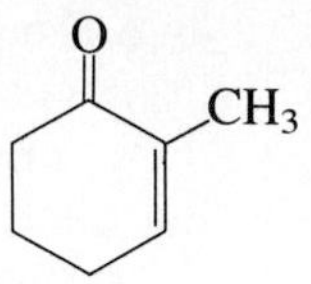

(e)

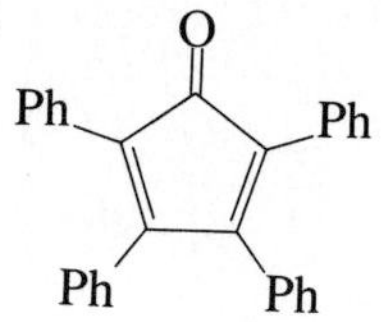

(g) $(CH_3)_2C{=}CH{-}CH{=}O$

(h) $Ph{-}CH{=}CH{-}C(=O){-}CH{=}CH{-}Ph$

20 Analyze the aldol condensation in Eq. 22.48 using the method given above. Show that there are four possible aldol condensation products that might in principle result from the starting material. Argue why the observed product is the most reasonable one.

22.5 CONDENSATION REACTIONS INVOLVING ESTER ENOLATE IONS

In this section we shall begin to use more compact abbreviations for several commonly occurring organic groups. These abbreviations, shown in Table 22.1, not only save space, but also make the structures of large molecules less cluttered and more easily read. Just as we have been using Ph— to symbolize the phenyl ring, we shall now use Me— for methyl, Et— for ethyl, Pr— for propyl, and so on. Thus, ethyl acetate is abbreviated EtOAc; sodium ethoxide ($NaOC_2H_5$) is simply written as NaOEt; and methanol is abbreviated as MeOH.

TABLE 22.1 Abbreviations of Some Common Organic Groups

Group	Structure	Abbreviation
methyl	$CH_3{-}$	Me
ethyl	$CH_3CH_2{-}$	Et
propyl	$CH_3CH_2CH_2{-}$	Pr
isopropyl	$(CH_3)_2CH{-}$	*i*-Pr
butyl	$CH_3CH_2CH_2CH_2{-}$	Bu
isobutyl	$(CH_3)_2CHCH_2{-}$	*i*-Bu
tert-butyl	$(CH_3)_3C{-}$	*t*-Bu
acetyl	$CH_3{-}C(=O){-}$	Ac
acetate (or acetoxy)	$CH_3{-}C(=O){-}O{-}$	AcO

Problem

21 Write out the structure that corresponds to each of the following abbreviations:
(a) Et$_3$C—OH (d) Pr—OH
(b) i-Pr—Ph (e) Ac$_2$O
(c) t-BuOAc (f) Ac—Ph

A. Claisen Condensation

In the last section we learned that the aldol condensation is a reaction that involves the enols or enolates of *aldehydes* and *ketones*. Now we shall learn about a reaction that involves the enolate ions of *esters*.

Ethyl acetate undergoes a condensation reaction in the presence of one equivalent of sodium ethoxide in ethanol to give a compound known commonly as ethyl acetoacetate.

$$2CH_3-\overset{\overset{\displaystyle O}{\|}}{C}-OEt \xrightarrow[\text{EtOH}]{\overset{\text{NaOEt}}{\text{(1 equiv)}}} \xrightarrow{H_3O^+} CH_3-\overset{\overset{\displaystyle O}{\|}}{C}-CH_2-\overset{\overset{\displaystyle O}{\|}}{C}-OEt + EtOH \quad (22.50)$$

ethyl acetoacetate
(75 76% yield)

This is the best-known example of the **Claisen condensation,** named for Ludwig Claisen (1851–1930), who was a Professor at the University of Kiel. (Do *not* confuse this reaction with the Claisen–Schmidt condensation in the previous section—same Claisen, different reaction!) The product of this reaction, ethyl acetoacetate, is an example of a β-**keto ester**: a compound with a ketone carbonyl group β to the ester carbonyl group.

$$CH_3-\underset{\beta}{\overset{\overset{\displaystyle O}{\|}}{C}}-CH_2-\underset{\alpha}{\overset{\overset{\displaystyle O}{\|}}{C}}-OEt$$

The first step in the mechanism of the Claisen condensation is formation of an *enolate ion* by reaction of the ester with the ethoxide base.

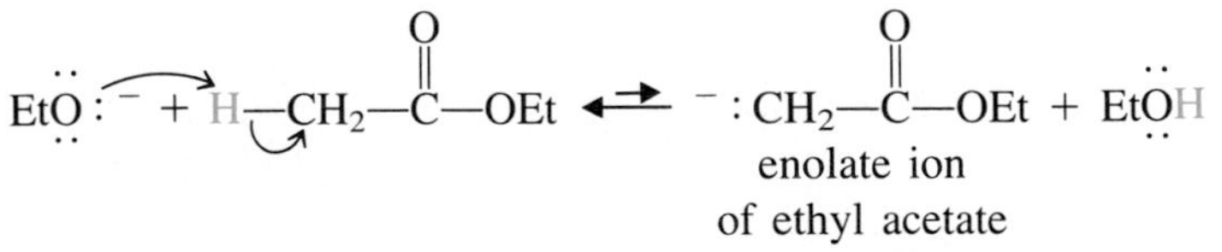

$$Et\overset{..}{\underset{..}{O}}:^- + H-CH_2-\overset{\overset{\displaystyle O}{\|}}{C}-OEt \rightleftarrows {}^-:CH_2-\overset{\overset{\displaystyle O}{\|}}{C}-OEt + Et\overset{..}{O}H \quad (22.51a)$$

enolate ion
of ethyl acetate

Since ethoxide ion is a nucleophile, we might also ask whether it can also attack the carbonyl group of the ester to give the usual carbonyl substitution reaction. This reaction undoubtedly takes place, but the products are the same as the reactants! This is why ethoxide ion is used as a base with ethyl esters in the Claisen condensation.

Although the ester enolate ion is formed in low concentration, it is a strong base and good nucleophile, and undergoes a *carbonyl substitution reaction* with a second molecule of ester (Eq. 22.51b).

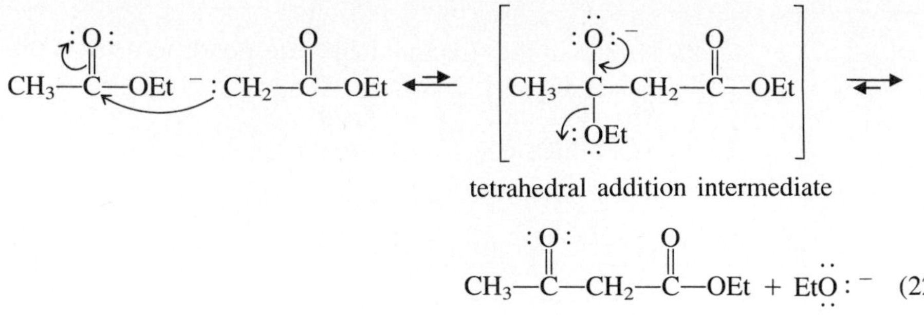
tetrahedral addition intermediate

(22.51b)

The overall reaction is readily reversible; in fact, the equilibrium favors starting materials. For this reason, the Claisen condensation has to be driven to completion by applying LeChatelier's principle. The most common technique is to use one equivalent of ethoxide catalyst. In the product β-keto ester, the hydrogens adjacent to both carbonyl groups (color in Eq. 22.51c) are especially acidic (why?), and the ethoxide removes one of these hydrogens to form quantitatively the conjugate base of the product.

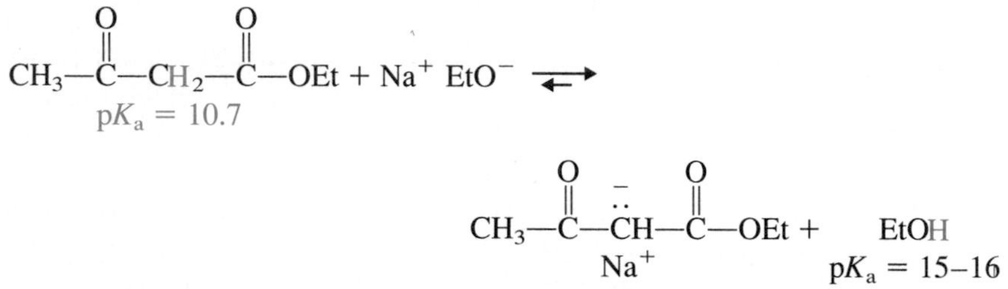
$pK_a = 10.7$
$pK_a = 15–16$

(22.51c)

The un-ionized β-keto ester product in Eq. 22.50 is formed when acid is added subsequently to the reaction mixture.

The removal of a product by ionization is the same strategy employed to drive ester saponification to completion (Sec. 21.7A). The importance of this strategy in the success of the Claisen condensation is evident if we attempt reaction with an ester that has only one α-hydrogen: *little or no condensation product is observed.* In this case, the desired condensation product has no α-protons acidic enough to react completely with ethoxide.

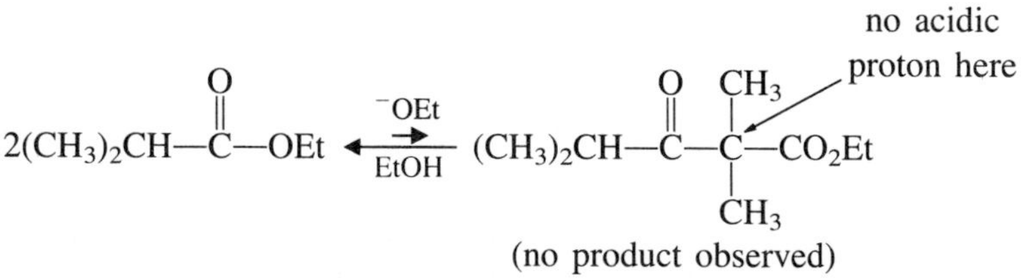
no acidic proton here
(no product observed)

(22.52)

Furthermore, if the desired product of Eq. 22.52 (made by another method) is subjected to the conditions of the Claisen condensation, it readily decomposes back to starting materials because of the reversibility of the Claisen condensation.

We can see that the Claisen condensation is one more example of *nucleophilic substitution at a carbonyl group.* In this reaction, the nucleophile is an enolate ion derived from an ester. Although the reaction may seem complex because of the number of carbon atoms in the product, it is not conceptually different from other carbonyl substitutions—for example, ester saponification:

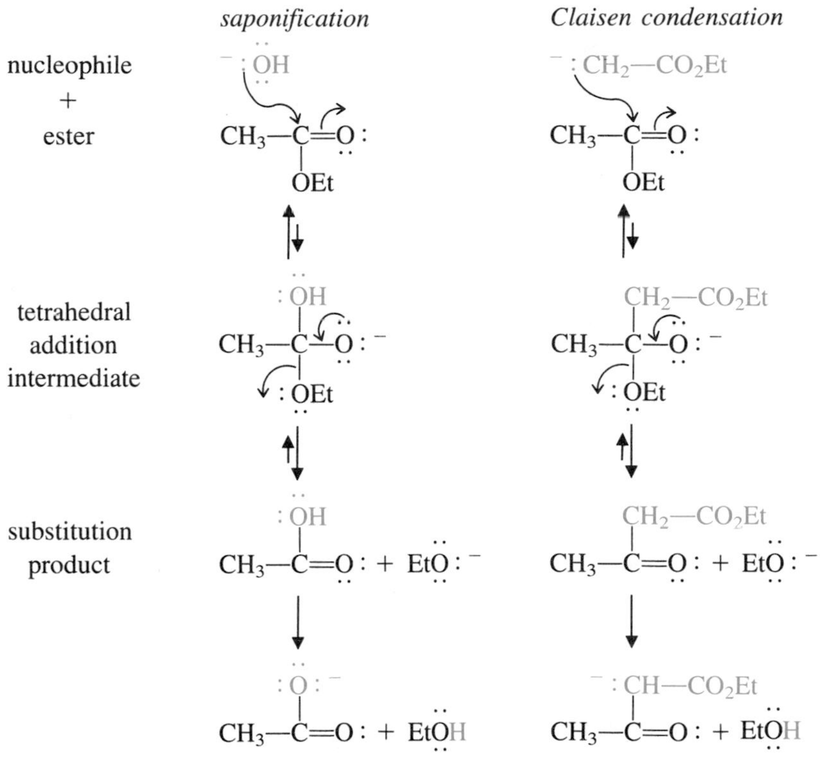

(22.53)

We have now studied two types of condensation reactions: the aldol condensation and the Claisen condensation. Although both condensations in basic solution involve reactions of enolate ions with carbonyl groups, they are quite different processes and should not be confused. To compare:

1. The aldol condensation is an *addition* reaction of an enolate ion with an aldehyde or ketone. The Claisen condensation is a *substitution* reaction of an enolate ion with an ester group.

2. The aldol condensation can be carried out in acidic as well as basic solution, because the reaction can be driven to completion by acid-catalyzed dehydration of the aldol product. The Claisen condensation requires a full equivalent of base, and does not occur in acid.

3. Because ionization of the product β-keto ester is crucial to the success of the Claisen condensation, the starting ester must have at least *two* α-hydrogens: one for each of the ionizations shown in Eqs. 22.51a and 22.51c. There is no such restriction in the aldol condensation.

Problems

22 Give the product of the Claisen condensation of ethyl butyrate with one equivalent of NaOEt.

Problems (Cont.)

23 Hydroxide is about as basic as ethoxide. Would NaOH be a suitable base for the Claisen condensation of ethyl acetate? Explain.

B. Dieckmann Condensation

Intramolecular Claisen condensations, like intramolecular aldol condensations, take place readily when five- or six-membered rings can be formed. The intramolecular Claisen condensation reaction is called the **Dieckmann condensation.**

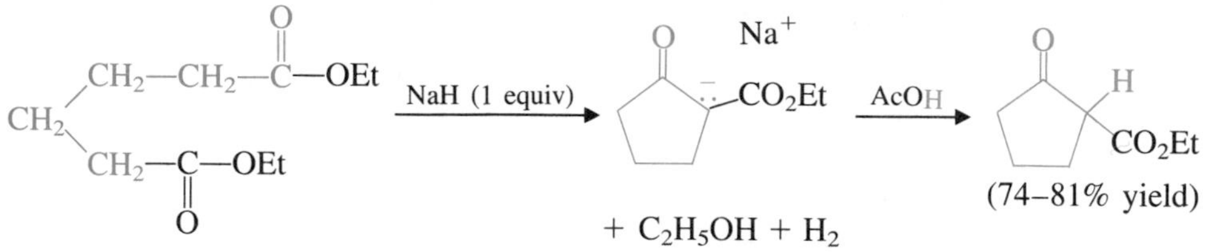

(22.54)

Notice that in this reaction, as in the Claisen condensation, the α-proton must be removed from the initially formed product in order for the reaction to be driven to completion. (The base used in Eq. 22.54, sodium hydride, was discussed in Eq. 8.11 of Sec. 8.6A.)

Problem

24 (a) Explain why compound *A*, when treated with one equivalent of NaOEt, followed by acidification, is completely converted into compound *B*.
(b) Give the structure of the only product formed when diethyl α-methyladipate (*C*) reacts in the Dieckmann condensation. Explain your reasoning.

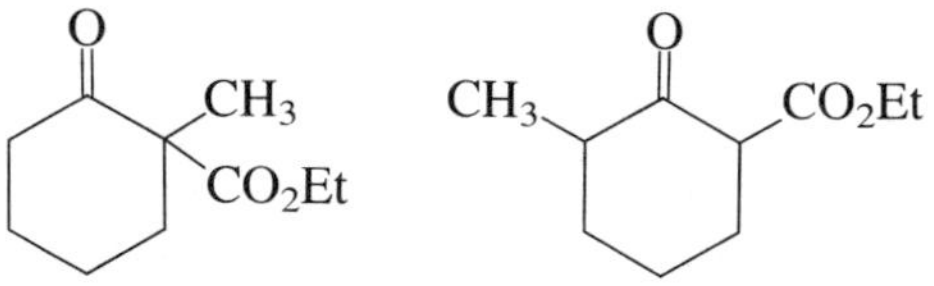

A *B*

$EtO_2C(CH_2)_3CH(CH_3)CO_2Et$

C

C. Crossed Claisen Condensation

The Claisen condensation of two *different* esters is called a **crossed Claisen condensation.** The crossed Claisen condensation of two esters that both have α-hydrogens gives a mixture of four compounds that are typically difficult to separate. Such reactions in most cases are not synthetically useful.

$$CH_3{-}CO_2Et + C_2H_5{-}CO_2Et \xrightarrow{NaOEt} \xrightarrow{H_3O^+}$$

$$CH_3{-}\overset{O}{\overset{\|}{C}}{-}CH_2{-}\overset{O}{\overset{\|}{C}}{-}OEt + CH_3CH_2{-}\overset{O}{\overset{\|}{C}}{-}CH_2{-}\overset{O}{\overset{\|}{C}}{-}OEt$$

$$CH_3{-}\overset{O}{\overset{\|}{C}}{-}\underset{CH_3}{\underset{|}{CH}}{-}\overset{O}{\overset{\|}{C}}{-}OEt + CH_3CH_2{-}\overset{O}{\overset{\|}{C}}{-}\underset{CH_3}{\underset{|}{CH}}{-}\overset{O}{\overset{\|}{C}}{-}OEt \quad (22.55)$$

However, crossed Claisen condensations are possible if one ester is especially reactive, or has no α-hydrogens. (We found a similar situation in crossed aldol condensations; Sec. 22.4C.) For example, formyl groups (—CH=O) are readily introduced with esters of formic acid such as ethyl formate:

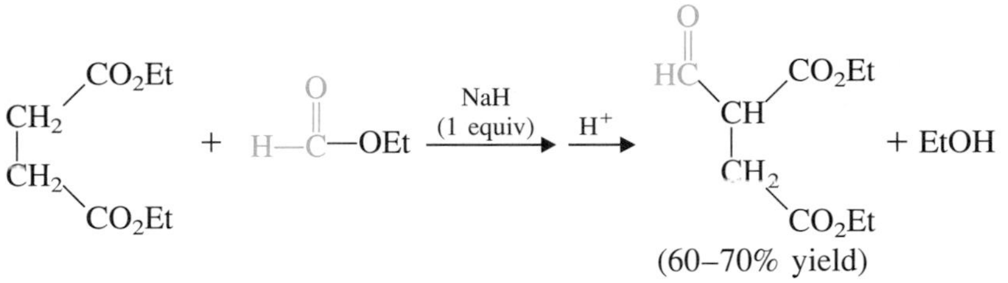

(22.56)

Formate esters fulfill both of the criteria for a crossed Claisen condensation. First, they have no α-hydrogens; second, their carbonyl reactivity is considerably greater than that of other esters. The reason for the greater reactivity is that the carbonyl group in formate esters is "part aldehyde," and aldehydes are particularly reactive toward nucleophiles (Eq. 21.65).

A less reactive ester without α-protons can be used if it is present in excess. For example, an ethoxycarbonyl group can be introduced with diethyl carbonate.

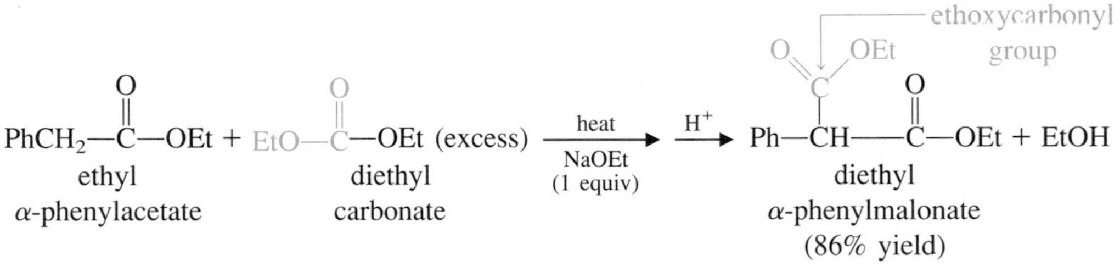

(22.57)

In this example, the enolate ion of ethyl α-phenylacetate condenses preferentially with diethyl carbonate rather than with another molecule of itself, because it "sees" a much higher concentration of diethyl carbonate, which is used in large excess. Of course, the excess diethyl carbonate must be separated from the product by fractional distillation.

Another type of crossed Claisen condensation is the reaction of ketones with esters. In this type of reaction the enolate ion of a ketone attacks the carbonyl group of an ester.

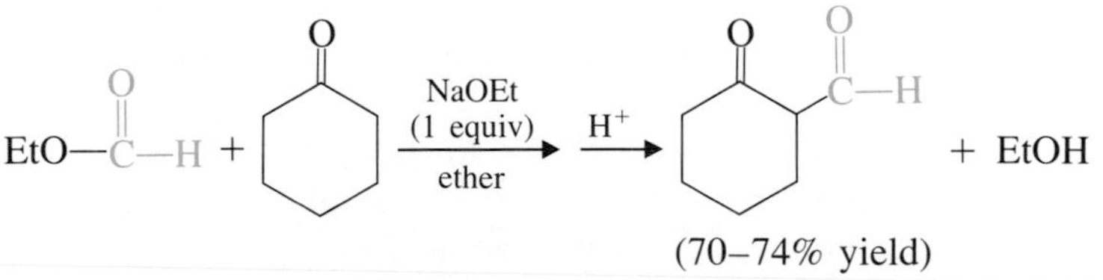

(22.58)

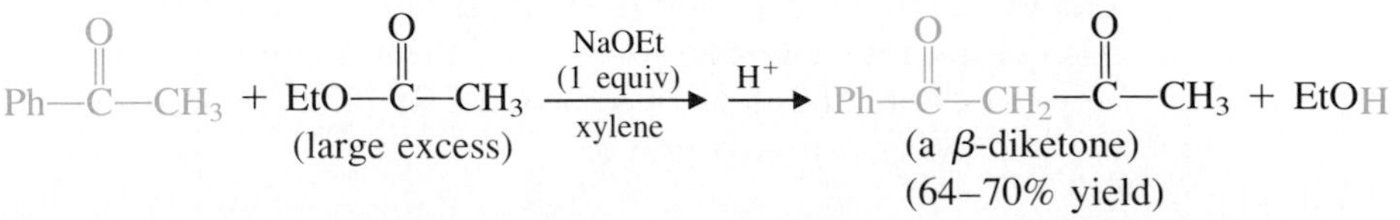

(22.59)

In Eq. 22.58, the enolate of the ketone cyclohexanone is acylated by ethyl formate. Let us examine the side reactions that might have occurred and try to understand why they in fact pose no problem. First, we might imagine that the aldol condensation of cyclohexanone with itself could be a possible side reaction. However, the aldol condensation of two ketones is reversible, whereas the Claisen condensation is irreversible *because one equivalent of base is used to form the enolate ion of the product.* Second, the ester has no α-hydrogens, and thus cannot condense with itself.

However, in Eq. 22.59, the ester *does* have α-hydrogens and is known (Eq. 22.50) to condense with itself. Why is such a condensation not an interfering side reaction? The answer is that ketones are far more acidic than esters (by about 5–7 pK_a units). Thus the enolate of the ketone is formed in much greater concentration than the ester enolate. The ketone enolate reacts with the large excess of ethyl acetate. The product is a β-diketone, which, like a β-keto ester, is especially acidic at the α-position, and ionizes completely under the basic conditions to drive the reaction to completion. The only reaction, then, that can occur is the crossed Claisen condensation.

We see from these examples that the crossed Claisen condensation of ketones and esters can be used for the synthesis of a wide variety of β-diketones.

Problem

25 Complete the following reactions. Assume that one equivalent of NaOEt or NaH is present in each case.

(a)

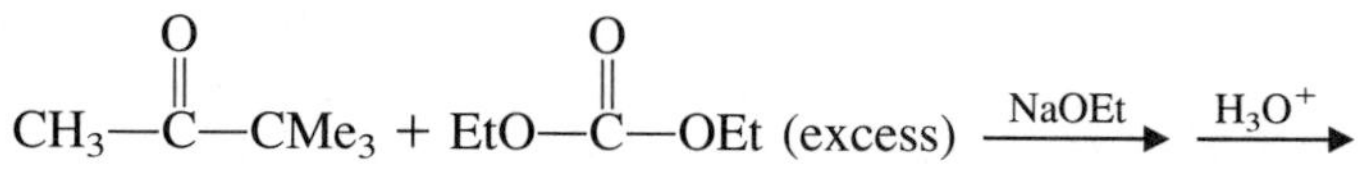

(b)

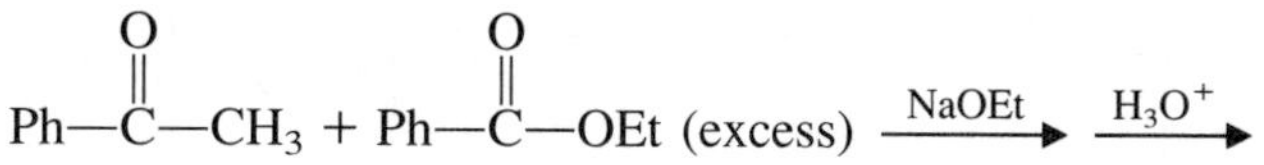

(c)

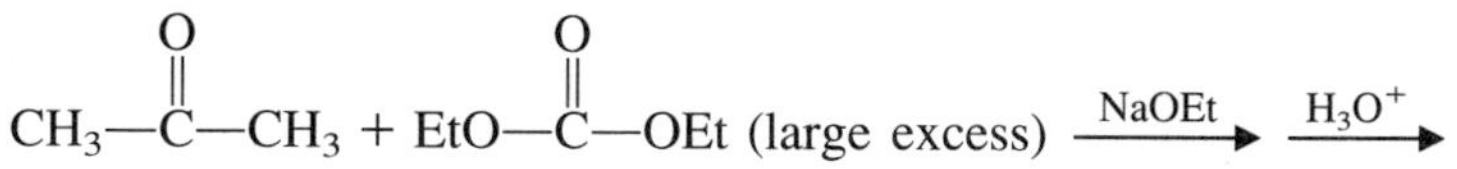

(d)

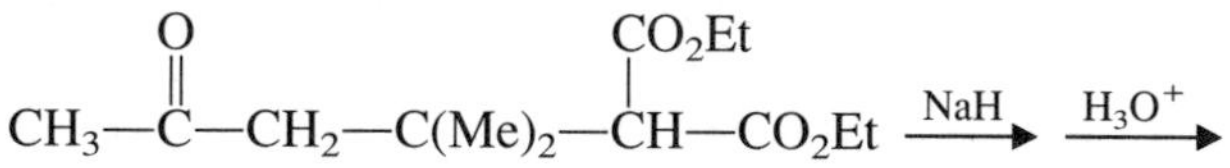

D. Synthesis with the Claisen Condensation

Let us now consider application of the Claisen condensation and related reactions to organic synthesis. As usual, we must examine the desired target molecule and work backwards to reasonable starting materials. We should think of the Claisen condensation when we desire to synthesize a β-dicarbonyl compound: a β-keto ester, a β-diketone, or the like. Compare these types of compounds with those prepared by the aldol condensation, and observe carefully the difference.

Given a β-dicarbonyl compound, we ask first what the required starting materials should be if the compound is to be prepared by a Claisen condensation. To answer this question we mentally reverse the Claisen condensation. We can do this in two ways by adding the elements of ethanol (or other alcohol) across the carbon–carbon bond at either carbonyl group:

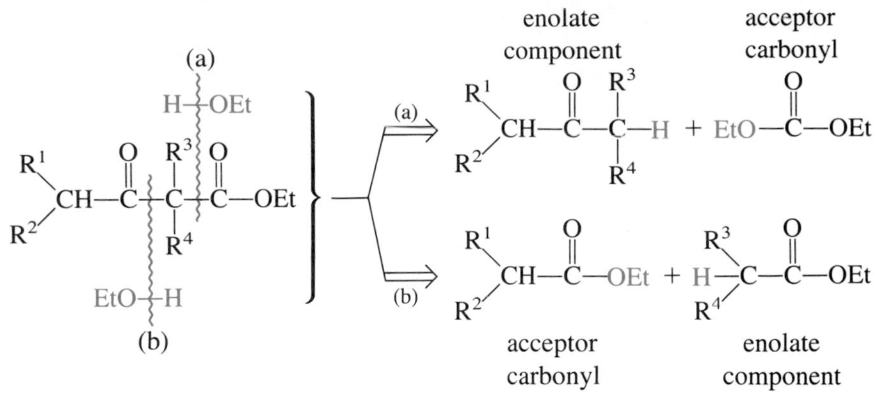

(22.60)

A β-diketone can also be mentally reversed in two different ways:

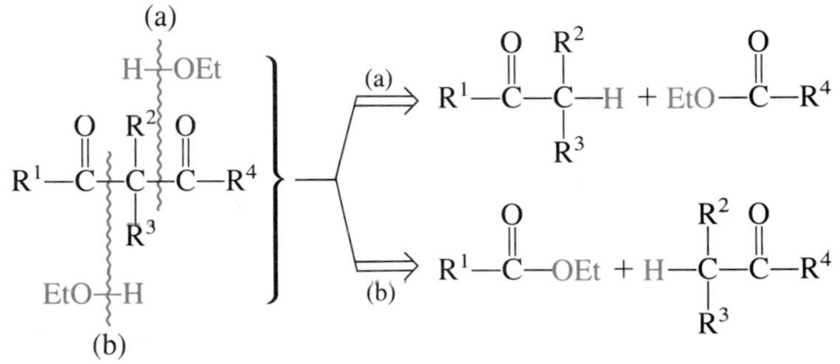

(22.61)

Having determined the starting materials required in the Claisen condensation, we then ask whether the Claisen condensation of the required starting materials is reasonable. If a crossed Claisen condensation is necessary, we must ask whether it is one of those that gives mostly one product, or whether it would be expected to give a difficult-to-separate mixture. (This is the same general procedure used in analyzing the aldol condensation in Sec. 22.4E.)

Problems

26 Analyze each of the following compounds and determine what starting materials would be required if they were to be made by a Claisen condensation. Then decide which of these reactions would be a reasonable route to starting material.

(a) 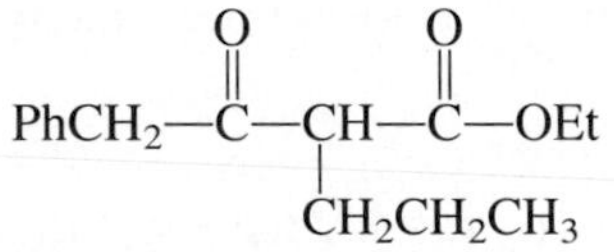

(c) 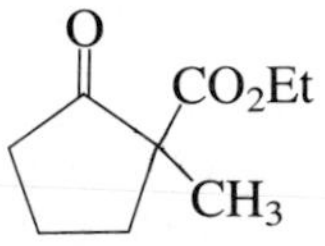

(b) 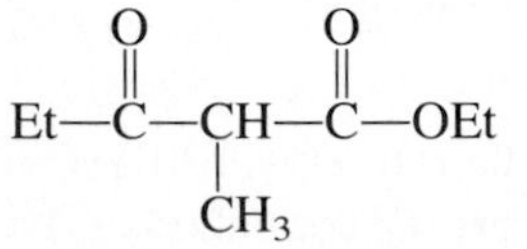

(d) 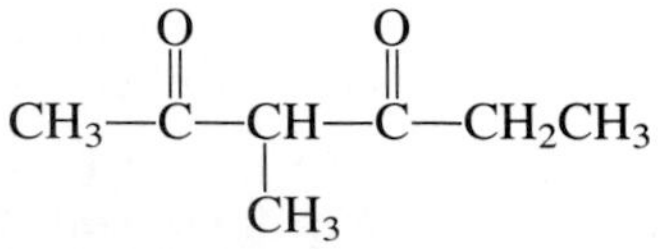

Problems (Cont.)

27 Give the starting material required for the synthesis of each of the following compounds by a Dieckmann condensation.

(a)

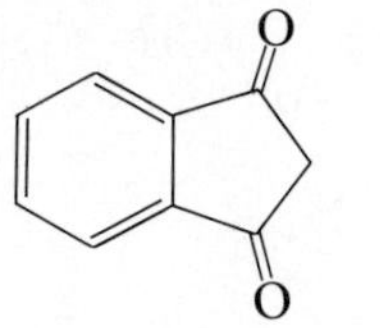

(b)

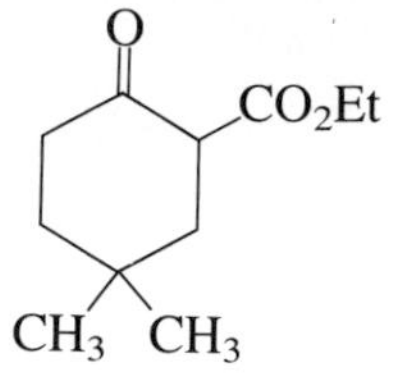

(c)

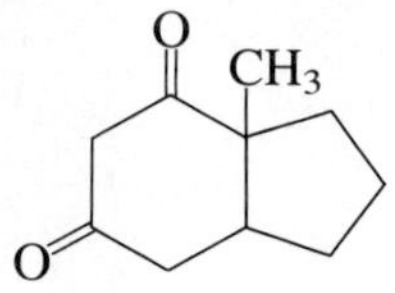

22.6 ALKYLATION OF ESTER ENOLATE IONS

In the previous two sections, we studied reactions in which enolate ions react as nucleophiles with carbonyl groups. In this section, we shall consider two reactions in which enolate ions behave as nucleophiles in S_N2 reactions with alkyl halides and sulfonate esters.

A. Malonic Ester Synthesis

Diethyl malonate (malonic ester), like many other β-dicarbonyl compounds, has unusually acidic α-protons (why?). Consequently, its conjugate-base enolate ion can be formed quantitatively with alkoxide bases such as sodium ethoxide.

$$\text{Et}\ddot{\text{O}}\text{:}^- + \text{EtO}-\overset{\overset{\text{O}}{\|}}{\text{C}}-\text{CH}_2-\overset{\overset{\text{O}}{\|}}{\text{C}}-\text{OEt} \rightleftharpoons \text{Et}\ddot{\text{O}}-\text{H} + \text{EtO}-\overset{\overset{\text{O}}{\|}}{\text{C}}-\overset{-}{\ddot{\text{C}}}\text{H}-\overset{\overset{\text{O}}{\|}}{\text{C}}-\text{OEt} \quad (22.62a)$$

diethyl malonate $pK_a = 12.9$ — enolate ion of diethyl malonate

The conjugate-base anion of diethyl malonate is nucleophilic, and reacts with alkyl halides and sulfonate esters in typical S_N2 reactions. This reaction can be used to introduce alkyl groups at the α-position of malonic ester.

$$\text{Na}^+ \ {}^-\text{:CH(CO}_2\text{Et)}_2 + \text{CH}_3\underset{}{\overset{\overset{\text{C}_2\text{H}_5}{|}}{\text{C}}}\text{H}-\text{Br} \xrightarrow{\text{EtOH}} \text{CH}_3\overset{\overset{\text{C}_2\text{H}_5}{|}}{\text{C}}\text{HCH(CO}_2\text{Et)}_2 + \text{Na}^+\ \text{Br}^- \quad (22.62b)$$

(83% yield)

Of course, this reaction is limited to alkyl halides and sulfonates that can undergo S_N2 reactions (which ones are these?). Monoalkylated derivatives of diethyl malonate can be treated again with ethoxide and alkylated a second time.

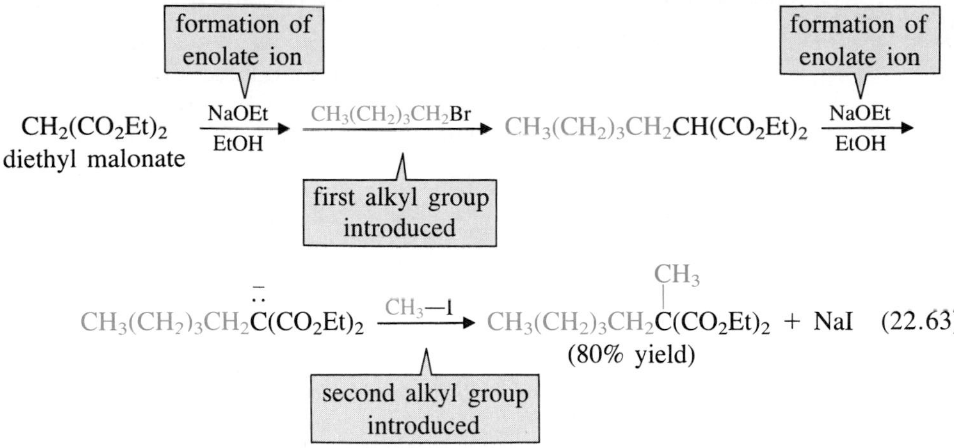

The synthesis of alkylated derivatives of malonate esters is called the **malonic ester synthesis.** Notice that the malonic ester synthesis also results in the formation of carbon–carbon bonds. The importance of this synthesis is that it can be extended to the preparation of carboxylic acids. Saponification and acidification of the diester gives a substituted malonic acid derivative (Sec. 21.7A). As we learned in Sec. 20.11, heating any malonic acid derivative causes it to *decarboxylate*. The result of the alkylation, saponification, and decarboxylation is a carboxylic acid that is formally a disubstituted acetic acid—an acetic acid molecule with two alkyl groups on its α-carbon.

An ester of malonic acid bearing only one alkyl group at its α-carbon can also be hydrolyzed and decarboxylated to give a "monoalkylated acetic acid."

When we wish to prepare a given carboxylic acid, how do we know whether a malonic acid synthesis can be used? First, we determine whether the desired carboxylic acid is formally an acetic acid with one or two alkyl groups on its α-carbon. If so, we mentally reverse the decarboxylation, hydrolysis, and alkylation steps to arrive at the structures of the alkyl halides (or sulfonate esters) that must be used in the malonic ester synthesis.

$$\text{acetic acid unit}$$

$$R-\underset{\underset{R'}{|}}{CH}-CO_2H \implies R-\underset{\underset{R'}{|}}{C}(CO_2Et)_2 \implies CH_2(CO_2Et)_2, \ R-Br, \ R'-Br \quad (22.65)$$

If the alkyl halides R—Br and R′—Br are among those that will undergo the S_N2 reaction, then the target carboxylic acid can in principle be prepared by the malonic ester synthesis.

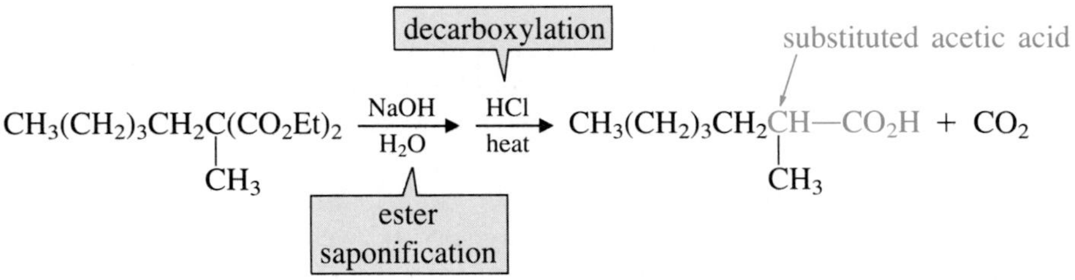

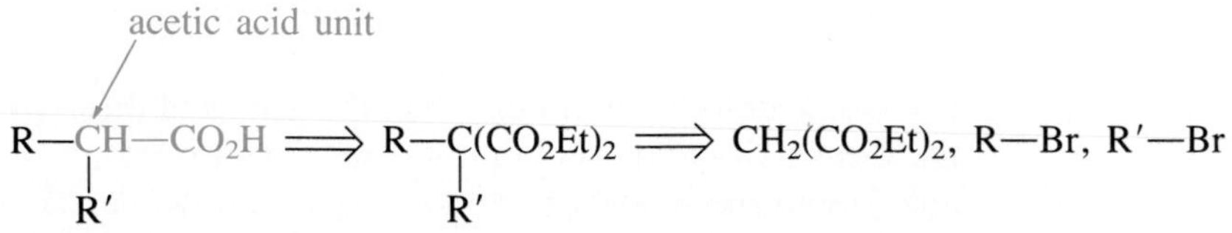

Problems

28 Indicate whether each of the following compounds could be prepared by the malonic ester synthesis. If so, outline a preparation from diethyl malonate and any other reagents. If not, explain why.
(a) 2-ethylbutanoic acid (b) 3,3-dimethylbutanoic acid

29 Complete the following reaction and explain your answer:

$$CH_2(CO_2Et)_2 + BrCH_2CH_2CH_2Cl \xrightarrow[\text{EtOH}]{\text{2NaOEt}} \xrightarrow{\text{NaOH}} \xrightarrow[\text{heat}]{\text{HCl}} (C_5H_8O_2)$$

B. Direct Alkylation of Enolate Ions Derived from Simple Monoesters

In the synthesis of carboxylic acids by malonic ester alkylation, a —CO_2Et group is "wasted" because it is later removed. Why not avoid this altogether and alkylate directly the enolate ion of an acetic acid ester?

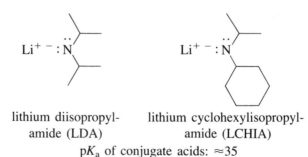

$$\underset{\text{(a base)}}{B:^-} + CH_3—\overset{O}{\overset{\|}{C}}—OR \longrightarrow \underset{+\ B—H}{^-:CH_2—\overset{O}{\overset{\|}{C}}—OR} \xrightarrow{CH_3CH_2CH_2CH_2—I}$$

$$CH_3CH_2CH_2CH_2—CH_2—\overset{O}{\overset{\|}{C}}—OR + I^- \quad (22.66)$$

In the past this idea could not be used in practice because ester enolates, once formed, undergo another, faster reaction: Claisen condensation with the parent ester. The direct alkylation shown in Eq. 22.66 is so attractive, however, that chemists have continued efforts to find conditions under which it will work.

It was discovered in the early 1970s that a family of very strong, hindered nitrogen bases, two of which are shown below, can be used to form stable enolate ions at −78° from esters.

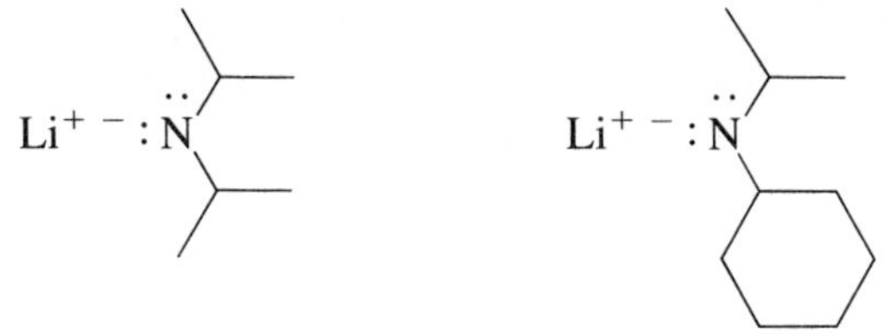

lithium diisopropylamide (LDA) lithium cyclohexylisopropylamide (LCHIA)

pK_a of conjugate acids: ≈35

(Do not confuse the term *amide* in the names of these bases with the carboxylic acid derivative. This term has a double usage. As used here, an *amide* is the conjugate-base anion of an amine.) Since esters have pK_a values near 25, these amide bases are strong enough to convert esters completely into their conjugate-base enolate ions. The ester enolate anions formed with these bases can be alkylated directly with alkyl halides. Notice that esters that are *trisubstituted* at the α-position can be prepared with this method. (These compounds cannot be prepared by the malonic ester synthesis; why?)

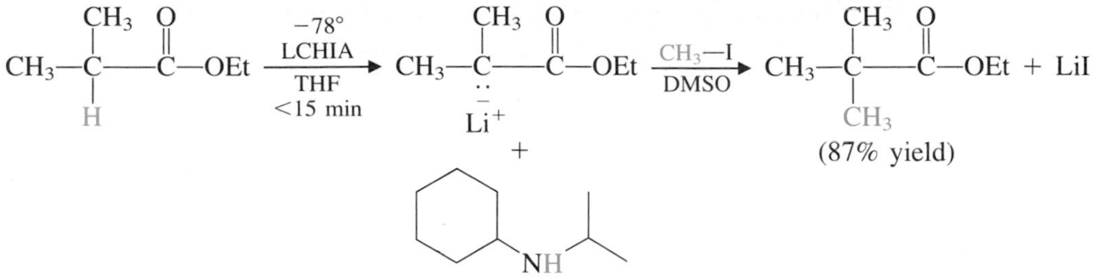

The nitrogen bases themselves are generated from the corresponding amines and butyllithium at $-78°$ in tetrahydrofuran (THF).

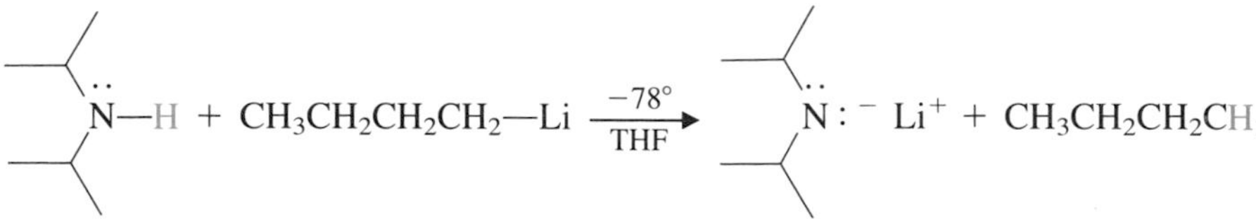

$$N—H + CH_3CH_2CH_2CH_2—Li \xrightarrow[THF]{-78°} N:^- Li^+ + CH_3CH_2CH_2CH_3 \quad (22.68)$$

This method of ester alkylation is considerably more expensive to use than the malonic ester synthesis. It also requires special inert-atmosphere techniques, because the strong bases used in the method react vigorously with both oxygen and water. For these reasons, the malonic ester synthesis remains very useful, particularly for large-scale syntheses. However, for the preparation of laboratory samples, or for preparation of compounds not available from the malonic ester synthesis, the preparation and alkylation of enolate ions with amide bases is particularly valuable.

The use of strong bases to form ester enolate ions looks simple enough. However, there are a number of potential problems with this method that evidently do not interfere. An analysis of these problems, and why they do not occur, reveals that this method, although simple, is also quite elegant in concept. First, we have already mentioned the Claisen condensation as a potential side reaction. This is avoided by *adding the ester to the amide base.* Thus, as a molecule of ester "hits" the solution, it has a choice: it can react with the very strong amide base, or it can react with an enolate ion in the Claisen condensation. The amide bases are so strong, and their reaction with α-protons is so rapid (even at $-78°$!) that the enolate ion is formed. No condensation is observed because the ester and its enolate are never present simultaneously—except for an instant—in the reaction flask.

Another potential side reaction is attack of the amide base (or even its conjugate acid amine, which is, after all, still a base) on the ester. We know that amines react with esters to give products of aminolysis (Sec. 21.8C), and it would not be unreasonable to expect the *conjugate bases* of amines—very strong bases indeed—to react even more rapidly with esters. That this does not happen is once again the result of a competition. When an amide base reacts with the ester, it has a choice: it can remove a proton, or it can attack the carbonyl group. The large branched groups on a hindered amide base retard its attack on the carbonyl group of an ester. For such a hindered amide to attack the carbonyl carbon is somewhat akin to our trying to put a dinner plate in the coin slot of a soft drink machine. (Such an effect of hindrance in a nucleophile has been aptly termed *F-strain,* or "front-strain.") If the amide base could be in contact with the ester long enough, it would eventually react at the carbonyl group; but the base instead reacts rapidly in a different way: it abstracts an α-proton. Reaction with a tiny proton is much less subject to steric hindrance than attack on a carbonyl group.

Hence, the amide base takes the path of least resistance: it forms the enolate ion. Thus, steric hindrance is used productively to avoid an undesired reaction.

Problems

30 Outline a synthesis of each of the following compounds from either diethyl malonate or ethyl acetate. Since the hindered amide bases are expensive, you may use them in only one reaction.

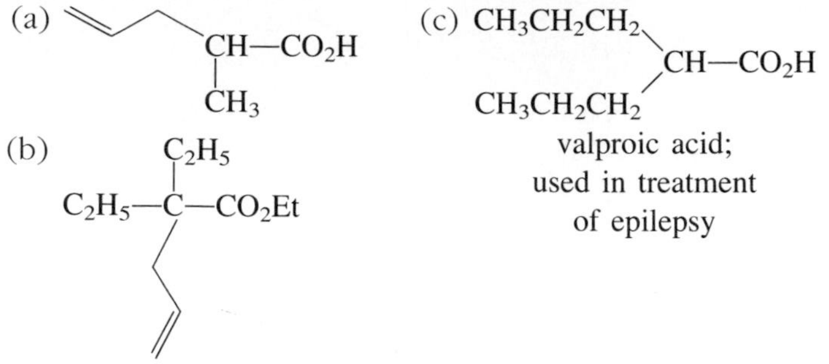

(c) valproic acid; used in treatment of epilepsy

31 The reactions of ester enolates are not restricted to alkylation. Complete the following reaction:

$$\textit{t}\text{-butyl acetate} \xrightarrow[-78°]{\text{LCHIA}} \xrightarrow[\text{(1 equiv)}]{\text{acetone}} \xrightarrow{\text{HCl (dil)}}$$

C. Acetoacetic Ester Synthesis

As we have learned (Eq. 22.51c), β-keto esters, like malonic esters, are substantially more acidic than ordinary esters, and are completely ionized by alkoxide bases.

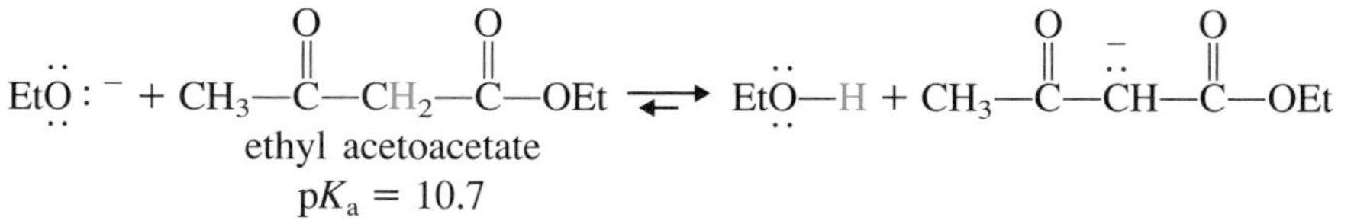

$$\ddot{\text{EtO}}:^- + CH_3-\overset{O}{\overset{\|}{C}}-CH_2-\overset{O}{\overset{\|}{C}}-OEt \rightleftharpoons \ddot{\text{EtO}}-H + CH_3-\overset{O}{\overset{\|}{C}}-\overset{-}{\overset{..}{C}}H-\overset{O}{\overset{\|}{C}}-OEt \quad (22.69)$$

ethyl acetoacetate
$pK_a = 10.7$

The resulting enolate ions, like those derived from malonate ester derivatives, can be alkylated by alkyl halides or sulfonate esters.

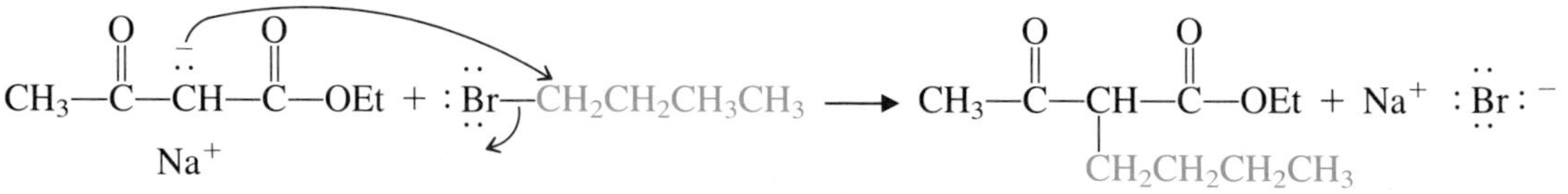

$$CH_3-\overset{O}{\overset{\|}{C}}-\overset{..}{\overset{-}{C}}H-\overset{O}{\overset{\|}{C}}-OEt + :\ddot{\text{Br}}-CH_2CH_2CH_3CH_3 \longrightarrow CH_3-\overset{O}{\overset{\|}{C}}-\overset{CH_2CH_2CH_2CH_3}{\overset{|}{C}H}-\overset{O}{\overset{\|}{C}}-OEt + Na^+ \ :\ddot{\text{Br}}:^- \quad (22.70)$$

Na^+

Dialkylation of β-keto esters is also possible.

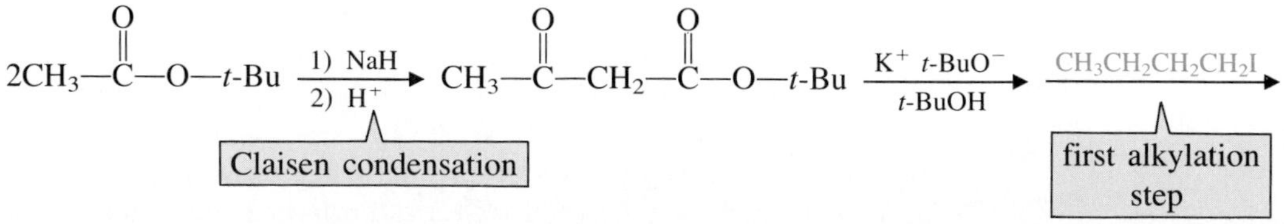

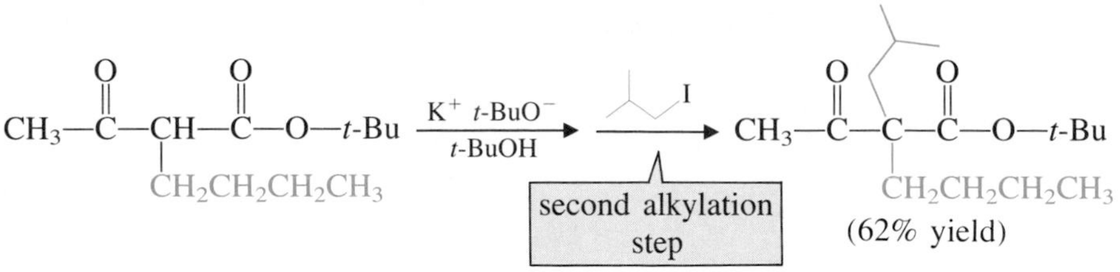

(22.71)

Alkylation of a Dieckmann condensation product is the same type of reaction:

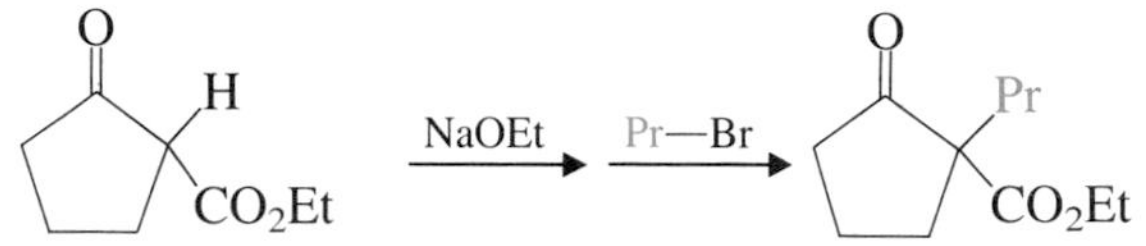

(from a Dieckmann condensation)

(85% yield) (22.72)

The preparation of alkylated β-keto esters is called the **acetoacetic ester synthesis.** Its value is that β-keto esters can be saponified or hydrolyzed, and the resulting β-keto acids can be easily decarboxylated (Sec. 20.11). Decarboxylation of a substituted β-keto acid yields a ketone.

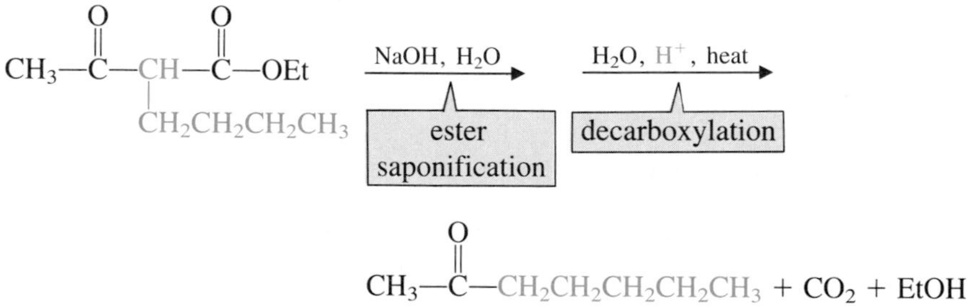

(22.73)

Thus, the acetoacetic ester synthesis can be extended to a preparation of ketones in the same sense that the malonic ester synthesis can be extended to a preparation of substituted acetic acids. The alkylation part of this sequence, like the alkylation of diethyl malonate, involves the construction of new carbon–carbon bonds.

If the desired β-keto ester is disubstituted at the α-position, alkaline hydrolysis is usually avoided since, as we learned in Eq. 22.52, alkaline conditions promote reversal of the Claisen or Dieckmann condensations of disubstituted β-keto esters.

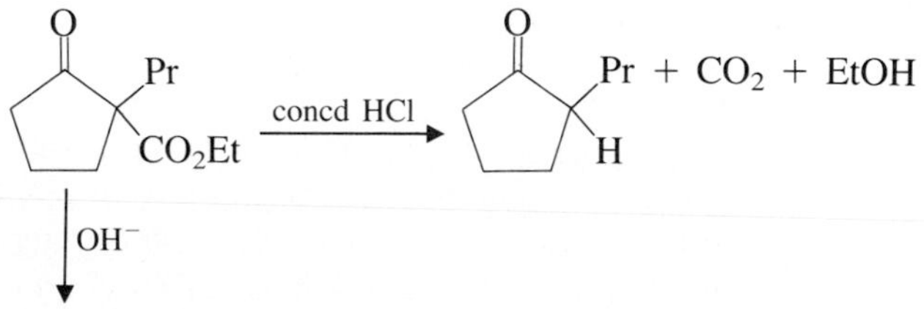

(22.74)

reverse Dieckmann

The use of the *t*-butyl ester in Eq. 22.75, for example, permits a particularly mild ester cleavage under acidic conditions (Sec. 21.7A).

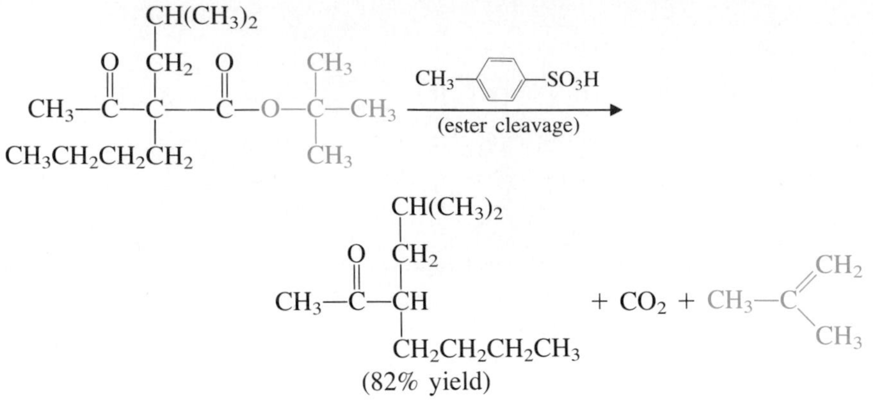

(22.75)

If we wish to prepare a target ketone, we can tell whether the acetoacetic ester synthesis can be used by our usual process of mentally reversing the synthesis.

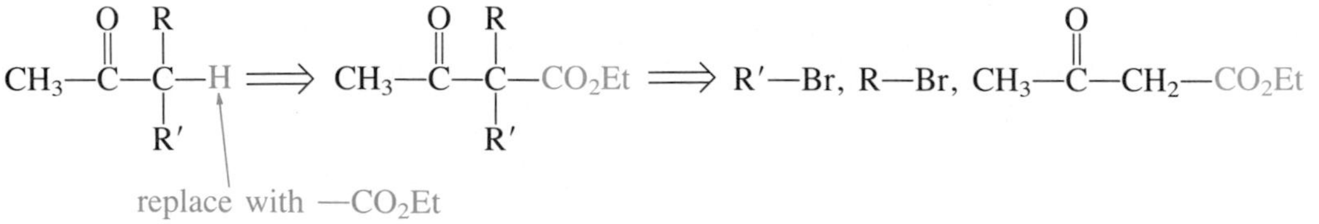

$$CH_3-\overset{O}{\underset{}{C}}-\overset{R}{\underset{R'}{C}}-H \implies CH_3-\overset{O}{\underset{}{C}}-\overset{R}{\underset{R'}{C}}-CO_2Et \implies R'-Br, R-Br, CH_3-\overset{O}{\underset{}{C}}-CH_2-CO_2Et \quad (22.76)$$

replace with —CO₂Et

As we can see from this example, this analysis involves replacing an α-hydrogen of our target ketone with a —CO₂Et group. This process unveils the β-keto ester required for the synthesis. The β-keto ester itself is then analyzed in terms of the alkyl halides required for the alkylation.

As we might imagine, ethyl acetoacetate itself is not the only β-keto ester that can be used in this type of synthesis. Once we have deduced a satisfactory β-keto ester starting material, we can then apply the analysis shown in Eq. 22.60 to determine whether the β-keto ester is one that can be made by the Claisen condensation.

Problems

32 Outline a synthesis of (a) 5-methyl-2-hexanone and (b) 4-phenyl-2-butanone from ethyl acetoacetate and any other reagents.

33 What β-keto ester would be required for the preparation of 4-methyl-3-pentanone? Could it be prepared by a Claisen condensation?

As we study the various alkylation and condensation reactions in this chapter, we should not allow the large number of reactions to obscure a very important central theme: *enolate ions are nucleophilic,* and they do many of the things that other nucleophiles do: addition to carbonyl groups, carbonyl substitution, attack on alkyl halides, and so on. Our study of enolate ions can only scratch the surface of their many reactions. Yet if we grasp this central idea—that enolate ions are nucleophiles—we can easily understand a new reaction when we see it, and can even predict the outcome of many reactions with confidence.

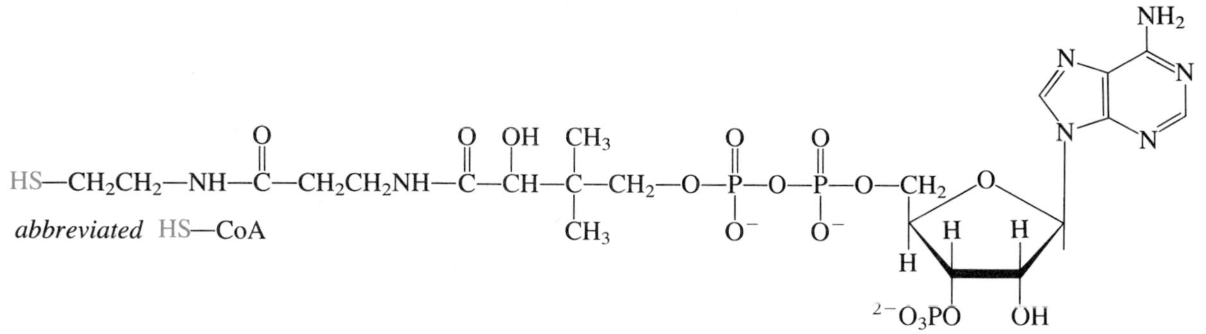

Figure 22.3 Structure of coenzyme A (CoA). Coenzyme A and other thiols are esterified to acetic acid and other straight-chain carboxylic acids in biological acyl-group transfer reactions.

Problem

34 Predict the outcome of the following reaction by identifying *A*, then *B*, then the final product:

$$\text{diethyl malonate} \xrightarrow{\text{NaOEt/EtOH}} A$$

$$A + \text{...} \xrightarrow{\text{EtOH}} [B] \longrightarrow (C_9H_{14}O_4)$$

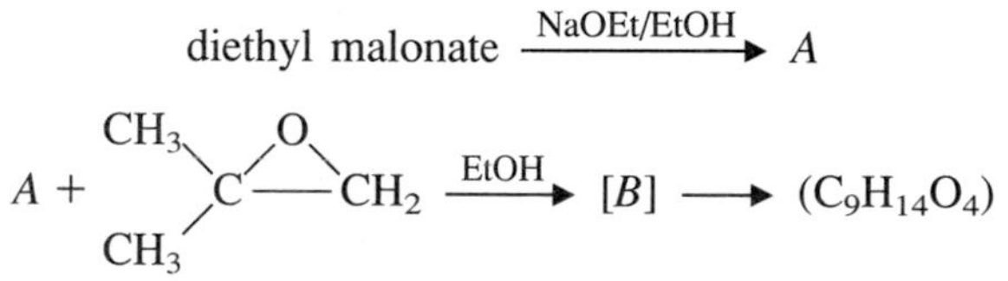

22.7 BIOSYNTHESIS OF COMPOUNDS DERIVED FROM ACETATE

The utility of the Claisen and aldol condensations is not confined to the laboratory: these reactions are important also in the biological world. Let us briefly survey some examples of these reactions in nature. The molecular starting point for the biosynthesis of a wide variety of biomolecules is a thiol ester of acetic acid called *acetyl-CoA*.

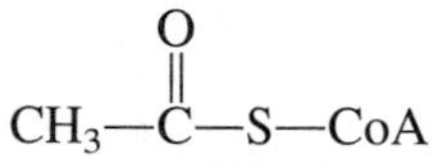

$$\underset{\text{acetyl-CoA}}{CH_3-\overset{\overset{\displaystyle O}{\|}}{C}-S-CoA}$$

The thiol CoA-SH, or **coenzyme A,** from which this ester is derived, is a complex-looking molecule; its complete structure is shown in Fig. 22.3. The complex functionality in this molecule is required for its proper binding to enzymes. However, in order to appreciate nature's chemistry with acetyl-CoA, we can ignore the complexity; all we need to understand is the fact that acetyl-CoA is an esterified thiol.

A. Biosynthesis of Fatty Acids

Acetyl-CoA is the basic building block for the biosynthesis of *fatty acids* (Sec. 20.5). Acetyl-CoA is first converted into a closely related derivative, malonyl-CoA:

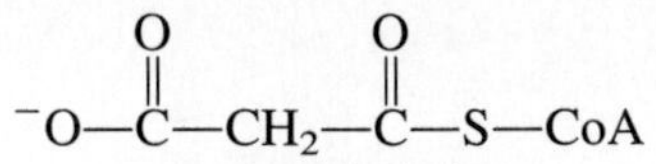

malonyl-CoA

The acetyl- and malonyl-CoA derivatives are transesterified to give different thiol esters, and, in a reaction closely resembling the Claisen condensation, these compounds react in an enzyme-catalyzed reaction to give an acetoacetyl thiol ester.

(22.77a)

acetoacetyl thiol ester

The nucleophilic electron pair (color in Eq. 22.77a) is made available not by proton abstraction, but by loss of CO_2 from malonyl-CoA. The loss of CO_2 also serves another role: to drive the Claisen condensation to completion. Recall that in the laboratory, a Claisen condensation is driven to completion by ionization of the product with a strong base like ethoxide. Such a strong base cannot be used in nature, where all reactions must occur near neutral pH.

The product of Eq. 22.77a, an acetoacetyl thiol ester, then undergoes successively a carbonyl reduction, a dehydration, and a double-bond reduction, each catalyzed by an enzyme.

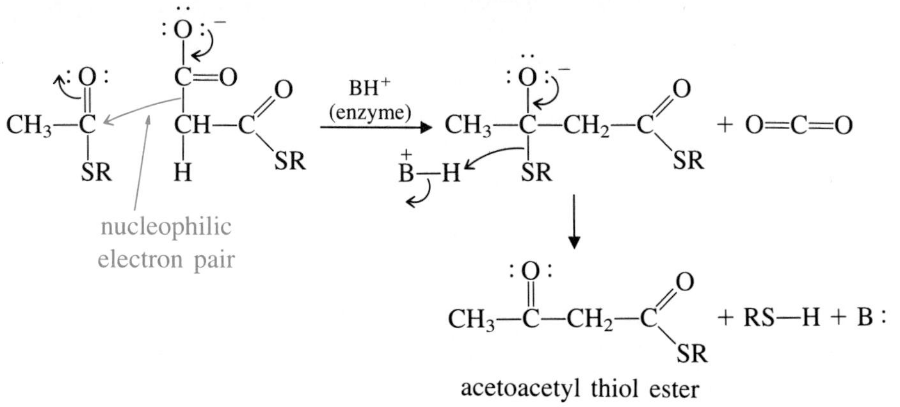

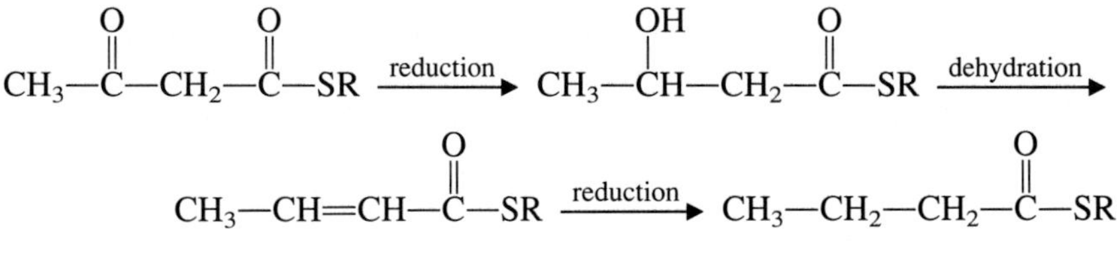

$$CH_3-\overset{O}{\overset{\|}{C}}-CH_2-\overset{O}{\overset{\|}{C}}-SR \xrightarrow{\text{reduction}} CH_3-\overset{OH}{\overset{|}{C}H}-CH_2-\overset{O}{\overset{\|}{C}}-SR \xrightarrow{\text{dehydration}}$$

$$CH_3-CH=CH-\overset{O}{\overset{\|}{C}}-SR \xrightarrow{\text{reduction}} CH_3-CH_2-CH_2-\overset{O}{\overset{\|}{C}}-SR \quad (22.77b)$$

The net result of these three reactions is that the acetyl thiol ester is converted into a thiol ester with *two additional carbons*. This sequence of reactions is then repeated, thus adding yet another two carbons to the chain.

$$CH_3CH_2CH_2\overset{O}{\overset{\|}{C}}-SR + {}^-O-\overset{O}{\overset{\|}{C}}-CH_2-\overset{O}{\overset{\|}{C}}-SR \xrightarrow{H^+}$$

$$CH_3CH_2CH_2\overset{O}{\overset{\|}{C}}-CH_2-\overset{O}{\overset{\|}{C}}-SR + R-SH + CO_2 \quad (22.77c)$$

$$CH_3CH_2CH_2\overset{O}{\overset{\|}{C}}-CH_2-\overset{O}{\overset{\|}{C}}-SR \xrightarrow[\text{of Eq. 22.77b}]{\text{three reactions}} CH_3CH_2CH_2CH_2CH_2-\overset{O}{\overset{\|}{C}}-SR \quad (22.77d)$$

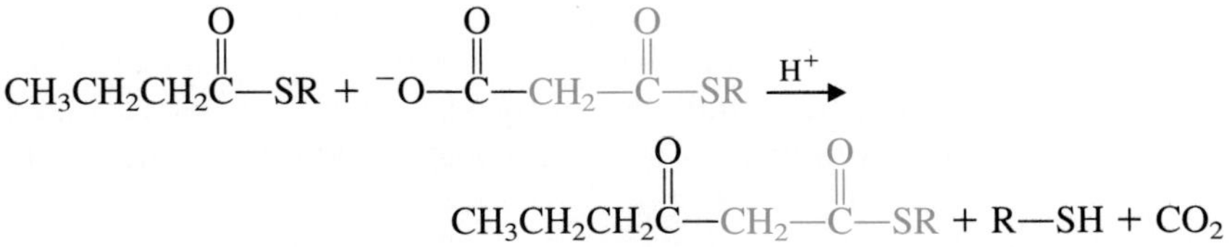

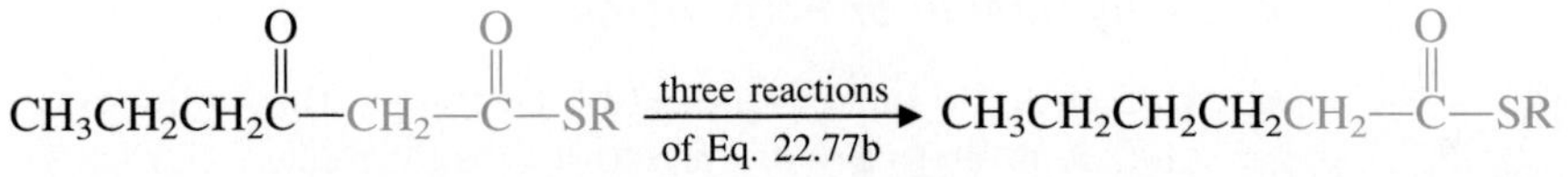

These four reactions are repeated with the addition of two carbons to the carbon chain at each cycle until a fatty acid with the proper chain length is obtained. The fatty acid thiol ester is then transesterified to glycerol to form fats and phospholipids (Sec. 21.12B).

The biosynthetic mechanism outlined above shows clearly why the common fatty acids have an *even number of carbon atoms:* they are formed from the successive addition of two-carbon acetate units. Fatty acids with an odd number of carbon atoms, although not unknown, are relatively rare.

B. Biosynthesis of Aromatic Compounds: Polyketides

One of the two major pathways for the biosynthesis of aromatic compounds involves ester condensations. In this pathway, acetoacetyl thiol ester does not undergo subsequent reductions and dehydration, as it does in fatty-acid biosynthesis. Instead, it undergoes successive condensations with more malonyl thiol ester molecules to yield **polyketides:** keto esters with carbonyl groups at alternate carbons.

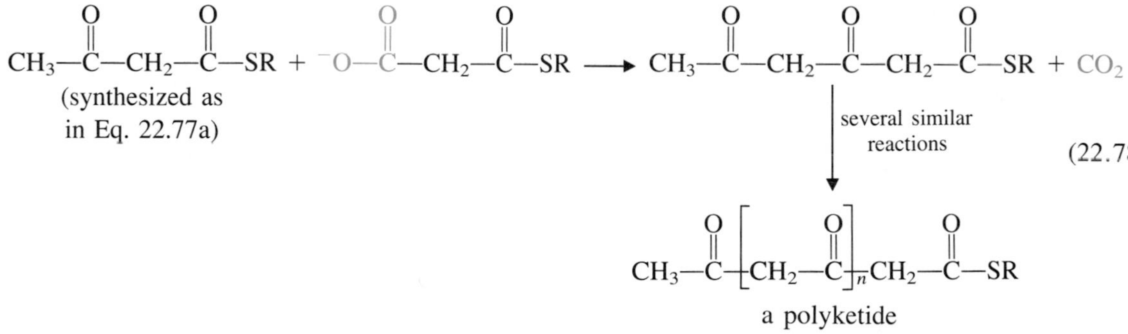

(22.78a)

The polyketide then cyclizes. The cyclization in the following example is essentially a Dieckmann condensation. The enol form of the product triketone is an aromatic compound.

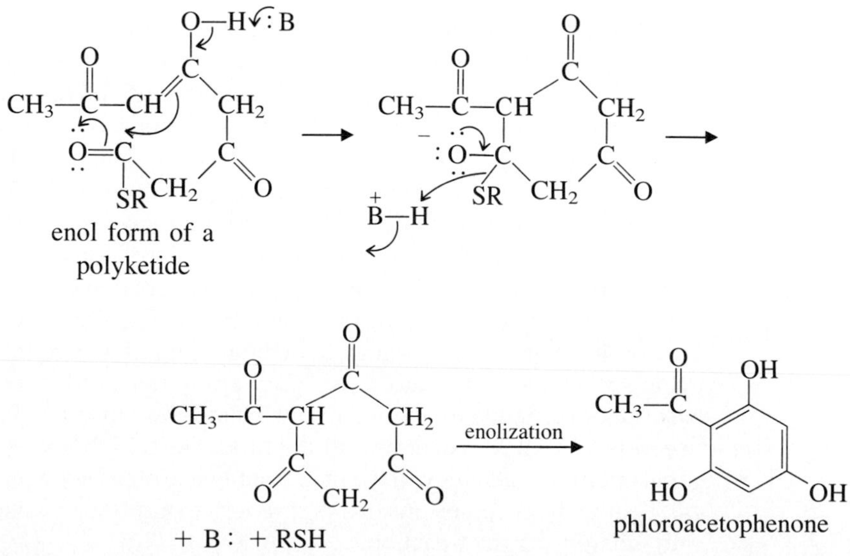

(22.78b)

Other modes of cyclization are also known. For example, the biosynthesis of orsellinic acid (Problem 35) can be rationalized as an intramolecular aldol condensation.

Problems

35 Show how cyclization of a polyketide can account for the biosynthesis of orsellinic acid.

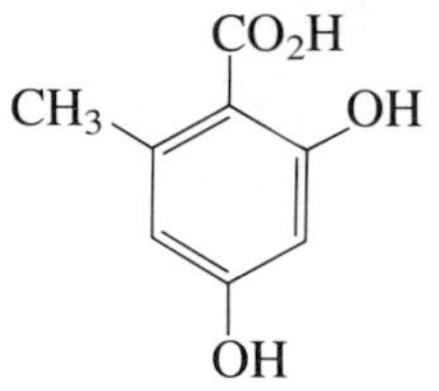

orsellinic acid

---

36 Hydroxymethylglutaryl-CoA (HMG-CoA) is the biosynthetic precursor of isopentenyl pyrophosphate (Sec. 17.6B), which, in turn, is the precursor for isoprenoids and steroids. Show how HMG-CoA can be built up entirely from acetyl-CoA.

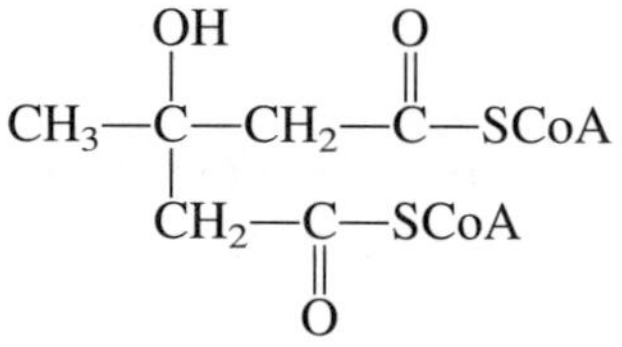

hydroxymethylglutaryl-CoA
(HMG-CoA)

If you study biochemistry, you will see many of the reactions in this section again. The purpose of this section is to illustrate two points about the organic syntheses carried out in nature. First is the *economy of starting materials.* Fatty acids, isoprenoids, steroids, and many aromatic compounds (as well as many compounds we have not considered) are all derived from a common building block: acetyl-CoA. Second, most of the reactions involved are transformations for which we have *close laboratory analogy*. With benefit of hindsight, it may seem obvious to us that natural chemistry and laboratory chemistry should be closely related. However, this point was far from obvious to early chemists, many of whom considered the chemistry of living things to be beyond the scope of laboratory investigation (Sec. 1.1B). The synthesis of urea by Wöhler (Eq. 1.1) signaled the beginning of an age in which the chemistry of living systems and laboratory chemistry are understood to be branches of the same basic science. Students who can understand the close mechanistic relationship between biochemical reactions and the fundamental laboratory transformations of organic chemistry generally have little trouble learning biochemical pathways, because each pathway becomes, instead of a series of unrelated conversions, a series of logically related organic reactions.

22.8 CONJUGATE-ADDITION REACTIONS

We now turn to our last topic in carbonyl chemistry: the reactions of α,β-unsaturated carbonyl compounds. Conjugation of the C=C and C=O bonds endows α,β-unsaturated carbonyl compounds with unique reactivity, which we shall consider in the remainder of this chapter.

A. Conjugate Addition to α,β-Unsaturated Carbonyl Compounds

The unusual reactivity of α,β-unsaturated carbonyl compounds is illustrated by the reaction of an α,β-unsaturated ketone with HCN.

$$\text{Ph—CH=CH—}\overset{\overset{\text{O}}{\|}}{\text{C}}\text{—Ph} + \text{NaCN} \xrightarrow[\text{EtOH}]{\text{35°, HOAc}} \text{Ph—}\underset{\underset{\text{CN}}{|}}{\text{CH}}\text{—CH}_2\text{—}\overset{\overset{\text{O}}{\|}}{\text{C}}\text{—Ph} \quad (22.79)$$

(93–96% yield)

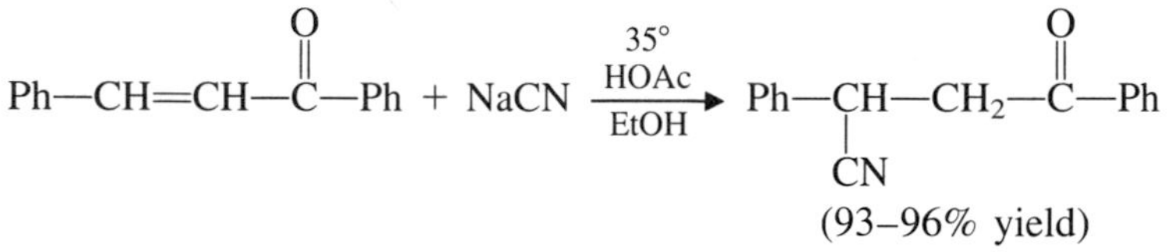

In this reaction, the elements of HCN appear to have added across the C=C bond. Yet this is not a reaction of ordinary double bonds:

$$CH_3CH{=}CH_2 + NaCN \xrightarrow{H_2O} \text{no reaction} \quad (22.80)$$

Nucleophilic addition to the double bond in an α,β-unsaturated carbonyl compound occurs because it gives a resonance-stabilized enolate ion intermediate:

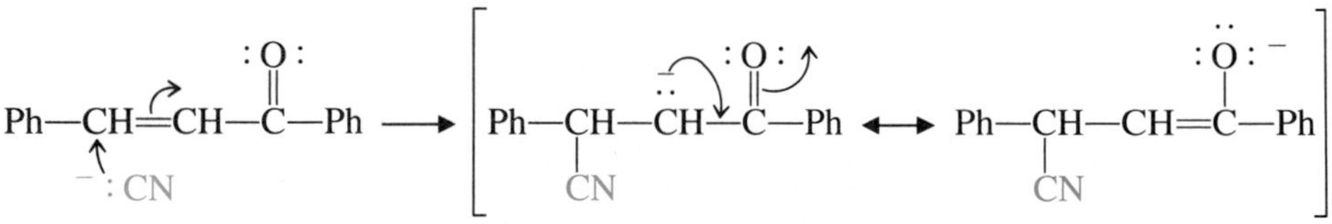

(22.81a)

(Nucleophilic addition to the alkene in Eq. 22.80, in contrast, would give a very unstable alkyl anion.) The enolate ion can protonate on either oxygen or carbon. In either case a carbonyl group is eventually regenerated. The *overall* result of the reaction is net addition to the double bond.

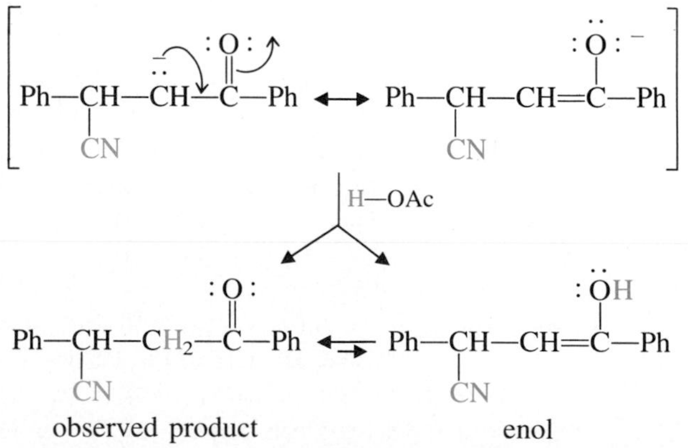

(22.81b)

Nucleophilic addition to the carbon–carbon double bonds of α,β-unsaturated aldehydes, ketones, esters, and nitriles is a rather general reaction that can be observed with a variety of nucleophiles. Some additional examples follow; try to write the mechanisms of these reactions.

$$CH_3—CH{=}CH—CO_2Et + NaCN \xrightarrow[H_2O]{EtOH} \xrightarrow{H^+}$$

$$\underset{\underset{CN}{|}}{CH_3—CH—CH_2—CO_2Et} \xrightarrow[heat]{H_2O, H^+} \underset{\underset{CO_2H}{|}}{CH_3—CH—CH_2—CO_2H} \quad (22.82)$$
$$\text{(66–70\% yield)}$$

$$\text{>}{-}SH + CH_2{=}CH—CO_2Me \xrightarrow[MeOH]{NaOMe} \text{>}{-}S—CH_2—CH_2—CO_2Me \quad (22.83)$$
$$\text{(97\% yield)}$$

$$\underset{\underset{O}{\overset{\|}{}}}{CH_3—C—CH{=}CH—Ph} + HN\text{<} \longrightarrow CH_3—\overset{O}{\overset{\|}{C}}—CH_2—\overset{N\text{<}}{CH}—Ph \quad (22.84)$$
$$\text{(85\% yield)}$$

$$CH_3SH + CH_2{=}CH—CN \xrightarrow[MeOH]{NaOMe} CH_3S—CH_2—CH_2—CN \quad (22.85)$$
$$\text{acrylonitrile} \qquad\qquad \text{(91\% yield)}$$

Notice that the addition of cyanide in Eq. 22.82 forms a new carbon–carbon bond, and that the nitrile group can then be converted into a carboxylic acid group by acid hydrolysis. The addition of a nucleophile to acrylonitrile (as in Eq. 22.85) is a useful reaction called **cyanoethylation.**

Quinones (Sec. 18.7A) are α,β-unsaturated carbonyl compounds, and also undergo similar addition reactions.

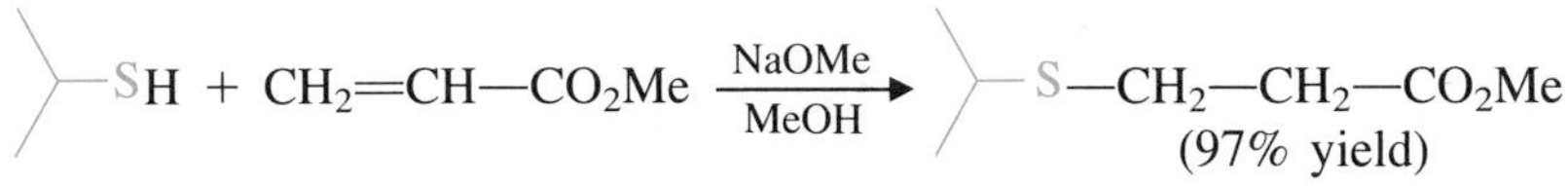

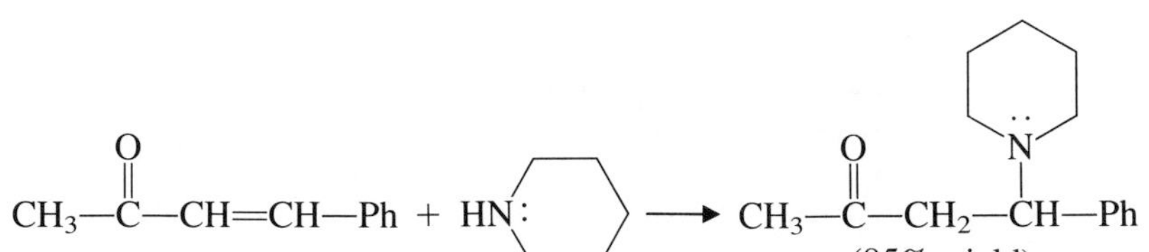

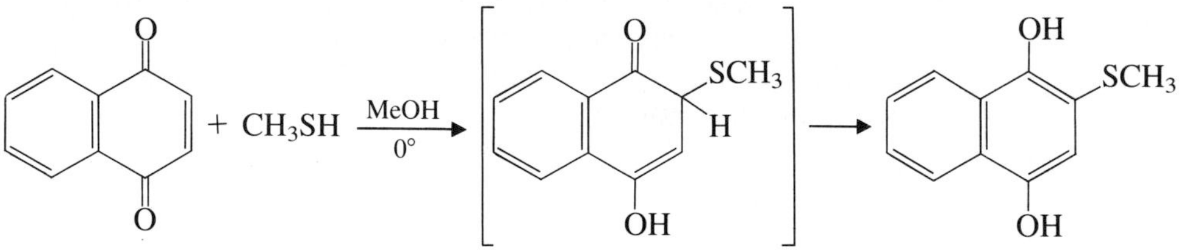

$$(22.86)$$

The examples above occur in base, but acid-catalyzed additions to the carbon–carbon double bonds of α,β-unsaturated carbonyl compounds are also known.

$$CH_2{=}CH—CO_2Me + HBr \xrightarrow{Et_2O} Br—CH_2CH_2—CO_2Me \quad \text{(80–84\% yield)} \quad (22.87)$$

$$CH_2{=}CH—CH{=}O + HCl \xrightarrow{-15°} Cl—CH_2CH_2—CH{=}O \quad (22.88)$$

Although such reactions appear to be nothing more than simple additions to the carbon–carbon double bond, this is not the case. The more basic position of an

α,β-unsaturated carbonyl compound is not the double bond, but the carbonyl oxygen (why?). Protonation on the carbonyl oxygen yields a cation in which some of the positive charge is present on the β-carbon.

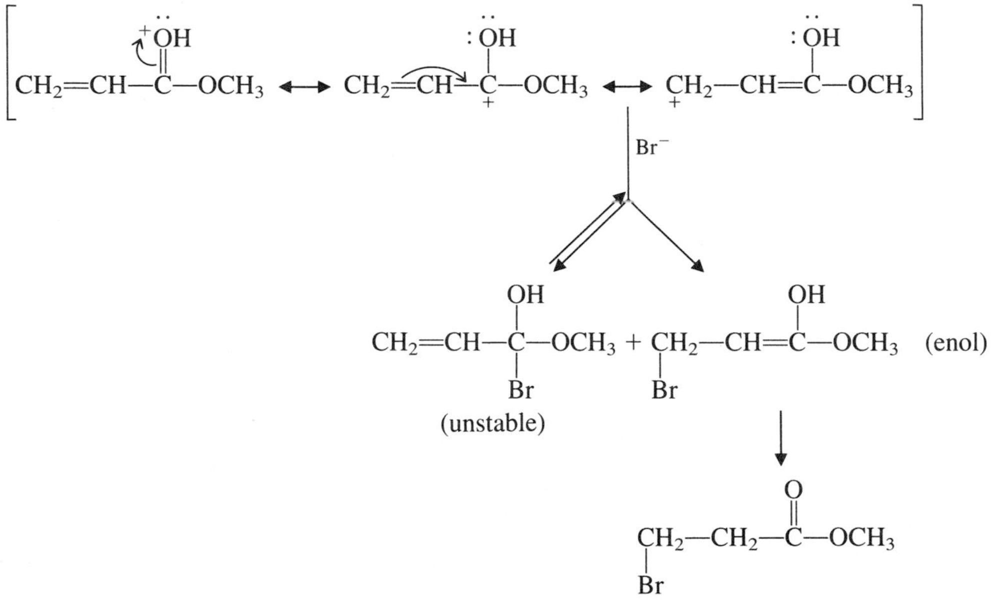

(22.89)

Attack of Br^- can occur at either the carbonyl carbon or the β-carbon. Attack at the carbonyl carbon yields an unstable compound; attack at the β-carbon yields an enol, which rapidly reverts to the observed carbonyl product.

An addition to the double bond of an α,β-unsaturated carbonyl compound is a *conjugate addition.* We have seen other examples of conjugate addition; for instance, the addition of HBr to butadiene (Sec. 15.4) occurs by essentially the same mechanism as the addition of HBr shown in Eq. 22.89. The Diels–Alder reaction (Sec. 15.3) is another example of a conjugate addition. However, the *nucleophilic conjugate addition,* such as the addition of cyanide in Eq. 22.79, has no parallel in the reactions of simple conjugated dienes such as butadiene. Such nucleophilic additions are sometimes called **Michael additions,** because of their resemblance to the Michael reaction, which we shall discuss in Sec. 22.8C.

Conjugate additions are sometimes called *1,4-additions* (Secs. 15.3A and 15.4A). This terminology may be somewhat confusing when used with α,β-unsaturated carbonyl compounds. If we consider the *initial* product in the addition of HBr to the ester in Eq. 22.87, we see that the H and the Br add in a 1,4-relationship, just as they do in the 1,4-addition to butadiene (Eq. 15.19).

1,4-addition to an α,β-unsaturated ester:

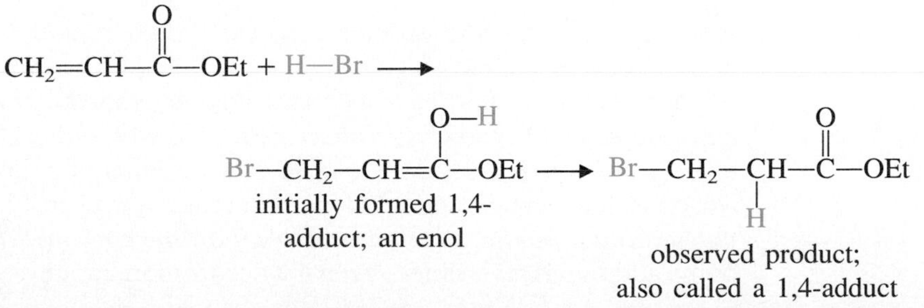

(22.90a)

1,4-addition to butadiene:

$$CH_2{=}CH{-}CH{=}CH_2 + HBr \longrightarrow \underset{\text{1,4-addition product}}{Br{-}CH_2{-}CH{=}CH{-}CH_2{-}H} + \underset{\text{1,2-addition product}}{CH_2{=}CH{-}CH(Br){-}CH_2{-}H} \quad (22.90b)$$

The product of the addition in Eq. 22.90a is an enol, which reverts to the corresponding carbonyl compound, in which the 1,4-relationship of the H and the Br no longer exists. The observed product is nevertheless called a 1,4-addition product because it is formed spontaneously from the actual 1,4-addition product. (The 1,4-relationship specifies the number of atoms between the H and the Br after the initial addition; it has nothing to do with the numbering of the carbonyl compound for nomenclature purposes.)

B. Conjugate Addition vs. Carbonyl-Group Reactions

Conjugate addition competes with familiar carbonyl-group reactions. In the case of aldehydes and ketones, conjugate addition competes with *addition to the carbonyl group*. (Nuc = nucleophile; for example, in cyanide addition, H—Nuc = H—CN.)

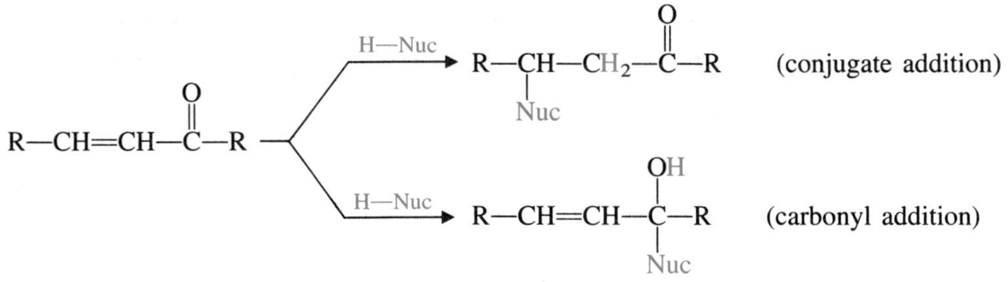

(22.91)

In the case of esters, conjugate addition competes with *substitution at the carbonyl group*.

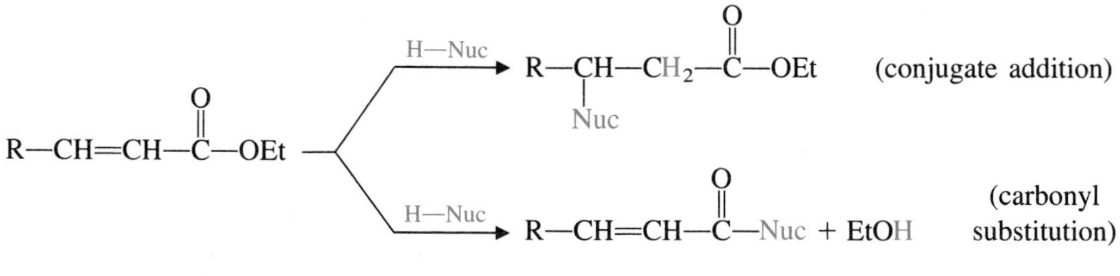

(22.92)

When can we expect to observe conjugate addition, and when can we expect carbonyl-group reactions?

Let us consider first the reactions of aldehydes and ketones. Relatively weak bases that give *reversible* carbonyl-addition reactions with ordinary aldehydes and ketones tend to give conjugate addition with α,β-unsaturated aldehydes and ketones. The reason is that conjugate addition, in most cases, is irreversible. In other words, *conjugate-addition products are more stable than the carbonyl-addition products.* If carbonyl addition is reversible—even if it occurs more rapidly—then conjugate addition drains the carbonyl compound from the addition equilibrium, and the conjugate-addition product is formed ultimately.

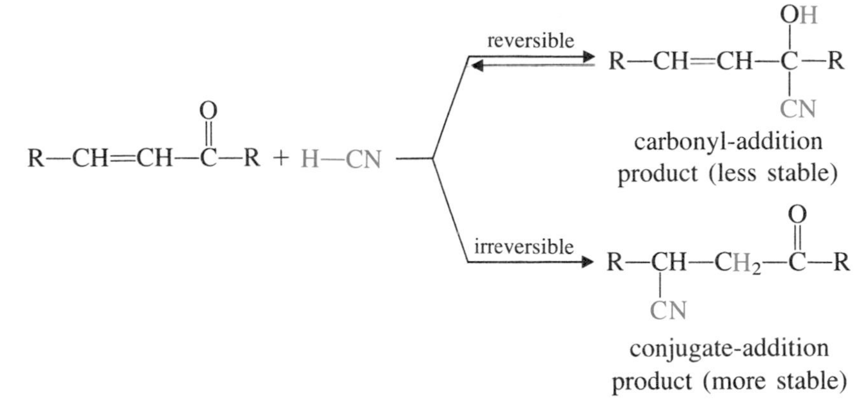

$$R-CH=CH-\overset{\overset{\displaystyle O}{\|}}{C}-R + H-CN \quad\begin{cases} \xrightarrow{\text{reversible}} R-CH=CH-\overset{\overset{\displaystyle OH}{|}}{\underset{\underset{\displaystyle CN}{|}}{C}}-R \\ \text{carbonyl-addition} \\ \text{product (less stable)} \\ \\ \xrightarrow{\text{irreversible}} R-\overset{\underset{\underset{\displaystyle CN}{|}}{|}}{CH}-CH_2-\overset{\overset{\displaystyle O}{\|}}{C}-R \\ \text{conjugate-addition} \\ \text{product (more stable)} \end{cases} \quad (22.93)$$

This, then, is another case of *kinetic vs. thermodynamic control of a reaction* (Sec. 15.4B). The conjugate-addition product is the thermodynamic product of the reaction.

Why is the conjugate-addition product more stable? The answer lies in a simple bond energy argument. Conjugate addition retains a carbonyl group at the expense of a carbon–carbon double bond. Carbonyl addition retains a carbon–carbon double bond at the expense of a carbonyl group. Since a C=O bond is considerably stronger than a C=C bond (Table 5.2), conjugate addition gives a more stable product. (Of course, other bonds are broken and formed as well, but the major effect is the relative strengths of the two kinds of double bonds.) We can see these factors reflected in the relative heats of formation of the isomers allyl alcohol and propionaldehyde:

$$\begin{array}{ccc} & CH_2=CH-CH_2-OH & CH_3-CH_2-CH=O \\ \Delta H_f^0 & -31.55 \text{ kcal/mol} & -45.90 \text{ kcal/mol} \end{array}$$

The reasoning used here suggests that if a carbonyl addition reaction were *not* reversible, and if it were faster than conjugate addition, it would be the observed mode of addition. As we shall see in Secs. 22.9 and 22.10A, this is precisely what happens in the addition of LiAlH$_4$ or organolithium reagents to α,β-unsaturated carbonyl compounds.

In the case of esters, we also find that relatively weak bases, such as $^-$CN, thiolate anions ($^-$SR), and amines (RNH$_2$ and R$_2$NH) give conjugate-addition products. In contrast, stronger bases that react irreversibly at the carbonyl group give carbonyl substitution products. Thus, hydroxide ion reacts with an α,β-unsaturated ester to give products of saponification, a carbonyl substitution reaction, because saponification is not reversible. Likewise, LiAlH$_4$ reduces α,β-unsaturated esters at the carbonyl group, because attack of hydride ion on the carbonyl group is irreversible.

In summary: We usually observe conjugate addition with nucleophiles that are relatively weak bases. Stronger bases give irreversible carbonyl addition or carbonyl substitution reactions.

Problems

37 Give the product expected when methyl methacrylate (methyl 2-methylpropenoate) reacts with each of the following reagents:
(a) $^-$CN in acidic EtOH (c) HBr
(b) C$_2$H$_5$SH and NaOEt (1 equiv) in EtOH (d) NaOH

38 Give a mechanism for each of the following reactions:

(a)

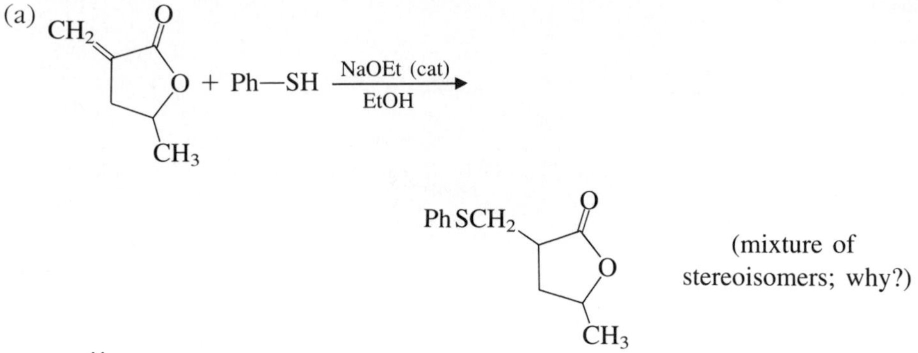

NaOEt (cat)
EtOH

(mixture of
stereoisomers; why?)

(b) $Me\ddot{N}H_2 + 2CH_2{=}CH{-}CO_2Me \longrightarrow$

$$MeO_2C{-}CH_2CH_2{-}\overset{..}{\underset{Me}{N}}{-}CH_2CH_2{-}CO_2Me$$

(c) EtO₂C, CO₂Et C=C H, H

$\xrightarrow[\text{(longrightleftharpoons)}]{\text{Et}_2\text{NH (catalytic amount)}}$

EtO₂C, H C=C H, CO₂Et

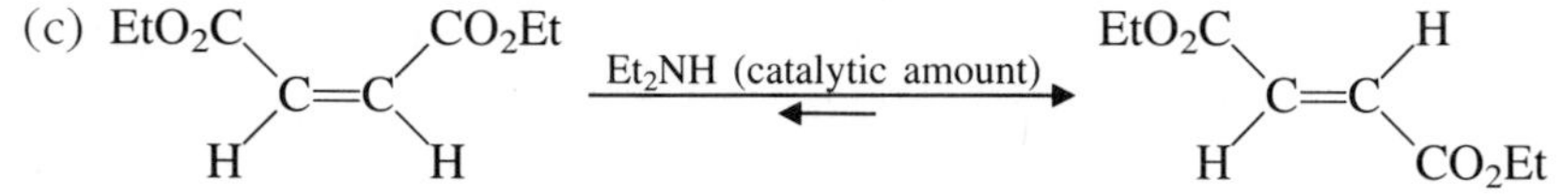

C. Conjugate Addition of Enolate Ions: Michael Addition

In Secs. 22.3–22.6 of this chapter we studied several reactions in which enolate ions act as nucleophiles. It should therefore be no surprise to learn that enolate ions, like other nucleophiles, undergo conjugate-addition reactions with α,β-unsaturated carbonyl compounds, as in the following example:

$$CH_3{-}\overset{O}{\overset{\|}{C}}{-}CH{=}CH_2 + CH_2(CO_2Et)_2 \xrightarrow[\text{EtOH}]{\text{NaOEt (cat)}} CH_3{-}\overset{O}{\overset{\|}{C}}{-}CH_2CH_2CH(CO_2Et)_2 \quad (22.94)$$

$$(65\text{–}71\% \text{ yield})$$

The mechanism of this reaction follows exactly the same pattern established for other nucleophilic conjugate additions; the nucleophile is the enolate ion formed in the reaction of ethoxide with diethyl malonate (Eq. 22.62a).

enolate ion from
diethyl malonate
(see Eq. 22.62a)

$CH_2(CO_2Et)_2$

$$EtO_2C \diagdown CH{-}CH_2{-}CH_2{-}\overset{O}{\overset{\|}{C}}{-}CH_3 + {}^-{:}CH(CO_2Et)_2 \quad (22.95)$$
EtO₂C

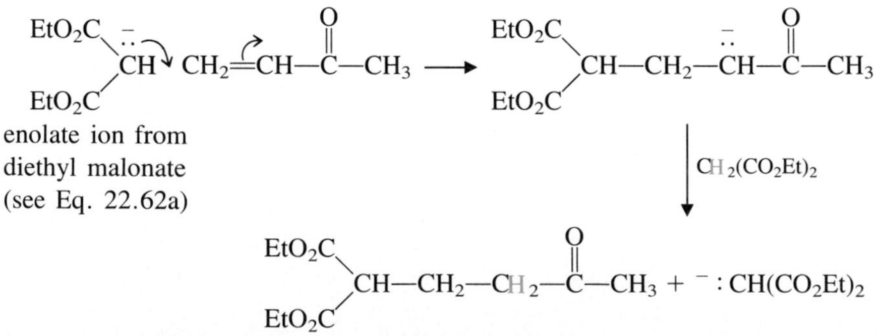

In contrast to the Claisen ester condensation (Sec. 22.5A), this reaction requires only a catalytic amount of base. The reaction does *not* rely on ionization of the product to drive it to completion. It goes to completion because a carbon–carbon π-bond in the starting α,β-unsaturated carbonyl compound is replaced by a stronger carbon–carbon σ-bond.

Conjugate additions of enolate ions to α,β-unsaturated carbonyl compounds are called **Michael additions,** after Arthur Michael (1853–1942), a Harvard professor who investigated these reactions extensively. (As noted in Sec. 22.8A, some chemists refer to all nucleophilic conjugate additions as Michael additions.)

Let us analyze the Michael addition as if we were planning to use it in a synthesis. The product in many Michael additions could originate from either of two pairs of reactants. For example, in the reaction shown in Eq. 22.94, the same product (in principle) might be obtained by the Michael addition of either of the following pairs of reactants (convince yourself of this point):

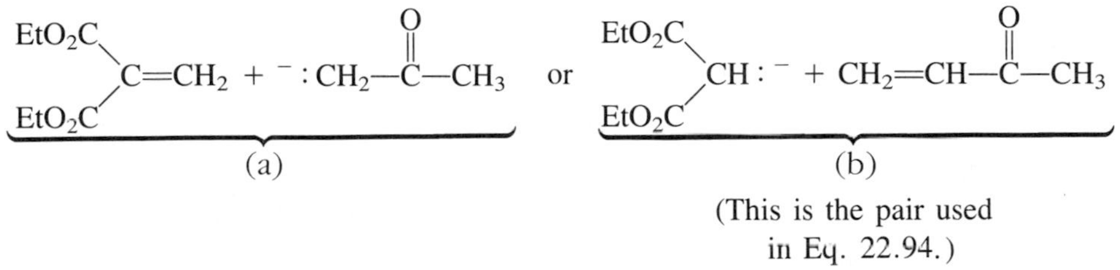

Which pair of reactants should be used? To answer this question, we use the result in Sec. 22.8A: Weaker bases tend to give conjugate addition, and stronger bases tend to give carbonyl-group reactions. Hence, to maximize conjugate addition, *we choose the pair of reactants with the less basic enolate ion*—pair (b) in the case above.

In one useful variation of the Michael addition, the immediate product of the reaction can be subjected to an aldol condensation that closes a ring.

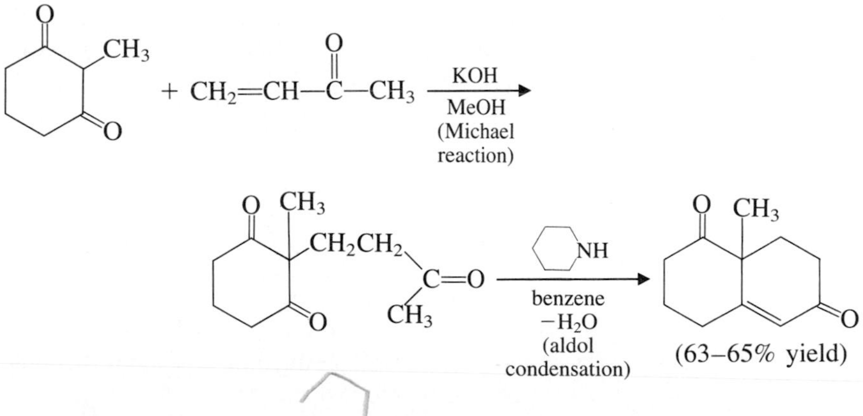

(22.96)

(Write out the detailed mechanisms of these reactions.) This sequence is an example of a **Robinson annulation,** named for Sir Robert Robinson, a British chemist who pioneered its use. (An annulation is a ring-forming reaction, from Latin *annulus,* ring.)

Problems

39 Complete the following reactions:

(a) $CH_2(CO_2Et)_2 + CH_2{=}C(CH_3){-}CO_2Et \xrightarrow[EtOH]{NaOEt} \xrightarrow[heat]{H_3O^+}$

(b) $CH_3{-}\overset{O}{\overset{\|}{C}}{-}CH_2{-}\overset{O}{\overset{\|}{C}}{-}OEt + CH_2{=}CH{-}CN \text{ (excess)} \xrightarrow[EtOH]{NaOEt} (C_{12}H_{16}O_3N_2)$

40 Give a rational mechanism for each of the following reactions. In each reaction identify the bracketed intermediate.

(a)

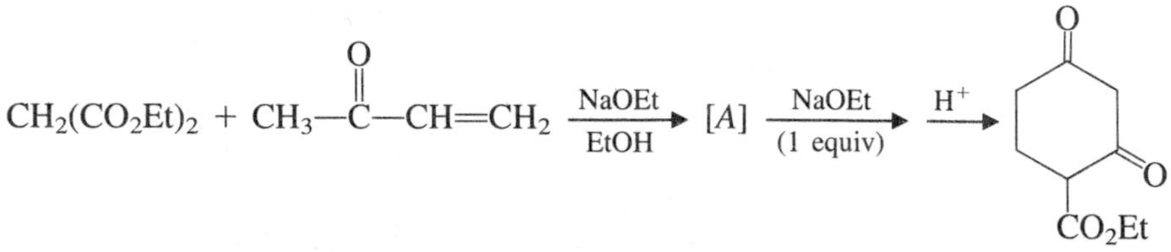

(b)

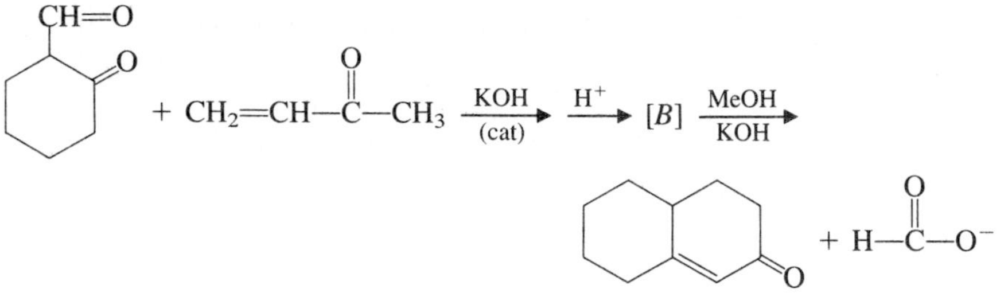

22.9 REDUCTION OF α,β-UNSATURATED CARBONYL COMPOUNDS

The carbonyl group of an α,β-unsaturated aldehyde or ketone, like that of an ordinary aldehyde or ketone (Sec. 19.8), is reduced to an alcohol with lithium aluminum hydride.

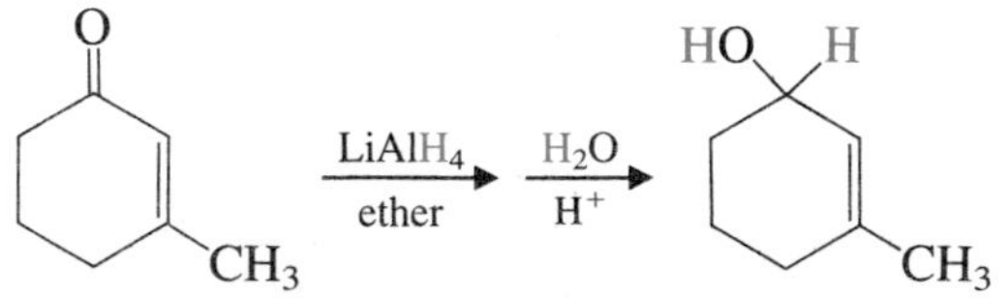

(98% yield) (22.97)

This reaction, like other $LiAlH_4$ reductions, involves the attack of hydride at the carbonyl carbon, and is therefore a carbonyl addition.

Why is carbonyl addition, rather than conjugate addition, observed in this case? The answer is that carbonyl addition is not only *faster* than conjugate addition but, in this case, also *irreversible*. It is irreversible because hydride is a very poor leaving group. Since carbonyl addition of $LiAlH_4$ is irreversible, conjugate addition never has a chance to occur, and is therefore not observed.

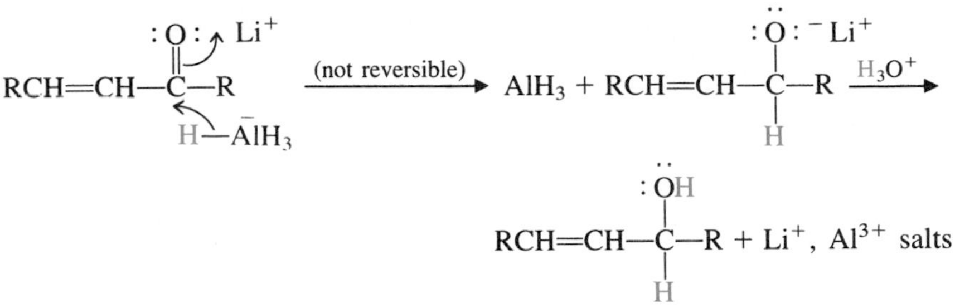

In other words, reduction of the carbonyl group with LiAlH$_4$ is a *kinetically controlled* reaction.

Many α,β-unsaturated carbonyl compounds are reduced by NaBH$_4$ to give mixtures of both carbonyl-addition products and conjugate-addition products. Because mixtures are obtained, NaBH$_4$ reductions of α,β-unsaturated ketones are not useful. Why conjugate addition is observed with NaBH$_4$ is not well understood. Although some cases of conjugate addition with LiAlH$_4$ are known, this reagent usually reduces carbonyl groups without affecting double bonds.

The carbon–carbon double bond of an α,β-unsaturated carbonyl compound can in most cases be reduced selectively by catalytic hydrogenation. (See also Eq. 19.32.)

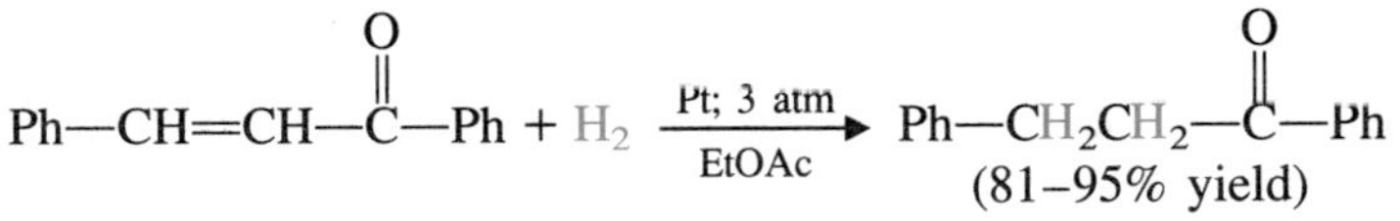

$$\text{Ph}-\text{CH}=\text{CH}-\overset{\text{O}}{\overset{\|}{\text{C}}}-\text{Ph} + \text{H}_2 \xrightarrow[\text{EtOAc}]{\text{Pt; 3 atm}} \text{Ph}-\text{CH}_2\text{CH}_2-\overset{\text{O}}{\overset{\|}{\text{C}}}-\text{Ph} \quad \text{(22.99)}$$
$$\underset{\text{(81–95\% yield)}}{}$$

Problem

41 Show how ethyl 2-butenoate can be used as a starting material to prepare (a) ethyl butanoate; (b) 2-buten-1-ol.

22.10 REACTIONS OF α,β-UNSATURATED CARBONYL COMPOUNDS WITH ORGANOMETALLIC REAGENTS

A. Addition of Organolithium Reagents to the Carbonyl Group

Organolithium reagents react with α,β-unsaturated carbonyl compounds to yield products of carbonyl addition.

$$\text{(22.100)}$$

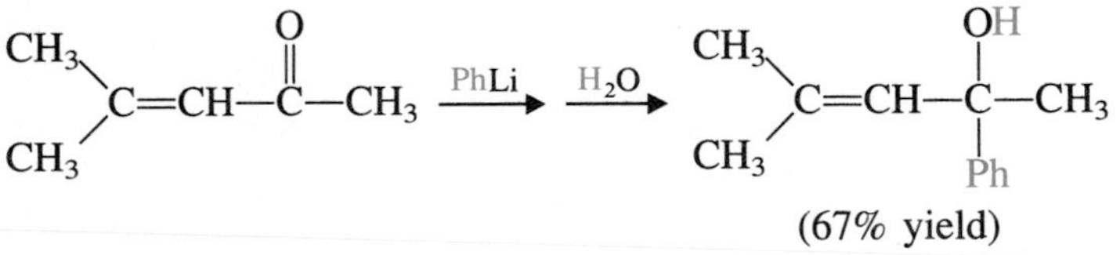

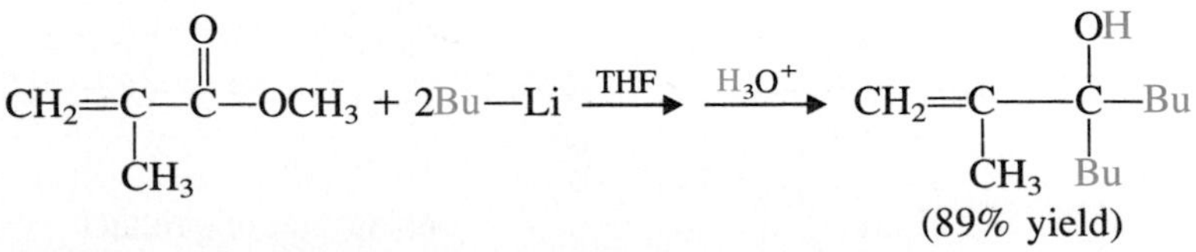

The reason for carbonyl addition, rather than conjugate addition, is the same as in the case of LiAlH$_4$ reduction (Sec. 22.9): since carbonyl addition is rapid and irreversible, the product is the result of kinetic control.

B. Conjugate Addition of Lithium Dialkylcuprate Reagents

Lithium dialkylcuprate reagents (Sec. 21.10B) give exclusively products of conjugate addition when they react with α,β-unsaturated esters and ketones.

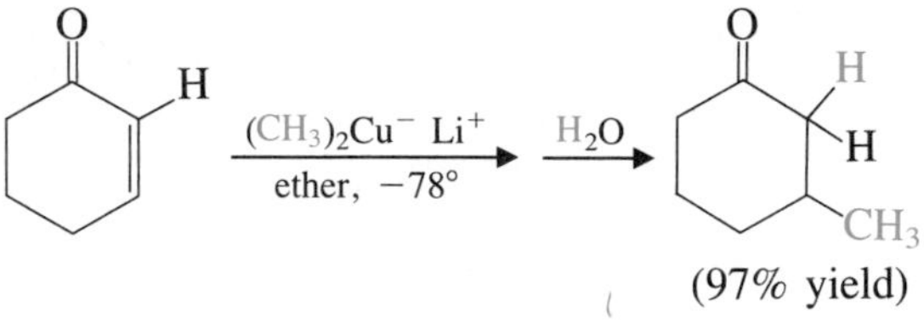

$$(22.102)$$

(97% yield)

Conjugate addition is also observed with Grignard reagents to which small amounts of CuCl have been added. Undoubtedly magnesium dialkylcuprate reagents are formed under these circumstances, and these react much like the corresponding lithium dialkylcuprates.

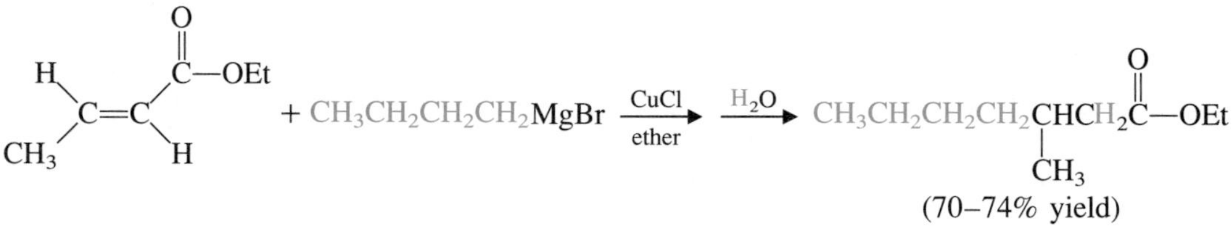

$$\underset{\substack{CH_3}}{\overset{\substack{H}}{C}}=\underset{\substack{H}}{\overset{\substack{O\\\parallel\\C-OEt}}{C}} + CH_3CH_2CH_2CH_2MgBr \xrightarrow[\text{ether}]{CuCl} \xrightarrow{H_2O} CH_3CH_2CH_2CH_2\underset{\substack{|\\CH_3}}{CH}CH_2\overset{\substack{O\\\parallel}}{C}-OEt \quad (22.103)$$

(70–74% yield)

Even α,β-unsaturated aldehydes, which are normally very reactive at the carbonyl group, give all or mostly products of conjugate addition, especially at low temperature.

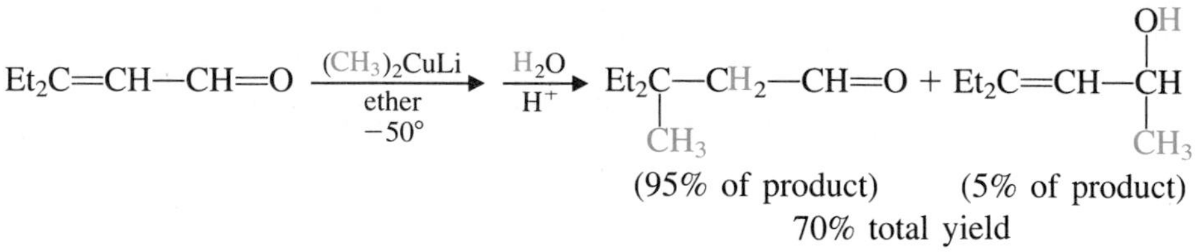

$$Et_2C=CH-CH=O \xrightarrow[\substack{\text{ether}\\-50°}]{(CH_3)_2CuLi} \xrightarrow[H^+]{H_2O} Et_2\underset{\substack{|\\CH_3}}{C}-CH_2-CH=O + Et_2C=CH-\underset{\substack{|\\CH_3}}{\overset{\substack{OH\\|}}{CH}} \quad (22.104)$$

(95% of product) (5% of product)
70% total yield

We can envision the conjugate addition of a lithium dialkylcuprate mechanistically as we do other conjugate additions: attack of an anion—in this case the "alkyl anion" of the dialkylcuprate reagent—on the double bond to give a resonance-stabilized enolate ion.

$$\underset{CH_3-Cu-CH_3}{\overset{\substack{Li^+ \qquad :O:\\|\\R-CH=CH-C-R}}{}} \longrightarrow \left[\underset{\substack{|\\CH_3}}{\overset{\substack{Li^+ \quad :O:\\\parallel\\R-CH-CH-C-R}}{}} \longleftrightarrow \underset{\substack{|\\CH_3}}{\overset{\substack{Li^+ \qquad :O:^-\\|\\R-CH-CH=C-R}}{}} \right] + CH_3Cu \quad (22.105)$$

enolate ion of product

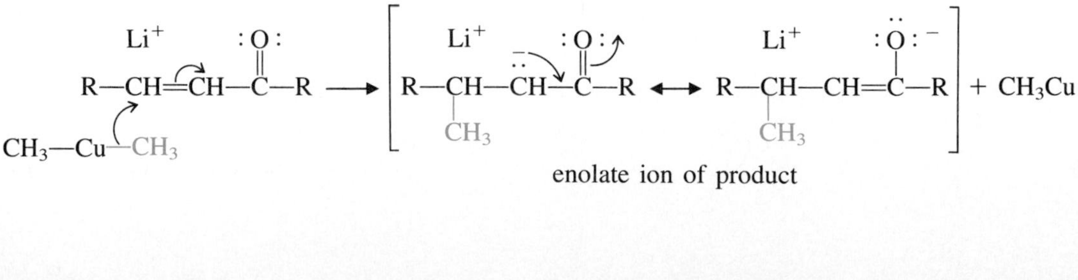

When water is added to the reaction mixture, protonation of this enolate ion gives the conjugate-addition product.

Although this mechanism is conceptually useful, it does not explain the effect of copper in promoting conjugate addition. Since carbonyl addition is observed in the *absence* of copper, there is obviously a special effect of the copper that is responsible for the conjugate addition. A substantial body of evidence suggests that a unique mechanism involving radicals is operative in this case, although we shall not consider this mechanism here.

Since Grignard reagents and organolithium reagents usually undergo similar types of reactions, we might expect Grignard reagents to give products of carbonyl addition when they react with α,β-unsaturated carbonyl compounds. However, it is found that mixtures of carbonyl addition products and conjugate-addition products are often obtained. The following example is typical:

$$\text{EtMgBr} + CH_3CH{=}CH{-}C({=}O){-}CH_3 \xrightarrow{\text{ether}} \xrightarrow{H_2O}$$

$$\underset{(39\%\ \text{yield})}{CH_3{-}CH(Et){-}CH_2{-}C({=}O){-}CH_3} + \underset{(41\%\ \text{yield})}{CH_3{-}CH{=}CH{-}C(OH)(Et){-}CH_3} \qquad (22.106)$$

In some cases the relative amounts of carbonyl addition and conjugate addition vary *in the same reaction* from investigation to investigation. Because mixtures are obtained, the addition of ordinary Grignard reagents to α,β-unsaturated carbonyl compounds is not very useful. It seems likely that the conjugate addition observed in many reactions of Grignard reagents with α,β-unsaturated carbonyl compounds is catalyzed by small amounts of transition-metal impurities in the magnesium. If we wish to carry out a carbonyl addition with an α,β-unsaturated carbonyl compound and avoid conjugate addition, it is usually best to use a lithium reagent (Sec. 22.10A). Lithium dialkylcuprates, or Grignard reagents intentionally doped with Cu(I) salts, are used when products of conjugate addition are desired.

The reaction of lithium dialkylcuprates with α,β-unsaturated carbonyl compounds adds yet another method to our repertoire of reactions that can be used for the formation of carbon–carbon bonds.

Problems

42 Outline a synthesis for each of the following compounds from acetone. Use a Grignard reagent, organolithium reagent, or lithium dialkylcuprate reagent in at least one step of each synthesis.

(a) $(CH_3)_3CCH_2{-}C({=}O){-}CH_3$

(b) $(CH_3)_2C{=}CH{-}C(OH)(CH_3)_2$

(c) $C_2H_5{-}C(CH_3)_2{-}CH{=}C(CH_3)_2$

(*Hint:* 4-Methyl-3-penten-2-one (mesityl oxide) can be prepared from acetone.)

Problems (Cont.)

43 Complete the following reactions. If more than one product seems reasonable, explain why and try to make a rational decision about which product should predominate.

(a) 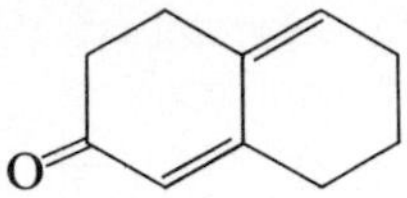 $+\ Me_2CuLi \longrightarrow \xrightarrow{H_2O}$

(b) $CH_3{-}C{\equiv}C{-}CO_2Me \quad + \quad Me_2CuLi \longrightarrow \xrightarrow{H_2O,\ H^+}$
(one equivalent)

22.11 ORGANIC SYNTHESIS WITH CONJUGATE-ADDITION REACTIONS

How do we know when to use a conjugate-addition reaction in an organic synthesis? One useful way to think of this problem is that any group at the β-position of a carbonyl compound (or nitrile) can *in principle* (although not always in practice) be delivered as a nucleophile in a conjugate addition. Hence we can mentally reverse a conjugate addition by subtracting a nucleophilic group from the β-position of our target molecule, and a positive fragment (usually a proton) from the α-position:

$$R{-}CH_2{-}\underset{\displaystyle H}{\underset{|}{CH}}{-}\overset{\displaystyle O}{\overset{\|}{C}}{-}R' \Longrightarrow \text{“}R^-,\ H^+\text{”} + CH_2{=}CH{-}\overset{\displaystyle O}{\overset{\|}{C}}{-}R' \qquad (22.107)$$

For example, suppose we want to prepare the following ketone, 2-octanone.

$$CH_3CH_2CH_2CH_2CH_2CH_2{-}\overset{\displaystyle O}{\overset{\|}{C}}{-}CH_3$$

If we count from the carbonyl group to the β-carbon, we see that there are two groups attached to that carbon: hydrogen, and an *n*-butyl group.

$$CH_3CH_2CH_2CH_2{-}CH_2{-}\overset{\displaystyle H}{\overset{|}{CH}}{-}\overset{\displaystyle O}{\overset{\|}{C}}{-}CH_3$$

We could introduce the *n*-butyl group as a “butyl anion” by the reaction of lithium dibutylcuprate with methyl vinyl ketone; the proton is provided in the subsequent protonolysis step:

$$Li^+\ (CH_3CH_2CH_2CH_2)_2Cu^- + CH_2{=}CH{-}\overset{\displaystyle O}{\overset{\|}{C}}{-}CH_3 \longrightarrow \xrightarrow{H_3O^+} \text{2-octanone} \qquad (22.108)$$

But we could just as easily think of the hydrogen as the group to be added. We have not learned any ways of introducing "H :$^{-}$" in a conjugate addition (although there are some!), but we certainly could hydrogenate the corresponding α,β-unsaturated ketone:

$$CH_3CH_2CH_2CH_2CH{=}CH{-}\overset{\overset{\large O}{\|}}{C}{-}CH_3 \xrightarrow{H_2/cat} \text{2-octanone} \qquad (22.109)$$

Problem

44 Show how a conjugate addition could be used to prepare each of the following compounds:

(a) 3,4-dimethyl-2-hexanone (2 ways)

(b) $N{\equiv}C{-}CH_2CH(CH_3){-}\overset{\overset{\large O}{\|}}{C}{-}OEt$

(c) $HO{-}\overset{\overset{\large O}{\|}}{C}{-}CH_2CH_2CH_2{-}\overset{\overset{\large O}{\|}}{C}{-}CH_3$ (*Hint:* See Eq. 22.94.)

Many of the reactions discussed in this chapter add to our list of those that can be used for forming carbon–carbon bonds. Let us once again review the ones we know. The first eleven methods are given in Sec. 21.11. To these we now add the following:

12. Aldol condensation (Sec. 22.4)
13. Claisen condensation (Sec. 22.5)
14. Malonic ester synthesis (Sec. 22.6A)
15. Alkylation of ester enolates with amide bases and alkyl halides (Sec. 22.6B)
16. Acetoacetic ester synthesis (Sec. 22.6C)
17. Addition of cyanide and enolate ions to α,β-unsaturated carbonyl compounds (Michael addition; Sec. 22.8C)
18. Reaction of lithium dialkylcuprates with α,β-unsaturated carbonyl compounds (Sec. 22.10B)

(A complete list of methods for forming carbon–carbon bonds is given in Appendix V.)

This chapter completes our study of what might be called "the organic chemistry of oxygen." That is, we have now learned the general methods for the preparation and the reactions of the oxygen-containing functional groups: alcohols, ethers, aldehydes and ketones, carboxylic acids, and carboxylic acid derivatives.

In the following chapter, we conclude our study of fundamental organic chemistry by focusing on the organic chemistry of nitrogen.

KEY IDEAS IN CHAPTER 22

- Hydrogens on carbon atoms α to carbonyl groups are acidic. Ionization of these hydrogens gives enolate ions.
- Most carbonyl compounds with α-hydrogens are in equilibrium with small amounts of enol (vinylic alcohol) tautomers. Generally, carbonyl–enol equilibria favor carbonyl tautomers, but there are important exceptions, such as phenols and β-diketones, in which enol tautomers are the major forms.
- Enolization is catalyzed by acids and bases.
- Enolate ions can act as nucleophiles in a number of reactions. In this chapter we have learned that enolate ions can

  1. Undergo α-halogenation (haloform reaction)
  2. Add to carbonyl groups (aldol condensation)
  3. Substitute at carbonyl groups (Claisen condensation)
  4. React with alkyl halides (enolate alkylation)
  5. Add to α,β-unsaturated carbonyl compounds (Michael addition)

  Enols undergo α-halogenation and aldol condensation reactions in acid solution.
- The aldol and Claisen condensations are reversible reactions. The aldol condensation is generally favorable for aldehydes, but unfavorable for ketones. It can be driven to completion by dehydration of the intermediate β-hydroxy carbonyl compound. The Claisen condensation is generally driven to completion by ionization of the product.
- Two types of addition to α,β-unsaturated carbonyl compounds are possible: addition to the carbonyl group and addition to the double bond (conjugate addition, or 1,4-addition). When carbonyl addition is reversible, conjugate addition is observed because it gives the more stable product. When carbonyl addition is irreversible, it is observed instead because it is faster. Conjugate addition of lithium dialkylcuprate reagents is attributable to a special effect of copper that is not yet clearly understood.
- Reactions closely resembling enolate condensations are observed in nature. Many such reactions employ acetyl-CoA as a starting material.

ADDITIONAL PROBLEMS

45 Give the principal organic product expected when 3-buten-2-one (methyl vinyl ketone) reacts with each of the following reagents:

(a) HBr
(b) Br_2 in CCl_4
(c) $LiAlH_4$, then H_2O
(d) HCN in water, pH 10
(e) $Et_2Cu^-\ Li^+$, then H_3O^+
(f) diethyl malonate and NaOEt
(g) ethylene glycol, H^+ (cat)
(h) 1,3-butadiene

46 Give the principal organic products expected when ethyl *trans*-2-butenoate (ethyl crotonate) reacts with each of the following reagents:

(a) ^-CN in water, then H_2O/H^+, heat
(b) Me_2NH, room temperature
(c) NaOH, H_2O, heat
(d) CH_3Li (excess), then H_3O^+
(e) H_2, catalyst
(f) 1,3-cyclopentadiene

47 A mixture of the *E* and *Z* isomers of the following ester was carefully fractionated into its components:

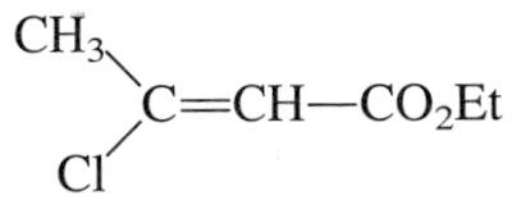

The dipole moment of isomer *A*, which has the lower boiling point, was found to be 1.43 D. The dipole moment of isomer *B*, which has the higher boiling point, was found to be 2.45 D. Which is the *E* isomer, and which is the *Z* isomer? How do you know?

48 Give the structure of a compound that meets each criterion:

(a) an optically active compound $C_6H_{12}O$ that racemizes in base
(b) an optically active compound $C_6H_{12}O$ that neither racemizes in base nor gives a positive Tollens' test
(c) an optically active compound $C_6H_{12}O$ that gives a positive Tollens' test and does not racemize in base
(d) an achiral compound $C_6H_{12}O$ that does not give a positive Tollens' test

49 Give the structure of a compound that meets each criterion:

(a) a compound C_4H_8O that gives a 2,4-DNP derivative but a negative haloform test
(b) a compound C_4H_8O that gives a 2,4-DNP derivative and a positive haloform test
(c) a compound C_4H_8O that gives neither a 2,4-DNP derivative nor a positive haloform test
(d) a compound C_4H_8O that gives a positive haloform test, but gives no 2,4-DNP derivative

Problems (Cont.)

50 Identify the compounds below that are unstable and either decompose to other compounds or exist as isomers. Explain your answers.

(a) $CH_3—C{\equiv}C—OH$

(b) HO—CH—OH (with Ph on CH)

(c) $Ph—O—CH_2—O—CH_3$

(d)

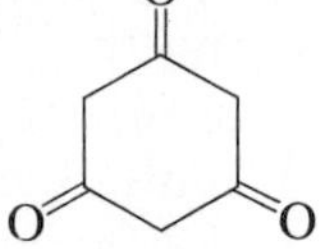

(e) $CH_3—CH(OH)—O—CH{=}CH—CH_3$

51 Which compound in each set is most acidic? Explain.

(a) $CH_3\overset{O}{\overset{\|}{C}}CH_2\overset{O}{\overset{\|}{C}}CH_3$, $CH_3\overset{O}{\overset{\|}{C}}\underset{Ph}{CH}\overset{O}{\overset{\|}{C}}CH_3$, or $CH_3\overset{O}{\overset{\|}{C}}CH_2\overset{O}{\overset{\|}{C}}CH_2Ph$

(b) $CH_2{=}CHCH_2—\overset{O}{\overset{\|}{C}}—CH_3$ or $CH_3CH_2CH_2—\overset{O}{\overset{\|}{C}}—CH_3$

(c)

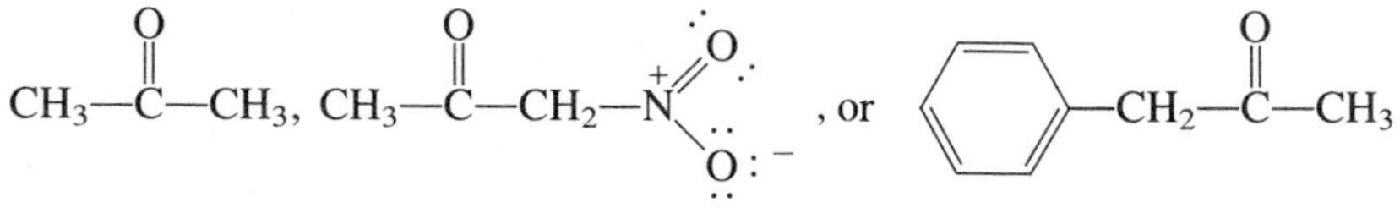

52 Arrange the compounds below in order of increasing acidity.

(a) isobutyramide
(b) octanoic acid
(c) toluene
(d) ethyl acetate
(e) phenylacetylene
(f) phenol

53 When 1,3-diphenyl-2-propanone is treated with Br_2 in acid, 1,3-dibromo-1,3-diphenyl-2-propanone is obtained in good yield. On further characterization, however, this product proves to have a very broad melting point (79–87°), a fact suggesting a mixture of compounds. Account for this observation.

54 Treatment of (*S*)-(+)-5-methyl-2-cyclohexen-1-one with lithium dimethylcuprate gives, after protonolysis, a good yield of a mixture containing mostly a dextrorotatory ketone *A* and a trace of an optically inactive isomer *B*. Treatment of *A* with zinc amalgam and HCl affords an optically active, dextrorotatory hydrocarbon *C*. Identify *A, B,* and *C,* and give the absolute configurations of *A* and *C.*

55 Cringe Labrack, a graduate student in his tenth year of study, has suggested each of the following synthetic procedures. Explain why each one cannot be expected to work.

(a) $CH_2(CO_2Et)_2 \xrightarrow[EtOH]{NaOEt} \xrightarrow{PhBr} Ph—CH(CO_2Et)_2$

(b) $CH_3CH_2CO_2Et \xrightarrow[EtOH]{NaOEt} \xrightarrow{CH_3I} (CH_3)_2CHCO_2Et$

(c) $CH_3\overset{OH}{\overset{|}{C}}HCO_2Et \xrightarrow[heat]{H_3O^+} CH_2{=}CH{-}CO_2Et$

(d) $CH_3CH_2CO_2H \xrightarrow[Br_2]{PBr_3\ (cat)} \xrightarrow[ether]{Mg} \xrightarrow{CH_3CH{=}O} CH_3{-}\overset{OH}{\overset{|}{C}}H{-}\overset{CH_3}{\overset{|}{C}}H{-}CO_2H$

(e) $C_2H_5{-}\overset{O}{\overset{\|}{C}}{-}CH_3 + CH_3{-}\overset{O}{\overset{\|}{C}}H \xrightarrow{OH^-} CH_3{-}CH{=}CH{-}\overset{O}{\overset{\|}{C}}{-}C_2H_5$

(f)

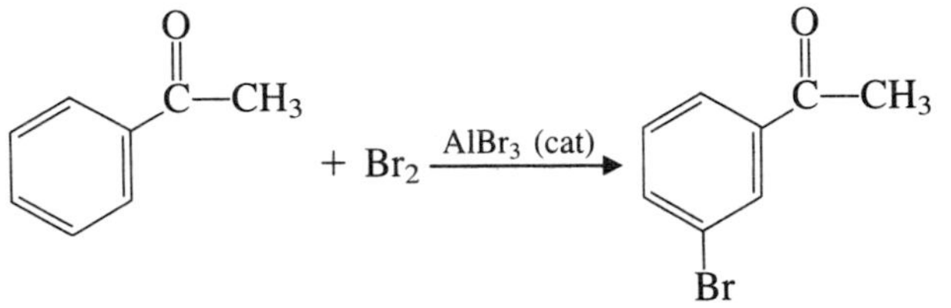

56 When compound *A* is treated with $NaOCH_3$ in CH_3OH, isomerization to *B* occurs; give a mechanism for this reaction. When compound *C* is treated in a similar fashion, no reaction takes place. Explain.

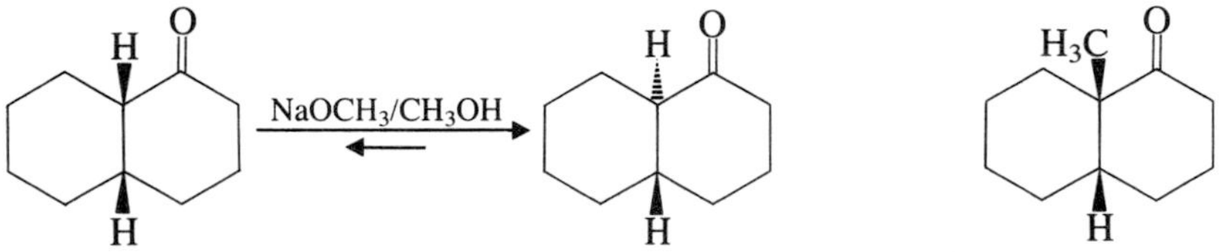

A *B* *C*

57 In either acid or base, 3-cyclohexen-1-one comes to equilibrium with 2-cyclohexen-1-one:

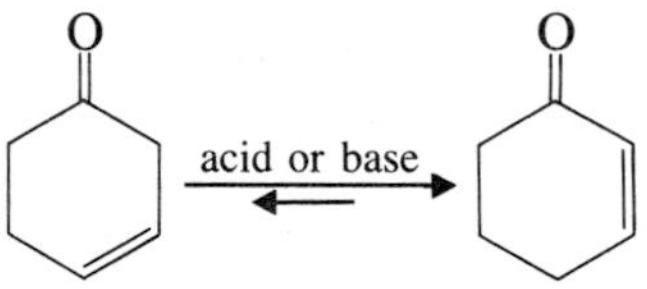

(a) Explain why the equilibrium favors the α,β-unsaturated ketone over the β,γ-unsaturated one.
(b) Give a mechanism for this reaction in aqueous NaOH.
(c) Give a mechanism for the same reaction in dilute aqueous H_2SO_4. (*Hint:* The enol 1,3-cyclohexadien-1-ol is an intermediate in the acid-catalyzed reaction.)
(d) Is the equilibrium constant for the analogous reaction of 4-methyl-3-cyclohexen-1-one expected to be greater or smaller? Explain.

Problems (Cont.)

58 In 3-methyl-2-cyclohexen-1-one the eight protons H^a, H^b, H^c, and H^d can be exchanged for deuterium in CH_3O^-/CH_3OD.

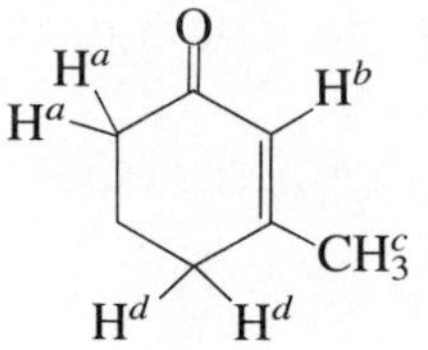

(a) Write a mechanism for the base-catalyzed exchange of protons H^a, H^c, and H^d.

(b) Explain why proton H^b is much less acidic than protons H^a, H^c, and H^d, even though it is an α-proton.

(c) Although proton H^b is not unusually acidic, it nevertheless exchanges readily in base. Write a mechanism for the exchange of H^b. (*Hint:* Use the results of Problem 57.)

59 (a) Compound *A* below, γ-pyrone, has a conjugate acid with an unusually high pK_a of -0.4. The pK_a of the conjugate acid of compound *B*, in contrast, is about -3.

A
γ-pyrone

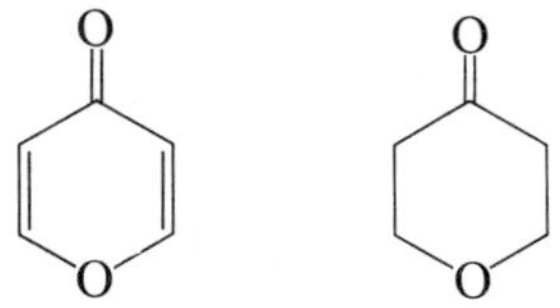

B

Draw the structures of the conjugate acids of both molecules, and explain why *A* is more basic than *B*.

(b) Tropone reacts with one equivalent of HBr to give a stable crystalline conjugate acid salt with a pK_a of -0.6, much higher than that of most protonated α,β-unsaturated ketones. Give the structure of the conjugate acid of tropone, and explain why tropone is unusually basic.

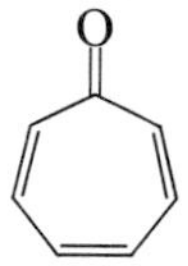

tropone

60 When *t*-butyl acetate was treated with one equivalent of lithium diisopropylamide (LDA) in THF at $-78°$, and the resulting solution was evaporated to dryness, a white powder was obtained that had the following NMR spectrum: δ 1.96 (9*H*, s); δ 3.14 (1*H*, d); δ 3.44 (1*H*, d). Addition of water to this powder gave back *t*-butyl acetate in good yield. What is the identity of the white powder? Rationalize its NMR spectrum.

61 (a) Give the structures of the three separable monobromo derivatives that could form when 2-methylcyclohexanone is treated with Br_2 in the presence of HBr.

(b) In fact, only one of these derivatives is formed. Assuming that the course of the reaction is governed by the relative stability of the possible enol intermediates, predict which product is formed and why. (*Hint:* See Sec. 4.4C.)

62 The compound *alternariol* is a metabolite isolated from a fungus.

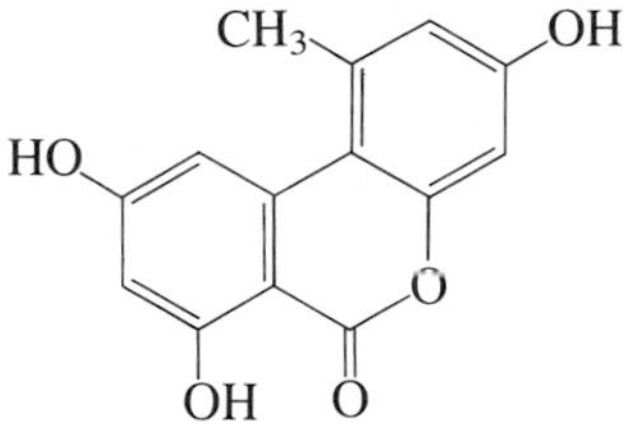

alternariol

(a) Suggest a pathway for the biosynthesis of alternariol from acetate thiol ester units.
(b) According to your biosynthetic scheme, if the fungus that produces this metabolite were fed acetic acid labeled at the carbonyl group with the radioactive isotope ^{14}C, which carbons of alternariol would carry the label?

63 A biochemist, Sal Monella, has come to you to ask your assistance in testing a brilliant biosynthetic hypothesis. She wishes to have two samples of α-methylsuccinic acid specifically labeled with ^{14}C as shown·

$$\underset{*}{\text{HO}-\overset{\overset{\text{O}}{\|}}{\text{C}}-\underset{\underset{\text{CH}_3}{|}}{\text{CH}}-\text{CH}_2-\overset{\overset{\text{O}}{\|}}{\text{C}}-\text{OH}} \qquad \text{HO}-\overset{\overset{\text{O}}{\|}}{\text{C}}-\underset{\underset{\text{CH}_3}{|}}{\text{CH}}-\underset{*}{\text{CH}_2}-\overset{\overset{\text{O}}{\|}}{\text{C}}-\text{OH}$$

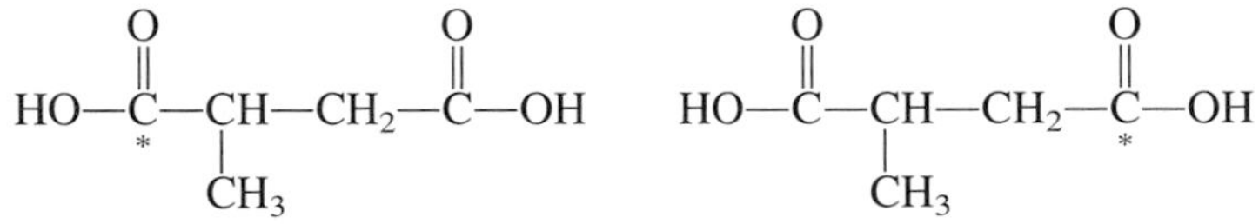

The source of the isotope, for reasons of expense, is to be Na^{14}CN. Outline syntheses that will accomplish the desired objective.

64 When acetoacetic acid is decarboxylated in the presence of bromine, α-bromoacetone is isolated.

$$\underset{\text{acetoacetic acid}}{\text{CH}_3-\overset{\overset{\text{O}}{\|}}{\text{C}}-\text{CH}_2-\overset{\overset{\text{O}}{\|}}{\text{C}}-\text{OH}} + \text{Br}_2 \xrightarrow{\text{H}_3\text{O}^+} \text{CH}_3-\overset{\overset{\text{O}}{\|}}{\text{C}}-\text{CH}_2-\text{Br} + \text{HBr} + \text{CO}_2$$

The rate of appearance of bromoacetone is described by the following rate law:

$$\text{rate} = k[\text{acetoacetic acid}]$$

(The reaction rate is zero order in bromine.) Suggest a mechanism for the reaction that is consistent with this rate law.

65 Account for the fact that 1,3-diphenyl-1,3-propanedione gives a positive iodoform test even though it is not a methyl ketone. (*Hint:* Besides iodoform, the product of the iodoform reaction, after acidification of the reaction mixture, is two equivalents of benzoic acid.)

Problems (Cont.)

66 Crossed aldol condensations can be carried out if one of the carbonyl compounds is unusually acidic.

(a) Give the structure of the α,β-unsaturated carbonyl compound that results from the crossed aldol condensation of diethyl malonate and acetone in NaOEt/EtOH.

(b) Explain why the aldol condensation of acetone with itself does not compete with the crossed aldol condensation in (a).

(c) Outline a sequence of reactions for the conversion of the product from (a) into 3,3-dimethylbutanoic acid.

67 Identify the intermediates *A* and *B* in the following transformation, and show how they are formed:

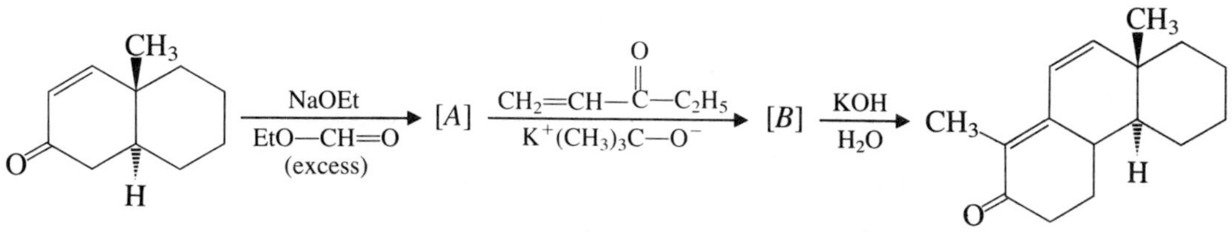

68 When the terpene *pulegone* is heated with aqueous NaOH, acetone is formed. Explain how this reaction occurs, and give the structure of the other product formed in the same reaction.

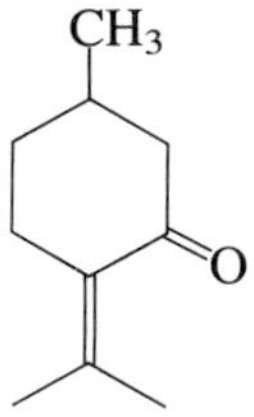

pulegone

69 Explain the following findings:

(a) One full equivalent of base must be used in the Claisen or Dieckmann condensation.

(b) Ethyl acetate readily undergoes a Claisen condensation in the presence of one equivalent of sodium ethoxide, but phenyl acetate does *not* undergo a Claisen condensation in the presence of one equivalent of sodium phenoxide.

70 Propose syntheses of each of the following compounds from the indicated starting materials and any other reagents:

(a) 3-ethyl-1-cyclopentanol from 2-cyclopenten-1-one
(b) 2-ethyl-1,3-hexanediol from butyric acid
(c) 4,4-dimethyl-3-hexanone from ethyl butyrate
(d) 2,2-dimethyl-1,3-propanediol from diethyl malonate
(e) 1,3,3-trimethyl-1-cyclohexanol from 3-methyl-2-cyclohexen-1-one
(f) $H_2NCH_2CH_2CH_2CH_2OH$ from ethyl acrylate (ethyl 2-propenoate)

(g) 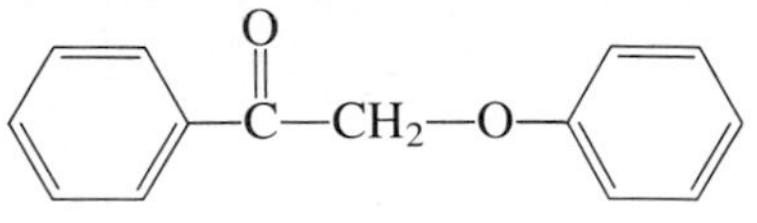 from acetyl chloride

(h) 2-benzylcyclohexanone from $EtO_2C{-}(CH_2)_5{-}CO_2Et$

(i) 1,3-diphenyl-1-butanone from acetophenone

(j) $Ph{-}CH(OH){-}CD_2{-}Ph$ 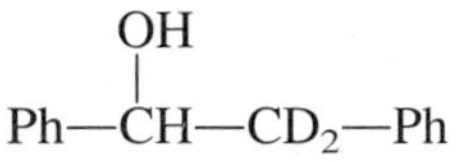 from $Ph{-}CH_2{-}CO_2H$

(k) 2-phenylbutanoic acid from α-phenylacetic acid
(Do not use hindered amide bases; see Eq. 22.57.)

(l) $CH_3CH_2CH_2CH_2\overset{O}{\overset{\|}{C}}CH(CH_3)(CH_2CH_2CH_3)$ 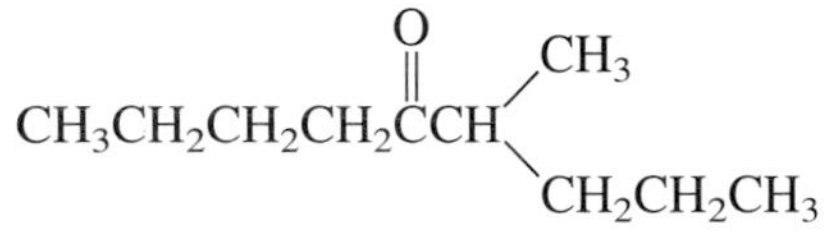 from diethyl malonate

(m) 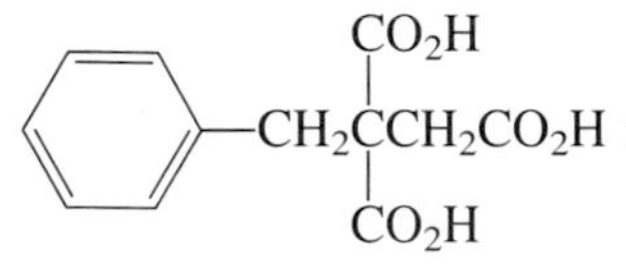 from diethyl malonate

(n) $(C_2H_5)_2N{-}CH_2CH_2CH_2NH_2$ 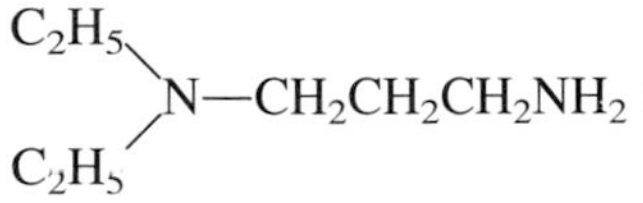 from acrylonitrile (propenenitrile)

(o) 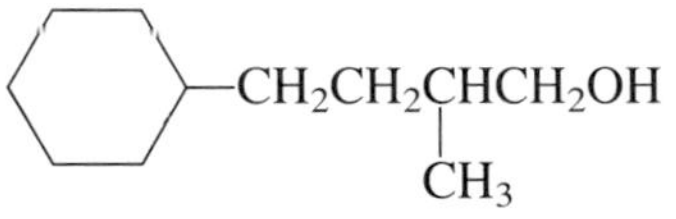 from bromocyclohexane

(p) 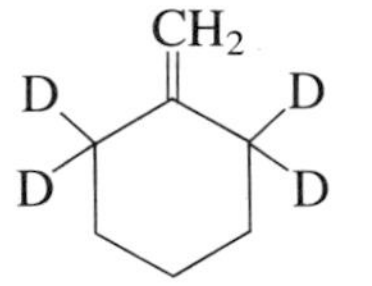 from cyclohexanone

(q) $Ph{-}\overset{O}{\overset{\|}{C}}{-}CH(CH_3){-}\overset{O}{\overset{\|}{C}}{-}Ph$ 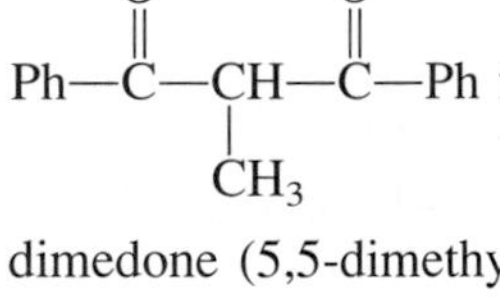 from propionic acid

(r) dimedone (5,5-dimethyl-1,3-cyclohexanedione)
from $CH_3{-}\overset{O}{\overset{\|}{C}}{-}CH{=}C(CH_3)_2$

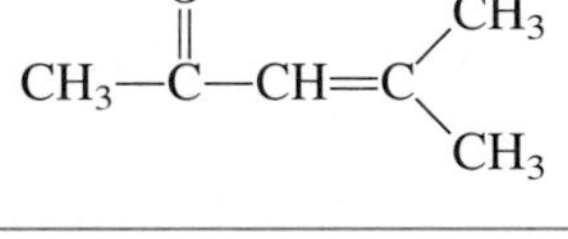

71 When the epoxide 2-vinyloxirane reacts with lithium di-*n*-butylcuprate, followed by protonolysis, a compound *A* is the major product formed. Oxidation of *A* with CrO_3 yields *B,* a compound that gives a positive Tollens' test and has an intense UV absorption around 215 nm. Treatment of *B* with Ag_2O, followed by catalytic hydrogenation, gives octanoic acid. Identify *A* and *B,* and outline a mechanism for the formation of *A*.

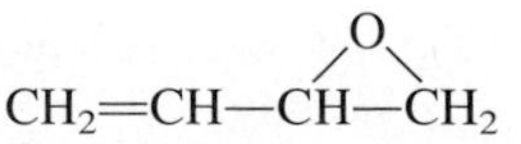

2-vinyloxirane

72 Ethyl vinyl ether, C_2H_5—O—CH=CH$_2$, hydrolyzes in weakly acidic water to acetaldehyde and ethanol. Under the same conditions, diethyl ether does not hydrolyze. Quantitative comparisons of the hydrolysis rates of the two ethers under comparable conditions show that ethyl vinyl ether hydrolyzes about 10^{13} times as fast as diethyl ether. The rapid hydrolysis rate for ethyl vinyl ether suggests an unusual mechanism for its hydrolysis. The acetaldehyde formed when the hydrolysis of ethyl vinyl ether is carried out in D_2O contains one deuterium in its methyl group (that is, D—CH$_2$—CH=O). Suggest a hydrolysis mechanism for ethyl vinyl ether consistent with these facts. (*Hint:* Vinyl ethers are also called *enol ethers*. Where do enols protonate? See Eq. 22.16b.)

73 The following resonance structures can be written for an α,β-unsaturated carboxylic acid:

$$R-CH=CH-\overset{\overset{\displaystyle :O:}{\|}}{C}-OH \longleftrightarrow R-\overset{+}{CH}-CH=\overset{\overset{\displaystyle :\ddot{O}:^-}{|}}{C}-OH$$

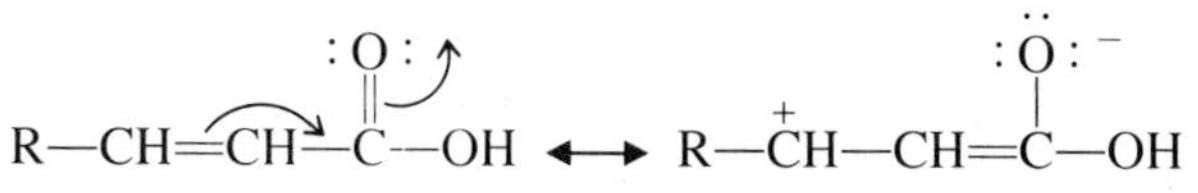

(a) Would this type of resonance interaction increase or diminish the acidity of a carboxylic acid relative to that of an ordinary carboxylic acid?
Consider the following pK_a data:

$$CH_2=CH-CH_2-CO_2H$$
$$pK_a = 4.37$$

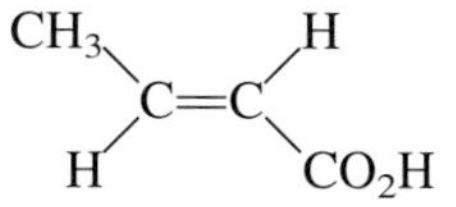

$$\underset{pK_a = 4.70}{\overset{\displaystyle \overset{CH_3}{}\;\;\overset{H}{}}{\underset{H}{\overset{\displaystyle}{C}}=\underset{CO_2H}{C}}}$$

$$CH_3CH_2CH_2-CO_2H$$
$$pK_a = 4.87$$

(b) According to these data, is the *inductive effect* of the double bond an electron-withdrawing or electron-donating effect?
(c) Is there any evidence in these data for the resonance effect on acidity discussed in (a)?

74 When 2,4-pentanedione in ether is treated with one equivalent of NaH, a gas is evolved, and a species A is formed.
(a) Give the structure of A. Which atoms of A should be nucleophilic? Explain.
(b) When A reacts with CH_3I, three isomeric compounds, B, C, and D ($C_6H_{10}O_2$), are formed. Suggest structures for these compounds.

75 Suggest a mechanism for the following reaction, an example of the Darzens glycidic ester condensation:

$$Ph-CH=O + Br-CH_2-\overset{\overset{\displaystyle O}{\|}}{C}-OEt \xrightarrow[t\text{-BuOH}]{K^+ \; t\text{-BuO}^-} Ph-\overset{\overset{\displaystyle O}{\triangle}}{CH}-CH-\overset{\overset{\displaystyle O}{\|}}{C}-OEt + KBr$$
$$\text{(mixture of stereoisomers)}$$

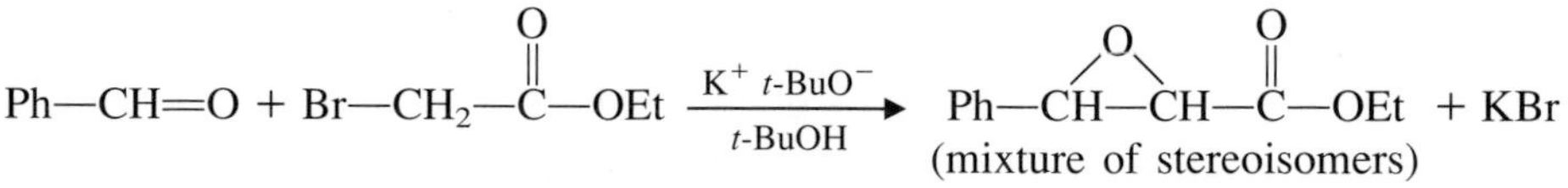

76 Complete the following reactions by giving the major organic products, and explain your reasoning:

(a)

$$CH_3-\overset{O}{\overset{\|}{C}}-CH_2-\overset{O}{\overset{\|}{C}}-OEt + Br(CH_2)_4Br \xrightarrow[\text{EtOH}]{\text{NaOEt (excess)}} \xrightarrow[\text{heat}]{H_3O^+} (C_7H_{12}O)$$

(b)

$$+ Na^+ \ ^- : CH(CO_2Et)_2 \xrightarrow{H^+}$$

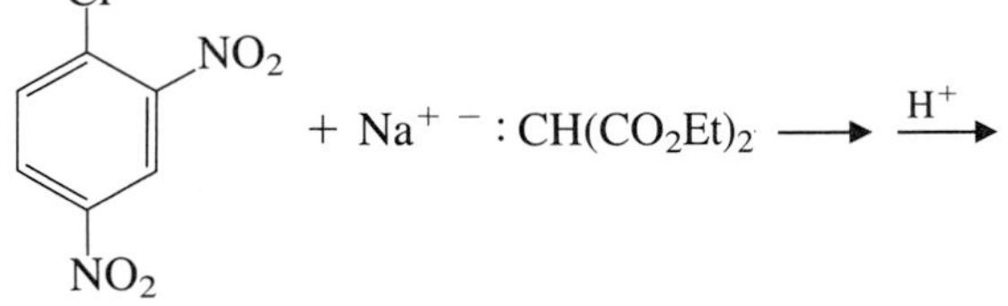

(c) γ-butyrolactone $\xrightarrow{Li^+ \ [(CH_3)_2CH]_2\ddot{N}:^-} \xrightarrow{CH_3I}$

(d)

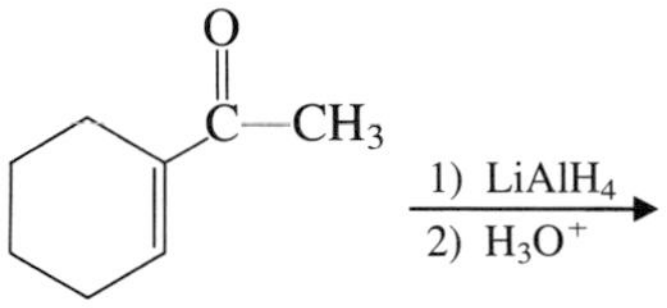

$$\xrightarrow[\text{2) } H_3O^+]{\text{1) LiAlH}_4}$$

(e)

$$\overset{CH_3}{\diagup} -CO_2Et + CH_3CH(CO_2Et)_2 \xrightarrow[\text{EtOH}]{\text{NaOEt}}$$

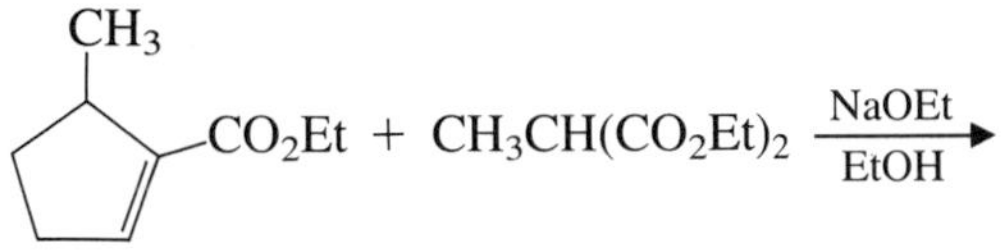

(f)

$$CH_3-\overset{O}{\overset{\|}{C}}-CH_2-CO_2Et + CH_2{=}CH-CO_2Et \xrightarrow{EtO^-} \xrightarrow{H^+}$$

(g)

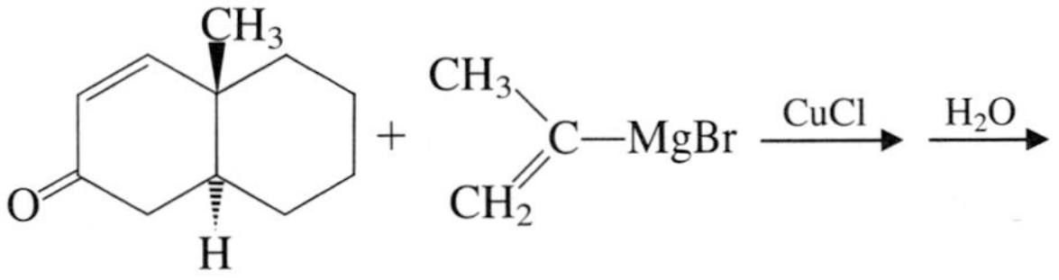

$$+ \overset{CH_3}{\underset{CH_2}{\diagup}}C-MgBr \xrightarrow{CuCl} \xrightarrow{H_2O}$$ (Give the stereochemistry of the product and explain.)

(h)

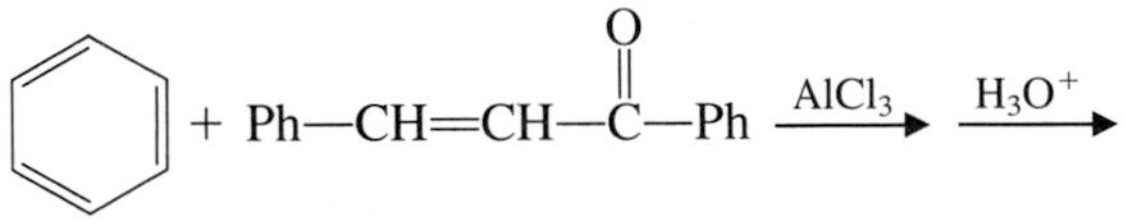

$$+ Ph-CH{=}CH-\overset{O}{\overset{\|}{C}}-Ph \xrightarrow{AlCl_3} \xrightarrow{H_3O^+}$$ (a ketone $C_{21}H_{18}O$)

(i)

$$Cl(CH_2)_3CH\overset{O-}{\underset{O-}{\diagdown}} \xrightarrow[\text{CuBr}]{\text{Mg}} \xrightarrow[\substack{\text{benzene} \\ -H_2O}]{H^+} (C_{10}H_{14}O)$$

77 (a) The pK_a of 2-nitropropane is 10. Give the structure of its conjugate base, and suggest reason(s) why 2-nitropropane has a particularly acidic C—H bond.

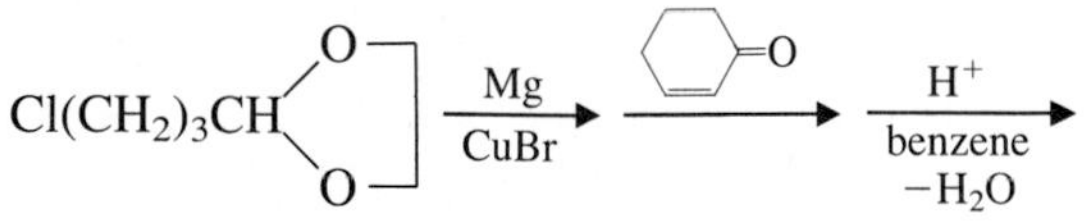

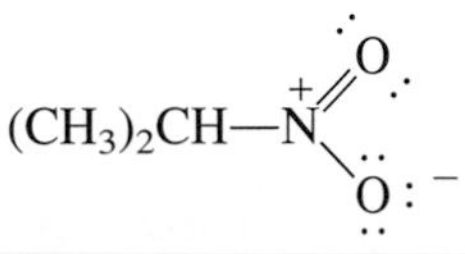

$$(CH_3)_2CH-\overset{+}{N}\overset{\overset{\cdot\cdot}{O}\cdot}{\underset{\overset{\cdot\cdot}{O}:^-}{}}$$ 2-nitropropane

(b) When the conjugate acid of 2-nitropropane is reprotonated, an isomer of 2-nitropropane is formed, which, on standing, is slowly converted into 2-nitropropane itself. Give the structure of this isomer.

(c) What product forms when 2-nitropropane reacts with ethyl acrylate ($CH_2{=}CH-CO_2Et$) in the presence of NaOEt in EtOH?

Problems (Cont.)

78 Using the arrow formalism where possible, give a reasonable mechanism for each of the following reactions:

(a)

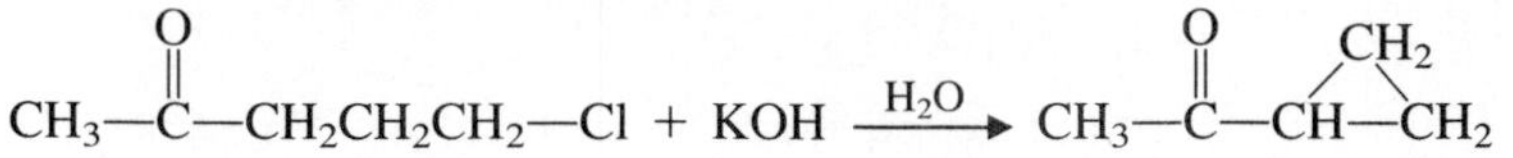

$$CH_3-\overset{O}{\underset{\|}{C}}-CH_2CH_2CH_2-Cl + KOH \xrightarrow{H_2O} CH_3-\overset{O}{\underset{\|}{C}}-\overset{CH_2}{\underset{|}{C}H}-CH_2$$
(77–83% yield)

(b)

$$\xrightarrow[\text{OH}^-]{H_2O}$$ (*Hint:* Aldol condensations are reversible.)

(c)

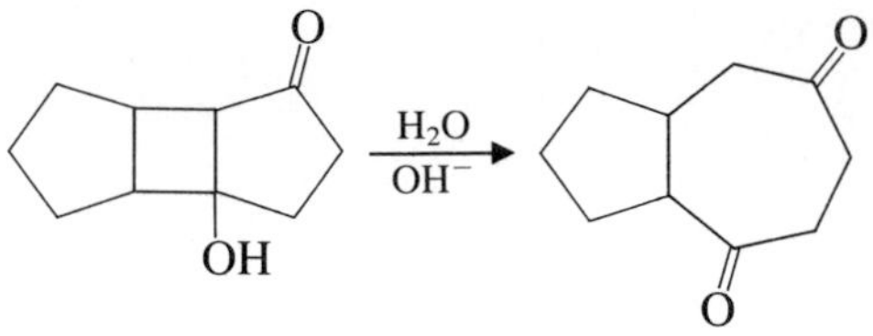

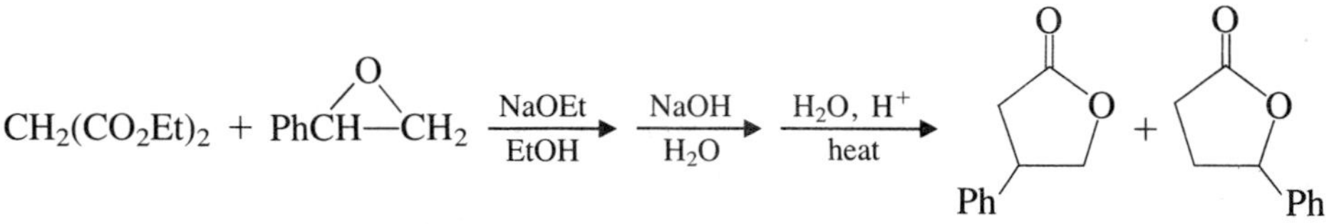

$$CH_2(CO_2Et)_2 + PhCH\overset{O}{\diagup}CH_2 \xrightarrow[\text{EtOH}]{\text{NaOEt}} \xrightarrow[\text{H}_2\text{O}]{\text{NaOH}} \xrightarrow[\text{heat}]{\text{H}_2\text{O, H}^+} \qquad +$$

(d)

$$\bigcirc-Br + 2CH_2(CO_2Et)_2 \xrightarrow[\text{NH}_3\text{(liq)}]{\text{4NaNH}_2} \xrightarrow{H^+} \bigcirc-CH(CO_2Et)_2 + \bigcirc-NH_2$$

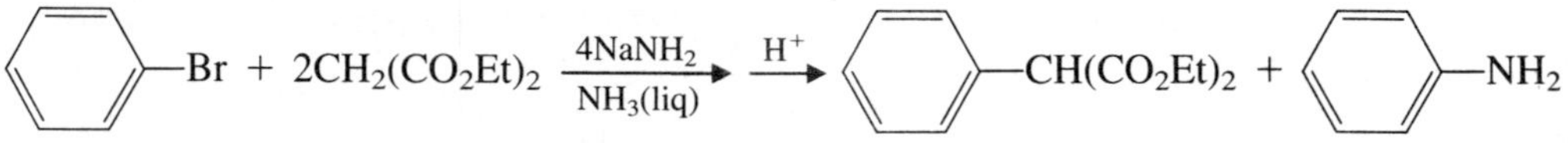

(This reaction does not occur if NaOEt/EtOH is used as the base.)

(e)

$$CH_3-\overset{O}{\underset{\|}{C}}-CH_2-CO_2Et + PhCO_2Et \xrightarrow{\text{NaOEt}}$$

$$PhÖ\overset{O}{\underset{\|}{C}}-CH_2-CO_2Et + CH_3CO_2Et$$
(removed as it is formed)

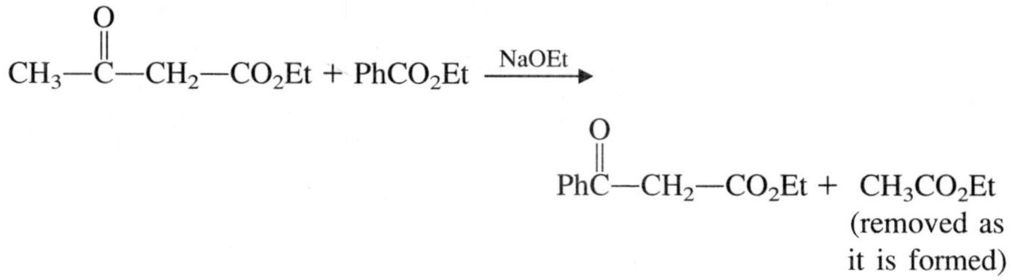

79 When the diester of a malonic acid derivative is treated with sodium ethoxide and urea, a *barbiturate* is formed. (Barbiturates are hypnotic drugs.)

$$H_2N-\overset{O}{\underset{\|}{C}}-NH_2 + EtO\overset{O}{\underset{\|}{C}}-\overset{Et}{\underset{\underset{Et}{|}}{C}}-\overset{O}{\underset{\|}{C}}OEt \xrightarrow[\text{EtOH}]{\text{NaOEt}} \xrightarrow{H^+}$$

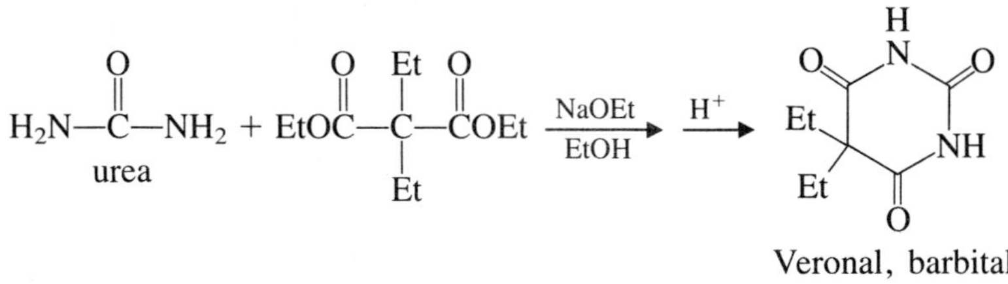

Veronal, barbital
(a barbiturate)

(a) Give the mechanism of the Veronal synthesis above.

(b) Outline a synthesis of the malonate ester derivative from diethyl malonate and any other reagents.

80 Using the arrow formalism, give a mechanism for each of the following condensation reactions:

(a)

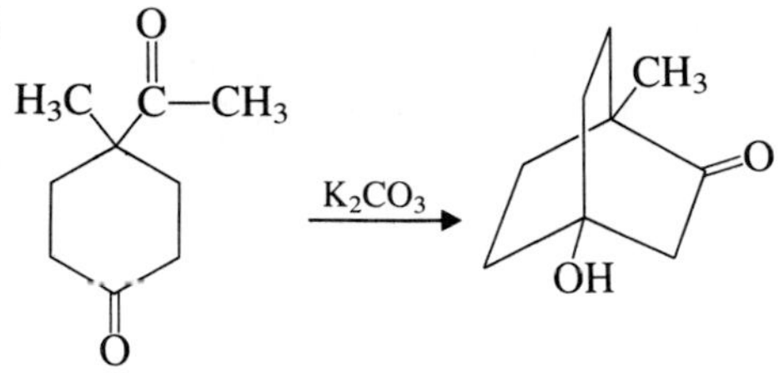

(b)

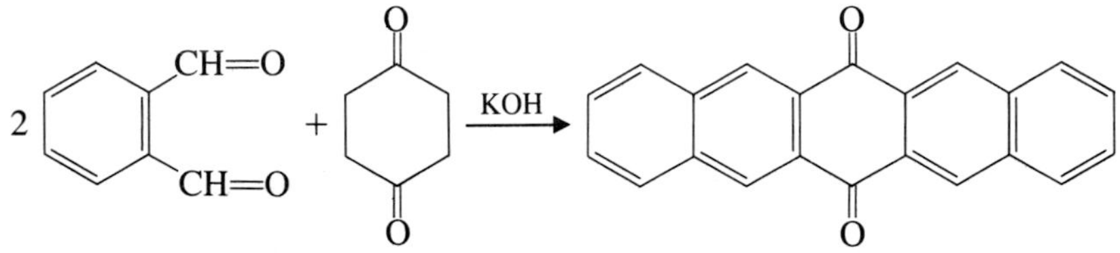

(90% yield)

(c)

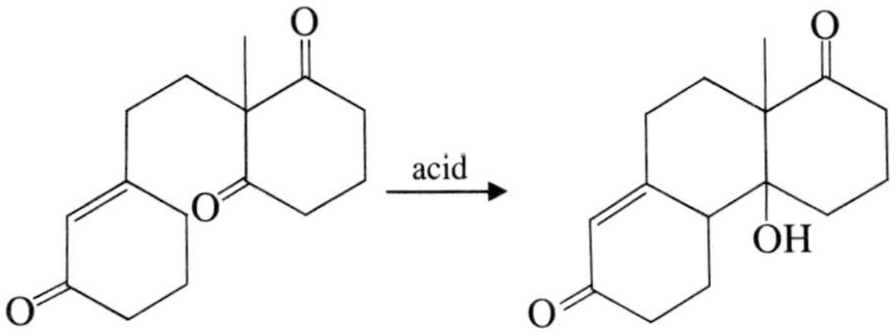

(d)

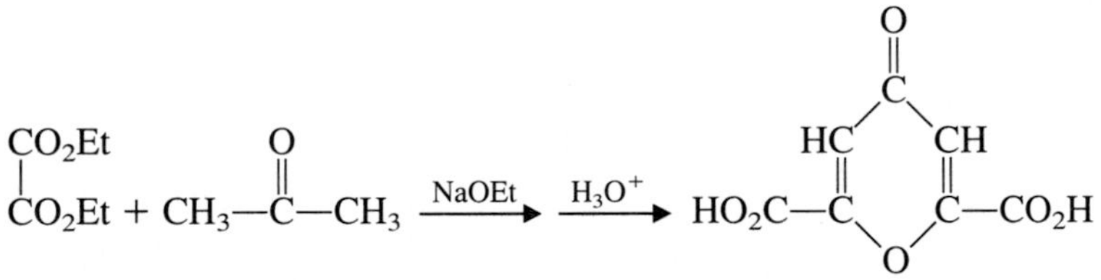

chelidonic acid

(e)

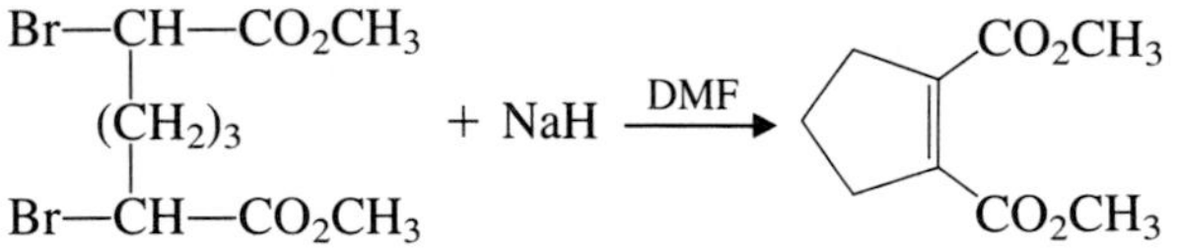

(f)

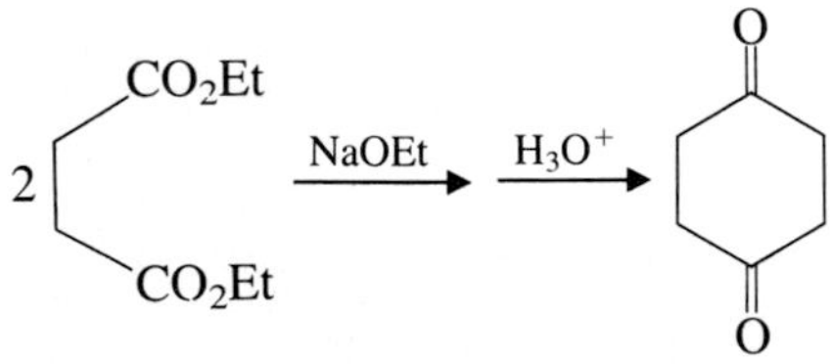

81 Suggest a mechanism for each of the following reactions, each of which involves an α,β-unsaturated carbonyl compound:

(a)

$$PhO{-}CH{=}CH{-}\overset{\overset{O}{\|}}{C}{-}CH_3 \xrightarrow{H_2O,\ OH^-} O{=}CH{-}CH_2{-}\overset{\overset{O}{\|}}{C}{-}CH_3 + PhOH$$

(b)

$$PhO{-}CH{=}CH{-}\overset{\overset{O}{\|}}{C}{-}CH_3 \xrightarrow{H_2O,\ H^+} O{=}CH{-}CH_2{-}\overset{\overset{O}{\|}}{C}{-}CH_3 + PhOH$$

Problems (Cont.)

(c)

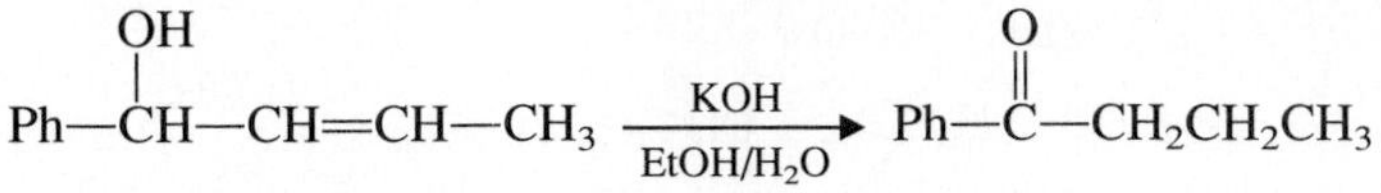

$$Ph-\underset{\underset{OH}{|}}{CH}-CH=CH-CH_3 \xrightarrow[\text{EtOH/H}_2\text{O}]{\text{KOH}} Ph-\underset{\underset{O}{||}}{C}-CH_2CH_2CH_3$$

(d)

$$CH_3-CH=CH-\underset{\underset{O}{||}}{C}-Cl + EtOH \longrightarrow CH_3-\underset{\underset{Cl}{|}}{CH}-CH_2-\underset{\underset{O}{||}}{C}-OEt$$

(e)

$$\text{(enoate)} + CH_3-NH-OH \longrightarrow [A] \longrightarrow \text{(isoxazolidinone)} \quad \text{(Identify } A.)$$

(f)

pyrylium $+ EtNH_2 \longrightarrow$ pyridinium $+ H_2O$

(g) $Ph(CH_2)_3CH=O + N\equiv C-CH_2-CO_2Et \xrightarrow[\substack{1)\ KCN/EtOH \\ 2)\ OH^- \text{ (1 equiv)} \\ 3)\ H_3O^+ \\ 4)\ \text{heat}}]{} Ph(CH_2)_3\underset{\underset{CH_2C\equiv N}{|}}{CHC\equiv N}$

(h)

bicyclic ester $\xrightarrow{^-CN} \xrightarrow[\text{heat}]{\text{H}_2\text{O, H}^+}$ cyclopentanone-CH_2CO_2H

(i)

maleic anhydride $+ Ph_3P: \longrightarrow [A] \xrightarrow[\text{H}_2\text{O}]{\text{CH}_3\text{CH}_2\text{CH}_2\text{CH}=\text{O} \quad \text{H}^+}$

$$CH_3CH_2CH_2CH=\underset{\underset{CO_2H}{|}}{\overset{\overset{CH_2CO_2H}{|}}{C}} + Ph_3\overset{+}{P}-\overset{..}{\underset{..}{O}}:$$

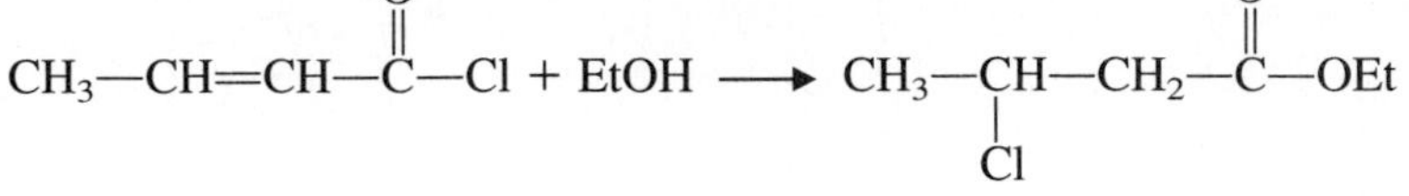

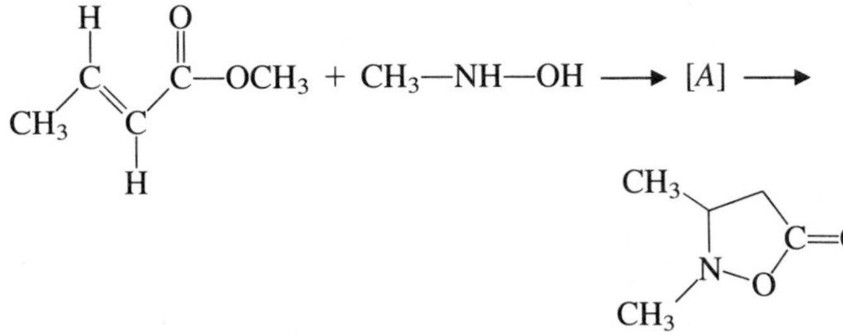

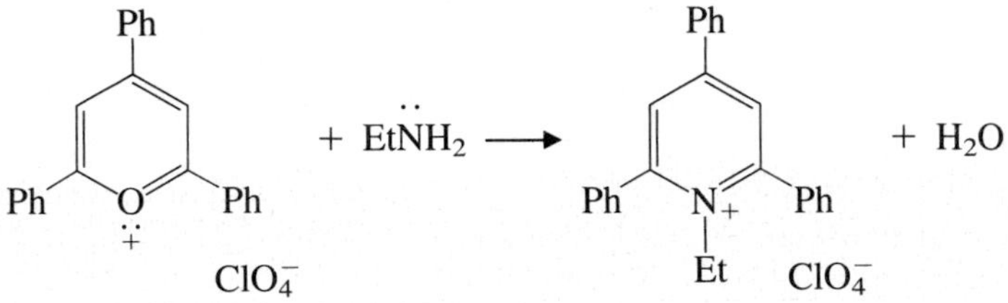

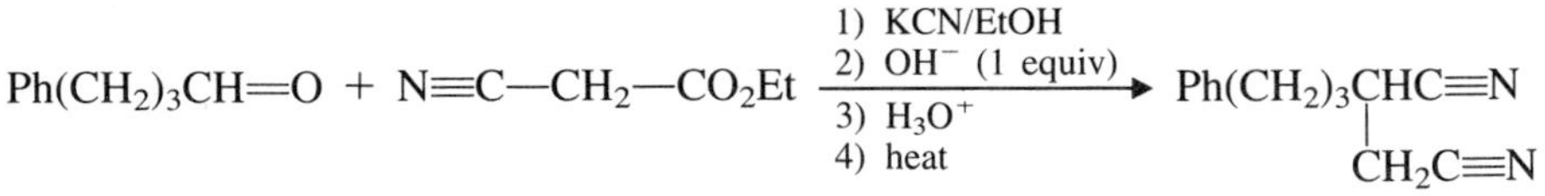

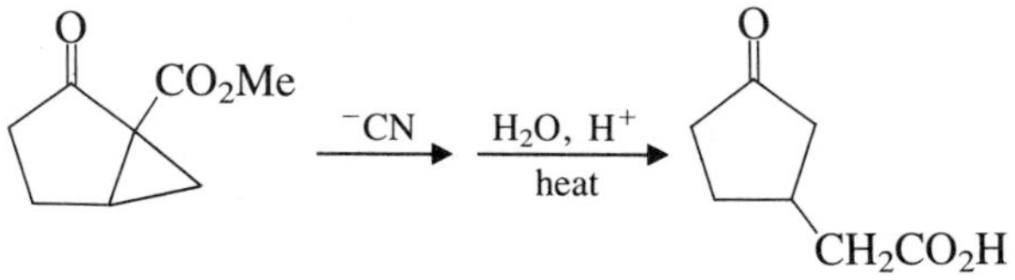

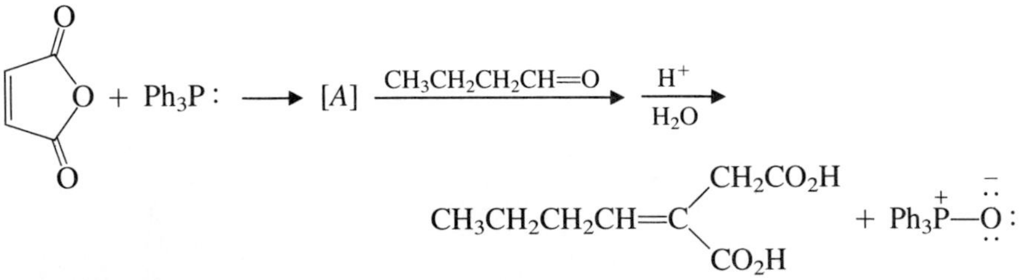

82 The following reaction is an example of the "abnormal Michael reaction." Give a mechanism for this reaction that is consistent with the isotopic labeling pattern shown. (* = ^{14}C)

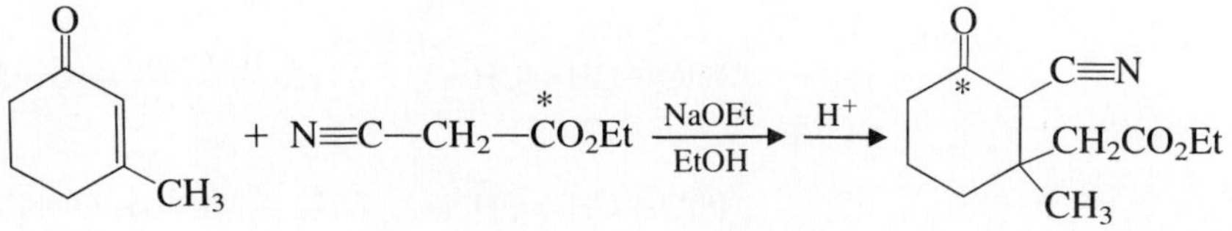

23 Chemistry of Amines

Amines are organic derivatives of ammonia in which the ammonia protons are formally substituted with organic groups. Amines are classified according to the number of hydrogens on the amine nitrogen. A **primary amine** has two N—H protons, a **secondary amine** has one N—H proton, and a **tertiary amine** has no N—H protons.

$R—\ddot{N}H_2$	$R—\ddot{N}H—R$	$R—\underset{\cdot\cdot}{N}(R)—R$
primary amine	secondary amine	tertiary amine

examples:

$C_2H_5—\ddot{N}H_2$	$C_2H_5—\ddot{N}H—C_2H_5$	$C_2H_5—\underset{\cdot\cdot}{N}(C_2H_5)—C_2H_5$
ethylamine	diethylamine (a secondary amine)	triethylamine (a tertiary amine)

(This classification is like that of amides; see Sec. 21.1E.) It is important to distinguish between the classifications of alcohols and amines; alcohols are classified according to the number of protons on the α-carbon, but amines (like amides) are classified according to the number of protons *on nitrogen.*

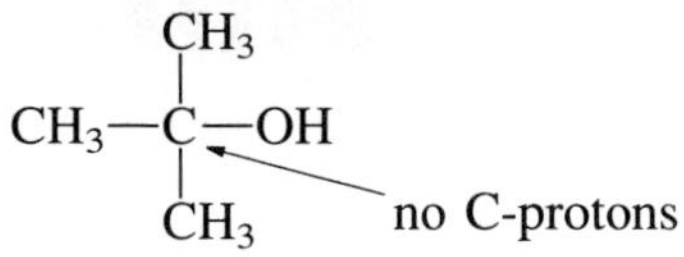

a tertiary alcohol

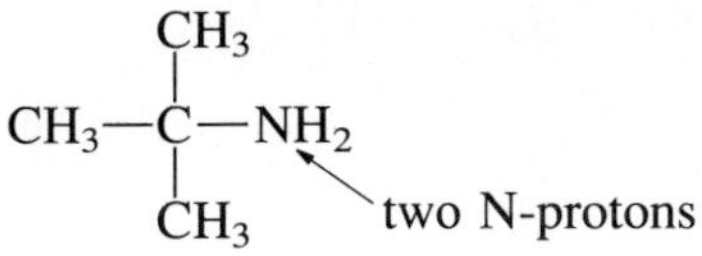

a primary amine (even though the α-carbon is tertiary)

Besides amines we shall also consider briefly in this chapter some other nitrogen-containing compounds that are formed from, or converted into, amines: quaternary ammonium salts, azobenzenes, diazonium salts, acyl azides, and nitro compounds. This overview of the organic chemistry of nitrogen will complete our study of fundamental organic chemistry.

23.1 NOMENCLATURE OF AMINES

A. Common Nomenclature

In common nomenclature an amine is named by appending the suffix *amine* to the name of the alkyl group; the name of the amine is written as one word.

$C_2H_5NH_2$ ethylamine $(CH_3)_3N$ trimethylamine

When two or more alkyl groups in a secondary or tertiary amine are different, the compound is named as an *N*-substituted derivative of the larger group.

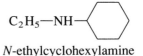

$(CH_3)_2N—CH_2CH_2CH_2CH_3$
N,N-dimethylbutylamine

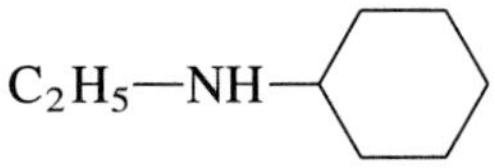
N-ethylcyclohexylamine

This type of notation is required to show that the substituents are on the amine nitrogen and not on an alkyl group carbon.

Aromatic amines are named as derivatives of aniline.

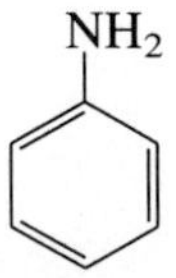
aniline

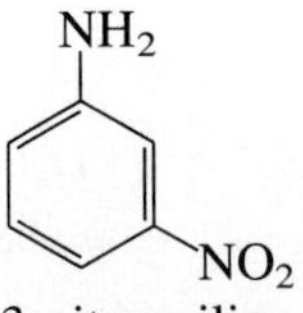
3-nitroaniline (*m*-nitroaniline)

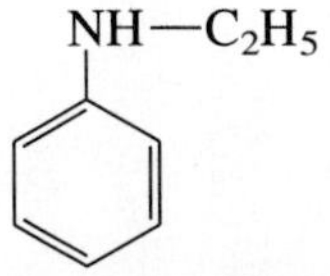
N-ethylaniline

The anilines with a methyl group on the benzene ring are called *toluidines*.

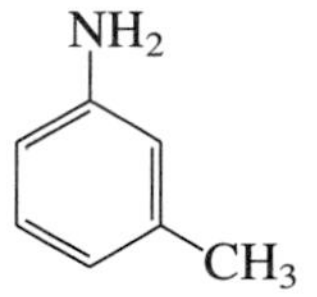

m-toluidine

B. Systematic Nomenclature

We shall not consider the IUPAC system for amine nomenclature because it is not logically consistent with the IUPAC nomenclature of other organic compounds. The most widely used system of amine nomenclature is that of *Chemical Abstracts,* a comprehensive index to the world's chemical literature. In this system we name an amine systematically much as we would name the analogous alcohol, except that the suffix *amine* is used.

$$\underset{\text{2-pentanol}}{\overset{\overset{\displaystyle OH}{|}}{CH_3CHCH_2CH_2CH_3}} \qquad \underset{\text{2-pentanamine}}{\overset{\overset{\displaystyle NH_2}{|}}{CH_3CHCH_2CH_2CH_3}}$$

$$\underset{\textit{N}\text{-methylcyclohexanamine}}{\bigcirc\!-NH-CH_3} \qquad \underset{\text{1,5-pentanediamine}}{H_2N-CH_2CH_2CH_2CH_2CH_2-NH_2}$$

$$\underset{\textit{N}\text{-ethyl-}\textit{N}'\text{-methyl-1,3-propanediamine}}{CH_3-NH-CH_2CH_2CH_2-NH-C_2H_5}$$

Notice that in diamine nomenclature, as in diol nomenclature, the final *e* of the hydrocarbon name is retained. In the last example, the prime is used to show that the ethyl and methyl groups are on two different nitrogens.

The priority for citation of amine groups as principal groups is just below that of alcohols:

$$\underset{\text{(carboxylic acid)}}{-\overset{\overset{\displaystyle O}{\|}}{C}-OH} \text{ and derivatives} > \underset{\text{(aldehyde)}}{-\overset{\overset{\displaystyle O}{\|}}{C}-H}, \ \underset{\text{(ketone)}}{-\overset{\overset{\displaystyle O}{\|}}{C}-} > -OH > -NR_2 \quad (23.1)$$

(A complete list of group priorities is given in Appendix I.) When cited as a substituent, the $-NH_2$ group is called the **amino** group.

$$\underset{\substack{\text{2-amino-1-ethanol}\\\text{(OH has priority)}}}{H_2N-CH_2CH_2-OH} \qquad \underset{\substack{\text{2-propen-1-amine (common: allylamine)}\\(NH_2 \text{ has priority})}}{H_2N-\overset{1}{C}H_2-\overset{2}{C}H=\overset{3}{C}H_2}$$

$$\underset{\substack{\\ \\ \text{5-chloro-}\textit{N}\text{-ethyl-}\textit{N},3\text{-dimethyl-1-pentanamine}}}{\overset{\overset{\displaystyle CH_3}{|}}{C_2H_5NCH_2CH_2CHCH_2CH_2Cl}\atop\underset{\displaystyle CH_3}{|}} \qquad \underset{\text{3-(dimethylamino)-1-propanol}}{(CH_3)_2N-CH_2CH_2CH_2-OH}$$

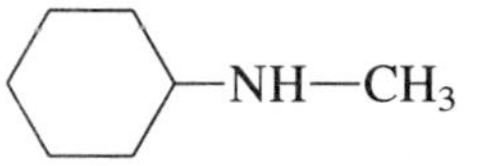

An *N* designation in the last example is unnecessary because the position of the methyl groups is clear from the parentheses.

Although *Chemical Abstracts* calls aniline *benzenamine,* we shall follow the conventional practice of using the common name *aniline* in systematic nomenclature.

The nomenclature of *heterocyclic compounds* was introduced in Sec. 8.1C in our discussion of ether nomenclature. There are many important nitrogen-containing heterocyclic compounds, most of which are known by specific names that must be learned. Some important saturated heterocyclic amines are the following:

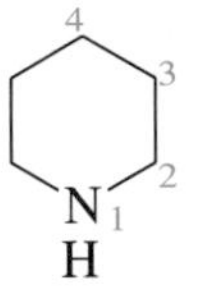
piperidine

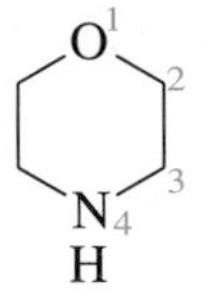
morpholine

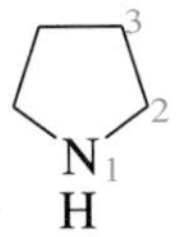
pyrrolidine

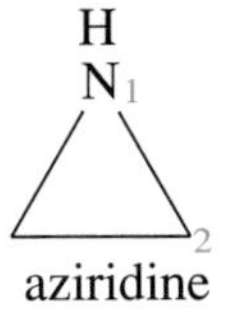
aziridine

As in the oxygen heterocyclics, numbering generally begins with the heteroatom. Following are examples of substituted derivatives:

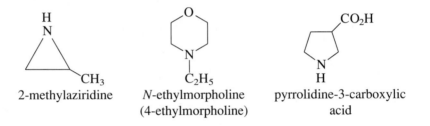

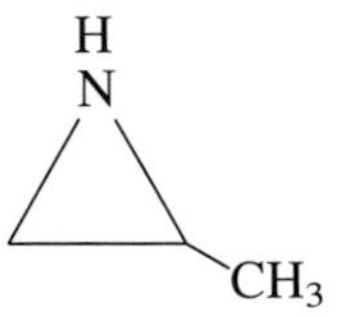
2-methylaziridine

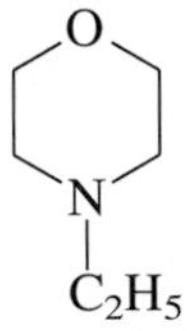
N-ethylmorpholine
(4-ethylmorpholine)

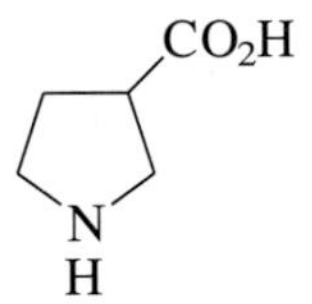
pyrrolidine-3-carboxylic acid

To a useful approximation, much of the chemistry of the saturated heterocyclic amines parallels the chemistry of the corresponding open-chain amines.

Problems

1 Draw the structure for each of the following compounds:
(a) *N*-isopropylaniline
(b) *t*-butylamine
(c) 3-methoxypiperidine
(d) 2,2-dimethyl-3-hexanamine
(e) ethyl 2-(diethylamino)pentanoate
(f) *N*,*N*-diethyl-3-heptanamine

2 Give an acceptable name for each of the following compounds:

(a)
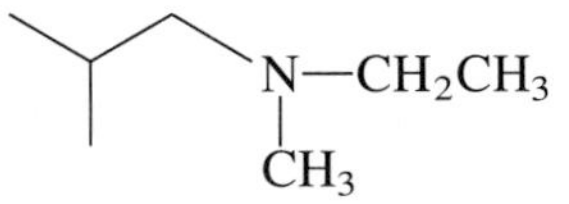

(b)
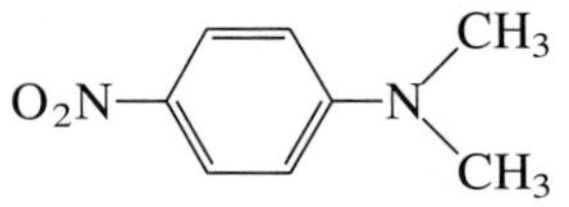

(c)
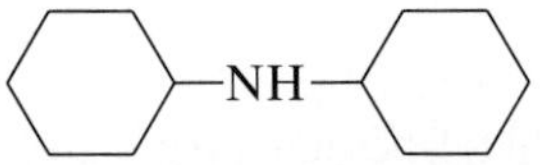

(d) $CH_3NH\underset{C_2H_5}{\underset{|}{C}}HCH_2CH_2OH$

23.2 STRUCTURE OF AMINES

The C—N bonds of aliphatic amines are longer than the C—O bonds of alcohols, but shorter than the C—C bonds of alkanes, as we would expect from the effect of electronegativity on bond length (Sec. 1.6B).

bond:	C—C	C—N	C—O	C—F
typical length:	1.54 Å	1.47 Å	1.43 Å	1.39 Å

Aliphatic amines have a pyramidal shape (or approximately tetrahedral shape, if we regard the electron pair as a "group"). The structures of some amines are shown in Fig. 23.1. Most amines undergo rapid *inversion* at nitrogen, which occurs through a planar transition state and converts an amine into its mirror image (Sec. 6.8B).

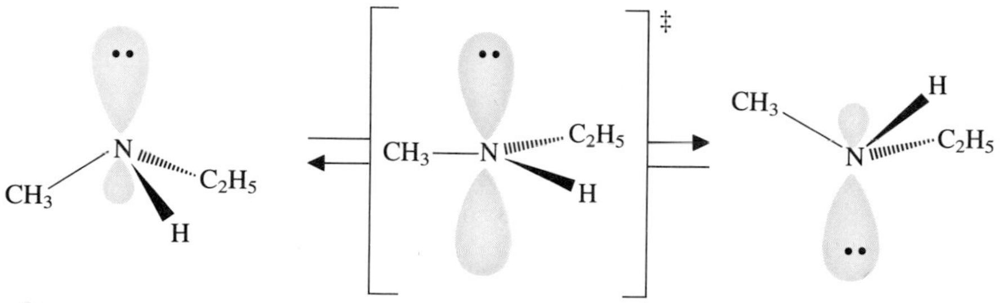

(23.2)

Because of this inversion, amines in which the only asymmetric atom is the amine nitrogen cannot be resolved into enantiomers.

The C—N bond in aniline is shorter (1.40 Å) than that in aliphatic amines. This reflects both the sp^2 hybridization of the adjacent carbon and overlap of the unshared electrons on nitrogen with the π-electron system of the ring. This overlap, shown by the following resonance structures, gives some double-bond character to the C—N bond.

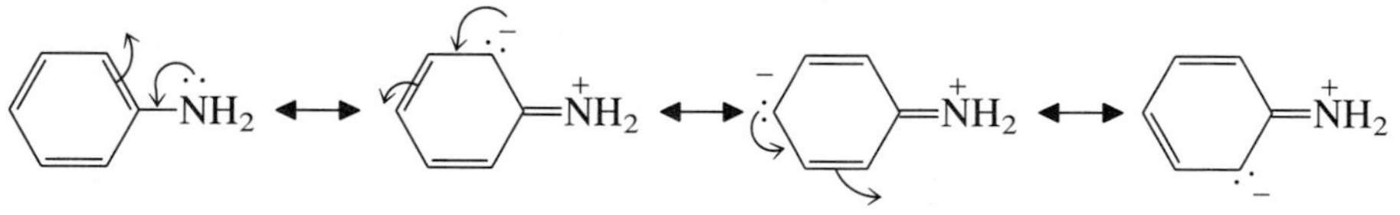

(23.3)

Figure 23.1 Structures of some amines.

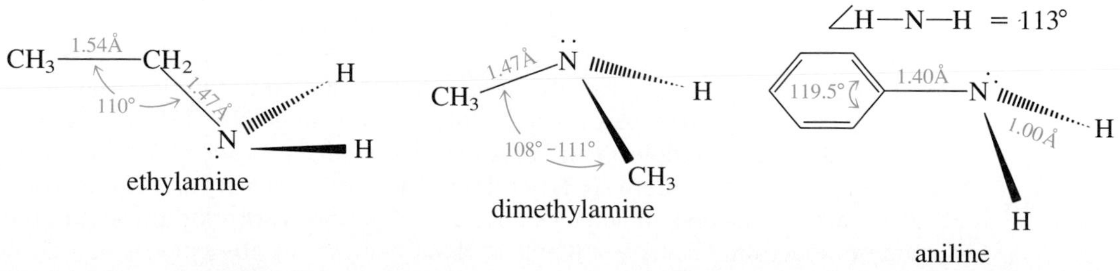

Problem

3 Arrange the following compounds in order of increasing C—N bond length. Explain your answer.
(a) $HN{=}CH_2$ (b) $CH_3{-}NH_2$ (c) $CH_2{=}CH{-}NH_2$

23.3 PHYSICAL PROPERTIES OF AMINES

Most amines are somewhat polar liquids with unpleasant odors that range from fishy to putrid. Primary and secondary amines, which can both donate and accept hydrogen bonds, have higher boiling points than isomeric tertiary amines, which cannot donate hydrogen bonds.

	$CH_3CH_2CH_2CH_2NH_2$	$(C_2H_5)_2NH$	$C_2H_5N(CH_3)_2$	$C_2H_5CH(CH_3)CH_3$
boiling point	77.8°	56.3°	37.5°	27.8°
dipole moment	1.4 D	1.2–1.3 D	0.6 D	0 D

———— decreased hydrogen bonding and polarity ————→

The fact that primary and secondary amines can both donate and accept hydrogen bonds also accounts for the fact that they have higher boiling points than ethers. On the other hand, alcohols are better hydrogen-bond donors than amines because they are more acidic. Therefore, alcohols have higher boiling points than amines.

	$(C_2H_5)_2NH$	$(C_2H_5)_2O$	$(C_2H_5)_2CH_2$
boiling point	56.3°	37.5°	36.1°

	$CH_3CH_2CH_2CH_2OH$	$CH_3CH_2CH_2CH_2NH_2$
boiling point	117.3°	77.8°

Most primary and secondary amines with four or fewer carbons, as well as trimethylamine, are miscible with water, a fact consistent with their hydrogen-bonding ability. Amines with large carbon groups have little or no water solubility.

23.4 SPECTROSCOPY OF AMINES

A. IR Spectroscopy

The most important absorptions in the infrared spectra of primary amines are the N—H stretching absorptions, which usually occur as two or more peaks at 3200–3375 cm^{-1}. Also characteristic of primary amines is an NH_2 scissoring absorption (see Fig. 12.7) near 1600 cm^{-1}. These absorptions are illustrated in the IR spectrum of butylamine (Fig. 23.2). Most secondary amines show a single N—H stretching absorption rather than the multiple peaks observed for primary amines, and the absorptions associated with the various NH_2 bending vibrations of primary amines are not present. For example, diethylamine lacks the NH_2 scissoring absorption present in the butylamine spectrum. Tertiary amines obviously show no absorptions associated with

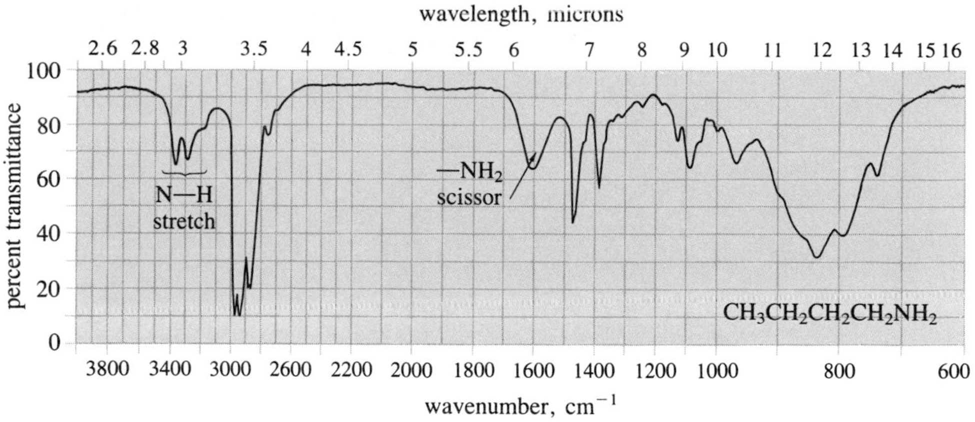

Figure 23.2 IR spectrum of butylamine.

N—H vibrations. The C—N stretching absorptions of amines, which occur in the same general part of the spectrum as the C—O stretching absorptions (1050–1225 cm^{-1}), are not very useful.

B. NMR Spectroscopy

The characteristic resonances in the NMR spectra of amines are due to the protons adjacent to the nitrogen (the α-protons) and the N—H protons. In alkylamines the α-protons are observed in the δ 2.5–3.0 region of the spectrum. In aromatic amines, the α-protons of *N*-alkyl groups are somewhat further downfield (why?), near δ 3. The following chemical shifts are typical:

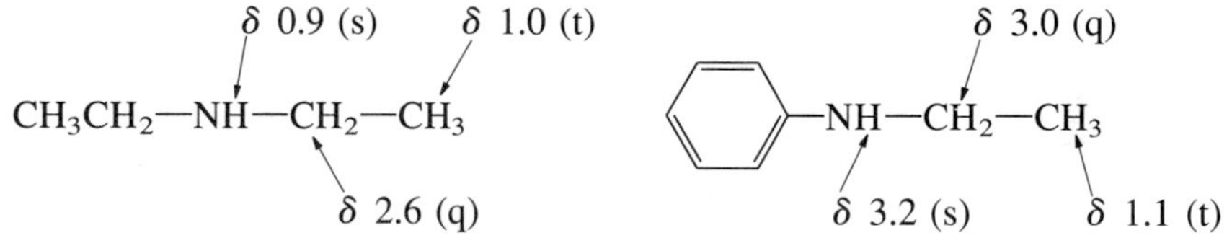

The chemical shift of the N—H proton, like that of the O—H proton in alcohols, depends on the concentration of the amine, and on other conditions of the NMR experiment. In alkylamines, this resonance typically occurs at rather high field—typically around δ 1. In aromatic amines, this resonance is at considerably lower field, as we can see in the example above.

Like the OH proton of alcohols, phenols, and carboxylic acids, the NH proton of amines under most conditions undergoes rapid exchange (Secs. 13.5D and 13.7). For this reason splitting between the amine N—H and adjacent C—H groups is usually not observed. Thus, in the NMR spectrum of diethylamine the N—H resonance is a singlet rather than the triplet we would expect if splitting by the adjacent —CH_2— protons were observed. In some amine samples the N—H resonance is broadened and, like the O—H proton of alcohols, it can be obliterated from the spectrum by exchange with D_2O.

In aromatic amines the absorptions of the ring protons *ortho* and *para* to the nitrogen are shifted to higher field than other aromatic-ring protons, as shown in the following example. This shift reflects the donation of electrons by resonance from the nitrogen onto the *ortho* and *para* carbons of the ring (Eq. 23.3).

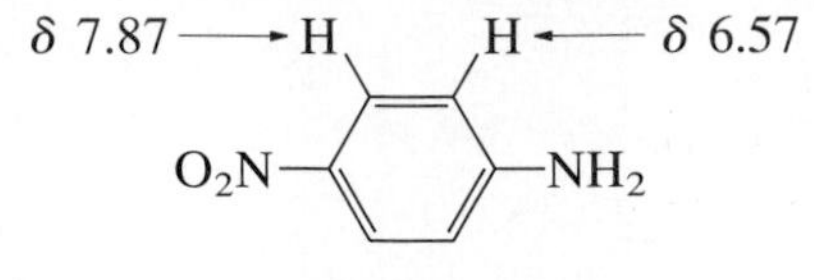

C. Mass Spectrometry

As we learned in Sec. 21.4C, compounds containing an odd number of nitrogens have odd molecular weights. It follows that in the mass spectrum of an amine, the parent ion occurs at an *odd mass* if the amine contains an *odd* number of nitrogens. Like the parent ion, odd-electron ions containing an odd number of nitrogens are observed at odd mass, but even-electron ions containing an odd number of nitrogens are observed at even mass.

α-Cleavage (Sec. 12.5C) is a particularly important fragmentation mode of aliphatic amines. For example, in the mass spectrum of butylamine, the only significant peaks are the parent at $m/e = 73$ (note the odd mass) and the base peak at $m/e = 30$. The even mass of the latter peak tells us that it is attributable either to an odd-electron ion containing no nitrogen or an even-electron ion containing one nitrogen. In fact this peak can be rationalized by an α-cleavage mechanism, and is thus attributable to an even-electron ion:

$$C_2H_5{-}CH_2{-}CH_2{-}\ddot{N}H_2 \xrightarrow{-e^-} C_2H_5{-}CH_2{-}CH_2{-}\dot{\overset{+}{N}}H_2 \xrightarrow{\alpha\text{-cleavage}} C_2H_5{-}\dot{C}H_2 + CH_2{=}\overset{+}{N}H_2 \quad (23.4)$$
$$m/e = 30$$

Problems

4 A compound has IR absorptions at 3400–3500 cm^{-1} and the following NMR spectrum:
δ 2.07 (6*H*, s), δ 2.16 (3*H*, s), δ 3.19 (broad, exchanges with D_2O), δ 6.63 (2*H*, s). To which one of the following compounds do these spectra belong? Explain.
(a) 2,4-dimethylbenzylamine
(b) 2,4,6-trimethylaniline
(c) *N,N*-dimethyl-*p*-methylaniline
(d) 3,5-dimethyl-*N*-methylaniline

5 Identify the compound that has the following spectra:
IR spectrum: 3279 cm^{-1}
NMR spectrum: δ 0.91 (1*H*, s), δ 1.07 (3*H*, t, $J = 7$ Hz), δ 2.60 (2*H*, q, $J = 7$ Hz), δ 3.70 (2*H*, s), δ 7.18 (5*H*, apparent s)
Mass spectrum: $m/e = 135$ (*p*, 17%), 120 (40%), 91 (base peak)

6 In an abandoned laboratory you have found two vials, *A* and *B*, each containing a very small amount of material. Near the vials are lying two labels; one bears the name "2-methyl-2-heptanamine" and the other "*N*-ethyl-4-methyl-2-pentanamine." There is only enough of each compound for a mass spectrum. The mass spectrum of sample *A* has a base peak at $m/e = 72$, and that of sample *B* has a base peak at $m/e = 58$. Which amine is contained in each vial?

23.5 BASICITY AND ACIDITY OF AMINES

A. Basicity of Amines

Amines, like ammonia, are good bases, and are completely protonated in dilute mineral acids.

$$CH_3\ddot{N}H_2 + H{-}Cl \rightleftharpoons CH_3{-}\overset{+}{N}H_3\ Cl^- + H_2O \qquad (23.5)$$

methylammonium chloride

Protonated amines are called **ammonium salts.** The ammonium salts of simple alkylamines are named as substituted derivatives of the ammonium ion. Other ammonium salts are named by replacing the final *e* in the name of the amine with the suffix *ium*.

$(CH_3)_2\overset{+}{N}H_2\ Cl^-$

dimethylammonium chloride

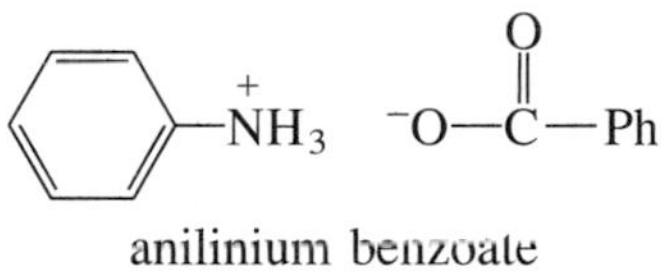

anilinium benzoate
(aniline + *ium*)

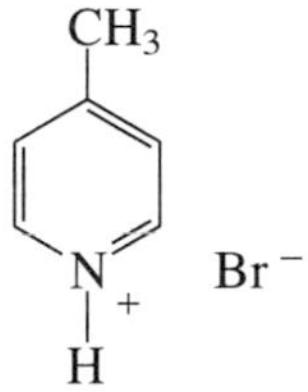

4-methylpyridinium bromide

It is important to understand that ammonium salts are fully ionic compounds. Although we often see ammonium chloride written as NH_4Cl, the structure is more properly represented as $^+NH_4\ Cl^-$. Although the N—H bonds are covalent, there is no covalent bond between the nitrogen and the chlorine. (A covalent bond would violate the octet rule.)

The basicity of an amine is conveniently expressed by the pK_a of its conjugate-acid ammonium salt. For the equilibrium

$$R\overset{+}{N}H_3 + H_2\ddot{O}: \rightleftharpoons R\ddot{N}H_2 + H_3\ddot{O}^+ \qquad (23.6)$$

the *dissociation constant* K_a is defined by the following expression:

$$K_a = \frac{[R{-}NH_2][H_3O^+]}{[R{-}\overset{+}{N}H_3]} \qquad (23.7)$$

and pK_a as usual, is $-\log K_a$. The higher the pK_a of the ammonium ion, the more basic is its conjugate-base amine.

In older literature the **basicity constant** K_b was used to measure amine basicity. This is essentially a measure of the ability of the amine to accept a proton from water rather than from H_3O^+.

$$R{-}NH_2 + H_2O \rightleftharpoons R{-}\overset{+}{N}H_3 + OH^- \qquad (23.8)$$

The equilibrium constant for this reaction is

$$K_{eq} = \frac{[R{-}NH_3^+][OH^-]}{[R{-}NH_2][H_2O]} \quad (23.9)$$

Since $[H_2O]$ is constant, it is convenient to incorporate it into the equilibrium constant. Multiplying through Eq. 23.9 by $[H_2O]$ gives the *basicity constant,* K_b:

$$\text{basicity constant} = K_{eq}[H_2O] = K_b = \frac{[R{-}NH_3^+][OH^-]}{[R{-}NH_2]} \quad (23.10)$$

By multiplying through numerator and denominator by $[H_3O^+]$, we can easily show that

$$K_b = \frac{K_w}{K_a}, \quad (23.11)$$

where $K_w = [H_3O^+][OH^-]$, which is the autoprotolysis constant of water ($= 10^{-14}$ $mol^2/liter^2$). (Verify Eq. 23.11.) Taking negative logarithms,

$$pK_b = 14 - pK_a \quad (23.12)$$

Although in this text we shall use ammonium-ion pK_a values as the quantitative measure of amine basicity, calculation of the amine pK_b from the pK_a of the ammonium salt is a simple matter with Eq. 23.12. A higher pK_a value for an ammonium salt implies a higher basicity (and lower pK_b) for its conjugate-base amine.

B. Substituent Effects on Amine Basicity

The pK_a values for the conjugate acids of some representative amines are given in Table 23.1. As we can see from this table, the exact basicity of an amine depends on the groups with which the nitrogen is substituted. Three types of effects influence the basicity of amines—the same effects that influence the acid–base properties of other compounds. They are:

1. The effect of alkyl substitution
2. The inductive effect
3. The resonance effect

To understand these effects we must realize that the pK_a of an ammonium ion, like that of any other acid, is directly related to the standard free-energy difference ΔG^0 between it and its conjugate base by Eq. 8.17b, which we repeat here:

$$\Delta G^0 = 2.3RT\ pK_a \quad (23.13)$$

The effect of a substituent group on pK_a can be analyzed in terms of how it affects the energy of either an ammonium ion or its conjugate-base amine, as shown in Fig. 23.3.

TABLE 23.1 Basicities of Some Amines (All pK_a values are for dissociation of the corresponding ammonium ion.)

Amine	pK_a	Amine	pK_a	Amine	pK_a
CH_3NH_2	10.62	$(CH_3)_2NH$	10.64	$(CH_3)_3N$	9.76
$C_2H_5NH_2$	10.63	$(C_2H_5)_2NH$	10.98	$(C_2H_5)_3N$	10.65
$PhCH_2NH_2$	9.34				
$PhNH_2$	4.62	$PhNHCH_3$	4.85	$PhN(CH_3)_2$	5.06
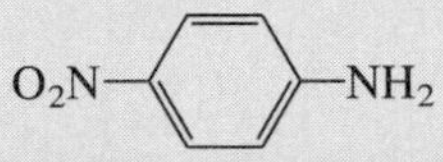	~1.0	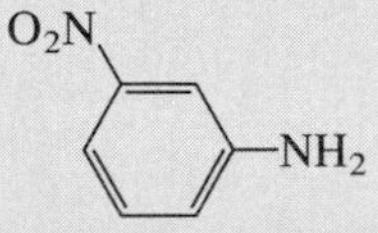	2.45		
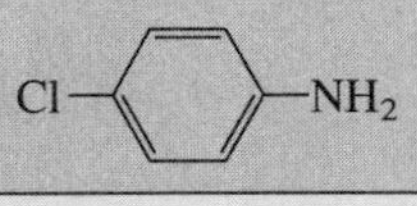	3.81	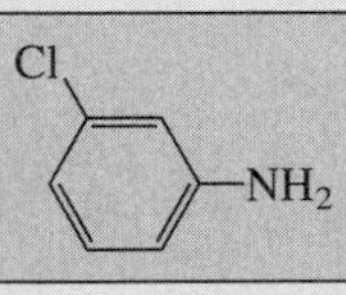	3.32	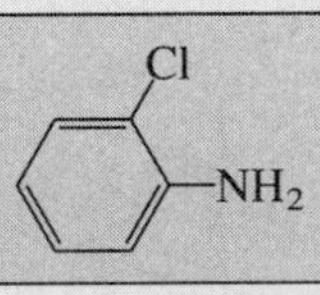	2.62
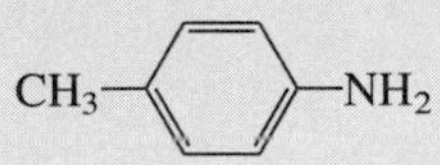	5.07	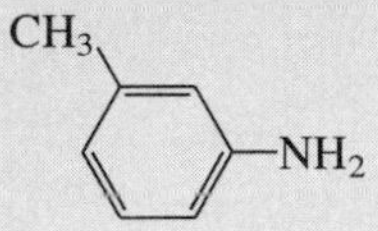	4.67	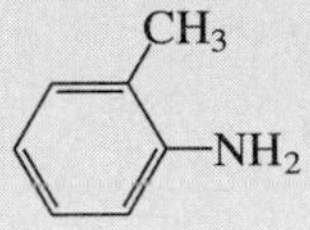	4.38

Figure 23.3 Effect of the relative standard free energies of ammonium ions and amines on the pK_a values of ammonium ions. Reading from the center to the right, for example, we see that destabilization of RNH_2 or stabilization of RNH_3^+ increases the ammonium-ion pK_a (increases the amine basicity).

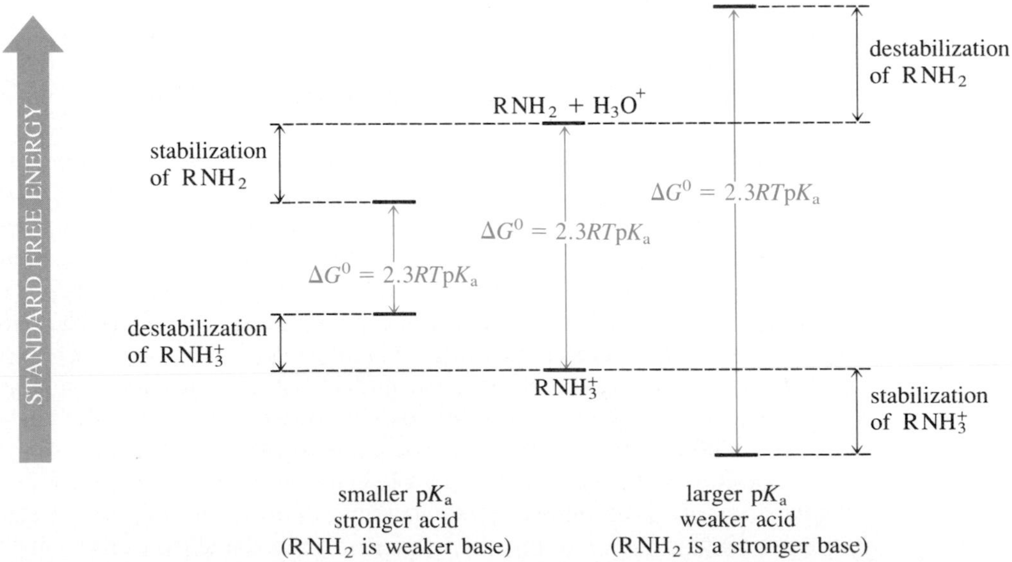

For example, if a substituent stabilizes the ammonium ion more than the amine, the free energy of the ammonium ion is lowered, ΔG^0 is increased, and pK_a is raised. If a substituent destabilizes the ammonium ion relative to the amine, the opposite effect is observed: the pK_a is lowered. Or, if a substituent group stabilizes the amine relative to the ammonium ion, the pK_a of the ammonium ion is also lowered.

Let us first consider the effect of alkyl substitution. Most common alkylamines are somewhat more basic than ammonia in aqueous solution:

pK_a *in aqueous solution:*

$$\begin{array}{cccc} \overset{+}{N}H_4 & Et\overset{+}{N}H_3 & Et_2\overset{+}{N}H_2 & Et_3\overset{+}{N}H \\ 9.21 & 10.63 & 10.98 & 10.65 \end{array} \qquad (23.14)$$

But if we look at the pK_a values along this series, it is interesting that there is neither a simple increase nor decrease with increasing substitution: the secondary amine is the most basic. Two factors are actually at work here. The first is the tendency of alkyl groups to stabilize charge through a *polarization* effect. The electron clouds of the alkyl groups distort so as to create a net attraction between them and the positive charge of the ammonium ion:

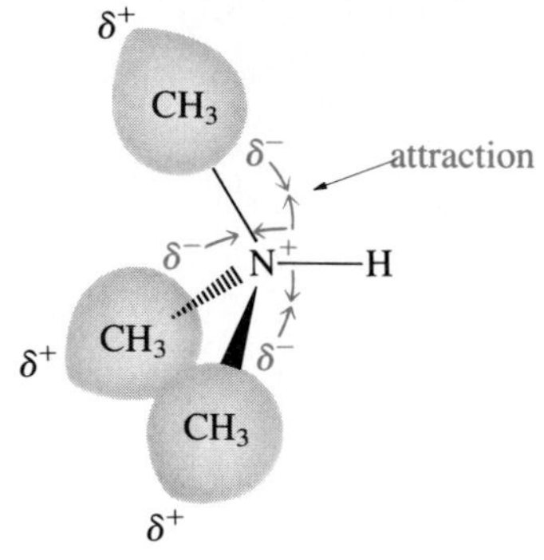

Because the ammonium ion is stabilized by this effect, its pK_a is increased. This effect is clearly evident in the *gas phase* basicity of amines. In the gas phase, the pK_a values of ammonium ions increase regularly with increasing alkyl substitution:

gas phase pK_a:

$$\overset{+}{N}H_4 < C_2H_5\overset{+}{N}H_3 < (C_2H_5)_2\overset{+}{N}H_2 < (C_2H_5)_3\overset{+}{N}H \qquad (23.15)$$

The polarization effect also operates in the gas-phase acidity of alcohols (Sec. 8.6C), except that it is in the opposite direction. In other words, the polarization of alkyl groups can act to stabilize *either* positive or negative charge. In the presence of a positive charge, as in ammonium ions, electrons polarize *toward* the charge; in the presence of a negative charge, as in alkoxide ions, they polarize *away from* the charge. We might say that electron clouds are like some politicians: they polarize in whatever way is necessary to create the most favorable situation.

The second factor involved in the effect of alkyl substitution on amine basicity must be a *solvent effect,* because the basicity order of amines in the gas phase (Eq. 23.15) is different from that in aqueous solution (Eq. 23.14). In other words, the solvent water must play an important role in the solution basicity of amines. A simple explanation of this solvent effect is that ammonium ions in solution are stabilized not only by alkyl groups, but also by hydrogen-bond donation to the solvent:

$$\begin{array}{c}
H\text{---}:\overset{..}{O}H_2 \\
| \\
CH_3\text{---}\overset{+}{N}\text{---}H\text{---}:\overset{..}{O}H_2 \\
| \\
H\text{---}:\overset{..}{O}H_2
\end{array}$$

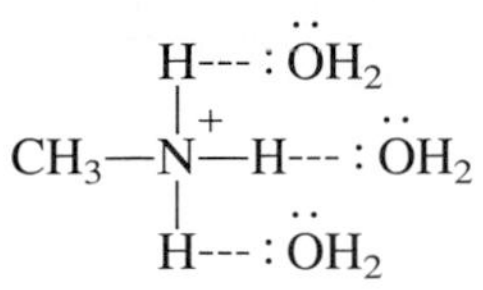

Primary ammonium salts have three hydrogens that can be donated to form hydrogen bonds, but a tertiary ammonium salt has only one. Furthermore, there is probably some steric hindrance to hydrogen bonding in a tertiary ammonium ion. Thus, primary ammonium ions are stabilized by hydrogen bonding more than tertiary ones.

The pK_a values of alkylammonium salts reflect the operation of both hydrogen bonding and alkyl-group polarization. Because these effects work in opposite directions, the basicity in Eq. 23.14 is maximum for the secondary amine.

Ammonium-ion pK_a values, like the pK_a values of other acids, are also sensitive to *inductive effects* of substituents.

$$\text{Et}_2\overset{+}{N}HCH_2C\equiv N \qquad \text{Et}_2\overset{+}{N}HCH_2CH_2C\equiv N \qquad \text{Et}_2\overset{+}{N}H(CH_2)_4C\equiv N \qquad \text{Et}_3\overset{+}{N}H \qquad (23.16)$$

pK_a 4.55 7.65 10.08 10.65

An electronegative (electron-withdrawing) group such as halogen or cyano destabilizes an ammonium ion because of a repulsive electrostatic interaction between the positive charge on the ammonium ion and the positive end of the substituent bond dipole.

$$\text{repulsive interaction}$$
$$\overset{\delta-}{N}\equiv C\overset{\delta+}{\text{---}}CH_2\text{---}CH_2\text{---}\overset{+}{N}H_3$$

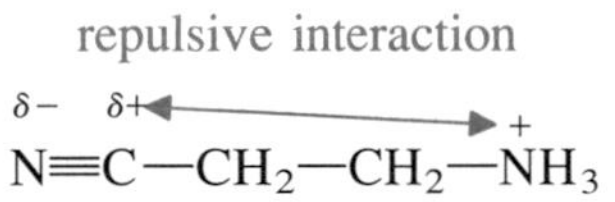

The data in Eq. 23.16 show that the base-weakening effect of electron-withdrawing groups, like all inductive effects, falls off rapidly with distance.

Resonance effects on amine basicity are illustrated by the difference in pK_a values between the conjugate acids of aniline and cyclohexylamine, two primary amines of almost the same shape and molecular weight.

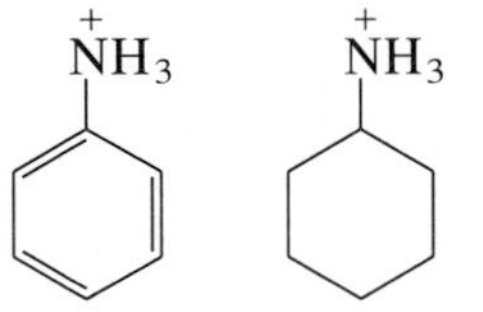

pK_a 4.62 10.64

(Notice that all the substituted anilinium ions in Table 23.1 have considerably lower pK_a values than the alkylammonium ions.) Aniline is stabilized by resonance interaction of the lone electron pair on nitrogen with the aromatic ring (Eq. 23.3). When aniline is protonated, this resonance stabilization is no longer present, because the lone pair is bound to a proton and is "out of circulation." The stabilization of aniline relative to its conjugate acid reduces its basicity by Eq. 23.13 and Fig. 23.3. In other words, the resonance stabilization of aniline reduces its energy and lowers its basicity relative to that of cyclohexylamine, in which the resonance effect is absent. (An electron-withdrawing inductive effect of the aromatic ring also contributes significantly to the reduced basicity of aromatic amines.)

Problems

7 Explain the basicity order of the following three amines:

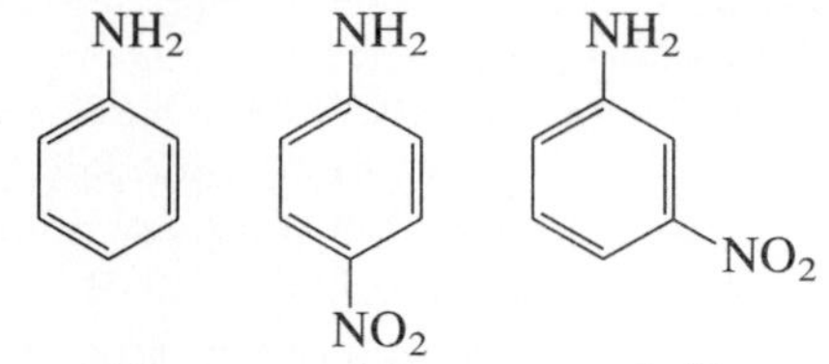

conjugate-acid pK_a 4.62 ~1.0 2.45

8 Arrange the amines within each set in order of increasing basicity, least basic first.
(a) propylamine, ammonia, dipropylamine
(b) methyl 3-aminopropanoate, *sec*-butylamine, $H_3\overset{+}{N}CH_2CH_2NH_2$
(c) aniline, methyl *m*-aminobenzoate, methyl *p*-aminobenzoate
(d) benzylamine, *p*-nitrobenzylamine, cyclohexylamine, aniline

C. Separations Using Amine Basicity

Because ammonium salts are ionic compounds, many have appreciable water solubilities. Hence, when a water-insoluble amine is treated with dilute aqueous mineral acid—for example, 5% HCl solution—the amine dissolves as its ammonium salt. Upon treatment with base, the ammonium salt is converted back into the amine. These observations can be used to design separations of amines from other compounds. Suppose, for example, that we wish to separate the following two compounds:

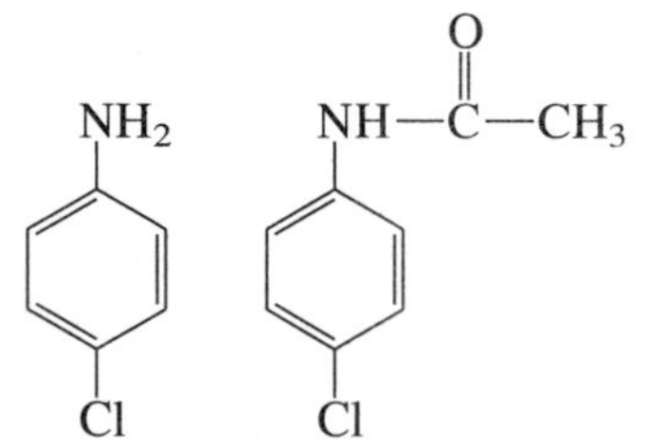

If the mixture is treated with 5% aqueous HCl, the amine will form the hydrochloride salt and dissolve. The amide, however, is not basic enough to be protonated in 5% HCl, and therefore does not dissolve. The amide can be filtered or extracted away from the aqueous solution of the ammonium salt, which can then be made basic to liberate the free amine.

D. Use of Amine Basicity in Optical Resolution

Obviously, any separation of the components of a mixture must be based on the different properties of the components. Separation by fractional distillation, for example, takes advantage of different boiling points. Separation by crystallization takes advantage of different solubilities. We can appreciate that the separation of a *racemate* into its enantiomeric components presents special problems when we recall that *a pair of enantiomers have identical physical properties* (Sec. 6.3). Yet such separations are routinely carried out. The separation of an enantiomeric pair of compounds is called an **optical resolution.** The very first optical resolution by human hands was carried out by Louis Pasteur (Sec. 6.10), who separated the enantiomeric crystals of a tartaric acid salt with tweezers under a microscope. Fortunately, we do not have to go to such lengths for most optical resolutions, since there are chemical methods that can be used. Let us see how we might go about designing an optical resolution of a chiral amine.

Suppose, for example, we wish to separate the racemate of α-phenethylamine, or (±)-1-phenyl-1-ethanamine, into its components.

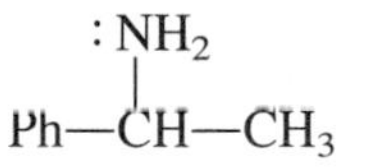

$$: NH_2$$
$$Ph-\overset{|}{C}H-CH_3 \qquad \text{α-phenethylamine}$$

The salt-forming characteristics of amines can be used in this resolution. Suppose we have available a pure enantiomer of a chiral carboxylic acid. Such optically active compounds are available from natural sources—for example, $(2R,3R)$-(+)-tartaric acid:

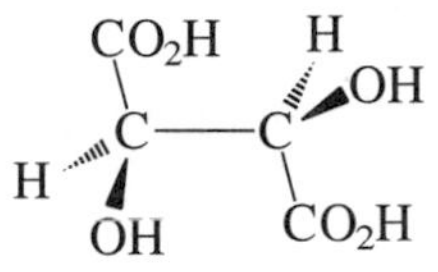

$(2R,3R)$-(+)-tartaric acid

The key to the optical resolution is that the tartaric acid is *optically pure*—it is a single enantiomer. If we allow (+)-tartaric acid to react with the racemic amine, a *mixture* of two salts is formed.

diasteromeric
salts

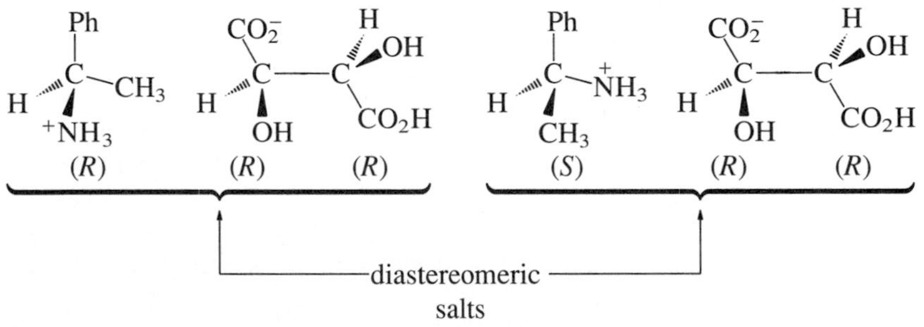

The salt is a mixture because the starting amine is a mixture—a mixture of enantiomers. However, the salt is a mixture of two *diastereomers* (why?). Now as we learned in Sec. 6.6, diastereomers have different physical properties. It turns out that the two salts we have just formed have significantly different solubilities in methanol. The (S,R,R) diastereomer crystallizes from methanol, leaving the (R,R,R) diastereomer in solution, from which it may be recovered. Once either pure diastereomer is in hand, the salt can be decomposed with base to liberate the water-insoluble, optically active amine, leaving the tartaric acid in solution as its conjugate-base anion.

$$\text{NaOH} + \qquad\qquad\qquad\qquad\qquad\qquad \longrightarrow \qquad\qquad\qquad\qquad\qquad + \text{H}-\text{OH} \qquad (23.17)$$

salt

(insoluble in water)

(soluble in water)

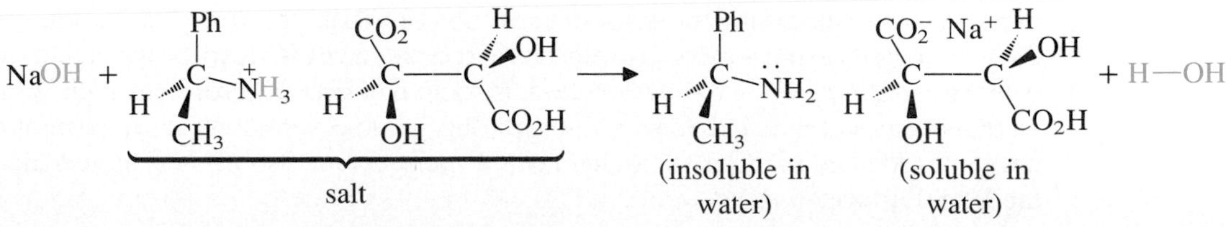

What we have done here is involve the amine *temporarily* in diastereomer formation to endow its two enantiomers with different physical properties. Salt formation is such a rapid, quantitative, and easily reversed reaction that it is often used for the optical resolution of carboxylic acids and amines.

Problems

9 Design a separation of a mixture containing *p*-chlorobenzoic acid, *p*-chloroaniline, and *p*-chlorotoluene into its pure components. Indicate what you would do and what you would expect to observe.

10 Suppose you have performed the optical resolution of α-phenethylamine described in the text above and have in hand the pure enantiomers of this compound. Given the racemate of 2-phenylpropanoic acid, outline the steps you would use to carry out an optical resolution of this compound.

11 Explain why a pure enantiomer of tartaric acid, rather than the racemate, must be used in the optical resolution discussed in the text above.

E. Acidity of Amines

Although we normally think of amines as bases, primary and secondary amines are also weakly acidic. The conjugate base of an amine is called an **amide** (not to be confused with amide derivatives of carboxylic acids).

$$2Na + 2\ddot{N}H_3 \xrightarrow{Fe^{3+}} \underset{\substack{\text{sodium amide} \\ \text{(sodamide)}}}{2Na^+ \; ^-\!:\ddot{N}H_2} + H_2 \qquad (23.18)$$

$$(Me_2CH)_2\ddot{N}{-}H + CH_3CH_2CH_2CH_2{-}Li \longrightarrow \underset{\substack{\text{lithium} \\ \text{diisopropylamide}}}{(Me_2CH)_2\ddot{N}:^- \; Li^+} + CH_3CH_2CH_2CH_3 \qquad (23.19)$$

The pK_a of a typical amine is about 35. Thus, amide bases are very strong bases—stronger than acetylide or enolate anions, which, as we learned in Secs. 14.7A and 22.6B, can themselves be formed with amide bases.

F. Summary of Acidity and Basicity

We have now surveyed the acidity and basicity of the most important organic functional groups. This information is summarized in the Tables in Appendix VI. Although the values given in these tables are typical values, we have also learned that acidity and basicity are affected by alkyl substitution, inductive effects, and resonance effects. As we have learned, the acid–base properties of organic compounds are important not only in predicting many of their chemical properties, but also in their industrial and medicinal applications.

23.6 QUATERNARY AMMONIUM SALTS

Closely related to ammonium salts are compounds in which all four protons of $^+NH_4$ are replaced by organic groups. Such compounds are called **quaternary ammonium salts.** The following compounds are examples:

$$(CH_3)_4N^+ \ Cl^-$$
tetramethylammonium
chloride

$$Ph\overset{+}{N}Et_3 \ Br^-$$
N,N,N-triethylanilinium bromide

$$PhCH_2\overset{+}{N}(CH_3)_3 \ OH^-$$
"Triton B"

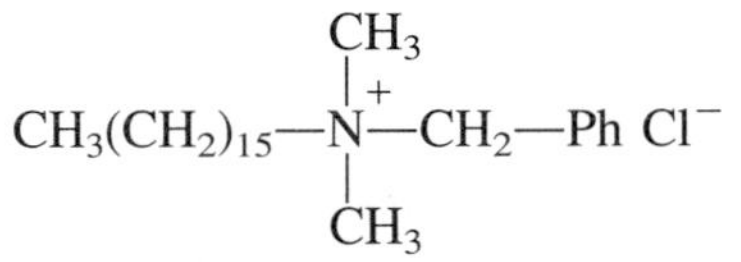

"benzalkonium chloride," a cationic
surfactant (Sec. 20.5)

Like the corresponding ammonium ions, quaternary ammonium salts are fully ionic compounds. Many quaternary ammonium salts containing large organic groups are soluble in nonaqueous solvents. Triton B, for example, is used as an ^-OH source that is soluble in organic solvents. Benzalkonium chloride, a common antiseptic, acts as a surfactant in water (Sec. 20.5) and is also soluble in several organic solvents. We can think of such compounds as "positive charges surrounded by greasy groups."

Problem

12 Explain why compound *A* can be isolated in optically active form, but compound *B* cannot.

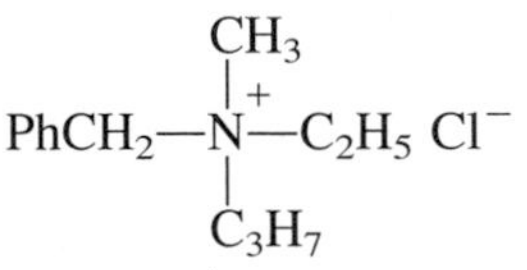

$$PhCH_2\overset{\underset{|}{CH_3}}{\underset{\underset{|}{C_3H_7}}{\overset{+}{N}}}C_2H_5 \ Cl^-$$
A

$$Ph-CH_2\overset{\underset{|}{CH_3}}{\underset{\underset{|}{H}}{\overset{+}{N}}}C_2H_5 \ Cl^-$$
B

23.7 ALKYLATION AND ACYLATION REACTIONS OF AMINES

In the previous section we learned that amines are *Brønsted bases*. Amines, like many other Brønsted bases, are also good *nucleophiles* (Lewis bases). Three reactions of nucleophiles we have studied are:

1. S_N2 reaction with alkyl halides, sulfonate esters, or epoxides

2. Addition to aldehydes and ketones and related reactions

3. Substitution at the carbonyl groups of carboxylic acid derivatives

In this section we shall study (or review) reactions of amines that fit into each of these categories.

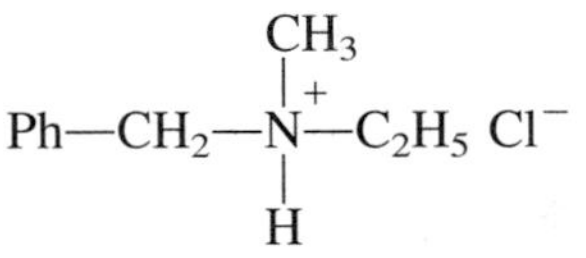

A. Direct Alkylation of Amines

Treatment of ammonia or an amine with an alkyl halide or other alkylating agent results in alkylation on the nitrogen.

$$R_3N\!: + CH_3\text{—}I \longrightarrow R_3\overset{+}{N}\text{—}CH_3\ I^- \qquad (23.20)$$

This process is an example of an S_N2 reaction in which the amine acts as the nucleophile.

The immediate product of the reaction shown in Eq. 23.20 is an ammonium ion. However, if this ammonium ion has N—H bonds, further alkylations can take place to give a complex product mixture, as in the following example:

$$\ddot{N}H_3 + CH_3I \longrightarrow CH_3NH_3^+\ I^- + (CH_3)_2NH_2^+\ I^- + (CH_3)_3NH^+\ I^- + (CH_3)_4N^+\ I^- \qquad (23.21)$$

A mixture of products is formed because the methylammonium ion produced initially is partially deprotonated by the ammonia starting material. Since the resulting methylamine is also a good nucleophile, it too reacts with methyl iodide.

$$\ddot{N}H_3 + CH_3I \longrightarrow CH_3\text{—}\overset{+}{N}H_3\ I^- \qquad (23.22a)$$

$$CH_3\text{—}\overset{+}{N}H_3\ I^- + :NH_3 \rightleftharpoons CH_3\text{—}\ddot{N}H_2 + \overset{+}{N}H_4\ I^- \qquad (23.22b)$$

$$CH_3\text{—}\ddot{N}H_2 + CH_3I \longrightarrow (CH_3)_2\overset{+}{N}H_2\ I^- \qquad (23.22c)$$

Analogous deprotonation–alkylation reactions give the mixture of products shown in Eq. 23.21. (To be sure you understand this point, write the full sequence of reactions leading to all of the products in Eq. 23.21.)

Epoxides, as well as α,β-unsaturated carbonyl compounds and α,β-unsaturated nitriles, also react with amines and ammonia. As the following results show, multiple alkylation can be a problem with these alkylating agents as well.

$$t\text{-Bu—}NH_2 + \overset{O}{CH_2\text{—}CH_2} \xrightarrow{H_2O} t\text{-Bu—NH—}CH_2CH_2\text{—OH} + t\text{-Bu—N}(CH_2CH_2OH)_2 \qquad (23.23)$$

$$\underset{}{NH_3\ (\text{excess})} + CH_2\!=\!CH\text{—}CN \longrightarrow \underset{(31\text{–}33\%\ \text{yield})}{H_2N\text{—}CH_2CH_2CN} + \underset{(57\%\ \text{yield})}{HN(CH_2CH_2CN)_2} \qquad (23.24)$$

In an alkylation reaction, the exact amount of each product obtained depends on the precise reaction conditions and on the relative amounts of starting amine and alkyl halide. Because a mixture of products results, the utility of alkylation as a preparative method for amines is limited, although in specific cases, conditions have been worked out to favor particular products. In Sec. 23.11 we shall discuss other methods that are more useful for the preparation of amines.

Quaternization of Amines Amines can be converted into quaternary ammonium salts with excess alkyl halide under forcing conditions. This process, called **quaternization,** is one of the most important synthetic applications of amine alkylation. The reaction is particularly useful when especially reactive alkyl halides, such as methyl iodide or benzyl halides, are used.

$$PhCH_2NMe_2 + MeI \xrightarrow{EtOH} PhCH_2\overset{+}{N}Me_3\ I^- \quad (94\text{–}99\%\ \text{yield}) \qquad (23.25)$$

$$CH_3(CH_2)_{15}NMe_2 + PhCH_2{-}Cl \xrightarrow{\text{acetone}} CH_3(CH_2)_{15}\overset{+}{N}Me_2(CH_2Ph)\ Cl^- \qquad (23.26)$$

$$CH_3CH(C_2H_5)NHMe + MeI\ (\text{excess}) \xrightarrow[\text{heat}]{\text{ether}} CH_3CH(C_2H_5)\overset{+}{N}Me_3\ I^- + HI \qquad (23.27)$$

Conversion of an amine into a quaternary ammonium salt with excess methyl iodide (as in Eqs. 23.25 and 23.27) is called **exhaustive methylation.**

B. Reductive Amination

When primary and secondary amines react with aldehydes and ketones, they form imines and enamines, respectively (Sec. 19.11). In the presence of a reducing agent, imines and enamines are reduced to amines.

$$EtNH_2 + CH_3{-}\overset{O}{\overset{\|}{C}}{-}CH_3 \xrightarrow{-H_2O} \underset{\substack{\text{imine} \\ \text{(not isolated)}}}{\left[CH_3{-}\overset{NEt}{\overset{\|}{C}}{-}CH_3\right]} \xrightarrow[\substack{30\ \text{psi} \\ EtOH}]{H_2,\ Pt} CH_3{-}\overset{NH-Et}{\overset{|}{C}H}{-}CH_3 \qquad (23.28)$$

Reduction of the C=N double bond is analogous to reduction of the C=O double bond (Sec. 19.8). Notice that the imine or enamine does not have to be isolated, but is reduced within the reaction mixture as it forms. Because imines and enamines are reduced much more rapidly than carbonyl compounds, reduction of the carbonyl compound is not a competing reaction.

The formation of an amine from the reaction of an aldehyde or ketone with another amine and a reducing agent is called **reductive amination.** Sodium borohydride, $NaBH_4$, and a related compound, sodium cyanoborohydride, $NaBH_3CN$, find frequent use as reducing agents in reductive amination.

$$PhCH{=}O + Ph{-}NH_2 \xrightarrow[EtOH]{NaBH_4} PhCH_2{-}NH{-}Ph \qquad (83\%\ \text{yield}) \qquad (23.29)$$

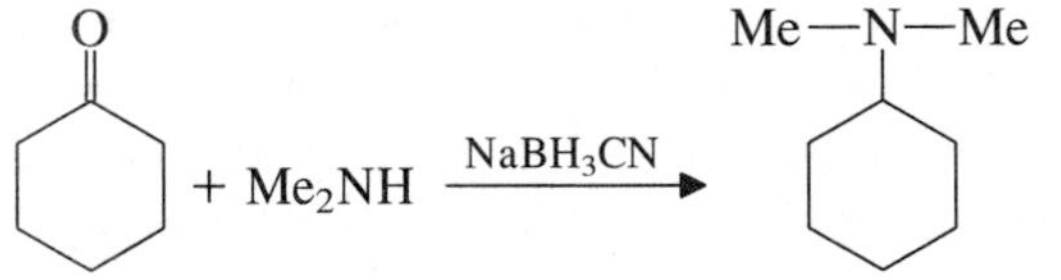

(52% yield) (23.30)

(Sodium cyanoborohydride is supplied as an easily handled, commercially available powder that is stable even in aqueous solution above pH 3.)

In the following synthesis, a secondary amine is reductively aminated with formaldehyde.

$$Et{-}NH{-}CH_2Ph + CH_2{=}O \xrightarrow{NaBH_3CN} Et{-}\overset{CH_3}{\overset{|}{N}}{-}CH_2Ph \qquad (23.31)$$

Neither an imine nor an enamine can be an intermediate in this reaction (why?). In this case a small amount of a cationic intermediate, an *immonium ion,* is formed in solution by ionization of the carbinolamine intermediate. The immonium ion is rapidly and irreversibly reduced.

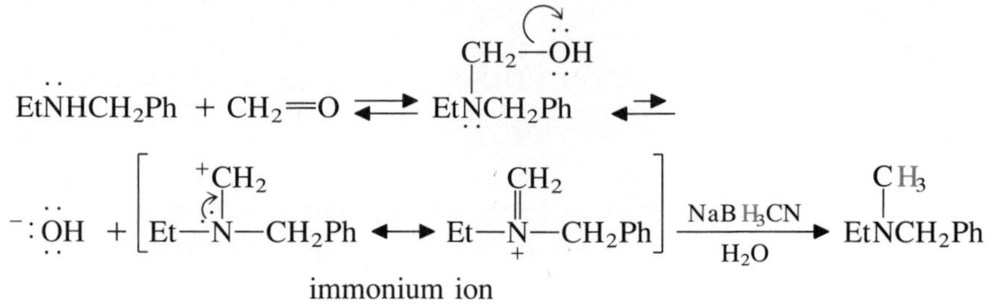

(23.32)

As Eq. 23.32 and the following equation show, the reaction of an amine with an excess of formaldehyde is a useful way to introduce methyl groups to the level of a tertiary amine.

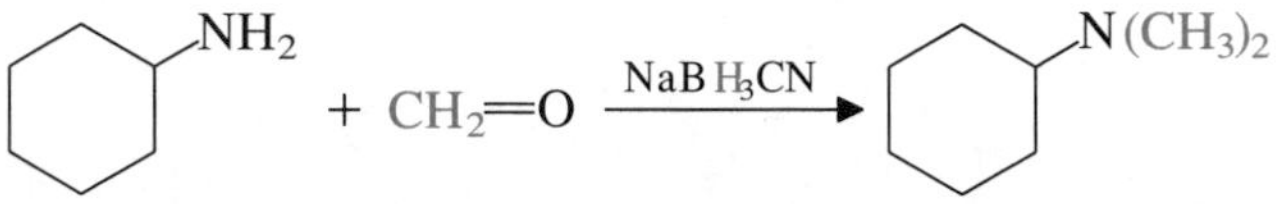

(84% yield) (23.33)

(Quaternization does not occur in this reaction; why?)

Suppose we wish to prepare a given amine and want to decide rationally whether reductive amination would be a suitable preparative method. How do we determine the required starting materials? As usual, we work in reverse, starting at the desired compound, and mentally reverse the reductive amination process. Let us use the synthesis of *N*-ethyl-*N*-methylaniline as an example. We mentally break the C—N bond and replace it on the nitrogen side with an N—H bond. On the carbon side, we drop a hydrogen from the carbon and add a carbonyl oxygen:

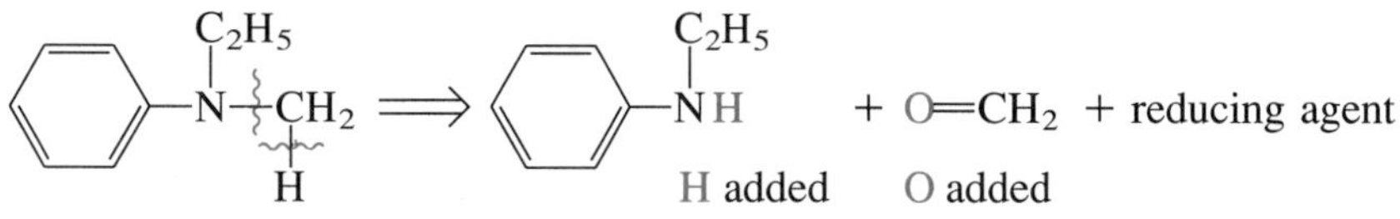

(23.34)

This analysis reveals that *N*-ethylaniline and formaldehyde, plus a suitable reducing agent, would be suitable starting materials.

Our decision to break the CH_3—N in this analysis is not the only possibility; we could break the C_2H_5—N bond instead to obtain a different set of starting materials. What starting materials are suggested by this analysis? Could other reasonable starting materials be discovered by mentally breaking the phenyl-N bond? Why or why not?

Problems

13 Suggest two syntheses for *N*-ethyldicyclohexylamine by reductive amination.

14 Outline a synthesis of the quaternary ammonium salt $(CH_3)_3\overset{+}{N}CH_2Ph\ I^-$ from dimethylamine and any other reagents.

15 A chemist, Mada Meens, treated ammonia with pentanal in the presence of H_2 and a catalyst expecting to isolate 1-pentanamine. Instead, she obtained a mixture of products. What competing reactions are likely to have occurred?

16 Another chemist, Caleb J. Cookbook, heated ammonia with bromobenzene expecting to isolate tetraphenylammonium bromide. Can Caleb expect this reaction to succeed? Why or why not?

C. Acylation of Amines

We have already learned (Sec. 21.7) that amines can be converted into amides by reaction with acid chlorides, anhydrides, or esters.

$$R'-\overset{\overset{O}{\|}}{C}-Cl + 2R_2NH \longrightarrow R'-\overset{\overset{O}{\|}}{C}-NR_2 + R_2\overset{+}{N}H_2\ Cl^- \qquad (23.35)$$

$$R'-\overset{\overset{O}{\|}}{C}-O-\overset{\overset{O}{\|}}{C}-R' + 2R_2NH \longrightarrow R'-\overset{\overset{O}{\|}}{C}-NR_2 + R-\overset{\overset{O}{\|}}{C}-O^- + R_2\overset{+}{N}H_2 \qquad (23.36)$$

$$R'-\overset{\overset{O}{\|}}{C}-OR' + R_2NH \longrightarrow R'-\overset{\overset{O}{\|}}{C}-NR_2 + R'OH \qquad (23.37)$$

In this type of reaction, a bond is formed between the amine and a carbonyl carbon rather than an alkyl carbon (as in the previous two sections). This is another example of an *acylation* reaction—the introduction of an acyl group (Sec. 16.4E). The reactions of amines with carboxylic acid derivatives should be reviewed in Sec. 21.8 if necessary.

23.8 HOFMANN ELIMINATION OF QUATERNARY AMMONIUM HYDROXIDES

In the previous section, we learned about ways to *make* carbon–nitrogen bonds. In these reactions, amines react as *nucleophiles*. In this section, we shall study an elimination reaction used to *break* carbon–nitrogen bonds. In this reaction, which involves *quaternary ammonium hydroxides* ($R_4N^+\ ^-OH$) as starting materials, amines act as *leaving groups*.

When a quaternary ammonium hydroxide is heated, a β-elimination reaction takes place to give an alkene, which distills from the reaction mixture.

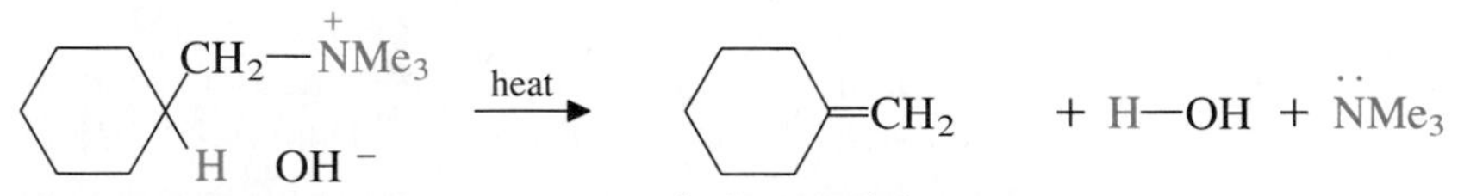

(23.38)

a quaternary ammonium hydroxide — methylenecyclohexane (74% yield)

This type of elimination reaction is called a **Hofmann elimination,** after August Wilhelm Hofmann (1818–1895), a German chemist who became professor at the Royal College of Chemistry in London. Hofmann was particularly noted for his work on amines.

How do we form the quaternary ammonium hydroxides used as the starting materials in Hofmann eliminations? A quaternary ammonium hydroxide is formed by treating a quaternary ammonium salt with silver hydroxide (AgOH), which, in turn, is formed from water and silver oxide (Ag_2O).

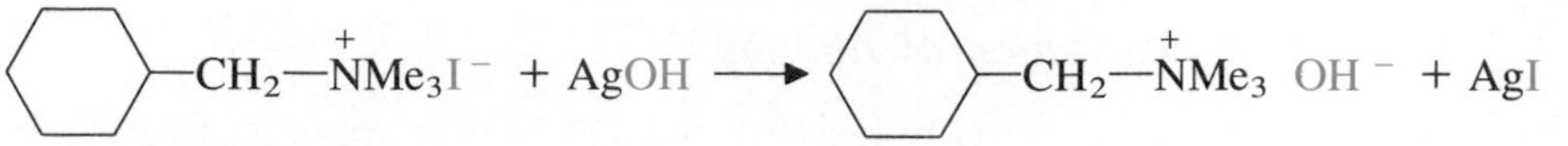

$$\text{---CH}_2\text{---}\overset{+}{\text{N}}\text{Me}_3\text{I}^- + \text{AgOH} \longrightarrow \text{---CH}_2\text{---}\overset{+}{\text{N}}\text{Me}_3 \; \text{OH}^- + \text{AgI} \quad (23.39)$$

It follows that alkenes can be formed from amines by a three-step process: exhaustive methylation (see Eq. 23.27), conversion of the ammonium salt to the hydroxide (Eq. 23.39), and Hofmann elimination.

The Hofmann elimination is formally analogous to the E2 reaction of alkyl halides (Sec. 9.4), in which a proton and a halide ion are eliminated; in the Hofmann elimination, a proton and a tertiary amine are eliminated. Because the amine leaving group is very basic, and therefore a relatively poor leaving group, the conditions of the Hofmann elimination are typically harsh.

Like the analogous E2 reaction of alkyl halides, the Hofmann elimination generally occurs as an *anti* elimination (Sec. 9.4E).

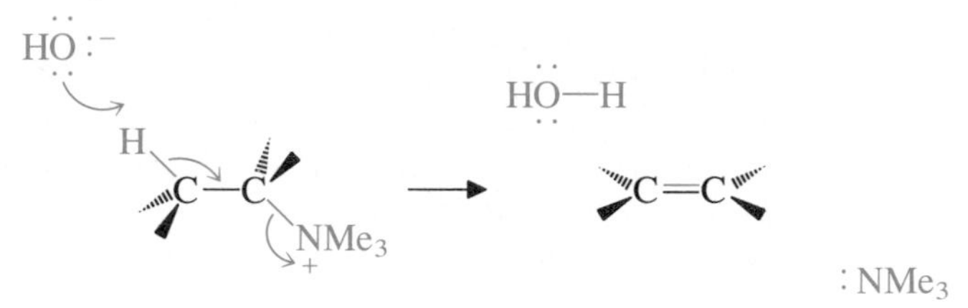

$$(23.40)$$

17 What product (including stereochemistry) is expected from the Hofmann elimination of each of the following stereoisomers?

(a)

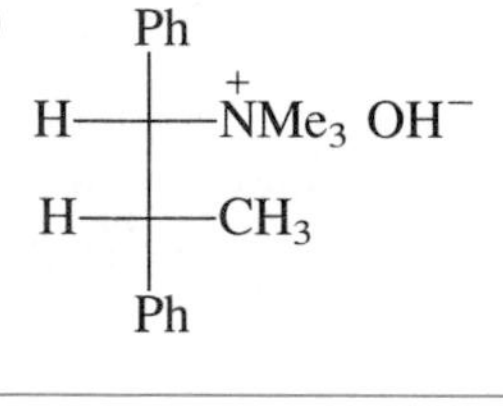

```
        Ph
        |
H------|----NMe3 OH⁻
        |
H------|----CH3
        |
        Ph
```

(b)

```
         Ph
         |
H-------|----NMe3 OH⁻
         |
CH3------|----H
         |
         Ph
```

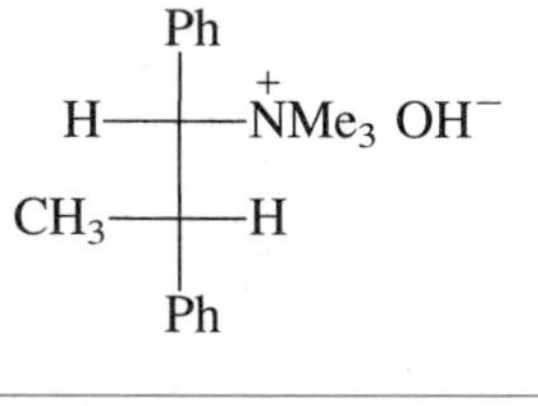

Despite their similarities, the elimination reactions of alkyl halides and quaternary ammonium salts show distinct differences in *regiochemistry*—that is, in the position of the double bond in the product. Recall that E2 elimination of most alkyl halides gives a predominance of *the alkene with the greatest amount of branching at the double bond* (Saytzeff rule; Sec. 9.4D). Thus, E2 elimination from 2-bromobutane promoted by sodium ethoxide gives mostly 2-butene.

$$\underset{\underset{\text{Br}}{|}}{\text{CH}_3\text{CH}_2\text{CHCH}_3} \xrightarrow{\text{NaOEt}} \underset{\substack{(81\% \text{ of alkene} \\ \text{formed; mostly } trans)}}{\text{CH}_3\text{CH}\text{=}\text{CHCH}_3} + \underset{\substack{(19\% \text{ of alkene} \\ \text{formed})}}{\text{CH}_3\text{CH}_2\text{CH}\text{=}\text{CH}_2} + \underset{\underset{\text{OEt}}{|}}{\text{CH}_3\text{CH}_2\text{CHCH}_3} \quad (23.41)$$

In contrast, Hofmann elimination of the corresponding trimethylammonium salt gives mostly 1-butene.

$$\underset{\displaystyle {}^{+}NMe_3\ {}^{-}OH}{CH_3CH_2CHCH_3} \xrightarrow{\text{heat}} \underset{(95\%)}{CH_3CH_2CH{=}CH_2} + \underset{(5\%;\ cis \text{ and } trans)}{CH_3CH{=}CHCH_3} \quad (23.42)$$

This observation has been generalized as a second elimination "rule," called the **Hofmann rule.** A general statement of the Hofmann rule is that *elimination of a trialkylammonium salt generally occurs in such a way that the base abstracts a proton from the β-carbon atom with the least branching.*

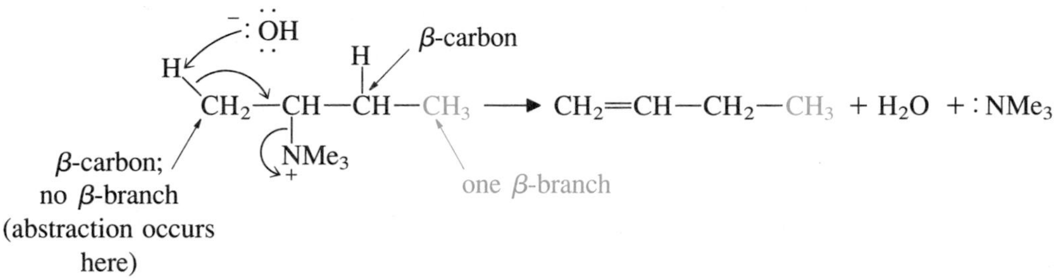

(23.43)

Chemists have disagreed about the reasons for the Hofmann rule. In some cases the explanation may lie, at least partly, in both the preference for *anti* elimination and the conformation of the ammonium salt in the transition state of the reaction. Taking the elimination shown in Eq. 23.42 as an example, let us first consider the possible conformations of the transition state for elimination at the C2–C3 bond—the reaction that is *not* observed.

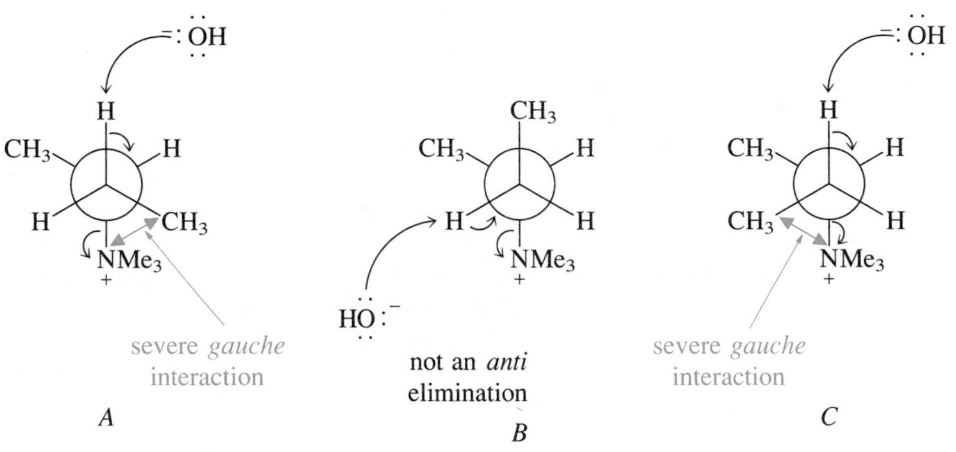

(23.44)

There is an energetically unfavorable feature in each of these transition-state conformations. Although transition states *A* and *C* are *anti* eliminations, they contain severe *gauche* interactions, because a trimethylammonium group is about the same size as a *t*-butyl group. Transition state *B* avoids the *gauche* interaction between the trimethylammonium group and a methyl group, but is not an *anti* elimination. Thus no transition state for elimination across the C2–C3 bond incorporates both *anti* elimination and absence of significant van der Waals repulsions. In contrast, since elimination at the C1–C2 bond—Hofmann elimination—can occur with *anti* stereochemistry and without severe *gauche* interactions, this is the reaction that occurs:

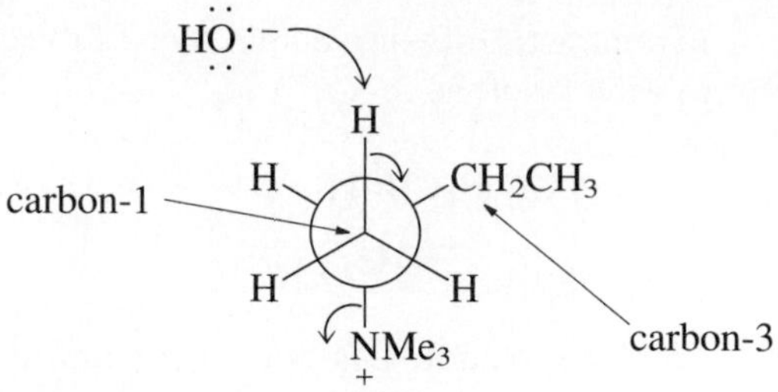

In the case of elimination from 2-bromobutane (Eq. 23.41), the *anti*-transition states leading to 2-butene have much less severe *gauche* interactions, because the bromine leaving group is much smaller than a trimethylammonium group—it is about the same size as a methyl group. Hence, elimination in this case involves formation of the more stable alkene isomer.

When there are especially acidic β-hydrogens, violation of the Hofmann rule can occur.

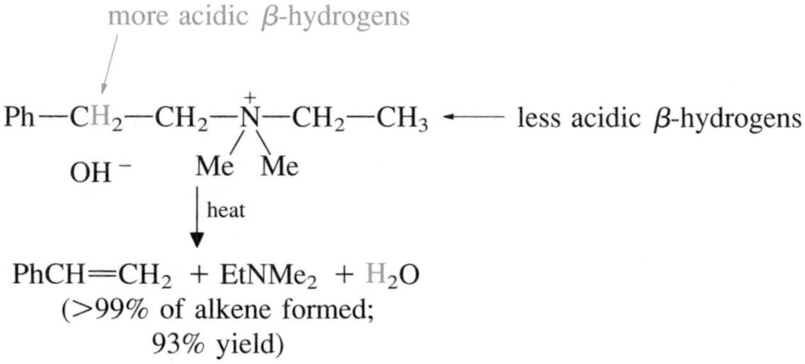

(23.45)

In this example, the hydrogens next to the phenyl group are more acidic (why?). Since base must abstract a proton in the elimination, it is reasonable that the acidity of the β-hydrogen should play an important role in determining the direction of elimination. For a similar reason, protons adjacent to carbonyl groups are also preferentially abstracted in the Hofmann elimination.

Although the Hofmann elimination can be used to prepare 1-alkenes reasonably free of isomers (as in Eq. 23.38), it has been largely supplanted by the Wittig reaction (Sec. 19.13) as a method of alkene synthesis. The importance of the Hofmann elimination is due to the key role it has played in the development of the theory of elimination mechanisms, and its use in determining the structures of amine-containing natural products in the older literature (Problem 20).

Problems

18 Predict the predominant alkene formed when each of the following quaternary ammonium hydroxides are heated:

(a)

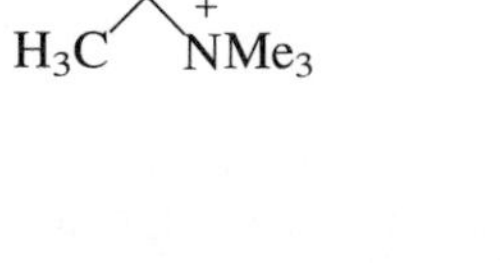

(b)

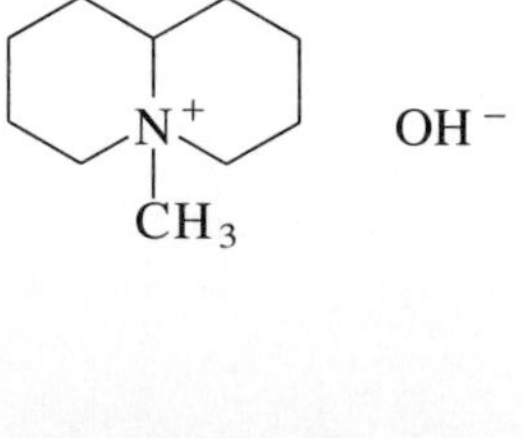

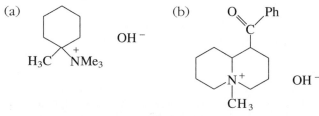

19 Give the structures of three alkenes that could in principle form when the following compound is heated. Rank these products in order of the relative amounts produced, greatest first.

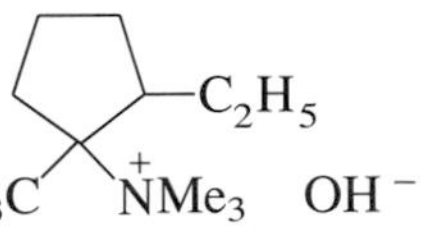

20 *Coniine* is the toxic component of hemlock, the poisonous plant believed to have killed Socrates. Coniine is a secondary amine containing a piperidine ring. Exhaustive methylation of coniine yields a salt that, when treated with Ag_2O, then heat, gives a mixture of alkenes from which a compound *A* ($C_{10}H_{21}N$) is isolated. Exhaustive methylation of *A* and treatment with Ag_2O and heat gives a mixture of 1,4-octadiene and 1,5-octadiene. Propose a structure for coniine.

23.9 AROMATIC SUBSTITUTION REACTIONS OF ANILINE DERIVATIVES

Aromatic and aliphatic amines share many reactions. However, only aromatic amines can undergo *electrophilic aromatic substitution* reactions (Sec. 16.4). Although these reactions occur on the aromatic ring, their outcome is strongly influenced by the activating and directing effects of the amino nitrogen.

The amino group is one of the most powerful activating groups in electrophilic aromatic substitution. It is also an *ortho, para*-directing group (why?). For example, aniline, like phenol, brominates three times under mild conditions.

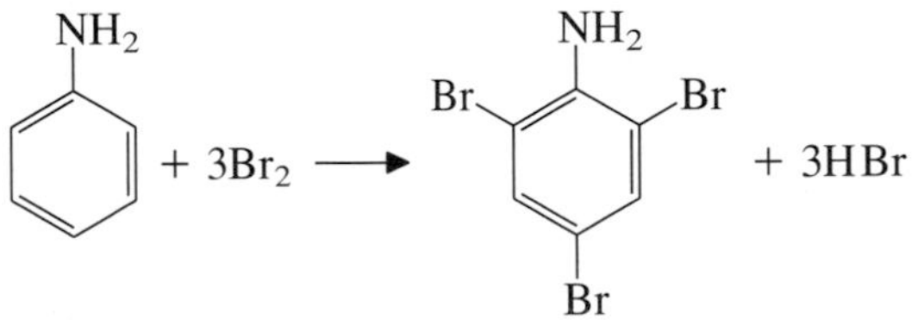

(23.46)

In view of the activating and *ortho, para*-directing effects of the amino group in many aromatic substitution reactions, the results of the following nitration at first seem strange: only one nitro group is introduced into the ring, and a substantial amount of *m*-nitroaniline is formed.

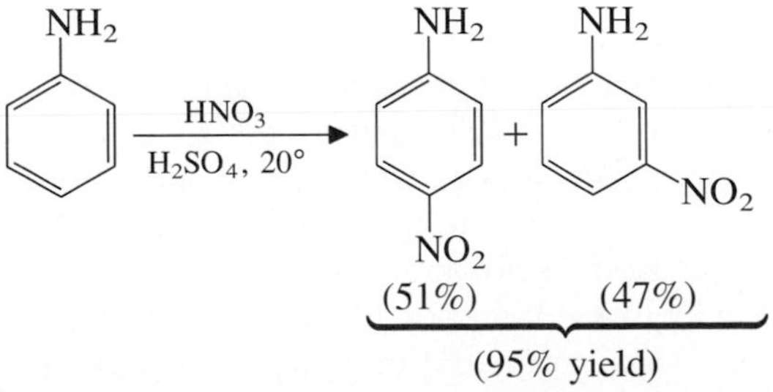

(+ 2% *o*-isomer) (23.47)

One reason for these results is that the amine is protonated under the reaction conditions; remember that nitration takes place in a very acidic solution. *Protonated* aniline no longer has the free nitrogen electron pair responsible for the activation of the *ortho* and *para* positions, and ammonium salts are largely *meta*-directing groups.

We would actually expect the —NH_3^+ group to act as a *meta*-directing group. Yet a substantial amount of *para* isomer is formed in Eq. 23.47, even though aniline must be completely protonated under the reaction conditions. It is possible that the *p*-nitroaniline arises from nitration of the very small amount of highly reactive unprotonated aniline in the reaction mixture. In such a case the unprotected aniline would be replenished immediately as it reacts, by LeChatelier's principle.

We can nitrate aniline at only the *para* position if we *protect* the amino nitrogen so that it cannot be protonated so easily. One way of accomplishing this objective is to convert the aniline into an amide, which is much less basic than the amine. In Eq. 23.48, the requisite amide is formed by treating the amine with acetyl chloride to give *N*-acetylaniline (acetanilide). Following the nitration reaction, the acetyl group is removed to afford the nitrated aniline, *p*-nitroaniline.

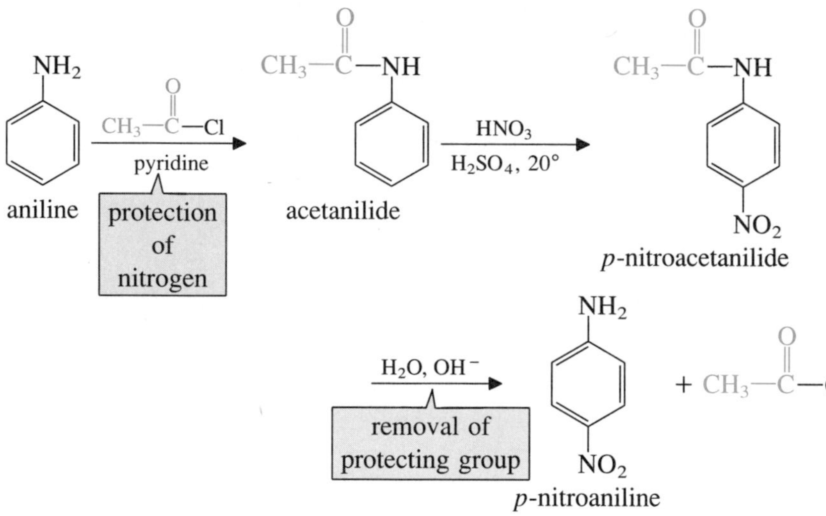

(23.48)

The amide group of acetanilide is a *protecting group* (Sec. 19.10A) in this sequence of reactions. Because it is considerably less basic than the amino group of aniline, the amide group is only partially protonated under the reaction conditions. Hence, the amide group is an *ortho, para*-directing group even under the strongly acidic conditions of nitration. Notice that nitration (and other electrophilic substitution reactions) of acetanilide can be stopped after a single substitution has occurred, because the acetamido group is not so strong an activating group as the free amino group (why?).

A widely used undergraduate laboratory experiment that relies on a protection strategy of this sort is the preparation of sulfanilamide, one of the *sulfa drugs*.

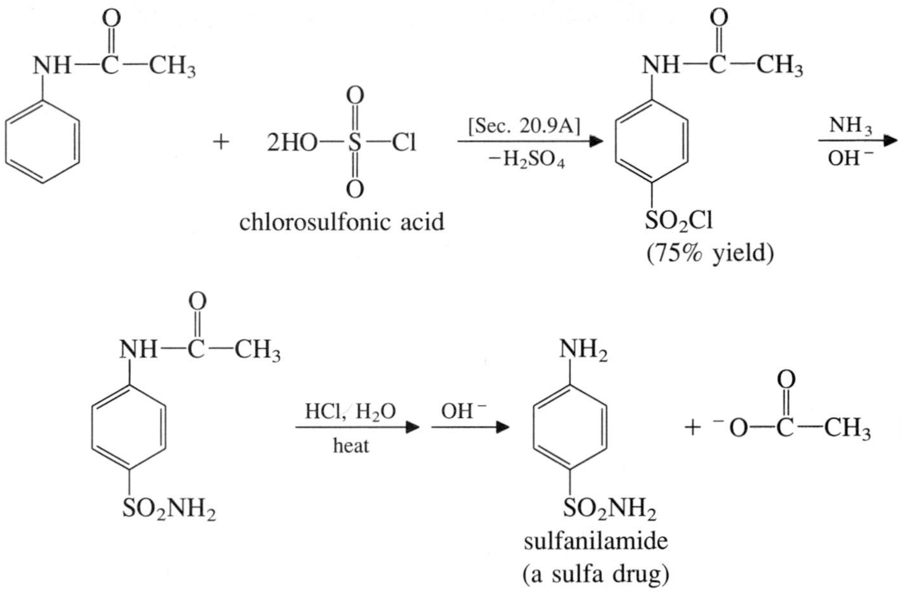

(23.49)

Problems

21 Suppose the *N*-acetyl protecting group were not used in the first step of Eq. 23.49. Besides the problem of *meta* substitution, what other serious side reaction would occur?

22 Outline syntheses of each of the following compounds from aniline and any other starting materials:
(a) 2,4-dinitroaniline
(b) sulfathiazole, a sulfa drug

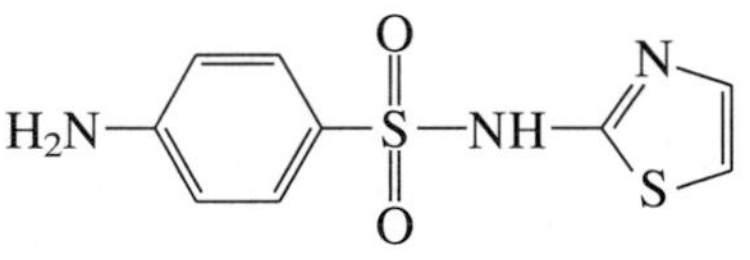

sulfathiazole

23.10 REACTIONS OF AMINES WITH NITROUS ACID. REACTIONS OF DIAZONIUM IONS

From the reactions we have considered so far, we can see that the chemistry of the amino group bears a vague resemblance to the chemistry of the hydroxy group. Thus, amino groups, like hydroxy groups, can both donate and accept hydrogen bonds; amino groups, like hydroxy groups, are basic and nucleophilic (only more so); and amino groups, like hydroxy groups, activate aromatic rings toward electrophilic aromatic substitution.

In contrast, when it comes to oxidation reactions, there is no parallel between amines and alcohols or phenols. Oxidation of amines usually occurs at the amino-group nitrogen, whereas oxidation of alcohols and phenols occurs at the α-carbon. In this section, we consider one of the important amine oxidations—the conversion of primary amines into **diazonium salts,** a reaction that is brought about by nitrous acid (HNO_2):

$$HCl + CH_3CH_2CH_2CH_2NH_2 + HNO_2 \longrightarrow \underset{\text{1-butanediazonium chloride (a diazonium salt)}}{CH_3CH_2CH_2CH_2\overset{+}{N}{\equiv}N\!:\ Cl^-} + 2H_2O \qquad (23.50)$$

Because nitrous acid is unstable, it is generated as it is needed from the reaction of sodium nitrite ($NaNO_2$) and a mineral acid such as HCl or H_2SO_4, as in the following example:

$$Ph{-}NH_2 \xrightarrow[0\text{–}5^\circ]{NaNO_2,\ HCl} \underset{\text{benzenediazonium chloride}}{Ph{-}\overset{+}{N}{\equiv}N\!:\ Cl^-} \qquad (23.51)$$

Nitrite esters (esters of alcohols and nitrous acid; see Sec. 10.3) can also be used to generate diazonium salts.

$$HCl + PhNH_2 + \underset{\text{amyl nitrite (a nitrite ester)}}{CH_3(CH_2)_4{-}O{-}N{=}O} \longrightarrow \underset{\text{(98\% yield)}}{Ph{-}\overset{+}{N}{\equiv}N\!:\ Cl^-} + CH_3(CH_2)_4{-}OH + H_2O \qquad (23.52)$$

The conversion of a primary amine into a diazonium salt is called **diazotization.**

Alkanediazonium salts, such as the one formed in Eq. 23.50, are very unstable and decompose under the reaction conditions, as we shall see in Sec. 23.10A. Aryldiazonium salts, such as the one formed in Eq. 23.51, are more stable and undergo several important reactions that we shall consider in Secs. 23.10B and 23.10C. However, before concerning ourselves with the reactions of these salts, let us try to understand the process of diazotization itself.

In solution nitrous acid is in equilibrium with its anhydride, dinitrogen trioxide N_2O_3.

$$2HNO_2 \rightleftarrows \underset{\text{dinitrogen trioxide}}{O{=}N{-}O{-}N{=}O} + H_2O \qquad (23.53)$$

Dinitrogen trioxide is the actual diazotizing reagent. We can understand its reactivity by recognizing that it contains a good leaving group, nitrite. (The pK_a of nitrous acid is 3.2; thus, nitrite is a fairly weak base and a good leaving group.) Although N_2O_3 is not an ionic compound, it reacts as if it were a salt of the *nitrosyl cation* (sometimes called a *nitrosonium ion*) and the nitrite anion:

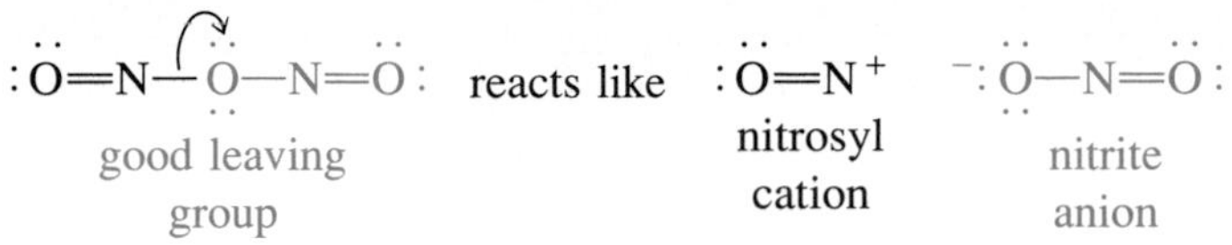
reacts like
good leaving
group
nitrosyl
cation
nitrite
anion
(23.54)

In other words, N_2O_3 contains a nitrogen that has some electron-deficient character.

The first mechanistic step in diazotization is the attack of the amine nitrogen on the electron-deficient nitrogen of N_2O_3 to form a **nitrosamine.** This process very much resembles a carbonyl-substitution reaction, with the —N=O group behaving much like a carbonyl group.

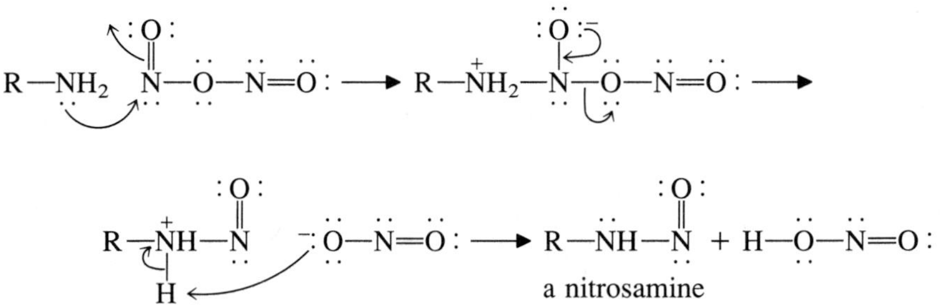
a nitrosamine
(23.55a)

The nitrosamine is in equilibrium with a tautomer called a *diazoic acid.*

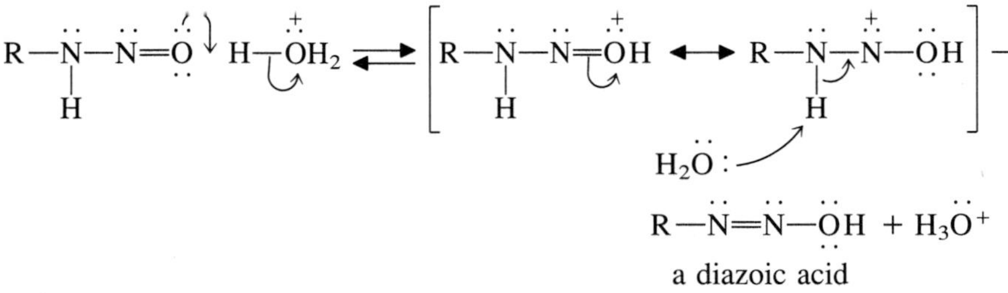
a diazoic acid
(23.55b)

Under the acidic conditions of diazotization, diazoic acids dehydrate to diazonium ions in much the same sense that alcohols dehydrate to carbocations:

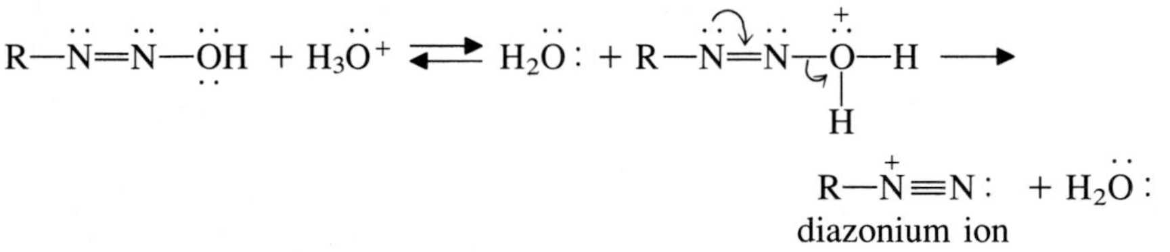
diazonium ion
(23.55c)

A. Reactions of Alkanediazonium Salts

Now let us consider the reactions of diazonium salts. Because diazonium ions incorporate one of the best leaving groups—molecular nitrogen—alkanediazonium ions are unstable and decompose spontaneously with the evolution of nitrogen gas to carbocations, which, in turn, react to give a mixture of products. (The evolution of nitrogen during diazotization is a good qualitative test for aliphatic primary amines.)

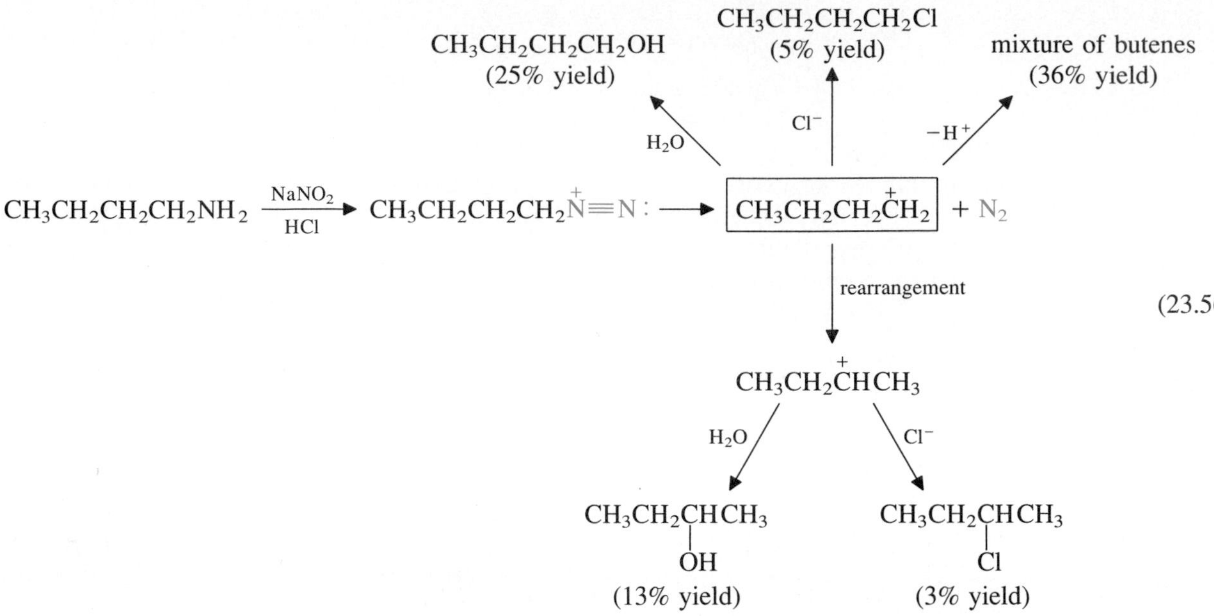

(23.56)

In this case an unstable primary carbocation intermediate can be formed because nitrogen is such a good leaving group. That is, the instability of the primary carbocation is offset by the stability of molecular nitrogen. Because a complex mixture of products is obtained, the diazotization of most primary amines is not generally a useful synthetic procedure.

Problem

23 The diazotization of 2-amino alcohols, as in the last step of the reaction sequence below, is an unusual case of a useful diazotization reaction of an alkylamine. Propose reagents for reactions (*a*) and (*b*) of this sequence, which is an example of the *Tiffeneau–Demjanov ring expansion*. Then suggest a mechanism for reaction (*c*), which is somewhat related to the pinacol rearrangement (Sec. 10.1B).

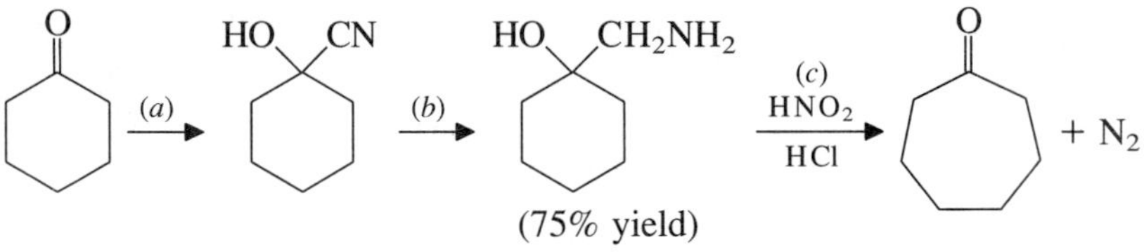

B. Substitution Reactions of Aryldiazonium Salts

Aryldiazonium salts decompose much more slowly than alkanediazonium salts, because loss of N_2 from an aryldiazonium salt would yield a very unstable *aryl cation* (Sec. 18.3). Consequently, aryldiazonium salts can be prepared in solution and used in subsequent reactions. (Although aryldiazonium salts can be isolated, they rarely are isolated because in the dry state they are dangerously explosive.) What are some of the reactions of aryldiazonium salts?

Addition of cuprous chloride or cuprous bromide to a solution of an aromatic diazonium salt gives products in which the nitrogen is replaced by halide.

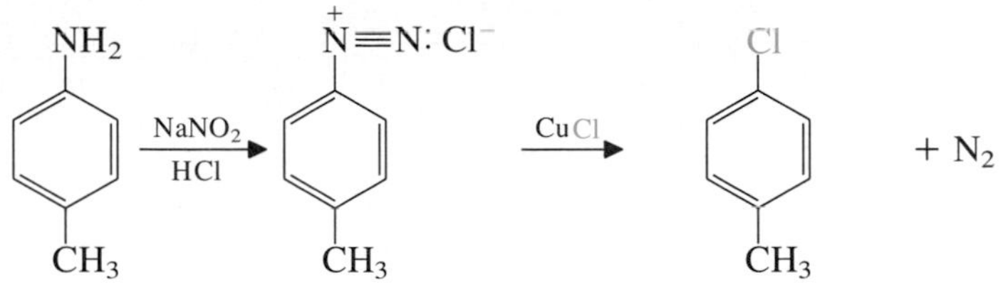

(70–71% yield) (23.57)

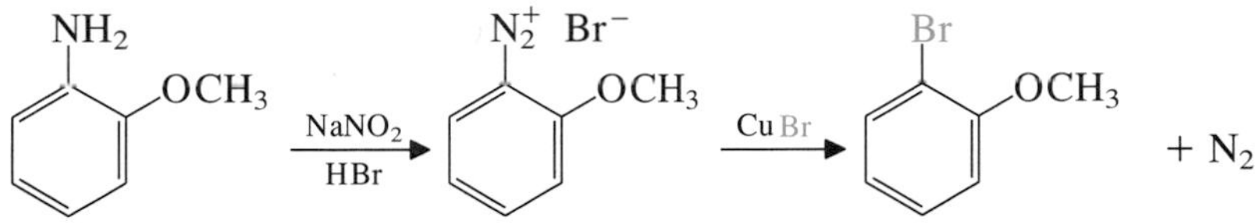

(88–93% yield) (23.58)

An analogous reaction occurs with cuprous cyanide, CuCN.

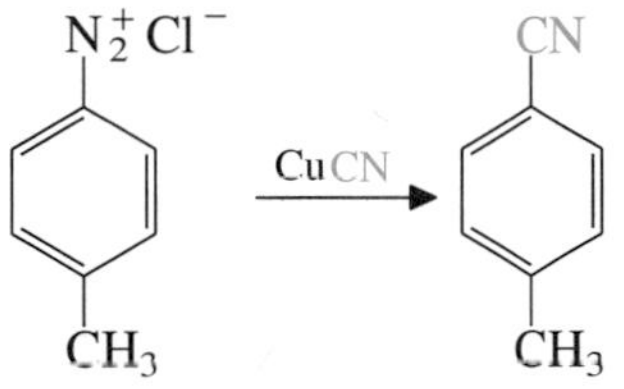

(64–70% yield) (23.59)

This reaction is another way of forming a carbon–carbon bond, in this case to an aromatic ring (see Appendix V). The resulting nitrile can be converted into a carboxylic acid, which can, in turn, serve as the starting material for a variety of other types of compounds. The reaction of an aryldiazonium ion with a cuprous salt is called the **Sandmeyer reaction.** This reaction is an important method for the synthesis of aryl halides and nitriles.

Aryl iodides can also be made by the reaction of diazonium salts with the potassium salt KI.

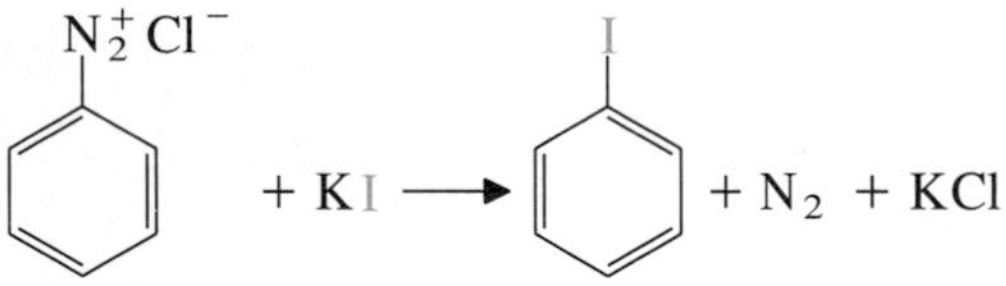

(74–76% yield) (23.60)

Notice that the analogous reactions with KBr and and KCl do not work; cuprous salts are required.

Replacement of the diazonium group with fluorine can be effected by heating the diazonium salt with fluoroboric acid (HBF_4); this reaction is called the **Schiemann reaction.**

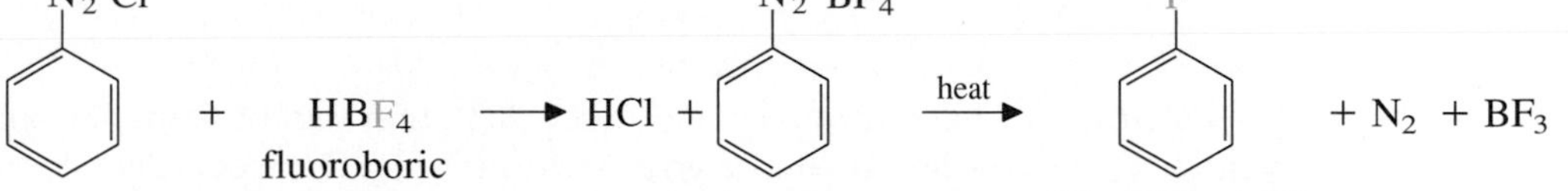

(51–57% yield) (23.61)

Diazonium salts can also be hydrolyzed to phenols by heating them in water:

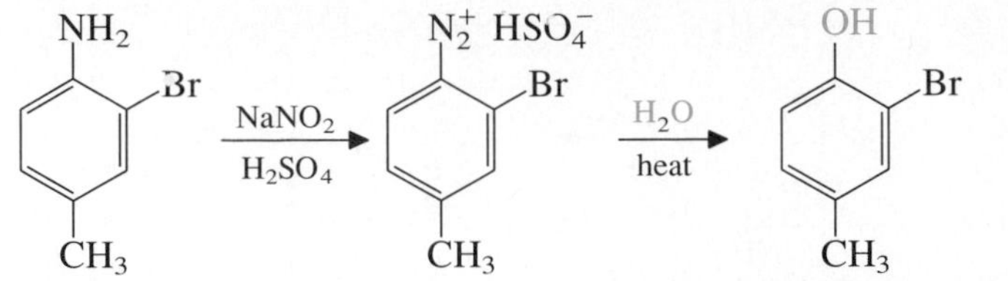

$$+ N_2 + H_2SO_4 \quad (23.62)$$

(80–92% yield)

Finally, the diazonium group is replaced by hydrogen when the diazonium salt is treated with hypophosphorous acid, H_3PO_2.

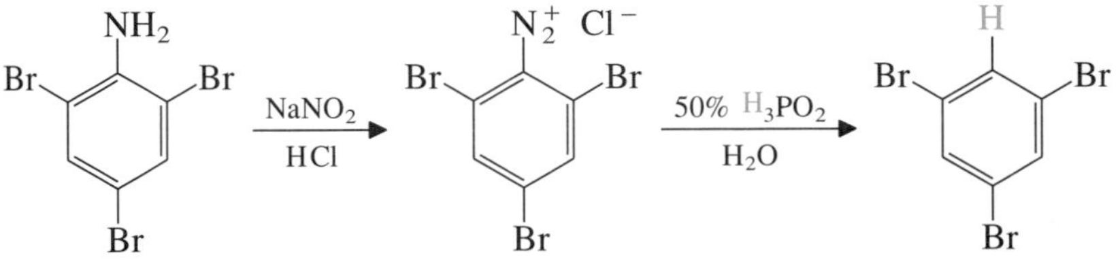

$$+ N_2 + H_3PO_3 + HCl \quad (23.63)$$

(70% yield)

All of the diazonium salt reactions just outlined are substitution reactions; however, they occur by a variety of mechanisms. The Sandmeyer reaction cannot be an S_N2 reaction of Br^- with the diazonium ion, because aromatic compounds do not undergo S_N2 reactions (Sec. 18.1). Furthermore, other bromide salts, such as NaBr, which should be equally effective if the reaction were an S_N2 reaction, do not work. It turns out that copper(I) has a catalytic effect in the Sandmeyer reaction not unlike its effect in lithium dialkylcuprate additions to α,β-unsaturated carbonyl compounds. Although we shall not consider the details of the mechanism here, the Sandmeyer reaction is believed to involve radical or radical-like species. The reaction of diazonium salts with KI (Eq. 23.60) does not involve copper(I), but is also believed to take place by an unusual mechanism. The reaction of diazonium salts with H_3PO_2 (Eq. 23.63) has been shown definitively to be a free-radical chain reaction.

The Schiemann reaction and the hydrolysis of diazonium salts to phenols are not free-radical reactions; they probably involve aryl cations as intermediates.

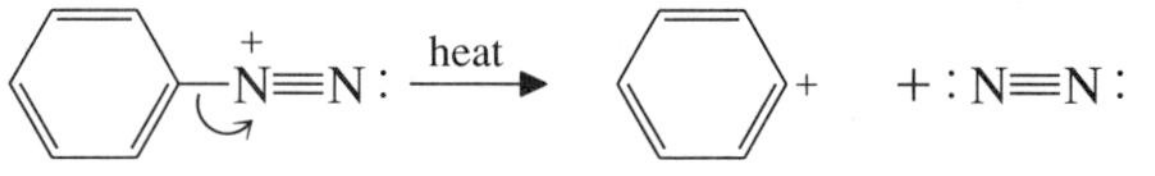

$$\overset{+}{\underset{}{\text{Ph–N}}} {\equiv} N : \quad \xrightarrow{\text{heat}} \quad \text{Ph}^+ \quad + \quad +:N{\equiv}N : \qquad (23.64a)$$

phenyl cation

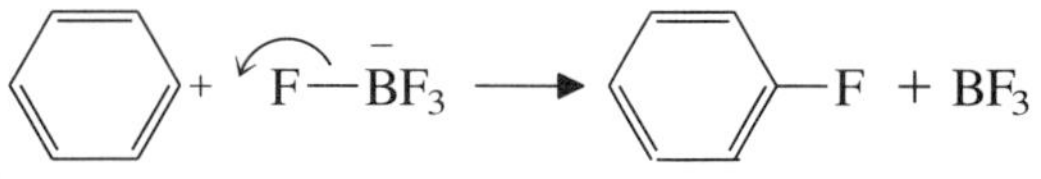

$$\text{Ph}^+ \ + \ \overset{-}{F}{-}BF_3 \quad \longrightarrow \quad \text{Ph–}F \ + \ BF_3 \qquad \text{(Schiemann reaction)} \quad (23.64b)$$

$$\text{Ph}^+ \ + \ :\overset{..}{O}H_2 \quad \longrightarrow \quad \text{Ph–}\overset{\overset{+.}{|}}{\underset{H}{O}}{-}H \quad \xrightarrow{H_2O} \quad \text{Ph–}\overset{..}{O}H \ + \ H_3O^+ \qquad \text{(hydrolysis)} \quad (23.64c)$$

Despite the instability of aryl cations (Sec. 18.3), they can be formed from diazonium salts because nitrogen is such a good leaving group. However, notice that both the Schiemann reaction and the hydrolysis reaction require heating.

All of the diazonium ion reactions discussed above have an important role in the laboratory synthesis of organic compounds. Let us consider, for example, the preparation of *p*-bromotoluene. Because the methyl group of toluene is an activating, *ortho,*

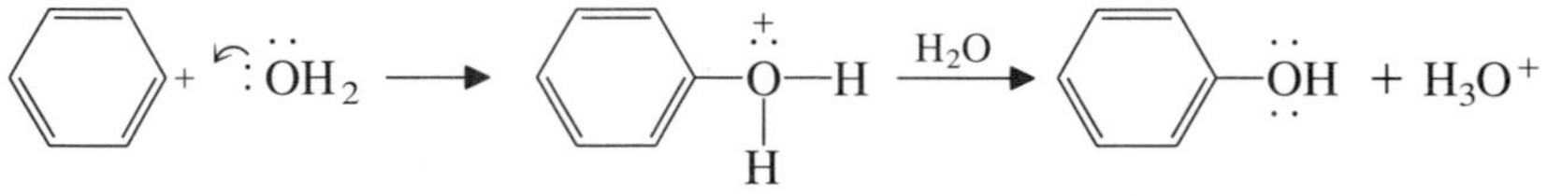

para-directing group, bromination of toluene would seem to be a satisfactory method for preparing *p*-bromotoluene. Unfortunately, bromination of toluene affords a mixture of the *ortho*- and *para*-bromotoluene isomers. These compounds are very difficult to separate because both have low melting points and nearly the same boiling point. The *o*- and *p*-nitrotoluenes, however, can be separated by fractional distillation and, as we shall learn in Sec. 23.11B, reduced to the corresponding amines. The Sandmeyer reactions of these amines provide a good laboratory synthesis of the corresponding aryl halides.

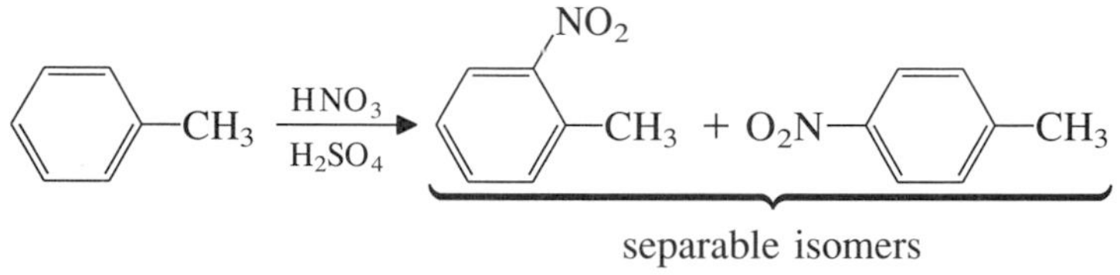

(23.65a)

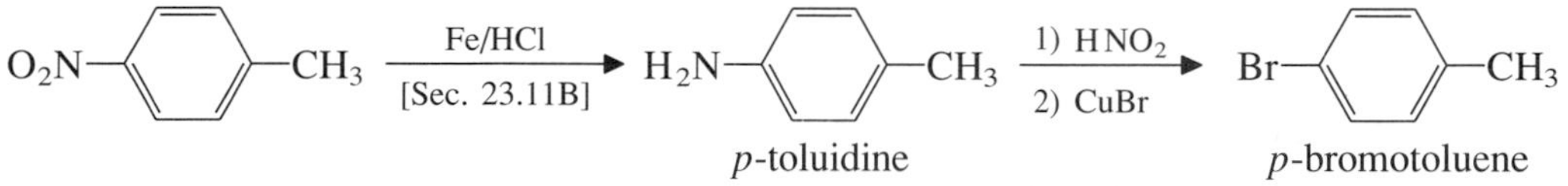

(23.65b)

The extra steps involved in nitrating toluene and reducing *o*- or *p*-nitrotoluene to *o*- or *p*-toluidine are worth the effort, because pure compounds are obtained.

The substitution reactions of diazonium salts also enable us to achieve ring substitution patterns that cannot be obtained in other ways. For example, 1,3,5-tribromobenzene could *not* be obtained by direct bromination of benzene (why?). However, it can be made by bromination of aniline, followed by diazotization of the tribromoaniline and reaction of the resulting diazonium salt with H_3PO_2.

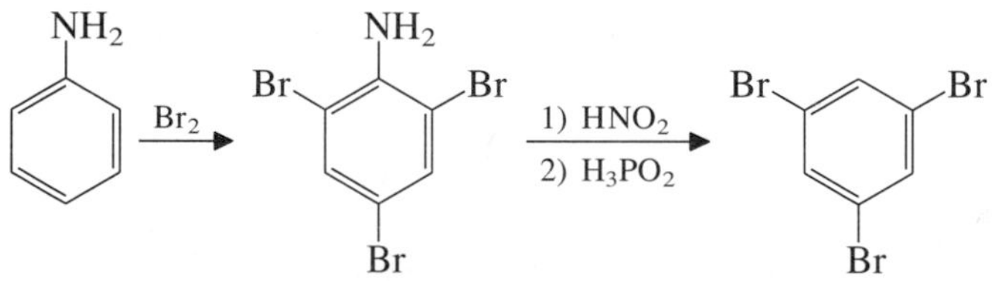

(23.66)

In this reaction sequence, the powerful directing effect of the amino nitrogen is used to place the bromines in *ortho* and *para* ring positions. The amino group is then removed by diazotization and replaced with hydrogen by the reaction of the resulting diazonium salt with hypophosphorous acid.

C. Aromatic Substitution with Diazonium Ions

Aryldiazonium ions react with aromatic compounds containing strongly activating substituent groups, such as amines and phenols, to give substituted *azobenzenes*. (Azobenzene itself is Ph—N=N—Ph.)

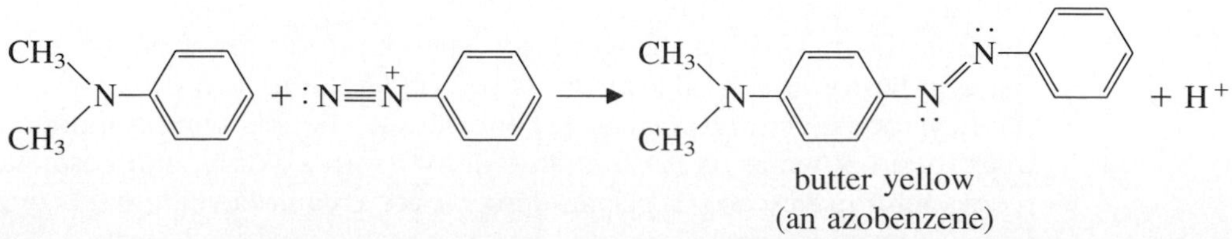

(23.67)

This is an electrophilic aromatic substitution reaction in which the terminal nitrogen of the diazonium ion is the electrophile. The mechanism follows the usual pattern of electrophilic substitution. First the electrophile is attacked by the π-electrons of the aromatic compound to give a resonance-stabilized carbocation:

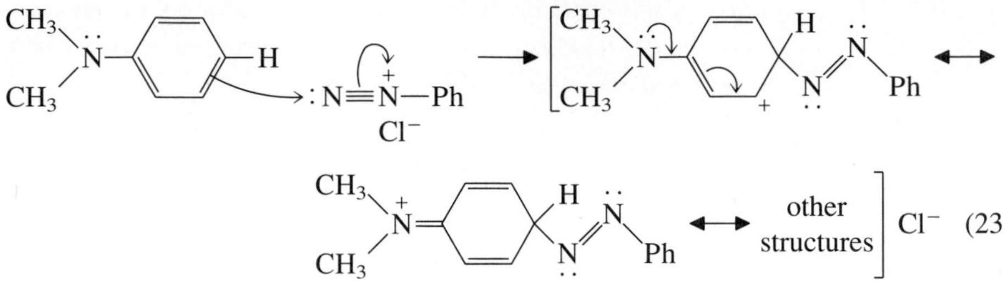

(23.68a)

This carbocation then loses a proton to give the substitution product.

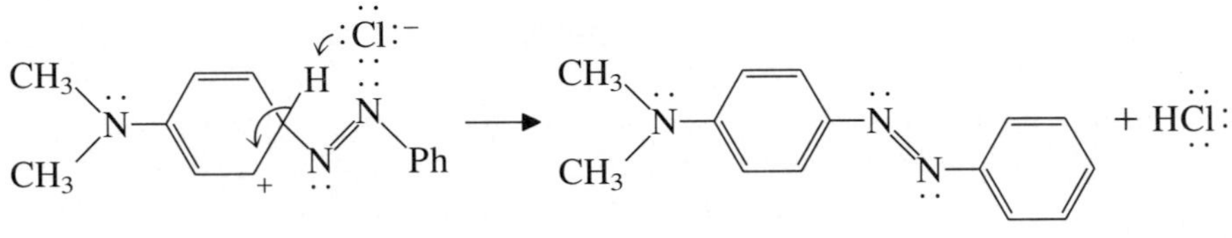

(23.68b)

(Why does substitution occur at the *para* position?)

The azobenzene derivatives formed in these reactions have extensive conjugated π-electron systems, and most of them are colored (Sec. 15.2C). Some of these compounds are used as dyes and indicators; as a class they are known as **azo dyes.** For example, the azo dye methyl orange is a well-known indicator.

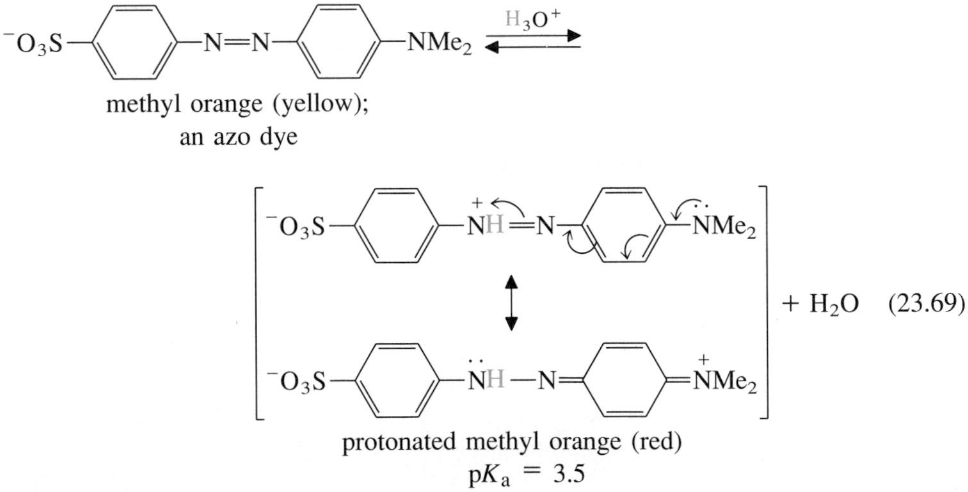

(23.69)

Because methyl orange changes color when it protonates, it can be used as an acid–base indicator at pH values near its pK_a of 3.5. Some azo dyes are used in dyeing fabrics, foodstuffs, and cosmetics. For example, the following compound is a widely used dye known as as FD & C #6 (FD & C = food, drug, and cosmetic), a yellow compound used to color gelatin desserts, ice cream, beverages, candy, and so on.

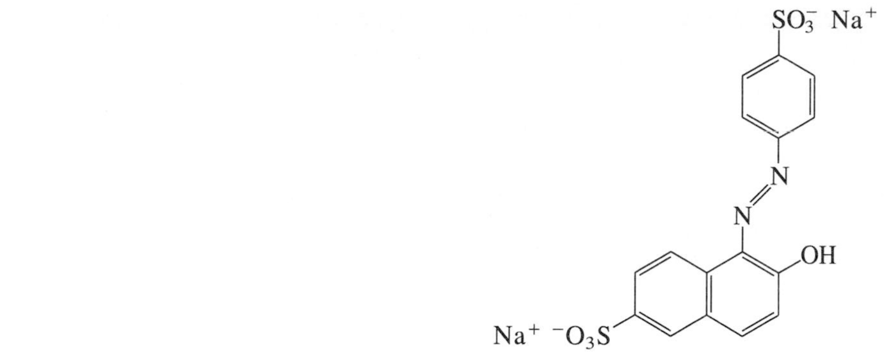

FD & C #6 (yellow)

D. Reactions of Secondary and Tertiary Amines with Nitrous Acid

We have seen that primary amines react with nitrous acid to give diazonium salts. How do secondary and tertiary amines react under the same conditions? Secondary amines react with nitrous acid to yield *N*-nitrosoamines, usually called simply **nitrosamines.**

$$Me_2NH + HNO_2 \longrightarrow \underset{\substack{N,N\text{-dimethyl} \\ \text{nitrosamine} \\ (89\text{–}90\%\ \text{yield})}}{Me_2N{-}N{=}O} + H_2O \qquad (23.70)$$

$$Ph{-}NH{-}CH_3 \xrightarrow[HCl]{NaNO_2} \underset{N\text{-methyl-}N\text{-nitrosoaniline}}{Ph{-}N(N{=}O){-}CH_3} \qquad (23.71)$$

The mechanism of this reaction is almost identical to that of the first step in the diazotization of primary amines (Eq. 23.55a). Since the nitrosamine derivative of a secondary amine has no N—H protons, it cannot react further, as a primary amine can. Therefore, secondary nitrosamines can be isolated as stable compounds.

Nitrosamines and Cancer The fact that many nitrosamines are known to be potent carcinogens has created a conflict over the use of sodium nitrite ($NaNO_2$) as a meat preservative. The meat-packing industry has argued that sodium nitrite is important in preventing the botulism that results from meat spoilage. But because sodium nitrite is, in combination with acid, a diazotizing reagent, it has the capacity for producing nitrosamines. For example, the frying of bacon generates nitrosamines that concentrate in the fat. (Well-drained bacon contains fewer nitrosamines, and nitrosamines are destroyed by ascorbic acid (vitamin C), which is present in fruit and vegetable juices. Perhaps this is a good reason for drinking orange juice when we have bacon for breakfast!) The potential hazards of sodium nitrite led to a long campaign by consumer groups to have it banned as a meat preservative; the campaign was fought by the meat-packing industry. Then researchers found that we humans carry our own source of nitrite, for nitrite is produced by our intestinal bacteria! It became questionable whether the risk from nitrite in meat is any greater than the risk we have faced all along from our own intestinal flora. These new findings caused the Food and Drug Administration in 1980 to back away from banning sodium nitrite as a preservative, recommending only that it be kept to a minimum. Here again (see Sec. 8.9B) we have a situation in which risk (risk of cancer from nitrosamines) is weighed against benefit (ability to preserve foods; freedom from toxic effects of food spoilage) over which reasonable persons may differ.

There have been some involuntary human tests on the effects of large amounts of nitrosamines in the diet. In one rather bizarre case, a chemistry teacher in Ulm, West Germany, was sentenced to life imprisonment for systematically injecting dimethylnitrosamine into his wife's favorite dessert. The woman became suspicious and had the food analyzed. She died later of severe liver dysfunction; precancerous lesions were found on her liver during autopsy.

The nitrogen of tertiary amines does not react under the strongly acidic conditions used in diazotization reactions. (Complex reactions occur under less acidic conditions.) However, *N,N*-disubstituted aromatic amines react with the electrophilic reagent N_2O_3 generated during diazotization (Eq. 23.53) in a conventional electrophilic substitution reaction, just as these compounds react with other electrophiles (see Eq. 23.67).

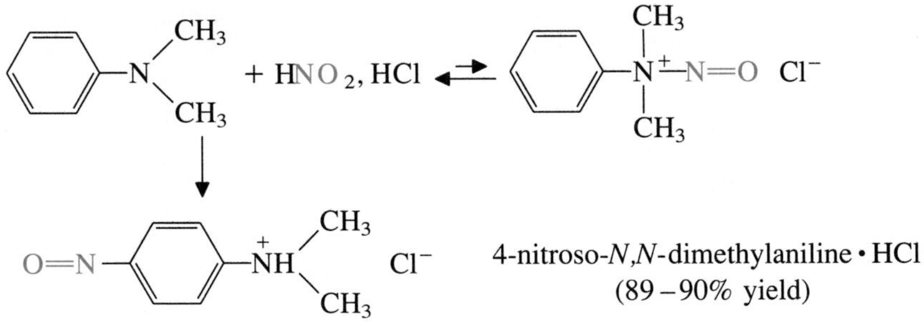

(23.72)

Problems

24 Outline a synthesis for each of the following compounds from the indicated starting materials using a reaction sequence involving a diazonium salt.
(a) 2-bromobenzoic acid from *o*-toluidine
(b) 2,4,6-tribromobenzoic acid from aniline

25 Design a synthesis of methyl orange (Eq. 23.69) using aniline as the only aromatic starting material.

26 What two compounds would react in a diazo coupling reaction to form FD & C dye #6?

23.11 SYNTHESIS OF AMINES

A. Gabriel Synthesis of Primary Amines

We learned in Sec. 23.7A that direct alkylation of ammonia is generally not a good synthetic method for the preparation of amines because multiple alkylation takes place. This problem can be avoided by protecting the amine nitrogen so that it can react only once with alkylating reagents. One approach of this sort begins with the imide *phthalimide*. The conjugate base of phthalimide ($pK_a \approx 9$) is easily formed with KOH or NaOH. This anion is a good nucleophile, and is alkylated by alkyl halides or sulfonate esters in an S_N2 reaction.

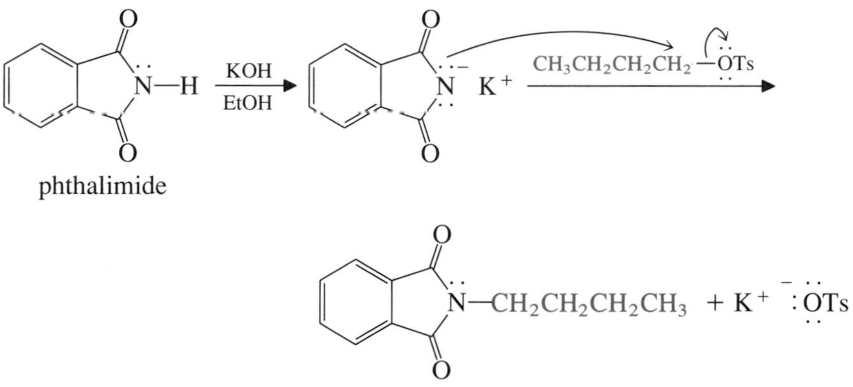

phthalimide

$$N\!\!-\!\!CH_2CH_2CH_2CH_3 + K^+ \; :\!\ddot{O}Ts \quad (23.73a)$$

The alkyl halides and sulfonates used in this reaction are primary or unbranched secondary (why?). Since the *N*-alkylated phthalimide formed in this reaction is really a double amide, it can be converted into the free amine by amide hydrolysis in either strong acid or base.

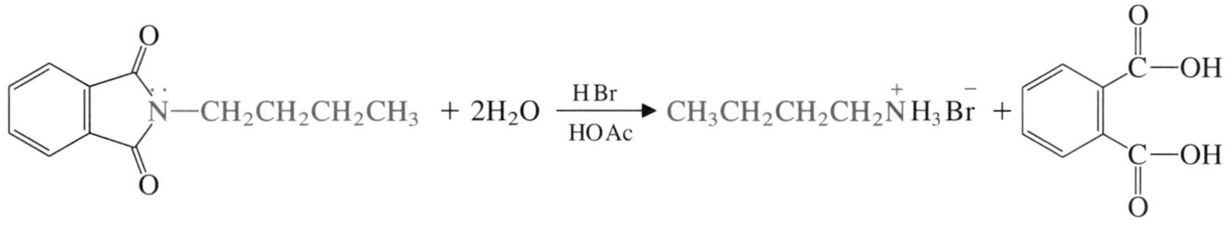

$$N\!\!-\!\!CH_2CH_2CH_2CH_3 + 2H_2O \xrightarrow[\text{HOAc}]{\text{HBr}} CH_3CH_2CH_2CH_2\overset{+}{N}H_3\overset{-}{Br} + \quad (23.73b)$$

In this example, acidic hydrolysis gives the ammonium salt, which can be converted into the free amine by neutralization with base.

The alkylation of phthalimide anion followed by hydrolysis of the alkylated derivative to the primary amine is called the **Gabriel synthesis.** Notice that the nitrogen in phthalimide has only one acidic hydrogen, and can thus be alkylated only once. Although *N*-alkylphthalimides also have a pair of unshared electrons on nitrogen, they do not alkylate further, because neutral imides are much less basic, and therefore less nucleophilic, than phthalimide anions. Hence, the problem of overalkylation encountered in the direct alkylation of ammonia is avoided.

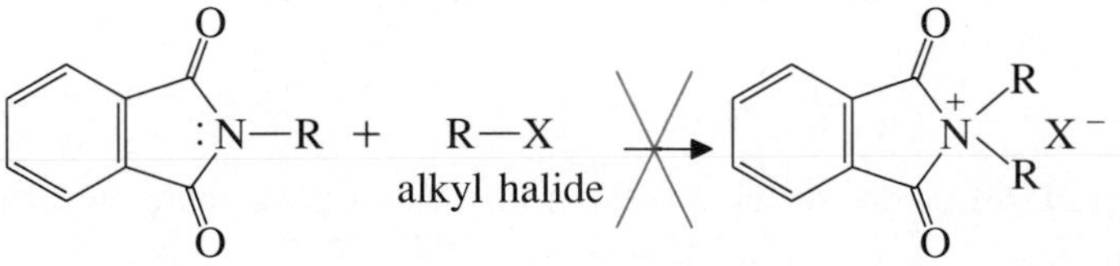

$$:\!N\!\!-\!\!R + \quad R\!\!-\!\!X \quad \xrightarrow{\quad\quad} \quad \overset{+}{N}\overset{R}{\underset{R}{\diagup}} \; X^- \quad \text{(does not occur)} \quad (23.74)$$

alkyl halide

B. Reduction of Nitro Compounds

Nitro compounds can be reduced to amines under a variety of conditions. The nitro group is usually reduced very easily by catalytic hydrogenation (Eq. 23.75).

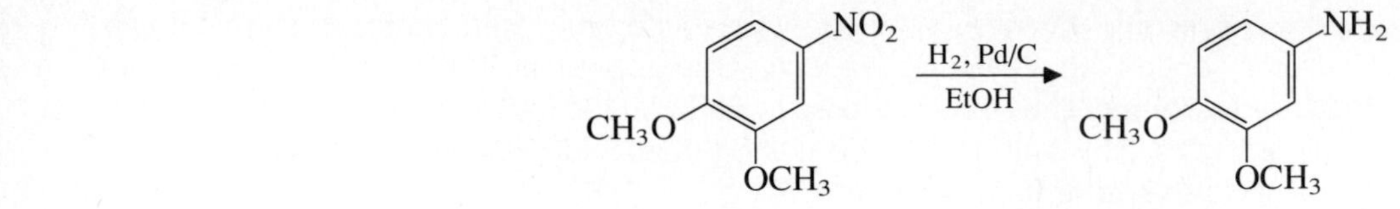

(97% yield) (23.75)

An older but effective method for reducing the nitro group is the reduction of aromatic nitro compounds to primary amines with finely divided metal powders and HCl; iron or tin powder is frequently used.

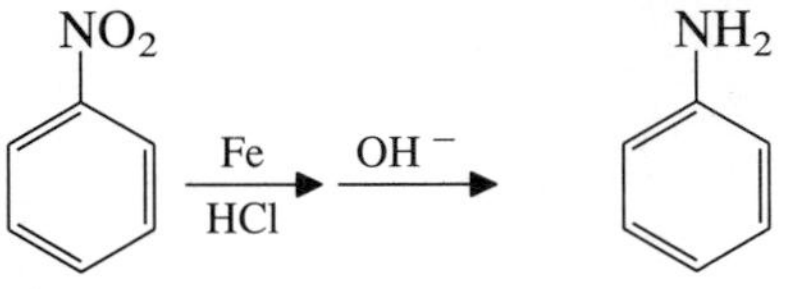

+ Fe^{3+} salts (23.76)

(82–86% yield)

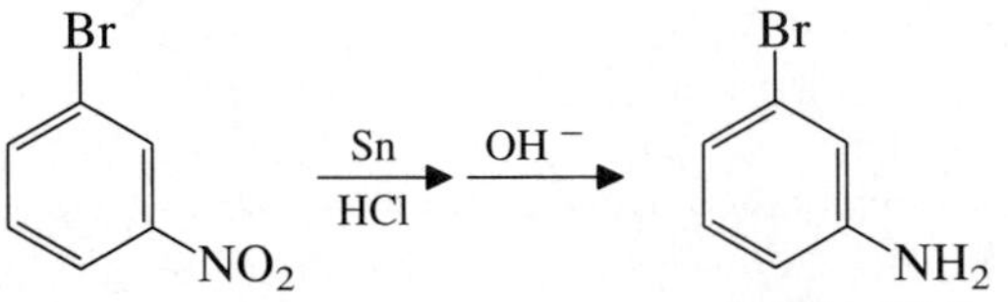

+ Sn^{2+} salts (23.77)

(80% yield)

Although we shall not consider the mechanisms of these reactions, we can recognize that the nitro compound is reduced *at nitrogen,* and the metal, which is oxidized, is the reducing agent.

We have learned that $LiAlH_4$ and $NaBH_4$ can be used instead of catalytic hydrogenation in many reductions. How do these reagents react with nitrobenzenes? Aromatic nitro compounds do react with $LiAlH_4$, but the reduction products are azobenzenes, not amines:

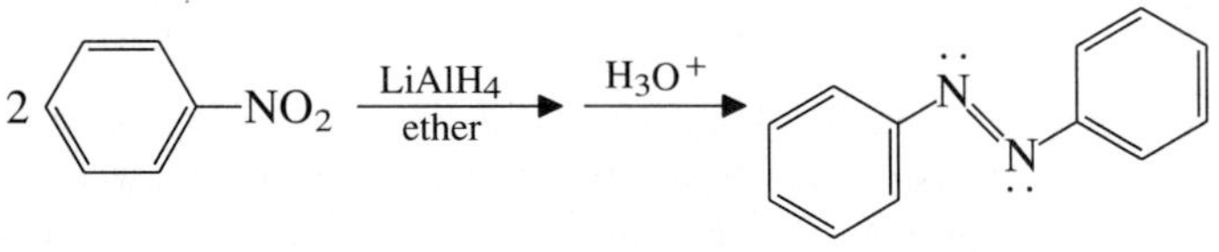

(23.78)

Nitro groups do not react at all with sodium borohydride under the usual conditions.

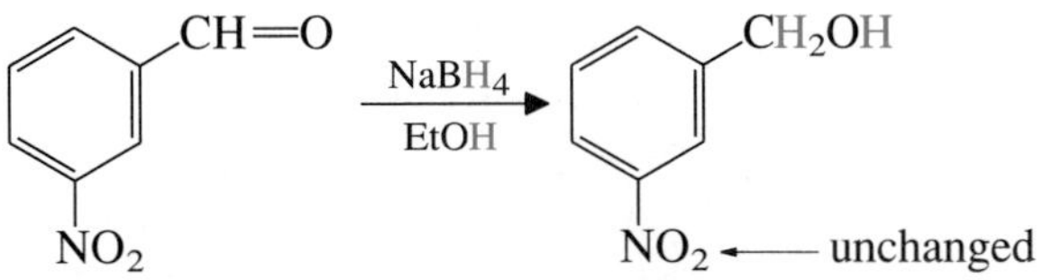

(23.79)

Hence, $LiAlH_4$ and $NaBH_4$ are *not* useful in forming aromatic amines from nitro compounds.

Problem

27 Outline a synthesis of *p*-iodoanisole from phenol and any other reagents.

C. Hofmann and Curtius Rearrangements

When a primary amide is treated with bromine in base, a very unusual reaction takes place: the amide is converted into an amine with one fewer carbon atom, and the carbonyl carbon of the amide is lost as CO_2.

$$Br_2 + 2NaOH + (CH_3)_3CCH_2\overset{\displaystyle O}{\overset{\|}{-C}}-NH_2 \longrightarrow$$

$$(CH_3)_3CCH_2-NH_2 + O-C-O + 2NaBr + H_2O \quad (23.80)$$
$$\text{(94\% yield)}$$

(Notice the fate of the amide carbonyl group, shown in color.) We can see that a rearrangement has taken place in this reaction because the alkyl group originally attached to the carbonyl carbon in the amide is bound to the amine nitrogen of the product. This reaction is called the **Hofmann rearrangement** or **Hofmann hypobromite reaction.**

Although the reaction looks complex, it can be dissected mechanistically into a sequence of individual steps, some of which have close analogy in chemistry that we have studied already. The first step in the mechanism is ionization of the amide N—H proton (Sec. 22.1A). The resulting anion is then brominated.

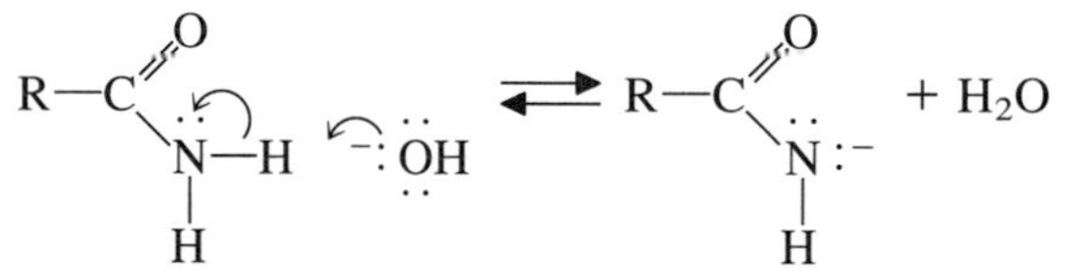

$$(23.81a)$$

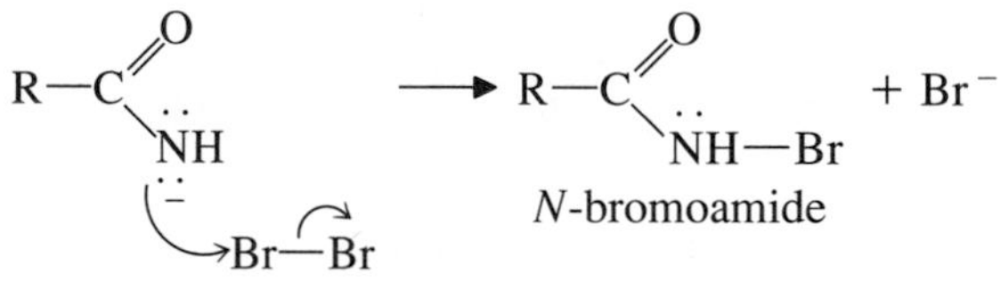

$$(23.81b)$$

N-bromoamide

(This reaction is analogous to α-bromination of a ketone in base; Sec. 22.3B.) The *N*-bromoamide product is even more acidic than the amide starting material (why?), and it too ionizes.

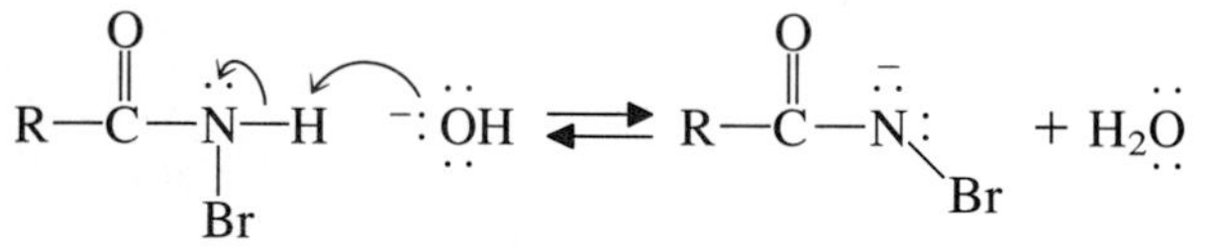

$$R-\overset{\displaystyle O}{\overset{\|}{C}}-\overset{\displaystyle }{\underset{\underset{Br}{|}}{N}}-H \quad \overset{..}{:}OH \rightleftharpoons R-\overset{\displaystyle O}{\overset{\|}{C}}-\overset{..}{\underset{Br}{N}:}{}^{-} + H_2\overset{..}{O} \quad (23.81c)$$

The *N*-bromo anion does not brominate again, because it undergoes an even faster reaction: a rearrangement.

$$\underset{R}{\overset{:O:}{\overset{\|}{C}}}\overset{..}{N}\!-\!\overset{..}{Br}: \longrightarrow \overset{..}{O}=C=N-R + :\overset{..}{Br}:{}^{-} \quad (23.81d)$$
$$\text{an isocyanate}$$

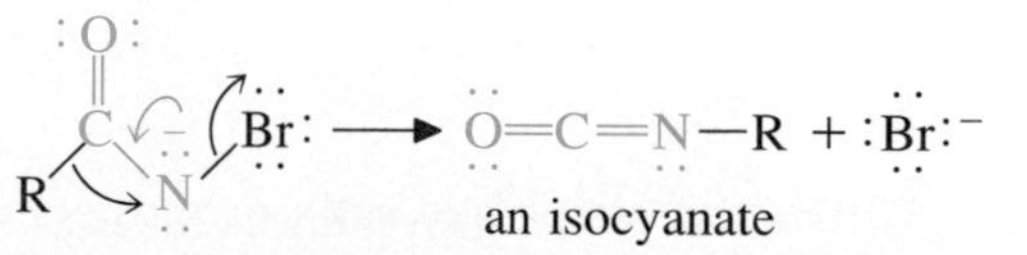

It is in this rearrangement that the alkyl–nitrogen bond is established. The product of this rearrangement is called an **isocyanate.** Isocyanates are very reactive, and are attacked by nucleophiles at the carbonyl carbon. For example, when a basic aqueous solution is used in the Hofmann rearrangement, hydroxide, a good nucleophile, is present. Attack of hydroxide on the carbonyl group leads to the formation of a carbamate ion, the conjugate base of a *carbamic acid* (see Sec. 21.1G).

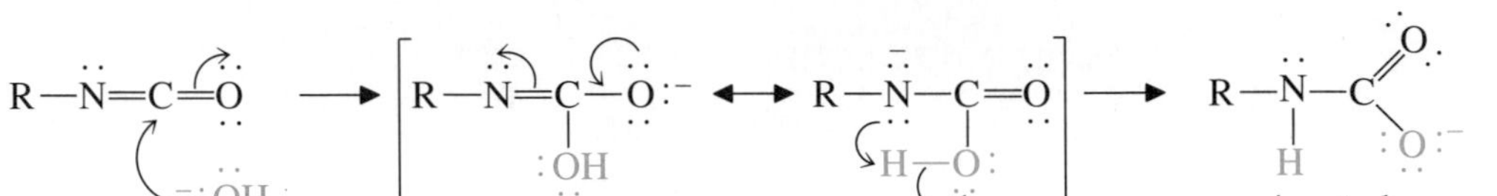

conjugate base of a carbamic acid (23.81e)

Notice that the overall reaction with the isocyanate is an addition across the *carbon–nitrogen bond*. The carbonyl group is left intact because the C=O bond is a stronger bond than the C=N bond.

When the reaction mixture is acidified, the carbamate decarboxylates (Sec. 20.11) to give the amine.

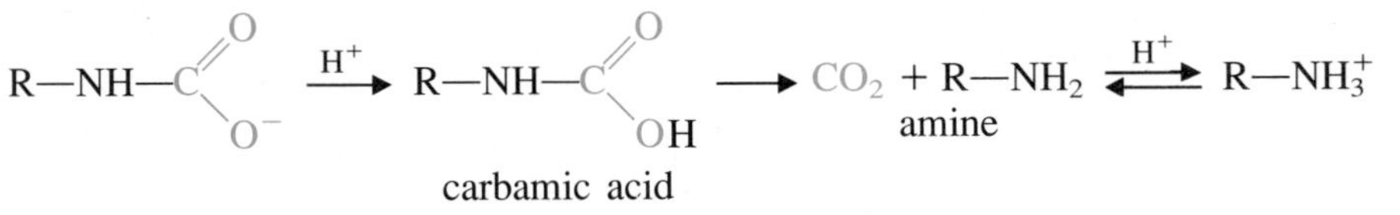

(23.81f)

The Hofmann rearrangement can be summarized by the following steps:

amide ⟶ *N*-bromoamide ⟶ isocyanate ⟶ carbamic acid ⟶ amine + CO_2 (23.82)

amide carbonyl group removed as CO_2

When an alcohol solvent is used for the Hofmann rearrangement, the nucleophile that attacks the isocyanate is the conjugate-base anion of the alcohol, and a *carbamate ester* is formed, as in the following example:

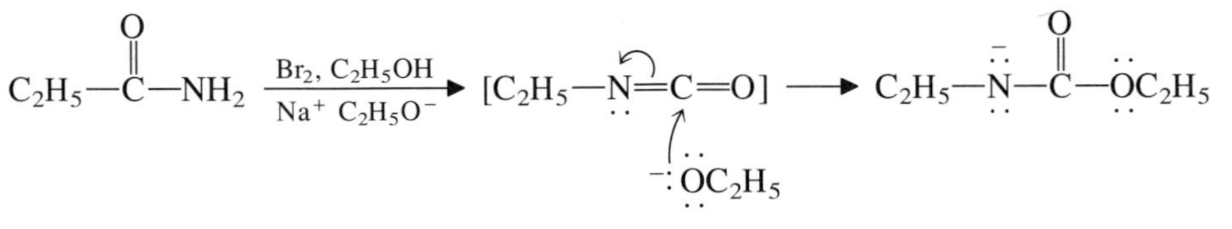

(23.83a)

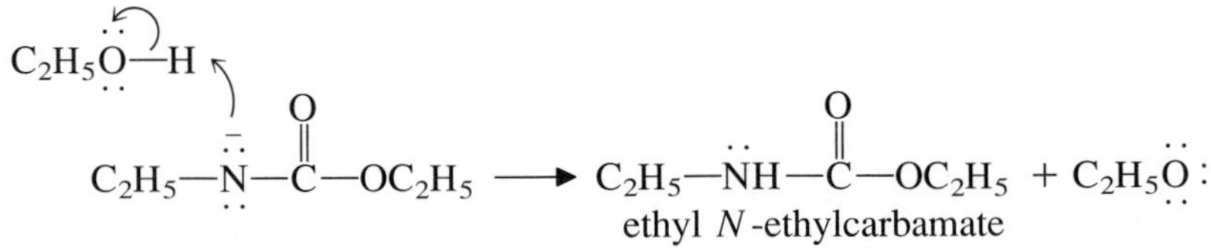

(23.83b)

ethyl *N*-ethylcarbamate
(a carbamate ester;
80% yield)

Carbamate esters, like carbonate esters (Sec. 20.11), are stable compounds; they do not decarboxylate.

Closely related to the Hofmann rearrangement is an analogous reaction of *acyl azides*. An acyl azide has the following general structure:

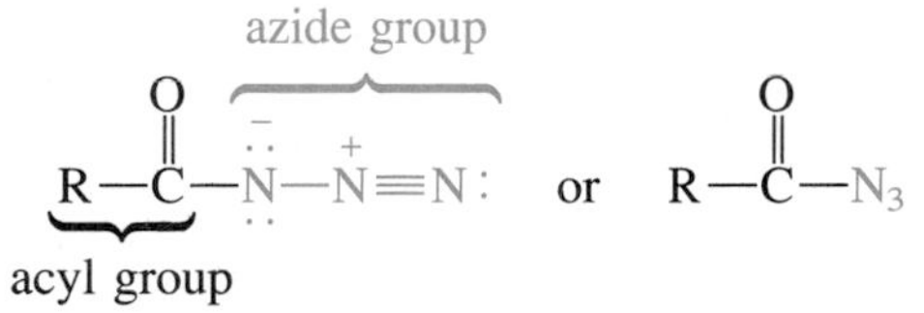

an acyl azide

When acyl azides are heated, they are transformed, with loss of nitrogen, into isocyanates.

$$CH_3(CH_2)_{10}\text{—}\overset{\displaystyle O}{\overset{\|}{C}}\text{—}N_3 \xrightarrow[\text{benzene}]{\text{heat}} CH_3(CH_2)_{10}\text{—}N\text{=}C\text{=}O + N_2 \qquad (23.84)$$

undecyl isocyanate
(81–86% yield)

This reaction is called the **Curtius rearrangement.** The mechanistic similarity between the Curtius and Hofmann rearrangements can be seen by the following comparison. In the Hofmann rearrangement, bromide ion is the leaving group; in the Curtius rearrangement, molecular nitrogen is the leaving group.

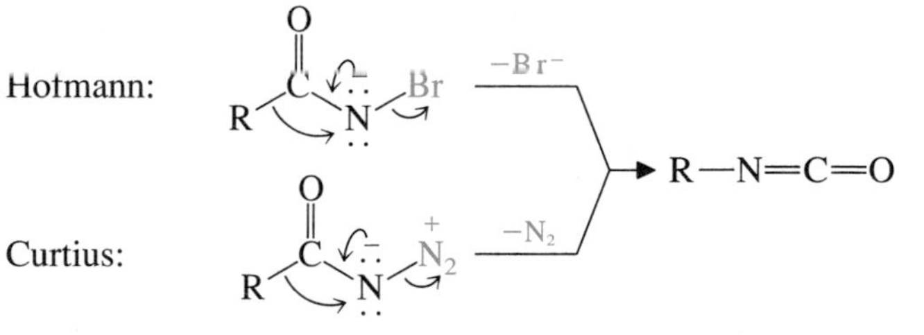

(23.85)

Other than this difference, the two rearrangement reactions are formally identical.

Despite the similarity between the Hofmann and Curtius rearrangements, there are important practical differences. Because the Hofmann rearrangement is carried out in protic solvents, and because isocyanates react with these solvents, the isocyanate intermediates cannot be isolated in the Hofmann rearrangement. The Curtius rearrangement, on the other hand, can be run in inert aprotic solvents such as benzene or toluene, and can be used to prepare pure isocyanates, as shown in Eq. 23.84. (If isolation of the isocyanate is not desired, the Curtius rearrangement may also be run in protic solvents.) The isocyanate formed in the Curtius reaction may be treated with other nucleophiles such as amines, phenols, alcohols, and so on, to give carbamate derivatives.

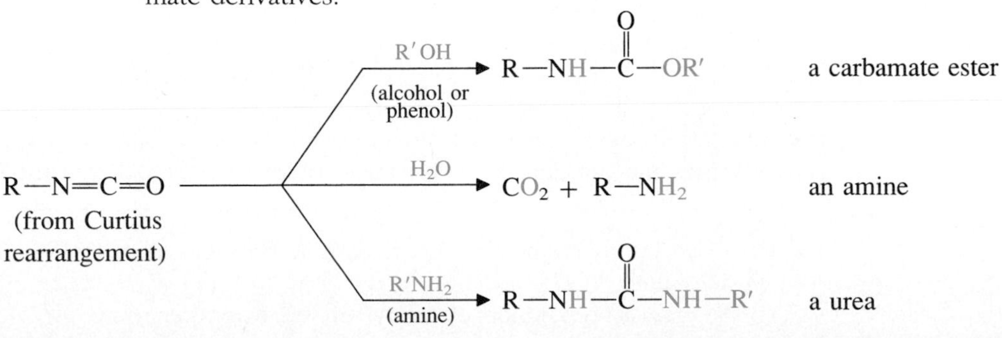

(23.86)

How do we prepare the acyl azides used in the Curtius rearrangement? The key is to recognize that these compounds are carboxylic acid derivatives. The most straight-forward preparation is the reaction of an acid chloride with sodium azide.

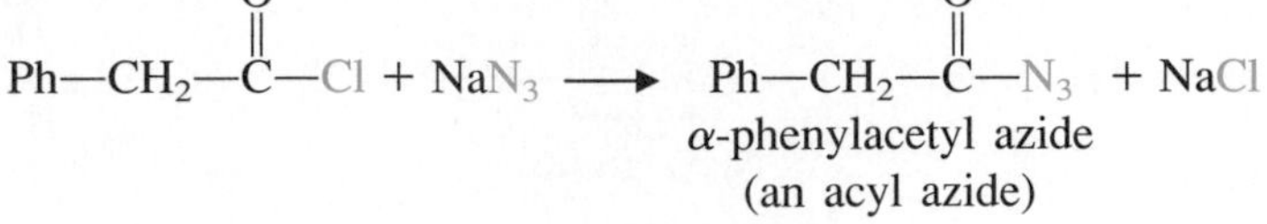

$$\underset{\text{α-phenylacetyl azide}}{\text{Ph—CH}_2\text{—}\overset{\displaystyle O}{\overset{\|}{\text{C}}}\text{—Cl}} + \text{NaN}_3 \longrightarrow \underset{\substack{\text{α-phenylacetyl azide} \\ \text{(an acyl azide)}}}{\text{Ph—CH}_2\text{—}\overset{\displaystyle O}{\overset{\|}{\text{C}}}\text{—N}_3} + \text{NaCl} \qquad (23.87)$$

Another widely used method is to convert an ethyl ester into an acyl derivative of hydrazine ($\text{NH}_2\text{—NH}_2$) by aminolysis (Sec. 21.8C). The resulting amide, an *acyl hydrazide,* is then diazotized with nitrous acid to give the acyl azide.

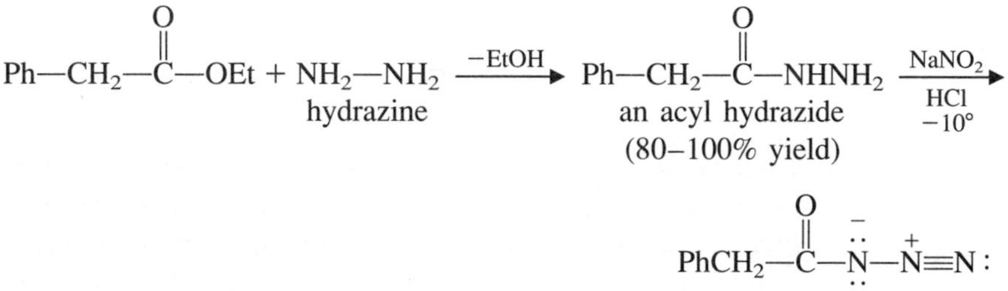

$$\underset{\text{hydrazine}}{\text{Ph—CH}_2\text{—}\overset{\displaystyle O}{\overset{\|}{\text{C}}}\text{—OEt} + \text{NH}_2\text{—NH}_2} \xrightarrow{\text{—EtOH}} \underset{\substack{\text{an acyl hydrazide} \\ \text{(80–100\% yield)}}}{\text{Ph—CH}_2\text{—}\overset{\displaystyle O}{\overset{\|}{\text{C}}}\text{—NHNH}_2} \xrightarrow[\substack{\text{HCl} \\ -10°}]{\text{NaNO}_2}$$

$$\underset{}{\text{PhCH}_2\text{—}\overset{\displaystyle O}{\overset{\|}{\text{C}}}\text{—}\overset{..}{\underset{..}{\text{N}}}\text{—}\overset{-}{\overset{..}{\text{N}}}\overset{+}{\equiv}\text{N}:} \quad (23.88)$$

We can see what is going on in this diazotization if we compare it to the diazotization of alkylamines:

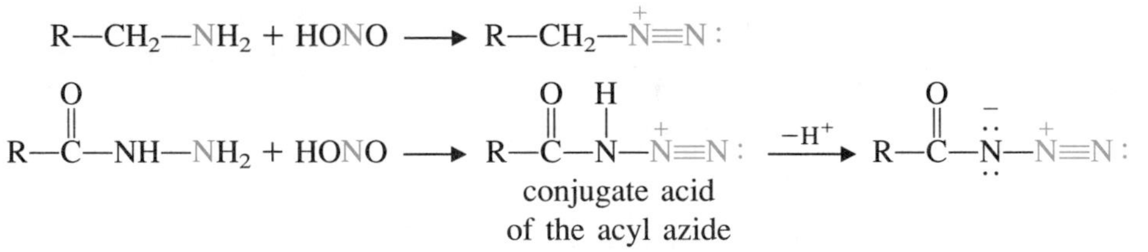

$$\text{R—CH}_2\text{—NH}_2 + \text{HONO} \longrightarrow \text{R—CH}_2\text{—}\overset{+}{\text{N}}\equiv\text{N}:$$

$$\underset{\substack{\text{conjugate acid} \\ \text{of the acyl azide}}}{\text{R—}\overset{\displaystyle O}{\overset{\|}{\text{C}}}\text{—NH—NH}_2 + \text{HONO} \longrightarrow \text{R—}\overset{\displaystyle O}{\overset{\|}{\text{C}}}\text{—}\overset{\text{H}}{\underset{}{\text{N}}}\text{—}\overset{+}{\text{N}}\equiv\text{N}:} \xrightarrow{\text{—H}^+} \text{R—}\overset{\displaystyle O}{\overset{\|}{\text{C}}}\text{—}\overset{..}{\underset{..}{\text{N}}}\text{—}\overset{+}{\text{N}}\equiv\text{N}: \quad (23.89)$$

Because the conjugate acid of the acyl azide is quite acidic (why?), it loses a proton from the adjacent nitrogen to give the neutral acyl azide.

An interesting and useful aspect of both the Hofmann and Curtius rearrangements is that they take place with *retention of stereochemical configuration* in the migrating alkyl group.

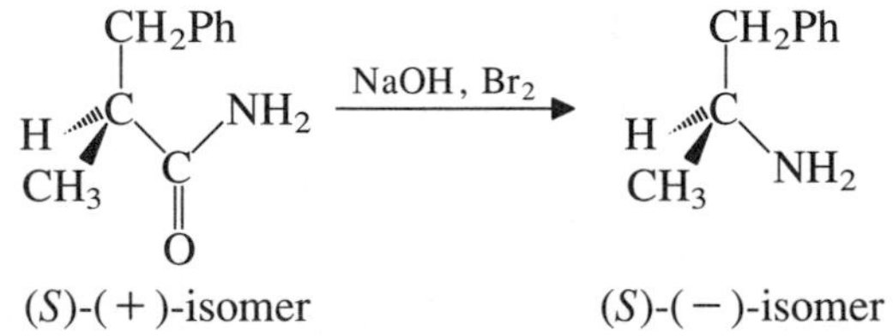

(23.90)

Hence, optically active carboxylic acid derivatives can be used to prepare optically active amines and carbamic acid derivatives of known stereochemical configuration.

Problems

28 (a) Could *t*-butylamine be prepared by the Gabriel synthesis? If so, how? If not, why not?

(b) Suggest a preparation of *t*-butylamine by another route.

29 Outline the preparation of the following compounds from 3-methylbenzoic acid and any other reagents:

(a)

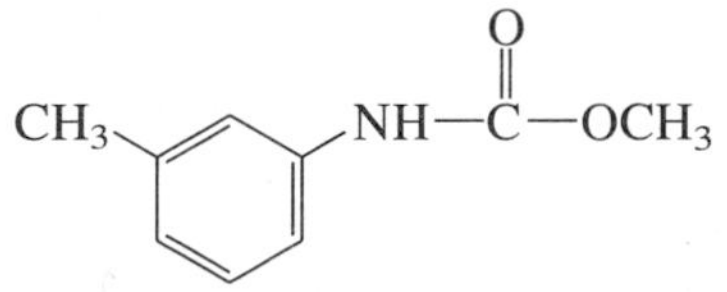

(c)

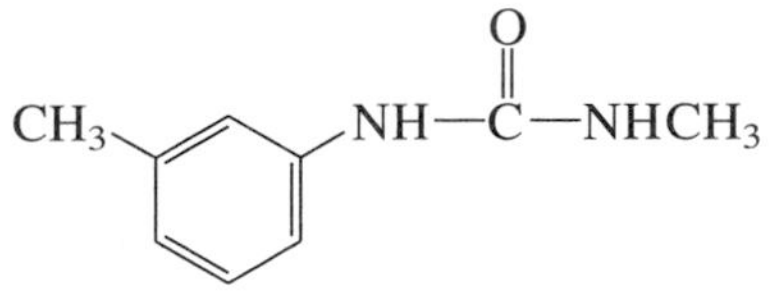

(b)

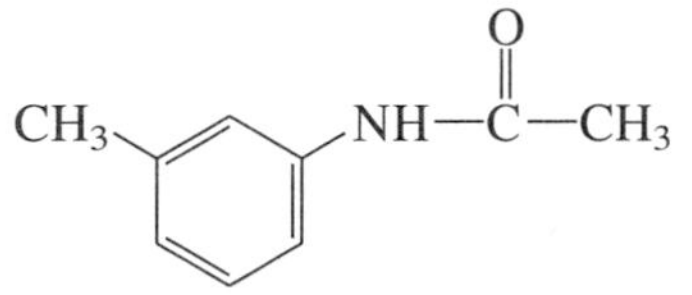

D. Synthesis of Amines: Summary

In addition to the methods discussed in this section, we have also learned about other ways to prepare amines. Let us summarize the methods we have studied:

1. Reduction of amides and nitriles with $LiAlH_4$ (Secs. 21.9B, 21.9C)
2. Direct alkylation of amines (Sec. 23.7). This reaction is of limited utility, but is useful for preparing quaternary ammonium salts.
3. Reductive amination (Sec. 23.7B)
4. Aromatic substitution reactions of anilines (Sec. 23.9)
5. Gabriel synthesis of primary amines (Sec. 23.11A)
6. Reduction of nitro compounds (Sec. 23.11B)
7. Hofmann and Curtius rearrangements (Sec. 23.11C)

Methods 2, 3, and 4 represent methods of preparing amines from other amines. To the extent that an amide used in method 1 can be prepared from an amine, this method, too, represents a method for obtaining one amine from another. Method 5 is limited to the preparation of primary amines, and methods 1, 6, and 7 can be used for obtaining amines from other functional groups.

Problem

30 Show how 2-cyclopentyl-*N,N*-dimethyl-1-ethanamine could be synthesized from each of the following starting materials.

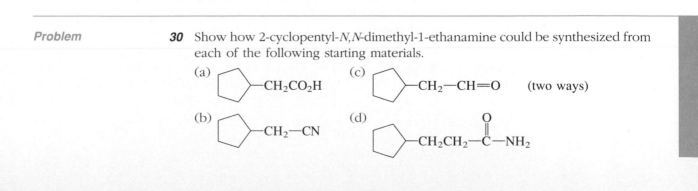

(a)

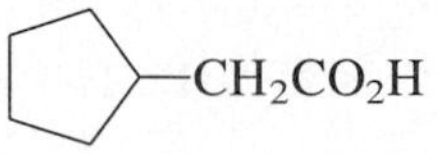

(c)

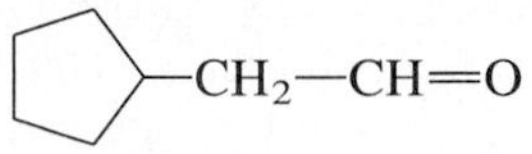

(two ways)

(b)

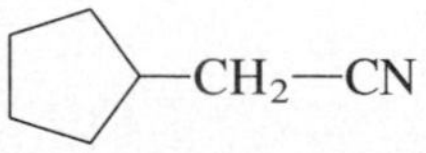

(d)

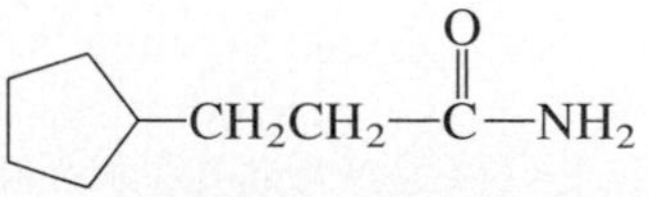

23.12 OCCURRENCE AND USE OF AMINES

A. Industrial Use of Amines and Ammonia

Among the relatively few industrially important amines is hexamethylenediamine, $H_2N(CH_2)_6NH_2$, used in the synthesis of nylon-6,6 (Sec. 21.12A). Ammonia is also an important "amine," and is a key source of nitrogen in a number of manufacturing processes. In agricultural chemistry, for example, liquid ammonia itself and urea, which is made from ammonia and CO_2, are important nitrogen fertilizers. Ammonia is manufactured by the hydrogenation of N_2. Although it might not seem that the industrial synthesis of ammonia has anything to do with organic chemistry, the hydrogen used in its manufacture in fact comes from the cracking of hydrocarbons (Sec. 5.9B). Thus, the availability of ammonia is tied inexorably to the availability of hydrocarbons.

Many biologically important compounds are amines. Some of the most important of these—amino acids—are the subject of Chapter 26. Many drugs are amines. Although these are not classified as major industrial organic chemicals, their manufacture has had a profound impact on life on this planet. In the next section, we shall consider a few of the biologically important amines that occur in nature.

B. Naturally Occurring Amines

Alkaloids Alkaloids are nitrogen-containing bases that occur naturally in plants. This simple definition encompasses an incredibly diverse group of compounds; the structures of a few alkaloids are shown in Fig. 23.4. Since amines are the most common organic bases, it is not surprising that most alkaloids are amines, including heterocyclic amines. It is believed that the first alkaloid ever isolated and studied is morphine, discovered in 1805. Many alkaloids have biological activity (See Fig. 23.4); others have no known activity, and their functions in the plants from which they come are, in many cases, obscure. Investigations dealing with the isolation, structure, and medicinal properties of alkaloids continue to be major research activities in organic chemistry.

Problem

31 Identify the structural feature(s) of (a) morphine and (b) cocaine that are responsible for their basicity. (See Fig. 23.4.)

Hormones and Neurotransmitters Epinephrine (adrenaline) is an amine secreted by both the adrenal medulla and sympathetic nerve endings; it is an example of a **hormone**—a compound that regulates the biochemistry of multicellular organisms, particularly vertebrates.

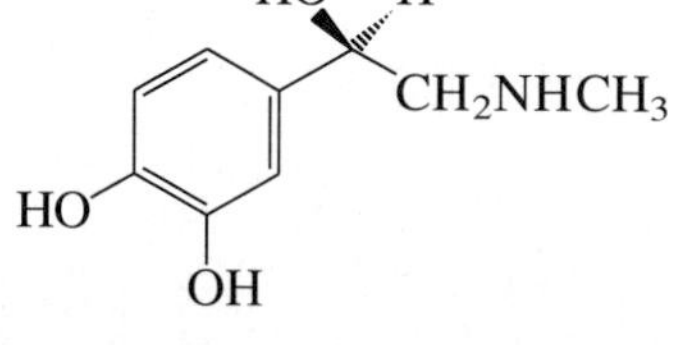

epinephrine

Figure 23.4 Structures of some alkaloids.

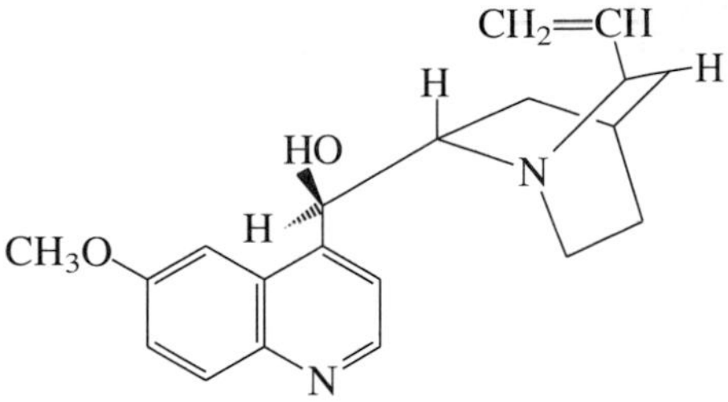

quinine
(antimalarial drug)

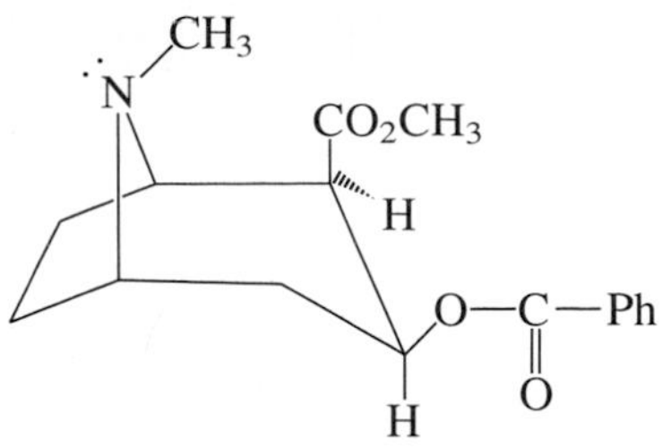

cocaine
(stimulant of central nervous system; induces euphoria; widely abused)

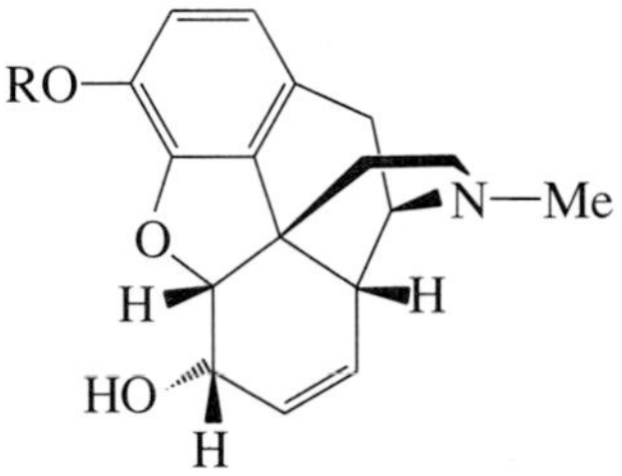

morphine (R = H)
codeine (R = CH_3)
(medically important analgesics)

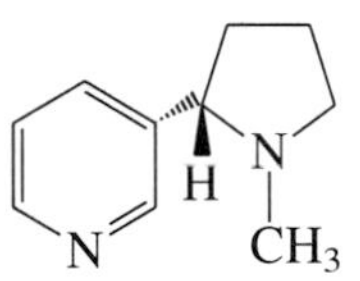

nicotine
(principal alkaloid from tobacco)

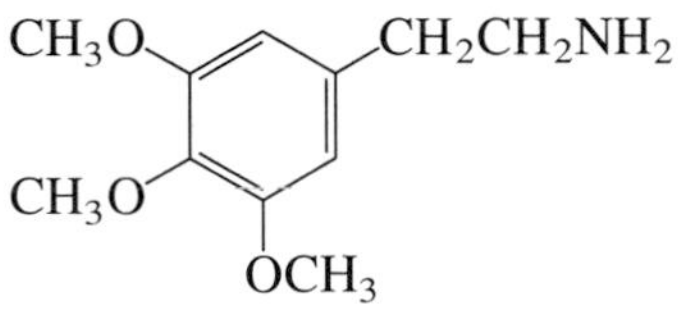

mescaline
(from peyote cactus; used as hallucinogen by American Indians during religious observances)

Epinephrine, for example, is associated with the "fight or flight" response to external stimuli; you might feel the effects of epinephrine secretion when you walk unprepared into your organic chemistry class and your instructor says, "Pop quiz today." The mechanisms by which hormones exert their effects is an important research area in contemporary biochemistry.

Norepinephrine, another amine, and acetylcholine, a quaternary ammonium ion, are examples of *neurotransmitters.*

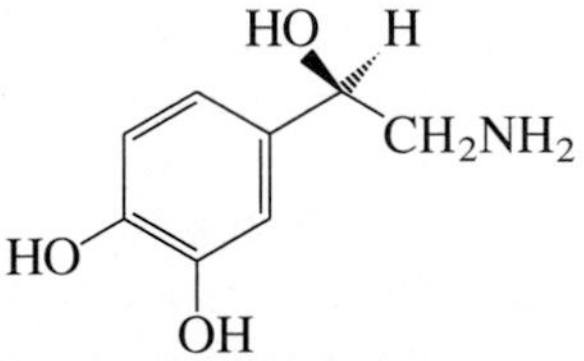

norepinephrine

$$\overset{+}{Me_3N}—CH_2—CH_2—O—\overset{\overset{\displaystyle O}{\|}}{C}—CH_3$$

acetylcholine

Neurotransmitters are molecules used for communication between nerve cells, or between nerve cells and their target organs. This communication occurs at junctions called *synapses.* A nerve impulse is transmitted when a neurotransmitter is released from a nerve cell on one side of the synapse, moves by diffusion across the synapse, and binds to a protein receptor molecule on the other side. This binding triggers

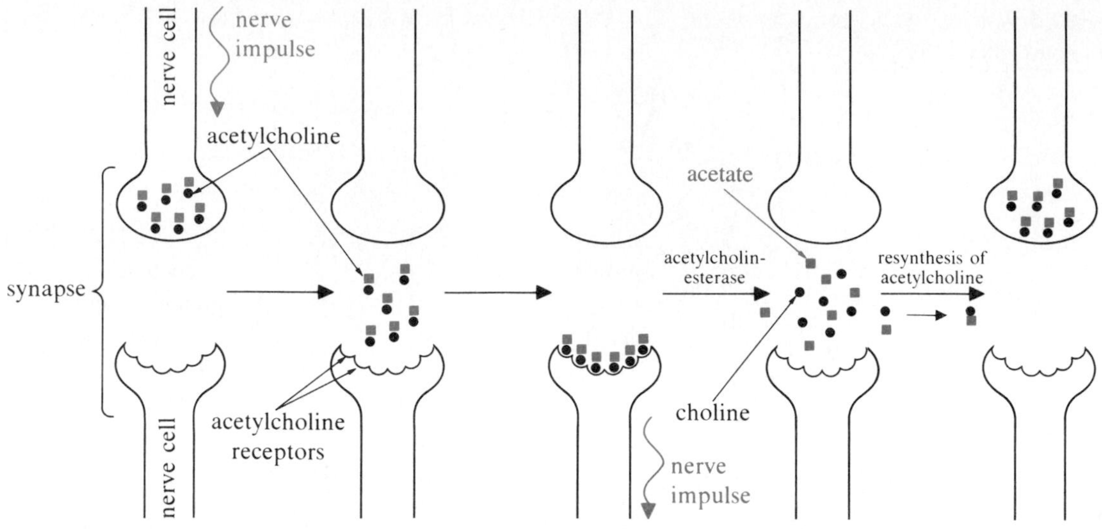

Figure 23.5 A diagram showing the transmission of nerve impulses across a synapse by acetylcholine.

either the transmission of the impulse to the next synapse (Fig. 23.5) or a response of the target organ. Different neurotransmitters are used in different parts of the nervous system. When acetylcholine, for example, is used as a neurotransmitter, its effects must be removed once the nerve impulse has been transmitted. This is accomplished by the hydrolysis of acetylcholine to acetate and choline, catalyzed by the enzyme *acetylcholinesterase*. The choline is then used in the resynthesis of acetylcholine on the other side of the synapse.

$$H_2O + Me_3\overset{+}{N}{-}CH_2CH_2{-}O{-}\overset{\overset{O}{\|}}{C}{-}CH_3 \xrightarrow[\text{(enzyme)}]{\text{acetylcholinesterase}}$$

$$\underset{\text{choline}}{Me_3\overset{+}{N}{-}CH_2CH_2{-}OH} + CH_3{-}\overset{\overset{O}{\|}}{C}{-}O^- + H^+ \qquad (23.91)$$

Some **neurotoxins** (nerve poisons) act by interfering with this series of events. For example, one type of neurotoxin binds to the enzyme acetylcholinesterase because it has a structure very similar to that of acetylcholine, and cannot be differentiated from acetylcholine by the enzyme—something of a molecular "wolf in sheep's clothing." If the enzyme cannot hydrolyze the toxin, the nerve cell is retained in a continuously excited state. Such neurotoxins have been developed as agents of chemical warfare, but also have other uses as insecticides and medicinals. Neostigmine, for example, is used to treat glaucoma; can you see any structural features of neostigmine that might make it a good molecular mimic of acetylcholine?

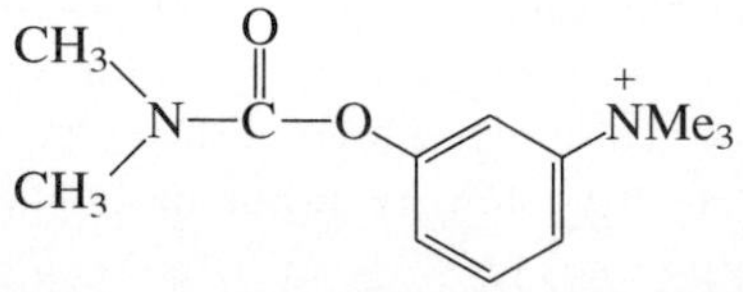

neostigmine

KEY IDEAS IN CHAPTER 23

- Amines are classified as primary, secondary, or tertiary. Quaternary ammonium salts are compounds in which all four protons of the ammonium ion are formally replaced with alkyl or aryl groups.
- Most amines undergo rapid inversion at nitrogen. This inversion interconverts an amine and its mirror image.
- Simple amines are liquids with unpleasant odors. The lower amines are miscible with water.
- The N—H stretching absorption is the most important infrared absorption of primary and secondary amines. In the NMR spectrum, the protons on the carbon adjacent to nitrogen have chemical shifts in the δ 2.5–3.0 range.
- The basicity of amines is one of their most important chemical properties. The basicity of an amine is measured by the pK_a of its conjugate-acid ammonium ion. Ammonium-ion pK_a values are affected by alkyl substitution on the nitrogen, the inductive effects of nearby substituent groups, and resonance interaction of the amine's unshared electrons with an adjacent aromatic ring.
- Because amines are basic, they are also nucleophilic. The nucleophilicity of amines is important in many of their reactions, for example, alkylation, imine formation, acylation, and diazotization.
- Because amines are basic, they are relatively poor leaving groups in substitution and elimination reactions. Thus the elimination reaction of a quaternary ammonium hydroxide requires heat, but does proceed to give an alkene (Hofmann elimination). In this reaction the proton is lost from the least branched β-carbon atom.
- The unprotonated amino group is *ortho, para*-directing in electrophilic aromatic substitution reactions. However, in many reactions of this type, such as nitration and sulfonation, protection of the nitrogen as an amide is required to prevent protonation of the amine as well as other side reactions.
- Treatment of primary amines with nitrous acid gives diazonium ions. Alkanediazonium ions decompose under the diazotization conditions to give a mixture of many products. Aryldiazonium ions, however, can be used in a number of substitution reactions: the Sandmeyer reaction (substitution of N_2 by halide or cyanide groups), the Schiemann reaction (substitution with fluoride), or hydrolysis (substitution with —OH). Because aryldiazonium ions are electrophiles, they react with activated aromatic compounds, such as aromatic amines and phenols, to give substituted azobenzenes, some of which are dyes. Secondary amines react with nitrous acid to give nitrosamines. Under acidic conditions, tertiary amines do not react. (Aromatic tertiary amines give ring-nitrosated products.)

- Amines can be synthesized from amides, nitriles, nitro compounds, and other amines, as summarized in Sec. 23.11D. The Hofmann and Curtius rearrangements yield amines with one fewer carbon atom. The Curtius rearrangement can also be used to prepare isocyanates, which react with nucleophiles by addition across the C=N bond to give carbamic acid derivatives.

ADDITIONAL PROBLEMS

32 Give the principal organic product(s) expected when *p*-chloroaniline reacts with each of the following reagents:
(a) dilute HBr
(b) C_2H_5MgBr
(c) $NaNO_2$, HCl, 0°
(d) *p*-toluenesulfonyl chloride
(e) product of (c) with H_2O and heat
(f) product of (c) with CuBr
(g) product of (c) with H_3PO_2
(h) product of (c) with CuCN
(i) product of (c) with HBF_4, heat
(j) product of (d) + NaOH, 25°

33 Give the principal organic product(s) expected when *N*-methylaniline reacts with each of the following reagents:
(a) Br_2
(b) benzoyl chloride
(c) benzyl chloride, then dilute OH^-
(d) *p*-toluenesulfonic acid
(e) $NaNO_2$, HCl
(f) excess CH_3I, heat, then Ag_2O
(g) $CH_3CH{=}O$, $NaBH_3CN$

34 Give the principal organic product(s) expected when isopropylamine reacts with each of the following reagents:
(a) dilute H_2SO_4
(b) dilute NaOH solution
(c) butyllithium, −78°
(d) acetyl chloride, pyridine
(e) $NaNO_2$, aqueous HBr, 0°
(f) acetone, H_2/Pd/C
(g) excess CH_3I, heat
(h) benzoic acid, 25°
(i) formaldehyde, $NaBH_4$
(j) product of (g) + Ag_2O, then heat
(k) product of (d) + $LiAlH_4$, then H_3O^+

35 Give the structure and name of a compound that fits each description. (There may be more than one correct answer for each.)
(a) a chiral primary amine $C_4H_{11}N$
(b) a chiral primary amine C_4H_7N with no triple bonds
(c) two secondary amines, which, when treated with CH_3I, then Ag_2O and heat, give propylene and *N,N*-dimethylaniline
(d) a compound C_4H_9N that reacts with $NaBH_4$ to give *N*-methyl-2-propanamine (Do not attempt to name this compound.)

36 Behold the molecule *grossamine,* extracted from the base imagination of an organic chemist.

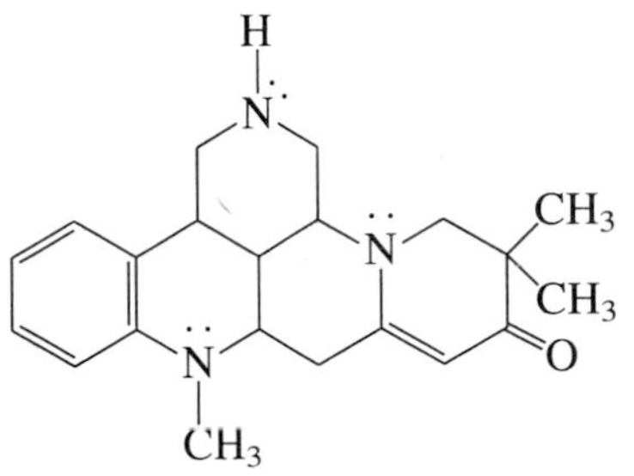

Give the structure of this molecule as it would exist when allowed to react with one molar equivalent of HCl.

37 (a) When anthranilic acid (below) is treated with $NaNO_2$ in aqueous HCl solution, and the resulting solution is treated with *N,N*-dimethylaniline, a dye called *methyl red* is formed. Give the structure of methyl red.

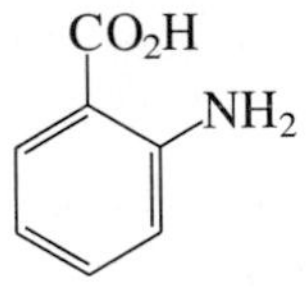

anthranilic acid

(b) When an acidic solution of methyl red is titrated with base, the dye behaves as a diprotic acid with pK_a values of 2.3 and 5.0. The color of the methyl red solution changes very little as the pH is raised past 2.3, but as the pH is raised past 5.0, the color of the solution changes dramatically from red to yellow. Explain.

38 Explain why base must be used in the Hofmann rearrangement but not in the Curtius rearrangement.

Problems (Cont.)

39 Edy Amine, professional organic chemistry student, has proposed the following reactions. Indicate the problems, if any, that would occur in each case.

(a)

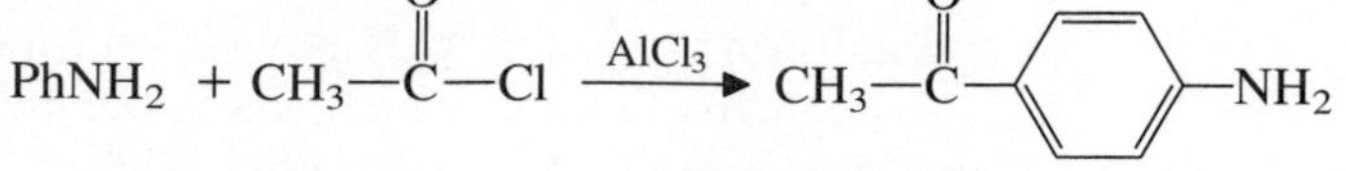

(b) $Me_3C{-}NH_2 + CH_3I \longrightarrow \xrightarrow{OH^-} Me_3C{-}NH{-}CH_3 + I^-$

(c)

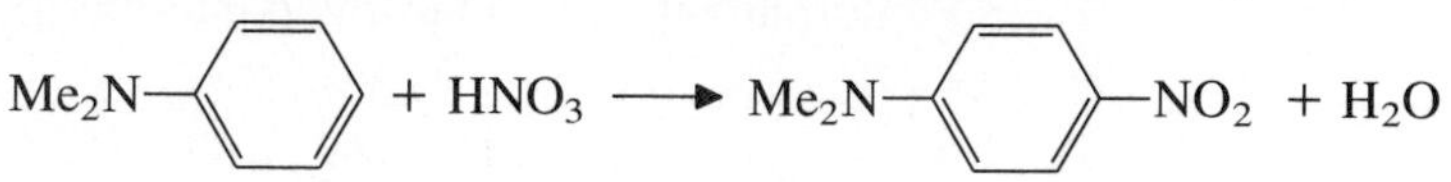

(d) $(CH_3)_2NH + CH_3CH_2CH_2OH \xrightarrow{H_2SO_4} CH_3CH_2CH_2N(CH_3)_2 + H_2O$

(e)

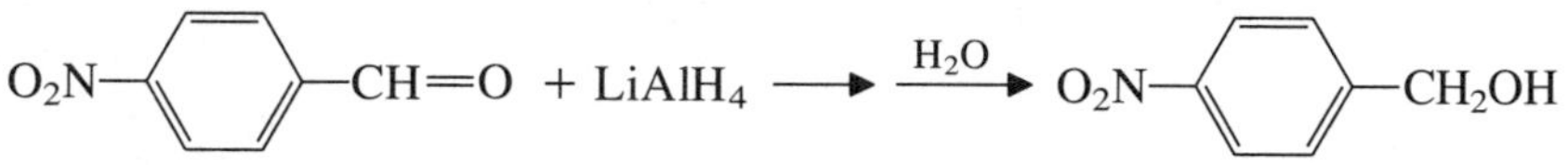

(f)

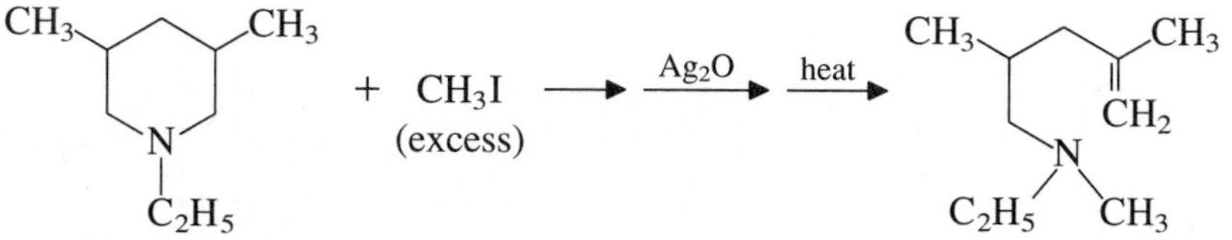

(g)

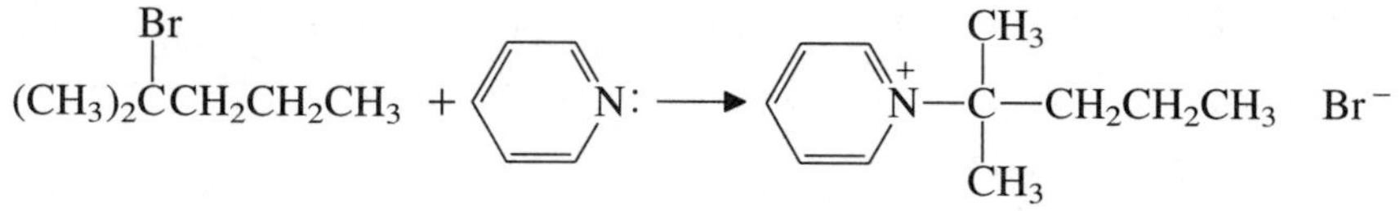

40 Explain how you would distinguish the compounds within each set by simple chemical tests with readily observable results.

(a) *N*-methylhexanamide, 1-octanamine, *N,N*-dimethyl-1-hexanamine
(b) *p*-toluidine, benzylamine, *p*-cresol, anisole

41 When *p*-aminophenol reacts with one molar equivalent of acetic anhydride, a compound acetaminophen, $C_8H_9NO_2$, is formed that dissolves in dilute NaOH. What is the structure of this product? Explain your reasoning.

42 Design a separation of a mixture containing the following four compounds into its pure components. Describe exactly what you would do and what you would expect to observe.

(a) nitrobenzene (c) *p*-chlorophenol
(b) aniline (d) *p*-nitrobenzoic acid

43 Outline a sequence of reactions that would effect the conversion of aniline into each of the following compounds:

(a) benzylamine
(b) benzyl alcohol
(c) 2-phenyl-1-ethanamine
(d) *N*-phenyl-2-butanamine
(e) *p*-chlorobenzoic acid
(f) *p*-fluoronitrobenzene

44 Show how the insecticide *carbaryl* can be prepared from methyl isocyanate, $CH_3—N{=}C{=}O$.

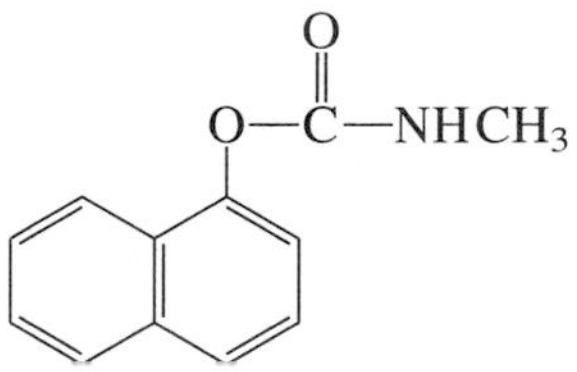

carbaryl

45 Three bottles have been found in an abandoned laboratory. Each bottle contains a liquid and is labeled "amine $C_8H_{11}N$." As an expert in amine chemistry, you have been hired as a consultant and asked to identify each compound. Compounds *A* and *B* give off a gas when they react with $NaNO_2$ and HCl at 0°; *C* does not. However, when the aqueous reaction mixture from the diazotization of *C* is warmed, a gas is evolved. Compound *A* is optically inactive, but when it reacts with (+)-tartaric acid, two isomeric salts with different physical properties are obtained. Titration of *C* with aqueous HCl reveals that its conjugate acid has a $pK_a = 5.1$. Oxidation of *C* with H_2O_2 (a reagent known to oxidize amines to nitro compounds), followed by vigorous oxidation with $KMnO_4$ gives *p*-nitrobenzoic acid. Oxidation of *B* in a similar manner yields 1,4-benzenedicarboxylic acid (terephthalic acid), and oxidation of *A* yields benzoic acid. Identify compounds *A*, *B*, and *C*.

46 Imagine that you have been given a sample of racemic 2-phenylbutanoic acid. Outline steps that would allow you to obtain pure samples of each of the following compounds from this starting material and any other reagents. (You should be aware that optical resolutions are time consuming. One resolution that would serve all four syntheses would be most efficient.)

(a)

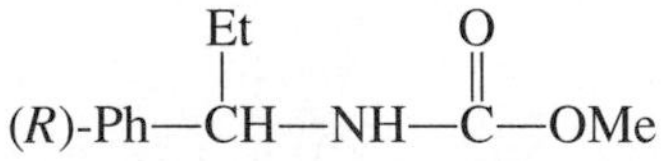

(b)

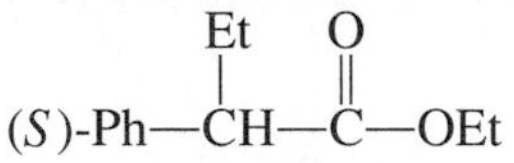

(c)

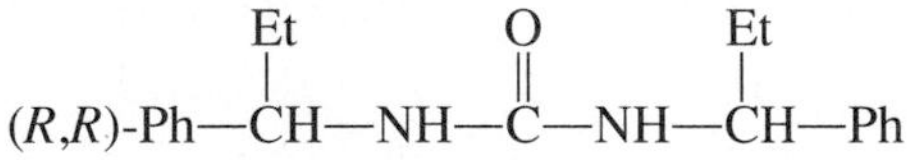

(d)

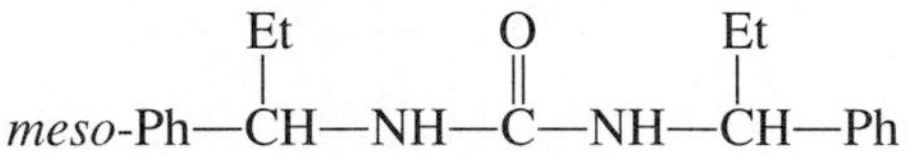

47 When 1,5-dibromopentane reacts with ammonia, among the several products isolated is a water-soluble compound *A* that rapidly gives a light yellow precipitate with acidic $AgNO_3$ solution. This compound is unchanged when treated with dilute base, but treatment of *A* with concentrated NaOH and heat gives a new compound *B* ($C_{10}H_{19}N$) that decolorizes Br_2 in CCl_4. Compound *B* is identical to the product obtained from the following reaction sequence:

Problems (Cont.)

$$\text{4-pentenoic acid} \xrightarrow{SOCl_2} \xrightarrow{\text{piperidine}} \xrightarrow[\text{2) } H_3O^+]{\text{1) } LiAlH_4} B$$

Identify *A* and explain your reasoning.

---

48 Give a rational explanation for each of the following facts.

(a) The barrier to rotation about the *N*-phenyl bond in *N*-methyl-*p*-nitroaniline is considerably higher (10–11 kcal/mol) than that in *N*-methylaniline itself (about 6 kcal/mol).

(b) *cis*- and *trans*-1,3-dimethylpyrrolidine cannot be separately isolated.

(c) $H_3CNH—CH_2—NHCH_3$ is unstable in aqueous solution.

(d) The following compound exists as the enamine tautomer shown rather than as an imine:

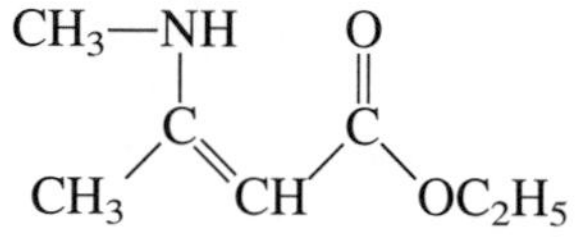

(e) Diazotization of 2,4-cyclopentadien-1-amine gives a diazonium salt that does not decompose to a carbocation.

---

49 Complete the following reactions.

(a) $NH_3 + HNO_2 \longrightarrow$

(b)

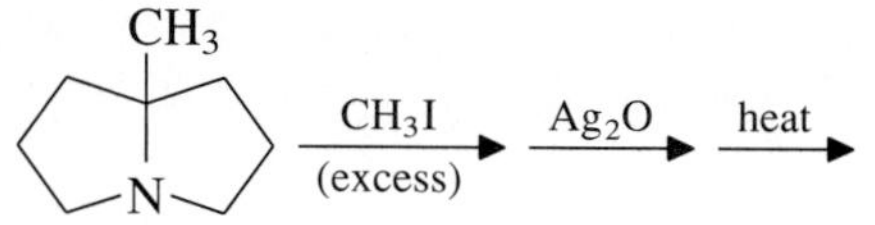

(c)

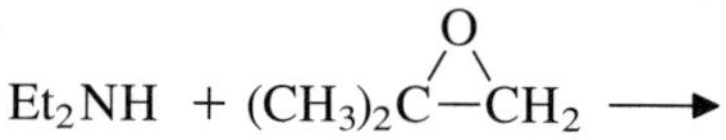

(d) $[(CH_3)_2CH]_2NH + CH_3CH_2CH_2CH_2Li \xrightarrow{C_2H_5C\equiv CH} \xrightarrow{(CH_3)_2CHCH_2I}$

(e)

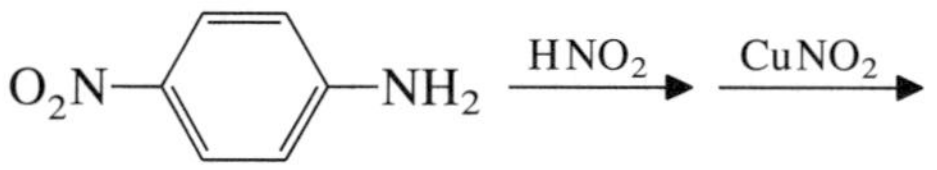

(f)

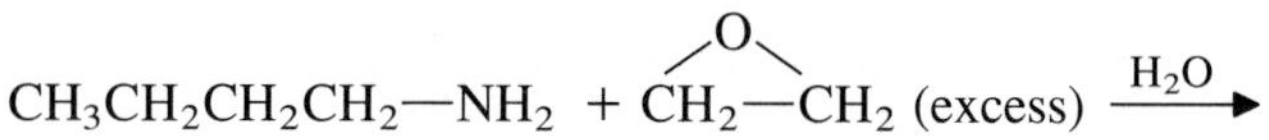

(g)

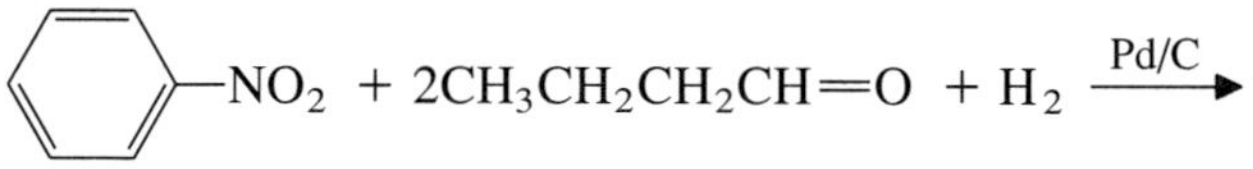

(h)

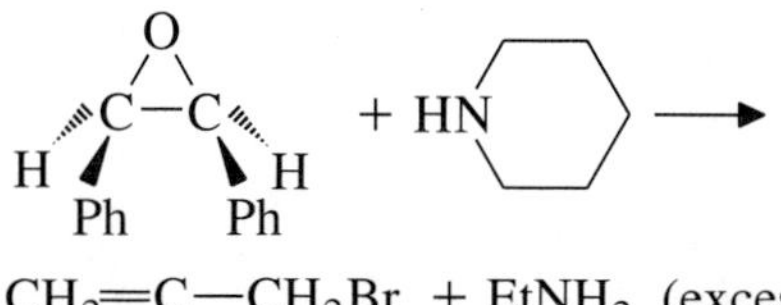

(i) $CH_2{=}C(Br){-}CH_2Br + EtNH_2 \text{ (excess)} \longrightarrow$ (a compound with 5 carbons)

(j)

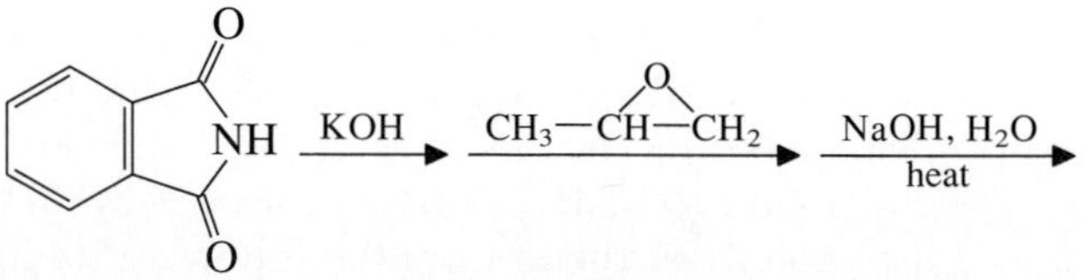

(k) 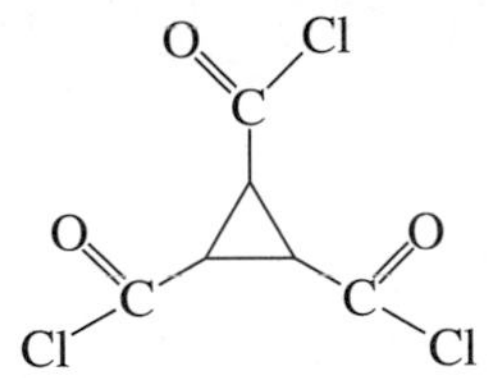 $\xrightarrow[\text{excess}]{(CH_3)_2NH}$ $\xrightarrow[2)\ H_2O]{1)\ LiAlH_4}$ $\xrightarrow{MeI}$ $\xrightarrow{Ag_2O}$ $\xrightarrow{heat}$ (an isomer of benzene)

(l) 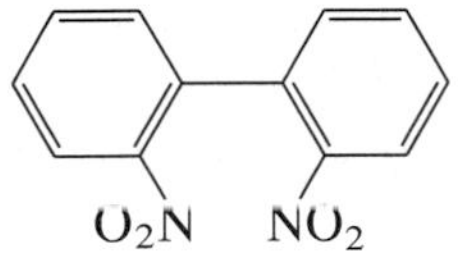 $\xrightarrow{LiAlH_4}$ $\xrightarrow{H_2O}$ (a compound with twelve carbons)

(m) 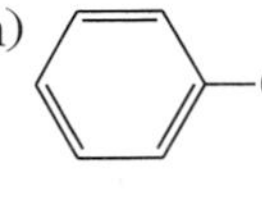 —OH + HCl + $NaNO_2$ ⟶

50 Outline a synthesis for each of the following compounds from the indicated starting materials and any other reagents. The starting material for (a) through (e) is pentanoic acid.

(a) *N*-methyl-1-hexanamine
(b) *N,N*-dimethyl-1-pentanamine
(c) pentylamine
(d) butylamine
(e) hexylamine
(f) *N*-ethyl-3-phenyl-1-propanamine from toluene
(g) $PhCH_2—\overset{+}{N}(CH_3)_2—CH_2CH_2CH_2CH_3\ Br^-$ from butyraldehyde
(h) 2-pentanamine from diethyl malonate
(i) 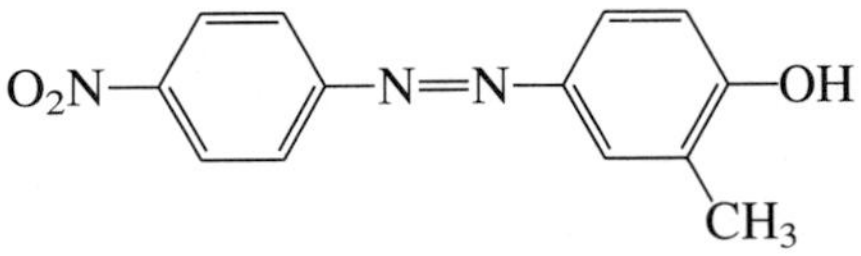 from aniline
(j) isobutylamine from acetone
(k) isopentylamine from acetone
(l) *m*-chlorobromobenzene from nitrobenzene
(m) *p*-chlorobromobenzene from nitrobenzene
(n) *p*-methoxybenzonitrile from phenol
(o) 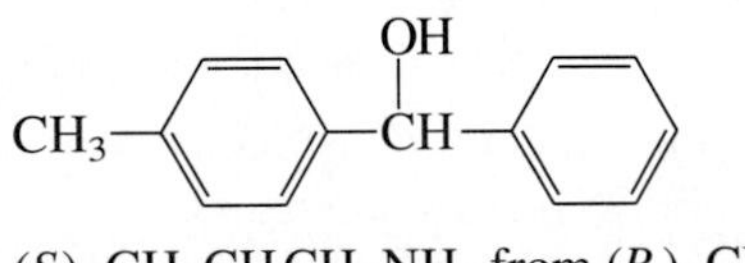 from *p*-toluidine
(p) (*S*)-$CH_3CH(D)CH_2NH_2$ from (*R*)-$CH_3CH(D)OH$
(q) 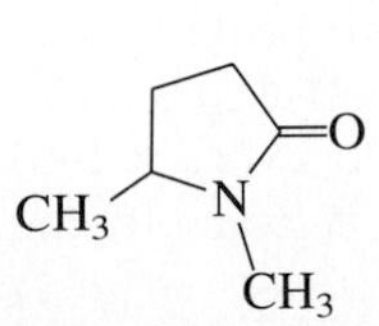 from $CH_3—\overset{O}{\overset{\|}{C}}—CH_2CH_2—CO_2H$ (levulinic acid)

Problems (Cont.)

51 Rank the following compounds according to increasing relative rates of Hofmann elimination, and explain your answers carefully:

$$R{-}CH_2{-}CH_2{-}\overset{+}{N}Me_3$$

compound *A*	$R = {-}C(CH_3)_3$
compound *B*	$R = {-}H$
compound *C*	$R = {-}CH_3$

52 (a) The conjugate acids of enamines have pK_a values near 3. Give the structure of the conjugate acid of the enamine derived from cyclopentanone and piperidine.

(b) When the enamine from (a) reacts with CH_3I, a mixture of two compounds, *A* and *B*, is formed.

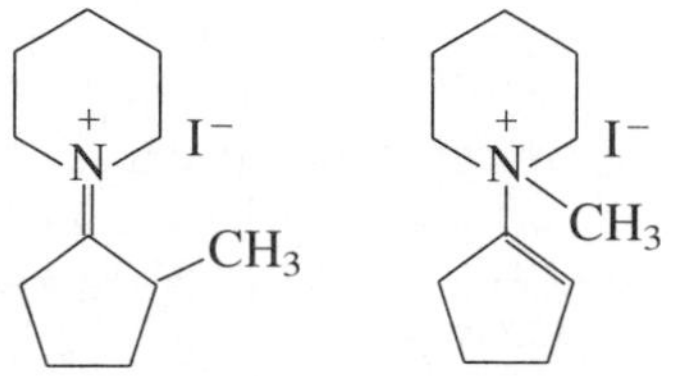

A *B*

Give a mechanism for the formation of each product.

(c) When *B* is heated, an equilibrium between *B* and *A* is established that strongly favors *A*. Give a mechanism for the formation of *A* from *B* that is consistent with the observation that the interconversion is catalyzed by added iodide ion.

53 A compound *A* ($C_{22}H_{27}NO$) has been found in an abandoned laboratory, and you have been called in as a consultant to identify it. Compound *A* is insoluble in acid and base but reacts with concentrated aqueous HCl and heat to give a clear aqueous solution from which, on cooling, benzoic acid precipitates. When the supernatant solution is made basic, a liquid *B* separates. Treatment of *B* with benzoyl chloride in pyridine gives back *A*. Evolution of gas is not observed when *B* is treated with an aqueous solution of $NaNO_2$ and HCl. Compound *B* is achiral. Treatment of *B* with excess CH_3I, then Ag_2O and heat, gives a compound *C*, $C_9H_{19}N$, plus styrene, $Ph{-}CH{=}CH_2$. Compound *C*, when treated with excess CH_3I, then Ag_2O and heat, gives an alkene *D* that is identical to the compound obtained when cyclohexanone is treated with the ylid $^{-}{:}CH_2{-}\overset{+}{P}(Ph)_3$. Give the structure of *A* and explain your reasoning.

54 Aniline has a UV spectrum with peaks at $\lambda_{max} = 230$ ($\epsilon = 8600$) and 280 ($\epsilon = 1430$). In the presence of dilute HCl, the spectrum of aniline changes dramatically: $\lambda_{max} = 203$ ($\epsilon = 7500$) and 254 ($\epsilon = 160$). This spectrum is nearly identical to the UV spectrum of benzene. Account for the effect of acid on the UV spectrum of aniline.

Figure 23.6 Spectra for problem 55.

(a)
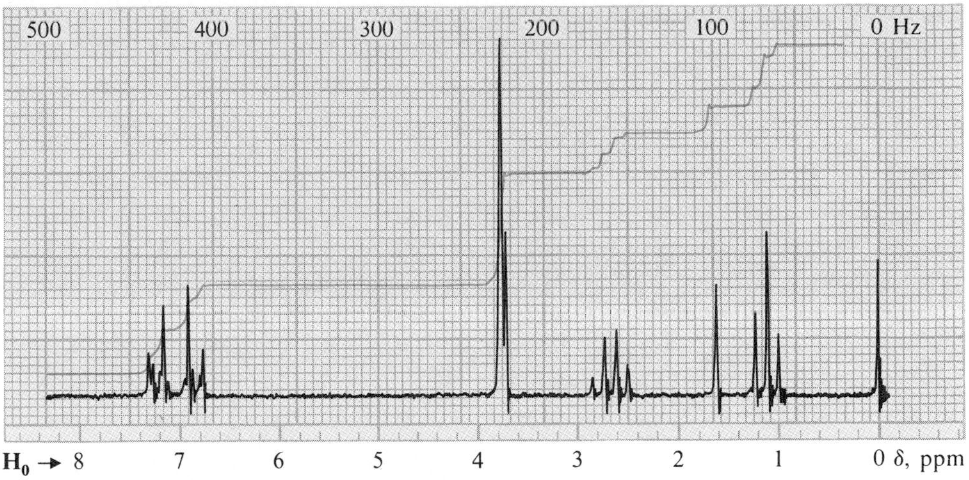

(b)
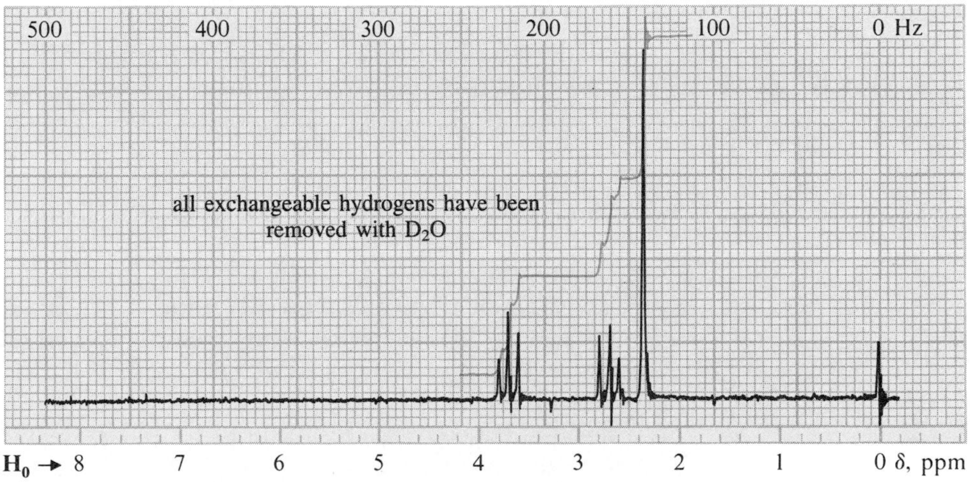

(b)
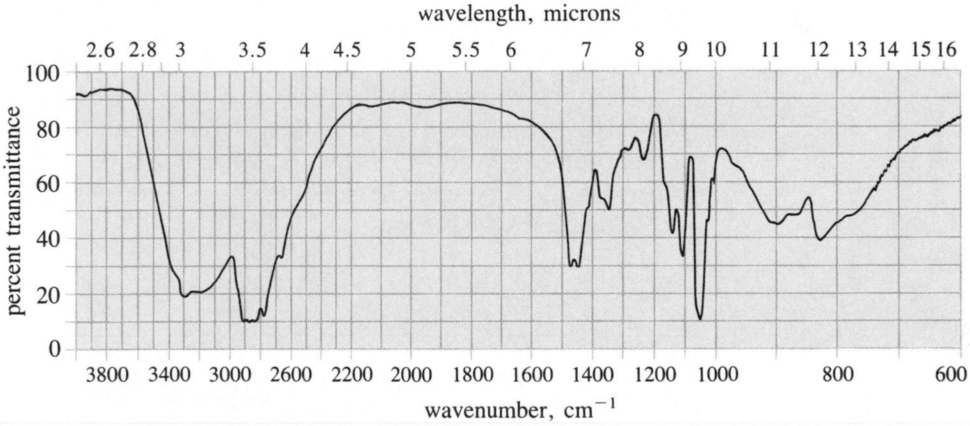

55 In the warehouse of the company Tumany Amines, Inc., two unidentified compounds have been found. The president of the company, Wotta Stench, has hired you to identify them from their spectra:

(a) $C_{10}H_{15}NO$: IR spectrum 3355, 1618 cm^{-1}, no carbonyl absorption. NMR spectrum in Fig. 23.6a

(b) C_3H_9NO: NMR and IR spectra in Fig. 23.6b

56 Give a mechanism for each of the following rearrangement reactions:

(a)

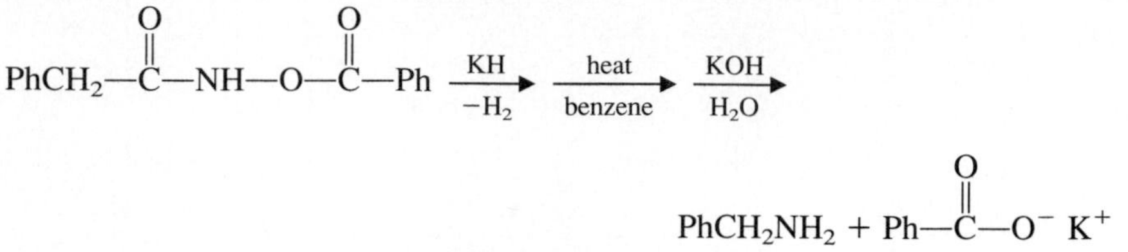

$$PhCH_2-\overset{O}{\underset{\|}{C}}-NH-O-\overset{O}{\underset{\|}{C}}-Ph \xrightarrow[-H_2]{KH} \xrightarrow[benzene]{heat} \xrightarrow[H_2O]{KOH}$$

$$PhCH_2NH_2 + Ph-\overset{O}{\underset{\|}{C}}-O^-\ K^+$$

(b)

$$\text{(succinimide)}N-Br + KOH \longrightarrow \xrightarrow{H^+} H_2N-CH_2CH_2-CO_2H$$

(c)

$$H_2N-\overset{O}{\underset{\|}{C}}-CH_2CH_2-\overset{O}{\underset{\|}{C}}-NH_2 + Br_2 \xrightarrow{NaOH} \text{(dihydrouracil ring structure)}$$

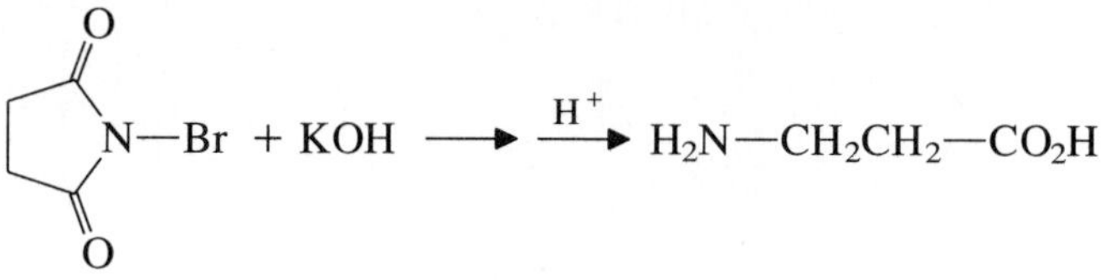

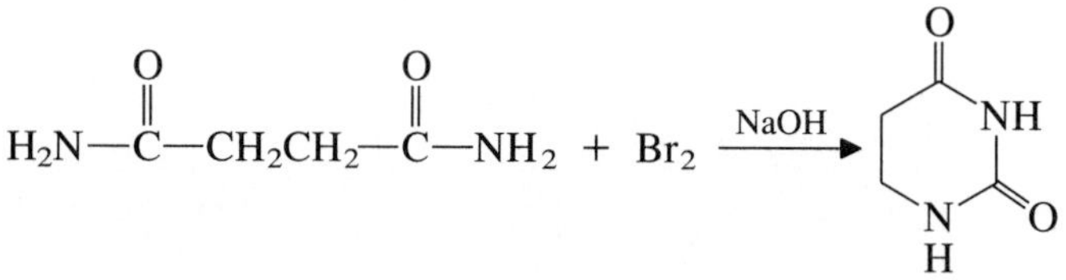

57 You have been hired as a consultant by the Chicago police, who have brought to you a compound A confiscated in an important drug raid. Compound A is a water-soluble white solid that rapidly gives a white precipitate with acidic $AgNO_3$ solution. Treatment of the solid with 5% NaOH solution liberates a water-insoluble liquid B that is found to be optically active and dextrorotatory; a gas is evolved when B is treated with $NaNO_2$ and HCl at 0°. The elemental analysis laboratory returns the following analysis for B: 79.9% carbon, 9.7% hydrogen, and 10.4% nitrogen, and the mass spectrum of B suggests a molecular weight of 135. Treatment of B successively with excess CH_3I, Ag_2O, and then heat gives a mixture of compounds C. Catalytic hydrogenation of C yields one compound D that is identical to the compound obtained when propiophenone is heated with hydrazine (NH_2NH_2) and KOH. The NMR spectrum of A is complex, but contains a clean doublet at δ 1.1 ($J = 7.5$ Hz). Identify compound A. (Do not try to determine its absolute configuration.) Conviction of a notorious drug syndicate may hinge on your successful analysis.

58 Suggest a rational explanation for each of the following reactions, including a detailed mechanism where possible.

(a)

$$H_2N-\overset{O}{\underset{\|}{C}}-NH_2 + HNO_2 \xrightarrow[H_2O]{HCl} N_2 + CO_2 + H_2O$$

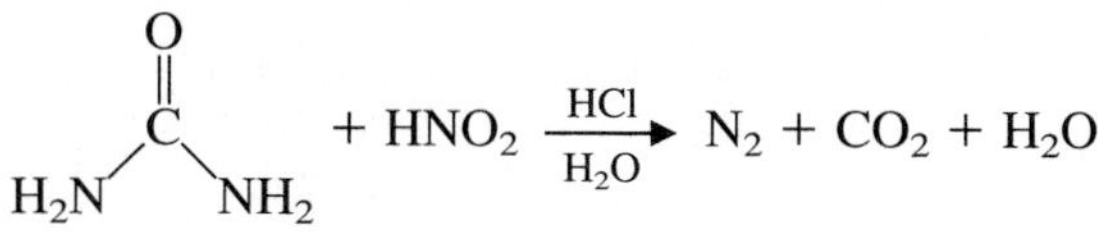

(This reaction can be used to scavenge unwanted nitrous acid.)

(b)

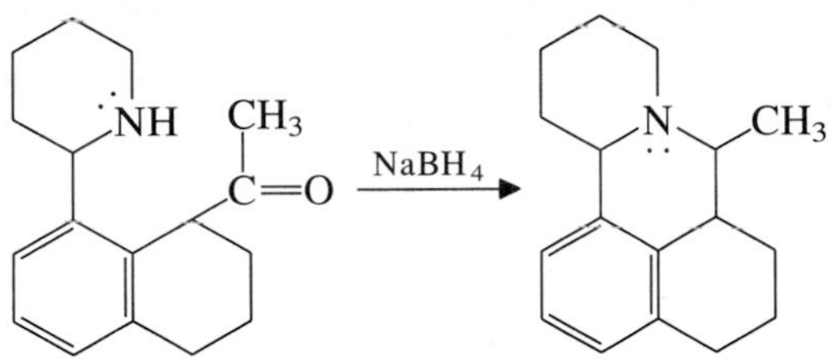

(c) $(CH_3)_2CHCH_2NH_2 \xrightarrow[\text{H}_2\text{SO}_4]{\text{NaNO}_2} (CH_3)_3C—OH + (CH_3)_2C=CH_2$

(d)

$$Ph—\overset{O}{\overset{\|}{C}}—CH_2CH_2—\overset{+}{N}Me_3\ Cl^- \xrightarrow[\text{H}_2\text{O}]{\text{KCN, heat}} Ph—\overset{O}{\overset{\|}{C}}—CH_2CH_2—CN$$

(e)

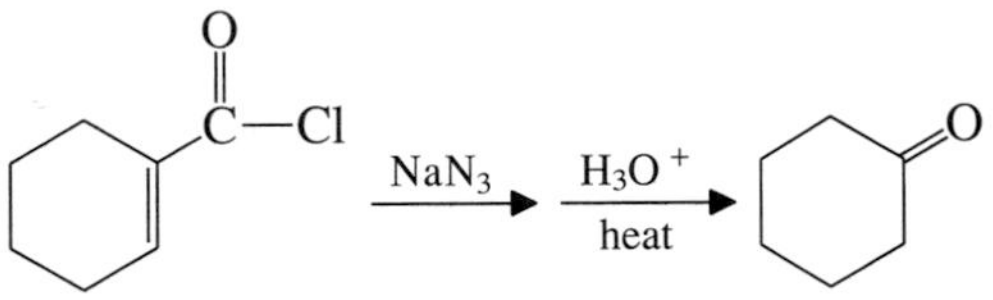

$$\xrightarrow[\text{heat}]{\text{NaN}_3\quad \text{H}_3\text{O}^+}$$

(f)

$$\xrightarrow[\text{heat}]{CH_3(CH_2)_4—O—N=O} \quad + N_2 + CO_2$$

(g)

$$CH_3—\overset{OH}{\underset{CH_3}{\overset{|}{\underset{|}{C}}}}—\overset{NH_2}{\underset{CH_3}{\overset{|}{\underset{|}{C}}}}—CH_3 \xrightarrow[\text{H}_2\text{O}]{\text{HNO}_2,\ \text{HCl}} CH_3—\overset{O}{\overset{\|}{C}}—\overset{CH_3}{\underset{CH_3}{\overset{|}{\underset{|}{C}}}}—CH_3$$

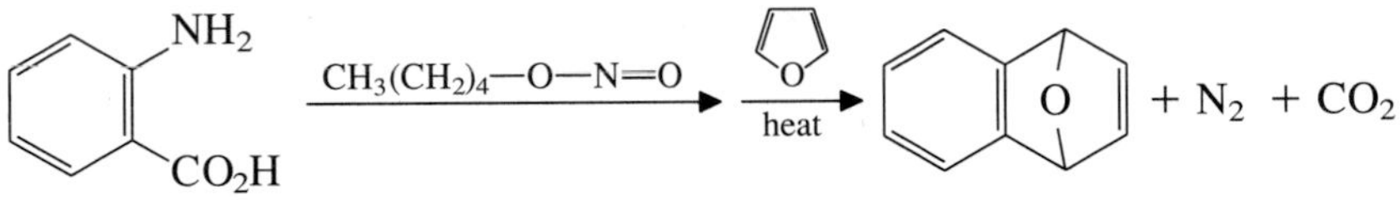

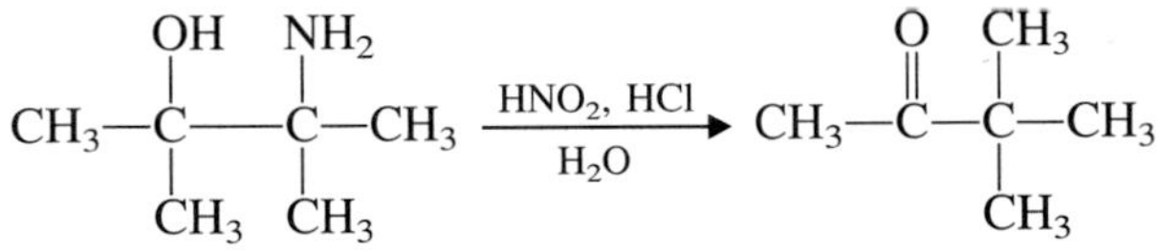

59 In the NMR spectrum of a concentrated (4.5M) aqueous solution of methylamine, the methyl group appears as a quartet when the solution pH is 1. At intermediate pH, the methyl group appears as a broad line. At pH = 9 the methyl group is observed as a single sharp line. Explain these observations.

60 Explain the fact that the following amines, despite obvious similarities in structure, have considerably different basicities. (*Hint:* Make a model of compound (*B*). Look at the relationship of the nitrogen unshared electron pair to the *p* orbitals of the benzene ring.)

	A	*B*	*C*
conjugate-acid pK_a:	5.20	7.79	10.58

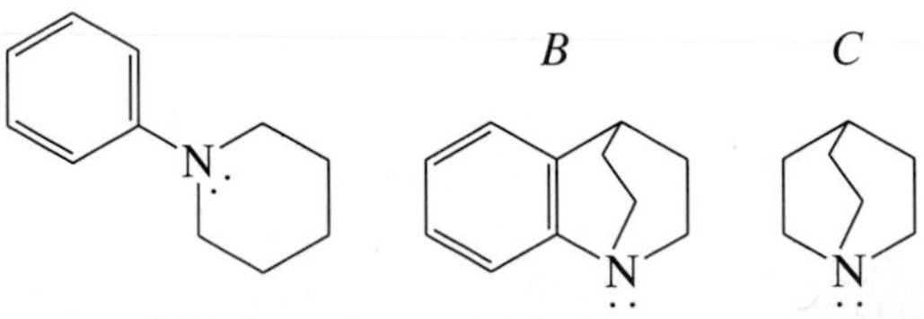

Problems (Cont.)

61 Amide *A* below, δ-valerolactam, is a typical amide with a conjugate-acid pK_a of 0.8. The two cyclic tertiary amines *B* and *C* also have typical conjugate-acid pK_a values.

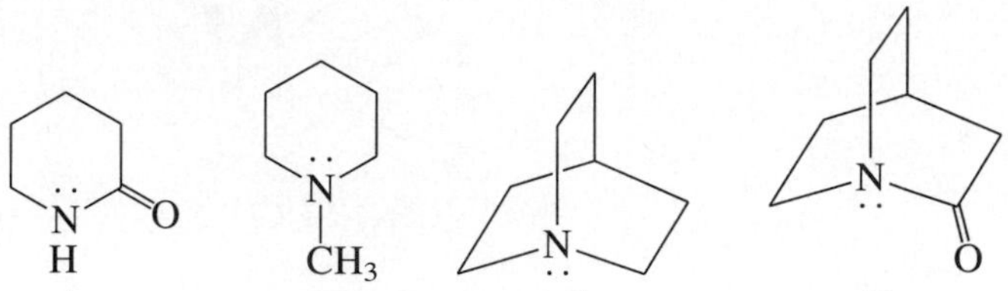

	A	*B*	*C*	*D*
conjugate-acid pK_a	0.8	10.1	10.65	5.33

In contrast, the conjugate-acid pK_a of amide *D* is unusually high for an amide, and it hydrolyzes much more rapidly than other amides. Draw the structure of the conjugate acid of amide *D*, and suggest a reason for both its unusual pK_a and its rapid hydrolysis.

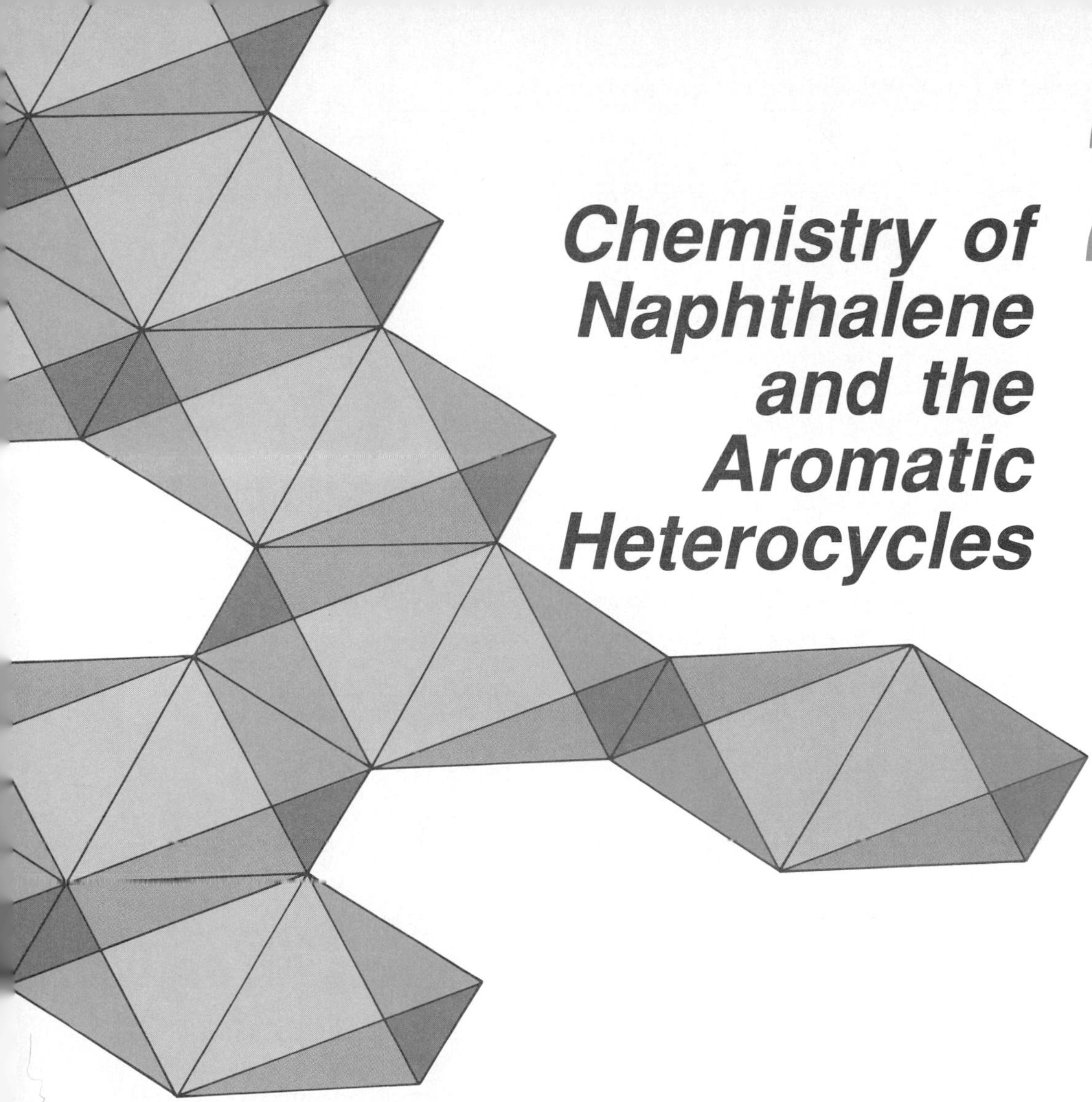

24 Chemistry of Naphthalene and the Aromatic Heterocycles

Our study of aromatic compounds has focused mainly on derivatives of benzene. In Sec. 15.7D, however, we learned that many other compounds exhibit aromatic behavior. After benzene, **naphthalene** is the most important aromatic hydrocarbon, and is the simplest of the **polycyclic aromatic hydrocarbons**—hydrocarbons with more than one aromatic ring.

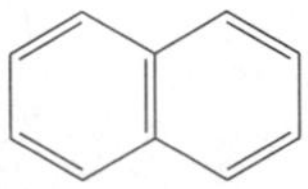

naphthalene

In this chapter we shall consider some of the chemistry of naphthalene. In particular, we shall contrast the chemistry of naphthalene with that of benzene.

Heterocyclic compounds are compounds with rings that contain more than one type of atom. The heterocyclic compounds of greatest interest to organic chemists have carbon rings containing one or two **heteroatoms** (atoms other than carbon).

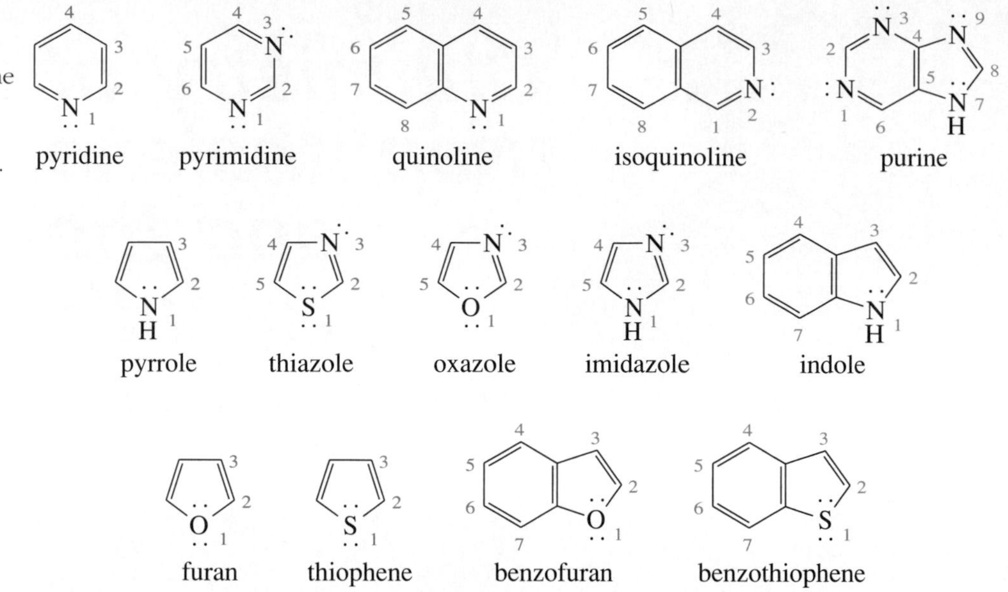

Figure 24.1 Common aromatic heterocyclic compounds. The numbers are used in systematic nomenclature, discussed in Sec. 24.2A.

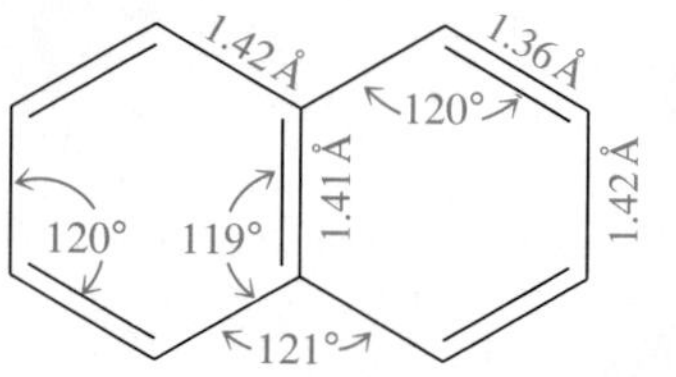

Figure 24.2 Structure of naphthalene.

Although the chemistry of many *saturated* heterocyclic compounds is rather analogous to that of their open-chain counterparts, a significant number of *unsaturated* heterocyclic compounds exhibit aromatic behavior. Some of these are shown in Fig. 24.1. In this chapter, we shall focus for the most part on the unique chemistry of the aromatic heterocycles. Since many of these compounds occur widely in natural products of biological importance, their chemistry is of substantial practical significance.

24.1 CHEMISTRY OF NAPHTHALENE

A. Physical Properties and Structure

Although naphthalene is a solid (mp 80.5°), it has a high vapor pressure and is readily steam distilled. The familiar odor of moth balls is due to naphthalene.

Naphthalene can be represented by three resonance structures; two are equivalent, and one is unique. The structure of naphthalene is shown in Fig. 24.2.

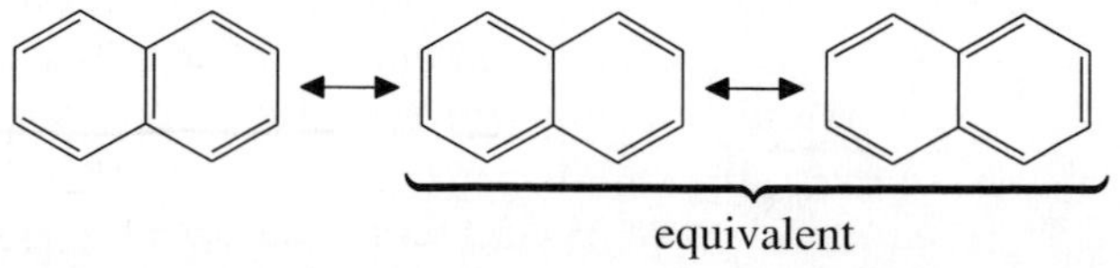

(24.1)

Figure 24.3 Some polycyclic aromatic hydrocarbons.

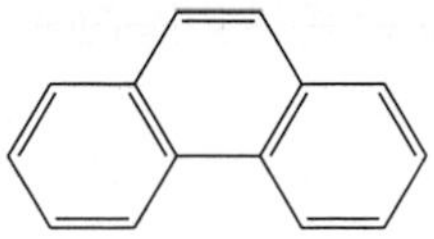

phenanthrene

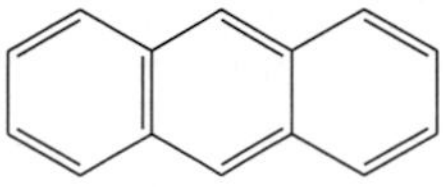

anthracene

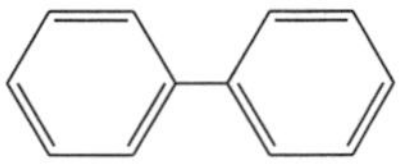

biphenyl

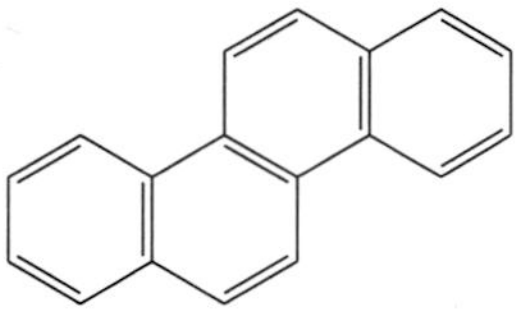

chrysene

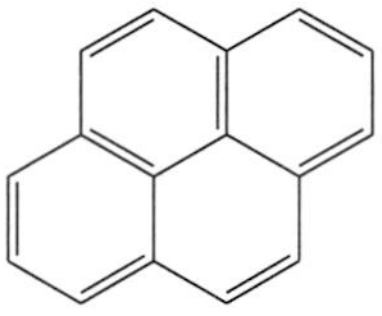

pyrene

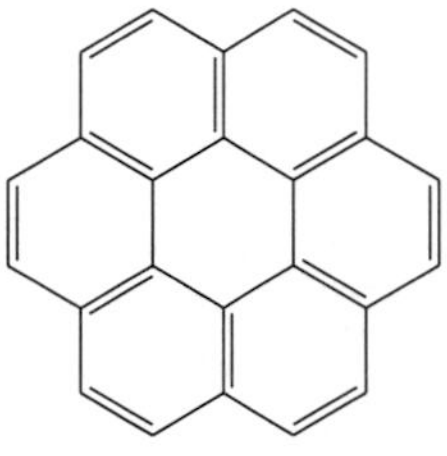

coronene

Problem

1 Examine the resonance structures above and predict which bond in naphthalene should be the shortest. Check your answer by looking at the structure of naphthalene in Fig. 24.2.

Some other polycyclic aromatic hydrocarbons are shown in Fig. 24.3. All consist of planar six-membered rings with formal alternating single and double bonds. Perhaps the ultimate polycyclic compound is graphite, a carbon polymer that consists of layers of fused aromatic rings (Fig. 24.4). The softness of graphite and its capacity to act as a lubricant can be attributed to the ability of its layers to slide with respect to each other. The distance between the layers of graphite is taken as the approximate "height" of a π-orbital; can you see why?

Figure 24.4 Structure of graphite. Graphite consists of fused aromatic rings in layers separated by 3.35 Å. (The formal double bonds in the aromatic rings are not shown.)

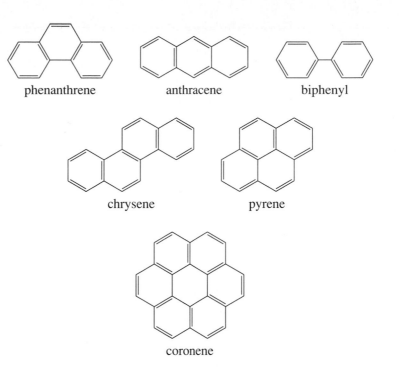

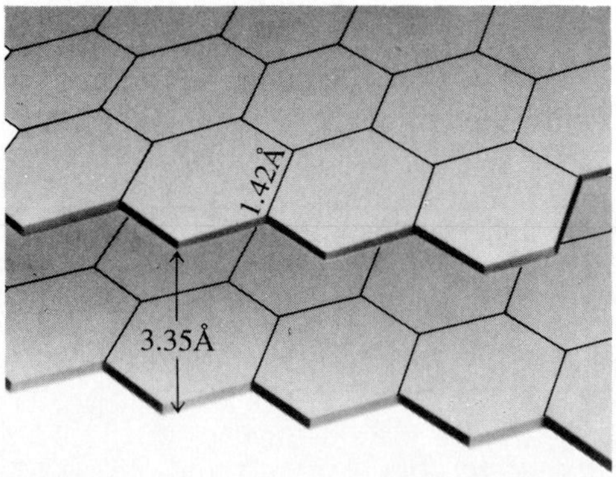

B. Nomenclature of Naphthalene Derivatives

In systematic nomenclature, carbon-1 of naphthalene is the carbon adjacent to a *bridgehead* carbon (a vertex at which the rings are fused). Substituents are given the lowest numbers consistent with this scheme and their relative priorities. The following examples illustrate this idea:

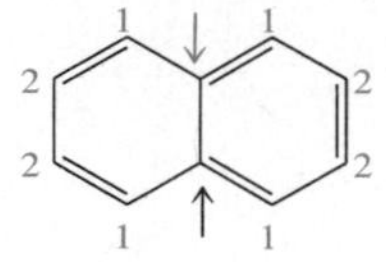

bridgehead carbons (→)

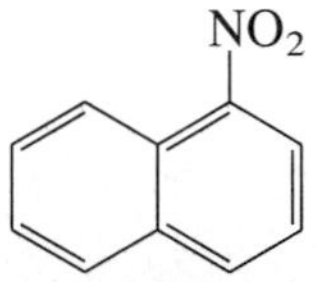

1-nitronaphthalene

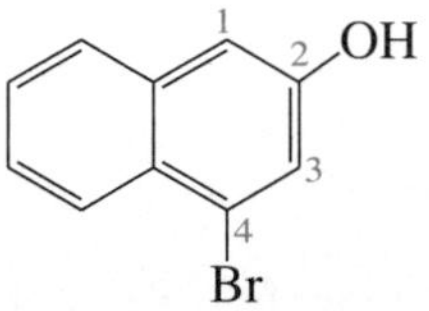

4-bromo-2-naphthol
(*not* 1-bromo-3-naphthol;
—OH has priority)

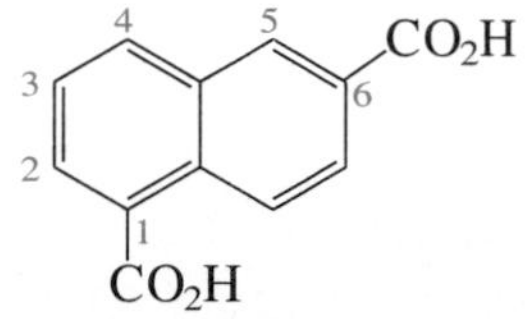

1,6-naphthalenedi-
carboxylic acid

Naphthalene also has a common nomenclature that uses Greek letters. In this system the 1-position is designated as α and the 2-position as β. Common and systematic nomenclature should not be mixed.

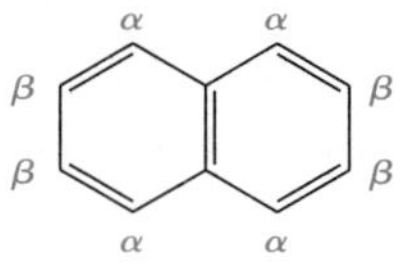

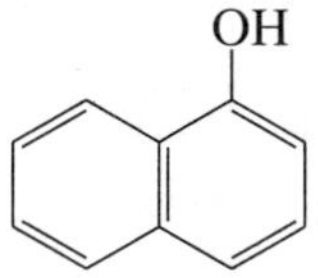

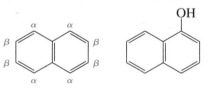

systematic: 1-naphthol
common: α-naphthol

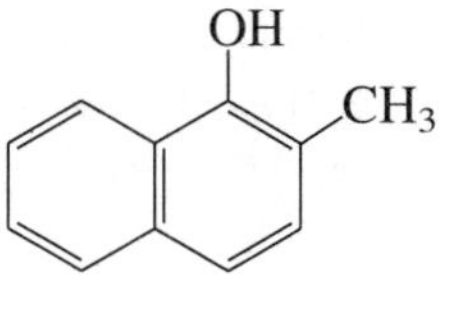

correct: 2-methyl-1-naphthol
incorrect: 2-methyl-α-naphthol
(mixes common and systematic
nomenclature)

The naphthalene ring can also be named as a substituent group, the *naphthyl* group. The term *naphthyl* for a naphthylene ring is analogous to the term *phenyl* for a benzene ring.

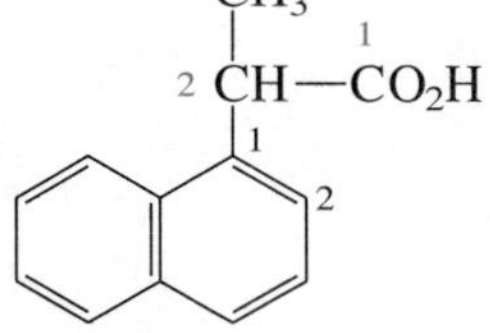

2-(1-naphthyl)propanoic acid

↑ position of attachment on naphthalene ring

↑ position of substitution on the parent propanoic acid

Problem

2 Name the following compounds:

(a)

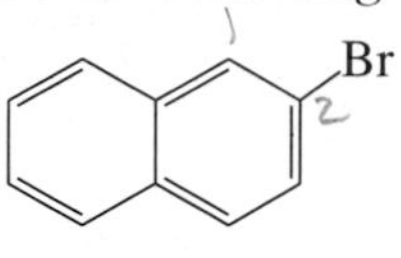

(b)

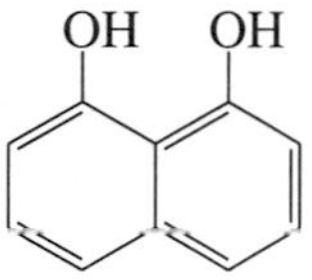

(c)

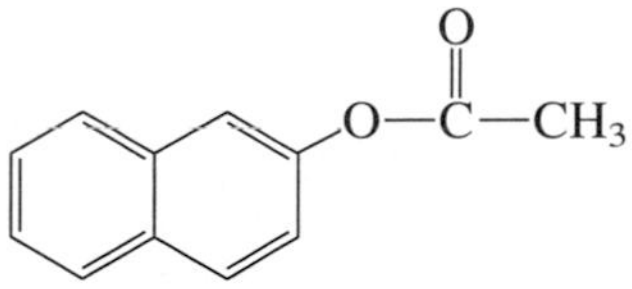

(d)

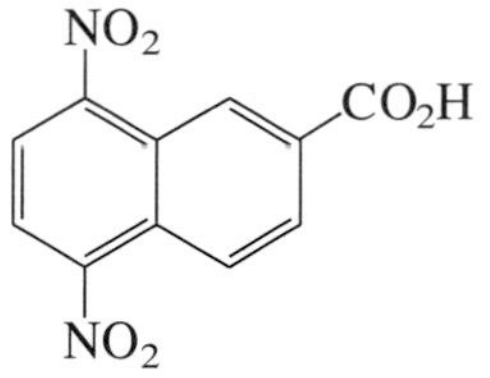

C. Electrophilic Aromatic Substitution Reactions of Naphthalene and Its Derivatives

Electrophilic Aromatic Substitution Reactions of Naphthalene Because naphthalene is an aromatic compound, it is not surprising to find that it undergoes electrophilic aromatic substitution reactions much like those of benzene. (You should review the electrophilic aromatic substitution reactions of benzene found in Secs. 16.4, 16.5, 18.7B, and 23.9.)

We shall consider two questions about electrophilic substitution in naphthalene: (1) At what position does naphthalene undergo substitution? and (2) how reactive is naphthalene in electrophilic aromatic substitution reactions?

Electrophilic aromatic substitution of naphthalene generally occurs at the 1-position.

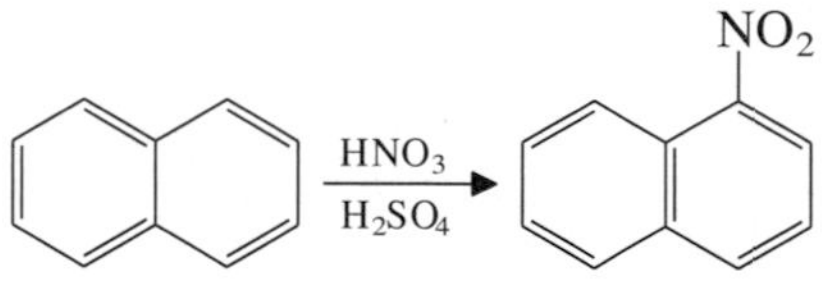

(95% yield) (24.2)

We can see why 1-substitution rather than 2-substitution occurs by comparing resonance structures for the carbocation intermediates in the two processes. We can draw seven resonance structures for the carbocation intermediate in electrophilic substitution at the 1-position.

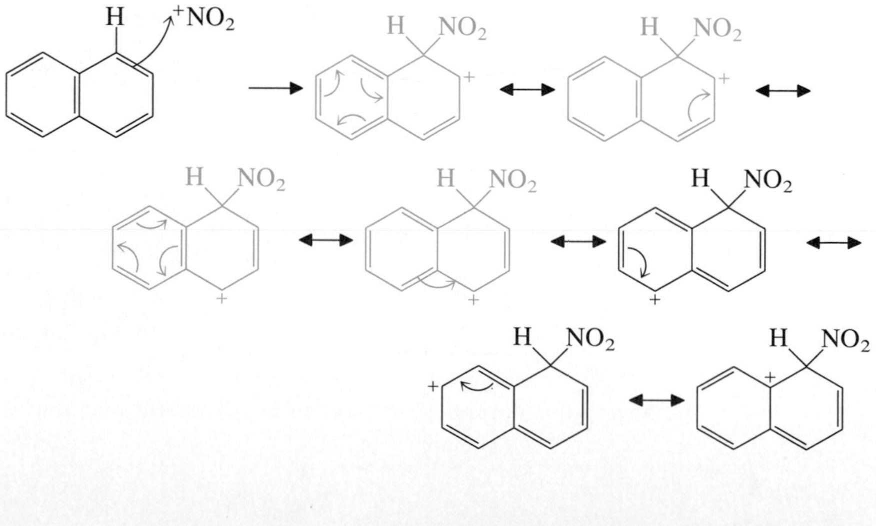

(24.3)

Furthermore, four of these (color) contain intact benzene rings. This is an important point because *structures in which benzene rings are left intact are more important than those in which the formal double bonds are moved out of the ring.* The reason this is so is that structures lacking the intact benzene rings are formally not aromatic, and are thus less stable. (Recall from Sec. 15.6B that the importance of resonance structures is determined by pretending they are separate molecules, even though they really aren't.) Substitution in the 2-position, in contrast, gives a carbocation intermediate for which we can draw six resonance structures (draw these!). Of these six, only two contain intact benzene rings (identify these). Therefore substitution at the 1-position gives the more stable carbocation intermediate and, by Hammond's postulate (Sec. 4.7C), occurs more rapidly. Notice that there is nothing wrong with 2-substitution—1-substitution is simply more favorable.

An interesting result occurs in the sulfonation of naphthalene, a result which hinges on the fact that *aromatic sulfonation is a reversible reaction.* This reversibility is apparent if a benzenesulfonic acid derivative is heated in the presence of acid: the sulfonic acid group is replaced by hydrogen.

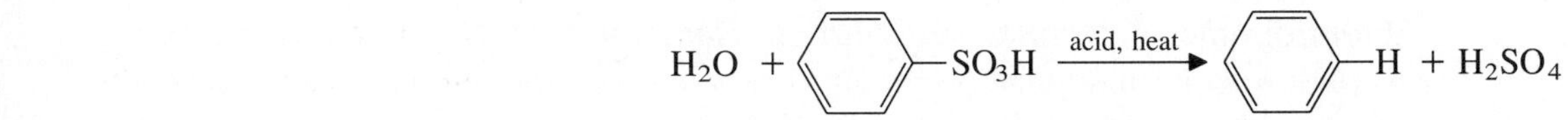

$$H_2O + \underset{}{\bigcirc}\!\!-SO_3H \xrightarrow{\text{acid, heat}} \underset{}{\bigcirc}\!\!-H + H_2SO_4 \qquad (24.4)$$

Sulfonation of naphthalene under mild conditions gives mostly 1-naphthalenesulfonic acid. Substitution at the 1-position is expected by analogy to give other electrophilic substitutions of naphthalene. However, under more vigorous conditions, sulfonation yields mostly 2-naphthalenesulfonic acid.

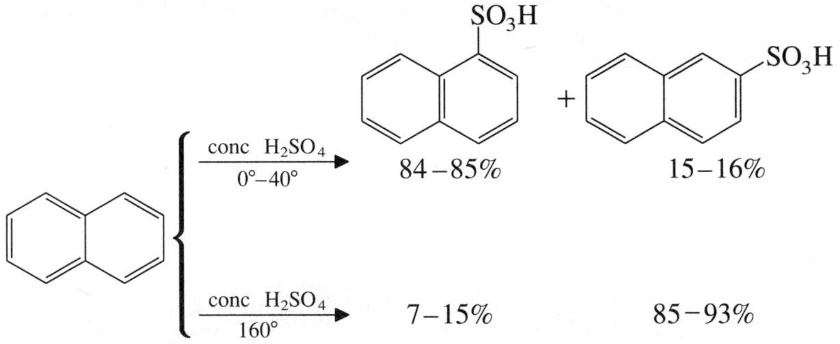

$$(24.5)$$

This is another case of *kinetic vs. thermodynamic control* of a reaction (Sec. 15.4B). At low temperature, substitution at the 1-position is observed because it is faster. At higher temperature, formation of 1-naphthalenesulfonic acid is reversible, and the *more stable,* but more slowly formed, 2-naphthalenesulfonic acid is observed. This reversibility is confirmed when pure 1-naphthalenesulfonic acid is subjected to high-temperature conditions:

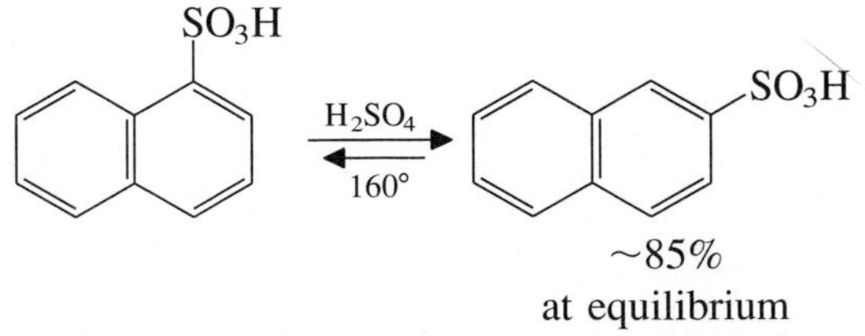

$$(24.6)$$

~85%
at equilibrium

Why is the 2-isomer more stable? The answer lies in a unique steric effect that occurs in the naphthalene ring system. The sulfonic acid group is large—about as large as a *t*-butyl group. In the 1-position, this group interacts unfavorably with the adjacent hydrogen on the 8-position. This interaction, called a *peri* interaction, destabilizes the 1-isomer.

peri interaction

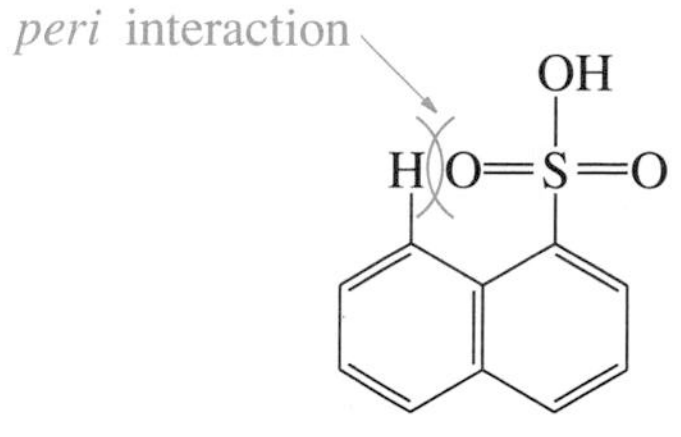

A *peri* interaction is much more severe than the interaction of the same two groups in *ortho* positions because bonds in *ortho* positions diverge from each other, whereas the bonds in *peri* positions are parallel. Thus, 1-naphthalenesulfonic acid, if allowed to equilibrate, is converted into the 2-isomer to avoid this unfavorable steric interaction. In contrast, such an equilibrium is not observed in other electrophilic substitution reactions that are not reversible.

Now let us consider the reactivity of naphthalene. *Naphthalene is considerably more reactive than benzene in electrophilic aromatic substitution.* For example, recall that bromination of benzene requires a Lewis acid catalyst such as $FeBr_3$ (Sec. 16.4A). In contrast, naphthalene is readily brominated in CCl_4 without catalysts.

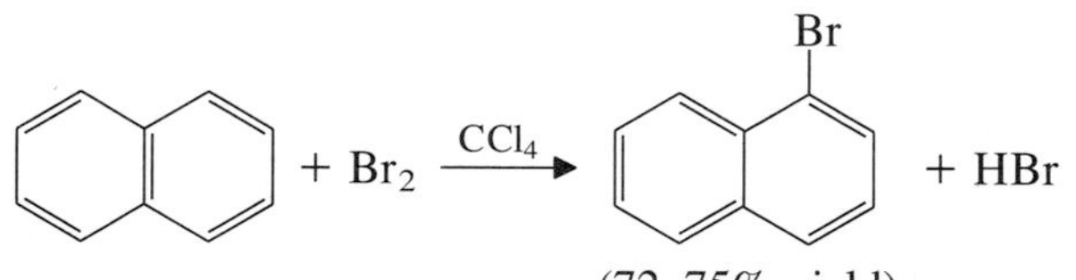

$$\text{naphthalene} + Br_2 \xrightarrow{CCl_4} \text{1-bromonaphthalene} + HBr \qquad (24.7)$$

(72–75% yield)

The greater reactivity of naphthalene in electrophilic aromatic substitution reactions reflects the considerable resonance stabilization of the carbocation intermediate (see Eq. 24.3).

Electrophilic Aromatic Substitution Reactions of Substituted Naphthalenes

In our study of electrophilic aromatic substitution reactions of substituted benzenes, we learned that substituents on the ring either *accelerate* or *retard* a substitution reaction relative to benzene itself; that is, substituents may *activate* or *deactivate* further substitution. We also know that substituents exert very definite *directing effects* in the aromatic substitution reactions of benzene derivatives; that is, the nature of the substituent determines the ring position(s) at which further substitution takes place. (See Table 16.2.) The same effects are observed in napthalene chemistry. However, an additional question arises in naphthalene chemistry that has no counterpart in the chemistry of benzene: does the second substitution occur on the ring that is already substituted, or on the unsubstituted ring?

The following trends are observed in most cases:

1. When one ring of naphthalene is substituted with deactivating groups (such as $-NO_2$ or $-SO_3H$), further substitution occurs in the *unsubstituted* ring at an open α-position (if available).

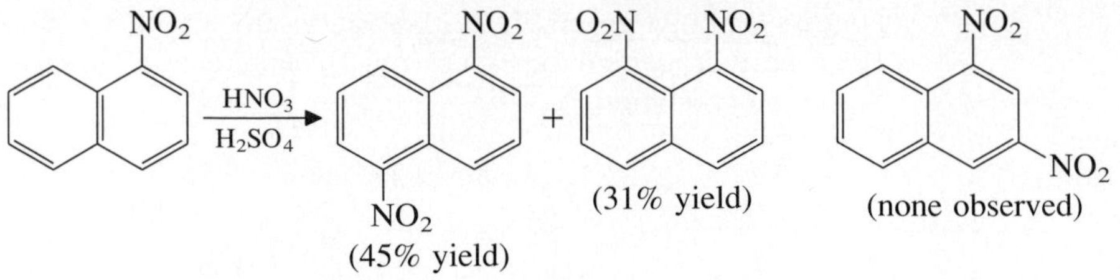

(24.8)

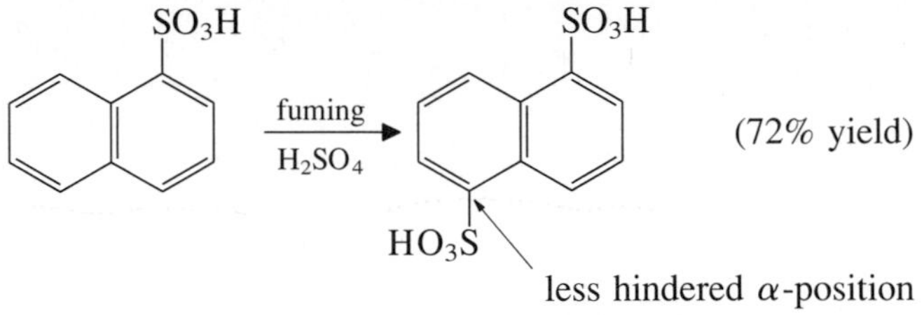

(24.9)

2. When one ring of naphthalene is substituted with activating groups (such as $—CH_3$ or $—OCH_3$), further substitution occurs in the substituted ring at the *ortho* or *para* positions.

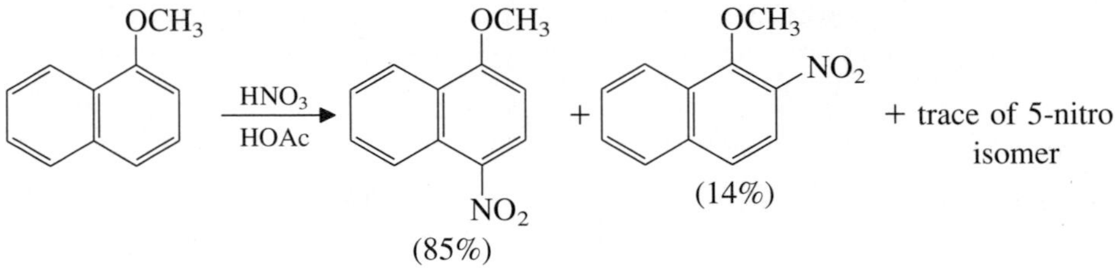

(24.10)

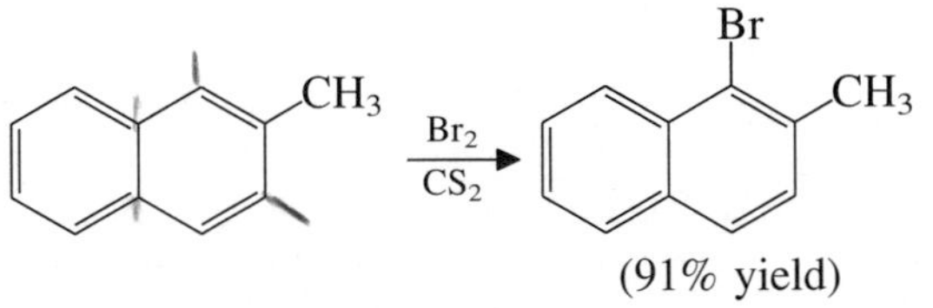

(24.11)

In the last example, we see that since no *para* position is open, substitution occurs at an *ortho* position. Of the two *ortho* positions open (identify them), substitution at an α-position is preferred to substitution at a β-position (why?).

Problems

3 Complete the following reactions by giving the structure(s) of the major organic product(s). Explain your answers.

(a)

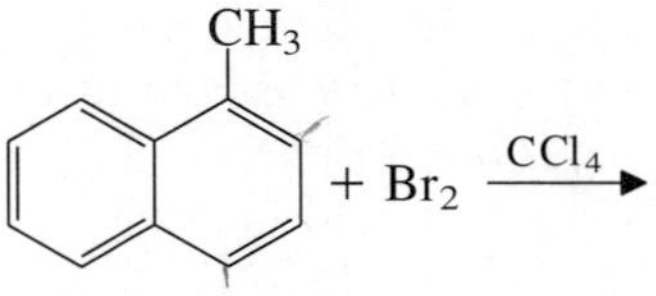

(b)

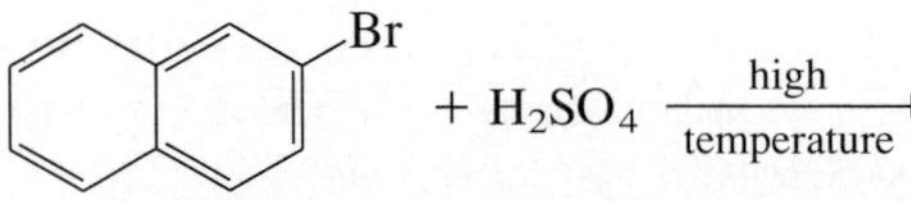

(c)

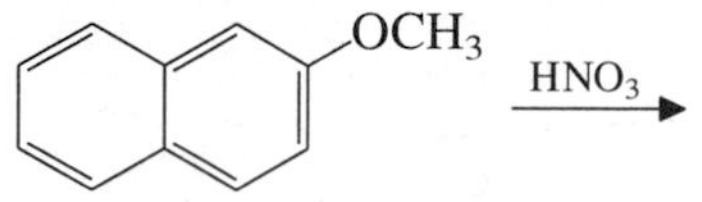

(d)

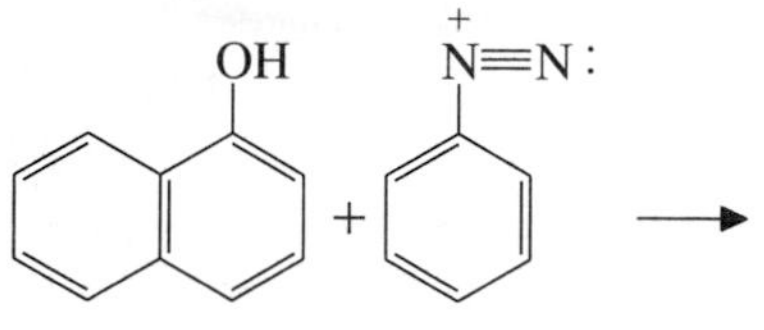

4 Propose a synthesis of the following compound from naphthalene. (The Friedel–Crafts reaction cannot be used because it gives a mixture of 1- and 2-acetylnaphthalene that is difficult to separate.)

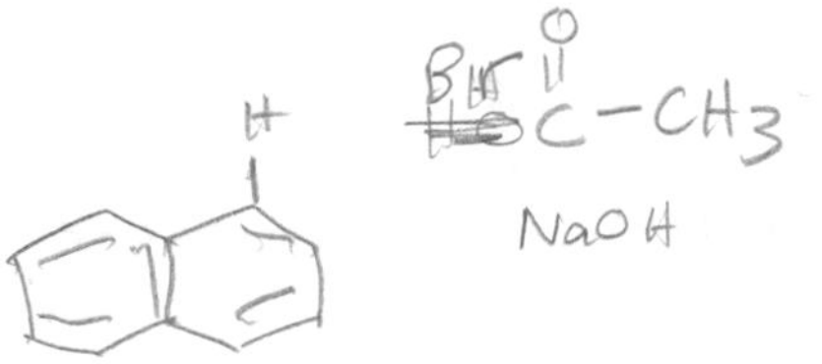

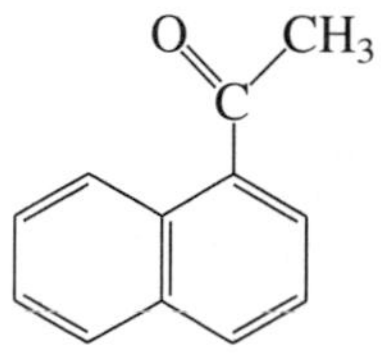

24.2 INTRODUCTION TO THE AROMATIC HETEROCYCLES

A. Nomenclature

The names and structures of some common aromatic heterocyclic compounds are given in Fig. 24.1. This figure also shows how the rings are numbered in systematic nomenclature. In all but a few cases, a heteroatom is given the number 1. (Isoquinoline is an exception.) As we see in thiazole and oxazole, oxygen and sulfur are given a lower number than nitrogen when a choice exists. Substituent groups are given the lowest number consistent with this scheme.

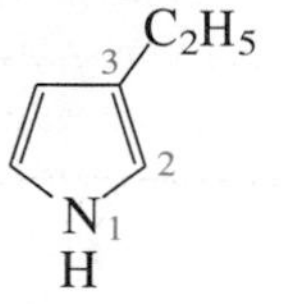

3-ethylpyrrole

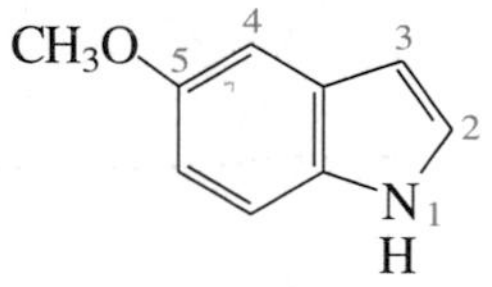

5-methoxyindole

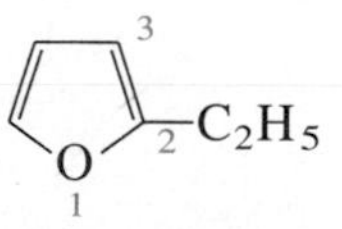

2-ethylfuran

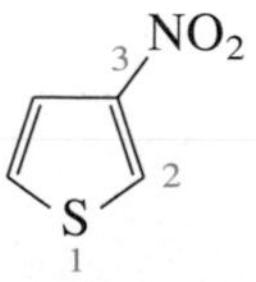

3-nitrothiophene

(These are the same rules used in numbering and naming saturated heterocyclic compounds; see Secs. 8.1C and 23.1B.)

Problems

5 Draw the structure of (a) 4-(dimethylamino)pyridine; (b) 4-ethyl-2-nitroimidazole.

6 Name the following compounds:

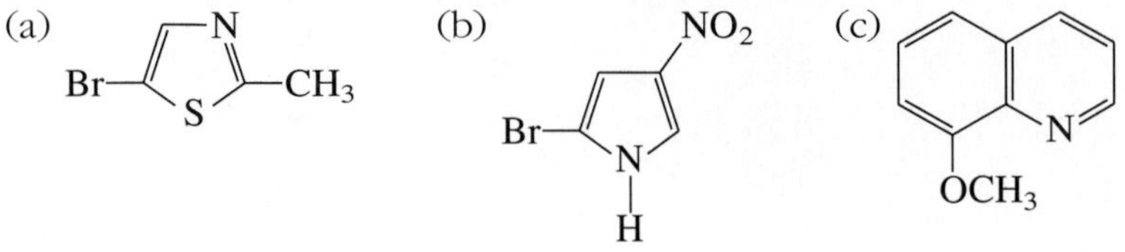

B. Structure and Aromaticity

The aromatic heterocyclic compounds furan, thiophene, and pyrrole can be written as resonance hybrids, illustrated here for furan.

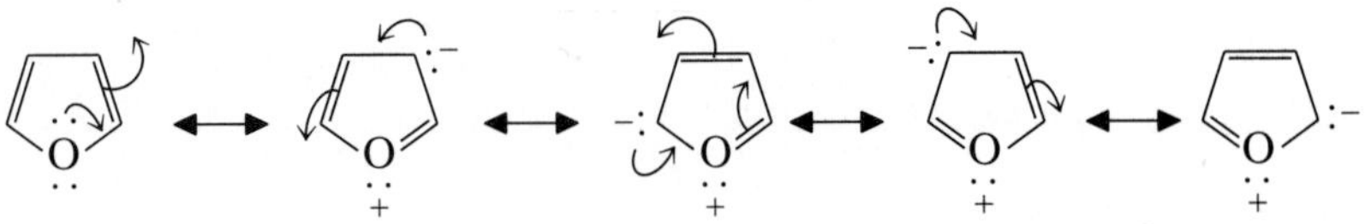
(24.12)

Since separation of charge is present in all but the first structure, the first structure is considerably more important than the others. Nevertheless, the importance of the other structures is evident if we compare the dipole moments of furan and tetrahydrofuran, a saturated heterocyclic ether.

	tetrahydrofuran	furan
dipole moment	1.7 D	0.7 D
boiling point	67°	31.4°

The dipole moment of tetrahydrofuran is attributable mostly to the bond dipoles of its polar C—O single bonds. That is, electrons in the σ-bonds are pulled toward the oxygen because of its electronegativity. This same effect is present in furan, but in addition there is a second effect: the resonance delocalization of the oxygen unshared electrons into the ring shown in Eq. 24.12. This tends to push electrons away from oxygen into the π-electron system of the ring.

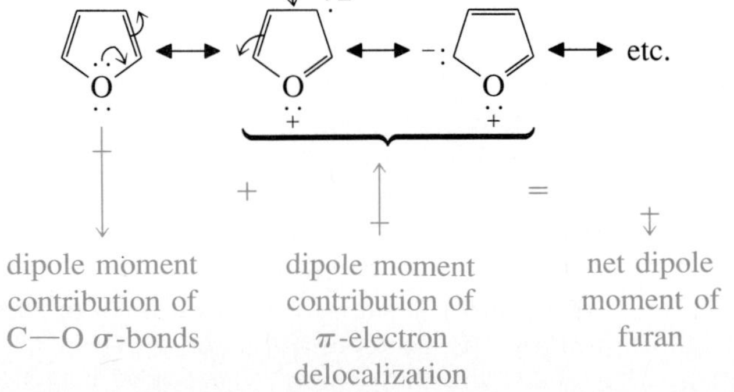

(24.13)

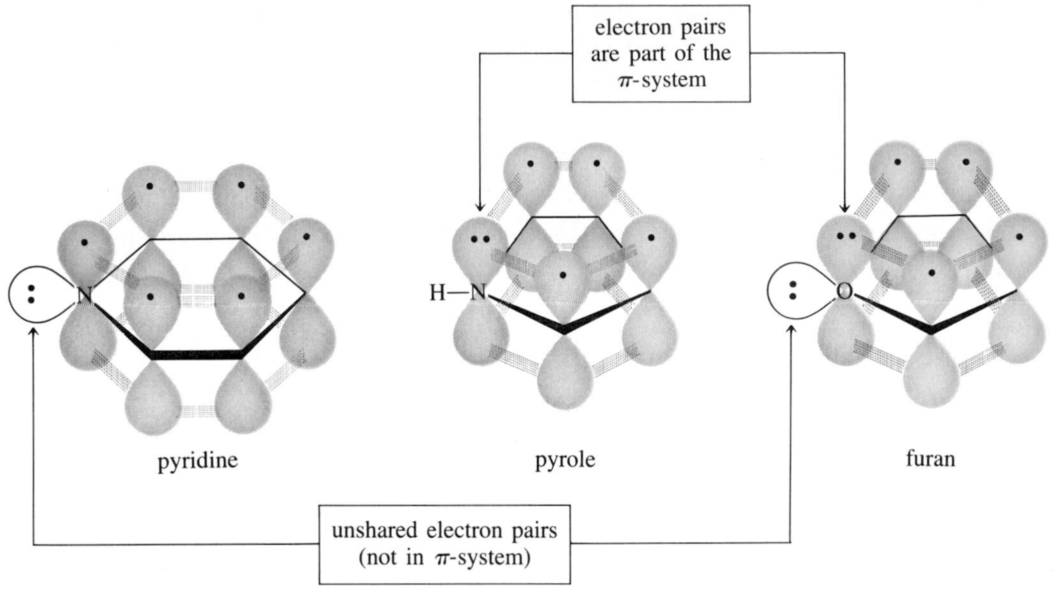

Figure 24.5 The configurations of the unshared electron pairs and π-electrons in pyridine, pyrrole, and furan. The orbitals in each $4n + 2$-electron π-system are shown in grey; π-interactions are shown in color. Unshared electron pairs not in the π-system are shown in white.

Because these two effects in furan nearly cancel, furan has a very small dipole moment. We can see the effect of dipole moment on the relative boiling points of tetrahydrofuran and furan.

Pyridine, like benzene, can be represented by two equivalent neutral resonance structures. Three additional structures, although involving separation of charge, have some importance because they reflect the relative electronegativity of nitrogen.

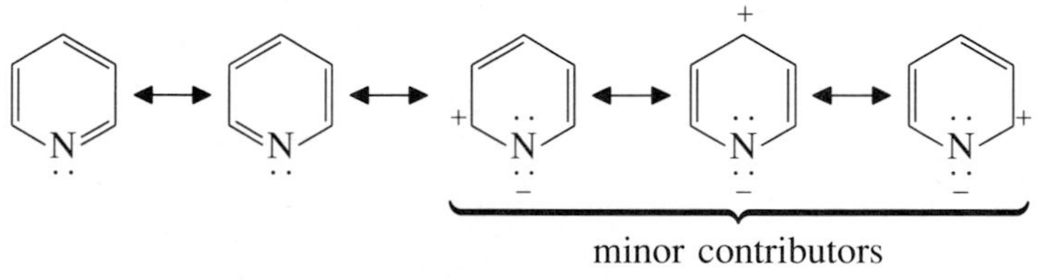

(24.14)

The aromaticity of some heterocyclic compounds was considered in our discussion of the Hückel $4n + 2$ rule (Sec. 15.6D). It is important to understand which unshared electron pairs in a heterocyclic compound are part of the $4n + 2$ aromatic π-electron system, and which are not. Heteroatoms involved in formal double bonds—such as the nitrogen of pyridine—contribute one π-electron to the six π-electron aromatic system, just like each of the carbon atoms in the π-system. The orbital containing the unshared electron pair of the pyridine nitrogen is perpendicular to the p orbitals of the ring and is therefore not involved in π-bonding (Fig. 24.5a). An unshared electron pair on a heteroatom in a formally allylic position—such as the un-

TABLE 24.1 Empirical Resonance Energies of Some Aromatic Compounds

Compound	Resonance energy, kcal/mol	Compound	Resonance energy, kcal/mol
benzene	34–36	furan	16
pyridine	23–28	pyrrole	21–22
naphthalene	61	thiophene	29

shared pair on the nitrogen of pyrrole—is part of the aromatic π-system. Thus, the hydrogen of pyrrole lies in the plane of the ring (Fig. 24.5b). The oxygen of furan (Fig. 24.5c) contributes one unshared electron pair to the aromatic π-electron system, and the other unshared electron pair occupies a position analogous to the hydrogen of pyrrole—in the ring plane, perpendicular to the p orbitals of the ring.

How much stability does each heterocyclic compound owe to its aromatic character? In Sec. 15.7C we learned that the *empirical resonance energy* can be used to estimate this stability. (Remember that this is the energy a compound "doesn't have" because of its aromaticity—that is, its aromatic stability.) The empirical resonance energies of benzene, naphthalene, and some heterocyclic compounds are given in Table 24.1. To the extent that resonance energy is a measure of aromatic character, we can see that furan has the least aromatic character of the heterocyclic compounds in the table.

Problems

7 (a) The dipole moments of pyrrole and pyrrolidine are similar in magnitude but have opposite directions. Explain, indicating the direction of the dipole moment in each compound.

μ = 1.80 D μ = 1.57 D

(b) Explain why the dipole moments of furan and pyrrole have opposite directions.

8 Each of the following chemical shifts goes with a hydrogen at carbon-2 of either pyridine, pyrrolidine, or pyrrole. Match each chemical shift with the appropriate heterocyclic compound, and explain your answer. δ 8.51; δ 6.41; and δ 2.82.

C. Basicity and Acidity of the Nitrogen Heterocycles

Basicity Pyridine and quinoline act as ordinary aromatic amine bases; they are about as basic as aniline.

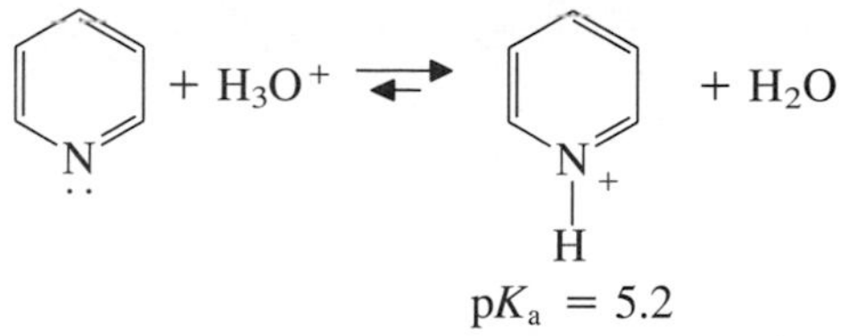
(24.15)

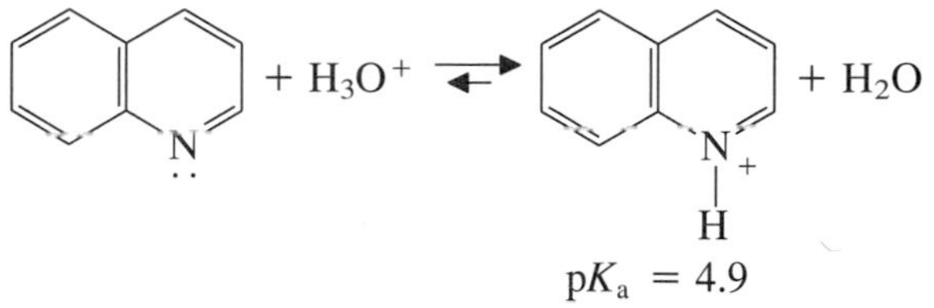
(24.16)

Pyridine and quinoline are less basic than aliphatic tertiary amines because of the sp^2 hybridization of their nitrogen unshared electron pairs. (Recall from Sec. 14.7A that basicity of an unshared electron pair decreases with increasing *s*-character.)

Since pyrrole and indole look like amines, it may come as a surprise that the nitrogens of these two heterocycles are not basic at all! These compounds are protonated only in strong acid, and protonation occurs on carbon, not nitrogen.

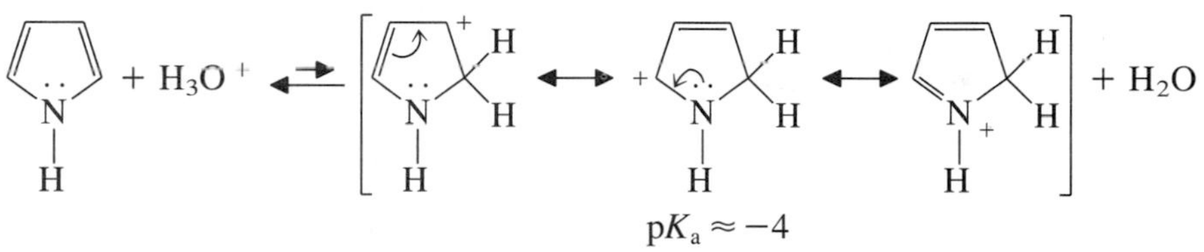
(24.17)

We can understand the marked contrast between the basicities of pyridine and pyrrole by considering the role of nitrogen's unshared electron pair in the aromaticity of each compound (Fig. 24.5). Protonation of the pyrrole nitrogen disrupts the aromatic six π-electron system by taking the nitrogen's unshared pair "out of circulation."

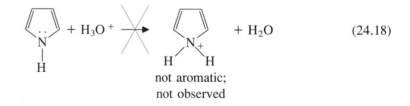

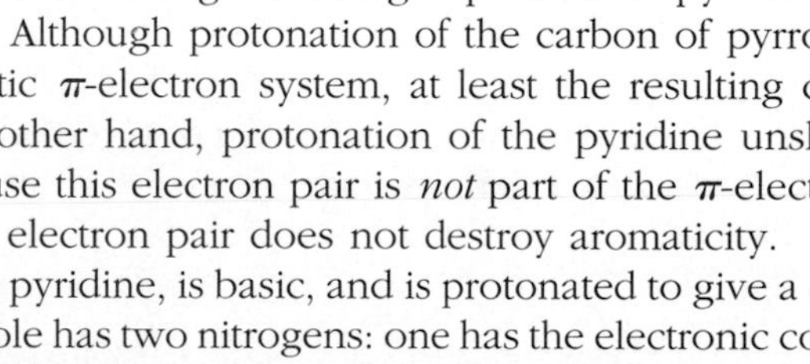
(24.18)

Furthermore, the positive charge in nitrogen-protonated pyrrole cannot be delocalized by resonance. Although protonation of the carbon of pyrrole (Eq. 24.17) also disrupts the aromatic π-electron system, at least the resulting cation is resonance stabilized. On the other hand, protonation of the pyridine unshared electron pair occurs easily because this electron pair is *not* part of the π-electron system. Hence protonation of this electron pair does not destroy aromaticity.

Imidazole, like pyridine, is basic, and is protonated to give a conjugate acid with $pK_a = 6.95$. Imidazole has two nitrogens: one has the electronic configuration of pyridine, but the other is like the nitrogen of pyrrole. Protonation of imidazole occurs on the pyridine-like nitrogen—the one whose electron pair is not part of the aromatic sextet.

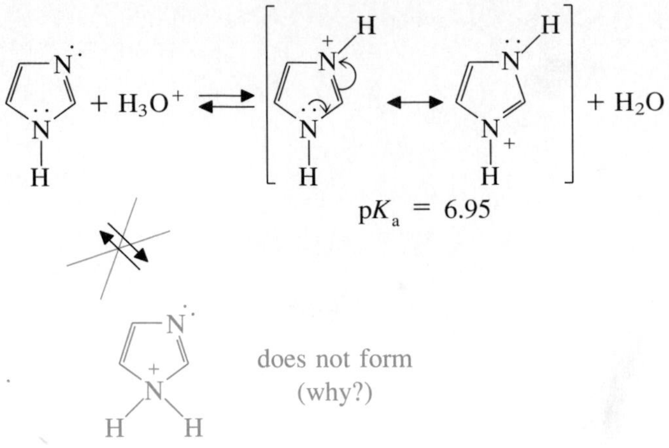

$$pK_a = 6.95 \qquad (24.19)$$

does not form
(why?)

However, once imidazole is protonated, its two nitrogens are indistinguishable because the resonance structures of the protonated form are equivalent (Eq. 24.19). Furthermore, because the protonated form is resonance-stabilized, imidazole is more basic than pyridine.

Acidity Pyrrole and indole are weak acids. The N—H protons of pyrrole and indole ($pK_a \approx 17.5$) are about as acidic as alcohol O—H protons. Thus, these protons are much more acidic than those of amines (Sec. 23.4E). The greater acidity of these compounds, relative to the acidity of amines, is a consequence of the resonance stabilization of the conjugate-base anions.

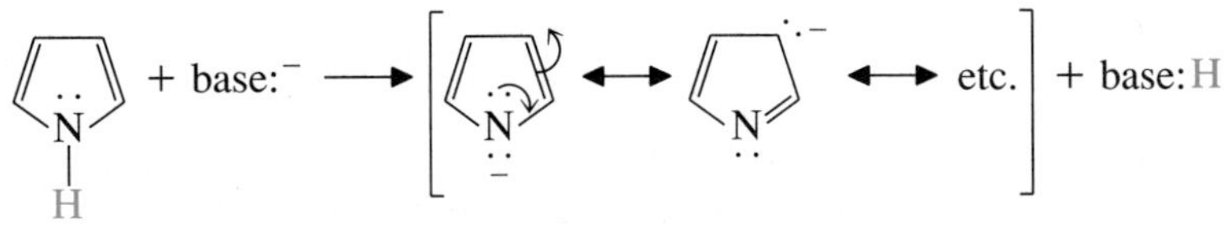

$$\qquad + \text{base:}^- \longrightarrow \left[\cdots \longleftrightarrow \cdots \longleftrightarrow \text{etc.} \right] + \text{base:H} \quad (24.20)$$

(Draw the other resonance structures of the pyrrole conjugate-base anion.) Pyrrole and indole behave as acids toward basic organometallic compounds such as Grignard or organolithium reagents.

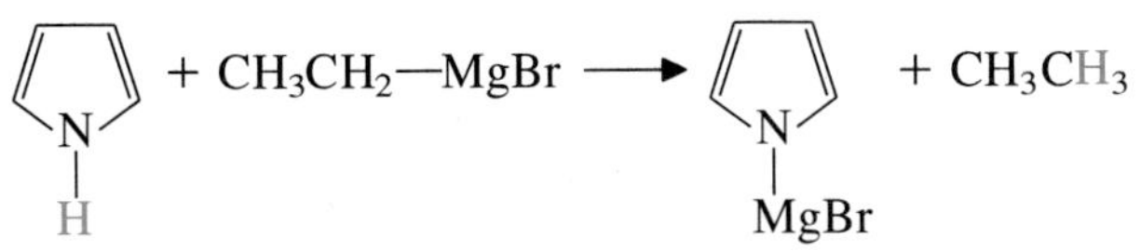

$$\qquad + \text{CH}_3\text{CH}_2\text{—MgBr} \longrightarrow \qquad + \text{CH}_3\text{CH}_3 \qquad (24.21)$$

It is interesting that pyrrole, a nitrogen acid, is somewhat *less* acidic than 1,3-cyclopentadiene, a carbon acid.

$pK_a \approx 15$ $pK_a \approx 17.5$

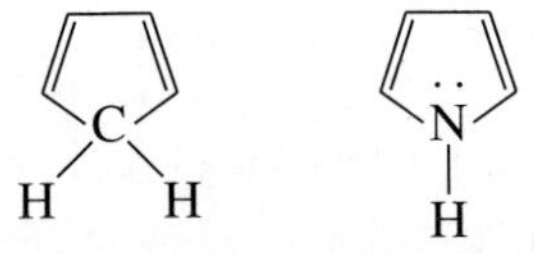

This contrasts with the more common observation that protons on more electronegative atoms are more acidic than protons on less electronegative atoms; compare, for example, the acidities of carboxylic acids, amides, and ketones. The reason for the greater acidity of 1,3-cyclopentadiene is that the cyclopentadienyl anion is aromatic, but 1,3-cyclopentadiene itself is not. Thus, when cyclopentadiene ionizes, it gains aromaticity. However, both pyrrole and its conjugate-base anion are aromatic; ionization of pyrrole does not gain any additional aromatic stabilization for the molecule.

Problems

9 (a) Suggest a reason why pyridine is miscible with water, whereas pyrrole has little water solubility.

(b) Indicate whether you would expect imidazole to have high or low water solubility, and why.

10 (a) The compound 4-(dimethylamino)pyridine protonates to give a conjugate acid with a pK_a value of 9.9. This compound is thus 4.7 pK_a units more basic than pyridine itself. Draw the structure of the conjugate acid of 4-(dimethylamino)pyridine, and explain why it is much more basic than pyridine.

(b) What product is expected when 4-(dimethylamino)pyridine reacts with CH_3I?

11 Protonation of aniline causes a dramatic shift of its UV spectrum to lower wavelength, but protonation of pyridine has almost no effect on its UV spectrum. Explain.

24.3 CHEMISTRY OF FURAN, PYRROLE, AND THIOPHENE

A. Electrophilic Aromatic Substitution

Since furan, pyrrole, and thiophene are aromatic, they, like benzene, undergo aromatic substitution reactions. We can ask the same questions about the aromatic substitution reactions of heterocyclic aromatic compounds that we asked about naphthalene: (1) In what ring position do these compounds substitute? (2) How reactive are these compounds?

Furan, pyrrole, and thiophene all undergo electrophilic substitution predominately in the 2-position of the ring, as the following examples illustrate:

$$\text{(furan)} + CH_3-\overset{O}{\underset{\|}{C}}-O-\overset{O}{\underset{\|}{C}}-CH_3 \xrightarrow[CH_3CO_2H]{BF_3} \text{(2-acetylfuran)}\ \overset{O}{\underset{\|}{C}}-CH_3 + CH_3-\overset{O}{\underset{\|}{C}}-OH \quad (24.22)$$

(a Friedel–Crafts reaction)　　　　　　(75–92% yield)

$$\text{(thiophene)} + HNO_3 \xrightarrow{\text{acetic anhydride}} \text{(thiophene)}-NO_2 + \text{(thiophene-NO}_2\text{)} + H_2O \quad (24.23)$$

(70% yield)　　(5% yield)

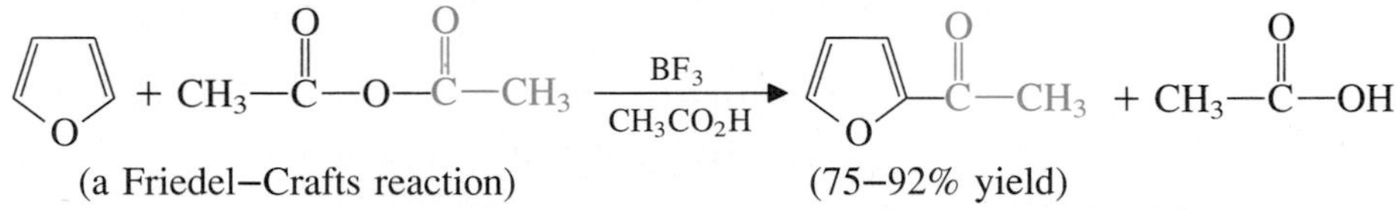

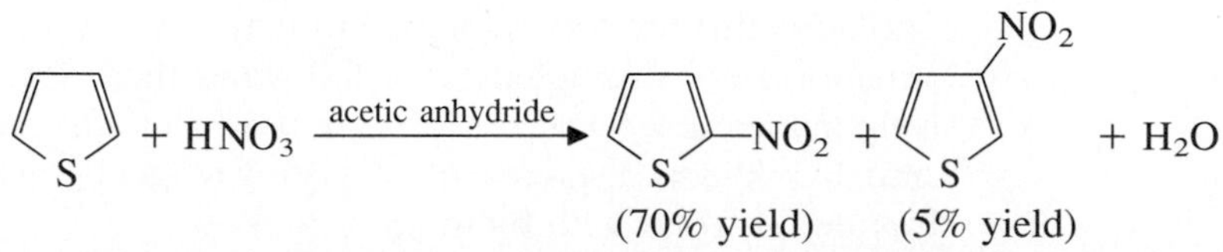

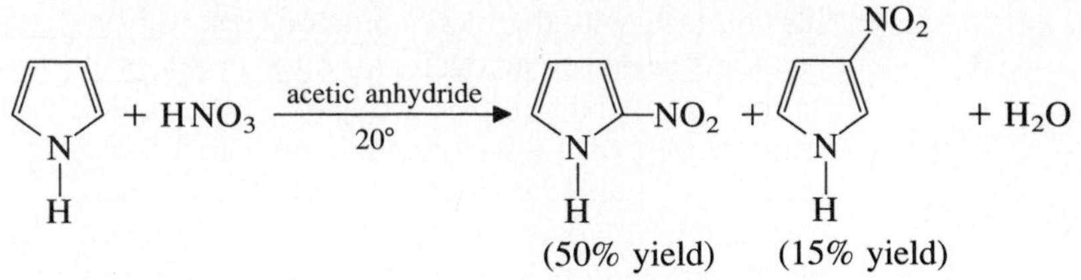

(24.24)

To understand why substitution at the 2-position is preferred, let us compare the carbocation intermediates involved in the two modes of substitution.

substitution at carbon-2:

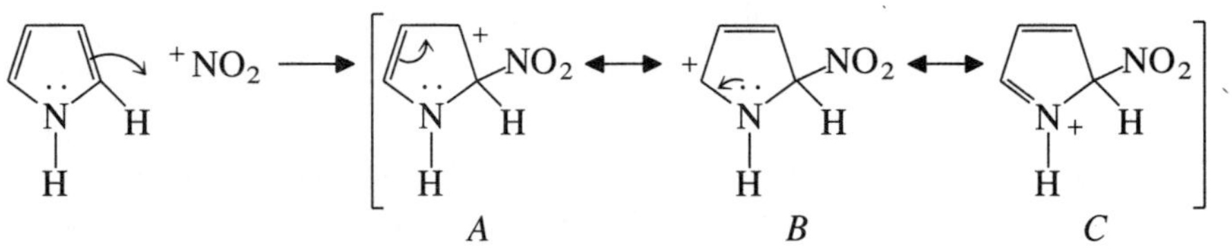

(24.25a)

substitution at carbon-3:

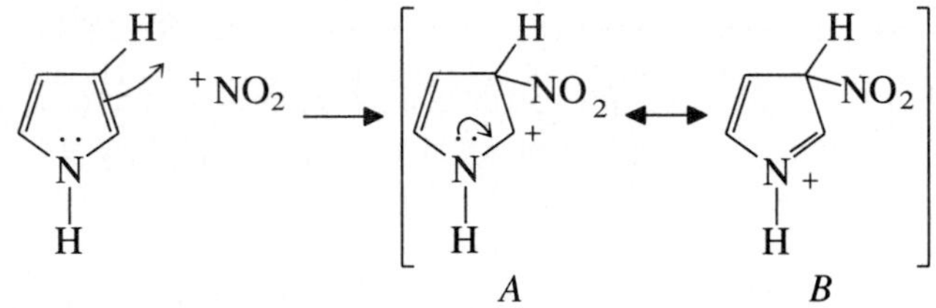

(24.25b)

This comparison shows that the carbocation resulting from attack of the electrophile at carbon-2 has more important resonance structures, and is therefore more stable, than the carbocation resulting from attack at carbon-3. Hammond's postulate (Sec. 4.7C) suggests that the reaction involving the more stable intermediate should be faster. There is nothing wrong with reaction at carbon-3; reaction at carbon-2 is simply more favorable. In fact, careful examination of reaction mixtures (see Eq. 24.23) has shown that in many cases some carbon-3 substitution product accompanies the major carbon-2 substitution product. Furthermore, if both carbon-2 and carbon-5 are substituted, substitution at carbon-3 then occurs.

Furan, pyrrole, and thiophene are all much more reactive than benzene in electrophilic substitution reactions. Although precise reactivity ratios depend on the particular reaction, the relative rates of bromination are typical:

pyrrole >	furan >	thiophene ≫	benzene
3×10^{18}	6×10^{11}	5×10^{9}	1

(24.26)

Milder reaction conditions must be used with more reactive compounds. (Reaction conditions that are too vigorous in many cases bring about polymerization and tar formation.) For example, in the following three Friedel–Crafts reactions, progressively milder Lewis-acid catalysts are used along the series benzene, thiophene, and furan. In addition, a less reactive acylating reagent—acetic anhydride instead of acetyl chloride—is used with furan.

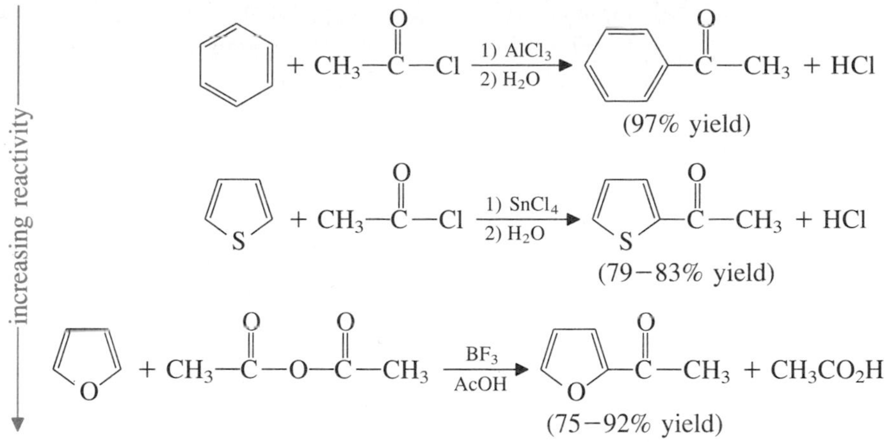

increasing reactivity

The reactivity order of the heterocycles (Eq. 24.26) is a consequence of the relative abilities of the heteroatoms to stabilize positive charge in the intermediate carbocations (structure C in Eq. 24.25a). Both pyrrole and furan have heteroatoms from the first row of the periodic table. Because nitrogen is better than oxygen at delocalizing positive charge (it is less electronegative), pyrrole is more reactive than furan. The sulfur of thiophene is a second row element and, although it is less electronegative than oxygen, its $3p$ orbitals overlap less efficiently with the $2p$ orbitals of the aromatic π-electron system (see Fig. 16.10). In fact, the reactivity _order_ of the heterocycles in aromatic substitution parallels the reactivity order of the correspondingly substituted benzene derivatives:

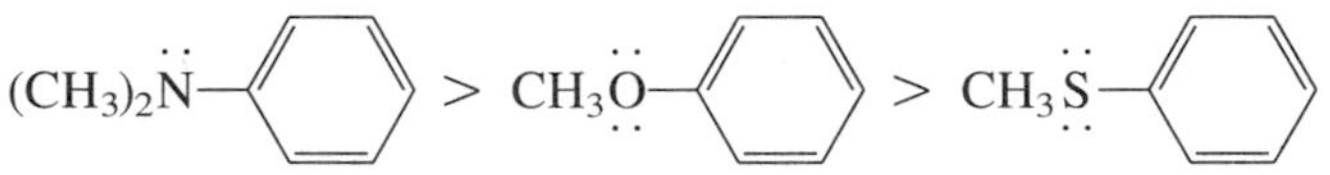

$$(CH_3)_2\overset{..}{N}- \hspace{-0.3em} \bigcirc \hspace{-0.3em} > CH_3\overset{..}{O}- \hspace{-0.3em} \bigcirc \hspace{-0.3em} > CH_3\overset{..}{S}- \hspace{-0.3em} \bigcirc \hspace{2em} (24.28)$$

What happens when we try to introduce a second substituent into a furan, pyrrole, or thiophene ring that already contains one substituent group? As we might expect from what we know about benzene and naphthalene chemistry, the usual activating and directing effects of substituents in aromatic substitution apply (see Table 16.2). Superimposed on these effects is the normal effect of the heterocyclic atom in directing substitution to the 2-position. The following example illustrates these effects:

"_meta_" (1,3)

(69% yield;
satisfies directing
effect of both the
heteroatom and —CO$_2$H)

(not observed; satisfies
directing effect of
heteroatom only)

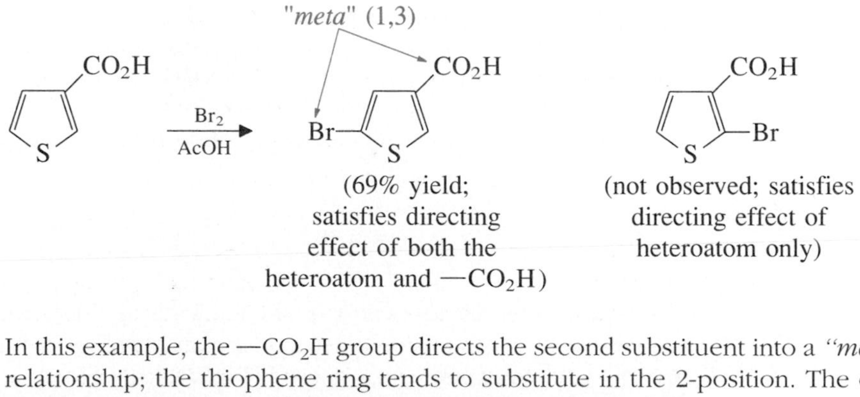

$$(24.29)$$

In this example, the —CO$_2$H group directs the second substituent into a _"meta"_ (1,3) relationship; the thiophene ring tends to substitute in the 2-position. The observed product satisfies both of these directing effects. (Notice that we count around the *carbon* framework of the heterocyclic compound, not through the heteroatom, when using this *ortho, meta, para* analogy.) In the following example, the chloro group is an *ortho, para*-directing group. Since the position *"para"* to the chloro group is also a 2-position, both the sulfur of the ring and the chloro group direct the incoming nitro group to the same position.

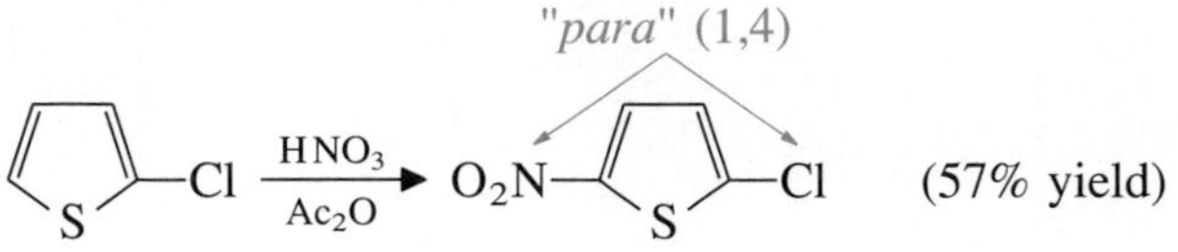

(24.30)

When the directing effects of substituents and the ring compete, it is not unusual to observe mixtures of products.

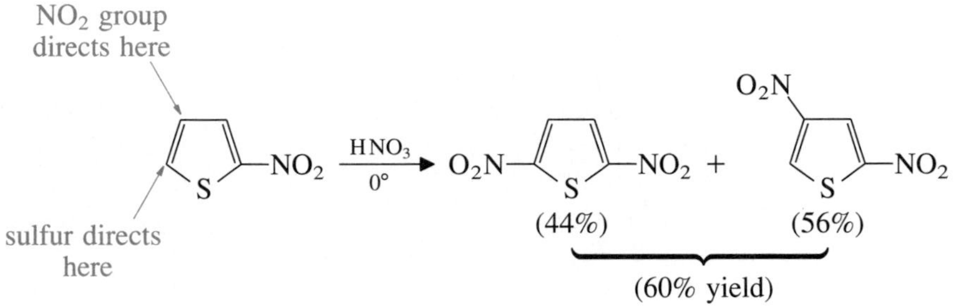

(24.31)

Finally, if both 2-positions are occupied, 3-substitution takes place.

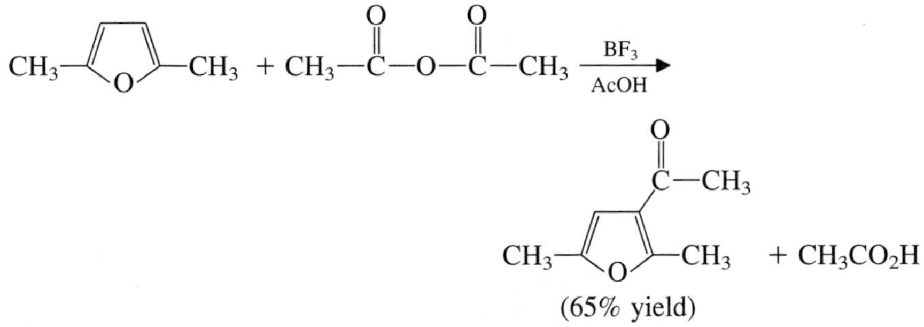

(24.32)

B. Addition Reactions of Furan

In the previous sections we focused on the aromatic character of furan, pyrrole, and thiophene. We could, however, also view furan, pyrrole, or thiophene formally as a 1,3-butadiene with its terminal carbons "tied down" by a heteroatom bridge.

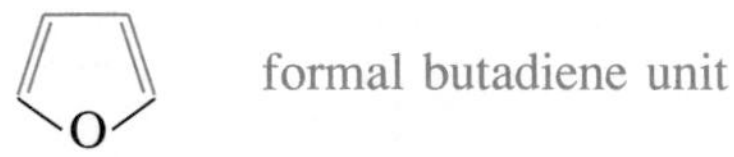

Do the heterocycles ever behave chemically as if they are conjugated dienes? The answer is yes. Of the three heterocyclic compounds furan, pyrrole, and thiophene, furan has the least resonance energy (Table 24.1) and, by implication, the least aromatic character. That is, of the three compounds, furan has the greatest tendency to behave like a conjugated diene.

One characteristic reaction of conjugated dienes is 1,4-addition, or conjugate addition (Sec. 15.4A). Indeed, furan does undergo some 1,4-addition reactions. One example of such a reaction occurs in bromination. For example, furan undergoes conjugate addition of bromine and methanol in methanol solvent; the 1,4-addition product reacts further in a rapid S_N1 displacement with the methanol. (Write mechanisms for both parts of this reaction; refer to Sec. 15.4A if necessary.)

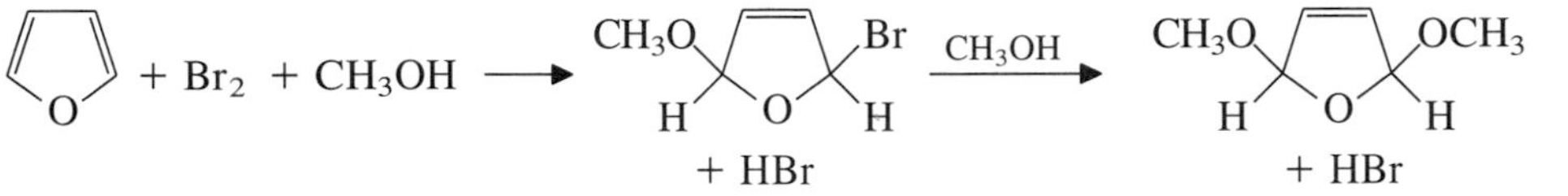

mixture of stereoisomers
(72–76% yield)

(24.33)

Another manifestation of the conjugated-diene character of furan is that it undergoes Diels–Alder reactions (Sec. 15.3) with reactive dienophiles such as maleic anhydride.

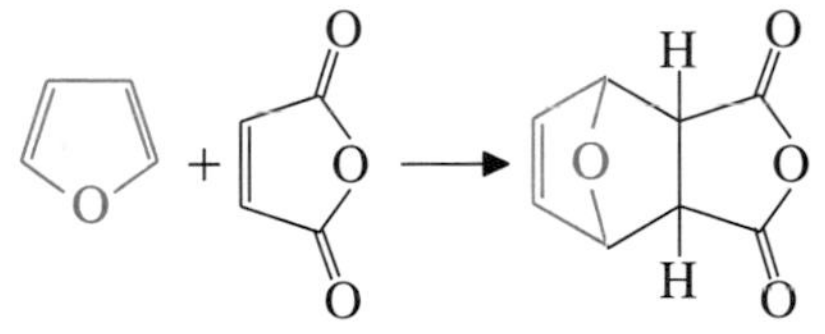

(>90% yield) (24.34)

C. Side-Chain Reactions

Although we have been considering the reactivity of heterocyclic rings themselves, many reactions occur at the side chains of heterocyclic compounds without affecting the rings—in the same sense that some reactions occur at the side chain of a substituted benzene (Secs. 17.1–17.5).

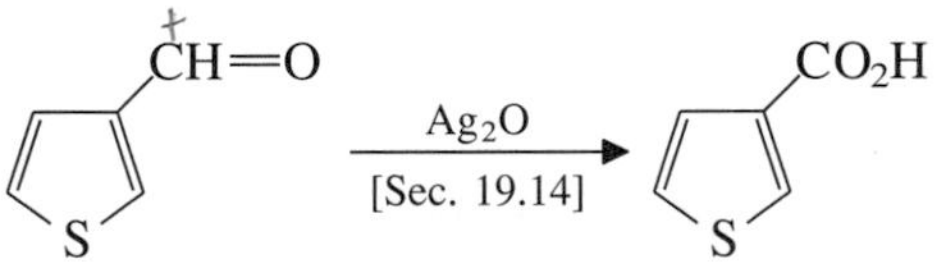

(95–97% yield) (24.35)

A particularly useful example of a side-chain reaction is removal of a carboxy group directly attached to the ring (*decarboxylation*). This reaction, which is effected by strong heating (in some cases with catalysts), is important in the synthesis of some unsubstituted heterocyclic compounds.

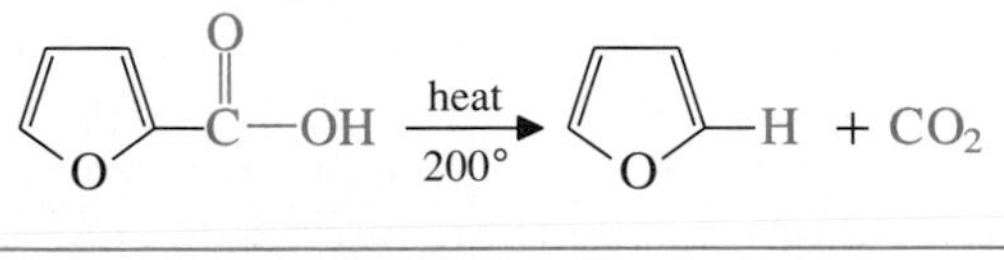

(24.36)

Problems

12 Complete each of the following reactions by giving the principal organic product(s):

(a)

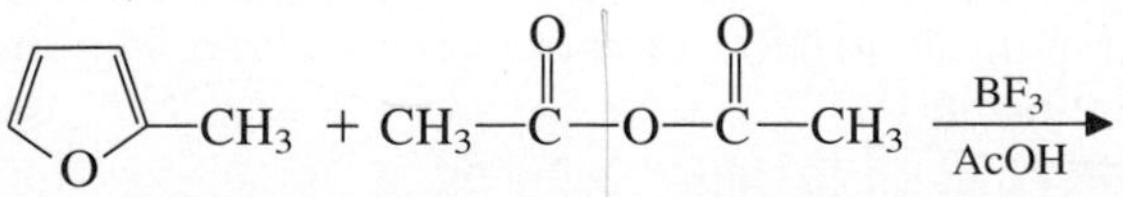

Problems (Cont.)

(b)

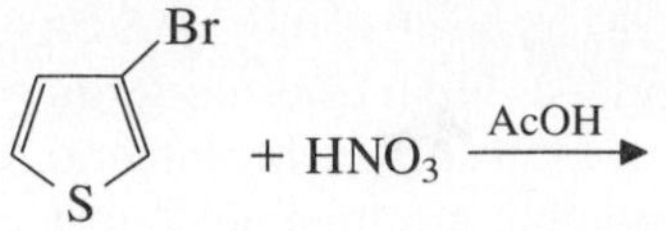

(c)

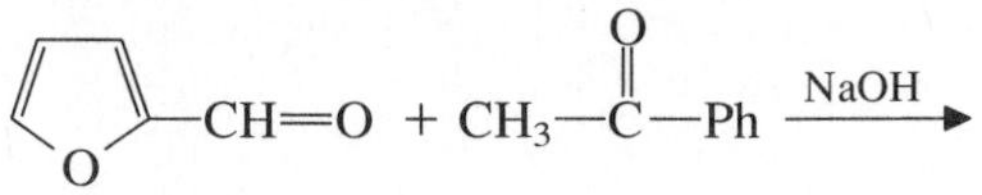

(d)

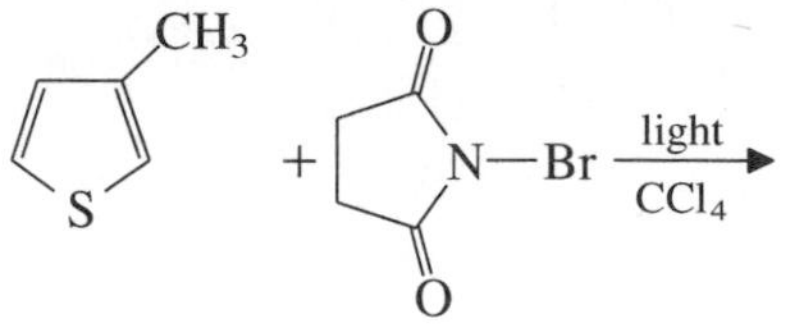

13 Write a mechanism for each of the following reactions:

(a)

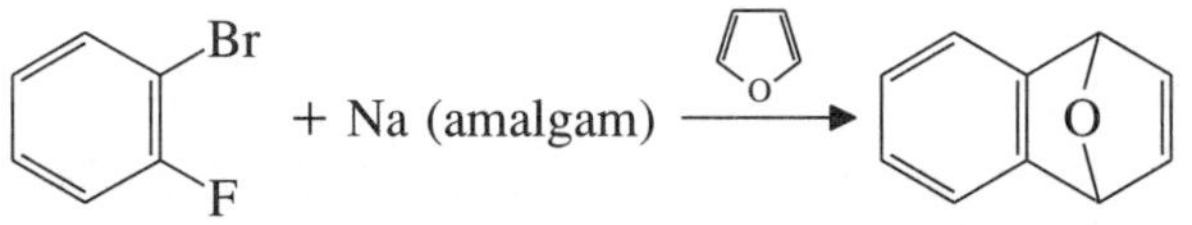

(*Hint:* See Sec. 15.3D.)

(b)

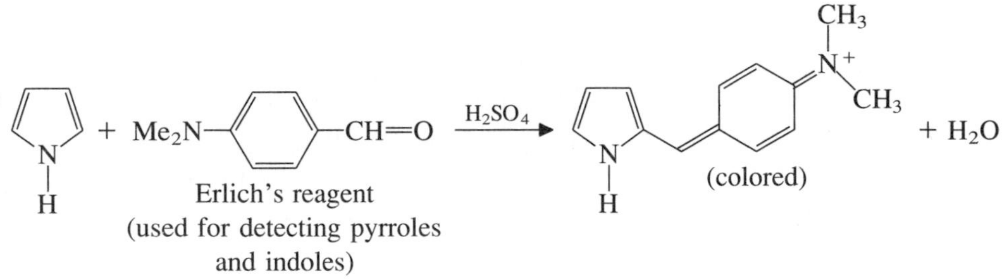

24.4 CHEMISTRY OF INDOLE, BENZOFURAN, AND BENZOTHIOPHENE

A. Electrophilic Aromatic Substitution

When we consider the electrophilic aromatic substitution reactions of indole, benzofuran, and benzothiophene, we again ask two familiar questions: (1) What is the preferred position of substitution? (2) How reactive are these compounds relative to other aromatic compounds we have studied?

Substitution occurs on the heterocyclic ring rather than on the benzene ring of benzofuran, benzothiophene, and indole. It is found experimentally that the relative reactivity of the 2- and 3-positions is more closely balanced in indole, benzofuran, and benzothiophene than it is in pyrrole, furan, or thiophene. As a generalization, benzothiophene and indole, unlike thiophene and pyrrole, show predominant substitution in the 3-position. However, benzofuran, like furan, substitutes predominately at the 2-position.

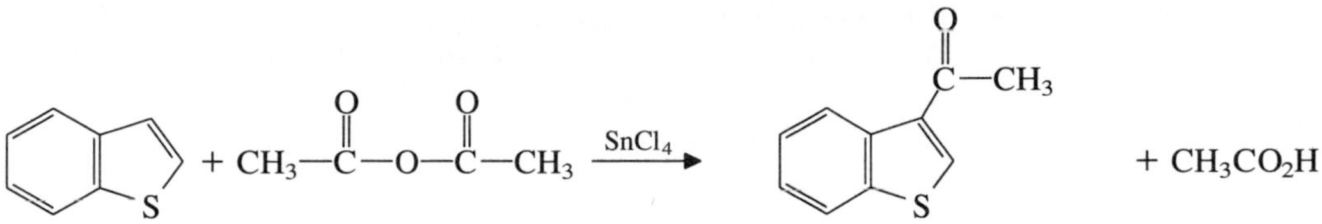

6.7 : 1
3-substitution : 2-substitution

(24.37)

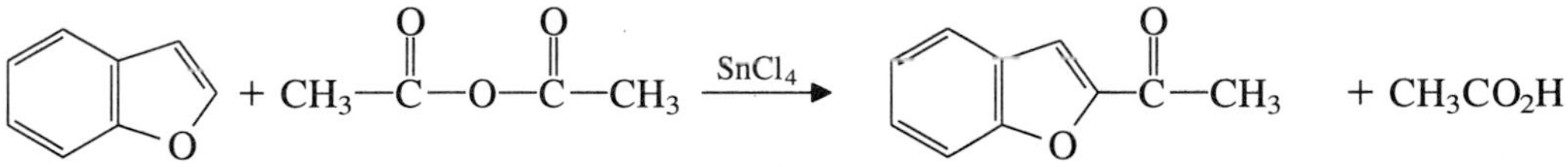

7.25 : 1
2-substitution : 3-substitution

(24.38)

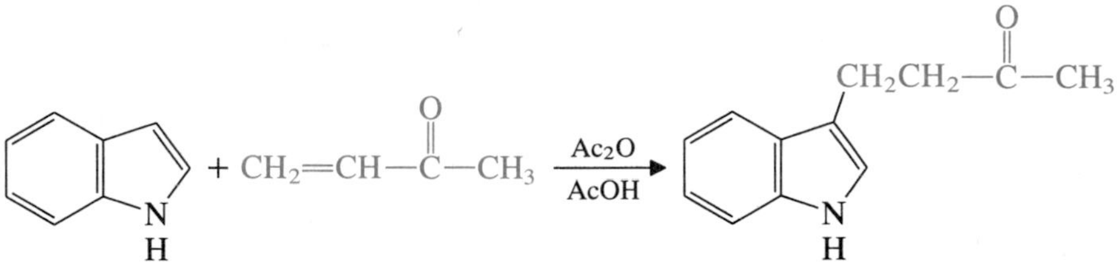

(75% yield)

(24.39)

(Write the mechanism of the reaction in Eq. 24.39.)

Benzofuran, benzothiophene, and indole are less reactive than furan, thiophene, or pyrrole, respectively, but more reactive than benzene. For example, benzothiophene is about 5–10% as reactive as thiophene, and benzofuran is about 0.5% as reactive as furan.

B. Synthesis of Indoles

Conceptually, the simplest approach to the synthesis of a heterocyclic compound is to introduce the substituent into an existing ring. The electrophilic aromatic substitution reactions we have been considering are useful in this strategy. The second approach is to form the substituted ring system itself in some type of cyclization reaction. Both types of approaches are used in the synthesis of heterocyclic compounds, but the latter approach is particularly important in the synthesis of indole and quinoline derivatives. Let us consider two well-established approaches to the construction of substituted indoles.

One of the best-known methods for preparing indoles is the **Fischer indole synthesis,** named for the great German chemist Emil Fischer (see Sec. 27.9A). In this reaction, an aldehyde or ketone with an α- —CH_2— group is condensed with a phenylhydrazine derivative in the presence of an acid catalyst and/or heat.

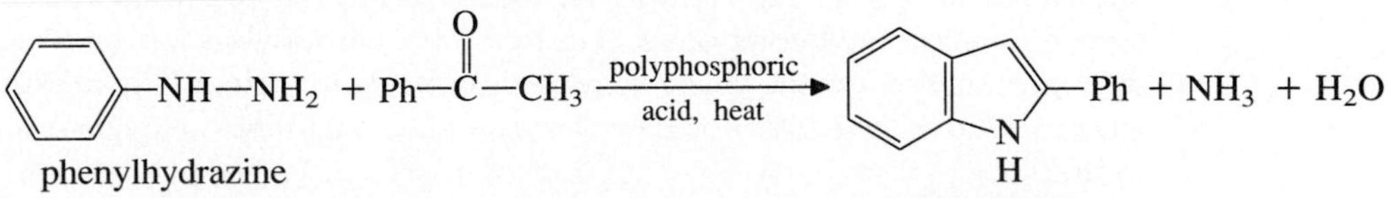

phenylhydrazine

(76% yield)

(24.40)

A variety of Brønsted or Lewis acid catalysts can be used: H_2SO_4, BF_3, $ZnCl_2$, and others. (The acid used in Eq. 24.40, *polyphosphoric acid,* is a syrupy mixture of P_2O_5 and phosphoric acid.) The reaction also works with many different substituted phenylhydrazines and carbonyl compounds. However, acetaldehyde, which could in principle be used to give indole itself, does not work in this reaction, probably because it polymerizes under the reaction conditions.

The mechanism of the Fischer indole synthesis begins with a familiar reaction: conversion of the carbonyl compound into a *phenylhydrazone,* a type of imine (Table 19.3).

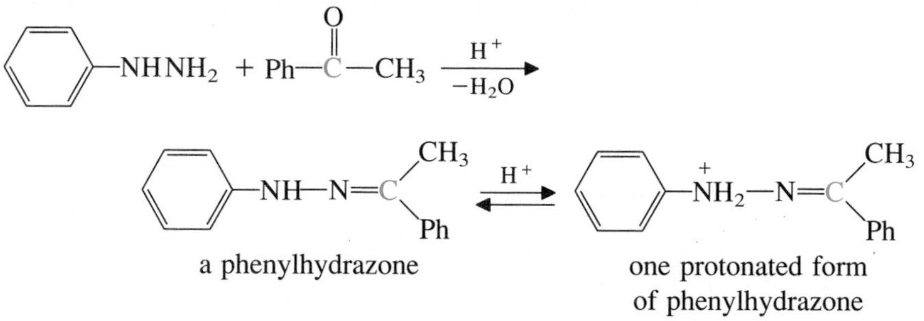

(24.41a)

The phenylhydrazone, which is protonated under the reaction conditions, is in equilibrium with a small amount of its enamine tautomer, which is also protonated.

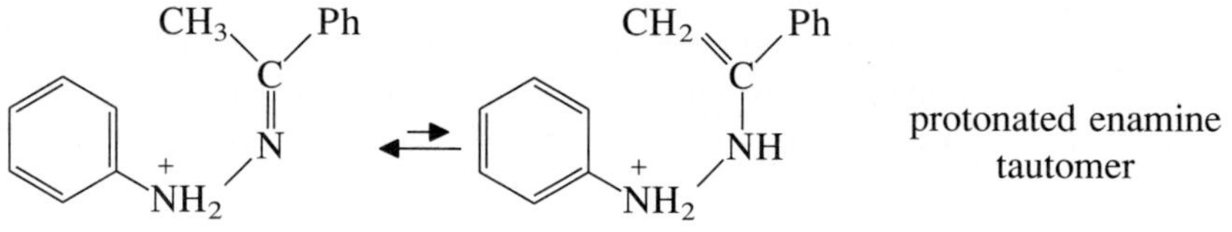

(24.41b)

The latter species undergoes a cyclic shift of *three electron pairs* (six electrons) to give a new intermediate in which the N—N bond of the phenylhydrazone has been broken.

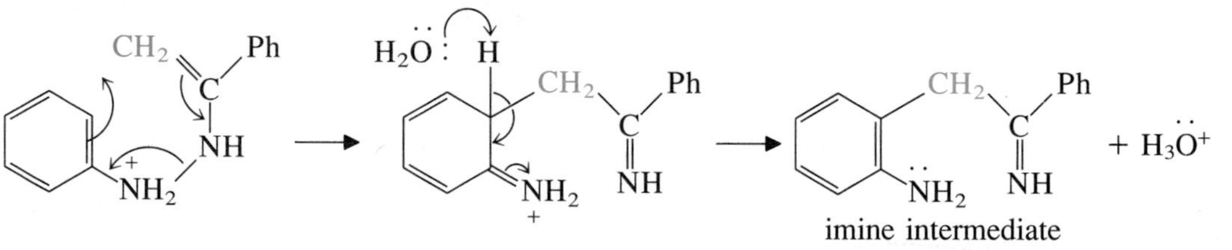

(24.41c)

This step occurs by the cyclic flow of electrons—a *pericyclic reaction*. The intermediate formed in Eq. 24.41c is an imine (Sec. 19.11A). Since an imine is the nitrogen analog of a carbonyl compound, it reacts like a carbonyl group. Indeed, the imine, after protonation on the imine nitrogen, undergoes nucleophilic addition with the amine group in the same molecule. The resulting "enamine" derivative is the product indole.

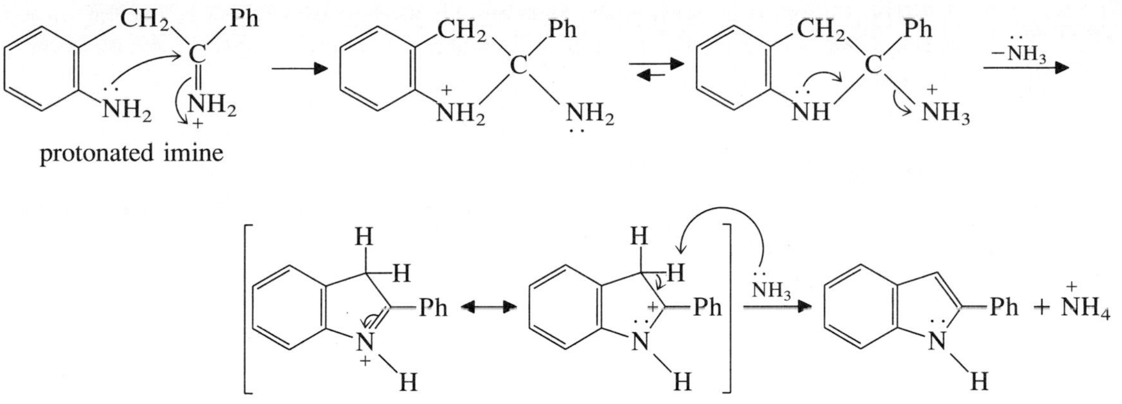

(24.41d)

Problems

14 What starting materials are required for the synthesis of each of the following compounds by the Fischer indole synthesis?

(a)

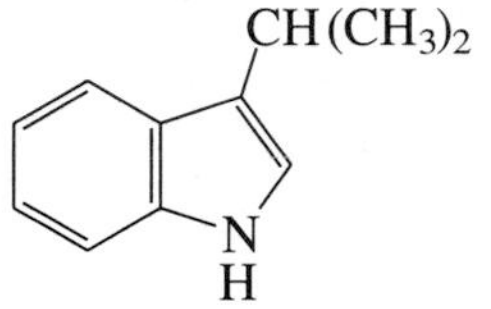

(c)

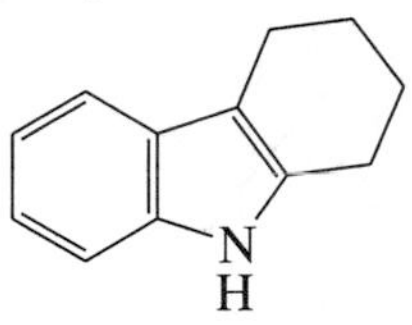

(b)

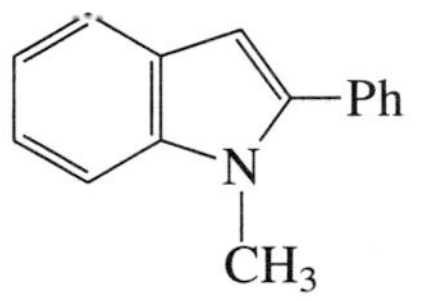

15 When phenylhydrazine is condensed with 2-butanone, a mixture of two isomeric indoles is obtained. Explain.

As we can see, the Fischer indole synthesis involves acidic conditions. Another synthesis of indoles, the **Reissert synthesis,** occurs under basic conditions. The key starting materials for this synthesis are diethyl oxalate and *o*-nitrotoluene or a substituted derivative.

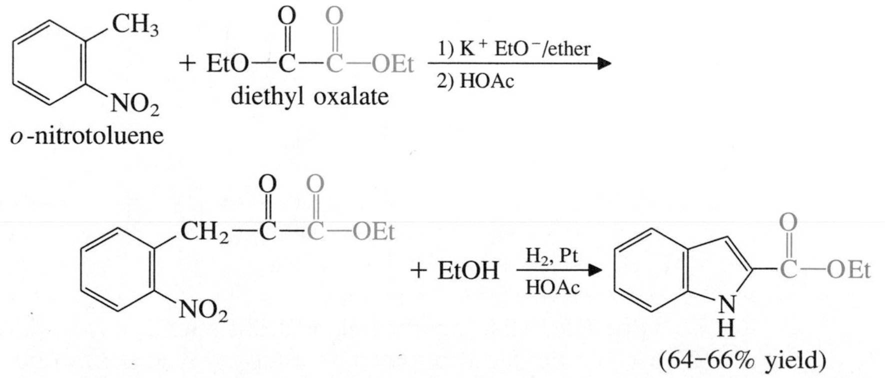

(24.42)

The *o*-nitro group is an essential element in the success of this reaction because it stabilizes by resonance the anion formed when a proton is removed from the methyl group:

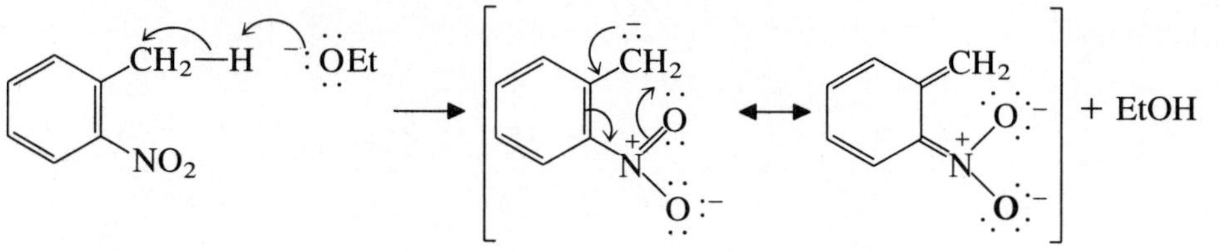

(24.43a)

This nucleophilic anion attacks a carbonyl group of diethyl oxalate, displacing ethanol; the reaction is a variation of the Claisen condensation (Sec. 22.5A). Like the Claisen condensation, this reaction is driven to completion by ionization of the product. For this reason, at least one equivalent of the base must be used.

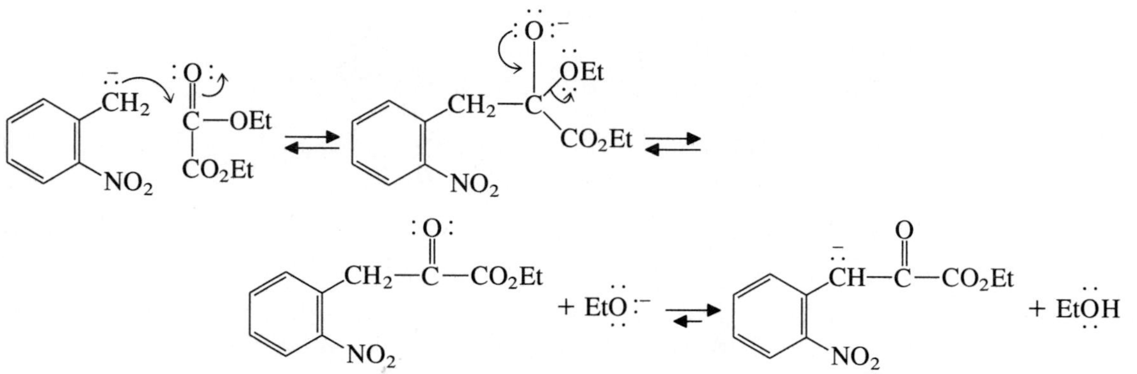

(24.43b)

The anion is neutralized by protonation in acetic acid, and the nitro group is converted into an amino group in a separate reduction step. (Catalytic hydrogenation is the reduction method used in Eq. 24.42.) The amino group thus formed reacts with the neighboring ketone to yield, after acid–base equilibria, an "enamine"—that is, the aromatic indole. (Fill in the mechanistic details for formation of the product from the amine.)

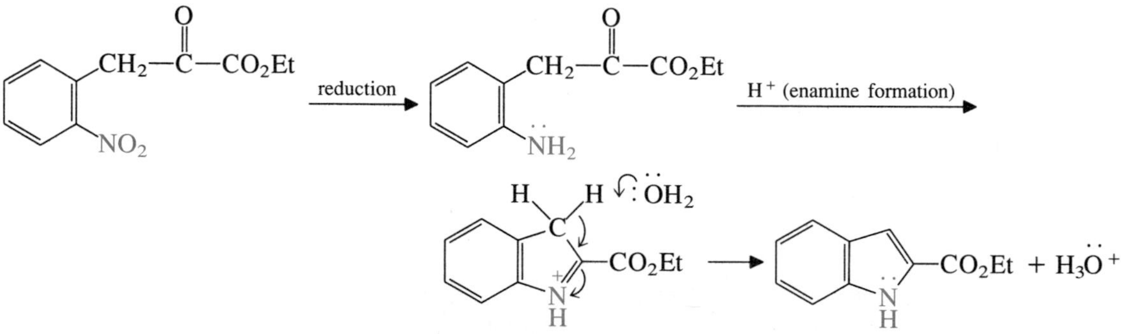

(24.43c)

Because indole-2-carboxylic acid, like other heterocyclic carboxylic acids (see Eq. 24.36), can be decarboxylated, the Reissert reaction can be used to prepare indole itself.

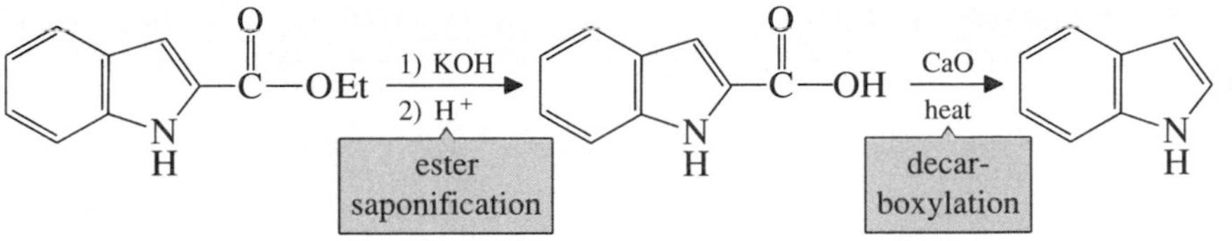

(24.44)

If substituted nitrotoluenes are used in the Reissert reaction, this reaction, in conjunction with the final decarboxylation step, can be used to prepare indoles that are substituted in the benzene ring and unsubstituted at the 2- or 3-positions. In this sense, the Reissert synthesis is complementary to the Fischer synthesis. Although many substituted phenylhydrazines work in the Fischer indole synthesis, some are difficult to prepare; thus, the Fischer synthesis is most often used to prepare indoles that are substituted at the 2- or 3-positions.

Problem

16 Outline a synthesis of 5-bromoindole from *m*-toluidine.

24.5 CHEMISTRY OF PYRIDINE AND QUINOLINE

A. Electrophilic Aromatic Substitution

In general, pyridine has very low reactivity in electrophilic aromatic substitutions; it is much less reactive than benzene. This behavior stands in direct contrast to that of the other heterocyclic compounds we have discussed. When substitution in pyridine does occur, it takes place at the 3-position, and the yields are in some cases poor.

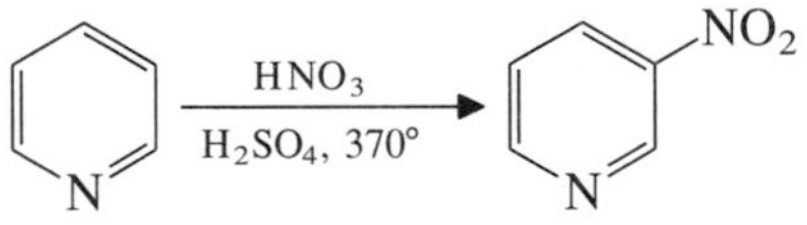

(5% yield) (24.45)

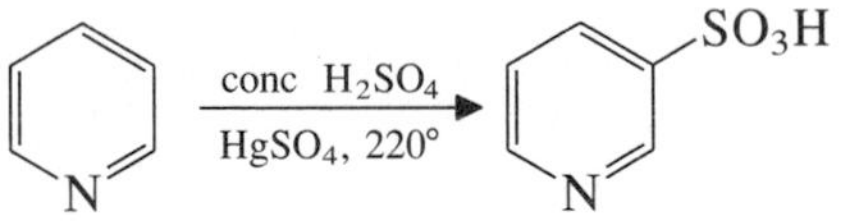

(70% yield) (24.46)

Even a pyridine derivative with three activating methyl groups requires very severe conditions for nitration.

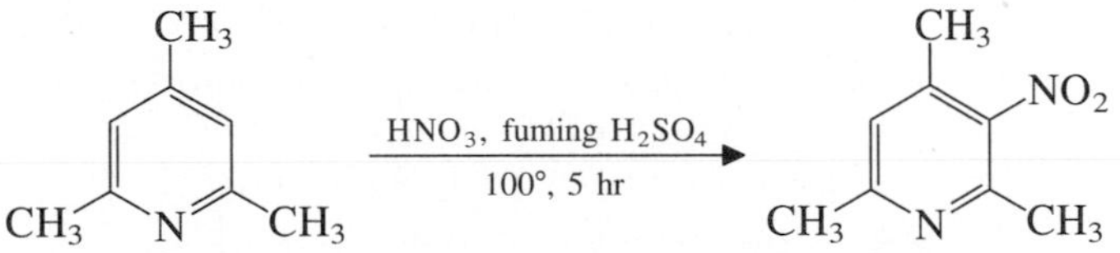

(93% yield) (24.47)

We can understand the preference for 3-substitution in pyridines by considering the resonance structures for the possible carbocation intermediates. Substitution in the 3-position gives a carbocation with three resonance structures:

3-substitution:

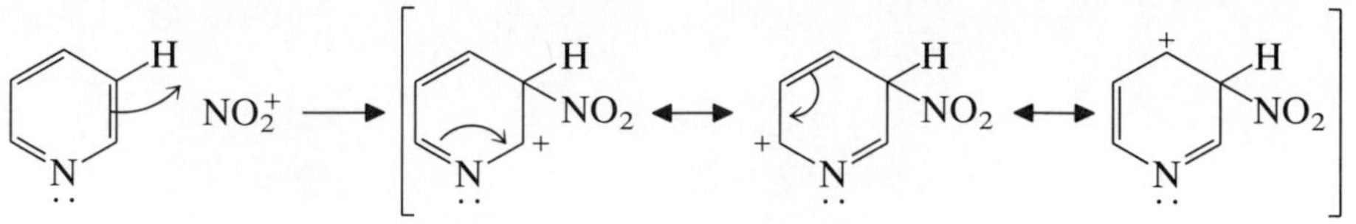

(24.48a)

Substitution at the 2-position also involves an intermediate with three resonance structures, but the one shown in color is particularly unfavorable because *the nitrogen is not only positively charged, but also electron deficient*—that is, it does not have an electronic octet.

2-substitution:

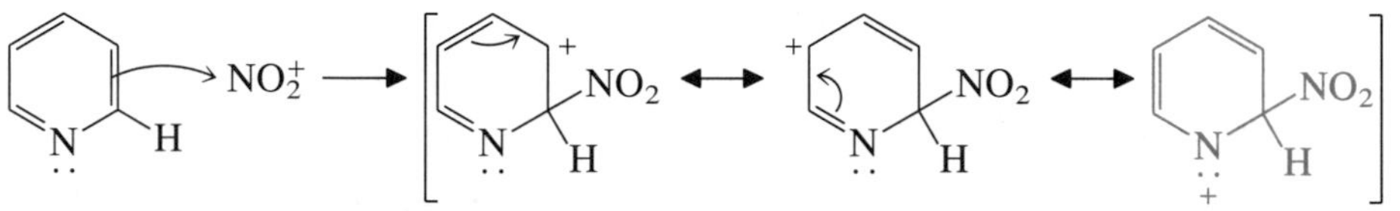

(24.48b)

We must be sure to understand that the nitrogen in the colored structure has a very different electronic configuration from the nitrogen in pyrrole (Eq. 24.25a, structure *C*). The pyrrole nitrogen is also positively charged, but because it has a complete octet, it is not electron-deficient. Electron deficiency on an electronegative atom is particularly unfavorable energetically.

Why is pyridine so unreactive? One reason is that the electronegative nitrogen destabilizes the positive charge in the structures of Eq. 24.48a. A more important reason has to do with the basicity of pyridine. Most methods of electrophilic aromatic substitution involve highly acidic conditions (either Brønsted or Lewis acids). An acid can coordinate with the unshared electron pair of pyridine to give a positively charged nitrogen. Introducing a further positive charge into the ring by electrophilic substitution is then especially difficult.

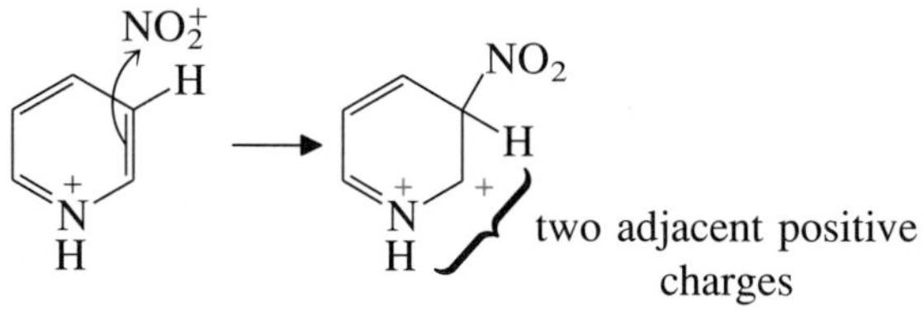

(24.49)

To understand electrophilic substitution in quinoline, we can apply the principles we learned for substitution on naphthalene (Sec. 24.1C). The pyridine ring of quinoline is deactivated; as with naphthalene, substitution occurs in the more activated ring—in this case, the benzene ring—at the α-positions. Because two nonequivalent α-positions are available, a mixture of compounds is obtained.

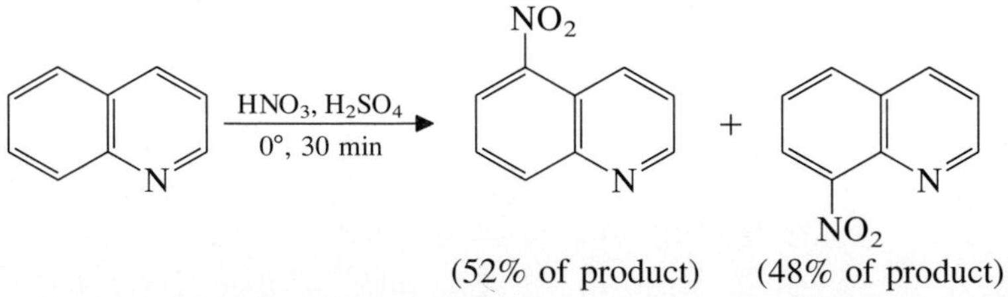

(24.50)

Because most electrophilic substitution reactions of quinoline give mixtures, substituted quinolines are generally synthesized from acyclic compounds, as we shall see in Sec. 24.5E.

If electrophilic substitution in pyridine occurs at the 3-position, how can we obtain pyridine derivatives substituted at other positions? One compound used to obtain 4-substituted pyridines is pyridine-*N*-oxide, formed by oxidation of pyridine with 30% hydrogen peroxide.

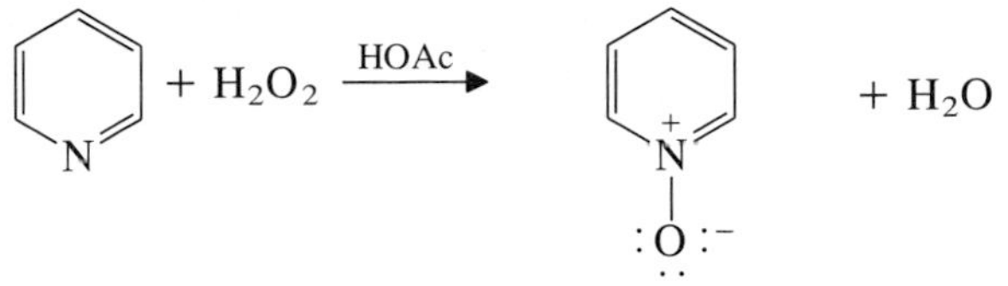

pyridine-*N*-oxide
(90% yield)

(24.51)

An analogy to pyridine-*N*-oxide from benzene chemistry is phenoxide—the conjugate base of phenol. Just as phenol or phenoxide is much more reactive in electrophilic aromatic substitution than benzene (Sec. 18.7B), pyridine-*N*-oxide is much more reactive than pyridine. Of course, since there is a positive charge on the nitrogen of pyridine-*N*-oxide, this compound is *much* less reactive than phenol or phenoxide. Nevertheless, pyridine-*N*-oxide undergoes useful aromatic substitution reactions, and substitution occurs in the 4-position.

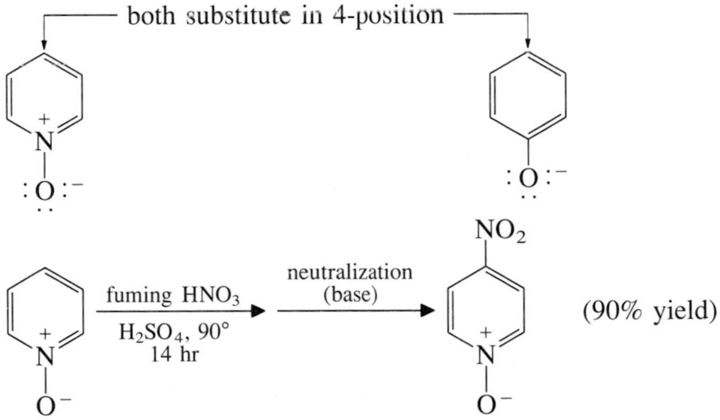

(24.52)

Once the *N*-oxide function is no longer needed, it can be removed by catalytic hydrogenation; this procedure also reduces the nitro group. Reaction with trivalent phosphorus compounds, such as PCl_3, removes the *N*-oxide function without reducing the nitro group.

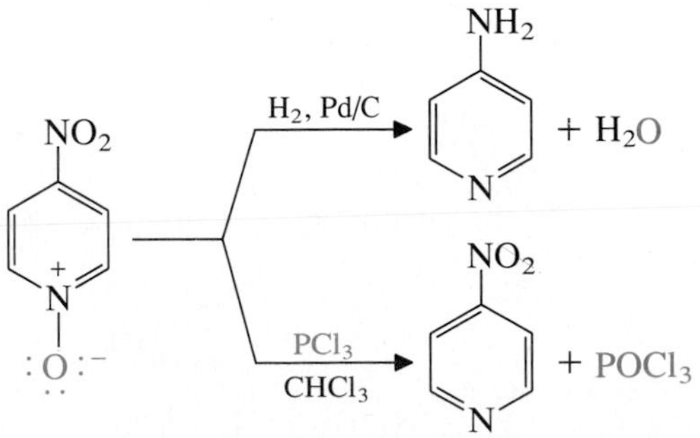

(24.53)

Similar reactions are possible with quinoline.

Problem

17 By drawing the resonance structures for the carbocation intermediate, show why aromatic substitution in pyridine-*N*-oxide occurs at the 4-position rather than at the 3-position.

B. Nucleophilic Aromatic Substitution

We have learned ways to prepare some 3-substituted and 4-substituted pyridines. Is there a way to synthesize 2-substituted pyridines? One reaction used for this purpose is a rather unusual one, called the **Chichibabin reaction.** In this reaction, treatment of a pyridine derivative with the strong base sodium amide ($Na^+ \ ^-NH_2$; Sec. 23.5E) brings about the direct substitution of an amino group for a ring hydrogen.

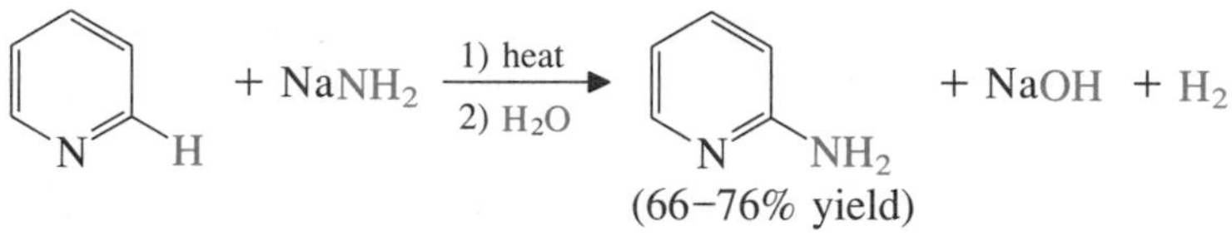

(24.54)

This reaction is a *nucleophilic substitution* in which the nucleophile is the strongly basic amide ion, $^-NH_2$.

In the first step of the mechanism, the amide ion attacks the 2-position of the ring to form a *tetrahedral addition intermediate.*

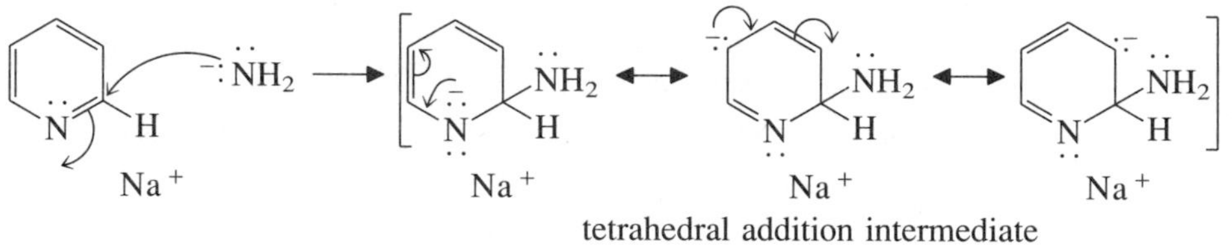

(24.55a)

It helps to understand this step of the mechanism if we recognize that the C═N linkage of the pyridine ring is somewhat analogous to a carbonyl group; that is, carbon at the 2-position has some of the character of a carbonyl carbon, and can be attacked by nucleophiles. The C═N group of pyridine is, of course, *much* less reactive than a carbonyl group because it is part of an aromatic system.

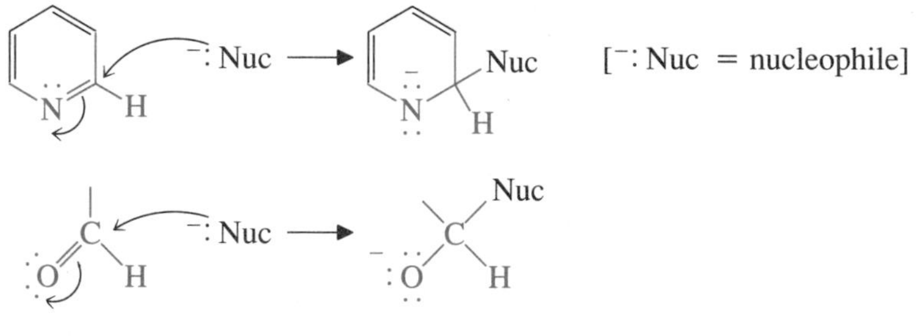

(24.55b)

In the second step of the mechanism, the leaving group, a *hydride ion,* is lost.

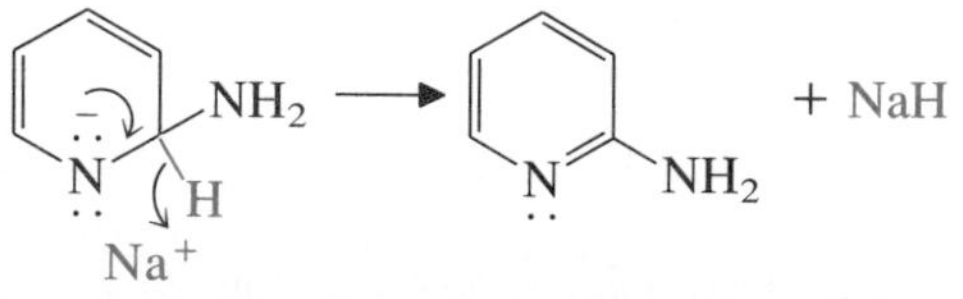

(24.55c)

Hydride ion is a very poor, and thus very unusual, leaving group because it is very basic. There are two reasons why this reaction is driven to completion. First, the aromatic pyridine ring is re-formed; aromaticity, lost in the formation of the tetrahedral addition intermediate, is regained when the leaving group departs. Second, the basic sodium hydride produced in the reaction reacts with the —NH_2 group irreversibly to form hydrogen gas and the resonance-stabilized conjugate-base anion of 2-aminopyridine—once again, LeChatelier's principle at work.

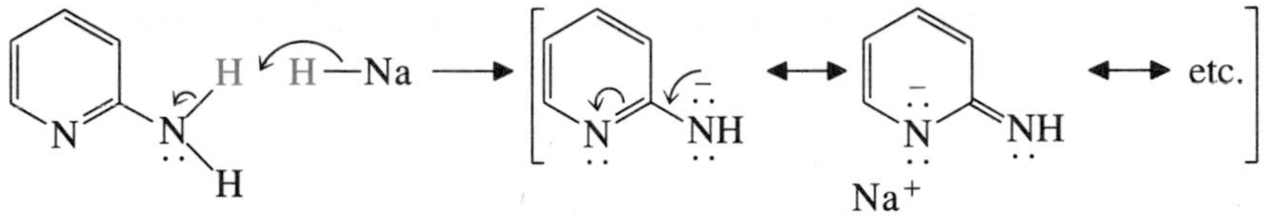

(24.55d)

The neutral 2-aminopyridine is formed when water is added in a separate step.

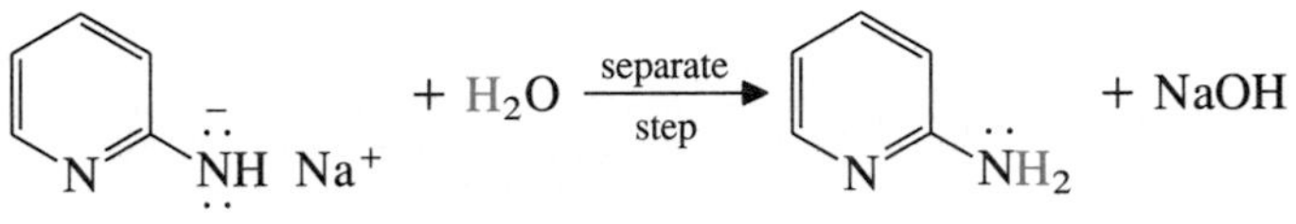

(24.55e)

A reaction similar to the Chichibabin reaction occurs with organolithium reagents.

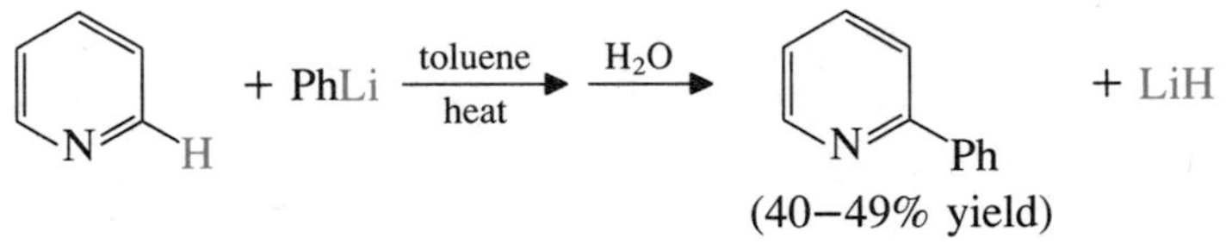

(24.56)

As we might imagine, when pyridine is substituted with a better leaving group than hydride at the 2-position, it reacts more rapidly with nucleophiles. The 2-halopyridines, for example, readily undergo substitution of the halogen by other nucleophiles under conditions milder than those used in the Chichibabin reaction.

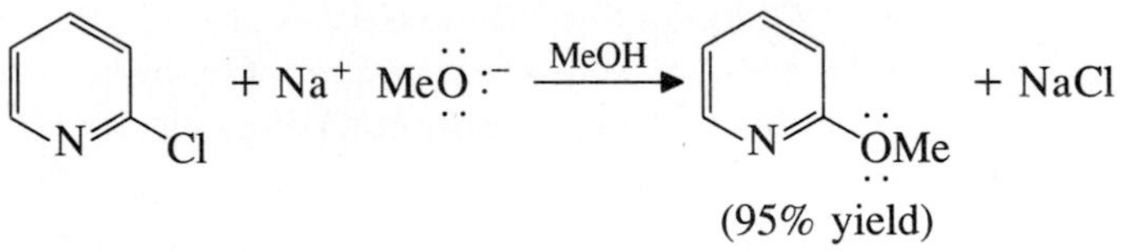

(24.57)

We can also relate this nucleophilic displacement to the analogous reaction of a carbonyl compound. This reaction of a 2-chloropyridine formally resembles the displacement reaction of an acid chloride—except that acid chlorides are *much* more reactive than 2-halopyridines.

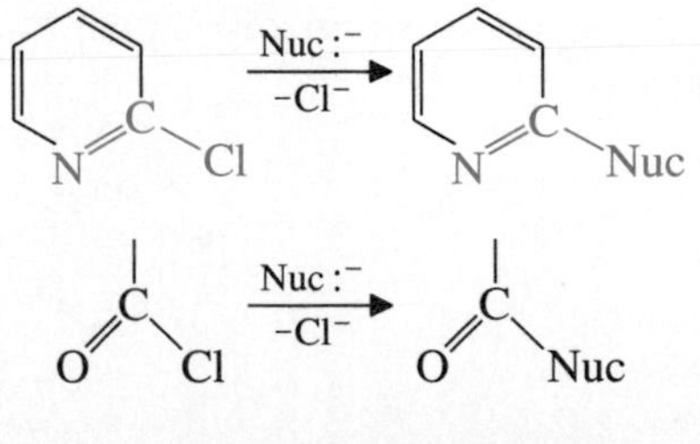

(24.58)

Be sure you understand that nucleophilic substitution reactions on pyridine rings are *not* S_N2 reactions; aryl halides cannot undergo S_N2 reactions (why? Sec. 18.1). We have already seen the analogy between these reactions and carbonyl substitution reactions. These reactions are also analogous to *nucleophilic aromatic substitution* reactions. Recall that aryl halides undergo nucleophilic aromatic substitution when the benzene ring is substituted with electron-withdrawing groups (Sec. 18.4A). The "electron-withdrawing group" in the reactions of pyridines is the pyridine nitrogen itself. The tetrahedral addition intermediate (Eq. 24.55a) is analogous to the Meisenheimer complex of nucleophilic aromatic substitution (Eq. 18.15). Thus we see a mechanistic parallel between three types of reactions: (1) carbonyl substitution, a typical reaction of carboxylic acid derivatives; (2) nucleophilic aromatic substitution; and (3) nucleophilic substitution on the pyridine ring.

The 2-aminopyridines formed in the Chichibabin reaction serve as starting materials for a variety of other 2-substituted pyridines. For example, diazotization of 2-aminopyridine gives a diazonium ion that can undergo substitution reactions (see Sec. 23.10B).

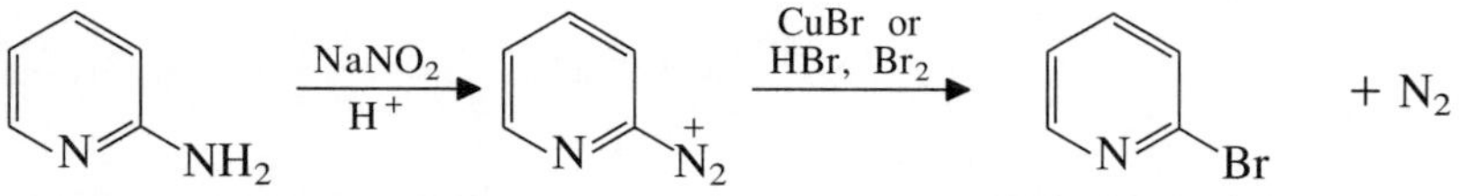

(86–92% yield)

(24.59)

When the diazonium salt reacts with water, it is hydrolyzed to 2-hydroxypyridine, which in most solvents exists in its carbonyl form, 2-pyridone, despite the aromaticity of the 2-hydroxy form. In water the ratio of 2-hydroxypyridine to 2-pyridone is about 1 : 340.

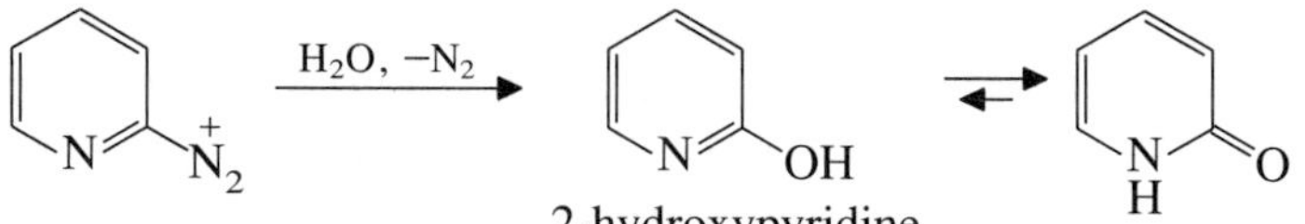

2-hydroxypyridine 2-pyridone

(24.60)

Although 2-pyridone exists largely in the carbonyl form, it nevertheless undergoes some reactions reminiscent of hydroxy compounds. For example, when 2-pyridone reacts with phosphorus pentachloride, the 2-hydroxy group is substituted by chloride.

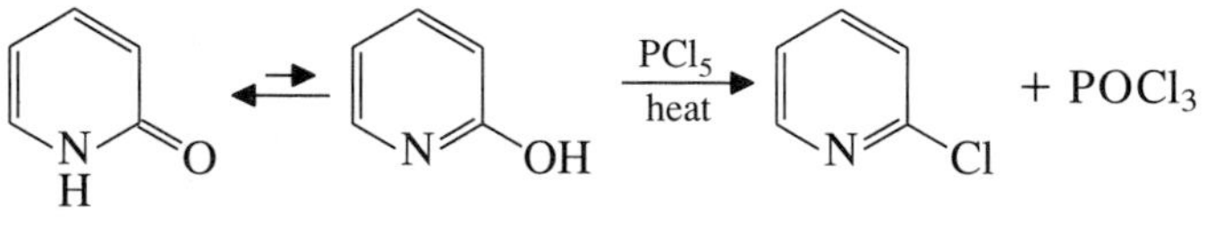

(24.61)

This reaction is formally similar to the preparation of acid chlorides from carboxylic acids.

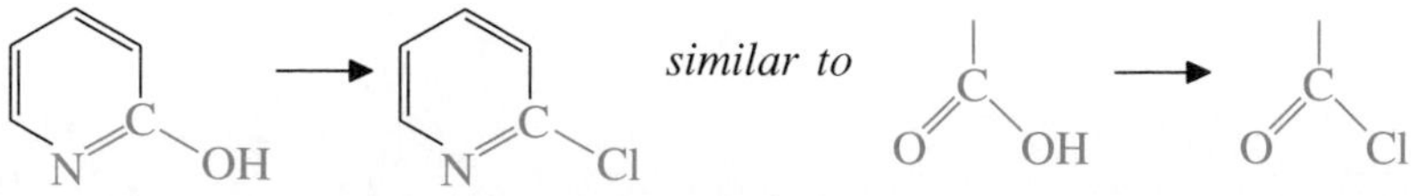

Here again we see the analogy between pyridine chemistry and carbonyl chemistry.

Pyridines with leaving groups in the 4-position also undergo nucleophilic substitution reactions.

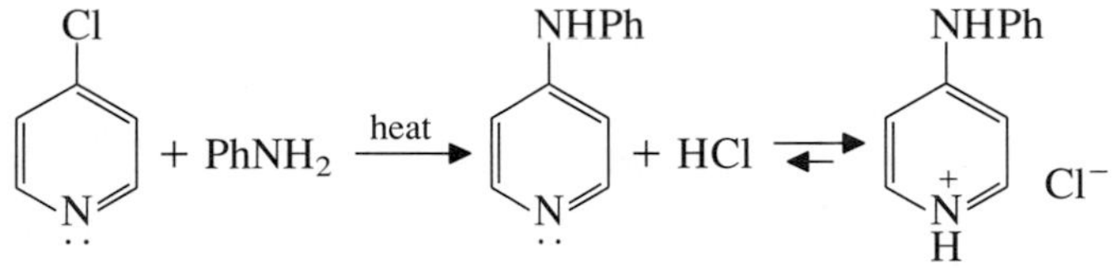

(24.62)

As we have seen, nucleophilic substitution at the 2- and 4-positions of a pyridine ring is particularly common. The reason is clear from the mechanism of this type of reaction: negative charge in the addition intermediate is delocalized onto the electronegative pyridine nitrogen.

substitution at carbon-2: (Y = leaving group)

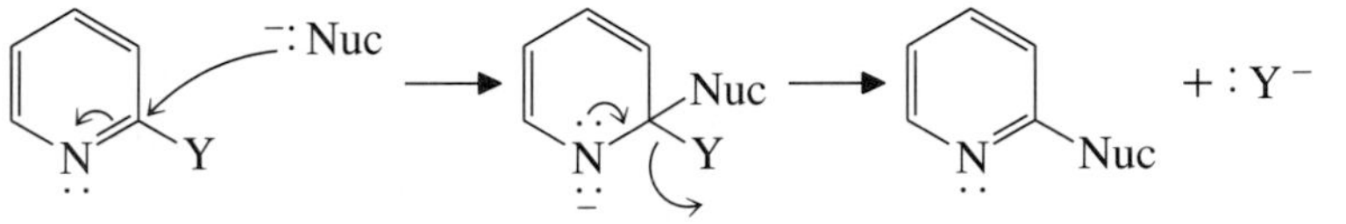

(24.63a)

substitution at carbon-4:

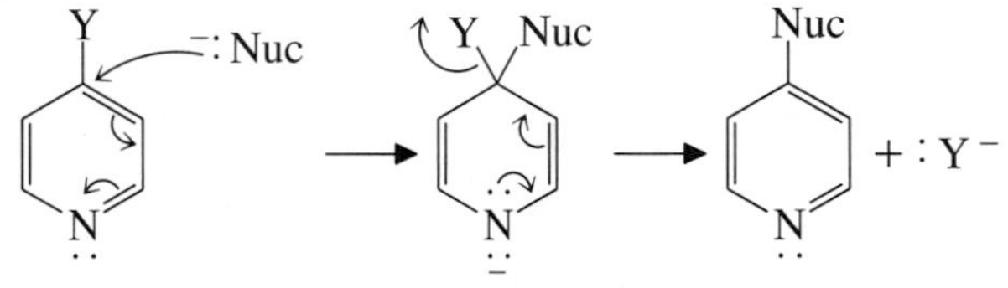

(24.63b)

What about substitution at carbon-3? 3-Substituted pyridines are *not* reactive in nucleophilic substitution because negative charge in the addition intermediate *cannot* be delocalized onto the electronegative nitrogen:

substitution at carbon-3:

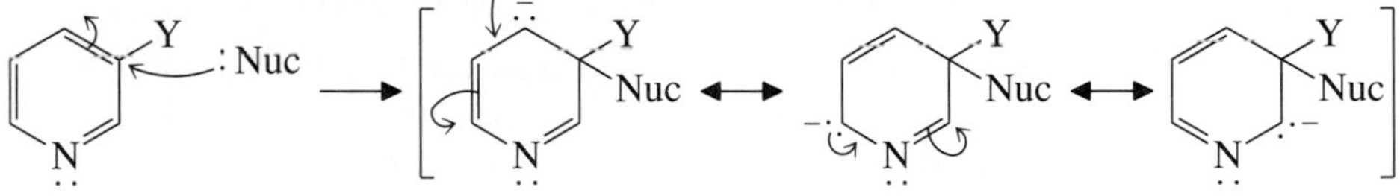

(24.63c)

Problem

18 Using the arrow formalism, give the mechanism for the reaction of 4-chloropyridine with sodium methoxide; draw all important resonance structures for the addition intermediate.

C. Pyridinium Salts and Their Reactions

Pyridine, like many Lewis bases, is a nucleophile. When pyridines react in S_N2 reactions with alkyl halides or sulfonate esters, quaternary ammonium salts, called *pyridinium salts,* are formed.

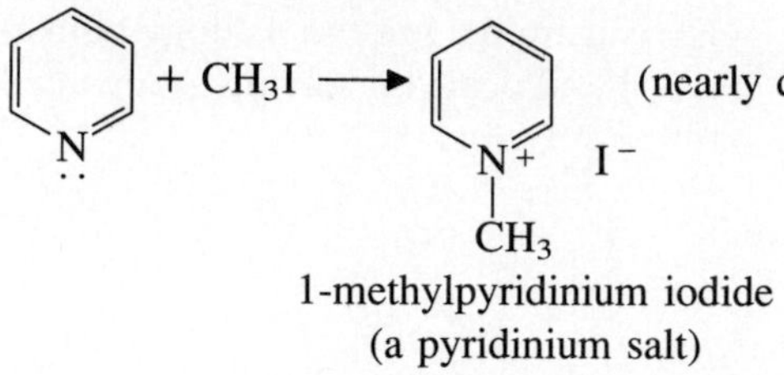

1-methylpyridinium iodide
(a pyridinium salt)

Pyridinium salts are activated toward nucleophilic displacement of groups at the 2- and 4-positions of the ring much more than pyridines themselves, because the positively charged nitrogen is much more electronegative than the neutral nitrogen of a pyridine. When the nucleophiles in such displacement reactions are anions, charge is neutralized. In the following reaction, for example, the pyridinium salt is attacked at the 2-position by hydroxide ion; the resulting hydroxy compound is then oxidized by potassium ferricyanide [$K_3Fe(CN)_6$] present in the reaction mixture.

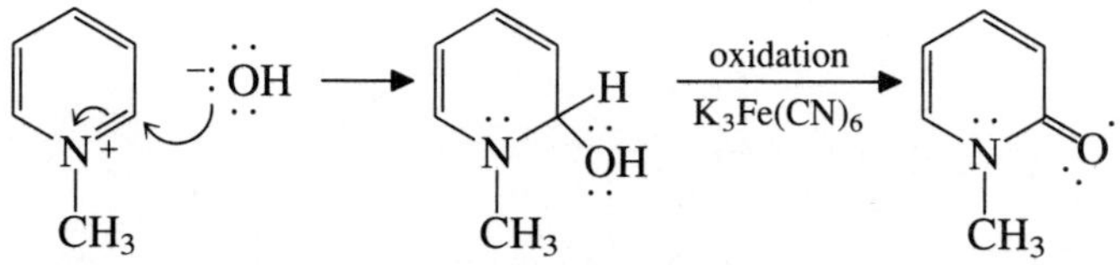

A biological example of nucleophilic addition to the 4-position of a pyridinium ring is found in biological oxidations with NAD^+ (Sec. 10.7).

Pyridine-*N*-oxides are formally pyridinium ions, and they react with nucleophiles in much the same way as quaternary pyridinium salts:

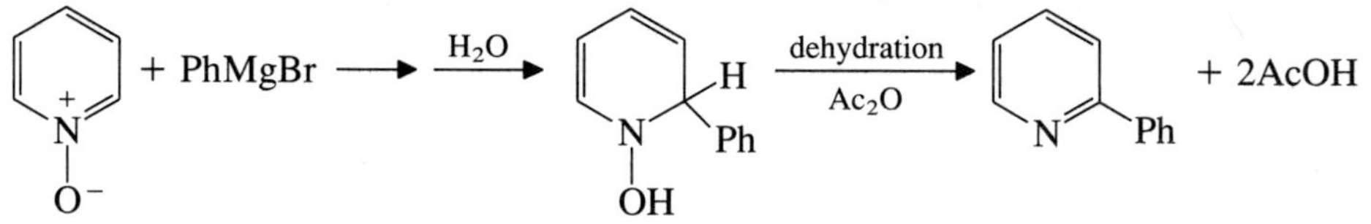

D. Side-Chain Reactions of Pyridine Derivatives

The "benzylic" protons of an alkyl group at the 2- or 4-position of a pyridine ring are especially acidic because the electron pair (and charge) in the conjugate-base anion is delocalized onto the electronegative pyridine nitrogen.

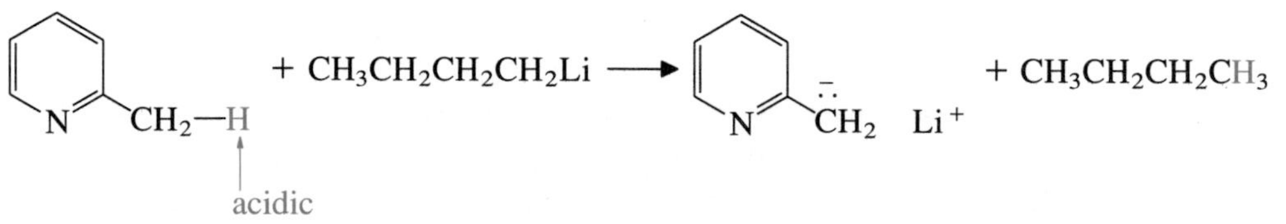

$$\text{(structure)} + CH_3CH_2CH_2CH_2Li \longrightarrow \text{(structure)} + CH_3CH_2CH_2CH_3 \quad (24.67)$$

(Write the resonance structures of this ion and verify that charge is delocalized onto the pyridine nitrogen.) As the example in Eq. 24.67 illustrates, strongly basic reagents such as organolithium reagents or $NaNH_2$ abstract the "benzylic" proton from 2- or 4-alkylpyridines. The anion formed in this way has a reactivity much like that of other organolithium reagents. In Eq. 24.68, for example, it adds to the carbonyl group of an aldehyde to give an alcohol (Sec. 19.9).

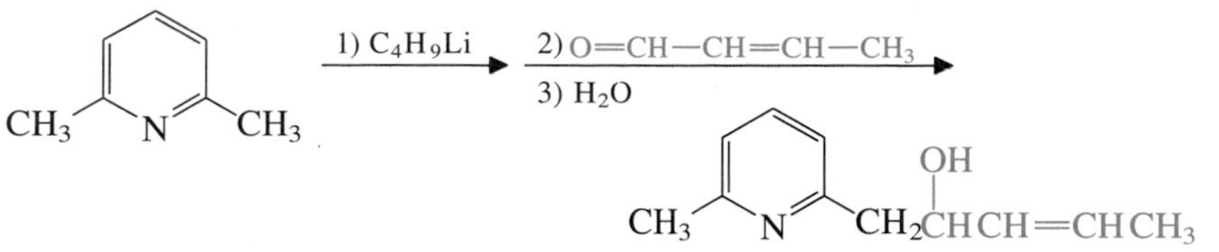

(24.68)

In this example we see again the analogy between pyridine chemistry and carbonyl chemistry. If the C=N linkage of a pyridine ring is analogous to a carbonyl group, then the benzylic anion is analogous to an enolate anion.

We should not be surprised, then, to find that these anions undergo some of the reactions of enolate anions, such as the aldol condensation in Eq. 24.68.

The "benzylic" protons of 2- or 4-alkylpyridinium salts are much more acidic than those of the analogous pyridines because the conjugate base "anion" is actually a neutral compound, as the following resonance structures demonstrate:

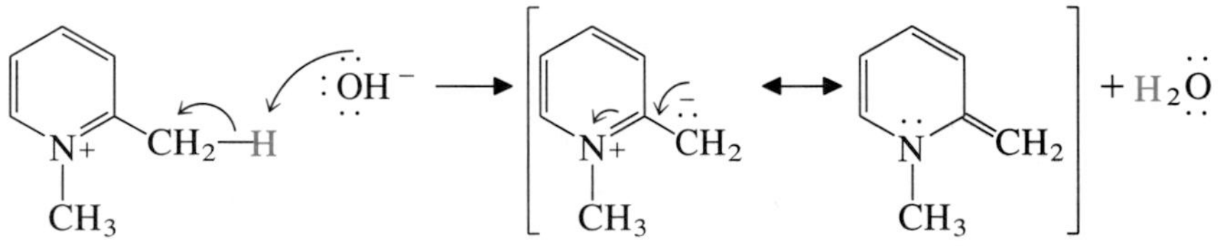

(24.69)

The "benzylic" protons of 2- or 4-alkylpyridinium salts are acidic enough that the conjugate-base "anions" can be formed in useful concentrations by aqueous NaOH or amines. In the following reaction, which exploits this acidity, the conjugate base of a pyridinium salt is used as the enolate component in a variation of the Claisen–Schmidt condensation (Sec. 22.4C).

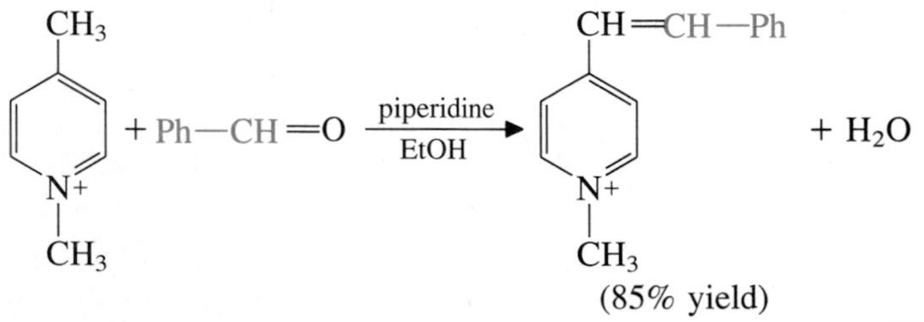

(24.70)

Many side-chain reactions of pyridines are analogous to those of the corresponding benzene derivatives. For example, side-chain oxidation (Sec. 17.5D) is a useful reaction of both alkylbenzenes and alkylpyridines. Nicotinic acid (pyridine-3-carboxylic acid) is readily obtained by side-chain oxidation of either 3-picoline (3-methylpyridine), a component of coal tar, or nicotine, an alkaloid that is abundant in tobacco.

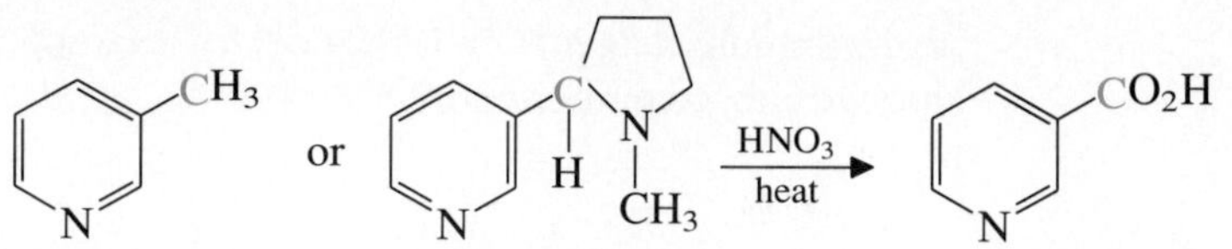

3-picoline nicotine nicotinic acid (50–60% yield) (24.71)

Problems

19 Give the principal organic product in the reaction of quinoline with each of the following reagents. (*Hint:* Consider the similar reactions of pyridine.)
(a) 30% H_2O_2
(b) $NaNH_2$, heat; then H_2O
(c) product of (a), then HNO_3, H_2SO_4

20 Outline a synthesis for each of the following compounds from the indicated starting material and any other reagents. (A *picoline* is a methylpyridine.)
(a) 4-methyl-3-nitropyridine from 4-picoline
(b) 3-methyl-4-nitropyridine from 3-picoline
(c) 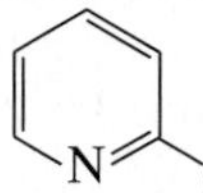 CH_2—CO_2H from 2-picoline (*Hint:* See Sec.20.6.)
(d) 3-aminopyridine from 3-picoline

21 Predict the predominant product in each of the following reactions. Explain your answer.
(a) 3,4-dimethylpyridine + butyllithium (1 equiv), then CH_3I ⟶ ($C_8H_{11}N$)
(b) 3,4-dibromopyridine + NH_3, heat ⟶ ($C_5H_5BrN_2$)

E. Synthesis of Quinolines

A number of reasonably versatile syntheses of quinolines from acyclic compounds are known. This is fortunate, since many direct substitution reactions of the quinoline nucleus give mixtures (Eq. 24.50). One of the best known syntheses of quinolines is the **Skraup synthesis,** in which glycerol reacts with an aniline derivative under acid catalysis.

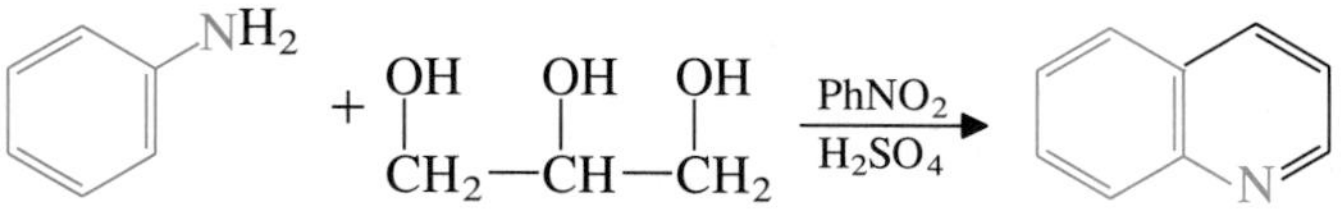

glycerol (84–91% yield) (24.72)

In this reaction, glycerol undergoes an acid-catalyzed dehydration to provide a small but continuously replenished amount of acrolein, an α,β-unsaturated aldehyde. (If acrolein itself were used as a reactant at high concentration, it would polymerize.)

$$\underset{\text{OH}}{CH_2}—\underset{\text{OH}}{CH}—\underset{\text{OH}}{CH_2} \xrightarrow[-H_2O]{H^+} O{=}CH—CH_2—CH_2OH \xrightarrow[-H_2O]{H^+} \underset{\text{acrolein}}{O{=}CH—CH{=}CH_2} \quad (24.73a)$$

Aniline undergoes a Michael-type conjugate addition (Sec. 22.8A) with the acrolein and, under the influence of acid, the resulting aldehyde protonates. The protonated aldehyde has carbocation character, and its electron-deficient carbon serves as the electrophile in an intramolecular electrophilic aromatic substitution reaction. Dehydration of the resulting alcohol yields a 1,2-dihydroquinoline. (You should fill in the details of the mechanism outlined in Eq. 24.73a and 24.73b.)

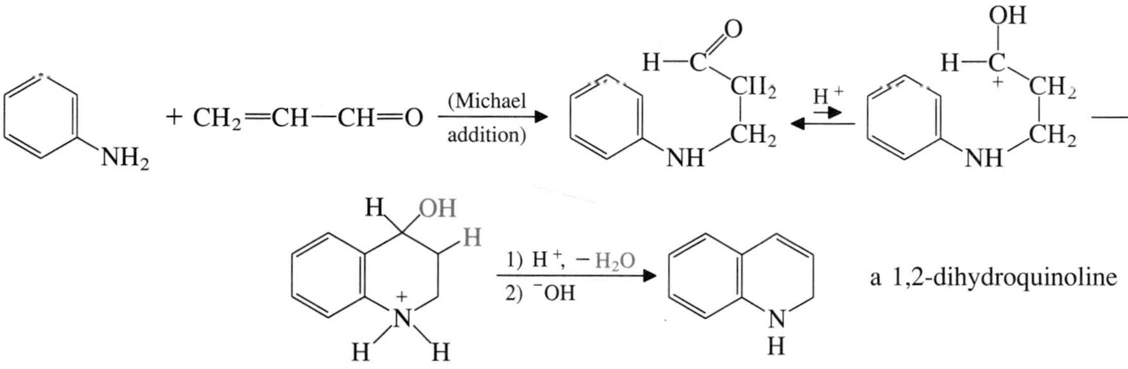

a 1,2-dihydroquinoline (24.73b)

The 1,2-dihydroquinoline product differs from quinoline by only one degree of unsaturation, and is readily oxidized to the aromatic quinoline by mild oxidants. Nitrobenzene, As_2O_5, or Fe^{3+} are commonly used oxidants in the Skraup synthesis; these are included in the reaction mixture.

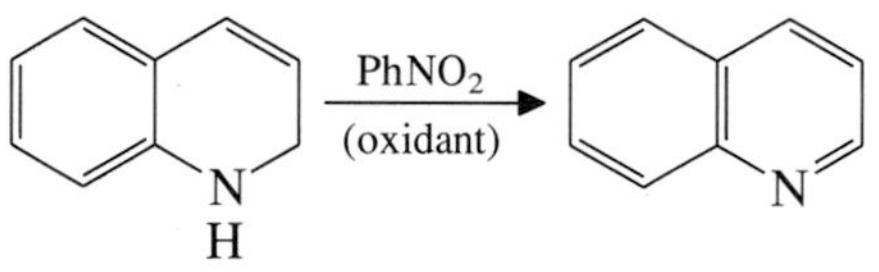

(24.73c)

α,β-Unsaturated aldehydes and ketones that are less prone to polymerize than acrolein can be used instead of glycerol in the Skraup synthesis to give substituted quinolines.

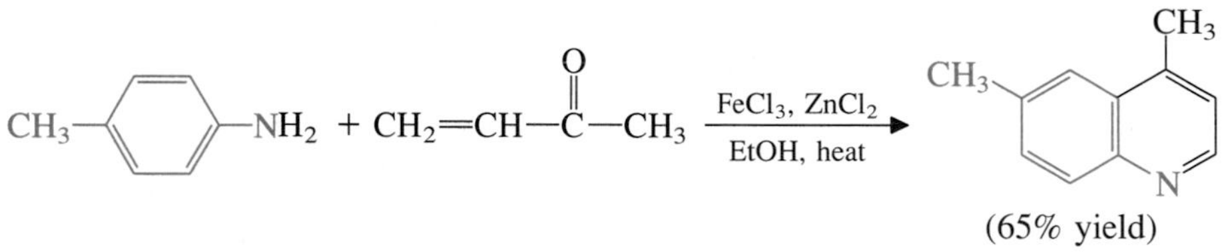

(24.74)

Problems

22 What product is expected when *p*-methoxyaniline reacts with each of the following compounds under the conditions of the Skraup synthesis?
(a) glycerol (b) 1-phenyl-2-buten-1-one

23 What reactants are required for a Skraup synthesis of 6-chloro-3,4-dimethylquinoline?

24.6 OCCURRENCE OF HETEROCYCLIC COMPOUNDS

Nitrogen heterocycles occur widely in nature. We have already discussed the *alkaloids* (Fig. 23.4), many of which contain heterocyclic ring systems. The naturally occurring amino acids proline, histidine, and tryptophan, about which we shall learn more in Chapter 26, contain respectively a pyrrolidine, imidazole, and indole ring (Fig. 24.6). A number of vitamins are heterocyclic compounds; without these compounds, many important metabolic processes could not take place. For example, vitamin B_6 is a substituted pyridine; NAD^+ (Sec. 10.7) is a pyridinium ion; and vitamin B_1 (thiamin) contains a pyrimidine ring and an *N*-substituted thiazolium salt. As we shall learn at the end of Chapter 27, the nucleic acids, which carry and transmit genetic information in the cell, contain purine and pyrimidine rings (Fig. 24.1) in combined form.

Heterocyclic compounds are involved in some of the colors in nature that have intrigued mankind from the earliest times. Why is blood red? Why is grass green? The color of blood is due to an iron complex of heme, a heterocycle composed of pyrrole units. This type of heterocycle is called a **porphyrin.**

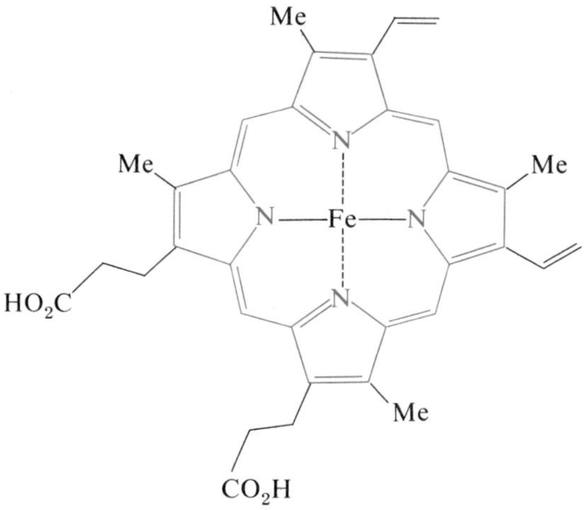

heme
(occurs in the protein complex hemoglobin)

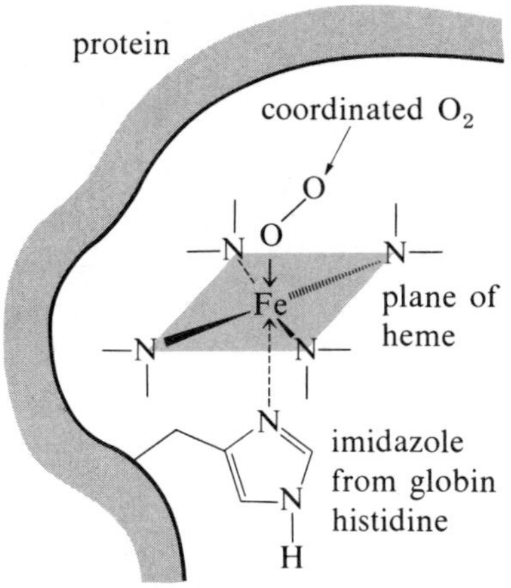

schematic view of oxygenated heme in hemoglobin

Heme is an aromatic compound (why?), and is found in red blood cells as a tight complex with a protein called *globin;* the complex is called **hemoglobin.** The iron, held in position by coordination with the nitrogens of heme and an imidazole of globin, complexes reversibly with oxygen. Thus, hemoglobin is the oxygen carrier of blood, and the red color of blood is due to oxygenated hemoglobin. Carbon monoxide and cyanide, two well-known respiratory poisons, also complex with the iron in hemoglobin as well as with iron in the heme groups of other respiratory proteins.

The green color of plants is caused by a compound closely related to the porphyrins, called **chlorophyll.**

naturally occurring amino acids:

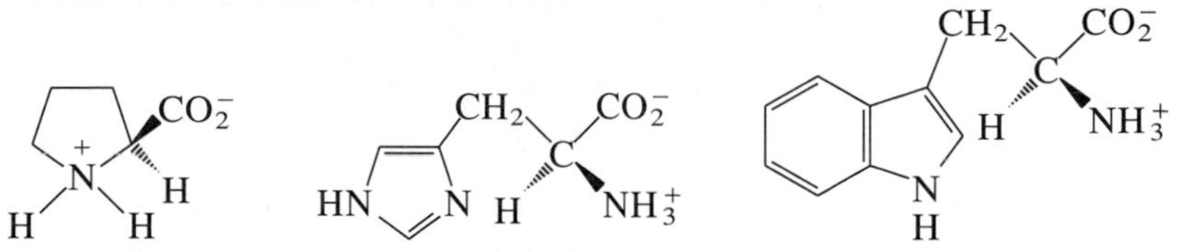

proline histidine tryptophan

vitamins:

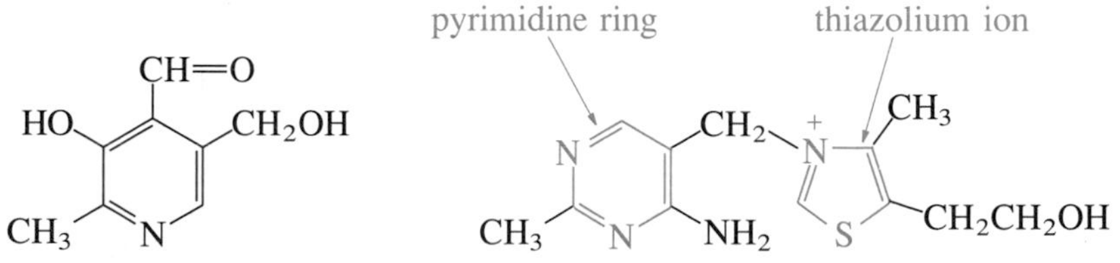

pyridoxal (a form of vitamin B_6)

thiamin (vitamin B_1)

chlorophyll *a*

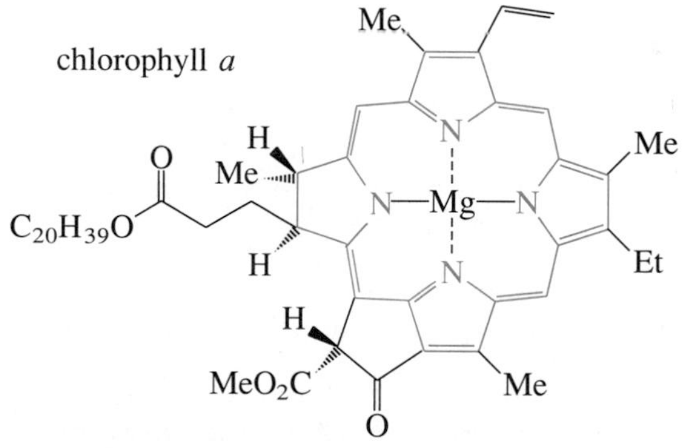

Figure 24.6 Some naturally occurring heterocyclic compounds. (See also Fig. 23.4 for some representative alkaloid structures, and Figures 10.1 and 22.1 for the structures of NAD^+ and coenzyme A.)

The absorption of sunlight by chlorophylls is the first step in the conversion of sunlight into useable energy by plants. Thus the chlorophylls are nature's "solar energy collectors."

KEY IDEAS IN CHAPTER 24

- Electrophilic aromatic substitution of naphthalene generally occurs at the 1-position. Sulfonation occurs most rapidly at the 1-position, but the most stable sulfonation product, observed at higher temperature, comes from sulfonation at the 2-position.

- Naphthalene and the fused-ring heterocycles benzofuran, benzothiophene, indole, and quinoline react in electrophilic substitution at the more activated ring. That is, the two rings act as if they are more or less independent. If a naphthalene derivative contains electron-donating, activating substituents, further substitution occurs on the substituted ring. If the substituted ring bears deactivating substituents, further substitution occurs on the unsubstituted ring. Benzofuran, benzothiophene, and indole undergo substitution on the heterocyclic ring, but quinoline substitutes on its benzene ring.
- Pyridine reacts very slowly in electrophilic aromatic substitution, and substitution occurs at the 3-position. Electrophilic substitution of pyridine-*N*-oxides, however, occurs at the 4-position.
- The aromatic heterocycles containing nitrogen atoms as part of a double bond—for example, imidazole, pyridine, and quinoline—are basic toward protons. Those with a nitrogen in which the unshared electron pair is part of the π-electron system—for example, pyrrole and indole—are not basic, because protonation of the nitrogen would disrupt the aromatic π-electron system.
- Many side-chain reactions proceed normally without disrupting heterocyclic rings.
- Pyridine derivatives undergo nucleophilic substitution reactions at the 2- and 4-positions. Thus pyridines react in the Chichibabin and related reactions; 2- and 4-chloropyridines undergo nucleophilic aromatic substitution reactions. Pyridinium salts are even more reactive than pyridines in these reactions. The chemistry of the pyridine C=N linkage has some similarity to that of the carbonyl group.
- The "benzylic" hydrogens of pyridines and especially pyridinium salts are acidic enough to be removed by bases. The resulting anions can act as nucleophiles in substitution and condensation reactions.
- Among the syntheses of indoles is the reaction of arylhydrazines with aldehydes or ketones that have an α- —CH_2— group (the Fischer indole synthesis), and the reaction of *o*-nitrotoluene and its derivatives in base with diethyl oxalate (the Reissert synthesis).
- The reactions of α,β-unsaturated aldehydes or ketones with aniline derivatives give 1,2-dihydroquinolines, which are oxidized to the corresponding quinolines by mild oxidants present in the reaction mixture (Skraup synthesis).

ADDITIONAL PROBLEMS

24 Give the principal organic product(s) expected when 1-methylnaphthalene reacts with each of the following reagents:

(a) concentrated HNO_3
(b) H_2SO_4, 40°
(c) *N*-bromosuccinimide, CCl_4, light
(d) Br_2, CCl_4

25 Give the principal organic product(s) expected when 2-methylthiophene reacts with each of the following reagents:
(a) acetic anhydride, BF_3, H^+
(b) *N*-bromosuccinimide, CCl_4, light
(c) HNO_3 in acetic anhydride
(d) dilute aqueous HCl
(e) dilute aqueous NaOH
(f) product of (b) + Mg/ether, then CO_2, then H_3O^+
(g) product of (a) + Ph—CH=O and NaOH

26 Give the principal organic product(s) expected when 2-methylpyridine reacts with each of the following reagents:
(a) dilute aqueous HCl
(b) dilute aqueous NaOH
(c) HNO_3, H_2SO_4, heat; then ^-OH
(d) $CH_3CH_2CH_2CH_2$—Li
(e) $NaNH_2$, heat, then H_2O
(f) CH_3I
(g) 30% H_2O_2
(h) product of (c) + H_2, catalyst
(i) product of (d) + PhCH=O, then H_3O^+
(j) product of (e) + $NaNO_2$/HCl, then H_2O

27 Rank each of the following compounds in order of increasing reactivity toward HNO_3/H_2SO_4, and explain your choices:
(a) naphthalene (b) pyridine (c) quinoline

28 Bromination of 1,6-dimethylnaphthalene gives a mixture of three isomeric monobromo derivatives. Two of these are formed in major amount, and one in very small amount. Give the structures and names of the three isomers, and indicate which are the major products. Explain your reasoning.

29 Rank the compounds within each of the following sets in order of increasing basicity, and explain your reasoning:
(a) pyridine, 4-methoxypyridine, 5-methoxyindole, 3-methoxypyridine
(b) pyridine, 3-nitropyridine, 3-chloropyridine
(c)
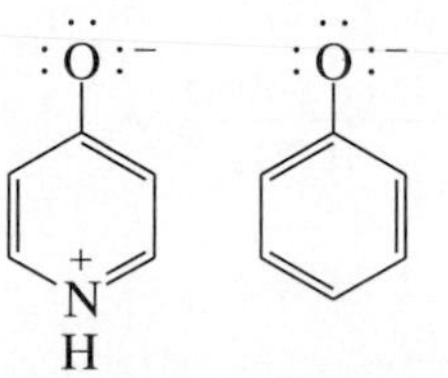
(d) imidazole and oxazole

30 Explain each of the following observations.

(a) Naphthalene undergoes electrophilic aromatic substitution at the 1-position.

(b) Quinoline undergoes electrophilic aromatic substitution at the 5- and 8-positions.

(c) Pyridine undergoes electrophilic aromatic substitution at the 3-position.

(d) 2-Bromopyridine and 4-bromopyridine undergo nucleophilic aromatic substitution, but 3-bromopyridine does not.

(e) Pyridine is protonated on nitrogen in dilute acid, but pyrrole is protonated on carbon, and only in concentrated acid.

31 Complete the following reactions by giving the major organic product(s):

(a)

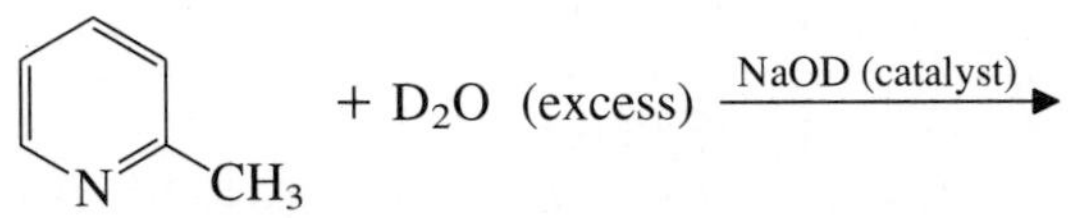

 + D$_2$O (excess) $\xrightarrow{\text{NaOD (catalyst)}}$

(b)

 + HNO$_3$ $\longrightarrow$

(c) SO$_3$H

 $\xrightarrow[40°]{\text{fuming H}_2\text{SO}_4}$

(d) SO$_3$H

 $\xrightarrow[180°]{\text{fuming H}_2\text{SO}_4}$ (two products)

(e)

 + Br$_2$ $\xrightarrow[\text{dark}]{\text{CCl}_4}$

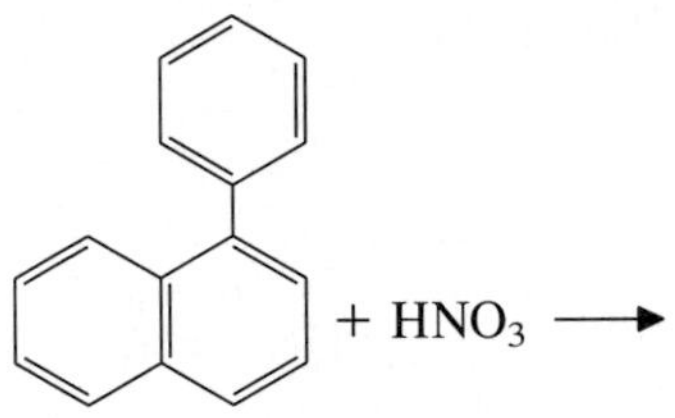

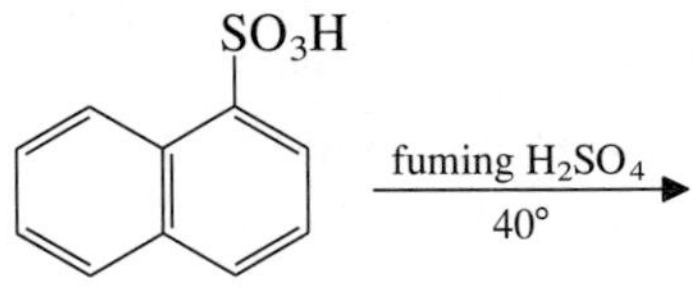

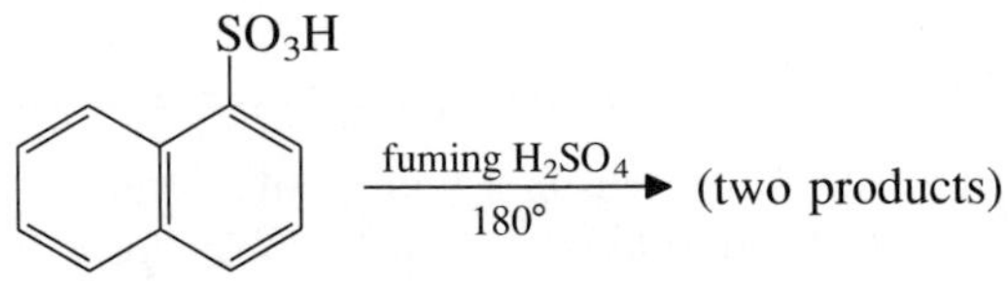

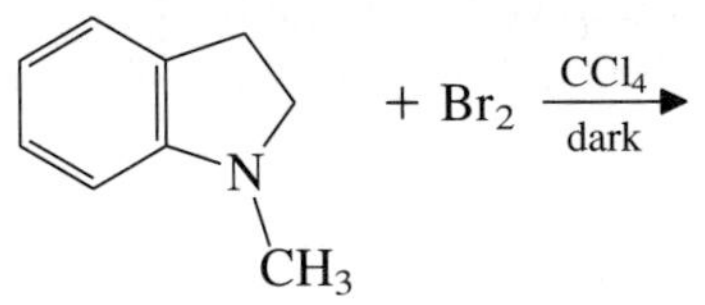

(*Note:* the starting material has no double bond between positions 2 and 3; that is, it is not an indole.)

(f)

 $\xrightarrow[\text{H}_2\text{SO}_4]{\text{fuming HNO}_3}$

(g)

 + PhLi $\longrightarrow$

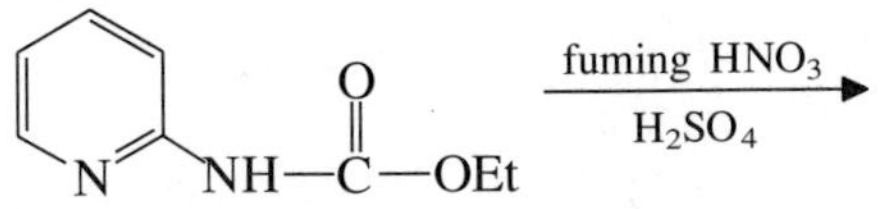

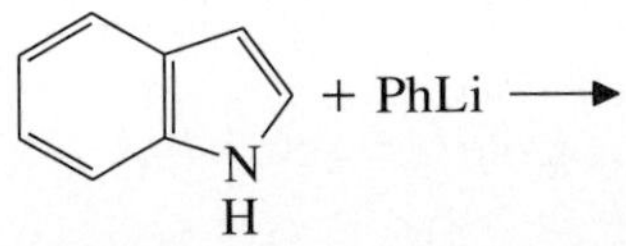

(h)

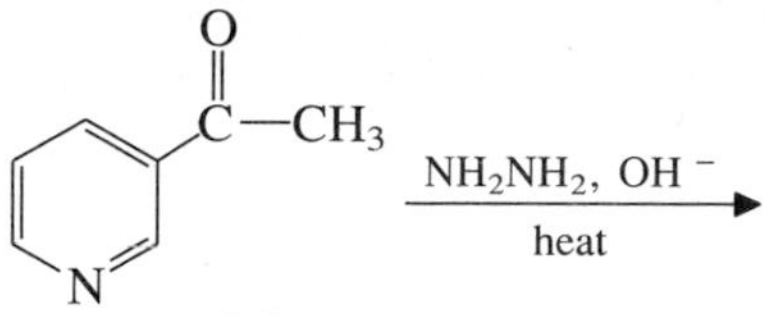

(i)

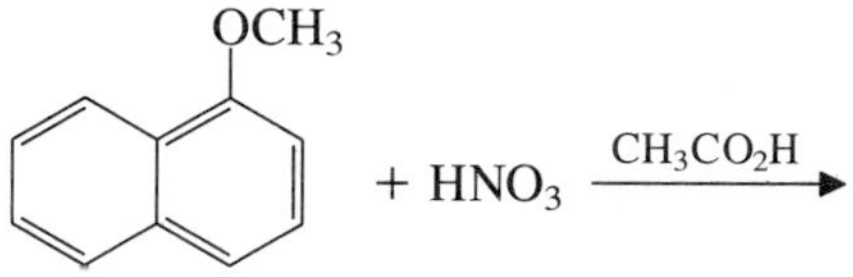

(j)

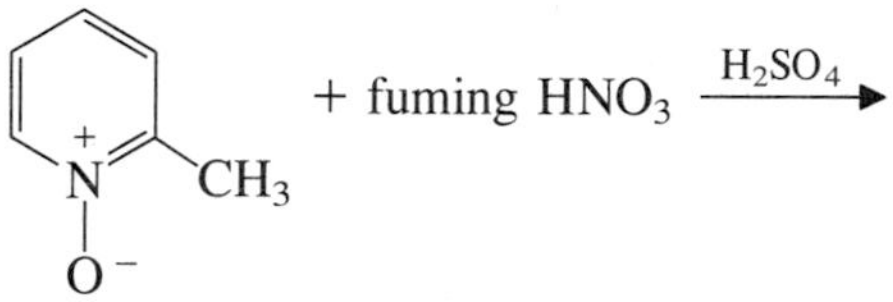

(k)

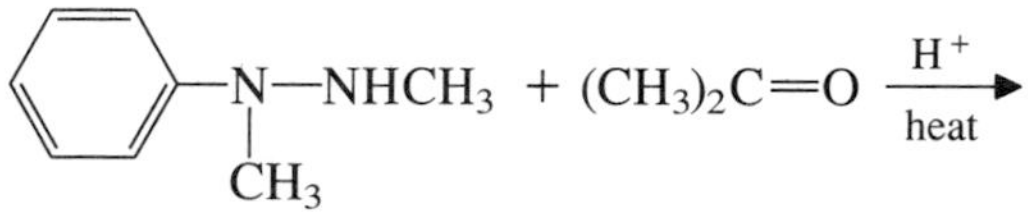

(l)

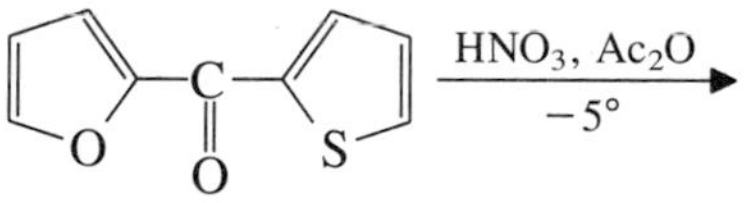

(nitration occurs in a 5-position—but in which ring?)

(m)

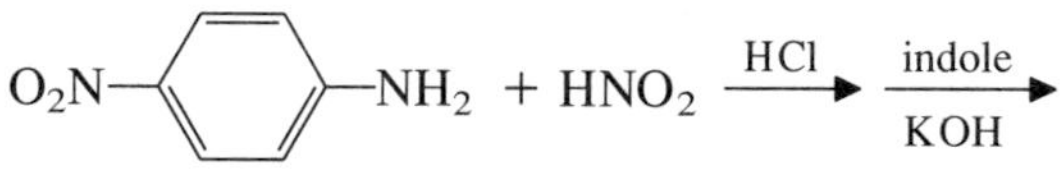

(an azo dye)

(n)

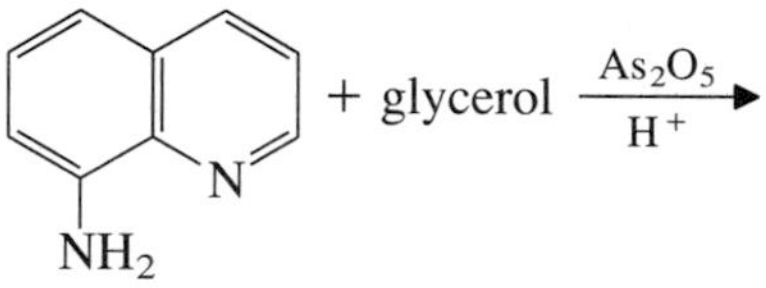

32 Outline a synthesis of each of the following compounds from naphthalene and any other reagents:
(a) 1-bromonaphthalene
(b) 1-naphthalenecarboxylic acid
(c) 1-chloro-4-nitronaphthalene
(d) 2-(1-naphthyl)-1-ethanol
(e) 1-bromo-4-chloronaphthalene
(f) 1-naphthyl acetate
(g) 1,4-naphthalenediamine
(h) 5-amino-1-naphthalenesulfonic acid

33 Doreen Dimwhistle has paused from her usual pursuit of testing the effect of H_2SO_4 on her clothing long enough to propose the following variations on the Chichibabin reaction:
(a) indole + $NaNH_2$ $\longrightarrow$ 2-aminoindole
(b) 2-chloropyridine + $NaNH_2$ $\longrightarrow$ 2-amino-6-chloropyridine

She is shocked to find that neither of these reactions works as planned and has come to you for an explanation. Explain what reaction, if any, occurs instead in each case.

Problems (Cont.)

34 Draw the structure of the major form of each of the following compounds present in an aqueous solution containing initially one molar equivalent of 1*N* HCl:

(a)

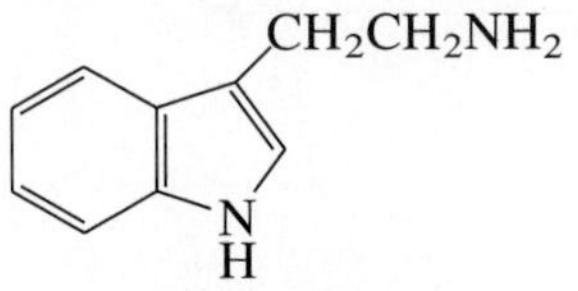

tryptamine

(b) quinine (Fig. 23.4)

(c) nicotine (Eq. 24.71 or Fig. 23.4)

(d)

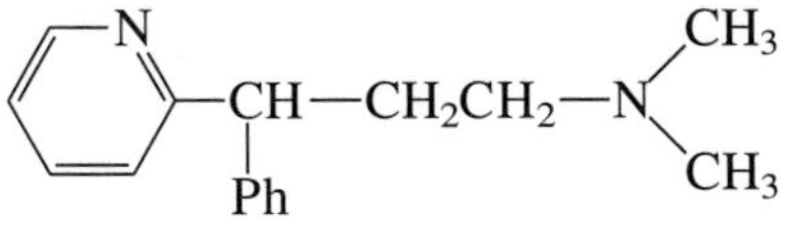

pheniramine (an antihistaminic drug)

(e) 3,4-diaminopyridine

(f)

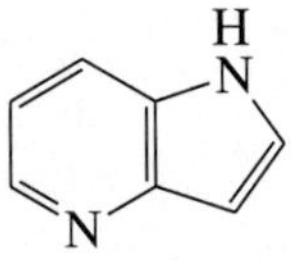

1,4-diazaindene

---

35 When 3-methyl-2-butanone reacts with phenylhydrazine and a Lewis acid, the following compound, an example of an *indolenine,* is formed. Give a mechanism for the formation of this compound.

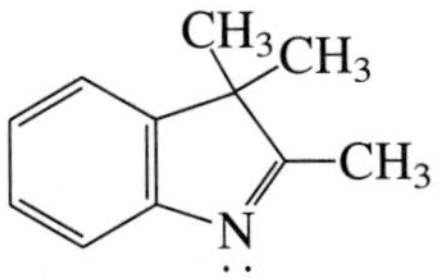

---

36 Outline a synthesis for each of the following compounds from the indicated starting material and any other reagents:

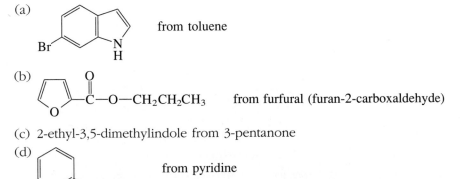

(a)

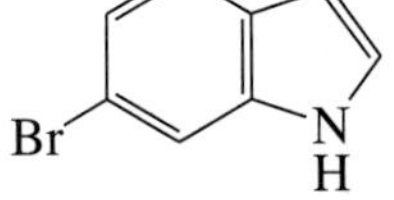

from toluene

(b)

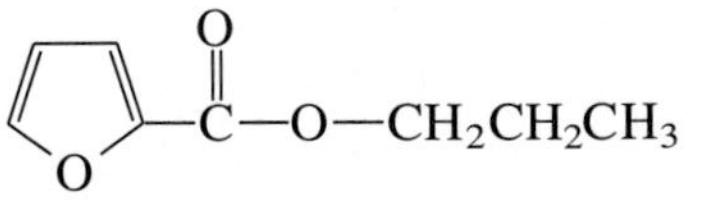

from furfural (furan-2-carboxaldehyde)

(c) 2-ethyl-3,5-dimethylindole from 3-pentanone

(d)

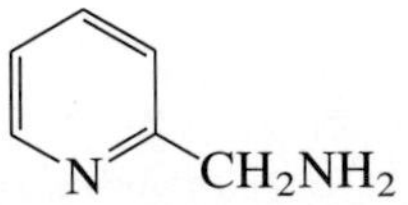

from pyridine

(e) 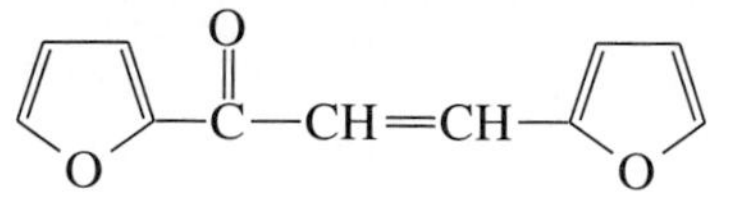 from furfural (furan-2-carbaldehyde) as the only source of furan rings

(f) 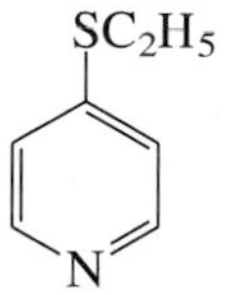 from pyridine

(g) 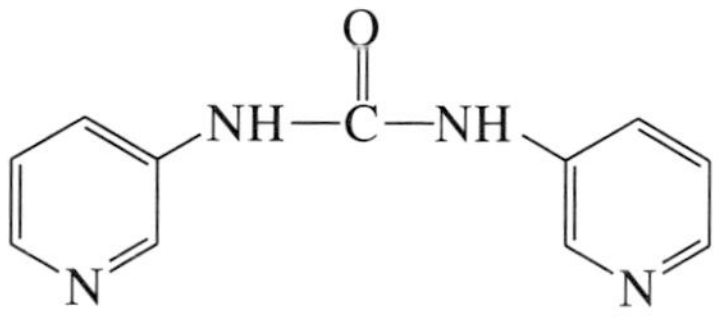 from 3-methylpyridine

(h) 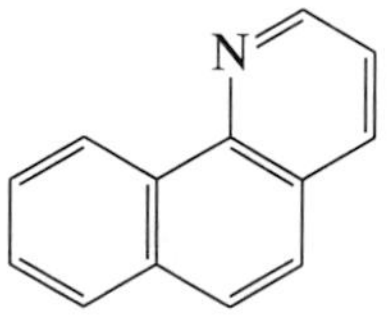 from naphthalene

(i) 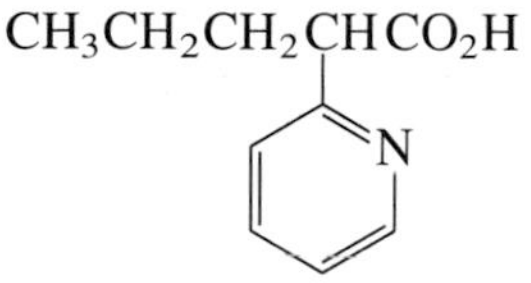 from 2-methylpyridine

(j) 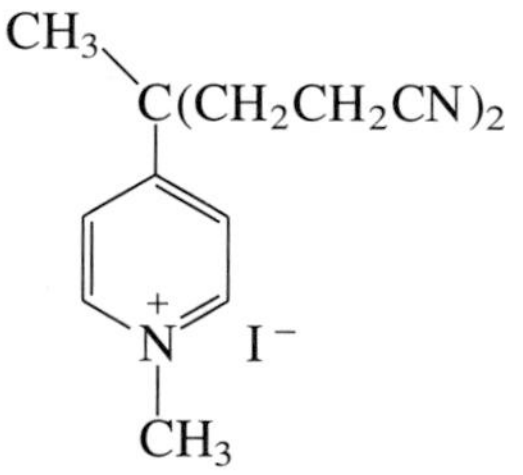 from 4-ethylpyridine

37 Rank the following compounds in order of increasing reactivity in an S_N1 solvolysis reaction in ethanol, and explain your choices. Then propose a synthesis for compound (d) from thiophene and any other reagents.

(a) 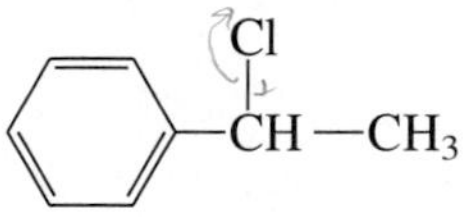

CH3CH2OH SN1

p.688

(b) 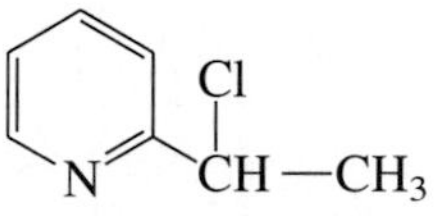

(c) 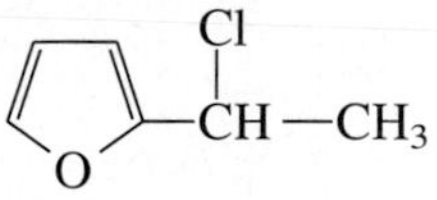

(d) 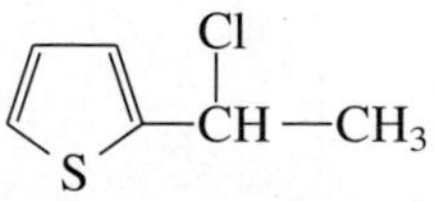

Problems (Cont.)

38 Outline rational mechanisms for each of the following reactions. Give the structure for the bracketed intermediate in part (e).

(a)

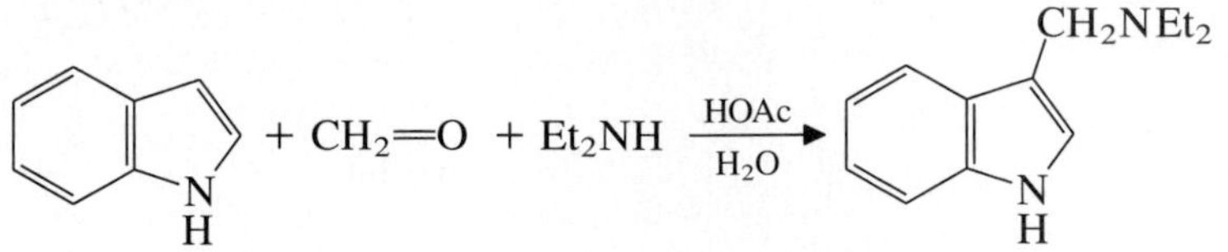

(b)

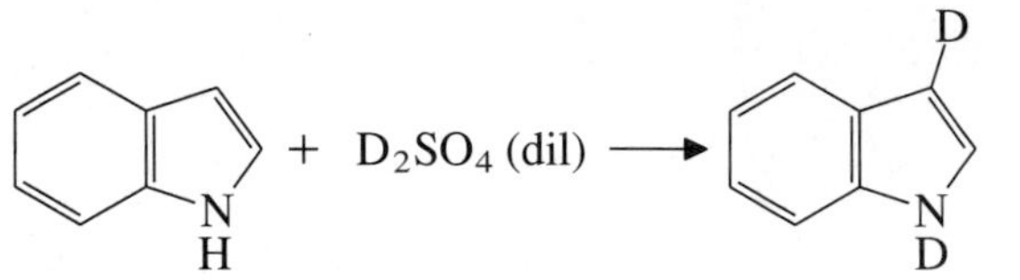

(c)

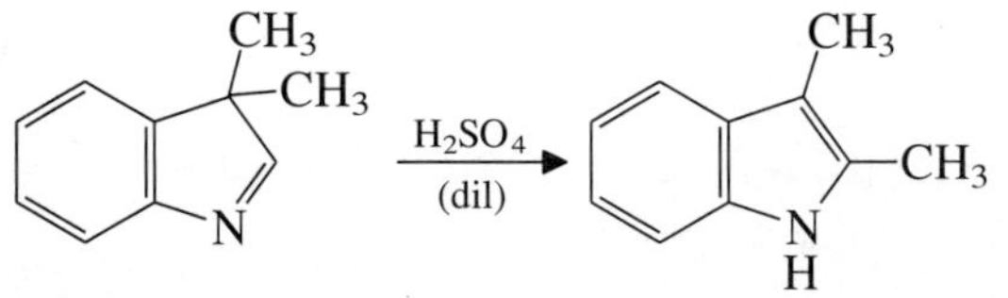

(d)

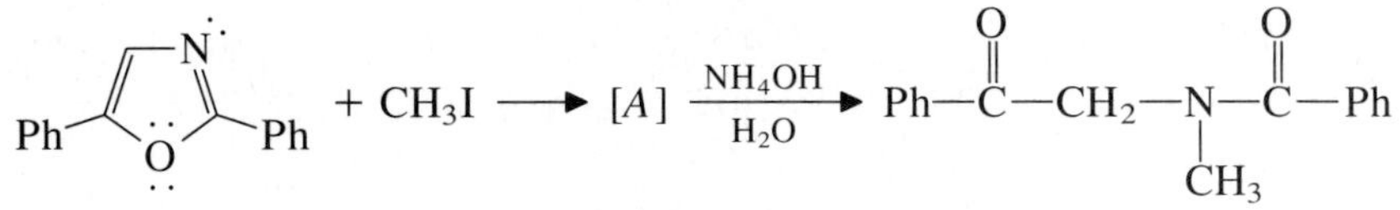

(e)

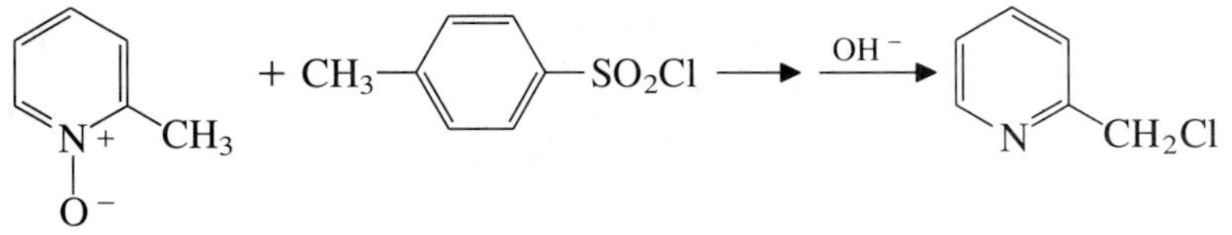

(f)

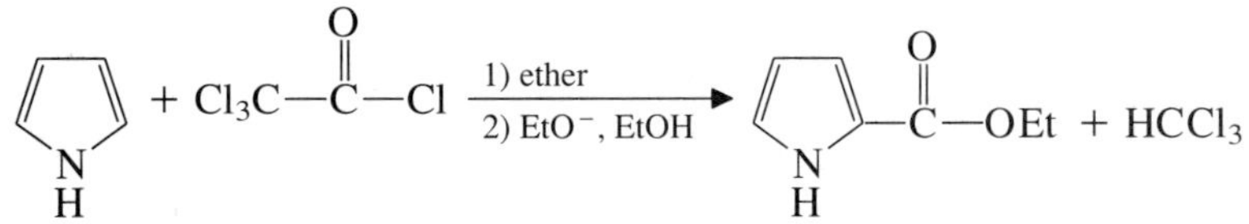

39 Identify *A, B,* and *C* in the following scheme. Explain why *C* cannot be synthesized in one step from thiophene.

$$\text{thiophene} \xrightarrow[-15°]{\text{ClSO}_3\text{H}} [A] \xrightarrow[40°]{\text{conc HNO}_3} [B] \xrightarrow[\text{heat}]{\text{H}_2\text{O}} [C] \ (\text{C}_4\text{H}_3\text{NSO}_2)$$

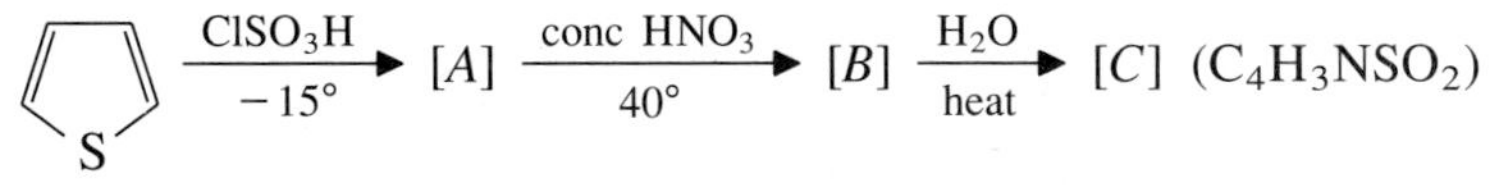

40 The following compound is a very strong base; its conjugate acid has a pK_a of about 13.5. Give the structure of its conjugate acid and suggest a reason why the compound is such a strong base.

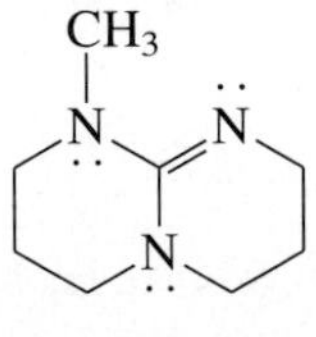

41 In an abandoned laboratory you have found a bottle containing a compound *A*, $C_8H_{11}NO$, that smells as if it might have been isolated from an extract of dirty socks. This compound can be resolved into enantiomers and it dissolves in 5% aqueous HCl. No gas is evolved when *A* reacts with $NaNO_2$ and HCl. Reaction of *A* with concentrated HNO_3 and heat gives nicotinic acid (3-pyridinecarboxylic acid), but *A* is recovered unchanged after treatment with dilute nitric acid followed by neutralization with NaOH. When *A* reacts with CrO_3 in pyridine, a compound *B* (C_8H_9NO) is obtained. Compound *B*, when treated with dilute NaOD in D_2O, incorporates five deuterium atoms/molecule. Identify *A* and explain your reasoning.

42 The following compound is isolated as a by-product in the Chichibabin reaction of pyridine and sodium amide. Give a mechanism for its formation.

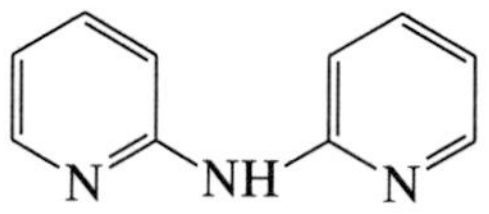

43 Each of the following reactions is an example of a heterocyclic ring synthesis that was not discussed explicitly in the text. Using the arrow formalism, give a mechanism for each reaction. As usual, begin by analyzing the relationship of atoms in the reactants and products.

(a)

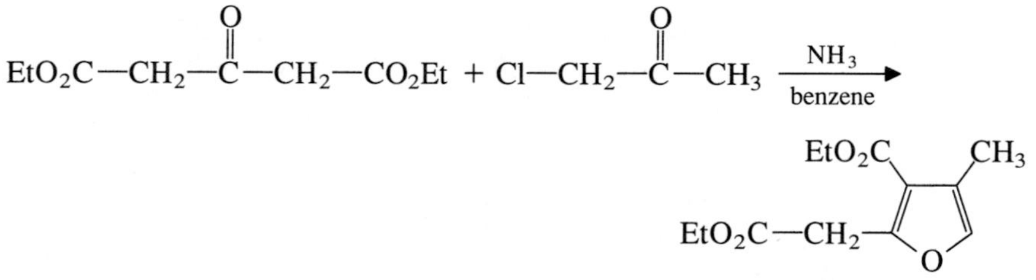

(b)

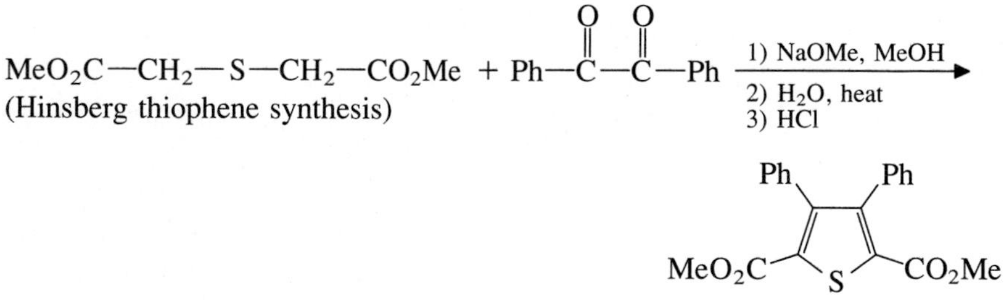

(Hinsberg thiophene synthesis)

(c)

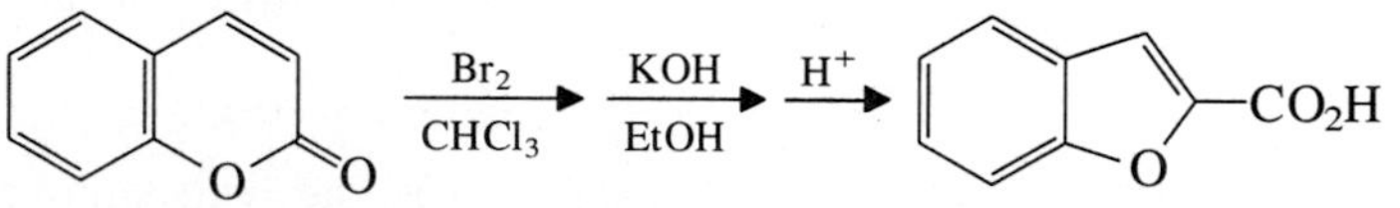

(d)

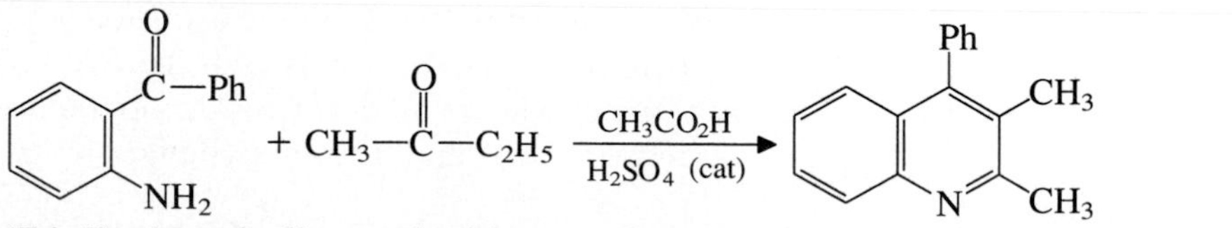

(Friedlander quinoline synthesis)

Problems (Cont.)

(e)

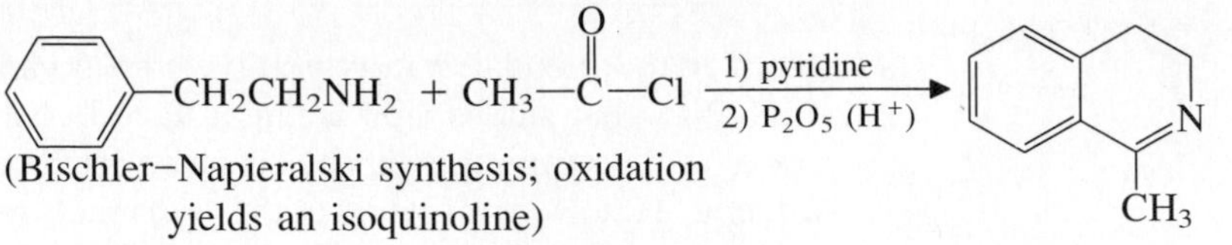

(Bischler–Napieralski synthesis; oxidation
yields an isoquinoline)

(f)

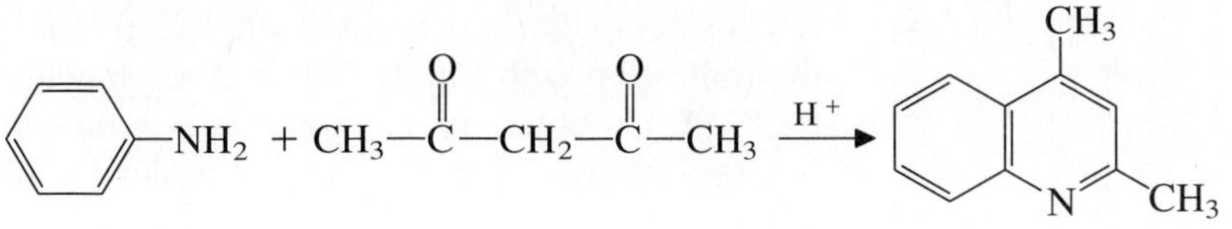

(Combes quinoline synthesis)

(g)

$$2CH_3-\overset{O}{\underset{\parallel}{C}}-CH_2-\overset{O}{\underset{\parallel}{C}}-OEt + NH_3 + CH_2{=}O \xrightarrow{Et_2NH}$$

$$+ 3H_2O$$

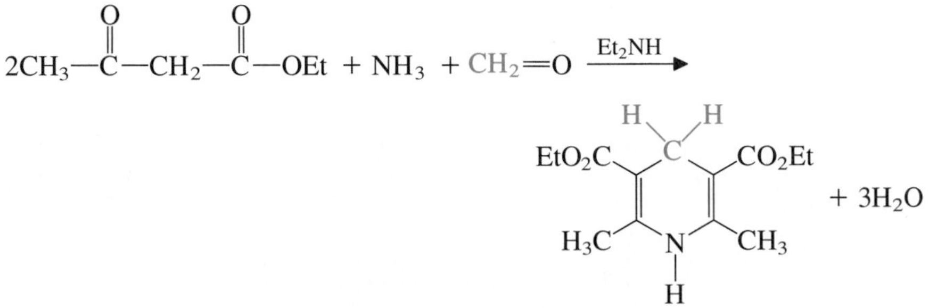

(Hantzsch dihydropyridine synthesis; oxidation gives a pyridine.)

44 Identify the bracketed compounds in the reaction sequence below. Note that there are two reasonable possibilities for compound *D*. The correct structure can be deduced from the fact that the dipole moment of compound *D* is zero.

3-methylpyridine $\xrightarrow{\text{KMnO}_4,\ \text{heat}}$ [*A*] $\xrightarrow{\text{SOCl}_2}$ $\xrightarrow{\text{NH}_3}$ [*B*] $\xrightarrow{\text{Br}_2,\ \text{NaOH}}$

[*C*] $\xrightarrow{\text{glycerol, H}^+,\ \text{As}_2\text{O}_5}$ [*D*]

45 Many furans are unstable in strong acid. Hydrolysis of 2,5-dimethylfuran in aqueous acid gives a compound *A*, $C_6H_{10}O_2$, that has an NMR spectrum consisting entirely of two singlets at δ 2.1 and δ 2.6 in the ratio 3 : 2, respectively. On treatment of compound *A* with very dilute NaOD in D_2O, both NMR signals virtually disappear. Treatment of *A* with zinc amalgam and HCl gives hexane. Propose a structure for *A* and give a mechanism for its formation.

46 (a) Treatment of 4-chloropyridine with ammonia gives 4-aminopyridine. However, when 3-chloropyridine is treated with ammonia, nothing happens. Explain the different results with the two chloropyridines.
 (b) Explain why treatment of 3-chloropyridine with sodamide (NaNH$_2$) gives a mixture of 3- and 4-aminopyridines.

47 Explain each of the following facts:

(a) The compound 2-pyridone does not hydrolyze in aqueous NaOH using conditions that bring about the rapid hydrolysis of δ-butyrolactam.

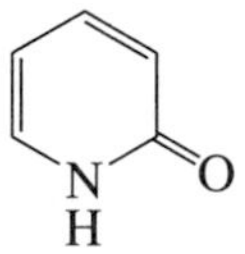

2-pyridone

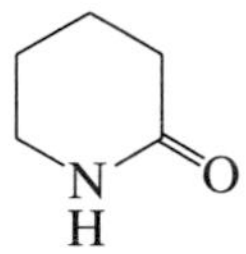

δ-butyrolactam

(b) In the ion below, the protons of the methyl group shown in color are most acidic, even though the other methyl group is directly attached to the positively charged nitrogen.

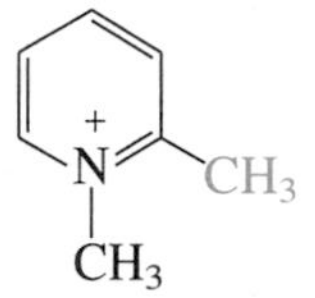

(c) The following reaction takes place in aqueous base:

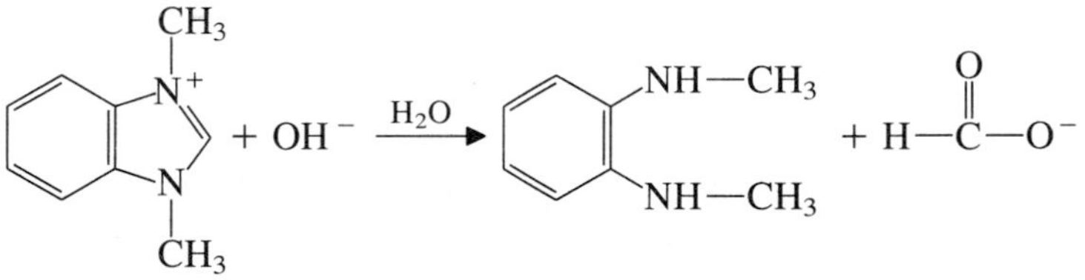

(d) Arsabenzene undergoes electrophilic aromatic substitution at the 2- and 4-positions, but pyridine undergoes electrophilic aromatic substitution at the 3-position. (Note that arsenic is in the same group of the periodic table as nitrogen.)

arsabenzene

(e) The protons of the methyl group shown in color, as well as the imide proton, in the compound below are readily exchanged for deuterium by dilute NaOD in D_2O, but those of the other methyl group are not.

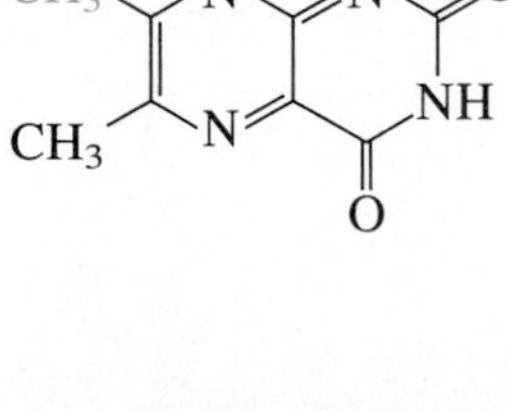

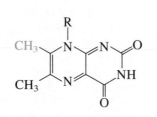

Problems (Cont.)

48 You work for VITA-men, a pharmaceutical company, whose management has decided to produce synthetic vitamin B_6. The company is in possession of some fragmentary notes from one of their early chemists that outline the following synthesis of pyridoxine (a form of vitamin B_6). Unfortunately, reagents for each of the numbered steps have been omitted. They have hired you as a consultant; suggest the reagents that would accomplish each step.

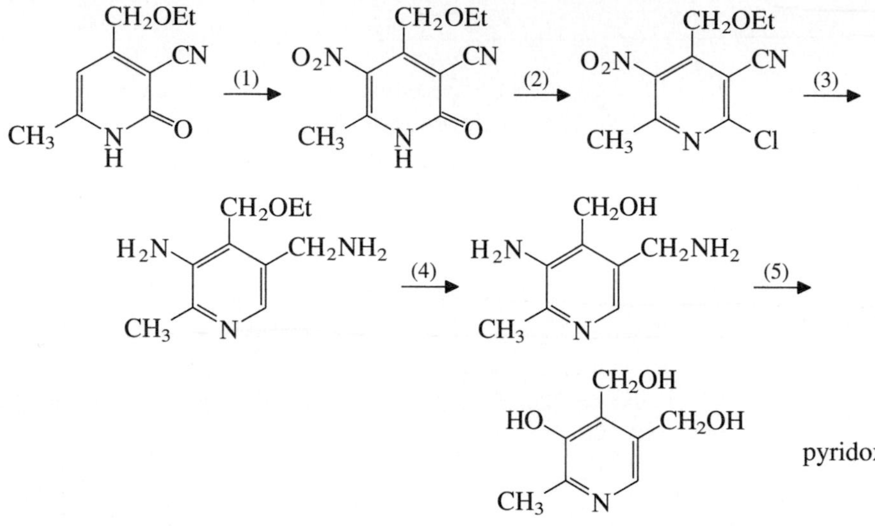

pyridoxine

49 One problem with the Fischer indole synthesis is the lack of many substituted phenylhydrazines or phenylhydrazones to use as starting materials. The following reaction (Japp–Klingemann synthesis) provides one solution to this problem.

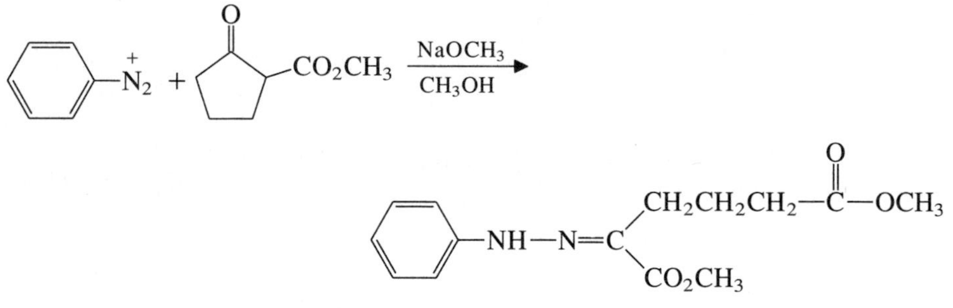

(a) Give a mechanism for this reaction.
(b) What indole would be formed when the product of the synthesis above is subjected to the conditions of the Fischer indole synthesis?
(c) Show how to prepare the following compound from toluene, diethyl pimelate [$EtO_2C—(CH_2)_5—CO_2Et$], and any other reagents.

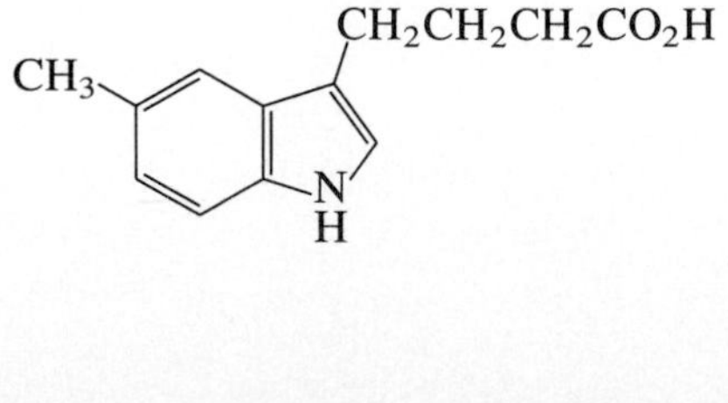

25 Pericyclic Reactions

Pericyclic reactions are defined as reactions that occur by a concerted cyclic shift of electrons. This definition states two key elements that characterize a pericyclic reaction. First, the reaction is *concerted*. In a *concerted reaction,* reactant bonds are broken and product bonds are formed at the same time, without intermediates. Second, a pericyclic reaction involves a *cyclic shift of electrons*. (The word *pericyclic* means "around the circle.") The Diels–Alder reaction (Sec. 15.3) and the S_N2 reaction (Sec. 9.3) are both concerted reactions, but only the Diels–Alder reaction occurs by a *cyclic electron shift*. Hence the Diels–Alder reaction is a pericyclic reaction, but the S_N2 reaction is not.

In this chapter, we shall concern ourselves with three major types of pericyclic reactions, although there are others. The first type is the **electrocyclic reaction:** a reaction in which a ring is closed at the expense of a conjugated double (or triple) bond:

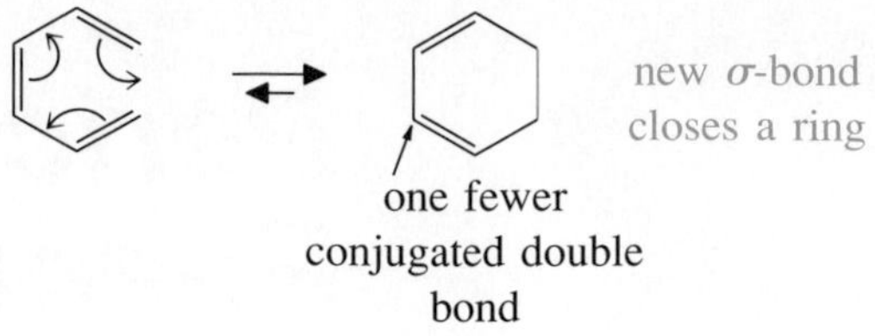

(25.1)

The second type of reaction is **cycloaddition:** a reaction in which two or more π-electron systems react to form a ring at the expense of one double (or triple) bond in each of the reacting partners.

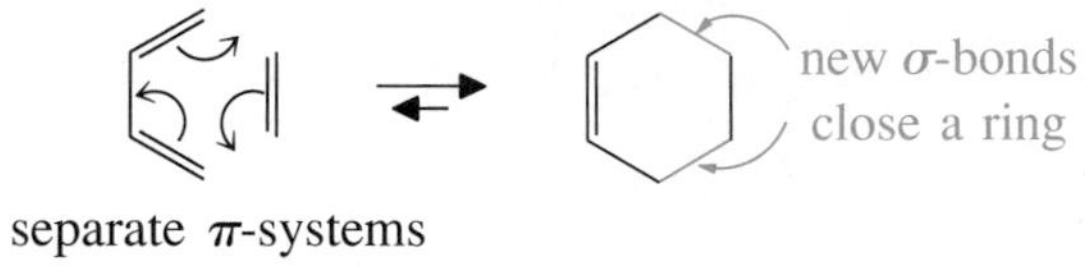

(25.2)

The third type of pericyclic reaction is the **sigmatropic reaction:** a reaction in which a σ-bond formally migrates from one end to the other of a π-electron system and the net number of double (or triple) bonds remains the same.

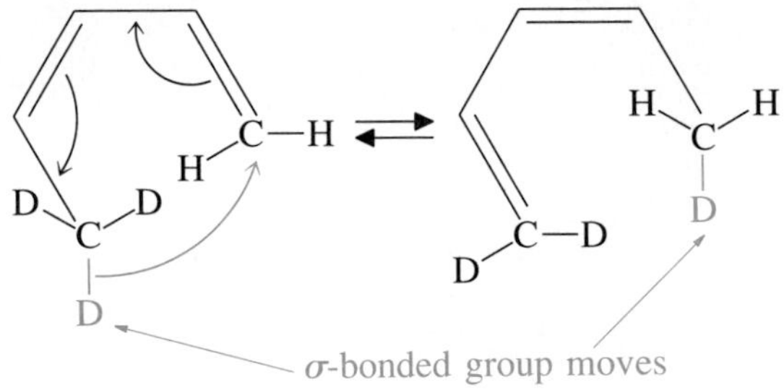

(25.3)

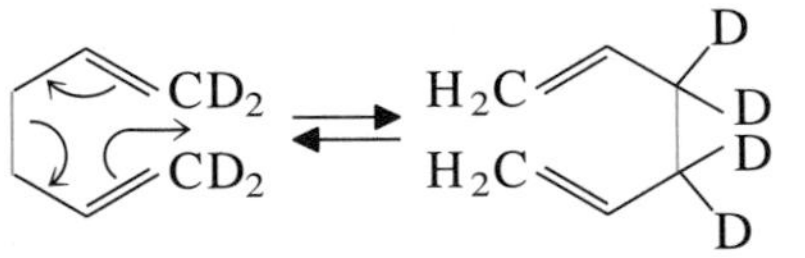

(25.4)

Three features of any pericyclic reaction are intimately interrelated. These are:

1. The way the reaction is activated (heat or light)

Many pericyclic reactions require no catalysts or reagents other than the reacting partners. They take place merely on heating or irradiation with ultraviolet light. However, many reactions that require heat are not initiated by light, and vice versa.

2. The number of π-electrons involved in the reaction
3. The stereochemistry of the reaction

These points are illustrated by the following electrocyclic reactions.

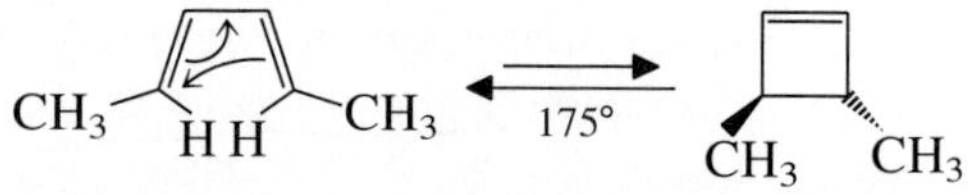

(25.5a)

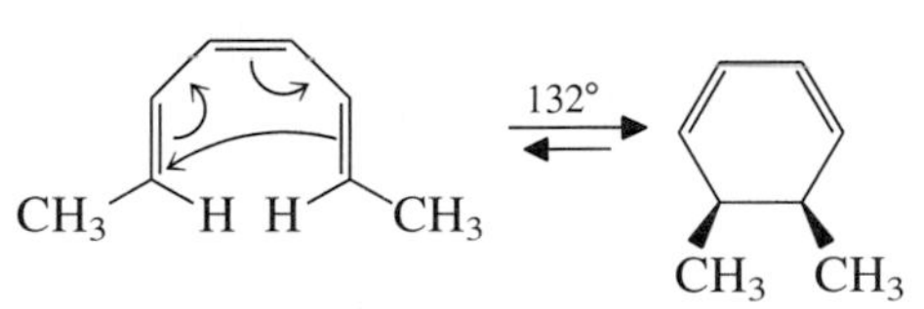

(25.5b)

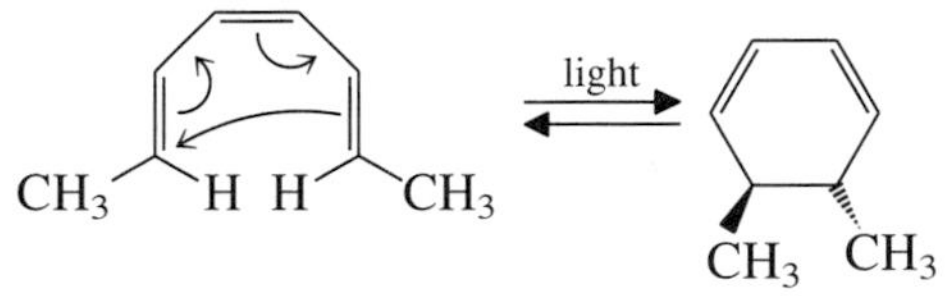

(25.5c)

The relationship between the number of π-electrons involved in the reaction and its stereochemistry is illustrated by a comparison of Eq. 25.5a and 25.5b. Both are activated by heat; however, the former reaction, involving four π-electrons, gives only the *trans*-disubstituted isomer of the cyclic product, whereas the latter reaction, involving six π-electrons, gives only the *cis* isomer.

The relationship between the mode of activation and the stereochemistry is exemplified by a comparison of Eqs. 25.5b and 25.5c. When the starting material is heated, only the *cis*-disubstituted isomer of the cyclic product is obtained. When the starting material is irradiated with ultraviolet light, the only product obtained is the *trans* isomer.

Relationships such as these had been observed for many years, but the reasons for them were not understood. In 1965 a theory that clearly explained these observations and successfully predicted many new ones was put forth by R. B. Woodward (1917–1979), then a professor of chemistry at Harvard University, and Roald Hoffmann, at the time a Junior Fellow at Harvard and presently Professor of Chemistry at Cornell University. For this theory, called *Conservation of Orbital Symmetry,* Hoffmann received the 1981 Nobel Prize in Chemistry. He shared the prize with Kenichi Fukui, a professor of chemistry at Kyoto University in Japan, who had advanced a related theory, called *Frontier–Orbital Theory.* (The two theories make the same predictions; they are alternate ways of looking at the same reactions.) Woodward undoubtedly would have also shared the Nobel Prize had he not died prior to its announcement. (The terms of Nobel's bequest require that the prize only be awarded to living scientists.) Woodward had, however, received an earlier Nobel Prize for his work in organic synthesis. Our goal in this chapter is to learn elements of the Woodward–Hoffmann theory that will enable us to understand and predict the outcome of pericyclic reactions.

Problem

1 Classify each of the following pericyclic reactions as an electrocyclic, cycloaddition, or sigmatropic reaction. Give the curved-arrow formalism for each.

(a)

(b)

$$CH_3—CH—\underset{\underset{H}{|}}{\overset{\overset{CH_3}{|}}{C}}—CH_3 \longrightarrow CH_3—CH_2—\underset{+}{\overset{\overset{CH_3}{|}}{C}}—CH_3$$

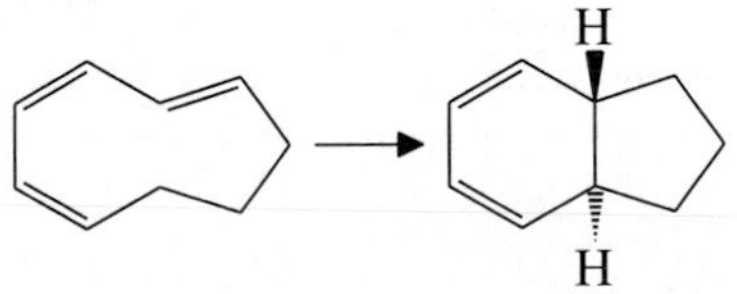

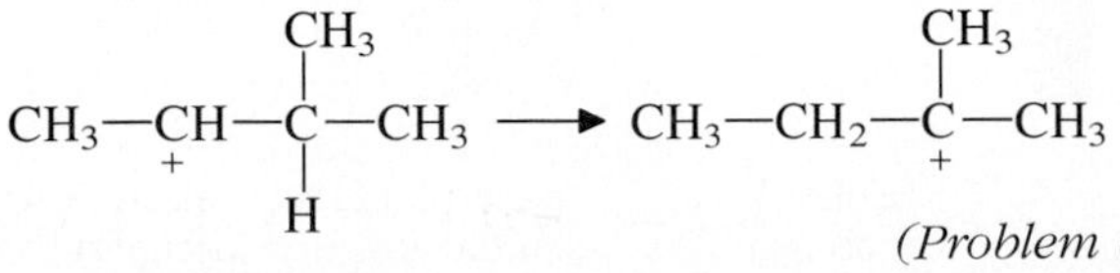

(Problem continued on page 1094.)

Problems (Cont.)

(c)

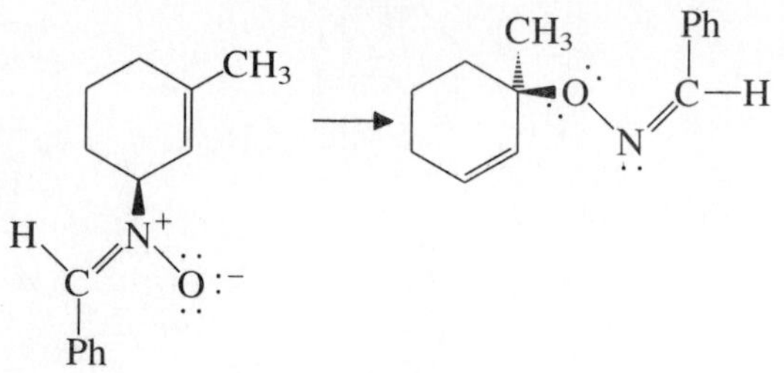

(d)

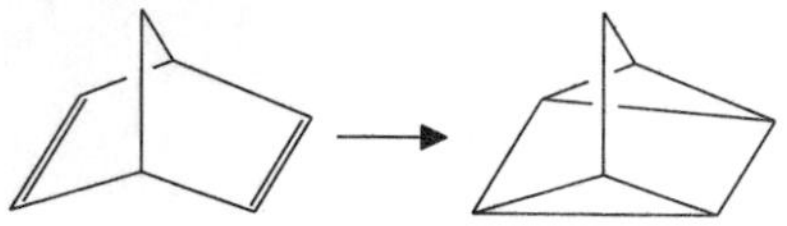

(e)

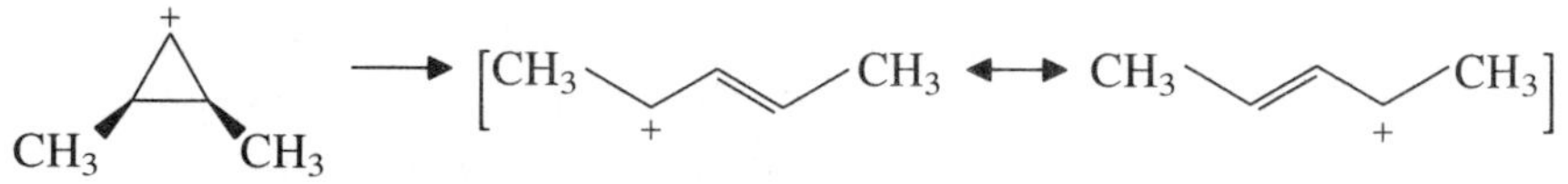

25.1 MOLECULAR ORBITALS OF CONJUGATED π-ELECTRON SYSTEMS

In order to understand the theory of pericyclic reactions, we must first understand some rudiments of *molecular orbital theory,* particularly as it applies to molecules containing π-electrons. This theory was introduced in Secs. 2.4 and 4.1A, which should be reviewed carefully.

A. Molecular Orbitals of Conjugated Alkenes

The overlap of *p* orbitals to give π-molecular orbitals is described by the mathematics of quantum theory. However, we do not need to consider the mathematical aspects of this theory to appreciate the results. We shall restrict our attention here to ethylene and ordinary conjugated alkenes. The π-molecular orbitals for these molecules can be constructed according to the following generalizations, which are applied to ethylene and 1,3-butadiene in Figs. 25.1 and 25.2, respectively.

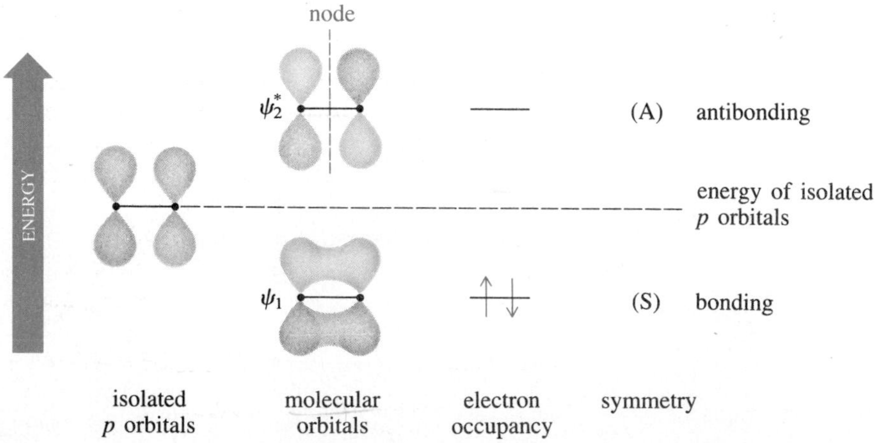

Figure 25.1 π-Molecular orbitals of ethylene.

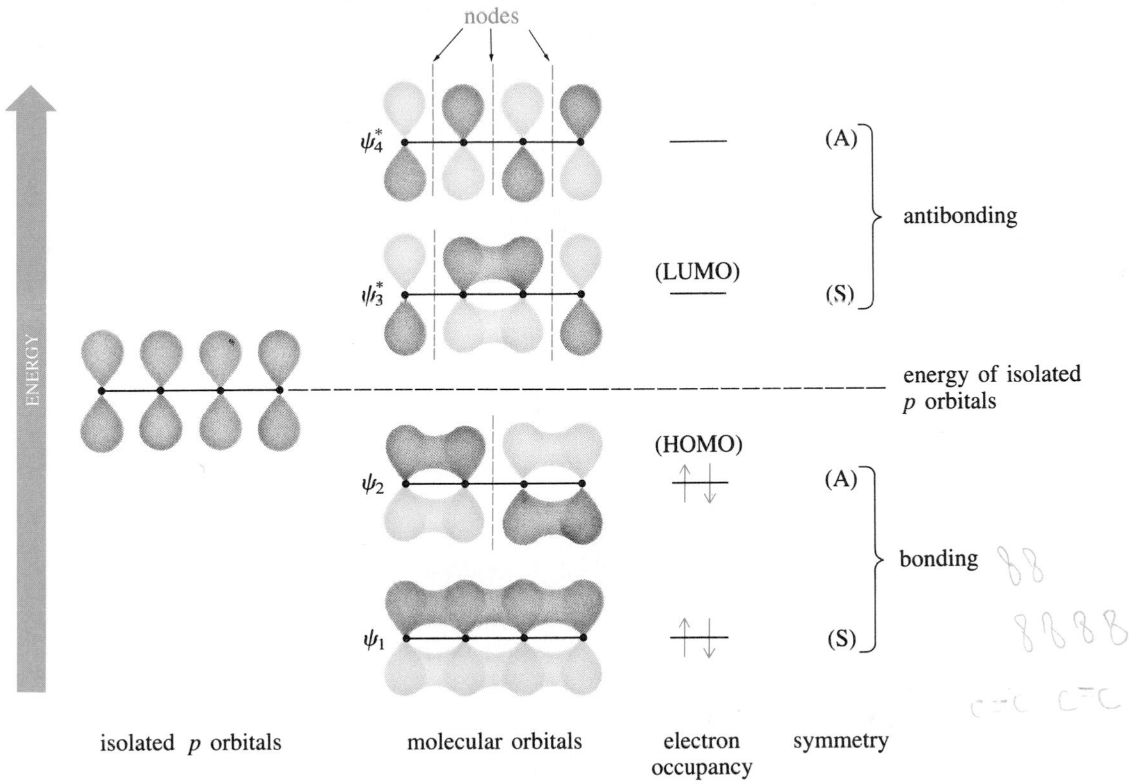

Figure 25.2 π-Molecular orbitals of 1,3-butadiene.

1. A π-electron system derived from the interaction of a number *m* of *p* orbitals contains *m* molecular orbitals (MOs) that differ in energy.

Since two *p* orbitals contribute to the π-electron system of ethylene, this molecule has two π-MOs, which we shall abbreviate as ψ_1 and ψ_2. Similarly, the four *p* orbitals of 1,3-butadiene combine to form four MOs, ψ_1, ψ_2, ψ_3, and ψ_4.

2. Half of the molecular orbitals have lower energy than the isolated *p* orbitals. These are called **bonding molecular orbitals.** The other half have energy higher than the isolated *p* orbitals. These are called **antibonding molecular orbitals.**

To emphasize this distinction, antibonding MOs will be indicated with asterisks. Thus, ethylene has one bonding MO (ψ_1) and one antibonding MO (ψ_2^*); 1,3-butadiene has two bonding MOs (ψ_1 and ψ_2) and two antibonding MOs (ψ_3^* and ψ_4^*).

3. The bonding molecular orbital of lowest energy, ψ_1, has no nodes (except, of course, the plane of the molecule, which is a node of the component *p* orbitals). Each molecular orbital of increasingly higher energy has one additional node.

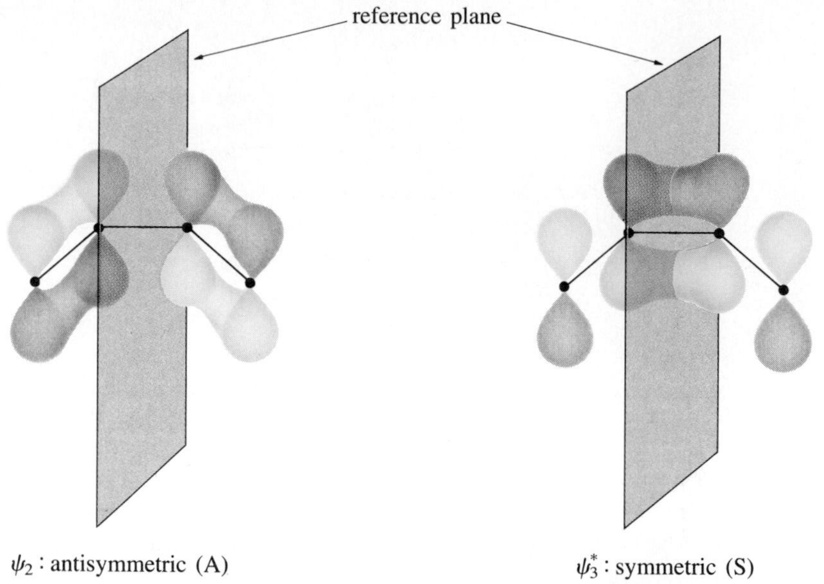

Figure 25.3 Symmetry classification of ψ_2 and ψ_3^* of 1,3-butadiene with respect to a reference plane through the center of the molecule and perpendicular to the plane of the molecule. The symmetry classifications of the MOs in Figs. 25.1 and 25.2 are indicated by the abbreviations (A) and (S).

Recall from Sec. 2.2 that a *node* is a plane at which any wave, including an electron wave (orbital), is zero; that is, when an electron is in a given MO there is zero *probability of finding the electron,* or zero *electron density,* at the node. A particularly important feature of the node for understanding pericyclic reactions is that the electron wave has a *peak* on one side of the node (color in Figs. 25.1 and 25.2) and a *trough* on the other side. (See also Fig. 2.3.) We say that the orbital *changes phase* at the node.

Thus ψ_1 of ethylene has no nodes, and ψ_2^* has one node. In 1,3-butadiene, ψ_1 has no nodes, ψ_2 has one, ψ_3^* has two, and ψ_4^* has three.

4. The nodes occur *between* atoms and are arranged symmetrically with respect to the center of the π-electron system.

The node in ψ_2^* of ethylene is between the two carbon atoms, in the center of the π-system. The node in ψ_2 of 1,3-butadiene is also symmetrically placed in the center of the π-system. The two nodes in ψ_3^* are placed between carbons 1 and 2, and between carbons 3 and 4, respectively—equidistant from the center of the π-system. Each of the three nodes in ψ_4^*, the orbital of highest energy, must occur between carbon atoms.

The next generalization relates to the *symmetry* of the molecular orbitals. In particular, let us concentrate on the symmetry with respect to a plane through the center of the π-electron system and perpendicular to the plane of the molecule. This plane, which we shall call the *reference plane,* is shown for the 1,3-butadiene molecule in Fig. 25.3. If peaks of an MO reflect into peaks, and troughs into troughs, the MO is **symmetric,** abbreviated (S); if peaks reflect into troughs, the MO is **antisymmetric,** abbreviated (A). Notice that symmetric MOs, such as ψ_1 and ψ_3^* of 1,3-butadiene, have the *same* phase on the terminal carbons of the π-system, and antisymmetric MOs, such as ψ_2 and ψ_4^*, have *opposite* phase at these two carbons. This phase relationship will prove to be crucial in our study of pericyclic reactions.

The symmetry and phase relationships of MOs can be summarized by the following generalization:

5. The first MO (ψ_1) is symmetric with respect to the reference plane; MOs of progressively higher energy alternate in symmetry.

We can see this alternating symmetry quite clearly in Figs. 25.1 and 25.2.

Finally, we concern ourselves with how π-electrons are distributed in the MOs.

6. Electrons are placed pairwise into each molecular orbital, beginning with the orbital of lowest energy.

This point is illustrated in Figs. 25.1 and 25.2 in the column labeled "electron occupancy." An alkene with some number *m* of *p* orbitals has *m* π-electrons. Thus ethylene, with two *p* orbitals, has two π-electrons. These are both placed (with opposite spin) into ψ_1 (Fig. 25.1). 1,3-Butadiene, with four *p* orbitals, has four π-electrons. Two are placed in ψ_1 and two in ψ_2 (Fig. 25.2). We can see from these examples that the bonding MOs are fully filled in conjugated alkenes and the antibonding MOs are empty.

The π-electron contribution to the energy of a molecule is determined by the energies of its *occupied* MOs. Since the bonding MOs have lower energy than the isolated *p* orbitals, there is an energetic advantage to π-molecular orbital formation; this is why π-bonds exist.

Two MOs are of particular importance in understanding pericyclic reactions. One is the occupied molecular orbital of highest energy, which we call the **highest occupied molecular orbital (HOMO).** The other is the unoccupied molecular orbital of lowest energy, which we call the **lowest unoccupied molecular orbital (LUMO).** These are labeled in Figs. 25.1 and 25.2. In ethylene, ψ_1 is the HOMO and ψ_2^* the LUMO; in 1,3-butadiene, ψ_2 is the HOMO and ψ_3^* the LUMO. Notice that the HOMO and LUMO of any given compound have opposite symmetries.

Problem

2 Construct a diagram like Fig. 25.2 for 1,3,5-hexatriene, the conjugated triene containing six carbons. Indicate the bonding and antibonding molecular orbitals, the HOMO, the LUMO, and the nodes in each molecular orbital. Classify each MO as symmetric (S) or antisymmetric (A) with respect to a plane through the center of the π-system.

B. Molecular Orbitals of Conjugated Ions and Radicals

Conjugated unbranched ions and radicals have an odd number of carbon atoms. For example, the allyl cation has three carbon atoms and three *p* orbitals—hence, three MOs.

$$[CH_2{=}CH{-}\overset{+}{C}H_2 \longleftrightarrow \overset{+}{C}H_2{-}CH{=}CH_2] \qquad \text{allyl cation}$$

The MOs of such species follow many of the same patterns as those of conjugated alkenes. The MOs for the allyl and 2,4-pentadienyl systems are shown in Figs. 25.4 and 25.5, respectively. These figures show two important differences between these MOs and those of conjugated alkenes. First, one MO is neither bonding nor antibonding, but has the same energy as the isolated *p* orbitals; this MO is called a **nonbonding**

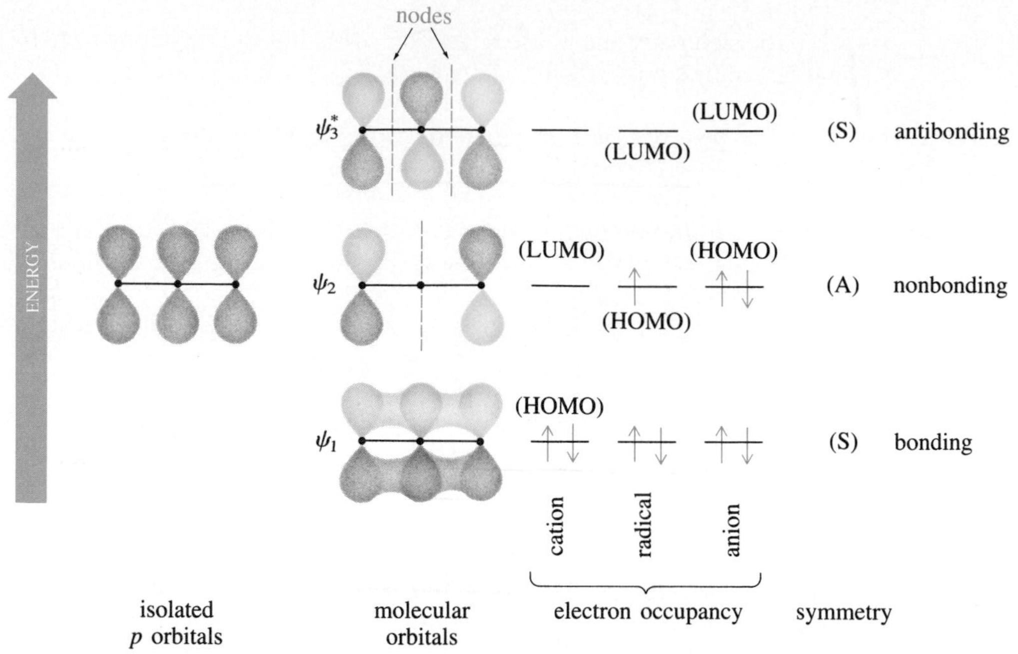

Figure 25.4 π-Molecular orbitals of the allyl system.

molecular orbital. The nonbonding MO in the allyl system is ψ_2. The remaining orbitals are either bonding or antibonding, and there are an equal number of each type. Second, in some of the MOs, nodes pass through carbon atoms. For example, in the allyl system, there is a node on the central carbon of ψ_2. This means an electron in ψ_2 has no electron density on the central carbon. This is why, for example, charge in the allyl anion resides only on the terminal carbons, something we already knew from resonance arguments:

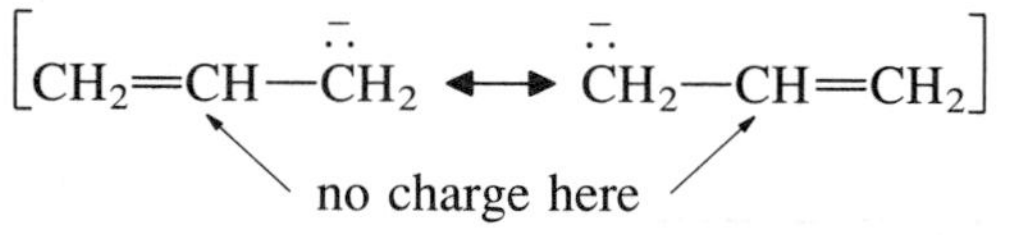
$CH_2{=}CH{-}\ddot{C}H_2^{-}$ ⟷ $\ddot{C}H_2^{-}{-}CH{=}CH_2$
no charge here

allyl anion

Just as we associate the charge in an atomic anion with an excess of valence electrons, we can associate the charge in a conjugated carbanion with the electrons in its HOMO.

Notice that cations, radicals, and anions involving the same π-system have the same molecular orbitals. For example, the MOs of the allyl system apply equally well to the allyl cation, allyl radical, and allyl anion, because all three species contain the same *p* orbitals. These species differ only in the *number* of π-electrons, as shown in the "electron occupancy" column of Fig. 25.4. Thus, the HOMO of the allyl cation is ψ_1, and the LUMO is ψ_2. In contrast, the HOMO of the allyl anion is ψ_2, and the LUMO is ψ_3^*.

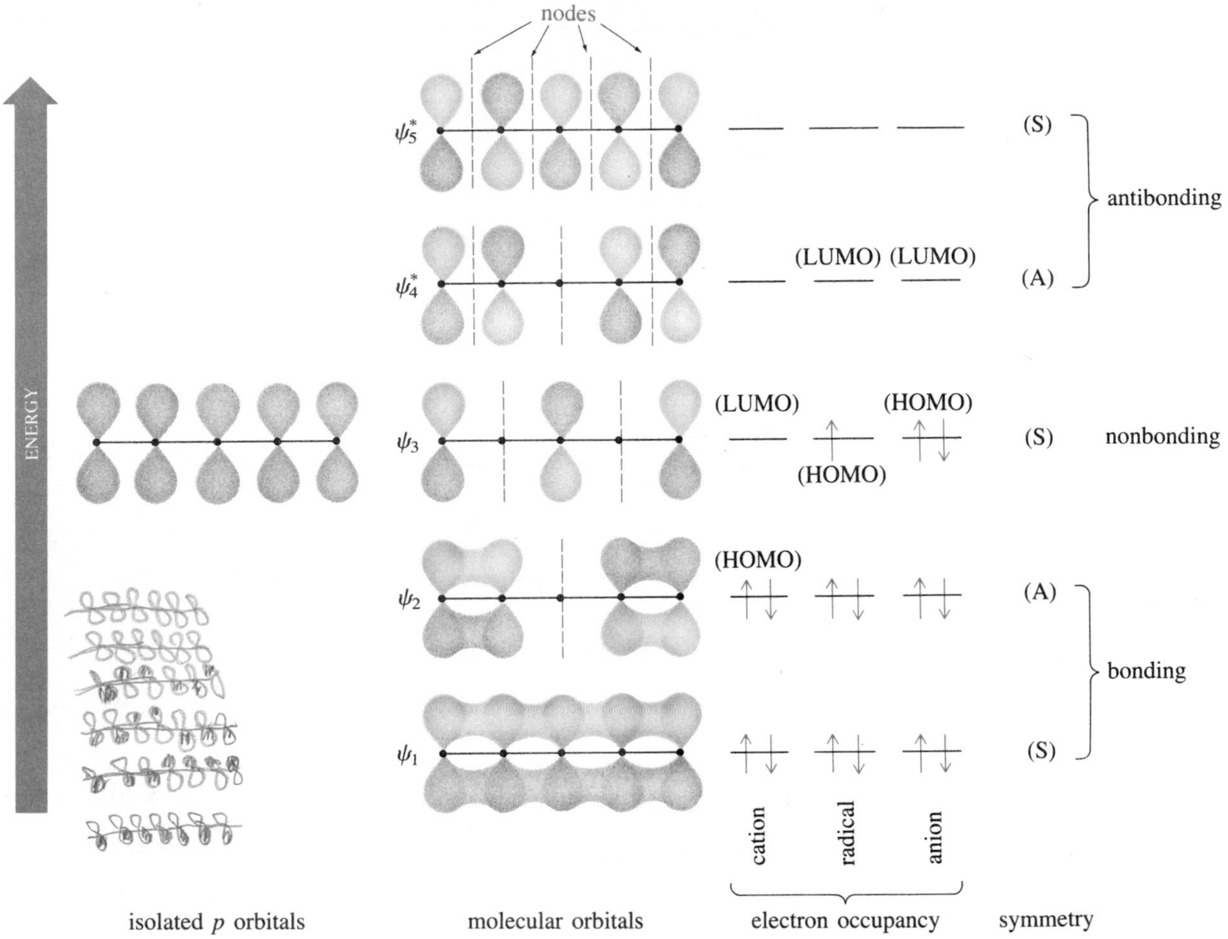

Figure 25.5 π-Molecular orbitals of the 2,4-pentadienyl system.

Problem

3 (a) Generalizing from the patterns in Figs. 25.4 and 25.5, construct the MOs of the 2,4,6-heptatrienyl cation. Label the nonbonding MO and indicate the nodes in each MO. Classify each MO as symmetric (S) or antisymmetric (A).

(b) To which carbons is the positive charge in this cation delocalized? Explain.

C. Excited States

The molecules and ions we have been discussing can absorb energy from light of certain wavelengths. This process, which is also responsible for the UV spectra of these species (Sec. 15.2A), is shown schematically in Fig. 25.6 for 1,3-butadiene. Let us refer to the normal electronic configuration of butadiene as the **ground state.** Energy from absorbed light is used to promote an electron from the HOMO of ground-state 1,3-butadiene (ψ_2) into the LUMO (ψ_3^*). The species with the promoted electron is an **excited state** of 1,3-butadiene. The orbital ψ_3^* becomes the HOMO of the excited state. Notice that *the HOMOs of the ground state and the excited state have opposite symmetries.*

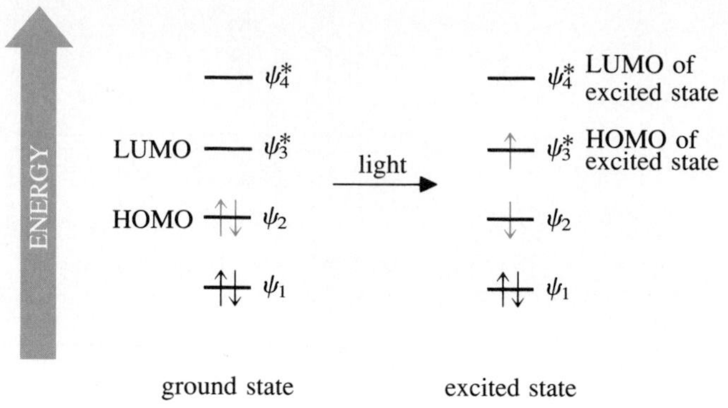

Figure 25.6 Light absorption by 1,3-butadiene promotes an electron from the HOMO to the LUMO and produces an excited state.

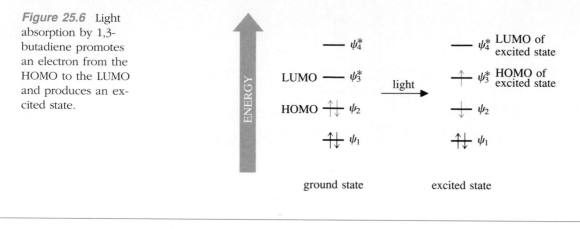

Problem

4 Give the electronic configuration for the excited state of the allyl anion.

25.2 ELECTROCYCLIC REACTIONS

A. Stereochemistry of Electrocyclic Reactions

We are now ready to apply what we have learned about MO theory to pericyclic reactions; we begin with *electrocyclic reactions,* in which a ring is closed at the expense of a double or triple bond (see Eq. 25.1). When an electrocyclic reaction takes place, the carbons at each end of the conjugated π-system must turn in a concerted fashion so that the *p* orbitals can overlap (and rehybridize) to form the σ-bond that closes the ring. Let us illustrate this point with the reaction shown in Eq. 25.5a, the electrocyclic closure of (2*E*,4*E*)-2,4-hexadiene to give 3,4-dimethylcyclobutene. This turning can occur in two stereochemically distinct ways. In a **conrotatory** closure the two carbon atoms turn in the same direction. (The colored arrows show the direction of motion, not electron flow.)

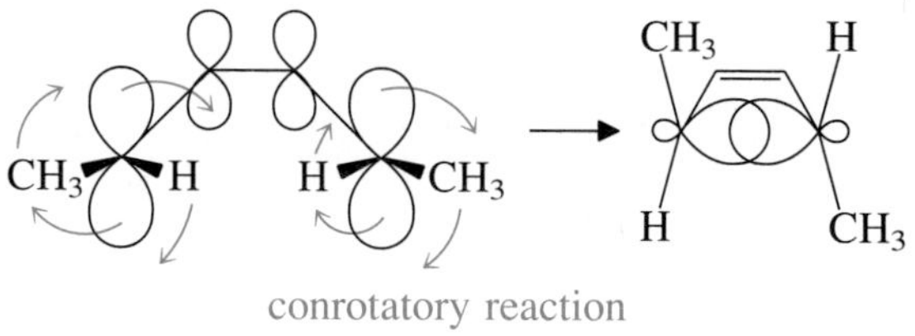

(25.6)

(There are, of course, two conrotatory modes, clockwise and counterclockwise; the clockwise mode is shown, but the counterclockwise mode in this case is equally probable, since it gives the enantiomer.) In the second mode of ring closure, called a **disrotatory** mode, the carbon atoms turn in opposite directions:

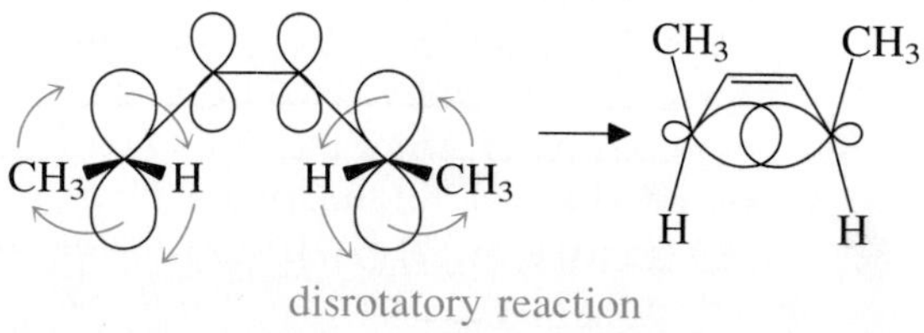

(25.7)

As we can see from Eq. 25.5a, *the conrotatory reaction is observed*. *Trans*-, not *cis*-3,4-dimethylcyclobutene is the product.

Molecular orbital theory explains this result. A simple way to look at the reaction is to focus on the *HOMO of the diene*. This molecular orbital contains the π-electrons of highest energy. We can think of these π-electrons in the same way that we think of the valence electrons of an atom, which are the atomic electrons of highest energy. Just as the atomic valence electrons are the ones involved in most chemical reactions, the electrons in the HOMO are the ones that govern the course of pericyclic reactions. When the ring closure takes place, the two *p* orbitals on the ends of the π-system must overlap. But simple overlap is not enough: they must overlap *in phase*. That is, the wave peak on one carbon must overlap with the wave peak on the other, or a wave trough must overlap with a wave trough. If a peak were to overlap with a trough, the electron waves would cancel and therefore no bond would form.

Let us see what it takes to provide the required bonding overlap. We first make the approximation that the nodal properties in the MOs of 2,4-hexadiene are essentially identical to those of 1,3-butadiene (Fig. 25.2). In other words, the methyl groups at each end of the molecule can be largely ignored when considering the MOs of the system. Examining the HOMO of a conjugated diene (ψ_2 in Fig. 25.2), we see that because of the antisymmetric nature of ψ_2, conrotatory ring closure is required for in-phase, or bonding, overlap, as observed:

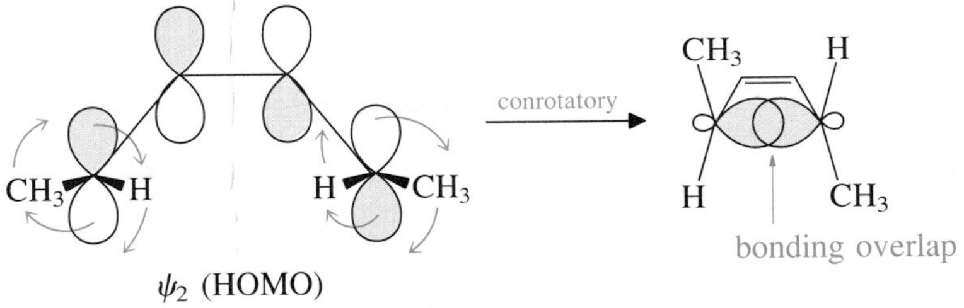

(25.8)

In contrast, disrotatory ring closure gives out-of-phase overlap, or an antibonding (and hence unstable) situation:

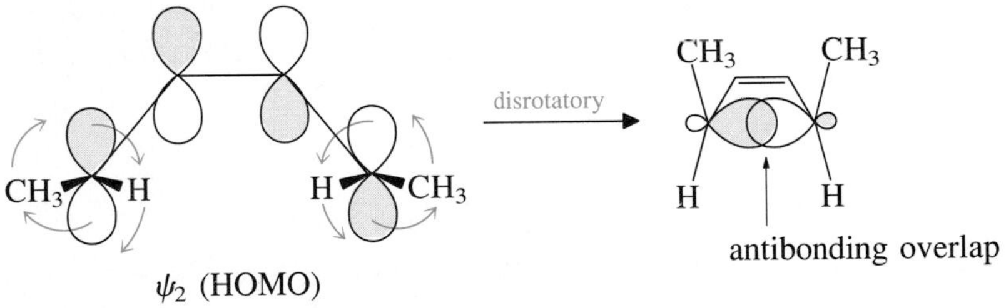

(25.9)

Thus it is the relative orbital phase at the terminal carbon atoms of the HOMO—in other words, the *orbital symmetry*—that determines whether the reaction is conrotatory or disrotatory. This observation suggests that *all* conjugated polyenes with *antisymmetric* HOMOs will undergo conrotatory ring closure, and indeed, such is the case.

What about polyenes with *symmetric* HOMOs? The triene (2*E*,4*Z*,6*E*)-2,4,6-octadiene is just such a compound. Equation 25.5b shows that the electrocyclic reaction of this triene is a *disrotatory* closure. Once again we ignore the substituent groups and examine the HOMO of the simpler triene, 1,3,5-hexatriene (Problem 2). Because the HOMO of this triene (ψ_3) is *symmetric,* the HOMO has the *same phase* at each end of the π-system. Hence, bonding overlap can occur only if the ring closure is *disrotatory,* as observed.

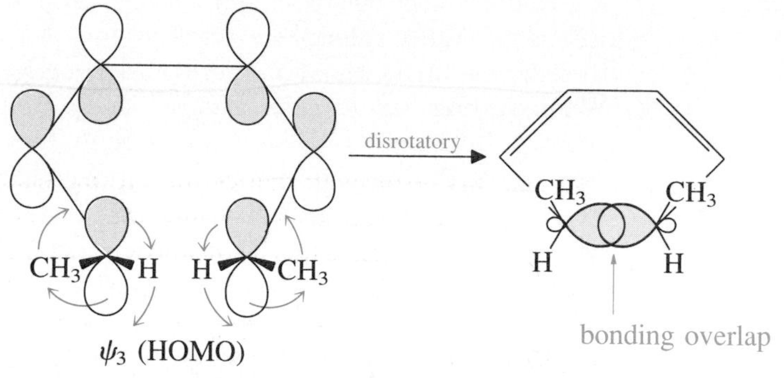

(25.10)

To summarize: electrocyclic closure of a conjugated diene is conrotatory, and that of a conjugated triene is disrotatory. The reason for the difference is the phase relationships within the HOMO at the terminal carbons of the π-system. In the diene the HOMO has opposite phase at these two carbons; in the triene the HOMO has the same phase. A different type of rotation is thus required in each case for bonding overlap.

This result can be generalized. Conjugated alkenes with $4n$ π-electrons (n = any integer) have antisymmetric HOMOs, and undergo conrotatory ring closure; those with $4n + 2$ π-electrons have symmetric HOMOs and undergo disrotatory ring closure. That is, conrotatory ring closure is *allowed* for systems with $4n$ π-electrons; it is *forbidden* for systems with $4n + 2$ π-electrons. Conversely, disrotatory ring closure is *allowed* for systems with $4n + 2$ π-electrons; it is *forbidden* for systems with $4n$ π-electrons.

B. Excited-State Electrocyclic Reactions

When a molecule absorbs light, it reacts through its *excited state* (Sec. 25.1C). The HOMO of the excited state is different from the HOMO of the ground state, and has different symmetry. For example, as Eq. 25.5c shows, the *photochemical* ring closure of (2*E*,4*Z*,6*E*)-2,4,6-octatriene is *conrotatory.* This is understandable in terms of the symmetry of the excited state HOMO:

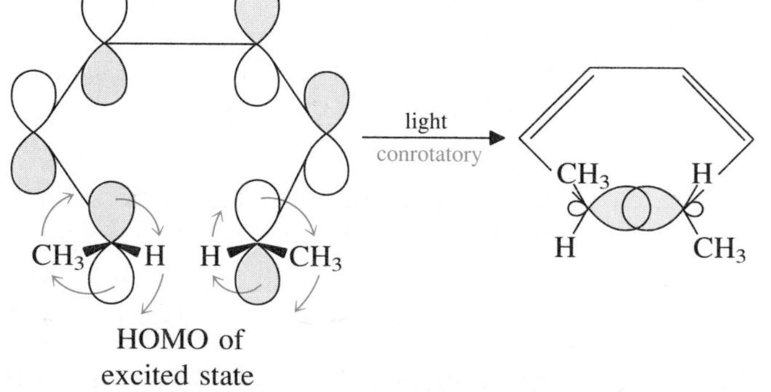

(25.11)

tac

TABLE 25.1 Selection Rules for Electrocyclic Reactions

Number of electrons	Mode of activation	Allowed stereochemistry
$4n$	thermal	conrotatory
	photochemical	disrotatory
$4n + 2$	thermal	disrotatory
	photochemical	conrotatory

Contrast this result with that of the ground-state reaction in Eq. 25.5b. The stereochemical result is opposite because the symmetry of the HOMO is opposite.

We can generalize this result and say that the mode of ring closure in *photochemical* electrocyclic reactions—reactions that occur through electronically excited states—is opposite to that of electrocyclic reactions that occur through electronic ground states. (Ground-state reactions are also called *thermal* reactions because most heat-promoted electrocyclic reactions occur through ground states.) We can summarize these results with a series of *selection rules* for allowed pericyclic reactions, given in Table 25.1. The number of electrons involved in the reaction (column 1 of the table) is determined by counting the electron pairs when we write the reaction mechanism in the curved-arrow notation, as in the following example:

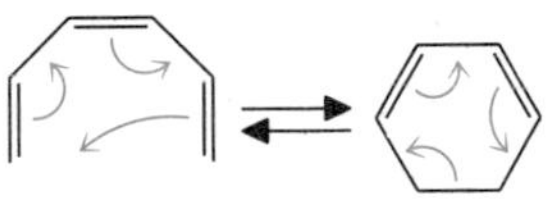

(25.12)

three curved arrows;
six electrons

C. Selection Rules and Microscopic Reversibility

It is important to understand that the selection rules in Table 25.1 (as well as all others we shall consider) refer to the *rates* of pericyclic reactions and not the *position of the equilibria* involved. Thus the electrocyclic reaction of the diene in Eq. 25.5a to give a cyclobutene favors the diene at equilibrium because of the strain in the cyclobutene, but the electrocyclic reaction of the conjugated triene in Eq. 25.5b favors the cyclic compound because σ-bonds are stronger than π-bonds, and because six-membered rings are relatively stable.

It is also common for a photochemical reaction to favor the less stable isomer of an equilibrium because the energy of light is harnessed to drive the equilibrium energetically "uphill." For example, in the following reaction, the conjugated alkene absorbs UV light, but the bicyclic compound does not; hence, the photochemical reaction favors the latter.

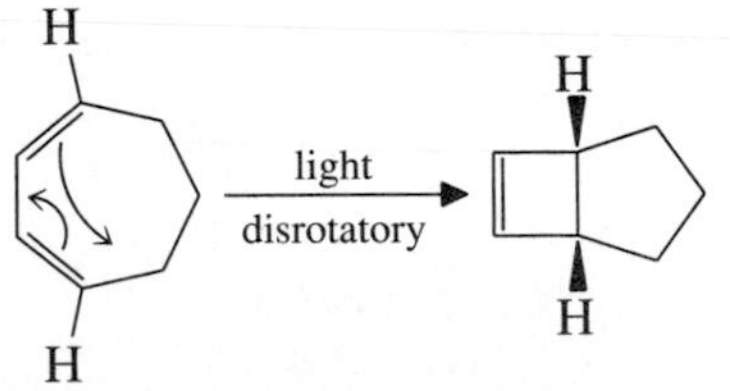

(42% yield) (25.13)

Thus, the selection rules do not tell us which component of an equilibrium will be favored—only whether the equilibrium will be established in the first place at a reasonable rate.

The *principle of microscopic reversibility* (Sec. 10.1A) assures us that selection rules apply equally well to the forward and reverse of any pericyclic reaction, because the reaction in both directions must proceed through the same transition state. Hence, an electrocyclic ring *opening* must follow the same selection rules as its reverse, an electrocyclic ring *closure*. Thus, the thermal ring-opening reaction of the cyclobutene in Eq. 25.14, like the reverse ring-closure reaction, must be a conrotatory process, by Table 25.1.

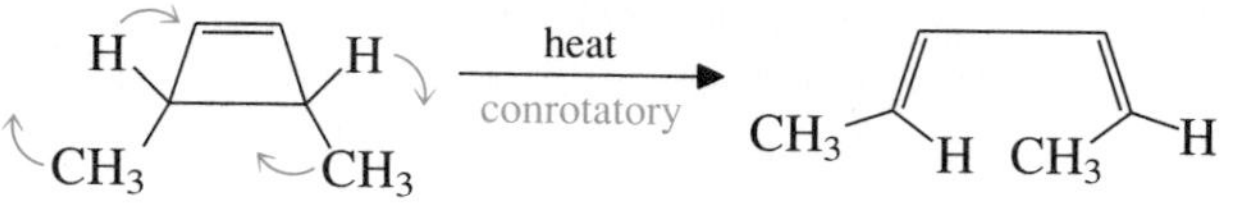

(25.14)

Another interesting example of an electrocyclic ring-opening reaction is the following one, in which the allowed thermal conrotatory process would give a highly strained molecule containing a *trans* double bond within a small ring.

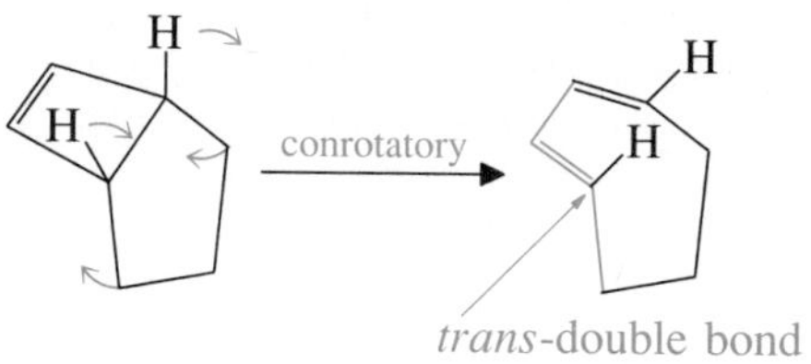

(25.15)

Although the selection rules allow the reaction, it does not occur because of the strain in the product. In other words, "allowed" reactions are sometimes prevented from occurring for reasons having nothing to do with the selection rules. Concerted ring opening to the relatively unstrained all-*cis* diene is forbidden, because this would be a disrotatory process—a process forbidden by the selection rules in Table 25.1 for a concerted $4n$-electron reaction.

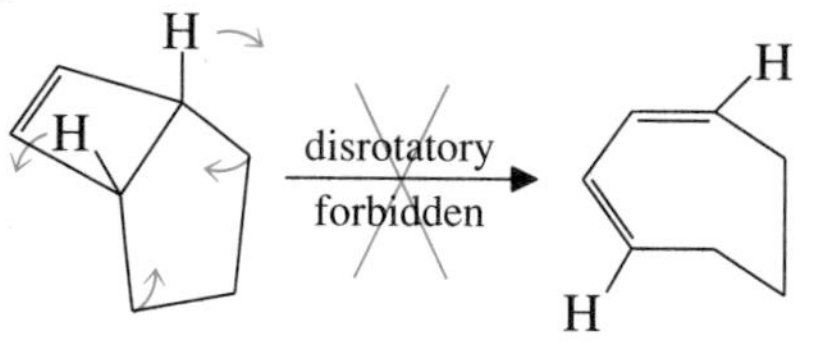

(25.16)

Hence, the bicyclic compound is effectively "trapped into existence"; that is, there is no concerted thermal pathway by which it can reopen. (Recall that it is formed *photochemically* from the all-*cis* diene; Eq. 25.13.)

Problems

5 Which of the following electrocyclic reactions should occur readily by a concerted mechanism?

(a)

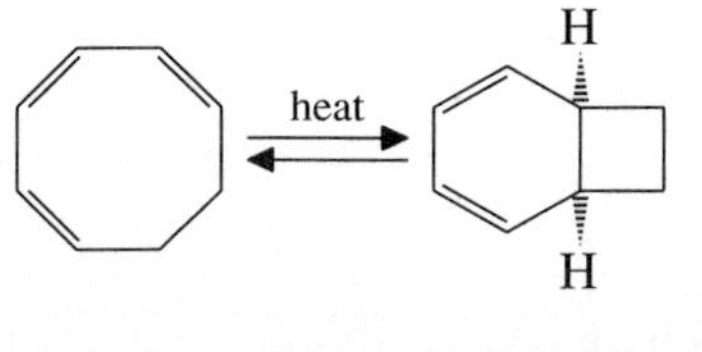

(b)

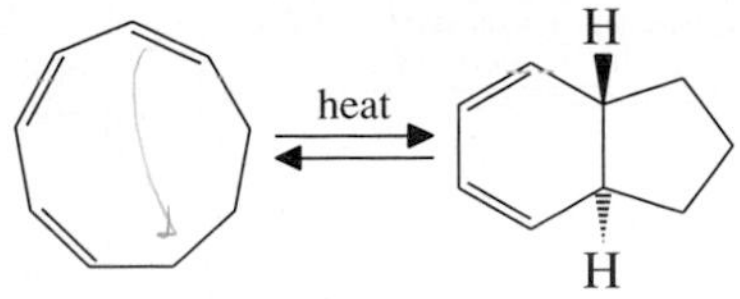

6 In the thermal ring opening of *trans*-3,4-dimethylcyclobutene, *two* products could be formed by a conrotatory mechanism, but only one is observed. Give the two possible products. Which one is observed, and why?

7 After heating to 200°, the following compound is converted in 95% yield into an isomer *A* that can be hydrogenated to cyclododecane. Give the structure of *A*, including its stereochemistry.

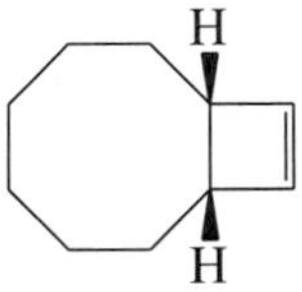

25.3 CYCLOADDITION REACTIONS

Let us now consider *cycloaddition reactions,* in which two or more π-electron systems form a ring at the expense of one double (or triple) bond in each of the reacting partners (see Eq. 25.2). Cycloadditions are classified, first, by the number of electrons involved in the reaction. The reaction in Eq. 25.17a is a [4 + 2] cycloaddition because the reaction involves four electrons from one reacting component and two electrons from the other. The reaction in Eq. 25.17b is a [2 + 2] cycloaddition.

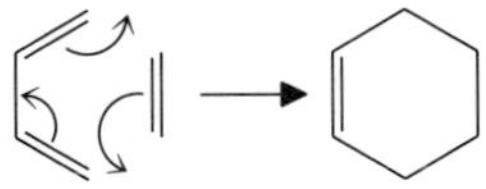

[4 + 2]

(25.17a)

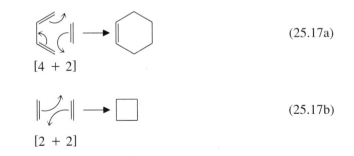

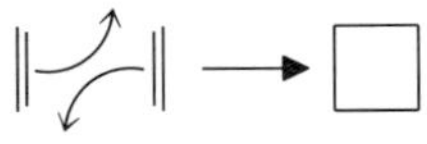

[2 + 2]

(25.17b)

As in electrocyclic reactions, the number of electrons involved is determined simply by writing the reaction mechanism in the curved-arrow notation. The number of electrons contributed by a given reactant is equal to twice the number of curved arrows originating from that component.

Cycloadditions are also classified by their stereochemistry with respect to *the plane of each reacting molecule.* (Recall that the carbons and their attached atoms in π-electron systems are coplanar.) This classification is shown for a [4 + 2] cycloaddition in Fig. 25.7. A cycloaddition may in principle occur either across the same face, or

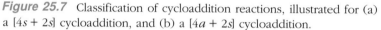

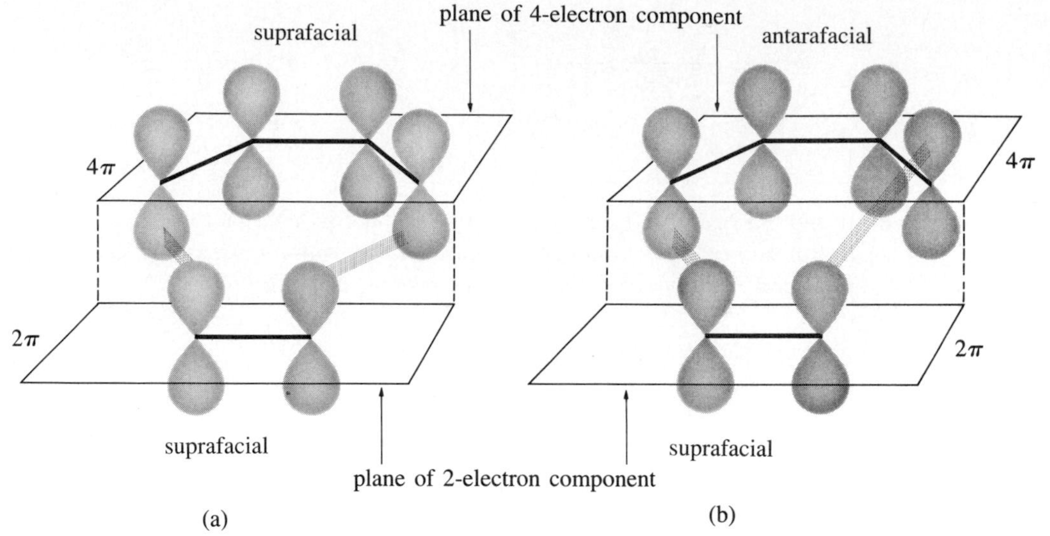

Figure 25.7 Classification of cycloaddition reactions, illustrated for (a) a [4*s* + 2*s*] cycloaddition, and (b) a [4*a* + 2*s*] cycloaddition.

across opposite faces, of the planes in each reacting component. If the reaction occurs across the same face of a π-system, the reaction is said to be **suprafacial** with respect to that π-system. A suprafacial addition is nothing more than a *syn* addition (Sec. 7.10A) that occurs in a single mechanistic step. If the reaction bridges opposite faces of a π-system, it is said to be **antarafacial.** An antarafacial addition is just an *anti* addition that occurs in one mechanistic step. Thus a [2*s* + 4*s*] cycloaddition is one that occurs suprafacially (or *syn*) on both the 2π component and the 4π component. A [2*s* + 4*a*] cycloaddition occurs suprafacially on the 2π component, but antarafacially (or *anti*) on the 4π component.

In order for a cycloaddition to occur, there must be bonding overlap between the *p* orbitals at the terminal carbons of each π-electron system; this is where the new bonds are formed. Let us illustrate this point with a [4 + 2] cycloaddition. We can think of this reaction in somewhat the same way that we think of a Lewis acid–Lewis base reaction: the Lewis base donates a pair of valence electrons to an empty orbital in the Lewis acid. By analogy, let us think of the diene (the 4π component) as an electron donor and the 2π component as the electron acceptor. (The result would be the same if we reversed the roles of the two components.) What electrons will the 4π component donate? Logically, these will be its "valence electrons"—the electrons in its HOMO (ψ_2 in Fig. 25.2). The 2π component will accept these electrons to form the new bonds. The MO used to accept these electrons must be *empty,* since an MO cannot contain more than two electrons. The empty MO used will be the one of lowest energy—the LUMO (ψ_2^* in Fig. 25.1). It follows that the two MOs used—the HOMO of the 4π component and the LUMO of the 2π component—must have *matching phases* if bonding overlap is to be achieved.

This phase match is achieved when a [4 + 2] cycloaddition occurs *suprafacially* on each component—that is, when the cycloaddition is a [4*s* + 2*s*] process:

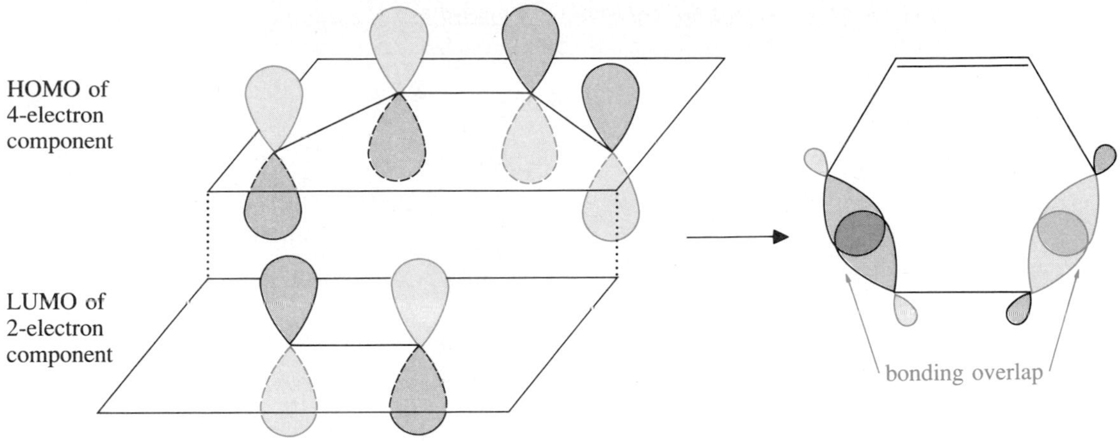

(25.18)

As we learned in Sec. 15.3, the Diels–Alder reaction, the most important example of a [4*s* + 2*s*] cycloaddition, indeed occurs suprafacially on each component. This can be seen from the retention of stereochemistry observed in both the diene and dienophile in the following examples:

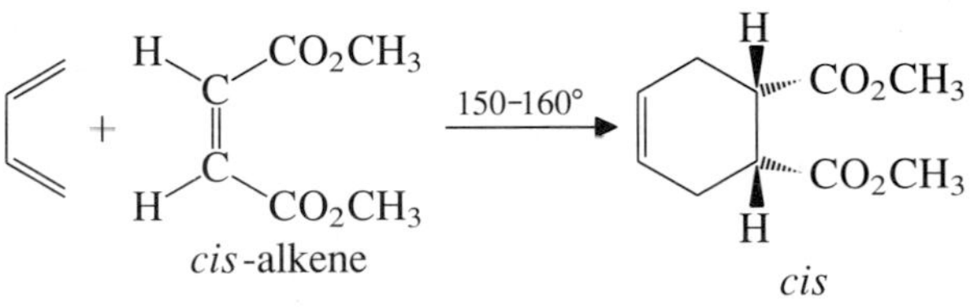

(25.19)

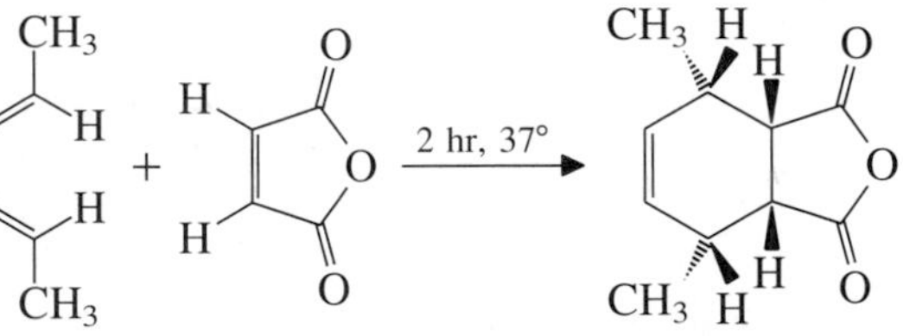

(25.20)

(A detailed discussion of the stereochemistry of the Diels–Alder reaction is given in Sec. 15.3C; this should be reviewed.) You should convince yourself that the [4*s* + 2*a*] and [4*a* + 2*s*] modes of cycloaddition do *not* provide bonding overlap at both ends of the π-electron systems.

The situation is different in a [2 + 2] cycloaddition. Again we use the HOMO of one component and the LUMO of the other. We can see that the orbital symmetries will *not* accommodate a cycloaddition that is suprafacial on both components:

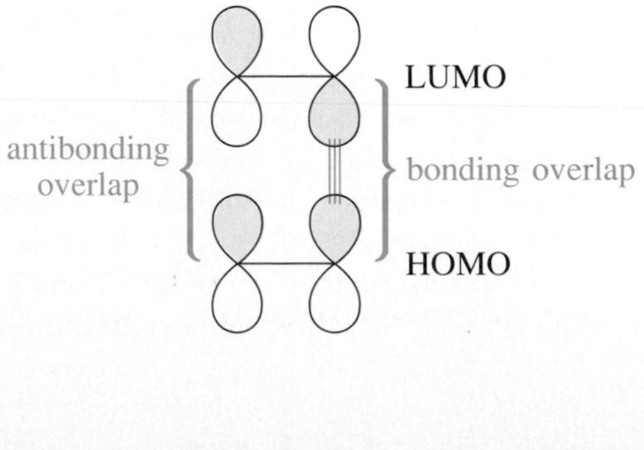

TABLE 25.2 Selection Rules for Cycloaddition Reactions

Number of electrons	Mode of activation	Allowed stereochemistry
$4n$	thermal	supra–antara antara–supra
	photochemical	supra–supra antara–antara
$4n + 2$	thermal	supra–supra antara–antara
	photochemical	supra–antara antara–supra

However, an addition that is suprafacial on one component but antarafacial on the other is allowed by orbital symmetry, but is geometrically more difficult. Indeed, the thermal [2 + 2] cycloaddition is a much less common reaction than the Diels–Alder reaction. Many of the known [2 + 2] additions occur by nonconcerted mechanisms, and therefore do not fall under the purview of the rules for pericyclic reactions.

Although the [$2s + 2s$] cycloaddition is forbidden by orbital symmetry under thermal conditions, it is allowed under photochemical conditions. Under these conditions it is the *excited state* of one alkene that reacts with the other alkene. The HOMO of the excited state has the proper symmetry to interact in a bonding way with the LUMO of the reacting partner:

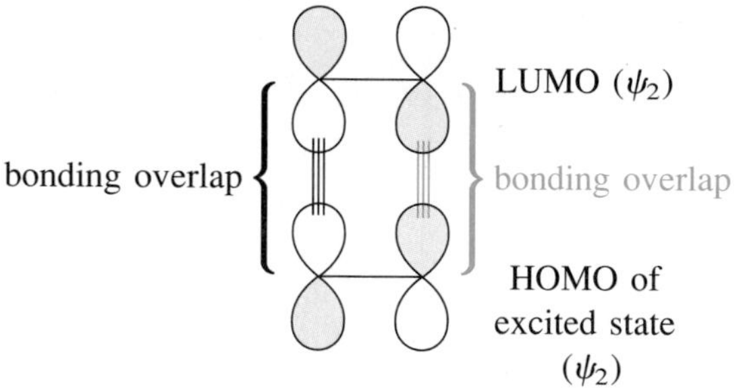

Indeed, many examples of photochemical [$2s + 2s$] cycloadditions are known. Such processes are widely used for making cyclobutanes.

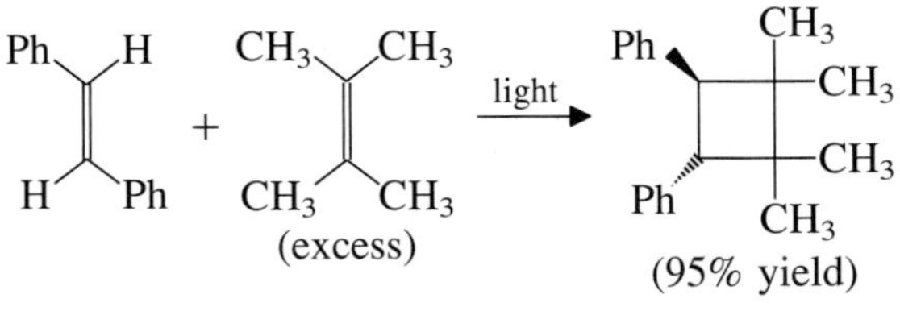

(25.21)

The results of this section can be generalized to the cycloaddition selection rules shown in Table 25.2. We can see that all-suprafacial cycloadditions are allowed thermally for systems in which the total number of reacting electrons is $4n + 2$, and photochemically for systems in which the number is $4n$.

A consequence of the selection rules is that suprafacial cycloadditions should be allowed for certain systems with more than six π-electrons. For example, the following all-suprafacial cycloaddition is a [4*s* + 6*s*] process involving ten electrons (five curved arrows). Notice that $4n + 2$ equals 10 for $n = 2$.

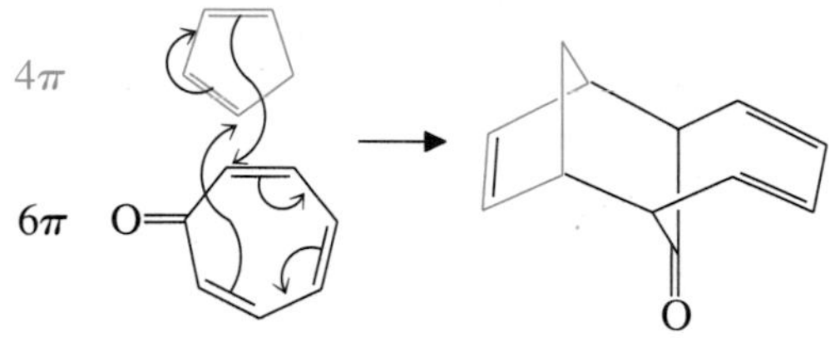

(25.22)

Problems

8 Give the product of the following reaction, which involves an [8*s* + 2*s*] cycloaddition:

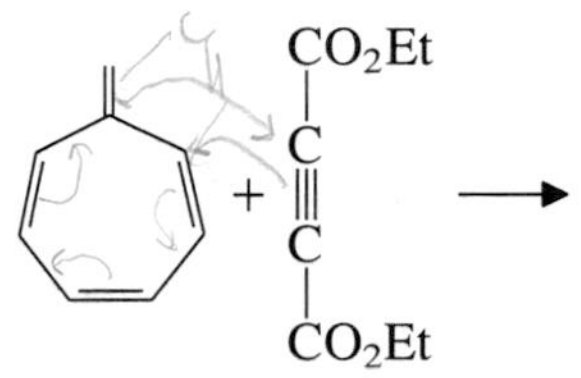

9 The photochemical cycloaddition of two molecules of *cis*-2-butene gives a mixture of two products, *A* and *B*. The analogous photochemical cycloaddition of *trans*-2-butene also gives a mixture of two products, *B* and *C*. The photochemical reaction of a *mixture* of *cis*- and *trans*-2-butene gives a mixture of *A, B,* and *C,* along with a fourth product, *D*. Propose structures for all four compounds.

25.4 SIGMATROPIC REACTIONS

A. Classification and Stereochemistry of Sigmatropic Reactions. [1,3] and [1,5] Sigmatropic Rearrangements

In a *sigmatropic reaction,* a σ-bond formally migrates from one end of a π-system to the other, and the number of double or triple bonds remains the same (Eqs. 25.3 and 25.4). Sigmatropic reactions are classified by using bracketed numbers to indicate the number of atoms over which a σ-bond formally migrates. In some reactions, both ends of a σ-bond migrate. In the following reaction, for example, each end of a σ-bond formally migrates over three atoms. (We count the point of original attachment as atom #1.) The reaction is therefore a [3,3] sigmatropic reaction.

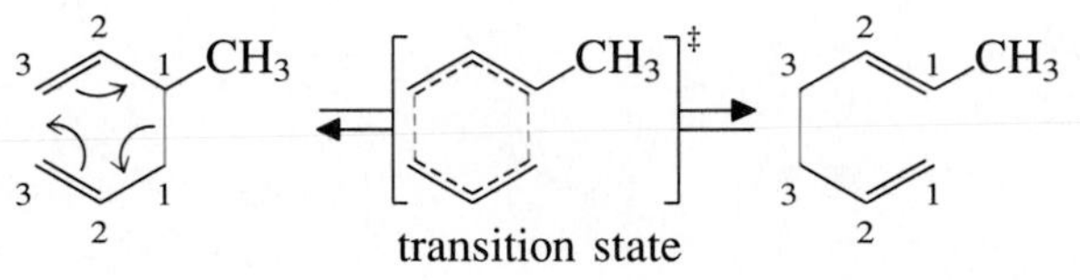

(25.23)

In other reactions, one end of a σ-bond remains fixed to the same group and the other end formally migrates. For example, the following reaction is a [1,5] sigmatropic reaction because one end of the bond "moves" from atom #1 to atom #1 (that is, it doesn't move), and the other end moves over five atoms.

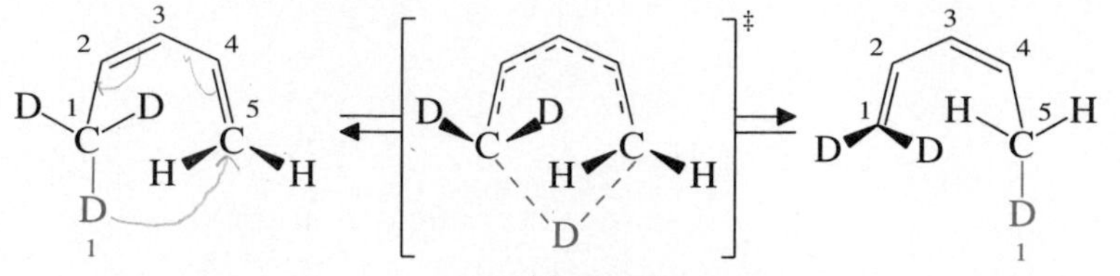

(25.24)

Sigmatropic reactions, like other pericyclic reactions, are also classified by their stereochemistry. In this classification, we consider whether the migrating bond moves over the same face, or between opposite faces, of the π-electron system. If the migrating bond moves across one face of the π-system, the reaction is said to be *suprafacial.* For example, if the [1,5] sigmatropic reaction of Eq. 25.24 were suprafacial, it would occur in the following manner:

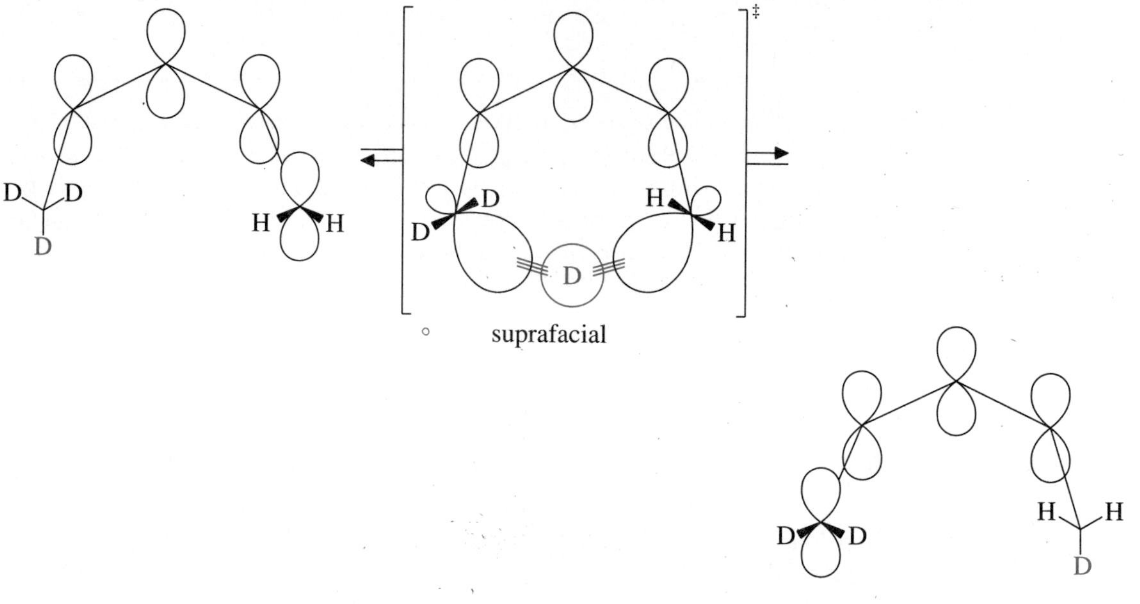

(25.25)

If the reaction were antarafacial, it would occur instead as shown in Eq. 25.26:

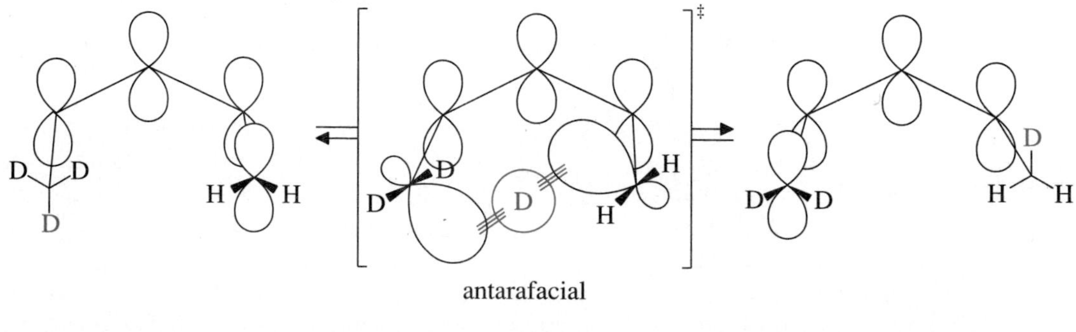

(25.26)

When both ends of a σ-bond migrate, the reaction can be suprafacial or antarafacial with respect to either π-system. For example, if the [3,3] sigmatropic reaction in Eq. 25.23 were suprafacial on both π-systems, it could occur as follows:

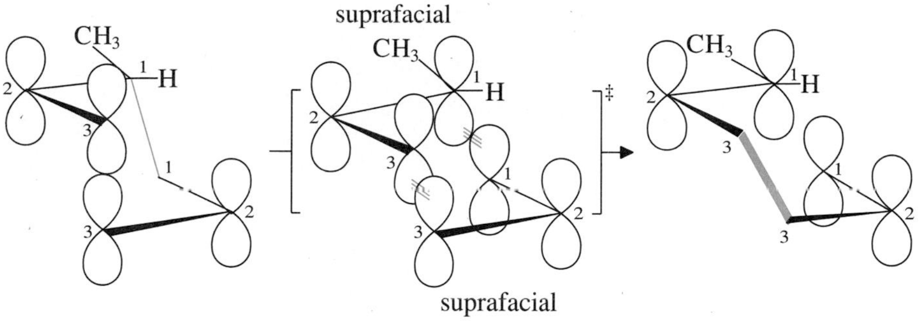

(25.27)

Problems

10 Classify the following sigmatropic reaction by giving its bracketed-number designation and its stereochemistry with respect to the plane of the π-electron system.

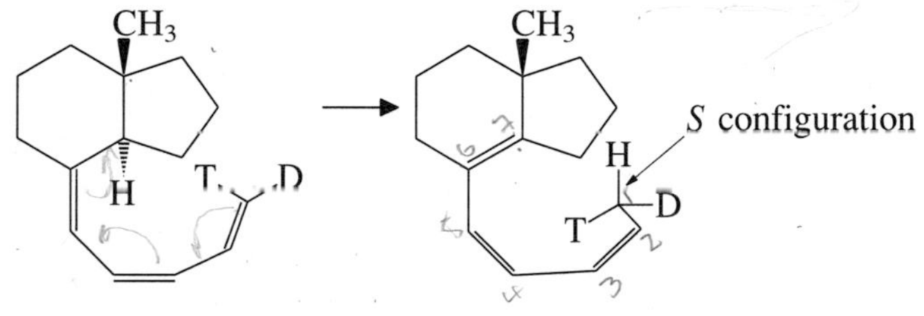

11 Classify the following sigmatropic reactions by bracketed number:

(a)

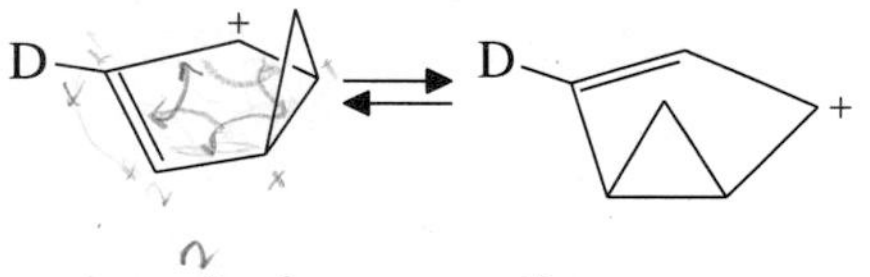

(b)

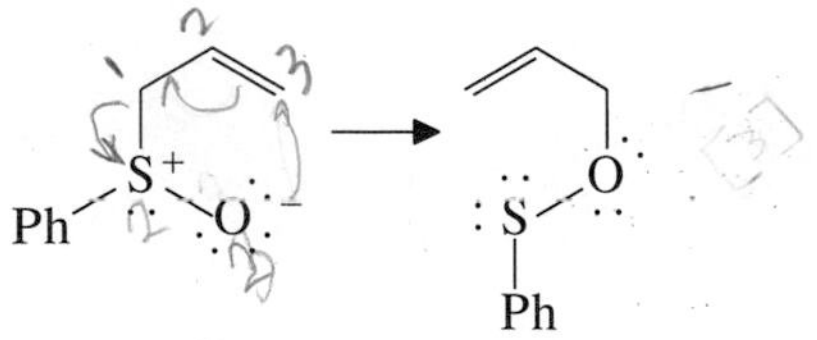

(c)

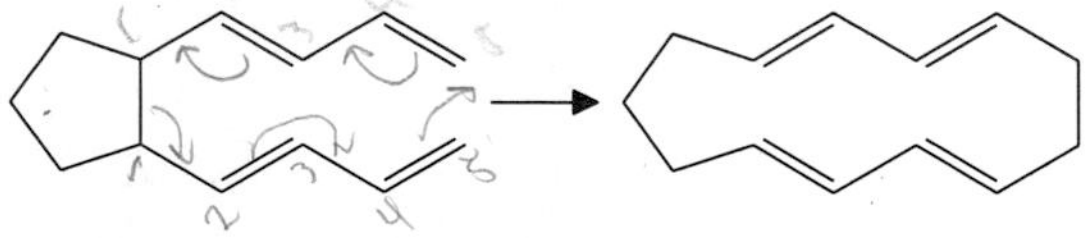

Molecular orbital theory provides the connection between the type of sigmatropic reaction and its stereochemistry. Consider, for example, a [1,5] sigmatropic migration of hydrogen across a π-electron system. We can think of this reaction as the migration of a proton from one end of the 2,4-pentadienyl anion (Fig. 25.5) to the other:

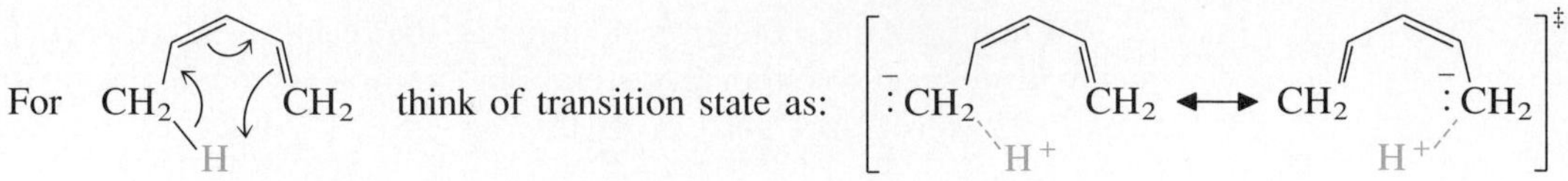

Sigmatropic reactions are concerted and do not involve free ions, but the orbital relationships are nevertheless easily understood this way. We could equally well think of the reaction as the migration of a hydrogen atom across a 2,4-pentadienyl radical, or a hydride anion across a 2,4-pentadienyl cation; the result would be the same.

The interaction of the proton LUMO—an empty 1*s* orbital—with the HOMO of the π-system controls the stereochemistry of the reaction. For the 2,4-pentadienyl anion (Fig. 25.5), the HOMO is symmetric. This means that bonding overlap can occur if the migration occurs suprafacially:

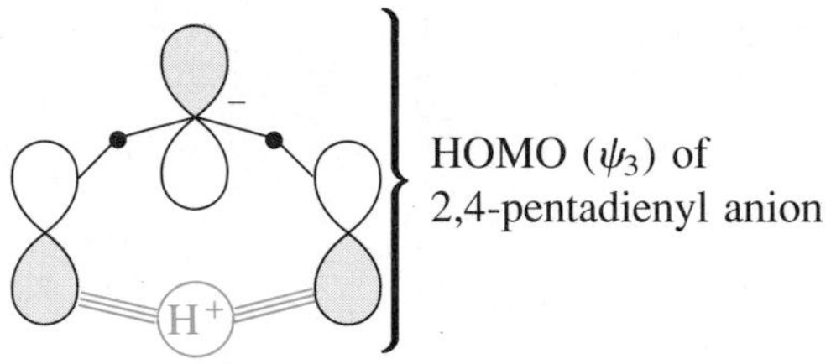

The suprafacial [1,5] migration of hydrogen causes no serious distortions or strains in most molecules and is in fact a relatively common reaction.

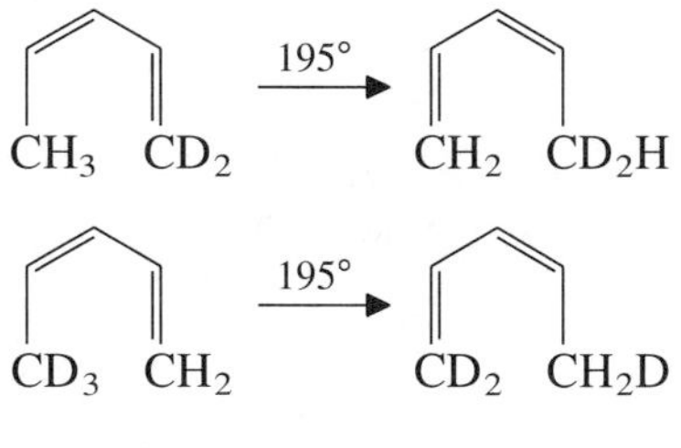

(25.28a)

(25.28b)

In contrast, the [1,3] hydrogen shift involves the HOMO of the allyl anion, an antisymmetric orbital:

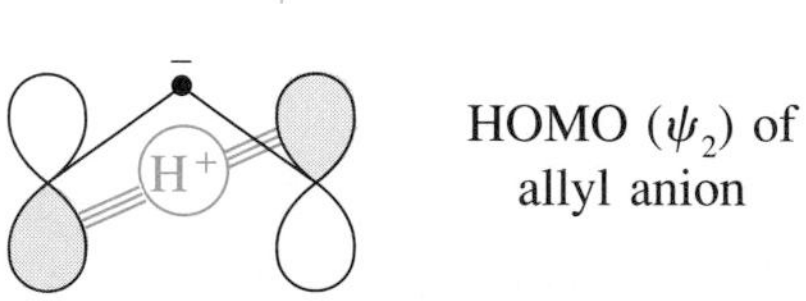

In order for a [1,3] hydrogen shift to occur, the migrating hydrogen must pass from one face of the allyl π-system to the other. Despite the fact that this reaction is "allowed" by MO theory, it requires that the migrating proton bridge too long a distance for adequate bonding. Alternatively, the terminal lobes of the allyl π-system could twist; but then a new problem arises: these lobes cannot overlap with the *p* orbital of the central carbon. The resulting loss of conjugation also raises the energy of the transition state. As we might anticipate from these arguments, the concerted sigmatropic [1,3] hydrogen shift is virtually nonexistent in organic chemistry. For example, the following [1,3] sigmatropic reaction was shown *not* to occur even under extreme temperatures and pressures:

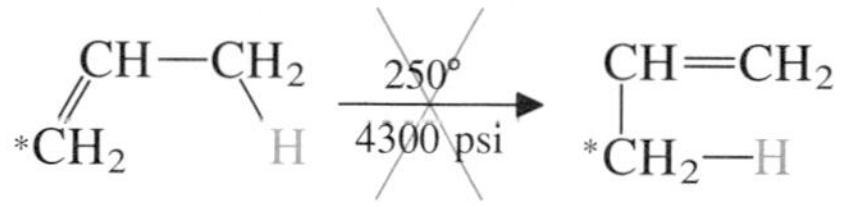

$$\overset{\displaystyle CH-CH_2}{\underset{\displaystyle *CH_2\quad\quad H}{\|}} \xrightarrow[\substack{4300\text{ psi}}]{250°} \overset{\displaystyle CH=CH_2}{\underset{\displaystyle *CH_2-H}{|}} \tag{25.29}$$

Several interesting experiments have been conducted in which a molecule is offered a choice between [1,5] and [1,3] hydrogen shifts. In one such experiment—an experiment of elegant simplicity—carried out in 1964 by Wolfgang Roth at the University of Cologne, 1,3,5-cyclooctatriene was labeled at carbons 7 and 8 with deuterium and then allowed to undergo many hydrogen shifts for a long period of time. When the molecule undergoes [1,5] hydrogen shifts, the D should migrate part of the time, and the H part of the time. However, if one waits long enough, the D should eventually scramble to all positions that have a 1,5-relationship. In such a case, only carbons 3, 4, 7, and 8 would be partially deuterated:

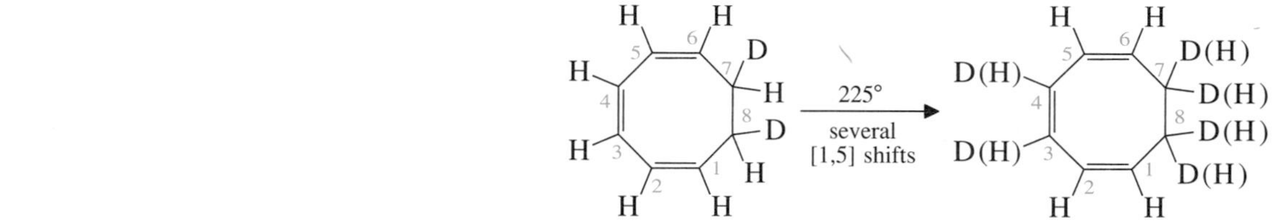

$$\tag{25.30a}$$

(You should write a series of steps for this transformation to convince yourself that it is the predicted result.) On the other hand, if the molecule undergoes successive [1,3] hydrogen (or deuterium) shifts, the deuterium should be scrambled eventually to all positions.

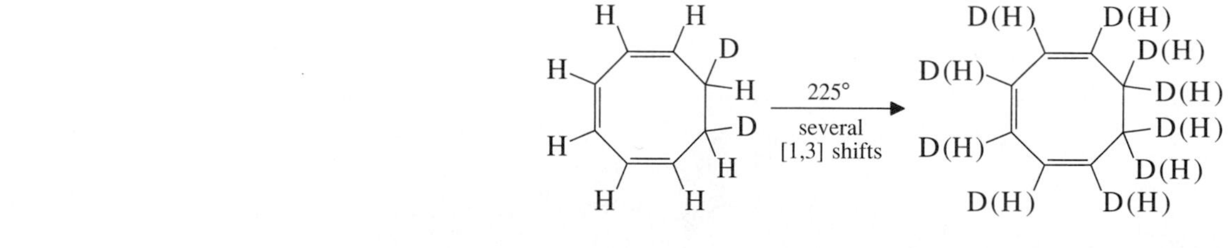

$$\tag{25.30b}$$

The experimental result was that even after very long reaction times, deuterium appeared only in the positions predicted by the [1,5] shift.

Roth and his collaborators also carried out another ingenious experiment that determined the stereochemistry of the [1,5] sigmatropic hydrogen shift. In the isotopically labeled, optically active alkene shown in Eq. 25.31, two suprafacial [1,5] hydrogen shifts are possible, one from each of the torsional isomers shown:

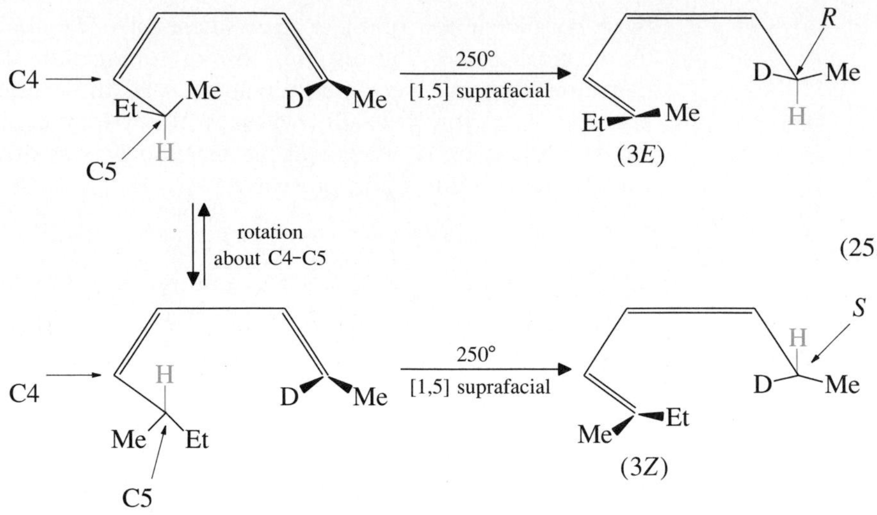

(25.31)

We can see that if the migration is suprafacial, the 3*E* isomer of the product must have the *R* configuration, and the 3*Z* isomer of the product must have the *S* configuration. (What is the predicted result if the migration were antarafacial?) The experimental result is shown in Eq. 25.31: the suprafacial migration was confirmed.

Although the suprafacial [1,3] shift of a hydrogen is *not* allowed, the corresponding shift of a carbon atom *is* allowed, provided that certain stereochemical conditions are met. Suppose that an alkyl group—suitably substituted so that we can trace its stereochemistry—were to undergo a suprafacial [1,3] sigmatropic shift. This shift could occur in two stereochemically distinct ways. In the first way, the carbon migrates with *retention* of configuration.

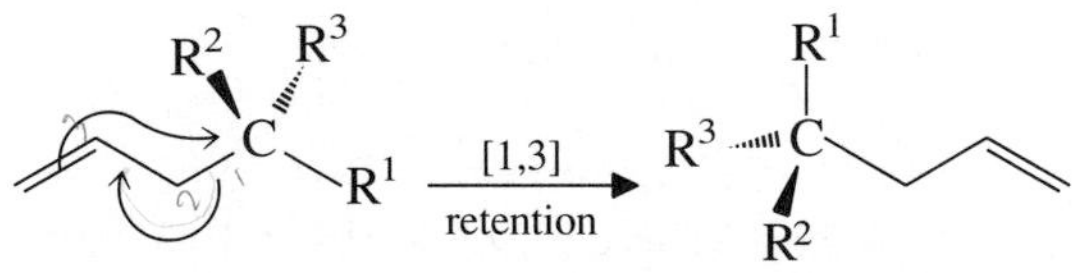

(25.32)

In the second way, the carbon migrates with *inversion* of configuration.

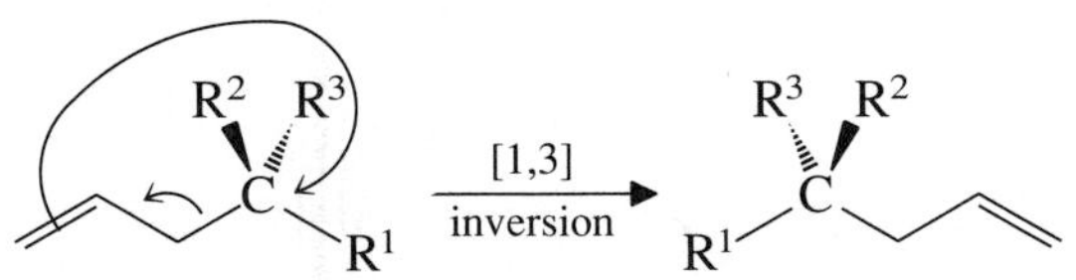

(25.33)

Let us look at the orbital symmetry relationships in these two modes of reaction. We can think of the migrating group as an alkyl cation migrating between the ends of an allyl anion. The LUMO of the alkyl cation—an empty *p* orbital—interacts with the HOMO of an allyl anion (ψ_2 in Fig. 25.4). In the case of migration with *retention,* the phase relationships between the orbitals involved lead to antibonding overlap:

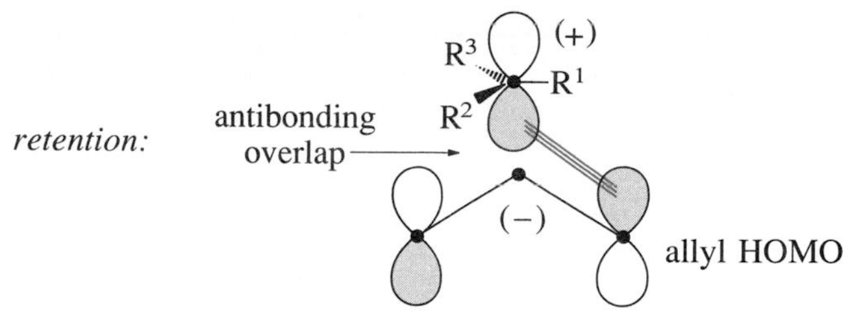

Hence, suprafacial carbon migration with retention is forbidden by orbital symmetry, in the same sense that hydrogen migration is forbidden. If migration occurs with *inversion,* however, there is bonding overlap in the transition state.

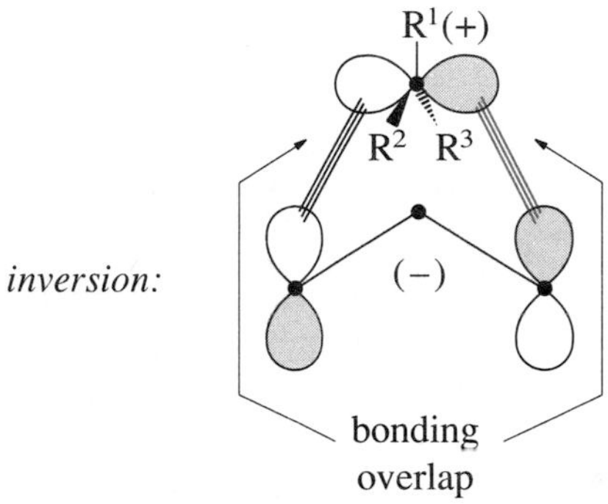

Thus, carbon migration with inversion is allowed by orbital symmetry.

We can see from this analysis that it is the node in the *p* orbital that makes the [1,3] suprafacial migration of carbon possible; each of the two lobes of the *p* orbital, which have opposite phase, can overlap with each end of the allyl π-system. Because a bond is broken at one side of the migrating carbon and formed at the other side, inversion of configuration is observed. In the migration of a hydrogen, the orbital involved is a 1*s* orbital, which has no nodes. Hence, [1,3] suprafacial migration of hydrogen is not allowed.

Orbital symmetry, then, makes a very straightforward prediction: the suprafacial [1,3] sigmatropic migration of carbon must occur with inversion of configuration. The following result confirms this prediction:

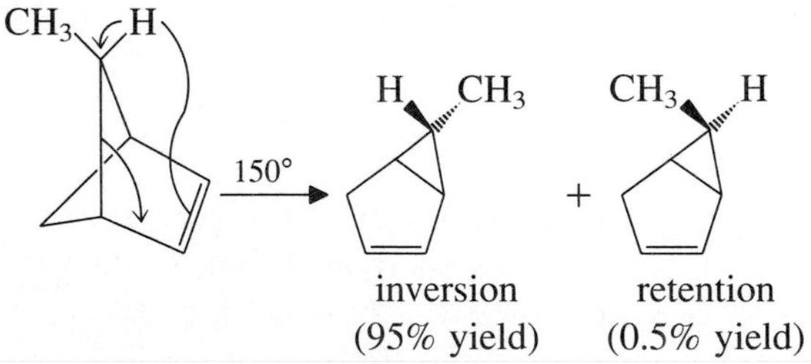

(25.34)

Without a knowledge of orbital symmetry, we might have guessed that migration with retention of configuration would have been the most straightforward, least contorted pathway that the rearrangement could take; yet the theory predicts otherwise. One of the remarkable things about the theory of orbital symmetry is that it correctly predicts so many reactions that otherwise would have appeared unlikely.

As we might expect, orbital symmetry dictates a reversal of stereochemistry for [1,5] migrations. Carbon, like hydrogen, undergoes suprafacial [1,5] migrations with retention of configuration (Problem 14).

B. [3,3] Sigmatropic Rearrangements: Cope and Claisen Rearrangements

Let us now turn to a sigmatropic rearrangement in which two groups migrate: the [3,3] sigmatropic rearrangement. We can think of the transition state for this rearrangement as the interaction of two allylic systems, one a cation and one an anion.

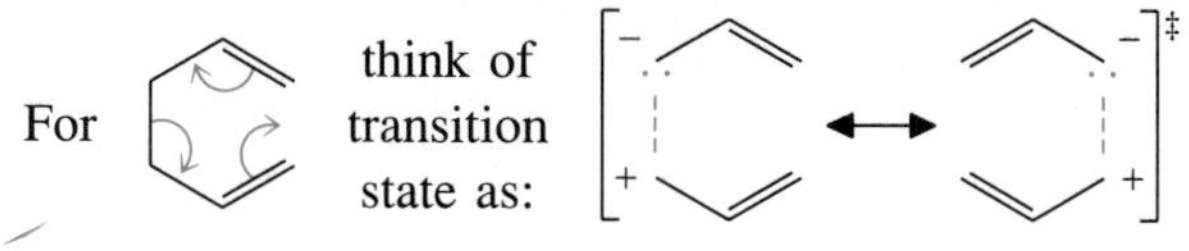

The "valence orbitals" involved are the HOMO of the anion and the LUMO of the cation, which, as we can see from Fig. 25.4, are the same orbital (ψ_2) of the allyl system.

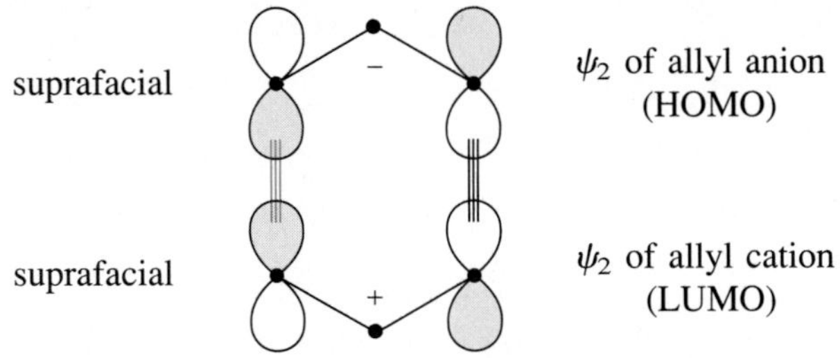

The two MOs involved achieve bonding overlap when the [3,3] sigmatropic reaction occurs suprafacially on both components. (You should convince yourself that a reaction that is antarafacial on both π-systems is also allowed, but one that is suprafacial on one component and antarafacial on the other is forbidden.)

There are many examples of [3,3] sigmatropic reactions. One of the best known is the **Cope rearrangement,** which was extensively investigated by Arthur C. Cope of the Massachusetts Institute of Technology long before the principles of orbital symmetry were known. The Cope rearrangement is simply a hydrocarbon isomerization:

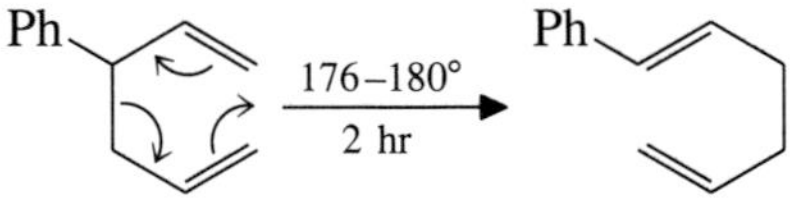

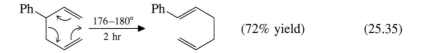

(72% yield) (25.35)

An interesting variation of the Cope rearrangement is the "oxyCope" reaction. In this type of reaction, an enol is formed initially; tautomerization of the enol into the corresponding carbonyl compound is a very favorable equilibrium that drives the reaction to completion.

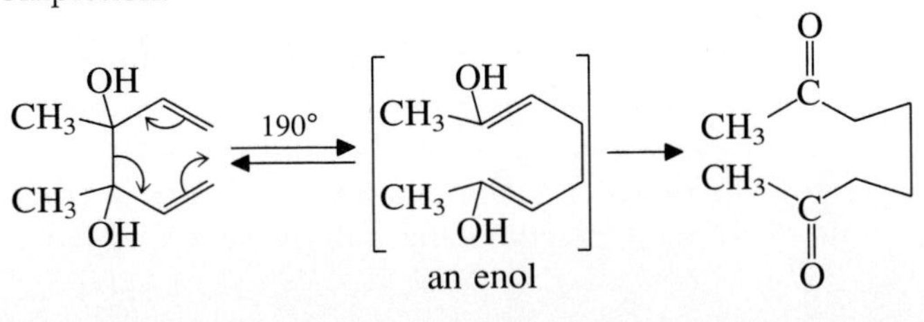

(25.36)

In the **Claisen rearrangement,** an allylic ether undergoes a [3,3] sigmatropic rearrangement.

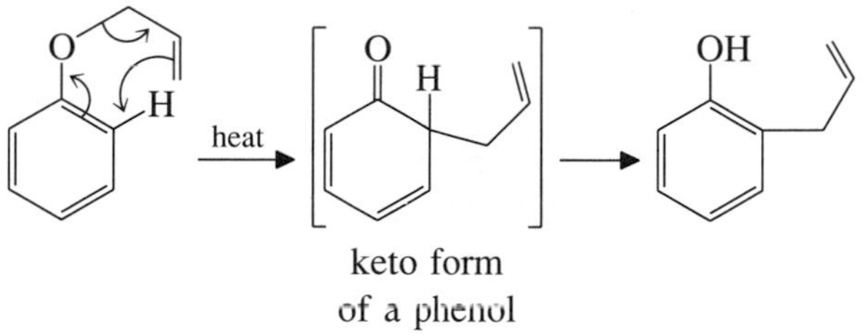

(25.37)

If both *ortho* positions are blocked by substituent groups, the *para*-substituted derivative is obtained:

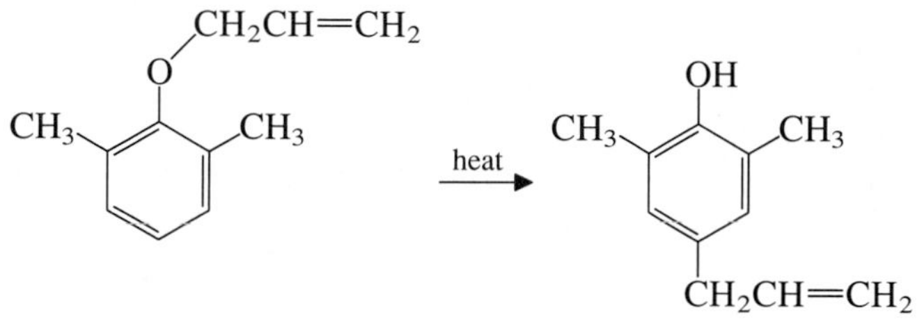

(95% yield) (25.38)

This reaction occurs by a sequence of two Claisen rearrangements, followed by tautomerization of the product to the phenol:

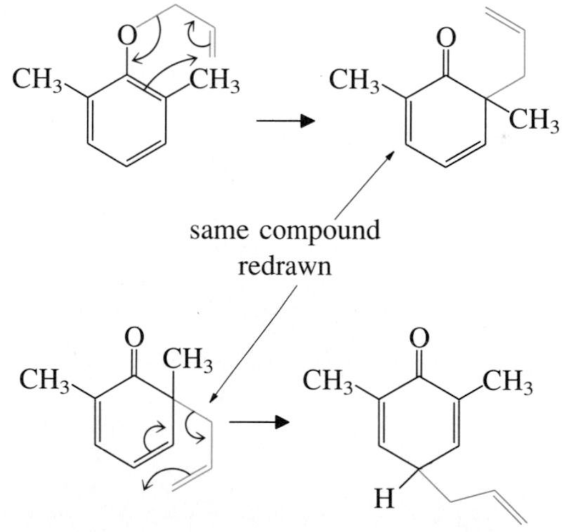

(25.39a)

(25.39b)

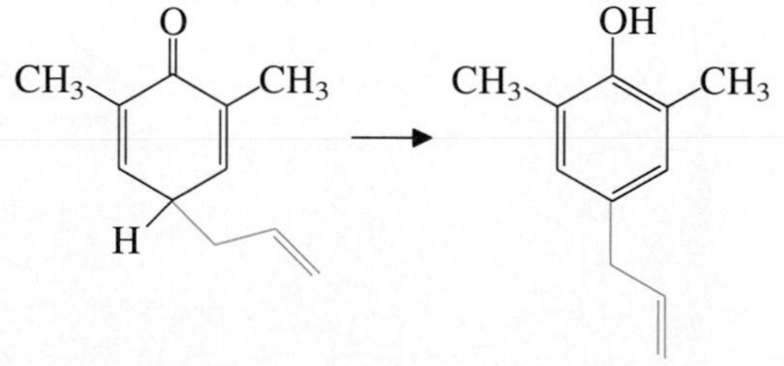

(25.39c)

Claisen rearrangements of aliphatic ethers are also well known (Problem 13).

The Claisen rearrangement is the third reaction we have studied that is named after Ludwig Claisen. The other two are the Claisen–Schmidt condensation, a variation of the aldol condensation (Sec. 22.4C); and the Claisen condensation of esters (Sec. 22.5A).

C. Summary: Selection Rules for Sigmatropic Reactions

The stereochemistry of sigmatropic reactions is a simple function of the number of electrons involved. (As with other pericyclic reactions, the number of electrons involved is easily determined from the curved-arrow formalism: simply count the curved arrows and multiply by two.) All-suprafacial sigmatropic reactions occur when there are $4n + 2$ electrons involved in the reaction—that is, an odd number of electron pairs, or curved arrows. In contrast, a sigmatropic reaction must be antarafacial on one component and suprafacial on the other when $4n$ electrons (an even number of electron pairs, or curved arrows) are involved. When a single carbon migrates, the term "suprafacial" is taken to mean "retention of configuration," and the term "antarafacial" is taken to mean "inversion of configuration."

These generalizations are summarized in Table 25.3 as selection rules for sigmatropic reactions. These selection rules hold for thermal reactions—that is, reactions that occur from electronic ground states. As with other pericyclic reactions, there are selection rules for photochemical sigmatropic reactions—reactions that occur from excited states. However, since these are rather uncommon, we shall not concern ourselves with them here.

Problems

12 (a) What allowed and reasonable sigmatropic reaction(s) can account for the following transformation?

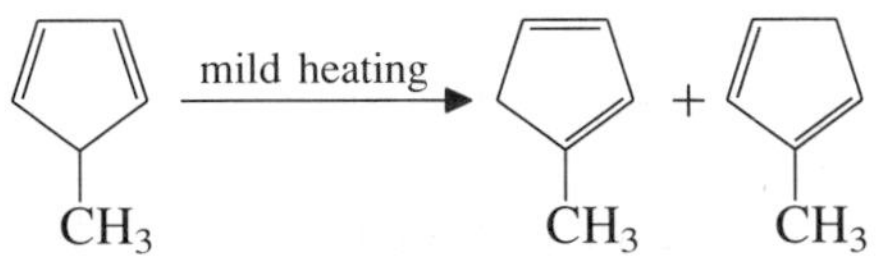

(b) What product(s) are expected from a similar reaction of 2,3-dimethyl-1,3-cyclopentadiene?

TABLE 25.3 Selection Rules for Thermal Sigmatropic Reactions

Number of electrons	Allowed stereochemistry: Generalized stereochemistry	Allowed stereochemistry: Stereochemistry of single-atom migrations
$4n$	supra–antara	supra–inversion
	antara–supra	antara–retention
$4n + 2$	supra–supra	supra–retention
	antara–antara	antara–inversion

13 (a) Aliphatic vinylic allylic ethers undergo the Claisen rearrangement. Complete the following reaction:

$$(CH_3)_2C{=}CH{-}CH_2{-}O{-}CH{=}CH_2 \xrightarrow{\text{heat}}$$

(b) What starting material would give the following compound in an aliphatic Claisen rearrangement?

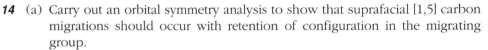

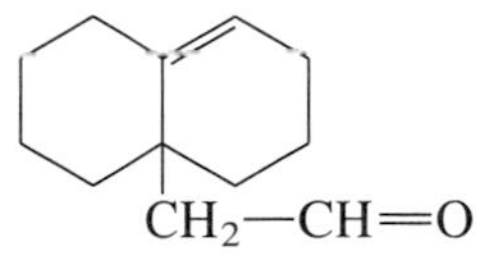

14 (a) Carry out an orbital symmetry analysis to show that suprafacial [1,5] carbon migrations should occur with retention of configuration in the migrating group.

(b) Indicate what type of sigmatropic reactions are involved in the following transformation. Is the stereochemistry of the first step in accord with the predictions of orbital symmetry?

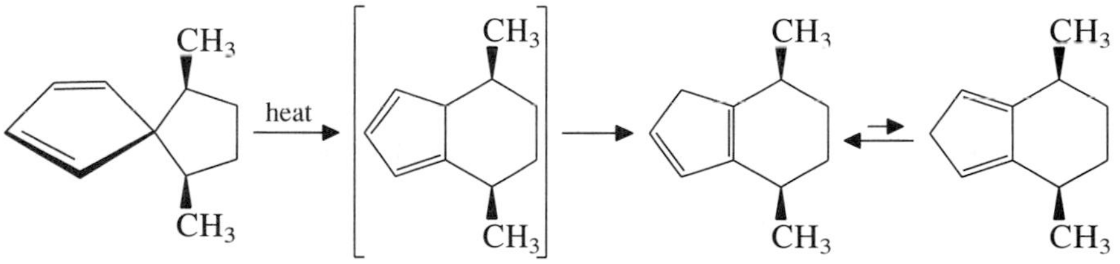

25.5 SUMMARY OF THE PERICYCLIC SELECTION RULES

It is always possible to remember the allowed stereochemistry of a pericyclic reaction by using the phase relationships within the molecular orbitals involved. However, there is a convenient way to remember the selection rules without re-deriving them each time they are needed. This involves assigning either a $+1$ or a -1 to each of the following aspects of the reaction: thermal or photochemical; $4n + 2$ or $4n$ reacting electrons; and stereochemistry. First, we assign $+1$ to thermal reactions and -1 to photochemical reactions. Next, we assign $+1$ to systems with $4n + 2$ reacting electrons, and -1 to systems with $4n$ reacting electrons. Finally, we assign $+1$ to disrotatory or suprafacial stereochemistry, and -1 to conrotatory or antarafacial stereochemistry; and we make this last assignment for each stereochemical component of the reaction. Then we multiply together the resulting numbers. If we get a $+1$, the reaction is allowed; if we get a -1, the reaction if forbidden.

Let us illustrate with some examples.

For the thermal $(+1)$ disrotatory $(+1)$ electrocyclic ring closure of 1,3,5-hexatriene: $4n + 2$ electrons $(+1)$.
Result: $(+1)(+1)(+1) = +1$: allowed.

For the photochemical (−1) [4*s* + 2*s*] cycloaddition: $4n + 2$ electrons (+1). There are two stereochemical components, one on each reacting π-system, and both are suprafacial (+1).
Result: (−1)(+1)(+1)(+1) = −1: forbidden

For the thermal (+1) [1,3] suprafacial migration of carbon with inversion of configuration: $4n$ electrons (−1); supra-inversion, (+1)(−1).
Result: (+1)(−1)(+1)(−1) = +1: allowed

However we choose to remember the selection rules, we can see that orbital symmetry is a powerful tool for understanding and predicting the course of pericyclic reactions.

Problem

15 Without consulting tables in this chapter, classify the stereochemistry (+1 or −1) of each of the following reactions. Indicate whether each is allowed or forbidden.
(a) [4*s* + 4*a*] thermal cycloaddition
(b) suprafacial [1,7] thermal hydrogen migration
(c) disrotatory photochemical electrocyclic ring closure reaction of (2*E*,4*Z*,6*Z*,8*E*)-2,4,6,8-decatetraene

25.6 FLUXIONAL MOLECULES

A number of compounds continually undergo rapid sigmatropic rearrangements at room temperature. One such compound is *bullvalene,* which was first prepared in 1963.

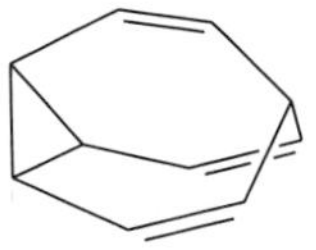

bullvalene

The [3,3] sigmatropic rearrangements in bullvalene rapidly interconvert *identical forms of the molecule.*

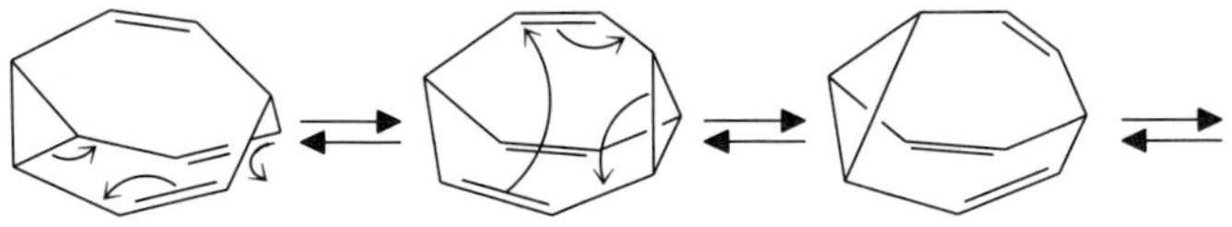

many others (25.40)

If the carbons could be individually labeled, there would be 1,209,600 different structures of bullvalene in equilibrium! Each one of these forms is converted into another at a rate of about 2000 times per second at room temperature.

Molecules such as bullvalene that undergo rapid bond shifts are called **fluxional molecules.** Their atoms are in a continual state of motion associated with the rapid changes in bonding.

Since all the forms of bullvalene are identical, we might well ask how it is known that the equilibria in Eq. 25.40 are actually taking place. The solution to this problem was obtained by the NMR methods discussed in Sec. 13.7. (See Problem 43.)

The fluxional behavior of bullvalene was predicted by Professor William von E. Doering of Harvard University prior to its synthesis. Doering and his collaborators carried out much of the early research on fluxional molecules, as well as important research on the stereochemistry of sigmatropic rearrangements.

Problem

16 Each of the following compounds exists as a fluxional molecule, interconverted into one or more identical forms by the sigmatropic process indicated. Draw one structure in each case that demonstrates the process involved and explain why each process is an allowed pericyclic reaction.

(a)

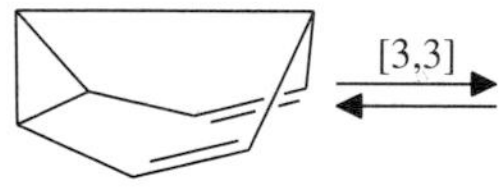

(b)

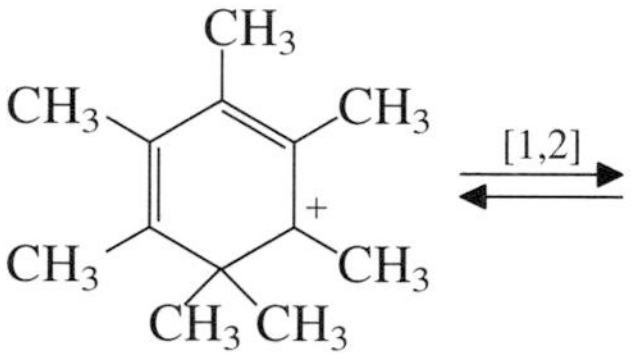

25.7 PERICYCLIC REACTIONS IN BIOLOGY: FORMATION OF VITAMIN D

It has long been known that, in areas of the world where winters are long and there is little sunlight, children suffer from a disease called *ricketts* (from old English, *wrickken,* to twist). This disease is characterized by inadequate calcification of bones. A similar disease in adults, *osteomalacia,* is particularly prominent among Bedouin Arab women who must remain completely covered when they are outdoors.

Ricketts can be prevented by administration of any one of the forms of vitamin D, a hormone that controls calcium deposition in bone. The human body manufactures a chemical precursor to vitamin D called 7-dehydrocholesterol. This is converted into vitamin D_3, or *cholecalciferol,* only when the skin receives adequate ultraviolet radiation from the sun or other source.

The reaction by which 7-dehydrocholesterol is converted into vitamin D_3 is a sequence of two pericyclic reactions. The first is an electrocyclic reaction:

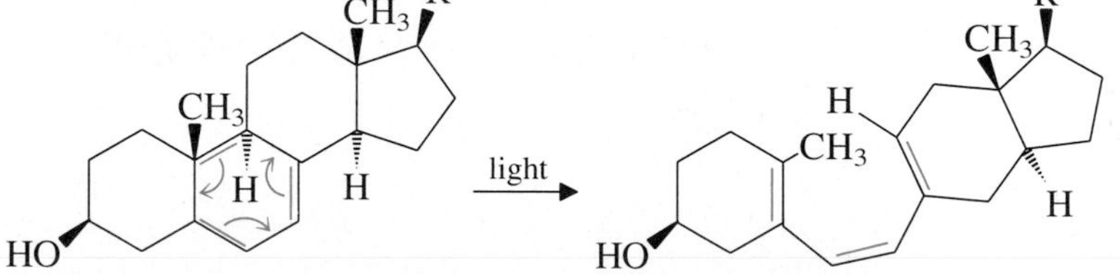

(25.41)

7-dehydrocholesterol previtamin D_3

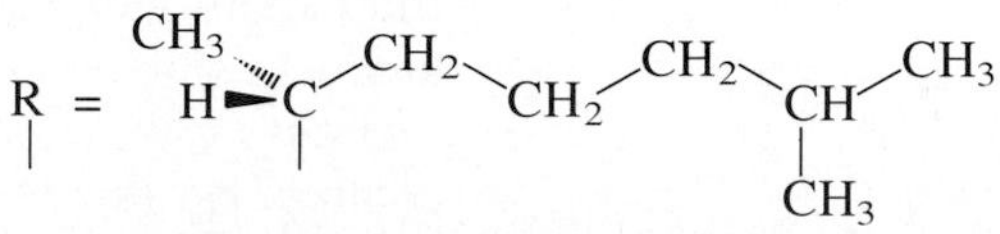

This reaction is clearly a *conrotatory* process. (Be sure you see why this is so; examine the reverse reaction if necessary.) This is precisely the stereochemistry required for a *photochemically allowed* electrocyclic reaction involving $4n + 2$ electrons. Sunlight ordinarily provides the UV radiation necessary for this reaction to occur in the human organism.

The final step in the formation of vitamin D_3 is a [1,7] sigmatropic hydrogen shift.

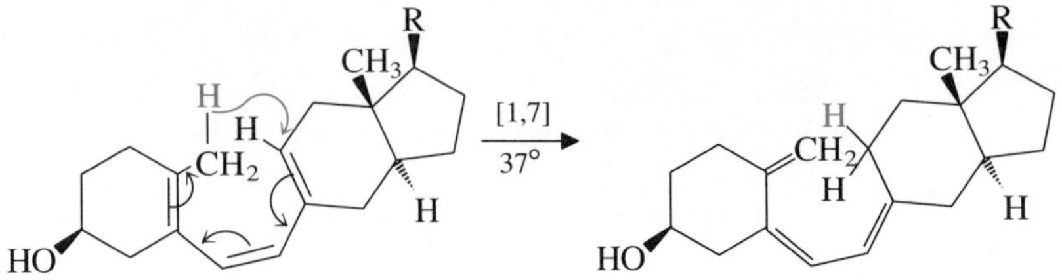

cholecalciferol
(vitamin D_3)

(25.42)

Vitamin D_3 exists in the more stable *s-trans* form, attained by internal rotation about the bond shown in color:

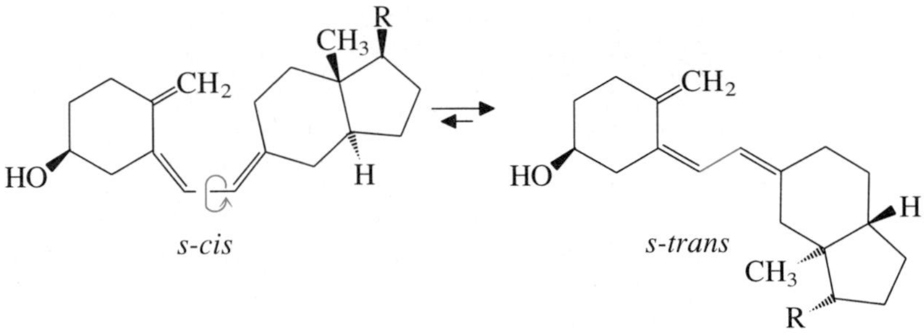

(25.43)

Vitamin D_2 (ergocalciferol), a compound closely related to vitamin D_3, is formed by irradiation of a steroid called *ergosterol*. The only difference between ergosterol and 7-dehydrocholesterol is the side-chain R:

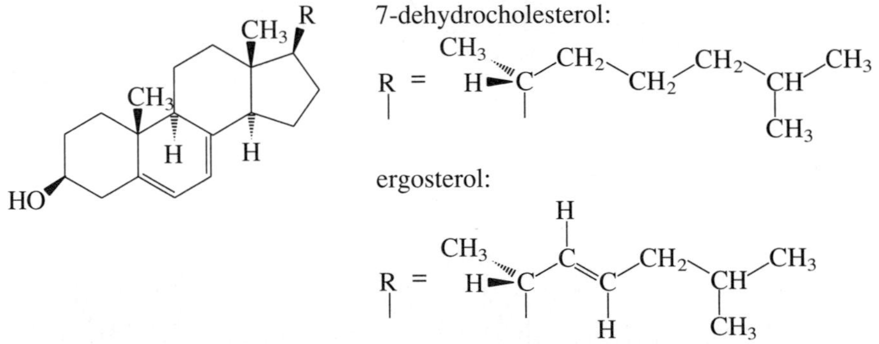

Irradiation of ergosterol gives successively previtamin D_2 and vitamin D_2, which are identical to the products of Eq. 25.41 and 25.42, respectively, except for the R-group. Vitamin D_2, sometimes called "irradiated ergosterol," is the form of vitamin D that is commonly added to milk and other foods as a dietary supplement.

Problems

17 When previtamin D_2 is isolated and irradiated, ergosterol is obtained along with a stereoisomer, *lumisterol.*

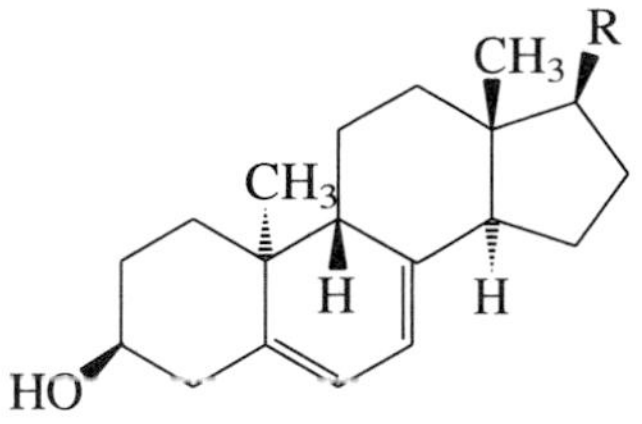

lumisterol

Explain mechanistically the origin of lumisterol.

18 When previtamin D_2 is *heated,* two compounds *A* and *B* are obtained that are stereoisomers of both ergosterol and lumisterol. Suggest structures for these compounds and explain mechanistically how they are formed.

19 When the compounds *A* and *B* in the previous problem are *irradiated,* two stereoisomeric compounds *C* and *D,* respectively, are obtained, each of which contains a cyclobutene ring. Suggest structures for *C* and *D* and explain mechanistically how they are formed. Explain why irradiation of *A* and *B* does *not* give back previtamin D_2.

KEY IDEAS IN CHAPTER 25

- Pericyclic reactions are concerted reactions that occur by cyclic electron shifts. Electrocyclic reactions, cycloadditions, and sigmatropic rearrangements are important pericyclic reactions.
- Electrocyclic reactions are stereochemically classified as conrotatory or disrotatory; cycloadditions and sigmatropic rearrangements are classified as suprafacial or antarafacial.
- The stereochemical course of a pericyclic reaction is governed largely by the symmetry of the reactant HOMO (highest occupied molecular orbital), or, if there are two reacting components, by the relative symmetries of the HOMO of one component and the LUMO (lowest unoccupied molecular orbital) of the other.
- Considerations of orbital symmetry lead to selection rules for pericyclic reactions. Whether a pericyclic reaction is allowed or forbidden depends on the number of electrons involved, the mode of activation (thermal or photochemical), and the stereochemical course of the reaction. The selection rules are summarized in Tables 25.1–25.3 and Sec. 25.5.

ADDITIONAL PROBLEMS

20 Without consulting tables or figures, answer the following questions:

(a) Is a thermal disrotatory electrocyclic reaction involving twelve electrons allowed or forbidden?

(b) Is a [4*s* + 8*s*] photochemical cycloaddition allowed?

(c) Is the HOMO of (3*Z*)-3,4-dimethyl-1,3,5-hexatriene symmetric or antisymmetric?

21 What do the pericyclic selection rules have to say about the *position of equilibrium* in each of the following reactions? Which side of each equilibrium is favored and why?

(a)

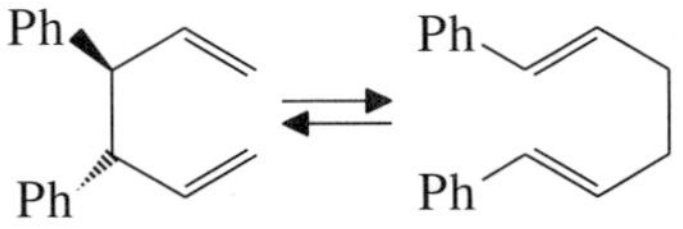

(b)

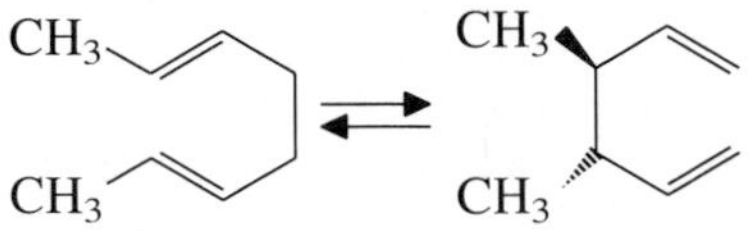

(c)

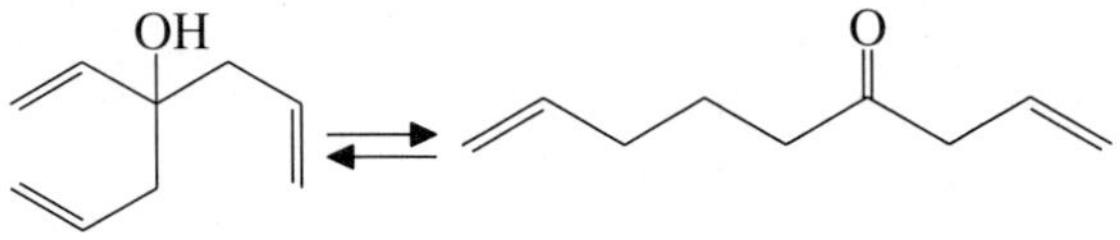

(d)

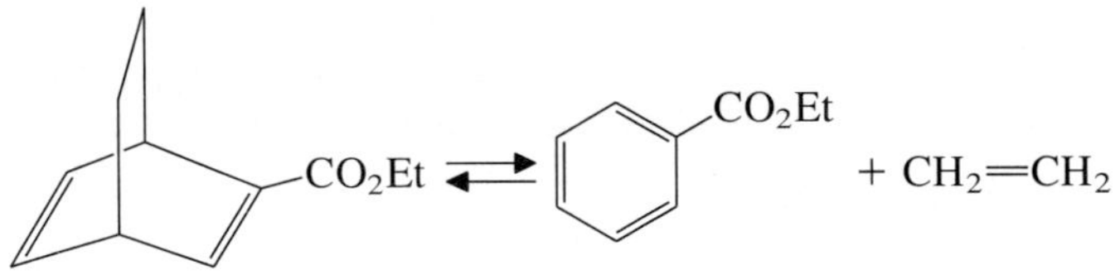

(e)

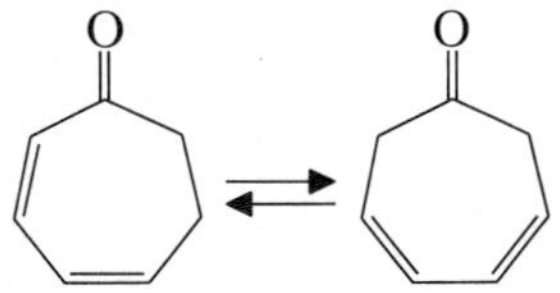

22 (a) Predict the stereochemistry of compounds *B* and *C*.

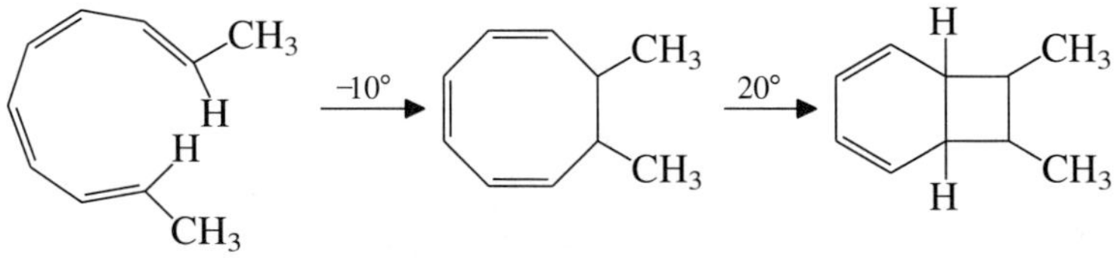

A *B* *C*

(b) What stereoisomer of *A* also gives compound *C* on heating?

23 Classify the following pericyclic reaction. Is it allowed or forbidden by orbital symmetry?

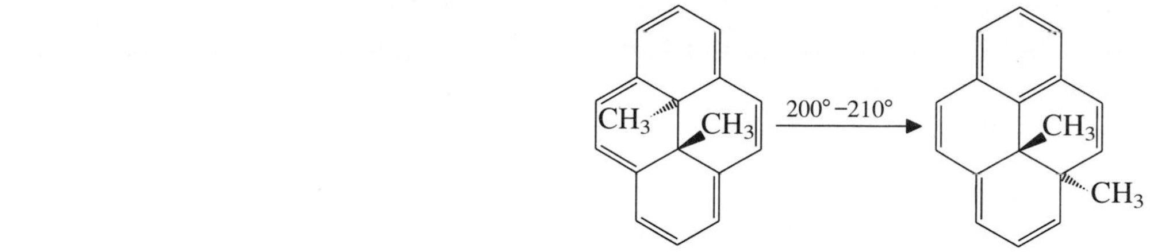

(80% yield)

24 Complete the following reactions by giving the major organic product(s):

(a) 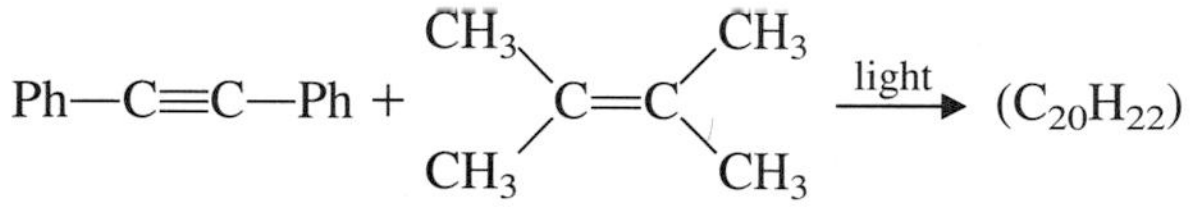

(b) 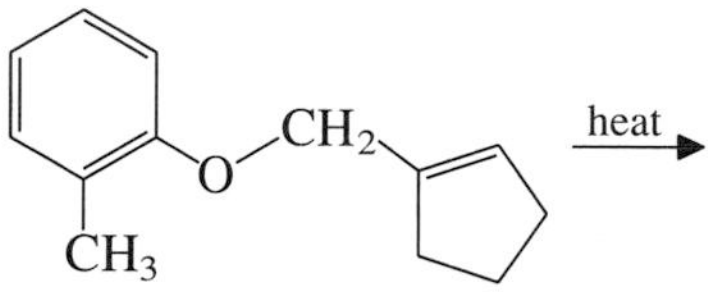

(c) 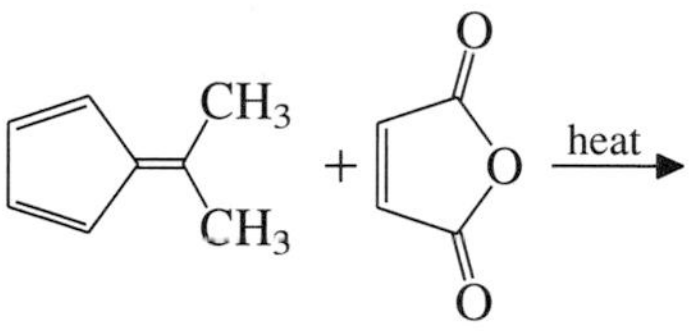

25 When compound *A* is irradiated for 115 hours in pentane, an isomeric compound *B* is obtained that decolorizes bromine in CCl_4 and reacts with ozone to give, after the usual workup, a compound *C*.

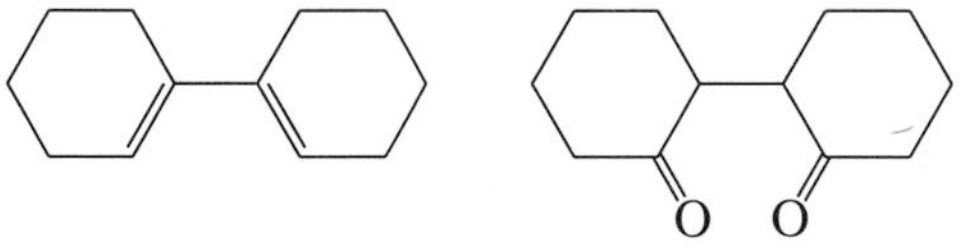

A *C*

(a) Give the structure of *B* and the stereochemistry of both *B* and *C*.

(b) On heating to 90°, compound *D*, a stereoisomer of *B*, is converted into *A*, but compound *B* is virtually inert under the same conditions. Identify compound *D* and account for these observations.

26 *Heptafulvalene* undergoes a thermal reaction with tetracyanoethylene (TCNE) to give the adduct shown below. What is the stereochemistry of this adduct? Explain.

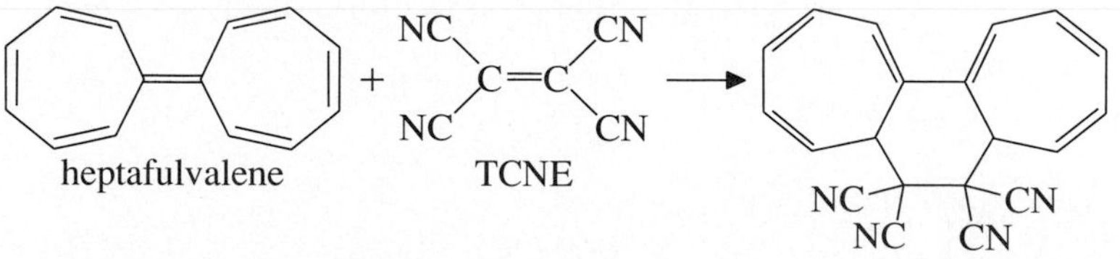

Problems (Cont.)

27 Suggest a mechanism for each of the following transformations. Some involve pericyclic reactions only; others involve pericyclic reactions as well as other steps. Invoke the appropriate selection rules to explain any stereochemical features observed.

(a)

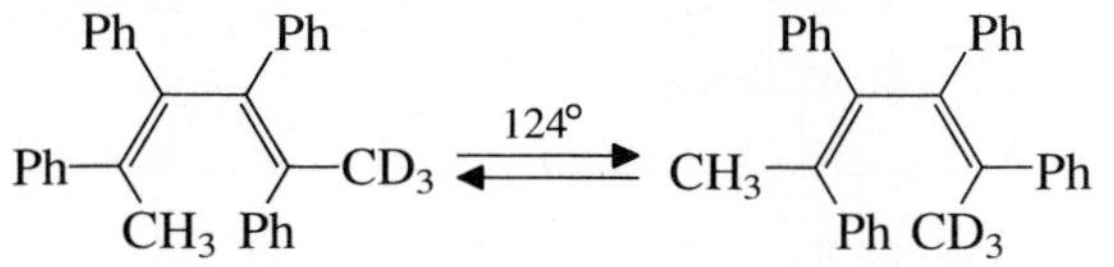

(b)

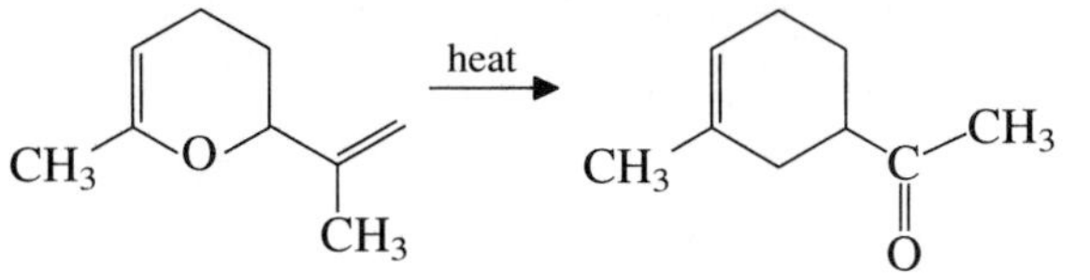

(c)

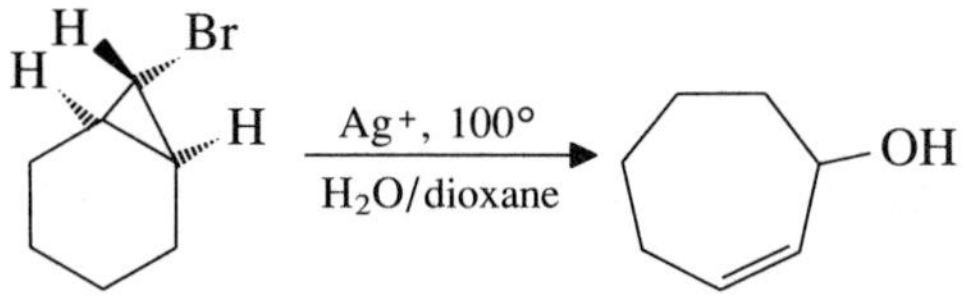

(d)

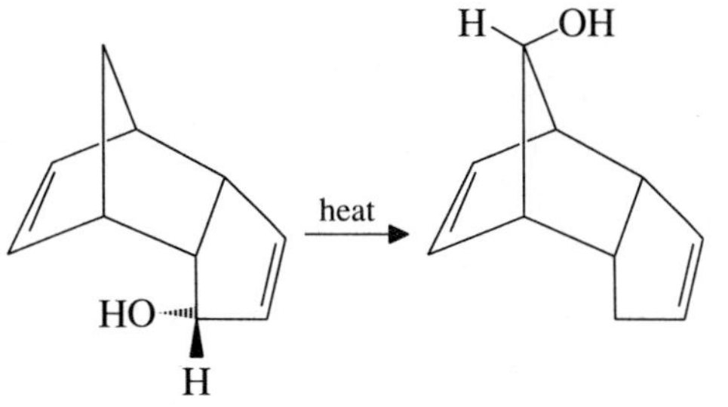

(e)

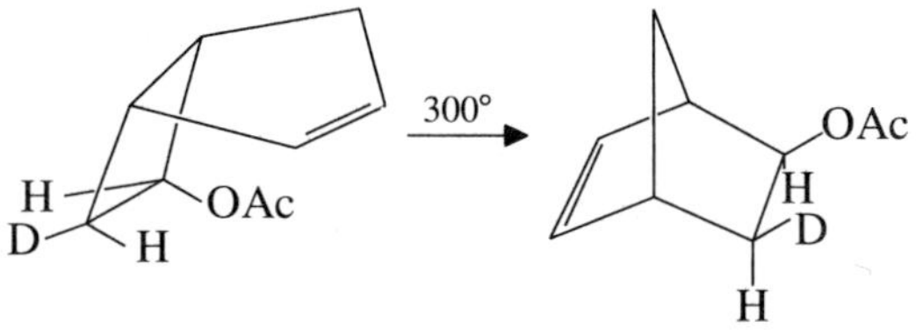

28 The following reaction occurs as a sequence of two pericyclic reactions. Identify the intermediate *A* and describe the two reactions.

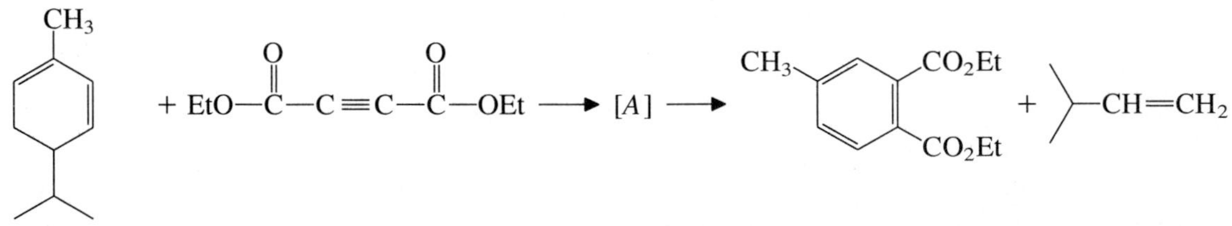

α-phellandrene

29 When 1,3,5-cyclooctatriene, *A*, is heated to 80–100°, it comes to equilibrium with an isomeric compound *B*. Treatment of the mixture of *A* and *B* with $CH_3O_2C{-}C{\equiv}C{-}CO_2CH_3$ gives a compound *C*, which, when heated to 200° for twenty minutes, gives dimethyl phthalate and a volatile alkene *D*. Identify compounds *B*, *C*, and *D*, and explain what reactions have occurred.

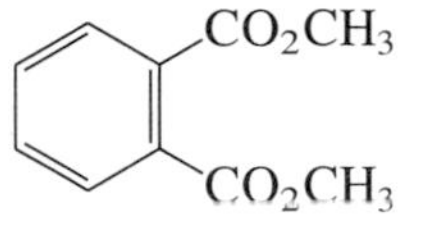

dimethyl phthalate

30 Black Hemptra bugs, generally observed in the tropical regions of India immediately after the rainy season, give off a characteristic nauseating smell whenever they are disturbed or crushed. Substance *A*, the compound causing the odor, can be obtained either by extracting the bugs with petroleum ether (which no doubt disturbs them greatly), or it can be prepared by heating the compound below at 170–180° for a short time. Give the structure of compound *A*.

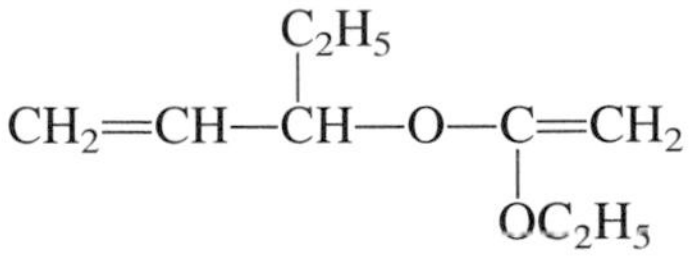

31 Chemist Perry Psychlick has found that an optically active sample of (+)-α-phellandrene loses optical activity when it is distilled, and has come to you for an explanation. Provide one, using what you know about pericyclic reactions.

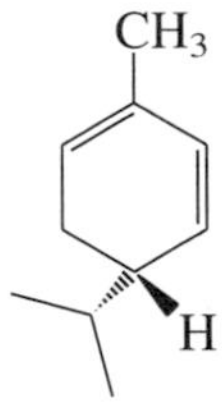

(*S*)-(+)-α-phellandrene

32 A compound *A* ($C_{11}H_{14}O_3$) is insoluble in base and gives an isomeric compound *B* when heated strongly. Compound *B* gives a sodium salt when treated with NaOH. Treatment of the sodium salt of *B* with dimethyl sulfate gives a new compound *C* ($C_{12}H_{16}O_3$) that is identical in all respects to a natural product *elemicin*. On ozonolysis followed by oxidation, elemicin gives the carboxylic acid *D* below. Propose structures for compounds *A*, *B*, and *C*.

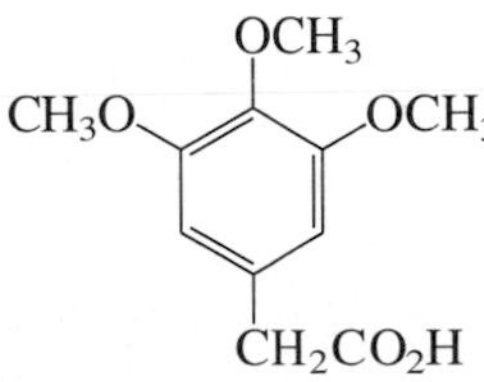

D

Problems (Cont.)

33 Using phenol and any other reagents as starting materials, outline a synthesis of each of the following compounds:

(a) 1-ethoxy-2-propylbenzene

(b)

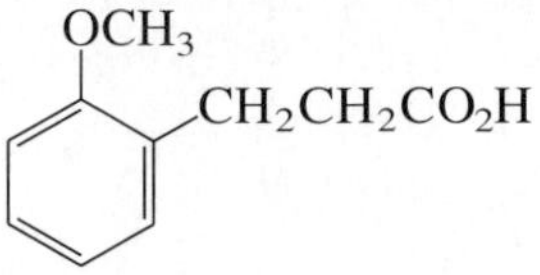

34 When 2-methyl-2-propenal is treated with allylmagnesium chloride (CH_2=CH—CH_2—MgCl) in ether, then with dilute aqueous acid, a compound *A* is obtained, which, when heated strongly, yields a compound *B*. Give the structures of compounds *A* and *B*.

35 Each of the following reactions involves a sequence of two pericyclic reactions. Identify the intermediate *X* involved in each reaction, and describe the pericyclic reactions involved.

(a)

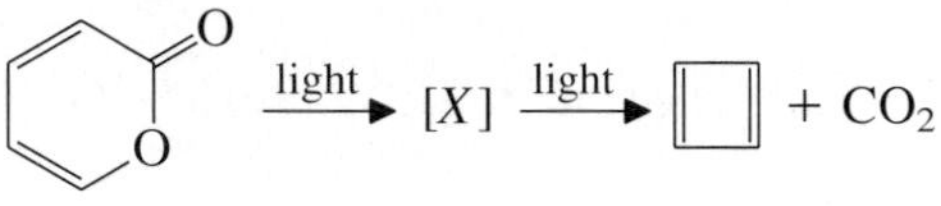

(b)

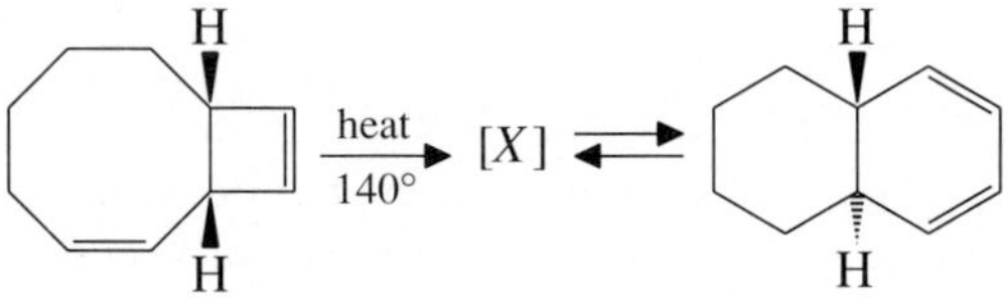

36 When each of the compounds below is heated in the presence of maleic anhydride, an intermediate is trapped as a Diels–Alder adduct. What is the intermediate formed in each reaction, and how is it formed from the starting material?

(a)

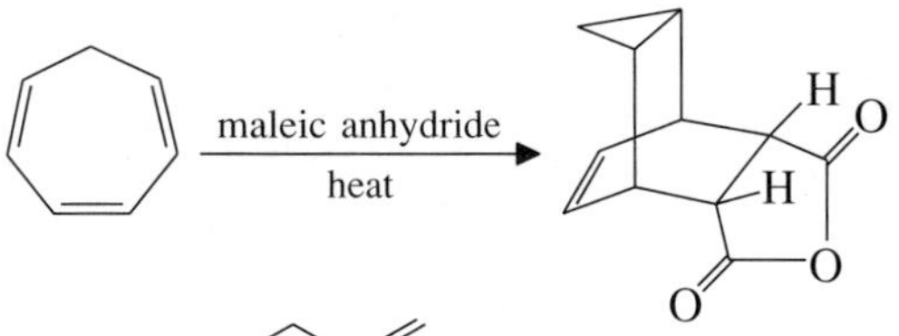

(b)

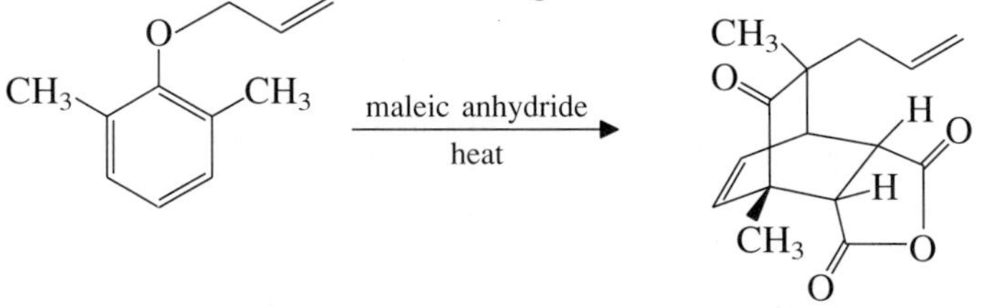

(mixture of stereoisomers)

(c)

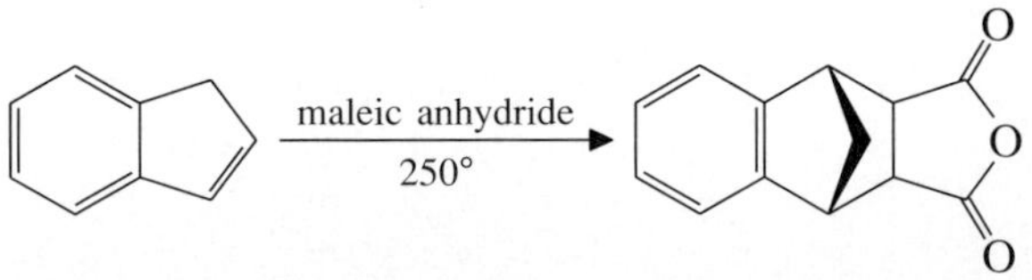

37 Ions as well as neutral molecules undergo pericyclic reactions. Classify the pericyclic reaction of the cation involved in the transformation below. What is the relative stereochemistry of the methyl groups in the product?

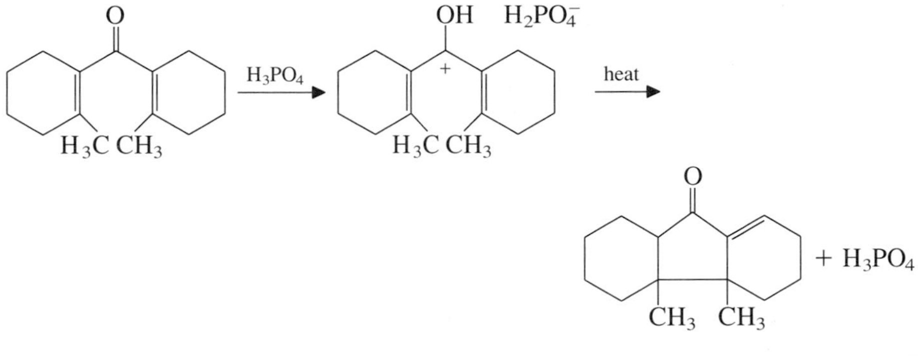

38 (a) The following transformation, involving a sequence of two pericyclic reactions, was used as a key step in a synthesis of the sex hormone estrone. Identify the unstable intermediate *A* and give the mechanism for both its formation and destruction.

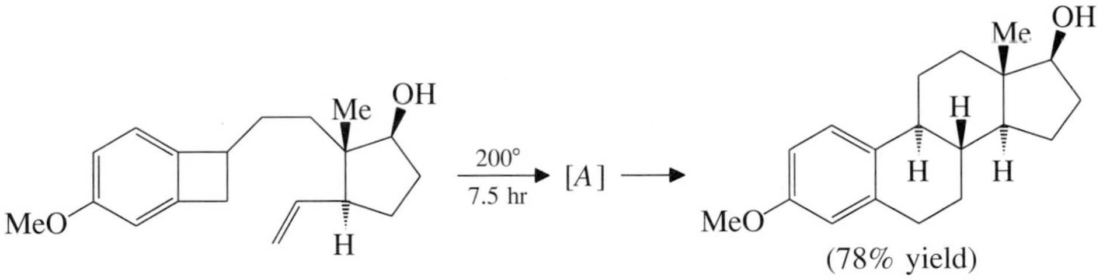

(b) Show how the product of this transformation can be converted into estrone.

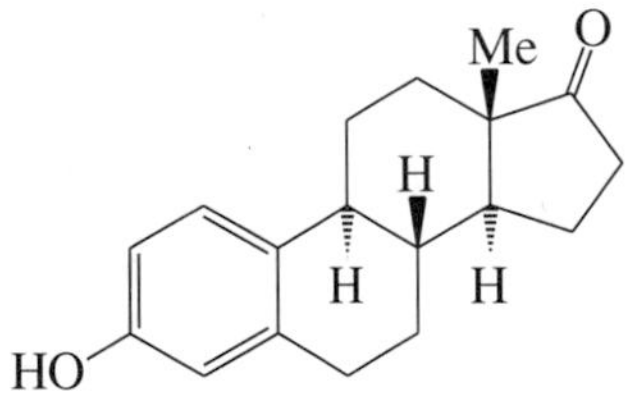

39 An all-suprafacial [3,3] sigmatropic rearrangement could in principle take place through either a chair-like or a boat-like transition state:

(a) According to the following result, which of these two transition states is preferred?

Problems (Cont.)

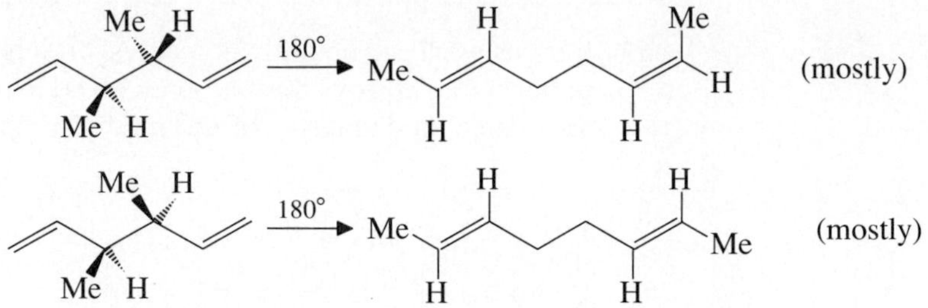

(b) When the terpene germacrone is distilled under reduced pressure at 165° it is transformed to β-elemenone by a Cope rearrangement. Give the structure of germacrone, including its stereochemistry.

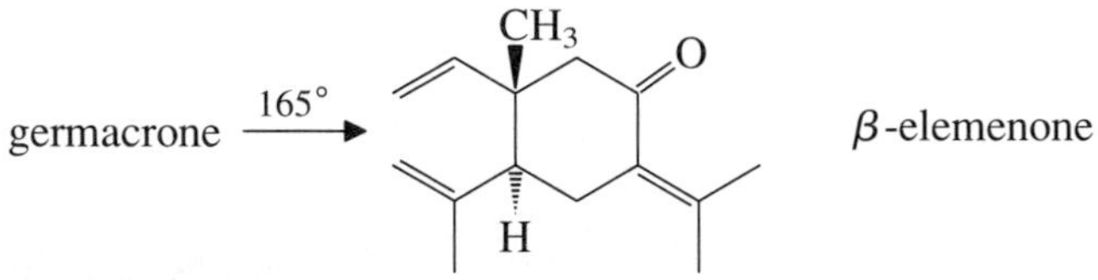

---

40 An interesting heterocyclic compound *C* was prepared and trapped by the following sequence of reactions. Give the structure of all missing compounds, and explain what happens in each reaction.

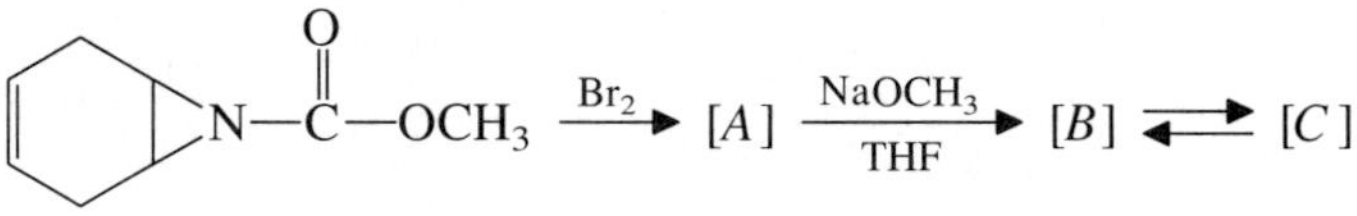

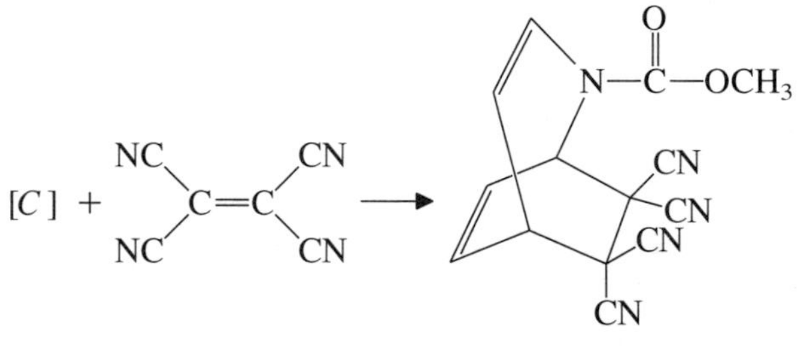

---

41 Anticipating the isolation of the potentially aromatic hydrocarbon *A*, a group of chemists irradiated compound *B* below. Compound *C* was obtained as a product instead of *A*.

(a) Explain why compound *A* might be expected to be unstable in spite of its cyclic array of $4n + 2$ π-electrons.

(b) Explain why the proposed synthesis of compound *A* was reasonable.

(c) Account for the formation of the observed product *C*.

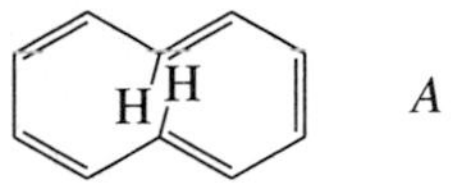

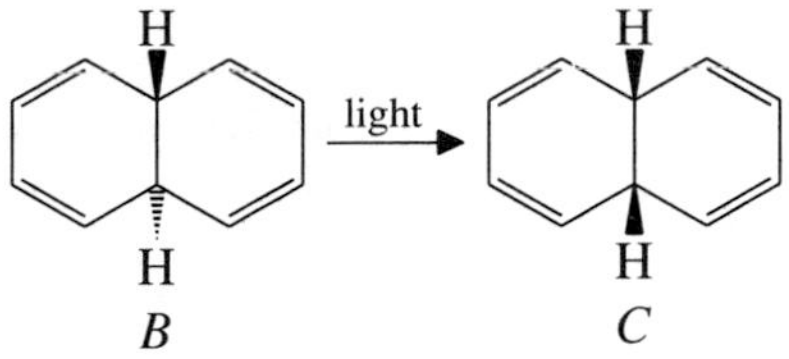

42 Identify the hydrocarbon *B* and the intermediate *A* (an isomer of *B*) in the reaction sequence below. Compound *B* is formed spontaneously from *A* in a pericyclic reaction. The NMR spectrum of *B* consists of a complex absorption at δ 7.1 (8*H*) and a singlet at δ (−0.5) (2*H*). Account particularly for the absorption of *B* at negative chemical shift.

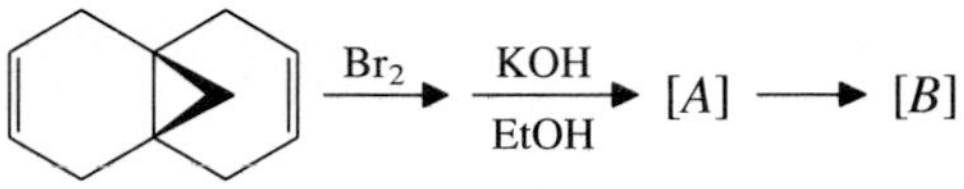

43 *Bullvalene* (Sec. 25.6) has a complex NMR spectrum at low temperature, consisting of resonances in both the allylic and vinylic region of the spectrum. As the temperature is raised, the NMR spectrum changes, and at 100°, consists of one singlet. Account for this change in the NMR spectrum with temperature in terms of what is known about the structure of bullvalene.

26 Amino Acids, Peptides, and Proteins

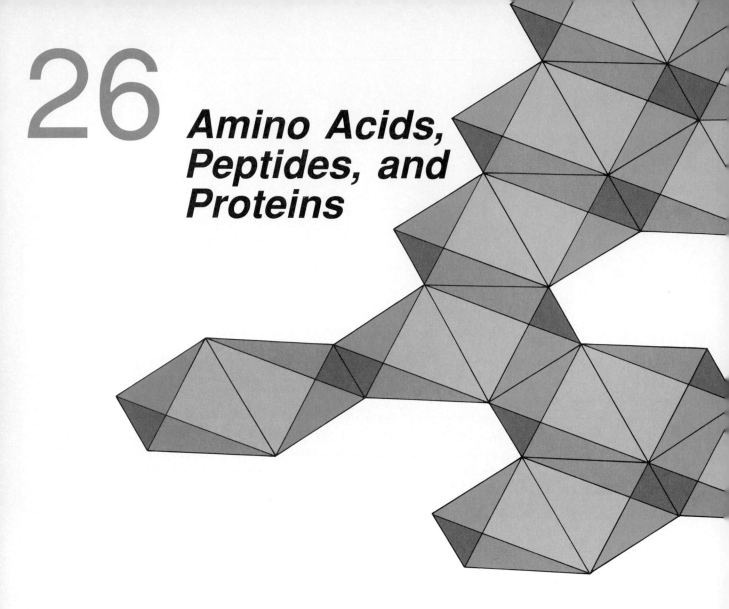

Amino acids, as the name implies, are compounds that contain both an amino group and a carboxylic acid group.

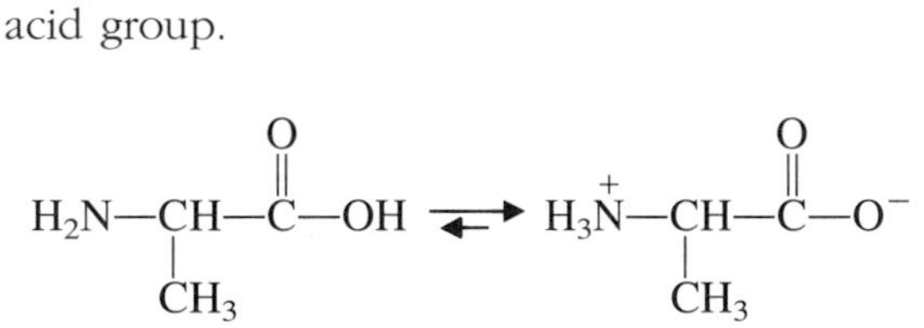

alanine
(an α-amino acid)

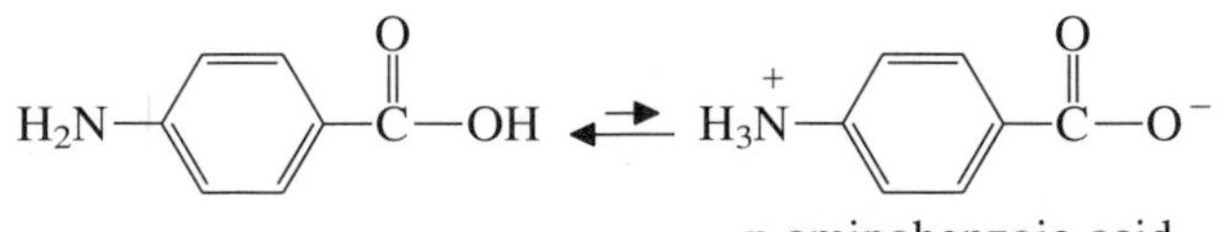

p-aminobenzoic acid
(PABA, a component
of folic acid, a vitamin)

As these structures show, an amino acid can contain within the same molecule two groups of opposite charge. Molecules containing oppositely charged groups are known as **zwitterions,** (from the German: "hybrid ion"). A zwitterionic structure is possible for an amino acid because the neutral amino group is basic, and can formally accept a proton from the acidic carboxylic acid group. The ***α*-amino acids,** of which alanine above is an example, have an amino group at the α-carbon—the carbon adjacent to the carboxylic acid group.

Peptides are biologically important polymers in which α-amino acids are joined into chains through amide bonds, called **peptide bonds.** A peptide bond is derived from the amino group of one amino acid and the carboxylic acid group of another.

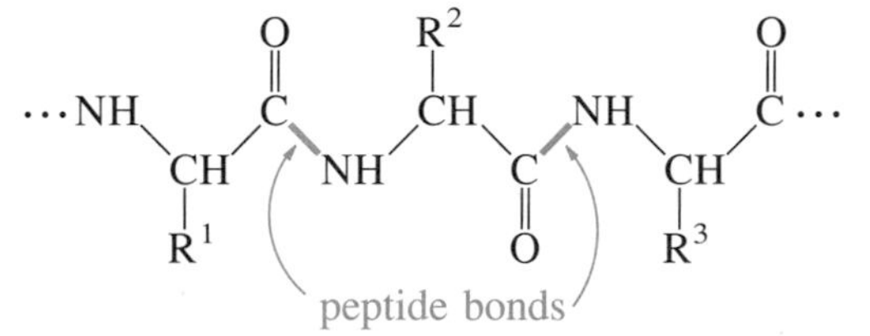

general peptide structure

Proteins are simply large peptides. (The distinction between peptides and proteins is not always clear.) The name *protein* (from the Greek: "of first rank") is particularly apt because peptides and proteins serve many important roles in biology. For example, enzymes (biological catalysts) and some hormones are peptides or proteins.

In our study of amino acids, peptides, and proteins we shall encounter again the principles of stereochemistry. Much of the chemistry of amines and carboxylic acids will also be relevant to this study—especially their acid–base behavior—but we shall also see how the occurrence of these two functional groups together in the same molecule leads to some unusual molecular properties and chemistry.

26.1 NOMENCLATURE OF AMINO ACIDS AND PEPTIDES

A. Nomenclature of Amino Acids

Some amino acids are named systematically as carboxylic acids with amino substituents.

$$H_3\overset{+}{N}-CH_2CH_2CH_2-CO_2^- \rightleftarrows H_2N-CH_2CH_2CH_2-CO_2H$$

4-aminobutanoic acid
(γ-aminobutyric acid)

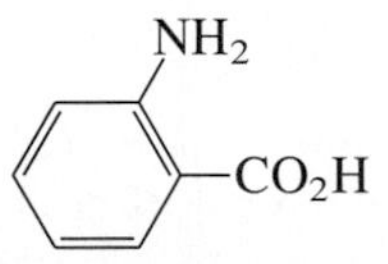

2-aminobenzoic acid
(*o*-aminobenzoic acid or anthranilic acid)

$$(CH_3)_2\overset{+}{N}H-CH_2CH_2-\overset{\displaystyle O}{\overset{\|}{C}}-O^-$$

3-(dimethylamino)-
propanoic acid

Twenty α-amino acids, however, are known by widely accepted traditional names. These are the amino acids that occur commonly as constituents of most proteins. The names and structures of the twenty naturally occurring amino acids are given in Table 26.1 (pp. 1136–1137).

Two points about the structures of the α-amino acids will help you to remember them. First, with the exception of proline, all α-amino acids have the same general structure, differing only in the identity of the side chain R.

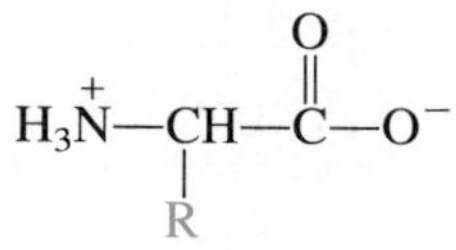

general structure

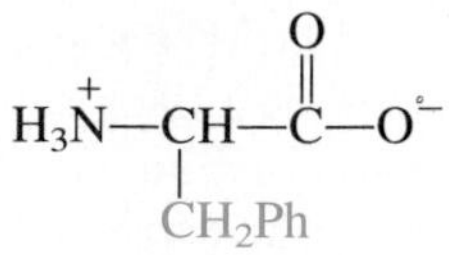

phenylalanine (R = CH_2Ph)

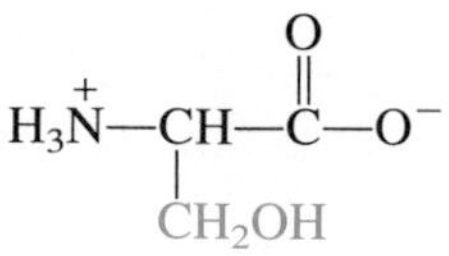

serine (R = CH_2OH)

Proline is the only naturally occurring amino acid with a secondary amino group. In proline the —NH— and the side chain are "tied together" in a ring.

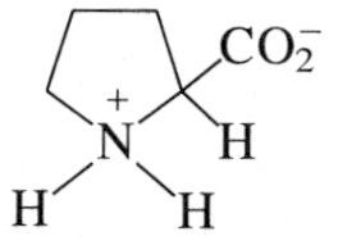

proline

Second, the amino acids can be grouped according to the nature of their side-chains. There are six groups:

1. Amino acids with —H or aliphatic hydrocarbon side chains
2. Amino acids with side chains containing aromatic groups
3. Amino acids with side chains containing —SH, —SCH_3, or alcohol —OH groups
4. Amino acids with side chains containing carboxylic acid or amide groups
5. Amino acids with basic side chains
6. Proline

The amino acids are arranged in this way in Table 26.1.

A student in the author's organic chemistry course proposed the following mnemonics for remembering the amino acids.

Group 1 and proline: LIAGVP (*L*oudon *I*s *A* *G*hastly *V*olleyball *P*layer)
Group 2: HTTP (*H*ail *T*o *T*errific *P*urdue)
Group 3: CTMS (*C*hemistry *T*akes *M*uch *S*tudy)
Groups 4 and 5: AGAGLA (*A*ll *G*irls *A*nd *G*uys *L*ove *A*rtichokes)

The α-amino acids are often designated by abbreviations. The generally accepted three-letter abbreviations for the naturally occurring amino acids are given in Table 26.1. Standard single-letter abbreviations, which are being used more frequently, are also listed in the table.

B. Nomenclature of Peptides

The terminology and nomenclature associated with peptides is best illustrated by an example. Consider the following peptide formed from the three amino acids alanine, valine, and lysine.

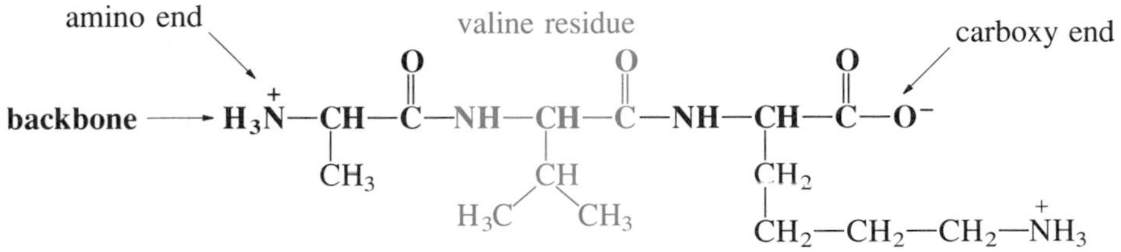

alanylvalyllysine
(abbreviated Ala-Val-Lys or A-V-K)

The **peptide backbone** is the repeating sequence of nitrogen, α-carbon, and carbonyl groups. The characteristic amino acid side chains are attached to the peptide backbone at the respective α-carbon atoms. Each amino acid unit in the peptide is called a **residue.** For example, the part of the peptide derived from valine, the *valine residue,* is shown in color in the structure above. The ends of a peptide are labeled as the **amino end** or **amino terminus** and the **carboxy end** or **carboxy terminus.** A peptide can be characterized by the number of residues it contains. For example, the peptide above is a *tripeptide* because it contains three amino acid residues.

A peptide is conventionally named by giving successively the names of the amino acid residues, *starting at the amino end.* The names of all but the carboxy terminal residue are formed by dropping the final ending (*ine, ic,* or *an*) and replacing it with *yl.* Thus, the peptide above is named alanylvalyllysine. In practice, this type of nomenclature is cumbersome for all but the smallest peptides. A simpler way of naming peptides is to connect with hyphens the three-letter (or one-letter) abbreviations of the component amino acid residues beginning with the amino-terminal residue. Thus, the peptide above is also written as Ala-Val-Lys or A-V-K.

Large peptides of biological importance are known by common names. Thus, *insulin* is an important peptide hormone that contains fifty-one amino acid residues; *ribonuclease,* an enzyme, is a protein containing 124 amino acid residues (and a rather small protein at that!).

Problems

1 Draw the structures of the following peptides:
(a) tryptophylglycylisoleucylaspartic acid
(b) Glu-Gln-Phe-Arg (or E-Q-F-R)

2 Using three-letter abbreviations for the amino acid residues, name the following peptide:

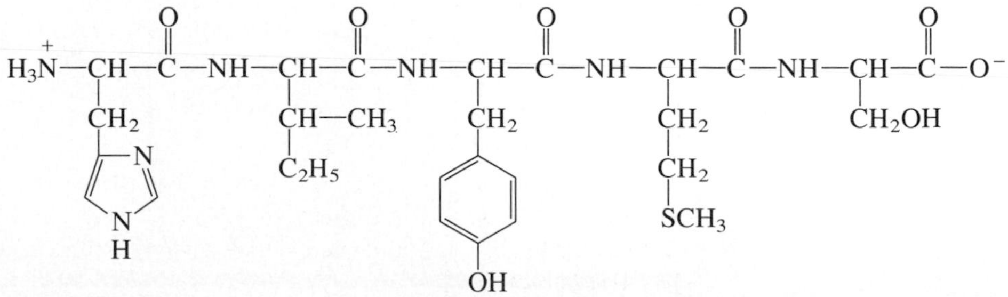

TABLE 26.1 Names, Structures, Abbreviations, and Properties of the Twenty Common Naturally Occurring Amino Acids

General structure: $H_3\overset{+}{N}-CH(R)-C(=O)-O^-$

Name and Abbreviations	R*	Optical rotation of L enantiomer in H_2O	pK_{a1}	pK_{a2}	pK_{a3}	Isoelectric point, p*I*	Water solubility, wt % at 30°
AMINO ACIDS WITH SIMPLE ALIPHATIC SIDE CHAINS							
glycine, Gly, G	—H		2.34	9.60	–	5.97	22
alanine, Ala, A	$-CH_3$	(+)	2.35	9.69	–	6.02	15
valine, Val, V	$-CH(CH_3)_2$	(+)	2.32	9.62	–	5.97	7
leucine, Leu, L	$-CH_2CH(CH_3)_2$	(–)	2.36	9.60	–	5.98	2.4
isoleucine, Ile, I	$-CH(CH_3)-C_2H_5$ [*S* configuration]	(+)	2.36	9.68	–	6.02	4.0
AMINO ACIDS WITH AROMATIC SIDE CHAINS							
phenylalanine, Phe, F	$-CH_2Ph$	(–)	1.83	9.13	–	5.48	3.1
tryptophan, Trp, W	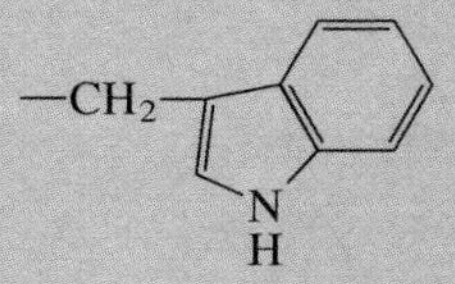	(–)	2.38	9.39	–	5.88	1.2
tyrosine, Tyr, Y	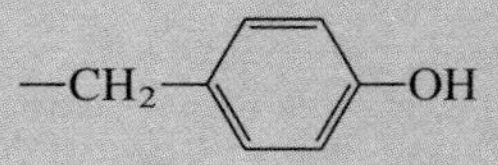	(–)	2.20	9.11	10.07	5.65	0.05
histidine, His, H†	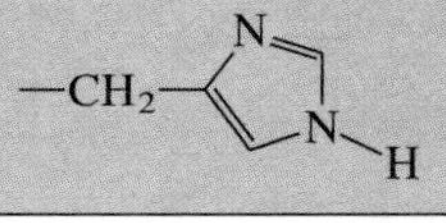	(–)	1.82	6.00	9.17	7.58	

Amino Acids with Side-chain Alcohol $-OH$, $-SH$, and $-SCH_3$ Groups							
serine, Ser, S	$-CH_2-OH$	(−)	2.21	9.15	–	5.68	32
threonine, Thr, T	$-CH(CH_3)-OH$ [*R* configuration]	(−)	2.09	9.10	–	5.59	
methionine, Met, M	$-CH_2CH_2-SCH_3$	(−)	2.28	9.21	–	5.75	3.7
cysteine, Cys, C‡	$-CH_2-SH$	(−)	1.71	8.33	10.78	5.02	
Amino Acids with Side Chains Containing Carboxylic Acid or Amide Groups							
aspartic acid, Asp, D	$-CH_2-CO_2H$	(+)	2.09	3.86	9.82	2.97	0.6
glutamic acid, Glu, E	$-CH_2CH_2-CO_2H$	(+)	2.19	4.25	9.67	3.22	1.0
asparagine, Asn, N	$-CH_2-CO-NH_2$	(−)	2.02	8.80	–	5.41	3.6
glutamine, Gln, Q	$-CH_2CH_2-CO-NH_2$	(+)	2.17	9.13	–	5.65	
Amino Acids with Side Chains Containing Strongly Basic Groups							
lysine, Lys, K	$-(CH_2)_4-NH_2$	(+)	2.18	8.95	10.53	9.74	44 (as HCl salt)
arginine, Arg, R	$-(CH_2)_3NH-C(=NH)-NH_2$	(+)	2.17	9.04	12.48	10.76	48 (as HCl salt)
Cyclic (Secondary) Amino Acid							
proline, Pro, P	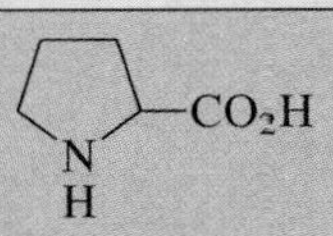	(−)	1.99	10.60	–	6.10	63

Side chains are shown in their uncharged form.

Histidine is considered a weakly basic amino acid.

Cysteine often occurs in proteins as a disulfide dimer, called cystine: $H_2N-CH(CO_2H)-CH_2-S-S-CH_2-CH(CO_2H)-NH_2$. For this reason, cysteine is sometimes called ***half-cysteine*** and abbreviated Cys/2.

26.2 STEREOCHEMISTRY OF THE α-AMINO ACIDS

With the exception of glycine, the α-amino acids have at least one asymmetric carbon atom. The naturally occurring chiral amino acids in Table 26.1 all have the *S* configuration at the α-carbon atom.

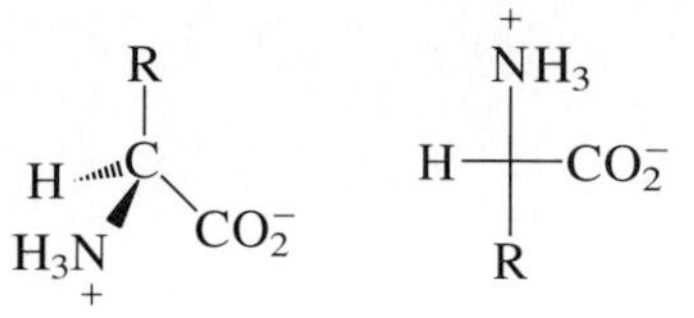

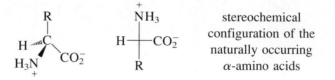

stereochemical configuration of the naturally occurring α-amino acids

The stereochemistry of α-amino acids is often described with an older system, the **D,L system.** In this system the naturally occurring amino acids are said to have the L configuration. Thus, (*S*)-serine could also be called L-serine; its enantiomer is D-serine. The L designation refers to the configuration of a reference carbon—the α-carbon—in each amino acid, regardless of the number of asymmetric carbons in the molecule. Thus L-threonine, which has two asymmetric carbons, is the (2*S*,3*R*) stereoisomer. Its enantiomer, D-threonine, is the (2*R*,3*S*) stereoisomer.

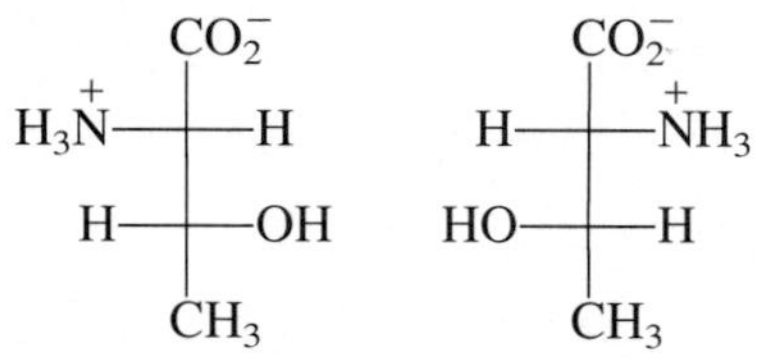

L-threonine D-threonine

Since threonine has two asymmetric carbons, it also has two other stereoisomers—its diastereomers. In the D,L system, *diastereomers are given different names*. The diastereomers of threonine are called *allothreonine*. Thus L-allothreonine is the (2*S*,3*S*) stereoisomer of threonine, and D-allothreonine is its enantiomer. We need not dwell further on this older system except to be aware that it is still used for amino acids and sugars. (The correspondence between *S* and L is not general for other compounds.)

Problem

3 (a) L-Isoleucine has two asymmetric carbons and has the (2*S*,3*S*) configuration. Draw a Fischer projection of L-isoleucine.

(b) Alloisoleucine is the diastereomer of isoleucine. Draw Fischer projections of L-alloisoleucine and D-alloisoleucine.

26.3 UV SPECTRA OF AMINO ACIDS AND PEPTIDES

An important physical property of amino acids used especially in the detection of peptides is their UV spectra. Most amino acids have negligible UV absorption, but the amino acids tyrosine and tryptophan have an appreciable absorption near 280 nm, and phenylalanine near 260 nm, attributable to their aromatic chromophores (Table 26.2). The absorbance at 280 nm, called "A_{280}" by protein chemists, is an important physical

TABLE 26.2 UV Absorptions of Amino Acids

Amino acid	λ_{max}	ϵ
Phe	259	200
Tyr	278	1,100
Trp	279	5,200

characteristic of a protein. This number obviously will be large for a protein or peptide if it contains a substantial number of tryptophan and/or tyrosine residues.

Problem

4 The enzyme *lysozyme* is a protein that consists of a single peptide chain of 129 amino acids (MW 14,100) containing three tyrosine and six tryptophan residues. Estimate the A_{280} of a solution that contains 0.1 mg of lysozyme per mL; assume the light beam passes through a 1.0 cm thickness of the sample.

26.4 ACID–BASE PROPERTIES OF AMINO ACIDS AND PEPTIDES

A. Zwitterionic Structures of Amino Acids and Peptides

We stated in the introduction to this chapter that the neutral forms of the α-amino acids are *zwitterions*. How do we know this is so? Some of the evidence is as follows:

1. Amino acids are insoluble in apolar aprotic solvents such as ether. On the other hand, most neutral amines and carboxylic acids dissolve in ether. Although the water solubility of the different amino acids varies from case to case, all are more soluble in water than they are in ether.

2. Amino acids have very high melting points. For example, glycine melts at 262° (with decomposition), and tyrosine melts at 310° (also with decomposition). Both hippuric acid, a much larger molecule than glycine, and glycinamide, the amide of glycine, have much lower melting points. The former compound lacks the amino group, the latter lacks the carboxylic acid group, and neither can exist as a zwitterion.

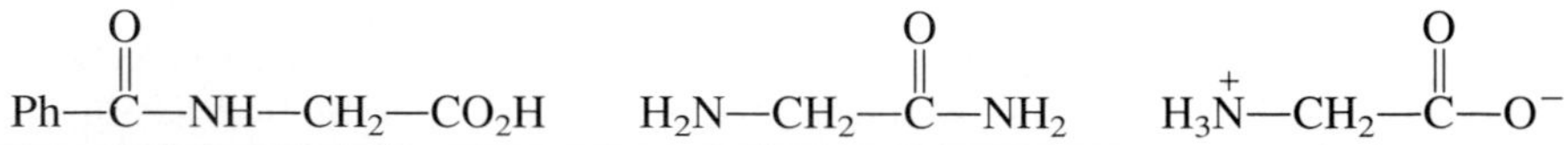

$$Ph-\overset{\overset{\displaystyle O}{\|}}{C}-NH-CH_2-CO_2H \qquad H_2N-CH_2-\overset{\overset{\displaystyle O}{\|}}{C}-NH_2 \qquad \overset{+}{H_3N}-CH_2-\overset{\overset{\displaystyle O}{\|}}{C}-O^-$$

$$\text{\textit{N}-benzoylglycine} \qquad\qquad \text{glycinamide} \qquad\qquad\qquad \text{glycine}$$
$$\text{(hippuric acid)} \qquad\qquad\quad \text{mp 67–68°} \qquad\qquad\qquad \text{mp 262° (d)}$$
$$\text{mp 190°}$$

The high melting points and greater solubility in water than in ether are characteristics we have come to expect from salts—not organic compounds. These saltlike characteristics are, however, what we *would* expect from a zwitterionic compound. The strong forces in the solid states of the amino acids that result from the attraction of full positive and negative charges on *different molecules* are much like those between the ions in a salt. These attractions stabilize the solid state and resist conversion of the solid into a liquid—whether a pure liquid melt or a solution. Water is the best solvent for most amino acids because it solvates the ionic groups of the amino acid, much as it solvates the ions of a salt (Sec. 8.4B).

3. The dipole moments of the amino acids are very large—much larger than those of similar-sized molecules with only a single amine or carboxylic acid group.

$H_3\overset{+}{N}—CH_2—CO_2^-$	$CH_3—CH_2—CO_2H$	$CH_3CH_2CH_2CH_2—NH_2$
glycine	propionic acid	butylamine
$\mu \sim 14$ D	$\mu = 1.7$ D	$\mu = 1.4$ D

A large dipole moment is expected for molecules that contain a great deal of separated charge (Eq. 1.5).

4. The pK_a values for amino acids are what we would expect for the zwitterionic forms of the neutral molecules.

Suppose we titrate a neutral amino acid with acid. When one equivalent of acid is added, we will have protonated the *basic* group of the amino acid. When we carry out this experiment with glycine, the pK_a of the basic group is found to be 2.3. If glycine is indeed a zwitterion, this basic group can only be the carboxylate ion. If glycine is not a zwitterion, this basic group has to be the amine.

$$H_3O^+ + H_3\overset{+}{N}—CH_2—\overset{\overset{\large O}{\|}}{C}—\ddot{\underset{..}{O}}:^- \rightleftharpoons H_3\overset{+}{N}—CH_2—\overset{\overset{\large O}{\|}}{C}—\ddot{\underset{..}{O}}H + H_2O \qquad (26.1a)$$

basic group of the zwitterionic form

$$H_3O^+ + H_2\ddot{N}—CH_2—\overset{\overset{\large O}{\|}}{C}—OH \rightleftharpoons H_3\overset{+}{N}—CH_2—\overset{\overset{\large O}{\|}}{C}—OH + H_2O \qquad (26.1b)$$

basic group of the nonzwitterionic form

How do we know which is the correct description of the titration? The pK_a of 2.3 is what we would expect for a carboxylic acid in a molecule containing a nearby electron-withdrawing group (in this case, the $—\overset{+}{N}H_3$ group). In contrast, the conjugate acids of amines have pK_a values in the 8–10 range. This analysis suggests that the zwitterion, not the uncharged form, is being titrated.

Along the same lines, if we add NaOH to neutral glycine, we titrate a group with pK_a = 9.6. This is a reasonable pK_a value for an ammonium ion, but would be very unusual for a carboxylic acid. This comparison also suggests that the neutral form of glycine is a zwitterion.

The acid–base equilibria for glycine can be summarized as follows:

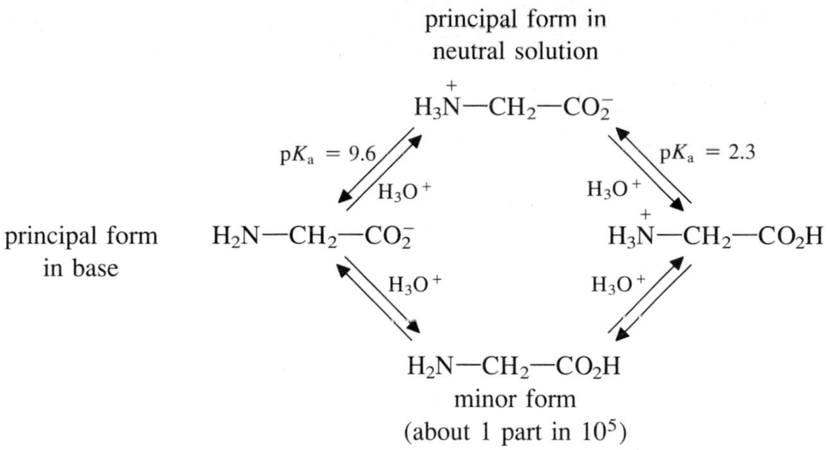

The major neutral form of any α-amino acid is the zwitterion. In fact, it can be estimated that the ratio of the uncharged form of an α-amino acid to the zwitterion form is about one part in 10^5! (See Problem 52.)

Peptides also exist as zwitterions; that is, at neutral pH, the terminal amino group is protonated and the terminal carboxylic acid group is ionized.

Problem

5 Draw the structure of the major neutral form of the peptide Ala-Lys-Val-Glu-Met.

B. Isoelectric Points of Amino Acids and Peptides

An important measure of the acidity or basicity of an amino acid is its **isoelectric point,** or **isoelectric pH.** This is the pH of a dilute aqueous solution of the amino acid at which amino acid is exactly neutral. At the isoelectric point, two conditions are met. First, the concentration of positively charged amino acid molecules equals the concentration of negatively charged ones. Second, the relative concentration of neutral amino acid molecules is greater than at any other pH.

To illustrate, consider the ionization equilibria of the amino acid alanine, which has an isoelectric point of 6.0.

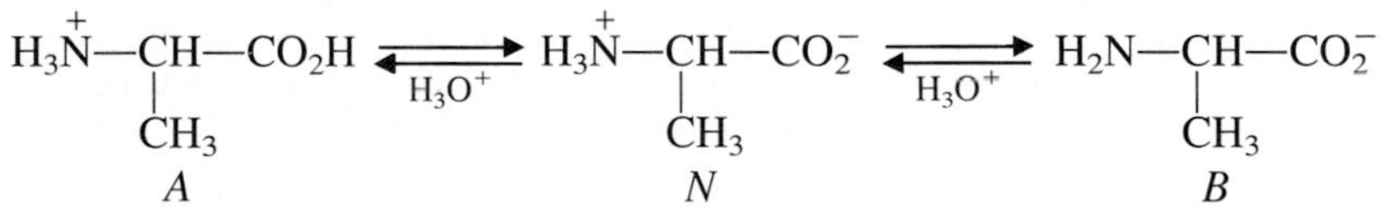

$$\overset{+}{H_3N}-CH-CO_2H \underset{H_3O^+}{\overset{}{\rightleftharpoons}} \overset{+}{H_3N}-CH-CO_2^- \underset{H_3O^+}{\overset{}{\rightleftharpoons}} H_2N-CH-CO_2^- \quad (26.3)$$
$$\overset{|}{CH_3} \qquad\qquad \overset{|}{CH_3} \qquad\qquad \overset{|}{CH_3}$$
$$A \qquad\qquad\qquad N \qquad\qquad\qquad B$$

In a dilute aqueous solution of alanine in which the pH has been adjusted to 6.0—the isoelectric point of alanine—a small amount of the alanine is in form A, an exactly equal amount is in form B, and most of the alanine is in the neutral form N. The isoelectric points of the α-amino acids are given in Table 26.1.

Since the charge on an amino acid at any pH is determined by its state of protonation, we can deduce the isoelectric point of an amino acid from its pK_a values. Let K_{a1} be the dissociation constant of the carboxylic acid group of form A, and K_{a2} the dissociation constant of the ammonium group in form N (Eq. 26.3).

$$K_{a1} = \frac{[N][H_3O^+]}{[A]} \qquad K_{a2} = \frac{[B][H_3O^+]}{[N]} \qquad (26.4)$$

It follows that

$$\frac{K_{a1}}{[H_3O^+]} = \frac{[N]}{[A]} \quad \text{and} \quad \frac{K_{a2}}{[H_3O^+]} = \frac{[B]}{[N]} \tag{26.5}$$

Letting $pK_{a1} = -\log K_{a1}$, and $pK_{a2} = -\log K_{a2}$,

$$pK_{a1} - pH = \log\left(\frac{[A]}{[N]}\right) \tag{26.6a}$$

and

$$pH - pK_{a2} = \log\left(\frac{[B]}{[N]}\right) \tag{26.6b}$$

If the pH of an alanine solution is lowered so that pH $\ll$ pK_{a1}, Eq. 26.26a shows that most of the alanine molecules will be in form *A*, which carries a positive charge. If the pH is raised so that pH $\gg$ pK_{a2}, Eq. 26.6b shows that most of the alanine molecules will be in form *B*, which carries a negative charge. It is then reasonable that the pH at which $[A] = [B]$—the isoelectric point—should be at a pH that is between pK_{a1} and pK_{a2}. In fact, it is not difficult to show (Problem 8) that the isoelectric point, p*I*, is the average of the two pK_a values of an amino acid:

$$\text{isoelectric point} = pI = \frac{pK_{a1} + pK_{a2}}{2} \tag{26.7}$$

Thus, alanine, with pK_a values of 2.3 and 9.7 (Table 26.1), has an isoelectric point of $(2.3 + 9.7)/2 = 6.0$.

The significance of the isoelectric point is that it tells us not only the pH at which the amino acid has zero charge, but also the net charge on an amino acid at *any* pH. For example, at a pH less than its isoelectric point, more molecules of an amino acid in solution have a positive charge than have a negative charge; we say that under these conditions the amino acid is positively charged. At a pH greater than its isoelectric point, more molecules of an amino acid have a negative charge than have a positive charge; we say that the amino acid is negatively charged. We shall see in Sec. 26.4C how a knowledge of the net charge of an amino acid can be useful.

When an amino acid has a side chain containing an acidic or basic group, the isoelectric point is markedly changed. The amino acid lysine, for example, has a very basic side-chain amino group as well as its α-amino and carboxy groups. It turns out that the isoelectric point of lysine is 9.74, halfway between its two highest pK_a values. At this pH one amino group is largely protonated, one is largely unprotonated, and the carboxy group is ionized, for a net charge of zero. Thus, at pH 6 (the pH at which alanine has a net charge of zero), lysine carries a net positive charge. This is reasonable because pH 6 is well below the pK_a values of both protonated amino groups and well above the pK_a value of the carboxylic acid group, for a net charge of $(+2 - 1) = +1$.

The amino acid aspartic acid has a carboxylic acid group on its side chain and a very low isoelectric point of 2.87, halfway between its two lowest pK_a values. Aspartic acid carries a negative charge at pH 6 because, at this pH, both carboxylic acid groups are ionized and the amino group is protonated.

Amino acids with high isoelectric points are classified as *basic amino acids*. Lysine and arginine are by far the two most basic amino acids (see Table 26.1); and as we can see from its isoelectric point, arginine is the most basic amino acid of all. Its side chain carries the very basic guanidino group, the conjugate acid of which has a pK_a of 12.5. The basicity of this group is a consequence of the fact that its conjugate acid is resonance stabilized.

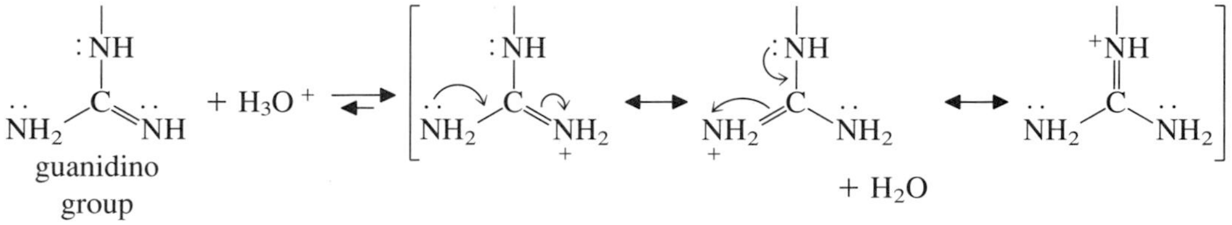

(26.8)

Amino acids with low isoelectric points are classified as *acidic amino acids.* Aspartic acid and glutamic acid are the two most acidic amino acids.

Amino acids with isoelectric points near 6, such as glycine or alanine, are classified as *neutral amino acids.* Of course, pH 6 is not *exactly* neutral; neutral amino acids are actually slightly acidic because the carboxylic acid group is somewhat more acidic than the amino group is basic.

principal forms at pH 6:

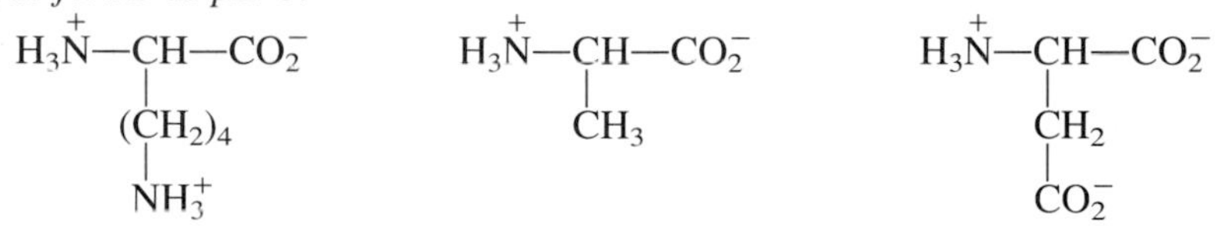

net +1 charge, lysine (a *basic* amino acid) | net 0 charge, alanine (a *neutral* amino acid) | net −1 charge, aspartic acid (an *acidic* amino acid)

Amphoteric peptides (peptides with both acidic and basic groups) also have isoelectric points. We can tell by inspection whether a peptide is acidic, basic, or neutral by examining the number of acidic and basic groups that it contains. A peptide with more amino and guanidino groups than carboxylic acid groups, for example, will have a high isoelectric point.

Problems

6 (a) Point out the ionizable groups on the amino acid *tyrosine* (Table 26.1).
(b) What is the net charge on tyrosine at pH = 6? How do you know?

7 Classify the following peptides as acidic, basic, or neutral. What is the net charge on each peptide at pH = 6?
(a) Gly-Leu-Val
(b) Leu-Trp-Lys-Gly-Lys
(c) *N*-acetyl-Asp-Val-Ser-Arg-Arg (*N*-acetyl means that the terminal amino group of the peptide is acetylated.)

8 By definition the isoelectric point of an amino acid is that pH at which [*A*] = [*B*] (Eq. 26.3). Use this condition to derive Eq. 26.7.

C. Separations of Amino Acids and Peptides Using Acid–Base Properties

The isoelectric point is a useful criterion on which we can design separations of amino acids and peptides. Consider, for example, the water solubilities of amino acids and peptides. Most peptides and amino acids, like carboxylic acids or amines, are more soluble in their ionic forms, and least soluble in their neutral forms. Thus some peptides, proteins, and amino acids precipitate from water if the pH is adjusted to their isoelectric points. These same compounds are more soluble at pH values far from their isoelectric points.

A separation technique used a great deal in amino acid and peptide chemistry is **ion-exchange chromatography.** This method, too, depends on the isoelectric points of amino acids and peptides. In this technique, a hollow tube, or column, is filled with a buffer solution in which is suspended a finely powdered, insoluble polymer called an *ion-exchange resin*. The resin bears acidic or basic groups. One popular resin, for example, is a sulfonated polystyrene—a polystyrene in which the phenyl rings contain strongly acidic sulfonic acid groups. If the pH of the buffer is such that the sulfonic acid groups are ionized, the resin bears a negative charge.

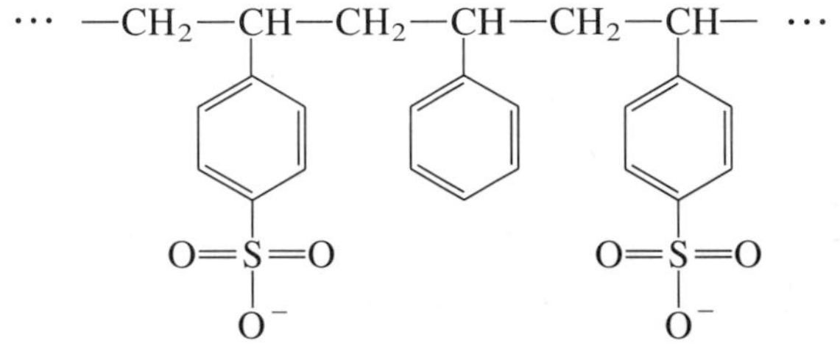

structure of sulfonated polystyrene with ionized sulfonic acid groups

The way ion-exchange chromatography works is shown in Fig. 26.1. Suppose the buffer in the column has a pH of 6, and we add to the top of the column a solution that contains a mixture of two amino acids in the same buffer: Val and Lys. We then allow the buffer to flow through the column; a frit keeps the resin from washing out. Since valine has zero charge at this pH, it is not attracted by the ionic groups on the column, and is washed through the column with a relatively small volume of buffer. Lysine, on the other hand, bears a net positive charge at pH 6, and is strongly attracted to the negatively charged resin. Because of this attraction, lysine is retained on the column and emerges from the column only after a considerably larger amount of buffer has passed through the column. The two amino acids are thus separated. Clearly, whether an amino acid or peptide is adsorbed by the column depends on its charge—which, in turn, depends on the relationship of its isoelectric point to the pH of the buffer.

In the experiment shown in Fig. 26.1, the ion-exchange resin is negatively charged and adsorbs cations; it is therefore called a *cation-exchange resin*. Resins that bear positively charged pendant groups (typically ammonium ions) adsorb anions, and are called *anion-exchange resins*.

Ion exchange has very important commercial applications—for example, in water treatment. Commercial water softeners contain cation-exchange resins much like the one used in this example, which adsorb the more highly charged calcium and magnesium ions in hard water and replace them with sodium ions with which the column is supplied. When the supply of sodium ions is exhausted, the column has to be flushed extensively with concentrated NaCl solution to replace the adsorbed calcium and magnesium ions with sodium ions.

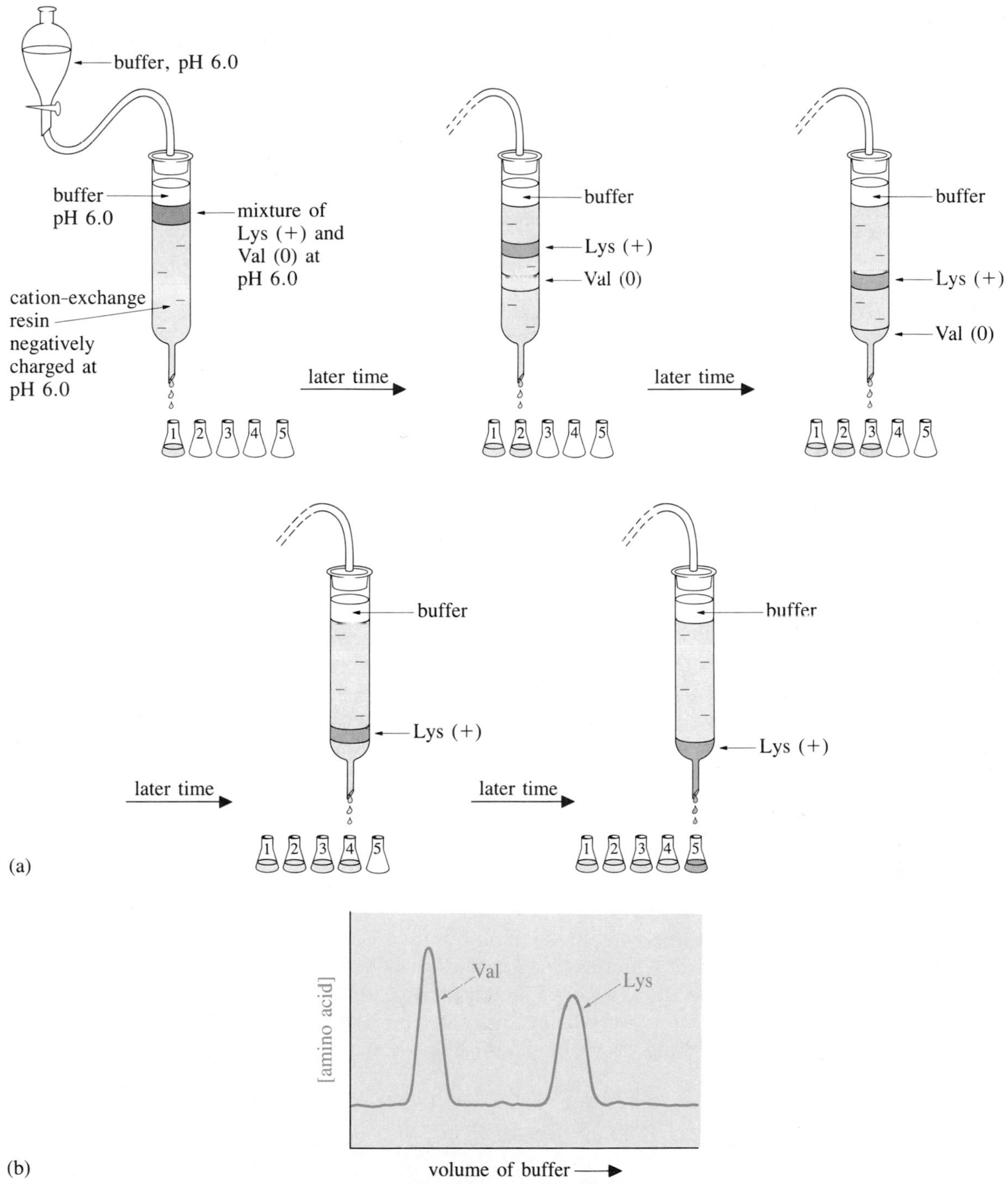

Figure 26.1 (a) Diagram of the cation-exchange separation of Val and Lys as described in the text. (b) Amino acid concentration in the effluent as a function of the volume of buffer passed through the cation-exchange column.

Problem

9 A mixture of *N*-acetyl-Leu-Gly, Lys-Gly-Arg, and Lys-Gly-Leu are applied to a sulfonated polystyrene cation-exchange column at a buffer pH of 6.0. Predict the order in which these three peptides will elute from the column and explain your reasoning. (See Problem 7c for an explanation of the *N*-acetyl nomenclature.)

26.5 SYNTHESIS AND OPTICAL RESOLUTION OF α-AMINO ACIDS

A. Alkylation of Ammonia

Some α-amino acids can be prepared by alkylation of ammonia with α-bromocarboxylic acids.

$$CH_3CH_2CH-\overset{Br}{\underset{CH_3}{\underset{|}{CH}}}-CO_2H + \underset{\text{(large excess)}}{2NH_3} \longrightarrow CH_3CH_2CH-\overset{^+NH_3}{\underset{CH_3}{\underset{|}{CH}}}-CO_2^- + \overset{+}{N}H_4 \, Br^- \quad (26.9)$$

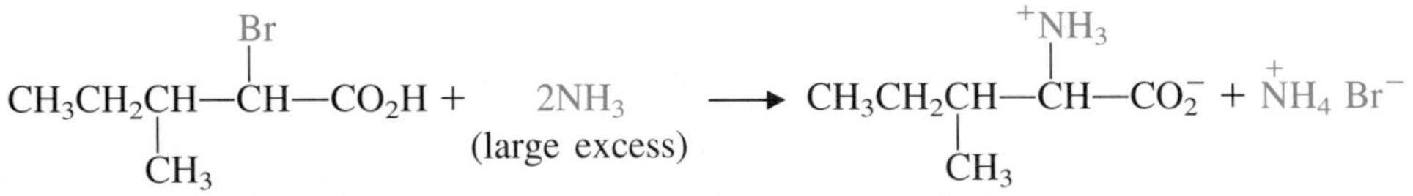

isoleucine
(49% yield)

This is an S_N2 reaction in which ammonia acts as the nucleophile. (Recall from Sec. 22.3D that α-halo carbonyl compounds are very reactive in S_N2 reactions.) We learned in Sec. 23.7A that alkylation of ammonia is generally not an acceptable method for preparing primary amines because ammonia is alkylated more than once, giving complex mixtures. However, the use of a large excess of ammonia in this synthesis favors monoalkylation. Furthermore, amino acids are less reactive toward alkylating agents than simple alkylamines because amino acids are less basic and more sterically hindered. Thus, reasonable yields of some amino acids can be obtained by ammonia alkylation.

B. Alkylation of Aminomalonate Derivatives

One of the most widely used methods for preparing α-amino acids is a variation of the malonic ester synthesis (Sec. 22.4A). The malonic ester derivative used is one in which a protected amino group is already in place: diethyl α-acetamidomalonate. This derivative is treated with sodium ethoxide in ethanol to form the ester–enolate ion, which is then alkylated with an alkyl halide—benzyl chloride in this example.

diethyl
α-acetamidomalonate

$$CH_3-\overset{O}{\overset{\|}{C}}-NH-\overset{CO_2Et}{\underset{CO_2Et}{\underset{|}{C}}}-CH_2Ph + Cl^- \quad (26.10a)$$

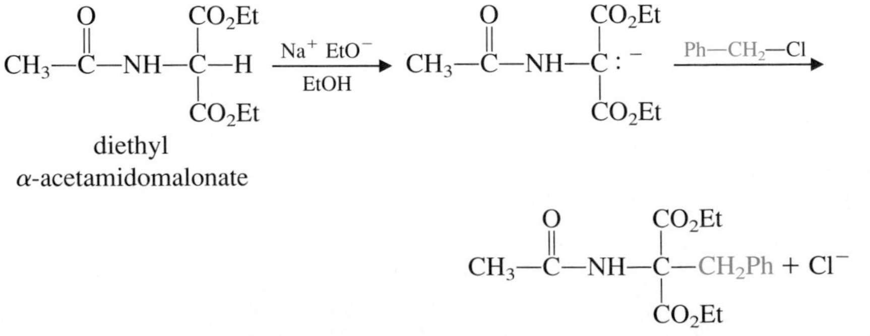

The resulting compound is then treated with hot aqueous HCl or HBr. This treatment accomplishes three things: First, the ester groups are hydrolyzed to carboxylic acids (Sec. 21.7A), yielding a substituted malonic acid (draw it!). Second, the malonic acid derivative decarboxylates under the reaction conditions (Sec. 20.11). Third, the acetamido group, an amide, is also hydrolyzed (Sec. 21.7B). Neutralization affords the α-amino acid.

$$CH_3 \!-\! \overset{\overset{\displaystyle O}{\|}}{C} \!-\! NH \!-\! \underset{\underset{\displaystyle CO_2Et}{|}}{\overset{\overset{\displaystyle CO_2Et}{|}}{C}} \!-\! CH_2Ph \xrightarrow[\text{heat}]{H_2O,\ HBr} CH_3CO_2H + 2EtOH + CO_2 + H_3\overset{+}{N} \!-\! \underset{\underset{\displaystyle CO_2H}{|}}{CH} \!-\! CH_2Ph\ Br^- \qquad (26.10b)$$

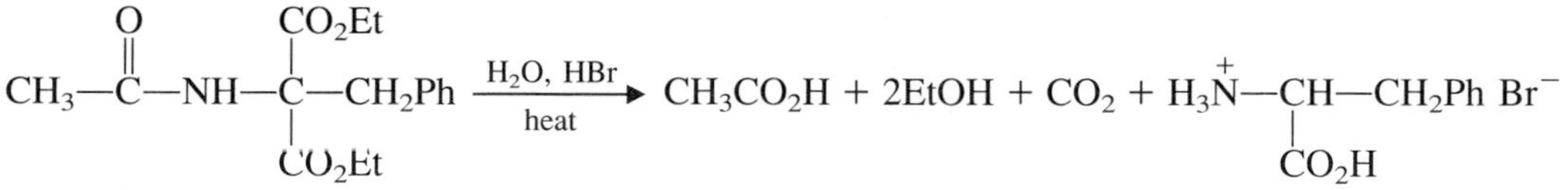

phenylalanine
(65% yield after
neutralization)

C. Strecker Synthesis

An important method for the synthesis of carboxylic acids is the hydrolysis of nitriles (Sec. 21.7D, 21.11). It follows that an α-amino nitrile might be used to prepare α-amino acids.

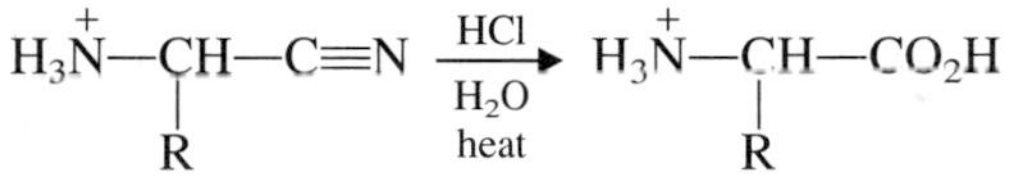

$$H_3\overset{+}{N} \!-\! \underset{\underset{\displaystyle R}{|}}{CH} \!-\! C\!\equiv\!N \xrightarrow[\substack{H_2O \\ \text{heat}}]{HCl} H_3\overset{+}{N} \!-\! \underset{\underset{\displaystyle R}{|}}{CH} \!-\! CO_2H \qquad (26.11)$$

an α-amino nitrile

α-Amino nitriles can be prepared by treatment of aldehydes with ammonia in the presence of cyanide ion. Hydrolysis of α-amino nitriles indeed gives α-amino acids.

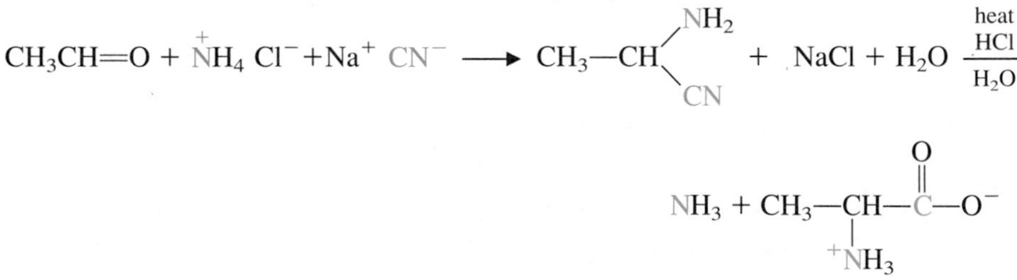

$$CH_3CH\!=\!O + \overset{+}{N}H_4\ Cl^- + Na^+\ CN^- \longrightarrow CH_3 \!-\! CH\!\!\underset{\diagdown CN}{\overset{\diagup NH_2}{}} + NaCl + H_2O \xrightarrow[\substack{HCl \\ H_2O}]{\text{heat}} \xrightarrow{OH^-}$$

$$NH_3 + CH_3 \!-\! \underset{\underset{\displaystyle {}^+NH_3}{|}}{CH} \!-\! \overset{\overset{\displaystyle O}{\|}}{C} \!-\! O^- \qquad \text{alanine} \qquad (26.12)$$
(52–60% yield)

This preparation of α-amino acids is called the **Strecker synthesis.**

The mechanism of this reaction probably involves the transient formation of an imine.

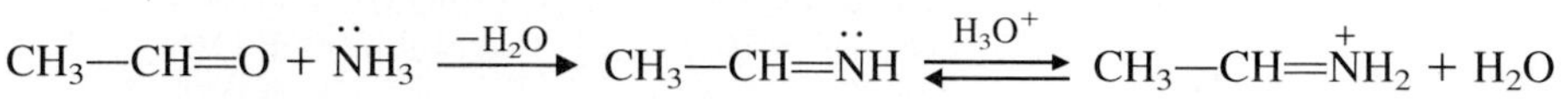

$$CH_3 \!-\! CH\!=\!O + \overset{..}{N}H_3 \xrightarrow{-H_2O} CH_3 \!-\! CH\!=\!\overset{..}{N}H \underset{}{\overset{H_3O^+}{\rightleftharpoons}} CH_3 \!-\! CH\!=\!\overset{+}{N}H_2 + H_2O \qquad (26.13a)$$

imine

The conjugate acid of the imine reacts with cyanide to give the α-amino nitrile.

$$CH_3 \!-\! CH\!\!\overset{+}{=}\!NH_2 + {}^-\!:CN \longrightarrow CH_3 \!-\! \underset{\underset{\displaystyle CN}{|}}{CH} \!-\! \overset{..}{N}H_2 \qquad (26.13b)$$

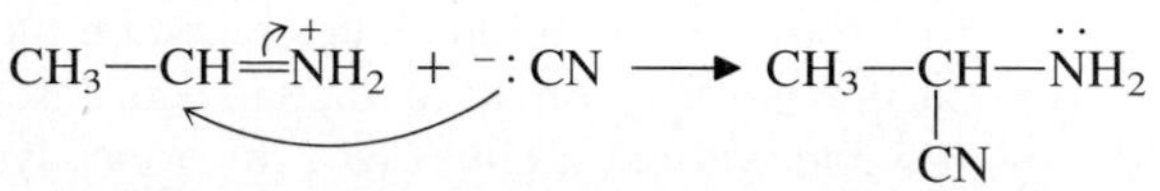

The addition of cyanide to an imine is analogous to the formation of a cyanohydrin from an aldehyde or ketone (Sec. 19.7).

$$HCN + CH_3-CH=O \longrightarrow CH_3-\underset{\underset{CN}{|}}{CH}-OH \qquad (26.14)$$

Problems

10 Suggest a synthesis of
(a) leucine from diethyl α-acetamidomalonate.
(b) α-phenylglycine from benzaldehyde.

11 Why is the *N*-acetyl group necessary in the α-acetamidomalonate synthesis? What side reaction might occur if diethyl α-aminomalonate were used instead?

D. *Optical Resolution of α-Amino Acids*

Amino acids synthesized by common laboratory methods are *racemic*. Since many applications require the pure enantiomers, the racemic compounds must be optically resolved. As useful as the diastereomeric salt method is (Sec. 23.5D), it can be tedious and time consuming. An alternative approach to the preparation of optically pure amino acids, and one that is used industrially, is the synthesis of amino acids by microbiological techniques—that is, by fermentation. Some cultures of microorganisms can be used to produce industrial quantities of certain amino acids in the natural L form.

Certain enzymes—biological catalysts—can also be used to resolve racemic amino acids into enantiomers. For example, a preparation of the enzyme *acylase* from hog kidney selectively catalyzes the hydrolysis of *N*-acetyl-L-amino acids and leaves the corresponding D isomers unaffected. Thus treatment of the *N*-acetylated racemate with this enzyme affords the free L-amino acid only.

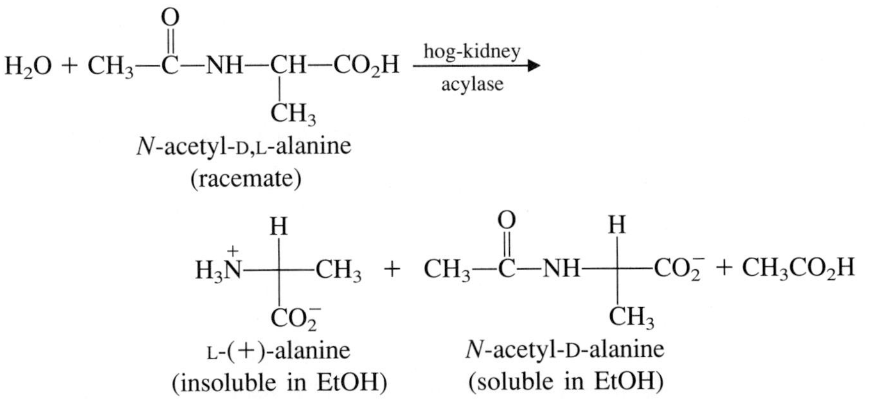

$$H_2O + CH_3-\overset{O}{\overset{||}{C}}-NH-\underset{\underset{CH_3}{|}}{CH}-CO_2H \xrightarrow[\text{acylase}]{\text{hog-kidney}}$$

N-acetyl-D,L-alanine
(racemate)

$$H_3\overset{+}{N}\overset{H}{\underset{CO_2^-}{\vert}}CH_3 + CH_3-\overset{O}{\overset{||}{C}}-NH\overset{H}{\underset{CH_3}{\vert}}CO_2^- + CH_3CO_2H \quad (26.15)$$

L-(+)-alanine *N*-acetyl-D-alanine
(insoluble in EtOH) (soluble in EtOH)

In this example the liberated L-alanine is precipitated from ethanol; the *N*-acetyl-D-alanine remains in solution, from which it can be recovered and hydrolyzed in aqueous acid to D-alanine.

The enzyme differentiates between the two enantiomers of *N*-acetylalanine because it is *chiral*. As we learned in Sec. 7.8C, such a selective hydrolysis would not be possible without a chiral reagent or catalyst.

26.6 REACTIONS OF AMINO ACIDS

A. Acylation and Esterification

Amino acids undergo many of the characteristic reactions of both amines and carboxylic acids. A typical reaction of amines that is also very important in amino acid chemistry is *acylation* by acid chlorides or anhydrides.

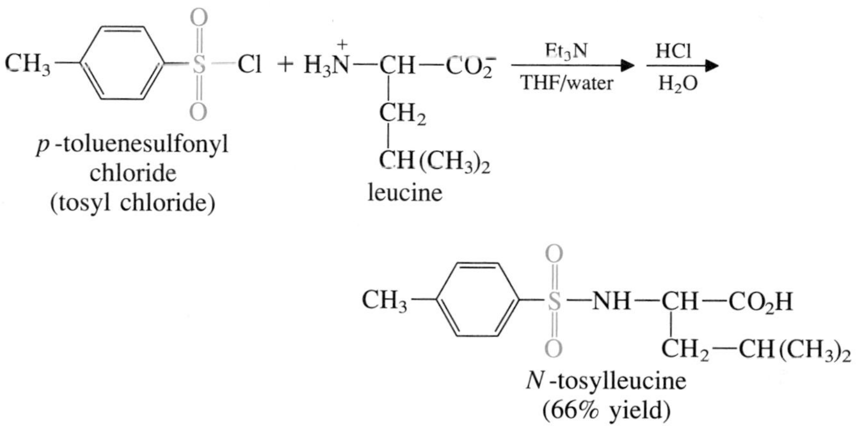

(26.16)

$$\overset{+}{H_3N}-\underset{\substack{|\\CH_2\\|\\CH(CH_3)_2}}{CH}-CO_2^- + CH_3-\overset{O}{\overset{\|}{C}}-O-\overset{O}{\overset{\|}{C}}-CH_3 \xrightarrow{CH_3CO_2H} \xrightarrow{H_2O} CH_3-\overset{O}{\overset{\|}{C}}-NH-\underset{\substack{|\\CH_2\\|\\CH(CH_3)_2}}{CH}-CO_2H \quad (26.17)$$

N-acetylleucine (85–95% yield)

Amino acids, like ordinary carboxylic acids, are easily esterified by heating with an alcohol and a mineral acid catalyst (acid-catalyzed esterification; Sec. 20.8A).

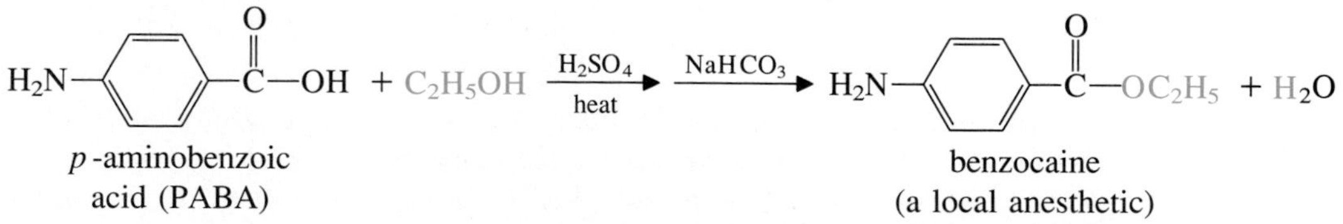

(26.18)

$$\overset{+}{H_3N}-\underset{\substack{|\\CH_3}}{CH}-CO_2^- + CH_3OH \xrightarrow[heat]{HCl} \overset{+}{H_3N}-\underset{\substack{|\\CH_3}}{CH}-CO_2CH_3\ Cl^- \quad (26.19)$$

alanine methyl ester hydrochloride

Problem

12 If the hydrochloride of glycine methyl ester is neutralized and allowed to stand in solution, a polymer forms. If the hydrochloride itself is allowed to stand, nothing happens. Explain these observations.

B. Reaction with Ninhydrin

α-Amino acids react with *ninhydrin* to form a dye called *Ruhemann's purple.*

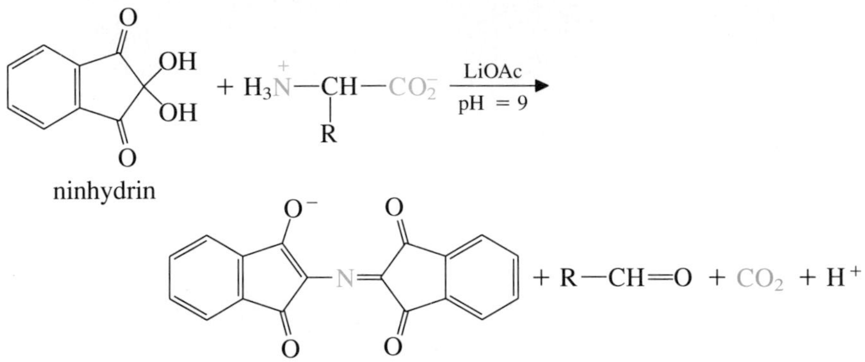

ninhydrin

Ruhemann's purple
λ_{max} = 570 nm

(26.20)

This is an important reaction that is used to detect small amounts of amino acids.

We can see from the structure of the product in Eq. 26.20 that the same dye is formed from all α-amino acids with primary amino groups, since the amino acid side chain R is lost as an aldehyde. This dye has an intense bluish-purple color because it strongly absorbs visible light at 570 nm, and it can be used for the quantitative analysis of amino acids and peptides in the laboratory (Sec. 26.7A). Because both hydrogens of the amino group are lost, it is clear that secondary amines cannot react in this way with ninhydrin. Thus, proline reacts with ninhydrin to give a different adduct, which has a different color and absorbs light at a different wavelength. (See Problem 13.)

A startling demonstration of the sensitivity of ninhydrin can be observed if we handle a piece of clean filter paper and spray it with ninhydrin. A blue fingerprint rapidly develops from the presence of small amounts (about 10^{-10} mol) of amino acids on the fingers.

The mechanism of the ninhydrin reaction involves a combination of simpler reactions with which we are already familiar. First, notice that *two* molecules of ninhydrin react per molecule of amino acid. Second, you should recognize that ninhydrin is the hydrate of a tricarbonyl compound.

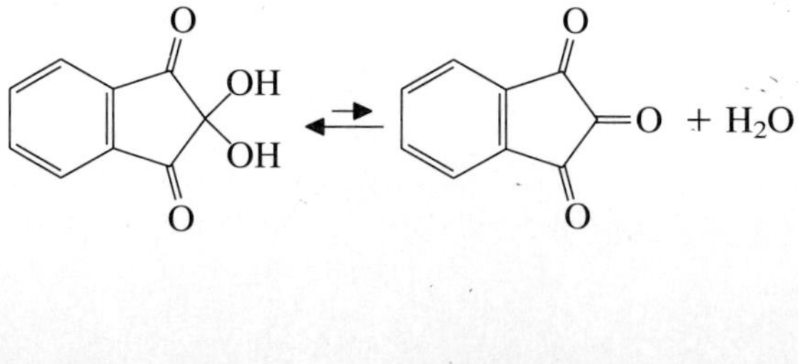

(26.21a)

The first step in the ninhydrin reaction is the formation of an imine (Sec. 19.11A) between the tricarbonyl form of ninhydrin and the amino group of the amino acid. (Why is the central carbonyl group of ninhydrin more reactive?)

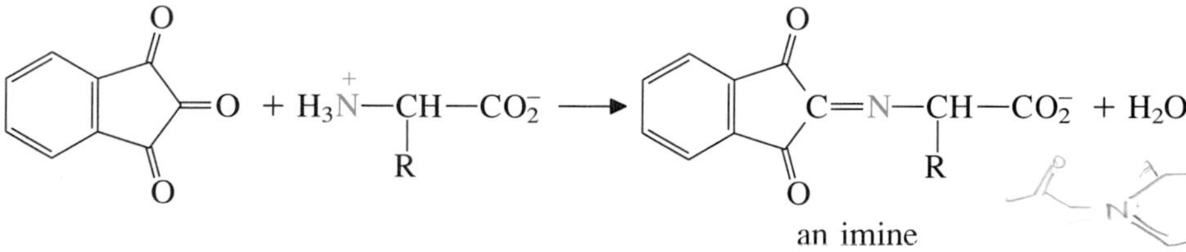

(26.21b)

This imine is converted into a different imine by the loss of CO_2 (decarboxylation). The new imine hydrolyzes to an aldehyde and an amine. (Recall, imine formation is reversible.) The aldehyde carries the amino acid side chain.

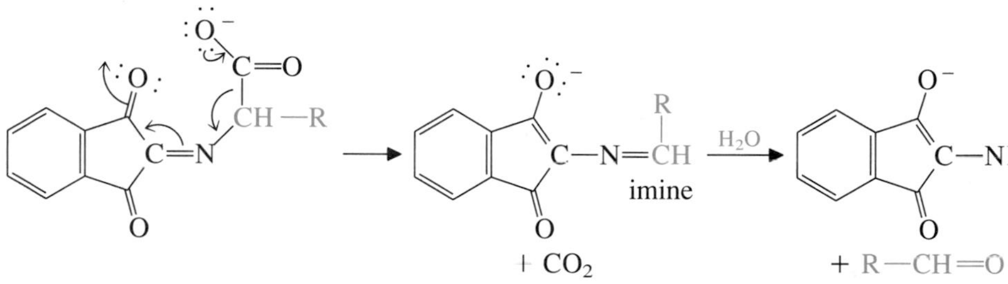

(26.21c)

The resulting amine then forms yet another imine with a second molecule of ninhydrin; this is the final product.

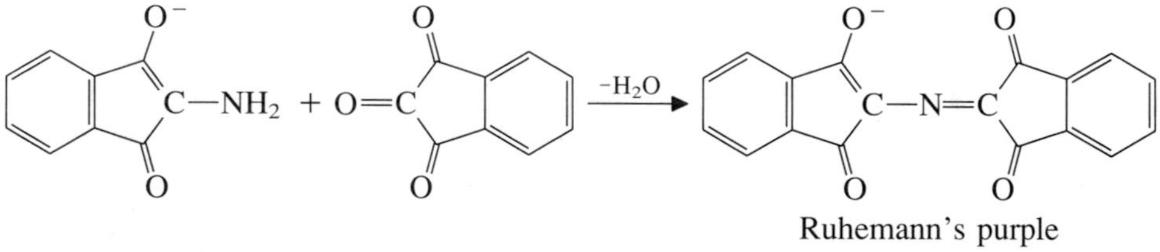

(26.21d)

Simple primary amines also react with ninhydrin to give the same dye (see Problem 13b), but the reaction with α-amino acids is particularly efficient.

Problem

13 (a) As noted above, proline does not give Ruhemann's purple when it reacts with ninhydrin. The following structure has been postulated for the adduct of proline and ninhydrin. Give a mechanism for the formation of this compound.

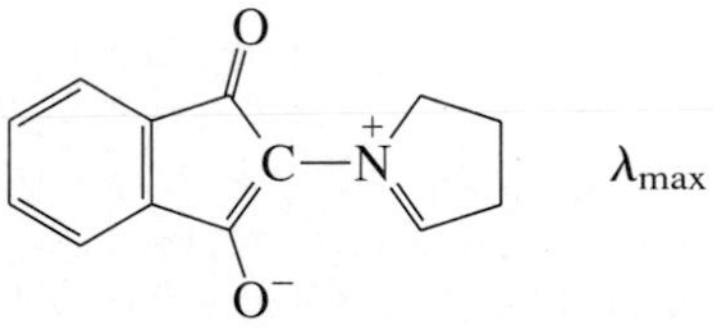

$\lambda_{max} = 440$ nm

(b) Ninhydrin also reacts with primary amines to give Ruhemann's purple. Using the reaction of ninhydrin with ethylamine, give a mechanism for this reaction.

26.7 ANALYTICALLY IMPORTANT REACTIONS OF PEPTIDES

In this section we consider some laboratory reactions of peptides that are important in their analysis and structure determination.

A. Hydrolysis of Peptides. Amino Acid Analysis

Since the peptide bonds that connect amino acid residues in peptides are amide bonds, it is not surprising that peptides can be hydrolyzed under the conditions of amide hydrolysis: moderately concentrated aqueous acid or base and heat. Hydrolysis with $6N$ aqueous HCl for 20–24 hr at 110° effects complete conversion of most peptides into their constituent amino acids.

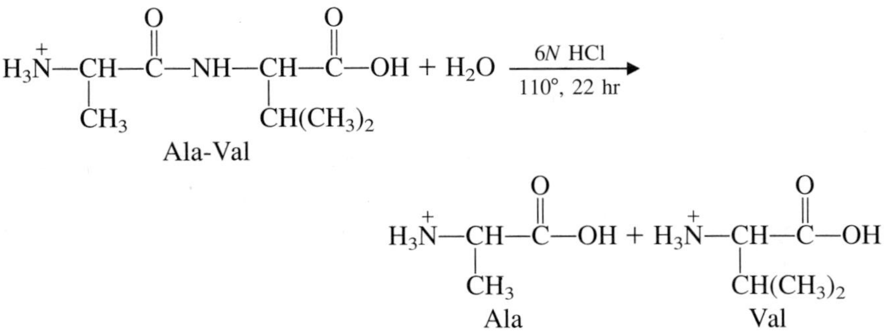

$$
\underset{\text{Ala-Val}}{\overset{+}{H_3N}-\underset{\underset{CH_3}{|}}{CH}-\overset{\overset{O}{\|}}{C}-NH-\underset{\underset{CH(CH_3)_2}{|}}{CH}-\overset{\overset{O}{\|}}{C}-OH} + H_2O \xrightarrow[110°,\ 22\ hr]{6N\ HCl}
$$

$$
\underset{\text{Ala}}{\overset{+}{H_3N}-\underset{\underset{CH_3}{|}}{CH}-\overset{\overset{O}{\|}}{C}-OH} + \underset{\text{Val}}{\overset{+}{H_3N}-\underset{\underset{CH(CH_3)_2}{|}}{CH}-\overset{\overset{O}{\|}}{C}-OH} \quad (26.22)
$$

When a peptide or protein is hydrolyzed, the product amino acids can be separated, identified, and quantitated in a technique called **amino acid analysis,** shown schematically in Fig. 26.2. The amino acids in the hydrolyzed mixture are separated by passing them through a cation-exchange column (see Fig. 26.1) under very carefully defined conditions. The time at which each amino acid emerges from the column is accurately known. As each amino acid emerges, it is mixed with ninhydrin (Sec. 26.6B). The intensity of the resulting color is proportional to the amount of the amino acid present. The color intensity, and therefore the amount of each amino acid, is recorded as a function of time. Figure 26.3 shows the recorder trace obtained from a mixture of 10 nanomoles (10^{-8} mole) of each α-amino acid. The trace obtained from the analysis of a peptide of unknown composition will contain a peak corresponding to each amino acid present; the area of the peak is proportional to the amount of the amino acid. Thus, by hydrolysis, reaction with ninhydrin, and quantitation of the color produced, the *identities* and *relative amounts* of the amino acid residues in a peptide can be determined.

For example, let us imagine that a hypothetical peptide P has been hydrolyzed and subjected to amino acid analysis, and that the results are as follows:

P: Ala$_3$,Arg,(Asp or Asn),Gly$_2$,His,Ile,Leu,Lys,Met,NH$_3$,Phe,Pro,Trp,Tyr,Val

According to this analysis, the peptide contains three times as much Ala and twice as much Gly as Arg, His, or Lys. The absolute number of each amino acid residue is not known unless, in addition, the molecular weight of the peptide is known. Notice also that the relative order of the amino acid residues within the peptide is also not known. In this sense, amino acid analysis is to the amino acid composition of a peptide as elemental analysis is to the molecular formula of an organic compound.

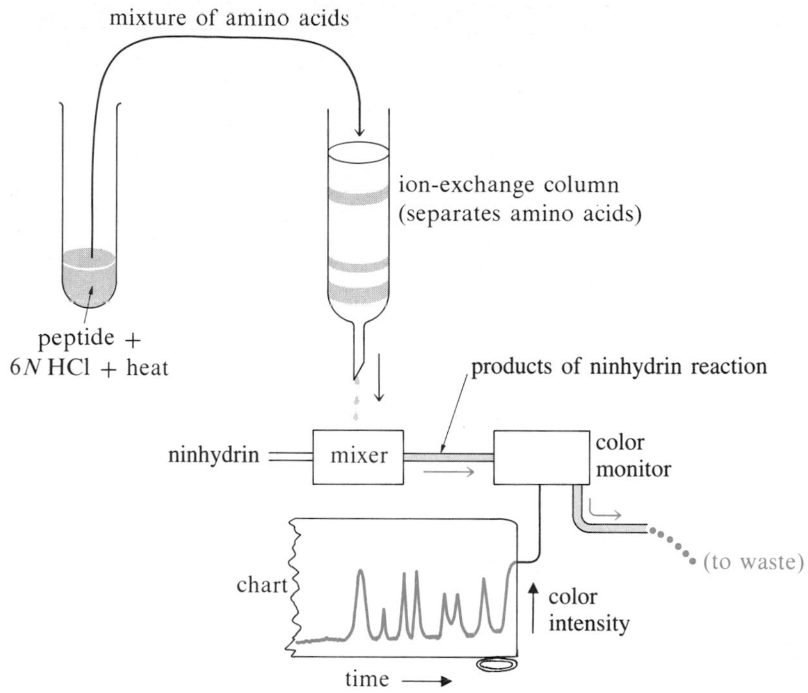

Figure 26.2 Diagram of amino acid analysis. A chart trace from an actual analysis is shown in Fig. 26.3.

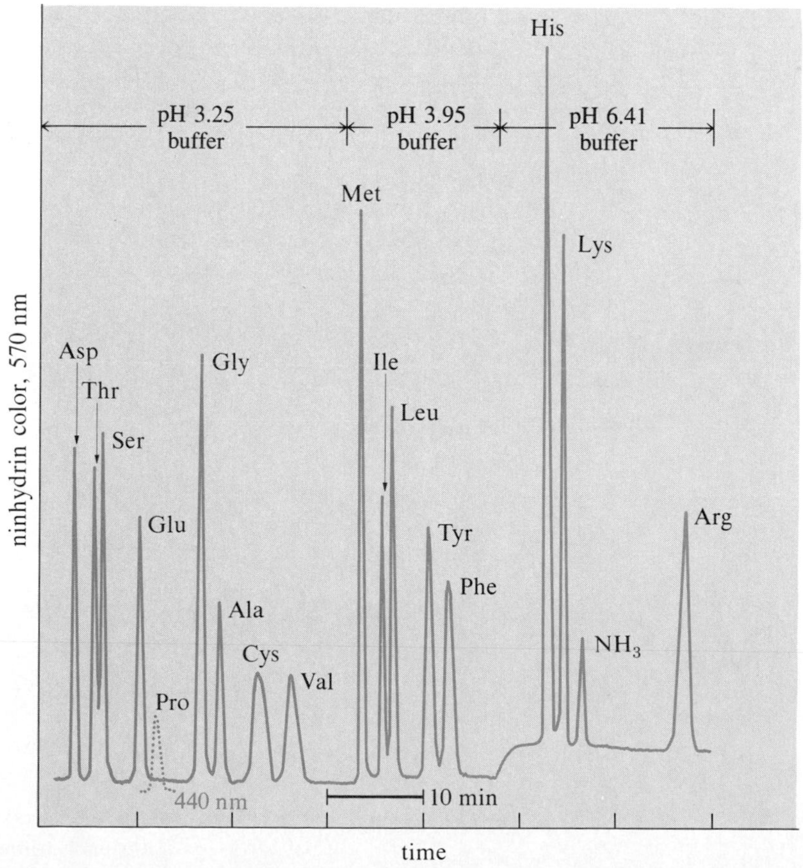

Figure 26.3 Chart trace from the amino acid analysis of a mixture containing 10 nanomoles (10^{-8} mol) of each amino acid. The pH values at the top of the chart show the pH of the buffers used to elute (wash) the amino acids from the ion-exchange column.

The ion-exchange column used in the separation shown in Fig. 26.2 contains negatively charged groups much like the one shown in Fig. 26.1. We can see that the amino acids with acidic side chains emerge from the column earlier and those with basic side chains emerge from the column later, as we expect from our discussion of ion exchange. To a first approximation, we expect all neutral amino acids to emerge from the column at the same time; obviously, separation of the amino acids succeeds because this is *not* the case. Each neutral amino acid interacts differently with the ion-exchange resin. The ion-exchange resin contains, in addition to charged groups, aromatic rings that have an affinity for other hydrocarbonlike groups. The adsorption process is much like the process of dissolving a solute, with the resin acting as a "solvent." Thus, phenylalanine, with its aromatic side chain, is adsorbed to the resin more tightly than serine, with its alcohol-like, $—CH_2OH$ side chain (like dissolves like); therefore phenylalanine emerges much later from the column than serine.

Problem

14 Explain why acid hydrolysis followed by amino acid analysis does not distinguish between Asp and Asn. (*Hint:* why is ammonia present in the amino acid analysis of the peptide *P* above?) What other pair of amino acids are not differentiated by amino acid analysis?

B. Sequential Degradation of Peptides

The actual arrangement, or sequential order, of amino acid residues in a peptide is called its **amino acid sequence** or **primary sequence.** There are a large number of possible sequences for even a relatively small peptide with a given amino acid composition. For example, more than 10^{14} sequences are possible for peptide *P* described in the previous section! How do we determine the amino acid sequence of a peptide or protein?

This apparently complex problem is actually solved rather easily. It is possible to remove one residue at a time from the amino end of a peptide, identify it, and then repeat the process sequentially on the remaining peptide. Schematically, the strategy is as follows:

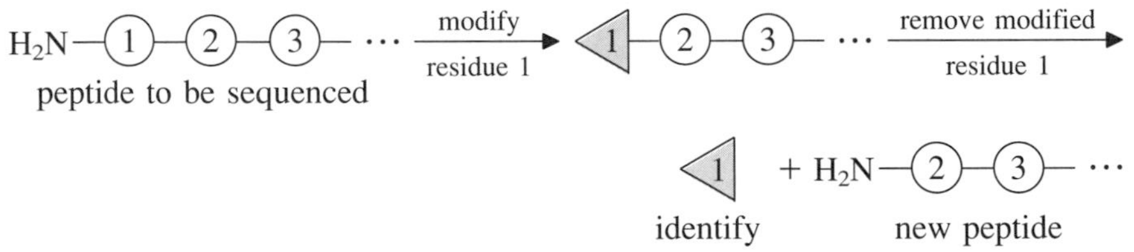

(26.23a)

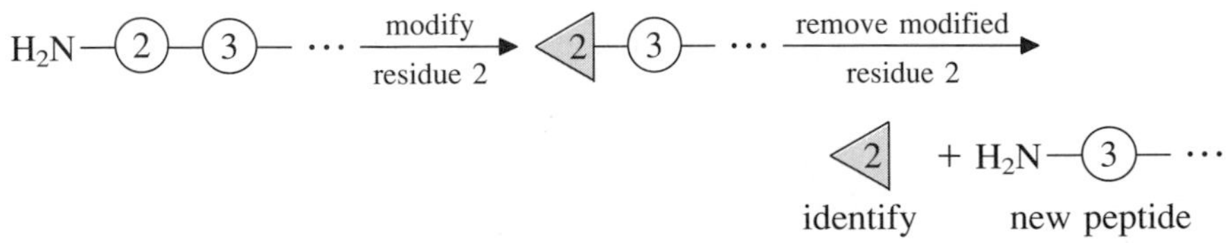

(26.23b)

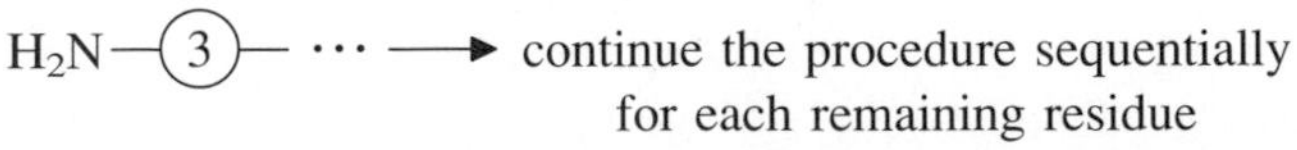

(26.23c)

The standard method for implementing this strategy is called the **Edman degradation,** after Per Edman, a Swedish biochemist who devised the method in 1952. In this method, the peptide is treated with *phenyl isothiocyanate* (often called the **Edman reagent**). The peptide reacts with the Edman reagent at its amino terminal amino group to give a thiourea derivative. (In this and subsequent equations, we shall use the abbreviation PepN for the amino terminal part of a peptide and PepC for the carboxy terminal part of a peptide.)

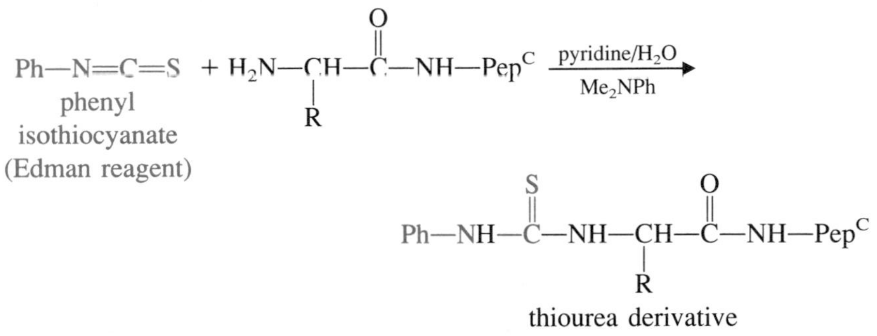

$$Ph-N=C=S \quad + \quad H_2N-\underset{\underset{R}{|}}{CH}-\overset{\overset{O}{\|}}{C}-NH-Pep^C \quad \xrightarrow[Me_2NPh]{pyridine/H_2O}$$

phenyl
isothiocyanate
(Edman reagent)

$$Ph-NH-\overset{\overset{S}{\|}}{C}-NH-\underset{\underset{R}{|}}{CH}-\overset{\overset{O}{\|}}{C}-NH-Pep^C \quad (26.24a)$$

thiourea derivative

This reaction, which is the modification of residue 1 shown schematically in Eq. 26.23a, is exactly analogous to the reaction of amines with *isocyanates,* the oxygen analogs of *isothiocyanates* (Eq. 23.86). Any remaining phenyl isothiocyanate is removed, and the modified peptide is then treated with anhydrous trifluoroacetic acid. As a result of this treatment, the sulfur of the thiourea, which is nucleophilic, displaces the amino group of the adjacent residue to yield a five-membered heterocycle called a *thiazolinone;* the other product of the reaction is *a peptide that is one residue shorter.*

thiourea

thiazolinone
derivative

$+ \quad H_3\overset{+}{N}-Pep^C \quad (26.24b)$
new peptide, one
residue shorter

When treated subsequently with aqueous acid, the thiazolinone derivative forms an isomer called a **phenylthiohydantoin.** This probably occurs by reopening of the thiazolinone to the thiourea, followed by ring formation involving the thiourea nitrogen. Notice in this and the previous equation the intramolecular formation of five-membered rings.

thiazolinone

thiourea

phenylthiohydantoin
(PTH) derivative
of amino terminal residue

side chain
of amino terminal
residue

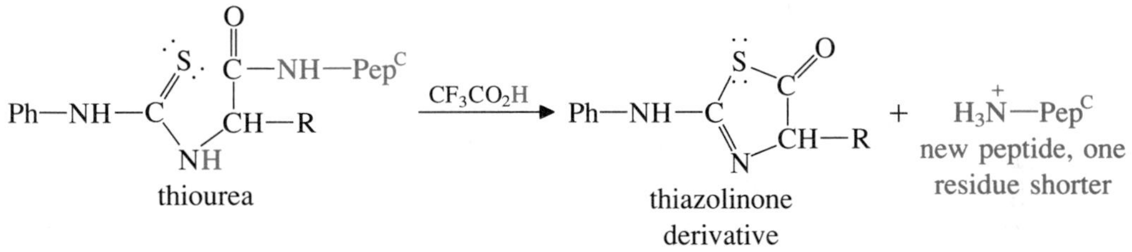

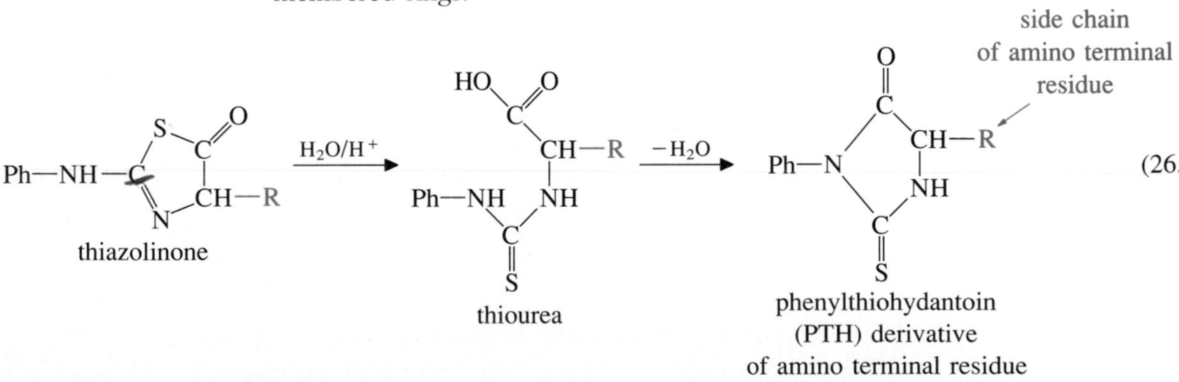

(26.24c)

(You should fill in the details of these reactions using the arrow formalism.)

Because the phenylthiohydantoin (PTH) derivative carries the characteristic side chain of the amino terminal residue, the structure of the PTH derivative tells us immediately which amino acid residue has been removed. Methods for identifying the PTH derivatives of all amino acids are well established. The peptide liberated in Eq. 26.24b can be subjected in turn to the Edman degradation again to yield the PTH derivative of the next amino acid and a new peptide that is shorter by yet another residue.

In principle, the Edman degradation can be continued indefinitely for as many residues as necessary to define completely the sequence of the peptide. In practice, because the yields at each step are not perfectly quantitative, an increasingly complex mixture of peptides is formed with each successive step in the cleavage, and after a number of such steps the results become ambiguous. Hence, the number of residues in a sequence that can be determined by the Edman method is limited. Nevertheless, instruments are now in use that can apply Edman chemistry to structure determination of peptides in a highly standardized, automated, and reproducible form. In such instruments the sequential degradation of twenty residues is common, and the degradation of as many as sixty or seventy amino acid residues is sometimes possible.

Let us illustrate application of the Edman degradation to two peptides of unknown structure. Suppose peptide *A* is a pentapeptide with the composition (Gly,Ile,Leu,Lys,Phe), and peptide *B* is also a pentapeptide with the composition (Ala,Gly,His,Met,Pro). Suppose further that application of the Edman degradation to these peptides gives the following results:

	PTH derivative found in repetition number			
Peptide	1	2	3	4
A	Leu	Gly	Ile	Phe
B	Ala	His	Pro	Gly

The sequence of each peptide is read directly from the identity of the PTH derivative liberated in each cycle of the degradation:

A: Leu-Gly-Ile-Phe-Lys
B: Ala-His-Pro-Gly-Met

The Lys in peptide *A*, and the Met in peptide *B*, must lie at the carboxy termini of their respective peptides because they are the only residues not determined in the Edman degradation.

Problems

15 Using the arrow formalism, write out in detail the mechanisms for the reactions shown in Eq. 26.24a, b, and c.

16 Some peptides found in nature have an amino-terminal acetyl group (color):

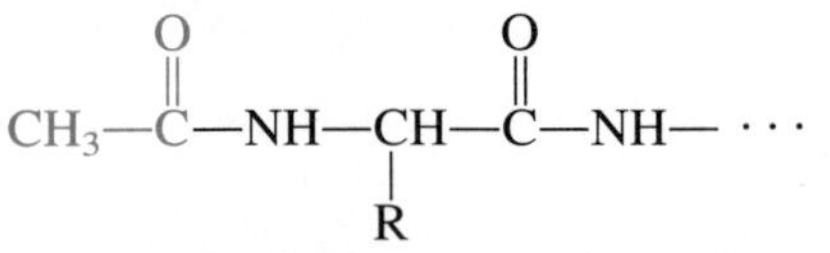

Can these peptides undergo the Edman degradation? Explain.

C. Specific Cleavage of Peptides

Most common proteins contain hundreds of amino acid residues. Hence the Edman chemistry, which is limited by yield to 20–60 consecutive residues, cannot be used to determine the structure of such proteins in a single set of sequential degradations. The amino acid sequence of most large proteins is determined by breaking the protein into a number of smaller peptides, and sequencing these peptides individually. Then the sequence of the protein is reconstructed from the sequences of the peptides. In other words, the large sequencing problem is divided into a series of smaller sequencing problems. When breaking a larger protein into smaller peptides, it is desirable to use reactions that cleave the protein in high yield at well-defined points so that a relatively small number of peptides are obtained. (Cleavages at random points in the protein chain would give complex, difficult-to-separate mixtures.) In this section, we shall consider two methods used frequently by protein chemists to cleave peptides at specific amino acid residues into smaller fragments. One method uses ordinary reagents; another method involves the use of enzymes to catalyze peptide-bond hydrolysis.

Peptide Cleavage at Methionine with Cyanogen Bromide When a peptide reacts with *cyanogen bromide* (Br—C≡N) in aqueous HCl, a peptide bond is cleaved specifically at the carboxy side of each methionine residue.

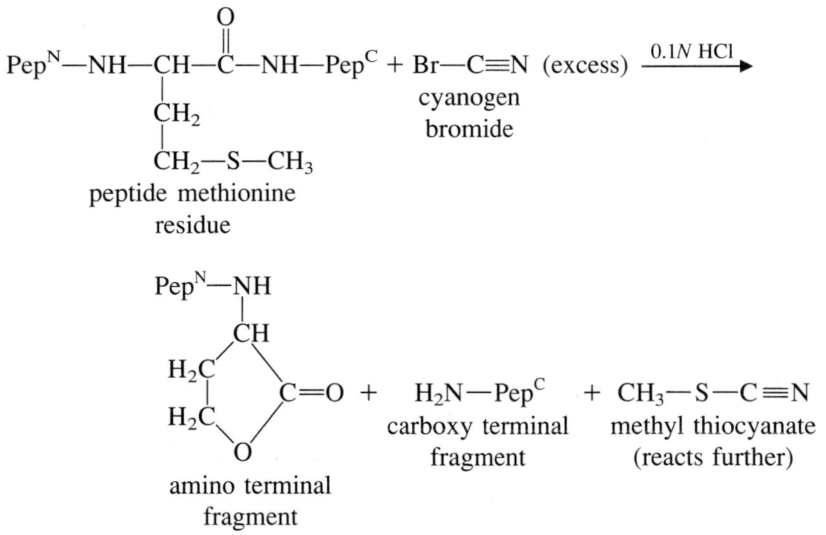

(26.25)

It is important to observe that the amino terminal fragment from the cleavage shown in Eq. 26.25 has a carboxy terminal *homoserine lactone* residue instead of the starting methionine.

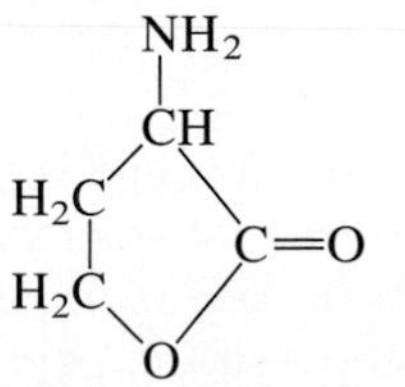

homoserine lactone

Methionine is a relatively rare amino acid; hence, when a typical protein is cleaved with CNBr, relatively few cleavage peptides are obtained, and all of them are derived from cleavage at methionine residues.

Why does this cleavage work? Cyanogen bromide has something of the character of an acid halide. Although it can in principle react with any nucleophilic amino acid side chain, under the conditions of the reaction only its reaction at methionine leads to a peptide cleavage, for reasons that will become apparent in the mechanism outlined below. (Fill in the details of each step using the arrow formalism.)

The sulfur in the methionine side chain acts as a nucleophile, displacing bromide from cyanogen bromide to give a type of *sulfonium ion* (Sec. 11.5A).

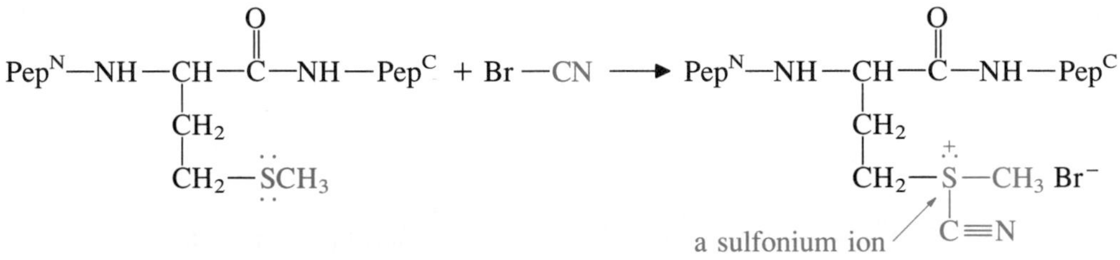

(26.26a)

The sulfonium ion, with its electron-withdrawing cyanide, is an excellent leaving group, and is displaced by the oxygen of the neighboring amide bond to form a five-membered ring. (The amide oxygen is normally not a very good nucleophile, but as we have seen before, reactions that occur with the formation of five- or six-membered rings are particularly rapid; see Sec. 11.6.)

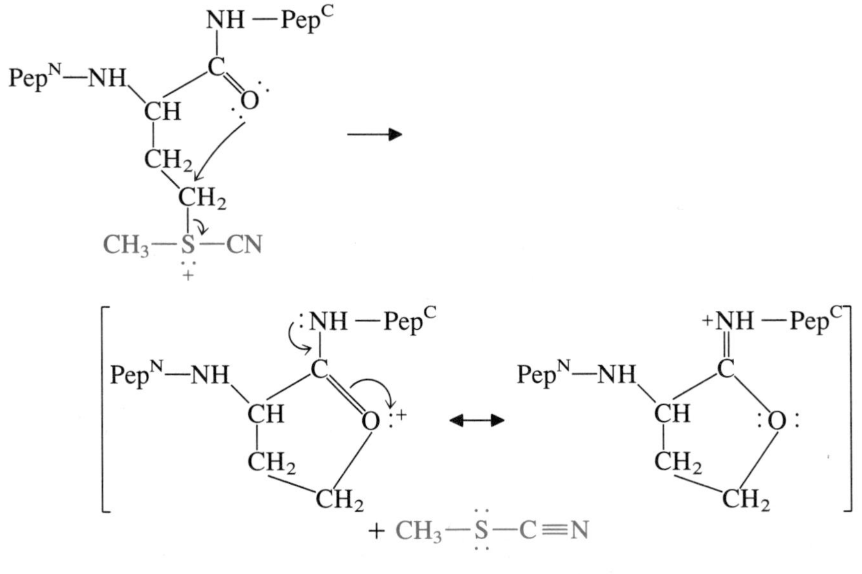

(26.26b)

This step shows why the cleavage is specific for methionine: only methionine has a side chain that can form a five-membered ring by such a mechanism.

The last mechanistic step in the cleavage is hydrolysis of the ion formed in Eq. 26.26b. It is in this step that the peptide bond is actually broken:

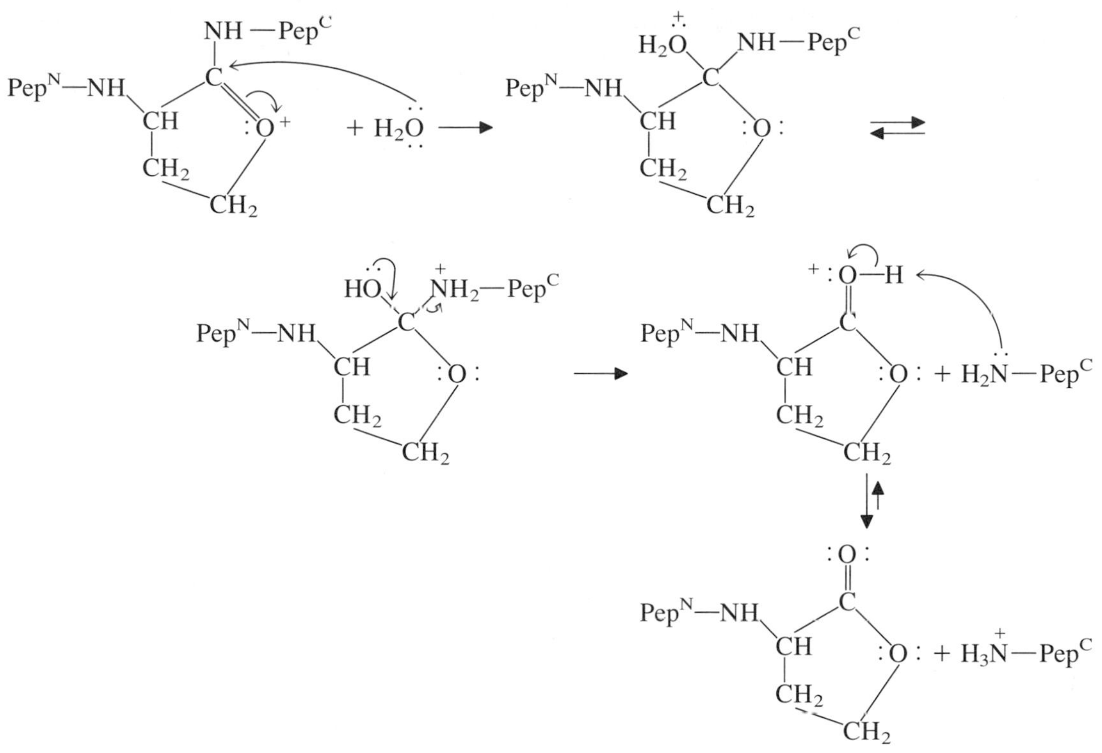

(26.26c)

Consider how this cleavage might be used in determining the amino acid sequence of the peptide *P* introduced on p. 1152. (Actually peptide *P* is small enough that it could be completely sequenced by the Edman method; but for illustrative purposes we shall imagine that is too long for a single set of Edman degradations.) Suppose cyanogen bromide treatment of *P* gives two cleavage peptides, *A* and *B*:

composition of A: Ala,Gly_2,His,Ile,Leu,Lys,Phe, Pro,Tyr,homoserine lactone

composition of B: Ala_2,Arg,(Asp or Asn),Val,Trp

From these data we can construct a partial sequence of *P*:

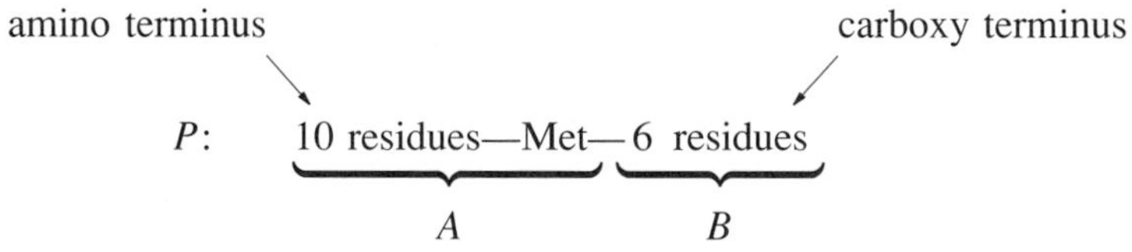

We know that *A* precedes *B* in the sequence because peptide *A* contains the homoserine lactone residue. Each of the peptides *A* and *B* could be sequenced individually with the Edman method. Knowing the order of *A* and *B* in the sequence of peptide *P*, the sequence of the full peptide *P* would then be known.

Peptide Cleavage with Proteolytic Enzymes The cyanogen bromide reaction is one of the relatively few useful nonenzymatic cleavages of peptides. In contrast, there are a number of enzymes that catalyze the hydrolysis of peptide bonds at specific points in an amino acid sequence. Such peptide-hydrolyzing enzymes are called **proteases, peptidases,** or **proteolytic enzymes.** One of the most widely used proteases is the enzyme *trypsin*. This enzyme catalyzes the hydrolysis of peptides or proteins at the carbonyl group of arginine or lysine residues, provided that these residues (a) are not at the amino end of the protein, and (b) are not followed by a proline residue.

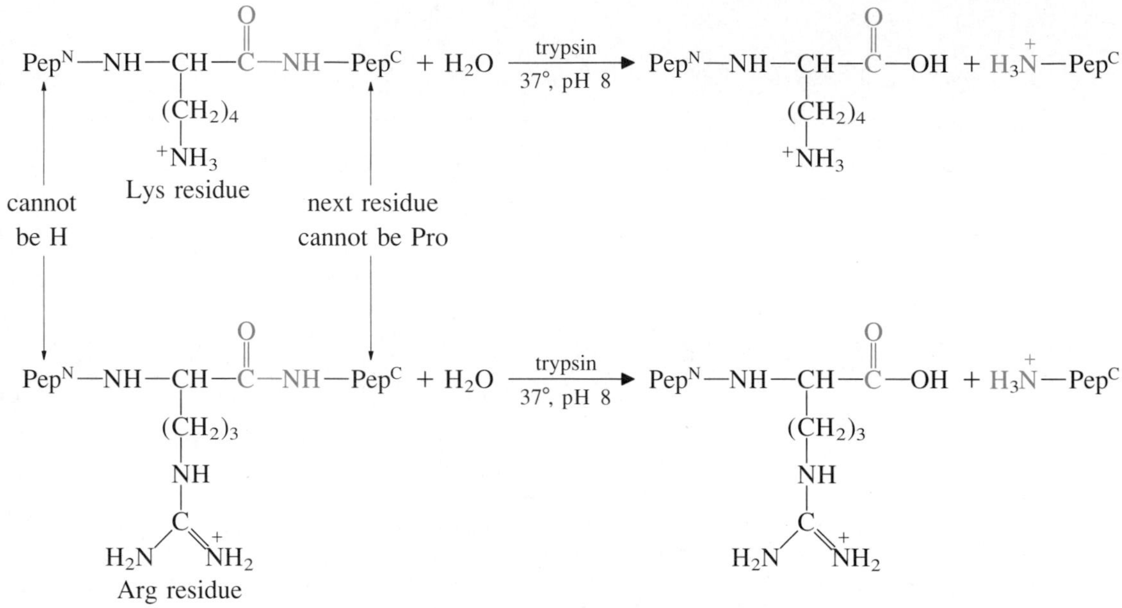

(26.27a)

(26.27b)

(We shall consider the mechanism of trypsin-catalyzed hydrolysis in Sec. 26.10.) Because trypsin catalyzes the hydrolysis of peptides at internal rather than terminal residues, it is called an **endopeptidase.** (Enzymes that cleave peptides only at terminal residues are termed **exopeptidases.**)

Suppose we were to treat each of the peptides *A* and *B* above with trypsin and obtain the following data. See how much information about the sequence of the original peptide *P* you can deduce from these data. (The answer follows.)

Treatment of *A* with trypsin gives *C* + *D*.
composition of C: Gly,Ile,Leu,Phe,Tyr,Lys
composition of D: Ala,Gly,His,Pro,homoserine lactone

Treatment of *B* with trypsin gives *E* and *F*.
composition of E: Ala,Trp,Arg
composition of F: Ala,(Asp or Asn),Val

Therefore a partial sequence of *P* is:

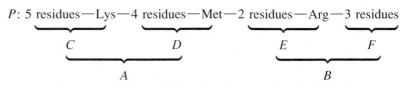

P: 5 residues—Lys—4 residues—Met—2 residues—Arg—3 residues

C (5 residues—Lys); *D* (4 residues—Met); *E* (2 residues—Arg); *F* (3 residues)

A (*C* + *D*); *B* (*E* + *F*)

The sequence could be completed by applying the Edman method to the short peptides *C*-*F*.

Problem

17 Explain the logic used in constructing the partial sequence above by answering the following:
(a) Why is the order of peptides C-D-E-F known?
(b) How do we know the position of the Lys, Arg, and Met residues?

Several enzymes other than trypsin are also used to cleave peptides. Chymotrypsin, a protein related to trypsin, is used to cleave peptides at amino acid residues with aromatic side chains and, to a lesser extent, residues with large hydrocarbon side chains. Thus, chymotrypsin cleaves peptides at Phe, Trp, Tyr, and occasionally Leu and Ile residues. Chymotrypsin and trypsin are mammalian digestive enzymes; their biological role, understandably, is to catalyze the hydrolytic breakdown of dietary proteins in the intestine. Not long ago an important endopeptidase was discovered in a microorganism, *Staphylococcus aureus*. This enzyme catalyzes the hydrolysis of peptides at glutamic acid residues. Thus, biochemists have an arsenal of different proteases that can be used to break proteins into peptides at specific sites.

Problems

18 (a) What product would you expect from the reaction of cyanogen bromide with free amino groups in a protein, such as the side-chain amino group of lysine?
(b) What conditions of the reaction with CNBr prevent such a reaction? Explain. (See Eq. 26.25.)

19 A peptide *Q* has the following amino acid composition.

$$Q\text{: Arg}_2\text{,Ile}_2\text{,Glu,Gly}_2\text{,Leu,Lys,Phe,Pro,Ser,Trp}$$

When *Q* was subjected to a cycle of the Edman degradation, the PTH derivative of leucine was formed, along with a new peptide. Treatment of *Q* with trypsin gave the following peptides. (Their individual sequences were determined by the Edman degradation.)

$$Q \xrightarrow{\text{trypsin}} \underset{A}{\text{Gly-Arg}},\ \underset{B}{\text{Ile-Trp-Phe-Pro-Gly-Arg}},\ \underset{C}{\text{Leu-Lys}},\ \underset{D}{\text{Ser-Glu-Ile}}$$

Cleavage of *Q* with chymotrypsin gave the following peptides:

$$Q \xrightarrow{\text{chymotrypsin}} \begin{array}{l} E\text{: partial sequence Leu-Lys-Gly . . . ; and} \\ F\text{: partial sequence Phe-Pro-Gly-Arg-Ser . . .} \end{array}$$

From these data construct the amino acid sequence of *Q*. Explain why the additional cleavage data from chymotrypsin are necessary to define the sequence. (This problem illustrates the use of *overlapping peptides,* a technique frequently used in the sequencing of large proteins.)

26.8 SYNTHESIS OF PEPTIDES

A. Strategy of Peptide Synthesis

In the previous section we learned how to analyze peptides by carrying out reactions that literally "take peptides apart." In this section, we shall learn how we can approach the problem of constructing peptides synthetically.

Superficially, the strategy of peptide synthesis is simple. Because peptides are linear sequences of amino acid residues, in principle a peptide can be built up from one end to the other by simply forming successive peptide bonds between the amino group of one amino acid and the carboxy group of another.

Let us consider a simple problem in peptide synthesis: the preparation of the dipeptide Ala-Val from the amino acids alanine and valine. How can we bring about the formation of the peptide bond? Since the peptide bond is really just an amide bond, we can imagine allowing the acid chloride of the amino terminal amino acid, alanine, to react with the amino group of valine.

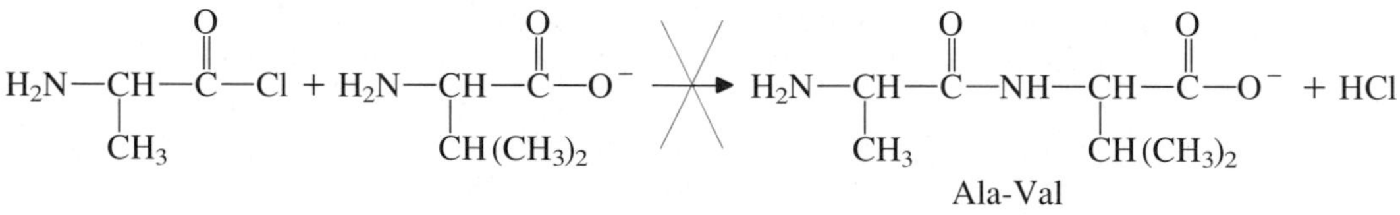
(26.28)

However, a little reflection will show that this cannot work very well. One important reason for the failure of this approach is that the amino group of one Ala-acid chloride can attack the acid chloride group of another to form polyalanine.

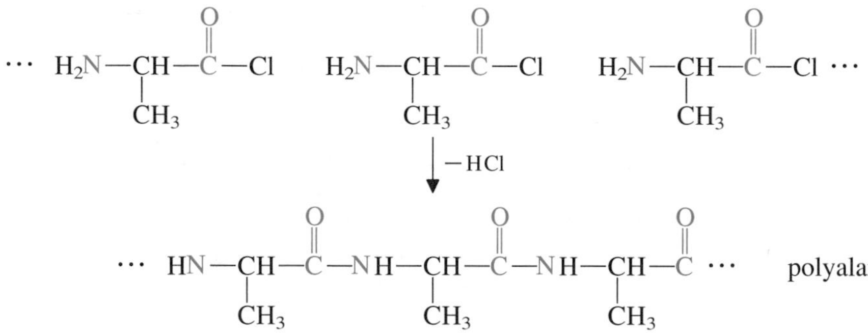
(26.29)

polyalanine

Another undesired reaction is attack of the carboxylate group of valine on the acid chloride of alanine to form an anhydride (Sec. 21.8A).

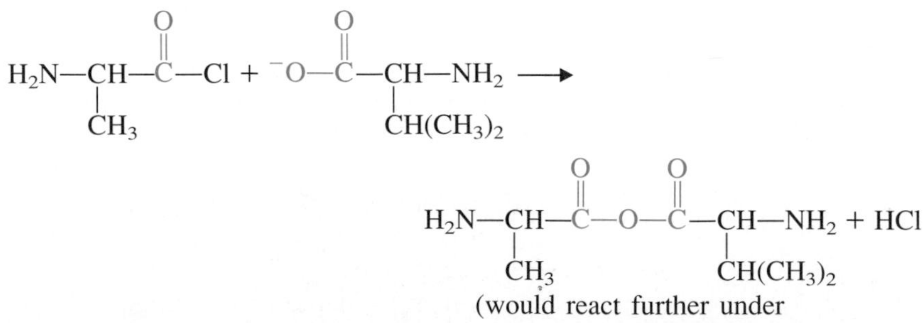
(26.30)

(would react further under reaction conditions)

These considerations show that *we can form the desired peptide bond only if the amino group of Ala and the carboxy group of Val are blocked with protecting groups* so that they cannot take part in undesired reactions. (The philosophy of protecting groups is discussed in Sec. 19.10B.) Once the protecting groups are no longer needed, they are removed. We shall see this strategy in practice in the next section.

B. Solid-Phase Peptide Synthesis

Many procedures can be used for the synthesis of peptides. The most widely used method is an ingenious technique called **solid-phase peptide synthesis.** In this method the carboxy terminal amino acid is covalently anchored to an *insoluble* polymer, and the peptide is "grown," one residue at a time, on this polymer. Solutions containing the appropriate reagents are allowed to contact the polymer with shaking. At the conclusion of each step, the polymer containing the peptide is simply filtered away from the solution, which contains soluble by-products and impurities. The completed peptide is removed from the polymer by a reaction that cleaves its bond to the resin, just as a plant is harvested by cutting it away from the ground. The advantage of this method is the ease with which the peptide is separated from soluble by-products of the reaction. The reactions used in solid-phase peptide synthesis also illustrate some important amino acid and peptide chemistry.

Suppose we wish to prepare the tripeptide Phe-Gly-Ala. The solid-phase peptide synthesis begins with a derivative of the carboxy terminal residue Ala. In this derivative the amino group is protected with a special acyl group, the *tert*-butyloxycarbonyl group, which is abbreviated *t*-Boc, or simply Boc.

Boc group

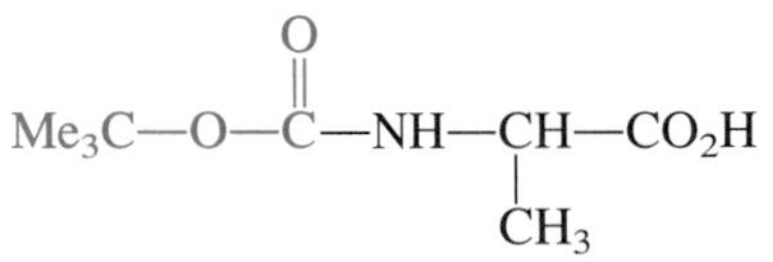

$$\underset{\text{Boc-protected alanine (Boc-Ala)}}{Me_3C-O-\overset{\displaystyle O}{\overset{\|}{C}}-NH-\underset{\underset{CH_3}{|}}{CH}-CO_2H}$$

The Boc group is introduced by allowing alanine to react with the anhydride di-*t*-butyldicarbonate. (The corresponding acid chloride is not stable.)

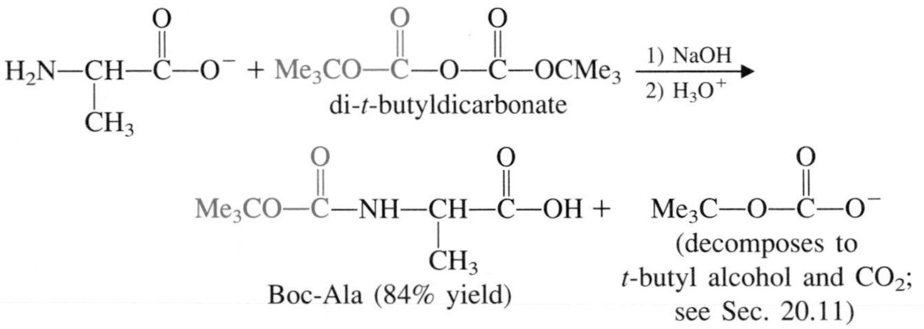

$$
\underset{\underset{CH_3}{|}}{H_2N-CH}-\overset{\overset{\displaystyle O}{\|}}{C}-O^- \;+\; Me_3CO-\overset{\overset{\displaystyle O}{\|}}{C}-O-\overset{\overset{\displaystyle O}{\|}}{C}-OCMe_3 \quad \xrightarrow[\text{2) } H_3O^+]{\text{1) NaOH}}
$$

$$
\underset{\underset{CH_3}{|}}{Me_3CO-\overset{\overset{\displaystyle O}{\|}}{C}-NH-CH}-\overset{\overset{\displaystyle O}{\|}}{C}-OH \;+\; Me_3C-O-\overset{\overset{\displaystyle O}{\|}}{C}-O^- \qquad (26.31)
$$

Boc-Ala (84% yield) (decomposes to *t*-butyl alcohol and CO_2; see Sec. 20.11)

The amino group rather than the carboxylate group reacts with the anhydride because the amino group is the more basic, and therefore more nucleophilic, group.

The Boc-Ala formed in Eq. 26.31 is then anchored to the insoluble solid support using the reactivity of its free carboxylic acid group. The insoluble solid support, or resin, is supplied as a chloromethyl derivative of polystyrene called a *Merrifield resin*:

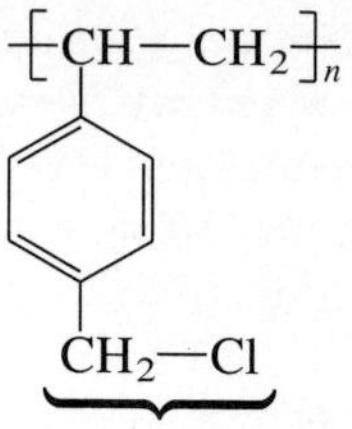

reactive in nucleophilic displacements

abbreviated

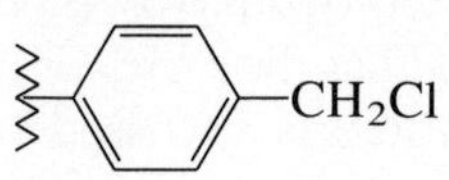

chloromethylated polystyrene (Merrifield resin)

This is, in effect, an "insoluble benzyl chloride," and has the enhanced reactivity generally associated with benzylic halides (Sec. 17.4). An S_N2 reaction between the cesium salt of Boc-Ala and the resin results in the formation of an ester linkage to the resin. (See Sec. 20.8B for related chemistry.)

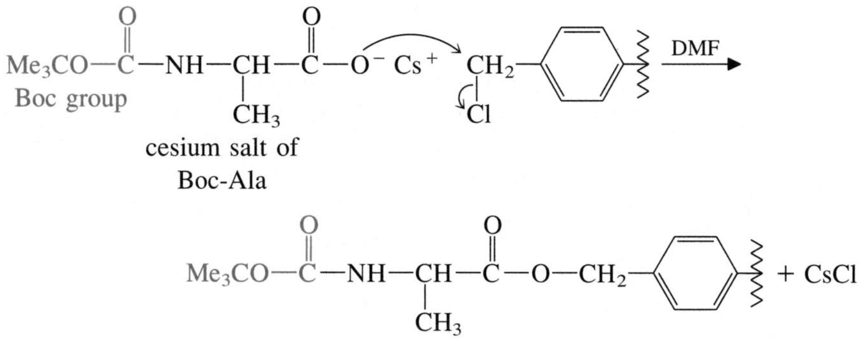

(26.32)

Boc-Ala linked to resin

The Boc-protecting group is necessary in this reaction because if it were not present, the amino group would compete with the carboxylate group as a nucleophile for the benzylic halide group on the resin.

Once the Boc-amino acid is anchored to the resin, the Boc-protecting group is removed. Notice that the Boc-protected derivative is not only an amide, but also a *t*-butyl ester. We learned in Sec. 21.7A that *t*-butyl esters are rapidly cleaved with anhydrous acid by a special carbocation mechanism, and so it is with the Boc group:

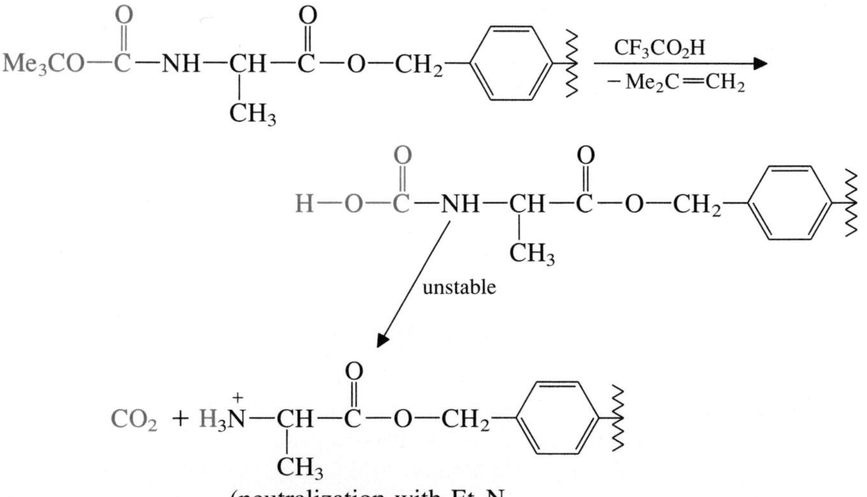

(26.33)

(neutralization with Et_3N gives free amino group)

There are two important points to notice about this reaction. First, the carboxylic acid group exposed in this reaction (color) is of the carbamic acid type, and carbamic acids decarboxylate under acidic conditions (Eq. 20.48). This decarboxylation exposes the free amino group. Second, since the ester linkage holding the Ala residue to the resin is *not* a *t*-butyl ester, it is *not* cleaved under the anhydrous acidic conditions. (Water would have to be present for acid-catalyzed ester hydrolysis to occur.)

This deprotection step, after neutralization of the ammonium salt, exposes the free amino group of the resin-bound amino acid, which is used as a nucleophile in the next reaction.

We are now ready to couple a second residue to the resin-bound alanine. Coupling of Boc-glycine to the free amino group of the resin-bound Ala is effected by the reagent *N,N'*-dicyclohexylcarbodiimide, or DCC. The result of this reaction is that the resin-bound Ala is acylated by the Boc-Gly.

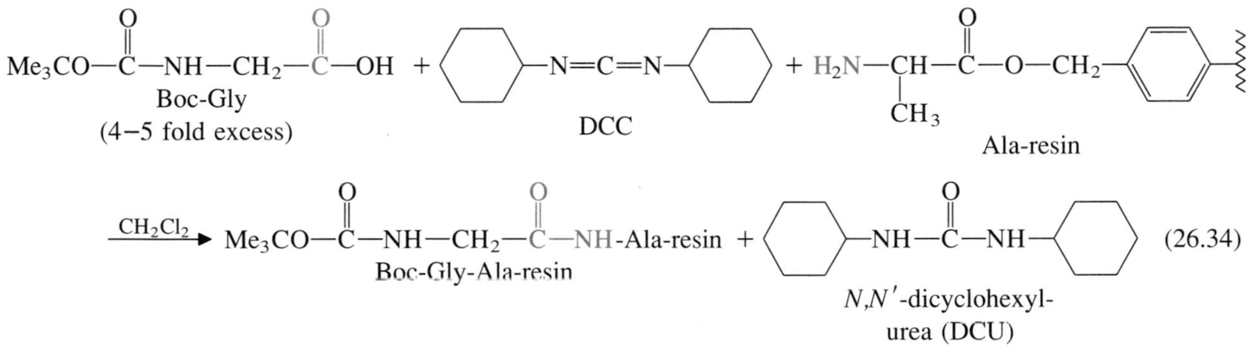

(26.34)

What is the role of DCC in this reaction? Markovnikov addition of the amino acid carboxylic acid group to a double bond of DCC gives a derivative called an *O*-acylisourea.

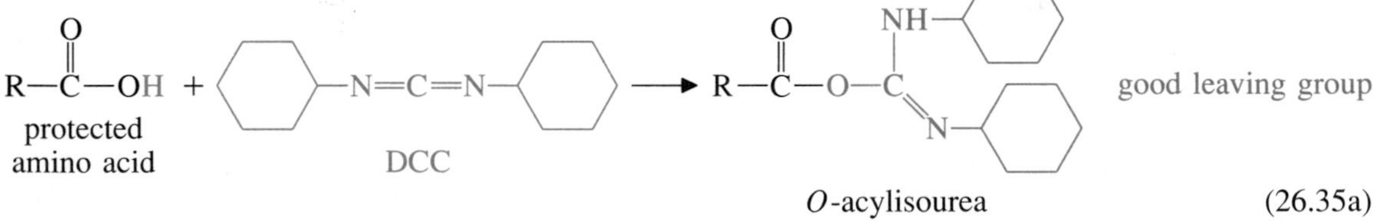

(26.35a)

This derivative behaves somewhat like an anhydride, and is an excellent acylating agent. It reacts in either of two ways. First, it can react directly with the amino group of the resin-bound amino acid to form the peptide bond. This reaction is an example of ester aminolysis (Sec. 21.8C).

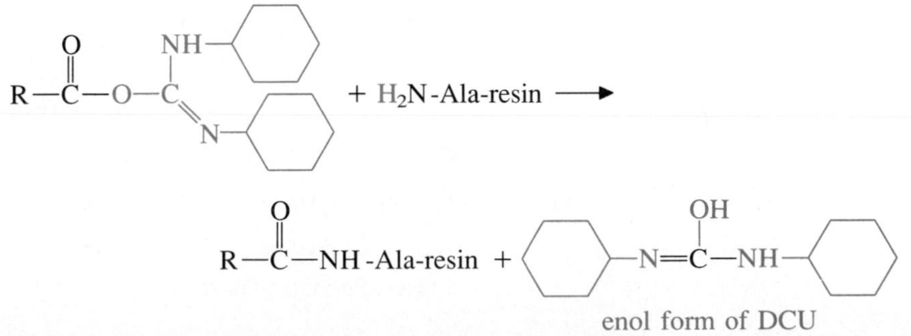

(26.35b)

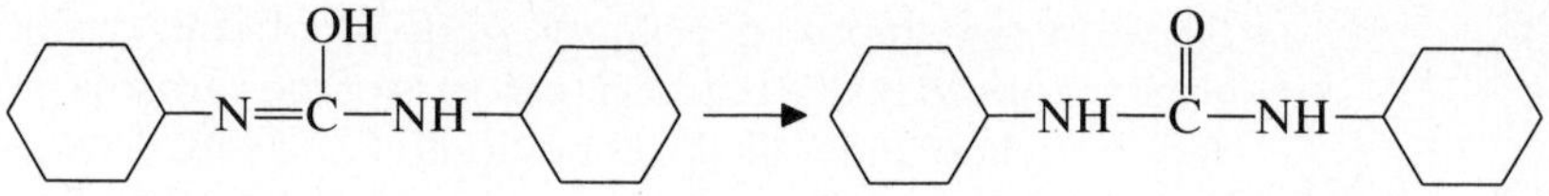

DCU

(26.35c)

The by-product of this reaction is the enol form of DCU, which is spontaneously transformed into DCU itself. (You should write the mechanistic details of each of these reactions using the arrow formalism.)

The second way in which the *O*-acylisourea can react is with the carboxy group of another equivalent of the protected amino acid, giving an anhydride.

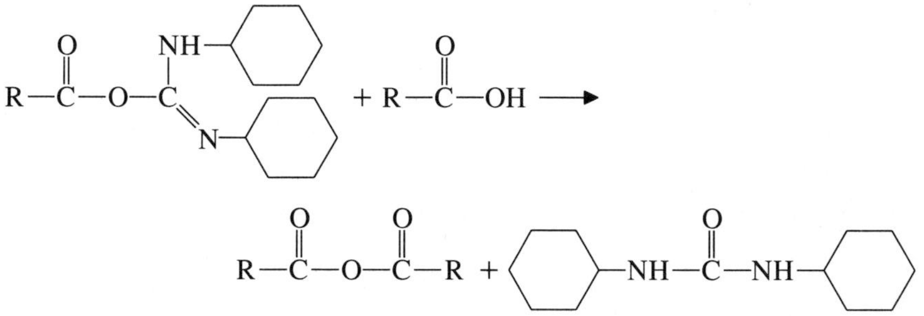

(26.36a)

This anhydride can also acylate the amino group of the resin-bound amino acid to complete the formation of the peptide bond.

$$\text{R}-\overset{\text{O}}{\overset{\|}{\text{C}}}-\text{O}-\overset{\text{O}}{\overset{\|}{\text{C}}}-\text{R} + \text{H}_2\text{N-Ala-resin} \longrightarrow \text{R}-\overset{\text{O}}{\overset{\|}{\text{C}}}-\text{NH}-\text{Ala-resin} + \text{R}-\overset{\text{O}}{\overset{\|}{\text{C}}}-\text{OH} \quad (26.36\text{b})$$

Both of these acylation reactions occur in practice, depending on the conditions actually used in the solid-phase peptide synthesis. In either case, the desired peptide bond is formed. The by-product DCU is simply washed away from the peptide because it is not attached covalently to the resin.

Completion of the peptide synthesis requires deprotection of the resin-bound dipeptide in the usual way and a final coupling step with Boc-Phe and DCC:

$$\text{Me}_3\text{CO}-\overset{\text{O}}{\overset{\|}{\text{C}}}-\text{NH-Gly-Ala-resin} \xrightarrow[-\text{Me}_3\text{COH},\ -\text{CO}_2]{\text{CF}_3\text{CO}_2\text{H}} \xrightarrow{\text{Et}_3\text{N}} \text{H}_2\text{N-Gly-Ala-resin} \xrightarrow[\underset{\text{(Boc-Phe)}}{\text{Me}_3\text{CO}-\underset{\text{O}}{\underset{\|}{\text{C}}}-\text{NH}-\underset{\text{CH}_2\text{Ph}}{\underset{|}{\text{CH}}}-\text{CO}_2\text{H}}]{\text{DCC}}$$

$$\underset{\text{Boc-Phe-Gly-Ala-resin}}{\text{Me}_3\text{CO}-\overset{\text{O}}{\overset{\|}{\text{C}}}-\text{NH}-\underset{\text{CH}_2\text{Ph}}{\underset{|}{\text{CH}}}-\overset{\text{O}}{\overset{\|}{\text{C}}}-\text{NH-Gly-Ala-resin}} + \underset{\text{(in solution)}}{\text{DCU}} \quad (26.37)$$

All the peptide bonds in the desired tripeptide are now assembled, but the completed tripeptide must be removed from the resin. The ester linkage that connects the peptide to the resin, like most esters, is more easily cleaved than the peptide (amide) bonds (Sec. 21.7F), and is typically broken by liquid HF or by a mixture of CF_3SO_3H (trifluoromethanesulfonic acid) and trifluoroacetic acid. (These acidic reagents also remove the Boc group from the product peptide.)

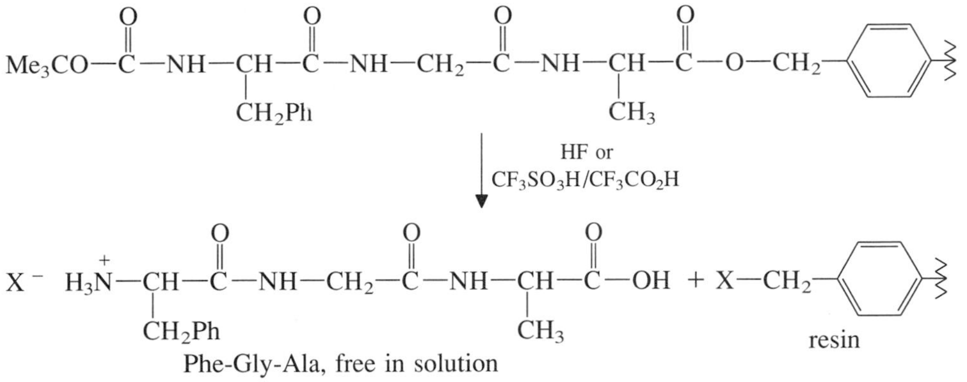

(26.38)

In this equation, the X— group that ends up on the resin is either trifluoroacetate or fluoride, depending on the conditions used for cleavage.

The same reagents used in solid-phase peptide synthesis can also be used in solution, but removal of the DCU from the product peptide is sometimes difficult. The advantage of the solid-phase method, then, is the ease with which dissolved impurities and by-products are removed from the resin-bound peptide by simple filtration.

Despite its advantages, solid-phase peptide synthesis also has some problems. Suppose, for example, that a coupling reaction is incomplete, or that other side reactions take place to give impurities that remain covalently bound to the resin. These are then carried along to the end of the synthesis, when these are also removed from the resin and must be separated (in some cases tediously) from the desired peptide product. (This situation is something like what might occur if a flight attendant on a flight from San Francisco to New York discovers over Denver a passenger without a ticket; the offending party cannot be removed until the end of the line.) In order to avoid impurities, then, each step in the solid-phase synthesis must occur with virtually 100% yield. Remarkably, this ideal is approached closely in practice (Problem 20).

Solid-phase peptide synthesis was devised by R. Bruce Merrifield (1921–   ), a Rockefeller University chemist, and first reported in the early 1960s. A particularly impressive achievement of the method was the synthesis of an active enzyme by Merrifield's research group in 1969 using a homemade machine in which the various steps of the method were preprogrammed. (Modern instruments for automated solid-phase peptide synthesis are now commercially available.) The enzyme that was synthesized, ribonuclease, contains 124 amino acid residues; the synthesis required 369 separate reactions and 11,931 individual operations, yet was carried out in 17% overall yield. For his invention and development of the solid-phase method, Merrifield was honored with the 1984 Nobel Prize in Chemistry.

Problems

20 Calculate the average yield of each of the 369 steps in the synthesis of ribonuclease by the solid-phase method (discussed above), assuming a 17% overall yield.

21 (a) An aspiring peptide chemist, Mo Bonds, has decided to attempt the synthesis of the peptide Gly-Lys-Ala using the solid-phase method. To the Ala-resin he couples the following derivative of lysine:

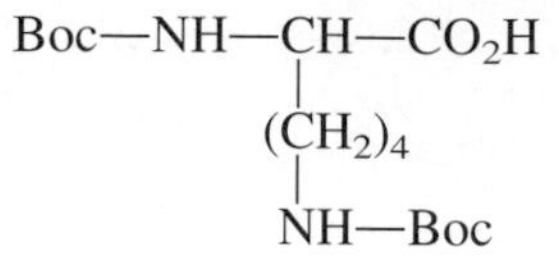

$$Boc-NH-\underset{\underset{\underset{NH-Boc}{|}}{\underset{(CH_2)_4}{|}}{CH}-CO_2H \qquad \alpha,\epsilon\text{-diBoc-lysine}$$

Why are *two* Boc groups necessary for the protection of lysine?

(b) After the coupling, he deprotects his resin-bound peptide with anhydrous CF_3CO_2H, and then completes the synthesis in the usual way by coupling Boc-Gly, deprotecting the peptide, and removing it from the resin. He is shocked to find a mixture of several peptide products. Two of them give the amino acid analysis (Ala,Gly,Lys), and one gives the amino acid analysis (Ala,Gly$_2$,Lys). Suggest a structure for each product and explain what happened.

(c) On the suggestion of a colleague, Uri Thane, Mo Bonds repeats the chemistry in (b) using the lysine derivative below, and obtains a good yield of the desired tripeptide. Explain.

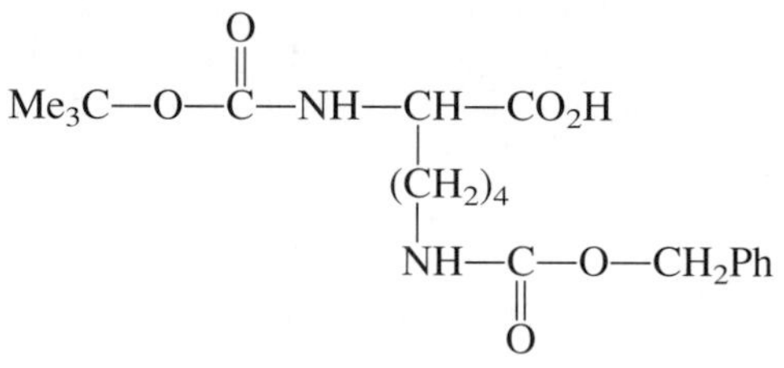

Problem 21 shows that certain amino acid side chains can also react under the conditions of peptide synthesis. Special protecting groups must be introduced on these side chains. These protecting groups must survive the entire synthesis, including the removal of the amino-protecting group at each stage, yet themselves be removable at the end of the synthesis. The design of protecting groups that can meet these exacting requirements is an important aspect of peptide synthesis that is under active investigation.

26.9 STRUCTURES OF PEPTIDES AND PROTEINS

A. Primary Structure

The structures of molecules as large as peptides and proteins are complex and can be discussed at different levels. The simplest description of a peptide or protein structure is its covalent structure, or **primary structure.** The most important aspect of any

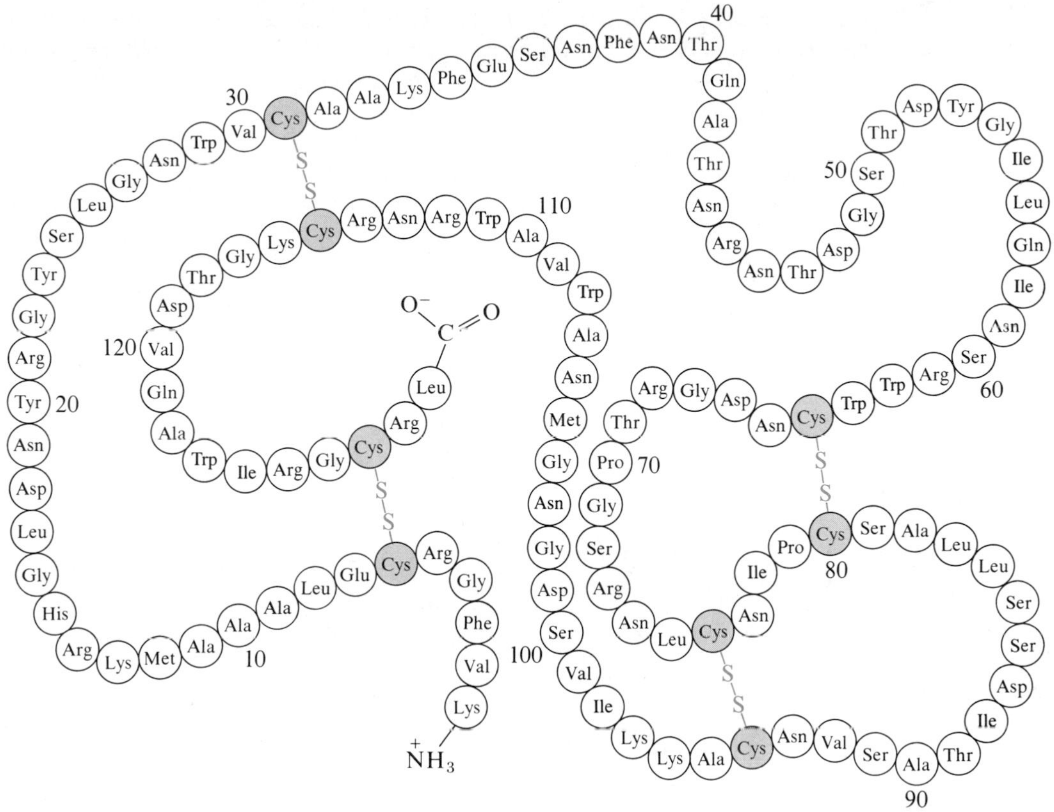

Figure 26.4 Primary structure of the enzyme lysozyme from hen egg white. The disulfide bonds are shown in color.

primary structure is the amino acid sequence (Sec. 26.7). However, peptide bonds are not the only covalent bonds that connect amino acid residues. In addition, **disulfide bonds** (Sec. 10.9) link the cysteine residues in different parts of a sequence.

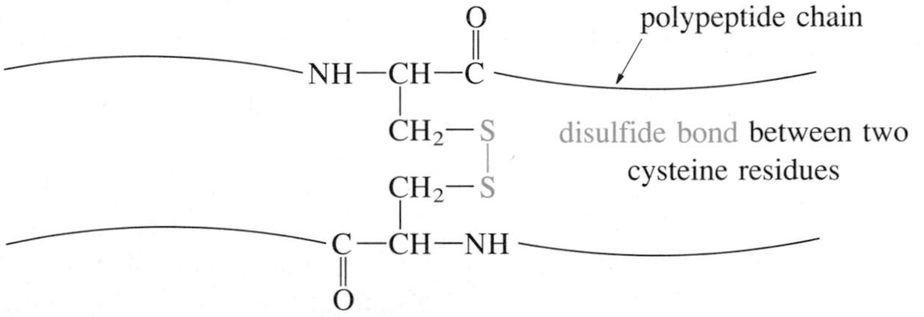

Disulfide bonds thus serve as crosslinks between different parts of a peptide chain. Many proteins contain several peptide chains; disulfide bonds hold these chains together. The primary structure of a peptide or protein, then, includes its amino acid sequence and its disulfide bonds. The primary structure of *lysozyme,* a small enzyme that is abundant in hen egg white, is shown in Fig. 26.4. Lysozyme is a single polypeptide chain of 129 amino acids, including eight cysteine residues linked together into four disulfide bonds. (Physiologically, lysozyme catalyzes the hydrolysis of bacterial cell walls. Different variants of this enzyme are present in tears, nasal mucus, and even viruses—anywhere antibacterial action is important. Lysozyme is one of the smallest known enzymes!)

The disulfide bonds of a protein are readily reduced to free cysteine thiols by other thiols. Two commonly used thiol reagents are 2-mercapto-1-ethanol ($HSCH_2CH_2OH$), and dithiothreitol (known to biochemists as DTT, or Cleland's reagent).

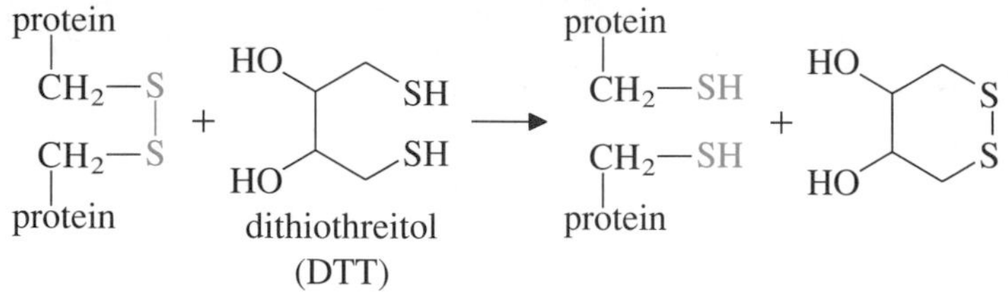

(26.39)

This reaction is simply a biological example of the thiol-disulfide equilibrium shown in Eq. 10.67. Typically, when the extraneous thiols are removed, the thiols of the protein spontaneously reoxidize in air back to disulfides.

An interesting example of the biological effects of disulfide-bond reduction occurs in the ordinary hair permanent. Hair (protein) is treated with a thiol solution; this solution is responsible for the unpleasant smell of permanents. The thiol solution reduces the disulfide bonds in the hair. With the hair in curlers, the disulfides are allowed to reoxidize. The hair is thus set by disulfide bond reformation into the conformation dictated by the curlers. Only after a long period of time do the disulfide bonds rescramble to their normal configuration, when another permanent becomes necessary.

An industrial example of the use of disulfide bonds is the process of *vulcanization* (Sec. 15.5), which introduces disulfide bonds into synthetic polymers.

B. Secondary Structure

The description of the primary structure of a protein gives no indication of how the molecule might actually appear in three dimensions. The structural characteristics of a typical peptide bond are shown in Fig. 26.5. With some exceptions, the amide units in most peptides are planar. In Sec. 21.2 we learned that slow rotation occurs about the carbonyl-nitrogen bond of most amides, and that the preferred conformation about this bond is *Z;* the same is true for the amide bonds in a peptide. There are two other single bonds in a typical peptide residue: the two bonds to the α-carbon. In principle, rotation can occur about these bonds as well, allowing the peptide to adopt a variety of conformations. Because of these rotations, it might seem that a very large number of conformations could occur in a protein or large peptide. However, it turns out that only three conformations occur commonly.

In one type of peptide conformation, the peptide backbone adopts the conformation of a **right-handed alpha (α)-helix,** shown in Fig. 26.6. "Right-handed" means that the helix turns in a clockwise fashion along the helical axis. In this conformation,

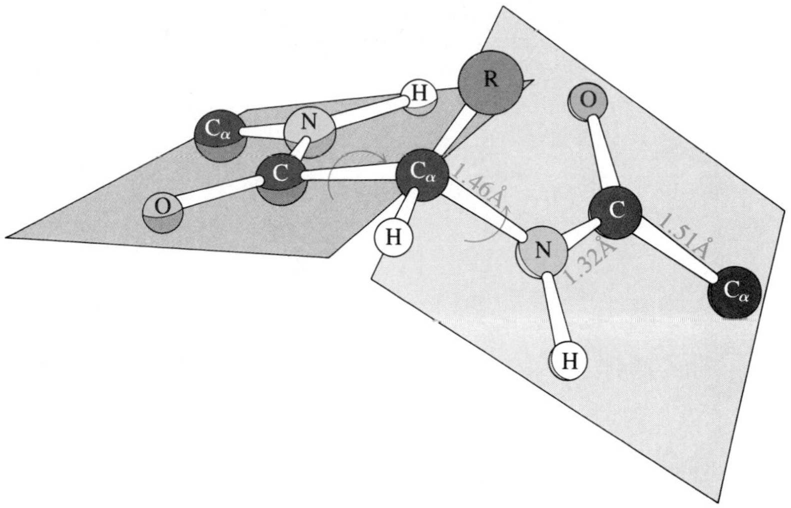

Figure 26.5 Typical dimensions of a peptide bond. The plane is that of an amide group, and amino acid side chains are represented by R. In principle, rotation about the bonds to the α-carbons (marked with arrows) is possible.

Figure 26.6 The protein α-helix. (a) All atoms are shown, with side chains represented by R. Note that the side chains extend away from the helix on the outside. (b) Backbone atoms only are shown.

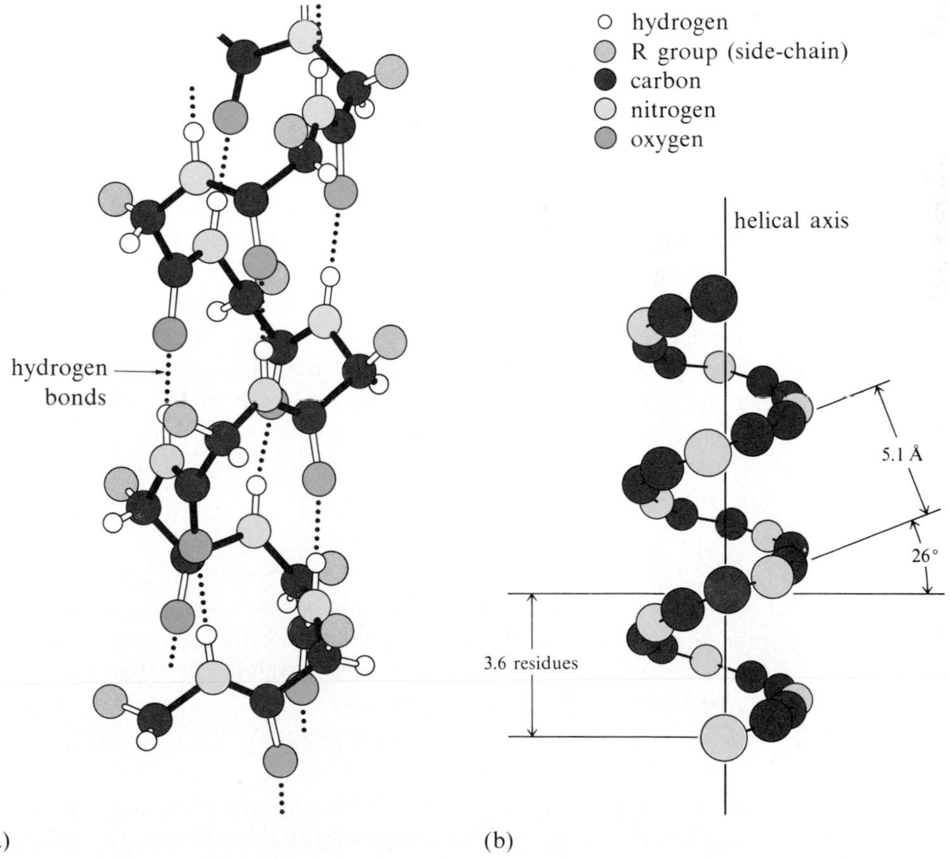

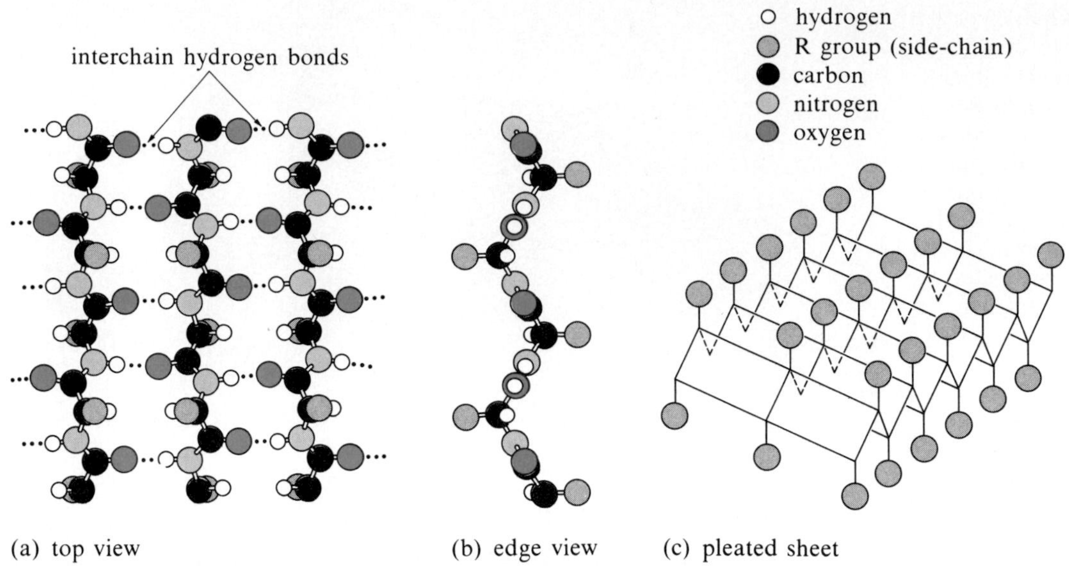

Figure 26.7 The β-antiparallel pleated-sheet structure of proteins. In (a) and (b) are shown top and edge views, respectively, with the nitrogen atoms in color. Part (c) shows the pleated-sheet surface described by the backbone atoms.

the side-chain groups are positioned on the outside of the helix, and the helix is stabilized by *hydrogen bonds* between the amide N—H of one residue and the carbonyl oxygen four residues away.

Another type of peptide conformation is called **beta (β)-structure** or **pleated sheet.** In this type of structure, a peptide chain adopts an open, zigzag conformation, and is engaged in hydrogen bonding with another peptide chain (or a different part of the same chain) in a similar conformation. The successive hydrogen-bonded chains can run (in the amino terminal to carboxy terminal sense) in the same direction (**parallel pleated sheet**) or in opposite directions (**antiparallel pleated sheet**). The antiparallel pleated sheet structure is shown in Fig. 26.7. The name "pleated sheet" is derived from the fanlike surface described by the aggregate of several hydrogen-bonded chains (Fig. 26.7c). Notice that the side-chain R-groups alternate between positions above and below the sheet.

The terms α and β refer to two characteristic X-ray diffraction patterns observed for certain proteins before chemists fully understood their structures. The α type of pattern was eventually shown to be associated with the right-handed helix, and the β pattern with the pleated sheet.

A third peptide conformation is actually a nonconformation: the **random coil.** As the name implies, peptides that adopt a random coil show no discernible pattern in

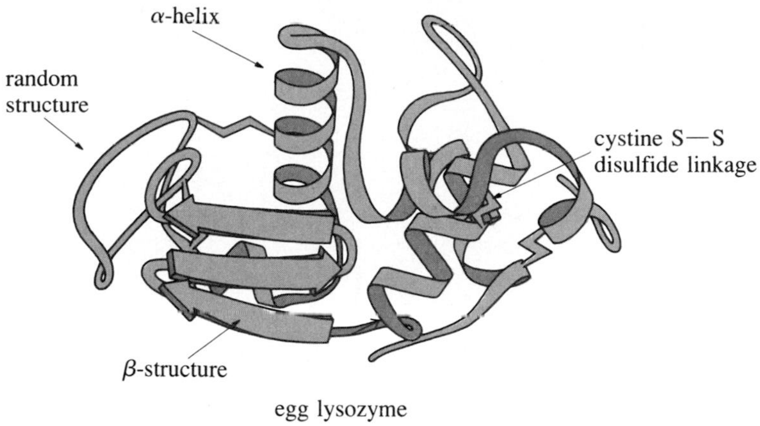

Figure 26.8 Conformation of lysozyme. In this figure the peptide backbone is traced as a ribbon.

their conformation. The best analogy for the random coil "conformation" of peptides is perhaps a large plate of cooked spaghetti.

There are other conformations known in peptides, but the α-helix, β-structure, and random coil are the major ones. Some peptides and proteins exist entirely in one of these three conformations. For example, the α-keratins, major proteins of hair and wool, exist in the α-helical conformation. In these proteins, several α-helices are coiled about each other to form "molecular ropes." These structures have considerable physical strength. In contrast, silk fibroin, the fiber secreted by the silkworm, adopts the β-antiparallel pleated sheet conformation. The process of steaming wool breaks the hydrogen bonds of the α-helix and allows the keratin molecules to adopt temporarily a β-structure much like that in silk. It is the disulfide bonds between chains in α-keratin that help maintain the curl of hair (Sec. 26.9A).

The description of peptide or protein structure in terms of α-helix, β-structure, and random-coil conformations is called **secondary structure.** Despite the examples above, proteins that contain a single type of secondary structure are relatively rare. Rather, *most proteins contain different types of secondary structure* in different parts of their peptide chains. This point is illustrated for the enzyme lysozyme in Fig. 26.8. In this illustration, the backbone of the peptide chain is traced as a ribbon. The regions of α-helix, β-structure, and random coil are clearly evident. (Lysozyme is the protein whose primary sequence is given in Fig. 26.4.)

C. Tertiary and Quaternary Structures

The complete three-dimensional description of protein structure at the atomic level is called **tertiary structure.** The tertiary structures of proteins are determined by X-ray crystallography; each crystallographic structure analysis represents many person-years of labor. Nevertheless, since the first protein crystallographic structure was determined in 1960, more than a hundred protein structures have been elucidated, and more determinations are continually appearing.

The tertiary structure of any given protein is an aggregate of α-helix, β-sheet, random-coil, and other structural elements whose precise arrangements are governed by noncovalent interactions of groups within the protein molecules, and by noncovalent interactions of protein groups with surrounding solvent. What are these noncovalent forces? There are three general types (Fig. 26.9):

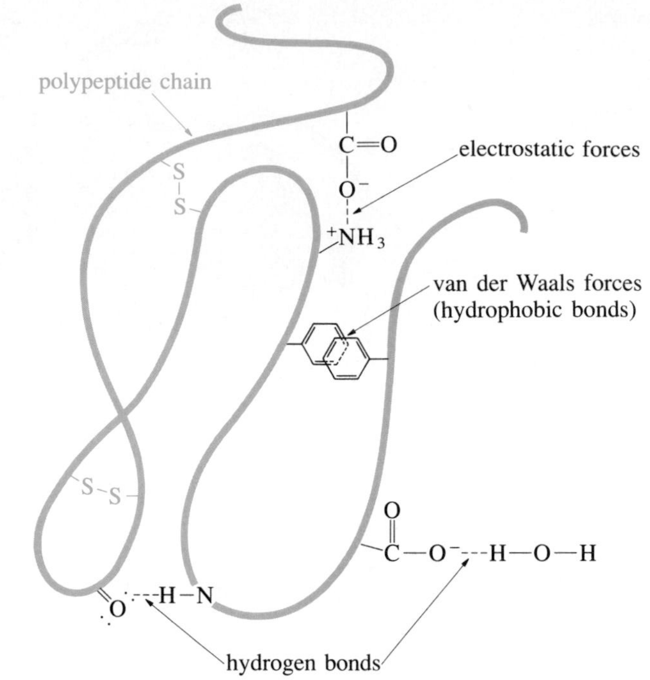

Figure 26.9 Protein conformation is maintained by disulfide bonds and by noncovalent forces: van der Waals interactions (hydrophobic bonds), hydrogen bonds, and electrostatic attractions and repulsions.

1. Hydrogen bonds
2. Van der Waals interactions
3. Electrostatic interactions

We have already seen how *hydrogen bonds* stabilize both the α-helix and the β-structure. All protein structures appear to contain many stable hydrogen bonds, not only within regions of helix or β-structure, but also in other regions. Protein conformations are also stabilized in part by hydrogen bonding of certain groups to solvent water.

Van der Waals interactions, or dispersion forces, are the same interactions that provide the cohesive force in a liquid hydrocarbon (see Fig. 3.3), and can be regarded as examples of the "like-dissolves-like" phenomenon. For example, we know that benzene does not dissolve in water, but dissolves readily in hexane. In the same sense, the benzene ring of a phenylalanine side chain would prefer to be near other aromatic or hydrocarbon side chains rather than the "waterlike" side chain of a serine, the charged, polar side chain of an aspartic acid, or solvent water on the outside of a protein. (This does *not* mean that *all* phenylalanine residues are buried next to other hydrocarbonlike residues; it does mean, however, that phenylalanine residues, more often than not, will be found in hydrocarbonlike environments.) The van der Waals interactions of this type between hydrocarbonlike residues are sometimes called *hydrophobic bonds,* because the hydrocarbon groups would rather be near each other than water. Residues such as the side chain of phenylalanine or isoleucine are sometimes called *hydrophobic residues.* In contrast, polar residues, such as aspartic acid, glutamic acid, or lysine, are often found to interact with other polar residues or with the aqueous solvent. Such residues are sometimes termed *hydrophilic residues.*

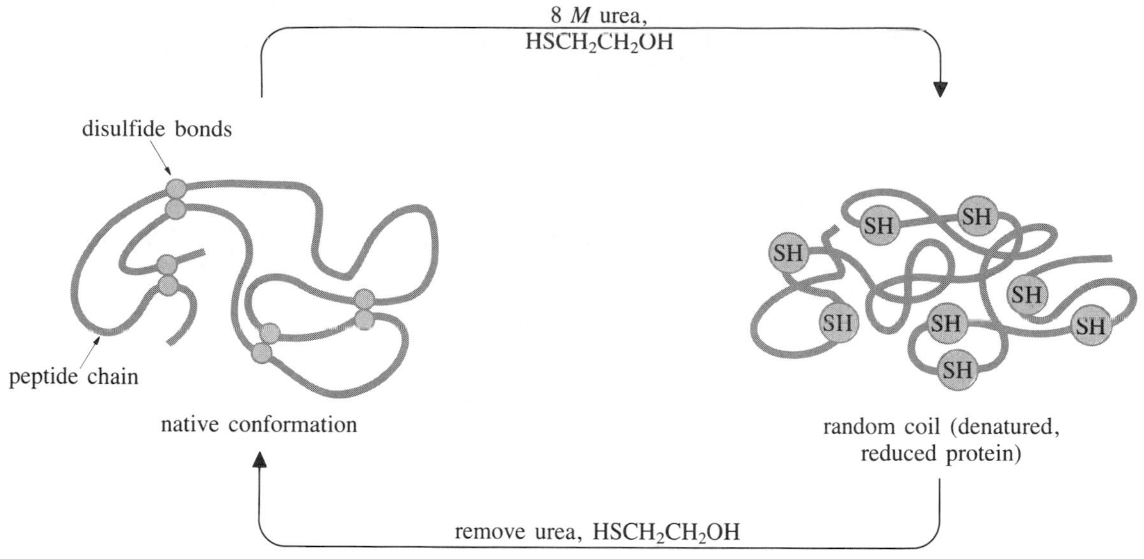

Figure 26.10 Protein denaturation.

Electrostatic interactions are noncovalent interactions between charged groups governed by the electrostatic law (Eq. 1.2). A typical stabilizing electrostatic interaction is the attraction of a protonated, positively charged amino group and a nearby ionized, negatively charged carboxylate ion.

A protein adopts a tertiary structure in which favorable interactions are maximized and unfavorable interactions are minimized. Although there are exceptions, most soluble proteins are globular and compact rather than extended. For example, lysozyme (Fig. 26.8) is a globular protein. Globular, nearly spherical, shapes are a consequence of the fact that proteins expose to aqueous solvent as small a surface as possible. (A sphere is the geometrical object with the minimum surface-to-volume ratio.) The reason for minimizing the exposed surface is that the majority of the residues in most proteins are hydrophobic, and the interaction of hydrophobic side chains with solvent water is unfavorable. The conformations of proteins are probably as much a consequence of these *unfavorable* interactions with water as the *favorable* interactions of the amino acid residues with each other. Indeed, to a first approximation, proteins are large "grease balls" (with a few charged or polar groups on their surfaces) floating about in aqueous solution. In this respect, proteins somewhat resemble micelles (Sec. 20.5).

Suppose we were to synthesize a protein. Would the finished protein automatically "know" what conformation to assume, or is some external agent required to direct the protein into its naturally occurring conformation? This question has been answered in two elegant experiments with the enzyme ribonuclease. First of all, synthetic ribonuclease was prepared by the solid-phase method and found to be an active enzyme. (See the small print in Sec. 26.8B.) Since the enzyme must have its natural, or native, conformation in order to be active, it follows that ribonuclease, once synthesized, spontaneously folds into this conformation.

The second type of experiment involved **denaturation** of ribonuclease. Denaturation is illustrated schematically in Fig. 26.10. When a protein is denatured, it is converted entirely into a random-coil structure. Denaturation of a protein is brought about typically by breaking its disulfide bonds with thiols, such as DTT or 2-mercapto-ethanol (Eq. 26.39), and treating it with 8*M* urea, detergents, or heat. Ribonuclease was denatured by treatment with 2-mercaptoethanol and 8*M* urea. After the urea was removed, and the cysteine —SH groups allowed to reoxidize back to disulfides, the protein spontaneously reassumed its original, native conformation. This experiment shows that *the amino acid sequence of ribonuclease somehow specifies its conformation; that is, the native structure is the most stable structure.* If this were not so, another, more stable structure would have formed when the protein was allowed to refold after the urea was removed.

There is a great deal of evidence that the ribonuclease result is general: proteins spontaneously assume their native conformations at the time of their biosynthesis. That is, *primary structure dictates tertiary structure.*

This result is a significant one, for it means that chemists and physicists can in principle predict protein tertiary structure starting from only the amino acid sequence and the laws governing the appropriate noncovalent forces. Because of the size of proteins and the large number of interactions involved, such predictions require extensive "number-crunching" that can only be carried out with large computers. That these efforts have met with, at best, only modest success is probably attributable more to the complexity of the problem than to the fundamental soundness of the underlying physical principles.

To summarize:

1. Proteins appear to fold spontaneously into their native conformations.
2. The noncovalent forces that determine conformation are: (a) hydrogen bonds; (b) van der Waals forces (hydrophobic bonds); and (c) electrostatic forces.
3. Most soluble proteins appear to be compact, globular structures. Although there are exceptions, hydrocarbonlike amino acid residues tend to be found on the interior of a protein, away from solvent water, and the polar residues tend to be on the exterior of a protein.

Some proteins are aggregates of other individual proteins. The best-known example of such proteins is *hemoglobin,* which transports oxygen in the bloodstream. Hemoglobin (Fig. 26.11) contains two different types of polypeptide chains that differ somewhat in sequence, but are fundamentally similar to each other. One type of chain is called the alpha-chain, and the other is called the beta-chain (not to be confused with α- and β-secondary structure). Hemoglobin is an aggregate of four polypeptide chains, or *subunits,* two alpha and two beta. These subunits are held together solely by noncovalent forces—hydrogen bonds, electrostatic interactions, and van der Waals interactions. Notice in Fig. 26.11 that the individual subunits lie more or less at the vertices of a regular tetrahedron. This is no accident; this shape is the most compact arrangement that can be assumed by four objects. Many important proteins are aggregates of individual polypeptide subunits. In some proteins the subunits are identical; in other cases they are different. The description of subunit arrangement in a protein is called **quaternary structure.**

Problem

22 What would you expect to happen when hemoglobin is treated with a denaturant such as 8*M* urea? Explain.

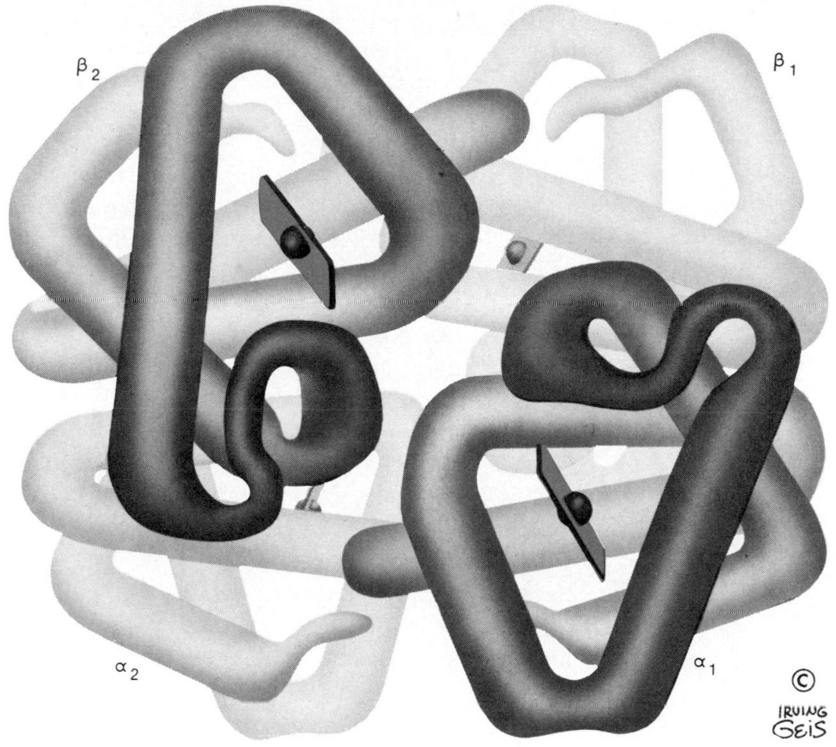

Figure 26.11 Quaternary structure of hemoglobin showing the general outline of each polypeptide chain. The rectangles represent the heme groups (Sec. 24.6) and the spheres represent the iron at which oxygen is bound to hemoglobin. Notice the tetrahedral orientation of the subunits.

26.10 ENZYMES: BIOLOGICAL CATALYSTS

Many of the proteins that occur naturally are **enzymes.** Enzymes are the catalysts for biological reactions. Through detailed studies of certain enzymes, chemists have come to understand some of the reasons why these proteins are efficient catalysts. To illustrate enzyme catalysis, let us consider the mechanism by which the enzyme *trypsin* catalyzes the hydrolysis of peptide bonds. Trypsin, it will be recalled, is the enzyme used in the sequencing of proteins (Sec. 26.7C). With a molecular weight of about 24,000, trypsin is an enzyme of modest size. It is a globular protein containing three polypeptide chains held together by disulfide bonds. The following comparison gives us some idea of the catalytic effectiveness of trypsin. Peptides in the presence of trypsin are rapidly hydrolyzed at 37° and pH = 8. In the absence of trypsin, the same peptides are indefinitely stable under the same conditions, and hydrolysis requires boiling them in 6*N* HCl for several hours. However trypsin, in contrast to hot HCl solution, does not catalyze hydrolysis at just *any* peptide bonds. It is specific for hydrolysis of the peptide bonds at lysine and arginine residues (Eq. 26.27). These two aspects of trypsin catalysis—**catalytic efficiency** and **specificity**—are characteristic of catalysis by all enzymes. These are the phenomena that must be understood if we are to understand enzyme catalysis in general.

If an enzyme catalyzes a reaction of a certain compound, the compound is said to be a **substrate** for the enzyme. Enzymes act on their substrates in at least three stages, shown schematically in the following equation.

$$\underset{\text{enzyme}}{\mathrm{E}} + \underset{\text{substrate}}{\mathrm{S}} \rightleftarrows \underset{\substack{\text{noncovalent}\\ \text{enzyme–substrate}\\ \text{complex}}}{\mathrm{E\cdot S}} \rightleftarrows \underset{\substack{\text{noncovalent}\\ \text{enzyme–product}\\ \text{complex}}}{\mathrm{E\cdot P}} \rightleftarrows \mathrm{E} + \underset{\text{product}}{\mathrm{P}} \qquad (26.40)$$

First, the substrate binds to the enzyme in a noncovalent **enzyme–substrate complex.** The binding occurs at a part of the enzyme called the **active site.** Within the active site are groups that attract the substrate by interacting favorably with it. The noncovalent interactions that cause a substrate to bind to an enzyme are typically the same ones that stabilize protein conformations: electrostatic interactions, hydrogen bonding, and van der Waals interactions, or hydrophobic bonds.

In the second stage of enzyme catalysis, the enzyme promotes the appropriate chemical reaction(s) on the bound substrate to give an enzyme-product complex. The necessary chemical transformations are brought about by groups in the active site of the enzyme. In most enzymes, these groups are certain amino acid side chains of the enzyme itself. However, in some cases, other molecules, called **coenzymes,** are also required. (An example of a coenzyme is NAD^+; Sec. 10.7. Most vitamins are coenzymes.)

In the last step of enzyme catalysis, the product(s) depart from the active site, leaving the enzyme ready to repeat the process on a new substrate molecule.

Enzymes are true catalysts; their concentrations are typically much lower than the concentrations of their substrates. They do not affect the equilibrium constants of the reactions they catalyze. Therefore, by the principle of microscopic reversibility (Sec. 10.1A), they catalyze both the forward and reverse reactions of an equilibrium.

Let us see how the trypsin-catalyzed hydrolysis of peptide bonds fits this general picture of enzyme catalysis. The active site of trypsin containing a bound peptide substrate is shown in Fig. 26.12. The active site of trypsin consists of a cavity, or "pocket," that just accommodates the amino acid side chain of a lysine or arginine residue from the substrate; an arginine is shown in Fig. 26.12. Several hydrophobic residues line this cavity. At the bottom of the cavity is the side-chain carboxylic acid group of an aspartic acid residue (Asp-189 in the trypsin sequence). This group is ionized, and therefore *negatively charged,* at neutral pH. The amino group of a lysine side chain and the guanidino group of an arginine side chain are both protonated, and therefore *positively charged,* at neutral pH. The favorable electrostatic attraction between the ionized Asp-189 side chain of the enzyme and the positively charged side chain of the substrate helps stabilize the enzyme-substrate complex. This complex is also stabilized by the van der Waals interactions between the —CH_2— groups of the substrate side chain and the hydrophobic residues that line the cavity. Here, then, we can see some of the reasons for the *specificity* of trypsin. The active site just "fits" the substrate (or vice versa), and contains groups that are noncovalently attracted to groups on the substrate.

Figure 26.12 The enzyme trypsin. Only the α-carbons are shown. Disulfide bridges are shown in outline, and a portion of the polypeptide chain is shown in color. A substrate peptide containing an arginine residue is shown in the active site; the positively charged side chain lies in the "specificity pocket," sketched in shading. The negatively charged side-chain carboxylate group of Asp-189 lies at the bottom of this pocket. The catalytically important His-57 and Ser-195 residues are prepared for cleavage of the peptide bond marked with an arrow.

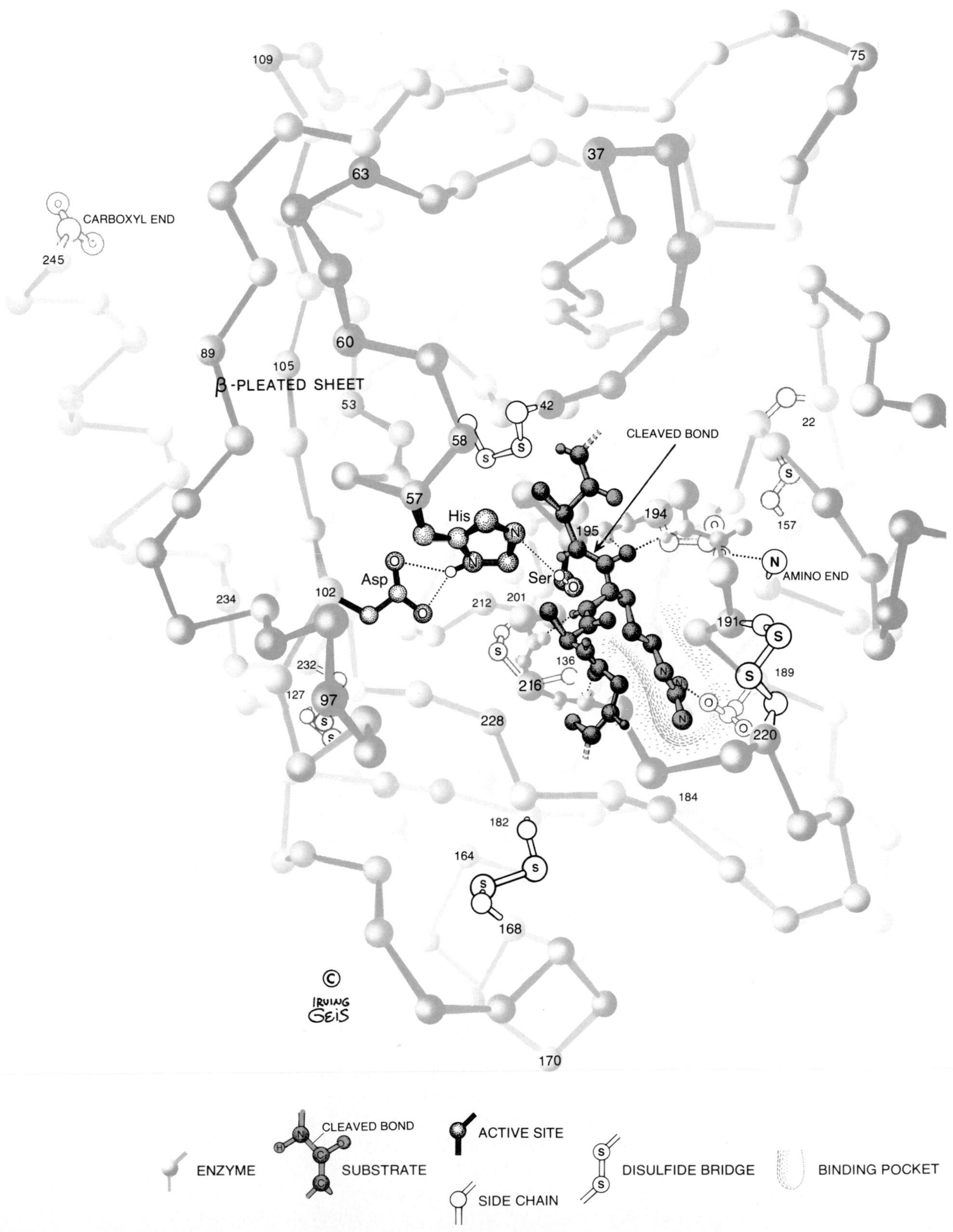
109
75
37
63
CARBOXYL END
245
60
89
105
β-PLEATED SHEET
53
42
22
58
CLEAVED BOND
57
His
194
195
157
N
AMINO END
Asp
Ser
102
234
212
201
191
S
S
189
232
136
216
127
97
228
220
184
182
164
S
S
168
IRVING GEIS
170
CLEAVED BOND
ACTIVE SITE
ENZYME
SUBSTRATE
DISULFIDE BRIDGE
BINDING POCKET
SIDE CHAIN

Near the mouth of the active site are two amino acid residues that serve a critical catalytic function: a serine (Ser-195) and a histidine (His-57). The way that these residues act to catalyze peptide bond hydrolysis is shown in Fig. 26.13. The —OH group of the serine side chain acts as a nucleophile to displace the peptide leaving group from the carbonyl group of the substrate. The resulting product is an **acyl-enzyme;** in this covalent complex, the residual peptide substrate is actually esterified to the enzyme! The imidazole group of histidine-57 serves as a base catalyst to remove the proton from the attacking serine. When water enters the active site, it is deprotonated by the histidine as it attacks the carbonyl group of the acyl-enzyme, giving the free carboxy group of the substrate and regenerating the enzyme. After the product leaves the active site, the enzyme is ready for a new substrate molecule.

The *catalytic efficiency* of trypsin, as well as that of other enzymes, is attributable mostly to the fact that all of the necessary reactive groups are positioned in proximity within the enzyme–substrate complex: the substrate carbonyl, a nucleophile (the serine —OH group), and an acid–base catalyst (the imidazole of the histidine). These groups do not have to "find" each other by random collision, as they would if they were all free in solution. Notice that the reactions shown in Fig. 26.13 are not particularly unusual. As we have observed before, enzyme-catalyzed transformations find close analogy in common organic reactions. As we can see from this example, understanding the chemistry of life does not require the idea of a "vital force" so prevalent in Wöhler's day; it is simply good organic chemistry! We should hasten to add that the rationality of this chemistry makes it no less remarkable and elegant.

We can also now appreciate, at least in a general way, why most enzymes act on one enantiomer of a chiral substrate but not on another (see, for example, Eq. 26.15). Enzymes are chiral because they are peptides derived solely from L-amino acids. A chiral enzyme reacts with a chiral substrate to form first an enzyme–substrate complex. The enzyme–substrate complex derived from the enantiomer of the substrate is a *diastereomer* of the one derived from the substrate itself.

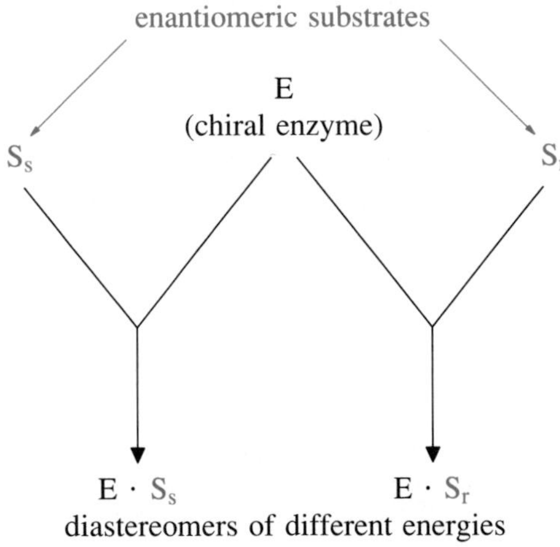

(26.41)

Since the two complexes are diastereomers, they have different energies, and one is more favorable than the other. For the same reason, the rates at which these complexes are converted into products also differ. This is an important example of the *differentiation of enantiomers by a chiral reagent* (Sec. 7.8C). In fact, most enzymes do not catalyze reactions involving the enantiomers of their natural substrates to any significant extent. Putting a substrate enantiomer into the chiral active site is like putting a right hand into a left-handed glove: things simply don't fit!

Figure 26.13 Mechanism of the trypsin-catalyzed hydrolysis of an arginyl-peptide bond, beginning with the enzyme-substrate complex and ending with the enzyme-product complex. Notice how the imidazole in the side chain of the His-57 residue acts alternately as an acid and as a base to facilitate the necessary proton transfers. Pep^N and Pep^C denote the amino and carboxy terminal parts of the peptide substrate, respectively.

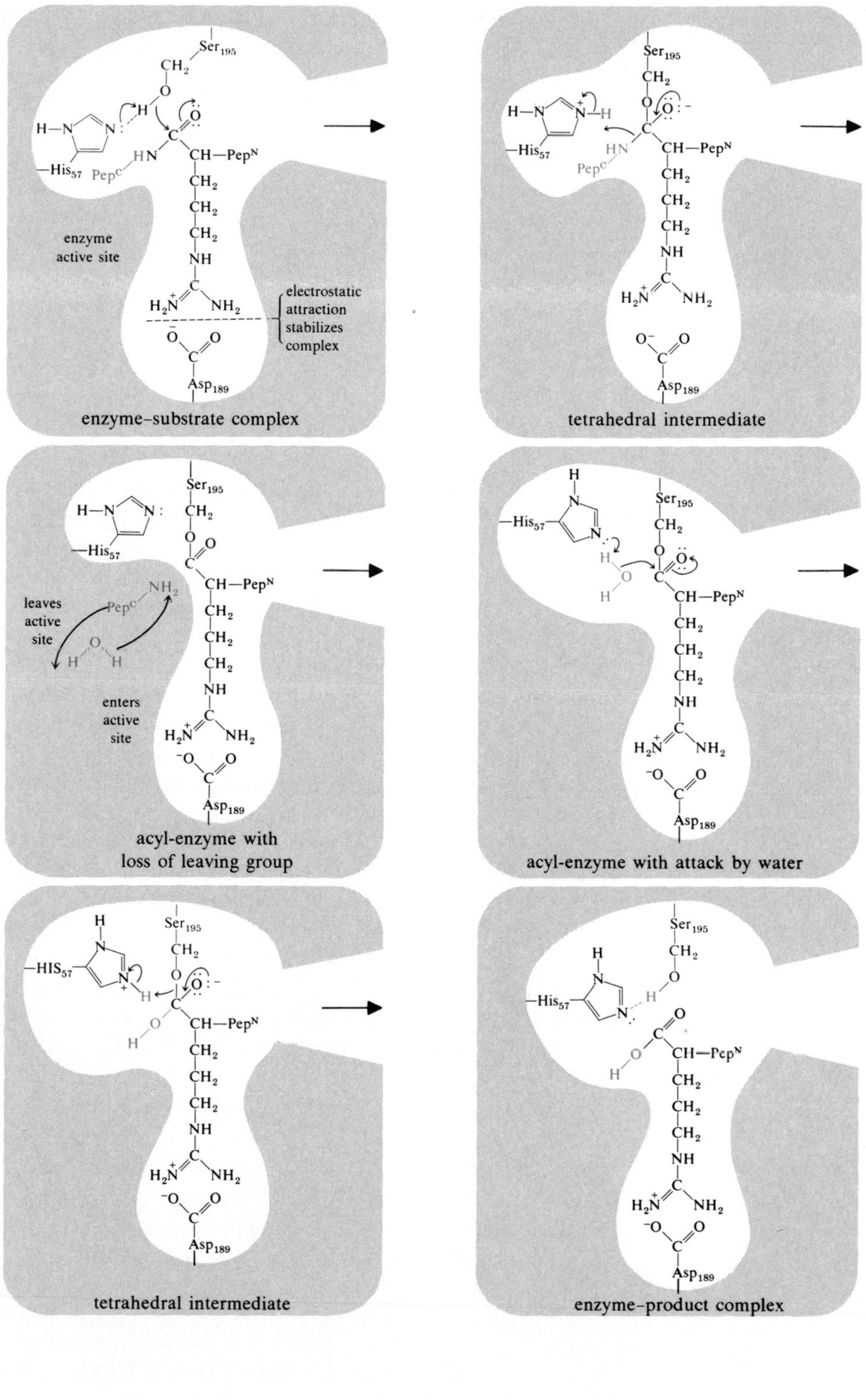

If we understand the details of enzyme catalysis, we should be able to design and synthesize artificial enzymes that can bind specific chemical compounds and act on them catalytically. With a few exceptions, this sort of activity is yet to meet with great success. The rational synthesis of molecules with the *catalytic efficiency* and *specificity* of enzymes is an area of research that will occupy chemists of the future.

26.11 OCCURRENCE OF PEPTIDES AND PROTEINS

We have just seen that proteins serve an important role as enzymes—biological catalysts. Peptides and proteins also serve other important biological roles. For example, proteins serve as transporters; thus, hemoglobin transports oxygen in the bloodstream. Some proteins have a structural role: collagen, a protein, is the major component of connective tissue. Proteins and peptides act as a line of defense: antibodies, or immunoglobulins, are the proteins that protect higher animals from invasion by foreign substances, including infectious agents. In some cases the active components present in venoms and toxins (for example snake and bee venoms) are proteins. One of the most exciting developments in peptide chemistry is the discovery that many hormones are peptides. Two classical examples of peptide hormones are insulin, which, among other things, controls the uptake of glucose into cells, and glucagon, which counterbalances the action of insulin. Gastrin is a peptide that controls the release of stomach acid. A peptide has been discovered that induces drowsiness (Δ-sleep-inducing peptide).

An important recent discovery in the peptide field was prompted by the curiosity of scientists to find out why morphine, a compound that does not occur naturally in the human body, relieves pain. They reasoned that the body must contain another natural substance that might have the same effects as morphine. Certain peptides, called *enkephalins,* and β-endorphin, a longer peptide from which the enkephalins are derived, were discovered and found to have morphinelike effects. These may be the substances ultimately responsible for tolerance to pain. (It has been theorized that "joggers' high," the state of quasi-addiction to long-distance running, may be attributable to the release of these peptides.) It is now clear that peptide hormones control a large number of biological functions. Who knows: perhaps your ability to learn organic chemistry is regulated by one or more peptides!

Chemists have, on occasion, taken a cue for synthetic substances from protein chemistry. Nylon (Sec. 21.12A), with its many amide bonds, might be regarded as "synthetic silk"; silk is a protein fiber. Not long ago peptide chemists at G. D. Searle Co. accidentally discovered a sweet-tasting peptide, L-aspartyl-L-phenylalanine methyl ester (aspartame), which is now an important artificial sweetener sold under the trade name NutraSweet®.

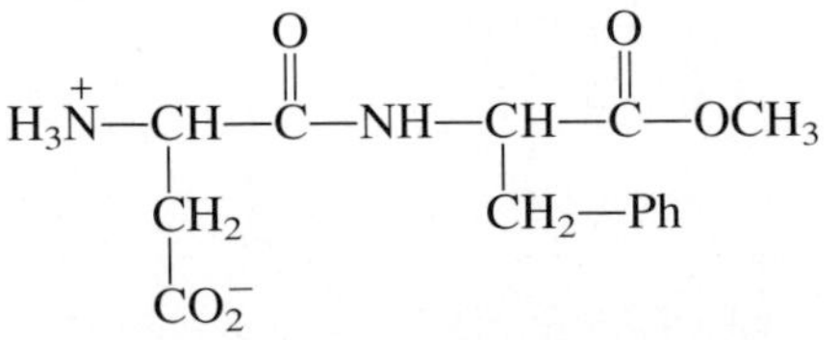

aspartame (an artificial sweetener)

Peptide chemists are working to develop new, metabolically stable analogs of physiologically important peptides that can be used in medicine. It is clear that peptide and protein chemistry is an important branch of organic chemistry in which many exciting developments can be anticipated in the future.

KEY IDEAS IN CHAPTER 26

- Amino acids are compounds that contain both an amino group and a carboxylic acid group. Peptides and proteins are polymers of the α-amino acids.
- The naturally occurring α-amino acids have the *S* configuration at their α-carbon atoms. (2*S*)-Amino acids are called L-amino acids in an older system of stereochemistry.
- Amino acids (as well as peptides containing both acidic and basic groups) exist at neutrality as zwitterions.
- The acidic or basic character of an amino acid, peptide, or protein is measured by its isoelectric point.
- Common methods for the synthesis of α-amino acids include the alkylation of ammonia with α-halo acids; alkylation of α-acetamidomalonate esters, followed by hydrolysis and decarboxylation; and the Strecker synthesis.
- Amino acids react both as amines and carboxylic acids. Thus, the amino group can be acylated; the carboxylic acid group can be esterified.
- Reaction with ninhydrin is an analytically important transformation of α-amino acids, although primary amines also undergo the same reaction.
- Peptides can be hydrolyzed to their constituent amino acids by heating them in aqueous acid or base. Hydrolysis of a peptide, separation of the resulting amino acids by ion-exchange chromatography, and quantitation give the relative number of each amino acid in the peptide. This procedure is called amino acid analysis.
- Peptides and proteins can be cleaved specifically at methionine residues with cyanogen bromide, and at arginine and lysine residues by trypsin-catalyzed hydrolysis.
- The primary sequence of a peptide or protein can be determined by the Edman degradation: treatment of the peptide with phenyl isothiocyanate, followed by acid. Each cycle of this degradation affords a PTH derivative of the amino terminal amino acid plus a new peptide that is one residue shorter.
- Peptide synthesis strategically involves attaching amino acids, one at a time, to a peptide chain beginning at the carboxy terminus. Protecting groups such as the Boc group must be used to prevent competing reactions that would occur in their absence. In solid-phase peptide synthesis, the Boc-protected carboxy terminal

amino acid is covalently bound to an insoluble resin; deprotected; condensed with another Boc-protected amino acid using DCC; deprotected; and the cycle continued until the peptide chain has been assembled, at which point the peptide is removed from the resin by cleavage with HF or other acidic reagents.

- The primary structure of a protein is a description of its covalent bonds, including disulfide bonds. Its secondary structure is a description of its content of α-helix, β-structure, and random coil; and its tertiary structure is a complete description of its three-dimensional structure. The manner in which smaller proteins aggregate into larger ones is termed quaternary structure.
- The noncovalent forces that determine the three-dimensional structure of a protein include hydrogen bonds, van der Waals forces, and electrostatic interactions.
- Enzymes are biological catalysts. A substrate is bound noncovalently at the enzyme active site before it is transformed into products.

ADDITIONAL PROBLEMS

23 Write the products expected when valine reacts with each of the following reagents:

(a) C_2H_5OH, H_2SO_4
(b) benzoyl chloride, Et_3N
(c) HCl solution
(d) NaOH solution
(e) benzaldehyde, heat, NaCN
(f) ninhydrin
(g)

$$Me_3CO-\overset{\overset{\large O}{\|}}{C}-O-\overset{\overset{\large O}{\|}}{C}-OCMe_3,\ \text{DMF},\ Et_3N$$

(h) product of (g) + DCC + glycine *t*-butyl ester
(i) product of (h) + anhydrous CF_3CO_2H, then neutralize with NaOH
(j) product of (i) + 6*N* HCl, heat

24 Referring to Table 26.1, identify the amino acid(s) that satisfy each of the following criteria:

(a) the most acidic amino acid
(b) the most basic amino acid
(c) the amino acids that can exist as diastereomers
(d) the amino acid that does not react with ninhydrin to give Ruhemann's purple
(e) the amino acid that has zero optical rotation under all conditions
(f) the amino acids that are converted into other amino acids on treatment with concentrated NaOH solution

25 In repeated attempts to synthesize the dipeptide Val-Leu, aspiring peptide chemist Polly Styreen performs each of the following operations. Explain what, if anything, is wrong with each procedure.

(a) The cesium salt of leucine is allowed to react with chloromethylated polystyrene (Merrifield resin). The resulting derivative is treated with Boc-Val and DCC, then with liquid HF.
(b) The cesium salt of Boc-Leu is allowed to react with the Merrifield resin. The resulting derivative is then treated with Boc-Val and DCC, then with liquid HF.

26 According to its amino acid composition (Fig. 26.4), lysozyme has an isoelectric point that is (choose one and explain):
(a) ≪6 (b) about 6 (c) ≫6

27 A previously unknown amino acid, γ-carboxyglutamic acid (Gla), was discovered in the amino acid sequence of the blood-clotting protein prothrombin.

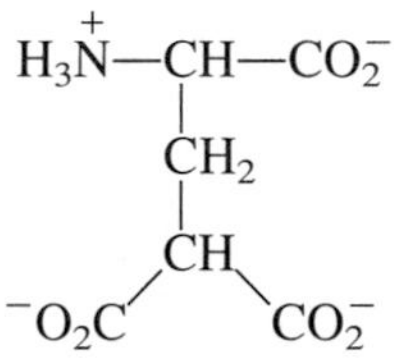

γ-carboxyglutamic acid (Gla)

This amino acid escaped detection for many years because, on acid hydrolysis, it is converted into another common amino acid. Explain.

28 Most L-amino acids have a flat to slightly bitter taste; most D-amino acids are sweet. What does the different taste of enantiomeric amino acids indicate about the taste-sensing apparatus of the body?

29 When bovine insulin is treated with the Edman reagent followed by anhydrous CF_3CO_2H, then by aqueous acid, the PTH derivatives of *both* glycine and phenylalanine are obtained in nearly equal amounts. What can be deduced about the structure of insulin from this information?

30 Which of the statements below would correctly describe the isoelectric point of *cysteic acid*, an oxidation product of cysteine? Explain your answer.

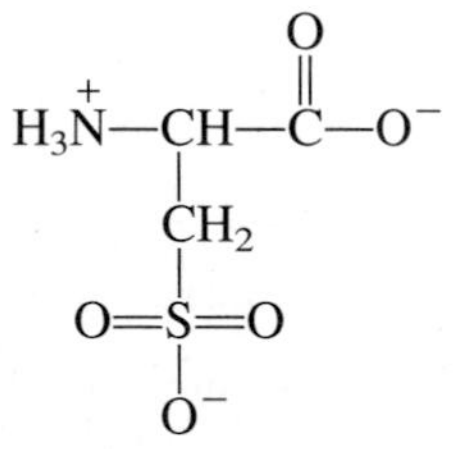

cysteic acid

(a) lower than that of aspartic acid
(b) about the same as that of aspartic acid
(c) about the same as that of cysteine
(d) about the same as that of lysine
(e) higher than that of lysine

Problems (Cont.)

31 The peptide hormone glucagon has the following amino acid sequence:

His-Ser-Gln-Gly-Thr-Phe-Thr-Ser-Asp-Tyr-Ser-Lys-Tyr-Leu-
Asp-Ser-Arg-Arg-Ala-Gln-Asp-Phe-Val-Gln-Trp-Leu-Met-
Asn-Thr

Give the products that would be obtained when this protein is treated with
(a) trypsin at pH 8
(b) cyanogen bromide in HCl
(c) Ph—N=C=S, then CF_3CO_2H, then aqueous acid

32 *Dansyl chloride* (5-dimethylamino-1-naphthalenesulfonyl chloride) reacts with amino groups to give a fluorescent derivative.

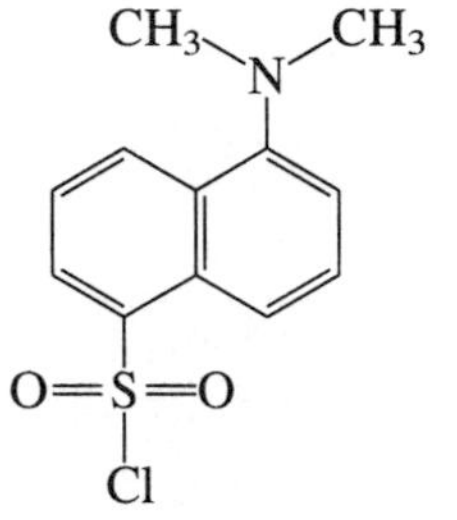

dansyl chloride

After a peptide with the composition (Arg,Asp,Gly,Leu_2, Thr,Val) reacts with dansyl chloride at pH 9, it is hydrolyzed in 6*N* aqueous HCl. The following derivative, detected by its fluorescence, is isolated, along with the free amino acids Arg, Asp, Gly, Leu, and Thr. What conclusion can be drawn about the structure of the peptide from this result?

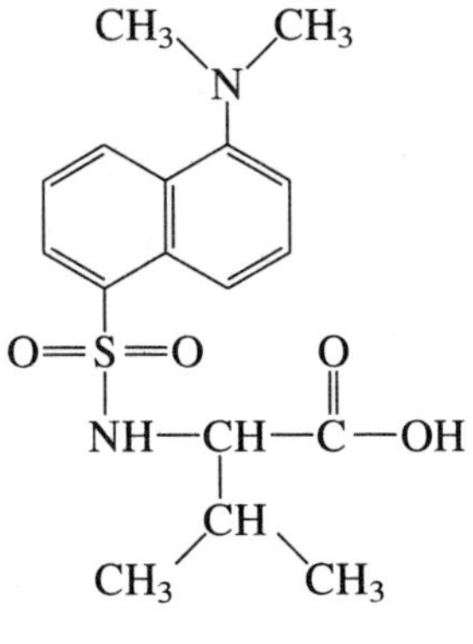

33 Complete the following reactions by giving the structure of the major organic product(s):

(a) ethylamine + Ph—N=C=S $\longrightarrow$

(b) PhCH=O + KCN + CH_3NH_2 $\longrightarrow \xrightarrow[\text{heat}]{H^+/H_2O}$

(c) $H_3\overset{+}{N}—\underset{\displaystyle CH_3}{\underset{|}{CH}}—CO_2^- \xrightarrow[\text{(1 equiv)}]{\text{NaOH}} \xrightarrow[H_2O]{(CH_3)_2CH—CH=O} \xrightarrow{NaBH_4}$

(d) 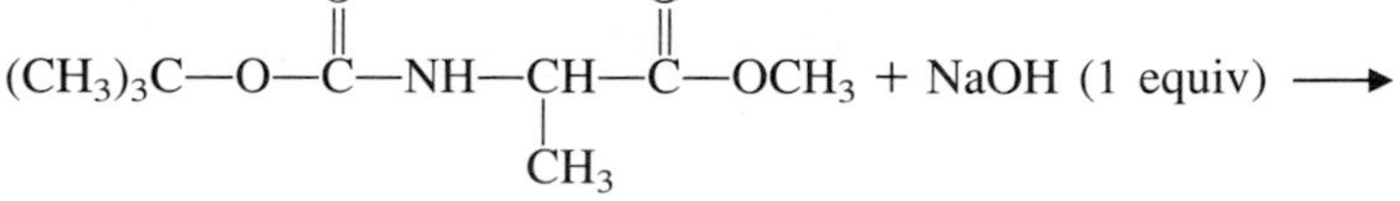

(e) 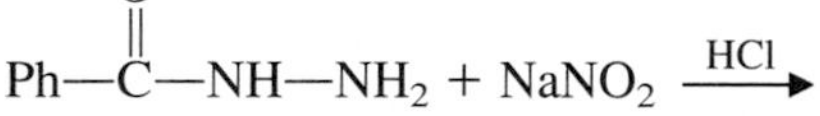

(f) product of (e) $\xrightarrow{\text{heat in benzene}}$

(g) product of (f) + valine methyl ester $\longrightarrow$

(h)

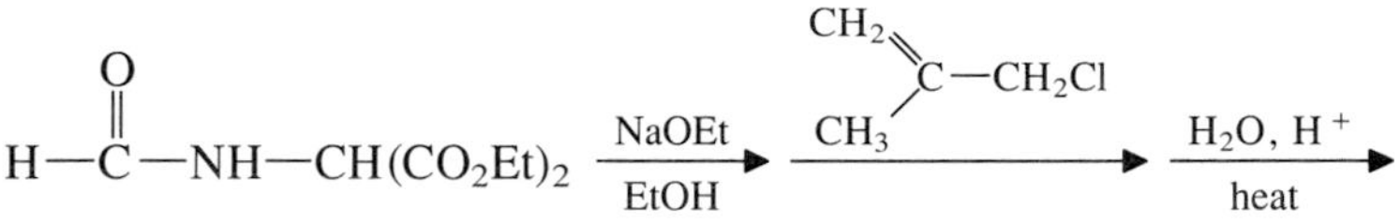

(i) 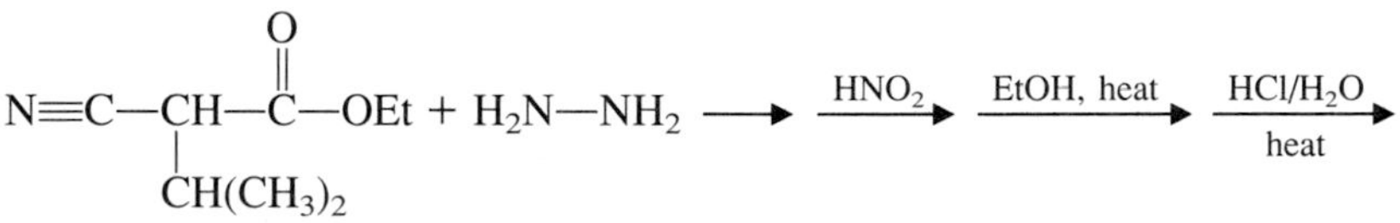

34 A peptide was subjected to one cycle of the Edman degradation, and the following compound was obtained:

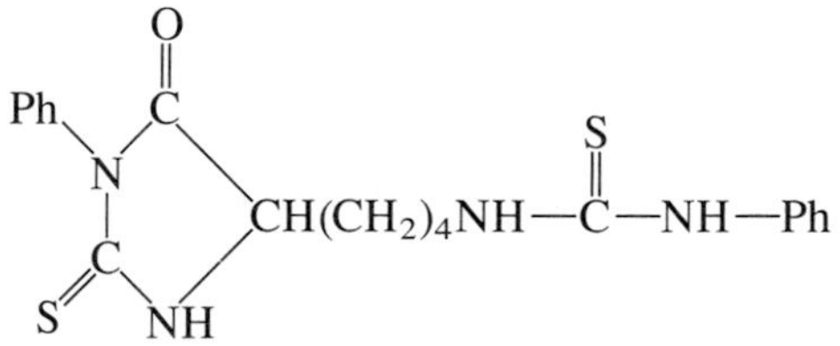

What is the amino-terminal residue of the peptide?

35 Explain each of the following observations:

(a) The optical rotations of alanine are different in water, 1*N* HCl, and 1*N* NaOH.

(b) Two mono-*N*-acetyl derivatives of lysine are known.

(c) The peptide Gly-Ala-Arg-Ala-Glu is readily cleaved by trypsin in water at pH = 8, but is inert to trypsin in 8*M* urea at the same pH.

(d) After peptides containing cysteine are treated with $HSCH_2CH_2OH$, then with aziridine, they can be cleaved by trypsin at their (modified) cysteine residues.

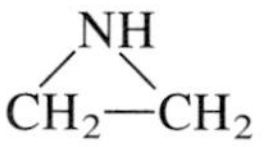

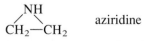

aziridine

(e) When L-methionine is oxidized with H_2O_2, two separable methionine sulfoxides with the following structure are formed:

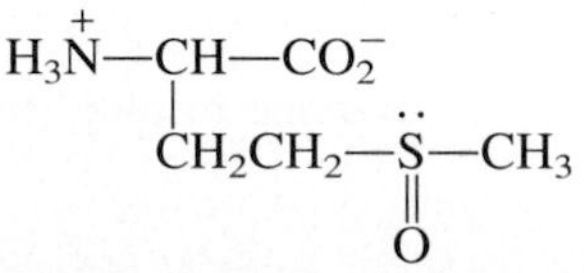

Problems (Cont.)

(f) When benzamidine hydrochloride is present in solution, trypsin no longer catalyzes peptide hydrolysis. (Benzamidine hydrochloride is not a denaturant at the concentrations used.)

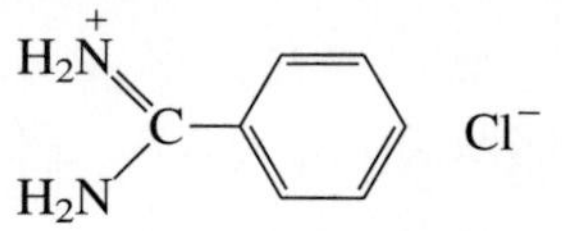

benzamidine hydrochloride

36 A peptide *P* has the following composition by amino acid analysis:

P: Ala,Arg,Asp,Gly_2,Glu,Leu,Val_2,NH_3

Treatment of *P* once with the Edman reagent followed by anhydrous acid gives a new peptide *Q* with the following composition by amino acid analysis:

Q: Ala,Arg,Asp,Gly_2,Glu,Val_2,NH_3

Treatment of *P* and *Q* with the enzyme *dipeptidylaminopeptidase* (DPAP) yields a mixture of the following peptides:

$$P \xrightarrow{\text{DPAP}} \text{Arg-Gly, Gln-Ala, Leu-Val, Val-Asp, Gly}$$

$$Q \xrightarrow{\text{DPAP}} \text{Ala-Gly, Asp-Gln, Gly-Val, Val-Arg}$$

What is the amino acid sequence of *P*?

37 Poly-L-lysine (a peptide containing only lysine residues) exists entirely in an α-helical conformation at pH > 11. Below pH 10, however, the peptide assumes a random-coil conformation. Poly-L-glutamic acid, on the other hand, exists in the α-helical conformation at pH < 4, but above pH 5 it assumes a random-coil conformation. Explain the effect of pH on the secondary structure of both polymers. Your explanation should indicate why the helical conformation predominates at high pH in one case and low pH in the other. (*Hint:* Look carefully at the location of the amino acid side chains in Fig. 26.6.)

38 Complete the following reactions, assuming the amino acid residue is part of a peptide and is at neither the amino nor the carboxy terminus:

(a)

lysine residue +

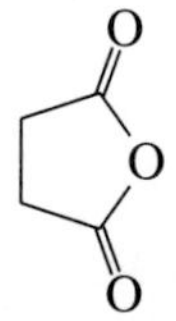

$\xrightarrow{\text{pH 8}}$

succinic anhydride

(b)

cysteine residue + $I—CH_2—\overset{O}{\overset{\|}{C}}—O^-$ $\xrightarrow{\text{pH 8-9}}$

iodoacetate

(c)

cysteine residue + [maleimide] $\xrightarrow{pH > 5}$ (a conjugate addition)

maleimide

(d) lysine residue + $CH_2{=}O$ + $NaBH_4$ $\xrightarrow{pH\ 9}$
(excess of both)

(e) aspartic acid residue + [cyclohexyl]—$N{=}C{=}N$—CH_2CH_2—$\overset{+}{N}$(morpholine)H_3C + Cl^-
a water-soluble carbodiimide that
reacts much like DCC

H_2N—CH_2—CO_2CH_3 $\longrightarrow$

(f) tyrosine residue + $Ph\overset{+}{N}{\equiv}N\ Cl^-$ $\xrightarrow{pH\ 9}$

(g) cysteine residue + $HgCl_2$ $\longrightarrow$ (This reaction is one reason why mercury
and other heavy metals are toxic.)

(h) lysine residue + $NaNO_2$ + HCl $\xrightarrow{H_2O}$

39 Show how the acetamidomalonate method can be used to prepare the following
unusual amino acids from the indicated starting material and any other reagents:

(a) $(CH_3)_2CDCH_2$—$\underset{\overset{|}{\overset{+}{N}H_3}}{CH}$—$CO_2^-$ from isobutylene

(b) Ph—CHD—$\underset{\overset{|}{\overset{+}{N}H_3}}{CH}$—$CO_2^-$ from benzaldehyde

(c)

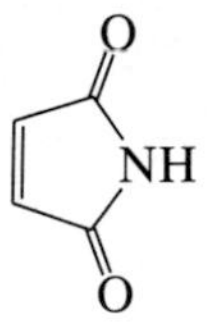

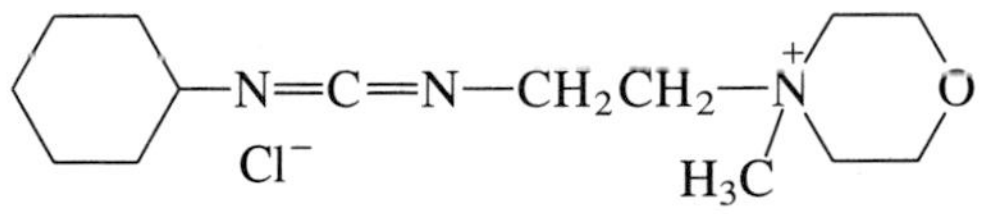

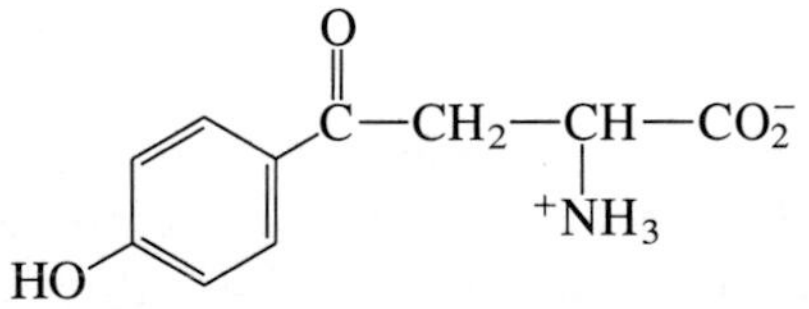

HO—[C_6H_4]—$\overset{\overset{O}{\|}}{C}$—$CH_2$—$\underset{\overset{|}{\overset{+}{N}H_3}}{CH}$—$CO_2^-$ (γ-oxohomotyrosine)
from anisole

40 The artificial sweetener *aspartame* was withheld from the market for several
years because, on storage for extended periods of time in aqueous solution, it
forms a *diketopiperazine*.

aspartame $\longrightarrow$ a diketopiperazine + CH_3OH

(Extensive biological testing was required to show that this by-product was safe
for consumers.) Give a mechanism for the formation of the diketopiperazine.

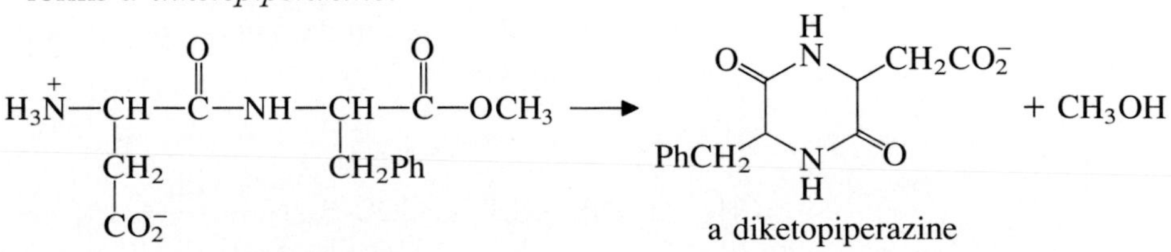

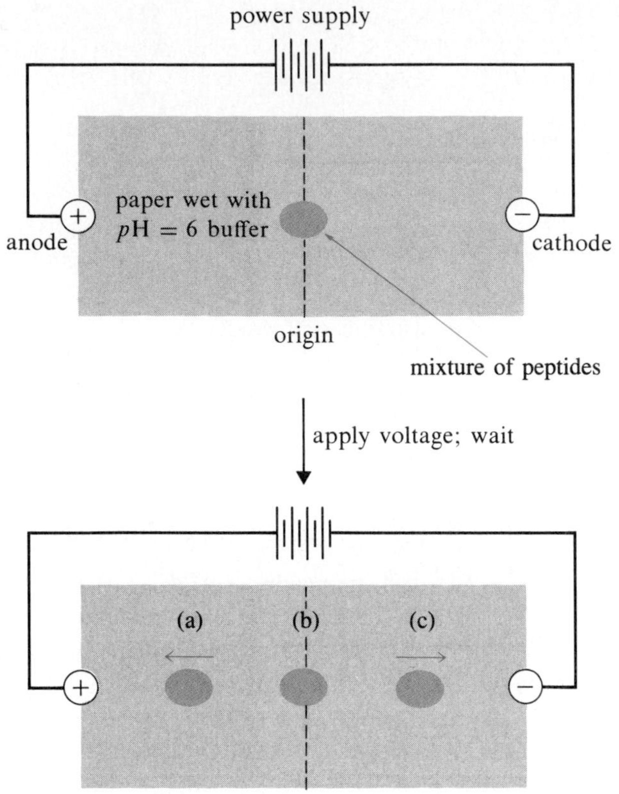

Figure 26.14 Separation of three amino acids by electrophoresis at pH = 6.

Problems (Cont.)

41 In *paper electrophoresis,* amino acids and peptides can be separated by their differential migration in an electric field. To the center of a paper, wet with buffer at pH 6, is applied a mixture of the following three peptides: Gly-Lys, Gly-Asp, and Gly-Ala. (See Fig. 26.14.) A voltage is applied across the ends of the paper for a period of time, after which the peptides have migrated into three distinct spots. Which peptide is in each spot? Explain.

42 Identify each of the bracketed compounds in the following reaction scheme. Explain your answers.

$$CH_3-\overset{\overset{O}{\|}}{C}-NH-\underset{\underset{CH(CH_3)_2}{|}}{CH}-\overset{\overset{O}{\|}}{C}-OC_2H_5 \xrightarrow{H_2N-NH_2} [A] \xrightarrow{NaNO_2/HCl} [B] \xrightarrow[HCl, H_2O]{heat}$$

$$[C] \xrightarrow[OH^-]{H_2O, \ heat} (CH_3)_2CH-CH=O + [D]$$

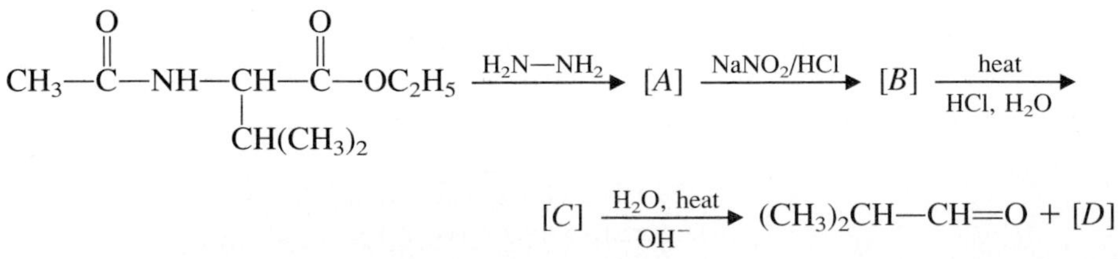

43 (a) With what reagent would the Merrifield chloromethylated polystyrene resin react to give the following resin?

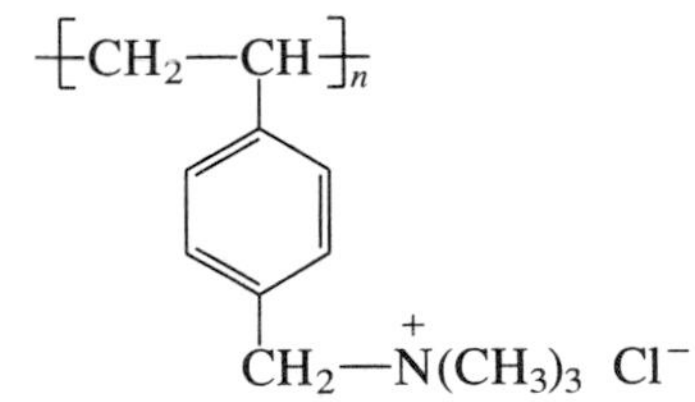

(b) To a column containing this resin suspended in a pH 6 buffer is added a mixture of the amino acids Arg, Glu, and Leu, and the column is eluted with the same buffer. In what order will the amino acids be washed from the column? Explain.

44 Outline a synthesis for each of the following compounds from the indicated starting material and any other reagents:

(a)

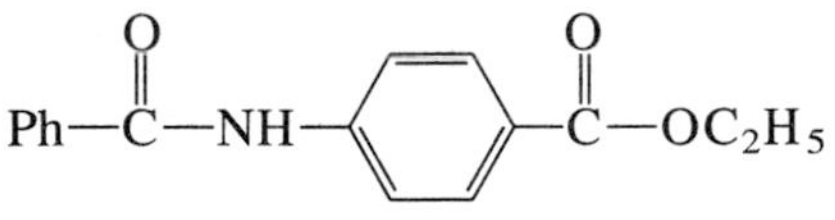

$$Ph—\overset{O}{\underset{||}{C}}—NH—\bigcirc—\overset{O}{\underset{||}{C}}—OC_2H_5 \qquad \text{from } p\text{-aminobenzoic acid}$$

(b) $Ph—\underset{\underset{^+NH_3}{|}}{CD}—CO_2^-$ from benzoic acid

(c) $CD_3—\underset{\underset{^+NH_3}{|}}{CH}—CO_2^-$ from $CD_3—CH{=}O$

(d) L-Lys-L-Ala-L-Pro from L-proline, L-alanine, and the following compound using a solid-phase peptide synthesis:

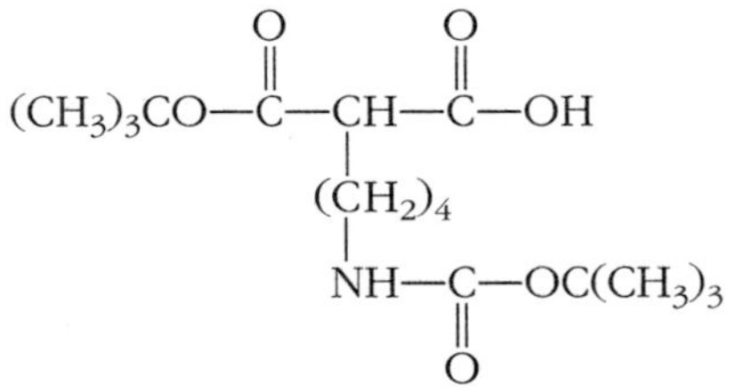

α,ϵ-diBoc-L-lysine

(e)

$$C_2H_5O—\overset{O}{\underset{||}{C}}—CH\overset{\displaystyle NH—\overset{O}{\underset{||}{C}}—CH_3}{\underset{\displaystyle C{\equiv}N}{\Big\langle}} \qquad \text{from } C_2H_5O—\overset{O}{\underset{||}{C}}—CH{=}O$$

(f)

$$\bigcirc\!\!\!-\!\!\!-\underset{\underset{NH_3^+}{|}}{CH}—\overset{O}{\underset{||}{C}}—O^-$$

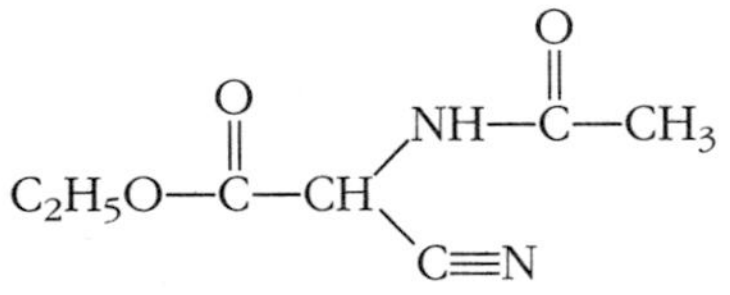

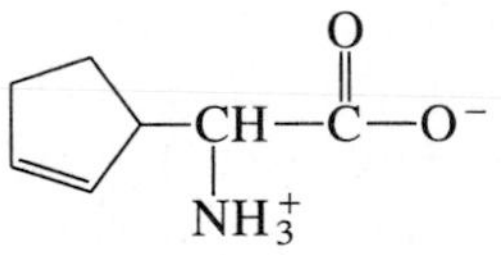

from the product of (e) and cyclopentene. (Cyclopentenyl amino acids are produced by certain plants. *Hint:* Protons α to a cyano group are about as acidic as those α to an ester group.)

Problems (Cont.)

(g) The polymer *p*-aramid from terephthalic acid (1,4-benzenedicarboxylic acid).

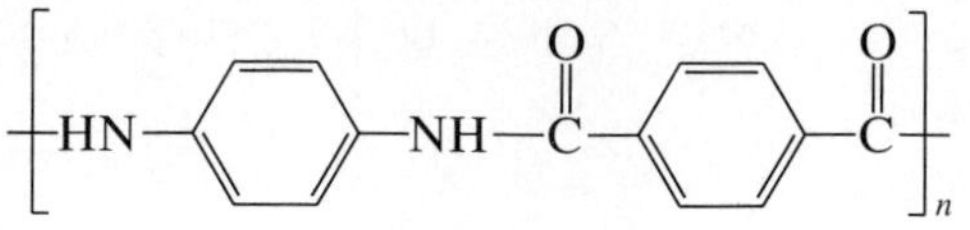

p-aramid

(This polymer is used in tire cord and other applications that require rigidity and strength.)

---

45 A peptide *C* was found to have a molecular weight of about 1100. Amino acid analysis of *C* revealed its composition to be (Ala$_2$,Arg,Gly,Ser). The peptide was unchanged on treatment with the Edman reagent, then CF_3CO_2H. Treatment of *C* with trypsin gave a single peptide *D* with an amino acid analysis identical to that of *C*. Three cycles of the Edman degradation applied to *D* revealed the partial sequence Ala-Ser-Gly. . . . Suggest a structure for peptide *C*.

---

46 The (9-fluorenyl)methoxycarbonyl (Fmoc) group is a protecting group that can be readily removed with amines or dilute NaOH.

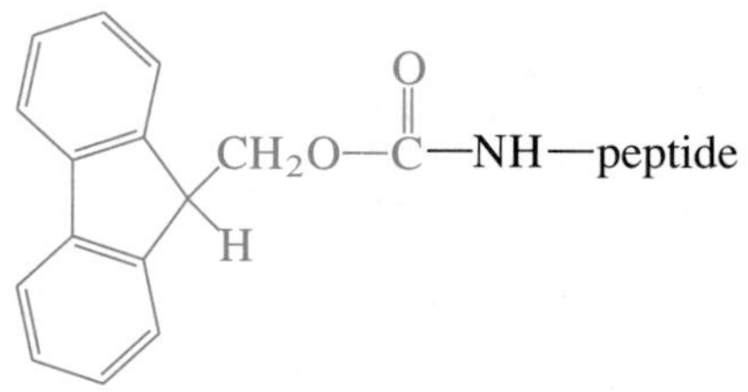

Fmoc group

Explain why this group is readily removed by base. Give the products and the mechanism for their formation when the peptide above reacts with the base triethylamine.

---

47 An amino acid *A*, isolated from the acid hydrolysis of a peptide antibiotic, gave a positive test with ninhydrin and had an optical rotation (HCl solution) of +37.5°. Compound *A* was not identical to any of the amino acids in Table 26.1. In amino acid analysis, compound *A* emerged from the ion-exchange column with the basic amino acids, and it could be prepared by the reaction of L-glutamine with Br_2 in NaOH. Suggest a structure for *A*.

---

48 When *N*-acetyl-L-aspartic acid is treated with acetic anhydride, an optically active compound *A*, $C_6H_7NO_4$, is formed. Treatment of *A* with the amino acid L-valine yields two separable, isomeric peptides, *B* and *C*, that are both converted into a mixture of L-alanine and L-aspartic acid by acid hydrolysis. Suggest structures for *A*, *B*, and *C*.

---

49 Using the arrow formalism, outline a reasonable mechanism for each of the following reactions:

(a)

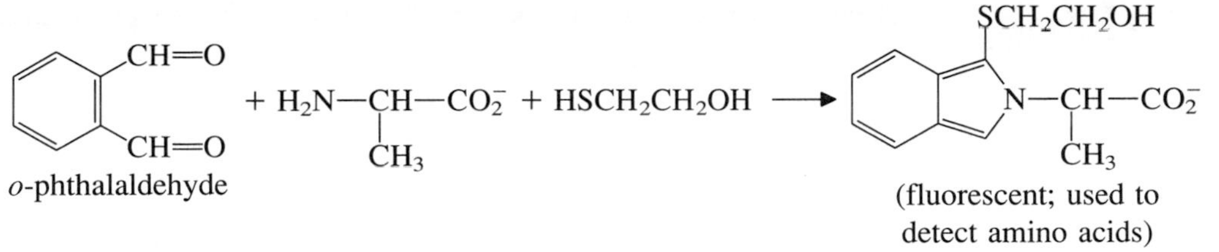

(b)

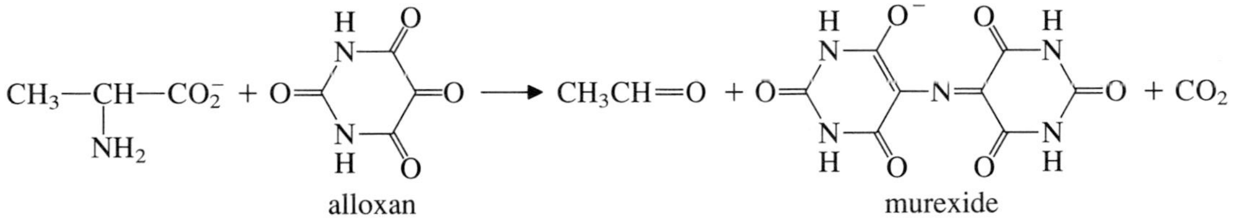

(c)

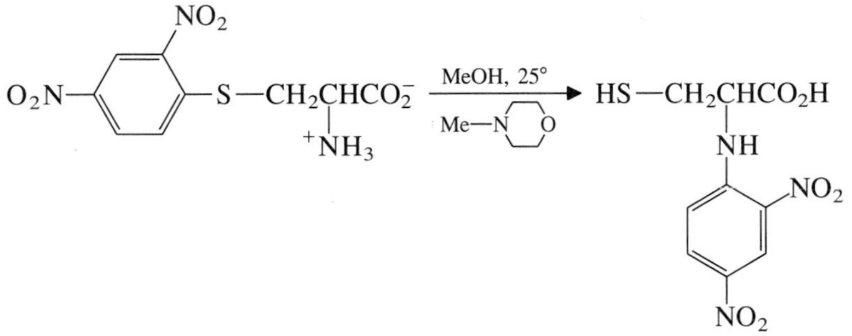

(d)

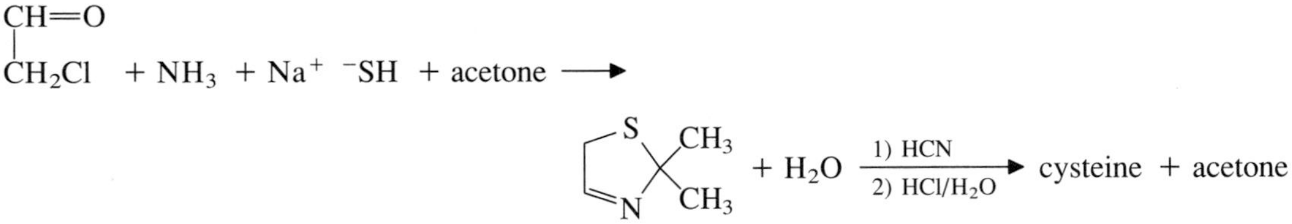

(e)

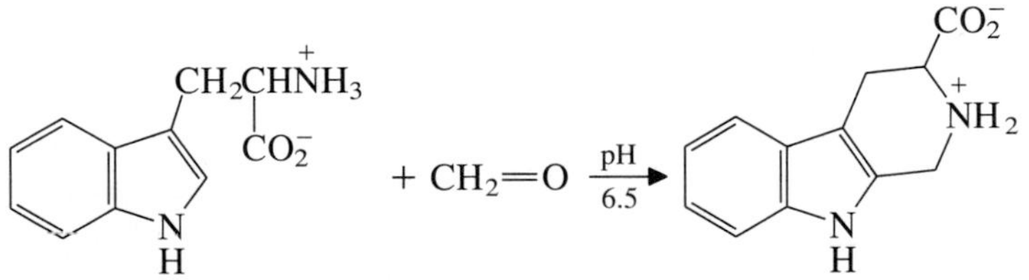

(f)

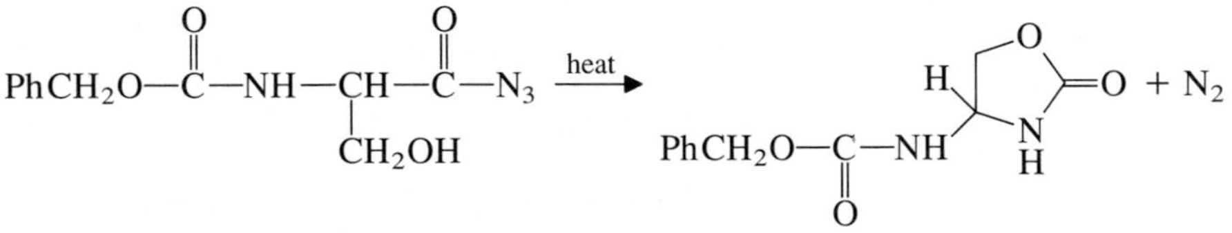

50 The thiol ester below is an excellent acylating agent for amines.

$$RNH_2 + CF_3-\overset{O}{\overset{\|}{C}}-SEt \longrightarrow RNH-\overset{O}{\overset{\|}{C}}-CF_3 + EtSH$$

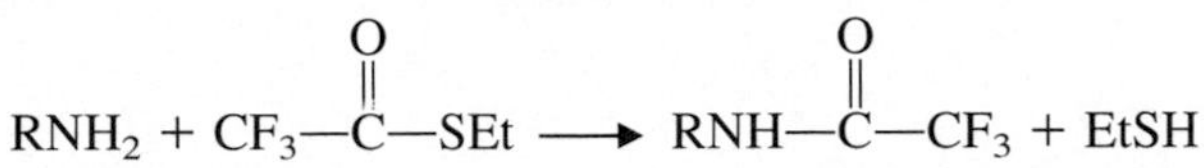

(Problem continues on page 1194.)

Problems (Cont.)

When lysine is treated with one equivalent of this ester under basic conditions, only one of its two amino groups is acylated. Which amino group of lysine is acylated? Explain.

51 (a) Explain why two monomethyl esters of *N*-acetyl-L-aspartic acid are known. Draw their structures.

(b) Explain why a mixture of these two compounds can be separated by anion-exchange chromatography at pH = 3.5, but not at pH = 7. (*Hint:* Use the pK_a values of aspartic acid in Table 26.1.)

(c) Which of the two compounds would emerge first from an anion-exchange column at pH = 3.5? Explain.

52 Assume that the pK_a of protonated glycine methyl ester is about the same as that of the protonated amino group of glycine:

$$\overset{+}{H_3N}—CH_2—CO_2CH_3 \qquad \overset{+}{H_3N}—CH_2—CO_2H$$

$$\text{assume the same } pK_a = 7.8$$

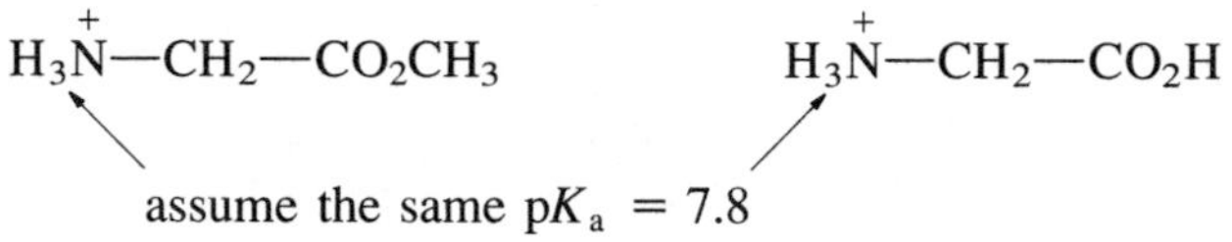

Under this assumption, use the scheme in Eq. 26.2 to estimate the fraction of neutral glycine that exists in the nonzwitterionic form.

53 *Dehydro-α-amino acids* are α,β-unsaturated α-amino acids. (Peptides that contain these unusual amino acids are believed to be intermediates in the biosynthesis of certain natural products.) The following peptide contains a dehydroalanine residue:

$$
\begin{array}{ccc}
& O & O \\
& \| & \| \\
\overset{+}{H_3N}—CH—C—NH—C—C—NH—CH_2—CO_2H \\
| & \| \\
CH_2Ph & CH_2 \\
\text{Phe} & \text{dehydroAla} & \text{Gly}
\end{array}
$$

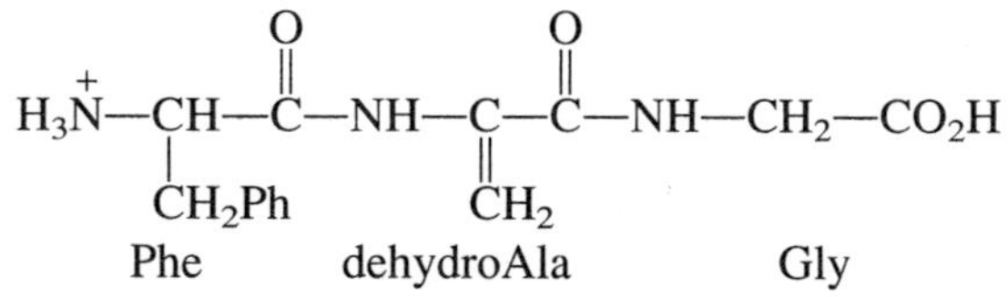

Explain each of the following observations:

(a) When this peptide is hydrolyzed in 6*N* HCl, *pyruvic acid,* ammonia (as NH_4^+), phenylalanine, and glycine are formed.

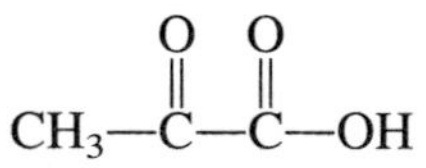

$$
\begin{array}{c}
O \ \ O \\
\| \ \ \| \\
CH_3—C—C—OH \qquad \text{pyruvic acid}
\end{array}
$$

(b) If the peptide is treated with NaB^3H_4 prior to hydrolysis, radioactive alanine is formed in addition to phenylalanine and glycine, and neither ammonia nor pyruvic acid are formed. (3H = tritium, the radioactive isotope of hydrogen.)

54 Treatment of the enzyme trypsin with radioactive phenylmethanesulfonyl fluoride gives a new protein T_1 in which one equivalent of the reagent is covalently bound to the enzyme.

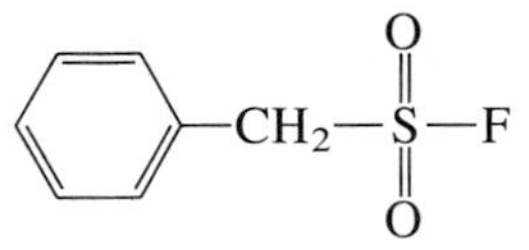

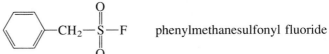

phenylmethanesulfonyl fluoride

After standing briefly in 0.05*N* KOH, all radioactivity is lost from T_1 and a new trypsin derivative T_2 is formed. Both T_1 and T_2 are catalytically inactive; that is, they do not catalyze the hydrolysis of trypsin substrates. However, T_2, but not T_1, binds trypsin substrates noncovalently; that is, T_2 forms an enzyme-substrate complex. Amino acid analysis of T_2 reveals that it contains one fewer serine than is present in the native enzyme. When T_2 is treated with NaB^3H_4 prior to acid hydrolysis, amino acid analysis indicates the presence of ^{3}H-alanine.

Using the chemistry in Problem 53, interpret these results in terms of the mechanism of trypsin catalysis shown in Fig. 26.13 and explain all observations. (This is one of the definitive experiments that established a catalytically important residue in the active site of trypsin.)

27 Carbohydrates and Nucleic Acids

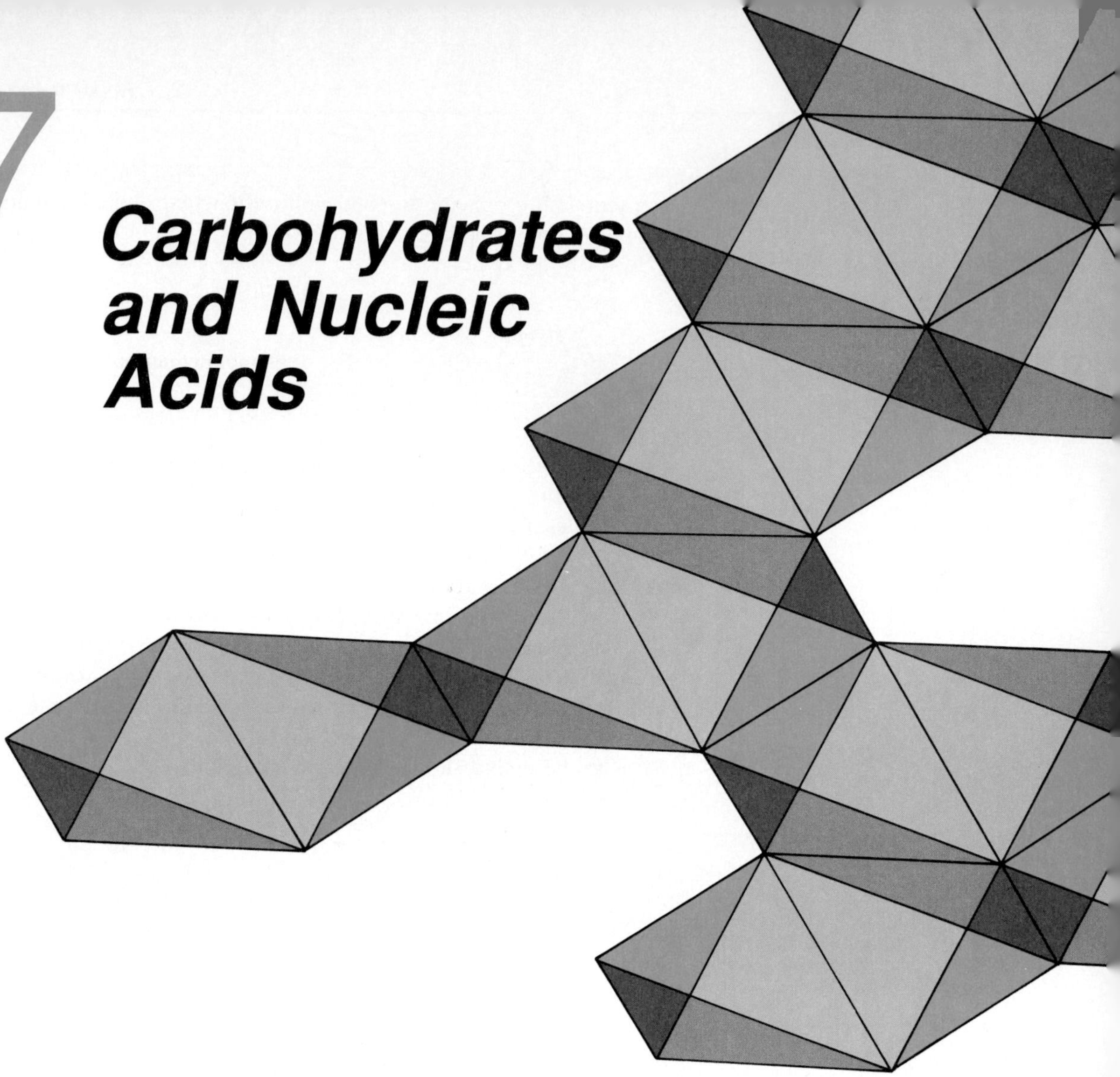

Carbohydrates are the sugars and their derivatives. Most common sugars contain an aldehyde or ketone carbonyl group as well as a number of hydroxy groups on an unbranched carbon chain. For example, glucose, the most common six-carbon sugar, is one of the stereoisomers with the following structure:

$$\mathrm{HO{-}CH_2{-}\underset{\displaystyle OH}{\underset{|}{CH}}{-}\underset{\displaystyle OH}{\underset{|}{CH}}{-}\underset{\displaystyle OH}{\underset{|}{CH}}{-}\underset{\displaystyle OH}{\underset{|}{CH}}{-}CH{=}O}$$

The name *carbohydrate* is derived from the fact that many common sugars have formulas that are formally what we would expect for a "carbon hydrate."

$$\left.\begin{array}{ll}\text{glucose} & C_6H_{12}O_6 = C_6(H_2O)_6 \\ \text{sucrose} & C_{12}H_{22}O_{11} = C_{12}(H_2O)_{11}\end{array}\right\} C_m(H_2O)_n$$

This relationship is, in fact, more than a formal one. Anyone who has added H_2SO_4 (a strong dehydrating agent) to sucrose (table sugar) has observed the violent formation of carbon and steam:

Thus charcoal, which possesses almost the property of ubiquity, is to be found as the very cornerstone in the fabric of the fairest faces of Eve's daughters. It lurks in the sweetened cup of tea; and a quarter pound of nice white lump sugar put into a breakfast cup with the smallest possible dash of boiling water and then the addition of plenty of oil of vitriol [H_2SO_4] is a truly wonderful spectacle, and more instructive than much reading, to see the white sugar turn black, then boil spontaneously, and now, rising out of the cup in solemn black, it heaves and throbs as the oil of vitriol continues its work in the lower part of the cup, emitting volumes of steam . . . until the acid has spent its fury. The elements forming water in the sugar have been attracted, and are now united with the oil of vitriol; a divorce has taken place between the water and the charcoal, which latter now tumbles over the side of the cup." [J. W. Pepper, "Scientific Amusements for Young People," 1863]

Although carbohydrate chemistry is centered around the common sugars, modern carbohydrate chemistry also encompasses related compounds. Thus, the following compounds, which have empirical formulas that do not fit the "hydrate of carbon" pattern, are also regarded as carbohydrates:

$$HOCH_2—CH—CH—CH—CH—CH_2OH \qquad HO_2C—CH—CH—CH—CH—CO_2H$$
$$\quad\;\; | \quad\; | \quad\; | \quad\; | \qquad\qquad\qquad\quad | \quad\; | \quad\; | \quad\; |$$
$$\quad\;\; OH \;\; OH \;\; OH \;\; OH \qquad\qquad\qquad OH \;\; OH \;\; OH \;\; OH$$

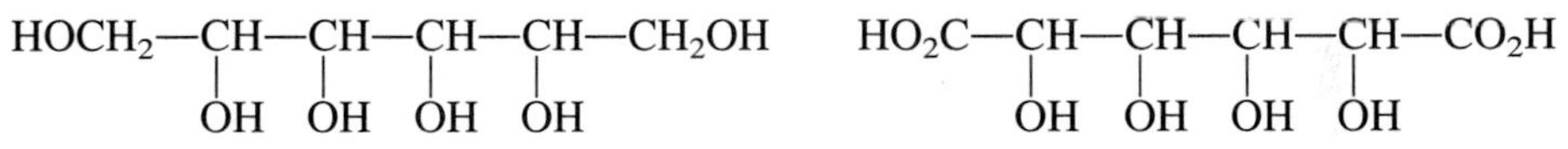

Carbohydrates are among the most abundant organic compounds on the earth. In polymerized form, carbohydrates account for 50–80% of the dry weight of plants. Sugars are a major source of food; sucrose (table sugar) and lactose (milk sugar) are examples. Even the shells of arthropods such as lobsters consist largely of carbohydrate.

In our study of carbohydrates we shall rely heavily on our knowledge of stereochemistry (Chapter 6) and the conformational aspects of cyclohexane rings (Chapter 7). Do not hesitate to use molecular models when necessary.

Nucleic acids are so named because they are a principal component of the cell nucleus (although they also occur elsewhere). Nucleic acids are of two types: **deoxyribonucleic acid** (DNA) and **ribonucleic acid** (RNA). DNA is the storehouse of genetic information in the cell. Your hair color, your sex, and your eye color are all determined by the structure of the DNA in your cells. RNA serves various roles in translating and processing the information encoded in the structure of DNA. DNA and RNA are both polymers of the basic building building blocks called **nucleotides.**

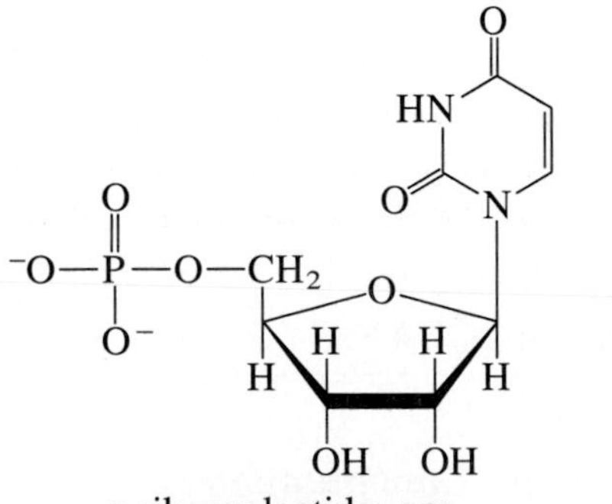

a ribonucleotide; one
of the nucleotide building
blocks of RNA

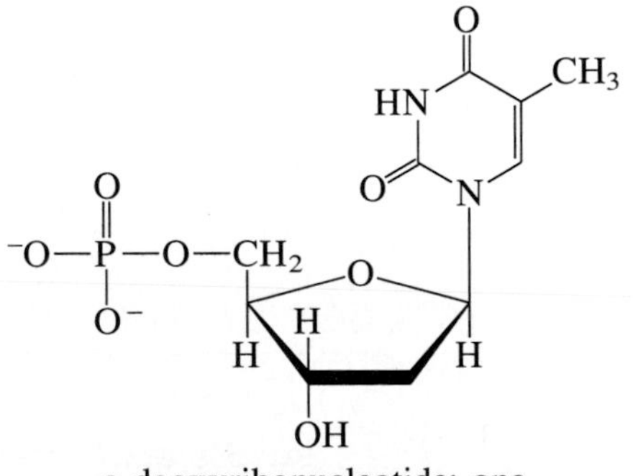

a deoxyribonucleotide; one
of the nucleotide building
blocks of DNA

Just as proteins are polymers of amino acids, RNA and DNA are polymers of nucleotides.

The determination of the structure of DNA was a key development in the field of **molecular biology:** the study of biological systems at the molecular level. This field has given the world dramatic and, at times, controversial new technologies characterized by words like "clone," "genetic engineering," and "recombinant DNA." The newspaper headlines alone engendered by these terms attest to the importance of nucleic acids. Their chemistry, with which the text concludes, is related to that of both sugars and heterocyclic compounds.

27.1 CLASSIFICATION AND PROPERTIES OF SUGARS

The classification of sugars is illustrated in the following examples:

$$HOCH_2-CH-CH-CH-CH-CH=O$$
$$\quad\quad\quad |\quad\ |\quad\ |\quad\ |$$
$$\quad\quad\quad OH\ \ OH\ \ OH\ \ OH$$

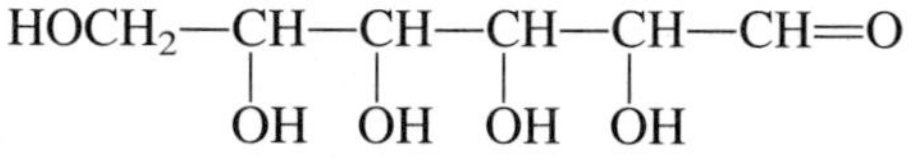

an *aldose* (aldehyde carbonyl group)
a *hexose* (six carbon atoms)
an *aldohexose* (combination of the above classifications)

$$HOCH_2-CH-CH-C-CH_2OH$$
$$\quad\quad\quad |\quad\ |\quad\ ||$$
$$\quad\quad\quad OH\ \ OH\ \ O$$

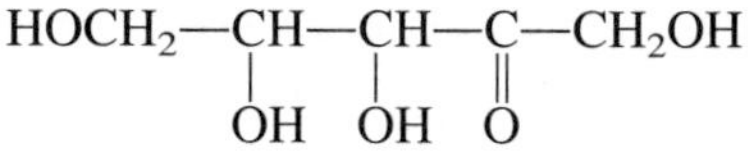

a *ketose* (ketone carbonyl group)
a *pentose* (five carbon atoms)
a *ketopentose* or *pentulose* (combination of the above classifications)

One type of classification is based on the type of carbonyl group in the sugar. A sugar with an aldehyde carbonyl group is called an **aldose;** a sugar with a ketone carbonyl group is called a **ketose.** Sugars can also be classified by the number of carbon atoms. A six-carbon sugar is called a **hexose,** and a five-carbon sugar is called a **pentose.** These two classifications can be combined: an **aldohexose** is an aldose containing six carbon atoms, and a **ketopentose** is a a ketose containing five carbon atoms. A ketose can also be indicated with the suffix *ulose*; thus, a five-carbon ketose is also termed a **pentulose.**

Sugars can also be classified on the basis of their hydrolysis to simpler sugars. Lactose, for example, is classified as a **disaccharide** because it can be hydrolyzed in acid or with certain enzymes to one molecule each of the simpler sugars, glucose and galactose. Because glucose and galactose cannot be hydrolytically decomposed to other sugars, they are examples of **monosaccharides.** Likewise, **trisaccharides** can be hydrolyzed to three molecules of monosaccharides, and **polysaccharides** to many molecules of monosaccharides.

Because of their many hydroxy groups, sugars are very soluble in water. The ease with which table sugar dissolves in water is an example from common experience of sugar solubility. Sugars are virtually insoluble in nonpolar solvents.

27.2 STRUCTURES OF THE MONOSACCHARIDES

A. Stereochemistry and Configuration

We shall consider stereochemistry of sugars by focusing primarily on the aldoses with six or fewer carbons. The aldohexoses have four asymmetric carbons and exist as 2^4

or sixteen possible stereoisomers. These can be divided into two enantiomeric sets of eight diastereomers.

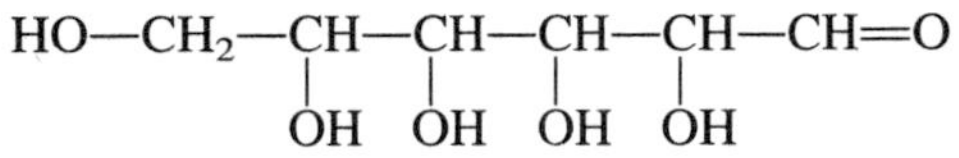

aldohexoses
four asymmetric carbons
$2^4 = 16$ stereoisomers

Similarly, there are two enantiomeric sets of four diastereomers (eight stereoisomers total) in the aldopentose series. Each diastereomer is a *different sugar* with *different properties,* known by a *different name.* The diastereomeric aldoses are given in Fig. 27.1.

Each of the monosaccharides in Fig. 27.1 has an enantiomer. For example, the two enantiomers of glucose have the following structures:

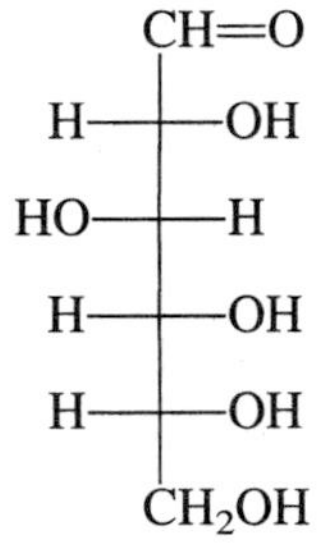

enantiomers of glucose

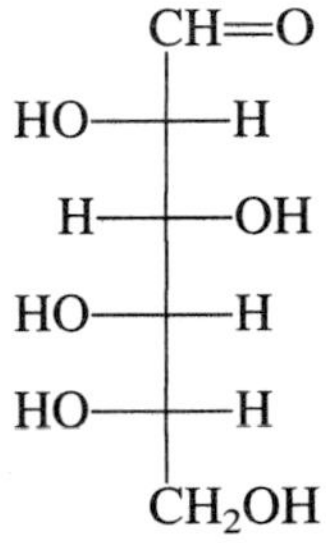

It is important for us to be able to name the enantiomers of sugars in a simple way. Suppose you had a model of one of these glucose enantiomers in your hand; how would you explain to someone who cannot see the model (for example, over the telephone) which enantiomer you were holding? You could, of course, use the *R,S* system to describe the configuration of one or more of the asymmetric carbon atoms. A different system, however, was in use long before the *R,S* system was established. The **D,L system** was proposed in 1906 by a New York University chemist, M. A. Rosanoff, and is still used. In this system, the configuration of a sugar enantiomer is specified by applying the following conventions:

1. The following stereoisomer of the aldotriose glyceraldehyde (the 2*R* enantiomer) is arbitrarily said to have the D configuration; its enantiomer is then said to have the L configuration.

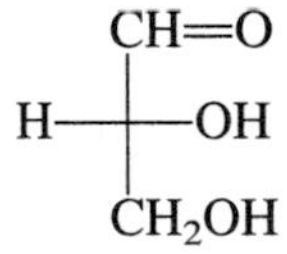

D-glyceraldehyde

2. The other aldoses or ketoses are written in a Fischer projection with their carbon atoms in a straight vertical line, and the carbons are numbered consecutively as they would be in systematic nomenclature, so that the carbonyl carbon receives the lowest number.

3. We now designate as a *reference carbon* the *asymmetric carbon of highest number.* If this carbon has the H, OH, and CH_2OH in the same relative configuration as the same three groups of D-glyceraldehyde, the sugar is said to have the D configuration. If this carbon has the configuration opposite to D-glyceraldehyde (that is, the same as L-glyceraldehyde), the sugar is said to have the L configuration.

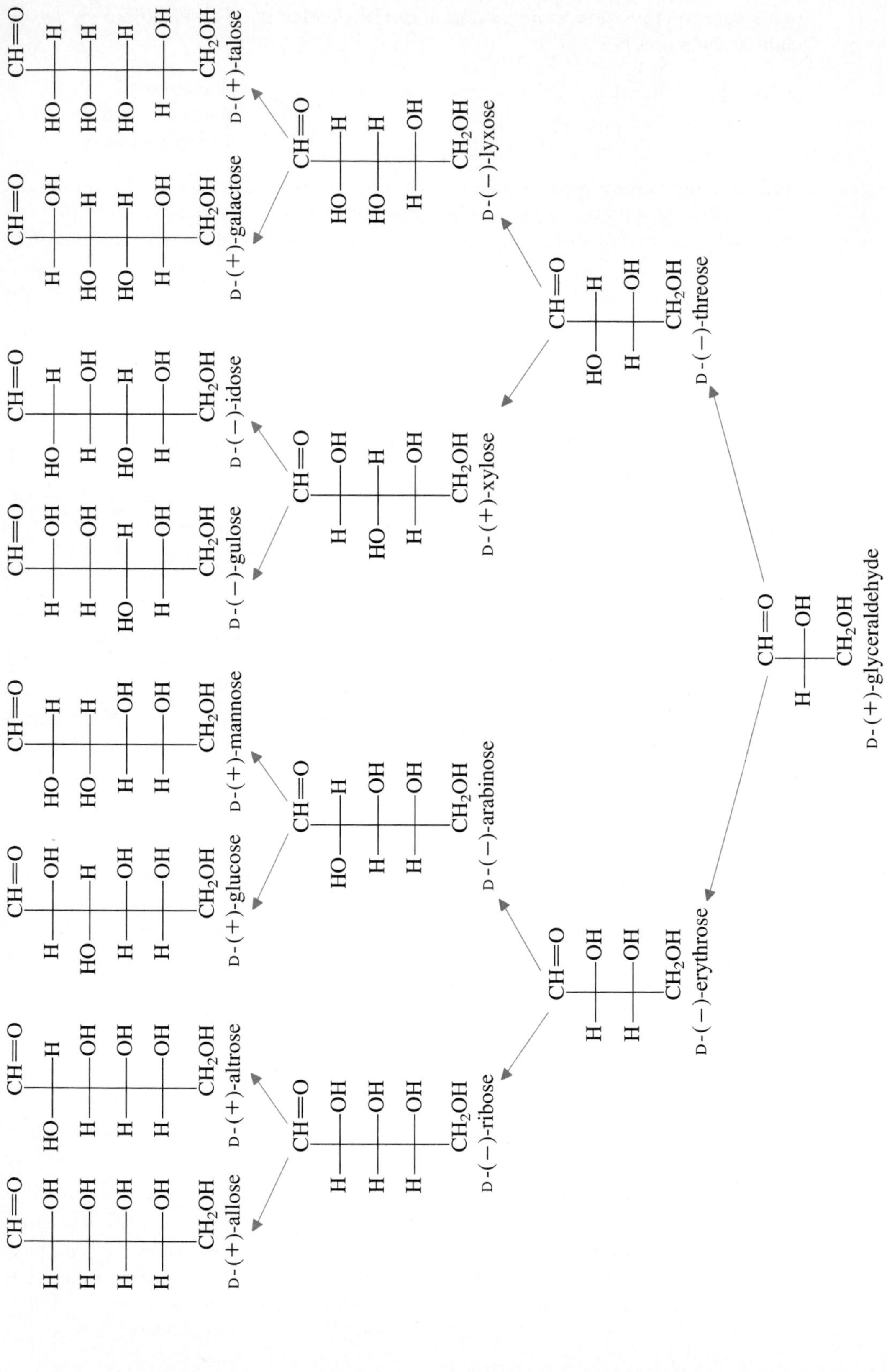

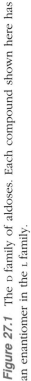

Figure 27.1 The D family of aldoses. Each compound shown here has an enantiomer in the L family.

Applying these rules to one of the glucose structures above, we have the following analysis:

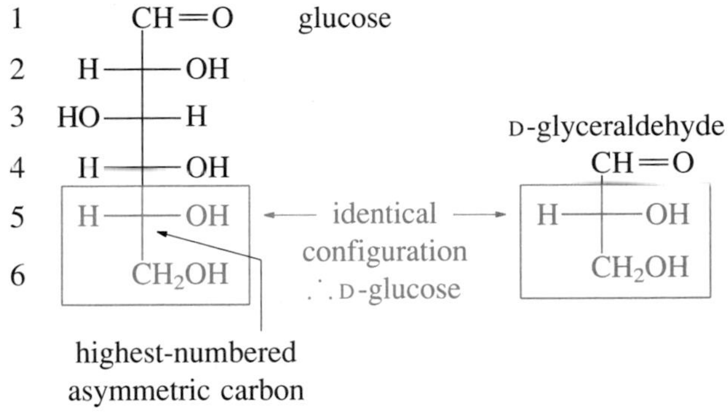

(27.1)

Carbon-5 is the asymmetric carbon with the highest number. Since it has the same configuration as D-glyceraldehyde, this aldohexose is D-glucose. Its enantiomer is L-glucose.

The monosaccharides shown in Fig. 27.1 are the D family of enantiomers. Each of the compounds in this figure has an enantiomer with the L configuration. Recall that *there is no general correspondence between configuration and the sign of the optical rotation* (see Sec. 6.3B). We see, for example, that some D sugars have positive rotations, but others have negative rotations. Also *there is no simple relationship between the D,L system and the R,S system.* The *R,S system is used to specify the configuration of each* asymmetric carbon atom in a molecule, but *the D,L system specifies a particular enantiomer of a molecule that might contain many asymmetric carbons.* Although use of the D,L system is straightforward for sugars and amino acids, it has been virtually abandoned for other compounds.

One of the annoyances of the D,L system is that each diastereomer is given a different name. This is one reason why the D,L system has been generally replaced with the *R,S* system, which can be used with systematic nomenclature. Nevertheless, the common names of many sugars are so well entrenched that they remain important.

A few of the aldoses in Fig. 27.1 are particularly important, and their structures should be learned. D-Glucose, D-mannose, and D-galactose are the most important aldohexoses because of their wide natural occurrence. The structures of the latter two sugars are easy to remember once the structure of glucose is learned, because their configurations are related to the configuration of glucose. D-Glucose and D-mannose differ in configuration only at carbon-2; D-glucose and D-galactose differ only at carbon-4. Compounds that differ in configuration at only one of several asymmetric carbons are called **epimers.** Hence, D-glucose and D-mannose are epimeric at carbon-2; D-glucose and D-galactose are epimeric at carbon-4.

D-Ribose is a particularly important aldopentose; its structure is easy to remember because all of its —OH groups are on the right in the standard Fischer projection. D-Fructose is an important naturally occurring ketose:

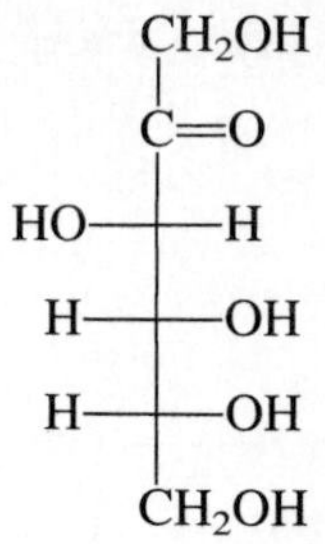

D-fructose

Carbons 3, 4, and 5 of D-fructose have the same stereochemical configuration as carbons 3, 4, and 5 of D-glucose.

Problems

1 By using the rules for manipulating Fischer projections (Sec. 6.9), tell which pair of the following sugars are epimers, and which pair are enantiomers:

(a)

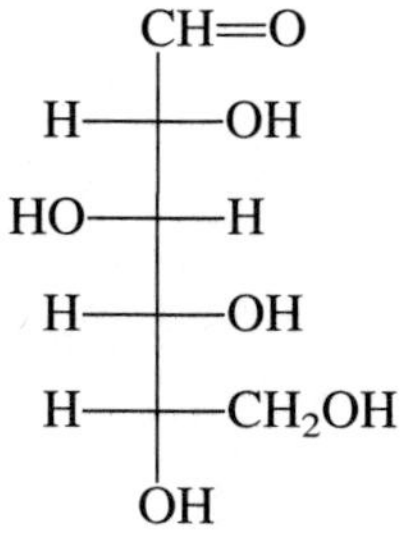

(b)

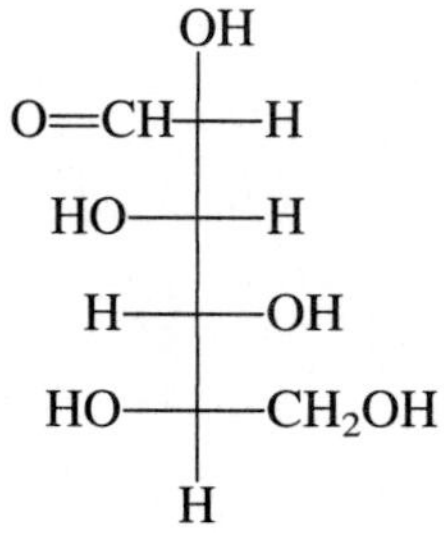

(c)

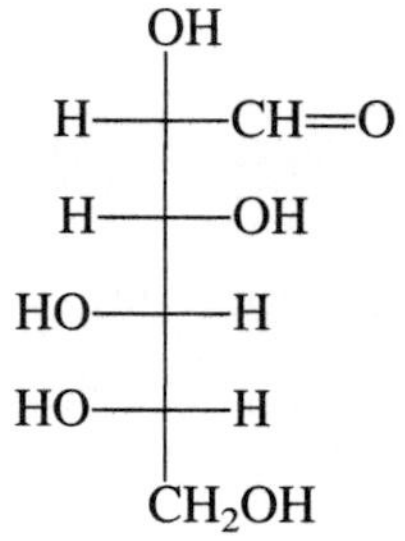

2 Classify each of the following sugars as D or L:

(a)

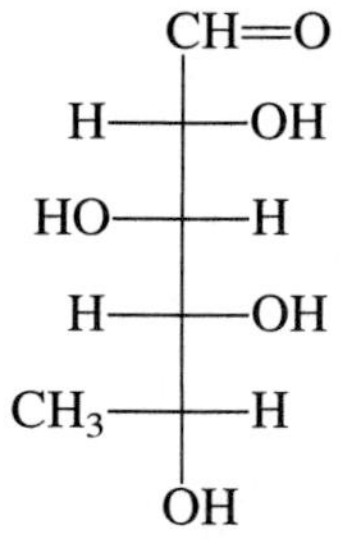

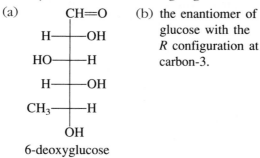

6-deoxyglucose

(b) the enantiomer of glucose with the *R* configuration at carbon-3.

B. Cyclic Structures of the Monosaccharides

In Sec. 19.10A we learned that γ- or δ-hydroxy aldehydes exist predominately as *cyclic hemiacetals:*

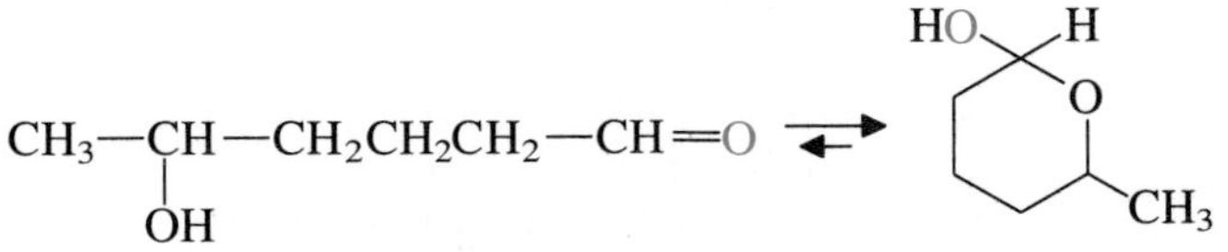

(27.2a)

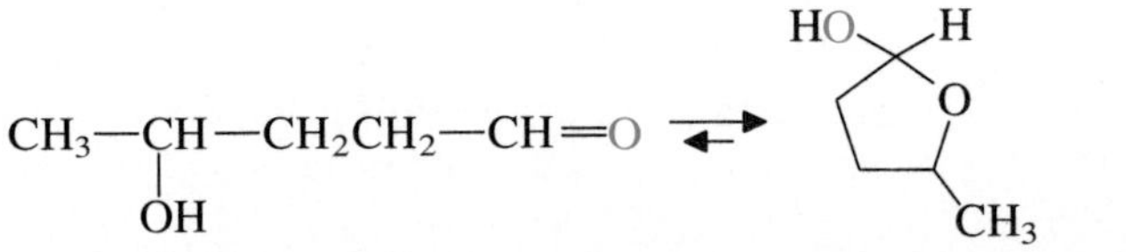

(27.2b)

The same is true of sugars. Although sugars are often written by convention as acyclic carbonyl compounds, they exist predominately as cyclic hemiacetals. An aldohexose, for example, can exist as either a five-membered or six-membered cyclic hemiacetal, depending on which hydroxy group undergoes cyclization.

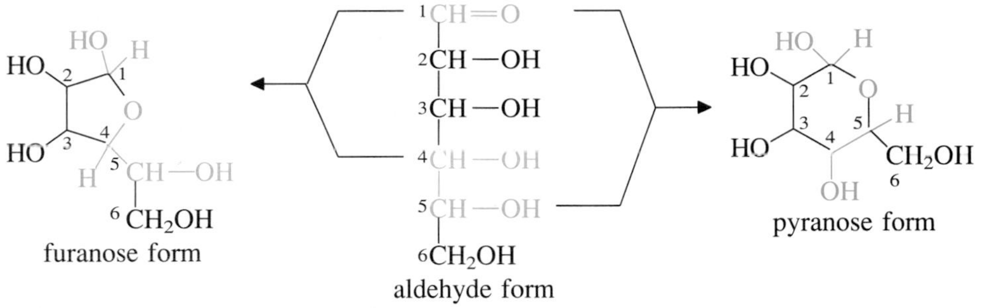

(27.3)

The five-membered cyclic acetal form of a sugar is called the **furanose** form (after furan, a five-membered oxygen heterocycle); the six-membered cyclic acetal form of a sugar is called the **pyranose** form (after pyran, a six-membered oxygen heterocycle).

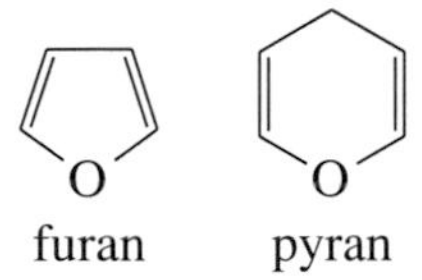

Glucose and the other aldohexoses exist predominately in the pyranose form, but the furanose forms of some sugars are important.

When we use a name such as *glucose,* we are referring to any or all of the many forms of the sugar. When we wish to refer specifically to a cyclic form, we use a different name that is derived from the size of the hemiacetal ring. For example, the six-membered cyclic hemiacetal form of D-glucose is called D-glucopyranose; the five-membered cyclic hemiacetal form is called D-glucofuranose.

Anomers It is important to notice that *the furanose or pyranose form of a sugar has one more asymmetric carbon than the open chain form*—carbon-1. Thus there are two possible diastereomers of D-glucopyranose:

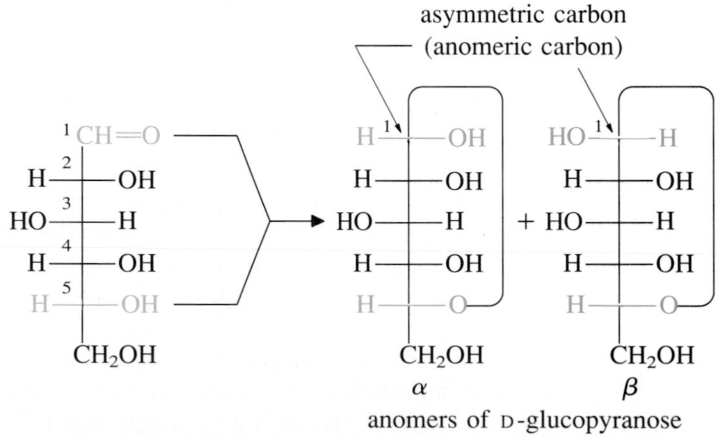

(27.4)

(The rings in the Fischer projections of these cyclic compounds are closed with a rather strange-looking long bond.) Both of these compounds are forms of D-glucopyranose, and in fact, glucose in solution exists as a mixture of both. They are diastereomers, and are therefore separable compounds with different properties. When two cyclic forms of sugars differ in configuration only at the hemiacetal carbon (carbon-1), they are said to be **anomers.** Thus, the two forms of D-glucopyranose are anomers of glucose.

As the structures above show, anomers are named with the Greek letters α and β. This nomenclature refers to the Fischer projection of the cyclic form of the sugar, written with all carbon atoms in a straight vertical line. *In the α-anomer the hemiacetal —OH group is on the same side of the Fischer projection as the oxygen at the configurational carbon.* (The configurational carbon is the one used for specifying the D,L designation—carbon-5 for the aldohexoses.) Conversely, in the β-anomer the hemiacetal —OH group is on the side of the Fischer projection opposite the oxygen at the configurational carbon. The application of these definitions to the nomenclature of the D-glucopyranose anomers is shown below.

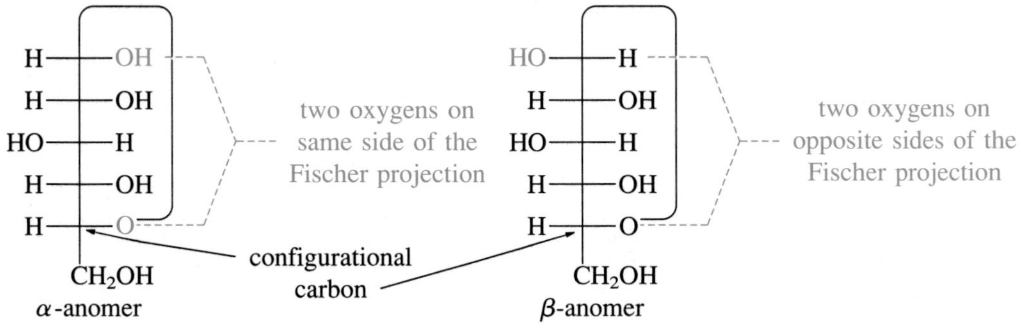

Conformational Representations of Pyranoses Fischer projections of sugars are convenient for specifying the *configuration* at each asymmetric carbon, but they contain no information about the *conformations* of the sugars. It is important to be able to relate Fischer projections to more realistic structural formulas for the sugars.

The six-membered ring of a pyranose, like that of cyclohexane, exists in two chair conformations related by the chair-chair interconversion (Sec. 7.2). How do we relate a Fischer projection to a chair conformation? The process is shown in Fig. 27.2 for α-D-glucopyranose. We first redraw the molecule in an equivalent Fischer projection in which the ring oxygen is in a down position. This is easily done using an allowed manipulation of Fischer projections, the cyclic permutation of groups on a terminal carbon (Sec. 6.9). The rest is simple: when we draw the chair conformation with the ring oxygen *in a right rear position* (as shown in Fig. 27.2), the groups on the *left* of the Fischer projection are up and those on the right are down. We can see from this procedure that the —OH group at the anomeric carbon is down in an α-anomer, and up in a β-anomer. (It is important that these relationships be verified with the use of models; see Fig. 6.18.)

Frequently we shall deal with sugars that either are mixtures of anomers or have uncertain anomeric compositions. In such cases, the configuration at the anomeric carbon is represented with a "squiggly bond."

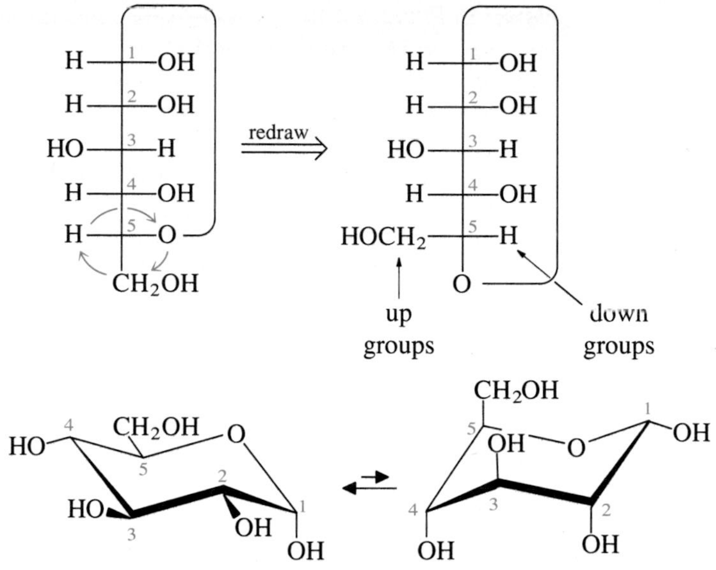

Figure 27.2 Conversion of a Fischer projection of α-D-glucopyranose into its chair conformations. Notice the position of the ring oxygen in the chair conformations.

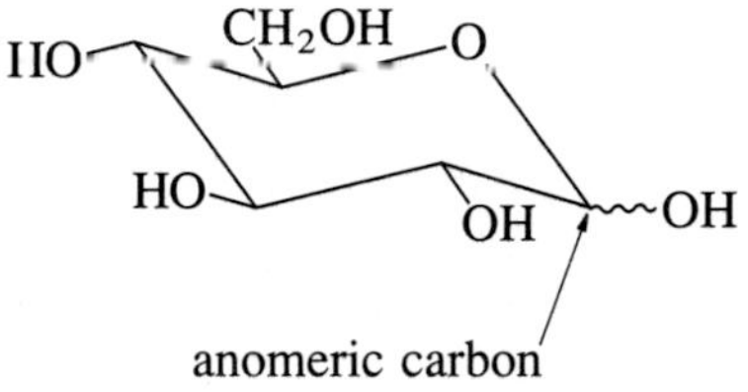

D-glucopyranose of mixed or uncertain anomeric composition

Although we can use the procedure shown in Fig. 27.2 for any sugar, it is also easy to derive the chair forms of D-mannopyranose and D-galactopyranose from their epimeric relationships to D-glucopyranose. For example, since galactose and glucose are epimeric at carbon-4, the chair form of D-galactopyranose can be drawn simply by interchanging the —H and —OH groups).

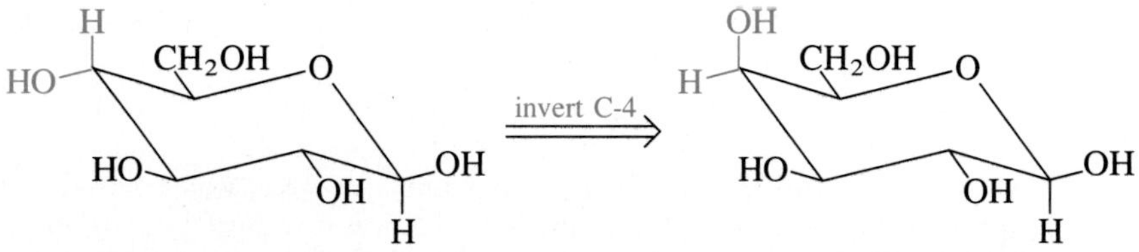

(27.5)

β-D-glucopyranose β-D-galactopyranose

Once we understand that the pyranose forms of sugars have the chair cyclohexane structure, we can ask which of the two chair forms is more stable. In the case of glucopyranose (Fig. 27.2) the answer is obvious. In one form most of the substituent groups are equatorial, and in the other most are axial. Since equatorial substitution avoids the unfavorable *gauche* and 1,3-diaxial interactions associated with axial substitution, the conformational isomer with more equatorial groups is strongly favored at equilibrium.

Haworth Projections A widely used convention for representing the cyclic forms of sugars is the **Haworth projection.** In this convention, the cyclic form of a sugar is represented as a *planar ring* at right angles to the page. This projection simply shows whether the groups on the ring are up or down without indicating a specific conformation. The Haworth projection is simple to draw from either of the corresponding chair forms, as we can see in the following representation of β-D-glucopyranose.

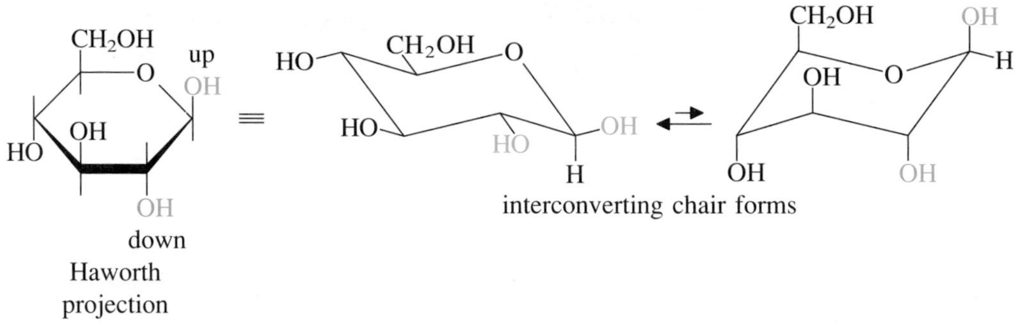

(27.6)

In the Haworth projection, the heavy, shaded lines of the ring are meant to be visualized in front of the page; the lighter lines are in back. The bonds with nothing on the ends bear hydrogen atoms. (Notice that this convention differs from the one used in ordinary skeletal formulas, in which such bonds represent methyl groups.)

Although cyclopentane rings are not planar, they are close enough to planarity that the Haworth projections for furanose sugars are good approximations to their actual structures. Haworth projections are frequently used to represent furanoses for this reason. The Haworth projections for furanoses can also be derived using the method shown in Fig. 27.2, except that a five-membered ring is formed instead. Thus, the Haworth projection for β-D-ribofuranose is derived as follows:

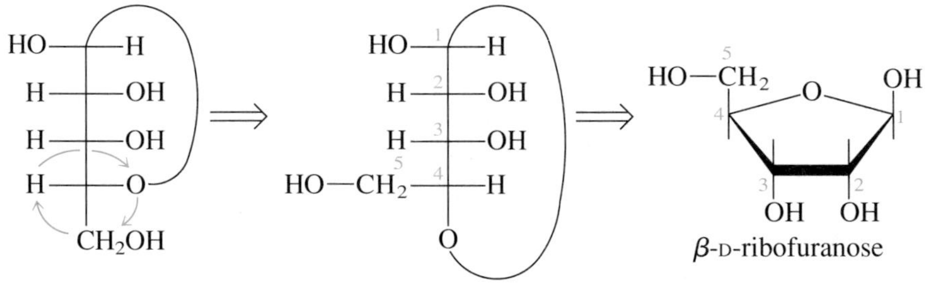

(27.7)

The Haworth formula is named after Sir Walter Norman Haworth (1883–1950), a noted British carbohydrate chemist who carried out important research on the cyclic structures of sugar derivatives. Haworth received the Nobel Prize in Chemistry in 1937 and was knighted in 1947.

Problems

3 Draw the Fischer projections, Haworth projections, and, for the pyranoses, the chair structures of each of the following compounds:

(a) β-D-mannopyranose (c) β-D-xylofuranose

(b) α-D-fructopyranose (d) α-L-glucopyranose

4 Name each of the following sugars. In some cases, you may be able to relate the sugar to a structure with which you are already familiar. In other cases, you may have to work back to the Fischer projection and consult Fig. 27.1.

(a)

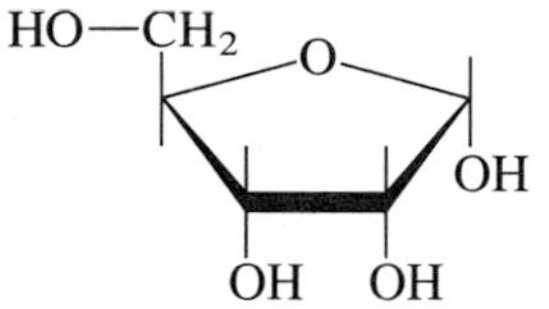

(c)

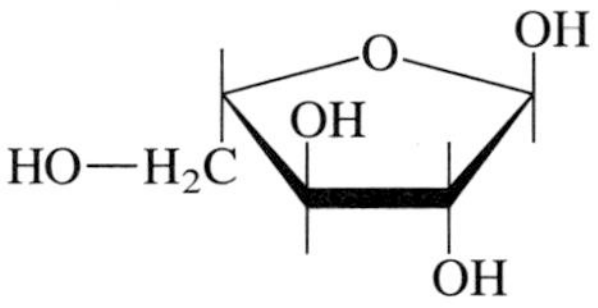

(b)

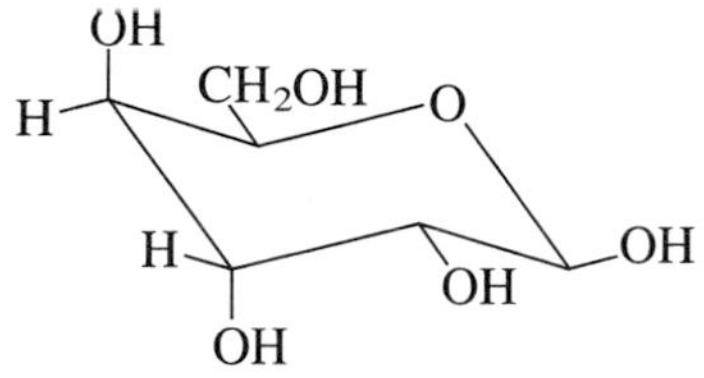

27.3 MUTAROTATION OF SUGARS

When pure α-D-glucopyranose is dissolved in water, its specific rotation is found to be $+112°$. With time, however, the rotation of the solution falls, ultimately reaching a stable value of $+52.7°$ (Fig. 27.3). When pure β-D-glucopyranose is dissolved in water, it has a specific rotation of $+18.7°$. The rotation of this solution increases with time, also to $+52.7°$. This change of optical rotation with time is called **mutarotation** (*muta* = change). A similar phenomenon occurs when pure anomers of other sugars are dissolved in aqueous solution.

The mutarotation of glucose is caused by the conversion of the α- and β-glucopyranose anomers into an equilibrium mixture of both. The same equilibrium

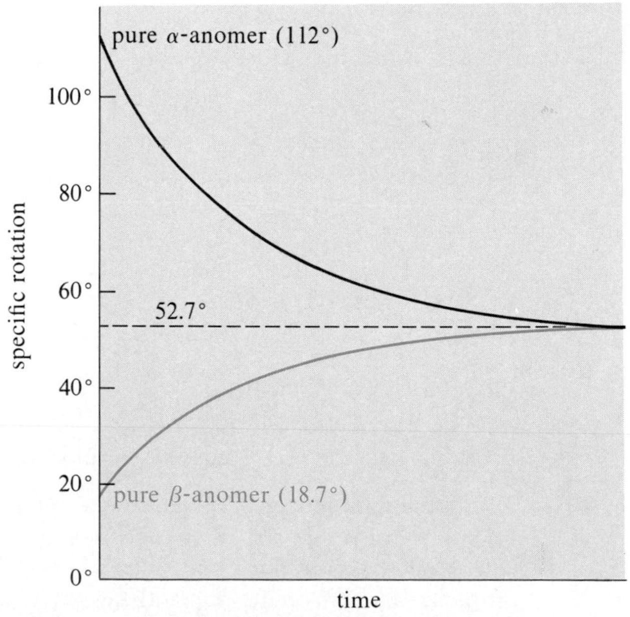

Figure 27.3 Mutarotation of D-glucose. Equimolar aqueous solutions of pure α- or β-glucopyranose gradually change their specific optical rotations to the same final value that is characteristic of the equilibrium mixture.

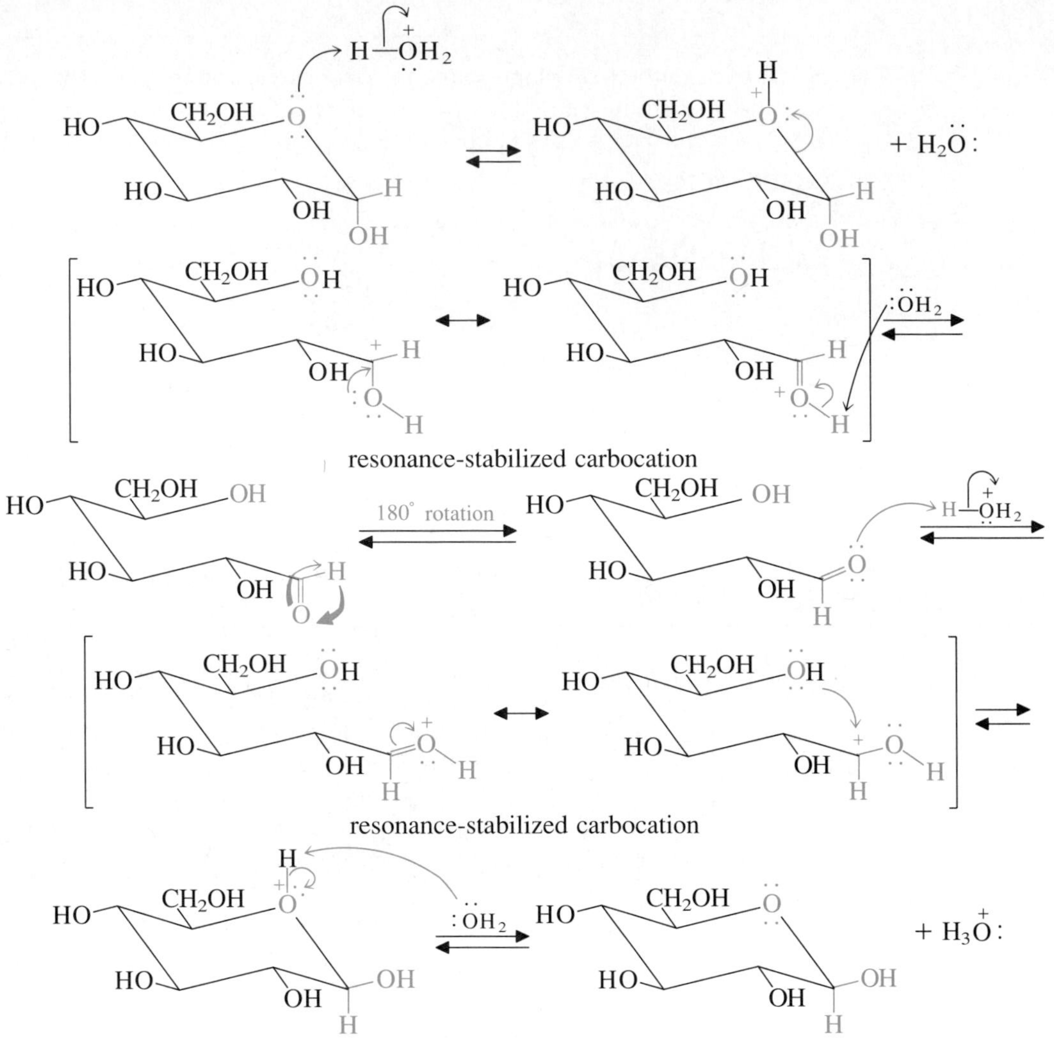

Figure 27.4 Mechanism of mutarotation for D-glucopyranose.

mixture is formed, as it must be, from either pure α-D-glucopyranose or β-D-glucopyranose. Mutarotation is catalyzed by both acid and base.

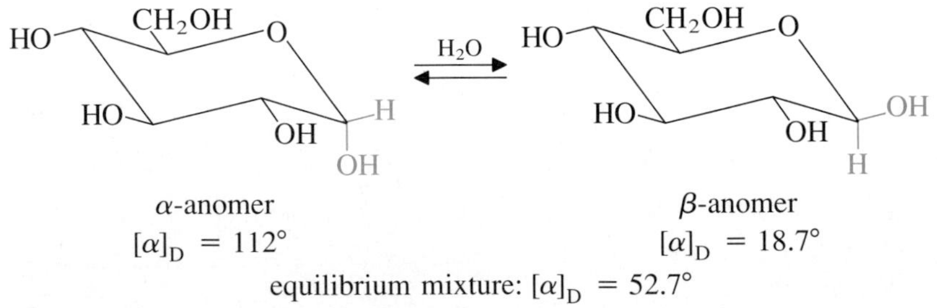

(27.8)

Mutarotation occurs by opening of the pyranose ring to the free aldehyde form. The mechanism, shown in Fig. 27.4, begins as the *reverse* of hemiacetal formation. A 180° rotation about the bond to the carbonyl group permits attack of the hydroxy group at carbon-5 on the opposite face of the carbonyl carbon. Hemiacetal formation then gives the other anomer.

The mutarotation of glucose is due almost entirely to the interconversion of its two pyranose forms. Other sugars undergo more complex mutarotations. For example, the crystalline form of D-fructose, a 2-ketohexose, is β-D-fructopyranose. When crystalline D-fructose is dissolved in aqueous solution, it equilibrates to both pyranose and furanose forms.

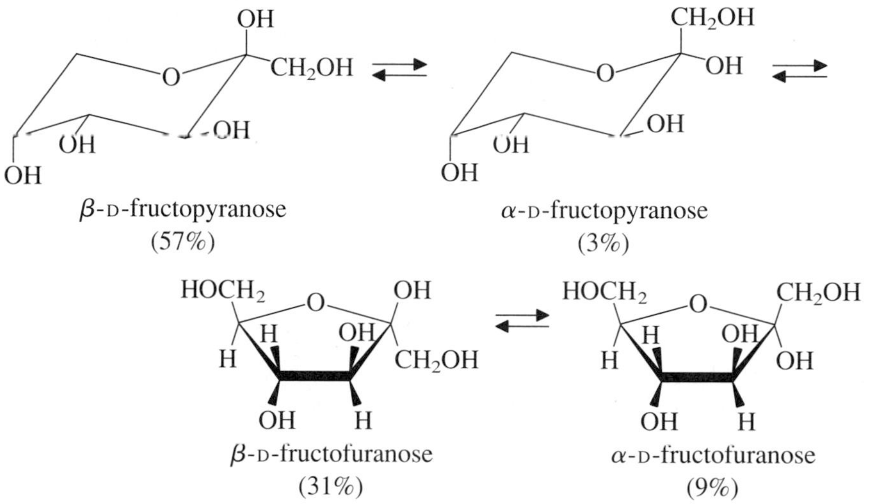

(27.9)

Glucose in solution undoubtedly also contains furanose forms, but these are present in amounts too small to be detected.

We have seen in the foregoing discussion that a single hexose can exist in no fewer than five forms: the acyclic aldehyde or keto form, the α- and β-pyranose forms, and the α- and β-furanose forms. Mutarotation allows these forms to come to equilibrium in aqueous solution. How do we know which forms are preferred by a given sugar? The answer is that, until relatively recent times, we didn't know! However, several modern techniques, particularly NMR spectroscopy, have enabled chemists to determine which of the various forms are preferred by sugars in aqueous solution. The results for some sugars are summarized in Table 27.1.

TABLE 27.1 Conformations of Sugars at Equilibrium in Aqueous Solution at 40°

	Percent at equilibrium				
	pyranose		furanose		aldehyde
Sugar	α	β	α	β	or ketone
D-glucose	36	64	trace		0.003
D-galactose	27–36	64–73	trace		trace
D-mannose	68	32	0.7	0	trace
D-allose	18	70	5	7	
D-altrose	27	40	20	13	
D-idose (24°)	38	38	10	14	
D-talose	40	29	20	11	11
D-arabinose	63	34	3		
D-xylose	37	63			
D-ribose	20	56	6	18	0.02
D-fructose	0–3	57–75	4–9	21–31	0.25

Some general conclusions from this table are:

1. Most aldohexoses exist primarily as pyranoses, although a few have substantial amounts of furanose forms.
2. There are relatively small amounts of open-chain carbonyl forms of most sugars.
3. Mixtures of α- and β-anomers are usually found, although the exact amounts of each vary from case to case.

The fraction of any form in solution at equilibrium is determined by its stability relative to that of all other forms. In order to predict the data in Table 27.1 for a given sugar, we would have to understand all the factors that contribute to the stability or instability of *every* one of its isomeric forms, and an analysis of these factors is beyond the scope of our discussion. In some cases, though, the principles of cyclohexane conformational analysis (Sec. 7.3, 7.4) can help us to understand the relative stabilities of pyranosides, as Problem 7 illustrates.

Problems

5 Write a mechanism for the *base-catalyzed* mutarotation of glucose, using OH^- as the catalyzing base.

6 From the specific rotations shown in Fig. 27.3, calculate the amount of α- and β-D-glucopyranose present at equilibrium. (Assume the amounts of aldehyde and furanose forms are negligible.) Compare your answer with the data in Table 27.1.

7 Consider the β-D-pyranose forms of (a) glucose and (b) talose. Suggest one reason why talose contains a smaller fraction of β-pyranose form than glucose.

27.4 BASE-CATALYZED ISOMERIZATION OF SUGARS

In base, aldoses and ketoses rapidly equilibrate to mixtures of sugars.

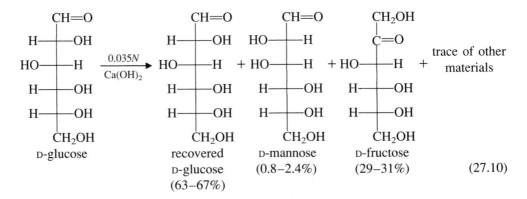

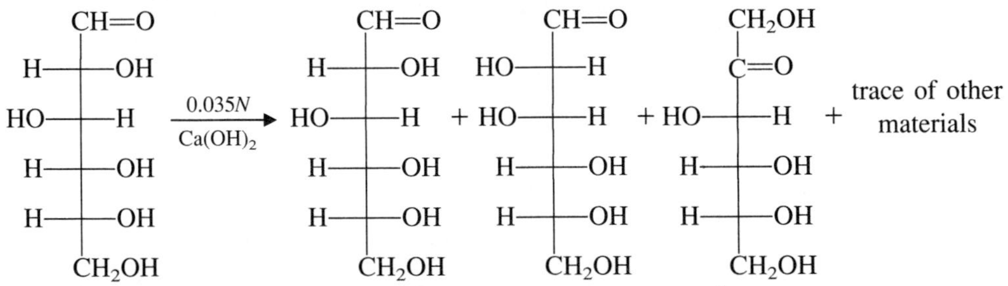

D-glucose → recovered D-glucose (63–67%) + D-mannose (0.8–2.4%) + D-fructose (29–31%) + trace of other materials (27.10)

This transformation is an example of the **Lobry de Bruyn–Alberda van Ekenstein reaction,** named for two Dutch chemists, Cornelius Adriaan van Troostenbery Lobry de Bruyn (1857–1904) and Willem Alberda van Ekenstein (1858–1907). Despite its rather formidable name, this reaction is a simple one, and is closely related to processes we have already studied.

Although glucose in solution exists mostly in its cyclic hemiacetal forms, it is also in equilibrium with its acyclic aldehyde form. This aldehyde, like other carbonyl compounds with α-hydrogens, ionizes to give small amounts of its enolate ion in base. Protonation of this enolate ion at one side gives back glucose; protonation at the other side gives mannose. This is much like the process shown in Eq. 22.7.

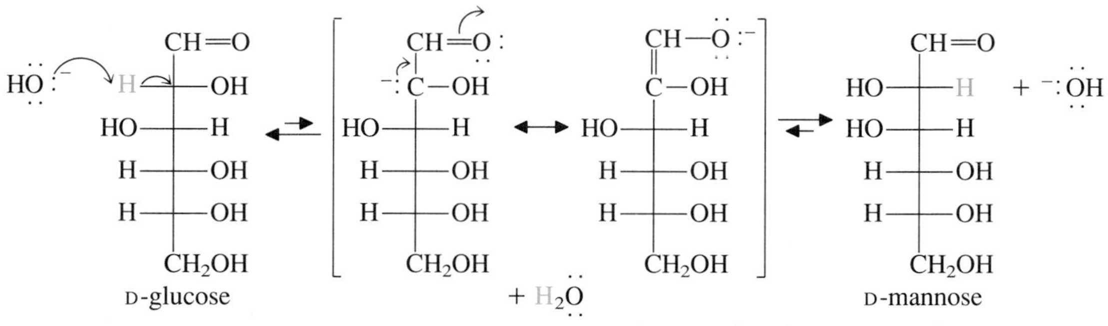

(27.11)

The enolate ion can also protonate on oxygen to give a new enol, called an **enediol.** Because the enediol has a hydroxy group at *both* ends of the double bond, it is the enol of not only the two aldoses glucose and mannose, but also fructose, a ketose.

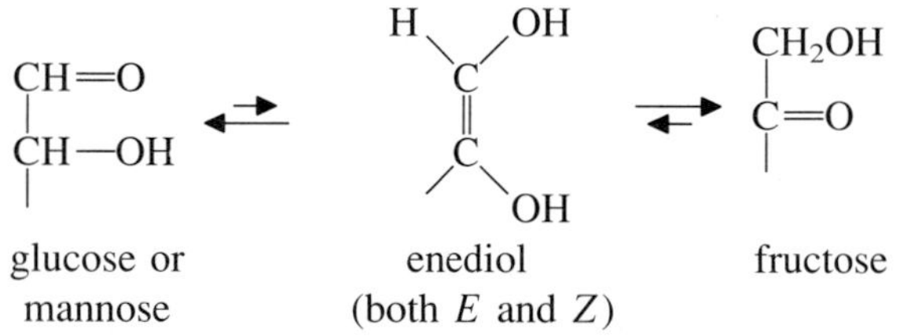

(27.12)

Such base-catalyzed epimerizations and aldose-ketose equilibria need not stop at the 2-carbon. For example, fructose epimerizes on prolonged treatment with base (why?).

Several transformations of this type are important in metabolism. One such reaction, the conversion of glucose-6-phosphate into fructose-6-phosphate, occurs in the breakdown of glucose (glycolysis), the series of reactions by which glucose is utilized as a food source. Since biochemical reactions occur near pH 7, there is too little hydroxide ion present to catalyze the reaction. Instead, the reaction is catalyzed by an enzyme, glucose-6-phosphate isomerase.

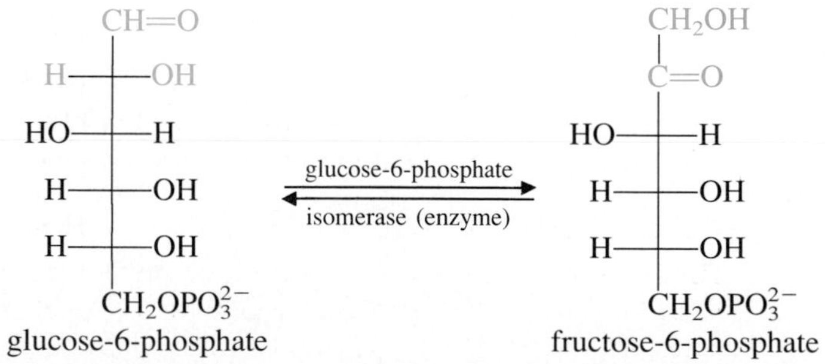

(27.13)

Problem

8 Into what aldose and 2-ketose would D-galactose first be transformed on treatment with base? Name the aldose (see Fig. 27.1) and give the structure of the ketose.

27.5 GLYCOSIDES

Most sugars react with alcohols under acidic conditions to yield cyclic acetals.

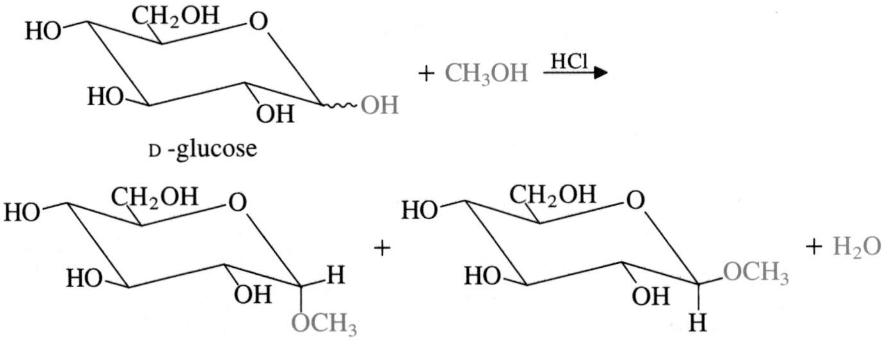

(27.14)

methyl α-D-glucopyranoside   methyl β-D-glucopyranoside
(83–85% yield; separated by fractional crystallization)

Such compounds are called **glycosides.** They are special types of acetals in which one of the oxygens of the acetal linkage is the ring oxygen of the pyranose or furanose.

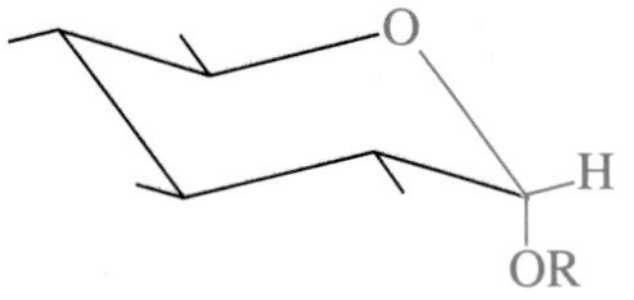

As illustrated in Eq. 27.14, glycosides are named as derivatives of the parent sugar. The term *pyranoside* indicates that the glycoside ring is a six-membered ring. The term *furanoside* is used for a five-membered ring.

Glycoside formation, like acetal formation, is catalyzed by acid and involves carbocation intermediates.

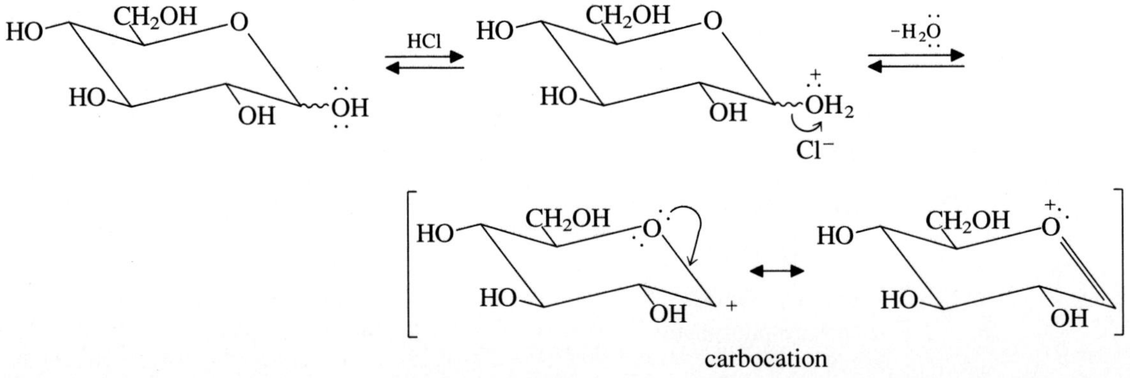

(27.15a)

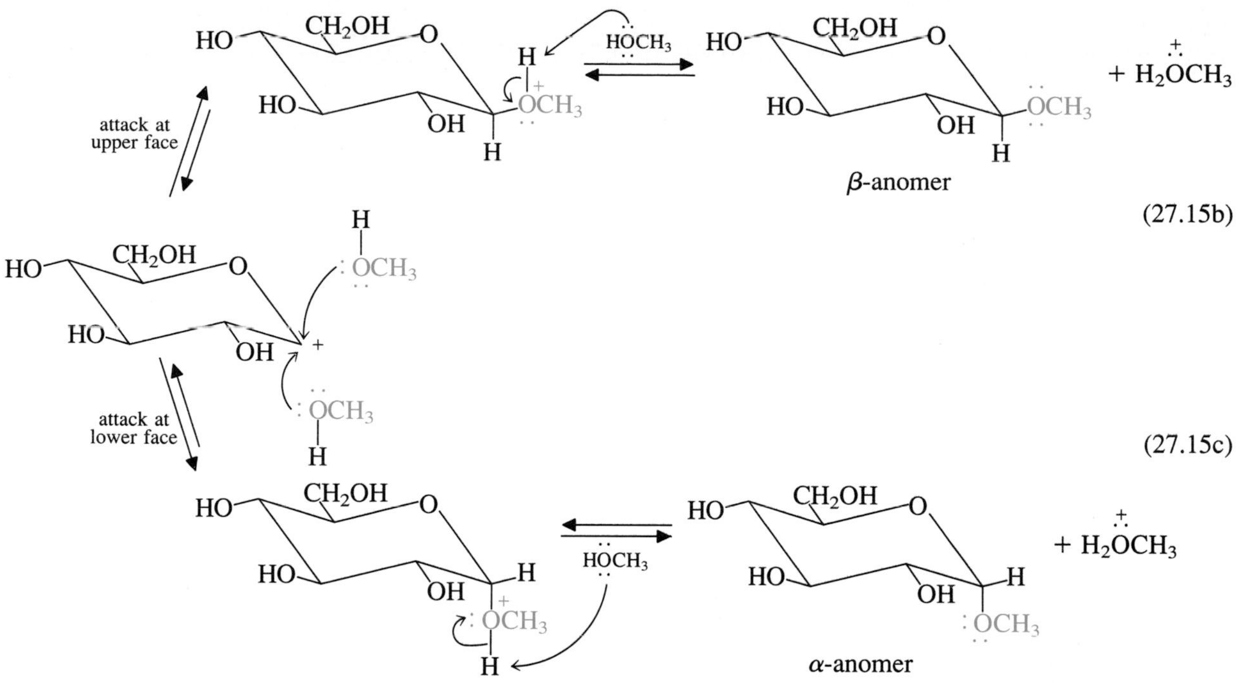

(27.15b)

(27.15c)

Although glycoside formation involves protonation of the —OH oxygen at carbon-1, and mutarotation (Fig. 27.4) involves protonation of the ring oxygen, it is important to understand that protonation of *any one* of the sugar oxygens can take place in acid. The different protonated forms are simply in rapid equilibrium, and are all present whether they are shown explicitly or not.

It is useful to contrast what happens to a sugar and what happens to an ordinary aldehyde or ketone in the presence of alcohol and acid. When an aldehyde reacts with methanol to form an acetal (Sec. 19.10A), two additional carbon atoms are incorporated into the molecule. When glucose forms a methyl glycoside, only one additional carbon atom is incorporated; the other part of the acetal group is derived from the sugar itself.

$$\mathrm{R{-}CH{=}O} + 2\,\mathrm{CH_3OH} \overset{\mathrm{H^+}}{\rightleftharpoons} \underset{\text{acetal}}{\mathrm{R{-}CH(OCH_3)_2}} + \mathrm{H_2O} \qquad (27.16a)$$

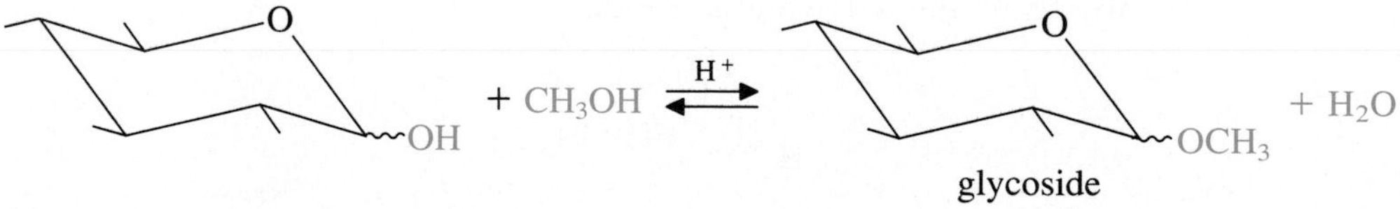

(27.16b)

This difference between aldoses and ordinary aldehydes is one piece of evidence that led early chemists to suspect the cyclic (pyranose or furanose) structure of sugars.

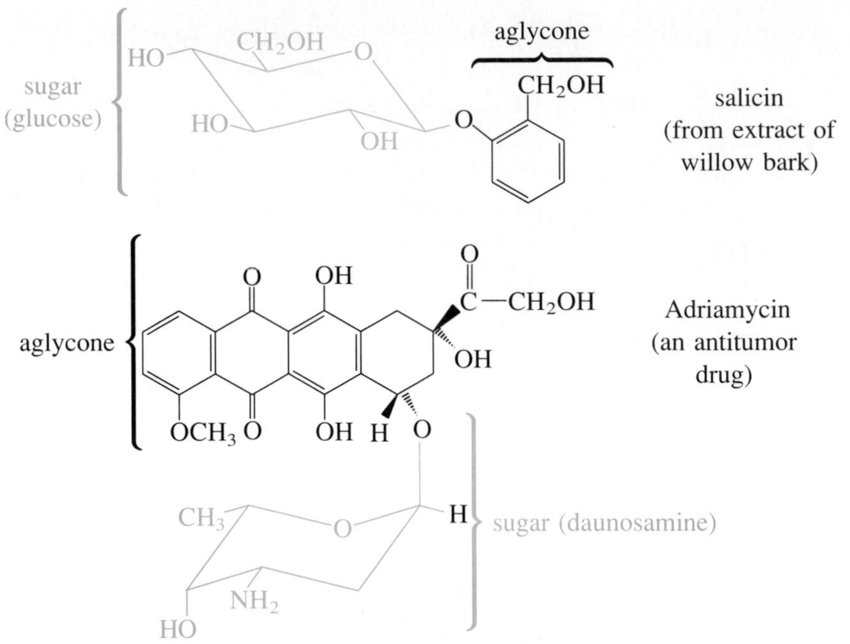

Figure 27.5 Two naturally occurring glycosides of medicinal interest.

Like other acetals, glycosides are *stable to base,* but are hydrolyzed in dilute aqueous acid back to their parent sugars.

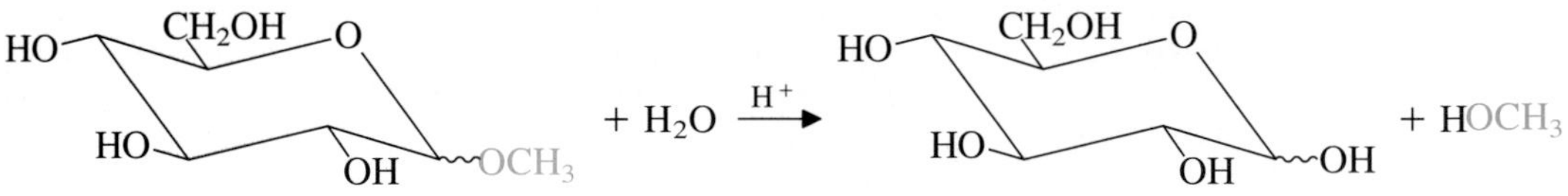

(27.17)

Many compounds occur naturally as glycosides; two examples are shown in Fig. 27.5.

Like simple methyl glycosides, the glycoside of a natural product can be hydrolyzed to its component alcohol, called an **aglycone,** and a sugar.

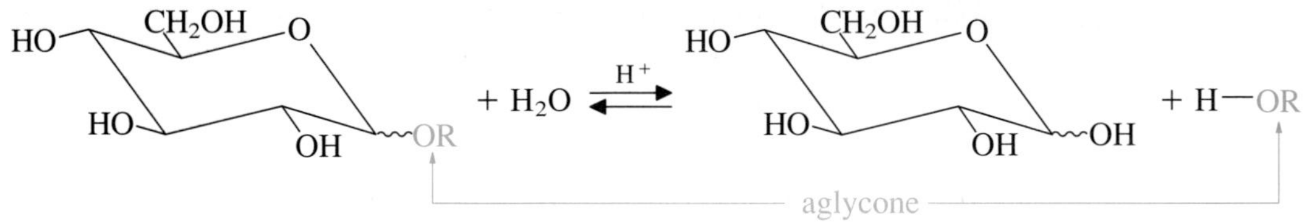

(27.18)

Problems

9 (a) Draw the structure of methyl β-D-fructofuranoside.
(b) Name the following glycoside:

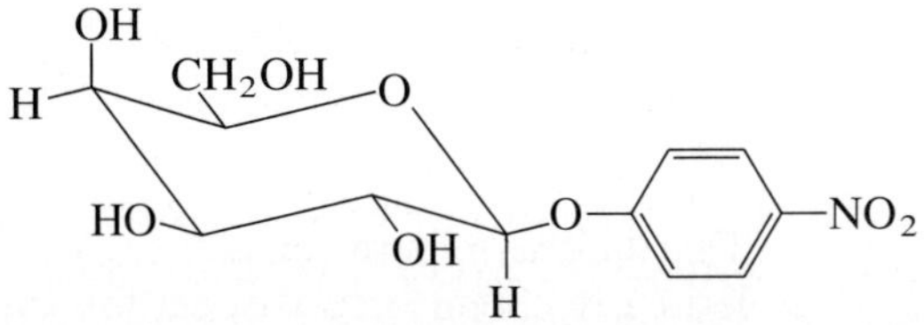

(c) Into what products will this glycoside be hydrolyzed in aqueous acid?

10 (a) Draw the structure of adriamycinone, the aglycone of adriamycin. (See Fig. 27.5.)
(b) Vanillin (the natural vanilla flavoring) is found in nature as a *β*-glycoside of glucose. Suggest a structure for this glycoside.

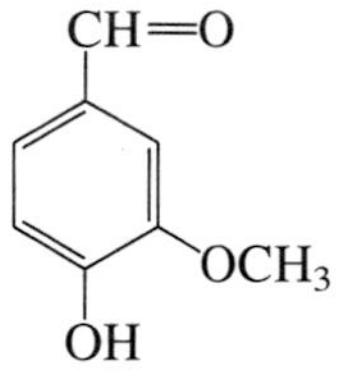

vanillin

27.6 ETHER AND ESTER DERIVATIVES OF SUGARS

Since sugars contain many —OH groups, it should not be surprising that sugars undergo many of the reactions of alcohols. One such reaction is ether formation. In the presence of concentrated base, sugars are converted into ethers by reactive alkylating agents such as dimethyl sulfate, methyl iodide, or benzyl chloride.

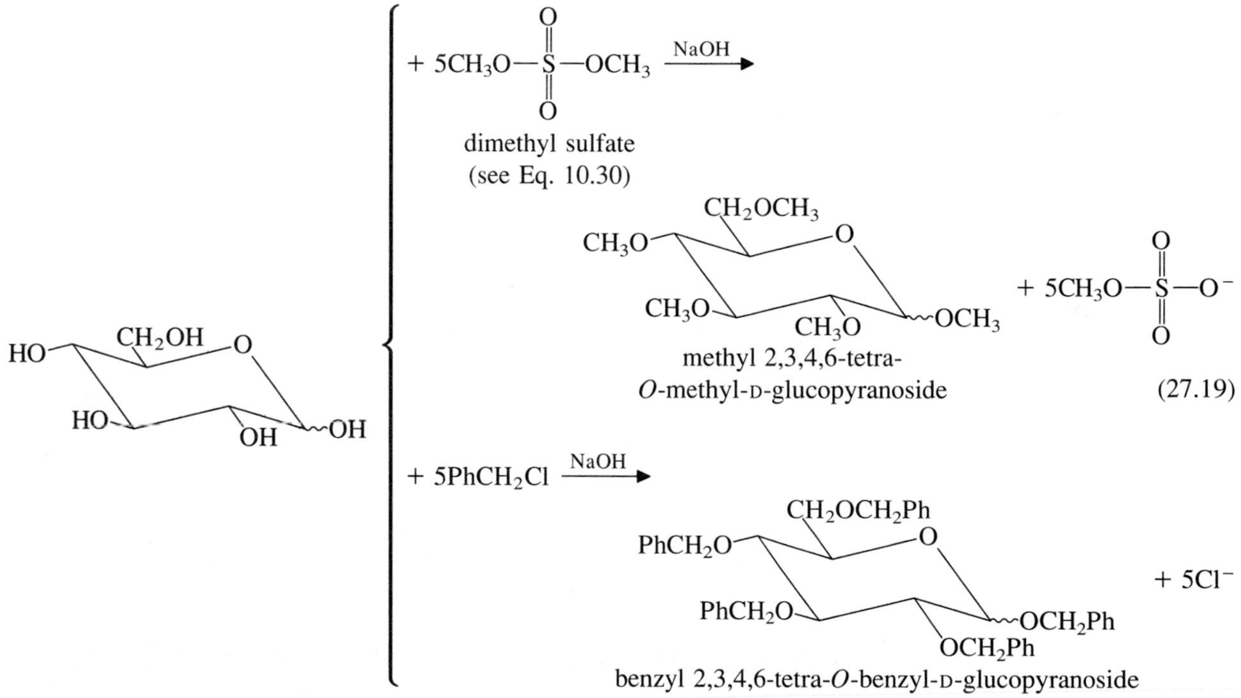

(27.19)

(Note that the ethers are named as *O*-alkyl derivatives of the sugars.) These reactions are examples of the Williamson ether synthesis (Sec. 11.1A). This synthesis with most alcohols requires a base stronger than ^{-}OH to form a conjugate-base alkoxide. The hydroxy groups of sugars, however, are more acidic ($pK_a \approx 12$) than those of ordinary alcohols. (The higher acidity of sugar hydroxy groups is attributable to the inductive effect of the many neighboring oxygens in the molecule.) Consequently, substantial concentrations of their conjugate-base alkoxide ions are formed in concentrated NaOH. A large excess of the alkylating reagent is used because hydroxide itself, present in large excess, also reacts with alkylating agents. It is interesting that little or no base-catalyzed epimerization (Sec. 27.4) is observed in this reaction, despite the strongly basic conditions used. Evidently, alkylation of the hydroxy group at carbon-1 is much faster than epimerization. Once this oxygen is alkylated, epimerization can no longer occur (why?).

Other reagents used to form methyl ethers of sugars include CH_3I/Ag_2O at high temperature, and the strongly basic $NaNH_2$ (sodium amide) in liquid NH_3 followed by CH_3I.

It is important to distinguish the ether at carbon-1 from the other ether groups in the alkylated sugar. The ether at carbon-1 is part of the glycosidic linkage. Since it is an acetal, it can be hydrolyzed in aqueous acid under mild conditions:

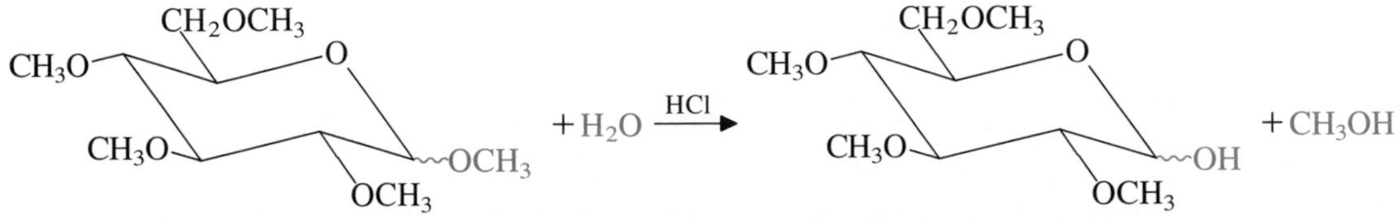

(27.20)

The other ethers are ordinary ethers and do not hydrolyze under these conditions. They require *much* stronger conditions for cleavage (Sec. 11.3A).

Problem

11 Why do acetals hydrolyze more readily than ordinary ethers? (*Hint:* Look at the carbocation intermediate.)

Another reaction of alcohols is esterification; indeed, the hydroxy groups of sugars, like those of other alcohols, can be esterified.

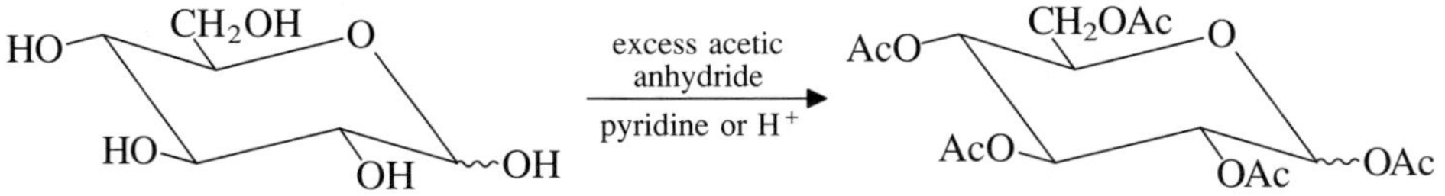

(27.21)

1,2,3,4,6-penta-*O*-acetyl-D-glucopyranose
(83% yield)

These esters can be saponified in base or removed by transesterification with an alkoxide such as methoxide:

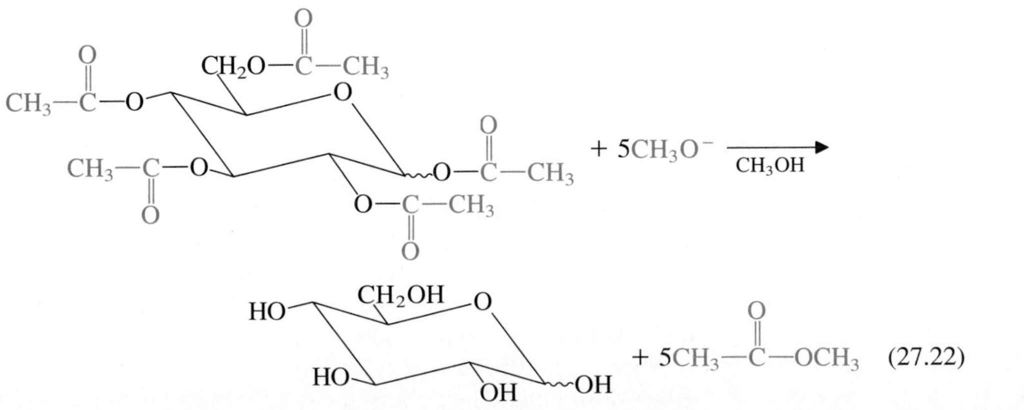

(27.22)

Ethers and esters are used as protecting groups in organic synthesis with sugars. Because ethers and esters of sugars have broader solubility characteristics and greater volatility than sugars themselves, they also find use in the characterization of sugars by chromatography and mass spectrometry.

Problem

12 Outline a sequence of reactions that will effect the conversion of D-galactopyranose to 1-*O*-ethyl-2,3,4,6-tetra-*O*-methyl-D-galactopyranoside.

27.7 OXIDATION AND REDUCTION REACTIONS OF SUGARS

A. Oxidation to Aldonic Acids

Treatment of an aldose with bromine water oxidizes the aldehyde group to a carboxylic acid. The oxidation product is known as an **aldonic acid.**

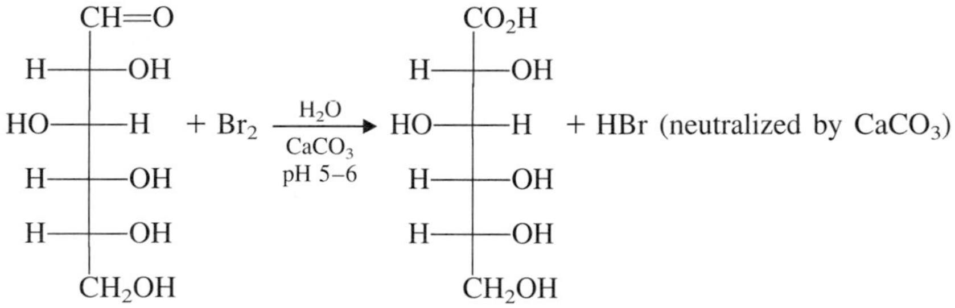

(27.23)

D-glucose

D-gluconic acid (an aldonic acid; 77–96% yield as Ca^{2+} salt)

Although it is customary to represent aldonic acids in the free carboxylic acid form, they, like other γ- and δ-hydroxy acids (Sec. 21.6A), exist in acidic solution as lactones called **aldonolactones.** The lactones with five-membered rings are somewhat more stable than those with six-membered rings.

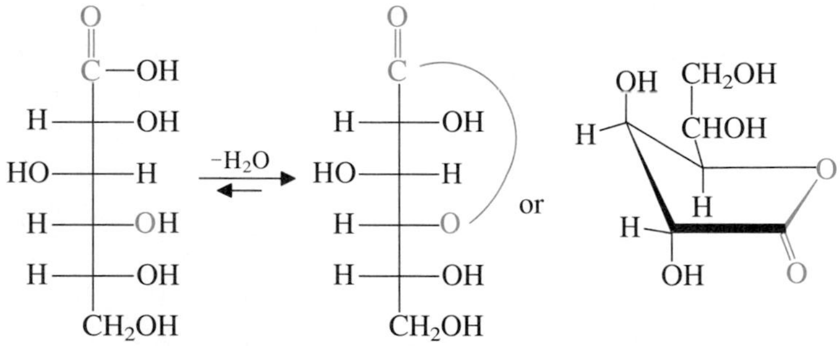

(27.24)

D-γ-gluconolactone (an aldonolactone)

Oxidation with bromine water is a useful test for aldoses. Aldoses can also be oxidized with other reagents—for example, Tollens' reagent ($Ag^+(NH_3)_2$; Sec. 19.14). However, because Tollens' reagent is alkaline and causes base-catalyzed epimerization of the aldose (Sec. 27.4), it is less useful synthetically. The alkaline conditions of Tollens' test also promote the equilibration of ketoses and aldoses; thus ketoses also give positive Tollens' tests. Glycosides are *not* oxidized by bromine water, because the aldehyde carbonyl group is protected as an acetal.

B. Oxidation to Aldaric Acids

We learned in Sec. 10.6C that dilute nitric acid selectively oxidizes primary alcohols to carboxylic acids. Both ends of an aldose—the primary alcohol and the aldehyde group—are oxidized to carboxylic acid groups by dilute HNO_3, but the secondary alcohol groups are not affected. The oxidation product is an **aldaric acid.** (More vigorous HNO_3 oxidation breaks carbon–carbon bonds.)

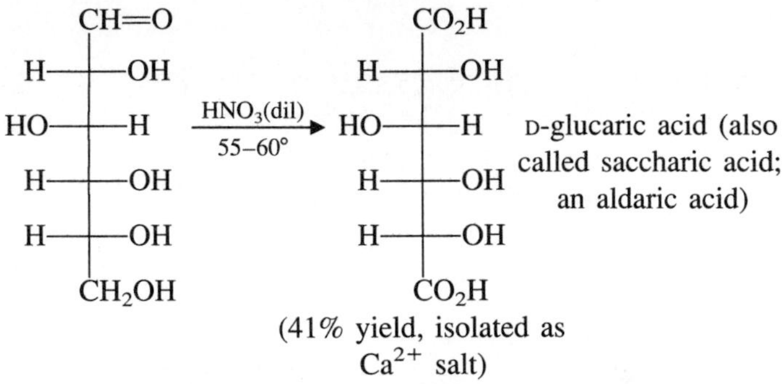

(27.25)

Like aldonic acids, aldaric acids in acidic solution form lactones. Two different five-membered lactones are possible, depending on which carboxylic acid group undergoes lactonization. Furthermore, under certain conditions, some aldaric acids can be isolated as dilactones, in which both carboxylic acid groups are lactonized.

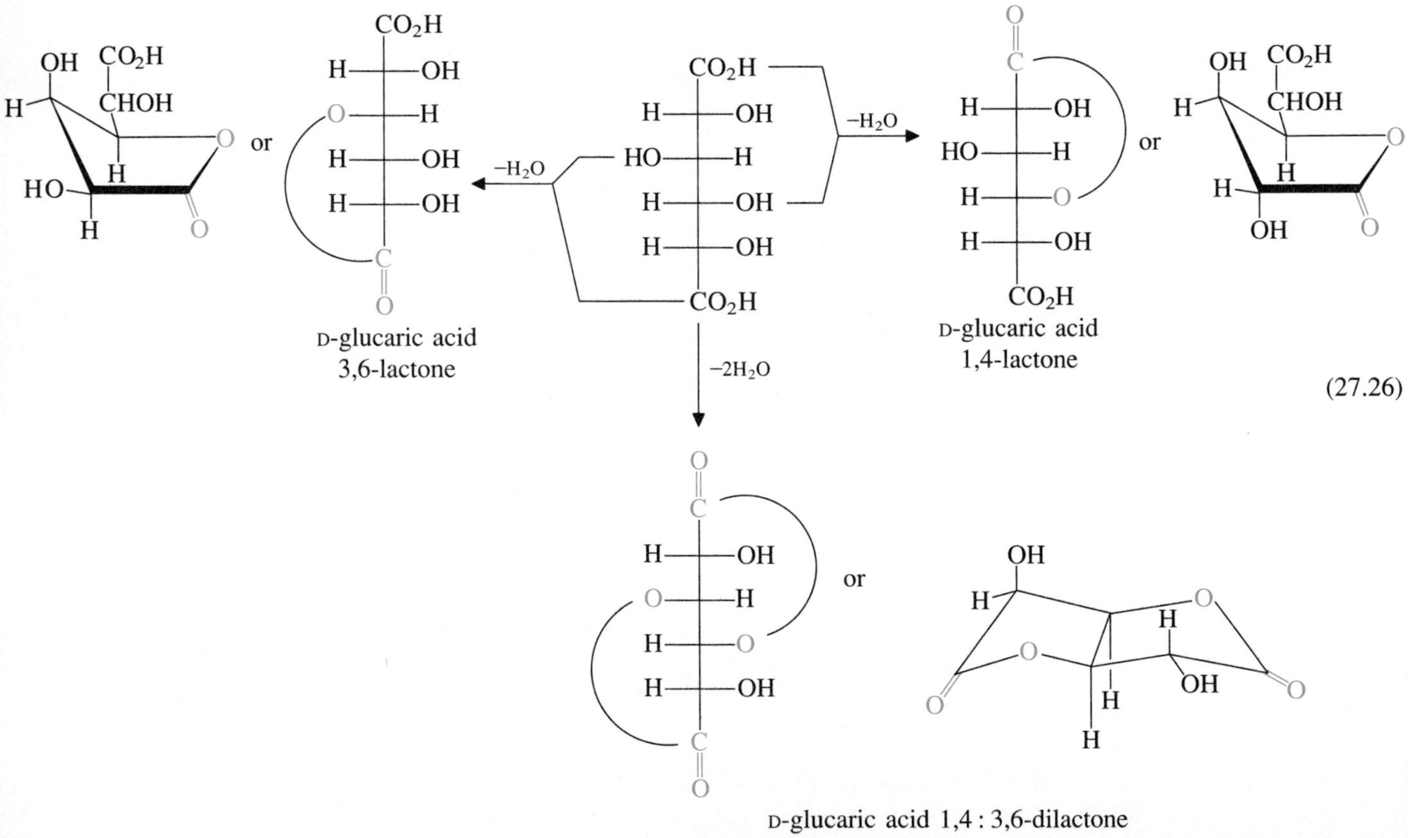

(27.26)

Problem

13 Give the structure of another aldose (it may not have the D configuration!) that would give the *same* aldaric acid as D-glucose when oxidized with dilute HNO_3.

Problem 13 illustrates a subtle yet important aspect of the D,L system of configuration. Some compounds—for example, aldaric acids—that have identical end groups can be derived from either a D or an L sugar.

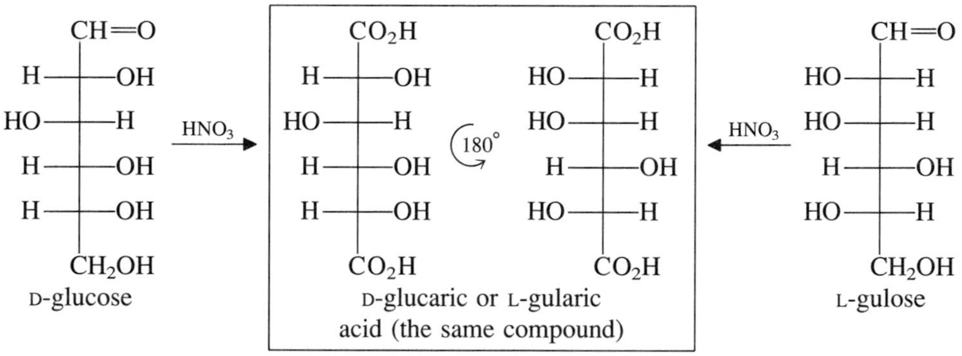

D-glucose — D-glucaric or L-gularic acid (the same compound) — L-gulose (27.27)

In this case, whether we classify the aldaric acid as D or L depends on which carbon we call carbon-1. Since the two end groups of the molecule are equivalent, the choice is completely arbitrary! The names D-glucaric acid and L-gularic acid are both correct names for this aldaric acid.

This situation arises because the —OH groups on the endmost asymmetric carbons are on the same side of the Fischer projection. When these —OH groups are on opposite sides, the configuration is unambiguous. For example, D-tartaric acid also has functionally identical ends, but it has the D configuration no matter how it is turned:

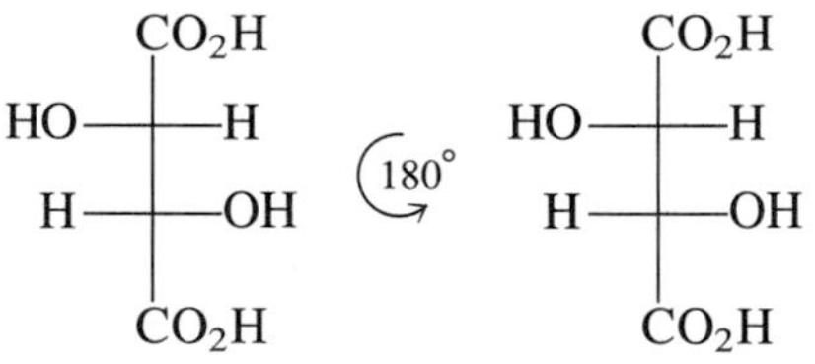

D-tartaric acid (27.28)

Problem

14 Consider the following aldaric acids:

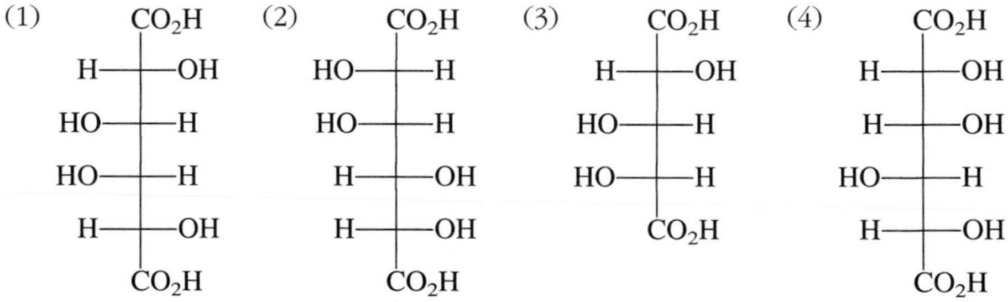

(a) Which can be classified as both D and L? Which are only D or only L? Which are *meso?*

Problem (Cont.)

(b) Which of the sugars above, if any, fulfill each of the following criteria? Explain.
(i) Can be obtained by oxidation of enantiomeric aldoses.
(ii) Can be obtained by oxidation of either of two diastereomeric aldoses with opposite (D and L) configurations.
(iii) Can be obtained by oxidation of either of two diastereomeric aldoses with the same (D or L) configuration.
(iv) Can be obtained by oxidation of a single aldose only.

C. Periodate Oxidation

Many sugars contain vicinal glycol units and, like other 1,2-glycols, are oxidized by periodate (Sec. 10.6D). A complication arises when, as in many sugars, more than two adjacent carbons bear hydroxy groups. When one of the oxidation products is an α-hydroxy aldehyde, as in the following example, it is oxidized further to formic acid and another aldehyde.

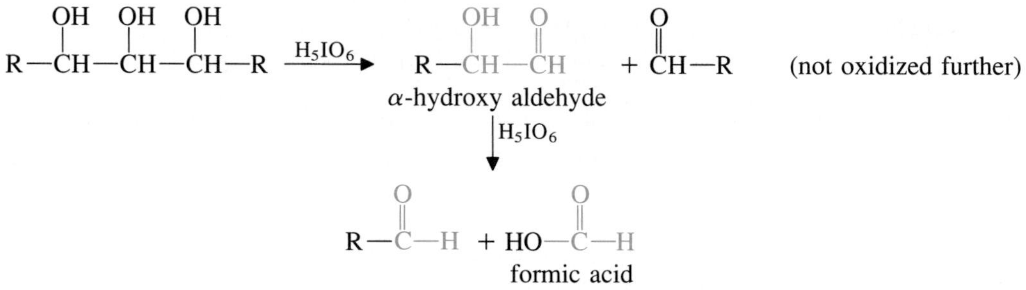

(27.29)

By analogy, an α-hydroxy ketone is oxidized to an aldehyde and a carboxylic acid.

Because it is possible to determine accurately both the amount of periodate consumed and the amount of formic acid produced, periodate oxidation can be used to differentiate between pyranose and furanose structures for saccharide derivatives. For example, the fact that periodate oxidation of methyl α-D-glucopyranoside liberates one equivalent of formic acid shows that it must be a pyranose (Eq. 27.30a) rather than a furanose (Eq. 27.30b).

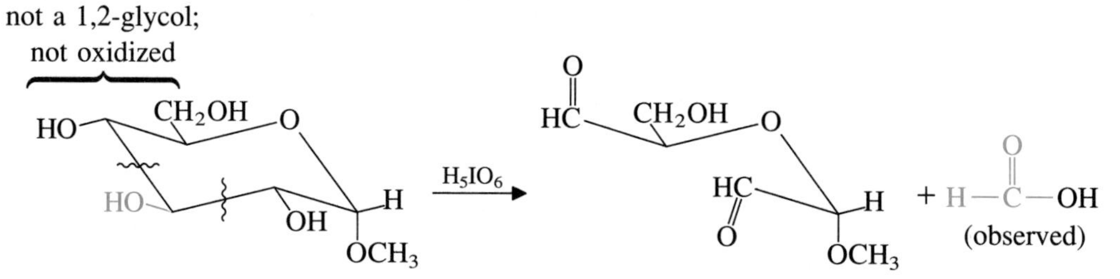

(27.30a)

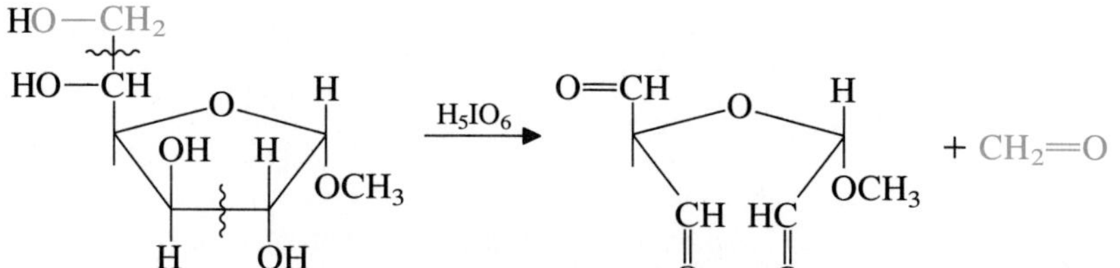

(27.30b)

The periodate oxidation of sugars was developed by C. S. Hudson (1881–1952), a noted American carbohydrate chemist. It was used extensively to relate the anomeric configurations of many sugar derivatives. How this was done is suggested by the following problem:

Problem

15 Explain why the methyl α-D-pyranosides of all D-aldohexoses give, in addition to formic acid, the same compound when oxidized by periodate. Assuming you knew the properties of this compound (including its optical rotation), show how you could use periodate oxidation to distinguish between methyl α-D-galactopyranoside from methyl β-D-galactopyranoside.

D. Reduction to Alditols

Aldohexoses, like ordinary aldehydes, undergo many of the usual carbonyl reductions. With H_2/Ni or $NaBH_4$, for example, an aldose is reduced to a primary alcohol (Sec. 19.10) known as an **alditol.**

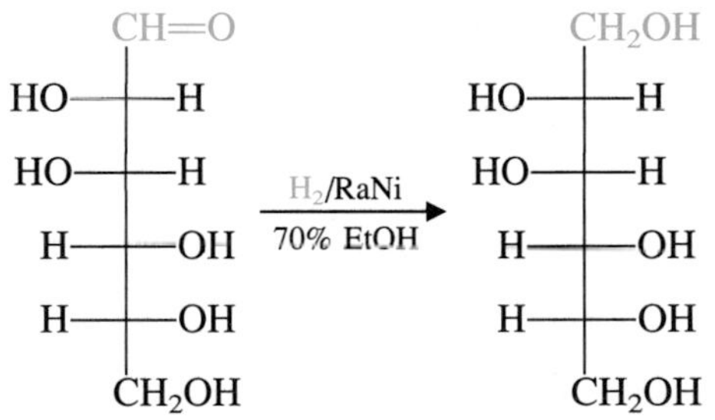

D-mannose → D-mannitol (an alditol) (70% yield) (27.31)

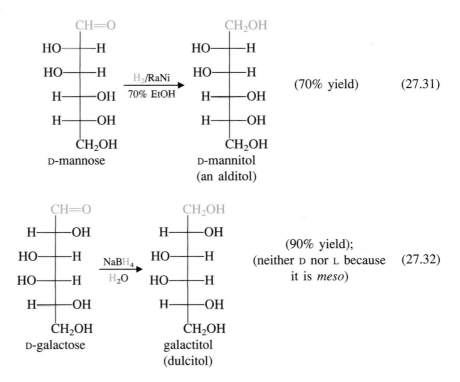

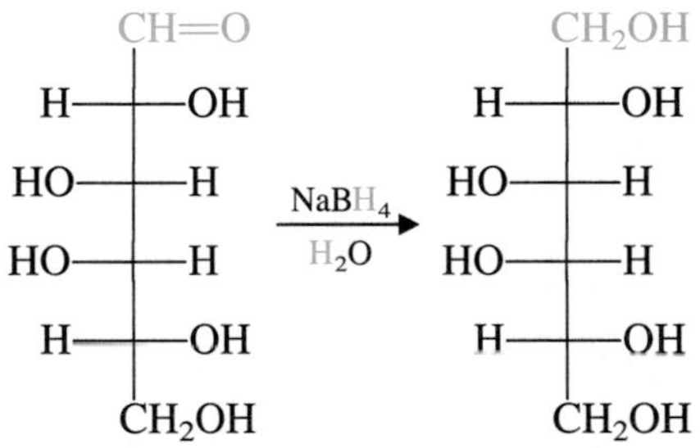

D-galactose → galactitol (dulcitol) (90% yield); (neither D nor L because it is *meso*) (27.32)

27.8 SYNTHESIS OF SUGARS FROM OTHER SUGARS

A. Kiliani–Fischer Synthesis: Increasing the Length of an Aldose Chain

Aldoses, like other aldehydes, add hydrogen cyanide to give cyanohydrins (Sec. 19.7). Notice that Reaction 27.33, like several others we have discussed, involves the aldehyde form of a sugar:

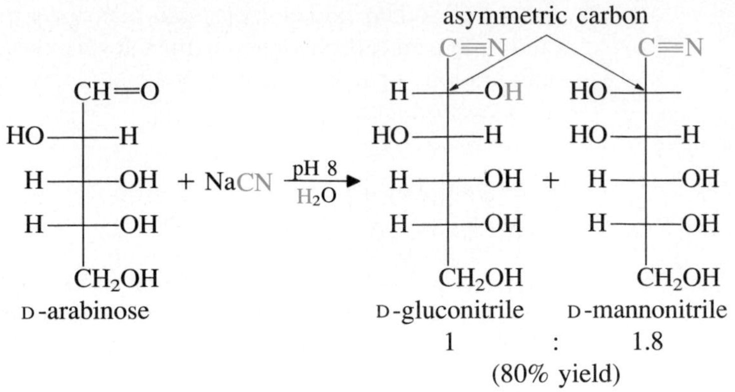

(27.33)

Because the cyanohydrin product has an additional asymmetric carbon, it is formed as a mixture of two epimers. Since these epimers are diastereomers, they are typically formed in different amounts. Although the exact amounts of each are not easily predicted, in most cases significant amounts of both are obtained. The mixture of cyanohydrins can be converted into a mixture of aldoses by catalytic hydrogenation, and these aldoses can be separated.

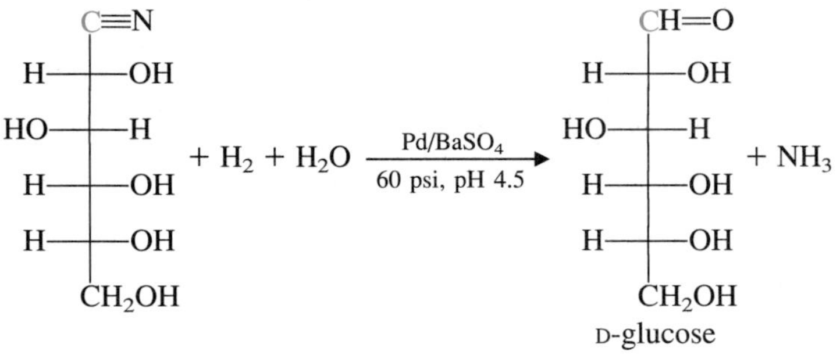

(27.34)

The hydrogenation reaction involves reduction of the nitrile to the imine (or a cyclic derivative), which, under the aqueous reaction conditions, hydrolyzes to the aldose and ammonia.

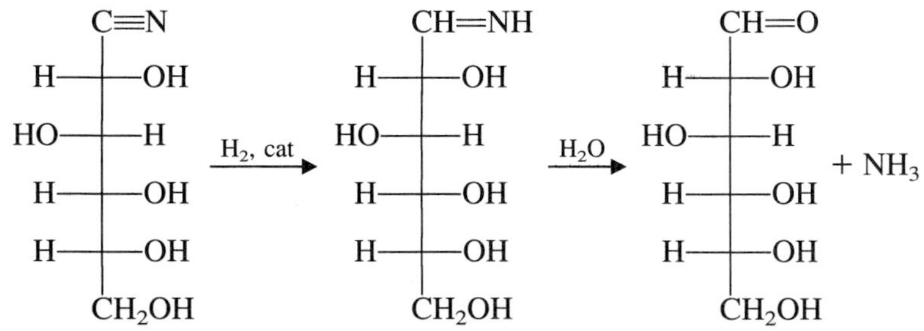

(27.35)

We can see from this example that the sequence of cyanohydrin formation–reduction converts an aldose into two epimeric aldoses with one additional carbon. That is, two aldohexoses, epimeric at carbon-2, are formed from a pentose. Notice particularly that this synthesis does not affect the stereochemistry of carbons 2, 3, and 4 in the starting material.

This process is known as the **Kiliani–Fischer synthesis.** The colored arrows in Fig. 27.1 indicate how sugars can be related by the Kiliani–Fischer synthesis. (The version of the Kiliani–Fischer synthesis discussed here is different from the original; it involves modern reagents, and occurs in better yield.)

B. Ruff and Wohl Degradations: Decreasing the Length of an Aldose Chain

Using either of two reactions, an aldose can be degraded to another aldose with one fewer carbon atom. In both degradations the aldehyde carbon is removed, and carbon-2 of the original sugar becomes the aldehyde carbon of the lower sugar.

In the **Ruff degradation,** the calcium salt of an aldonic acid is oxidized with hydrogen peroxide in the presence of Fe^{3+}.

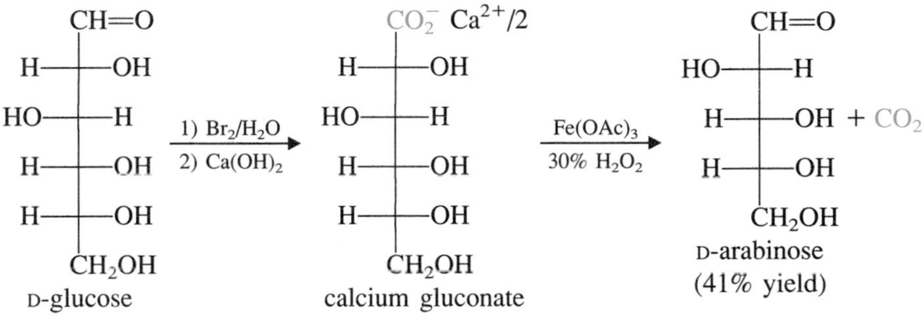

(27.36)

(The mechanism of this reaction, which appears to involve free radicals, is not well understood.)

The following sequence is an example of the **Wohl degradation.**

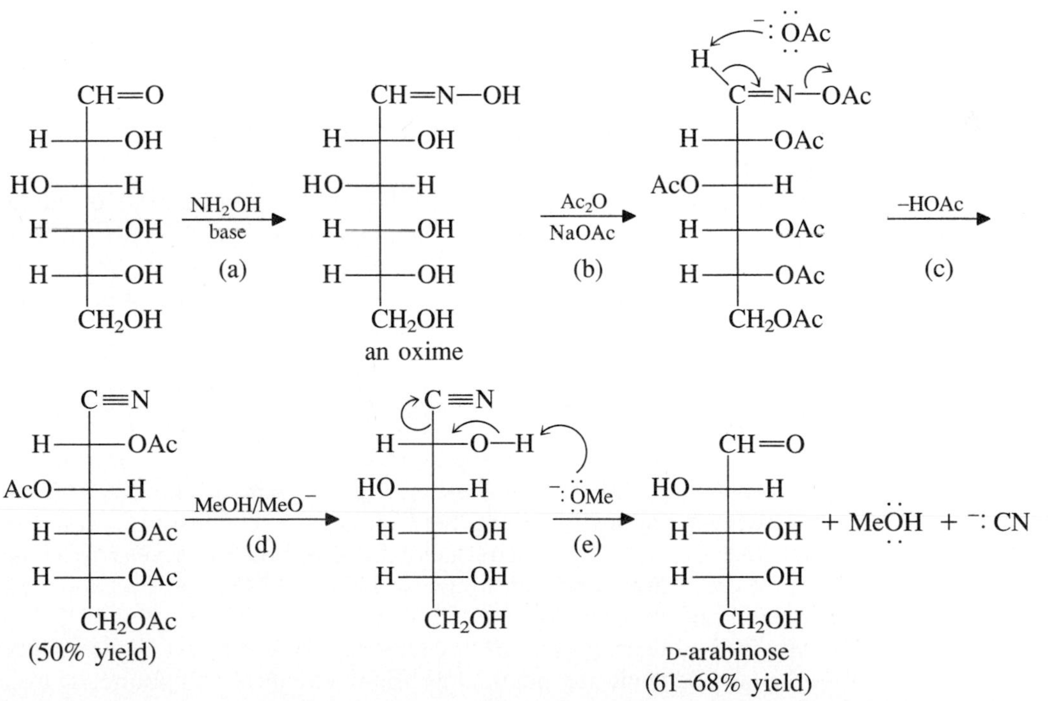

(27.37)

In this degradation an aldose is first converted into its oxime (step a; see Table 19.3). The oxime is then converted into its hexa-acetate ester with acetic anhydride (step b). The sodium acetate present in the reaction mixture acts as a base to convert the oxime into a nitrile by elimination (step c). The nitrile is the cyanohydrin of the next lower sugar. When treated with methoxide ion, the acetate groups are removed by transesterification (step d; see Eq. 27.22), and the cyanohydrin loses the elements of HCN to form the aldose. (Cyanohydrin formation is reversible; Sec. 19.7B.)

Problems

16 An optically active aldose *A* can be oxidized by dilute HNO_3 to give an optically inactive dicarboxylic acid. When subjected to the Wohl degradation, compound *A* gives a compound *B* that can be oxidized to D-(−)-tartaric acid. Identify aldoses *A* and *B*.

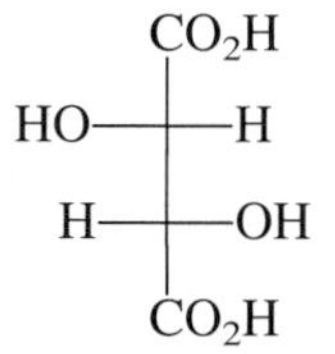

D-(−)-tartaric acid

17 Identify *A* and *B*:

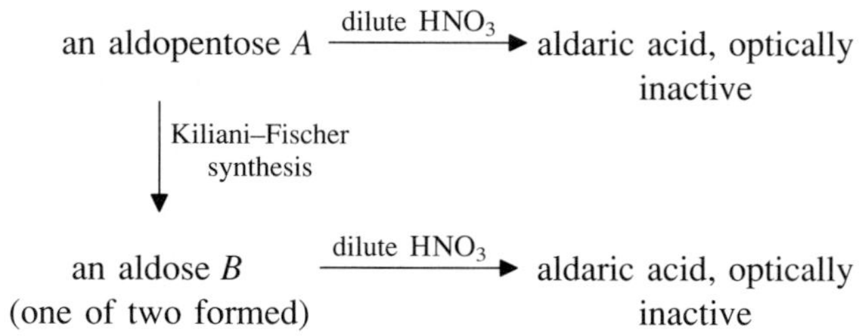

27.9 PROOF OF GLUCOSE STEREOCHEMISTRY

The aldohexose structure of (+)-glucose was established around 1870. The van't Hoff–LeBel theory of the tetrahedral carbon atom, published in 1874 (Sec. 6.10), suggested the possibility that glucose and the other aldohexoses could be stereoisomers. The problem to be solved, then, was: which one of the 2^4 possible stereoisomers is glucose? This problem was solved in two stages.

A. Which Diastereomer? The Fischer Proof

The first (and major) part of the solution to the problem of glucose stereochemistry was published in 1891 by Emil Fischer, a great German chemist who carried out landmark investigations in several fields of organic chemistry. It would be reason enough to study Fischer's proof as one of the most brilliant pieces of reasoning in the history of chemistry. However, it also will serve to sharpen our understanding of stereochemical relationships.

We must understand that in Fischer's day there was no way to determine the absolute stereochemical configuration of any chemical compound. Thus, although

Fischer would determine which *diastereomer* of the aldohexoses corresponds to (+)-glucose, he realized that he had no way of determining which *enantiomer* of this compound he was dealing with. Thus, Fischer arbitrarily *assumed* that carbon-5 (the configurational carbon in the D,L system) of (+)-glucose has the —OH on the right in the standard Fischer projection; that is, Fischer assumed that (+)-glucose has what we now call the D configuration. No one knew whether this assumption was correct; the solution to this problem had to await the development of special physical methods not available in Fischer's day. If Fischer's guess were wrong, then it would be necessary to reverse all of his stereochemical assignments. Fischer, then, proved the stereochemistry of (+)-glucose *relative* to an assumed configuration at carbon-5. The remarkable thing about his proof is that it allowed him to assign relative configurations in space using only chemical reactions and optical activity. The logic involved is direct, simple, and elegant, and can be summarized in four steps:

Step 1 (−)-Arabinose is converted into both (+)-glucose and (+)-mannose by a Kiliani–Fischer synthesis. From this fact (see Sec. 27.8A), Fischer deduced that (+)-glucose and (+)-mannose are epimeric at carbon-2, and that the configuration of (−)-arabinose at carbons 2, 3, and 4 is the same as that of (+)-glucose and (+)-mannose at carbons 3, 4, and 5, respectively.

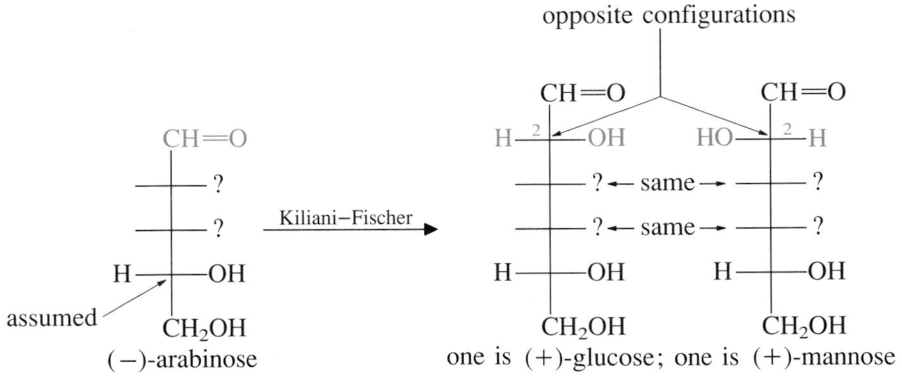

(27.38)

Step 2 (−)-Arabinose can be oxidized by dilute HNO_3 (Sec. 27.7B) to an optically active aldaric acid. From this, Fischer concluded that the —OH group at carbon-2 of arabinose must be on the left. If this —OH group were on the right, then the aldaric acid of arabinose would have to be *meso,* and thus optically inactive, *regardless of the configuration of the —OH group at carbon-3.* (Be sure you see why this is so; if necessary, draw both possible structures for (−)-arabinose to verify this deduction.)

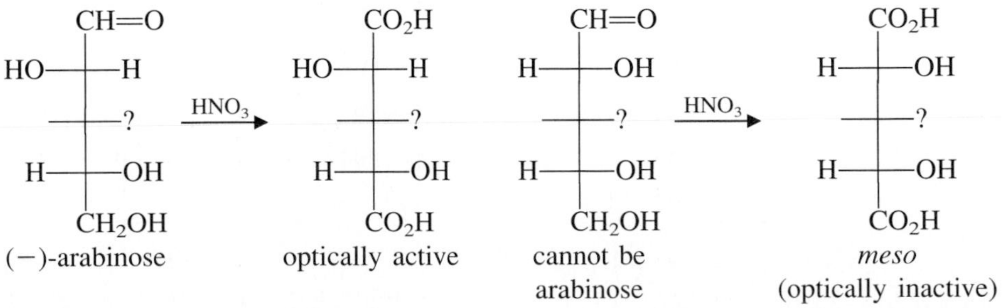

(27.39)

The relationships among arabinose, glucose, and mannose established in step 1 require the following partial structures for (+)-glucose and (+)-mannose.

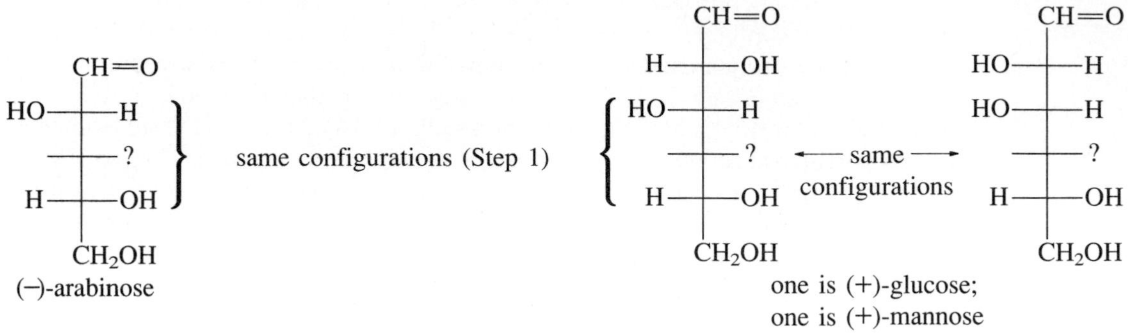

(27.40)

Step 3 Oxidations of *both* (+)-glucose and (+)-mannose with HNO_3 give optically active aldaric acids. From this, Fischer deduced that the —OH group at carbon-4 is on the right in both (+)-glucose and (+)-mannose. Recall that whatever the configuration at carbon-4 in these two sugars, it must be the same in both. Only if the —OH is on the right will *both* structures yield, on oxidation, optically active aldaric acids. If the —OH were on the left, *one* of the two sugars would have given a *meso*, and hence, optically inactive, aldaric acid.

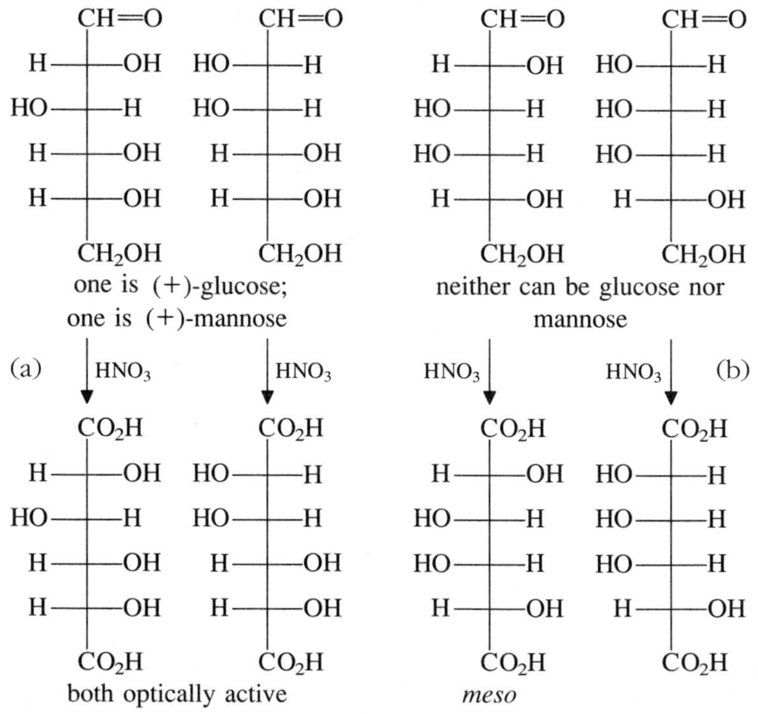

(27.41)

The configuration at carbon-4 of (+)-glucose and (+)-mannose is the same as that at carbon-3 of (−)-arabinose (Step 1); thus, at this point Fischer could deduce the complete structure of (−)-arabinose.

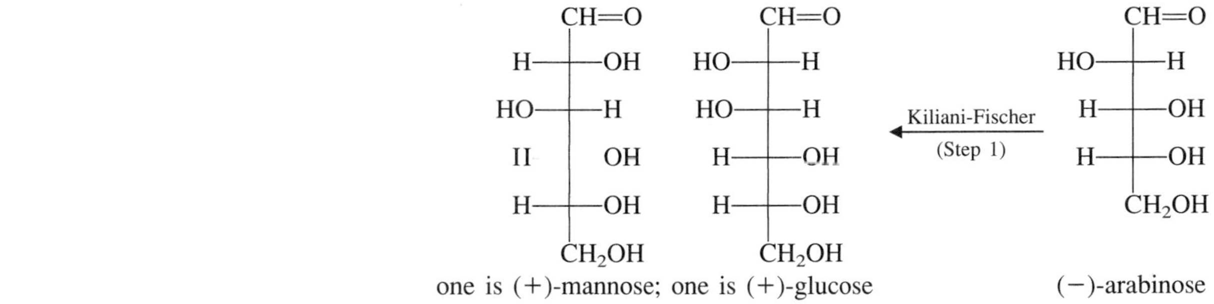

(27.42)

Step 4 The previous steps had established that (+)-glucose had one of the two structures in Eq. 27.42 and (+)-mannose had the other, but Fischer did not yet know which structure goes with which sugar. This point is confusing to some students. Fischer's situation is analogous to that of the young man who has just met two sisters; he wants to ask one of them to dinner. So he asks a friend: "What are their names?" The friend says, "Oh, they are Mannose and Glucose; I just wish I knew which is which!" Just because the young man knows *both* names doesn't mean that he can associate *each* name with *each* face. Similarly, although Fischer knew the structures associated with both (+)-glucose and (+)-mannose, he did not yet know how to correlate *each* sugar with *each* structure.

This problem was solved when Fischer discovered that another sugar, (+)-gulose, can be oxidized with HNO_3 to the same aldaric acid as (+)-glucose. How does this fact differentiate between (+)-glucose and (+)-mannose? Two *different* sugars can give the same aldaric acid only if their —CH=O and —CH_2OH groups are at opposite ends of an otherwise identical molecule (Problems 13 and 14). When we switch the —CH_2OH and —CH=O groups in one of the aldohexose structures in Eq. 27.42, we get the *same* sugar. (We can verify that these two structures are identical by rotating either one 180° in the plane of the page and comparing it to the other.)

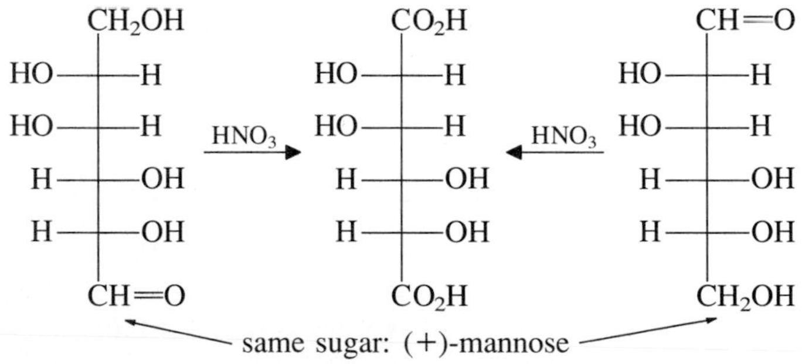

(27.43)

Since only one aldose can be oxidized to this aldaric acid, this aldose cannot be (+)-glucose; therefore it must be (+)-mannose. If we switch the end groups of the other aldohexose structure in Eq. 27.42, we obtain a *different* aldose:

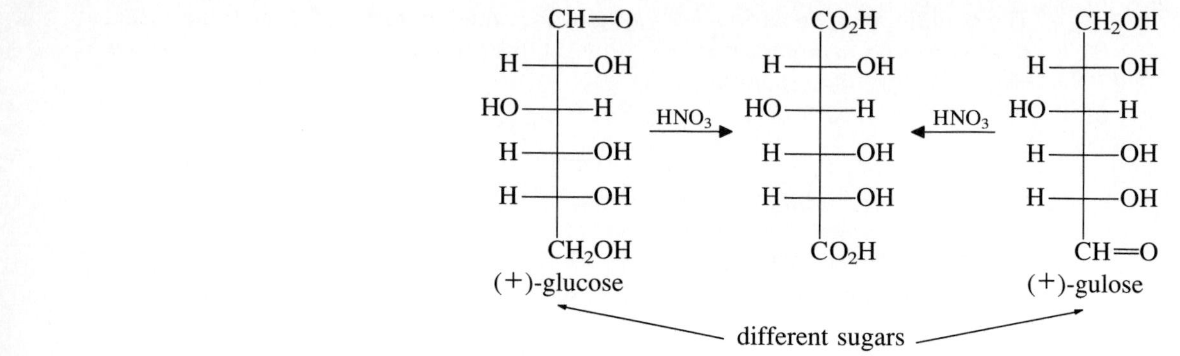

(27.44)

Since there are two aldohexoses that can be oxidized to this aldaric acid, one of them—the one in Eq. 27.44—must be (+)-glucose.

Emil Fischer (1852–1919) was one of the greatest organic chemists who ever lived. Fischer studied with Adolph von Baeyer, and ultimately became Professor at Berlin in 1892. Fischer carried out important researches on sugars, proteins (he devised the first rational syntheses of peptides), and heterocycles (for example, the Fischer indole synthesis). Fischer was a technical advisor to Kaiser Wilhelm. The following story gives some indication of the authority that Fischer commanded in Germany. It is said that one day he and the Kaiser were arguing questions of science policy, and the Kaiser sought to end debate by pounding his fist on the table, shouting, "Ich bin der Kaiser!" (I am the King!) Fischer, not to be silenced, responded in kind: "Ich bin Fischer!" Another story, perhaps apocryphal, attributes an important laboratory function to Fischer's long flowing beard. It was said that when a student had difficulty crystallizing a sugar derivative (some of which are notoriously difficult to crystallize), Fisher would shake his beard over the flask containing the recalcitrant compound. The accumulated seed crystals in his beard would fall into the flask and bring about the desired crystallization. Fischer was awarded the Nobel Prize in 1902.

Problem

18 An aldopentose *A* can be oxidized with dilute HNO_3 to an optically active aldaric acid. A Kiliani–Fischer synthesis starting with *A* gives two new sugars *B* and *C*. Sugar *B* can be oxidized to an achiral, and therefore optically inactive, aldaric acid, but sugar *C* is oxidized to an optically active aldaric acid. Assuming the D configuration, give the structures of *A*, *B*, and *C*.

B. Which Enantiomer? The Absolute Configuration of Glucose

Fischer died not knowing whether his arbitrary assignment of the absolute configuration of (+)-glucose was correct. That is, it was not known whether the —OH group at carbon-5 of glucose was really on the right in the Fischer projection. The groundwork for solving this problem was laid when glucose was related to both D-tartaric acid and D-glyceraldehyde by the following sequence of reactions. Two cycles of the Ruff degradation (Sec. 27.8B) converted (+)-glucose into (−)-erythrose.

$$\text{(+)-glucose} \xrightarrow{\text{Ruff}} \text{(−)-arabinose} \xrightarrow{\text{Ruff}} \text{(−)-erythrose} \qquad (27.45)$$

D-Glyceraldehyde, in turn, was related to (−)-erythrose by a Kiliani–Fischer synthesis:

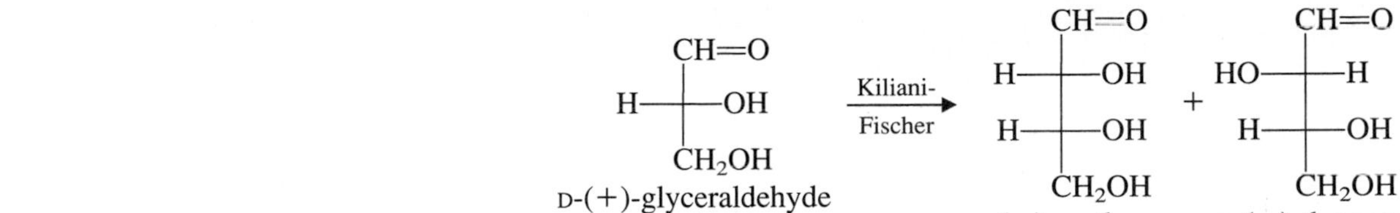

D-(+)-glyceraldehyde (absolute configuration assumed by convention)

D-(−)-erythrose D-(−)-threose (configurations at carbon-3 assumed by convention)

(27.46)

This sequence of reactions showed that (+)-glucose, (−)-erythrose, (−)-threose, and (+)-glyceraldehyde were all of the same stereochemical series—the D series. Oxidation of D-(−)-threose with dilute HNO_3 gave D-(−)-tartaric acid.

In 1950 the absolute configuration of naturally occurring (+)-tartaric acid (as its potassium rubidium double salt) was determined by a special technique of X-ray crystallography called *anomalous dispersion*. This determination was made by J. M. Bijvoet, A. F. Peerdeman, and A. J. van Bommel, Dutch chemists who worked, appropriately enough, at the van't Hoff laboratory in Utrecht. If Fischer had made the right choice for the D configuration, the assumed structure for D-(−)-tartaric acid and the experimentally determined structure of (+)-tartaric acid determined by the Dutch crystallographers would be enantiomers. If Fischer had guessed wrong, the assumed structure for D-tartaric acid would be the same as the experimentally determined structure of L-tartaric acid, and would have to be reversed. To quote Bijvoet and his colleagues: *"The result is that Emil Fischer's convention* [for the D configuration] *appears to answer to reality."*

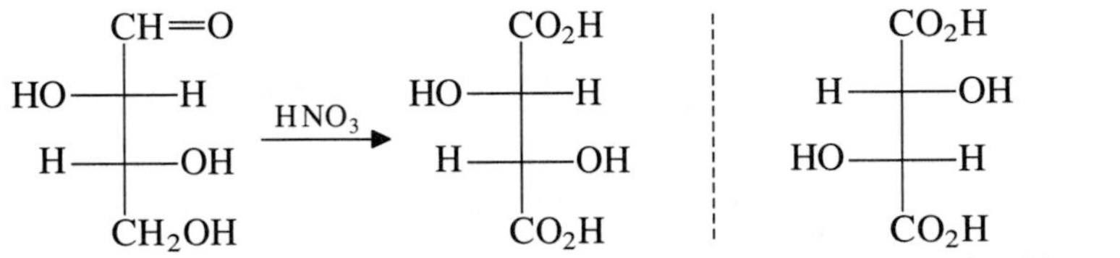

D-(−)-threose D-(−)-tartaric acid L-(+)-tartaric acid (by X-ray crystallography)

(27.47)

Problems

19 Given the structure of D-glyceraldehyde, how would you assign a structure to each of the two sugars obtained from it by Eq. 27.46, assuming that these compounds were previously unknown?

20 Suppose that a scientist were to reexamine the crystallographic work that established the absolute configuration of L-(+)-tartaric acid and finds that the structure of this compound was the mirror image of the one given in the text above. What changes would have to be made in Fischer's structure of D-(+)-glucose?

27.10 DISACCHARIDES AND POLYSACCHARIDES

A. Disaccharides

Disaccharides consist of two simple sugar residues, or *monosaccharides,* connected by a glycosidic linkage. **(+)-Lactose** is an example of a disaccharide. ((+)-Lactose is present to the extent of about 4.5% in cow's milk and 6–7% in human milk.)

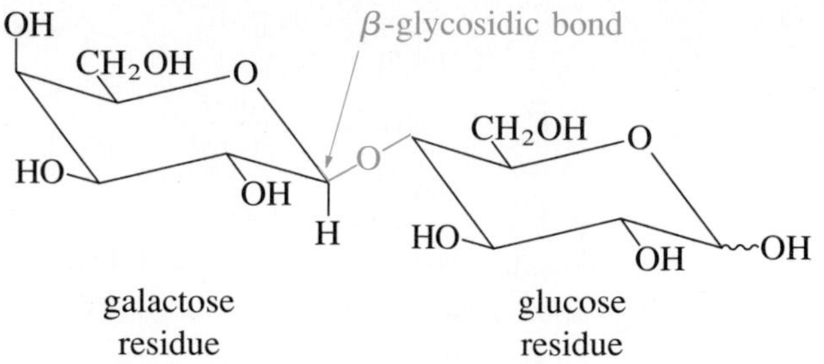

(+)-lactose

In (+)-lactose, a D-glucopyranose molecule is linked by its oxygen at carbon-4 to carbon-1 of D-galactopyranose. In effect, (+)-lactose is a glycoside in which galactose is the sugar and glucose is the "aglycone." As we have learned, the glycosidic linkage is an acetal, and therefore hydrolyzes under acidic conditions. Hence it is not surprising that (+)-lactose can be hydrolyzed in acidic solution to give one molecule each of D-glucose and D-galactose, in the same sense that a methyl glycoside can be hydrolyzed to give methanol and a sugar:

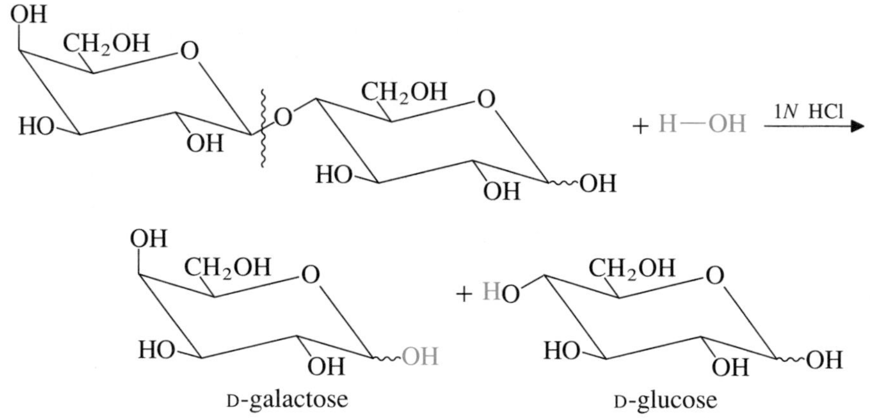

(27.48a)

compare:

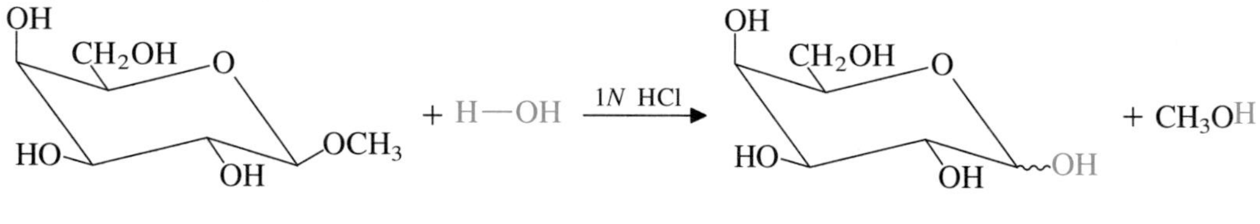

(27.48b)

Certain enzymes also catalyze the same reaction at pH values near neutrality.

The stereochemistry of the glycosidic bond in (+)-lactose is **β**. That is, the stereochemistry of the oxygen linking the two sugar residues in the glycosidic bond corresponds to that in the β-anomer of D-galactopyranose. This stereochemistry is very important in biology, because higher animals possess an enzyme, β-galactosidase, that catalyzes the hydrolysis of this β-glycosidic linkage; this hydrolysis allows lactose to act as a source of glucose. α-Glycosides of galactose are inert to the action of this enzyme.

Because carbon-1 of the galactose residue in (+)-lactose is involved in a glycosidic linkage, it cannot be oxidized. However, carbon-1 of the glucose residue is part of a hemiacetal group, which, like the hemiacetal group, of monosaccharides, is in equilibrium with the free aldehyde, and can undergo characteristic aldehyde reactions. Thus, oxidation of (+)-lactose with bromine water (Sec. 27.7A) effects oxidation of the glucose residue:

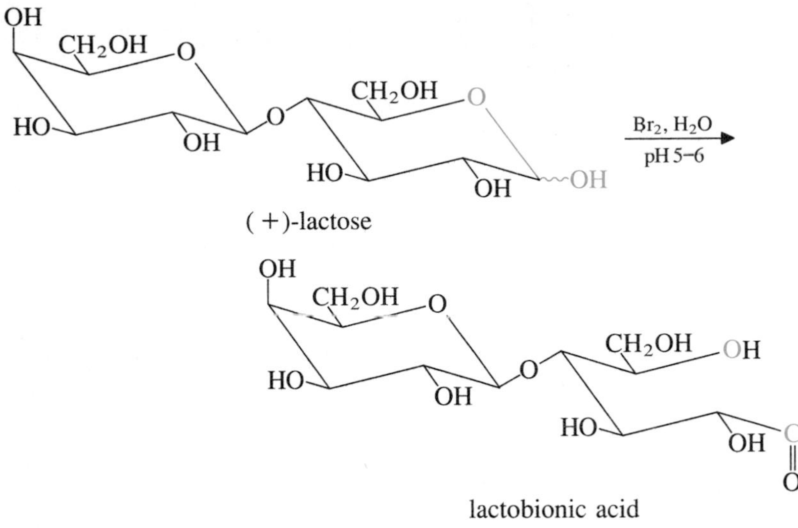

(27.49)

Sugars such as (+)-lactose that can be oxidized in this way are called **reducing sugars.** The glucose residue is said to be at the *reducing end* of the sugar, and the galactose residue at the *nonreducing end.* Because of its hemiacetal group, (+)-lactose also undergoes many other reactions of sugar hemiacetals, such as mutarotation.

Problem

21 What products are expected from each of the following reactions?
(a) lactobionic acid (Eq. 27.49) + 1*N* aqueous HCl ⟶
(b) (+)-lactose + dimethyl sulfate, NaOH ⟶
(c) product of (b) + 1*N* aqueous H_2SO_4 ⟶

(+)-Sucrose, or table sugar, is another important disaccharide. About ninety million tons of sucrose are produced annually in the world. Sucrose consists of a D-glucopyranose residue and a D-fructofuranose residue connected by a glycosidic bond (color) at the anomeric carbons of *both* monosaccharides.

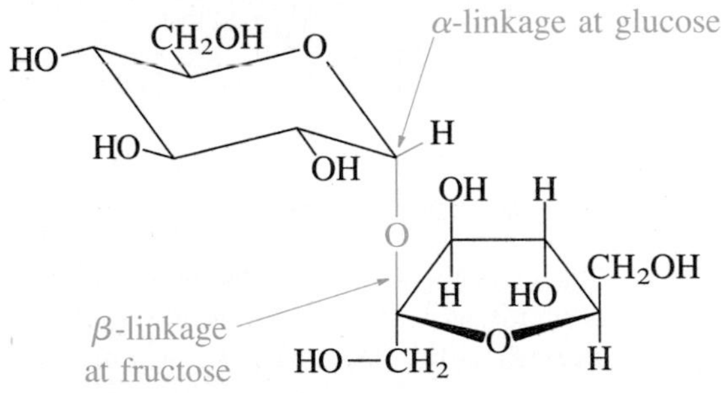

(+)-sucrose

The glycosidic bond in (+)-sucrose is different from the one in lactose. Only one of the residues of lactose—the galactose residue—contains an acetal (glycosidic) carbon. In contrast, *both* residues of (+)-sucrose have an acetal carbon. The glycosidic bond in (+)-sucrose bridges carbon-2 of the fructofuranose residue and carbon-1 of the glucopyranose residue. These are the carbonyl carbons in the open-chain forms of the individual sugars; remember that the carbonyl carbons become the acetal or hemiacetal carbons in the cyclic forms.

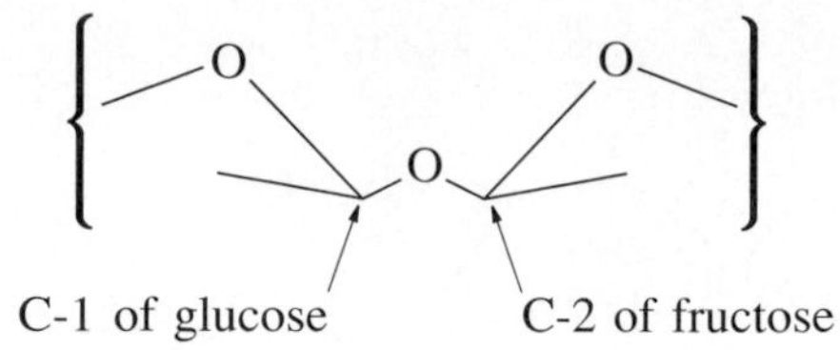

double glycoside linkage

Thus, neither sugar residue has a free hemiacetal group. Hence, (+)-sucrose cannot be oxidized by bromine water, nor does it undergo mutarotation. Sugars such as (+)-sucrose that cannot be oxidized by bromine water are classified as **nonreducing sugars.**

Like other glycosides, (+)-sucrose can be hydrolyzed to its component monosaccharides. Sucrose is hydrolyzed by aqueous acid or by enzymes (called *invertases*) to an equimolar mixture of D-glucose and D-fructose.

The mixture of D-glucose and D-fructose obtained on hydrolysis of sucrose is called *invert sugar*. (Honey is mostly invert sugar.) The name arises from the fact that (+)-sucrose (specific rotation +66°) is hydrolyzed to (+)-glucose (formerly called *dextrose:* specific rotation +52.7°) and (−)-fructose (formerly called *levulose:* specific rotation −92°). The rotation of the mixture is dominated by the negative rotation of fructose. Thus, as hydrolysis proceeds, the initially positive rotation of the sucrose solution "inverts" to a negative rotation. Fructose, which is the sweetest of the common sugars (about twice as sweet as sucrose), accounts for the greater sweetness of invert sugar.

Problem

22 The structure of cellobiose, a disaccharide obtained from the hydrolysis of the polysaccharide cellulose, is given below. Into what monosaccharide(s) is cellobiose hydrolyzed by aqueous HCl?

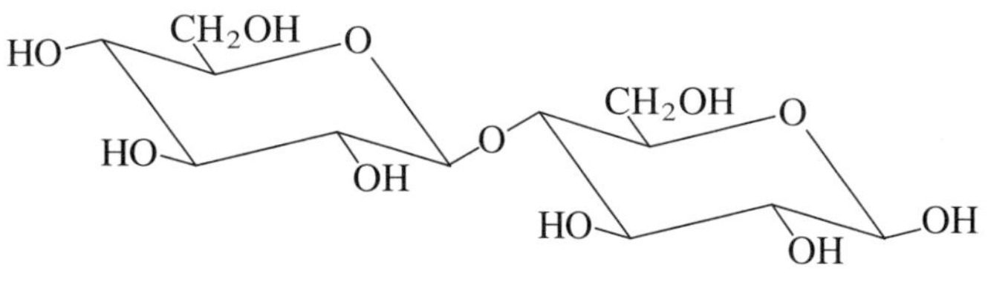

cellobiose

B. Polysaccharides

In principle, any number of monosaccharide residues can be linked together with glycosidic bonds to form chains. When such chains are long, the sugars are called **polysaccharides.** Let us survey a few important polysaccharides.

Cellulose Cellulose, the principal structural component of plants, is the most abundant organic compound in the world. Cotton is almost pure cellulose; wood is cellulose combined with an aromatic polymer called *lignin*. About 5×10^{14} kg of cellulose are biosynthesized and degraded annually on the earth.

Cellulose is a regular polymer of D-glucopyranose residues connected by β-1,4-glycosidic linkages.

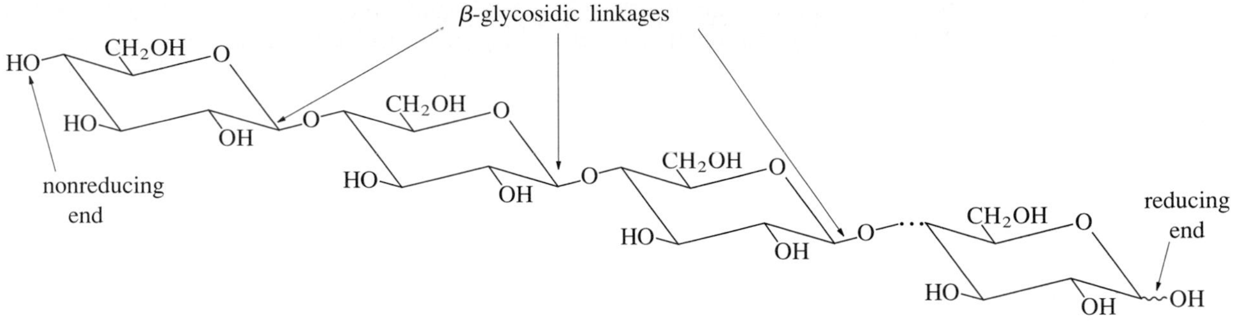

cellulose

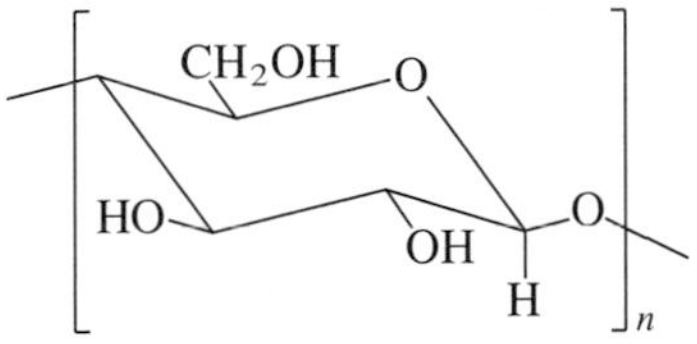

general structure

Like disaccharides, polysaccharides can be hydrolyzed to their constituent monosaccharides. Thus, cellulose can be hydrolyzed to D-glucose residues. Mammals lack the enzymes that catalyze the hydrolysis of the β-glycosidic linkages of cellulose; this is why humans cannot digest grasses, which are principally cellulose. Cattle can, of course, derive nourishment from grasses, but this is because the bacteria in their rumen provide the appropriate enzymes that break down plant cellulose to glucose.

Processed cellulose (cellulose that has been specially treated) has many other uses. It can be spun into fibers (rayon) or made into wraps (cellophane). The paper on which this book is printed is largely processed cellulose. Nitration of the cellulose hydroxy groups gives nitrocellulose, a powerful explosive. Cellulose acetate, in which the hydroxy groups of cellulose are esterified with acetic acid, is known by the trade names Celanese, Arnel, etc., and is used in knitting yarn and decorative household articles.

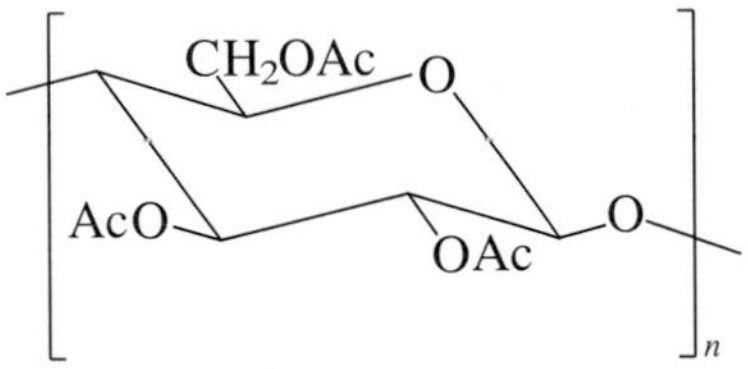

cellulose acetate

Cellulose is potentially important as an alternative energy source. As we have seen, biomass is largely cellulose, and cellulose is merely polymerized glucose. The glucose derived from hydrolysis of cellulose can be fermented to ethanol, which can be used as a fuel (as in gasohol). And plants obtain the energy to manufacture cellulose from the sun. Thus, the cellulose in plants—the most abundant source of carbon on the earth—can be regarded as a storehouse of solar energy.

Starch Starch, like cellulose, is also a polymer of glucose. In fact, starch is a mixture of two different types of glucose polymers. In one, **amylose,** the glucose residues are connected by α-1,4-glycosidic linkages. Formally, the only chemical difference between amylose and cellulose is the stereochemistry of the glycosidic bond.

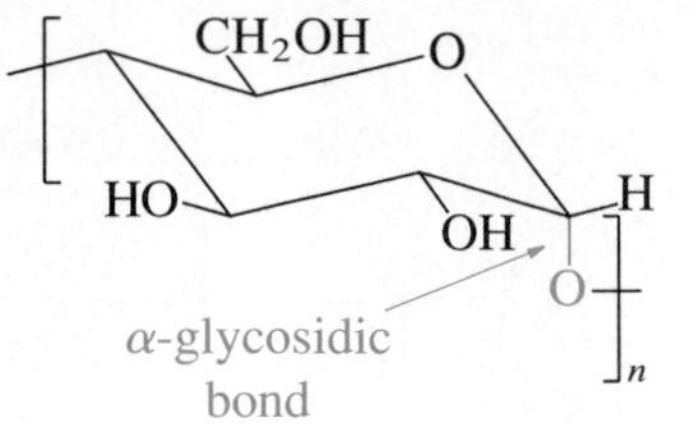

amylose ($n \cong 400$)

The other constituent of starch is **amylopectin,** a branched polysaccharide. Amylopectin contains relatively short chains of glucose residues in α-1,4-linkages. In addition, it contains branches that involve α-1,6-glycosidic linkages. Part of a typical amylopectin molecule might look as follows:

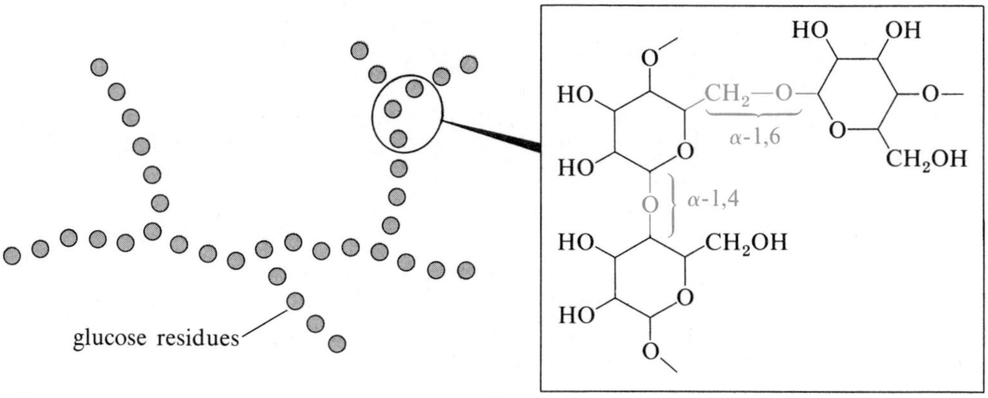

amylopectin branch

Starch is the important storage polysaccharide in corn, potatoes, and other starchy vegetables. Humans have enzymes that catalyze the hydrolysis of the α-glycosidic bonds in starch, and can therefore use starch as a source of glucose.

Chitin Chitin is a polysaccharide that also occurs widely in nature—notably, in the shells of arthropods (for example, lobsters and crabs). Crab shell is an excellent source of nearly pure chitin.

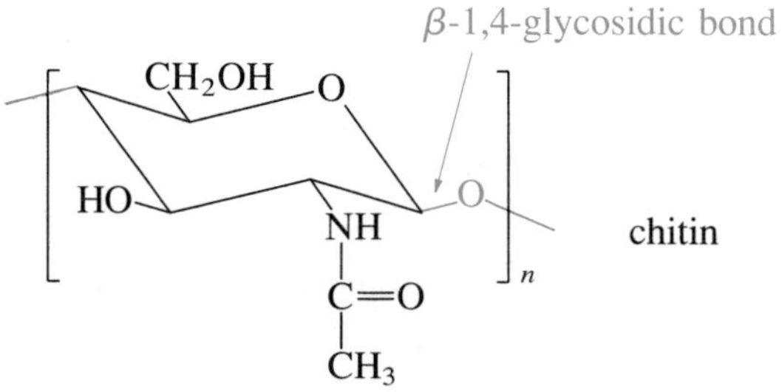

chitin

Chitin is a polymer of *N*-acetyl-D-glucosamine (or, as it is known systematically, 2-acetamido-2-deoxy-D-glucose). Residues of this sugar are connected by β-1,4-

glycosidic bonds within the chitin polymer. *N*-Acetyl-D-glucosamine is liberated when chitin is hydrolyzed in aqueous acid. Stronger acid brings about hydrolysis of the amide bond to give D-glucosamine hydrochloride and acetic acid.

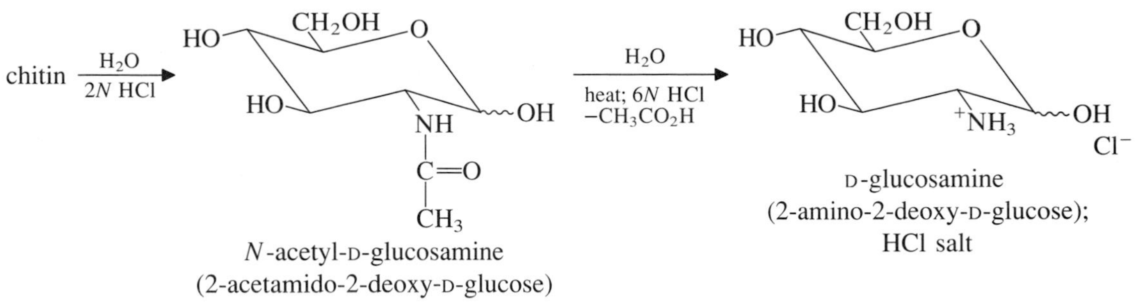

(27.50)

Glucosamine and *N*-acetylglucosamine are the best-known examples of the **amino sugars.** A number of amino sugars occur widely in nature. Amino sugars linked to proteins (glycoproteins) are found at the outer surfaces of cell membranes, and some of these are responsible for blood-group specificity.

New uses are being found for chitin. One promising application is the use of chitin fibers as surgical sutures, since chitin is nonallergenic and can be biologically absorbed.

The discovery of D-glucosamine is the subject of an amusing story. In 1876 Georg Ledderhose was a premedical student working in the laboratory of his uncle, Friedrich Wöhler (the same chemist who first synthesized urea; Sec. 1.1B). One day, Wöhler had lobster for lunch, and returned to the laboratory carrying the lobster shell. "Find out what this is," he told his nephew. History does not record Ledderhose's thoughts on receiving the refuse from his uncle's lunch, but he proceeded to do what all chemists did with unknown material—he boiled it in concentrated HCl. After hydrolysis of the shell, crystals of the previously unknown D-glucosamine hydrochloride precipitated from the cooled solution (see Eq. 27.50).

Principles of Polysaccharide Structure Studies of many polysaccharides have revealed the following generalizations about polysaccharide structure:

1. Polysaccharides are mostly long chains with some branches; there are no highly cross-linked, three-dimensional networks. Cyclic polysaccharides are known.
2. The linkages between monosaccharide units are in every case glycosidic linkages; thus, monosaccharides can be liberated from all polysaccharides by acid hydrolysis.
3. A given polysaccharide incorporates only one stereochemical type of glycoside linkage. Thus, the glycoside linkages in cellulose are all β; those in starch are all α.

Problem

23 What product(s) would be obtained when cellulose is treated first exhaustively with dimethyl sulfate/NaOH, then 1*N* aqueous HCl?

27.11 NUCLEOSIDES, NUCLEOTIDES, AND NUCLEIC ACIDS

A. Nucleosides and Nucleotides

A **ribonucleoside** is a β-glycoside formed between D-ribose and a heterocyclic nitrogen compound referred to generally as a *base.* A **deoxyribonucleoside** is a β-glycoside of D-2-deoxyribose and a heterocyclic base. The prefix *deoxy* means "without oxygen"; thus the sugar 2-deoxyribose is simply a ribose that lacks the—OH group at carbon-2 of the sugar.

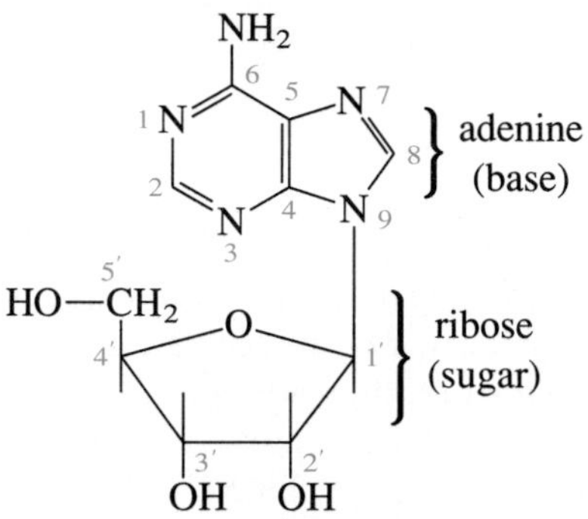

adenosine
(a ribonucleoside)

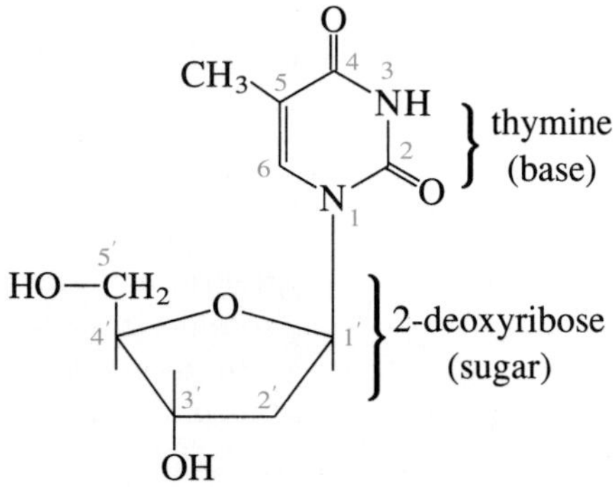

2′-deoxythymidine
(a deoxyribonucleoside)

The bases that occur most frequently in nucleosides are derived from two heterocyclic ring systems: **pyrimidine** and **purine.** Notice particularly the numbering of these rings. Three bases of the pyrimidine type and two of the purine type occur most commonly.

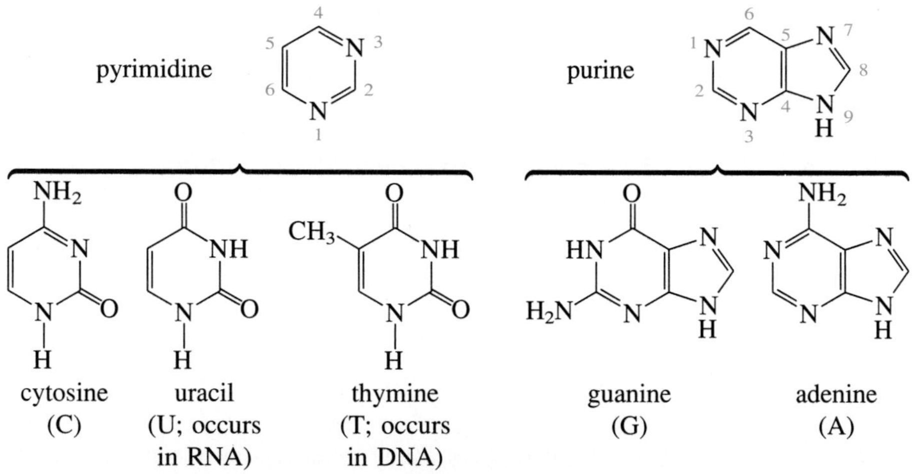

The base is attached to the sugar at N-9 of the purines and N-1 of the pyrimidines, as in the examples above. Notice that the base and the sugar ring systems are numbered separately. To differentiate the two systems, primes (′) are used to refer to the sugar carbon atoms. Thus the 2′ (pronounced *two-prime*) carbon of adenosine is carbon-2 of the sugar ring.

TABLE 27.2 Nomenclature of Nucleic Acid Bases, Nucleosides, and Nucleotides*

Base	Nucleoside	Nucleotide monophosphate	Abbreviation
adenine (A)	adenosine	adenylic acid	AMP
uracil (U)	uridine	uridylic acid	UMP
thymine (T)	thymidine	thymidylic acid	TMP
cytosine (C)	cytidine	cytidylic acid	CMP
guanine (G)	guanosine	guanidylic acid	GMP

*The deoxyribonucleosides and deoxyribonucleotides are named by appending the prefix *deoxy*:

deoxyadenosine deoxyadenylic acid *d*AMP

The prefix *deoxy* means 2′-deoxy unless stated otherwise.

The 5′ —OH group of the ribose in a nucleoside is often found esterified to a phosphate group. A 5′-phosphorylated nucleoside is called a **nucleotide.** A **ribonucleotide** is derived from the sugar ribose; a **deoxyribonucleotide** is derived from 2′-deoxyribose. Some nucleotides contain a single phosphate group; others contain two or three phosphate groups condensed in phosphoric anhydride linkages.

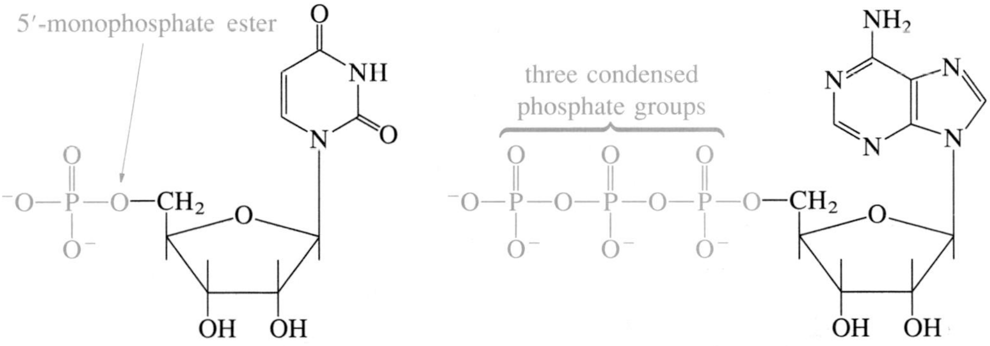

uridylic acid or UMP
(for *u*ridine *m*ono*p*hosphate)

adenosine triphosphate
or ATP

Although the ionization state of the phosphate groups depends on pH, these groups are written conventionally in the ionized form.

Nomenclature of the five common bases and their corresponding nucleosides and nucleotides is summarized in Table 27.2. This table gives the names of the ribonucleosides and ribonucleotides. To name the corresponding 2′-deoxy derivatives, the prefix *2′-deoxy* (or simply *deoxy*) is appended to the names of the corresponding ribose derivatives. For example, the 2′-deoxy analog of adenosine is called 2′-deoxyadenosine or simply deoxyadenosine. In addition, the names of the mono-, di-, and triphosphonucleotide derivatives are often abbreviated. Thus, adenylic acid is abbreviated AMP (for adenosine monophosphate); the di- and tri-phosphorylated derivatives are called ADP and ATP, respectively (see example above). The abbreviations for the corresponding deoxy derivatives contain a *d* prefix. Thus, 2′-deoxythymidylic acid can be abbreviated *d*TMP.

Nucleotides, as we shall see, are important because they are the building blocks of ribonucleic acid (RNA) and deoxyribonucleic acid (DNA), polymeric molecules that are responsible for the storage and transmission of genetic information. Ribonucleotides also have other important biochemical functions, some of which we have already studied. NAD^+, one of nature's important oxidizing agents (Fig. 10.1, Sec. 10.7), and coenzyme A (Fig. 22.7, Sec. 22.5A) are both ribonucleotides. One of the most ubiquitous nucleotides is ATP (adenosine triphosphate), which serves as the fundamental energy source for the living cell. The hydrolysis of ATP to ADP, shown in Eq. 27.51, liberates 7.3 kcal/mol of energy at pH 7; living systems harness this energy to drive energy-requiring biochemical processes.

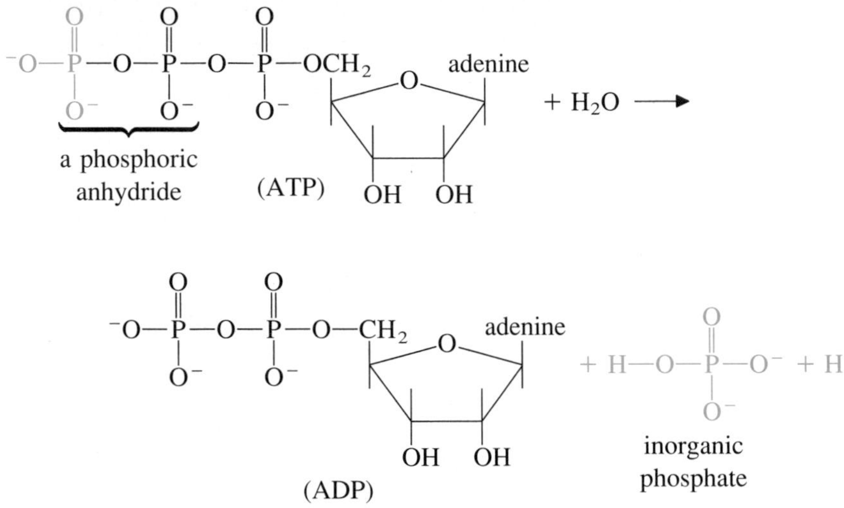

(27.51)

Muscle contraction—obviously an energy-requiring process—is an example of the biological use of ATP hydrolysis to provide energy. Abbreviating inorganic phosphate as P_i, the overall process for muscle contraction can be summarized as follows:

$$\begin{array}{rrcl} & \text{energy} + \text{muscle} & \longrightarrow & \text{contracted muscle} \\ & \text{ATP} + H_2O & \longrightarrow & \text{ADP} + P_i + \text{energy} \\ \hline \textit{sum:} & \text{ATP} + \text{muscle} + H_2O & \longrightarrow & \text{ADP} + P_i + \text{contracted muscle} \end{array} \qquad (27.52)$$

(A human might use 0.5 kilogram of ATP/hr during strenuous exercise!) *How* living organisms use the energy from ATP hydrolysis is a subject that we leave for your study of biochemistry; the important point is that overall ATP hydrolysis is invariably involved in any biological process that requires energy. The energy for making ATP from ADP is ultimately derived from the foods we eat—for example, sugars; and the sugars that we use as foods are produced by plants, using solar energy harnessed by the processes of photosynthesis.

Problem

24 Draw the structure of (a) GDP; (b) deoxythymidine monophosphate (*d*TMP).

B. Structures of DNA and RNA

Deoxyribonucleic acid (DNA) is a polymer of deoxyribonucleotides, and is the storehouse of genetic information throughout all of nature (with the exception of some viruses). A typical section of DNA is shown in Fig. 27.6. As we can see from this figure,

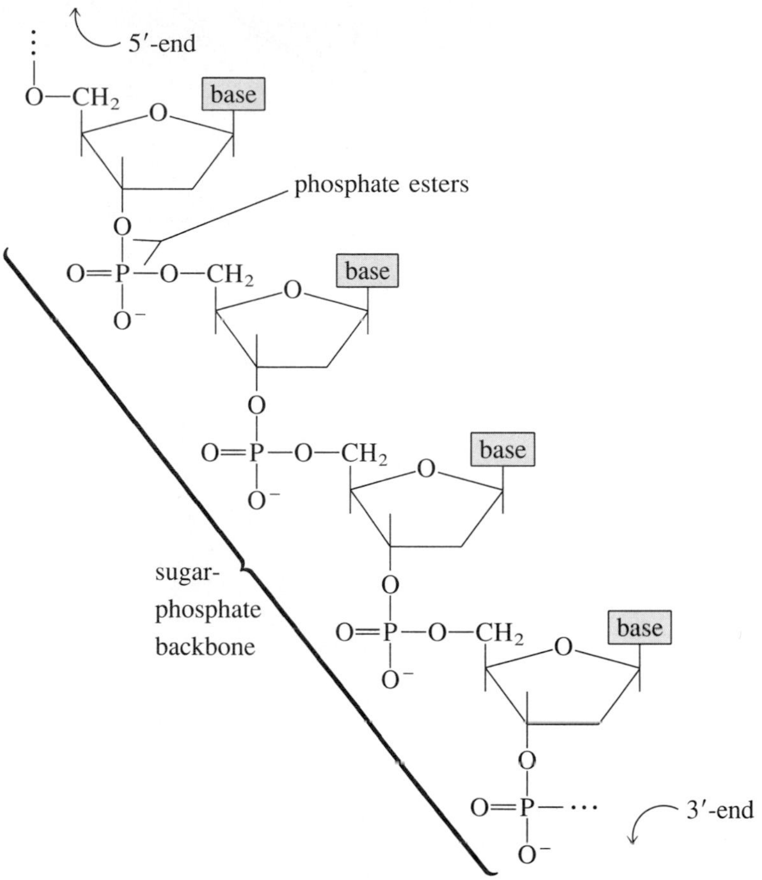

Figure 27.6 General structure of DNA. (base = A, T, G, or C; Table 27.2)

the nucleotide residues in DNA are interconnected by phosphate groups that are esterified both to the 3′—OH group of one ribose and the 5′—OH of another. The DNA polymer incorporates adenine, thymine, guanine, and cytidine as the nucleotide bases. Although only four residues are shown in Fig. 27.6, a typical strand of DNA might be thousands of nucleotides long. Just as each amino acid residue in a peptide is differentiated by its amino acid side chain, *each residue in a polynucleotide is differentiated by the identity of its base.* The DNA polymer is thus a **backbone** of alternating phosphates and 2′-deoxyribose groups to which are connected **bases** that differ from residue to residue. The ends of the DNA polymer are labeled 3′ or 5′, corresponding to the deoxyribose carbon on which the terminal hydroxy group is attached.

Ribonucleic acid (RNA) polymers are formally much like DNA polymers, except that ribose, rather than 2′-deoxyribose, is the sugar. RNA also incorporates essentially the same bases as DNA, except that uracil occurs in RNA instead of thymine, and some rare bases (which we shall not consider here) are found in certain types of RNA.

It was known for many years before the detailed structure of DNA was determined that DNA carries genetic information. It was also known that DNA is *replicated,* or copied, during cell reproduction. In 1950, Erwin Chargaff of Columbia University showed that the ratios of adenine to thymine, and guanosine to cytosine, in DNA are both 1.0; these observations are called *Chargaff's rules.* How these facts relate to the storage and transmission of genetic information, however, remained a mystery. It

Figure 27.7 Space-filling model of the DNA double helix. The sugar-phosphate backbone is traced with dark lines.

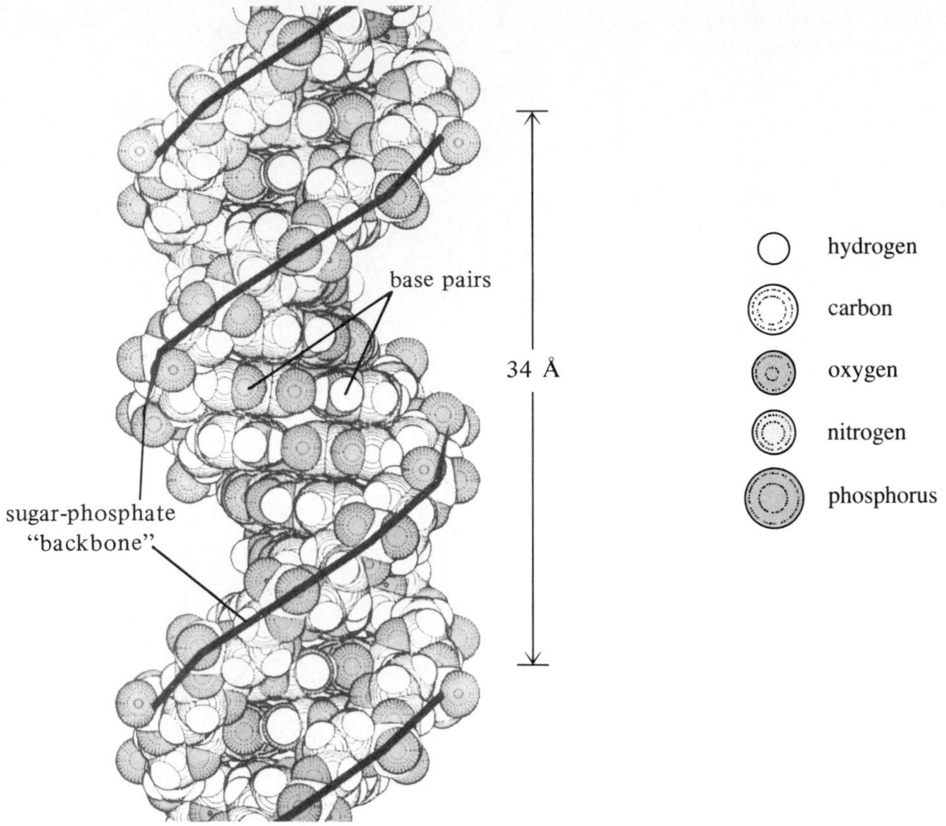

became clear to a number of scientists that a knowledge of the three-dimensional structure of DNA would be essential in order to understand how DNA functions as it does. The importance of this problem was sufficiently obvious that several scientists worked feverishly to be the first to determine the three-dimensional structure of DNA. In 1953, James D. Watson, now at Cold Spring Harbor Laboratory, and Francis C. Crick, now at the Salk Institute, proposed a structure for DNA. Their proposal was based on X-ray diffraction patterns of DNA fibers obtained by their colleagues at the Medical Research Council laboratory in England, Rosalind Franklin and Maurice Wilkins. (For an intriguing account of the race for the DNA structure, see "The Double Helix," by James D. Watson; Atheneum, 1968.) For their work on the structure of DNA, Watson and Crick received the Nobel Prize for Medicine and Physiology in 1962.

The Watson–Crick structure of DNA is shown in Fig. 27.7. The structure has the following important features:

1. The structure contains *two* right-handed helical polynucleotide chains that run in opposite directions, coiled around a common axis; the structure is therefore that of a **double helix.** The helix makes a complete turn every ten nucleotide residues. (Other conformations of DNA have been found subsequently; current research is aimed at elucidating the biological roles of different DNA conformations.)

2. The sugars and phosphates, which are rich in —OH groups and charges, are on the outside of the helix, where they can interact with solvent water; the bases,

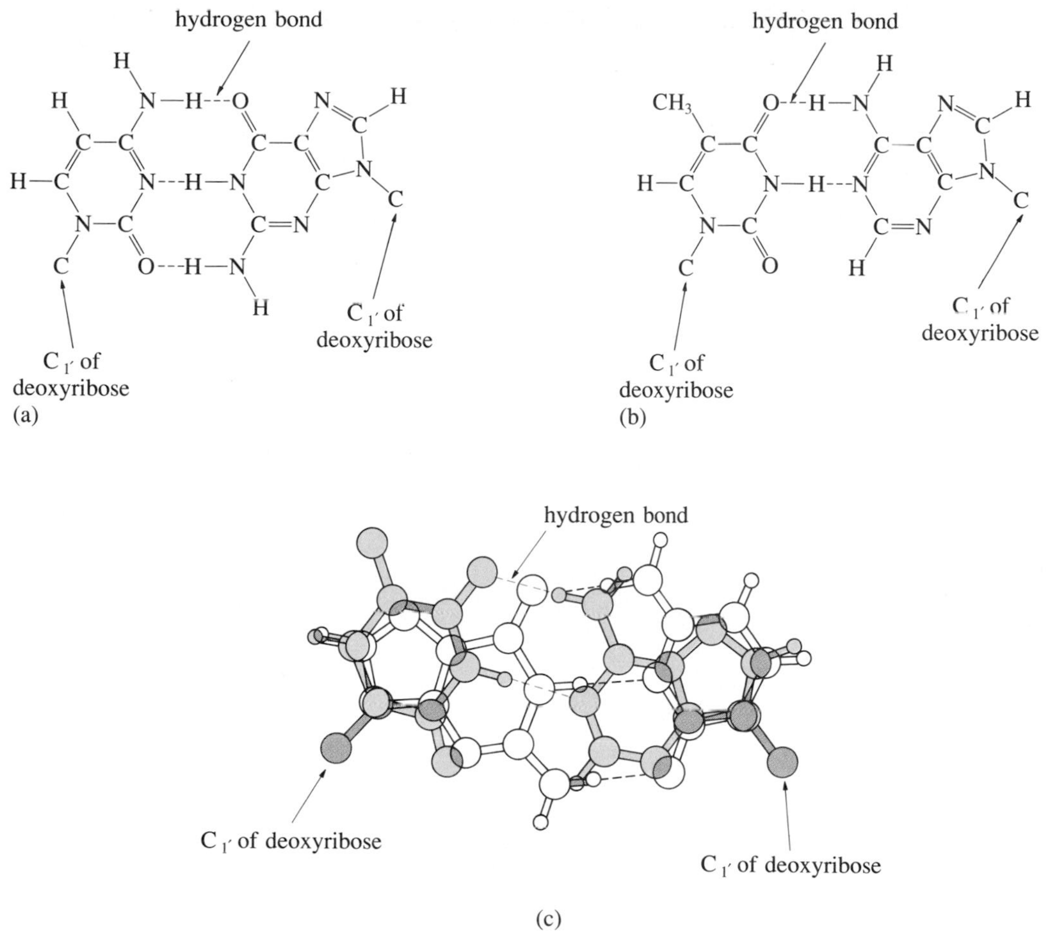

Figure 27.8 A closer look at the complementary base pairing in DNA. (a) A cytosine–guanine (C–G) base pair involves three hydrogen bonds. (b) A thymine–adenine (T–A) base pair involves two hydrogen bonds. (c) Superposition of the C–G and T–A base pairs shows that the two occupy the same space.

which are more hydrocarbonlike, are largely buried in the interior of the double helix, away from water.

3. The chains are held together by hydrogen bonds between bases. *Adenine (A) in one chain always hydrogen-bonds to thymine (T) in the other, and guanosine (G) in one chain always hydrogen-bonds to cytosine (C) in the other.* Thus in each pair a purine is always hydrogen-bonded to a pyrimidine. For this reason, A is said to be **complementary** to T, and G complementary to C. A closer look at these hydrogen-bonded *Watson–Crick base pairs* is shown in Fig. 27.8. Notice that the A–T pair has about the same spatial dimensions as the G–C pair.

4. The planes of successive complementary base pairs are stacked, one on top of the other, and are perpendicular to the axis of the helix. The distance between each successive base-pair plane is 3.4 Å. Since the helix makes a complete turn

every ten residues, this means that there is a distance of 10 × 3.4 = 34 Å along the helix per complete turn.

5. There is no restriction on the sequence of bases in a polynucleotide; however, because of the hydrogen bonding described in point 3, the sequence of one polynucleotide strand in the double helix is complementary to that in the other strand. Thus everywhere there is an A in one strand, there is a T in the other; everywhere there is a G in one strand, there is a C in the other.

We can see that the hydrogen-bonding complementarity in DNA accounts nicely for Chargaff's rules: if A always hydrogen bonds to T and G always hydrogen bonds to C, then the number of As must equal the number of Ts, and the number of Gs must equal the number of Cs. This structure also suggests a reasonable mechanism for the duplication of DNA during cell division: the two strands can come apart, and a new strand can be grown as a complement of each original strand. In other words, *the proper sequence of each new DNA strand during cellular reproduction is assured by hydrogen-bonding complementarity* (Fig. 27.9).

Problems

25 Suppose you could separate the strands of double-helical DNA. Would you expect Chargaff's rules to apply to the individual strands? Explain.

26 Draw in detail the structure of a section of RNA four residues long, which, from the 5′-end, has the following sequence of bases: A, U, C, G.

C. DNA, RNA, and the Genetic Code

The sequence of nucleotides in DNA forms a linear code for every protein and RNA molecule. To understand this point, let us see how the following strand of DNA, which we can imagine as part of a gene in some organism, could be used biologically to direct the synthesis of a section of a specific protein. If this were a DNA strand from a cell, it would be one of the two strands of the double helix; each letter in the sequence identifies a residue of DNA by its particular base.

← 3′-end 5′-end →
. . .–A–A–A–G–A–T–T–C–A–C–C–C–C–T–C–A–T–C–. . .

First, a strand of DNA directs the synthesis of a complementary strand of RNA. This RNA is called **messenger RNA** (mRNA) and the process by which it is assembled is called **transcription.** The sequence of the mRNA transcript is complementary to one DNA strand of the gene. For example, everywhere there is a G in DNA, there is a C (the complementary base) in the mRNA transcript. An adenine (A) in DNA is transcribed into a uracil (U). (Messenger RNA contains uracil rather than thymine (T); uracil is just a thymine without its methyl group.) Thus our gene fragment above would be transcribed as follows:

DNA, a polydeoxyribonucleotide
3′-end 5′-end
. . .– A–A–A–G–A–T–T–C–A–C–C–C–C–T–C–A–T–C–. . .
. . .– U–U–U–C–U–A–A–G–U–G–G–G–G–A–G–U–A–G–. . .
5′-end 3′-end
mRNA, a polyribonucleotide

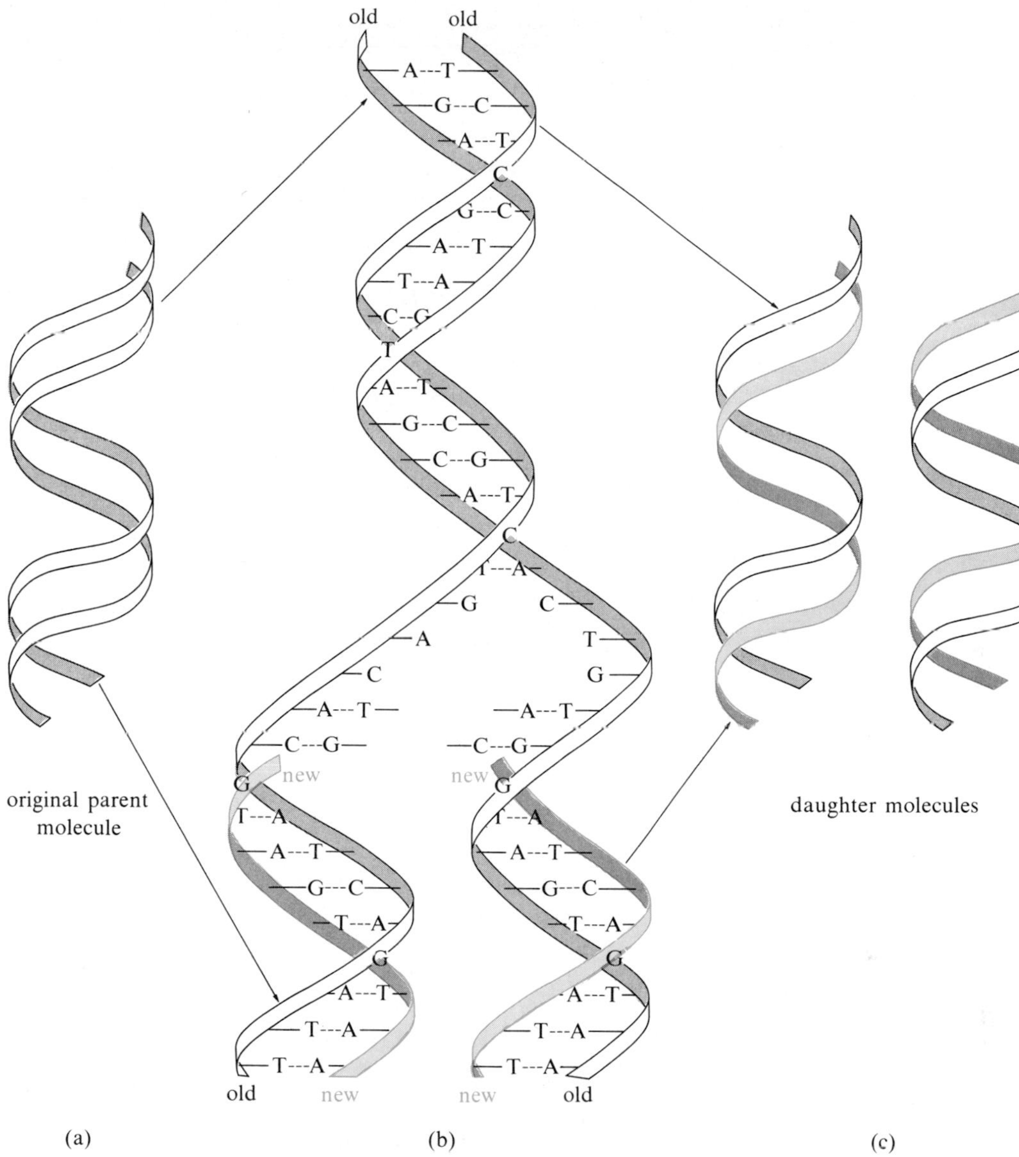

Figure 27.9 Complementary base pairing in DNA is crucial to its faithful replication. (a) A typical DNA double helix. (b) In the replicating DNA a new strand grows on each of the original strands. (c) Two new molecules of DNA, each containing one old strand and one new strand.

Notice that the complementary sequence of mRNA runs in the opposite direction to that of its parent DNA—the 3′-end of RNA matches the 5′-end of DNA, and vice versa.

Once the mRNA synthesis is complete, the mRNA sequence is used by the cell to direct the synthesis of a specific protein from its component amino acids. This process is called **translation.** The result is very simple. Each successive three-residue triplet in the sequence of mRNA is translated as a specific amino acid in the sequence of a

TABLE 27.3 The Genetic Code

5′-OH Terminal base	Middle base				3′-OH Terminal base
	U	C	A	G	
U	Phe	Ser	Tyr	Cys	U
	Phe	Ser	Tyr	Cys	U
	Leu	Ser	(Stop)	(Stop)	A
	Leu	Ser	(Stop)	Trp	G
C	Leu	Pro	His	Arg	U
	Leu	Pro	His	Arg	C
	Leu	Pro	Gln	Arg	A
	Leu	Pro	Gln	Arg	G
A	Ile	Thr	Asn	Ser	U
	Ile	Thr	Asn	Ser	C
	Ile	Thr	Lys	Arg	A
	Met*	Thr	Lys	Arg	G
G	Val	Ala	Asp	Gly	U
	Val	Ala	Asp	Gly	C
	Val	Ala	Glu	Gly	A
	Val*	Ala	Glu	Gly	G

*Sometimes used as "start" codons.

protein according to the **genetic code** given in Table 27.3. Thus, the particular stretch of mRNA shown above would be translated into a protein sequence as follows:

mRNA

5′-end 3′-end

. . .–U–U–U –C–U–A–A–G–U–G–G–G–G–A–G–U–A–G–. . .

. . .–Phe – Arg – Ser – Gly – Glu STOP!

amino end carboxy end

peptide chain

Just as a sequence of dots and dashes in Morse code can be used to form words, *the precise sequence of bases in DNA (by way of its complementary mRNA transcription product) codes for the successive amino acids of a protein.* Morse code has two coding units—the dot and the dash. In DNA or mRNA, there are four: A, T (U in mRNA), G, and C, the four nucleotide bases. Notice that the sequences of DNA and mRNA contain no "commas." The protein-synthesizing "machinery" of the cell knows where one amino acid code ends and another starts because, as Table 27.3 shows, there is a specific "start" signal—either of the nucleotide sequences AUG or GUG—at the appropriate point in the mRNA. Because mRNA also contains "stop" signals (UAA, UGA, or UAG), protein synthesis is also terminated at the right place.

Some amino acids have multiple codes. For example, we can see from Table 27.3 that glycine, the most abundant amino acid in proteins, is coded by GGU, GGC, GGA, and GGG.

It is possible for the change of only one base in the DNA (and consequently in the mRNA) of an organism to cause the change of an amino acid in the corresponding protein. A dramatic example of such a change is the genetic disease *sickle-cell anemia.* In this painful disease, the red blood cells take on a peculiar sickle shape that causes them to clog capillaries. The molecular basis for this disease is a single amino acid substitution in hemoglobin, the protein that transports oxygen in the blood (Fig. 26.11). In sickle-cell hemoglobin, glutamic acid at position 6 in one of the protein chains of normal hemoglobin is changed to valine. That is, sickle-cell disease results from a change in but one of the 141 amino acids in this hemoglobin chain! The mRNA genetic code for Glu is GAA and GAG; that for Val is GUA and GUG (among others). Clearly, a change of only one nucleotide (A $\rightarrow$ U) of the (3×141), or 423, nucleotides that code for this chain of hemoglobin is responsible for the disease.

Although we have focused on the structure and function of DNA, the structure of RNA is also important. A detailed discussion of this topic is beyond the scope of this text. However, you should be aware that there are many different types of RNA besides messenger RNA, each with a specific function in the cell.

Very powerful methods have been developed for sequencing DNA and RNA. With these methods, DNA and RNA sequences containing several hundred nucleotides can be determined in a relatively short time. These nucleotide sequences can be used to check, or even predict, protein sequences by applying the genetic code. Similarly, excellent methods for the synthesis of DNA and RNA fragments also exist. These methods strategically resemble peptide synthesis in that a strand of DNA or RNA is "grown" from individual nucleotides, using a series of protection, coupling, and deprotection steps. Molecular biologists have also discovered ways in which foreign DNA can be incorporated into, and expressed by, host organisms. All of these techniques used together have led to new biotechnologies that have been termed collectively "genetic engineering." One major pharmaceutical house employs a lowly bacterium—*Escherichia coli*—for the commercial production of *human* insulin using these techniques. Formerly all insulin used for the treatment of diabetes came from horses and pigs, and shortages of this important hormone occurred. The new process promises an abundant supply of human insulin, and foretells a revolution in the ways that many complex biological materials will be produced. It is also possible that, using the techniques of genetic engineering, we might one day be able to cure genetic diseases such as sickle-cell anemia.

D. DNA Modification and Chemical Carcinogenesis

We have seen how the double-helical structure of DNA, DNA replication, and the fidelity of DNA transcription into RNA involve very specific base-pairing complementarity. Other important processes we have not discussed, such as the recognition of the three-base triplet code of mRNA during protein biosynthesis, also involve this type of complementarity. The molecular basis of this complementarity is the specific hydrogen bonding between a pyrimidine and a purine base. We can imagine that, if this hydrogen bonding were upset, the base-pairing complementarity would also be upset, and with it, some or all of the biological processes that rely on this phenomenon. There is strong circumstantial evidence that chemical damage to DNA can interfere with this hydrogen-bonding complementarity and can trigger the state of uncontrolled cell division that we call cancer.

One type of chemical damage is caused by alkylating agents—compounds that have significant reactivity with nucleophiles in S_N2 reactions. Certain types of alkylating agents react with DNA by alkylating one or more of the nucleotide bases. These same alkylating agents are also carcinogens (cancer-causing compounds). A few such compounds are shown below.

$$CH_3-O-\overset{\overset{\displaystyle O}{\|}}{\underset{\underset{\displaystyle O}{\|}}{S}}-CH_3 \qquad CH_3-O-\overset{\overset{\displaystyle O}{\|}}{\underset{\underset{\displaystyle O}{\|}}{S}}-O-CH_3$$

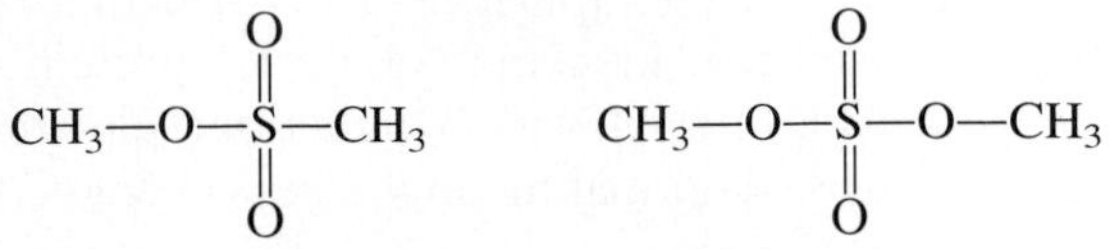

methyl methanesulfonate dimethyl sulfate
(weak carcinogen) (weak carcinogen)

$$H^+ + \quad CH_3-\overset{\overset{\displaystyle O=N}{|}}{N}-\overset{\overset{\displaystyle O}{\|}}{C}-NH_2 \xrightarrow{\text{living cell}} \quad CH_3-\overset{+}{N}\equiv N: \qquad + CO_2 + NH_3 \quad (27.53)$$

N-methyl-N-nitrosourea methyldiazonium ion
(potent carcinogen) (the actual alkylating agent)

When such alkylating agents (abbreviated CH₃—X below) react with DNA, among the products are alkylated guanosines. The major product is alkylated on N-7 of the guanine base, but an important minor product is alkylated on the oxygen at C-6 (called the O-6 position).

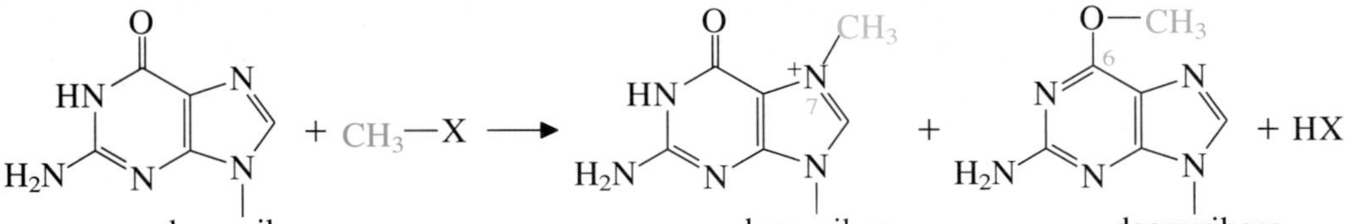

G residue of
DNA

N-7 alkylation
product
(major)

O-6 alkylation
product
(minor)

$$(27.54)$$

Notice that the alkylation at O-6 prevents the N-1 nitrogen from acting as a hydrogen-bond donor in a Watson–Crick base pair (Fig. 27.8) because the proton is lost from this nitrogen as a result of alkylation.

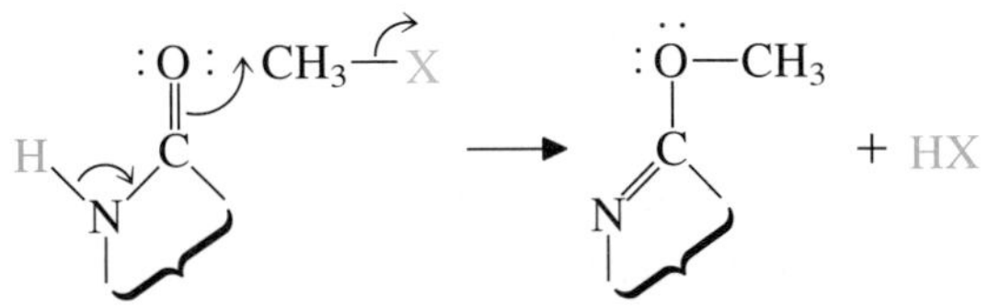

$$(27.55)$$

The N-7 alkylation does not directly affect any of the atoms involved in the hydrogen-bonding complementarity. It has been found that the alkylating agents that are the most potent carcinogens also yield the greatest amount of the product alkylated at O-6. Although this correlation does not prove that O-6 alkylation is a primary event in carcinogenesis, it provides strong circumstantial evidence in this direction.

In Sec. 16.8 we discussed the way in which aromatic hydrocarbons are converted into carcinogenic epoxides by enzymes in living systems. These epoxides have been shown to react with DNA; among the products of this reaction is a guanosine residue alkylated on the nitrogen at carbon-2 of the guanine base.

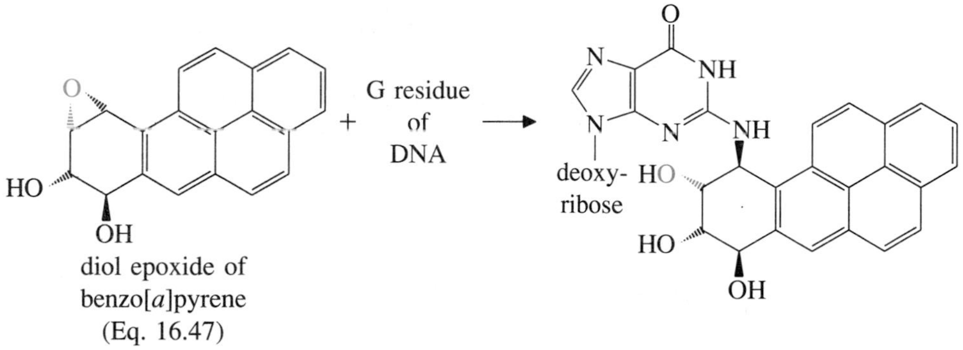

(27.56)

This nitrogen is also involved in the hydrogen-bonding interaction of G with C (Fig. 27.8). Thus, it may be that alkylation by aromatic hydrocarbon epoxides also triggers the onset of cancer by interfering with the base-pairing complementarity.

It is well known that excessive exposure to ultraviolet radiation can cause skin cancer. An interesting example of this phenomenon occurs when two pyrimidines occur in adjacent positions on a strand of DNA. Ultraviolet light promotes the [2 + 2] cycloaddition of the two pyrimidines (Sec. 25.3). In the example below, a thymine dimer is formed from two adjacent thymines.

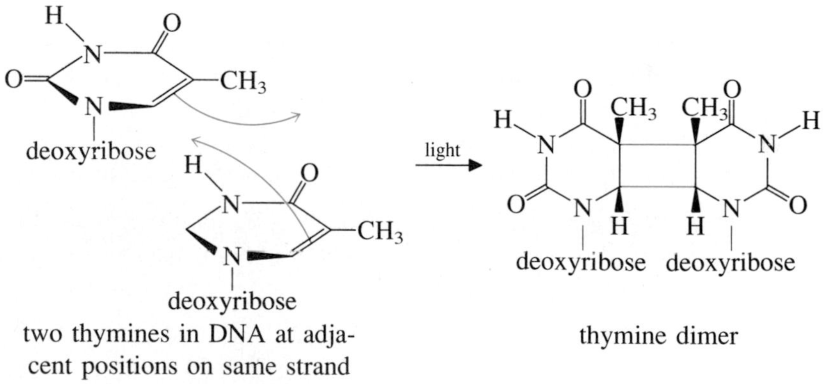

(27.57)

Most people have a biological repair system that effects the removal of the modified pyrimidines and repairs the DNA. People with a rare skin disease, *xeroderma pigmentosum,* have a genetic deficiency in the enzyme that initiates this repair. Most of these people contract skin cancer and die at an early age. Here, then, is a situation in which chemical modification of DNA has been clearly associated with the onset of cancer.

An understanding of the molecular basis of diseases such as cancer is clearly in its infancy. It should be clear, however, that progress is being made—progress, which, in many cases, stems from an understanding of the fundamental organic chemistry of the living cell.

KEY IDEAS IN CHAPTER 27

- Carbohydrates are polyhydroxy aldehydes and ketones and their derivatives.
- The D,L system is an older, but still widely used, method for specifying carbohydrate enantiomers. The D enantiomer is the one in which the asymmetric carbon of highest number has the same configuration as 2*R*-glyceraldehyde (D-glyceraldehyde).
- Sugars exist in cyclic furanose or pyranose forms in which a hydroxy group and the carbonyl group of the aldehyde or ketone have reacted to form a cyclic hemiacetal.
- The cyclic forms of sugars are in equilibrium with small amounts of their respective aldehydes or ketones, and can therefore undergo a number of aldehyde and ketone reactions. These include oxidation (bromine water or dilute nitric acid); reduction with sodium borohydride; cyanohydrin formation (the first step in the Kiliani–Fischer synthesis); oxime formation (the first step in the Wohl degradation); and base-catalyzed enolization and enolate-ion formation (the Lobry de Bruyn–Alberda van Eckenstein reaction).
- The —OH groups of sugars undergo many typical reactions of alcohols and glycols, such as ether formation, ester formation, and glycol cleavage with periodate.
- Because the hemiacetal carbon of sugars is asymmetric, the cyclic forms of sugars exist as diastereomers called anomers. The equilibration of anomers is why sugars undergo mutarotation.
- In a glycoside the —OH group at the anomeric carbon of a sugar is substituted with an ether (—OR) group. In disaccharides or polysaccharides, the —OR group is derived from another sugar residue. The —OR group of glycosides can be replaced with an —OH group by hydrolysis. Thus, higher saccharides can be hydrolyzed to their component monosaccharides in aqueous acid.
- Disaccharides, trisaccharides, etc., can be classified as reducing or nonreducing sugars. Reducing sugars have at least one free hemiacetal group. In nonreducing sugars all anomeric carbons are involved in glycosidic linkages.
- Ribonucleotides and deoxyribonucleotides, which are phosphorylated derivatives of ribonucleosides and deoxyribonucleosides, are the building blocks of RNA and DNA, respectively. These compounds are β-glycosides of either ribose or 2′-deoxyribose, respectively, and a purine or pyrimidine base. Adenine, guanine, and cytosine are bases in both DNA and RNA; thymine is unique to DNA, and uracil to RNA.
- An important conformation of DNA is the double helix, in which two strands of DNA running in opposite directions wrap around a common helical axis. The sugars and phosphate groups lie on the outside of the helix, and the bases are stacked in parallel planes on the inside. The two strands of the double helix are held together by hydrogen bonds between a purine base on one strand and a

pyrimidine base on the other (C or T with G or A, respectively). A number of known carcinogens apparently interfere with this hydrogen-bonding complementarity.

ADDITIONAL PROBLEMS

27 Give the product(s) expected when D-mannose reacts with each of the following reagents. (Assume that cyclic mannose derivatives are pyranoses.)
(a) $Ag^+(NH_3)_2$
(b) dilute HCl
(c) dilute NaOH
(d) Br_2/H_2O, then H_3O^+
(e) CH_3OH, HCl
(f) acetic anhydride
(g) product of (d) + $Ca(OH)_2$, then $Fe(OAc)_3$, H_2O_2
(h) product of (e) + $(CH_3)_2SO_4$ and NaOH

28 Give the products expected when D-ribose reacts with each of the following reagents.
(a) dilute HNO_3
(b) CN^-, H_2O
(c) product of (b) + $H_2/Pd/BaSO_4$
(d) CH_3OH, HCl (four compounds)
(e) products of (d) + $NaNH_2/NH_3$, then CH_3I
(f) NH_2OH
(g) product of (f) + $Ac_2O/NaOAc$, then NaOMe/MeOH

29 Draw the indicated type of structure for each compound below:
(a) (+)-lactose (Haworth projection)
(b) α-D-talopyranose (chair)
(c) CDP (sugar ring in Haworth projection)
(d) propyl β-L-arabinopyranoside (chair)

30 Name each aldohexose shown below. Be sure to give the appropriate name for the specific form shown for each sugar.

(a) 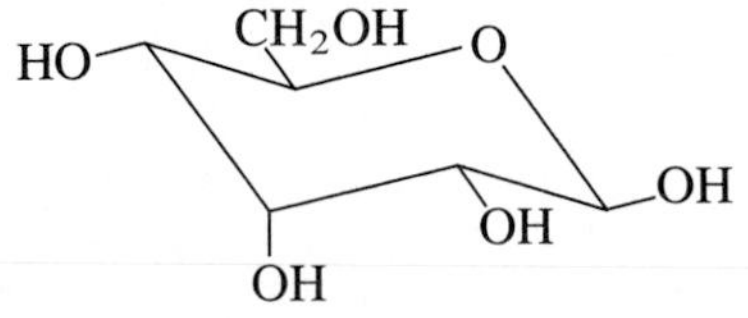

(b) 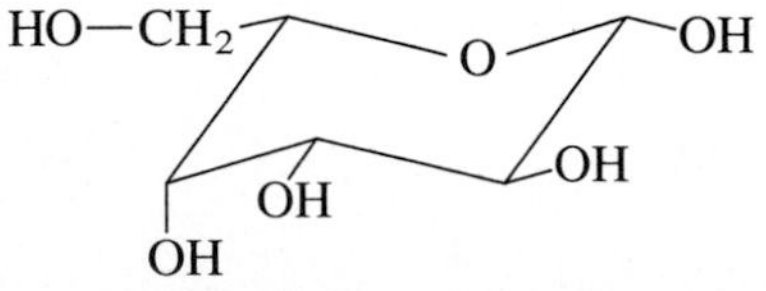

(Problem continues on page 1250.)

Problems (Cont.)

(c)

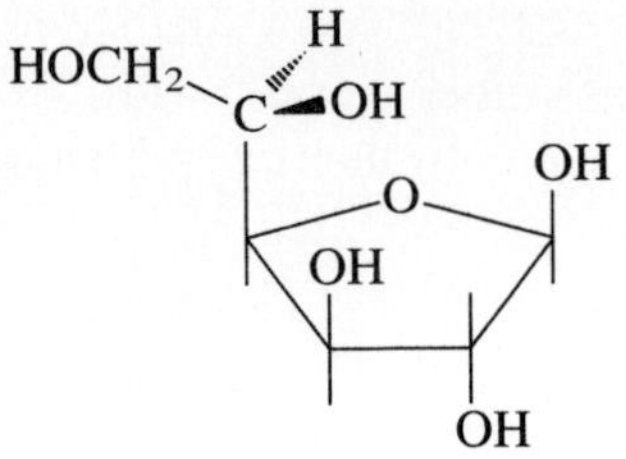

31 Draw the structure(s) of
(a) all the 2-ketohexoses.
(b) an achiral ketopentose $C_5H_{10}O_5$.
(c) α-D-galactofuranose.

32 Specify the relationship(s) of the compounds in each of the following sets. Choose among the following terms: epimers, anomers, enantiomers, diastereomers, structural isomers, none of the above. (More than one answer may be correct.)
(a) α-D-glucopyranose and β-D-glucopyranose
(b) α-D-glucopyranose and α-D-mannopyranose
(c) β-D-mannopyranose and β-L-mannopyranose
(d) α-D-ribofuranose and α-D-ribopyranose
(e) aldehyde form of D-glucose and α-D-glucopyranose
(f) methyl α-D-fructofuranoside and 2-*O*-methyl α-D-fructofuranose

33 Indicate the compounds or terms below that correspond to a correct description or structure of L-sorbose. (A structure is correct even if it is present in small amount.)

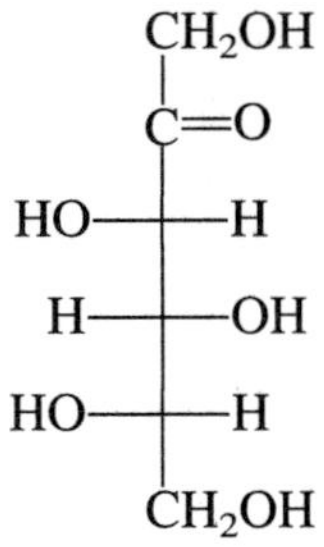

L-sorbose

(a) a hexose

(b) a ketohexose

(c) a glycoside

(d) an aldohexose

(e)

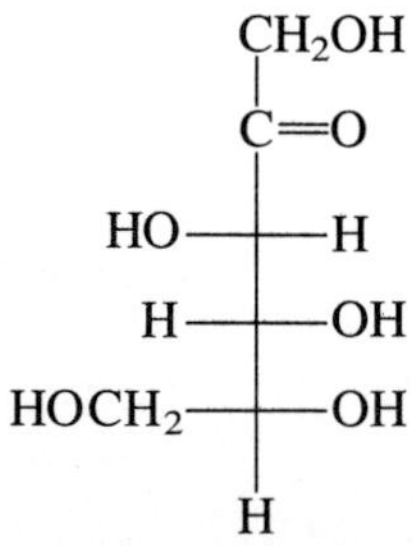

(f)

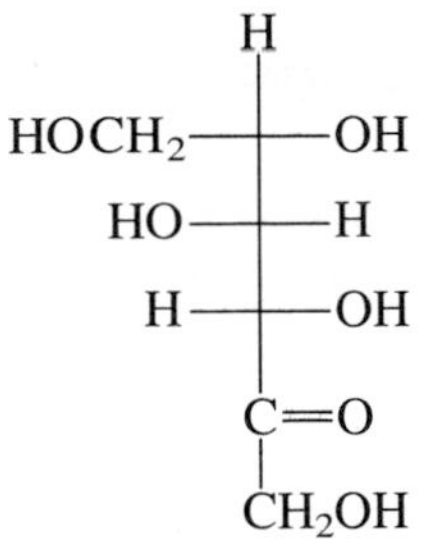

(g)

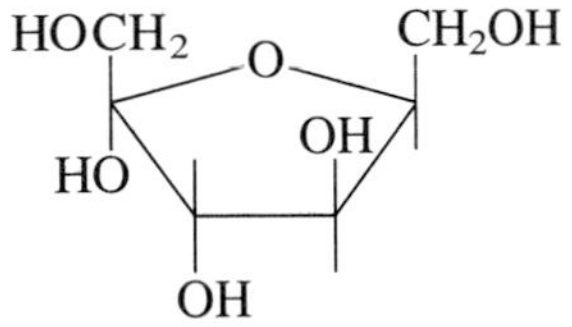

(h)

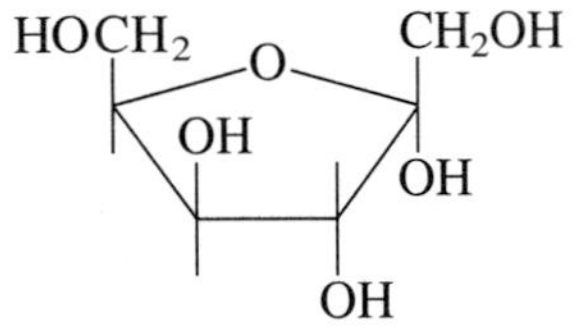

(i)

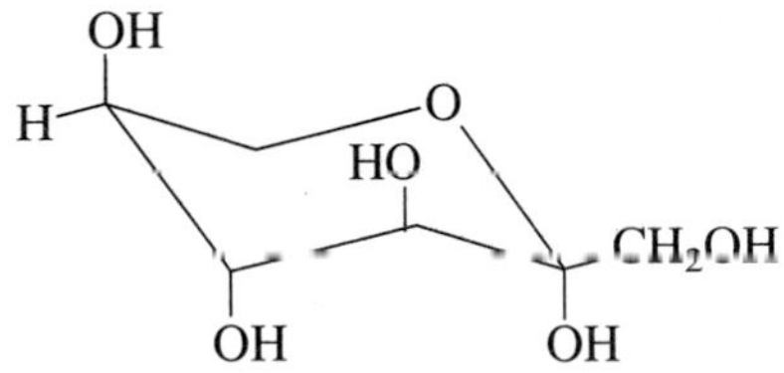

(j)

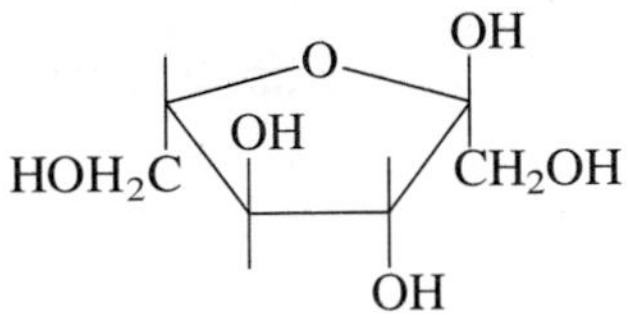

(k)

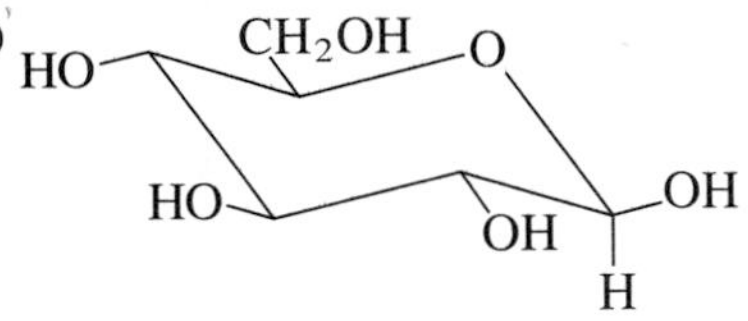

(l)

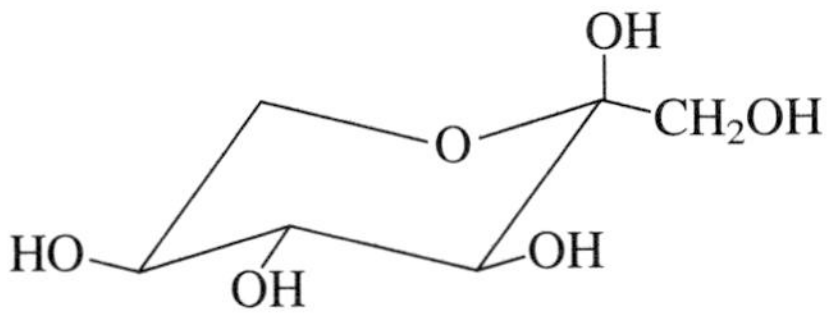

34 Consider the structure of *raffinose,* a trisaccharide found in sugar beets and a number of higher plants.

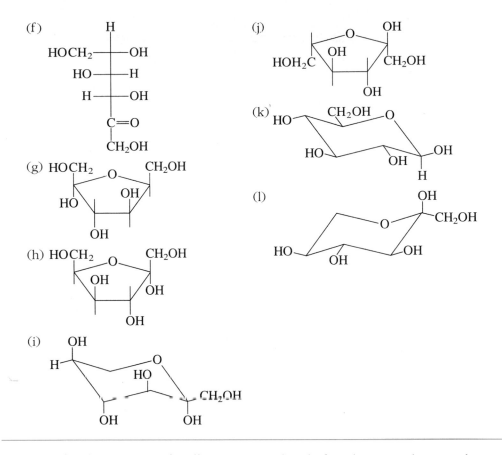

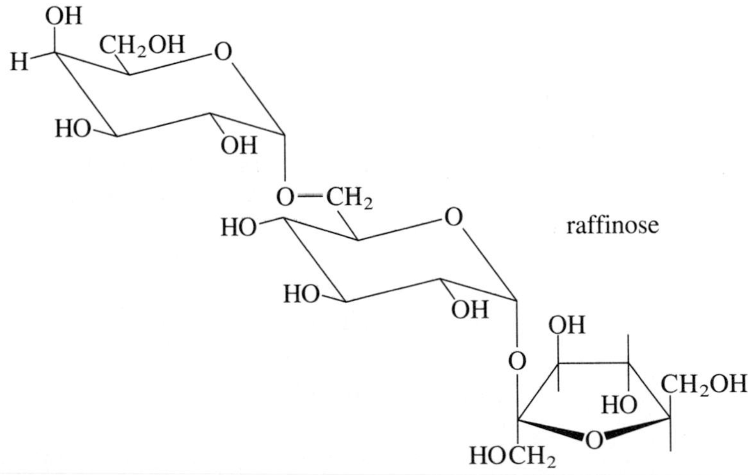

raffinose

(a) Classify raffinose as a reducing or nonreducing sugar, and tell how you know.
(b) Identify the glycoside linkages in raffinose, and classify each as either α or β.
(c) Name the monosaccharides formed when raffinose is hydrolyzed in aqueous acid.
(d) What products are formed when raffinose is treated with dimethyl sulfate in NaOH, and then with aqueous acid and heat?

Problems (Cont.)

35 The *inositols* are the 1,2,3,4,5,6-cyclohexanehexols. Of the nine stereoisomeric inositols, two are an enantiomeric pair and the rest are *meso*. Draw stereochemical formulas of all the inositols using wedges and dashed wedges. Identify the enantiomeric pair and show the internal mirror plane in each of the *meso* compounds.

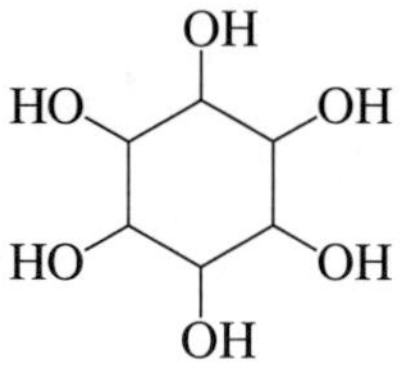

general structure of the inositols

36 (a) Lactose in its common crystalline form has the structure shown below. When crystalline lactose is dissolved in water, the optical rotation of the solution is found to increase slightly. Explain why the optical rotation changes.

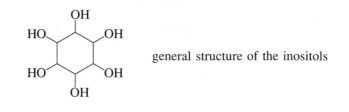

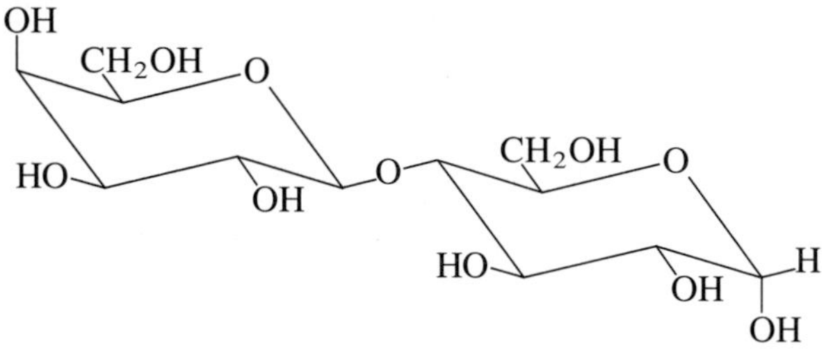

crystalline lactose

(b) After the increase in optical rotation described in part (a) has stopped, the sugar is heated in 1*N* aqueous HCl, and the optical rotation increases again. Explain this second change.

37 Explain each of the following observations:
(a) D-Fructose gives a positive Tollens' test.
(b) D-Glucose epimerizes in base but D-glucitol (Sec. 27.7D) does not.
(c) Treatment of 2-amino-2-deoxy-D-glucose (D-glucosamine; Eq. 27.50) with aqueous NaOH liberates ammonia.
(d) Treatment of cellulose with periodic acid followed by mild acid hydrolysis yields D-erythrose.
(e) $NaBH_4$ reduction of D-fructose gives *two* alditols, but the same reduction of D-glucose gives only one.
(f) Some sugars liberate formaldehyde when treated with periodic acid and others do not.

38 You have just prepared a methyl glycoside of D-fructose, but you are not sure whether it is a pyranoside or a furanoside. Show how you could use periodic acid to distinguish between the two possibilities.

39 Complete the following reactions by giving the major organic product(s):

(a) phenyl β-D-glucopyranoside + CH_3OH $\xrightarrow{H_2SO_4}$

(b)

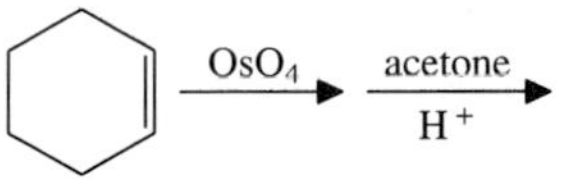

(c)

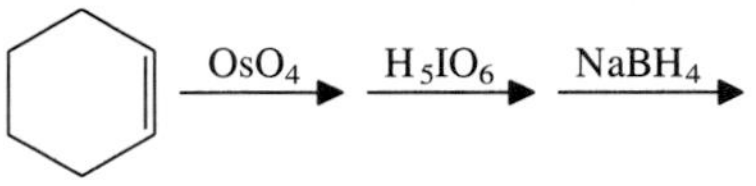

(d) $HOCH_2CH_2CH_2CH_2CH{=}O$ + CH_3OH $\xrightarrow{HCl}$

(e) (+)-lactose + C_2H_5OH $\xrightarrow[\text{heat}]{H_2SO_4}$

(f) (+)-sucrose + CH_3I (excess) $\xrightarrow{Ag_2O}$

(g)

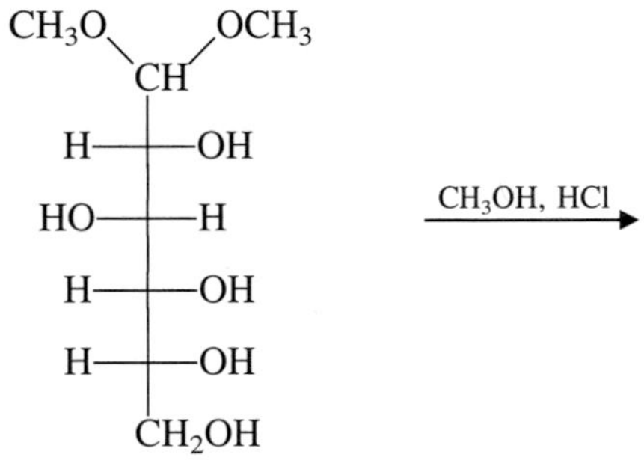

D-glucose dimethyl acetal
(prepared by special methods)

40 A biologist, Simone Spore, needs the following isotopically labeled sugars for some feeding experiments. Realizing your expertise in the saccharide field, she has come to you to ask whether you will synthesize these compounds for her. She has agreed to provide an unlimited supply of D-(−)-arabinose as a starting material. (* = ^{14}C, T = ^{3}H = tritium)

(a)

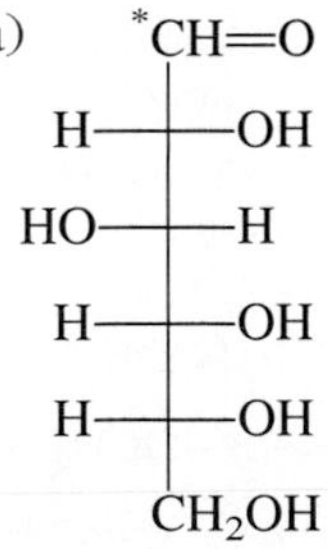

(b)

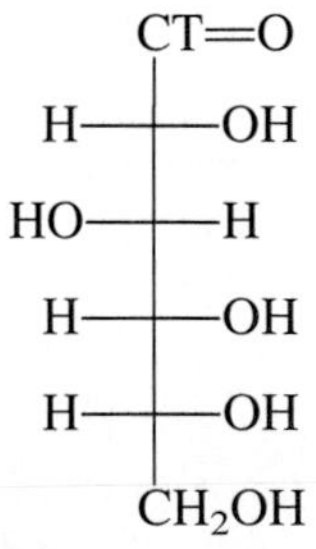

(c)

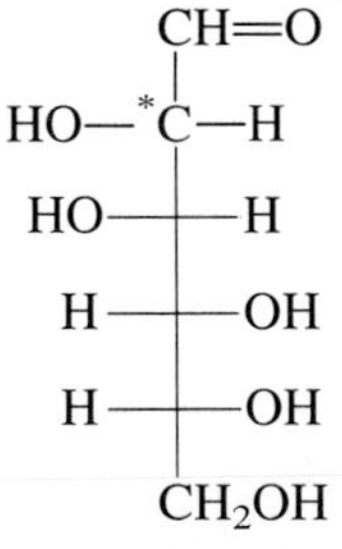

Available commercial sources of isotopes include $Na_2{}^*CO_3$, Na^*CN, $^3H_2$, and ^3H_2O. Outline a synthesis of each isotopically labeled compound.

Problems (Cont.)

41 Compound *A*, known to be a monomethyl ether of D-glucose, can be oxidized with bromine water. When *A* is subjected to the Wohl degradation, another sugar is obtained that can also be oxidized with bromine water. When *A* is subjected to the Kiliani–Fischer synthesis, two new methylated sugars are obtained. Both are optically active, and one of them can be oxidized with dilute HNO_3 to an optically inactive compound. Suggest a structure for *A*.

42 When D-ribose-5-phosphate was treated with an extract of mouse spleen, an optically *inactive* compound *X*, $C_5H_{10}O_5$, was produced. Treatment of *X* with $NaBH_4$ gave a mixture of ribitol and xylitol. (These are the alditols obtained when D-ribose and D-xylose, respectively, are reduced with $NaBH_4$.) Treatment of *X* with periodic acid produced two molar equivalents of formaldehyde. Suggest a structure for *X*.

43 After starch was fed to a culture of the bacterium *Bacillus macerans*, a saccharide *A* was isolated and its molecular weight determined. Hydrolysis of *A* gave six equivalents of D-glucose. Compound *A* was not oxidized by bromine water. Reaction of *A* with periodic acid consumed six moles of periodate per mole of *A*. When *A* was treated exhaustively with dimethyl sulfate and sodium hydroxide, and the product hydrolyzed in aqueous acid, 2,3,6-tri-*O*-methyl-D-glucose was obtained in excellent yield as the only ether product. Compound *A* was hydrolyzed to glucose by the enzyme α-amylase, an enzyme that hydrolyzes starch. Suggest a structure for *A*. This is one of an interesting group of compounds called *Schardinger dextrins*.

44 L-Rhamnose is a 6-deoxysugar with the following structure:

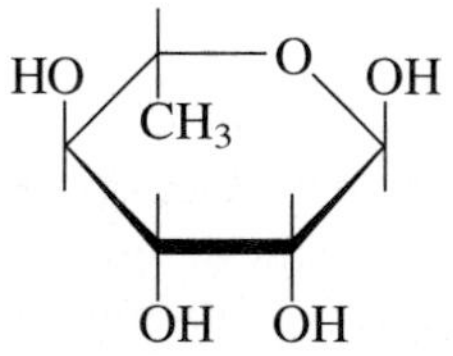

L-rhamnose

When a methyl glycoside of L-rhamnose, methyl α-L-rhamnopyranoside, is treated with $NaIO_4$, a compound *A*, $C_6H_{12}O_5$, was obtained that showed no evidence of a carbonyl group in its IR spectrum. Treatment of *A* with CH_3I/Ag_2O gave a derivative *B*, $C_8H_{16}O_5$. Treatment of *A* with H_2/Ni or $NaBH_4$ gave the following compound, *C*, shown in Fischer projection.

C

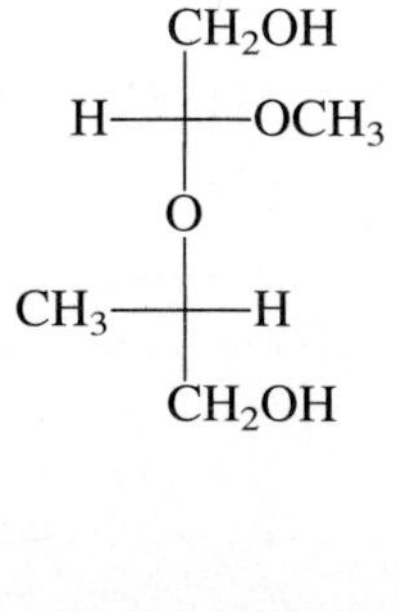

Give the structure of *A*. Explain why *A* gives no detectable carbonyl absorption in its IR spectrum, yet reacts with $NaBH_4$.

45 When RNA is treated with periodic acid, and the product of that reaction treated with base, only the nucleotide residue at the 3′-end is removed.
(a) Explain why the degradation is reasonable by showing its chemistry.
(b) Would the same degradation work with DNA? Explain.

46 The stability of a DNA double helix can be measured by the *melting temperature*, T_m, defined as the temperature at which the helix is 50% dissociated into random chains.
(a) Explain why the double helix formed between polydeoxyadenylic acid (polyA) and polydeoxythymidylic acid (polyT) has a considerably lower T_m (68°) than that of the double helix formed between polydeoxyguanidylic acid (polyG) and polydeoxycytidylic acid (polyC) (91°).
(b) Which of the following viruses has the higher ratio of (G + C)/(A + T) in its DNA? Explain.

Viral DNA source	T_m, °C
Human adenovirus I	58.5
Fowl pox	35

47 Planteose, a saccharide isolated from tobacco seeds, can be hydrolyzed in dilute acid to yield one equivalent each of D-fructose, D-glucose, and D-galactose. Almond emulsin (an enzyme preparation that hydrolyzes α-galactosides) catalyzes the hydrolysis of planteose to D-galactose and sucrose. Planteose does not react with bromine water. Treatment of planteose with $(CH_3)_2SO_4$/NaOH, followed by dilute acid hydrolysis, yields, among other compounds, 1,3,4-tri-*O*-methyl-D-fructose. Suggest a structure for planteose.

48 Maltose is a disaccharide obtained from the hydrolysis of starch. Maltose can be hydrolyzed to two equivalents of glucose, and can be oxidized to an acid, maltobionic acid, with bromine water. Treatment of maltose with dimethyl sulfate and sodium hydroxide, followed by hydrolysis of the product in aqueous acid, yields one equivalent each of 2,3,4,6-tetra-*O*-methyl-D-glucose and 2,3,6-tri-*O*-methyl-D-glucose. Hydrolysis of maltose is catalyzed by α-amylase, an enzyme known to affect only α-glycosidic linkages. Give *two* structures of maltose consistent with these data, and explain your answers.

Treatment of maltobionic acid with dimethyl sulfate and sodium hydroxide followed by hydrolysis of the product in aqueous acid gives 2,3,4,6-tetra-*O*-methyl-D-glucose and 2,3,5,6-tetra-*O*-methyl-D-gluconic acid. (See Eq. 27.23 for the structure of D-gluconic acid.) Give the structure of maltose.

Problems (Cont.)

49 The following sequence of reactions, called the *Weerman degradation,* can be used to degrade an aldose to another aldose with one fewer carbon atom.

$$\text{aldose} \xrightarrow{Br_2/H_2O} \xrightarrow{H_3O^+} [A] \xrightarrow[\text{heat}]{NH_3} [B] \xrightarrow[NaOH]{Cl_2} \text{aldose with one fewer carbon}$$

Explain what is happening in each step of this sequence. Your explanation should include the identity of each compound in brackets.

50 Saccharides of the following type are obtained from the partial hydrolysis of starch amylopectin:

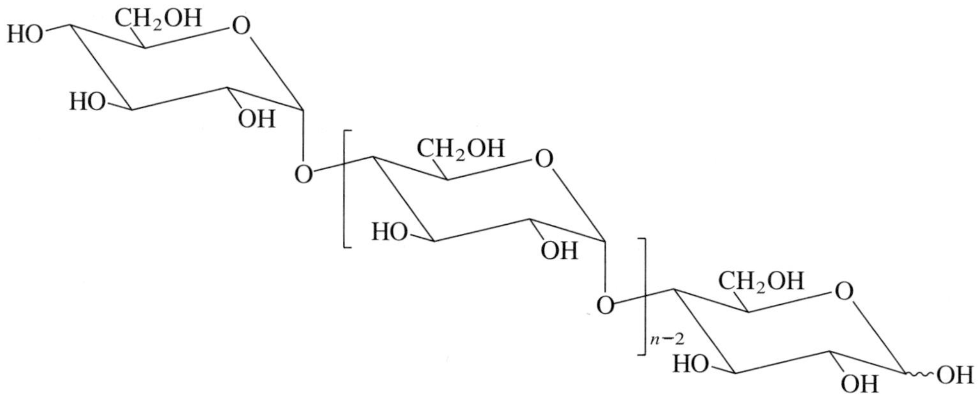

(a) When such a saccharide is treated with periodic acid followed by $NaBH_4$, and the resulting compound is hydrolyzed in aqueous acid, what products are obtained from

1. the residue at the nonreducing end?
2. the residue at the reducing end?
3. the residues in the middle?

(b) After a certain saccharide of the type shown above was treated successively with $NaIO_4$, $NaBH_4$, and then hydrolyzed in aqueous acid, some of the products were quantitated. A 1 : 11 ratio of glycerol to erythritol (below) was obtained. From this result what can you say about the structure of the saccharide?

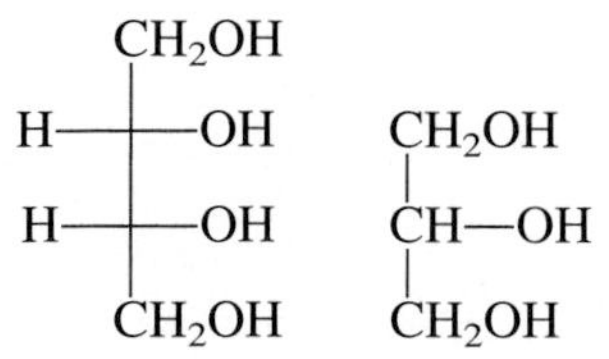

erythritol glycerol

51 What is the amino acid sequence of a peptide coded by a strand of RNA with the following base sequence?

5′ AUGAAACAAGAUUUUUAAUGGGG 3′

52 Suggest a reasonable mechanism for the following reaction, an example of the *Amadori rearrangement.*

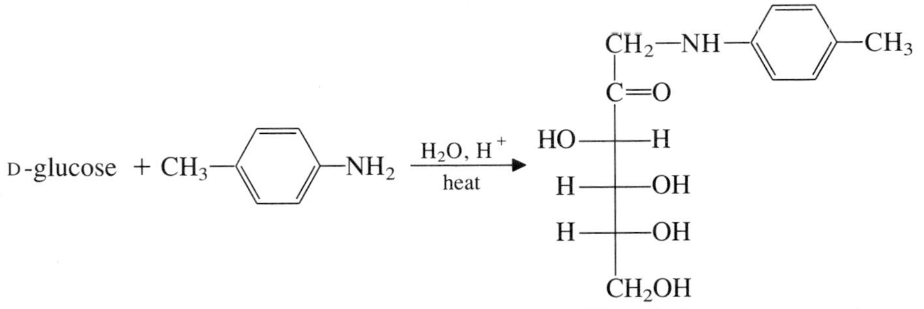

(Amadori rearrangement)

53 Furfural (furan-2-carboxaldehyde) is manufactured by heating oat hulls with acid. The source of furfural is pentose sugars in the hulls. Using the arrow formalism, give a mechanism for the acid catalyzed conversion of any aldopentose (for example, ribose or arabinose) into furfural.

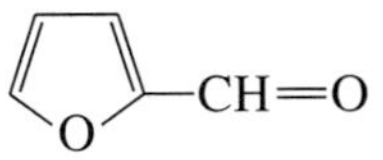

furfural

54 L-Ascorbic acid (vitamin C) has the following structure:

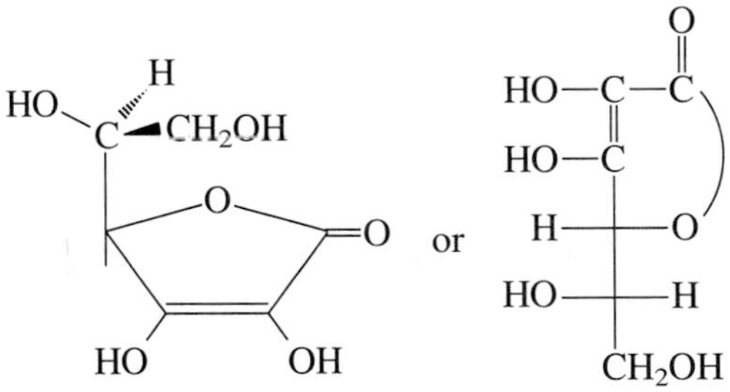

(a) Ascorbic acid has pK_a = 4.21, and is thus about as acidic as a typical carboxylic acid. Explain why ascorbic acid is so acidic.

(b) Thousands of tons annually of L-ascorbic acid are made commercially from D-glucose. In the synthesis below give the structures of the missing compounds.

(Problem continues on page 1258.)

Problems (*Cont.*)

D-glucose $\xrightarrow{\text{NaBH}_4}$ [A] $\xrightarrow[\text{oxidation}]{\text{biological}}$ [B] $\xrightarrow{[C]}$
D-glucitol L-sorbose
(L-sorbitol) (a keto sugar)

$\xrightarrow{\text{KMnO}_4,\text{OH}^-}$ $\xrightarrow{\text{H}^+/\text{H}_2\text{O}}$ [D] $\xrightarrow{\text{heat, H}^+}$ L-ascorbic acid

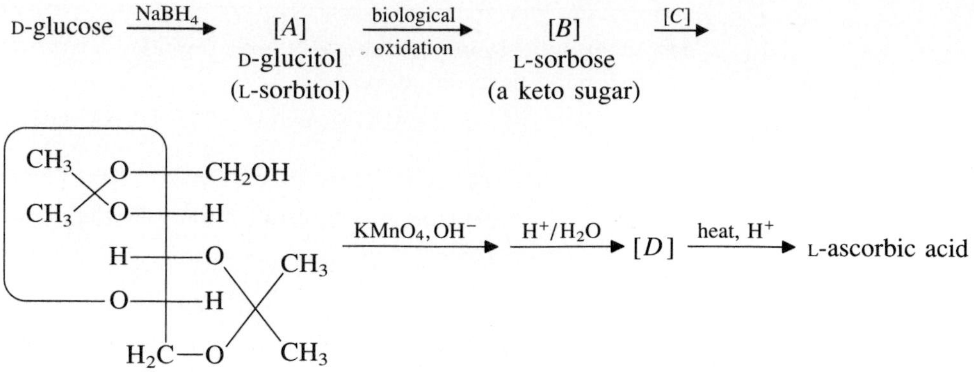

55 The optical activity of aldaric acids is a key element in the Fischer proof of glucose stereochemistry.
 (a) Is the aldaric acid that results from the oxidation of D-galactose with dilute HNO$_3$ chiral?
 (b) Is the γ-lactone formed from this aldaric acid chiral? Explain.
 (c) The aldaric acids are typically isolated from acidic solution as their γ-lactones (see, for example, Eq. 27.26). The γ-lactone in part (b) is found to be optically *inactive*. How do you reconcile this fact with your answer to (b)?

56 When DNA is treated with 0.5*N* NaOH at 25°, no reaction takes place, but when RNA is subjected to the same conditions, it is rapidly cleaved into mononucleotide 2- and 3-phosphates. Explain. (*Hint:* what is the only formal structural difference between RNA and DNA? How can this difference promote the observed behavior? See Sec. 11.6.)

57 At 100°, D-idose exists mostly (about 86%) as a 1,6-anhydropyranose:

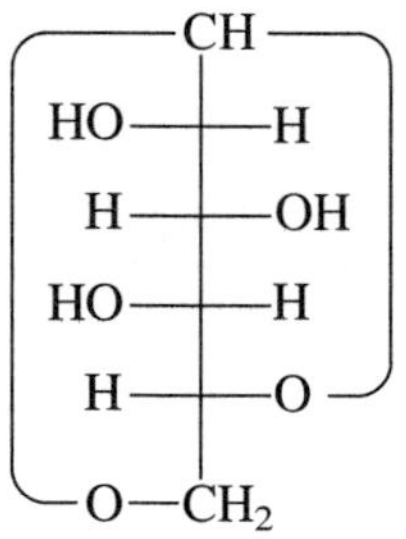

1,6-anhydro-D-idopyranose

 (a) Draw the chair conformation for this compound.
 (b) Explain why D-idose has more of the anhydro form than D-glucose. (Under the same conditions glucose contains only 0.2% of the 1,6-anhydro sugar.)

58 Consider the following equilibrium:

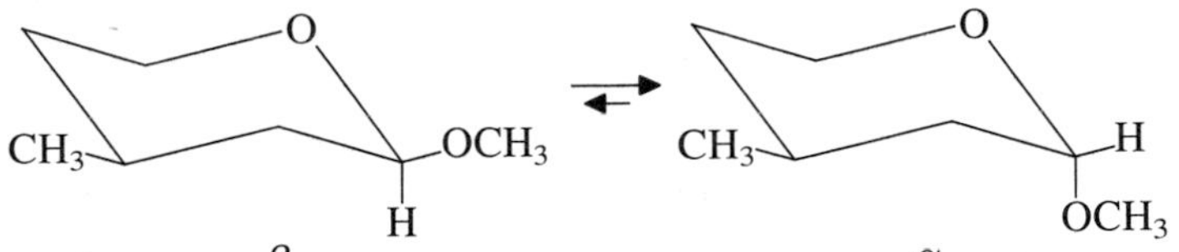

β (35%) α (65%)

(a) Is the predominant anomer the one you would expect from the conformational analysis of cyclohexanes? Explain.

The preference for the anomer with an axial substituent at carbon-1 is called the *anomeric effect*. It has been explained by considering the alignment of the C—O bond dipoles and their interaction, as shown.

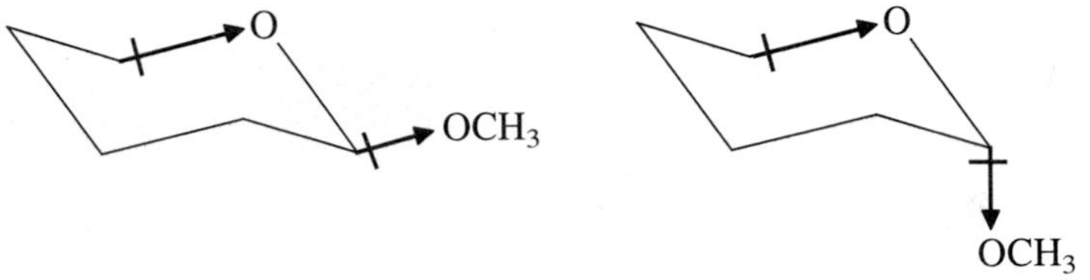

β α

(b) Explain how the interaction between these bond dipoles accounts qualitatively for the predominance of the α-anomer.

(c) Explain why D-mannopyranose contains more α-anomer at equilibrium than D-glucopyranose, even though there are more axial hydroxy groups in the mannose derivative. (See the data in Table 27.1.)

Appendices

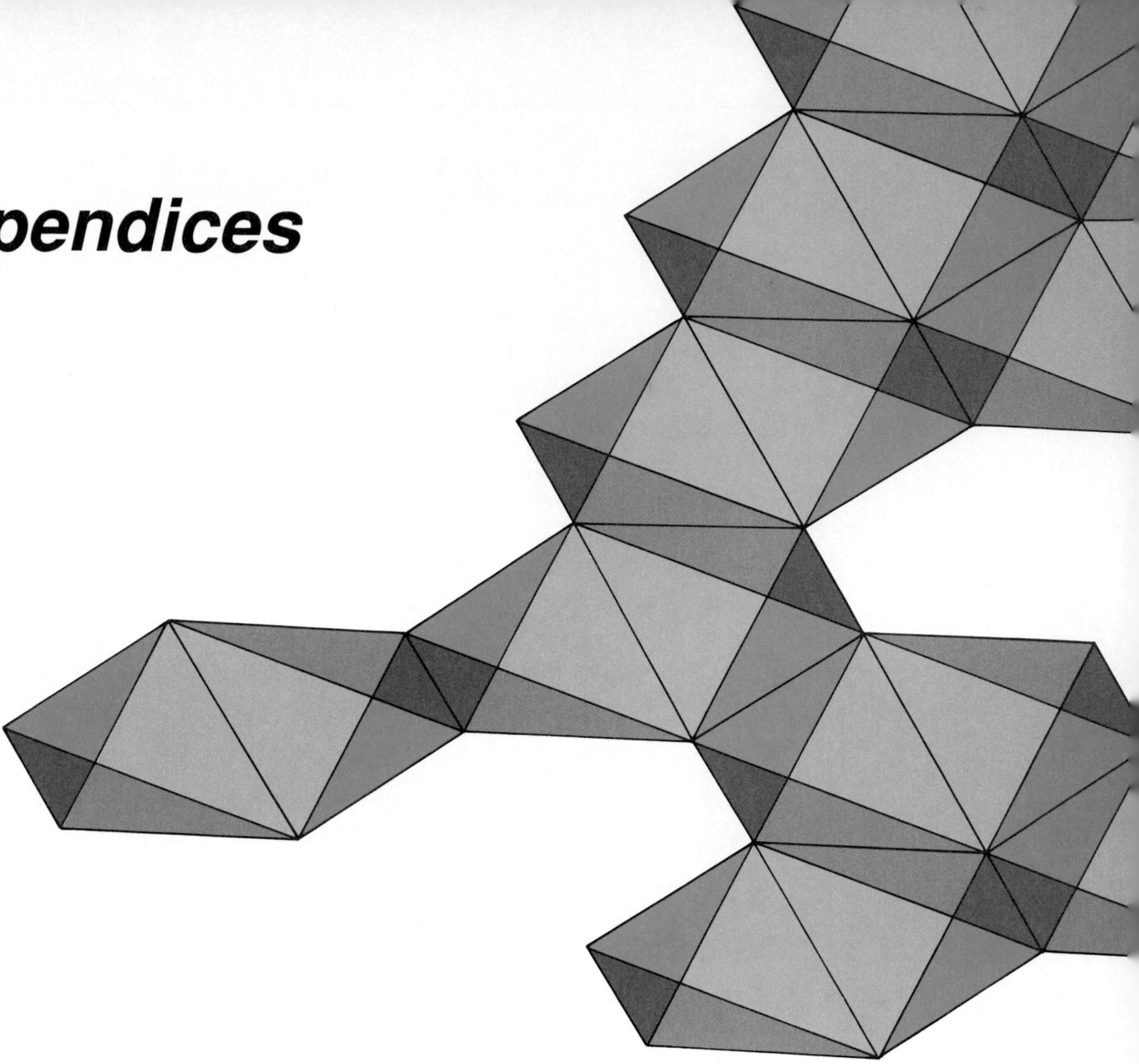

APPENDIX I. NOMENCLATURE OF ORGANIC COMPOUNDS

The name of an organic compound is based on its *principal group* and *principal chain.*

The *principal group* is *assigned* according to the following priorities:

$$-\overset{\text{O}}{\overset{\|}{\text{C}}}-\text{OH} \text{ (carboxylic acid)} > -\overset{\text{O}}{\overset{\|}{\text{C}}}-\text{O}-\overset{\text{O}}{\overset{\|}{\text{C}}}- \text{ (anhydride)} > -\overset{\text{O}}{\overset{\|}{\text{C}}}-\text{OR} \text{ (ester)} >$$

$$-\overset{\text{O}}{\overset{\|}{\text{C}}}-\text{X} \text{ (acid halide)} > -\overset{\text{O}}{\overset{\|}{\text{C}}}-\text{NR}_2 \text{ (amide)} > -\text{C}{\equiv}\text{N} \text{ (nitrile)} >$$

$$-\overset{\text{O}}{\overset{\|}{\text{C}}}-\text{H} \text{ (aldehyde)} > -\overset{\text{O}}{\overset{\|}{\text{C}}}- \text{ (ketone)} > -\text{OH} \text{ (alcohol, phenol)} >$$

$$-\text{SH} \text{ (thiol)} > -\text{NR}_2 \text{ (amino)} > -\text{OR} \text{ (ether)} > -\text{SR} \text{ (sulfide)}$$

The *principal chain* is *identified* by applying the following criteria in order until a decision can be made.

1. Maximum number of substituents corresponding to the principal group
2. Maximum number of double and triple bonds considered together
3. Maximum length
4. Maximum number of double bonds
5. Maximum number of substituents cited as prefixes

A *principal chain* is *numbered* by applying the following criteria in order until there is no ambiguity. Where multiple numbers are possible, comparisons are made at the first point of difference.

1. Lowest number for the principal group cited as a suffix—that is, the group on which the name is based. (Ethers and sulfides are never cited as suffixes and therefore do not fall under this rule.)
2. Lowest numbers for multiple bonds, with double bonds having priority over triple bonds.
3. Lowest numbers for all other substituents.
4. Lowest number for the substituent named as a prefix that is cited first in the name.

These are partial lists that cover most of the cases in this text. For a more complete discussion of nomenclature, see "Nomenclature of Organic Chemistry, 1979 Edition," by the International Union of Pure and Applied Chemistry, published by Pergamon Press.

APPENDIX II. IMPORTANT INFRARED ABSORPTIONS

Type of Absorption	Frequency, cm^{-1}	Comment
	Alkanes	
C—H stretch	2850–3000	occurs in all compounds with aliphatic C—H bonds
	Alkenes	
C=C stretch		
$—CH=CH_2$	1640	
$>C=CH_2$	1655	
others	1660–1675	not observed if alkene is symmetrical

(Table continues on page A-3.)

APPENDIX II (CONT.) Important Infrared Absorptions

Type of Absorption	Frequency, cm^{-1}	Comment
=C—H bend		
—CH=CH_2	910–990	
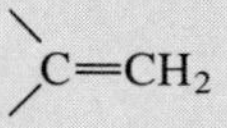	890	
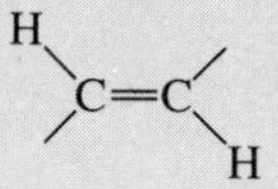	960–980	
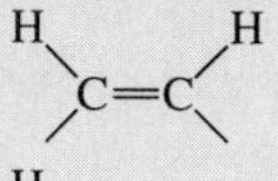	675–730	highly variable
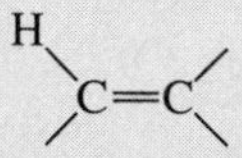	800–840	
	Alcohols and Phenols	
O—H stretch	3200–3400	
C—O stretch	1050–1250	also present in other compounds with C—O bonds: ethers, esters, etc.
	Alkynes	
C≡C stretch	2100–2200	not present in many internal alkynes
≡C—H stretch	3300	1-alkynes only
	Aromatic Compounds	
C=C stretch	1500, 1600	two absorptions
C—H bend	650–750	
overtone	1660–2000	
	Aldehydes	
C=O stretch		
ordinary	1720–1725	
α,β-unsaturated	1680–1690	
benzaldehydes	1700	
C—H stretch	2720	
	Ketones	
C=O stretch		
ordinary	1710–1715	increases with decreasing ring size (Table 21.3)
α,β-unsaturated	1670–1680	
aryl ketones	1680–1690	
	Carboxylic Acids	
C=O stretch		
ordinary	1710	
benzoic acids	1680–1690	
O—H stretch	2400–3000	strong and broad
	Esters and Lactones	
C=O stretch	1735	increases with decreasing ring size (Table 21.3)

APPENDIX II (CONT.) Important Infrared Absorptions

Type of Absorption	Frequency, cm^{-1}	Comment
	Acid Chlorides	
C=O stretch	1800	second weaker band sometimes observed at 1700–1750
	Anhydrides	
C=O stretch	1760, 1820	two bands; increases with decreasing ring size in cyclic anhydrides
	Amides and Lactams	
C=O stretch	1650–1655	increases with decreasing ring size (Table 21.3)
N—H bend	1640	
N—H stretch	3200–3400	doublet absorption observed for some primary amides
	Nitriles	
C≡N stretch	2200–2250	
	Amines	
N—H stretch	3200–3375	several absorptions sometimes observed, especially for primary amines

APPENDIX III. IMPORTANT NMR CHEMICAL SHIFTS

A. Protons within Functional Groups

Group	Chemical shift, ppm	Group	Chemical shift, ppm
—C—C—H	0.7–1.5	—C(=O)—H	9–11
>C=C<H	4.6–5.7	—C(=O)—N—H	7.5–9.5
—O—H	varies with solvent and with acidity of O—H	—C—NH—	0.5–1.5
—C≡C—H	1.7–2.5		
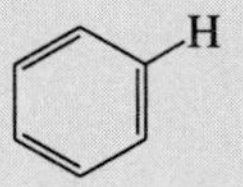	6.5–8.5	—NH—	2.5–3.5

(Table continues on page A-5.)

B. Alkyl Protons Adjacent to Functional Groups

Group G	δ for CH_3—G, ppm	G*
—H	0.2	—
—CR_3 or —CR_2— (R = H, alkyl)	0.9	0.5
—F	4.3	3.3
—Cl	3.0	2.5
—Br	2.7	2.3
—I	2.2	2.2
—CR=CR_2 (R = H, alkyl)	1.8	1.3
—C≡C—R (R = H, alkyl)	2.0	1.4
—OH	3.5	2.6
—OR (R = alkyl)	3.3	2.4
—OR (R = aryl)	3.7	2.9
—SH	2.4	1.6
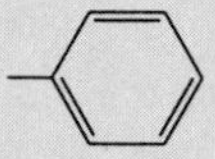	2.3	1.8
—C(=O)—R	2.1 (R = alkyl) 2.6 (R = aryl)	1.5
—C(=O)—OR (R = H, alkyl)	2.1	1.5
—O—C(=O)—R	3.6 (R = alkyl) 3.8 (R = aryl)	3.0
—C(=O)—NR_2 (R = H, alkyl)	2.0	1.5
—NR—C(=O)—R (R = H, alkyl)	2.8	
—NR_2 (R = H, alkyl)	2.2	1.6
—NR (R = aryl)	2.9	
—C≡N	2.0	1.6

*For use with Eq. 13.5

Some ^{13}C chemical shifts are given in Table 13.7, page 531.

APPENDIX IV. SUMMARY OF SYNTHETIC METHODS

The numbers in parentheses following each method refer to the sections in which the method is discussed. These methods are listed in order of occurrence in the text. Thus, a review at any point in the text is possible by considering the methods listed prior to that point.

Care should be taken in applying this summary, since syntheses classified for one purpose can in some cases be used for another. For example, phenols can be converted into alcohols by catalytic hydrogenation; but catalytic hydrogenation is listed under "Synthesis of Alkanes" (method no. 6) because the actual transformation is the formation of —CH_2—CH_2— groups from —CH=CH— groups; the presence of the —OH group is incidental.

Synthesis of Alkanes and Arenes

1. Catalytic hydrogenation of alkenes (5.7A)
2. Protonolysis of Grignard or related organometallic reagents (8.8B)
3. Cleavage of sulfides with Raney nickel (11.3B)
4. Catalytic hydrogenation of alkynes (14.6A)
5. Friedel–Crafts alkylation of benzene derivatives (16.4F)
6. Catalytic hydrogenation of benzene derivatives (16.6)
7. Wolff–Kishner or Clemmensen reductions of aldehydes or ketones (19.12)
8. Reaction of aryldiazonium ions with hypophosphorus acid (23.10A)

Synthesis of Alkenes

1. E2 elimination of alkyl halides or sulfonates (9.4, 10.3A)
2. Acid-catalyzed dehydration of alcohols (10.1)
3. Catalytic hydrogenation of alkynes to *cis*-alkenes (14.6A)
4. Reduction of alkynes to *trans*-alkenes with alkali metals in liquid ammonia (14.6B)
5. Diels–Alder reaction of dienes and alkenes to give cyclic alkenes (15.3)
6. Wittig reaction of aldehydes or ketones (19.13)
7. Hofmann elimination of quaternary ammonium hydroxides (23.8)

Synthesis of Alkynes

1. S_N2 reaction of acetylide anions with alkyl halides or sulfonate esters (14.7B)

Synthesis of Alkyl and Aryl Halides

(Syntheses apply only to alkyl halides unless noted otherwise.)

1. Addition of hydrogen halides to alkenes (4.5, 15.4A)
2. Addition of HBr to alkenes in the presence of peroxides or other free-radical initiators (5.8)
3. Addition of halogens to alkenes (for dihalide or halohydrin formation; 5.1)
4. Synthesis of dihalocyclopropanes by addition of dihalomethylene to alkenes (9.7)
5. Reaction of alcohols with hydrogen halides, thionyl chloride, or phosphorus tribromide (10.2, 10.4)

6. S_N2 reaction of sulfonate esters with halide ion (10.3A)
7. Halogenation of benzene derivatives to aryl halides (16.4A)
8. Allylic and benzylic bromination of alkenes or arenes (17.2)
9. α-Halogenation of aldehydes, ketones, or carboxylic acids (22.3A,C)
10. Reaction of aryldiazonium ions with cuprous halides or fluoroboric acid to give aryl halides (Sandmeyer and Schiemann reactions; 23.10B)

Synthesis of Grignard and Related Organometallic Compounds

1. Reaction of alkyl or aryl halides with metals (8.8A, 18.5)
2. Synthesis of lithium dialkylcuprates by reaction of lithium reagents with cuprous halides (21.10B)

Synthesis of Alcohols and Phenols

(Syntheses apply only to alcohols unless noted otherwise.)

1. Hydration of alkenes (used industrially, but usually not a good laboratory method; 4.8)
2. Hydroboration–oxidation of alkenes (5.3A)
3. Oxymercuration–reduction of alkenes (5.3B)
4. Reaction of ethylene oxide with Grignard reagents (11.4B)
5. Reduction of aldehydes or ketones (19.8, 22.9, 27.7D)
6. Reaction of Grignard or related organometallic reagents with aldehydes or ketones (19.9, 22.10B)
7. Reduction of carboxylic acids to primary alcohols (20.10)
8. Reduction of esters to primary alcohols (21.9A, 22.9)
9. Reaction of esters with Grignard or related organometallic reagents (21.10A)

Synthesis of Ethers

1. Williamson synthesis (S_N2 reaction of alkyl halides with alkoxides; 11.1A, 27.6)
2. Alkoxymercuration–reduction of alkenes (11.1B)

Synthesis of Epoxides

1. Oxidation of alkenes with peroxyacids (11.2A)
2. Cyclization of halohydrins (11.2B)

Synthesis of Glycols

1. Reaction of alkenes with osmium tetroxide or potassium permanganate (5.4)
2. Hydrolysis of epoxides (11.4A)

Synthesis of Aldehydes

1. Ozonolysis–reduction of alkenes (of limited utility, since carbon–carbon bonds are broken; 5.5)
2. Oxidation of primary alcohols (10.6)
3. Oxidative cleavage of glycols with sodium periodate (of limited utility, since carbon–carbon bonds are broken; 10.6D; 27.7C)
4. Hydroboration–oxidation of alkynes (14.5B)
5. Reduction of acid chlorides (21.9D)
6. Aldol condensation of aldehydes to give α,β-unsaturated aldehydes (22.4)

Synthesis of Ketones

1. Ozonolysis of alkenes (of limited utility, since carbon–carbon bonds are broken; 5.5)
2. Pinacol rearrangement (10.1B)
3. Oxidation of secondary alcohols (10.6)
4. Oxidative cleavage of glycols with sodium periodate (of limited utility, since carbon–carbon bonds are broken; 10.6D)
5. Hydration of alkynes (14.5A)
6. Friedel–Crafts acylation of benzene derivatives or aromatic heterocycles (16.4E, 24.3A)
7. Reaction of acid chlorides with lithium dialkylcuprates (21.10B)
8. Aldol condensation of ketones to give α,β-unsaturated ketones (22.4)
9. Crossed Claisen condensation of esters to give β-diketones (22.5C)
10. Acetoacetic ester synthesis (22.6C)
11. Reaction of lithium dialkylcuprates with α,β-unsaturated ketones (22.10B)

Synthesis of Carboxylic and Sulfonic Acids

(Syntheses apply to carboxylic acids unless noted otherwise.)

1. Ozonolysis–oxidation of alkenes (of limited utility, since carbon–carbon bonds are broken; 5.5)
2. Oxidation of primary alcohols (10.6A,C; 27.7B)
3. Oxidation of thiols to sulfonic acids (10.9)
4. Sulfonation of benzene derivatives to arylsulfonic acids (16.4D)
5. Side-chain oxidation of alkylbenzenes (17.5)
6. Oxidation of aldehydes (19.14, 27.7A)
7. Reaction of Grignard reagents with carbon dioxide (20.6)

8. Acid- or base-promoted hydrolysis of carboxylic acid derivatives, especially nitriles (21.7, 26.5C)
9. Haloform reaction of methyl ketones or methyl carbinols (of limited utility, since this reaction breaks carbon–carbon bonds; 22.3B)
10. Malonic ester synthesis (22.6A, 26.5B)

Synthesis of Esters

1. Acid-catalyzed reaction of carboxylic acids with alcohols (20.8A)
2. Alkylation of carboxylic acids or carboxylates (20.8B)
3. Reaction of acid chlorides, anhydrides, or esters with alcohols (21.8A,B,C; 27.6)
4. Claisen and Dieckmann condensations (22.5A,B)
5. Alkylation of ester-enolate ions (22.6B)
6. Conjugate addition of enolate ions to α,β-unsaturated carbonyl compounds (Michael addition; 22.8C)

Synthesis of Anhydrides

1. Reaction of carboxylic acids with dehydrating reagents (20.9B)
2. Reaction of acid chlorides with carboxylate salts (21.8B)

Synthesis of Acid Chlorides

1. Reaction of carboxylic acids with thionyl chloride, phosphorus pentachloride, or related reagents (20.9A)

Synthesis of Amides

1. Reaction of acid chlorides, anhydrides, or esters with amines (21.8)
2. Condensation of amines and carboxylic acids with dicyclohexylcarbodiimide (26.8B)

Synthesis of Nitriles

1. Cyanohydrin formation from aldehydes or some ketones (19.7, 21.11, 27.8A)
2. S_N2 reaction of cyanide ion with alkyl halides or sulfonate esters (21.11)
3. Conjugate addition of cyanide to α,β-unsaturated carbonyl compounds (22.8A)
4. Reaction of aryldiazonium ions with cuprous cyanide (23.10B)

Synthesis of Amines

1. Reduction of amides (21.9B)
2. Reduction of nitriles (21.9C)

3. Direct alkylation of amines (of limited utility because overalkylation can occur; useful for forming quaternary ammonium salts; 23.7A; 26.5A)
4. Reductive amination of aldehydes or ketones (23.7B)
5. Aromatic substitution reactions of aniline derivatives (23.9)
6. Gabriel synthesis of primary amines (23.11A)
7. Reduction of nitro compounds (23.11B)
8. Hofmann or Curtius rearrangements (23.11C)

APPENDIX V. REACTIONS USED TO FORM CARBON–CARBON BONDS

Reactions are listed in order of occurrence in the text. The numbers in parentheses following each method refer to the sections in which the reaction is discussed.

1. Cyclopropane formation by addition of carbenes to alkenes (9.7)
2. Reaction of Grignard reagents with ethylene oxide (11.4B)
3. Reaction of acetylide ions with alkyl halides or sulfonate esters (14.7B)
4. Diels–Alder reactions (15.3)
5. Friedel–Crafts acylation (16.4E) and alkylation (16.4F)
6. Cyanohydrin formation (19.7, 27.8A)
7. Reaction of Grignard reagents with aldehydes or ketones (19.9)
8. Reaction of Grignard reagents with carbon dioxide (20.6)
9. Reaction of Grignard reagents with esters (21.10A)
10. Reaction of lithium dialkylcuprates with acid chlorides (21.10B)
11. Reaction of cyanide ion with alkyl halides or sulfonate esters (21.11).
12. Aldol condensation (22.4)
13. Claisen condensation (22.5)
14. Malonic ester synthesis (22.6A)
15. Alkylation of ester enolates with amide bases and alkyl halides (22.6B).
16. Acetoacetic ester synthesis (22.6C).
17. Addition of cyanide or enolate ions to α,β-unsaturated carbonyl compounds (Michael addition; 22.8C).
18. Reaction of lithium dialkylcuprates with α,β-unsaturated carbonyl compounds (22.10B).
19. Reaction of aryldiazonium salts with cuprous cyanide (23.10B)

APPENDIX VI. TYPICAL ACIDITIES AND BASICITIES OF ORGANIC FUNCTIONAL GROUPS

A. Acidities of Groups that Ionize to Give Anionic Conjugate Bases (Acidic Protons in Color)

Functional group	Structure	Typical pK_a
sulfonic acid*,†	$R-S(=O)_2-O-H$	<1 (strong acid)
carboxylic acid*,†	$R-C(=O)-O-H$	3–5
phenol†	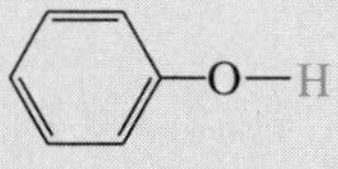	10
thiol†	$R-S-H$	10
sulfonamide†	$R-S(=O)_2-N(H)-R$	10
amide	$R-C(=O)-N(H)-R$	15–17
alcohol	$R-O-H$	15–19
aldehyde, ketone	$R-C(=O)-C(H)R_2$	17–20
ester	$R_2C(H)-C(=O)-OR$	20–25
nitrile	$R_2C(H)-C\equiv N$	25
amine	R_2N-H	32–35
alkane	R_3C-H	50–60 (?)

*Acidic enough to dissolve in aqueous 5% $NaHCO_3$ solution.
†Acidic enough to dissolve in aqueous 5% NaOH solution.

B. Basicities of Groups that Protonate to Give Cationic Conjugate Acids (Acidic Protons in Color)

Functional group	Structure of conjugate acid	Typical pK_a of conjugate acid
alkylamine*	$R_3\overset{+}{N}-H$	9–11
aromatic amine*	$Ar-\overset{+}{N}R_2-H$	4–5
amide	$R-C(=\overset{+}{O}-H)-NR_2$	−1
alcohol, ether (R,R′ = alkyl, H)	$R-\overset{+}{O}(H)-R$	−2 to −3
phenol, aromatic ether	$Ar-\overset{+}{O}(H)-R$	−6 to −7
thiol, sulfide (R,R′ = alkyl, H)	$R-\overset{+}{S}(H)-R$	−6 to −7
ester, acid	$R-C(=\overset{+}{O}-H)-OR$	−6
aldehyde, ketone	$R-C(=\overset{+}{O}-H)-R$	−7
nitrile	$R-C\equiv\overset{+}{N}-H$	−10

*Basic enough to dissolve in 5% aqueous HCL.

Credits

Fig. 1.5: Photo courtesy of John Kozlowski.

Fig. 2.7: Adapted from *University Chemistry,* 3rd Edition by Bruce H. Mahan. © 1975 by Addison-Wesley Publishing Company and used with permission.

Fig. 3.9: Courtesy of Dow Chemical Company, Texas Division.

Fig. 3.10(a): T. J. Beveridge, University of Guelph and G. Sprott, National Research Council of Canada/BPS.

Fig. 3.10(b): T. J. Beveridge and C. Forsberg, University of Guelph/BPS.

Fig. 4.16: Reprinted from R. Ember, P. S. Layman, Wil Lepkowski, and P. S. Zurer, *Chemical and Engineering News,* November 24, 1986.

Fig. 5.2: Photo courtesy of Kyle Loudon.

Fig. 6.4: From the Nobel Lecture of Vladimir Prelog, *Science,* **193:**17 (1976). Used by permission of Elsevier Scientific Publishing Company.

Fig. 6.19: From G. B. Kauffman and R. D. Myers, *Journal of Chemical Education,* **52:**777 (1975) and used with permission.

Figs. 12.4, 12.8, 12.9, 12.10, and 12.11: Adapted from *Aldrich Library of FT-IR Spectra,* Charles J. Pouchert, editor. Aldrich Chemical Company, © 1985, and used with permission.

Figs. 12.14, 12.15, 12.16, 12.17, and 12.18: From *NPA-NIH Mass Spectral Data Base,* published by the National Bureau of Standards, United States Department of Commerce.

Fig. 12.19: Reprinted by permission from F. W. McLafferty, *Interpretation of Mass Spectra.* © 1977. Benjamin/Cummings Publishing Company, Inc.

Fig. 12.20: Adapted from *Aldrich Library of FT-IR Spectra,* Charles J. Pouchert, editor. Aldrich Chemical Company, © 1985, and used with permission.

Fig. 12.21: From *NPA-NIH Mass Spectral Data Base,* published by the National Bureau of Standards, United States Department of Commerce.

Fig. 12.22: Adapted from *Aldrich Library of FT-IR Spectra,* Charles J. Pouchert, editor. Aldrich Chemical Company, © 1985, and used with permission.

Fig. 12.23: From *NPA-NIH Mass Spectral Data Base,* published by the National Bureau of Standards, United States Department of Commerce.

Figs. 12.24 and 12.25: Adapted from *Aldrich Library of FT-IR Spectra,* Charles J. Pouchert, editor. Aldrich Chemical Company, © 1985, and used with permission.

Figs. 13.2, 13.3, 13.5, 13.9, 13.10, and 13.13: © Sadtler Research Laboratories, Division of Bio-Rad Laboratories, Inc.

Fig. 13.14: Spectrum courtesy of John Kozlowski.

Figs. 13.15 and 13.18: © Sadtler Research Laboratories, Division of Bio-Rad Laboratories, Inc.

Fig. 13.19: From Daniel J. Pasto and Carl R. Johnson, *Organic Structure Determination,* © 1969, pp. 192 and 200. Reprinted by permission of Prentice-Hall, Inc., Englewood Cliffs, NJ.

Fig. 13.21: Reprinted with permission from F. A. Bovey, *Chemical and Engineering News,* August 30, 1965. © 1965 by the American Chemical Society.

Figs. 13.22 and 13.23: © Sadtler Research Laboratories, Division of Bio-Rad Laboratories, Inc.

Fig. 13.24: Spectrum courtesy of John Kozlowski.

Fig. 13.25: Adapted from *Aldrich Library of FT-IR Spectra,* Charles J. Pouchert, editor. Aldrich Chemical Company, © 1985, and used with permission. From *NPA-NIH Mass Spectral Data Base,* published by the National Bureau of Standards, United States Department of Commerce. © Sadtler Research Laboratories, Division of Bio-Rad Laboratories, Inc.

Fig. 13.27: Courtesy of Dr. Paul J. Keller, Barrow Neurological Institute.

Fig. 13.28: © Sadtler Research Laboratories, Division of Bio-Rad Laboratories, Inc.

Fig. 13.29: Adapted from *Aldrich Library of FT-IR Spectra,* Charles J. Pouchert, editor. Aldrich Chemical Company, © 1985, and used with permission. © Sadtler Re search Laboratories, Division of Bio-Rad Laboratories, Inc.

Fig. 13.31: © Sadtler Research Laboratories, Division of Bio-Rad Laboratories, Inc.

Fig. 13.32: Reprinted by permission from G. A. Olah, *et al., J. Am. Chem. Soc.,* **86:**1362 (1964). Copyright © 1964 by the American Chemical Society.

Fig. 13.33: Adapted from *Aldrich Library of FT-IR Spectra,* Charles J. Pouchert, editor. Aldrich Chemical Company, © 1985, and used with permission. © Sadtler Research Laboratories, Division of Bio-Rad Laboratories, Inc.

Fig. 13.34: Adapted from *Aldrich Library of FT-IR Spectra,* Charles J. Pouchert, editor. Aldrich Chemical Company, © 1985, and used with permission. From *NPA-NIH Mass Spectral Data Base,* published by the National Bureau of Standards, United States Department of Commerce. © Sadtler Research Laboratories, Division of Bio-Rad Laboratories, Inc.

Figs. 14.3 and 14.5: Adapted from *Aldrich Library of FT-IR Spectra,* Charles J. Pouchert, editor. Aldrich Chemical Company, © 1985, and used with permission.

Fig. 14.7: Photo courtesy of Professors Tom Eisner and Jerrold Meinwald, Cornell University.

Fig. 14.8: Adapted from *Aldrich Library of FT-IR Spectra,* Charles J. Pouchert, editor. Aldrich Chemical Company, © 1985, and used with permission.

Figs. 14.9 and 15.3: © Sadtler Research Laboratories, Division of Bio-Rad Laboratories, Inc.

Fig. 15.5: From T. B. Brill, *J. Chem. Education,* **57:**259 (1980). Used with permission.

Fig. 16.1: Adapted from *Aldrich Library of FT-IR Spectra,* Charles J. Pouchert, editor. Aldrich Chemical Company, © 1985, and used with permission.

Figs. 16.3, 16.4, 16.5, 16.6, 16.11, and 16.12: © Sadtler Research Laboratories, Division of Bio-Rad Laboratories, Inc.

Fig. 16.13: © Sadtler Research Laboratories, Division of Bio-Rad Laboratories, Inc. From *NPA-NIH Mass Spectral Data Base,* published by the National Bureau of Standards, United States Department of Commerce.

Fig. 18.7: © Sadtler Research Laboratories, Division of Bio-Rad Laboratories, Inc.

Fig. 19.3: Adapted from *Aldrich Library of FT-IR Spectra,* Charles J. Pouchert, editor. Aldrich Chemical Company, © 1985, and used with permission.

Fig. 19.4: Copyright © 1962 and 1963 by Varian Associates, Palo Alto, California. Reprinted by permission of Varian Associates.

Figs. 19.5 and 19.6: © Sadtler Research Laboratories, Division of Bio-Rad Laboratories, Inc.

Fig. 19.7: From *NPA-NIH Mass Spectral Data Base,* published by the National Bureau of Standards, United States Department of Commerce.

Fig. 19.10: Adapted from *Aldrich Library of FT-IR Spectra,* Charles J. Pouchert, editor. Aldrich Chemical Company, © 1985, and used with permission. © Sadtler Research Laboratories, Division of Bio-Rad Laboratories, Inc.

Figs. 19.11 and 19.12: © Sadtler Research Laboratories, Division of Bio-Rad Laboratories, Inc.

Fig. 20.2: Adapted from *Aldrich Library of FT-IR Spectra,* Charles J. Pouchert, editor. Aldrich Chemical Company, © 1985, and used with permission. © Sadtler Research Laboratories, Division of Bio-Rad Laboratories, Inc.

Figs. 20.2 and 20.6: © Sadtler Research Laboratories, Division of Bio-Rad Laboratories, Inc.

Fig. 20.7: Adapted from *Aldrich Library of FT-IR Spectra,* Charles J. Pouchert, editor. Aldrich Chemical Company, © 1985, and used with permission. © Sadtler Research Laboratories, Division of Bio-Rad Laboratories, Inc.

Figs. 21.2 and 21.3: Adapted from *Aldrich Library of FT-IR Spectra,* Charles J. Pouchert, editor. Aldrich Chemical Company, © 1985, and used with permission.

Figs. 21.4 and 21.7(a–c): © Sadtler Research Laboratories, Division of Bio-Rad Laboratories, Inc.

Fig. 21.7(d): From *NPA-NIH Mass Spectral Data Base,* published by the National Bureau of Standards, United States Department of Commerce. © Sadtler Research Laboratories, Division of Bio-Rad Laboratories, Inc.

Fig. at bottom of page 906: From *Biochemistry,* by Geoffrey Zubay. © 1983 by Addison-Wesley Publishing Company. Used with permission.

Fig. 23.2: Adapted from *Aldrich Library of FT-IR Spectra,* Charles J. Pouchert, editor. Aldrich Chemical Company, © 1985, and used with permission.

Fig. 23.6: Adapted from *Aldrich Library of FT-IR Spectra,* Charles J. Pouchert, editor. Aldrich Chemical Company, © 1985, and used with permission. © Sadtler Research Laboratories, Division of Bio-Rad Laboratories, Inc.

Fig. 24.4: From *Chemistry, Matter, and the Universe,* by R. E. Dickerson and I. Geis, Benjamin/Cummings Publishing Company, 1969. Illustrations copyright by Irving Geis. Used with permission.

Fig. 26.8: From *Biochemistry,* by Geoffrey Zubay. © 1983 by Addison-Wesley Publishing Company. Used with permission.

Fig. 26.11: From *Structure and Action of Proteins,* by R. E. Dickerson and I. Geis. Benjamin/Cummings Publishing Company, 1969. Illustrations copyright by Irving Geis. Used with permission.

Fig. 26.12: From *Chemistry, Matter, and the Universe,* by R. E. Dickerson and I. Geis, Benjamin/Cummings Publishing Company, 1969. Illustrations copyright by Irving Geis. Used with permission.

Fig. 27.7: From *Biochemistry,* by Geoffrey Zubay. © 1983 by Addison-Wesley Publishing Company. Used with permission.

Fig. 27.9: From James D. Watson, *Molecular Biology of the Gene,* © 1976 by Benjamin/Cummings Publishing Company. Used with permission.

Index

Key: Page numbers in **boldface** give locations of exact definitions. When the page number is followed by *f, p,* or *t,* the item is found in a *figure, problem,* or *table,* respectively.

A

F

G

H

P

Q

T

W

X

Y

Z

Organic Functional Groups
(R = alkyl or aryl unless indicated otherwise)

	Aliphatic Hydrocarbons			
	Alkane	**Alkene**	**Alkyne**	**Aromatic Hydrocarbons**
General formula	R—H (R = alkyl)	$R_2C{=}CR_2$ (R = alkyl, aryl, H)	R—C≡C—R (R = alkyl, aryl, H)	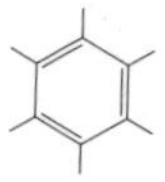
Functional group	—C—H or —C—C—	>C=C<	—C≡C—	benzene or other aromatic ring
Specific example	CH_3—CH_3	$CH_2{=}CH_2$	HC≡CH	
Common name	ethane	ethylene	acetylene	benzene
Systematic name	ethane	ethene	ethyne	benzene

	Alkyl Halide	**Alcohol**	**Phenol**	**Ether**
General formula	R—X (R = alkyl; X = halogen)	R—OH (R = alkyl)	R—OH (R = aryl)	R—O—R
Functional group	—C—X	—C—OH	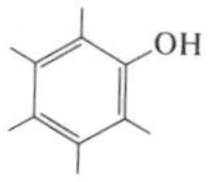	—C—O—C—
Specific example	CH_3—CH_2—Cl	CH_3—CH_2—OH	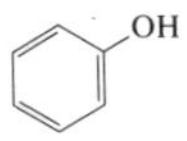	CH_3—O—CH_2—CH_3
Common name	ethyl chloride	ethyl alcohol	phenol	ethyl methyl ether
Systematic name	chloroethane	ethanol	phenol	methoxyethane